American College of Veterinary Internal Medicine (ACVIM) Forum 2012

New Orleans, Louisiana, USA
30 May - 2 June 2012

Editors:

Dianne McFarlane **Rance Sellon**

ISBN: 978-1-62276-153-1

Printed from e-media with permission by:

Curran Associates, Inc.
57 Morehouse Lane
Red Hook, NY 12571

Some format issues inherent in the e-media version may also appear in this print version.

Printed by Curran Associates, Inc. (2012)

American College of Veterinary Internal Medicine
1997 Wadsworth Boulevard Suite A
Lakewood CO 80214-5293

Phone: (800) 245-9081
Fax: (303) 231-0880

acvim@acvim.org

Additional copies of this publication are available from:

Curran Associates, Inc.
57 Morehouse Lane
Red Hook, NY 12571 USA
Phone: 845-758-0400
Fax: 845-758-2634
Email: curran@proceedings.com
Web: www.proceedings.com

ACVIM Specialty Symposium (Pre-Forum) Proceedings

Cardiology

Neurology

Oncology

Large Animal Internal Medicine - Equine

Large Animal Internal Medicine - Food Animal

Small Animal Internal Medicine

Veterinary Scientific Program Proceedings

AAEP

AAVN/ACVN

AAVN/ACVN (Small Animal)

ABVP

ACVCP/AAVPT

ACVIM Business Management

ACVIM Cardiology

ACVIM Large Animal Internal Medicine - Equine

ACVIM Large Animal Internal Medicine - Food Animal

ACVIM Neurology

ACVIM Oncology

ACVIM One Medicine

ACVIM Small Animal Internal Medicine

ASVCP

AVHTM

CGS

ISCAID

Liver Study Group

SCE

SVN/U

VCRS (Food Animal)

VECCS

Session#	Title / Author	Page#

Technician Program Proceedings

Technician

WELCOME TO THE 2012 ACVIM FORUM IN NEW ORLEANS, LA

Welcome to New Orleans! I am pleased and honored that you have chosen to attend the 2012 ACVIM Forum, the premier veterinary internal medicine continuing education meeting. Much of the credit for this distinction goes to the ACVIM Forum Program Committee as well as the many specialists who present state-of-the-art clinical science and the investigators who present the very latest advances in veterinary research to our attendees.

This year's agenda is full, with a core program of approximately 270 scientific lectures, more than 360 research abstracts, and a technician lecture series. All sessions of the core program are open to every attendee. Thursday through Saturday you'll find a variety of special interest group (SIG) sessions (workshops, case discussions, and labs) that address diverse topics in veterinary medicine. Many of these sessions are available in the evenings. Veterinary technicians will be pleased to see a scientific program and workshops designed specifically for them, although veterinary technicians are also welcome to attend other scientific sessions.

Make plans to attend the Opening Night Reception, sponsored by Hill's Pet Nutrition and the ACVIM on Thursday evening. On Friday morning, all attendees are invited to the Keynote Address: Veterinarians in Space: The Continuing Evolution of the Profession. Dr. Richard Linnehan, BS, DVM, MPA was nominated to deliver this keynote by ACVIM Diplomates who believe that he will inspire attendees with the story of his unusual and exciting career path. After the Keynote address, please stay for the Small Animal or Large Animal Consensus Statement DRAFT session.

Research posters are on display throughout the meeting. Content may be discussed with the authors on Friday evening during the Wine and Cheese Reception sponsored by Nestlé Purina.

There is an abundance of quality session topics on the daily scientific sessions grid. However, if you are unable attend a session of interest the Proceedings notes will help to fill in the gaps.

Make sure to leave plenty of time to meet with our many exhibitors in the Exhibit Hall. You will be surprised by the number and diversity of companies that come to bring you the very latest in veterinary technology, pharmaceutical advances, superior nutritional products, educational materials, and career opportunities.

The ACVIM is indebted to the 2012 ACVIM Forum Program Committee (chaired by Dr. Susan White) for the incredible effort it takes to put together a program of this magnitude. We also want to reiterate our thanks to our many speakers, facilitators, abstract presenters, and moderators who are willing to share their knowledge, talent, and time. We are very grateful to our sponsors for their recognition, presence, and support of our efforts to provide an unequaled educational experience for all attendees. The ACVIM Forum continues to grow substantially in both size and quality due to a wide range of sponsor support.

The dedicated ACVIM staff members have handled all of the organizational details needed to bring this meeting to you, and are available at the registration desk to help you with any special needs or requests. Please take a few minutes to introduce yourself and to personally thank them for a job well done.

The members of the ACVIM Board of Regents realize that you have choices regarding where to spend your continuing education dollars and your valuable time, and we appreciate the fact that you have chosen to join us in New Orleans. We hope that you have an outstanding educational experience, and that your attendance at the 2012 ACVIM Forum will facilitate your continued commitment to the highest quality of service in veterinary practice, academia, or industry. Mark you calendars now to join us for the 2013 ACVIM Forum, June 12 – 15, in Seattle, Washington.

Wishing you a stimulating and productive meeting,

Dr. Leah A. Cohn, DVM, PhD, DACVIM (SAIM)
President, ACVIM Board of Regents

SPEAKER DISCLOSURE INDEX
TECHNICIAN PROGRAM

Session	Speaker	Disclosure
PFT01a	Hally Niehaus	Nothing Disclosed
PFT01b	Pamela Knowles	Nothing Disclosed
PFT01c	Kate Brumbaugh	Nothing Disclosed
PFT02a	Gina Galyon	Nothing Disclosed
PFT02b	Victoria Lee Mudge	Nothing Disclosed
PFT02c	Ed Carlson	Nothing Disclosed
PFT03a	Erin Wright	Nothing Disclosed
PFT03b	Julie Nettifee Osborne	Nothing Disclosed
T05	Allison Jean Stewart	Nothing Disclosed
T06	Allison Jean Stewart	Nothing Disclosed
T07	Stephanie Nicole Gilliam	Nothing Disclosed
T08	Stephanie Nicole Gilliam	Nothing Disclosed
T09	David Liss	Nothing Disclosed
T10	Amy Breton	Nothing Disclosed
T11	Jessica M. Quimby	Independant consultant for Aratana Therapeutics: Clinical Trial design.
T12	Chad M. Johannes	Formerly employed by Pfizer Animal Health (2006-2010); own stock in Pfizer Inc; product names - Cerenia, Palladia, Vanguard L4.
T13	Tracy Lyn Gieger	Nothing Disclosed
T14	Chad M. Johannes	Formerly employed by Pfizer Animal Health (2006-2010); own stock in Pfizer Inc; product names - Cerenia, Palladia, Vanguard L4.
T15	Gregory K. Ogilvie	Nothing Disclosed
T16	Gregory K. Ogilvie	Nothing Disclosed
T17	Vicky Ograin	Nothing Disclosed
T18	Kara Burns	Nothing Disclosed
T19	Jessica M. Quimby	Independant consultant for Aratana Therapeutics: Clinical Trial design.
T20	Kara Burns	Nothing Disclosed
T21	Kim M. Albin	Nothing Disclosed
T22	Chelsea D. Tripp	Nothing Disclosed
T23	Laura D. Garrett	Consultant for Pfizer/Palladia (ELITE group).
T24	Meri Frances Hall	Nothing Disclosed
T25	Joshua Estrin	Nothing Disclosed
T25	Lisa Allison	Nothing Disclosed
T26	Mark G. Papich	Nothing Disclosed
T27	Romain Pariaut	Nothing Disclosed
T28	Henry W. Green	Nothing Disclosed
T29	Deborah M. Fine	Nothing Disclosed
T30	H. Edward Durham	Nothing Disclosed

T31	Richard J. Hepburn	Sanofi Merial are part funding my PhD at the University of Liverpool.
T32	Shirley I. Sandoval	Nothing Disclosed
T33	Thomas Passler	Nothing Disclosed
T34	Anna May Firshman	Nothing Disclosed
T35	H. Edward Durham	Nothing Disclosed
T36	Charlotte Pace	Nothing Disclosed
T37	Charlotte Pace	Nothing Disclosed
T38	Bonnie J. Heatwole	Nothing Disclosed
T39	Caroline Mansfield	Nothing Disclosed
T40	J. Scott Weese	Nothing Disclosed
T41	Bea Biddinger	Nothing Disclosed
T42	Bea Biddinger	Nothing Disclosed
T43	Jason M. Berg	Nothing Disclosed
T44	R. Timothy Bentley	Nothing Disclosed
T45	Kristin Kirkby	Nothing Disclosed
T46	Jessica Marie Weber	Nothing Disclosed
T47	Anne Myers	Nothing Disclosed
T48	Kristin Kirkby	Nothing Disclosed
T49	Colleen Suzanne Ruderman	Nothing Disclosed
T50	Amy Breton	Nothing Disclosed

SPEAKER DISCLOSURE INDEX
VETERINARY PROGRAM

Session	Speaker	Disclosure
Keynote	Richard M. Linnehan	Nothing Disclosed
PFC01	Jens Haggstrom	Nothing Disclosed
PFC02	Kathryn Montgomery Meurs	Co- investigator on 3 patents associated with the work presented.
PFC03	James W. Buchanan	Nothing Disclosed
PFC04	Margaret M. Sleeper	Nothing Disclosed
PFC05	Robert Roberts	Nothing Disclosed
PFC06	John D. Bonagura	Nothing Disclosed
PFC07	Mark D. Kittleson	Nothing Disclosed
PFC08	Brendan Corcoran	Nothing Disclosed
PFC09	Hannah Mary Hodgkiss-Geere	Nothing Disclosed
PFL01	Alexander de Lahunta	Nothing Disclosed
PFL02	Gerald Kent Carter	Nothing Disclosed
PFL02	Mark T. Donaldson	Nothing Disclosed
PFL02	Abby M. Sage	Nothing Disclosed
PFL03	Kenneth W. Hinchcliff	Nothing Disclosed
PFL04	Paul Lunn	Nothing Disclosed
PFL05	Margaret Bynoe	Nothing Disclosed
PFL06	Margaret Bynoe	Nothing Disclosed
PFL07	Paul S. Morley	Nothing Disclosed
PFL08	Sébastien M.C. Buczinski	Nothing Disclosed
PFL09	David C. Van Metre	Nothing Disclosed
PFN01	Michael S. Kent	Nothing Disclosed
PFN02	Ian D. Duncan	Nothing Disclosed
PFN03	W. Douglas Sheffield	Nothing Disclosed
PFN04	Simon Ronald Platt	Nothing Disclosed
PFN05	Peter J. Dickinson	Nothing Disclosed
PFN06	Peter J. Dickinson	Nothing Disclosed
PFO01	Amy K. LeBlanc	Nothing Disclosed
PFO02	Susan L. Kraft	I do a very small amount of consulting part-time as a clinical teleradiologist for PetRays, which occasionally includes interpreting MRI or CT studies performed at various hospitals. However, I am not at all involved in marketing for that company, recruitment of clientele nor continuing education events. My MRI and PET/CT research is completely independent and unrelated to those teleradiology activities, but I am happy to list this on my disclosure slide.
PFO03	Robert T. O'Brien	Nothing Disclosed
PFO04	Chand Khanna	Nothing Disclosed
PFO05	Anne Avery	Nothing Disclosed

PFO06	Rodney L. Page	Nothing Disclosed
PFO07	Anne Barger	Nothing Disclosed
PFS01a	Steven W. Dow	Nothing Disclosed
PFS01b	Steven W. Dow	Nothing Disclosed
PFS02	Brendan C. McKiernan	Nothing Disclosed
PFS03a	Daniel Lacorazza	Nothing Disclosed
PFS03b	Daniel Lacorazza	Nothing Disclosed
PFS04a	Domenico Bianco	Nothing Disclosed
PFS04b	Oliver A. Garden	Nothing Disclosed
PFS05a	Laura D. Garrett	Consultant for Pfizer/Palladia (ELITE group).
PFS05b	Anne Barger	Nothing Disclosed
PFS06a	Michael Podell	Nothing Disclosed
PFS06b	Michael Podell	Nothing Disclosed
PFS07a	Mark A. Oyama	Serve on veterinary advisory board for IDEXX Laboratories, have received funding for clinical studies from IDEXX Laboratories, have received programmatic support from IDEXX Laboratories.
PFS07b	John M. MacGregor	Clinical investigator (Pimobendan) 11/11 to present Speaker (Pimobendan) 10/10 to present.
PFS08a	Lauren A. Trepanier	Nothing Disclosed
PFS08b	Lauren A. Trepanier	Nothing Disclosed
PFS09a	Mark Mitchell	I consult for them. I give lectures once monthly via webinars.
PFS09b	Mark Mitchell	I consult for them. I give lectures once monthly via webinars.
PFS10	James Korndorffer	Nothing Disclosed
PFS11	Thomas K. Graves	Nothing Disclosed
1	Lisbeth H. Olsen	Nothing Disclosed
2	Jens Haggstrom	Nothing Disclosed
2	Ingrid Ljungvall	Nothing Disclosed
3	Mark A. Oyama	Serve on veterinary advisory board for IDEXX Laboratories, have received funding for clinical studies from IDEXX Laboratories, have received programmatic support from IDEXX Laboratories.
4	Sofia Cerda-Gonzalez	Nothing Disclosed
5	Ronaldo Casimiro da Costa	Nothing Disclosed
6	Ronaldo Casimiro da Costa	Nothing Disclosed
7a	Douglas H. Thamm	Dr. Thamm owns stock options in VetDC, the company that holds the North American rights to VDC-1101 as an animal cancer therapeutic.
7b	Douglas H. Thamm	Dr. Thamm owns stock options in VetDC, the company that holds the North American rights to VDC-1101 as an animal cancer therapeutic.
8a	Dominic Marino	Nothing Disclosed
8b	Dominic Marino	Nothing Disclosed
9a	Nicola Mason	Founder and CEO of Vetagenics LLC. The company was founded in 2011.

9b	Jenna Hart Burton	Nothing Disclosed
10	Kari V. Lunsford	Nothing Disclosed
11	Marilyn E. Dunn	Nothing Disclosed
12	John M. Thomason	Nothing Disclosed
13	Elizabeth Anne Rozanski	Nothing Disclosed
14	Carol R. Reinero	Nothing Disclosed
15	Jonathan F. Bach	Nothing Disclosed
16	David C. Twedt	Nothing Disclosed
16	Penny J. Watson	Nothing Disclosed
17	Cynthia Leveille-Webster	Nothing Disclosed
18	Sharon A. Center	Nothing Disclosed
19	Mark G. Papich	Nothing Disclosed
20	Dawn M. Boothe	Nothing Disclosed
21	David Lloyd	Nothing Disclosed
22	Bettina M. Dunkel	Nothing Disclosed
23	Kelsey A. Hart	Nothing Disclosed
24	Veronique Anne Lacombe	Nothing Disclosed
25	Sarel R. Van Amstel	Nothing Disclosed
26	Jan Shearer	Nothing Disclosed
27	Jan Shearer	Nothing Disclosed
28	David Scott McVey	Nothing Disclosed
29	Katherine M. Kocan	Nothing Disclosed
30	Johann F. Coetzee	Nothing Disclosed
31	E. Christopher Orton	Co-Inventor of MitralSeal (transcatheter mitral valve device). Patent rights are assigned to Colorado State University and have been licensed to Avalon Medical (veterinary application) and Tendyne Medical (human application). I recieve a portion of royalites paid to CSU.
32	Mark A. Oyama	Serve on veterinary advisory board for IDEXX Laboratories, have received funding for clinical studies from IDEXX Laboratories, have received programmatic support from IDEXX Laboratories.
33	Joao Sacadura Orvalho	Nothing Disclosed
34	Gerard Le Bobinnec	Nothing Disclosed
35	Christine E. Thomson	I don't think this is significant but mention it for completeness. The material I am presenting at ACVIM comes from the textbook that I have written with coauthor Caroline Hahn. The textbook 'Veterinary Neuroanatomy: A Clinical Approach" is being released by Elsevier in May 2012.

36	Christine E. Thomson	I don't think this is significant but mention it for completeness. The material I am presenting at ACVIM comes from the textbook that I have written with coauthor Caroline Hahn. The textbook 'Veterinary Neuroanatomy: A Clinical Approach" is being released by Elsevier in May 2012.
37	Charles H. Vite	Nothing Disclosed
38	Phillip Lerche	Nothing Disclosed
39	Deborah W. Knapp	Nothing Disclosed
40	Susan M. LaRue	Nothing Disclosed
41	William Culp	I teach laboratory sessions that are sponsored by a company that produces interventional radiology equipment.
42	Melissa C. Paoloni	Nothing Disclosed
43	Linda Kidd	Nothing Disclosed
44	Alfred M. Legendre	Nothing Disclosed
45	Audrey Karen Cook	Nothing Disclosed
46	Julia K. Veir	Nothing Disclosed
47	Steven W. Dow	Nothing Disclosed
48	J. Scott Weese	Nothing Disclosed
49	Todd Marlow Archer	Nothing Disclosed
50	Clare R. Gregory	Nothing Disclosed
51	Lowell Anthony	Nothing Disclosed
52	Richard J. Campeau	Nothing Disclosed
53	Gregory B. Daniel	Nothing Disclosed
54	Claudia E. Reusch	Nothing Disclosed
55	Kenneth W. Hinchcliff	Nothing Disclosed
56	Emmanuelle van Erck	Nothing Disclosed
57	Emmanuelle van Erck	Nothing Disclosed
58	Erica C. McKenzie	Nothing Disclosed
59	Melissa R. Mazan	Nothing Disclosed
60	Melissa R. Mazan	Nothing Disclosed
61	Rose Dunbar Nolen-Walston	Nothing Disclosed
62	Mark G. Papich	Nothing Disclosed
63	Barry J. Maron	Nothing Disclosed
64	Barry J. Maron	Nothing Disclosed
65	Bermans J. Iskandar	Nothing Disclosed
66	Bermans J. Iskandar	Nothing Disclosed
67	Lauren A. Trepanier	Nothing Disclosed
68	Lauren A. Trepanier	Nothing Disclosed
69	Joseph M. Blondeau	Nothing Disclosed

70	Joseph M. Blondeau	Nothing Disclosed
71a	Karin Weber	Nothing Disclosed
71b	Markus Rick	Nothing Disclosed
72a	Penny J. Watson	Nothing Disclosed
72b	Jonathan D. Dear	Nothing Disclosed
73a	Carol R. Reinero	Nothing Disclosed
73b	Oscar Cortadellas	Nothing Disclosed
74a	Hasan Albasan	Nothing Disclosed
74b	Jessica M. Quimby	Independant consultant for Aratana Therapeutics: Clinical Trial design.
75	Eleanor C. Hawkins	Receive financial compensation for participation in Pfizer Speakers Bureau (providing CE opportunities) and consultation.
76	Vanessa Rosemary Barrs	Nothing Disclosed
77	Dwight Bowman	Research over the years with most veterinary associated pharma and diagnostic corporations.
78	Dwight Bowman	Research over the years with most veterinary associated pharma and diagnostic corporations.
79	Meryl P. Littman	Nothing Disclosed
80	Richard E. Goldstein	Serve of advisory councils, speaker bureaus, and assist with R and D, have received funding for research from multiple companies. Was an employee of Cornell University until 12/31/2011.
81	Michael R. Lappin	Nothing Disclosed
82	Severine Tasker	Nothing Disclosed
83	Ashley G. Boyle	Nothing Disclosed
84	J. Scott Weese	Nothing Disclosed
85a	Dianne McFarlane	Nothing Disclosed
85b	Harold C. Schott	Consultant for Boehringer-Ingelheim Vetmedica, Inc.
86a	Molly E. McCue	Nothing Disclosed
86b	Raymond J. Geor	Nothing Disclosed
87	Thomas M. Craig	Nothing Disclosed
88	Thomas M. Craig	Nothing Disclosed
89a	Rosangela Alves	Possess financial support for research of the National Council for Scientific and Technological Development (CNPq) of the Brazilian Government.
89b	Jessie Rose Payne	IDEXX Laboratories provide costs associated with my PhD studies (October 2009 - September 2013). They are running NT-proBNP and troponin levels of blood samples collected as part of my PhD project. They have not had any direct input on the research report to be presented.

90a	Bruce W. Keene	Nothing Disclosed
90b	Adrian Boswood	Nothing Disclosed
91	Virginia Luis Fuentes	IDEXX Laboratories are part funding our clinical research in HCM in shelter cats.
91	Kristin A. MacDonald	Nothing Disclosed
91	Kathryn Montgomery Meurs	Co- investigator on 3 patents associated with the work presented.
91	Christopher Paige	Nothing Disclosed
92	Philip R. Fox	Nothing Disclosed
92	Mark D. Kittleson	Nothing Disclosed
92	Karsten Schober	Nothing Disclosed
92	Daniel F. Hogan	Nothing Disclosed
93	Sofia Cerda-Gonzalez	Nothing Disclosed
94	Dennis O'Brien	Nothing Disclosed
95	Fred A. Wininger	Nothing Disclosed
96	Joan R. Coates	Nothing Disclosed
97	Erika L. Krick	Nothing Disclosed
98	Michael Owen Childress	Nothing Disclosed
99a	Kim A. Selting	I am both collaborating with VDI and am paid as a consultant.
99b	Laura E. Selmic	Nothing Disclosed
100a	Deborah W. Knapp	Nothing Disclosed
100b	Deborah W. Knapp	Nothing Disclosed
101	Nicola Mason	Founder and CEO of Vetagenics LLC. The company was founded in 2011.
102	Jaime F. Modiano	Nothing Disclosed
103	Kenneth Marshall Rassnick	Nothing Disclosed
104	Timothy M. Fan	Nothing Disclosed
105	James G. Fox	Nothing Disclosed
106	Jeffrey N. Bryan	Nothing Disclosed
107	Maria Julia Bevilaqua Felippe	Nothing Disclosed
108a	Laurent L. Couetil	Nothing Disclosed
108b	Megan Duckett	Nothing Disclosed
109a	Tracy L. Hill	Nothing Disclosed
109b	Patrick Meeus	Employed by Pfizer Animal Health.
110a	Leslie Lyons	Nothing Disclosed
110b	Michael R. Lappin	Nothing Disclosed

111a	Leah Ann Cohn	Nothing Disclosed
111b	Giacomo Rossi	Nothing Disclosed
112a	Arleigh Reynolds	I do basic research for Nestle Purina; I am directly employed by them.
112b	Paul R. Hess	Nothing Disclosed
113	Rose E. Raskin	Nothing Disclosed
114	Rose E. Raskin	Nothing Disclosed
115	James H. Meinkoth	Nothing Disclosed
116	James H. Meinkoth	Nothing Disclosed
117	Ann E. Hohenhaus	Nothing Disclosed
118	Sarah E. Helmond	Nothing Disclosed
119	J. Corey Orava	Full time employee of company involved with stem cells. One of the mechanisms of action of stem cells is immunomodulation.
120	Sarah E. Helmond	Nothing Disclosed
120	Ann E. Hohenhaus	Nothing Disclosed
120	J. Corey Orava	Full time employee of company involved with stem cells. One of the mechanisms of action of stem cells is immunomodulation.
121	Jane E. Sykes	Nothing Disclosed
121	Joseph Wheat	Nothing Disclosed
122	Jane E. Sykes	Nothing Disclosed
123	Katrin Hartmann	Nothing Disclosed
124	Michael R. Lappin	Nothing Disclosed
125	J. Glenn Songer	Nothing Disclosed
126	J. Scott Weese	Nothing Disclosed
127	Stanley L. Marks	Nothing Disclosed
128	Stanley L. Marks	Nothing Disclosed
128	J. Glenn Songer	Nothing Disclosed
128	J. Scott Weese	Nothing Disclosed
129a	David J. Polzin	Nothing Disclosed
129b	Mary B. Nabity	Nothing Disclosed
130a	Carrie A. Palm	Nothing Disclosed
130b	Gilad Segev	Nothing Disclosed
131a	George E. Lees	Nothing Disclosed
131b	Rosanne E. Jepson	Nothing Disclosed
132a	Jodi L. Westropp	I have spoken for Bayer animal health at various CE meetings. They provided funds to perform the high dose short duration enrofloxacin study I mention in the proceedings. I have no relations with Uricult Vet system, another product we evaluated in a scientific study.

132b	Dennis J. Chew	Bayer financially supported the clinical study that compared the use of 14 days Clavamox to 3 days of high dose enrofloxacin (Baytril) while I was still employed full time at OSU. Bayer has sponsored sessions at TNAVC, WVC, and AAHA at which I spoke on the general topic of urinary tract infections in dogs- they paid my honoriarium.
133	Manuel Boller	Nothing Disclosed
134	Daniel John Fletcher	Nothing Disclosed
135	Daniel John Fletcher	Nothing Disclosed
136	Manuel Boller	Nothing Disclosed
137	Paul Marik	Nothing Disclosed
138	Paul Marik	Nothing Disclosed
139a	Elizabeth M. Tadros	Nothing Disclosed
139b	Mary M. Durando	Nothing Disclosed
140a	Sandra D. Taylor	Nothing Disclosed
140b	Christopher Riley	Nothing Disclosed
141	Modest Vengust	Nothing Disclosed
142	Bettina M. Dunkel	Nothing Disclosed
143	Renaud Leguillette	Nothing Disclosed
144	Morgan Keith Chaffin	Nothing Disclosed
145a	Christie E. Mayo	Nothing Disclosed
145b	Johanna Rebecca Elfenbein	Nothing Disclosed
146a	Maria E. Prado	Nothing Disclosed
146b	Manuel F. Chamorro	Dr. Tom Earleywine is one of the authors of the reserach report abstract submitted for this meeting and Land O' Lakes Animal Milk products provided the Colostrum replacer for this study. The rest of the authors do not have any financial or commercial relationship with Land O' Lakes Animal Milk Products.
147a	Derek M. Foster	Nothing Disclosed
147b	Elizabeth Dore	Nothing Disclosed
148a	Stacey Renee Byers	Nothing Disclosed
148b	Sebastien M.C. Buczinski	Nothing Disclosed
149	Joanne Hewson	These organizations have provided funding for research in the area of diagnosing bovine respiratory disease in 2011-2012. There is not an associated product.

150	Tim Potter	Nothing Disclosed
151	Paul H. Walz	Nothing Disclosed
152	Thomas Passler	Nothing Disclosed
153a	Ruthnea Muzzi	There are governmental funding agencies that give us support to present our research abroad, such as CAPES (Coordination of Improvement of Researchers), FAPEMIG (Agency of Sientific and Technological Development of Minas Gerais) and CNPQ (The National Council for Scientific and Technological Development). All these agencies are linked to thc Ministry of Science and Technology (MCT), dedicated to the promotion of research and formation of human resources, being nonprofit and noncommercial.
153b	Ruthnea Muzzi	There are governmental funding agencies that give us support to present our research abroad, such as CAPES (Coordination of Improvement of Researchers), FAPEMIG (Agency of Sientific and Technological Development of Minas Gerais) and CNPQ (The National Council for Scientific and Technological Development). All these agencies are linked to the Ministry of Science and Technology (MCT), dedicated to the promotion of research and formation of human resources, being nonprofit and noncommercial.
154a	Deborah M. Fine	Nothing Disclosed
154b	John M. MacGregor	Clinical investigator (Pimobendan) 11/11 to present Speaker (Pimobendan) 10/10 to present.
155a	Edward Cooper	Nothing Disclosed
155b	Clarke E. Atkins	Nothing Disclosed
156a	Gerard Le Bobinnec	Nothing Disclosed
156b	Katja Hoglund	Nothing Disclosed
157	Andrea C. Vollmar	Boehringer Vetmedica GmbH, Ingelheim, provided financial support for cardiovascular examinations and medications used in this study of preclinical DCM in Irish wolfhounds.
158	Clarke E. Atkins	Nothing Disclosed
159	Brian A. Scansen	Nothing Disclosed

160	Mark Rishniw	Nothing Disclosed
161a	Karen M. Vernau	Director Veterinary Affairs, CanCog Technology CanCog Technologies developed and validated these protocols in conjunction with other researchers. Therefore likely not a conflict of interest but I will be presenting our research findings.
161b	Alireza A. Gorgi	Nothing Disclosed
162a	Karen R. Munana	Nothing Disclosed
162b	Byeong-Teck Kang	Nothing Disclosed
163a	Ronaldo Casimiro da Costa	Nothing Disclosed
163b	Thomas Flegel	Nothing Disclosed
164a	Nicolas Granger	Nothing Disclosed
164b	Christine E. Thomson	I don't think this is significant but mention it for completeness. The material I am presenting at ACVIM comes from the textbook that I have written with coauthor Caroline Hahn. The textbook 'Veterinary Neuroanatomy: A Clinical Approach" is being released by Elsevier in May 2012.
165	Peter J. Dickinson	Nothing Disclosed
166	Marjorie Milne	Nothing Disclosed
167	Nicolas Granger	Nothing Disclosed
168	Andy Shores	Nothing Disclosed
169	Barbara J. Biller	Nothing Disclosed
170	Nicola Mason	Founder and CEO of Vetagenics LLC. The company was founded in 2011.
171	Michael Owen Childress	Nothing Disclosed
172	Dennis Burt Bailey	During the initial release of Palladia, I was a member of Pfizer's ELITE training program and gave CE talks to other veterinarians regarding the use of Palladia.
173	Mark E. Peterson	Nothing Disclosed
174	Thomas Schermerhorn	Nothing Disclosed
175	Audrey Karen Cook	Nothing Disclosed
176	Patty A. Lathan	Nothing Disclosed
177	Michael D. Willard	Nothing Disclosed
178	Adam George Gow	Nothing Disclosed
179	Silke Schmitz	Nothing Disclosed

180	Caroline Mansfield	Nothing Disclosed
181	Jessica M. Quimby	Independant consultant for Aratana Therapeutics: Clinical Trial design.
182	Jacqueline Sinclair Rand	Nothing Disclosed
183	Lisa Moses	Nothing Disclosed
183	Lisa Moses	Nothing Disclosed
184	Marnin A. Forman	Nothing Disclosed
185	Marie Holowaychuk	Nothing Disclosed
186	Marie Holowaychuk	Nothing Disclosed
187	Armelle M. De Laforcade	Nothing Disclosed
188	Elizabeth Anne Rozanski	Nothing Disclosed
189	Daniel L. Chan	Nothing Disclosed
190	Daniel L. Chan	Nothing Disclosed
191	Nick Cave	Nothing Disclosed
192	Nick Cave	Nothing Disclosed
193	Raymond J. Geor	Nothing Disclosed
194	Raymond J. Geor	Nothing Disclosed
195	Sarah L. Ralston	The Owner of the company, Dr. Harlan Anderson, has provided the rations free of cost for the research since 2003. Dr. Ralston also receives 0.1% of the net income from sales of the product in the form of feed for her own horses (no monetary income). Patent pending on the diagnostic "tool" and feed formulation.
196	Sarah L. Ralston	The Owner of the company, Dr. Harlan Anderson, has provided the rations free of cost for the research since 2003. Dr. Ralston also receives 0.1% of the net income from sales of the product in the form of feed for her own horses (no monetary income). Patent pending on the diagnostic "tool" and feed formulation.
197a	Lutz Steffen Goehring	Nothing Disclosed
197b	Guy D. Lester	Nothing Disclosed
198a	Sarah M. Reuss	Nothing Disclosed
198b	Stacy Tinkler	Nothing Disclosed
199a	Renaud Leguillette	Nothing Disclosed
199b	Andy E. Durham	Nothing Disclosed
200a	Andrew William van Eps	Nothing Disclosed
200b	Steeve Giguere	Nothing Disclosed
201	Sylvain Nichols	Nothing Disclosed
202	Sylvain Nichols	Nothing Disclosed

204	Marie-Eve Fecteau	Nothing Disclosed
203	Simon Francis Peek	Nothing Disclosed
205	Lisa H. Williamson	Nothing Disclosed
206	Christopher K. Cebra	Nothing Disclosed
207	Christopher K. Cebra	Nothing Disclosed
208	Erica C. McKenzie	Nothing Disclosed
209	Shawn G. McVey	Nothing Disclosed
210	Shawn G. McVey	Nothing Disclosed
211	Shawn G. McVey	Nothing Disclosed
212	Shawn G. McVey	Nothing Disclosed
213	Heidi Britt Kellihan	Nothing Disclosed
214	Heidi Britt Kellihan	Nothing Disclosed
215	Michael R. O'Grady	Nothing Disclosed
216	Lynne O'Sullivan	Nothing Disclosed
217a	Filippo Adamo	Nothing Disclosed
217b	Sarah Anne Moore	Nothing Disclosed
218a	Gary Landsberg	Nothing Disclosed
218b	Sam Long	Nothing Disclosed
219a	Karen M. Vernau	Director Veterinary Affairs, CanCog Technology CanCog Technologies developed and validated these protocols in conjunction with other researchers. Therefore likely not a conflict of interest but I will be presenting our research findings.
219b	Christopher Mariani	Nothing Disclosed
220a	Arianna Maiolini	Nothing Disclosed
220b	R. Timothy Bentley	Nothing Disclosed
221	Debra Kamstock	Nothing Disclosed
222a	Debra Kamstock	Nothing Disclosed
222B	Dennis Burt Bailey	During the initial release of Palladia, I was a member of Pfizer's ELITE training program and gave CE talks to other veterinarians regarding the use of Palladia.
223	Dennis Burt Bailey	During the initial release of Palladia, I was a member of Pfizer's ELITE training program and gave CE talks to other veterinarians regarding the use of Palladia.
224	Laura D. Garrett	Consultant for Pfizer/Palladia (ELITE group).
225	Claire Rebecca Sharp	Nothing Disclosed
226	Claire Rebecca Sharp	Nothing Disclosed
227	Claire Rebecca Sharp	Nothing Disclosed
228	Claire Rebecca Sharp	Nothing Disclosed

229	Keith P. Richter	Nothing Disclosed
230	Cynthia R. Leveille-Webster	Nothing Disclosed
231	Sean P. McDonough	Nothing Disclosed
232	Sharon A. Center	Nothing Disclosed
233	Larry G. Adams	Nothing Disclosed
234	Stephanie A. Smith	Nothing Disclosed
235	Stanley L. Marks	Nothing Disclosed
236	Ellen Nicole Behrend	Nothing Disclosed
237	Ashley Baker Saunders	Nothing Disclosed
238	Joseph William Bartges	Nothing Disclosed
239	Claudia Ann Kirk	Nothing Disclosed
240	Jacqueline C. Whittemore	Nothing Disclosed
241	Kendal Harr	Nothing Disclosed
242	Bente Flatland	Nothing Disclosed
243	Kendal Harr	Nothing Disclosed
244	Bente Flatland	Nothing Disclosed
245	Catherine E. Langston	Nothing Disclosed
246	Mary H. Bowles	Nothing Disclosed
247	Jennifer Larsen	Nothing Disclosed
247	Jodi L. Westropp	I have spoken for Bayer animal health at various CE meetings. They provided funds to perform the high dose short duration enrofloxacin study I mention in the proceedings. I have no relations with Uricult Vet system, another product we evaluated in a scientific study.
248	Edward Cooper	Nothing Disclosed
249	Erica C. McKenzie	Nothing Disclosed
250	Stephanie J. Valberg	Nothing Disclosed
251	Carrie Jacqueline Finno	Nothing Disclosed
252	Sharon Witonsky	Nothing Disclosed
253	Frank M. Andrews	Nothing Disclosed
254	Frank M. Andrews	Nothing Disclosed
255	Richard J. Hepburn	Sanofi Merial are part funding completion of my PhD study at the University of Liverpool looking into Equine Glandular Gastritis-Erosion Syndrome
256	Richard J. Hepburn	Sanofi Merial are part funding completion of my PhD study at the University of Liverpool looking into Equine Glandular Gastritis-Erosion Syndrome
257	H. Morgan Scott	Nothing Disclosed
258	Christine B. Navarre	Nothing Disclosed

259	Gilles Fecteau	Nothing Disclosed
260	Allen J. Roussel	Nothing Disclosed
261	Nancy D. Kay	Nothing Disclosed
262	Nancy D. Kay	Nothing Disclosed
263	Peter Weinstein	Nothing Disclosed
264	Peter Weinstein	Nothing Disclosed
237	Kenneth James Drobatz	Nothing Disclosed
237	Erica Reineke	Nothing Disclosed
238	Kenneth James Drobatz	Nothing Disclosed
238	Erica Reineke	Nothing Disclosed
239	Kenneth James Drobatz	Nothing Disclosed
239	Erica Reineke	Nothing Disclosed
240	Kenneth James Drobatz	Nothing Disclosed
240	Erica Reineke	Nothing Disclosed
241	Allyson C. Berent	Nothing Disclosed
242	Charles Weisse	Consultant and minority partner of Infiniti Medical
243	Larry G. Adams	Nothing Disclosed
244	Larry G. Adams	Nothing Disclosed
244	Allyson C. Berent	Nothing Disclosed
244	Charles Weisse	Consultant and minority partner of Infiniti Medical
245	Joanne R. Smith	Nothing Disclosed
246	Kate E. Creevy	Nothing Disclosed
247	Joanne R. Smith	Nothing Disclosed
248	Kate E. Creevy	Nothing Disclosed
249	Rebecca L. Davies	Nothing Disclosed
250	Rebecca L. Davies	Nothing Disclosed
251	Kirstin F Barnhart	Nothing Disclosed
252	Kirstin F Barnhart	Nothing Disclosed
253	Anthony Pease	Nothing Disclosed
254	Shannon P Holmes	Nothing Disclosed
254	Fred A. Wininger	Nothing Disclosed
255	Shannon P Holmes	Nothing Disclosed
255	Fred A. Wininger	Nothing Disclosed
256	Anthony Pease	Nothing Disclosed
257	Melissa Trogdon Hines	Nothing Disclosed
258	K. Gary Magdesian	Nothing Disclosed
259	Samuel David Alan Hurcombe	Nothing Disclosed
260	Harold C. Schott	Nothing Disclosed
261	Jeffrey Carl Phillips	Some of the work described was partially funded by Merial Limited.
262	Jeffrey Carl Phillips	I received support from Merial Limited for some of the clinical studies that I will describe.
263	Robert H. Mealey	Nothing Disclosed

263	Angela M. Pelzel	Nothing Disclosed
263	Josie L. Traub-Dargatz	Nothing Disclosed
264	Robert H. Mealey	Nothing Disclosed
264	Angela M. Pelzel	Nothing Disclosed
264	Josie L. Traub-Dargatz	Nothing Disclosed
265	Kevin E. Washburn	Nothing Disclosed
266	Paul S. Morley	Nothing Disclosed
266	Paul J. Plummer	Nothing Disclosed
266	Allen J. Roussel	Nothing Disclosed
267	Gilles Fecteau	Nothing Disclosed
268	Allen J. Roussel	Nothing Disclosed
269	Karyn Gavzer	Nothing Disclosed
270	Karyn Gavzer	Nothing Disclosed
271	Karyn Gavzer	Nothing Disclosed
272	Karyn Gavzer	Nothing Disclosed

ACVIM FORUM KEYNOTE ADDRESS

Friday, June 1, 2012 · 10:30 am – 11:20 am · Ernest N. Morial Convention Center, New Orleans

Sponsored by

Hill's®

EXTRA-ORDINARY VETERINARIAN: DR. RICHARD LINNEHAN

"VETERINARY MEDICINE IN SPACE: A NATURAL EVOLUTION OF THE PROFESSION"

Dr. Linnehan graduated from the University of New Hampshire with a Bachelor of Science in Animal Sciences and a minor in Microbiology and subsequently received the Doctor of Veterinary Medicine degree from The Ohio State University. Initially entering private veterinary practice, Dr. Linnehan was later commissioned as a Captain in the U.S. Army Veterinary Corps and served as chief clinical veterinarian for the U.S. Navy's Marine Mammal Program. Selected by NASA in March 1992, Dr. Linnehan first flew as a mission specialist aboard Columbia in 1996 on STS-78, the Life Sciences and Microgravity Spacelab (LMS) mission. A veteran of four space flights, Dr. Linnehan has logged over 58 days in space, including 6 EVAs (spacewalks) and traveled approximately 25 million miles in space over the course of his NASA career.

In 2009 Dr. Linnehan received the degree Master of Public Administration (MPA) from the Kennedy School of Government at Harvard University in Cambridge, MA.

Please join us in hearing Dr. Linnehan's remarkable story.

ACVIM
As far as you want to go.
www.ACVIMForum.org

MOLECULAR GENETIC RESEARCH IN DOGS FROM A VETERINARY CARDIOLOGIST'S PERSPECTIVE

Jens Häggström, DVM, PhD, DECVIM-CA (Cardiology), Ingrid Ljungvall, DVM, PhD and Katja Höglund, DVM, PhD
Uppsala, Sweden

INTRODUCTION

The advances in molecular genetics have been considerable during the last decade.[1] The number of publications in this area in medical literature has soared, and in more recent years, there has been an increasing number of publications in this area concerning non-laboratory dogs and cats. Indeed, several publications reporting genetic molecular research findings in dogs and cats have been published in the most prestigious scientific journals. [2-5] The immediate result from these studies is that more genetic tests have become available. However, findings in these studies offer more in the future than genetic tests; these findings have the potential to offer a deepened knowledge and understanding of fundamental pathophysiologic processes, which, in turn, may lead to improved treatment strategies. Veterinary cardiology is not isolated from this process. Indeed, several successful studies have been published,[6-8] and several projects are currently in the process of being published or still ongoing. The advances in molecular genetics in veterinary cardiology may in the future impact how veterinary cardiologists and general practitioners practice. Thus, it may be of value to understand how such research is currently conducted and how results may be interpreted. The authors of this proceeding are not experts in molecular genetics, but mere veterinary cardiologists involved in this type of research. The purpose of this proceeding and presentation is to share some of the experiences gained in this process from a veterinary cardiologist's perspective.

WHY DOGS?

During the last 5 years, several comparably large research grants have been allocated to groups working with molecular genetics in dogs. For example, the European Union granted 12 million Euros to a project with the acronym "LUPA" (this is the main project the authors have been involved in).[9] This EU-project spanned over several different diseases/traits in dogs, including cardiovascular disease.[9] These large grants were not allocated to dog projects to primarily improve pet health or provide breeders with genetic tests; they were granted to use naturally occurring diseases/traits in dogs as models for similar human diseases/traits. A rightful question is: Why dogs? Why not another mammalian species? The answer to this question lies in the history of dogs. The dog species has passed through two "genetic bottlenecks" in history.[2,3] The first was thousands of years ago, when a few wolves founded the dog species. The second was only a little more than 100 years ago, when the dog breeds were created (by inbreeding to promote certain traits). These two bottlenecks have led to a great morphologic variation between breeds, but extensive genetic homogeneity within breeds, which, genetically, is characterized by long haplotypes and high linkage disequilibrium.[2,3] As a consequence of the genetic homogeneity within breeds, certain breed specific diseases have developed, including cardiovascular disease (e.g. dilated cardiomyopathy (DCM) in Doberman Pinschers). This implies that the same causative mutations exist within one breed. Because of the genetic similarity within breeds, fewer individuals are needed to identify disease causing genomic sequences, compared to people. Therefore, domestic dogs represent excellent models for human genetic disease because dogs and people develop diseases that are strikingly similar in etiology and the two species share the vast majority of their genes. Dogs also share the same environment as people.

WHO ARE "THEY"?

This research fueled by the allocation of large grants would not be possible without the engagement of world leading molecular geneticists. These scientists have embraced the potential of studying genetic causes of naturally occurring disease in primarily pure-bred dogs, and lead comparably large research teams that consist of a broad range of expertise including molecular genetics, biomedical chemistry, proteomics, bioinformatics/biostatistics, and veterinary specialists. This wide range of expertise is needed in this type of research because creation of study populations, handling of samples, managing the databases (which are usually quite large) and analyzing the data is challenging in more ways than just being time consuming and expensive. Thus, the final study report is usually the result of an extensive joint effort.

WHAT IS OUR ROLE AS VETERINARY CARDIOLOGY SPECIALISTS IN THIS RESEARCH?

With a few exceptions, it is uncommon for molecular genetic scientists to have veterinary specialists training. In fact, many of the molecular geneticists have no veterinary training at all. Much, if not all, of the research is heavily dependent on correct characterization of the phenotype (correct diagnosis) and this is where veterinary expertise is vital. To correctly diagnose a disease, adequate diagnostic methods must be chosen, diagnostic criteria must be established, and the diagnostic workup must be conducted in an acceptable manner by trained staff. Furthermore, many of the problems in the dog molecular genetic projects have been associated with definition of the control population. Many of the projects have been aimed at studying naturally occurring acquired diseases, and normal findings in a dog at a young age does not guarantee that the dog does not develop disease in the future, potentially leading to inclusion of false negative cases. In fact, adequate definition of the control population requires much background knowledge on the disease to be studied, such as breed predilection, disease prevalence/incidence, and natural history.

WHAT IS GWAS AND WHAT INFORMATION CAN BE OBTAINED?

Until a little more than a decade age, a common method to study an inherited disorder was through genetic linkage studies. However, the results from these studies were shown to poorly replicate for common and complex disorders.[10] The Genome Wide Association Study (GWAS) approach was brought forward as an alternative to genetic linkage studies. This type of study basically answers the question if an allele of a genetic variant is more commonly found in individuals with the phenotype of interest (e.g. the disease being studied).[10] The dog genome consists of approximately 2.4 billion base-pairs[2,3] and the genome between two individuals of the same specie may differ in millions of ways: there may be deletions, insertions, copy number variations, and there may be small variations in the individual nucleotides of the genomes (single nucleotide polymorphisms (SNPs)). In the dog, millions of these SNPs have been identified throughout the genome (currently 2.5 million)[2] and arrays have been developed capable of analyzing thousands of SNPs in one single sample (the most recent SNP array has a resolution of 170 000 SNPs). Each of these SNPs may serve as a marker in the genome, and if presence of disease may be statistically associated to one (or several neighboring SNPs), so-called genome wide significance has been obtained.[10] Although the SNPs used in the SNP arrays are many, they do not cover the entire genome. Therefore, a GWAS study is primarily used to find areas in the genome that are of interest to continue to study in a higher resolution, either by using SNPs located closer to each other in the DNA segment of interest (so-called fine mapping), and/or by sequencing the entire DNA segment.

HOW IS GWAS USED?

The most common approach of GWA studies is the case-control design, which compares two large groups of individuals, one group with disease and one healthy group (control).[10] All individuals in each group are genotyped for the majority of common known SNPs. For each of these SNPs, it is then investigated if the allele frequency is significantly different between the two groups using the chi-squared test (because of the thousands of comparisons in GWAS, the *P*-value is usually corrected for multiple comparisons). Examples from the veterinary cardiology literature of such studies are recent reports of DCM in Doberman Pinschers.[8,11] One of these two studies is included in the LUPA project, but other studies of DCM with a similar study design are also included in the LUPA project and they include Great Danes, Newfoundlands, Boxers, Irish Wolfhounds. The results from these studies have been presented at genetic conferences and are currently in the process of being published. However, there are many different approaches how to design and analyze the data. The GWAS approach may also be used to study continuous variables, such as a biomarker concentration or height.[12] In the LUPA project, there was one large prospective study aimed at identifying genetic factors that impact blood pressure, neuroendocrine variables and metabolic variables in normal dogs.[9] Results from this study are currently being analyzed, and many of the outcome variables are continuous. One other example from the veterinary cardiology literature, where a continuous trait was investigated using GWAS is a recent LUPA project report concerning Cavalier King Charles Spaniels (CKCS) and myxomatous mitral valve disease (MMVD).[7] Age of onset of MMVD in this breed is a continuous variable.[13] In this study CKCS were grouped depending on the age of onset, either into a group of early onset (<4 years with MMVD or <8 years of age with CHF caused by MMVD), or into a group of late (or no) onset of MMVD, defined as either no or mild MMVD at an age >10 years. This study reported of two haplotypes associated with an early onset of MMVD. Because of the design of this study, it did not provide information of genetic variants causing MMVD as such (the disease is common in many other breeds), but it was suggestive of genetic variants that caused the disease to manifest at a younger age (indicating that these variants impacted disease progression).

WHICH ARE THE PITFALLS?

In addition to certain technical problems with handling of samples, extraction of DNA, and microarrays, there are some issues associated with GWAS that should be observed. The GWAS are sensitive to group size. In fact, there are human studies including up to 300 000 people for that reason. The number required to obtain genome significance is considerable lower for studies in pure-bred dogs (which is an important reason why this type of research is conducted in dogs). It has been shown that reasonable power may be achieved for group sizes as small as 20+20 dogs for recessive traits, 50+50 dogs for dominant traits, and 100+100 dogs for complex traits.[2] These estimated sizes of the study populations have been corroborated in published studies in dogs.[4,14,15] However, larger group sizes are often required. There may be many reasons for this: suggested group sizes may result in border-line genome wide significance. In some instances, the cause for this may be that the GWAS results indicate that the trait may be polygenic rather than monogenic (multiple associations in the GWAS may indicate this). Furthermore, more samples are often needed or replicating GWAS results in other study populations. The need for additional samples may be a problem, depending on the disease that is being studied. The less common it is, the more difficult it may be to obtain additional samples. Conversely, the more common it is, the more difficult it may be to obtain reliable controls. Results from the GWAS are never reliable if the included individuals were poorly characterized. Here lies our greatest challenge in this type of research; because at the same time samples from many individuals are required for the GWAS we are responsible for a thorough characterization of included dogs. The GWAS are sensitive for stratification,[16] meaning that dogs from the same breed, but from different sub-populations, that differ genetically are included in the same analysis. The problem is that, if the frequency of disease differs between these two sub-population (for example only cases in one sub-population and only controls in the other), there is an increased risk that the significant GWAS associations were false.[17] The stratification may be geographic, where dogs from the same breed differ between countries, or it may be sub-populations that have been created within one breed. One example of the latter in Europe is the Labrador, where the hunting type differs genetically from the show type. Another problem that may arise is if individuals in one of the two groups are closely related, but not individuals in the other. For this reason, it has been recommended not to include two closely related dogs (siblings, sires, dams and offspring in the same family) in the LUPA projects. To some extent the problem with stratification may be overcome in the statistical analyses,[17] but there may be instances where the stratification is unknown.

SUMMARY

The advances in molecular genetic research in dogs and cats are likely to impact small animal cardiovascular medicine in the future in more ways than genetic tests. Increased knowledge of fundamental pathophysiology of diseases opens new possibilities for targeted treatments and personalized medicine. The GWAS is a powerful tool for associating regions on the genome to disease, but, in addition to methodological issues of the laboratory work, care must be taken to control for systematical errors and confounding factors associated with the included study population. Although GWAS is currently frequently used in genetic research, it is likely only one step on the way to more refined methods for identifying causative genetic factors for disease.

REFERENCES

[1] O'Donnell CJ, et al. **N Engl J Med** 2011;365:2098-2109, [2] Lindblad-Toh K, et al. **Nature** 2005;438:803-819, [3]Karlsson EK, et al. **Nat Rev Genet** 2008;9:713-725, [4]Drogemuller C, et al. **Science** 2008;321:1462, [5]Parker HG, et al. **Science** 2009;325:995-998, [6] Meurs KM, et al. **Hum Genet** 2010;128:315-324, [7]Madsen MB, et al. **J Hered** 2011;102 Suppl 1:S62-67, [8]Mausberg TB, et al. **PLoS One** 2011;6:e20042, [9]Lequarre AS, et al. **Vet J** 2011;189:155-159, [10]Manolio T, et al. **N Engl J Med** 2010;363:166–176, [11]Meurs K, et al. **J Vet Intern Med** 2010;24:693, [12]Strachan T, et al. **Human Molecular Genetics (4 ed.)**: Garland Science, 2010, [13]Haggstrom J, et al. **Vet Rec** 1992;131:549-553, [14]Karlsson EK, et al. **Nat Genet** 2007;39:1321-1328, [15]Drogemuller C, et al. **PLoS Genet** 2009;5:e1000579, [16]Nsengimana J, et al. **Methods Mol Biol** 2012;850:237-262, [17]Sun L, et al. **Genet Epidemiol** 2006;30:519-530.

INTERPRETATION AND USE OF GENETIC TESTING FOR CARDIAC DISEASE

Kathryn M. Meurs, DVM, PhD
Raleigh, North Carolina

INTRODUCTION

At this time, the available genetic tests are for cardiac disease are limited to some feline and canine cardiomyopathies. Genetic testing can be performed by submitting a DNA sample to a reputable screening laboratory. Good quality DNA samples can be obtained either from a blood sample in an EDTA tube or by brushing the inner cheek of the cat or dog with a special buccal swab, although many labs will even accept samples submitted on a cotton swab. In general, buccal swabs work fine as long as careful attention is paid to brushing the inner cheek of the animal. In some cases, particularly with an owner who has not used these swabs before an insufficient amount of DNA will be obtained and the sample will need to be repeated. Therefore, if the owner is in a hurry to get the results it is often more efficient to use a blood sample.

FELINE CARDIOMYOPATHY

Hypertrophic cardiomyopathy is the most common heart disease in the cat. It is known to be inherited in some breeds of cats including the Maine Coon and Ragdoll.1 A separate causative genetic mutation has now been identified in these two breeds and testing is available for these mutations. Both mutations appears to be fairly breed specific.2,3

There is no evidence that these mutations are causative for the disease in other breeds.

Maine Coon

A substitution mutation has been identified in the Myosin Binding Protein C gene.2 It is single base pair change that alters the amino acid produced from alanine to proline. Additionally the alteration of the amino acid is predicted to change the structure of myosin binding protein C in this region from its normal helix shape and alters the ability of the protein to interact with other sarcomeric proteins. The trait is inherited in an autosomal dominant fashion and homozygous (2 copies of the mutation) cats are thought to develop a more severe form of the disease. Heterozygous (one copy of the mutated gene) cats may develop the disease at a later age and possibly a less severe form. Not all cats with the mutation will ever show the disease. Genetic modifiers that affect the expression of the disease are likely but are not yet known.

The presence of the mutation in the Maine Coon population appears to be fairly common with over 30% of the samples submitted from all over the world showing the mutation. Prospective studies are needed to identify other factors that might be important in expression of disease.

Ragdoll

A substitution mutation has been identified in the same gene as the Maine Coon, the Myosin Binding Protein C gene in the Ragdoll.3 However, in the Ragdoll the mutation is in a different position and changes the amino acid produced from arginine to tryptophan. It is extremely unlikely that the Maine Coon and Ragdoll mutations were inherited from a common ancestor since the mutations are located in such different regions of the gene. It is more likely that these mutations developed independently. Although the mode of inheritance of this mutation in the Ragdoll has not been identified, it is most likely an autosomal dominant trait as well. In the Ragdoll, homozygous cats appear to be very severely affected with development of heart failure and thromboembolic episodes often before 2 years of age. Heterozygous cats appear to have a much more mild form of the disease that may include mild papillary muscle hypertrophy.

The presence of the mutation in this breed appears to be slightly less common than that of the Maine Coon, with mutations present in about only 20% of the submissions.

Breeding recommendations

Due to the fairly high prevalence of the mutation in both breeds, it would seem to be unwise to recommend that all cats with the mutation be removed from the breeding programs since this could really alter the gene pool.

Our current recommendations for both breeds are to remove cats that are homozygous for the mutation from the breeding pool since they will certainly pass on the mutation and they have the highest risk of developing the disease. Heterozygous cats should be carefully evaluated. Cats that have many strong positive breed attributes and are disease negative at time of breeding could be bred to a mutation negative cat. The offspring of that mating should be screened, and if possible, a mutation negative kitten with desirable traits selected to replace the mutation positive parent in the breeding pool. Over a few generations this will decrease the prevalence of the disease mutation in the population, hopefully without greatly altering the gene pool. Finally, disease negative but mutation positive cats should be evaluated annually for presence of disease.

CANINE DILATED CARDIOMYOPATHY

Portuguese Water Dog

A juvenile form of familial DCM has been reported in the Portuguese Water Dog. It is thought to be inherited as an autosomal recessive trait that is linked to a region on canine chromosome 8. 4 Affected puppies were from seemingly unaffected parents and typically died between 2 and 32 weeks of age, either from sudden collapse and death without any preceding signs or congestive heart failure. 5 A test for a genetic marker highly associated with this disease is available through PennGen, http://research.vet.upenn.edu/penngen. Since this is a recessive trait, dogs that have two copies of the mutation are most likely to develop the disease. Silent carriers exist since dogs that only have one copy of the trait may pass the gene on, but will not show the trait. The risk of an individual inheriting the mutation from both parents and having clinical evidence of the trait can generally be decreased by out breeding and not breeding to related dogs.

Doberman pinscher

Dilated cardiomyopathy in the Doberman pinscher appears to be inherited as an autosomal dominant trait with incomplete penetrance. 6 Several genes known to cause the human form of the disease have been evaluated in the Doberman pinscher and have not identified a mutation. Recently a genetic deletion has been identified in a gene that helps regulate mitochondrial energy in at least some Doberman pinschers; however this is not likely to be the cause in all cases since a second chromosomal region of interest has been identified in a population of German Doberman pinschers. 6,7 Therefore clinical screening on an annual basis is strongly recommended.

Since this is an autosomal dominant trait, breeding of affected individuals is generally not recommended. If a dog is homozygous it would certainly pass on the trait. If a dog is heterozygous it would have a fifty percent chance of passing on the trait, and a fifty percent chance of not passing on the trait. Therefore, unaffected offspring of affected dogs that have been carefully cleared of the disease annually by echocardiography and Holter should be allowable for breeding since silent carriers should not exist. Additionally, with an autosomal dominant trait all affected dogs should have at least one affected parent. Therefore, it is recommended that all parents of affected dogs be carefully evaluated with echocardiography and Holter monitoring.

Boxer Arrhythmogenic Right Ventricular Cardiomyopathy

Arrhythmogenic right ventricular cardiomyopathy (ARVC) is an inherited disease in the boxer and appears to be an autosomal dominant trait with reduced penetrance.8 A deletion mutation associated with the development of canine ARVC has been identified in the untranslated region of a desmosomal gene (striatin). Homozygous dogs appear to have a more severe form of the disease with a statistically greater number of ventricular premature complexes/24 hours and in some cases, more severe systolic dysfunction than heterozygous dogs. 9 It is not yet known if this is the only causative mutation for canine ARVC since in human beings there are at least 9 known causative genes (genetic heterogeneity) for the disease. Since there are very likely other causes for the disease clinical screening is still recommended.

It is not yet known what percentage of Boxers have the ARVC deletion but it is likely to be high enough that removal of all dogs with the mutation from the breeding pool could have a deleterious effect on the breed. Since this is a disease of incomplete penetrance and variable expressivity not all affected dogs will ever develop clinical signs and many may live a normal lifespan. It is likely that there are multiple factors that may influence which dogs become symptomatic for the disease. We recommend that dogs that are positive heterozygous for the mutation should be annually evaluated for signs of disease (Holter monitor, echocardiogram). Adult dogs without signs of disease and that have other positive breed attributes could be bred to mutation negative dogs, particularly if they do not appear to have a strong family history of disease. These may be dogs that have a reduced penetrance form of the disease. Puppies from this mating may be screened for the mutation and over a few generations, mutation negative puppies may be selected to replace the mutation positive parent and gradually decrease the number of mutation positive dogs in the population. Boxers that are homozygous for the mutation should probably not be used for breeding unless exceptional circumstances occur and in these cases they should only be bred to negative dogs. They have more significant disease and since this is an autosomal dominant trait they will certainly pass on the mutation.

Since there is likely more than one cause of the disease, further evaluations should be performed for additional causes of this disease.

REFERENCES

1. Kittleson MD. Textbook of Veterinary Internal Medicine. 2005; 1082. 2. Meurs KM, et al. Human Molecular Genetics 2005;14:3587 3. Meurs KM et al.Genomics, 90, 2007,261-264. 4. Werner P, et al. Genomics 2008;91:517-21. 5. Dambach DM, et al: J Vet Intern Med 1999;13:65-71. 6. Meurs KM, et al. *J Vet Intern Med* 2010;23:1023 7. Mausberg TB et al. PLOS 2011;6:Epub 8. Meurs KM, et al. J Vet Intern Med 1999;13:437-9. 9. Meurs KM et al. *Human Genetics* 128, 2010, 315-324.

VETERINARY CARDIOLOGY HISTORY

James W. Buchanan, DVM, MMedSci, DACVIM
Philadelphia PA

Dr David Detweiler (1920-2009) is generally considered the Father of Veterinary Cardiology which is a well deserved accolade for his role in the development of clinical and comparative cardiology over a period of 50 years. However, researching history on this topic revealed fascinating reports of cardiovascular studies in animals in Europe long before Detweiler devoted himself to the field in 1948.[1] Highlights include Willem Harvey's treatise in 1628 describing cardiovascular studies in dogs and other animals which proved that blood circulated instead of going to and fro in arteries and veins. Stephen Hales reported measurements of blood pressure, stroke volume and cardiac output in horses, dogs and other animals in 1733. Rene Laennec, inventor of the stethoscope reported its usefulness in animals in his 1819 treatise on mediate auscultation. In 1831, James Hope clarified the etiology of heart sounds and murmurs in donkeys by applying his stethoscope directly on the heart while palpating the heart in open chest experiments.

J.B, Auguste Chauveau, a French veterinarian reported pulse velocity in horses and determined heart sound origin and timing by open chest experiments in1855. He and Etienne Marey, a physician collaborator became world famous for performing the first hemodynamic cardiac catheterizations and proving the etiology of the apex beat in 1861. Augustus D Waller introduced the term electrocardiogram and performed the first ECG on his own dog and a man in 1886. Franz Hutyra and Josef Marek wrote a widely used Hungarian textbook on the pathology, diagnosis and treatment of domestic animals in 1895. The expanded 1905 third edition included a chapter on circulatory diseases with examples of percussion to determine heart size and pleural effusion, correct descriptions of heart valve areas, and auscultatory features of valve stenosis and insufficiency. Thomas Lewis began experimental studies in dogs in 1908 using a string galvanometer obtained from Willem Einthoven and visually demonstrated the nature of atrial fibrillation in open chest experiments in horses. He then reported 3 horses with the arrhythmia in 1911. Heinrich Jakob was a professor of internal medicine at the University of Utrecht in Holland. In 1913 he wrote a textbook with 331 illustrations emphasizing physical examination findings in dogs. Included in the book was a 30 page chapter on the diagnosis and therapy of cardiovascular diseases in dogs in which he described and illustrated most of the features detectable by inspection, palpation, percussion, and auscultation of four valve areas.

Louis Desliens, a French veterinarian was the first to employ cardiac catheterization in the study of diseases in horses and cattle and developed a carotid artery injection method and a "blood throw" recording technique. He developed a direct blood pressure gauge to measure venous and arterial pressure and called his approach "Hémodynamométrie". In 1935 he published a 553 page book describing his methods and hemodynamic changes in arrhythmias and valve disease as well as studies of the circulatory effects of arterial and venous injections of cardioactive drugs and therapeutic agents.

Detweiler became aware of these studies during a sabbatical year in Switzerland in 1956. After returning to America he developed a screening protocol to determine the types and prevalence of heart disease in dogs and obtained a million dollar grant to establish the Comparative Cardiovascular Studies Unit in 1960.With the assistance of Donald Patterson and others they examined all dogs presented to the veterinary hospital at the University of Pennsylvania and found reliable evidence of heart disease in 11% of them. Initial productivity of this unit included clinical studies, improved diagnostic methods, epidemiology, cardiac pathology, surgery, electrophysiology, and hemodynamics. The results of some of these studies were summarized in a NY Academy of Science monograph on Comparative Cardiology in 1965.[2] The CCSU received continued funding for 17 years and yielded many research reports as well as training for many veterinary cardiologists. Discovery of a higher prevalence of congenital heart disease in certain breeds led to breeding experiments characterizing diagnostic and pathologic features and the genetics of congenital heart disease which Patterson expanded into a new discipline of comparative medical genetics. Chief among the heart studies were tetralogy of Fallot in Keeshunds, patent ductus arteriosus in a mixed poodle strain of dogs, subaortic stenosis in Newfoundland dogs, pulmonic stenosis in Beagles, and persistent right aortic arch in German Shepherd dogs.[3]

Other centers of veterinary cardiology research that began in 1960 include the Biology of the Heart program at the Ohio State University led by Robert Hamlin and C Roger Smith and the Mary Caspary Research Institute at the Animal Medical Center in New York under the direction of Robert Tashjian. Equine cardiology was the focus of research by James Holmes and his graduate students beginning in 1960 at the University of Bristol in Great Britain. Many authors of subsequent veterinary cardiology research papers and chapters in textbooks from 1960-70 trained at one of these programs. The first full textbook devoted solely to cardiology in animals was published in 1970.[4] Since then hundreds of outstanding articles about cardiovascular diseases, diagnostic methods, medical and surgical treatment in animals have been published and there are over 50 other textbooks on large and small animal cardiology.

Veterinary cardiology organizations were formed in the 1960s and 1970s to advance education, research and specialty certification. These include the Academy of Veterinary Cardiology (AVC) in 1968, the American College of Veterinary Internal Medicine (ACVIM) in 1972 and the American Heartworm Society (AHS) in 1974. The Veterinary Cardiovascular Society in England (VCS) was formed in 1975 followed by the European Society of Veterinary Cardiology

(ESVC) in1981. The ESVC and other specialties established the European School for Advanced Veterinary Studies (ESAVS) in 1991 to provide training and certification as specialists in various organ systems. In 1994 the European College of Veterinary Internal Medicine –Companion Animals (ECVIM-CA) was created to promote higher standards of cardiology training, research, communication and examination in order to award diplomate credentials as a veterinary cardiologist. The first ACVIM cardiology certifying examination was given in 1974. The Journal of Veterinary Cardiology was started in 1998 as the official journal of the European Society of Veterinary Cardiology and has become a premier journal for publication of veterinary cardiology papers along with the Journal of Veterinary Internal Medicine.

The knowledge base and diagnostic methods for heart disease in the 1960s were rather simple compared to current standards. Diagnoses were made by skillful physical examinations, radiography and electrocardiography. Training a resident usually took about 6 months and then they got practical experience in the next 6 months. Later, cardiac catheterization and angiography became important if the equipment was available. This was followed by computed tomography and later by contrast CT angiography which yielded definitive information about thoracic masses and vascular ring abnormalities. Body surface electrocardiography grew to include transmitters, 24 hour Holter monitors and event recorders over several days with interpretation aided by automatic playback equipment. Transthoracic and later transvenous cardiac pacemaker implantation became a standard procedure with sophisticated programs, monitors and programmers. Intracardiac electrocardiography for bundle of His recording and endocardial mapping yielded site specific diagnoses and enabled therapeutic ablation of arrhythmogenic foci, dual chamber pacing and improved synchrony of contraction. Echocardiography mostly replaced angiocardiography to diagnose morphologic abnormalities and is unsurpassed in assessing cardiac morphology, function, flow patterns and obstruction to flow. Transcatheter therapy using balloons for dilation of stenotic valves and occluding devices for intracardiac and vascular shunts provided ways to treat common cardiac diseases without reliance on surgery. Digitalis and diuretics that were mainstays of treatment for congestive heart failure were replaced by vasodilators, various inotropic agents and better diuretics. Drugs for heartworm prevention and treatment proved effective in controlling this disease. Observation led to the major discovery of taurine deficiency as the cause of dilated cardiomyopathy in cats. Biomarkers and other laboratory tests to identify and quantify disease agents and their effects have opened the door to improved diagnosis and treatment. Advances in genetics provided insights into hereditary diseases and show promise of eliminating some diseases by detection and elimination of affected animals as well as the promise of gene therapy to treat genetic diseases. All of the above advances over the past 50 years make the training of a cardiologist in 6 months only a distant memory.

REFERENCES

[1]Buchanan JW, **J Vet Card** 2012; in press. [2]Detweiler DK and Patterson DF. **Ann NYAcad Sci** 1965;127:471. [3]Patterson DF, **Circ Res** 1968;23:171. [4]Ettinger SJ and Suter PF. *Canine Cardiology* WB Saunders, 1970

THERAPEUTIC GENE DELIVERY: HOW CLOSE IS IT TO PRIME TIME?

Meg M. Sleeper, VMD, DACVIM (Cardiology)
Philadelphia, PA

INTRODUCTION

Therapeutic gene transfer is one molecular based option for heart disease patients. Gene therapy has traditionally been used to transfer a gene that encodes a functional protein into a diseased patient, resulting in long-term expression of the protein that was deficient (1,2). This strategy is often referred to as "gene replacement therapy", and it requires that the mutated gene be previously identified. This approach has been used effectively to treat canine hemophilia, lysosomal storage diseases, and inherited retinal diseases (3-9). However, gene transfer can also be performed when the causative mutation is unknown or in acquired diseases, with the goal of increasing the concentration of a therapeutic gene product in a tissue or organ. When used in this manner, gene transfer results in a "drug effect," and multiple therapeutic gene products can be considered.

When naked DNA is injected directly into cells, it is largely degraded. Therefore, most gene transfer techniques package the genetic material so that it can be more efficiently introduced into the cells of interest. Most commonly, viruses are used for this packaging purpose. Viruses bind to their host cells and introduce their genetic material (and directions for producing more copies of the virus) into the host cell as part of their replication process. Therapeutic gene transfer using viral vectors for gene delivery re-engineers the virus to replace the viral disease causing genes with the gene of interest (the therapeutic gene). Multiple viruses have been used for this packaging purpose including retroviruses, adenoviruses, and adeno-associated viruses. All viral vectors have positive and negative aspects to their use in gene transfer. Adeno-associated viruses (AAV) are small viruses from the parvovirus family, The engineered vector (recombinant AAV or rAAV) does not insert the viral gene into the host genome and the virus is non-pathogenic so an inflammatory response should not occur after the therapy which allows long term production of the gene product. However, because of the small viral size, AAV can only carry small therapeutic transgenes.

We have developed a system whereby intramyocardial injections (40-60) of rAAV are delivered throughout the left ventricle using a cardiac injection catheter and a carotid artery approach. We believe this technique will be better tolerated by patients with heart disease than direct injections via thoracotomy or other more invasive techniques such as the procedure described above (10,11).

GENE THERAPY APPROACHES FOR MYOCARDIAL FAILURE

There is substantial evidence that Ca^{2+} handling in the failing heart is impaired, and that abnormalities of calcium cycling represent a final common pathway in the pathogenesis of heart disease and failure (12-14). Furthermore, the reduced rate at which cytosolic Ca^{2+} is returned to the sarcoplasmic reticulum results in impaired myocardial relaxation and a decrease in the amount of Ca^{2+} released via the ryanodine receptor. As our understanding of the molecular Ca^{2+} handling pathways has improved, various target proteins for gene-transfer based therapeutics have become possible. The intracellular calcium gradient is partially maintained by the cardiac sarcoplasmic reticulum Ca^{2+} ATPase (SERCA2a), an energy-dependent molecular pump that transports Ca^{2+} from the cytosol across the membrane of the sarcoplasmic reticulum. The activity of SERCA2a is modulated by several proteins including phospholamban (15). By altering expression of the proteins which move calcium between the cytosol and the sarcoplasmic reticulum, Ca^{2+} handling can be normalized in diseased myocardial cells resulting in improved cardiac function regardless of the underlying disease. Results from a phase 2 clinical trial were published in 2011 demonstrating safety and suggesting some clinical improvement in a population of humans with advanced heart failure whom were treated percutaneously with an rAAV vector expressing SERCA2a compared to those treated with placebo (16). Another group has used an alternate transgene (s100A1) to normalize myocardial Ca^{2+} pathways, and has demonstrated improved heart function in a preclinical large animal model and improved function in human failing cardiomyocytes (17,18). We hope to commence a placebo controlled, double blinded clinical trial in dogs with dilated cardiomyopathy using a gene therapy approach in late 2012 or early 2013.

GENE THERAPY APPROACHES FOR X-LINKED MUSCULAR DYSTROPHY

Cardiac

X-linked muscular dystrophy is an inherited disease caused by loss of dystrophin in striated muscle. It occurs in 1 out of every 3500 male births and occurs intermittently in various mammal species including the Golden Retriever dog. Restoration of dystrophin expression via exon skipping in both cardiac and skeletal muscle has been successfully demonstrated in the murine model of X-linked muscular dystrophy, but restoration of cardiac dystrophin expression in large animal models has proven to be a challenge. We sought to optimize restoration of cardiac dystrophin expression in the Golden Retriever muscular dystrophy (GRMD) model using transendocardial delivery of an rAAV vector to deliver a modified U7 snRNA carrying antisense sequence to target the exon splicing enhancers of exons 6 and 8. The goal of this technique is to correct the disrupted reading frame. We demonstrated efficient restoration of cardiac dystrophin expression at 13 months confirmed by RT-PCR and immunoblot as well as membrane localization by immunohistochemistry. We also reported restoration of membrane-localized nNOS and accompaning improved cardiac function as assessed by cardiac MRI (19).

Systemic

The above cardiac approach does not address the skeletal muscle disease in these animals or children and a novel strategy that has therapeutic potential for DMD is inhibition of myostatin, a negative regulator of skeletal muscle mass that may also promote fibrosis. Systemic myostatin inhibition (using a dominant negative myostatin transgene) in the GRMD was attempted with an intravenous administration of adeno-associated virus vector encoding a dominant negative myostatin transgene. Treated dogs were compared with age-matched, untreated GRMD controls. Dogs were followed with serial magnetic resonance imaging (MRI) for 13 months to assess cross-sectional area and volume of skeletal muscle, then euthanized so that tissue could be harvested for morphological and histological analysis. We found that systemic myostatin inhibition resulted in increased muscle mass in GRMD dogs as assessed by MRI and confirmed at tissue harvest. We also found that hypertrophy of type IIA fibers was largely responsible for the increased muscle mass and that reductions in serum creatine kinase and muscle fibrosis were associated with long-term myostatin inhibition in GRMD (20).

CONCLUSION

Therapeutic gene transfer is a technique which has been proven to be effective for treating various inherited canine diseases. Small animal and human studies suggest it is likely to also be useful for canine heart failure. As the costs associated with vector production have come down, the barrier to beginning a canine clinical trial to assess its efficacy in dogs with naturally occurring disease is also becoming more feasible.

REFERENCES

1 Lyon AR, et al. **Heart** 2008; 94:89-99. 2 Vinge LE, et al. **Circ Res** 2008; 102:1458-70. 3 Haskins M. **ILAR J**. 2009;50:112-21. 4 Tessitore A, et al. **Mol Ther** 2008; 16 :30-7. 5 Margaritis P, et al. **Blood** 2009; 113:3682-9. 6 Niemeyer GP, et al. **Blood** 2008; 113:797-806. 7 Acland GM, et al. **Mol Ther** 2005; 12:1072-82. 8 Manno CS, et al. **Nat Med** 2006; 12(3):342-7. 9 Ponder KP, et al. **Proc Natl Acad Sci USA** 2002; 99(20): 13102-7. 10 Vera Janavel G, et al. **Gene Ther** 2006; 13:1133-1142. 11 Li JJ, et al. **Cardiovasc Res** 1995; 30:97-105. 12 Manning BS, et al. **Circulation** 2000; 102(22):2751-7. 13 Most P, et al. **J Clin Invest** 2004; 114(11):1550-63. 14 Pleger ST, et al. **Circulation** 2007;115(19):2506-15. 15 Mattiazzi A, et al. **Cardiovasc Res** 2005; 68(3):366-75. 16 Jessup M, et al. Circulation 2011; 124:304-313. 17 Pleger ST et al. **Sci Transl Med** 2011;3:92ra64. 18 Brinks H et al. **J Am Coll Cardiol.** 2011 58;966-73. 19 Bish LT et al. **Mol Ther** advance online publication 6 December 2011. 20 Bish LT, et al. **Hum Gene Ther** 2011;22:1499-509.

ECHOCARDIOGRAPHIC ASSESSMENT OF VENTRICULAR SYSTOLIC FUNCTION

John D. Bonagura, DVM, DACVIM
Columbus, OH

INTRODUCTION

Quantification of left ventricular (LV) systolic function is one of the most fundamental components of the echocardiographic examination. Deterioration in systolic performance carries prognostic implications and leads to therapeutic recommendations. LV systolic dysfunction is most often observed with idiopathic dilated cardiomyopathy in dogs and cats, but can also develop from myocarditis, drug toxicosis, taurine deficiency, relentless tachycardia, or as a "cardiomyopathy" of chronic volume or pressure overload. Conversely, heart failure in the setting of normal systolic function suggests cardiac disease dominated by diastolic dysfunction, such as hypertrophic cardiomyopathy, or severe volume overload with preserved systolic function, as might be encountered in chronic mitral regurgitation (MR) of small-breed dogs or in a compensated left-to-right shunt. The prognosis and treatment of these disorders can be quite different.

The ventricles can be assessed globally – as with the determination of LV ejection fraction (EF) and systolic time intervals – or regionally with myocardial tissue velocities and myocardial deformation (strain). The LV also wrings and untwists during the cardiac cycle: these rotations can be captured by advanced echocardiographic methods.[1-18] Both the LV and the right ventricle (RV) can be evaluated with an array M-mode, 2D, 3D, Doppler-based, and tissue-focused examinations.

This lecture offers an overview of the echocardiographic methods used to assess left ventricular systolic function, considering both standard and investigational forms of ultrasound imaging. Some methods for assessing the RV are also considered. Some of the current issues and practical difficulties encountered when assessing ventricular systolic function will be highlighted during the presentation. The role of 3D (4D) imaging for assessment of LV function is addressed by others at this meeting. The difficult problem of assessing the LV in the setting of primary mitral regurgitation is addressed by the author in a separate presentation at this conference.

OVERVIEW & COMMENTARY

Modeling of the LV is generally more straightforward when compared to the complicated anatomy of the right ventricle. Most clinicians quantify LV systolic function using the underlying assumption of the ventricle as a prolate ellipsoid with a one major and two equally-sized minor dimensions. Indeed this model is applied to most of the common calculations used in clinical practice. These include calculation of shortening fraction (across one minor axis between the ventricular septum and LV free wall); fractional area shortening (using multiple cross-sectional diameters at the high papillary muscle level); and methods that calculate volumes from long axis images of the LV using Simpson's rule or an area-length method with a single or bi-plane approach. Since the heart functions as a pump, it is also possible to assess the timing of systolic phases and to estimate the actual volume ejected by the ventricles during an average beat.

However, these methods may be insensitive to mild global dysfunction or to regional abnormalities of ventricular function. Detecting mild global dysfunction is one of our greatest difficulties; whereas, most experienced examiners can "eyeball" moderate to severe ventricular dysfunction. This lack of sensitivity has fostered the development of tissue Doppler based methods, 2D myocardial speckle strain imaging, and vector imaging for the assessment of regional and segmental function. These advanced methods evaluate a small portion of the LV (or RV) and may consider longitudinal, radial, circumferential, or rotational movements or deformation as well as subendocardial and subepicardial differences in contraction. The challenges in employing these new techniques include a lack of standardization in veterinary patients of varying size and heart rate; the need to use "black-box" algorithms developed for human patients; the lack of equally sensitive gold standards (e.g., invasive or MRI-based) needed for validation; and the near absence of longitudinal studies of spontaneous disease that prove these methods are actually superior to conventional Echo variables.

Echocardiographic studies are affected by biological variation, disease, drug therapy, and technical factors; these can significantly influence the results of an evaluation and the ultimate assessment of ventricular function. Some pertinent technical issues include obtaining consistent images; employing repeatable measurement techniques; addressing the timing of systolic and diastolic measurements within the cardiac cycle; obtaining optimal endocardial border recognition; appreciating the potential for translational or rotational movements across the imaging plane; and understanding the issue of tethered regional motion (as opposed to true contractile activity or deformation).

The influence of cardiac rhythm on function cannot be underestimated. Abnormal heart rhythms (premature complexes and post-extrasystolic potentiation), varying cycle lengths, conduction disturbances (bundle branch block), and ventricular dyssynchrony can affect systolic function, sometimes underestimating actual performance. Protracted tachyarrhythmias can create transient LV systolic dysfunction that may persist for some time, even with reestablishment of sinus rhythm.

Changes in ventricular preload and afterload also influence estimates of ventricular function, which may seemingly improve (or deteriorate) as a consequence of altered loading. Drug intervention may also affect heart function, by directly influencing contractility, altering heart rate, or changing loading conditions.

Inter-observer issues and day-to-day repeatability are still important concerns for many Echo measurements. Considering that day to day variation of a single examiner often approaches 10% (or more) for many Echo and Doppler variables, it may be very difficult to identify subtle changes in ventricular function within a single patient.

Additional problems occur when assessing systolic ventricular function. Methods that are touted during a laboratory study of ventricular function may be of little use when applied to a clinical population where the population variance is much greater and the important outcomes are clinically relevant, not statistically-significant, differences in cardiac function. For example, ambiguous findings of LV function are identified very often in the practice setting. These may be dismissed as insignificant when there is no clinical context to explain the abnormality, or supplemented with additional functional estimates and planned follow up examinations at suitable intervals (which is usually the most practical approach). It is quite common to perform 8 or 10 different estimates of LV function in one dog only to find that half of these variables point in one direction and the other half the opposite way. Owing to species and breed differences, as well as different methods used by examiners when measuring cardiac function, it can be difficult to apply rigid benchmarks and cutoffs for "normal" systolic function. These problems are especially common in "screening clinics", where subtleties are sought and are too often found. Decisions in these situations are very often based on multiple factors that extend beyond the echocardiographic findings (and this is probably the most prudent approach). However, when dealing with patients (as opposed to breeding stock), the importance of an "abnormal" designation cannot be underestimated, because a well-defined decrease in systolic function will prompt additional investigations, more frequent follow-up examinations, and life-long therapy. Of importance is the need for specialty working groups to standardize the interpretation of echocardiographic findings that define likelihoods for having or developing systolic dysfunction.

SPECIFIC FUNCTION INDICES

Most commonly used M-mode, 2D, and real-time 3D estimates of LV systolic function provide an estimate of LV ejection fraction (EF) and are therefore considered ejection phase indices. These are grounded in linear, area, or volumetric estimates of EF or stroke volume. Some examiners prefer to estimate global LV function based on a single diameter end-systolic diameter (or calculated volume) index, but the author believes this index is too influenced by chamber remodeling and extrapolated cut-offs for normal volumes from human studies to be definitive (see *Echocardiographic Assessment of Mitral Regurgitation*, elsewhere in these Proceedings). Global function also can also be quantified with time intervals, such as pre-ejection period and ejection time, as well as hybrid estimates of global function like the Tei index of myocardial performance. These are calculated from M-mode, spectral Doppler, or tissue Doppler modalities. However, there are considerable overlaps between normals and affected dogs and a single point calculation can be ambiguous or misleading. In particular, the value of systolic time indices in valvular heart disease is very limited and in the author's experience, can be unduly influenced by the duration of the QRS complex and the heart rate even in dogs with suspected DCM. Additional Doppler approaches include quantifying stroke volume and cardiac output by measurement of velocity time integrals. This approach is of greatest utility when tracking trends in the intensive care setting or when applied to monitor anesthesia. Stroke volume estimates are typically insensitive for detecting mild LV or RV dysfunction.

Regional estimates of LV function are gaining more attention and include measuring tissue velocities or deformation with tissue Doppler imaging (TDI); calculating myocardial velocity gradients using color Doppler imaging (and proprietary analysis algorithms); and calculating cardiac deformation of cardiac tissue, or strain, using 2D speckle-tracking technologies. These can be assessed regionally or combined automatically (or manually averaged) to obtain a "global" index of longitudinal, radial, or circumferential strain/strain rate. This same technology can be used for quantitation of cardiac rotation and to calculate ventricular torsion. Frequently tissue or regional assessments combine both systolic and diastolic function indices.

M-mode Estimates

One of the oldest methods for assessing LV systolic function is still among the most popular, namely, the linear estimate of shortening fraction (SF), typically measured at the high papillary muscle or chordal level. This is a minor axis approximation of global function. When the SF is recorded within a healthy reference range for species and breed, LV systolic function is generally considered "normal" and most examiners will be satisfied with the result. However, there are situations when global or regional myocardial function may be quite abnormal in the setting of normal SF. Furthermore, reference values for healthy dogs are not straightforward, and there is no single cut-off value that has been proven to have optimal sensitivity/specificity for predicting advancing heart failure across canine breeds. Many healthy dogs have SF values between 20-25% (or lower), a factor that may be related to relatively greater shortening in other planes, especially in the apical-basilar axis. In breeds that are predisposed to dilated cardiomyopathy, the problem becomes still more difficult when the septum and free wall contract dyssynchronously due to regional wall abnormalities or conduction issues. Even in dogs with dilated cardiomyopathy (DCM), marked dyssynchrony of the LV can produce a very low SF, while other estimates such as shortening area might show only mild to moderate LV dysfunction. Other potential problems with SF (as well as with shortening area and EF) is the masking of myocardial failure by increased preload or decreased afterload, a finding commonly observed in smaller breed dogs with severe mitral regurgitation (see elsewhere in these Proceedings). Similarly, dogs with severe aortic stenosis and concentric hypertrophy may have normal values for SF even when myocardial failure develops after the "supernormal" compensatory phase of hypertrophy. This phenomenon has been observed in both spontaneous and laboratory-induced cases of canine LV outflow tract obstruction. Thus, while the SF is efficient, the clinician must appreciate the limitations, particularly in healthy dogs. The M-mode situation in cats seems less confusing at least in terms of global function estimates. Most normal cats exceed a SF of 35%, though some sedatives can markedly depress these values. Horses are typically >28%, but this can be affected by stress or sedation.

Other reported M-mode indices include percentage wall thickening, amplitude of wall excursion, E-point to septal separation (EPSS), mitral annular motion, and calculation of EF. The normal range for EPSS varies with breed and size, but this value increases in LV failure (in the absence of severe mitral regurgitation). False positives can arise in the setting of aortic regurgitation or mitral stenosis. Mitral annular motion tracks longitudinal contraction and is recorded by an M-mode study or tissue-tracking techniques. This is one of the few M-mode methods considering apical-basilar shortening, but it is not widely measured. Most reports automatically use M-mode study data to generate volume-estimates by diameter-cubed or Teicholz formulae; these numbers are used to calculate EF and end-systolic volume (index). There are inherent limitations to these minor axis methods, and the volumetric results may simply multiply the linear errors. The end-systolic volume index is a theoretically sound approach to identifying LV dysfunction. However, in the author's opinion, there are too many limitations of this approach as used currently in veterinary medicine, specifically: the calculations for estimating LV volume; reliance on an M-mode estimates of LV volume; ventricular remodeling that increases LV systolic dimension, regardless of functional changes; the reference standards used for any given breed; and the non-validated 30-60-90-120 ml per m2 “benchmarks” used to classify grades of dysfunction (that have been directly extrapolated from human angiographic studies).

Subjectively watching the systolic excursion of the LV free wall and the systolic basilar to apical movement of the mitral annulus can provide insight regarding LV systolic function and are especially useful when septal motion is flat or dyssynchronous.

Two-dimensional Estimates

The 2D echocardiogram can be used to estimate global and regional myocardial function. One of the more practical area-based measurements of systolic function is the fractional area shortening of the LV. This measurement is derived from the 2DE short axis image of the left ventricle, and is calculated (like SF) as the percentage reduction in left ventricular cross-sectional area between diastole and systole (with areas traced across the bases of the papillary muscles). It has advantages over SF in that it represents wall motion across the entire short axis area. Therefore, it is less sensitive to asynchronous wall motion. However, it does not address shortening in the apical-basilar direction, and off-angle images may create inconsistent results. Values between 50 to 72% are normal in dogs in our lab, with values of 43 to 49% representing a more ambiguous "borderline" contraction across the minor axis planes.

The gold standard for estimation of global LV function in human laboratories is calculation of LV EF using a bi-plane 2D (or 3-D reconstruction) method. Many different formulae have been used for calculation of LV volumes, but calculations based on a modified Simpson's rule seem least affected by abnormal chamber geometry. Unfortunately, left apical tomograms can yield truncated views of the LV in dogs and cats. Right parasternal long-axis views provide better length, and have been applied for single plane, EF estimation in dogs using a modified Simpson's rule or length-area method of volume estimation. Normal values in dogs for a single plane EF are typically >45 to 48%. A major concern relates the capture of true apical to basilar motion as opposed to translational movement leading to foreshortening of LV length.

Regional and global estimates of LV function also can be obtained with 2D echo, using "speckle" imaging modalities. Unique signatures of 2D images are used to calculate regional myocardial deformation or strain. This approach may have some benefits--including less angle dependency--when compared to tissue Doppler methods of strain estimation.

Intervals of Systole

Measurements of isovolumetric periods, pre-ejection intervals, and ejection times have been used for assessment of systolic or global ventricular function. The pre-ejection period and the ejection period can be recorded using either M-mode or spectral Doppler methods. Whilst systolic time intervals are influenced by loading conditions, they provide an alternative means of assessing global systolic function, provided the QRS duration is not prolonged or heart rate extreme. The ratio of PEP/LVET may be less heart rate dependent, and as myocardial dysfunction deteriorates the PEP to LVET ratio increases.

Normal values from our laboratory in healthy dogs of varying sizes measured by PW Doppler have identified values of LV PEP/ET between 0.24-0.42, with small dogs generally <0.3 and large dogs generally < 0.42. For the RV PEP/ET values of 0.23 to 0.34 have been reported. Other laboratories have reported different values, demonstrating again the importance of standardizing methods or better defining reference populations. Another problem is a prolonged QRS duration. This is a major limitation in applying systolic time intervals for assessment of LV function.
A less often-used function variable is the mean velocity of circumferential fiber shortening. This is an extension of SF, calculated by dividing the shortening fraction by the LV ejection time, and subject to many of the same limitations as SF. Finally, a combined index of ventricular systolic and diastolic myocardial performance has been reported by Tei using PW-Doppler {IMP = (ICT + IRT) /ET; where ICT and IRT are isovolumetric contraction and relaxation times and ET is the ejection time}. In one study of Newfoundland dogs, IMP was 0.28 +/- 0.12 in the normal dogs, and higher in dogs with DCM. We have found generally greater values for IMP in healthy dogs of different breeds, with a wide range of 0.32 to 0.50 for the left ventricle and with lower values for the RV (in every dog).

Doppler Methods for Assessment of LV Function

There are additional Doppler-derived variables used to quantify global and regional ventricular function. For example, Doppler estimates of ventricular stroke volume or cardiac output represent indices of global ventricular function. These calculations are based on the area under the aortic or pulmonic flow signal--the velocity-time integral or stroke distance--which correlate to stroke volume. The actual stroke volume (in cc) can be calculated by multiplying the stroke distance (VTI) by the calculated cross-sectional area of the aorta or pulmonary artery. Measurements of stroke volume index and cardiac index have been related favorably to invasive indicator dilution techniques.

Doppler correlations also have been demonstrated between traditional invasive indices of left ventricular function, such as the rate of LV pressure rise (LV dP/dt). LV dp/dt can be estimated non-invasively from a mitral regurgitation jet by recording the time interval in seconds between one and 3 meters per second, and dividing the value into 32 mm Hg. This is an "averaged" dp/dt over that period and is generally lower than peak dp/dt measured with a high-fidelity manometer catheter.

Peak aortic velocity has been found to be abnormal in dogs with dilated cardiomyopathy, but generally, myocardial dysfunction must be very severe for peak flow velocity to be affected. Instantaneous peak aortic acceleration (dv/dt max) relates to inotropy, but is difficult to calculate (as only averaged acceleration is typically measured using conventional Doppler systems).
TDI records velocities of myocardial or annular segments in longitudinal (apical-basilar) or radial (short axis) planes. A variety of proprietary methods are used to generate these data including color M-mode imaging, color 2D imaging with velocity signals averaged from regions of interest, and single pulse wave Doppler methods. The values obtained from a specified segment are not always equivalent. The same principles apply for TDI as for conventional Doppler, including the importance of sample volume placement and alignment with movement. TDI or TDI velocity gradients have been used to assess LV systolic function by comparing epicardium to endocardium. These techniques are quite sensitive, and when standardized in an echo laboratory, can be used to identify early signs of ventricular dysfunction. Apical to basilar TDI velocities in healthy dogs are typically between 12 to 20 cm/sec near the annulus. At least for the LV, a decreasing peak tissue velocity is evident as the sample is moved from the base to the apex. Unfortunately, the author has been repeatedly disappointed by longitudinal TDI velocities, which are often relatively normal (12 to 20 cm/s) even in the setting of reduced EF.

Myocardial Strain & Twist

Calculation of regional myocardial deformation – strain and strain rate – is a recent application of TDI (as well as 2D speckle imaging). Longitudinal strain is calculated as a negative value because segments shorten during systole, while radial strain is positive (as the wall thickens). Circumferential strain and segmental rotation and ventricular twist are also available from a number of vendors. Strain by this method is purported to overcome the TDI limitation of tethering. Values for strain obtained by TDI and speckle technology are similar, but standardization of both the methods and reference values is needed before widespread veterinary application can be recommended.

Rotation of the heart at the base and apex can be done to allow estimation of ventricular twisting or torsion, yet another index of LV function. It should be noted that the values obtained are related to non-simultaneous peak twisting.

REFERENCES

[1]Sagawa K, Circ 1981; 63:1223. [2]Ross J. J Am Coll Cardiol 1985; 5:811. [3]Kittleson MD, et al. J Am Vet Med Assoc 1984;184. [4]Borgarelli M, et al. J Vet Intern Med 2007;21:61. [5]Bonagura JD et al. J Small Anim Pract 2009;50:12. [6]Spinale FG, et al. J Thorac Surg 1993;106:1147. [7]Urabe Y, et al. Circ Res 1992;70:131. [8]Edvardsen T, et al. Prog Cardiovasc Dis 2006;49:207. [9]Mazur W, et al. Curr Opin Cardiol 2001;16:246. [10]Young AA, et al. Am J Physiol 1996;271:H2689. [11]Takano H, et al. Am J Vet Res 2010;71:421. [12]Chetboul V, et al. J Vet Intern Med 2008;22:342. [13]Griffiths LG, et al. J Vet Cardiol 2011;13:115. [14]Culwell NM, et al. Am J Vet Res 2011;72:650. [15]Chetboul V, et al. J Vet Intern Med 2007;21:719. [16]Tidholm A, et al. J Vet Intern Med 2009;23:1197. [17]Smith DN, et al. J Vet Cardiol 2012, doi:10.1016/j.jvc.2011.11.002. [18]Chandra S, et al. Heart Failure Rev 2011;16:327.

FELINE HYPERTROPHIC CARDIOMYOPATHY: SCREENS AND GENES

Mark D. Kittleson, DVM, PhD, DACVIM (Cardiology)
Davis, CA

SARCOMERIC GENE MUTATIONS IN HUMANS WITH HCM

Hypertrophic cardiomyopathy (HCM) is a primary myocardial disease characterized by myocardial wall thickening (almost always confined to the left ventricle). In humans HCM is a genetic disease and is usually inherited in an autosomal dominant pattern. The first mutation (in the β-myosin heavy chain gene [β-MHC]) responsible for HCM in humans was identified in 1989 and since then over 600 mutations in 10 genes that encode for sarcomeric proteins have been identified in human families with HCM.[1-5] They include the β-myosin heavy chain, α-tropomyosin, cardiac troponins I, C, and T, myosin binding protein C, essential and regulatory light chains, titin, and actin genes.[1,6-10] The genes with the most mutations described and the ones that most commonly produce disease are the β-MHC and cardiac myosin binding protein C (MYBPC3) genes, which account for roughly 60% of the mutations identified and for 60% of HCM cases. It is now known that sarcomeric gene mutations actually cause HCM since several mutations that have been identified in human families with HCM have been placed in transgenic mice and the disease reproduced (at least partially), thus fulfilling Koch's postulates.[11-13]

A minority of HCM cases in humans are caused by mutations in non sarcomeric genes. These include the gene that codes for LIM protein and the gene coding for AMP-activated kinase. The latter produces glycogen accumulation in the myocardium that mimics HCM.[14,15] Interestingly mutations in β-myosin heavy chain, α-tropomyosin, titin, and troponin T genes also cause dilated and restrictive cardiomyopathies along with left ventricular noncompaction. Mutations in the LIM protein gene also cause DCM in mice.[16] However, the mutations that cause other forms of cardiomyopathy are at different sites on these genes than those that cause HCM.

Mutations in the MYBPC3 gene in humans with HCM were first identified in 1995 and it was the fourth causal gene to be identified.[17] To date approximately 200 mutations in MYBPC3 that cause HCM have been identified in humans. Compared with mutations in β-MHC, a larger number of the MYBPC3 mutations are associated with a more benign clinical course (later average age at onset of symptoms and a lower incidence of sudden death). Nonetheless MYBPC3 mutations are a significant cause of morbidity and mortality for millions of people worldwide. For instance, a single MYBPC3 founder mutation found in people in South Asia (India, Pakistan, Sri Lanka, Indonesia and Malaysia; close to 5% of this population and upwards of 40 million people) makes them up to 7 times more at risk for development of cardiac dysfunction and heart failure than normal individuals.[18] Typically individuals affected with this mutation are free of HCM through the third decade of life but 90% of older individuals eventually develop cardiac complications.

The majority of MYBPC3 mutations are splice site donor/acceptor or other insertion/deletion mutations that are predicted to lead to reading frame shifts, premature stop codons, and truncated proteins.[17] However, approximately 40% of the mutations in MYBPC3 are due to point mutations (single base pair changes resulting in single amino acid substitutions). When compared to truncation/frame shift/splice variants, point mutations more commonly result in a more benign clinical course although there are those that produce severe disease.

The functional pathophysiology of missense mutations in MYBPC3 is not yet fully known, but potentially they can alter cMyBP-C structure, disrupt cardiac myosin binding C (cMyBP-C) protein function or alter interactions with other proteins.[17] Missense mutations could result in so-called haploinsufficiency where one allele is rendered nonfunctional resulting in only half the protein product being produced and so in sarcomere dysfunction, as can occur with truncation mutations. Alternatively, a protein that is produced by a mutated gene can be so altered that it is lysed within the cell before it ever is incorporated into a sarcomere. A missense mutation can also just result in a dysfunctional protein that is incorporated into a sarcomere. For example, the N775K mutation in humans allows the C5 domain of the protein to unfold rendering it dysfunctional. In other instances a mutation alters the ability of the protein (cMyBP-C) to interact with other proteins. For example, domains C7 to C10 anchor cMyBP-C to myosin and titin and disruptions of this interaction could impair the function of cMyBP-C. Lastly, missense mutations can result in a dominant-negative effect where an altered protein produced by a mutated gene has a detrimental on the normal protein.

The manner by which HCM is produced by sarcomeric mutations is relatively unknown and still controversial. One theory is that the abnormal protein produced by a mutated gene results in dysfunctional sarcomeres through any of the means discussed in the previous paragraph. Those dysfunctional sarcomeres contract poorly and this either forces functional sarcomeres to bear a larger work load (if, for example half of the sarcomeres are rendered dysfunctional) or all of the sarcomeres together can do less work (generate less force; if, for example, all of the sarcomeres are working at 50% capacity). The myocardium then has to compensate by replacing the dysfunctional sarcomeres with additional functional ones or has to add in more dysfunctional sarcomeres to help carry the load.[7,19-21] The net result is that the myocardium adds in new sarcomeres within myocytes resulting in the cells growing wider. Because myocytes make up a large part of the heart muscle, the thicker myocytes result in the wall of the left ventricle increasing in thickness.

FAMILIAL FELINE HCM

The first "family" of cats with an inherited form of HCM was identified in a research colony of Maine coon cats in 1992 and first reported in 1999.[22] The disease appeared to be inherited as a simple autosomal dominant trait in this breed with 100% penetrance (all cats affected with HCM). However, this colony was particularly inbred and so genetic modifying factors were probably uniform and a number of cats were probably homozygous for the subsequently identified mutation. In Maine coon cats in the real world it is now known that the disease is not 100% penetrant. Two lines of evidence support this. First, it is known that when an affected cat is identified (proband) that it is possible that neither parent may have echocardiographic evidence of the disease. Secondly, Maine Coon cats screened for the identified mutation and for HCM often have the mutation but no HCM, especially when the screening clinic includes only young cats.

The disease has been reproduced in Maine coon cats in the original research colony by mating affected to unaffected and affected to affected cats as well as by breeding affected Maine coon male cats to domestic shorthair female cats. The fact that HCM can be produced in offspring from mating a Maine Coon cat with HCM to a normal domestic shorthair cat proves that the disease is inherited as an autosomal dominant trait. The course of the disease can be accelerated by mating affected to affected cats, probably because cats that are homozygous for the mutation are produced. Penetrance is age related and the disease is progressive over months to years. In many affected Maine coon cats in the research colony, HCM is not apparent during the first year of life but becomes apparent by 2 years of age in males. Females tend to get the disease later, with many manifesting the disease by 3-4 years of age but some not showing evidence of HCM until 6 or 7 years of age and, others not until 10 and 13 years of age (most recently identified in two cats with the identified mutation in the research colony). When both parents have HCM, an affected Maine coon kitten may have echocardiographic evidence of the disease as early as 6 months of age and have severe disease by one year of age, whether male or female. Again these are assumed to be cats that are homozygous for the subsequently identified mutation.

In 2005, Dr. Kate Meurs identified the first gene mutation responsible for HCM in Maine Coon cats from our research colony.[23] This mutation is in exon 3 of MYBPC3. The exact location is codon 31 of the gene where a single point base pair (missense) mutation changes alanine to proline in the encoded protein (A31P). This region of the gene is highly conserved across species and the resultant amino acid change results in a change in the computed protein structure. Several studies have now shown that this mutation is only found in Maine Coon cats (not in other purebred cats). Prevalence of the mutation across continents is generally in the 30-40% range (highly prevalent).

In 2007 a mutation in the same gene (MYBPC3) was identified in Ragdoll cats with HCM (R820W).[24] This mutation is at a completely different location on the gene but it again occurs in a highly conserved region. Ragdoll cats have long been known to have a particularly malignant form of the disease where they often die before reaching one year of age. This now appears to be due to the presence of a larger number of homozygous cats in this highly inbred population.

A family of American shorthair cats, primarily with SAM of the mitral valve, but with other evidence of HCM as well, has also been identified.[25] The disease in this breed also appears to be inherited as an autosomal dominant trait. In addition to these breeds, there is anecdotal evidence of HCM being inherited in numerous other breeds, including Persian, British shorthair, Norwegian forest, Ragdoll, Turkish van, Sphynx, Siberian, and Scottish fold cats, along with others. HCM is most likely inherited when it is identified in a specific breed. However, HCM is most commonly identified in domestic (mixed-breed) cats. Whether the disease is inherited in these cats, is due to a de novo mutation, or is associated with a different disease process is unknown although suspicion of inheritance has been reported even in mixed breed cats.[26-28] Since the disease is inherited as an autosomal dominant trait in some purebred cats, it is not hard to imagine how these mutations could make their way into the mixed breed population and disseminate.

Several studies have now been done to look at the penetrance of the A31P mutation in Maine Coon cats (determine what percent of cats with the mutation have HCM). To date these have all been done as part of echocardiographic clinics to screen for HCM. Unfortunately, this type of study is inherently biased since the cats being screened are chosen by the breeder, not the investigator. This has resulted in populations of younger and female-predominant cats.

The first study looked at a group of Maine Coon cats during a screening clinic and correlated the genetic findings with echocardiographic findings.[29] The investigators found that many Maine Coon cats with the A31P mutation did not have echocardiographic evidence of HCM and a few cats without the A31P mutation had HCM. The veiled conclusion by the authors was that the A31P mutation might not be causal and that genetic testing for the mutation in this breed may not be warranted. However, what was shown instead is that the penetrance of this mutation is relatively low, at least in young cats, which is not surprising since most MYBPC3mutations in humans have a low and an age-related penetrance. They also showed that there appears to be at least one more cause of HCM in this breed. This has been apparent in the colony at UCDavis for a number of years.

Another study made observations on the echocardiographic appearance of hearts from Maine coon cats with the A31P MYBPC3 mutation.[30] In this study, echocardiography was performed on 96 Maine coon cats presented for screening for HCM. Both two-dimensional and tissue Doppler imaging echocardiography were performed. Cats had to have an LV wall thickness greater than 6 mm to make the diagnosis of HCM. Of the 96 cats, 44 of the cats had the A31P mutation (38 were heterozygous and 6 were homozygous). Unfortunately, 45 of the 96 cats were less than 2 years of age and so too young to have evidence of HCM if they were heterozygous. These young cats probably should have been excluded from analysis if they were heterozygous. Of the 38 heterozygous cats, four had clear evidence of at least moderate HCM. However, only 10 of the 34 heterozygous cats that did not have HCM were over 4 years of age. Of the 44 cats that had the mutation, 13 were male and 31were female. This produced an additional bias since female Maine coon cats get HCM at a later age and get less severe disease. Still, this study clearly suggests that the A31P mutation is not nearly 100% penetrant in Maine coon cats when a cat is heterozygous for the mutation (11% in this study). As expected, homozygous cats appear to be a different story. Of the 6 cats that were homozygous for the A31P MYBPC3 mutation in this same study, 4 had clear echocardiographic evidence of HCM and the other two had abnormal diastolic function as assessed by TDI, an abnormality previously shown to be present in cats that go on to develop HCM.30 Consequently, it appears that all 6 of these cats had HCM, which once again proves the mutation to be causal. Two cats in this study without the known mutation also had HCM. This once again documents that there is at least one more cause of HCM in Maine coon cats. Lastly, some of the cats heterozygous for the A31P mutation had evidence of regional diastolic dysfunction, showing that there myocardium was not normal even when they had no evidence of hypertrophy.

The most recent study into this subject looked at 332 cats.[31] They similarly showed that penetrance for Maine Coon cats heterozygous for the mutation is low (6%), that most but not all cats homozygous for the mutation get HCM (penetrance is high) and that most of the Maine Coon cats that get severe HCM prior to 6 years of age do so because they are homozygous for the mutation. The prevalence of HCM was 6%. Eighteen cats were homozygous and 89 cats were heterozygous for the mutation. The odds ratio for having HCM for homozygous cats was 21.6 (95% confidence interval 7-66). Overall, 50% of the cats that were homozygous for the mutation had HCM. Only two cats over four years were homozygous and both had HCM.

The previously referenced studies have only made a preliminary attempt to look at penetrance of the A31P mutation in Maine coon cats that are heterozygous for the A31P mutation. What really needs to be done is a longitudinal study looking at the same mutated cats until they are at least 10 years of age to tell what the true penetrance is for heterozygous cats rather than looking at a group of cats once at one point in time.

We are currently examining the cellular effects of this mutation and have preliminary evidence to show clear abnormalities produced by this mutation (e.g., myocardium from cats homozygous for the A31P mutation has no myosin binding protein C). Our preliminary findings clearly show that the A31P mutation causes intracellular derangement. It has also been shown that the mutated protein in cats with the A31P mutation is incorporated into the sarcomere, even in cats homozygous for the mutation.[17] This means the mutation does not result in haploinsufficiency.

There are currently several labs that test for the A31P MYBPC3 mutation in Maine Coon cats (one in the USA and at least two in Europe). This service has been set up so that breeders can try to rid their breed of this mutation and the HCM caused by this mutation. Many breeders have been reluctant to do testing and even more reluctant not to breed mutated cats. To be fair, the method of dealing with this problem is controversial (discussed below). In Dr. Meurs' lab, approximately 35% of the DNA samples submitted have had the mutation.[32]

Concern has been expressed about what would happen if all of these cats were removed from the breeding pool. The concern has to do with decreasing the size of the gene pool and so producing more recessive traits in this breed. In other words, if 35% of the cats could no longer be bred the remaining cats would have to make up the breeding pool and by shrinking that pool the breed may be worse off because of increased inbreeding. The author's counter to that argument is as follows. Breeders commonly only use around 10% of cats in a purebred population for breeding. That means they already exclude 90% of cats from being bred. So if we recommend that they don't breed any cat with a mutation that means we're recommending that they not use 35% of 10% or 3.5% of the entire population. Yes, that does mean that 35% of the cats they would normally breed (the good-looking 10%) would be no longer eligible for breeding. But it also means that they could still use the 65% of the cats they normally don't breed (the less than good-looking 90%) to breed. What would they sacrifice? Maybe their breed wouldn't be quite as visually attractive as before – but it also might be healthier without the mutation. Of course, getting breeders to breed cats they don't think look quite as good as some others is probably an impossible task. So one recommendation has been made that if a cat is heterozygous for the A31P mutation and it is very good looking, it can be bred once. Its kittens then should be tested for the mutation and only those without should be bred. The question then comes, what do you do with the mutant kittens that have been produced in this process? And this recommendation was first made 3 years ago meaning breeders have had plenty of time to do this and should no longer be relying on this method.

Lastly, it should be mentioned that the Ragdoll mutation (R820W) has also now been identified in humans. In humans, it also produces HCM but can also produce left ventricular noncompaction in some individuals.[33] Another mutational variant affecting the same codon (R820Q) also produces HCM and one that progresses to the so-called burn-out phase (i.e., DCM) more frequently.[34] The A31P mutation has not been identified in humans.

REFERENCES

1. Marian AJ, et al. **Curr Opin Cardiol** 2002;17(3):242. 2. Jarcho JA, et al. **N Engl J Med** 1989;321:1372. 3. Rosenzweig A, et al. **Mol Biol Med** 1991;8:159 4. Tanigawa G. et al. **Cell** 1990;62:991. 5. Geisterfer-Lowrance AA, et al. **Cell** 1990;62:999. 6. Thierfelder L, et al. **Cell** 1994;77:701. 7. Marian AJ, et al. **Circulation** 1995;92(5):1336. 8. Watkins H, et al. **Nat Genet** 1995;11:434. 9. Bonne G, et al. **Nat Genet** 1995;11:438. 10. Poetter K, et al. **Nat Genet** 1996;13:63. 11. Geisterfer Lowrance AA, et al. **Science** 1996;272:731. 12. Vikstrom KL, et al. **Mol Biol Cell** 1993;4:39a. 13. Wang Y, et al. **J Biol Chem** 2012;287(3):2156. 14. Geier C, et al. **Circulation** 2003;107(10):1390. 15. Blair E, et al. **Hum Mol Genet** 2001;10:1215. 16. Morimoto S. **Cardiovasc Res** 2008;77(4):659. 17. Harris S, et al. **Circ Res** 2011;108:751. 18. Dhandapanny PS, et al. **Nat Genet** 2009;41:187. 19. Lankford EB, et al. **J Clin Invest** 1995; 95:1409. 20. Cuda G, et al. **J Clin Invest** 1993; 91:2861. 21. Sweeney HL, et al. **J Biol Chem** 1994;269:1603. 22. Kittleson MD, et al. **Circulation** 1999:99(24):3172. 23. Meurs KM, et al. **Hum Mol Genet** 2005; 14(23):3587. 24. Meurs KM, et al. **Genomics** 2007;90(2):261. 25. Meurs KM, et al. **J Vet Intern Med** 1997;11(2):138. 26. Nakagawa K, et al. **J Vet Med Sci** 2002; 64(7):619. 27. Bary CJ, et al. **J Vet Intern Med** 2001;15(6):595. 28. Fujii Y, et al. **J Vet Med Sci** 2001;63(5):538. 29. Wess G, et al. **J Vet Intern Med** 2010;24(3):527. 30. Sampedrano CC, et al. **J Vet Intern Med** 2009;23:91. 31. Godiksen MT, et al. 2011; **Acta Vet Scand**; 53:7. 32. Fries R, et al. **J Vet Intern Med** 2008; 22(4):893. 33. Ripoll Vera T, et al. **Int J Cardiol** 2010;145(2):405. 34. Konno T, et al. **J Am Coll Cardiol** 2006;41:781.

TISSUE ENGINEERED CANINE MITRAL VALVES; VALUE IN VALVE DISEASE RESEARCH

Brendan M. Corcoran, MVB, DPharm, PhD, MRCVS
Edinburgh, Scotland, UK

INTRODUCTION

Myxomatous mitral valve disease (MMVD) is the single most common acquired cardiac disease of the dog and its ubiquity is such that all dogs are likely to develop the disease if they live long enough. However, the disease is characteristically associated with small dogs, particularly pedigree breeds, and becomes apparent during middle age and progresses over an extended time (years). While the disease in pedigree dogs is likely to be inherited, it is also commonly found in out-bred dog populations. The heritability of the disease probably affects the manner in which MMVD presents in a particular in-bred population, but not the end-stage pathological consequences of the disease. Furthermore, there is some anecdotal evidence that the pathological changes might begin in early adulthood.

PATHOLOGY

MMVD is characterised by myxoid changes to the valve stroma which results in damage and loss of the collagen rich fibrosa layer and expansion of the loose connective tissue spongiosa layer. The deformation of the valve leaflet and loss of mechanical integrity contribute to its failure of co-aptation at the end of ventricular systole. The stromal changes seem to occur in association with damage to and loss of the valve endothelium (VEC).

The exact mechanisms that initiate myxomatous degeneration and valve extra-cellular matrix remodelling are unknown. The cellular and molecular mechanisms involved in these processes appear to involve a combination of valve interstitial cell phenotypic alteration (activated myofibroblasts), proliferation and migration of activated myofibroblasts, disturbance of the valve endothelium, and changes in the balance of ECM remodelling processes[1-4]. Evidence for aberrant remodelling has been identified using powerful imaging modalities, such as nuclear diffraction, and by demonstrating altered expression of genes encoding for matrix metalloproteinase and their tissue inhibitors[5-9]. At a structural and morphological level, there is a clear decrease in total collagen and connective tissue content in diseased valves[10-11]. This is accompanied by an increase in glycosaminoglycan content and the extent and severity of ECM change is associated with disease development and progression. More interestingly, there is a significant change in collagen fibril alignment in overtly myxomatous areas of diseased valves and a concomitant failure of maturation of newly produced collagen. In the most severely affected valves, there is clearly significant loss of collagen, but this is still predominantly localised to the distal zone of the leaflet.

The investigation of pathobiology of MMVD and our understanding to date relies on access to canine valves and these are typically harvested at various stages of disease. This means that studies investigate mechanisms at time points in disease pathogenesis over which the investigator has no control making it difficult to separate cause from effect, and age-matched controls can be nearly impossible to obtain. The maintenance of a MMVD dog colony could be an approach to overcome this problem, but would be very expensive and raise ethical issues. Investigating isolated valve cells in culture can give insight into cellular mechanisms, but is hampered by the unknown effects of culturing conditions itself on cell function. Mouse knock-out models of MMVD are another option, but have limitations. The fibrillin-1deficient mouse has been proposed as a model of MMVD, but this is questionable as it is in fact a model of Marfan syndrome for which mitral valve prolapse is only one part of the clinical phenotype. An alternative approach is to manufacture tissue-engineered mitral valves from cultured cells and thereby attempt to mimic (but not replicate exactly) the normal mitral valve, and to then maintain valve constructs in a bioreactor system that permits manipulation of the constructs environment under quasi-physiological conditions. This is one research strategy being explored by the Cardiology Group at the Royal (Dick) School of Veterinary Studies and Roslin Institute, the University of Edinburgh.

TISSUE ENGINEERED CANINE MITRAL VALVES (CONSTRUCTS)

Tissue engineered (TE) constructs consists of cells embedded in a scaffold material. In the case of heart valves there is a need to have a central stromal matrix populated with valve interstitial cells and encased in an endothelial cell layer. The first step in TE manufacture is the accurate culturing of a near-as-pure VIC and VEC populations. This can be difficult to achieve, particularly in the case of VECs. Careful native valve preparation for cell recovery and use of appropriate cell culture techniques and cell specific media are needed. Accurate phenotyping of the cultured cell population is achieved using a panel of cell specific markers; assessed by both RT-PCR and immunostaining. However, culture passage might itself result in cell phenotypic changes or culture conditions might grow out the unwanted cell type. Lastly VICs are very easy to culture and can over-whelm VECs, which are much harder to grow.

The panel of markers used for phenotyping are different for VICs and VECs. In our laboratory we have used vimentin, SM22, αSMA, vWF, CD146 and CD31 and cell sorting techniques to identify VEC and VIC populations. While, with the right conditions, a reasonably pure population of cells can be obtained, there tends to be some contamination of the culture by the other cell type. Tissue engineers tend to be unconcerned about this factor as their main aim is to produce constructs that are compatible as prosthetic implants rather than true representations of native tissues. For constructs that are needed for research it would make sense to have the greatest purity. However, it transpires that with valve conditioning in a culture system, and selection of specific scaffolds, that cells can differentiate to the phenotype required.

Various manufacturing strategies can be applied for engineering valve constructs. We currently use a fibrin gel scaffold (fibrinogen reacting with thrombin in the presence of calcium ions) to suspend VICs to form a 3-D construct (Figure 1). The intention is to then coat the conditioned construct with VECs producing a sandwich-like structure with valve features, but not necessarily valve-like in shape (our preference is a tube structure).

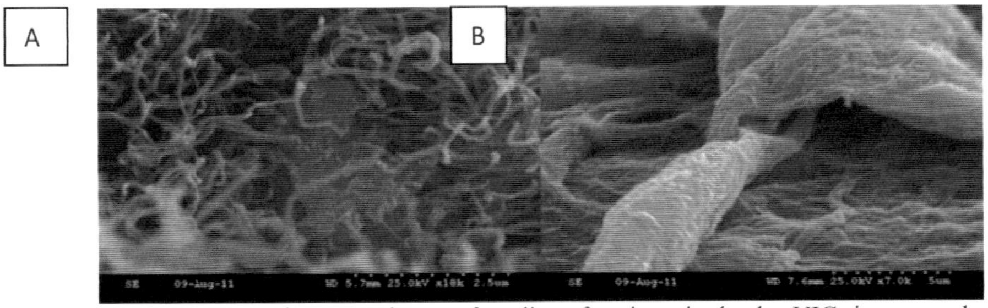

Figure 1. Scanning electron microscopy image of seeding of canine mitral valve VICs into a standard 3-D culture system using a fibrin scaffold. The images show some similarity to what is seen with native valves. A: VIC embedded in the fibrin scaffold. B: Higher power image showing a VIC adherent to the scaffold.

An alternative approach used in prosthesis manufacture is to use a mould (Figure 2a). This gives a structure that looks like a valve and would allow testing under mimicked physiological conditions in a bioreactor. A bioreactor is basically a chamber that allows the tissue construct to be maintained for relatively prolonged periods[12]. The aim is to maintain the construct under physiological conditions. In the case of heart valves this can include controlling pressure and flow fluctuations. Bioreactors can be simple or complex devices; from a simple tray arrangement (Figure 2b) to a fully integrated device where pressure in the chamber and flow of media can be regulated over significant periods of time (Figure 2c).

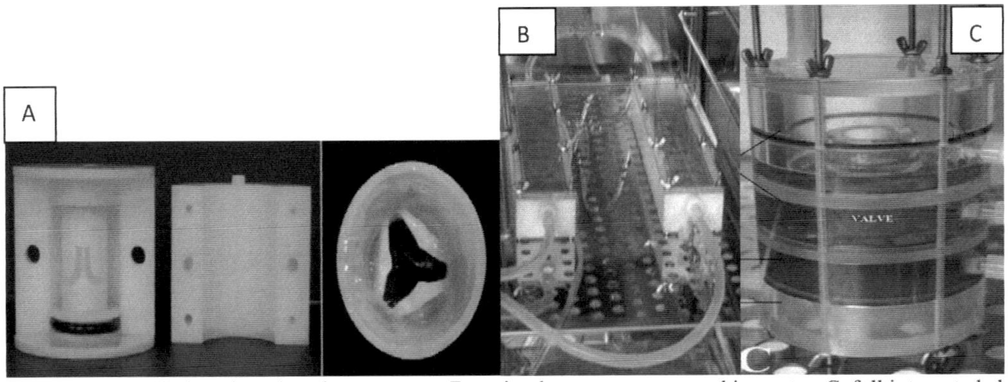

Figure 2. A: Fibrin gel aortic valve construct. B: a simple tray arrangement bioreactor. C: full integrated chamber bioreactor (with kind permission Dr T Flanagan & Prof S Jockenhovel).

Complex bioreactors also allow the investigator to access the construct and manipulate it as necessary. For example, damaging the endothelial covering or introducing agents into the culture media or directly into the construct. The ability to manipulate an artificial valve over a period of time and to sample at set time points is a powerful experimental stratagem which is not feasible with live animals.

Despite the potential for bioreactor maintained canine mitral valves in cardiology research, the question remains as to how comparable a TE valve is to a native valve? The mitral valve is a complex structure with clearly defined layers that can be readily identified on histological examination. Tissue engineered valves tend to lack the layered structure of a native valve, but this can be improved through conditioning; which is a combination of a properly selected scaffold and bioreactor maintenance (Figure 3)[13]. TE constructs with a mature phenotype are still in development, and such an end-point might not be achievable. However, valve constructs seem to have an embryonic mesenchymal appearance and this also appear to have some similarities with myxomatous degeneration. Mitral valve constructs might, therefore, allow investigation of valve biology in early development and this may have some bearing on MMVD pathogenesis. Constructs will be of use in evaluating biochemical mechanisms in valve function, with the further potential to evaluate new therapeutic strategies. While canine TE mitral valve constructs can provide a nearly inexhaustible supply of tissue for experimental work that would not be possible using native valves alone, they will not replace native valves as the ideal tissue source necessary to achieve understanding MMVD pathogenesis.

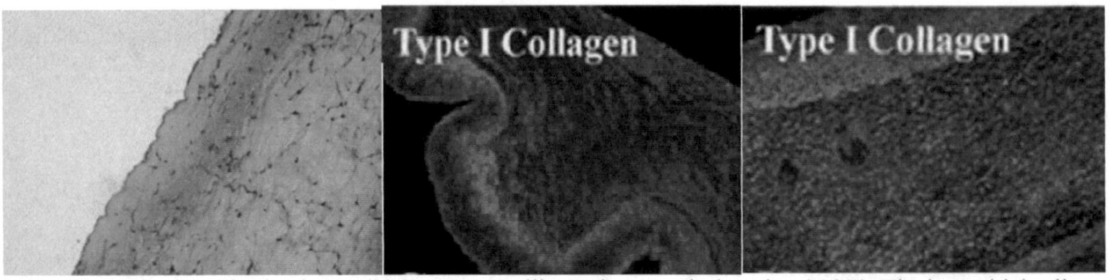

Figure 3. A: Canine mitral valve VIC 3-D construct illustrating pseudo-layering (H&E). The interstitial cells can be seen within the stroma and at the construct edge there is a distinct layer developing. B: Porcine native aortic valve (left) and TE conditioned valve (right) illustrating layering as a consequence of the conditioning process (with permission Dr T Flanagan).

REFERENCES

[1] Disatian S, et. al. **J Heart Valve Dis** 2008;17:402. [2] Han R, et.al. **AJVR** 2008; 69:763. [3]Corcoran BM, et.al. **Am J Vet Res** 2004;65:198. [4]Aupperle H, et.al. **Res Vet Sci** 2009;87:277. [5]Hadian M, et.al. Cardiovasc Pathol 2010;19:e141. [6]Aupperle H, et. al. **J Comp Pathol** 2009;140:271. [7]Aupperle H, et. al. **Vet J** 2009;180:88. [8]Lacerda CMR, et.al. **Proteomics Clinical Applications** 2009;3:1422. [9]Oyama MA & Chittur SV. **Am J Vet Res** 2006;67:1307. [10]Han RI, et. al. **J Heart Valve Dis** 2009;19:60. [11]Hadian M, et. al. **Biophys J** 2007;93:2472. [12]Wendt D et. al., **Adv Mater**, 2009; 21:3352. [13]Flanagan TC, et. al. **Biomaterials** 2007;28: 3388.

CANINE ADULT STEM CELLS; THEIR BIOLOGY AND THERAPEUTIC POTENTIAL IN HEART DISEASE

Hannah M. Hodgkiss-Geere, BVM&S, MSc, MRCVS
Edinburgh, UK

INTRODUCTION

In pet pedigree dogs cardiac disease is the third biggest cause of mortality after cancer and 'old age' and is therefore of significant importance in Veterinary practice. Acquired cardiac diseases in dogs are predominantly seen in two major categories; cardiomyopathies and valvular degeneration, with congenital causes of disease generally being very rare. [1,2]. Risk of death from cardiac disease increases with the age of the patient, and is also highly breed specific. For example, large breed dogs, such as Great Danes and Irish Wolfhounds are at major risk of developing dilated cardiomyopathy (DCM) whereas the Cavalier King Charles Spaniel has a significant risk of developing myxomatous mitral valve disease (MVMD).[1] Managing cardiac disease morbidity in the dog has been the focus of significant pharmaceutical research, with the key goal being the improvement of quality of life first and foremost, followed by increased longevity. The need for the study of cardiac diseases in the dog is evident, and the use of stem cells for this is an exciting prospect, not only for study of disease processes *in vitro*, but also for their future therapeutic potential *in vivo*.

The pet dog contributes to a significant amount of cases seen in small animal veterinary practice. The use of dogs as a subject allows us to study canine specific diseases more closely, advancing the understanding of the biology of veterinary disease. Furthermore, canine research may contribute to the development of novel therapeutics which could improve treatment regimes for other veterinary species. The dog is not only a patient in its own right, but also inadvertently an ideal model for the study of human diseases. Large animal models of disease are a requirement for any translation of basic science into therapeutics and practice[3]. Many diseases are known to affect both species, and in the western world the dog and human are exposed to many similar environmental factors of disease. The dog is a larger species than many commonly used rodent models, and also has greater longevity thereby allowing for age influence. Furthermore the pedigree dog is a relatively inbred model allowing for genetic influences on disease to be studied. Isolating stem cell populations from the dog will allow for *in vitro* disease modeling and may also act as an animal sparing process in drug discovery. Dog lifespan allows for longer term follow up on both cellular transplantation and gene therapy, and similarities can be seen in the molecular and physiological basis of disease, including cancer.

STEM CELLS

Stem cells are defined by having two characteristic properties: the ability for self-renewal, giving rise to more stem cells and the ability to differentiate into cell lineages from all three germ layers of the embryo under appropriate conditions (the endoderm, mesoderm and ectoderm).[4] This differentiation has lead to the cells being termed pluripotent - with the ability to divide into all cell types, or termed multipotent - with the ability to differentiate into multiple cell types of a limited number. There are three key types of stem cells described; embryonic, adult and induced pluripotent stem cells. These differ in both origin and differentiation potential, but all have the possibility for usage in the future of regenerative medicine. This study focuses on adult stem cells.

Stem cells hold great promise for the field of regenerative medicine and tissue engineering, however currently there are very few true clinical applications, largely due to difficulties in optimising cell culture, and targeting cultured cells to specific tissue sites *in vivo*. Ultimately the study of stem cell differentiation and the production of viable new tissue specific cells would enable the treatment of many diseases, such as neurological disorders, diabetes mellitus, orthopaedic disease and cardiac disease.

Defining Stem Cells

Stem cell marker profiles examining gene expression have been used to try to define stem cells *in vitro*. This technique has also proven useful in determining what stage of differentiation the cell may have reached. Cellular morphology, such as ability to form spheres, and behaviour in culture, including clonal expansion have also been used in combination with marker expression panels to attempt to define stem cells. However, this is a developing and challenging area; defining the marker or groups of markers which can be reliably used to define a cell as a stem cell is still being investigated. This is made more complicated by the possibility of interspecies differences and differences between different stem cell populations and stages of differentiation.

Adult Stem Cells

Although ES cells were the first truly pluripotent stem cell to be identified and characterised, it was hypothesised that a mechanism existed whereby organs could repair and replenish cells as they died. The hypothesis was confirmed when it was subsequently demonstrated that many developed tissues possessed a reservoir of somatic stem cells, which still retained some of the characteristics of their ES cell predecessors, including the capacity to self-renew but with a more limited differentiation capacity, termed multipotent.[5] Non-embryonic stem cells are lower in the stem cell hierarchy. Their multipotent function allows them to differentiate into a limited number of cell types generally common to the tissue of origin and generally assigned to one of the three germ cell layers of the embryo. Many tissues within the body have been found to contain ASCs, including liver, bone marrow, brain and heart.

Cardiomyocyte Renewal

The heart was long thought to be a terminally differentiated organ with no capacity for renewal and repair. Using Ki-67 staining and genetic fate mapping it has been shown that regeneration in the heart is increased following injury, and therefore suggests that cardiac renewal is possible.[6,7] Furthermore it has been shown using carbon-14 dating that renewal is constantly occurring, and is at its highest level in humans under 25, and decreases to under half the original level by the age of 75.[8] It would be fair to extrapolate from these results and assume that the dog also has a level of cardiomyocyte turnover, and that the presence of stem cells within the heart may contribute to normal remodelling and repair.

Following work demonstrating mitosis and cytokinesis in human hearts under normal conditions it was found that the heart did indeed have a natural population of stem-cell like cells.[6, 9, 10] Messina *et al* identified and characterised a cell population from both the mouse and human heart with properties of CSCs. The isolated cells grew as self-adherent clusters, which they termed 'cardiospheres'. These cells were capable of both self-renewal and could differentiate *in vitro* and *in vivo* into the major specialised cell types of the heart; cardiomyocytes, endothelial cells and smooth muscle cells.[11]

CARDIAC STEM CELLS

Adult stem cells derived from human endomyocardial biopsies cultured to form cardiospheres have been used to examine the therapeutic use of CSCs *in vivo*. The cardiosphere derived cells injected into the border zone of acute myocardial infarcts in immunodeficient mice migrate into the infarct zone and remain for at least 20 days. The percentage of viable myocardium within the infarct zone was higher in the cardiosphere derived cell treated group and this group also had a higher left ventricular ejection fraction when compared to the non-treated animals.[12] In a mouse study it was shown that CSCs home to the myocardium when injected intravenously and differentiated to express α-MHC – however it was impossible to discern if this expression was caused solely by differentiation of the injected cells, or by some level of cell fusion.[13] Human CSCs, and cardiomyocytes derived from CSCs injected into the border zone of infarcts in mice found that both cell populations improved cardiac function equally over three months, indicating that pure stem cell populations may be used therapeutically, avoiding the lengthy *in vitro* differentiation procedures which may increase cellular mutation. Crucially it was seen that these cells did not fuse with the murine cells, thereby suggesting that it was a differentiation effect rather than cell fusion causing the improvements seen.[14] Realistically the therapeutic use of CSCs is going to be limited due to the difficulty in isolation from cardiac tissue and the limited expansion potential *in vitro*. However, activation of the endogenous stem cell population *in situ* would bypass the need to remove the CSCs. In a canine model the activation of resident CSCs in chronically diseased hearts was demonstrated using the growth factors Insulin like Growth Factor (IGF-1) and Hepatocyte growth factor (HGF).[15]

Canine Adult Cardiac Stem Cells and Cardiomyocyte Differentiation

During our experiments, atrial cardiac explants were taken from dogs *post-mortem* and cultured to isolate CSCs. These cells were able to survive successive passages in serum free media, forming cardiospheres and were capable of clonal expansion under controlled culture conditions, demonstrating their ability for self-renewal. Characterisation of the CSCs demonstrated expression of the commonly used CSC markers c-Kit, GATA 4 and Flk-1 with no expression of cardiac lineage markers. Cardiac directed differentiation was performed based upon published literature and subsequent analysis demonstrated that partial cardiac directed differentiation was achieved with an increase in the expression of Nkx2.5 and cardiac troponin T. This data indicated that there was a resident population of CSCs found in the canine heart, and that these cells have an inherent ability to differentiate towards cardiac lineages.

An alternative differentiation experiment was performed using indirect co-culture of CSCs with mesenchymal stem cells isolated from the bone marrow which appeared to drive the CSCs toward a more endothelial lineage, and away from the cardiac lineages seen previously based upon cellular morphology and up-regulation of Flk-1 and vWF.

The data in this study suggested that canine adult CSCs are directly comparable to human stem cells and are multipotent, and therefore may be a source of cells for *in vitro* cardiac research.[16]

MESENCHYMAL STEM CELLS

Mesenchymal stem cells (MSCs) have a capacity for multi-lineage differentiation thus providing a promising source of stem cells for regeneration. MSCs are derived from the mesodermal lineage, and offer an opportunity to study an alternative adult stem cell source *in vitro*. Furthermore MSCs have been shown to have a paracrine effect upon damaged tissues which may be of great therapeutic potential. Using rodent models there is a suggestion that bone marrow derived stem cells can be driven toward cardiac lineages *in vitro* using various techniques, and therefore may present a more accessible source of stem cells for cardiac differentiation than CSCs.[17,18, 19] This work has been repeated using human MSCs demonstrating successful cardiac directed differentiation with increased expression of cardiac specific genes such as β-MHC and α-cardiac actin and although no spontaneous beating was seen a rhythmical calcium fluctuation was demonstrated.[20,21]

Due to their ease of isolation from a variety of anatomical locations and the proposed ability for cardiac differentiation there has been extensive investigation of MSCs as a therapeutic agent and as a source of cellular precursors. Human bone marrow derived stem cells were found to differentiate into cardiomyocytes when transplanted into the adult mouse heart and green fluorescent protein (GFP) labeled mouse bone marrow cells injected intravenously into mice with an induced myocardial infarction were found to assist in restoration of cardiac function parameters and cardiac remodeling.[22, 23]

Animal models and clinical trials in humans using MSCs in cardiac disease have suggested a short term improvement in some cases. Injection of MSCs into porcine infarcted hearts differentiated into cardiomyocytes and vascular structures, and also upregulated the presence of c-Kit positive CSCs.[24] The current rodent models are poor substitutes for direct comparisons to humans in terms of both size and cardiac physiology and as previously discussed larger animal models are being utilized including the pig and the dog, and data generated from these studies has the potential to be used in veterinary medicine. Clinical trials using MSCs for cardiac disease in humans are already underway and overall demonstrate a trend for generalized short term improvement. However these trials have been undertaken following limited basic stem cell science research making it difficult to ascertain at this stage whether the improvements seen are caused by a paracrine effect promoting scar remodeling versus a replacement/renewal effect, or perhaps a combination of both. Trials include the REGENERATE, TOPCARE, ASTAMI and BOOST trials. These trials all suggest a generalised improvement in measurable outcomes in patients with cardiac disease, such as increased ejection fraction and decreased end-systolic volume. Long term follow up of these MSC treated patients is still ongoing, and results of these studies will be critical in deciding the future of MSC therapy in cardiac disease. Further work elucidating direct and indirect effects of MSCs within the damaged heart must be performed.

Canine Mesenchymal Stem Cells and Cardiac Disease

Canine MSC isolation was first described by Kadiyala *et al* in the mid 1990's. These cells were isolated using standard human techniques and were capable of osteochondrogenic differentiation, a key determinant of successful MSC isolation.[25] Following initial isolation, a large amount of research has been performed investigating their use for joint and cartilage diseases; but more recently the use of canine MSCs for cardiac diseases has been investigated. Currently, dogs are being used as models for human disease research, and the results from these studies may be directly translated into veterinary medicine. Bone marrow derived MSCs need to be investigated for their safety when used for cardiac therapy in dogs, and route of delivery, cell type and cell number are all critical factors. Currently conflicting literature surrounds the use of MSCs in canine heart disease, with an indication that myocardial damage may occur following injection of MSCs either directly or via the coronary artery.[26, 27]

In a chronic ischemia study using dogs as a model, allogeneic bone marrow derived MSCs were injected into chronically infarcted myocardium and were found to improve vascularisation and cardiac function over 60 days.[28] Human MSCs transfected with pacemaker genes were transplanted into the heart of dogs with induced heart block and in these dogs pacemaking function was seen for up to 6 weeks of study, providing a possible alternative to the use of electrical pacemakers.[29] These results may suggest a positive effect of MSCs in longer term repair of the heart, counteracting the acute damage previously described.

In two further studies utilizing human MSCs injected into the canine heart, it was found that MSCs home to the site of ablation following radio-ablation therapy and differentiate into cardiomyocyte-like cells[30] and that MSCs laced onto an extra-cellular matrix scaffold can be used to repair full-thickness right ventricular defects.[31] No data is available on the clinical benefits of MSCs injected into canine hearts for the treatment of naturally occurring congenital or acquired canine cardiac diseases but based on the preliminary data from experimental studies and findings from human clinical trials this is an area worthy of further study.

Canine Mesenchymal Stem Cells and Cardiomyocyte Differentiation

In our study canine MSCs were isolated from bone marrow aspirates and cultured using traditional techniques. Canine MSCs behaved in culture and appeared morphologically similar to previously described MSCs from other species, including humans and mice. Furthermore, canine MSCs demonstrated a marker panel similar to other MSC populations including expression of CD44 and STRO-1, and did not express CD34 and CD45. Canine MSCs were subsequently exposed to a standard cardiac differentiation protocol described for murine MSCs. We saw no significant alteration in marker profile which would be indicative of cardiac differentiation, therefore successful differentiation was not achieved using this technique in our canine cells. In a follow up study analyzing baseline murine MSCs we demonstrated inherent expression of cardiac lineage markers in MSCs from this species, which increased in expression level over time in culture alone. This may explain previously reported cardiac differentiation in murine MSCs.[32]

CONCLUSION

This study highlights the ability to obtain adult stem cells from multiple sources in the dog which are readily isolated and cultured. Differences are seen in cardiac differentiation potential between CSCs and MSCs which may be relevant in downstream applications of these cells. The need to analyse a stem population extensively at baseline prior to differentiation is evident, and also the limitations in extrapolating data from experiments performed in different species such as the mouse, highlighting the importance of investigation using cells directly from the species of interest.

REFERENCES

[1]Egenvall, A. et al. **J Vet Intern Med** 2006; 20: 894. [2]Adams, V. J. et al. **J Small Anim Pract** 2010; 51: 512. [3]Dixon, J. A. & Spinale, F. G. **Circ Heart Fail** 2009; 2: 262. [4]Bajada, S. et al . **J Tissue Eng Regen Med** 2008; 2: 169. [5]Beltrami, A. P. et al.. **Blood** 2007; 110: 3438. [6]Beltrami, A. P. et al. **N Engl J Med** 2001; 344: 1750. [7]Hsieh, P. C. et al. **Nat Med** 2007; 13: 970. [8]Bergmann, O. et al. **Science** 2009; 324: 98. [9]Anversa, P. & Kajstura, J. **Circ Res** 199; 83: 1. [10]Nadal-Ginard, B. et al. **Circ Res** 2003; 92: 139. [11]Messina, E. et al. **Circulation Research** 2004; 95: 911. [12]Smith, R. R. et al. **Circulation** 2007; 115: 896. [13] Oh, H. et al. **PNAS USA** 2003; 100: 12313. [14]Smits, A. M. et al. **Cardiovasc Res** 2009; 83: 527. [15] Linke, A. et al. **PNAS USA** 2005; 102: 8966 [16] Hodgkiss-Geere, H. M. et al. **Vet J** 2011; epub Feb 15. [17] Makino, S. et al. **J Clin Invest** 1999; 103: 697. [18] Fukuda, K. **Art Organs** 2001; 25: 187. [19] Kruglyakov, P. V. et al. **Bull Exp Bio Med** 2006; 142: 503. [20]Xu, W. et al. **Exp Bio Med (Maywood)** 2004; 229: 623. [21]Antonitsis, P. et al. **Int Cardiovasc Thor Surg** 2007;6: 593. [22] Toma, C. et al. **Circulation** 2002; 105: 93. [23] Orlic, D. et al. **Nature** 2001; 410: 701. [24] Hatzistergos, K. E. et al. **Circ Res** 2010; 107: 913. [25] Kadiyala, S. et al. **Cell Transplant** 1997; 6:125. [26] Liu, T. M. et al. **Stem Cells** 2007; 25: 750. [27] Vulliet, P. R..et al. **Lancet** 2004; 363: 783. [28] Silva, G. V. et al. **Circulation** 2005; 111: 150. [29] Plotnikov, A.N. et al. **Circulation** 2007;116:706. [30] Kim, U. et al. **Int J Cardiol** 2011; 146: 371. [31] Potapova, I. A.. et al. **Am J Physiol Heart Circ Physiol** 2008; 295: H2257. [32] Hodgkiss-Geere, H.M. et al. **Int J Stem Cells** 2011; 4(2):143.

STEREOTACTIC RADIOSURGERY FOR BRAIN TUMORS: CLINICAL TRIAL UPDATE

Michael S. Kent, MAS, DVM, DACVIM (Oncology), DACVR (Radiation Oncology)
Davis, CA

INTRODUCTION

Radiation therapy is playing an increasing role in the treatment of brain tumors in cats and dogs. With the advent of improved advanced imaging techniques, improvements in the accuracy of patient positioning, advances in linear accelerator equipment and the use of imaging on linear accelerators the targeting of tumors has become more precise. In combination these have allowed the advent of stereotactic radiosurgery. Stereotactic radiosurgery (SRS) is a radiation therapy technique whereby a high dose of very conformal radiation is delivered to the target in a single fraction. This allows for sparing of surrounding normal tissue better than other radiotherapy techniques. While there is actually no surgery or cutting taking place with this procedure, the dose given is high enough to ablate the area treated, hence the use of the word "surgery". Stereotactic radiotherapy on the other hand is defined the same as SRS, except the treatment is fractionated into several or many fractions. Stereotactic radiosurgery most commonly refers to intracranial treatments while stereotactic body radiosurgery refers to a stereotactic radiotherapy procedure where the radiation dose is delivered outside of the head.

STEREOTACTIC RADIOSURGERY

Equipment

Most commonly gamma radiation or photons are used in stereotactic radiosurgery and radiotherapy, although protons or other particles can be used. There are several different types of radiation producing machines that can be used to perform SRS procedures, including linear accelerators and the Gamma Knife®, although there are no veterinary facilities using a gamma knife to treat patients currently. The principal of stereotaxis uses the Cartesian coordinate system to localize the target (tumor) to an external frame to allow increased accuracy and precision in targetting[1]. In the case of the gamma knife and some linear accelerator based SRS sytems the frame is screwed into the patient's head to allow rigid fixation. The system utilized at UC Davis uses the BrainLab frameless system. This frameless system uses a triple layer thermoplastic mold that is attached to the frame, eliminating the need to attach the frame directly to the skull.

Linear accelerator based radiotherapy uses a standard linear accelerator fitted with either stereotactic cones or a micro-multi leaf collimator to shape the field. A Gamma Knife® unit is a machine with over 200 cobalt-60 sources in a hemisperical array allowing all beams to cross at a single point. Treatments with this system are limited to intracranial disease of less than several centimeters in diameter. Another newer type of machine used in SRS is the Cyberknife®. The Cyberknife® is a compact 6MV linear accelerator mounted on a jointed robotic arm allowing 6 degrees of freedom. The couch can also be controlled in multiple planes allowing further flexibility in the system. The Cyberknife® is not limited to planning around isocenters as are other systems and due to the real time imaging system does not require a patient frame like other systems. No one system has shown superiority in terms of patient outcome in human medicine.

Treatment planning and quality assurance

It was not until the advent of computed tomography (CT) and magnetic resonance imaging (MRI) however that the treatment of intracranial tumors with targeted form of radiotherapy became possible. Planning for SRS is done on imaging data obtained from the CT scan. If the tumor targeted is of soft tissue origin then (MRI) can be used for better tumor delineation and the treatment planning software can fuse the images to the CT scan to assist in planning.

There are multiple different planning systems available. Generally image sets are imported and fused in the software. Treatment planning is then begun. Beam configurations are made using multiple static beams or arcs. In some systems inverse planning is used to make an intensity modulated radiation therapy plan for SRS. Dose can be viewed on planar images or in three dimensions. Dose volume histograms can then be compared ensuring maximal dose to the tumor while limiting dose to the surrounding critical structures. The goals of treatment planning are to create high dose gradients at the edge of the planning target volume of the tumor.

Precision and accuracy are required in patient positioning to ensure that the planned treatment is delivered correctly. The patient is imaged and positioned for treatment in the frame. Depending on the system used, portions of the frame may be removed after positioning for treatment. The patient should be imaged on the treatment machine just prior to or during treatment to confirm positioning.

The quality assurance required for SRS is more rigorous than for other types of radiation therapy given the small volumes of tissue that are added to create the planning target volume (PTV). This added volume around the tumor is to account for inaccuracies in the imaging equipment, the machine, the treatment planning system, and the treatment plan itself. The treatment plan can be tested on a phantom to ensure the calculated dose is the dose that will be delivered to the patient during the procedure. It is essential to limit the PTV as much as possible without sacrificing accuracy to limit the amount of normal tissue receiving a high radiation dose.

In humans SRS has been used to treat malignant and benign tumors, primary and metastatic lesions and either single or multiple lesions. Benign lesions treated in people include arteriovenous malformations, Parkinson's disease, trigeminal neuralgia and refractory obsessive-compulsive disorder. In humans treatment of meningiomas, gliomas and pituitary tumors, which represent the majority of intracranial tumors found in veterinary medicine have been well described. With non-resectable skull base meningiomas, 5 year local control rates are reported between 85-97%[2]. The use of SRS to control pituitary tumor growth, resulting in Cushing's disease vary widely in the human literature ranging from 66-100%, with endocrine remission rates ranging from 17-100%[3]. Median survival times for gliomas which are notoriously difficult to treat were recently reported at 11.5 months[4].

Clinical Trial Update

As this data is unpublished only a brief summary is being provided in these notes. Since November 2009 we have treated 94 cases in dogs and cats including pituitary tumors, meningiomas, gliomas, and peripheral nerve sheath tumors. The Brainlab frameless stereotactic system for patient positioning, Brainlab iplan for treatment planning and a linear accelerator using a cone based system with treatment delivered in multiple arcs are used. The majority of animals did not have pretreatment biopsies. Dogs with masses in the area of the pituitary were prescribed a single 15Gy fraction. Cats with a mass in the area of the pituitary were prescribed either a single 15Gy fraction or two 10Gy fractions. Dogs with masses consistent with the diagnosis of a meningioma were prescribed 15Gy in a single fraction and dogs with masses consistent on imaging with a glioma or peripheral nerve sheath tumor were prescribed 24Gy in three 8Gy fractions. All plans were tested on a canine phantom using a microchamber and gafchromic film to confirm dose prior to treatment. Efficacy data is still pending with updated preliminary results to be presented. Dog and cats are being re-imaged at 3 and 6 months after treatment to determine tumor volume using MRI for all non-pituitary tumors and CT for all tumors in the area of the pituitary. Further in a subset of patients the CT scan included a dynamic acquisition over 90 seconds with non-ionic iodinated contrast given at a rate of 5 mL/s intravenously to evaluate tumor perfusion.

CONCLUSIONS

To date there are only several published papers in the veterinary literature regarding the clinical use of SRS including one on the treatment of brain masses in 3 dogs[6], one on the treatment of 11 dogs with appendicular osteosarcoma[7] and one on the treatment of pituitary tumors in 11 cats[8]. Data on the efficacy of these types of treatments are still preliminary but promising.

REFERENCES

1. Leksell L. The stereotaxic method and radiosurgery of the brain. Acta Chir Scand. 1951;102: 316-319. 2. Minniti G, Amichetti M, Enrici RM. Radiotherapy and radiosurgery for benign skull base meningiomas. Radiat Oncol. 2009;4: 42. 3. Starke RM, Williams BJ, Vance ML, Sheehan JP. Radiation therapy and stereotactic radiosurgery for the treatment of Cushing's disease: an evidence-based review. Curr Opin Endocrinol Diabetes Obes. 2010;17: 356-364. 4. Villavicencio AT, Burneikiene S, Romanelli P, Fariselli L, McNeely L, Lipani JD, et al. Survival following stereotactic radiosurgery for newly diagnosed and recurrent glioblastoma multiforme: a multicenter experience. Neurosurg Rev. 2009;32: 417-424. 5. Lester NV, Hopkins AL, Bova FJ, Friedman WA, Buatti JM, Meeks SL, et al. Radiosurgery using a stereotactic headframe system for irradiation of brain tumors in dogs. J Am Vet Med Assoc. 2001;219: 1562-1567, 1550. 7. Farese JP, Milner R, Thompson MS, Lester N, Cooke K, Fox L, et al. Stereotactic radiosurgery for treatment of osteosarcomas involving the distal portions of the limbs in dogs. J Am Vet Med Assoc. 2004;225: 1567-1572, 1548. 8. Sellon RK, Fidel J, Houston R, Gavin PR. Linear-accelerator-based modified radiosurgical treatment of pituitary tumors in cats: 11 cases (1997-2008). J Vet Intern Med. 2009;23: 1038-1044.

CONVECTION ENHANCED DELIVERY OF CETUXIMAB FOR THE TREATMENT OF CANINE INTRACRANIAL GLIOMAS: THE UGA BRAIN TUMOR TRIAL

SR Platt, M. Kent, AC Freeman, N. Northrup, G Hadjipanayis
Athens, GA

Glioblastoma (GBM) is the most common and deadly human brain tumor for which there is no cure. Virtually all GBM tumors recur at the site of their initial treatment due to the presence of infiltrating cancer cells in the surrounding normal brain that resist therapy or go untreated. Infiltrating cancer cells include a subpopulation of glioblastoma stem cells (GSCs) shown to be integral to tumor development, perpetuation, and therapy resistance. Imaging and targeted therapy of infiltrating GBM cells, including GSCs, within the normal brain remains limited. The biocompatible, magnetic iron-oxide nanoparticle (IONP) has emerged as a potential multifunctional clinical tool that can provide cancer cell detection by magnetic resonance imaging (MRI) contrast enhancement as well as therapy by cancer cell surface targeting. The epidermal growth factor receptor (EGFR) represents the most common GBM alteration and a candidate for targeting of GBM cells for imaging and therapeutic purposes. A major barrier in the use of IONPs for the targeted imaging and therapy of brain tumors is the delivery of IONPs to the intracranial tumor. Systemic delivery of IONPs is difficult due to the nonspecific uptake of IONPs and difficulty penetrating the blood-brain barrier (BBB). Convection-enhanced delivery (CED), or fluid convection by a pressure gradient in the brain, can bypass the BBB and deliver therapeutic agents in high concentration to tumors and infiltrating cancer cells in the surrounding brain. The small size of IONPs (10 nm core size) make them promising agents for distribution in the brain by CED.

Our preliminary studies have confirmed EGFRvIII-antibody bioconjugated IONPs (core size of 10 nm) can provide MRI contrast enhancement of GBM cells and can therapeutically target GBM, including patient derived GSCs. IONPs can distribute well in the rodent brain after CED and can disperse over days to potentially target infiltrating GBM cells and prolong animal survival. IONPs can also be distributed in the larger canine brain after CED and have no toxic effects on animals. Our group is using IONPs bioconjugated to antibodies specific to the wild type (wt) EGFR and/or the EGFRvIII deletion mutant (Cetuximab or an EGFRvIII deletion mutant antibody) which we believe will provide MRI contrast enhancement of GBM cells and targeted therapy of GBM tumors after CED. GBM cell toxicity, neuronal/astrocyte toxicity, EGFR signaling, and mechanisms of cell death will be studied after treatment with bioconjugated IONPs.

CED is an approach developed to overcome the obstacles associated with current CNS agent delivery. Recently, multiple clinical trials have been performed utilizing CED for the treatment of recurrent GBM. In CED, a small hydrostatic pressure differential imposed to distribute infusate directly to small or large regions of the brain with a catheter is used in a safe, reliable, targeted, and homogeneous manner. Both small (rodent) and larger (dog) animal models have been used to study CED. Due to the small size of IONPs (10 nm core size), CED is an attractive method for targeting brain tumors. Penetration of IONPs through the extracellular matrix (ECM) in the brain is possible due to the larger effective pore size of the ECM (50 nm).

Difficulty imaging CED of agents is one major limitation of this approach. Visualizing distribution of infused agents is necessary to ensure accurate delivery into target sites and provides feedback on catheter placement and control of agent delivery. Ineffective drug distribution is one major criticism of CED which may compromise brain tumor targeting and therapy. Agent surface properties (cationic charge) and size, catheter positioning, and high interstitial tumor pressures can result in poor agent distribution. Dilating the extracellular matrix by co-infusion of the hyperosmolar agent, mannitol, can enhance the distribution of agents by CED. Poor drug distribution due to improper catheter placement has recently been shown to be a possible explanation for the lack of overall survival increase in the Phase III PRECISE trial. Groups have radio-labeled their therapeutic agent, co-infused their agent with radio-labeled albumin (^{123}I-labeled albumin), or used liposomes containing an MRI contrast agent (e.g., gadoteridol) for real-time CED imaging. Radio-labeling of therapeutic agents relies on low resolution single-photon emission computerized tomography (SPECT) for agent imaging. Recently, CED of 80 nm maghemite magnetic nanoparticles have been depicted by MRI in a normal rat brain model. Direct imaging of magnetic nanoparticles by MRI can permit distribution studies of nanoparticles in the brain after CED. CED represents a magnetic nanoparticle delivery method that can avoid non-specific uptake of nanoparticles, bypass the BBB, minimize systemic toxicity, and provide for the effective targeted therapy of infiltrating malignant glioma cells with MRI guidance.

Accurate targeting and better delivery efficiency are two major goals in the development of therapeutic agents for the treatment of GBM. Ideally, a therapeutic agent would be able to overcome the BBB and selectively be enriched in tumors with minimal toxicity to normal tissues. Conjugation of agents to antibodies or other ligands that bind to antigens or receptors that are usually abundantly or uniquely expressed on the tumor surface represent a promising approach for treatment of GBM and have been taken to clinical trial. The use of multifunctional IONPs that may target these infiltrating cells, including GSCs, for imaging and therapy purposes may represent a new paradigm shift in the treatment of GBM.

Initial CED and Distribution of IONPs in the Canine Brain.

Two groups of 4 healthy dogs underwent CED of free IONPs or bioconjugated Cetuximab-IONPs at the University of Georgia College of Veterinary Medicine. Group 1 (Dogs 1-4) underwent CED of free IONPs (0.2 mg/ml) for 60 minutes at 2 different doses; (Group 2 (Dogs 5-8) underwent CED of bioconjugated Cetuximab-IONPs (0.2 mg/ml) at two different rates for 6, 12, or 24 h. All dogs underwent CED after placement of a single open-tip catheter (Hermetic Catheter from Integra NeuroSciences) through a right frontal burr hole into the frontal lobe after general anesthesia. Infusions were performed utilizing an external programmable reservoir pump (Medtronic Synchromed II) strapped to each animal. Animals did not require anesthesia for their CED procedures. These pumps have been used for human intrathecal infusions and a prior human GBM CED trial. CED parameters (I_R and V_i) were programmed utilizing a handheld device. Brain MRI scans (3T) were performed on all dogs prior and 1 h after their CED procedure. Group 1 dogs underwent a brain MRI scan 7 d after CED. Group 2 animals underwent MRI scans at 2 time points after CED to determine IONP volume distribution, dispersion, and clearance. Distribution of the IONPs in the canine brain was achieved for the first time by CED and direct IONP imaging by T2WI was possible. T_1-weighted imaging (T1WI) confirmed no presence of hemorrhage after each CED procedure. In addition, infusate leak-back along the catheter was found at the higher infusion rates. Use of a slower rate provided for more uniform distribution of IONPs and low infusate leakback. We did find that V_D is linearly proportional to the V_I. Greater dispersion was found with use of the slowest infusion rate.

Toxicity and Clearance of IONPs.

All dogs showed no signs of toxicity after IONP CED. All dogs showed no abnormalities on cerebrospinal fluid (CSF), hematology, and serum biochemistry analysis 7 d after CED. No neurologic deficits were noted in any of the dogs during the 30 d observation period after their CED procedure and all dogs maintained their normal weight. Detection of IONPs was found on T2WI at 30 d at a much lower volume suggestive of IONP clearance from the brain. Brains were harvested from all the animals 30 d after their CED procedure and histopathologic analysis was performed. Lymphoplasmacytic infiltration was found around the tracts as well as gemistocytic astrocytosis and gitter cells (phagocytizing microglia) were also present. Both the astrocytes and microglia contained small spherical iron-positive particles. The intracellular presence of these iron-positve particles confirm the uptake, degradation, and clearance of the IONPs from the brain by astrocytes and microglia.

Canine Glioma Cells and EGFR expression.

Cells harvested from a dog with a spontaneous glioma have been grown in culture. Detection of EGFR expression was confirmed after Western blot analysis with a rabbit polyclonal antibody specific to the human wt EGFR.

REFERENCES

Lustig R. Cancer Biol Ther. 2006;5:1242-1243. Neyns B, et al. Ann Oncol. 2009;20:1596-1603. Patel D, Lahiji A, Patel S, et al. Anticancer Res. 2007;27:3355-3366. Balyasnikova IV, et al. PLoS One.5:e9750. Sampson JH, et al. Proc Natl Acad Sci U S A. 2000;97:7503-7508. Wikstrand CJ, Hale LP, Batra SK, et al. Monoclonal antibodies against EGFRvIII are tumor specific and react with breast and lung carcinomas and malignant gliomas. Cancer Res. 1995;55:3140-3148. Mishima et al.. Cancer Res. 2001;61:5349-5354.

TOCA-511 GENE THERAPY CLINICAL TRIAL UPDATE

Peter. J. Dickinson. BVSc, PhD, DACVIM
Davis, CA

INTRODUCTION

Toca 511 is replication competent retrovirus based on a cloned Moloney MLV, with the original ecotropic envelope replaced with the amphotropic envelope from the 4070A virus. The inserted internal ribosome entry site (IRES) from the Encephalomyocarditis Virus (EMCV) and the cytosine deaminase gene is inserted downstream of the envelope gene. This virus can only replicate in dividing cells, as it has no nuclear localization signal and can only integrate with the host cell's genome during mitosis, when the nuclear membrane has dissolved. Preclinical studies have been published demonstrating that in multiple murine tumor models, including malignant gliomas, intratumoral injection of a replication competent MLV vector containing a cytosine deaminase gene followed by administration of 5-Fluorocytosine (flucytosine, 5-FC) significantly improves survival when compared to controls. [1] Viremia or measurable viral dissemination to non-target tissues in immune competent animals has not been reported. Toca 511 is currently in Phase I clinical trials in humans.

GENE THERAPY CONSTRUCT

Toca 511 is a novel replication-competent amphotrophic moloney murine leukemia virus (MLV) based vector carrying the yeast cytosine deaminase gene capable of converting 5-fluorocytosine (5-FC) to the antineoplastic drug 5-fluorouracil (5-FU).

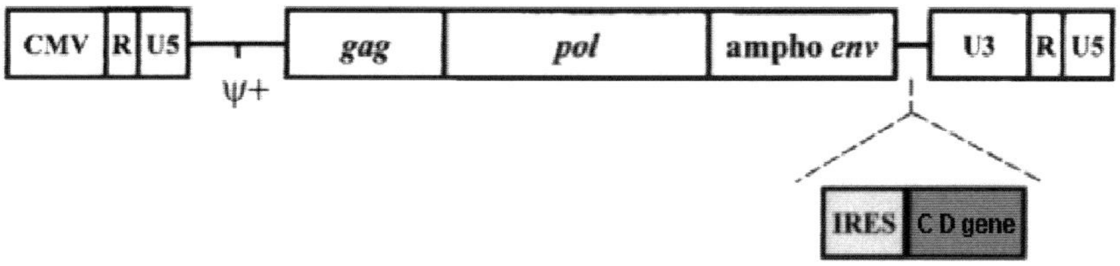

Pro-drug

Flucytosine, an immediate release formulation of 5-fluorocytosine is an approved anti-fungal agent. A gastric retention tablet formulation of 5-fluorocytosine, Flucytosine XR, has been developed that requires dosing only three times a day (TID) rather than four. Flucytosine XR, is a white, film coated oval tablet containing 500mg 5-fluorocytosine formulated with Carbopol to increase gastric retention. Since the 5-FU is generated intracellularly in the tumor and has a very short half life, none of the toxicities seen with systemic administration of 5-FU are anticipated.

Target validation

MLV entry into cells requires specific binding to the Pit2 transmembrane receptor protein (a Na-dependant inorganic phosphate transporter). Semi-quantitative RT-PCR demonstrated that Pit2 receptor mRNA is strongly expressed in both canine astrocytic cell lines and canine glioma tissue. *In vitro* transduction of MLV was assessed using a GFP/FACS based reporter system in 3 canine glioma cell lines (J3t-Bg, G06A, SDT-3G) and a human fibrosarcoma control cell line (HT1080). Percentage transduction of G06A was similar to HT1080 with 62% and 95% transduction at 3 and 6 days respectively. Transduction of J3T-Bg and SDT-3G was 81% and 31% at 6 days.

Preclinical safety:

A laboratory dog was infused with Toca 511 (1.8 x 108 TU/mL; total volume 200uL) using convection enhanced delivery targeting the thalamus and corona radiata. Infusions were monitored and confirmed using real-time imaging and gadoteridol as a surrogate marker. 5-FC was given at a dose of 125mg/kg/day for 7 days starting 14 days after infusion, and the dog was killed 7 days following completion of the 5-FC administration. Minimal inflammatory responses were seen histologically at infusion sites. All non-brain tissues were negative for virus (at 10 copies/ug) by PCR, and blood, saliva and urine were negative for virus by RT-PCR throughout the study period.

CLINICAL DOGS (THERAPEUTIC TRIAL)

Two client-owned dogs with spontaneous brain tumors (oligodendroglioma III) were treated in a prospective open label non-controlled experimental protocol with Toca 511 and Flucytosine XR following intratumoral delivery of Toca-511. Percentage tumor coverage was 10 and 40%. 5-FC administration was initiated 7 and 21 days following vector delivery.

Dog 1: 2 cycles (150mg/kg/day) 5 days, 4 weeks apart (total dose 1.5g/kg)
Dog 2: 2 cycles (115mg/kg/day) 6 days, 4 weeks apart (total dose 1.3g/kg)

No decrease in tumor volume based on sequential MRI was seen in either dog. Blood, urine, saliva and feces were negative for virus by PCR and RT-PCR in both dogs (up to 3 months for dog 2). Dog 2 had a surgical resection 15 weeks following vector delivery. Integrated vector sequence was detected in resected tumor.

5-FC Toxicity

Both clinical cases developed limiting adverse events following multiple courses of 5-FC characterized predominantly by pustular dermatitis. Previous reports of toxicity associated with flucytosine reported skin lesions developing 13 to 41 days (median 20 days) after commencing flucytosine (105 to 188 mg/kg/day divided and given every 8 h; median dose rate 150 mg/kg/day). The cumulative dose of flucytosine given prior to the first signs of the drug eruption ranged from 1.7 to 6.8 g/kg (median 2.3 g/kg). [2]

Further characterization of pharmacokinetics and pharmacodynamics of Flucytosine XR was done in experimental dogs. Exposure of to Flucytosine XR at 250 or 500 mg TID (mean body weight 9kg) resulted in relatively stable levels of 5-FC in the blood between 15 and 40ug/mL and 20 and 70ug/mL respectively over four dosing cycles. However both dose levels resulted in skin lesions, mainly confined to the scrotum, after 3 cycles of 4 days on 10 days off drug (minimum total dose ~1.4g/kg). Consistent with published reports, there was considerable variation in time and total dose exposure before lesions developed. Lesions were in general less severe and took longer to appear in animals exposed to the lower dose. Development of lesions was not associated with higher peak or trough values of plasma 5-fluorocytosine at a particular cycle. Re-exposure to drug after a longer washout (17 days) did not exacerbated existing lesions or result in new lesions. Exposure to 500 mg TID for 5 days or 250 mg TID for 8 days did not result in new lesions after a one month washout.

The idiosyncratic appearance of lesions both during and after drug exposure and lack of induction after re-challenge suggests a non-immune mediated reaction perhaps due to accumulation of a long lived metabolite. Lower cumulative doses and longer washout periods between courses of Flucytosine XR appear to be tolerated. A 5 day cycle at 500 mg TID and 8 day cycle at 250 mg TID were well tolerated in beagle dogs. Re-exposure after sufficient wash out is possible.

Recommendations

≤150 mg/kg/day
≤ 5 days drug per cycle
≥ 3 week wash out between cycles
Cumulative dose ≤ 1.5 g/kg per 2 month period.

CLINICAL DOGS (DEVELOPMENTAL TRIAL)

To further define factors associated with viral integration, spread and expression, four dogs with surgically resectable gliomas were infused with Toca 511 following CT-guided stereotactic biopsy. Infusions were confirmed using co-infusion of gadoteridol and post infusion/real-time MRI. Tumors were resected 2-3 weeks following infusion and tumor tissue analyzed for vector integration and expression. All animals were negative by PCR and RT-PCR for viral shedding following infusion. No adverse events were seen following infusion. Toca 511 was detected in multiple samples from spatially distant regions of tumor in 3 of 4 tumors. Positivity indicative of both integrated vector (qPCR) and replicating virus (qrtPCR) was found, although levels of expression varied.

CONCLUSIONS

1) Delivery of Toca 511 is feasible in canine glioma and is associated with minimal adverse events.
2) Vector is detectable in tumor tissue up to 3 months following delivery, even in the presence of circulating anti-viral antibodies.
3) Vector is only detectable in tumor tissue with no evidence of systemic dissemination.
4) Analysis of resected tumor tissue is suggestive of spread within the tumor.
5) Adverse events associated with multiple cycles of 5-FC may be limiting in the clinical setting.
6) Current delivery and therapeutic protocols have not resulted in clinical benefit as defined by decrease in tumor volume.

Viral integration and spread, as seen in rodent preclinical models is essential for therapeutic efficacy. Ongoing studies will investigate strategies to increase dissemination of the vector within the tumor and optimize timing of therapeutic drug dosing.

REGULATORY

All experimental protocols were approved by the Institutional Animal Care and Use Committee (IACUC) and Biological Use Authorization committee (BUA). All clinical protocols were approved by IACUC, BUA, the VMTH Clinical Trial Review Board (CTRB), and FDA.

FUNDING

The clinical and research studies were funded by Tocagen Inc., The Paul C and Borghild T Petersen Foundation, and the American Brain Tumor Association.

REFERENCES

[1]Tai C, **Mol Therapy** 2005;12(5):842-851. [2]Malik R, **Aust Vet J** 1996;74:285-88.

PET IMAGING IN VETERINARY MEDICINE: ONCOLOGY AND BEYOND

Amy K. LeBlanc, DVM, DACVIM (Oncology)
Knoxville, TN

INTRODUCTION

A growing body of knowledge and experience relating to Position Emission Tomography fused with Computed Tomography (PET/CT) is allowing veterinarians to use this powerful molecular imaging technique for clinical and research applications. This lecture provides an overview of how commercially available and investigational PET radiopharmaceuticals can be applied to the management of veterinary diseases.

PET/CT BASICS: EQUIPMENT, FACILITIES AND RADIOPHARMACEUTICALS

Access to PET scanners is generally limited to selected academic centers and to those with access to a human PET facility. Many PET radiopharmaceuticals may be purchased via commercial distribution networks such as PETNET or Cardinal Health, as in the case of ^{18}FDG and ^{18}F-NaF. ^{18}FDG (^{18}F-fluorodeoxyglucose) is an ^{18}F-labelled glucose analogue that is widely used for human nuclear medicine studies. Although other position-emitting isotopes such as I-124, Rb-82 , Cu-64, and C-11 are available, this lecture focuses on ^{18}F-labelled radiopharmaceuticals as they are most user-friendly and accessible through existing distribution networks. Other PET radiopharmaceuticals may be synthesized on-site with access to a cyclotron/reactor and radiochemistry expertise, which are currently limited to a few medical centers in the US, mainly for research endeavors.

Previous literature describing whole-body biodistribution of ^{18}FDG in cats and dogs, and the use of ^{18}FLT and ^{18}FDG in detection of malignancies and monitoring response to antineoplastic therapy have been published.[1-5] Additional data from our lab and others relating to ^{18}FDG, ^{18}F-NaF, ^{18}FLT, and other investigational tracers described below is forthcoming. In moving PET expertise and knowledge forward in veterinary medicine, consideration must be given to adequate facilities for holding radioactive animals and protocols for transport to off-site facilities, monitoring exposure to staff, and release limits for pets to return home to their owners. Compliance with applicable state regulations and good communication with IACUC and Radiation Safety offices is crucial.

PET IMAGING PROTCOLS FOR COMPANION ANIMALS

Patient preparation involves careful planning for general anesthesia, which is generally required for positioning during the scan procedure. State-specific radiation safety guidelines must be followed to limit exposure to the animals and their excreta once injected with the radiopharmaceutical under study. In the case of ^{18}FDG, artifactual uptake within skeletal muscle can occur if the animal is physically active in the uptake period prior to the scan, so use of a sedative/premedicant is recommended. Also in the case of ^{18}FDG, animals should be normoglycemic so as not to interfere with tissue uptake of the radiopharmaceutical, leading to erroneous scan results.

As PET imaging becomes more common in veterinary medicine, a need for standardized imaging protocols, post-image processing and data analysis arises. The most commonly used clinical indicator of radiopharmaceutical uptake within tissues/lesions is the Standardized Uptake Value (SUV) which may be influenced by many factors (scanner design and reconstruction parameters, patient body condition, scan protocol, etc) Use of a uniform reporting method to minimize factors affecting the generation of a PET image is needed so that data can be reproduced at other PET imaging centers, particularly in support of multi-site clinical trials that involve PET imaging. Finally and most importantly, as PET/CT is still very much a novel technique for veterinary medicine, validation studies are needed for the large variety of veterinary diseases and the value of PET for the many applications discussed here must be determined. The convenience and 'one stop shopping' of a single scan procedure must be weighed against the cost, exposure to ionizing radiation, and paucity of data collected to date on the wide variety of veterinary diseases to know exactly where PET/CT should be considered routinely in veterinary medicine. In the case of oncology, guidelines for objective evaluation of PET/CT-based responses to antineoplastic therapy (PERCIST: Positron Emission tomography Response Criteria In Solid Tumors) were recently reported for human patients; the adoption of this approach should be considered by those routinely involved in oncologic PET imaging.[6]

COMMONLY AVAILABLE PET RADIOPHARMACEUTICALS AND THEIR APPLICATIONS

^{18}F-fluoro-D-deoxyglucose (^{18}FDG)

^{18}FDG is the standard PET radiopharmaceutical for cancer imaging. ^{18}FDG is widely used for lesion characterization, tumor staging and re-staging during therapy, and monitoring patients for disease relapse. ^{18}FDG PET/CT is also valuable for prognostication in certain diseases e.g. ^{18}FDG uptake as measured via serial SUVs can predict patient outcomes with standardized therapy and can be used to tailor chemotherapy protocols early in the course of treatment.[7] We should also recognize the potential of whole-body PET/CT as an efficient staging technique that can identify lesions not detected on physical examination that may represent additional sites of metastasis and/or additional unrelated malignancies. Veterinary applications of ^{18}FDG-PET/CT in oncology are growing with increasing access to scanners and experience with ^{18}FDG. Descriptions of ^{18}FDG biodistribution in normal cats and dogs are available to assist the clinician in lesion interpretation. Avidity of several canine and feline tumors for ^{18}FDG has been reported and characterized.(Figure 1) Using serial ^{18}FDG scans is attractive for monitoring response to therapy in a single procedure, investigations of this approach in canine cancers treated with Palladia™ showed discordance between RECIST and PERCIST criteria applied , emphasizing the need for larger clinical studies of serial ^{18}FDG PET/CT in this area.(LeBlanc et al, data in press)

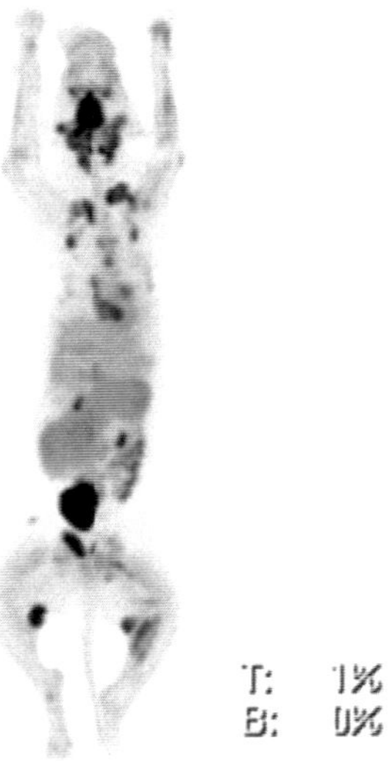

Figure 1 (left): Whole-body MIP of a dog with high-grade multicentric lymphoma imaged with ^{18}FDG. Intense radiopharmaceutical uptake is seen in affected lymph nodes, and physiologic uptake is seen within the brain, salivary glands, and urinary tract due to ^{18}FDG excretion.

Figure 2 (right): Whole-body MIP of a dog with multiple migrating thorn foreign bodies, where sites of inflammation are represented by areas of intense ^{18}FDG uptake within the skin and subcutis.

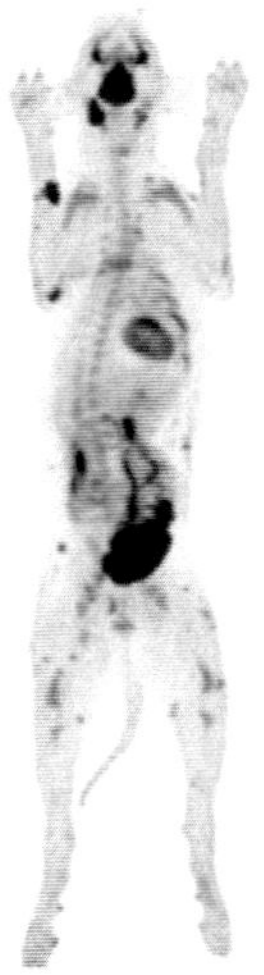

Imaging of inflammatory processes with ^{18}FDG is well established in humans and the application of this technique to veterinary medicine is ongoing. This application takes advantage of ^{18}FDG uptake by inflammatory cells, which is a limitation of ^{18}FDG PET/CT of undiagnosed lesions or cancers that are associated with inflammatory reactions. Use of ^{18}FDG to characterize localized inflammation associated with multiple migrating thorn foreign bodies was helpful in characterizing location of these thorns and planning surgical intervention (Figure 2). Even with this data in our literature, insufficient information exists to establish thresholds for SUVs to characterize lesions as inflammatory vs. benign vs. malignant. Biopsy and/or cytology to confirm lesion etiology is highly recommended prior to definitive treatment for ^{18}FDG-avid lesions detected with PET.

^{18}F-fluoro-thymidine (^{18}FLT)

3'-Deoxy-3'- [18F] fluorothymidine, or ^{18}FLT, is a substrate of the mammalian thymidine kinase, an integral part of cellular DNA synthesis machinery. After mono-phosphorylation, the compound is trapped in cells proportional to cellular thymidine kinase 1 (TK1) activity and thus reflects the cell's rate of proliferation.[8] Many studies demonstrate the strong correlation between ^{18}FLT uptake and cell proliferation as assessed by Ki-67 immunohistochemistry. Although ^{18}FLT uptake in reactive lymphoid hyperplasia, mediated by B lymphocyte proliferation in germinal centers does occur, several studies demonstrate superiority of ^{18}FLT over ^{18}FDG for differentiation of tumor from inflammation and indicate that FLT is a useful adjunct to ^{18}FDG in order to improve diagnostic specificity. The complementary strengths of ^{18}FDG and ^{18}FLT are currently under investigation for many tumor types where accurate tumor staging carries significant prognostic weight.[9,10] A study in canine lymphoma demonstrated utility of ^{18}FLT imaging in determining response to and predicting relapse following chemotherapy.[5] Our group has studied the whole-body biodistribution of ^{18}FLT in the normal cat and dog and has applied serial ^{18}FLT imaging in a canine model of chemotherapy-induced myelosuppression. This study is intended to provide validation of the ^{18}FLT signal as it relates to marrow function. Uptake of ^{18}FLT within bone marrow lends itself to PET imaging of leukemias and other primary bone marrow disorders (MDS, aplastic anemia, idiopathic marrow fibrosis). ^{18}FLT PET/CT may also be applied to identify bone marrow in external beam radiotherapy planning, so as to limit dose in chemoradiation schemes, which carry a significant risk for myelosuppression.

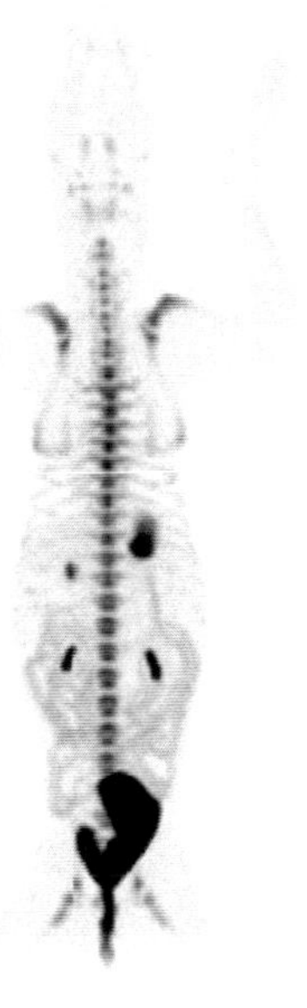

Figure 3 (left): Whole-body MIP of ^{18}FLT biodistribution in a normal dog. Note strong physiologic ^{18}FLT uptake within the proximal humerii, vertebrae, and proximal femurs. ^{18}FLT metabolites are also noted within the urinary bladder, gall bladder/GI tract due to hepatobiliary excretion.

Figure 4 (right): Whole-body MIP of ^{18}FLT biodistribution in a normal cat. Physiologic uptake of the radiopharmaceutical is noted within the bone marrow as noted in the dog; excretion is noted within the urinary tract and gall bladder/GI tract as above.

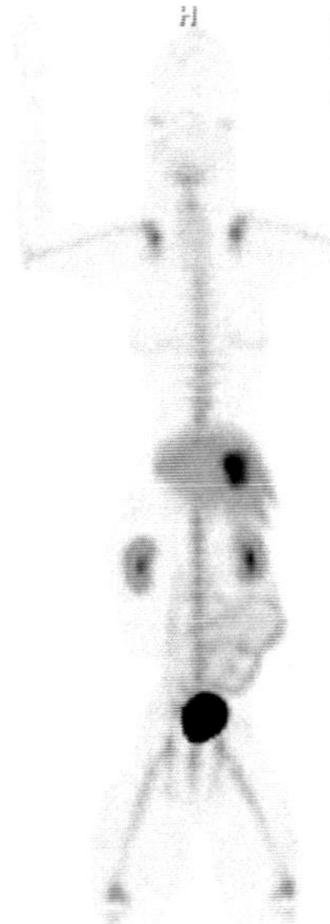

^{18}F-Sodium fluoride (^{18}F-NaF)

^{18}F-sodium fluoride was first used for skeletal scintigraphy in the 1960s, but was replaced by ^{99m}Tc diphosphonates due to cost and wide availability of clinical planar and SPECT gamma cameras in both human and veterinary medicine. Recently ^{18}F-NaF has been reevaluated and is gaining in popularity for bone imaging due mainly to the technological advances made possible with PET/CT. This technique provides greater sensitivity, resolution and efficiency for whole-body skeletal studies. ^{18}F-NaF has a high and rapid uptake in the bone with rapid blood-pool clearance, producing high target:background ratios in a short time after injection.[11] Based on the human experience, ^{18}F-NaF should be useful for characterization of primary and metastatic bone lesions, but data to support this technique over traditional bone scans in veterinary nuclear medicine is yet unknown.

INVESTIGATIONAL PET RADIOPHARMACEUTICALS AND THEIR APPLICATIONS

^{18}F-fluoro-thia-heptadecanoic acid (^{18}FTHA)

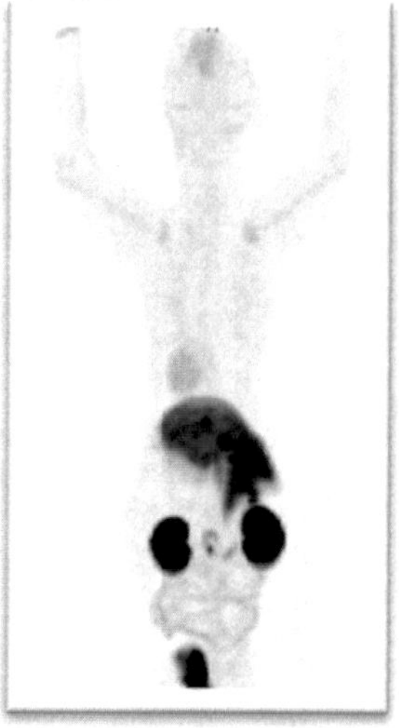

As a complement to ^{18}FDG for cardiac studies, a palmitate analogue 14(R,S)-[^{18}F]Fluoro-6-thia-heptadecanoic acid (^{18}FTHA) has been shown to accumulate preferentially within the myocardium via beta-oxidative metabolic trapping. Intense myocardial uptake, long myocardial retention, and rapid clearance from the bloodstream make ^{18}FTHA a useful PET imaging agent, and a vast improvement over previous efforts to interrogate the FAO pathway with C-11 labeled fatty acids, which were difficult to model and required on-site cyclotron facilities due to the 20 minute half-life of C-11. [12-14] Our group has studied the whole-body biodistribution of ^{18}FTHA in the normal cat and dog, in addition to a dynamic myocardial imaging protocol in which the kinetics of uptake of both ^{18}FTHA and ^{18}FDG were studied for 60 minutes post-injection. This study was performed to assess normal feline myocardial metabolic kinetics, and future work with cardiac diseases is planned. A whole-body MIP of 18FHTA biodistribution is presented (**Figure 5, right**), demonstrating intense uptake within the liver/gall bladder, renal cortices and urinary bladder. Further, our group is currently using ^{18}FTHA to evaluate whole-body FAO in study of a nutraceutical product in obese and lean dogs. Potential for ^{18}FTHA as a tracer for hepatic and/or renal function seems plausible.

^{18}F-labelled peptide imaging agents

A research focus of our lab is development of heparin-binding peptides target hypersulfated heparan sulfate proteoglycans on the surface of tumor cells. In contrast, this form of heparan sulfate is not observed within healthy organs. Peptides were labeled with biotin for histological staining of tumors removed surgically or radioiodinated/radiofluorinated for *in vivo* SPECT and PET/CT imaging. Biotinylated peptides were shown to bind preferentially to human and canine malignant melanoma in formalin-fixed paraffin-embedded tissue sections. By using small animal SPECT/CT imaging, we detected pulmonary colonies of B16F10 cells in C57Bl/6 mice and demonstrated co-localization of ^{125}I-labeled p5 peptides but not control peptide with the tumor lesions by microautoradiography. Binding of the ^{125}I-labeled p5 peptide was not uniform throughout the tumor mass; however both pulmonary and metastatic nodal tumor cells were targeted. In addition, a variant of peptide p5 was shown in similar experiments to associate specifically with tumor vasculature. Investigation of ^{18}F-labelled peptides into imaging of *in vivo* canine malignancies is underway.

REFERENCES

[1] LeBlanc AK et al. **Vet Radiology & Ultrasound** 2009; 50(4): 435-40. [2] Hansen AE, et al. **Vet Radiology & Ultrasound** 2011; 52(2): 201-206. [3] LeBlanc AK et al. **Vet Radiology & Ultrasound** 2009; 50(2): 215-223. [4] LeBlanc AK et al. **Vet Radiology & Ultrasound** 2008; 49(2): 182-188. [5] Lawrence J et al. **Vet Radiology and Ultrasound** 2009; 50(6): 660-668. [6] Wahl RL, et al. **Journal of Nuclear Medicine** 2009; 50 suppl 1: 122S. [7] Nahmias C, et al. **Journal of Nuclear Medicine** 2007; 48(5): 744-51. [8] Vesselle H et al. **Journal of Nuclear Medicine** 2003; 44(9): 1482-8 [9] Lee TS et al. **Nucl Med Biol** 2009; 36(6): 681. [10] Van Waarde A, et al. **Curr Pharm Des** 2008; 14 (31): 3326-3339. [11] Cook GJ. **Cancer Imaging** 2010; 10: 1-8. [12] DeGrado TR et al. **Journal of Nuclear Medicine** 1991; 32: 1888-96. [13] Ebert A et al. **Journal of Nuclear Medicine** 1994; 35: 51–56. [14] DeGrado TR et al. **Journal of Nuclear Medicine** 2006; 47: 173-81

ADVANCED MRI FOR CANCER DIAGNOSIS, STAGING AND EVALUATION OF TREATMENT RESPONSE

Susan L. Kraft, DVM, PhD, DACVR, DACVR-RO
Fort Collins, CO

INTRODUCTION

Magnetic resonance imaging (MRI) is unquestionably the best method for imaging neurological disease including neoplasia, and has become well established and fairly routine for veterinary medical neuro- oncology and other neuroradiological applications. The use of MRI for non-neurological veterinary oncologic imaging has progressed more slowly due to practical considerations such as cost and availability; nevertheless, MRI for veterinary oncology patients is increasingly being performed. Intense interest in MRI for human oncologic imaging is related to MR's high spatial and contrast resolution and the absence of ionizing radiation. There are many potential applications including primary tumor diagnosis, evaluation of margins for surgical or radiation therapy, cancer staging for metastases, sentinel lymph node identification and assessing treatment response.[1-4] In addition to anatomic and pathological imaging, advanced MRI techniques provide information about tumor functional biological and molecular properties. These can be quantitated and have potential value as biomarkers for diagnosis, prognosis and evaluation of treatment response. The role of these specialized MR applications is in transition from translational cancer research to human clinical investigations. In veterinary diagnostic imaging we are still in the early stages of evaluating these advanced MR techniques for oncology.

POWER OF MRI FOR CANCER IMAGING

In general, neoplasia involving the soft tissues is more conspicuous with MR imaging compared to computed tomography (CT). This advantage potentially translates into more accurate determination of tumor margins, better chances of detecting early tumor recurrence and/or residual disease as well as metastases. This is because of MRI's superior soft tissue contrast resolution (the ability to discriminate tissues and disease processes on the basis of their properties). MRI is based upon the ability to detect and differentiate the resonance characteristics of tissue nuclei, which are inherently influenced by the local microenvironment including biochemistry, proton content and diffusion, and blood flow. Those factors change during disease progression, resulting in altered MR signal intensity on the images. In addition to disease detection, the nature of a pathological process can be characterized non-invasively with MRI. Features such as its water and fat content, presence of hemorrhage, edema, proteinaceous fluid and necrosis can often be deduced.[1]

Although there were initial hopes that the tissue resonance characteristics detected by MRI would be highly diagnostic and specific for cancer, that has proven to not be the case.[4] Biopsy or cytology still remain necessary for definitive diagnosis. However, the combined pattern of altered MR signal characteristics of a suspected tumor, along with its geographic distribution, morphology, margins, degree and pattern of paramagnetic contrast enhancement and functional biological information from advanced MR techniques can be very informative and greatly assist in narrowing and better ranking differential diagnoses.

In comparison, the grey scale images derived from CT are based on essentially a single factor, tissue electron density, which becomes altered only after a larger degree of pathology has occurred. Soft tissue infiltrate and tumor margins can therefore be more challenging to distinguish on CT. This is partially compensated for by the use of iodinated contrast media during CT imaging, however many tumors are poorly enhancing and therefore can still be difficult to distinguish from surrounding normal anatomy on CT. An advantage of CT is that it provides more detail than MRI regarding cortical bone anatomy and lysis, as well as regions of mineralization and early periosteal response, however cortical bone lysis is evident with MRI by visualization of the associated invasive soft tissues.

PRACTICAL AND TECHNICAL CONSIDERATIONS

Although in many ways MRI is more informative, practical considerations still often favor CT for non-central nervous system cancer imaging. As MRI develops in veterinary oncology, it will be important to determine those clinical scenarios in which its use might improve the treatment and outcome of veterinary oncologic patients compared to other modalities such as CT or diagnostic ultrasound. Scan time is an important difference between MRI and CT. A full MRI examination involves multiple scans or pulse sequences, and can take significantly longer than a CT scan (especially multislice CT). This is a critical factor for veterinary patients due to the need for general anesthesia. The longer MR scan times can also lead to artifact from respiratory and cardiac motion on certain scans, and although gating can circumvent this problem, it adds technical complexity.

CT is traditionally used for radiation treatment planning. This is because CT provides tissue attenuation data for dosimetric heterogeneity corrections. This need may be overemphasized for some forms of radiation therapy however. The use of CT heterogeneity correction was found to be insignificant during dosimetric calculations for 6 Mv treatment plans for canine brain tumors.[5] In general, treatment simulation is more conveniently done with CT due to the ready availability of laser triangulation systems for landmarking to isocenter and CT radiation therapy immobilization tables, although repositioning accuracy is becoming more possible with MRI especially through the use of positioning aids. Last but not least, MRI is often more costly and may be less available to some hospitals. The purchase/ installation costs and operating costs of newer MR systems capable of fastest scans and the advanced MR methods described here remain substantial.

Imaging techniques and protocols for neuro-oncology have been well described in veterinary medicine and will not be reviewed here.[6] An MRI examination for suspected neoplasia of the head and neck, torso or musculoskeletal system is less standardized. Typically, this would include a minimum of either STIR or alternatively T2 weighted imaging, plus T1 weighted imaging before and after gadolinium contrast enhancement.[1,2] Malignant neoplasia will be hyperintense on the STIR or T2 pulse sequences because of its higher water content. The STIR pulse sequence has the added advantage of negating and darkening the signal from surrounding fat, making neoplasia quite conspicuous.[7] A fat saturation technique can be used during T2 imaging that has similar effects, although the suppression of fat signal can be less complete than for STIR images. On T1 weighted images, neoplastic tissues are typically hypointense, or darker, than surrounding fat which is inherently hyperintense. T1 images have excellent signal-to-noise and anatomic detail, and the gadolinium-based paramagnetic contrast enhancement used during T1 imaging results in tumor signal hyperintensity. The use of fat saturation with the post-contrast T1 weighted scan is again helpful and makes subtle contrast enhancement and tumor margins more visible especially when there is surrounding fat.

ADVANCED MR TECHNIQUES FOR IMAGING BIOMARKERS

One of MRI's most powerful advantages over CT is the ability to assess cellular, molecular and functional aspects of tumors. This is done using more advanced MR techniques that are available on contemporary high-field strength MR instruments. These parameters can serve as biomarkers or signatures of neoplasia and provide evidence to support the noninvasive diagnosis of cancer.[8] These features are more useful and relevant indicators of therapeutic response than traditional gross tumor volume or RECIST measurements because they can change prior to tumor size. Reliable methods to determine early therapeutic response would make it possible to tailor the most appropriate course of management for individual cancer patients. Some examples of imaging biomarkers from advanced MR techniques include tumor diffusion, metabolic biomarkers derived from MR spectroscopy and dynamic contrast enhancement as an estimate of tumor perfusion.

Diffusion-weighted imaging has been adopted from its applications for stroke in neuroradiology. The basis of MR diffusion imaging is its ability to probe the Brownian motion of water molecules in a tissue. Densely packed cells in viable tumor tissue restrict water diffusion, whereas regions where cells are dying due to effective therapy will lead to increased water diffusion. This can be measured as the ADC, apparent diffusion coefficient. Although singular ADC measurements are not diagnostically specific[9], serial changes in tumor ADC have been shown to be an indicator of therapeutic effect and to have prognostic significance in a variety of human cancers.[4,8] Proton MR spectroscopy (MRS) has been extensively studied in research on experimental preclinical cancers. There are also a relatively large number of clinical investigations evaluating MRS for human brain, breast and prostate cancer, but its exact clinical role and benefit has as yet been determined.[4] There are few reports using MRS from clinical veterinary medicine.[10] MR spectra are acquired in vivo (similar to ex vivo NMR or nuclear magnetic resonance) for the relative quantitation of the more abundant endogenous metabolites. The spatial resolution of MR spectroscopy is coarse and a fairly large tissue volume (typically 1 cm^3) is needed to obtain adequate signal for spectra. The relative amount of choline, a marker of membrane proliferation and turnover, is often increased in neoplastic tissue and correlates with malignant behavior and tumor grade. We have been evaluating the use of MR spectroscopy for canine normal brains and brain lesions to evaluate its diagnostic utility. The small size of the feline and canine brain relative to thick adjacent skull bone and sinuses presents special technical challenges, and acquiring MR spectra can add 15 to 20 minutes onto imaging time. We have found elevated choline levels in veterinary brain tumors but whether this provides significantly increased diagnostic specificity remains to be determined.

Dynamic contrast enhanced MR is performed by rapidly repeating scans of a tumor during and after gadolinium contrast injection. This results in signal intensity time curves that can be analyzed pharmacokinetically to derive wash in and washout rate constants and tissue compartment volumes. Those data is then used to infer degree of tumor angiogenesis, perfusion and vascular permeability. Changes in these parameters have prognostic value in longitudinal human and preclinical oncology studies.[11] Again, these techniques are in the early stages of exploration for veterinary patients with brain and body tumors.[12-15]

Novel contrast agents are being developed for MRI cancer applications but none are yet used routinely in clinical work. Gadolinium nanoparticles and other gadolinium constructs, various sizes of paramagnetic and superparamagnetic iron oxide particles (some of which are coupled with targeting molecules and reporter systems) are being devised for molecular imaging.[16]

Whole body MR using STIR or diffusion weighted pulse sequences, has great potential for staging cancer in veterinary patients, comparable to the more expensive and less available positron emission tomography imaging (PET).[17-18] Whole body MR has been found to be superior to CT or 99mTc-MDP bone scintigraphy for detecting skeletal metastases and also to bone marrow infiltrate, and in some situations can obtain comparable or even better results than PET.

MRI FOR PRIMARY TUMOR DIAGNOSIS AND EVALUATION

The MRI appearance and characteristics of the more common canine and feline brain and spinal tumors have been well described over the last 2 decades and will not be reviewed here. MRI has been described for veterinary soft tissue tumors of the nasal passages and sinuses, skull, ears, orbit and neck, and its multiplanar capabilities are valuable for treatment planning in the context of the complex anatomy of the head. MR is particularly sensitive to meningeal and/or brain invasion from locally aggressive skull and nasal tumors.[19] MRI has been shown to provide more accurate diagnoses of orbital neoplasia compared to radiography and ultrasound.[20] The abilities of MRI to characterize disease processes is helpful in distinguishing neoplasia in the nasal and/or paranasal sinuses from exudate accumulation, as well as rhinitis and sinusitis because of differing combinations of MR signal characteristics, and invasion through the cribriform plate is seen well with MRI.[1,21] Thin slice volume MR scans are best to see the full extent of invasion of malignant cancer into the middle and inner ear, intralabyrinthine structures, cranial nerves and brainstem due to lack of artifact when imaging the brainstem region.[22]

MR is the modality of choice for musculoskeletal cancer because of the ability to delineating the full extent of tumor margins from neurovascular and skeletal structures.[23-24] It has been shown to be better than CT and/or radiography for identifying extraosseous and intramedullary tumor margins, joint involvement and skip lesions.[1,25] These are critical factors for radiation therapy and limb spare procedures. MR has less associated artifact than CT when imaging in the presence of non-ferrous metal orthopedic implants.

As previously stated, MR's sensitivity to skeletal metastases and bone marrow infiltrate exceeds that of CT and 99m-Tc-MDP bone scintigraphy. The MR appearance of canine bone marrow changes with age and varies geographically due to the relative distribution of yellow (fatty) versus red (cellular) bone marrow.[26] An adult dog's yellow fatty bone marrow is hyperintense relative to musculature on T1 and T2-weighted images, and hypointense on fat-suppressed images such as STIR. Primary or metastatic lesions reverse those signal intensities and appear as mottled to uniformly hypointense regions on T1 images and hyperintensity on STIR images. Although the STIR pulse sequence is sensitive to bone marrow infltrate, the hyperintense signal is not specific and can also be caused by edema, inflammation or conversion of yellow to red bone marrow during certain diseases.

MR of thoracic neoplasia may increase because of improving technical abilities to decrease artifact from respiratory and cardiac motion, including faster scan sequences that allow acquisitions during breath-hold, respiratory compensation, and respiratory and cardiac gating.[27] Multiplanar imaging made possible with MRI allows curved anatomy such as the rib cage or diaphragm to be accurately depicted. MRI is useful for imaging tumors of the mediastinum, thoracic wall, cardiovascular structures and diaphragm.[28] Detailed lung imaging is possible especially with the newest techniques, although radiography and CT are likely to remain the methods of choice for some time.[1]

The widespread availability and relative low cost of radiographs and diagnostic ultrasound has slowed the development of MRI for abdominal cancer but there are many potential applications.[29-30] Newer fast MR scans (and general anesthesia of our veterinary patients) minimize artifacts from bowel motion. A prospective MR study in dogs found that MRI had 90% specificity to distinguish benign hyperplastic from malignant nodules of the liver or spleen, indicating a potential strength over ultrasound imaging.[31] MR's gastrointestinal applications have been slower to develop due to lack of an effective intraluminal contrast media and efficacy of existing techniques, although its use is now recognized for staging of human gastrointestinal cancers. MRI's usefulness in distinguishing canine mesenteric lymphoma from adjacent adhesions and bowel has been demonstrated.[32]

The appearance of normal and neoplastic lymph nodes has now been described for veterinary patients and the value of MRI for identifying sentinel and metastatic lymph nodes is being explored.[17, 32-34] Lymph nodes are hyperintense and conspicuous on STIR images. Metastatic lymph nodes are enlarged and often have strong heterogeneous contrast enhancement on T1 images.[34]

REFERENCES

1. Kraft SL; Pract Sm Anim MRI 2009:333. 2. LeBlanc A et. al.; Vet Clin NA: Sm An Pract 2007;37:1059. 3. Wisner ER et. al.; VCOG 2004; 2:49. 4. Sorenson A; J Clin Oncol 2006;24:3274. 5. Lyons J et.al.; VRUS 48, 2007,292-296. 6. Robertson I; VRU&S 2011;52:S15. 7. Dwyer AJ et al; Radiology 1988;168:827-836. 8. Yankeelov TE et.al.; Semin Oncol 2011; 38:16. 9. Sutherland-Smith J et.al.;VRU&S 2011;52:142. 10. McConnell JF et.al.; Proc Europ Assoc VDI 2005;47:421. 11. Turkbey B et.al.; Diagn Interv Radiol 2010;16:186. 12. Kraft S et.al.; Proc Annual ISMRM; 2008:. 13. Drees R et.al;, VCOG 2008;6:151. 14. Viglianti BL et.al.; Clin Cancer Res 2009;15:4993. 15. Zhao Q et.al; VRU&S 2010;51:122. 16. Bogdanov A et.al.; Semin Oncol 2011:38:42. 17. Kraft SL et.al.; VRU&S 2007; 48: 212. 18. Kwee TC et.al.; Europ J Radiol 2009;70:409. 19. Lipsitz D et.al.; VRU&S 2001;42:14. 20. Morgan RV et.al.; VRU&S 1996;37:185. 21. Moore MP et.al.; VRU&S 1991;32:19. 22. Garosi LS et.al.; VRU&S 2003;44:137. 23. Kraft SL et.al.; VRU&S 2007;48:1. 24. Mellanby RJ et.al.; VRU&S 3002;44:522. 25. Wallack ST et.al.; VRU&S 2002;43:432. 26. Armbrust LJ et.al.; VRU&S 2004;45:393. 27. Gilbert SH et.al.; Vet Jour 2010;183:124. 28. Mai W et.al.; VUR&S 2010;51:428 29. Spall B et.al.; VRU&S 2011;52:534. 30. Muleya JS et.al. VRU&S 1997;38:444. 31. Clifford CA et.al.; JVIM 2004;18:330. 32. Yasuda D et.al.; J Vet Med Sci 2004;66:1447. 33. Kneissl S et.al. VRU&S 2006;47:538. 34. Pokorny E et.al. VRU&S 2011;00:1.

METHODS FOR DIAGNOSIS AND PROGNOSIS IN CANINE LYMPHOMA AND LEUKEMIA

Anne C. Avery, VMD, PhD
Fort Collins, CO

INTRODUCTION

Lymphoproliferative disorders are a complex group of diseases that arise from neoplastic lymphocytes at various stages of development. In people, more than 40 different classifications of lymphoproliferative disorders are recognized – each arises at a different point in the life of a B or T cell. Originally, these diseases were characterized by their histologic appearance, as well as clinical attributes (age at presentation, location of disease). Contemporary methods, however, use primarily immunophenotype (beginning by classifying disorders as B or T cell) combined with histologic appearance and/or expression of a characteristic constellation of antigens as assessed by flow cytometry, and detection of consistent genetic aberrations such as translocations (Jaffe, 2009). The goals of such detailed subclassification are multiple: different types of mature B cell lymphomas, for example, have significantly different prognoses. Small cell lymphoma/chronic lymphocytic leukemia is a neoplasm of mature B cells with a prolonged indolent clinical course, and many cases can remain untreated for years. On the other hand, mantle cell lymphoma, also a neoplasm of mature B cells, requires aggressive therapy. Therefore these distinctions are important to make for making therapy decisions and guiding prognosis.

Another benefit of extensive subclassification of lymphoproliferative disorders is it allows researchers to better focus clinical trials and epidemiologic investigations to a more homogeneous population of patients. Today, clinical trials do not target "B cell lymphoma", but rather "follicular lymphoma" (a B cell disorder), or "diffuse large B cell lymphoma"

The contemporary histologic classification of lymphoproliferative disorders has been applied to dogs (Valli, 2011, Ponce, 2010), although since it requires determining immunophenotype as well as specialized knowledge on the part of the pathologist, the classification scheme is not in widespread use in a clinical setting. In addition, there are almost no studies correlating histology with outcome, although such studies would be highly beneficial. The one study available (Ponce, 2004) demonstrates the utility of this kind of correlation – these investigators found that the lymphoma with the best outcome is small cell T cell lymphoma (T zone lymphoma, which is not a WHO classification), and the lymphoma with the worst outcome was Burkitt's lymphoma which is a B cell disorder. Since numerous studies have shown that overall, T cell lymphomas tend to do worse than B cell lymphomas, the findings of Ponce et al demonstrate that further subclassification of lymphoma is important for accurate prognostic information.

Beyond B versus T cell, the constellation of antigens expressed by many human lymphomas and leukemias is well enough established that in some cases, they can be classified by flow cytometry alone. For example flow cytometry can distinguish mantle zone lymphoma, follicular lymphoma and small cell lymphoma/leukemia – all of which consist of mature B cells. In veterinary medicine, we do not have the broad range of antibodies available for this level of discrimination, nor do we have studies correlating histologic diagnosis with immunophenotype beyond the distinction between B and T cells. Nonetheless, until such studies have been carried out, flow cytometry can provide good prognostic and diagnostic information in many cases.

MAKING A DIAGNOSIS OF A LYMPHOPROLIFERATIVE DISORDER – IS BIOPSY NECESSARY?

Cytologic examination is the first step toward a diagnosis of lymphoma or leukemia. There is no data in the literature to support the use of FNA in determining prognosis in canine lymphoma, but studies in the human literature suggests that it is highly specific for simply making a diagnosis – a positive diagnosis by an experienced pathologist can be considered definitive (Meda, 2000). If the owners are simply looking to determine if their dog has neoplasia, then FNA may be sufficient. If, however, the cytology results are ambiguous, or the owners are interested in prognosis, additional testing is necessary. Three options are available for further characterization/confirmation of the neoplastic process: PARR (clonality), flow cytometry, and histology/immunohistochemistry. Each offers different information.

PARR

The PARR assay determines if the lymphocytes present are derived from a single precursor (if they are clonal) or if they are polyclonal. The latter suggests a reactive process, and the former, a neoplastic disorder. A positive PARR result together with a suggestive cytology can be considered highly specific for lymphoproliferative disease – different laboratories report different levels of sensitivity and specificity, but overall the specificity of a positive result is high (Burnett, 2002). In most cases, the PARR assay also will establish whether the neoplastic disorder is B cell in origin (including plasma cell tumors) or T cell. Beyond phenotype, however, the assay does not provide any prognostic information, nor is there evidence that this assay can be used to stage lymphoma. The main two advantages of the PARR assay is that it can be carried out on small amounts of material from almost any source, (including fixed, stained cytology specimens), and that it is sensitive to approximately 1:100 neoplastic cells. Thus, it can be very useful in situations where the cytologist identifies a few suspicious cells in an otherwise reactive looking node, or a slight excess of plasma cells in a bone marrow raising suspicion for myeloma, or in a cytologic specimen from a visceral organ that would otherwise require invasive biopsy.

Histology

Despite the limited data available on prognostic significance of different histologic subtypes, biopsy together with immunophenotyping read by a pathologist with training in the contemporary classification systems can provide a great deal of useful information. Biopsies read by a pathologist who use consistent diagnostic criteria, even if they are not those currently used in human lymphoma, can help clincians in predicting prognosis, particularly if the oncologists have extensive clinical experience with a single pathologist or classification scheme. In addition, biopsy is the only method available that will allow us to directly compare canine lymphoma with human lymphoma. Thus a histologic diagnosis is particularly important for studies in which the dog is being used as a pre-clinical animal model of human lymphoma, and for epidemiologic studies of breed predisposition. As mentioned above, even if a clinical trial of a novel therapy, for example, is conducted only for the benefit of dogs without any human relevance, histologic diagnoses will allow investigators to restrict their study to a more homogeneous group of patients – only dogs with diffuse large B cell lymphoma, for example. This will almost certainly result in less variation in the data, and smaller studies required to obtain significant results.

Flow cytometry

If the goal of diagnosis and prognosis, however, is for the purpose of advising an individual owner and treating an individual patient, there is not a compelling reason in the current literature to biopsy when less invasive methods are available and provide similar information. Flow cytometry is now routinely available, and is performed at multiple institutions and diagnostic laboratories. Flow cytometry involves staining cells in suspension from an aspirate or blood with a panel of antibodies specific for B cells, T cells and other lineage markers. In one respect, flow cytometry is similar to cytology and histology in that there are no published objective criteria on which to base a definitive diagnosis of lymphoma or leukemia in dogs using flow, although such studies would be very important. Nonetheless, there is general agreement that homogeneous expansion of a single cell type (usually, but not always, a clearly defined population of large cells) can be considered diagnostic for lymphoid neoplasia. Another feature which indicates neoplasia is the presence of cells with an aberrant phenotype. The most common example is the presence of T cells that have lost expression of the pan-leukocyte antigen CD45. Even in small numbers, the presence of these cells can be used for a definitive diagnosis.

When there is peripheral lymphocytosis present, regardless of the presence of lymphadenopathy, flow cytometry on peripheral blood is an extremely useful method for establishing a definitive diagnosis of neoplasia. In the analysis of peripheral blood, absolute numbers of different lymphocyte subsets per ul of blood can be determined and compared with normal ranges. With absolute numbers, it is easier to establish if there is expansion of a single lymphocyte subset, and such numbers can also be used as a baseline for treatment.

It has been repeatedly demonstrated that T cell lymphomas have a better prognosis than B cell lymphomas (Dobson, 2001, Ruslander, 1997). Thus flow cytometry can give prognostic information because it reveals the immunophenotype. In addition, a small number of reports are now available which link additional immunophenotypic information to prognosis. Most recently we demonstrated that, as with human B cell lymphoma, canine B cell lymphomas that express low levels of class II MHC have a worse prognosis than B cell lymphomas with high class II MHC (Rao, 2011). Similarly, when the cells comprising the lymphoma are identified as large by light scatter properties, the median survival is closer to that of T cell lymphoma. Two studies (Rao, 2011, Comazzi, 2011) linked immunophenotype in cases of leukemia with survival – Comazzi et al focused on chronic lymphocytic leukemia specifically, and found that T cell CLL had a significantly better prognosis than B cell CLL. Williams et al showed that CD34+ leukemias have a terrible prognosis (median survival 16 days) when compared with all other types of lymphoproliferative disorder involving peripheral blood. CD34 expression is considered diagnostic for acute leukemia, and although acute leukemia can often be diagnosed by cytologic criteria and clinical presentation, there are rare cases than are originally misclassified as CLL in the early stages. Thus flow cytometry on peripheral blood can provide very useful diagnostic and prognostic information if blood is involved in the neoplastic process.

IS THIS LYMPHOMA OR LEUKEMIA?

Does a patient with enlarged lymph nodes and lymphocytosis have lymphoma or leukemia? In human patients, this question is no longer relevant, since several categories of lymphoproliferative disorders have been grouped, and others have such clearly defined histologic criteria than the presence of circulating cells and bone marrow involvement is not a part of the diagnosis (but may have prognostic significance). Small cell lymphoma/chronic lymphocytic leukemia are now considered one disease of mature B cells. Precursor B lymphoblastic leukemia/lymphoma, and precursor T lymphoblastic leukemia/lymphoma are now two entities instead of four. On the other hand, mature B cell disorders are known to involve blood and marrow in a subset of cases. Diffuse large B cell lymphoma, for example, can have circulating neoplastic B cells in a subset of cases, but this doesn't change the diagnosis (de Leval, 2009).

Currently, with the exception of CD34+ acute leukemias, we do not have a way to determine if a process is primarily a leukemia or a lymphoma with peripheral blood involvement by flow cytometry. There appears to be overall agreement that chronic lymphocytic leukemia can be diagnosed by the cytologic appearance of the cells in the peripheral blood (Comazzi), and that phenotype in these cases may offer insight into the prognosis, but since CLL in dogs can either be T cell or B cell, the immunophenotype itself does not lead to the diagnosis of CLL. Williams et al demonstrated that small cell size, as determined by flow cytometry, predicted that a leukemia would behave like a CLL (median survival >1000 days) but the same did not appear to be true for T cell leukemias. Comazzi et al showed in an analysis restricted to CLL as determined by cytologic criteria, T cell CLL had a better prognosis than did B cell CLL.

Looking ahead, clinicians, pathologists and basic scientists should strive for better classification of lymphoproliferative disorders. Future studies that attempt to better classify different lymphoproliferative disorders using histology, immunophenotyping and cytogenetics, must *correlate these with prognosis*, or they will not have any clinical utility. Such studies will have the added benefit of allowing oncologists to treat based on experience in human medicine with different types of well characterized lymphomas, as well as better positioning the dog as a model for the human disease. Most importantly, better characterization and understanding of the natural history of different types of lymphoproliferative disorders will allow owners to make more informed choices. While the development of advanced diagnostics (such as genetic analysis for chromosomal aberrations, or complex immunophenotyping) may be expensive, once it is established the tests themselves are not complicated or expensive. In the long term, they will help owners receive more timely and accurate information, and likely save money spent on diagnostics.

REFERENCES

Jaffe ES **Hematology** 523-531, 2009. Valli VE, et.al. **Vet Path** 48:198-211, 2011. Ponce F, et.al. **Vet Path** 47:414-433, 2010. Ponce F, et.al. **Vet Journal** 167:158-166, 2004. Meda BA, et.al. **Am J Clin Path** 113:688-699, 2000. Burnett RC, et.al. **Vet Path** 39:617, 2002. Dobson JM, et.al. **J Sm An Prac** 42:377-384, 2001. Ruslander DA, et.al. **In vivo** 11:169-172, 1997. Rao S, et.al. **J Vet Int Med** 25:1097-1105, 2011. Comazzi S, et.al. **J Vet Int Med** 25:100-106, 2011. Williams MJ, et.al. **J Vet Int Med** 22:596-601, 2008. De Leval L, et.al. **Hematology - Oncology Clinics of North America** 23:791-827, 2009.

CYTOLOGY REVIEW FOR DIPLOMATES AND CANDIDATES

Anne Barger, DVM, MS, DACVP
Urbana, Illinois

INTRODUCTION

Cytology is a useful diagnostic tool for evaluation of solid tissues, masses and fluids. It is a noninvasive and fairly inexpensive tool that can often yield diagnostic results. Fine needle aspirations, fenestrations, imprints and fluid analysis are techniques useful to obtaining a cytologic specimen. Slides are prepared and once the sample has dried on the slide, it is ready to stain. No fixation is necessary. Romanowsky stains are used most commonly for cytology in veterinary medicine.

MICROSCOPIC EVALUATION

Once the slide is stained, it is ready for microscopic evaluation. Scan the slide first at low power to evaluate the quality of the sample and interpret the overall cellularity. Inflammatory lesions usually consist of a pleomorphic population of cells and neoplasms often have a monomorphic population of cells. Low power evaluation is also useful in evaluating the cellular arrangement (clusters, individualized, sheets, etc.) The sample can then be evaluated at higher power to evaluate the cell populations more closely. Identification of cell types, evaluation for uniformity of cells and presence of criteria of malignancy.

The sample is evaluated for the overall process, inflammation or neoplasia and then further classified based on the specific cell type identified. Inflammatory lesions are classified by the cell types observed. If neutrophils predominate and represent >85% the sample is classified as suppurative. Neutrophils should be further examined for the condition of the neutrophils, particularly the condition of the nucleus and for the presence of intracellular bacteria. Pyogranulomatous inflammation is still predominated by neutrophils with lesser numbers of macrophages and multinucleated giant cells. This type of inflammation is commonly associated with fungal infection or foreign body response. Granulomatous inflammation is predominated by macrophages with multinucleated giant cells and the low numbers of other types of inflammatory cells such as lymphocytes, plasma cells and neutrophils. Eosinophilic inflammation is unique because this classification only requires 12-15% eosinophils. Causes of this type of inflammation include allergic or hypersensitivity response, parasitic infestation and paraneoplastic response. Tumors commonly associated with a paraneoplastic eosinophil infiltrate include mast cell tumor and T cell lymphoma.

The first step to evaluating an aspirate of a neoplasm is to try and identify the tissue if origin. Tissue types are differentiated using a variety of features. The first step is to classify the cells as epithelial, mesenchymal or round cell. These categories can have very different characteristics. Epithelial cells are often arranged in clusters which can be distinguished even at low power. The cells are often round although columnar epithelial cells can appear spindle shaped. The nuclei are usually round. Mesenchymal cells are usually individualized and can be round or elongate or spindle shaped. Spindle shaped cells often have wispy cytoplasmic borders and long projections. Round cells, not surprisingly are round. These cells have distinct cell borders and round nuclei. It is not always clear what the tissue of origin is not always clear, especially with poorly differentiated neoplasms and melanoma. Cytologic criteria of malignancy include anisocytosis, anisokaryosis, prominent and multiple nucleoli, irregularly shaped nucleoli, multiple nuclei and nuclear molding. These are the primary criteria used but this is an incomplete list. Distinguishing a hyper plastic response is very important but also can be very challenging. Not all malignant neoplasms exhibit obvious criteria of malignancy (apocrine gland adenocarcinoma of the anal sac for example). When differentiating hyperplastic from neoplastic responses, it is important to consider what tissue is being sampled. For example, a dog with an enlarged prostate, if the cells appear uniform it is a hyperplastic response however if the cells exhibit obvious criteria of malignancy it is a malignant neoplasm like a prostatic adenocarcinoma or transitional cell carcinoma.

Finally, the final step to evaluating a slide is identifying background material. Blood is commonly observed in the background, and is generally a contaminant. The important thing to remember is the red blood cells will be accompanied by white blood cells so it is important to carefully distinguish peripheral blood contamination from inflammation. Background material may also consist of cellular products such as protein, osteoid or mucin. If granulated cells such as mast cells or melanocytes rupture, these structures can appear extracellularly and must be differentiated from bacteria.

Diagnosis of neoplasia with cytology can be very useful but is not without limitations. Many neoplasms such as mast cells and soft tissue sarcomas rely on grading to predict the behavior on the tumor. A grading system for these neoplasms has not been established for cytology. The diagnosis for the tumors listed above can be successful with cytology but a follow-up of histopathology for appropriate grading is essential. It is unrealistic to include cytologic descriptions of all tumors however below is a brief description of some of the easier tumors to diagnose with cytology. Round cell tumors can often be diagnosed with cytology. This grouping of tumors includes mast cell tumor, plasma cell tumor, lymphoma, histiocytoma and transmissible venereal tumor. Of these tumors, mast cell tumors and histiocytomas are the most common. Mast cell tumors are more common in dogs than cats. Cytologically, they are cellular and consist of a population of round cells with a moderate amount of basophilic cytoplasm, filled with metachromatic (purple) granules. These granules do not stain consistently with Diff-Quik and special staining may be necessary to clearly visualize these granules. Eosinophils often accompany mast cell tumors in dogs but do so less commonly in cats. Well-differentiated mast cell tumors in cats usually do not contain eosinophils. Histopathologic assessment is necessary for grading in dogs but grading has not proven useful in cats. In dogs the grading is important to attempt to predict the behavior of these tumors.

Histiocytomas are unique to dogs. They are benign tumors that will often resolve on their own after several weeks. Grossly they appear as a round, red, raised mass. Cytologically this tumor consists of a population of round cells with a moderate amount of basophilic cytoplasm and a centrally placed nucleus. The cytoplasm often has a clearing of the basophilia at the periphery. As the tumor begins to resolve, increasing numbers of small lymphocyte (cytotoxic T lymphocytes) can be observed, sometimes outnumbering the neoplastic cells.

Lipomas are a benign mesenchymal tumor and probably the most common benign tumor in dogs. Generally they are of low cellularity. Large droplets of lipid are observed on the unstained slide and often wash away during staining. Adipocytes are large cells with abundant, pale staining cytoplasm packed with lipid and a small round nucleus. Benign spindle cells can also be observed in these samples. Lipomas, cytologically, are often indistinguishable from mature subcutaneous adipose tissue.

Soft tissue sarcomas exfoliate extremely well for a connective tissue tumor. The cells have voluminous cytoplasm with long wispy projections. The nuclei are round and multinucleation is common. Marked anisocytosis and anisokaryosis are prominent features of the sample. The most remarkable and distinguishing feature of these cells are the long cytoplasmic projections. They are often bipolar and the cells, with their cytoplasmic projections included can appear quite large.

Squamous cell carcinomas (SCC) are a frequently diagnosed epithelial tumor. This tumor is the most common head and neck tumor diagnosed in cats. Samples often contain a mixture of cell types because the neoplastic squamous cells are frequently accompanied by inflammatory cells. The presence of inflammation can make diagnosis challenging because epithelium can become dysplastic secondary to inflammation. The neoplastic squamous epithelial cells are arranged in clusters with varying amounts of basophilic, occasionally keratinized cytoplasm. The nuclei are large and round with a stippled chromatin pattern and prominent nucleoli. Occasionally cells are coated in neutrophils. Grossly the masses are often ulcerated which contributes to the inflammatory component of the tumor.

In conclusion, there are many benign and malignant processes that can be successfully diagnosed with cytology. The veterinarian's clinical impression is critical in the appropriate interpretation of a cytologic specimen. Neoplastic processes are more likely to have a monomorphic population of cells, inflammatory responses are more likely to be pleomorphic and be painful to the patient. If there is ever any question about the cytologic interpretation, a pathologist should be contacted and/or a biopsy performed.

HAVEMEYER HIGHLIGHTS: FOCUS ON EHV-1 WORKSHOP

D. Paul Lunn BVSc, MS, PhD, MRCVS, DACVIM
Raleigh, North Carolina
Julia H Kydd, BSc, PhD
Nottingham, UK
Josh D. Slater, BVM&S, PhD
London, UK
Nicklaus Osterrieder Dr. Med Vet. (Habil.)
Berlin, Germany

The third international workshop on Equine Herpesvirus-1 (EHV-1) was held in Steamboat Springs, Colorado, USA on 19th – 22nd September 2011, and included presentations from university and biological company researchers from around the world. The meeting occurred shortly after the so-called "Ogden" EHV-1 outbreak of 2011, which was geographically the largest recorded epidemic of equine herpesvirus myeloencephalopathy (EHM) to date, with confirmed cases in 10 states in the USA and several Canadian provinces. New research was presented in epidemiology, infection models, virulence factors, immunity and therapeutics.

In the field of epidemiology, analysis of 31 peer reviewed publication (Hugh Townsend, Saskatchewan, & Lutz Goehring, Colorado) identified a series of probable increased risk factors for EHM including barn and stable confinement, increased age, exposure to infected animals, and failure to segregate. While season, viral strain, reproductive status and limb edema were suspected risk factors, they were confounders in each case. The impact of factors such as vaccination, previous exposure, treatment and stress could not be determined.

Genetic analysis of an extensive series of EHV-1 isolates (Udeni Balasuriya, Kentucky) demonstrated significant variation in ORF30 between EHV-1 isolates, but most nonsynonymous substitutions were isolated from non-EHM cases; the G2254 (D752) genotype was most commonly but not exclusively associated with EHM cases. Caution must be exercised in interpreting allelic discrimination PCR assays for the A or G2254 genotype in terms of clinical risk. Presentations on virulence factors explored the recently described role of MHC I as an entry receptor for EHV-1, with additional importance in cell to cell spread of the virus (Arthur Frampton, North Carolina).

A variety of *in vitro* infection models were described. The use of whole mucosal explants (Hans Nauwynck, Ghent) was valuable for describing the ability of EHV-1 but not EHV-4 to infect cells below the basement membrane of the respiratory mucosa. This system is also well adapted to the study of anti-viral agents. The phenotype of these infected cells appears to be monocytic (Gerlinde Van de Walle, Ghent). A 3-dimensional equine respiratory epithelial cell culture system was described (Gisela Hussey, Colorado), which allows for study of EHV-1 immunomodulation of innate immunity, and the impact of specific viral gene deletions. This system demonstrates the typical MHC-I down-regulation resulting from EHV-1 infection, and its dependence on the ORF-1 gene, which was independently demonstrated in other in vivo cell infections with deletion mutants (Van de Walle).

Additional infection model studies conducted *in vivo* utilized the aged mare model to study EHM and the use of anti-viral agents (Lara Maxwell, Oklahoma). These studies demonstrated the possible value of therapy after the onset of fever in preventing or ameliorating EHM, and the drug ganciclovir showed potentially greater promise in this regard. A new model of EHV-1 endothelial infection that focuses on ocular blood vessel infection was described (Gisela Hussey & Lutz Goehring, Colorado), which can occur at a high prevalence, and may be useful for studying immune protection from this important intermediary step in the development of EHM. Finally, EHM was recorded in 10 to 37% of horses infected with Ab4 in two different conventional challenge experiments (Gisela Hussy, Colorado, & Gillian Perkins, Cornell, respectively).

Presentations on virulence factors explored the recently described role of MHC I as an entry receptor for EHV-1, with additional importance in cell to cell spread of the virus. The impact of EHV-1 infection on pro-inflammatory cytokine responses was described by a number of investigators (Bettina Wagner, Cornell, & Gisela Hussey), and the specific effects of ORF1 gene deletion were described including reducing fever and specific cytokine increases. These results can be reproduced *in vitro* as described above. Immunological responses to infection may be predicted by IgGb:IgG(T) rations (IgG4/7:IgG3/5) (Bettina Wagner), which may also predict the outcome of vaccination. There were extensive discussions of the use of bioinformatics approaches to determine the molecular basis of neurovirulence (Klaus Osterrieder & Walid Azab, Berlin, Peggy Brosnahan, Cornell, & Debs Elton, Newmarket), although results were pending. Similarly, the need for novel reagent systems for studying cytotoxic T-lymphocyte (CTL) responses and MHC I typing were discussed (Julia Kydd, Nottingham & Doug Antczak, Cornell). While efforts to further define CTL epitopes in the ORF64 IE gene product have progressed (Gisela Hussey), tetramers have still not been produced. New research strategies using dermal responses to study CTL epitopes in EIA were described (David Horohov, Kentucky), which offer considerable promise for streamlining this process.

Therapeutic agents for EHM included anti-virals, and silencing-interfering RNA (siRNA). The value of anti-viral agents is referred to above. Experiments using siRNA (Gillian Perkins, Cornell) were able to demonstrate anti-viral effects *in vitro* and in murine infection experiments, but failed to reduce clinical disease, viral shedding or viremia in equine challenges.

The group conducted extensive discussions of research priorities, and refinements of current paradigms and critical questions which will be presented at the forum. In addition, it was identified as a priority to approach the OIE to update the standards for current EHV-1 diagnosis.

RESIDENT TRAINING: CLINICAL EXPERIENCES OUTSIDE OF THE TEACHING HOSPITAL

Sébastien Buczinski, Dr Vét, DÉS. MSc, DACVIM
Saint-Hyacinthe, Quebec, Canada

INTRODUCTION

The ACVIM Food-animal emphasis residency program has a worldwide reputation. One of the big challenges of this residency program is that it is frequently perceived by many graduated veterinarians as a program that will uniquely serve to the formation of specialists that will remain in the academia and hospital settings. This may lead to difficulty to recruit new residents as previously mentioned in ACVIM annual forums. Another challenge that needs to be addressed is to improve the perception by the veterinarians that this program can be used in a practically field approach of the ruminant health approach and is not only preparing future faculties.

WHAT NEED THE VETERINARY PROFESSION AND FOOD ANIMAL INDUSTRY?

Before entering in the heart of the talk it is important to look on the industry needs because most of time the industry is driven by the producers' needs. Higher productivity and efficiency is the ultimate goal of a dairy or beef producer. This allows them to maintain the sustainability of their business. This statement also drives the Mission Statement of important food animal veterinary associations. This may be illustrated by the mission statement of the AABP:[1] "The American Association of Bovine Practitioners is an international association of veterinarians organized to enhance the professional lives of its members through relevant continuing education that will improve the well-being of cattle and the economic success of their owners, increase awareness and promote leadership for issues critical to cattle industries, and improve opportunities for careers in bovine medicine." An ACVIM Diplomate could be a little concerned of this statement since there seems to be little place given to internal medicine in this sentences.

Two different types of producers can be found in the industry: the pure-breed producers who hardly work on animal genetic selection and have high-valuable animals (for dairy or beef breeds). However, both beef and dairy production are more commonly done in commercial farms in which the individual value of each animal is mainly driven by the market. This second type of farms represents more than 90% of daily veterinary practice, whereas pure breed-farms animals are an important part of veterinary teaching hospitals caseload. Since most of the interventions done by a veterinarian in practice need to be economically feasible (on a cost-effectiveness basis) it is important to keep in mind the real economic world. This may explain why each specialist in food-animal medicine should be fully aware of these challenges especially when wanting to practice in the field after the residency program.

The aim of the ACVIM residency program is to prepare the resident to the highest standards of food animal internal medicine, this objective of the program is actually well done and give to the resident strong knowledge of individual case management..

To the question : what should I expect during a visit with a Board-certified Large Animal Internist?, the ACVIM website answer is[2]: "The internist will perform a complete and thorough physical examination of your animal, and based on this findings, additional tests will be discussed. Depending on your animal's condition, diagnostic testing or treatment may include:

- Advanced laboratory testing of various tissue and blood samples
- Biopsies of masses, internal organs, or bone marrow
- Cerebral spinal fluid (CSF) taps for neurological diseases
- Diagnostic Imaging – ultrasound, radiographs (x-rays), CT scans, MRIs
- Echocardiography – ultrasound of the heart
- Electrocardiography (ECGs) – electrical reading of the heart's rhythm
- Endoscopy – bronchoscopy (lungs), cystoscopy (bladder & urethra), gastroscopy (stomach), rhinoscopy (nasal cavity & guttural pouches of horses, & laparoscopy (minimally invasive surgery for biopsies of internal organs)"

More specifically, we can see that any aspect of the animal as belonging to a larger group unit (a herd) is missing with that definition. It is the author's opinion that the Food-Animal internist should also be aware of the impact of the disease on the animal's productivity (meat, milk, or reproduction) as well as on the herd (especially for infectious diseases). An in depth knowledge of prevention of the disease (from vaccination to biosafety program) should also be known since they are an important part of the cattle industry especially to produce highly safe products.

STRENGTHS OF THE RESIDENCY PROGRAM THAT CAN BE FULLY APPLIED IN A FARM SETTING

One of the strengths of an ACVIM Food-Animal emphasis residency program is that this 3 year program is directed by one mentor and other ACVIM and non ACVIM Diplomates that have a broad knowledge of individual animal care. Having these highly specialized ressources create a highly stimulating environment for the resident which help to provide a high quality program. Another advantage to this program at least in some of the veterinary colleges offering that program is that surgical and medical cases are not separated by specialty. Internists can do (and needs to continue to do) most of the common surgeries performed in cattle. It is of tremendous importance to continue to offer the possibility to the resident to perform common surgical procedures (such as soft tissue surgeries, casts techniques for limb fractures). This lead to an important practical experience that is highly valuable for Candidates that will not stay in Academia and will go back to private practices. It is the author's opinion that it is quite impossible in the real field life to not perform surgeries and other common « generalist » procedures (pregnancy checking, minimal herd health data analysis…). This is an important difference when compared with equine internists who can then work in private referral centers exclusively doing internal medicine.

All that is learned in a veterinary teaching hospital is of potentially of great interest for the resident that will not practice in hospital conditions. Learning from referred cases with different ancillary tests greatly help to understand the interest of the different ancillary tests that can be done, how they can be interpreted and how they can be proposed and used in field practice. This may at least partially explain why in large veterinary clinics, the specialists are frequently asked for interpretations of CBC or biochemistry panels when the changes are not typical. On the other side, when the ACVIM-boarded specialist has been exposed to complex clinical cases management or unusual surgeries, he(she) is considered as the expert resource for "individual management" of sick animals. This is especially true in farms where the producers want to invest money on case management without referring the animal to a veterinary teaching hospital.

WEAKNESSES OF THE ACVIM-FOOD ANIMAL EMPHASIS PROGRAM FOR RESIDENTS WHO WON'T STAY IN HOSPITAL SETTINGS

On the opposite one of the big lacks of the program is the lack of population medicine approach such as basic reproduction as a mandatory part of the residency. The lack of practical epidemiology learning, a key component of the preventive medicine program, is also an important gap to consider since daily practice needs to be based on data management and collection. These concepts may have been used by the candidate that has a former experience in private practice but maybe not for the candidate coming from an internship done in hospital conditions.

These gaps of knowledge need to be covered at least their general concepts before coming back to the field. Since food animal industry is mostly based on cost-effectiveness medicine it is important for the resident to be able to know common economic data and concerning her/his recommendations. For example, the ACVIM board certified in practice can do in the same farm a transthoracic echocardiography for a cow with suspicion of heart disease; dehorn some dairy calves and then performing pregnancy examination suggesting hormonal protocols for open cows.

For all these reasons and since ACVIM residency program should prepare the candidate (at least those who expressed the desire to not stay in academia) to the more common procedures done in the field even if it is not purely considered as internal medicine from an academic point. Rotations in large veterinary practices, exposition to different situations using epidemiology and population medicine approach should be integrated in the residency program. This may increase the interest of the candidate for choosing a residency program that is more adapted to field needs.

INTEGRATION OF AN ACVIM-FA DIPLOMATE IN A FIELD SETTING

For this part of the talk I will use my own example and examples from other ACVIM-FA board certified veterinarians working in private practice. My example is biased since I am working in a field setting for the Faculté de medicine vétérinaire, which is mainly a dairy practice (90% dairy cows, 5% cow-calf operations and feedlots and 5% small ruminants (mainly meat lamb and dairy goats). The annual caseload is 4500 visits for individual cases (sick cows or calves) and 1500 visits for preventive medicine (monthly or bimonthly herd visits), with 4-5 veterinarians doing individual cases and 4 veterinarians doing mainly preventive medicine.

First rule: come back to the basic!

I would like to perform internal medicine every time of the day but unfortunately (or sometime fortunately) in my clinical time in the farm I do internal medicine represent less than 5 to 10% of my clinical time. My main activity represents classical bovine practitioner activity. I will treat a downer hypocalcemic periparturient cow with intravenous calcium, then doing a right-flank omentopexy in a Holstein cow with a left displaced abomasum, make some foot trimming in lame cows, doing a couple of pregnancy checking and transrectal examinations for anoestrus cows, finally treating a uterine prolapse in a beef cow. Having this said that, one can easily understand that with that type of day, internal medicine is not an important component of daily bovine practice.

However, practicing these daily interventions with ease is of primary importance to be fully trusted by the producers. These basic interventions could be seen as an investment for the internist to be able to propose more complex ancillary tests and procedures in more complicated cases. I consider these simple technical procedures as a very interesting part of my practice since it is always interesting to see that we can with simple procedure make a dramatic change in animal condition. For example, calving a cow with dystocia and deliver a live calf is one of the most beautiful sensations I have in my daily activity at least as good as doing an echocardiography to reach an endocarditis diagnosis in a cow with fever of unknown origin.

Second rule: Serve as a specialist opinion for tricky cases

As we discussed earlier, in large practices the ACVIM-board certified veterinarian is considered as a reference for unusual medical and surgical cases. For these reasons it is important for the internist to be able to serve as a second opinion veterinarian for more complicated cases. In these situations, the ACVIM board-certified veterinarian may have opportunities to use part of the skills that he learned to use during his/her residency program. This is particularly true for performing and interpreting ancillary tests such as different blood tests, CSF or other body fluid puncture, medical and surgical ultrasonography…Since the Diplomate has a unique experience of unusual and complex cases, he/she is the most qualified person in the veterinary clinic to do that type of work.

This can be seen as a valuable argument to encourage people to perform a residency program.

And… Be prepared for other opportunities!

The residency program and the ACVIM-FA training provide a unique background coming from the caseload seen during the program and the great teachers that are mentoring the resident (ACVIM Board-certified veterinarians as well as non ACVIM Board-certified veterinarians). Since most of the residents also have performed a clinical research project during their residency program they are a valuable resource for performing clinical research trials conducted by the industry or in relation with academia.

On the other side, they can actively be involved in continuing education in their practice or area of practice as well as to design small clinical trials to increase evidence-based veterinary medicine practice in the farm. For these reasons they are a unique resource for Hospital faculty to help in clinical research.[3]

Finally, other opportunities also mean exotic or uncommon ruminants medicine which can be very attractive for the Board-certified veterinarian. Small ruminant medicine, wild ruminant, New World Camelids, or water buffaloes can all be various examples in which the internist background in comparative medicine may help a lot to offer a better service for these clients.

CONCLUSIONS

Since recruiting resident in food animal internal medicine may be more and more challenging in a context where cost-effectiveness medicine is a reality in cattle industry, it is important to insist on the fact that ACVIM residency program can be used outside of the classical hospital setting. Integration of residents' rotations in veterinary clinics that may be susceptible to improve the resident's skills in preventive and population medicine as well as in structures employing specialist could be an interesting way to maintain a high interest for the resident. It is important to have Food-animal internists working in private practice since it may help to advance clinical research and finally medical care of Food Animals. They are also important maintaining the link between classical Hospital Internists and bovine practitioners.

REFERENCES

[1]AABP website, Mission statement http://www.aabp.org/ assessed in Januarry 30th, 2012, [2]ACVIM Website, http://www.acvim.org/websites/acvim/index.php?p=294 assessed in January 30th 2012, 2012. [3]Vandeweerd JM, et al., **Vet Clin North Am Food Anim Pract 2012;28.**

ZOONOTIC DISEASES: CASE SCENARIOS

David C. Van Metre, DVM, DACVIM
Fort Collins, CO

INTRODUCTION

Zoonotic diseases (zoonoses) are defined as infectious diseases that can be shared by humans and other vertebrate animals. Well over half of all currently known infectious diseases of humans are considered to be zoonoses.[1] Zoonoses comprise and even larger proportion of novel, emerging infectious disease threats, with three-quarters of all emerging infectious disease pathogens classified as having zoonotic potential.[2] In addition, endemic zoonotic diseases continue to pose a public health threat in North America; these range in occurrence and clinical consequences from common and mild (e.g. dermatophytosis) to relatively uncommon and severe (e.g. plague). In the past decade, there has been a resurgence of awareness of the interdependence of human and animal health, due in part to the potential for many zoonotic pathogens to serve as bioterrorism agents.

Farmers and ranchers, pet owners, veterinarians, and veterinary technicians often are the first persons to encounter infected animals; therefore, these individuals may serve as sentinels for incursion of zoonoses into human populations, bridges for disease incursion into human populations, or a first line of defense against such incursions.[3,4] These individuals have been demonstrated to be at risk of exposure to a spectrum of disease agents, including emerging pathogens such as monkeypox and endemic pathogens such as methicillin-resistant *Staphylococcus aureus* and *Salmonella enterica*.[4] For example, the seroprevalance of exposure to *Coxiella burnetii,* the causative agent of Q fever, was found to be nearly 10-fold higher in individuals with frequent contact with livestock than in the general public.[5] The probability of animal owners and animal health care providers serving as a sentinel of infection, a bridge of incursion into human populations, or a line of defense depends heavily upon awareness of disease risks and symptoms and prompt, clear communication among veterinarians, veterinary staff, those who work with livestock, and physicians. More frequent and efficient communication among physicians, veterinarians, and public health agencies is frequently cited as a vital component of effective zoonotic disease prevention.[3,4,6]

In a survey of veterinary practitioners in Washington, the majority of whom were involved in small animal practice, 28% of respondents reported having been diagnosed with a zoonosis during their careers.[6] Assuming that not all zoonotic infections are reported or diagnosed, the true prevalence of zoonoses in veterinarians is likely to be higher than what has been reported. A sound working knowledge of common zoonoses is essential for the veterinary practitioner. In this seminar, several zoonotic diseases will be reviewed, using a (hypothetical) case-scenario format. The zoonoses covered will be predominantly those considered to be commonly transmitted to humans from livestock, horses, and/or the farm /ranch environment. Data on the signalment, presenting complaint(s), and where available, historical data disease signs will be derived from published reports of human cases and applied to the hypothetical patient(s) in each scenario.[7] The working assumption for this seminar will be that large animal internists are generally well aware of the potential for exposure to such zoonotic disease agents as *Salmonella enterica*, methicillin-resistant *Staphylococcus aureus*, and *Cryptosporidium parvum*, owing to recent publication of multiple reports of disease outbreaks involving these agents in large animals at veterinary referral hospitals. This seminar is intended to focus on the salient clinical and epidemiologic features of less common zoonoses that are endemic to the U.S.[8]

ZOONOTIC DISEASE SCENARIOS

Scenario 1

Alabama - A 56-year old male veterinarian attended to a periparturient adult Holstein cow that was recumbent in a flooded pasture. The cow responded to intravenous calcium administration. Heavy rain had recently caused flooding of a nearby stream. During examination and treatment of the cow, the veterinarian's lower legs and feet were submerged. Eleven days later, the veterinarian became ill with fever, vomiting, and chills. Tremor, poor balance, and visual flashes of light began a few days later. The attending physician confirmed the disease as leptospirosis through serologic tests, and the veterinarian improved with a course of oral doxycycline.

Scenario 2

California - A 51 year-old rancher was referred to an infectious disease physician with a chief complaint of mildly pruritic but non-painful vesicles on his left hand. The vesicles had an erythematous base with a surrounding pale halo. Ten days previously, the rancher had shorn weaned lambs that had been purchased at an auction market. He had abraded his hands on thistles and burs embedded in the wool. Fluid aspirated from the lesions was submitted for testing for contagious ecthyma (orf) virus, which was detected by real-time PCR. Spontaneous healing occurred over the following 2 weeks.

Scenario 3

Montana – A 41 year-old, non-pregnant female veterinarian presented to a physician with a history of variable fever, coughing, myalgia, and fatigue of 2 weeks duration. The veterinarian reported having assisted in the births of several lambs, calves, and goat kids over the previous 2 months. One week previously, another physician had examined the veterinarian and made a diagnosis of influenza-related bronchitis. Treatment with 7 days of oral amoxicillin had not alleviated the patient's symptoms. The second physician initiated a 3-week course of doxycycline for presumed Q fever and submitted serum from the veterinarian to assess for Q fever exposure. Acute and convalescent serum samples from the veterinarian were tested for IgM and IgG antibodies to the phase I and II antigens of *Coxiella burnetii,* which confirmed the diagnosis. The veterinarian reported feeling better within 3 days after initiation of treatment and recovered.

Scenario 4

Iowa – A married couple, consisting of a 25 year-old husband and a 23 year-old wife, presented a 1 year-old Vietnamese pot-bellied pig to a veterinarian. The couple had noticed gradually progressive generalized pruritis, skin crusts, and erythema in the pig over the previous 4 weeks. The veterinarian confirmed the presence of *Sarcoptes scabei* var *suis* adult mites upon microscopic examination of deep skin scrapings of the axillae and external pinnae. When the veterinarian informed the owners that the mite can cause pruritic lesions in humans, the owners disclosed that they both had pruritic, red skin lesions on the arms and thorax. The owners were advised by the veterinarian to seek medical attention, and the veterinarian offered to consult with the physician if needed. The lesions in the owners gradually resolved after the pig was treated with injectable ivermectin.

Scenario 5

North Dakota – A 48 year-old male veterinarian presented to his physician with fever, headache, malaise and a non-painful, dime-sized cutaneous ulcer covered by a black eschar on his right wrist. Edema was marked in the right wrist, extending from the fingers to the elbow. The cutaneous lesion had begun as a pruritic papule approximately 7 days before presentation. The physician's diagnosis was cutaneous anthrax, which was confirmed by Gram stain and culture. The case was reported to state public health authorities. Subsequently, investigation of the patient history revealed no contact with suspected or confirmed cases of anthrax in animals. However, the veterinarian had changed a tire on his vehicle while traveling through a country where previous cases of anthrax had been reported. The lesion location on his arm suggested inoculation through a break in the skin caused by a steel watch band.

Scenario 6

New Mexico – During a farm visit, the attending veterinarian was informed that the farm owner, a 62 year-old female, was unable to participate in restraint and handling of livestock, owing to a febrile illness of 2 days duration. The farm owner's son reported that the woman had complained of arm pain that morning and had returned to bed after a meager breakfast; she had been disoriented and weak when her son had checked her later that day. Aware of the existence of plague in the region, the veterinarian discussed her concerns with the farmer's son, who then consented to seek medical care for his mother. The veterinarian telephoned the local physician, who dispatched an ambulance to the farm. The woman was hospitalized, and aspirates obtained from an enlarged axillary lymph node were positive for the F1 antigen of *Yersinia pestis*. She recovered following a course of intravenous gentamicin. Investigation by the state health department revealed plague infection in local ground squirrels and prairie dogs.

PREVENTION

Veterinarians are considered more likely to encounter zoonotic diseases in practice than are physicians.[3] In a Wisconsin survey, physicians other than infectious disease specialists professed to be relatively unfamiliar with the role of animal exposure and associated risks in the transmission of zoonoses to people.[3] Surveyed physicians in that study reported that they believed veterinarians should serve a greater role in educating animal owners about zoonoses and that veterinarians should provide information directly to physicians to aid in diagnosis, prevention, and patient education. Notably, it appears that animal owners may not consistently self-identify as being at higher risk for infectious diseases when communicating with their veterinarian. In a recent study, only 12 of 368 (3%) of surveyed veterinarians encountered clients on a daily basis who identified themselves or family members as having a higher risk of infection.[6] If veterinarians are to serve as effective educators on zoonoses risk, it appears that these conditions of unique disease susceptibility may need to be presented by the veterinarian, rather than relying on clients to initiate that discussion.

While prevention of exposure to all zoonotic agents is likely impossible, is worthwhile to recognize certain scenarios wherein zoonotic pathogens could be involved, and to take appropriate precautions while encouraging similar precautions in livestock owners. Situations that should be empirically considered as higher risk for exposure to zoonotic disease agents include the following: 1) diarrhea in neonatal or adult livestock; 2) obstetrical cases; 3) cases of abortion, stillbirth, or birth of weak or ill neonates; 4) unfamiliar signs of illness and/or unexplained death in livestock; and 5) progressive neurologic disease in livestock. Barrier precautions such as protective outerwear, boots, and gloves should be considered as routine when such situations are encountered.[9] Additionally, when the potential exists for aerosol exposure to particular zoonotic pathogens – for example, when handling tissues or bedding of livestock abortions - eye protection and respiratory particulate filters (e.g. masks rated at N95 or higher) should be worn.[9] Strict adherence to rigorous hand hygiene and careful handling and disinfection of soiled clothing and equipment are also essential preventive measures. Particular attention to hygienic measures is necessary if the veterinarian or his or her soiled work wear should come into contact with children, the elderly, pregnant women, or immune-suppressed individuals.

THE VETERINARIAN OR RANCHER AS A PATIENT

Communication among veterinarians and physicians may be particularly important when the veterinarian or the livestock owner assumes the role of the patient, because individuals within these two occupations can be considered to represent a very small proportion of the adult patient population seen by physicians in the U.S. When labor and census data are examined, it becomes clear that the veterinarian and the farmer / rancher represent very small proportions of the adult (>21 yr) population in the U.S. According to data from 2008-2010, veterinarians comprise approximately 0.02% of the approximately 221 million adults in the U.S.; ranchers, farmers, and agricultural managers together represent roughly 1% of that population,[10-12] with unknown fractions of each occupation encountering livestock on a daily to weekly basis. Simply put, one can use these crude estimates to hypothesize that the unique occupational risks encountered by these individuals may not be commonly considered by most physicians, particularly those outside of a rural setting or those without advanced training in infectious diseases. If that hypothesis were to be true, it then falls to the patient – the veterinarian and the farmer / rancher – to clearly communicate his or her occupational risks to the health care provider, including details on recent exposures that may be relevant to the illness at hand. To illustrate: Large animal veterinarians are likely aware that the springtime is a time period with higher potential for exposure to zoonotic agents shed by periparturient ruminants; handling of aborted fetuses and placentas, stillborn neonates, and weak neonates warrants particular caution. Diseases such as Q fever, toxoplasmosis, chlamydophilosis, leptospirosis, and listeriosis may occur after such exposures. Disclosure of any exposure may be particularly important in such cases, as the incidence of seasonal influenza in humans typically peaks in the late winter months. Accordingly, if flu-like signs comprise the presenting complaint, differentiation of these zoonoses from influenza can be potentially problematic for the physician.

REFERENCES

1. Woolhouse MEJ, et al. **Emerg Infect Dis** 2005;11:1842. 2. Taylor LH, et al. **Philos Trans R Soc Lond Biol Sci** 2001; 356:983. 3. Grant S, et al. **Emerg Inf Dis** 1999;5:159. 4. Wright JG, et al, **JAVMA** 2008; 232:1863. 5. McQuiston JH, et al. **Vector Borne Zoonotic Dis** 2002;2:179. 6. Lipton BA, et al. **JAVMA** 2008; 233:1242. 7. **Centers for Disease Control and Prevention**; http://www.cdc.gov/mmwr/ 8. **Centers for Disease Control and Prevention**, 2011; http://www.cdc.gov/mmwr/preview/mmwrhtml/mm6102md.htm?s_cid=mm6102md_w 9. Morley PS, Weese S; 2009. Biosecurity and infection control for large animal practices. In Smith BP, ed. **Large Animal Internal Medicine**, 4th Ed., 1524. 10. **Census 2010**; http://www.census.gov/prod/cen2010/briefs/c2010br-03.pdf 11. **US Bureau of Labor Statistics** 2008; http://www.bls.gov/oco/ocos176.htm#outlook 12. **US Bureau of Labor Statistics** 2010; http://www.bls.gov/oes/current/oes291131.htm#nat 10

MESENCHYMAL STEM CELLS: NATURE'S ALTERNATIVE TO STEROIDS?

Steven Dow, DVM, PhD, DACVIM
Ft. Collins, Colorado

BIOLOGY OF MESENCHYMAL STEM CELLS

Despite the current intense interest in stem cell therapy, the only stem cells likely to be used clinically in the near future are mesenchymal stem cells (MSC). Mesenchymal stem cells consist of a small population of self-renewing cells present in virtually all organs that can be enriched and induced to proliferate following *in vitro* culture. Recent studies have suggested that these cells should be considered mesenchymal stromal cells rather than mesenchymal stem cells because they lack the capacity to differentiate into all cell lineages. However, the MSC subpopulation does have stem cell-like qualities in the sense that they can be induced in vitro to differentiate into at least 3 cell types, including adipocytes, osteoblasts, and chondroblasts [1]. However, MSC cannot give rise to all cell lineages and are thus more appropriately considered multipotent rather than pluripotent cells.

Nonetheless, the ability of MSC to differentiate into a number of different mesenchymal tissues (eg, muscle, tendon, cartilage) makes MSC extremely attractive for use in orthopedic conditions [2]Le Blanc, 2006 #267}[3]. For this reason, stem cell therapy in veterinary medicine currently involves the use of MSC rather than true stem cells. The primary indications for use of MSC therapy in companion animal practice include treatment to enhance healing of acute orthopedic injuries, to stimulate bone healing and graft incorporation, and to reduce the severity of inflammation and discomfort in chronic osteoarthritis. While the use of MSC for orthopedic conditions in veterinary medicine is increasing, at present few well-controlled studies have been performed to substantiate the clinical benefits of MSC therapy.

It is also difficult to compare MSC studies because the cell preparations used in individual studies are often very different. For example, the most widely used MSC preparation in veterinary medicine currently consists of a very heterogenous population of cells (stromal vascular fraction cells) that are collected by enzymatic digestion of adipose tissues and then immediately injected, without undergoing *in vitro* culture or selection [4]. In vitro propagation of MSC results in selection for a much more homogeneous population of cells with well-defined properties. For example, mechanistic studies of MSC are typically conducted using cells obtained following *in vitro* propagation, and with cells that have been well-characterized phenotypically and through use of functional assays.

Fig 1. Feline adipose-derived mesenchymal stem cells. Following enzymatic digestion and culture of plastic-adherent cells, MSC were expanded from adipose tissues of healthy donor cats.

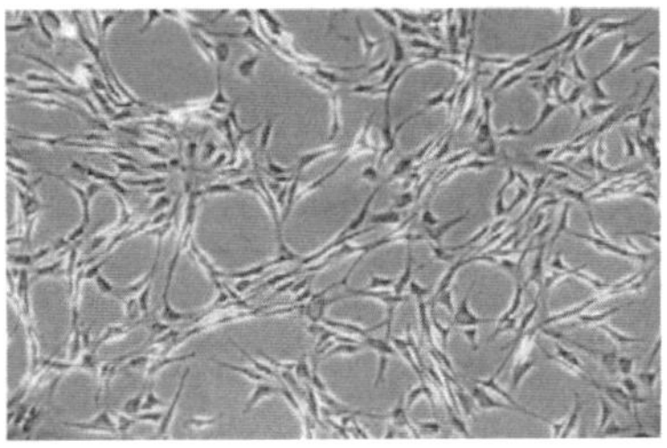

WHY USE MSC TO SUPPRESS IMMUNE RESPONSES?

Experimental studies in rodent models (though not necessarily in clinical trials in human or veterinary patients) have established the effectiveness of MSC therapy for stimulating musculoskeletal healing and regeneration. However, studies over the past 5 years have also uncovered another remarkable property of MSC, which is their ability to suppress immune responses and inflammation. Thus, it is now apparent that MSC propagated *in vitro* and administered to animals can potently suppress a variety of different inflammatory responses [5-9]. For example, MSC have been shown to suppress experimental colitis, arthritis, immune-mediated encephalitis, asthma, chronic kidney disease and graft-versus-host disease in rodents [10-12]. Moreover, small human clinical trials have also demonstrated apparent benefit from systemically administered MSC in suppressing inflammatory bowel disease and refractory graft-versus-host disease [13]. Indeed, the immune modulatory properties of MSC is now an important factor that must now be taken into account when interpreting the results of experimental or clinical studies involving stem cell therapy.

HOW DO MSC SUPPRESS IMMUNITY?

While administration of MSC suppresses immune responses, even in animals with established inflammatory diseases, the mechanisms responsible for these MSC properties have not been fully elucidated. Most mechanistic studies have been conducted primarily using *in vitro* propagated MSC. While the immune modulatory properties of MSC are well-established, there have been few published studies assessing the immunological properties of uncultured stromal vascular fraction cells isolated from adipose tissues. What has emerged from the MSC immune function studies conducted to date is the concept that MSC can suppress immune responses by a variety of distinct and non-redundant mechanisms (**Table 1**). The fact that MSC exert a variety of different immune modulatory or suppressive effects has made it difficult to identify a single predominant mechanism *in vivo*. Nonetheless, the emerging consensus is that MSC generally suppress immune responses by 3 primary immunological mechanisms: 1) release of broadly immune suppressive cytokines (IL-10, TGF-β); 2) production of reactive nitrogen or oxygen metabolites that target primarily T cells; and 3) release of soluble inhibitors such as galectins and soluble IL-1 receptor antagonist [14-15]. While *in vitro* assays can identify production of immune suppressive factors, recent studies also suggest that the interaction of MSC with their tissue environment *in vivo* can significantly modify their immune suppressive properties [16]. Thus, it is important to keep in mind the limitations of *in vitro* assays when assessing the true immune modulatory potential of MSC.

Mediator	Cellular targets	Mechanism
IL-10	T cells, APC	Suppress cytokine production
TGF-β	T cells, APC	Suppress cytokine production
Reactive nitrogen	T cells	Block T cell signaling
Reactive oxygen	APC	Block APC activation
Indoleamine-deoxygenase	APC	Block APC activation
PGE2a	APC	Block APC activation
Galectins	T cells	Immune suppressive receptors
Solu[illegible]	[illegible]	[illegible]tor

Table 1. Mechanisms of immune suppression by MSC

UNDERSTANDING MSC IMMUNE MODULATION FROM ANIMAL MODELS

The net effects of MSC immune modulation are best appreciated from studies of animal models of inflammation. For example, when mice with inflammatory bowel disease were treated with MSC, it was found that the MSC homed to the site of inflammation and once there began to produce immune suppressive factors including IL-10 and TGF-β [11]. The effects of MSC administration were profound, with marked suppression of intestinal inflammation and improvement in intestinal function. Similar results have also been obtained in mouse models of immune-mediated arthritis, autoimmune encephalitis, and asthma. In some cases, even more impressive suppression of inflammation can be achieved when MSC are co-administered with immunosuppressive drugs. It should also be noted that intravenous administration of MSC to humans with refractory inflammatory bowel disease or with severe graft-versus-host disease has been associated with significant clinical improvement, though these results have not yet been confirmed in randomized clinical trials.

CAN MSC THERAPY BE USED TO TREAT INFLAMMATORY DISORDERS IN ANIMALS?

The evidence accumulated from experimental studies in rodents, together with results from small pilot studies in human patients, indicate that MSC therapy can significantly downmodulate inflammation associated with immune-mediated disorders. Therefore, there is strong rationale to believe that MSC therapy could also demonstrate significant clinical benefit in companion animals with immune-mediated diseases. To date, results of clinical trials of MSC therapy for modulation of immunity in companion animals have not been published. However, our research group at the Center for Immune and Regenerative Medicine at Colorado State University, along with colleagues at the University of Missouri, is currently conducting clinical trials assessing the potential effectiveness of MSC therapy for immune modulation in animals.

In ongoing studies supported by the Winn Foundation and the Morris Animal Foundation, the effects of MSC therapy on intra-renal inflammation in cats with chronic kidney disease are being assessed [17]. In these studies, repeated MSC therapy is being administered to cats with CKD based on the hypothesis that MSC can suppress the marked intra-renal inflammatory response that is characteristic of tubulo-interstitial nephritis and CKD in cats. The effects of MSC therapy on renal inflammation are being assessed primarily by measuring the impact on cytokine concentrations in urine, particularly urinary IL-8 and MCP-1 concentrations, which have been shown in other studies to closely correlate with the degree of intra-renal inflammation. In preliminary studies, we have found that a series of 3 intravenous administrations of allogeneic MSC could significantly improve renal function in cats with naturally-occurring CKD (Quimby J and Dow S, manuscript in preparation).

In a second clinical trial being conducted by Dr. Carol Reinero at the University of Missouri, the effects of intravenous MSC therapy on the severity of asthma in cats are being evaluated. These studies are based on compelling data from our lab and others that MSC can potently suppress airway inflammation and resistance in mouse models of asthma. In preliminary studies, it was found that repeated MSC administration could significantly improve airway resistance in cats with experimentally induced asthma (Reinero C, unpublished data). Thus, there is promising preliminary evidence that MSC therapy can suppress inflammation in cats with immune-mediated diseases.

OPPORTUNITIES FOR IMMUNE MODULATION WITH MSC THERAPY IN DOGS AND CATS?

Based on studies in rodent disease models and early data from trials in companion animals, there is reason to believe that MSC therapy may be useful for treatment of selected immune-mediated disorders in companion animals. Indeed, the immune modulatory effects of MSC therapy have been observed in equines treated with cultured MSC for treatment of chronic joint disease. For example, in horses receiving intra-articular injections of MSC, it was found that injection of MSC alone resulted in increased intra-articular production of PGE2 and transient cellular infiltrates [18-19]. Injection of cultured MSC was also reported to be superior to injection of uncultured stromal vascular fraction cells in another study [18].

In dogs, there is a significant need for alternative therapies to improve control of the major immune-mediated disorders, including IMHA, IMTP, and immune-mediated polyarthritis. Other disorders potentially amenable to MSC immune modulatory therapy include inflammatory bowel disease, atopic dermatitis, and chronic allergic disorders. While it may not be realistic to expect MSC therapy alone to completely control these immune-mediated disorders, it may be more reasonable to expect that MSC therapy may reduce the requirement for high doses of immune suppressive drugs to control these disorders. For immune-mediated disorders in cats (in addition to asthma), there would be strong rationale for administration of MSC to control inflammatory bowel disease, and potentially to help ameliorate inflammation in chronic stomatitis and chronic sinusitis, both of which have a strong immune component.

SUMMARY AND IMPLICATIONS

In summary, the use of biological therapy with MSC holds considerable appeal as a novel and potentially effective means of controlling immune-mediated diseases, using an approach that has fewer side-effects than conventional therapy. However, there are still considerable hurdles to overcome. For one, the culture and expansion of MSC in laboratories is still very expensive and not yet widely available. However, the ability to safely utilize allogeneic MSC, which we have used in our studies, helps to alleviate some of the logistical hurdles. Other questions include the duration of the MSC immune modulatory effects, the number of cells that should be administered, and the optimal frequency of administration. Moreover, *in vitro* activation of MSC prior to administration may also considerably improve their effectiveness [20].

REFERENCES

1. Caplan AI. Adult mesenchymal stem cells for tissue engineering versus regenerative medicine. J Cell Physiol 2007;213:341-347. 2. Uccelli A, Moretta L, Pistoia V. Mesenchymal stem cells in health and disease. Nat Rev Immunol 2008;8:726-736. 3. Wang S, Qu X, Zhao RC. Mesenchymal stem cells hold promise for regenerative medicine. Front Med 2011;5:372-378. 4. Black LL, Gaynor J, Adams C, et al. Effect of intraarticular injection of autologous adipose-derived mesenchymal stem and regenerative cells on clinical signs of chronic osteoarthritis of the elbow joint in dogs. Vet Ther 2008;9:192-200. 5. Abumaree M, Al Jumah M, Pace RA, et al. Immunosuppressive Properties of Mesenchymal Stem Cells. Stem Cell Rev 2011. 6. Cho KS, Roh HJ. Immunomodulatory effects of adipose-derived stem cells in airway allergic diseases. Curr Stem Cell Res Ther 2010;5:111-115. 7. Griffin MD, Ritter T, Mahon BP. Immunological aspects of allogeneic mesenchymal stem cell therapies. Hum Gene Ther 2010;21:1641-1655. 8. Jones BJ, McTaggart SJ. Immunosuppression by mesenchymal stromal cells: from culture to clinic. Exp Hematol 2008;36:733-741. 9. Rasmusson I. Immune modulation by mesenchymal stem cells. Exp Cell Res 2006;312:2169-2179. 10. Park HK, Cho KS, Park HY, et al. Adipose-derived stromal cells inhibit allergic airway inflammation in mice. Stem Cells Dev 2010;19:1811-1818. 11. Gonzalez MA, Gonzalez-Rey E, Rico L, et al. Adipose-derived mesenchymal stem cells alleviate experimental colitis by inhibiting inflammatory and autoimmune responses. Gastroenterology 2009;136:978-989. 12. Semedo P, Correa-Costa M, Antonio Cenedeze M, et al. Mesenchymal stem cells attenuate renal fibrosis through immune modulation and remodeling properties in a rat remnant kidney model. Stem Cells 2009;27:3063-3073. 13. Wernicke CM, Grunewald TG, Juenger H, et al. Mesenchymal stromal cells for treatment of steroid-refractory GvHD: a review of the literature and two pediatric cases. Int Arch Med 2011;4:27. 14. Nemeth K, Keane-Myers A, Brown JM, et al. Bone marrow stromal cells use TGF-beta to suppress allergic responses in a mouse model of ragweed-induced asthma. Proc Natl Acad Sci U S A 2010;107:5652-5657. 15. Bunnell BA, Betancourt AM, Sullivan DE. New concepts on the immune modulation mediated by mesenchymal stem cells. Stem Cell Res Ther 2010;1:34. 16. Waterman RS, Tomchuck SL, Henkle SL, et al. A new mesenchymal stem cell (MSC) paradigm: polarization into a pro-inflammatory MSC1 or an Immunosuppressive MSC2 phenotype. PLoS One 2010;5:e10088. 17. Quimby JM, Webb TL, Gibbons DS, et al. Evaluation of intrarenal mesenchymal stem cell injection for treatment of chronic kidney disease in cats: a pilot study. J Feline Med Surg 2011;13:418-426. 18. Frisbie DD, Kisiday JD, Kawcak CE, et al. Evaluation of adipose-derived stromal vascular fraction or bone marrow-derived mesenchymal stem cells for treatment of osteoarthritis. J Orthop Res 2009;27:1675-1680. 19. Frisbie DD, Smith RK. Clinical update on the use of mesenchymal stem cells in equine orthopaedics. Equine Vet J 2010;42:86-89. 20. DelaRosa O, Lombardo E. Modulation of adult mesenchymal stem cells activity by toll-like receptors: implications on therapeutic potential. Mediators Inflamm 2010;2010:865601.

MONOCYTES REDISCOVERED: THEIR CRITICAL ROLE IN CANCER AND INFECTION

Steven Dow, DVM, PhD, DACVIM
Ft. Collins, CO

RECENT ADVANCES IN UNDERSTANDING MONOCYTE BIOLOGY

Most clinicians can readily generate a list of functions performed by neutrophils, but the odds are most will be stumped when it comes to generating a similar list for monocytes. Monocytes are largely ignored in the veterinary curriculum, and few veterinary researchers study monocytes. Who can remember the last time they got excited about finding monocytosis on a CBC? Or worried when the monocyte count was low? The reality is that monocytes are virtually a forgotten cell type when it comes to clinical medicine.

However, things have begun to change in the monocyte arena and there has recently been a marked resurgence of interest in monocytes and monocyte biology, spanning disciplines as diverse as infectious diseases, autoimmunity, cancer, and developmental biology. What has sparked this renewed interest? In part, improved availability of reagents has made it possible to identify specific monocyte subsets with diverse functions in health and disease in rodents and humans. For example, in human medicine there is now a new understanding of the critical role of inflammatory monocytes in atherosclerosis, stroke, and myocardial infarction and in the metabolic inflammatory syndrome associated with obesity. Identifying the chemokines and receptors that regulate monocyte mobilization and recruitment has also sparked interest in development of new drugs to block the deleterious effects of monocytes. In addition, new animal models of monocyte depletion have generated remarkable new insights into the often contradictory roles of monocytes in infectious diseases and cancer.

TWO POPULATIONS OF MONOCYTES IN HEALTH AND DISEASE

Monocytes are derived from progenitor cells in the bone marrow (common myeloid progenitors) from which neutrophils are also derived. The generation of monocytes in the bone marrow is governed by a complex of set of cytokines and growth factors, including GM-CSF, M-CSF, IL-3, and SCF (stem cell factor) [1]. In the bloodstream, there are normally two distinct subpopulations of monocytes, known as steady state monocytes (also known as non-classical monocytes) and inflammatory monocytes (also known as classical monocytes)[2]. The two monocyte subpopulations can be distinguished in mice and humans by using flow cytometry and a panel of at least 3-4 antibodies recognizing specific cell surface determinants. In health, a small population of inflammatory monocytes exists in the bloodstream, while the majority of circulating monocytes are steady state monocytes [1,3]. After circulating for several days, steady state monocytes leave the bloodstream at a relatively constant rate and differentiate into tissue macrophages in all tissues except the CNS. In contrast, during inflammation, which can include infection, trauma, or even vaccination, the numbers of inflammatory monocytes in circulation increases rapidly. Most inflammatory monocytes leave the circulation and only enter tissues where inflammation is present. Here, the inflammatory monocytes differentiate into short-lived inflammatory macrophages or inflammatory dendritic cells (DC). Once the inflammatory stimulus is removed, the numbers of inflammatory monocytes in circulation rapidly returns to normal.

The primary signal for mobilization of inflammatory monocytes from the bone marrow and into circulation is delivered by the chemokine MCP-1 (also known as CCL2), which is produced in response to inflammation by many different cell types, including epithelial cells, endothelial cells, and bone marrow stromal cells [4]. MCP-1 also serves a critical role in recruiting monocytes from the bloodstream and into sites of tissue inflammation. Monocytes respond to MCP-1 primarily via the receptor CCR2, a G protein coupled receptor. Strategies to block monocyte-mediated inflammation have therefore focused on blocking MCP-1 production, neutralizing its activity, or blocking the CCR2 receptor [5].

The number of steady state monocytes in circulation is governed primarily by the chemokine CX3CL1, also known as fractaline. The only receptor for CX3CL1 is CX3CR1, which is expressed primarily on steady state monocytes and thus serves as a useful marker for this subpopulation of monocytes in the bloodstream. Much less is known about the biology of steady state monocytes and their role in health and disease.

Regulation/Function	Steady state monocyte	Inflammatory monocyte
Receptor expression	CX3CR1, CD16 (hu), CD43 (m)	CCR2, CD14 (hu), Ly6C (m)
Circulating in health	Constant basal level	Low
Circulating in inflammation	Unchanged	Rapidly increased
Role in infection	None?	Control acute bacterial infection
Role in cancer	None?	Suppress tumor immunity
Role in inflammation	None?	Promote vascular injury, organ damage

ROLE OF MCP-1 AND INFLAMMATORY MONOCYTES IN INFECTION

Studies in mice lacking the CCR2 receptor (CCR2$^{-/-}$ mice), or the ability to produce MCP-1 (CCL2$^{-/-}$ mice), have revealed a key role for inflammatory monocytes in generating protective immunity against bacterial infections[6]. For example, CCR2$^{-/-}$ mice were significantly impaired in their ability to control a sublethal challenge with *Listeria monocytogenes*, due in part to a marked defect in the ability of the mice to produce NO[7]. Similarly, decreased resistance to infection was also observed in CCR2$^{-/-}$ mice infected with Salmonella or with Mycobacterium. In studies in our laboratory, we observed that CCR2$^{-/-}$ mice were markedly impaired in their ability to withstand low-dose pulmonary inoculation with the gram-negative pathogen *Burkholderia pseudomallei*[8]. Notably, the defect in protective immunity to infection was associated with a marked reduction in the numbers of inflammatory monocytes recruited to the lungs, which could not be overcome by greatly increased numbers of neutrophils. The $CCR2^{-/-}$ mice were also significantly impaired in their ability to mount an IFN-γ response, which may therefore represent the critical immune defect. Thus, mouse models of infection reveal a key role for rapidly mobilized monocytes in controlling acute infections with both $Gram^-$ and $Gram^+$ pathogens. It is quite likely therefore that inflammatory monocytes play a similar role in bacterial infections in humans and companion animals.

PARADOXICAL IMMUNE SUPPRESSIVE ROLE OF INFLAMMATORY MONOCYTES IN CANCER

The preceding results would lead one to believe that monocytes also play a protective role in cancer. While for years it has been assumed that monocytes and macrophages help control tumor growth, more recent studies suggest that just the opposite is true. Indeed, increased numbers of tumor-associated macrophages and circulating monocytes have been found to be a consistent negative prognostic finding in a number of different studies in humans with multiple different tumor types. Tumor associated macrophage suppress tumor immunity by multiple independent mechanisms, including suppression of T cell immunity, stimulation of tumor metastasis, enhancement of tumor angiogenesis, and promotion of tumor genetic instability. More recently, an important role for inflammatory monocytes in promoting tumor metastasis was also identified [9]. In those studies, it was found that inflammatory monocytes were recruited to sites of tumor metastases, where the monocytes promoted tumor growth and survival. Blocking monocyte recruitment to sites of tumor metastases significantly suppressed the growth and spread of tumor metastases. Thus, it is apparent that monocytes play very different roles in cancer and infections.

MONOCYTE ABNORMALITIES IN COMPANION ANIMALS AND THE SIGNIFICANCE OF MONOCYTOSIS

The literature regarding monocytes and their clinical significance in veterinary medicine is sparse. Monocytosis has been reported to occur in dogs as a paraneoplastic syndrome, and also following rebound from parvovirus infection[10]. Clinically, persistent monocytosis has been reported to be a reliable biomarker for the presence of persistent fungal infection in dogs with cocidiodomycosis [11]. Persistent monocytopenia as a clinical entity has not been reported previously in dogs, although dogs with pronounced bone marrow suppression due to chemotherapy or canine parvovirus suppression will develop monocytopenia concurrently with neutropenia[12]. In animals treated with chemotherapy that develop significant myelosuppression, monocyte recovery often precedes neutrophil recovery and the appearance on monocytes in the bloodstream may therefore be the earliest indication of bone marrow recovery.

MCP-1 CONCENTRATIONS IN DOGS AND CATS WITH INFECTIOUS AND AUTOIMMUNE DISEASES.

Several recent studies have evaluated MCP-1 concentrations in dogs with infectious or immune-mediated disorders. We reported recently that serum MCP-1 concentrations were significantly increased in dogs with sepsis, IMHA and severe trauma [13]. Moreover, we found that serum MCP-1 concentrations were significantly correlated with outcomes in dogs with both sepsis and IMHA, wherein dogs with the highest MCP-1 concentrations were significantly more likely to succumb to their disease. A study in human critically-ill patients found that MCP-1 concentrations were the single best predictor of treatment outcomes when 22 different cytokines were compared [14]. Another recent study in dogs found that MCP-1 concentrations were significantly elevated in dogs with IMHA, and that this cytokine (along with IL-18) was significantly correlated with decreased survival [15]. Thus, MCP-1 concentrations, and by extension inflammatory monocytes, appear to play critical roles in both infectious and immune-mediated diseases of dogs. In cats with chronic kidney disease (CKD), we have found recently that urinary concentrations of MCP-1 were significantly increased compared to healthy cats, which we have attributed to monocyte/macrophage infiltration due to tubulointerstitial nephritis (Quimby, et al, manuscript submitted).

INCREASED NUMBERS OF MONOCYTES ASSOCIATED WITH POOR PROGNOSIS IN DOGS WITH OSTEOSARCOMA AND LYMPHOMA.

We recently evaluated the role of monocytes in dogs with cancer, based on accumulating clinical evidence that increased numbers of monocytes are associated with more rapid tumor progression and metastasis in humans with cancer [16]. In one study, we evaluated that association between circulating numbers of monocytes and time to development of metastases in dogs with osteosarcoma. Increased numbers of circulating monocytes were strongly associated with decreased time to development of metastasis, and were the only leukocyte parameter that predicted disease outcome. Interestingly, most animals at the time of initial diagnosis did not have overt monocytosis. Instead, we found that animals with monocyte counts in the upper range of normal had a significantly shorter time to development of metastases than those with a monocyte count in the lower range of normal. Though reagents to identify canine monocyte subsets are not yet available, we would predict that the animals with the highest numbers of inflammatory monocytes would have the shortest disease-free intervals.

In a second study of dogs with B cell lymphoma, the relationship between monocyte count and serum MCP-1 concentration and duration of first remission following chemotherapy was evaluated [17]. Here again we also found that higher monocyte counts were significantly associated with poorer treatment outcomes, as reflected by significantly shorter disease remission. In addition, there was also a significant correlation between increased MCP-1 concentrations and shorter remission times. Therefore, this second study also suggests that increased numbers of monocytes may suppress tumor immunity or promote tumor growth and metastasis, or both.

SUMMARY, IMPLICATIONS, AND AREAS FOR FUTURE INVESTIGATION.

While monocytes have been ignored for years, it is apparent that now may be a good time to revisit their role in diseases of companion animals. However, it is also worth noting that investigating the role of monocytes may not be as simple as looking at CBCs to identify animals with monocytosis. Indeed, the presence of two functionally distinct monocyte subsets suggests that it will be much more fruitful to examine monocyte subsets and their correlations with various disease states. The large interest by pharmaceutical companies in development of chemokine receptor antagonists for treatment of inflammatory disorders in humans also suggests that there may soon be drugs available for suppressing the numbers of inflammatory monocytes in animals with cancer or other diseases characterized by pathological inflammation.

REFERENCES

1. Ingersoll MA, Platt AM, Potteaux S, et al. Monocyte trafficking in acute and chronic inflammation. Trends Immunol 2011;32:470-477. 2. Ziegler-Heitbrock L, Ancuta P, Crowe S, et al. Nomenclature of monocytes and dendritic cells in blood. Blood 2010;116:e74-80. 3. Gautier EL, Jakubzick C, Randolph GJ. Regulation of the migration and survival of monocyte subsets by chemokine receptors and its relevance to atherosclerosis. Arterioscler Thromb Vasc Biol 2009;29:1412-1418. 4. Garin A, Proudfoot AE. Chemokines as targets for therapy. Exp Cell Res 2011;317:602-612. 5. Proudfoot AE, Power CA, Schwarz MK. Anti-chemokine small molecule drugs: a promising future? Expert Opin Investig Drugs 2010;19:345-355. 6. Serbina NV, Jia T, Hohl TM, et al. Monocyte-mediated defense against microbial pathogens. Annu Rev Immunol 2008;26:421-452. 7. Serbina NV, Salazar-Mather TP, Biron CA, et al. TNF/iNOS-producing dendritic cells mediate innate immune defense against bacterial infection. Immunity 2003;19:59-70. 8. Goodyear A, Jones A, Troyer R, et al. Critical protective role for MCP-1 in pneumonic Burkholderia mallei infection. J Immunol 2010;184:1445-1454. 9. Pollard JW. Tumour-educated macrophages promote tumour progression and metastasis. Nat Rev Cancer 2004;4:71-78. 10. Petterino C, Luzio E, Baracchini L, et al. Paraneoplastic leukocytosis in a dog with a renal carcinoma. Vet Clin Pathol 2011;40:89-94. 11. Kerl ME. Update on canine and feline fungal diseases. Vet Clin North Am Small Anim Pract 2003;33:721-747. 12. Marinaro M, Mari V, Bellacicco AL, et al. Prolonged depletion of circulating CD4+ T lymphocytes and acute monocytosis after pantropic canine coronavirus infection in dogs. Virus Res 2010;152:73-78. 13. Duffy AL, Olea-Popelka FJ, Eucher J, et al. Serum concentrations of monocyte chemoattractant protein-1 in healthy and critically ill dogs. Vet Clin Pathol 2010;39:302-305. 14. Bozza FA, Salluh JI, Japiassu AM, et al. Cytokine profiles as markers of disease severity in sepsis: a multiplex analysis. Crit Care 2007;11:R49. 15. Kjelgaard-Hansen M, Goggs R, Wiinberg B, et al. Use of serum concentrations of interleukin-18 and monocyte chemoattractant protein-1 as prognostic indicators in primary immune-mediated hemolytic anemia in dogs. J Vet Intern Med 2011;25:76-82. 16. Sottnik JL, Rao S, Lafferty MH, et al. Association of blood monocyte and lymphocyte count and disease-free interval in dogs with osteosarcoma. J Vet Intern Med 2010;24:1439-1444. 17. Perry JA, Thamm DH, Eickhoff J, et al. Increased monocyte chemotactic protein-1 concentration and monocyte count independently associate with a poor prognosis in dogs with lymphoma. Vet Comp Oncol 2011;9:55-64.

MOUSE MODELS TO STUDY NORMAL AND MALIGNANT HEMATOPOIESIS

Daniel Lacorazza, PhD
Houston, Texas

Normal hematopoiesis

Homeostasis of tissues relies on the multi-potentiality of a rare population of tissue-specific stem cells. The blood-forming hematopoietic stem cells (HSCs) are the most well characterized adult stem cells. Hematopoiesis is the formation of blood cells from primitive hematopoietic progenitor cells. Two important features of HSCs are their capacity to self-renew and to form all types of blood cells (multi-lineage potential) [1]. Self-renewal is an important property of HSCs, it is the ability to generate daughter cells that are functionally identical to the stem cell that gave rise to them. HSCs are lodged in specialized microenvironments (niches) that support their dormancy (quiescence) and self-renewal, more than 80% of HSCs remain in this reversible cell cycle arrest [2,3]. A balanced control of proliferation and differentiation maintains the number of HSCs in the bone marrow throughout an organism's lifespan, ensuring a continuous supply of blood cells during homeostasis without leading to stem cell exhaustion. Only few thousands HSCs can generate billions of blood cells every day at expense of the extraordinary proliferative capacity of hematopoietic progenitor cells. HSCs confront several choices upon commitment to proliferate (Figure 1): differentiation into multiple lineages, self-renewal, cell death, mobilization to peripheral tissues or re-entry into a quiescent state. The initial entry to the cell cycle is regulated by cell intrinsic and extrinsic factors. HSCs first differentiate into multipotent progenitor cells (MPP) that can commit to either myeloid (common myeloid progenitor cells: CMP) or lymphoid (common lymphoid progenitor cells: CLP) lineages (Figure 2). Cell cycle toxicity (i.e. chemotherapy, radiation) eliminates most of progenitor cells and mature cells leading to a lymphopenic environment that promotes proliferation of quiescent HSC to re-establish the entire hematopoietic system. Similarly, bone marrow transplants take advantage of this property of HSCs to reconstitute an ablated host.

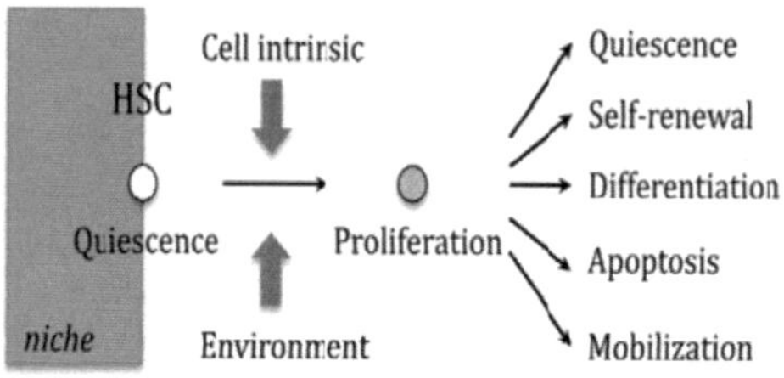

Figure 1: Hematopoietic stem cells are blood-forming stem cells located in the bone marrow. Release from their quiescent state and entry into the cell cycle exposes HSCs to various cell fates.

Isolation and study of HSCs

Multi-parameter flow cytometry and the use of monoclonal antibodies conjugated to fluorochromes ignited HSC research by allowing isolation of a wide variety of stem cells. There is no single cell surface marker that can define a stem cell population; however, a combination of markers of differentiated cells, to eliminate mature blood cells, and markers of primitive hematopoietic progenitor cells can delineate cell populations enriched in HSCs. For example, murine HSCs can be defined as negative for lineage markers (Lin negative), positive for Sca-1 and c-kit, and negative for Flk2 and CD34 (Lin^- $Sca\text{-}1^+$ $c\text{-}kit^+$ $Flk2^-$ $CD34^-$) [4,5] or using the SLAM markers as Lin^- $Sca\text{-}1^+$ $c\text{-}kit^+$ $CD150^+$ $CD48^-$ $CD41^-$ cells [6]. Other groups have taken advantage of the capacity of stem cells to extrude the fluorescent dye Hoechst 33342 to define HSCs as those bone marrow cells with low retention of Hoechst 33342 (side population) [7]. This differential dye efflux activity can be also combined with cell surface markers.

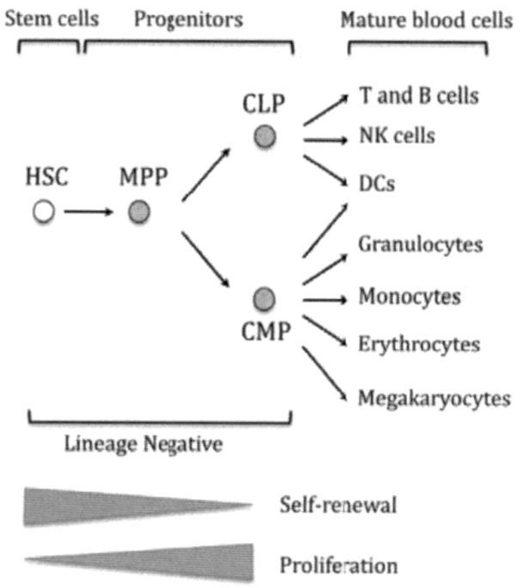

Figure 2: HSCs differentiate into multipotent progenitor cells (MPP) that then commit to either lymphoid (common lymphoid progenitors: CLP) or myeloid (common myeloid progenitors: CMP) lineages. Self-renewal potential decreases with differentiation.

Although the immunophenotypic analysis by flow cytometry has been useful to define hematopoietic populations enriched in HSCs, the best way to define stem cells is through functional assays. Most of the HSC studies are based on a retrospective definition of stem cell characteristics using functional studies both *in vitro* and *in vivo*. Various functional assays have been developed that test different aspects of hematopoietic progenitor and stem cells. The colony forming cell (CFC) assay consists in plating cells in a semi-solid medium in the presence of cytokines (i.e. IL-3, IL-6, thrompoietin, stem cell factor) and counting the number of clonal colonies formed after incubation [8,9]. Erythroid, granulocyte-macrophage and multi-potential granulocyte-erythroid-macrophage-megakaryocyte progenitors can be enumerated in total bone marrow cells using this assay. The gold standard to define stem cells is to test their capacity to restore all blood lineages (multi-lineage potential) in lethally irradiated recipient mice for long time (stem cell function) [4,10]. In bone marrow transplants, a test population (bone marrow cells from knockout mice) is transplanted into recipient mice with different congenic markers to easily distinguish donor from host to provide long-term bone-marrow engraftment. Bone marrow transplants can be performed in non-competitive (using dilutions of donor cells to enumerate HSCs), competitive (to test proliferative potential) and serially (to test self-renewal capacity). To assess the number of stem cells, limiting dilution transplantation studies are performed and the number of long-term HSCs is calculated using Poisson statistics.

Genetic control of HSC quiescence

Our group focuses on genetic studies using mouse models to define the role of stem cell quiescence in the maintenance of stem cell function. These studies also provide insights into the molecular mechanisms that regulate entry and progression through the cell cycle in normal stem cells and during leukemogenic transformation. Using homozygous deletion of the murine gene, we identified the transcription factor ELF4 as a negative regulator of HSC quiescence (Figure 3) [11,12]. *Elf4*-null mice exhibit an accumulation of $Lin^- Sca\text{-}1^+ c\text{-}kit^+ Flk2^- CD34^-$ cells in the bone marrow and enhanced competitive reconstitution in secondary transplants. The increased number and function of HSCs coincided with increased quiescence compared to wild type HSCs and thus *Elf4*-null mice recovered more rapidly from treatment with cytoablative agents that target dividing cells such as chemodrugs (i.e. 5-fluorouracil) or radiation. Even though *Elf4*-null HSCs are more quiescent during homeostatic hematopoiesis, HSCs can rapidly proliferate after chemotherapy, implying that HSC proliferation is differentially regulated during steady state and regenerative hematopoiesis. Collectively, our data suggest that ELF4 is a novel negative regulator of HSC quiescence and its deletion results in increased quiescence and expansion of HSCs (Figure 3). Modulation of ELF4 expression in HSCs may have clinical impact by enhancing engraftment efficiency in bone marrow transplants. In addition, the ability to target ELF4 function in HSCs can be important for leukemic stem cells, a rare population of leukemic cells proposed to maintain blood cell cancers. The majority of therapies currently used for leukemia treatment have been designed based on the highly proliferative properties of malignant blasts although a rare subset of leukemic cells with stem cell features are largely quiescent and thus they are refractory to current treatments leading to relapses.

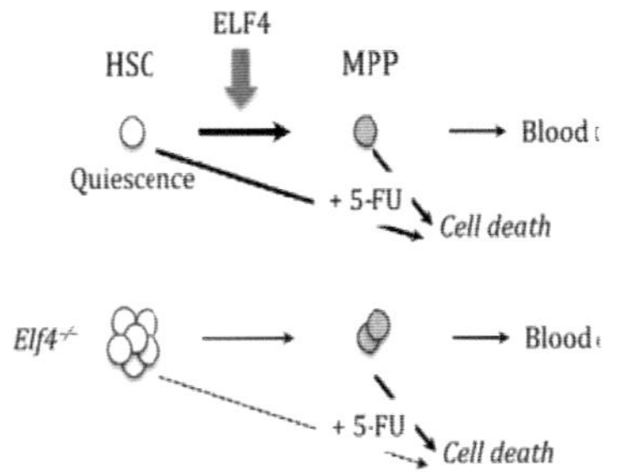

Figure 3: ELF4 negatively regulates stem cell quiescence by promoting cell cycle entry of HSC during homeostasis.

Malignant hematopoiesis and mouse models of acute lymphoblastic leukemia

Many leukemias arise by transformation of primitive hematopoietic progenitor cells leading to deregulated expansion and survival of a specific blood lineage. The discovery of mutations or chromosome translocations in human patients allowed the development of leukemic mouse models by introducing in murine bone marrow cells those genetic alterations found in patients and transplanting them in host mice to recapitulate the leukemogenic process. As described above, we identified a transcription factor that promotes cell cycle entry in HSCs [12]. Interestingly, same molecule induces cell cycle arrest in naïve CD8+ T cells by activating the expression of the tumor suppressor KLF4 [13]. Therefore, we are currently testing the hypothesis that inactivation of this 'proliferation brake' can contribute to the development and progression of lymphoblastic leukemias.

Acute lymphoblastic leukemia (ALL) is the most common hematological malignancy in children. Even though most of these patients have good therapeutic outcome, a fifth of them relapse after standard treatment, which is a common cause of cancer-related deaths in children. Insufficient information on ALL pathogenesis is a major roadblock in the development of alternative therapies for these high-risk patients. In spite of the identification of specific chromosomal translocations and mutations in ALL, the generation and outgrowth of pre-leukemic cells are likely caused by a combination of additional genetic aberrations and environmental pressure. A well-established mouse model of T-cell ALL (T-ALL) is based on the expression of activating mutants of NOTCH1, found in more than 50% of T-cell ALL patients, in murine bone marrow cells using retroviral gene transfer. NOTCH1 encodes for a transmembrane receptor involved in differentiation, proliferation and survival. Ligand-induced proteolytic processing releases the intracellular NOTCH1 domain which then translocates to the nucleus to regulate gene expression [14,15]. Retroviral-driven expression of intracellular NOTCH1 leads to strong NOTCH1 signaling and consequently to an aggressive T-ALL in mouse models. In contrast, the expression of NOTCH1 mutants that leads to ligand-independent activation can induce T-ALL with longer latency and lower penetrance (Figure 4). To study the role of specific tumor suppressors, a multi-step leukemogenic process can be recapitulated by using for transformation bone marrow cells from mice that carry deletion of the tumor suppressor. Homozygous deletion can be used if null mice are viable. In case of lethality, a floxed allele is used to induce deletion with lineage specific (i.e. vav-cre-recombinase) or inducible (ER-ROSA-recombinase). In this way, transformation can be manipulated before or after gene deletion.

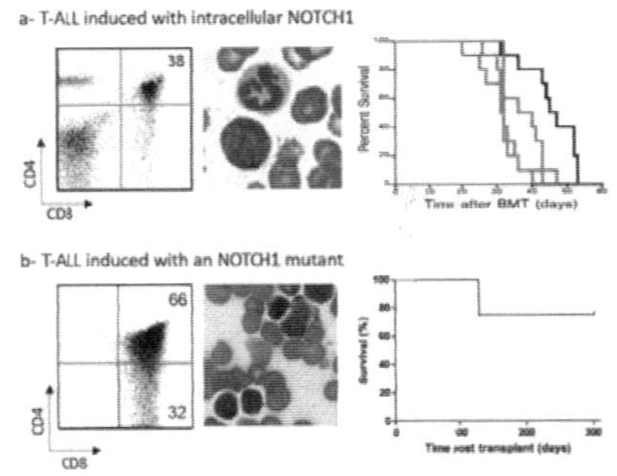

Figure 4: Mouse models of T-cell Acute Lymphoblastic Leukemia (T-ALL). Bone marrow cells are transduced with retrovirus carrying either intracellular NOTCH1 (a) or an activated NOTCH1 mutant (b) and transplanted into recipient mice. Flow cytometry, blood smears and survival curves are shown.

REFERENCES

1. Spangrude, G.J., Heimfeld, S. & Weissman, I.L. Purification and characterization of mouse hematopoietic stem cells. *Science* **241**, 58-62 (1988). 2. Passegue, E., Wagers, A.J., Giuriato, S., Anderson, W.C. & Weissman, I.L. Global analysis of proliferation and cell cycle gene expression in the regulation of hematopoietic stem and progenitor cell fates. *J Exp Med* **202**, 1599-1611 (2005). 3. Cheshier, S.H., Morrison, S.J., Liao, X. & Weissman, I.L. In vivo proliferation and cell cycle kinetics of long-term self-renewing hematopoietic stem cells. *Proc Natl Acad Sci U S A* **96**, 3120-3125 (1999). 4. Osawa, M., Hanada, K., Hamada, H. & Nakauchi, H. Long-term lymphohematopoietic reconstitution by a single CD34-low/negative hematopoietic stem cell. *Science* **273**, 242-245 (1996). 5. Christensen, J.L. & Weissman, I.L. Flk-2 is a marker in hematopoietic stem cell differentiation: a simple method to isolate long-term stem cells. *Proc Natl Acad Sci U S A* **98**, 14541-14546 (2001). 6. Yilmaz, O.H., Kiel, M.J. & Morrison, S.J. SLAM family markers are conserved among hematopoietic stem cells from old and reconstituted mice and markedly increase their purity. *Blood* **107**, 924-930 (2006). 7. Goodell, M.A., Brose, K., Paradis, G., Conner, A.S. & Mulligan, R.C. Isolation and functional properties of murine hematopoietic stem cells that are replicating in vivo. *J Exp Med* **183**, 1797-1806 (1996). 8. Leary, A.G., Ogawa, M., Strauss, L.C. & Civin, C.I. Single cell origin of multilineage colonies in culture. Evidence that differentiation of multipotent progenitors and restriction of proliferative potential of monopotent progenitors are stochastic processes. *J Clin Invest* **74**, 2193-2197 (1984). 9. Suda, T., Suda, J. & Ogawa, M. Single-cell origin of mouse hemopoietic colonies expressing multiple lineages in variable combinations. *Proc Natl Acad Sci U S A* **80**, 6689-6693 (1983). 10. Harrison, D.E. Competitive repopulation: a new assay for long-term stem cell functional capacity. *Blood* **55**, 77-81 (1980). 11. Lacorazza, H.D., *et al.* The ETS protein MEF plays a critical role in perforin gene expression and the development of natural killer and NK-T cells. *Immunity* **17**, 437-449 (2002). 12. Lacorazza, H.D., *et al.* The transcription factor MEF/ELF4 regulates the quiescence of primitive hematopoietic cells. *Cancer Cell* **9**, 175-187 (2006). 13. Yamada, T., Park, C.S., Mamonkin, M. & Lacorazza, H.D. Transcription factor ELF4 controls the proliferation and homing of CD8+ T cells via the Kruppel-like factors KLF4 and KLF2. *Nat Immunol* **10**, 618-626 (2009). 14. Grabher, C., von Boehmer, H. & Look, A.T. Notch 1 activation in the molecular pathogenesis of T-cell acute lymphoblastic leukaemia. *Nat Rev Cancer* **6**, 347-359 (2006). 15. Aster, J.C., Pear, W.S. & Blacklow, S.C. Notch signaling in leukemia. *Annu Rev Pathol* **3**, 587-613 (2008).

DIFFERENTIATION OF EFFECTOR AND MEMORY T CELLS UPON IMMUNE RESPONSE

Daniel Lacorazza, PhD
Houston, Texas

Differentiation of Effector T cells

Mono-functional and non-self reactive T cells are generated in the thymus from bone marrow progenitor cells. Mature naïve T cells circulate in secondary lymphoid organs, a microenvironment where T cells find signals of homeostatic proliferation (maintenance) and antigen-mediated activation (antigen presentation). The use of mouse models, and more recently non-human primates, has significantly aid the advancement of our knowledge on T cell development and differentiation into effector T cells.

The frequency of naïve T cells in the total repertoire able to recognize specific pathogen epitopes is low. Naïve T cells circulate through blood and secondary lymphoid organs surveying antigens displayed on dendritic cells (DC). During infection or immunization, activated DCs initiate in the lymph nodes an immune response tailored to a specific pathogen by "presenting" short peptides processed from the pathogen. DCs express on their surface pathogen peptides in the context of the major histocompatibility complex (MHC) that can be recognized by a T-cell receptor (TCR). Only those naïve T cells with right specificity are activated during this process known as antigen presentation (clonal selection). T cell activation requires a minimum of 3 signals for proper activation: interaction of MHC:peptide and TCR (signal 1), co-stimulation (signal 2), and pro-inflammatory cytokines (signal 3) [1-3]. The frequency of pathogen-specific T cells is extremely low (less than 100 naïve cells with same specificity), thus T cell activation leads to a massive expansion of these rare pathogen-specific T cells ('effectors') to ensure effective clearance of the infectious agent (> 10,000 effector T cells per naïve T cell) [4]. However, these large number of effector T cells generated from few naïve T cells can pose a pathological risk if no controlled after infection has been resolved and can dominate the T cell pool limiting future responses to different pathogens. Therefore, terminally differentiated effector T cells are short lived and die after eliminating pathogen via cytolysis of infected cells and secretion of IFNg[5] (Figure 1).

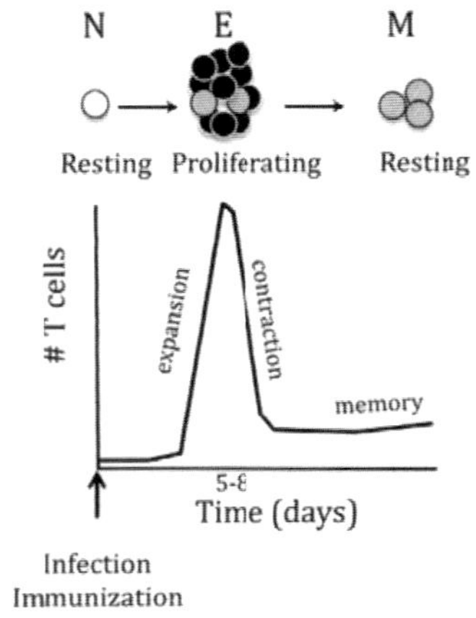

Figure 1: Differentiation of naïve (N) to effector (E) T cells and development of memory (M) T cells upon immune activation by infection or immunization.

The expansion of naïve T cells (naïve to effector transition) is regulated by intrinsic and environmental factors present in the site of activation [6]. It has been recognized that inflammation plays a critical role in the differentiation of effector T cells, particularly the inflammatory cytokines IFNg and IL-12 [7]. The transcription factor T-bet modulates the degree of inflammation and consequently the magnitude of differentiation during T cell activation. In addition to these cell extrinsic and intrinsic factors, two landmark reports showed that the metabolism of T cells is critical for the transition of naïve into effector T cells. Naïve T cells obtain energy from fatty acid oxidation (FAO) whereas effector T cells produce energy via glycolysis, thus, a 'metabolic switch' from catabolism to anabolism must occur for proper expansion and differentiation of effector T cells [8,9]. The transition of naïve (N) to effector (E) T cells is modulated by the nutrient-sensing mammalian target of rapamycin (mTOR) and stress-sensing adenosine monophosphate kinase (AMPK). Hence, drugs that inhibit these two targets (i.e. metformin and rapamycin) can potentially enhance long-lasting immune protection during infection or vaccination.

Development and maintenance of Memory T cells

Following the expansion phase of effector T cells that peaks between 5-8 days after infection, most effector T cells die during a contraction phase to maintain homeostasis of the T cell pool. The immune response 'adapt' to their environment by saving a subset of T cells (immunological memory), which have increased frequency compared to naïve T cells specific for that pathogen. Similarly to humans, the number of memory T cells formed during immune response in laboratory mice is maintained constant for the life of the host. Memory CD8+ T cells are heterogeneous, at least two subsets can be distinguished based on the expression of CD62L (L-selectin): effector memory T cells lack the expression of CD62L whereas central memory T cells express this homing molecule and thus localize predominantly in lymph nodes [10]. The longevity of memory T cells is maintained by cytokine-driven proliferation independently of pathogen [11]. In a second encounter with same pathogen that elicited primary immune response, memory T cells can rapidly expand and eliminate pathogen. The development of memory T cells has been controversial for many years, whether they are effector T cells that survived the

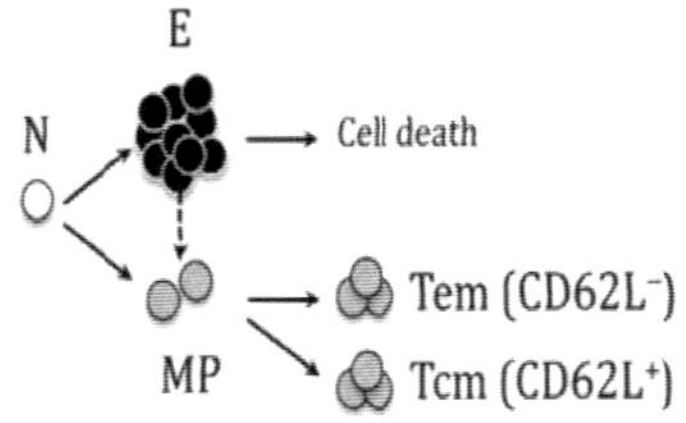

Figure 2: Activation of naïve T cells generates precursors of memory (MP) and terminally differentiated effector cells (E). Consolidation of memory leads to two subsets of memory T cells based on the expression of CD62L: effector memory T cell (Tem) and central memory T cell (Tcm).

contraction phase or whether lineage determination to memory T cells occurs concomitantly to effector formation (Figure 2) [12]. The capacity of the adaptive immune response to establish an immunological memory is the basis of immunization to protect an organism from infections during their lifetime.

Mouse models to study activation of naïve CD8 T cells

Research on the differentiation of effector T cells and development of memory T cells has taken advantage of TCR transgenic mice, MHC class I tetramers, multiparameter flow cytometry and intracellular staining, which made possible to identify pathogen specific T cells both *in vivo* and *ex vivo*. TCR transgenic mice express an abTCR of known specificity in most CD8+ T cells; the most commonly used is the OT-1 transgenic mice that carry a TCR with specificity against ovalbumin (OVA) in CD8+ T cells. MHC class I tetramers are a soluble complex of four MHC class I molecules with a peptide of interest and b2-microglobulin linked to fluorochrome-coupled avidin for detection by flow cytometry. Intracellular staining allows flow cytometric detection of cytokines produced by T cells in response to a specific stimulus. Naïve CD8+ T cells isolated from TCR transgenic mice are adoptively transferred into recipient mice that are then immunized with viruses or bacteria that express antigens specific for the transgenic TCRs used (Figure 3). Most groups use the intracellular bacteria *Listeria monocytogenes* or *Lymphocytic choriomeningitis virus* (LCMV) carrying ovalbumin to induce OVA specific immune response. The different phases of immune response (expansion, contraction, memory formation) can be followed in peripheral blood or tissues (i.e. spleen). Because of the abundance of antigen specific T cells in transgenic mice, typically millions of naïve T cells were used for adoptive transfer for many years. However, it has been demonstrated recently that in order to mimic endogenous immune response only few hundreds or thousands of transgenic T cells must be used in the murine model [13,14]. Current research focuses on deciphering the molecular mechanism of their longevity and homing aimed to improve immune protection against pathogens that elicit a poor immune response.

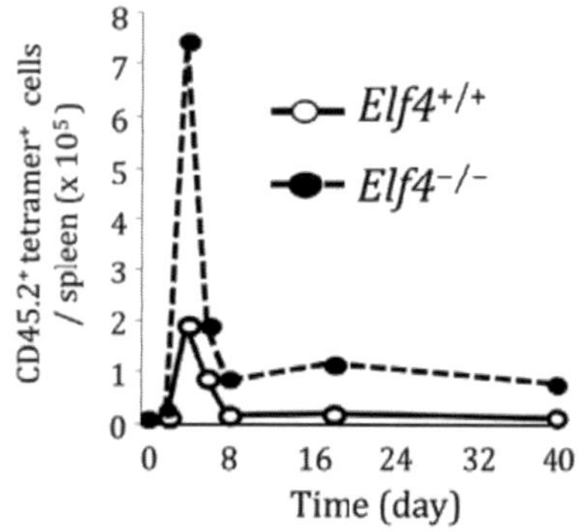

Figure 3: Homozygous deletion of the Elf4 gene leads to increased expansion and memory formation of CD8+ T cells upon infection with *Listeria monocytogenes*.

Transcriptional control of T cell quiescence

The intrinsic factors that regulate quiescence, proliferation, and homing of lymphocytes are critical for immune function. Our group hypothesizes that naïve T cells regulate their quiescence using similar mechanisms as hematopoietic stem cells (HSC). Quiescence is a reversible cell cycle arrest poised to enter cell cycle in response to cell intrinsic and extrinsic mechanisms. We identified that the transcription factor ELF4 inhibits proliferation of naïve CD8+ T cells by activating the expression of the tumor suppressor Krüppel-like factor 4 (KLF4) downstream of the TCR and mTOR pathways [15,16]. We demonstrated that the transcription factor ELF4 directly activates the tumor suppressor KLF4 downstream of TCR signaling to induce cell cycle arrest in naïve $CD8^+$ T cells (Figure 4). *Elf4* and *Klf4*-deficient mice accumulated $CD8^+$ $CD44^{hi}$ T cells during homeostasis and produced more memory T cells upon immunization. The homeostatic expansion of $CD8^+$ $CD44^{hi}$ T cells in *Elf4*-null mice resulted in a redistribution of cells to non-lymphoid tissue due to the loss of KLF2, CCR7 and CD62L expression. This work describes the combinatorial role of intrinsic factors in the control of T cell homeostasis, activation and homing [16]. TCR activation inhibits expression of the ELF4 and KLF4 genes to release naïve $CD8^+$ T cells from their quiescent state. Here we show that ELF4 controls the ERK-mediated proliferative response by maintaining normal levels of dual specificity phosphatases (DUSP1 and DUSP5) in $CD8^+$ T cells. In activated $CD8^+$ T cells, the mTOR pathway inhibits ELF4 and KLF4 expression downstream of ERK and PI3K signaling. These findings demonstrate that rapamycin could be used to modulate expression of this transcriptional network involved in cell cycle regulation [15].

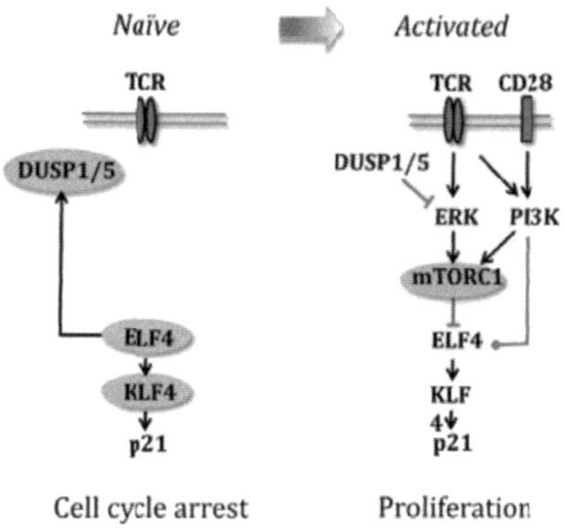

Figure 4: Diagram depicting mechanism of ELF4/KLF4 'proliferation brake' in naïve CD8+ T cells.

REFERENCES

1 Kaech, S.M., Wherry, E.J. & Ahmed, R. Effector and memory T-cell differentiation: implications for vaccine development. *Nat Rev Immunol* **2**, 251-262 (2002). 2. Haring, J.S., Badovinac, V.P. & Harty, J.T. Inflaming the CD8+ T cell response. *Immunity* **25**, 19-29 (2006). 3. Mescher, M.F., *et al.* Signals required for programming effector and memory development by CD8+ T cells. *Immunol Rev* **211**, 81-92 (2006). 4. Badovinac, V.P. & Harty, J.T. Programming, demarcating, and manipulating CD8+ T-cell memory. *Immunol Rev* **211**, 67-80 (2006). 5. D'Cruz, L.M., Rubinstein, M.P. & Goldrath, A.W. Surviving the crash: transitioning from effector to memory CD8+ T cell. *Semin Immunol* **21**, 92-98 (2009). 6. Cui, W. & Kaech, S.M. Generation of effector CD8+ T cells and their conversion to memory T cells. *Immunol Rev* **236**, 151-166 (2010). 7. Joshi, N.S., *et al.* Inflammation directs memory precursor and short-lived effector CD8(+) T cell fates via the graded expression of T-bet transcription factor. *Immunity* **27**, 281-295 (2007). 8. Pearce, E.L., *et al.* Enhancing CD8 T-cell memory by modulating fatty acid metabolism. *Nature* **460**, 103-107 (2009). 9. Araki, K., *et al.* mTOR regulates memory CD8 T-cell differentiation. *Nature* **460**, 108-112 (2009). 10. Sallusto, F., Lenig, D., Forster, R., Lipp, M. & Lanzavecchia, A. Two subsets of memory T lymphocytes with distinct homing potentials and effector functions. *Nature* **401**, 708-712 (1999).

11. Surh, C.D. & Sprent, J. Homeostasis of naive and memory T cells. *Immunity* **29**, 848-862 (2008). 12. Ahmed, R., Bevan, M.J., Reiner, S.L. & Fearon, D.T. The precursors of memory: models and controversies. *Nat Rev Immunol* **9**, 662-668 (2009). 13. Wirth, T.C., Pham, N.L., Harty, J.T. & Badovinac, V.P. High initial frequency of TCR-transgenic CD8 T cells alters inflammation and pathogen clearance without affecting memory T cell function. *Mol Immunol* **47**, 71-78 (2009).

14. Badovinac, V.P., Haring, J.S. & Harty, J.T. Initial T cell receptor transgenic cell precursor frequency dictates critical aspects of the CD8(+) T cell response to infection. *Immunity* **26**, 827-841 (2007). 15. Yamada, T., Gierach, K., Lee, P.H., Wang, X. & Lacorazza, H.D. Cutting edge: Expression of the transcription factor E74-like factor 4 is regulated by the mammalian target of rapamycin pathway in CD8+ T cells. *J Immunol* **185**, 3824-3828 (2010). 16. Yamada, T., Park, C.S., Mamonkin, M. & Lacorazza, H.D. Transcription factor ELF4 controls the proliferation and homing of CD8+ T cells via the Kruppel-like factors KLF4 and KLF2. *Nat Immunol* **10**, 618-626 (2009).

HUMAN INTRAVENOUS IMMUNOGLOBULIN: CURRENT INDICATIONS IN SMALL ANIMAL MEDICINE

Domenico Bianco, DVM, PhD, DACVIM
Woodland Hills, CA

INTRODUCTION

Human intravenous immunoglobulin (hIVIg) is prepared from pooled plasma of a large number of healthy blood donors.[1] It consists mainly of intact immunoglobulin G (IgG) molecules, and trace of IgA, IgM, soluble CD4, CD8, human leukocyte antigen molecules, and certain cytokines.[1,2] The proposed immunodulatory effects of hIVIg are not completely understood and include: fucose (Fc) receptor-mediated effects, modulation of complement, modulation of cytokine production, superantigen neutralization, neutralization antibodies by idiotype network, increased catabolism of IgG, and possible role in apoptosis.[2-5] Binding of hIVIg to Fc receptors has also been demonstrated to in canine lymphocytes and monocytes *in vitro*.[4] It has been recently demonstrated that most of the hIVIg properties are associated with a small fraction of Fc fragment-sialylated IgG, which offered the possibility of creating more clinically effective solutions by increasing the Fc-sialylated IgG fraction.[6-7] The half-life of hIVIg is 21-33 days in humans and 7-9 days in dogs.[2,5] Currently, there are only six conditions approved by the United States Food and Drug Administration for "on-label" use of hIVIg, including primary immunoglobulin deficiency, idiopathic thrombocytopenic purpura (ITP), Kawasaki disease, human immunodeficiency virus infection in children, B-cell chronic lymphocytic leukemia, and bone marrow transplantation.[5] In human medicine, hIVIg is administered for "off-label" use in more than 50% of cases, including many hematologic, neurologic, and dermatologic conditions; furthermore, hIVIg is often used in patients with sepsis, septic shock, and nosocomial infections.[8-16] The administration of hIVIg is well tolerated in people with less than 5% of patients suffering from adverse reactions, including most commonly acute hypersensitivity, and less commonly thromboembolic disease, acute kidney injury, hypotension, pseudohyponatremia, aseptic meningitis, and fluid overload.[3,5,17-23] Only 3.97% of reported dogs treated with hIVIg therapy administered intravenously (IV) experienced adverse reactions, including severe non-fatal anaphylactic reaction, mild immediate reactions, such as vomiting and facial swelling, and fatal delayed reaction, such as protein-losing nephropathy.[5,24-26] No adverse effects were noted in the two reported cats treated with systemic hIVIg therapy.[27,28] Thrombocytopenia was noted in 3 healthy dogs and 6 dogs with immune-mediated hemolytic anemia (IMHA) who received hIVIg infusion.[5,29] Thromboembolic disease was detected on necropsy in 5 hIVIg-treated dogs with IMHA, a known predisposing condition for thrombosis based on autopsy examination.[28] Increased concentration of circulating thrombin-antithrombin complexes and fibrinogen degradation products was found in healthy beagles receiving hIVIg, but there was no statistical significant difference in D-dimer concentrations between dogs with IMHA receiving hIVIg and dogs with IMHA receiving placebo.[30-31] Serial hIVIg infusions are part of most treatment protocols in human medicine, and have been reported in animals with different immune-mediated conditions without significant adverse effects.[2,3,5,8-16,27,29-34] On the other hand, a recent study reported a severe anaphylactic reaction in a dog with myasthenia gravis (MG) after multiple hIVIg infusions.[24] In 5 dogs with immune-mediated retinitis (IMR) and sudden acquired retinal degeneration syndrome (SARDS), membranous glomerulonephritis was found on necropsy; consequently, evaluation of urine microalbumin level and protein/creatinine ratio, is recommended before hIVIg administration; furthermore, mild vitreal hemorrhage was reported in only 1out of 16 dogs with SARDS or IMR treated with intraocular hIVIg therapy, using the Iowa State University (ISU) protocol (Dr. Grozdanic, unpublished data). A preliminary study of 20 dogs with IMHA receiving equine immunoglobulin G (eIgG) reported a higher rate of severe adverse reactions and mortality related to eIgG therapy when compared to a control group.[35] The optimal dose of hIVIg in veterinary medicine is unknown, and the available literature reports doses ranging from 0.25 to 2.2 g/kg over 4 to 12 hours.[3,5,24-42] There is no consensus regarding the indications of hIVIg therapy in small animal medicine, but only limited documentation of its use in some immune-mediated hematologic, dermatologic, neurologic and ophthalmic diseases.[3,5,24-42]

Immune-mediated cytopenias

The mechanism of action of hIVIg in the treatment of hematologic conditions is unknown.[5] Infusion of hIVIg in human patients with idiopathic thrombocytopenic purpura (ITP) resulted in a prolonged clearance of autologous antibody-coated red blood cells; other evidence for the role of Fc-receptor blockade is that antibodies directed against Fc-receptor blockade have similar effects to those of hIVIg in patients with ITP; it is also possible that bindings of hIVIg to Fc receptors has other immunomodulatory effects such as changes in synthesis or release of cytokines from mononuclear cells; another proposed mechanism of action of hIVIg is decreased serum concentrations of autoantibodies or decreased autoantibody production by B cells; $F(ab)_2$ fragments of hIVIg inhibit the binding of autoantibodies to autoantigens *in vitro* and neutralize the functional activity of autoantibodies in a dose-dependent manner; anti-idiopathic antibodies in the hIVIg may also decrease autoantibody production by modulating the function of the idiothype-anti-idiotype network; other mechanisms that have been hypothesized include functional modulation of T lymphocytes, decreased natural cell killer activity, blockade of complement-mediated cell damage, and modulation of the release and function of proinflammatory cytokines.[5,29] Because of its high mortality rate, IMHA was one of the first diseases where the use of hIVIg has been evaluated in veterinary medicine.[29,31,37,38] Initial studies reported some encouraging results regarding the efficacy of hIVIg (Gammar, 1 g/kg IV over 6-12 hours; Gammacard, 0.5 g/kg IV over 4 hours) in the acute management of canine IMHA.[29,37,38] A recent prospective randomized clinical trial reported no significant difference in length of hospitalization, survival time, or initial transfusion requirements between dogs with newly diagnosed IMHA treated with corticosteroids and hIVIg (Gammaguard SD) at the dose of 0.5 g/kg IV over 6 hours on 3 consecutive days, compared to those only treated with corticosteroids.[31] In human hematology, the general consensus is that hIVIg is not recommended for routine use in either acute or chronic treatment of IMHA, but it may be considered for the treatment of severe life-threatening IMHA.[10] At this time, in the veterinary literature there is no evidence in favor or against the use of hIVIg in the acute management of canine IMHA.[29,31,37,38] Further larger randomized controlled clinical trials are required to fully elucidate the role of hIVIg in the management of dogs with IMHA.[3] An initial case report and a case series had promising results regarding the use of a single hIVIg (Polygam SD) infusion at the dose of 0.44 g/kg (0.28-0.76 g/kg) IV over 6 hours in five dogs with presumed primary immune-mediated thrombocytopenia (pIMT) and severe hemorrhage.[5,39] Evidence of gastrointestinal hemorrhage, as presence of melena or increased blood urea nitrogen, is a negative prognostic indicator in dogs with presumed pIMT.[43] A subsequent randomized prospective clinical trial reported a significant reduction in time of resolution of thrombocytopenia (median of 3.5 vs. 7.5 days) and duration of hospitalization (median of 4 vs. 8 days) in dogs with presumed pIMT treated with corticosteroids and a single hIVIg (Carimune NF) infusion in a 6% solution at a dose of 0.5 g/kg IV over 6-12 hours compared to dogs treated only with corticosteroids in the acute phase of the disease.[40] The preliminary results of a recent prospective clinical trial found no difference in regard of platelet count recovery time (median, 2.5 days for both groups) and length of hospitalization (median, 4 days for both groups) when comparing the effects of corticosteroids and a single hIVIg (0.5 g/kg IV) infusion vs. corticosteroids and vincristine (0.02 mg/kg IV) in dogs with presumed pIMT.[42] In human hematology, current guidelines recommend the use of hIVIg at the dose of 1 g/kg in adults with acute ITP and hemorrhage or if not responsive to corticosteroids or if corticosteroids are contraindicated.[10,44] Only a single case of a diabetic dog with Evans' syndrome (ES) successfully treated with a combination of a single hIVIg (Polygam SD) infusion at a dose of 1.3 g/kg IV in a 5% solution over 8 hours and oral leflunomide has been reported.[41] The general consensus for the use of hIVIg in people with ES is that the primary symptoms should be the focus of treatment: for patients with predominantly ITP symptoms, use of hIVIg is recommended as described in ITP guidelines; for patients with primarily IMHA symptoms, hIVIg is recommended as outlined in IMHA section.[10] In a case series of five dogs with myelofibrosis and presumed immune-mediated non-regenerative anemia that fail to respond to conventional therapy, the use of hIVIg at a dose of 1 g/kg IV over 12 hours caused a transient complete response in 2 dogs for 37 and 49 days, respectively; the remaining 3 dogs had only a partial response.[5,32] In human hematology, the general recommendations are to do not use hIVIg as first-line therapy for pure red cell aplasia (PRCA), but to consider it as a reasonable option for patients with PRCA who failed other therapies, such as corticosteroids or cyclosporine.[10] A case report of a cat with immune-mediated erythroid and megakaryocytic aplasia described the successful use of a single hIVIg (Redimune) infusion at the dose of 1.21 g/kg IV over 6 hours after corticosteroid therapy failed to control the disease.[28] Because of limited data and availability of other effective treatment options, such as bone marrow transplantation, the use of hIVIg is not recommended in people with bone marrow aplasia.[10]

Dermatologic and miscellaneous diseases

Adverse cutaneous drug reactions, such as toxic epidermal necrosis (TEN), Stevens-Johnson syndrome (SJS), and erythema multiforme (EM), can cause severe dermatologic and systemic signs: they are rare diseases, carry a poor prognosis, and may not respond well to standard immunosuppressive therapy.[27] They are taught to represent a host-specific cell-mediated hypersensitivity reaction induced by various antigens, including infiltrating CD8+ lymphocytes, tumor necrosis factor α, and activation of a "death receptor" on the keratinocyte surface, such as Fas and its ligand (FasL) causing keratinocyte apoptosis or necrosis; because at least some preparations of hIVIg contain anti-Fas antibodies that are able to block Fas-FasL binding *in vitro* and thereby inhibit the death signal of apoptosis, hIVIg has been proposed as a promising treatment of adverse cutaneous drug reactions.[27,34] A recent study suggest that the pathogenesis of TEN is directly related to drug-specific, major histocompatability complex class 1-restricted, perforin/granzyne-mediated cytotoxicity; therefore, this would suggest that an alternate mechanism of action of hIVIg for the treatment of cutaneous drug reactions is possible.[34] The beneficial effects of hIVIg therapy (Sandoglobulin), in a 6% solution in 0.9% NaCl at a dose of 1 g/kg IV over 4 hours on two consecutive days, were initially described in a case report of a cat diagnosed with EM.[27] Subsequently, a case of SJS in a dog had complete resolution of systemic and dermatologic signs following a single 0.5 g/kg hIVIg infusion (Flebogamma).[36] More recently, two dogs with severe cutaneous drug reactions were successfully treated with hIVIg therapy (Sandoglobulin at the dose of 1 g/kg IV over 4 hours in a 6% solution in 0.9%NaCl on two consecutive days) .[34] No adverse effects related to hIVIg administration were noted in any of those cases.[27,34,36] Pemphigus foliaceus (PF) is the most common autoimmune disease of dogs and is characterized by the presence of autoantibodies to desmoglein 1, a keratinocyte desmosome component, resulting in breakdown of epithelial adhesions, epidermal detachment (acantholysis), subcorneal blister formation, and pastules; the mechanism of action of hIVIg in controlling PF is not entirely clear: increased concentration of

IgG may result in rapid catabolism of autoantibodies to desmoglein 1; other potential mechanisms of action include an influence on the production of autoantiboides by B cells, complement fixation, T-cell cytokine release, and anti-idiotype autoantibody binding.[33] There is a case report of a dog with PF unresponsive to conventional therapy successfully treated with a total of 11 hIVIg (Polygam SD) infusions at a dose of 0.5 g/kg over a 31-week period, and no adverse effects related to multiple hIVIg infusions were noted.[33] In human dermatology, guidelines support the use of IVIG in people with severe or complicated PF, and the general consensus is to recommend the use of hIVIg in patients with autoimmune blistering cutaneous disease not responding to conventional treatments.[12,45,46] Acquired MG is and immune-mediated neuromuscular disorder resulting from auto-antibody-mediated destruction of acetylcholine receptors (AChR) on the postsynaptic neuromuscular junction.[9,11] The immunologic properties of hIVIg modulate AChR antibody effects temporarily.[11] Temporary improvement of clinical signs was reported in two dogs with MG after a single hIVIg (Gammaguard) infusion at the dose of 0.5 g/kg without observed adverse reactions in one dog, but the other dog experienced a non-fatal anaphylactic reaction during the fourth hIVIg infusion.[24] In human neurology, hIVIg therapy is recommended only for patients with acute exacerbations of MG and for short-term treatment of severe MG, while it is not recommended as maintenance therapy for patients with chronic MG.[9,11] The effects of systemic hIVIg therapy at the dose of 0.5 g/kg in a 5% solution over 7 hours in dogs with SARDS has been initially evaluated in 8 dogs at ISU, and all dogs showed improvement in visual maze behavior, with limited visual navigation skills and no improvement in menace response; subsequently, an additional 11 dogs with SARDS were treated with hIVIg and only 5 regained navigational skills based on visual maze.[26] Although the ISU protocol appears to be relatively safe, systemic hIVIg therapy and intraocular injection of hIVIg in dogs with SARDS or IMR, administered not following the ISU protocol, resulted in fatal anaphylactic reaction in 1 dog, hepatic failure in 1 dog, renal failure in 1 dog, and uncontrolled uveitis and glaucoma in 6 dogs; furthermore, the use of hIVIg is discouraged in dogs with concurrent glaucoma, retinal degeneration, cataracts, or duration of blindness of over 2 months (Dr. Grozdanic, unpublished data).[26] Information regarding the use of hIVIg in people with antibody-mediated retinopathy is also scarce.[47,48]

REFERENCES

1. Kazatchinke MD, et al. **N Engl J Med** 2001;34:747-755. **2.** Emmi L, et al. **Neurol Sci** 2002;23:S1-S8. **3.** Spurlock NK, et al. **J Vet Emerg Crit Care** 2011;21(5):471-483. **4.** Reagan WJ, et al. **Am J Vet Res** 1998;59(12):1536-1574. **5.** Scott-Moncrieff JCR, et al. **Semin Vet Med Surg (Small Anim)** 1997;12:178-185. **6.** Kaneko Y, et al. **Science** 2006;313:670-673. **7.** Anthony RM, et al. **Science** 2008;320:373-376. **8.** Foster R, et al. **Transfus Med** 2010;20(6):403-408. **9.** Elovaara, et al. **Eur J Neurol** 2008;15(9):893-908. **10.** Anderson D, et al. **Transfus Med Rev** 2007;21(2S1):S9-S56. **11.** Feasby D, et al. **Transfus Med Rev** 2007;21(2S1):S57-S107. **12.** Enk A, et al. **Eur J Dermatol** 2009;19(1):90-98. **13.** Laupland KB, et al. **Crit Care Med** 2007;35(12):2686-2692. **14.** Turgeon et al. **Ann Intern Med** 2007;146(3):193-203. **15.** Werdan K, et al. **Crit Care Med** 2008;36(3):716-723. **16.** Douzinas EE, et al. **Crit Care Med** 2000;28(1):8-15. **17.** De Albuquerque CR, et al. **J Clin Immunol** 2000;20:70-82. **18.** Nydegge UE, et al. **Drug Saf** 1999;21(3):171-185. **19.** Knezevic-Maramica I, et al. **Transfusion** 2003;43(10):1460-1480. **20.** Duhem C, et al. **Clin Exp Immunol** 1994;97(S1):79-83. **21.** Sekul E, et al. **Ann Intern Med** 1994;121:259-262. **22.** Orbach H, et al. **Clin Rev Allergy Immunol** 2005;29:173-184. **23.** Kroez M, et al. **Biologicals** 2003;31(4):277-286. **24.** Abelson A, et al. **J Vet Emerg Crit Care** 2009;19(4):369-374. **25.** Augusto M, et al. **J Vet Intern Med** 2009; 23(6):1340 (abstract). **26.** Grozdanic S. **Vet Clin N An Small Anim Pract** 2008;38(2):361-387. **27.** Byrne KP, et al. **J Am Vet Med Assoc** 2002;220:197-201. **28.** Zini E, et al. **J Am Vet Med Assoc** 2007;203(7):1024-1027. **29.** Scott-Moncrieff JC, et al. **J Am Vet Med Assoc** 1997;210(11):1623-1627. **30.** Tsuchiya R, et al. **J Vet Intern Med** 2009;23:1164-1169. **31.** Whelan M, et al. **J Vet Emerg Crit Care** 2009;19(2):158-164. **32.** Scott-Moncrieff JC, et al. **J Am Vet Med Assoc** 1995;206:1895-1900. **33.** Rahilly L, et al. **J Vet Intern Med** 2006; 20:1483-1486. **34.** Trotman TK, et al. **J Am Anim Assoc** 2006;42:312-320. **35.** O'Brian MA, et al. **J Vet Emerg Crit Care** 2007;17(3S1):S7 (abstract). **36.** Nuttall TJ, et al. **J Small Anim Pract** 2004;45:357-361. **37.** Kellerman DL, et al. **J Vet Intern Med** 1997;11(6):327-332. **38.** Gerber B, et al. **Schweiz Arch Tierheilkd** 2002;144(4):180-185. **39.** Bianco D, et al. **J Vet Intern Med** 2007;21:694-699. **40.** Bianco D, et al. **J Vet Intern Med** 2009;23:1071-1078. **41.** Bianco D, et al. **J An Anim Hosp Assoc** 2009;45:147-150. **42.** Balog K, et al. **J Vet Intern Med** 2011; 25(6):1503 (abstract). **43.** O'Marra SK, et al. **J Am Vet Med Assoc** 2011;238(3):346-352. **44.** Neunert C, et al. **Blood** 2011;117(16):4190-4207. **45.** Akerman L, et al. **Clin Rev Allergy Immunol** 2005;29(3):289-294. **46.** Ahmed AR, et al. **Arch Dermatol** 2003;139:1051-1059. **47.** Saoudi A, et al. **Int Immunol** 1993;5(12):1559-1567. **48.** Guy J, et al. **Arch Ophthalmol** 1999;117:471-477.

ALL CREATURES, GREAT AND SMALL: A REVIEW OF REGULATORY T CELLS IN MICE, HUMANS, DOGS AND OTHER DOMESTIC SPECIES

Oliver A. Garden, BSc, BVetMed, PhD, FHEA, MSB, MRCVS, DACVIM, DECVIM-CA
London, UK

INTRODUCTION

The rise, fall and enthusiastic resurrection of the suppressor T cell – latterly re-named the regulatory T cell (Treg) – represent defining events in modern immunology. Tregs have once again ascended into mainstream immunology following the seminal discovery of a population of suppressive $CD4^+$ T cells characterized by high constitutive expression of the IL-2Rα chain (CD25) and growing realization of their exciting therapeutic potential. A number of interacting T cells with regulatory function are now recognized, including $CD4^+CD25^{+/high}$, T regulatory 1 (Tr1) and T helper 3 (Th3) cells, to name a few.

REGULATORY T CELLS – CURRENT CONCEPTS IN MOUSE AND MAN

Regulatory T cells may be broadly divided into naturally occurring subsets (nTregs), which develop in the thymus along a regulatory lineage – for example, the canonical $CD4^+CD25^{+/high}$ Treg; and induced – or 'adaptive' – subsets, which arise in the periphery following activation of conventional (non-regulatory) T cells (Tcons) in a microenvironment rich in regulatory cytokines, or following the interaction of Tcons with nTregs in a process called infectious tolerance – for example, the Tr1, Th3 and 'induced (i) Treg' cells. Naturally occurring Tregs have traditionally been thought to possess a broad repertoire of autoreactive TCRs, supported by evidence in various murine models, though recent studies suggest that there is considerable overlap between the repertoires of nTregs and Tcons. Regardless, nTregs make an important contribution to both peripheral tolerance of autoantigens and the regulation of immune responses to non-self antigens, such as those associated with pathogens or the microbiota of the gut. The inter-relationships and relative importance of the various Treg subsets remain to be fully characterized.

The transcription factor Forkhead box p3 (Foxp3) – designated in upper case letters (FOXP3) in humans – serves a crucial role in stabilising the transcriptome of Tregs and is a faithful marker of peripheral Tregs in the mouse. Originally dubbed the 'master regulator' of Tregs, Foxp3/FOXP3 is essential for the maintenance of functional competence of Tregs, as suggested by the severe immunodeficiency diseases associated with its absence or defective function in mice and humans – but whether it is strictly necessary for Treg lineage commitment has been questioned in recent years. As well as being expressed by nTregs, Foxp3 may also be induced in murine Tcons by their stimulation in the presence of exogenous TGF-β, yielding T cells – labelled 'iTregs' – with a basic phenotype ($CD4^+CD25^+Foxp3^+$) that is identical to that of nTregs; however, regulatory function following stimulation of Tcons in the presence of TGF-β has not been observed by all authors, while others have observed only transient regulatory function associated with unstable Foxp3 expression. Human Tcons, in contrast to those of the mouse, show upregulation of FOXP3 with activation in the *absence* of exogenous TGF-β, the very low concentrations of this cytokine found in complete medium – coupled with endogenous synthesis and activation – being sufficient to induce FOXP3; however, the induced FOXP3 expression is transient and the resulting $CD4^+CD25^+FOXP3^+$ T cells lack or rapidly lose regulatory function in most studies. A picture is thus emerging that robust regulatory function requires stable and high-level Foxp3/FOXP3 expression.

Mechanisms of suppression mediated by Tregs include those operating in a contact-dependent or close range manner and those dependent on soluble cytokines; both Tcons and antigen-presenting cells (APCs) – including macrophages and B cells – may be the target of regulation, though Tregs are also thought to regulate NK cells, neutrophils and other innate immune cells [1]. The potential plasticity of Tregs has recently been recognized, with reports of their conversion to a number of the Th subsets in appropriate pro-inflammatory microenvironments.

REGULATORY T CELLS IN THE DOG – WHAT DO WE KNOW?

The concept of peripheral tolerance in the dog is not new: Weiden and colleagues postulated that a suppressor T cell population was responsible for preventing graft-versus-host disease in stable canine radiation chimaeras challenged by the infusion of donor lymphocytes in 1976. Furthermore, circulating suppressor cells were reported in German shepherd dogs with progressive myelopathy in 1980 and a short-lived suppressor population – thought to be neutrophils – were documented in the peripheral blood of healthy dogs in 1989. A study performed in the early 1990s described the inhibitory activity of spontaneous and induced suppressor lymphocyte populations – assumed to be T cells – in peripheral blood of dogs with atopic dermatitis. Further studies documented the ability of the azaspirane drug SK&F 105685 to elicit suppressor cell activity in the spleen and bone marrow of healthy beagle dogs, putatively by the induction of 'suppressive T cells'. The role of immunosuppressive cytokines was also recognized in this early work – for example, levels of TGF-β mRNA were assessed in canine atopic skin by semi-quantitative reverse transcriptase PCR – but further substantive advances in the field of canine peripheral tolerance were not made until more than a decade later, when we and others noted the ability of the antimouse / rat Foxp3 monoclonal antibody (mAb) FJK-16s to stain a population of $CD4^+$ T cells in the peripheral blood and lymph nodes of dogs, which we speculated were Tregs [2, 3]. Stimulation of both peripheral blood and lymph node cells with ConA yielded impressive increases in both the proportion and absolute numbers of $CD4^+$ and $CD8^+$ FJK-16s$^+$ T cells in the absence of exogenous TGF-β, recapitulating similar findings in human but not murine peripheral lymphoid cells. The specificity of FJK-16s for canine FOXP3 was subsequently confirmed by examining cell lines over-expressing the canine *FOXP3* gene [4]. Cross-reactivity of the anti-human CD25 mAb ACT-1 with the canine protein was also confirmed using the same strategy [4]. Subsequent studies have verified cross-reactivity of ACT-1 for canine CD25 [5], but have suggested that the affinity of this antibody for the canine protein is low [6]; indeed, Abrams and colleagues have developed a novel, higher-affinity, canine-specific anti-CD25 mAb (P4A10), which is now commercially available [6].

A number of recent studies have examined the changes in proportions of $CD4^+FOXP3^+$ T cells occurring in canine cancer, following exploratory work performed by Biller and colleagues [3]. When compared to healthy control animals, the proportions of $CD4^+FOXP3^+$ cells in dogs with a variety of neoplasms – including round cell tumors, carcinomas and sarcomas – were greater in both the blood and, when assessed, tumor-draining lymph nodes [3, 7]; when individual tumor types were analysed, those animals with carcinomas had significantly higher proportions of these cells than the control dogs [7]. Similarly, neoplastic dogs showed an increase in the ratio of peripheral $CD4^+FOXP3^+$: $CD8^+$ T cells [7]. The presence of metastasis further increased the proportions of peripheral blood $CD4^+FOXP3^+$ T cells in dogs with a range of tumors [8]. Further work performed by the same group demonstrated increased proportions of $CD4^+FOXP3^+$ T cells in the peripheral blood and tumor beds of canine oral malignant melanoma patients [9]; furthermore, the proportion of peripheral blood $CD4^+FOXP3^+$ T cells showed a positive correlation with tumor stage and negative correlations with the proportions of Th1 and cytotoxic T cells in these animals [10]. However, not all studies of canine tumors have yielded such a clear message: for example, Rissetto and colleagues found no difference in the proportions of $CD4^+FOXP3^+$ cells in the peripheral blood or lymph nodes – either regional or distant – in dogs with osteosarcoma [5], yet Biller and colleagues documented increased proportions of these cells in the blood, but not lymph nodes, of dogs with the same type of tumor, with an associated decrease in both the proportion of $CD8^+$ T cells and the ratio of $CD8^+$: $CD4^+FOXP3^+$ cells [11]; indeed, these authors also demonstrated that $CD8^+$: $CD4^+FOXP3^+$ ratios lower than the mean value in the patient group were associated with significantly shorter survival times, raising the intriguing possibility that the Tregs in these dogs were implicated in the immune evasion of the cancer [11]. A recent study has shown that low-dose cyclophosphamide selectively depletes Tregs and inhibits angiogenesis in dogs with soft tissue sarcomas, providing a mechanistic basis for metronomic administration of this drug [12]. Studies interrogating $CD4^+FOXP3^+/CD25^+$ T cells or *FOXP3* expression in other areas of canine medicine are growing in number. Keppel and colleagues provided evidence for an increase in the proportion of peripheral blood $CD4^+FOXP3^+$ T cells and concentration of serum IL-10 in dogs with atopic dermatitis receiving allergen-specific immunotherapy, associated with a decrease in serum IgE concentration [13]; while Veenhof and colleagues demonstrated increased *FOXP3* expression in both lesional and non-lesional skin of dogs with cutaneous adverse food reactions [14]. In contrast, de Lima and others showed diminished proportions of $CD4^+CD25^+$ T cells and increased concentrations of supernatant IFN-γ in cultures of peripheral blood mononuclear cells from dogs immunized with a vaccine against *Leishmania spp* [15], but Menezes-Souza and colleagues demonstrated high copy numbers of transcripts encoding FOXP3, TGF-β1 and IL-10 in the skin of asymptomatic dogs infected with *L. chagasi* [16] – consistent with murine models of leishmaniasis, in which Tregs are thought to allow the persistence of low numbers of parasites. Recent immunohistochemical studies of spirocercosis-induced oesophageal nodules revealed no evidence of accumulation of $FOXP3^+$ cells in lesional tissue, suggesting that Tregs may not be involved in the neoplastic transformation characteristic of this parasite [17].

We and others have recently provided direct evidence of the regulatory function of canine $CD4^+CD25^{high}FOXP3^{high}$ T cells. Developing a novel canine-specific anti-CD25 mAb, Abrams and colleagues showed that immunomagnetically-selected $CD4^+CD25^+$ T cells were able to inhibit the proliferation of responder T cells in one-way mixed leukocyte reactions [6]. Using the cross-reactive anti-CD25 mAb ACT-1, we demonstrated that $CD4^+$ T cells with the highest CD25 expression were enriched for FOXP3 and stained positively with the newly-developed anti-murine/human Helios mAb 22F6 [18], suggesting that they expressed Helios, an Ikaros transcription factor family member that has recently been associated with nTregs in both mice and humans. Following activation of $CD4^+$ T cells derived from either peripheral blood or lymph nodes with ConA and subsequent withdrawal of the stimulus, we demonstrated the emergence of two populations of T cells defined on the basis of FOXP3 and IFN-γ expression [18] – $CD4^+FOXP3^{high}IFN\text{-}\gamma^-$ T cells, which we speculated were activated Tregs, and $CD4^+FOXP3^{intermediate}IFN\text{-}\gamma^{-/+}$ T cells, which we speculated were a more heterogeneous population containing predominantly activated Tcons [18]. We used flow-assisted cell sorting (FACS™) to isolate $CD4^+CD25^{high}$ T cells, demonstrating their regulatory function *in vitro* [18]. More recent studies by Knueppel and colleagues have recapitulated these findings *in vitro* [19] and have shown that denileukin diftitox, a recombinant cytotoxic fusion protein, is able to neutralize Tregs *in vitro* and to deplete Tregs *in vivo* [20].

REGULATORY T CELLS IN OTHER DOMESTIC SPECIES – AN EXPANDING PANOPLY!

The existence of Tregs has also been established in a number of domestic animal species other than the dog. 'Suppressor cells' were first described in the cat in the early 1980s, but it was not until the pivotal work of Wayne and Mary Tompkins and colleagues that $CD4^+CD25^+$ Tregs were characterized in detail, in the context of the pathogenesis of feline immunodeficiency virus (FIV) infection. These cells expressed FOXP3 [21]; mediated contact-dependent suppression of proliferation and IL-2 synthesis *in vitro* [22]; were present in FIV-positive cats in a pre-activated state, providing a cellular 'sanctuary' for productive FIV infection [23-26]; could be recruited from the Tcon pool by stimulation with Con-A and TGF-β, depending on TGF-β/TGF-β-RII signalling for their regulatory function [27]; and induced anergy in $CD8^+$ T cells in cats infected with FIV [28], while inhibiting their synthesis of IFN-γ by means of membrane-bound TGF-β [29]. More recent work has described the depletion of Tregs *in vivo* [30] and its impact on anti-viral responses mounted in cats chronically infected with FIV [31]; however, partial Treg depletion prior to *acute* FIV infection had no impact on viraemia or numbers and proportions of peripheral $CD4^+$ T cells [32]. The presence of endometrial Tregs was suggested in a study documenting transcripts encoding FOXP3 and CTLA-4 in placental tissues of FIV-infected cats [33]. Tregs have also been described in the pig, initially in the context of fetal tolerance and renal and cardiac allotransplantation. More recently, $CD4^+CD25^{high}$ Tregs [34] and porcine FOXP3 [35, 36] have been specifically characterized, and further studies have described the induction of $CD4^+CD25^+FOXP3^+$ [37, 38] and Th3 cells [39] by porcine reproductive and respiratory syndrome virus; however, European strains of this virus were not able to induce Tregs [40], suggesting strain dependence of this phenomenon. Retinoic acid was implicated in the augmentation of a diverse Th1, Th2 and Treg response in pigs infected with *Ascaris suum* [41], while suckling piglets infected with *Isospora suis* showed decreased proportions of Tregs in the mesenteric lymph nodes but not the spleen. Preliminary studies have provided evidence for the induction of Tregs by dendritic cells treated with the immunosuppressive drug mycophenolic acid [42], suggesting that Tregs may contribute to this drug's suppressive action.

Early reports of bovine 'suppressor cells' described a population of T cells capable of suppressing the proliferation of autologous responder cells by the release of a soluble factor; these cells, generated by the activation of lymphocytes with ConA, did not require cell: cell contact and resembled human ConA-induced suppressor T cells documented in the preceding few years. Whether these cells were *bona fide* Tregs or activated Tcons, which in some assay systems are capable of mediating non-specific suppression, remains unclear. A number of subsequent studies presented indirect evidence for populations of $CD4^+$, $CD8^+$ and γδ T cells with regulatory activity [43-48], but T cells expressing FOXP3 were first documented in 2009: thus, Seo and colleagues developed novel mAbs against bovine FOXP3, demonstrating that its expression was limited to $CD4^+CD25^+$ T cells [49], while Gerner and colleagues documented cross-reactivity of the anti-mouse/rat Foxp3 mAb FJK-16s, showing that the majority of $FOXP3^+$ T cells were $CD4^+$ [50]. However, the regulatory properties of the $CD4^+CD25^{high}FOXP3^+$ population were called into question by work performed by Hoek and colleagues, who suggested that these cells were neither anergic nor suppressive; rather, $WC1.1^+$ and $WC1.2^+$ γδ T cells, and $CD14^+$ monocytes, showed regulatory properties in their experiments, supported by the observation that these cells transcribed IL-10 [51]. Ovine ConA-induced 'suppressor cells' were described in 1985. Unlike the bovine suppressor cells generated in a similar manner, the ovine cells appeared not to mediate their regulatory function *in vitro* by means of a soluble factor, but by the depletion of IL-2. Interestingly, these findings reconcile with recent work describing IL-2 deprivation by murine and human nTregs. Subsequent ovine studies presented indirect evidence for both $CD4^+$ and $CD8^+$ Tregs [52, 53], as well as endometrial suppressor cells that were unlikely to be conventional T, B or NK cells [54, 55]. Ovine FOXP3 has recently been sequenced, showing high homology with FOXP3 from other species [56], and peripheral $CD4^+CD25^{high}FOXP3^+$ T cells – thought to be Tregs – have been identified [57]. Immunohistochemical studies have revealed infiltration of $FOXP3^+$ T cells into the skin of sheep infested with *Psoroptes ovis* [56] and recent work has documented the kinetics of Th1, Th2 and Treg immune response genes in lambs resistant to the abomasal parasite *Teladorsagia circumcincta* [58].

Work on equine Tregs is gathering pace. Following the generation of novel mAbs against equine IL-10, Wagner and colleagues defined a population of $CD4^+IL\text{-}10^+IFN\text{-}\gamma^+$ cells that were thought to represent Tr1 cells [59]. Further indirect evidence for the role of immunosuppressive cytokines was generated by cultures of peripheral blood mononuclear cells derived from Icelandic horses, which suggested that heavy helminth burdens imposed a suppressive effect on IL-4 production by inducing IL-10 and TGF-β synthesis [60]. Further work by the same group revealed that the immune response of young foals is Th1-biased, but that regulatory IL-10 production by T cells is developmentally mature within the first three months of life [61]. Recent work has documented the expression of FOXP3 by equine $CD4^+$ T cells using cross-reactive anti-human FOXP3 and anti-murine Foxp3 mAbs [62], as well as equine $CD4^+CD25^{high}$ T cells with regulatory function dependent on close contact and cytokine-dependent mechanisms *in vitro* [63]. Furthermore, a deficit of skin-infiltrating Tregs was thought to contribute to the pathogenesis of insect bite hypersensitivity in Icelandic horses [64]. Other species in which regulatory cells have been described include the guinea pig [65-69], baboon [70, 71], macaque [72-74] and chimpanzee [75]. Furthermore, a recent study described the generation of $IL\text{-}10^+$ T cells from chicken $CD4^+CD25^-$ T cells *in vitro* [76]; additional work has characterized mAbs against chicken CD25 [77], defined the regulatory properties of chicken $CD4^+CD25^+$

T cells [78] and their responsiveness to lipopolysaccharide [79], and demonstrated temporary depletion of Tregs *in vivo* by the administration of anti-CD25 mAb [80]. The anti-human FOXP3 mAb 236A/E7 yielded positive immunohistochemical staining of the mesenteric lymph nodes of harbour seals and a walrus [81], while a number of studies have characterized Foxp3 in teleost fish, including zebrafish [82], Atlantic salmon [83] and rainbow trout [84], with evidence of Foxp3-based regulation in zebrafish [85] and Treg-like cells in pufferfish [86].

REFERENCES

[1] Garden OA et al Int Immunopharmacol 2011;11:576-88; [2] Sacchini F et al Immunology 2007;120:11; [3] Biller BJ et al Vet Immunol Immunopathol 2007;116:69-78; [4] Mizuno T et al J Vet Med Sci 2009;71:1561-8; [5] Rissetto KC et al Vet Immunol Immunopathol 2010;135:137-45; [6] Abrams VK et al Vet Immunol Immunopathol 2010;135:257-65; [7] O'Neill K et al J Vet Intern Med 2009;23:875-81; [8] Horiuchi Y et al Microbiol Immunol 2009;53:468-74; [9] Tominaga M et al J Vet Diagn Invest 2010;22:438-41; [10] Horiuchi Y et al Microbiol Immunol 2010;54:152-9; [11] Biller BJ et al J Vet Intern Med 2010; 24:1118-23; [12] Burton JH et al J Vet Intern Med 2011;25:920-6; [13] Keppel KE et al Vet Immunol Immunopathol 2008;123:337-44; [14] Veenhof EZ et al Vet J 2010; 187:320-4; [15] de Lima VM et al Vet Immunol Immunopathol 2010;135:296-302; [16] Menezes-Souza D et al Vet Parasitol 2011;177:39-49; [17] Dvir E et al Parasite Immunol 2011;33:545-53; [18] Pinheiro D et al Immunology 2011;132:111-22; [19] Knueppel A et al Exp Anim 2011;60:471-9; [20] Knueppel A et al Vet Immunol Immunopathol 2011; doi10.1016/j.vetimm.2011.11.009; [21] Lankford S et al Vet Immunol Immunopathol 2008;122:159-66; [22] Vahlenkamp TW et al. J Immunol 2004;172:4752-61; [23] Joshi A et al. J Virol 2005;79:4965-76; [24] Joshi A et al Virol 2005;335:212-21; [25] Joshi A et al Virol 2004;321:307-22; [26] Mexas AM et al Vet Immunol Immunopathol 2008;126:263-72; [27] Petty CS et al J Acquir Immune Defic Syndr 2008;47:148-60; [28] Fogle JE et al Retrovirol 2010;7:97; [29] Fogle JE et al AIDS Res Human Retrovir 2010;26:201-16; [30] Smithberg SR et al J Immunol Methods 2008;329:81-91; [31] Mikkelsen SR et al Virol 2010;403:163-72; [32] Mikkelsen SR et al PLoS One 2011;6:e17183; [33] Lockett NN et al Placenta 2010;31:796-802; [34] Kaser T et al Vet Immunol Immunopathol 2008;122:153-8; [35] Kaser T et al Vet Immunol Immunopathol 2008;125:92-101; [36] Bolzer K et al Vet Immunol Immunopathol 2009;132:275-81; [37] Wongyanin P et al Vet Immunol Immunopathol 2010;133:170-82; [38] LeRoith T et al Vet Immunol Immunopathol 2011;140:312-6; [39] Silva-Campa E et al Virol 2009;387:373-9; [40] Silva-Campa E et al Virol 2010;396:264-71; [41] Dawson H et al Infect Immun 2009;77:2576-87; [42] Pilon C et al. Transplant Proc 2009;41:700-2; [43] Shafer-Weaver KA et al Vet Immunol Immunopathol 1997;56:53-64; [44] Chang BS et al J Vet Sci 2005;6:247-50; [45] Park YH et al J Vet Sci 2000;1:39-48; [46] Weiss DJ et al Vet Pathol 2006;43:127-35; [47] Seo KS et al Infect Immun 2007;75:260-9; [48] de Almeida DE et al Vet Immunol Immunopathol 2008;125:234-45; [49] Seo KS et al Vet Immunol Immunopathol 2009;128:30-6; [50] Gerner W et al Vet Immunol Immunopathol 2010;138;154-8; [51] Hoek A et al Vet Res 2009;40:6; [52] Chen X et al J Autoimmun 1995;8:539-59; [53] Gunnes G et al Res Vet Sci 2004;77:115-22; [54] Segerson EC et al Biol Reprod 1998;58:397-406; [55] Miles JR et al Theriogenol 2002;58:1097-109; [56] McNeilly TN et al Parasite Immunol 2010;32:361-9; [57] Rocchi MS et al Vet Immunol Immunopathol 2011;144:172-7; [58] Hassan M et al Vet Res 2011;42:45; [59] Wagner B et al Vet Immunol Immunopathol 2008;122:57-64; [60] Hamza E et al Vet Immunol Immunopathol 2010;133:40-50; [61] Wagner B et al Vet Res 2010;41:47; [62] Robbin MG et al Vet Immunol Immunopathol 2011;140:90-101; [63] Hamza E et al Immunology 2011;134:292-304; [64] Heimann M et al Vet Immunol Immunopathol 2011;140:63-74; [65] Katayama I et al Immunology 1982;47:709-16; [66] Parker D and Turk JL Immunology 1982;47:61-6; [67] Wadee AA et al Clin Exp Immunol 1981;45:427-32; [68] Rudnicka W Immunology 1980;39:435-40; [69] Shang S et al Tuberculosis (Edin) 2011;91:378-85; [70] Porter CM et al Xenotransplant 2007;14:298-308; [71] Singh AK et al Transplant Proc 2009;41:418-21; [72] Hartigan-O'Connor DJ et al J Exp Med 2007;204:2679-92; [73] Green AM et al J Infect Dis 2010;202:533-41; [74] Allers K et al J Virol 2010;84:3259-69; [75] Manigold T et al Blood 2006;107:4424-32; [76] Shanmugasundaram R and Selvaraj RK Vet Immunol Immunopathol 2010;137:161-5; [77] Lee SH et al Vet Immunol Immunopathol 2011;144:396-404; [78] Shanmugasundaram R and Selvaraj RK J Immunol 2011;186:1997-2002; [79] Shanmugasundaram R and Selvaraj RK Vet Immunol Immunopathol 2011;144:476-81; [80] Shanmugasundaram R and Selvaraj RK Dev Comp Immunol 2012;36:578-83; [81] Seibel H et al Vet Immunol Immunopathol 2010;137:305-9; [82] Mitra S et al FEBS J 2010;277:128-47; [83] Zhang Z et al Fish Shellfish Immunol 2011;30:902-9; [84] Wang T et al Mol Immunol 2010;47:2563-74; [85] Quintana FJ et al PLoS One 2010;5:e9478; [86] Wen Y et al Cell Mol Life Sci 2011;68:2615-26

ONCOLOGY REVIEWS: CURRENT CONTROVERSIES IN CANINE MAST CELL TUMORS

Laura D. Garrett, DVM, DACVIM (Oncology)
Urbana, IL

INTRODUCTION

Mast cell tumors (MCT) are the most common malignant skin cancer in dogs, and significant variability exists in their biological behavior. Most of the tumors are cured with appropriate local therapy, but a subset shows very malignant behavior and the potential to spread to lymph nodes, liver, spleen, and other areas and to thus become a systemic cancer. Because of this variable behavior, it is difficult to predict how an individual tumor is going to behave, and thus difficult to know if staging tests are needed and what treatments to offer in many cases. In addition to controversies over systemic disease, controversies also exist over what treatment is needed to best attain local control of these tumors.

PROGNOSTIC FACTORS FOR SYSTEMICALLY AGGRESSIVE BEHAVIOR

HISTORY AND PHYSICAL EXAMINATION

Some factors that can be obtained from a history and physical examination that are considered to carry a more guarded prognosis in dogs with MCTs include recent, rapid tumor growth; certain locations (mucocutaneous, muzzle, pinnae, inguinal, nail bed); and fixed, ulcerated tumors.

Mucocutaneous Location

In limited published cases eyelid margin MCT appeared to have relatively benign behavior and were effectively treated with local therapy, although one dog was reported to have regional lymph node metastasis.[1-3] MCT of the conjuctiva may be of concern only locally, without reported metastasis in three dogs.[4, 5] In contrast, two case reports of MCT of the lip showed aggressive behavior and local metastatic disease at diagnosis, with survival time of 6 months or less.[6, 7] Five dogs were reported with MCT of the tongue, two presented with lymph node and/or systemic metastasis, with two of the remaining three having post-operative local recurrence leading to euthanasia.[8] In a paper evaluating chemotherapy for high risk MCT patients, 11 dogs with mucous membrane MCT (vulva, prepuce, conjunctiva, oral cavity) had significantly shorter MST than 50 dogs with MCT of haired skin.[9] A recent paper of 32 dogs with 33 conjunctival MCT treated with surgery showed prolonged survival times, with only 2 dogs have local recurrence despite incomplete margins in 25 cases, and no dogs dying of mast cell related disease.[10]

Muzzle/Perioral/Oral Location

Two studies have confirmed a more aggressive behavior for muzzle MCT, with one showing similar behavior for perioral and oral cases as well.[11, 12] The rate of documented metastasis to local (mandibular) lymph nodes was similar in both papers, at 58 and 59%, compared to <10% rate for other cutaneous sites. The median survivals in the studies were 30 and 52 months, which shows that despite a more aggressive tumor behavior, dogs with these tumors may have prolonged survival times. In both studies dogs with lymph node metastasis had significantly shorter survival times than dogs without nodal metastasis, with a median of 14 months in one paper, and an unreported median (less than 20 months) in the other.[11, 12]

Inguinal Location

Historically, the "back half" of the dog was considered to carry a worse prognosis, and while this is generally not supported, there are still concerns about MCT located in the inguinal areas, especially involving the prepuce or scrotum. One study reported on dogs with MCT in the perineal and/or inguinal region treated aggressively, most receiving tri-modality therapy with surgery, radiation and chemotherapy. The median survival was 37 months, and the authors concluded that with appropriate therapy dogs with perineal or inguinal MCT can do well.[13] It is important to note that these dogs received a more aggressive treatment protocol than do the majority of dogs with MCT at other cutaneous sites. A later study compared the outcome of dogs with inguinal or perineal MCT to dogs with MCT in other cutaneous locations; these dogs received a range of therapies. While there was no statistically identified difference between the locations, preputial or scrotal MCT, included in the inguinal category, were also analyzed separately from other inguinal or perineal tumors. The disease-free interval (DFI) for these 12 dogs was significantly shorter (4.2 months) than for the 84 dogs with tumors in other cutaneous locations (33.9 months).[14] Dogs with inguinal or perineal tumors were also significantly more likely to have received chemotherapy, thus potentially biasing the results. The authors state that the power of the study may not have been adequate to identify a difference between the locations. Overall, many oncologists still feel that scrotal and preputial MCT can have a more malignant biological behavior.

Subcutaneous Location

Recently, several papers reported information regarding MCT in the subcutaneous (SQ) location.[15-17] As the grading system (see below) was developed only with cutaneous tumors, it cannot necessarily be applied to SQ tumors with prognostic accuracy. Survival times for the SQ tumors was found to be prolonged, with the majority of dogs in both papers (53 dogs and 306 dogs) not succumbing to mast cell disease.[15, 16] Out of the 306 dogs, only 8% had local recurrence, and only 4% had metastasis.[16] Risk factors identified for local recurrence and metastasis included mitotic index, infiltrative vs circumscribed histologic pattern, the presence of multinucleation, and age at diagnosis as well as Ki67, Ki67+AgNOR, and KIT cellular localization pattern.[16, 17]

HISTOLOGIC INFORMATION

Grade

A histologic grading scheme (I to III) was developed for classification of MCTs affecting haired skin and is still one of the most accepted prognostic indicators of canine MCT behavior. However, histologic assessment is prone to operator subjectivity, giving rise to extremely variable grading results for the same tumor among different pathologists.[18, 19] In one study with10 pathologists evaluating the same 60 MCT, there was agreement on grade by all pathologists for only 4 tumors, and 6 of the MCT had all 3 grades assigned. Differences in the references used to grade the MCT was the suspected reason for the disparity, and despite 6 different references being used in this study group, at least 80% of the pathologists agreed on the grade of 50% of the tumors.[19] In a follow-up study involving the same tumors and pathologists, the Patnaik grading scheme was used by all. While mean agreement improved to 62.1%, only 16/60 tumors received the same grade by all, and 3/60 still received all three grades.[18] The subjectivity and variability between pathologists bring into question the heavy reliance placed on the grade of a MCT to predict its behavior. In addition, most MCTs are designated grade II and most of these are cured with surgery. However, a subset of grade II tumors show aggressive behavior, and the question of how best to identify this subset of high risk cases is a matter of great debate.

Recently, a two-tier (high vs low grade) grading scheme was proposed in a study of 95 dogs with MCT treated with surgical excision alone.[20] High grade tumors had one of the following criteria: MI (see below) of ≥ 7, ≥ 3 mutinucleated cells or cells with bizarre nuclei in 10 hpf, or karyomegaly. Areas of highest mitoses or anisokaryosis were evaluated. Median survival time for high grade (10 dogs) was 3.6 months, vs median not reached (>2 years) for low grade (85 dogs). Further studies validating this proposed system are warranted; currently the author's university pathology department provides the standard grade, a MI, and the two-tier grade for every sample.

Other Assessments on Biopsies

As grade is so variable and subjective, many other prognostic factors, including DNA aneuploidy, c-kit–staining pattern, presence of c-kit mutations, microvessel density, Ki67, PCNA, and mitotic index (MI), have been evaluated in an attempt to better predict the behavior of canine MCTs and to pick out the "bad" grade II tumors. Two groups have evaluated the MI (total number of mitotic figures counted in 10 high-power fields – areas with the highest mitoses are the ones to be counted) and found that it is predictive of survival time, even within the grade II tumor category. The groups did identify different values for the index, with the first paper showing a median survival of > 70 months for an index score ≤5 and a survival < 2 months for a score > 5.[21] The second group wrote a letter to the editor in response to the first paper and confirmed that high scores have very short survival times, with index scores broken down into three groups: MI=0, median survival time not reached ; MI=1 to 7, median survival time = 18 months; and MI > 7, median survival = 3 months.[22]

STAGING INFORMATION

In general, staging tests for asymptomatic dogs with cutaneous MCT are extremely low yield. The test most often positive is regional lymph node aspiration. Cytologic assessment of the locoregional LN is important, even if the node is not enlarged. In the study of dogs with muzzle MCT, 4/11 lymph nodes with metastases were of normal size.[11] Lymph node involvement has been associated with a worse prognosis in a number of studies.[12, 23-25] FNA is ideally performed prior to excision of the primary MCT, as surgical treatment of the tumor can produce confusing LN results due to local inflammation. Due to the fact that mast cells are a normal feature in LNs, some MCT cases will not have a definitive answer on LN cytology. Criteria for LN involvement has been proposed[23] and morphometrical analysis seemed useful in a preliminary study.[24]

Other staging tests are rarely positive and may have false positive results as well. In general, staging tests other than LN aspiration are recommended in patients that have negative prognostic factors associated with their MCT. Thoracic radiographs are indicated to evaluate the sternal lymph node if the mass is on the ventral abdomen, or to rule out other non-MCT diseases. MCT metastasize so rarely to the lungs that radiographs are not indicated to evaluate for pulmonary spread. Buffy coat preparation to look for circulating mast cells is a quick and easy test, but it is both insensitive and non-specific. The bone marrow may still be infiltrated in spite of a normal buffy coat, and dogs with skin disease, parvo virus, and non-mast cell illnesses often have positive buffy coats despite not having a MCT. Bone marrow evaluation may show infiltration, but appears to be positive in <5% of cases, even in cases with poorly differentiated tumors.[26] This author rarely assesses the bone marrow for MCT cases anymore.

Abdominal ultrasound (US) also shows low positive results. I strongly recommend US of the abdomen to evaluate the sublumbar LNs if the MCT is on the back half of the patient. The benefit of fine needle aspiration and cytology of ultrasonographically normal-appearing liver and spleen for staging of dogs with mast cell tumors (MCT) is a topic of ongoing debate. Mast cells can be found in moderate numbers in normal canine liver and spleen, and while increased numbers were found in splenic aspirates in 51 MCT-bearing dogs with normal spleens on US compared to 32 unaffected dogs, the splenic cytology did not correlate with systemic behavior.[27] In another study, 52 dogs with MCT underwent ultrasound evaluation and FNA with cytology of their liver and spleen.[28] Cytologic criteria of MCT infiltration included clustering of well-differentiated mast cells, large numbers of well-differentiated mast cells, or mast cells with atypical morphology (pleomorphic and poorly granulated). The dogs were separated into 2 groups; those without MCT infiltration into either organ (n=42) and those with infiltration into one or both organs (n=10); survival time between these groups was significantly different at 733 vs 34 days, respectively (p<.0001). This dramatic difference in survival time supports that the cytologic findings were consistent with systemic MCT. Survival time based on US appearance alone in the dogs was not significant, suggesting that cytologic evaluation of the spleen and liver may be indicated for complete staging of dogs with MCT regardless of the ultrasonographic findings.[28] More recently, 19 dogs with aggressive grade II or III MCT treated with vinblastine/CCNU chemotherapy were evaluated to determine the specificity and sensitivity of ultrasound findings in determination of MCT infiltration of the liver and/or spleen, using cytologic assessment to determine infiltration.[29] Seven dogs had MCT infiltration, and the sensitivity of US for detection of infiltration was 43% for the spleen and 0% for the liver. Dogs with cytologic determination of infiltration of the spleen or liver had significantly shorter survival than dogs without infiltration (100 vs. 291days and 100 vs 276 days, respectively, p<.0001), this clinical survival thus corroborating the cytologic finding of systemic disease. Due to the poor sensitivity of ultrasound in detection of cytologically confirmed organ infiltration and the decreased survival associated with identified infiltration, FNA and cytology of the spleen and liver is recommended for staging of dogs with MCT thought to be at high risk for metastasis.[29]

LOCAL CONTROL

Surgical Margins

While surgery is the mainstay of therapy for MCT, and is curative in a large percentage of cases, there is still much controversy over surgical margins. Removing the tumor with gross margins of 2-3 cm lateral margins and a deep margin one tissue plane beyond the tumor has been recommended. Twenty-three MCT in 21 dogs were evaluated for completeness of excision at 1, 2, and 3 cm from the tumor edge; 100% of tumors had complete excision at the 2 cm margin mark. Only grade 1 (3) and grade 2 (20) MCTs were in the study, so it is not clear if 2 cm margins will be adequate for grade 3 tumors.[30] In a subsequent study, 91% (21/23) of grade 1 and 2 tumors were completely excised with 2 cm margins and had no local recurrence for >538 days.[31] Thus, 2 cm margins provide good results for grade 1 and 2 MCT with decreased morbidity. More recently, a report on 100 dogs with 115completely resected MCT showed no recurrence or metastasis in grade 1 and 2 tumors when excised with lateral margins ≥ 10mm and deep margins ≥ 4 mm. The margin around the gross tumor was assessed on the histologic sample, so some shrinking would have occurred, but it does appear that even 2 cm margins may be more than is needed for successful removal of these tumors.[32]

Histologic Margins

Another controversy rages over what to do about clean vs dirty margins. MCT will recur both with clean and dirty margins, and the % of such recurrence varies widely between papers. With recurrence defined as a new mass within 2 cm of previous incision developing up to 1 year post-op, two papers stated 19-37.5% recurrence with clean margins.[33, 34] Other papers site lower recurrence with clean margins, from 0-11%.[32, 35, 36, 37] With incomplete resection, one paper found a 63.6% recurrence rate.[34] This is in contrast to a number of papers which report a much lower recurrence rate for dirty margins, ranging from 12-30%.[16, 35, 38-41] In one of these papers, only 23% of 28 dogs with incompletely resected grade II tumors had local regrowth. The combination of Ki67 and PCNA scores were prognostic for local recurrence, and dogs with recurrence had significantly decreased survival times.[39] Thus, in cases where further local therapy is being discussed, these proliferation indices may be beneficial. Also, SQ MCT recurrence was associated with a number of histologic markers (see above). With such low recurrence rates for "incomplete" resections, further therapy to treat the "residual" disease becomes a controversial topic and one for further discussion with pet owners.

SYSTEMIC THERAPY

Additional therapy (chemotherapy, tyrosine kinase inhibitor (TKI) therapy) may be considered in dogs with poor prognostic indicators (e.g., grade III, high MI, poor location) after excision of the MCT. Dogs with nonexcisable tumors may also be considered for chemotherapy or TKI therapy; both of these therapies show better success against smaller tumors, but certainly can be beneficial in cases with bulky tumors.

Vinblastine/prednisone and lomustine are commonly used chemotherapeutics for MCT. Kinavet and Palladia, new small molecule inhibitors of TKs, have activity against bulky canine MCT. These drugs are likely to also provide benefit in the microscopic disease setting, including treatment in dogs with MCT with poor prognostic factors such as high grade, high mitotic index, poor location, etc. Masitinib (Kinavet--AB Science) is a TKI that mainly works against Kit. A publication described 202 dogs with grade II or III MCTs in a randomized study evaluating Kinavet.[42] Dogs treated with Kinavet at 12.5 mg/kg PO once daily had a longer time to disease progression (178 days) versus dogs receiving placebo (75 days). Kinavet did not improve overall survival time compared with placebo, except in the subset of dogs with KIT mutations. In that group, survival with Kinavet was 417 days versus 182 days with placebo. [42] A subsequent paper examined the 12 and 24 month survival rates in 132 dogs that continued on their respective treatments from the earlier study..[43] Survival rates at 12 and 24 months were significantly improved in the dogs receiving Kinavet, with 62% and 39.8% of Kinavet dogs alive at 12 and 24 months, versus 36% and 15% of placebo dogs. However, the median survival time of 617 days for Kinavet dogs was not significantly different from the median of 322 days for placebo.

Toceranib phosphate (Palladia--Pfizer) has activity against the split kinase family members (VEGFR, PDGFR, and Kit) and is believed to have both direct antitumor and antiangiogenic activity. An early study showed 54% of dogs with a variety of tumors having some response to the drug.[44] A subsequent study in 149 dogs with bulky (nonresectable) grade II or III MCT showed a 37% response rate in dogs receiving Palladia (3.25 mg/kg PO EOD, $n = 86$) vs an 8% response in dogs treated with placebo ($n = 63$).[45] After 6 weeks, the study was unblinded and all dogs were eligible to receive Palladia. In the 58 placebo dogs that subsequently received Palladia 40% showed a response. The response rate of 145 dogs that received Palladia was thus 43% (21 CR, 41 PR). If dogs that maintained SD for ≥ 10 weeks (12%) are considered to have shown a biological response, the overall RR increased to 59.5%. While dogs with KIT mutations were more likely to have a response (82%), dogs without mutations still showed a response (54.5%).

REFERENCES

[1.] Northrup NC, et al., **JAHAA** 2004; 40(4): 309.[2.] Roberts SM, et al., **JAVMA** 1986; 189(10): 1355.[3.] Jacobi S, et al., **Vet Ophth** 2008; 11(6): 395.[4] Barsotti G, et al., **Vet Ophth** 2007; 10(1): 60.[5.] Johnson BW, et al., **JAHAA** 1988; 24: 439[6] Herman LH, et al., **JAVMA** 1967; 151(10): 1322.[7] Patnaik AK, et al., **Vet Path** 1982; 19(6): 608.[8.] Beck ER, et al., **JAHAA** 1986; 22(4): 525.[9.] Thamm DH, et al., **J Vet Med Sci** 2006; 68(6): 581[10.] Fife M, et al., **Vet Ophth**2011; 14(3): 153.[11.] Gieger TL, et al., **JVIM** 2003; 17(5): 687.[12.] Hillman LA, et al., **JAVMA** 2010; 237(8): 936.[13.] Cahalane AK, et al., **JAVMA** 2004; 225(3): 401.[14.] Sfiligoi G, et al., **JAVMA** 2005; 226(8): 1368.[15.] Newman SJ, et al., **J Comp Path** 2007; 136(4): 231.[16.] Thompson JJ, et al., **Vet Path** 2011; 48(1): 156.[17.] Thompson JJ, et al., **Vet Path** 2011; 48(1): 169.[18.] Northrup NC, et al., **J Vet Diagn Invest** 2005; 17(6): 561.[19.] Northrup NC, et al., **J Vet Diagn Invest** 2005; 17(3): 245.[20.] Kiupel M, et al., **Vet Path** 2011; 48(1): 147.[21.] Romansik EM, et al., **Vet Path** 2007; 44(3): 335.[22.] Elston L, et al., **Vet Path** 2009.[23.] Krick EL, et al., **Vet Comp Oncol** 2009; 7(2): 130.[24.] Marconato L, et al., **Vet Comp Oncol** 2008; 6(3): 162.[25.] Murphy S, et al., **Vet Rec** 2006; 158(9): 287.[26.] Endicott MM, et al., **Vet Comp Oncol** 2007; 5(1): 31.[27.] Finora K, et al., **Vet Comp Oncol** 2006; 4(3): 178.[28.] Stefanello D, et al., **JVIM** 2009; 23(5): 1051.[29.] Book AP, et al., **Vet Rad US** 2011; 52(5): 548.[30.] Simpson AM, et al., **JAVMA** 2004; 224(2): 236.[31.] Fulcher RP, et al., **JAVMA** 2006; 228(2): 210.[32.] Schultheiss PC, et al., **JAVMA** 2011; 238(11): 1464.[33.] Jaffe MH, et al., **JSAP** 2000; 41(1): 7.[34.] Grier RL, et al., **JSAP** 1995; 36(9): 385.[35.] Michels GM, et al., **JAHAA** 2002; 38(5): 458.[36.] Seguin B, et al., **JAVMA** 2001; 218(7): 1120.[37.] Weisse C, et al., **JAHAA** 2002; 38(1): 71.[38.] Misdorp W, **Vet Q** 1987; 9(3): 279.[39.] Seguin B, et al., **JVIM** 2006; 20(4): 933.[40.] Brocks BA, et al., **Vet Surg** 2008; 37(5): 472.[41.] Murphy S, et al., **Vet Rec** 2004; 154(24): 743.[42.] Hahn KA, et al., **JVIM** 2008; 22(6): 1301.[43.] Hahn KA, et al., **AJVR** 2010; 71(11): 1354.[44.] London CA, et al., **Clin Cancer Res** 2003; 9(7): 2755.[45.] London CA, et al., **Clin Cancer Res** 2009; 15(11): 3856.

DIAGNOSTIC UTILITY OF IMMUNOCYTOCHEMISTRY AND CYTOCHEMSITRY FOR CLINICAL CYTOLOGY

Anne Barger, DVM, MS, DACVP
Urbana, IL

INTRODUCTION

Immunocytochemistry (ICC) is a technique used to evaluate the presence of specific cellular proteins (cellular antigens) by use of monoclonal or polyclonal antibodies. The primary antibody binds to the antigen and is produced primarily in rabbits, mice or goats. An antirabbit, goat or mouse secondary antibody conjugated to a labeling mechanism is then used to identify the binding of the primary antibody to the antigen. The label can be an enzyme, fluorochrome or metal. Enzymes are used most commonly in ICC and immunohistochemistry (IHC). The sample is treated with the enzyme substrate and a reaction occurs at the site of antibody binding which results in a color change. This can be visualized with light microscopy.

GENERAL PRINCIPLES

An understanding of the antigen-antibody interaction is essential for successful use of immunocytochemistry. Immunoglobulin structure consists of the variable region or Fab portion and a constant region or Fc. The Fab binds the antigen of interest. The Fc portion is specific to the species used to generate the antibody. This antibody is referred to as the primary antibody. The secondary antibody is species specific and binds to the Fc portion of the primary antibody. The cellular antigen may be a glycoprotein, lipoprotein or just a protein. The antibody is also a protein, more specifically a glycoprotein. Interaction of these two proteins is essential for this technique to work. The cellular antigen is fixed on the cytology slide or for immunohistochemistry, in a paraffin embedded formalin fixed tissue. The antibody is diluted and free in the buffer and using Brownian kinetics, finds the antigen[1]. By controlling the pH, buffer concentration, salt content and temperature, the process is greatly enhanced. The bonds involved in the antigen antibody binding are weak and are primarily hydrophobic and electrostatic. Van der Waals forces are enhanced by high increases in temperature and electrostatic bonds by neutral pH, low ionic strength of buffer and low incubation temperature[2].

Antigens have multiple binding sites, also referred to as epitopes. A monoclonal antibody recognizes only one epitope of a particular antigen whereas a polyclonal antibody recognizes multiple epitopes. Use of a polyclonal antibody will increase the chance that the antibody will recognize a particular antigen in a different species however it also increases the possibility of cross-reactivity with other antigens.

A mechanism of detection is necessary to quantify the antigen antibody reaction. A reporter mechanism or label is conjugated to the primary or secondary antibody. Several types of labels are available and include fluorescent, enzyme and particulate labels[3]. Fluorescence relies on light excited from a fluorochrome. This method requires a special microscope. Enzymatic labels rely on an interaction between enzyme and substrate which results in a color change at the point of antibody binding. These are commonly used in many immunohistochemical and immunocytochemical reactions. They require use of a light microscope and are fairly inexpensive. The particulate labels are heavy metals such as gold. These are used most commonly for electron microscopy[3].

Immunocytochemistry can be used as a diagnostic tool to increase the specificity of diagnosis, identify infectious organisms, assist in therapeutic decisions or is commonly used in research as a phenotyping tool. Some neoplasms are so poorly differentiated that the cells cannot be morphologically identified as epithelial, mesenchymal or round cell. Antibodies which recognize intermediate filaments vimentin and cytokeratin can help differentiate mesenchymal and epithelial cells. Other antibodies such as melan A are more specific and positive staining can lead to a diagnosis of melanoma. Panels of antibodies are often necessary to gain the information needed to more specifically identify a particular cell type as lymphoid, myeloid, histiocytic or more specifically T lymphocyte, B lymphocyte, macrophage or dendritic cell. A panel of antibodies is very useful in these situations. A typical lymphoid panel in blood in human medicine includes CD2, CD3, CD4 and CD8 for T lymphocytes, CD19, CD20 and CD79a for B cells and CD56 and CD57 for natural killer cells[4,5]. Evaluation of l and k light chains can also be evaluated in an attempt to establish clonality[6]. CD34 is used as a stem cell marker and CD45 is a general leukocyte marker. In veterinary medicine, many different panels are used to evaluate neoplastic cells. Common markers used to diagnose and differentiate histiocytic neoplasms include: CD1c, CD3, CD11b, CD11c, CD11d, CD18, CD45, CD79a, CD80, CD86, CD90 and MHCII[7].

SAMPLE PREPARATION

ICC can be performed on multiple sample types (Table 1)[8]. Sample handling and preparation are critical when performing panels of ICCs. Limitations for ICC of fine needle aspirations include appropriate numbers of diagnostic slides. Some panels include 10 or more antibodies so ensuring there are enough diagnostic quality slides can be challenging. If the clinician considers ICC prior to aspiration, the sample can be placed into transport media. The transport media recipe we use in our lab consists of 85% saline, 10% protein and 5% acetone. Advantages of this are many. Many uniform quality slides can be prepared from one sample. Ideally, the cytospin will consist of 250-300 cells/μl so the cells can spread out and lay flat allowing for thorough microscopic examination[9]. Additionally, effusions are excellent sources for ICC because again, uniform cellularity samples can be prepared. Slides from FNAs can still be used but the number of antibodies will likely be limited. For research purposes, cytospin preparations of cell cultures can also be used. Additionally these can serve as positive controls for diagnostic purposes. The cells are washed with PBS to eliminate the high protein growth media and resuspended in PBS. A cell count should be performed and cell preparations of 250-300 cells/μl should be used. The samples for all preparations should be fairly fresh, ideally less than 1 week old. Antigen retrieval techniques similar to those used for IHC can be used on older samples[10]. However antibodies directed against extracellular epitopes such as CD4 and CD8 may not work as well with antigen retrieval. In our laboratory, we can perform a minimal lymphoma package on a lymph node if 5 diagnostic quality slides are submitted but we will only perform CD3 and CD79a on those slides. Cells can be fixed in formalin, embedded in agarose and then sectioned as they would for IHC. This technique also allows for even distribution of the cells on the slide. Cells fixed in formalin require antigen retrieval before the immunostaining.

Table 1: Pros and Cons of Slide preparation for ICC

	Pros	Cons
Direct Smear	Slides from the original sample can be used	Increased background artifact Prior staining may affect results Use of panels is limited
Cytospins	Useful with limited material Panels possible	Background artifact Extra sampling of patient required
Cell Block	IHC lab can handle like routine material Material easily stored	Low cellular samples cannot be used Antigen retrieval may be necessary

Adapted from Fowler LJ and Whitney AL. Application of immunohistochemistry to cytology. Arch Pathol Lab Med 2008;132:373-383.

CONTROLS

Many papers discuss several levels of controls but often they are actually referring to antibody validation. Validation of the primary antibody for antigen recognition and specificity is essential. It is thought that most antibodies that will work for histopathology will work for cytology. However, the conditions of the sample, antigen retrieval and formalin fixation can change the condition of the epitope; therefore validation for the sample type you are intending to use is strongly recommended. The FDA definition of antibody validation is "the process of demonstrating through the use of specific laboratory investigations, that the performance characteristics of an analytical method are suit able for its intended analytical use[10]." This means the laboratory must prove the antibody is specific, selective and reproducible. This is actually more straightforward for ICC compared to IHC because the tissue is not formalin fixed. Type and time of fixation can greatly impact the reproducibility of the assay. Antibody specificity is very important, especially with the common use of polyclonal antibodies. Antibodies produced against synthetic antigens will not have the same 3-D conformation and therefore may not be as useful in IHC and ICC. Also, it is important to recognize where the antigen is located within the cell and therefore the location of the anticipated staining (cytoplasm, membranous or nuclear). Nonspecific nuclear or cytoplasm staining can occur with this technique. A positive and negative control run with each sample is critical for appropriate interpretation of the test. A cell type or cell line known to express the antigen serves as a positive control. Two conceptually different negative controls should be evaluated. A patient sample treated only with the secondary antibody is used to detect nonspecific binding and false positive staining. In the process of validating the antibody, a known negative cell line or type should be evaluated to show the antibody is recognizing the appropriate antigen. This does not need to be run with every sample, just done as part of the antibody validation.

The antibodies must also be optimized for each laboratory. Several dilutions of the antibody are prepared, usually a minimum of three dilutions are evaluated to determine the optimal staining of the sample. It is best to perform this process on a known positive cell line or tissue type.

CONCLUSION

Immunocytochemistry is a technique that is useful for research and in the diagnostic setting. It can be used in conjunction with cytology to improve specificity or assist with prognostic information. However care should be taken to establish that the antibodies are appropriate for the tissue and the species being evaluated.

REFERENCES

[1]Montero C. The antigen-antibody reaction in immunohistochemistry. J histochem and cytochem 51:1-4, 2003. [2]Ramos-Vara JA. Technical aspects of immunohistochemistry. Vet Pathol 42:405-426, 2005.[3]Burry RW. Controls for Immunocytochemistry: An update. J Histo &Cyto 59(1):6-12,2011.[4]Robson, A. Immunocytochemistry and the diagnosis of cutaneous lymphoma. Histopath 56:71-90, 2010.[5]Oertel J, Huhn D. Immunocytochemical methods in haematology and oncology. J Can Res Clin Oncol 126:425-440, 2000.[6]Venkatraman L, Catherwood MA, Patterson A, et al. Role of polymerase chain reaction and immunocytochemistry in the cytological assessment of lymphoid proliferations.[7] Moore PF, Affolter VK, Vernau W. Canine hemophagocytic histiocytic sarcoma: a proliferative disorder of CD11d+ macrophages. Vet Pathol 43:632-645, 2006.[8]Fowler LJ and Whitney AL. Application of immunohistochemistry to cytology. Arch Pathol Lab Med 2008;132:373-383.[9]Barger AM, Fan TM, et al. Expression of receptor activator of nuclear factor kappa-beta ligand (RANKL) in neoplasms of dogs and cats. J Vet Intern Med 21:133-140, 2007.[10]Bordeaux J, Welsh AW, et al. Antibody validation. Biotech 48:197-209, 2010.

Epilepsy 101 for Internists

Michael Podell MSC, DVM, DACVIM (Neurology)
Chicago, IL

The approach and treatment of seizure disorders in small animals is similar in many respects to the treatment of various other ailments in veterinary medicine: an antecedent historical problem arises, a proper diagnosis is made to confirm the condition, and therapy is initiated to treat the underlying disease and/or signs of the disease. Important differences arise, however, when approaching the diagnosis and treatment of seizures in the cat and dog. First, a specific underlying etiology is often not identified. Second, the clinician must often make a therapeutic decision based on historical accounts alone. Third, treatment is often initiated when the pet is normal, with little ability to predict frequency of seizure recurrence. Finally, both the pet's and owner's quality of life during the interictal period must be balanced with the ability to limit the severity, frequency and duration of future seizure events. This presentation is designed to help clinicians understand the variables for consideration in the treatment of seizure disorders.

CLINICAL SIGNS AND DIAGNOSIS

Classification of seizures and epilepsy into a universally accepted, coherent, and relevant scheme for clinicians has been an ongoing dynamic process in human epilepsy over the past two decades. The standardized classification scheme for seizures and epilepsy established by the International League Against Epilepsy (ILAE) in the 1980's, (Commission for ILAE 1981; 1989) provided the first basis for a taxonomic foundation for an analytic approach in the diagnosis and treatment of epilepsy. This classification scheme, however, is restricted by the following limitations: 1. The reliance on the clinician's ability to classify seizure types based on the presence of "impaired consciousness;" 2. The reliance on electroencephalographic (EEG) features to classify seizure type; and 3. The difficulty in distinguishing between an "idiopathic" disorder of confirmed undetermined etiology versus a "cryptogenic" cause of highly suspect morphologic disease of the brain.(Engel, 2001) Considering that as veterinarians we rarely perform EEG analysis on affected pets, rely heavily on owner perception of the event to determine if consciousness is "impaired," and may not always have the opportunity to perform a complete battery of neurodiagnostic tests (including brain imaging), it is no wonder that we are all left in a confused state of trying to describe, classify, and categorize seizures and epilepsy in our patient population.

My goal is to attempt to piece together a rationale categorization for use in small animal epileptic patients adapted on the recent recommendations of the ILAE Task Force on Classification and Terminology (Engel, 2001). The purpose is to establish a common ground mode of communication to allow diagnostic and therapeutic data to be tabulated for clinical outcome measures. The proposed new diagnostic scheme consists of five levels or axes as follows as proposed by Engel (Engel, 2001):

Axis 1: Ictal Phenomenology

A **seizure** can be defined as a non-specific, paroxysmal, abnormal event of the body. An **epileptic seizure** is the clinical manifestation of excessive and/or hypersynchronous abnormal neuronal activity in the cerebral cortex.(Podell, 1996) Thus, an epileptic seizure has a specific neural origin. Absolute confirmation that a seizure is epileptic may be difficult, as it requires simultaneous visualization of behavioral and EEG changes. As a result, historical information is often used to diagnose an epileptic seizure. The clinical features of epileptic seizures can be separated into four components.(Engel, 1989), (Podell, 1996) The *prodrome* is the time period prior to the onset of seizure activity. Owners report that they can "predict" the onset of their pet's seizures by behaviors exhibited during this time, such as increased anxiety related behaviors (attention-seeking, whining, etc.), reluctance to do normal activity patterns, or increased hiding (in cats). The *aura* is the initial manifestation of a seizure. During this time period, which can last from minutes to hours, animals can exhibit stereotypic sensory or motor behavior (e.g., pacing, licking), autonomic patterns (e.g., salivating, urinating, vomiting) or even unusual psychic events (e.g., excessive barking, increased or decreased attention seeking). The *ictal* period is the actual seizure event manifested by involuntary muscle tone or movement, and/or abnormal sensations or behavior lasting usually from seconds to minutes. Following is the *post-ictal* period, which can last from minutes to days. During this time, an animal can exhibit unusual behavior, disorientation, inappropriate bowel/bladder activity, excessive or depressed thirst and appetite, and/or actual neurologic deficits of weakness, blindness, and sensory and motor disturbances. The latter problems are known as *Todd's paralysis*, and are often an indicator of a focal, contralateral cortical epileptic focus. Often owners only observe the post-ictal period as evidence that their pet has had a seizure. Thus, careful questioning is needed by the clinician to decide if an animal did experience a seizure.

Regardless of cause, a patient's epileptic seizures may be recurrent over time or may occur as a single event. If the patient has a chronic brain disorder characterized by recurrent epileptic seizures, then that patient has **epilepsy**. Note that neither the term seizure nor epilepsy connotes the underlying etiology of the disorder.

Status epilepticus can be defined as a state of continuous seizure activity lasting for 10 minutes or longer or repeated seizures with failure to return to normalcy within 30 minutes. **Epilepsia partialis continua** is a continuous focal seizure involving the motor cortex.(Engel, 2001) Although not well documented by EEG in animals, the typical manifestations include facial muscle movements with "chewing gum" activity, repetitive eye/lip twitching, or myoclonic jerking of limb muscles.

Axis 2: Seizure type

Seizure types are first classified as either being self-limiting (isolated) or clustered (2 or more within 24 hours) and/or continuous (status epilepticus). Within each category, seizures are divided into being either focal or generalized. Focal seizures are the manifestation of a discrete, epileptogenic event in the cerebral cortex. The focal nature of this seizure type is associated with a higher incidence of focal intracranial pathology. (Podell et al, 1995) Focal seizures can be elementary motor seizures, commonly seen as facial muscle twitching, or manifested by more abnormal behavioral disorders. Progressive involvement of the facial, neck/shoulder, and/or limb muscles is known as a "Jacksonian" march seizure event. More complex behavior patterns with focal seizures will include impaired consciousness often with bizarre behavioral activity. Previously termed complex partial or psychomotor seizures, these events are now classified as automatisms, or automotor seizures.(Engel, 2001) Animals may show "fly-biting" behavior patterns, become aggressive without provocation, howl incessantly, become restless or exhibit a variety of motor disturbances. Cats may show a variety of abnormal behaviors and/or motor signs, to include drooling, hippus, excessive vocalizations, or random, rapid running behaviors in house. Whenever a focal seizure is suspected, the clinician should be suspicious of a focal cerebral disturbance and plan the diagnostic work up accordingly.

Generalized seizures are subdivided into tonic-clonic, clonic, myoclonic, atonic, or absence types. The terms convulsive ("grand mal") and non-convulsive ("petit mal") seizures are no longer in use. Generalized seizures originate from both cerebral hemispheres from the start, or progress secondarily from focal seizures. (Engel, 1989),(Berendt and Gram, 1998) Unlike focal seizures, generalized seizures are not necessarily associated with focal cerebrocortical disease.

Axis 3: Syndrome

A syndrome by definition is a group of signs or characteristics that define a particular abnormality. Epilepsy syndromes are not well defined in veterinary medicine, although familial epilepsies are now being identified with segregation analysis. A number of purebred dogs have been identified with either proven or highly suspect familial epilepsy, to include Belgian tervurn, Vizslas, and a variety of other breeds. The vast majority of epileptic syndromes in dogs will be idiopathic in nature.

Axis 4: Etiology

The differential diagnosis of epileptic seizures due to underlying brain disease can be divided into three main etiologic categories: Idiopathic, symptomatic, and probably symptomatic (previously termed "cryptogenic") epileptic seizures. Idiopathic epilepsy implies that no underlying structural brain lesion is present and is presumed to be genetic in origin. Symptomatic epilepsy is the result of an identifiable structural lesion of the brain. Probably symptomatic epilepsy, or "cryptogenic," is believed to be the result of a structural lesion of the brain but not identified. Reactive epileptic seizures are due to metabolic disease and therefore are not classified as an etiology for epilepsy, as the brain returns to normal once the underlying inciting change in metabolism is corrected. The differential diagnoses for these categories are presented below.

Axis 5: Impairment from the epilepsy

Inclusion of signs that are associated with epilepsy allows one to evaluate for persistence of functional and/or structural neurologic changes with associated seizures. The majority of signs in cats and dogs are transient, to include disorientation, visual impairment, salivation, incontinence, altered behaviors, etc. Dogs have been found to demonstrate transient structural changes with cerebral edema of the temporal lobe on MRI scans of the brain,(Mellema et al., 1999) along with altered cerebral metabolism with proton MR spectroscopy after seizures.(Neppl R, 2001) Symptomatic temporal lobe epilepsy with associated hippocampal neuronal loss, however, appears not to be present in idiopathic epileptic dogs.(Buckmaster P.S., 2002)

PATHOPHYSIOLOGY

Epilepsy represents a heterogeneous disease consisting of diverse etiologies, electrophysiologic and behavioral seizure patterns, and responses to pharmacological intervention. As such, the pathogenesis of epilepsy is multifactorial. Genetically determined "seizure susceptibility factors" play a crucial role in the brain's response to triggering or precipitating factors, also known as the seizure threshold. The seizure threshold in people has been shown to lower during sleep (in particular stage 2 sleep), where the hypersynchrony of sleep facilitates both the initiation and propagation of partial seizures in the parietal and occipital lobes.(Herman, 2001) Seizures in these individuals may be activated from unrecognized changes in neuronal activity, intrinsic neurochemical transmission, or by environmental stimuli that do not cause seizures in the normal brain.

A basic tenet in the mechanism of epilepsy is the presence of an imbalance in excitatory and inhibitory neurotransmission. A seizure develops when the balance shifts towards excessive excitation. Recently, much research has been focused on the role of glutamate, the principle excitatory neurotransmitter in the brain, and its receptor complex, the N-methyl-D-aspartate receptor.(Lipton and Rosenburg, 1994) Conditions leading to excessive excitation or loss of inhibition will lead to depolarization of neurons without normal regulatory feedback mechanisms. The result is a paroxysmal depolarization shift of a neuronal aggregate. Epileptic foci will then develop with the potential to spread throughout the brain. Recently, ion channel mutations have been linked to a variety of epilepsies considered idiopathic in humans. The majority of genes identified to date for human idiopathic epilepsy are inherited disorders of ion channels, known as channelopathies.(Noebels, 2003) Excessive influx of sodium, blockade of efflux of potassium or altered calcium flux can lead to repetitive neuronal firing. Specific functional genetic mutations have been identified for each of these ion channels in people (Noebels, 2003), but not in animals.

DIFFERENTIAL DIAGNOSIS

The differential diagnosis of epileptic seizures can be divided into four main etiologic categories: Idiopathic, symptomatic (or secondary), probably symptomatic (cryptogenic), and reactive epileptic seizures. Idiopathic epilepsy (IE) is diagnosed if no underlying cause for the seizure can be identified and presumed to be of genetic origin. This diagnosis is most common in dogs with the first onset of seizures between the ages of 1 to 5 years, a normal inter-ictal neurologic examination, in purebred dogs, and if there is a longer initial inter-ictal period (> 4 weeks)(Podell et al., 1995) True idiopathic epilepsy is much less common in the cat due to the more diverse genetic background of most cats(Quesnel et al., 1997) As such, all cats should be evaluated for underlying reactive or secondary seizures with appropriate diagnostic testing before a diagnosis of idiopathic epilepsy is made in the cat.

Symptomatic, or secondary epileptic seizures, are the direct result of structural forebrain pathology. Any age or breed of dog may develop this type of epileptic seizures. Younger dogs are more prone to developmental and encephalitic diseases, while older dogs (> 7 years of age) are more likely to develop intracranial neoplasia. As expected with underlying cerebral pathology, these animals are more likely to exhibit focal or multifocal neurologic deficits. However, focal lesions in "silent" cortical areas of the brain (e.g., olfactory, pyriform, and occipital lobes) may have seizures as the only neurologic problem.

TEN GUIDING PRINCIPLES OF THERAPY

1. THE BRAIN IS SIMILAR TO OTHER ORGANS IN THE BODY. The brain is an organ that can get sick and recover, similar to other body systems. Epileptic seizures are a sign of a sick brain. Thus, all epileptic seizures should be taken seriously, and approached in a similar fashion.

2. BE CERTAIN THAT EPILEPTIC SEIZURES HAVE OCCURRED. It is essential to determine a positive diagnosis to ascertain that an epileptic seizure has occurred. Many events may mimic epileptic seizures, such as syncope. The majority of seizures are paroxysmal in onset, have a finite duration, and are followed by a change in behavior from seconds to hours.

3. IDENTIFY THE SEIZURE ETIOLOGY. The most important aspect of seizure management is establishing the underlying cause. The differential diagnoses in terms of prevalence of diagnosis differ according the age of onset of the seizure. Overall, dogs that present at less than 1 or greater than 7 years of age, have an initial interictal interval less than 4 weeks, and/or have a partial seizure as the first observed seizure should carry a high suspicion for an underlying identifiable etiology for their seizures with appropriate diagnostic tests done, to include MRI scanning of the brain. All cats without metabolic cause for seizures should undergo MRI scanning of the brain and possible cerebrospinal fluid analysis.

4. ALWAYS TREAT THE UNDERLYING DISEASE. Symptomatic or reactive seizures can be viewed as a sign of the overall disease. This sign can potentially be eliminated or greatly reduced by directing specific therapy to the primary diagnosis.

5. START AED TREATMENT EARLY IN THE COURSE OF DISEASE. The earlier antiepileptic drug (AED) therapy is started, the better the potential outcome may be for seizure control. The decision to initiate AED therapy is based on the underlying etiology, seizure type and frequency, and diagnostic evaluation. Reasons to start monotherapy include: 1) Identifiable structural lesion present; 2) Status epilepticus has occurred; 3) More than 3 generalized seizures occurred within a 24 hour period; 4) Two or more cluster seizure events (2 or more seizures) occur within an 1 year period; 5) Two or more isolated seizure events occur within 6 month period; 6) The first seizure was within one week of head trauma; 7) Prolonged ictal events (> 5 minutes), regardless of frequency; and 8) Prolonged, severe, or unusual post-ictal periods occur.

6. START WITH THE APPROPRIATE AED. Limiting total AED intake is advantageous and a goal in treating epileptic dogs. With any AED drug, there is a balance between seizure control and quality of life afforded. As such, monotherapy is recommended to reduce adverse effects, allow better owner compliance, and reduce overall costs. Phenobarbital (PB) and bromide (BR) are the most widely used initial AEDs. The appropriate starting dose of phenobarbital is 2.5 mg/kg PO BID for the dog and 2.5 mg/kg/day for the cat. If BR is to be used as monotherapy the initial maintenance dose is 40 mg/kg/day. Intravenous loading dose is described under emergency treatment of seizures. Bromide is ***NOT*** recommended for use in cats due to the potential for allergic bronchitis to develop.

7. MONITOR SERUM CONCENTRATIONS. Serial serum trough PB concentrations should be evaluated at 14, 45, 90, 180, and 360 days after the initiation of treatment, at 6 months intervals thereafter, if a dog has more than two seizure events between these times, and/or at 2 weeks after a dosage change. BR serum concentrations should be taken at 3 weeks (the first elimination half-life), and at the first steady state concentration (3 months). Blood samples are best taken prior to dosing and in a fasted dog. Blood samples should be submitted in non-serum separation tubes to avoid factitious decreases in drug concentration.

8. KNOW HOW AND WHEN TO ADJUST THE DOSAGE. The use of therapeutic drug levels is based on population statistical information to provide a probability guide on the potential for successful treatment without adverse effects for an individual patient. Important criteria in determination of these values include larger uniform patient sample size, appropriate use of AED for the diagnosis, and adequate patient monitoring over longer periods of time. Unfortunately, these criteria are rarely available for the standardization of therapeutic AED ranges. In general, the main goal of the use of the values is to prevent toxic reactions . Therapeutic decision making may be better served by setting therapeutic target values for each patient, as not all patients will respond to the same drug dose.

The most effective initial trough therapeutic PB range for the dog is 15-25 μg/ml and 10-20 ug/ml for the cat. Five μg/ml increments are beneficial if seizures are occurring at an equal frequency or worsening after 30 days of therapy. Adjustments of the trough phenobarbital concentration can be calculated with the following formula:

(Desired concentration/ Actual concentration) X (# mg phenobarbital per day) = Total # mg phenobarbital/day

The optimal therapeutic range for monotherapy BR treated patients is 1500-2500 mg/L. The new maintenance dose can be calculated with the following formula: **(Css=steady-state concentration)**

(Target Css – Actual Css) X (Clearance/Bioavailability) = (2000 mg/L – Actual Css) X 0.02 = added mg/kg/day to existing dose

9. MAKE SURE THERE IS GOOD OWNER COMPLIANCE IN DOSING AND MONITORING. All owners should be sent home with a specialized calendar to record seizure events, adverse effects, and changes in drug dosing.

10. KNOW WHEN AND HOW TO ADD OR CHANGE MEDICATIONS. Antiepileptic drug therapy should be changed when either no improvement in seizure control is seen despite maximal trough therapeutic serum concentration and/or when toxic effects are developing.

REFERENCES

Berendt, M., and Gram, L. (1998): *J Vet Internal Med* **13**, 5-8.;Buckmaster P.S., et al. (2002). *J Vet Intern Med* **16**, 95-99.; Commission on Classification and Terminology of the International League Against Epilepsy. (1989): *Epilepsia* **30**, 389-399.; Engel, J. (1989): *Seizures and epilepsy*. FA Davis. Philadelphia.; Engel, J. J. (2001): *Epilepsia* **42**, 796-803.; Mellema, L. M., Koblik, P. D., Kortz, G. D., et al. . (1999): *Vet Radiol Ultrasound* **40**, 588-95.; Neppl R, N. C., Bowen W, Al-Saadi T, et al. (2001*Am J Neuroradiol* **22**, 1933-1943.; Noebels, J. L. (2003):. *Epilepsia* **44 (supplement)**, 16-21.Podell, M. (1996) *Vet Clincs NA- Small Animal Practice* **26**, 779-809.; Podell, M., Fenner, W. R., and Powers, J. D. (1995):. *Journal of the American Veterinary Medical Association* **206**, 1721-8; Podell M. (2005). BSAVA Manual of Canine and Feline Neurology, Platt SR and Olby NJ (eds).

ADVANCED TREATMENT OPTIONS FOR EPILEPSY

Michael Podell MSc, DVM, DACVIM (Neurology)
Chicago, IL

Over the past two decades, the number of antiepileptic drugs (AEDs) available to veterinarians has quadrupled. Coupled with this increase is the ability to rapidly and accurately diagnose underlying brain disease with readily accessible MRI scanning. As a result, the veterinary community is attuned to the need for early treatment intervention. As more treatment choices become available, the unrelenting questions still arise are when should treatment begin, which initial drug therapy is best for our patients, when should treatment changes be considered, and finally, what advantages do newer drugs provide for our patients. The *purpose* of this presenta-tion is three-fold: 1. Review decision making strategies for AED therapy; 2. Provide an overview of the potential applicability of several newer AEDs currently available; and 3. Present information on the therapeutic advances in epilepsy.

DECISION MAKING STRATEGIES FOR AED THERAPY

When to start treatment

The decision to start AED treatment is based on a number of factors, including etiology, risk of recurrence, seizure type and effect on patient, and risk of treatment. Risk factors for seizure recurrence are not well-established for cats and dogs. A number of relative risk factors have been identified in epileptic people, to include a diagnosis of current or previous defined cerebral lesions or trauma, presence of interictal EEG epileptic discharges (up to 90 % recurrence rate) and a history of marked post-ictal adverse effects (Todd's paralysis).[1] Evidence-based guidelines through several international groups are well established for people based on risk: benefit ratio and predictability factors of drug effect.[1,2] From these guidelines, several commonalities exist in guiding clinical practice to include confirmation of an epileptic seizure event and seizure type, obtaining a definitive diagnosis, knowledge that recurrent seizure activity is correlated with poorer long term treatment success, and the influence of treatment on quality of life factors.[3] Thus, the decision to treat is a reflection of the treatment goals to reduce or eliminate epileptic events, reduce seizure severity, avoid adverse effects, and reduce seizure-related mortality and morbidity.[4]

While similar information is not as readily available in our patient population, extrapolation is possible to provide rationale treatment guidelines. *Overwhelming evidence exists in people that there is no benefit to start treatment after a single unprovoked event.*[5] The earlier AED therapy is started, however, the better the potential outcome may be for seizure control. Reasons to start AED include: 1) Identifiable structural lesion present or prior history of brain disease or injury; 2) Status epilepticus has occurred (ictal event $\geq$ 10 minutes; 3) Two or more generalized seizures occur within a 24 hour period; 4) Two or more isolated seizure events occur within a 6 month period; and 5) Prolonged, severe, or unusual post-ictal periods occur.

Drug selection

AED selection is based on a number of factors, including seizure type, efficacy, and tolerability. Again, evidence-based guidelines established by the International League Against Epilepsy (ILAE), American Academy of Neurology, and Standard and New AED Trials (SANAD) provide probability based recommendations.[1,2,5] Despite these guidelines, *no evidence exists that any single AED provides a better outcome for adults with unprovoked epilepsy when early treatment is started.*[3] Drug selection, therefore, is often based on tolerability. Adverse effects can be divided into transient, persistent and life-threatening categories (idiosyncratic or predictable). Most transient adverse effects are avoidable with titration dosing and dissipate within several weeks. Persistent effects are either CNS dose dependent effects associated with sedation, ataxia, vertigo or cognitive impairment or metabolic related with hormonal imbalances, metabolic syndromes and degenerative effects (e.g., osteoporosis). Severe life-threatening effects are mainly associated with either idiosyncratic bone marrow disease (e.g., aplastic anemia) or predictable organ damage over time (e.g., hepatotoxicity).

Success parameters

Epilepsy treatment should be goal-oriented and approached in an objective fashion. Seizure elimination or significant seizure reduction, reducing seizure severity and maintaining normal lifestyle for patient and owner are all important considerations. While many drugs may provide initial improvement in seizure control, long-term efficacy is dependent on many factors. With only approximately 60 – 80% of both human and canine epileptic patients responsive to treatment,[6,7] evaluation of reasons for recurrent seizure activity (refractory epilepsy) is important. These factors can be broken down into three main dependent variables: disease related, drug-related, and patient-related problems. Disease related factors include presence of an undiagnosed underlying brain disease, such as cortical malformation, prior trauma, or active disease process. Occult conditions can lead to localization related epilepsy where epileptic foci develop drug resistance due to architectural brain changes. Drug-related mechanisms include ineffective mechanism of action, development of tolerance, and alteration of drug target or uptake over time.[8] Seizure specific therapy targets a drug for a specific seizure type, and inappropriate drug selection may result in poor control. Patient-related issues are now being discovered in relation to gene polymorphisms that affect AED pharmacokinetic or pharmacodynamic properties. As a result, altered drug metabolism or action is no longer predictable as compared to the general patient population.

Drug-Resistant Epilepsy

The ILAE recently developed a consensus statement on defining drug-resistant epilepsy based on a 2 level structure. In level 1, a patient is either classified as seizure free, undetermined outcome, or a treatment failure based on a standard monotherapy approach. Undetermined outcome is reserved for situations where an inadequate time period or drug level was documented as cause for failure. In level 2, patients have recurrent seizures on 2 or more appropriately chosen and used AED.[23] Moreover, definitive evidence exists that a third AED is unlikely to provide further benefit in people.[24] A variety of mechanisms of drug resistant have been postulated, to include altered neuronal membrane ion channels, enhanced drug efflux through the blood-brain barrier, and other immune-related interactions.[25]

THE GENERATION GAP

History of AED Development

From the introduction of bromide in 1857 by Sir Charles Locock to treat "hysterical" epileptic fits in women, to the development of phenobarbital in early 1900's and the introduction of phenytoin, valproate and carbamazepine in the early 1970s, the use of these first generation AEDs was limited by mechanism, benefit, and adverse effects. Not until the early 1990's did a revolution in drug development occur with the introduction of felbamate. The enhancement of rodent models for drug discovery screening was a driving factor in this expansion. The succession of the second generation drugs over the next two decades resulted in greater treatment success with significant improvement in tolerability. As the second decade of the 21st century starts, a third generation of AED therapy enters with more advances in improving seizure control and patient quality of life. **(Table 1)**

Mechanisms of action

Antiepileptic drugs can be classified into three broad mechanistic categories **(Table 2)**: 1. Enhancement of inhibitory processes via facilitated action of gamma amino-butyric acid (GABA); 2. Reduction of excitatory transmission; and 3. Modulation of membrane cation (Na or Ca^{+2}) conductance. Unfortunately, several limitations exist in the selection of AEDs for use in veterinary medicine.[9] In the past, many of the AEDs useful in people could not be prescribed to small animals either due to inappropriate pharmacokinetics (too rapid of an elimination), ineffectiveness, or the potential hepatotoxicity. The result was that the most commonly used AEDs in veterinary medicine were from the same mechanistic category, that of enhancing brain inhibition. The advent of new AEDs with alternative mechanisms of action is now available to allow a broader selection of treatment options.

Table 1. Categorization of Antiepileptic Drugs by Generation of Drug Development

FIRST GENERATION	SECOND GENERATION	THIRD GENERATION	NEXT GENERATION
Bromide	Felbamate	Lacosamide	Flurofelbamate
Phenobarbital	Gabapentin	Rufinamide	Brivacetam
Benzodiazepines	Oxcarbazine		Carisbamate
Phenytoin	Zonisamide		Retigabine
Carbamazepine	Lamotrigine		Tiagabine
Valproate	Levetiracetam		Losigamone
	Pregabalin		Remacemide
	Topiramate		Seletracetam

Clinical AED pharmacology

The efficacy and safety profiles of AEDs are determined in large part by their pharmacokinetic properties. Drugs that are the easiest to use by the general population are ones that have the most favorable pharmacokinetic properties. For this reason, phenobarbital (PB) has been widely used in veterinary medicine for generations. Ultimately, the most desirable pharmacokinetic profile of an AED is one that has complete bioavailability, availability as a parenteral formulation, an elimination half-life suitable for daily or twice daily dosing, linear elimination kinetics, no autoinduction of enzymatic biotransformation, no pharmacokinetic interactions with other drugs, rapid brain penetration, a volume of distribution with a single compartment, low and nonsaturable protein binding, and no active metabolites. Unfortunately, the ideal AED has not been formulated for any species.

Table 2. Summary of the Mechanism of Actions of Currently Available Antiepileptic Drugs.

	MECHANISM OF ACTION				
	Decrease seizure onset		Decrease seizure spread		
Drug	**Enhanced Na+ channel inactivation**	**Enhanced GABA activated Cl- conductance**	**Reduced Ca^{+2} channel current**	**Reduced glutamate excitation**	**Novel mechanism**
Phenobarbital		++	+	?	
Bromide		++*			
Felbamate	+	+		+	
Benzodiazepines		++			
Gabapentin	?	+	++		
Topiramate	+	+		+	
Zonisamide	+		++		
Levetiracetam		?	?	?	SV2a receptor
Pregabalin			++		
Lacosamide	+				CRMP
Rufinamide	++				

*Competitive displacement of Cl- through activated GABA receptors; ? = Possible mechanism; SV2a=synaptic vesicle 2A binding; CRMP=collapsin response mediator protein 2 binding.

First Generation AEDs

Phenobarbital is well-tolerated and is still the standard of care for first generation AED use in the cat and dog, as it is highly efficacious and well tolerated. The drug has high bioavailability, is over 50% protein bound, has a long elimination half-life, with predominant hepatic metabolism and autoinduction of hepatic microsomal enzymes (p-450 system), which can progressively reduce the elimination half-life with chronic dosing. *Bromide* is an inorganic salt with a prolonged elimination half-life (15-20 days), is solely renal excreted unchanged, has proven efficacy, but does have greater dose-dependent adverse effects.[10]

Second Generation AEDs

Felbamate is believed to increase seizure threshold and prevent seizure spreading by reducing excitatory neurotransmission in the brain and increasing GABA activity. Felbamate is most useful as monotherapy in the treatment of uncontrolled partial epilepsy and Lennox-Gastaut syndrome with atonic seizures. In dogs, the drug has a high bioavailability and protein binding capability and is metabolized by the hepatic microsomal p450 enzymes.[11] Effective control of focal seizure activity (automatisms) with documented therapeutic serum concentrations has been shown with felbamate therapy in dogs.[12] Felbamate is a non-sedating drug, but has been correlated with a higher incidence of idiosyncratic aplastic anemia and liver toxicity in people. Similar, but rare and reversible blood cytopenias and hepatotoxcity have been observed in the dog. Serial monitoring of the complete blood count and chemistry panel is recommended at 1 month and every 3 months during treatment. Initial dose is 5-10 mg/kg q 12 hours with titration to q 8 hr dosing. Trough serum drug concentration is typically done 1 –2 weeks after initiation of treatment, with a therapeutic range between 25-100 mg/L.

Topiramate is a sulphamate-substituted monosaccaride with multiple mechanisms of action of rapidly potentiated GABA activity, sodium channel blockade and post-synaptic glutamate receptor blockage that is efficacious against a wide variety of seizure types.[12] In people, topiramate is well-absorbed, is primarily renal excreted as an unchanged drug and has a relatively long-half life of 20-30 hours. In the dog, however, topiramate has a relatively short elimination half-life in the dog of 2-4 hours,[13] but may exhibit prolonged pharmacodynamic activity in the brain due to high affinity receptor binding. Initial dosing is 2- 5 mg/kg q 12 hours with gradual adaption to 10 mg/kg q 12 hr. Loss of appetite, irritability, GI upset and sedation are potential adverse effects.

Zonisamide is a substituted 1,2-benzisoxazole derivative that works via multiple mechanisms to block the propagation of epileptic discharges and suppress focal epileptogenic activity through sodium and calcium channel blockade and GABA activity potentiaion .[14] In the dog, zonisamide is well-absorbed, has a relatively long-half life (15-20 hours), has high protein-binding affinity and has both hepatic and renal metabolism.[15] The drug is highly concentrated in red blood cells due to high binding to carbonic anhydrase and other red cell protein components. Broad-spectrum antiepileptic activity has been reported against a variety of seizure types, with particular improvement in the treatment of adult myoclonus epilepsy.[16]. Since serum drug levels are lowered by phenobarbital, phenobarbital doses should be decreased by 25%. . Initial dose is 5 mg/kg q 12 hr with a therapeutic range of 10-40 mcg/ml, with greater efficacy at 20 mcg/ml.[17] Major adverse effects include a higher incidence of sedation, GI upset, metabolic acidosis and renal calculi formation.

Levetiracetam is the S-enantiomer of the ethyl analogue of piracetam with a unique mechanism of action mediated by binding to the synaptic vesicular protein, SV2A, that decreases neurotransmitter release. In dogs, the drug is well-absorbed, is rapidly metabolized with an elimination half-life estimate of 4-8 hours and is predominantly renal excreted (> 80%).[18] Wide fluctuations of drug metabolism, however, occur in the dog. Initial dose is 10- 20 mg/kg q 12 hours with gradual increment to 20 or higher mg/kg q 8 hour dosing. The therapeutic range is not well-defined and drug monitoring is recommended only to establish individual patient pharmacokinetic pattern. The drug is well-tolerated, with sedation noted as the most common adverse effect. Parenteral formulation is available for IV loading at 40-60 mg/kg over 10 minutes.

Lamotrigine is a novel drug that is chemically unrelated to current AEDs. Although efficacious in human epileptic patients, the drug is converted to a cardiotoxic 2-N methyl metabolite in dogs,[19] not found in people, and is not recommended for use in dogs.

Pregabalin is a structural analogue of GABA with a mechanism of action of voltage-gated calcium channel modulation that decreases depolarization induced calcium influx at nerve terminals, ultimately, to reduce excitatory neurotransmitter release. The mean elimination half-life is estimated at 7 hours in dogs.[20] Metabolism appears to be predominantly renal excretion with minimal protein binding and drug interaction. Efficacy as an add-on for refractory partial seizures was found in several studies in people and in a recent study in dogs.[21] A definitive therapeutic range has yet to be determined for people or dogs. Initial dose recommendation is 4 mg/kg q 8 to 12 hours. Adverse effects appear limited to sedation and ataxia.

Third Generation AEDs

Lacosamide (Vimpat®) is a functionalized amino acid proven to decrease neuronal discharge frequency and synaptic excitability.[22] The postulated mechanisms of action includes selective slow inactivation of sodium channels and novel binding to collapsin response mediator protein-2. In people, the drug is well-absorbed, has minimal first pass effect with predominant renal excretion, low protein binding, favorable drug-drug interactions with other AEDs, and is well-tolerated. Clinical trials demonstrate comparable decrease in seizure frequency as that of levetiracetam and zonisamide at a dose of 100-200 mg q 12 hours. A parenteral formulation is available for IV loading. No data was found regarding clinical use in dogs or cats.

Rufinamide (Banzel®) is a novel AED that is structurally unrelated to any other AED. The main mechanism of action is related to prolongation of the inactive state of the sodium channel to prevent neuronal depolarization. In people, the drug is slowly absorbed with low bioavailability. Renal excretion is high and no induction of the hepatic p450 system was found, although other hepatic metabolized drugs decrease serum concentration. Of a total of 9 double-blinded studies in people, 5 revealed a positive effect of rufinamide to treat refractory partial seizures but not generalized seizures.[23] Initial dosing ranged from 10 to 40 mg/kg/day, with dose dependent adverse effects of sedation, fatigue and dizziness noted. A parenteral formulation is not available. No data was found regarding clinical use in dogs or cats.

REFERENCES

1.SIGN Guideline 70. (2003) Scottish Intercollegiate Guidelines Network. Edinburgh . www.sign.ac.uk; 2. American Academy of Neurology, (2004) www.aesnet.org/go/practice/guidelines; 3. Stephen LJ and Brodie MJ (2009*), Neurol Clinics* 27:967-992; 4. Perucca E. (2009) In: The Treatment of Epilepsy (ed, Shorvon, Perucca, Engel), p 121; 5. Glauser T, et al. ILAE Guidelines (2006), *Epilepsia* 47:1094-1020; 6. Brodie MJ, et al (2007) *Neurology* 68:402-408; 7. Podell M (2005) Manual of Canine and Feline Neurology, Platt SR and Olby N. BSAVA, 97-112; 8. Loscher and Schmidt D. (2006) *Epilepsia* 47:1253-1284; 9. Podell, M. (1998): *Clin Tech Small Animal Practice* **13**, 185-192; 10. March, P. A., et al. (2002):. *J Vet Pharmacol Ther* **25**, 425-432; 11. Adusumalli V. A., et al. (1992). *Epilepsia* **33**, 955-960; 12. Ruehlmann, D. S et al (2001):. *J Small Animal Practice* **42**, 403-408.; 13. Petroff, Oa, et al (2001). *Epilepsia* **42**, 543-548; 14. Streeter AJ et al (1995). *Drug Metab Dispos* 23:90-93; 15. Ito, T., et al.. (1980):. *Arzneimittel-Forschung* **30**, 603-609; 16. Walker, R. M., et al. (1988): *Fundam Appl Toxicol* **11**, 333-342; 17. Dewey C (2004) *JAAHA* 40:2850291; 18. Volk HA et al. (2008) *Vet J* 176:310-319; 19. Wong I., and Lhatoo S. (2000):. *Drug Safety* **23**, 35-56; 20. Salazar V, et al. (2009) *Vet Anesth Analg*, 36:574-580; 21. Dewey C et al. (2009) *JAVMA* 235:1442-1449); 22.Halford JJ and Lapointe M (2009) *Epilepsy Currents* 9: 1-9; 23. Biton V. (2009) In: The Treatment of Epilepsy (ed, Shorvon, Perucca, Engel), p. 651.; 22. Kwan P et al, Epilepsia (2010) 51:1069-1077; 24. Poolos NP et al Neurologu (2012) 78:62-68; 25. Kwan P et al,. (2011) NEJM 365:919-926.

CARDIOLOGY REVIEWS: DIAGNOSIS, PROGNOSIS AND RISK STRATIFICATION OF HEART DISEASE USING NT-PROBNP ASSAY

Mark Oyama, DVM, DACVIM (Cardiology)
Philadelphia, PA

Testing of cardiac biomarkers is an avenue of investigation that is relatively recent veterinary medicine. NT proBNP or N-terminal proB-type natruretic peptide is a marker with both great promise and great confusion surrounding its clinical use. This presentation will review the current indications for NT pro BNP testing in both dogs and cats, as well as present data suggesting that NT proBNP can provide novel information that is not obtainable by current diagnostic or prognostic means.

Expansion of plasma volume, vasoconstriction, and activation of the renin-angiotensin aldosterone system as well as the sympathetic nervous system occur in animals with heart disease. The natruretic peptide system has evolved in parallel to provide a counterbalance to the systems. The atrial and ventricular myocardium produced 2 types of natruretic peptide ANP and BNP. Both are secreted as pro-hormones that are subsequently cleaved into an active molecule and an inactive N-terminal fragment. It is well established the in both dogs and cats with variety of primary as well as secondary heart diseases circulating NT proBNP is elevated. In general, circulating concentrations are proportional to indices of disease severity obtained from physical, radiographic, and echocardiographic examination; thus the great interest in evaluating this marker over the last 5 years.

The clinical utility of NT proBNP to diagnose heart disease is dependent on both breed and specific disease. In some instances such as in dogs with mitral valve disease biomarkers play a negligible role in diagnosis such that the detection of mitral valve disease is easily accomplished by auscultation. Whereas in other situations such as in cats with occult cardiomyopathy, NT proBNP has adequate sensitivity and specificity to help identify cats that are likely to have significant underlying heart disease. Biomarker testing does not provide a definitive diagnosis but simply increases the likelihood that significant disease will be found if further and more definitive diagnostics are pursued. Biomarker testing is not a substitute for genetic testing in evaluating animals of young age for breeding soundness.

There have been a number of studies evaluating use of NT proBNP in the differentiation of respiratory disease versus congestive heart failure in dogs and cats with respiratory signs. Similar to results in humans NT proBNP is best applied as a rule out diagnosis for heart failure, that is, in animals with normal NT proBNP congestive heart failure is highly unlikely. As a rule in diagnosis NT proBNP is affected by concurrent morbidities such as pulmonary hypertension and renal insufficiency as well as a relatively low upper reference limit for the NT proBNP assay.

Perhaps the most intriguing potential use of biomarkers is to help stratify risk of morbidity and mortality. Previous studies have shown that elevated NT proBNP is associated with cardiovascular and all cause mortality in dogs with symptomatic mitral valve disease. Studies have also shown that NT proBNP can help assess risk for first onset of congestive heart failure in dogs with asymptomatic mitral valve disease. The use of biomarkers for risk stratification is particularly attractive such that the information gained is novel and not readily obtainable by current diagnostic methods. It should be noted that in order for biomarkers to achieve maximal clinical utility, interventional strategies that then mitigate high risk need to be discovered and proven.

REFERENCES

Fox PR, et al. J Vet Cardiol 2009;11Suppl1:S51-61. Oyama MA et al. Vet Clin North Am Small Anim Pract 2010;40:545-558. Wess G et. al. Vet Clin Pathol 2011;40:237-44. Wess et al. Am J Vet Res 2011;72:642-9. Fox PR et. al. J Vet Intern Med 2011;25:1010-6. Moonarmart W et. al. J Small Anim Pract 2010;51:84-96. Boswood A et al. J Small Anim Pract 2008;49:26-32.

RATIONAL USE OF PIMOBENDAN

John M. MacGregor, DVM, DACVIM (Cardiology)
Portsmouth, NH

INTRODUCTION

Pimobendan has many attributes that make it a potentially useful medication in the treatment of cardiovascular disorders. Pimobendan is a benzimidazole-pyridazinone derivative that combines positive inotropic and vasodilator effects. It acts as a calcium-sensitizer by increasing the binding affinity of calcium to its binding site in the myocyte, troponin C, leading to positive inotropy. Pimobendan acts as a potent balanced vasodilator by inhibiting peripheral phosphodiesterase III. Pimobendan has also been shown to produce coronary vessel dilation in several species ensuring the myocardium is supplied with oxygen and nutrients. Neurohormonal effects of pimobendan include the reduction of certain pro-inflammatory cytokines and catecholamines. In cats and other species, pimobendan is an inhibitor of platelet aggregation. These qualities have led to pimobendan's use in a variety of different clinical situations. The rationale and evidence for its use will be discussed.

USE IN CONGESTIVE HEART FAILURE

Pimobendan has been shown to be safe and useful in treatment of congestive heart failure in dogs. This has been demonstrated in dogs with congestive heart failure due to dilated cardiomyopathy and dogs with congestive heart failure due to mitral valve disease. Additionally, pimobendan has been shown to be superior to an ACE inhibitor as monotherapy (in addition to Lasix) for treatment of congestive heart failure. Pimobendan has become part of the standard therapy for most dogs with congestive heart failure but there are certain conditions for which its use is not appropriate and it does not replace medications that have proven efficacy for treating congestive heart failure. A treatment plan for dogs with congestive heart failure will be enumerated that incorporates the use of pimobendan in concert with other medications.

USE IN DOGS WITH PRECLINICAL HEART DISEASE

Use of pimobendan in dogs with preclinical heart disease is controversial at best. In an experimental model of mild mitral valve disease, pimobendan was found to reduce norepinephrine, increase renal blood flow, decrease blood pressure and decrease left heart size over a four-week period. However, in a longer study in dogs with naturally occurring mitral valve disease, there was an initial decrease in left ventricular size at 30 days relative to a control group that disappeared at 180 days. Pimobendan was reported to increase regurgitant fraction in 2 dogs and resulted in histologic damage to the mitral valve and left ventricle in beagles with asymptomatic mitral disease. Therapy for asymptomatic dogs with mitral valve disease will be discussed relative to the ACVIM consensus statement and the existing evidence. There are no current reports about the effects of pimobendan in asymptomatic dogs with dilated cardiomyopathy, however use of positive inotropes such as amrinone and milrinone is associated with increased mortality in human patients. Several studies are ongoing and will begin to answer the questions about use of pimobendan in preclinical patients.

USE FOR PULMONARY HYPERTENSION

Pimobendan has been used successfully to treat pulmonary hypertension in dogs with pulmonary hypertension caused by myxomatous mitral valve disease and is approved for treatment of pulmonary hypertension from a variety of causes in some countries for human patients. Beyond the documented efficacy of pimobendan and sildenafil, very little published information is available about the treatment of pulmonary hypertension, despite its increased diagnosis in canine patients.

USE IN CATS

Very little information is published about the use of pimobendan in cats for the treatment of congestive heart failure. A recent retrospective study documented its use in cats with naturally occurring heart disease and concluded that it was likely safe when administered to cats. However, conclusions could not be drawn about the efficacy of pimobendan in treating cats with congestive heart failure as the study lacked a control population.

SUMMARY

Pimobendan is a safe and useful drug for treatment of many causes of congestive heart failure and pulmonary hypertension in dogs. Use of pimobendan in cats with congestive heart failure has not been demonstrated to be efficacious at the current time but it appears safe. Use of pimobendan in dogs that are not in congestive heart failure and that do not have pulmonary hypertension has no proven benefit at this time and may cause harm.

REFERENCES

1. Fuentes VL, et al. Journal of Veterinary Internal Medicine. 2002; 255-261. 2. O'Grady et al. Journal of Veterinary Internal Medicine. 2008. 22(4): 897-904. 3. Haggstrom J et al. Journal Veterinary Internal Medicine. 2008; 22: 1124-1135. 4. Smith PJ et al. Journal of Small Animal Practice. 2005. 46:121-130. 5. Lombard CW et al. Journal of the American Animal Hospital Association. 2006. 42: 249-261. 6. Kanno N et al. Journal of Veterinary Medical Science. 2007. April 69(4): 373-7. 7. Ouellet M et al Journal of Veterinary Internal Medicine. 2009. 23(2): 258-63. 8. Tissier R et al. Cardiovascular Toxicology. 2005. 5(1):43-51. 9. Chetboul et al.Journal of Veterinary Internal Medicine. 2007. 21(4): 742-53. 10. Atkinson KJ et al. Journal of Veterinary Internal Medicine 2009 Nov-Dec;23(6):1190-6. 11. Sahara M, et al. Cardiovasc Drugs Ther. 2006 Oct;20(5):377-86. 12. MacGregor et al, Journal of Veterinary Cardiology, 2011; 13(4): 251-260.

TURNING A CLINICAL QUESTION INTO A TESTABLE HYPOTHESIS

Lauren A. Trepanier, DVM, PhD, DACVIM, DACVCP
Madison, WI

As residents or new faculty members, we are faced with the expectation to carry out clinical research. As clinicians, we encounter frustrating clinical questions that we may want to resolve. However, many of us as clinicians have not received formal research training, and research may feel outside of our comfort zones. When trying to identify a good research question, trust your own clinical experience. Where are the controversies that are impairing your ability to manage your cases? What types of cases are *commonly* seen in your clinic? What is new and interesting about those diseases in the human field?

Clinical research questions may address new diagnostic tests or treatment options, adverse outcomes, prognostic indicators, or underlying etiology. [1,2] Review your journal club papers and re-read papers that have interested you – is there a logical follow-up to a published study, or a study that could be repeated with a better design? Look through proceedings and research abstracts from recent ACVIM Forums; where are the knowledge gaps? Talk with your mentors about their ideas for good clinical research questions. Your question should be clinically relevant and interesting to both you and your research mentor.

The next step in approaching a clinical research question is to perform a thorough literature search to determine **what is known** and **what needs to be answered.** Start with the PubMed site (http://www.ncbi.nlm.nih.gov/sites/entrez), and use multiple search terms (scan relevant papers to see what words show up in the titles and abstracts). Supplement your PubMed search with additional publications from reference lists of papers you are reading. *Read the entire papers, not just the abstracts.* When reviewing research abstracts from conferences, check PubMed to see whether older abstracts have reached the peer-reviewed and published stage (exercise caution if they have not). Summarize the results of your literature search in a bullet point outline. This summary should tell a logical story about why the problem is important, what is known, and what still needs to be discovered. This outline can later form the basis for the background section of a grant application, or the introduction portion of a manuscript. From your outline, ask yourself whether answering your original clinical question still makes sense, or whether it needs to be refined based on your literature review.

Once you have decided on a research question, generate a hypothesis about the likely outcome/answer to your question. Most granting agencies and many journals will ask you to overtly state your hypothesis(es). Your hypothesis should be concise; more detail can be added in your specific research objectives and study design. For example, your original clinical question might have been how to manage proteinuria that is refractory to ACE inhibitors in dogs. After your literature review, your research question is to determine whether spironolactone is an effective adjunctive treatment to ACE inhibitors in dogs with persistent proteinuria. Your hypothesis is that spironolactone will be more effective than placebo in reducing proteinuria in dogs that are refractory to ACE inhibitors.

The next step is to frame a specific research approach that will test your hypothesis in a clear and specific manner. This will flesh out your research question into full study objectives. Study objectives often start with active descriptors such as to "characterize," "determine," "evaluate" or "compare." One practical approach to refining your study objectives is to use the PICOT method to identify five components for your particular study, to include Population, Intervention, Comparator(s), Outcomes, and Time frame. [3,4]

POPULATION

The study population should be as clearly defined as possible, and should encompass specific inclusion and exclusion criteria. If you are studying a pre-existing disease, your **inclusion criteria** should include a gold standard for diagnosis. Make sure that the gold standard is achievable in sufficient patients. If the gold standard is fairly invasive (for example, pancreatic biopsy for pancreatitis or intracranial biopsy for CNS tumors), consider whether there is a reasonably valid surrogate marker. You may be able justify your surrogate marker based on previous studies, or you may need to validate it first. This might shift the focus of your project to marker validation itself as your research objective. Think not only about diagnosis, but also about specific clinical presentations, age ranges, and specific breeds versus all breeds. For example, a study of serologic markers of chronic hepatitis in dogs might be more powerful if focused on a single breed, while breed might be less important in a study on medical management of pyothorax. Consider chronicity of disease – will you enroll only newly diagnosed patients, only patients with established stable disease, or both? Remember that the more heterogeneous your groups, the more patients you will need to demonstrate a difference.

Exclusion criteria should also be carefully defined at the onset. Consider severity of disease – will you exclude animals not expected to survive 24 hours? If so, is this valid for the question you are trying to answer? Will you allow prior treatments, and if so, what minimum washout periods are reasonable? Think about co-morbidities. Is it valid to include some FeLV or FIV positive cats in a study of antioxidant status in renal disease? The more homogeneous your groups, the fewer patients you will need to demonstrate a difference. However, you need to make sure that your study groups are representative of the larger clinical population that you are trying to understand. Finally, there are practical and safety considerations in exclusion criteria. For example, if three 10 ml blood samples are needed, small or anemic dogs may need to be excluded, and breeding animals may need to be excluded from a drug intervention study in a purebred kennel.

INTERVENTION

Most prospective studies have an intervention that is being evaluated, such as test drug versus placebo, a new procedure versus a standard of care, or a new diagnostic assay versus a commonly used test. Interventions should be well defined as to drug dosage, duration, and washout; specific procedure (and clinicians involved); and timing of testing relative to stage of disease. Decide which concurrent treatments will be allowed. Leaving much of case management to "clinician discretion" is not ideal and can cloud your results. You will also need to decide whether your study needs to be blinded (the client or investigator) or double-blinded (both the client and investigator). Anyone providing *any subjective evaluation* for the study should be blinded to the intervention group assignments; this includes owners, managing clinicians, vet techs, radiologists, cytologists, and pathologists.

COMPARATORS (CONTROLS)

Clinically relevant controls are key to the structure of your research question. The control group(s) should reflect the accepted standard of care or a placebo. For most studies, concurrent controls are necessary, since they help to minimize bias associated with the passage of time. Such biases could include the development of better diagnostic tests or interventions during your study period compared to ten years ago, the adoption of a new medical records system with better search capabilities for certain adverse outcomes, or inherent differences in the consistency of evaluations between two time periods. Historical controls may be acceptable in some instances based on ethical grounds, such as when studying new interventions for well-established, uniformly fatal outcomes.

OUTCOMES

All studies should have at least one defined primary outcome. This is often the most important question that needs to be answered. Primary outcomes are strongest when they are objective, such as hematocrit, blood pressure, days in hospital, or survival. If surrogate markers are used, they should be validated against a clinically relevant outcome, such as force plate analysis for lameness in polyarthropathy, or cardiac troponin I concentrations for myocardial disease. Primary outcomes should be easily measured, clinically available, and validated for your species. [4] When subjective assessments are needed, seek out a validated scoring system, such as the Glasgow Composite Measure Pain Score, [5] or the Survival Prediction Index (SPI2) for illness severity. [6] Alternatively, you might develop your own scoring system. Even if subjective scoring systems are chosen, it is best to complement these outcomes with objective or surrogate markers to strengthen your study. If blinding is not possible (e.g. for an invasive procedure), use only objective outcome measures.

To gain the maximal amount of information from the study, secondary outcomes can also be included. These may be of lesser importance, or more difficult to prove, but provide depth to the study. For example, the primary outcome of a study on IV IgG in dogs with immune thrombocytopenia was days to a platelet count > 40,000 with IV IgG versus placebo. [7] The secondary outcomes were the incidence of acute or delayed reactions to IV IgG, cost of hospitalization between groups, and 6-month morality rates. [7]

Once you establish your primary and secondary outcomes, you need to determine the number of subjects you will need to enroll. This is a **sample size calculation** based on the desired power of the study to detect a clinically relevant difference between groups. This step can be approached in one of two ways: 1) start with a known sample size (e.g. number of retrospective cases available in each group, or number of cases likely to be recruited or that the budget will allow), and determine what power you will have to detect the difference you are interested in; or 2) start with a minimum power threshold (typically 80% or higher), and determine how many patients you will need in each group to reach that. The information that you will need for a sample size calculation is: the expected mean outcome, with SD, in your control group (either from a pilot study, or previous veterinary or human studies); and the difference that you want to detect in your experimental/treated group (e.g. 20 mmHg of blood pressure, 50 umol/L of fructosamine, or 1 less day of hospitalization). P (a) is typically set at 0.05. Power is either calculated from your available sample size or set at a minimum of 0.8 (i.e. 80% chance to detect a true difference if it is present). Then use an on-line sample size calculator (e.g. http://www.stat.uiowa.edu/~rlenth/Power/) to calculate your sample size/power. Follow up with a statistician (*before* you start your study) if your study design is complex. Sample size/power calculations are essential for any research question that involves the statistical comparison of outcomes in two or more groups, and most granting agencies require a clear description defending the proposed sample size.

TIME FRAME

Finally, the time frame for study observation and follow-up should also be established. What time points for evaluation are clinically relevant? How much follow-up would you as a clinician find adequate? Is the proposed duration of follow-up feasible, or are many cases likely to drop out over time? If your recruitment time frame is short (less than one year), does it take into consideration possible seasonal variables such as tick exposure (which could bias a fever-of-unknown-origin study) or seasonal vaccines (which could affect a study of immune function)?

Once you have refined your clinical research question, formulated your hypothesis, and framed your study in PICOT format, your study design will fall into place quickly. Make sure that your study design still addresses your hypothesis, and has not "drifted" to address a slightly different question as you have formulated your PICOT. This can happen as you make practical decisions and concessions in your study design, and you may need to reframe or narrow your hypothesis at this stage. Remember that after your PICOT efforts, your research question should still be relevant to clinical practice and interesting enough to warrant your time and effort. It is even more critical that your research question be feasible, given your clinical expertise, caseload, support staff, available funds, and career time frame. The best advice for new researchers is to **choose a disease that you see commonly, and start with a simple focused research question!**

Common roadblocks to new researchers in formulating successful research questions [8]

- Question is unfocused
- Question is not clinically interesting
    - "Because no one has looked at it" is not sufficient!
- Disease diagnosis is uncommon
- Outcome to be studied is rare (e.g. uncommon side effect or complication)
- Data collection is invasive, labor intensive, or off-putting to owners
- Validated assays are not in place
- New pieces of equipment are needed
- Case identification is outside of your sphere of control
- Collaborators are not motivated to complete the study
- Prolonged follow-up is required

Helpful websites

http://consortiumlibrary.org/aml/researchaids/handouts/PICOSworksheet.pdf

- Provides a worksheet to help define a research question and determine search terms to provide background research.

http://www.stat.uiowa.edu/~rlenth/Power/

- A useful on-line sample size calculator from Dr. Russell Lenth at the University of Iowa. Sigma = SD; n = sample size; difference of means = clinical relevant difference that needs to be detected (in same units as SD)

REFERENCES

1. Geddes J. Asking structured and focused clinical questions: essential first step of evidence-based practice. *Evidence-Based Mental Health* 1999; 2:35-36. 2. Heddle NM. The research question. *Transfusion* 2007;47:15-17. 3. Richardson WS, et al. The well-built clinical question: a key to evidence-based decisions. *ACP J Club* 1995;123:A12-13. 4. Thabane L, et al. Posing the research question: not so simple. *Can J Anaesth* 2009;56:71-79. 5. Holton L, et al. Development of a behaviour-based scale to measure acute pain in dogs. *Vet Rec* 2001;148:525-531. 6. King LG, et al. Evaluation of the survival prediction index as a model of risk stratification for clinical research in dogs admitted to intensive care units at four locations. *Am J Vet Res* 2001;62:948-954. 7. Bianco D, et al. A prospective, randomized, double-blinded, placebo-controlled study of human intravenous immunoglobulin for the acute management of presumptive primary immune-mediated thrombocytopenia in dogs. *J Vet Intern Med* 2009;23:1071-1078. 8. Chulay M. Good research ideas for clinicians. *AACN Adv Crit Care* 2006;17:253-265.

FELLOWSHIP OPPORTUNITIES AND FUNDING SOURCES FOR INTERNISTS

Lauren A. Trepanier, DVM, PhD, DACVIM, DACVCP
Madison, WI

POST-RESIDENCY FELLOWSHIPS WITH NO DEGREE REQUIREMENTS				
Sponsor	**Funding**	**Graduate degree?**	**Allows clinic duty?**	**More information and submission deadlines**
NIH National Research Service Award (NRSA) Postdoctoral Fellowship (Individual F32)	Stipend at NIH scale ($46K plus), tuition, some research funds	Often used for PhD training, but degree program not required	Not formally	http://grants.nih.gov/training/nrsa.htm Funding for up to 3 years Must engage in health-related biomedical research upon completion of fellowship, for time equivalent to duration of fellowship support (typically 1-3 years).
Morris Animal Foundation Fellowship Training Grant	Stipend for 2 years (~$50K), tuition not covered	Can be used for PhD training, but not required; 90% research effort	No	http://www.morrisanimalfoundation.org/for-grant-seekers/fellowships.html Applicant cannot be more than 2 years into a postdoctoral research training program.
ACVIM Foundation	Stipend for 2 years (NIH scale), no tuition	Not required	Yes, up to 20% of total effort	http://www.acvimfoundation.org/fellowships/
Cornell University Clinical Fellowship	Annual stipend ($65K) plus $15K for research	Not eligible for PhD program support	Yes, up to 20% of total effort	http://www.vet.cornell.edu/oge/training/clinicalF/
FDA Commissioner's Fellowship program	Stipend for 2 years at the FDA	No, but coursework provided	No	http://www.fda.gov/AboutFDA/WorkingatFDA/FellowshipInternshipGraduateFacultyPrograms/CommissionersFellowshipProgram/default.htm Recent 2011 FDA Fellow projects: mechanisms of antibiotic resistance; hepatotoxicity of herbal supplements; adenovirus vectors for gene therapy; renal function and drug dosing; screening of p-glycoprotein substrates
PHD TRAINING FELLOWSHIPS				
Sponsor	**Funding**	**Graduate degree?**	**Allows clinic duty?**	**More information and submission deadlines**
NIH Institutional Postdoctoral Training Grants (Institutional T32)	Stipend for up to 3 years	Often requires PhD program	Not formally	Search for T32, postdoctoral and "DVM" to find schools of interest (not all are veterinary schools) The University of Pennsylvania School of Veterinary Medicine also maintains an active list of T32 postdoctoral programs for veterinarians (with links) on their website: Stanford ; University of California, Davis; Colorado State; Tulane University; The Jackson Laboratory; Johns Hopkins Harvard; Massachusetts Institute of Technology; Michigan State; University of Michigan; University of Minnesota; University of Missouri; Wake Forest; Ohio State; University of Pennsylvania; Virginia-Maryland; University of Wisconsin
NRSA postdoctoral fellowship (Individual F32)	Stipend at NIH scale plus tuition, research funds	Often used for PhD training	Not formally	http://grants.nih.gov/training/nrsa.htm Must engage in health-related biomedical research upon completion of fellowship, for time equivalent to duration of fellowship support (typically 1-3 years).
Pfizer Animal Health-Morris Animal Foundation	$60K stipend (for salary and tuition) for 4 years	PhD	Not formally	http://www.morrisanimalfoundation.org/for-grant-seekers/pfizer-fellows.html
Cornell University Research Fellowship	Stipend for 3 years	PhD	No	http://www.vet.cornell.edu/oge/training/researchF/

University of Pennsylvania Research Fellowship in Comparative Medical and Molecular Genetics	Stipend and tuition	PhD encouraged	No	http://www.vet.upenn.edu/EducationandTraining/Training/ResidenciesInternshipsandFellowships/ResearchFellowshipGraduateTrainingForVets/tabid/1029/Default.aspx
Institute for Clinical and Translational Research, University of Wisconsin-Madison	Stipend provided by research advisor	MS or PhD in Clinical Investigation	Informally	https://ictr.wisc.edu/GraduateProgram

TRANSITION TO FACULTY FELLOWSHIPS

Sponsor	**Funding**	**Research Effort**	**More information and submission deadlines**
NIH Mentored Clinical Scientist Career Development Award (Individual K08)	Stipend (NIH pay scale), plus research funds, for 3 to 5 years	Designed for transition to new faculty position; requires 75% research effort	http://grants1.nih.gov/grants/guide/pa-files/PA-11-193.html Must have full time academic appointment before award activation Applications are judged on candidate, mentor, academic environment, and research plan
KL2 Scholar Career Development Award (also called K12 awards)	75% of faculty salary, plus research funds, for 3 years	Designed for transition to new faculty position; requires 75% research effort	Offered by institutions with CTSA (Clinical and Translational Science) awards from NIH University of Wisconsin-Madison https://ictr.wisc.edu/KL2Scholars Ohio State http://ccts.osu.edu/content/career-development-program-kl2-0 University of Pennsylvania http://www.itmat.upenn.edu/ctsa/ctsa_k12_program.shtml University of California at Davis http://www.ucdmc.ucdavis.edu/ctsc/area/education/k12/ Tufts University http://www.tuftsctsi.org/Education-and-Career-Development.aspx University of Florida http://www.ctsi.ufl.edu/education/programs/junior-faculty/ Other institutions with CTSA awards also offer KL2 awards to DVMs, even if not affiliated with a vet school; for list of CTSA sites, see https://www.ctsacentral.org/ctsa-consortium
NIH Pathway to Independence (Individual K99/K00)	Salary plus $20K in research funds for up to 5 years	Designed for transition from clinical faculty or research scientist position into clinician-scientist faculty position; requires 75% research effort	http://grants1.nih.gov/grants/guide/pa-files/PA-11-197.html K99 = mentored research phase; K00 = independent research phase
NIH Loan Repayment Program (LRP)	$35K in loan repayment per year of service (minimum 2 years)	DVM with a PhD working in human clinical research (at least 50% effort)	**http://www.lrp.nih.gov/**

To check your eligibility for various NIH awards, use the web-based "Career Award Wizard" questionnaire: http://grants.nih.gov/training/kwizard/index.htm

RESEARCH DOLLARS FOR EARLY-CAREER INTERNISTS				
Agency	**Focus**	**Applications**	**Funding**	**Website**
Internal SVM grants	Seed grants for small projects	Typically annually	Typically <$10,000	See your SVM research office
Morris Animal Foundation Pilot Study	Companion animal, equine, and wildlife; strongly discourage induction of disease	Deadlines vary with grant emphasis; see website.	$10,000 in direct costs	http://www.morrisanimalfoundation.org/for-grant-seekers/pilot-studies.html **Residents are eligible as PIs**
AKC Canine Health Foundation Acorn Grants	Dog diseases, especially breed related and genetic	Continuous submission, rapid turn around	One year pilot project, up to $12,000	http://www.akcchf.org/research/application-process/acorn-grant-program/
Morris Animal Foundation First Investigator award	Companion animal, equine, and wildlife; strongly discourage induction of disease	Deadlines vary with grant emphasis; see website.	Maximum of $100,000 for 2 years	http://www.morrisanimalfoundation.org/ for-grant-seekers/first-award.html
AKC Canine Health Foundation Oak Grants	Dog diseases, especially breed related and genetic.	Annual applications with a pre-proposal; typically Feb-March deadline	Average of $100,000 for 2 years	http://www.akcchf.org/research/application-process/oak-grant-program/ Disease priorities are listed on the website Must appeal to lay as well as scientific reviewers
ACVIM Foundation	Internal medicine diseases	Check the website	Urgent study and annual competition $10-30K is typical	http://www.acvimfoundation.org/grants/info.html
Winn Feline Foundation	Feline health problems	Typically December	Maximum of $25,000	http://www.winnfelinehealth.org/Pages/Researchers.html **Residents are eligible as PIs**
Waltham Foundation	Health and welfare of companion animals	Typically February	Maximum of $15,000	http://www.waltham.com/foundation.htm
International Feline Foundation	Feline health problems	Typically December	Maximum of $7,500	http://www.theinternationalfelinefoundation.com/ **Residents are eligible as PIs**
AAVPT Veterinary Pharmacology Research Foundation	Therapeutics, animal models, and food safety	Check the website	Typically $10-20,000 for 1-2 years	http://www.aavpt.org/VeterinaryPharmacologyResearchFoundation.shtml#grant
Comparative Gastroenterology Society	Gastrointestinal diseases	October	See website	http://vetmed.tamu.edu/cgs/opportunities

MAKING A CONTRIBUTION TO EVIDENCE-BASED MEDICINE: HOW TO DESIGN A STUDY

Mark A. Mitchell, DVM, MS, PhD, DECZM (Herpetology)
Urbana, IL

Success with research starts with the design of a study. Many of the failures clinicians experience can be directly related to a lack of planning during the original design and development of a study. The types of studies performed in veterinary medicine are varied and may include basic bench top research to clinical trials. With such a broad range of topics and hypotheses to be tested, it should not be unexpected that the types of studies required to answer these scientific questions can vary. Developing a sound understanding of the different types of studies available and when/how to use them to answer specific research questions is essential to the success of an individual(s) planning a research project.

There are four basic study designs that will be discussed: cross-sectional studies, case-control studies, cohort studies, and experimental studies. To decide which study design is best, first consider the types of data to be collected and the specific hypotheses being proposed. Other considerations when planning a study include the feasibility of collecting the data, any temporal components associated with the study, and the resources (both financial and personnel) available. If you can answer these questions in the early stages of planning a study, it will help ultimately determine which study design will be best fit your needs.

Cross-sectional studies are an excellent first step for attempting to ascertain the occurrence of some outcome (e.g., disease process) and any associated independent variables in a population. This type of study is commonly referred to as a prevalence study. Cross-sectional studies evaluate a study population at a given point ("snapshot") in time. These types of studies are routinely pursued in veterinary clinical medicine because the data can be collected from patients as they present to determine the prevalence of some occurrence. These types of studies are well suited for those research questions attempting to identify how common or uncommon a disease might be or whether there are any inherent risk factors associated with a specific disease. A deficiency of these types of studies is that they do not provide information relative to the temporal nature of a disease in a case/population. For example, if you are interested in estimating the prevalence of blastomycosis in dogs by evaluating PCR results, you will be able to determine which dogs are positive or negative and which, if any, risk factors (e.g., breed, sex, activity) predispose them to an outcome of being positive or negative, but you will be unable to determine when dogs were exposed or for how long. Of course once you have completed a cross-sectional study and determined such a disease was prevalent, follow up studies (e.g., cohort studies) could be designed to further evaluate a temporal question.

Case-control studies are commonly done in veterinary clinical medicine because they are often designed to be retrospective in nature and can utilize historic (already available) information. These types of studies can provide a great deal of information regarding risk factors for a disease or outcome without needing a large amount of research dollars. With case-control studies the disease outcome is known and the relative effect different risk factors have on that outcome can be assessed by matching the patients to conspecifics that are presented for different outcomes. This type of study is well suited for preliminary investigations (pilot studies) that hope to collect sufficient results to pursue research grants for more developed studies or as a stand-alone study that can contribute to evidence based medicine.

Cohort studies are not commonly done in veterinary clinical medicine because of their longitudinal design and cost. Whereas a case-control study evaluates a known outcome and seeks to determine the effect of specific risk factors on that outcome, cohort studies seek to longitudinally evaluate the potential outcome(s) that occur as a result of various risk factors. This type of study should not be used for rare disease events, as it might not be possible to measure a specific outcome within a reasonable amount of time. These types of studies have contributed significantly to human medicine, and would likely do the same in veterinary medicine if the resources were available to pursue them.

Experimental or intervention studies are one of the most common types of studies pursued in veterinary clinical medicine. This type of study, when done properly, places the control of the research in the hands of the investigators. Experimental studies can range from being as simple as evaluating a simple outcome while controlling a single independent variable (e.g., one group receiving a treatment and another a placebo) to more complex cross-over studies evaluating multiple independent variables and outcomes.

Regardless of the study design being used, it is imperative to consider the sample size required for the study, the methods used for the randomization of the study subjects, and the potential for introducing and controlling different types of bias that could have negative effects on the results. Something that is expected for most study proposals through granting agencies and institutional animal care and use committees is the pre-determination of the sample size required to answer specific hypotheses. In the past, investigators might get by using examples of previous studies to estimate their sample size for a study; however, this is not considered sufficient and pre-determining sample size using a priori information specific to the sample population and the data collected are now required. Being able to have a sound estimate for a sample size for a study can go a long way in limiting problems later (e.g., introduction of a type II error). Random allocation of subjects into test groups should be done using appropriate methods, such as random number generators. Software to perform this is readily available, so it is now possible to be thorough and random when implementing a study. Finally, bias is a major contributor to study failure. Pre-evaluating a study for these types of error can help minimize the likelihood of them occurring, which will increase the overall power in the results.

There is always a great deal of work required to complete a research study. To ensure that the results of the study will have value and contribute to evidence-based medicine, it is important to design a study that will answer the specific hypotheses being proposed. Taking time to consider the potential pitfalls that might occur in a study during the initial planning stages will greatly improve an investigator's likelihood of success

MAKING A CONTRIBUTION TO EVIDENCE-BASED MEDICINE: HOW TO ANALYZE A STUDY

Mark A. Mitchell, DVM, MS, PhD, DECZM (Herpetology)
Urbana, IL

Clinicians depend on the peer-reviewed literature to develop their foundation of knowledge. However, it is this author's opinion that most clinicians do not fully comprehend or appreciate all the material that is presented in these articles. More specifically, it is in the statistical methods that most clinicians skim or skip that which is written. Why is this? The most common responses I hear are that statistics are "hard", "difficult to understand", "make no sense", or are "a collection of confusing formulas and numbers". While I can agree that statistical analysis of data does pose its challenges, clinicians are trained to comprehend "hard" and "difficult to understand" concepts; after all, understanding the dynamic intricacies of physiology is not without its challenges. Also, as clinicians, we rely on formulas and numbers to interpret test results, physiologic data, and therapeutics for our patients, so the reality is that although we might argue against it, we are quite capable of understanding what statistics are and how we can use them to answer the specific research questions we have or are reviewing in an article. It is important to note that reviews of the literature suggest that 50% or more of the articles we rely on to "develop our knowledge base" have inappropriate statistical methods. This means that we our building our foundation of knowledge using a less than stable base. Knowing whether the methods are correct or not is essential to our ability to understand whether what is presented is useful or not. Blindly accepting that something is "correct" because it has been peer reviewed (remembering the previous reference regarding the mis-analysis in publications) should not be considered acceptable. Remember that the articles you rely on to build your foundation of knowledge were reviewed by individuals such as yourself who may not have been comfortable reviewing the statistical methods; in many cases it appears the reviewers assume this is being done at some point during the editorial review. Unfortunately, (again, referring to the previous references) this is not always the case.

So how is it that we can overcome these deficiencies? To start, we need to realize we can't know and be experts in all things. To truly become "experts" in statistics we need to utilize these methods on a regular basis. Nobody became a specialist overnight. You needed to practice your craft to develop your skill. The goal of this lecture is not to turn you into a statistician, but instead make you comfortable with how statistics can be used to answer specific research questions or the process of reviewing statistics in an article to maximize the value of the material for you. If you walk away thinking they are "less scary", than that is a great start.

Clinicians are process oriented. We collect data (history, physical examination) to answer a specific question(s) (e.g., why does a dog have chronic diarrhea) we have regarding our cases. We then formulate specific hypotheses (e.g., this dog has chronic diarrhea because of parasitism, neoplasia, etc.) that we test using diagnostic tests. The results of these tests provide us information that we can use evaluate our specific hypotheses and draw final conclusions regarding the case and institute therapy (which comes with a new set of hypotheses). This same process is what statistical methods can provide for research data and specific research hypotheses. If we look at the statistical methods using this approach, it should be easier for us to process.

The following is a review of the process I use to train house officer and graduate students. Note that it requires us to start at a beginning and follow a process to draw any conclusions. A common problem with statistical interpretation as it relates to publications is that clinicians are told which test is used, but they don't understand why it was selected (sometimes the authors themselves don't know why they selected the test) (e.g., they don't appreciate how you determine which test to use). This is the same feeling new clinicians feel when looking at diagnostic tests. Until they understand the process, it is confusing.

THE DATA

It all starts with the data. Knowledge about the different types of data is essential to our understanding of statistics. There are three basic types of data: categorical, ordinal, and continuous. Categorical data, also called nominal data, is that which is based on categories without any measurement. There is no numerical value to the data. Common categories include: sex (male, female) or age (adult, juvenile). Ordinal data is rank data. This type of data is assigned into distinct groups, but the numerical value assigned to the group is not continuous. A pain scoring system would be an example of a type of ordinal data. Continuous data is that which is measureable on a continuous scale. Examples of this type of data include body temperature and heart rate. You can appreciate with this type of data that there is a big difference between a body temperature of 100.7°F and 106.8°F.

Parametric vs. Non-Parametric Statistics

The type of data ultimately dictates the statistic to be used. Parametric statistics have a requirement that the data be normally distributed. Because categorical and ordinal data do not adhere to this type of distribution, they are evaluated using non-parametric statistics. Continuous data can be normally distributed, but it needs to be evaluated to determine whether it is or not. If continuous data is normally distributed, parametric statistics are best. These methods are more robust than the non-parametric methods. If the continuous data is not normally distributed, non-parametric methods are best or the data must be transformed to meet the assumption of normality. The good news is that for every parametric test there is a non-parametric test that can be used to answer the same type of question.

Tests of Normality

There are a number of different methods for assessing the distribution of continuous data. The most common methods tend to be using statistics such as the Shapiro-Wilk test or Kolmogorov-Smirnoff test. Both of these statistics test the null hypothesis that the distribution of the data is normal. The Shapiro-Wilk test is more commonly recommended for smaller samples sizes, which are common in veterinary clinical medicine study designs. In addition to these tests, the skewness and kurtosis can be used to assess the distribution of the data.

Dependent and. Independent Variables

Once you have identified the type of data and the general classification of statistics to be used (parametric or non-parametric), the type of variables being assessed in the research are used to determine which specific test should be used. If there is only a single dependent variable (outcome), the statistical test selected will be based on the number of responses in the independent variables being evaluated. Table 1 is an example of the different statistics that can be used to evaluate data when there is a single dependent variable. Note that the different tests are organized in parallel based of the type of data used. If there are two or more samples for a dependent variable (paired or repeated measures), than the statistical test is selected based on this information. Table 2 is an example of the different statistics that can be used in these cases. For studies where multiple variables are assessed, regression models can be developed that enable the researcher to evaluate multiple pieces of data simultaneously. Table 3 provides an example of these types of statistics.

CONCLUSIONS

Developing a basic understanding of statistics is essential to the veterinary clinician. We rely on this type of knowledge to interpret results reported in articles used to develop our foundation of knowledge, as well as analyze the results of studies we perform that can be used to contribute to the body of evidence-based medicine.

Table 1. Statistical tests to be used for a single dependent variable.

Independent variable	**Categorical data (non-normally**	**Ordinal data (normally distributed continuous data)**	**Continuous data distributed)**
One category	Binomial test	Wilcoxon rank test	Independent t-test
Two categories	Chi-square test/ Fisher exact test	Mann-Whitney U test	Student's t-test
Three or more categories	R x C chi-square test	Kruskal-Wallis test	ANOVA

Table 2. Statistical tests to be used for dependent variables with multiple samples.

Dependent variable	**Categorical data (non-normally**	**Ordinal data (normally distributed continuous data)**	**Continuous data distributed)**
Paired	McNemar's test	Wilcoxon paired rank test	Paired t-test
Serial (>2 samples)	Cochran's Q test	Friedman's repeated measures Test	Repeated measures ANOVA

Table 3. Statistical tests to be used to evaluate multiple variables simultaneously.

Categorical data (non-normally	**Ordinal data (normally distributed continuous data)**	**Continuous data distributed)**
Logistic regression	Ordinal regression	Linear regression

Movement Disorders in Horses: Shivers, Stringhalt and other conundrums

S.J. Valberg, DVM, PhD, DACVIM, ACVSMR, A. Draper, BVetMed
St. Paul, MN

In human medicine, movement disorders comprise any type of disorder that causes abnormal movement or control of a part of the body. In veterinary medicine, movement disorders are better characterized in small animal compared to equine medicine, however, a number of equine movement disorders have been described. Although diseases can be characterized into distinct categories, it is important to appreciate that overlap may occur with dysfunction of both the sensory and motor systems. From an anatomic point of view, equine movement disorders may occur as a result of;

1. Abnormal muscle cell depolarization. Disorders in the sarcolemma often produce muscle fasciculations or sustained contraction, however, this rarely results in abnormal limb movement. Equine examples include;
    - Myotonia congenita = sarcolemmal chloride channels
    - Hyperkalemic periodic paralysis = sarcolemmal sodium channels
2. Abnormal firing within peripheral nerves. Sustained firing of alpha motor neurons can produce painful sustained contractions. Equine examples include;
    - Muscle cramps due to electrolyte derangements
    - Neuromyotonia caused by ear tick otobius megnini
3. Dysfunction of sensory systems. Includes abnormalities of the spinocerebellar and spinothalamic tracts. Abnormalities of the spinocerebellar tracts often result in ataxia and potentially tremor if the cerebellum is involved. Equine examples include;
    - Neuroaxonal dystrophy/equine degenerative myeloencephalopathy, cervical vertebral compressive myelopathy, equine protozoal myeloencephalitis, cerebellar abiotrophy
4. Dysfunction of pyramidal system. Abnormalities of the corticospinal tracts (information to motor neurons of the spinal cord) and corticobulbar tracts (information to motor neurons of the cranial nerves) Equine examples include;
    - Cervical vertebral compressive myelopathy, equine degenerative myeloencephalopathy, equine protozoal myeloencephalitis
5. Dysfunction of extrapyramidal system. Dystonia characterized by sustained muscle contractions, usually producing twisting and repetitive movements or abnormal body positions with a sustained directional quality. Equine examples include
    - Fluphenazine reactions
6. Dysfunction of interneurons within the spinal cord. Abnormalities disrupt smooth coordination of agonist and antagonist muscle contractions often causing myoclonus. Myoclonus is characterized by variable short irregular bursts of muscle contraction. Equine examples include;
    - Tetanus, myoclonus and stiff horse syndrome which affect glycine or GABAergic receptors

Horses have other unique forms of movement disorders that are triggered by a particular form of locomotion and are stylized and repeatable. Stringhalt for example is triggered by walking forward and Shivers is triggered by walking backward or by manually picking up the limb. These abnormal movement patterns disappear when the horse trots or canters. Such movement disorders fall into a grey area between the lameness focus of surgeons and the neurologic focus of internal medicine and remain poorly understood. The purpose of our research is to develop a clinical characterization of locomotor-induced movement disorders in horses and to identify pathophysiologic bases for these conditions. In this proceeding, we will provide a clinical characterization of locomotor induced movement disorders, an overview of the neural control of locomotion and some potentially novel hypotheses for the cause of locomotor-induced movement disorders.

Clinical characterization

The Neuromuscular Diagnostic Laboratory at the University of Minnesota solicited owners of horses with movement disorders loosely termed Shivers to fill in an on-line questionnaire and submit videos of their horses walking, turning, backing and lifting their limbs. Over 200 owners described their horse's movement and many provided videos of their horses. Based on examination of this material several distinct categories of movement disorders were apparent.

1. Dystonia of the head and neck
    - Resistance to manual flexion of either forelimb
    - Trembling of the forelimb once flexed
    - Patterned extension and twisting of the neck to one side with movement of the lips and eyelids when forelimb is manually flexed
    - Normal backwards and forwards walking

2. Hyperflexion of a hindlimb
    - Occurs when manually picking up the limb
    - Normal backwards and forwards walking
3. Hitch
    - Hyperflexion of the hindlimb, a pause and then placement at a normal speed back to the ground
    - Occurs intermittently for a few strides when walking forward.
    - Backwards walking occurs normally and willingly
4. Shivers
    - Hyperflexion and abduction of the hindlimb, muscle trembling of the flexed hindlimb, a pause and then placement at a normal or slightly increased speed back to the ground when walking backwards
    - Alternatively hyperextension of the hindlimb with difficulty flexing to move backwards
    - Occurs with backwards walking especially if horse has been standing still, transported or stressed
    - Backwards walking occurs with resistance and reluctance
    - Forward walking is normal
    - Difficulty with farrier work due to inability to hold limb flexed when standing
5. Advanced Shivers
    - Hyperflexion and abduction of the hindlimb, muscle trembling of the flexed hindlimb, a pause and then placement at a rapid speed back to the ground when walking backwards
    - Alternatively, hyperextension of hindlimbs and inability to flex the limb or to walk backwards
    - Occurs consistently with backwards walking, when pushed sideways, when turned sharply or for a few strides walking forward
    - Backwards walking occurs with resistance and reluctance
    - Inability to trim hind feet because horse cannot hold the limb flexed and will slam the foot to the ground
    - Occasional tremor of upper lip
6. Stringhalt
    - Hyperflexion of the limb forward and under the belly while walking forward followed by a rapid return of the limb to the ground
    - Abnormal movement occurs for many strides when walking forward and may occur for a few strides walking backward.
    - Horse walks backward willingly
    - May be bilateral or unilateral

Control of locomotion

Locomotion involves the precise activation and inhibition of many muscles, the pattern of which varies between gaits. One of the perplexing features of the locomotor-induced movement disorders is why they occur only within a specific gait. For example, why is Shivers only apparent in horses when they walk backwards whereas Stringhalt occurs when horses are walking forwards? In both cases, the locomotor-induced abnormality disappears when horses canter. A potential explanation for this might lie in the neural mechanisms that control locomotion. The information presented in this section is derived from studies performed in the cat which might well be similar in horses.

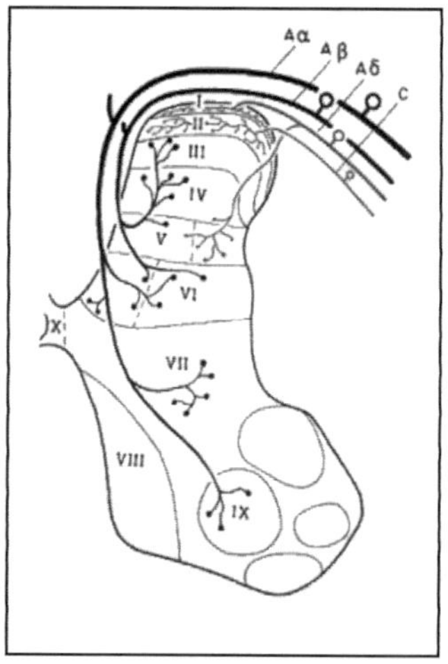

The key features of walking forwards and backwards are the rhythm, the ipsilateral coordination of flexors and extensors across the same or different joints in a limb, as well as left/right and fore/hind coordination. These functions are integrated in central pattern generators (CPG) located within the thoracic and lumbar intumescence in laminae VII, VIII, X of each limb. CPGs provide much of the timing and pattern of complex, rhythmic, coordinated muscle activities. When locomotion starts, activity in the CPG is turned on and maintained by inputs from descending locomotor commands originating from neurons in the brainstem and midbrain. Rhythm of locomotion during walking is generated by a subpopulation of CPG excitatory interneurons that provide excitatory drive to segmental, ipsilaterally located motor neuron pools. Transition from stance to swing phase happens when the hip is extended to an 80–90° angle and is governed by midlumbar inhibitory interneurons that receive multisensory inputs from muscle spindles near the hip and joint afferents. Flexor and extensor motor neurons receive rhythmic glycinergic inhibition alternating with rhythmic glutamatergic excitation during locomotion.

The extensor related interneurons lose their rhythmicity and become tonically active when flexor motor neurons momentarily drop out of the locomotor rhythm (owing to a decreased or absent drive from the flexor CPG). Left-right coordination is mediated by ventrally located commissural interneurons (CINs) whose axons cross the midline via the ventral commissure Rhythmic, ascending lumbar CINs, some of which are cholinergic, have been described in the cat. Many of these neurons have long, ascending fibers and might therefore, in addition to being involved in hindlimb coordination, provide rhythmic signals to the forelimb region of the spinal cord or the brainstem. Intrasegmental CINs is may also play an important role in gait that switches from alternation to synchrony. This may occur when crossed dual inhibitory systems are active during walking, whereas the excitatory systems are active during segmental synchronous activity such as hopping or gallop. Another possibility is that gait changes happen because of a slight change in the balance between inhibitory and excitatory crossed actions. This implies both systems are active simultaneously, but one or the other system dominates depending on the gait. How such a change of balance could take place during normal gait changes is, of course, unknown. One possibility is that descending modulatory fibers have specific inhibitory and/or excitatory actions on the crossed inhibitory and excitatory CIN pathways.

Thus there are several specific groups of interneurons that function to control locomotion and disruption in the function of one of these groups may well produce a locomoto- induced movement disorder.

Application to Shivers and Stringhalt

We hypothesize that CPGs for backwards vs forwards walking are comprised of a largely separate group of interneurons in horses. Forward walking CPGs might be activated by muscle spindle Ia afferents that fire when the hip flexes whereas with backwards walking CPGs may be activated by spindles stretched when the hip extends. Dysfunction in separate CPGs might provide an explanation for why Stringhalt horses are affected when walking forward and Shivers when walking backwards. Further separate CPGs might explain why horses with both of these conditions can trot and canter normally but cannot back or lift a hindlimb while standing.

There are only two reported postmortem studies of horses with Shivers and no neurologic lesions were identified in H&E sections. We are currently evaluating in great detail the musculature and central nervous system of 6 horses with Shivers and results will be presented at the forum. Findings are supportive of an abnormality in the interneurons in the lumbosacral intumescence.

Several studies describe various causes for stringhalt including; 1) peripheral axonopathy due to toxicity (false dandelion), 2) trauma to the lateral aspect of the lower hindlimb (unilateral), 3) painful conditions of the limb (causing false stringhalt) and 4) unknown. Phenytoin will alleviate clinical signs in some horses with stringhalt and not others. Similarly cutting the lateral digital extensor tendon may alleviate signs in some but not all cases. Axonopathies in humans usually present with sensory loss or tingling and motor weakness without the typical movement disorder described for stringhalt. The reason why a peripheral axonopathy might produce signs of stringhalt may be that either axons in muscles spindles or the CPG may be affected. Muscle spindles are stretch receptors enervated by gamma motor neurons to adjust spindle length, and with afferent fibers that directly increase alpha motor neuron firing in the same muscle and indirectly inhibit motor neuron firing in antagonist muscles. Normally, stretching of a muscle stimulates muscle spindle afferents which produces a contraction in that muscle to resist stretching and at the same time induces relaxation in agonist muscles. Dysfunction in the spindles, their enervation or in CPGs could all potentially contribute to signs of Stringhalt. Further, cutting and separating the lateral digital extensor tendon might change the input from muscles spindles to the CPGs regarding the degree of muscle stretch and thus alter the locomotor-induced movement disorder.

Thus it appears that there are many unexplored neural pathways involved in gait that might be implicated in locomotor-induced movement disorders in horses.

REFERENCES

(1) Grillner S, Wallen P. Central pattern generators for locomotion, with special reference to vertebrates. *Annu Rev Neurosci* 1985;8:233-261. (2) Grillner S, Rossignol S. On the initiation of the swing phase of locomotion in chronic spinal cats. *Brain Res* 1978;146(2):269-277. (3)Kjaerulff O, Kiehn O. Distribution of networks generating and coordinating locomotor activity. *J Neurosci* 1996;16(18):5777-5794. (4) Kjaerulff O, Kiehn O. Crossed rhythmic synaptic input to motoneurons during selective activation of the contralateral spinal locomotor network. *J Neurosci* 1997;17(24):9433-9447. (5) Armengou L, Anor S, Climent F, Shelton GD, Monreal L. Antemortem diagnosis of a distal axonopathy causing severe stringhalt in a horse. *J Vet Intern Med* 2010;24(1):220-223. (6) Cahill JI, Goulden BE, Jolly RD. Stringhalt in horses: a distal axonopathy. *Neuropathol Appl Neurobiol* 1986;12(5):459-475. (7) Crabill MR, Honnas CM, Taylor DS, Schumacher J, Watkins JP, Snyder JR. Stringhalt secondary to trauma to the dorsoproximal region of the metatarsus in horses: 10 cases (1986-1991). *J Am Vet Med Assoc* 1994;205(6):867-869. (8) Huntington PJ, Seneque S, Slocombe RF, Jeffcott LB, McLean A, Luff AR. Use of phenytoin to treat horses with Australian stringhalt. *Aust Vet J* 991;68(7):221-224. (9) Baird JD., Firshman AM, Valberg S.J. Shivers (shivering) in the horse: A review. *Proc Am Assoc Equine Pract* 2006;52:359-354. (10) Firshman AM, Baird JD, Valberg SJ. Prevalences and clinical signs of polysaccharide storage myopathy and shivers in Belgian draft horses. *J Am Vet Med Assoc* 2005;227(12):1958-1964. (11) Valentine BA, de LA, Divers TJ, Ducharme NG, Orcutt RS. Clinical and pathologic findings in two draft horses with progressive muscle atrophy, neuromuscular weakness, and abnormal gait characteristic of shivers syndrome. *J Am Vet Med Assoc* 1999;215(11):1661-5, 1621.

GENETIC AND NUTRITIONAL ASPECTS OF EQUINE NEUROAXONAL DYSTROPHY

Carrie J. Finno, DVM, DACVIM; Monica Aleman, MVZ, PhD, DACVIM; John Madigan, DVM, DACVIM, Birgit Puschner, DVM, PhD, DABVT and Danika Bannasch, DVM, PhD
Davis, CA

INTRODUCTION

Neuroaxonal dystrophy (NAD) is a common neurologic disorder in numerous equine breeds characterized by symmetric ataxia, abnormal base-wide stance at rest and proprioceptive deficits of all limbs, which usually develop by six to twelve months of age. Affected foals often have low serum vitamin E concentrations. NAD is clinically indistinguishable from equine degenerative myeloencephalopathy (EDM). The only difference between these disorders is the location of axonal degeneration within the central nervous system. NAD/EDM appears to have a genetic basis with clinical expression in genetically predisposed foals being influenced by dietary vitamin E. The cause of NAD/EDM is unknown.

Significance

NAD/EDM was rated the second most common cause (24% of cases) of equine spinal cord disease at Cornell University in 1978[1] and ranked second in the causes of spinal ataxia at the University of Montreal from 1985 to 1988,[2] yet there is no antemortem test for the disease. Currently, a clinical diagnosis is made based on normal cervical radiographs/myelography, negative equine protozoal myeloencephalitis (EPM) testing and low serum vitamin E. In such cases, empirical treatment with vitamin E is attempted but is ineffective with horses remaining unfit for any performance activity. Because definitive diagnosis requires a postmortem exam conducted shortly after euthanasia and an experienced pathologist to identify characteristic lesions in the brainstem and spinal cord, many cases go undiagnosed. NAD/EDM easily goes unrecognized because an antemortem diagnostic test is not available and postmortem histopathological lesions can be subtle.

Nomenclature

Clinically, NAD and its closely related counterpart EDM, are indistinguishable and they also resemble other symmetrical causes of spinal cord ataxia such as cervical vertebral compressive myelopathy (CVCM). The distinction between NAD and EDM depends mainly on the distribution of the histologic lesions. Although the lesions in both NAD and EDM consist of neuroaxonal degeneration, historically the disease is classified as NAD if the histological lesions are confined to the lateral accessory cuneate, medial cuneate and gracilis nuclei[3-5] whereas a diagnosis of EDM is assigned when axonal necrosis and demyelination additionally involve the dorsal and ventral spinocerebellar tracts and ventromedial funiculi of the cervicothoracic spinal cord.[6-10] Histologic lesions consistent with both NAD and EDM can occur in the same animal;[11] thus, in all likelihood, NAD and EDM are the same disease.

Inheritance:

NAD/EDM occurs in Standardbreds, Paso Finos, Quarter horses, Mongolian horses, Appaloosas, Haflingers, Arabians, Morgans, Lusitanos, Thoroughbreds, Paint horses, Tennessee Walking Horses, Norwegian Fjord Horses, Welsh Pony, Pony of the Americas and various mixed breeds. Although clinical signs of NAD/EDM are similar across breeds, it is not known whether all breeds share the same etiologic basis. Both a breeding trial[12] and clusters of cases of related horses provide strong evidence for a genetic basis with an autosomal dominant mode of inheritance with variable expression suggested in Morgan horses.[12] One study found that foals from dams that had an NAD/EDM-affected foal were 25x more likely to develop NAD/EDM than foals from other dams further supporting a genetic basis.[13] Although the clinical disease appears similar across breeds, there is currently no evidence that the underlying etiology is shared.

Role of vitamin E

Vitamin E appears to play an important role in the development of NAD/EDM in genetically predisposed foals. It has been demonstrated that, during the first year of life, vitamin E supplementation of genetically susceptible foals (from a dam/sire that produced affected foals), lowers the overall incidence (40% to 10%) and severity of NAD/EDM.[10,13] We have studied the effects of vitamin E supplementation of pregnant mares and their foals in a herd of Quarter horses with a high incidence of NAD and low serum vitamin E concentrations. Supplementation decreased the overall number of severely (neurologic grade ≥ 3 on a scale of 0 to 5) affected cases from 6% to 0%; however, the incidence of NAD/EDM was unaffected.[11] Others have shown that low serum vitamin E in affected foals is not related to abnormal vitamin E absorption,[14] suggesting a difference in vitamin E requirements or metabolism. Empirical treatment with vitamin E once clinical signs are apparent is often attempted but is ineffective and these horses exhibit lifelong, stable neurologic deficits that prevent use for any performance activity. Although vitamin E clearly plays a role in these susceptible families, low serum vitamin E is not consistently reported in all cases of NAD/EDM. Overall, there is very strong evidence that NAD/EDM is an inherited disorder and, in susceptible families, dietary vitamin E during early development modifies the severity of the phenotype.

ELECTROPHYSIOLOGIC TESTING

In humans, mice, and dogs, dietary vitamin E deficiencies have also been associated with retinal degeneration. In horses with equine motor neuron disease (EMND), another neurodegenerative condition associated with vitamin E deficiency, lipofuscin deposits can be observed both in the motor neurons of the ventral horns of the spinal cord and in the retina. Flash electroretinogram (ERG) of

EMND-affected horses has demonstrated decreased B-wave amplitudes.[15] Additionally, a unique feature of NAD/EDM in the American Quarter horse is an inconsistent menace response and abnormal mentation, ranging from quiet to obtunded. Electroencephalograms (EEGs) can be used to evaluate this abnormal mentation.

ERG and EEG testing of NAD/EDM affected horses has revealed no significant abnormalities when compared to normal control and published values.[11,16,17] An inconsistent menace response is noted in many NAD/EDM-affected horses; however, ophthalmologic examination reveals no evidence of lipofuscin deposition or retinal degeneration.[16,17] The quiet to obtunded mentation found in Quarter horses with NAD/EDM does not appear to be associated with any EEG changes.[17]

GENETIC INVESTIGATION

Currently, candidate genes discovered by a genome-wide association analysis using single nucleotide polymorphism (SNP) markers are being evaluated for NAD/EDM in the Quarter horse. Elucidation of the genetic mechanism underlying NAD/EDM will provide an antemortem diagnostic method in clinical cases while assisting breeders to decrease the overall incidence of the disease. Genetically susceptible foals may be supplemented with vitamin E during the first two years of life to prevent severe ataxia.

VITAMIN E SUPPLEMENTATION

Healthy horses

The Nutritional Research Council set the upper safe diet concentration at 20 IU/kg body weight (10,000 IU/500 kg horse) based on 2% dietary intake.[18] Although complications and potential toxicity resulting from high dose-supplementation of vitamin E are thought to be minimal in horses, coagulopathy and impaired bone mineralization have been reported when horses are administered dosages in excess of this upper limit.[18] Recently, in human medicine, high dose supplementation with vitamin E has been associated with increased all-cause mortality.[19-21] Therefore, in humans, indiscriminate supplementation of high dose vitamin E is not recommended.[22] Supplementation with vitamin E may alter drug metabolism and disposition since the same isoforms that metabolize vitamin E metabolize >50% of therapeutic drugs.[23] In healthy exercising horses, high dosage of vitamin E supplementation (10x NRC requirements) was shown to be potentially detrimental to beta-carotene absorption and thus not recommended.[24] Therefore, it is the authors' recommendation that *healthy* horses receive the baseline 2007 NRC dietary recommendation of vitamin E (1000 IU/500 kg horse).

NAD/EDM

In equine families where previous NAD/EDM cases have been diagnosed, supplementation above the baseline 2007 NRC requirement is advised to pregnant broodmares and foals during the first two years of life. This is based on the findings that supplementation appears to decrease the overall severity of neurologic defects in genetically susceptible foals. It is important to realize, however, that the foal may still be mildly affected with neurologic deficits due to NAD/EDM DESPITE adequate vitamin E supplementation. Breeding changes are required to completely prevent new NAD/EDM cases.

Vitamin E supplements

The term vitamin E actually describes a closely related family of eight fat-soluble naturally occurring compounds. The family consists of two subgroups: tocopherols (saturated) and tocotrienols (unsaturated). Within each subgroup, there are four individual isoforms (α, β, γ and δ). There are eight different stereoisomers of α-tocopherol. Naturally-occurring α-tocopherol (d-α-tocopherol) has the R-configuration at the three positions (2, 4, 8') in the chromanol ring and is therefore termed RRR-α-tocopherol. Due to preferential hepatic uptake and subsequent secretion into plasma for transport to the peripheral tissues, RRR-α-tocopherol is the most biologically active isoform of vitamin E and RRR-α-tocopherol also possesses the most potent antioxidant properties.[25]

Types of α-tocopherol supplements

Synthetic: There are two types of synthetic α-tocopherol, all-rac-α-tocopherol and dl-α-tocopherol. All-rac-α- tocopherol is an equimolar mixture of all eight possible stereoisomers. All-rac-α-tocopherol acetate is accepted as the International Standard (1 mg = 1 international unit). Forms of synthetic vitamin E are available as powdered or pelleted supplements. The all-rac-α-tocopherol acetate form is highly dependent on adequate amount of bile salts to generate hydrolysis and subsequent absorption of α-tocopherol into the plasma.[26]

Natural: RRR-α-tocopherol is available as a micellized liquid form and as an esterified form (acetate). In order for α-tocopherol acetates to be utilized in the body, the ester has to be removed and the α-tocopherol made water-soluble by the action of bile salts (micellization). These additional steps may limit α-tocopherol absorption in the horse.[27]

It is important to account for the biopotency ratio of 2:1 (RRR-α-tocopherol to all-rac-α-tocopherol acetate), when considering α-tocopherol supplementation as most recommendations for supplementation were based on using synthetic α-tocopherol. For example, if pure RRR-α-tocopherol is the supplement provided to a 500 kg adult sedentary horse (recommended dose of 1 IU/kg), only 500-660 IU should be administered whereas if a synthetic all-rac-α-tocopherol acetate supplement is provided, 1000 IU should be administered. These respective dosages will provide the same active amount of α-tocopherol. The difference between RRR-α-tocopherol and synthetic vitamin E (all-rac-α-tocopherol acetate) in the horse was recently demonstrated by Pusterla et al.,[28] where cerebrospinal fluid concentrations of α-tocopherol were significantly elevated above baseline values after supplementation with the natural vitamin E but not following supplementation of a synthetic all-rac-α-tocopherol acetate at equivalent high dosages (10,000 IU/500kg horse/day). In addition, serum α-tocopherol concentrations were significantly higher in the group supplemented with the RRR-α-tocopherol compared to the all-rac-α-tocopherol acetate, with values approximately 2x that of concentrations obtained using all-rac-α-tocopherol acetate. This is in agreement with other studies showing that the activity of RRR-α-tocopherol is 1.5-2 IU/mg higher than all-rac-α-tocopherol.[29-32]

It is therefore recommended by the authors that natural RRR-α-tocopherol (non-acetate) form of vitamin E be used to supplement horses from families with previously diagnosed NAD/EDM cases as this form is the most biologically available, most readily absorbed, and has the most potent antioxidant activity.[33]

Dose of α-tocopherol: It is important to measure serum α-tocopherol concentrations BEFORE implementing a supplementation program to determine if there is an underlying deficiency.

Case		Dose (IU/kg/d)	Amount fed as all-rac-α-tocopherol acetate (synthetic)	Amount fed as RRR-α-tocopherol (natural) (1.5-2x activity of synthetic)
Healthy horse		1.6-2	1000 IU (500 kg horse)	500-660 IU (500 kg horse)
NAD/EDM	Pregnant broodmare: 1st two trimesters	8.5-10.5	5000 IU (500 kg horse)	2500-3300 IU (500 kg horse)
	Pregnant broodmare: last trimester	17-21	10,000 IU (500 kg horse)	5000-6,600 IU (500 kg horse)
	*Foals: 1st year of life	8.5-10.5	500 IU (50 kg foal) 1500-1900 (200 kg weanling)	250-330 IU (50 kg foal) 750-1260 IU (200 kg weanling)
	>1 year of age	4-5	1350-2250 IU (300-500 kg)	675-1500 IU (300-500 kg)
	**Affected cases	8.5-10.5	5000 IU (500 kg horse)	2500-3300 IU (500 kg horse)

* Recommended to provide an injectable dose of vitamin E at birth (E-Se, Schering Plough, 1 ml/100 lbs)
** Supplementation is unlikely to improve neurologic status once abnormalities are evident

It is recommended to have hay and grain analyzed to determine the concentration of α-tocopherol and to then factor in these amounts to the total α-tocopherol.

REFERENCES

[1]Mayhew IG, et al. **Cornell Vet** 1978;68 Suppl 6:1-207. [2]Nappert G, et al. **Can Vet J** 1989;30:802-806. [3]Adams AP, et al. **Can Vet J** 1986;37:43-44. [4]Beech J. **Vet Pathol** 1984;21:383-393. [5]Baumgartner W, et al. **J Comp Pathol** 1990;103:114-119. [6]Blythe LL, et al. **J Am Vet Med Assoc** 1991;198:1005-1013. [7]Siso S, et al. **Vet J** 2003;166:238-243. [8]Liu SK, et al. **J Am Vet Med Assoc** 1983;183:1266-1268. [9]Gandini G, et al. **Equine Vet J** 2004;36:83-85. [10]Mayhew IG, et al. **J Vet Intern Med** 1987;1:45-50. [11]Aleman M, et al. **J Am Vet Med Assoc** 2011;239(6):823-33. [12]Beech J, et al. **Am J Vet Res** 1987;48:109-113. [13]Dill SG, et al. **Am J Vet Res** 1990;51:1300-1305. [14]Blythe LL, et al. **Am J Vet Res** 1991;52:908-911. [15]Riis RC, et al. **Equine Vet J** 1999;31(2):99-110. [17]Finno CJ, et al. **Vet Ophthalmol** *in press.* [18]NRC, 6th Ed. National Research Council, 2007. [19]Miller ER, et al. **Ann Intern Med** 2005;142:37-46. [20]Waters EE, et al. **J Am Med Assoc** 2002; 288:2432-2440. [21]Bjelakovic G, et al. **J Am Med Assoc** 2007; 297:842-857. [22]Dotan Y, et al. **Biofactors** 2009;35:469-473. [23]Kliewer SA, et al. **Endocr Rev** 002;23:687-702. [24]Williams CA, et al. **Equine Vet J Supplem** 2006:617-621. [25]Mustacich DJ, et al. **Vitam Horm** 2007;76:1-21. [26]Knarreborg A, et al. **J Nutr** 2004;134:1487-1492. [27]Pagan JD, et al. **Pferdeheikunde** 2005;21:101-102. [28]Pusterla N, et al. **Vet Rec** 2010;166:366-368. [29]Weber P, et al. **Nutrition** 1997;13:450-460. [30]Burton GW, et al. **Lipids** 1988;23:834-840. [31]Acuff RV, et al. **Am J Clin Nutr** 1994;60:397-402. [32]Traber MG, et al. **Atherosclerosis** 1994;108:27-37. [33]Kayden HG, et al. **J Lipid Res** 1993;34:343-358

IMMUNITY TO *SARCOSYSTIS NEURONA*
Sharon Witonsky DVM, PhD, DACVIM
Blacksburg, VA

INTRODUCTION

Sarcocystis neurona is the most common cause of Equine Protozoal Myeloencephalitis. Incidence of disease is between 0.-5-1% of all horses that will get disease during their lifetime (NAHMS). National seroprevalence is >50% with some regional variation (MacKay, 1997). With high exposure rates, but low incidence of disease, it would suggest that the immune system would be critical in prevention of disease. Some studies report stressful factors or steroid treatment can increase incidence of disease in horses (Saville et al., 2001). Other studies in horses that develop EPM, both from natural and experimental exposure, have altered immune responses (Tornquist et al., 2001; Spencer et al., 2004; Yang et al., 2006; Witonsky et al., 2008). Additionally, some immunocompromised mouse models infected with *S. neurona* develop disease (Sellon et al., 2004; Witonsky et al., 2003, 2005). This talk with provide a brief review of the pathophysiology of disease in horses and mice, as well as review what is known about the protective immune response. The focus of the talk will be on the knowledge we have gained regarding the immune response to *S. neurona*, both from studies in mice and horses.

LIFE CYLE OF *S. NEURONA*

The definitive host for *S. neurona* is the opossum. Intermediate hosts are skunk, cat, raccoon, armadillo, and sea otters. *S. neurona's* life cycle includes both sexual and asexual stages. In the lumen of the gut of the opossum, sarcocysts release bradyzoites. These enter the small intestine and transform into male and female gamonts; the male gamonts divide into gametes. Fertilization by the male gamete of the female gamont causes an oocyst to be formed. Within the lamina propria, the oocysts sporulate to produce two sporocysts. Each sporocyst has four sporozoites. Opossum feces contain fully sporulated oocysts or sporocyst (Dubey et al., 2001a; Sellon and Dubey, 2007).

INFECTION

Horses ingest the sporozoites, which invade the intestinal epithelium and undergo asexual replication in many tissues. The mechanism of dissemination to the tissues is not well understood. *S. neurona* replicates in the peripheral tissues and central nervous system (CNS) by schizogony (endopolygeny). During this process, the nucleus becomes lobulated, with chromatin strand connecting the strands. The merozoites can undergo replication, and can also develop to become mature schizonts. The mechanism by which the merozoites enter the central nervous system (CNS) is unknown, but it may involve mononuclear cells (Ellison et al., 2004; Lindsay et al., 2006). Some studies demonstrate that *S. neurona* can infect endothelial cells as well as leukocytes (Witonsky et al., 2005). Therefore, both endothelial cell invasion and entrance via infected mononuclear cells are possible mechanisms by which *S. neurona* invades the CNS. Studies in gamma interferon knock-out mice (KO), demonstrate that *S. neurona* can infect the peripheral tissues prior to invasion of the CNS. It is not certain whether this applies in immunocompetent animals (Dubey et al., 2001a; Sellon and Dubey, 2007).

Typical lesions in the CNS, which often affect predominantly brain stem and spinal cord, and grossly include multifocal hemorrhagic foci, which on histopathology, consist of hemorrhage, nonsuppurative inflammation, small areas of necrosis, and perivascular cuffing with mononuclear cells. Organisms are identified less than 50% of the time (Sellon and Dubey, 2007). A small study demonstrated an increased number of T-cells vs. B-cells present in the lesions (Scott et al., 2005).

ALTERED IMMUNE RESPONSE INCREASING RISK OF DISEASE

In initial efforts to establish a valid experimental model for *S. neurona* infection, several methods have been attempted to create a reproducible model, all of which involve some sort of "stress and/or immune compromise" (Saville et al., 2001). Shipping and glucocorticoids were used in some models (Cutler et al., 2001; Saville et al., 2001). In another model, immunodeficient horses were used (Sellon et al., 2004). Finally, a model has been established which involves inoculating horses using infected autologous leukocytes (Ellison et al., 2004). This method may allow for evasion of some of the host immune response early in the disease because cells are infected directly. Mononuclear cell infection may also represent the normal method of entry into the CNS. Lindsay et al., (2006) and Heid et al., (2008), both demonstrated that *S. neurona* can infect leukocytes. As a correlate for the role of the immune response in disease, Saville et al., (2000), compiled data from multiple EPM vs. non neurologic controls and determined that risk factors for disease include racing or showing as well as adverse health incidents. This study would support that "stress" may increase incidence of disease. Additionally, there are several studies in mouse models demonstrating that *S. neurona* infected immunocompetent mice are resistant to disease, whereas infected immunocompromised mice develop disease (Dubey et al., 2001; Witonsky et al., 2003a, b; 2005; Sellon et al., 2004a). Collectively, these data support a role for a dysfunctional immune response associated with disease.

PROTECTIVE IMMUNE RESPONE TO *S. NEURONA*

The protective immune response to many intracellular agents is a cell mediated CD4 T-helper-1 (CD4 Th1) with CD8 T-cytotoxic-1 (CD8 Tc1) response, with associated interferon gamma production. Data from studies in horses and mice, which will be discussed below, suggest that the protective immune response to *S. neurona* is a CD4 Th1 CD8 Tc1 response.

MOUSE MODELS OF S. NEURONA: WHAT HAVE WE LEARNED?

To better understand the pathology of *S. neurona* infection, the development of EPM, and the role of the immune response, investigators performed some studies in immunodeficient mice to determine the pathology and the role of each immune component in disease. Data in immunocompetent mice demonstrated that C57BL/6 mice infected with *S. neurona* develop early activation and memory CD8 response (Witonsky et al., 2003a). Furthermore, additional studies in CD8KO, interferon (IFN)-gamma KO, and B-cell deficient mice, all of which were on a C57BL/6 background, confirmed the most significant role for CD8 cells (Witonsky et al., 2003b, 2003c, 2005). *S.neurona* infected IFN-gamma KO mice developed an increased number of memory CD4 and CD8 lymphocytes in the spleen (Witonsky et al., 2003b). Infected CD8 KO mice develop splenomegaly and lymphadenopathy, histopathological changes including follicular hypoplasia, endothelitis and perivascular inflammation in the liver, and meningoencephalitis in the brain. Brains from infected CD8 KO mice were positive by PCR for *S. neurona* (Witonsky et al., 2005a). B-cell deficient mice also did not develop encephalitis (Witonsky et al., 2005b). *S. neurona* infected mice also developed significant CD4 and CD8 effector (CD45RBhigh) and memory (CD44high) cells in lymph nodes and spleen. These data support the role for both a CD4 and CD8 cell mediated immune response, and in particular, the role for CD8 cells (Witonsky et al., 2005a).

One of the main cytokines that both CD4 and CD8 cells produce is IFN-gamma. The studies in CD4 and CD8KO mice as well as the studies in IFN-gamma KO mice, all confirm the critical role for IFN-gamma. Besides CD4 and CD8 cells, NK cells, macrophages and dendritic cells also make IFN-gamma. In studies performed by Sellon et al., (2004a), with immunodeficient mice, severe combined immunodeficient (SCID) mice did not prevent parasitemia, but did prevent encephalitis. However, when the SCID mice were treated with antibodies to interferon-gamma, the mice developed encephalitis. Similar studies in SCID foals demonstrated that SCID foals developed parasitemia, but not neurologic disease (Sellon et al., 2004b). These data suggest that one or more IFN-gamma producing cell populations likely prevents *S. neurona* from causing encephalitis, possibly by preventing *S. neurona* from infecting specific cell types and crossing the blood brain barrier. Additional studies are warranted to investigate these mechanisms.

HORSES INFECTED WITH S. NEURONA HAVE ALTERED IMMUNE RESPONSES

Tornquist et al., (2001) was one of the first studies demonstrating a decrease in percentage of CD4 cells in EPM horses compared to asymptomatic seropositive and seronegative horses. Asymptomatic horses had increased Concanavalin A (Con A), a marker of T-cell proliferation, and antigen specific proliferation compared to symptomatic and seronegative horses. Spencer et al., also showed that there EPM affected horses had decreased Surface antigen 1 (SAG-1) mediated proliferation as well as IFN-gamma vs. IL-4 expression (Spencer et al., 2004; 2005). In contrast to Tornquist's study, Yang et al., (2006) identified a significantly increased number of CD4 cells and neutrophils in naturally occurring cases of EPM, but that non-antigen specific proliferation to mitogens phorbol myristate acetate and ionomycin (PMA/I) was decreased in cells incubated overnight. The decreased PMA/I proliferation response was confirmed in experimentally infected horses. Experimentally infected horses also demonstrated a trend towards decreased antigen specific interferon gamma responses (Witonsky et al., 2008). In a subsequent study to define the mechanism of suppressed responses, the antigen specific proliferation to merozoites was decreased compared to non-specific PMA/I induced proliferation. These data suggest that antigen presenting cell function (i.e. dendritic cells) may have decreased function (Lewis et al., 2011). In addition to the roles for cell mediated immunity for protection, Ellison and Witonsky (2009) demonstrated that there was a correlation between antibody titer and protection in SAG-1 vaccinated horses against disease. Collectively, these data suggest that while the cell mediated immune response may be critical for protection, multiple components of the immune response, innate, CD4, CD8 and B-cell, are likely needed for optimal immunity.

REFERENCES

Cutler TJ, et al. **Vet Parasitol** 2001. 95:197-210, 2001. Dubey JP, et al., **Vet Parasitol** 2001a: 95(2-4):89. Dubey JP, et al., **J Parasitol** 2001b; 87(5): 1171-3. Ellison SP, et al., **Intern J Appl Res Vet Med** 2003: 1: 318. Ellison SP, et al., **Intern J Appl Res Vet Med** 2004: 2: 79. Ellison SP, Witonsky SG: **Can J Vet Res** 2009: 73: 176. Ellison SP, et al., **Intern J Appl Res Vet Med** 2003:1: 318. Heid B., et al, **ACVIM EPM SIG,** 2008. Lewis R, et al., **ACVIM EPM SIG,** 2011. Lindsay DS, et al., **Vet Parasitol** 2006: 138:371. Long MT et al, **Exp Parasitol** 2001: 100:150. MacKay RJ: **J Am Vet Med Assoc** 1997: 210:482. National Animal Health Monitoring System: *Equine protozoal myeloencephalitis (EPM) in the U.S.* 2001, USDA:APHIS:VS, CEAH, NAHMS. Saville WJ et al., **J Am Vet Med Assoc** 2000: 217:1174. Scott et al., **J Parasitol** 2005. 91(6): 1499. Sellon DC, et al., **Clin Diagn Lab Immunol** 2004b: 11:1134. Sellon DC, et al:, **J Parasitol** 2004a: 90:782. Sellon DC and Dubey JP in. **Equine Infectious Disease**, p453. Spencer JA, et al: **J Parasitol** 2004: 90:428. Spencer et al., **Clin Diag Lab Immunol** 2005. 12(5): 644. Tornquist SJ et al., **Eq Vet J** 2001: 33: 726. Witonsky SG, et al., **J Parasitol** 2003b: 89:932. Witonsky SG, et al, **Int J Parasitol** 2005a:35:113. 116. Witonsky SG, et al., **J Parasitol** 2003a: 89:924. Witonsky SG et al., **J Parasitol** 2005b: 91(4): 830. Yang J, et al., **Vet Parasitol** 2006:138: 200. Witonsky S, et al., **J Parasitol** 2008: 94(5):1047.

GASTRIC ULCERS IN HORSES: PATHOPHYSIOLOGY AND DIAGNOSIS

Frank M. Andrews, DVM, MS, DACVIM
Baton Rouge, LA

INTRODUCTION

Gastric ulcers are common in horses, with prevalence estimated from 53 to 93%, depending on populations surveyed and type of athletic activity.[1-3] Gastric ulcers in horses are primarily caused by stomach acids. However, the anatomy of the stomach, diet, restricted feed intake, exercise, stress (stall or transport), and the use of non-steroidal anti-inflammatory agents (NSAIDs) are risk factor in development of gastric ulcers in horses. Because many factors are involved in the cause of ulcers, the term Equine Gastric Ulcer Syndrome (EGUS) was coined in 1999 to describe the condition of erosions and ulcerations occurring in the distal esophagus, non-glandular (NG) and glandular stomach, and proximal duodenum of horses.[4] This paper reviews the pathophysiology and diagnosis of EGUS, with an emphasis on NG gastric ulcers.

ANATOMY AND BARRIER FUNCTION

The proximal portion of the equine stomach is covered by stratified squamous epithelium and approximately 80% of ulcers occur in this region.[5] Histologically, the NG squamous mucosa consists of four distinct zones made up of cell layers.The outermost cell layer (stratum corneum) is cornified and functions as a barrier to diffusion of strong electrolytes, such as hydrochloric acid (HCl) and sodium (Na). The cells of the middle (stratum transitionale) and deeper (stratum spinosum) layers contain sodium-potassium ATPase, which functions in the transcellular transport of Na.[6-7] These layers are principle barriers to the diffusion of strong acids in the deeper layers and consist of glycoconjugate substances containing bicarbonate, which are secreted by the superficial cells in the stratum spinosum. A minimal barrier of tight junctions between cells exists in the stratum corneum. The deepest layer is the stratum basale, which is one cell thick and presents no barrier to diffusion. In total, these epithelial layers present a minimal physical barrier to gastric acid diffusion compared with the glandular mucosa. For example, in a recent study, HCl (60 mmol/L; pH 1.5) exposed to the NG mucosa for 60 minutes in an Ussing chamber system disrupted barrier function.[8-9] However, disruption of barrier occurred almost immediately when volatile fatty acids (VFAs) were added to the acid solution. Thus, VFAs in the stomach may act synergistically with HCl *in vivo* to disrupt the NG mucosa and cause acid damage. The distal portion of the stomach is covered by glandular mucosa, which is responsible for secreting mucus, bicarbonate, HCl, and pepsinogen.[10] Approximately, 20% of gastric ulcers occur in this region of the stomach. The glandular mucosa has extensive mechanisms for protection, including a bicarbonate-rich mucus layer, an extensive capillary network, and a rapid ability for restitution or healing. Gastric ulcers occurring in this region are likely due to the reduction of prostaglandins and loss of these protective mechanisms as occurs with treatment of NSAIDs or stress-induced release of endogenous cortisol.

PATHOGENESIS

Horses are continuous gastric HCl secretors, and acid exposure is thought to be the primary cause of NG gastric ulcers in horses. Also, performance horses are typically fed relatively low-roughage, high grain diets containing water-soluble carbohydrate (WSCs). A diet high in WSCs provides substrates for gastric fermentation by resident bacteria. Gastric fermentation by-products, such as VFAs, alcohol, and lactic acid (LA), may damage the NG squamous mucosa. Several species of bacteria (*Lactobacillus, Streptococcus, E. coli)* were isolated from the stomach of horses, adding credence to this theory.[11] Previously, an *in vivo* study in cannulated horses showed that a diet high in WSCs and protein (alfalfa hay/sweet-feed grain diet) produced high VFA concentrations in the stomach.[12] A stepwise model generated from those data showed that the presence of VFAs (butyric, propionic, and valeric acids) and low gastric juice pH were important predictors of ulcer severity. Furthermore, an *in vitro* study with Ussing chambers showed that a VFAs (60 mmol/L; acetic, propionic, butyric, or valeric acids) led to decreased chloride-dependent Na transport across the NG mucosa and histologic changes of cell swelling.[8-9] Also in that previous study, HCl alone and in combination with VFA caused inhibition of cellular chloride-dependent Na transport, cell swelling, and eventual sloughing of mucosa when exposed to the NG mucosa at pH < 4.0. The ulcerogenic effects of the VFAs in combination with HCl were pH, dose and chain-length dependent. In addition to VFAs, bile acids were shown to increase NG mucosal cell permeability to hydrogen ions, which eventually lead to ulceration.[13] However, the effects of bile acids in EGUS is questionable because they usually come from less acidic duodenal reflux and are non-ulcergenic at a pH > 4. Pepsinogen, which is cleaved to pepsin at a pH < 4, may play a role in the development of NG ulcers. Pepsin may act with HCl to produce acid damage, but they are not synergistic.[14]

RISK FACTORS

Although HCl, VFAs and other organic acids have been implicated in the cause of NG ulcers, several risk factors for ulcer development have been identified.[3,5,15]

Exercise Intensity

Horses in training and racing are at high risk of developing NG gastric ulcers.[16] Previously, horses running on a high-speed treadmill showed increased abdominal pressure and decreased stomach volume.[17] The authors speculated that compression of the stomach allowed acids from the glandular mucosa to reflux into the NG region ("acid splash"), leading to NG mucosal injury. Intense exercise may increase the exposure of the NG mucosa to acid, explaining the increased prevalence of gastric ulcers in horses in race training and racing. Furthermore, an increase in serum gastrin concentration occurs during exercise[18] which stimulates HCl secretion, resulting in a lower stomach pH.

Intermittent vs. Continuous Feeding

Horses grazing on pasture have a decreased prevalence of EGUS. During grazing, there is a continuous flow of saliva and ingesta that buffers stomach acid and maintain stomach pH > 4 for a large portion of the day. Conversely, when feed is withheld from horses, before racing or in managed stables, gastric pH decreases rapidly and the NG mucosa is exposed to an acid environment. Intermittent feeding has been shown to cause and increase the severity of NG gastric ulcers, and an alternating feed-deprivation model was developed to produce EGUS experimentally.[19] The NG mucosa is the most susceptible to ulceration in horses subjected to intermittent feeding due to its lack of mucosal protective factors. Studies have shown that stomach pH drops 6 h after feeding and DM content decreases 12 h after feeding a mixed-feed diet compared with horses fed a hay diet.[12,20] Thus, horses should be fed hay continuously or every 5 to 6 h to buffer stomach acids.

Diet

Diet has been implicated as a risk factor for EGUS. Serum gastrin concentrations increase significantly in horses fed high-concentrate diets. In addition, as noted previously, concentrate diets are high in WSCs and are fermented by resident bacteria, resulting in the production of VFAs, which, in the presence of low stomach pH (≤4), cause damage to the NG mucosa.[8-9] In another study, horses fed a high-protein and high calcium (alfalfa hay/grain) diet showed higher stomach pH than horses fed a low protein and low-Ca (brome grass hay) diet. Horses on the alfalfa hay diet (high-protein, high-Ca) had fewer and less severe gastric ulcers.[12] Feeding alfalfa hay may have some protective effect on the NG mucosa in horses. The protective effects of alfalfa hay fed to exercising horses was confirmed in a more recent study.[21] Furthermore, horses fed mixed feed (128 g of CP and 175 g of crude fiber/kg of DM) for at least 14 d showed increased gastric ulcers in the NG mucosa localized along the margo plicatus compared with horses fed a hay diet.[20] A diet high in grains (WSCs [sweet feeds]) have been implicated in causing NG gastric ulcers in horses, however alfalfa hay has a protective effect by buffering gastric juice pH and decreasing the effects of VFAs on NG mucosa.

Stall Confinement

Stall confinement has been implicated as a risk factor for EGUS.[23] In that study, 6 of 7 horses housed in stalls had gastric ulcers, whereas no horse had gastric ulcers after 7 days turn out to pasture. However, in another study, the prevalence of ulcer severity did not differ significantly between horses stabled full-time, horses kept in a stable part-time or horses kept in a pasture full-time .[24] In one study evaluating housing in horses showed that neither proximal stomach nor ventral stomach pH changed significantly in horses housed in stalls alone, housed in stalls with a companion or housed in a grass paddock.[25] However, pH in the proximal stomach was lower during the early morning hours regardless of the housing and feed intake was lowest during these hours. Thus, other factors may play role in stabled horses that increase the risk of EGUS. Stabled horses are typically fed two large meals daily. These meals are traditionally high in grains and consumed rapidly, which lead to a decrease in saliva production and less buffering of stomach contents.

Non-steriodal Anti-inflammatory Agents

The use of NSAIDs is common in horses presenting with acute abdominal pain. Phenylbutazone or flunixin meglumine are typically given to control pain during a colic episode. These agents have been shown to cause gastric ulcers in horses, but usually at higher than therapeutic doses.[26] Also, the use of NSAIDs in racehorses has not been shown to be a risk factor for EGUS in other epidemiologic studies. Furthermore, NSAIDs are thought to cause more severe ulcers in the glandular mucosa because of their effect on prostaglandin inhibition. Prostaglandin inhibition results in decreased mucosal blood flow, decreased mucus production, and increased HCl secretion. Although prostaglandins are also important in the regulation of acid production and Na transport, it may be their effect on mucosal blood flow that is the most important. Adequate blood flow is necessary to remove hydrogen ions that diffuse through the mucus layer covering the glandular mucosa. Gastric mucosal ischemia may lead to a hypoxia-induced cellular acidosis, release of oxygen-free radicals, phospholipase, and proteases, which may damage the cell membrane leading to necrosis. NSAIDs have the potential to exacerbate EGUS in horses with colic and one should use them with caution.

Helicobacter spp. and other Bacteria

Helicobacter spp. have been isolated from a man and a variety of animals suffering from gastric ulcers and gastritis. Recently a new enterohepatic *Helicobacter* species, *Helicobactor equorum,* was isolated from fecal samples of two clinically healthy horses [35].[27,28] Also, *Helicobacter equorum* DNA was demonstrated in the feces of 2 of 7 (28.6%) foals less than 1 month of age and 40 of 59 (67.8%) foals 1 to 6 months of age. Furthermore, Helicobacter-like DNA was detected in the stomach of 10 Thoroughbred horses in Venezuela.[29] In this study, Helicobacter-like DNA was detected in two out of seven horses with gastric ulcers, three out of five horses with gastritis and five out of six horses with both pathologies and one horse with normal gastric mucosa. Furthermore, 10/11 of the horses infected with Helicobacter had either gastric ulcers or gastritis or both pathologies. However, 39% of the horses in that study did not have gastric lesions, so multiple causes are likely.

Other bacteria have been implicated in inhibiting ulcer healing. Bacteria, including *E. coli*, were cultured from the stomach of horses.[11] In rats, which have a compound stomach similar to horses, *E. coli* administered orally, rapidly colonized acetic acid-induced gastric ulcers and impaired healing.[30] Oral antibiotic treatment with streptomycin and/or penicillin suppressed bacterial colonization of the ulcer and accelerated ulcer healing. Also, oral administration of lactulose resulted in an increase *Lactobacillus spp.* growth and colonization of the ulcer bed. Accelerated ulcer healing was seen in the rats compared to placebo treated controls. Thus, bacterial colonization of gastric ulcers in the stomach of horses may delay ulcer healing and in this case treatment with antibiotics may be indicated.

DIAGNOSIS

In cases where EGUS is suspected, a thorough history, physical examination, and a minimum data base should be obtained to rule out other differential diagnoses. Identifying risk factors and clinical signs are helpful in making a diagnosis of EGUS. However, gastroscopy is the only definitive diagnosis for gastric ulcers currently available. Standing gastroscopy procedures have been described in detail elsewhere in the literature and require an endoscope of least 2 meters in length to visualize the NG mucosa and margo plicatus and a 2.5m to 3m endoscope to visualize the pylorus and proximal duodenum in most adult horses.[4] Use of a gastric ulcer scoring system allows clinicians to compare gastroscopic findings and monitor healing of ulcers and evaluate efficacy of treatment.

Currently, there are no hematologic or biochemical markers to diagnose EGUS. However, a recent report showed that horses with gastric ulcers had lower RBC counts and hemoglobin concentrations than horses that did not have gastric ulcers.[31] Some horses with EGUS may be slightly anemic or hypoprotenemic, but in this author's experience the RBC and hemoglobin concentrations may be low, but are rarely outside normal reference ranges.

Other presumptive diagnostic techniques include a sucrose absorption test.[32] Urine sucrose concentrations were significantly higher for horses with gastric ulcer scores > 1. Using a urine sucrose concentration cutoff value of 0.7 mg/ml or higher, revealed an apparent sensitivity of 83% and specificity of 90% to detect ulcers in horses tested using the sucrose permeability test. Thus, this test may provide a simple, noninvasive test to detect and monitor gastric ulcers.

In a recent study, a fecal occult blood test was found to be helpful in diagnosis of gastric ulcers.[33] The positive predictive value of the fecal occult blood test (FOBT) in horses with EGUS was 90%; however, the negative predictive value was only 17%, which suggests that a horse with a positive FOBT is likely to have a gastric ulcer. In an attempt to improve the negative predictive value of the FOBT, investigators developed another test (SUCCEED™ Equine Fecal Blood Test, Freedom Health LLC., Aurora, OH) that utilizes specific equine monoclonal antibodies to both albumin and hemoglobin in easy to use kit. The package insert contained with the kits and information provided by the manufacturer showed an improved predictive value of a negative test (72%), but the predictive value of a positive test was slightly lower (77%). Thus, this new test may be helpful in diagnosing EGUS in horses, but should be used as part of a complete work-up. A false positive FOBT may occur following a recent rectal examination, rectal biopsy or other rectal trauma has occurred or if the horse has a protein losing enteropathy.

Unfortunately, laboratory techniques provide only presumptive diagnostic evidence of EGUS, whereas a definitive diagnosis can only be made by endoscopic examination. Therefore, if gastroscopy is not available and ulcers are strongly suspected, it may be worthwhile to start empirical treatment and observe for resolution of clinical signs. If the horse does not respond to treatment, further evaluation is indicated.

REFERENCES

[1]Vatistas NJ, et al. Proc Ann Conv AAEP Vancouver, British Columbia, Canada 1994;125–126. [2]Hammond CJ, et al. EVJ 1996;18:284–287. [3]Murray MJ, et al. Equine Vet. J. 1996;28:368–374. [4]Anon. 1999. Eq Vet Educ 1999;1:122–134. [5]Andrews FM, Nadeau JA. EVJ Suppl 1999;29:81–86. [6]Schnorr B, Willie KH. Z Zellforsch 1971;114:482–492. [7]Argenzio RA. EVJ Suppl 1999;29:19–23. [8]Nadeau JA, et al. AJVR 2003;64:404–412. [9]Nadeau JA, et al. AJVR 2003;64:413–417. [10]Murray MJ. Vet. Med. 1991;8:815–819. [11]Al Jassim RAM, et al. FEMS Microbiology Letters 2006;248:75-81. [12]Nadeau JA, et al. AJVR 2000;61:784–790. [13]Berschneider HM, et al. EVJ Suppl. 1999;29:24–29. [14]Widenhouse T, et al. AJVR 2002;63:744–749. [15]Rabuffo TS, et al. JAVMA 2002;221:1156–1159. [16]Vastistis NJ, et al. EVJ Suppl 1999;29:40–44. [17]Lorenzo-Figueras M, Merritt AM. 2002;63:1481-1487. [18]Furr M, et al, Cornell Vet. 1994;84:41–45. [19]Murray, MJ. Dig. Dis. Sci. 1994;12:2530–2535. [20]Coenen M. Schweiz Arch Tierheilkd 1990;132:121–126. [21]Lybbert T et al. Proceedings of the 54th Annual AAEP, Orlando. 2007;525-526. [22]Feige K, et al. Schweiz Arch Tierheilkd 2002;144 (7):348-55. [23]Bell RJW, et al. NZVJ 2007;55(1):13-18. [24]Murray MJ, et al. EVJ 1993;25:417-421. [25]Husted L, et al. EVJ 2008;40:337-341. [26]Andrews FM, et al. J Pharm Therap 2009;10(3):113-120. [27]Moyaert H, et al. Int J Syst Evol Microbiol 2009;57, 213-218. [28]Moyaert H et al. Vet Microbiol. 2009;1:133(1-2):190-192. [29]Contreras M., et al. Lett Appl Microbiol. 2007;45(5):553-557. [30]Elliott SN, et al. AJP-GI 1998;275:425-432. [31]McClure SR, et al. JAVMA 2005; 226(10):1681-1684. [32]O'connor MS et al. AJVR 2004; 65(1):31-39. [33]Pellegrini FL. JEVS 2005;25 113-117.

INTRALIPIDS THERAPY: FROM NUTRITION TO ANTIDOTE

Daniel L. Chan, DVM, DACVECC, DACVN, FHEA, MRCVS
North Mymms, Hertfordshire, UK

INTRODUCTION

The use of intralipids in parenteral nutrition (PN) has been controversial for many years. Determining whether this practice should be continued merits careful consideration and three central questions must be evaluated: 1) What benefit does the inclusion of intralipids confer to patients? 2) What are the risks associated with their use? 3) What alternatives are there to intralipids for formulating PN?

INTRALIPIDS IN NUTRITION

As the practice of providing nutritional support to critically ill veterinary patients is increasingly becoming more accepted, the use of PN in veterinary medicine has increased in recent years. However, routine use of PN in general veterinary practice is hindered by the relatively high cost of this form of nutrition, the logistical requirements for safely compounding PN and the perceived high rate of complications associated with its use. Classically, total parenteral nutrition (TPN) was defined as the provision of all of the patient's protein, calorie and micronutrient requirements intravenously, whereas partial parenteral nutrition (PPN) referred to a formulation that provided only a part of this requirement (typically 40-70% of the energy requirement). These terms are becoming less used and PN solutions are more or less defined by their composition. For example, PN containing amino acids, dextrose and intralipid are more commonly referred to as total nutrient admixtures (TNA). All three components of PN (i.e. lipid, amino acid and glucose) provide energy, yet the lipid component is the most energy dense solution. In addition to this, the lipid component being iso-osmolar serves an important role to reduce the tonicity of the resultant solution, thereby allowing peripheral administration of PN solutions. For these two reasons, namely high energy density and low osmolarity, intralipids offer important advantages in the formulation of PN solutions.

However, despite the benefits of intralipids in PN solutions, the inclusion of intralipids has raised some concerns with regard to possible immune-suppression, pro-inflammatory effects, pro-oxidant effects, increased risk for hyperlipidaemia, lipid embolization, increased risk of contamination, and negative hemodynamic effects.[1] For example, several *in vitro* studies have shown that neutrophil chemotactic and phagocytic function is compromised after lipid infusions.[2,3] A veterinary study investigating the effect of lipid infusions on the neutrophil phagocytic responses of dogs revealed reduced function when the lipid was administered at a rate of 200% of the basal energy requirement (BER).[4] However neutrophil function normalized 24 hours following cessation of the infusion. One could argue, however, that as the administration of lipids at 200% of BER is far in excess of what is normally administered in clinical practice (normally up to 50% RER), the risk of clinically-relevant immunosuppression maybe quite small. Nevertheless, this study did demonstrate that the infusion of intralipid was able to significantly and adversely impact the immune system of healthy dogs. The impact of the immune-suppressive effects of high-dose intralipid therapy in critically ill animals warrants urgent evaluation.

As the principal type of lipid used in TNAs is derived from soybean oil (a long chain triglyceride containing linoleic acid) its affect on the inflammatory system via activation of the omega-6 family of inflammatory mediators has also been investigated. *In vivo* studies in people have shown an exaggerated inflammatory response to endotoxin following a long-chain triglyceride infusion. Another source of concern with the use of lipid emulsions has been the tendency for lipids to support bacterial and fungal growth due to its isotonicity and neutral pH, although compounding standards have greatly reduced risk for contamination. Highlighting this concern is reflected by the fact that the Center of Disease Control have issued guidelines requiring that lipids should only be administered over a period no longer than 12 hours (also referred to as '12 hour hang time.') It should be pointed out however, that the risk of bacterial and fungal growth is highest when intralipids are administered as stand alone solutions.[5] Various studies have demonstrated that when intralipids are delivered as part of TNAs, there is no increased risk of infections.[6]

Experimental studies performed on dogs have also demonstrated possible hemodynamic effects of intralipids including reduced ventricular function and decreased systemic vascular resistance.[1] Finally, toxicities due to leaching of lipid soluble phthalates (used as plasticizers in polyvinyl-chloride based delivery systems) is another concern in long-term PN, particularly in infants. Although, other components in PN solutions may also contribute to leaching of phthalates from polyvinyl-chloride materials, the lipid component has been recently implicated as the principal component mediating toxicity.[1] Recent shortages in the supply of intravenous lipid solutions have also highlighted the need to seek alternatives forms of TNAs other than 3-in-1 solutions.[1]

The lipid component of TNAs provides an energy dense solution and is a source of essential fatty acids, which are necessary for maintenance health. Being iso-osmolar to blood (approximately 300 mOsm/L), the lipid suspension also serves an important role in reducing the relative osmolarity of the amino acid and glucose solutions (both of which have osmolarities > 1000 mOsm/L). However, despite these benefits there is a requirement for aseptic compounding of lipid-containing PN solutions and their shelf life is relatively short once compounded.

Given the benefits and risks of intralipids in PN solutions, it is reasonable to explore possible alternatives to intralipids in PN or re-evaluate the role of lipids in PN in general. Some authors already do not include any form of lipids when formulating PN for small animals. Lipid-free PN formulations (eg, ProcalAmine) is sometimes the preferred method of PN as they are extremely easy to implement as these solutions require no further manipulation and can be administered directly out of the original bottle. Lipid-free PN solutions are inexpensive, provide a relative high amino acid content and lack the problems associated with intralipids. However, the major disadvantages of lipid-free solutions are the relative low energy content and the high osmolarity.

Simply eliminating the lipid component is not, however, the only possible solution. There are now several lipid-based solutions, which have been specifically designed to reduce complications and to even confer a benefit to patients. As previously mentioned, intralipids are mostly derived from soybean oil and consists of long-chain triglycerides. In contrast to this, medium-chain triglycerides have been considered as an alternative lipid source as these lipids are metabolized much more rapidly, do not accumulated in tissues and do not even require carnitine for oxidation. However, increased plasma concentrations of medium-chain free fatty acids may lead to various adverse reactions including ketosis, increase energy expenditure and in dogs, severe neurologic toxicity. In order to improve the safety profile of medium-chain triglycerides, chemical manipulation has yielded the so-called structured triglycerides. These structured lipids are produced by first hydrolyzing soybean (source of long-chain triglycerides) and coconut oil (source of medium chain triglycerides) and then re-esterifying them into structured triglycerides with medium- and long-chained fatty acids at randomly attached to positions within the same glycerol molecule. The use of these emulsions appears to result in several beneficial effects including, lower plasma triglyceride concentrations, faster plasma clearance, improved nitrogen balance, increased hepatic protein synthesis, and improved albumin status.[7,8]

More recently, there has been a great interest in the use of omega-3 fatty acids to modulate disease. Well characterized benefits of omega-3 fatty acids include modulation of the inflammatory cascade, beneficial effects on plasma lipid and lipoprotein profiles, antiarrhythmic effects, and even anti-epileptic effects. With the advent of intravenous omega-3 fatty acids designed for PN administration has opened a new door for possible therapeutics. It is interesting to note that enteral administration of omega-3 fatty acids results in some significant loss due to intestinal remodeling, which is bypassed when administered parenterally. The incorporation of omega-3 fatty acids into cellular membranes also appears to occur much more rapidly after intravenous administrations (ie days) compared to the several weeks reported with enteral intake.[9] Some of the documented benefits of parenteral omega-3 fatty acid administration includes improved microcirculatory and splanchnic perfusion in the face of sepsis.[9] The major problem with access to such products (eg, Omegaven) is that is not currently licensed in the US; it has been used in Europe of some time and many studies support the use of such fatty acids in specific conditions. Newer formulations of lipids for PN incorporate medium chain triglycerides, omega-3 and omega-9 fatty acids and there are promising results in clinical trials with these new lipid formulations.

INTRALIPID AS ANTIDOTE THERAPY

Although there has been growing concerns regarding the benefits on intralipids in PN formulations, the use of intralipids as an antidote for fat-soluble drug toxicity in emergency medicine has become more established. The genesis of the use of intralipid as an antidote began with treatment of local anesthestic overdoses in people (i.e., bupivacaine) that resulted in cardiopulmonary arrest.[10] Since then, there have been a number of case reports and experimental studies describing the use of intralipids in the treatment of patients intoxicated with other drugs, including verapamil, calcium-channel blockers, and beta-blockers. In veterinary medicine the use of intralipids to treat fat-soluble drug toxicity has also been reported.[11-13] The exact mechanism by which intralipids reverse toxicty is unknown. Identification of the mode of action of intralipids remain elusive. Hypothesis range from enhancing myocardial performance in the face of toxicity, to myocardial protection due to intracellular calcium influx to the most plausible hypothesis, whereby the intravenous administration of intralipids creates a "lipid sink." Fat-soluble substances are trapped within this lipid compartment and is eliminated through the normal metabolic pathways, ie, macrophage and reticuloendothelial systems.[12]

The dosage scheme for use of intralipids as an antidote was empirically determined and then simply extrapolated for use in animals. The most commonly quoted dosage scheme is as follows: Using a 20% intralipid solution, a 1.5 mL/kg IV is bolused over 1 minute, followed by 0.25 mL/kg/min for 30-60 minutes.[12] The initial bolus could be repeated 1-2 times if no response to the initial bolus is obtained, with a goal to not exceed 8 mL/kg/day.[12] Based on experience with the use of intralipids for PN support, an upper limit of 2g/kg/day could be safely employed, which would mean 10 mL/kg/day. Some nutritionists have exceeded this upper limit and do not report over complications. This would suggest that further investigation evaluating the optimal dose of intralipid as antidotes are warranted. It has also become apparent that intralipid therapy does not work in all cases and that further work is necessary to elucidate the mechanism of action, define situations the merit intralipid therapy and establish effective protocols.[13]

SUMMARY

In summary, given the various concerns regarding intralipids (especially those derived from soybean and safflower oils) in PN formulations, it is entirely reasonable to explore alternatives to their inclusion in PN. Currently, the only viable option in the US appears to simply use PN formulations composed of amino acids and dextrose. Such formulations are inexpensive, widely available and simple to implement. Their major disadvantages include to the inability to supply enough calories to meet RER in the majority of patients, relative high osmolarity (necessitating central venous administration) and relatively high dextrose content which could result in persistent hyperglycemia which we now know to be pathological. Upcoming innovation in this area are likely to come in the form combining medium-chain tryglycerides, omega-3 and omega-9 fatty acids in PN. These forms of lipids been associated with positive clinical results in Europe and will hopefully be available in the US soon. Meanwhile, the use of conventional intralipids as antidotes is increasing in popularity and reports of its use in various lipid-soluble toxicities are increasing. Despite apparent successes, our understanding of its mode of action and optimal protocols remain elusive and require further evaluation.

REFERENCES

1. Calder PC, et al. Intens Care Med 2010; 36:735; 2. Jarstrand C, et al. JPEN J Parenter Enteral Nutr 1978;2(5):663; 3. Wiernik A, et al. Am J Clin Nutr 1983;37:256; 4. Kang JH, et al. J Vet Intern Med 2008;22:1166; 5. D'Angio R, et al. JPEN J Parenter Enteral Nutr 1987;11:394; 6. Kuwahara T, et al. Int J Med Sci 2010; 7:43; 7. Kruimel JW, et al. JPEN J Parenteral Enteral Nutr 2001;25(5):237; 8. Chambrier C, et al. Nutr Clin Pract 2006;21(4):342; 9. Chen WJ, et al. J Nutr 2003;19(2):275; 10. Weinberg, *et al. Reg Anesth Pain Med* 2003;28(3):198; 11. Crandell DE, *et al. J Vet Emerg Crit Care* 2009;19(2):181; 12. Fernandez AL, et al. J Vet Emerg Crit Care 2011;21(4):309; 13. Wright HM, et al. J Vet Emerg Crit Care 2011;21(6):666

DIETARY ISOFLAVONES AND THEIR BIOLOGICAL EFFECTS

Nick Cave, BVSc, MVSc, PhD, DipACVN
Palmerston North, New Zealand

INTRODUCTION

Molecules found in or derived from plants that have estrogenic activity are termed phytoestrogens, and they include structurally diverse compounds such as isoflavones (e.g. genistein), coumestans (e.g. coumestrol), sterols (e.g. silosterol), lignans (e.g. enterolacton), and macrolides (e.g. zearalenone). The isoflavone genistein is principally found in plants of the family Leguminosae including soy and clover.[1] Leguminous plants form symbiotic associations with certain soil bacteria (rhizobia) that "fix" atmospheric nitrogen to form ammonia, which the plants can use. Infection of legume hairs by rhizobia requires transcription of so-called nodulation (nod) genes by the bacteria. The gene product of the regulatory nod gene, nodD, requires the presence of a plant signal to convert it to an "active form", which then induces the expression of other nodulation genes allowing infection of the plant and establishment of symbiosis. The plant signal in soybeans and other legumes is the production of flavanoid compounds such as genistein.[2]

Genistein is structurally similar to 17 β-estradiol as depicted in figure 1 below.

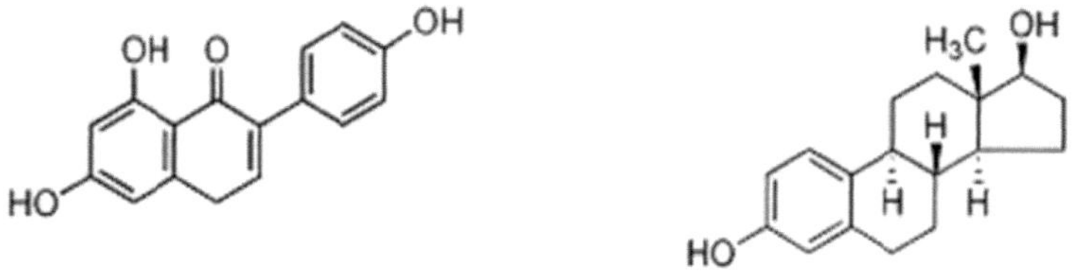

Genistein 17 β-estradiol

Figure 1. Structural diagrams of genistein and estradiol

Soy-based ingredients are common in commercial diets fed to dogs and cats; the soy plant is a relatively cheap source of protein, fiber, and polyunsaturated oil. The isoflavone content of several pet foods has been assayed and concentrations have been found that would result in a dog ingesting up to 3 mg/kg BW, and a cat ingesting up to 8 mg/kg BW.[3,4]

GENISTEIN AS AN ESTROGENIC COMPOUND

Genistein has been confirmed as a phytoestrogen *in vivo* through its ability to increase uterine weight, mammary gland development, and stimulation of prolactin secretion in ovariectomized rats.[5] It has been shown to function as an estrogen *in vitro* in some estrogen-dependant cell lines.[6,7] The biological actions of estrogens are mediated by binding to one of two estrogen receptors (ERs), termed ER-α and ER-β, which are ligand-regulated nuclear transcription factors. Binding of a ligand to the nuclear receptor induces a conformation change in the receptor that is influenced by the ligand[8] The conformational change leads to receptor dimerization, recruitment of coregulator proteins and other transcription factors and ultimately protein–DNA interaction with the formation of the transcription initiation complex. Both the ER-α and ER-β bind estrogen responsive elements (EREs) with similar specificity and affinity. The ER-α is expressed in the ovaries, uterus, kidneys, and mammary glands, whereas the ER-β has been detected in the prostate, ovaries, lungs, and peripheral nerves, and only very low expression in mammary tissue.[8] Lymphocytes express both ER-α and ER-β, which appear to regulate different responses.[9]

In addition to the ligand binding and DNA-response element binding by the two receptor types, estrogen receptors also recruit coactivators or corepressors; molecules that enhance or repress the transcription of target genes by interacting with transcription machinery, or by remodeling the chromatin structure without themselves binding DNA.[8,10] Different cells express different coactivator/repressors, thus the "estrogenicity" of a compound, and the tissues in which it acts, depends not only on its structural ability to bind to ERs, but also on the particular set of molecules that interact with the ER after it binds to ligand. This means that different compounds can even have opposite effects in different tissues depending on the particular set of auxiliary proteins expressed by that particular tissue.[11] This diversity of estrogenic activity displayed by various molecules capable of binding to ERs has lead to the term "selective estrogen response modulators" or SERMs. Different SERMs may bind to ER-α and ER-β with differing affinities, they may lead to the recruitment of different corepressor or coactivator proteins, which may differ for receptor types, and its activity may depend on the nature of the response element in that particular cell type resulting in tissue specificity.[12,13] Genistein binds to both the alpha and beta forms of the estrogen receptor (ER-α and ER-β), but the relative estrogenic potency of genistein on ER-β is about 30-fold higher compared with the potency on ER-α. However, the *in vivo* estrogenic activity of isoflavones is impossible to predict because of the complexity and tissue specificity of expression of co-activator and co-repressor molecules *in vivo*.[13] Thus in differing cells, perhaps even at varying times, a compound such as genistein can have estrogenic activity, no activity, or anti-estrogenic activity, and thus satisfies the definition of a SERM.[14]

We have shown that a large dose of genistein (100mg/kg/day) is required to have any effect on the cytological appearance of the vaginal mucosa in cats, but there may be a subtle effect on estrogen receptor expression in the uterus after long term dietary ingestion.[15] In neutered male and female cats, the same dose increased lean body mass.[16] This effect has been recognized in rodents, where ER-β agonists increase muscle mass.[17] This effect has some potential in preserving lean mass during weight loss programs, and may improve the recovery of muscle following injury or atrophy.[18] In dogs, a dose of 500mg/kg was required to have a significant effect on the reproductive tract, where it increased uterine, and decreased testicular and prostatic weights.[19] Thus, when administered orally, either as a dietary "ingredient", or as a high dose treatment, genistein (and daidzein) are not potent estrogenic compounds in dogs and cats. This is explained in part by their inherantly weak agonist effects, but mostly by the poor oral bioavailability. This suggests that a significant estrogenic effect might occur within the intestinal mucosa, but not systemically. This mucosal effect has been demonstrated in mice, where the stimulation of ER-β by genistein protects against intestinal carcinoma by promoting epithelial apoptosis.[20]

TYROSINE KINASE INHIBITION

In addition to genistein's estrogenic activity is its ability to inhibit tyrosine kinases.[21] Genistein inhibits free cytosolic, and receptor kinase kinase activity by forming nonproductive enzyme-substrate complexes. In contrast, genistein does not significantly inhibit the activity of serine and threonine kinases. Genistein is now utilized widely as a tool for elucidating the involvement of tyrosine kinases in many different cellular signaling pathways.

Of particular importance is the ubiquity of tyrosine kinase signaling cascades in leukocyte responses. These include T-cell activation and proliferation following engagement of T-cell antigen receptor (TCR), natural killer cell activation, neutrophil activation and superoxide production, bacterial phagocytosis by macrophages, antibody responses, and delayed-type hypersensitivity responses. In addition, most growth factor receptors (e.g. epidermal growth factor, platelet-derived growth factor, insulin-like growth factor-1) and many other receptors (e.g. insulin, leptin) have tyrosine kinase activity to their cytoplasmic portions that autophosphorylate tyrosine residues providing docking sites for the recruitment of specific accessory proteins with phosphotyrosine-binding sites.

TOPOISOMERASE INHIBITION

Genistein has also been found to DNA topoisomerase II, resulting in DNA topoisomerase II-mediated double strand breaks in DNA.[22] This activity is associated with G2-phase arrest in human Jurkat and murine P388 leukemia cells lines and has been linked to efficacy as a cancer chemotherapeutic, and as a disrupter of lymphocyte proliferation.[23-25] Genistein can also inhibit the proliferation of an established B-cell lymphocyte line *in vitro*, but only at concentrations that are 5-10 times more than the plasma concentration that is achievable through oral dosing, and probably 100 times more than what occurs following dietary exposure.[26]

CALCIUM CHANNEL BLOCKER

Finally, genistein has been shown to inhibit the function of certain calcium channels, notably functioning as a calcium channel blocker in human platelets demonstrating a 51% reduction in intracellular Ca_i^{2+} at 10mM concentration.[27] The inhibitory effect was immediate and appeared to directly affect the calcium influx channels and did not involve inhibition of tyrosine phosphorylation. In another study utilizing Guinea pig myocytes, inhibition of calcium L-channel function was dependant upon genistein's ability to inhibit tyrosine kinase activity.[28]

THYROID HORMONE METABOLISM

Several flavanoids can interfere with thyroid hormone metabolism and may increase the risk of hypothyroisism. Although the exact mechanism has not been clarified, possibilities include inhibition of T4 synthesis in the thyroid gland, inhibition deiodinase to reduce T3 synthesis from T4, and hepatic induction of phase II enzymes, such as UDP-glucuronosyltransferase, which are responsible for T3 and T4 clearance.[29] Lastly, in iodide deficient states, genistein forms a terminal inactive by-product with thyroid peroxidase.[30] In cats at least, any effect is probably mediated via inhibition of the deiodinase enzyme. When cats were fed a diet containing 0.198mg/g genistein for 3 months, their serum T3 was decreased, and fT4 and TT4 were increased, such that 4/18 cats had a serum T4 greater than the reference range.[31]After 12 months of feeding a soy-based diet that contained 68mg/kg DMB, there was no clear effect on thyroid hormone status (thyroid panel) compared with dogs fed a low genistein diet.[32] There is some evidence that another soy isoflavone, daidzein, is also capable of interfering with thyroid hormone production, and thus the two may need to be considered together when evaluating diets.[33] Dogs treated with 500mg/kg daily for 1 year, did not show any clinical signs of hypothyroidism, nor experienced any significant changes on serum biochemistry or haematology, although thyroid function was not specifically assayed.[19]

IMMUNOMODULATION

Given the number of mechanisms by which compounds such as genistein can have an effect, there is potential for their use as immunomodulatory compounds. In ovariectomized mice, there is a dose-dependant reduction in thymic, splenic, and circulating, CD4+ thymocytes and lymphocytes, and reduced IgG in response to an adjuvanted antigen with genistein treatment comparable with doses ingested by babies fed soy milk.[34] We have shown that daily oral genistein treatment can decrease circulating CD8+ cells, increase neutrophil respiratory burst, and decrease delayed-type hypersensitivity responses in cats.[35] However, this required a dose of 100mg/kg per day. As noted above for dogs, it does not appear possible to easily achieve plasma concentrations that would affect circulating lymphocyte function. However, we did observe a significant inhibition of the proliferation of lymphocytes isolated from the mesenteric lymph nodes of cats fed a purified soy-protein based diet.[36] This suggests that soy isoflavones might have a significant effect on mucosal immunity where cells are exposed to high concentrations of the free forms, whilst efficient intestinal and hepatic conjugation dramatically reduces systemic exposure to the biologically active, unconjugated forms. Thus, although systemic immunomodulation is unlikely to occur at dietary concentrations, it is likely that currently available, soy-based diets might have a significant inhibitory effect on intestinal mucosal lymphocytes that could be beneficial in chronic enteritis.

The oral bioavailability of free genistein is low when administered as a gelatin encapsulated powder, which explains the requirement for a high dose to elicit an estrogenic response. This is partly because of poor oral absorption, but also because of efficient conjugation to glycosylated or sulphated moieties.[19,37] It is suspected that the large intestine is an important site for conjugation as in other species. It is also apparent that the long elimination half-life is due to the slow release of genistein from peripheral - presumably liphophilic – compartments, such as fat and neural tissue. When genistein is incorporated into diets, accumulation in peripheral compartments such as fat and neural tissue may occur to a degree that is not predicted by plasma concentrations.

CONCLUSION

Dogs and cats consuming commercial diets are frequently exposed to isoflavones, mostly from soy. Genistein, and to a lesser extent other isoflavones, can act as weak estrogens, and genistein also affects thyroid and skeletal metabolism, and inhibits carcinogenesis in some cells. Genistein's low bioavailability as the result of poor absorption and efficient conjugation, limits these effects at normal dietary concentrations. However, there is still potential for therapeutic and adverse effects. In particular, the intestinal mucosa can be exposed concentrations that are expected to have significant immunomodulatory, predominantly immunosuppressive effects, which may have a role in the management of chronic enteritis.

REFERENCES

[1]Dixon RA, Ferreira D. **Phytochemistry** 2002;60:205. [2]Rolfe BG. **Biofactors** 1988;1:3. [3]Bell KM, et al. **N Z Vet J** 2006;54:103. [4]Court MH, Freeman LM. **Am J Vet Res** 2002;63:181. [5]Santell RC, et al. **J Nutr** 1997;127:263. [6]Morito K, et al. **BiolPharmBull** 2001;24:351. [7]Gutendorf B, Westendorf J. **Toxicology** 2001;166:79. [8]Diel P. **Toxicol Lett** 2002;127:217. [9]Hill L, et al. **Mol Med**;17:211. [10]Glass CK, Rosenfeld MG. **Genes Dev** 2000;14:121. [11]Geistlinger TR, et al **Chem Biol** 2004;11:273. [12]Dutertre M, Smith CL. **J Pharmacol Exp Ther** 2000;295:431. [13]Wong CW, et al. **Biochemistry (Mosc)** 2001;40:6756. [14]Diel P, et al. **Planta Med** 2001;67:510. [15]Cave NJ, et al. **N Z Vet J** 2007;55:113. [16]Cave NJ, et al. **J Anim Physiol Anim Nutr (Berl)** 2007;91:400. [17]Weigt C, et al. **Mol Cell Endocrinol**. 2012, Epub. [18]Velders M, et al. **FASEB J**. 2012, Epub. [19]McClain RM, et al. **Food Chem Toxicol** 2005;43:1461. [20]Schleipen B, et al. **Carcinogenesis**;32:1675. [21]Akiyama T, et al. **JBiolChem** 1987;262:5592. [22]Markovits J, et al. **Cancer Res** 1989;49:5111. [23]McCabe MJ, Jr., Orrenius S. **BiochemBiophysResCommun** 1993;194:944. [24]Finlay GJ, et al. **OncolRes** 1994;6:33. [25]Salti GI, et al. **EurJCancer** 2000;36:796. [26]Jamadar-Shroff V, et al. **Clin Cancer Res** 2009;15:1269. [27]Dobrydneva Y, et al. **JCardiovascPharmacol** 2002;40:399. [28]Ji ES, et al. **Sheng Li XueBao** 2004;56:466. [29]Doerge DR, Sheehan DM. **Environ Health Perspect** 2002;110 Suppl 3:349. [30]Divi RL, et al. **BiochemPharmacol** 1997;54:1087. [31]White HL, et al. **Am J Vet Res** 2004;65:586. [32]Cerundolo R, et al. **Am J Vet Res** 2004;65:592. [33]Sosic-Jurjevic B, et al. **Exp Biol Med**;235:590. [34]Yellayi S, et al. **Proc Natl Acad Sci USA** 2002;99:7616. [35]Cave NJ, et al. **Vet Immunol Immunopathol** 2007;117:42. [36]Cave NJ, **Am J Vet Res** 2004;65:1427. [37]Cave NJ, et al. **J Vet Pharmacol Ther** 2007;30:327.

EQUINE METABOLIC SYNDROME – NUTRITIONAL MANAGEMENT

Raymond J. Geor, BVSc, PhD, DACVIM (LAIM), DACVN (Honorary)
East Lansing, MI

INTRODUCTION

The term equine metabolic syndrome (EMS) has been used to describe the clustering of obesity (generalized and/or regional), insulin resistance (IR), and prior or current laminitis.[1] Diet appears to play an important role in the triggering of laminitis in horses or ponies with this phenotype, particularly the ingestion of pasture forage or other feeds that are high in nonstructural carbohydrates (simple sugars, starches, and/or fructans).[2,3] The associations between obesity, IR, diet and laminitis have focused attention on strategies that reduce the risk for laminitis, especially feeding management for weight loss and improvement in insulin sensitivity. This paper reviews dietary countermeasures to obesity, insulin resistance and hyperinsulinaemia in EMS animals.

WEIGHT LOSS IN OBESITY

Most cases of obesity reflect an imbalance between energy intake and expenditure – and, as in humans, 'eating less' and 'exercising more' are the key strategies to achieve a more ideal bodyweight and condition in horses and ponies. Important components in the development of a weight loss and management program include:

- Documentation of current body weight (BW), body condition score (BCS), and other morphometric measurements e.g. neck circumference, abdominal circumference. These baseline data are important for subsequent assessments of the effectiveness of any weight loss program.
- Owner/trainer recognition that the horse or pony is overweight or obese – different equestrian disciplines and breeds have adopted different accepted "norms" in body condition. Nonetheless, the effectiveness of any weight loss program is critically dependent upon the willingness of the owner or caregiver to comply with the plan. Achieving weight loss and maintaining a more ideal BW and BCS is a long term commitment.
- Evaluation of the current feeding program and housing – this will include a thorough evaluation of what feed is being provided (including supplementary feed, hay, pasture quality and time allowed for grazing) and in what quantities. Proximate analysis of feeds and forage will enable estimation of true digestible energy (DE) and nutrient intake. Note that some animals, when allowed free access to pasture, preserved forage or forage replacer feeds, may ingest up to 5% of BW per day.[4-6] Ration components (including haynets) should be weighed for accurate assessment of feed intake. Maintenance energy requirements for horses typically range between 30 and 35 kcal digestible energy (DE) per kg BW per day, i.e. 15.0 to 17.5 Mcal DE/day for a 500-kg horse – the low end of this range has been used to estimate maintenance needs of 'easy keeper' horses or ponies with a tendency to be overweight or obese. However, the amount and type of feed actual needed to maintain bodyweight varies between individuals and depends on several factors, including age, temperament, environmental conditions, level of physical activity and metabolic efficiency.
- Assessment of the weekly workload and soundness for exercise – how many hours per week is the horse or pony engaged in structured physical activity? There is a tendency for horse owners and caregivers to overestimate the DE needs of horses performing structured activities, with resultant overfeeding and weight gain. The impact of exercise level on DE requirements varies between animals and adjustments in feeding should be based on regular monitoring of BCS and BW. Information on current activity level and soundness for exercise will form the basis for development of recommendations for physical activity.
- Set realistic goals for weight loss and regularly monitor progress – there can be wide variation in the response to weight loss treatment programs. In some, there is a substantial loss of body weight and adiposity after 2-3 months of diet restriction and increased physical activity. In others, progress can be frustratingly slow and further adjustments to diet and the level of physical activity may be needed for satisfactory improvement. However, a target weight loss of up to 0.5% per week (relative to BW after the first week of dietary restriction) is reasonable based upon data from studies in ponies and other species. Bodyweight and BCS should be monitored regularly (i.e. every 2-4 weeks) under standardized conditions; note *that appropriate levels of weight loss may not always be accompanied by a detectable change in BCS in the first few months.*[7]
- Make all dietary changes gradually and avoid prolonged periods of feed withholding. Abrupt starvation in obese ponies, donkeys and miniature horses (especially pregnant animals) carries the risk of hyperlipemia and is not recommended. In addition, severely restricting access to forage may increase risk of gastric ulceration and stereotypic behaviors.
- Develop, and continually update, an appropriate *weight maintenance* program once the target weight and body condition have been achieved. This will include monthly assessment of BW and BCS to ensure that the feeding program is appropriate to the current level of physical activity and other environmental influences on energy requirements (e.g. ambient conditions).

Feeding Recommendations for Weight Loss

Caloric restriction is of paramount importance in the management of obese equids – creation of a state of negative energy balance is needed to achieve loss of BW. Several different dietary strategies can be applied depending on the present and desired BW. In some situations, the removal of calorie-dense manufactured feeds from the diet, restriction from access to 'lush' pasture, and an increase in structured physical activity may be sufficient to promote weight loss (e.g. overweight but not obese animals). For obese animals, however, more severe energy restriction may be needed for effective weight loss. The quality/nutritional value of the current forage should be evaluated, ideally by proximate analysis. Early maturity hay with a high leaf-to-stem ration should be replaced by later maturity hay with lower energy content. Forage-based, low-calorie feeds complete with vitamins and minerals are now available commercially. This type of feed offers convenience and may be used as a substitute for hay or fed as a component of the ration along with hay. This forage or substitute can be fed at 1.5% to 2.0% of BW depending on previous level of feeding and current BCS. When mature grass hays are fed, intake of vitamin E, vitamin, copper, zinc, selenium and other minerals may not meet requirements and provision of a vitamin-mineral supplement is therefore recommended. Many feed companies sell a low calorie 'ration balancer' feed for this purpose. In addition to vitamins and minerals, these products contain sources of high-quality protein and are usually designed to be fed in small quantities (e.g. 0.5 to 1.0 kg/day, fed as is or mixed with hay chop). Studies in humans and other species have indicated that maintaining adequate indispensable amino acid is important for mitigation of loss of lean (muscle) body mass during severe energy restriction.[8]

At least partial restriction of pasture grazing is recommended, especially on pastures that have abundant, rapidly growing forage. It is preferable to maintain turnout, either in a large dry-lot or at pasture with the horse wearing a 'grazing' muzzle (attached to a breakaway halter) that restricts but does not eliminate grazing. Longland et al.[6] have shown that grazing muzzles will significantly reduce the rate of dry matter (DM) intake on autumn pasture (sward height 8-15cm). Without muzzles ponies consumed an average of 0.08 BW in a 3 h period vs. 0.014 BW with muzzles. It is important to ensure that horses wearing grazing muzzles are able to consume adequate water. Some horses do not tolerate grazing muzzles, and they can alter herd dynamics. Alternate approaches that may restrict pasture forage intake include strip grazing behind other livestock; mowing and removal of clippings before grazing activity is permitted; and application of a deep layer of woodchips over a small paddock. It should be noted that simply restricting the time allowed for grazing may not be an effective strategy for weight loss. A study in obese pony mares reported no change in body weight when ponies were provided 12 h/day of access to pasture (during daytime or night time), perhaps due to increased forage consumption during the restricted grazing period.[9] Ponies have been observed to consume 40% of their daily DM intake during 3 hours of pasture turnout[10] and can consume up to 1% of BW as DM within 3 h of pasture turnout.[5]

A more severe reduction in feed (caloric intake) is indicated for more obese animals or when the above approaches prove ineffective. Grain and commercial, energy-dense feeds must be should be removed from the diet. Pasture intake should be severely limited or eliminated. Forage with low-to-medium energy content ($\leq$ 2 MCal per kg DM) should be the primary energy-providing component of the ration. In a study of obese ponies provided a free choice forage (chaff) diet during winter and summer, voluntary intake (DM basis) was about 2% of bodyweight and BCS was unchanged during the study period.[4] As a general guide, therefore, hay or hay substitute should initially be provided at no more than 1.5% of current BW per day (DM basis, with most hays around 90% DM), i.e. 3.7 kg for a 250 kg pony; 7.5 kg for a 500 kg horse. Horse owners/clients should be instructed to weigh the ration. If there has been minimal weight loss after 6-8 weeks, the feeding rate should be decreased to 1.25% BW and then to 1.0% BW. Providing lower amounts of forage (<1.0% BW, DM basis) as this may increase risk of hindgut dysfunction, gastric ulcers, and stereotypical behaviors (e.g. wood chewing). Substituting straw for up to 50% of the hay ration is one way to lower the energy density diet and maintain a reasonable level of DM intake. On average, the DE content of straw is lower than that of grass hay. Straw that is clean and with minimal cereal head should be selected; thoroughly shaking the straw prior to feeding will remove any loose cereal. As above, provide a ration balancer or vitamin-mineral-protein product along with forage. The ration should be divided into 3 or 4 feedings per day. Strategies to prolong feed intake time should also be considered, such as the use of hay nets with multiple small holes.

Monitoring Weight Loss

The BCS system is useful for estimation of subcutaneous fat mass in equids but may not be a sensitive indicator of weight loss, at least during the early phase. Dugdale et al.[4] recorded BCS, heart and belly girth and ultrasound-derived measures of subcutaneous fat depth in obese ponies during weight loss over a 12-week period. Mean BW decreased by approximately 12% and was associated with significant decreases in heart and belly girth measurements as well as rump and rib subcutaneous fat depth. However, a change in BCS was not detected. Similarly, neck circumference – a measure of neck crest adiposity – did not change during the study. Other studies of longer duration and with a greater magnitude of weight loss (18%) when compared to the study of Dugdale et al. (2010) have reported a significant decrease in BCS in association with weight loss.[11] It is therefore recommended that, in addition to BCS, heart and belly girth (and, if feasible, ultrasound fat depth) be recorded during a weight loss program. These measures should be recorded at monthly intervals during the weight loss program, and every 6-8 weeks during weight maintenance.

COUNTERING INSULIN RESISTANCE AND HYPERINSULINEMIA

In obese horses with EMS, weight loss can be accompanied by improvement in insulin sensitivity. However, IR and exaggerated insulin responses to consumption of feeds rich in nonstructural carbohydrates (NSC; starches, sugars and/or fructans) can persist after weight loss. Furthermore, some EMS animals at initial diagnosis are not obese but do have evidence of IR and abnormal insulin dynamics. Marked postprandial increases in circulating insulin are a concern due to the potential for high insulin concentrations to cause laminitis.[12,13] Restriction in dietary NSC is therefore important in the long-term management of EMS animals. Key approaches in this regard are: 1) elimination of grains and sweet feeds from the diet; 2) restricted access to pasture forage that, under certain conditions, can have high NSC content (~30-40% DM); and 3) the feeding of forage with low NSC content. Ideally, the results of proximate nutrient analysis, including direct measurement of starch and sugars, should be reviewed before selection of the hay. A NSC content of less than 10-12% DM is currently recommended. In the absence of data on hay NSC content, some nutritionists have recommended soaking hay in water for 30 to 60 min before feeding to leach water soluble carbohydrates (WSC; sugars and fructans). However, recent work has suggested that under typical management conditions this practice is variable in outcome and may not result in substantial change in the WSC content of some hays.[14] Soaking hay should therefore be used as an adjunct to choosing low NSC-containing hay. Although scientific evidence is not currently available to determine whether the 10-12% DM upper limit of NSC content is optimal, clinical experience suggests that this recommendation is reasonable. Caution is required in using the ethanol soluble carbohydrate (ESC) fraction as a measure of the simple sugar fraction of the feed in question (and based on this, the assumption that WSC – ESC = fructan content) as some smaller molecular weight fructans and other carbohydrate oligomers may also be soluble in 80% ethanol.[15]

As mentioned, some EMS animals are not obese and a ration of mostly hay may not meet DE requirements particularly when some weight gain is desired or the animal is competing in athletic events. One approach is to add non-molassed sugar beet pulp to the ration, e.g. 0.5 to 1.5 lb/day. Beet pulp is rich in highly-digestible fibres, provides more digestible energy when compared to most hay types, and does not elicit a marked glycemic or insulinemic response unless molasses is added at the time of processing. Beet pulp shreds should be soaked (in a volume of water 3- to 4-fold higher than that of the beet pulp) prior to feeding. The energy density of the ration also can be increased by feeding vegetable oil, for example mixed with sugar beet pulp shreds or with hay cubes that have been softened in water. Corn and soy oils are commonly used in equine rations, but need to be fresh, non-rancid and introduced gradually to the ration. One standard cup (about 225 ml or 210 g) of vegetable oil provides 1.7 Mcal of digestible energy. Depending on energy requirements, ½ to 1 cup of oil can be fed once or twice daily (up to a maximum of ~1 ml oil/kg bodyweight). Smaller amounts (e.g. ¼ cup once daily) should initially be fed, with a gradual increase over a 7-10 day period. Supplemental antioxidant (100-200 IU vitamin E per 100 ml of added oil) should be provided. Stabilized rice bran (~20% fat) is another option for increasing the energy density of the diet, providing the calcium:phosphorus ration of the final ration is considered.

Another approach to dietary management of the lean or working insulin resistant horse is to provide a commercial feed along with hay. Most feed companies now offer products with lower starch and sugar content (<20-25% NSC, DM basis) when compared to traditional sweet feeds (40-50% NSC) or cereal grains. Digestible fibers (sugar beet pulp and/or soy hulls) and vegetable oils are included in place of starch-rich ingredients and energy density is similar or even higher when compared to sweet feeds. In theory, these lower NSC feeds result in lower post-feeding glycemic and insulinemic responses and carry lower risk for disturbances in hindgut function associated with the rapid fermentation of starch and/or sugars. However, glycemic/insulinemic responses cannot be reliably predicted from the formulation (or assessment of NSC content) and actual measurement of post-feeding glucose and insulin concentrations in horses is currently required to verify that the feed is 'low-glycemic'.[16] These products are often marketed for use in horses and ponies at high risk for laminitis, including those with IR or PPID, although minimal published data is available to support or refute these medical claims. Nonetheless, these feeds are convenient for clients, the principle is sound, and providing these feeds does actually result in low post-feeding glucose and insulin responses which are likely to be useful.

DIETARY SUPPLEMENTS

A number of supplements are marketed with claims for improved insulin sensitivity or reduced risk of laminitis but evidence of efficacy is scant.[17] Many products contain magnesium, chromium and/or cinnamon. Chromium is thought to potentiate insulin action via activation of insulin receptor kinase and/or inhibition of insulin receptor tyrosine phosphatase.[18] Studies in people have shown that suboptimal intake of chromium (trivalent form) contributes to IR in type 2 diabetes and metabolic syndrome. Moreover, some but not all studies have shown improvement in glucose tolerance in insulin resistant human patients.[19] In one equine study, however, the feeding of a supplement containing 5 mg/day chromium (as yeast), 8.8 g/day magnesium and other unspecified nutraceuticals for 16 weeks did not alter morphometric measurements, resting serum glucose and insulin concentrations, or insulin sensitivity in obese horses with a history of laminitis.[20] Additional research is required to determine the efficacy of chromium supplements in the management of IR in horses. There also is interest in the effects of magnesium nutrition on insulin sensitivity and other aspects of metabolism. Some studies in humans have demonstrated an association between magnesium status and IR in non-insulin dependent diabetes mellitus and type 2 diabetes.[21] A review of randomized double-blind controlled trials that evaluated the effects of magnesium supplementation in patients with type 2 diabetes concluded that supplementation may be effective in reducing fasting plasma glucose concentrations and increasing HDL cholesterol but only in patients with actual magnesium deficiency which was determined by measurement of intra-erythrocytic magnesium concentrations.[21] There are no published data on the magnesium status of obese and/or insulin resistant horses or ponies. The aforementioned study by Chameroy et al.[20] reported that magnesium supplementation at 8.8 g/day (an amount which exceeds daily requirements) did not impact insulin sensitivity in obese laminitic horses.

Daily supplementation with 45 g short-chain fructo-oligosaccharide (scFOS) for 6 weeks resulted in a modest improvement in insulin sensitivity and decrease in resting serum insulin concentrations of obese Arabian horses fed a 50:50 grass hay and sweet feed diet.[22] The mechanism of this apparent improvement in insulin sensitivity is unknown although it warrants mention that scFOS supplementation has also been shown to improve insulin sensitivity and modulate the transcription of genes involved in adipose tissue glucose and lipid metabolism in obese dogs.[23]

REFERENCES

[1]Frank N, et al. **J Vet Intern Med** 2010;24:467. [2]Treiber KH, et al. **J Am Vet Med Assoc** 2006;228:1538. [3]Carter RA, et al. **Equine Vet J** 2009;40:171. [4]Dugdale AH, et al. **Vet J** 2011;190:329. [5]Longland AC, et al. **J Eq Vet Sci** 2011;31:275. [6]Longland AC, et al. **J Eq Vet Sci** 2011;31:282. [7]Dugdale AH, et al. **Equine Vet J** 2010;42:600. [8]Arathuzik GG, et al. **Curr Diab Rep** 2011;11:1106. [9]Buff PR, et al. **Vet Therap** 2007;7:64. [10]Ince JE, et al. **Brit Soc An Sci** 2005;109. [11]Van Weyenberg S, et al. **J An Nutr An Phys** 2008;92:538. [12]Asplin KE, et al. **Vet J** 2007;174:530. [13]De Laat MA, et al. **Equine Vet J** 2010;42:129. [14]Longland AC, et al. **Vet Rec** 2011;168:618. [15]Logland AC, et al. **J Sci Food Agric** 2011 (in press). [16]Harris PA, et al. **Vet Clin Nth Am Equine Pract** 200925:23. [17]Tinworth KD, et al. **Vet J** 2010;186:282. [18]Lau FC, et al. **Mol Cell Biochem** 2008;317:1. [19]Anderson RA. **Diab Metab** 2000;26:22. [20]Chameroy KA, et al. **Equine Vet J** 2011;43:494. [21]Song Y, et al. **Diab Med** 2006;23:1050. [22]Respondek F, et al. **J An Sci** 2011;89:77. [23]Respondek F, et al. **J Nutr** 2008;138:1712.

METABONOMIC STUDIES: POTENTIAL FOR IDENTIFICATION AND PREVENTION OF OCD IN FOALS

S.L. Ralston, VMD, PhD, DACVN[1], L. Pappalardo, PhD[2], I. Pelczer, PhD[3], P.F. Spears, MD[4], Y. Chen, PhD[5]
[1]New Brunswick, NJ; [2]Sharjah, United Arab Emirates; [3]Princeton, NJ; [4]Hanover, PA; [5]Durham, NC

INTRODUCTION

Metabonomic studies utilize nuclear magnetic resonance (NMR) of biological fluids and tissues to identify distinct metabolic profiles (metabonomes) associated with metabolic states, diseases, responses to toxins or drugs and genetic variations (Fonville et al., 2010, Nicholoson et al., 1999), using a "top down" approach. By use of principle component analysis (PCA), partial least squares with discriminant analyses (PLS-DA) and orthogonal PLS-DA (OPLS-DA) on the integral values of the peaks, metabolite components which vary systematically can be identified both between groups and over time (Pelczer, 2005; Beckonert et al., 2007). These discrete metabonomes permit identification of diagnostic metabolic markers and frequently provide novel understanding of the underlying disease processes (Trygg et al., 2007; Fonville et al., 2010, Keller et al., 2012). For example, metabonomic analyses of serum from insulin resistant humans (Zhang et al., 2009) and horses (Hodavance et al., 2007) revealed distinct metabolic differences when compared with individuals with normal insulin sensitivity, not only in glucose metabolites but also in lipoproteins, cholines and certain amino acids that had not been previously recognized as significant contributors to the syndrome. Metabonomic studies have been utilized with increasing frequency to follow changes with drug treatments and disease conditions in human medicine (Trygg et al., 2007; Fonville et al., 2010).

Abnormal calcification of cartilage matrices leading to fragmentation of articular cartilage, joint effusion and lameness (osteochondrosis dissecans (OCD)) is a growth-related metabolic disease in horses (McIlwraith, 2004). The radiographic lesions become apparent in young horses between 3 to 12 months of age, although clinical signs may not surface until the animals are put into training (McIlwraith, 2004). Defective cartilage development can be caused by improper nutrition (Savage et al., 1993a; 1993b) and trauma (NRC, 2007), however it has been well documented that especially tibiotarsal lesions are heritable (Hoppe and Phillipson, 1985; Piermattei et al., 1993; Lykkjen et al., 2011). The incidence of lesions in Standardbred horses was reported to be 10 and 26% in European Standardbred trotters (Hoppe and Phillipson, 1985; Grondahl, 1992) and was up to 43 and 58% in more recent USA Standardbred studies, (Kapraun et al., 2008; Hilliken et al., 2011). Altered gene expression in the chondrocytes of affected animals has been documented in genomic studies (Mirams et al., 2009), however the associated metabolic defects have not been defined. It was hypothesized that NMR-based metabonomic analyses would reveal discrete and consistent metabolic profiles associated with the development of hock OCD lesions in young Standardbred horses. It was further hypothesized that these differences would be present in 2- to 3-month-old Standardbred foals and would be predictive of development of lesions. A series of three experiments were conducted over 4 years to test these hypotheses.

MATERIALS AND METHODS

Experiments 1 and 2

To test the hypothesis that NMR-based metabonomic analyses would reveal discrete and consistent metabolic profiles associated with the development of hock OCD lesions blood samples were collected from yearling Standardbreds in 2007 (n = 40) and 2008 (n = 71). In 2007 the horses chosen for the study were paired on the basis of sire (9 different sires represented in 20 pairs) and dam's breeding. One member of each pair had undergone hock OCD surgery 2 to 9 months before the samples were taken (OCD); the other had had no clinical evidence of OCD (Control). Status of the control horses was based only on the absence of clinical signs and was not verified radiographically. In 2008 Standardbred yearlings again were used from the same farm. Fifty of the horses were paired as before, with one having had surgery to treat hock OCD lesions 1 to 9 months before the samples were taken (OCD) and their matched sibling (Control) had never exhibited clinical signs of hock OCD. The other horses were either half or full siblings to the yearlings used in Experiment 1. The absence of hock OCD lesions was radiographically verified in 22 of the Control horses. In both years all of the horses had been raised on the same farm. They had been maintained on pastures with additional free access to alfalfa hay plus group feeding of 2 to 3 kg of Purina Strategy® feed (Purina Mills, St. Lois, MO) per head per day since weaning the previous fall. They had been transferred to individual stalls at a training facility for sale preparation 2 weeks prior to sampling. At the training facility they were all fed 1.5 to 2 kg of Strategy® feed and 0 to 0.5 kg of Ultium® feed (Purina Mills) twice daily (0730h and 1600h) with free access to alfalfa hay, salt and water. In the 2007 study all blood samples were collected by venipuncture on a single day (September 21, 2007) within a 3 hour period (12:45–14:30). The serum was processed within 6 hours and stored at –80 °C pending NMR analysis. In 2008 blood samples were taken by venipuncture on August 21, 2008 (n = 41) and September 10, 2008 (n = 39). Serum and plasma were processed and stored at –80 °C pending the NMR, glucose and insulin analyses as described previously (Ralston et al., 2011).

Experiment 3

To test the hypothesis that the differences observed in yearlings in Experiments 1 and 2 would be present in 2- to 3-month-old Standardbred foals and would be predictive of development of lesions, 18 nursing foals (1 to 3 months of age) on the same farm were brought with their dams into individual box stalls 3 to 4 hours before blood samples were drawn by venipuncture on May 15, 2009. The mares and foals had free access to alfalfa hay in the stalls and no attempt was made to control nursing activity by the foals. Blood samples were drawn by venipuncture and sample handling was as in the previous experiments. All subsequent management and feeding practices were the same for these foals as in previous years. At 12 to 20 months of age the presence or absence of OCD lesions was documented either radiographically or by surgery in 15 of the horses. The other three foals had been sold before radiographs could be taken and follow up was not possible.

NMR data acquisition and statistical analyses

The techniques for NMR analysis have been described previously (Ralston et al., 2011). Briefly NMR experiments were conducted using a Bruker Avance-II spectrometer (Bruker-Biospin, Billerica, MA) set at 500 MHz and equipped with an inverse

TCI CryoProbe (Bruker-Biospin, Billerica, MA) at a controlled temperature of 25 °C. The water suppressed, relaxation filtered spectra were processed using MNova v.6.1.1 software (Mestrelab Research S.L., Santiago de Compostela, Spain). The ^{1}H-NMR spectra were referenced to the alpha-glucose anomeric proton resonance at δ 5.233 ppm. Phasing, baseline corrections and peak alignment were performed in MNova. Multivariate cluster analyses (SIMCA-P+, v.12.0.1, Umetrics, Umeå, Sweden) were performed using unsupervised principle component analysis (PCA), as well as supervised projection to latent structure discriminant analysis (PLS-DA) and orthogonal PLS-DA (OPLS-DA) on the spectral data. Validation was conducted with PLS-DA and OPLS-DA analyses to minimize spurious correlations (Trygg et al., 2007). Peaks identified as contributing ($P < 0.05$) to the validity of the models were identified based on their chemical shift position on the spectra, using previously published data (Nicholson et al., 1995). This was further validated by spiking experiments using alanine and valine to verify congruence with data from other species (Pelczer and Ralston, unpublished data).

RESULTS

Experiment 1

Alcohol (a mixture of ethanol and methanol) was identified as a contaminant in the majority of the samples from 2007, presumably from the alcohol swabs used to prepare the venepuncture site that year. These peaks were suppressed in the statistical analyses. This type of contamination has been observed in other studies (Lykkejen et al., 2010) and emphasizes the sensitivity of the NMR analyses.

The PLS-DA scores plot for the data revealed good class (Control vs OCD) separation and highly significant model validation. With only two components in the analysis, the R^2 was 0.662 and the Q^2 was 0.32. Duplicate aliquots were located close to each other in the plots, showing excellent reproducibility and there were no significant outliers. The OPLS-DA scores plots of the OCD horses versus the Controls showed even greater and more distinct separation. With only one component in the analysis, the R^2 was 0.82 and the Q^2 was 0.83.

Experiment 2

In horses sampled in September, insulin concentrations did not differ between OCD and control groups (OCD group: 11.1 ± 0.9 µIU/dL, n = 9; Control group: 17.0 ± 1.6 µIU/dL, n = 30; $P > 0.05$). Corresponding plasma glucose also did not differ between groups (OCD: 85.4 ± 2.6 mg/dL vs Control: 82.4 ± 3.9 mg/dL, $P > 0.5$). Insulin concentrations also did not differ between horses when grouped by date (August: 13.7 ± 0.7 µIU/dL; September: 15.6 ± 1.4 µIU/dL; $P > 0.2$).

There was distinct separation with strong predictability (R^2 (Y) = 0.64, $Q^2 = 0.75$) between the data sets for the August and September samples which, based on the loadings plots (not shown), was due primarily to higher sugar peaks and a variety of lipid peaks in the September samples that are under investigation.

There was no clear separation between the OCD and Control groups when the complete data set was used, presumably due to the strong confounding influence of the metabolic differences between the sampling days. However, when only data from animals with radiographically-verified OCD status (presence or absence) were analysed, regardless of the month of sampling, using only the peaks of metabolites that were significant for separation in Experiment 1, the OPLS-DA separation of the OCD and Control groups was clear ($R^2 = 0.98$) but had poor predictability ($Q^2 < 0.0$).

Experiment 3

Despite the relatively small number of samples (15 for which OCD status had been verified radiographically), separation was highly significant (OPLS-DA; R^2(Y) = 0.955, $Q^2 = 0.90$) between two distinct clusters. Only two individuals were "misidentified" by the clusters, with one yearling that had clustered with the controls having OCD, and the other grouped with the OCD had radiographs read as clean. The model was extremely well validated when the data for one individual was excluded. However, when that sample was included in the set identified as a control, the R^2(Y) and Q^2 were significantly reduced since the data point appeared in the center of the cluster of the OCD samples. When its status was changed to OCD, the R^2(Y) and Q^2 were again very high. As the individual in question had been sold by the time the apparent discrepancy was discovered, it was not possible to determine if there had been a misdiagnosis.

The loadings/coefficient plot for Experiment 3 OPLS-DA revealed that, as in previous years, concentrations of alpha glucose and other sugars were lower in the serum of horses with OCD than in the controls, whereas those of certain unsaturated lipids and amino acids, creatinine and choline were higher.

DISCUSSION

This series of experiments demonstrates that NMR-based metabonomic analyses can detect metabolic differences between groups of young horses that do or do not develop OCD. However, these analyses are sensitive to many different influences and it is very important that the resultant models are rigorously validated to eliminate extraneous correlations. In the "forced" OPLS-DA analyses, in which group identifiers are used a priori, spurious correlations can arise (Fonville et al., 2010) that give importance to metabolites that are not truly significant factors in the target disease metabonome but are implicated as important in the statistical model. For example, this was observed when the loadings plots for OPLS-DA analyses were compared between experiments. A strong urea peak observed in Experiment 1 loadings did not appear in Experiment 2 or 3 as significant factor in OCD versus control groupings. Also the strength of the distinct clustering ($R^2 = 0.98$) of the radiographically verified OCD versus Control yearlings in Experiment 2 was offset by the poor predictability ($Q^2 < 0.0$) of the model, probably due to the confounding factors associated with the two collection day differences, whereas there was good predictability in Experiments 1 and 3. Comparison of the loadings plots across the data sets will permit identification of metabolites that are consistently associated with OCD, hopefully eliminating spurious correlations. This is currently underway, using more sophisticated orthogonal signal correction (OSC) analyses that identify the metabolites most correlated with the condition of interest across data sets (Trygg et al., 2007; Fonville et al., 2010). Various manipulations of the data sets, such as suppression of peaks that are known contaminants (such as the alcohol peak in Experiment 1) or known to be non-contributing factors (such as water) also will enable greater refinement of the resultant OCD metabonome.

The validated distinct difference between the August and September groups of yearlings in Experiment 2 was surprising. There were no significant differences between these groups in age, management or sample handling, but the loadings plots revealed apparent differences in the peaks of certain sugars, lipids and urea between the two sample times that may have obscured the previously observed OCD vs Control metabonomic differences when the data sets were combined for statistical analysis. The origin of the differences between the two sample times is currently being investigated using chemometric analyses (Trygg et al., 2007; Fonville et al., 2010).

The three NMR experiments used relaxation filtered, water suppressed spectra. As plasma and serum contain large proportions of water and the peak for water is large, it can overwhelm and obscure the signals of interest (Hwang et al., 1995; Pelczer, 2005). If the water peak is not suppressed, the baselines of the spectra become distorted and can hide peaks produced from various metabolites. A simple water-suppressed experiment reflects the unbiased, full molecular content of the sample in a quantitative manner. The relaxation-filtered version combines water suppression with a pulse method; this induces controlled loss of signal intensity based on relaxation, which is differential and more efficient for detecting the large and slow-moving molecules in the mixture. As a result, the relaxation filtered spectrum highlights the small molecules, whereas large biopolymers, such as proteins, are highly suppressed (Pelczer, 2005). The direct quantitative information is lost, but variation in relative concentrations is consistent when the same experimental parameters are used for successive samples. In subsequent analyses of these data, we will also investigate the relative importance of large molecules.

CONCLUSIONS

NMR-based metabonomic analysis is a sensitive method for identifying differences between distinct groups of individuals. However, the statistical analyses are influenced by a variety of factors such as sample collection techniques, and spurious correlations caused by non-relevant variations need to be identified to maximize the utility of the resultant models. Young Standardbred horses that develop hock OCD differ metabolically from those that do not develop the disease despite similar management and genetic backgrounds. These metabolic differences can be detected using NMR-based metabonomic analysis before lesions appear in nursing foals. The apparent correlation with glucose metabolism and the appearance of OCD lesions in young horses has been previously reported (Ralston, 1996, Pagan et al., 2001) but the influence of amino acids, lipids, cholines and creatinine are novel and currently under further investigation. The better understanding of the underlying metabolic deficits associated with a predisposition to OCD will allow targeted formulation of nutritional supplements that may prevent or at least reduce the incidence in affected young horses.

Acknowledgments

The authors thank Hanover Shoe Farm for the samples and acknowledge the assistance and invaluable input from Mr. J. Simpson, Dr. B Jablonsky, Dr. J Petersen, Dr. C Pink and Mr. D Meirs III, and the assistance with laboratory and data analysis of Dr. L O'Byrne, and students C Shatynski, K Zeigler, B LaBarre, E Baez, J Bansal and S Cimbol. LP wishes to thank the American University of Sharjah for supporting her participation in this research.

REFERENCES

Beckonert O, et al. **Nature Prot** 2007: 2(11): 2692. Fonville JM, et al. **J Chemometrics** 2010: 24: 636. Grondahl AM. (1992). **J Eq Vet Sci** 1992:12(2): 81. Hilliken K, et al. **J Eq Vet Sci** 2011: 31 (5-6):236. Hodavance MS, et al**. Analytical Bioanalyt Chem** 2007: 387(2): 533. Hoppe F, et al (1985) **Eq Vet Prac**. 1985: 7: 7. Hwang TL, et al. **J Magnetic Resonance Series A** 1995:112:275. Kapraun PR, et al. **Proc Australian Eq Sci Soc Symp** 2008:2:38. Keller MD, et al. **Eq Vet J**. 2011: 43(6):737. Lykkejen S, et al. **International Fdn Anim Genetics**. 2010: 41(2):111. McIlwraith, CW. **J Eq Vet Sci**. 2004: 24(11), 475. Mirams M, et al, **J Orthopaedic Res**. 2009: 27(4):452. **National Research Council (NRC).** National Academy Press, Washington, DC.2007. Nicholson JK, et al**. Analyt Chem** 1995: 67:793. Nicholson JK, et al. **Xenobiota** 1999: 29:1181. Pagan JD, et al. **Proc Am Assoc Eq Prac.** 2001: 7: 322. Pelczer I, **Current Op Drug Res Dev**. 2005: 8(1):127–133. Phillipsson J, et al. **Eq Vet J**. 1993: 16: 38–41. Piermattei C, et al. **Livestock Prod Sci** 2003: 79: 249. Ralston SL. **Pferdeheilkunde** 1996:12(3):320. Ralston SL, et al. **Recent Adv Anim Nutr** 2011:18:197. Savage CJ, et al. **Eq Vet J Suppl.** 1993a: 16: 74. Savage CJ, et al. **Eq Vet J Suppl.** 1993b:16: 80. Tang H, et al, **Analyt Biochem** , 2 004:325:260. Trygg J, et al. **J Proteome Res** 2007: 6(2): 469. Zhang X, et al. **J Proteome Res** 2009: 8:5188.

FORAGE-BASED TOTAL MIXED RATIONS: AN ECONOMIC ALTERNATIVE?

Sarah Ralston, VMD, PhD, DACVN and Harlan Anderson, DVM
New Brunswick, NJ and Cokato, MN

INTRODUCTION

Horse rations should be based primarily on forages (NRC, 2007). The most recent recommendations are that horses receive at least 2% of their body weight in forage dry matter daily (Coenen et al, 2011) to optimize gastrointestinal health and well-being. Traditionally this has been accomplished by feeding dried long stem hay or maintaining horses on pastures. However, in many regions pastures are limited, if available at all. Adverse weather conditions in recent years have decimated hay crops, making good quality not only hard to find in many regions but also very expensive. For example, in January, 2012 in Texas 20 to 25 kg square bales of alfalfa and Coastal Bermuda grass, if available, were selling for $10 to $12 (USDA Market News, Jan 27, 2012) whereas the same size bales were going for <$5.00 in northern states that were not affected. Commercial grain mix prices have also increased to over $15.00 for an average quality 25 kg bag, with premium products exceeding $20/bag. Horse owners in densely populated, weather challenged areas are finding it increasingly difficult to provide adequate, consistent forage-based rations. Most horse feed companies offer "complete" pelleted or textured mixes of grains and fiber sources that are designed to be fed without forage. However, these products meet or exceed nutrient needs at <1.5% BW daily intake. Limit feeding horses to this degree to avoid obesity is associated with an increase in not only behavioral vices such as wood chewing (NRC, 2007) but also gastric ulceration (Flores et al, 2009, Warren et al, 2011).

Forage-based total mixed rations (TMR) are made from mixtures of chopped and/or processed hays with minimal additives. They can be formulated to meet or exceed the nutritional needs of any class of animal. The TMR rations are stored in 20 to 22 kg bags for ease of handling and storage. The advantages of TMRs over traditional long stem hay/concentrate rations are the ease of handling, reduced waste, ability to completely control/regulate nutrient intake and improved nutrient availability. They are designed to be offered at a minimum of 2% BW daily, and preferably free choice. Another benefit is that TMR formulations obviate the need for supplements of any sort, being designed to be the sole source of nutrition aside from water and salt. Horse owners often have no knowledge of how to evaluate hay quality and tend to add concentrates and supplements on a purely empirical basis. This frequently results in over- or underfeeding nutrients, especially nitrogen and protein. Over feeding nitrogen (protein) and phosphorus are impending environmental concerns (Williams et al, 2011, Westendorf et al, 2011).

TMR TRIALS

The safety and efficacy of TMRs with or without added grain has been evaluated for over 7 years in young horses (Ralston, 2011). In a series of trials draft cross weanlings and yearlings (n=6 per year) were paired by sex and type and fed either 3% BW TMR cubes (basically free choice) or a hay/grain (HG) ration in which the concentrates provided 50% of the calories recommended for growth (NRC, 1989) with ad libitum grass/alfalfa mix hay supplying the remainder of the ration. The trials were all done in a crossover design. Feed intakes were recorded daily and height and weight were measured weekly. An experienced observer blind to the dietary treatments evaluated the horses before and at the end of each trial for clinical signs of developmental orthopedic disease (DOD). A low dose oral dextrose challenge test (0.25 gm dextrose/kg BW, Ralston, 2002) was used to evaluate insulin sensitivity before and after each trial, as was the glycemic index of each of the feeds. In the first two trials (2004, 2005) the TMR cubes had 10-15% cracked corn added to the formulation, in the third (2006) ground oats were used instead of corn. In 2007-2010 grain-free forage-based TMR rations were compared to the traditional hay/grain (HG) rations using draft crosses and BLM mustangs yearlings and two year olds.

Feed efficiency (Kg gain/Mcal DE consumed) was higher ($P<0.05$) in the TMR fed horses relative to the HG fed horses in all trials(Ralston, 2011). Growth rates tended ($P<0.1$) to be higher with TMR than HG and were higher (1.2 kg/day as weanlings in the fall, 0.8kg/day as yearlings in the spring) than predicted by NRC (1989) in virtually every trial. Weanling voluntary intakes were between 2.5 and 3.2% BW in the fall trials, which in spring decreased to 2.2 to 2.5% BW on both treatments (Ralston et al, 2007b, 2011). Incidence of DOD was low (0 to 2 per year) and not affected by ration. Plasma glucose responses to the dextrose challenges did not differ between treatments in the first two years but insulin responses tended ($P<0.1$) to be higher in HG fed horses,(Ralston et al, 2007a,c) suggesting somewhat reduced insulin sensitivity. There were no feed-related adverse events observed with the TMRs. Reactivity to stimuli and trainability were not affected by the addition of grain to the TMRs (Ralston, unpublished data, 2010)

The results obtained with the draft crosses and mustangs were corroborated in a study at another university, using Standardbred yearlings fed either TMR cubes with 25% oats at the rate of 3% BW/day versus equivalent amounts of hay cubes and oats fed separately as discrete meals. The same hay was used in both types of cubes. The horses on the TMR rations again had higher ($P<0.05$) growth rates (1.69 kg gain/d) and gain to feed ratio (0.9kg/kg feed intake) than the meal fed animals (0.95kg gain/day, 0.05 kg gain/kg feed). No gastric ulcerations were found on endoscopic examination in either treatment group (Warren et al, 2011).

In 2009 it cost an average of $4.26/day for the TMR ration versus $4.18/day for HG. However the HG horses also wasted 0.8 to 1 kg of hay per day, incurring an average loss of about $0.30/day in wasted hay that should be added to the daily cost. There was no waste with the cubes. The cost differential between the two rations is further reduced if labor costs are taken into consideration. It also was observed that it took less time to feed the TMR and that the bags were easier to store and handle than hay bales.

CONCLUSIONS

Feeding TMR cubes formulated for growth with or without added grain concentrate free choice is an efficient, economic alternative to traditional HG rations under the conditions of these trials. This type of ration may be especially appropriate for young horses that are predisposed to DOD. Anecdotal reports of use of TMR rations formulated for adult horses have been equally encouraging. The use of bagged TMR rations that can easily be transported and stored worldwide may be the most economic and practical answer for horse owners in tough economic and environmental conditions.

REFERENCES

Flores RS, et al. **Proc 21rst Eq Sci Soc** 2009:152. National Research Council (NRC). 1989. **Natl Acad Press**. National Research Council (NRC). 2007. **Natl Acad Press**. Ralston SL. **Vet Clin NA:Eq**. 2002:18:295. Ralston SL, et al. **Am Soc Anim Sci** 2007:61. Ralston SL, **Proc 20th Eq Sci Soc Symp**.2007b:235. Ralston SL, et al. ***Appl Eq Nutr Train***. Lindner A (ed), 2007c:233. Ralston SL, et al. **J Eq Vet Sci** 2009:29:386. **Ralston, S.L.**, ***Appl Eq Nutr Train***. Lindner A (ed), 2011:185. Warren SN, et al. J Eq Vet Sci2011:31(5-6):330. Williams CA, et al J Eq Vet Sci 2011:31(5-6):305. Westendorf ML, et al J Eq Vet Sci 2011:31(5-6):306.

197a

ANTI-INFLAMMATORY DRUGS DECREASE THE RATE OF ENDOTHELIAL CELL INFECTION WITH EHV-1 *IN VITRO*. LS Goehring, KM Brandes, L Wittenburg, F Olea-Popelka, G Soboll Hussey. College of Veterinary Medicine and Biomedical Sciences, Colorado State University, Fort Collins, CO.

Central nervous system (CNS) endothelial cell (EC) infection with EHV-1 is most likely the initiating cause of EHV-1 Myeloencephalopathy (EHM). Cell-associated viremia in peripheral blood mononuclear cells (PBMC) transports virus to the CNS EC; however, EC infection likely requires a cell-to-cell contact between EC and PBMC. Intercellular contact is thought to be facilitated by inducible adhesion molecules. We hypothesize that anti-inflammatory drugs will decrease the interaction between PBMC and EC, and, therefore, will decrease EC infection.

Carotid artery EC monolayers and *in vitro* EHV-1 infected (strain Ab4; MOI 0.01) PBMC were incubated (24 hrs) with either media alone (NTx), or with equine therapeutic plasma concentrations (1xTx) or 10x concentrations (10xTx) of one of the following drugs: flunixin meglumine, firocoxib or dexamethasone. A virus suspension (EHV-1 Ab4; 100 PFU/mL) without or NTx/Tx PBMC in media containing EHV-1 neutralizing (VN) antibody (titer 1: 200) were added to monolayers for 4 hours. Then, monolayers were washed 3x times, and media NTx, 1xTx or 10xTx -various drugs- were re-applied for additional 48 hrs. All media contained a VN antibody titer of 1:200. Monolayers were stained with crystal violet solution to allow a plaque count. Generalized linear and latent mixed models were used to evaluate differences in plaque counts. Statistical significance was assumed with a p-value of ≤ 0.05.

In the cell contact model all drugs significantly decreased the plaque count at 1xTx and 10xTx ($p<0.001$); however, there was no statistical difference between the effects of 1xTx and 10xTx ($p=0.163$) on plaque reduction. With direct virus inoculation there was no statistical difference between groups.

These results provide a rationale for the use of anti-inflammatory drugs during early phases of EHV-1 infection. Moreover, the use of anti-inflammatory drugs during viremia may aid in preventing EC infection *in vivo*.

THE HUMAN FACTOR TO FEEDING PETS

Nick Cave, BVSc, MVSc, PhD, DACVN
Palmerston North, New Zealand

INTRODUCTION

By far the greatest problem associated with feeding pets in the developed world is obesity. Obesity has been simply defined as an excessive body fat mass. A more complete, if cumbersome definition is that obesity is "an excess of body fat that results in disease or reduced life span". Numerous studies have documented the adverse effects of obesity in multiple species, but it is not known at what percentage of body fat mass, for a given individual, a clinically significant adverse effect will occur. Current statements of "ideal" bodyweights, or body condition scores (BSC) are not firmly based on evidence. In fact if longevity is seen as an appropriate measure of the effects of bodyfat mass, then an ideal BCS is likely to be considerably less than the currently touted 5/9. These are not idle questions, since success rates for weight management remains relatively poor, and the prevalence of obesity remains extremely high. So objectively identifying those animals that are truly adversely affected by their fat mass may be important for motivating clients, and focusing veterinary efforts. Obesity occurs when energy intake exceeds expenditure. Although this is often termed a simplistic statement, it remains true, and however complex the metabolic derangements associated with obesity may be, the quintessence remains that an owned animal that is obese, has been allowed access to an amount of food that is excessive. At this time, too little emphasis has been placed on the human variables that influence the development of obesity and its management, and some recent studies have illustrated the need for a much deeper understanding.

RISK FACTORS FOR OBESITY

Several studies have identified different risk factors, which differ between studies such that opposite results have been reported. For instance, the provision of ad lib feeding has been shown to increase the risk of feline obesity in some studies, whilst in others it was either not significant or protective.[1,2] Similarly, the feeding of scraps to dogs was shown to be associated with a decreased risk if fed frequently in one study, whilst an increased risk if fed infrequently.[3]However, it appears true that no particular feeding practice is a strong risk factor for cats, whilst dogs are less accurate at regulating intake, and ad lib feeding cannot be considered a sensible approach. It is tempting to suggest that this species difference relates to their ancestral dietary approaches, whereby wild canids will often consume large meals infrequently, whilst small felids eat small meals often throughout the day. For both cats and dogs, the most consistent risk factors for obesity other than age, are being neutered, low activity, and the owners underestimation of their pet's body condition. As mentioned, there are conflicting results between studies, caused mostly by differences in sampling strategies, survey questions, geography, and sample populations. None-the-less, when specifically asked, there appears to be a strong association between owner underestimation of body condition and obesity.

Low activity as a risk seems intuitive: reduced activity decreases energy expenditure and therefore if food intake remains the same, weight gain will result. However, this statement requires that food intake remains the same for it to be true. Food intake is surprisingly accurately matched to fluctuations in energy expenditure in normal animals, thus for inactivity to lead to obesity, there still needs to be a dysregulation in the control of food intake. Alternatively, for activity to prevent obesity, there has to be an effect of activity on the regulation of food intake. Although not studied mechanistically in either dogs or cats, it is likely that inactivity is associated with inappropriately high intake, and that a certain amount of activity leads to a reduction in intake relative to the increased requirements, and thus it appears that sustained, "regular activity" decreases the risk of obesity. The amount and type of activity necessary has not been defined.

A reasonable hypothesis for an increase in the prevalence of obesity in pets is that highly palatable, energy dense dry food diets have lead to an inappropriate energy intake by pets. The energy density of the edible portion of a whole mouse, whole rabbit, or whole bird is between 5 and 10 kJ/g as fed, whilst most premium dry diets are around 16-18 kJ/g. This is certainly consistent with the hypothesis. However, in several studies, home prepared diets, or canned diets have not been associated with a decreased prevalence of obesity.[4] What does rapidly result in obesity, at least in rodents, is the provision of a variety of foods (cafeteria diets). A similar effect is predicted to occur in cats and dogs, especially when associated with variations in energy density (see below).

THE CONTROL OF FOOD INTAKE

Appropriate food intake requires control of both meal initiation (signalled by hunger) and meal cessation (signalled by satiety). In most species, oestrogen is a key negative regulator of food intake that functions by increasing the sensitivity to CCK released during a meal, and leading to satiety.[5] Gonadectomy of both male and female cats results in a significant and rapid increase in food intake, which is reversible with exogenous estrogen, and the effect is very likely to occur in bitches as well.[6] With a sustained increase in food intake as body fat mass (BFM) increases, more of the hormone leptin is released from adipocytes, which is supposed to function as a controller of food intake promoting early meal cessation. However, this supposes that mammals have evolved control mechanisms to avoid obesity. Rather it is likely that leptin functions less as an inhibitor of food intake, but rather when BFM decreases, the decreased leptin acts as a facilitator of increased hunger to restore a healthy BFM.

Most animals can regulate food intake appropriately within a wide range of energy densities, and a high energy density per se is not predicted to lead to obesity. However, when changing from one to another energy density, there is much better compensation when changing to a low density food, than when changing to a high density food. Thus when switched from a high to low energy density food, a cat or dog might lose weight briefly, then adapt and put it back on to restore its original mass. However, when switched from a low to a high energy density diet, there is a period of over-intake, weight gain, followed by a normalisation in energy intake, but *not a compensatory decrease to lose the weight gained.*[7] This emphasizes that most mammals have evolved in an environment where access to nutrients is limiting, and the evolutionary drive is to be efficient at securing food.

There has not been evolutionary pressure to protect against obesity, thus there is much more accurate defense against energy deficiency than energy excess. This probably explains the effect of feeding scraps to dogs.

Because modern pet foods are frequently erroneously blamed for increasing the prevalence of obesity, the common perception is that obesity prevention can be solved through formulation of a magical, "ideal diet". As an animal eats "the ideal diet", it would receive satiety signals at exactly the right point in a meal to meet its energy requirements for that period. Certainly some dogs consuming a high fiber and high protein diet will consume less energy than when fed either high protein or high fiber alone, despite comparable palabablities.[8] However, long term trials that have allowed true ad libitum access to food have not been conducted. Thus no diet, including the large number of diets marketed as "low energy density", has been shown to reduce the risk of obesity.

THE OWNER FACTOR

Neutering cats and dogs disturbs the normal control of food intake, and no diet at present is magically capable of providing that control. However, since in most pets, the only food that they consume is provided by their owners, obesity should be completely preventable. Owners ultimately have control over the intake of their pet, and thus our efforts should be directed towards assisting owners to feed the correct amount. It really is that simple. Feeding guides provided by pet food companies are the bane of their existence. A good manufacturer is expected to provide a feeding guide with their food but consider the challenges: for all 20kg dogs, their resting energy requirements (RER) varies by at least ±25%, and the energy requirement for "normal" activity varies such that the maintenance energy requirements (MER) ranges from 1.1 – 2x RER. This excludes working or sporting dogs. This means that at the lowest, there are 20kg dogs that require 1800kJ, and dogs at the same weight that require 4358kJ per day. Low = $[20kg^{0.67} \times 220] \times 1.1 = 1800$ kJ / day = 111 g of Royal Canin Medium Adult dry (16.2 kJ/g). High = $[20kg^{0.67} \times 366] \times 1.6 = 4358$ kJ / day = 336 g Royal Canin Medium Adult dry. Although the majority of dogs lie within a few percentages of the mean, it is sobering to remember the consequences of an animal consuming 5% extra than the MER. For a 20kg dog with an MER of 3052kJ, that equates to 152kJ per day, which will accrue body fat at a rate of approximately 5g/day, or 1.8kg/year, or 9kg in 5 years.

Even if an owner knew the exact requirements of their pet, if they use a measuring cup to measure the food, they may vary inadvertently measure as much as 80% more than intended.[9] To make matters worse, the amount owners feed is influenced by the size of the food bowl and scoop used to dispense dry food.[10] So, the exact amount of food an animal needs is not known, animals of the same weight can vary enormously in their requirements, a neutered animal is unable to regulate its food intake normally, and there is great inaccuracy in the normal methods of food allocation. Thus, the emphasis from veterinarians must shift from the dietary factors, to feeding to an "ideal" body condition, or to maintain an established ideal weight.

If an owner does not perceive their cat as overweight, he/she will not be motivated to reduce their cat's calorie intake. Unfortunately, owners are unreliable in assessing body condition. Underestimation of the body condition of their pet by owners, is a common risk factor for obesity, and was the strongest explanatory independent risk factor for obesity in cats in a study in New Zealand, and has been confirmed in Scotland and France.[2,4,11] Owners with overweight dogs underestimate their dog's BCS, although as the BCS increases, the odds of underestimation decrease (Courcier et al).[12] In the UK, owners underscored their dogs when judged overweight, but better concordance when very obese.[13] In that study, it was also found that in most cases, the dog's obesity had been discussed with a veterinarian, but they still underscored their dog. The use of a body condition scoring chart, such as that developed by Laflamme (1997), may be a useful tool in establishing a body condition benchmark.[14] However, one study has suggested that even when owners of overweight cats are shown silhouettes of cats ranging from emaciated to obese, many chose the silhouette of a cat that was considerably thinner than that of an objective independent assessor.[15] This suggests that a number of psychological approaches may need to be in place to change an owner's perception of their pet's body condition.

In the Netherlands, there is an association between canine and human obesity, but not feline obesity, that appears to be explained by the amount of time spent walking with the dog.[16] A similar positive relationship between the weight of dogs and their owners has been demonstrated in other European countries.[3,17] It seems highly likely that this finding can be extrapolated across cultures. Socioeconomic class seems to be an important factor in the development of canine obesity. The risk of obesity in dogs is also associated with owner income in the UK, whereby those in the highest income bracket are much less likely to have obese dogs.[18]

An owner's perception of their pet's body condition will be influenced not just by their understanding of what constitutes "normal", but also what is "healthy". An owner is unlikely to ascribe the definition of "obese" to their pet if they do not associate that with disease. In a recent phone survey of pet owners in the US and Australia, 356 of 1,104 (32%) of owners in USA and AUS reported their pet was overweight or obese.[19] However, *only three* of those owners (0.8%) believed it was a health problem. Owners of obese dogs more likely to believe exercise and balanced dog nutrition were less important than were owners of normal weight dogs.[3] Social contagion is likely to be an important factor for the appearance of obesity as a contagious, infectious disease in humans when viewed from an epidemiological pandemic. Similarly, owner perceptions, and perhaps veterinary perceptions, might be shifting towards acceptance of obesity as a normal state. Thus educational strategies to address owner perception must educate owners on what is normal, and what is unhealthy.

Weak intentions to feed appropriately are associated with pets being overweight. Owners who believe they lack knowledge of appropriate feeding behaviour and that feeding appropriately is unimportant are less likely to intend to feed their dogs appropriately, whereas owners who believe they have control over feeding appropriately are more likely to hold strong intentions to do so.[20] The more owners value exercise, the more likely they are to hold strong intentions to exercise their dogs, whereas the more owners think that exercising their dogs is hampered by barriers such as time, dislike of exercise, and little access to appropriate areas, the less likely their intent to exercise their dogs appropriately is. To increase an owner's motivation for exercise, it is important not just to increase their awareness of the health benefits, but to assist them in overcoming the perceived barriers. Improving feeding and exercising behaviours in owners requires modification of both their beliefs and intentions.

In an online survey of more than 30,000 cat owners in New Zealand, it was found that there were significant associations between how an owner perceives their cat, what type of information was important to them, and to what food they routinely fed. In addition, an owner's description of how fussy their cat is, was strongly correlated with their choice of food. Owners for whom objective, veterinary-sourced information was important, and who had a strong attachment with their pet were more likely to feed premium diets purchased at a veterinary clinic. Owners with a strong attachment, but for whom the perception of what their cat liked most was more important than objective information, were more likely to feed multiple flavours of canned food, and to describe their cat as fussy. Although these latter clients were still highly motivated and attached to their cat, they are probably less likely to adhere to strict dietary recommendations, and they need to be motivated in very different ways to other clients.

THE VETERINARIAN FACTOR

When veterinarians were recently surveyed in Australia as to what they considered the "cause of obesity" in dogs was, 36% replied dietary factors (e.g. overfeeding 12%), 21% cited insufficient exercise, 34% said it was human related. Of those that cited "human factors, 11% related the factor to exercise such as the sedentary or overly busy lifestyle of owners, but only 3% thought the owner perception of ideal weight was an important factor.[21] When asked what their preferred management strategy for obesity was, the most popular were: 1) instruct owner to reduce food, 2) reduce treats 3) change to a low energy diet, 4) natural remedies, 5) obesity clinic. Owners saw veterinarians as the preferred source of help for weight-loss help by owners. However it seems that the majority of animal health professionals recommend simple nutritional management and nothing else. It is now well known that successful prevention and management of human obesity requires multifaceted management strategies that change the environment in a way that causes sustainable behavioural changes in diet and exercise habits.[22] It is clear that we need a similar approach.

CONCLUSION

A simple consult-room observation that an animal is overweight and advice to reduce its intake will not lead to weight loss. Offering low energy dense foods and generic information is equally unlikely to prevent or manage obesity. We need to make owners believe that obesity is a health issue, and that healthy feeding and appropriate activity requires informed, active engagement by them. We need to understand that an owner's current feeding practice gives us a clue as to which, multifaceted weight management strategy is most likely to be successful. One size does not fit all.

REFERENCES

[1]Russell K, et al. **J Small Anim Pract** 2000;41:12. [2]Colliard L, et al. **J Feline Med Surg** 2009;11:135. [3]Kienzle E, et al. **J Nutr** 1998;128:2779S. [4]Allan FJ, et al. **Prev Vet Med** 2000;46:183. [5]Geary N. **Peptides** 2001;22:1251. [6]Cave NJ, et al. **N Z Vet J** 2007;55:113. [7]Backus RC, et al. **Br J Nutr** 2007:1. [8]Weber M, et al. **J Vet Intern Med** 2007;21:1203. [9]German AJ, et al. **J Anim Physiol Anim Nutr (Berl)**;2011; 95:368. [10]Murphy M, et al. **J Anim Physiol Anim Nutr (Berl)**. 2012, Epub. [11]Courcier EA, et al. **J Feline Med Surg** 2011;12:746. [12]Courcier EA, et al. **Prev Vet Med** 2012;102:66. [13]White GA, et al. **J Small Anim Pract**;52:622. [14]Laflamme D. **Feline Practice** 1997;25:1. [15]Kienzle E, Bergler R. **J Nut** 2006;136:1947S. [16]Nijland ML, et al. **Public Health Nutr**;13:102. [17]Mason E. **Vet Rec** 1970;86:612. [18]Courcier EA, et al. J Small Anim Pract 2011;51:362. [19]Laflamme DP, et al. **J Am Vet Med Assoc** 2008;232:687. [20]Rohlf VI, et al. **J Appl Anim Welf Sci**;13:213. [21]Bland IM, et al. **Prev Vet Med**;94:310. [22]Wolf AM, Woodworth KA. **Am J Med** 2009;122:S19.

EQUINE METABOLIC SYNDROME – NEW INSIGHTS ON PATHOPHYSIOLOGY

Raymond J. Geor, BVSc, PhD, DACVIM (LAIM), DACVN (Honorary)
East Lansing, MI

INTRODUCTION

Over the last 15-20 years there has been an epidemic surge in the incidence of obesity in human populations.[1,2] This trend has been triggered by a lifestyle that emphasizes overeating and inadequate physical activity. Similarly, the incidence of obesity appears to be on the rise in horses and pony populations, potentially associated with modern husbandry practices that are characterized by limited physical activity and over-feeding. In both species, there is evidence that obesity can be accompanied by a number of metabolic and endocrine disturbances that increase risk of disease – cardiovascular disease and type 2 diabetes in humans; laminitis in horses and ponies.[3,4] The term *metabolic syndrome* has been used to describe the clustering of these metabolic abnormalities that, together, signal increased risk of disease. This paper briefly reviews information on the pathophysiology of equine metabolic syndrome (EMS), with a focus on the potential mechanisms of EMS-associated laminitis.

EQUINE METABOLIC SYNDROME PHENOTYPE

The EMS phenotype is characterized by increased adiposity (generalized obesity and/or regional fat accumulations), insulin resistance (IR) with associated abnormal glucose and insulin dynamics, and predisposition to laminitis.[5] In an inbred herd of Welsh and Dartmoor ponies, the clustering of hyperinsulinemia, obesity and hypertriglyceridemia was associated with predisposition to pasture-associated laminitis.[6] This metabolic phenotype predicted 11 of 13 cases of clinical laminitis observed in May of the same year, with an odds ratio of 10.4, i.e. the EMS ponies were at approximately 10-times higher risk for development of laminitis. A subsequent study of this population of ponies demonstrated that the presence of obesity, hyperinsulinemia (insulin >32 mU/L when sampled on winter pasture) or hyperleptinemia (>7.3 ng/ml) were useful predictors of laminitis episodes when ponies were exposed to high carbohydrate pasture.[7] A study of out-bred ponies in the UK confirmed the association between IR and predisposition to pasture laminitis, and provided evidence of hypertension in the high risk ponies.[8] Interestingly, signs of this metabolic syndrome (i.e. IR and hypertension) were evident in summer but not winter, suggesting environmental factors such as consumption of summer pasture forage affected expression of the phenotype. Ongoing studies that involve collaboration between researchers at the University of Minnesota, Michigan State University and Tufts University have a goal of characterizing the EMS phenotype in a larger, more breed-diverse population (~300 horses and ponies with suspected EMS). Preliminary results are that baseline insulin and triglyceride concentrations as well as insulin values 75 min after a modified oral glucose challenge are good predictors of EMS. On the other hand, subjective (body condition score) and objective (girth and neck circumference measurements) assessments of obesity are less useful for identifying EMS animals. Across several breeds, dysregulation of insulin dynamics – manifest as high basal insulin concentrations and/or an exaggerated increase in insulin post-oral glucose challenge – appears to be a central feature of EMS.

EMS-ASSOCIATED LAMINITIS

Recognition that horses and ponies with the metabolic syndrome phenotype are prone to repeated episodes of laminitis has stimulated much discussion and some research on possible mechanisms underlying laminitis in these animals. Clinical experience with EMS animals suggests that episodes of laminitis often occur in pastured animals coincident with an increase in the pasture forage nonstructural carbohydrate (NSC, which includes sugars, fructans and starches) content, i.e. rapidly growing pastures in the spring, after summer and fall rains, or after frost. The association between grazing NSC-rich pasture and incident laminitis suggests the involvement of hindgut carbohydrate overload, with one or more factors associated with EMS potentially lowering the threshold for the development of laminitis in the face of carbohydrate challenge. An alternate theory is that laminitis in EMS animals arises due to the direct effects of hyperinsulinaemia (*'insulin-induced laminitis'*). High serum insulin is sometimes detected in EMS animals when laminitis is diagnosed, and this increase in circulating insulin may be highly relevant to the development of laminitis because the condition has been experimentally induced in healthy horses and ponies by infusing insulin intravenously.[9,10] Both theories are discussed in more detail below.

Intestinal Carbohydrate Overload-Associated Laminitis

Much of our current knowledge of the pathogenesis of laminitis has been extrapolated from experimental models of alimentary carbohydrate overload that involve the bolus administration of a mixture of cornstarch and wood flour (~17 g/kg bwt) or oligofructose (5-12 g/kg bwt).[11,12] In both situations there is marked disturbance to the hindgut environment secondary to rapid fermentation of the carbohydrate, including death of gram-negative bacteria, proliferation of gram-positive bacteria, a sharp increase in hindgut acidity (with pH values as low as 4.0), death and lysis of large numbers of bacteria and release of endotoxins, exotoxins and other microbial components, and damage to hindgut epithelial cells with a resultant increase in intestinal permeability.[13,14] The increase in intestinal permeability presumably favors absorption of exotoxins, endotoxins, other bacterial components and vasoactive amines into the circulation. There also is clinical and laboratory evidence of a systemic inflammatory response[15] and it has been argued that these models are germane to sepsis- and pasture-associated laminitis.[11,12]

One or more factors associated with the EMS phenotype could enhance susceptibility to laminitis under conditions of alimentary carbohydrate overload. One possibility is that a pro-inflammatory state in EMS amplifies lamellar inflammation and destruction associated with carbohydrate overload. Similar to findings in other species,[16] there is some evidence linking obesity with inflammation. Blood TNF-α and IL-1β mRNA expression was higher in obese horses and identified as an independent risk for IR,[17] while increased serum TNF-α concentrations have been observed in a well characterized population of laminitis-prone ponies.[18] Given the possibility that inflammatory mechanisms are central to the development of carbohydrate overload-induced laminitis, it is conceivable that EMS animals are more susceptible because of their pre-existing pro-inflammatory state (or that quantity of carbohydrate required to trigger events that result in clinical laminitis may be lower in EMS vs. normal animals). However, recent studies by our group have not supported this hypothesis. Groups of lean and obese, insulin resistant ponies were fed a low or high NSC diet for 7 days prior to harvest of lamellar and adipose tissues for evaluation of inflammatory events. The high NSC challenge mimicked exposure to spring pasture. Although some of the obese, insulin resistant ponies fed the high NSC diet developed clinical laminitis toward the end of the feeding period, there was no evidence of induction or exacerbation of inflammatory signaling in adipose or lamellar tissues (unpublished data). These findings suggest that other mechanisms contributed to development of laminitis under these carbohydrate challenge conditions.

Vascular endothelial dysfunction associated with IR is another factor that could alter susceptibility to laminitis after carbohydrate challenge. Insulin is a vasoregulatory hormone, invoking vasodilatation through pathways similar to those of insulin-mediated glucose metabolism.[19] In insulin resistant states, insulin's ability to counteract endothelin-1 associated vasoconstriction may be compromised due to decreased nitric oxide synthesis, while compensatory hyperinsulinemia might stimulate increased endothelin-1 production.[19] The resulting imbalance between the production of nitric oxide and secretion of endothelin-1 favors vasoconstriction and also contributes to platelet activation and leukocyte adhesion, all of which have been proposed as pathophysiological mechanisms in the development of carbohydrate overload-induced laminitis.[20,21] Bailey et al.[8] documented hypertension in insulin resistant, laminitis prone ponies at summer pasture, suggesting that vascular endothelial dysfunction also is a component of the metabolic syndrome phenotype in equids. EMS animals may therefore be more susceptible to digital vasoconstriction, platelet aggregation and neutrophil adherence and emigration into lamellar tissues under conditions of alimentary carbohydrate overload. Further studies are needed to explore this possibility.

Insulin-Induced Laminitis

In healthy ponies and horses, the administration of a prolonged (~48 h) euglycaemic hyperinsulinaemic clamp (EHC) via simultaneous infusions of insulin and glucose induces Obel grade 1-2 laminitis.[9,10] In this model, the induction of supraphysiological hyperinsulinaemia (serum insulin ~1000-1100 mU/L) is accompanied by pronounced digital pulses and a significant increase in hoof wall surface temperature.[10] Pathological features are similar to those described in the oligofructose overload model, although marked leukocyte infiltration and inflammation do not appear to be features of the insulin model.[22,23] There also is no global separation of the lamellar dermal/epidermal interface, as occurs in carbohydrate overload laminitis.[22] Instead, there is elongation of secondary epidermal lamellae and an apparent increase in mitotic activity in epidermal basal cells, with speculation that aberrant lamellar mitosis may lengthen and weaken lamellar-distal phalanx attachment and contribute to clinical signs of laminitis.[22] The mechanism by which high circulating insulin concentrations (and perhaps also increased glucose flux) induce laminitis has not been elucidated and is the subject of current research. The involvement of increased matrix metalloproteinase activity is unlikely.[24] Insulin-like growth factor (IGF)-1 receptor has been identified on lamellar epithelial cells[25] and it has been proposed that insulin signaling via these receptors may be involved in the pathogenesis of insulin-induced laminitis.

Findings from the insulin model have important implications for EMS-associated laminitis, providing a potential explanation for episodes of laminitis following consumption of feeds and forages that elicit pronounced insulinemic responses. Studies of grazing horses have shown a strong, positive relationship between pasture NSC content and circulating insulin concentrations,[26] and marked exacerbation of hyperinsulinemia has been observed in EMS ponies when they are grazing spring pasture (NSC ~15-18% DM).[6,27] In healthy, non-obese Thoroughbred mares grazing spring (April) pasture in northern Virginia, serum insulin concentrations followed a circadian pattern that mirrored changes in forage NSC content, with peak insulin concentrations approaching 100-110 mU/L.[26] In a herd of Welsh and Dartmoor ponies kept at pasture, some of which were insulin resistant and prone to recurrent pasture laminitis, serum insulin concentrations markedly increased during the months of April and May (values as high as 700 mU/L in some ponies), and this occurrence coincided with an increase in pasture grass NSC content and incident laminitis.[27] Additionally, feeding inulin (to simulate intake of fructan from spring grass) to ponies can elicit an exaggerated insulin response in animals predisposed to laminitis.[28] Seasonal influences on insulin dynamics also may influence risk of pasture laminitis. One study demonstrated exaggerated post-dexamethasone insulin responses in previously laminitic ponies in April and July when compared to responses in December. Additionally, the increase in serum insulin concentrations in response to dexamethasone administration was significantly higher in previously laminitic than in normal ponies.[29] It is therefore possible that episodes of laminitis in pasture kept animals, especially those with an EMS phenotype, are directly linked to increases in circulating insulin associated with season and/or the consumption of NSC-rich forage.

Whether or not hyperinsulinemia plays an essential or exclusive role in the development of EMS-associated laminitis remains to be determined. Clinical observations have indicated that not all animals with profound and persistent hyperinsulinemia develop clinical laminitis while, conversely, some EMS animals are not hyperinsulinemic during episodes of laminitis. Circulating insulin concentrations can vary markedly in EMS animals, so detection of an association between hyperinsulinemia and incident laminitis is challenging. It is also possible that the threshold for induction of lamellar damage and clinical laminitis (i.e. extent of hyperinsulinaemia) is lower in susceptible animals that have had previous episodes.

SUMMARY

Knowledge regarding the pathophysiology of EMS, including the mechanisms of laminitis, is incomplete. The association between grazing NSC-rich pasture and incident laminitis suggests the involvement of hindgut carbohydrate overload, with one or more factors associated with the EMS phenotype potentially lowering the threshold for the development of laminitis in the face of carbohydrate challenge. Evidence from recent studies suggests that inflammatory mechanisms are not primary to the development of laminitis in EMS animals when they subject to clinically relevant carbohydrate challenge. However, oxidant or vascular insults under these conditions may be contributing factors. Alternatively, exacerbations of hyperinsulinemia in response to diet or other factors may trigger laminitis ('insulin-induced laminitis'), supported by the observation that the condition can be induced in healthy ponies and horses by maintaining supraphysiological circulating insulin concentrations for 2–3 days.

REFERENCES

[1]Hossain P, et al. **N Engl J Med** 2007;356:213. [2]Janssen I, et al. **Obesity Rev** 2010;12:397. [3]Alberti KG, et al. **Diabetic Med** 2006;23:469. [4]Geor R, et al. **Vet Immunol Immunopath** 2009;129:151. [5]Frank N, et al. **J Vet Intern Med** 2010;24:467. [6]Treiber KH, et al. **J Am Vet Med Assoc** 2006;228:1538. [7]Carter RA, et al. **Equine Vet J** 2009;40:171. [8]Bailey SR, et al. **Am J Vet Res** 2008;69:122. [9]Asplin KE, et al. **Vet J** 2007;174:530. [10]De Laat MA, et al. **Equine Vet J** 2010;42:129. [11]van Eps AW and Pollitt CC. **Equine Vet J** 2006;38:203. [12]Pollitt CC, et al. **Vet Clin Nth Am Equine Pract** 2010;26:65. [13]Milinovich GJ, et al. **Vet Clin Nth Am Equine Pract** 2010;26:79. [14]Milinovich GJ, et al. **Environ Microbiol** 2006;8:885. [15]Bailey SR, et al. **Vet Immunol Immunopath** 2009;129:167. [16]Hutley L, et al. **Am J Med Sci** 2005;330:280. [17]Vick MM, et al. **J Anim Sci** 2007;85:1144. [18]Treiber K, et al. **Vet Immunol Immunopathol** 2009;129:216. [19]Kim JK, et al. **Circulation** 2006;113:1888. [20]Eades SC, et al. **Am J Vet Res** 2007;68:87. [21]Eades SC. **Vet Clin Nth Am Equine Pract** 2010;26:51. [22]De Laat MA, et al. **J Comp Pathol** 2011;145:399. [23]Asplin KE, et al. **Equine Vet J** 2010;42:700. [24]De Laat MA, et al. **Vet Immunol Immunopathol** 2011;140:275. [25]Bailey SR, et al. **Proc AAEP Workshop** 2009. [26]Byrd BM, et al. **J Anim Sci** 2006;84(Suppl 1):330. [27]Treiber KH, et al. **J Vet Int Med** 2008;22:735. [28]Bailey SR, et al. **J Am Vet Med Assoc** 2007;231:1365. [29]Borer KE, et al. **Proc AAEP Workshop** 2009.

DRUG INDUCED LIVER INJURY IN DOG

Cynthia R.L. Webster, DVM, DACVIM
North Grafton, MA

INTRODUCTION

The liver is uniquely susceptible to drug injury. After absorption from the gastrointestinal tract, drugs are delivered directly to the liver via the portal circulation. Thus the liver is consistently exposed to high drug concentrations. The liver also has a central role in drug metabolism which can lead to the generation of toxic intermediates. Drug-induced liver injury (DILI) is an increasing problem in human medicine and presents a challenge for physicians, as well as regulatory bodies and the pharmaceutical companies.[2,4] As veterinarians begin to use more and more human approved medications in dogs, it is likely that the veterinary profession will have to join in with this challenge and become more diligent in its approach to the recognition, prevention and treatment of DILI.

HEPATIC DRUG METABOLISM

Hepatic drug biotransformation is accomplished by Phase 1 and Phase 2 reactions. Phase 1 reactions are catalyzed by a family of enzymes called cytochrome P450 (CP450) that oxidize, reduce or hydrolyze drugs. Phase 1 reactions either activate or detoxify a drug. Some phase 1 products can be excreted, while others are acted upon by Phase II enzymes. Phase II reactions conjugate drugs or their metabolites to glucuronic acid, sulfate, glutathione, or acetyl or methyl groups. While the products of phase II reactions are always non-toxic, phase I reactions may lead to the generation of reactive intermediates. Toxicity from Phase 1 products is normally kept in check by prompt Phase II conjugation or by the myriad of anti-oxidant defence mechanisms present in the liver including superoxide dismutase, catalase and glutathione. The CP450 enzymes exhibit significant genetic polymorphism and thus an individual's genetic complement of CP450 enzymes can profoundly influence drug metabolism.

A phase III drug metabolism system is now acknowledged. These reactions primarily involve the transport of the products of phase II metabolism.[6] Phase III transport proteins include the multiple drug resistance gene MDR1 (p-glycoprotein), the phosphoflippase, MDR3, the bilirubin transporter, multidrug resistance associated protein, MRP-2 and the bile salt export pump, BSEP. The expression and function of these transporters can be altered under the influence of several inducers and inhibitors and thus they can profoundly influence drug uptake, metabolism and excretion. Some drug toxicities are associated with interference with these phase III transporters. In addition, humans with genetic alteration in transport proteins can have increased drug toxicity.

PATTERNS OF INJURY

Drug toxicity can either be intrinsic or idiosyncratic. Intrinsic toxins exhibit predictable dose-dependent toxicity. Idiosyncratic toxins cause unpredictable primarily dose-independent toxicities that are difficult to reproduce. They are associated with an individual's unique sensitivity to a drug and occur due to a metabolic aberration in the host. This aberration could be a genetic polymorphism in drug metabolism, disposition, transport and/or interaction at receptors. Recent genetic studies also suggest that susceptibility may idiosyncratic reactions may be immunologically controlled at the level of HLA expression. Some idiosyncratic reactions will show a degree of dose-dependency.

The histopathological reaction to drug mediated toxicity is quite varied. Many reactions are acute, however, some drugs cause chronic hepatotoxicity. Hepatocellular necrosis and/or apoptosis are common. Immunologically mediated reactions are characterized by inflammatory infiltrates that may be granulomatous in nature. Some drugs preferentially destroy bile ducts or interfere with bile formation and histopathological lesions reflect this. Other drugs cause lipidosis, fibrosis or damage the sinusoidal endothelium.

DIAGNOSIS AND TREATMENT OF DILI

The methods used to diagnose DILI for the most part rely on either expert judgment or specially designed algorithms or scales. An example of the latter is the Council for International Organizations of Medical Sciences (CIOMS) scale.[7] Scales are useful ways of enhancing objectivity, grading the strength of probability among categories of evidence and act as a checklist to remind clinicians of the information needed to make the determination. There has been limited use of these scales in veterinary medicine although a recent case report of zonisamide toxicity in the dog made use of a variation of the CIOMS scale.[21]
The expert judgment determination should make use of the following criteria to make a diagnosis of DILI. 1.) The animal becomes ill while taking the drug. Typically the latency period is within the first 3 months of therapy, but some toxicities are associated with an accumulated dose over a longer time period. 2.) The animal recovers after drug withdrawal. Some reactions may take months to resolve particularly if they involve an immune reaction since offending allergens may remain on the hepatocyte surface despite drug discontinuation. 4.) An awareness of the hepatotoxic potential of the drug including the typical clinical presentation and histologic reaction. 5.) The animal gets sick upon re-challenge, which in reality is almost never done for ethical reasons. 6.) A careful exclusion of other causes of liver disease.

Treatment for hepatic ADR's involves immediate drug withdrawal. Most ADR's do not have specific antidotes, the exception being acetaminophen in which N-acetylcystiene is administered to increase glutathione levels and aid in de-toxification of the reactive intermediates. In the absence of a specific antidote, treatment of drug reactions involves intensive management of the complications of acute and/or chronic liver failure to give the damaged liver time to repair and regenerate. Corticosteroids are contra-indicated when acute liver failure is present. The one exception may be with severe allergic reactions.

PREVENTION OF DILI

The key to limiting disease from hepatic ADR is rapid recognition and discontinuation of the drug. One of the keys to the recognition of hepatotoxic drug reactions is the drug's track record for causing hepatic ADR. Newly released drugs, however, do not have track records. In addition, pre-market approval drug testing is generally not sensitive enough to detect idiosyncratic hepatic reactions. Since most idiosyncratic reactions occur at a frequency of 1:10000, Phase 3 studies would have to enroll 30,000 patients to identify 1 reaction. Thus most idiosyncratic ADR's become apparent post-marketing.

If a patient is on a potentially hepatotoxic drug, monitoring of serum liver enzymes is warranted. Hepatic function must be documented prior to the administration of these drugs. This is especially important in geriatric animals who are predisposed to DILI. Other risk factors for DILI in human patients include female sex, high body mass index, the presence of underlying renal or liver disease and concurrent administration of potentially hepatotoxic drugs or CP450 enzyme inhibitors or inducers. Recent reports in humans with genome wide association studies on patients taking drugs that did and did not have DILI have identified HLA variants as risk factors for hepatotoxicity.[2,4] Some breeds of dogs appear to be at higher risk for hepatotoxicity with certain drugs. For instance, Doberman pinschers and Labrador retrievers appear to be predisposed to hepatotoxocity from potentiated sulfonamides and carprofen, respectively.[15,23]

COMMON DILI IN DOG

Table 1: Pharmacological Agents Associated With DILI in the Dog

Acetaminophen	Corticosteroids	Methotrexate	Tetracycline/ doxycycline[12,20]
Amiodarone[11,13]	CCNU [9,10,14]	Mitotane[25]	Zonisamide[17,21]
Anabolic steroids	Halothane	Nitrofurantoin	
Azathioprine	Ketoconazole	Phenobarbital [1,5,8,17]	
Carprofen[15]	Mebendazole/oxibendazole	Potentiated sulphonamides [23]	

SELECTED TOXICITIES

Phenobarbital Phenobarbital is an idiosyncratic hepatotoxic drug in dogs.[1,5,8,17] The most common histological pattern is that of chronic inflammatory disease, although acute hepatic necrosis has been reported. The clinical signs of toxicity are non-specific and suggestive of chronic liver failure with anorexia, ascites and encephalopathy noted in some dogs. The reaction is somewhat dose-dependent. With drug termination and appropriate supportive care, many dogs can recover. The mechanism of hepatotoxicity has not been well characterized. It may be associated with defective epoxide hydrolase activity leading to the build up of toxic intermediates. Since phenobarbital induces CP450 enzymes, it may potentiate the toxicity of other drugs, pesticides or environmental toxins. In one report on phenobarbital induced hepatotoxicity, 44% of the dogs were on other potentially hepatotoxic drugs. Alternatively, CP-450 induction itself may play a role in toxicity since increased CP450 activity by itself can lead to increased free radical generation. Phenobarbital can also predispose the liver to oxidant damage by depleting hepatic glutathione stores and activating macrophages. Monitoring patients on phenobarbital for hepatotoxicity is complicated by the fact that it can induce hepatic enzymes in the absence of hepatic injury. Nonetheless abnormal hepatic function tests such as increases in serum bilirubin, total serum bile acids or hypoalbuminemia should prompt discontinuation of Phenobarbital, if possible.

Chronic therapy with phenobarbital has been implicated in the development of superficial necrolytic dermatitis (SND).[16] SND in these dogs was consistently associated with hepatic changes distinct from those seen in dogs with phenobarbital hepatotoxicity, but typical of that seen in dogs with non-phenobarbital associated SND. Histologically these changes include severe parenchymal collapse with glycogen laden hepatocytes and moderate fibrosis sharply demarcated by nodules of normal hepatic parenchyma. Most dogs with phenobarbital associated SND are not in overt liver failure and prognosis is related more to the severity of the skin lesions.

Potentiated sulphonamides Potentiated sulfonamides cause a variety of adverse drug reactions in dogs.[19,23,24] A recent study suggests that DILI to potentiated sulfonamides carries a worse prognosis (46% recovery with hepatopathy vs 89% recovery without hepatopathy) than adverse reactions in other organs. The histopathologic lesion is one of massive centrilobular and mid-zonal necrosis with minimal centrilobular lymphoplasmacytic inflammation. A less fulminant cholestatic pattern with more inflammation on histopathology and a destructive eosinophlic cholangitis have also been reported in the dog. An increased incidence of reactions is seen with Dobermans and has been suggested in Miniature schnauzers and Samoyeds. The reaction typically occurs within 5-36 days (mean of 12 days). Acute liver failure with encephalopathy and coagulopathy may occur. The dogs can have other immune manifestations such as arthritis, thrombocytopenia, and immune mediated haemolytic anemia. There is no association with prior exposure to potentiated sulfonamides. There is a trend for the hepatotoxicity to occur with higher doses. In humans, the toxicity is a hypersensitivity reaction. In those genetically predisposed a reactive oxidative metabolite, a sulfamethoxazolenitroso forms and binds to tissue which results in generation of an immune response. Although it is not known if this occurs in dogs as well, the reaction in the dog is presumed to be immunologic as anti-platelet antibodies have been demonstrated in dogs with sulfonamide induced thrombocytopenia. As the toxic nitroso intermediate can be reduced nonenyzymatically by anti-oxidants, sulfonamide toxicity can be treated with glutathione precursors (S-adenylmethionine or n-acetylcystiene) and vitamin C supplementation.

Recently, acute hepatotoxicity has been reported with therpapeutic dosese of zonisamide dogs.[17,21] Zonisamide is a sulfonamide based anti-convulsant that does not contain the sulfonylarylamine structure that is known to generate the toxic nitrosyl molecule in other sulfonamides. Thus the toxicity is likely mediated by an alternative mechanism. In one report the dog presented 10 days after starting zonisamide with severe acute panlobular hepatic necrosis and was euthanized due to nonresponsive coma and bleeding diathesis. Hepatic biopsy showed severe necrosis and marked microvesicular hepatic lipidosis and moderate lymphoplasmacytic and neutrophilic hepatitis. Sulfa crystals were seen in the urine. In a second report a dog on the mediation for 3 weeks returned for vomiting, inappetance and icterus and was found to have severe elevations in serum liver enzymes. The drug was immediately stopped and the dog made a full recovery.[21]

Non steroidal anti-inflammatory agents (NSAIDS) All NSAIDS should be considered potential hepatotoxins.[3] Acetaminophen and aspirin are intrinsic toxins albeit at doses far above those necessary for therapeutic effects (except in the case of the cat who poorly glucuronidates reactive acetaminophen metabolites). Hepatocellular toxicosis with carprofen has been reported in the dog.[15] A breed predisposition in Labrador retrievers has been noted. Clinical signs (anorexia, vomiting) typically develop within the first 3 weeks of therapy. All dogs have increases in serum liver enzymes and many are hyperbilirubinemic. A spectrum of histopathologic changes occurs characterized by ballooning degeneration, vacuolar change, necrosis and apoptosis with mild to moderate periportal and interacinar lymphoplasmacytic inflammation. Most dogs recover with prompt drug discontinuation. Although there is considerable evidence in human patients suggesting that DILI to NSAID's are due an immune hypersensitivity there is also evidence that NSAIDS may directly damage hepatic mitochondria.

CCNU (1-(2-chloroethyl)-3-cyclohexyl-1-nitrosurea, lomustine) CCNU is an orally active alkylating agent used to treat lymphoma, mast cell tumor, histiocytic sarcoma and CNS tumors. The dose limiting side effect in dogs (not cats) is hepatotoxicity.[9,10,14] Retrospective studies have reported a prevalence of any elevation in serum ALT enzyme activity to be from 60 to 80%. Elevations >1.5 times the upper limit of normal have been reported in 48% of dogs on CCNU, while 29% of dogs had "major" elevations in ALT (defined as ALT 5 times the upper limit of normal). Typically the elevations occur after 1 to 3 doses. The toxicity was somewhat dependent on cumulative dose in one study, but dose independent in 2 others. The percentage of dogs that develop a clinical hepatopathy is much lower (1-6%) than those that get elevations in serum liver enzymes. Apparently many dogs develop a subclinical hepatopathy that resolves away when the drug is stopped. In one study young age (< 5 yrs) and Boxer breed were associated with a higher incidence of ALT elevation.

Dogs receiving CCNU should have routine evaluation of serum liver enzymes (ALT and ALP) throughout the course of their therapy. The lack of elevations early in therapy does not seem to preclude the development of increases at a later date. The results of a recent study suggest that pre-treatment with a silymarin/S-adenosylmethionine product attenuates the increase in ALT seen after CCNU therapy. More importantly dogs on these hepatoprotective agents were less likely to have a delay in treatment and more likely to complete their chemotherapy cycles.[22]

The exact mechanism of hepatotoxicity with CCNU is not fully understood, but may involve oxidant damage with glutathione depletion and lipid peroxidation. Only limited hepatic histological assessment of clinical cases of hepatoxicity has been reported. In one study multiple changes including lipogranulomas, hepatocellular vacuolization, periportal lymphocytic plasmacytic inflammation, fibrosis and bile duct hyperplasia were seen.

REFERENCES

1 Aitken MM, et. al. **Vet Rec**. 2003;153:13-6. **2** Au JS, et. al. **Aliment Pharmacol Ther** 2011;34:11-20. **3** Bessone F. **World J Gastroeneterol** 2010 ;16:5651-61. **4** Chalasani N. et. al. **Gastroenterology** 2010;138: 2246-59. **5** Dayrell-Hart B. et. al. **J Am Vet Med Assoc**. 1991;199:1060-8. **6** Funk C. **Expert Opin Dr Metab Toxicol.** 2008;89:109-16. **10** Hosoya K, et al. Vet Comp Oncol 2009;7:244-55. **11** Jacobs G, et. Al. J Vet Intern Med 2000;14:96-9. **12** Kaufman AC, et. Al **Aust Vet J**. 2011;89:109-16. **10** Hosoya K et. al. **Vet Comp Oncol**. 2009;7:244-55. **11** Jacobs G et. al. **J Vet Intern Med**. 2000;14:96-9. **12** Kaufman AC et. al. **J Am Vet Med Assoc**. 1993;202:628-30. **13** Kraus MS et. al. **J Vet Intern Med**. 2009;23(1):1-10. **14** Kristal O et. al. **J Vet Intern Med**. 2004;18:75-80. **15** MacPhail CM, et. al. **J Am Vet Med Assoc**. 1998;212:1895-901. **16** March PA, et.al. **J Vet Intern Med**. 2004;18:65-74. **17** Miller ML, et. al. J Vet Intern Med. 2011;25:1156-60. **18** Müller PB, et.al. **J Vet Intern Med**. 2000;14:165-71. **19** Rowland PH, et. al. **J Am Vet Med Assoc**. 1992;200:348-50. **20** Schulz BS, et. al. **Vet Rec**. 2011;169:229. **21** Schwartz M, et. al. **J Vet Med Sci**. 2011;73:1505-8 **22** Skorupski KA, et. al. **J Vet Intern Med**. 2011;25:838-45 **23** Trepanier LA. **J Vet Pharmacol Ther**. 2004;27:129-3. **24** Twedt DC, et. al. **J Vet Intern Med**. 1997;11:20-3. **25** Webb CB, et. al. **J Am Anim Hosp Assoc**. 2006;42:298-301.

Diagnostic and Managerial Pearls for Selected Hepatobiliary Disorders

S.A. Center, DVM, DACVIM
Ithaca, NY

The Bedlington Terrier (B T) is the only breed for which a genetic cause of Copper Associated Hepatopathy (CuAH) is confirmed. This autosomal recessive trait derives from a 13.9 kb deletion in exon 2 of COMMD1. However, there are some affected BT kindreds lacking this mutation. This is not unprecedented as >300 mutations in ATP7β are causal to Wilson's disease in human beings. Initially, a microsatellite marker (C04107) adjacent to the COMMD1 mutation predicted BT phenotype; this test remains in use by some BT breeders with historical kindreds who use it tandem with the restriction fragment PCR test detecting the COMMD1 mutation. Published case series verify that CuAH also occurs in other pure breeds of dogs (e.g. Dalmatian, Corgi, Labrador Retriever, Doberman Pincher, West Highland White Terrier). While an inherited trait has been proposed in some of these, none are confirmed. Since CuAH also occurs in small pure breed and mixed breed dogs, the syndrome deserves consideration in any dog with chronically increased liver enzymes particularly if fold increase in ALT>ALP. The diagnosis is conflicted by a tendency for Cu to accumulate secondary to severe necroinflammatory or cholestatic lesions. Status of WHWT is confused as many dogs with high hepatic Cu concentrations lack clinical signs, clinicopathologic features, and histologic lesions indicative of CuAH, even into geriatric years.

Canine liver has a higher Cu concentration than many other species (e.g.≤ 400 ppm, ~8 fold greater human beings). Hepatic Cu concentrations in dogs have progressively increased since 1930 coinciding with introduction of commercial dog foods. Recent study of hepatic Cu in Labrador Retrievers with and without chronic hepatitis (1985 to 2011) demonstrated significant increase in tissue Cu concentrations over time, further implicating a role for dietary Cu exposure. Use of premixes additional to source food and change in Cu supplements (1997 AFFCO recommendation) may underlie the recent surge in canine CuAH. In health, Cu homeostasis is primarily maintained by self-adjusted biliary excretion such that hepatic Cu concentrations remain constant even with wide intake variation. Thus current intake levels may be challenging tolerable limits for certain dogs. Excess hepatic Cu is injurious because it is a redox-reactive transition metal. Cellular accumulation results in lysosomal loading and eventual oxidative injury to structural and functional cell components. Quiescent Cu loading increases hepatocyte vulnerability in Zone 3 to other injurious events (e.g. NSAIDs, hypoxia or hypoperfusion during anesthesia). Definitive diagnosis of CuAH requires liver biopsy although CuAH can sometimes be deduced from liver aspirates stained with Rhodanine. Diagnosis optimally involves examination of an H&E and Cu- specific stained tissue sections (Rhodanine most commonly used) and quantification of tissue Cu concentration. CuAH primarily involves Zone 3 (centrilobular area= periacinar region) hepatocellular Cu retention and is characterized by individual hepatocyte necrosis and a pyogranulomatous inflammatory response (small foci of macrophages, neutrophils, fewer lymphocytes). Pink cytosolic granules on H&E stained sections are confirmed as Cu-binding protein with Cu specific stains. Additional application of special stains allows early recognition of Zone 3 lobular collapse (reticulin, Type III collagen), sinusoidal collagen deposition (Masson's Trichrome, Type I collagen in sinusoidal chicken wire pattern), and Kupffer cell/macrophage iron aggregation (Prussian Blue). Examination of young hepatocytes within regenerative nodules shows Cu "+" granules when CuAH is the primary disorder. Quantification of tissue Cu concentrations (atomic absorption spectroscopy & other methods) requires submission of a separate tissue sample for analysis. If a regenerative fibrotic nodule is sampled, Cu quantification may be discordant and lower than expected. This problem has been eliminated by use of a digital scanning system (Aperio Scanscope system) that allows Cu quantification from Rhodanine stained sections. The system was validated with a spectrum of low to high tissue Cu concentrations and a predictive linear regression derived.

Treatment: two patient subsets: 1) dogs with necroinflammatory injury associated with Zone 3 Cu accumulation and tissue Cu > 850 ppm, and 2) asymptomatic high risk large breed dogs with serendipitous discovery of Zone 3 Cu retention (e.g. Labrador Retrievers, Doberman pinchers, Dalmatians) with future anticipated NSAID exposure.

Treatment: **1)** removal of tissue bound Cu by chelation, **2)** restricted Cu intake, **3)** antioxidants, and **4)** chronic restricted Cu uptake by dietary restriction or lower dose or intermittent (every-other-day [EOD] chelation or oral zinc to restrict alimentary Cu uptake.

1. *Initial Chelation:* **d-Penicillamine**- considered the first line chelator as it mobilizes Cu from all tissues and has an anti-inflammatory effect. In man, excess Cu mobilized during the first treatment year is eliminated in urine. Yet, health status improves before substantial Cu mobilization suggesting that tissue detoxification precedes Cu chelation. d-Pen sequesters Cu in an innocuous state either through direct binding or by induction of Cu binding proteins (e.g. metallothioneins). It also imparts an immuno-modulatory and antifibrotic effect (inhibits collagen cross-linking, stimulates collagenase activity), and restores cellular GSH and function of the mononuclear-macrophage system (increasing host defense against antigenic/ endotoxic insults). *Dose*-15 mg/kg PO BID given 30-minutes before feeding for 2 mths in animals with asymptomatic hepatocyte Cu and for 6 mths in animals with necroinflammatory injury due to CuAH. Serum ALT usually normalizes within 6-8 weeks. *Pyridoxine (Vitamin B_6)* is given adjunctively: 10–25 mg total dose/day, to thwart deficiency secondary to chelating agent inhibition of pyridoxine-dependent enzymes (however, there is no experimental data that substantiate this recommendation). Chronic or maintenance therapy after de-coppering can be done with treatment EOD or by using a 50% dose reduction given daily along with continued Cu restriction. The canine dose of d-Pen has been successful used in a few cats with CuAH. *d-Pen Side Effects* (dogs): vomiting is common but may decline by dividing total dose into 2-3 treatments/day or dose administration with a small amount of food. d-Pen also chelates zinc & other essential metals and can lead to Fe deficiency anemia (reported once in a dog on chronic therapy). Co-treatment for these possible scenarios reduces its Cu chelating efficacy. Co-treatment with zinc is ill advised as this may negate benefit of each therapy. If zinc is

concurrently given, the interval between drug administration must be as wide as possible (dosing numerous times). Other side effects of d-Pen rarely observed in dogs: acute hypersensitivity reaction, immune mediated renal injury and Nephrotic syndrome, a Lupus-like syndrome, polyarthritis, anorexia (altered taste sensation, nausea), cytopenias, peripheral neuropathy or a myasthenia-like syndrome, dermatitis and hair loss. In patients intolerant of d-Pen, alternative chelation is done using *2,3,2 Tetramine* (Syprine™, Cuprid) at a dose of 5-7 mg/kg PO BID, given 1-2 hr before meals (enhances bioavailability) for 2-6 months and then titrated for chronic administration (as for d-Pen, previously). Unlike d-Pen, Tetramine does not remove CNS Cu. Work in human Wilson's patients suggests it may increase free circulating Cu that can precipitate neurologic signs (leading to death or permanent disability). *Side Effects*: similar to d-Pen. No adverse effects occurred in a clinical trial in healthy dogs. However, acute renal injury occurred in 2 chelated dogs using initial dose recommendations of10-15 mg/kg/day PO BID, perhaps due to free Cu induced renal tubular injury.

2. *Zinc Reduction of Alimentary Cu Uptake:* increased enterocyte metallothionein creates a Cu "sink" restricting uptake when Zn is given 30-60 mins before meals. Cu is subsequently lost in feces with effete enterocytes. Efficacy of Zn for CuAH remains controversial; some dogs cannot tolerate oral Zn at any dose. Recent retrospective studies in man do not recommend Zn as initial therapy, demonstrating dismal outcome in Wilson's patients. *Loading Phase:* 5-10 mg/kg of elemental Zn/day given in 2 divided doses (separated by at least 8 hrs), 30-60 minutes before meals. In case of vomiting, Zn is given with a small piece of food (reduces bioavailability). Although Zn gluconate, sulfate or acetate can be used, the latter is best tolerated. Encapsulated Zn acetate can be acquired as Galzin™ from GATE Pharmaceuticals. *Maintenance Phase:* 50% of the effective loading phase dose after 2-6 months. Theoretical advantage of Zn in CuAH is that interrupted therapy (up to 24 days) is associated with sustained Zn release from acquired body stores providing some protective coverage. Zn therapy has allowed Wilson's patients to diversify their normally restricted diet (e.g. allowing chocolate !). However, Wilson's Disease patients are monitored for dose adequacy by measuring 24 hour urinary Cu elimination or enteric uptake of ^{64}Cu because standard Zn dosing has failed to control enteric Cu uptake in ~10-15% of patients. Chronic Zn therapy may cause iron deficiency, similar to chelation. *Elemental Zinc Dose:* based on Zn contribution to molecular weight: Zn acetate = 30%, Zn gluconate= 14%, Zn sulfate = 23%. Plasma Zn concentrations are monitored at baseline, then periodically, during the first month and then q2 -3 months during the initial 6 months to avoid Zn toxicity (>800 ug/dL [8-10 ppm] promotes hemolysis). Plasma Zn concentrations do not correlate with tissue Zn concentrations or prove treatment efficacy.

2. *Antioxidants:* Vitamin E- (10 U/kg/day a- tocopherol) and/or SAMe-(20 mg/kg/day-bioavailable product. Vitamin C-avoided as *in vitro* work suggests it augments transition metal oxidative reactivity.

3. *Restricting Cu Intake:* several commercial diets designed to maintain CuAH affected BT in neutral Cu balance (Hills L/d, Royal Canin Hepatic Support) deliver ~5-6.8 ug Cu/kg dry wt. food. Because these diets provide ~1.8-2.5 gm protein/kg body wt we add supplemental low Cu protein to increase protein intake to 3.5-4.0 gm/kg/day. Most CuAH cases are diagnosed well before onset of hepatic insufficiency and do not require protein restriction. A home prepared diet can also be used but formulation by a clinical nutritionist is recommended. Web applications useful for adjusting nutritional support: http://www.ars.usda.gov/Services/docs.htm?docid=17477 directs users to food Cu content. Further diet adjustments (e.g. protein, energy, sodium content) can be done using the NAT-2 program at the department of human nutrition, U. Illinois (also linked to USDA food tables) http://nat.crgq.com/mainnat.html (no Cu concentrations). Home-made diets should avoid organ meats, shellfish, nuts, and certain cereals rich in Cu. It is important also to avoid vitamins or treats containing Cu. These may not be specified requiring an investigative call to the manufacturer. A nutritionist should advise which vitamin/mineral supplement to use. Source house water Cu concentration should also be investigated, as it is advised to limit water Cu intake: <0.1 ppm. If Cu plumping fixtures exist, running tap water for several minutes should adequately flush the system.

Uncommon Clinical Features of CuAH: transient Fanconi glucosuria, hemolytic anemia (massive hepatic necrosis syndrome).

Long Term Survival (surgical vs medical) in Dogs with Portosystemic Shunts (n=612, 31 yr experience at Cornell University)

While it is inarguable that best treatment for portosystemic vascular anomalies (PSVA) is surgical attenuation, a subset of dogs cannot tolerate complete ligation and an expanding number of clients cannot afford the expense. Further, outcome and morbidity associated with PSVA surgery varies among individual patients, surgeons, and hospitals. All symptomatic PSVA dogs benefit from medical management, yet not all dogs require all forms of medical intervention. As a result of improved condition after initiation of medical management, many clients have questioned the need to purse surgical intervention. Because no single treatment plan fits all cases, individuation of care is the best course of action. We have successfully managed longterm a spectrum of PSVA patients medically. It is difficult to judge outcome of these patients against those undergoing surgical shunt attenuation because medical management is often used for: 1) minimally symptomatic patients, 2) dogs failing surgical intervention, and 3) dogs with a degenerative zone 3 hepatic lesion having dismal surgical outcome. With this in mind, we retrospectively determined survival of dogs with PSVA (extrahepatic [EPSVA, n=378], intrahepatic [IPSVA, n=105]) and anicteric APSS (n=114) over 31 contiguous yrs with an ~92% follow-up. (see poster presented at this meeting) Medical management included dietary protein-restriction/modification, ± lactulose, ± metronidazole and surgical PSVA attenuation was by suture ligation (n=225/239) to tolerance. Comparison of survival (50th percentile, [75th-25th percentiles] of EPSVA and IPSVA treated surgically (n=189 and 50, respectively) and medically (n=166 and 46, respectively), and APSS were compared. Median age (yrs) at diagnosis was significantly younger for PSVA (1[0.1-12]) vs APSS (4[0.12-12]), and IPSVA (0.5[0.1-5]) vs EPSVA (1.5[0.12-12]). As expected, EPSVA survival (all dogs: n=334, 10[6.0-13]; Yorkshire Terriers: n=91; 12[6-13]) was significantly longer than IPSVA: n=104, 4.5(1.2-9). However, there was no survival difference between surgically- or medically-treated EPSVA and surgically- or medically-treated IPSVA. Of surgically treated dogs,

63 % EPSVA and 58% IPSVA also received medical treatment. APSS survival (9.0[4.6-11]) was significantly shorter than medically-treated EPSVA (all dogs: 10 [6-13]; Yorkshire Terriers:12.0 [6.0-13]). So how did we accomplish success with medical intervention ?

Tailored HE Management: important considerations

1) *Prompt recognition & intervention of precipitating factors.* Do not assume deterioration in liver function, rather carefully examine the patient, the recent history, & medications. **2)** *Prevent GI formation of encephalogenic substrates.* **3)** *Evacuate enteric substrates & toxins contributing to HE* (especially in abrupt onset or recalcitrant HE). Most HE manifestations are reversible with appropriate intervention and consistent management.

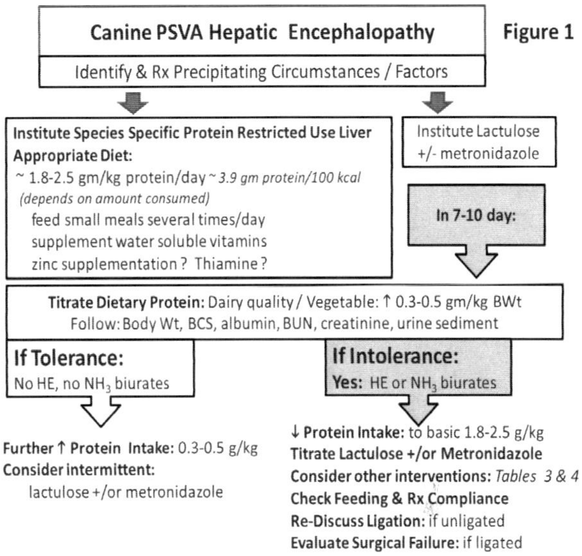

Figure 1

Therapeutic Interventions

1. **Dietary Modification:** Nutritional support must be individually tailored as different magnitudes of shunting seemingly influence dietary protein tolerance. Standard strategy for nutritional management of PSVA dogs in our hospital is shown in Fig 1. Overzealous protein restriction can impose a negative nitrogen balance leading to muscle catabolism and increased risk for HE. Skeletal muscle is the second largest site of NH_3 metabolism & transient NH_3 detoxification i(glutamine/glutamate cycle) and muscle catabolism contributes to nitrogenous waste. We first feed an appropriate species-specific protein restricted commercial diet formulated for hepatic insufficiency (either L/d [Science Diets] or Hepatic Support [Royal Canin, Waltham]) , initially combined with lactulose +/- metronidazole. This provides a dietary protein allowance of ~1.8-2.5 g/kg body wt. per day (~3.9 g protein/100 kcal food with estimated daily energy intake [e.g. (30 x body wt kg) + 70]). Depending on the quantity of food consumed, a lower gm/kg protein intake may occur. Starting estimates for protein intake are based on study of surgically created canine PSS. A modified protein quality (focused on vegetable and dairy quality protein) is used if a homemade diet is developed. While several small meals are recommended/day to optimize digestion/assimilation (reducing enteric debris that generates HE toxins, some dogs do well with a single large evening meal and a small morning snack. Protein content is thereafter adjusted using supplementary dairy (cottage cheese, feta cheese, hard cheese, yogurt) or tofu sources. The patient is sequentially assessed for change in: body weight and condition, clinicopathologic markers of protein sufficiency or excess (e.g. total protein, albumin, BUN, creatinine [reflecting muscle mass] and ammonium biurate crystalluria) initially at 7 to 10 day intervals. If all goes well (HE controlled, good body condition, absence of crystalluria) additional protein supplements are explored using 0.3 gm/protein/kg body wt. followed-by sequential assessments to determine patient benefit and protein tolerance. Selection of food sources and amount of supplemental protein is facilitated using the NAT-2 website (cited previously). Yogurt delivers several beneficial constituents including lactose, milk protein, and live "beneficial" bacteria that together augment HE control with regular use. With low levels of methionine, aromatic amino acids, and ammoniagenic amino acids, vegetable proteins are used for snacks. Certain dry cereals are used as treats (e.g. frosted bite size shredded wheat, etc). Dietary fiber has value in increasing enteric nitrogen incorporation /elimination, decreasing intestinal transit time, and intraluminal enteric pH, and increases fecal NH_3 elimination. Human liver patients consuming a high protein vegetable-based diet demonstrate fewer HE episodes and maintain higher plasma arginine and citrulline concentrations which may facilitate NH_3 detoxification. Personal experience with daily maintenance of PSVA dogs (> 10 years) confirms that even dogs with portal atresia can tolerate milk-bone like dog cookies, supervised recreation with rawhide chews (chews assist with maintaining dental health as PSVA dogs form prodigious tarter/calculus), popcorn, yogurt, cheese, cottage cheese, ice cream, fruit (not raisins), carrots, broccoli, and almonds with no ill effects. If acceptance of a modified liver friendly diet needs coaxing with some form of "meat" protein, only small amounts of white meat chicken is allowed. Owners are warned to AVOID hot dogs, Vienna sausages, hamburger, steak, organ meats, fish, and shell fish. Rare episodic HE has developed in a few PSVA dogs fed eggs and in one dog fed hydrolyzed protein diet for IBD.

2. *Enemas:* for episodic HE and are best provided by owners at home at first recognition of early HE. Enemas also thwart constipation that can provoke HE. Cleansing and retention enemas are routinely used for PSVA dogs hospitalized for HE.

3. *Colonic Catharsis*: colonic evacuation is usually achieved with lactulose, lactitol, or lactose (in lactose intolerant animals) +/- dietary fiber. Catharsis reduces colonic retention of HE toxins (e.g. ammonia, SCFA) and colonic flora generating toxic products.

4. *Non-Absorbable Fermentable Carbohydrates:* acidify the enteric lumen trapping NH_3 as NH_4^+, reduce activity of bacterial ureases and enteric peptidases, increase bacterial nitrogen fixation (increasing nitrogen tolerance), and promote colonic catharsis secondary to osmotic effects of generated organic acids. The prototype medication is *lactulose*, a synthetic disaccharide (galactose & fructose) not digested by mammalian enzymes that is fermented to organic acids by enteric flora and is accepted by most dogs. *Lactitol,* a similar product, is more palatable to humans (less sweet) but is more costly. Dose-limiting adverse effects include cramping, diarrhea and flatulence. Hypernatremia and hypercalcemia may result from lactulose administration; the former reflects fluid loss and the latter drug assisted calcium uptake. Large overdoses of lactulose/lactitol may result in diarrhea, electrolyte disturbances, and hypovolemia severe enough to provoke HE. Lactulose and Lactitol dosing is started low (0.5 mL/kg PO BID) and

gradually increased to achieve passage of several soft stools/day. While there is some evidence of a synergistic effect between lactulose and a co-administered oral antimicrobial, lactulose efficacy remains controversial. This controversy should be considered when designing long term medical management schemes as some dogs do well with tailored dietary management and lactulose reserved for episodic HE.

5. *Enteric-Non-Absorbable Antibiotics*- these potentially reduce SIBO and modifies enteric flora contributing to HE (suppressing urease producers and production of encephalogenic products) and has potential for reducing bowel flora overall. There is some evidence that neomycin may improve HE by inhibiting intestinal glutaminase. While systemic absorption of oral Neomycin is minimal at ~4%, both irreversible ototoxicity and renal toxicity have been witnessed in several PSVA dogs receiving chronic neomycin. Thus, neomycin is not recommended as first line chronic therapy. Recent information (human studies) suggests that Rifaximin is highly effective in controlling episodic HE even when administered as single therapy. Rifaximin is a poorly absorbed synthetic antimicrobial with broad spectrum antibacterial activity (both aerobic and anaerobic Gram-positive and Gram-negative organisms). Yet, it also preserves beneficial host enteric microbiota. Trials in humans show a favorable tolerability profile with a lower overall risk of adverse effects compared to neomycin. However, the trade-off comes at a considerable cost, with rifaximin being >4 fold more expensive than other agents (e.g. lactulose or neomycin).

6. *Enteric Absorbable Antibiotics given Orally*- may thwart episodic HE by adjusting enteric flora and reducing formation of encephalogenic products contributing to HE. Clavamox or metronidazole, alone or in conjunction for only a few doses can interrupt an HE episode in some patients. Clavamox is especially useful in cats for this effect.

7. *Oral Glucose Supplementation* (Karo syrup-home therapy) useful in pediatric toy breed dogs at risk for episodic neuroglycopenia. Synergistic effect of neuroglycopenia with other encephalogenic toxins increases risk for episodic or protracted HE.

8. *Fluid Therapy* (subcutaneous & oral supplementation): plays a pivotal role in averting episodic HE and is useful in the home environment. Episodic HE leads to dehydration through reduced normal water and food intake. Thus, both oral and SQ fluid administration helps resolve pre-renal azotemia . Considering that ~25% of urea fluxes into the intestines where it converts to NH_3 via bacterial ureases, increased renal urea elimination is important.

9.*Probiotics, Prebiotics, Synbiotics:* Probiotics (beneficial living microorganisms), prebiotics (indigestible carbohydrates that stimulate growth and activity of beneficial enteric bacteria), and synbiotics (a combination of these two) also may help attenuate HE. A common application is use of probiotic yogurt but consistency of treatment is important. *Gut Re-Population with a Probiotic* (supplementing "beneficial" bacteria) may modulate bowel flora to a less encephalogenic state (colonization with non-urease-producing bacteria), and may reduce SIBO that is thought to contribute to HE. Probiotics have been shown to reduce blood NH_3 in humans prone to HE. Yet gut repopulation is difficult to achieve and some believe is impossible, requiring frequent agent administration. Gastric acidity reduces efficacy of this treatment.

10. *Adjust Drug Dosages Responsibly*, anticipate reduced first pass hepatic extraction and minimize or avoid use of drugs known to aggravate HE or that produce prolong depressive effects.

11. *Avoid Vitamin Deficiencies:* provide a daily water soluble vitamin supplement. Thiamine deficiency can provoke metabolic encephalopathy confused with HE.

12. *Zinc supplementation:* a trace metal essential for function of >300 metalloenzymes, including two urea cycle enzymes and glutamine synthetase (temporary muscle NH_3 detoxification and astrocyte NH_3 shuttling). While we have shown subnormal hepatic zinc concentrations in ~45% of dogs with PSVA, its role in HE remains speculative. Among mechanisms contributing to zinc deficiency are: reduced dietary intake (low protein diets deliver lower zinc concentrations), reduced intestinal absorption, reduced hepatointestinal extraction, portosystemic shunting (increases urinary loss), and altered protein and amino acid metabolism. An inverse relationship between serum zinc and NH_3 has been described in humans with hepatic insufficiency, however, plasma zinc cannot predict tissue zinc status so this cannot be interpreted as evidence supporting zinc supplementation . Little information is available regarding the clinical efficacy of zinc supplementation in HE in humans or animals. Some evidence suggests improved amino acid metabolism and some results are conflicting. Nevertheless, patients with subnormal tissue zinc concentrations and perhaps patients with recalcitrant HE for which an underlying precipitating cause remains undiscovered, might benefit from oral zinc supplementation. A safe dosing strategy is 1-3 mg elemental zinc/kg body weight per day, measuring plasma zinc before and several weeks after treatment initiation to document **1)** a lack of toxicity and **2)** a slight increase in plasma zinc concentration.

13. *Seizure control:* in our database, dogs with PSVA have increased risk for epilepsy. Diazepam intervention should be avoided.. Zonisamide (sulfa drug) and Keppra are preferred anticonvulsants. We do not prophylactically treat PSVA patients lacking a seizure history with anticonvulsants before ligation surgery. Postoperative seizures are exceeding rare (< 1/150 cases).

14. *Flumazenil:* the once popular benzodiazepine theory of HE is now refuted. Flumazenil, a benzodiazepine antagonists, was theorized to reverse benzodiazepine modified neurotransmission associated with HE. With increased clinical assessment and research, it is clear that flumazenil does not reliably alter HE unless benzodiazepines have been administered.

15: *BCAA vs AAA Adjustment:* In patients with PSS and hepatic insufficiency, an increased ratio of aromatic amino acids (AAA, phenylalanine, tryptophan, tyrosine):branched chain amino acids (BCAA, leucine, isoleucine, valine) reflects preferential metabolic (hepatic and peripheral) utilization of BCAA. Since BCAA and AAA share a common BBB transporter, the altered ratio favors AAA transport into the CNS where they function as false neurotransmitter. While adjustment of the AAA:BCAA ratio has been proposed to modulate HE, controlled studies fail to justify a consensus. Specific studies in Eck fistula dogs (model of PSVA) demonstrated that oral BCAA provide no benefit. Some evidence suggests that parenteral BCAA might provided an HE crisis intervention benefit.

Conditions Antecedent to Gallbladder Mucoceles (GBM)

Considering GBM cases in the author's hospital (n=100) and published case series, >200 canine cases have been reviewed with most from 2 Teaching Hospitals (Cornell- unpublished, U. Pennsylvania, published). This information provides the format for this discussion (n=176 cases). Risk factors for GBM formation include: 1.*Breed:* Shetland Sheepdog (most common), mixed breed dogs, Bichon frise, Miniature and Toy poodle, West Highland White Terrier, Dachshund, German Shepherd, and Cocker Spaniel. A mutation in ABCB4, a phospholipid translocater, flipping phosphatidylcholine across hepatocyte canalicular membranes into biliary canaliculi, has been defined in Shetland sheepdogs and other dogs with GBM. Median weight of GBM dogs is: 11.7 and 13.2 kg (3-17). *2. Age*: ~11 years (3-17) with no gender predisposition. *3. Endocrinopathies* (% in each hospital, Cornell first, U. Pennsylvania second): hyperadrenocorticism in 21% & 21%, hypothyroidism in 12% & 14%. Cornell case files only for the remainder- *4: Diabetes mellitus* in 6%. *5: Pancreatitis* in 12%. *6:GB Dysmotility i*n 10%. *7:High fat diet* in 10-%. *8: Protein Losing Nephropathy* in 6% (fed high fat food). *9: GB Neoplasia* in 5%. *10: Hyperlipidemia* in 21% (includes dogs in other categories). There are only 6 dogs in our data base lacking one of these predisposing factors. GB Dysmotility is assessed prior to mature GBM formation, using 3 dimensional US measurements and the equation for volume of a pronate ellipsoid, with 1.2 ml/kg volume considered normal and a $\geq$ 25% change in GB volume with feeding or feeding with erythromycin (1.0 mg/kg) as a motilin agonist. At GBM resection, a liver biopsy should be collected distant to the GB to determine the presence of associated liver disease or neoplasia. Importantly, this may disclose the presence of a diffuse vacuolar hepatopathy that may implicate an unidentified but associated steroidogenic endocrinopathy. Refrain from collecting liver immediately adjacent to and attached to the GB as this tissue is not representative of typical hepatic parenchyma having a greater number of unusual Von Meyenberg-like biliary complexes. Early cholecystectomy after recognition of GB dysmotility and early stage GBM formation has seemingly improved survival in dogs with this disorder. Infection may occur in GBM secondary to tissue ischemia, cholestasis, and perhaps underlying endocrinopathies or diabetes mellitus but in our hospital infection is uncommon.

Managing Hepatobiliary Infections

Predisposition of the hepatobiliary system to infection derives from its intimate contact with the splanchnic circulation, active macrophage functions, and the enterohepatic-biliary route of bacterial translocation. In health, bacterial invasion is prevented by the livers innate immunological defenses (Kupffer cell, biliary IgA), physical barriers (anatomical tight junctions), the mechanical effects of bile flow and mucus, and the chemical effect of bile acids. Disorders disrupting bile flow or causing hepatocellular necrosis or hypoxia increase risk for infection, especially when the enteric mucosal barrier function is altered (e.g. as in IBD). Another high risk scenario for hepatic infection occurs in dogs with a centrally necrotic hepatic adenoma or hepatocellular carcinoma (HCCA) in which a large hepatic abscess may form. These impose high risk for systemic infection during surgical manipulation. However, extracorporeal drainage and parenteral antimicrobial therapy can be ineffectual if HCCA is the underlying cause. These tumors remain local to the liver, range widely in size and histologic features and often support polymicrobial and anaerobic infections. Biliary tree obstruction increases risk for bacterial infection. Consequently, pre-operative antimicrobial therapy is important for animals undergoing biliary diversion or decompressive procedures. Several studies have shown that bile from normal animals may culture positive, especially when subcultured in enrichment broth. This finding complicates interpretation of cholecentesed bile cultures and underscores the importance of simultaneous cytologic assessment. Several studies have likewise shown greater likelihood of finding organisms in bile compared to liver in clinical patients with liver disease. The similarity of these observations complicates bacterial culture interpretation. While bacterial organisms are difficult to identify in histologic sections of liver even with special stains, the presence of bacter in bile or tissue may be confirmed using PCR of 16S rRNA primers or by using a universal eubacterial FISH probe (EUB-338). However, molecular probing for bacteria has not been practically proven to be more sensitive than optimally sampled and cultured specimens. Cats with nonsuppurative cholangiohepatitis usually do not harbor biliary or tissue infections and have been negative using molecular methods of bacterial detection. However, use of a universal eubacterial FISH probe (EUB-338) on canine samples lacking adequate culture evaluations, has proven fruitful for bacterial identification.

Cats with lymphocytic cholangitis or lymphoma rarely culture positive for bacteria and in a recent study were also shown to be bacterial negative with a universal eubacterial FISH probe applied to fixed sections. Conversely, cats with suppurative cholangiohepatitis often demonstrate bacterial organisms on cytologic samples of tissue, bile, or biliary structures, and have positive cultures derived bile and liver (bile samples are more commonly culture positive). In one report, bacterial culture (aerobic, anaerobic) of 190 canine samples (liver n=166, bile n=46; >80% US guided tru-cut) was more often "+" in cholecystitis or biliary tree obstruction (20% [1/5] tissue, 62% [8/13] bile) as compared to "inflammatory" liver disease (6% [6/103] liver, 23% [7/30] bile). In the author's hospital, bacterial cultures (aerobic, anaerobic) of liver/bile from >425 dogs (Cornell, surgical/laparoscopic samples) were "+" in ~25% for aerobic and in ~9% for anaerobic organisms. Of "+" cultures, ~25% were polymicrobial with primary GB or bile duct disorders predominating and with liver disease complicated by IBD. Most "+"cultures involved Enterobacteriaceae spp. Since "+" results also may occur in healthy animals (with samples sub-cultured in enrichment media), results must be interpreted in light of histopathology, cytology, and patient clinical signs. Bacterial cultures from 324 feline cases with hepatobiliary disease were positive in 11% for aerobic and in 2% for anaerobic organisms. Of "+" cultures, ~5.0% were polymicrobial.

Bacterial Susceptibility-Hepatobiliary Infections

Enterobacterial isolates have variable response to first generation amino-penicillins or cephalosporins, and >80% susceptibility to fluorinated quinolones, amikacin, or gentamicin; most enterococci are susceptible to penicillin or a fluorinated quinolone; most anaerobes to metronidazole. Thus, combination of a potentiated penicillin, fluorinated quinolone, and metronidazole is recommended as standard therapy before cultures are available and has had a significant positive influence on surgical patient survival in the author's hospital. Treatments should ultimately be modified in light of sensitivity results, patient response, and in respect to re-evaluation of bile/liver aspirates and cultures. Antibiotics should be given within 1-hr before surgical interventions, especially when biliary tree obstruction is suspected. GB bile and liver tissue should be collected for culture at surgery. Because bacteria are rarely visible on histopathology sections, cytologic imprints of bile and tissue should be inspected and bacterial morphology characterized by gram stain. If infection remains suspected but unproven, non-culture molecular method of bacteria detection can be used to investigate for infectious agents.

Consequences of Biliary Enteric-Anastomosi

Iimposes life-long risk for retrograde septic cholangitis/choledochitis. Dogs tolerate this procedure with fewer clinical signs than cats. Owners should monitor these patients for fever, inappetence, and vomiting, and seek veterinary assistance if cyclic illness is recognized (may indicate hepatic microabscessation). A CBC,serum biochemistry profile (liver enzymes) should be monitored quarterly. Chronic or pulsatile (intermittent) antimicrobial administration can be used to control ascending infection of biliary structures. Usually illness is antibiotic responsive and transient.

REFERENCES

Mealey KL, et al: Comp Hepatol. 2010;9:6 doi: 10.1186/1476-5926-9-6. Mesich ML, et al: J Small Anim Pract. 2009;50(12):630-635. Ramstedt KL, et al: Am J Vet Res 2008;69(5):647-651. Sepesy LM, et al: J Am Vet Med Assoc 2006;15;229:246-252. Kook PH, et al: J Vet Intern Med 2010;24:224-228. Wagner KA, et al: J Vet Intern Med. 2007;21:417-24.

CONTROVERSIES IN THE MANAGEMENT OF URETEROLITHS IN CATS

Larry G. Adams, DVM, PhD, DACVIM
West Lafayette, IN

INTRODUCTION

Nephroliths are uroliths located in the renal pelvis or collecting diverticula of the kidney and ureteroliths are uroliths in the ureter that result from migration of nephroliths or nephrolith fragments into the ureter. Nephroliths in cats are not treated unless they migrate into the ureter and cause ureteral obstruction. Ureteroliths often require treatment if they are causing partial to complete ureteral obstruction. Ureteroliths may pass down the ureter into the urinary bladder; however, we previously reported spontaneous retrograde movement of ureteroliths in 2 dogs and five cats.[1] We have also documented uroliths moving from the renal pelvis into the proximal ureter, and back into the renal pelvis in some animals.

For feline nephroliths and ureteroliths submitted to urolith centers at the Minnesota Urolith Center and the University Of California-Davis, calcium oxalate is the most common urolith composition (>90%), followed by dried solidified blood calculi, struvite, calcium phosphate, and urate.[2] Dried solidified blood calculi may be radiolucent and approximately 40% of the dried solidified blood calculi in cats were from the upper urinary tract in one study.[2] Feline ureteroliths are commonly are much more common than 20 years ago with one study revealing a 50-fold increase in the frequency of calcium oxalate uroliths in the upper urinary tract of cats from 1981 to 2000.[3]

Acute ureteral obstruction secondary to ureteroliths has become a common clinical syndrome in cats.[4] Many cats with acute uremia from ureteral obstruction have one enlarged kidney and one small kidney. The small kidney may be fibrotic and non-functional from prior ureteral obstruction. The cat may be asymptomatic until an acute uremic crisis develops from ureteral obstruction of the contralateral enlarged kidney. Renal enlargement may be from compensatory hypertrophy, hydronephrosis or a combination of the two. Cats may present with bilateral ureteral calculi, but relative renal function of both kidneys should be determined because both kidneys may not be functional. Urine output in patients with ureteroliths may be anuric, oliguric, polyuric or normal urine output depending on severity of ureteral obstruction. In one retrospective study, 9% of cats with ureteral obstruction also had signs suggestive of lower urinary tract disease (e.g., dysuria and pollakiuria).[4]

Ureteroliths are frequently diagnosed by a combination of abdominal radiography and ultrasonography. In one study ultrasonographic confirmation of ureteroliths was not possible in 23% of cats with ureteroliths.[4] Cats with moderate to severe azotemia from ureteral obstruction often have poor excretion of dye during excretory urography, and identification of the cause of ureteral obstruction may be facilitated by antegrade pyelography via nephropyelocentesis or contrast CT scan.

CONTROVERSIES IN MANAGEMENT – WHAT ARE THE OPTIONS?

Calcium oxalate and matrix nephroliths, which are the most common types of nephroliths in cats, are not amenable to dissolution. Similarly, ureteroliths are not amenable to medical dissolution because they are only intermittently bathed in urine. Therefore, medical dissolution is not a viable option for management of ureteroliths. Some authors have advocated benign neglect for ureteroliths that are not causing complete ureteral obstruction based on case examples where the ureteroliths eventually passed over weeks to months without intervention.[5] In veterinary medicine, the most common method of treatment of ureteroliths has been surgical removal by ureterotomy.[6, 7] Temporary diversion of urine via nephrostomy tube placement is may be used pre-operatively and post-operatively to assist with management of ureteral obstruction in cats. Because of the small ureteral lumen diameter of approximately 0.4 mm, even minor mucosal edema may cause ureteral obstruction after ureterotomy; so urinary some authors recommend diversion by placement of nephrostomy tubes. Movement or displacement of nephrostomy tubes is a common problem that can be minimized by utilization of locking loop design nephrostomy catheters. Alternatively, hemodialysis if available may be used for medical management of oliguric patients prior to surgical intervention.

Ureterotomy and primary closure of the feline ureter requires microsurgical expertise unless the ureter is grossly distended.[6, 8] For ureteroliths in the middle and distal ureter in cats, some authors recommend transection of the ureter and reimplantation of the healthy proximal ureter into the bladder (ureteroneocystostomy).[6, 7] Other authors prefer to attempt ureterotomy for removal of ureteroliths regardless of location. Success of surgical management (or ureteral stent placement) is dependent on the microsurgical skills of the surgeon and generalization of published results is not consistently applicable to all cases.

Medical management of ureteral obstruction

Conservative medical management by serial monitoring of ureteroliths along with administration of intravenous fluids and diuretics has been recommended provided there is minimal renal function compromise, and no infection, renal colic or progressive ureteral dilation. Typically, ureteroliths with evidence of complete obstruction, worsening azotemia, or evidence of pyelonephritis should be treated by either some interventional procedure. One suggested approach is that if a cat with partial to complete ureteral obstruction is monitored for two weeks and the ureterolith is not moving distally, intervention should be considered. An alternative approach is to be much more aggressive with quicker intervention to prevent renal damage from obstructive uropathy. Which of these approaches is the best approach is not definitively known.

Expulsive therapy may have a role in management of obstructing ureteroliths in cats provided the cat is not severely azotemic or metabolically compromised. Suggested therapy includes use of IV fluids with diuretics (e.g., mannitol), calcium channel blocking agents, alpha-adrenergic antagonists, amitriptyline, and glucagon. The medical therapy that is agreed on by most authors is the administration of IV fluids and diuretics to increase urine flow to facilitate ureterolith passage with serial monitoring of the ureterolith position by radiographs or ultrasonography.

In one recent report, 52 cats were treated by medical management (IV fluid administration and diuretics) alone and serial radiographs were monitored in 14 of these cats. The ureterolith passed spontaneously to the urinary bladder in 9 of these 14 cats.[6] However, 17 of the 52 medically managed cats did not respond to medical management and were euthanatized or died within 1 month.[6] In this study, the time interval from initial diagnosis to surgical intervention decreased from 19 days in the initial 11 cats to only 2 days for cats in the later part of the study. The optimal time to allow for passage of ureteroliths by medical management prior to surgical intervention is not known.

Although there are anecdotal clinical reports of glucagon facilitating passage of ureteral calculi in cats, the only clinical study of glucagon administration in cats did not demonstrate any benefit for management of obstructive ureteroliths and glucagon had an unacceptably high incidence of side-effects.[9] Given the unacceptable number of side effects and lack of documented efficacy, administration of glucagon is not recommended at this time.

Amitriptyline facilitated passage of urethral plugs in male cats with urethral obstruction.[10] Although the article title implies this study involved ureteral calculi, these cats all appeared to have urethral plugs rather than actual uroliths. Amitriptyline also inhibited contraction of urinary tract smooth muscle from rats, pigs and humans *in vitro* and was suggested as a potential therapy for promoting passage of ureteroliths. There are no other reports supporting the use of amitriptyline for expulsive therapy of ureteroliths.

In humans, multiple clinical studies have confirmed that the alpha-adrenergic antagonist tamsulosin (Flowmax®) facilitates passage of ureteral calculi compared to other medications.[11-14] Calcium channel blocking agents like nifedipine also facilitate passage of ureteroliths in humans.[15] Tamsulosin appears to be superior to calcium channel blocking agents such as nifedipine.[13] There are also other alpha-adrenergic antagonists and beta-adenoreceptor agonists that may facilitate expulsion of ureteroliths but none have been evaluated clinically in dogs and cats.[16, 17]

Impaction of ureteroliths causes inflammation and edema of the ureteral mucosa, which further impairs passage of the ureterolith. Therefore the use of steroids or non-steroidal anti-inflammatory medications is sometimes advocated as an addition to other expulsive therapy in humans.[12, 15] There are no trials examining the potential role of these medications in dogs and cats with ureteroliths. The author suggests avoiding NSAIDs in cats with ureteral obstruction due to the risk of nephrotoxicity in cats.

Treatment of nephroliths and ureteroliths in cats using shock wave lithotripsy

Lithotripsy is crushing or fragmenting uroliths by shock waves or laser energy. Types of lithotripsy include extracorporeal shock wave lithotripsy, electrohydraulic lithotripsy and laser lithotripsy. Shock wave lithotripsy (SWL) is used for treatment of nephroliths or ureteroliths. SWL uses repeated shock waves generated outside the body to fragment the uroliths into many small fragments that can pass spontaneously through the excretory system. Preliminary research with SWL in cats revealed that the feline kidney appears to be more susceptible to SWL induced renal injury than dogs and humans. Therefore, if the kidney is within the treatment area, shock wave dose must be reduced in cats compared to dogs to prevent shock wave induced injury to the kidney. We treated 5 cats with spontaneously occurring calcium oxalate ureteroliths and chronic renal failure with SWL. Four of these cats also had nephroliths. Treatment with SWL only resolved the ureterolith in one of these 5 cats. Partial fragmentation of the nephroliths occurred in two of four cats. Serum creatinine concentrations were transiently or permanently increased after SWL treatment in 3 of 5 cats.

We also demonstrated that feline calcium oxalate uroliths are significantly more difficult to fragment using SWL than were paired canine calcium oxalate uroliths.[18] Shock wave lithotripsy is unlikely to be as effective for treatment of nephroliths or ureteroliths in cats compared to dogs because the small diameter of the feline ureter (approximately 0.4 mm diameter) requires that all fragments be fine sand. Even with sand-like fragments, ureteral obstruction may occur in some cats.

Ureteral stents

Ureteral stents are a new interventional radiology procedure for management of ureteral stents in cats. Initial reports indicated that only 25% of human pediatric ureteral stents could be placed retrograde via cystoscopy using fluoroscopic guidance and up to 80% stents could be successfully placed surgically.[19] Recently 2.5 French diameter ureteral stents designed for cats have become available. In a recent report, these ureteral stents were successfully placed in 94% of cats.[20] In this report, 90% of the cats were azotemic before stenting and after stent placement, only 40% remained azotemic. The vast majority of cats had improvement in severity of azotemia with only 5% of cats not showing any improvement.

The benefits of ureteral stent placement include drainage of urine past obstructive ureteroliths and passive dilation of the ureter around the ureteral stent creating room for urine to pass both through and around the stent. Complications of ureteral stents include ureteral stent migration, pollakiuria, tissue reaction and inflammation to the stent material and UTI. The lumen of the stent may become encrusted resulting in obstruction if there is inadequate room for urine to pass around the outside of the stent. Stent encrustation appears to be uncommon in cats compared to humans. In humans, ureteral stents are either removed or exchanged within 6 months to prevent stent encrustation. Stent exchange does not appear to be necessary in cats.

Ureteral stents are the preferred interventional procedure for cats with multiple ureteroliths. In the one study of cats that received ureteral stents, surgery was not considered a viable alternative because most cats had multiple ureteroliths and many had concurrent nephroliths.[20] More than 80% of the cats would have required multiple ureterotomies and thus were better candidates for ureteral stent placement. For cats treated by conventional surgery, movement of remaining nephroliths into the ureter may cause ureteral obstruction weeks to months after successful ureterotomy.[6] The morbidity and mortality associated with ureteral stent placement are less than conventional ureteral surgery such as ureterotomy.

CONCLUSIONS

Medical therapy of ureteroliths may permit passage of ureteroliths into the urinary bladder and avoid surgical techniques. Optimal medical management to facilitate passage of ureteroliths in cats needs to be further defined through prospective, placebo-controlled clinical trials. Ureterotomy is a viable option for single ureteroliths, although success rate is dependent on the surgeon's microsurgical experience and skill. Ureteral stent placement is an interventional option for management of ureteroliths in cats and may be preferred over conventional ureteral surgery.

REFERENCES

1. Dalby AM, et al. *J Am Vet Med Assoc* 2006, 229:1118-1121. 2. Westropp JL, et al. *J Vet Itern Med* 2006, 20:828-834. 3. Lekcharoensuk C, et al. *J Am Anim Hosp Assoc* 2005, 41:39-46. 4. Kyles AE, et al. *J Am Vet Med Assoc* 2005, 226:932-936. 5. Lulich JP, et al. In *Consultations in Feline Internal Medicine*, 2005, 399-406. 6. Kyles AE, et al. *J Am Vet Med Assoc* 2005, 226:937-944. 7. Hardie EM, et al. *Vet Clinics N Am Small Anim Pract* 2004, 34:989-1010. 8. Kyles AE, et al. *J Am Vet Med Assoc* 1998, 213:1150-1156. 9. Forman MA, et al. *J Vet Itern Med* 2004, 18:417(Abstract). 10.Achar E, et al. *Kidney Int* 2003, 64:1356-1364. 11. Autorino R, et al *Urol Res* 2005, 33:460-464. 12. Dellabella M, et al. *Urology* 2005, 66:712-715. 13. Dellabella M, et al. *J Urol* 2005, 174:167-172. 14. Dellabella M, et al. *J Urol* 2003, 170:2202-2205. 15. Borghi L, et al. *J Urol* 1994, 152:1095-1098. 16. Wanajo I, et al. *Urol Res* 2005, 33:409-414. 17. Yilmaz E, et al. *J Urol* 2005, 173:2010-2012. 18. Adams LG, et al. *Am J Vet Res* 2005, 66:1651-1654. 19. Berent A, et al. *J Vet Itern Med* 2009, 23:688-689 (Abstract). 20. Berent AC. *J Vet Emerg Crit Care* 2011, 21:86-103.

THE CHALLENGES OF EMPIRICAL ANTIMICROBIAL THERAPY: A FOCUS ON UTI AND PYODERMA

Mark G. Papich, DVM, MS, DACVCP
Raleigh, North Carolina

INTRODUCTION

Empirical treatment of common infections in small animals has been developed by several authors and reviewed in the veterinary textbooks. A committee of experts has also recently developed consensus-based guidelines for empirical treatment of urinary tract infections (UTI) and pyoderma (superficial bacterial folliculitis). These guidelines were developed by the International Society for Companion Animal Infectious Diseases (ISCAID) and published on-line (Weese et al, 2011) as well as on their internet site (http://www.iscaid.org/). The guidelines for pyoderma treatment are in development (see presentation by Lloyd & Hillier) and will be published shortly.

These guidelines provide recommendations for empirical treatment of urinary tract infections and pyoderma. The empirical choice for initial treatment is based on the assumption that the infection is not complicated and the infection is caused by wild-type bacteria. It is critically important at this state to define what is meant by *wild-type* and *non-wild type* bacteria. *Wild-type* strains of bacteria are those that have an absence of acquired and mutational resistance mechanisms. Whereas *Non-wild-type* strains of bacteria are those that have the presence of an acquired or mutational resistance mechanism to the drug in question. Wild-type strains may include bacteria that have inherent resistance to antimicrobials. For example, wild-type anaerobic bacteria are inherently resistant to aminoglycosides by virture of a lack of an oxygen-dependent drug entry to the bacteria. Gram-negative bacteria of the Enterobacteriaceae and *Pseudomonas aeruginosa* are inherently resistant to macrolide antibiotics. Wild-type strains of bacteria may or may not respond clinically to antimicrobial treatment. Likewise, non-wild type strains may or may not respond clinically to antimicrobial treatment. The prediction of whether the bacteria will, or will not respond to treatment is commonly referred to as the "90/60 rule" (Doern & Brecher, 2011).

The 90/60 rule was derived from the observation that, in general, bacteria treated with antimicrobials to which the strain is sensitive will have a favorable therapeutic response in approximately 90% of the patients. On the other hand, when the bacteria is resistant to the antimicrobial administered, despite the susceptibility result, approximately 60% of patients will respond to therapy. In veterinary medicine, we have no data to confirm or challenge the 90/60 rule. The authors (Doern & Brecher, 2011) emphasize that these observations apply to immunocompetent patients with infections caused by a single bacteria, when the drug is expected to penetrate to the site of infection adequately. Most clinicians would agree that these cases do not comprise all of their patients. Many patients (especially those seen by specialists) have polymicrobial infections treated with more than one antibiotic, have pathologic changes that may affect drug distribution (eg, protein binding changes), have received oral antibiotics that are insufficiently absorbed, are immune compromised patients, or have infections at sites that are either poorly penetrated or diluted, or for which antibiotics are concentrated (for example from topical treatment or by tubular concentration prior to clearance by the kidneys).

EMPIRICAL TREATMENT

Susceptibility of the most common isolates has been documented well enough to make empirical antimicrobial drug choices. Drug manufacturers have produced several important drugs to treat the most common infections encountered in small animals. Most of the medications that can be used for empirical treatment are approved for use in small animals.

Bacterial Pyoderma

As discussed in another presentation at this forum (see Lloyd & Hillier), empirical treatment of pyoderma consists of administration of cephalosporins, amoxicillin-clavulanic acid (also known as potentiated amoxicillin), clindamycin, or trimethoprim-sulfonamides (which can include ormetoprim-sulfadimethoxine, trimethoprim-sulfamethoxazole, or trimethoprim-sulfadiazine). If the lesion is amenable to topical treatment, ointments with medications active against these pathogens also can be used. Other systemic drugs to which wild-type strains are usually susceptible include the β-lactam antibiotics cloxacillin, dicloxacillin, or oxacillin, the lincosamide lincomycin, and the macrolide erythromycin. However these alternatives are rarely used today because of poor oral absorption, lack of known efficacy, better alternatives within the same class, or lack of available formulations for animals.

Urinary Tract Infections

Empirical treatment of UTI assumes that the infection is uncomplicated. In the document produced by the ISCAID working group (Weese et al, 2011) uncomplicated UTI was defined as a bacterial infection of the bladder "in an otherwise healthy individual with normal urinary tract anatomy and function". In these patients, the urine is concentrated and normal patient defense mechanisms exist that will assist in eradicating the infection. Ordinarily, the low pH of the urine, high osmolarity, and presence of salts, urea, and organic acids inhibit bacterial colonization and growth. Empirical drug selection in these patients can consist of oral treatment with amoxicillin, amoxicillin-clavulanate, or trimethoprim-sulfonamides.

THE CHALLENGE OF EMPIRICAL ANTIMICROBIAL THERAPY

What is the veterinarian to do when empirical antimicrobial therapy is not appropriate or sufficient? This is the challenge for many specialists. The drugs and approaches to therapy are more limited, and less successful, when the infection is more refractory, resistant, or is associated with another complicating factor. The complicating factors for treating pyoderma include endocrinopathy, immune-mediated disorder, immune-deficiency (or drugs that suppress the immune response), and resistant bacteria. For UTI the complicating factors include diabetes mellitus, excess glucocorticoids, urinary or reproductive tract conformational abnormalities, prostate involvement, urinary calculi, kidney involvement, or recurrent episodes of UTI. In these instances other strategies and drugs may be necessary. When empirical antibiotic therapy fails for treatment of pyoderma it is most often because of antibacterial drug resistance. But when empirical therapy fails for treatment of UTI drug resistance is just one of the many possibilities that cause failure. The complicating factors of prostate involvement or kidney infection (pyleonephritis) also require special consideration. As with many new treatments, there are few veterinary clinical studies to support a recommended use and dose and many of these details have been extrapolated from human medicine.

TREATING UTI WHEN THERE ARE COMPLICATING FACTORS

Urinary tract infections that are refractory, recurrent, and/or caused by resistant bacteria can be some of the most frustrating cases to treat in veterinary medicine. Some of these patients are immunocompromised, diabetic, receiving corticosteroids, or have underlying neurological disease (McGuire, et al, 2002; Torres et al, 2005; Ihrke et al, 1985). Many of these patients have chronic kidney disease. Inability to concentrate urine produces inadequate urinary drug concentrations. Treatment of complicated UTI should be based on urine culture and susceptibility testing, but even with this information, it can be very difficult to eradicate the infection (Sequin et al, 2003). The first step is to obtain a culture, via cystocentesis and perform a urinalysis. Culture and susceptibility testing should be performed by a reliable laboratory that uses appropriate CLSI standards (CLSI, 2008). If the patient is symptomatic and the urinalysis indicates presence of an infection, treatment should be initiated. If the patient has asymptomatic bacteriuria (subclinical bacteriuria) treatment can be withheld and the patient simply monitored. Patients with subclinical bacteriuria can often resolve the infection spontaneously.

After initiating antimicrobial therapy, follow-up urine culture should be considered after approximately 5-7 days to ensure antimicrobial efficacy *in vivo*. Whereas duration of treatment for uncomplicated UTI is typically 7-14 days, the duration of treatment for a complicated infection is less established. Some clinicians treat for as long as 6 weeks, but the duration ultimately depends on whether or not the underlying complications can be resolved. With either complicated or uncomplicated infections, culture should be repeated approximately one week after antibiotics are discontinued, and again several weeks later if there is a history of recurrent infection.

Even though high urine concentrations of many drugs can be achieved in the urine owing to tubular concentration of the drugs (Ling 1984), in many cases high urine concentrations are not sufficient. Uropathogenic *Escherichia coli* can adhere to the bladder wall, invade the bladder mucosa, and survive for weeks (Mulvey et al, 2000; Schilling & Hultgren, 2002). In the bladder mucosa, the bacteria can either replicate and produce clinical disease, or enter a quiescent state and remain dormant and unresponsive to antibiotics. These bacteria can potentially serve as a source for recurrent infections. Effective therapies to target these bacteria have not been established.

Treating UTI when there is resistance

The most common resistance problem encountered in UTI are the gram-negative bacilli of the Enterobacteriaceae – most commonly *Escherichia coli*. Some can be producers of extended-spectrum β-lactamase (ESBL). Other bacteria that occur in UTI and can be resistant are *Klebsiella pneumoniae*, *Enterobacter*, *Pseudomonas aeruginosa*, *Enterococcus* species, and *Staphylococcus* species. (The treatment of methicillin-resistant strains of *Staphylococcus* is discussed with the treatment of pyoderma below.) After a susceptibility report is available, one may find that the only drugs to which some gram-negative bacilli are sensitive are extended-spectrum cephalosporins, carbapenems, or amikacin. Cefotaxime and ceftazidime are the injectable cephalosporins usually used to treat bacteria resistant to other β-lactam drugs, although individual veterinary hospitals have utilized others in this group. These drugs are expensive, injectable, and must be administered frequently. The activity of the oral third-generation cephalosporin, cefpodoxime proxetil (Simplicef) is variable among gram-negative bacteria. Cefpodoxime is more active than 1st-generation cephalosporins against gram-negative bacteria, but not as active as the injectable drugs in this class. It is not active against resistant *E. coli*, (eg, ESBL producers), *Pseudomonas aeruginosa*, *Enterococcus*, or methicillin-resistant *Staphylococcus*. The injectable 3rd-generation cephalosporin, cefovecin (Convenia) has greater activity than 1st -generation cephalosporins (Stegemann et al, 2007), but comparisons to other 3rd-generation cephalosporins are not available. Concentrations of free (unbound) cefovecin achieved systemically in plasma are not above the MIC of wild-type strains of non-urinary *E. coli* (Stegemann et al, 2006, 2007); therefore, the activity against resistant strains is doubtful.

The carbapenems are β-lactam antibiotics that include imipenem-cilastatin sodium (Primaxin), meropenem (Merrem), ertapenem (Invanz), and doripenem (Doribax). All have activity against the enteric gram-negative bacilli, including ESBL-producing strains. All except for ertapenem have good activity against *Pseudomonas aeruginosa*. Resistance (carbapenemases) among veterinary isolates has been very rare. Some disadvantages of imipenem are the inconvenience of administration, short shelf-life after reconstitution, and high cost. Meropenem is safer, and has antibacterial activity greater than imipenem against some isolates. One important advantage over imipenem is that it is more soluble and can be administered in less fluid volume subcutaneously with almost complete absorption (Bidgood & Papich, 2002).

Aminoglycosides can be valuable for treating gram-negative bacilli that are resistant to other drugs. They are rapidly bactericidal, less expensive than injectable drugs listed above, and can be administered once-daily. Among these, amikacin is the most active and is often the first choice in small animal medicine. It has been administered once-daily (Drusano, et al, 2007) for systemic infections IV, IM, or SC. One of the important disadvantages to systemic use of aminoglycosides: There is a risk of nephrotoxicosis, which increases with longer duration of treatment. Risk of nephrotoxicosis is much greater in patients with kidney disease and administration to these patients is discouraged.

Resistant Enterococcus

Drug-resistant enterococci can be isolated from UTI that can be persistent and difficult to treat empirically. The most common species identified are *Enterococcus faecalis* and *E. faecium*. *Enterococcus faecalis* is more common, but *E. faecium* is usually the more resistant. Wild-type strains of enterococci may still be sensitive to penicillin G and ampicillin, or amoxicillin. However, the enterococci have an inherent resistance to cephalosporins and fluoroquinolones. These strains also are usually resistant to trimethoprim-sulfonamide combinations, clindamycin, and erythromycin. Susceptibility test results for cephalosporins, β-lactamase resistant penicillins (eg, oxacillin), trimethoprim-sulfonamide combinations, and clindamycin can give misleading results (CLSI, 2008). Even if isolates are shown to be susceptible to a fluoroquinolone, this class of drugs may not be a good alternative for treatment. Treatment of *Enterococcus* is frustrating because there are so few drug choices. If the *Enterococcus* isolated is sensitive to penicillins, one should administer amoxicillin or ampicillin at the high-end of the dose range. Combination therapy with an aminoglycoside and a β-lactam antibiotic has not been shown to have an advantage for infections other than infective endocarditis in people. Occasionally, one of the carbapenems (imipenem-cilastatin) or an extended-spectrum penicillin (eg, piperacillin) can be considered for treatment of *E. faecalis* (but not *E. faecium*). When enterococci are present in lower urinary tract infections, the organism may exist with other bacteria such as gram-negative bacilli. In these cases, there is evidence that treatment should be aimed at the gram-negative bacilli and not directed at the enterococcus. Treatment cures are possible if the other organisms are eliminated without specific therapy for enterococcus (Bartlett et al 1978). In some cases, the only active drug will be a glycopeptide or the oxazolidinone linezolid, which are discussed below.

TREATMENT OF PYODERMA WHEN THERE ARE COMPLICATING FACTORS

Like the UTI, treatment of uncomplicated pyoderma is usually successful. Success rates published in clinical studies approach or exceed 90%. When there are complicating factors such as endocrinopathy, immune-mediated disorder, mite infection, immunosuppression, or resistant bacteria therapy is more challenging. Managing the other underlying problems will go a long way toward successful treatment of the bacterial infection and preventing recurrence. Despite adequate management of the primary problems, a common cause for antibacterial failure is drug resistance.

Resistant Staphylococcus

A review on the treatment of methicillin-resistant *Staphylococcus* was recently published by the author (Papich, 2012). Specific treatments and medications are provided in more detail in that review. Staphylococcus isolated from small animals is most likely to be *S. pseudintermedius* rather than *S. aureus*. (Note that previously identified *Staph intermedius* are now referred to as *S. pseudintermedius* by most laboratories.) β-lactamase production among *Staphylococcus* spp. is common and can easily be overcome with administration of β-lactamase resistant drugs (eg, cephalosporins), which are included in empirical regimens. The most important resistance mechanism for *Staphylococcus* is methicillin-resistance. Methicillin-resistance presents a problem for veterinarians because, in addition to resistance to β-lactam antibiotics, most of these bacteria are also multi-drug resistant (Bemis et al, 2009). The increased emergence of methicillin-resistant *Staphylococcus* in animals has been discussed in several publications and review articles (eg, Weese & van Duijkeren, 2010; Bemis et al, 2009).

If staphylococci are resistant to oxacillin or methicillin, they should be considered resistant to all other β-lactams, including cephalosporins and amoxicillin-clavulanate (eg, Clavamox), regardless of the susceptibility test result. Unfortunately, these bacteria often carry co-resistance to many other non- β-lactam drugs, including lincosamides (clindamycin, lincomycin), fluroquinolones, macrolides (erythromycin), tetracyclines, and trimethoprim-sulfonamides. In the report by Bemis et al (2009), more than 90% of the methicillin-resistant isolates of *S. pseudintermedius* also were resistant to > 4 other drugs.

Because susceptibility to non-β-lactam antibiotics is unpredictable, a susceptibility test is needed to identify the most appropriate drug to administer for these infections. Chloramphenicol, tetracyclines, aminoglycosides (gentamicin) and rifampin, are drugs to consider for these infections if a susceptibility test can confirm activity. Unlike the human strains of community-associated *Staphylococcus aureus* (CA-MRSA), the veterinary strains of methicillin-resistant *Staphylococcus pseudintermedius* (MRSP) are usually not susceptible to trimethoprim-sulfonamides, clindamycin, or fluoroquinolones (Perreten et al, 2010; Bemis et al, 2009). However, a susceptibility test should always be used to confirm whether or not these drugs may have activity against isolates from animals. Topical drugs also should be considered for treatment of localized infections. These drugs (eg, mupirocin or fusidic acid) are available in topical ointments and have been used in dogs and cats.

Drugs used in people for these infections, but rarely used in veterinary medicine, are linezolid (Zyvox), the first in the class of oxazolidinones, and the glycopeptides vancomycin. Both drugs are consistently active against methicillin-resistant strains of *Staphylococcus* species. However, both drugs have significant disadvantages. Linezolid can be administered orally to dogs and cats, and has been well tolerated, but is extremely expensive. Vancomycin is less expensive, but must be administered via IV infusion to maintain plasma concentrations above the minimally effective level (eg, > 10 μg/mL). Therefore, the use of these drugs in veterinary patients will probably remain rare.

REFERENCES

Bemis DA, et al. J Vet Diagn Invest. 2009 Jan;21(1):53-8. Bidgood T, & Papich MG. Am J Vet Res 2002;63(12):1622-1628. Clinical Laboratory Standards Institute. CLSI document M31-A3. Wayne, PA: Clinical and Laboratory Standards Institute; 2008 *(new edition to be published in late 2012)*. Doern GV & Brecher SM. J Clin Microbiol 49(9 suppl): S11-S14, 2011. Drusano GL, et al. Clin Infect Dis 2007; 45, 753-760. Ihrke PJ, et al. J Am Vet Med Assoc 1985; 186(1):43-46. Ling GV. J Am Vet Med Assoc. 1984;185(10):1162-1164. McGuire NC, et al. J Am Anim Hosp Assoc. 2002;38:541-544. Mulvey MA, Proc Natl Acad Sci U S A. Proc Natl Acad Sci U S A;97(16):8829-8835. Papich MG. Vet Derm (in press), 2012. Perreten V, et al. J Antimicrob Chemother 65: 1137-1144, 2010. Seguin MA, et al. J Vet Intern Med 17: 622-631, 2003. Stegemann MR, et al. Antimicrob Agents Chemother. 2006c;50(7):2286-92. Stegemann MR, et al. JVPT 29 : 501-511, 2006. Torres SM, et al. J Am Vet Med Assoc. 2005;227:239-243. Weese JS, et al. Vet Med Inter, Vol 2011, Article ID 263768. Weese JS, & van Duijkeren E. Vet Microbiol. 2010 Jan 27;140(3-4):418-29.

ANTIMICROBIAL THERAPY OF URINARY TRACT INFECTIONS IN DOGS AND CATS

D.M. Boothe, DVM, PhD, DACVIM, DACVCP for the * International Society for Companion Animal Infectious Disease (ISCAID), Antimicrobial Guidelines Working Group
Auburn, AL

INTRODUCTION

E. coli serves as an example of the pathophysiology of UTI. Ascending infection from the urethra is by far the most common route of infection in UTI. In both humans and dogs, the origin of bacteria infecting the urinary tract is generally fecal, with the frequency of infection by a particular strain depending on the virulence of the organism. Pathogens generally travel along the urethra to the bladder. Anatomic deformity and turbulent urine flow may facilitate antegrade movement of organisms toward the bladder. Female patients are more predisposed to ascending infection due to the shorter length of the tract and increased risk of contamination. Once in the bladder, infection can continue to ascend up the ureter to the kidney, particularly if vesiculoureteral refux is present. Uropathogenic *E coli (UPEC)* are capable of causing extraintestinal infection because they contain specialized virulence factors and as such, differ from commensals. *Escherichia coli* contains adhesins that bind to the glycolipid receptors. Mannose-containing receptors also have been documented in humans; *E. coli* (and other Enterobacteraicea) carries mannose-specific adhesions which, upon entry into the lower urinary tract, appear to be the primary mechanism of adherence for *E. coli* and thus are important for colonization of the bladder or lower structures. In contrast, mannose-resistant fimbriae and other adhesins appear to be critical for colonization of the renal structures. Binding between receptor and adhesin changes the receptor-bearing cell. The severity of a UTI may be correlated with the degree of adherence to uroepithelial cells. Organisms causing acute pyelonephritis in human patients are characterized by higher adherence compared with organisms causing asymptomatic bacteria. These *E. coli* produce a number of virulent factors, including hemolysin and aerobactin, and are able to resist the antibacterial effects of serum. Toxins such as endotoxin may act in conjunction with adherence to induce inflammation. Eventually, identification of virulence factors may facilitate detection of infection in tissues other than the uroepithelium, providing guidance regarding treatment.

SIMPLE UNCOMPLICATED URINARY TRACT INFECTION

Simple uncomplicated UTI is a sporadic bacterial infection of the bladder in an otherwise healthy individual with normal urinary tract anatomy and function. Clinically significant infection implies the presence of a clinical abnormality and is characterized by dysuria, pollakiuria, and/or increased urgency of urination along with the presence of bacteria in urine. None of these signs is considered pathognomonic for infection. Further, the committee did not confirm the need to treat covert bacteriuria/subclinical bacteriuria present in the absence of clinical signs. The minimum database for evaluation of suspected UTI should include complete urinalysis, including urine-specific gravity, urine glucose level determination, and examination of the sediment for crystalluria is considered. Underlying causes of infection must be indentified and treated if possible. Bacterial C&S also should be performed in order to confirm the presence of infection, and equally important, to identify the presence of resistant bacteria that may not respond to initial therapy. C&S is particularly important in light of patterns of increasing resistance particularly for uropathogenic *E. coli.* (see Figure 1 and Table 1). Increasing resistance limits empirical use of drugs that even a decade ago might have been considered effective. Culture may also help differentiate reinfection from relapse should a UTI return. Routine culture offers the added advantage of provide the clinician with cumulative data regarding the most common bacteria (and their susceptibility) causing infection in their

Figure 1. An antibiogram for Dogs collected at an academic teaching hospital (2007-2010). The number of isolates is indicated in the first column; the percent of isolates susceptible to each drug is indicated in the cell. A blank cell indicates that the drug is inappropriate for that organism (e.g., cephalosporins and *Enterococcus*). Based on this data, close to 50 % of *E. coli* isolates are resistant to first choice drugs (e.g,

Canine	No. of Isolates	Amikacin	Amoxicillin/CA	Ampicillin	Cefazolin	Cefoxitin	Ceftiofur	Cephalothin[b]	Chloramphenicol	Clindamycin	Enrofloxacin	Erythromycin	Gentamicin	Marbofloxacin	Orbifloxacin	Oxacillin	Penicillin	Tetracycline	Ticarcillin	Ticarcillin/CA	Trimethoprim/Sulfa
Enterococcus faecalis	30		97	97					83		40	28 (29)		40 (15)	10		93	63			
E. faecium	13		15	13					100		0	15		0 (11)	0		15	23			
Escherichia coli	120	98	63	48	69	65	76 (103)	46	81		63		83	65	62 (119)			68	52	64	67 (119)
Klebsiella pneumoniae	30	80	70	0	70	70	63	70	80		77		77	77	73			73	0	70	77
Proteus mirabilis	32	100	97	94	94	97	100	94	94		97		91	100	91				94	100	81
Pseudomonas aeruginosa	61	100							0		64		97	76 (50)	25				97	98	
Staphylococcus aureus	8	100	50	13	50	50	29 (7)	50	100	50	75	50	88	75	75	50	13	88			88
S. intermedius	102	92	66	17	64	64	60 (89)	64	97	43 (101)	55	43	87	63	60 (99)	72	18	42			60 (101)
S. schleifen ss coagulans	20	95	65	45	65	65	59 (17)	65	100	95 (19)	30	100	70	50	40	65	50	95			100
S. pseudintermedius	4	75	50	0	50	50	33 (3)	50	100	50	75	50	75	75	75	50	0	50			75

practice. Samples should be collected by cystocentesis; catheterized samples are acceptable (particularly for male dogs, if properly collected), but free-catch (midstream voiding or manual expression) samples are not. Results of samples that take 24 hours or more to reach the laboratory should be interpreted with caution; "urine paddles" may facilitate accuracy. Whereas in-clinic testing can minimize the impact of sample deterioration and can be cost-effective, this should be for screening purposes; bacterial isolation should only be attempted in clinics with appropriate laboratory facilities. Susceptibility testing should be performed according to accepted standards, such as those published by the CLSI or EUCAST, or another internationally recognized public standard. Culture should be quantitative such that colony counts can facilitate interpretation. of results. For samples collected by cystocentesis, any level of bacterial growth may be significant, although samples from a UTI typically contain ≥10^3 colony forming units (CFU)/mL. Small numbers of minimally pathogenic skin commensals (i.e., coagulase negative staphylococci) likely represent contamination. For samples collected via catheter, bacterial counts ≥10^4 CFU/mL in males and ≥10^5 in females are typically considered significant. Samples with lower counts should be interpreted with caution. The interpretive criteria for susceptibility testing and breakpoints for systemic infections also apply to UTIs. Those drugs for which urinary tract-specific breakpoints have been provided are limited to ampicillin or amoxicillin in dogs (≤8 µg/mL), amoxicillin-clavulanate in dogs and cats <8/4 µg/mL), and nitrofurantoin. Antimicrobial therapy should begin while awaiting C&S results to relieve patient discomfort. Appropriate choices include amoxicillin (11–15 mg/kg PO q8h), or trimethoprim-sulfonamide (15mg/kg PO q12h) (Tables 1 and 2); Amoxicillin/clavulanic acid (12.5–25 mg/kg PO q8h) is an acceptable option but is not recommended initially because of the lack of evidence regarding the need for clavulanic acid and the desire to use the narrowest spectrum that is possible while maintaining optimal efficacy. Antibiograms might be consulted. Objective data regarding the prevalence of resistance that indicates a need to change initial therapy is lacking. If baseline resistance ratesto a given drug from non biased sample collection exceed 10%, the drug chosen for initial therapy should be changed to another of the recommended initial choices (see Table 1). If C&S indicates the presence of an isolate that is resistant *in vitro* to initial therapy but the patient has responded to therapy, the chosen protocol might be maintained assuming a follow-up urinalysis and C&S confirms efficacy. If the patient has not responded, an alternative drug should be chosen. The lack of adequate evidence regarding duration of treatment precludes specific recommendations. Typically, uncomplicated UTIs are treated for 7–14 days. However, a shorter treatment time (≤7 days) may be effective. Clinical trials supporting shorter durations for treatment of UTIs in dogs and cats are strongly encouraged. There is currently insufficient evidence supporting the use of other adjunctive treatment measures (e.g., cranberry extract) for uncomplicated UTIs in animals. Specific follow up (e.g, reculture) is not indicated if response is sufficient.

Table 1: Resistance of *E. coli* canine and feline uropathogens (n=1500) in the US (2008-2010).

Drugs	% Resistance	95 % CI
Meropenem	1.26	0.76-1.96
Gentamicin	7.94	6.62-9.30
Cefoxitin	8.47	7.11-9.98
Trimeth-Sulfa	8.86	7.48-10.41
Cefotazidime	8.93	7.54-10.48
Ciprofloxacin	9.32	7.72-11.13
Enrofloxacin	11.64	10.07-13.36
Chlormphenicol	12.63	11.00-14.41
Cefotaxime	12.76	11.12-14.55
Cefpodoxime	12.95	11.31-14.76
Doxycycline	15.61	13.81-17.54
Ticarc-clav	19.44	17.48-21.53
Amoxicillina	41.53	39.04-4.02
Ampicillin	49.07	46.53-51.63
Cephalothin	63.1	60.61-65.53

COMPLICATED URINARY TRACT INFECTION

A UTI is considered complicated if it occurs in the presence of an anatomic or functional abnormality or a comorbidity
that predisposes the patient to persistent infection, recurrent infection, or treatment failure. Concurrent disease complicating UTI might include prostatitis, urinary calculi, a neurogenic bladder, pregnancy, diabetes mellitus, or immunocompromising disorders (humans). *Recurrent UTIs* are defined by the presence of 3 or more episodes of UTI during a 12-month period. *Reinfection* is recurrence of a UTI within 6 months of cessation of previous, apparently successful treatment and isolation of a different microorganism. Evaluation of antimicrobial susceptibility pattern can be helpful, although unrelated organisms can have the same susceptibility pattern. *Relapse* is recurrence of a UTI within 6 months of cessation of previous, apparently successful treatment and isolation of an indistinguishable organism from the one that was present previously; failure to completely eliminate the pathogen is assumed. In general, relapses tend to occur earlier than reinfections (i.e., within weeks rather than months) and are characterized by a period of apparent bladder sterility during treatment. Isolation of the same bacterial species, even if the same antimicrobial susceptibility pattern is present, does not necessarily indicate that the organism was not successfully eliminated. A *refractory* infection is similar to a relapse except that it is characterized by persistently positive results using culture during treatment (despite *in vitro* susceptibility to the antimicrobial), with no period of eliminated of bacteriuria during or after treatment.

Diagnosis is as described for uncomplicated UTI, with the addition of further diagnostic work to determine the extent of illness. Culture of a bladder wall biopsy and is recommended if cystotomy is performed. Failure to find and correct the underlying cause may result in therapeutic failure. A complete physical examination, including rectal palpation and examination of the vulva, is required. Referral to a specialist in internal medicine or surgery (particularly in females) should be considered to further investigate underlying causes (e.g., cystoscopy). Ideally, therapy should be withheld until C&S

results are received, if permitted based on patient needs. If treatment must be initiated immediately, a "first tier" drug – that is, one recommended for initial treatment of simple uncomplicated UTI – should be chosen; ideally, the drug will be different from that used to treat the prior UTI(s). After treatment has been initiated, continued treatment should be based on the results of culture and susceptibility testing. Preference should be given to drugs that are excreted in urine predominantly in an active form. If more than one bacterial species is identified on initial C&S, the relevance of the each organism should be considered, based on the bacterial counts and the pathogenicity of the organisms. Anecdotal evidence suggests that infection by *Enterococcus* spp. will often resolve when the other organism is successfully treated. Ideally, antimicrobial therapy should be directed against both organisms. In some instances, an antimicrobial effective against both organisms will not be available. Combination therapy that would be potentially effective against both organisms should be considered. Targeting therapy towards the organism perceived as the most clinically relevant is a reasonable approach. There is no supporting evidence for efficacy of drugs that decrease bacterial biofilm nor that direct instillation of antimicrobials, antiseptics, or DMSO directly into the bladder via a urinary catheter is effective for treatment of recurrent UTIs. As with uncomplicated UTI, evidence supporting the duration of therapy for complicated UTI does not exist. Typically, 4 weeks of treatment has been recommended although it is likely that shorter courses of appropriate treatment might be effective in some or all situations, particularly if the underlying disease can be controlled. Followup monitoring should consist of urine C&S 5–7 days after initiation of therapy, particularly in patients with a history of relapsing or refractory infection, or those considered at high risk for ascending or systemic infection. Any bacterial growth during treatment indicates potential treatment failure and should prompt immediate reevaluation. Referral or consultation with a specialist is recommended. If a positive urine culture is obtained after treatment, more in-depth investigation of predisposing factors for relapse or reinfection should be performed. Referral is strongly recommended. There is insufficient evidence to recommend use of either pulse (intermittent) or chronic low dose therapy for prevention of UTIs or methanamide or other antiseptics. Although there is some evidence for a benefit of nutritional supplements (e.g., cranberry juice extract) for this indication in humans, the evidence is not strong and some studies have shown no effect.

.MISCELLANEOUS

Subclinical bacteriuria is the presence of bacteria in the urine as determined by positive bacterial culture, in the absence of clinical and cytological evidence of UTI. Treatment may not be necessary in animals that have no clinical signs of UTI and no evidence of UTI based on examination of urine sediment. In some circumstances, treatment may be considered if there is concern that there is a particularly high risk of ascending or systemic infection (e.g., immunocompromised patients, patients with underlying renal disease) or that the bladder may be a focus of extraurinary infection.

Urinary Catheters. UTI and subclinical bacterial colonization of the bladder are commonly identified in dogs with indwelling urinary catheters. ***Without clinical signs of UTI.*** Culture in the absence of clinical signs consistent with infection is not recommended in the absence of clinical signs. Treatment is not necessary for catheterized animals with bacteriuria in the absence of clinical or cytological evidence of infection supporting the presence of an infection. Prophylactic antimicrobial therapy for prevention of UTI in catheterized animals is never indicated. Culture of the catheter tip at time of catheter removal is not indicated since catheter tip culture results are not predictive of development of catheter-associated UTI. However, culture of the urine (ideally collected by cystocentesis after catheter removal) may be reasonable in patients in which the risk and implications of UTI are high (e.g., recently obstructed male cat with at high risk for reobstruction). ***With Clinical Signs:*** UTI should be suspected in catheterized animals exhibiting clinical signs of infection. If a catheter must remain in place and signs of UTI are present, the catheter should be replaced and a urine sample collected through the new catheter for culture. Several milliliters of urine should be withdrawn to clear the catheter and discarded prior to obtaining the sample for culture. When possible, catheters should be removed and urine sampled by cystocentesis after an appropriate period of time has passed for the bladder to fill with urine. Less ideal is a sample collected from the urinary catheter prior to removal, in which case positive cultures should ideally be confirmed with a subsequent cystocentesis sample. In all instances, cultures should be quantitative. Urine culture should never be performed from the collection bag. Treatment is more likely to be successful if the catheter can be removed.

Upper UTI: Pyelonephritis: C&S (cystocentesis) interpretive criteria should be based on serum. Treatment should be initiated immediately. If regional data are supportive, treatment with a fluoroquinolone excreted in urine in the active form (e.g., not difloxacin) is a reasonable first choice. If infection is ascending, urine culture results obtained for diagnosis of lower UTI might be considered; if hematogenous, initial therapy should be based on cultures of blood or the infected site. Treatment of 4–6 weeks is often recommended although a shorter duration of therapy might be effective. Urinalysis and culture should be performed 1 week from the start of treatment because of potential severity of disease and long treatment duration. If the same organism is isolated, one should consider adding an antimicrobial to which the organism is susceptible *in vitro*, if possible. Consultation with a specialist is recommended

Multidrug-resistant Infections. Organisms (e.g, various *Enterobacteriaceae*, staphylococci, and enterococci) may represent a public health concern with regard to the potential for zoonotic transmission of resistant pathogens. Judicious use of drugs important to humans is supported if appropriate criteria are met (refer to published guidelines).

* The Antimicrobial Use Working Group, sponsored unconditionally by Bayer Animal Health, is comprised of a spectrum of representatives from specialty and non specialty groups supporting veterinary medicine and human microbiologists. Their purpose was to discuss and reach consensus recommendations for the treatment of bacterial infections in selected body systems, whether empirical or culture based. For each system, pathophysiologies of disease were delineated, potential target organisms identified and aspects of antimicrobial clinical pharmacology relevant to therapeutic interventions were collated such that recommendations in support of judicious antimicrobial use were promulgated. These recommendations for treatment of UTI as summarized here are published at http://www.hindawi.com/journals/vmi/2011/263768/ (public access). Recommendations for treatment of superficial pyoderma and respiratory tract infections are yet to be published; synopses are published elsewhere in this proceedings. During the course of guideline development, significant limitations in objective, published information were recognized. The guidelines thus are based on available data, expert opinion, principles of infectious diseases, antimicrobial therapy, antimicrobial resistance, pharmacology, and internal medicine. The guidelines are intended as general recommendations and should be applicable to the majority of cases. They are not intended to substitute for standards of care, but rather are offered as the basis of decision-making. Because these guidelines are intended for international dissemination, regional differences (e.g., antimicrobial resistance rates, antimicrobial availability, prescribing regulations) must be considered. **The members of the ISCAID Working Group were Scott Weese (chair), Joseph Blondeau, Dawn Boothe, Edward Breitschwerdt, Luca Guardabassi, Andrew Hillier, Michael Lappin, David Lloyd, Mark Papich, Shelley Rankin, Jane Sykes, and John Turnidge.

REFERENCES

Guardabassi L, et al. Guidelines for antimicrobial use in dogs and cats. In: Guardabassi L, Jensen LB, Kruse H (eds.), Guide to Antimicrobial Use in Animals, Blackwell Publishing, Oxford, 2008, pp.182-206. Devriese LA, et al. Boothe DM : Principles of Antmicorbial Use ; Antimicrobial Drugs and Treatment of Bacterial Infections ; in Boothd DM (ed) Small Animal Clinical Pharmacology and Therapeutics ; Elseviere, 2012 ; Wiles TJ, Kulesus RR, Mulvey MA. Origins and virulence mechanisms of uropathogenic Escherichia coli. Exp Mol Pathol. 2008 Aug;85(1):11-9. ; Soto SM, Smithson A, Horcajada JP, Martinez JA, Mensa JP, Vila J. Implication of biofilm formation in the persistence of urinary tract infection caused by uropathogenic Escherichia coli. Clin Microbiol Infect. 2006 Oct;12(10):1034-6 ; Shaheen BW, Boothe DM, Oyarzabal OA, Wang C, Johnson CM. Evaluation of the contribution of gyrA mutation and efflux pumps to fluoroquinolone and multidrug resistance in pathogenic Escherichia coli isolates from dogs and cats. Am J Vet Res. 2011 Jan;72(1):25-32. Shaheen BW, Boothe DM, Oyarzabal OA, Smaha T. Antimicrobial resistance profiles and clonal relatedness of canine and feline Escherichia coli pathogens expressing multidrug resistance in the United States. J Vet Intern Med. 2010 Mar-Apr;24(2):323-30. Platell JL, Cobbold RN, Johnson JR, MA, Trott DJ. Commonality among fluoroquinolone-resistant sequence type ST131 extraintestinal Escherichia coli isolates from humans and companion animals in Australia. Antimicrob Agents Chemother. 2011 Aug;55(8):3782-7. J. W. Bartges, "Diagnosis of urinary tract infections," *Veterinary Clinics of North America—Small Animal Practice*, vol. 34, no. 4, pp. 923–933, 2004.

ANTIMICROBIAL THERAPY OF SUPERFICIAL BACTERIAL FOLLICULITIS IN DOGS

David H. Lloyd, PhD, BVetMed, FRCVS, DECVD, Andrew Hillier, BVSc, MACVSc, DACVD and the International Society for Companion Animal Infectious Disease (ISCAID) Antimicrobial Guidelines Working Group
London, UK

INTRODUCTION

Dogs are highly susceptible to cutaneous infection and particularly to superficial bacterial folliculitis (SBF), the most frequently encountered form of canine pyoderma. This disease is routinely treated with systemic antimicrobials and is probably the principal reason for the use of such drugs in small animal practice. The cause of infection in the great majority of cases is *Staphylococcus pseudintermedius*, which is able to invade the skin as a consequence of reduced immunity associated with a variety of other conditions. Such conditions can be difficult to diagnose and resolve, particularly in general small animal practice. Thus SBF is frequently a recurrent disease and affected animals receive repeated courses of systemic antimicrobial drugs. In the past, *S. pseudintermedius* (formerly known as *S. intermedius*) infections could be routinely treated successfully with beta-lactam antimicrobials, to which this species did not show a propensity to develop resistance. However, methicillin-resistant *S. pseudintermedius* (MRSP), which is highly multi-resistant, is now rapidly spreading throughout the world. This has focused attention on the need to ensure that diagnosis and treatment of SBF is managed in an optimal way so as to resist the spread of this organism, the generation of even more resistant clones, and the wasteful and ineffective use of antimicrobial drugs to which this organism is resistant.

DEVELOPMENT OF GUIDELINES FOR THERAPY

The emergence of MRSP has occurred at a time when there is already worldwide development of multiresistance amongst a variety of pathogens causing disease both in humans and in animals. In the human field the employment of guidelines for antimicrobial use has been shown to be effective in improving antimicrobial stewardship and prescribing practices, particularly in hospitals. It is increasingly recognised that there is a need for such guidelines to be developed in the veterinary field and that these should be, as far as possible, applicable on an international basis. This awareness lead to the formation of a Working Group on Antimicrobial Guidelines by the International Society for Companion Animal Infectious Disease (ISCAID) in 2009. One of the priorities that the group recognised was for guidelines on the therapy of SBF and this paper summarises the concepts behind the development of such guidelines and outlines the recommendations which have been developed. It should be noted that these guidelines are aimed primarily at general veterinary practitioners; they will be published in detail elsewhere. Recommendations for the treatment of other canine bacterial infections of the skin will be developed also and published separately. The members of the ISCAID Working Group were Scott Weese (chair), Joseph Blondeau, Dawn Boothe, Edward Breitschwerdt, Luca Guardabassi, Andrew Hillier, Michael Lappin, David Lloyd, Mark Papich, Shelley Rankin, Jane Sykes, and John Turnidge. Input was also obtained from panels of diplomates of the American College of Veterinary Dermatology and the European College of Veterinary Dermatology (ECVD).

PATHOGENESIS, CLINICAL SIGNS AND DIAGNOSIS OF SBF

Although SBF is normally caused by follicular infection with *S. pseudintermedius*, other pathogenic staphylococci such as *S. aureus*, and coagulase variable (e.g. *S. schleiferi*) or coagulase negative species may be occasionally involved. Lesions compatible with SBF may also be rarely caused by other bacterial species, which may also act as secondary invaders, particularly in more serious cases. Staphylococci are not able to invade normal skin and thus SBF only occurs when there is some kind of underlying problem. This may be related to diseases such as hyperadrenocorticism, hypothyroidism, demodicosis and allergic dermatoses. The clinical signs of these underlying diseases need to be differentiated from those of SBF.

The early cutaneous signs of SBF are typically erythematous papules and pustules associated with the hair follicles; however, the pustules can be quite short lived and may only be found on careful examination. Subsequently, annular areas of alopecia, scaling, erythema and hyperpigmentation may appear, commonly surrounded by thin crusts (epidermal collarettes). Thicker crusts may be present over infected areas in more severely affected skin. Follicular infection generally causes hair loss which may be apparent as alopecic patches particularly over the trunk and in short coated breeds.

Diagnosis of SBF is based on the characteristic clinical signs but should be supported by cytological examination and the demonstration of coccoid bacteria associated with the presence of inflammatory cells and within phagocytes. However, failure to demonstrate such cytological features does not rule out bacterial infection. Inflammatory infiltrates and phagocytosis may be reduced or absent in patients with immunosuppression associated with underlying diseases or therapy. Cytology is essential if pustules are absent or rare, and if typical lesions are present but there is a poor response to empirical antimicrobial therapy. Positive cytology in the face of failure to culture bacteria from lesions should prompt repeated culture rather than the diagnosis of sterile pustular disease. Cytology is also very important for identification of co-infection with other microorganisms, particularly the yeast, *Malassezia pachydermatis* and, rarely, rod-shaped bacteria. Cytology is a powerful technique in the diagnosis of SBF but is underused in veterinary practice. This is due, in part, to lack of familiarity with cytological techniques and lack of confidence in their efficacy. The guidelines will not attempt to describe cytological techniques but practitioners will be encouraged to study and become proficient in such diagnostic procedures.

SBF needs to be distinguished from other causes of folliculitis and pustular disease. It should be routinely differentiated from demodicosis by deep skin scrapings and from dermatophytosis by fungal culture and such tests are required when history and clinical signs are suggestive but not typical of SBF, or when antibacterial treatment fails. Sterile pustular diseases must also be considered although these conditions are relatively rare e.g. pemphigus foliaceus, and sterile neutrophilic or eosinophilic pustulosis. These diseases are differentiated by absence of bacteria on cytology and the presence of acantholytic cells (pemphigus foliaceus), by failure to culture bacteria and lack of response to antimicrobial drugs, and by histopathology.

THE NEED FOR BACTERIAL CULTURE AND ANTIMICROBIAL SUSCEPTIBILITY TESTING

Bacterial culture in cases of SBF is essential if there is a poor response to two weeks of appropriate systemic antimicrobial therapy, emergence of new lesions two weeks or more after the initiation of such therapy, presence of residual lesions after six weeks of therapy combined with cytology demonstrating infection with cocci, and when cytology demonstrates intracellular bacterial rods. Culture and susceptibility testing is strongly encouraged in cases of recurrent SBF infection and multiple antimicrobial drug therapy. However, where initial treatment of SBF has been limited to topical antimicrobials alone, empirical choice of systemic antimicrobial drugs is acceptable. Bacterial culture in cases of SBF is, however, never contraindicated. Ideally, evaluation of response to therapy should always be done by the clinician but often this is not the case and owners are relied upon to report progress. Thus owners need to be informed of the changes that they should expect in response to antimicrobial therapy, namely reduction and disappearance of papules, pustules and crusts. They need to be able to understand how to differentiate these from clinical signs of any underlying diseases, which may include alopecia and seborrhoeic changes, and symptoms such as pruritus. It is important for owners to be able to recognise when progress is not being made so that they may seek veterinary advice prior to the end of a course of treatment, which in the case of SBF is recommended to be three weeks or more initially, allow changes to be made if necessary.

It is particularly important that samples for bacterial culture should be appropriately obtained. Ideally, samples should be taken aseptically from pustules, and careful examination of the skin with good illumination and removal of the coat in hairy areas is necessary for this process. If pustules are not found samples may be taken from pus beneath crusts, and from papules or epidermal collarettes. No surface disinfection should be carried out. To maximise survival of significant organisms, transport medium (e.g. Amies Transport Medium, Oxoid CM0425) should always be used and delivery of specimens to laboratories should occur without delay. There are no published reports showing that current use of antimicrobial drugs affects isolation of causative bacteria in canine SBF and samples may be collected for culture and sensitivity testing whenever indicated. Laboratories used for such testing should be those that follow protocols, including updated breakpoints for animal species, published by the Clinical and Laboratory Standards Institute (CLSI) and other internationally recognised institutions. Tests should include representatives of all antimicrobial drugs which submitting clinicians intend to consider for therapy, taking into account regional and national restrictions relating to the use of specific drugs in animals, where these exist.

Laboratories should carry out appropriate tests to differentiate coagulase positive and coagulase negative staphylococci and to distinguish *S. aureus*. This is important because breakpoints for antimicrobials may differ between *S. aureus* and other veterinary coagulase positive staphylococci, and because of public health issues. It cannot be assumed that coagulase positive staphylococci isolated from canine lesions are *S. pseudintermedius*. It should be noted that automated systems used in human medicine to speciate staphylococci may not be reliable for veterinary species, particularly in the identification of *S. pseudintermedius* and *S. schleiferi*. Microbiology reports need to be interpreted carefully by clinicians, bearing in mind public health considerations as well as the patient's clinical status and therapeutic history.

ANTIMICROBIAL THERAPY OF SBF

Selection of the modes of therapy and antimicrobials to be used in SBF always needs to be related to the nature of the disease in the affected dog. In particular, the severity and extent of lesions, concurrent disease and any other drugs which are being administered, and the ability to perform topical therapy, will influence the choice of antimicrobial drugs, in addition to known or expected antimicrobial resistance. Reliance on empirical choice of systemic antimicrobial drugs is now being challenged by multiresistance amongst the causative staphylococci, particularly in certain regions, and thus practitioners need to become familiar with the prevalence of multi-resistant staphylococci in the areas which they serve.

Drugs registered for use in the treatment of SBF should be used whenever possible and appropriate, and "on label" products should be used rather than generics. Clinicians should ensure that owners are aware of the requirements of the selected therapy and are able to complete the selected treatment protocols appropriately, particularly where topical therapy is involved. Contact with owners should be maintained to promote effective compliance and to determine if and when failure to respond to treatment, or recurrence, occurs. In recurrent infection, an effective diagnostic plan should be formulated to identify and resolve any underlying disease as this is the best means of controlling recurrence and reducing antimicrobial drug use. Owners should be made aware of the implications of the diagnostic plan and the necessity for dealing with the underlying disease.

Topical therapy offers considerable benefit in the treatment of SBF. It can promote more rapid lesion resolution and reduce the duration of antimicrobial drug therapy when combined with systemic antimicrobials. It can also be effective as a sole therapy, particularly in mild or moderate infections. There is very little evidence suggesting that the bacteria involved in canine SBF, even those which are multi-resistant, are not susceptible to the antimicrobial agents available for topical application in the dog at the concentrations expected at the skin surface when they are used. Topical therapy has been used successfully to treat difficult cases of MRSP infection both as a single therapy and in combination with systemic antimicrobial drugs. Topical products which have been shown to be effective include antimicrobial ointments, gels and creams, which can be applied to localised lesions particularly in relatively hairless areas, 2 to 3 times daily. Such products may contain antibiotics (mupirocin, fusidic acid, novobiocin, bacitracin, pristinamycin), silver sulfadiazine, benzoyl peroxide and hydroxyl acids e.g. acetic acid, lactic acid, malic acid. More extensive areas of infection can be treated with shampoos, lotions, sprays and rinses. These contain antiseptics such as chlorhexidine, benzoyl peroxide, ethyl lactate, povidone iodine, triclosan and hydroxyl acids; they are commonly used 2 to 3 times weekly and until seven days after lesions resolve, with contact times of 10 minutes if possible before rinsing or conditioners. They can then be used for prophylaxis on a weekly or less frequent basis depending on response.

Systemic antimicrobial therapy is very effective in SBF. Selection of appropriate drugs will depend on availability, safety, cost, local prevalence of resistant staphylococci and patient specific factors. Empirical choices can be made in non-recurrent cases and in animals which have not been exposed to multiple courses of antimicrobial drugs. Otherwise, drug selection should be based on culture and sensitivity tests. First-tier empirical drugs which can be used in cases of confirmed or suspected SBF include clindamycin, first generation cephalosporins, potentiated sulphonamides, erythromycin, lincomycin and doxycycline. When first-tier drugs are not appropriate and topical therapy cannot be used, second tier drugs may be chosen; these can include cefovecin and cefpodoxime, fluoroquinolones, doxycycline, chloramphenicol and rifampin. Use of third tier drugs, which include vancomycin, teicoplanin and linezolid, may be considered when first and second tier drugs are not appropriate and when culture and sensitivity results indicate susceptibility but their use is strongly discouraged, and may be prohibited in some countries, as these are preferential antimicrobial drugs for treatment of life-threatening infections in humans. Recommended dosages of systemic drugs will be presented when the SBF guidelines are published in full.

PREVENTION AND CONTROL OF SBF

If the diseases underlying SBF can be eliminated or controlled, the disease will not recur. However this can be difficult and recurrent disease is often a problem. Systemic antimicrobial drugs given at subtherapeutic doses or as pulse therapy have been used to prevent or delay recurrence but such protocols are likely to promote the development of antimicrobial resistance and their use is strongly discouraged. In cases where these approaches appear to be necessary it is recommended that the patient be referred to a specialist for further evaluation and treatment. There is evidence that injection of autogenous bacterins and commercial bacterial antigens can be used to stimulate immunity and prevent SBF or make it much easier to control e.g. with shampoos and other forms of topical antiseptic therapy; these modalities are recommended but further research is required in this area.

INFECTION CONTROL AND PUBLIC HEALTH

Staphylococci are readily transferred between humans and pets and veterinarians, especially those recently in contact with methicillin-resistant staphylococci, tend to carry these organisms with much higher frequency than the general public. Infections in humans by pathogenic staphylococci acquired from pets have also been documented in a number of different circumstances. Such infections can have serious consequences particularly with MRSA and MRSP which may be very difficult to treat effectively. Veterinary staff members need to be aware of this potential hazard, and owners should be informed of the potential risks. Nosocomial infection of other animals by such organisms in veterinary hospitals is also well documented.

Control of such hazards depends on rigorous adherence to good practice in hygiene and disinfection within veterinary practices and isolation or barrier nursing of animals carrying such organisms, and particularly MRSA. Specific recommendations for hygiene and disinfection do not form part of these guidelines but posters describing appropriate measures and guideline documents are available from veterinary associations in North America and in Europe and these should be followed.

REFERENCES

Guardabassi L, et al. Guidelines for antimicrobial use in dogs and cats. In: Guardabassi L, Jensen LB, Kruse H (eds.), Guide to Antimicrobial Use in Animals, Blackwell Publishing, Oxford, 2008, pp.182-206. Devriese LA, et al. **Int J Syst Evol Microbiol** 2005; 55: 1569-73. Lloyd D H, et al. **Vet Dermatol** 1996; 7: 171-4. Weese JS, et al. **Vet Microbiol** 2010; 27:140 (3-4):418-29. Deuster S, et al. **J Clin Pharm Ther** 2010;35:71-8. Cain CL, et al. **Am J Vet Res** 2011; 72 (1): 96-102. Hill PB, et al. **J Am Vet Med Assoc** 1994; 204(3): 334-40. White SD, et al. **J Am Vet Med Assoc** 2005; 226(6): 904-8. Loeffler, A, et al. **J Antimicrob. Chemother** 2008; 62: 1301-4. Loeffler, A, et al. **Vet Dermatol** 2007: 18; 412-421. Soares Magalhaes, RJ, et al. **Vet Res** 2010; 41: 55. Nienhoff U, et al. **Vet Microbiol** 2011; 150 (1-2) :191-7. Curtis CF, et al. **Vet Dermatol** 2006; 17: 163-8. DeBoer, D. J. et al. **Am. J. Vet. Res.** 1990; 51: 636-9. Stegmann R, et al. **J Antimicrob Chemother** 2010; 65(9): 2047-8. Riegel P, et al. **Int J Med Microbiol** 2011; 301(3): 237-9. Cohn LA, et al. **J Vet Emerg Crit Care (San Antonio)** 2010; 20(1): 31-45. Federation of European Companion Animal Veterinary Associations (FECAVA).
ttp://www.fecava.org/sites/default/files/files/Hygiene%20poster.pdf. January 2012. British Small Animal Veterinary Association (BSAVA).
http://www.bsava.com/LinkClick.aspx?fileticket=FZpsnz6kDEU%3D&tabid=171 January 2012.

T CELL TARGETS FOR IMMUNE MODULATION

Steven Dow, DVM, PhD, DACVIM
Ft. Collins, CO

CENTRAL ROLE OF CD4 T CELLS IN AUTOIMMUNITY

All autoimmune diseases are ultimately driven by pathogenic CD4 T cells. While CD4 T cells themselves cannot damage tissues directly, the cytokines produced by pathogenic CD4 T cells lead to a number of downstream effects. For example, cytokines they produce can drive the production of certain subclasses of autoantibodies by B cells, which in turn leads to direct and indirect tissue injury. In addition, cytokines such as IL-2 and IFN-γ are critical for the effective generation and function of cytotoxic CD8 T cells, which can damage tissues directly. T cell cytokines such as IL-17 lead to neutrophil mobilization from the bone marrow and recruitment into tissues, while GM-CSF can promote the generation of monocytes and macrophages.

The mechanisms responsible for development of autoimmune disease are complex, but all healthy individuals have potentially autoreactive CD4 T cells present at all times in their bodies. The reasons why some individuals go on to develop autoimmune disease are not fully understood, but at some point antigen presentation to autoreactive CD4 T cells triggers disease. Activation and proliferation of autoreactive CD4 T cells leads to production of cytokines that activate effector T cells and trigger tissue injury. Individuals with autoimmune disease may also fail to develop effective counter-regulatory responses to suppress the ongoing T cell activation process.

MAJOR CD4 T CELL SUBSETS AND THEIR FUNCTIONAL RELEVANCE IN AUTOIMMUNITY AND INFLAMMATION

Abnormal cytokine production ultimately drives the tissue injury and pathology characteristic of autoimmune diseases and associated inflammatory responses. Immunological studies have revealed that the CD4 T cell population can be subdivided into several different categories based on their cytokine production profile [1]. Since specific groups of cytokines will have very different effects on the immune system, dividing T cells in this manner provides important insights into the pathogenesis of T cell mediated diseases. Currently, CD4 T cells can be subdivided into 4 major subgroups, though undoubtedly more subgroups will be identified in the future. The **Th1 subset** of CD4 T cells is associated with cellular immunity to chronic infections (bacterial, fungal, viral) as well as to cancer. The Th1 subsets of T cells are also major mediators of autoimmune diseases. The major cytokines produced by Th1 T cells include IFN-γ and IL-2, as well as TNF-α and GM-CSF. These cytokines, especially IFN-γ, lead to activation of cells of the innate immune system (macrophages, NK cells) and to induction of cellular immune responses by CD8 T cells. The **Th2 subset** of CD4 T cells is associated with allergic responses and is characterized by production of the cytokines IL-4, IL-5, and IL-13 that drive antibody production, particularly production of IgE antibodies. Immune responses dominated by Th2 T cells typically are characterized by strong humoral immunity, with weak cellular immune responses. Immune responses to infections at mucosal surfaces, especially the gut and lung, often have a strong Th17 T cell component. The **Th17 cells** produce high levels of several different IL-17 molecules and a related cytokine IL-22. These cytokines in turn drive the release of G-CSF and mobilization of neutrophils to sites of infection and inflammation. In situations of chronic, sustained inflammation, **T regulatory cells** (Tregs) are expanded in numbers, and these T cells produce high levels of the immune suppressive cytokines IL-10 and TGF-β, which serves as a potent brake on overall T cell responses and inflammation [2].

CD4 T cell subset	Cytokine profile	Characteristic inflammatory response	Disease associations
Th1 T cells	IFN-γ, IL-2, TNF-α, GM-CSF	Lymphocyte, macrophage infiltration	Chronic infections, autoimmunity, cancer
Th2 T cells	IL-4, IL-5, IL-13	IgE, eosinophils, mast cells	Allergy, asthma, atopy
Th17 T cells	IL-17, IL-22	Strong neutrophil infiltrate	Bacterial infections, autoimmunity
T regulatory cells	IL-10, TGF-β	Reduced T cell response	Chronic inflammation

EFFECTOR CELLS IN AUTOIMMUNITY AND INFLAMMATION

While CD4 T cells initiate and perpetuate immune responses in autoimmunity, the ultimate effector cells that generate direct tissue damage include activated CD8 T cells, neutrophils, and macrophages. In addition, B cells participate indirectly by producing autoantibodies, which in turn can damage tissues by targeting them for destruction by macrophages or NK cells, or by activating complement locally. It is therefore important to keep in mind that even when CD4 T cells can be blocked from proliferating or functionally inactivated by immunosuppressive drugs, many of the effector cell-mediated processes can still continue for weeks later. Thus, for rapid control of immune-mediated diseases, these effector cell populations must also be suppressed or eliminated. At present, the only practical and effective way to accomplish this is by administration of broadly acting corticosteroids, though experimental methods of targeting specific subpopulations of effector T cells are being developed[3]. Thus, corticosteroids remain an important component of the initial treatment approach to immune-mediated diseases, though it should be noted that memory and effector T cells are relatively resistant to suppressive effects of steroids. .

TARGETED THERAPY FOR T CELLS

While high-dose corticosteroid therapy is usually effective for rapid control of immune-mediated disorders, it is also clear that the long-term administration of high-dose steroid therapy is associated with many unacceptable side-effects. In addition, some severe immune mediated disorders (eg, IMHA, IMTP, inflammatory bowel disease) may not be adequately controlled with corticosteroid therapy alone. Therefore, therapy targeted specifically at CD4 effector T cells is often used in conjunction with steroid therapy for companion animals[3]. These T cell-targeted drugs can be broadly grouped into drugs that inhibit T cell cytokine production (cyclosporine, tacrolimus, sirolimus) and those that block T cell proliferation (azathioprine, mycophenolate, leflunomide). Cyclosporine is widely used for the management of immune-mediated disorders, particularly as a steroid-sparing medication. Key considerations for cyclosporine therapy are cost, potential side-effects, and the potential delay in onset of activity until T effector populations are depleted. Nonetheless, cyclosporine is currently the most reliably effective drug for suppressing T cell mediated diseases in companion animal medicine. Tacrolimus is much more potent than cyclosporine, but high cost and high incidence of side-effects limits its use to topical administration. Somewhat paradoxically, Rapamycin (sirolimus) is currently being evaluated in companion animals as a potential cancer therapeutic, based on its ability to inhibit mTOR activity [4].

Agent	Cellular target	Mechanism	Adverse effects
Corticosteroids	All immune cells	Multiple	Significant
Azathioprine	T cells	Inhibition T cell replication	Mild (occas. Severe)
Cyclosporine	T cells	Inhibit T cell cytokine production	Moderate
Mycophenolate	T cells	Inhibition T cell replication	Moderate
Leflunomide	T cells	Inhibition T cell replication	Moderate
Immune globulins	Macrophage, DC	Inhibitory signaling to macrophage	Mild (occas Severe)
Anti-CD21	B cells	Depletion of B cells	Experimental (Unknown)
Liposomal clodronate	Macrophage, DC	Depletion of macrophages	Experimental (Mild-Moderate)
Alter gut microflora	Regulatory T cells, Th17 T cells	Expand Tregs, reduce Th17	Experimental (Mild)
Mesenchymal stem cells	T cells, DC	Suppress T cell and DC function	Experimental (Mild)

Azathioprine, which blocks T cell proliferation, is currently the most widely used adjunctive immune suppressive drug for a large number of immune-mediated disorders. It is somewhat therefore surprising that azathioprine has never been shown in a properly designed prospective study to be effective in the management of immune-mediated disorders in companion animals. Indeed, a recent meta-analysis failed to find any benefit from azathioprine therapy for IMHA in dogs[5]. Mycophenolate also suppresses T cell proliferation by a similar mechanism to azathioprine, but is much more potent and rapidly acting [3,6]. Moreover, the immune suppressive effectiveness of mycophenolate has been repeatedly demonstrated in bone marrow transplantation studies in dogs [6]. Leflunomide is another inhibitor of T cell proliferation that has received increasing attention recently because the generic version of the drug has become available at significantly reduced cost. However, the benefits of leflunomide therapy have yet to be demonstrated in properly designed clinical trials[3].

TARGETED THERAPY FOR B CELLS

At present, there is no effective method of depleting or suppressing B cell function, other than treatment with corticosteroids. None of the T cell targeted drugs have strong, direct effects on B cells, and therefore would not be expected to rapidly affect antibody production in antibody-mediated diseases such as IMHA. In human medicine, currently the first line of therapy for IMHA and IMTP is now B cell depletion, accomplished by administration of monoclonal antibodies to CD20 (eg, rituximab) that specifically target B cells for immune destruction. There is a large need for a similar B cell therapeutic in companion animal medicine, and there are reports that canine versions of a CD21 depleting antibody are being developed.

TARGETED THERAPY FOR MACROPHAGES

Particularly in diseases such as IMHA and IMTP that are largely mediated by macrophage destruction of antibody-coated RBC or platelets, an alternative approach to therapy is to target macrophages for suppression or removal. Administration of high doses of purified, monomeric immunoglobulins is extremely immunosuppressive for macrophages. The mechanism appears to involve triggering of inhibitory receptors on the surface of macrophages by contaminating sialyl groups present on the Fc portion of the immunoglobulin molecules [7]. The net result of intravenous administration of pooled human immune globulins is rapid suppression of macrophage phagocytosis, and this treatment was shown to be beneficial recently in dogs with IMTP [8]. Alternatively, macrophages can be depleted by administration of a macrophage toxin such as a bisphophonate drug, encapsulated in liposomes. We reported earlier that administration of liposomal clodronate (LC) depleted splenic macrophages and prolonged survival times in dogs with IMHA [9]. The effectiveness of LC therapy for IMHA in dogs is currently being investigated in a randomized clinical trial.

UNCONVENTIONAL THERAPY FOR IMMUNE-MEDIATED DISEASES

New insights into mechanisms of immune counter-regulation have suggested new strategies for modulating immune responses. Mesenchymal stem cells (MSC) have been investigated primarily for their ability to stimulate healing of connective tissues (see MSC discussion elsewhere in the 2012 ACVIM Proceedings). However, MSC also have potent immune suppressive effects in vivo and have been used to suppress inflammation associated with a number of different immune-mediated diseases, including inflammatory bowel disease, asthma, and arthritis. Thus, administration of MSC may be one means of enhancing the efficacy of conventional immune suppressive therapy.

A number of very interesting recent studies have revealed an important link between the gut microbiota and the development of autoimmune diseases [10]. In particular, certain species of endogenous gut bacteria are associated with immune protective effects (eg, *Bacteroides thetaiotoamicron*), or with enhancement of autoimmunity by stimulation of Th17 T cells (eg, segmented filamentous bacteria) [11]. Thus, it may be possible in the future to selectively alter the gut flora by antimicrobial therapy or transfaunation to suppress ongoing immune-mediated inflammatory responses.

REFERENCES

1. Leung S, et al. The cytokine milieu in the interplay of pathogenic Th1/Th17 cells and regulatory T cells in autoimmune disease. Cell Mol Immunol 2010;7:182-189. 2. La Cava A. Tregs are regulated by cytokines: implications for autoimmunity. Autoimmun Rev 2008;8:83-87. 3. Whitley NT, Day MJ. Immunomodulatory drugs and their application to the management of canine immune-mediated disease. J Small Anim Pract 2011;52:70-85. 4. Paoloni MC, Mazcko C, Fox E, et al. Rapamycin pharmacokinetic and pharmacodynamic relationships in osteosarcoma: a comparative oncology study in dogs. PLoS One 2010;5:e11013. 5. Piek CJ, van Spil WE, Junius G, et al. Lack of evidence of a beneficial effect of azathioprine in dogs treated with prednisolone for idiopathic immune-mediated hemolytic anemia: a retrospective cohort study. BMC Vet Res 2011;7:15. 6. Lupu M, McCune JS, Kuhr CS, et al. Pharmacokinetics of oral mycophenolate mofetil in dog: bioavailability studies and the impact of antibiotic therapy. Biol Blood Marrow Transplant 2006;12:1352-1354. 7. Teeling JL, Jansen-Hendriks T, Kuijpers TW, et al. Therapeutic efficacy of intravenous immunoglobulin preparations depends on the immunoglobulin G dimers: studies in experimental immune thrombocytopenia. Blood 2001;98:1095-1099. 8. Bianco D, Armstrong PJ, Washabau RJ. A prospective, randomized, double-blinded, placebo-controlled study of human intravenous immunoglobulin for the acute management of presumptive primary immune-mediated thrombocytopenia in dogs. J Vet Intern Med 2009;23:1071-1078. 9. Mathes M, Jordan M, Dow S. Evaluation of liposomal clodronate in experimental spontaneous autoimmune hemolytic anemia in dogs. Experimental Hematology 2006;34:1393-1402. 10. Kosiewicz MM, Zirnheld AL, Alard P. Gut microbiota, immunity, and disease: a complex relationship. Front Microbiol 2011;2:180. 11. Jarchum I, Pamer EG. Regulation of innate and adaptive immunity by the commensal microbiota. Curr Opin Immunol 2011;23:353-360.

PROBIOTICS: INFLUENCING THE IMMUNE RESPONSE 'NATURALLY'

J. Scott Weese, DVM, DVSc, DACVIM
Guelph, Ontario, Canada

INTRODUCTION

Probiotics can be defined as 'living microorganisms, that when ingested in certain numbers, confer health benefits beyond that of their inherent nutritional value.[1] Probiotics are becoming increasingly popular for treatment and prevention of disease in humans, as well as companion and food animals for various reasons, including the desire to reduce the use of antimicrobials, the perception of probiotics as 'natural' or 'alternative' therapies and the potential for efficacy. Unfortunately, despite the widespread use of probiotics in these species, there has been minimal objective investigation of efficacy or safety, particularly in companion animals.

The actual mechanism of action of probiotics is not well understood and is likely complex and multifactorial. Initial theories revolved around 'competitive exclusion' where 'good' bacteria eliminated or suppressed 'bad' bacteria. While competitive exclusion may play a role in probiotic therapy, it is a rather simplistic approach and is unlikely to account for the wide array of beneficial effects that have been reported in various animal species. Alternative proposed mechanisms include antimicrobial factor production, immunoregulatory effects, anti-inflammatory effects, anticarcinogenic effects and direct effects on the intestinal mucosa. It is these complex actions that probably play a key role in potential efficacy of probiotics and create the potential for efficacy against a wide range of conditions, yet our understanding of this area is extremely limited.

Microbiomes

An understanding of the intestinal microbiome (the microbial population of the intestinal tract) in both health and disease is critical to understanding the potential effect of probiotics. Certainly, it is necessary to understand the microbiome and its interaction with the body to determine when and how to manipulate it for preventive or therapeutic purposes. However, the complexity of the microbiome is a major limiting factor in the development, assessment and use of probiotics. Only recently have next generation sequencing and bioinformatics methods developed to the point where they appear to provide a good representation of the intestinal microbiome. A revolution in understanding of the gut microbiome is now underway, including studies of companion animals,[2-6] although it is reasonable to consider that we are still in our infancy in terms of truly understanding how to use and interpret the vast data that are obtained. Additionally, reliance on fecal samples, something that is a practical necessity, has limitations since the intestinal microbiome varies throughout its length and feces only provide some indication of the state of more proximal areas.

Microbial Interactions with the Immune System

There is overwhelming evidence that the gut microbiome has a complex and continuous interaction with the immune system. This would have been an evolutionary necessity, as the body needed to learn how to handle the potentially overwhelming antigen exposure from the gut, with it's trillions of bacteria from hundreds of different species. Learning how to tolerate some organisms while still maintaining an ability to address pathogenic species was a critical step. Beyond being a way to live with the complex gut microbial population, development of tolerance may also be a key factor in prevention of various inflammatory or immunological disorders. There is much interest in humans in the 'Hygiene Hypothesis", whereby there are concerns that decreased antigen exposure early in life has lead to faulty immunoregulation and an immune system that is more apt to respond abnormally over time.[7] While controversial, it focuses needed attention on the complex interaction between the body and its microbial partners, and the positive and negative impacts of this important (and unavoidable) interaction. Much attention has been focused on helminths, since they have essentially been eliminated from individuals in developed countries and development of parasitic infection has been associated with some positive impacts such as decreased multiple sclerosis exacerbations,[8] and improvement in ulcerative colitis and Crohn's disease has been achieved with ingestion of *Trichuris suis*.[9-11] While large and better-controlled studies are needed, these studies provide further support to the notion that alteration of the local intestinal environment can have significant effects on inflammation.

Mechanisms by which the body interacts with the microbiome are complex, only partially understood and beyond the scope of this review, but a few general aspects should be considered. The gut-associated lymphoid tissue (GALT) plays a key role in immunoregulation in the intestinal tract, something that is mitigated in large degree by the intestinal microbiome.[12] Dendritic cells in the lamina propria project dendrites through tight junctions of intestinal epithelial cells, allowing direct contact (and sampling) with the intestinal microflora.[13] These dendritic cells express toll-like receptors that can have marked and variable effects on the immune system. Overall, exposure to the commensal microbiome drives T_{reg} vs Th1 or Th2 effector cells.[12]

One landmark study indicating a role of the gut microbiome in inflammatory and immune disorders demonstrated that the susceptibility of rats to adjuvant-associated arthritis depended on their gut microflora.[14] Subsequent studies in conventional and laboratory animals expanded on this initial association and demonstrated a complex (and sometimes contradictory) relationship. The wealth of laboratory animal studies that have evaluated microbial/immune system interactions are beyond the scope of this review, however it is important to consider insights that have been derived from study of germ free animals. At a basic level, it is known that germ free animals develop a relatively 'immature' immune system with decreased lymphoid cell number and function, decreased circulating immunoglobulins and limited mucosal immunity.[12] One of the earlier studies demonstrating the impact of the gut microflora compared germ free and specific pathogen free (SPF) mice and showed that a bacterial microflora was required for oral tolerance induction.[15] In that study, germ free mice maintained Th2-mediated responses such as IgE, IgG and IL-4 production in response to ovalbumin administration while the Th2 responses of SPF mice (who had a commensal microflora) allowed them to develop tolerance. Further, treatment of germ free mice with *Bifidobacterium infantis*, a major component of the mouse microbiome, restored the ability of these mice to develop tolerance. Germ free animals are also relatively (or completely) resistant to a variety of disorders, indicating a role of the gut flora in their pathophysiology.

Various observations in humans provide further evidence, such as the fact that IBD tends to occur in areas with the highest bacterial populations, and remission of disease following fecal stream diversion and recurrence after restoration of normal intestinal flow.

CLINICAL DATA IN HUMAN MEDICINE

The probiotic field of one filled with equal parts promise, skepticism and disappointment. Numerous trials of varying quality have been performed on a range of allergy and immune disorders. Some evidence is weak or controversial, but there is strong suggestion that certain probiotics can have beneficial effects for certain conditions. Administration of *Lactobacillus rhamnosus* GG to pregnant women resulted in a decrease in the incidence of atopic dermatitis in children at 2 and 4 yrs of age.[16,17] At the 7yr follow-up, eczema rates were still significantly lower in the treatment group, yet a trend (bordering on statistically significant) towards increased incidence of allergic rhinitis and asthma was identified.[18] A similar but larger study involving different organisms reported no impact of probiotics on allergic diseases (food allergy, eczema, asthma, allergic rhinitis) but a reduction in IgE-associated (atopic) diseases.[19] These demonstrate the complexity of this field and the need for longterm studies (something that are almost completely lacking in veterinary medicine). There is also reasonable evidence for efficacy against a range of disorders such as chronic pouchitis and ulcerative colitis, with weaker evidence for inflammatory bowel disease and Crohn's disease.[20,21]

While encouraging pockets of data are present, it is clear that information indicating an effect of the gut microbiome is much stronger than evidence of clinical efficacy of probiotic therapy, and further research is needed to understand the role of probiotics in these and other disorders.

CLINICAL DATA IN VETERINARY MEDICINE

In veterinary medicine, most studies of the immune response and probiotics have involved food animals, mainly poultry. Administration of various probiotics has been shown to enhance adaptive and innate defenses in chickens, as demonstrated through *in vitro* or *ex vivo* studies, investigations of germ free chicks, modulation of various immune parameters (e.g. IgA production) in healthy chickens and clinical efficacy studies.

Very limited information is available for companion animals and the studies that have been reported offer little insight into potential immunological benefits of probiotics. This is in part because of the nature of studies, often being small and relatively superficial. Another aspect is the lack of any clear and objective selection process for commercial probiotics that would suggest they might have positive immunological benefits, or development of probiotics with a specific immunomodulatory target (rather than random testing and seeing what happens immunologically). Further, determining what constitutes a positive immunological benefit is challenging, and probably varies greatly with the disease that is being targeted. Indeed, for some conditions, enhancement of immune responses may be desired (e.g. infectious enteritis) while in others, immunomodulation may be the key (e.g. inflammatory intestinal disease, allergy…). Selection of strains with desirable survival (e.g. acid tolerance, bile tolerance) and antibacterial properties is much easier than screening of strains for immunological effects, providing further hindrances to commercial development. Some basic *in vitro* testing may be useful to obtain relevant information, such as chemokine or cytokine release in response to exposure of intestinal cell lines to individual organisms,[22] however the true relevance of such testing, performed with cell lines and in the absence of other microorganisms, is still of some question.

Enterococcus faecium SF68, a commonly used commercial probiotic, was reported to induce higher plasma IgA and IgA in puppies in response to CDV vaccination, with no differences in CD4+ and CD8+ T cells or CD4/CD8 ratios.[23] Additionally, a significant effect on fecal IgA was claimed, yet a non-significant P value was reported. The same organism was also shown to have no impact on fecal IgA concentration of leukocyte phagocytic activity as part of a giardiasis treatment study.[24] *Lactobacillus acidophilus* DSM13241 administration to healthy adult dogs resulted in increased concentrations of neutrophils, monocytes and serum IgG, but no impact on a variety of other indicators[25] and some gaps in study design and data presentation hamper interpretation of the results. An *ex vivo* study of the impact of a combination of lactobacilli on harvested small intestinal cells demonstrated increased expression of IL-10 in samples from healthy dogs, and increased IL-10 along with decreased TNFalpha/IL-10, IFN-gamma/IL-10 amd IL12p40/IL10 ratios.[26]

In cats, *E. faecium* SF68 supplementation increased the percentage of circulating CD4+ lymphocytes but had no effect on other measured parameters,[27] and in another study, the same strain had no effect on lymphocyte response in cats infected with FIV.[28] *Lactobacillus acidophilus* DSM13241 administration resulted in increased eosinophil counts during and after probiotic therapy, along changes in monocyte and granulocyte populations (apparently predominantly after cessation of probiotic therapy) in cats.[29] Overall, the clinical relevance of these changes is hard to determine and some vagarities in the design of some studies may interpretation of the results rather difficult. Simply finding changes, while indicating something is probably happening with the immune system, does not necessarily indicate whether those changes are good or bad, and how they may be related to the potential for clinical efficacy. Nonetheless, they indicate the potential for probiotics to have at least some effect on the immune system. This is best considered to be at the 'proof of principle' stage, with large and more specific studies needed to investigate this further.

MANIPULATION OF MICROORGANISMS TO MANIPULATE THE IMMUNE SYSTEM

Modification of bacteria to enhance potentially useful properties is another approach, although one fraught with controversy based more on emotion than data. Modification of a *Lactobacillus casei* strain to produce active canine granulocyte macrophage colony stimulating factor was shown to produce higher monocyte counts and serum IgA in treated puppies versus controls in response to canine coronavirus vaccination, along with a greater serum IgG response.[30]

COMMERCIAL VETERINARY PROBIOTICS

A limiting factor in assessment and clinical use of probiotics is variability in the quality of commercial products. As nutritional supplements, not drugs, probiotics undergo limited regulatory scrutiny. Studies of veterinary probiotics have shown numerous deficiencies in product contents, with a minority of products containing anywhere near the claimed number of viable organisms and/or the absence of many claimed species.[31,32] Even something as proper description, including spelling, of the contents is not always adequate, as evidenced by studies identifying misspelling of bacterial species and failure to properly identify product contents.[32,33] Fortunately, recent work has identified the presence of some commercial products with stated contents and proper labeling,[32] although there is still a large component of 'buyer beware' in clinical application of probiotic therapy.

THE LAST AND NEXT GENERATION: FECAL TRANSPLANTATION

Methods to restore the intestinal microflora using rumen contents or feces are certainly not new to veterinary medicine. They probably represent the earliest specific measures to manipulate the intestinal microbiome and are still used empirically in different conditions in different species. However, the approach has recently been taken up in human medicine, most notably using fecal transplantation (administration of feces by enema or colonoscopy) as a treatment of severe recurrent *Clostridium difficile* infection.[34-37] Astounding preliminary success has sparked interest in using this approach for a range of conditions, including ulcerative colitis, and it has been suggested that this procedure should be evaluated for a range of conditions such as IBD, irritable bowel syndrome, metabolic syndrome, neurodevelopmental disorders, autoimmune diseases and allergic diseases,[34] encompassing much of the range of disorders potentially associated with altered interactions between the immune system and microbiome. From a basic standpoint, this makes more sense than traditional probiotic therapy, as it provides very large numbers of the core components of the microbiome not a relatively small dose of organisms like lactic acid bacteria that represent a very small fraction of the normal microbiome.

A logical extension of fecal transplantation, and one that merges it with more 'mainstream' probiotic therapy is development of refined cocktails of organisms, something referred to as 'synthetic stool'. This logical (and more palatable) extension is based on the concept that if fecal transplantation can be shown to be effective, selected components of feces can probably be used in lieu of the entire fecal sample. This allows for laboratory production, standardization and no risk of transfer of undesirable organisms, and may be the next (and potentially the most successful) wave of probiotic therapy. Preliminary study of a cocktail of fecal organisms in a small number of human patients with CDI has shown promising results (unpublished data), and extension of this approach to veterinary medicine is appealing.

REFERENCES

[1] Guarner F, et al. ***Int J Food Microbiol*** 1998;39:237-238. [2] Chaban B, et al. ***Microb Ecol*** 2011. [3] Middelbos IS, et al. ***PLoS ONE*** 2010;5:e9768. [4] Swanson KS, et al. ***The ISME journal*** 2010. [5] Handl S, et al. ***FEMS microbiology ecology*** 2011. [6]Costa M, et al. Proc Annual Forum of the American College of Veterinary Internal Medicine 2011. [7] Rook GAW. ***Clin Rev Allergy Immunol*** 2012;42:5-15. [8] Correale J, et al. ***Ann Neurol*** 2007;61:97-108. [9] Summers RW, et al. ***Gut*** 2005;54:87-90. [10] Summers RW, et al. ***Gastroenterology*** 2005;128:825-832. [11] Summers RW, et al. ***Arch Pathol Lab Med*** 2006;130:1753. [12]Rook GAW, et al. Probiotics and other organisms in allergy and autoimmune disorders In: Versalovic J,Wilson M, eds. *Therapeutic Microbiology*. Washington, DC: ASM Press, 2008;231-247. [13] Rescigno M, et al. ***Nat Immunol*** 2001;2:361-367. [14] Kohashi O, et al. ***Microbiol Immunol*** 1985;29:487-497. [15] Sudo N, et al. ***J Immunol*** 1997;159:1739-1745. [16] Kalliomäki M, et al. ***Lancet*** 2001;357:1076-1079. [17] Kalliomäki MA, et al. ***Immunol Allergy Clin North Am*** 2004;24:739-752, viii. [18] Kalliomäki M, et al. ***J Allergy Clin Immunol*** 2007;119:1019-1021. [19] Kukkonen K, et al. ***J Allergy Clin Immunol*** 2007;119:192-198. [20] Hammer HF. ***Dig Dis*** 2011;29:550-553. [21] Meijer BJ, et al. ***J Clin Gastroenterol*** 2011;45 Suppl:S139-144. [22] Vizoso Pinto MG, et al. ***J Food Prot*** 2007;70:125-134. [23] Benyacoub J, et al. ***J Nutr*** 2003;133:1158-1162. [24] Simpson KW, et al. ***J Vet Intern Med*** 2009;23:476-481. [25] Baillon M-LA, et al. ***Am J Vet Res*** 2004;65:338-343. [26] Sauter SN, et al. ***Domest Anim Endocrinol*** 2005;29:605-622. [27] Veir JK, et al. ***Vet Ther*** 2008;8:229-238. [28] Lappin MR, et al. ***J Feline Med Surg*** 2009;11:650-654. [29] Marshall-Jones ZV, et al. ***Am J Vet Res*** 2006;67:1005-1012. [30] Chung JY, et al. ***J Microbiol Biotechnol*** 2009;19:1401-1407. [31] Weese JS. ***J Am Vet Med Assoc*** 2002;220:794-797. [32] Weese JS, et al. ***Can Vet J*** 2011;52:43-46. [33] Weese J. ***The Canadian Veterinary Journal*** 2003. [34] Borody TJ, et al. ***Nature reviews Gastroenterology & hepatology*** 2011. [35] Russell G, et al. ***Pediatrics*** 2010;126:e239-242. [36] Silverman MS, et al. ***Clin Gastroenterol Hepatol*** 2010;8:471-473. [37] Yoon SS, et al. ***J Clin Gastroenterol*** 2010;44:562-566.

THE ROLE OF CYCLOSPORINE IN IMMUNOMODULATION

Todd M Archer, DVM, DACVIM
Mississippi State, MS

INTRODUCTION

Cyclosporine is a potent immunosuppressive agent with treatment applications in both human and veterinary medicine. It is a cyclic polypeptide derived from the soil fungus *Tolypocladium inflatum*. Cyclosporine was discovered in the Sandoz laboratory by Borel in Switzerland in the 1970s, who demonstrated *in vitro* inhibition of human immune reactions as well as prolonged survival of skin grafts.[1] This data was presented in 1976 at the British Society of Immunology, paving the pathway towards utilization of its immunosuppressive abilities with organ transplantation in human medicine.[1,2] Cyclosporine was first described in human medicine by Calne in 1978 with published data documenting prevention of organ rejection in those receiving renal allografts.[2] Once these results were shown to be repeatable, cyclosporine became and still remains a cornerstone of immunosuppression in human transplant medicine. Cyclosporine was approved by the U.S. Food and Drug Administration in 1983 for the treatment and prevention of transplant rejection in human medicine. Novartis Animal Health received FDA approval in 2003 for oral cyclosporine capsules marketed as Atopica® for the treatment of atopic dermatitis in dogs. Cyclosporine is currently being applied to a spectrum of inflammatory and immune-mediated diseases in dogs, including but not limited to atopic dermatitis, keratoconjunctivitis sicca, autoimmune skin disorders, perianal fistula, inflammatory bowel disease, granulomatous meningoencephalitis, and immune-mediated blood disorders such as immune-mediate hemolytic anemia and immune-mediated thrombocytopenia. Cyclosporine has also been used in canine organ transplantation.

ORAL FORMULATIONS

Two oral formulations of cyclosporine exist. The initial formulation of cyclosporine was a vegetable-oil based preparation called Sandimmune®. Sandimmune® demonstrates significant variation in absorption from patient to patient, causing wide intra-individual and inter-individual differences in pharmacokinetics. Because of this, a newer formulation was introduced into the human market by Novartis in 1996. The newer formulation is an ultramicronised formulation which forms a microemulsion upon contact with aqueous fluids and has a much more consistent and predictable absorption.[3] Novartis manufactures the approved human ultramicronised formulation (Neoral®) as well as the veterinary approved ultramicronised formulation product (Atopica®). There are now also generic formulations available, both as the ultramicronised formulation and the non-ultramicronised formulation. The generic ultramicronised formulations demonstrate equivalent product bioavailability to the FDA approved product, but only the FDA approved veterinary product (Atopica®) has undergone extensive pharmacokinetic testing in dogs to demonstrate its clinical utility, and Atopica® is therefore the recommended product. Because of the poor absorption and extreme variability of the non-ultramicronised Sandimmune®-type preparations, they are not recommended in dogs.

MECHANISM OF ACTION

Cyclosporine is a calcineurin inhibitor, thus modulating the adaptive immune system and specifically inhibiting T-cell function, potently inhibiting cell-mediated immunity with a relatively lesser effect on humoral immunity.[4] Calcineurin inhibitors such as cyclosporine begin their biological effects by binding to intracellular proteins called cyclophilins. The most abundant cyclophilin found in T cells is cyclophilin A. The binding of cyclosporine to cyclophilin A facilitates the complex's affinity for calcineurin. When this complex binds to calcineurin, inhibition of calcineurin occurs by inhibition of its phosphatase activity, thus preventing its binding to and subsequent activation of nuclear factor of activated T-cells (NFAT). This facilitates decreased cytokine gene expression, with the most notable cytokine being IL-2. Other cytokines shown to have decreased expression in response to exposure to cyclosporine include IL-4, INF-γ, and TNF-α. The role of IL-2 in the inflammatory process is complex and multifactorial, involving both pro-inflammatory as well as regulatory functions.[5] In T-cells, IL-2 recognition by the IL-2 receptor initiates a signaling cascade which ultimately promotes cytokine transcription, cell survival, cell-cycle entry, and proliferation. IL-2 works within T-cells to further induce the expression of IL-2 receptors. IL-2 remains the most effective cytokine for T-cell clonal expansion. By decreasing IL-2 expression in CD4+ Th1 cells, proliferation and activation of both T-helper and T-cytotoxic lymphocytes is inhibited and the immune response is blunted.

PHARMCOKINETICS

The pharmacokinetics of cyclosporine have been well described in both humans and dogs. After oral dosing with Atopica® in dogs, there is a sharp upswing in blood concentrations, with a peak seen in most patients at approximately 2 hours post-dosing. The blood levels then slowly fade across the dosing interval.[6]

Absorption

With oral dosing of cyclosporine, absorption occurs through the gut epithelium in the small intestine. The Sandimmune® formulation has an extremely variable absorption profile, being dependent on factors such as bile flow and the presence or absence of food and gastrointestinal motility.[3] The newer ultramicronised formulation does not depend on biliary action nor enzymes or small intestinal secretions for absorption, and therefore has much more consistent and predictable absorption. Recommendations are to administer Atopica® on empty stomach, either two hours before or after feeding, because a decreased oral bioavailability as well as an increased variability of individual blood concentrations have been shown when oral ultramicronised Atopica® is given to dogs concurrently with food.[6]

Distribution

Cyclosporine has an increased binding affinity for red blood cells as well as plasma lipoproteins. Once in the systemic circulation, cyclosporine distributes widely, accumulating in the skin, liver, kidneys, and fat of dogs.[3] Tissue levels often exceed levels in serum. Peak plasma concentration times are variable, but usually occur approximately 2 hours after oral administration. Blood levels then fade away over the dosing interval.

Metabolism

Cyclosporine is extensively metabolized by hydroxylation and/or N-demethylation, with many metabolites being produced.[7] Metabolism occurs in the liver, small intestine, and kidneys, with the liver being the major site of metabolism. In the liver, the P-450 3A microsomal enzymes serve as the key pathway of metabolism. Several drugs have been reported to or are theorized to decrease or increase the metabolism of cyclosporine in both human and veterinary medicine through the hepatic P-450 enzyme system, and allow for increased or decreased cyclosporine blood concentrations, respectively. The drug most commonly utilized concurrently with cyclosporine therapy in dogs in order to decrease the daily cyclosporine dosage requirements is ketoconazole.

Excretion

Most cyclosporine metabolites are excreted through the biliary system, with minimal renal excretion.

ADVERSE EFFECTS

A number of adverse side effects have been associated with the use of cyclosporine in dogs. The most notable have been gastrointestinal-related, including vomiting, diarrhea, and inappetence. Many other side effects have been reported in the literature, including but not limited to lethargy, weight loss, gingival hyperplasia, hypertrichosis, coat shedding, cutaneous papillomatosis, concurrent infections, hepatotoxicity, nephrotoxicity, tremors, seizures, and the emergence of neoplasia. Clinicopathologic changes in dogs associated with the use of Atopica® have included lymphopenia, eosinopenia, transient thrombocytosis, anemia, leukocytosis, elevated creatinine, elevated BUN, hyperglobulinemia, hypoalbuminemia, hypocalcemia, hypernatremia, hyperkalemia, elevated ALT, elevated ALP, hypercalcemia, and hyperchloridemia, but rarely are these reported changes of any real clinical significance. One advantage of this drug in dogs with immune-mediated blood disorders is that, in contrast to many immunosuppressive agents, cyclosporine is not myelosuppressive. Recent findings have suggested cyclosporine may increase platelet procoagulant activity in dogs.[8]

MEASUREMENT OF BLOOD LEVELS

The assay used to measure cyclosporine blood levels as well as the sample type must be considered when interpreting results. Plasma values will be lower than whole blood values due to the concentration of cyclosporine in erythrocytes and leukocytes. While either whole blood or plasma cyclosporine concentrations can be measured, most advocate measurement of whole blood concentrations. Cyclosporine concentrations can be measured by high-pressure liquid chromatography (HPLC), fluorescence polarization immunoassay (TDx method), and specific monoclonal antibody radioimmunoassay (RIA) methods. HPLC measures only the parent drug without measuring metabolite levels. Both TDx and RIA methods measure parent drug as well as metabolites, and the blood concentrations will therefore be higher compared to the same sample being analyzed using HPLC. Although HPLC is labor intensive and not routinely offered for patient monitoring, it is considered the gold standard for measuring cyclosporine blood concentrations. TDx and RIA are most often clinically employed, with the laboratory providing the assay typically providing recommendations regarding ideal target blood drug levels.

THERAPEUTIC DRUG MONITORING

Although cyclosporine is gaining favor for treating a wide number of inflammatory and immune-mediated diseases in veterinary medicine, optimal oral dosing protocols in dogs are still unclear. In dermatologic applications such as canine atopic dermatitis, cyclosporine is often started at a low once daily oral dosage (5 mg/kg) that has been shown in many studies to be effective in the treatment of skin disease, and used at that dosage until remission of disease is achieved.[9,10] Cyclosporine therapy is then slowly weaned down until the lowest effective dosage to maintain disease remission is found.

In more critical and life-threatening diseases such as immune-mediated hemolytic anemia and immune-mediated thrombocytopenia, the best way to utilize cyclosporine therapy is not known. Cyclosporine is often started at a much higher dosage than that used to treat canine atopic dermatitis. Multiple possible side effects can sometimes affect our patients at these higher doses, and the drug can be cost-prohibitive for many clients. In contrast to dogs with skin diseases such as atopy, in dogs with more serious illnesses therapeutic drug monitoring is recommended. Cyclosporine blood concentrations are measured during the initial treatment period to ensure that drug levels are within a recommended therapeutic window, being high enough for disease control but still no higher than needed in order to minimize adverse drug side effects and unnecessary expense. Monitoring drug blood concentrations is recommended in more critical patients because, in the individual patient, cyclosporine pharmacokinetics are inherently variable and unpredictable even with the ultramicronised formulations, largely due to issues with oral bioavailability. A number of studies demonstrate a wide variability in cyclosporine pharmacokinetics, with differing blood levels in individual dogs administered the same dosage. Therapeutic drug monitoring attempts to optimize the dosage for individual patients with varying oral bioavailability. However, there are only a limited number of clinical studies in veterinary medicine which address ideal target therapeutic cyclosporine blood concentrations. Unfortunately, target blood concentrations do not necessarily correlate with clinical outcome or response to therapy. Many of our canine patients respond to therapy although their blood concentrations are less than the recommended cyclosporine target range, and some have not responded even when their cyclosporine blood concentrations were within the recommended target range.

Veterinary pharmacology laboratory recommendations vary regarding the type of assay used to determine cyclosporine concentrations, the ideal target cyclosporine concentration range for controlling disease, and whether the target should be based on trough blood concentrations, peak blood concentrations or both. In veterinary medicine, measurement of trough cyclosporine concentrations prevailed for many years based on initial work done in renal transplant studies. However, in human medicine, peak cyclosporine concentrations are now well-recognized to provide information regarding therapeutic efficacy that is superior to trough levels.[11] Recent recommendations for therapeutic drug monitoring in dogs receiving cyclosporine therapy are based on the evaluation of both peak and trough levels, although specific recommended target levels still depend on the laboratory reference ranges as well as the assay used to measure cyclosporine concentrations. Unfortunately, when using cyclosporine at higher doses, strictly relying on blood cyclosporine concentrations to monitor therapy may not be sufficient to adequately assess the immune system or to determine if there is sufficient immunosuppression to attain disease control.

PHARMACODYNAMICS

Therapeutic drug monitoring provides evidence that cyclosporine levels in patients may be in the estimated therapeutic range. For some patients, this estimation may be valid while, for others, there may not be clinical efficacy and response to therapy despite blood cyclosporine concentrations within 'therapeutic range'. This variation in clinical responses to comparable blood levels can be attributed to individual variability in biological responses to cyclosporine. Studies in human transplant medicine are now emerging that demonstrate the extra benefits of pharmacodynamic monitoring above those provided by monitoring of blood cyclosporine concentrations alone for adjusting immunosuppressive therapy in individual patients.[12,13]

Pharmacodynamic monitoring allows direct quantitation of the effects of cyclosporine on target T-cells, allowing adjustment of therapy based on measurable immunosuppressive effects even when individual patient responses to comparable blood levels are variable. Several pharmacodynamic biomarkers have been studied in human medicine, including lymphocyte proliferation, enzyme (calcineurin) activity, lymphocyte surface antigens, and intracellular cytokine quantification. Through pharmacodynamic monitoring, human studies have shown individually distinct degrees of calcineurin inhibitor sensitivity in patients. These methods show great promise for optimizing cyclosporine therapy as well as delivering individualized therapy.

Studies performed by our research group have documented measurable suppression of select biomarkers of immunosuppression in T-cells of normal dogs, particularly IL-2 and INF-γ, when T-cells are exposed to cyclosporine. An initial *in vitro* study documented suppression of a panel of biomarkers when T-cells were exposed to cyclosporine.[14] During a subsequent *in vivo* study, measurable suppression of T-cell cytokine expression was noted to occur not only at high dosages of cyclosporine but also in some dogs at the low dosage which is used for the treatment of canine atopic dermatitis.[15] Some dogs receiving the low dosage had suppression of T-cell function despite very low trough blood cyclosporine concentrations that were in fact well below the target immunosuppressive ranges recommended by most reference laboratories. Our findings call into question the ability of pharmacokinetic monitoring of blood drug concentrations to be the sole technique for assessing the efficacy of cyclosporine therapy in individual clinical patients. Our initial studies utilized flow cytometry to evaluate T-cell expression of cytokines. Flow cytometry, which requires rapid and specialized sample processing, may not however be the ideal tool for the development of pharmacodynamic monitoring in clinical patients, and a quantitative real-time PCR assay may in the long term hold more promise as a method for clinical pharmacodynamic monitoring of patients. Pharmacodynamic assays may offer additional information to enable further assessment of the status of the immune system. With further investigation into the pharmacokinetics and pharmacodynamics of cyclosporine in veterinary medicine, a better assessment of the immune status may be gained when combining these two techniques and allow for optimization of therapy and delivery of individualized treatment.

REFERENCES

[1]Calne R. **Transplant Proc** 2004; 36:13S. [2]Graeb C, et.al. **Transplant Proc** 2004; 36:125S. [3]Robson D. **Vet Rec** 2003;152:739. [4]Robson D. **Vet Rec** 2003; 152:768. [5]Lan RY, et.al. **Journal of Autoimmunity** 2008; 31:7. [6]Steffan J, et.al. **J Vet Pharmacol Ther** 2004; 27:231. [7]Whalen RD, et.al. **Xenobiotica** 1999; 29:3. [8]Thomason JM, et.al. **J Vet Intern Med** 2009; 23:692. [9]Steffan J, et.al. **J Am Vet Med Assoc** 2005; 226:1855. [10]Olivry T, et.al. **J Am Vet Med Assoc** 2002; 221:370. [11]Cantarovich M, et.al. **Transplant Proc** 1998; 30:1462. [12]Barten MJ, et.al. **Cell Prolif** 2007, 40, 50. [13]Kuzuya T, et.al. **Biol. Pharm. Bull**. 2009, 32, 604. [14]Fellman CL, et.al. **Vet Immunol Immunopathol** 2011; 140:175. [15]Archer TM, et.al. **J Vet Intern Med** 2011, 25:1391.

NON-GLUCOCORTICOID IMMUNMODULATORS

Clare R. Gregory, DVM, DACVS
Santa Rosa, California

MYELOTOXIC AGENT

Azathioprine is a purine analog that is metabolized to ribonucleotide monophosphates. Poor conversion to diphosphates and triphosphates leads to an intracellular accumulation of monophosphates that produces a feedback inhibition of the enzymes required for the biosynthesis of purine nucleotides. The triphosphate analogs that do form become incorporated into DNA and result in ribonucleic acid miscoding and faulty transcription. Azathioprine has a greater effect on humoral than on cell-mediated immunity. For the treatment of immune-mediated diseases in dogs, azathioprine is generally administered in conjunction with a corticosteroid and/or cyclophosphamide.

Azathioprine has been used for the treatment of IMTP, IMHA, autoimmune skin diseases, chronic hepatitis, myasthenia gravis, immune-mediated glomerulopathy, chronic atrophic gastritis, systemic lupus erythematosus, and inflammatory bowel disease. In recent studies azathioprine combined with prednisolone was no more effective than prednisolone administered alone in the treatment of IMHA, and azathioprine monotherapy was only moderately effective in the treatment of perianal fistulas. Azathioprine monotherapy had insufficient efficacy to recommend its general use for the treatment of atopic dermatitis in dogs. Although very myelotoxic in cats, azathioprine has been used for the treatment of feline autoimmune skin diseases. Azathioprine and prednisolone, when administered at maximally tolerated levels, do not effectively suppress the rejection response against canine major histocompatibility complex (MHC)–nonmatched renal allografts. However, when administered on an every-other-day schedule (1 to 3 mg/kg orally [PO]) with cyclosporine, azathioprine has been used to successfully maintain canine MHC-matched renal allografts. In feline renal transplantation patients with suspected inflammatory bowel disease or following an allograft rejection episode, azathioprine is administered with cyclosporine at a dose of 0.3 mg/kg every 3 days PO. The dose is decreased if the total white blood cell count falls below 3000 cells/μl or there is biochemical evidence of hepatotoxicity. The primary complication encountered with the administration of azathioprine is bone marrow suppression that can result in leukopenia, anemia, and thrombocytopenia. Acute pancreatitis and hepatotoxicity may also occur.

CALCINEURIN INHIBITORS

Cyclosporine is bound in the cytosol of lymphocytes by cyclophilins (cyclosporine-binding proteins). The cyclosporine-cyclophilin complexes associate with calcium-dependent calcineurin-calmodulin complexes to impede calcium-dependent signal transduction. Transcription factors that promote cytokine gene activation are either direct or indirect substrates of the serine-threonine phosphatase activity of calcineurin. This enzymatic activity is reduced by association of the cyclosporine-cyclophilin bimolecular complex with calcineurin. Via this mechanism of action, cyclosporine inhibits early T-cell activation (G_0 phase of the cell cycle) and prevents synthesis of several cytokines, in particular interleukin-2 (IL-2). Without stimulation by IL-2, further T-cell proliferation is inhibited, and T-cell cytotoxic activity is reduced. Cyclosporine may also exert an immunosuppressive effect as it stimulates mammalian cells to secrete transforming growth factor–β (TGF-β) protein. TGF-β is a potent inhibitor of IL-2–stimulated T-cell proliferation and generation of antigen-specific cytotoxic lymphocytes. Cyclosporine is not cytotoxic or myelotoxic and is specific for lymphocytes. This specificity spares other rapidly dividing cells and allows nonspecific host defense mechanisms to continue to function.

Cyclosporine has gained widespread use in veterinary medicine. Combination cyclosporine and prednisolone immunosuppression has maintained normal function of MHC-nonmatched feline renal allografts for more than 12 years. Cyclosporine in combination with azathioprine, prednisolone, and antithymocyte serum has been used to maintain MHC-nonmatched canine renal allografts. Cyclosporine has also been used to control corticosteroid-resistant IMHA and IMTP in dogs. Cyclosporine is available in an ophthalmic preparation (Optimmune, Schering-Plough, Kenilworth, NJ) for the control of keratoconjunctivitis sicca in dogs. An ophthalmic preparation of cyclosporine was shown to be an effective treatment for feline eosinophilic keratitis. Cyclosporine (10 to 20 mg/kg q24h PO) was found to significantly reduce the size and depth of perianal fistulas in dogs[1]. Most dogs did not require further therapy, either medical or surgical, after 6 to 8 weeks of therapy. In a more recent study cyclosporine resolved or reduced the lesions in 25 of 26 dogs[2]. However, residual or recurrent lesions were encountered that required long-term medical therapy or surgical resection. Cyclosporine (5mg/kg q24h PO induction dose, reduced by 50% to 75% over weeks to months, depending on response) has been shown to be effective in the reduction of skin lesions and pruritus in dogs with atopic dermatitis[3]. Cyclosporine has also been shown to be effective in the treatment of atopic dermatitis in cats, being an alternative treatment to prednisolone[4].

Cyclosporine has been used for the treatment of other dermatologic diseases such as pemphigus foliaceus[5]. Cyclosporine has also been shown to be effective for the treatment of granulomatous meningoencephalitis and with prednisolone for the treatment of necrotizing meningoencephalitis[6]. Cyclosporine has been used to treat steroid-refractory inflammatory bowel disease (5 mg/kg q24h PO for 10 weeks[7], and has recently been shown, in combination with prednisolone, to control inflammatory colorectal polyps in miniature dachshunds. Cyclosporine appears to be effective in the management of myasthenia gravis in dogs, especially when anticholinesterase medication fails and other immunosuppressive medications such as corticosteroids lead to deterioration in clinical signs. Very recently, feline idiopathic pure red cell aplasia was responsive to combination therapy with cyclosporine and glucocorticoids[8]. Most cats required long-term low dose therapy. Cyclosporine is available in two oral formulations: Sandimmune and Neoral (Novartis). Both contain cyclosporine in a concentration of 100 mg/ml, but the two solutions are not biologically equivalent. Sandimmune consists of an olive oil base, and adsorption of cyclosporine requires emulsification of the agent by bile salts and digestion by pancreatic enzymes. The absorption percentage can be as little as 4%, and there is a tremendous variation in dose-trough whole-blood levels among individuals of the same species. Neoral, (and now various generics) is a microemulsion preconcentrate of cyclosporine that becomes a microemulsion when in contact with gastrointestinal fluids. The microemulsion is directly absorbed through the gut epithelium, resulting in more sustained and consistent blood levels of the drug. The various generic forms of the microemulsion form of cyclosporine do not appear to be interchangeable; i.e. the same dose does not predict the same trough blood concentrations in patients. If one microemulsion form is exchanged for another, it is important to check whole blood trough concentrations to be sure adequate levels are achieved or that the concentration of the drug in the blood has not become too high.

To achieve immunosuppression with cyclosporine in dogs, I recommend attaining a 12-hour whole-blood trough level (measured just before the next oral dose) of at least 500 ng/ml. With Sandimmune, achieving this level requires an oral dosage of 10 to 25 mg/kg over 24 hours divided into two doses. Neoral can be initiated at 5 to 10 mg/kg over 24 hours divided into two doses. With either formulation gastrointestinal inflammation increases the dosage requirements, and blood levels of the agent must be measured starting 24 to 48 hours after initiation of therapy to ensure that adequate blood levels are achieved. Blood levels of cyclosporine should be measured at periodic intervals during the time of therapy. To achieve immunosuppression with cyclosporine in cats, I recommend attaining a 12-hour whole blood trough level of 250 to 500 ng/ml. With Sandimmune, obtaining this level requires an oral dosage of 4 to 15 mg/kg over 24 hours divided into two doses. Neoral can be initiated at 1 to 5 mg/kg over 24 hours divided into two doses. Again it is imperative to measure blood levels 24 to 48 hours after initiation of therapy to ensure that adequate blood levels have been achieved. Blood levels must also be measured periodically during the course of therapy[9].

Unlike the situation in humans, cyclosporine does not appear to be hepatotoxic in dogs unless extremely high blood levels are maintained (>3000 ng/ml). Although not as frequently encountered in humans, cyclosporine can be nephrotoxic in the cat. Nephrotoxicity in the cat does not seem to be related to the level of the drug in whole blood and can occur at relatively low plasma concentrations. Cats with extremely high cyclosporine whole-blood concentrations (>4000 ng/ml) often show no toxicity at all. In my experience cyclosporine nephrotoxicity has only developed in one dog. Levels higher than 1000 ng/ml can cause inappetence in cats. If levels of 1000 ng/ml are maintained for several weeks or months, opportunistic bacterial and fungal infections can occur. As in humans, cyclosporine could promote the development of neoplasia, particularly lymphomas, in cats and dogs. The administration of high levels (1 to 2 mg/kg over 24 hours) of prednisolone with cyclosporine increases the likelihood of tumor formation. As in humans, cyclosporine has resulted in a marked increase in hair growth in several feline renal transplant recipients. When administered to dogs, cyclosporine can result in severe gingival hyperplasia, fibropapillomatosis, and severe or fatal pyodermas, especially when combined with azathioprine[10].

Cyclosporine is also available in an intravenous solution (Sandimmune IV) that must be diluted in 0.9% sodium chloride or 5% dextrose in water. For the treatment of acute renal allograft rejection, I administer a dose of 6 mg/kg over 4 hours in the calculated maintenance fluid requirement. Intravenous cyclosporine is administered to control organ rejection episodes, for an acute hemolytic crisis, or during periods when a patient cannot tolerate oral medications.

Although tacrolimus, or FK-506 (Prograf®, Fujisawa), is structurally different from cyclosporine, it shares a similar mechanism of action. Tacrolimus binds in the cytosol of lymphocytes with an immunophilin, FK-binding protein (FKBP). As with the cyclosporine-cyclophilin complex, the tacrolimus-FKBP complex binds to calcineurin and inhibits its phosphatase activity. This inhibition directly and indirectly inhibits de novo expression of nuclear regulatory proteins and T-cell activation genes. The transcription of cytokines (IL-2, IL-3, IL-4, IL-5, interferon-γ, tumor necrosis factor-α [TNF-α], and granulocyte-macrophage colony-stimulating factor) responsible for lymphocyte activation is suppressed, as is the expression of IL-2 and IL-7 receptors. Tacrolimus in vitro is 50 to 100 times more potent as an inhibitor of lymphocyte activation than cyclosporine. Tacrolimus also inhibits B-cell proliferation and production of antibody by unknown mechanisms.

Little if any use of tacrolimus has been applied systemically to veterinary patients, but based on its effectiveness in experimental animal trials, it could be very useful in controlling a wide range of immune-mediated conditions. Tacrolimus may be particularly effective in controlling IMHA, IMTP, and immune-mediated arthritis because of its inhibition of antibody synthesis T-cell proliferation. Despite the potential benefits of tacrolimus for treating diseases in dogs, a major concern is the possible toxicity of the drug. A dose of 0.16 mg/kg every 24 hours intramuscularly or 1 mg/kg every 24 hours PO has been reported to be effective in prolonging renal allograft survival in beagle dogs. The side effects included anorexia, vasculitis, and intestinal intussusception. In a study using mongrel dogs the same doses were not effective in prolonging renal allograft survival, and most of the dogs developed severe vasculitis leading to fatal myocardial infarction, hepatic failure, and intussusception. Blood levels of tacrolimus are assayed at human medical centers with the use of monoclonal immunoassays. The effective serum trough level of tacrolimus in dogs is approximately 0.1 to 0.4 ng/ml, about 100 times lower than that of cyclosporine. Trough levels of 2 ng/ml or greater can result in death. Used topically in a 0.1% solution, tacrolimus has safely controlled the lesions of discoid lupus erythematosus and pemphigus foliaceus in dogs[11]. The International Task Force on Canine Atopic Dermatitis included topical tracrolimus (Protopic®, Astellas Pharma) in the group of medications with good evidence of high efficacy in the treatment of that disease. Tacrolimus in a topical 0.02% - 0.03% aqueous suspension is effective for the treatment of dogs with keratoconjunctivitis sicca and chronic superficial keratitis[12]. Tracrolimus has been shown to effectively increase the Schirmer tear test in dogs nonresponsive to cyclosporine therapy. Topical tacrolimus is also very effective in the control of perianal fistulas in dogs.

INHIBITORS OF CYTOKINE AND GROWTH FACTOR ACTION

Sirolimus, or rapamycin (Rapamune, Wyeth-Ayerst) and everolimus (Afinator®, Novartis) are a macrocyclic antibiotic with a structure similar to that of tacrolimus that also binds in the cell cytosol to FKBP. However, sirolimus and tacrolimus affect different and distinct sites in the signal transduction pathway. The immunosuppressive activity of sirolimus appears to be a consequence in part of the sirolimus-FKBP complex blocking the activation of the mammalian target of rapamycin, (mTOR). mTOR is a serine/threonine protein kinase and is involved in the regulation of cell proliferation through the initiation of gene translation in response to amino acids, growth factors, cytokines, and mitogens. The kinase activity of additional cell cycle regulatory proteins cyclin-dependent kinase-2 and cyclin-dependent kinase-4 is also inhibited by sirolimus. Sirolimus blocks IL-2 and other growth factor–mediated signal transduction (signal 3 of the allograft rejection response) and the calcium-independent CD28/B7 (CD80/CD86) costimulatory pathway. Cyclosporine and tacrolimus block T-cell cell cycle progression at the G_O to G_1 stage; sirolimus prevents cells from progressing from G_1 to the S phase. Sirolimus blocks T-cell activation by IL-2, IL-4, and IL-6 and stimulation of B-cell proliferation by lipopolysaccharide. Sirolimus directly inhibits B-cell immunoglobulin synthesis caused by interleukins.

It has been shown that signaling through the mTOR pathway contributes to the growth, progression and chemoresistance of several cancers. Accordingly, mTOR inhibitors are being studied as potentially valuable therapeutics. It has been shown that rapamycin may be safely administered to dogs and can yield therapeutic exposures . In vitro, rapamycin has been effective in decreasing tumor cell survival in both canine osteosarcoma and melanoma cell lines. Investigations into the efficacy of treatment with mTOR inhibitors for naturally occurring osteosarcomas in dogs are currently underway.

Mycophenolate mofetil (MMF), also known as RS-61443 or mycophenolic acid (Cellcept, Roche), is a prodrug hydrolyzed by liver esterases to mycophenolic acid. MMF is cytostatic for lymphocytes, owing to its inhibition of inosine monophosphate dehydrogenase, an enzyme necessary for de novo purine biosynthesis. MMF is a relatively selective inhibitor of T- and B-cell proliferation during the S phase of the cell cycle via its ability to prevent guanosine and deoxyguanosine biosynthesis. MMF has been shown to reduce allograft rejection in multiple animal models, demonstrating the most effect when combined with cyclosporine, tacrolimus, and/or sirolimus. MMF was developed in part as a nonmyelotoxic replacement for azathioprine in human allograft patients. Early clinical trials in human renal allograft recipients showed a decrease in biopsy-proven acute rejection episodes in patients receiving MMF in place of azathioprine. At therapeutic doses MMF can be toxic to animals. The primary dose-limiting effects are anemia and weight loss in rats; leukopenia, diarrhea, and anorexia in monkeys; and gastrointestinal hemorrhage, anorexia, and diarrhea in dogs. To reduce the toxic effects, the dose can be lowered, or mycophenolic acid can be given in combination with other immunosuppressive agents. MMF (10 mg/kg q12h PO) has been used in combination therapy in veterinary medicine to control renal allograft rejection in dogs[13].

There are anecdotal reports of the use of MMF for the treatment of RA, and renal allograft rejection in the dog. In a study involving 27 dogs with myasthenia gravis, MMF was not found to be an effective treatment. MMF, (10 mg/kg, PO, q 12 hr) was used to control primary immune-mediated hemolytic anemia in two cats that had a lack of response to corticosteroid therapy[14]. MMF(20 mg/kg, PO q 12 hr, then reduced to 10mg/kg, q 12 hr), with prednisolone was used to control subepidermal blistering autoimmune disease in a dog. In this dog, the dose of 20 mg/kg q 12 hours was well tolerated. In humans MMF has been used to treat scleroderma lung disease and Evans syndrome.

Leflunomide is a synthetic organic isoxazole that the intestinal mucosa metabolizes to the active form, A77 1726. Leflunomide mediates at least part of its antiproliferative activity during the S phase of the cell cycle by inhibiting the de novo pathway of pyrimidine biosynthesis. The target of A77 1726 in this pathway is the enzyme dihydroorotate dehydrogenase. At higher concentrations leflunomide is also an inhibitor of tyrosine kinases associated with growth factor receptors. In addition to T and B lymphocytes, leflunomide also has an antiproliferative effect on smooth muscle cells and fibroblasts, which is also caused by the result of inhibition of the pyrimidine biosynthetic pathway in these cells. Leflunomide is currently approved for the treatment of RA in humans. It has been shown to be an effective disease-modifying antirheumatic drug free from the side effects commonly associated with currently approved immunosuppressants.

In addition to its efficacy in humans and animal models with autoimmune diseases, leflunomide has been found to control acute, ongoing, and chronic allograft rejection of the kidney, skin, heart, vessels, and lung in small- and large-animal models. I have used leflunomide to successfully treat steroid-resistant autoimmune hemolytic anemia, Evan's syndrome, immune-mediated thrombocytopenia, multifocal nonsuppurative encephalitis/meningomyelitis, polyarthritis, pemphigus folicaceous and systemic histiocytosis in dogs[15]. In combination with cyclosporine, leflunomide has completely prevented the rejection of canine MHC-nonmatched renal allografts in both experimental and clinical studies. At doses used in humans, leflunomide causes gastrointestinal toxicity in dogs as a result of the accumulation of a metabolite trimethylfluoroanaline (TMFA). Fortunately the canine lymphocyte is far more sensitive than the human lymphocyte to the effects of the active agent A77 172, and much lower oral doses are equally effective in achieving immunosuppression. I currently use a dose of 4 mg/kg PO over 24 hours and adjust the dose as needed to obtain a 24-hour serum trough level of 20 mg/ml. Early studies in the cat suggest that TMFA does not present the toxicity problem encountered in dogs. The recommended dose for cats is 2 mg/kg, PO q 24 hours for two consecutive days and then a maintenance dose of 2 mg/kg, PO, q 48 hours. This dose produces a mean blood concentration of 16 µg/ml, and produces a 50% reduction in mitogen-stimulated lymphocyte proliferation[16]. Leflunomide serum concentrations can be measured by simple high-performance liquid chromatography assays.

Leflunomide (70 mg/cat once weekly PO) in combination with methotrexate (7.5 mg/cat once weekly PO) has been used to treat feline RA[17]. Leflunomide (3-4 g/kg, PO, q 24 hours was effective in controlling immune-mediated polyarthritis in 14 dogs[18].

Leflunomide is showing effectiveness in treating viral diseases and certain cancers. Leflunomide has been used successfully to treat primary cytomegalovirus infection and ganciclovir-resistant cytomegalovirus infection in human renal transplant patients. In vitro, leflunomide inhibits herpesvirus-1 replication in feline kidney cell cultures. Leflunomide may be useful in the treatment of reactivation of latent FHV-1 that occurs secondary to the stress of illness, surgery, or pharmacologic immunosuppression. Leflunomide promotes cytostatic, antioxidant and apoptotic effects in human transformed epithelial cells, and may play a role in prostate cancer chemoprevention. In mouse models investigating the treatment of melanoma, leflunomide significantly inhibited tumor growth. When combined with PLX4720, a new anti-melanoma agent in clinical trials, tumor growth was almost completely inhibited. Both cats and dogs with diminished renal function may be subject to TFMA toxicity since it is excreted by the kidneys.

REFERENCES

1Mathews KA et. al. J Am Vet Med Assoc 1997; 21 1 :124 . 2Hardie RJ et. al. J Small Anim Pract 2005; 6 : 3. 3Steffan J et. al. Vet Dermatol 2006; 7 : 3.
4Wisselink MA et.al. Vet J 2009; 80: 55. 5Guaguere E et. al. Vet Dermatol 2004;5 : 61 . 6Adamo FP et. Al. J Am Vet Med Assoc 2004; 225 : 1211. 7Allenspach K
et al. J Vet Intern Med 2006; 20: 239 .8Viviano KR et.al. J Feline Med Surg 2011; 13:885. 9Gregory et. al. In Bonagura JD, ed: Current Vet Therapy XIV 2009; 254
10 Gregory CR et. al. Vet Surg 2006; 35 : 105. 11 Rosenkrantz WS. Vet Dermatol 2004;15 :90. 12 Berdoulay A et al. Vet Ophthalmol 2005; 8 : 225. 13 Broaddus KD
et al . Vet Surg 2006; 35 : 125. 14 Beck LM et al. J Vet Emerg Crit Care 2011; 21: 45. 15 Gregory CR et. al. Transplant Proc 1998; 30: 4143. 16 Mehl ML et. al. J Vet
Pharmacol Therp doi: 10.111/j. 1365-2885.2011.01306.x (epub ahead of print). 17 Hanna, FY. Vet Comp Orthop Traumatol 2005; 18 : 94. 18 Colopy SA et. al. J Am
Vet Med Assoc 2010; 236: 312.

Hiring Winners: EI Based Hiring

Shawn G. McVey, MA, MSW
Chicago, IL

INTRODUCTION

Research shows that no more than 25% of one's overall success is attributable to general intelligence or IQ. A good portion of that remaining 75% is related to emotional intelligence competencies: self-awareness, self-management, social awareness, and relationship management. After this session, you will understand the components of emotional intelligence (EI), understand the importance of EI as a competency in need of measurement when selecting for leaders, assess your own EI in order to have a frame of reference for selecting for EI in others, possess tools to identify ways to improve your own emotional intelligence, and be able to identify EI characteristics of effective leaders to measure in a selection process. You will leave the session with resources or tools for measuring EI, such as interview questions, written exercises, behavioral tools, and EI inventories.

THE DIFFERENCE BETWEEN IQ AND EI

General intelligence (measured as IQ) is typically focused on analytic reasoning, verbal skills, and spatial ability. It's a murky concept with many definitions and not the best predictor of work success. Emotional intelligence (EI) is the capacity for recognizing our own feelings and those of others, for motivating ourselves, and for managing emotions well in ourselves and in our relationships.

It's critical that our team members and leaders have emotional intelligence in order to maximize our own personal success and the success of the practice. For every 1% improvement in the service climate, there is a 2% increase in "profit."

EI AND THE BRAIN

Emotions begin in the limbic area. There is an open loop relying on external cues. The main purpose of the innermost part of the brain is survival. Open loops mean that we affect and are affected by someone else's emotion (e.g., ICU patients get better if someone is present with them). A leader transmits signals that can alter hormone levels and immune and cardiovascular functioning of others. In fact, after a 15-minute conversation, body rhythms and physiology match—a concept called mirroring. Leaders "manage meaning" for us. About 60% of perceptions about organizational climate can be traced directly to the leader. Workers are more likely to be overtaken by negative emotions (hijacked) after interactions with the boss rather than with peers or clients. Distress from hijacking lowers performance.

THE COMPONENTS OF EI

The four components of EI are self-awareness, self-management, social awareness, and relationship management. Self-awareness includes emotional self-awareness, accurate self-assessment, and self-confidence. Self-management includes emotional self-control, transparency, adaptability, achievement, initiative, and optimism.

Social awareness includes empathy, organizational awareness, and service. Relationship management includes inspirational leadership, influence, developing others, change catalyst, conflict management, building bonds, and teamwork and collaboration.

When it comes to emotional self-awareness, the inability to notice our true feelings leaves us at their mercy. People with greater self-awareness are better pilots of their lives and are more confident about personal decisions. As a person matures, emotions begin to shape and improve thinking by directing a person's attention to important changes. For example, a teacher becomes concerned about a lesson that needs to be completed for the next day. The teacher moves on to complete the task before concern takes over enjoyment.

You can develop social awareness through empathy. Empathy is the ability to recognize another's emotional state. In research on married couples, empathy appears to include matching the physiological changes of the other person. Developing empathy leads to greater emotional stability, greater interpersonal sensitivity, and better performance at school, work, volunteer jobs, as well as with family.

The art of social relationships involves managing emotions in others. To excel at people skills means having and using the competencies to be an effective friend, negotiator, and leader. One should be able to guide an interaction, inspire others, make others comfortable in social situations, and influence and persuade others. The subtle and complex abilities that underlie people skills are being attuned to others' emotions, promoting comfort in others through the proper use of display rules, and using your own emotional display to establish a sense of rapport.

USING EI TO HIRE LEADERS

So, now that you understand EI, how do you hire people with an abundance of it so that they and your organization maximize success? First, let's talk about traditional interviews. Have you ever hired someone who seemed fabulous in the interview and bombed on the job? Why does this happen so often? Interviews fail because we over-rely on our intuition and our "gut feelings." People just aren't good at predicting future behavior. If we were, would the divorce rate be 50%? We more often select for general intelligence or technical competencies rather than emotional intelligence.

When hiring for EI, the successful selection methods include the structured interview, both situational and behavioral; written exercises, written exercises, including reactions to case studies and EI assessment; and behavioral assessments, including presentations and role plays.

In structured interviews, types of interviewing questions include:

- Credentials/Technical: "Why did you pursue the undergraduate degree that you did?" (Self-Awareness)
- Experience: "Describe a time when you were presented with a problem about which you did not have all of the information to solve the problem. How did you proceed?" (Social Awareness, Self-Management)
- Opinion: "Coming in as a new employee, what would you do to get oriented to the organization?" (Social Awareness, Relationship Management) and "Choose one or the other: I have little trouble taking prudent risks or I am particularly cautious about change. Explain your answer." (Self-Awareness)
- Situational/ Hypothetical: "Assume you are confronted by a donor who is unhappy with something the organization has done. The donor is calling to let you know that they are going to discontinue paying on their pledge. What do you do?" (Relationship Management)
- Behavior Description: "Tell me about a time when you felt least effective or were most frustrated in your efforts to deal with a conflict between yourself and a co-worker." (Self-Management, Relationship Management)

In behavior-based interviews, the best predictor of future behavior is past behavior in a similar circumstance. Here are some guidelines:

- DON'T ask questions that can clearly be answered by reading the application or the resume.
- DO balance the types of questions you ask.
- DO make the majority of your questions behavioral description questions.
- DO determine the "best" answer BEFORE the interviewing process begins.

In written exercises, you want the candidate to identify critical incidents. Use job descriptions, challenges, successes, and essay questions. Develop case studies, or have the candidate write media plans, case statements, or grant proposal outlines.

Three types of behavioral exercises are interpersonal role plays, tests of people skills, and knowledge application. Another is having the candidate give a presentation, which assesses communication, public speaking, networking, and training. Yet another behavioral exercise is mock meetings, which assess facilitation skills, leadership ability, teambuilding skills, and group process skills.

In order to assess the results of the EI-based exercises, you have to know what the correct "answer" is before you pose the question or elicit behavior. Narrative description is not enough. Match the most desirable answer to the dimensions of EI, including self-awareness, self-management, social awareness, and relationship management. Develop rating or scoring rubrics ahead of time so that you can easily evaluate the responses and compare the candidates.

EXAMPLES OF EI-BASED INTERVIEW QUESTIONS

Here are some examples of EI interview questions that test self-awareness.

- Impact on others
    - Tell me about a time when you were surprised about the positive impact your behavior or words had on a coworker, a customer, or an employee. How did you learn this information? What did you do when you learned this information?
    - Tell me about a time when you did or said something and it had a negative impact on a coworker, a customer, or an employee.
- Emotional and inner awareness
    - Tell me about a time when you were distracted or preoccupied about something. How did you know? What impact did that have on your performance? What impact did it have on others at work?
    - Tell me about a time when you were in a good mood at work. How did that affect your performance? What impact did your mood have on others at work?
- Accurate self-assessment
    - Describe a time when you received feedback about your performance and were in agreement. What did you agree with?
    - Describe a time when you received feedback about your performance and you disagreed with that feedback. What did you disagree with?
    - Was there ever a time when you initially disagreed with feedback you received and later came to accept it? Tell me about that.

Here are some examples of EI interview questions that test self-management.

- Emotional expression
    - Tell me about a time when you were angry with someone at work. What did you do?
    - Has there ever been a situation at work where you said something and later regretted saying it? Tell me about that.
    - Describe a situation at work when you were very enthusiastic about something. How did your enthusiasm affect others?
- Courage or assertiveness
    - Tell me about a time when you spoke up about something in the workplace. What was the issue? Why did you speak up about it? What did you say? What did others think?
    - Has there ever been a situation at work where you wish you had said something in a meeting or encounter but didn't? Tell me about that.
    - Describe what you did the last time someone blamed you for something at work that wasn't your fault. What did you do?
- Resilience
    - Tell me about a time when you felt that you were defeated at work. What did you do?
    - Tell me about a time when you were distracted by or preoccupied about something. What did you do?
    - Tell me about a time when you felt like giving up on something. What did you do?

- Awareness in the moment
    - Tell me about a time when you realized that a conversation wasn't going very well. What did you do?
    - Tell me about a time when you realized that you weren't speaking up during a meeting. What did you do?
    - Tell me about a time when you realized that something was best left unsaid. What did you do?
- Planning tone
    - Tell me about a time when you deliberately planned the tone of a particular conversation. How did you do that? What result did it have?
    - Tell me about a time when you missed an opportunity to set the tone in a discussion. What happened?

  Here are some examples of EI interview questions that test relationship management.
- Respectful listening
    - Think about a time when you didn't understand something in the workplace. What did you do?
    - Describe a situation when you didn't understand why someone was acting a certain way or taking a certain position on some issue? What did you do?
    - Describe a time when you jumped to conclusions.
- Feeling the impact on others
    - Tell me about a situation when you sensed something was bothering a peer or coworker. How did you know? What did you do?
    - Describe a situation when you knew that something was wrong with a relationship you had with a peer, customer, or supervisor. What did you do?
    - Relate a situation in which you determined that something that you did or said didn't go over very well. How did you know?
- Service orientation
    - Tell me about a time when you offered assistance to someone without being asked. What did you do?
    - Describe a situation when you offered assistance to someone even though it was outside of your job description. What did you do?
    - Was there ever a time when you resented helping someone at work? Tell me about that.

  Here are some examples of EI interview questions that test social awareness.
- Building relationships
    - Who are some key people in your organization who you currently must work with on a regular basis to get your work done? Describe your relationship with these people.
    - Describe your present responsibility for building and maintaining relationships at work. Whom do you build relationships with? How? Why?
- Collaboration
    - Describe a time when you had to solve a problem that involved or affected other people within the company. How did you solve it?
    - Have you ever implemented an idea or solved a problem and had your solution met with resistance? What do you think you could have done to avoid the resistance?
    - Describe a time when you sought someone's ideas or opinions about a project or idea you were working on.
- Conflict resolution
    - Tell me about a dispute with a peer. What was it about? What did you do? How did it end up?
    - Tell me about a time when someone suggested something that you disagreed with. What did you say?
    - How have you resolved differences with peers or others?
    - Tell me about the process you use to resolve your difference.
- Organizational savvy
    - Did you ever have an opportunity to advance a new idea at your last job? How did you go about doing that?
    - Tell me about a time when you gained support for an idea that you had. How did you do that? Why was this idea important to you?
    - Describe a time when you couldn't get support for an idea that you had. What happened? Why was this idea important to you?

Managing Change & Growth: Key Performance Indicators

Shawn G. McVey, MA, MSW
Chicago, IL

INTRODUCTION

Change compels us to get out of a place of complacency, but managing growth can be a difficult and emotional process. Here we look at the signs of complacency, how to get change-ready, and how to implement a ten-step process for consistently handling change and growth with ease and grace.

HOW TO GAUGE WHEN YOUR PRACTICE IS COMPLACENT

You know that management is complacent and letting the practice get stuck when there is no highly visible crisis, the practice measures itself against low standards, and planning and control systems are rigged (or non-existent) to make it easy for everyone to reach goals.

Other signs of complacency are that the practice is not losing money—no big threats are on the horizon; the practice compares itself to the industry average, not the industry leaders; and the typical manager or employee can work for months without encountering resistance or questioning.

HOW TO MAKE YOUR PRACTICE CHANGE-READY

First, push decision making down to the lowest possible level. Begin sharing information.
Talk, but listen too. It is a two-way street. Encourage participation, and get into the trenches with front-line employees. Help people see the "why" of change, and work with them to discover the "what."

Step 1: Mobilize Energy and Commitment

Do this by sharing values and a clear vision. People do not argue with what they help create. You will get buy-in through participation. The group who will implement change must agree on what the problems are and why they need to change.

People resist having solutions imposed upon them by individuals who lack familiarity with day-to-day operations. Resistance is expressed through lack of motivation and commitment to the change. You MUST have top-level support, not control.

Step 2: Develop a Shared Vision

Do you have a mission, vision, and values? Is it just a sign on the wall, or do the owners and key managers inspire others in the organization toward a vision? Vision is "what can be seen," so focus on specifics about how to improve the business and how those changes will benefit the employee and patients/clients.

Price Pritchett at Pritchett and Associates says that 20% of employees embrace change from the start, 50% are fence-sitters, and 30% resist any kind of change or growth. You may notice that resistance comes up in the form of these questions, so be prepared to answer them:

- "What's in it for me?"
- "Why is this important?"
- "How do these people even know what the problems are? They don't even bother to ask us."
- "Do they really think they can change the entire practice at once?"
- "How much of our time and their money will they sink into this dry hole?"

To create a shared vision, describe a desirable future. Be compelling. Make it much better than the current state of things. Be realistic. Make it within the grasp of the hard-working folks who will make it happen. Focus on a manageable set of goals. Be flexible. Make it easy to communicate with all levels of employees.

Step 3: Hire Qualified Employees

The biggest decisions are to get the right people on the bus and the wrong people off! Answer the question: What isn't getting done well? What is holding the practice back?

Plan for employee turnover in this process. Would a key employee's departure upset the applecart? As the volume of business increases, some people will not keep up. Without clearly defined roles, new employees will leave. Creating systems is the answer to this problem.

Identify leaders. Good leaders have three characteristics:

- A persistent belief that revitalization and change is the key to staying competitive
- Can articulate their conviction in the form of a credible and compelling vision
- Have people and organizational skills to implement their vision

Step 4: Focus on Results, Not Activities

If you can't measure it, you can't manage it. Operations, experience, and expertise are critical. Identify short-term goals. Discuss them with employees with regular, good communication.

Step 5: Implement Systems

HR, project management, and quality-control systems give you the infrastructure for growth. Systems provide checks and balances for your practice. Systems reduce stress and make workloads predictable and manageable. They help reduce turnover.

Step 6: Start Change at the Periphery:

Your likelihood of success is greatest when change is implemented in small, autonomous units. Success breeds success. When you implement change and grow in this way, there are a few key benefits:

- Clear advantage over the status quo
- Compatibility with peoples' desires and values
- Requirements that are understandable
- Option for people to experience the change model at a small scale first
- The possibility of people observing the suggested change in another setting

Step 7: Train Employees and Train Again

Employees cannot read your mind. Invest in training and supervision of the employee. There is a direct correlation between training and productivity. Our goal is continuous improvement and change. This is a process, not an event.

Step 8: Implement Technology to Support Operations

Give people the tools they need to do their jobs. You cannot run a business on a wing nut and a prayer. Strategically plan and budget for technological improvements.

Step 9: Improve Quality

Deliver a consistent client experience. Develop checklists and scripts to recover from service failures. Map your processes.

Increase client satisfaction by increasing value. Listen more than ever to what employees and clients say about their experience with your business. If clients are negatively impacted by a system (or lack thereof), fix it quick! Complaints and surveys are great tools.

Increase competitiveness. Stay current with what's important to your clients, what's happening in the marketplace, and what your competition is doing well. What is your practice doing well? What do you need to improve?

Step 10: Expand Services, Not Just Service

Add new procedures and profit centers. 80/20: Do you know the rule?

THE PROBLEMS WITH IMPLEMENTATION

The single biggest impediment to growth is the inability of the culture to evolve. We are often failures at implementation. There are six common problems:

- It takes more time than we thought (76%).
- Major problems surfaced that we did not predict (74%).
- Coordination of activities was not effective enough (66%).
- Competing activities and crises distracted us from implementing the strategic decisions (64%).
- The capabilities (skills and abilities) of employees involved with change were not sufficient (63%).
- Training and instructions were not clear to lower-level employees (63%).

HOW PEOPLE REACT TO CHANGE

The typical employee spends eight hours a day doing, in general, fairly routine tasks. The completion of those tasks becomes the culture. Change creates anxiety! People will adapt, but there is a predictable series of responses. The stages of emotion in reaction to change are: shock, defensive retreat, acknowledgement, and acceptance and adaptation.

People get stuck for a couple of reasons. Change is not a single event. Everyone comes to the table with a different level of experience relative to change and different coping skills. People who are emotionally fragile are particularly vulnerable during times of change.

To minimize the negatives associated with change and growth, follow some key steps. Keep your cool. Handle pressure smoothly, and respond non-defensively. Develop creative and innovative solutions to problems. Be willing to take risks and try out new ideas. Be willing to adjust priorities. Demonstrate enthusiasm for long-term goals. Be open and candid. Participate. When a decision needs to be made, make it.

Managing in an Economic Downturn

Shawn G. McVey, MA, MSW
Chicago, IL

INTRODUCTION

In tough economic times, you must reinvent yourself and the practice to stay viable. The traditional ways of doing business and measuring performance are outdated and do not help the team or the practice move forward. If you want to grow the practice and stay competitive, you need to be lighter on your feet, which means becoming a learning organization. Here we look at how you can create a learning culture in your practice.

WHAT IS A LEARNING CULTURE?

In the early 90s, the term "learning culture" was coined by Peter M. Senge in his book titled *The 5th Discipline*. The goal was to create flexibility and rapid response to business climate change. It was, and is, a good theory, but difficult or impossible to implement in veterinary practices, because we spend all of our time working in the business instead of on the business.

"Supportive learning environments allow time for a pause in the action and encourage thoughtful review of the practice's processes." A learning practice is one where employees excel at creating, sharing, and transferring knowledge. In such a practice, the team pinpoints areas to foster knowledge sharing. Management develops new ideas and teaches everyone how to learn from mistakes.

HOW CAN YOU CREATE A LEARNING CULTURE?

To create a learning culture, you must address three key elements:

- A supportive learning environment
- Concrete learning processes and practices
- Leadership behavior that reinforces learning

Practices do not perform evenly across the three building blocks. Each building block can be viewed as a separate developmental stage, so each unit may require specific tools to improve. Even in a "small" practice, people learn differently and respond to different stimuli.

Building Block 1: Supportive Learning Environment

A supportive learning environment has four distinctive characteristics:

- Psychological safety
- Appreciation of differences
- Openness to new ideas
- Time for reflection

To learn, team members cannot fear being belittled or marginalized when they disagree with peers or authority figures, ask naïve questions, own up to mistakes, or present a minority viewpoint. They must be comfortable with their thoughts about the work at hand. This is defined as psychological safety.

Recognizing the value of competing functional outlooks and alternative solutions increases energy and motivation. The result of appreciation of differences is fresh thinking and the prevention of lethargy and drift.

Veterinary practice leaders often do not reward reflection. Owners and DVMs (by nature) are seldom curious about human nature and rely too much on science to achieve success.

Team members are measured by archaic standards, such as number of hours worked, gross revenue, or tasks accomplished. Or they are not measured at all, but rather subjectively and randomly evaluated. When people are overly stressed and over-scheduled, their ability to think creatively is diminished. They are less able to diagnose problems and learn from their experience.

A case in point is Children's Hospital and Clinic of Minnesota, which has a "blameless reporting" policy. They have replaced threatening terms such as "error" and "investigation" with less emotional descriptions like "accidents" and "analysis." The hospital culture was changed to one of everyone working together to understand process and priorities and report on them without fear of blame. The result was increased productivity and engagement, along with improved patient outcomes and satisfaction.

Building Block 2: Concrete Learning Processes and Practices

A learning practice is not achieved effortlessly, but arises from a series of concrete steps and widely distributed activities. In our business, examples would be: logistics, billing, compliance, and healthcare plan delivery. For maximum impact, knowledge must be shared in systematic and clearly defined ways.

Knowledge must flow laterally and vertically. You can be internally focused, with an eye toward corrective action, or externally focused, as in focus groups or employee surveys. All of this should result in information moving quickly and efficiently into the hands and heads of people who must implement change.

A case in point is that after-action reviews (AARs) are now commonly used in service organizations. AARs encompass a systematic review of every service, activity, mission, or critical mission. They ask four simple questions:

- What did we set out to do?
- What actually happened?
- Why did it happen?

- What do we do next time?

Building Block 3: Leadership that Reinforces Learning

Organizational learning is strongly influenced by leadership: DVMs, managers, and team leads. When leaders actively question and listen to employees (prompting dialogue and debate), people in the practice feel encouraged to learn.

What message do you send? Do they know that you know what's important? For example, do they know that you value problem identification versus symptom reporting, knowledge transfer versus CYA/no training, and autopsy versus blame?

The result of unconventional wisdom can be emboldened employees, higher productivity, and new options. Do you want them to have the right answer (be right), or do you want them to think about the problem the right way (be correct)? Questions you can use to generate discussion:

- Which criteria did you use?
- Why did you come to that conclusion?
- What alternatives did you consider?
- What was your primary motivation?
- Which value/mission is supported by your thinking?

HOW CAN YOU MEASURE YOUR LEARNING CULTURE?

What if you could score your practice and compare it to other learning organizations? What if you could compare the learning environment of different work units in your practice? And what if you could have a broader view of how your practice reacts to change and challenges? Use the free tool at learning.tools.hbr.org, or the complete tool at los.hbs.edu. The tool is designed to help you answer two questions: 1) To what extent is your practice (team, employee, DVMs) functioning as a learning organization? and 2) What are the relationships among the factors that affect learning in your practice?

Data is stratified into quartiles. Identify areas of excellence and those that you need to improve. Do group and/or individual assessments to pinpoint cultural/habit differences and commonalities and what you can learn from one another. Address specific problems.

Ask each participant to score themselves and the team in each area on a scale of 1 to 100 and then compare to means below. What will you discover? Mean scores for service businesses:

- Leadership: 76
- Concrete learning process: 74
- Supportive learning environment: 71

This process helps to clear the fog between intention, cultural myth, and reality. It helps employees better understand what critique and analysis really look like. A high score does not indicate a competitive advantage, but rather an awareness of the concept and a belief that these things are paid attention to in your hospital.

Test assumptions in a quantifiable way. Create the background for having productive debates and seek out dissenting views. Identify critical development areas for the business. Use it to initiate conversation and self-reflection, but not as the sole cure for what ails your practice.

Leadership alone is insufficient. The cultural and process dimensions of learning require more explicit, targeted interventions. In veterinary medicine, you must do more than modify or change the leadership structure. Practices (in spite of a common mission) are not monolithic. Groups or individuals vary in their learning maturity and focus. A one-size-fits-all approach for building "team" is not likely to work. Learning is multi-dimensional. You can enhance learning by addressing one or more of a multitude of factors.

Pulmonary Hypertension: Pathophysiology and Diagnosis

Heidi B. Kellihan, DVM, DACVIM (Cardiology)
Madison, WI

DEFINITIONS

In veterinary medicine, pulmonary hypertension (PH) had been described as echocardiographically estimated pulmonary arterial systolic pressure greater than ~30mmHg.[1-5] Pulmonary hypertension can be further classified as pulmonary arterial hypertension (PAH) (i.e. pre-capillary) or pulmonary venous hypertension (PVH) (i.e. post-capillary). Pulmonary arterial hypertension is influenced by pulmonary blood flow, pulmonary vascular resistance (PVR), and pulmonary venous pressure. Pulmonary arterial pressure and pulmonary capillary wedge pressure (PCWP) can be directly measured, invasively, and the PVR can be derived from these two measurements. Pulmonary vascular resistance can then be useful in distinguishing pre-capillary PH from post-capillary PH. PVR is calculated from the following equation and expressed as dynes*sec*cm^{-5}/m^2.

PVR=[mPAP – PCWP] X 80/CO

By definition, PAH or pre-capillary PH occurs when there is an elevation in PAP (i.e. increased mPAP) as a result of an increased PVR with a normal left atrial pressure (i.e. normal PCWP). Diseases that result in PAH or pre-capillary PH include: idiopathic PAH, heartworm disease, congenital systemic-to-pulmonary shunts (i.e. atrial septal defects, ventricular septal defects and patent ductus arterious), necrotizing vasculitis/arteritis, pulmonary disease and pulmonary thrombotic and/or embolic disease. In contrast, the definition of PVH or post-capillary PH is an elevated PAP as a result of increased PCWP and a normal PVR. The phenomenon of PVH or post-capillary PH occurs with mitral valve disease and myocardial disease (i.e. dilated cardiomyopathy). Lastly, there is a phenomenon called reactive precapillary PAH that typically occurs with chronic left-sided heart disease. In this situation, the increase in PAP is a result of increased PCWP and increased PVR. The increased PVR can be transient and reversible from PA vasoconstriction secondary to hypoxia, as with pulmonary edema, or permanent and irreversible secondary to chronic vascular remodeling secondary to hypoxia. Disproportionate reactive PH is seen when the mPAP is disproportionately increased above what the PCWP would be with left-sided heart disease alone. This implies that there is intrinsic disease of the pulmonary arterial system from chronic remodeling changes secondary to chronic left-heart disease or from others diseases resulting in pulmonary arterial hypertension.

PATHOLOGY AND PATHOPHYSIOLOGY

General Pulmonary Hypertension Pathology

The normal pulmonary circulation is comprised of a vast network of thin-walled arteries, veins and capillaries characterized by a very low vascular resistance, low pressure and high capacitance. Pulmonary hypertension occurs secondary to an imbalance of endogenous and exogenous pulmonary arterial vasoconstrictors and vasodilators resulting in pulmonary arterial intimal proliferation, medial hypertrophy, vasoconstriction and thrombosis.

Endothelin Pathway

One of the major mediators of PH is endothelin-1 (ET-1) which is released by pulmonary vascular endothelium resulting in potent pulmonary artery and vein vasoconstriction, smooth muscle cell proliferation and increased collagen synthesis. A primary stimulus for increased ET-1 concentrations is heart failure due to increased local production, decreased pulmonary clearance and upregulation of endothelin-A receptors (ETA). In addition to powerful affects influencing PH, ET-1 also results in sodium and water retention, potentiation of other neurohormonal systems and promotion of myocardial hypertrophy, fibrosis and remodeling.

Prostanoid Pathway

Prostacyclin and thromboxane A2 are endogenous arachidonic acid metabolites produces from pulmonary artery endothelial cells and platelets involved in maintenance of vascular tone in normal physiologic and pathyphysiologic states. Prostacyclin is a vasodilator, inhibitor of platelet activation and has antiproliferative effects on the pulmonary artery. In contrast, thromboxane A2 is a potent vasoconstrictor and promotor of platelet activation. In cases of pulmonary hypertension, the balance of protacyclin and thromboxane A2 formation favors thromboxane A2 leading to vasoconstriction, cellular proliferation and thrombosis.

Nitric Oxide Pathway

Nitric oxide (NO) is synthesized endogenously from L-arginine and oxygen by nitric oxide synthase isoenzymes in the pulmonary artery vascular endothelium. Nitric oxide activates guanylate cyclase and subsequently cyclic guanosine monophosphate (cGMP) which results in a reduction of intracellular calcium concentration. The NO pathway enhances smooth muscle cell relaxation, inhibits smooth muscle cell proliferation and hypertrophy, and inhibits platelet aggregation and adhesion. cGMP is rapidly inactivated by phosphodiesterase (PDE), particularly PDE-5 isoenzymes. In experimental models and clinical studies of heart failure in people, a deficiency in basal NO synthesis as well as a decrease in endothelial release and response to NO has been found, suggesting that the loss of NO-dependant vasodilation and anti-cellular proliferation may contribute to the production of PH through vascular remodeling.

Intimal Proliferation

Intimal lesions associated with PH consist of eccentric intimal thickening and concentric fibrotic plexiform lesions comprised of myofibroblasts or endothelial cell in an "onion-skin" morphology that account for most of the reduction in luminal area of the small pulmonary arteries. Deposition of an acellular extracellular matrix in the lumen of pulmonary arteries also contributes to luminal obstruction and an increase in PVR. The pathogenesis of these intimal lesions is largely undetermined, although it is thought that these lesions may result from endothelial injury, smooth muscle cell migration, extracellular matrix deposition and endothelial proliferation. Microthrombi may also serve as the scaffolding for theses lesions.

Medial Hypertrophy

The pathologic feature of medial smooth muscle hypertrophy is characteristic of PH and involves muscularization of pulmonary arteries. Normally, the medial smooth muscle layer of pulmonary arteries represent 10-15% of the total outside vessel diameter. In PH, medial hypertrophy can result in an increase in pulmonary artery outside vessel diameter of up to 30-60%. The increase in the tunica media layer occurs as a result of both smooth muscle cell hypertrophy and hyperplasia. The mechanism by which this hypertrophy and hyperplasia occurs is thought to be primarily a result of cell proliferation and also inhibition of cell apoptosis.

Adventitial Remodeling

More recently, the fibroblast-rich adventitia has been postulated to play a role in PH by regulating vascular function from the "outside-in". In idiopathic PH in people, the adventitial layer, which is normally 15% of the external diameter of pulmonary arteries larger than 50 microns in diameter has been shown to be increased to 28% of the artery diameter because of collagen deposition. Inflammatory cells are present in the adventitial layer and may affect fibroblasts contributing to PH when vascular stress, hypoxia, high flow states and toxins (i.e. monocrotaline) are present.

DIAGNOSTIC STRATEGY

Invasive hemodynamic assessment

Right heart catheterization can provide multiple hemodynamic parameters that aide in the diagnosis and etiologic classification of PH. Right heart catheterization provides hemodynamic information regarding the right atrial, right ventricular, pulmonary artery and pulmonary capillary wedge pressures.

Echocardiography

Echo can estimate the systolic, diastolic and mean PAP. The peak systolic TR velocity and gradient reflects the estimated peak systolic right ventricular pressure and is used to classify PH as mild (≥ 2.8 to <3.5 m/s, ≥ 31.4 to <50 mmHg), moderate (3.5-4.3 m/s, 50-75 mmHg) or severe (>4.3 m/s, >75 mmHg). Other parameters that echo can evaluate include: pulmonary artery systolic flow profiles, Tei index of myocardial performance, systolic time intervals and right ventricular tissue Doppler.

Cardiac biomarkers

NT-proBNP has been shown to be elevtaed in dogs in the presence of PAH and PVH.[6,7]

CT scan and MRI

In people, CT scanning and MRI provide valuable diagnostic information regarding PH.

REFERENCES

[1]Kellum HB, et.al. **J Vet Intern Med.** 2007;21(6):1258. [2]Johnson L, et.al. **J Vet Intern Med.** 1999;13(5):440. [3]Bach JF, et.al. **J Vet Intern Med.** 2006;20(5):1135. [4]Schober KE, et.al. **J Vet Intern Med.** 2006;20(4):912. [5]Pyle RL, et.al. **Intern J Appl Res Vet Med** 2004;2(2):99. [6]Atkinson KJ, et.al. **J Vet Intern Med.** 2009;23(6):1190. [7]Kellihan HB, et.al. **J Vet Cardiol** 2011;13(3):171.

EMAIL COMMUNICATION WITH YOUR CLIENTS: THE GOOD, THE BAD, & THE UGLY

Nancy D. Kay, DVM, DACVIM
Hendersonville, NC

Access to the Internet has changed our clients' expectations dramatically over the last several years. We've entered what I like to refer to as the "Age of the Empowered Client". One expectation that is changing is our clients' desire to communicate with us via email. Email communication with doctors is not a new concept. The Kaiser Permanente "My Health Manager" program with an "email my doctor" feature has been wildly successful. More and more patients are using the program and a study conducted by Kaiser between 2004 and 2007 documented email access to physicians resulted in fewer telephone contacts and office visits, and proved to be more convenient for non-urgent concerns[1]. A follow-up study on more than 30,000 Kaiser Permanente patients with high blood pressure and/or diabetes mellitus documented that those who communicated via email with their physicians enjoyed better health outcomes[2].

I recently provided a written survey to 120 of my northern California colleagues to learn more about email communication between veterinary hospital staff members and their clients (unpublished data). Seventy-two veterinarians responded (23 board certified specialists and 49 general practitioners). Here is what I learned.

1. Fifty-eight percent of respondents reported that they do communicate with their clients via email.
   - -55% general practitioners
   - -65% specialists
2. Of those vets who use email 62% are selective about which clients are allowed email access.
   - -59% general practitioners
   - -66% specialists
3. Of those vets who use email 26% set ground rules with their clients pertaining to email communications. Many commented that they feel the need to set email ground rules, but have been too "wimpy" to do so.
   - -23% general practitioners
   - -33% specialists
4. Nurses/technicians use email to communicate with clients in 21% of practices.
   - -20% general practices
   - -22% specialty practices
5. Receptionists use email to communicate with clients in 37% of practices.
   - -39% general practices
   - -35% specialty practices
6. Of those vets who use email 95% viewed it to be a mostly positive experience.

GROUND RULES

No urgent matters
One to two-day turnaround time
Email is not instant messaging
No response on doc's day off
For established clients only
No "cutesy" emails
No "What's your diagnosis?"

THE POSITIVES

Clients like it
Mutually convenient
Eliminates phone tag
Provides written documentation for medical record
Succinctness/clarity
Better for clients who are "visual learners"
Client perceives 24-hour availability
Allows more thoughtful responses
Useful with time zone differences
Good for simple questions
A good way to deliver
- Cost estimates
- Discharge instructions
- Ancillary information
- Mass mailings
- Vaccine reminders
- Appointment reminders

Patient photos

THE NEGATIVES

Urgent emails missed
Inconvenient transfer to medical record
Can follow you anywhere
Expectation of constant access/delayed response may imply lack of concern
Doesn't eliminate phone calls/adds to workload
Clients want "on line diagnosis"
Too time consuming
- Inadequate word processing skills
- Need to edit
- Verbose clients
- Too many questions
- Need to check multiple times daily

Messages lost to spam
"Accidental sends"
Miss out on nonverbal communication
Miss out on voice inflections

WHAT WOULD IT TAKE FOR YOU TO USE EMAIL

More time
More efficient email system
Connectivity with medical record
Separate work and personal email addresses
Decrease in the amount of other email workload
Voice recognition software
My own computer
Client email addresses

REFERENCES

1. Chen C, et al. **Health Affairs** 2009; 28:323-333. 2. Zhou YY, et al. **Health Affairs** 2010; 0048:1370-1375.

SYSTEMATIZING THE MONEY TALK TO CLIENTS

Peter Weinstein, DVM, MBA
Irvine, CA

INTRODUCTION

You leave one exam room on the way to the next when Kathy slinks up to you with a look of terror in her eyes and says, "Dr. Weinstein, Mrs. Noble wants to talk with YOU!"

All of a sudden your mind starts to race. Oscar has been hospitalized for the last three days with pancreatitis. He has been doing great and, in fact, it is about his scheduled pick up time. What happened, did Oscar relapse? Did he not get the clean up bath? Worse...did he die? As your blood pressure goes up and Kathy waits for you to get the glazed look off your face, panic begins. "Yes, Kathy", you say concerned. "What are Mrs. Noble's concerns?"

"Dr. Weinstein, Oscar is fine. She is thrilled that he did so well in light of the prognosis that you gave. However, she can't afford the bill! She knows you have been keeping her up to date the whole time on both Oscar's medical progress and her total bill. But she can't pay it."

Your eyes go from glazed to rolled up in your head. You run into the office, close the door, climb under your desk, pull the chair in behind you and hope...against all odds, that somebody else will handle this one. As you pull the few remaining hairs from your head and start to bang it against the wall, a little voice whispers in your ear...you could have become an accountant!!

Obviously, this scenario is a bit on the exaggerated side...Or is it?

What does your practice do when this scenario occurs? Does it vary from client to client, event to event? Is there a standardized approach to the 'money talk'?

SYSTEMS, CHECKLISTS, PROCESSES, RECIPES

Call it what you want, a SYSTEM is a formal, documented approach to handling an event. The event could be checking in a client or prepping a surgery or performing a cardiac ultrasound. It is a consistent manner by which something is approached. It can be delineated as a line by line check list; a formal process; or a recipe. A SYSTEM says this is HOW we do it here, each time; every time; until somebody finds a better way. Do you have a system for talking about money with clients?

THE MONEY SYSTEM includes:

1. Financial Policy
2. Financial Policy Form
3. Payment options
4. Estimates/Healthcare Plans
5. Communicating Your Money System

FINANCIAL POLICY

If a clearly defined financial policy doesn't exist in your practice your freezer will be receiving more deposits than your bank. If the barrier to Getting to Yes is the cost of care, then the pet owner must be made aware of the options available to allow for decisions to be made from the heart rather than from the wallet.

You NEED a written and clearly understood and discussed financial policy to make it much easier for the entire hospital team to discuss payment options with the client WITHOUT having to have a different policy for every client. From their very first visit, you need a pet owner to understand what the options that they have for payment are. Besides cash, check, money order, what credit cards are accepted? Do you offer third party payment programs such as Care Credit? What are your feelings on pet insurance? What about if the client has no cash, has maxed out credit cards, doesn't qualify for Care Credit? What are your 'final/alternate' financial options? What are your practice's options to help clients 'afford' treatment? How are you "giving patients more ways to say YES to treatment"? All of these questions are clearly answered in a written Financial Policy.

Your Financial Policy is clearly expressed in your Financial Policy Form-- a non-confrontational and non-stressful way to inform clients of the options that they will have to take care of the bill so they know BEFORE they get the bill.

THE "OUR FINANCIAL POLICY" FORM

By putting your financial policies in writing you are documenting and validating what you believe for your staff and your clients. The policy/form can be included on your website, posted in the reception area, summarized on your brochures, and presented to clients new and old. The premise: you believe in clear communication, mutual understanding and respect, and thus your clients would appreciate knowing in advance what your financial policies are.

It is your goal to make sure that financial considerations are not an obstacle to a client providing important, life-saving care to their four-legged family members. So, clearly define that in your Form. Not until they understand how much you care do you discuss your payment options.

Use the form to tell clients about deposits, how much and when. Let them know when you will update them about changes to their estimate. And clearly define what options exist to pay for the care that is being provided. The obvious choices are: Cash, Check, Money Order, Visa, MasterCard, American Express, etc. In most cases, these options are sufficient.

On the other hand, on those rare occasions when your client has no cash and no credit cards, what can you do? The recent advent of "Client Payment Plans" through a variety of companies (for example, Care Credit©), has opened up a whole new modality for client financing. Third Party Payment Plans are explained as an option in the Form.

Finally, and as a last resort, you can be the bank. With all of the options available, the days of being your client's bank can be over if you want them to be. On the other hand, if you enjoy sending out monthly statements, reminder notes, late payment statements, dunning notes, etc., at least have a "Financial Arrangement" Form (FAF) that let's your clients know that you are serious about getting paid. The FAF should outline the Treatment Plan total, the deposit and the amount financed. It should delineate how long the balance will be financed for, what the amount of each installment will be, what the late fee (after 30 days without payment) will be, and what the annual percentage rate for delinquent accounts is. This is similar to any financial arrangement form or credit card form that you have filled out for yourself and should be reviewed by your attorneys for enforceability and have a place for signatures and witnesses at the bottom. Don't make this the first option and make sure clients know that is not your first choice but a last resort to ensure that they can provide the necessary care.

We know that most pet owners are not that pro-active. So, what is your plan for the more common, 'I didn't know it would be that expensive, I can't afford that' client? This form will should have a statement about pet insurance. This may be as simple as "Pet Insurance is available to assist you in the cost of care for your pet, for additional information, ask one of our healthcare team". It may be more complicated, such as: "We believe that every pet that is eligible for pet insurance coverage should enroll as young as possible with XYZ Insurance Company. The insurance will be provide a real benefit to you and your pet's long-term healthcare'. This addresses the big picture part.

OFFERING FINANCIAL OPTIONS

Besides the more mundane cash, check and credit cards, you need options to discuss or offer to clients to help with the cost of care? By providing options to clients you are letting them know that you care about their pet's health and want them to be able to financially afford the best care.

In a system you will provide payment options in the following order:

- Cash, check, credit cards
- Third party payment options—e.g. Care Credit
- Internal banking—essentially YOU act as the bank and 'loan' money to the client who will pay you back over time
- Barter
- Discount it or just give it away.
- Purchasing pet insurance policies for young healthy pets to provide coverage when needed.

Since there is no cost to the practice until it is utilized, all clients should apply for approval for third party payment such as Care Credit. Care Credit and others in the third party financing world provide low or no interest financing for veterinary healthcare services. Clients can apply for it and have it in their back pocket for the unexpected.

A client with third party payment options allows you to:

Discuss the patient's needs from the optimal level

Have prepared clients for the unexpected illness/ injury/ accident, etc.

Charge a fair and reasonable fee for your services

The following forms of financing are your practice acting as a bank for the client: held checks, delayed credit card payments, sending a bill, etc. These are best avoided because most of the time they only lead to ill will.

Pet insurance has been available in the US and Canada for over 25 years but only one percent of all American dogs and cats are insured. Insured pets see their veterinarians more frequently and sooner when they have a problem. And pet owners with insurance will pay more for veterinary care than those without insurance. Of course, if you don't have insurance and a problem arises, it is too late. Pet owners should be encouraged to buy policies for their young healthy pets and retain it for those emergencies, diseases or accidents where the wallet and the heart both weigh heavily. Besides current insurance endorsements will help pay for basic wellness needs until the crisis mode hits.

At the same time a pet owner is signing up for care credit, they can sign up for pet insurance and have two systems to assist them in the payment for care.

Create a very simple list clearly showing all of the payment options you offer and make it visible for the client in the exam room, reception area, web site, etc. Offering options really does show you care.

HEALTHCARE PLANS

Estimates are just one way of communicating the cost of services to a client. Estimates that used to be given only to new clients or for higher bills have become more and more a necessity for financial and legal reasons. I have worked with practices that give written estimates for everything from a nail trim to a splenectomy. And there are others that just reach into the back of the minds and say, "Mrs. Jones, it will cost about $XXXX for today's work." Only to find out that they grossly underestimated the costs and were faced with having to explain that or grossly overestimated the cost and the services weren't accepted.

Take estimates seriously. Actually, if you think about estimates, they are what your repairman provides you for fixing your car or clothes washer. What we are actually providing is a cost of services based upon a clinical plan that we have ascertained would be best for the patient. Consider changing the terminology in your practice to a HEALTH CARE PLAN (HCP) or Treatment Plan.

With the HCP, you are now able to focus on the left hand column of information that has the services that are needed delineated. The exam room nurses or receptionists can focus on the right hand column that is the costs. In a systematized practice, the doctor focuses on the needs of the pet (left column). The staff focuses on the cost for the care (right column).

To get past euthanasia as option, euthanasia can't be provided as the first option. Healthcare plans (Estimates) must provide optimal care for the case in question. These plans must be clearly explained and clarified for the pet parent. When hesitancy is noted, it is imperative that the reason for the hesitancy be identified: is it the risk of the procedure? The prognosis for a positive outcome? The place of the pet in the family hierarchy? Or the cost of care?

Should the pet's needs change and the HCP change, advise the client ASAP. Clients are much, much, much happier AND the Staff is much, much, much happier when they are forewarned about what their bill might be.

COMMUNICATION

Positive and confident body language and verbiage are important both when discussing HCPs and when eventually collecting the bill. Practiced scripts are very effective in this arena and an integral part of the system.

Having the Financial Policy and Financial Guidelines Forms in conjunction with the FAF provides you with the tools needed to communicate clearly that you really want to help them provide the optimal level of care to their pets. You have the knowledge, skills, and desire to provide the healthcare and NOW you have communicated to them that not only can you solve their pet's problems but you can also help them with the financial problem that might force them to make a decision between life and death.

The combination of pet insurance, Care Credit, healthcare plans, and a well-trained team will work together to give patients the level of care they deserve. The combination of pet insurance, Care Credit, healthcare plans, and a well trained team will work together to provide clients with the peace of mind that their four legged children can be given they need. The combination of pet insurance, Care Credit, healthcare plans, and a well-trained team will give your practice a reputation as a caring and compassionate hospital. The combination of pet insurance, Care Credit, healthcare plans, and a well-trained team will provide you with the additional revenue to make the entire healthcare team happy and proud of the work that they do.

SUMMARY

Having a step by step program for each client based upon agreed upon values and beliefs allows for consistency within your practice when it comes to the money talk. This area is extremely sensitive for the client and the healthcare team. By creating a system you are providing standards that are openly shared and communicated and that the client will appreciate. As will your staff.

Look closely how you are handling this area of your practice today and determine what steps you need to institute to put a system into place.

HOW TRANSPARENT COMMUNICATION CAN HELP THE BOTTOM LINE

Peter Weinstein, DVM, MBA
Irvine, CA

INTRODUCTION

A system is successful based upon two premises—the consistency in which it is performed and the manner in which it is communicated. Having created a system to talk to clients about money, communicating it is the next step. In healthcare, and in my opinion, decision-making must be made in the following order:

Pet's best interest
Client's best interest
Staff's best interest
Hospital's best interest
Doctor's best interest
Owner's best interest

You need to make decisions with this order in mind and thus offer the best care possible that the client can afford and that the staff and hospital can deliver. However, more importantly, you need to communicate this decision-making hierarchy to the clients both in words and the way you act..

How you handle the money talk sends a HUGE message to the pet owner. Just read some on line reviews on Yelp if you want to find out what clients think about practices that have not learned how to talk about money. But it isn't easy...

DOLLAR AND CENTS TAKES COMMON SENSE

As a practitioner, I found it easier to discuss life and death than dollars and cents. What is it about money, fees, costs, etc. that is so intimidating, daunting, threatening, and downright scary?

Many of us were brought up with very negative connotations of money. Look at some of these quotes:

"money is the root of all evil"

"Money is dirty"

Is it any wonder that in a caring and compassionate profession, such as veterinary medicine, talking about money has its own negative connotations?

Is the barrier to charging more and collecting more your clients or YOU?

The cost of providing care for pets has grown significantly over the last decade. The amount of money spent on pets both for veterinary care and for other pet related services has continued to grow in spite of the recessionary economy.

In the 21st century, where pets have taken on the role of four-legged children in my families, we have an increasing obligation to both the client and the pet. Our obligation, by oath, will always be to the pet's best interest and health. However, we can't forget the human holding the leash or carrier.

A challenge that we face as a profession is to educate the public on the level of care that is available to their pets. It is amazing, still, how many pet owners really have NO idea about the amount of education that veterinarians have, the technology that is readily available to us, and the knowledge base with which we work. For pet owners to appreciate the value of the care that we provide, we must educate them about the value of the care that they receive. Until we can do that, the cost of veterinary care will always be questioned by those not in the know.

To be able to feel comfortable in discussing your fees with clients, you must be comfortable that clients are getting value for the level of care provided. You must make sure that they are a part of the decision making process. You must educate, educate, educate and then educate some more so that they are comfortable in the decisions that they made. And you need to start the education at the first visit and never stop.

TALKING ABOUT MONEY...

How many people do you think have a line item on their personal budget for their pet care? For that matter, how many people do you think actually have a personal budget? How many people who purchase a pet have any idea what the basic veterinary costs will be over the life of the pet? And this doesn't include food, treats, toys, grooming, etc.

DVM Newsmagazine, Veterinary Economics, Consumer Reports, Smart Money, USA Today, Forbes, etc. have addressed a variety of notable trends in the financial levels within the profession. Some of the truths noted:

*Level of skills, technology, and care have all increased astronomically
*The actual cost to the practice to deliver the above has increased.
*The reflected cost to the consumer to receive the above has increased
*The relationship of pets within families has changed.
*Expectations of clients for the services provided by veterinarians have increased.

The two most difficult things to talk about in practice are euthanasia and money. Interestingly, in both conversations, we are taking something very near and dear away from a client.

In practice, most of the non-clinical discussions are handled by the para-professional and management staff and NOT by the doctor. Thus, it is IMPERATIVE that the staff (and associates) believe in what you are doing, the level of care that you offer, the standards that you follow, and the value that you provide. The greatest barrier to a successful money talk can be a staff member that feels you are over-charging or gouging a client. Thus, communication starts in the clinic and must have everybody on the same page.

Most importantly, when it comes to both medicine and money, clients hate surprises.

Transparency comes from not only communicating about your financial policy but also the manner in which you discuss it and the timing. Discussing it up front gets it out in the open, clears the air and hopefully minimizes surprises.

Talking to clients up front about their budget, what they can afford, and then describing payment options they may not have known about should mitigate surprises.

PUTTING THE SYSTEM INTO PLAY

You have the following tools created:

- Financial Policy
- Financial Policy Form
- Healthcare Plan
- Payment Options

Now how do you communicate this to the client? Let's look at this somewhat chronological order.

FINANCIAL POLICY AND PAYMENT OPTIONS

This are the initial areas communicated and in many cases they are not communicated directly. The Financial Policy and Payment options should be shared and in a public forum. Website, exam rooms, waiting area, brochures, etc. that discuss how you want payment handled and the payment options should all be readily available and not hidden away. New clients should be encouraged to look at them before they come in. Existing clients should be encouraged to review the policies and MUST be kept up to date on any changes.

These provide the foundation for all future talk.

COMMUNICATING HEALTHCARE PLANS? –WHO SHOULD DO IT?

Once the client is in the practice and clinical relationship started, the opportunity to create an estimate (or HealthCare Plan) comes to the fore. When should you create one? Every time and anytime especially if a client says they don't need one. Now, once it is printed or posted on the monitor, who discusses what?

In my practice, the doctor's role was to be a clinician. The doctor ensures that the client understands the care the pet needs and the treatment options available. They make sure all questions are answered about the nature of the care and then they say, "do you have any questions at all about what Sugar needs today?"

And the usual response is, "what is this all going to cost?" At which point the doctor asks, " Before I ask Kathy to go over the Plan with you, I just want to make sure you understand both what Sugar needs and why?" Once there is clarity, a receptionist or exam room nurse can go over the healthcare plan from a cost standpoint and collect a deposit if appropriate.

Focus on the pet's needs first. Ensure they are understood. Then focus on the client's needs (cost).

What happens if this is a major medical case or one that might change based upon some initial testing? Then, think…

NO SURPRISES

Updating healthcare plans twice per day and including an update on the costs accrued AFTER and ONLY AFTER you update the client on the status of their pet helps minimize surprises. Promise your clients that if their pet is NOT at home with them that you will keep them updated on a regular basis. By phone, by e-mail , by text, by owl, let them know both the clinical and financial update.

If you find yourself about to exceed the costs projected, don't do anything until you advise the client by how much and WHY the costs have escalated. Keep clients in the loop as things develop. Given an honest prognosis of the case. Give a worst-case scenario of the progression both clinically and financially.

Think about it, is there anything worse than telling a client that everything will be just fine or things are going great and then everything tanks?

Clear ongoing communication will prevent the explosion at the front desk that so disrupts your day. But when that happens, it is all about….

LOCATION, LOCATION, LOCATION

The reception area is NOT the best place to be talking about money. If at all possible, move money conversations into an exam room or business office. Letting a client vent is a good way to start. Don't interrupt. Make eye contact. Take notes. And repeat back what you identify as the greatest concern/issue. This is anger management 101 in most cases. Giving the client the opportunity to present their side without interruption is all that is needed sometimes.

Once the issue is identified. Solve it. Then and there! How you solve it depends upon if you are a....

BIG BAD MONEY LADY or FINANCIAL COUNSELOR

A big, bad money lady solves it by getting in the client's face, using the Financial Policy as a crutch and looking to create a win for the practice. In a nutshell, if you are determined to win...you lose...the client...forever.

Thus, be a financial counselor and look for compromise. Sometimes, when you lose your win Is a client that pays your practice $30,000 over the lifetime of the pet worth losing over a $15 prescription that their pet was given and they couldn't get the pet to take the pills.

This level of communication is THE most important if a client is teetering on the edge of going somewhere else. Learning how to read a client, speaking in a smooth, quiet, calming tone, and gaining trust and most importantly, solving the problem, is the goal of transparent communication.

COMMUNICATION TIP: NEEDS vs. RECOMMENDS

As a healthcare provider, a client and their pet come to you to identify a condition, cure the condition, or prevent the condition OR provide peace of mind that there is no condition. A recommendation does not emanate the same level of confidence that a need does. A recommendation is more of a suggestion, a need is more or a requirement. As a clinician, your role should be to communicate the REQUIRED care not just suggested care.

SUMMARY

The financial implications of Veterinary Practice are a matter of:

1) Avoiding surprises by making pet owners aware of their pet's needs and other associated costs.
2) Providing high quality care and higher quality service to give the impression of high value
3) Anticipating a pet's needs and discussing these needs with clients
4) Providing options for clients so that the decisions for their pet's health care are made with their hearts NOT with their wallets.

So, the financial implications of what we do everyday are huge!! I challenge you to increase the level of client service that you provide, increase the standards of care that you operate under, and astronomically increase the type and nature of your communications with your clients.

Clients want and need to be a part of the decision making process. They came to you to have their problems solved. To enhance the value of the services you provide they must leave not only having had their problems solved but also having expectations for future care clearly delineated. They must have a greater appreciation of the value that you have to offer for them and their pet and be willing to find ways, with you assistance, to pay for that care.

Talking about money is not easy. But being able to talk about money is good for business.

MYXOMATOUS MITRAL VALVE DISEASE IN DOGS

Lisbeth H. Olsen, DVM, DVSc
Frederiksberg, Denmark

INTRODUCTION

Myxomatous mitral valve disease (also known as chronic valvular disease, mxomatous valve degeneration or degenerative myxomatous mitral valve disease) is a thickening and/or enlargement of the mitral valve leading to mitral regurgitation in dogs. The mitral valve can be affected alone or in combination with the tricuspid valve; however the aortic and pulmonary valves are seldom affected. The disease is the most common heart disorder in dogs and the prevalence is age- and breed-related. The disease is most commonly seen in small to medium-size breeds, especially Cavalier King Charles Spaniels (CKCS). The course of the disease can be divided into three stages: 1. Early signs of mitral valve prolapse (MVP) and/or thickening of the valve without any or with only minimal mitral regurgitation, 2. Presence of mild to severe mitral regurgitation but no other clinical signs of heart failure and 3. Clinical signs of heart failure. Mild stages of myxomatous mitral valve disease are seen as relative benign conditions; however, it is challenging to define dogs at high risk of progression.[1]

AETIOLOGY

The major question is what causes the mitral valve changes in dogs with myxomatous mitral valve disease. The primary initiating factors are not known. Genetic factors are known to play a major role; however, what exactly initiates this disease process is unclear. There are many possible initiating factors such as defects in collagen and/or other extracellular components.

A hereditary background of myxomatous mitral valve disease has long been suspected on the basis of the higher prevalence of disease in certain breeds of dogs. More direct evidence for a genetic basis has been found.[2-5] Significant correlations between cardiac status in parents and offspring both with regard to regurgitant murmur intensity[2] and MVP severity[3] have been found. Based on these two studies a polygenic inheritance was suggested. In a recent genome-wide association study, two loci have been associated with early onset of myxomatous mitral valve disease in CKCS.[5]

PATHOPHYSIOLOGY

Different pathophysiological mechanisms have been suggested to be involved including changes in platelet and endothelial function, neuroendoendocrine regulation and myocardial function. However, it is not clear whether these changes play a primary role or they are secondary changes in relation to the disease process. In addition, it can be difficult to rule out whether pathophysiological changes seen in dog breeds predisposed to the disease, are disease related, or represent breed differences not related to the disease.

It is hypothesized that altered motion of the mitral valve in relation to MVP may be due to a defect in the connective tissue, and mitral regurgitation leads to changed haemodynamic forces which act on the surface of the valve. This may cause repetitive minor damage to the valve, which in turn mediate subendothelial changes that further weaken the valve. This might mediate a vicious circle in the mitral valve tissue that might be influenced by other cardiovascular factors including changed platelet function, circulating cardiovascular mediators and myocardial function.

Platelet function

Different results have been found concerning relations between myxomatous mitral valve disease and platelet function. The results seem to depend on the assessment technique used and the stage of disease. An increased platelet aggregation response is found in CKCS compared to dogs of other breeds.[6,7] In addition, a decreased platelet life span[8] indicating increased platelet activity has been shown in dogs with severe mitral regurgitation. However, other studies have showed a decreased platelet-aggregation response in CKCS and other dogs with severe mitral regurgitation.[9,10] In addition CKCS and Dachshunds with mild to severe mitral regurgitation have been found to have prolonged closure times using a platelet function analyzer (PFA100) reflecting decreased ability to form a platelet plug in a hole in a membrane at high shear rates.[11,12] In the CKCS, these findings were associated with a selective loss of von Willebrand factor (vWf) high molecular weight multimers which might have been caused by high shear stress of the blood in relation to the mitral regurgitation.[13] Therefore the prolonged closure time measured with PFA100 in dogs with mitral regurgitation might be explained by the fact that platelet aggregation at high shear rates depend on vWf.

Macrothrombocytopenia (i.e. enlarged platelets together with thrombocytopenia) is seen in about one third of CKCS.[14,15] An autosomal recessive inheritance of the macrothrombocytopenia has been shown; however, clinical signs of bleeding are rare in these dogs despite the presence of low manual platelet counts, typically between 30,000 and 90,000 platelets/μL.[14-16] The macrothrombocytopenia in CKCS is not related to the severity of MVP or the degree of mitral regurgitation.[6,14,16] On the other hand, CKCS with moderate to severe mitral regurgitation show a tendency to have higher platelet counts compared to CKCS with no or only minimal mitral regurgitation.[11,17] Shear stress due to the mitral regurgitation might cause platelet fragmentation leading to a higher platelet count.[18] However, further studies are warranted to study the possible relations between mitral regurgitation and platelet count in dogs, and these studies should include other breeds than CKCS.

The clinical consequences of platelet dysfunction in dogs with myxomatous mitral valve disease are not clear. In theory, the increased platelet aggregation response found in CKCS could be associated with development of the vascular changes in femoral and pulmonary arteries and with pathological changes in the mitral valve as well as thromboembolic events.[6,19,20] On the other hand, the decreased platelet function found in CKCS with mitral regurgitation might be associated with an increased bleeding tendency during surgery.[10,11,21]

Endothelial function

Different circulating biomarkers associated with endothelial function have been evaluated in relation to myxomatous mitral valve disease. Some studies have indicated that dogs with the disease have endothelial dysfunction due to decreased nitric oxide (NO) availability similar to human patients with heart disease. The plasma concentration of the NO metabolites nitrate and nitrite is used as an estimate of the ability of the endothelium to release NO. A decreased plasma concentration of nitrate and nitrite has been found in CKCS with clinical silent mild to severe mitral regurgitation, when compared with CKCS without or with only minimal mitral regurgitation and dogs of other breeds.[17] The finding suggests, in line with histopathological findings, a systemic endothelial dysfunction in CKCS with mild to severe myxomoatous mitral valve disease without clinical signs.[17,19,20] However, increased levels of plasma NO metabolites have been found in dogs with clinical signs due to advanced disease.[22] In addition, plasma levels of asymmetric dimethylarginine (ADMA), an endogen inhibitor of nitric oxid synthase (NOS), that play an important role for the regulation of endothelial NO production and the metabolite of the vasodilator prostacyclin (6-keto-prostaglandin (PG) $I_{2\alpha}$) were not found to be associated with clinical silent myxomatous mitral valve disease.[23,24] Plasma von Willebrand factor decreased in dogs with myxomatous mitral valve disease, as mentioned previously, probably due to a shear stress induced destruction in relation to mitral regurgitation.[13]

Evaluation of plasma markers of endothelial function in dogs requires highly standardised study conditions. A study showed that the plasma markers were influenced by several factors such as dog breed, exercise and clinical settings.[25] Functional measurements evaluating endothelial function are desirable to support findings suggesting endothelial dysfunction in dogs with myxomatous mitral valve disease. A recent study reports decreased flow-mediated vasodilation (FMD) response in CKCS with myxomatous mitral valve disease using a functional non invasive measurements of endothelial function.[26]

Local valvular mediators

Alterations in the extracellular matrix such as deposition of mucopolyschaccarides and changes in the collagen and elastic fibres are central to the pathogenesis of myxomatous mitral valve disease. Several local mediators seem to be involved in the disease process. Changes in NO availability are suggested to play a role in the pathogenesis of myxomatous mitral valve disease in dogs. In diseased canine mitral valves, an increased NADPH-diaphorase (NADPH-d) activity probably reflected increased expression of NOS is found mainly in the endothelium.[27] Haemodynamic and/or mechanical forces in certain areas of the valve might cause focal increase in NOS expression in the endothelium which in turn can contribute to the induction of focal pathological valvular changes such as apoptosis. This hypothesis could explain the slow development over years of myxomatous mitral valve disease. In addition, the hypothesis is supported by *in vitro* studies showing that mechanical injury, shear stress and pressure influence the synthesis and release of mediators including NO from endothelial cells.

A disturbance in the regulatory balance between NO and endothelin could also be of key importance in the local disease process in the myxomatous mitral valve. In the canine myxomatous mitral valve leaflet, an increased expression of endothelin-receptors was found both in the endothelium and in the stroma of the leaflet.[28] It has been shown that both NO and endothelin can influence the production of extracellular matrix in cell cultures with cardiac fibroblasts and smooth muscle cells. Valvular interstitial cells (VIC) are the most common type of cells in the valve and are responsible for the production of extracellular matrix including collagen and mucopolysaccharides. Matrix metalloproteinases (MMPs) are proteases that are inhibited by specific tissue inhibitors (TIMPs). Both MMPs and TIMPs are expressed in canine valvular interstiel cells (VICs) and some isoforms are up regulated in severe myxomatous mitral valve disease.[29-31] In addition, MMPs and TIMPs are both involved in the regulation of transforming growth factor-β isoforms (TGF-β1, β2 and β3). TGF-β1 and TGF-β3 are suggested to be involved in the myofibroblast-like differentiation of VICs.[29-31] Moreover, serotonin signaling pathways are suggested to play a major role for VIC differentiation and myxomatous valve damage.

Myocardial function

Intramyocardial fibromuscular arteriosclerosis and fibrosis have been found in dogs with advanced stages of myxomatous mitral valve disease. They may be part of a generalised disease process and play a role in the development of heart failure and sudden death. A significant association between intramyocardial arteriosclerosis and myocardial fibrosis has been found. Also there was an association between intramyocardial arteriosclerosis and decreased systolic function.[32] Despite the myocardial changes, the global pump function is apparently maintained for a long period of time in most dogs with advanced myxomatous mitral valve disease. However, a myocardial dysfunction might be masked by favorable loading conditions since reduced load-independent contractile reserve has been observed in dogs with experimental heart disease.[33] Tissue Doppler imaging reflecting systolic and diastolic myocardial function has not been found to be changed in dogs with myxomatous mitral valve disease.[34] It is interesting that circulating level of markers of myocardial function have been found to change in dogs with myxomatous mitral valve disease. Increased plasma levels of cardiac troponin I and decreased plasma levels of MMP-9 and some cytokines are associated with decreased myocardial function.[35-37]

Electrocardiographic changes

Cardiac arrhythmias are not common in dogs with early stages of myxomatous mitral valve disease; however supraventricular arrhythmias has been associated with early signs of the disease (mitral valve prolapse) in young Dachshunds[38] and with left atrium enlargement in older dogs with advanced disease.[39] The occurrence of ventricular arrhythmias is higher in dogs with advanced myxomatous mitral valve disease compared to dogs without heart disease.[39] It is interesting that ventricular arrhythmias tended to be associated with intra myocardial arterial narrowing, myocardial interstitial fibrosis and decreased systolic function in dogs with advanced disease. [32] Holter derived electrocardiographic variables change with disease severity in CKCS and the changes are most pronounced in advanced stages of the disease.[40] Heart rate variability decreases with increasing disease severity, whereas minimum and mean heart rates measured over 24-hours increase.[40] In a recent study, advanced electrocardiographic parameters (5-minute 12-lead electrocardiography) have been shown to predict the severity of mitral regurgitation in CKCS with myxomatous mitral valve disease.[41]

CONCLUSIONS AND PERSPECTIVES

New knowledge regarding disease mechanisms may contribute to further clarification of the complex pathophysiology of myxomatous mitral valve disease, and thus hopefully leads to improved prognostic, diagnostic and therapeutic strategies in the future.

REFERENCES

[1]Borgarelli M, et al. **J Vet Intern Med** 2012;26(1):69. [2]Swenson L, et al. **J Am Vet Med Assoc** 1996;208(12):2009. [3]Olsen LH, et al. **J Vet Intern Med** 1999;13(5):448. [4]Lewis T, et al. **Vet J** 2011;188(1):73. [5]Madsen MB, et al. **J Hered** 2011;102(suppl 1):S62. [6]Olsen LH, et al. **J Vet Intern Med** 2001; 15(3):209. [7]Nielsen L, et al. **Vet Clin Pathol** 2007;36(3):267. [8]Tanaka R, et al. **J Vet Intern Med** 2002;16(4):446. [9]Cowan SM, et al. **Exp Hematol** 2004;32(4):344. [10]Tanaka R, et al. **Am J Vet Res** 2000;61(10):1248. [11]Tarnow I, et al. **J Vet Intern Med** 2003;17(5):680. [12]Moesgaard SG, et al. **Res Vet Sci** 2009;86(2):320. [13]Tarnow I, et al. **J Vet Intern Med** 2005;19(4):515. [14]Eksell P, et al. **J Small Animal Pract** 1994;35:153. [15]Olsen LH, et al. **J Vet Diagn Invest** 2004;16(2):167. [16]Smedile LE, et al. **J Am Anim Hosp Assoc** 1997;33(5):411. [17]Pedersen HD, et al. **J Vet Intern Med** 2003;17(2):178. [18]Brown CH 3rd, et al. **J Lab Clin Med** 1975;86(3):462. [19] Buchanan JW, et al. J Am Vet Med Assoc 1997;211(7):872. [20]Karlstam E, et al. Vet Rec. 2000;147(6):166. [21]Shiroshita Y,et al. **Vet Rec** 2000;146(1):16. [22]de Laforcade, et al. **J Vet Intern Med** 2003;17(3):315. [23]Pedersen LG, et al. **Res Vet Sci** 2006 Jun;80(3):336-42 [24]Rasmussen CE, et al. **Vet J** 2010;184(2):241. [25]Moesgaard SG, et al. **Res Vet Sci** 2007;82(3):409. [26]Moesgaard SG, et al. **J Vet Intern Med** 2012;26(1):61. [27]Olsen LH, et al. **J Comp Pathol** 2003;129(2-3):120. [28]Mow T, et al. **J Cardiovasc Pharmacol** 1999;34(2):254. [29]Aupperle H, et al. **J Comp Pathol** 2009;140(4):271. [30]Aupperle H, et al. **Res Vet Sci** 2009;87(2):277. [31]Obayashi K, et al. **Am J Vet Res** 2011;72(2):194. [32]Falk T, et al. **Vet J** 2010 Jul;185(1):68. [33]Urabe Y, et al. **Circ Res** 1992;70(1):131. [34]Tidholm A, et al. **J Vet Intern Med** 2009;23(6):1197. [35]Ljungvall I, et al. **J Vet Intern Med** 2010 ;24(1):153. [36]Ljungvall I, et al. **Am J Vet Res** 2011;72(8):1022. [37]Zois NE, et al. **Vet J** 2011 (in press). [38]Olsen LH, et al. **J Vet Cardiol** 1999;1(2):7. [39]Crosara S, et al. **Aust Vet J** 2010;88(10):386. [40]Rasmussen CE, et al. **J Vet Intern Med** 2012 Jan;26(1):76. [41]Spiljak Pakkanen M, et al. **J Vet Intern Med** 2012 (in press).

CONTROVERSIES IN MYXOMATOUS MITRAL VALVE DISEASE

Ingrid Ljungvall, DVM, PhD and Jens Häggström, DVM, PhD, DECVIM-CA (Cardiology)
Uppsala, Sweden

INTRODUCTION

Myxomatous mitral valve disease (MMVD) is by far the most common heart disease in dogs. Comparably extensive research has been conducted concerning different aspects of this disease in dogs and much has been learned from these studies. However, many issues are still unresolved, or available data are conflicting, or, in some instances, confusion prevails amongst veterinary clinicians, even veterinary cardiology specialists. This presentation is aimed at identifying some of these controversies and to comment on them from our perspective, but we do not aim at providing definite answers.

TERMINOLOGY

Consistent terminology in the scientific literature is important for defining a disease. Unfortunately, this disease has many names: chronic valvular disease, degenerative mitral valve disease, mitral valve disease, endocardiosis, and more recently myxomatous mitral valve disease. If the condition is simply referred to as "chronic mitral valve disease", or "mitral valve disease", this could include mitral stenosis, mitral dysplasia or even endocarditis, because the term "chronic" denotes a temporal association, not a pathophysiological one. The same applies to the term "chronic valvular disease" – the term "chronic" does not define the pathology and therefore does not help in characterizing the disease process. The term "myxomatous" describes a characteristic histologic feature of this disease. Conversely, the term "myxomatous" excludes most, if not all, other forms of mitral valve disease. Although the myxomatous degeneration most commonly affects the mitral valve, any of the four intracardiac valves can be affected. However, the most commonly used descriptor "mitral" does not necessarily exclude involvement of other valves. Thus, it is our opinion that "Myxomatous mitral valve disease" accurately and succinctly defines the pathophenotype of this condition.

ARE CAVALIER KING CHARLES SPANIEL DOGS DIFFERENT THAN DOGS OF OTHER BREEDS?

A search in MedLine shows that approximately 10% of all published articles on mitral valve disease in dogs have "Cavalier King Charles Spaniel (CKCS)" in the title. No other breed is nearly as frequently specifically mentioned in article titles. It has been suggested that CKCS have a very different form of myxomatous mitral valve disease compared to other breeds. It has been suggested that CKCS have a more rapid disease progression into congestive heart failure (CHF), and that they respond differently in comparison to other breeds in response to certain pharmacological therapy.[1,2] In our opinion, many of these suggestions lack support in the form of scientific evidence. Furthermore, results suggesting a different outcome for CKCS are often not easily interpreted and may have been confounded by other factors. For example, the QUEST trial suggested a more favorable outcome (longer survival) CKCS dogs than for dogs of other breeds; even after adjustment for all other baseline variables. The authors of this article avoided making this conclusion. Clearly, CKCS dogs remained longer in the study before they reached the primary endpoint (which was cardiac related death/euthanasia or treatment failure) and this result could reflect a true difference in survival as a consequence of breed, but could also reflect other factors that are associated with CKCS dogs; such as owner attitude to their pet and to heart failure medications. It has, to our knowledge, never been shown that the pathology of the mitral valve apparatus in MMVD is different in CKCS than in other breeds. We think that such a difference does not exist. Still, some reports have clearly shown that CKCS dogs, in some other aspects, are different from dogs of other breeds. It has been shown that CKCS have an onset of MMVD at a younger age than certain other breeds. Consequently, CKCS dogs have an alarmingly high cause specific mortality before the age of 10 years in actuarial data (only the Irish Wolfhound have a higher cause specific mortality).[3] Two loci has recently been reported to be associated with the age of onset of MMVD in CKCS,[4] but it is unknown if the same loci are similarly associated in other breeds. A recent report at ACVIM forum described a mutation in the beta X receptor in CKCS dogs.[5] Furthermore, it has been shown that CKCS have a different plasma renin to aldosterone ratio and higher plasma aldosterone concentrations,[6] and higher plasma serotonin concentrations[7] than other breeds. CKCS dogs have also recently been shown to have a higher heart rate compared to other breeds of comparable size.[8]

ARE FINDINGS FROM LABORATORY DOGS WITH EXPERIMENTALLY INDUCED MITRAL REGURGITATION RELEVANT FOR DOGS WITH NATURALLY ACQUIRED MMVD?

The mechanisms involved in the remodeling process of MMVD remain poorly characterized and understood; partly owing to the slow progression from early degenerative changes to development of CHF; which complicates studies of disease progression. Cardiac remodeling processes in experimental settings have been more extensively documented.[9,10] These cardiovascular events differ fundamentally from naturally acquired chronic valvular disease in origin and progression, which emphasizes the importance of investigating the left ventricular (LV) remodeling process also in dogs with naturally acquired MMVD. However, it is difficult to obtain cardiac tissue from a large group of client-owned dogs with differing severities of naturally occurring MMVD, and as a consequence; comprehensive histologic and biochemical evaluation of cardiac tissue from dogs in various MMVD severity stages is lacking. Myocardial changes seen in experimentally induced MR dogs which have been followed during a prolonged study-period following MR induction,[11] can probably better reflect some of the myocardial processes in dogs with naturally acquired MMVD. Nevertheless, the following-up period in these dogs fails to mimic the naturally acquired MMVD progression. Even though results from experimental settings, as well as from acute vascular diseases in other species, cannot be automatically applicable for dogs with naturally acquired MMVD, such study results might stimulate further investigations in dogs with naturally occurring disease progression. Potential existence of similar pathophysiological mechanisms can be investigated using other, less invasive methods; such as circulating biomarker analyses.

IS FIBROSIS A PROBLEM?

The existence of significant myocardial fibrosis in dogs with MMVD has been questioned. Even though the valvular apparatus is the site of origin in MMVD, secondary changes also occurs in the myocardium in response to the mitral regurgitation (MR). The remodeling process of the extracellular matrix (ECM) has been suggested to occur as a mechanically mediated response to stretch caused by volume overload. Normal collagen chains are fractured and replaced by poorly cross-linked collagen, resulting in loss of normal structural support and myocyte slippage.[11,12] The structural changes in the cardiac ECM and myocytes allow for chronic dilatation without overstretching of the myocytes, thereby minimizing the effects on the myocardium from the increased volume overload.[13] The process of cardiac dilatation is presumably accompanied by additional myocardial processes in dogs with MMVD, e.g. myocyte death and replacement fibrosis.[14] Replacement fibrosis is mainly oriented close to vessels, and through serial sectioning an association between arteriosclerosis and replacement fibrosis/infarcts has been demonstrated in dogs with MMVD.[14,15] An arteriosclerosis-induced myocyte death might be a causative factor in development of myocardial fibrosis in MMVD in dogs. Based on these findings, we think that strong evidence exists regarding the presence of myocardial fibrosis in MMVD dogs. In our opinion, myocardial fibrosis might affect both systolic function, and indirectly survival. Myocardial fibrosis, perhaps in combination with inadequate myocyte hypertrophy, might contribute to the loss of force of contractility through the LV in MR dogs;[16,17] An increased amount of myocardial fibrosis might furthermore influence the Frank-Starling mechanism thereby disable an optimal transduction of contractile force generated by the myocytes in systole. It is not known at which stage of MMVD these changes occur.

CAN SYSTOLIC DYSFUNCTION BE ASSESSED BY ECHOCARDIOGRPAHY IN DOGS WITH MMVD?

The systolic function has been suggested to decline in dogs with naturally acquired MMVD and in dogs with experimentally induced MR.[16,17] However, identification of systolic dysfunction is challenging in dogs with MMVD. A reduced afterload and an increased preload lead to normal to hyperdynamic LV contraction, even in the presence of intrinsic myocardial dysfunction. Many of the commonly used echocardiographic indices for evaluating systolic function, such as the ejection phase indices (ejection fraction and shortening fraction) are, besides being dependent on intrinsic contractility, also known to be influenced by hemodynamic load and sympathetic tone, which potentially mask significant myocardial dysfunction in dogs with MR. Left ventricular internal dimension in systole corrected for body weight ($LVIDs_n$or $LVIDs_{inc}$), which has been suggested to be a more load-independent variable, has been shown to increase in dogs with more severe MMVD; indicating systolic dysfunction.[17,18] Use of this echocardiographic variable as an indicator of systolic function has been questioned using the argument that the left ventricular internal dimension in systole might increase only as a consequence of increased preload in MMDV dogs, and thereby not actually reflect systolic dysfunction. However, a fully preserved contractile function in severely affected MMVD dogs could be assumed to shorten the LV to an almost normal end-systolic dimension as the LV stroke volume can be ejected both forward as well as retrograde into the low resistance LA; thereby facilitating LV ejection. Hence, in our opinin; increased $LVIDs_{inc}$, indicates systolic dysfunction in MMVD dogs. The more recently introduced tissue doppler imaging (TDI) technique has been considered comparably independent of loading conditions, but recent studies suggest that this technique is affected by loading condition and sympathetic tone activity to a greater extent than previously expected, which might limit the value obtained from this technique, compared to using $LVIDs_{inc}$.[19] Although myocardial systolic function declines with progression of the disease, the remodeling process allows the LV to retain a relatively well preserved forward cardiac pump function even in advanced MMVD.[20] Increased pulmonary blood volume, and not decreased forward stroke volume, has been shown to be the main cause of abnormal cardiopulmonary function in dogs with MMVD.[21]

CAN THE ONSET OF CLINICAL SIGNS OF CONGESTIVE HEART FAILURE BE PREDICTED?

Because MMVD in small dogs is characterized by chronic progression, the owner and veterinarian often have no other alternative than to observe how the valvular lesions and MR progress, often over years. At a certain point, the heart can no longer compensate for the MR, so the forward cardiac output decreases, and the pulmonary capillary pressure exceeds the threshold for pulmonary edema. Thus, there is a need for clinical variables that have prognostic value. Several physical examination- echocardiographical- and hormonal variables have been evaluated in these aspects. At present, left atrial dimension (LA/AO ratio, LA/predicted AO etc.), body weight normalized left ventricular end-diastolic dimensions ($LVIDd_{inc}$, $LVIDd_{n}$) , vertebral heart score index (VHS), and natriuretic peptides (Atrial and B-type natriuretic peptides) stand out as being reliable in predicting the onset of decompensated CHF. Significant changes occur in all of them 0.5 to 1 year before the onset of clinical signs of decompensated heart failure. Recent studies show that these indicators change in a non-linear manner over time.[22-25] Small changes are seen over the comparably long period during which a dog progresses from very mild to moderate and severe MR, but the rate of change of the variables increases as the dog progresses into CHF.[22-25] It is our opinion that a number of these variables have the acceptable predictive value for groups, but less so for individuals. Furthermore, the performance of the test improves considerably if more than one observation point is available, and if the observation point is closer to the time when the dog goes into CHF.

CAN THE PROGRESSION OF MMVD BE CHANGED BY DRUG INTERVENTION?

Several studies have shown that clinical signs of CHF and survival in dogs with MMVD may be improved as a consequence of drug therapy. The improved survival is likely to be caused by an alleviation of the heart and the circulation, not by a fundamental change in the pathophysiology of valve degeneration. Much work has been performed in order to investigate if certain drugs could change the progression of disease in asymptomatic MMVD dogs; as this group of dogs is very large. Clearly, dogs with cardiomegaly, secondary to MMVD, demonstrate changes in hemodynamic status compared to dogs with normal heart size, as the left side of the heart is undergoing a remodeling process (eccentric hypertrophy) in response to increasing MR and volume overload.[12,17] Furthermore, it has recently been shown that heart rate normalized to pulmonary transit time (the number of heart beats required for a unit of blood to travel from the main pulmonary artery to the left atrium) increases with worsening MR, even in asymptomatic dogs.[26] This increase is associated with increased pulmonary blood volume rather than reduced forward stroke volume.[21] Thus, it would appear that drugs that improve the hemodynamic situation would have the potential to change progression. Unfortunately, we think that no pharmacological, or any other form of management, has convincingly been shown to substantially change the rate of progression in this group of dogs. There are currently two large clinical trials ongoing investigating the effect of two different substances on progression in asymptomatic MMVD dogs.

REFERENCES

[1]Haggstrom J, et al. **J Vet Intern Med** 2008;22:1124-1135, [2]Pouchelon JL, et al. **J Vet Intern Med** 2008;22:905-914, [3]Egenvall A, et al. **J Vet Intern Med** 2006;20:894-903, [4]Madsen MB, et al. **J Hered** 2011;102 Suppl 1:S62-67, [5] Nelson OL, et al. **Proc ACVIM Forum, Denver**, *USA* 2011, [6]Pedersen HD, et al. **J Vet Med Ser A** 1995;42:435-441, [7] Arndt JW, et al. **J Vet Intern Med** 2009;23:1208-1213, [8] Rasmussen CE, et al. **J Vet Intern Med** 2011;26:76-84, [9] Dell'Italia LJ, et al. **Am J Physiol** 1995;269:H2065-2073, [10] Stewart JA, Jr, et al. **J Mol Cell Cardiol** 2003;35:311-319, [11]Zheng J, et al. **Circulation** 2009;119:2086-2095, [12] Dell'Italia L, et al. **Am J Physiol** 1997;273:H961-H970, [13] Komamura K, et al. **Am J Physiol** 1993;265:H1119-1131, [14]Falk T, et al. **Vet J** 2010;185:68-74, [15] Jonsson L. **Acta Vet Scand** 1972;38, [16] Urabe Y, et al. **Circ Res** 1992;70:131-147, [17] Borgarelli M, et al. **J Vet Intern Med** 2007;21:61-67, [18] Ljungvall I, et al. **Am J Vet Res** 2011;72:1022-1028, [19] Tidholm A, et al. **J Vet Intern Med** 2009;23:1197-1207, [20] Kittleson MD, et al. **J Am Vet Med Assoc** 1984;184:455-459, [21] Eriksson A, et al. **J Vet Intern Med** 2010;24:1393-1399, [22] Häggström J, et al. **Am J Vet Res** 1997;58:77-82, [23]Lord P, et al. **J Small Anim Pract** 2010;51:210-218, [24]Tarnow I, et al. **Vet J** 2009;180:195-201, [25] Ljungvall I, et al. **J Vet Intern Med** 2011, [26]Lord P, et al. **J Vet Intern Med** 2003;17:824-829.

PREDICTING FIRST ONSET OF CONGESTIVE HEART FAILURE IN DOGS WITH MITRAL VALVE DISEASE

Mark A. Oyama, DVM, DACVIM
Philadelphia, PA

Degenerative mitral valve disease (DMVD) is a common disease of adult dogs. Approximately 3/4 of dogs with heart disease will have mitral valve disease, and evaluating severity of heart disease and risk for congestive heart failure is an important feature of this condition. This presentation will review current and emerging methods used to diagnosis and assess risk for heart failure, with the goal of predicting the first onset of heart failure in dogs with asymptomatic mitral valve disease.

Methods involve the physical examination, ECG, thoracic radiographs, biomarker testing, and echocardiography. The goal of these methods is to identify the subset of dogs with progressive mitral valve disease, volume overload, and impending congestive heart failure. Once identified, these dogs might benefit from medical therapy or increased vigilance that would potentially prevent an unexpected and catastrophic episode of heart failure.

The diagnosis of mitral valve disease can be presumptively made on the basis of the classic murmur in an adult small breed dog. As mitral regurgitation worsens, the intensity of the left basilar heart murmur increases; however there is a great deal of individual variation, particularly in dogs with mild disease. Due to the orientation of the mitral regurgitant jet, it is not uncommon for dogs with mild mitral valve disease to have relatively loud murmurs. On echocardiographic examination disease of the anterior mitral valve leaflet can be seen to typically produce a mitral regurgitant jet orientated towards the lateral atrial wall, thus creating a relatively loud focal murmur. While it can be said that dogs with a grade 4 or 5 murmur will typically have more disease than dogs with a murmur of lesser intensity, it is difficult to base individualized treatment plans on the absolute intensity of the heart murmur.

The ECG is a relatively insensitive diagnostic test for severity of heart enlargement or risk of congestive heart failure. The ability to detect left atrial and left ventricular enlargement is as low as 50%. Dogs with advanced disease can demonstrate atrial or ventricular premature beats, or in cases of the worst severity, atrial fibrillation. Activation of the sympathetic nervous system is one of the characteristic responses to mitral regurgitation and subsequent decreased cardiac output. Dogs with congestive heart failure typically demonstrate sinus tachycardia whereas dogs with signs due to respiratory disease often demonstrate sinus arrhythmia. Therefore persistent tachycardia or loss of sinus arrhythmia can be consistent with progressive disease.

Thoracic radiography is perhaps the single best tool to assess severity of disease, progression of disease, and risk of heart failure. Radiographs are relatively sensitive and specific for left heart enlargement, pulmonary vein enlargement, and the caudal dorsal and perihilar interstitial pattern--features that are typical of pulmonary edema. Due to the high degree of individual variation in dogs of differing weights and chest confirmation, standardized radiographic measurements, such as the vertebral heart size, can be very useful. In the author's opinion it is relatively uncommon for dogs to experience heart failure with vertebral heart sizes less than 11.5-12.0. Thus in dogs with respiratory signs, the finding of a normal or only mildly enlarged cardiac silhouette and radiographs can help rule out heart failure. In addition to absolute size, previous studies (Lord, 2010; Lord, 2011) have shown that change in vertebral heart size is greatest in the 2-3 months directly preceding onset of heart failure. The performance of serial thoracic radiographs is one of the most useful means to monitor disease in the dog with moderate to advanced, yet still asymptomatic, mitral valve disease.

Echocardiography is a useful modality to confirm diagnosis, evaluate heart size and contractility, and to detect current heart diseases such as tricuspid regurgitation and pulmonary hypertension. Sound radiographic skills lessen the importance and impact of echocardiography, and in many if not most cases, mitral valve disease can be successfully diagnosed and managed without echocardiography. With respect to the presence of left heart failure, mitral inflow velocity indexed to isovolumic relaxation time is significantly elevated in dogs with left heart failure versus dogs with asymptomatic disease. Other parameters such as left atrial size can also be useful.

Recently, testing of blood based biomarkers such as NTproBNP has been studied. NT proBNP is a natriuretic peptide produced by myocardial tissue. Its circulating concentration is proportional to severity of heart disease. In both humans and dogs, NT proBNP is used to help rule out congestive heart failure in patient with respiratory signs. Recently, NT proBNP was shown to predict the first onset of congestive heart failure in dogs with asymptomatic mitral valve disease. In this multicenter study NT proBNP along with radiographic and echocardiographic heart size was able to detect dogs that would experience their first episode of heart failure within the subsequent 3 months. (Reynolds et al., 2012)

REFERENCES

Lord PR et al. J Small Anim Pract 2010;51:210-8. Borgarelli M et al. J Vet Intern Med 2012;26:69-75. Lord PR, et al. J Vet Intern Med 2011;25:1312-19. Reynolds CA et al. J Vet Cardiol;2012:in press.

TRANSCATHETER MITRAL VALVE IMPLANTATION (TMVI) FOR DOGS

Christopher Orton, DVM, PhD, DACVS
Fort Collins, Colorado

Transcatheter cardiac therapies have been developed and continue to emerge for several congenital and acquired cardiac conditions that previously could only be corrected by traditional open heart surgery. These therapies involve catheter delivery of a cardiac device into a beating heart. Transcatheter delivery of cardiac devices can utilize a **transvascular** or **transcardiac** approach. Transvascular approaches typically employ **percutaneous** access to a major artery or vein utilizing commercially available systems for gaining access to the vessel and controlling bleeding after catheter removal. Transvascular access to the right heart is gained by a **transvenous** approach (i.e. jugular or femoral vein). Access to the left heart generally employs either a **transarterial** (**retrograde**) approach or a transvenous approach combined with an **atrial transseptal** (atrial puncture) approach (**antegrade** approach). The antegrade approach is generally favored for access to the mitral valve. Percutaneous transvascular approaches to transcatheter cardiac therapies have the advantage of being less invasive compared to traditional cardiac surgery. However percutaneous approaches limit the size of delivery catheter and device that can be delivered. This can be particularly true in small animal patients such as small- and toy-breed dogs. Depending on the complexity of the procedure and cardiac structure needing intervention, delivery catheters may require complex systems to provide steerability. Such systems can further limit the size of the patient that can be treated. Transcardiac approaches require a minimally-invasive thoracotomy to gain access to the outside of the heart, and are thus referred to as **hybrid** procedures. Currently, the most common transcardiac approach is through the apex of the left ventricular wall, generally referred to as a **transapical** approach. Access for a transapical approach is gained by a caudal sternotomy with the patient in dorsal recumbency or a left thoracotomy with the patient in lateral recumbency. The principle advantage of a transapical approach is that it allows use of larger and simpler delivery catheters and systems. The use of transcardiac approaches has expanded the possibilities for delivery of new generation cardiac devices such as heart valve prostheses. They have also necessitated the establishment of hybrid operating rooms equipped with cardiac imaging such as C-arm fluoroscopy and transesophageal echocardiography (TEE) or intracardiac echocardiography (ICE). Lastly they have led to the development of **multidisciplinary teams** consisting of interventional cardiologists and cardiothoracic surgeons.

Current Heart Valve Interventions

The most common surgical heart valve interventions performed in humans are aortic valve replacement (AVR), mitral valve replacement (MVR), and mitral valve repair (MVRe). Each requires open heart surgery to accomplish. AVR and MVR require replacement of the native valve with a valve prosthesis. Valve prostheses are either mechanical prostheses or bioprostheses. Current transcatheter heart valve interventions available for clinical use in human patients include **transcatheter aortic valve implantation (TAVI)** and **percutaneous edge-to-edge mitral repair**. TAVI involves the catheter delivery of a bioprosthesis into the aortic position without removing the aortic valve and hence is most correctly referred to as valve implantation rather than valve replacement. Two TAVI systems are currently available or under development for clinical use in humans: the **Edwards-SAPIEN valve** (http://www.youtube.com/watch?v=N7o6K4onPxI&feature=related) and the **Medtronic CoreValve** (http://www.youtube.com/watch?v=Qg9Kvq-nTN4&NR=1). The Edwards-SAPIEN valve is a trileaflet bovine pericardial valve mounted on a balloon-expandable stainless steel stent. The Medtronic CoreValve is trileaflet porcine pericardial valve mounted on a self-expanding nitinol stent. Both are designed for treatment of calcific aortic valve disease, generally after balloon valvuloplasty to correct aortic stenosis. Fixation for both devices is dependent on radial force against a fixed diameter (calcified) aortic annulus. The Edwards-SAPIEN valve can be delivered using a percutaneous transarterial (transfemoral) retrograde approach or a hybrid transapical antegrade approach. The CoreValve is designed for delivery using a transfemoral approach.

Percutaneous edge-to-edge mitral repair is based on the open surgical edge-to-edge mitral valve repair technique described by Alfieri in 1998. **MitraClip** (http://www.youtube.com/watch?v=GwDgPDYf3Qo), the commercial system for percutaneous edge-to-edge mitral repair, is currently approved for use in humans in Europe and is an investigational device in the USA. The system utilizes a 24 F delivery system and an atrial transseptal approach to the mitral valve. The MitraClip is placed under TEE guidance. The emerging indication for MitraClip in humans is treatment of functional mitral regurgitation (MR).

Transcatheter Mitral Valve Implantation

Transcatheter mitral valve implantation (TMVI) is the catheter delivery of a valve prosthesis into the mitral position without the removal of the native mitral valve. Several TMVI designs are currently being explored, however none have yet reached the stage of first-in-human clinical trial. Design of a TMVI prosthesis presents significant challenges that are different from the challenges associated with the currently successful TAVI devices. Chief among these challenges are achieving reliable fixation of a prosthesis in the mitral position. Unlike the calcified aortic annulus, the mitral annulus is pliable and not fixed in diameter. Therefore fixation or anchoring of a mitral prosthesis must be based on something other than radial traction between a fixed-size annulus and the sides of the prosthesis. Further, the design must be such that it does not interfere with the left ventricular outflow tract (LVOT). Lastly, transcatheter access to the mitral valve is inherently more difficult than access to the aortic valve.

MitralSeal (Avalon Medical) is a TMVI prosthesis and delivery system being developed for dogs. The prosthesis consists of a trileaflet porcine pericardial bioprosthesis mounted on a self-expanding nitinol stent. The stent body consists of an isodiametric cylinder that houses the bioprosthetic leaflets and an approximately 90° atrial cuff at the inflow end of the cuff (Figure 1). The atrial cuff is covered by Dacron on the ventricular side and pericardium on the atrial side to minimize perivalvular leak around the prosthesis. The outflow end of the stent body has 3 attachment points for 3 suture tethers. The tethers are used to manipulate the prosthesis into the mitral orifice during deployment and then to fix the prosthesis to the ventricular apex. Fixation of the prosthesis is thus based on longitudinal (rather than radial) traction with the atrial cuff preventing displacement into the ventricle during atrial filling (valve open) and the tethers preventing displacement into the atrium during ventricular ejection (valve closed). The MitralSeal delivery utilizes a hybrid transapical approach and a straight non-steerable delivery catheter (Figure 2). Access to the left ventricular apex is gained via a minimally-invasive left thoracotomy. The pericardium is opened and 2 pledget-buttressed purse-string sutures are placed in the left ventricular apex to control bleeding from the cardiotomy site. A needle puncture is made in the left ventricular (LV) apex. An LV angiogram is performed to identify the mitral annulus using C-arm fluoroscopy with a lateral thoracic projection. The cardiotomy site is progressively dilated with obturators. A guide-wire is passed across the mitral annulus under fluoroscopic guidance. The delivery catheter is passed over the guide-wire across the mitral annulus into the left atrium. The guide-wire and catheter are removed. The prosthesis is back loaded into the delivery catheter and deployed into the left atrium under fluoroscopic guidance. The delivery catheter is removed, externalizing the prosthesis tethers. The prosthesis is manipulated into the mitral annulus by placing traction on the tethers under fluoroscopic and TEE guidance. The tethers are temporarily fixed on the outside of the heart using vascular clamps. The tethers are passed through a buttressing pledget. Tension on the tethers is individually adjusted to minimize perivalvular leak using TEE. The controlling purse-string sutures are tied. The tethers are permanently fixed and tied. A sealing cap is placed over the cardiotomy and tether fixation site to prevent bleeding. The thoracotomy is closed.

The MitralSeal device has undergone successful chronic implantation in normal pigs and dogs. These studies have identified problems including perioperative bleeding, perioperative pulmonary edema, prosthesis displacement, prosthesis leak, perivalvular leak, and thrombosis of the prosthesis. Each of these problems have been multifactorial and addressed by a combination of ongoing changes to the prosthesis design and changes to perioperative protocols. Perioperative bleeding was addressed by altering perioperative anticoagulation protocols and by adding a sealing cap to the cardiotomy site. Perioperative pulmonary edema was addressed by improving the hemodynamic flow properties of the prosthesis and avoidance of over-administration of fluids during and immediately after the procedure. Only one animal experienced displacement of the prosthesis and that problem was addressed by increasing both the diameter and stiffness of the atrial cuff. These design changes also addressed the problem of perivalvular leak around the prosthesis. Improvements in the bioprosthetic leaflet design, thickness, and coaptation have largely addressed problems with prosthesis leak and have increased the long-term competence of the prosthesis. The last remaining challenge is thrombosis of the prosthesis in dogs. Whereas pigs have gone multiple months without thrombosis or a need for chronic antithrombotic therapies, dogs have experienced significant thrombosis of the prosthesis within days to weeks of implantation despite antithrombotic strategies (aspirin and dose-adjusted warfarin) that have previously worked in dogs undergoing open MVR. Anecdotal evidence suggests that dogs are likely more prone to thrombosis of intravascular devices compared to other species. This may be to high activity levels of coagulation factors in dogs compared to other species.[1] Possible solutions to this problem in dogs include: 1) design changes to the prosthesis intended to decrease areas of disturbed flow around the prosthesis, and 2) robust antithrombotic protocols based on dual antiplatelet therapy (aspirin and clopidogrel[2]) and new generation anticoagulant drugs. New generation anticoagulation drugs could include direct thrombin inhibitors (e.g. dabigatran[3]) or oral factor Xa inhibitors (e.g. rivaroxaban, apixaban[4]). Lastly, it is an industry standard for new heart valve prostheses to undergo *in vitro* accelerated fatigue testing for at least up to 200 million cycles (5 year equivalent durability). Accelerated fatigue testing for the MitralSeal device is ongoing.

Design of a Canine TMVI Clinical Trial

Once prosthesis design has been optimized and perioperative protocols established for normal animals, a clinical trial of the MitralSeal device in dogs with naturally-occurring mitral valve disease will become appropriate. Such a clinical trial would likely represent a milestone in veterinary medicine in that the MitralSeal device will have been conceived and developed, and will be manufactured specifically for dogs by a veterinary medical device company. It is possible that TMVI will be successfully accomplished in a canine patient before it is accomplished in a human patient. At the time of this writing, a clinical trial of client-owned dogs with severe mitral regurgitation had not yet begun. Issues that will need to be considered before undertaking a canine clinical trial:

- Clinical trial design
- Inclusion criteria
- Establishment of centers
- Endpoints
- Establishment of device registry

The initial veterinary trial will be a phase I-type pilot study in client-owned dogs based on the principles of "compassionate use" and informed consent. The principal purpose of a phase I-type trial is to establish safety/feasibility rather than efficacy. Compassionate use is defined as a treatment for a disease for which no effective alternative therapy exists. Because open heart surgery is not generally available/feasible for dogs with mitral valve disease, the most appropriate treatment group for this first study will be dogs with medically-refractory (stage D) mitral valve disease. The specific goal of this study will be to demonstrate that the device can be successfully deployed and will function in a patient with severe MR. A secondary goal will be to demonstrate efficacy based on reversal of diuretic-dependent CHF for a meaningful period of time. It should be anticipated that this early study will identify necessary modifications to the prosthesis design and/or perioperative protocols based on differences between normal animals and animals with severe MR. A possible outcome is that dogs with stage D mitral valve disease may not tolerate the procedure well because of secondary myocardial dysfunction associated with end-stage MR. This early study will be performed at a single center.

It is anticipated that the primary veterinary study will be a phase IIB-type trial on client-owned dogs to demonstrate efficiency and device durability. Efficiency in dogs would be based on comparison to standard-of-care treatment for dogs with advanced mitral valve disease, i.e. medical therapy for CHF. Because the comparison is between a device and medical therapy, animals would not be randomized to a treatment group. The control would be historical or concurrently collected data on outcome to standard medical therapy in an appropriate cohort. Endpoints for this study could include change in disease stage level, quality of life, freedom from cardiovascular-related events, cardiac mortality, and all-cause mortality. Because of a lack of criteria for surgical intervention for MR in dogs, entry and exclusion criteria for this study would likely be extrapolated from ACC/AHA guidelines for surgical management of severe MR in human patients.[5] It is anticipated that this study would be multi-institutional.

REFERENCES

1. DODD WJ. CLINICAL BIOCHEMISTRY OF DOMESTIC ANIMALS, 5TH EDITION, 1997; 241. 2. BRAINARD BM, ET AL. AM J VET RES 2010; 71:822-830. 3. MCKELLAR SH, ET AL. J THORAC CARDIOVASC SURG 2011; 141:1410-1416. 4. GRANGER CB, ET AL. N ENGL J MED 2011; 365:981-992.
5. BONOW RO, ET AL. CIRCULATION 2008;118:E523-E661.

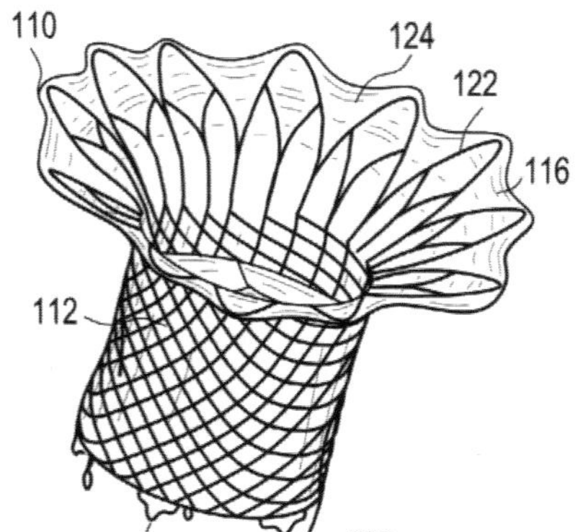

Figure 1

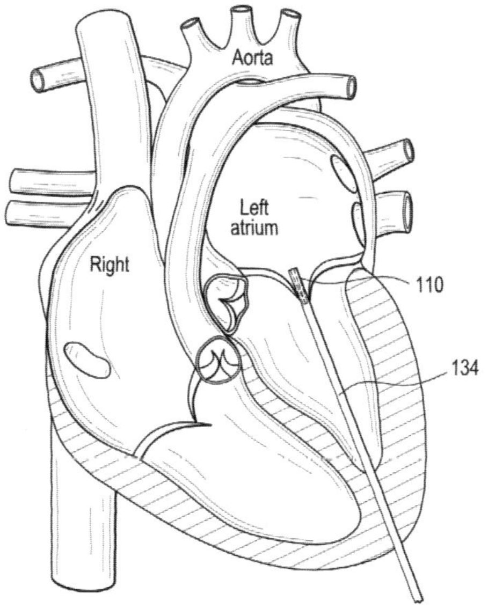

Figure 2

DIURETIC RESISTANCE IN DOGS TREATED FOR HEART FAILURE

Mark A. Oyama, DVM, DACVIM
Philadelphia, PA

INTRODUCTION

Diuretics are a mainstay of treatment in animals with congestive heart failure. Diuretics reduce plasma volume, hydrostatic pressure, and edema or effusion. Diuresis, however, is not without adverse effects such as activation of the renin-angiotensin - aldosterone and sympathetic nervous systems, electrolyte imbalances, and renal side effects. These and other mechanisms can reduce the efficacy of long-term oral diuretic use in animals with advanced heart disease. This presentation will present and discuss the mechanisms behind and clinical importance of diuretic resistance in dogs treated for heart failure.

Furosemide is the most commonly used diuretic for the treatment of canine heart failure. Furosemide, a loop diuretic, acts within the nephron to inhibit chloride resorption in the ascending portion of the loop of Henle. Furosemide is a relatively potent diuretic with good oral bioavailability, a rapid onset of action, and short half-life. These characteristics make for effective treatment of heart failure, however as disease progresses increasing doses of furosemide are needed to alleviate clinical signs. Data from experimental models as well as our own clinical experience indicates that chronic diuretic use encounters resistance. Thus recurrent heart failure signs often occur in dogs receiving aggressive medical management that includes higher and higher doses of furosemide.

The mechanisms behind diuretic resistance are varied and relatively poorly understood. Potential mechanisms include activation of neurohormonal systems, hypertrophy of nephron segments, and increased renal solute absorption in portions of the nephron other than the loop of Henle. Diuretic resistance has been shown in dogs receiving doses of furosemide as low as 2 mg/kg and for durations as short as 14 days. Resistance could be lessened by administration of diuretics with different pharmacologic and pharmacokinetic properties. These mechanisms, in addition to causing diuretic resistance, might also have direct effects on the pathophysiology of disease and morbidity and mortality. In humans, the dose of furosemide is a marker of heart failure severity and independently predicts long term mortality. Thus, strategies to prevent diuretic resistance and minimize diuretic dose might improve outcome.

Strategies to minimize resistance in dogs with heart failure include, administration of multiple diuretics, each of which act at a different part of the nephron, replacement of furosemide with other loop diuretics, such as torsemide, or delivering diuretics via subcutaneous injection rather than orally. During the course of this presentation, we will discuss the practical application of these and other clinical strategies.

REFERENCES

Uechi M. et al. J Vet Med Sci 2003;65:1057. Eshaghian S et al. Am J Cardiol 2006;97:1759-64. Metra M et al. Exp Opin Pharmacother 2001;12:1369-79. Oyama MA et al. J Vet Cardiol 2011:13:287-92. Peddle GD et al. J Vet Cardiol 2012:in press.

REAL TIME THREE-DIMENTIONAL ECHOCARDIOGRAPHY: FROM DIAGNOSIS TO INTERVENTION

Joao S. Orvalho, DVM, DACVIM
San Diego, CA

INTRODUCTION

Real time three-dimensional echocardiography (RT3DE) is a new ultrasound modality that provides comprehensive views of the cardiac valves and congenital heart defects. RT3D is a potentially more accurate echocardiographic means of evaluating chamber volumes and a more precise interventional and postoperative tool.

RT3D transesophageal imaging may overcome technical and image quality limitations of transthoracic techniques.

HISTORICAL PERSPECTIVE

Advances in ultrasound techniques have made echocardiography one of the most clinically relevant diagnostic modalities in veterinary cardiology. The evolution occurred from M-mode, pulsed and continuous wave Doppler, color flow Doppler, 2-dimensional echocardiography and transesophageal echocardiography to real time 3-dimensional echocardiography and more recently real time three-dimensional transesophageal echocardiography.

TECHONOLOGY

New discoveries in electronic and computer technology created a fully sampled matrix array transducer and workstation for three-dimensional display of acquired images, with the possibility of post-processing and quantification.

The matrix array transducer provides a full volume and 360° samples, with real time acquisition and excellent image quality. This technology results in surface-rendered or wire-framed reconstructions of the cardiac chambers that allow more accurate calculations of the ventricular volumes and mass.[1] Visualization and quantification of color flow jets in three dimensions is also possible using this technique.

CLINICAL APPLICATIONS

Since its invention, echocardiography has become one of most important diagnostic tools in veterinary cardiology. The most significant development of this century has been the introduction of real time three-dimensional imaging.

RT3DE allows direct evaluation of the cardiac chambers without the need for geometric assumptions, preventing mistakes caused by foreshortened views ultimately creating a more accurate and reproducible measurement.[2] Left ventricular wall motion and left ventricular volume are acquired at the same time resulting in the complete dynamic information on LV contraction, which can be used not only to calculate the standard LV function parameters, such as ejection fraction but also to assess asynchrony. Measurement of LV mass is also possible and it is generated based on endocardial and epicardial visualization.

Non-invasive realistic views of the valves can be obtained with unlimited image plane orientation permitting a better understanding of the geometry of the valves and valvular apparatus.[3] In myxomatous valvular degeneration of the mitral and tricuspid valves, accurate identification of a chordae tendineae and area of prolapse of the leaflet can usually be achieved. In mitral or tricuspid valve stenosis, as well as aortic and pulmonic stenosis an "en face" view generates a more accurate measurement of the valve area.[4]

Full volume acquisitions and especially three-dimensional zoom views reveal a precise display of the spatial relationship between the cardiac structures, which is particularly useful in congenital abnormalities.[5] RT3DE has been shown to reliably define anatomic details of patent ductus arteriosus, atrioventricular valvular dysplasias, septal defects, aortic and pulmonic valve anomalies, cor triatriatum dexter and other complex cardiac defects.

Identification and characterization of cardiac masses and thrombi is facilitated by the addition of a new dimension, which can help improve the accuracy of the diagnosis.

RT3DE can also be used to assess surgical/transcatheter interventions such as patent ductus arteriosus occlusion, atrial septal defect closure, balloon valvuloplasty, heartworm removal, and pacemaker lead or catheter localization. Evaluations pre- and post-catheterization as well as during the procedure can confirm the success of the intervention as well as enhance decision making.[4]

REAL TIME THREE-DIMENSIONAL TRANSESOPHAGEAL ECHOCARDIOGRAPHY

Real time three-dimensional transesophageal echocardiography (RT3D TEE) is a recently developed imaging tool that is particularly useful to define the location and morphology of the cardiac structures.[6]

The diagnosis of congenital heart disease is highly dependent on imaging, and RT3D TEE provides a superior comprehensive evaluation of the cardiac defect(s) prior to and during interventional procedures. The main advantages of this technique are the ability to visualize the catheters (including the tip), and the balloons or devices that they carry, as well as the possibility to image the structure that is undergoing intervention with unprecedented quality. These advantages may reduce fluoroscopy time, and increase procedure safety and efficacy.[7] RT3D TEE may become the technique of choice for guidance of percutaneous catheter procedures.

In human cardiology RT3D TEE has already been extensively used to evaluate a wide variety of cardiac conditions and to assist in interventional cardiac procedures, as well as to guide and follow up intracardiac surgery. The assessment of complex septal anatomy, cor triatriatum sinister membrane morphology, transseptal puncture protocols and guidance of electrophysiologic studies are some of the most challenging applications of RT3D TEE recently described in human medicine.[8,9]

In veterinary medicine, the relevance of two-dimensional transesophageal imaging has been proven and its use in interventional cardiac procedures, especially at academic institutions has increased in the last decade. The main limitations for the use of the three-dimensional version are still the price of the equipment and the need for general anesthesia.

CONCLUSIONS AND FUTURE DIRECTIONS

Real time three-dimensional echocardiography has become a clinically relevant imaging technique. Its main advantages are more accurate quantification of cardiac chamber size and function and unlimited image plane orientation for better evaluation of valvular and congenital heart diseases.

In human cardiology RT3DE is considered to be superior to the traditional 2-dimensional techniques in quantification of left ventricular volume, ejection fraction and mass, as well as in the evaluation of mitral valve morphology and function. Future developments in transducer technology will likely allow full volume acquisitions with and without color Doppler in one cardiac cycle.Real time three-dimensional transesophageal imaging is currently used in human interventional cardiology and will likely have similar applications in veterinary cardiology. It has the potential to overcome some technical and image quality limitations of transthoracic techniques.

REFERENCES

[1]Lang RM, et al, JACC 2006; 48:2053–2069. [2]Lang RM, et al. Eur J Echocardiography 2006;7:79–108. [3]Marsan NA, et al. Heart 2009; 95:1881–1890. [4]Shiota T, J Cardiology 2008; 52:169-185. [5]Hage F, et al. Echocardiography 2012;29:220-231. [6]Perk G, et al. J Am Soc Echocardiogr 2009;22:865-82. [7]Pui-Wai Lee A, et al. Heart 2010;96:1485e1493. [8]Faletra F, et al. J Am Soc Echocardiogr 2011;24:593-9. [9]Mercer-Rosa L, et al. Congenit Heart Dis. 2006;1:335–339.

AUSCULTATION OF MURMURS IN THE DOG AND CAT: HOW COLOR DOPPLER FLOW MAPPING IMPROVES OUR UNDERSTANDING

Gerard Le Bobinnec, DVM, DECVIM-Cardiology
Bordeaux, France

Auscultation remains the cornerstone of the physical examination in Cardiology: there is not any machine to replace clinical sense and experience in one of the last strongholds of the veterinary art. However, the appreciation of cardiac frequency, rhythm, and especially murmurs, includes an important part of subjectivity, and auscultation is still a thorny challenge for the generalist. Echocardiography, essentially through the doppler modes, has considerably improved our understanding of murmurs. Thus, in the light of Color Flow Mapping (CFM), the difficult art of auscultation can acquire an increased efficacy and reliability.

In the context of congenital heart diseases (CHD), the interest of CFM is double: to explain innocent murmurs, in order to distinguish them from pathologic ones, and to explain complex auscultations in case of associated congenital anomalies.

In young animals, innocent murmurs constitute a real pitfall for the clinician, because the epidemiologic range is exactly the same for CHD. The most demonstrative example of this ambiguity is provided byaortic stenosis (AS) or subaortic stenosis (SAS) in the Boxer breed ; the partition between physiologic and pathologic murmurs in the breed remained elusive for a long time, until echocardiographic studies revealed the smaller aorta of this breed, and CFM demonstrated the resultant turbulence in the ascending portion of the aorta. The elevated sympathetic tone and smaller left ventricular outflow tract found in the Boxer breed further complicates the situation. With the help of CFM, some tricks may help to identify innocent murmurs only with the stethoscope: soft, 2/6, mesosystolic, strictly left mid-thoracic, and transient murmurs (disappearing with bradycardia) are more likely to be innocent in nature. With the same functional mechanism of high velocities in the ascending part of the aorta, innocent murmurs are now also described in other breeds includingthe Whippet & Greyhound, Cavalier King Charles Spaniel (CKCS), and the miniature Bull Terrier.

When more than one CHD coexists, the general rule of auscultation is that only the stronger murmur is audible. The most common exemple is the frequent association between subaortic stenosis (SAS) and pulmonic stenosis (PS) in young Boxers. Before the generalization of echo examination, some of them seemed to have PS at 2 or 3 months and SAS at one year old. Doppler echo explains this ambiguity ; PS is a real congenital anomaly (present at birth), while SAS is not really, because it often develops or worsens with growth. As a result, it is possible to identify first a PS murmur, and a few months after clearly an SAS murmur.

Ventricular septal defect (VSD) is the most frequently associated CHD in dogs, so much so that when a VSD is detected, it's important to systematically look for other possible defects such as AS, PS, patent ductus arteriosus (PDA), atrial septal defect (ASD), and mitral or tricuspid dysplasia (MD & TD). A meticulous auscultation can detect these associations, as in the presented example of a dog with a harsh systolic mid-thoracic murmur, 4/6 on the left, but 5/6 on the right. CFM demonstrated a VSD associated with AS. On the contrary, a VSD associated with PS generates a right-to-left shunt through the VSD (Eisenmenger syndrom) and an ambiguous holosystolic murmur on the left hemithorax.

In young Boxers, once more, ASD may be underestimated, because the specific auscultation area (left basilar) of this soft murmur is often not explored, or is drowned out by an SAS murmur.

In young Golden Retrievers, TD and its right cranial systolic murmur often obscures an associated MD murmur (and the contrary is possible too, especially in CKCS). But a murmur louder in the tricuspid area can also be heard in the rare case of atrioventricular canal defect.

In the context of acquired heart diseases, the interest of CFM is to visualize the murmurs of DCM and endocarditis, and to clear up some aspects of the evolution of mitral valve disease (MVD).

In canine DCM, CFM shows these mitral and tricuspid regurgitation jets, generating soft systolic functional murmurs, because of mitral and tricuspid annulus dilation. These murmurs become variable in cases of concurrent atrial fibrillation, and disappear in settings of very low cardiac output when, in fact, CFM shows a persistent but non-turbulent regurgitation jet.

Diastolic murmurs are uncommon in canine and feline cardiology (except the continuous murmur of PDA) ; a majority of them are not loud enough to be detected with a stethoscope, and can only be recorded with phonocardiography. The main exception is aortic vegetative endocarditis, which produces a protosystolic murmur, confirmed in CFM with a spectacular jet of diastolic aortic insufficiency.

In MVD

- CFM first confirmed the direct relationship between intensity of the murmur and volume of the regurgitation, but in the early stages of the disease, this intensity can be variable in case of respiratory sinus arrhythmia: the regurgitation – and then the murmur – increases with increased heart rate (then in inspiration), and may disappear during the expiratory bradycardia.
- But it also demonstrated that, in the beginning of the evolution of the disease, several small mitral leaks coexist, making delusive all the colour doppler evaluations (PISA, RJA).
- CFM combined with 2D imaging underlined that rupture of chordæ tendinæ is often overestimated: in cases of

tangential leak of the parietal leaflet, vibrations can generate a 5/6 or a 6/6 murmur without associated rupture of chordæ tendinæ.

- Pulmonary hypertension (PHT) as a complication of advanced MVD is also often overestimated. CFM shows that tricuspid valve disease often coexists with the MVD from the beginning, without any PHT. In this case, auscultation is able to detect a systolic soft murmur, with a location on the right hemithorax much more cranial than the mitral area.

In feline cardiology, the value of auscultation is much less interesting than in dogs, but CFM allowed us to elucidate the origin of murmurs in the case of cardiomyopathies (essentially aortic ejection and mitral regurgitation murmurs, and then a left parasternal location); and also to understand the frequent right parasternal murmurs, with the demonstration of dynamic right ventricular obstruction (DRVO).

Lastly, arrhythmias may also generate auscultation ambiguity, for example atrial fibrillation (AF) associated with AS results in a mid-thoracic systolic ejection murmur, variable in intensity with each beat.

REFERENCES

Abbott J. **J Vet Intern Med** 2001 ; 15 : 505. Amberger C et al. **J Vet Cardiol** 2002 ; 4 (2) : 37. Bavegems VC et al. **J Am Vet Med Assoc** 2011 ; 238 (4) : 468. Chetboul V. **J Vet Med A Physiol Pathol Clin Med** 2006 ; 53 (4) : 179. Fabrizio F. **J Vet Intern Med** 2006 ; 20 : 78. Jenni S. **J Vet Intern Med** 2009 ; 23 : 81. Koplitz SL et al. **J Vet Intern Med** 2006 ; 20 : 904. Linde A et al. **J Vet Cardiol** 2006 ; 8 : 79. Naylor JM et al. **J Vet Intern Med** 2001 ; 15 : 507. Olsen LH et al. **J Vet Intern Med** 2006 ; 20 : abstract 136. Rishniw et al. **J Vet Intern Med** 2002 ; 16 : 547.

89a

THE ACTION OF A NEW PROTOTYPE CARDIOTONIC DRUG (LASSBio 294) ON THE CARDIOVASCULAR SYSTEM OF DOGS. A. C. B. Tôrres, P. R. Nasciutti, R. C. O. L. Brasil, P. N. Oliveira, M. C. Souza, I. P. Bittar, V. De Oliveira, E.J. Barreiro, R. Oliveira Alves . School of Veterinary and Animal Science - Federal University of Goiás - Brazil.

LASSBio 294 is a new prototype cardiotonic drug produced from safrole substrate, found in Brazilian plants like "canela-branca" (*Ocotea pretiosa*), which showed positive inotropic and vasodilatory effects in *in vitro* tests. Due to its attractive pharmacological profile, we carried out an investigation of its probable toxic effects, which exhibited no acute toxicity. Thus, the present study carried out *in vivo* tests in dogs in order to determine the effects of this molecule on the cardiovascular system.

We used five adult Beagles, female and healthy. In the first stage, all dogs received a 2mg/kg dose of LASSBio294 orally. In the second and third phases of experimentation, they received 5mg/kg and 10mg/kg respectively. Underwent blood and urine sampling; doppler echocardiographic examinations, ECG and blood pressure screening. Evaluations were performed two, four, six, twelve and twenty four hours before and after administration of the drug.

The animals showed no clinical, hematological, serum biochemistry, or behavior changes, confirming the preliminary toxicology findings by BARREIRO et al., 2002. The Doppler echocardiographic evaluations showed no changes during the time of assessment, or between the groups, unlike the findings SUDO et al. (2001). No dog had rhythm abnormalities, differing from other compounds with similar therapeutic effects, such as digitalis, which can cause arrhythmias. In evaluation of systolic blood pressure was observed its decrease in a 2mg/kg dose, at four ($p = 0.010$), six ($p = 0.018$) and twenty four ($p = 0.045$) hours, demonstrating the vasodilator potential of the molecule besides its safety, since there was no hypotension. Similar effects were observed by THOMASON & FALLOW (2007) using pimobendan. For doses of 5mg/kg and 10mg/kg, there was no dose-dependent effect, unlike that observed by SUDO et al. (2001).

Considering the results obtained and the conditions under which this experiment was conducted, it can be concluded that the oral administration of LASSBio294 in doses of 2mg/kg, 5mg/kg and 10mg/kg did not cause side effects in healthy Beagles, demonstrating that the prototype drug is safe. In occasional doses, it causes a systolic blood pressure decrease only at a dose of 2mg/kg. Despite the fact that it showed no increase in cardiac contractility, it is believed that continuous daily use of the molecule might allow us obtain more promising results.

89b

PROGNOSTIC INDICATORS IN FELINE HYPERTROPHIC CARDIOMYOPATHY.

J.R. Payne, V. Luis Fuentes, K. Borgeat, D. Connolly, A. Boswood, D. Brodbelt. Dept of Veterinary Clinical Sciences, Royal Veterinary College. U.K.

Left atrial (LA) enlargement, congestive heart failure (CHF) and aortic thromboembolism (ATE) are known to be associated with reduced survival in feline hypertrophic cardiomyopathy (HCM). We sought to investigate the prognostic value of additional factors in cats diagnosed with HCM.

Clinical and echocardiographic records were reviewed of feline HCM cases seen at the Royal Veterinary College from 2004-2009. Only cats with echocardiographic confirmation of left ventricular (LV) diastolic wall thickness ≥6mm were included. Outcome was obtained from clinical records or referring veterinarians/owners.

Of 285 cats meeting the inclusion criteria, 146 had died (99 cardiac deaths). Median survival was 2,153 (0-2,374) days.

Univariable analysis showed cardiac death was associated with absence of murmur, a gallop, arrhythmia, fractional shortening (FS) ≤35%, increasing LA size, septal or free wall diastolic thickness ≥8.0mm, regional LV wall hypokinesis, spontaneous echo contrast and/or thrombus, absence of dynamic LV outflow tract obstruction, mitral inflow E:A ≥2, left auricular appendage velocity ≤0.25m/s and E:E' ≥20.

Multivariable analysis showed the following to be independent predictors of outcome: having a greater LA:Ao (HR for a one unit increase in LA:Ao = 4.04, 95% CI 2.83-5.77), FS ≤35% (HR if FS ≤35% 3.41, 95% CI 1.80-6.45) and an arrhythmia on ECG (HR if arrhythmia present = 1.86, 95% CI 1.10-3.15) increased the risk of cardiac death.

Although LA size is an important prognostic indicator in feline HCM, other factors should also be taken into account, such as FS and presence of arrhythmias.

90a

EFFICACY OF BAY 41-9202 (BISOPROLOL ORAL SOLUTION) FOR THE TREATMENT OF CHRONIC VALVULAR HEART DISEASE (CVHD) IN DOGS. BW Keene[1], PR Fox[2], RL Hamlin[3], GF Beddies[4], TJ Keefe[5], T Settje[4], LS Treml[4] for the HECTOR investigators. 1. North Carolina State University, Raleigh, NC; 2. The Animal Medical Center, NY, NY; 3. The Ohio State University, Columbus, OH; 4. Bayer Healthcare LLC, Shawnee, KS; 5. Colorado State University, Ft. Collins, CO

This prospective, randomized, blinded, placebo-controlled trial was designed to determine whether BAY 41-9202 would slow the progression of CVHD and mitral regurgitation (MR) in dogs, thus delaying the onset of clinical signs of heart failure (HF, the primary endpoint).

This study conformed to Good Clinical Practice and New Animal Drugs for Clinical Investigation guidelines, with a single protocol followed at 14 sites. Inclusion criteria included signalment, history, physical, radiographic, ECG, echo, and lab parameters designed to identify dogs with hemodynamically significant CVHD (MR and left atrial enlargement), but no clinical signs of HF. Included dogs received placebo or BAY 41-9202, with scheduled dosage up-titration. A power analysis suggested that a total enrollment of 200 subjects provided an 80% chance of identifying a 30% benefit in the primary endpoint. Subjects were monitored regularly until reaching the primary endpoint. Planned safety analyses were performed by independent monitors. Statistical differences between groups were sought via Kaplan-Meier analysis.

172 dogs received treatment; 178 placebo. No differences were identified between groups at baseline. The study was stopped because a prespecified safety analysis revealed a remote likelihood of identifying a significant difference between groups. Final analysis revealed no significant difference in the estimated mean time-to-HF between treatment and placebo (p=0.917, 528.8 [SE=21.5] and 653.4 [SE=24.4] days, respectively).

These results do not support the clinical efficacy of BAY 41-9202 to delay HF in dogs with CVHD.

90b

PIMOBENDAN IMPROVES SURVIVAL IN DOBERMAN PINSCHERS WITH PRECLINICAL DCM: THE RESULTS OF PROTECT A DOUBLE-BLIND, PLACEBO-CONTROLLED, RANDOMIZED CLINICAL TRIAL. N.J. Summerfield[1], A. Boswood[2], M.R. O'Grady[3], S.G. Gordon[4], J. Dukes McEwan[5], M.A. Oyama[6], S. Smith[7], M. Patteson[8], A.T. French[9], G. Culshaw[9], L. Braz-Rivo[10], A. Estrada[11], M.L. O'Sullivan[3], J. Loureiro[1], R. Willis[12], P. Watson[13] 1 North Downs Specialist Referrals, Bletchingley, Surrey, UK 2 Department of Veterinary Clinical Sciences, The Royal Veterinary College, Hertfordshire, UK 3 Ontario Veterinary College, University of Guelph, Ontario, Canada 4 College of Veterinary Medicine and Biomedical Sciences, Texas A&M University, College Station, Texas, USA 5 School of Veterinary Science and Department of Musculoskeletal Biology, University of Liverpool, Neston, UK 6 Matthew J. Ryan Veterinary Hospital of the University of Pennsylvania, School of Veterinary Medicine, Philadelphia, USA 7 Sarah Smith Cardiology, Etwall, Derbyshire, UK 8 HeartVets @ Vale Referrals, The Animal Hospital, Dursley, Gloucestershire, UK 9 Royal (Dick) School of Veterinary Studies, University of Edinburgh and Roslin Institute, Scotland, UK 10 Dogs & Cats Veterinary Referral and ER, Bowie, Maryland, USA 11 Small Animal Clinical Sciences, University of Florida College of Veterinary Medicine, Florida, USA 12 Holter Monitoring Service, Dollar, Clackmannanshire, Scotland, UK 13 Boehringer Ingelheim Animal Health, Ingelheim am Rhein, Germany

The potential benefit of pimobendan therapy in delaying the progression of preclinical dilated cardiomyopathy (DCM) in Dobermans has not previously been evaluated.

Pimobendan is a benzimidazopyridazinone with a potent positive inotropic and a vasodilatory effect. Two separate mechanisms produce the positive inotropic action, an increase in calcium sensitivity of cardiac contractile proteins and an inhibition of cardiac phosphodiesterase activity. Pimobendan also results in selective peripheral phosphodiesterase inhibition leading to systemic vasodilatation and improvements in coronary and renal perfusion. This combined effect of preload and afterload reduction, together with positive inotropic support, could result in a reduction of cardiac size and filling pressures in dogs with preclinical DCM. We hypothesized that pimobendan would prolong the preclinical phase of DCM in Dobermans and therefore delay the onset of clinical signs.

The objective of this study was to prospectively evaluate if chronic oral administration of pimobendan in Dobermans with preclinical DCM would delay the onset of congestive heart failure (CHF) or sudden death. The trial was a randomized, blinded, placebo-controlled, parallel group multicenter study. Dogs were allocated in a 1:1 ratio to receive pimobendan (Vetmedin capsules) or visually identical placebo. Seventy-six client-owned Dobermans were recruited at 10 centers in the UK and North America. The composite primary endpoint was defined as either onset of CHF or sudden death. Time to death from all causes was a secondary endpoint.

Pimobendan administered to Dobermans with pre-clinical DCM prolongs the time to the onset of CHF or sudden death and increases survival time.

153a

LONGITUDINAL STRAIN AND STRAIN RATE BY TWO-DIMENSIONAL SPECKLE TRACKING IN FELINES. AC Silva, RAL Muzzi, LAL Muzzi, G Oberlender, MM Mantovani, GFM Reis, RB Nogueira. Federal University of Lavras, Minas Gerais, Brazil.

Two-dimensional Speckle Tracking (2D-STE) is a new tool that has been used in veterinary medicine to evaluate myocardial strain (St) and strain rate (StR). This technique is already validated for dogs, showing to be useful in the diagnosis of cardiomyopathies. There is little information regarding 2D-STE strain and strain rate in cats. In addition, to our knowledge there is no data about longitudinal St and StR by use of 2D-STE in awake clinically healthy cats.

Thus, the aim of this study was to evaluate myocardial longitudinal St and StR in awake clinically healthy cats, as well as, myocardial longitudinal velocity by means of 2D-STE.

For these purpose fifteen healthy cats were evaluated clinically and Doppler systolic blood pressure (SBP), electrocardiography (ECG) and conventional and Doppler echocardiography were performed to obtain the healthy status of all animals. All variables were into normal range.

All 2D-STE images were obtained in a Mylab® 40 ultrasound machine (Esaote) and analyzed off line with the software Xstrain ™. Cineloops were acquired from left apical four chamber view, which were used to perform the analyzes and 6 measurements were obtained in two different images (3 for each image) for all variables, with a total of 540 measurements and the mean ± standard deviation were calculated. In addition, after one week 20% of animals (n=3) were analyzed again and compared with previous values to obtain the intraobserver variability.

The group of cats studied was composed by 5 females and 10 males, and the mean age was 26.13 ± 14.09 months and mean weight of 4.16 ± 0.86 kg. Heart rate was also obtained and its value was 164.60 ± 38.28 beats for minute, being elevated due to the fact that cats were awake.

Regarding endocardial longitudinal St and StR was observed values of -15.95 ± 7.35% and -0.92± 0.92 s^{-1}, respectively. On the other hand, for the epicardial region, the values obtained were -16.52 ± 6.67% for St and a StR of -1.84 ± 0.76 s^{-1}. It was noted that the mean values were similar in both myocardial regions. Myocardial longitudinal velocity was also obtained from endocardial and epicardial regions, and were observed values of 1.38 ± 0.98 cm/s and 1.47 ± 1.19 cm/s, respectively, demonstrating also similar values in both regions. Intraobserver variability was about 10% for St and lower than 2% for StR in both regions, showing acceptable variability percentage.

2D-STE demonstrated to be useful to determine longitudinal St and StR in cats. Given that, there is an absence of data about longitudinal St and StR in felines, this study could be of great value for future studies with 2D-STE in feline cardiomyopathies.

Support by CAPES, CNPQ, FAPEMIG

153b

FLOW PROPAGATION VELOCITY ON THE EVALUATION OF LEFT VENTRICULAR DIASTOLIC FUNCTION IN DOGS WITH CHRONIC MITRAL VALVE DISEASE WITH AND WITHOUT CONGESTIVE HEART FAILURE. RAL Muzzi[1], RM Resende[1], MM Mantovani[1], AC Silva[1], C Pace[2], RB Nogueira[1], G Reis[1], LAL Muzzi[1]. 1. Federal University of Lavras, MG, Brazil/CNPq. 2. Royal Veterinary College, Hatfield, UK

Chronic mitral valve disease (CMVD) is one of the most prevalent acquired valvular diseases in dogs. Patients with CMVD may present with increased preload and heart rate causing pseudonormalization of conventional echocardiographic indices, which makes it difficult to evaluate diastolic function in these animals. Thus, the use of mitral flow propagation velocity (Vp) and tissue Doppler (TD) Em and Am waves velocity, have been recently implemented with the purpose to assess more effectively diastolic function. The relationship of mitral flow E wave with Vp (E/Vp), isovolumetric relaxation time (E/IVRT) and peak early diastolic tissue velocity (E/Em) has also been used successfully to detect congestive heart failure (CHF) in dogs.

Therefore, the aim of this study was to assess left ventricular diastolic function in dogs affected by CMVD through the mitral flow propagation velocity and TD, as well as verify that this relationship can predict CHF. Thirty-six dogs were clinically evaluated and divided into three groups, which were composed of a control group (healthy dogs, n = 10), CMVD group without CHF (n = 13) and CMVD group with CHF (n = 13). CHF was based upon clinical and radiographic findings. The ratio between left atrium/aorta (LA/Ao) was increased in all diseased animals, presenting with a mean ratio greater than 1.7 in animals with CHF. All groups demonstrated a preserved systolic function assessed by ejection fraction (EF) and left ventricular end-systolic volume index (LVESVI) obtained through the Simpson method.

Vp was not significantly different between the three evaluated groups, not being able to detect diastolic dysfunction in dogs with CMVD in this study. However, the E/Vp ratio was shown to be effective in diagnosing CHF on symptomatic animals (> 1.5), as well as the E/IVRT, which had values greater than 2.5 and E/Em that presented values over than 12.

A transmitral flow pseudonormalization pattern was detected in the animals of the group with CHF, but the TD Em/Am ratio found impaired diastolic pattern in the control group and in the dogs with CHF. The abnormal relaxation pattern detected in the Em/Am ratio was decisive in the diagnosis of diastolic dysfunction in older dogs with CMVD.

Support by CAPES, CNPQ, FAPEMIG

154a

HEMODYNAMIC AND BIOCHEMICAL ALTERATIONS IN CANINE LYMPHOMA.

DM Fine, K Selting, RC Backus, HE Durham. University of Missouri, Columbia, MO.

Cancer patients are frequently evaluated by echocardiography (Echo) prior to receiving doxorubicin, a chemotherapeutic with cardiotoxic effects. Dogs with lymphoma (LSA) often show changes consistent with volume depletion on Echo: small left ventricular (LV) chamber dimensions and thickened walls (referred to as pseudohypertrophy). Concomitantly, systolic function appears depressed or dyskinetic. This can impair the ability to determine the risk of using doxorubicin. The purpose of this study was to characterize the hemodynamic and biochemical alterations in dogs with LSA and compare the findings to experimental volume depletion.

Dogs presenting with chemotherapy-naïve LSA were enrolled (n=13). Dogs were evaluated at 1 week post-chemotherapy induction. Drugs used in induction included: prednisone, l-asparaginase, and vincristine. Baseline evaluation included: weight, heart rate, systolic blood pressure (BP), Echo, glucose, BUN, creatinine, PCV, T.P., USG, plasma renin activity, aldosterone, ionized calcium, plasma and urine concentrations of sodium, potassium, and osmolarity. LSA dogs were then given LRS+16 mEq KCl (6 ml/kg/hr for 24h) and reevaluated at 4 and 24h. Healthy control (CON) dogs (n=12) were similarly evaluated and then given 2.5 mg/kg of IV furosemide; water was withheld and parameters were reevaluated at 4 and 8 h. In a separate group (LSA, n=8; CON, n=6), total body water and extracellular water space were assayed using deuterium oxide and bromide dilution respectively. Data was analyzed using one-way or repeated measures ANOVA. Post-hoc analysis was performed using the Tukey Kramer test. $P<0.05$ was accepted as significant.

At baseline, LSA dogs were significantly older, but not different in weight or gender. The following were significantly decreased in LSA: creatinine 0.7±0.2 v 0.9±0.1 mg/dl, PCV 38.5±5.6 v 49.8±4.5%, USG 1.021±8.5 v 1.031±1.5, urine potassium 60.2±39.2 v 117.9±19.6 mEq/L, and bromide dilution space 25.6±4.4 v 30.9+4.3. On Echo, LSA had decreased M-mode LV systolic 2.2±.7 v 2.5±.5 cm, diastolic 3.49±.7 v 3.8±.6, and left atrial dimensions 2.44±.05 v 2.64±.04, and lower cardiac index 3.68±1.2 v 4.2 ±1.1 L/min/M^2 among others. LSA dogs had significantly increased BP 147.2±17 v 125±14 mmHg, glucose 112.2±12.2 v 97.5±12.2 mg/dl, urine sodium 133.5±93.8 v 117.9±64.9 mEq/L, and BUN/creatinine 26±10 v 15±4. After fluid administration, there were no differences between LSA and CON's baseline Echos. Similarly, furosemide treatment eliminated most of the differences between CON and LSA's baseline echo.

Results of this study support the hypothesis that the Echo changes in dogs with LSA are related to volume depletion. Bromide dilution space, a measure of extracellular water was decreased in LSA, and fluid therapy reversed the Echo changes. However, volume depletion alone does not appear to fully explain the changes as no evidence of neurohormonal activation or hemoconcentration was seen in dogs with LSA. Further, in spite of reversal of Echo changes, there was no weight change after 24h of fluid therapy in LSA dogs. Further study is necessary to fully elucidate the mechanisms of these observations.

154b

SURVEY OF TREATMENT STRATEGIES OF CANINE TACHYARRHYTHMIAS BY CARDIOLOGIST. JM MacGregor, DVM, DACVIM (Cardiology)1,2 M Rishniw, BVSc, MS, PhD, DACVIM (Cardiology)3,4. 1. PortCity Veterinary Referral Hospital, Portsmouth, NH, 2. New England Veterinary Cardiology, Durham NH, 3. Veterinary Information Network, Davis, CA.

We conducted an online survey of 468 cardiology specialty listserve members in April 2011 to assess treatment strategies of 3 tachyarrhythmias – atrial fibrillation (AF), supraventricular arrhythmia/tachycardia (SVA/T), and ventricular arrhythmia/tachycardia (VA/T). 130 responses were obtained.

46% of respondents attempted chemical conversion of AF; of these, 54% chose amiodarone, 38% chose sotalol or lidocaine as first-line antiarrhythmics, with a 5% median success rate (range: 0-100%) for a median duration of sinus rhythm of 8 weeks (range: 0-100 weeks). 30% of respondents attempted electrical cardioversion with a median success rate of 1% (0-100%) with 1% staying in sinus rhythm for more than one month (0-100%). 100% of respondents attempted rate control for some dogs with AF – 88% attempted it in all dogs with AF; 12% only attempted it in dogs with abnormal hearts. Respondents chose a combination of digoxin and diltiazem (29%) most frequently.

Short acting diltiazem was chosen most commonly for treatment of SVA/T (24%), with sotalol (19%), long acting diltiazem (19%), or beta blockers (12%) being utilized frequently.

Treatment choices for VA/T depended on breed and clinical status. Boxer dogs without myocardial dysfunction were most often treated with sotalol (81% of respondents); Doberman Pinschers with subclinical DCM were most commonly treated with sotalol (54% of respondents) or mexiletine (24% of respondents); Doberman Pinschers with CHF were most commonly treated with mexiletine (55% of respondents) or sotalol (26% of respondents).

Our study shows that cardiologists adopt similar treatment strategies for managing AF and VA/T, but are less consistent in managing SVA/T.

155a

MICROCIRCULATORY IMAGING: A NEW FRONTEIR IN HEMODYNAMIC MONITORING. ES Cooper. Ohio State University, Columbus, OH.

Macrovascular measurements (heart rate, blood pressure, cardiac output, etc) may serve to indicate global hemodynamic status but local changes in microcirculatory flow (and thereby tissue perfusion) may not be accurately reflected. This becomes especially important in shock states or when region perfusion differences may occur (e.g. hemorrhagic shock and sepsis). Direct assessment of the microcirculation is possible through the use of sidestream darkfield microscopy (SDF). The MicroScan© is a handheld videomicroscope that uses SDF technology to directly assess capillary density and blood flow. Consensus criteria have been established for determination of microcirculatory parameters including total vessel density, proportion of perfused vessels, perfused vessel density, and microvascular flow index.

Microcirculatory imaging has been used extensively in human patients. A brief overview of the existing literature will be presented, highlighting applications in sepsis, heart failure and fluid therapy in human patients. These studies have consistently revealed a poor correlation between macro- and microcirculatory parameters. Applications in veterinary medicine have been limited thus far. Published studies in dogs have established normal microcirculatory parameters for anesthetized dogs and documented microcirculatory changes in experimental hemorrhagic shock.

Data from recently completed studies will be presented. Determination of normal microcirculatory parameters in healthy anesthetized cats undergoing ovariohysterectomy has revealed significantly different values when compared to dogs and people. It was further found that there were no significant microcirculatory changes brought about by surgical manipulation. Microcirculatory assessment in hemorrhagic shock has revealed successful restoration of baseline parameters after resuscitation with a hemoglobin-based oxygen carrier (HBOC). This is contrary to previously published information which has suggested that HBOCs promote microvascular vasoconstriction and impaired tissue perfusion.

A major benefit of the Microscan is its portable nature conferring the potential for bedside use. There are limitations to this technology, including a steep learning curve for acquiring quality microvascular videos, as well as the tedious and time consuming nature of the vascular analysis. However, given the limitless potential applications, microvascular imaging presents an exciting avenue for hemodynamic monitoring in veterinary medicine.

155b

INCIDENCE OF ALDOSTERONE BREAKTHROUGH (ESCAPE) FROM ACE-INHIBITION IN DOGS WITH EXPERIMENTAL RENIN-ANGIOTENSIN-ALDOSTERONE SYSTEM ACTIVATION. C. Atkins, A. Lantis, M. Ames. North Carolina State University, Raleigh, NC.

Approximately 40-50% of human heart-failure-patients experience a recrudescence of aldosterone secretion, despite chronic (at 1-year) ACE-inhibition. This finding, termed "aldosterone escape" or, more appropriately, "aldosterone breakthrough" (ABT) is associated with a poorer prognosis.

The authors utilize a subacute (≤17day) model of furosemide- and amlodipine-induced renin-angiotensin-aldosterone (RAAS) activation to study drug effects on RAAS, as indicated by the urinary aldosterone:creatinine ratio (A:C). A:C was assessed in 75 normal dogs from 5 studies (mean resting A:C=0.56+/-0.41).

ABT incidence (defined as A:Cr ≥100%>baseline or ≥1.37 ug/g with ACE-I) was determined by evaluation of 31 of these dogs, which were receiving ACE-I and either amlodipine or furosemide. A:C increased from baseline in 28/31 (87%). The increase was ≥25% in 84%, ≥50% in 81%, and ≥100% in 61%; A:C was >1.37ug/g in 26%. Overall, ABT was documented in all (5) studies in 71% of dogs. ABT occurred regardless of ACE-I (enalapril/benazepril), RAAS-activator (amlodipine/furosemide), whether ACE-I was begun with, or added to, RAAS stimulants, and whether pimobendan was administered concurrently.

In this model, the incidence of ABT approximates that seen in human cardiac patients. However, ABT was recognized early (from 1-17 days post-institution) in this study, as compared to observations in human cardiac patients. If corroborated in clinical patients, these data argue for the use of additional RAAS suppression in dogs receiving ACE-I for cardiac disease, glomerular disease, or hypertension.

156a

INTEREST OF FLECAINIDE FOR THE TREATMENT OF VENTRICULAR ARRHYTHMIAS. G. Le Bobinnec, DECVIM-Cardiology. Bordeaux, France.

Flecaïnide is a Vaughan-Williams Class 1C AAR, marketed since 1983 in France (100 mg tablets, 50-100-150-200 mg long-acting capsules). In human cardiology, this drug has a bad notoriety because of its implication in the CAST Trial (increased mortality after infarction, with pro-arrhythmogenic effects). For these reasons, it's now used in human medicine with caution, and with many contra-indications : Myocardial Infarction, CHF, Conduction troubles (blocks), and betablockade treatment.

We do not confirm this datas in Veterinary cardiology : Using Flecaïnide for 5 years (since Mexiletine has been removed from the french market), on the 45 cases treated (9 DCM, 10 Doberman CM, 5 Boxer CM , 8 Ischemic myocarditis, 5 SAS, 2 MVD, 1 German Sheperd V-Tach, 5 shocks with free-radicals), we had 11 failure (24,4%) in the control of ventricular arrhythmias (5 of the 11 afterwards well-controlled with imported Mexiletine), but no one case of pro-arrhythmogenic effect and none worsened CHF. None other side-effects (as seen with Tocaïnide or Amiodarone) were detected, even after several weeks (up to 6 months).We start the treatment with 3-4 mg/kg/day (given if possible in 3 oral intakes) and increase to 5-6 mg/kg/day if insufficient result after 7-10 days (empirically : to date, there are no datas about pharmakokinetics of flecaïnide in dogs). Use and interest of Flecaïnide is illustrated with 3 different cases : 2 successful treatments (a paroxysmal V-tach on a occult Doberman Cardiomyopathy, and an Accelerated Idioventricular Rhythm on an ischemic myocarditis), and one failure (the German Sheperd V-Tach).

156b

BLOOD PRESSURE, HEART RATE AND CATECHOLAMINES IN GENOTYPED HEALTHY DOGS IN A CLINICAL SITUATION. K. Höglund[1], S. Hanås[2], C. Carnabuci[3], I. Ljungvall[3], A. Tidholm[4], J. Häggström[3]. [1]Dept. of Anatomy, Physiology and Biochemistry, Swedish University of Agricultural Sciences, Uppsala, Sweden, [2] Strömsholm Specialist Animal Hospital Dog and Cat, Sweden, [3]Dept. of Clinical Sciences, Swedish University of Agricultural Sciences, Uppsala, Sweden, [4]Albano Animal Hospital, Danderyd, Sweden.

Correct interpretation of blood pressure (BP) and heart rate (HR) recordings is important in dogs in the clinical situation. In a potentially stressful situation, renal function may also become affected. Phenotypic differences between breeds in the dog population have become evident during recent years. The aim of the study was to investigate BP, HR, and kidney function in healthy, genotyped dogs from three different breeds; Labrador retrievers, Cavalier King Charles spaniels (CKCS) and Dachshunds. Eighty-nine dogs were examined, and BP and HR were measured in 4 clinical settings with or without veterinarian and owner present. Urine samples, obtained before and after examination, were analysed by ELISA assays for epinephrine, norepinephrine, creatinine, cystatin-C, vasopressin and aquaporin-2.

BP and HR were higher when measured by veterinarian alone without owner present ($P<0.020$), indicating a white-coat effect. Urinary catecholamine concentrations increased after examination in all dogs ($P<0.0001$) as did creatinine ($P<0.0022$) and cystatin-C ($P<0.0003$). Urine density decreased ($P<0.0001$), although vasopressin/creatinine and aquaporin-2/creatinine ratios were unchanged. Thus, BP seems to have increased glomerular filtration rate, overruling the hours without access to water during examinations.

Labrador retrievers had lower diastolic BP than Dachshunds in 2 settings ($P\leq 0.041$), lower HR than CKCSs in 3 settings (all $P<0.0001$) and lower catecholamine concentrations than both other breeds ($P\leq 0.035$).

In conclusion, both circumstance and breed should be taken into consideration when interpreting BP and HR recordings in a clinical situation. Stress-induced changes in renal function might affect the results of diagnostic tests and could potentially implicate clinical trials.

UPDATE AND REVIEW OF CARDIOMYOPATHY IN THE IRISH WOLFHOUND, EARLY THERAPEUTIC INTERVENTION

Andrea C. Vollmar[1], Dr.med.vet., DECVIM (Cardiology); Nicole A. Mohren, Dr.med.vet; Philip R. Fox, DVM, MS, DACVIM (Cardiology), DECVIM (Cardiology), DACVECC
[1]Bonn and Wissen, Germany

INTRODUCTION

The Irish wolfhound (IW) dog has an apparently high prevalence of heart disease, particularly dilated cardiomyopathy (DCM) and atrial fibrillation (AF).[1-8] Many questions remain unresolved regarding cardiovascular morbidity, mortality and the influence of cardiovascular therapies upon outcome. Echocardiographic parameters have been reported for normal and cardiomyopathic Irish wolfhounds.[9,10] Criteria most indicative of dilated cardiomyopathy include LV end-diastolic dimension > 60mm, LV end-systolic dimension >41mm, ESVI >41 ml/m², and fractional shortening < 25%. Left atrial and right ventricular dilatation may be present (LA dimension at end-systole >56mm, and RV end-diastolic dimension >35mm).

PREVALENCE AND CHARACTERIZATION OF DCM IN IRISH WOLFHOUNDS

Several investigators have reported arrhythmias in IW dogs. In a study of 496 IWs in Great Britain, arrhythmias were detected in 22 %. Atrial fibrillation was the most common rhythm abnormality, recorded in 10.5% of the dogs which were 10 months to 9.25 years of age.[1] Atrial fibrillation was also detected in 25 IWs from New England, USA, (mean age 4.5 years, median age, 5 years) with and without overt DCM.[2] In a study of IWs from northwestern Europe,[11] sequential echocardiographic and EKG examinations of 232 IWs detected AF in 30 dogs (12.9%) affected with DCM and 19 dogs (8.2%) with relatively normal echocardiograms. Nine additional dogs (3.9%) with DCM were in sinus rhythm, and 12 dogs (5.2%) had sinus rhythm with intraventricular conduction abnormalities. In another study evaluating heart screening examinations performed in 500 IWs from northwestern Europe, the prevalence of dogs with AF associated with DCM was 21.2%, while 3% of dogs had DCM and sinus rhythm, and 8 dogs (1.6%) had AF without evidence of DCM.[12]

Based upon prospective longitudinal evaluations of a larger population of Irish wolfhounds (1038 dogs),evaluated between 1990 and 2004, DCM and/or AF were diagnosed in 29.4% of these animals.[13] Males were overrepresented. At the time of diagnosis the mean age was approximately four years of age. Animals with decompensated heart failure were most often characterized by pleural and pericardial effusions while pulmonary edema occured mostly in dogs with moderate to severe secondary mitral regurgitation due to mitral annular dilatation. Ascites was not commonly detected. Although AF was detected from some IWs without echocardiographic evidence of DCM initially, most of these animals tended to develop overt DCM later on. These findings suggested that AF may be useful clinical marker for development of DCM. Most IWs with AF had very mild elevations of ventricular heart rate.

EFFECT OF DCM ON LIFE EXPECTANCY

IWs are subject to a number of reported hereditary diseases which reduce life expectancy. Intrahepatic portosystemic shunts, fibrocartilarginous embolism, and rhinitis/bronchopneumonia have been reported to affect juvenile IW dogs, while other severe disease complexes (DCM, osteosarcoma, gastric dilation volvulus, bronchopneumonia, soft tissue cancer, renal failure) develop in adult dogs.[14] In a study comparing lifespan and causes of death in the IW, in dogs born in the 1960s vs dogs born after 1980, a highly significant decrease in mean lifespan of about 1.5 years was detected.[14] An evaluation of survival information from 752 dogs recorded a median life expectancy of about 6 years. DCM had a statistically significant adverse effect on survival. However, due to a predisposition of Irish wolfhounds to a variety of other disease complexes that reduce longevity, the magnitude of the DCM effect on survival was about 5 months.[12, 13]

GENETIC ASPECT

IWs have been reported to be extremely inbred, with one ancestor explaining over 25% and 3 ancestors explaining over 50% of the genetic variability of the population born between 1965 and 2005.[14] This circumstance makes all molecular genetic evaluations extremely difficult. Concerning the familial distribution of DCM in the IW, analyses of smaller pedigrees initially had suggested an autosomal dominant pattern of inheritance.[8] After unsuccessful results with candidate gene approaches, a complex segregation analysis based on results of cardiovascular examinations carried out between 1987 and 2003 by the author was performed.[15] Data of 878 dogs from 531 litters in 147 different kennels were used for complex segregation analysis. Pedigree information was available for more than 15 generations. Male dogs were affected significantly more often by DCM than female dogs. The segregation analysis indicated different quantitative trait loci (QTL) involved in pathogenesis of DCM. A simple monogenic mode of inheritance could be excluded.[15] Recently, a genome-wide association study using 106 DCM cases and 84 controls identified QTL on 13 chromosomes with highest peaks on 3 different dog chromosomes. Within three QTL, highest associated single nucleotide polymorphisms (SNPs) were localized for the first time within or proximate to genes connected to lipid metabolism. [16]

DCM AND TREATMENT

For many years cardiovascular screening examinations have become obligatory for breeding dogs and recommended for non breeding dogs by IW breeding clubs in several northwestern European countries. Dogs affected with DCM or AF are excluded from breeding and early start of medical therapy is expected to improve outcome. However, due to the relatively long time course of DCM and the high number of predispositions to other fatal disease affecting this breed, many IW dogs affected by DCM die from non-cardiac causes.[12] This makes the influence of early therapeutic intervention on outcome difficult to evaluate.A number of drugs have been suggested for treating occult DCM or AF in IW dogs.

Angiotensin-converting-enzyme inhibitors (ACEIs) are part of the recommended standard treatment of chronic congestive heart failure (CHF) in people[17]. The American College of Cardiology and the American Heart Association guidelines recommend the use of ACEI therapy also in people with asymptomatic LV dysfunction.[17] There is one study in veterinary medicine which demonstrated that initiation of ACEI therapy (benazepril hydrochloride) during the occult phase of DCM significantly delayed the progression to overt DCM in Doberman Pinschers.[18]

Digoxin has been used in people and dogs with DCM and heart failure for decades.Today, the role of digoxin in contemporary therapy is controversial. Randomized controlled trials conducted in people in the pre-ß-blocker era suggested that digoxin could reduce symptoms and improve exercise capacity; however, a large trial that enrolled patients mostly with mild symptoms suggested no overall effect on mortality. A small increase in sudden death was balanced by a reduction in death from worsening heart failure.[19] In a more recently published study analysing the outcome in a population of relatively young patients with severe HF assessed for heart transplantation, digoxin did not improve outcomes.[20] In yet another study carried out in patients with AF and HF, digoxin alone without concomitant ß-blocker therapy was associated with worse survival, similar to that of patients without any heart rate control.[21] In veterinary medicine there is only one published study which assessed the efficacy of digoxin in 10 dogs with dilated cardiomyopathy. Of these, 4 dogs showed an increase in % SF of more than 5.5%.[22]

Pimobendan significantly extends survival time and improves the quality of life in dogs with congestive heart failure (CHF) due to dilated cardiomyopathy.[23,24] Studies investigating the effect of pimobendan in dogs initiated during the occult phase of DCM are ongoing.

REFERENCES

[1]Brownlie SE. **Vet Rec** 1991;129:470. [2]Harpster NK **Proceed. 12th ACVIM Forum** 1994;1:17. [3]Sisson DD et al. In: **Ettinger SJ, Feldman** EC. eds.Textb. of Vet. Int. Med. Philadelphia: WB Saunders, 1995:995. [4]Vollmar AC. **Kleintierpraxis** 1996;41:393 [5]Vollmar AC. **J Am Anim Hosp Assoc** 2000; 36:125. [6]Brownlie SE. **Sci Proc BCVS**, Birmingham UK, 2003. [7]Brownlie SE et al. **J Small Anim Pract** 1999;40:371. [8]Cobb MA et.al. **Proc BSAVA,** Birmingham UK 1996; 215. [9]Vollmar A. **J Am Anim Hosp Assoc** 1999;35:271. [10]Vollmar A. **J Am Anim Hosp Assoc** 1999;35:279. [11]Vollmar A. et al. **J Vet Int Med** 2001:15;279. [12]Vollmar AC et al. **Proc. 22nd ACVIM Forum 2004**: 811. [13]Vollmar AC et.al. **Proc. 23rd ACVIM-Forum**, Baltimore 2005:66. [14] Urfer SR. **Diss Vetsuisse Univ Bern** 2007. [15]Distl O et.al. **Heredity adv online pub** 2007,Jul4,doi:10.1038/sj.hdy.6801024. [16]Philipp U et al. **PLoS** 2012, publication submitted. [17]Hunt SA **J Am Coll Cardiol** 2005;(46):81. [18]O´Grady MR et.al. **J Vet Int Med** 2009;23:977. [19]Cleland JGF et.al. **Circ Heart Fail**,2009;(2):81. [20]Georgiopoulou VV et.al. **Circ Heart Fail** 2009,(2);90. [21]Fauchier L et.al. **Am J Cardiol** 2009;103;248. [22]Kittleson MD et.al. **JAVMA** 1985;(186):162. [23]Fuentes VL et.al. **J Vet Int Med** 2002;(16):255. [24]O´Grady MR et.al. **J Vet Int Med** 2008;(22):897.

CONTROVERSIES AND DILEMMAS IN MANAGEMENT OF CANINE HEARTWORM INFECTION

Clarke Atkins, DVM, DACVIM (SAIM, Cardiology)
Raleigh, NC

PROPHYLAXIS

Prevention of HWI is an obvious and attainable goal for the veterinary profession. Prevention failure results from ignorance on the part of owners as to the presence or potential severity of HWI, lack of owner compliance, or from inadequate instruction on preventative measures by the attending veterinarian.[1,32-34] Studies of owner compliance have revealed that approximately 55% of dog owners that use veterinary care purchase heartworm preventative, and enough medication is dispensed only to meet the needs of approximately 56% of those dogs. Hence the proportion of "cared for" dogs in the population that receive adequate heartworm prophylaxis is less than one third.[33] If one takes into consideration doses purchased but not administered and dogs that are never taken to a veterinarian, the percentage of protected dogs falls drastically. This was emphasized in North Carolina in 1999, when Hurricane Floyd caused extensive flooding and disruption in the poorest part of the state. Of dogs rescued from the floodwaters, 67% were infected with heartworms (personal communication, Dr. Kelli Ferris, North Carolina State University, 2003). In addition, evidence suggests that the veterinary profession is failing in its education of clients. New and colleagues,[35] upon questioning veterinary clients purchasing macrolide preventatives, found that 38% did not realize that their prescribed drug's spectrum was broader than solely preventing HWI.

Macrocyclic Lactone (Macrolide) Antibiotics

The introduction of the macrocyclic lactone endectocides (macrolides), ivermectin (Heartgard®, Iverhart ®, TriHeart®), ivermectin with pyrantel pamoate (Heartgard® Plus, Iverhart Plus®, TriHeart® Plus), ivermectin with pyranel pamoate and praziquantel (Iverhart Max®), milbemycin oxime (Interceptor®), milbemycin with lufenuron (Sentinal®) and with spinosad (Trifexis®), selamectin (Revolution™), and moxidectin (ProHeart®, ProHeart® 6), and moxidectin with imidicoprid (Advantage/Multi™) has provided the veterinary profession with highly effective, incredibly safe heartworm preventatives in a variety of formulations and with a variety of spectra. These agents, because they interrupt larval development (L3 and L4) during the first 2 months after infection, have a large temporal window of efficacy and are administered monthly. These products have enjoyed great efficacy, "virtually 100%", when used as directed. Recently, a single isolate (MP3) from north-eastern Georgia has shown restistance/tolerance to some macrocyclic lactones, when administered <u>once</u> 30 days after heavy experimental challenge.

All are safe in collies when used as directed at preventive dosages. They each have microfilaricidal efficacy and render female heartworms sterile. Hence microfilarial tests for HWI cannot be reliably used in dogs receiving these products. Prophylaxis should be commenced no later than 6 to 8 weeks of age in endemic areas or as soon thereafter as climatic conditions dictate.[19,49] Macrolides should be administered precisely as indicated by the manufacturer. If accidental lapses of more than 10 weeks occur, the preventative should be reinstituted at recommended doses and maintained for at least12 consecutive months.[25] In the event of a lapse in preventative administration during a time of known exposure risk, an antigen test should be performed 7-8 months after the last possible exposure to determine if infection has occurred. It is recommended by the AHS and by CAPC that these agents be used year-around in all areas of the U.S.

<u>Off-Label Use.</u> The macrolides are effective microfilaricides, with varying microfilarial kill rates, but microfilariae, in reduced numbers, are often found in the circulation for months after treatment has begun. Recently it has been shown that some macrolides have adulticidal activity if used continuously for prolonged periods.[39-41]

Macrocyclic Lactone "Resistance/Tolerance"

In 2005, the FDA-CVM reported an increase in the reports of LOEs (Lack of Effectiveness) for macrocyclic lactones and required that such agents no longer be labelled as "perfect" in terms of efficacy. This failure of complete rapid microfilarial clearing, coupled with concern in the Mississippi River delta region (areas of LA, AR, MS, TN), has caused concern that resistance to this class of drugs may be developing.[41a] The proof of this is small, but taken together, the data argue that a small percentage of microfilariae, isolated from dogs in this region have characteristics suggesting tolerance to the drug group. A joint consensus of the AHS and CAPC stated the following (excerpts). "There is evidence in some HW populations for genetic variations that are associated with decreased in vitro susceptibility to the macrocyclic lactones. Whether the observed genetic variations constitute heritable resistance is being investigated. Most credible reports of LOE that are not attributable to compliance failure are geographically limited at this time. The extent of the problem is obscured by demonstrated lack of owner and DVM compliance, possible changes in environmental/vector factors, and more effective antigen testing. The potential for resistance is *not* a reason to abandon use of approved preventive products."

The concern relative to the presence of circulating microfilariae in dogs that are started and maintained on monthly preventives is that they could be a source of propagation of microfilariae that are preselected for resistance to macrocyclic lactones. In 2005, Prichard wrote "Consideration of the proportion of the *D. immitis* population in refugia, the life cycle stage targeted, and the anthelmintic dosages used suggest that it is unlikely that significant avermectin/milbemycin [macrolides] resistance will be selected in *D. immitis* with current treatment strategies."[118] However, this belief was based upon the assumption that people were using the preventives as per label instructions, not using them as adulticides and microfilarial suppressants. The prudent approach is simply to administer the products as approved by the FDA: as preventives that should be given to microfilaria-negative dogs. This means that the "soft" or "slow" kill approach to adulticidal therapy should be avoided. Likewise, one could argue against the use of macrocyclic lactones in microfilaria-positive dogs prior to beginning them on adulticidal therapy (ie, a method advocated by this author; see below) as 10% to 20% of these dogs will have circulating microfilariae for months after they start this regimen – microfilariae that have seen a macrocyclic lactone. If this approach is utilized, the clinician must ensure that microfilaria are eradicated in the first months of macrolide therapy. All current heartworm preventives belong to the same class of molecule, the macrocyclic lactones, and thus, we need to be very prudent in our long-term stewardship of these drugs.

THERAPY

Adulticidal therapy

An important breakthrough in the management of heartworm infection (HWI) is the adulticide melarsomine, an organoarsenical superior in safety and efficacy to thiacetarsemide.[17] This product, which is administered twice, at 2.5 mg/kg q24h, has a mean retention time 5 times longer than thiacetarsemide and its metabolites are free in the plasma, on which HW feed.[18] In a study of 382 dogs with HWI receiving melarsomine, none required cessation of therapy due to hepato-renal toxicity, as compared to 15-30% with thiacetarsemide, the agent previously used.[18] With 2 doses, the efficacy is over 90% (FDA pivotal study) with the useful flexibility of a 50% worm kill with 1 dose. This then allows "split-dose" protocol to be utilized in severely afflicted individuals or in those in which pulmonary thromboembolism (PTE) is a concern. This method allows destruction of only one-half the worms initially (1 IM injection of 2.5 mg/kg), thereby lessening the chance for embolic complications. This single dosage is followed by a 2 dose regimen in 1-3 months, if clinical conditions permit. While the manufacturer recommends this protocol (Figure 1) for severely affected dogs, the author employs it in **all cases** unless there is financial constraint or underlying concern for arsenical toxicity (for example, preexistent severe renal or hepatic disease).[18a] One disadvantage to the "split-dose" method, in addition to the expense, is the need for 2 months' exercise restriction.

In 55 dogs, with severe heartworm disease (HWD) and treated in this 3-dose manner, 96% had a good or very good outcome with >98% negative for antigenemia 90 days post-therapy.[18] Although symptomatic and even fatal PTE can result from treatment with melarsomine, no case of severe PTE was seen in the 382 dogs of this series.[19] Of the 55 severely affected dogs, 31% had "mild or moderate PTE"; no fatalities resulted. The most common sign was fever, cough, and anorexia 5-7 days post-treatment. This was associated with mild perivascular caudal lobar pulmonary radiographic densities and subsided spontaneously or after corticosteroid therapy.

The most common complication to melarsomine therapy is the local inflammatory reaction at the injection site. This can be minimized by following the manufacturer's directions explicitly (change needles before injecting, choose deep IM site with care, put pressure on site after injection, and alternate sites). In addition, corticosteroids (e.g. dexamethasone) or NSAIDs can be given at the time melarsomine is administered to lessen the reaction.

"Soft" or "Slow" Kill

It is now known that certain macrolides have adulticidal properties.[6,16,20] Ivermectin, when administered for 31 months continuously has nearly 100% efficacy in young heartworm infections.[20] It has been shown, however, that lung and pulmonary vascular manifestations of HWD still result when ivermectin "prophylaxis" is begun 5.5 and 6.5 months post-infection and continued for 1 year.[21] Selamectin, when administered continously for 18 months killed approximately 40% of transplanted worms.[16] Sustained release moxidectin also appears to have some adulticidal efficacy.[12] Recent data suggests that an aggressive macrolide protocol (ivermectin, given at 6 ug/kg weekly instead of monthly), coupled with a complex regimen of doxycycline (10 mg/kg/day) will hasten worm destruction, with worm eradication with approximately 9 months' therapy.[21a] Furthermore, microfilariae are eradicated more quickly in this manner. This has caused many to invoke the use of doxycycline routinely in the management of heartworm infection in dogs. While there may be a role for this therapeutic strategy (Slow Kill) in cases in which patient age, financial constraints or concurrent medical problems prohibit melarsomine therapy, the current recommendations are that **marcolides <u>not</u> be adapted as the primary adulticidal approach.**

Surgery

Surgical removal of HW can minimize PTE, as compared to pharmacologic adulticides, such as melarsomine.[22,23] This procedure, however, requires specialized training and instrumentation, including fluoroscopic imaging capabilities. Nevertheless, it remains an alternative for the management of high risk patients.

Ancillary therapy

As mentioned above, aggressive macrolide therapy (ivermectin, given at 6 ug/kg weekly instead of monthly), coupled with a complex regimen of doxycycline (10 mg/kg/day) hastens worm destruction and quickly eradicates microfilariae. This has resulted in increasing use of doxycycline in the management of HWI in dogs.

Cage rest is an important aspect of the management of HWD after adulticidal therapy, after PTE, or during therapy of heart failure. This can often be best, or only, accomplished in the veterinary clinic. If financial constraints preclude this, crating at home and/or tranquilization are useful alternatives.

Microfilaricidal therapy

Despite the fact that no agent is FDA-approved for the elimination of microfilaria, microfilaricidal therapy is traditionally instituted 4-6 weeks after adulticide administration. The macrolides offer a new and effective alternative to levamisole and dithiazanine.[5,5a] Microfilariae are efficiently and rapidly cleared with ivermectin at 50 ug/kg (approximately 8 times preventative dose) or milbemycin at 500 ug/kg (preventative dose), although this represents an extra-label use of these drugs.[5,5a] The off-label use of livestock formulations of ivermectin is discouraged because of the possibility of dosing errors and resultant toxicity. Adverse reactions, the severity of which is likely related to microfilarial numbers, were observed in 6% of 126 dogs receiving ivermectin at the microfilaricidal dose (50 ug/kg).[30] Signs included shock, depression, hypothermia, and vomiting. With fluid and corticosteroid (dexamethasone at 2-4 mg/kg IV) therapy, all dogs recovered within 12 hours. One fatality was observed 4 days after microfilaricidal therapy. Similar findings and frequency have been reported with milbemycin at the preventative dosage.[31] Dogs so treated should be hospitalized and carefully observed for the day. Dogs <16 kg, harboring >10,000 microfilaria per ml blood, are more apt to suffer adverse reactions.[32] Benadryl (2 mg/kg IM) and dexamethasone (0.25 mg/kg IV) can be administered prophylactically to prevent adverse reactions to microfilaricidal doses of macrolides.

A 90% microfilaricidal success rate can be expected with ivermectin[25], while milbemycin at 500 mcg/kg cleared 6/8 (75%) dogs which had received adulticide therapy and did not harbor male and female adults; microfilarial numbers were reduced by 99% on the day after after treatment.[31] A slower microfilarial kill rate can also be achieved with ivermectin, moxidectin, moxidectin-imidacloprid, and selamectin at preventative doses.[1-4]

The time-honored approach to ridding the patient of microfilariae involves macrolide therapy (50 mcg/kg for ivermectin or 500 mcg/kg milbemycin) instituted 3-6 weeks after adulticide.[5,5a] In 2-3 weeks, a second microfilaria concentration test is performed and, if negative, preventative started. If still positive, the treatment is repeated or alternatively, chemoprophylaxis begun (assuming that no adverse reaction occurred on the initial treatment).[5,5a] Persistent antigenemia (after 6-7 months) indicates continued patent infection.

This author chooses an alternative approach (Figure 1[18a]), beginning the administration of a macrolide preventative at the time of diagnosis, often days to weeks prior to adulticidal therapy. With the "slow microfilaricides" (ivermectin, moxidectin, or selamectin), there is little chance of an adverse reaction, but the owner is warned and advised to administer the medication on a day when he/she will be at home. If Milbemycin (a superior microfilarial agent) is used, it is administered in the hospital and/or preceded by administration of dexamethasone and benadryl, as described above. If this approach is used, the dog must be rendered microfilariae-free by 1-3 months post-diagnosis. Recent evidence demonstrates that concurrent usage of a macrolide and doxycycline reduces microfilarial numbers more rapidly, rendering dogs negative in less than 3 months. It is imperative that dogs on macrocyclic lactones be rendered microfilaria-free.

REFERENCES:

[1]McCall, JW, et al. Amer J Vet Res 57:1189-1192; 1996. [2]Bowman, DD et al. Proc Amer Heartworm Soc '92, p151, 1992. [6] McCall, JW, Proc Amer Heartworm Sym '98. 209-215, 1998. [7]Blagburn et al. Proc Amer Heartworm Sym '92. 159-164, 1992. [7]Blagburn et al. Proc Amer Heartworm Sym '92. 159-164, 1992. [18a]Atkins, CE, et al. Vet Med 2003; April:310-317. [21a]McCall, JW et al. Vet Parasit. 204-214, 2008. [41a]Geary TG, et al. Topics Comp Anim Med. 186-192, 26:2011.

Figure 1. The author's preferred approach to adulticidal therapy invirtually *all* (severely affected or not) dogs infected with heartworms includes 3 doses of melarsomine. Macrolide prophylaxis is begun at the time of diagnosis, if not already in use. *If microfilaremic, care should be taken to prevent or observe and treat adverse reactions, based on microfilarial numbers and macrolide used. It is imperative to rid patients receiving macrolides of microfilariae to reduce the chance of resistance.

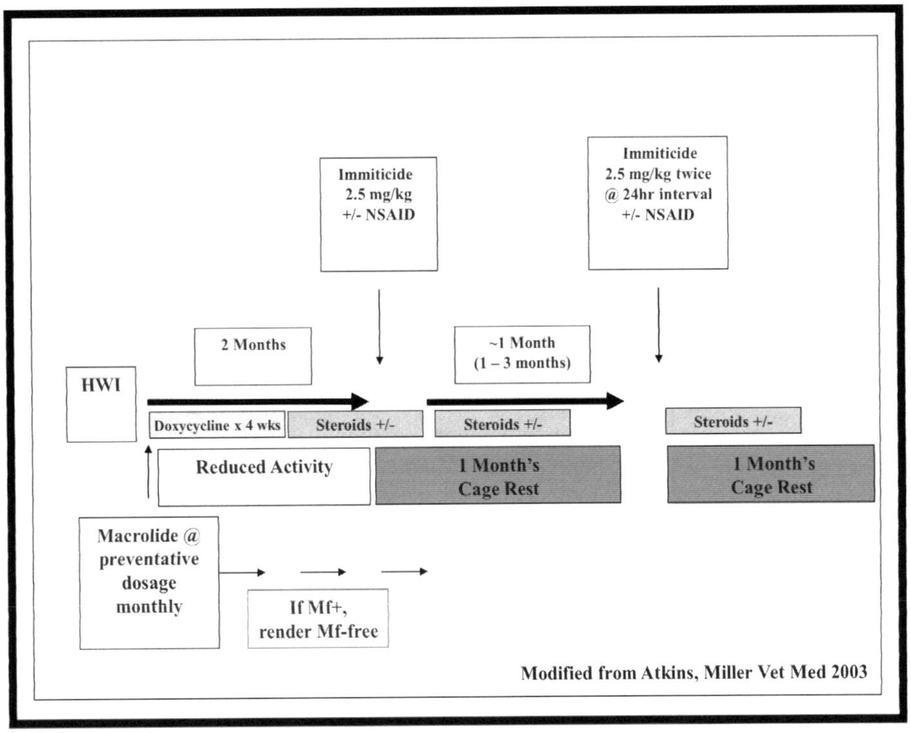

ECHOCARDIOGRAPHIC ASSESSMENT OF RV FUNCTION IN 150 DOGS WITH PULMONIC STENOSIS

Brian A. Scansen, DVM, MS, DACVIM (Cardiology) and Tara A. Amburgy
Columbus, Ohio

INTRODUCTION

Congenital pulmonary valve stenosis (PS) has traditionally been considered the third most commonly diagnosed congenital heart defect of dogs.[1] However, recent studies suggest this condition may be the most common congenital cardiac malformation in the canine species.[2] As in children,[3] valvular PS in dogs is described in three principal forms – (1) a dome-shaped valve with commissural fusion, (2) a dysplastic and markedly thickened valve with or without hypoplasia of one or more leaflets, and (3) hypoplasia of the pulmonary annulus.[4,5] While the human form of valvular PS is predominately that of commissural fusion,[3] the canine form is more commonly characterized by thickened, dysplastic valves.[4,6,7]

The obstruction to blood flow that results from PS causes a compensatory hypertrophy of the right ventricle (RV), resulting in a thick, high-pressure chamber in contrast to the normal RV that is thin-walled and low pressure. The long-term consequence of pathologic RV remodeling is the development of RV ischemia, dysfunction, dysrhythmias, and heart failure. Dogs with PS are asymptomatic early in life, but may progress to signs of lethargy, collapse, cyanosis, right-heart failure, or sudden death.[5,8,9] These symptoms are alleviated or delayed with therapy (adrenergic receptor antagonists and/or balloon pulmonary valvuloplasty), but the morbidity and mortality of dogs with severe PS remains substantial. Balloon pulmonary valvuloplasty (BPV) was first performed in a dog in 1980 and reported in a child in 1982.[10,11] There is evidence that BPV improves the clinical outcome of human and canine patients with valvular PS, both with a reduction in clinical symptomatology and an improvement in survival.[6,9,12,13] The procedure is now routinely performed in clinical canine practice with low morbidity and mortality for those patients with a severe gradient or the presence of clinical signs referable to their disease.

Although effective for many patients, human and canine studies suggest that patients with valvular dysplasia and a hypoplastic annulus show less improvement following BPV than those with purely valvular fusion.[7,14] This stands to reason as the principal benefit of BPV is tearing of commissural fusion; it is unable to alleviate obstruction secondary to annular hypoplasia or thick, redundant valve tissue. In a study of 30 dogs with valvular PS, 94% of the 18 dogs with a normal pulmonary annular diameter and minimal valve thickening had resolution of their clinical signs and survived more than 1 year after the procedure; however, for the 12 dogs with marked leaflet thickening and a hypoplastic pulmonary annulus, only 66.6% had a favorable outcome and only 50% showed a resolution in their clinical signs.[7]

While it is known that the valve morphology of PS impacts response to BPV, there is little data available to predict which dogs with PS will suffer clinical symptoms of their disease nor is there information available to prognostic when such symptoms may develop. The focus of this study was to retrospectively evaluate signalment, Doppler echocardiographic measures of disease severity, 2-dimensional echocardiographic indices of RV function, treatment offered, and response to treatment in a population of dogs with PS. In so doing, our aim was to define the breeds at risk, evaluate secondary markers of treatment efficacy, evaluate basic echocardiographic estimates of RV function, and assess change in RV function over time.

MATERIALS & METHODS

Medical records from the Health Information Database of The Ohio State University Veterinary Medical Center were retrospectively analyzed and all canine cases with pulmonic stenosis from January 2004 to June 2011 were selected. Cases were reviewed for the appropriate inclusion criteria (PS with no other concurrent cardiac lesions and at least one echocardiogram recorded). Selected cases were tabulated to document signalment, severity of PS, interventions performed, medications administered, change in echocardiographically-derived pulmonary valve pressure gradient after beta-blockade (BB) or BPV, and change in RV pressure measured during cardiac catheterization.

A single trained observer measured each echocardiogram and averaged the following variables over 3-4 cardiac cycles: the maximum (PG_{max}) and mean (PG_{mean}) trans-pulmonary valve pressure gradient, the velocity time integral of the aortic (Ao_{VTI}) and pulmonic (PV_{VTI}) flow signals, the RV freewall thickness in diastole ($RVFW_d$), the RV fractional shortening ($\%FS_{RV}$), the RV fractional area change ($\%FAC_{RV}$), and the tricuspid annular plane systolic excursion (TAPSE). These values allowed calculation of the pulmonary valve effective orifice area (EOA_{PV}) and the ratio of the Ao_{VTI} to the PV_{VTI} ($VTI_{Ao/PV}$).

Statistical analyses were performed using GraphPad Prism 5 software and those with a normal distribution (as assessed by the D'Agostino & Pearson omnibus normality test) are presented as mean with standard deviation (SD). Those not meeting criteria for normalcy are presented as median and range. The presence of a correlation between groups was assessed with either a Pearson correlation or a Spearman's rank correlation for parametric and non-parametric data, respectively. Simple comparisons between groups were assessed with either a Student's t test or Mann Whitney U test for parametric and non-parametric data, respectively. Repeated measures ANOVA and regression analyses were utilized to evaluate changes in a given variable over time. Significance was set at $P < 0.05$.

RESULTS

The medical record search yielded 203 dogs with a diagnosis of PS. Of these, 151 dogs met the entry criteria.

Subjects

The dogs included in the analysis were composed of 95 males (63%) and 56 females (37%). The breed with the greatest prevalence was the English Bulldog, comprising 17% (n=26) of the cases. Other breeds (n=# of cases) included mixed breed (23), Boxer (12), Pomeranian (7), Cavalier King Charles Spaniel (6), Yorkshire Terrier (6), Bull Mastiff (5), French Bulldog (5), Chihuahua (4), Miniature Pinscher (4), American Pit Bull Terrier (3), German Shepherd (3), and the Scottish Terrier (3). Other breeds with only 1 or 2 dogs represented 44 additional cases. The median age at diagnosis was 6.8 months (range, 1.7 to 164 months).

Echo Findings

Several dogs had multiple imaging studies (range, 1 to 8 echocardiograms during the study period). Unless specified otherwise, the data are given based on the first echocardiogram recorded for a given dog. The PG_{max} at diagnosis averaged 89 mmHg ± 44 mmHg (range, 15 to 209 mmHg). The PG_{mean} averaged 50 mmHg ± 26 mmHg. The mean pulmonary valve annulus measured 12.2 mm ± 3.7 mm. 78 dogs had sufficient images during the first echocardiographic study to calculate variables of RV systolic function. The mean $\%FS_{RV}$ was 38% ± 15%, the median $\%FAC_{RV}$ was 63% (range, 37% to 82%), and the mean TAPSE was 1.7 cm ± 0.5 cm. There was no correlation between $\%FS_{RV}$, $\%FAC_{RV}$, or TAPSE and the PG_{max}. The $\%FS_{RV}$ and $\%FAC_{RV}$ were positively correlated ($r = 0.66$); however, there was no correlation between TAPSE and either $\%FS_{RV}$ or $\%FAC_{RV}$. There was a significant negative correlation ($r = -0.72$) between the $VTI_{Ao/PV}$ and PG_{max}. There was a weak correlation ($r = 0.41$) between $RVFW_d$ and PG_{max}.

Balloon Pulmonary Valvuloplasty

71 dogs underwent BPV, at a median age of 7.4 months (range, 2.6 to 136.5 months). The mean balloon to annulus ratio (BAR) was 1.25 ± 0.19. For the 63 dogs with available data, the right ventricular systolic pressure decreased significantly after ballooning from a median of 85 mmHg (range, 39 to 150 mmHg) to 45 mmHg (range, 23 to 150 mmHg). By echocardiography (Figure 1), the estimated pulmonary valve pressure gradient decreased significantly from 125 mmHg (range, 52 to 228 mmHg) to 52 mmHg (range, 15 to 180 mmHg).

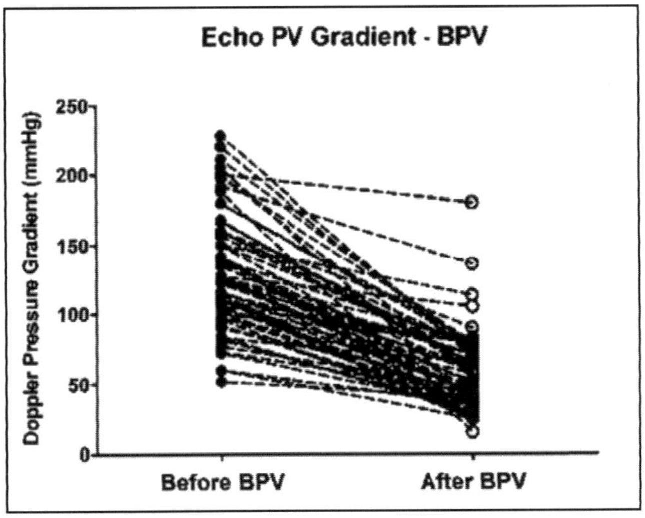

Figure 1. Estimated pulmonary valve gradient in 63 dogs before and after BPV. The post-BPV gradient was significantly less than the pre-BPV gradient.

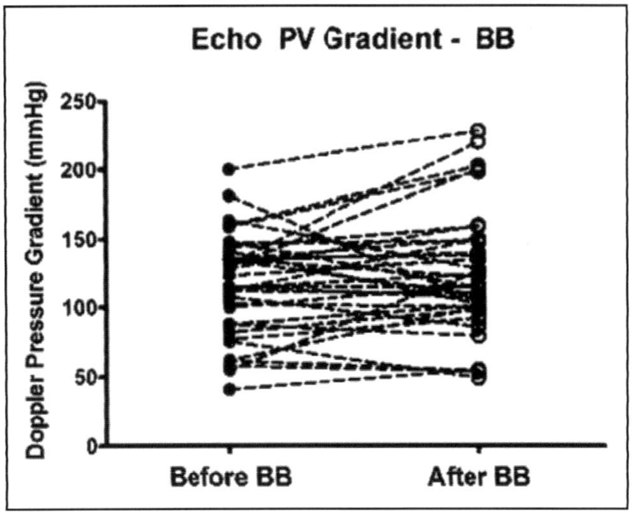

Figure 2. Estimated pulmonary valve gradient in 36 dogs before and after receiving beta-blocker (BB) therapy. There was no significant difference between the two time points ($P = 0.17$).

Beta-blockade

93 dogs received beta-blockade (atenolol). For the 36 dogs with available data, there was no significant change in the pulmonary valve pressure gradient after starting beta-blockade with a mean of 117 ± 37 mmHg before and 124 ± 45 mmHg after (Figure 2).

Congestive Heart Failure

Although follow-up data were not available for all dogs, at least 13 dogs experienced right-sided congestive heart failure (CHF). The English Bulldog was over-represented among dogs developing heart failure as 5 of the 13 (38%) dogs in heart failure were English Bulldogs, compared to 17% of the total population of dogs with PS. There was insufficient echocardiographic data on these 13 dogs to draw conclusions about echocardiographic estimates of right heart function in the setting of CHF.

CONCLUSIONS AND CLINICAL RELEVANCE

In this study, an almost 2:1 male predominance was seen in dogs presenting with congenital pulmonary valve stenosis. Earlier studies suggested that there was no gender predilection in this condition except for in the English Bulldog.[1] However, more recent studies also find a male predominance for PS.[6] As noted before, the English Bulldog is over-represented with this condition. In addition, these results further suggest that the English Bulldog may be more likely to develop CHF than other breeds, comprising 38% of dogs that went into CHF. The higher proportion of CHF cases in the English Bulldog may be related to an inability to perform BPV in these dogs due to the common presence of a coronary artery anomaly that wraps around the pulmonary annulus. Without the benefit of BPV, there is little that can be done to alleviate the severe elevation in RV afterload in these dogs.

Echocardiography to assess the severity of PS showed that BPV provided a significant reduction in pulmonary valve pressure gradient, while BB did not alter the gradient. This supports earlier studies that found reduced symptoms and prolonged survival in dogs that underwent BPV compared to those that did not.[9]

The correlation between $VTI_{Ao/PV}$ and PG_{max} is anticipated from the continuity equation, which states that the EOA_{PV} can be estimated by the stroke volume of the left heart divided by the VTI of the pulmonary outflow signal. Indeed, the EOA_{PV} is likely a better estimate of PS severity than PG_{max} because it is less dependent on flow than the maximum velocity through the stenosis. As the stroke volume of the left heart is given by the aortic VTI multiplied by the cross-sectional area of the aortic valve, one can presume that the ratio of the aortic VTI to the pulmonic VTI is a comparable estimate of PS severity. While we have shown that the $VTI_{Ao/PV}$ is proportional to the current clinical standard of PS severity (PG_{max}), whether this ratio reflects a preferred means of estimating PS severity that is less flow-dependent than the peak Doppler velocity requires further study.

There is unfortunately a paucity of data concerning echocardiographic evaluation of RV function in dogs. In recent years the importance of right ventricular (RV) function in the clinical status and outcome of patients suffering from numerous cardiac and non-cardiac diseases has become increasingly apparent in human medicine. In this study we describe the technique of RV functional assessment in the setting of PS by 2-dimensional estimates. However, there was no correlation apparent between RV function and PS severity. The echocardiographic estimation of RV function in dogs with PS would be appealing if additional prognostic information could be obtained – eg, risk for CHF. In this retrospective series, there is insufficient data to draw any conclusions as to the prognostic utility of RV function. Further studies evaluating RV function in dogs with PS through both conventional 2-dimensional echocardiography as well as advanced Doppler and speckle-tracking techniques are warranted to further characterize this condition.

REFERENCES

1. Buchanan JW. In: Kirk RW, Bonagura JD, eds. *Current Veterinary Therapy XI: small animal practice*. 1992:647-654. **2.** Oliveira P, et al. *J Vet Intern Med.* 2011;25(3):477-483. **3.** Waller BF, et al. *Clinical Cardiology.* 1995;18(1):45-50. **4.** Patterson DF, et al. *The American Journal of Cardiology.* 1981;47(3):631-641. **5.** Fingland RB, et al. *Journal of the American Veterinary Medical Association.* 1986;189(2):218-226. **6.** Ristic J, et al. *Journal of Veterinary Cardiology.* 2001;3(2):13-19. **7.** Bussadori C, et al. *J Vet Intern Med.* 2001;15(6):553-558. **8.** Lombard CW, et al. *Journal of the American Veterinary Medical Association.* 1989;194(1):71-75. **9.** Johnson MS, et al. *J Vet Intern Med.* 2004;18(5):656-662. **10.** Buchanan JW, et al. *J Vet Intern Med.* 2002;16(1):116-117. **11.** Kan JS, et al. *The New England Journal of Medicine.* 1982;307(9):540-542. **12.** Fawzy ME, et al. *American Heart Journal.* 2007;153(3):433-438. **13.** Rao PS. *Catheterization and Cardiovascular Interventions.* 2007;69(5):747-763. **14.** McCrindle BW. *Circulation.* 1994;89(4):1751-1759.

CARDIOLOGY MYTHBUSTERS: DISPELLING DIAGNOSTIC DOGMA IN VETERINARY CARDIOLOGY

Mark Rishniw, BVSc, MS, PhD, DACVIM
Ithaca, NY

INTRODUCTION

Clinicians are taught a range of diagnostic tests during their veterinary training and after graduation, which are expected to help them better diagnose cardiac disease in their patients. These include historical and physical examination findings, and findings from specific diagnostic procedures, such as electrocardiography (ECG), radiography and echocardiography. More recently, various biochemical analyses have been added to this list of diagnostic procedures.

Many of these diagnostic procedures are taught as dogma, with little evidence to support their validity. Examples are carefully selected to underscore the teacher's claims, but whether these findings hold true in the broader population of cardiac patients is often not specified. These procedures and findings are often based in out-dated or historical literature, or assume a level of skill from the examiner that might not be appropriate for the person performing the examination.

The following is a collection of personal observations based in experience that should help clinicians better determine the value of specific diagnostic observations and findings, and allow them to adjust their diagnostic expectations, interpretations and ultimately, their diagnosis.

HISTORY & PHYSICAL EXAMINATION

A cough is a cough is a cough

One of the most common cardio-pulmonary complaints of clients about their pets is the presence of coughing. Clinicians are expected to identify the cause of the cough and treat it appropriately. It is important to recognize that coughing is not a consistent feature of congestive heart failure (CHF), and the finding of a murmur in a coughing dog should not result in the automatic prescription of a diuretic.

$$\text{MURMUR} + \text{COUGH} \neq \text{CHF}$$

Coughing IS a sign of CHF in humans. Generally, this is described as a moist soft cough, often at night. The nocturnal nature of coughing in people is due to the change in body axis from vertical to horizontal when sleeping. Soft coughing CAN BE a sign of CHF in dogs. However, protracted, loud, and chronic coughing (lasting weeks or months) is almost never due to CHF. Additionally, coughing without any other clinical signs of CHF is strongly indicative of non-cardiac causes. Thus, in a dog that is coughing due to CHF, one would expect to observe tachypnea, tachycardia, and potentially dyspnea. Coughing from CHF is not chronic. CHF does not "fester" for weeks or months without therapy. Thus, history of a chronic, non-productive, and vigorous or frequent coughing should encourage clinicians to exclude CHF as a cause.

Coughing in cats is almost always associated with non-cardiac causes. Thus, unless proven otherwise, a history of coughing in a cat should prompt an investigation of airway disease, rather than cardiac disease.

Wheezy is not one of the seven dwarves

Wheezing is a feature of small airway constriction. Thus, cats that wheeze (and often cough) almost invariably have non-cardiac disease (respiratory disease) underlying their clinical signs.

Snaps, crackles, pops

The most common pulmonary sounds associated with CHF in animals are accentuated bronchovesicular sounds. Pulmonary crackles are described in humans with CHF, and have been reported to occur in dogs and cats with CHF. However, in humans, the crackles are generally restricted to the ventral tips of the caudal lung lobes, and are due to dependent alveolar edema in those portions of the lung. Since quadrupeds have a horizontal pulmonary axis, they do not develop dependent pulmonary edema. Consequently, the rare occasions when CHF causes crackles can be easily identified – these animals have fulminant CHF, with generalized alveolar edema. However, this degree of CHF is uncommon – most cases present with primarily interstitial edema, rather than alveolar edema. Without alveolar fluid, there are no crackles. Therefore, the vast majority of pets with audible crackles have primary airway disease, rather than CHF.

Pinky blue and <2 seconds

Mucous membrane color and capillary refill time are commonly assessed. However, no studies exist showing any association with these variables and presence or absence of common cardiac diseases or CHF. With the exception of cyanotic heart defects, membrane color is of little value. Similarly, I have observed CRT<2 seconds in recently deceased patients – hardly a useful test! Examination of these variables is straight-forward, gives the appearance of "doing medicine" to the client, and provides "thinking space" for the clinician while they assess their findings. However, clinicians should not anticipate finding abnormalities with these variables even in severely affected cardiac patients, but use them for what they really are – window-dressing.

Murmur, murmur…

Classical cardiac texts describe characteristics of cardiac murmurs – timing, loudness, location, phasic changes. These are based on descriptions in humans, where the heart sounds are (A) of a lower frequency than those observed in dogs and cats and (B) occur at a much lower heart rate. Thus, findings such as crescendo-decrescendo vs plateau, base vs apex, etc are often easily identified in people with heart rates of 60-70 bpm, but are almost impossible to identify in animals with heart rates >120 bpm.

Furthermore, small patients, like cats and small dogs have hearts that are often as small as the diaphragm of the clinician's stethoscope – in such cases, localization becomes virtually impossible.

In most cases, characterization of feline murmurs is not possible beyond the intensity (soft, loud, very loud), and general location (parasternal, or lateralized). Since systole and diastole are of approximately similar duration in cats, timing becomes difficult to determine – it suffices to say that 99.9% of murmurs in cats are systolic, so a clinician is safe betting on this!

RADIOGRAPHY

Sternal schmernal!

"Increased sternal contact" is taught as a finding of cardiomegaly to veterinary students by most radiologists. This finding was proposed in the 60s and 70s, when radiography and angiography were used exclusively to identify cardiomegaly antemortem. However, the advent of echocardiography allowed cardiologists to more accurately examine the size of cardiac chambers in awake patients and compare these findings with radiographic impressions. This led to several observations – marked right heart enlargement can, occasionally be associated with increased sternal contact, but is associated with other signs of cardiomegaly; sternal contact is commonly seen in aged cats that have no heart disease; sternal contact in dogs is more commonly a feature of thoracic conformation than cardiomegaly. Thus, clinicians would be well served to forget they ever heard of "sternal contact."

The reverse D

Similarly, clinicians have been taught about the "reverse D" finding, which is claimed to be suggestive of right heart enlargement, and often associated with heartworm disease. However, like sternal contact, the specific definitions of what constitutes a true "reverse D" vs a variant of normal anatomy are, not surprisingly, lacking from the literature. This finding becomes like the definition of pornography – "I can't define it but I know it when I see it." In my experience, the "reverse D" is far more likely to be a misinterpreted normal variant than to be a true finding. Thus, clinicians would also do well to forget this term.

Coupled with the "reverse D" is the very common observation by clinicians of "right heart enlargement and clinical history of coughing." Physiologically speaking, there is no cardiac disease that would cause right heart enlargement and result in coughing. The converse is potentially feasible – diseases that cause coughing can cause pulmonary hypertension and subsequent right cardiomegaly; however, these are relatively uncommon.

Mild heart enlargement

Cardiac-wise, the only thing that can be described with any accuracy as being mildly enlarged on thoracic radiographs is the left atrium. The beat-to-beat variability, positional variability, and inter-individual variability of the other chambers radiographically speaking is such that "mild" changes are not detectable. Again, cardiologists, with the benefit of echocardiographic hindsight have found that their ability to accurately and consistently identify mild enlargement of cardiac chambers is limited to the left atrium. Indeed, Suter and Gomez stated that "The differences among normal hearts of varying canine breeds or the differences between hearts radiographed at inspiration or expiration are often greater than the differences among normal and diseased hearts." [8]

Perihilar edema

Pulmonary edema is not uniform in its distribution. As pulmonary edema develops in quadrupeds, the peripheral pulmonary lymphatics open and shunt interstitial fluid towards the regional (hilar) lymph nodes. This results in preferential accumulation of interstitial fluid at the hilus. In humans, gravity results in mild alveolar edema developing in the ventral lung lobes – a finding that is particular to bipeds. However, clinicians tend to grossly over diagnose "hilar edema,"especially when interpreting radiographs of coughing patients. With left-sided CHF, the vast majority of patients have to have severe chronic left heart disease. Thus, the finding of "mild perihilar edema" in the absence of strong physical and radiographic evidence of severe heart disease is almost always a false positive observation. From personal experience with clinicians' interpretations of "perihilar edema,"this radiographic finding is of virtually no clinical value and is much more likely to mislead a clinician than to guide them to a correct diagnosis.

VHS and heart disease

About 15 years ago, Jim Buchanan published a cardiac mensuration method known as the vertebral heart scale.[3] This system allowed clinicians to identify cardiomegaly, and, more importantly, to compare changes in VHS within a patient over time.

What can the VHS do? The VHS system is a relatively specific (few false positives) means of identifying cardiac volume overload. Thus, a big heart, secondary to volume overload, as determined by a big VHS, is likely to be big. On the other hand, minor increases in cardiac size might well go undetected as increases in VHS. Subsequent studies of VHS in various breeds showed that this method of cardiac measurement was not breed-independent, but that some breeds appeared to have VHS that exceeded the initial limits. However, even in these breeds, changes in cardiac size over time can still be detected.

What can the VHS NOT do? The VHS system was not developed to identify "heart disease." There are many diseases, both those that cause volume overload and those that don't, which cannot be identified by VHS. Either the volume overload disease is too mild (low sensitivity of the test), or there is no marked change in cardiac size from the disease (e.g. concentric hypertrophy). Thus, a normal VHS does not rule out heart disease. However, it does reliably rule out severe left-sided volume overload.

Costal counts

Counting rib spaces covered by a cardiac silhouette has also been proposed as a means of estimating heart size. Again, no studies have demonstrated that, against an echocardiographic standard of heart size, this system reliably or accurately identifies cardiomegaly. Indeed, with "horizontalization" of hearts in older patients, the cardiac axis changes sufficiently to cover a greater span of ribs. This means of assessing heart size is non-sensical.

ELECTROCARDIOGRAPHY

Big R, big heart

Measurements of the R-wave amplitude and duration have been used as indicators of left-heart enlargement. However, studies have shown that patient position, electrode position, and body condition/conformation can all affect the amplitude of the R-wave. Thus, an R-wave that is smaller than the reference limit does not exclude left-heart enlargement, and an R-wave larger than the reference limit does not confirm left-heart enlargement.

Tick Tock-Tock

Therapy of chronic ventricular arrhythmias is often recommended in cardiac patients. Efficacy of such therapy cannot be determined by routine in-house ECG recording. A dog has an estimated 140K heart beats per day. A 2-minute ECG recording identifies approximately 200 of these – or 0.13% of the total cardiac events. Thus, the probability of accurately determining therapeutic response from such a small sample approaches zero. Even 24 ambulatory monitoring has been questioned in humans and dogs with ventricular arrhythmia – evidence suggests that multi-day sampling provides the best estimate of arrhythmia control.

CONCLUSIONS

These selected tidbits are aimed at helping clinicians understand the basis on which they make a diagnosis of cardiac disease or CHF in their patients, and to optimize their diagnostic armamentarium, develop skills that matter, and to ignore those findings which are more likely to confuse than aid the diagnosis.

REFERENCES

1. Rishniw M, *et al.* J Vet Intern Med, 2002;16(1):69-73. 2. Rishniw M. Vet Clin North Am Small Anim Pract. 2000;30(2):395-425. 3. Buchanan JW, Bucheler J. JAVMA 1995;206:194. 4. Lamb CR, *et al.*, Vet Rec 2001;148:707. 5. Spier AW, Meurs KW. J Am Vet Med Assoc 2004;224(4):538-41. 6. Demling RH, *et al.*, New Horiz. 1993;1(3):371-80. 7 Fabiani M. IVECC Symposium, 2010. 8. Suter P, Gomez JA (eds). Diseases of the Thorax: radiographic diagnosis. Iowa State University Press, 1987.

Pulmonary Hypertension: Clinical Applications and Treatment

Heidi B. Kellihan, DVM, DACVIM (Cardiology)
Madison, WI

TREATMENT

Pulmonary hypertension treatment focuses on targeting the derangements that result in medial hypertophy, intimal proliferation and decreased vascular compliance. Pulmonary vasodilating drugs currently in use target the pathophysiologic abnormalities associated with the pulmonary arterial endothelin pathway(endothelin receptor antagonists), prostanoid pathway (prostacyclin analogues), and nitric oxide pathway (specific or non-specific phosphodiesterase inhibitors). More recently, calcium sensitizing agents with phosphodiesterase-3 inhibiting actions have also been used for clinical PH, especially when left heart disease is a contributing cause.

Sildenafil

Sildenafil (Viagra→ , Revatio→) is an orally active, highly selective PDE-5 inhibitor. Multiple studies have demonstrated the benefits of sildenafil in people and dogs with PH.[1-3] Sildenafil appears to produce beneficial effects in PH by multiple mechanisms, but the primary mechanism operative in PH patients appears to be direct pulmonary artery vasodilation.

Pimobendan

Pimobendan and levosimendan are calcium sensitizing agents and PDE-3 inhibitors. PDE-3 has activity at the level of large and small (resistance)PAs, whereas PDE-5 exerts its activity in primarily large pulmonary arteries. PDE-3 inhibitors promote PA vasodilation via enhancement of adrenergic relaxation. The dual action of PDE-3 inhibition and the positive inotropic effects of calcium sensitization may provide some attenuation of PH, especially in PH secondary to left-sided heart disease. Pimobendan has been used in dogs with PH associated with mitral valve disease, and there was a significant decrease in estimated PAPs and an improved quality of life in the short term.[4]

REFERENCES

[1]Kellum HB, et.al. **J Vet Intern Med.** 2007;21(6):1258. [2]Bach JF, et.al. **J Vet Intern Med.** 2006;20(5):1135. [3]Brown, et.al. **J Vet Intern Med.** 2010;24(4):850. [4]Atkinson KJ, et.al. **J Vet Intern Med.** 2009;23(6):1190.

PATHOPHYSIOLOGY OF DILATED CARDIOMYOPATHY

Dr. Michael O'Grady and Dr. M. Lynne O'Sullivan
Guelph, Ontario, Canada

Definition

Dilated cardiomyopathy (DCM) has traditionally been defined based on a functional and structural phenotype; that is dilation of the left ventricle with or without right ventricular dilation accompanied by systolic dysfunction. While this definition continues to be useful we increasingly recognize that DCM, as thus defined, can be the final common pathway to a variety of myocardial injuries. Additionally this final common pathway perspective is supported by the findings that altered signaling pathways and compensatory mechanisms are observed in the failing heart regardless of the underlying etiology. Nevertheless DCM continues to be reserved for heart disease that primarily affects the myocardium thus in the absence of primary valve disease, pericardial and epicardial disease, congenital heart disease, parasitic disease, secondary to systemic or pulmonary artery hypertension, and coronary artery disease (in people). Cardiomyopathy that primarily affects the right ventricle (arrhythmogenic right ventricular cardiomyopathy/dysplasia) while an important cause of heart disease will not be discussed in this presentation.

Much of the data available concerning DCM in the dog is derived from work with the Doberman Pinscher. To what extent this data is transferable to other breeds with DCM is still to be determined. It has been demonstrated that DCM in the Doberman has a much more rapid progression than that observed in other dogs[1].

Etiologies

Identifying an etiology is particularly important from a management perspective. While many forms of DCM induced heart failure will benefit from neurohormonal blockade, volume contraction, and inotropic support it is primarily important to identify those uncommon cases that represent a potentially reversible form of DCM capable of significant improvement when specific treatment for the underlying disorder is employed.

I will spend considerable time presenting genetic etiologies discovered in the pursuit of the abnormal gene(s) or mechanisms that underlie breed specific DCM. These data provide potential treatment targets to arrest the progression of DCM or provide the opportunity for reverse remodeling.

1. Genetic

Doberman - Recent work by Dr. Meurs et al[2] established that DCM in the Doberman Pinscher is inherited as an autosomal dominant trait. Additionally she has demonstrated a gene deletion (PDK4, 16 base pairs) on chromosome 14 that codes for a mitochondrial protein that is involved with cardiac regulation of glucose metabolism. It appears that there are many Dobermans with DCM that do not demonstrate this abnormality, hence other genetic defects must be involved in some Dobermans with DCM. German and UK Dobermans[3] demonstrated a locus on chromosome 5 with a strong association for DCM. A specific candidate gene was not identified. This allelic association accounted for 50% of affected Dobermans hence the other 50% require a different genetic etiology some of which will probably be the PDK4 deletion.

Irish wolfhounds – A complex mode of inheritance of DCM in the Irish wolfhound has been demonstrated consisting of a dominant major gene model with a sex-dependent allele effect[4]. Studies investigating isolated candidate genes as causal for DCM in the Irish wolfhound have failed to find an association with DCM in this breed[5,6].

Great Danes – Oyama et al[7] demonstrated a downregulation of the triadin gene and an upregulation of the calstabin2 gene, both involved with regulation of the cardiac ryanodine receptor (RyR2), in 3 Great Dane dogs with end-stage CHF due to DCM. RyR2 is a Ca release channel, spans the sarcoplasmic reticulum (SR), and is responsible for Ca release from the SR into the cytosol. Opening of sarcolemmal L-type Ca channels trigger the release of SR Ca. This SR Ca binds to cardiac contractile proteins initiating systolic contraction. Abnormalities in RyR2 function produce reduced systolic Ca release with reduced contractility, and diastolic Ca leakage (Ca sparks) promoting cardiac arrhythmias. RyR2 function is closely regulated by a host of complementary molecules, including triadin, calstabin2, junctin, sorcin, calsequestrin, and phosphatases 1 and 2A. Thus regulation of RyR2 is critical to proper myocardial function. Triadin is a transmembrane protein closely bound to RyR2, and in conjunction with junctin, functions as a sensor of Ca concentration within the SR. Triadin affects RyR2 function by binding to calsequestrin. Binding of triadin to calsequestrin reduces the probability of channel opening; less bind to calsequestrin increase probability of channel opening. When SR Ca concentration is low triadin tends to bind to calsequestrin; when high triadin-calsequestrin binding is lessened. Downregulation of triadin promotes diastolic Ca leakage (promoting arrhythmias and impaired diastolic function) and diminished systolic Ca flux (reducing contractility).

Calstabin2 is another important component of the RyR2 regulatory complex. It binds to the RyR2 receptor and stabilizes it in the closed configuration during diastole. Downregulation of calstabin2 expression and protein is purported to promote diastolic Ca leakage and development of arrhythmias in arrhythmogenic right ventricular cardiomyopathy of Boxers. Oyama[7] observed an increased transcription of calstabin2 in Great Danes with DCM. There is support for the notion that increased calstabin2 reduces the propensity for arrhythmias. It is possible that the increased expression of calstabin2 is responsible for the reduced frequency of ventricular arrhythmias observed in Great Danes with DCM. These findings may only reflect the consequences of heart failure and not the initiating cause. The use of a control group with heart failure due to another disorder such as chronic mitral valve disease may have helped to sort this out.

German Shepherds – In German Shepherds with DCM SERCA1 was observed to be over expressed in ventricular and interventricular septal myocardial samples[8]. SERCA1 is mainly expressed in skeletal muscle, SERCA2 is expressed primarily in the heart and SERCA3 is expressed in a wide variety of tissues. Studies in dogs and people have demonstrated reduced expression of SERCA2a in DCM. SERCA1a can functionally substitute for SERCA2a. Therefore it is tempting to suggest that the increased expression of SERCA1 is a compensation for reduced expression of SERCA2a attempting to restore otherwise impaired excitation-contraction coupling. No other studies address the nature of the defect in German Shepherd DCM.

Newfoundlands – DCM is inherited in the NFLD as an autosomal dominant disorder with reduced penetrance[9]. Additional studies attempted to isolated candidate genes as causal for DCM in the Newfoundland dog failed to find an association with DCM in this breed[9]. Newfoundlands appear to synthesize less taurine compared with Beagles and this was associated with lower plasma taurine concentrations[10]. DCM was observed in several of these dogs that was reversible with taurine supplementation. Low sulfur amino acid dietary intake may be responsible for the low taurine synthesis observed in Newfoundland dogs.

Labrador retrievers – a juvenile form of DCM has been observed in a Labrador Retriever[11]. No further data is available to address the mechanism/cause.

Spaniels – American cocker spaniels with DCM are frequently characterized as taurine deficient and supplementation with either taurine or the combination of taurine and l-carnitine can markedly improve outcome[12]. It is uncertain whether these dogs required just taurine supplementation. In a review of all forms of canine DCM, Cockers supplemented with taurine had the best outcome[1]. This was likely due to the fact that DCM in this group of dogs is very responsive to taurine supplementation. Myocardial carnitine deficiency has been estimated to occur in 40% of dogs with DCM[12]. It appears this is likely a consequence of heart failure rather than causal. Colonies of English cocker spaniels have been studied suggesting a genetic predisposition[13].

Portuguese Water dogs – Juvenile DCM is described as a fully penetrant autosomal recessive inherited trait in the Portuguese water dog[14]. It was associated with a locus on chromosome 8[15]. The authors identified a previous unknown region of the canine and homologous human genome as an area responsible for DCM. There are 28 genes within the identified critical region on chromosome 8 for this juvenile DCM, some of which can be considered very good candidate genes based on known function and tissue expression patterns. Among the most intriguing candidate genes is Nesprin 2 (SYNE2), it codes for a giant protein that connects the nucleus and the actin cytoskeleton and has been shown to interact with ACTC1 (actin, , cardiac muscle 1), mutations in which have been associated with idiopathic DCM and familial hypertrophic cardiomyopathy[16]. Alterations in the regions of SYNE1 and SYNE2 were found in families segregating Emery–Dreifuss muscular dystrophy, some of these patients suffered from severe DCM[17].

2. Toxic

The anthracyclines, doxorubicin, daunorubicin, epirubicin, and idarubicin are key components of many chemotherapeutic regimens. This class of drugs represents the most common form of toxic cardiomyopathy observed in dogs. They can produce arrhythmias, left ventricular dilation and systolic dysfunction (DCM), and pericarditis. This toxicity is dose related. These anthracyclines cause myocardial necrosis with myofibrillar loss. Cellular damage appears to be due to oxidative stress with iron oxidation and production of free radicals. With chronic administration myocardial changes develop at a cumulative dose of doxorubicin of from 90 to 180 mg/m^2.[18] Epirubicin, a steroisomer of doxorubicin, is believed to be less cardiotoxic than doxorubicin at comparable doses[19].

Monitoring LV function is helpful to assess the risks of patients beginning and in treatment protocols, particularly as the accumulative dose rises. In people a normal troponin I level was highly predictive of patients that would not develop LV dysfunction. Additionally in people prophylactic antioxidant or angiotensin converting enzyme inhibitor therapy with elevated troponin I levels prevented progression to heart failure[19].

3. Infectious/inflammatory

Myocarditis is not well described in the dog. Myocarditis may result from viral (or other pathogen) infection, immune mediate disease, or both. Genetic factors may predispose to cardiac disease after a viral infection or to an immune mediated process. It has been difficult in people to establish a link between viral infections and the subsequent development of cardiomyopathy. Inflammatory myocarditis can occur in the absence of an infectious agent. Additionally viral myocardial disease could be the result of direct injury or immune mediated. The authors have observed several examples of young dogs with an echocardiographic phenotype of DCM that reverse remodeled over time with the help of supportive therapy. It is presumed a viral insult could have been involved in these cases and the insult was self-limiting. The role of immune mediated myocarditis in people has been suggested as causal for some cases of DCM. Autoantibodies to myosin heavy chain, muscarinic receptors, beta adrenoceptors, sarcolemmal sodium-potassium adenosine triphosphatase, laminin and mitochondrial proteins have been reported in people[20]. Few studies have investigated the role of autoimmunity in canine cardiomyopathy and none have been able to demonstrate an association.

Chagas myocarditis is a well recognized form of myocarditis in the dog. It is caused by the hemoflagellated protozoan *Trypanosoma cruzi*. This parasite is common in South and Central America. It causes arrhythmias and myocarditis. Three phases of myocarditis are observed: acute, intermediate, and chronic. The chronic form manifests as typical DCM and serology is usually required to identify the underlying parasite[21].

4. Metabolic

Hypothyroidism. Thyroid hormone (T3) inhibits the expression of phospholamban by increasing the phosphorylation of phospholamban. SERCA2a activity promotes the entry of cytosolic Ca into the SR. Phospholamban inhibits SERCA2a activity. Thyroid hormone thereby promotes SERCA2a activity. However diastolic function is also increased in the hypothyroid state[22]. Hypothyroidism is associated with an increase in systemic vascular resistance, reduction in cardiac output, reduced contractility, lower heart rate, reduced circulating blood volume and reduced energy efficiency (increased oxygen cost of work) due to the increased afterload[22].

A report of 2 Great Danes with DCM and hypothyroidism described the benefit of thyroid supplementation[23]. Thyroid supplementation in addition to conventional therapy resulted in reverse remodeling and improved systolic function.

5. Tachycardia induced

Both experimental and endogenous recurrent or persistent tachycardias can induce DCM in the dog. Control of the arrhythmia usually permits complete reverse remodeling with restoration of normal function. This etiology represents another of the potentially reversible forms of DCM. Left ventricular dysfunction was more marked experimentally with chronic ventricular tachycardia than in chronic supraventricular tachycardia[24].

6. Idiopathic

Idiopathic is a diagnosis of exclusion and accounts for the majority of cases of canine DCM.

Pathology

Histologic changes noted in dogs with DCM are generally considered modest and non-specific evidence of injury. These include scattered foci of myocardial degeneration, necrosis, and myocyte atropy. Modest fibrosis or fatty replacement is noted. Attenuated, wavy myofibers are commonly observed. The importance of wavy fibers to the progression of this disease remains unclear[25].

Pathophysiology

Oyama et al[26] examined the genomic expression patterns of cardiac tissue from Dobermans with DCM. They observed that 478 of 23,851 analyzed transcripts were differentially expressed in dogs with severe DCM. The authors summarized there findings to describe genes that served 8 different cellular functional categories such as; major energy production pathways [specifically glycolysis, the tricarboxylic acid cycle, and oxidative phosphorylation (this could be instrumental in the progressive loss of global systolic and diastolic function)]; cell signaling and communication pathways [downregulation of SERCA2a and beta adrenergic signaling (important in Ca cycling with Ca transients that are smaller and last longer promoting arrhythmias and systolic dysfunction) and upregulation of protein kinase C (important in pathologic remodeling) and adrenomedullin (a counter regulator to the RAAS)]; cell structure involved in the cytoskeleton [downregulation of titin and nebulin (this could alter systolic and diastolic function)] and extracellular matrix [upregulation of proinflammatory matrix components, activation of matrix metalloproteinases, abnormal collagen content, turnover and distribution and specifically osteopontin (all factors that promote remodeling of the diseased heart)]; cell defense function [upregulation of genes associated with cellular stress (metallothionein 1), upregulation of genes associated with the cathepsin protease family, complement cascade, and heat-shock proteins (indicate catabolic and inflammatory activity)]; shift to fetal genes [upregulation of skeletal muscle alpha actin (believed to represent a compensatory effort to increase contractility) and the Kickkopf-2 gene [regulates cardiac valve formation in embryos]; and downregulation of transcriptional proteins [fused in liposarcoma-like protein].

The specific genetic derangements that are now coming to light with breed specific DCM help us begin to understand the myocardial insult that initiates/promotes the morphologic and hemodynamic changes inherent in heart failure and the cascade of compensatory changes both functional and neurohumoral that accompany heart failure in the setting of DCM. Understanding these mechanisms introduces potential therapeutic targets to arrest disease progression.

And finally when considering the final common pathway hypothesis as a means to explain why a variety of myocardial injuries can eventually cause the phenotype of DCM, stem cell therapy has demonstrated that depletion or dysfunction of endogenous cells with capacity to divide and to differentiate into cardiac cellular constituents may be a central pathophysiologic contributor to the cardiomyopathic processes[27].

REFERENCES

1. Bronsoiler J, et al. abstract JVIM 2005;19:456. 2. Meurs K, et al. abstract JVIM 2010;24:693. 3. Mausberg TB, et al. PLOS One 2011;6:e20042. 4. Distl O, et al. Heredity 2007;99:460-5. 5. Philipp U, et al. Anim Genet 2008;39:88-9. 6. Philipp U, et al. Anim Biotech 2008;19:231-6. 7. Oyama M, et al. JVIM 2009;23:1014-9. 8. Summerfield N, et al. JVC 2010;12:17-23. 9. Wiersma AC, et al. J Heredity 2008;99:73-80. 10. Backus RC, et al. J Nut 2006;136:2525-33. 11. Schneider C, et al. Tierärztl Prax Ausg Kleintiere Heimtiere 2010;39:39-44. 12. Kittleson MD, et al. JVIM 1997;11:204-11. 13. Day MJ. Vet Immunol Immunopath 1996;53:207. 14. Sleeper MM, et al. JVIM 2002;16:52-62. 15. Werner P, et al. Genomics 2008;91:517-21. 16. Zhang Q, et al. J Cell Sci 2005;118:673-87. 17. Zhang Q, et al. Hum Mol Genet 2007;16:2816-33. 18. De Souza RCA, et al. AJVR 2006;67:1319-25. 19. Force T et al. In Braunwald's Heart Disease: A textbook of cardiovascular medicine. 9th ed'n 2012; p.1896. 20. Yu X, et al. J Am Soc Hyperten 2009;3:133-40. 21. Barr SC. VCNA Sm An 2009;39:1055-64. 22. Klein E. In Braunwald's Heart Disease: A textbook of cardiovascular medicine. 9th ed'n 2012; p.1833. 23. Phillips DE. et al. JAAHA 2003;39:133-7. 24. Zupan I et al. Int J Card 1996;56:75-81. 25. Tidholm A et al. JAVMA 1998;212:1732-4. 26. Oyama MA, et al. AJVR 2005;66:1140-55. 27. Hare JM. In Braunwald's Heart Disease: A textbook of cardiovascular medicine. 9th ed'n 2012; p.1561.

TREATMENT OF DILATED CARDIOMYOPATHY

M. Lynne O'Sullivan, DVM, DVSc, DACVIM (Cardiology); Michael R. O'Grady, DVM, MS, DACVIM (Cardiology)
Guelph, ON, Canada

Dilated cardiomyopathy (DCM) is a common cause of acquired cardiac disease in dogs, likely second only to degenerative mitral valve disease (MVD). It is defined by its morphologic and functional description, which includes dilation and systolic dysfunction of the left or both ventricles.[1] Etiologies described in the veterinary literature that may be responsible for this phenotype include genetic (breed specific mutations), toxic (e.g. doxorubicin-induced), infectious/inflammatory (e.g. Chagas disease), metabolic (e.g. hypothyroidism), nutritional (e.g. taurine deficiency), and tachycardia-induced (tachycardiomyopathy). [2] Idiopathic is the term used when no etiology is known or identified, and is often a diagnosis of exclusion. It stands to reason that the treatment of any case with an identified underlying cause would address reversing that underlying cause, in addition to addressing the hemodynamic alterations that result. For the purposes of this presentation, discussion will focus mostly on the genetic and idiopathic cases, with only brief discussion of etiology-specific therapy.

CLASSIFICATION OF STAGE OF DCM

It is important to consider both the stage of cardiac disease and the level of evidence for therapeutic efficacy when considering treatment. These are the overriding principles that guide the consensus recommendations of the American College of Cardiology Foundation (ACCF) and the American Heart Association (AHA), two authorities on human cardiac disease.[3] Levels of evidence for treatment efficacy may vary from randomized prospective clinical trials, to non-randomized trials, to case studies, to consensus opinion. While prospective randomized clinical trials are becoming more common in veterinary medicine, and wherever possible we base therapeutic recommendations on well-designed trials, the reality is that many of our therapeutic recommendations are still based on extrapolation from the human literature, small non-randomized studies, or opinion.

DCM is a progressive disease that changes along a continuum, and being as therapeutic goals may differ depending on the stage of disease, it is essential to consider this disease spectrum when making therapeutic choices. One classification system recently used in the ACVIM Consensus statement on the treatment of canine chronic MVD was a modification of a scheme used in the ACCF/AHA human guidelines.[4] This scheme attempts to link the severity of signs to appropriate treatments at each stage of disease.

Class	Description
Stage A	Patient at high risk for developing heart disease but that currently does not have evidence of structural disease
Stage B	Patient with structural heart disease but that has never had clinical signs of heart failure **B1**: Asymptomatic patient with no radiographic or echocardiographic evidence of cardiac enlargement or remodeling **B2**: Asymptomatic patient with radiographic or echocardiographic evidence of cardiac enlargement or remodeling
Stage C	Patient with past or current clinical signs of heart failure associated with structural heart disease
Stage D	Patients with advanced heart failure having clinical signs that are refractory to standard therapy

TREATMENT OF STAGE A

Patients in this class may include breeds at high risk of developing DCM (e.g. Doberman Pinschers, Irish Wolfhounds, Great Danes), Dobermans testing positive for the genetic mutations identified to date, [5, 6] or patients at risk of DCM due to an underlying associated condition such as anthracycline chemotherapy or primary tachyarrhythmia, but all of whom do not yet express the DCM phenotype. Therapies at this stage would therefore be aimed at preventing DCM. Apart from treating underlying predisposing conditions mentioned above, there are no therapies currently indicated to prevent the development of canine DCM. For patients with the potential for a genetic form of DCM, this is perhaps a stage where future development of gene therapy may be useful. [7] Breeding recommendations may also form part of the plan at this stage, including not breeding homozygous dogs, and breeding heterozygous dogs only to negative dogs for the case of Doberman DCM which has autosomal dominant inheritance. [8] A number of cardioprotective therapies are being investigated for the prevention of anthracycline-associated DCM in humans, including probucol (antioxidant effects), erythropoietin, statins, dietary nitrate supplementation, and dexrazoxane (iron chelator and antioxidant). Only the latter is approved for this use in humans in North America. [9] Dexrazoxane is reportedly cardioprotective in canine models of anthracycline cardiotoxicity, [10] and its safety has been demonstrated, [11] however its efficacy in the canine cancer patient has not been investigated.

TREATMENT OF STAGE B

The asymptomatic Stage B DCM patient is typically identified through abnormalities detected on physical exam (murmur, gallop, arrhythmia) or through active screening of high-risk populations. The B1 stage describes the asymptomatic patient that has no clinical evidence of cardiac enlargement. While this stage is commonly identified in MVD patients, it is unknown how likely it is to identify a DCM patient with detectable systolic dysfunction in the absence of cardiac dilation. Most commonly we identify stage B2 cases. In either case, therapies would be directed at slowing systolic dysfunction, preventing ventricular remodeling and disease progression, and delaying the onset of congestive heart failure (CHF) or sudden death (SD). The two classes of drugs currently known to achieve these goals in humans include angiotensin converting enzyme inhibitors (ACEi) and β-blockers (BB).

The role of the renin angiotensin aldosterone system (RAAS) in maladaptive remodeling and progression of systolic dysfunction is well known, through promotion of myocardial and vascular fibrosis, endothelial and baroreceptor dysfunction, sympathetic nervous system (SNS) activation, and myocyte apoptosis. [12] There is definitive evidence in the human literature for the benefit of ACEi in asymptomatic DCM, with a 29% risk reduction of heart failure or mortality with the use of enalapril. [13] Consequently ACEi are recommended in all human stage B DCM patients. [3] Only one study in the veterinary literature has examined the utility of ACEi in stage B canine DCM. This retrospective study in Dobermans similarly found that time to CHF or SD was significantly prolonged in the benazepril treated group, with a 55-64% reduction of risk on multivariate analysis compared to no treatment. [14] While the Doberman may not be representative of all cases of canine DCM, I advocate the use of ACEi in any stage B DCM patient on the basis of the collective evidence in the human and veterinary literature, unless there is a specific contraindication to this class of drug. Evaluation of renal function one week after initiation and periodically thereafter is recommended.

It is well established that progressive and long-term SNS activation characterizes many forms of heart disease and heart failure. Adrenergic overactivity leads to myocyte necrosis and apoptosis, fibrosis, decreased β-receptor responsiveness, RAAS activation, and arrhythmias. While there is ample evidence in the human literature that BB therapy reverses left ventricular (LV) remodeling, improves LV systolic function in the long term, improves patient symptoms, and prolongs life in the symptomatic patient (stage C), controlled clinical trial data is lacking for the stage B patient. [15] Nevertheless, BB use is often recommended in the asymptomatic human patient.[3] The only published prospective study of BB use in naturally-occurring canine DCM evaluated the effect of carvedilol on LV volume and function over a three month period. [16] While no effect was noted, this study was significantly limited by low patient numbers, low carvedilol dose, and short term follow-up. Furthermore, it is unclear how many dogs were in stage B vs stage C. Therefore it remains to be seen whether BBs have utility in altering disease progression or outcome in stage B canine DCM. Concerns with BB use typically include provocation of hypotension or even CHF. Current BB use in stage B is veterinary cardiologist dependent and based on opinion. While I do not routinely use BBs in stage B patients independent of treating ventricular arrhythmias (see below), I believe this is likely an underutilization of a potentially beneficial class of medication, particularly in stage B where patients are more likely to tolerate β-blockade and live long enough to benefit.

The stage B patient with ventricular arrhythmias is worthy of separate consideration. Who are the patients for consideration of antiarrhythmic therapy and whether antiarrhythmic therapy can decrease the risk of SD are the central questions. In humans with DCM, the risk of SD increases with declining systolic function and with a history of syncope.[17] Yet there is no clear evidence that any antiarrhythmic strategy is effective in preventing SD in stage B DCM, with the exception of the post-myocardial infarction patient with systolic dysfunction in whom BBs or implantable cardioverter-defibrillators (ICDs) reduce arrhythmic death. [3] There are no prospective studies addressing the efficacy of antiarrhythmic therapy in canine DCM. A retrospective case-control study in occult Dobermans suggested that antiarrhythmic drug therapy may have retarded SD compared to non-treated dogs, however this data must be interpreted with great caution as it is retrospective, includes small numbers, and involved a number of different antiarrhythmic drugs including combinations (tocainide, mexiletine, procainamide, quinidine, amiodarone, BBs). [18] I tend to consider antiarrhythmic therapy in any patient with significant systolic dysfunction demonstrating frequent ventricular premature contractions (VPCs > 1 per minute) or VPCs with higher complexity (couplets, triplets, or runs with instantaneous heart rate > 180 bpm) independent of systolic function. My therapy of choice tends to be sotalol, a class III antiarrhythmic with both β-blockade and K^+ channel blockade properties. Initiating low doses and titrating upwards is often advised due to the β-blockade. A second choice tends to be mexiletine, either alone or in combination with sotalol.

In terms of other potential therapies that have been evaluated in veterinary medicine, the multicenter PROTECT trial enrolled 76 Dobermans with stage B DCM. The dogs were prospectively randomized to receive pimobendan or placebo, and treatment was double-blinded. Preliminary analysis of the results suggests that the dogs treated with pimobendan had a significantly prolonged time to the onset of CHF or SD as compared with the placebo dogs. Additionally the dogs administered pimobendan had a significantly longer time to all cause mortality when compared with the placebo dogs. The final results of this trial are anticipated this year.

TREATMENT OF STAGE C

Therapeutic goals in the CHF patient include resolution of clinical signs and improvement in quality of life in addition to delay in disease progression and prolongation of survival. Strategies include intravascular volume reduction to resolve pulmonary edema or decrease cavitary effusions, improvement in hemodynamic performance through decreased systemic vascular resistance and improved contractility, neurohumoral modulation, arrhythmia management, and nutritional management.

The mainstay of treatment of edema and effusions remains diuretics, with furosemide continuing to be the most commonly used agent. As a loop diuretic, it possesses potent diuretic properties and may be administered orally or intravenously (as bolus or constant rate infusion), depending on the severity of congestion and whether the patient is being treated as an inpatient or outpatient. The goal is to eventually achieve the lowest effective dose that maintains ease of breathing and elimination of respiratory signs. While there are no clinical trials that specifically address the benefit of diuretics in CHF, it is generally accepted that they are a requirement of the stage C patient.

ACEi continue to be indicated for the RAAS modulating effects alluded to above, particularly in the setting of furosemide use which will heighten RAAS stimulation, and for their hemodynamic effects (mixed vasodilation and reduction in Na^+ and water retention). The results of veterinary clinical trials using ACEi in the treatment of canine CHF have been available for quite some time and as a result the use of ACEi is well entrenched as standard therapy. Both enalapril [19, 20] and benazepril [21] were shown to improve quality of life in DCM patients with CHF. The effects on survival, while suggestive of improvement, did not reach statistical significance. [21, 22] However, there is ample evidence in the human literature that ACEi are effective in reducing mortality.[3, 15]

Pimobendan is a calcium sensitizer and phosphodiesterase inhibitor with positive inotropic and balanced vasodilating properties. It is the third medication that would now be considered standard therapy in dogs. Three placebo-controlled trials have found significantly longer time to treatment failure or longer survival time in dogs treated with pimobendan in addition to diuretics,

ACEi, +/- digoxin. [23-25] While positive inotropes, as a drug class, typically raise concern of increased myocardial oxygen consumption (MVO_2) and potential provocation of arrhythmias, pimobendan appears to improve contractility without increasing MVO_2 due to its mechanisms of action and did not increase VPC frequency in the one study mentioned above that evaluated repeated Holter data in its subjects.[23]

Treatment of the stage C human DCM patient continues to focus heavily on neurohormonal blockade. In addition to ACEi, BBs, aldosterone (aldo) antagonists, and angiotensin receptor blockers (ARBs) are used. According to the most recent ACCF/AHA guidelines, BBs are indicated in all stable stage C patients, an aldo antagonist is indicated in all with moderate to severe HF and without pre-existing renal or electrolyte abnormalities, and ARBs are indicated in patients that are intolerant of ACEi.[3] As alluded to above, there is minimal published work in the veterinary literature with respect to BB use in naturally-occurring DCM. One study failed to find a benefit of carvedilol but was underpowered to do so, [16] while two retrospective studies have shown that BBs are generally well tolerated in dogs with CHF. [26, 27] This area awaits more much-needed research in dogs, given the profound positive effect of this class of drug in humans. It is worth noting that BBs should not be initiated in the unstable patient with active pulmonary edema, rather should be reserved until the patient is stabilized on initial CHF therapy.

The aldo antagonist spironolactone has been shown to significantly reduce morbidity and mortality in humans with advanced CHF and LV systolic dysfunction.[28] The only published study on the use of spironolactone in dogs that included a small number with naturally occurring DCM failed to find a difference in survival time between the treatment and placebo groups.[29] However this trial, like many small veterinary studies, was underpowered and may have used less than adequate doses. A randomized, placebo-controlled trial was undertaken in Dobermans with DCM at our institution, and data analysis is underway. Currently, I use spironolactone in the patient who is necessitating additional management despite use of triple therapy with furosemide, ACEi, and pimobendan. Spironolactone is licensed for use in dogs with CHF in Europe.

Management of the stage C patient with ventricular arrhythmias focuses first on optimal management of congestion and hemodynamics to reduce SNS stimulation and improve perfusion. Once this is accomplished, then a judgment call is necessary to determine if the patient needs additional antiarrhythmic aid. Patients could be managed as outlined above for the stage B patient. Atrial fibrillation represents another common arrhythmic disorder encountered in the stage C DCM patient. While cardioversion to sinus rhythm may be physiologically ideal, rate control with antiarrythmics is much more accessible and feasible. Diltiazem tends to be my first choice for rate control, however having said that, recent evidence suggests that the combination of diltiazem with digoxin is more efficacious than either drug alone.[30] Having personally instituted this combination more frequently since that publication, my clinical experience would corroborate these findings.

In terms of nutritional management, cardiac cachexia is identified commonly in DCM CHF patients. Through anorexia, increased energy requirements, and production of inflammatory cytokines, lean muscle mass is lost. Management involves providing a good quality diet with adequate calories and protein, and there is evidence that supplementation with n-3 fatty acids is beneficial. [31] While diets restricted in sodium are commonly advocated, the effects on quality of life or survival have not been assessed in dogs. [31]

TREATMENT OF STAGE D

Refractory CHF can be a very difficult condition to manage. Initial strategies include confirming owner compliance with respect to current medications, and identification of any complicating factors that may contribute to difficulty in managing CHF (poorly controlled or new arrhythmias, concurrent systemic diseases). Ultimately a change in medication is often needed, and may include strategies such as an increase in the frequency of furosemide administration, change in the route of furosemide administration (IV bolus or CRI for short period), addition of diuretics such as spironolactone or hydrochlorothiazide (which should be accompanied by a reduction in furosemide dose by half), stricter sodium restriction, increase in inodilator dose and/or ACEi dose, parenteral positive inotrope administration (dobutamine), or entertainment of the use of drugs not yet commonly used in dogs (e.g. nitrates, ARBs, ADH antagonists). A recent case series describing the use of torsemide (a more potent loop diuretic with a longer duration of action than furosemide) in patients with refractory CHF showed promising results. [32]

FUTURE THERAPIES

The treatment of DCM continues to be an evolving area both in human and veterinary medicine. Some therapies that may be particularly exciting for the future include stem cell therapy, gene therapy, and interventional therapies such as epicardially-placed support devices. Research in these areas is ongoing but may be closer to use than one might believe. [7, 33, 34]

REFERENCES

1. Hare JM. *Braunwald's Heart Disease* 2012:1561. 2. O'Grady MR, et al. *Vet Clin North Am Small Anim Pract* 2004;34:1187. 3. Hunt SA, et al. *J Am Coll Cardiol* 2009;53:e1. 4. Atkins C, et al. *J Vet Intern Med* 2009;23:1142. 5. Mausberg TB, et al. *PLoS One* 2011;6:e20042. 6. Meurs KM. *Proceedings of the ACVIM Forum*. Anaheim, CA: 2010. 7. Sleeper MM, et al. *Vet Clin North Am Small Anim Pract* 2010;40:717. 8. Meurs KM. *Vet Clin North Am Small Anim Pract* 2010;40:701. 9. Ludke AR, et al. *Can J Physiol Pharmacol* 2009;87:756. 10. Herman EH, et al. *Semin Oncol* 1998;25:15. 11. FitzPatrick WM, et al. *Vet Comp Oncol* 2010;8:273. 12. Mann DL. *Braunwald's Heart Disease* 2012:487. 13. SOLVD investigators. *N Engl J Med* 1992;327:685. 14. O'Grady MR, et al. *J Vet Intern Med* 2009;23:977. 15. Mann DL. *Braunwald's Heart Disease* 2012:543. 16. Oyama MA, et al. *J Vet Intern Med* 2007;21:1272. 17. Myerburg RJ, et al. *Braunwald's Heart Disease* 2012:845. 18. Calvert CA, et al. *J Am Anim Hosp Assoc* 2004;40:24. 19. COVE study group. *J Vet Intern Med* 1995;9:243. 20. IMPROVE study group. *J Vet Intern Med* 1995;9:234. 21. BENCH study group. *J Vet Cardiol* 1999;1:7. 22. Ettinger SJ, et al. *J Am Vet Med Assoc* 1998;213:1573. 23. O'Grady MR, et al. *J Vet Intern Med* 2008;22:897. 24. Fuentes V, et al. *J Vet Intern Med* 2002;16:255. 25. Lombard CW, et al. *World Small Animal Association Congress*. Prague:2006:23. 26. Tidholm A. *J Vet Cardiol* 2006;8:41. 27. Rush JE, et al. *J Vet Cardiol* 2002;4:23. 28. Pitt B, et al. *N Engl J Med* 1999;341:709. 29. Schuller S, et al. *J Vet Pharmacol Ther* 2011;34:322. 30. Gelzer AR, et al. *J Vet Intern Med* 2009;23:499. 31. Freeman LM, et al. *Curr Heart Fail Rep* 2007;4:84. 32. Oyama MA, et al. *J Vet Cardiol* 2011;13:287. 33. Dimmeler S, et al. *Braunwald's Heart Disease* 2012:627. 34. Estrada A. *Proceedings of the ACVIM Forum*. Montreal, QC:2009.

217a

CERVICAL DISC ARTHROPLASTY IN 12 DOGS AFFECTED BY DISC ASSOCIATED WOBBLER SYNDROME. PRELIMINARY RESULTS. PF Adamo[1], RC da Costa[2], C Giovannella[3], M Podell[4], R Kroll[5]. [1]East Bay Area Veterinary Specialists, Walnut Creek, CA. [2]Department of Veterinary Clinical Sciences, The Ohio State University, Columbus, OH. [3]Gulf Coast Veterinary Neurology & Neurosurgery, Houston, TX. [4]Chicago Veterinary Specialty Group, Chicago, IL. [5]VCA Northwest Veterinary Specialists, Clackamas, OR.

The objective of this study was to evaluate the immediate post-operative recovery, and the short- and intermediate term follow-up of dogs with disc associated wobbler syndrome (DAWS) treated with cervical disc arthroplasty.

Twelve dogs with over 2 months history of DAWS diagnosed by MRI were evaluated. 8 dogs were treated for a single lesion and 4 dogs for a double lesion.

The cervical prosthesis and the technique were similar to that described in the preliminary clinical study (JAVMA, 239(6), 2011). However, additional tools were used to facilitate implantation and the external surface of the prosthesis was treated to promote bone-implant-incorporation. All dogs were evaluated neurologically shortly after surgery and with immediate and serial post-operative radiographs.

All dogs had immediate post-operative recovery with good degree of distraction in the immediate post-operative radiographs. In the majority of dogs, the distraction was moderately lost over time; this except for one dog was not clinically relevant. In this one dog, neurological signs recurred with tapering of NSAID after surgery. In all dogs, the mobility at the treated sites decreased over time, and in a few dogs became not detectable; except for the one dog, this did not appear to be clinically significant. Median clinical follow-up was 15 months (2-40 months). All dogs (except one) showed improvement in neurological status during the observation period.

Cervical disc arthroplasty is a well-tolerated surgical technique and it might be a valuable method to treat DAWS in dogs. Short- and intermediate-term results are promising. Long-term follow-up studies are underway.

THE ROLE OF PLATELETS IN EQUINE INFLAMMATION AND IMMUNITY

Bettina Dunkel, DVM, PhD, DACVIM, DECEIM, DACVECC, MRCVS
North Mymms, UK

The role of platelets was long thought to be limited to their hemostatic functions, namely control of bleeding and wound healing. With expanding knowledge of the inflammatory cascade the close interactions between inflammation and coagulation have become increasingly evident. In recent years, platelets have been recognized or suspected to play a fundamental part in multiple acute and chronic inflammatory processes and diseases in man and other species, such as myocardial infarction, atherosclerosis, stroke, asthma, adult respiratory distress syndrome (ARDS) and cystic fibrosis[1-7]. The role of platelets in equine diseases is also increasingly recognized[8,9]. Platelets contribute to inflammatory conditions directly by release and synthesis of pro-inflammatory mediators and indirectly via interaction with leukocytes and endothelial cells[10].

PLATELET PHYSIOLOGY:

Platelets or thrombocytes are anucleate blood cells derived from polyploidy precursors, megacaryocytes, by budding off from the boarders of the large precursors in the bone marrow[11]. In other species, platelet generation has also been reported in the peripheral blood, particularly in the pulmonary circulation[14]. Platelet numbers in horses vary between 100-200 x 10^9/L (100-200 x 10^3/μL) with a size of 5.1 fL and a protein content of 99 mg/10^{11} platelets[12]. Similar values have been reported for foals up to one year of age[13]. The main platelet storage site in humans is the pulmonary vasculature; this might also be true for horses[15]. Platelets from most species contain an invaginated, tubular membranous structure called the surface-connected open canalicular system (OCS), which serves as a site for granule fusion and as a reservoir of membrane for platelet spreading[16] The OCS is absent in equine platelets[16,17]. Based on comparison of platelet aggregation, it has been suggested that the absence of an OCS might delay platelet responses to stimuli[16]. Equine platelets contain dense bodies, which are similar in appearance and number to human dense bodies, alpha granules, which are slightly larger compared to their human counterparts[17] and lysosomes. Although by far not all contents of the alpha granules in equine platelets have been determined, it is assumed that they contain similar mediators as found in other species, including CD62P (P-selectin), an adhesion molecule and marker of platelet activation, mitogenic factors (platelet derived growth factor (PDGF), transforming growth factor-ß (TGF), vascular endothelial growth factor and insulin-like growth factor have been demonstrated in equine platelets[18]) and an array of pro-inflammatory enzymes and cytokines. Mitogenic factors play an important part in tissue healing and revascularization of the injured tissue; these properties are currently intensely explored for treatment of various orthopedic injuries in horses and man[6,19-21]. Dense granules contain high levels of adenine and guanine nucleotides, divalent cations and serotonin (5-HT), while lysosomes contain acid proteases, glycohydrolases and matrix metalloproteinases (MMPs) that exhibit potent cytotoxic and proteolytic activities[2,4,10]. Alpha granules can fuse with one another and with the external surface upon platelet activation which releases their content into the external milieu[17]. Evidence of differential release of presumably alpha granule content, previously described in other species, has recently been demonstrated in equine platelets: While growth factors are rapidly and amlost completely released upon coagulation, CCL5, an inflammatory mediator, is released independent of coagulation in a more protracted fashion[18]. Similarly, release of 5-HT from dense granules has been shown to occur independent of equine platelet aggregation[22].

Due to the absence of a nucleus, platelets have long been thought incapable of synthesizing new proteins[23]. It is now known that mature human platelets carry mRNA originating from the parent megacaryocyte that can be rapidly translated in response to activation in a process termed signal-dependent translation. This is used for example for the production and release of interleukin 1ß (IL-1ß)[10,24]. There is some, although indirect, evidence that equine platelets may also be capable of *de novo* synthesis[18,25].

Human platelets posses a cytoskeletal framework allowing active cell locomotion and have been observed to role on activated endothelium *in vivo*, similar to leukocytes. Release of MMPs facilitates degradation of extacellular matrix and allows transmigration through vascular endothelial cells and basement membranes[4]. Gradient-driven chemotaxis in response to bacterial peptides[26] and allergenic stimuli in human asthma patients[4] has also been demonstrated. To the author's knowledge these capacities have not yet been investigated in equine platelets.

ROLE OF PLATELETS IN INFALMMATION AND INFECTION

Besides their well described functions as hemostatic cells, platelets initiate and perpetuate inflammation at sites of vascular injury[27]. Platelets remain functional in the thrombotic milieu for an extended period of time and continue to express pro-inflammatory cytokines while encast in the fibrin network for at least 18h after clot formation[10]. Expression of CD62P upon stimulation mediates adherence to leukocytes and endothelial cells and enhances expression of tissue factor on monocytes[28], one of the most important activators of the coagulation cascade. As mentioned above platelet granules contain an array of enzymes and cytokines, that allow modification of the inflammatory and coagulation system while platelet lysosomal contents have strong cytotoxic and proteolytic properties[2,4,10]. Platelet aggregates have been observed to accumulate in inflammatory lesions either concomitantly with[6] or preceding leukocytes[4] and platelets are believed to be important, if not essential, for leukocyte recruitment into an inflamed area. They are suspected to significantly contribute to inflammatory organ injury[29]. Modulation of dendritic cell activation, initiation of IgG production by B cells and enhancement of T cell responses by platelets have been described[30]. Based on observations in animal models, platelets may modulate T-cell mediated immune responses including adaptive immune responses to infection, delayed-type hypersensitivity and contact sensitivity[24].

There is evidence that platelets not only augment the antimicrobial functions of leukocytes, but are able to inhibit or kill a variety of bacterial species *in vitro* and *in vivo*[31]. Antimicrobial activity against *E.coli*, *Staphylococcus aureus*, *Candida albicans* and others has been demonstrated[32,33] and interaction with viruses, fungi and protozoal parasites has been described. Being among the most rapid and abundant inflammatory cells accumulating at sites of endothelial damage or microbial colonialization platelets contribute to the clearance of microorganisms by binding, aggregating and internalizing invaders in phagosome-like vacuoles and by release of antimicrobial peptides[31]. Platelets share structural and functional features with phagocytic cells which include receptors for IgE, thrombospondin, complement C3 and C-reactive protein and expression of toll-like receptors 1, 2, 4, 6 and 9[34,35]. Platelets are capable of releasing cytotoxic oxygen metabolites including superoxide anions, hydroxyl radicals, hydrogen peroxide, cationic peptides and other potent antimicrobial peptides[31,32]. LPS amplifies the interaction between platelets, IgG, microorganisms and immune complexes and may contribute to the development of thrombocytopenia accompanying endotoxemia and gram negative infections. Rapid platelet sequestration into lungs and liver has been demonstrated in LPS-induced and anaphylactic shock in mice and platelet degranulation in these organs is believed to contribute to the systemic signs observed[36]. Furthermore, the severity of organ injury and inflammation following hemorrhagic or LPS-induced shock is closely associated with the presence of platelets and can, under experimental conditions, be significantly ameliorated or prevented by induction of thrombocytopenia or inhibition of platelet CD62P[29,37,38].

Equine platelets up-regulate CD62P expression in response to LPS and lipoteichoic acid (LTA) but formation of platelet-neutrophil aggregates was not observed (Dunkel and Cunningham, unpublished data). Furthermore, equine platelets are *in vitro* 1000 times more sensitive to LPS in terms of thromboxane A_2 (TXA_2) production and p38 MAP kinase phosphorylation than equine leukocytes[39]. Activated equine platelets appear to be able to inhibit bacterial growth *in vitro*; however, effects were less pronounced compared to reports from other species and similar effects were achieved with plasma alone. Although equine platelets produce superoxide when stimulated with strong stimuli, this was not observed following exposure to whole killed bacteria or bacterial products (Aktan and Cunningham, unpublished data). Equine platelets also contain functional myeloperoxidase (MPO) but its role in antimicrobial defense has not been established (Dunkel and Cunningham, unpublished data). Other effects of LPS on equine platelets, such as aggregation and 5-HT release, are dependent on the presence of leukocytes. Both processes are mediated by platelet activating factor (PAF), released from activated leukocytes, but not cyclooxygenase products and can be inhibited by PAF-inhibitors but not aspirin[22,40].

PLATELETS IN HUMAN DISEASES:

Platelets are implicated in contributing to the development of cardiovascular diseases, namely stroke, artherosclerosis and septic shock. There is growing evidence, although mainly derived from animal models, that platelets play a role in the initiation and progression of artherosclerotic plaques[30]. Platelets adhere to intact, but activated endothelial cells prior to the development of manifest artherosclerotic lesions before leukocytes invade. There they become activated and release inflammatory and chemotactic mediators, leading to an altered endothelial phenotype and leukocyte recruitment which ultimately favors plaque formation[1].

In many respiratory diseases, such as asthma, cystic fibrosis and ARDS, platelets are recognized to be an integral part of the inflammatory pathogenesis of the diseases. Circulating platelet-leukocyte aggregates have been observed in human allergic asthma patients during the late asthmatic response and the presence of autologous platelets augmented the adhesion of human leukocytes from asthmatic patients to endothelial cells *in vitro*. Depletion of platelets in a mouse model of allergic airway disease prevented leukocyte migration into the airways, but airway leukocytosis could be reestablished by reinfusion of the platelets[6]. As mentioned, platelets from asthmatic human patients and sensitized mice have the ability to migrate *in vitro* towards the specific allergen the patient is sensitized to[4,6]. Histology of tissue obtained from sensitized mice revealed platelets migrating out of blood vessels and localizing underneath the airway walls[4].

Circulating platelets of inflammatory bowel disease patients are activated and are thought to trigger an inflammatory response in the microcirculation of these patients. In ulcerative colitis, activated platelets aggregate in affected colonic tissues in close association with neutrophils and may contribute to and amplify colonic inflammation[41].

The severity of organ injury and inflammation following hemorrhagic or LPS-induced shock, ischemia and reperfusion is closely associated to the presence of platelets and can, under experimental conditions, be significantly ameliorated or prevented by induction of thrombocytopenia or inhibition of platelet CD62P[29,37,38] indicating the significance of platelets in the pathogenesis of these conditions. Without doubt platelet involvement will be demonstrated for many if not most inflammatory diseases in the future.

PLATELETS IN EQUINE DISEASES:

Gastrointestinal diseases: Coagulation disorders, including subclinical and clinical disseminated intravascular coagulation (DIC), are commonly observed in horses with gastrointestinal disease and thrombocytopenia is occasionally observed[42-44]. However, platelet biology has not been extensively investigated in equine gastrointestinal disease. Indirect evidence for platelet activation and degranulation has been identified in horses with strangulating intestinal lesions where increased concentrations of 5-HT were identified in the circulation and, to a lesser degree, peritoneal fluid. The authors speculated that this increase was due to systemic 5-HT release from platelets; a β-TG/PF4 ratio >2 provided further evidence for systemic platelet activation[45].

Laminitis: Hyperaggreability, increased numbers of circulating platelet-neutrophil aggregates and decreased platelet survival have been identified in ponies with induced alimentary laminitis[46,47]. Inhibition of platelet aggregation prevented carbohydrate-induced laminitis in one study[48]. Digital hypoperfusion following LPS administration was temporally associated with systemic increases in TXA_2 and 5-HT[49]; both mediators, but not PAF, appeared to play a role in LPS-induced digital hypoperfusion in horses[50].

Airway diseases: Five hours following initiation of an antigen challenge, platelets from horses with RAO show decreased aggregation in response to PAF compared to healthy controls, potentially suggesting local *in vivo* exposure to PAF and desensitization of platelets[51]. Further indirect evidence for platelet activation in RAO was provided by demonstration of increased levels of TXA_2 in horses with active RAO[52]. Direct platelet activation was demonstrated by an increased adhesion of platelets to serum coated plastic in horses with RAO, 24h after initiation of a natural antigen challenge, but not in healthy controls[8]. Furthermore, differences in phosphodiesterase isoenzyme inhibition were documented[53].

Miscellaneous: Increased numbers of platelet-neutrophil aggregates were also detected in situations of increased shear stress[54] and in horses undergoing near maximal treadmill exercise[55]. PDGF BB stimulates proliferation of equine corneal epithelial cells and keratocytes *in vitro*[56] and treatment of wounds in the lower limb region of horses with platelet rich plasma resulted in accelerated epithelial differentiation[57] indicating that platelets play an active part in equine cell proliferation and differentiation. Epithelial injury leading to platelet activation has been suggested to participate in clinical and experimentally induced glomerulonephritis. PDFG stimulates DNA synthesis in equine glomerular mesangial cells and the authors suggested a role for platelets in equine glomerular disease[58]. No difference in platelet aggregation was detected in ponies with and without *Culicoides* hypersensitivity[59], in contrast to findings in humans with atopic dermatitis[60].

ASSESSMENT OF PLATELET NUMBERS AND FUNCTION:

Platelet numbers: Thrombocytosis is rare, occurring in approximately 1% of horses presenting to a referral hospital and has been associated with infectious or inflammatory conditions[61]. *Rhodococcus equi* infection in weanlings has also been linked to severe thrombocytosis[62] but to the author's knowledge this has not been confirmed in larger studies. Pseudothrombocytopaenia, caused by clumping of platelets is commonly observed in horses, particularly in ethylenediaminetetraacetic acid (EDTA)-anticoagulated blood[63]. It is easily recognized by scanning blood smears for platelet clumps or performing a platelet count using citrate- or heparin-anticoagulated samples. Although most clumping is due to *in vitro* platelet aggregation, occasionally true pseudothrombocytopenia due to the presence of EDTA-dependent platelet agglutinating antibodies might be observed[63,64]. Thrombocytopenia can be associated with systemic diseases and coagulopathies, immune-mediated destruction of platelets or insufficient production and has been linked to a worse prognosis[65].

Platelet aggregation: Measuring platelet aggregation is a frequently employed method to evaluate platelet function and can be determined by performing platelet counts before and after stirring of samples using a magnetic stir bar and calculation of the percentage decrease[66] or by use of aggregometers[51,67]. In comparison to other species, equine platelets are more sensitive to the aggregating effect of platelet-activating factor (PAF)[66].

Flow cytometry: Segura *et al.* (2006) demonstrated cross reactivity of monoclonal anti-human antibodies for the platelet membrane glycoproteins (GP) IIb-IIIa (CD42), GPIV, CD62P and lysosomal integral membrane protein (LIMP) but not GPIbα[17]. Binding of annexin V and fibrinogen to equine platelets has also been measured[17,68]. Platelet-leukocyte aggregates have been detected in equine blood using microscopic and flow cytometric detection systems[46,54,55]. *In vitro* experiments suggest that concurrent platelet and neutrophil activation is necessary for aggregate formation to occur[9] and that attachment is CD62P mediated[54].

Measurement of platelet adhesion under static and dynamic conditions: Methods to detect platelet adhesion to coated microtiter plates under static conditions and to subendothelium under flow conditions have been described[8,17]; equine platelets are reported to adhere more readily to subendothelial collagen compared to their human counterparts.

Measurement of platelet release products: Several platelet release products can be measured in plasma and other biological fluid samples as an indication of platelet degranulation. Growth factors, 5-HT and TXA_2 are among the most frequently used markers[18,45,49]. Due to the short half-life of PF4, a β-TG/PF4 ratio >2 has been used to distinguish between *in vivo* from accidental *in vitro* platelet activation[45].

REFERENCES

[1]Gawaz M. **Blood Cells Mol Dis** 2006;36(2):206. [2]Gawaz M. **Cardiovasc Res** 2004;61(3):498. [3]O'Sullivan BP, et al. **Am J Respir Crit Care Med** 2006;173(5):483[4]Pitchford SC, et al. **Am J Respir Crit Care Med** 2008;177:604[5]Pitchford SC, et al. **Blood** 2004;103(2):639. [6]Pitchford SC, et al. **J Allergy Clin Immunol** 2003;112(1)109.[7]Gries A, et al. **Crit Care Med** 2003;31(6):1697.[8]Dunkel B, et al. **Equine Vet J** 2007;39(6):557.[9]Dunkel B, et al. **Vet Immunol Immunopathol**, 2009;131(1-2):25.[10]Weyrich AS, et al. **J Thromb Haemost** 2003;1(9):1897.[11]Tschudi P, et al. **Equine Vet J** 1975;7(3):141.[12]Meyers KM. **Adv Vet Sci Comp Med** 1985;30:131.[13]Harvey JW, et al. **Equine Vet J** 1984;16(4):347.[14]Lunetta P, et al. **Eur J Haematol** 1997;59(1):63.[15]Lepherd EE. **Vet Rec** 1977;101(24):488.[16]Choi W, et al. **Platelets** 2010;21(3):167.[17]Segura D, et al. **J Vet Intern Med** 2006;20(3):581.[8]Dunkel B, et al. **Equine Vet J** 2011; Epub ahead of print.[19]Fortier LA, et al. **Clin Orthop Relat Res** 2011; Epub ahead of print.[20]Fortier LA, et al. **Vet Clin North Am Equine Pract** 2008;24(1):191.[21]Huo Y, et al. **Trends Cardiovasc Med** 2004;14(1):18.[22]Bailey SR, et al. **Equine Vet J** 2000;32(6):497. [23]Evans RJ. **Equine Vet J Suppl** 1992;13:31.[23]Weyrich AS, et al. **Trends Immunol** 2004;25(9):489.[25]McCarrel T, et al. **J Orthop Res** 2009;27(8):1033.[26]Czapiga M, et al. **Exp Hematol** 2005;33(1):73.[27]Yaguchi A, et al. **J Thromb Haemost** 2004;2(12):2096.[28]Levi M. **Crit Care Med** 2005;33(12 Suppl.):S523.[29]Tweardy DJ, et al. **Shock** 2006;26(4):386.[30]Elzey BD, et al. **Cell Immunol** 2005;238(1):1.[31]Yeaman MR. **Clin Infect Dis** 1997;25(5):951.[32]Tang YQ, et al. **Infect Immun** 2002;70(12):6524.[33]Krijgsveld J, et al. **J Biol Chem** 2000;275(27):20374.[34]Shiraki R, et al. **Thromb Res** 2004;113(6):379.[35]Cognasse F, et al. **Immunol Cell Biol** 2005;83(2):196.[36]Yoshida A, et al. **Br J Pharmacol** 2002;137(2):146.[37]Yu Z, et al. **Int Immunopharmacol** 2005;5(3):571.[38]Khandoga A, et al. **Hepatology** 2006;43(2):306.[39]Brooks AC, et al. **Inflamm Res** 2007;56(4):154.[40]Jarvis GE, et al. **Res Vet Sci** 1994;57(3):317.[41]Kayo S, et al. **Inflamm Bowel Dis** 2006;12(8):727.[42]Dolente BA, et al. **J Am Vet Med Assoc** 2002;220(7):1034.[43]Cotovio M, et al. **J Vet Intern Med** 2007;21(2):308.[44]Monreal L, et al. **Equine Vet J Suppl** 2000;32:19.[45]Delesalle C, et al. **Equine Vet J** 2008;40(4):326.[46]Weiss DJ, et al. **Am J Vet Res** 1997;58(12):1376.[47]Weiss DJ et al. **Am J Vet Res** 1995;56(8):986.[48]Weiss DJ, et al. **Am J Vet Res** 1998;59(7):814.[49]Menzies-Gow NJ, et al. **Equine Vet J** 2004;36(3):273.[50]Menzies-Gow NJ, et al. **Am J Vet Res** 2008;69(2):199.[51]Ablett JM, et al. **Equine Vet J** 1997;29(5):382.[52]Gray PR, et al. **Am Rev Respir Dis** 1989;140(1):154.[53]Dunkel B, et al. **Am J Vet Res** 2010;71(5):534.[54]Weiss DJ, et al. **Am J Vet Res** 1998;59(10):1243.[55]Weiss DJ, et al. **Am J Vet Res** 1998;59(4):393.[56]Haber M, et al. **Vet Ophthalmol** 2003;6(3):211.[57]Carter CA, et al. **Exp Mol Pathol** 2003. 74(3):244.[58]Ennulat D, et al. **Am J Vet Res** 199; 58(11):1308.[59]Bailey SR, et al. **Inflamm Res** 2000;49(5):224.[60]Neuber K, et al. **Int Arch Allergy Immunol** 1992;98(4):339.[61]Sellon DC, et al. **J Vet Intern Med** 1997;11(1):24.[62]Leadon D, et al. **Vet Rec** 1988;123(10):279.[63]Hinchcliff, KW, et al. **J Am Vet Med Assoc** 1993;203(12):1715.[64]Casonato A, et al. **J Clin Pathol** 1994;47(7):625.[65]Sellon DC,. **J Vet Intern Med** 1996;10(3):127.[66]Weiss, DJ, et al. **Am J Vet Res** 1997;58(8):823.[67]Foster AP, et al. **Equine Vet J** 1992;24(3):203. [68]Kingston JK, et al. **Equine Vet J Suppl** 2002;34:502.

CURRENT RECOMMENDATIONS REGARDING THE USE OF STEROIDS IN SEPTIC FOALS

Kelsey A. Hart, DVM, PhD, DACVIM (LAIM)

INTRODUCTION

The hypothalamic-pituitary-adrenal (HPA) axis regulates circulating concentrations of the corticosteroid hormone cortisol, which is vital for the stress response to critical illness in both people and animals. Despite the importance of HPA activity in the endocrine response to illness, HPA axis dysfunction resulting in cortisol insufficiency occurs frequently in humans and animals with critical illness such as sepsis and septic shock. This syndrome, coined relative adrenal insufficiency or critical illness-related corticosteroid insufficiency (CIRCI), is correlated with both increased disease severity and increased mortality in humans and animals. In people, cortisol replacement therapy with low, physiologically-relevant doses of hydrocortisone appears to improve shock reversal and survival in septic patients with CIRCI. The purposes of this review are to describe current understanding of the incidence and pathogenesis of CIRCI in septic foals and discuss current recommendations for cortisol replacement therapy in septic foals with CIRCI.

CRITICAL ILLNESS-RELATED CORTICOSTEROID INSUFFICIENCY (CIRCI)

Definition

During critical illness, the HPA axis is activated when stress signals from the periphery are integrated in the hypothalamus and induce the release of corticotropin-releasing hormone (CRH). CRH stimulates the adjacent anterior pituitary gland to secrete adrenocorticotropic hormone (ACTH) into the systemic circulation, which then acts on the adrenal cortices to induce the synthesis and secretion of the glucocorticoid hormone cortisol. Cortisol has diverse systemic effects throughout the body; during critical illness, the sum of these effects serve to help maintain blood pressure, coordinate energy metabolism, and regulate the immune and inflammatory response.

However, a number of studies suggest that up to 40-60% of people, dogs, and horses with sepsis and septic shock suffer from transient HPA axis dysfunction (CIRCI).[1-4] In these patients, the cortisol response to illness is inadequate for the existing degree of physiologic stress. Circulating cortisol concentrations may be increased from basal levels found in healthy unstressed individuals, but are still not adequate to meet demands to support and regulate cardiovascular, metabolic and inflammatory responses.[3,5] In some patients with CIRCI, cortisol concentrations may be appropriately increased, but peripheral glucocorticoid resistance results in impaired cortisol activity in target tissues.[3,5] Clinical manifestations of CIRCI are consistent with glucocorticoid deficiency, including persistent hypotension in the face of adequate volume resuscitation and vasopressor therapy, metabolic derangements such as hypoglycemia, and a dysregulated or overzealous immune and inflammatory response.[3,5] Signs associated with specific mineralocorticoid deficiency such as hyponatremia and hyperkalemia are not typical. Inadequate glucocorticoid concentrations or activity in patients with CIRCI limit appropriate stress responses and result in unregulated inflammation and poor tissue perfusion, ultimately damaging tissues and organs and leading to increased disease severity, organ failure, and death in many instances.[6,7]

Pathogenesis

The specific pathogenesis of CIRCI is poorly understood and is likely multifactorial. Pathophysiologic manifestations of cortisol deficiency can result from dysfunction at one or more levels of the HPA axis, including: (1) inadequate regulatory hormone (CRH, ACTH) secretion; (2) impaired adrenal sensitivity to ACTH due to receptor antagonism or down-regulation, (3) impaired adrenal cortisol synthetic capacity due to loss of substrate or enzymatic failure ("adrenal exhaustion" or "loss of adrenal reserve"); or (4) impaired cortisol activity in the target tissues due to decreased glucocorticoid receptor (GR) expression or activity or intracellular conversion of cortisol to cortisone, which cannot bind the GR.[3,5]

This HPA axis dysfunction may be a result of direct injury to one or several components of the HPA axis caused by hemorrhage or ischemia associated with the primary illness.[3,5] In addition, the HPA axis can be suppressed by specific drugs (e.g. etomidate),[8] bacterial components such as endotoxin,[9] or the patient's own immune response.[5,10] Sepsis-induced inflammatory cytokines can stimulate and/or suppress the HPA axis at several levels.[5,11] For example, tumor necrosis factor-α (TNF-α), which is increased in people and animals with sepsis, impairs both ACTH secretion from the pituitary gland and ACTH activity at the level of the adrenal gland.[5,10,11] In addition, increased leukocyte gene expression and/or increased serum concentration of interleukin-6 (IL-6), another pro-inflammatory cytokine, is correlated with both HPA axis dysfunction and a poor prognosis for survival in septic human neonates and in septic foals.[12-15] Impaired GR binding affinity has been documented in an ovine model of acute lung injury and in a rodent burn model, and this effect was partially ameliorated in the rodent model when the pro-inflammatory cytokines TNF-α and IL-1β were reduced by neutralizing antibodies.[16,17] Finally, in an *ex vivo* study in which leukocytes from healthy people were exposed to plasma from patients with acute respiratory distress syndrome, nuclear localization of cortisol-GR complexes was impaired in cells exposed to plasma from non-survivors as compared to plasma from survivors.[18]

LOW-DOSE HYDROCORTISONE THERAPY FOR THE MANAGEMENT OF CIRCI IN PEOPLE

One key aspect of CIRCI is that the HPA axis dysfunction is transient, in contrast to absolute adrenal insufficiency or Addison's Disease, which are characterized by permanent HPA axis dysfunction and cortisol insufficiency.[3] Thus, patients with CIRCI that recover from their primary disease regain normal HPA axis function and don't require lifelong steroid supplementation. However, because of the important immunomodulatory and anti-inflammatory effects of cortisol, HPA axis dysfunction during critical illness such as sepsis can ultimately result in immunoendocrine dysregulation, allowing an initially appropriate inflammatory response to become dangerously exaggerated.[10,11] Dysregulation of the immune/inflammatory response due to HPA axis failure may impair the patient's ability to combat infection and trigger overwhelming inflammation that further damages host tissues and organs.[6] This further perpetuates septic shock and SIRS, often culminating in the death of the patient.[6] Thus, appropriate management of CIRCI in conjunction with treatment of the primary illness is critical.

Several human studies,[19-22] including 3 meta-analyses,[23-25] have demonstrated improved outcomes in septic patients with CIRCI when they are provided with cortisol replacement in the form of low-dose hydrocortisone supplementation. The general premise for hydrocortisone therapy during critical illness is to provide synthetic cortisol in the form of hydrocortisone in an amount comparable to an appropriate cortisol response in a patient with intact HPA axis function.[3,26] It is important to note that this is *low-dose* corticosteroid therapy based on physiologically appropriate cortisol concentrations, not the *high-dose* corticosteroid therapy previously and convincingly shown to be detrimental in septic patients.[23-25]

However, it is difficult to predict the specific physiologic cortisol response to critical illness that would be ideal for individual patients. Hydrocortisone dosage recommendations used in human intensive care units thus vary widely, although current recommendations for hydrocortisone replacement regimens for septic adults consist of a 50-mg IV bolus every 6 hours (approx. 2.5 to 3 mg/kg/d) or an 100-mg loading dose followed by a continuous rate infusion of 10 mg/h IV (approx. 4 to 6 mg/kg/d)[26]. Hydrocortisone replacement protocols for use in neonatal and pediatric septic shock are less well defined; current recommendations range from 2 to 50 mg/kg/d as a continuous rate infusion or intermittent bolus dosing.[27,28] Such dosing strategies are based on the maximal cortisol secretory rate in healthy people in response to exogenous HPA-axis stimulation with cosyntropin[29] or calculated by multiplying the daily endogenous cortisol production rate in healthy, unstressed individuals by a variable "illness factor" to take into account appropriate increases in cortisol in response to the stress of severe illness[30]. To date, strategies for cortisol replacement therapy in animal species have only been reportedly anecdotally and have not been evaluated critically to date.

EVIDENCE FOR CIRCI IN SEPTIC NEONATAL FOALS

The neonatal foal may be especially susceptible to the development of CIRCI for 2 reasons: (1) HPA axis function matures very late in the perinatal period in the foal in comparison to other mammalian species and has been shown in a number of studies to be somewhat impaired during the neonatal period;[31-34] and (2) sepsis, the critical illness most frequently associated with CIRCI in people,[3,5] is a common and important problem in neonatal foals.[35] Indeed, a number of recent studies provide evidence that CIRCI occurs in critically ill and septic neonatal foals with comparable frequency and impact as in septic people and infants. One case report of transient HPA axis dysfunction, as evidenced by both a low basal cortisol concentration and an impaired cortisol response to ACTH stimulation, has been described in a septic neonatal foal.[36] In addition, two independent studies that measured basal ACTH and cortisol concentrations in healthy and septic neonatal foals found significantly increased ACTH:cortisol ratios in non-surviving septic foals.[37,38] Such high ACTH:cortisol ratios, with high ACTH concentrations and low corresponding cortisol concentrations, suggest HPA axis dysfunction at the level of the adrenal gland may occur in the septic foal.

HPA axis function has been further characterized in hospitalized foals using ACTH stimulation tests in two studies.[2,39] Neither study identified a significant difference in cortisol responses between groups of age-matched healthy and ill foals in response to a low-dose ACTH stimulation test (0.1 μg/kg)[39] or in response to a paired low-dose (10 μg) / high-dose (100 μg) ACTH stimulation test.[2] However, when human diagnostic criteria for CIRCI[26] were adapted and applied to a group of 72 hospitalized foals, approximately 50% of all hospitalized foals and approximately 40% of septic foals met these criteria.[2] Furthermore, increased disease severity and poorer prognoses were correlated with decreased cortisol responses to ACTH stimulation in foals in both studies.[2,39] Specifically, non-surviving foals had significantly lower cortisol responses to low-dose ACTH stimulation as compared to survivors,[39] and foals that met criteria for CIRCI (as characterized by an inadequate cortisol response to a high-dose ACTH stimulation test) had a significantly greater incidence of shock, multiple organ dysfunction syndrome, and non-survival than foals with an adequate cortisol response to ACTH.[2]

Recent data also provide preliminary evidence for interactions between HPA axis function and immune and inflammatory responses in horses and foals, as have been described in other species. Gold *et al.* recently investigated links between HPA axis function and inflammatory cytokine responses in septic neonatal foals, and demonstrated positive associations between HPA axis activation and gene expression of the pro-inflammatory cytokine IL-6 in peripheral blood leukocytes[40]. We have also recently demonstrated direct endogenous production of the pro-inflammatory cytokines TNF-α and IL-6 and the anti-inflammatory cytokine IL-10 by equine adrenocortical tissue after exposure to bacterial endotoxin in an *ex vivo* model (Hart et al., unpublished data), similar to findings in rodent, bovine and human models.[41-43] Together, these data suggest that local and systemic inflammatory responses may be important for regulation of both adrenal and immune responses to sepsis in the horse and foal, as in other species.

DEVELOPMENT OF A LOW-DOSE HYDROCORTISONE REGIMEN FOR POTENTIAL USE IN FOALS WITH CIRCI

Although septic neonatal foals with CIRCI may potentially benefit from cortisol replacement therapy in the form of low-dose hydrocortisone, such treatment has not been critically evaluated in foals. Given the marked differences in HPA-axis function and cortisol dynamics in foals, compared with adult horses and with humans,[31,32,34,44] hydrocortisone replacement regimens used in people may not be directly applicable to foals.

In order to determine a physiologically appropriate low-dose hydrocortisone regimen for potential use in foals, we measured daily endogenous cortisol production and hydrocortisone pharmacokinetics in healthy horses and foals. Both adult horses and foals exhibited pulsatile cortisol secretion over a 24 hour period, but a circadian rhythm was only evident in horses.[45] Mean 24-hour cortisol concentration was slightly but significantly lower in foals (20 ± 1 ng/mL) than adult horses (26 ± ng/mL), despite a significantly higher daily endogenous cortisol production rate in foals.[45] This is likely due to substantially increased hydrocortisone/cortisol clearance rates in foals,[45] presumably associated with much higher percentages of free cortisol in foals as compared to adult horses.[44] These data suggest that approximately 1.3 mg/kg/day, or two times the daily endogenous cortisol production rate in healthy foals, might be an appropriate hydrocortisone dose in septic foals with CIRCI. Division of this daily dose into q. 4-hour IV boluses would best mimic cortisol secretion patterns in healthy foals.

Corticosteroid-related immunosuppression is an important concern when considering the use of steroids in septic patients with CIRCI. In healthy 2-to-6-day-old foals, a 3.5 day tapering course of hydrocortisone at this initial dose of 1.3 mg/kg/day IV divided q. 4 hours has potentially beneficial anti-inflammatory effects without significantly impairing innate immune function[46]. Specifically, peripheral blood leukocytes isolated from foals treated with hydrocortisone exhibited decreased expression of the pro-inflammatory cytokines TNF-α, IL-6, and IL-1β after *ex vivo* exposure to LPS as compared to leukocytes from untreated foals. Oxidative burst activity and phagocytic function were not significantly different between neutrophils from treated and untreated foals. Further study is needed, however, to confirm these effects *in vivo* and in septic foals in a clinical setting.

At present, we are conducting a multicenter double-blinded placebo-controlled trial of low-dose hydrocortisone in neonatal foals with septic shock. In this trial, HPA-axis function as determined by cortisol responses to a paired low-dose/high-dose ACTH stimulation test and inflammatory cytokine gene expression in peripheral blood leukocytes will be compared between hydrocortisone- and placebo-treated foals before, during, and after a 3.5 day course of hydrocortisone as described above. Disease severity and outcome data will also be compared between groups. All foals with septic shock will be enrolled at admission, and effects in foals with and without a diagnosis of CIRCI will be compared to determine if currently available diagnostic testing for CIRCI adequately predicts foals that might benefit from hydrocortisone therapy. Data collection is still underway and study participants remain blinded to group assignments, precluding data analysis at this stage. However, preliminary cytokine gene expression data does illustrate changes in inflammatory cytokine gene expression during hospitalization, characterized by a shift from pro-inflammatory to anti-inflammatory cytokine expression in some foals. Whether or not this effect is associated with hydrocortisone therapy or is simply related to disease progression and response to other therapy remains to be seen.

CURRENT RECOMMENDATIONS REGARDING THE USE OF STEROIDS IN SEPTIC FOALS

It is important to note potential risks and benefits of low-dose hydrocortisone replacement therapy in septic foals have not yet been evaluated. Our currently proposed hydrocortisone dose of 1.3 mg/kg/day IV divided q. 4 hours is based on physiologically-relevant cortisol secretion in healthy foals and was calculated in a similar fashion as hydrocortisone doses used in septic adult and pediatric human patients with CIRCI. This dose does not appear to cause undue immunosuppression in healthy foals,[46] but its effects in critically ill foals are not known. In addition, the ideal hydrocortisone dose and length of treatment may vary substantially between individual patients, so it may be impossible to determine a universally appropriate hydrocortisone treatment regimen applicable to all septic foals, or to all septic people for that matter. Continued study is needed to evaluate the effects of this proposed hydrocortisone dose in septic neonatal foals.

Furthermore, the ideal diagnostic test for CIRCI in foals – to identify which foals might be most likely to benefit from hydrocortisone replacement therapy – has not been defined. In people, patients with persistent hypotension and septic shock that fail to improve with fluid resuscitation and vasopressor support are likely to have CIRCI and often demonstrate shock reversal in response to hydrocortisone.[21] Hydrocortisone treatment may be warranted in foals with similar clinical presentations without the need for specific testing to diagnose CIRCI. However, upon provocative HPA-axis testing, many other patients meet criteria for CIRCI and may benefit from hydrocortisone therapy as well.[26] To date, the best diagnostic test for CIRCI in foals is not known, but assessment of cortisol responses to ACTH stimulation may offer the most comprehensive means for evaluation of HPA axis function in the critically ill foal available to date. Specific cortisol response cut-off values to diagnose CIRCI in foals have not been established, but healthy foals during the first week of life demonstrate at least a 2- to 4-fold increase in cortisol concentration at 90 minutes after IV administration of 100μg ACTH (cosyntropin),[31] so dampened cortisol responses in critically ill foals may indicate foals with CIRCI that could also benefit from short-term hydrocortisone replacement therapy.

In summary, a short tapering course of hydrocortisone at 1.3 mg/kg/day IV divided q. 4-hours may limit detrimental systemic inflammatory responses and may provide beneficial cortisol support in foals with CIRCI, though further study is needed. At present, universal treatment of most or all septic foals with low-dose hydrocortisone therapy is not recommended. Hydrocortisone therapy may be indicated in foals that exhibit clinical signs consistent with CIRCI, including persistent hypotension after fluid and vasopressor support, persistent hypoglycemia, or persistent systemic inflammatory responses despite standard anti-inflammatory therapy. Foals with inadequate cortisol responses to ACTH stimulation testing may also benefit from a short tapering course of hydrocortisone, though further study is needed to better define cut-off values for classification of such cases.

REFERENCES

1. Burkitt J, et al., **J Vet Intern Med.** 2007;21:226. 2. Hart K, et al., **J Vet Intern Med.** 2009; 23:901. 3. Marik P. **Chest.** 2009;135:181. 4. Pizzaro C, et al., **Crit Care Med.** 2005; 33:855. 5. Marik P. **Curr Opin Crit Care.** 2007; 13:363. 6. Loisa P, et al., **Acta Anaesthesiol Scand.** 2002; 46:145. 7. Wang H, Ma S. **Am J Emerg Med.** 2008; 26:711. 8. De Jong F, et al., **J Clin Endocrin Metab.** 1984; 59:1143. 9. Beisuizen A, Thijs L. **J Endotoxin Res.** 2003; 9:3. 10. Turnbull A, Rivier C. **Physiol Reviews.** 1999; 79:1. 11. Haddad J, et al. **J Neuroimmunol.** 2002; 133:1. 12. Gibson S, et al. **Emerg Med Clin North Am.** 2005: 23:909. 13. Gold J, et al.,**J Vet Intern Med.** 2007;21:482. 14. Harris M, et al., **J Pediatr.** 2005; 147:462. 15. Kozyra W, et al., **Ann Pharmacother.** 2005; 39:691. 16. Liu D, et al. **Crit Care Med.** 2002; 30:623. 17. Liu L, et al., **Am Rev Respir Dis.** 1993; 148:878. 18. Meduri G, et al., **Neuroimmunomodulation.** 2005; 12:321. 19. Annane D., et al., **J Am Med Assoc.** 2002; 288:862. 20. Keh D, et al., **Am J Respir Crit Care Med.** 2003; 167:512. 21. Oppert M, et al., **Crit Care Med.** 2005; 33:2457. 22. Sprung C, et al., **N Engl J Med.** 2008; 358:111. 23. Annane D, et al., **J Am Med Assoc.** 2009; 301:2362. 24. Keh D, Sprung C. **Crit Care Med.** 2004; 32:S527. 25. Minneci P, et al., **Ann Intern Med.** 2004; 141:47. 26. Marik P, et al., **Crit Care Med.** 2008; 36:1937. 27. Brierley J, et al., **Crit Care Med.** 2009; 37:666. 28. Fernandez E, Watterberg K. **J Perinatol.** 2009; 29: S44. 29. Briegel J, et al., **Clin Investig.** 1994; 72:782. 30. Lamberts S, et al., **N Engl J Med.** 1997; 337:1285. 31. Hart K, et al., **J Vet Intern Med.** 2009; 23:344. 32. Rossdale P, Ousey J, **Equine Vet J.** 1984; 16:300. 33. Silver M, Fowden A., **J Endocrinol.** 1994; 142:417. 34. Silver M, et al., **Equine Vet J.** 1984; 16:278. 35. Cohen N. **J Am Vet Med Assoc.** 1994; 204:1644. 36. Couteil L, Hoffman A. **J Am Vet Med Assoc.** 1998; 212:1594. 37. Gold J, et al., **J Vet Intern Med.** 2007; 21:791. 38. Hurcombe S, et al., **J Vet Intern Med.** 2008; 22:639. 39. Wong D, et al., **J Am Vet Med Assoc.** 2009; 234:126. 40. Gold J, et al., **Proc. ACVIM Annual Forum**, June 9-12, 2010, Anaheim, CA. 41. Call G, et al., **Gen Comparative Endocrinol.** 2000; 118:249. 42. Judd A, et al., **Ann NY Acad Sci.** 2000; 917:628. 43. Kanczkowski W, et al., **Mol Cell Endocrinol.** 2009; 300:57. 44. Hart K, et al., **J Vet Intern Med.** 2011; 25:345. 45. Hart K, et al., **Am J Vet Res.** 2012; 73:68. 46. Hart K, et al., **Pediatr Res.** 2011; 70:72.

REGULATION OF GLUCOSE TRANSPORT IN INSULIN-SENSITIVE TISSUE: FROM HEALTH TO DISEASE

Véronique A. Lacombe, DVM, PhD, DACVIM, DECVIM
Stillwater, OK

Insulin resistance (IR) is one of the hallmarks of equine metabolic syndrome and is the most common metabolic predisposition for laminitis in horses. It is characterized by the impaired ability of insulin to stimulate glucose disposal into insulin-sensitive tissues, namely muscle and adipose tissue. Therefore, the goal of this presentation is to review the main processes regulating glucose uptake, which is, as for other species, the rate-limiting step in glucose utilization into insulin-sensitive tissues.

REGULATION OF GLUCOSE TRANSPORT IN HEALTHY STATE

Regulation of glucose uptake at the molecular level

Glucose uptake from the bloodstream into the cell is the rate-limiting step in glucose utilization in insulin-sensitive tissue (i.e., adipose tissue and striated muscle). Glucose transport occurs primarily by facilitated diffusion via a family of structurally related proteins (GLUT1 to GLUT12) as glucose carriers in all species. The glucose transporter family is expressed in a tissue-specific manner. For instance, GLUT4 is the major isoform expressed in insulin-sensitive tissues and as such plays a crucial role in whole-body glucose homeostasis. Whereas the GLUT5 isoform is mainly associated with the cell surface and is not insulin-stimulated, the translocation of GLUT4 protein from an intracellular (non active) pool to the sarcolemma (active site) increases glucose transport and is largely regulated by insulin-dependent processes. In addition, GLUT12 has been recently identified as a potential novel second insulin-sensitive GLUT, although its role is not well known in any species.[1,2]

GLUT4 is located in an intracellular pool (*non active site*) under basal condition. Upon insulin release, insulin binds to the insulin receptor, thereby triggering its intrinsic protein-tyrosine kinase activity. The subsequent autophosphorylation of several insulin receptor tyrosine residues promotes the recruitment and tyrosine phosphorylation of insulin receptor substrates. That action is followed by activation of phosphatidylinositide 3-kinase (PI3K) and serine/threonine protein kinase (Akt), which in turn promote the downstream effects of insulin and ultimately GLUT4 translocation to the cell surface.[1,3] However, despite intensive research for over 50 years, the insulin-activated downstream signaling pathways that mediate GLUT4 translocation to the cell surface have remained elusive. An Akt substrate protein of 160 kDa (AS160) has recently emerged as a key downstream signaling molecule. Under basal conditions, AS160 is primarily unphosphorylated and retains GLUT4 vesicles intracellularly, while activation of AS160 by phosphorylation permits GLUT4 exocytosis.[2]

Contribution of glucose transport to whole-body insulin sensitivity

There is evidence that glucose transport across the sarcolemma is the rate-limiting step for glucose uptake and subsequent glycogen synthesis *in equine skeletal muscle*. It is well documented that horses have a slower rate of glycogen synthesis compared to other species. In parallel, it has been shown that, in contrast to reports in humans and rats, there is no increase in whole-body insulin sensitivity following carbohydrate feeding during 24 h of recovery from moderate or high-intensity exercise in horses.[4] Of importance, total protein content of GLUT4 in muscle is unchanged or only mildly increased immediately after exercise.[5,6] In addition, IV glucose infusion after exercise or starch-rich meals failed to enhance GLUT4 gene and protein expression in the equine skeletal muscle within the first 24 h after strenuous exercise.[5,7,8] Finally, our laboratory and other groups have demonstrated that *in vitro* insulin stimulation minimally increases GLUT4 translocation to the cell surface in the equine skeletal muscle, the increase only being 13.5% and 15.9% after physiological and supraphysiological insulin concentrations in healthy ponies, respectively (Duehlmeir et al. 2010, Waller et al. 2011a).[2,9] The observed lack of substantial increase in GLUT4 translocation in healthy skeletal muscle in response to insulin may be the result of the insulin-dependent signaling pathways regulating GLUTs already being at near maximal physiologic limits for the species.[2] Alternatively, it has also been suggested that the relatively high resting muscle glycogen concentrations (typical of the equine species) may partially prevent further GLUT4 translocation to the plasma membrane via negative feedback.[8] Taken together, these studies demonstrated that glucose uptake across the sarcolemma is the rate-limiting in glucose utilization and could be a major mechanism for the slow rate of muscle glycogen replenishment observed after exercise.[1,8]

In addition, *the central role of adipose tissue glucose transport in regulating whole-body insulin sensitivity* is becoming increasingly evident, as adipose tissue-selective reduction of GLUT4 in mice results in impaired glucose tolerance and peripheral IR.[10] In addition, human omental adipocytes have higher GLUT4 expression and basal- and insulin-stimulated glucose uptake rate.[11] Similar to these findings, we reported that total GLUT4 protein expression was significantly increased in visceral compared to subcutaneous adipose sites and skeletal muscle of insulin-sensitive horses, with the highest content present in the omental site.[12] Since GLUT4 is abundant in visceral adipose tissue, it is therefore likely to play a substantial role in the regulation of visceral glucose transport and in the maintenance of peripheral insulin sensitivity in horses.

REGULATION OF GLUCOSE TRANSPORT IN INSULIN-RESISTANT HORSES

Regulation of glucose transport: a major pathogenic factor during insulin resistance?

Recently, there has been increased clinical recognition of insulin resistance (**IR**) and diabetes in horses,[13] as well as an Equine Metabolic Syndrome phenotype consisting of a general clustering of obesity, IR, hyperinsulinemia, hypertriglyceridemia, and an increased risk for laminitis.[14] Although the proposed mechanisms involved in the pathogenesis of laminitis are numerous and interrelated, IRis a factor at the forefront of current investigations.[15]

Insulin resistance is characterized by an impaired ability of insulin to stimulate glucose uptake into muscle and adipose tissue in all species. However, a clear understanding of the pathophysiology of IR is still missing both in horses and in other species. Cellular defects associated with IR can include reduced insulin receptor number, reduced expression and activation of insulin signaling cascade proteins, and an impaired glucose transport process.[16] Alterations in GLUT4 trafficking appear to be a key pathogenic factor in IR in multiple species. For example, decreased insulin-stimulated GLUT4 translocation in skeletal muscle is a hallmark of IR associated with type II diabetes in humans and rodents.[17,18] However, in contrast to the extensive research mostly done in rodents and humans, there have been limited studies on the regulation of muscle glucose transport in horses with naturally-occurring IR. Our laboratory recently demonstrated that naturally-occurring IR decreases translocation of GLUT4 to the cell surface in equine skeletal muscle, without a change in its total protein expression.[2] It is worthy to note that these horses exhibited early stage IR (with normal baseline blood glucose concentrations), suggesting that the observed alteration in muscle glucose transport is an early pathogenic event.[2] Under these conditions of reduced insulin sensitivity, increase pancreatic insulin release is required to activate GLUT4-mediated glucose transport in order to maintain euglycaemia, a phenomenon referred as compensated IR.[1]

It is well accepted that preferential deposition of fat into visceral deposits instead of subcutaneous deposits increases the risk of IR and diabetes in humans. The "Portal Theory" suggests that IR and many of its related features could arise from visceral adipose tissue delivering free fatty acids at a high rate to the liver via the portal vein into which visceral adipose tissue directly drains. This in turn would increase hepatic glucose production, reduce hepatic insulin clearance, and ultimately lead to hyperinsulinemia, and hyperglycemia and IR. In contrast, in horses, adiposity and expansion of fat deposits in the crest and tailhead regions are associated with IR.[1] However, the importance of these regional fat sites to the pathogenesis of IR and glucose transport in horses has not been well investigated. We recently reported a decrease in GLUT4 translocation to the cell surface in both visceral and subcutaneous adipose tissue in IR compared to IS horses, suggesting an impairment of glucose transport in both of these adipose depots.[12]

Inflammation & regulation of glucose uptake: novel link and pathogenic factor?

Numerous studies in humans and animal models have demonstrated that obesity induces an inflammatory state characterized by increased expression and secretion of proinflammatory cytokines such as tumor necrosis factor (TNF)-α and interleukins (IL)-1 and-6, and which likely plays a key role in the etiology of IR.[19] In addition, suppressors of cytokine signaling (SOCS) proteins are increased by proinflammatory cytokines such as TNF-α, and have been suggested to be crucial key link between inflammation and IR in the liver. The feedback regulation of SOCS-3 is complex, and there are conflicting reports describing it's relationship with other components of the inflammatory signaling pathways such as toll-like receptors (TLR), and ultimately the pathways linking inflammation to altered glucose transport remain elusive in all species.

Recent data from our laboratory suggested that TLR4 could be a mechanism by which impaired fatty acid metabolism induces and propagates inflammation, and ultimately decreases insulin signaling and glucose transport in IR horses. Activation of TLR4 signaling in adipocytes could activate JNK and NFkB signalling leading to increased production of proinflammatory cytokines which in turn increases SOCS3. Increased SOCS-3 in muscle and adipose may lead to decreased signal transduction through IRS-1 and thus impairment of the insulin signaling pathway, which ultimately may result in dysregulation of glucose homeostasis and whole-body insulin sensitivity. Overall these findings provide new insights into the crucial link between a proinflammatory state and dysregulated insulin and glucose metabolism during IR.

TRANSLATIONAL RESEARCH ON GLUCOSE TRANSPORT

Diabetes mellitus, a global epidemic health problem affecting over 200 million people worldwide, is characterized by dysfunctional glucose uptake into insulin-sensitive tissues, due to a lack of insulin production and/or action (type 1 and 2 diabetes, respectively). However, despite intensive research for the last 50 years, primarily in rodents and humans, the downstream signaling pathways, which regulate glucose transport remain are not well defined for all insulin-sensitive tissues. In addition, exercise is vital for the maintenance of glucose homeostasis in diabetic patients, suggesting that insulin-independent pathways are a major mechanism to regulate glucose entry into striated muscles.[20] Indeed, GLUT4 translocation to the cell surface is not only largely regulated by insulin but also by contraction-dependent processes.[21] For instance, contractile activity during exercise increases glucose uptake and the number of glucose transporters at the cell surface of the skeletal muscle, in the absence of insulin. Holloszy and Narahara, in 1967, suggested the possibility that exercise-induced contraction and the resultant increase in cytosolic calcium in frog skeletal muscle provides the signal to activate glucose transport.[22] However, since the first major discovery in 1967, the molecular signaling mechanisms by which contraction/calcium regulates glucose metabolism have remained elusive in striated muscle. Therefore our laboratory is also investigating how intracellular Ca2+ level regulates glucose uptake independently of insulin using transgenic diabetic rodent model.

CONCLUSION AND CLINICAL RELEVANCE:

Glucose uptake is the rate-limiting step in glucose utilization and is regulated by a family of proteins, the glucose transporter. Insulin resistance is characterized by an impaired ability of insulin to stimulate glucose uptake into insulin sensitive tissues (i.e., muscle and adipose tissue) in all species. As for other species, the functional capability of the insulin-responsive glucose transporter GLUT4 is impaired in skeletal muscle and adipose tissue during naturally-occurring IR in horses. Therefore, these insulin-responsive glucose transporters constitute a major pathogenic factor and potential therapeutic targets for equine patients affected by IR. However, despite intensive research for over 50 years, the molecular mechanisms of altered glucose transport remain elusive, and thus clear understanding of the pathophysiology of IR is still missing in all species. Therefore, further research is required to characterize impairments of the downstream signaling pathways regulating GLUT4 translocation.

REFERENCES

[1]Lacombe VA. **J Equine Vet Sciences**. 2011;31: 578; [2]Waller AP, et al. **J Vet Intern Med** 2011; 25: 315; [3]Bryant NJ, et al. **Nat rev Mol Cell Biol** 2002; 3: 267;[4]Pratt SE, et al. **J Appl Physiol** 2007;103:1063; [5]Lacombe, V.A., et al. **Am J Vet Research** 2003; 64:1500. ; [6]McCutcheon LJ, et al. **Equine Vet J Suppl** 2002;34: 199;[7]Jose-Cunilleras EJ, et al. **Am J Vet Research** 2005; 66: 379; [8]Lacombe VA, et al. **Am J Vet Research** 2004;65:916; [9]Duehlmeier R, et al. **Vet J** 2010;184:176; [10]Abel ED, et al. **Nature** 2001; 409:729; [11]Virtanen KA, et al. **J Clin Endocrinol Metab** 2002; 87: 3902; [12]Waller AP, et al. **BBA**- Molecular basis of diseases 2011;1812(9):1098; [13]Durham AE, et al. **Equine Vet J** 2008; 40:493; [14]Frank N, et al. **J Vet Intern Med** 2010; 24:467; [15]De Laat MA, et al. **Vet Clin Equine** 2010; 26 : 257; [16]Kitamura Y, et al. **Rev Endrocr Metab Disord** 2004;5:143; [17]Handberg A, et al. **Diabetologia** 1990;33:625;[18]Zierath JR, et al. **Diabetes** 1997;46:215; [19]Kim JK. **Cell Metab.** 2006;4(6):417; [20]Goodyear LJ, et al. **Am J Physiol** 1991;261: E795; [21]Brozinick JT Jr, et al. **Biochem J**. 1994;297: 539; [22]Holloszy JO, et al**. J Gen Physiol.** 1967;50: 551.

PHYSIOLOGICAL RESPONSES TO EXERCISE – CAUSES OF POOR PERFORMANCE

Kenneth W. Hinchcliff, BVSc, PhD, DACVIM
Williamstown, Victoria, Australia

The horse is an extraordinary athlete, a characteristic that is the result of evolution of horses as grazing animals on the ancient prairies of North America. Survival in these open lands was enhanced by speed, to escape predators, and endurance, required to travel long distances in search of feed and water. These attributes are shared by prong horn antelopes, another species that evolved on the prairies. The equid characteristics of speed and endurance were subsequently modified or enhanced by selective breeding by humans. A large number of physiological and anatomical features act in concert to endow the horse with extraordinary athletic capacity. Optimal athletic performance is dependent upon optimal integrated functioning of these physiological and anatomical features.(*1, 2*)

The athletic capacity of horses is attributable to a number of physiological adaptations. In some cases these adaptations are not affected by training, for example lung size, whereas others change in response to training, for example blood volume.(*3, 4*) The superior athletic ability of horses is attributable to their high maximal aerobic capacity, large intramuscular stores of energy substrates and in particular glycogen, high mitochondrial volume in muscle, the ability to increase oxygen carrying capacity of blood at the onset of exercise through splenic contraction, efficiency of gait, and efficient thermoregulation. A recent review provides an excellent consideration and description of this topic.(*5*)

The maximal aerobic capacity ($\dot{V}O_{2max}$) of horses is approximately 2.6 times that of similarly sized cattle.(*6*) The larger aerobic capacity in horses is permitted by a larger maximum cardiac output and stroke volume and higher hemoglobin concentration.(*6*) Maximum heart rate is not different between horses and cattle. In addition to the cardiovascular differences between cattle and horses, horses also have lungs that are twice as large as those of cattle with gas exchange surfaces 1.6 times that of cattle.(*7*) Thus, horses have structural adaptations that enhance oxygenation of blood in the lungs, oxygen transport capacity of blood and the ability to deliver oxygen to tissues. The oxygen transport chain, from air to muscle, of horses is suited to transportation of the large volumes of oxygen required to support the high metabolic rate of strenuously exercising horses.

Substrate is required to support these high metabolic rates during exercise. Substrate to support exercise is either carbohydrate or fatty acids. Oxidation of fatty acids is limited and reaches maximal values in other species at a work intensity of approximately 40% to 60% of $\dot{V}O_{2max}$.(*8, 9*) It is likely that a similar phenomenon occurs in horses. Additional work above this exercise intensity is fueled solely by oxidation of carbohydrates, predominantly glycogen.(*10*) Horses have high intramuscular concentrations of glycogen, as do other athletic species such as dogs.(*11*) Muscle concentrations of glycogen in horses are approximately 140 mmol per kg of muscle (wet weight) compared with 80-100 mmol/kg in humans.(*12*) High intramuscular concentrations of substrate are important for fueling muscle contractions during exercise. The flux of glucose from blood into muscle and subsequently to the mitochondria provides only a small amount (< 10%) of the energy used during intense exercise,(*13*) likely because of limits to the rate of transportation of these compounds during exercise.(*11*) The presence of large amounts of readily available substrate in close proximity to mitochondria is therefore essential for horses to undertake strenuous exercise.

Mitochondria provide the energy for muscle contraction. The greater the quantity of mitochondria per unit of muscle weight, the greater is the oxidative capacity of muscle. Muscle of horses contains approximately twice the concentration of mitochondria as does muscle of cattle, a similarly sized animal but with a much lower aerobic capacity.(*14*) This greater aerobic capacity in muscle, when supported by adequate substrate availability and oxygen delivery, permits a higher whole animal maximal aerobic capacity.

Oxygen transport from the lungs to exercising muscle is achieved by the circulation. In addition to cardiac output, oxygen delivery is limited by the oxygen carrying capacity of blood. Horses achieve rapid increases in the oxygen carrying capacity of blood by increasing hemoglobin concentration through splenic contraction. Splenic contraction in anticipation of exercise and during exercise increases the circulating red cell mass without concomitant increases in plasma volume.(*15*) The resulting increase in hemoglobin concentration increases the oxygen carrying capacity of arterial blood by up to 50% during intense exercise. The beneficial effect of this autoinfusion of red cells at the start of exercise is apparent in horses from which the spleen has been removed.(*16-18*) Splenectomized horses have lower hematocrits during exercise, altered systemic hemodynamics including lower right atrial and pulmonary artery pressures, and reduced capacity to perform strenuous exercise.

Energetically efficient gait is challenging for large animals because of the slow rate of contraction and low power output of their muscles.(*19*) However, the gait of horses is energetically efficient,(*20*) with the muscular work of galloping being halved by elastic storage of energy in muscle and tendon units.(*21*) For the forelimb, this use of stored energy and the subsequent catapult action mean that the biceps and brachiocephalicus muscles are less than one one-hundredth the size that they would need to be were there no use of stored energy.(*19*)

Maximal performance involves the coordinated optimal functioning of almost all body systems. In most cases, maximal performance requires that these body systems operate at or close to their maximum capacity. Conceptually, this integrated maximal function may be viewed as a pipeline. The maximum overall flow through the pipe is limited by the narrowest segment of the pipe. This analogy is often employed for oxygen transport during exercise and the system is viewed as one of tuned resistors, with no one individual element limiting the capacity of the system.(*22*) While this analogy is appropriate for healthy animals, it may not be so for animals with performance limiting abnormalities, such as lameness or airway obstruction. In this instance, a single abnormality is sufficient to impair performance. Specific aspects of poor performance are dealt with elsewhere in this book.

For healthy horses the actual performance limiting factor depends on the type of exercise and its duration. Standardbred or Thoroughbred race horses running at top speed are likely limited by oxygen transport. In these animals the malleable components of the oxygen transport chain (red cell mass, mitochondrial volume, muscle capillarity) have adapted to the extent that the capacity of these components approaches or exceeds the capacity of the non-malleable components, such as lung volume or tracheal diameter. A reduction in the capacity of the non-malleable component, for example a reduction in laryngeal diameter secondary to laryngeal hemiplegia, will reduce the capacity of the whole system. This has important consequences for a horse performing at maximal intensity. However, if the capacity of the non-malleable components exceeds those of the malleable components, then a reduction in capacity of the non-malleable component may not reduce performance, for instance in the case of a dressage horse with laryngeal hemiplegia. In this case, the disorder will likely not limit the physiological capacity of the horse to perform its task (although the associated respiratory noise may detract from the performance).

For other types of performance, other factors are limiting. Three day event horses may be limited by their capacity to thermoregulate, endurance horses by their capacity to maintain fluid and electrolyte homeostasis and draft horses by the strength of their muscles. Clearly, the factors limiting exercise capacity of horses varies with the type and duration of exercise. However, an understanding of what likely limits performance for each breed and use of horse is important to understanding not only the physiology of that form of exercise, but also in determining the likely causes of poor performance in animals with clinical disease.

There has been considerable interest in investigating the relationship between readily measured physiological variables and performance in horses. A clear relationship exists between V_4 (the speed at which a blood lactate concentration of 4 mmol/L is achieved) and, to a lesser extent, V_{200} (the speed at which a heart rate of 200 beats per minute is measured) and level of performance for both healthy trotting harness horses and trotters with sub-clinical disease.(*23-26*) Horses with superior performance have higher speeds at V4 and V200, whereas horses with poor performance have lower values. These results are generalizations across a number of horses and application to individual animals can be challenging unless this information is collected as part of the regular training program.

Horses with superior performance also have, as a group, higher aerobic and anaerobic capacity and associated greater workload capacity during high speed exercise tests to fatigue run on a treadmill.(*27*) The greater aerobic capacity of horses with superior performance is considered to be related at least in part to their greater heart size and hence capacity for oxygen delivery.(*28*) Male Standardbreds with a lower percentage of body fat (a relatively higher fat free mass) have a shorter race time, perhaps because of a greater capacity to generate power per kilogram of body weight.(*29*)

Overtraining syndrome (OTS)

Overtraining is a well recognized phenomenon in both human and equine athletes and is an important cause of poor performance.(*30*) Humans affected by overtraining are excessively fatigued, have lower plasma lactate concentrations during exercise and have chronic depletion of muscle glycogen.(*30*) Race horse trainers and veterinarians refer to affected horses as being "stale", or "sour". The term OTS should be used for horses in heavy training losing performance without an obvious clinical reason and which is sustained for >2 weeks.(*31*) It is suggested that the diagnosis of overtraining syndrome is imprecise and that in the absence of a scoring system to assess the psychological status of horses, the most effective method for diagnosis is the thorough assessment of the history and presenting complaint.(*31*) Signs of overtraining include (*30, 32*):

* Decreased performance, such as slower split times
* Unexplained poor performance or adverse persistent change in performance
* A decrease in body weight $\geq$ 6 kg
* Reduced appetite
* Decreased enthusiasm to train
* Behavioral changes in which the horse might become moody and lethargic or nervous(*33*)
* Increased heart rate during exercise and possibly slower recovery heart rate
* An increase in blood lactate concentration between 8%-16% after pace and fast work(*32*) (this is not a consistent finding in studies of overtraining in horses(*30*))
* Reduced plasma cortisol response to maximal exercise(*32, 34*)

There are not reliable changes in hemogram or leucogram of overtrained horses.(*33*) Changes in serum activity of muscle derived enzymes are minor in overtrained horses and of limited diagnostic value.(*33*) Conversely, serum gamma-glutamyl transferase activity is markedly elevated in overtrained horses and can exceed 100 IU/L.(*30, 33*)

The mechanism of overtraining in horses is poorly understood although it is hypothesized that disturbances of hypothalamic function are central to the disorder rather than a primary effect on muscle function.(*35*) Overtrained Standardbreds have abnormal pulsatile nocturnal growth hormone secretion which is speculated to be associated with longer phases of somatostatin withdrawal.(*36*) Standardised exercise tests are suggested to provide a way to detect subtle changes in hormonal responses in the individual, which may make an important contribution to the detection of early overtraining.(*31, 36*) Red cell hypervolemia is not an underlying mechanism associated with overtraining,(*37*) and chronic muscle glycogen depletion is not a feature of the experimentally induced syndrome.(*38*)

REFERENCES

1. K. W. Hinchcliff, R. J. Geor, in *Equine Sports Medicine and Surgery: Basic and clinical sciences of the equine athlete,* K. W. Hinchcliff, A. Kaneps, R. J. Geor, Eds. (WB Saunders, London, 2004), pp. 3-8. 2.D. L. Evans, *Equine Veterinary Journal* **39**, 373 (2007). 3. K. McKeever, in *Equine Sports Medicine and Surgery: basic and clinical sciences of the equine athlete,* K. W. Hinchcliff, A. Kaneps, R. J. Geor, Eds. (Elsevier, London, 2003). 4.D. Ainsworth, in *Equine Sports Medicine and Surgery: basic and clinical sciences of the equine athlete,* K. W. Hinchcliff, A. Kaneps, R. J. Geor, Eds. (Elsevier, London, 2003) 5. D. C. Poole, H. H. Erickson, *Compr Physiol* **1**, 1 (2011). 6. J. H. Jones *et al.*, *Journal of applied physiology (Bethesda Md : 1985)* **67**, 862 (1989).7.M. Constantinopol *et al.*, *Journal of applied physiology (Bethesda Md : 1985)* **67**, 871 (1989). 8.J. M. Weber *et al.*, *The Journal of experimental biology* **199 (Pt 8)**, 1667 (1996). 9. J. Achten, Gleeson, M, Jeukendrup, AE. x, *Med Sci Sports Exerc* **34**, 92 (2002). 10. J. M. Weber, T. J. Roberts, R. Vock, E. R. Weibel, C. R. Taylor, *The Journal of experimental biology* **199 (Pt 8)**, 1659 (1996). 11. E. R. Weibel *et al.*, *J Exp Biol* **199**, 1699 (1996).12. B. Essen-Gustavsson, D. McMiken, K. Karlstrom, A. Lindholm, S. D. G. Persson, *Equine Vet J* **21**, 27 (1989).13. R. Geor, K. Hinchcliff, R. Sams, *J Appl Physiol* **88**, 1765 (2000). 14. S. R. Kayar *et al.*, *Pflugers Archiv : European journal of physiology* **413**, 343 (1989). 15. S. D. G. Persson, *Acta Physiol Scan* **Suppl. 19**, 9 (1967). 16.K. H. McKeever, K. W. Hinchcliff, S. M. Reed, R. L. Hamlin, *Amer J Physiol* **265**, R409 (1993). 17. S. G. Persson, L. Ekman, G. Lydin, G. Tufvesson, *Zentralblatt fur Veterinarmedizin Reihe A* **20**, 456 (1973). 18. S. G. Persson, G. Lydin, *Zentralblatt fur Veterinarmedizin Reihe A* **20**, 521 (1973). 19. A. Wilson, J. Watson, G. Lichtwark, *Nature* **421**, 35 (2003). 20. D. Hoyt, C. R. Taylor, *Nature* **292**, 239 (1981). 21. A. Minetti, L. Ardigo, E. Reinach, F. Saibene, *J Exp Biol* **202**, 2329 (1999). 22. S. L. Lindstedt, D. J. Wells, J. Jones, H. Hoppeler, H. A. J. Thronson, *Int J Sports Med* **9**, 210 (1988). 23. A. Courouce, J. C. Chatard, B. Auvinet, *Equine Vet J* **29**, 365 (Sep, 1997). 24. C. Leleu, *Veterinary record* **156**, 339 (2005). 25. A. E. Lindner, *Journal of Animal Science* **88**, 950 (2009). 26. E. A. Richard, *The veterinary journal* **184**, 282 (2009). 27. G. M. Gauvreau, H. Staempfli, L. J. McCutcheon, S. S. Young, W. N. McDonell, *J Appl Physiol* **78**, 1447 (Apr, 1995). 28. R. Buhl, *Journal of the American Veterinary Medical Association* **226**, 1881 (2005). 29. C. F. Kearns, K. H. McKeever, K. Kumagai, T. Abe, *Vet J* **163**, 260 (May, 2002). 30. C. M. McGowan, *Comparative exercise physiology* **5**, 57 (2008). 31. J. L. L. Rivero, *Equine Veterinary Journal* **40**, 611 (2008). 32. M. J. Hamlin, *Equine Veterinary Journal* **34**, 383 (2002). 33. C. M. TylermcGowan, *Equine Veterinary Journal* **31**, 621 (1999). 34. L. C. Golland, *Pflügers Archiv* **439**, 11 (1999). 35. I. D. Wijnberg, *Journal of applied physiology: respiratory, environmental and exercise physiology* **105**, 1746 (2008). 36. E. de Graaf-Roelfsema, *American journal of physiology. Regulatory, integrative and comparative physiology* **297**, R403 (2009). 37. L. C. Golland, *The Veterinary Journal* **165**, 228 (2003). 38. C. M. McGowan, *Equine Veterinary Journal* **34**, 257.

FIELD EXERCISE TESTING OF HORSES

Emmanuelle van Erck, DVM, PhD, DECEIM
Belgium

INTRODUCTION

Successful performance in a sport horse is the sum of many factors inherent to individual qualities determined by genetic potential and temperament, good health, soundness and conformation. The full expression of the horse's potential also requires a finely-tuned management: the quality of the horse's environment and nutrition, the way it is physically prepared to reach a satisfactory level of fitness for a specific type of work, the careful preparation of competitions, as well as the follow-up of short and long term recovery. These measures likewise, regular veterinary check-ups, contribute to the prevention of illness. Exercising testing is a powerful tool to monitor a horse's health and progress in fitness during a competitive season as well as to investigate causes of poor performance. Exertion may bring forth problems otherwise impossible to reveal upon resting examination.

The investigation into equine exercise physiology has been initially and extensively based on the evaluation of horses on a treadmill. Treadmill exercise enables to perform comprehensive measurements in a controlled environment thereby ensuring a higher degree of standardization and repeatability of the procedures. These studies have shown the pertinence of measuring specific parameters to evaluate the athletic capacity and metabolic response of horses during exercise and their relation to performance. Among these parameters, maximal oxygen consumption (VO2max), arterial blood gases (PaO2, PaCO2), heart rate (HR) or lactate (LA) against speed have served as reference values in both physiological and pathological conditions. Because investigation of a horse on a treadmill requires transport to a specialized centre and the intervention of a qualified team of several persons, the implementation of treadmill exercise tests have never become routinely integrated to fitness or clinical evaluations of equine athletes. Furthermore, exercising on a treadmill has sometimes been wrongly perceived as unsafe (Franklin et al., 2010). This has encouraged the development of alternative techniques allowing similar evaluations but in the field. These different techniques have benefited from the miniaturization of measuring and recording equipment and techniques as well as telemetric transmission of the data. A choice of measurement can now be done with the horse exercising in normal working conditions or conditions simulating competition, with the effect of its rider, tack, track type, etc… taken into account.

DESIGNING AN APPROPRIATE FIELD TEST

Most trainers rely on empirical training regimes to train a horse for a specific event. Their assessment of fitness often relies on subjective evaluation. While in human athletes a number of reliable exercise tests are available to assess both stamina and strength and carried out routinely under standardized conditions, validated testing protocols are more scarce in equine sports medicine, especially for sports horses.

Because the requirements of racing and equestrian sports can be significantly different, exercise tests should aim at testing if the horse can meet the physical requirements of his specific discipline. Although it is often unnecessary, and sometimes contraindicated, to reproduce the level of strain reached during competition, the exercise test should be of sufficient duration and intensity to yield pertinent information. Because tests are conducted in less repeatable conditions, extra care should be taken to ensure as much standardization of the exercise and procedures as possible. Exercising in the horse's usual environment and with its usual rider is recommended as it helps reduce stress for both horse and rider. Tests should not be conducted if the environmental conditions are unsuitable or unsafe (bad weather, frozen track, …). The horse should not be lame and should not have competed the previous days.

Field exercise test characteristically consist of consecutive incremental steps, which may be interrupted by short periods of recovery during which blood samples may be taken. Tests to fatigue are possible in racehorses but more hazardous and difficult to program during a competition season. The test can be implemented to a routine work, as long as it can be compared to other tests intra-individually over time or inter-individually. Different testing procedures have been described according to the horse's activity. To facilitate meaningful interpretation, velocity should be kept as constant as possible during each step and the duration of each step should be sufficient to provide a steady-state in the parameters evaluated (heart rate, lactates,…). The number, intensity and duration of the steps vary accordingly.

Some examples are given in the table below.

Horses	Distance/ step	Number of steps	Speed range	Reference
Thoroughbred	600-800 m	4	400-660 m/min	Koyabashi et al, 1999
Standardbred	1200 m	3	500-630 m/min	Courouce et al 1999
Eventers	1000 m	4	400-700 m/min	Munoz et al 1998 Van Erck et al 2005
Endurance	27 km / 1.5 km / 1.5 km	3	6 – 8.8 m/s	Fraipont et al, 2011

MEASUREMENTS ACHIEVABLE IN THE FIELD

In the field, heart rate (HR) and velocity (v) are the simplest and easiest parameters to monitor. There is a choice of commercially available equipment to record both parameters simultaneously. The GPS based systems may also record the course over which the horse has been ridden at give additional information such as total distance covered, average speed and incline at different steps, etc... Recordings can be downloaded and stored for later comparative analysis. When investigating a poorly performing horse, the recording of an ECG not only allows to record heart rate but also allows to detect cardiac arrhythmias occurring only during exercise. Some systems permit direct telemetric monitoring of the ECG tracings.

Blood lactate concentrations (LA) are also an interesting parameter to assess during a stepwise incremental test. Samples can be readily analyzed in the field with portable systems validated for the equine species. Their main limitation is the accuracy of the data in the low and high lactate concentration range. LA measurements give information on the adaptation of the aerobic metabolism to exercise and training.

Indices may be calculated from HR, v and LA parameters that facilitate comparison of the test results, especially when they are subject to more variability. Heart rates can be plotted against speed, and parameters expressing velocity at any given heart rate can be extrapolated from the curve. Some of the most common parameters used are V200, the velocity at which heart rate is 200 bpm. Expression of the velocity at lower HR is equally valid (V140, V160 and V170) although more variable as potentially more influenced by factors such as stress. The same principle is applied to LA measurements, which can be expressed according to speed. Comparison of the velocity at a LA values of 4 mmol/l (VLA4 or V4) is a commonly derived parameter. A LA concentration of 4 mmol/l is thought to correspond to onset of blood lactate accumulation (OBLA), considered as the aero-anaerobic metabolic threshold.

Other more sophisticated metabolic measurements have been tried in the field such as measurement of oxygen consumption (VO2) and carbon-dioxide production (VCO2). However these techniques are more tedious and time consuming to implement in practice and have limitations when used in the field, especially at maximal levels of exercise (Van Erck et al, 2009).

In the clinical evaluation of poor performers, other more accessible diagnostic tools can be implemented, such as overground endoscopy and accelerometers to analyze the horse's gait. Overground upper airway endoscopy has become an essential diagnostic tool when doing a field exercise test to establish the cause of poor performance. In sport horses, dynamic upper airway obstruction limiting performance may occur without producing any respiratory noise (see cardio-respiratory causes of poor performance in the proceeding).

Post-exercise measurements may also give interesting fitness and clinical information. Post-effort HR recovery is a simple and widely used measurement. Post-effort muscle enzyme activity, such as evaluation of CK even 1 hour post-effort is sufficient to diagnose an episode of exercise-induced rhabdomyolysis, even in the absence of other specific clinical symptoms.

INTERPRETATION OF FIELD EXERCISE TESTS

The results of a field exercise test depend on many factors, some intrinsic to the horse's physical status, other extrinsic such as environmental factors. Taking into account these factors will avoid misinterpretation. The horse's fitness status can depend on age, gender, breed, level of training and level of performance in competition. Reference values should be recorded in other horses with a similar level to allow a significant comparison of the results. Repeated monitoring of the same horse during a season will also allow more pertinent interpretation of consecutive tests. Extrinsic factors should also be accounted for such as climate, humidity, track softness. For instance higher outdoor temperatures will increase energy expenditure due to thermoregulation and increase both LA and HR. Similarly, deeper sand tracks will increase muscular effort and similarly increase HR and LA.

When following horses participating in competitions, their weekly working program and competition schedule should be initially carefully reviewed with all persons involved: the horse's caretaker, trainer and rider. The objectives missed and the level of performance expected should also be questioned, as an impression of poor performance can originate from divergence between the owner's hopes and the horse's true capacities or training status. Changes in the horse's usual work sessions, increases in the number of events run, changes in diet or environment may also influence the outcome of tests.

The HR response increasing speed or exercise intensity is linear until maximal HR is reached. The regression HR(v) plot is highly reproducible when measured during a standardized exercise (Evans and Rose, 1988). Training adaptations are associated with a decreased HR at an identical submaximal step and an increase in V200. But the relationship between HR and speed is may be negatively influenced by concomitant disease and are helpful tools to detect subclinical problems causing poor performance. Loss of fitness, increase in body mass, excessive stress, lameness, cardiovascular and respiratory disease and overtraining influence the response of HR to exercise and may cause a decrease in V200 (Courouce-Malblanc et al., 2002; Richard et al., 2010; Fraipont et al. 2011).

VLA4 is considered as an indirect indicator of aerobic capacity. High VLA4 indicates superior athletic capacity and has been related to performance in racing (Evans et al., 1993; Couroucé et al., 1997; Leleu et al., 2005). LA at a given sub-maximal step will decrease with training, VLA4 will increase whereas LA measured after a test to fatigue will be higher in well trained horses. Cardio-respiratory disease will cause an earlier OBLA, increase LA at any given submaximal step and decrease VLA4. These parameters have been used as predictors of health in racehorses and sport horses. VLA4 is negatively influenced by lack of training, cardio-respiratory and muscular diseases (Van Erck et al, 2006).

To conclude, field exercise tests are well accepted procedures which help the practitioner to follow a horse's clinical and fitness status regularly throughout a competitive season. Even simple parameters may help detect occurrence of subclinical diseases or overtraining that may be detrimental to performance.

REFERENCES

Couroucé A. et al., **Equine Vet J** 1999; Suppl 30:528; Evans D. et al., **Equine Vet J** 1993; 25:44; Franklin S. et al **Equine Vet J. Suppl** 2010; 42,38:70; Leleu C. et al., **Vet Rec**. 2005;12,156:339; Munoz A. et al, **J. Equine Science**, 9 (1), 1; Van Erck E. et al., **Pferdeheilkunde** 22 (5), 625; Van Erck E. **Equine Vet J Suppl**. 2011; 43 Suppl 40:18

CARDIO-RESPIRATORY CAUSES OF POOR PERFORMANCE

Emmanuelle van Erck, DVM, PhD, DECEIM
Belgium

INTRODUCTION

In the field of exercise physiology, horses demonstrate exceptional athletic capability in comparison to many Mammalians studied. There superior performances originate from various physiological adaptations, namely an outstanding aerobic metabolism. Maximal oxygen consumption or VO2max is the gold standard parameter to estimate aerobic capacity in human athletes. Values of VO2max reached in Thoroughbred racehorses during maximal exertion reach values that go beyond any measurements normally expected in other species. The importance of the aerobic pathway in the muscular energetic supply, whatever the equestrian discipline, is well recognized in horses. Oxygen transport from the ambient air to the exchange zones in the lung, then from the exchanging zone into the muscles mitochondria, is achieved by the coordinate action of the respiratory and cardiovascular systems. During exercise, the increase in the metabolic rates imposes an increased load on all physiological steps involved in what is called the "oxygen chain": increase in minute ventilation and in alveolar ventilation and diffusion, in cardiac output and oxygen-carrying capacity (splenic contraction) and ultimately in perfusion to the muscles. The efficiency of each element and link in this chain may be modified positively by training or negatively by illness. The soundness and intrinsic capacity of each of these systems will therefore play a substantial role in determining the performance potential of a horse, whatever the equestrian discipline he is involved in. Conversely, any limitation in of one of the chain links will limit the efficiency of the entire chain and may result in poor performance.

SOME KEY ELEMENTS OF CARDIO-RESPIRATORY PHYSIOLOGY

Racehorses are probably mainly limited by the oxygen supply at the muscle level. Indeed, despite a large pulmonary exchange area, a dramatic increase in pulmonary ventilation, an increase in cardiac output and packed cell volume, horses exercising at intensities higher than 60% of VO2max become hypoxemic, and even hypercapnic. The severity of hypoxaemia increases with training and disease.

There is plentiful research demonstrating that horses have remarkably high muscle mass. The enormous metabolic requirements of this muscular mass results in O2 uptake and CO2 production levels that exceed the ventilation capacities of the respiratory system, especially in well-trained horses. Some other anatomical and physiological peculiarities in horses explain the fact that the cardio-respiratory systems are a weak link in the oxygen chain.

(1) The length, section and shape of the upper airway contribute to a high anatomical dead-space, and associated to the compulsory nasal breathing, result in an increased resistance to airflow and energetic cost of breathing, which probably "downregulate" the increase in pulmonary ventilation during intense exercise.

(2) Mixed venous blood coming back from the peripheral tissues (muscles) has a partial pressure in O2 as low as 15 mmHg, which is in relation to the dramatic O2 removal by the working muscles.

(3) Impairment of gas diffusion at the level of the alveolo-capillary membrane because of the dramatic increase of cardiac output reduces the time for gas exchange. A relative impairment of the ventilation/perfusion ratio has been also suggested as a possible reason for the gas exchange impairment. These observations indicate that the respiratory system may be the ultimate limiting factor to high intensity exercise in horses.

RESPIRATORY PROBLEMS CAUSING POOR PERFORMANCE

The respiratory system is the major limiting factor in exercising horses (Art et al., 2002) and respiratory diseases can affect both exercise capacity and athletic performance. Because diseases of the respiratory system are highly prevalent in horses, ensuring respiratory health should be a priority in equine athletes. Clinical imaging of the respiratory tract and sampling of the airways are useful diagnostic tools in assessing the severity of disease and level of dysfunction.

1. Epidemiology of respiratory problems

Respiratory diseases can be considered as the second most common cause of wastage in sport horses and racehorses after musculoskeletal problems. The overall prevalence of respiratory diseases is probably underestimated due to the fact that some chronic forms of diseases may develop subclinically, although they still impede performance. The high prevalence of respiratory diseases in the equine population is due to a number of risk factors naturally present in the horses' environment as well as management and work practices. Housing in dusty environment and feeding with dry food, training, transportation, mingling during competition, all factors that favor the development of infectious, inflammatory and allergic diseases.

There are only 4 published retrospective studies reporting causes of poor performance on at least 50 cases, on the basis of multiple clinical and ancillary exams, including echocardiography, exercise tests with video-endoscopy and electrocardiography, as well as pre- and post-exercise biochemistry (Morris & Seeherman, 1991; Martin et al, 2000; Richard et al, 2009; Fraipont et al,

2011). Some other studies report tests on a little number of horses in laboratory or in the field. Most of the horses examined in these latter studies were standardbred and/or thoroughbred horses (Couëtil and Denicola, 1999; van Erck et al, 2006). On the other hand, information for jumping, event and dressage horses are sparse or nonexistent. In the study of Martin et al (2000), a definite cause of poor performance was indentified in 256/348 (73.5%) horses. The causes of poor performance were distributed as follows: 148 upper airway (UA) dynamic obstructions (mainly dorsal displacement of the soft palate, dynamic pharyngeal collapse, idiopathic laryngeal hemiplegia), 22 combinations of upper airway obstruction and cardiac arrhythmias, 19 cases of cardiac dysfunction (decreased fractional shortening), 10 subclinical exertional rhabdomyolysis, 15 lameness and 9 other disorders. These results were close to the results of the report of Morris & Seeherman in 1991. Surprisingly, none of these 2 laboratories performed lower airway examination or sampling. In the other aforementioned studies, cytologic examination of airway samples (tracheal wash (TW) and/or broncho-alveolar lavage (BAL)) was included in the string of tests. The major incidence of lower airway subclinical problems (including Inflammatory Airway Disease (IAD), Recurrent Airway Obstruction (RAO), Exercise-Induced Pulmonary Hemorrhage (EIPH) on exercise tolerance was reported (Couëtil and Denicola, 1999; van Erck et al, 2006; Richard et al 2009; Fraipont et al, 2011). So respiratory problems (upper or lower airway diseases or a combination of both conditions) are reported as being a major cause of poor performance and exercise intolerance in competition horses. The lack of adaptation to training of the respiratory system and the high prevalence of respiratory disease explain the important role of this system in the occurrence of poor performance and the necessity of diagnosing problems early in their process.

2. Dynamic obstruction of the upper airways

To provide the lowest resistance due to inertance, friction and turbulence, the horse's upper airway should ideally have been be short, large, straight and rigid. However, they are long, narrow, curved and collapsible. The nasopharynx is a soft tissue structure that is sustained solely by the activity of intrinsic and extrinsic muscles. This particular area of the upper airways is therefore susceptible to dynamic collapse when submitted to high negative transmural pressures, as those encountered during exercise. According to Poiseuille's law, the resistance to airflow in a tube is inversely dependent upon the radius to the fourth power. A small reduction in airway section has thus a dramatic effect on the resistance to airflow. Any abnormality that would reduce the airway section would be a potential factor of impaired ventilation and consequently impaired performance. This reduction or sub-obstruction may be due to functional, inflammatory or infectious problems.

Dynamic UA obstruction is a common cause of respiratory noise during exercise and may lead to poor performance in equine athletes (Tan et al., 2005). The noise generated by an obstruction in the airway, be it functionally disabling or not, is often of concern to the horse's rider, can count against it in competition and can decrease the commercial value of the horse. A clinical exam should include close observation of the respiratory pattern, palpation of the larynx and induction of cough, evaluation of airflow through both nostrils, head conformation, ... During a prepurchase examination, observation of the horse in usual working conditions is essential if the horse is destined for sports. Upper airway obstruction may be caused by multiple conditions, most of which remain undetectable or misdiagnosed upon resting examination (Parente and Martin, 1995, Tan et al., 2005; Lane et al., 2006).

The recent advent of overground endoscopes has allowed the evaluation of upper airway mechanical behavior in normal exercising conditions. The technique has already been used to evaluate racehorses at the track, as well as ridden sport horses (Franklin et al., 2008; Desmaizières et al., 2009; Pollock et al., 2009). Head flexion has been recognized as a factor contributing to an increase in upper respiratory tract resistance. Rider equitation and head flexion have a combined effects on upper airway morphology and aggravate UA dynamic obstruction (Van Erck, 2011). In sport horses, ridden examination is of particular importance, as equitation-related maneuvers or movements are a factor in the development of dynamic airway instability, both at the level of the nasopharynx and of the larynx. Dynamic obstructive conditions usually occurred when the intensity of exercise increased, i.e. at the canter or with increased collection and vertical movements, or when the horse starts to tire. Similarly, when audible, the intensity of noise generated by obstructive conditions increases with increasing gait, fatigue, head flexion or other forms of rider intervention.

Underlying airway inflammation has been suggested as a possible etiology or predisposing factor to UA obstruction. Pharyngeal instability, under the form of palatal instability or nasopharyngeal collapse, is significantly affected by the presence of upper (pharyngeal) or lower airway inflammation (Van Erck, 2011). The pathophysiology of nasopharyngeal collapse is not fully understood but has been associated to neuromuscular dysfunction of the upper airways (Holcombe *et al.* 1999). Inflammatory conditions such as pharyngeal lymphoid hyperplasia could result in nasopharyngeal instability and possibly displacement of the soft palate (Blythe *et al.* 1983; Holcombe *et al.* 2001; Sullivan and Parente 2003).

3. Lower airway issues

Whatever the type of sport, most lower airway diseases cause a significant decrease in athletic parameters and delayed cardio-respiratory recovery (Sanchez et al, 2006, Richard et al, 2009, Fraipont et al 2011). The main lower respiratory diseases involved in poor performance reported are inflammatory airway disease (IAD), recurrent airway obstruction (RAO), exercise induced pulmonary haemorrhage (EIPH) and viral or bacterial infections. Several factors inevitably promote the occurrence of inflammatory diseases, such as age (higher prevalence in older individuals) and environmental conditions. Horses are often housed indoors in buildings were ventilation within the boxes is limited, indoor storage of hay and straw can be encountered and doors and windows are often closed during the cold months. This type of management promotes chronic respiratory inflammation. Intensive training, repeated transports and mingling during competitions may also promote the development and transmission of infectious diseases.

The quality of environmental management is not the only explanation to the high prevalence of lower airway diseases in horses: competition itself is also a key factor. Competition may trigger metabolic and hormonal alterations which disturb immunity. Depression of the immune system and the occurrence of repeated inflammatory and oxidative processes during training and competitions seem to create favorable conditions to the outbreak of infectious respiratory diseases (Tyler et al., 1996; Robson et al., 2003). These respiratory problems are often subclinical but have an impact on performance and may impede the pursuit of their physical preparation as well as post-race short and long-term recovery.

For the practitioner, the diagnosis of subclinical lower airway diseases is difficult without ancillary examinations. Most horses with chronic IAD typically do not show respiratory symptoms such as cough or nasal discharge but display more subtle signs such as premature fatigue during exercise, prolonged recovery or decreased energy during competitions. Tracheal endoscopy is useful to evaluate the amount of mucus or even blood present in the lower airways. When possible, post-exercise endoscopy is often more relevant. A significant correlation has been shown between the presence of tracheal mucus and poor performance in racehorses (Holcombe et al. 2006). In showhorses, although IAD does not seem to affect some defined factors of performance (Gerber et al. 2003), it has been associated to decreased willingness to perform (Widmer et al. 2009). This is in accordance with other works, where lower airway inflammation, including inflammatory airway disease (IAD), is shown to be detrimental to upper airway stability and may accelerate the onset of fatigue (Richard et al., 2009, Courouсé et al, 2010, Fraipont et al., 2011). Definitive diagnosis of lower airway disease can be established through airway sampling. A tracheal or trans-tracheal wash are mostly useful to diagnose infectious diseases whereas broncho-alveolar lavage is indicated in cases of chronic problems such as IAD, RAO or EIPH. Cytologic analysis allows evaluating the severity of the disease process, evaluating environmental quality and of course orientating adequate therapy. Evaluating the functional repercussions of a disease may also be indicated if permanent pathological damage is suspected. There are currently commercially available non-invasive respiratory function tests which may be useful in the future (impedance measurements, impulse oscillometry). Prevention remains a guarantee for the maintenance of lower airway health. A dust-free and hygienic environment both at home and on shows limits the risks of lower airway inflammation.

CARDIAC CAUSES OF POOR PERFORMANCE

Although cardiac problems are relatively rare in sport horses, they have significant repercussions on exercise tolerance and may be dangerous for the rider. The most common cardiac problems causing poor performance are cardiac arrhythmias, valvular regurgitations and less commonly congenital defects or overtraining.

1. Exercising arrhythmias

Physiological cardiac arrhythmias are more frequently seen in athletic horses, both at rest and during post-effort recovery. They are usually related to a high vagal tone and should disappear with exercise. There are various types of cardiac arrhythmias, primary or secondary in nature, which may potentially limit performance, the most frequent ones being supraventricular or ventricular premature beats and atrial fibrillation. It is not uncommon that these arrhythmias occur exclusively during exercise. They will be easily missed if the exam is performed only at rest, so performing an exercising ECG is an easy way of diagnosing the presence and type of arrhythmia present as well as checking for the value of heart rate for a given standard exercise. Small recording units are now available to record and eventually transmit ECG signals throughout an exercise session. Arrhythmias may originate from a primary cardiac disease (valvular regurgitation, myocarditis, endocarditis) or secondary to a systemic disease or acid-base/ electrolyte imbalance. The presence of underlying cardiac disease should be assessed by careful bilateral auscultation, blood work including measurement of electrolytes, LDH and/or CK iso-enzymes and troponine I.

Atrial fibrillation is probably the most frequent arrhythmia associated with poor performance. In a mixed population of horses, the prevalence of atrial fibrillation was reported to be around 2.5 %. It can be permanent or paroxysmal and resolve within minutes or up to 48 hours after exercise. It occurs sometimes without obvious underlying cardiac disease or can be secondary to atrial enlargement. It decreases diastolic ventricular filling and may cause a disruption in hemodynamics during strenuous exercise, leading to poor muscle perfusion, premature fatigue and in some cases epistaxis, distress, ataxia or collapse during exercise. If cardioversion treatment is not considered (because of risk factors inherent to the use of quinidine or underlying cardiac hypertrophy for instance), an exercising ECG is mandatory to check for heart rate and R on T phenomena as these may potentially predispose to fatal ventricular tachycardia.

2. Pathological cardiac murmurs

Cardiac dimensions and left ventricular volume and mass are correlated to performance in race in adult horses (Young et al, 2006) and can be estimated by morphological measurements obtained by echocardiography. Training for racing induces changes in cardiac morphology and function characterized by an eccentric hypertrophy and defined as "athlete's heart". This phenomenon may promote the pathogenesis of atrio-ventricular valve regurgitation in horses. Consequently, the prevalence of heart murmurs is higher in more athletic individuals and in older horses (Young & Wood, 1998, 2001). Evaluating the pathological character of valvular regurgitations and their functional consequences is important in competition horses, both to determine if it can be the cause of poor performance and if the horse can be safely kept at work. Doppler echocardiography not only allows to establish a diagnosis but is also helpful to have both a morphological and functional appreciation of the heart. Assessing the rate of progression of cardiac disorders (whatever their degree of importance) by regular echocardiographic follow-up, is important if the horse is to continue work and competition.

Generally, compensated valvular regurgitations are well tolerated in performance horses and appear to have a slow progression. Factors influencing prograssion may be the level of work performed by the horse, the rhythm and strain of competition and concurrent diseases. During exercise, very high levels of pulmonary arterial pressures may be reached, predisposing horses to episodes of EIPH. It is thought that chronic pulmonary hypertension may influence the development of tricuspid valve insufficiency in athletic horses, whatever their discipline (Young, 2003). Tricuspid valve incompetence seems to have less consequences on performance in comparison to mitral valve regurgitation. These atrio-ventricular valve dysfunctions evolve, they are usually associated with progressive clinical signs, beginning with abnormal fatigue and poor post-effort recovery. Aortic regurgitation is more commonly found in horses from the age of 10, as there is a progressive age-related degeneration of the valves with age. Aortic incompetence is more frequently associated with arrhythmias during exercise and may generate more acute clinical signs when the condition progresses.

REFERENCES

Art et al, **Equine Vet J Suppl.** 2002 Sep;(34):442; Couëtil, L. and Denicola, **Equine vet. J.** 1999; 31:77.; Couroucé et al, **Equine Vet J.** 2010 Nov;42 Suppl 38:246; Desmaizières L-M. et al., **Equine vet. J.** 2009;41:347.; Fraipont et al, **Vet Rec.** 2011 Aug 6;169(6):154; Franklin S. et al., **Equine vet. J.** 2008;40:712; Gerber et al. **Equine vet. J.** 2003; 35:491. ; Holcombe et al. **Equine vet. J.** 1999;31:45; Holcombe et al, **Am. J. vet.** 2001 Res. 62, 1706; Lane et al., **Equine vet. J.** 2006 38, 393; Martin et al, **J. Am. vet. Med. Ass.** 2000; 216, 554; Morris & Seeherman, **Equine vet. J.** 1991; 23, 169; Parente and Martin, **J Am Vet Med Assoc.** 1998,1;212(3):399

Pollock et al., **Equine vet. J.** 2009; 41, 354; Richard et al, **vet. J.** 2009; 184, 282; Robson et al., Equine **Vet J.** 2003 Mar;35(2):133; Sullivan and Parente **vet. Clin. North. Am. Equine Pract.** 2003 19, 159; Tan et al., **vet. J.** 2005, 170, 243; Tyler et al., **J Appl Physiol.** 1996 Nov;81(5):2244; **Van Erck E. et al.,** Pferdeheilkunde **2006, 22 (5), 625; Van Erck E.** Equine Vet J Suppl. **2011; 43 Suppl 40:18,** Widmer et al. **Vet J.** 2009, 182(3):430, Young, **Exp Physiol.** 2003, 88(5):659; Young et al, **Equine Vet J Suppl.** 2006, 36:193.

MUSCULAR CAUSES OF POOR PERFORMANCE IN ATHLETIC HORSES

Erica C. McKenzie BSc, BVMS, PhD. DACVIM, DACVSMR
Corvallis, OR

INTRODUCTION

Muscular causes of poor performance are somewhat under-recognized phenomena in athletic horses, and are often confused with other conditions causing lameness or suboptimal performance. However, a progressively greater number of research investigations are focusing on the biomechanics of equine athletes, spinal function, and other features that relate directly or indirectly to the muscular system, such as saddle fit.[1,2] Equine athletes have proportionally greater muscle mass than comparably athletic species, with muscle tissue comprising approximately 45% of bodyweight in most equine breeds, and up to 55% of bodyweight in the Thoroughbred. It is also the organ with the greatest energy demand. Despite their massive aerobic capacity, horses have a relatively limited number of slow twitch muscle fibers (≤ 40%) even after considerable endurance training. This feature contributes to their ability for phenomenal speed, but may also increase vulnerability to traumatic and metabolic muscle injury during intense or prolonged efforts. In addition, equine athletes are participating in ever more varied and demanding athletic activities which can promote strain in specific regions of the muscular system during certain activities. Significant muscular issues in athletic horses include post-exercise muscle soreness and muscle strain, traumatic muscle injury, spinal muscle pain and inflammation, and exertional rhabdomyolysis.

Muscle strain and soreness

Depending on the activity in question, athletic horses may participate in brief , intense athletic events, prolonged endurance events, events that require a combination of activities (jumping and dressage), or events that require the horse to perform athletic activities while sustaining a demanding body position such as working in a collected frame or sliding stops. Different situations obviously give rise to different demands in regard to muscular effort or strain, in addition to influencing regulation of cellular energy metabolism and muscle blood flow and oxygenation during exercise. Significant demands on the muscular system of the inappropriately prepared horse can readily create local muscle strain or tears, or may provoke more generalized rhabdomyolysis, depending on the activity in question. In many cases low grade muscle strain is subtle enough clinically to go unnoticed. For example, higher serum muscle enzyme concentrations have been reported in normal horses training two times a week compared to three times a week.[3]A somewhat similar phenomenon was noted in Thoroughbred horses with exertional rhabdomyolysis - even when clinical signs and creatine kinase values were very well controlled in these horses via dietary manipulations, post-exercise serum CK activity was significantly higher after two days of stall rest when compared to values obtained after five consecutive days of exercise.[4] This phenomenon can be extended to a seasonal scale also, with a study of polo ponies identifying more frequent and significant muscle damage early in the polo season compared to later in the season.[5] Collectively, these findings demonstrate the important influence that day to day training management can have in regard to potentially preventing or promoting muscle strain in athletic horses. Additional factors that may contribute to the occurrence of muscle strain include inadequate warm-up procedures, pre-existing lameness, exercising to fatigue, and the performance of novel activities to which the horse is not accustomed. The lumbar and gluteal muscle groups in jumpers, harness horses and dressage horses are commonly affected. The adductor, gluteal, gastrocnemius, biceps femoris, gracilis and semimembranosus/tendinosus muscles of the hind limb and shoulder muscles of the forelimb may also be injured during intense athletic activities such as jumping and barrel racing, or during a fall or contact with other animals or objects. Even with adequate preparation, some degree of muscle strain and necrosis is expected in healthy horses undertaking prolonged exercise events, as evidenced by the fairly prevalent tendency for moderate increases in serum CK activity (of several thousand U/L) in a significant proportion of apparently conditioned endurance horses.[6]

As described previously, muscle strain may be a subtle phenomenon evident only on biochemical investigation. More significant strain however may manifest as subtle or non-specific lameness, tenderness on palpation of the affected muscle group and unwillingness to engage the hindquarters during subsequent activity. Acute severe strain can be expected to result in pain, heat and swelling of the muscle, and possibly lameness if limb musculature is involved. Repetitive injury over time can eventually result in fibrosis and even ossification of abused muscles, which may become abnormally firm on palpation. In severe and chronic cases of repeated injury, mechanical lameness may eventually arise (and will be most prominent at slow gaits), particularly if the semitendinosus muscle is involved, reflecting the development of fibrotic myopathy.

Diagnosis of muscle strain can be difficult since clinical signs are often subtle, and localizing the source of the pain may prove challenging. Physical examination should focus on assessment of muscle symmetry, response to palpation and identification of any other abnormalities of the musculature. Serum CK and AST activity may be normal or mildly increased. Ultrasonography and thermography can assist in detecting inflammation, damage and ossification using the contralateral limb for comparison in unilateral disease.

Treatment of muscle strain usually requires reduced activity or rest, potentially combined with short term anti-inflammatory therapy. Ice therapy of severe acute injuries can provide analgesia and reduce inflammation. A host of accessory modalities are readily available to assist in the rehabilitation of athletic injuries in horses, including application of physical therapy techniques, therapeutic ultrasound and cold laser, acupuncture, electromagnetic therapy, massage and spinal manipulation.[7] Severe muscle strain or tearing may benefit from a combination of these therapies. In addition, a blinded placebo controlled study of dantrolene sodium in human athletes demonstrated substantial reductions in pain and muscle tension and 100% functional recovery in athletes with traumatic muscle contracture.[8] Therefore horses with severe muscle injuries associated with significant pain and inhibited motion (such as with gracilis muscle tears) may benefit from administration of dantrolene at 4 to 6 mg/kg of bodyweight PO twice daily to induce muscle relaxation and to prevent additional damage. Dantrolene is poorly absorbed in fasted horses and therefore should be given no longer than 4 hours after a most recent meal.[9] Other pharmaceuticals that can promote muscle relaxation or vasodilation may be useful including methocarbamol, acepromazine and diazepam. Horses returning to training after muscle strain should be gradually reintroduced to exercise and a thorough and gentle warm-up routine instituted that focuses on gradually increasing the intensity towards the specific activity for the day (e.g. after appropriate warming up with walk, trot and canter, riders preparing for jumping practice can progress through cavaletti work to low jumps to warm up the spinal and hind limb muscles appropriately before commencing more rigorous jumping activity). The prognosis for full return to function is usually good for mild injuries but more guarded for chronic recurrent injuries or in cases where horses will continue to perform an activity that carries a high probability of repeated aggravation.

Spinal muscle pathology

Back pain has received growing attention as a frequent cause of abnormal hind limb motion in performance horses, since progressive research investigations appear to be identifying very high prevalences of muscular and osseous spinal pathology in specific performance groups such as Thoroughbred racehorses.[1] Indeed there are now entire texts devoted to back pathology of the horse, and it is becoming a frequent area of management for veterinarians, physiotherapists and chiropractors that treat performance horses.[10] The degree and detail of available information in regard to back pathology in the horse is now extensive and describes a very complicated relationship between the epaxial musculature and spine and limb function.[1] If detailed examination of the spinal region is consistent with findings of muscular or osseous pathology of the back, further investigations using ultrasonography, thermography, radiography and nuclear scintigraphy can be employed.[10] A remarkable number of rehabilitative modalities are now available as described, and back pain is one of the most likely phenomena to benefit from their application.[1,7]

Exertional rhabdomyolysis

Exertional rhabdomyolysis currently remains the most important and prevalent muscular cause of poor performance in athletic horses.[12] Given the recently enhanced understanding of the varied etiology of this phenomenon, it is now often helpful to consider exertional rhabdomyolysis in light of specific performance or breed groups.

In Thoroughbred horses, exertional rhabdomyolysis was attributed to a heritable disorder of intramuscular calcium regulation after multiple intensive *in vitro* and *in vivo* studies, and epidemiological investigations. The performance horse most likely to be clinically affected with this condition is the young female in race training, however, mild or subclinical disease is a prevalent phenomenon in both genders and may be accompanied by sometimes profound increases in serum CK.[12] Anxiety, pre-existing lameness, high starch diets, and intermittent training patterns are significant precipitating factors for clinical episodes of rhabdomyolysis in genetically predisposed Thoroughbreds. Unfortunately to date, definitive diagnosis of this condition remains challenging since the associated genetic mutation has not been identified, and standard muscle biopsy is frequently unrewarding. Although *in vitro* contracture testing of intercostal muscle bundles is abnormal in affected Thoroughbreds, this is neither a practical nor safe means of diagnosis for clinical cases. In general, affected Thoroughbreds can usually be successfully managed by substituting a large part of the dietary carbohydrate fraction with fat sources, instituting measures to reduce stress and anxiety, and by establishing a regular exercise routine with no more than 24 consecutive hours of stall rest. Dantrolene sodium is commonly used in Thoroughbred training yards in the United Kingdom to control clinical episodes in vulnerable horses, and clinical investigation of this agent in Thoroughbred horses definitively diagnosed with RER supported its efficacy for this purpose.[13]

Quarter horses, Appaloosas and Paint horses usually have exertional rhabdomyolysis associated with Polysaccharide Storage Myopathy (PSSM), a hereditary disorder of glycogen metabolism. Approximately 80% of Quarter horses and a high proportion of Appaloosa and Paint horses are heterozygous for a mutation of the glycogen synthase gene (*GYS1*). Horses that are homozygous for this mutation or concurrently carrying a known mutation of the ryanodine receptor gene (*RYR1*) tend to have more severe clinical disease. Of particular concern amongst performance groups in this classification is the US Halter horse group, in which a recent study incorporating a high number of elite animals identified a particularly high prevalence of the PSSM and HYPP genetic alleles. In addition 20% of halter horses were found to be carrying alleles for two or more of the known hereditary equine disorders.[14]

The last performance group that also has a substantial prevalence of muscular disease is the Warmbloods. The most common muscular disorder in this group is PSSM, however, distinct differences are evident between Warmbloods and the Quarter horse group. In contrast to the Quarter horse group, a relatively low percentage (~20%) of Warmbloods with PSSM carry the *GYS1* mutation, in spite of classic histopathologic findings consistent with this disease in affected animals. In addition, Warmbloods with PSSM frequently have normal muscle enzyme activities, and only approximately 20% of affected horses display typical signs of exertional rhabdomyolysis. Clinical signs of muscular disease in this breed are more likely to consist of reluctance to move forward or perform in a collected frame, or inability to back without transiently elevating and flexing the hind limb (shivers) which may complicate diagnosis. A low percentage of Warmbloods with muscular disease have also been diagnosed with recurrent exertional rhabdomyolysis.[15] In spite of these differences however, both Quarter horses and Warmbloods with PSSM can show good clinical responses to similar dietary and exercise recommendations provided for Thoroughbreds, with the exception that clinical signs consistent with shivers are likely to persist.

REFERENCES

[1]Stubbs, N. **AAEP Proceedings** 2011;57:153; [2]von Peinen K, et al. **Equine vet. J Suppl** 2010;42:650; [3]Lindner A, et al. **Equine Vet J Suppl** 2006;36:88; [4]McKenzie EC, et al. **J Vet Intern Med** 2003;17(5):693; [5]McGowan CM, et al. **Vet Rec** 2002;150(17):53; [6]Barton MH, et al. **Am J Vet Res** 2003;64(6):746; [7]Haussler KK. **Equine vet. J Suppl** 2010;42(38):695; [8]Flacco L, et al. **Clin Ther** 1989;11(5):623; [9] McKenzie E, et al. **Equine Vet J Suppl** 2010;42(38):613; [10]Henson MD (ed). Equine Back Pathology. Wiley Blackwell, 2009; [11]Haussler KK, et al. **AAEP Proceedings** 2011;57:141;[12]Martin BB Jr, et al. **J Am Vet Med Assoc** 2000;216(4):554; [13]McKenzie, et al. **Am J Vet Res** 2004;65(1):74; [14] Tryon RC, et al. **J Am Vet Med Assoc** 2009;234(1):120;[15]Hunt LM, et al. **Equine Vet J** 2008;40(2):171.

DETECTION OF *S. EQUI* IN WASH SAMPLES AND SEM TITERS FOLLOWING STRANGLES OUTBREAKS

Ashley G. Boyle, DVM, DACVIM
Kennett Square, PA

INTRODUCTION

Strangles is a highly contagious disease of the respiratory tract of horses caused by *Streptococcus equi* subsp *equi*. Vaccination against *S equi* is common, but the prevalence of secondary complications associated with currently available intramuscular and intranasal *S equi* vaccines is perceived to be higher than prevalence for other vaccines commonly used in horses. A rare but serious possible sequela to vaccination against *S equi* is purpura hemorrhagica, an immune-mediated vasculitis that leads to a type III hypersensitivity reaction.[1,2]The true incidence of purpura hemorrhagica secondary to strangles or vaccination against *S equi* is not known.[1,3] However, in a retrospective study[4] of 53 horses with purpura hemorrhagica, 4 reportedly had received an *S equi* M protein extract vaccine IM during the 2 weeks prior to admission. In addition, a study[5] from 1999 reported an incidence of 2 cases of purpura-like disease for every 100,000 doses of attenuated-live intranasal *S equi* vaccine sold.

Various authors have suggested that horses with high serum SeM-specific antibody titers may be predisposed to developing purpura hemorrhagica when vaccinated against *S equi*,[1,5] and the 2005 consensus statement[1]for the American College of Veterinary Internal Medicine recommended both that horses with SeM-specific antibody titers ≥ 1:1,600 and that horses with SeM-specific antibody titers ≥ 1:3,200 not be vaccinated against *S equi*. It has been suggested that SeM-specific antibody titers be measured prior to instituting a vaccination program,[1] but the cost of testing an individual horse is approximately 4.5 times the cost of vaccinating the horse. Testing also requires an additional farm call fee.[6]

SEM TITERS IN HEALTHY HORSES

A 2009 study[6] identified factors associated with an increased likelihood that horses would have SeM-specific antibody titers ≥ 1:1,600. This was a cross-sectional study in which we obtained a single serum sample from 188 healthy, client-owned animals which was tested for SeM-specific antibody titer with an ELISA. Multivariate logistic regression was used to identify factors associated with having a titer ≥ 1:1,600. Age, breed, and vaccination status were significantly associated with the likelihood of having a titer ≥ 1:1,600. The odds of having a titer ≥ 1:1,600 increased by a factor of 1.07 with each 1-year increase in age. Quarter Horses and horses of other breeds were 4.08 times as likely to have a titer this high as were Thoroughbreds and Warmbloods. Horses that had previously received an intranasal *S equi* vaccine were 4.7 times as likely to have a titer this high as were horses without any history of vaccination. Results indicated that older horses, horses other than Thoroughbreds and Warmbloods, and horses that had been vaccinated with an attenuated-live intranasal *S equi* vaccine between 1 and 3 years previously had an increased likelihood of having a serum SeM-specific antibody titer ≥ 1:1,600. We also found that 26/35 (74%) horses that had been vaccinated with IN strangles when their SeM titer was equal to 1:1,600 did not have any complications associated with purpura.[6] Since this study, the common recommendation by the laboratories that provide the SeM ELISA is to not vaccinate when the titer is ≥ 1: 3,200.[7]

SEM TITERS POST *S. EQUI* OUTBREAK

We then followed the SeM-specific antibody titers of horses post a natural *Streptococcus equi* infection in order to make a more accurate, clinical recommendation of when to vaccinate horses after a *Streptococcus equi* infection outbreak. This was a cross-sectional study in which we obtained serum samples 6 weeks, and 3, 6, 9, 12 months after the resolution of the last clinical case to test for *S. equi* SeM-specific antibody titer using ELISA in three outbreaks consisting of 57 horses. If *S. equi* SeM-specific antibody titers were not < 1:3,200 by 12 months, additional samples were obtained at 18 and 24 months. All titers were ≤1:1,600 by 9 months post outbreak A, ≤ 1:1,600 by 27 months post outbreak B, and ≤ 1:1,600 by 16 months post outbreak C. We also evaluated the risk factors that resulted in titers ≥ 1:3,200.[8]

DETECTION OF *S. EQUI* IN WASH SAMPLES

As veterinary clinical practices have become more refined, it has become necessary to test both symptomatic animals and potential carrier animals for the etiological agent: *S. equi*. Bacterial culture and polymerase chain reaction (PCR) of nasopharyngeal washes and guttural pouch lavages have been used for both diagnostic testing and for the detection of *S. equi* in carrier animals, but currently have low sensitivities.[9-12] Flocked swabs have been used in human medicine to increase the sensitivity of tests.[12-16] Currently, there is no standard laboratory method being used to perform either bacterial culture or PCR on nasopharyngeal wash and guttural pouch lavage samples.

Using 7 known serial dilutions of *S. equi* in 0.9% NaCl, the sensitivities for 4 methods of processing samples for bacterial culture were compared (BBL™ CultureSwab™, flocked swab, centrifugation and sampling with a loop, and centrifugation and sampling with BBL™ CultureSwab™) and for 3 methods of processing samples prior to DNA amplification (BBL™CultureSwab™ immersed in 1 ml NaCl, centrifugation of a 1 ml aliquot, and flocked swab immersed in 1 ml NaCl).

The DNA amplification assays were the most sensitive in their ability to detect low numbers of *S. equi*. The flocked swab method was superior to all other methods tested at the higher organism concentrations (>100 cfu/ml). Flocked swab culture, flocked swab/PCR, and 1 ml aliquot PCR methods will be compared in a future field trial using strangles animals and outwardly healthy animals screened for strangles carrier status.[17]

REFERENCES

[1]Sweeney C, et al. **J Vet Intern Med** 2005;19:123. [2]Galan JE, et al. **J Immunol** 1985;135:3134. [3]Sweeney CR et al. **J Am Vet Med Assoc** 1989;9:1281. [4]Pusterla N et al. **Vet Rec** 2003;153:118. [5]Timoney JF ***Proceedings*. 45th Annu Meet Am Assoc Equine Pract** 1999;31. [6] Boyle AG et al. **J Am Vet Med Assoc** 2009;235: 973. [7]**IDEXX Web site**. *S. equi* ELISA testing recommendations. Available at: www.idexx.com/equine/laboratory/sequi_elisarecommend.jsp. 2008. [8]Boyle AG, et al. unpublished data. [9]Timoney JF et al. **Vet Rec** 1997;141: 446. [10]Gronbaek, LM et al. **Equine Vet J** 2006;38:59. [11] Newton JR et al. **Vet Rec**1997;140:84. [12] Newton JR et al. **Equine Vet J** 2000;32:515. [13] Chernesky M et al. **J Clin Microbiol** 2006;44:1084. [14] Goldfarb DM et al. **J Clin Microbiol** 2009;47:3029. [15] Goldfarb D et al. **An Clin Microbiol Antimicrob** 2011;10:1. [16] Van Horn KG et al. **Diagn Microbiol Infect Dis** 2008; 62:471. [17] Boyle AG et al. Abstr. 53rd **Ann meet Am Ass Vet Lab Diagn**, 2010; abstr. 35.

METHICILLIN-RESISTANT *STAPHYLOCOCCUS AUREUS* (MRSA) IN HORSES

J. Scott Weese, DVM, DVSc, DACVIM
Guelph, Ontario, Canada

INTRODUCTION

Staphylococcus aureus has historically been an important opportunistic pathogen in a number of animal species, including humans and horses. In the late 20th century, methicillin-resistant *S. aureus* (MRSA) emerged as a leading cause of hospital-associated infection in humans, and it is now among the most important infections in hospitalized individuals.[1] Since then, MRSA has emerged as a significant community-associated pathogen, causing disease in people in the general population, including those that would previously be considered at low risk for infection.[2] Subsequently, there has been increasing recognition of MRSA in various animal species, in large part because of spread of MRSA from humans. While MRSA has emerged in animals at a slower rate, this pathogen is now a significant concern in veterinary medicine, involving both animal and public health issues.

Methicillin-resistance

Staphylococci, particularly *S. aureus*, have been notable in their tendency to acquire resistance to antimicrobials, and introduction of new drugs (often to treat resistant *S. aureus*) has lead to further development of resistance. Resistance to penicillin emerged not long after penicillin was in widespread use. This resistance, based on beta-lactamase production, is now very common.[3-5] Subsequent introductions of new antimicrobials were typically met with rapid identification of new resistance, a pattern that included recognition of methicillin-resistance shortly after the drug was released for use in humans. Methicillin resistance in MRSA is mediated by production of an altered penicillin-binding protein (PBP2a) that possesses low affinity for beta-lactam antimicrobials. Accordingly, MRSA strains are resistant to virtually all beta-lactam antimicrobials. Resistance is mediated by the *mecA* gene, which resides on a chromosomal cassette (SCC*mec*). This cassette can also harbour various other resistance genes, and MRSA isolates are often resistant to numerous additional antimicrobial classes, thereby further limiting treatment options.

Emergence of MRSA in horses

Initial reports of MRSA infection in horses were largely sporadic infections or small clusters, mainly associated with veterinary hospitals,[6-8] with the general thought that horses were directly infected by colonized humans. This progressed to reports of larger numbers of MRSA infections in horses internationally, involving both veterinary hospitals and primary community-onset infections, and with evidence of horse-horse transmission. Additionally, transmission between humans and horses, in both directions, was observed in equine hospitals and on farms.

Clinical Presentation

Clinical MRSA infections can occur as sporadic cases or in outbreaks involving a wide range of opportunistic infections. Joint, incision and skin/soft tissue infections are most common in community-onset cases, with surgical site infections predominating in hospitalized horses,[7-11] but a variety of other infections have been reported, including pneumonia, omphalophlebitis, sinusitis, bloodstream infection, invasive device infection, osteomyelitis, tenosynovitis, metritis and mastitis.[6,10,12-14]

The severity of disease can be quite variable, from mild and superficial infection to aggressive and life-threatening. There is no evidence that MRSA infections have a different clinical presentation than infections by methicillin-susceptible *S. aureus* (MSSA) or other opportunists. A multicentre study did not identify a significant difference in mortality of MRSA versus MSSA infections, with 84% of horses with MRSA infections surviving until the time of discharge.[11] This is an important aspect as SCC*mec* is a resistance determinant, not a virulence factor. The inherent virulence of MRSA and MSSA should be the same and the potential for more serious illness and worse outcome with MRSA relates to inadequate or ill-timed treatment. Therefore, prompt diagnosis and implementation of appropriate treatment is likely the most important determinant of outcome in MRSA vs MSSA infections.

Colonization

Only a small percentage of horses that are exposed to MRSA develop a clinical infection. Others either eliminate the organism completely or become colonized (most often in the nares). The prevalence of MRSA colonization has been investigated in various horse populations, with rates of 0-11% reported in horses in the community and upon admission to veterinary hospitals, and it is likely that MRSA is endemic in the horse population throughout much of the world. MRSA can cluster on horse farms, with colonization rates approaching 50% occasionally being found.[15] High rates can also be found in some hospitalized populations.[16] The nasal passages seem to be the primary site of colonization, however MRSA can be found at other sites such as the intestinal tract and skin.[16,17] While the majority of colonized horses do not develop clinical infection, colonization at the time of admission was identified as a risk factor for development of a clinical MRSA infection in hospitalized horses.[18]

Molecular Epidemiology

Most initial reports of MRSA in horses involved a sequence type (ST) 8 strain called Canadian MRSA-5 or USA500.[9,10,19,20] This is a human epidemic clone however it only accounts for a small percentage of human infections,[21] and the predominance of this clone in horses and horse personnel in many regions suggests that this human-origin clone is horse-adapted. Other human epidemic clones have also been identified, and recently there have been reports of colonization and infection with sequence type (ST) 398 in horses in Europe[22,23] and Canada,[24] including a study that identified ST398 MRSA colonization in 11% of horses admitted to a Belgian teaching hospital.[22] This is a livestock-associated strain that is endemic in pigs internationally but which can also infect other species, including humans. A large outbreak of MRSA predominantly caused by ST398 was recently reported horses in a Dutch veterinary teaching hospital.[25] Considering the recent identification of ST398 in horses and its prevalence in food animals, it is reasonable to suspect that ST398 has entered the horse population directly or indirectly from food animals. It is currently unclear whether this may be an emerging problem in horses.

Risk factors for MRSA colonization and infection have been only superficially investigated. Administration of ceftiofur or aminoglycosides was associated with acquisition of MRSA during hospitalization.[18] Previous colonization of the horse, previous identification of colonized horses on the farm, antimicrobial administration within 30 days, admission to the neonatal intensive care unit, and admission to a service other than the surgical service were risk factors for community-associated colonization in another study.[26] Factors associated with clinical infection are unclear.

Diagnosis

Diagnosis of MRSA infection involves isolation of *S. aureus* from an infected site followed by determination of methicillin resistance. A key aspect for proper case management is timely and proper collection of diagnostic specimens. There are no clinical or historical factors that indicate MRSA over any other opportunistic infection, so factors like a lack of antimicrobial exposure, hospitalization or contact with a person with MRSA infection or colonization should not be taken to indicate that testing is not required.

Staphylococci are relatively easy to isolate and standard protocols are well described for detection of methicillin resistance. Cefoxitin has replaced oxacillin as the *in vitro* marker of methicillin resistance, and cefoxitin testing should be performed on all *S. aureus* isolates. Care should be taken with reports of isolates that are multidrug resistant but not identified as methicillin-resistant, since while multidrug resistance can occur in the absence of methicillin-resistance, it is rather uncommon. Questionable isolates should be tested for the presence of *mecA* by PCR or PBP2a by latex agglutination test. Veterinarians should try to ensure that the laboratory they use tests all *S. aureus* for methicillin-resistance by an accepted method.

Treatment

Treatment of MRSA is no different than treatment of susceptible staphylococcal infections, besides the antimicrobial that is used. The principles of MRSA therapy are no different than the principles of treatment of any other infection, and the simple presence of MRSA does not mean that longer or more intensive treatment is indicated.

Some superficial infections may be treated topically with antibacterials or antiseptics such chlorhexidine digluconate, silver sulfadiazine, fusidic acid, mupirocin, povidone iodine or other topical antimicrobials. Miconazole, while an antifungal, has excellent anti-staphylococcal properties. Honey may be another effective option, particularly with large wound infections.

While topical therapy alone can be highly effective for superficial infection, systemic antimicrobial therapy will be required in most cases, alone or in conjunction with local therapy. Drug choices should be based on *in vitro* antimicrobial susceptibility testing in combination with other patient and drug factors (e.g. drug penetration, concentration at the site of infection, activity at the site of infection, drug interactions, etc.) The level of resistance possessed by MRSA isolates can be quite variable, ranging from resistance to only beta-lactams to resistance to most antimicrobials. It is critical to remember that commonly used antimicrobials can be highly effective if the isolate is sensitive *in vitro*, and if the chosen drug is able to achieve therapeutic levels at the site of infection. Therefore, identification of an MRSA infection does not mean that the latest generation or most potent antimicrobials are required, particularly in non-life-threatening infections. Fluoroquinolones tend not to be good options for treatment because resistance can emerge very quickly and because *in vitro* susceptibility does not always correlate well with clinical response. Extended susceptibility testing may need to be requested from diagnostic laboratories if there are no viable options among the antimicrobials that are routinely tested. Ideally, drugs that are important for treatment of critically ill humans should be avoided and it is extremely rare to encounter a situation where these need to be considered.

In addition to systemic antimicrobials, local therapy consisting of regional limb perfusion, intra-articular or intra-osseous adminstration of antimicrobials, or implantation of antimicrobial-impregnated materials may be useful in some cases.

Decolonization

Colonized horses are a potential source of infection for other horses and humans, and are also at risk of MRSA infection in certain circumstances. Fortunately, MRSA colonization in horses appears to be transient in most cases, with decolonization occurring naturally if reinfection is prevented.[15] This is an important aspect for control of MRSA and management of infected or colonized horses. Because of the transient nature of colonization and lack of any evidence demonstrating that active decolonization (e.g. topical or systemic antimicrobials) is effective, active decolonization is not current recommended.

Public Health Concerns

There are significant public health concerns about MRSA in animals, most of which have only been preliminarily investigated. Most of the concern involves the potential for animals to act as reservoirs of MRSA with subsequent transmission to humans, irrespective of whether MRSA originated in humans. Zoonotic infections have also been reported in people working with colonized or infected horses [10,27] and it is clear that MRSA can be transmitted between people and both healthy and infected horses. Regardless, while information from most studies should be considered preliminary and circumstantial, it is very likely that interspecies transmission can occur in households and on horse farms, in both directions.

There have been numerous studies of MRSA colonization in veterinary personnel, indicating abnormally high rates in veterinarians, particularly equine and swine veterinarians. Studies of equine veterinarians have reported a predominance of clones associated with horses (i.e. CMRSA-5, CC8),[10,13,20,28] despite those clones being relatively uncommon in the general population, providing support to the hypothesis that infection is acquired from horses. Having diagnosed a horse with MRSA or having been personally diagnosed with MRSA over the past year were risk factors for colonization of equine veterinarians in another study.[29] An important result from the latter study was a statistically significant protective effect of routine hand hygiene following contact with infectious cases and between farms, demonstrating the importance of routine infection control practices such as hand hygiene on MRSA control. Zoonotic transmission of ST398 has also been reported.[30]

Equine Hospital MRSA Control

MRSA is a potential concern in equine hospitals from many standpoints, including sporadic disease, outbreaks and transmission to hospital personnel. While there are few published reported of MRSA in veterinary hospitals, anecdotal information abounds and MRSA was the 2nd most commonly reported cause of outbreaks in a survey of veterinary teaching hospitals.[31] Challenges encountered in MRSA control in equine facilities include the presence of MRSA colonization in horses at admission, colonization of veterinary personnel and lack of objective data regarding control measures. Information from human medicine, where abundant research has been performed, can provide some information but the many differences hamper direct extrapolation of human data. In general, MRSA control in equine hospitals revolves around good general hygiene practices (especially hand hygiene), routine cleaning and disinfection practices, cohorting of risk groups, isolation of infected animals, judicious antimicrobial use, passive surveillance (monitoring routine culture results) and ensuring prompt culture of infected sites. MRSA screening is widely used in humans as a key control measure, but it is uncommonly used in horses. Active surveillance of horses at admission (+/- further screening during hospitalization and discharge) is used in some facilities and can identify colonized horses to allow for prompt isolation, facilitate identification of transmission of MRSA in the hospital and identify at-risk horses (e.g. horses from specific farms with high MRSA rates), but the cost is often prohibitive. Surveillance is more often used as a transient measure during outbreaks, but the cost-benefit of routine screening should be considered by all facilities, taking into account the infection pressure, facility issues (e.g. isolation capacity), risk aversity and cost. Very rarely is screening of personnel indicated or useful. Facilities should assume that one or more of their personnel are colonized and structure routine infection control practices to reduce the implications of colonized personnel, since logistical, privacy, legal and efficacy challenges preclude the use of personnel screening as an effective and useful tool.

REFERENCES

[1] Klevens RM, et al. *JAMA* 2007;298:1763-1771. [2] Frazee B, et al. ***Ann Emerg Med*** 2005;45:311-320. [3] Sieber S, et al. ***Microb Drug Resist*** 2011;17:471-478. [4] Moore CP, et al. ***J Am Vet Med Assoc*** 1995;207:928-933. [5] Clark C, et al. ***Can Vet J*** 2008;49:153-160. [6] Shimizu A, et al. ***J Vet Med Sci*** 1997;59:935-937. [7] Hartmann FA, et al. ***J Am Vet Med Assoc*** 1997;211:590-592. [8] Seguin JC, et al. ***J Clin Microbiol*** 1999;37:1459-1463. [9] O'Mahony R, et al. ***Vet Microbiol*** 2005;109:285-296. [10] Weese J, et al. ***Emerging Infect Dis*** 2005;11:430-435. [11] Anderson M, et al. ***Equine Veterinary Journal*** 2009;41:401-405. [12] Baptiste K, et al. ***Emerg Infect Dis*** 2005;11:1942-1944. [13] Cuny C, et al. ***Euro Surveill*** 2006;11:44-47. [14] Middleton J, et al. ***J Clin Microbiol*** 2005;43:2916-2919. [15] Weese J, et al. ***Equine Vet J*** 2005;37:510-514. [16] Van den Eede A, et al. ***Veterinary journal*** 2012. [17] Weese JS, et al. ***Emerging Infect Dis*** 2005;11:430-435. [18] Weese J, et al. ***J Vet Intern Med*** 2006;20:182-186. [19] Anderson M, et al. ***Vet Microbiol*** 2007;122:185-189. [20] Weese J, et al. ***J Am Vet Med Assoc*** 2005;226:580-583. [21] Christianson S, et al. ***J Clin Microbiol*** 2007;45:1904-1911. [22] Van den Eede A, et al. ***Veterinary Microbiology*** 2008;ePub ahead of print. [23] Witte W, et al. ***Emerging Infect Dis*** 2007;13:255-258. [24] Tokateloff N, et al. ***Can Vet J*** 2009;50:1177-1180. [25] van Duijkeren E, et al. ***Vet Microbiol*** 2009. [26] Weese JS, et al. ***Can Vet J*** 2007;48:921-926. [27] Weese J, et al. ***Vet Microbiol*** 2006;114:160-164. [28] Hanselman B, et al. ***Emerg Infect Dis*** 2006;12:1933-1938. [29] Anderson ME, et al. ***Vet Microbiol*** 2007;129:410-417. [30] van Duijkeren E, et al. ***Emerging Infect Dis*** 2011;17:1137-1139. [31] Benedict KM, et al. ***J Am Vet Med Assoc*** 2008;233:767-773.

85a

COMPARISON OF PLASMA ACTH CONCENTRATION, PLASMA α-MSH CONCENTRATION, AND OVERNIGHT DEXAMETHASONE SUPPRESSION TEST FOR TEST FOR DIAGNOSIS OF PPID. D McFarlane, MA Breshears, M Cordero, K Hill, R Carmichael, LK Maxwell. Oklahoma State University, Stillwater, OK.

Despite the suggestion that the overnight dexamethasone suppression test (DST) is the gold standard for the diagnosis of equine pituitary pars intermedia dysfunction (PPID), its performance has never been directly compared to that of plasma ACTH or α-MSH concentration for identification of PPID. We hypothesized that measurement of plasma ACTH or α-MSH concentration would more accurately predict PPID than would a DST.

One hundred and twenty equids (mean age ± SD: 21.5 ± 5.2 yrs) that presented for euthanasia were included. Horses were categorized as clinically positive if they had an overtly abnormal haircoat consistent with PPID. Plasma ACTH and α-MSH concentration were measured and a DST performed. Following euthanasia, an H&E stained sagittal section of the pituitary gland was scored (1-5) and the pars intermedia to total pituitary (PI/TPit) area was determined using morphometric analysis. Test results were considered positive for PPID at α-MSH > 35 pmol/L, ACTH >50 pg/ml or cortisol ≥1 ug/dl 19 hours after dexamethasone.

There were 105 clinically normal horses and 15 (12.5%) with signs of PPID. PI/TPit area significantly increased with an increase in pituitary score (P<0.0001). PI/TPit area was positively correlated to plasma α-MSH (n=120, r=0.5, P<0.0001), plasma ACTH (n=120, r=0.2, P=0.01) and serum cortisol post dexamethasone (n=93, r=0.4, P=0.0002), but not serum cortisol prior to dexamethasone. DST results differed from ACTH in 21.5% of cases and from α-MSH in 16% of cases. ACTH and α-MSH differed in 25% of cases. All animals with grade 1 or 2 pituitary glands (n=48) were clinically normal; mean plasma ACTH and α-MSH concentration did not differ between grades 1 and 2. There were 2 horses with a positive α-MSH test, 13 horses with a positive ACTH test (2 in fall) and one horse that failed to suppress following DST (in fall) in grade 1 and 2. All animals with grade 5 pituitaries (n=11) were positive on all 3 diagnostic tests. All animals with a PI/TPit area <0.3 were clinically normal, all horses with clinical PPID had a PI/Pit area >0.3. Using a PI/Pit area > 0.3 as a positive PPID diagnosis, sensitivity and specificity in non-fall samples for the ACTH test was 71% and 81%, for α-MSH was 63% and 90%, and for DST was 65% and 98%, respectively. The best results were obtained when both DST and α-MSH were performed and a positive result by either test was used to diagnosis PPID (sensitivity= 80%, specificity= 90%). In conclusion, all three diagnostic tests were poor at correctly identifying non-end stage cases of PPID. Improved diagnostic tests are needed for early diagnosis of PPID.

85b

EVALUATION OF THE SAFETY AND CLINICAL EFFICACY OF PRASCEND® (PERGOLIDE MESYLATE) TABLETS FOR THE CONTROL OF CLINICAL SIGNS ASSOCIATED WITH PITUITARY PARS INTERMEDIA DYSFUNTION IN EQUIDS. H Schott II[1], D McFarlane[2], F Andrews[3], J Marteniuk[1], A Stokes[3], R Bimes[4], J Hunt[5], D Kolb[6], S Toppin[7], G White[8]. 1. MSU, East Lansing, MI; 2. OSU, Stillwater, OK; 3. LSU, Baton Rouge, LA; 4. Quakertown, PA; 5. Carthage, IN; 6. Lodi, WI; 7. Littleton, CO; 8. Sallisaw, OK .

To investigate the safety and efficacy of Prascend® (pergolide mesylate, 1 mg tablets) in controlling clinical signs of pituitary pars intermedia dysfunction (PPID) or Equine Cushing's Disease (ECD), 122 equids (59 male, 63 female, 10-35 yr, 137-623 kg, and 16 breeds) were evaluated in a multi-site, clinical field efficacy study. Equids were enrolled between 11/1/08 and 1/31/09 by physical examination and endocrine testing. Animals were scored (0-3) for hirsutism, hyperhidrosis, polyuria-polydipsia, abnormal fat distribution, and muscle wasting. Inclusion criteria were a hirsutism score ≥1 AND either a plasma adrenocorticotropin (ACTH) concentration ≥50 pg/ml or failure of endogenous cortisol concentration to suppress (<1.0 μg/dl) 19 h after dexamethasone administration (40 μg/kg, IM). Treatment with pergolide mesylate (2 μg/kg, PO, q 24 h) was initiated at study enrollment. Animals were re-evaluated (physical exam and endocrine testing) after 90 d (n=113) and 180 d (n=111) of treatment. When endocrine test results remained abnormal at 90 d (n=47/113) the dose was increased to 4 μg/kg, PO, q 24 h. Treatment success after 180 d was defined as either normalization of dexamethasone suppression test results (<1.0 ug/dl) or a decrease in plasma ACTH concentration by 50% (or return to normal [<50 pg/ml] by reduction of at least 5 pg/ml) AND improvement by a score of ≥1 in at least one clinical sign, without worsening of other clinical signs. Treatment was also considered successful when the sum of clinical scores improved by 3 or more, regardless of endocrine test results, as long as there was no worsening of any clinical signs.

In total, 76% (86/113) equids were treatment successes at Day 180. Of the 122 equids enrolled, eight (8) animals died and one (1) was euthanized due to worsening of pre-existing conditions or colic and these horses were excluded from evaluations. Both median scores of clinical signs and mean concentrations of ACTH and cortisol decreased during the study period. Transient inappetance was the most common adverse event observed in 40/122 (33%) equids, mostly during the initial 30 d of treatment. Treatment with Prascend® resulted in improvement in clinical signs and endocrine test results and is efficacious in treatment of PPID.

86b

CHARACTERIZATION OF THE EQUINE METABOLIC SYNDROME PHENOTYPE – PART I: BIOCHEMISTRY, MORPHOMETRICS AND BREED COMPARISONS.

Schultz NE[1], McCue ME[1], Martinson K[2], Frank N[3], Geor RJ[4]. [1]Colleges of Veterinary Medicine and [2]Food, Agricultural and Natural Resources, University of Minnesota, St. Paul MN [3]Cummings School of Veterinary Medicine, Tufts University, North Grafton, MA [4]College of Veterinary Medicine, Michigan State University, East Lansing MI.

Equine Metabolic Syndrome (EMS) is characterized by increased adiposity, insulin resistance, hyperinsulinemia and laminitis. While adiposity and insulin resistance have been reported as defining features of EMS, physical and biochemical measures have varied between studies, leading to uncertainty in diagnostic criteria. The purpose of this study was to identify differences in morphometric and biochemical measurements in EMS suspects with (EMS-L) and without (EMS-NL) a clinical history of laminitis, when compared to control horses and to identify differences in phenotypic criteria among breeds. As part of an ongoing effort to establish phenotypic criteria in ~300 EMS cases and ~300 controls; we have collected epidemiologic, morphometric, biochemical, and environmental data from 181 EMS suspects (132 EMS-L, and 49 EMS-NL) and 205 controls. Horses classified as EMS suspects had generalized or regional adiposity; were considered "easy keepers" by owners, and were defined as EMS suspects by a referring DVM. Samples for biochemical measurements were taken before and after an oral sugar test (OST). After an ~8 hour fast, baseline blood samples were collected for measurement of insulin, glucose, triglycerides and NEFA concentrations. After baseline sample collection, 0.25 ml/kg corn syrup was administered per os; at 75 min post OST, samples were collected for measurement of insulin and glucose concentrations. Differences in morphometric and biochemical measurements between clinical groups (control, EMS-NL and EMS-L) and breeds (Welsh, Tennessee Walking-Horse, Quarter Horse, Arabian, Paso Fino and Morgan) were determined using ANOVA or Kruskal-Wallis rank sum test, where appropriate. Baseline glucose concentrations were not significantly different between groups. Post-OST glucose values were significantly higher in EMS-L horses, but not EMS-NL horses when compared to controls. Baseline insulin and post-OST insulin were significantly higher in EMS-NL and EMS-L horses compared to controls. Triglyceride but not NEFA concentrations were significantly higher in EMS-NL and EMS-L horses. Breed differences were detected in girth- and neck-to-height ratios, baseline insulin, post-OST glucose and insulin, triglyceride and NEFA concentrations. Across breeds, baseline insulin, insulin-OST and triglycerides were good predictors of EMS status, although breed differences in these measurements suggest that breed-specific cut-off values may improve diagnostic accuracy. In conclusion, our results confirm that dysregulation of insulin dynamics is a central feature of the EMS phenotype across horse breeds. Further, the size and across-breed nature of this sampling cohort should clarify uncertainties arising from phenotypic variability in previous studies and lead to improved diagnostic criteria for EMS.

86b

CHARACTERIZATION OF THE EQUINE METABOLIC SYNDROME PHENOTYPE – PART II: IMPACT OF DIETARY COMPONENTS ON PHENOTYPIC MEASUREMENTS. Schultz, NE[1], McCue[1], ME, Martinson[2], K, Frank[3], N, Geor RJ[4]. [1]Colleges of Veterinary Medicine and [2]Food, Agricultural and Natural Resources, University of Minnesota, St. Paul MN [3]Cummings School of Veterinary Medicine, Tufts University, North Grafton, MA [4]College of Veterinary Medicine, Michigan State University, East Lansing MI.

Previous studies in healthy horses have demonstrated that dietary adaptation has an impact on many of the phenotypic variables measured and used as diagnostic criteria for EMS. However, less is understood about the relationship between dietary intake and measured responses in EMS horses. Our objectives were to evaluate the relationships between dietary components and phenotypic variables used as EMS criteria and to determine if the relationship differs between control horses and EMS suspects. We have collected biochemical and dietary data from 181 EMS suspects with (EMS-L, n=132) and without (EMS-NL, n=49) a clinical history of laminitis and 205 controls. Biochemical variables were measured in the fasted state (insulin, glucose, triglycerides and NEFA concentrations) and after an oral sugar test (OST, glucose and insulin concentrations). For each individual, total daily dietary consumption was calculated by weighing the daily rations of hay and concentrate and estimating pasture consumption based on hours of pasture access/day. Concentrate, hay and pasture samples were analyzed by Equi-Analytical (Ithaca, NY). Total daily digestible energy (DE), crude protein (CP), acid detergent fiber (ADF), nonstructural carbohydrates, water and ethanol soluble carbohydrates, fructans and starch were calculated as a % body weight in dry matter intake. DE was significantly different across clinical groups, with DE highest in controls, followed by EMS-NL and EMS-L. Therefore, correlations between dietary intake and phenotypic measurements were determined using multivariate linear regression including all dietary components, while controlling for the effects of EMS status. Significant correlations occurred between fasting glucose and DE, CP, ADF and fructan; whereas post-OST glucose was correlated with CP and starch. Both fasting and post-OST insulin were significantly correlated with fructan concentration; post-OST insulin was also correlated with DE, CP and ADF. Triglyceride concentrations were not correlated with dietary components. In contrast NEFA concentration was correlated with DE and CP. In addition to dietary components, EMS status accounted for significant variation in fasting insulin, post-OST insulin and glucose, and triglyceride concentrations. Differences in the correlations between these measurements and dietary components in different clinical groups were identified by modeling the interaction between dietary component(s) and clinical status in the regression equation. Significant interactions were detected for both post-OST insulin and triglycerides. These data suggest that the response to dietary adaptation may be different between EMS and control horses. Further, the impact of dietary adaptation may need to be taken into account when interpreting diagnostic tests in EMS horses.

CRITICAL ILLNES RELATED CORTICOSTEROID INSUFFICIENCY

Paul E. Marik, MD, FCP(SA), FRCP(C), FCCP, FCCM, FACP
Norfolk, VA

Exposure of the host to diverse noxious stimuli results in a stereotypic and coordinated response, referred to by Hans Selye as the "*general adaption syndrome*" (or stress response) which serves to restore homeostasis and enhance survival. (1) The stress response is mediated primarily by the hypothalamic-pituitary-adrenal (HPA) axis as well as the sympathoadrenal system (SAS). Activation of the HPA axis results in increased secretion from the paraventricular nucleus of the hypothalamus of corticotropin-releasing hormone (CRH) and arginine vasopressin (AVP). CRH plays a pivotal integrative role in the response to stress. CRH stimulates the production of ACTH by the anterior pituitary, causing the zona fasciculata of the adrenal cortex to produce more glucocorticoids (cortisol in humans). AVP is a weak ACTH secretagogue and vasoactive peptide that acts synergistically with CRH to increase secretion of ACTH. The increase in cortisol production results in multiple effects (metabolic, cardiovascular, and immune) aimed at restoring homeostasis during stress. In addition, the HPA axis and immune system are closely integrated in multiple positive and negative feed-back loops. Activation of the SAS results in the secretion of epinephrine and norepinephrine from the adrenal medulla and to an increased production of inflammatory cytokines such as interleukin-6 (IL-6).

CORTISOL PHYSIOLOGY

Cortisol (hydrocortisone) is the major endogenous glucocorticoid secreted by the adrenal cortex. Over 90% of circulating cortisol is bound to corticosteroid-binding globulin (CBG) with less than 10% in the free, biologically active form. CBG is the predominant binding protein with albumin binding a lesser amount. During acute illness, particularly sepsis, CBG levels fall by as much as 50%, resulting in a significant increase in the percentage of free cortisol. The circulating half-life of cortisol varies from 70 to 120 minutes, with a biological half-life of about 6-8 hours. The adrenal gland does not store cortisol; increased secretion arises due to increased synthesis under the control of ACTH. Cholesterol is the principal precursor for steroid biosynthesis in steroidogenic tissue. At rest and during stress about 80% of circulating cortisol is derived from plasma cholesterol, the remaining 20% being synthesized *in situ* from acetate and other precursors. High-density lipoprotein (HDL) is the preferred cholesterol source of steroidogenic substrate in the adrenal gland.(2)

The activity of glucocorticoids are mediated by both the glucocorticoid (GR) and mineralocorticoid receptor (MR). The GR and MR share both functional and structural homology. Both aldosterone and glucocorticoid hormones bind to both the GR and MR. At low basal levels cortisol binds to the high affinity, low capacity MR. However, with increased cortisol secretion the MR are saturated and cortisol then binds to the low affinity, high capacity glucocorticoid receptor (GR). In addition, the 11β-hydroxysteroid dehydrogenase (11β-HSD) enzymes plays an important role in preventing glucocorticoid access to cells that express the MR. This enzyme has two isoforms, a NAD+-dependent form (11β-HSD-2) and a NADP+ dependent form (11β-HSD-1). 11β-HSD-2 is found in tissues with high levels of MR activity such as the kidney, sweat and salivary glands, placenta and colon. 11β-HSD-2 converts cortisol to cortisone, its inactive reduced metabolite which is unable to bind to the GR and MR. 11β-HSD-1 which is found in glucocorticoid target tissues, catalyzes the conversion of cortisone to the active glucocorticoid cortisol. Pro-inflammatory cytokines modulate the activity of the 11β-HSD enzymes, with interleukin-1β (IL-1β) and tumor necrosis factor-α (TNF-α) increasing the activity of 11β-HSD-1 while decreasing that of 11β-HSD-2.

Cortisol diffuses rapidly across cell membranes binding to the GR. Two isoforms of the GR have been isolated, namely GR-α and GR-β. The GR-β isoform fails to bind cortisol and activate gene expression and thus functions as a negative inhibitor of GR-α Seven isoforms of GR-α have been reported; these isoforms may be selectively expressed by different tissues with each isoform eliciting a distinct response.(3) Through the association and disassociation of chaperone molecules the glucocorticoid-GR-α complex moves into the nucleus where it has both genomic and non-genomic (protein-protein) effects.

CRITICAL ILLNESS RELATED CORTICOSTEROID INSUFFICIENCY

There has recently been a great deal of interest regarding the assessment of "adrenal function" and the indications for corticosteroid therapy in critically ill patients. While the use of high-dose corticosteroid (10 000 to 40 000 mg of hydrocortisone equivalent over 24 hours) in patients with severe sepsis and ARDS failed to improve outcome and was associated with increased complications,(4,5) an extended course of "stress-dose" corticosteroids (200-350 mg hydrocortisone equivalent/day for up to 21 days) has been demonstrated to increase ventilator and hospital free days and improve short-term survival in select groups of ICU patients.(6-9) These patients typically have an exaggerated pro-inflammatory response and are considered to be "relatively" corticosteroid insufficient.

Until recently the exaggerated pro-inflammatory response that characterizes patients with systemic inflammation has focused on suppression of the HPA axis and "adrenal failure." However, experimental and clinical data suggest that corticosteroid tissue resistance may also play an important role. This complex syndrome is referred to as "*Critical Illness Related Corticosteroid Insufficiency* (CIRCI)." (10) CIRCI is defined as inadequate corticosteroid activity for the severity of the patients illness. CIRCI manifests with insufficient corticosteroid mediated downregulation of inflammatory transcription factors.

TISSUE CORTICOSTEROID RESISTANCE DURING CRITICAL ILLNESS

Tissue corticosteroid resistance is a well known manifestation of chronic inflammatory diseases such as chronic obstructive pulmonary disease (COPD), severe asthma, systematic lupus erythematosus (SLE), ulcerative colitis and rheumatoid arthritis. Emerging data suggests that corticosteroid tissue resistance may develop in patients with acute inflammatory diseases such as sepsis and acute lung injury (ALI). In an *ex vivo* model Meduri and colleagues compared the cytoplasmic to nuclear density of the GR-complex in patients with ARDS who survived as compared to those who died. (11) These authors demonstrated a markedly reduced nuclear density of the GR-complex in those who died as compared to the survivors.

HPA AXIS FAILURE IN ACUTE ILLNESS

HPA axis failure appears to be a common problem in patients with systemic inflammation. Patients at risk for developing tissue glucocorticoid resistance are similarly at risk for HPA axis failure. The incidence of HPA axis failure varies widely depending on the criteria used to make the diagnosis and the patient population studied. The overall incidence of "adrenal insufficiency" in critically ill patients approximates 20%, with an incidence as high as 60% in patients with severe sepsis and septic shock.(12) The mechanisms leading to inadequate cortisol production during critical illness are complex and poorly understood and likely include decreased production of CRH, ACTH and cortisol. A subset of patients may suffer structural damage to the adrenal gland from either hemorrhage or infarction and this may result in long term adrenal dysfunction. In addition, a number of drugs are associated with adrenal failure. However, reversible HPA dysfunction is increasingly being recognized in critically ill patients with systemic inflammation associated with the sepsis, ALI, liver disease and following cardiopulmonary bypass.

CLINICAL MANIFESTATIONS OF CIRCI

Patients with chronic adrenal insufficiency (Addison's Disease) usually present with a history of weakness, weight loss, anorexia and lethargy with some patients' complaining of nausea, vomiting, abdominal pain and diarrhea. Clinical signs include orthostatic hypotension and hyperpigmentation (primary adrenal insufficiency). Laboratory testing may demonstrate hyponatremia, hyperkalemia, hypoglycemia and a normocytic anemia. This presentation contrasts with the features of CIRCI. The clinical manifestations CIRCI are consequent upon an exaggerated pro-inflammatory immune response. Hypotension refractory to fluids and requiring vasopressors is a common manifestation of CIRCI. CIRCI should therefore be considered in all ICU patients requiring vasopressor support as well as those with severe progressive ARDS. Patients usually have a hyperdynamic circulation which may compound the hyperdynamic profile of the patient with sepsis/systemic inflammation. However, the systemic vascular resistance, cardiac output, and pulmonary capillary wedge pressure can be low, normal, or high. The variability in hemodynamics reflects the combination of CIRCI and the underlying disease. CIRCI should also be considered in patients with progressive ALI. Laboratory assessment may demonstrate eosinophilia and hypoglycemia. Hyponatremia and hyperkalemia are uncommon.

DIAGNOSIS OF ADRENAL INSUFFICIENCY AND CIRCI

The diagnosis of adrenal insufficiency in the critically ill is fraught with difficulties. Furthermore we have no test that quantifies corticosteroid activity at the tissue level. Traditionally the diagnosis of adrenal insufficiency in the critically ill has been based on the measurement of a random total serum cortisol ("stress" cortisol level) or the change in the serum cortisol in response to 250 ug of synthetic ACTH (cosyntropin) the so called "delta cortisol". Both of these tests have significant limitations in the critically ill. Despite the limitations, Annane and colleagues have reported that a delta cortisol of less than 9 mg/dl was the best predictor of adrenal insufficiency (as determined by metyrapone testing) in patients with severe sepsis/septic shock.(12) A cortisol of less than 10 mg/dl was also highly predictive of adrenal insufficiency (PPV of 0.93), however the sensitivity of the test was poor (0.19).

One approach to resolving the question of whether too little glucocorticoid signal ultimately "gets through" is to examine target tissues whose function is regulated in part by glucocorticoids. Through their inhibitory effects on nuclear factor-κβ signaling pathways, glucocorticoids are the most potent anti-inflammatory hormones in the body and thereby serve to suppress the production and activity of proinflammatory cytokines during exposure to stress. Inadequate glucocorticoid-mediated feedback inhibition of the immune response will result in excess circulating levels of pro-inflammatory mediators. Kwon and colleagues measured the levels of pro-inflammatory mediators in a cohort of 82 patients, most of whom had sepsis.(13) Thirty six patients (43%) met the above cited criteria for adrenal inefficiency. The authors divided the patients with adrenal inefficiency into two groups, namely i) those with a low basal cortisol (basal cortisol < 10 ug/dl) and ii) those with a basal cortisol > 10 ug/dl and a delta cortisol < 9 ug/dl. In the group of patients with a low delta cortisol the serum levels of pro-inflammatory mediators were markedly elevated compared to group of patients' with a low basal cortisol. In the low basal cortisol group the levels of proinflammatory mediators were similar to those of the non- adrenal inefficiency control patients. This data suggests that the adrenal insufficiency sub-group with a low delta cortisol may truly have too little glucocorticoid signaling (glucocorticoid resistance) while the low baseline sub-group has inadequate glucocorticoid production (adrenal insufficiency) but without glucocorticoid resistance. From a pathophysiological and therapeutic standpoint it may therefore be useful to divide adrenal inefficiency /CIRCI into two subgroups, namely; Type I, characterized by a random (stress) cortisol < 10 ug/dl and Type II, characterized by a random cortisol ≥10 ug/dl and a delta cortisol less than 9 u/g/dl. The clinical implications and the response to treatment of this sub-classification require further study.

TREATMENT WITH CORTICOSTEROIDS. WHO AND HOW?

Over the last 3 decades approximately 20 randomized controlled trials (RCTs) have been conducted evaluating the role of glucocorticoids in patients with sepsis, severe sepsis, septic shock and ARDS. Varying doses (37.5 - 40 000 mg/hydrocortisone eq/day), dosing strategies (single bolus/ repeat boluses/continuous infusion/ dose taper) and duration of therapy (1-32 days) were used in these studies.(4,5) Similarly approximately 20 meta-analysis have been published in an attempt to better understand the role of glucocorticoids in the treatment of critically ill patients. The results of these studies together with our current understanding of CIRCI allow us to make a number of general recommendations. It should be appreciated that the non-stressed daily production of cortisol (hydrocortisone) in adults is approximately 15-25 mg/day while the maximal stressed daily production of cortisol (hydrocortisone) is about 200-350 mg/day.(14) Based on this data, a daily dose of hydrocortisone (or equivalent) of 25-200 mg/day can be considered "low-dose", 200-350 mg/day a physiologic "stress-dose", 351-1000 mg/day as a "supra-physiologic" dose and > 1000 mg/day as mega-dose corticosteroid. A number of RCTs investigated the clinical outcomes of mega-dose, short course corticosteroid treatment in patients with ARDS and sepsis. Doses of methylprednisolone as high as 20 to 30 mg/kg body weight (10 000 to 40 000 mg of hydrocortisone) over 24 hours were investigated.(4,5) These studies were unable to demonstrate improved outcomes and there was a higher incidence of complications in the patients who received high-dose corticosteroids.(4,5) This dosing strategy is at odds with our current understanding of CIRCI. Ideally the dose of corticosteroid should be sufficient to downregulate the pro-inflammatory response without causing excessive immune-paresis and interfering with wound healing. Similarly, the duration of glucocorticoid therapy should be guided by the duration of CIRCI and the associated duration of systemic inflammation.

Schneider and Voerman were the first investigators (in 1991) to suggest that *"physiologic and not pharmacologic doses of glucocorticoids [be administered] in the course of septic shock"*. (15) These authors demonstrated reversal of shock in 3 of 8 patients given "*100 mg of hydrocortisone intravenously followed by 100 mg every 8 hours with dose tapering with improvement.*" The use of extended course, stress-dose corticosteroids has been evaluated in 10 RCTs in critically ill patients with sepsis, septic shock and ARDS. Overall, this dosing strategy has been reported to be associated with a significant reduction in 28 day all-cause mortality, more rapid weaning of vasopressor agents (septic shock), a reduction in ICU length of stay and an increase in ventilator free days (ARDS).(4,5,10,16) It is however important to realize that the analysis and meta-analyses of these studies are largely influenced by the largest two studies; namely the study by Annae et al and the CORTICUS study.(8,17) Both of these studies have important limitation in that 24% patients received etomidate in the Annane study while 19 received etomidate in the CORTICUS study. Furthermore, only patients with "refractory septic shock" were enrolled into the Annane study while an overwhelming selection bias resulted in approximately only 5% of eligible patients being enrolled into the CORTICUS study. A more recent study found no benefit from a 7 day course of 40 mg prednisolone in patients hospitalized with community acquired pneumonia.(18) Based on the current data it is not possible to make strong recommendations regarding the use of corticosteroids in the critically ill. The risk/benefit ratio should be determined in each patient and prolonged course (10-14days) of low dose (200 mg/day hydrocortisone or equivalent) should be considered in patients with septic shock who have responded poorly to fluids and vasopessor agents as well as in patients with ARDS who have progressive disease (after 48h ours) despite supportive care. Infection surveillance is critical in patients treated with corticosteroids and to prevent the rebound phenomenon the drug should be weaned slowly (after 10-14days) and never stopped abruptly.

The use of a continuous infusion of hydrocortisone has been reported to result in better glycemic control with less variability of blood glucose concentration. This may be important as it has been demonstrated that oscillating blood glucose is associated with greater oxidative injury than sustained hyperglycemia.A number of reports indicate that glucose variability may be an independent predictor of outcome in critically ill patients. However, a continuous infusion of glucocorticoid may result in greater suppression of the HPA axis. In the "sepsis" studies patients were treated with hydrocortisone, while methylprednisolone was the corticosteroid of choice in the ARDS studies. Different corticosteroids differentially effect gene transcription and have differing pharmacodynamic effects. Consequently, the preferred corticosteroid as well as the optimal dosing strategy in critically ill patients with sepsis, ARDS and other inflammatory states remains to be determined.

The complications associated with the use of corticosteroids are dependent upon the dose, the dosing strategy and the duration of therapy. In the ICU setting (short term treatment of CIRCI) the most important complication include immune suppression with an increased risk of infections (typical and opportunistic), impaired wound healing, hyperglycemia, myopathy, hypokalemic metabolic acidosis, psychosis and HPA axis and GR suppression. The effect of glucocorticoids on immune suppression may be critically dose dependent. It is well know from the organ transplant experience that high-dose corticosteroids effectively abolish T-cell mediated immune responsiveness and are very effective in preventing/treating graft rejection. Recent studies, however, suggest that while stress-doses of corticosteroids inhibit systemic inflammation with decreased transcription of pro-inflammatory mediators, they maintain innate and Th1 immune responsiveness and prevent an overwhelming compensatory anti-inflammatory response (CARS). While the effects of corticosteroids on IL-10 and soluble TNF receptors (TNFsr) is conflicting, corticosteroids should be avoided in chronically critically ill ICU patients with presumed CARS. Similarly, while myopathy is common in patients treated with high-dose corticosteroids this complication is uncommon with stress dose of corticosteroids.

CONCLUSION

Critical Illness Related Corticosteroid Insufficiency (CIRCI) is a complex disease our understanding of which continues to evolve. In critically ill patients with catecholamine refractory septic shock and patients with persistent severe ARDS for more than 48 hours after supportive therapy a course of stress-dose corticosteroids (200-350 mg hydrocortisone/day or 40–70 mg methylprednisolone/day) should be considered. Treatment for at least 7 days (and up to 14 days) is suggested, followed by a slow taper. These recommendations are based on limited data and addition studies are therefore required.

REFERENCES

1.Selye H. A syndrome produced by diverse nocuous agents. Nature 1936; 136:32 2.Yaguchi H, Tsutsumi K, Shimono K et al. Involvement of high density lipoprotein as substrate cholestrol for steroidogenesis by bovine adrenal fasciculo-reticularis cells. Life Sci 1998; 62:1387-1395. 3. Duma D, Jewell CM, Cidlowski JA. Multiple glucocorticoid receptor isoforms and mechanisms of post-translational modification. J Steroid Biochem Mol Biol 2006; 102:11-21. 4. Annane D, Bellissant E, Bollaert PE et al. Corticosteroids for severe sepsis and septic shock: a systematic review and meta-analysis. Br Med J 2004; 329:480-489. 5. Minneci PC, Deans KJ, Banks SM et al. Meta-analysis: the effect of steroids on survival and shock during sepsis depends on the dose. Ann Intern Med 2004; 141:47-56. 6. McKee JI, Finlay WE. Cortisol replacement in severely stressed patients [letter]. Lancet 1983; 1:484. 7. Confalonieri M, Urbino R, Potena A et al. Hydrocortisone infusion for severe community-acquired pneumonia: a preliminary randomized study. Am J Respir Crit Care Med 2005; 171:242-248. 8. Annane D, Sebille V, Charpentier C et al. Effect of treatment with low doses of hydrocortisone and fludrocortisone on mortality in patients with septic shock. JAMA 2002; 288:862-871. 9. Meduri GU, Golden E, Freire AX et al. Methyprednisolone infusion in patients with early severe ARDS: results of a randomized trial. Chest 2007; 131:954-963. 10. Marik PE, Pastores SM, Annane D et al. Recommendations for the diagnosis and management of corticosteroid insufficiency in critically ill adult patients: Consensus statements from an international task force by the American College of Critical Care Medicine. Crit Care Med 2008; 36:1937-1949. 11. Meduri GU, Muthiah MP, Carratu P et al. Nuclear factor-kappaB- and glucocorticoid receptor alpha- mediated mechanisms in the regulation of systemic and pulmonary inflammation during sepsis and acute respiratory distress syndrome. Evidence for inflammation-induced target tissue resistance to glucocorticoids. Neuroimmunomodulation 2005; 12:321-338. 12. Annane D, Maxime V, Ibrahim F et al. Diagnosis of adrenal insufficiency in severe sepsis and septic shock. Am J Respir Crit Care Med 2006; 174:1319-1326. 13. Kwon YS, Suh GY, Jeon K et al. Cytokine levels and dysfunction in the hypothalamus-pituitary-adrenal axis in critically-ill patients. Intensive Care Med 2010; 36:1845-1851. 14. Thomas JP, el-Shaboury AH. Aldosterone secretion in steroid-treated patients with adrenal suppression. Lancet 1971; 1:623-625. 15. Schneider AJ, Voerman HJ. Abrupt hemodynamic improvement in late septic shock with physiological doses of glucocorticoids. Intensive Care Med 1991; 17:436-437. 16. Meduri GU, Marik PE, Chrousos GP et al. Steroid treatment in ARDS: a critical appraisal of the ARDS network trial and the recent literature. Intensive Care Med 2008; 34:61-69. 17. Sprung CL, Annane D, Keh D et al. Hydrocortisone therapy for patients with septic shock. N Engl J Med 2008; 358:111-124. 18. Snijders D, Daniels JM, de Graaff CS et al. Efficacy of corticosteroids in community-acquired pneumonia: a randomized double-blinded clinical trial. Am J Respir Crit Care Med 2010; 181:975-982.

139a

ENDOTOXEMIA AS A PREDISPOSING FACTOR FOR LAMINITIS IN HORSES.

EM Tadros[1], N Frank[1,2]. 1. Department of Large Animal Clinical Sciences, College of Veterinary Medicine, University of Tennessee, Knoxville, TN. 2. Department of Clinical Sciences, Cummings School of Veterinary Medicine, Tufts University, North Grafton, MA.

Endotoxemia is associated with intestinal disease in horses and laminitis develops in some affected animals, but not others. It is our overall hypothesis that laminitis develops when repeated inflammatory events exacerbate systemic inflammation and sepsis-associated organ damage. This study was performed to test whether exaggerated inflammation increases the risk of laminitis in an equine "two-hit" sepsis model.

Twenty-four light-breed horses randomly received an 8-hour intravenous infusion of either lipopolysaccharide (5 ng/kg/hr) or saline beginning at -24h, followed by oligofructose (OF; 5 g/kg) via nasogastric tube at 0h. Euthanasia and tissue collection occurred at Obel grade 2 laminitis, or at 48h if laminitis had not developed. Liver biopsies were performed at 24h in laminitis non-responders. Blood cytokine gene expression was measured throughout the study. Mixed model ANOVA and Fisher's exact tests were performed.

Lipopolysaccharide and OF administration independently increased mean rectal temperature ($P<0.001$), heart rate ($P=0.003$), respiratory rate ($P<0.001$), and blood interleukin (IL)-1β gene expression ($P<0.0016$), but responses to OF were not exaggerated in endotoxin-pretreated horses. The overall laminitis induction rate of 63% did not differ between treatment groups. When horses were classified as laminitis responders and non-responders, area under the blood IL-1β expression curve ($P=0.010$) and liver and lung gene expression of IL-1β, IL-6, IL-8, IL-10, and tumor necrosis factor-α ($P<0.05$) were higher in responders following OF administration.

Endotoxin pretreatment did not enhance responses to OF, so our hypothesis was not supported. However, systemic inflammation was more pronounced in laminitis responders, and tissues were a source of inflammatory mediators.

139b

COMPARISON OF ECHOCARDIOGRAPHIC MEASURMENTS IN ELITE AND NON-ELITE ARABIAN ENDURANCE HORSES.

MM Durando[1], T Holbrook[2], MM Sleeper[3], M Payton[2], EK Birks[1]. 1 Landenberg, PA. 2. Oklahoma State University College of Veterinary Medicine, Stillwater, OK. 3. University of Pennsylvania School of Veterinary Medicine, Philadelphia, PA.

The purpose of this study was to determine if performance ability is related to echocardiographic indices of cardiac dimensions or function. In Thoroughbred racehorses, an association has been made between caliber of racehorse, maximal oxygen consumption and certain echocardiographic variables in longer, more aerobic races. Because of the aerobic nature of the sport of endurance racing, we postulated that elite endurance Arabians would have different echocardiographic variables than less successful competitors.

Arabians between 7 and 17 years old were considered. Horses were assigned to either an elite or a non-elite group based on results from specified competitions. The competitions used to define elite competitors were AERC-sanctioned 100-mile rides. Horses were considered elite competitors if they won or finished within 10% of the winning time or within 30 minutes of the winner. Horses were considered non-elite if they had never won a 100-mile ride or placed within 2 hours of the winning time, or were not actively competing in endurance rides. Horses were evaluated in their home environment within 1-4 weeks after the competitions chosen to determine groups. 23 horses were considered elite and 18 horses non-elite. A standardized 2-dimensional, M-mode and Doppler echocardiographic exam was performed with a portable ultrasound machine, and cardiac dimensions and function were recorded. Differences in measured variables between the 2 groups were compared using a T test. Logistic regression was performed to determine which variables were significantly associated with performance. After backward elimination to remove insignificant variables, a model was developed using the significant variables. Significance was set at $P<0.05$.

Four horses in the non-elite group had cardiac abnormalities that could negatively affect performance and/or safety (significant valvular regurgitation and/or ventricular dysrhythmias). Two of these were eliminated from the study because of their cardiac abnormalities. No horses in the elite group had significant abnormalities. Left ventricular internal diameter in systole and diastole, interventricular septal thickness in diastole, left ventricular mass and stroke volume were significantly greater in the elite group. A model developed using logistic regression was applied to these horses. It correctly predicted 96% of the elite horses, and 83% of the non-elite horses.

Elite endurance Arabians had larger left ventricular cardiac dimensions and stroke volume than non-elite and non-competitive Arabians in this group of horses. This could correspond to greater cardiovascular ability and aerobic capacity, and thus superior performance, in this group of horses, although many other factors contribute to successful racing ability. These findings should be tested on an independent group of horses.

140a

FACTORS ASSOCIATED WITH SURVIVAL IN 50 HORSES WITH POTOMAC HORSE FEVER. FR Bertin[1], A Reising[2], NM Slovis[2], PD Constable[1], SD Taylor[1]. 1. Purdue University, West Lafayette, IN. 2. Hagyard Equine Medical Institute, Lexington, KY.

Neorickettsia risticii is the etiologic agent of Potomac Horse Fever (PHF). The objective of this retrospective study was to describe the signalment, clinical signs, clinicopathologic values, antimicrobial use, and survival status in horses affected with PHF, and to identify variables for prediction of survival. Horses diagnosed with PHF at two referral institutions over a 15-year period were included in the study. A diagnosis of PHF was based on the presence of positive blood or fecal PCR, an IFA titer greater than 1:2,560 in non-vaccinated horses with compatible clinical signs, or a four-fold increase between acute and convalescent titers.

Horses ranged from 4 months to 29 years of age. The month of presentation ranged from June to December with a peak incidence in August. The most common clinical signs included diarrhea (66%), fever (48%), anorexia (42%), depression (40%), colic (38%), and lameness (18%). The median duration of hospitalization was 6 days and 76% of horses survived to discharge. Laminitis was confirmed in 26% and suspected in 36% of cases. Serum urea nitrogen and creatinine concentrations, anion gap and hematocrit were significantly ($P<0.01$) higher in nonsurvivors, while serum chloride, sodium, calcium, and magnesium concentrations were significantly lower in nonsurvivors.

Logistic regression indicated that serum chloride concentration on admission and initial antimicrobial treatment with oxytetracycline were the only independent factors associated with survival. We conclude that the severity of colitis in horses with PHF is the best predictor of survival and that oxytetracycline is efficacious in the treatment of horses with PHF.

140b

IDENTIFICATION OF MARE-ASSOCIATED PREDICTORS OF PASSIVELY ACQUIRED IMMUNOGLOBULIN CONCENTRATIONS IN THOROUGHBRED FOALS. C.B. Riley, C.J. Jenvey, M. Baltussen, and C. Caraguel. University of Adelaide, Adelaide, SA, Australia.

Critical investigations into the identification of mare-associated prepartum predictors of foal immunoglobulin (Ig) concentrations are limited. The objective of this study was to determine if periparturient mare-associated factors were useful as predictors of neonatal serum Ig levels.

Blood, milk, and medical records were collected for 124 Thoroughbred mare-foal pairs from 1 month before to 2 months after birth. Samples were tested using radial immunodiffusion assay and enzyme-linked immunoassays for concentrations of Ig classes. Multiple linear regression was used to identify significant interactions ($P<0.05$).

Significant differences were present between successive sampling points for IgG, IgGa, IgA and IgM for mare serum, and for IgG and IgA in foal serum and colostrum. Subclasses IgG(T), IgGa, IgGb and IgM were not significantly associated with any mare-associated factors of passive transfer. Mare age was associated with colostral and foal IgG concentrations at 12 h. Mare parity was associated with mare serum IgA concentrations up to 1 month before, and foal serum IgA concentrations 12 h after birth. Supplementation with colostrum positively interacted with colostral and foal serum IgA concentrations from 19-60 days. A temporal relationship was observed among IgA concentrations of mare serum, foal serum, colostrum and milk samples. Mare serum IgA measured 1 month before foaling was predictive of colostral IgA concentration at parturition and foal serum IgA concentration 12 h after birth.

The peripartum forecasting of Ig concentrations in the foal may enable immediate neonatal intervention to significantly improve foal survival and reduce risk of failure of the transfer of passive immunity.

STARLING FORCES IN THE LUNG OF HORSES

Modest Vengust, DVM, DVSc, PhD, DACVIM
Slovenia

Ernest H. Starling was an English physiologist, whose discoveries into circulatory, endocrine, renal and gastrointestinal physiology are the foundation for understanding several basic mechanisms of disease in humans and animals. His work, known as Starling principle (Starling's hypothesis) of microvascular fluid exchange, provides a means to quantitatively and/or qualitatively explain the balance of fluid movement between the interstitial and circulating compartments. The tool to review the Starling principle is the Starling equation, which balances osmotic pressure of proteins (colloid osmotic pressure) against hydrostatic pressure in the capillaries.[1,2]

STARLING PRINCILE AND STARLING EQUATION

The Starling equation determines fluid flux *(Jv)* per unit time and unit area/length as the difference between hydrostatic *(Pc – Pi)* and oncotic *(pc – pi)* pressure gradients: $Jv = K [(Pc - Pi) - S (pc - pi)]$, where *Pc* (mmHg) is the intravascular hydrostatic pressure, *Pi* (mmHg) is the perimicrovascular interstitial hydrostatic pressure, pc (mmHg) is the intravascular oncotic pressure, and p*i* (mmHg) is the perimicrovascular interstitial oncotic pressure. Albumin is the major determinant of oncotic forces in plasma. *K* (ml $sec^{-1}mmHg^{-1}100g^{-1}$) is the filtration coefficient, a product of the hydraulic conductivity and the filtration surface area. *S* is the osmotic reflection coefficient, which represents the permeability of the membrane or pore size. *S* = 1 when the full osmotic potential of the solute is present across an impermeable membrane, and shifts towards 0 when the membrane is becoming less discriminative to the passage of larger molecules (albumin).

Forces governing microvascular fluid exchange, however, are not as symmetrical in nature as it is proposed by the Starling principle. They often fail to explain the non-sustained behavior of *Jv* along the capillary, and the nonlinearity of *Jv* to the lymph flow. The error is caused by the osmotic asymmetry of the capillary endothelium. To minimize the error the oncotic pressure gradient included in the Starling equation should be derived across the endothelial glycocalyx and not across the whole microvascular wall. The osmotic forces present in a paracellular cleft in the microvascular subglycocalyx space, which are significantly less than *pi*, codetermine *Jv*. Therefore, the osmotic pressure gradient is then shown as *(pc – pg)*, where *pg* is the subglycocalyx fluid oncotic pressure.[3,4]

PULMONARY TRANSVASCULAR FLUID DYNAMICS

The capillary structure of the lung is designed for gas exchange with a large surface area and an ultrathin (1 μm) diffusion barrier. Despite such architecture of the blood gas barrier the alveolar space remains dry even over the extreme range of pulmonary artery pressures (>100 mmHg), pulmonary capillary pressures (~42 mmHg), pulmonary wedge pressures (38.5 mmHg) and blood flow (>300 $Lmin^{-1}$), which develop in exercising horses. This is because the pulmonary microvascular endothelium is at least 10x less permeable to solutes (electrolytes, proteins) and fluid than capillaries in other tissues and the pulmonary macrovasculature. Low pulmonary capillary hydraulic conductivity and a powerful pulmonary lymphatic pump limit accumulation of interstitial fluid in the lung parenchyma and generate a subatmospheric interstitial pressure (~ -8 mmHg). Even if there is excessive fluid extravasation out of pulmonary circulation, which cannot be removed by the lymphatic pump and/or bronchial circulation, the fluid accumulates perivascularly and peribronchialy where it forms cuffs, which have little effect on respiratory gas exchange at rest. Alveolar flooding, however, produces a rapid drop in arterial PO_2.[5,6]

Other mechanisms prove to be equally or even more important in pulmonary fluid dynamics in healthy lungs than Starling forces, especially during exercise. Erythrocytes form a tight unit with the pulmonary capillary endothelium analogous to a single semi-permeable membrane, which assumes the osmotic characteristics of the erythrocyte membrane. Such apposition of membranes enhances gas exchange between the pulmonary capillary and alveolus. It also provides means for strong ion, CO_2 and pulmonary transvascular fluid fluxes that follow the Jacobs-Stewart cycle and are independent of Starling forces.[7,8,9] The popular belief that pulmonary transvascular fluid fluxes during high intensity exercise are solely dependent on Starling forces, mostly hydrostatic forces, which are conducted by 1) pulmonary capillary recruitment and/or dilatation coupled with the increase in the pulmonary surface area and induced by increased cardiac output; 2) increase in pulmonary macro- and microvascular pressures; and 3) redistribution of blood flow across the lung[10], was challenged by a recent furosemide study performed in Standardbreds during submaximal exercise where furosemide failed to modify pulmonary transvascular fluid fluxes despite a significant decrease in cardiac output.[11]

STARLING FORCES AND LUNG FLUID ACCUMULATION (PULMONARY EDEMA DEVELOPMENT)

The pressure acting across the pulmonary microvasculature (transmural forces) is positive in the normal lung (~8 mmHg) favoring filtration of fluid into perivascular interstitial space. At the level when fluid removal from pulmonary interstitium cannot match fluid efflux from the pulmonary vasculature, perivascular and peribronchial cuffs develop followed by alveolar flooding. Moreover, when *Pi* exceeds pleural pressure, fluid moves across the visceral pleura, and forms pleural effusions. Several adaptations in the lung prevent such events from developing. They are named "Starling safety factors against edema" and include:

- low pulmonary *Pc*
- low *Pi* when compared to pulmonary alveolar pressure
- lower *pg* compare to *pc*
- high pulmonary interstitial compliance
- lymph drainage
- alveolar epithelial barrier substantially less permeable than the pulmonary capillary endothelial barrier.[12]

Starling safety factors against edema can be overwhelmed by forces that have hydrostatic (high pressure or cardiogenic) or permeability (non-cardiogenic) characteristics. Other less common causes for excessive lung fluid accumulation in the lung is primary lymphatic obstruction. Clinically evident pulmonary edema is most often the result of a combination of high pressure and permeability problems in the pulmonary circulation.[13]

Hydrostatic causes for lung fluid accumulation

The precise level of pressure in the left atrium that could be deemed as "edemagenic pressure" (pressure level with great likelihood of causing excessive pulmonary fluid accumulation with intact pulmonary microvascular permeability) and would be associated with the development of hydrostatic edema has not been determined in horses. However, pulmonary artery occlusion pressures (an estimate for the left atrial pressure) substantially > 18 mmHg would be relevant.[14]

The lungs of healthy horses can normally accommodate the extremely large *Pc* that is present during high intensity exercise. *Pc* is determined by elevated pulmonary artery pressure and consequently influences transient increase in transvascular fluid filtration from the pulmonary circulation.[6,15]

A similar scenario leading to excessive fluid extravasation is pathognomonic for cardiogenic and volume-overload edema. These are mostly non-transient in nature, and usually arise from the pulmonary venous compartment and less likely from the pulmonary artery. Elevated pulmonary venous pressure is caused by increased left ventricular end-diastolic pressure and/or left atrial pressure.[13]

Volume-overload edema may occur in horses with impaired renal function. On the other hand, even in the face of normal renal function, a clinically significant increase in lung fluid accumulation due to volume-overload edema with crystalloid solution infusion is possible, if the pulmonary circulation has preexisting permeability problems. Excessive crystalloid solution infusion could lower *pc* independently or acts as a contributing factor to changes in hydrostatic pressure or capillary permeability. A substantial decrease of *pc* may lower the upper limit of hydrostatic pressure tolerance.[16] The importance of *pc* in pulmonary transendothelial forces is minor; therefore, clinically evident respiratory limitations due to the crystalloid solution overload are not likely to occur in adult horses. Colloid fluid therapy may decrease extravasation because of their capacity to increase *pc*. The potentially protective role of *pc* may diminish when pulmonary capillary permeability is increased, which may aggravate *pg* and promote the accumulation of lung fluid. Studies in humans could not detect a significant difference between crystalloid or colloid treatment on pulmonary fluid loading and pulmonary permeability.[17]

In the initial stages of hydrostatic edema the capillary endothelium remains normal. Extravasated fluid has low protein content; therefore, *pg* gradually decreases and counterbalances *Pc*. However, chronically challenged blood-gas barrier (i.e. chronic heart failure) leads to pulmonary parenchymal inflammation and disruption of the alveolar-capillary membrane. Damage to the alveolar-capillary membrane raises the permeability to water and solutes, including albumin, which is followed by the loss of the pulmonary interstitial compliance and alveolar flooding.[18] Fluid extravasation from pulmonary circulation, sustained initially only by overwhelming *Pc,* gradually benefits from deteriorated *K* and *S*, with consequent worsening of *Pi* and ($pc - pg$). Further aggravation of *Pc* is then compounded by circulatory adaptations through renin-angiotensin-aldosterone system and catecholamine production.[19]

Permeability causes for lung fluid accumulation

Permeability or non-cardiogenic pulmonary edema fluid has high protein content because the vascular membrane is more permeable to the outward movement of plasma proteins and is associated primarily with pneumonia, sepsis, aspiration of gastric contents, iatrogenic insult, causes classified under Vet ALI/ARDS.[14] To classify lung fluid accumulation under permeability pulmonary edema, the pulmonary artery occlusion pressure must not exceed 18 mmHg, the threshold suggested by Vet ALI/ARDS.[14]

Marked increase in fluid extravasation from the pulmonary circulation and protein exiting the vascular space are due to increased *K* and decreased *S*, and changed *(Pc – Pi)* and *(pc – pg)*. This is caused and sustained by injury and/or mediator-activated disruption of the alveolar-capillary membrane. Inflammation is critical for innate immunity and host defense; however, it can injure the alveolar-capillary membrane. [13]

Diseases causing permeability edema are specific with regards to the origin of forces governing fluid accumulation in the lungs. Particularly after inhalation lung injury (chemical, burn and smoke inhalation), bronchial circulation primarily contributes to edema formation (permeability nature of edema formation), which is followed by the fluid flux from the pulmonary circulation. There is a dramatic increase in *K*, a reduction in *S* and an increase in *Pc* in pulmonary microcirculation (hydrostatic nature of edema formation). Ablation of the bronchial artery significantly attenuates lung fluid accumulation, minimizing the effect of disrupted Starling forces.[20]

Unclassified causes for lung fluid accumulation

Post-obstructive pulmonary edema is an uncommon complication of upper airway obstruction in horses, which is characterized by large fluid extravasation from the pulmonary vasculature caused by changes in intrathoracic pressure. In most cases this is the post-anesthetic complication.[21] The pathogenesis of the condition is not well determined. It may include causes that are similar to hydrostatic edema formation and/or causes that are involved in permeability pulmonary edema.[13]

Pulmonary limitation to exercise performance in horses, among other factors, includes exercise-induced pulmonary hemorrhage (EIPH). EIPH is caused by extreme transmural pulmonary capillary pressures during exertion, which damage the alveolar-capillary membrane. Such events are aggravated by high *Pc,* increased *K*, decreased *S*, sustained or decreased *Pi*, sustained *pc* and increased *pg*. Therefore, hydrostatic and permeability factors are involved. Damage due to the injury and inflammation can compromise pulmonary interstitial and barrier compliance, especially in recurrent cases of EIPH.[22]

Pulmonary fluid accumulation due to primary lymphatic obstruction in horses is also possible and is mostly associated with neoplastic processes. Lung fluid accumulation develops retrograde to pulmonary circulation, with *Pi* being initially the most prominent causative Starling force.

CONCLUSION

The Starling principle is a useful tool for comparing the permeability characteristics of different tissues, including lungs. Some physiological inconsistencies, however, raise questions on the role of some Starling forces in pulmonary circulation transvascular fluid dynamics. In clinical settings in horses the Starling equation does not have quantitative value but can be logically used to qualitatively explain clinical observations. This is because most of Starling forces are impossible to measure in real time in a clinical setting. However, quantitative determination of only *Pc* indirectly through the measurement of circulation pressures benefits the horse that is critically ill or undergoes invasive procedures, with regards to evaluation of the severity of shock, to evaluate/monitor success of the volume expansion treatment and to monitor responses in circulatory system. It also enables deviatiation from predetermined treatments (doses based on patient's weight) and proactively meets and adapts to the physiological endpoints of the patient.

REFERENCES

1. Starling EH, J Physiol 1896; 19: 312 –326. 2. Hu X & Weinbaum S, Microvasc Res 1999; 58:281-304. 3. 2. Adamson RH, et al. J Physiol 2004; 557:889-907.3. Levick JR & Michel CC, Cardiovasc Res 2010; 87:198-210.4. Effros RM & Parker JC, Microvasc Res 2009;78:71-77.5. Sinha AK, et al. J Appl Physiol 1996; 80:1792-1798.6. Hansen AT, Nature 1961; 190: 504–508.7. Vengust M, et al. Equine Vet J Suppl. 2006; 36: 508-515.8. Brechuea WF & Hammel HT, Comparative Exercise Physiology 2011; 1-19. DOI: 10.1017/S1755254011000092.9. Hlastala, MP, et al. J Appl Physiol 1996; 81: 1051–1061.10. Vengust M, et al. Equine Vet J 2011; 43: 451-459.11. Sartori C, et al. Respir Physiol Neurobiol 2007;159:338-349.12. Michel RP & Goldberg P, Eds: Hamid Q, et al. Hamilton, Ont., Canada, BC Decker, 2005.13. Wilkins PA et al. J Vet Emerg Crit Care 2007; 17: 333–339.14. Vengust M, et al. J Physiol 2006; 570: 397-405.15. Stein L, et al. JAMA 1974; 229: 65-68.16. van der Heijden M, et al. Crit Care Med 2009; 37: 1275-1281.17. Miserocchi G, et al. News Physiol Sci 2001; 16: 66-71.18. Cotter G, et al. Eur J Heart Fail 2008; 10: 165-169. 19. Hamahata A, et al. Burns 2009; 35:802-810.20. Senior M, Vet Anaesth Analg 2005; 32: 193-200.21. Derksen FJ, et al. Equine Vet J 2007; 39: 334-339.

EQUINE INTERSTITIAL LUNG DISEASES: CLINICAL AND RADIOGRAPHIC ASSESSMENT

Bettina Dunkel, DVM, PhD, DACVIM, DECEIM, DACVECC, MRCVS
North Mymms, UK

INTRODUCTION

Interstitial pulmonary diseases comprise a heterogenous group of disorders characterised by inflammation and/or fibrosis of the pulmonary interstitium. In the absence of a lung biopsy, the diagnosis is frequently based on the presence of a dense diffuse or miliary to nodular interstitial radiographic pattern. The following discussion will briefly summarise equine interstitial pneumonias that have been reported in foals and adult horses.

FOALS

Acute broncho-interstitial pneumonia/ acute lung injury and acute respiratory distress syndrome (VetALI/VetARDS)

A condition similar to acute lung injury (ALI) and acute respiratory distress syndrome (ARDS) in humans has previously been described in foals under the name of proliferative interstitial lung disease[1] and acute broncho-interstitial pneumonia[2,3]. Recently, the term veterinary ALI/ARDS (VetALI/VetARDS) has been introduced for this condition in adult horses and foals and criteria to define VetALI/VetARDS have been determined[4]. The criteria include acute onset (<72h) of tachypnea and laboured breathing, presence of a known risk factor, evidence of pulmonary capillary leak without an increase in pulmonary capillary pressure (such as left-sided heart failure), evidence of inefficient gas exchange and evidence of diffuse pulmonary inflammation. To identify compromised gas exchange and to differentiate VetALI from VetARDS, the ratio of partial arterial oxygen pressure (PaO_2) to fraction of inspired oxygen (FiO_2) is used, similar to the human definition. A $PaO_2/FiO_2 < 300$mmHg is compatible with VetALI while a ratio of <200 indicates the more severe form of VetARDS[4]. Different cut off values have been introduced for neonatal foals. The condition appears to most commonly affect foals up to weanling age. Out of 84 reported cases, 66 were 1-8 months and 18 were <14d old[1-3,5-8]. Clinically, the syndrome is characterized by rapid onset of severe respiratory distress with tachycardia, tachypnea and rapidly worsening respiratory function, often minimally responsive to intranasal oxygen insufflation. Radiographic findings vary from a heavy interstitial to coalescing alveolar pattern and rapid change from the former to the latter has been observed[2], presumably as edema fluid is released from the interstitium into the alveoli. Based on the human disease pathogenesis and animal models, the clinical signs are paralleled by acute, uncontrolled interstitial and alveolar inflammation provoked by a pulmonary or systemic insult (risk factors). Endothelial and epithelial damage cause an increased alveolo-capillary permeability and rapid accumulation of protein-rich alveolar edema leading to severe hypoxemia. In most cases in foals the initial insult can be attributed to an underlying bacterial or viral pulmonary infection. *R. equi* infection has been reported in 14/60 (23%) of cases in which culture was performed followed by β-hemolytic *Streptococcus* spp. in 10/60 cases (16%). Whether this simply reflects high prevalence of these pathogens in the particular age group or whether these pathogens are more likely to trigger an exaggerated pulmonary immune response remains to be established. Treatment consists of intranasal oxygen supplementation, anti-inflammatories and antimicrobials, if an underlying bacterial infection is suspected. It has been suggested that corticosteroids might be useful to decrease pulmonary inflammation,; however, the use of steroids remains controversial in human medicine and recommendations in foals are solely based on clinical impressions[2,8]. Based on limited available reports in foals, the prognosis is guarded for weanlings and potentially worse for neonates[2,8]; encouragingly, the few foals for which follow up information was available did not seem to suffer from lasting pulmonary impairment[8].

Chronic broncho-interstitial pneumonia in foals

The term chronic broncho-interstitial pneumonia has been used to describe a diverse group of weanling foals with a chronic (weeks-months) history of respiratory disease[9]. The uniting feature in all of these foals was a diffuse, interstitial pattern of varying severity on thoracic radiographs, often in combination with areas of consolidation. No common infectious organism was identified and foals responded to symptomatic treatment consisting of antimicrobials and corticosteroids. Whether or not these foals represent a distinct disease entity or lingering pulmonary inflammation secondary to previous infection/inflammation remains to be established.

Interstitial pneumonia secondary to viral infection

VetALI/VetARDS secondary to influenza infection has been described in individual neonatal foals[1,5,6,8] and in an outbreak following an outbreak of equine influenza infection in a naive population[7]. Whether the viral infection triggers VetALI/VetARDS or directly causes destruction of pulmonary tissue remains unclear. Severe and often fatal respiratory distress following natural infection with equine arteritis virus (EAV) has also been described in neonatal foals[10-12]. Histopathological findings demonstrated alveolar epithelial type II cell hyperplasia, hyaline membrane formation and severe arteritis. Aerosol, but also transplacental infection appeared to be responsible for transmission. Although most commonly a cause of mild and upper respiratory track signs, severe interstitial to necrotizing broncho-interstitial pneumonia has been reported in neonatal and slightly older foals infected with equine herpesvirus 1(EHV-1) *in utero* or *post partum*, occasionally in association with disseminated EHV-1 infection[13-15].

ADULT HORSES

Acute lung injury and acute respiratory distress syndrome (VetALI/VetARDS)

A condition compatible with VetALI/VetARDS has been reported occasionally in adult horses. However, identification of cases in the literature so far involves screening of individual and often incomplete case reports and a certain amount of guessing whether or not VetALI/VetARDS might have been present. Potential risk factors leading to VetALI/VetARDS in the described cases include smoke inhalation[16], near drowning[17,18], experimental intravenous *E. coli* infusion[19], natural equine influenza infection[20] and no known underlying cause[21,22]. Pulmonary vasculotropic EHV-1 infection in young adult horses can lead to severe multisystemic vasculitis, diffuse alveolar damage and pulmonary edema resembling VetALI/VetARDS[23-25].

Equine multi-nodular pulmonary fibrosis (EMPF)

Recently, an association between multi-nodular pulmonary fibrosis and equine herpes virus 5 (EHV-5) infection has been established[26,27]. The disease presents as numerous, small (<1-5cm) coalescing or, less commonly, multiple, discreet, larger (up to 10cm) pulmonary nodules. The interstitial spaces are expanded by well organised, mature collagen and infiltrating inflammatory cells. The predominant cell types are lymphocytes but macrophages, neutrophils and eosinophils are also observed. Frequently, airway spaces reminiscent of alveoli (described as "alveolar-like") lined with cuboidal cells and filled with neutrophils and macrophages can be noted. However, in other cases broad collagen bundles replacing the pulmonary architecture are encountered. Eosinophilic nuclear inclusion bodies might be observed in intralesional marcophages and the presence of EHV-5 in tissues has been confirmed by PCR or virus isolation[26,27]. Suggested treatment included long term corticosteroid administration and, in some cases, acyclovir or valacyclovir. However, recovery of cases has been reported without the use of either drug[27] (Schwarz *et al*, unpublished data) and in some cases, repeated testing showed persistence of EHV-5 in the airways despite treatment with valacyclovir at recommended doses and improvement of the clinical condition. The best treatment regime and the exact role of EHV-5 for EMPF therefore still need to be established. The prognosis is guarded but return to full athletic function and resolution of the radiographic abnormalities has been reported in a few cases[27] (Schwarz *et al*, unpublished data).

Idiopathic eosinophilic pneumonia in adult horses

Acute and chronic respiratory compromise associated with nodular eosinophilic pulmonary interstitial infiltrates, eosinophilic airway inflammation evident on broncho-alveolar lavage and/or histology and occasional peripheral eosinophilia has been described. Other organ systems were not involved in these cases and improvement with corticosteroids treatment has been documented[28-32]. Severe pulmonary disease due to multisystemic eosionophilic epitheliotrophic disease has also been reported[33]. It is not known whether the two conditions are related or represent different diseases.

Idiopathic granulomatous interstitial pneumonia

Idiopathic granulomatous interstitial pneumonia is a rare, chronic respiratory disease. Radiographically, individual or multiple diffuse military or nodular opacities can be identified. In a single case, mineralization of the infiltrates was observed. Histologically, granulomas consist of a central focus of necrosis or degeneration surrounded by multinucleated cells and epitheloid macrophages. The lesions might be separated by fibrotic tissue. Occasionally, predominately fibrosis or histiocytes and multinucleated giant cells were observed. As no underlying cause has been identified, the term idiopathic has been used[34]. A similar picture of pulmonary inflammation with a predominately perivascular distribution has also been described in a case of systemic granulomatous disease in a pony[35].

Other military or nodular interstitial pneumonias

Fungal, and occasionally bacterial or parasitic infections or neoplasia might present radiographically as military or multinodular interstitial pneumonia[36,37]. As with other interstitial conditions, cytology and/or histology are necessary to achieve a diagnosis. Silicate pneumoconiosis with accompanying pulmonary fibrosis has been identified predominately in horses from certain areas in California[38,39]. Osteoporosis and vague chronic lameness might also be seen in these cases and silicate has been speculated to play a role in this bone fragility syndrome[40].

Toxic interstitial pneumonia

Perilla mint (*Perilla fructenscens)* ketones induce diffuse alveolar injury and pulmonary edema in horses and have been used to provoke ARDS experimentally[41,42]. Cofton weed (*Eupatorium adenophorum*) ingestion by horses has been associated with focal fibrosis, pulmonary abscessation, and pulmonary infarcts with vascular thrombosis. The principle toxins of *Crotalaria juncea* are pyrrolizidine alkaloids; diffuse fibrosing alveolitis and concurrent hepatic damage have been observed following experimental feeding of a diet containing 40% of tritured *Crotalaria juncea* seeds[43]. Horses develop severe and occasionally fatal bronchiolitis following oral or intravenous administration of 3-methylindole while alveolar changes are limited to over-inflation. This is in contrast to cattle where severe diffuse alveolitis and pulmonary edema occur[44]. Paraquat is a potent pulmonary toxin in many species[45] and has been suggested to cause acute pulmonary edema and subsequent fibrosis in horses. Grazing of recently sprayed pasture induced mouth lesions and increased mucous secretions which quickly resolved[46]. The author is not aware of a report detailing the effects of paraquat on the equine lung.

REFERENCES

[1]Buergelt CD, et al. **Vet Pathol** 1986;23(6):750-6.[2]Lakritz J, et al. **J Vet Intern Med** 1993;7(5):277-88.[3]Prescott JF, et al. **Can Vet J** 1991;32:421-425.[4]Wilkins PA, et al. **J Vet Emerg Crit Care** 2007;17(4):333-339.[5]Peek SF, et al. **J Vet Intern Med** 2004;18(1):132-4.[6]Britton AP, et al. **Can Vet J** 2002;43(1):55-6.[7]Patterson-Kane JC, et al. **Equine Vet J** 2008;40(3):199-203.[8]Dunkel B, et al. **Equine Vet J** 2005;37(5):435-40.[9]Nout YS, et al. **Equine Vet J** 2002;34(6):542-8.[10]Del Piero F, et al. **Equine Vet J** 1997;29(3):178-85.[11]Wilkins PA, et al. **Equine Vet J** 1995;27(5):398.[12]Vaala WE, et al. **Equine Vet J** 1992;24(2):155-8.[13]McCartan CG, et al. **Vet Rec** 1995;136(1):7-12.[14]Murray MJ, et al. **J Vet Intern Med** 1998;12(1):36-41.[15]Platt H. **In Pract** 1983;5(2):41-9.[16]Kemper T, et al. **J Am Vet Med Assoc** 1993;202(1):91-4.[17]Humber KA. **J Am Vet Med Assoc** 1988;192(3):377-8.[18]Austin SF, et al. **Equine Vet Science** 1988;8(4)313.[19]Sembrat R, et al. **Am J Vet Res** 1978;39(7):1147-54.[20]Begg AP, et al. **Aust Vet J** 2011;89 Suppl 1:19-22.[21]Kelly DF, et al. **Equine Vet** J 1995;27(1):76-8.[22]Turk JR, et al. **Vet Pathol** 1981;18(4):560-2.[23]Del Piero F, et al. **Vet Pathol** 2001;38(4):474.[24]Del Piero F, et al. **Vet Pathol** 2000;37(6):672-6.[25]Hamir AN, et al. **J Vet Diagn Invest** 1994;6(4):493-6.[26]Williams KJ, et al. **Vet Pathol** 2007;44(6):849-62.[27]Wong DM, et al. **J Am Vet Med Assoc** 2008;232(6):898-905.[28]Bell SA, et al. **J Vet Intern Med** 2008;22(3):648-53.[29]Dixon PM, et al. **Vet Rec** 1992;130(17):367-72.[30]Uhlhorn M, et al. **Equine Vet Educ** 2006;18(4):178-181.[31]Dixon PM, et al. **Equine Vet J** 1995;27(6):416-21.[32]Dixon PM, et al. **Equine Vet J** 1995;27(6):428-35.[33]Singh K, et al. **Vet Pathol** 2006;43(2):189-93.[34]Pusterla N, et al. **Vet Rec** 2003;153(21):653-5.[35]Sellers RS, et al. **J Comp Pathol** 2001;125(2-3):214-8.[36]Ziemer EL, et al. **J Am Vet Med Assoc** 1992;201(6):910-6.[37]Sweeney CR et al. **J Am Vet Med Assoc** 1999;214(6):808-11.[38]Schwartz LW, et al. **Chest** 1981;80(1 Suppl):82-5.[39]Berry CR, et al. **J Vet Intern Med** 1991;5(4):248-56.[40]Arens AM, et al. **Vet Pathol** 2011;48(3):593-615.[41]Breeze RG, et al. **Equine Vet J** 1984;16(3):180-4.[42]Schmidbauer SM, et al. **J Comp Pathol** 2004;131(2-3):186-98.[43]Nobre D, et al. **Vet Hum Toxicol** 1994;36(5):445-8.[44]Breeze RG et al. **Equine Vet J** 1984;16(2):108-12.[45]Dinis-Oliveira RJ, et al. **Crit Rev Toxicol** 2008;38(1):13-71.[46]Calderbank A, et al. **J Sci Food Agric** 1968;19(5):246-250.

UPPER AND LOWER AIRWAY INFLAMMATION IN HORSES: ENDOSCOPIC SCORING, DIAGNOSIS AND TREATMENT

Renaud Léguillette, DVM, MSc, PhD, DACVIM
Calgary, AB, Canada

INTRODUCTION:

Endoscopy is often used in equine medicine to investigate respiratory disorders. Upper and lower airways are both integral parts of the respiratory tract and human studies brought the concept of "one airway, one disease".[1] Indeed, there is evidence that allergic rhinitis and sinusitis co-exist in a high proportion of patients with asthma and chronic obstructive pulmonary disease (COPD).[2] Based on these findings, the World Health Organization (WHO) published recommendations for physicians to assess the lower airway condition in patients with rhinitis and conversely to perform upper airway evaluation in patients with asthma.[3] To the contrary, there is evidence in equine medicine that pharyngeal inflammation is not associated 1) with changes in tracheal cytology[4], or 2) with BAL cytology when done in a small number of 18-month-old healthy horses with low mucus scores.[5] Since upper airway endoscopy is routinely performed by equine practitioners, we will assess if the finding of upper airway inflammation also suggests the presence of lower airway inflammation in a patient, thus indicating the utility of further diagnostic tests. We will also review the diagnostic tests available to diagnose non-infectious airway inflammatory pathologies. Lastly, we will review the treatment options available once a diagnosis of airway inflammation is established.

ENDOSCOPIC EXAMINATION AND SCORING:

In an attempt to establish the relationship between mucus accumulation, inflammation observed by endoscopy, and bronchoalveolar lavage (BAL) cell type percentages in a large population of horses, we present here the results of a field study using mucus and inflammation scores of the upper and lower airways on 127 horses with and without lung inflammation.[6] We are essentially evaluating whether i an association exists 1) between inflammation scores and mucus accumulation scores and 2) between endoscopic scores and the lower airway inflammation measured with BAL cytology.
Endoscopic scores have been proposed for pharyngitis[7], accumulation of mucus in the trachea[8], and tracheal septum thickness[8,9], but they have not been combined altogether in one study to establish a comprehensive endoscopic score of both the upper and lower the airways. We assembled the scoring system shown in the Figure 1 and now use it systematically in a clinical setting.
Comprehensive endoscopy of the upper and lower airways is best reported using validated scoring systems. Parameters that are of clinical value are the amount of mucus and the degree of inflammation observed.

Mucus scores:

Mucus scoring has been well validated[8] but was limited to the trachea. With a comprehensive scoring system (Figure 1) the amount of mucus can be scored reliably from the pharynx down to the bronchi with an excellent inter-reader agreement (ICC>0.87).
-The pharyngeal mucus score did not show any correlation with any BAL cell type percentage, but was significantly greater in the group of horses with severe BAL inflammation. Pharyngeal mucus may not be a very reliable sign of airway inflammation possibly because horses can swallow it any time before or during the endoscopic examination.
-The tracheal mucus score is well correlated with the bronchial mucus score, which indicates that the use of a long endoscope brings little benefit in terms of mucus scoring. Both scores were correlated with BAL neutrophilia.
Altogether the presence of mucus in the trachea and bronchi indicates that the BAL will more likely show neutrophilia. It cannot be predicted if the inflammation will be severe or moderate, which is why a BAL is interesting to do and provides good information.

Inflammation scores:

The pharyngeal inflammation score is routinely used and is based on a Grade 1-4[7]. Tracheal inflammation is difficult to score and to our knowledge, is limited to the scoring of the septum thickness of the carina. There is some controversy on the usefulness of this score.[8,9] On 127 horses, we found a weak correlation between the tracheal septum score and lower airway mucus scores, but no positive correlation with the BAL neutrophils or mast cells. However, the tracheal septum score was significantly greater in horses with severe BAL inflammation than in horses with moderate or no BAL inflammation.
Overall, we found that the mucus and inflammation scores provide useful and clinically relevant information and that the "one airway, one disease" concept is not true in horses, meaning practically that if some mucus or pharyngitis is observed in a horse, it is not correlated with lung inflammation. A systemic endoscopic examination of the lower airways and/ or a BAL is thus recommended in cases with clinical signs and history suggestive of lower airway inflammation.

DIAGNOSIS:

Upper airway:

The significance of inflammation in the oropharynx is not clear. It is well accepted that young horses have some pharyngeal lymphoid hyperplasia and it is considered as normal. In more mature horses, pharyngitis develops most likely secondary to infection (viral or bacterial) or an allergic process. The diagnosis is often limited to endoscopic observation. When a primary infection is suspected, it is possible to perform pharyngeal swabs and submit them for PCR analysis (viral, *Strep. equi equi*).

Lower airway:

- Inflammatory airway disease (IAD) vs recurrent airway obstruction (RAO)

Three non-infectious lower airway diseases have been defined in horses: Inflammatory airway disease (IAD), recurrent airway obstruction (RAO) and summer pasture-associated obstructive disease (SPAOD). It is important to note that a diagnostis for these diseases cannot be made simply based on a BAL cytology report but also needs to take into account clinical signs and patient history. There is often some confusion in practice on the differences between the definitions of IAD and RAO. Table 1 provides a summary of clinical and laboratory results helping differentiate between IAD and RAO. Basically, the most important is to know is if the horse has clinical signs of labored breathing at rest. This symptom (caused by lower airway obstruction, hence the name RAO), combined with an abnormal BAL cytology is diagnostic of RAO. Based on a consensus paper[10] and workshop[11] IAD is defined as a noninfectious inflammatory lung disease with 1) Respiratory clinical signs at work and/or exercise intolerance, without clinical signs of labored breathing at rest, even after exposure to moldy hay, 2) Airway hyperresponsiveness, 3) BAL inflammation. Consequently, if a horse has any respiratory clinical signs (coughing, increased mucoid discharge at the nose or in the trachea/ pharynx), combined with an abnormal BAL cytology, this is diagnostic of IAD. Depending on the type of usage and severity of the inflammation, it is important to note that not all of these horses have exercise intolerance. The prevalence of IAD has been shown to be high in several studies and can affect horses of any age and discipline.[12, 13]

- Lower airway cytology:

Based on the above, a definitive diagnosis of RAO or IAD is made after lower airway cytology is obtained. There are 3 available tests:tracheal cytology, BAL, and lung biopsy.BAL is most accurate and practical. -

- Lung biopsies are not without risks and should be limited to research or when neoplasia or lung fibrosis are suspected.
- Although easy to perform, it is now accepted that tracheal ctytology is not the most reliable for diagnosis of lung inflammation, probably because the sample is taken remotely from the alveoli. Tracheal cytology is commonly performed at racetracks through the endoscope because it is quick and practical. Another option for obtaining tracheal fluid for analysis is a transtracheal wash (TTW), which is more invasive and necessitates more sedation.
- Lastly, BAL cytology is a reliable test to diagnose lung inflammation, probably because it samples cells closer to the alveoli. The technique used to perform a BAL can be performed easily in the field with limited equipment.. The BAL technique along with requirements for sample shipping, processing and interpretation will be presented including staining[14] and cell counting techniques[15] necessary for reliable interpretation.

-Lower airway bronchoprovocation:

Bronchoprovocation tests are available in specialized centers and allow measurement of parameters indicative of airway hyperresponsiveness. Airway hyperresponsiveness is a valuable parameter for both research and clinical practice because it can be detected before more obvious clinical signs develop.[16, 17]

TREATMENT:

Pharyngitis:

Very little information is available on the efficacy of treatments for oropharynx inflammation. Treatments are routinely based on solutions containing corticosteroids, glycerin, DMSO and other anti-inflammatory drugs. We also use inhaled corticosteroids to treat these cases but our clinical impression is that they are not more effective than traditional treatments.

RAO and IAD:

- Management:

- The key to treatment of RAO is environmental management, through limiting exposure to organic dust. This is easier said than done, since horses live as a group in a dusty environment. Usual recommendations are to avoid straw bedding, turn the horse outside as much as possible especially when sweeping in the barn. Round hay bales should be avoided as they are a risk factor for severe lung inflammation[18]; interestingly, we found that many horses in the Calgary area have RAO when they are outside because of the round hay bales. Recommendations are usually to replace hay with cubes or pellets, or to use hay steamers.
- The exact etiology of IAD is still unknown. Several studies have demonstrated a link between IAD and a poor environment.[17,19]Although the role of the management in IAD has not been described, it is recommended to pay attention to obvious exposure to dust or allergens (pollen for example) in these cases.

- Medical treatment of RAO and AID:

RAO:

Table 3 summarizes the protocol we use for dexamethasone or fluticasone (administered with the AeroHippus device). Both dexamethasone and fluticasone have been proven to be effective in relieving clinical signs and altering cytologic findings of bronchoalveolar lavage fluid in horses with RAO.[20- 24]

IAD:

Although many studies have been published on the diagnosis and characterization of the phenotype of IAD in horses, the scientific evidence for treatments of IAD is extremely limited. It is however common practice to treat IAD with glucocorticoids. We reported that both dexamethasone and fluticasone treatments inhibit airway hypersensitivity and hyperreactivity in horses with IAD, but do not improve clinical signs or the number of most inflammatory cells in the BAL.[25] We also found that the treatments have no residual effect 3 weeks after discontinuation.

REFERENCES

1- Bachert C, et al, Immunol Allergy Clin North Am 2004, 24 (1):19- 43; **2-** Bousquet J, et al, Allergy 2003;58:691-706; **3-** Bachert C, et al, WHO workshop report in Allergy 2002;57:841-855; **4-** Holcombe SJ, et al, Equine Vet J 2006;38:300-304; **5-** Holcombe SJ, et al, Equine Vet J 2001;33:244-249; **6-** Koblinger K., et al., J Vet Intern Med. 2011 Sep;25(5):1118-26; **7-** Raker CW. The Nasopharynx. In: Mansmann R.A. MAES, ed. Equine Medicine and Surgery, 3rd Ed, 1982; **8-** Gerber V, et al., Equine Vet J 2004;36:576-582; **9-** Koch C, et al., Equine Vet J 2007;39:107-112; **10-** Couetil L.L., et al.,J Vet Intern Med 2007, 21, 356-361; **11-** V. Buechner-Maxwell, et al., Havemeyer Foundation Workshop 2009; **12-** Allen, K.J. et al Equine Vet J Suppl, 2006 529-534.; **13-** Wasko, A., et al. Equine Vet J, 2011 43 (2), p.145-152; **14-** Leclere, M., et al., J Vet Intern Med 20, 377-381; **15-** Fernandez N.J., et al.Veterinary Clinical Pathology 2012, in press; **16-** Couetil, L.L., et al., Am J Vet Res 2001 62, 538-546; **17-** Hoffman, A.M., et al. Am J Vet Res 1998 59, 176-181; **18-** Wasko A., et al.: Veterinary Comparative Respiratory Society, 26th symposium, 2008 p72; **19-** Holcombe, S.J., et al. Equine Vet J 2001 33, 244-249; **20-** Giguere, S. et al Vet Immunol Immunopathol 2002 85, 147-158; **21-** Picandet, V. et al Equine Vet J 2003 35, 419-424; **22-** Robinson, N.E., et al. Equine Vet J 2002 34, 17-22; **23-** Rush, B.R., et al. Am J Vet Res 1998 59, 1033-1038; **24-** Viel, L. et al. AAEP annual convention 1999; **25-** Tohver T. et al., American College of Veterinary internal Medicine Forum, 2010.

Table 1: Diagnostic criteria for IAD and RAO

History	Clinical signs	Diagnostic tests	Other
RAO			
Recurrent/ seasonal clinical signs	**MUST HAVE**: Labored breathing at rest. **MAY HAVE**: Cough, mucoid to purulent discharge.	**BAL:** Neutrophilia (usually marked) **Lung mechanics at rest:** Increased DP_L and R_L **Bronchoprovocation**: Variable	Positive "moldy hay challenge"
IAD			
No history specific to IAD.	**MUST <u>NOT</u> HAVE**: Labored breathing at rest. **MUST HAVE at least one of**: - Cough (often at beginning or during exercise) - Mucoid discharge - Exercise intolerance	**BAL:** Increased neutrophils or mast cells (usually Mild inflammation). **Lung mechanics at rest:** Normal **Bronchoprovocation**: Positive	Negative "moldy hay challenge"

Table 2: Treatment protocol RAO (and IAD): These doses are for a horse not suspected of having PPID, Equine metabolic syndrome, or laminitis. Lower doses are also used on poneys, fat horses and horses with small feet.

Systemic corticosteroids	
Dexamethasone (IV or IM)	20mg SID in the morning for 10 days, then 15mg SID in the morning for 10 days, then 10mg SID in the morning for 10 days, then 10mg every other day in the morning for 10 days
Inhaled corticosteroids, with Aerohippus device (Trudell)	
Fluticasone (Flovent 250 ug) or Ciclesonide (Alvesco 200ug)	15 puffs BID for 15 days, then 10 puffs BID for 15 days, then 10 puffs SID for 15 days, then 5 to 8 puffs as needed Also used in prevention: For example when changing barn or exposure to dust is noticed or predictable
Bronchodilators (only in very severe cases and when exposure to organic dust is under control)	
Clenbuterol (Ventipulmin)	8 or 12 pump strokes BID (i.e. 1.6 or 2.4ug/kg) orally

Figure 1: Comprehensive endoscopic scoring system of upper and lower airway mucus and inflammation.

1. Scoring system for upper airways:

Pharyngitis: Grade I - IV

Pharyngeal Mucus Accumulation: Score 0 - 2

2. Scoring system for lower airways:

Trachea Mucus Accumulation: Score 0 - 5

Bronchial Mucus Accumulation: Score 0 - 5
(scored from carina to bronchus 2.4)

0	0.5	1	1.5	2	2.5	3	3.5	4	4.5	5
None		Little		Moderate		Marked		Large		Extreme
Clear, singular		Multiple small blobs		Larger blobs	Confluent		Stream-forming	Pool-forming		Profuse amounts

Septum Thickness: Score 0 - 4

THORACOTOMY FOR TREATMENT OF RECURRENT PLEURAL ABSCESSES ASSOCIATED WITH EQUINE PLEUROPNEUMONIA

M. Keith Chaffin, DVM, MS, DACVIM
College Station, TX

Bacterial pleuropneumonia is a common disorder that results in sizable economic losses to the equine industry. Pathophysiology of this disease involves bacterial colonization of the lung, development of pneumonia and/or lung abscesses, and subsequent extension to the visceral pleura and pleural space, resulting in pleural effusion.[1-4] Horses of any age, breed, sex, or occupation may be affected. Risk factors include long-distance transport, strenuous exercise (such as racing), viral infections, laryngeal and/or pharyngeal dysfunction, recent general anesthesia, esophageal obstruction, and aspiration.[1-4] Bacteria can colonize the lung due to compromised or overwhelmend pulmonary mucosal defense mechanisms. Polymicrobial and mixed aerobic/anaerobic infections are common.[5] Bacterial pleuritis may also arise in the absence of pneumonia. Thoracic puncture wounds and esophageal perforation can lead to pleural empyema.[1] Primary pleuritis has also been described.[6]

Clinical signs may include fever, lethargy, anorexia, cough, tachypnea, tachycardia, nasal discharge, exercise intolerance, halitosis, sternal edema, pleurodynia and weight loss.[2-4,7] Diagnosis of pleuropneumonia is based upon physical examination, hematology, thoracic auscultation, thoracic percussion, thoracic radiography, thoracic ultrasonography, tracheobronchial aspiration, thoracocentesis, and cytologic and microbiologic evaluation of tracheal aspirates and pleural fluid samples.[7] Treatment of bacterial pleuropneumonia varies depending on the severity of disease, the causative organisms, the presence and character of pleural effusion, the amount of fibrin deposition, the microenvironment of the lung and pleural space and the development of sequelae.[8] Treatment modalities include antimicrobials, anti-inflammatory agents, pleural drainage, pleural lavage, intra-pleural antimicrobial agents, rest, intravenous fluids and supportive care.[8-9] Intra-pleural administration of fibrinolytic agents has been described.[10] Frequent imaging of the thoracic cavity, via ultrasonographic scanning, is crucial to monitor the horse's response to therapy, and to determine the need for alteration of therapy.

Sequelae of pleuropneumonia include pulmonary and pleural abscesses, bronchopleural communications, pleural fibrin deposition, pleural loculation, cranial thoracic abscesses, laminitis, thrombophlebitis, pericarditis, and pneumothorax.[8,11] Pleural absesses are one of the more common sequela of pleuropneumonia. Abscesses may be single or multiple and can vary from small to very large in size.[8] Ultrasonography is required to detect and localize pleural abscesses.[8] The character of the abscess contents can often be determined songraphically. Large bright internal echoes may represent inspissated debris and/or necrotic lung and debris. A fibrotic wall can often be imaged surrounding the axial, dorsal, ventral and abaxial borders of the abscess. In some cases, the fibrotic wall will appear hyperechoic relative to the abscess contents, whereas, in other cases, the wall may be relatively hypoechoic. Some abscesses may have sonographically-visible free gas present in the dorsal region of the abscess cavity.[12]

Although some pleural abscesses may respond to prolonged antimicrobial therapy, I prefer to drain large abscesses (>10 cm diameter) via large-bore trocars or chest tubes placed directly into the abscess.[8] One should exercise caution to avoid rupturing the abscess wall, which could result in spillage of abscess contents into the pleural space. Suction can be applied, if needed, and the tube can be repositioned to remove as much inspissated pus and debris as possible. Lavage of the abscess cavity with warmed, sterile, isotonic fluids may help dilute thick pus and facilitate drainage. Indwelling tubes in the abscess may be necessary to allow repeated drainage and lavage. Some abscesses contain inspissated exudate, fibrino-necrotic debris, and necrotic lung. Such material is too large to drain via chest tubes,[8] and these abscesses are either persistent or recur after repeated efforts at drainage and lavage.

In horses with recurrent pleural abscesses that are refractory to medical therapy, surgical intervention is needed to facilitate manual debridement and lavage of the abscess cavity.[8,13-17] Thoracotomy may promote healing, and ultimately curtail the overall expense of prolonged hospitalization and antimicrobial therapy. I do not use surgical intervention as a primary form of therapy for pleuropneumonia, but instead reserve it for chronic pleural abscesses that have failed to respond to repeated drainage and lavage.

In my experience, proper case selection is critical for a successful outcome. Indications for thoracotomy include failure to respond to aggressive antimicrobial therapy, pleural drainage and lavage, and the presence of significant volumes of sonographically-visible organized fluid and debris within a compartment of the pleural space.[8] Other requirements that I consider for a surgical candidate include a stabilized systemic condition despite the presence of severe, localized and persistent pleural abscess(es). It is critical that the pleural abscess be walled-off from the remainder of the ipsilateral hemithorax. The mediastinum should be complete or a visceral-to-parietal seal must be present around the abscess.

Prior to thoracotomy, the location and nature of the recurrent pleural abscess must be thoroughly characterized with ultrasonographic imaging.[8] The anatomic borders of the abscess are mapped out in advance. It is very helpful to place a chest tube into the center of the abscess and infuse sterile, isotonic fluid into the abscess and to perform simultaneous ultrasound examination (contrast pleurosonography) to assure that the abscess does not communicate with the remainder of the ipsilateral or contralateral pleural space. If the infused fluid remains compartmentalized within the abscess cavity, then the infused fluid serves as a contrast agent and helps the sonographer to identify the dorsal, ventral and axial margins of the abscess cavity. In some horses, the infused fluid will be coughed up by the horse indicating a bronchopleural communication. In most horses, the chronic pleural abscess will be in the cranioventral thorax just behind the heart and just in front of the diaphragm, but occasionally, more dorsal or cranial sites will be affected. The anatomic boundaries of the abscess are marked on the horse's chest and the optimum surgical site is identified. Occasional cases may require multiple thoracotomies.

At this point, the indwelling chest tube is opened to atmospheric air, and the horse is monitored closely for the following 1-2 hours. If the abscess is adequately walled off, the horse will show no untoward effects. If the abscess is not adequately walled off, unilateral or bilateral pneumothorax will develop and the horse may show tachypnea. Because unilateral pneumothorax often does not result in clinical signs, I recommend sonographically imaging the dorsal lung fields to make sure that subclinical pneumothorax has not developed. If the horse does not develop pneumothorax and the infused fluid remains compartmentalized, then I consider the horse a candidate for standing thoracotomy. If these conditions are not met, then I recommend aspirating any air from the dorsal pleural space to allow reexpansion of the lung, delaying surgery, and continue medical therapy until the abscess cavity is better walled off.[16,17]

Several surgical procedures have been described including thoracotomy via an intercostal approach,[13] thoracotomy via intercostal muscle myectomy,[13] thoracotomy and rib resection under general anesthesia,[15] and thoracotomy and rib resection in the standing position.[16-17] I have most frequently utilized thoracotomy and rib resection in the standing position because it eliminates the risks of general anesthesia and allows superior exposure for intrathoracic exploration, debridement, post-operative endoscopic assessment and lavage. Also, many large recurrent pleural abscesses have long, tortuous "communicating tunnels" that are not recognized pre-operatively, and may not be recognized during surgery unless adequate exploration of the abscess cavity is possible. These tunnels require careful and complete lavage for successful resolution. For smaller abscesses, I often elect the intercostal approach. For cranial mediastinal abscesses, I prefer general anesthesia to facilitate positioning of the forelimb, and an intercostal approach.

For standing thoracotomy and rib resection, the patient is placed in restraint stocks and sedated with intravenous detomidine combined with butorphanol. I recommend having ultrasound equipment and materials to remove pleural air readily available should the horse develop pneumothorax during surgery. The surgical site is infiltrated locally with lidocaine. A 10-15 cm skin incision is made over the chosen rib and through the soft tissues down to the periosteum of the rib.[8] The periosteum is carefully reflected using periostial elevators and obstetrical wire is used to resect a section of rib. Often the ventral limit of resected rib is simply disarticulated at the costochondral junction. The remaining periostium and parietal pleura are then incised, the pleural cavity is explored, all fluid and necrotic debris are manually removed and the cavity is thoroughly lavaged. I recommend caution to avoid overzealous debridement and resultant damage to the abscess wall because unilateral or bilateral pneumothorax can inadvertently develop. The incision is left open to heal by second intention.[9] A sterile towel is generally sutured over the incision to prevention introduction of foreign material into the abscess cavity. Non-steroidal anti-inflammatory agents are used for first few days after surgery. Postoperatively, the abscess cavity is periodically debrided via gentle massage performed by a clinician wearing a sterile obstetrical sleeve. Any manually loosened debris is removed via lavage. An endoscope is often introduced into the abscess cavity to visually explore the cavity for remaining necrotic debris and communicating tunnels. In the event that communicating tunnels are identified, they should be thoroughly lavaged via a sterile nasogastric tube placed directly in the tunnel(s). Usually, granulation tissue closes the incision over a period of 2-3 weeks, but some cases will continue to drain for a more extended period.[9]

In previous reports, clinical improvement has been reported in 3 of 3,[16] 3 of 5,[17] 2 of 2,[14,18] and 1 of 3[9] horses treated with thoracotomy. In a more recent report, Hilton et al[13] described the indications, complications and outcomes of 16 horses that were treated with standing lateral thoracotomy. Fourteen of the horses had recurrent pleural abscessation. Intercostal thoracotomy was used in 13 horses, intercostal muscle myectomy was used in 5 horses, and thoracotomy and rib resection was used in 1 horse. Perioperative complications included hemipneumothorax in 2 horses, and cellulitis and abscessation of the incision site in 6 horses.[13] Median healing time for incisions was 2.5 months (range, 1-48 months). Fourteen horses survived to discharge and 46% of the survivors returned to their previous level of athletic activity.[13]

REFERENCES

[1]Chaffin MK, et. al. **Comp Cont Ed Pract Vet** 1993;15(12):1642. [2]Raphel CF, et. al. **J Am Vet Med Assoc** 1982;181(8):808. [3]Byars TD, et. al. **Vet Clin of North America (Equine)** 1991;7(1):63. [4]Sweeney CR **Current Therapy in Equine Medicine, ed 3**. 1992:327. [5]Sweeney CR, et. al. **J Am Vet Med Assoc** 1991;198(5):839. [6]Smith BP **J Am Vet Med Assoc** 1977;170(2):208. [7]Chaffin MK, et. al. **Comp Cont Ed Pract Vet** 1994;16(3):362. [8]Chaffin MK, et. al. **Comp Cont Ed Pract Vet**1994;16:1585. [9]Schott HC III, et. al. **Comp Cont Ed Pract Vet** 1990;12(2):251. [10]Hilton H, et. al. **Vet Record,**2009;164:558-559. [11]Byars TD, et. al. **Equine Vet J** 1991;23(1):22. [12]Reef VB **Ultrasonographic evaluation. In Beech J (ed)** 1991;69. [13]Hilton H, et. al. **Vet Surgery** 2010;39:847. [14]Henninger R, et. al. **Equine Pract** 1988;10(6):13. [15]Colahan PT, et. al. **J Am Vet Med Assoc** 1979;174(11):1231. [16]Shearer DC, et. al. **Proc 31st Ann Conv Am Assoc Equine Pract** 1986;31:39. [17]Carter GK, et. al. **Vet Surg** 1986;15(1):115. [18]Dechant J **Can Vet J** 1997;38:499.

197b

WEIGHT GAIN IN HOSPITALISED NEONATAL FOALS. GD Lester[1], LC Sanchez[2], RC Hill[2]. 1. Murdoch University School of Veterinary and Biomedical Sciences, Murdoch, Australia, 2. University of Florida College of Veterinary Medicine, Gainesville, FL.

Nutrition is an important component of neonatal foal care. The average daily weight gain (ADG) of healthy foals is around 1.6 kg/d from 1-17d of age, but there are limited data regarding ADG in sick foals. The study objective was to evaluate the impact of disease on ADG in a population of hospitalized newborn foals.

The medical records of 210 foals presenting before 7d of age and weighing >18kg were reviewed. Information recorded included estimates of volume and caloric delivery when intake was controlled, bodyweight, diagnosis, survival, and basic laboratory data. Parenteral nutrition (PN; IV glucose+amino acids ± lipids) was used in 71 foals. Average daily weight gain was also calculated when foals had unlimited access to their dam. Fifty four foals had a day 1 increase in bodyweight >5%, thus their day 2 value was used as initial bodyweight.

During periods of controlled caloric intake ADG was 0.50 kg/d, and was significantly greater in surviving foals (0.61 kg/d) than non-survivors (0.33 kg/d). On average 6459 kcal were required for 1kg increase in bodyweight. Non-survival in septic foals was associated with significantly reduced ADG and increased caloric intake per kg increase in weight (4962±546 survivors and 7463±982 non-survivors). Non-survival in PN foals was significantly associated with reduced ADG, increased caloric demand, and increased blood glucose, creatinine, and triglycerides. The ADG in foals on mares was 1.00 kg/d.

Weight gain in neonatal foals is impacted by disease and is highlighted by a reduced efficiency of converting calories to body mass.

198a

EVALUATION OF EMPLOYMENT OPPORTUNITIES, CAREER SATISFACTION, AND FINANCIAL COMPENSATION OF RECENTLY-BOARDED ACVIM-LAIM DIPLOMATES.

SM Reuss[1], MK Chaffin[2], BM Waldridge[3], JH Wilson[4], J Gold[5]. 1. University of Florida College of Veterinary Medicine, Gainesville, FL. 2. Texas A&M University College of Veterinary Medicine, College Station, TX. 3. Georgetown, KY. 4. Stillwater, MN. 5. Dubai, UAE.

An online survey was conducted to determine the employment opportunities, career satisfaction, and financial compensation encountered by recent ACVIM-LAIM diplomates. The survey was designed to collect data to identify methods to more effectively market current internists, as well as to identify methods to alter residency training programs to promote future marketability.

The survey consisted of 67 questions, beginning with demographic and background data. Participants were then asked a series of questions regarding their employment opportunities since achieving board certification, financial compensation and practice ownership, job satisfaction and perceived benefits of being LAIM-boarded, and ways in which residency training programs could be altered to better meet employment needs.

The survey was sent to 123 LAIM diplomates who achieved board certification between 2006 and 2011 of which 90 responded (73% response rate). Respondents were evenly distributed between academia and private practice employment. 31/90 respondents enrolled in a graduate program after completion of their residency. The majority (55%) of respondents in clinical practice treat predominantly horses, while 14% treat predominantly food animal, and 13% treat mixed large animals. 42% of respondents found it difficult to find satisfactory employment post-residency, and 59% rated the current employment opportunities for newly-boarded large animal internists as poor. 49% stated that their current job does not meet the employment expectations that they had when starting a residency program, and only 34% would recommend pursuing a large animal medicine residency to current students or interns. 93% of respondents think that the general public is unaware of the specialized skills and depth of training of internists, and 54% believe general practitioners are also unaware. 78% think that owners do not know the difference between an "intern" and an "internist." 51% believe that having had an opportunity to get advanced training in disciplines other than medicine during their residency would have improved their employment opportunities with the most popular disciplines being lameness, theriogenology, general surgery, ophthalmology, and cardiology.

These data will be shared with the residency training committee and will be used to design a marketing strategy for large animal internal medicine specialists.

198b

SODIUM BICARBONATE ADMINISTRATION EXACERBATES POST EXERCISE HYPOKALEMIA IN STANDARDBREDS UNDERGOING A SIMULATED RACE PROTOCOL S Tinkler, L Couëtil, PD Constable. Purdue University School of Veterinary Medicine, West-Lafayette, IN.

Administration of alkalinizing agents is a major concern to the horseracing industry because of the potential for performance enhancement or masking of other performance altering drugs by changing urine pH or urine concentration. We hypothesized that the pre-race administration of sodium bicarbonate would result in clinically relevant changes in acid-base status and plasma electrolyte concentrations before, during, and after a simulated race.

Eight Standardbreds (2 to 6 years of age; 4 females, 4 geldings) were fed a typical training diet (DCAD $\approx$ 200 mEq/kg DM) and trained 5 times a week for 6 weeks using a high-speed treadmill. Sodium bicarbonate was administered by nasal intubation at 0, 0.5, and 1.0 g/kg BW in 3 L of distilled water in a randomized cross-over design 4 hours before horses underwent a simulated race protocol (SRP) on a high-speed treadmill. Jugular venous blood samples were collected via an indwelling catheter immediately before treatment and periodically up to 4 hours after the SRP. Plasma Na, K, and Cl concentrations were determined using a Nova-4 analyzer, blood pH and gas tensions were determined using an ABL-5 blood gas analyzer, and plasma L-lactate concentration was measured using an Accusport unit. Data was analyzed using repeated measures analysis of variance and $P<0.05$ was significant.

Oral sodium bicarbonate administration induced a clinically significant dose-dependent alkalemia, strong ion (metabolic) alkalosis, hypernatremia, hyperlactatemia, and profound hypokalemia. Plasma potassium concentrations were lowest at 90 minutes after the end of the SRP, being significantly lower ($P<0.0001$) than the baseline value (3.4±0.2 mEq/L, mean±SD) at 2.6±0.3, 2.2±0.4, and 1.9±0.3 mEq/L in the 0, 0.5 and 1.0 g/kg sodium bicarbonate groups, respectively. Plasma potassium concentrations had returned to baseline values by approximately 4 hours after the end of the SRP in horses in the 0 g/kg group, but not the 0.5 and 1.0 g/kg groups. We conclude that the prerace administration of oral sodium bicarbonate induces clinically significant and persistent electrolyte and acid-base changes in Standardbreds. Furthermore, our results suggest that post race hypokalemia may play a role in the development of cardiac arrhythmias in exercising racehorses.

199a

EFFECTS OF HYPERBARIC OXYGEN THERAPY ON ARTERIAL BLOOD GASES AND LUNG IMFLAMMATION IN HEALTHY HORSES. M. Looijen[1], D. New[1], M. Scott[2], C. Berezowiski[2], R. Léguillette[1,2]. University of Calgary, Faculty of Veterinary Medicine, Calgary, AB, Canada. 2. Moore Equine Veterinary Centre, Calgary, AB, Canada.

No data is available on the effects of hyperbaric oxygen (HBO) therapy on lung physiology in equines. Because of oxygen toxicity, HBO could induce lung inflammation. We hypothesized that (1) HBO therapy induces lung inflammation in healthy horses and (2) HBO induces a transient increase in arterial oxygen levels with no other effect on arterial blood parameters.

Eight healthy thoroughbred horses were used in a randomized controlled cross-over design; Treated horses were exposed to 100% oxygen at 3 ATA for 20 minutes for 10 days whereas the chamber was not pressurized for control horses. A bronchoalveolar lavage (BAL) was performed at baseline and on day 10. Arterial blood parameters were measured before and right after unloading horses from the HBO chamber on day 1 and 10. Four additional horses were used to measure arterial blood gases collected through an arterial line during the HBO sessions at baseline, 3 ATA (for 0, 10 and 20 minutes), during and 10 minutes after decompression.

Results show that HBO induced a significant decrease in total BAL cell counts but had no effect on the differential cell counts. Arterial blood parameters before vs after unloading the horses from the HBO chamber and on day 1 vs day 10 were not significantly different. However, during HBO PaO_2 increased (>800mmHg) rapidly and decreased to baseline values within 10 minutes after HBO. The increase in $PaCO_2$, HCO3- and lactate was not significant.

These results suggest that HBO reaches extremely high blood oxygenation levels very transiently and does not induce inflammation in the lungs of horses.

199b

THE EFFECTS OF METFORMIN HYDROCHLORIDE ON INTESTINAL GLUCOSE ABSORPTION AND USE OF TESTS FOR HYPERINSULINAEMIA.

AE Durham[1], DI Rendle[1,2], F Rutledge[2], J Heller[2], KJ Hughes[2]. [1]Liphook Equine Hospital, Hampshire, UK; [2]Charles Sturt University, New South Wales, Australia.

Metformin has been proposed as a treatment for Equine Metabolic Syndrome. Whereas some studies have found decreased plasma insulin and glucose following metformin therapy in horses, other studies have found poor bioavailability and no change in insulin sensitivity following treatment. This study was designed to test the hypothesis that metformin impairs carbohydrate absorption in the equine intestine as a possible explanation of the apparently contrasting findings of previous studies.

Nine horses were subject to an oral glucose challenge test with plasma glucose and serum insulin measured before and for up to 240 minutes after consumption of a feed containing glucose. The same test was repeated 30 minutes after an oral dose of 30 mg/kg metformin. Tests were performed using 0.5 or 1.0 g/kg bodyweight glucose and were also performed with and without prior treatment with dexamethasone.

Pre-treatment with metformin significantly decreased absorption of glucose and the subsequent insulin response in all horses (figure 1).

Given that previous evidence indicates that hyperinsulinaemia (without insulin resistance) may cause laminitis, tests examining dietary insulinaemic response and strategies that moderate hyperinsulinaemia, such as low NSC diets and metformin therapy, are logical approaches to controlling laminitis risk.

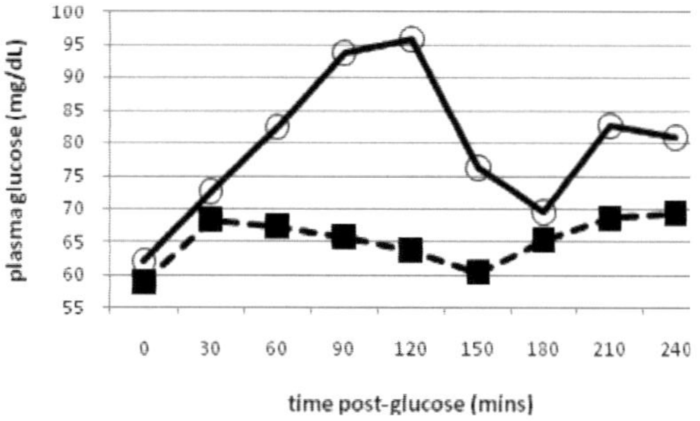

Figure 1. Glycaemic response to 0.5 g/kg glucose in feed with (dashed line) and without (solid line) pretreatment with 30 mg/kg metformin.

200a

DIGITAL HYPOTHERMIA APPLIED AFTER THE ONSET OF LAMENESS PREVENTS LAMELLAR FAILURE IN THE OLIGOFRUCTOSE LAMINITIS MODEL

A. Van Eps[1], C. Pollitt[1], C. Underwood[1], C. Medina-Torres[1], W. Goodwin[1], J. Belknap[2].
1.University of Queensland, Gatton, Queensland, Australia 2.Ohio State University, Columbus, Ohio.

Several studies have demonstrated that digital hypothermia reduces the severity of laminitis lesions when initiated early in the developmental stage of oligofructose (OF)-induced laminitis. Digital hypothermia is now commonly used clinically in the horse at risk of laminitis; however, it remains unclear whether there is still a beneficial effect when cryotherapy application is delayed until the detection of lameness. Thus, the aim of this study was to test the efficacy of digital hypothermia for decreasing the severity of lamellar injury in acute laminitis when treatment is initiated at the onset of detectable lameness.

Eight horses were administered 10g/kg OF by nasogastric tube. Upon the first observed signs of lameness at the walk, one forelimb (CRYO) of each horse was continuously cooled using ice and water. The other forelimb (NON-RX) was maintained at ambient temperature. Continuous analgesia of the palmar nerves of both forelimbs (via perineural catheters) was also initiated at the onset of lameness for humane reasons. Sections of dorsal lamellae from both forelimbs were harvested 36 h after the initial development of lameness (36 h of hypothermia for treated limb) for paraffin embedding. Sections were stained with H&E and PAS, randomized and coded for histological analysis performed by 2 blinded observers. The severity of laminitis pathology was scored (1 to 4) for each section using a system based on that previously described by Pollitt (1996) and morphometric analyses were also performed on 10 individual primary (PEL) and secondary (SEL) epidermal lamellae per section. Data were analyzed using Wilcoxon signed ranks tests.

Histopathology scores for the CRYO limbs (1 [1-1]) (median [interquartile range]) were significantly ($p<0.05$) less than those of the NON-RX limbs (3.5 [2-4]). There was total separation and dislocation of epidermis from dermis in the NON-RX limb (score of 4) in four horses, prohibiting morphometric analysis and comparison; the corresponding score of the CRYO limb in these four horses was dramatically decreased (score of 1 in three horses and a score of 2 in one horse). Morphometric analysis was thus limited to 4 horses in which the NON-RX lamellae remained intact; there was a trend of reduced PEL length (3112μm [2934-3248] vs. 3318μm [3146-3667]) and SEL length (207 μm [159-320] vs. 262 μm [192-316]), and increased SEL width (19.9μm [18.2-23.9] vs. 14μm [9.1-17.9]) in the CRYO limbs.

These data indicate that digital hypothermia effectively prevents the progression of lamellar injury (and even structural failure) when initiated at the detection of lameness in an acute laminitis model. This study provides the first evidence to support the use of digital hypothermia in clinical acute laminitis cases.

200b

EMERGENCE OF WIDESPREAD MACROLIDE AND RIFAMPIN RESISTANCE IN RHODOCOCCUS EQUI ISOLATES FROM A HORSE BREEDING FARM. S. Giguère,[1] A.J. Burton,[1] T.L Sturgill,[1] L.J. Berghaus,[1] N.M. Slovis,[2] J. Whitman,[3] K.R. Kuskie,[4] and N.D Cohen[4] 1. University of Georgia, Athens, GA; 2. Hagyard Equine Medical Institute, Lexington, KY; 3. Equine Medical Associates, Lexington, KY; 4. Texas A & M University, College Station, TX.

The combination macrolide-rifampin has been the mainstay of therapy in foals infected with *Rhodococcus equi* for nearly 30 years, with only rare isolated reports of resistance. In recent years, mass antimicrobial treatment of subclinically affected foals has become common practice on endemic farms in an attempt to prevent mortality. The objective of the present study was to describe the emergence of macrolide and rifampin resistance on a farm where widespread use of these drugs was instituted.

In 2010, 45 of 138 (33%) foals born on the farm were treated for subclinical ultrasonographic pulmonary lesions attributed to infection caused by *R. equi.* Culture of a tracheobronchial aspirate (TBA) was performed on 28 foals and yielded macrolide- and rifampin-resistant *R. equi* in 9 of 22 (41%) foals from which *R. equi* was cultured. Subsequently, air samples were collected from multiple sites on the farm. In 2011, TBAs were collected from each foal with subclinical lesions prior to initiation of therapy and 2 weeks later. *R. equi* isolates were tested for susceptibility to macrolides and rifampin, and genotyped by rep-PCR.

Two of 83 (2%) *R. equi* isolates from air samples were resistant. Resistant isolates were recovered from pre-treatment TBA in 7 of 21 (33%) foals from which *R. equi* was cultured. The proportion of resistant isolates from the post-treatment TBA was 7/12 (58%). Resistant isolates formed 2 distinct genotypic clusters whereas there was considerable genotypic heterogeneity among susceptible isolates. These results underscore the potential risk of widespread use of macrolides and rifampin.

WHAT'S NEW IN GASTROINTESTINAL SURGERY
Sylvain Nichols, DMV, MS, DACVS
St-Hyacinthe, QC, Canada

ABOMASAL SURGERY: LAPAROSCOPIC FIXATION

Surgical techniques to correct or prevent abomasal displacement have changed in the past 30 years. Today, in an era of minimally invasive surgery in human and veterinary medicine, laparoscopic techniques of abomasal fixation are gaining in popularity. Laparoscopic abomasal fixation can be divided in 2 categories: laparoscopic guided toggle fixation (LGTF) and laparoscopic abomasopexy (LA). The later is more challenging and it is only applicable to a hospital setting. The LGTF has been developed for field conditions.

Laparoscopic guided toggle fixation (LGTF)

Three different techniques have been developed: A) Janowitz's technique[1] (2 steps), B) Christiansen's technique[2] (standing 1 step), and C) Newman's technique[3] (recumbent 1 step).

A) The Janowitz's technique[1] is the mother of all the other laparoscopic techniques. It has been developed for LDA but it could be used for RDA. With the cow standing, the upper part of the left flank, from the 10th ICS to the mid paralumbar fossa, is surgically prepared. Local anesthesia is performed just caudal to the last rib, proximal to the transverse process of the lumbar vertebrae (laparoscope portal) and at the 11th ICS dorsally (instrument portal). First, a 10mm in diameter cannula and its trocar are introduced blindly in the abdomen. Air is allowed to enter the abdomen. The trocar is replaced by the scope and the abdomen is explored. The abomasum is visualized and the instrument portal is created. Under laparoscopic guidance, a 5mm in diameter and a 32.5cm long cannula and its trocar are introduced in the abdomen and in the abomasum at the level of the greater curvature. The trocar is removed and a specialized toggle pin (double suture toggle) is introduced in the abomasum. The toggle is pushed with a blunt trocar. The abomasum is deflated by the long cannula and the suture ends of the toggle are left in the abdomen. The scope and all the cannulas are removed and the portals are closed with skin sutures.

The cow is placed in dorsal recumbency and two 10cm x10cm squares are shaved and prepared for the introduction of the laparoscope and instruments. The laparoscope is introduced 10 cm lateral (right) to midline and 30 cm cranial to the umbilicus. The instrument portal is introduced 10 cm lateral (right) to midline and 10 cm cranial to the umbilicus. The toggle ends are brought out of the instrument portal with laparoscopic forceps introduced through the instrument portal. The instruments are removed, the portals are closed with skin sutures, and the cow is placed in right lateral recumbency. The sutures are tied over a roll of gauze until the dye marker is visualized (5 cm from the pin).

B) The Christiansen's technique[2] has been developed to eliminate the need for placing the cow in dorsal recumbency. The approach to the abomasum is similar. However, the instrument portal is much larger to accommodate the spiker trocar. The toggle is placed within the abomasum using the previous equipment (5mm cannula and trocar). However, the ends of the sutures are kept outside the animal until they are attached to a specialized "spiker" trocar and cannula (1 meter long trocar and cannula). The spiker is directed in the right cranio-ventral part of the abdomen. The trocar within the cannula is felt externally prior to being pushed through the abdominal wall, where the toggle ends are retrieved and tied over a roll of gauze.

C) The Newman's technique[3] is a one step technique with the cow in dorsal recumbency. The ventral approach is similar to the second step of the Janowitz's technique. The greater curvature of the abomasum is localized (abomasum still gas distended) and the toggle pin is introduced as mentioned earlier on the greater curvature. The abomasum is deflated and the ends of the sutures are kept outside the cow. The portals are closed with skin sutures and the cow is rolled on her side to tie the toggle over a roll of gauze. To be successful with this technique, it is important to be efficient when the cow is in dorsal recumbency. If too much time is taken to penetrate the abdomen, the abomasum can deflate and the introduction of the toggle is much more difficult.

The main advantage of LGTF is the minimally invasive nature of the surgery which is reported in all species to decrease morbidity and to speed up recovery. Also because it is a minimally invasive surgery, therapeutic and even prophylaxis use of antibiotics is not warranted except if other pathologies are diagnosed at the time of the surgery. Therefore, decreased use of antibiotics equals less milk withdrawal.

The main disadvantages of LGTF are the specialized equipment needed for the surgery, the cost of the equipment, the challenge of learning a new surgical technique, even if the learning curve is steep, and the necessity of placing the cow in dorsal recumbency with the Janowitz and the Newman techniques. At this point, we do not recommend the Christiansen's technique (one step standing). This technique is interesting because the procedure is performed with the cow standing. However, the second step of the procedure, which basically is a blind tack of the abomasum, can be dangerous. We have seen punctures of the omentum, the abomasum, the jejunum and even the uterus with this technique.

Recently, the Janowitz's technique was compared to the omentopexy[4-5]. The success rate was similar in regard to milk production and the relapse rate. The main advantages of the laparoscopic technique versus the conventional omentopexy are that fewer antibiotics are needed and that the surgery time is shorter.

Laparoscopic abomasopexy (LA)

This technique is a right paramediane abomasopexy realized by laparoscopy[6]. For this procedure, the cow is positioned in dorsal recumbency. A laparoscope (proximal to umbilicus), grasping laparoscopic forceps (right cranial paramedian) and a laparoscopic needle holder (right paramedian at the level of the umbilicus) are introduced into the abdomen. The abomasum is deflated prior to being grasped with the laparoscopic forceps. A needle of a USP 2 (PDS or Vicryl) is straightened and introduce in the abdomen through a stab incision in the skin at the pexy site (right paramedian). The needle is grasped with the needle holder and the suture is introduced in the abomasum wall on the greater curvature next to the greater omentum attachment. Then, the needle is pushed back out of the abdomen where it was first introduced. The needle is retrieved by an assistant. Those manipulations are repeated 2 or 3 times to create a strong abomasopexy. The sutures are tightened subcutaneously only when they are all in position. The skin portal and the stab incisions are closed with simple interrupted sutures.

This technique allows the formation of a strong pexy. However, this surgery is technically more challenging than the LGTF. The main disadvantages of this technique are that the cow is in dorsal recumbency and that the surgery is extremely difficult to do without the help of an assistant.

HEMORRHAGIC BOWEL SYNDROME (HBS)

Hemorrhagic bowel syndrome has been recognized in veterinary medicine for over a decade. The etiology of this syndrome remains unclear and treating cattle with this disease is always challenging. The first reports of HBS 10 years ago showed that the prognosis of survival after either surgical or medical treatment was poor[7,8]. After those reports, HBS was better recognized and seemed to become more prevalent. As we start understanding the disease better the prognosis has improved[9]. However, a couple of years after the encouraging report by Peek, it seemed that the prognosis, at least in our institution is back to what it was earlier, which is very poor. The way that we manage those cases has not changed much; if anything it got better as we gained a better understanding of the importance of pain control. We also know more about the uses of motility drugs However, it seems that we are seeing more advanced or more severe cases of HBS. The single clot obstruction does not seem to exist anymore. We are seeing enormous clots, multifocal clots, and a fair number of cases where the bowel has ruptured at the obstruction site. An important thing that the paper by Peek showed, and that we noticed as well, is that massaging the clot, when possible, improves the prognosis over a resection and anastomosis. If massage is unsuccessful, we wonder if a side-to-side anastomosis, without even removing the damaged bowel, is better than trying to do an end-to-end anastomosis in a less than healthy jejunum.

Cattle with HBS are difficult to keep standing during the surgery. Recently, we started using our conventional right flank approach on cattle in left sternal recumbency, if we felt that the odds of the animal going down during surgery were high. The position allows a similar exteriorisation of the jejenum than with the standing approach. Compared with the right flank approach with the cow in lateral recumbency, the exposition is much better. If you use this position on a cow under general anesthesia, you need to be careful on the position of the lower rear limb to avoid post-operative neuropathy and myopathy especially on a cow that is hypotensive.

DUODENAL SIGMOID FLEXURE VOLVULUS (DSFV)

Recently, we have recognized, in dairy cattle, a new condition that we called duodenal sigmoid flexure volvulus (DSFV). This pathology is caused by a dorsal displacement followed by a counter clockwise volvulus (as seen from the top of the cow) of the duodenal sigmoid flexure. The etiology of this condition remains to be evaluated.

Last year, we conducted a retrospective study of this condition and our results have been approved for publication in JAVMA[10]. In our paper, we had 29 cases (19 hospitalized and 10 treated on the farm) over a period of 4 years (2006-2010). Since the submission of the paper, we have seen many more cases in our hospital. This condition is diagnosed mainly on dairy cattle of any age and of any stage of lactation. The clinical signs are similar to those of an abomasal volvulus. However, the ping location and diameter are different. A high intensity ping associated with a positive succussion will be located at the 11th and 12th ICS. The classic hypochloremia, metabolic alkalosis and hypokalemia are present but much worse than with an abomasal volvulus as described by Garry[11]. At surgery, a fluid filled abomasum and cranial duodemun, and a gas distended duodenal sigmoid flexure are palpated. A twist is usually palpated at the sigmoid flexure. A constant finding with this pathology is an empty descending duodenum. The volvulus is resolved by flipping the gas distended duodenum ventrally and caudally. Upon resolution of the volvulus, the descending duodenum is filling with ingesta. No pexy of the duodenum is performed.

The abdomen is routinely closed without performing an omentopexy. Since more than half our cases had a previous omentopexy prior to having a DSFV, we believe that an omentopexy could be a risk factor for this pathology. The omentopexy positions the abomasum more dorsally than usual placing some slack on the double omental attachment of the sigmoid flexure. In all our cases, we had no recurrence of DSFV. However, in one case, we had a recurrence of an abomasal displacement.

In 2 cases, a possible cause of the DSFV could be identified. Those 2 cows had duodenal ulcers. We wonder if the ulcers were the primary disease and the secondary ileus, caused by the ulcers, caused the DSFV. One of those 2 cows had a side-to-side duodeno-duodenal anastamosis to bypass a compromised duodenum sigmoid flexure. Both cows survived.

With intensive fluid therapy, systemic antibiotic and NSAID, most of the cows survived. The cows that did not survive were the ones with extensive peritonitis at surgery. In those cows, the DSFV was often missed during the first laparotomy. They were then referred to us. The DSFV was found at the second laparotomy. Because of the increased duration of the condition, the duodenum was more compromised and the peritonitis was more extensive. Therefore, it is important to evaluate the area of the duodenal sigmoid flexure at surgery, especially when the surgical findings do not match the clinical signs.

REFERENCES

[1]Janowitz H. Tierarztl Prax Ausg G Grosstiere Nutztiere 1998;26(6):308, [2]Christiansen K. Tierärztl Praxis 2004;32:118, [3]Newman KD et al. J Am Vet Med Assoc 2005;227(7):1142, [4]Roy JP et al. J Am Vet Med Assoc 2008;232:1700, [5]Seeger T et al. AJVR 2006;67(3):472, [6]Babkine M et al. Can Vet J 2006;47(4):343, [7]Dennison AC et al. J Am Vet Med Assoc 2002;221(5) :686, [8]Abutarbush SM et al. Can Vet J 2005;46:711, [9]Peek SF et al. J Am Med Vet Assoc 2009;234(10):1308, [10]Vogel S et al. J Am Vet Med Assoc In press, [11] J Am Vet Med Assoc 1988;192(8):1107.

GASTRIC ULCERS IN HORSES: PHARMACOLOGIC AND MANAGEMENT STRATEGIES

Frank M. Andrews, DVM, MS, DACVIM
Baton Rouge, LA

INTRODUCTION

Equine gastric ulcer syndrome (EGUS) is characterized by ulceration in the terminal esophagus, stomach (non-glandular [NG] squamous and glandular mucosa), and proximal duodenum.[1] Diagnosis of EGUS is based on history, clinical signs, endoscopic examination, and response to treatment. All ages and breeds of horses are susceptible to EGUS. Intermittent mild colic signs, especially during and after eating, can be a feature of this disease. Current pharmacologic agents focus on blocking gastric acid secretion and increasing stomach pH ($\geq$ 4.0), which creates a permissive environment for ulcer healing. A sustained pH $\geq$ 4.0 for 18 hours has been shown to heal esophageal ulcers in people with gastroesophageal reflux disease (GERD). Also, since gastric acid is likely the cause of the pain associated with ulcers, effective acid control can decrease abdominal pain.

MANAGEMENT

The mainstay of pharmacologic treatment of EGUS is to suppress hydrochloric acid (HCl) secretion and increase stomach pH. Because of the high recurrence rate, effective acid control should be followed by nutritional and dietary management strategies to prevent ulcer recurrence.

Pharmacologic Therapy

Omeprazole

Omeprazole paste (4 mg/kg, PO, q24h; Gastrogard®, Merial Limited, Duluth, GA) significantly inhibits gastric acid secretion for 24 hours in horses.[2] In an acid environment, omeprazole is activated to a sulphenamide derivative, binds reversibly to the H+/K+ ATPase in parietal cells, and inhibits transport of hydrogen ions into the stomach.[3] Because of its effect on the cell, omeprazole is often called a proton-pump blocker and the effect on gastric acid secretion is dose and time dependent. Omeprazole is metabolized in the liver and excreted in urine and bile, and significant liver disease may affect the metabolism of the drug. Omeprazole has been shown to be an effective treatment for EGUS at a dose of 4 mg/kg orally once daily.[4] A recent study of 565 horses in race training found that omeprazole was superior to ranitidine in healing and preventing gastric ulcers in this population of horses.[5] Furthermore, there was improved performance, weight gain, attitude, appetite, and appearance after treatment. A second study found that 99% of spontaneous ulcers in adult horses and foals over 4 weeks of age were improved and 86.7% healed with omeprazole treatment.[6] Effectiveness of omeprazole was also shown to increase the rate of ulcer healing in horses removed from race training.[7]

Recently, omeprazole (0.5 mg/kg, IV; Premier Pharmacy Labs, Inc., Weeki Wachee, FL) treated horses had increased mean gastric juice pH of greater than 4.0, 1 hour after administration.[8] In a study performed to evaluate the efficacy of omeprazole for 5 days, gastric juice pH was significantly increased to >4, 1 hour and >5, 24 hours after the 4th daily dose, and 7, one hour after the 5th daily dose. Furthermore, omeprazole-treated horses had fewer ulcers after 5 days of treatment; however, none of the ulcers healed. This omeprazole compound might show promise for treatment of EGUS in horses with dysphagia, gastric reflux, or other conditions that restrict oral intake of omeprazole paste. However, due to the variability of acid suppression after the first dose, a loading dose (1.0 mg/kg, IV, Q24h) is recommended.

Histamine Type-2 Receptor Antagonists

Cimetidine

Cimetidine has been used since the early 1980s to treat and prevent ulcers in horses and foals, but there is little scientific evidence in the veterinary literature showing that it has efficacy in the treatment of EGUS. The author cannot recommend cimetidine for treatment or prevention for EGUS.

Ranitidine

Ranitidine hydrochloride is 4 times more potent than cimetidine.[9] When given orally (6.6mg/kg, PO, q8h), ranitidine suppresses acid output and maintains a median stomach pH of 4.6.[10] Ranitidine (6.6 mg/kg, PO, q8h) was able to successfully limit ulcer development in a feed deprivation model. Lower doses (4.4 mg/kg, PO, q8h) given orally are ineffective for treatment of EGUS. The recent availability of the generic ranitidine has made this drug popular and effective in treating EGUS. While ranitidine has been the most studied, other H2 antagonists have been evaluated experimentally and may allow for less frequent dosing and more effective acid suppression. Bioavailability and pharmacodynamic studies with famotidine (2.8 mg/kg, PO, q12h; 0.3 mg/kg, IV, q12h) in horses suggest that it can be used for treatment of EGUS, but may be cost prohibitive.[11]

Coating or Binding Agents

Sucralfate and bismuth subsalcyclate are two compounds that bind to stomach ulcers and promote healing. Sucralfate is a hydroxyl aluminum salt of sucrose octasulfate and binds to the negatively charged particles in the ulcer bed, buffering HCl by increasing bicarbonate secretion, stimulating prostaglandin production, and adhering to the ulcer bed.[12] In the stomach, sucralfate is converted to a sticky amorphous mass, which is thought to prevent diffusion of HCl into the ulcer. In a clinical trial in horses, sucralfate (22 mg/kg, PO, q8h) did not improve healing in 6 and 7 month old foals. Therefore, sucralfate alone may not be beneficial in treatment of EGUS. However, it may be effective when used in conjunction with acid suppressive therapy.

Bismuth containing compounds may have a coating effect similar to sucralfate. Additionally, these compounds will inhibit the activation of pepsin and increase mucosal secretion. A compound containing 26.25 g of bismuth failed to raise stomach pH in 5 horses.[13] Bismuth subsalicylate may be converted to sodium subsalicylate in the gastrointestinal tract, which may cause gastric irritation. Also, salicylates, similar to aspirin, decrease prostaglandin secretion and may further compromise an already damaged mucosa. Thus, compounds containing bismuth are not recommended for treatment of EGUS. However, bismuth is used as part of the therapy in humans with *Helicobacter pylori*-induced gastric ulcers and may be used in horses with chronic recurring gastric ulcers in which Helicobacter is suspected.

Synthetic hormones

Misoprostol, a synthetic PGE 1 analogue, is effective in the treatment of gastric and duodenal ulcers in man. Acid suppression, increased mucosal blood flow, increased bicarbonate secretion, and increased mucosal restitution are mechanisms of misoprostol. In one study, misoprostol (5 mcg/kg, PO) increased stomach pH and inhibited gastric acid secretion for 8 hours. Misoprostol is contraindicated in pregnant and nursing horses due to its effect on increasing uterine contractions. Although no reports of side effects have been reported in horses, side effects reported in other species include: diarrhea, cramping, flatulence, and uterine contraction. Side effects are dose dependent.[3]

A somatostatin analogue, octreotide acetate, has also been evaluated in horses.[14] Octreotide (0.5 to 5.0 mcg/kg) raised the gastric pH > 4 for approximately 5 hours, with no adverse effects noted. While octreotide appears to be very safe in human patients, it requires multiple daily dosing and is cost-prohibitive in horses. The benefit of using a somatostatin analogue is the prevention of hypergastrinemia associated with long-term use of acid suppressive drugs. The hypergastrinemia has a positive tropic effect on gastric cells and may result in proliferation.[15] Because somatostatin inhibits gastrin secretion, this hypertrophy is avoided; however, no case of gastric hypertrophy has been reported in horses following long term use of acid suppressive drug therapy.

Prokinetic Agents

Prokinetic agents may be valuable as an adjunct therapy in the treatment of EGUS and when there is adynamic ileus and gastroduodenal reflux. Bethanechol (0.25 mg/kg, IV) and erythromycin lactobionate (0.1 and 1.0 mg/kg, IV) increased solid phase gastric emptying time in horses.[16] No adverse effects were seen in healthy patients; however, other forms of erythromycin can cause fatal colitis in adult horses at antimicrobial doses. Both prokinetics increase gastric emptying versus saline, but bethanechol appeared to be superior. It increased gastric emptying rate of solid food versus erythromycin and increased gastric emptying rate of liquid versus saline. Bethanechol is a synthetic muscarinic cholinergic agent that is not degraded by acetylcholinesterases. The only side effect of the bethanechol administration was increased salivation. Other authors have recommended a dose of 0.025 – 0.030 mg/kg subcutaneously every 3-4 hours followed by oral maintenance therapy of 0.3 – 0.45 mg/kg 3-4 time daily. It is also possible that gastroduodenal reflux may worsen after treatment in patients with a proximal small intestinal obstruction.

Duration of Pharmacologic Treatment

It is difficult to predict how long NG or glandular gastric ulcer will take to heal, but the initial recommended treatment time for most antiulcer medications is at least 28 days. However, management changes, in addition to pharmacologic therapy, can affect healing ulcers. For example, in horses with gastric ulcers induced by feed-deprivation, ulcers were healed or nearly healed in horses after 9 days of pasture turnout,[17] whereas, omeprazole-treated Thoroughbred horses kept in training took longer to heal. Furthermore, pastured horses with spontaneous occurring ulcers showed 86% healing after 28 days of treatment.[6] We recommend endoscopic examination in horses with gastric ulcers after 14 days of omeprazole treatment. Over half of the gastric ulcer healed in 14 days during the FDA-Approval clinical trial.[4] This could save the client money and if ulcers are healed the dose (1 mg/kg, PO, q24h) could be reduced to prevent recurrence, while the horse remains in race training. In addition, adding alfalfa hay to the diet might potentiate the effect of omeprazole on increasing gastric juice pH and facilitate healing. If healing has not occurred after 28 days of omeprazole treatment, then further work up may be indicated. When endoscopy is not available, horses should be treated for at least 28 days. It should be noted that clinical signs might resolve before complete healing has taken place. Ranitidine should be used a full 28 days or longer (>40 days) as it is not as effective in ulcer healing compared to omeprazole treatment.[5]

Environmental, Nutritional and Dietary Management of EGUS

Pharmacologic therapy may be necessary to heal both NG and glandular gastric ulcers in horses. Once pharmacologic therapy is discontinued ulcers will return within several days if management changes are not instituted. Environmental, nutritional, and dietary management can be initiated during therapy to help facilitate ulcer healing and prevent ulcer recurrence. Adding alfalfa hay to the diet will buffer gastric acid which can be helpful. Also, pasture turn-out to facilitate continuous eating will help saliva production.

Antibiotics vs. Probiotics

Helicobacter pylori and other *Helicobacter spp.* have not been shown to cause of EGUS, although *Helicobacter equorum,* a urease positive bacteria, was recently isolated from feces and stomach mucosa of horses and foals.[18] Also, a large population of diverse acid-tolerant bacteria (*E. coli, Lactobacillus and Streptococcus spp.)* was isolated from the gastric contents of horses fed various diets.[19] In rats, which have a compound stomach similar to horses, bacteria (*E. coli*) rapidly colonized acetic acid-induced stomach ulcers and impaired ulcer healing.[20] In that study, oral antibiotic treatment suppressed bacterial colonization of the ulcer and markedly accelerated ulcer healing. Oral administration of lactulose resulted in an increased *Lactobacillus spp.* growth and colonization of the ulcer bed, which accelerated ulcer healing. Trimethoprim suphadimidine or a probiotic (*Lactobacillus agilis, L. salivarius, L. equi, Streptococcus equinus, S. bovis*) administered orally to horses decreased the number and severity of ulcers compared to untreated controls.[21] These data suggest that resident stomach bacteria are important in maintenance and progression of NG gastric ulcers in horses, with gram negative bacteria inhibiting healing and gram positive bacteria promoting healing. Treatment with antibiotic and/or probiotics might facilitate ulcer healing or prevent ulcers from recurring. Antibiotic or probiotic administration may be indicated in horses with chronic non-responsive gastric ulcers as an adjunct to therapy.

Seabuckthorn Berry Extract

There is an increasing interest in the use of herbs and berries that have therapeutic application in man and animals. Berries and pulp from the seabuckthorn plant (SBT; *Hippophae rhamnoides*) are high in vitamins, trace minerals, amino acids, antioxidants and other bioactive substances. SBT berries and pulp have been used successfully to treat mucosal injury. In addition, SBT have been shown to successfully treat and prevent acetic acid-induced gastric ulcers in rats.[22] In a recent study, a SBT preparation (3 ounces fed twice daily; SeaBuck™ Complete, Seabuck LLC, Midvale, UT) administered to horses, prevented an increase in number and severity of gastric ulcers after feed-deprivation.[23] Although, SBT did not significantly (P=0.06) decrease NG gastric ulcer scores, compared to untreated controls, gastric ulcer scores in 7/8 SBT-treated horses either stayed the same or decreased compared to just 2/8 of the untreated controls. SBT may be used as an adjunct therapy.

Calcium Carbonate Supplements

There are many supplements on the market containing calcium carbonate, a primary component of human antacids. The author (FMA) performed a small study with an antacid preparation containing calcium carbonate (4 ounces, top-dressed on grain, q12h; Neigh-Lox®, Kentucky Performance Products, LLC, Versailles, KY) and gastric ulcer scores were not significantly different compared to controls after 3 weeks of feeding.[24] However, gastric juice pH remained ≥ 4 for 2 hours after feeding, when compared to control horses. Also, in an *in vitro* Ussing chamber study, this supplement, when added to VFA-acid damaged NG mucosa, resulted in recovery of sodium transport. These data suggest that calcium carbonate preparations may have some efficacy in maintaining mucosal integrity, but may need to be fed more frequently than twice daily to improve gastric ulcer scores.

Oils (corn oil, rice bran oil)

Ponies, with gastric cannulae, fed corn oil (45 ml, orally, once daily) by dose syringe had significantly lower gastric acid output and increased prostaglandin concentration in gastric juice.[25] The authors conclude that corn oil supplementation could be an economical approach to the therapeutic and prophylactic management of glandular ulcers in horses, especially those associated with the use of NSAIDs. In contrast to the previous study, horse fed corn oil, refined rice bran oil and crude rice bran oil (8 ounces, once daily, mixed in grain) for six weeks had no statistical differences in NG ulcer scores.[26] In this model, dietary oils did not prevent NG ulcers in these horses, suggesting that dietary oils may be more useful in treatment or prevention glandular ulcers.

REFERENCES

[1]Anon et al. EVE 1999;1(2):122-134. [2]Daurio CP, et al. EVJ Suppl 1999;29:59-62. [3]Plumb DC. Veterinary Drug Handbook, 4th edition. Ames: Iowa State Press; 2002. [4]Andrews FM, et al. EVJ 1999;29:81-86. [5]Lester GD, et al. JAVMA 2005;227:1636-1639. [6]MacAllister CG, et al. EVJ Suppl 1999;29:77-80. [7]Murray MJ et al. EVJ 1997;29 (6):425-429. [8]Andrews FM, et al. JVIM 2006; 20:1202-1206. [9]Holland PS, et al. J Vet Pharmacol Therap 1997;20:145-152. [10]Murray MJ, et al. EVJ 1993;25(5):417-421. [11]Duran SH. Compend Cont Ed Pract Vet 1999;21 (5):424-425. [12]Borne AT, MacAllister CG. JAVMA 1993;202(9):1465-1468. [13]Clark CK, et al. JAVMA 1996;208 (10):1687-1691. [14]Sojka JE, et al. AJVR 1992;53(10):1818-1821. [15]Tielemans T, et al. Gastroenterology 1989;96:723-729. [16]Ringger NC, et al. AJVR 1996;57(12):1771-1775. [17]Murray MJ. Dig Dis Sci 1994;12:2530-2535. [18]Moyaert H et al. Vet Microbiol. 2009;1:133(1-2):190-192. [19]Al Jassim RAM, et al. FEMS Microbiology Letters 2006;248:75-81. [20]Elliott SN, et al. AJP-GI 1998;275:425-432. [21]Al Jassim RAM, et al. Rural Industries Research and Development Corporation, Final Report, Australia 2008:1-26. [22]Xing J, et al. Fitoterapia 2002;73:644-650. [23]Reese RE, et al. Proceed 9th International Equine Colic Research Symposium, Liverpool, UK 2008;125-126. [24]Andrews FM. Unpublished data, 2001.[25]Cargile JL, et al. JVIM 2004;18:545-549. [26]Frank N, et al. AJVR 2005;66:2006-2011.

EQUINE GLANDULAR ULCERATION: PATHOPHYSIOLOGY AND EPIDEMIOLOGY

Richard Hepburn, BVSc, MS, CertEM (IntMed), DACVIM (LAIM), MRCVS
Gloucestershire, UK

INTRODUCTION

There is minimal published information regarding the pathophysiology and epidemiology of equine gastric glandular mucosal disease. When compared to the weight of literature regarding equine gastric squamous ulceration, this difference is striking. Squamous ulceration represents increased exposure of the mucosa to hydrochloric acid and the synergistic effects of volatile fatty acids and pepsins. Prolonged pH<4 induces cell injury by damaging the outer cell barrier, which allows acid diffusion into the stratum spinosum, leading to altered ion transport, cell swelling and cell death. Volatile fatty acids also disrupt sodium transport, but due to their lipid solubility, enter the cells by diffusion. Squamous ulceration is multifactorial, and experimentally many risk factors have been identified: feed deprivation, high soluble carbohydrate diets, transport, stabling, and intense exercise. None of these experimental protocols has ever induced glandular lesions. Indeed, this tissue is adapted to withstand continuous exposure to the HCl it secretes into the gastric lumen. It is therefore unlikely that increased acid exposure *alone* is the cause of glandular lesions. This assertion is supported by the lack of correlation between the presence of lesions in the two regions of the equine stomach. In man, glandular ulceration is considered the end result of an imbalance between aggressive gastric luminal factors, and mucosal barrier defense. Several environmental and host factors contribute to lesion formation by either increasing gastric acid secretion or weakening the mucosal barrier, or both. Given the continuous nature of equine gastric acid secretion, versus the intermittent nature in man, it is most likely that weakened mucosal barrier function is the primary cause of equine glandular lesions. The potential etiologies include: bacterial infection, NSAID toxicity, and mucosal stress.

EQUINE GASTRIC GLANDULAR ANATOMY AND PHYSIOLOGY

The glandular mucosa is divided into three regions:[1] The **Cardia** is a narrow strip of mucosa immediately ventral to the *margo plicatus*. Large numbers of somatostatin immunoreactive cells are present in the equine cardia, suggesting a role in regulating gastric acid secretion. In other species, the primary secretion from the cardia is sodium bicarbonate. The **Fundus** is the largest part of the glandular mucosa and occupies the ventral stomach along the greater curvature and up the sides to the cardia. It contains straight gastric glands with the highest density of parietal cells and is responsible for the bulk of gastric acid secretion. The gastric glands also contain zymogen cells, enterochromaffin like cells (ECL) and mucus neck cells, which secrete thin mucus and are the only gastric cells capable of division. Pyloric mucosa lines the **Antrum** and the pyloric outflow. The gastric glands here are branched rather than the straight form seen in the fundus, and their predominant secretions are mucus and bicarbonate, creating the mucus curtain. Mucus neck cells and ECL cells are more numerous in the antrum than the fundus. Characterization of the leukocyte populations in the different regions of normal equine glandular mucosa has not been performed.

Gastric mucus is a complex mixture of glycoproteins, water, electrolytes, antibodies and lipids that forms a hydrophobic flexible gel that covers the glandular epithelium in a continuous layer 100-400µm thick. The formation of this tightly adherent gel is pH dependent; at pH 2, a firm viscid gel occurs, whereas a loose solution forms at pH 7. Mucus production is stimulated by acetylcholine (ACh), histamine (HIST), prostaglandin E2 (PGE_2), cholecystokinin (CCK), aastrin and nitric oxide (NO). Experimentally, NSAIDs (such as indomethacin in rats) result in thinning of the mucus layer. The pH in the ventral half of the equine stomach is stable at pH1-3 throughout a 24-hour period, with the antrum being slightly less acidic than the fundus or luminal gastric juice.[2]

WHAT ACTUALLY ARE EQUINE GLANDULAR ULCERS AND WHERE DO THEY OCCUR?

The first report describing the endoscopic appearance of glandular lesions described 162 clinical cases.[3] The majority of lesions (EGUS grade ≥2) were located in the antrum or adjacent to the pylorus (88%), with the remainder in the fundus. The cardia was not assessed as a separate location, but was included in the fundus. Thickening and hyperemia of the rugal folds was commonly observed. Interestingly, of the 66 horses that had a squamous mucosal grade of 0 or 1, 33 (50%) had antral lesions of grade ≥2. In cases that were re-examined, the author commented that: "thickened rugal folds were a persistent and perhaps permanent feature".

The term "ulcer" is usually erroneous as most glandular lesions do not involve erosion to the level of the lamina propria. Three more recent reports of post mortem findings from small numbers of horses identified different gross lesions and patterns of cellular change. Examination of 21 stomachs removed from unrelated clinical cases, showed gross change in the glandular mucosa in 14/21 (67%).[4] Three lesion types were identified: hyperemia, shallow erosion, and deep erosion. Hyperemia occurred most frequently in the antrum in the region of the pylorus. Histologically it presented as superficial capillary congestion with a mild mononuclear inflammatory infiltrate and normal epithelial integrity. Hyperemia was also associated with shallow and deep erosions, seen as depressions of increasing depth on the crest of rugal folds, and found most commonly in the antrum. Histologically, these were associated with varying degrees of fibrin exudation, hemorrhage and inflammatory cell infiltration. When looking at the same population, but assessing histology samples from 3 standardized glandular mucosal sites (cardia, fundus, antrum) as well as from lesions, gastritis was recorded if inflammatory cells were present and graded by the weight of infiltration.[5] Histologic gastritis was found in 19/21 (90%) animals, with the antral region again most commonly affected. This increased recognition of microscopic inflammation over gross lesions could indicate that low-grade inflammation may be a precursor to erosion development. The gastritis was termed chronic if mononuclear cells alone were found; and chronic active if they were accompanied by neutrophils. Solely neutrophilic infiltrates were not appreciated, although a solely mild eosinophilia was seen in 2 cases. Four types of architectural change were also described: Atrophy,defined as a reduction in the number of gastric glands; dysplasia, described as areas where the inflammatory infiltrate was associated with tortuous or irregular gastric glands; and metaplasia, defined as replacement of cells with different cell types. Rarely, glandular hyperplasia was also seen. A cross-sectional post mortem study of 63 horses from a Danish abattoir demonstrated glandular lesions in 36 animals (57%), with the majority of lesions (92%) within the antrum.[6] Three gross lesion types were described: polypoid masses (2/36); hyperplastic rugal folds (13/36); and hyperemic, erosive lesions (21/36). Hyperplastic rugae were all found within the antrum, and histologically the thickening was caused by gastric gland hyperplasia. The surface of these lesions ranged from intense hyperemia and exudate formation, to a normal epithelium. The hyperemic, erosive lesions were all small in size (1-2cm^2) and characterized by superficial luminal epithelial sloughing, fibrinopurulent exudate with cellular debris, and a mononuclear cellular infiltrate. Deeper erosive lesions (9/21) extended down into the gastric pits and glands and were associated with a localized gastritis.

This author has recognized six different endoscopic appearances of glandular lesions: flat and hemorrhagic; flat and fibrinous; raised and hemorrhagic; raised and fibrinous; depressed; depressed and exudative. The antrum (excluding the pylorus), and the pylorus are the most frequently affected sites. Lesions of the cardia are found infrequently and fundic lesions rarely (typically found on a rugal fold on the greater curvature side of the fundus). Recent application of histopathology to biopsy samples obtained trans-endoscopically has shown a great variety of changes, none of which appear to be specific to the different endoscopic lesion types. Patterns of inflammation consistent with acute, chronic, chronic active and suppurative gastritis have been described. Inflammatory infiltrates are most commonly mixed (lymphocytes, plasma cells and neutrophils), but can be individual and include eosinophils. Superficial fibrin exudation with a neutrophilic infiltrate is frequently described, sometimes in combination with minimal epithelial changes. Erosion and epithelial sloughing are common, but ulceration is rare. Architectural changes are also common and include gastric gland hyperplasia, hypertrophy, atrophy and dysplasia; mucus gland metaplasia; and mucosal fibrosis (often extending to the lamina propria). Bacterial colonies of variable size and appearance (coccoid or rod shaped, 1-2μm) are found infrequently within the mucosa. Samples of endoscopically normal glandular tissue taken from the same area (cardia, fundus, antrum) will either demonstrate normal mucosa or will show mild, diffuse hyperemia and a variable cellular infiltration. Serial endoscopy will frequently show resolution of the fibrinous or hemorrhagic exudate overlying an area of rugal hyeremia and hyperplasia, but persistence of the gross thickening. Repeated biopsy often shows normal overlying epithelium, but ongoing inflammatory and architectural changes in the underlying tissue. Clearly a larger sample size and better molecular techniques are required to further describe what is actually happening within glandular mucosal lesions. In order to promote a consistent future approach to this topic and avoid confusion with squamous disease, this author proposes the term equine glandular gastritis-erosion syndrome (EGGES) be used instead of EGUS.

EQUINE GASTRIC GLANDULAR BACTERIA

In man and most domestic species, bacteria are intimately involved in the generation of glandular lesions, yet the equine literature is sparse and often conflicting. It is estimated that 50% of the world's human population is chronically infected with *Helicobacter pylori,* yet only 5-10% develop disease, with a marked geographic variation in prevalence and pathogenicity.[7] Reports of the presence and the activity of different bacteria in equine gastric contents and on the gastric mucosa are becoming more numerous, but Koch's postulates have definitely not been fulfilled in the same way as they have for *H.pylori* in man! Culture, staining and molecular techniques have shown a diverse population of acid tolerant bacteria in equine gastric contents.[8] Two main genera have been identified: *Lactobacillus* and *Streptococcus*. Given the pH gradient that exists within the equine stomach, the majority of this bacterial activity is in the squamous lined dorsal fundus.[9] Transient bacteria have also been identified, including those belonging to the genera *Pseudomonas, Prevotella, Proprionibacterium, Escherichia, Clostridia, Legionella, Voraxella* and *Pasturella.* Horses ingest a range of bacteria from the environment. It is not known whether these merely pass through the stomach, or whether they are able to colonize ulcer beds or even induce primary injury. In rats, *Escherichia coli* can rapidly colonize and delay the healing of experimentally induced ulceration.[11] The ability of bacteria to delay ulcer healing in horses has been demonstrated by the observation that significant ulcer healing can occur with prolonged antibiotic therapy.[8]

Fluorescence in situ hybridization (FISH) demonstrated four morphological types of bacterial cell associated with the glandular mucosal surface in tissue samples taken at post mortem: small, short coccoid rods, distinct rods, long chained rods, and large dividing coccoid bacteria.[2] Identification of the bacteria by 16s rDNA gene cloning showed two different clones in normal tissue samples: *Lactobacillus salivarius* (99% similarity) and *Sarcina ventriculi* (99% similarity). *L.salivarius* has previously been reported to be the predominant lactobacilli in the equine gastric contents, and can also be cultured from colon and rectal contents, indicating a wide pH tolerance.[8] Its presence on the glandular mucosa most likely reflects squamous wash-off. The *L.salivarius* clone was also found in hyperplastic rugal lesions and in erosive lesions, but its involvement with lesion development or perpetuation is unknown.[2] Interestingly, the potential for some bacterial species to be protective has been shown by ulcer healing in horses given a live culture probiotic formulation involving several *Lactobacillus* species including *L.salivarius.*[8] *Sarcina ventriculi* is able to grow within the stomach, and has been associated with gastric disease in lambs and kids and with gastric dilatation in dogs, but has not been associated with gastric disease in the horse.[2] Two bacterial isolates were found in a single case of bacterial gastritis, where a large, focal antral erosion was associated with erythrocytes, neutrophils and a heavy population of adherent short coccoid rods.[2] A clone with 100% similarity to *Enterococcus faecium* was found superficially, and a clone with 100% similarity to *Escherichia fergusonii* was found invading the mucosa. Bacteria were also seen intracellularly within epithelial cells and neutrophils. Extension of this inflammation into the gastric pits, with a deeper mononuclear cell infiltrate was also found.

The greatest controversy with EGGES regards the presence of *Helicobacter* like organisms (HLOs). In samples taken at post mortem in Denmark, no evidence of *Helicobacter* genus organisms was found by FISH, no spiral shaped bacteria were seen histologically, and no evidence of localized urease activity (by rapid urease testing) was found.[2] This contrasts to the results of a Venezuelan post mortem study where HLOs were detected on Warthin-Starry stain sections in 11/20 horses and *Helicobacter*-like

DNA was found by *Helicobacter* genus specific PCR in 11/20 cases.[11] In cases where gastritis and erosions were found, the bacteria were associated with a lymphoplasmacytic inflammatory infiltrate. Two further studies have also shown evidence of *Helicobacter* specific DNA in gastric biopsy samples from 19 horses, and evidence of intra-gastric urease activity in 8 horses.[12,13] Enterohepatic *Helicobacter equorum* has been isolated from the feces of up to 7.9% of hospitalized horses,[14] so it is possible that cross reactivity lead to some of the positive PCR results. Given the importance of HLOs to gastric disease in other species, it is vital that its role, if any, in EGGES is defined fully. It is this author's opinion that equine gastric HLOs do exist but that their pathogenicity is minimal when compared to *H.pylori*. Other opportunistic bacteria are likely to be more important in the development and maintenance of glandular disease.

NSAIDs

Phenylbutazone and flunixin meglumine can induce ulceration when given at elevated doses[15] although recently no effect was seen after oral dosing at recommended dose rates,[16] and the use of NSAIDs in young racehorses has not been shown to be a risk factor for ulceration.[17,18] This author does not recognize recent NSAID administration as being a risk factor for glandular lesion development. Occasionally, horses that have been administered oral NSAIDs prior to gastroscopy will have single, small depressed exudative lesions on the ventral floor of the antrum. These have an appearance similar to the mucoid cap reported with NSAID erosion in man.[19]

MUCOSAL STRESS

"Stress is the sum of the biological reactions to any adverse stimulus, physical, mental, or emotional, internal or external, that tends to disturb the homeostasis of an organism. Should these reactions be inappropriate, they may lead to disease states"[20] An example of the effect of stress is the rise in bleeding gastric ulcers in elderly Japanese patients after a severe earthquake.[7] The etiology of stress related gastric disease is incompletely understood, but probably reflects splanchnic hypoperfusion, loss of host defenses and resultant gastric acid injury. In critically ill animals, the hypoperfusion may be severe. As splanchnic blood flow is reduced in fasted exercising ponies, could a similar, cumulative, lower grade effect potentially be contributing to glandular lesion development in performance horses?[21] Exercise may contribute to mucosal stress; however, there is much variation in type, frequency and duration of exercise horses undergo. Therefore, the single finding of no worsening in EGUS scores in overtrained Standardbreds should not stop study in this area.[22] Could it also be possible that much more prolonged, repeated exposure to the risk factors associated with squamous disease is causing neuro-humoral effects that result in glandular lesion development?

EPIDEMIOLOGY

Unfortunately, most papers regard squamous and glandular ulceration as a single entity (EGUS), so there is minimal literature looking at the epidemiology of glandular lesions alone. A recent report of 201 horses from throughout Denmark found ulceration in 168 animals, of which glandular ulceration alone occurred in 29 (17%) and in combination with squamous lesions in 86 horses (51%).[23] In 32 horses (19%), the most severe lesion was glandular, 82% of these were located in the antrum. In all age groups >3 years, horses with significant lesions in both mucosae (twin site) were more common than horses with single site lesions. All horses >18 years only had twin site ulceration. No significant effect of breed was found, although twin site ulceration was more likely in Sport ponies and Warmbloods. Horses in light work and mature individuals at pasture were more likely to have twin site lesions, whereas the proportion of horses having glandular lesions alone did not significantly differ with work intensity. A second paper by this author assessed risk factors for EGUS, but only looked at risks for any site ulceration and risks for squamous ulceration alone.[24] No differences existed between these analyses to suggest glandular specific risks. The effect of concurrent disease was described in a retrospective study of endoscopic images of the glandular mucosa where 92/162 cases had concurrent disease including medical intestinal colic (34), surgical intestinal colic (16) and colitis (7).[3] In this author's population of predominantly sport and leisure horses, glandular lesions are frequently found without squamous disease in middle aged and older animals. Twin site ulceration is less frequently appreciated. Squamous disease is predominantly seen in younger horses, typically without glandular involvement. No effect of age on lesion grade is currently appreciated, although data analysis is ongoing. Administration of NSAIDs is not recognized as a risk factor in this population, and concurrent disease is rarely present. No effect of stabling versus turnout is appreciated. Most cases present toward the end of the competition season, although the increasing frequency of survey studies affects this. Increased exercise or competition intensity is commonly reported in the 3-6 month period prior to gastroscopy. Changes in feeding are reported regularly in cases that have a variable appetite, but are probably effect rather than cause. In cases that respond to therapy, positive changes in behavior and attitude occur to suggest that ulceration could have been present for a very prolonged period. Ongoing epidemiologic study of a large population will hopefully aid in identifying risk factors for glandular lesion development.

REFERENCES

[1]Merritt A **Equine Vet J Suppl** 1999:7-13, [2]Husted L **BMC Microbiology** 2010;10:84, [3]Murray M **J Vet Intern Med** 2001;15:401, [4]Martineau H et al **EVJ** 2009;41(7):638, [5]Martineau H et al **EVJ** 2009;41(7):646, [6]Husted L **BMC Microbiology** 2010;10:84, [7]Malfertheiner P et al **Lancet** 2009;374(24):1449, [8]Al Jassim et al **RIRDC Publication No. 08/033** (2008), [9]Burton A et al **Proc ACVIM Forum** (2007), [10]Elliott S et al **Am J Physiol Gastrointest Liver Physiol** (1998):G425, [11]Contreras M et al **Lett Appl Microbiol** (2007);45(2):553, [12]Hepburn et al **MS Thesis VMRCVM** (2004), [13]Scott et al in **Proc Am Soc Microbiol** (2001), [14]Moyaert H et al **Vet Microbiol** 2007;121:378, [15]MacAllister C et al **J Am Vet Med** 1993; 202(1):71, [16]Andrews FM et al **Vet Ther** 2009;10(3): 113-20, [17]Murray M et al **Equine Vet J** 1996;28(5):368, [18]Vastita N et al **Proc Am Assoc Equine Pract** 1994;40:125, [19]Wallace J **Gastroenterology** 1997;112:1000, [20]Monnig A **J Vet Emerg Crit Care** 2011;21(5):484, [21]Duren S **PhD Thesis, University of Kentucky** (1990), [22]De Graaf-Roelfsema E et al **Equine Vet J** (2010);42(S38):56, [23]Luthersson N **Equine Vet J** 2009;41(7):619, [24]Luthersson N **Equine Vet J** 2009;41(7):625,

EQUINE GLANDULAR ULCERATION: COMPARATIVE ASPECTS AND MANAGEMENT

Richard Hepburn, BVSc, MS, CertEM (IntMed), DACVIM (LAIM), MRCVS
Gloucestershire, UK

INTRODUCTION

Equine glandular gastritis-erosion syndrome (EGGES) describes a spectrum of gross and histologic changes that occur in the compound columnar epithelium that lines the ventral half of the stomach. What we commonly call an ulcer is actually a combination of gastritis and mucosal erosion. Our understanding of why these pathological changes occur is incomplete, yet similar changes seen in other species may guide our further investigations. Obviously great anatomic, physiologic, environmental and genetic differences exist between the horse and other mammals, but some of the similarities are intriguing. All that is needed is an open mind and a critical eye!

Management of EGGES cases is poorly reported. As lesions do not represent straightforward acid injury, acid suppression alone is probably insufficient. A simple literature search shows the paradigm shift that occurred in the treatment of human peptic ulcer disease away from the "no acid, no ulcer" dictum to *H.pylori* eradication. The importance of "non-acid" etiologies and combination therapies (mucosal protectants, antibiotics) in other species makes our reliance on acid suppression in the horse seem antiquated. Could it is just be a matter of time before we catch up?

COMPARATIVE ASPECTS IN MAN

Until the end of the 20th century, morbidity and mortality from peptic ulcer disease had a huge global effect.[1] The discovery of effective acid suppressants, and the realization of the importance of eradication of *Helicobacter pylori* revolutionized clinical practice. Recently, the increased use of NSAIDs and low dose aspirin, as well as the increasing prevalence of *H.pylori*–negative, NSAID-negative ulceration has again made this an intriguing clinical problem. Potential etiologies of this type of ulcer include acid hypersecretory states, surgical conditions (gastric resections/bands), neoplasia, viral infections (particularly in immunocompromised patients), dietary associated inflammation and eosinophilic intestinal disease. The primary clinical symptom of any peptic ulcer is epigastric pain, often accompanied by fullness, bloating, early satiety and nausea. Chronic ulceration can be asymptomatic. Disease is described as uncomplicated or complicated.[2] Gastric bleeding is the most frequent problem seen in 50-170 per 100,000 cases andperforation resulting in peritonitis is the other major complication. Gastric outlet obstruction is rarer and often associated with malignancy.

H.pylori colonizes the entire gastric mucosa. Clinical outcome is dependent on host-pathogen interaction, the pattern of chronic gastritis induced, and its location within the stomach. In antrum-predominant *H.pylori* gastritis, basal and stimulated gastric acid output is increased; whereas fundic gastritis can lead to decreased acid secretion. In the horse, we know that the endoscopic prevalence of fundic lesions is underestimated,[3] so could it be possible that small lesions (those reported were 5mm ulcers) are reducing acid secretion leading to a more favorable environment for opportunistic bacteria to colonize lesions within the antrum? Or could antral lesions, regardless of etiology, become self-perpetuating due to increased acid secretion? In man, genetic variation in premorbid acid secreting status could be a key factor in determining the location of *H.pylori* gastritis. Acid reduction using proton pump inhibitors actually increases the intensity of *H.pylori* induced gastritis. So could variation in the constitutive acid secretory capacity in horses be important in development of glandular disease? Or could the spontaneous periods of gastric alkalinization, associated with duodenal reflux,[4] create antral areas of reduced acidity that also favor the development of bacterial gastritis? *H.pylori* can affect feedback control of gastric acid secretion, as its ability to use urease to produce ammonia can alter the juxtamucosal pH, preventing antral D-cells from sensing the true level of acidity, causing inappropriate acid secretion and hypergastrinemia. This hormonal change exerts a trophic effect inducing gastric gland hyperplasia. There is already evidence of urease activity within the equine stomach and the glandular mucosa.[5] Could measurement of serum gastrin concentrations document hypergastrinemia? Could this be associated with rugal thickening and gastric gland hyperplasia, which this author has found in EGGES cases? *H.pylori* derived phospholipases A and C are also important in lesion development as they affect the phospholipid rich layer within gastric mucus, so reducing the integrity of the gastric epithelial barrier. *E.fergusonii* clones have been found in equine glandular lesions. As this bacteria has phospholipase A1 activity *(http://www.genome.jp/dbget-bin/www_bget?efe:EFER_3681)*, could it be reducing the equine mucosal barrier?

Given the global importance of *H.pylori* and the lack of similar etiological evidence in the horse, assessment of the prevalence of a comparable human condition to EGGES is difficult. Chronic gastritis occurs in 2/10,000 and acute gastritis in 8/1,000 people. Great differences in prevalence are seen geographically. Factors such as smoking, excessive alcohol and drug use, and emotional stress all increase ulcer prevalence. As >50% of the world's population is infected with *H.pylori,* ulcer-like symptoms are often initially tested non-invasively (^{13}C-urea breath test, fecal antigen test). A peptic ulcer diagnosed at endoscopy is described as a mucosal break of >5mm diameter. It can be single or multiple (kissing lesions can occur), and is covered by fibrinous exudate. A mucosal break of <5mm is termed an erosion. The predilection site is the angulus of the lesser curvature, an area analogous to the dorsal wall of the antrum in the horse. In cases of gastritis, endoscopy can be a poor predictor of pathologic changes, with biopsy histology required for a more accurate diagnosis.[7] Poor correlation can occur in 1/3 patients, with cases of normal endoscopy/abnormal histology, and of abnormal endoscopy/normal histology reported. In light of this and previous findings,[8] it was proposed that diagnoses such as erosion and gastritis should not be made endoscopically, but that descriptive terms be used instead. In order to create future uniformity in the study of EGGES, this author believes that this approach should also be applied to the horse. *H.pylori* causes a mixed inflammatory infiltrate, with neutrophils and macrophages seen within the epithelium and gastric glands.[1] B and T lymphocytes regulate the local immune response, with the type of T-cell response determining the extent of mucosal damage that occurs (Th-1 predominant response results in greater damage). Previously, a solely lymphocytic gastritis has been described,[9] which is believed to be immune-mediated, gluten-associated or idiopathic. The need for standardization of histological interpretation for small animal GI disease led to an ACVIM consensus statement in 2010.[14] Greater use of endoscopic biopsy histology from both lesions and standardized locations in equine clinical cases is vital to increasing our understanding of this condition.[10]

The formation of gastric or duodenal ulceration whilst on NSAIDs is due to a combination of topical injury by ion trapping; reduced mucus hydrophobicity; reduced protective prostaglandin production by COX inhibition; and neutrophil adherence to the gastric microcirculation.[1] Gastric acid deepens this lesion to become an ulcer. Histologically, NSAID ulceration in man is characterized by a lack of inflammatory cells on top of and within the epithelium. The decreasing prevalence of *H.pylori* infection is leading to an increased appreciation of *H.pylori*-negative, NSAID-negative ulceration. Endoscopy with biopsy samples taken from the lesion and surrounding mucosa is recommended in these cases. None of the established causes in humans (e.g. radiation therapy, Zollinger-Ellison syndrome) are relevant to the horse except idiopathic ulcers. These patients typically have antral-predominant ulceration and are maintained on acid suppressants, which is similar to EGGES.

Prior to the appreciation of *H.pylori*, therapy targeted gastric acid suppression and reinforcement of the mucosal barrier.[1] Of these, the proton pump inhibitors (PPIs) were most successful; however, ulcer recurrence after finishing therapy was common, leading to the use of long-term maintenance acid suppression. Reinforcement of the mucosal barrier is most commonly used as protection against NSAIDs and aspirin. Sucralfate and bismuth salts promote mucosal repair, although both may have some intrinsic antibacterial (anti-*H.pylori*) effects.[1] Sucralfate may also reduce acid secretion and increase bicarbonate secretion via a prostaglandin effect. The prostaglandin analogue misoprostol has also been used for its ability to reduce acid secretion, increase mucosal blood flow and bicarbonate secretion, and enhance cellular restitution, but its use is complicated by abdominal side effects. *H.pylori* eradication therapy is now global standard for peptic ulcer disease. This is achieved using a combination of acid suppression and antibiotics. Sole antibiotic therapy does result in healing, but at a slower rate. The initial eradication phase is followed by a further 4-8 weeks of acid suppression alone. Eradication relies on a combination of antibiotics (clarithromycin+amoxicillin; clarithromycin+metronidazole; amoxicillin+levofloxacin), a proton pump inhibitor and bismuth subsalicylate in areas of increased antibiotic resistance. The particular antibiotic combination given is decided by geographic location and known local *H.pylori* resistance patterns. Failure of the first line therapy leads to second/third line choices. The duration of eradication therapy is controversial, with prolonged regimes (14 *vs* 7d) only increasing eradication rates by a small degree. Healing of gastric ulcers is confirmed endoscopically, with repeated biopsy to exclude malignancy. *H.pylori* negative NSAID gastric ulcers are usually successfully treated with 8 weeks of acid suppressive therapy alone, providing the NSAIDs are stopped. Healing in patients receiving continuous NSAID therapy is more problematic and may require a combination of a PPI and a mucosal protectant or misoprostol.

COMPARATIVE ASPECTS IN DOMESTIC SPECIES

Gastric disease is also appreciated in many domestic species, although none as frequently as the performance horse. At least 7 species of *Helicobacter* are prevalent in healthy and sick dogs and cats, but a direct causal relationship has not been established. *Helicobacter* like organisms (HLOs) have also been found in a wide range of large cats and dogs.[11] *Helicobacter* associated gastritis has a variable appearance and is characterized by lymphoplasmacytic infiltration of the lamina propria, lymphoid follicular hyperplasia and colonization of the gastric glands by HLOs (seen on special stain or identified by PCR or FISH). The combination of chronic gastritis and the presence of *Helicobacter,* is used to distinguish *Helicobacter*-gastritis from IBD-gastritis. Whilst IBD principally affects the intestines, chronic gastritis can be seen as a component in small animals. In this author's experience, the finding of an eosinophilic gastritis in the horse is often associated with concurrent intestinal disease (IBD) with similar histologic changes found in duodenal and terminal colon biopsies. There is increasing evidence in cats and in man that chronic lymphoplasmacytic gastritis is a precursor to development of *Helicobacter*-associated gastric mucosa associated lymphoid tissue lymphoma.[12] This condition is not seen in horses and the stomach is not a common lymphoma site.[13] Gastritis and ulceration also occurs in pigs, whales and marsupials. In all species, it has been associated with a HLO, although the pathogenic potential of these organisms is undefined. Gastritis can also be associated with dietary indiscretion, foreign body or toxin ingestion, NSAIDs, metabolic disease (renal, hepatobiliary) and gastrointestinal parasites.

MANAGEMENT OF EQUINE GLANDULAR GASTRITIS-EROSION SYNDROME

There is minimal literature regarding clinical signs specific to glandular ulceration, as opposed to squamous and glandular disease taken as a whole. In a retrospective study of 162 cases, the clinical signs that prompted gastroscopy were: reduced appetite (34%), recurrent or intermittent colic (20%), acute colic (14%), poor body condition/recent weight loss (12%), altered behavior/attitude (10%) and asymptomatic surveys (11%).15 Thirty-one horses (20%) had more than one clinical sign. These results are skewed by the fact that 64/162 (40%) were hospitalized cases, and 92/162 (57%) had concurrent disease, of which 57/162 (35%) had colic or colitis. Retrospective analysis by this author of several hundred cases of glandular disease in sport and leisure horses has shown colic (of any duration) to be a rare clinical sign. More frequently reported clinical signs are altered appetite (inappetence, slowed eating), altered behavior, girthing pain, weight loss/inability to maintain condition during a competition season and poor performance. Most lesions would be EGUS grade 2-3, and the association between severity of clinical signs and severity of ulceration is influenced by attitude and use. Asymptomatic cases are also found when doing population surveys, and positive attitude/performance changes can occur with resolution.

There are no prospective or retrospective reports of the treatment of EGGES using acid suppression or management change. Again, this is a striking difference to squamous ulceration. This author recommends a combined approach of acid suppression with agents that reinforce the mucus barrier, or treatments that address suspected primary causes. Therapy without acid suppression is not recommended, as acid-lesion contact is the origin of the clinical signs that occur. In an EGUS review paper, it was suggested that glandular lesions heal faster than squamous ones.16 It is this author's experience that glandular lesions do not heal as readily as squamous ones of a comparable EGUS grade, and that treatment failure or lesion worsening are more likely to occur. Similar to squamous disease, larger or more severe lesions can require more prolonged therapy. In general, flat lesions appear to heal faster than raised ones, and hemorrhagic lesions faster than those covered with fibrinous exudate.

This author's first line therapy is a combination of proton pump inhibitor (Omeprazole 4mg/kg PO q24h, GastroGard, Merial) and mucosal protectant (Sucralfate 20mg/kg PO q12h, Antepsin, Chugai Pharma) for 4 weeks. During therapy horses remain on their previous exercise program in order to asses a clinical response. Bismuth subsalicylate is not routinely used due to prohibited drug testing concerns, but when used, this author has not appreciated any negative effects of further mucosal compromise that have been raised elsewhere.16 The need for a further month of treatment or for second line therapy (antibiotics, bismuth salts, apolectol) is based upon biopsy histology results taken at initial examination, which are viewed in light of the endoscopic and clinical response to the initial treatment period. Thickened rugae will often never completely resolve, even when the overlying erosion and inflammation have. Biopsy of these can reveal mucosal fibrosis and glandular changes. The overlying epithelium remains normal whilst acid suppressive or mucosal protective therapy is ongoing. This author believes that these architectural changes reduce the efficacy of the mucosal barrier overlying these lesions, leading to increased ulcer recurrence risk. The use of daily administration of 300g of a pectin-lecithin compound (apolectol) has been reported to lead to a significant reduction in glandular lesion severity over a 10 day treatment period.17 Apolectol acts as a mucosal coating agent, contributing to the hydrophobic barrier overlying the glandular mucosa. No effect on preventing squamous lesion development was seen during an experimental starvation protocol, which is not surprising given Apolectol's action.18 This author uses Apolectol in cases where reduced mucus barrier function is suggested by biopsy histology results. This drug warrants further investigation, both during the initial period of therapy when given in combination with acid suppression, and as a long term preventative in cases of permanent rugal change. There are no reports of glandular or squamous ulcer healing using misoprostol alone, but anecdotally it has been used for non-responding glandular cases. This author has not experienced any positive gastric effects with misoprostol, and due to the side effects that can occur with overdosage, does not recommend it. There are no controlled studies detailing the use of antibiotics is glandular ulceration cases. The only study looking at antibiotic therapy used a mean gastric ulcer score from gastroscopic examination of the entire stomach.19 Four weeks of treatment with the oral antibiotic trimethoprim sulphadimidine (TMS) was shown to significantly reduce gastric ulcer score over untreated controls. This author would not routinely use antibiotics as part of first line therapy. There are, however, cases of persistent glandular lesions that do not change, or even worsen with first line therapy, where intralesional bacterial can be found histologically. The addition of antibiotic therapy can have dramatic effects, with complete resolution occurring within 2-4 weeks. As the only study to guide antibiotic choice used TMS for 4 weeks, this drug is this author's first choice. Susceptibility of E.fergusonii to TMS has also been described in another veterinary species.20 The use of probiotics is also intriguing. Four weeks of daily administration of a live culture Lactobacillus and Streptococcus spp probiotic has been associated with a trend toward improved ulcer healing. Unfortunately, this formulation is not commercially available, and most equine probiotics do not resemble it at all. This author has found the effects of commercial probiotic therapy on glandular disease to be variable.

The only reports detailing the dietary or management changes important to EGGES refer to dietary oil supplementation. In gastric cannulated ponies, five weeks of once daily supplementation with 45ml of corn oil (60%w/v linoleic acid), reduced stimulated gastric acid secretion, and increased PGE2 and sodium outputs, under basal and pentagastrin stimulation.21 This could be an effective and inexpensive way to enhance the glandular barrier. This author recommends corn oil supplementation as part of the therapy of glandular disease, but at a higher rate in light of the increased exposure to putative risk factors in clinical cases.22 In the absence of any specific management information to limit glandular lesion development, this author follows the advice given for squamous disease.23

REFERENCES

[1]Malfertheiner P et al **Lancet** 2009;374:1449, [2]Lau et al **Digestion** 2011;84:102, [3]Andrews F et al **Equine Vet J** 2002;34(5):475, [4]Murray M. **Equine Vet J** 1989;S6:73, [5]Hepburn R et al. **J Vet Intern Med** 2004;18(3):376, [6]Yamamoto S **Digestion** 2012;85:108 , [7]Carr N et al **Ann Diagn Pathol** 2012;16(1):13, [8]Maratka Z **Endoscopy** 1986;18:75, [9]Dixon M et al **J Pathol** 1988;154:125, [10]Matrineau H et al **Eq Vet J** 2009;41(7):646, [11]Harbour S **Vet Immunol Immunopath** 2008;122:191, [12]Bridgeford E et al **Vet Immunol Immunopath** 2008;123:106, [13]Meyer J **Vet Pathol** 2006;43(6):914, [14]Washabau et al **J Vet Intern Med** 2010 24:10, [15]Murray M **J Vet Intern Med** 2001;15:401, [16]Videla R **Vet Clin Equine** 2009;25:283, [17]Venner M **Equine Vet J** 1999;S29:91, [18]Murray M **Equine Vet J** 2002;34(2):195, [19]Al Jassim et al **RIRDC Publication No. 08/033** 2008, [20]Hariharan H et al **Can Vet J** 2007 48(6):630, [21]Cargile J et al **J Vet Intern Med** 2004;18:545, [22]Frank N et al **Am J Vet Res** 2006;66(11):2006, [23]Reese R **Vet Clin Equine** 2009;25(1):79

WELFARE ISSUES IN CATTLE (EMPHASIS ON BEEF CATTLE)

J.K. Shearer, DVM, MS
Ames, Iowa

WHAT IS ANIMAL WELFARE?

Animal welfare is a science that arose as a result of ethical concerns about animals. Its primary concerns typically include 3 basic questions: 1) is the animal functioning well (in other words, is it producing well), 2) does the animal have pain or is it distressed (affective states), and 3) is the animal able to express or perform natural behaviors? In production agriculture some assume that we can best determine an animal's welfare status by addressing only the first of these questions (i.e. is performance affected?). Indeed, this was the approach advocated by the late Stanley Curtis, legendary swine specialist at the University of Illinois, that welfare be assessed on the basis of objective measures of animal performance (production and reproduction) and not on how an animal feels. He argued that the use of subjective criteria (i.e. based on beliefs and values) for assessment of an animal's welfare were not appropriate for decision-making in how to improve the well-being of animals.

Opponents of this view would say that if good performance equals good welfare, then how does one reconcile high milk yield as desirable knowing that it is associated with an increased potential for disease (mastitis, lameness, milk fever and retained placenta)? Or, how would one assess the welfare of a cow that despite poor body condition, management and handling, may still manage to have relatively high levels of milk production and even become pregnant? In this scenario using performance parameters alone, one might characterize the animal's welfare as good. On the other hand, knowing the "state of the animal" with respect to its environment and human interactions, one might conclude that its welfare was poor.

"Animal welfare" refers to the "state of the animal". As defined by Dr. Donald M. Broom, a well-known animal behavior and welfare scientist from the University of Cambridge, animal welfare is best defined as *"an animal's state as regards its attempts to cope with its environment"*. It includes the animal's feelings, behavioral and physiological responses and its health.

PROBLEM AREAS IN WELFARE OF CATTLE

Broom lists the following as some of the major problem areas in animal welfare:

- Physical abuse
- Animal neglect: calculated, accidental or that arising from ignorance
- Inadequacies in design of housing facilities
- Poor husbandry practices and rough or careless animal handling
- Unnecessary or poorly executed physical alterations of animals for the benefit of farming practice (tail docking, ear notching and branding)
- Poor conditions and procedures during transport, at markets and at the packing plant

Physical abuse

The willful abuse of animals is uncommon and in its worst form conducted by those with sadistic views or tendencies. Unconscious forms of abuse are more common and generally occur as a result of carelessness and/or ignorance. Whereas, one might consider the controlled-use of an electric prod for the purposes of motivating a down animal to stand acceptable; the continual uncontrolled goading of an animal that is physically unable to stand is willful abuse and animal cruelty. Sometimes the extent of *injury* is misunderstood or the animal's *will* to stand misinterpreted. Working with animals takes patience, and particularly so when they are physically impaired.

Another area where physical abuse is common but rarely recognized as abuse *per se* is physical restraint. Controlling the head of a fractious or frightened bovine is important for both human and animal safety. Cows use their heads as a defense mechanism and will throw or swing them like a battering ram to fight or fend off advances by herdmates or other challengers. Restraint devices for stabilizing the head come in a variety of forms on cattle working chutes; but certainly one of the simplest and best are the rope halter. It provides safe and secure restraint of the cow's head for most purposes. However, out of ignorance or convenience one of the most common forms of head restraint used on farms and ranches is the nose-lead. Pick up any book on restraint of cattle and one will quickly learn that the nose-lead is intended to be a distraction device that should never be used without a halter. This is a simple, but very common form of *unconscious* abuse that is so commonplace as to be considered normal practice. There are many other examples of unintended abuse (most of them minor, but nonetheless significant) in bovines and oddly enough, the most frequent offenders are often those who are most familiar with cattle.

Animal Neglect

Neglect occurs primarily as a consequence of failure to provide basic needs for food, clean water, and prompt treatment of disease and injury. It is a reality that cattle in certain parts of the country must endure times where rainfall is short of that needed to provide adequate grass for grazing. When this occurs owners must rely on stored forages, grasses adapted to cooler temperatures or feed supplements to maintain their animals during the winter months. In situations where owners may fail to provide supplementation as necessary underfeeding may lead to severe weight loss particularly in cows with nursing calves. The question is how much weight loss is acceptable before one might consider this animal neglect? When weather conditions lead to periods where forage is unavailable and contributes to excessive weight loss and starvation, owners are guilty of neglect. Some producers might counter that low cattle prices coupled with high costs for feed, fertilizer and other farming supplies and equipment justifies not supplementing cattle under such conditions. Reality is that it is very difficult to turn a profit in such situations; nonetheless, a decision to reduce input costs by not supplementing feed or hay at the animals' expense is not justifiable either from a moral or legal standpoint. Producers are obligated to provide supplemental feed and hay for normal body maintenance.

In recent years, urbanites in search of the country lifestyle have started a migration back to the rural areas. For many, a 5 or 10 acre parcel of land is sufficient to start a small farm. They purchase a few animals, but don't have all of the background needed to understand their nutritional, housing or health needs. In most cases, they have even less understanding of pasture management or the cost of good hay and other nutritional supplements. Problems with malnutrition and parasitism are not uncommon as people learn that there is more to maintaining animals than they realized. And, as most veterinarians have experienced, when animals become ill or injured in these situations, the lack of facilities to handle them makes the tasks of examination and treatment nearly impossible to do safely. The neglect that occurs in these cases is most surely unintentional, but nonetheless important.

Beyond abuse and neglect are housing and environmental conditions, animal handling, painful or stressful management procedures, and distress and injury associated with transportation, marketing and handling at the packing plant. A group of cattle specialists from North America were surveyed and asked to identify their top 5 welfare issues. Below is a partial list of issues adapted from their comments (North American Food Animal Well-Being Commission – Beef):

- Environmental conditions for animals (feedlot cattle): heat stress, dust, and mud
- Painful procedures (castration, dehorning and branding)
- Weaning/Preconditioning/Respiratory disease
- Transportation Issues
- Cattle handling in feed yards, on ranches and dairies
- Management of chronically ill animals
- Timely euthanasia and euthanasia techniques
- Metabolic disease
- Electro-ejaculation of bulls
- Euthanasia of unmarketable dairy bull calves
- Handling of cattle in markets
- Dystocia of heifers in feedyards

ENVIRONMENTAL STRESS

The performance, health and welfare of beef cattle are significantly affected by climatic conditions. Beef cattle are exposed to extremes of hot or severe cold weather that result in significant losses in various regions across the country. USDA APHIS VS statistics on mortality of calves and cattle on US Beef Cow Calf operations from 2007 indicate that weather-related causes were the fourth leading cause of death. A report by Mader in 2002 estimated average losses to the feedlot industry of $10-20 million annually from adverse weather conditions. Environmental stress from extreme cold or hot temperatures predisposes to pronounced suffering in affected animals. In cold conditions cattle increase their dry matter intake, seek protection from wind blocks and orient their body toward the sun to take advantage of warming from solar radiation. Combined with rain or snow in times of wet weather, mud significantly increases maintenance requirements as the moisture and mud accumulation on cattle hair coats greatly reduces its natural insulating properties.

During the summer elevated ambient air temperatures, high humidity, periods of intense solar radiation and limited airflow combined with periods of heavy rainfall and mud can be deadly. As body temperature rises a cascade of behavioral and physiological changes take place designed to reduce and/or dissipate heat load. Respiratory rates increase, dry matter intake decreases, cattle seek shade and climb mounds searching for greater airflow. In times of hot and dry weather dust creates serious air quality problems predisposing to significant respiratory problems for both humans and animals.

Management strategies for protection of feedlot cattle during harsh winter weather include the provision of bedding, mounds (to minimize mud) and windbreaks. Contrary to traditional thinking, low roughage diets with higher starch levels are most beneficial in cold weather conditions. Similarly, high energy finishing-type diets increase heat load and are contraindicated during periods of hot weather. Heat stress is also complicated by breed (color), condition score (fatness), temperament (excitable or not), and a history of previous respiratory disease. Mitigation strategies beyond dietary manipulation in hot weather include shade, sprinkling and mounds (for greater airflow).

PAIN MANAGEMENT

Castration, dehorning, ear tagging and branding are necessary livestock management procedures. Benefits of castration and dehorning are improved meat quality and fewer injuries. Animal identification is an important management procedure and is usually accomplished through the use of an ear tag, tattoo, or freeze brand and hot brands are still required in some states for proof of ownership. All of these procedures are associated with variable degrees of acute discomfort and some such as castration and dehorning cause prolonged pain that may justify therapy when practicable. However, at the present time there are no compounds approved for pain relief in cattle. Non-steroidal anti-inflammatorics such as flunixin meglumine provide short-term benefit and are also complicated by label requirements which call for intravenous administration. Recent work by Coetzee et al. with meloxicam, an NSAID approved for use in Canada, has found that a single oral administration of meloxicam at the rate of 1 mg/kg prior to castration was able to mitigate pain and also reduce the number of calves requiring treatment for bovine respiratory disease in the post-castration period. Earlier work by Heinrich et al., found a reduction in the physiological stress responses of calves to pain dehorned by cautery dehorning methods when calves were pre-treated with a 0.5 mg/kg intramuscular dose of meloxicam. These and several similar studies suggest that meloxicam provides real benefit as a pain management tool in calves and cattle.

REDUCING THE STRESS OF WEANING IN BEEF CALVES

Calf weaning strategies include the following: natural (weaned by the cow), traditional (abrupt removal of the calf from the cow), fence-line weaning and two-step weaning (a combination of fence-line and nose-flap weaning rings). Of these, traditional weaning presents what many believe is the single greatest stress on calves. Specific stressors in the traditional abrupt weaning system include: 1) separation from the dam, 2) age at weaning is normally younger than would naturally occur, 3) calves must adapt to a new environment without the dam, 4) calves are mixed with unfamiliar animals and must form new social hierarchies, 5) calves must endure transportation and hauling, 6) calves must adapt to a new diet, 7) greater exposure to disease pathogens, and more.

Time required for natural weaning by the cow varies greatly, but generally follows the milk production curve of the cow. The age of the calf is much less important to the weaning process than the available supply of milk. One report on observations in Zebu cattle found that natural weaning required about 2 weeks with bulls weaning on average at 11.3 and heifers at 8.8 months of age. Further, observations of feral and domestic cattle demonstrate that despite weaning the cow and calf maintain a lifelong relationship. Some have thought that the use of a "trainer cow" (adult cow babysitter) might reduce the anxiety and stress associated with separation of the calf from the cow. Observations by Gibb et al. did not show a beneficial effect on weight gain, health or calf stress with the use of a "trainer cow" even when that animal was familiar to the calves.

Observations of fence-line weaning demonstrate that calves pace less and rest more. Although cows and calves call to each other frequently across the fence they call less when in visual contact with each other across the fence compared to those situations where calves are housed or maintained out of visual contact (as with traditional abrupt weaning). However, by implementing a 2-Stage weaning strategy one is able to separate the two most significant stressors (i.e. loss of milk and separation from the dam) at different times rather than all at once. The first step in 2-stage weaning is to prevent nursing by fitting the calf with an anti-sucking device (Quietwean). After 5-10 days the anti-sucking device is removed and calves are separated from their mothers. This strategy proved to greatly reduce the vocalization and other stress responses normally observed in traditional weaning systems.

LAMENESS

Griffin reported that a review of records from five western feedlots indicated that of the 13% of 1,843,652 animals treated for health problems, lameness accounted for 16% of health disorders treated and 5% of deaths. He further stated that lameness accounted for 70% of all sales of non-performing cattle and that the salvage value of these animals was only 53% of the original purchase price. Seventy percent of lameness disorders involved foot problems, 15% upper leg conditions, 12% septic joints and 3% injection site lesions. Foot problems were primarily due to mechanical injuries from handling equipment, footrot and traumatic lesions of the sole from excessive wear. Despite how common lameness is in the feedyard setting, few operations have a proper chute or equipment to manage and treat these conditions. Whereas most animals with claw disorders benefit from corrective trimming and application of a foot block to the healthy claw, feedlot cattle are rarely if ever offered such treatment for foot problems.

EUTHANASIA OF UNMARKETABLE DAIRY BULL CALVES

During those market conditions when demand for dairy bull calves is low or non-existent, dairy farmers must resort to euthanasia of these animals. In most cases, these calves are euthanized at birth. Many are killed by blunt force trauma, but research indicates that the calf's skull is too hard for successful euthanasia by blunt force trauma. A better method is controlled blunt force trauma as with a non-penetrating captive bolt followed by the use of an adjunctive method such as exsanguination, pithing, or the rapid intravenous administration of potassium chloride. Alternative methods are gunshot (a .410 or 28 gauge shotgun are very effective) or the administration of a barbiturate overdose.

SUMMARY

Our livestock industries need to understand the term "animal welfare" and distinguish it from the "animal rights" connotation. Done correctly – what they do results in good animal welfare. Welfare includes *animal function* (is it productive?), *affective states* (does it have pain or distress?) and the *animal's natural behavior* (does the animal express behavior similar to what it would if in the natural world). As production oriented folks, livestock producers and others tend to equate good welfare with high production. Experience has shown that high levels of production are not always indicative of good animal welfare. This occurs in part because these systems often ignore that fact that animals are experiencing discomfort or distress or that housing conditions are inhibiting their ability to express natural behaviors. Blatant animal abuse is rare, but unconscious forms of abuse are relatively common. Knowledge of proper animal handling with a fresh and objective look at restraint practices can help to reduce some of these unintended abuses. Animal neglect is rarely intentional and generally results from ignorance. Neglect in cattle is complicated by poor husbandry practices and by high input costs and low returns on investment. Finally, multiple issues have been identified as concerns in the US cattle industry. There remain many challenges for the livestock industry in finding a balance between what's right for animals and consistent for a sustainable future.

REFERENCES

Broom, DM. Welfare: in *Bovine Medicine*, 2004;955. Brown-Brandl, TM. Livestock Sci, 2006. 105;57. Coetzee, JF, et al. J Anim Sci, 2011, 4068. Coetzee, JF, et al. Vet Ther. 2009. 10:E1-E8. Curtis, SE, Prof Anim Scientist. 2007. 23:573. Gibb, DJ, et al. J Anim Sci, 2000. 78:1716. Griffin, D, et al. Neb Guide, G93-1159-A, 1993. 1-6. Haley, DB, et al, J Anim Sci, 2005, 83(9): 2205. Heinrich, A, et al. J Dairy Sci. 2009. 92:540. Hungerford, LL, et al. Proc of the 9th Sym on Vet Epi and Econ, 2000. 1. Mader, TL. J Anim Sci. 2003. 81:E110. ,Rollin, BE: Farm Animal Welfare: Social, Bioethical and Research Issues. Iowa State University Press, 1995, 1-168. Shearer, JK, Reynolds,J. *Dairy Production Medicine,* Wiley-Blackwell, 2011. 331. Von Keyserlingk, M.A.G.et al, J. Dairy Sci., 2009. 92:4101.

PROPER EUTHANASIA METHODS IN FOOD ANIMALS

J.K. Shearer, DVM, MS
Ames, Iowa

INTRODUCTION

The AVMA Panel on Euthanasia has completed its most recent update which is planned for publication sometime during the 1st quarter of 2012. More than 70 people, including veterinarians and non-veterinarians with expertise from a broad range of disciplines and species, have been convened to develop this version of the document which constitutes the *Eighth Edition of AVMA Guidelines on Euthanasia*. This edition of the Guidelines has been expanded to include more species and greater detail on specific methods of euthanasia than ever before. In the following I have tried to provide a glimpse of what is found in the Guidelines as it pertains to the euthanasia of cattle, swine and poultry.

METHODS OF EUTHANASIA FOR POULTRY

Euthanasia methods for poultry (domesticated birds used for egg, meat or feather production, [e.g., chickens, turkeys, quail, pheasants, ducks, geese]) include gas inhalation, manually applied blunt force trauma, cervical dislocation, decapitation, electrocution, gunshot, captive bolt and injectable agents. Embryonated eggs may be destroyed by prolonged exposure (20 min) to carbon dioxide, cooling (4 hr at 40^0F), or freezing. In some cases inhalant anesthetics can be administered through the air cell at the large end of the egg. Embryos that have been exposed can be decapitated.

Maceration, using specially designed mechanical systems with rotating blades or projections causes immediate fragmentation and death of newly hatched poultry and embryonated eggs. A review by the American Association of Avian Pathologists on the use of macerators for euthanasia of chicks, poults, and pipped eggs indicates that death by maceration in poultry up to 72 hours old occurs immediately with minimal pain and distress. It is important to note that maceration requires specialized equipment that must be properly maintained. And, newly-hatched poultry must be delivered to the macerator in a way and at a rate that prevents a backlog of chicks at the point of entry into the macerator and without causing injury, suffocation, or avoidable distress to the chicks before maceration.

METHODS OF EUTHANASIA FOR CATTLE AND SWINE

The physical methods for conducting euthanasia in cattle, swine or poultry include gunshot and captive bolt. For sows, boars, and grower/finisher pigs methods include gunshot, penetrating captive bolt, and electrocution. Euthanasia may also be accomplished by the parenteral administration of a barbiturate in an amount sufficient to cause death. This latter method normally requires a veterinarian to administer the drug, is more costly in large animals, may interfere with diagnostic investigations and creates residue problems that limit carcass disposal options. Despite what some people are inclined to believe, there are few methods as humane as gunshot or penetrating captive bolt combined with a secondary or adjunctive step to ensure death such as a second shot, exsanguination, the rapid intravenous injection of a saturated solution of potassium chloride, or pithing of the brain and upper spinal cord. When properly performed, both gunshot and penetrating captive bolt meet the objectives of inducing immediate loss of consciousness and rapid death without pain or distress to the animal.

EUTHANASIA USING FIREARMS

In most circumstances on the farm or ranch, gunshot is the most practical method of euthanasia. A 0.22 long rifle solid point bullet fired from either a pistol or rifle is sufficient for young animals, however higher caliber firearms are recommended for consistent results with adult animals. Handguns ranging from .32 to .45 calibers are usually recommended. In all cases these should be loaded with solid point lead bullets. Hollow-point bullets do more damage, but because of rapid expansion and fragmentation may not penetrate the skull in all cases.

Shotguns, in particular the 12, 16 and 20 gauge, are ideal for euthanasia of adult cattle and swine, whereas the .410 and 28 gauge are good choices for calves or feeder/grower animals. Birdshot (lead BBs – number 6 or lower) or slugs may be used. When birdshot is fired from within 1 to 3 feet, BBs leave the muzzle and enter the skull as a compact mass that easily penetrates the skull causing massive brain damage, immediate loss of consciousness and rapid death. Birdshot offers maximum penetration and brain destruction, but BBs rarely exit the skull where they may cause damage to bystanders.

Rifles, particularly the .223 or higher, are particularly useful when conditions require the operator to be some distance from the animal. A .22 LR, or better yet, a.22 magnum is a good choice for euthanasia of adult cattle and swine assuming the operator can get within 1 to 3 feet of the intended target (skull). Proper placement of the bullet is essential and best achieved by holding the firearm within 12 to 24 inches of the intended target. Unlike captive bolt, the muzzle of a firearm should never be held flush with the skull. Discharge of a firearm results in the development of enormous pressure within the barrel that is likely to cause explosion or rupture of the barrel if the muzzle end is obstructed or blocked.

CAPTIVE BOLT

There are two basic types of captive bolt: penetrating and non-penetrating. Both are intended to cause an immediate loss of consciousness. Styles include an in-line or cylindrical (KR Stunner) and pistol grip (resembling a handgun, "Cash Special" .25 Caliber, by Bunzl) versions. Pneumatic captive bolt guns (air powered) are limited to use in slaughter plant environments. Models using gunpowder charges are more often used in farm environments.

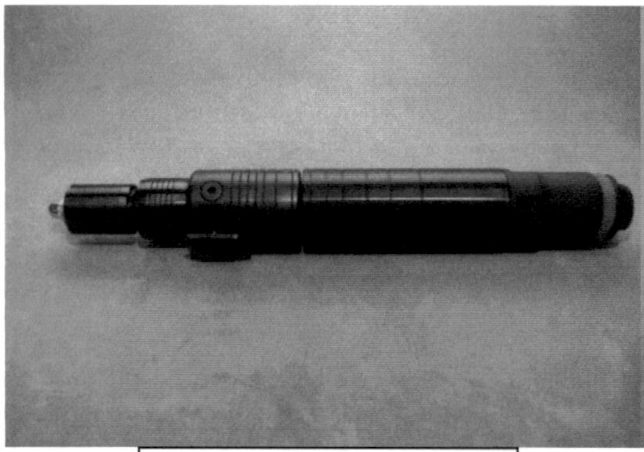

KR Stunner by Schermer

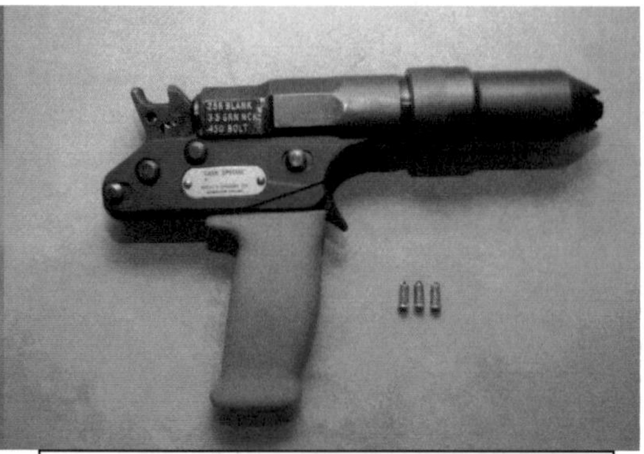

"Cash Special" .25 Caliber with extended bolt

Captive bolt guns should only be used by trained people. Even though the bolt protrudes only a short distance from the muzzle end, it should always be pointed towards the ground and away from the body or bystanders in case of accidental discharge. Protective gear for both ears and eyes are strongly recommended.

Death following use of the penetrating captive bolt is commonly achieved, but is not assured depending upon bolt length and depth of the frontal sinus in adult cows, bulls, sows and boars. Therefore, secondary steps to ensure death (e.g., a second application of the penetrating captive bolt, exsanguination, pithing and intravenous delivery of a saturated solution of KCl) should be applied as necessary. Breed differences result in variable skull shapes making determination of the best anatomic site for conducting euthanasia in mature sows and boars difficult. Similar problems may occur in bovines (buffalo and certain breeds of cattle).

Penetrating captive bolts offer safety advantages compared with firearms. Properly applied, the method is very effective and costs associated with its use are minimal. However, it is important that penetrating captive bolt guns be maintained regularly (cleaning and replacement of worn parts) and that charges be stored properly to ensure appropriate bolt velocity.

ELECTROCUTION

Electrocution as a sole method of euthanasia can achieve death via two-step or single-step processes. Electrical current must pass through the brain to achieve loss of consciousness, but then must cross the heart to cause fibrillation and cardiac arrest. As a two-step process, electrode placements are, head-head, followed by head to flank, for the appropriate time. For a single-step process for euthanasia, head to opposite flank is an appropriate placement. These methods are not commonly used in the on-farm environment. Proper training and special equipment are required to ensure adequate and safe euthanasia.

ANATOMICAL SITES FOR CONDUCTING EUTHANASIA IN MATURE CATTLE AND SWINE

In cattle, the ideal point of entry for a projectile (i.e. bullet or penetrating captive bolt) is on the intersection of two lines each drawn from the lateral canthus to the base of the horn on the contralateral side. If a firearm is used it should be positioned so that the muzzle is perpendicular to the skull to avoid the possibility of ricochet. Ideally, the bullet should travel toward the foramen magnum or in the direction of the tail of the animal (See diagram below). Proper positioning of the firearm or penetrating captive bolt is necessary to achieve the desired results since death can only be assured if there is sufficient damage to the brainstem. The preferred anatomic site for gunshot or penetrating captive bolt in swine is the frontal position of the forehead approximately a half inch above eye level, on the mid-line of the forehead and aiming toward the tail of the pig. An alternate site is behind the ear aiming or directing the bullet toward the opposite eye.

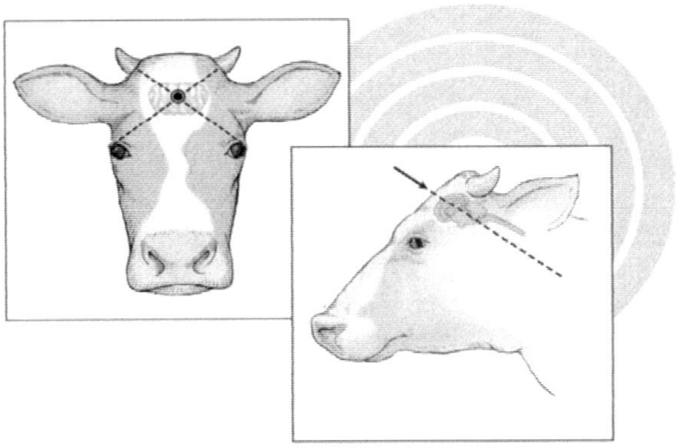

Anatomic site for gunshot or placement of a captive bolt and desired path of the projectile in bovids (Shearer, 2011) – On the intersection of 2 lines each drawn from the lateral canthus of the eyes to the base of the horn (or where the horn would be in hornless cattle) on the contalateral side.

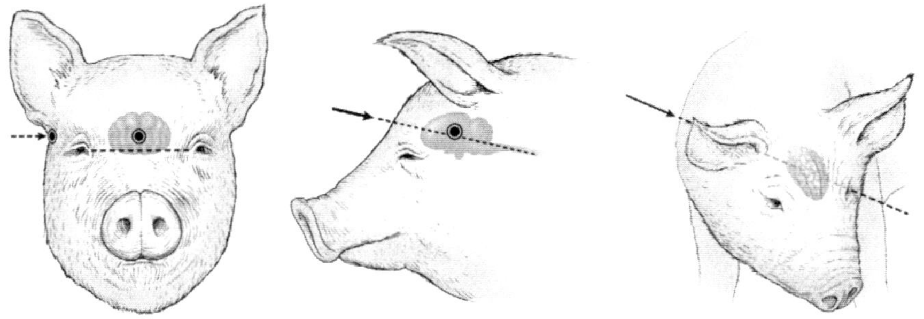

Anatomic site for gunshot and penetrating captive bolt application in swine. The ideal target is half an inch above eye level, on the midline of the forehead and aiming toward the tail of the pig. The projectile should enter the skull from behind the ear aiming toward the opposite eye.

REFERENCES

Althen TG, et al., J Anim Sci 1977;44:985-989; AAAP, Athens, Ga: American Association of Avian Pathologists, 2005; Baker HJ and Scrimgeour HJ,,Can Vet J 1995;36:160-165; Blackmore DK, et al., New Zealand Vet J 1995;43:134-137; Blackmore DK, Vet Rec 1985;116:36-40; Daly CC, Potters Bar, UK: Universities Federation for Animal Welfare, 1986a;15-20; Daly CC and Wittington PE, Res Vet Sci 1989;46:406-408; Erasmus MA, et al. Vet Rec 2010a;167(22):850-858; Erasmus MA, et al., Poultry Sci 2010b;89(7):1345-1354; Fulwider WK, et al., J Dairy Sci 2008;91: 1686-1692; Golab GC., J Dairy Sci, 2011(94):E-Suppl. 1, p. 156; Grandin T. Observations of cattle behavior applied to design of cattle-handling facilities. Appl Anim Ethol 1980;6:19-31; Gregory N, and Shaw F, J Appl Anim Welfare Sci 2000;3(3):215-230; Humane Slaughter Association. Captive bolt stunning of livestock, Guidance notes #2, 4th ed; 2006; Humane Slaughter Association. Electrical stunning of red meat animals, 2000; Humane Slaughter Association, Guidance notes #3, 2nd ed. Wheathampstead, UK: Humane Slaughter Association, 2005; Jaksch W, Int J Stud Anim Prob 1981;2:203-213; Leach TM, and Wilkins LJ, Meat Sci 1985;15:101-106; McKinstry JL and Anil MH, Meat Sci 2004;67:121-128; Millman S, Proc of AASV 2010;41:443-446; Muir W, Handbook of veterinary anesthesia, 2000; National Pork Board, AASP On farm euthanasia of swine, 2nd edition; 2009; Shearer JK, et al., J Dairy Sci, 2011(94):E-Suppl. 1, p. 156; Shearer JK, and Nicoletti P. Humane euthanasia of sick, injured and/or debilitated livestock, 2011. Available at www.vetmed.iastate.edu/vdpam/extension/dairy/programs/humane-euthanasia/anatomical-landmarks. Accessed June 24, 2011; Thurmon JC, Vet Clin North Amer Food Animal Practice 1986;2(3):743-756; USDA National animal health emergency management system guidelines, 2004, Available at www.dem.ri.gov/topics/erp/nahems_euthanasia.pdf. Accessed August 27, 2009; Van der Wal PP, Proc 2nd International Symposium on Conditions and Meat Quality of Pigs, 1971; Whiting T, et al., 2010; Widowski T, National Pork Board Research Report, 2008; Woods J, et al., In: Grandin T, ed. Improving animal welfare: a practical approach, Wallingford Oxfordshire, UK: CABI Publishing, 2010; 186-213; World Organization for Animal Health (OIE). Terrestrial animal health code, chapter 7.6: killing of animals for disease control purposes. Paris: OIE, 2010. Available at www.oie.int/en/international-standard-setting/terrestrial-code/access-online. Accessed May 16, 2011.

ARTHROPOD-BORNE FOOD ANIMAL DISEASES

D.S. McVey, B. Drolet, W. Wilson
Manhattan, KS

The Arthropod-Borne Animal Disease Research Unit (ABADRU), currently located in Manhattan, KS is part of the Agricultural Research Service (ARS). The mission of the ABADRU is to solve major endemic, emerging, and exotic arthropod-borne disease problems in U.S. livestock. Research emphasizes the molecular biology of pathogens and vectors, vector biology and competence, epidemiology, and animal pathogenesis. Arboviruses are the major focus of concern because they are especially difficult to control. This research effort is all in an effort to develop better detection, control, and prevention methods. Diseases of particular interest include:

- BSL-2 Research - domestic bluetongue, domestic vesicular stomatitis virus, Epizootic hemorrhagic disease virus, MP-12 Rift Valley fever virus (vaccine strain)
- BSL-3 Research - exotic bluetongue, vesicular stomatitis virus in animals, virulent Rift Valley fever virus

Bluetongue virus (BTV) is an orbivirus transmitted by biting midges in the genus *Culicoides* (Family: Ceratopogonidae) that infects both wild and domestic ruminants. Economic annual losses of $120 million U.S., $3 billion world-wide are attributed to effects on animal health and productivity [1-6], as well as non-tariff trade restrictions that affect the sale and movement of animals and germplasm.[7] Identified on all continents except Antarctica, the spread of BTV, and the resulting threat to North American livestock and wildlife, depends on many factors including climate [8, 9], altitude, vector competence (ability of the insect to acquire and transmit a specific serotype), and host availability. Over the past ten years, the epidemiological movement of BTV has changed dramatically from what was originally thought to have been enzootic areas for the disease.[10-12] Currently, there are 26 known serotypes worldwide [10-12], five of which are considered domestic to the U.S. (BTV-2, -10, -11, -13, -17).[8,14] Since 1999, 10 exotic serotypes have been isolated in southeastern U.S. (BTV-1, 3, 5, 6, 9, 12, 14, 19, 22, and 24).[11, 15-17] It is not clear from where these exotic serotypes originated. This raises questions of whether the epidemiology of BTV in the U.S. is changing and if these changes could result in more extensive livestock and wildlife disease.[18] Reported changes in climate, vector prevalence, and animal trade may provide the perfect storm for devastating BTV outbreaks in the U.S. similar to the most recent emergence of serotype 8 in Europe.[19, 20] Since 1998, BTV has spread from endemic areas of the Mediterranean Basin to previously BTV-free countries, including the UK. Infections in these geographic regions, previously believed to be free of BTV-competent vectors, have implicated climate changes, in conjunction with wind-borne vectors, as the culprits in emergent serotypes of the virus.[8, 9, 10, 21] In outbreaks of most BTV serotypes, the majority of disease was seen in sheep, therefore cattle were thought to only harbor virus with few clinical signs. However, the emergence of serotype 8 into the UK demonstrated that a new, more virulent strain can easily overwhelm naïve livestock populations, including cattle.[22-30] The rapid, devastating spread of BTV-8, and its atypical clinical manifestation in cattle, has highlighted the risks that face the U.S. of accidental or intentional introductions of serotypes with unknown virulence to our naïve livestock and wildlife.

The ABADRU is working to develop rapid, sensitive, specific diagnostic assays to detect and differentiate multiple serotypes of BTV and anti-BTV antibodies in cattle and sheep from a single blood or serum sample. The Unit is also evaluating the role of virus delivery routes (subcutaneous versus intradermal) and the role insect salivary proteins play in virus infection, pathogenesis and immune responses to BTV in order to develop better experimental models of pathogenesis and disease transmission via *Culicoides* spp. The long term goal is to develop a robust BTV infection and clinical disease animal model; a critical need for bluetongue infection, pathogenesis and vaccine research.

The potential for Rift Valley fever virus (RVFV) as a bioterrorism agent is widely recognized.[31-33] The intentional or unintentional introduction of RVFV into the U.S. is an agricultural and public health concern. RVFV is a Category A pathogen [34], and is therefore a high priority for biodefense. The disease is enzootic in sub-Saharan Africa with RVFV epidemics occurring at irregular intervals primarily in eastern and southern Africa.[35] Both RVFV and West Nile virus (WNV), are insect-transmitted zoonotic arboviruses. The introduction of WNV to the Western Hemisphere [36,37] and the expansion of RVFV from Africa to the Arabian Peninsula illustrate the natural ability of arboviruses to establish themselves in new ecosystems.[31-38] Although mosquitoes usually transmit RVFV, it is also very infectious by oral and aerosol routes.[39-40] Disease outbreaks can affect as few as 15% of susceptible animals to as many as 90% .[41] Infection of animals results in severe disease and abortion in mature cattle, sheep, and goats, with 70-100% mortality in young animals.[11] Endemic prevalence of antibody in humans ranges from 5-40% [40], but increases with risk factors such as livestock handling, exposure to aerosol during the slaughter of infected animals and consumption of raw milk.[39] Infection of humans can be subclinical or can cause four disease forms; 1) febrile, 2) encephalitis with confusion and coma, 3) ocular involvement, and 4) hemorrhagic, which appears to be the major cause of death.[3, 11] The number of human cases requiring hospitalization is usually less than 1% [42], but in the Saudi Arabia epidemic the mortality rate of infected individuals was 13.9% [43] demonstrating the potential severity of a RVF outbreak in a new region. Vision loss from retinitis occurs in approximately 10% of human patients either during acute febrile illness or up to 4 weeks after initial onset of illness.[43-46] Patients with ocular lesions (71%) remained legally blind for up to five years after initially acquiring disease.[32]

The potential introduction of Rift Valley fever (RVF) virus (RVFV) is considered to be the most significant arthropod-borne animal disease threat to U.S. livestock according to the USDA-APHIS National Veterinary Stockpile (NVS) Steering Committee. The ABADRU is investigating the epidemiological factors affecting disease outbreak and the inter-epizootic maintenance of RVFV which is necessary for the development of appropriate countermeasures strategies. This requires development of improved methods to detect and characterize emergent viruses (as RVFV is a multi-segmented RNA virus and could evolve to adapt to a new environment). Also, the proposed research will identify determinants of RVFV infection, pathogenesis and maintenance in mammalian and insect vector hosts. Information derived from these studies will also provide a better vaccine evaluation challenge model. The ABADRU is also working to improve RVFV vaccine formulations to improve immunogenicity, onset of immunity and in-use stability to provide better utility, compliance and effectiveness in the field.

REFERENCES

1. Bowne, J.G., *Bluetongue disease.* Adv. Vet. Sci. Comp. Med., 1971. **15**: p. 1-46. 2. MacLachlan, N.J., et al., *Detection of bluetongue virus in the blood of inoculated calves: comparison of virus isolation, PCR assay, and in vitro feeding of Culicoides variipennis.* Arch. Virol., 1994. **136**(1-2): p. 1-8. 3. Bonneau, K.R., et al., *Duration of viraemia infectious to Culicoides sonorensis in bluetongue virus-infected cattle and sheep.* Vet Microbiol, 2002. **88**(2): p. 115-25. 4. Luedke, A.J., *Effect of bluetongue virus on reproduction in sheep and cattle.* Prog. Clin. Bio. Res., 1985. **178**: p. 71-8. 5. MacLachlan, N.J., et al., *Orbivirus infection of the bovine fetus.* Prog Clin Biol Res, 1985. **178**: p. 79-84. 6. Kvasnicka, W.G., *Reproductive problems associated with bluetongue virus activity in Nebraska.* Prog Clin Biol Res, 1985. **178**: p. 109-19. 7. Roberts, D.H., M.H. Lucas, and R.A. Bell, *Animal and animal product importation and the assessment of risk from bluetongue and other ruminant orbiviruses.* Br Vet J, 1993. **149**(1): p. 87-99.
8. Maclachlan, N.J., *Global implications of the recent emergence of bluetongue virus in Europe.* Vet Clin North Am Food Anim Pract, 2010. **26**(1): p. 163-71, table of contents. 9. Purse, B.V., et al., *Climate change and the recent emergence of bluetongue in Europe.* Nature Reviews, Microbiology. 2005. 3:171-181. 10. Weaver, S.C. and W.K. Reisen, *Present and future arboviral threats.* Antiviral Res, 2010. **85**(2): p. 328-45. 11. Maclachlan, N.J. and A.J. Guthrie, *Re-emergence of bluetongue, African horse sickness, and other orbivirus diseases.* Vet Res, 2010. **41**(6): p. 35. 12. Maan, S., et al., *Novel bluetongue virus serotype from kuwait.* Emerg Infect Dis, 2011. **17**(5): p. 886-9. 13. Chaignat, V., et al., *Toggenburg Orbivirus, a new bluetongue virus: initial detection, first observations in field and experimental infection of goats and sheep.* Vet Microbiol, 2009. **138**(1-2): p. 11-9. 14. Tabachnick, W.J., *Culicoides variipennis and bluetongue virus epidemiology in the Unites States.* Annual Review of Entomology, 1996. **41**: p. 23-43. 15. Johnson, D.J., et al., *First report of bluetongue virus serotype 1 isolated from a white-tailed deer in the United States.* J Vet Diagn Invest, 2006. **18**(4): p. 398-401. 16. Johnson, D.J., et al. *Exotic Bluetongue Viruses Identified from Ruminants in the Southeastern U.S. from 1999-2006.* in *American Association of Veterinary Laboratory Diagnosticians.* 2007. Reno, NV. 17. Mertens, P.P.C., et al. *Sequencing and RT-PCR assays for genome segment 2 of the 24 bluetongue virus serotypes: Identification of exotic serotypes in the southe eastern USA (1999-2006).* in *World Association of Veterinary Laboratory Diagnosticians.* 2007. Melbourne, Australia. 18. Gibbs, E.P.J., et al., *U.S. concerns over bluetongue.* Science, 2008. **320**: p. 872. 19. Wilson, A.J. and P.S. Mellor, *Bluetongue in Europe: past, present and future.* Philos Trans R Soc Lond B Biol Sci, 2009. **364**(1530): p. 2669-81. 20. Slenning, B.D., *Global climate change and implications for disease emergence.* Vet Pathol, 2010. **47**(1): p. 28-33. 21. Gould, E.A. and S. Higgs, *Impact of climate change and other factors on emerging arbovirus diseases.* Trans R Soc Trop Med Hyg, 2009. **103**(2): p. 109-21. 22. Backx, A., et al., *Transplacental and oral transmission of wild-type bluetongue virus serotype 8 in cattle after experimental infection.* Vet Microbiol, 2009. **138**(3-4): p. 235-43. 23. Dal Pozzo, F., C. Saegerman, and E. Thiry, *Bovine infection with bluetongue virus with special emphasis on European serotype 8.* Vet J, 2009. **182**(2): p. 142-51. 24. Elbers, A.R., et al., *Field observations during the Bluetongue serotype 8 epidemic in 2006. II. Morbidity and mortality rate, case fatality and clinical recovery in sheep and cattle in the Netherlands.* Prev Vet Med, 2008. **87**(1-2): p. 31-40. 25. Menzies, F.D., et al., *Evidence for transplacental and contact transmission of bluetongue virus in cattle.* Vet Rec, 2008. **163**(7): p. 203-9. 26. Muller, U., et al., *Effects of bluetongue virus infection on sperm quality in bulls: A preliminary report.* Vet J, 2010. **186**(3): p. 402-3. 27. Santman-Berends, I.M., et al., *Bluetongue virus serotype 8 (BTV-8) infection reduces fertility of Dutch dairy cattle and is vertically transmitted to offspring.* Theriogenology, 2010. **74**(8): p. 1377-84. 28. Santman-Berends, I.M., et al., *Vertical transmission of bluetongue virus serotype 8 virus in Dutch dairy herds in 2007.* Vet Microbiol, 2010. **141**(1-2): p. 31-5. 29. Vercauteren, G., et al., *Bluetongue virus serotype 8-associated congenital hydranencephaly in calves.* Transbound Emerg Dis, 2008. **55**(7): p. 293-8. 30. Zanella, G., et al., *Bluetongue virus serotype 8: Abortion and transplacental transmission in cattle in the Burgundy region, France, 2008-2009.* Theriogenology, 2011. 31. Shoemaker, T., et al., *Genetic analysis of viruses associated with emergence of Rift Valley fever in Saudi Arabia and Yemen,*

2000-01. Emerg. Infect. Dis., 2002. **8**(12): p. 1415-20. 32. Al-Hazmi, A., et al., *Ocular complications of Rift Valley fever outbreak in Saudi Arabia.* Ophthalmology, 2005. **112**(2): p. 313-8. 33. Meegan, J.M., R.H. Watten, and L.W. Laughlin, *Clinical experience with Rift Valley fever in humans during the 1977 Egyptian epizootic.* Contr. Epidem. Biostatist., 1981. **3**: p. 114-123. 34. NIH. *NIH Biodefense Research,* http://www3.niaid.nih.gov/Biodefense. 2005 [cited 2005 February]. 35. Gerdes, G.H., *Rift valley fever.* Vet. Clin. North. Am. Food Anim. Pract., 2002. **18**(3): p. 549-55. 36. Lanciotti, R.S., et al., *Origin of the West Nile virus responsible for an outbreak of encephalitis in the northeastern United States.* Science, 1999. **286**(5448): p. 2333-7. 37. Anderson, J.F., et al., *Isolation of West Nile virus from mosquitoes, crows, and a Cooper's hawk in Connecticut.* Science, 1999. **286**(5448): p. 2331-3. 38. Miller, B.R., et al., *Isolation and genetic characterization of Rift Valley fever virus from Aedes vexans arabiensis, Kingdom of Saudi Arabia.* Emerg. Infect. Dis., 2002. **8**(12): p. 1492-4. 39. Smithburn, K., A. Mahaffy, and A. Haddow, *Rift Valley fever: accidental infections among laboratory workers.* J. Immunol., 1949. **49**: p. 141-173. 40. Gerdes, G.H., *Rift Valley fever.* Rev. Sci. Tech., 2004. **23**(2): p. 613-23.
41. EMPRES. *Emergency Prevention System, Transboundry Animal Diseases.* 2005 [cited February 2005]. 42. Flick, R. and M. Bouloy, *Rift Valley fever virus.* Curr. Mol. Med., 2005. **5**: p. 827-834. 43. Madani, T.A., et al., *Rift Valley fever epidemic in Saudi Arabia: epidemiological, clinical, and laboratory characteristics.* Clin. Infect. Dis., 2003. **37**(8): p. 1084-92. 44. Borio, L., et al., *Hemorrhagic fever viruses as biological weapons: medical and public health management.* J. Am. Med. Assoc., 2002. **287**(18): p. 2391-405. 45. Siam, A.L. and J.M. Meegan, *Ocular disease resulting from infection with Rift Valley fever virus.* Trans. R. Soc. Trop. Med. Hyg., 1980. **74**(4): p. 539-41. 46. Siam, A.L., J.M. Meegan, and K.F. Gharbawi, *Rift Valley fever ocular manifestations: observations during the 1977 epidemic in Egypt.* Brittish J. Ophthalmol., 1980. **64**(5): p. 366-74.

BOVINE ANAPLASMOSIS: AN OVERVIEW OF CURRENT CHALLENGES

Katherine M. Kocan, PhD, José de la Fuente, PhD, Douglas L. Step, DVM, DACVIM, Edmour F. Blouin, PhD, Johann F. Coetzee, BVSc, DVM, PhD, DACVCP, DVM, Katharine M. Simpson, DVM, Suzanne G. Genova, DVM, Melanie J. Boileau, DVM, MS, DACVIM
Stillwater, OK

INTRODUCTION

Bovine anaplasmosis, caused by the intracellular rickettsia *Anaplasma marginale*, has long been known to be endemic in several areas of the United States, including the south central and southeastern Gulf and northwestern states. However, more recently the disease has been reported in all states of the U.S. with the exception of Alaska and Hawaii. This wide and increasing distribution of *A. marginale* has most likely resulted from the unrestricted transport of asymptomatic but persistently infected cattle which serve as the major reservoir host for *A. marginale*. Because anaplasmosis is not a reportable disease in most states, the economic impact of this disease on U.S. cattle production has been difficult to accurately assess, and the lack of economic impact data has negatively impacted the need for continued research funding and the interest of industry in the development and marketing of new and novel control methods.

The primary site of *A. marginale* replication in cattle is within inclusion bodies in erythrocytes. Clinical disease is more likely to be observed in cattle over two years of age. The incubation period of infection (prepatent period) varies with the infective dose and ranges from 7 to 60 days, with an average prepatent period being approximately 28 days. The number of infected erythrocytes increases geometrically during acute infection. Infected erythrocytes are subsequently removed by the bovine reticulo-endothelial system which results in anemia and icterus, and the severity of anemia approximately correlates with the parasitemia. Cattle that recover from acute disease develop persistent, life-long infections. These persistently-infected or carrier cattle are the major reservoir host for *A. marginale*, which serve as a source of infection for mechanical transmission and for infection of ticks for biological transmission. Under conditions of stress such as poor nutrition, post-partum and lactation periods, concurrent infections and extreme weather conditions, carrier cattle may experience higher parasitemia which could cause mild to acute disease and also pose a higher risk of transmission (as reviewed by Kocan et al.[1,2]).

Bovine anaplasmosis is primarily a disease problem in cattle and its epidemiology is more complex than other tick-borne diseases because *A. marginale* can be transmitted in two in major ways: biological and mechanical. Biological transmission occurs when ticks become infected with *A. marginale* by feeding on an acutely or persistently infected animal. The first site of infection in ticks is in gut cells, after which a developmental cycle occurs that culminates in infection and multiplication of *A. marginale* in tick salivary glands from where the pathogen is transmitted to cattle during tick feeding. Mechanical transmission is effected by any means of transferring infected blood to susceptible cattle, and probably occurs more frequently than appreciated by most producers and veterinarians. Mechanical transmission occurs seasonally via the blood-contaminated mouthparts of biting flies or year-round when instruments frequently used in veterinary practice, such as needles, ear tag applicators, castration instruments, and dehorning equipment, are not thoroughly cleaned between individual cattle (as reviewed by Kocan et al.[1,2]). In some geographic areas where tick vectors are absent or refractive to infection with local *A. marginale* strains, mechanical transmission may be the only means of transmitting *A. marginale*. For example, strains of *A. marginale* from Florida studied thus far have been found to be non-infective for ticks[3]; therefore, outbreaks most likely result from mechanical transmission. *A. marginale* can also be transmitted *in utero* during pregnancy and results in healthy calves which are born with persistent infections of *A. marginale.* Anaplasmosis is a herd problem and when clinical disease is observed, other cattle will likely be in the incubation phase. Therefore, intervention on a herd basis will be required to prevent further clinical cases. In endemic areas, all three means of transmission are likely to occur, and cattle often become infected as calves when clinical symptoms would not be apparent. Therefore, these persistently-infected cattle contribute to endemic stability.

Even though bovine anaplasmosis is a well recognized disease, the epidemiology of anaplasmosis is complex and not well understood. The factors that lead to anaplasmosis outbreaks are ill defined and most likely vary with the geographic location. The findings of considerable research on *A. marginale* conducted over the past several decades have contributed to our understanding of the complexity of the pathogen's relationship with cattle and tick hosts, as well as the epidemiology of the disease. Overall, the goal of this review is to provide an understanding of the current status of knowledge on bovine anaplasmosis and provide veterinarians with answers to frequently asked questions. The information provided in this presentation is available in detail in the following review article published in the Bovine Practitioner: Kocan et al. 2010.[3]

FACTORS THAT HAVE CONTRIBUTED TO THE COMPLEXITY OF BOVINE ANAPLASMOSIS

Despite research conducted over the past 5 decades, control strategies for bovine anaplasmosis have advanced minimally since the first anaplasmosis vaccine was marketed in the 1960s. However, the complexity of *A. marginale* and bovine anaplasmosis has increased due to factors defined as a result of this research. Herein, we briefly review these factors because of their importance to the management of anaplasmosis and their impact on the design and application of future diagnostic and control strategies.

Diversity of *A. marginale* Strains or Genotypes

The diversity of *A. marginale* strains or genotypes has proved to be much greater than previously recognized[3,4,5,6]. This diversity has become apparent, in part, because molecular tools are able to define strains on the basis of the sequences of the major surface proteins (MSPs) and other genes. MSP1a has been shown to be a stable marker of strain identity but does not provide information on the geographic origin, while MSP4 was found to provide both strain identity and geographic information.[7,8] The overall result of these studies is that the diversity of *A. marginale* strains is extensive, especially in areas of intense cattle movement. Once introduced, individual *A. marginale* strains continue to be maintained in cattle and tick populations by independent transmission events. This increased strain diversity complicates anaplasmosis control strategies because widely genetically diverse strains may not be cross-protective when used as vaccine antigen and also may not have uniform susceptibility to antimicrobials. As mentioned previously, research has shown that some strains of *A. marginale* are not transmissible by ticks. In areas, such as Florida, where local isolates are not infective for ticks, outbreaks are most likely the result of mechanical transmission, rendering tick control an unnecessary component of a control program.

Persistent Infections in Cattle and Ticks.

All cattle that become infected with *A. marginale* (calves via *in utero* transmission, adults after acute infection or vaccinated or tetracycline-treated cattle) remain persistently infected for life. Persistently-infected cattle serve as the major reservoir of infection in nature, thereby being a source of infection for mechanical transmission and for biological transmission by ticks. Clearance of *A. marginale* is difficult both to effect and to unequivocally confirm and, importantly, if cattle are cleared of infection, they are fully susceptible to re-infection.[9,10] In addition, clearance of cattle in endemic areas may reduce endemic stability. When persistently infected cattle are transported into non-endemic areas, focal outbreaks of anaplasmosis are likely to occur and the disease may subsequently become apparent if infected cattle are not culled.

Wildlife Reservoirs of *A. marginale*

Wild ruminants have been reported to serve as reservoirs of *A. marginale,* including American bison (*Bison bison*), white-tailed deer (*Odocoileus virginianus*), mule deer (*Odocoileus hemionus hemionus*), black-tailed deer (*Odocoileus hemionus columbianus*) and Rocky Mountain elk (*Cervus elaphus nelsoni*) (as reviewed by Kuttler[11]). While wild ruminants, particularly mule deer and elk, have been implicated in the epizootiology of bovine anaplasmosis in some regions,[12,13,14,15] reinvestigation of wildlife hosts that serve as reservoirs of *A. marginale* using molecular diagnostic tools is needed to define the role of wild ruminants as reservoirs of *A. marginale*. For example, in two recent studies, *A. ovis* rather than *A. marginale* was detected in mule deer,[6,16] raising the question of whether mule deer actually serve as a reservoir host for *A. marginale*. In addition, future research is needed to determine whether mule deer could support co-infections of *A. ovis* and *A. marginale*. The status of these potential reservoir hosts for *A. marginale* can only be definitively demonstrated by use of molecular diagnostic tests because serologic tests based on the *A. marginale* major surface protein 5 (MSP5), a highly conserved MSP, may be cross reactive for *Anaplasma* spp. Although white-tailed deer in the southeastern U.S. were found to be susceptible to short-term experimental infection with *A. marginale*, they do not appear to be exposed naturally, even in enzootic areas. In addition, Keel et al.[17] demonstrated that experimentally-infected white-tailed deer did not develop parasitemia sufficient for mechanical transmission by biting flies. Therefore, while white-tailed deer populations often drive tick populations in much of the U.S., they do not appear to contribute to the epidemiology of bovine anaplasmosis. Serologic surveys have incriminated bison in Canada and the United States as a reservoir host for *A. marginale,*[18] and more recently this species was shown to be infected with *A. marginale* and to serve as a host for infection of *Dermacentor variabilis* ticks.[7,19,20] Therefore, co-mingling of susceptible cattle and persistently-infected bison may not be advisable.

The Role of Male Ticks in Transmission of *A. marginale*

The distribution of tick populations has been changing and may also contribute to outbreaks and development of endemic anaplasmosis in new areas of the U.S. Male ticks have been shown to be the major tick stage for biological transmission of *A. marginale*. Male ticks are intermittent feeders which transfer readily among cattle. Because of their persistent infections, these males can rapidly spread anaplasmosis throughout a herd of cattle.[1,2] In addition male ticks can survive in the environment and thereby also serve as a reservoir of *A. marginale*. The transmission of *A. marginale* by male ticks most likely contributes to the transmission dynamics of one-host ticks such as *Rhipicephalus* (*Boophilus*) sp. and *Dermacentor albipictus*.

Antigenic Variation in Persistently Infected Cattle and Ticks

Persistently infected cattle undergo 5-6 week cycles in which an antigenic variant of *A. marginale* arises, followed by an immune response against and clearance of the circulating *A. marginale* variant.[1] Subsequently, a population of a new antigenic variant arises creating another cycle of infection, followed again by immune control. Efficacy of vaccines or antimicrobials may be compromised by these constantly changing variants that may differ in treatment susceptibility. *A. marginale* in persistently-infected male ticks also undergoes antigenic variation, and transmission of these variants to cattle may further contribute to the overall antigenic diversity of *A. marginale* in any given region.[21]

Stress Impacts of Bovine Anaplasmosis

Stress factors impact the control of anaplasmosis, including the stresses of poor nutrition, adverse weather conditions, calving, lactation and infection of cattle with concurrent infections. Anaplasmosis may become apparent in endemic areas when one or multiple stressors are present. Mild to acute anaplasmosis results when the number of infected erythrocytes increases, and could contribute to enhanced mechanical transmission and biological transmission by ticks. While cattle often are infected with multiple hemoparasites that do not normally cause disease, concurrent infections may result in immunosuppression and therefore compromise the bovine immune response to *A. marginale* infection, especially during persistent infection.

Reorganization of the Classification of the Genus *Anaplasma*

The Genus *Anaplasma* that contains the causative agent of bovine anaplasmosis, *Anaplasma marginale*, was reorganized by Dumler et al.[22] based on the genetically-based similarity of organisms. This genus now contains organisms that infect other vertebrates, in addition to pathogens that infect ruminants (*A. marginale*, *A. centrale* and *A. ovis*). Notably, *A. phagocytophilum,* causative agent of human granulocytic anaplasmosis (HGA), is now included in the genus *Anaplasma*. *A. phagocytophilum,* formerly three pathogens (*Ehrlichia equi, E. phagocytophila*, and the previously unnamed agent of human granulocytic ehrlichiosis) is now recognized as the same agent and has a wide host range, including cattle. HGA is an emerging disease of humans in the U.S. and the pathogen could possibly enter the cattle population in the future. Clinical disease, if encountered, would be more likely to result with concurrent infections of *A. marginale* or other hemoparasites. However, at present *A. phagocytophilum* has not been reported in U.S. cattle, *A. marginale* and *A. phagocytophilum* cause separate and distinct diseases, and disease caused by *A. phagocytophilum* has only be recognized in humans.[23]

Emerging Pathogens

New and emerging tick-borne pathogens have been frequently reported in the U.S. A good example of an emerging tick-borne disease is HGA which was first described in the north central U.S. as human granulocytic ehrlichiosis (as reviewed by Goodman[23] and Woldehiwet[24]). In contrast to *A. marginale*, the etiologic agent of HGA, *A. phagocytophilum,* has a wide host range and has been shown to be experimentally infective for cattle. Since this pathogen is emerging in human populations of the U.S., it may also, in the future, emerge as a pathogen of cattle. While *A. phagocytophilum* may not cause clinical disease in cattle, its presence in cattle may complicate the diagnostic tools of bovine anaplasmosis because of the conservation of MSP5 which is the major antigenic component of the USDA approved serodiagnostic test for bovine anaplasmosis, the MSP5 cELISA[25]; therefore, a diagnostic test may be required in the future to differentiate between the two pathogens. Whenever cattle are seropositive by the *Anaplasma* cELISA without a history of bovine anaplasmosis or *A. marginale* infection, they should be tested by molecular diagnostics to determine the reason for the cross-reaction. An example of this type of cross reaction was reported recently in a cattle herd in British Columbia[26,27] which revealed that cattle that tested positive by the MSP5 cELISA were actually infected by a new and novel *Ehrlichia* sp.

SUMMARY

In this review presentation, we presented an update of the challenges of the management and epidemiology of bovine anaplasmosis. All of the factors reviewed herein contribute to the increasing complexity of the control and epidemiology of anaplasmosis in the U.S. The diversity of *A. marginale* strains has been more completely defined and is far greater than previously known. This genetic diversity, which has most likely resulted from extensive cattle movement, constitutes a major challenge for developing vaccines that are protective against a diverse range of *A. marginale* strains. For biological transmission of *A. marginale*, male ticks can contribute to the transmission of anaplasmosis because they become persistently infected and, as intermittent feeders, can transfer among cattle and transmit *A. marginale*. Because these males can remain in the environment, they also serve as a reservoir host for *A. marginale*. Male ticks are the most likely means of tick transmission by one-host ticks, *D. albipictus* and *Rhipicephalus* (*Boophilus*) spp. Finally, populations of tick species are constantly changing and may inhabit new geographic areas which may therefore contribute to the establishment of endemic bovine anaplasmosis in the future. Wildlife reservoirs of *A. marginale* may impact the epidemiology of anaplasmosis. The role of deer (white-tailed and mule) as reservoirs of *A. marginale* is not well defined and should be redefined using molecular diagnostic tools. However, bison from both Canada and the U.S. were found to be infected with *A. marginale* strains which proved to be infective for cattle and the *A. marginale* strain was shown to be transmissible by ticks. Therefore, the commingling of cattle and bison may not be advisable. The classification of *Anaplasma* has recently been reorganized and now includes several organisms in addition to those that are host-specific for ruminants. Most notably, *A. phagocytophilum,* an emerging tick-borne pathogen in the U.S., has a wide host range and in the future may include cattle which have been shown experimentally to be susceptible to infection. At present *A. phagocytophilum* has not been reported in cattle in the U.S. but diagnostic tests that differentiate between *A. marginale* and *A. phagocytophilum* may be needed in the future because the MSP5 cELISA cross reacts between these two *Anaplasma* species. While infections of *A. phagocytophilum* alone may not cause clinical disease in cattle, disease may be more likely to occur with concurrent infections of *A. marginale, A. phagocytophilum* and other hemoparasites frequently found in cattle. Importantly, HGA, caused by *A. phagocytophilum,* and bovine anaplasmosis do not share common features.

REFERENCES

[1]Kocan KM, et al. **Bovine Pract** 2010; 44:93. [2]Kocan KM, et al. **Clin Microbiol Rev** 2003; 16:698. [3]Kocan KM, et al. **Vet Parasitol** 2010; 167:95. [4]de la Fuente J, et al. **Int J Parasitol** 2001b; 31:1705. [5]de la Fuente J, et al.**Vet Microbiol** 2002; 88:275. [6]de la Fuente J, et al. **Vet Microbiol** 2007; 119:375. [7]de la Fuente J, et al. **J Clin Microbiol** 2003; 41:1609. [8]de la Fuente J, et al. **J Clin Microbiol** 2005; 43:1309. [9]Coetzee JF, et al. **Vet Ther** 2006; 7:347. [10]Reinbold JB, et al. **Vet Microbiol** 2010; 145:69. [11]Kuttler KL, J Wildl Dis 1984; 20:12. [12]Boyton WH, Woods GM. **Science** 1933; 78:559. [13]Howe DL, Hepworth WG, **Am J Vet Res** 1965; 26:1114. [14]Kuttler K, et al. **Am J Vet Res** 1984; 45:2223. [15]Zaugg JL, et al. **J Wildl Dis** 1996; 32:63. [16]Yabsley MJ, et al. **Vector Borne Zoonotic Dis** 2005; 5:351. [17]Keel MK, et al. **J Wildlife Dis** 1995; 31:378. [18]Taylor SK, et al. **J Wildl Dis** 1997; 33:308. [19]de la Fuente J, et al. **Vet Microbiol** 2004; 100:163. [20]Kocan KM, et al. **Annals NY Acad Sci** 2004; 1026:114. [21]de la Fuente J & Kocan KM, **Int J Parasitol** 2001a; 69(8):5151. [22]Dumler JS, et al. **Int J Sys Evol Microbiol** 2001; 51:2145. [23]Goodman JL, ASM Press, Washington D.C. 2005; pp. 218-238. [24]Woldehiwet, Z. 2010. **Vet Parasitol** 2010; 167:108. [25]Howden KJ, et al. **Canadian Vet** J 2010; 51:830. [26]Gajadhar AA, et al. **Vet Parasitol** 2010; 173:324. [27]Dreher, UM, et al. **J Clin Diagn Lab Immunol** 2005; 12:1177.

UPDATE ON THE DIAGNOSIS AND TREATMENT OF BOVINE ANAPLASMOSIS

Johann F. Coetzee, BVSc, Cert CHP, PhD, DACVCP and Katherine M. Kocan, PhD
Ames, IA

INTRODUCTION

Anaplasmosis, caused by the rickettsial hemoparasite *Anaplasma marginale,* is the most prevalent tick-transmitted disease of cattle worldwide and a major obstacle to profitable beef production in the continental United States.[1-3] Anaplasmosis is readily transmitted through biological and mechanical vectors such as ticks and biting flies and iatrogenically through needles and equipment contaminated with infected blood. Clinical anaplasmosis, characterized by anemia, icterus and fever, is associated with significant production losses, abortions and mortalities in cattle. It is estimated that the introduction of anaplasmosis into a previously naïve herd can result in a 3.6% reduction in calf crop, a 30% increase in cull rate and a 30% mortality rate in clinically infected adult cattle.[4] Furthermore, a study has shown that 16% of pregnant carrier cows will transmit anaplasmosis *in utero* producing persistently infected offspring.[5] The existence of both horizontal and vertical anaplasmosis transmission has important implications for disease control in endemic areas. The cost of a clinical case of anaplasmosis in the USA has been conservatively estimated to be over $400 per animal with some estimating the total cost to the beef industry to be over $300 million per year.[3,4,6] In addition to the costs associated with clinical anaplasmosis, animals recovering from acute anaplasmosis, including those treated with recommended doses of tetracyclines, remain lifelong *A. marginale* carriers. There are currently no antimicrobial compounds approved for elimination of persistent *A. marginale* infections in cattle, despite published reports of successful carrier clearance with tetracyclines. Carrier animals serve as reservoirs of infection for mechanical transmission and infection of ticks.[3,7,8] This restricts the export of cattle from endemic areas such as the United States to non-endemic territories such as Canada. Anaplasmosis is therefore a significant impediment to unrestricted international movement of cattle in North America. Successful measures to control and eradicate anaplasmosis are confounded by the absence of efficacious antimicrobial regimens to eliminate infections[9], inadequate information regarding the usefulness of newer diagnostic tests in determining the success of disease eradication and ineffective vaccines to protect against new infections.

Diagnosis of Anaplasmosis

Our research group conducted a study to compare the sensitivity of the complement fixation (CF) and a new competitive enzyme-linked immunosorbent assay (cELISA) tests for detection of *A. marginale* in experimentally infected steers.[10] Forty Angus X Simmental steers were experimentally infected with 2.6 x 10^9 *A. marginale* infected erythrocytes. Percent parasitized erythrocytes (PPE) were determined by microscopic examination and sera were tested by CF and cELISA using USDA approved methods from blood collected at 9, 13, 20, 28, 34, 41, 61, 96, 126 and 156 days post infection (DPI). At 9 DPI, sensitivity of the cELISA test was 47.5% whereas the CF test failed to identify positive animals. After 13 DPI, sensitivity of the cELISA and CF test were 100% and 20%, respectively. During peak parasitemia (20 DPI), each test had a sensitivity of 100%. Thereafter, sensitivity of the CF test fluctuated between 7.5% and 37.5% while the cELISA test remained at 100%. The overall sensitivity of the cELISA and CF tests was 94.8% and 26.5%, respectively with a kappa statistic of 0.039. These results indicate that the cELISA has superior sensitivity for the serological detection of *A. marginale.* It is however significant that both tests demonstrated a high percentage of false negatives during the prepatent period. For the purpose of identifying anaplasmosis carrier cattle, this new commercially available cELISA test is reported to have a sensitivity of 96% and specificity of 95%.

Microscopic examination of stained blood films is commonly used to detect *A. marginale* organisms in erythrocytes of infected animals. However, this diagnostic technique may be unreliable when cattle have minimal infections or in advanced cases of the disease when animals are severely anemic. In the study described previously, we observed that the cELISA accurately identified all infected cattle before the number of *A. marginale*–infected erythrocytes exceeded a PPE of 1%. This suggests that the cELISA may be more sensitive than examination of stained blood films for identifying early clinical cases. Furthermore, in instances in which the PPE is low, intraerythrocytic inclusions of *A. marginale* may easily be confused with Howell-Jolly bodies, basophilic stippling of reticulocytes, and stain contamination. This suggests that the cELISA may be a useful alternative to examination of stained blood films for the diagnosis of anaplasmosis, especially in situation in which experience of clinicians or the available facilities are inadequate for interpretation of blood films. Having said this, veterinarians should exercise caution before making a definitive diagnosis of acute anaplasmosis solely on the basis of a positive result for the cELISA and the presence of clinical signs such as fever, anemia, and icterus. *A. marginale* carrier cattle are cELISA positive but not rickettsemic and therefore do not develop anemia and icterus associated with erythrophagocytosis of parasitized erythrocytes. Differential diagnoses that should be ruled out based on these clinical signs include acute anthrax, leptospirosis, bacillary hemoglobinuria, oak poisoning, poisoning caused by ingestion of *Brassica* species and multicentric lymphosarcoma. In these circumstances, it would be advisable to detect intraerythrocytic inclusions of *A. marginale* by examination of blood films to assist in differentiating between anaplasmosis and these other diseases.

Molecular biological tests appear to be the future of definitive anaplasmosis identification and control strategies in very early stages of infection. Currently, polymerase chain reaction (PCR) is an area that is receiving the attention and focus of research efforts at K-State.[11] PCR utilizes biochemical and molecular biological processes to amplify the genetic material of an organism. DNA-based PCR for identification of *A. marginale* is presently being used based on previous publications.[12] Present research efforts at K-State are focused on developing a highly sensitive and specific diplex, RNA-based PCR diagnostic tool for identification of both *A. marginale* and *A. phagocytophilum* infections. The enhanced sensitivity of RNA-based versus DNA-based PCR is derived from the typical ratio of RNA: DNA molecules per organism being on the magnitude of 100:1. Torioni De Echaide (1998) and others report a sensitivity of

30 infected erythrocytes per milliliter of blood for the DNA-based PCR. This translates to 30 molecules of DNA and 3,000 molecules of RNA. Preliminary results for the RNA-based PCR test are projected to detect an infection with even fewer infected erythrocytes per milliliter of blood. Also, the RNA target within each respective organism is highly conserved and specific among isolates and provides for accurate and precise identification of infective organisms. RNA-based test results will provide a positive or negative diagnosis as well as an estimate of the number of infective organisms in the sample. The currently available DNA-based test result only yields a positive or negative test result.

Significance of iatrogenic anaplasmosis transmission

The significance of iatrogenic transmission has recently been demonstrated. This study compared iatrogenic transmission of *A. marginale* during simulated vaccination between needle and needle-free injection techniques and diagnostic method performance of light microscopy, cELISA, and an *A. marginale*-specific RT-PCR assay. 26 Holstein steers confirmed negative for anaplasmosis by cELISA and RT-PCR were infected with a Virginia isolate of *A. marginale* propagated to a circulating parasitemia of 2.0% in a splenectomized steer (SPS). A simulated vaccination of the infected steer was conducted by IM injection using a hypodermic needle fitted to a multi-dose syringe. The same needle and syringe was utilized to sham "vaccinate" a naïve steer. This two-step procedure was repeated until ten naïve steers (ND) were injected. Similarly, the right neck muscles of the SPS were injected by a needle-free injection system for a separate group of ten naïve calves (NF). Five calves remained non-injected, sentinel steers (CONT). Disinfectants were not used during the procedure. Disease status was monitored semi-weekly during a 61 day study by light microscopy, cELISA, and RT-PCR. Iatrogenic transmission occurred in 60% of steers in the ND group. No change in disease status occurred in the NF or CONT groups. Light microscopy, cELISA, and RT-PCR demonstrated 100% sensitivity on days 41, 41 and 20 post-vaccination, respectively; however, only cELISA and RT-PCR sustained 100% sensitivity thereafter. Needle-free injection was shown to be superior to needle injection for controlling iatrogenic transmission of *A. marginale*. The sensitivity of cELISA and RT-PCR were similar following the acute phase of infection.[13]

Treatment of Persistent Anaplasmosis infections

Chlortetracycline (CTC) and oxytetracycline (OTC) are the only compounds approved for use against acute anaplasmosis in the United States. In regard to the oral administration of oxytetracycline or chlortetracycline, there are currently no compounds approved for the <u>elimination of the carrier state</u> in the USA. Current label claims for chlortetracycline (Aureomycin 90, Alpharma) are as follows: Beef Cattle (over 700 lb): <u>Control of active infection</u> of anaplasmosis caused by *Anaplasma marginale* susceptible to chlortetracycline (0.5 mg/lb Chlortetracycline body wt/day); Beef and Non-Lactating Dairy Cattle (over 700 lb): <u>Control of active infection</u> of anaplasmosis caused by *Anaplasma marginale* susceptible to chlortetracycline when delivered in a free-choice feed (0.5 to 2.0 mg/lb Chlortetracycline body wt/day). Free-choice feed must be manufactured under a feed mill license utilizing an FDA approved formulation. Published studies that claim to have achieved successful clearance of carrier infections used the following variations of labeled dose regimens[14]: Chlortetracycline 2.2mg/kg (1 mg/lb) PO daily for 41 days; Chlortetracycline 1.1 mg/kg (0.5 mg/lb) PO for 120 days.

Chemosterilization has been reported in cattle fed chlortetracycline hydrochloride (CTC) at dosages ranging from 1.1 mg/kg orally for 120 days to 11 mg/kg orally for 30 to 60 days. The relationship between plasma CTC drug concentration and carrier clearance has not been described until recently.[15] In a study conducted by our research group, chronic carrier status was established in 21 steers with a Virginia isolate of *Anaplasma marginale* and confirmed by cELISA and the previously described *A. marginale* -specific RT-PCR. Four naïve, splenectomized steers served as active disease transmission sentinels. Steers were randomized to receive either 4.4 mg/kg/day (LD); 11 mg/kg/day (MD); or 22 mg/kg/day (HD) of oral chlortetracycline; or placebo (CONTROL) for 80 days. The LD, MD and HD treatment groups consisted of 5 infected steers and 1 splenectomized steer; CONTROL group had six infected steers and 1 splenectomized steer. The daily treatments and ration were divided equally and fed twice daily. Blood samples were collected semi-weekly for determining plasma drug concentration by ultrahigh performance liquid chromatography-mass spectrometry/mass spectrometry method and assessment of disease status by both cELISA and RT-PCR. Mean (CV%) chlortetracycline plasma drug concentrations in the LD, MD, and HD groups were 85.3 (28%), 214.5 (32%) and 518.9 (40%) ng/mL from day 4 to 53 of treatment. A negative RT-PCR assay result was confirmed in all CTC-treated groups within 49 days of treatment; however, cELISA required an additional 49 to 88 days before similar results. Subinoculation of splenectomized steers confirmed chemosterilization. These results demonstrate that CTC may be used to eliminate persistent *A. marginale* infections but cattle are susceptible to reinfection with anaplasmosis after clearance. These data are important for influencing future chemosterilization strategies and impacting free trade policy among countries and regions of contrasting endemicity.

REFERENCES

[1]Uilenberg, G., **Vet Parasit;** 1995; 57:19–41. [2]Dumler, JS, et al. **Int J Syst Evol Microbiol** 2001;51:2145-2165.[3] Kocan, KM., et al. **Clin Microbiol Rev** 2003:698-712.[4]Alderink, FJ. Et al. In Proceedings. *86th Annual Meeting of the United States Animal Health Association* 1982; 66-75. [5]Potgieter, FT. et al. **Onderstepoort J Vet Res** 1987; 54:557-560. [6] Goodger WJ et al. **J Am Vet Med Assoc;** 1979 174(2): 1333- 1336. [7]Eriks IS.et al. **J Clin Microbiol** 1993 31(8): 2091-2096.[8]Palmer, GH et al. **Microbes Infect;** 2000. 2:167-176.[9]. Coetzee JF et al.**Vet Parasitol**; 2005; 127:61-73.[10] Coetzee JF et al.,.**Am J Vet Res**. 2007; 68 (8). 872 – 878.[11] Sirigireddy, KR. et al. **J Mol Diag** 2005; 7(2):308-316.[12] Torioni De Echaide S et al., **J Clin Microbiol** 1998; 36(3):777-782. [13] Reinbold, JB et al. **Am J Vet Res** 2010; 71(10):1178-88 [14]Kuttler, KL. **J Am Vet Med Assoc**; 1980; 176(10):1103-1108.[15] Reinbold, JB et al. **Vet Micro** 2010. 145(1-2): 69-75

ANTHELMINTIC RESISTANCE IN SMALL RUMINANTS

Dr. Thomas M. Craig
College Station, TX

Nematode parasites are economically important in intensive grazing systems. Therefore livestock producers rely heavily upon anthelmintics to eliminate disease, and increase profitability. Approximately 50 years ago a revolution occurred in the development of anthelmintics that had a much broader spectrum of activity than anything previously used and the safety margin was sufficient that toxicity was unlikely. With the development of these safe effective products the approach to the control of parasitic disease shifted from one of treating "wormy" livestock to approaches which resulted in animals that ate more, grew faster, produced more milk or fiber and were able to exploit pastures, which formerly had been too dangerous to graze. The first modern anthelmintic Thiabendazole was introduced in in 1961 but almost immediately resistance by *Haemonchus contortus* was reported in North America (Drudge et al 1964) and rapidly spread to other drug classes of drugs such as levamisole / morantel and ivermectin / moxidectin increased to where anthelmintic resistance by gastrointestinal nematodes in small ruminants is the rule not the exception in North America (Kaplan 2004, Fleming et al 2006).

When parasite levels are reduced by use of anthelmintics if the drugs used are not 100% effective, then selection for resistance will inevitably occur as the survivors have characteristics which enable them survive. Resistance to each class of anthelmintic does not appear to be under the control of a single gene (Prichard 2001). The rapid onset and the widespread occurrence of resistance support the contention that selection by treatment not mutation is the driving force in the establishment of resistance. Factors in addition to the genetics of the parasite contribute to the high levels of resistance in goats and sheep. Under dosing by underestimating weight or by selecting a dose for the average size of animal in a flock, so the worms will be exposed to a less than lethal level of anthelmintic especially in larger animals. A form of under dosing is using the same dose to all host species; for instance goats metabolize anthelmintics efficiently so the exposure of ivermectin by worms in goats is 4 to 5 times lower in goats than cattle (Gonzalez et al 2006) Another practice that resulted in under dosage is the subcutaneous use of cattle dewormers in small ruminants, initially the blood levels are high then drop off. This gives the long residual levels, which kills incoming worms from a week to a month or more following its use, as the levels drop a few days to a week after treatment, the drug removes only those worms without alleles for resistance (Lespine et al 2005).

The molecular mechanisms of anthelmintic resistance vary with different worm populations. However, resistant worms accompany animals being introduced to different farms. Mechanisms associated with resistance; include gene variations in Beta-tubulin with benzimidazoles, and glutamate-gated chloride channels with ivermectin resistant worms. Resistance seen when multifamily anthelmintic resistance has been identified is associated with changes in drug transport such as P-glycoprotein efflux of lipophilic substances across cell membranes. This efflux is a mechanism used by worms to exclude noxious substances from their cells, similar to that utilized by cancer cells to inactivate chemicals used to control their growth. (James el al 2009) However, evaluation of several ivermectin resistant *Haemonchus* populations in Texas was unable to show that P-glycoprotein was a protective mechanism.

Biochemical analysis expressed by susceptible and resistant strains has led to confusion in the ability to consistently identify resistant populations. The confusion in genetic studies is likely due to the tremendous diversity seen in *Haemonchus* populations (Otsen et al. 2001, Prichard 2001). Tolerant or resistant worms accompany livestock being introduced into new areas. The worms accompanying their hosts are likely to be resistant and so clinical disease happens more rapidly than where there are still susceptible worms.

Determination of anthelmintic resistance is often made after a drug fails and animals die or develop severe clinical signs. A clinical response from an anthelmintic will occur with only 40 to 50% efficacy but this leads to massive pasture buildup of resistant worms and numbers win. The fecal egg count reduction test (FECRT) is the most common method of evaluating a herd or flock. When resistance is diagnosed the worms remain resistant at varying levels over the years (Lyons et al 1993). Resistant worms are purchased inside a host that has been treated with an anthelmintic that removed the competition. Maintaining newly purchased animals in a barn or dry lot until they have been evaluated two weeks after treatment with an effective combination of anthelmintics is an essential part of a herd health management (Fleming et al 2006).

When a premise has been identified where multiple anthelmintic resistances occurs combinations of anthelmintics from different families may be the only choice. Even though each individual anthelmintic is ineffective, combining drugs at full therapeutic levels may be clinically effective. Various combinations have been used and it is essential that full therapeutic dosages of each anthelmintic are used and they are used concurrently. Currently this approach is being used commercially in Australia where the product is a combination of albendazole, ivermectin and levamisole. Of course this approach may only have a limited life span of efficacy as worms with resistance to each compound are present and will mate to produce multi resistant worms.

Targeted deworming, that is treating only the animals at risk (Hoste et al 2002) or individuals with signs of disease (pale mucous membranes, high egg counts, hypoproteinemia), may lower the chances of selecting resistant parasite because the worms in the non-treated animals will survive whether possessing resistant genera or not. This approach is labor intensive and may require frequent observations. An approach, where ocular mucous membrane color is evaluated with a color chart FAMACHA to determine which individual animals are anemic. This methodology is effective where *Haemonchus* is the predominant helminth. The idea is: 1) Identify the individual animals with the largest numbers of worms and selectively treat them. This removes a major source of pasture contamination but puts no selective pressure on the worms in the remainder of the flock, the refugia worms. 2) Evaluate the flock to see how things are going before clinical signs are seen. Animals with varying levels of anemia will be identified and a flock score can be kept. Even if the producer decides to treat the entire flock the treatment will only be administered when needed. 3) The ability to improve the genetic resistance of flock. Resistance to infection with *H. contortus* is heritable. Therefore removing the most susceptible animals each year will have the long-term effect of improving the genetic resistance and/or resilience of the flock to *H. contortus*. The greatest problem with the FAMACHA approach in the southern United States is the lack of susceptible worms (in refugia) on most pastures.

Selection of individual animals with resistance to the parasite or those that have the capacity to rebound from the effects of parasitism seems to be a logical approach. The offspring of resistant hosts are likely to be resistant. Therefore resistance to *Haemonchus* or other locally important parasites should be a selection criterion used in flocks (Adams 1984, Notter et al 2003). Elevated numbers of CD4+ T cells, increases in IgG1 and IgA levels as well as the numbers of IgE bearing cells are associated with the ability of the host to respond to helminth infection (Pettit et al 2005). The genetic aspect of resistance to gastrointestinal parasites is not well understood. It appears that multiple genes are involved in the protective response and that different genes may be more important at various times in the course of infection and to infection by different species of worms (Stear et al 2001). To some extent some individual animals in all breeds possess resistance factors that make them more likely to survive exposure levels that severely affect other animals in the same flock. Even in resistant breeds all animals are not equal and factors such as lactation, age, or nutritional status may influence the course of disease. When animals are producing at their genetic maximum they are at greatest risk of parasitic disease because they are immunologically and or nutritionally challenged (Bricarello et al 2005). Desirable phenotypic characteristics such as production of milk, fiber, multiple births, and resistance to other agents may be at odds with resistance to helminths. However, research does not indicate that the traits that lead to increased production are linked to those of increased susceptibility or resistance.

Although anthelmintics are important in eliminating helminths from the host, good quality nutritious feed is equally important to the health and wellbeing of parasitized animals. None of the anthelmintics will increase the weight gain of an animal unless adequate rations are available. Providing sufficient dietary protein is vital during the periparturient period and during rapid growth so that animals will be able to better tolerate the worm burden as well as increase resistance to infection (Coop and Kyriazakis 2001). However the increased protein in legumes may or may not affect the establishment of gastrointestinal nematodes. When plants high in condensed tannins are grazed incoming larvae are adversely affected as well as providing bypass protein for the host and enhanced immune response (Min et al 2005). A nematode trapping fungus *Duddingtonia flagrans* has been utilized to lessen pasture survivability of larval nematodes. Non-organic chemicals, to control parasites, such as the use of copper oxide wire particles have been shown to protect ewes from *Haemonchus* infections but can only be used with care in sheep, as they are extremely susceptible to chronic copper poisoning. The concurrent use of copper oxide wire and *D. flagrans* if administered early in the grazing season may have prolonged effect in lowering pasture contamination and increased the safety margin (Burke et al 2005). Evading larvae in pastures by removal of livestock to render a pasture parasite safe may be as short as one month in a warm wet summer but may take as long as eight months in a cool humid climate. Management systems that lower the exposure of hosts to parasites by long term vacating of pastures may be counterproductive as the forage from the vacated pasture is overly mature with low palatability and protein content. On the other hand, high-density, short-duration rotation systems may increase the level of parasite exposure but optimize the value of forages resulting in healthier livestock because of a better nutritional plane. In humid tropical climates rapid rotation where pastures are vacated for as few as 30 days lower infective larval numbers significantly. Having livestock graze pastures that have been used for cropping, is an effective measure for reducing larval nematode challenge. Alternate or co-grazing by hosts of a different class or species may lower exposure when the target animals return to the pasture. In general *Haemonchus* of sheep and goats do not do well in cattle and vice versa. However, some populations of *H. contortus* may thrive in calves but not in adult cattle.

Because of the wide variation among the species, breeds, geographic locality, nutritional status, climate, age, sex and use of the animals that may be suffering from disease or freedom thereof and the variability in the populations of parasitic worms encountered there is no one approach that works universally. A working knowledge of the environment, host management and parasites is required. Perhaps the use of sophisticated computer generated programs will enable the livestock producer to select ways of avoiding anthelmintic resistance and have disease free profitable livestock (vanWyk, Reynecke 2011).

Anthelmintic resistance may become a problem in cattle. Two genera of nematodes, *Cooperia* and *Nematodirus* were tolerant of ivermectin from the beginning and selection for resistance by the use of only avermectins has enhanced resistance. Recently evidence of *Haemonchus* resistant to avermectins and benzimidazoles in calves has been documented (Gasbarre et al 2009). The worms probably did not originate from small ruminants and is cause for concern. All three genera are much more likely to be found in calves and therefore may become a management problem in replacement or stocker calf operations but are unlikely to be a problem in pastured adult cattle.

REFERENCES

Adams **International J Parasitol** 1994; 18:1071 Bricarello et al **Vet. Parasitol.** 2005; 134: 99 Burke et al **Vet Parasitol.** 2005; 134: 141 Coop, Kyriazakis **Trends Parasitol**. 2001; 325 Drudge et al **Am. J. Vet. Res**. 1964; 25: 1513 Fleming et al. **J. Vet. Intern. Med**. 2006; 20: 435 Gasbarre LC et al. **Vet Parsasitol** 2009; 166:275. Gonzalez et al **Am. J. Vet. Res**. 2006; 67: 323 Hoste et al **Vet. Parasitol.** 2002; 106: 345 James et al **Trends Parasitol** 2009; 25:328 Kaplan **Trends Parasitol.** , 2004; 20: 477 Lespine et al **Vet. Parasitol.** 2005; 128: 25 Lyons et al. **Am. J. Vet. Res**.1993; 54: 406 Min et al **Vet. Parasitol.** 2005; 130: 105 Notter et al **Small Rum. Res.** 2003; 47: 221 Otsen et al. **Int. J. Parasitol.** 2001; 31: 1138 Pettit et al. **Vet. Parasitol** 2005; 134: 131 Prichard **Trends Parasitol.** 2001; 17: 445 Stear et al. **Res Vet Sci** 2001; 71, 1 van Wyk, Reynecke **Vet. Parasitol.** 2011; 177: 212

CONTROL OF HELMINTH PARASITES IN COW CALF OPERATIONS

Dr. Thomas M. Craig
College Station, TX

To control parasitic disease in cattle one has to first determine which parasite(s) are present and if they are or could cause harm. Economic parasitism not clinical disease has become the gauge for parasite control programs. A primary economic effect of anthelmintics is increased milk production in grazing cattle and the increased growth rate of calves. Not only do cows produce more milk there is evidence they breed back more quickly after calving. The class of cattle, geography, time of year, weather conditions and management may help determine if the numbers of parasites are sufficient to cause clinical or economic disease. Each producer is likely to have a different array of parasites that infect his/her cattle and this will determine if the cost of parasitic infection will reach the economic threshold. Determining which cattle would benefit from treatment and when to do should be considered. The danger of selecting anthelmintic resistant worms may be important in some situations. What are the dangers and how may it be avoided?

Due to the widespread use of of broad spectrum, safe, effective drugs there has been a shift in the species of parasites a cattle producer is likely to encounter. Some parasites such as *Dictyocaulus viviparus*, the cattle lung worm, has nearly disappeared, others were, and still are, of little significance but remain. The remainders are as important now as they were then and, in some instances, much more important because of animals' failure to develop resistance due to low levels of exposure or increased exposure due to higher stocking density.

In general, parasitic nematodes are the most important profit-limiting group of parasites in the cattle industry at this time. Overall, the abomasal parasite *Ostertagia ostertagi* is the most important parasite of cattle in North America. Suggestions associated with control programs are directed toward this parasite. In the southern United States this parasite is transmitted to cattle during the cool season, October to May and in the north from green up in the spring to freeze up in the autumn. *Ostertagia* important as it is one of the few parasites that do well in older cattle. Cattle do become resistant to *Ostertagia* after being exposed over several years. However, if conditions are not conducive to the transmission of the parasite older cattle may react to the parasite similar to calves and cattle with Braham ancestry are less likely to develop resistance to *Ostertagia* (Suarez et al 1995).

The clinical disease caused by *Ostertagia* is associated with the emergence of the larval stages from deep in the crypts lining the abomasum. The emergence is associated with maturation and occurs 10 to 14 days after they were acquired by grazing, unless larvae go into hypobiosis. The abomasal glands are either destroyed or put in fast forward and do not convert the enzymes necessary for digestion. The animal becomes anorectic and in addition there is leakage of protein into the lumen of the gastrointestinal tract and a failure to absorb nutrients in the intestine. When large numbers of larvae emerge concurrently, as happens following hypobiosis, diarrhea is seen (Simpson 2000).

Because of a limited time during which successful transmission occurs on pasture many nematodes undergo hypobiosis or arrested development. Hypobiosis is a cessation of metabolic activity within the host where by the parasite survives by evading unfavorable conditions in the host or environment. If worms mature and reproduce at an unfavorable time the eggs or larvae encounter conditions too hot, dry, or cold for their successful gaining of a host. In North America *Ostertagia* over winters as hypobiotic larvae in the north and summers in the south and because they are inactive are not recognized by the host immune system. However, when the worms resume development they may successfully reproduce.

Host resistance to parasites is controlled by the immune system and requires exposure to the parasite in order to stimulate a protective response. This resistance is relative as animals become resistant to disease but not free of parasites. Within populations we see an extreme variability in parasite numbers even in resistant hosts. Approximately 20% of the herd will have 80% of the parasites some of which may be accounted for by factors such as stress, or lactation. For instance, first calf heifers are still trying to grow, feed a calf and compete with older and wiser cows for quality nutrients and tend to have a much higher parasite load than the remainder of the herd.

Early infection by gastrointestinal nematodes by single suckled beef calves is usually limited and the provision of high quality feed (milk) assures that the calf will probably not show signs of disease. Exposure to parasites is much less than when calves are run as a group and with adequate nutrition they are able to compensate for numbers of parasites that may otherwise cause disease. However, if replacement calves are not sufficiently exposed to parasites they may fail to gain sufficient immunity to give them a level of protection required later. Immunity to some species of parasitic nematodes is established during the first grazing year other take longer.

A common calf small intestinal nematode *Cooperia* spp. may be transmitted in the cool season along with *Ostertagia* or in the warm season. *Cooperia punctata,* a warm season parasite, may cause excessive inflammation of the small intestine, apparently due to the attempts of the body to rid itself of the parasite (Rodrigues et al 2004). This is also the major species implicated with anthelmintic resistance in cattle. The genus can cause diarrhea if it is present in large numbers and can be a problem in dairy or light weight stocker calves. As calves get older *Cooperia* stimulates a strong protective response (Kloosterman et al. 1991).

Haemonchus in young cattle can be a primary pathogen most problems have been associated with stocker or dairy calves on high density summer pastures. Suckling calves are infected but the resistance to the parasite by adult cattle is such that they act as a biological vacuum sweeper removing most of the larvae from the pasture so that the calves are not seriously affected. However calves grazing pastures with forages such as Coastal bermudagrass may encounter sufficient worms that they die of exsanguination. *Haemonchus* is a voracious blood sucker and calves will be sluggish and may develop bottle jaw but do not lose weight or are diarrheic because of this parasite. The losses caused by *Haemonchus* are clinical not just economic.

The large intestinal parasite *Oesophagostomum radiatum* is the only gastrointestinal nematode that can be transmitted by skin penetration as well as by grazing. The larval stages of the parasite stimulate the formation of nodules in the intestinal wall which are a cause for condemnation of intestines at slaughter. The adult worms are also blood suckers but pale in comparison with *Haemonchus*. The eggs of *Cooperia, Haemonchus, Ostertagia, Oesophagostomum,* and others look the same. So diagnosis of the worm species based on egg morphology may miss the correct diagnosis.

Strategic use of anthelmintics is important to clear the host of hypobiotic larvae before the worms do significant damage and lessen pasture contamination to prevent disease later. The seasonality of arrested development varies geographically. *Ostertagia* undergoes summer arrest in the south and thus becomes the strategic period. Tactical use of anthelmintics is the application of the drugs when worm levels are increasing in numbers or likely to do so. Such as two weeks following rains after a period of drought or a few weeks after green up of a pasture in the spring. Treatment based on egg counts, serum protein, hematocrit or pepsinogen levels can also be considered.

The classes of anthelmintics currently on the market in North America effective against cattle gastrointestinal nematodes are: Benzimidazoles (Albendazole, Fenbendazole, and Oxfendazole), are safe and effective against most adult nematodes and some arrested larvae, but have no residual effect. The Imidazothiazoles (Levamisole), and Tetrahydropyrimidines (Morantel), act on receptors on nematode muscle cells and cause spastic paralysis. Adequate doses are important because at low doses some worms recover and are able to reestablish. Overdoses may cause adverse reactions such as staggering, slobbering and increased urination. They have no effect on arrested larval worms nor is there any residual effect. The Macrocyclic lactones (Doramectin, Eprinomectin, Ivermectin and Moxidectin) vary in solubility and metabolism within cattle and the effects depend on the species and developmental stage of nematodes. Macrocyclic lactones cause flaccid paralysis, pharyngeal muscles are especially vulnerable to these compounds and the worm is unable to feed. The macrolides have residual effects against newly acquired larvae that are dispatched as they enter the host. The specific drug, method of administration, and species of worm determine the duration of residual activity. At first glance a product with residual activity has considerable advantages as cattle can remain in highly infested pastures without danger of disease. However, as the efficacy tails off it can be a powerful selection mechanism for anthelmintic resistance. Doramectin and moxidectin have longer residual activity than ivermectin. Moxidectin has the longest duration with 99% efficacy against *Ostertagia* at 35 days post injection.

Liver flukes in cattle are either present or absent from a premises. Maps showing their distribution are only an approximation, not the microhabitat of the snail intermediate host, which are low laying areas with clay soils that are periodically flooded, or seeps that keep the soil moist. The range of the intermediate host snails appears to be expanding. Seasonal transmission varies with geographic region and available moisture. Only a few cows infected with flukes will directly benefit from treatment. Treating adult cattle for *Fasciola hepatica* will have only minimal effect in their production. However, treating these cattle may lower the infection rate in the snail population sufficiently so that cattle on the pasture the next year will acquire levels of parasitism below the threshold of disease. Environmental manipulation of snail habitat may be feasible in some circumstances but is unlikely to have long-term effects. The formulations of drugs available against liver flukes in the U.S., Clorsulon and Albendazole, are only effective against adult *Fasciola* in the bile ducts. By the time the drugs are used the damage caused during fluke migration has already occurred but treating cattle before the snail activity season may lower transmission the following season. Treating cattle in the autumn appears to be the best approach in the south and in the early spring in the north.

The giant or deer fluke *Fascioloides magna* is present in the gulf coast, near the Great lakes and locally in other areas suitable for the intermediate host snails. This fluke is unable to complete its life cycle in cattle. A thick walled cyst, usually containing two or more flukes, is formed in the liver with no opening for eggs to escape. The developing flukes migrate extensively through the liver and occasionally in other organs destroying tissues as they migrate. A black pigment associated with the migration is pathognomonic for *F. magna*. Treatment with Albendazole at an elevated dose may kill the flukes but the damage has been done and treatment will not affect the pasture exposure which comes from white-tailed deer or wapiti infected with adult flukes and the eggs are passed in their feces. However, killing the adult flukes may improve the general health of a cow sufficiently that a producer may get an extra years production even though the liver and possibly other organs are condemned at slaughter.

From the time of anthelmintic development there have been a few worms out there that decided that they did not believe the label and were tolerant or resistant to the drug's effect. If the drug was used on that population of worms, after a while they were the only worms present. Because cattle ranchers do not tend to use anthelmintics as often as small ruminant or horse owners, the advent of detectable numbers of resistant worms took longer to become apparent in cattle. Is anthelmintic resistance a problem in cattle? It can be! Not every where all the time. *Cooperia* and *Nematodirus* were tolerant of the macrocyclic lactones from the get go. They are still resistant to this class of drugs but both genera stimulate an early robust resistance to disease. When light weight stocker or replacement calves are grazed on heavily contaminated pastures the use of a benzimidazole or combined with a macrocyclic lactone is warranted. A major concern is populations of *Haemonchus* that are resistant to both benzimidazoles and macrocyclic lactones just as in small ruminants (Gasbarre et al 2009). Did the resistant worms come from small ruminants? Probably not but? The resistant worms are found in the highest prevalence in the gulf coast and have been transported elsewhere. Properties where calves are back grounded or permanent summer pasture are used by one group of naïve calves after another are the major source of the resistant worms at this time. Resistant *Haemonchus* in cow calf operations is not common and if it were the cows cleaning up the pasture is likely to keep the level below that of any significance.

The bottom line is to treat the cow herd with a drug effective against arrested *Ostertagia* in the winter in the north or in late spring i.e. May or June in the south. If you want to treat treat only the animals with the highest worm load in the herd treat bulls, first and second calf heifers. If suckling calves are at least 200 lbs. body weight, treating them will usually result in heaver calves at weaning. Lighter calves, if nursing, will not likely be benefited by treatment. Deworm calves at or near weaning again using an injectable or oral product and there is considerable evidence that using both concurrently may improve parasite control. Treat for flukes in the early autumn (south) early spring (north) to protect snails. Evaluate the effectiveness of anthelmintics in replacement cattle on the farm every several years. The internet will not tell you which dewormer is best for a ranch, the cattle will.

REFERENCES

Gasbarre LC et al. **Vet Parsitol** 2009; 166:275. Kloosterman A et al. **Vet Parasitol** 1991; 39:101. Rodrigues RR et al. **J Helminthol** 2004 ;78:167. Simpson HV **Vet J** 2000; 160:177. Suarez VH et al. **Vet Parasitol** 1995; 58:263.

145a

EPIDEMIOLOGY OF BLUETONGUE VIRUS INFECTION AMONG CALIFORNIA RUMINANTS CE Mayo, DVM, MS and NJ MacLachlan, BVSc, MS, Ph.D., ACVP. Davis, CA.

Bluetongue virus (BTV) is the prototype virus in the genus *Orbivirus*, family *Reoviridae*, and the causative agent of bluetongue (BT), an economically important and re-emerging arboviral disease of ruminants that is transmitted by various species of *Culicoides* midges (gnats). [1]Thus, the global distribution of BTV infection of livestock and wildlife coincides with that of competent vector insect species.[2] Some twenty-six serotypes of BTV are now recognized globally, four (serotypes 10, 11, 13, 17) of which are endemic in much of the United States (US).[2, 3] Whereas BTV serotype 2 infection has been recognized in Florida (and some adjacent states) for at least 25 years, since 1998 ten additional previously exotic serotypes (serotypes 1,3,5,6,9,12,14,19,22,24) have been isolated in the southeastern US. Coincident with this invasion of novel BTV serotypes into the US, since 1998 eight serotypes of BTV (serotypes 1, 2, 4, 6, 8, 9, 11, 16) have also invaded and spread throughout extensive portions of Europe and even Scandinavia precipitating an economically devastating epidemic.[4] Although BT is typically a disease of sheep, infection with some European strains of BTV (serotype 8 in particular) has caused serious disease in cattle and other livestock, with substantial production impacts and lesions mimicking those of foot and mouth disease.[5] Climate change has been implicated as one cause of this dramatic change in the global distribution and nature of BTV infection of livestock, because of its potential impact on the activity, abundance, and vectorial capacity of populations of *Culicoides* insects resident in affected areas.[6, 7] The ongoing European BT pandemic involves novel vector species of the *C. obsoletus,C. pulicaris* complexes.[4,6] It is uncertain why these Palearctic *Culicoides* sp. have recently become efficient vectors of BTV but because some of these exist throughout the Holarctic region, which includes the US, there is an undefined risk that exotic and highly virulent strains of BTV could spread widely.[8, 9] The BTV transmission parameters are largely unstudied, especially within the US, and without such information it is impossible to accurately predict the potential for previously exotic BTV strains to invade and spread within the US. The economic consequences of such an epidemic would reflect direct losses from clinical disease, reproductive losses, animal movement restrictions, and the cost of vaccination programs.[4] Introduction or emergence of virulent BTV within the US could be catastrophic to the entire livestock industry, including the cattle industry (beef and dairy).[3] While recent focus has been directed towards the incursion of novel serotypes within previously unaffected regions such as Europe, few studies have focused on current epidemiology of BTV infection within endemic regions such as California. Therefore, we performed sentinel animal surveillance, as prescribed by the OIE, to determine 1. The current prevalence and distribution of BTV infection in cattle; 2. The seasonal patterns of abundance and BTV infection rates of *Culicoides* midges on individual dairy farms as determined by different insect trapping methods; and 3. The serotypes of BTV present at each farm and comparison of serotype diversity among *Culicoides* midges and cattle. Results of these studies have now been published.[10-12]

The initial epidemiological study began in January, 2009 with the enrollment of sentinel cattle from 10 commercial dairy farms, located in 5 different regions of California including northwestern California, an area historically BTV free.[13, 14] Thirty adult cows (ranging from 2 to 7 years of age) were enrolled from each farm for a serological survey. Serum collected from individual sentinel cattle was evaluated using a BTV-specific competitive ELISA (cELISA; VMRD Inc., Pullman, WA) for detection of virus-specific antibodies.[11] The second epidemiological investigation was an entomologic survey that focused on determining the abundance of *Culicoides sonorensis (C. sonorensis)* vector species on local dairy farms; *C. sonorensis* is the major vector of BTV in much of the US. A prospective cohort study to sample populations of vector *C. sonorensis* midges was initiated in May, 2010 with the enrollment of 4 dairy farms within the Central Valley.[11] These 4 farms were selected because of their proximity to one another (and presumed similar meteorological parameters) and on the basis of their individual BTV prevalence, as determined in the preceding cattle serological surveillance study that was conducted in 2009.[10] The trapping of *C. sonorensis* at each farm was performed until May, 2011, using 3 trapping methods: CDC suction traps baited with CO_2 and equipped with or without a 4W blacklight (UV) bulb, and mechanical aspiration directly from cows using a modified hand-held household vacuum (DC Insect Vac; BioQuip Products, Rancho Dominguez, CA). The CDC downdraft suction traps were placed within animal housing units, near dairy waste-water lagoons and watering troughs, and within open pasture sites. These traps were deployed 2 hours before sunset and removed 2 hours after sunrise.[11, 15, 16] Five seronegative adult (1-3 years) cows of similar size (as determined by girth measurement) and age were selected on each farm, and used as sentinels for mechanical aspiration of insects and for collection of blood and serum to detect BTV infection. Mechanical aspiration of insects was performed weekly, from 90 minutes before and 90 minutes after both sunset and sunrise, using a modification of previously described methods.[11, 15, 16] The harvested *C. sonorensis* midges were sorted independently based on collection method, recorded as to their relative abundance among other insects, and a subset of female midges on each farm were analyzed for viral RNA using group and serotype-specific quantitative reverse-transcriptase polymerase chain reaction (qRT-PCR) assays. Sera and whole blood collected from the sentinel cattle were analyzed for the presence of antibodies and viral RNA by cELISA and group/serotype specific qRT-PCR assays respectively.[12]

BTV seroprevalence among adult cattle (2-7 years of age) at the 10 study dairy farms ranged from 0 to 90% with lowest seroprevalence in northwestern California, a region historically free of BTV infection.[11] There was a non-uniform distribution of seroprevalence among older cattle throughout the remainder of the state. In particular, cattle at two farms within the Central Valley were all seronegative to BTV whereas two other farms in relatively close proximity had a high seroprevalence.[11] Due to the difference in BTV seroprevalence within a region of similar meteorological conditions, these 4 farms were selected for further analysis in an entomological survey. The data from the entomological survey indicates there was marked and significant variation ($p<0.001$) in the mean numbers of *C. sonorensis* midges that were trapped at each of the 4 study farms in the Central Valley and significantly greater numbers of *C. sonorensis* midges were collected using CO_2 baited suction traps equipped with or without light as compared with animal baited aspirations.[11] Seasonal BTV infection of the sentinel cattle occurred only at farms where substantial numbers of midges were harvested. Total numbers of midges collected in both the light and CO_2 traps peaked during August, whereas numbers of midges collected by animal-baited aspiration did not peak until September.[11] Importantly, infection of sentinel cattle coincided very closely with the first collection of midges by animal-baited aspiration, which was considerably later in the year than when midges were first detected using conventional light and CO_2traps. Furthermore, *C. sonorensis* midges were only rarely or never collected by aspiration from the sentinel cattle on the 2 farms where BTV infection did not occur. Seasonal mean daily maximum and minimum temperatures were similar among the 4 study farms.[11] However, both the number of waste-water lagoons and the total farmable acreage differed substantially between these farms, whereas other farm and herd management practices did not.[11] Further evaluation of midges for BTV infection rates indicated the number of serotypes circulating differed markedly among the individual farms.[12] More serotypes of BTV were present in midges than in sentinel cattle at individual farms where BTV circulated, and the virus was detected at each farm in midges prior to detection in cattle.[12] BTV infection rates were remarkably lower amongst female *C. sonorensis* midges collected by CO_2 traps with UV light than among midges collected by either animal-baited aspirations or in CO_2 traps without light.[12]

Findings from these studies confirm the importance of using sensitive surveillance methods for both midge collection and virus detection in epidemiological studies of BTV infection. Our results confirm those of prior studies showing regional variation in the prevalence of BTV infection of livestock throughout California, with defined areas of exposure.[14, 17-19] Sentinel cattle from coastal northwestern California were consistently negative whereas there was a non-uniform distribution of BTV infection amongst cattle on dairy farms in the Central Valley.[10, 11] The BTV-free status of the northwestern region is probably attributable to the cooler temperatures in this area and its effect on the vectorial capacity of resident insect vector populations. [11, 16, 20] Vector abundance and BTV infection prevalence at individual farms within the Central Valley was mirrored by the prevalence of BTV infection of sentinel cattle at each of the 4 farms studied. While mean maximum and minimum temperatures were similar at all of the 4 dairy farms, the variation in vector abundance most likely is the result of lagoon waste-water and irrigation management practices at individual farms.[11] Dairy wastewater lagoon ponds are present on many dairy farms throughout California and are a major larval habitat as larvae thrive within the muddy embankments of gently sloping and polluted lagoons.[11, 21]Lagoon wastewater provides a cost effective irrigation and nitrogen source for corn grown on dairy farms for silage. Planting of corn typically begins in late April and early May, and the waste lagoons are slowly drained for irrigation providing moist, muddy habitats at the margins for larval *C. sonorensis* populations.[11] Thus, extensive irrigation of crops coupled with the use of lagoon waste-water at individual farms likely promotes the expansion of resident populations of *C. sonorensis* midges and subsequent seasonal transmission of BTV to livestock.[11] BTV infection rates of *C. sonorensis* midges that were detected earlier in the season than sentinel cattle in addition to the plurality of serotypes on individual dairy farms might suggest that the midge vector constitutes a reservoir of genetically divergent BTVs that potentially sustain the virus in seasonally endemic areas.[12] The data does suggest infection prevalence may be misrepresented when UV light traps alone, a method most often used in routine vector surveillance, to collect female *C. sonorensis* midges.[12] In summary, our studies confirm that anthropogenic factors, and not just meteorological ones, can be important in determining the prevalence of BTV infection of livestock on individual farms in California.[10-12]

[1] Spreull J. **J Comp Pathol**. 1905; 18: 321-37. [2] Maclachlan NJ. **Vet Clin North Am Food Anim Pract**. 2010; 26(1): 163-71. [3] MacLachlan NJ, et al. **J Am Vet Med Assoc.** 2006; 228(9): 1346-9. [4] Gloster J, et al. **Vet Rec.** 2008; 162(10): 298-302. [5] Elbers AR, et al. **Prev Vet Med.** 2008; 87(1-2): 31-40. [6] Purse BV, et al. **Nat Rev Microbiol.** 005; 3(2): 171-81. [7] Purse BV, et al. **Med Vet Entomol.** 2006; 20(3): 335-44. [8] Meiswinkel R, et al. **Vet Ital.** 2004; 40: 151-9. [9] Purse BV, et al. **Rev Sci Tech.** 2008; 27(2): 427-42. [10] McKercher DG, et al. **J Am Vet Med Assoc.** 1953; 122(913): 300-1.[11] Metcalf HE, et al. **Am J Vet Res.** 1981; 42(6): 1057-61. [12] Mayo CE, et al. **Vet Microbiol.** 2011, in press. [13] Mayo CE, et al. **Transbound Emerg Dis.** 2010; 57(4): 277-81. [14] Gerry AC, et al. **J Med Entomol.** 2009; 46(3): 615-24. [15] Gerry AC, et al. **J Med Entomol.** 2001; 38(2): 197-209. [16] Mayo CE et al. **Vet Parasitol.** 2012; in press. [17] Osburn BI, et al. **Am J Vet Res.** 1981; 42(5): 884-7. [18] Stott JL, et al. **Prog Clin Biol Res.** 1985; 178: 571-82. [19] Uhaa IJ, et al. **Vet Res Commun.** 1990; 14(2): 99-112. [20] Mullens BA et al. **Vet Ital.** 2004; 40: 160-6. [21] Mullens BA. **Prog Clin Biol Res.** 1985; 178: 207-11

145b

EVALUATION OF CANDIDATE GENES IN SALMONELLA ENTERICA SEROTYPE TYPHIMURIUM REQUIRED FOR SURVIVAL DURING INTESTINAL INFLAMMATION.

JR Elfenbein[1], S Lawhon[1], TR Endicott-Yazdani[2], L Bogomolnaya[2], S Porwollik[3], M McClelland[3], HL Andrews-Polymenis[1,2]. 1. Texas A&M University College of Veterinary Medicine and Biomedical Sciences, College Station, TX; 2. Texas A&M Health Science Center, College Station, TX; 3. Vaccine Research Institute of San Diego , San Diego, CA.

Salmonella enterica serotype Typhimurium (STm) is a leading cause of human food-borne gastroenteritis and bovine enterocolitis and leads to production losses at all stages. A pool of ~1000 single-gene deletion (SGD) mutants was screened by microarray for survival defects during intestinal inflammation. The purpose of this study is to individually evaluate candidate genes in STm required for survival during intestinal inflammation.

Three Angus-cross calves (aged 21-30 days) were anesthetized prior to and for the duration of the experiment and euthanized by overdose of pentobarbital following completion. A right-flank laparotomy was performed in order to ligate 24-26 ileal loops (4-6cm long). Competitive infection experiments were performed in each loop. Infective dose was 10^9 colony-forming units (CFU) composed of equivalent numbers of wild type (WT) STm and SGD mutant. Experiments were performed in both inflammatory (WT) and non-inflammatory (by deletion of type-3 secretion system-1) conditions. At 12hr post-infection, loops were removed and fluid, mucus, and tissue homogenate cultured to enumerate CFU. Competitive index was determined by normalization of input WT:SGD mutant to output. Data were analyzed using Student's t-test. Significance was $p<0.05$.

Individual competitive infection experiments were performed for 4 SGD mutants in at least 2 calves under both conditions. Survival defects were present for the following genes: ΔSTM2303, a member of the polymyxin-resistance operon, in mucus during inflammation and tissue in the absence of inflammation and ΔSTM3602, a transcriptional regulator, in both fluid and mucus during inflammation. These genes are likely important for survival of STm during intestinal infection.

146a

CHARACTERIZATION OF *STREPTOCOCCUS UBERIS* BIOFILM FORMATION AND USE OF A BIOREPORTER TO DETERMINE ITS POTENTIAL IMPACT ON BOVINE MASTITIS.

M. E. Prado[1], O. Kerro-Dego[1], J. McGuire[1], S. Ripp[2], R. Almeida[1]. [1]Department of Animal Science and [2]The Center for Environmental Biotechnology, The University of Tennessee, Knoxville, TN.

Streptococcus uberis is an important environmental pathogen that causes mastitis in dairy cows. Furthermore, *S. uberis* can become established in the bovine mammary gland and cause chronic infections. The ability to cause recurrent infections has been attributed to biofilm formation in several bacterial pathogens. The objective of this study was to compare the ability of different *S. uberis* mastitis isolates to form biofilms *in vitro* and to further characterize biofilm formation by use of a *S. uberis* fluorescent bioreporter in a microfluidic model. *Staphylococcus epidermidis* was used as a positive control. Strains of *S. uberis* (n=38) were screened by qualitative and quantitative assays (microtiter plate [MP] and Congo Red agar [CRA]), and for presence of genes associated with biofilm formation (competence genes *comEA, comEC, comX, agrC, fasA* and the quorum sensing gene *luxS)* and genes responsible for hyaluronic acid capsule synthesis (*hasABC*) by PCR. The bioreporter was constructed by transforming *S. uberis* UT888 with the mCherry gene linked to the *luxS* promoter. Biofilm development was monitored under flow by observing attachment and aggregation of the *S. uberis* bioreporter to microfluidic channels with or without mammary epithelial cells. The ability of *S. uberis* to produce biofilms varied according to the method used. All strains produced biofilms when tested by the CRA method whereas most strains were positive by MP (30 of 38). Strains were further classified as negative (21%), weak (13%) and strong (66%) biofilm formers based on the MP assay. Competence and quorum sensing genes were amplified by PCR in 97% (37/38) of strains evaluated. Capsule formation did not appear to influence *S. uberis'* ability to produce biofilms (32/38 capsule-positive). Use of the bioreporter effectively demonstrated *S. uberis* biofilm development under flow conditions over time. The *S. uberis* bioreporter showed initial attachment (0.5-1 hour) followed by microcolony formation and development (4-6 hours) into mature biofilm (18 hours). Results suggest that *S. uberis* is capable of forming biofilms in vitro and genes associated with biofilms appear to be highly conserved in strains evaluated. In addition, use of a *S. uberis* bioreporter is an effective means for assessing biofilm formation under flow conditions. Future research plans include microfluidic optimization to try to mimic the bovine mammary gland conditions for conducting host-pathogen interaction *in vitro* studies using the bioreporter.

146b

COMPARISON OF ANTIBODY TITERS AGAINST BOVINE RESPIRATORY VIRUSES BETWEEN CALVES FED MATERNAL COLOSTRUM OR COLOSTRUM REPLACER AT BIRTH. M. Chamorro[1], P. Walz[1], D. Haines[2], T. Passler[1], T. Earleywine[3], Y. Zhang[1], K. Riddell[1], P. Galik[1], D. Givens[1]. [1]Auburn University College of Veterinary Medicine, Auburn, AL. [2]Western College of Veterinary Medicine, Saskatoon, Canada. [3]Land O' Lakes Animal Milk Products, Shoreview, MN.

The objective of this study was to compare the total concentration and decay of serum antibodies against BRSV, IBR, BVDV1, and BVDV2 between calves fed maternal colostrum (MC) or colostrum replacer (CR) at birth. Forty-three newborn male Holstein calves were removed from their dams at birth and assigned to colostrum replacer (CR) or maternal colostrum (MR) groups. Group CR calves (n=20) received 2 packets of Land O'Lakes-Bovine IgG® colostrum replacer (100 g of IgG per packet), while MC calves (n=23) received 3.8 L of high quality fresh or frozen maternal colostrum (> 50 g of IgG per liter). Pre-colostral serum samples were collected from each calf for viral antibody testing and BVDV virus isolation. An additional serum sample was collected for detection of viral antibodies at 2 and 7 days of life, and monthly until the calves became seronegative. Virus neutralization was performed for BVDV 1 and BVDV 2 testing, and indirect ELISA was performed for BRSV and IBR testing. Serum IgG concentrations were determined by radial immunodiffusion. Data were compared between treatment groups using the proc t-test function of SAS and repeated measures ANOVA.

Calves in the MC group had higher IgG concentrations at 2 days of life compared with CR calves but high individual variability was observed. Calves in the CR group had higher BRSV, IBR, BVDV1, and BVDV2 specific-antibody concentrations during the first 4 months of life compared with calves in the MC group ($P<0.05$); however high variability in antibody titers was detected in the MC group. With respect to BVDV, the mean time to become seronegative was not significantly different between groups ($P>0.05$).

Colostrum replacers provide adequate passive immunity against bovine viral respiratory pathogens. Compared with MC, colostrum replacers provide a more consistent concentration of specific virus-neutralizing antibodies and may offer a more effective protection against acute viral respiratory disease in young calves; however, more research is needed to evaluate potential application of these results in development of strategic vaccination. Individual variability of serum IgG concentration and viral antibody titers in MC calves could reflect variability of IgG mass in colostrum or differences in apparent efficiency of absorption (AEA).

147a

DETERMINING ENTEROCYTE DESTINY: CONTROL OF EPITHELIAL CELL LOSS IN *CRYPTOSPORIDIUM PARVUM* INFECTION. D. M. Foster, S. H. Stauffer, M. R. Stone, and J. L. Gookin. College of Veterinary Medicine, North Carolina State University, Raleigh, NC.

Cryptosporidium parvum (Cp), a parasite of the intestinal epithelium, leads to an untreatable malabsorptive and secretory diarrhea in neonatal calves. Further, cryptosporidiosis is emerging as one of the most significant causes of morbidity and mortality in children in developing nations. In order to better understand the pathophysiology of *Cp* infection, we used a piglet model that fully recapitulates *Cp* infection to explore a novel hypothesis that the host balances elimination of infected enterocytes with maintenance of barrier function.

Ileal mucosa was harvested from infected and age-matched control neonatal piglets at peak *Cp* infection. Enterocyte shedding and its effect on barrier function were studied ex vivo in Ussing chambers by measurement of transepithelial electrical resistance and mannitol flux. Location and activities of caspase-3, NFκB, and the inhibitor of apoptosis proteins (IAP) were assayed by ELISA, immunoblots, tissue immunoreactivity and the effect of specific pharmacological inhibitors. Enterocyte shedding was quantified by means light microscopy.

Cp infection precipitated widespread activation of apoptosis signaling pathways throughout the epithelium. Yet overt apoptosis was not seen amongst enterocytes residing in the epithelium while exfoliated cells were uniformly apoptotic. Enterocyte shedding was strictly regulated, and appeared to be due to NFκB activation. Inhibition of NFκB precipitated indiscriminant enterocyte shedding from the entire villus, loss of barrier function, and increased caspase-3 activity. These effects mediated by NFκB inhibition were rescued by specific caspase-3 inhibition. As IAP are the only known mechanism of inhibition of cleaved caspase-3, western blotting was used to assess the expression of this protein family. Upregulation of both X-linked inhibitor of apoptosis protein (XIAP) and survivin was demonstrated in the infected tissue. Expression of only XIAP and not survivin was dependent on NFκB. XIAP and not survivin binding to cleaved caspase-3 was demonstrated by co-immunoprecipitation. Specific pharmacologic inhibition of XIAP completely reproduced the effects of NFκB inhibition confirming its role as the mechanism of NFκB mediated control of cell shedding in *Cp* infection.

These studies identify a novel paradigm of epithelial defense in which loss of barrier function is ameliorated by active retention of infected enterocytes by NFκB-mediated expression of XIAP and its subsequent inhibition of cleaved caspase-3, which retains cells on the epithelium until they are physiologically shed at the villus tip. Thus, pharmacologic therapy to hasten clearance of CP may be best accomplished by increasing epithelial migration rate from crypt to villus tip rather than targeting infected epithelial cells for elimination.

147B

EVALUATION OF ENVIRONMENTAL SAMPLES TO PREDICT *MYCOBACTERIUM AVIUM* SUBSP. *PARATUBERCULOSIS* PREVALENCE IN QUÉBEC DAIRY HERDS. E. Doré[1], J. Paré[1], G. Côté[1], S. Buczinski[1], J-H. Fairbrother[2], O. Labrecque[2], J-P. Roy[1], G. Fecteau[1].
1. Faculté de médecine vétérinaire, Université de Montréal, Saint-Hyacinthe, Québec, Canada. 2. Laboratoire d'épidémiosurveillance animale du Québec, Saint-Hyacinthe, Québec, Canada.

Mycobacterium avium subsp. *paratuberculosis* (MAP) causes Johne's disease, a chronic and incurable enteric disease of ruminants. The majority of infected cows shows no apparent clinical signs but might be less productive and can be eliminated from the herd without being diagnosed with MAP. To detect presence of MAP in a herd, culture of environmental samples is a cost efficient diagnostic method.

The objective of this study was to estimate the relation between the prevalence of MAP in the environment estimated by the number of positive culture of environmental samples and the herd prevalence estimated by individual fecal cultures. Presence of MAP in the herd had to be previously confirmed by a positive environmental sample culture, a positive individual cow fecal culture or at histopathology. During a farm visit, an individual fecal culture was done on all cattle two years old and older and 6 environmental samples were taken according to the protocol used by the Québec Johne's disease control program. Briefly, three different sites were sampled twice: 1) a manure concentration area, 2) a manure storage area and 3) another manure concentration area. Linear regression with ArcSin transformation was done to study a possible association.

Thirteen tie-stall dairy herds were included in the study. Number of cows two years old or older in a herd varied from 26 to 140 (median: 67). Prevalence estimated with individual fecal culture varied from 0 to 31.1% (median: 2.3%). Five herds had no MAP detected in the environment and in individual cow fecal culture. Five herds had no MAP detected in the environment but positive individual fecal cultures (prevalence between 1.6 and 9%). The last three herds had respectiveley 2/6, 4/6 and 5/6 positive environmental samples and an estimated prevalence of 5.1%, 7.7% and 31.1%.

A positive significant association (P <0.001) was detected between the prevalence of MAP in the environment and the estimated prevalence by individual fecal cultures.

148a

A COMPREHENSIVE STUDY OF CARBOHYDRATE AND FAT METABOLISM IN OVINES THROUGHOUT GESTATION AND POSTPARTUM.

K Hornig[1], S Byers[1], R Callan[1], T Holt[1], M Field[2], H Han[2]. 1. Colorado State University, College of Veterinary Medicine and Biomedical Sciences, Fort Collins, CO. 2. Colorado State University, College of Agriculture Sciences, Fort Collins, CO.

Sheep are susceptible to metabolic derangements in late gestation, particularly when carrying multiple fetuses. One common metabolic problem is ketosis referred to as pregnancy toxemia (PT). The signs of PT can be subtle and easily missed which can lead to maternal and or fetal death if not treated appropriately. The purpose of this study was to analyze the effects of nutrient restriction on ewes carrying twin pregnancies through the analysis of glucose, beta-hydroxybutyrate (BHB), non-esterified fatty acids (NEFA), and urine ketone concentrations.

Thirty-six ewes were divided into 3 groups for this project: 1 (no feed restriction), 2 (partial feed restriction in second trimester) and 3 (partial feed restriction in second and third trimester). Blood and urine samples were collected weekly from mid gestation until parturition and periodically thereafter until lambs were weaned. Blood samples were tested on point of care BHB and glucose meters and a laboratory chemistry analyzer for BHB, NEFA, and glucose concentrations. Urine ketones were measured on commercial urine test strips.

Results indicated urine ketones, NEFAs, and BHB concentrations increased throughout pregnancy regardless of nutrition status. There were significant differences only for NEFAs and glucose in group 1 versus groups 2 and 3. In the third trimester, NEFAs and BHBs did not vary significantly between groups but glucose decreased significantly in groups 2 and 3. In conclusion, dietary restrictions in the second versus third trimester did not significantly change metabolic parameters in ewes predisposed to PT and both groups had an increased incidence of PT.

148b

THE DIAGNOSTIC CRITERIA USED IN BOVINE BACTERIAL ENDOCARDITIS: A META-ANALYSIS OF 460 PUBLISHED CASES FROM 1973 TO 2011.

S. Buczinski[1], T. Tsuka[2], T. Mohamed[3] 1. Faculté de Médecine Vétérinaire, Université de Montréal, St-Hyacinthe, QC, Canada. 2. Department of Clinical Sciences, Tottori University, Japan. 3. Faculty of Veterinary Medicine, Zagazig University, Egypt.

Bacterial endocarditis (BE) is the most common valvular disease in cattle. The diagnosis of this disease in the living animal remains a challenge for clinicians. The objective of the study was to report evidence-based veterinary medicine data concerning the clinical presentation and results of ancillary tests of necropsy-confirmed cases of bovine BE.

A systematic review and subsequent meta-analysis was performed using Medline and CAB abstracts of every article on bovine BE published in English, Japanese, German and French. The clinical criteria that were specifically assessed for diagnosis of BE were tachycardia, heart murmur, signs of congestive heart failure, presence of fever, evidence of lameness/polyarthritis, one or more positive blood cultures and positive echocardiograms for BE. A total of 34 studies (460 cases of BE) satisfied the inclusion criteria for the systematic review. The sensitivity (Se), specificity (Sp) and 95% confidence interval (CI) were then obtained using a random-effect meta-analysis for studies reporting 5 or more cases. The Se (95% CI) were 86.9% (39.1-98.6%) for positive haemoculture, 84.3% (60.4-95.0%) for echocardiography, 79.7% (70.1-86.8%) for the presence of tachycardia, 60.3% (51.8-68.3%) for the presence of a murmur, 45.7% (32.5-59.5%) for the presence of fever, 43.5% (25.6%-63.3%) for the presence of lameness/polyarthritis and 37.3% (21.6-57.0%) for the presence of clinical signs of heart failure. The Sp (95% CI) were 95.3% (93.3-96.8%) for lameness, 72.6% (45.8-89.2%) for the presence of murmur, 67.0% (55.5-76.7%) for the presence of fever and 27.1% (14.3-45.2%) for the presence of tachycardia. This meta-analysis confirmed that the diagnosis of BE is a difficult process. Echocardiography seems to be a sensitive tool for the diagnosis of BE, despite the absence of any consensus on the ultrasonographic definition of this disease. From the present study it is impossible to know whether or not the presence of multiple positive findings or clinical tests increases sensitivity for the diagnosis of bovine BE.

DIAGNOSING BRD: HAVE WE GOTTEN ANY BETTER OVER THE YEARS?

Joanne Hewson, DVM, PhD, DACVIM (LAIM)
Guelph, ON, Canada

INTRODUCTION

Bovine respiratory disease (BRD) is an important cause of animal suffering, and represents the major reason for preventative and therapeutic use of antibiotics in beef cattle. In spite of high prevalence of BRD during the first month after arrival at a feedlot, it remains an elusive diagnosis. It is difficult to accurately distinguish—particularly in recently weaned and transported beef calves—animals with pneumonia, those with non-respiratory diseases, and healthy animals that are distressed by recent weaning and change in environment. This review explores the effectiveness of our current methods to diagnose BRD, and identifies emerging technologies being investigated to more accurately diagnose respiratory disease in feedlot cattle in a clinically useful and cost effective manner.

CLINICAL DIAGNOSIS OF BOVINE RESPIRATORY DISEASE (BRD)

Review of past literature shows a broad array of criteria used to define clinical cases of BRD. Most commonly, a combination of clinical signs referable to the respiratory system and presence of an elevated rectal temperature has been used to define a clinical case of BRD in the feedlot setting. In reality, feedlot diagnosis of BRD first relies on trained personnel to detect subtle changes in individual animal behavior, feed intake, or clinical appearance, prompting removal of the calf from the home pen for more detailed assessment. An element of instinct is inherent to a skillful distance examination in the home pen, in order to pick up subtle behavioral changes in cattle with early BRD. In a study to assess BRD diagnosis by feedlot personnel[1], the sensitivity of BRD diagnosis by pen checkers was 94% and specificity was 77% in comparison with a diagnosis made by a veterinarian after clinical examination of the animal. In contrast, there was poor correlation between pulling a calf for suspected BRD and the presence of lung lesions at slaughter[2]. As a more consistent and reliable measure of animal health, remotely tracking behavioral changes has been demonstrated through use of accelerometers and pedometers on cattle with respiratory disease[3]. Remote recording of core body temperature using boluses in the rumen allowed for non-invasive monitoring of cattle for fever associated with *Mannheimia haemolytica* infection[4] and naturally-occurring BRD[5]. Near-infrared thermography has recently shown promise to screen radiant heat (as a measure of core body temperature) and watering activity in large numbers of cattle non-invasively, using a camera mounted at the pen water station[6,7]. Unique to this system was the moderate ability to anticipate impending BRD in cattle 4 – 5 days in advance of the onset of clinical signs[6].

Once a suspect animal has been pulled from the home pen, further examination and testing is performed to confirm the diagnosis of BRD. To this end, scoring systems using a combination of respiratory rate and character, rectal temperature, nasal or ocular discharge, appetite and demeanor are considered the most predictive clinical criteria to identify cattle with BRD[6,8,9]. Appetite is routinely assessed by gut fill, as determined by examining the left paralumbar fossa for a gaunt appearance in anorexic cattle. Monitoring of individual animal feed intake in a feedlot setting has been studied to more accurately detect reduced feed intake in the early phase of BRD[2].

Despite use of these clinical parameters, BRD continues to be difficult to detect under field conditions. Treatment for BRD during the feeding period correlates poorly with the incidence of lung lesions diagnosed at the time of slaughter[10-13]. In the study by Wittum *et al.* [10], treatment for respiratory disease either prior to weaning or during the feedlot period was not correlated with average daily weight gain (ADG). However, the presence of lung lesions at slaughter was negatively correlated with ADG; cattle with lung lesions at slaughter gained 0.076 kg less per day than animals with no lung lesions ($P < 0.01$). The lack of correlation between ADG and treatment for respiratory disease can be attributed to the disparity between clinical and pathological diagnoses of respiratory disease, with many subclinical cases of pneumonia going undetected throughout the feedlot period. A significant number of cattle defined as being clinically normal have been shown to have subclinical lung inflammation based on high neutrophil percentages in BAL fluid[14,15]. These studies highlight the clinical difficulty in accurately dividing cattle into "sick" versus "well" groups when making treatment decisions or for comparisons of response to various interventions.

Recently, White and Renter[16] estimated the sensitivity and specificity of diagnosing BRD based on clinical signs or by the presence of lung lesions at the time of slaughter in beef cattle. Since a Gold Standard test to diagnose BRD does not exist, they used Bayesian analysis to estimate the sensitivity and specificity of each test in comparison with "any diagnosis of BRD over the feedlot period". By performing a detailed review of prior literature, they were only able to identify 2 studies[12,13] that provided the appropriate data for this analysis – this reflects how little has been done to validate clinical versus pathological diagnosis of BRD or to define a universally accepted Gold Standard method to diagnose BRD. Like prior studies[10-13], White and Renter identified very poor agreement (Kappa statistic) between clinical diagnosis and the presence of lung lesions at slaughter; while these tests obviously measure different aspects of BRD and there is a substantial lag time between them, they still raise unsettling questions about our ability to accurately diagnose significant lung disease using clinical criteria. Using the sample populations from these two studies, the Bayesian method estimated that diagnosis based on clinical signs (depressed mentation, decreased feed intake, coughing, nasal discharge, +/- fever) yielded a test sensitivity of 62% and a specificity of 63%. In contrast, the presence of lung lesions at slaughter showed both improved sensitivity (77%) and specificity (90%) as a test for BRD. The obvious drawback to this test despite its improved Se/Sp is that it is not a diagnosis in "real time" and therefore cannot assist producers in making treatment decisions. It does suggest that surveillance of lung lesions at slaughter can be used as a measure of overall respiratory health in the herd and effectiveness of treatment protocols in managing BRD-affected cattle. However, retrieval of lungs at slaughter to compare sick versus healthy cattle presents the additional challenges of coordinating sampling from carcasses in commercial slaughter facilities where chain speeds may reach up to 400 head per hour[1].

CLINICOPATHOLOGY TO DETECT EARLY BRD

Various clinicopathologic indices have been assessed in an effort to predict presence or severity of the underlying lung lesions during the acute phase of BRD infection. Total white blood cell count and differential cell count were poor predictors of respiratory tract disease, and were not significantly different between "sick" and "well" animals throughout the week post-arrival at the feedlot[17] nor were they useful predictors of the severity of lung lesions at necropsy when measured in acutely sick calves[3]. Additionally, arterial blood gas analysis and blood lactate did not reliably predict disease severity in challenged calves with mild lung pathology[3]. In cattle with severe respiratory disease, a plasma lactate concentration of 4mmol/L as measured on a portable blood lactate analyser predicted impending (within 24 hours) animal death with a Se of 95% and a Sp of 96%[18].

Measurement of acute phase proteins (APPs)

The kinetics of the acute phase response have been detailed under experimental infection models of either viral[19-21] or bacterial[19,21,22] induction of disease. Various different proteins participate in the acute inflammatory response, with species variation in the predominant proteins involved. In cattle, APPs that have been assessed include haptoglobin, serum amyloid A (SAA), fibrinogen, lipopolysaccharide-binding protein (LBP), and alpha-1 acid glycoprotein (AGP).

In an early experimental study of bronchopneumonia induced by combined viral (BHV-1) and bacterial (*Mannheimia haemolytica*) aerosol exposure, a rise in serum haptoglobin concentration was observed following infection and there was a high correlation between the degree of haptoglobin elevation and likelihood of death following infection[19]. This pivotal study raised great interest in studying APPs as indicators of lung inflammation in bovine respiratory disease. In this study, the rise in haptoglobin concentration was not detected after BHV-1 aerosol exposure, but instead haptoglobin concentration rapidly rose in response to inhalation of *M. haemolytica* 3 days later. Based on this, it was theorized that the acute phase response was poorly elicited by viral infections but rather occurred with more severe tissue damage (as seen with bacterial pneumonia). It has since been shown that the APPs do become elevated in response to both viral and bacterial infections; although the *degree* of elevation is consistent regardless of viral versus bacterial infection[20], the *time* of rising APP after challenge varies with both the APP measured and the pathogen. Following viral infection with BRSV, haptoglobin and SAA were seen to increase starting approximately 5 days after challenge[20]. This correlated with the onset of fever in these calves. Haptoglobin, SAA and fibrinogen also increased at the onset of fever 4 – 8 days after intranasal exposure to BVDV[21]. In contrast, haptoglobin and SAA rose within 24 hours after experimental challenge with *Mannheimia haemolytica*, again correlating with the much earlier onset of fever starting within 6 hours after infection[19,21,22].

Recognizing that each acute phase protein may participate in different stages of the inflammatory cascade or be altered in response to differing stimuli, differences in their kinetics, diagnostic utility and therefore clinical applications are to be expected. In an experimental study of *M. haemolytica*-induced bronchopneumonia, both LBP and haptoglobin started to rise within 12 hours of challenge. Whereas haptoglobin peaked at 48 hours post-infection, LBP showed an earlier peak at 24 hours, suggesting that it is a marker of acute inflammation whereas haptoglobin is affected later in the inflammatory cascade[22].

Following the recognition that positive APPs are increased early (within days) of the onset of infection, measurement of these in cattle presents a unique opportunity to aid in the diagnosis of respiratory disease. Increased serum haptoglobin has been correlated with the presence of fever[23] and clinical diagnosis of BRD[24-26] in calves with naturally-occurring disease. Unlike haptoglobin, there was enough overlap between healthy and sick cattle for fibrinogen[24,25], SAA[24,26] and AGP[24] values to preclude use of these APPs to diagnose BRD under field conditions. Serum amyloid A and LBP were shown to rise more acutely in response to naturally-occurring BRSV infection, whereas the rise in haptoglobin was more gradual[27]. This suggests that SAA and LBP may have utility in assessing the early inflammatory response to viral infection, whereas haptoglobin would be a more robust APP to detect secondary bacterial infections.

Animals need to be sampled early in course of disease, at the onset of clinical signs, in order for haptoglobin and other APPs to have good diagnostic sensitivity and specificity. The sensitivity and specificity of detecting BRD are further improved by testing multiple APPs in parallel rather than use of a single protein (Humblet 2004). Since the rise in APPs corresponds with the onset of fever, APPs cannot be used to predict impending disease during the incubation period. By the time APPs rise, there are usually clinical signs detectable.

APPs may also be used to assess prognosis. Higher haptoglobin concentrations on day 0 were correlated with disease severity, as reflected by the need for more than one treatment for BRD seen more commonly in cattle with high haptoglobin at processing (Carter 2002). In another study where calves diagnosed with BRD were stratified based on severity of clinical disease, haptoglobin and fibrinogen elevations showed moderate sensitivity (71%) and specificity (83%) in predicting which calves required anti-inflammatory (flunixin or flumethasone) treatment in addition to an antimicrobial[25]. Serial monitoring of haptoglobin can also be used to track resolution of disease – untreated cattle continued to have an elevated haptoglobin concentration over time whereas the haptoglobin concentration returned to normal (low or undetectable concentration) in treated animals having clinical resolution of signs[10,24]. The association between elevations in APPs and presence of chronic lung lesions at the time of slaughter warrants further investigation given the poor sensitivity of clinically diagnosing BRD under field conditions; early studies have shown either no correlation[10] or increased prevalence of lesions at slaughter in cattle having had an elevated haptoglobin concentration at some time over the feeder period[28].

A peril in using APPs to detect BRD is that a rise in APPs reflects inflammation that is not specific to the respiratory system. Haptoglobin values were elevated to a similar degree as with BRD-affected calves when measured in dairy calves affected by diarrhea or other disease processes (eg. umbilical infections, oral abscesses)[23]. Thus a clinical assessment of the animal assists in the interpretation of elevated APP concentrations, by ruling out the presence of disorders of other body systems which may alter APP results.

SAMPLING OF THE AIRWAYS FOR PATHOGEN DETECTION OR INFLAMMATORY MARKERS

Sampling of the lower airways by nasotracheal aspirate with a guarded catheter, transtracheal aspirate, tracheobronchial lavage or bronchoalveolar lavage has been described. Beyond diagnosing BRD through cytology, specific pathogens can be detected using a variety of techniques including bacterial and Mycoplasma culture, PCR, ELISA and others. Specific tests to perform on samples vary with the suspected pathogen, and are not within the scope of this manuscript. The recommended sampling site for such tests has traditionally been the lower trachea or lung, although more recently a high correlation between deep nasopharyngeal swabs and lung pathogen culture results has been identified[29,30], allowing for less invasive sampling to be done. Detection of high numbers of bacteria or the presence of viral pathogens from these samples is consistent with BRD when obtained from clinically ill cattle. Bacterial culture and PCR yield an increased false positive rate, as demonstrated by the high percentage of clinically healthy cattle with positive culture or PCR results[26,31-33]. Test positive and negative predictive values are improved when used at the group level versus test application to diagnose BRD in an individual animal.

An additional method to sample the airways has recently been described through the measurement of constituents in exhaled breath. This non-invasive sampling method has been previously reported in young calves as well as feedlot cattle. To date, inflammatory markers tested in exhaled breath have not identified markers that reliably distinguish between healthy and sick cattle[5,34,35] although nitric oxide was correlated with disease severity as assessed by number of treatments in BRD affected cattle[35].

IMAGING MODALITIES TO DIAGNOSE BRD

The technique to obtain and interpret thoracic radiographs has been well described for adult cattle[36] and young (4 – 6 week old) calves[37]. Such populations are more likely to be transported to a referral facility where radiography of the chest would be possible. Through correlation between thoracic radiographs and necropsy results, these authors were able to more specifically interpret the significance of radiographic findings. Despite recognizing that radiography can improve our sensitivity in detecting bovine lung disease, thoracic radiography in a feedlot setting is neither cost-effective nor logistically feasible at present. Radiography may, however, have utility in improving our case definition in research studies on BRD through more correctly categorizing animals as "sick" versus "healthy" and by tracking pulmonary lesions over time when assessing treatment interventions.

Ultrasonography can be useful to detect a wide variety of pleural and pulmonary lesions provided that lung pathology extends to include the lung surface. Descriptions of common pathologic findings in cattle with a variety of respiratory disorders have been published[38-40]. Ultrasonographic changes were highly correlated with clinical signs (fever, tachypnea), reduced tidal volume, and lung pathology scores at necropsy in calves experimentally infected with *Pasteurella multocida*[41]. A simplified ultrasound protocol for rapid assessment of lung architecture in dairy calves has been developed that utilizes the 5 MHz, linear transducer commonly used for trans-rectal reproductive examinations; in this protocol, time and cost of the examination is greatly reduced through not clipping the haircoat and use of alcohol rather than acoustic coupling gel. When this protocol was applied to clinically healthy dairy calves, pulmonary abnormalities on ultrasound examination of the thorax were associated with reduced weight gain in comparison to calves without ultrasonographic abnormalities in the lung[42]. Rigorous validation of field ultrasonography in predicting underlying respiratory disease in dairy calves is presently ongoing using this imaging protocol. While ultrasonography is realistically feasible to perform in young dairy calves that can be manually restrained, it presents logistical challenges in the evaluation of intractable feedlot cattle restrained by a chute system. This includes safe access to the lung field and timely evaluation of both hemithoraces in a manner that does not impede feedlot efficiency. Ultrasonography to detect BRD in feedlot cattle has yet to be validated.

CONCLUSIONS

Despite historical inaccuracy in our ability to differentiate healthy cattle from those affected by BRD under field conditions, recent advances in technology hold promise in improving our case detection. The need for non-invasive test methods that can predict present or impending respiratory disease in cattle are desirable, with diagnosis in "real-time" to assist with management decisions.

REFERENCES

[1]Salman MD, et al. **J Am Vet Med Assoc** 1988;192(8):1069. [2] Buhman MJ, et al. **Am J Vet Res** 2000;61(10):1163. [3]Hanzlicek GA, et al. **Am J Vet Res** 2010;71(3):359. [4]Rose-Dye TK, et al. **J Anim Sci** 2011;89(4):1193. [5]Burciaga-Robles LO, et al. **Am J Vet Res** 2009;70(10):1291. [6]Schaefer AL, et al. **Res Vet Sci** 2007;83(3):376. [7]Schaefer AL, et al. **Res Vet Sci** 2011; [Epub ahead of print]. [8]Lekeux P, et al. **XXII World Buiatrics Congress** 2002;144. [9]Poulsen KP, et al. **Vet Clin North Am Food Anim Pract** 2009;25(1):121. [10]Wittum TE, et al. **J Am Vet Med Assoc 1996**; 209:814. [11]Bryant LK, et al. **Bovine Proc** 1996;147. [12]Gardner BA, et al. **J Anim Sci** 1999;77(12):3168. [13]Thompson PN, et al. **J Anim Sci** 2006;84(2):488. [14]Pringle JK, et al. **Can J Vet Res** 1988;52(2):239. [15]Allen JW, et al. **Can J Vet Res** 1992;56(2):122. [16]White BJ, et al. **J Vet Diagn Invest** 2009;21(4):446. [17]Thomson RG, et al. **Can J Comp Med** 1975; 39:194. [18]Coghe J, et al. **Vet J** 2000;160(2):139. [19]Godson DL, et al. **Vet Immunol Immunopathol** 1996;51(3-4):277. [20]Heegaard PM, et al. **Vet Immunol Immunopathol** 2000;77(1-2):151. [21]Ganheim C, et al. **J Vet Med B** 2003;50(4):183. [22]Knobloch H, et al. **Chem** 2010;144(1):81. [23]Svensson C, et al. **Vet J** 2007;174(2):288. [24]Carter JN, et al. **Am J Vet Res** 2002;63(8):1111. [25]Humblet MF, et al. **Res Vet Sci** 2004;77(1):41. [26]Angen O, et al. **Vet Microbiol** 2009;137(1-2):165. [27]Orro T, et al. **Comp Immunol Microbiol Infect Dis** 2011;34(1):23. [28]Young CR, et al. **Am J Vet Res** 1996;57(2):138. [29]DeRosa DC, et al. **J Clin Microbiol** 2000;38(1):327. [30]Godinho KS, et al. **Vet Rec** 2007;160(1):22. [31]Espinasse J, et al. **Vet Rec** 1991;129(15):339. [32]Virtala AM, et al. **J Am Vet Med Assoc** 1996;208(12):2035. [33]Autio T, et al. **Vet Microbiol** 2007;119(2-4):256. [34]Reinhold P, et al. **Am J Vet Res** 2000;61(7):742. [35]Roller CB, et al. **Appl Opt** 2007;46(8):1333. [36]Masseau I, et al. **Can Vet J** 2008;49(3):261. [37]Tegtmeier C, et al. **J Vet Med B** 2000;47(3):229. [38]Streeter RN, et al. **Vet Clin North Am Food Anim Pract** 2007;23(3):541. [39]Babkine M, et al. **Vet Clin North Am Food Anim Pract** 2009;25(3):633. [40]Tharwat M, et al. **Trop Anim Health Prod** 2011;43(4):803. [41]Reinhold P, et al. **Vet Rec** 2002;150(4):109. [42]Ollivett, T. L., et al. **AABP 44th Annual Meeting Proc** 2011; 148.

PK-PD INTEGRATION AND MODELING FOR ANTIMICROBIALS USED IN THE MANAGEMENT OF BRD

Tim Potter, BVetMed, PhD, MRCVS
Warnham, West Sussex, UK

INTRODUCTION

Whilst the positive benefits of antimicrobial drug use are readily apparent and include enhanced animal welfare, prevention or treatment of disease, contribution to food safety and major economic advantages; there are now significant concerns over the potential for emergence and spread of resistance. There have been major efforts by national and international organisations to monitor the incidence of resistance to antimicrobial drugs (both drug classes and individual drugs) in important pathogens and commensals of farm animals, and the veterinary profession is under increasing scrutiny as to how they are utilising antimicrobial drugs.

Given the large number of pathogens that have been implicated in BRD, treatment tends to be symptomatic with a broad spectrum of effect, with the main component of any protocol being the use of antimicrobial drugs. When considering which antimicrobial to use, there is a wide selection of products at the veterinary surgeons's disposal; in fact there are more antimicrobial drugs licensed for the management of respiratory disease than for any other disease of cattle. There are a number of key considerations when prescribing antimicrobial therapy:

1. Selecting an antimicrobial agent with an appropriate spectrum of activity.
2. Using a dosage protocol that attains and maintains an effective therapeutic concentration at the site of infection.
3. Treating for an appropriate duration.

Antimicrobial drugs are widely used in the treatment of bacterial infections in animals as well as humans. Their safe and effective use is critically dependent on selecting an appropriate drug, in an appropriately formulated product and its administration at an optimal dose rate and dose interval. The World Health Organization defines the appropriate use of antimicrobial drugs as "the cost effective use of antimicrobials which maximizes clinical therapeutic effect while minimizing both drug-related toxicity and the development and the development of drug resistance"[1]. Various strategies have been suggested to achieve appropriate use of antimicrobial drugs, one of which involves the use of PK-PD integration and modeling to optimize dosing regimes [2]. The objectives of this project were to evaluate PK-PD integration and modeling approaches to dose schedule design, by generating and correlating data on the PK and PD of five drugs; oxytetracycline, amoxicillin, florfenicol, marbofloxacin and tulathromycin commonly used to treat bovine respiratory disease (BRD).

PHARMACODYNAMICS

The efficacy and potency of antimicrobial drugs against each bacterial species and strain are most commonly defined by the MIC and MBC, both determined *in vitro*. The MIC is the lowest concentration of an antimicrobial drug (in µg/mL) that inhibits the visible growth of a microorganism after 24h incubation under standard conditions. The MBC is the concentration that reduces the population size of a given organism by 99.9%. When MIC and MBC are close, it suggests a bactericidal action, whereas a larger difference occurs with drugs that have a bacteriostatic action.

The standard method of determining the MIC using doubling dilutions has limitations with regards to accuracy, with the potential for almost a 100% error. It has been suggested that 5 sets of overlapping doubling dilutions can be used to increase the accuracy of the measurement [3]. This reduces the error to below 20%. The currently recommended standardised methodology for MIC and MBC determination [4] involves the artificial medium; Mueller-Hinton broth (MHB) as this provides optimal laboratory growth conditions for the majority of organisms. However, MHB is not representative of the *in vivo* situation. To provide a more biologically relevant environment for some organisms it is possible to determine the MIC and MBC in serum. The accurate determination of an antimicrobial's phamacodynamic properties (in particular MIC) is an important step in performing PK/PD integration to determine an effective dose.

MICs and MBCs were determined in MHB and serum *in vitro*. To obtain more accurate values, MICs were determined using 5 overlapping sets of doubling dilutions in contrast to the currently CLSI recommended methodology [4] which uses a single set of concentrations. For all drugs there were differences in the MIC and MBC values between MHB and serum. For amoxicillin, marbofloxacin and florfenicol, it is likely that the differences were due principally or solely to protein binding in serum, with only the unbound proportion of the drug having activity. For oxytetracycline and tulathromycin, however, the differences between MICs in MHB and serum were much greater. With oxytetracycline, the MIC values were much greater in serum than in MHB. The MIC values for tulathromycin were, in contrast, much lower in serum than in MHB.

PHARMACOKINETICS

For an antimicrobial agent to exert its action, it must first act on the cell wall or cross the bacterial cell wall and the cell membrane and then diffuse through the periplasmic space to the site of action. The passage of antimicrobial drugs from the extracellular fluid of the host animal into the cell is usually by passive diffusion and hence a concentration gradient dependent process although in some cases, most notably for the aminoglycosides, the process is energy dependent. The penetration of the antimicrobial drug into the bacterium and its subsequent movement to its site of action depends on delivery of the drug into the environment of the bacterial cell. Pharmacokinetics therefore plays a major role in determining the effectiveness of therapy and thus the outcome of treatment of disease. The exposure time (the time for which the antimicrobial agent remains at its site of action) is also influenced and controlled by a drug's pharmacokinetic properties.

Pharmacokinetics is the term used to describe the temporal and spatial distribution of a drug (or drugs) in a system. It encompasses the dissolution, absorption, distribution, metabolism and excretion of the drug. In general a drug's PK properties are determined by its simple physico-chemical properties (its molecular size and both water and lipid solubility). Molecular size is broadly similar for the majority of compounds; antimicrobial drugs are in general small enough to pass through capillary endothelial cells to enter interstitial fluid but are large enough to be prevented from entering most cells by simple filtration.

Data on serum concentrations of antimicrobial drugs are of great value in studies of their absorption and elimination, but they provide less information on the actual quantity of drug at sites of infection. Hence, establishing the PK of tissue penetration may be more valuable than serum data alone. To characterise the ability of amoxicillin, oxytetracycline, marbofloxacin, tulathromycin and florfenicol to penetrate tissue fluids, their distribution into carrageenan-inflamed tissue cage fluid (exudate) and non-inflamed tissue cage fluid (transudate) were investigated in calves.

THE PK-PD APPROACH TO OPTIMISING DRUG THERAPY

The effective use of drugs clearly depends on knowledge of both PK and PD properties. The initial concepts of PK-PD interrelationships were first identified in the 1940s and 1950s by Dr Harry Eagle, whose work in rodents identified the time dependent pattern of activity of benzyl penicillin [5, 6], the concentration dependent killing action of streptomycin and the mixed pattern of bactericidal activity of the tetracyclines [7]. Since then the utility of PK/PD integration has become greatly extended in human medicine. In veterinary medicine, the PK-PD approach provides a means of optimizing (and minimising) the use of chemotherapeutics in animals [8].

When examining PK-PD relationships the most useful PK variables are the maximum plasma concentration (C_{max}), the time (T) during which plasma concentrations exceed a specified PD threshold and the area under the plasma concentration time curve (AUC) from 0 to 24 h. The most commonly used PD parameter is the MIC. Based upon these parameters, the 3 most common integrated PK-PD measures are the time drug concentration remains above MIC (T>MIC), the ratio of the maximum plasma concentration to the MIC (C_{max}: MIC) and the ratio of AUC to the MIC ($AUC_{0\text{-}24}$:MIC) [9].

Toutain et al [10] proposed the use of AUC:MIC as a scaling factor without units by which the target MIC is multiplied to estimate the average in vivo plasma concentration for discrete 24 h periods. Once the targeted average concentration over discrete 24 h periods has been determined, the duration of a treatment for a single dose may be defined as the last 24 h partial area giving a value above a pre-determined scalar. Frequently, the scalar is 1, allowing identification of the 24 h periods for which the average daily plasma concentration is at least equal to MIC.

For drugs whose efficacy can be correlated with $AUC_{0\text{-}24}$:MIC, it is possible to predict the optimal dosage if appropriate PK and PD data are known. Furthermore, through utilisation of a sigmoidal E_{max} relationship for bacterial count vs ex vivo $AUC_{0\text{-}24}$:MIC, data can be modelled to provide an estimated dose for a specific levels (bacteriostasis, bactericidal activity or bacterial eradication) of activity [3, 11-14].

PK-PD integration and modeling were performed using data and samples from the tissue cage study. Serum samples collected during the tissue cage study were used in ex vivo time-kill curves with the aims of characterizing the nature of the killing action of each of the drugs and providing a spectrum of effects for a range of drug concentrations. For amoxicillin, florfenicol and marbofloxacin, a range of serum concentrations was obtained, which permitted characterization of the sweep of the concentration-effect relationship and the data were used to propose doses for 3 levels of effect; bacteriostatic, bactericidal and bacterial eradication. Further trial work is required to validate the proposed doses in a clinical setting and to investigate the potential use of MIC50 values in dose calculation in place of the traditionally used MIC90.

SUMMARY

The advantage of the PK/PD approach over the conventional dose-titration study for determining a dosage regimen results from the fact that all the PK/PD indices incorporate the MIC of the pathogen and can thus provide direct biological information about the susceptibility of the pathogen [2]. Whilst PK-PD integration and modeling provides a strong tool for the design of dosing regimens and a means of predicting efficacy in clinical practice, the findings of these investigations demonstrate the individual nature of the relationship for each different antimicrobial drug. This means for the approach to implemented appropriately, a detailed understanding of a drug's specific PK and PD characteristics is essential.

REFERENCES

1. *WHO global strategy for containment of antimicrobial resistance.* 2001, World Health Organisation. 2. Lees, P., F.S. Aliabadi, and P.L. Toutain, *PK-PD Modelling: An alternative to dose titration studies for antimicrobial drug dosage selection.* Regulatory Affairs Journal, 2004: p. 175-180. 3. Aliabadi, F.S. and P. Lees, *Pharmacokinetics and pharmacodynamics of danofloxacin in serum and tissue fluids of goats following intravenous and intramuscular administration.* Am J Vet Res, 2001. 62(12): p. 1979-89. 4. CLSI, *Performance Standards for Antimicrobial Disk and Dilution Susceptibility Tests for Bacteria Isolated From Animals; Approved Standard - Third Edition* M31-A3 2008. 28(8). 5. Eagle, H., R. Fleischman, and A.D. Musselman, *The effective concentrations of penicillin in vitro and in vivo for streptococci, pneumococci, and Treponema pallidum.* J Bacteriol, 1950. 59(5): p. 625-43. 6. Eagle, H., R. Fleischman, and A.D. Musselman, *Effect of schedule of administration on the therapeutic efficacy of penicillin; importance of the aggregate time penicillin remains at effectively bactericidal levels.* Am J Med, 1950. 9(3): p. 280-99. 7. Eagle, H., R. Fleischman, and M. Levy, "Continuous" vs. "discontinuous" therapy with penicillin; the effect of the interval between injections on therapeutic efficacy. N Engl J Med, 1953. 248(12): p. 481-8. 8. McKellar, Q.A., S.F. Sanchez Bruni, and D.G. Jones, Pharmacokinetic/pharmacodynamic relationships of antimicrobial drugs used in veterinary medicine. J Vet Pharmacol Ther, 2004. 27(6): p. 503-14. 9. Ambrose, P.G., et al., Pharmacokinetics-pharmacodynamics of antimicrobial therapy: it's not just for mice anymore. Clin Infect Dis, 2007. 44(1): p. 79-86. 10. Toutain, P.L., A. Bousquet-Melou, and M. Martinez, AUC/MIC: a PK/PD index for antibiotics with a time dimension or simply a dimensionless scoring factor? J Antimicrob Chemother, 2007. 60(6): p. 1185-8. 11. Sidhu, P.K., et al., PK-PD integration and modeling of marbofloxacin in sheep. Res Vet Sci, 2010. 88(1): p. 134-41. 12. Aliabadi, F.S. and P. Lees, Pharmacokinetics and pharmacokinetic/pharmacodynamic integration of marbofloxacin in calf serum, exudate and transudate. J Vet Pharmacol Ther, 2002. 25(3): p. 161-74. 13. Aliabadi, F.S., M.F. Landoni, and P. Lees, Pharmacokinetics (PK), pharmacodynamics (PD), and PK-PD integration of danofloxacin in sheep biological fluids. Antimicrob Agents Chemother, 2003. 47(2): p. 626-35. 14. Aliabadi, F.S., et al., Pharmacokinetics and PK-PD modelling of danofloxacin in camel serum and tissue cage fluids. Vet J, 2003. 165(2): p. 104-18.

EFFICACY OF VACCINATING CATTLE AGAINST BOVINE VIRAL DIARRHEA VIRUS

Paul H. Walz, DVM, MS, PhD, DACVIM
Auburn, AL

INTRODUCTION

Bovine viral diarrhea virus (BVDV) has long been recognized as a major cause of reproductive, respiratory, and gastrointestinal disease in cattle, resulting in significant economic loss to the beef cattle and dairy cattle industries throughout the world. BVDV is an RNA virus that may exist as either of 2 biotypes, noncytopathic or cytopathic, and 2 genotypes, BVDV-1 and BVDV-2.[1] BVDV infection in cattle may result in a wide spectrum of clinical manifestations ranging from subclinical to fatal disease. Clinical manifestations are dependent upon the interplay of host factors, environmental stress levels, and viral factors.[2]

The virus employs multiple strategies to ensure survival and successful propagation in cattle, and this includes suppression of the bovine immune system, transmission by various direct and indirect routes, and, perhaps most importantly, induction of persistently infected (PI) cattle that shed and transmit BVDV much more efficiently than other sources. Persistent infection is considered by many the most important aspect of BVDV infection as this is the key mode by which the virus maintains and perpetuates itself in the cattle population. PI calves are the result of an *in utero* BVDV infection during the period of fetal development from gestation day 45 to gestation day 125 which is the gestational period bracketed by the end of the embryonic stage and the development of fetal immunocompetence.[3] Infecting biotype is important to the development of persistent infection. While infection with either biotype is capable of causing fetal death, only NCP strains are associated with persistent infection.[4] All genotypes and subgenotypes of BVDV appear to be capable of causing PI's.

Successful control of BVDV in cattle herds requires a multidimensional approach, involving vaccination, biosecurity and identification of BVDV reservoirs.[5] Because PI cattle are the main source of transmission within and between beef and dairy cattle herds, most of recent research assessing BVDV vaccination efficacy has focused on prevention of fetal infection and the generation of PI offspring.

AVAILABLE VACCINES

Many BVDV vaccines are available, and the majority of USDA licensed vaccines contain BVDV in combination with other bovine respiratory and reproductive pathogens. Prior to 1995, most BVDV vaccines contained only BVDV1 strains, but because of antigenic diversity and outbreaks of severe clinical disease in association with BVDV 2 strains, modified-live and inactivated vaccines containing both BVDV1 and BVDV2 strains are now widely available and more routinely used. Advantages and disadvantages of BVDV modified-live viral vaccines and inactivated vaccines have been described.[6] Modified-live and inactivated vaccines have been shown to be safe when administered according to the manufacturer's label. In general, modified-live viral vaccines are believed to more effective since they induce both humoral and cell-mediated immune responses. One disadvantage of inactivated BVDV vaccines is that two doses are required for the initial immunization, and a major problem with programs using inactivated vaccines is the widespread lack of compliance among producers by failing to booster the primary series.[7]

BVDV VACCINE EFFICACY

When discussing BVDV vaccination efficacy, it is first important to discuss reasonable expectations following vaccination and to remember that disease and infection are not synonymous terms. Although vaccines are an important component to BVDV prevention and control, they are not 100% efficacious, meaning that no vaccine will prevent all infections from occurring.[8] Reasons for lack of efficacy of vaccination against BVDV are many, and include factors related to the administration of the vaccine and factors related to the ability of the host to respond to the vaccine.

MEASUREMENTS OF VACCINE EFFICACY

The requirements for vaccine licensure in the United States were first described within the Code of Federal Regulations (CFR) 113.311 for 'Bovine virus diarrhea vaccine" (MLV) and 113.215 "Bovine virus diarrhea vaccine, Killed virus." These licensing documents describe the requirements for immunogenicity. To summarize from the CFR113.311 guidelines for modified-live BVDV viral vaccines (http://edocket.access.gpo.gov/cfr_2003/9cfr113.311.htm), immunogenicity was determined by testing for neutralizing antibodies in 20 vaccinated calves as compared to 5 unvaccinated calves. Efficacy was determined by challenging the calves with virulent BVDV two to four weeks after vaccination. A BVDV vaccine is considered immunogenic if 19 of the 20 vaccinated develop an antibody response (>1:8 titer), with efficacy being defined as vaccinates not developing leukopenia where 4/5 unvaccinates did. These early licensing requirements did not address efficacy related to reproductive disease or reproduction-related label claims. The vaccine claims for protection of the fetus against BVDV were described in the Center for Veterinary Biologics Public Notice 02-19 (http://www.aphis.usda.gov/animal_health/vet_biologics/publications/notice_02_19.pdf). The label claims for BVDV reproductive effects are divided into claims for fetal protection and claims for abortion (maternal and/or fetal causes). Here, the label claims are type-specific, i.e. BVDV 1 or BVDV 2 protection. Supporting data for the label claim is performed according to Veterinary Services Memorandum 800.202. Three categories for label claims are: 1) aids in the prevention of abortion, 2) aids in the prevention of persistently infected calves, and 3) aids in the prevention of fetal infection or aids in the prevention of fetal infection including persistently infected calves. Most clinical trials evaluating vaccine efficacy are performed with the goal of achieving efficacy claims for the prevention of persistently infected calves. These studies are characterized by vaccination prior to breeding, then challenging the pregnant cattle between days 75-90 of gestation, and finally testing the fetuses on or after 150 days of gestation. Virus isolation procedures are performed on fetal tissues, and those fetuses from which BVDV is isolated are considered to be persistently infected.

Protection from clinical disease

Initially, the effectiveness of BVDV vaccination was focused on limiting clinical disease due to BVDV infection. Numerous studies have demonstrated the ability of vaccination to protect against overt clinical disease associated with BVDV, and many trials have investigated the ability of commercially available BVDV vaccines in protecting against the high virulent BVDV 2 strains isolated from outbreaks of severe peracute BVD in the late 80's and early 90's. Experimental[9,10] and field data[11] indicate vaccination with either inactivated or modified-live BVDV vaccines are effective at reducing or obviating clinical disease. In addition, modified-live viral vaccines containing BVDV 1 strains are effective at limiting or preventing clinical disease when vaccinated animals are subsequently challenged with a virulent BVDV 2 strain.[12,13] Protection from clinical disease is important for stocker/backgrounder and feedlot operations, and immunity to BVDV has been demonstrated to be protective against bovine respiratory disease complex. Preconditioning cattle by vaccinating cattle against BVDV prior to an expected exposure (commingling and shipping) reduces the effects of exposure of cattle to BVDV.

Prevention of fetal infection

For the reproductive herd, vaccination against BVDV should protect against viremia to prevent dissemination of virus throughout the host, including preventing infection of the reproductive tract and fetus. In the past decade, the focus for vaccine efficacy has shifted from protection against clinical disease to protection against fetal infection. Published studies indicate the protection against fetal infections following BVDV vaccination varies, with influences by use of inactivated or modified-live vaccine, the timing of challenge, and the degree of homology between vaccine and challenge strains. In general, most experimental studies indicate significant, although incomplete, protection against fetal infection using modified-live viral vaccines,[14-17] and partial protection using inactivated viral vaccines.[18,19] The genotype of the challenge strain is important, and fetal protection is superior when animals are challenged with strains from the same genotype.[17] An important observation regarding all of the fetal protection studies so far is that although protection may not be 100%, the level of protection is superior to that observed when proper vaccination is not utilized as evidenced by higher rates of PI animals in unvaccinated cattle. Recently, cattle PI with BVDV have been used for the challenge exposure of vaccinated and control cattle in BVDV vaccine efficacy studies evaluating fetal protection.[16,18-20] These types of studies provide a more natural and rigorous challenge method.

Within the United States cattle population, there are three major subtypes, BVDV1a, BVDV1b, and BVDV2a, with the BVDV1b subtype predominating from diagnostic laboratory submissions and PI prevalence studies, accounting for 78% of bovine persistent infections in one North American Study.[21] Because of the finding of higher isolation rates for BVDV 1b strains and the fact that commercial vaccines for BVDV contain BVDV 1a and/or BVDV 2a strains, logic would dictate that challenge experiments should be performed using cattle PI with BVDV 1b strains. Two recent clinical trials have evaluated commercial modified-live viral vaccines in their efficacy in preventing fetal infections.[14,16] Like previous studies, these studies have demonstrated that modified-live BVDV vaccines provide significant (85-96%), although incomplete, protection against fetal infection.

VACCINATION STRATEGIES

Designing a vaccine program is critical in helping to control BVDV associated losses and giving producers a sense of security. In general, vaccines do not fail; vaccination programs fail. Timing of vaccination is a critical issue. Providing young calves with immunity can help reduce disease and death associated with BVDV infections. Timing of vaccination has been performed to co-incide with the decay of colostral antibodies, which may occur as early as a few weeks to as long as 8 months of age. Recent research has demonstrated that vaccination of young calves that possess colostral antibodies can result in an immune response that provides protection against clinical BVD later on in life.

Vaccination programs aimed at preventing reproductive losses may have different timing than vaccination programs aimed at preventing losses associated with clinical disease, such as pneumonia in weaned calves. Maximizing immunity during the early periods of gestation is most likely to reduce BVDV-associated reproductive losses. This is achieved through the use of prebreeding vaccination and boostering. The use of a modified-live viral vaccine at 1 month prior to breeding has been recommended to the point of being indisputable. However, vaccinating or using booster vaccines in the early lactation period of dairy cows (15-45 days in milk) can be an immunological challenge for cows due to negative energy balance. Delaying vaccination until after this period may provide better immunity.

SUMMARY

Nothing seems to generate more opinions regarding BVDV than what vaccine is the best to control BVDV, and in fact, choices and opinions are many. Even though controversy exists, most everyone believes that when vaccines are given correctly, at appropriate times, to healthy cattle, they are better than not vaccinating at all. In general, currently available vaccines provide adequate protection against clinical disease. Prevention of infection in pregnant animals is the ideal measure of vaccine efficacy.

REFERENCES

1. Ridpath JF. Bovine viral diarrhea virus: Diagnosis, management and control. Ames, IA: Blackwell Publishing; 2005:65.2. Baker JC. **Vet Clin North Am Food Anim Pract** 1995;11:425.3. McClurkin AW, et al. **Can J Comp Med** 1984;48:156.4. Harding MJ, et al. **Am J Vet Res** 2002;63:1455.5. Walz PH, et al. **J Vet Intern Med** 2010;24:476.6. Fulton RW. Bovine viral diarrhea virus: Diagnosis, management and control. Ames, IA: Blackwell Publishing; 2005:209.7. Rauff Y, et al. **J Am Vet Med Assoc** 1996;209:1618.8. Grooms DL, et al. **Bovine Practitioner** 2009;43:106.9. Dean HJ, Leyh R. **Vaccine** 1999;17:1117.10. Cortese VS, et al. **J Am Vet Med Assoc** 1998;213:1312.11. Carman S, et al. **J Vet Diagn Invest** 1998;10:27.12. Kelling CL, et al. **Am J Vet Res** 2007;68:788.13. Makoschey B, et al. **Vaccine** 2001;19:3261.14. Xue W, et al. **Vaccine** 2011;29:4657.15. Brock KV, et al. **Vet Ther** 2006;7:27.16. Leyh RD, et al. **Am J Vet Res** 2011;72:367.17. Ficken MD, et al. **J Am Vet Med Assoc** 2006;228:1559.18. Grooms DL, et al. **Am J Vet Res** 2007;68:1417.19. Rodning SP, et al. **Theriogenology** 2010;73:1154.20. Ellsworth MA, et al. **Vet Ther** 2006;7:295.21. Fulton RW, et al. **J Am Vet Med Assoc** 2006;228:578.

BOVINE VIRAL DIARRHEA VIRUS IN WILDLIFE AND DOMESTIC SPECIES OTHER THAN CATTLE

Thomas Passler, DVM, PhD, DACVIM
Auburn, AL

INTRODUCTION

The genus Pestivirus of the family *Flaviviridae* is currently comprised of bovine viral diarrhea virus (BVDV) 1, BVDV 2, border disease virus (BDV), and classical swine fever virus (CSFV). This traditional classification of viral isolates according to mammalian host of origin may not always be accurate, as pestiviruses do not demonstrate strict host specificity.[1] Over recent years, pestiviral infections have been demonstrated in various species of the mammalian order Artiodactyla, and lack of host specificity may play an important role in the pestiviral survival strategy.[2] Among pestiviruses, BVDV has been the species most intensely studied as an infectious agent of heterologous hosts. Evidence of BVDV infection exists in over 50 species of artiodactyls in the families *Antilocapridae*, *Bovidae*, *Camelidae*, *Cervidae*, *Giraffidae*, *Suidae*, and *Tragulidae.* [3] The implications of heterologous pestivirus infections are twofold. First, heterologous BVDV infections may hamper the success of control strategies, as are in progress for BVDV in cattle and CSVF in swine. Second, heterologous BVDV infections may threaten the health and well-being of other mammalian hosts, such as free-ranging and captive wildlife. The role of heterologous infections for the epidemiology of BVDV is incompletely understood, but transmission and maintenance of BVDV within three host clusters: 1. domestic small ruminants and swine; 2. free-ranging and captive wildlife; 3. camelids was postulated.[4]

BOVINE VIRAL DIARRHEA VIRUS IN DOMESTIC SMALL RUMINANTS AND SWINE

Goats and Sheep

The close phylogenetic relatedness of small ruminants to cattle and resulting greater possibility of permissive infection may promote a unique role of sheep and goats in the ecology of BVDV. Additionally, sheep and goats often have closer contact with cattle as compared to other artiodactyls, especially in certain management conditions, such as subsistence agriculture, alpine farming, or hobby farming. Although one report did not identify a positive correlation between BVDV seropositivity of sheep flocks and presence of cattle,[5] other studies reported significantly higher seroprevalence rates in sheep on farms with cattle,[6] especially where communal alpine pasturing of sheep, goats, and cattle was practiced.[6]

Several serological surveys have demonstrated the occurrence of BVDV in small ruminants; however, cross-reactivity of pestiviral antibodies warrants caution when evaluating the results of such surveys. In recent reports on pestiviral seroprevalence in sheep, cross-neutralization assays identified highest reactivity against BVDV 1 rather than BDV.[5, 6] In Argentina, which is considered free from BDV, 79.7% (43/54 sheep) of samples were seropositive to BVDV 1 or BVDV 2, indicating BVDV circulation in Patagonian populations.[7] When antibodies against pestivirus are detected in small ruminants, the distinction of pestiviral species may hold critical information on epidemiology, prevention, and control.[8]

Infections with BVDV in domestic small ruminants result in clinical signs of border disease. Postnatal infections commonly cause mild clinical signs, including pyrexia and leucopenia.[9] Experimental inoculation of juvenile sheep resulted in anorexia, tachycardia, pyrexia, and lung lesions especially associated with the pulmonary vasculature.[10] Infections with BVDV in pregnant small ruminants may result in uteroplacental pathology and pregnancy loss by fetal resorption or abortions.[11] The outcome of transplacental infection depends on the length of gestation and biotype of infecting BVDV strain. In utero BVDV infections can cause fetal mummification, stillbirth, premature birth, and birth of offspring with congenital disorders. Congenital disorders of affected fetuses commonly involve the central nervous system, and neuropathogenicity may be more severe with cytopathic biotypes.[12] Although fetal death and non-viability of lambs are common sequelae of transplacental BVDV in sheep, reports of viable PI offspring exist.[13, 14] In contrast to cattle and sheep, viable PI kids appear to be a rare result of in utero BVDV infection in goats and reproductive failure or severe gross and histological pathology of affected fetuses are the common outcome of BVDV infection.[15, 11]

Swine

The close antigenic relationship between ruminant pestiviruses and CSFV has important implications on diagnostic testing and control of CSFV. Currently, examination of blood samples for CSFV antibodies relies on a combination of ELISA and virus neutralization techniques, and considerable efforts are necessary in establishing a CSFV-specific diagnosis.[16] During the CSFV outbreak in the Netherlands in 1997/1998, 26.5% of CSFV ELISA positive samples were demonstrated to be caused by presence of antibodies to ruminant pestiviruses.[16] After the first description of an antigenic relationship between CSFV and BVDV in 1960,[17] scientific interest revolved around the use of BVDV to immunize swine against CSFV.[18, 19] Results from these studies were discordant as to the immune protection upon CSFV challenge and amount of shedding of vaccine virus. A more recent study demonstrated that immunization with BVDV strain St. Oedenrode isolated from a naturally infected pig resulted in a self-limiting BVDV infection and protected vaccinates from clinical disease and CSFV transmission upon challenge.[20]

Only a limited number of serosurveys have been performed in domestic swine and reported seroprevalence rates are relatively low when compared to other domestic livestock. Risk factors associated with seropositivity of domestic swine include the presence of cattle on the same farm, high density of small ruminants near swine populations, vaccination with BVDV contaminated vaccines, and age of tested swine.[21-23] Studies on postnatal BVDV infections in non-pregnant swine have rarely described the occurrence of clinical signs based on observation of infected animals. When evaluated, an elevation of the rectal temperature after BVDV inoculation was not detected in most studies despite evidence of viremia and seroconversion.[19, 24] Experimental infection of eight-week-old pigs with either BVDV 1 or BVDV 2 did not result in clinical signs or increases in body temperature in any group, despite the use of a BVDV 2 isolate that was demonstrated to be virulent for cattle.[24] In contrast, inoculation of three-day-old piglets with either BVDV strain NY-1 or a cell-culture contaminant, BVDV-C, resulted in loss of appetite, diarrhea, and intestinal pathology for one to four days.[25] Importantly, the co-infection of either BVDV NY-1 or BVDV-C with attenuated transmissible gastroenteritis virus resulted in severe clinical signs that were worse with the cell-culture contaminant BVDV-C.[25] Experimental BVDV infection of pregnant gilts resulted in reproductive disease including intrauterine infection, pregnancy loss, and reduction in litter size.[26] While transplacental BVDV infection in pigs appears to less common than in ruminants, the birth of PI piglets has been reported.[27]

BOVINE VIRAL DIARRHEA VIRUS IN NEW AND OLD WORLD CAMELIDS

Bovine viral diarrhea virus infections were identified through detection of antibodies as early as 1975 in Old World camelids (OWC) and 1983 in New World camelids (NWC). [28, 29] However, most seroepidemiologic and experimental infection studies suggested that BVDV may cause infections without presenting a serious risk to NWC. Reports of BVDV isolation and identification of persistent infections in alpacas have prompted increased interest in BVDV infections in camelids, and BVDV is now considered an emerging pathogen of NWC.[30] The first description of a PI alpaca was made in Canada following natural exposure of a pregnant alpaca to a chronically ill cria.[31] From the persistently infected cria, a BVDV 1b strain was isolated from buffy coat cells on different occasions before euthanasia of this animal. Several cases of PI alpacas have since been reported in North America and Great Britain.[30, 32, 33] Persistently infected alpacas may survive for several months but are affected by low birth weight, failure to thrive, inappetence, lethargy, chronic diarrhea, and chronic recurrent infections especially of the respiratory tract. In OWC, reproductive disease including abortion, still births, weak calves, early neonatal death, and neonatal hemorrhagic disease was also reported.[34] Noncytopathic BVDV has also been isolated from tissues of camels in which histopathological changes were consistent with a BVDV infection.[35]

From NWC in Chile, both BVDV 1 and BVDV2 have been isolated.[36] In contrast, in reports from North America and the United Kingdom in which the subgenotype of the infecting isolate was determined, all isolates belonged to BVDV 1b.[30-32] A recent study analyzed 46 BVDV isolates from over 12,000 North American alpacas and classified all isolates as non-cytopathic strains of subgenotype 1b with $\geq$ 99% nucleotide homology in the 5'-UTR.[33] Kim et al. offer two explanations for this peculiarity, firstly that exposures of alpacas to BVDV are rare and spread of the existing BVDV 1b strain is by extensive movement of a few PI animals or secondly, only unique 1b subgenotypes are able to establish transplacental infections in alpacas.[33] Movement of alpacas (including dams with cria by foot) between farms, mainly for breeding purposes, are common practice and were described in reports of reproductive disease and birth of PI offspring, highlighting the importance of sound biosecurity practices.[31, 32, 37]

BOVINE VIRAL DIARRHEA VIRUS IN WILDLIFE

Although sources of BVDV infections in free-ranging wildlife are unknown, a likely source is contact with cattle. This is supported by the absence of antibodies in deer that were without contact with cattle for over 50 years.[38] Significantly higher seroprevalence rates were found in white-tailed deer on ranches where cattle were present as compared to ranches without cattle.[39] In contrast, an association between BVDV seroprevalence and cattle density was not found in another study.[40] Similar findings were described in a recent report from Minnesota where a greater percentage of deer were seropositive in a region with a lower cattle density.[41] The authors of that report concluded that cattle use and management (i.e. dairy or beef) may have an important impact on interspecific BVDV transmission, as there is likely less wildlife contact with housed dairy cattle as compared to pastured beef cattle. [41] Various factors likely influence the transmission of BVDV between cattle and wildlife as has been described for other pathogens such as bovine tuberculosis.[42] With presence of suitable environmental and management factors, transmission of BVDV from cattle to deer is possible as has been demonstrated in a cohabitation experiment.[43] In that study, BVDV was efficiently transmitted from PI cattle to white-tailed deer and resulted in seroconversion in all adult deer and birth of PI fawns from four of seven pregnant does.[43]

Maintenance of BVDV within a cervid population without presence of cattle was suggested by the presence of antibodies in over 60% of caribou that had not been in contact with domestic ruminants for over 25 years.[44] The presence of endemic BVDV infections as indicated by high seroprevalence rates was also suggested in reindeer in Norway and US cervid populations.[45-47] This is supported by identification of high seroprevalence rates in eland populations of Zimbabwe, in which a PI and a virus-isolation-positive eland were identified [48].

Only a limited number of studies have surveyed free-ranging wildlife populations for the presence of BVDV or BVDV antigen. Anderson and Rowe (1998) utilized an antigen capture ELISA to detect BVDV in a subset of 303 seronegative animals during their serosurvey in Zimbabwe and detected two BVDV-positive eland.[48] In Germany, cytopathic BVDV was detected in spleen samples from two of 203 deer and both animals were seronegative roe deer.[49] Three surveys utilizing the IHC or ELISA techniques, respectively, investigated the occurrence of BVDV in free-ranging cervids in the US and results suggest that PI cervids exist in wildlife populations.[50-52] In Alabama, one of 406 skin samples from white-tailed deer was positive on IHC and the antigen distribution resembled that of PI cattle.[51] The skin sample of one of 5597 deer in Colorado was positive on IHC and this result was confirmed by detection of viral RNA in skin and lymph node samples.[50] In Indiana, two of 745 white-tailed deer were positive for BVDV by antigen capture ELISA with subsequent isolation of cytopathic and non-cytopathic BVDV.[52] To date, the validation of BVDV assays for their use in wildlife has not been performed, and this may be critical as considerable variations were observed among the IHC and antigen capture ELISA on skin samples of white-tailed deer (Passler, T. and Walz, P.H., unpublished observations). In addition to surveys, isolation of BVDV was successful in free-ranging roe deer in Hungary, a mule deer in Wyoming, and two white-tailed deer in South Dakota.[53, 54, 47] In these reports, clinical illness including emaciation, weakness, and death prompted further investigations leading to the isolation of BVDV from tissues.

Few reports on the clinical outcome of BVDV infection in wildlife exist, but reported clinical signs are similar to BVDV infections in cattle. Experimental inoculation with BVDV NY-1 did not result in clinical signs in four mule deer and one white-tailed deer fawns, despite evidence of viremia and nasal shedding.[55] Similar findings were made in yearling elk that were inoculated either with BVDV 1 Singer or BVDV 2 24515.[56] In two young reindeer, loose, bloody, and mucoid feces, transient laminitis, or coronitis were observed after inoculation with BVDV Singer and mild lesions were detected at necropsy.[57] Naïve white-tailed deer fawns developed moderate pyrexia and marked to moderate decreases in lymphocyte populations in response to either BVDV 1 or BVDV 2; and lethargy or coughing was observed in individual fawns.[58, 59] Similar to other species, BVDV infections in wildlife may have the most important implications on reproductive health. Experimental inoculation of pregnant white-tailed deer may result in reproductive failure, including fetal resorption, fetal mummification, stillbirth, and abortion.[60, 61]

SUMMARY

Bovine viral diarrhea virus is one of the most important infectious agents affecting the cattle industry worldwide; however BVDV is not host-specific to cattle as numerous wild and domestic species have been reported susceptible to BVDV infection. Infection with BVDV in heterologous species, with the potential for persistently infected offspring may have important implications for BVDV control in cattle populations, as these species may become reservoirs and source of infection for cattle populations that are free from BVDV. In addition, BVDV-associated disease and mortality in heterologous species, including rare and endangered livestock or zoo animal species threatens natural resources and complicates laboratory testing strategies. Further research, including studies on interspecific interactions and BVDV transmission are warranted to fully understand the importance of BVDV in heterologous species.

REFERENCES

[1]Nettleton PF. **Rev Sci Tech** 1990;**9**:131.[2]Vilcek S, et al. **Vet Microbiol** 2006;**116**:1.[3]Passler T, et al. **Anim Health Res Rev** 2010;**11**:191.[4]Evermann JF. **Small Rumin Res** 2006;**61**:201.[5]Graham DA, et al. **Vet Rec** 2001;**148**:69.[6]Krametter-Frotscher R, et al. **Vet Rec** 2007;**160**:726.[7]Julia S, et al. **Prev Vet Med** 2009;**90**:274.[8]Konig M, et al. **Berl Munch Tierarztl Wochenschr** 2003;**116**:216.[9]Taylor WP, et al. **Trop Anim Health Prod** 1977;**9**:249.[10]Meehan JT, et al. **J Comp Pathol** 1998;**119**:277.[11]Loken T, et al. **J Comp Pathol** 1991;**105**:123.[12]Hewicker-Trautwein M, et al. **Zentralbl Veterinarmed [B]** 1995;**42**:557.[13]Carlsson U, et al. **Acta Vet Scand** 1994;**35**:79.[14]Scherer CF, et al. **Vet Microbiol** 2001;**79**:285.[15]Broaddus CC, et al. **Vet Pathol** 2009;**46**:45.[16]de Smit AJ, et al. **Prev Vet Med** 1999;**42**:185.[17]Darbyshire JH. **The Veterinary Record** 1960;**72**:331.[18]Baker JA, et al. **J Am Vet Med Assoc** 1969;**155**:1866.[19]Beckenhauer WH, et al. **Veterinary Medicine** 1961;**56**:108. [20]Wieringa-Jelsma T, et al. **Vet Microbiol** 2006;**118**:26.[21]Liess B, et al. **Rev Sci Tech** 1990;**9**:151.[22]Loeffen WL, et al. **Vet Microbiol** 2009;**136**:240.[23]Vannier P, et al. **Ann Rech Vet** 1988;**19**:283.[24]Walz PH, et al. **Can J Vet Res** 1999;**63**:119.[25]Woods RD, et al. **J Vet Diagn Invest** 1999;**11**:400.[26]Stewart WC, et al. **Am J Vet Res** 1980;**41**:459.[27]Terpstra C, et al. **Vet Q** 1997;**19**:97.[28]Burgemeister R, et al. **DTW - Dtsch Tierarztl Wochenschr** 1975;**82**:352.[29]Doyle LG, et al. **J Am Vet Med Assoc** 1983;**183**:1257.[30]Byers SR, et al. **J Vet Diagn Invest** 2009;**21**:145.[31]Carman S, et al. **J Vet Diagn Invest** 2005;**17**:589.[32]Foster AP, et al. **Vet Rec** 2007;**161**:94.[33]Kim SG, et al. **Vet Microbiol** 2008.[34]Hegazy AA, et al. in *International Symposium, Bovine Viral Diarrhea Virus: a 50 Year Review*. 1996. Ithaca, NY.[35]Wahab A, et al. **Int J Virol** 2005;**1**:45.[36]Celedon MO, et al. **Arch Med Vet** 2006;**38**:247.[37]Topliff CL, et al. in *AAVLD Annual Conference*. 2007. Reno, NV.[38]Sadi L, et al. **J Wildl Dis** 1991;**27**:569.[39]Cantu A, et al. **J Wildl Dis** 2008;**44**:1002.[40]Frolich K. **J Wildl Dis** 1995;**31**:247.[41]Wolf KN, et al. **J Wildl Dis** 2008;**44**:181.[42]Schmitt SM, et al. **Ann N Y Acad Sci** 2002;**969**:262.[43]Passler T, et al. **Vet Microbiol** 2009;**134**:362.[44]ElAzhary MA, et al. **J Wildl Dis** 1981;**17**:609.[45]Aguirre AA, et al. **Prev Vet Med** 1995;**21**:313.[46]Lillehaug A, et al. **J Wildl Dis** 2003;**39**:779.[47]Van Campen H, et al. **J Wildl Dis** 2001;**37**:306.[48]Anderson EC, et al. **Epidemiol Infect** 1998;**121**:441.[49]Frolich K, et al. **J Wildl Dis** 1995;**31**:243.[50]Duncan C, et al. **J Vet Diagn Invest** 2008;**20**:650.[51]Passler T, et al. **J Vet Diagn Invest** 2008;**20**:79.[52]Pogranichniy RM, et al. **J Vet Diagn Invest** 2008;**20**:71.[53]Chase CC, et al. **J Wildl Dis** 2008;**44**:753.[54]Romvary J. **Acta Vet Acad Sci Hung** 1965;**15**:451.[55]VanCampen H, et al. **J Wildl Dis** 1997;**33**:567.[56]Tessaro SV, et al. **J Wildl Dis** 1999;**35**:671.[57]Morton JK, et al. **Rangifer** 1990;**10**:75.[58]Mark CS, et al. in *Annual Meeting of the Conference of Research Workers in Animal Diseases*. 2005. St. Louis, MO.[59]Ridpath JF, et al. **J Wildl Dis** 2007;**43**:653.[60]Passler T, et al. **Vet Microbiol** 2007;**122**:350.[61]Ridpath JF, et al. **Am J Vet Res** 2008;**69**:1630.

CHALLENGING UMBILICAL SURGERY IN CALVES

Sylvain Nichols, DMV, MS, DACVS
St-Hyacinthe, QC, Canada

INTRODUCTION

Umbilical surgeries are frequent in cattle. The most frequent condition is the umbilical hernia. The principals of a herniorraphy are relatively simple. Other umbilical surgeries involving the embryonic internal structures or an incarcerated abdominal organ are technically more challenging. This presentation will focus on those challenging umbilical surgeries. First, the use of the ultrasound in the pre-operative period will be discussed. Second, anesthesia of those calves will be discussed in regards to their size and the length and difficulty of the procedure. Finally, the surgery and its post-operative management will be discussed.

ULTRASOUND EVALUATION

Every calf should undergo a complete physical exam prior to surgery for an umbilical pathology. A complete physical exam includes a deep palpation of the umbilical structures. Palpation could be difficult to perform on larger cattle. Ultrasound evaluation is an extension of the palpation. It will help determine the structures involved in the pathology (ex: urachus or arteries).[1] It will help determine the severity of the pathology (does the umbilical vein go into the liver). It will help determine the structure involved in the strangulated hernia (abomasum or jejunum) helping the veterinarian elaborate his surgical plan.

A 3.5 MHz curvilinear probe is ideal for umbilical evaluation. A linear probe, like a transrectal probe, can be useful, but lacks depth to evaluate some structures (ex: liver). The scan starts at the navel and goes toward the liver (right paracostal area) to evaluate the umbilical vein, and toward the bladder (caudal abdomen) to evaluate the urachus and the umbilical arteries. It is important to visualize the bladder, because oftentimes arteries and the urachus decrease in size going into the umbilicus but increase in size at the cranial aspect of the bladder.

ANESTHESIA

Some factors like the size of the animal and the structures involved in the pathology will change the protocol used to anesthetise the calf. Pathology, other than simple hernia, on larger cattle, should be tackled under general anesthesia. If available, gas anesthesia is ideal. If not, a double drip solution (guaifenesin and ketamine) given to effect allows the surgeon to perform the procedure without significant movement of the animal. We do not recommend giving large volume epidural anesthesia to cattle over 300kg to avoid complications during recovery (hip luxation). If used, it is recommended to recover the animal with hobbles on a non slippery floor (sand or thick bedding).

On smaller cattle with complicated umbilical surgery, we recommend using general anesthesia. Omphalophlebitis and deep omphaloarteritis should be tackled with gas anesthesia. During the manipulation of the arteries and vein, a lot of traction will be applied on the liver and on the aorta creating pain. Having the animal under gas anesthesia makes it easier to control the plane of anesthesia during those painful manipulations. Injectable general anesthesia is adequate for umbilical pathology where the traction is placed on the bladder (urachal surgery). A double drip solution (guaifenesin and ketamine) given to effect allows the surgeon to perform the procedure without significant movement of the animal.

On smaller cattle, except for simple hernia, a large volume epidural anesthesia is helpful to anesthetise the caudal abdomen and to immobilize the rear end of the calf. This is our protocol for a calf undergoing surgery for an infected urachus:

A. Premidication: Diazepam (0.2mg/kg IV).
B. Epidural anesthesia: Lidocaine 2% (0.15ml/kg sacro-coccygeal) +/- Xylazine (0.05mg/kg). The xylazine increases the sedation of the calves and increases the duration of action of the epidural (3-5 hours vs 1-2 hours if lidocaine is used alone). The calf is kept in sternal recumbency for 15 minutes after the epidural.
C. During surgical preparation, a nasotracheal tube, especially if a double drip is used, should be placed. The easiest way to perform a nasotracheal intubation is with the calf in dorsal recumbency. The neck is straightened and the tube is slid slowly into the trachea paying attention to the breathing pattern of the calf (advancing the tube during inspiration).
D. Local anesthesia: ring block around the umbilicus (especially at the cranial aspect, which is not anesthetized by the epidural).
E. Double drip: 5% guaifenesin mixed with ketamine (2mg/ml) given to effect

COMPLICATED URACHUS

Some urachal surgeries are more challenging than other. The challenge comes from the adhesion caused by the infected urachus. It is difficult, with the ultrasound, to detect those adhesions prior to the surgery. Most of the time, the omentum is adhered to the urachus. In this case, the omentum can be carefully dissected off or cut from the urachus. It is important to remember to suture any holes created in the omentum after the urachus is excised to avoid internal herniation. On other occasions, the abomasum, or even the jejunum can be adhered to the urachus. Rarely a resection and an anastomosis are needed. Most of the time, the adhesion can be dissected safely.

OMPHALOPHLEBITIS

Two clinical presentations of omphalophlebitis are encountered. The first one is a partially infected vein that does not penetrate into the liver. This umbilical pathology has only minimal repercussions on the health and the growth of the calves. The umbilical stump is usually enlarged and the owner often reports multiple episodes of purulent drainage from the stump. Antibiotic administration usually resolves the drainage temporarily. The diagnosis is confirmed by palpation and ultrasound. The surgery involves removing en block the infected vein. In some cases, liver adhesion around the vein might need to be dissected to gain access to an area where the vein tapers. The vein is double ligated. The ultrasound can be used intra-operatively to ensure that the vein is cut in a safe location. It is recommended to swab the stump and to submit it for bacterial culture and sensitivity.

The second presentation is a combination of extra and intra-hepatic infection of the vein. This umbilical pathology has significant repercussions on the health and the growth of the calves. Usually, they are poor doers and have multiple problems (septic arthritis). The diagnosis is made by palpation and ultrasound. In the liver, the vein could be the only structure infected. However, in some cases, multiple liver abscesses can be found, worsening the prognosis.

On presentation, those calves are usually in rough shape. It is important to stabilize them and start treating any concomitant diseases prior to surgery. The surgery can be delayed only if the infected vein is actively draining. In some cases, medical therapy combined with an efficient umbilical drainage can decrease the size of the infected structure to the point that it is possible to remove the vein en bloc. However, because of the angle of the vein in regards to the umbilical stump, it rarely drains efficiently. In those cases, the surgery cannot be postponed indefinitely. As soon as the animal can handle the surgery, it should be performed. Postponing the surgery for too long can cause the infection of other structures (especially joints).

Since the infected structure cannot be removed totally, it should be marsupialized. The marsupialization can be performed at the cranial aspect of the incision[2] or preferentially through an independent circular incision at the cranial and lateral aspect of the incision.[3] The goal of the marsupialization is to create a shorter, larger and straighter draining tract. The marsupialisation is created by suturing the vein in 2 or 3 plans to the body wall. Passing a large umbilical stump through the circular incision is impossible. The stump will have to be transacted during the surgery increasing the risk of contaminating the surgical field. It is usually performed in an area where the vein tapers just next to the body wall (they usually all do). A ligature is placed around the vein. A sterile glove (only a finger) is placed over the vein and is again ligated. The protected vein is then pulled through the body wall.

The day following the surgery, the stump is amputated. Often, the drainage coming out of the vein is disappointing. However, you need to resist the temptation of flushing actively the vein for the first couple of days. Frequently the wall, separating the abscess from the portal vein, is paper thin. Flushing actively might cause some bacteria to disseminate systemically.

The marsupialization site will heal by second intention after resolution of the infection. In a fair number of cases, a hernia will result and will need to be revised surgically. It is important to warn your client of the costs associated with a second procedure.

OMPHALOARTERITIS

Omphaloarteritis is uncommon. They are only few reports of this condition in the literature.[4-5] Like omphalophlebitis, the artery could be partially or almost completely involved. More systemic repercussions are seen with the later. When the artery is partially involved it is recommended to remove it en block. However, when it is almost completely involved, it could be extremely difficult to remove it completely. Again, if the artery is draining through the navel, it is recommended to treat the animal medically prior to surgery. In our hospital, I have seen a handful of cases where the artery regresses to the point where en block removal was possible. If the artery cannot be removed en block and it is not draining actively through the umbilicus, the artery can be marsupialized. However, this procedure is more challenging to do than with the vein. The marsupialization is created just cranial to the pubis and lateral to midline. Revision of the marsupilization incision is more difficult in this location if a hernia appears.

During surgery, the ultrasound can be used to find the safest location to ligate the artery. Usually, a lumen will be present in the abdominal stump. A lumen does not mean that the infected tissue is not totally excised. A swab of the stump should be submitted for bacteriological culture and sensitivity. If the lumen is large, an inverted suture should be performed on the stump.

After the surgery, I recommend dissecting the artery. I usually cut the artery at its widest point and I take a sample of the purulent material for bacteriological culture and sensitivity. Then, I feed a small gauge urinary catheter toward the amputation site to evaluate patency. If it does not exit at the amputation, I dissect the artery until I find the blind end following the catheter. If the catheter exits the amputation site, I will keep the animal on antibiotics for a longer period of time. I will readjust the therapy according to the results of culture and sensitivity.

INCARCERATED HERNIA

Incarcerated hernia is a common condition in calves. However, because the most frequent structures involved are the omentum, it does not usually cause severe clinical signs. After the omentum, the most frequent structure involved in an incarcerated hernia is the abomasum and the jejunum. Abomasal incarceration is usually found on poor doer calves with an enlarged umbilicus. Rarely, they have a complete obstruction of the abomasum. However, on bloodwork, they might have signs of proximal obstruction. The condition is confirmed with ultrasound. A neophyte ultrasonographer may misdiagnose an incarcerated abomasum for an umbilical abscess. It is important to always stick a needle in an enlarged umbilicus before cutting into it with the scalpel blade.

Prior to surgery, severely affected calves should be treated medically. The umbilicus is approached like an umbilical abscess. When the body wall is incised, the abdomen is protected and the abomasum is evaluated. The hernia ring is incised carefully and fibrinous adhesions between the abomasum and the hernia sac are broken carefully. Sharp dissection through fibrous adhesions is unrewarding. If the abomasum cannot be freed by blunt dissection, it will have to be incised. Therefore, an assistant is needed to hold the abomasum with its fingers to obstruct the flow of ingesta while it is incised. The hole created is closed with a double inverting suture pattern. In some cases, when I am worried about stricture, I close the incision with a simple interrupted pattern on the mucosa and submucosa and an inverting layer on the seromuscular layer.

REFERENCES

[1]Staller GS. J Am Vet Med Assoc 1995;206(1):77, [2]Steiner A. Vet Surg 1993;22(3):184, [3]Edwards RB. Vet Surg 1995;24:32, [4]Lopez MJ. Can Vet J 1996;37:170, [5]Hylton WE. J Am Vet Med Assoc 1985;186(10):1099.

IDIOPATHIC HEMORRHAGIC PERICARDIAL EFFUSION

Simon F. Peek, BVSc, PhD, DACVIM
Madison, Wisconsin

INTRODUCTION

There have been two previous reports in the literature concerning idiopathic hemorrhagic pericardial effusion in dairy cattle[1,2]. These were both small retrospective case series concerning mature dairy cattle presented for evaluation at referral teaching hospitals with clinical signs of heart failure associated with a hemorrhagic pericardial effusion and cardiac tamponade. Over the last few years we have seen an increasing number of similar cases presented for evaluation at the large animal hospital at the University of Wisconsin and this presentation will examine the condition and some of these cases in retrospective detail. The condition has become one of the two most common differentials for cattle with presenting signs of cardiac disease within our hospital in the last 5 years, and has exceeded septic pericarditis proportionately during that time period as the most common form of pericardial disease.

SIGNALMENT

Previous reports in the literature[1,2] and all of the cases that we have seen in our hospital have been female Holstein dairy cows over 2 years of age. There have been a variety of ages seen, and cattle can be either lactating or dry at presentation. Neither males, nor females less than 2 years of age have been documented to date. Both individually valuable pedigree cattle predominantly used as embryo transfer/oocyte donors or for showing, as well as grade cows from commercial confinement free stall dairies have been diagnosed with this condition in our hospital. In terms of reproductive status at the time of diagnosis, we have seen a spectrum from non-gravid early lactation animals through to cows in the last trimester of pregnancy.

PRESENTING CLINICAL SIGNS

Affected cattle present with classical outward signs of congestive heart failure, typically being noted by the producer as having brisket and ventral edema. Diminished milk production and appetite are anticipated and present and clients often indicate that they feel the condition has an acute to subacute onset (a few days to a week or two) with respect to the development of an appreciated change. On veterinary examination cattle are typically afebrile, with muffled heart sounds alongside tachycardia and tachypnea. Signs of congestive heart failure due to cardiac tamponade predominate. Other non-specific signs of illness, rumen hypomotility, moderate ketosis, etc., can be present somewhat according to signalment. Thoracic ultrasound examination is most rewarding and instantly diagnostic of a large volume of pericardial effusion. Ultrasonographically the fluid usually appears predominantly anechoic, but a more mixed echogenicity, with fronds of more fibrinous looking material can often be seen closer to the epicardial surface. Mild pleural effusion may also be noted. The pericardial sac is often so large that accurate echocardiographic evaluation of cardiac function is nearly impossible prior to drainage.

HEMATOLOGY AND BIOCHEMISTRY

We have not found complete blood counts to be of any diagnostic utility in cattle with this condition. Cattle with the condition that are seropositive for antibody to the bovine leukosis virus (BLV) (see below) do not reliably demonstrate a peripheral lymphocytosis. Similarly, cattle that either have lymphoblasts in the hemorrhagic effusion at the time of initial presentation or that develop a neoplastic effusion at a later date neither reliably demonstrate a peripheral lymphocytosis nor have blast forms in their circulation at any time. Routine biochemistry has been fairly unremarkable excepting the consistent finding of GGT elevations suggestive of hepatic congestion due to cardiac disease. This has been corroborated by gross and histologic examination of post mortem livers from affected cattle.

PERICARDIAL FLUID ANALYSIS

Pericardiocentesis and pericardial fluid analysis are critical parts of the evaluation of suspected cases from several perspectives. Firstly it is important to rule out other potential causes of pericardial disease and effusion; cytologic evaluation of pericardial fluid principally helping to exclude septic causes. We have repeatedly observed the pericardial fluid from cattle with this condition to appear hemorrhagic, with a low PCV (typically in the 5-15% range), variable total protein levels (often in the 3-5g/dl range), with highly variable total nucleated cell counts. The predominant cell type is small lymphocytes in the majority of cases, but some cases at first evaluation will already have obvious, phenotypically neoplastic lymphocytes present. Several cattle that we have followed demonstrate a small lymphocyte population initially but go onto exfoliate neoplastic cells at a later date, coincident with a recrudescence of the effusion and signs of tamponade. The lag time between initial presentation with a hemorrhaghic, non-neoplastic, small lymphocyte predominant pericardial effusion, and this recrudescence can be months to several years, and during that time frame cattle can return to expected production levels, and successful reproduction.

TREATMENT

Successful treatment in the published literature has revolved around the administration of anti-inflammatories, diuretics and pericardial drainage[1,2]. Our approach to treatment has been similar, and predictable, based upon the need for relief of tamponade due to a large volume pericardial effusion. Pericardial drainage using a 20-24 French trochar placed under ultrasound guidance, and local anesthesia, into the pericardial sac at the time of diagnosis, results in a perceptible improvement clinically, as well as ultrasonographically in terms of cardiac function, within hours. Chest trochar placement can be either right or left sided depending upon the most obvious window imaged in that individual, but is reliably performed in the 5th or 6th intercostal space at the approximate level of the elbow. Typically we attempt to remove as high a percentage of the effusion as possible, leaving a depth of 1-

2 cm from the epicardial surface. Given that the depth of effusion initially can be 10 cm or more there is an obvious and visible improvement in cardiac contractility and function following drainage. Concurrent with this we begin a short course of dexamethasone, initially administering the drug at 0.1 mg/kg for 1-3 days, followed by a tapering dose over the next 3-5 days. This combined steroid and drainage approach has proved remarkably successful in cattle with the classic non-neoplastic effusion, but is less effective should an individual already have neoplastic lymphocytes in the effusion. Under those circumstances we recommend palliative treatment for salvage (ovary harvest, etc.). In pregnant cattle, even those at a stage of gestation when dexamethasone is not be expected to cause abortion, we have experienced a high rate of pregnancy loss during therapy for this condition. We have limited experience with isoflupredone acetate as an alternative to dexamethasone, but subjectively do not feel the response is as good.

BOVINE LEUKOSIS STATUS

Although the BLV status of the cattle in the original reports was not reported we have examined each case admitted to our hospital for antibodies to BLV by either AGID or ELISA (and in many cases both) and have found all animals to be positive, indicating pre-existent infection at the time of initial presentation in all but one individual. Curiously that individual also became seropositive within 4 weeks of the initial diagnosis.

ASSOCIATION WITH LYMPHOSARCOMA

Although comprehensive post mortem and follow-up information was not available in the prior case series published by Jesty et al[1] and Firshman et al[2], we have been able to follow many cattle with this condition for months to years from the time of their initial diagnosis. As expected, grade cattle may be culled from commercial herds for many reasons and complete post mortem examination on many of these individuals is not possible. However, from close follow-up of several individually valuable cattle with this condition it appears likely that individuals will eventually suffer recrudescence, and that a high proportion of them will go on to develop cardiac lymphosarcoma, most commonly in an epicardial location. Those cattle that have gone onto develop neoplastic infiltrates in the effusion, or that demonstrate such an effusion at first presentation have most often demonstrated a particular form of epicardial lymphosarcoma at post mortem that is repeatable and quite consistent in its gross appearance. There is a single case report in the literature of a similar lesion[3].There have been variable degrees of lymphosarcoma within the rest of the body of these at post mortem, but most commonly there has been predominant involvement of just thoracic structures, especially local lymph nodes. Concomitant infiltration of other classic anatomic predilection sites[4,5], or generalized lymphadenopathy have been rare in affected cattle.

OUTCOME

As previously noted by Firshman et al[2], and Jesty et al[1], survival times with this condition can be extraordinarily long compared to other causes of cardiac disease with tamponade in cattle. In the original reports survival times of from 6 months to several years were reported and we have had similar experiences. In fact this particular condition appears to have the potential for far longer survival times than other clinically important causes of cardiac disease in cattle. Many of the individually valuable cows that we have treated have survived for several years during which time their individual genetic merit has been maximized through assisted reproductive technologies and/or they have continued to be productive lactating dairy cows. Several have conceived and carried normal pregnancies to term without recrudescence. Follow-up on a small series of cattle with the condition suggests that there is a very high likelihood that signs of cardiac tamponade and congestive heart failure will ultimately return and that a high proportion of cattle will ultimately develop cardiac lymphosarcoma. The form of lymphosarcoma that we have seen, as mentioned above, is different to the classical right atrial myocardial neoplastic infiltrate, instead tending to involve the epicardium.

REFERENCES

1.Jetsy S, et al. **JAVMA** 2005;226(9):1555. 2.Firshman A, et al. **J Vet Intern Med** 2006;20:1499. 3. Ivany JM, and Illanes OG. **Can Vet J** 1999, 40, 819. 4. Marshak RR et al. **Cancer Res** 1962, 22, 202. 5. Divers TJ and Peek SF. **Rebhun's Diseases of Dairy Cattle**, 2nd edition; 2008: 627-629.

NOVEL STRATEGIES FOR THE PREVENTION OF JOHNE'S DISEASE IN CATTLE

Marie-Eve Fecteau, DVM, DACVIM
Kennett Square, PA

INTRODUCTION

Because management practices, presently considered the best way to control the spread of *Mycobacterium avium* subsp. *paratuberculosis* (MAP), have not and most likely will not eliminate MAP from food animals, other preventive measures have been investigated. It is generally accepted that most calves become infected with MAP very early in life. For this reason, control programs are primarily based on preventing direct or indirect transmission of MAP organisms from adult cattle to young replacement stock on the farm. Vaccination, available on a limited basis in the United States, may reduce the incidence of clinical disease, and was associated with reduced colonization of intestinal tissues in experimental studies, but vaccinates are not fully protected from infection, and can still shed MAP in their feces.[1-2] Prophylactic administration of an antimicrobial agent to neonatal calves during the period of high susceptibility and high exposure may represent an additional approach to the prevention of infection.

MONENSIN

Monensin (Rumensin), an ionophore, has been widely used in the beef industry for years to enhance feed efficiency and to control coccidiosis. More recently, monensin has been used in dairy cattle to improve energy metabolism, milk production, and to alter milk components.[3] Monensin is a carboxylic polyether ionophore produced by *Streptomyces cinnamonensis* and is fed to cattle as a sodium salt. It binds to bacterial cell membranes, especially gram positive bacteria and mycobacteria, causing an efflux of potassium from the cell and an influx of hydrogen ions into the cell. The cell expends energy to maintain the electro-chemical gradient resulting in death or reduced growth of the bacterial cell.

Brumbaugh and collaborators first demonstrated a reduction in the number of colony forming units of MAP from the livers of experimentally infected mice treated with 15 and 30 mg/kg of monensin when compared to non-treated controls.[4] The MIC of monensin for 1 field isolate of MAP was investigated in an *in vitro* setting and was found to be 0.39 ug/mL.[5] In another *in vitro* experiment, monensin showed dose-dependent growth inhibition in all 5 strains of MAP tested.[6] Lastly, cattle naturally infected with MAP and fed monensin (450 mg daily) for 120 days had reduced severity of paratuberculosis-related histological lesions in the liver, ileum and rectal mucosa when compared to the control cows.[7] However, no significant difference was found when the mesenteric lymph nodes were compared.[7]

In a study of nearly 5,000 Canadian dairy cows in 94 herds monensin usage was associated with reduced occurrence of milk-ELISA positivity.[8] In 48 herds where Johne's disease (JD) had not been detected previously and in another group of 46 herds where paratuberculosis had been detected previously, monensin usage was associated with a reduced odds ratio of a cow testing positive for JD using the milk-ELISA test.[8] Monensin usage included both monensin premix and the controlled release capsule (CRC).[8] In another study that employed CRC boluses designed to release 335 mg monensin per day in culture positive dairy cows, treated cows had a moderate decrease in fecal MAP colony forming units during the 98 day study period.[9]

In an experiment that included 12 neonatal Holstein calves, 35 mg of monensin (n=6) or placebo (n=6) was added to the milk replacer at each twice-a-day feeding for a total of 70 mg of monensin/treated calf/day.[10] Both groups of calves were then administered two oral doses of viable MAP on two consecutive days between days seven and nine of the trial. At the time of challenge, calves were between 8 and 11 of age. Calves were euthanized between days 65 and 67 days after starting the trial. Feces and 50 individual tissues, including multiple sections of intestine, abdominal organs, and mesenteric and peripheral lymph nodes were harvested from each calf and processed for HEYM culture. Calves fed monensin had fewer culture-positive fecal samples and fewer MAP colony forming units detected in their manure compared to controls. Furthermore, monensin-fed calves had fewer culture-positive tissues and lower numbers of MAP colony forming units in their tissues compared to controls. Results of this study suggested that monensin effectively reduced tissue colonization with MAP following oral challenge, and also reduced fecal pass-through shedding of the organism.[10] Presumably, reduced tissue colonization in the short-term model would translate to lower mycobacterial burden and likelihood of shedding MAP in manure and clinical disease in adulthood.[10] In the study reported here, monensin was found to be relatively insoluble in milk. Therefore, the authors recommended proper and constant mixing if monensin is added to bulk milk intended to be fed to calves.

Monensin is safe when used in target species and at the recommended dosages. Intoxication may come after mixing errors that result in its inclusion in the diets of nontarget species or in excessive concentrations in the diets of target species. The toxicity of monensin in cattle and other species is known to be dose dependent. In cattle, the clinical signs of acute monensin toxicity are anorexia (24 to 36 hours post ingestion), diarrhea, dullness, weakness, ataxia, dyspnea, prostration, and death within 3 to 14 days post ingestion of the incriminated feed. Typical pathological findings in cattle suffering from monensin toxicity are cardiac and skeletal muscle degeneration and necrosis, with secondary lesions from acute cardiac failure or chronic cardiovascular insufficiency.

In conclusion, monensin added to cattle rations at all phases of their life, coupled with stringent implementation of biosecurity management practices at the farm level, may play a useful role both in the prevention of MAP infection in young cattle, and in diminishing MAP bioburden and fecal shedding in adults with established infections.

GALLIUM

Gallium (Ga) is a trivalent semi-metal that shares many similarities with ferric iron and functions as an iron mimic. Ga is preferentially taken up by phagocytes at sites of inflammation,[11] and its biologic effect on susceptible microorganisms appears to relate to its ability to substitute for ferric iron in many cellular metabolic pathways and disrupt them.[12] The *in vitro* antimicrobial activity of Ga has been shown against various microorganisms, including *Rhodococcus equi*, *Pseudomonas aeruginosa*, *Mycobacterium tuberculosis*, and *M. avium* complex.[12-14] More recently, the author demonstrated the antimicrobial efficacy of Ga against 10 MAP isolates in an *in vitro* setting.[15] In a follow-up experiment investigating the antimicrobial activity of Ga in neonatal calves experimentally infected with MAP, the author found that: (a) Ga treatment was associated with a significant reduction in MAP tissue burden when compared to control calves, and (b) no adverse effects related to Ga in the treated calves.[16] Further studies are needed to determine the optimal chemoprophylactic dose of Ga that should be employed in neonatal calves, as well as large on-farm clinical trial to further elucidate its effect on the prevention of MAP infection in cattle. Additional concerns regarding the use of Ga in cattle including drug residues, cost of prophylactic treatment, and environmental impact would also have to be considered.

SUMMARY

JD control programs are primarily based on preventing direct or indirect transmission of MAP organisms from adult cattle to young replacement stock on the farm. Prophylactic administration of an antimicrobial agent to neonatal calves during the period of high susceptibility and high exposure may represent an additional approach to the prevention of infection. Monensin, added to cattle rations at all phases of their life, may play a useful role both in the prevention of MAP infection in young cattle, and in diminishing MAP bioburden and fecal shedding in adults with established infections. More recently, the use of gallium as a chemoprophylactic agent against MAP has been investigated in neonatal calves and has promising results.

REFERENCES

[1]Kalis CH, et al. **Am J Vet Res** 2001; 62(2): 270. [2]Sweeney RW, et al. **Am J Vet Res** 2009; 70(4): 493. [3]Duffield TF, et al. **J Dairy Sci** 1999; 82(2): 272. [4]Brumbaugh GW, et al. **Am J Vet Res** 1992; 53(4): 544. [5]Brumbaugh GW, et al. **Can J Vet Res** 2004; 68(3): 175. [6]Greenstein RJ, et al. **Gut Path** 2009; 1(1): 4. [7]Brumbaugh GW, et al. **J Comp Path** 2000; 123(1): 22. [8]Hendrick SH, et al. **Prev Vet Med** 2006; 76(3-4): 237. [9]Hendrick SH, et al. **Prev Med** 2006; 75(3-4): 206. [10]Whitlock RH, et al. **Proceedings AABP**, Salt Lake City; 2005, p. 191. [11]Tsan MF. **J Nucl Med** 1985; 26(1): 88. [12]Olakanmi O, et al. **Infect Immun** 2000; 68(10): 5619. [13]Harrington JR, et al. **J Vet Pharmacol Therap** 2006; 29(2): 121. [14]Kaneko Y, et al. **J Clin Invest** 2007; 117(4): 877. [15]Fecteau ME, et al. **Am J Vet Res** 2011; 72(9): 1243. [16]Fecteau ME, et al. **J Vet Intern Med** 2011; 25(5): 1152.

ANTHELMINTIC RESISTANCE AND APPLICATION OF FAMACHA IN CAMELIDS

Lisa H. Williamson, DVM, MS, DACVIM
Athens, Georgia

INTRODUCTION

Anthelmintic resistance is a worldwide phenomenon that is rapidly progressing in many parasitic species of livestock.[1] Multiple-drug resistant parasites are currently a major threat to the health and productivity of small ruminants. Goat herds and sheep flocks with parasitic communities that are resistant to all available anthelmintics and now the status quo rather than a novel finding, particularly in temperate parts of the United States.[1] Evidence is mounting that camelid parasites are also acquiring resistance to commonly used anthelmintics.[2-3]

MULTI-DRUG RESISTANCE

A great deal of research has been conducted on small ruminants documenting resistance to various anthelmintics, and on methods to reduce the progession of resistance.[1] The knowledge acquired from these endeavors is applicable to camelids since they share several important parasites, most notably *Haemonchus contortus* and *Trichostrongylus colubriformis*. *Haemonchus contortus* is a blood feeding nematode that damages the third compartmental and abomasal mucosa in camelids and small ruminants, respectively. Heavy infections cause anemia, hypoproteinemia, loss of fecal consistency, and significant weight loss. Periparturient females, youngsters in their first grazing season, and animals debilitated from another disease are most at risk.. *Trichostrongylus colubriformis* is less virulent than *H. contortus*; morbidity occurs primarily through irritation of the small intestinal mucosa. Although *H. contortus* is most prevalent in warm, humid areas, it is not restricted to these areas. Life-threatening haemonchosis has been documented in small ruminants and camelids living in many northern states, and in Canada (B. Storey, personal communication). A recent study conducted on 45 sheep and goat farms in the southeastern United States demonstrated that *H. contortus* isolates were resistant to benzimidazoles, levamisole, and ivermectin on 98%, 54% and 24% of the farms, respectively.[4] Resistance to all three anthelmintics (hence, all 3 major anthelmintic classes) was evident on 48% of the farms, and total anthelmintic resistance (ie, resistance to benzimidazoles, levamisole, ivermectin and moxidectin) was identified on 17% of the farms.[4] The situation is not as dire in camelids, but the same trends are evident. Larval developmental assays conducted on 32 camelid farms showed that 97% of the farms had multi-anthelmintic resistant *H. contortus* isolates.[3] Resistance to the benzimidazoles was note on 100% of the farms, and resistance to ivermectin was noted on 88% of the farms. However, only 22% of the *H. contortus* isolates were resistant to levamisole, and 22% were resistant to moxidectin. Total anthelmintic resistance was identified on one camelid farm.[3] On 2 alpaca farms where initial testing demonstrated moxidectin susceptibility, conversion to moxidectin resistance occurred over a 1-2 year period in the *H. contortus* isolates (Williamson, unpublished data). Although the reason for the conversion from moxidectin susceptibility to resistance was not definitively established, both producers made management decisions that promote anthelmintic resistance: (1) they assimilated alpacas from other sources (with unknown parasite load) into their herds without appropriate quarantine, testing and treatment, and (2) they administered moxidectin to the whole herd on several occasions. Nonselective treatment (treat all the animals in the herd at the same time) is recognized as a major risk factor for escalating drug resistance in sheep nematodes.[5] Nonselective treatment protocols fail to preserve refugia, which is the proportion of the worm population spared selection pressure from exposure to anthelmintics. The refugia provide a reservoir of drug-susceptible genes that help dilute out resistant genes in the worm population. Refugia include eggs and larval stages in the environment, and parasites in animals that were not treated with an anthelmintic. When a large refugia pool is maintained, the evolution of parasite resistance is slowed.[5]

SELECTIVE TREATMENT

In healthy camelid herds, approximately 20-30% of the animals harbour 80 percent of the parasite population (Williamson, unpublished data). This uneven distribution of parasites among herd mates is seen in other classes of livestock, including small ruminants, cattle and horses. The majority of the animals have low parasite burdens, and they will not gain much benefit from anthelmintic treatment. Low-level parasitism is well tolerated by healthy hosts, and total elimination of parasites from grazing animals is an impractical and unsustainable goal. When minimally infected animals are left untreated, they provide a major source of refugia, which is beneficial to maintaining usefulness of anthelmintics. The benefits of adopting a selective treatment program cannot be over-emphasized. The consequence of frequent, nonselective use of anthelmintics will be loss of efficacy of current and future anthelmintics. The enormous cost associated with research and development of new drugs has made pharmaceutical companies less willing to invest in farm animal products.[6] As an example, monepantel is the first new class of anthelmintics to be launched in almost 3 decades.[7] Parasitologists predict that we are currently entering a phase where there will be even fewer new anthelmintic classes coming to the market to replace those that are failing.[6] Therefore, selective use of anthelmintics, and management practices aimed at breaking the parasitic life cycle, are critical aspects of current and future parasite control strategies. From a practical standpoint, the main obstacles to selective treatment is that it takes more time and effort to evaluate each individual in a herd, and treat on an as-needed basis. Body condition score (BCS) is a useful parameter to monitor in camelid herds, as it is a good barometer of overall health. Healthy-appearing camelids that have optimal-to high body condition score are less likely to be harbouring health-threatening worm burdens than are underweight camelids (Williamson, unpublished data). It is important to notice changes in body condition scores among individuals, and the overall herd, especially if the scores are declining. Selective treatment decisions can also be made using information obtained from quantitative fecal egg counts, Further, in herds where haemonchosis is an issue, physical indicators of anemia such as the FAMACHA ocular score can be used to make treatment decisions. These topics will be covered in the next 2 sections.

FECAL MONITORING FOR PARASITES

Fecal egg monitoring is vitally important for determining what kind of internal parasites are present on a farm. Most camelid farms have a variety of parasites present among herd members. *Haemonchus contortus* and *T. colubriformis*, are consistently high egg shedders, so quantitative fecal egg count are excellent indicators of the magnitude of infection. A female *H. contortus* typically produces approximately 5,000 eggs per day, and a *T. colubriformis* female produces about 500 eggs per day. Methods for fecal egg quantitation include the modified McMasters test (sensitivity of 25 eggs per gram), as well as centrifugation methods such as the Stoll's and the modified Wisconsin tests. These tests differ in their sensitivity end points as well as the likelihood of detecting certain types of parasites. Centrifugation techniques are more likely to detect low levels of *Trichuris* sp, Nematodirus sp eggs, and large coccidian oocysts than the McMasters tests. Fecal egg counts below < 900 epg of trichostrongyles were rarely associated with anemia or ill thrift in healthy adult llamas and alpacas when *H contortus* was the predominant trichostrongyle (Williamson, unpublished data). However, even if the fecal egg count is low, animals showing signs of haemonchosis should be treated with an effective anthelmintic, as prepatent *H contortus* infections can cause serious morbidity. Further, fecal egg counts are only as accurate as the methodology used to generate them, and "set points" for when therapeutic intervention is deemed necessary will differ accordingly. The fecal egg count reduction test (FECRT) is commonly used to test for phenotypic expression of anthelmintic resistance.[1] *In vitro* larval developmental assays are another way to assess this trait. The FECRT can be performed in herds of animals where there are sufficient numbers to dived them into groups of 10 (or more). The number of animals available will determine how many anthelmintics can be studied at one time. One group is left as an untreated control group. If smaller group sizes are used, then pre-treatment fecal egg counts are needed so that the egg counts can be balanced among the treatment groups. If *H. contortus* is the predominant worm, and there are sufficient animals to make large groups, camelids can be assigned to groups based on FAMACHA© scores, thereby eliminating the need for pre-treatment fecal egg counts. Camelids with FAMACHA scores of 3, 4 and 5 are the best test candidates, as they are most likely to be shedding sufficient eggs. Test groups should be balanced according to FAMACHA© scores so that infection is distributed more evenly among groups. Camelids that are extremely pale (FAMACHA© 5/5) should not be left in a control group. They should receive immediate treatment with an effective anthelmintic. Ten to 14 days after anthelmintic treatment, individual fecal samples are collected for quantitative fecal egg counts from the rectums of all the camelids participating in the FERCT. Air should be evacuated out of the plastic sandwich bag manually prior to sealing, or by using a Reynolds handi-vac system. Air exclusion and refrigeration delay larvation, thereby allowing samples to be stored for 5-7 days before processing them. Once individual post-treatment fecal egg counts are performed, the arithmetic mean FEC for each treatment group and the control group are determined. The means are compared using the formula: **FECR = (1- (mean of treatment group ÷ mean of control group) X 100** Based on guidelines established by the World Association for Advancement of Veterinary Parasitology, an efficacious treatment will result in a FECR of greater than 95%.[8] Reductions of 90-95% indicate low-level resistance, and reductions of less than 90% indicate resistance.[8] Varying clinical benefit will be perceived in the "gray zone" between 50 and 89%, but treatment does not bring about much improvement by the time efficacy ≤ 50%.

HAEMONCHUS CONTORTUS AND THE FAMACHA SYSTEM

Since anemia is a prominent feature of hemonchosis, pallor of the conjunctiva and mucous membranes is a good indicator of the severity of *H. contortus* infections. Dr. Francois Malan noticed the association between hematocrit and eyelid pallor in parasitized sheep in South Africa.[9] He subsequently developed a laminated card that depicted 5 illustrations of ocular membrane colors. The membrane colors are categorized as: 1 = deep red (nonanemic), 2= red-pink (nonanemic), 3 = pink (mild anemia), 4 white-pink = anemic, and 5 = white (severely anemic). The ocular colors correlated with mean hematocrit in parasitized sheep. The ocular chart was subsequently named the FAMACHA© chart. The acronym, "FAMACHA" is derived from Dr Malan's nickname (Faffa), his last name (Malan), and the word "chart". Studies conducted in the United States on sheep and goats with haemonchosis have demonstrated that the FAMACHA© eye score, packed cell volume, and fecal egg counts are all highly correlated.[10-12] Since introduction of the FAMACHA system in 2004, over 15,000 FAMACHA© cards have been distributed in the United States, and numerous workshops have been conducted to teach small ruminant producers how to implement the concepts in their herds and flocks.

The FAMACHA© system was recently evaluated on 17 llama and alpaca farms that had a high prevalence of *H. contortus*. Packed cell volume was highly correlated to FAMACHA© eye scores, fecal egg counts and body condition scores.[12] Based on these findings, the FAMACHA© system has been deemed a useful tool for management of *H. contortus* in llamas and alpacas. It is most accurate when the prevalence of anemia associated with haemonchosis is high. The FAMACHA© system is not intended as a management tool for other types of internal parasites, however. Ocular examinations should be conducted in bright natural light for the sake of accuracy, using the FAMACHA© card as a color reference. Alpacas and docile llamas can be restrained manually during the evaluation, but more fractious camelids need to be restrained in a chute. The normal tendency of camelids is to object to the presence of a hand on their forehead. Camelids are often more accepting of the scoring process if the examiner slowly slides one hand up the side of the animal's face, then uses the index finger to retropulse the globe through the upper eyelid. The thumb of the same hand is used to evert the lower lid. The FAMACHA© card is held beside the eye with the opposite hand. The examiner should expose as much conjunctiva as possible to accurately assess the color. The color of the lower eyelid conjunctiva is compared with the FAMACHA© card over a 1-3 second period, and assigned a score. Both eyelids are scored, and if variation is noted between the 2 eyelid scores, the more anemic (higher) score is assigned. Some camelids have heavily pigmented conjunctiva, which makes scoring more challenging. Animals scoring in categories 4 or 5 are almost always anemic, and should be treated with an effective anthelmintic. Animals scoring in categories 1 or 2 are not anemic so treatment for haemonchosis is not indicated. Many camelids that score in category 3 are not anemic, but they should be managed in light of prevailing circumstances. For instance, high risk animals (young, pregnant, nursing, elderly) in herds where the majority of the members have anemic scores and/or poor body conditions should be treated, particularly during periods of high *H. contortus* transmission.

Based on recommendations for sheep and goats, FAMACHA© examinations should be conducted every 1-2 weeks during periods of high *H. contortus* transmission, and less frequently during times of the year when transmission is typically low.[11] Although haemonchosis is a major cause of anemia in camelids, other causes should be considered, including ingestion of wilted red maple leaves, chronic disease, naturally occurring iron deficiency, and symptomatic Candidatus Mycoplasma haemolamae (CMhl) infections. Although most CMhl infections are asymptomatic, red blood cell destruction can occur if the animal becomes severely stressed or immunocompromised.[13]

REFERENCES

1. Kaplan, RM, et al. **Vet Parasitol** 2011; http://dx.doi.org/10.1016/j.vetpar.2011.11.048 2. Gillespie RM, et al. **Vet Parasitol** 2010;172:168-171.
3. Storey BE, et al. **Proc 23rd WAAVP** (abstract) 2011. 4. Howell S, et al. **J Am Vet Med Assoc** 2008;233:1913-1919. 5. Waghorn TS, et al. **NZ Vet J** 2008;56:158-163. 6. Geary TG, et al. **Tends Parasitol** 2004;20:449-455. 7. Kaminsky R, et al. **Parasitol Res** 2008;103:931-939. 8.Coles G, et al. **Vet Parasitol** 1992;44:35-44. 9. Malan FS, et al. **J Vet Res** 2001;68:165-174, 2001. 10. Kaplan RM, et al. **Vet Parasitol** 2004;123:105-120. 11. Burke JM, et al. **Vet Parasitol** 2007;147:89-95. 12. Williamson LH. Proc 21 **WAAVP** (abstract) 2009. 13. Tornquist SJ. **Vet Clin Food Anim** 2009;25:311-322.

COCCIDIOSIS IN NEW WORLD CAMELIDS

Christopher Cebra, VMD, MA, MS, DACVIM
Corvallis, OR

New World camelids appear to be susceptible to at least 5 species of *Eimeria*. These appear to be camelid-specific, unlike many worms, *Giardia*, and *Cryptosporidium*. Thus, transmission from ruminants or wildlife is not thought to be important. Camelid coccidia do, however, affect all species of New World camelid, and whether they also affect Old World camels is an open question. Whereas coccidiosis is primarily thought to be a disease of the young in many species, and is well publicized as a cause of illness and death in crias in South America, it is becoming more commonly recognized worldwide as a cause of adult morbidity and mortality as well.

Camelid coccidia can be lumped into categories of small and large. The small coccidia are relatively conventional in appearance and life cycle, and will be familiar to anyone acquainted with coccidial disease in domestic poultry or ruminants. The large coccidia of camelids are relatively unique. The small coccidia include, by increasing size of oocysts, *Eimeria punoensis* (17 to 22μ in length), *E. alpacae* (22 to 26μ), and *E. lamae* (30 to 40μ).[1] Oocysts become more ovoid as they get larger, so that *E. punoensis* is about 16% longer than it is wide without a clearly visible polar cap (by standard light microscopy), whereas *E. lamae* is about 60% longer than it is wide and has an obvious cap. These follow the general coccidial lifecycle. Oocysts sporulate and become infective after 4 to 12 days (for *E. lamae*) or more outside of the host.[2] Sporulation rate is highly dependent on environmental conditions and currently under investigation. After ingestion, sporulated oocysts usually release 4 sporocysts each containing 2 sporozoites, which penetrate epithelial cells. The host cell nucleus and organelles are marginalized by the parasitophorous vacuole. When this has matured, the cell ruptures and hundreds to thousands of merozooites are released. These repeat the cycle a number of times, creating more mucosal damage. Eventually, some merozoites develop into sexual forms and oocysts subsequently appear in the feces. The prepatent periods are approximately 10 days for *E. punoensis*, 15-16 days for *E. lamae*, and 16-18 days for E. alpacae.[3]

The small coccidia are best associated with hemorrhagic, watery diarrhea progressing to weakness, lethargy, weight loss or poor weight gain, feed refusal, dehydration, and eventually shock, coma, and death.[4] Colic, respiratory distress, and cerebral signs are uncommon or late findings. The gut, particularly the terminal jejunum and ileum, is occasional hemorrhagic or markedly edematous, and may have areas of mucosal hemorrhage, fibrinonecrotic pseudomembranes, or punctate white lesions, but is more commonly grossly non-remarkable. Histologically, lesions are most pronounced in the villi. There is mucosal loss and villus shortening. Immature and mature forms of the coccidia may be present. The submucosa is often filled with hemorrhagic or eosinophilic infiltrates. In severe cases, the mucosa is lost to the basement membrane. Protein loss is considerable, and hypoproteinemia is the most consistent blood abnormality. Anemia, hyponatremia, and hypochloremia are other common abnormalities.

The small coccidia primarily cause clinical disease in crias up to around 8 months of age. South American crias are usually shedding by around 23 days of age, with earliest shedding by 15 days, meaning they become infected shortly after birth. Shedding increases until 40 to 50 days of age, then gradually tapers off. Illness usually occurs in those first 2 months. Under rare circumstances, clinical disease is seen in older crias or adults. This usually reflects overwhelming exposure or a poor immune response.

The large *Eimeria* of New World camelids are *E. macusanensis* and *E. ivitaensis*. These are 3 to 4 times larger than small coccidia, and are approximately 80 to 100μ in length.[5-8] As such, they resemble *E. leuckarti* of horses, *E. camelli* of camels, and other large *Eimeria. E. ivitanensis* oocysts are elongated ellipses, whereas *E. macusaniensis* is ovoid and pyriform, resembling a cut avocado or watermelon seed in shape. Both have an obvious polar cap. There is some heterogeneity in size and shape of *E. macusanensis*, and it is possible that future research will reveal that distinct species exist. *E. macusanensis* also has a thick wall (approximately 8.5 to 11μ), which makes the cyst extremely durable; identifiable cysts have survived approximately 10,000 years in mummies.

The life cycles of the large coccidia resemble those of small coccidia, except that everything generally takes longer. The prepatent period for *E. macusaniensis* is from 32-43 days;[7] that from *E. ivitaensis* has not been reported. Sporulation times for *E. macusaniensis* range from 2 to 3 weeks, with faster times under warmer conditions. Sporulation appears to arrest at 7°C or below. Sporulation times for *E. ivitaensis* have not been reported, but appears to be in the 7 to 10 day range in our laboratory. The longer lifecycle means that patent infections appear later than with small coccidia, but not necessarily that disease occurs later. Severe disease and death appear to be able to occur within 3 weeks of initial exposure and 2 weeks before establishment of patency.[6] There is growing evidence that crias shedding small *Eimeria* oocysts or showing signs of enterotoxemia, may actually be dying of prepatent *E. macusaniensis* infection.[10] Additionally, there are increasing reports of prepatent or patent disease in adult camelids. Some of these are long-time herd residents, but most have a history of transportation and mixing with new groups of animals. Whereas shows, sales, and movement for breeding may cause stress and inhibit the immune response, the simplest explanation may lie in eating habits: new entrants in a herd are more likely to eat off the ground than out of feeders, or more likely to eat in the less desirable areas of pasture. Thus, they may ingest larger doses of the parasite and be more likely to show disease signs.

The characteristics of clinical disease associated with *E. macusaniensis* have been studied and reported more extensively than any other parasitic gastroenteritis in camelids.[8,11,12] It is likely that much of this information pertains to other coccidial infections and to parasitic gastroenteritis in general. In younger camelids, gastroenteritis is more likely to result in clinical diarrhea, whereas in older camelids, diarrhea is often absent or easy to miss. In fact, in adults, it is frequently counterintuitive to directly link the presenting signs to GI illness. In addition to or instead of diarrhea, general signs include weight loss or poor weight gain, ill-thrift, and increasing lethargy, weakness, and loss of appetite. It is one of the most common causes for weakness, weight loss, hypoproteinemia, or ill thrift in our area. Some animals show colic signs, probably more with *E. macusaniensis* than with any other parasitic gastroenteritis. As the disease progresses, hypoproteinemia worsens without commensurate anemia; mild anemia is common, but hypoproteinemia and hypoalbuminemia often become severe (such as 3.8 to 2.2 mg/dl for serum total protein and 1.8 to 0.6 mg/dl for serum albumin concentrations). There is also often some reduction in serum sodium and chloride concentrations, and hypokalemia become more marked with anorexia. With worsening disease, the animal becomes susceptible to translocation of bacteria or toxins through the damaged mucosa, the effects of hypoproteinemia, and circulatory shock. Signs of other organ systems are seen: camelids may develop ascites, hydrothorax, hydropericardium, pharyngeal edema, and cerebral edema, with ensuing abdominal distention, increased respiratory effort, lethargy, tachycardia, hypothermia, dysphagia, salivary loss, and cerebral signs. Eventually, there is also clinicopathologic evidence of the systemic disease including azotemia, metabolic (usually lactic) acidosis, high liver enzymes, hyperbilirubinemia, and increases in fat fractions. Abdominocentesis usually yields a transudate. Abdominal imaging is also inconclusive: colicky camelids may have ileus and fluid-distended intestine, though usually to a lesser degree than camelids with GI obstruction. Thickened bowel walls are rare (<10% of cases). In some cases, signs are insidious, and the animal is simply found dead.

Major differential diagnoses for the combination of ill-thrift, protein loss, and diarrhea include GI nematodes, subacute forestomach acidosis, salmonellosis, coronavirus infection, and in crias, cryptosporidiosis and giardiasis. In adults with this combination signs, particular if diarrhea and colic signs are evident, whipworm infestation most closely resembles coccidiosis; other gastrointestinal nematode infestations are rarely associated with diarrhea. The differential list for ill-thrift without diarrhea is extensive, with major additions including neoplasia, other forms of parasitism, chronic infections, and liver diseases.

Light microscopic examination of fecal samples is also the standard method of detecting coccidia. There are a number of pitfalls, namely prepatent or peripatent disease, intermittent shedding, and oocysts with a range of densities. Small *Eimeria* and *E. ivitaensis* oocyst appear to float readily in flotation solutions with a specific gravity of 1.18 (saturated saline, 33% $ZnSO^4$). *E. macusaniensis* requires denser solution (S.G. >1.025?).[13] Prepatent disease is more common in the coccidia with longer life cycles, namely *E. macusaniensis*. For *E. macusaniensis*, clinical disease and death can occur up to 2 weeks before the appearance of oocysts in the feces. For small coccidia, clinical disease usually appears around the same time as fecal cysts passage, but may precede it by a day or two. If prepatent disease is suspected, serial fecal analyses at two to four day intervals may be used to confirm the organism; presumptive treatments may be instituted in the interim. When looking for prepatent or peripatent shedding of protozoan parasites, it is important to keep the shedding curve in mind; *Eimeria* oocyst passage resembles a bell-shaped curve, with few at the earliest stages, peaking in 1 to 2 weeks before tapering off.

A number of other techniques can be used to detect or presume infections. The combination of hyponatremia and moderate to severe hypoalbuminemia with milder anemia is highly suggestive (more severe anemia suggests haemonchosis). A fecal PCR has been developed and shown promise in identifying prepatent infection.[14] Postmortem tests include gut mucosal impression smears or histopathology. The physical appearance of the gut is highly variable: from thin to thick, with a mucosal surface that can appear normal, appear hemorrhagic and inflamed, be coated with pseudomembranous plaques, and have multiple minute white foci embedded in it. Whenever examining intestines for coccidia, it is important to remember that the ileum is often the site of the most severe lesions. However, infections moves throughout the bowel during its course, and collection of multiple samples from the duodenum to the cranial loops of the ascending colon is recommended to increase the chance of finding it.

Treatment can be prophylactic or therapeutic, and can be based on a presumptive or a confirmed diagnosis. Dosage, frequency, agent, and population to be treated depend on the specific situation. The most commonly used anticoccidial medications for camelids in North America have historically been amprolium (10 mg/kg at the label dilution, PO, q24h for 5 days) and sulfa antibiotics for both prophylactic and therapeutic needs. Both are more effective against the immature forms of the parasite in ruminants and should not be expected to immediately reduce fecal shedding. Efficacy in camelids remains empirical, except for a small South American study investigating sulfas in crias, and pharmacokinetic trials on sulfa antibiotics suggests that uptake (and therefore efficacy) is negligible once the forestomach develops. Likewise, amprolium put in the water supply is often not drunk in sufficient quantities to be effective; individual animal dosing is preferred.

Treatment with amprolium during the prepatent period does appear to reduce subsequent shedding, and also decrease the viability of oocysts. There has been some concern that protracted treatment with amprolium may cause polioencephalomalacia, but I have not seen that with the 5-day course. Milling decoquinate (0.5 mg/kg for 28 days) into pelleted feeds may also be helpful, if the crias are old enough to eat the pellets.

The most commonly used anticoccidial medications for camelids in North America are amprolium (10 mg/kg at the label dilution, PO, q24h for 5 days) and sulfa antibiotics. Both are more effective against the immature forms of the parasite in ruminants and should not be expected to immediately reduce fecal shedding. Efficacy in camelids remains empirical, and pharmacokinetic trials on sulfa antibiotics suggests that uptake (and therefore efficacy) is negligible once the forestomach develops. Likewise, amprolium put in the water supply is often not drunk in sufficient quantities to be effective; individual animal dosing is preferred. Treatment with amprolium during the prepatent period does appear to reduce subsequent shedding, and also decrease the viability of oocysts. There has been some concern that protracted treatment with amprolium may cause polioencephalomalacia, but I have not seen that with the 5-day course. Milling decoquinate (0.5 mg/kg for 28 days) into pelleted feeds may also be helpful, if the crias are old enough to eat the pellets.

Benzeacetonitrile compounds have been gaining popularity as a coccidia treatment in the U.S. and are in heavy use in other countries. Ponazuril and toltrazuril (5 to 20 mg/kg, PO, q24h for up to 3 days) are effective against multiple stages of the parasite and rapidly stop shedding. The higher doses and longer courses are for treatment of individuals; the lower doses may be used for control. Diclazuril has seen limited use; in cattle, gastric absorption is lower than for toltrazuril.

With sick animals with presumptive or confirmed clinical coccidiosis, therapeutic courses are indicated. Contact animals may also be treated if considered at risk. This varies tremendously based on individual circumstances, especially in adults. Eating off the ground is a major risk factor. With crias, it is relatively ubiquitous, as they approach their surrounds with curiosity and ingest a variety of substances. Grazing adults or those fed on the ground are likewise at a universal level of risk. Adults or juveniles feed out of feeders may have dominant animals eat off the ground, whereas more submissive animals eat leavings off the ground, or are fed separately, increasing the risk of those individuals. In cases where the risk is universal, treatment of the entire group may be warranted on identification of a single sick animal. Where the risk is heterogeneous, selection of high risk animals for treatment is more efficient. If risk is unknown, simple tests such as body condition scoring, weight measurement, or blood protein determination can be used to identify animals suffering disease consistent with coccidiosis.

Prophylactic treatment strategies also depend on the circumstances. Younger animals have the lowest immunity and the greatest tendency to eat off the ground, including the ingestion of feces. They are also often born during or shortly before the onset of wet, mild weather (Spring or Fall), when conditions are most conducive to coccidiosis. Mass treatment 2 to 8 weeks after birth may be indicated; with the 2 or 3 week treatment reserved for operations at extremely high risk. Similar guidelines apply to groups of animals put on pasture or feed on the ground, especially during periods of mild, damp weather. Otherwise, mass medication should be based on tangible evidence of a widespread problem, such as property history or fecals with high oocyst counts.

REFERENCES

[1]Guerrero C. **J Protozool** 1967,14(4):613. [2]Rickard L, et al. **J Protozool** 1988;35:335. [3]Guerrero C, et al. *Proc 2nd Cong Nac Biol, Lima, Perú*, 1969. [4]Anccasi M, et al. **Rev Invest Sobre CSA Allpak'a UNA, Puno, Perú** 1985;1:78. [5]Guerrero C, et al. **J Protozool** 1971;18:163. [6]Palacios C, et al. **Vet Rec** 2006;158(10):344. [7]Leguia G, et al. **Revista Peruana de Parasitologia** 1998;13:59. [8]Cebra C, et al. **J Am Vet Med Assoc** 2007;230(1):94. [9]Rohbeck S, et al. In Gerken M, Renieri C: **South American camelids research--Volume 1**. The Netherlands, 2006, Wageningen Academic Publishers, pp 279. [10]Rosadio R, et al. **Vet Parasitol** 2010;168(1-2):116. [11]Rosadio R, et al. **Vet Parasitol** 2010;168(1-2):116. [12]Palacios C, et al. **Rev Inv Vet Perú** 2004;15:174. [13]Cebra CK, et al. **J Am Vet Med Assoc** 2008;232(5):733. [14]Cebra C, et al. **Am J Vet Res** 2012;73(1):13.

CAMELID DISEASES OF PUBLIC HEALTH SIGNIFICANCE

Christopher K. Cebra, VMD, MA, MS, DACVIM
Corvallis, OR

INTRODUCTION

The major role of a public health program is to protect and improve the health of the people in a community. This is different from the One Health Mission, which puts human health, animal health, and environmental health on more equal levels. In public health programs, the major roles that animals play are as hosts for diseases that also affect humans, as sources of nutrition for the human community, and to a lesser degree, as sources of human injury or allergy. The importance of individual species of animal relates mainly to the likelihood of them playing an important role in disease transmission, or their status in the community's food supply.

New World camelids have received little attention in North American public health programs, probably because their numbers were until recently low and their role in our food supply was, and still is, negligible. Llama numbers in the U.S. started to grow in the 1970's, and enjoyed steady growth through the 1990s. The current llama population in the U.S. is estimated to be around 170,000 animals, with some retraction in recent years. Numbers of alpacas have also been increasing steadily since the mid-1980's, with over 1,000 new alpacas registered each year since 1994, over 5,000 a year since 1999, and over 15,000 a year since 2005. The current alpaca population in the U.S. is estimated to be around 100,000 animals. With the rising numbers, there is a commensurate increase in human-camelid contact. Although the roughly 270,000 camelids reflect a population that is still small compared to 92 million cattle or 5.5 million sheep, many of these camelids are dispersed to small farms, the rapid growth of the population was fueled primarily by expanding the owner base, often including neophyte owners with minimal livestock experience, and camelids are often used in roles that involve substantial human-animal contact, including as show stock, 4H projects, fiber producers, and petting zoo animals. Thus, the small number of camelids is partially offset by the high level of handling, and the general lack of livestock experience in some of those owners, and non-owners exposed to the animals, means those handlers may lack basic knowledge of public health risks.

INFECTIOUS DISEASES

In the realm of infectious diseases, camelids can play roles either in transmission of pathogens to humans, or as sentinels of diseases which affect both camelids and humans, but which are unlikely to spread directly between the species. Camelids are especially useful in the sentinel role for diseases that they are thought to be more susceptible to or more exposed to than humans. Examples include infectious agents spread by arthropod vectors or wildlife, and those found on soil or contaminated pasture. Some of the infectious agents seen in recent years for which occurrence of disease in camelids appears to indicate heightened human risk include West Nile Virus,[1] Eastern Equine Encephalitis virus,[2] *Cryptococcus gattii*,[3] *Coccidioides immitis*,[4] and *Anaplasma phagocytophilum*.[5] For none of these infections is there any evidence for any innate susceptibility in camelids, but rather camelids are thought to have lifestyles that lead to frequent exposure. The encephalitis viruses are spread by mosquitoes during warmer weather, when camelids are frequently shorn and remain on pasture throughout the night. *Anaplasma* is spread by ticks, but is otherwise similar to the viruses. The fungi are thought to contaminate soil and are perhaps inhaled readily by herbivores. In all these cases, the occurrence of disease in camelids should alert health care professionals that humans in the community are also at risk.

Camelids can also harbor infectious agents which spread directly or indirectly to humans. As occurs in other livestock species, some of these infections are inapparent in camelids, but can cause severe human disease, especially in the very young, old, or infirm. With the increasing number of camelids and high level of handling, it should be no surprise that reports of zoonotic disease transmission are emerging. These, however, still represent the minority of actual events.

The major zoonotic organisms spread by casual contact with apparently healthy livestock include *E. coli* O157, *Salmonella*, and *Cryptosporidium*. Little is known about *E. coli* O157 in New World camelids. One small survey of camelids in animal attractions in the UK revealed that a high proportion of sampled camelids were infected.[6] A larger survey found no *E. coli* O157 in California llamas.[7] Salmonellosis appears to be rare, not showing up on 2 different surveys,[7,8] but showing up sporadically in clinical cases. *Cryptosporidium* may be responsible for around 10% of clinical enteritis cases in camelid crias,[8] and has also been found sporadically in sick and healthy adult camelids.[9] Thus, although these pathogens may be uncommon, camelids have shown to ability to shed all of them.

Transmissible infectious agents more commonly associated with clinical disease in camelid hosts include Contagious Ecthyma (and occasionally Camelpox), Tuberculosis and other mycobacterial infections, *Giardia*, Ringworm, Rabies, *Streptococcus equi* ssp. *zooepidemicus*, *Yersinia*, *Salmonella*, *Cryptosporidium* and others. Some of these are skin infections, whereas most spread via the fecal oral route. Direct contact or contact with contaminated facilities is also possible. Camelids propensity for spitting, particularly in regards to rabies, is occasionally voiced as a concern, but rapid camelids usually have some degree of pharyngeal compromise, and have not been reported to spit. Of the agents listed here, there are published reports on the passage of *Cryptosporidium*,[9] *M. bovis*,[10] and *M. microti*[11] from camelids to people. *Cryptosporidium* is a reasonably common isolate from outbreaks of diarrhea in neonatal camelids, during which contact adults often also shed and may have clinical signs, and also from sporadic cases of diarrhea.[9,12] *M. bovis* is a major concern in the UK[13] and some other countries, but it and other mycobacterial infections appear to be uncommon in US camelids. *Mycobacterium microti* is also reported in Europe,[11,14] and *M. avium* ssp. *paratuberculosis* has been reported from a variety of countries.[15,16] Most camelid mycobacterial infections are associated with wasting disease, with occasional diarrhea, but some cases of extensive respiratory infections with *M. bovis* raise concern of respiratory spread in countries where it is common. Potential spread through milk has also been identified.[17] Spread of contagious ecthyma and possibly camel pox, ringworm, and *Giardia* have all been reported anecdotally, and should be considered similar to these zoonoses in other species. *Giardia* is commonly found in the feces of crias with diarrhea in some parts of the country,[18] and similar to *Cryptosporidium*, has a variety of wild and domestic animal hosts, as well as man. Ironically, spread of *Giardia* from humans to camels is suspected in one Middle Eastern outbreak.[19] Human caregivers may likewise promote the spread of *Cryptosporidium* between camelids, either as fomites, or as hosts themselves. Rabies is uncommon in the US, but has been reported, and intraspecies transmission is suspected in camelids in South America. The other listed pathogens are infrequent in camelids, and the likelihood of their spread to humans is low.

Transmission of these infectious agents can occur whenever there is direct or indirect contact. Camelid handlers and health care professionals have the greatest contact, and may develop disease as individuals, but these people also have the greatest potential to be informed of the risks and take the necessary precautions. These precautions include personal protective gear, such as gloves and coveralls, frequent hand-washing, avoiding the juxtapositioning of animal handling and eating or food preparation, and the recognition that even casual contact with animals, animal wastes, or facilities bears risk of contaminating the skin or clothing.

Casual visitors to private establishments can also develop individual cases of disease. Outbreaks are more common when large numbers of casual visitors are exposed, such as at petting zoos, day farms, or other similar animal attractions.[20-22] These visitors appear to be at greater risk for infection because they are less aware of the risks, frequently are children, often wear street clothes which may become contaminated, often intermingle animal contact and eating, and may not follow ideal hand-washing protocols.[23] Facilities that allow causal contact, such as with animals over a single fence next to a human mingling area, represent greater risk than facilities where the animal contact is restricted to certain areas, and flow is directed to specific sanitation stations.

FOODBORNE PATHOGENS

An emerging area of public health interest involving camelids is the possibility of foodborne illnesses. As the growth of camelid ownership levels out and prices retract, some owners have investigated alternate uses for their unnecessary animals. Others have simply treated camelids as they are in South America, as food animals. A number of the common pathogens of food poisoning have been identified in camelids, including *Salmonella*, *E. coli* O157, *Listeria*, and *Toxoplasma*. The prevalence of these pathogens in U.S. camelids has not been investigated, nor has the effect of slaughtering these animals privately versus at government-inspected slaughter plants.

INJURIES

Bodily injuries are another possible consequence of camelid human interactions. Most injuries involve owners, handlers, or health care professionals interacting with camelids as part of their hobby or profession. These injuries can result from bites, particularly from camelids with developed fighting teeth, body slams, kicks, or any sort of struggling occurring during restraint or procedures. Serious injuries are believed to be rare, and involve llamas more commonly than alpacas, due to llamas' greater size. For the same reason, camels may represent an even greater health risk. Spitting is frequently mentioned as a possible health risk, but aside from transient eye irritation, I am not aware of a single complaint of a health complication associated with being spat upon by a camelid.

New World camelids have also been associated with a small number of random attacks on people. These include chance encounters, usually with llamas that have escaped from a pen or pasture, and at least one instance of a petting zoo llama attacking and injuring an elderly patron. In the minority of the attacks, the llama is thought to have 'Berserk male llama syndrome.' This behavioral abnormality is thought to develop in male llamas that are overly socialized at a young age, leading to possible aggression towards people once the animal attains sexual maturity. Most of the attacks are reported to involve otherwise docile llamas, and no cause for the incident of aggressive or violent behavior is identified. Videos of a number of less violent attacks are available online, and demonstrate the potential risk that is always present when dealing with relatively large animals.

The documented attacks and injuries highlight that there is risk in any human-animal interaction, and that risk increases with the number of interactions. Lack of experience, as is seen with new owners or the lay public, also increases risk.

ALLERGIES

Allergies are another area of public health concern. To date, there is little data concerning human allergic reactions to camelids.

REFERENCES

[1]Dunkel B, et al. **J Vet Intern Med** 2004;18(3):365. [2]Nolen-Walston R, et al. **J Vet Intern Med** 1007;21(4):846. [3]Stephen C, et al. **Can Vet J** 2002;43(10):792. [4]Fowler M, et al. **J Am Vet Med Assoc** 1992;201(10):1609. [5]Lascola K, et al. **J Vet Intern Med** 2009;23(2):379. [6]Pritchard G, et al. **Vet Rec** 2009;164(18):545. [7]Rulofson F, et al. **Am J Vet Res** 2001;62(4):637. [8]Cebra C, et al. **J Am Vet Med Assoc** 2003;223(12):1806. [9]Starkey S, et al. **J Am Vet Med Assoc** 2007;231(10):1562. [10]Twomey D, et al. **Vet Rec** 2010;166(6):175. [11]Horstkotte M, et al. J Clin Microbiol 2001;39(1):406. [12]Waitt L, et al. **J Am Vet Med Assoc** 2008;233(2):294. [13]Twomey D, et al. **Vet Rec** 2007;160(5):170. [14]Zanolari P, et al. **J Vet Intern Med** 2009;23(6):1266. [15]Belknap E, et al. **J Am Vet Med Assoc** 1994;204(11):1805. [16]Ridge S, et al. **Aust Vet J** 1995;72(4):150. [17]Richey M, et al. **Vet Rec** 2011;169(8):214. [18]Trout J, et al. **Vet Parasitol** 2008;153(3-4):203. [19]Al-Jabr O, et al. **Vet Rec** 2005;157(12):350. [20]LeJeune J, et al. **J Am Vet Med Assoc** 2004;224(9):1440. [21]Bender J, et al. **J Am Vet Med Assoc** 2004;224(7):1105. [22]National Association of State Public Health Veterinarians, Inc. **MMWR Recomm Rep** 2011;60(RR-04):1. [23]Anderson M, et al. **Epidemiol Infect** 2012;140(1):182.

EVALUATING AND MANAGING THE NEUROLOGIC OR RECUMBENT CAMELID

Erica C. McKenzie BSc, BVMS, PhD. DACVIM, DACVSMR
Corvallis, OR

INTRODUCTION

Recumbency is among one of the most frequent clinical signs of disease in New World camelids, and has been reported in up to 24% of camelids presenting to specific veterinary institutions.[1] Unfortunately recumbency can reflect any of a number of diseases affecting a variety of body systems. In addition to the relatively non-specific nature of this clinical sign, further frustration can arise due to interference with specific diagnostic procedures. Given these factors, collection of a complete and detailed history in addition to a thorough physical examination are critical first steps in assessing recumbent camelids and constructing a differential diagnosis list.

Initial Presentation

Collection of a history and performing a brief inspection of the recumbent camelid is often best achieved before moving them from the vehicle or trailer, so that appropriate precautions can be taken to move them safely without unknowingly jeopardizing significant lesions such as fractures. The collected history should be detailed and should seek to define the character of the recumbency as intermittent (animal voluntarily gets up and down), preferred (the animal can be coaxed to rise for short periods) or constant (the animal cannot physically rise) to assist construction of a differential diagnosis list, keeping in mind that recumbency status may have changed throughout the course of disease.[1] Once the current status of the animal's recumbency is determined, it can be prompted to walk into the clinic if capable and willing to rise, during which time assessment of gait and motion can be performed. However, if the animal is notably reluctant to rise or obviously impaired, transport should be organized using a tarp, cart or slide. After immediate lifesaving therapies are administered if necessary, every recumbent camelid should be subject to a thorough physical examination focused on identifying specific signs of causative disease and the likely system involved, without causing further injury via aggressive manipulations. Specific physical indications of many disorders associated with recumbency will be discussed in the following sections, which describe several broad categories of disease associated with recumbency in camelids.

Debilitation

Unfortunately, physical debilitation is a fairly frequent cause of recumbency in camelids presented for evaluation. The presence of a low body condition score usually indicates a chronic disease process, and the prognosis is often subsequently reduced. Severe debilitation may be encountered in camelids owned by inexperienced clients or those who do not perform routine palpation assessments of body condition in their animals. Debilitated animals will usually practice intermittent or preferred recumbency, but may progress to constant recumbency as a result of severe anemia or emaciation. Specific disorders associated with thin body condition in camelids include severe parasitism, megaesophagus, round cell neoplasia, peritonitis, malnutrition, abdominal abscessation (*S. zooepidemicus*), chronic renal disease, Johnes disease, and oral pathology.[2-4]

Examination of the debilitated camelid should commence with assessment of body condition via palpation of the spine, pelvis and ribs. Obtaining an actual bodyweight if feasible can be useful for monitoring small improvements over time. During the remainder of the physical examination the abdomen should be palpated for abnormal size or tenderness, and percussed gently with a hand on each side for evaluation of free abdominal fluid. Peripheral lymph nodes should be palpated for enlargement and the esophagus should be monitored visually for at least one minute for evidence of abnormal motility patterns or billowing which suggests esophageal dysfunction. The mouth should be thoroughly inspected if possible, though this may sometimes elicit great stress and should not be performed in animals with signs suggesting significant anemia such as pallor, weakness, and tachycardia. A complete blood count is indicated to identify anemia and/or hypoproteinemia in animals suspected of having significant parasitism, and may also provide indications of chronic infection such as neutrophilia. Fecal analysis via sugar centrifugation should be performed to identify problematic parasitism in the form of high numbers of coccidia, trichostrongyles or *Capillaria*, or the presence of more pathogenic parasites including *Trichuris spp.*, *Nematodirus battus*, *Haemonchus*, *Fasciola hepatica* and *Eimeria macusaniensis*.[2] Serum biochemistry can be helpful to assess renal and hepatic function and to direct further diagnostics.

Other specific diagnostic procedures of the debilitated animal should potentially include abdominal ultrasound with thorough inspection of all structures including the gastric compartments, kidneys and liver.[4] Abdominocentesis can be performed in the right paralumbar fossa if free abdominal fluid is identified (and is not thought to be transudate secondary to severe hypoproteinemia) or if gastrointestinal neoplasia is suspected. Megaesophagus can generally be confirmed via contrast radiography, and alpaca crias with megaesophagus should be carefully assessed for an underlying congenital vascular ring anomaly.[5]

Cardiorespiratory disease

Camelids are prone to a variety of cardiorespiratory conditions that can cause recumbency. Crias with bilateral choanal atresia will usually present with respiratory noise and open mouth breathing. Congenital cardiac defects also occur with some frequency and can be easily overlooked in the concurrently premature or septic cria. Hence thorough, bilateral cardiac auscultation and assessment of arterial oxygen content should be considered a routine diagnostic in weak or recumbent crias. Arterial samples are readily obtained from the median or saphenous arteries using a new 25 gauge needle attached to a very lightly heparinized syringe. Crias with cardiac anomalies frequently display tachypnea, weakness and a low arterial oxygen content (< 60 mmHg) that improves minimally with nasal oxygen insufflation if a shunting anomaly is present. The complexity of cardiac defects in camelids can range from simple septal defects to complex anomalies such as Tetralogy of Fallot. Echocardiography is essential to characterize the anatomy of cardiac defects and should be considered in any cria with an inappropriate murmur or hypoxemia that cannot be attributed to primary respiratory disease.[1] Premature lung syndrome can occur in crias with a gestational age less than 325 days. Signs of significant prematurity include floppy or tipped ears, an absence of erupted incisor teeth, low birth weight, and flexor laxity. Significantly premature crias may have hypoxemia and respiratory acidosis due to inadequate pulmonary surfactant, amongst the many other issues premature neonates face. In addition to nasal insufflation of oxygen and other supportive measures, administration of aminophylline (2 mg/kg subcutaneously every 4 to 8 hours) has been recommended for improving respiratory function in premature crias.[6]

Adult camelids will occasionally develop thrombotic endocarditis which can be easily overlooked since clinical signs referable to the cardiovascular system may be subtle. Recumbency is common in affected animals, and is frequently accompanied by signs of right sided heart failure including tachycardia, tachypnea and distended jugular veins. Cardiac murmurs are rarely appreciated in these cases and the absence of a pathologic murmur should not preclude a diagnosis of cardiac disease. Pericardial, pleural and peritoneal effusion is frequently detectable via auscultation, percussion and ultrasound, and a fibrinous mass is likely to be imaged in the right ventricle of the heart.[7] A complete blood count may be consistent with acute inflammation and serum chemistry often reveals increased hepatic enzyme activity. Euthanasia is usually indicated and clients should be instructed to consider anthelmintic therapy effective for liver flukes in their remaining animals (albendazole 10 mg/kg per os or clorsulon 7 mg/kg per os).

Diaphragmatic paralysis is a condition somewhat unique to camelids that can occur as an acute or chronic phenomenon in crias or adults.[8] Severely affected animals will present with recumbency, increased respiratory effort and a paradoxic respiratory pattern where the thorax and abdomen move in an asynchronous fashion. Blood gas analysis is consistent with hypoxemia and severe respiratory acidosis is expected in recumbent animals. Phrenic nerve degeneration is frequently present at necropsy, however the etiology of this condition remains unclear and there is no known therapy.[8] Pleuritis associated with *S. zooepidemicus* infection can also cause recumbency and increased respiratory effort in crias and mature camelids. Pleural effusion is usually evident on ultrasound and thoracocentesis can be easily performed using a vacutainer needle and tube. Pleural fluid will contain an increased nucleated cell count and chains of cocci are evident on routine staining. Gas analysis of the fluid is often consistent with septic pleuritis (low glucose concentration and pH, and high total protein and lactate concentrations). Early cases often respond successfully to parenteral penicillin therapy (22 000 IU/kg IM or SQ once to twice daily for 10 days or until CBC and ultrasound findings normalize). Owners should be queried in regard to any contact between camelids and horses on their property (shared grazing or feed bunks) which should be eliminated if identified. *S. zooepidemicus* has also been implicated in an outbreak of severe polyserositis in an alpaca herd.[9]

Colic

Colic is a common cause of recumbency in camelids. Apart from increased time spent in recumbency, additional signs of abdominal pain may vary from frequent shifting of position or lying in abnormal cush to rolling, thrashing and vocalizing. Colic can arise as a result of meconium impaction in crias, gastrointestinal obstruction with bezoars, torsion of the spiral colon, jejunal intussusception, intestinal adhesions, enteritis and other phenomena. Peritonitis may result from *S. zooepidemicus* infection or gastrointestinal perforation associated with third compartment ulceration, severe *Trichuris* infestation, or gastrointestinal neoplasia. Uterine torsion must be considered as one of the most likely differentials for colic in late pregnant females. In addition to signs of abdominal pain, other indications of gastrointestinal disease include reduced defecation, depression, abdominal distension and anorexia. Tachycardia is not a reliable sign of abdominal pain in camelids, and the presence of prominent tachycardia is more likely to indicate shock and severe underlying disease.

Physical examination should include abdominal palpation, particularly in young crias in which large bezoars may be externally palpable. Digital rectal examination can be attempted in young crias, and careful rectal palpation may be feasible in adults, with indications for this procedure including poor fecal passage, and the possibility of uterine torsion in the female or urinary obstruction in the male (feeling for pulsating urethra). CBC, serum biochemistry and/or blood gas analysis may help characterize severe abdominal disease such as peritonitis (neutropenia, hyperfibrinogenemia) or gastrointestinal obstruction in proximity to the pylorus (hypochloremic metabolic alkalosis). Fecal analysis should not be overlooked since severe parasitism can create substantial gastrointestinal disease and colic. Abdominocentesis in the right paralumbar fossa can be useful to identify gastrointestinal perforation or strangulation via cytology and blood gas analysis. Normal camelid peritoneal fluid should have a nucleated cell count below 1500 to 2000 cells/ul and a total protein concentration less than 2.5 mg/dl.[1] Glucose and lactate concentrations should mirror those of the peripheral blood. Abdominal ultrasonography can be invaluable in identifying gastrointestinal obstruction or peritonitis. In less clear cases of gastrointestinal disease, computed tomography performed three hours after orogastric contrast administration has been useful for identifying a variety of gastrointestinal lesions. Careful assessment of colic is critical to distinguish medical from surgical lesions since, in general, surgical investigation is rarely indicated to address colic in camelids, and is fraught with difficulty due to the variety of possible approaches to the abdomen of this species. Surgical investigation is therefore usually indicated for clearly defined cases of gastrointestinal obstruction, uterine torsions that cannot be manually corrected, or for other surgically amenable lesions.

Musculoskeletal defects

It is not uncommon for a painful skeletal lesion such as limb fracture to result in preferred or constant recumbency in the camelid, which can make such lesions more challenging to detect. Hence physical examination of recumbent camelids should include thorough palpation and manipulation of all joints and limbs. Particular caution should be applied when examining dark brown or black alpaca crias that are three to eight months of age and presenting in regions with long stretches of inclement weather, since rickets (hypovitaminosis D) can occur readily under these circumstances if preventative management is not practiced. Crias with rickets will usually have palpably enlarged costochondral junctions, swollen joints, and angular limb deformities. Spinal fractures are not uncommon. Laboratory findings consistent with rickets include low serum phosphorus (< 4.5 mg/dl) and vitamin D (<15 nmol/L), and elevated serum alkaline phosphatase. Radiographs of long bones will identify thin cortices and flared metaphyses. Treatment includes administration of vitamin D (1500-2000 IU/kg IM or SC) and supplementation with oral phosphorus at 0.5 to 1.0 gram per day and affected animals should be handled with great caution since iatrogenic fractures are possible.

Camelids developing recumbency and myopathy during periods of high environmental temperature are likely to have heat stress. This condition tends to arise in animals in full fiber during warm months in geographic regions that incur concurrent high humidity and high ambient temperatures. Affected animals may have tachycardia, tachypnea and very high rectal temperatures (up to 108°F) if examined in the acute stage. Males will often have scrotal swelling. Serum biochemistry usually reveals significant elevations of serum creatine kinase and aspartate transaminase, and azotemia may be present. Treatment includes intensive cooling via a combination of body clipping, alcohol or cold water baths, and cold air convection in addition to physical therapy aimed at reducing recumbency and improving mobility (float tanks or slings). Affected animals often lose their ability to thermoregulate indefinitely if they survive, and therefore require long term management to prevent recurrence.[1]

Neurologic disease

A wide variety of neurologic diseases occur in camelids and are frequently associated with recumbency. During physical examination of the animal, careful assessment of the individual cranial nerves should be performed, in addition to assessment of peripheral nerve function and spinal reflexes. Reflexes (patellar, withdrawal, biceps and triceps) can be carefully performed on the upper limb of the animal in lateral recumbency before rotating the animal to repeat on the other side, but can be difficult to evaluate in animals presenting after prolonged periods of recumbency with accompanying muscle trauma. The aim of the examination should be to differentiate if disease involves the brain (cranial nerve anomalies, obtundation) or spinal cord (ataxia and proprioceptive deficits in an animal that appears centrally normal), and as specifically as possible the probable location of the lesion within those regions. Camelids are susceptible to a variety of forms of meningitis and encephalitis including viral infections (equine herpes virus 1, rabies, eastern equine encephalitis, west nile virus), bacterial infections (*Listeria*, *Histophilus*, *Streptococcus*, *E. coli*), and fungal infections (*Cryptococcus*, *Coccidioides*), in addition to aberrant parasite migration (*Baylisascaris*, *Parelaphostrongylus*). In animals that appear to have ataxia or recumbency related to a spinal defect, traumatic injury should be considered (vertebral fractures, traumatic luxation of cervical vertebrae) in addition to intervertebral disk protrusion, spinal neoplasia and discospondylitis. Crias should also be assessed for the possibility of congenital brain or spinal anomalies.

Multiple diagnostic procedures are available for assessment of neurologic disease in camelids. Radiography ± myelography is often useful for traumatic injuries or suspected cervical spinal compression. Stressed views of the cervical spine may be required to detect luxating injuries. Cerebrospinal fluid collection is easily achieved from the lumbosacral space by palpating the obvious depression behind the 7th lumbar process and advancing a 20 gauge spinal needle 2-4 cm through the skin perpendicular to the spine. Increased CSF protein concentration (> 50 mg/dl) and nucleated cell count (> 3 cells/ul) are indicative of inflammation. Eosinophilic pleocytosis is suggestive of *Parelaphostrongylus* infestation, whereas monocytosis may be apparent with listeriosis. Suppurative inflammation is consistent with bacterial meningitis, and in particular geographic regions, such as the pacific northwestern USA, infection with *Cryptococcus* is readily apparent on routine cytological examination. Computed tomography can be performed quickly and easily in sedated or anesthetized camelids and is particularly useful for investigating possible intracranial lesions via contrast enhanced techniques. Other diseases that should be considered as possible causes of obtundation and abnormal neurologic function in camelids include tick paralysis and polioencephalomalacia (the latter particularly if opisthotonus is present). Recumbent animals in tick paralysis regions should be checked carefully since a single tick can cause profound but reversible disease if the tick can be identified and removed.

Managing recumbent camelids

Given the wide variety of conditions described above, management of individual cases will obviously vary widely depending on the cause of recumbency. However, specific considerations that should be awarded the recumbent patient include maintaining appropriate body temperature, protection from trauma and anxiety, providing adequate and appropriate nutrition, monitoring and maintaining urinary and fecal passage, providing respiratory system support, preventing infections, providing physical therapy and monitoring for and preventing miscellaneous complications. Appropriately managed, even prolonged recumbency can be associated with a successful return to function if the underlying disease can be identified and resolved, and substantial complications averted.

Camelids with a variety of illnesses often become anorexic and subsequently can develop significant peripheral fat mobilization which may promote further anorexia and result in hepatic lipidosis. Hence serum chemistry analysis should include measurement of betahydroxybutyrate, triglycerides and non-esterified fatty acid concentrations. Animals with significant anorexia or peripheral fat mobilization should receive forced nutrition (transfaunation and orogastric slurry diets) if not at risk of further injury (cervical lesions), and can be concurrently supported with partial parenteral nutrition solutions. Pain relief for significant abdominal or musculoskeletal pain can be provided using combinations of analgesics including non-steroidal anti-inflammatory agents (flunixin, meloxicam), opioids (butorphanol, fentanyl, buprenorphine, morphine), and anesthetics (infusion of lidocaine or ketamine). Oxygen insufflation is indicated for animals with significant hypoxemia, in addition to maintenance of appropriate body position (sternal recumbency) using straw bales or supports. Float tanks and transient slinging can be of great benefit in preventing or managing primary or secondary muscle damage. Eyes and limbs should be examined daily for evidence of ophthalmic or decubitus ulceration.

REFERENCES

[1]Belknap EB, et al. **Comp Cont Educ Pract Vet** 2000;22(1):36; [2]Ballweber LR. **Vet Clin North Am Food Anim Pract** 2009;25(2):295; [3]Watrous BJ, et al. **J Vet Intern Med** 1995;9(2):92; [4]Martin JM, et al. **Can Vet J** 2010;51(12):1379; [5]McKenzie EC, et al. **J Am Vet Med Assoc** 2010;237(3):311; [6]Whitehead CE. **Vet Clin North Am Food Anim Pract** 2009;25(2):353; [7]Firshman AM, et al. **J Vet Intern Med** 2008;22(2):456; [8]Byers S, et al. **J Vet Intern Med** 2011;25(2):380; [9]Jones M, et al. **J Vet Intern Med** 2009; 23(1):220; [10]Whitehead CE, et al. **Vet Clin North Am Food Anim Pract** 2009;25(2):385.

Money & Medicine: Getting Clients to Accept Your Fees

Shawn G. McVey, MA, MSW
Chicago, IL

INTRODUCTION

With an understanding of the client's perception of value, we can create client-centric pricing that the staff feels good about, too. Here we look at the factors that affect value, how to present estimates, and what to say in tough situations. We also address discounting and alternative payment options.

HOW WE GOT TO WHERE WE ARE RIGHT NOW

Pricing has been product-centric in that we have focused on the actual cost of services, the time we take to provide the service, and anything that can be quantified. We've considered our own feelings of guilt and inadequacy and stayed internally focused.

The internal focus can create consumer surplus. Everyone offers the same thing at the same price, teaching the customer that all of the products are worth the same thing. Value creation is getting into the heads of the client and finding out what they value. Extraction is the actual payment for services that leaves them satisfied.

Factors that affect value perception are product, status, loyalty, convenience, urgency, and quality. Price is a communication device between the client and the practice that reflects constantly changing market variables. Prices should reflect the answer to the question, "What is the client willing to pay for this service at this time"? If you guess too high, you risk the relationship. A 1% increase in prices can create an increase to the bottom line of 3%-12% annually.

The animal is never "free." Most people grossly underestimate the cost of taking care of a pet. In a May 2006 study, Dr. James Wilson found that the average lifetime cost of a medium-sized dog is $10,400, and cats cost slightly more at $10,600. For true pet lovers, the expense more than doubled. It is not uncommon for a pet owner to spend between $50,000 and $100,000 on a pet during its life, and 40% of that will be for healthcare.

THE BEST WAY TO PRESENT AN ESTIMATE TO THE CLIENT

We call this process presenting estimates, but we should call it a Health-Care Plan (HCP). Most employees would prefer to do almost anything else than present the HCP. Karen Felsted, DVM, CPA, wrote an article for *Firstline Live* that offers some helpful tips on how to present a Health-Care Plan.

Focus on the value you offer. Describe the service you're offering, along with its benefits and how it reduces risk. Consider the difference between, "This blood work will cost $112," and "First, we need to analyze some of Fluffy's blood to diagnose her problem and make sure we select the right anesthesia for the surgery." Is it necessary to state the cost when the client can see it on the HCP?

Off details that are thorough and easy to digest. If you say, "This surgery will cost about $2,000. Is that OK with you?", clients will think that sounds expensive and won't know what they are getting.

Understand your procedures. The AAHA/CareCredit/Pfizer Compliance Study showed us that the main reason clients are not compliant with their pet's care is that we do not follow up with the doctor's recommendations. Why? Consider the difference between, "The treatment plan includes a bile-acids test, and that will be $177," and "A bile-acids test will determine how well Fluffy's liver is functioning."

Encourage clients to ask questions. They are intimidated and overwhelmed. They are more likely to ask a staff member a pricing question than they are the doctor.

Don't get defensive. Clients will comment on cost. It is not necessarily a complaint about prices; rather, it is an initial questioning of value. Welcome the chance to explain, and if you can't, welcome the chance to train! Examples of good responses are, "Yes, that is a lot of money, but Fluffy is a sick kitty," and "Is there something specific on the HCP that you were concerned about?"

Be empathetic. You are a counselor, too! Allow the client time to get over the sticker shock of paying for health care. Provide them with contact name and information so that they can ask questions after they have gone home.

Don't use guilt trips. We don't appreciate it when a client guilts us into reducing our costs, so we should be respectful of their budgets. Always prepare a "second-best" alternative to the optimal plan.

Develop scripts for the practice team. Remember that it's too late to get it right when the HCP is presented in a stressful environment. Practice makes perfect, and you don't want to practice on the client. Allow the staff to help develop the script, and tie it to the practice mission statement. Practice the scripts in staff meetings. Here are some examples:

Tough Script #1

Client: "You charge more than ABC Animal Clinic!"

Team member: "I don't know what they charge for the same procedure, but I would want to be sure we are comparing apples to apples. An example is that our HCP includes pain management and pre-anesthetic testing."

Tough Script #2

Client: "Why does this cost so much?"

Team member: "I understand that this seems expensive, but your pet is very sick. We want to identify what is wrong and give her the best treatment, and we think these tests are necessary to discover what's wrong with Fluffy."

Tough Script #3

Client: "I pay more for my pet's veterinary care than I do for my own health care!"

Team member: "I completely understand why it seems that way, but insurance often keeps us from ever seeing the true cost of health care because you pay only a co-pay for the services. In reality, veterinary care is a fraction of the cost of human health care. Let's look at Fluffy's estimate and compare it to a human procedure."

Tough Script #4

Client: "I can't afford this. Is there anything you can do?"

Team member: First assess the client's intentions and see if it is a financial problem or a financial inconvenience. Are they asking for more information? "Let me talk to the doctor and see if we can come up with a treatment option that is less expensive."

WHEN IT IS RIGHT TO DISCOUNT

Veterinarians are more likely to do it than staff. There is always more of it happening than you are aware of. There is a direct correlation between lack of training and heavy discounting. Your competition is already trying to undercut you. Don't let your team do it, too!

Here are a few scenarios where discounting might happen. How do you handle each one in your practice? Scenario 1: A client complains about the cost of a $10 nail trim, so to avoid a confrontation, the receptionist removes the charge from the bill. $10 in profit requires $50 in revenue.

Scenario 2: Friends and family repeatedly receive under-the table-discounts because staff believes the fees are too high. Talk about the reality of the economics to your staff (many come to work at a practice because of the reduced health-care cost for their pets).

Scenario 3: When a popular and frequent visitor to the practice picks up over-the-counter products, the staff "forgets" to add all of them to the bill. In other businesses, this is called theft.

Scenario 4: Pity discounts go to the elderly and poorly dressed because they "obviously" cannot afford to pay full price. If this type of client is where your heart is, why would you work at a full-service, wellness-oriented veterinary practice?

Scenario 5: A team member overhears a doctor quoting an artificially low price and begins following that example.

How can you deal with discounting? Create protocols, and use a screening tool that requires authorization. Educate your team members and clients. Give tours to clients so that they can see what you do "in the back." Share costs and financials with your team. Provide leadership: If the "top" is discount happy, what will the staff do? Regular pricing audits help to uncover discounting that isn't approved and give some accountability to those who are giving the discounts.

Practice organized giving, and have your team drive the effort. Give in a smart way. First define the terms. There's a difference between mercy care and giving away the practice. There should be a planned budget for giving. And think more broadly. Do you have to give away services? In addition, you could conduct a culture audit to see if your efforts are linked to your mission. Train the staff, and track discounts. Give discounts judiciously.

WHEN CLIENTS WILL USE AN ALTERNATIVE PAYMENT METHOD

Yes, clients will use an alternative! They really do help. A recent Study done by Lynne Fondon, DVM, and John Volk of Brakke Consulting, Inc. looked at CareCredit card holders and found that the availability of alternative payment has a direct impact on the pet's health care. 79% of respondents said that they were more likely to follow the veterinarian's recommendations knowing they had access to this service.

76% of respondents would have opted for lesser treatment without the alternative option. More than 50% would have asked for a "veterinarian payment plan." Over half of the initial CareCredit transactions were for more than $1,000.

ISSUES IN ANTIMICROBIAL RESISTANCE

H. Morgan Scott, DVM, PhD
Manhattan, KS

HISTORICAL PERSPECTIVES (1960 to 2005)

Antimicrobials are perceived as an essential adjunct to both human and animal health systems worldwide. A range of antimicrobial products have been used to both treat and prevent infectious diseases of animals for over a half-century (Gustafson and Bowen, 1997). Resistance to certain classes of antimicrobials is an inherent feature of some microbes. In other cases, resistance traits can be acquired by microbial species through mutations, and transferred within and among species. For most bacterial species, it is widely acknowledged that the use of antimicrobials applies pressures that favor the selection and propagation of resistant strains. Indeed, some authors consider increasing resistance to be the inevitable outcome of the use of antimicrobials in both animal agriculture and human health (Levy, 1992). For over forty years, antimicrobial resistance has been a serious issue of concern to animal health pharmaceutical regulators in this country and elsewhere. This concern is often illustrated first in the issuance of the Swann Report in the United Kingdom in 1969, followed closely by FDA scrutiny in the US during the 1970's of feed grade antimicrobial use in agriculture. Later, outbreaks of multi-drug resistant Salmonella in countries such as Denmark helped push the development of resistance surveillance programs such as DANMAP and WHO activities in this area. While the Danish (see Aarestrup et al, 2010) and subsequently the E.U. bans on antimicrobials as growth promoters in 1996 and 2006, respectively, get much of the press, the Swedes had earlier discontinued their use in the mid 1980s. Generally speaking, foodborne outbreaks on a massive scale have tended to result in large scale policy shifts at the national and international levels, be they from organisms with no obvious connection to antimicrobial resistance (e.g., *E. coli* O157:H7) or of a multi- or pan-resistant phenotype (e.g., multi-resistant *Salmonella* in the U.K. and Denmark). In recent years, renewed attention has been focused on foodborne diseases. In particular, newly emerging and re-emerging enteric pathogens such as *Escherichia coli* O157:H7, non-typhoid *Salmonella* spp. and *Campylobacter* spp. have captured the interest of public health and veterinary science communities as well as the public at large. Concurrently, global recognition of the increasing resistance of many of these enteric pathogens to presently-available antimicrobials is rapidly developing (WHO, 2011). Many consider the development and spread of multiple antimicrobial resistances in pathogenic bacteria (e.g., *Salmonella* Typhimurium DT104) to be as important a health risk as the emergence of other more widely-recognized infectious diseases. As these "superbugs" continue to develop and acquire more resistance traits, some scientists (Levy 1992) and advocacy groups have begun to predict "doomsday scenarios", whereby the era of antimicrobials may almost be over.

There has been a growing chorus calling for wholesale change in the prescribing practices of antimicrobial agents licensed for both human and animal health. Some scientists and politicians point to antimicrobial use in animal agriculture as the primary culprit in promoting resistance of pathogenic organisms common to both animals and humans. These scientists often lead the call to either: 1) reduce the range of antimicrobial management practices used in animal agriculture (i.e., no growth promotant usages, no subtherapeutic levels in feedstuffs or water), 2) restrict certain classes of antimicrobials to last-line use in human medicine (e.g., 3rd and 4th generation cephalosporins, fluoroquinolones), or 3) remove virtually all uses of antimicrobials in animal agriculture except for strictly therapeutic purposes in individually-diagnosed clinical cases (WHO, 2000). Despite the many uncertainties surrounding the issue, livestock commodity groups have begun to include prudent-use guidelines for antimicrobials in their quality assurance programs. Consumer confidence in the safety of the foods of animal origin is paramount to the long term health and survival of the U.S. beef cattle, poultry and swine industries. Estimating the precise magnitude and importance of antimicrobial resistance as a food safety issue in animal agriculture is difficult. Any estimates need to account for the food safety risk uncertainties at each of multiple stages through production, consumption, and (potentially) human disease. Certainly, risk/benefit models must account for the tradeoffs of immediate cattle (and potentially, human) health and growth efficiency benefits from antimicrobial use in the short term, against potentially detrimental impacts on treatment of disease (animal and human) due to resistant organisms.

In general, there are four major uses of antimicrobials in food animals (with the nomenclature and taxonomy subject to change depending on the audience). First, and most recognizable, is the use of antimicrobials as therapy for clinically ill animals (therapeutic use). Such use can either be on-label, whereby the product is used in an FDA-approved manner or extra- or off- label where the dose, route, indication, or species of animal treated with the antimicrobial many vary from the approved label. It can be via parenteral (injection) or oral dosing. Prophylactic use (prevention) is generally accepted to mean treatment of individuals or groups of healthy, though 'at-risk', animals at or near therapeutic dose but before they become ill. This differs subtly from control (metaphylaxis) whereby treatment is effected when a certain critical threshold of ill animals during an epidemic is met. Thereafter, all animals in a pen or barn (for example), are treated. Finally, growth promotion is sometimes termed 'subtherapeutic' use of antimicrobials though the latter term is often disparaged by clinical pharmacologists. Generally, use of the antimicrobials is via feed and is provided to groups of animals to enhance growth rates, feed efficiency, and group uniformity, often through unknown or unclear biological mechanisms. Some also argue that the effects are likely to be through diminishment of subclinical disease and so the term growth promotion is a misnomer. Finally, it is generally accepted that the relative ranking of controversy concerning the use of the products is almost in reverse, with metaphylaxis likely to be slightly less problematic than prophylaxis.

THE RECENT PAST (2005-2012)

In 2005, a group of infectious disease physicians gathered in Canberra, Australia to develop the criteria for defining the relative importance of classes of antimicrobials for human medicine, and to apply those criteria to determine the status of antimicrobials of use in either or both of human and animal medicine. The resulting report (see WHO, 2011), which classifies antimicrobials as: critically important, very important, or important has been revisited and revised several times since (most recently in

Oslo, Norway in June 2011). The listing was controversial from the very start, and it spawned a series of copy-cat lists such as was developed by the world organization for animal health (OIE) as a listing of important antimicrobials for animal medicine. The WHO list is less controversial where there are products used solely in animals (e.g., ionophores, bambermycins) or humans (e.g., carbapenems). The controversy arises particularly for classes of antimicrobials such as the macrolides and even the tetracyclines. Whatever the problems with the WHO (and OIE, and other national lists) they are here to stay and they do provide a catalyst for discussing the prioritization of antimicrobials and their use in both human and veterinary circles.

In terms of the critically important antimicrobials, as listed by the WHO (WHO, 2011), the top three (fluoroquinolones, 3rd and 4th generation cephalosporins, and macrolides are of some concern to U.S. producers since their use spans all four categories as listed above. In general, the products approved in the U.S. as fluoroquinolones are limited to enrofloxacin and this product approval was revoked for use in poultry in 2005. Thus, it is used almost exclusively for therapeutic purposes in animal agriculture. For 3rd and 4th generation cephalosporins the product approved at this time in the U.S. is ceftiofur, available in short-, medium-, and long-acting formulations for injection. These facilitate use as therapeutic, metaphylactic and prophylactic situations under a 'control' label for some indications. This product is currently under some restrictions for extra-label use in agriculture. The third product on the top 3 listing is macrolides. The dominant macrolide in use in U.S. agriculture is tylosin, which while available in both parenteral and oral formulations is largely used in feed grade formulations.

The European ban on use of antimicrobials as growth promoters took effect on January 1, 2006. This followed other member country leads such as Sweden (mid 1980s) and Denmark (1996). The Danes in particular have been quite vigorous in tracking the effects of their restrictions on use based on both sales of antimicrobials, resistance among bacteria, and production of pigs in particular. Aarestrup and colleagues (Aarestrup et al, 2001; Aarestrup et al, 2010) have published results suggesting: 1) that the ban significantly reduced levels of resistance, particularly among the vancomycin resistant enterococci when avoparcin (never approved in US) and macrolides (tylosin) were removed. Their group also recently published work suggesting that while there was an early increase in treatment of clinically ill pigs, and a slight reduction in pig productivity, that overall the pig production efficiency has risen in the decade plus since the ban took effect. Of course there are many who disagree with the Danish interpretation of their own data. In particular, in many countries the use of antimicrobials in terms of tonnage hasn't changed much since 2006 suggesting that a shift to 'prophylaxis' or other has taken place. Obviously, this country is keeping a very close eye on the European experience as journalists, legislators, FDA commissioners, and others continue to advocate for changes in use of antimicrobials in the US.

Many in the veterinary and agriculture communities were surprised when the Food and Drug Administration (FDA, 2008) first issued a 'final rule' effectively banning the extra-label use of cephalosporins in food animals. The resultant backlash repeatedly delayed the implementation of this order for almost 4 years (a revised version went into effect in April, 2012). It was likely initiated by data originating out of Canada concerning *in ovo* injection of ceftiofur at low doses into eggs at broiler hatcheries. An FDA survey had found that the practice was also employed in the U.S. The data from Canada suggested a marked increase in *Salmonella* Heidelberg resistant to ceftiofur and a precipitous fall off in resistance when the practice was voluntarily ceased. However, since the ruling did not apply simply to that practice, the response to the request for comments was overwhelming and the rule changed substantially from the first draft.

The past 5 years have seen a plethora of scientific papers published exploring the presence/prevalence of a variety of different bacteria in retail meats and on-farm. The origin of most of these inquiries seems to relate to identifying non-hospital sources for the expansion of an otherwise hospital-restricted set of nosocomial infections such as *C. difficile* and MRSA. Indeed, we have investigated the prevalence of *C. difficile* in swine and in humans in a uniquely integrated system in Texas (Norman et al., 2011) and found that both healthy humans and swine harbor the organism within their intestines. However, it is rarely the hyper pathogenic strains that are present, and in the absence of risk factors such as hospital stay, antimicrobial use, and others that the organism causes problems. It is much more likely that there is a common environmental source for the anaerobic bacteria's spores that humans and animals alike are exposed to on a day-to-day basis. Methicillin-resistant *Staphylococcus aureus* (MRSA) has been a scourge of human hospitals for three decades. More recently, community acquired infections have led investigators to sample animals and food products and have identified many sequence types in the latter, and among companion animals, but only a restricted type among pigs: ST398. Despite initial fears that the organism could be spread through the food supply, it is generally accepted (EFSA, 2009) that the major risk to humans is through direct contact/colonization at the farm. It is very unclear at this point as to what role antimicrobials may play in selecting for the organism.

THE FUTURE (2012 AND BEYOND)

Existing empirical evidence supports theories suggesting that the use of antimicrobials in both human health and animal agriculture settings leads to the favorable selection (fitness advantage) of resistant strains of bacteria within treated humans/animals or groups of treated humans/animals. However, evidence pertaining to the further transmission of AMR from animals or animal populations to human populations has thus far been circumstantial; based largely on cross-sectional studies and qualitative data from case reports. At this point in time, it remains unclear whether AMR selection pressure in animal agriculture leads either to: 1) long-term maintenance of increased resistance levels in the animal population itself, 2) substantive transmission of resistant bacteria to human populations or, 3) a fitness advantage for any resistant bacteria transmitted to humans leading to long-term maintenance or propagation in the human population. While there is little doubt that cross-species (i.e., animal to human, human to animal) transmissions can occur, we presently have available little or no quantitative longitudinal data necessary for reliable risk assessments. Rapid technological advances in genetics, microbiology, and biochemistry have recently been integrated with the more field-based disciplines of bacterial population genetics, mathematical biosciences, ecology, evolutionary biology, and epidemiology. By combining these diverse fields, innovative research approaches to antimicrobial resistance in defined human populations have been.

In the meantime, animal agriculture remains responsive the consumer demands, and attuned to public health concerns. Prudent use guidelines as well as alternatives to antimicrobials as growth promoters have been developed. In the very near future, studies examining a variety of interventions against antimicrobial resistance will be conducted including: exploring potential for plasmid-curing via use of bambermycins, use of products that change lower bowel flora, changes in environments, and use of microminerals (e.g., Cu, Zn) instead of antimicrobials. Regarding the latter, we have found that the bacteria themselves adapt quickly to the presence of high levels of copper in feed and as a result can harbor resistance alongside that of antimicrobials on mobile genetic elements called plasmids (Amachawadi et al, 2011). Unfortunately, this means that bacterial adaptation to alternatives to antimicrobials can still co-select for resistance even when the antimicrobials are not being used. Despite some setbacks, pursuit of alternatives remains a worthwhile goal and there are likely to be some successes down the line that focus more on changing the host and environmental ecology to favor susceptible over resistant strains.

REFERENCES

Aarestrup, F.M., A.M. Seyfarth, H.D. Emborg, K. Pedersen, R.S. Hendriksen and F. Bager. 2001. Effect of abolishment of the use of antimicrobial agents for growth promotion on occurrence of antimicrobial resistance in fecal enterococci from food animals in Denmark. Antimicrob. Agents Ch. 45:2054-2059. Aarestrup, F.M., V.F. Jensen, H.D. Emborg, E. Jacobsen and H.C. Wegener. 2010. Changes in the use of antimicrobials and the effects on productivity of swine farms in Denmark. Am. J. Vet. Res. 71:726-733. Amachawadi, R.G., N.W. Shelton. X. Shi, J. Vinasco, S.S. Dritz, M.D. Tokach, J.L. Nelssen, H.M. Scott and T.G. Nagaraja. 2011. Selection of tcrB gene mediated copper resistant fecal enterococci in pigs fed diets supplemented with copper. Appl. Environ. Microb. 77:5597-5603. (EFSA) European Food Safety Agency. 2009. Assessment of the Public Health Significance of Meticillin Resistant *Staphylococcus aureus* (MRSA) in animals and foods: Scientific Opinion of the Panel on Biological Hazards. The EFSA Journal 993:3-3. (FDA) Food and Drug Administration. 2008. 21 CFR Part 530:[Docket No. FDA-2008-N-0326] New Animal Drugs; Cephalosporin Drugs; Extralabel Animal Drug Use; Order of Prohibition http://edocket.access.gpo.gov/2008/E8-15052.htm (accessed April 29, 2012) Gustafson, R.H. and R.E. Bowen. 1997. Antimicrobial use in animal agriculture. J. Appl. Microbiol. 83:531-541. Levy, S.B. 1992. The Antimicrobial Paradox. How Miracle Drugs are Destroying the Miracle. New York:Plenum Press, 1992. Norman, K.M., H.M. Scott, R.B. Harvey, B. Norby, M.E. Hume and K. Andrews. 2011. Comparison of the prevalence and genotypic characteristics of *Clostridium difficile* in a closed and integrated human and swine population. Appl. Environ. Microb. 77:5755-5760. (WHO) World Health Organization. 2000. WHO Global Principles for the Containment of Antimicrobial Resistance in Animals Intended for Food. Geneva, Switzerland:World Health Organization. http://whqlibdoc.who.int/hq/2000/WHO_CDS_CSR_APH_2000.4.pdf (Accessed April 29, 2012). (WHO) World Health Organization. 2011. Critically Important Antimicrobials for Human Medicine. Geneva:WHO. http://www.who.int/foodsafety/foodborne_disease/CIA_2nd_rev_2009.pdf (Accessed April 29, 2012).

DISASTER PREPAREDNESS FOR FOOD ANIMAL PRACTITONERS AND PRODUCERS

Christine B. Navarre, DVM, MS, DACVIM
Baton Rouge, LA

INTRODUCTION

In the fall of 2005, hurricanes Katrina and Rita were devastating to the Gulf Coast beef and dairy industries. In disasters of this magnitude, it is impossible to prevent all of the devastating losses. But whether it's flooding, fire, a hazardous material spill, or some other disaster, advanced planning can help veterinarians and producers minimize animal health problems associated with disasters. Three years after Katrina and Rita, when hurricanes Gustav and Ike hit the gulf coast, advanced planning proved invaluable.

Veterinarians play a key role in disaster preparation and mitigation. It is important for veterinarians to have plans for their families, their practices, and to be involved in animal emergency plans at the local, regional and state levels. Each geographic area is unique in the potential types of disasters. Plans should list the likely disasters for a particular area. There are many resources available to help prepare for different types of events, so there is no need to reinvent the wheel. Take plans from others and modify them.

Veterinarians are considered first responders and should be trained in the National Incident Management System (NIMS) and the Incident Command System (ICS). The ICS is a standardized all-hazards approach that enables a coordinated response among various jurisdictions and functional agencies, both public and private. ICS is flexible and can be used for incidents of any type and complexity (all hazards). In short, it allows everyone to speak the same language and use the same system so that duplicate efforts are minimized. It may also be useful to practice owners to organize practice responses to emergencies.

ADVANCED PLANNING AND COORDINATION

Planning at the local level will be the major determinant of successful disaster outcomes. In large disasters regional and state resources may be limited or exhausted so preparation at the local level is critical. Veterinarians should then help their clients and communities develop animal plans.

Plans should be coordinated with other local agriculture related groups such as county extension services, USDA's Natural Resource Conservation Services and Farm Service Agencies, local cattlemen's associations, livestock auction markets, feed stores, etc. Different tasks such as livestock hauling; feed, fuel and generator acquisition and distribution; and animal evacuation, rescue and treatment should be assigned to individuals or groups in advance. Plans should be made for mass evacuation of animals if warning of a disaster is given, and for dealing with rescue, treatment and feeding of animals following a disaster. Cattle, particularly beef cattle, pose special problems when it comes to mass evacuation, so plans should be made well in advance. Primary and contingent holding areas for evacuated and/or rescued cattle as well as staging areas for feed and fuel distribution should be identified in advance. Producers should partner with other farms to provide evacuation space, so other public holding areas can be used for rescued animals.

Portable working facilities and corrals and other equipment needs should be arranged. Tractors and forklifts to move feed and supplies, trucks and trailers for hauling cattle, and feed and water troughs may be needed. Following widespread flooding, airboats, barges, and cowboys on horseback may be able to reach areas where cattle are stranded. Local residents familiar with the area should be identified in advance to act as guides for out-of town volunteers. Anyone attempting to rescue or provide food and water to animals either by boat or on horseback should realize that the terrain might have been altered, and underwater debris and fences may not be visible and take proper precautions.

It's also important for agricultural groups to coordinate with the local offices of emergency preparedness. These organizations should be made aware of livestock evacuation plans so evacuation routes can be planned. Police and sheriff's deputies need to know who is authorized to enter areas that might be closed to the public for animal rescue operations. Following a disaster, many rescue organizations will offer help with animal rescue. However, due to the potential dangers of handling stressed cattle, only experienced persons should be involved directly in rescue operations.

Cooperation with state and federal veterinarians and state veterinary associations is a must to prepare for disasters that cover a wide area and involve multiple counties/parishes. It is prudent for veterinarians and other key agricultural leaders to have special training in the National Incident Management System. Setting up "central command" with phone banks and volunteers specifically for cattle related problems might be of benefit. Distribution of supplies and prioritization of rescue operations will be facilitated if all calls go through one place.

COMMUNICATION

The major widespread long-term destruction of communication systems following hurricanes Katrina and Rita posed major problems in getting information to those who needed it. Many areas were out of land and cell phone service for weeks. Without power, radios could not be recharged and soon were useless, and with many roads impassable, information by word of mouth was difficult. Even areas with phone service may experience very high call volumes leading to inability to get emergency calls through. Advanced notice of feed and fuel staging areas and acquisition of satellite phones for key places are imperative. Feed and hardware stores and milk processing plants are good places to post information. For the dairy producers, milk haulers can be valuable resources to distribute and collect information.

VETERINARY PREPARATION

Practices should plan for disasters in which staff remains local, and for the time when evacuation of the area is necessary. Practices should have copies of key papers at the clinic, in a safe place locally (safe deposit box for example) and in a remote town or state. Should a disaster destroy the whole community, these remote copies will still be available. Insurance needs, including business interruption insurance should be reviewed. An evacuation box of critical information such as insurance papers, staff contacts, licenses, DEA info, etc. should be prepared in advance.

A list of emergency contact information for staff, including where they will evacuate to and how you will get in touch with them is critical. Plan for power outages: have cash on hand and a plan for communications, pharmaceutical storage, etc. Let staff know where information will be made available in case of communication failures. Partner with a clinic in a remote area in case hospitalized patients have to be evacuated.

Veterinarians should maintain adequate supplies of emergency pharmaceuticals suitable to treat injured or diseased cattle. In large- scale disasters involving high numbers of cattle, providing food and fresh water is the first priority. If animals do need treatment, working facilities should be inspected before use as they may have been damaged. Access to portable working facilities should be arranged in advance. Stress induced respiratory disease outbreaks should be anticipated. Also, damage to chemical storage buildings and fences may allow cattle access to toxic chemicals or plants. Finally, severely injured animals may require euthanasia. Use of barbiturates should be limited to situations where the carcass can be properly buried or incinerated. Key personnel should be trained in the use and safety of firearms for euthanasia.

Hard copies of brand commission books and disaster recovery information should be stockpiled in a safe area for immediate distribution following a disaster. County extension offices and Red Cross Centers make good distribution points. Remember that with major power outages, web-based information is useless. Important phone numbers should be posted in several safe places. Numbers for emergency services, state and federal veterinarians, other local agricultural groups, volunteer organizations; information about where donations should be routed; and how to handle carcass disposals should be gathered in advance. Veterinarians should be familiar with county animal control regulations, as these may determine the disposition of rescued animals that go unclaimed.

It's important to remember that when disasters strike, not only are businesses and animals affected, but families and personal property are affected as well. Veterinarians and other first responders may be faced with the difficulty of juggling animal rescue operations with personal property and family issues. A backup "buddy" system should be in place for veterinarians and technicians to provide help in the immediate aftermath of a disaster, but to also provide some relief when rescue and cleanup operations continue for long periods of time. State veterinary medical associations can help practitioners set up these backup systems. Multiple backups are needed for each veterinarian/practice. Backups that are in the same area are needed for minor emergency situations, and backups in geographically separate areas are needed in case of widespread disasters.

PRODUCER PREPARATION

Producers should have safe, efficient penning and loading facilities ready in advance. If flooding or high winds are expected and animals cannot be evacuated, they should be left in large open pastures, and not put in barns.

Well in advance of a potential disaster situation, veterinarians should help producers evaluate their herd health programs. Cattle that undergo evacuation either before or after a disaster will be stressed and are likely to be commingled with other cattle. Herd biosecurity will be breeched, which makes increasing herd immunity imperative. Increased numbers of respiratory disease cases and abortions should be anticipated, and can be minimized with proper herd nutrition and vaccination. In addition, dairy herds are susceptible to increased mastitis problems due to interruption of normal milking schedules. One of the major problems for dairy producers following hurricane Katrina was prolonged power outages. Although most dairies had generators available, many were old and failed after several days of constant use. Fuel availability was also a problem. Having relatively new, well maintained generators and temporary sources of fuel will help minimize these problems.

Animal identification is also important. If cattle get evacuated and commingled, or escape and are later captured, it's essential to be able to identify the herd of origin through brands or tags. Individual animal identification will help owners evaluate which animals may be lost from the herd. Veterinarians should be aware of evacuation plans so health papers can be provided in situations such as hurricanes, where advanced warning is given. In some situations it may not be possible to evacuate or rescue all animals, so producers should prioritize animals so their most valuable stock gets attention first. Copies of herd records and registration papers should be stored in a safe place.

CONCLUSION

Whether it's at the veterinary practice level or state level, plans are always a work in progress and should be continually improved. Table top exercises helps to fine tune plans and identify weaknesses and holes. An experienced facilitator is a must for table top exercises. There is no way to prepare for every situation that arises in a disaster. However, veterinarians are well trained to be leaders in the preparation and implementation of plans for minimizing animal suffering and losses following disasters. By working closely with producers and other agricultural leaders, veterinarians can lesson the impact of a disaster on individual producers and the agricultural economy.

CHECKLISTS FOR PRODUCERS

Disaster Readiness Checklist for Beef Producers

- Herd health and vaccinations up-to-date
- Animal identification
- Health papers
- Prioritize herd
- Records stored in safe location
- Evacuation plan
- Cash available for emergency purchases (credit cards may not work)
- Stockpile food and water
- Emergency equipment and first aid supplies stored
- Partner with other producers/farms
- Coordinate plans with other local agricultural groups

Disaster Readiness Checklist for Dairy Producers

Farm Environment

- Do preventative trimming of trees around barns, driveway and fences
- Try to maintain as much free board in manure lagoon as possible to avoid potential overflow
- Have on hand 3 to 4 spools of barb wire plus 50 to 60 metal 'T' post to mend fence or to build a temporary fence
- Calf hutches that are not in use should be stacked together if possible or anchored
- Nail down all lose pieces of tin on barns
- Attach extra guide wires to augers on grain bins
- Have some extra tarps or shade cloth available in order to cover equipment if roof blown off or to provide temporary shade.
- Remove shade cloth from portable shade structures to prevent damage

Equipment

- Fill up all tractors, vehicles, generators and storage containers with fuel
- Service generator(s) and make sure they are operational
- Run the generator under a load for a couple of hours at least every two months
- Setup generator in place before storm
- If using a PTO-type generator, make sure the tractor being used has no fuel or oil leaks to prevent fire hazard

Feed and Water

- Have about a 2 week supply of all feedstuffs needed
- Have an emergency supply of water
- Put extra tires on top of silo to help hold down plastic
- Move round bales from low lying areas to an area that is readily accessible

Animals

- Evaluate herd health program
- Identify animals
- Move all animals to high ground if possible
- Remove calves from calf hutches made out of plastic or fiberglass if possible. If not anchor the hutches down
- Have a supply of intra-mammary mastitis treatment, broad spectrum antibiotics, electrolytes and calcium solutions, antiseptics, bandages, needles and syringes

Miscellaneous

- Have some cash on hand (often credit cards will not work)
- Try to work with milk hauler and marketing co-op to have the least amount of milk in bulk tank possible prior to potential hurricane
- Partner with other farms in remote areas for help

THE BEST PAPERS OF 2011

Allen Roussel, DVM, MS, DACVIM
College Station, TX

THE OTHER BEST PAPERS OF 2011

Gilles Fecteau, DMV, DACVIM
Saint-Hyacinthe, Quebec, Canada

DISCLOSURE

We think it important to describe the objective sought by the exercise of selecting 10 papers from the past year. Moreover, the terminology of "best papers" must be put into context. It would seem more appropriate to use the term "most interesting". It is not the pretention of the authors to have completely assessed all aspects of every manuscript published in the past year. These papers were selected because they were relevant to internal medicine; they brought important answers to clinical questions or new concepts. The intent of limiting the list to 10 studies was to focus attention and discussion on a small number of evidenced-based papers. These manuscripts caught the attention of both authors as being "interesting" and very relevant to the audience in 2012. We hope that the selected papers for discussion at the 2012 ACVIM Forum increase our understanding of the pathophysiology of disease or change the way we diagnose, treat, and control disease of food animals. Many excellent papers were not able to be listed because of the desire for diversity in species and organ system(s) affected, age of animals, causative agent, and geographic location.

SELECTION PROCESS

Peer-reviewed food animal medicine articles published in 2011 were examined. Only English-language articles were reviewed. The search was conducted in January 2012 in PubMed database, for articles published (paper or electronic format) in 2011. Search keywords used to scan title and abstract were: cow*, cattle, bovi*, calf, calves, heifer, sheep, goat; all combined with: medicine, surger*, treatment*, prognosis, diagnos*. The truncation * at the end of some words, include all possible terminations, i.e. surgery, surgeries, surgical, etc. in order to maximize the search. Each author searched independently with different combinations of the keywords above. The initial search result yielded 2427 references for one author and 3480 for the other with much obvious duplication. The first step involved reading only the titles. From that first selection, 74 and 256 papers were selected. For the second step, titles and abstracts were reviewed to restraint the list to 18 and 74 manuscripts. Then the entire manuscript was reviewed to obtain a final list of 11 and 27 articles. At this point the final list of the 2 authors was compared. Interestingly, five papers were common to both lists. After discussion, a final list of 10 papers were selected and agreed upon by both authors. Amongst criteria used to select a paper, the following were considered: relevance to internal medicine, geographical distribution, strength of study design and species diversity. Nonetheless, there was a perceived likelihood that the results of the study will impact the way we practice food animal medicine.

DESCRIPTIVE STATISTICS

Table 1 presents the diversity amongst the 10 selected papers in regard of the species studied, the journal of publication, the country of origin, the study design and the organ system involved. The different study designs include descriptive studies (case series) and analytic studies (observational studies; experimental studies such as laboratory studies and randomized clinical trials). Eight different journals were represented. Studies took place in 5 different countries. Six organ systems were involved in the presented results.

SELECTED PAPERS BY ALPHABETICAL ORDER

Ballou MA, Cobb CJ, Hulbert LE, Carroll JA. Effects of intravenous Escherichia coli dose on the pathophysiological response of colostrum-fed Jersey calves. *Vet Immunol Immunopathol.* May 2011;141(1-2):76-83.

Brisville AC, Fecteau G, Boysen S, Dorval P, Buczinski S, Blondin P, Smith LC. Respiratory disease in neonatal cloned calves. *J Vet Intern Med.* Mar-Apr 2011;25(2):373-9.1939-1676..

Feßler AT, Kaspar H, Lindeman CJ, et al. A proposal of interpretive criteria for cefoperazone applicable to bovine mastitis pathogens. *Vet Microbiol.* E-Pub Dec 2011.

Fecteau ME, Whitlock RH, Fyock TL, McAdams SC, Boston RC, Sweeney RW. Antimicrobial activity of gallium nitrate against Mycobacterium avium subsp. paratuberculosis in neonatal calves. *J Vet Intern Med.* Sep 2011;25(5):1152-5.

Mann S, Siler JD, Jordan D, Warnick LD. Antimicrobial susceptibility of fecal *Escherichia coli* isolates in dairy cows following systemic treatment with ceftiofur or penicillin..*Foodborne Pathog Dis.* Aug 2011;8(8):861-7.

Relun A, Guatteo R, Roussel P, Bareille N. A simple method to score digital dermatitis in dairy cows in the milking parlor. *J Dairy Sci.* Nov 2011;94(11):5424-5434.

Terrill TH, Miller JE, Burke JM, Mosjidis JA, Kaplan RM. Experiences with integrated concepts for the control of Haemonchus contortus in sheep and goats in the United States. *Vet Parasitol*. E-Pub Nov 2011.

Tharwat M, Oikawa S.Ultrasonographic evaluation of cattle and buffaloes with respiratory disorders. *Trop Anim Health Prod.* Apr 2011;43(4):803-10.

Timsit E, Bareille N, Seegers H, Lehebel A, Assie S. Visually undetected fever episodes in newly received beef bulls at a fattening operation: occurrence, duration, and impact on performance. *J Anim Sci.* Dec 2011;89(12):4272-4280.

Trefz FM, Lorch A, Feist M, Sauter-Louis C, Lorenz I. Metabolic acidosis in neonatal calf diarrhea-clinical findings and theoretical assessment of a simple treatment protocol. *J Vet Intern Med.* Jan 2012;26(1):162-170. E-Pub Dec 2011

Table 1: Characteristics of selected best papers of 2011

First Author	Animal studied	Journal	Country	Study design	Organ system
Ballou	Newborn calf	Veterinary Immunology and Immunopathology	USA	Experimental challenge	Systemic disease
Brisville	Newborn calf	Journal of Veterinary Internal Medicine	Canada	Retrospective study	Respiratory disease
Feßler	Dairy cattle	Veterinary Microbiology	Germany	Retrospective study	Udder health
Fecteau	Newborn calf	Journal of Veterinary Internal Medicine	USA	Randomized blind control experiment	Infectious disease
Mann	Dairy cattle	Foodborne Pathogens and Disease	USA	Longitudinal control trial	Gastroenterology
Relun	Dairy cattle	Journal of Dairy Science	France	Diagnostic validation study	Foot disease
Terrill	Sheep/goat	Veterinary Parasitology	USA	Review	Gastroenterology
Tharwat	Buffaloes	Tropical Animal Health Production	Egypt	Retrospective study	Respiratory disease
Timsit	Feedlot	Journal of Animal Science	France	Observational study	Systemic disease
Trefz	Newborn calf	Journal of Veterinary Internal Medicine	Germany	Prospective blind cohort study	Acid-base

This page intentionally left blank.

TRANSFORMING YOUR CLIENTS' INTERNET RESEARCH INTO A WIN-WIN SITUATION

Nancy D. Kay, DVM, DACVIM
Hendersonville, NC

Are you spending precious office visit time dissuading clients from the whackadoodle notions they glean from cyberspace? Do your clients cling to Dr. Google's recommendations more than they believe what you have to say? There's no question that Internet research is here to stay and a considerable amount has been published on this topic within the human literature[1,2,3]. The good news is there are a number of ways to ensure the result is a win-win situation for everyone involved without sacrificing quality of care or causing you to expend needless time and energy.

WHERE TO BEGIN

I encourage you to begin by teaching your clients how to avoid online "junk food" and become responsible surfers:

1. Create a list of recommended websites that you have already "vetted". If you see loads of patients with allergic skin disease, make sure your list covers this topic. If your specialty is surgery, be ready to provide a high quality website or two that address cruciate ligament disease. Help your clients avoid having to reinvent the wheel.
2. Veterinary college websites invariably provide reliable information. Advise your clients to search for them by entering "veterinary college" or "veterinary school" after the name of the disease or symptom they are researching. Try this yourself. Go online and compare the results of Googling "feline asthma" versus "feline asthma veterinary college." Quite a difference in the quality of information provided, don't you think?
3. If your patient has a breed-specific disease, encourage your client to visit the site hosted by that specific breed's national organization. For example, the national Bernese Mountain Dog website (www.bernergarde.org) provides a wealth of current information about malignant histiocytosis.
4. Teach your clients how to avoid business-sponsored websites, particularly those offering products for sale.
5. Teach your clients to recognize anecdotal information for what it is. As captivating as these *National Enquirer* type stories may seem (how Max's heartworm disease was cured with aromatherapy), it's not okay for such vignettes to gobble up precious office visit time or influence medical decision-making.
6. Encourage your clients to check out disease-specific online forums (ideally prevetted by you). Not only do they provide a wealth of educational information, members can be a wonderful source of emotional support (may translate into reduced need for emotional support from you). The best forums are those that limit themselves to a specific disease (kidney failure, diabetes, lymphoma, etc), have lots of members, utilize multiple moderators, and have been in existence for several years. Two examples of forums I readily refer clients to are AddisonDogs (www.addisondogs.com/support/helpers/guidelines) and Tripawds (tripawds.com/forums).

WHAT COMES NEXT

Now that you've taught your clients how to be responsible surfers, here are some tips for managing conversations about their Internet research:

1. Consider how your client may be feeling about approaching you with what she has learned online. She may be feeling:

   - Sheepish about obtaining a second opinion from "Dr. Google"
   - Worried that you will think she doesn't trust you
   - Anxious that you will judge her negatively
   - Apprehensive that there will be conflicting opinions
   - Concerned that there won't be time to discuss her research

As with any conversation, after putting yourself in your client's shoes, it's much easier to respond with genuine empathy rather than animosity or frustration.

2. Broach the topic first (with enthusiasm), just as soon as you see the stack of computer paper in your client's hands. This will help her know that you are interested in rather than put off by her research.
3. Let your client know that you are pleased that she is interested in learning more about her pet's health issues.
4. Reassure your client that you will reserve time to discuss what she has learned online. Chances are, once you've provided your thorough explanation the questions generated by the Internet research will have already been answered!

Transform your client's Internet research into a win-win situation. In addition to your clients feeling more supported and better informed there is always the possibility that you might learn a thing or two. Besides, a busy workday can always benefit from some comic relief- who knew that hip dysplasia is caused by global warming!

REFERENCES

[1]Schwartz Kendra L, et.al. **J of Am. Board of Family Med** 2006;39(19): 39-45.[2]McMullan, M. **Patient Educ Couns** 2006;63 (1-2):24-28.[3]Welch, RJ, et al. **Am Soc of Clinical Oncology** 2007;3(3):167-171.

SURGICAL MANAGEMENT OF CONGENITAL BRAIN AND CERVICAL SPINE DISORDERS IN DOGS

CW Dewey, DVM, MS, DACVIM (Neurology), DACVS
Ithaca, NY

INTRODUCTION

There are a number of congenital disorders of the brain and cervical spine encountered in dogs which are amenable to surgical treatment. Recognition of these abnormalities has increased in recent years due to increased availability of MR imaging in our profession. Whereas surgical treatment options are well described for some of these abnormalities, others have only recently been reported in the veterinary literature. The purpose of this presentation is to briefly describe the surgical procedures associated with a number of disorders: congenital hydrocephalus, intracranial arachnoid cyst, Chiari-like malformation (and similar disorders) and atlantoaxial instability.

CONGENITAL HYDROCEPHALUS

Ventriculoperitoneal shunt (VPS) placement is the typical method by which congenital hydrocephalus in dogs is surgically managed. There are a number of shunts available, all of which have the same basic design: a rostral end for placement into the lateral ventricle, a one-way valve, and a distal end for placement into the peritoneal cavity. There is some controversy regarding the appropriate opening valve pressure for shunts (i.e., low vs. high). For very small patients (e.g., Chihuahua, Yorkshire terrier), two incisions-one curvilinear incision over the caudodorsal region of the parietal bone and one vertical incison caudal to the last rib- are made. For larger patients (e.g., English Bulldog), an additional incision approximately midway between these 2 incisions on the lateral thorax is often required. A high speed drill is used to drill two holes in the parietal bone: one for the shunt and one for an anchoring suture. After insertion of the rostral end of the shunt into the lateral ventricle, it is secured with a Chinese finger trap suture via the anchor hole using 3-0 or 4-0 prolene. A grid approach to the peritoneal cavity is performed and the distal end of the shunt is tunneled from the cranial incision to the caudal incision using a long Carmalt or Doyen forceps. After placement of the distal end of the shunt tubing into the peritoneal cavity, it is secured to both the last rib and the abdominal musculature. Either post-operative radiographs or CT scans can be used to assess the accuracy of shunt placement postoperatively.

INTRACRANIALARACHNOID CYST

Surgical management of intracranial arachnoid cysts is generally accomplished via either fenestration or cystoperitoneal shunt (CPS) placement. The author prefers CPS placement after creating a wide craniotomy defect. With the exception of the placement of the rostral end of the shunt, the procedure is identical to that described for congenital hydrocephalus. For placement of the rostral end of the shunt in IAC cases, the author performs a wide caudolateral craniotomy incorporating the creation of both a rostrotentorial and suboccipital skull defect. The transverse sinus between these two defects is sacrificed and occluded with bone wax. As with congenital hydrocephalus, postoperative imaging is performed to assess accuracy of shunt placement. The success rate for surgical management of IAC appears to be high in humans and dogs, and whether fenestration or cystoperitoneal shunting is the preferred procedure remains controversial for both species.

CHIARI-LIKE MALFORMATION

Chiari-like malformation (CLM) is generally treated surgically via foramen magnum decompression (FMD). The author and colleagues developed a cranioplasty procedure to be used in conjunction with the FMD procedure. A dorsal midline incision is made extending from approximately 1 cm rostral to the external occipital protuberance cranially to the middle of the second cervical vertebra caudally. The superficial dorsal cervical musculature is separated at the median raphe, exposing the underlying biventer cervicis muscles. The paired biventer cervicis muscles are then separated on midline, exposing the rectus capitis dorsalis muscles. The caudal aspects of the rectus capitis dorsalis muscles are removed from the cranial half of C2 using sharp dissection and periosteal elevation and these muscles are then split on midline. The cranial aspects of the rectus capitis dorsalis muscles are sharply incised from the nuchal crest, exposing the caudal portion of the occiput and the arch of the atlas. Hemorrhage is controlled with bipolar electrocautery. A high-speed air drill with a 3- to 4-mm diameter round drill bit and Lempert rongeurs is used to remove a portion of the occiput and the dorsal aspect of the first cervical vertebra. Approximately 75% of the length of the dorsal arch of C1 is routinely removed. In cases in which there appears to be dorsal constriction at C1/C2, the entire arch of C1 as well as part of the dorsal lamina of C2 are also removed. Once the FMD is completed, the meninges (dura/arachnoid) are incised on midline and the meningeal tissue in the region of the FMD is resected. There is typically a fibrous band of tissue at the cervicomedullary junction that may be partially ossified. This band is transected carefully with a probe (Gross ear hook and spoon) and a sharp #11 blade. Four to 6 guide holes are drilled in the occipital bone around the periphery of the skull defect using a 1.1 mm drill bit. Self-tapping 6 mm length (1.5 mm width) screws are inserted into the guide holes for an approximate depth of 2-3 mm . The dog's head is then released from the flexed positon and repositioned at a normal resting angle. The skull plate is fashioned using titanium mesh (TM) and polymethylmethacrylate (PMMA) and fixed to the back of the skull, using the titanium screw heads as anchor posts for the PMMA. The plate is shaped somewhat like a guitar pick, the wide end of the pick toward the occiput. Only a thin layer of PMMA is applied to both surfaces of the plate, with some PMMA extending beyond the edges of the TM in order to adhere to the titanium screw heads. The caudal aspect of the plate is made to extend slightly over the dorsal defect of C1; the tail aspect of the plate is curved dorsally to avoid impinging on the medulla or cranial cervical spinal cord.

ATLANTOAXIAL INSTABILITY

The author prefers a ventral approach to the atlantoaxial (AA) joint for reduction and stabilization. A standard ventral cervical approach is made via splitting the sternohyoideus muscles and retracting both carotid sheaths laterally. When approaching the C1/C2 intervertebral space, the belly of the sternothyroideus muscle is transected to facilitate exposure of this joint space. Using a #11 blade, the small muscle (rectus capitis ventralis) over the ventral surface of C1 is incised and reflected laterally with Freer elevators to expose the C1 ventral arch. The longus colli muscles are elevated off of the ventral aspect of C2. With a #11 or #12 blade, the C1/C2 joint capsules bilaterally are incised, exposing the articular surfaces. Soft tissue attachment between C1 and C2 are incised, revealing the underlying dens (if present) and spinal cord. Redundant joint capsule tissue is sharply dissected with a #11 blade and Bishop-Harmon forceps. Using a small towel clamp at the C2/C3 disk space or a speed-lock reduction forceps on the body of C2, the C1/C2 joint is levered into reduction. Typically, the cranial aspect of C2 is displaced in a cranioventral direction from the surgeon's perspective (ventral approach), craniodorsally from an anatomic perspective. Mild but steady traction is applied dorsally and caudally on C2 while putting slight downward pressure on the ventral arch of C1 to attain reduction. The joint space is then distracted a small amount and the articular surfaces scarified with a #12 blade. While maintaining reduction, a pin or screw is placed through the articular processes (facets) on one side. In most toy and miniature dogs, the pin size will be 0.035 or 0.045 and the screw size will be 1.5 mm or 2.0 mm. The pin or screw is inserted in a medial to lateral direction, following the plane of the joint articulation, at an approximately 45° angle to the vertical plane. The towel clamp or reduction forceps is released. A pin or screw is inserted through the C1/C2 joint on the opposite side, using the same technique as the first side. Additional pins or screws are placed in the wings of C1 and body of C2, taking care to direct these implants laterally, away from the spinal cord. Depending on surgeon preference, the screws/pins can be reinforced with small K-wires and or cerclage wire prior to placing PMMA over the implants

REFERENCES

[1]Robinson S, et al. **Pediatr Neurosurg** 2002;37:287-294.[2]Dewey CW, et al. **Vet Surg** 2007; 36:404-414.[3]Dewey CW, et al.**Vet Surg** 2007;36: 416-421.

COMPARISON OF CERVICAL SPONDYLOMYELOPATHY IN DOGS AND CERVICAL SPONDYLOTIC MYELOPATHY IN PEOPLE

Ronaldo C. da Costa, DMV, MSc, PhD, DACVIM (Neurology)
Columbus, Ohio

INTRODUCTION

Cervical spondylomyelopathy in dogs (CSMd) is arguably the most common disease of the cervical spine of large and giant breed dogs.(1) Recent data from the Veterinary Medical Database indicates a prevalence of CSM in Doberman Pinschers of 5.5%, and of 4.2% in Great Danes. In most reports, these two breeds account for 60-70% of the cases. Canine CSM bears considerable similarities with the cervical spondylotic myelopathy of people, and the Doberman breed has been proposed as a natural model to study the disease in humans.(2) Cervical spondylotic myelopathy in people (CSMp) is the most common form of myelopathy in adults and it is considered the most common disorder in the United States causing dysfunction of the spinal cord.(3) CSMp refers to impaired function of the spinal cord caused by degenerative changes of the cervical spine typically resulting in spinal cord compression. Significant more information is available about CSMp than CSMd, therefore the goals of the review are to provide an overview of the disease in both species.

CERVICAL SPONDYLOMYELOPATHY IN DOGS

Etiology

The etiology of CSMd is still unresolved. Proposed etiologies include genetic, congenital, body conformation and nutritional. (4-8) The latter two appear to have a non-significant role in the development of CSM. Based on current evidence the diseaseappears to be congenital. (9) It is still unclear how much genetics contribute to the etiology of the disease, but the fact that two breeds are strongly over-represented, and the disease is congenital suggests a hereditary basis.

Pathophysiology

The pathophysiology of CSM involves both static and dynamic factors. The key static factor is vertebral canal stenosis. It may be an absolute vertebral canal stenosis (which then causes direct spinal cord compression and neurological signs) or a relative vertebral stenosis, which by itself does not lead to myelopathic signs, but predisposes the patient to develop myelopathy. (10) Despite some degree of overlap, the pathophysiology of the spinal cord compressions in CSMd can be divided into osseous or disc-associated compression. (11)

Disc-associated compression is typically seen in middle-aged large breed dogs (mostly Dobermans). It is caused by a combination of intervertebral protrusion with or without ligament hypertrophy (either the dorsal longitudinal ligament or ligamentum flavum). (11) The pathophysiology of osseous or bony associated CSMd is different. Osseous associated CSMd is seen predominantly in young adult giant breed dogs. Giant breeds usually have severe absolute vertebral canal stenosis secondary to proliferation of the vertebral arch (dorsally), articular facet joints (dorso-laterally) or articular facet joints and pedicles (laterally). (12) The cause of the compression appears to be a combination of vertebral malformation and osteoarthritic/osteoarthrotic changes at the level of the articular facet joints.(12)

Critical to the development of clinical signs in CSM-affected dogs is the concept of dynamic lesions. (13-16) Dynamic spinal cord compressions are present in both disc- and osseous-associated forms of CSMd. A dynamic lesion would be one that worsens or improves with different positions of the cervical spine.(17) This concept is very different from instability, which has been defined as the loss of ability of the cervical spine under physiologic loads to maintain its normal pattern of displacement so that there is no damage to the spinal cord or nerve roots. (18) Instability has not yet been documented in dogs with CSMd and it appears that with the severe disc and osseous degenerative changes, a restricted, rather than an excessive motion, takes place at the affected sites.(13)

Diagnosis

CSMd is primarily a disease of large and giant breed dogs.(1) It is occasionally seen in small breed dogs but it is unclear whether small dogs also have vertebral canal stenosis. (19, 20) The majority of large breed dogs (Weimaraners, Dalmatians) and Dobermans with CSMd are older than 3 years of age at presentation (mean age approximately 7 years). Great Danes and giant breeds (Mastiffs, Rottweilers, Bernese and Swiss Mountain dogs) with CSMd are seen usually at a younger age. The mean age of giant breed dogs with CSMd is 3.8 years, and the disease may be seen in dogs just a few months old. (12, 21-23)

History and Clinical Signs

A chronic progressive history (several weeks to months) is typical. Neck pain or cervical hyperesthesia is a common historical finding, but typically is not the main reason for presentation. Neck pain is part of the clinical findings in approximately 65-70% of Dobermans, and in 40-50% of other breeds, but it is the chief complaint in only 5 to 10% of dogs with CSMd.(20) Proprioceptive ataxia is seen in most dogs with CSM. Dogs with lesions in the cranial or mid cervical spine tend to present with ataxia affecting all four limbs more uniformly. Typically however, affected dogs have obvious pelvic limb ataxia with milder abnormalities in the thoracic limbs. In some cases, the thoracic limb ataxia or paresis may be very mild in comparison with the pelvic limbs signs, making the thoracic limb abnormalities go unnoticed. The thoracic limb gait can appear short-strided or spastic with a pseudo-hypermetric ("floating") appearance.

Imaging

The definitive diagnosis of CSMd requires either myelography, computed tomography or magnetic resonance imaging. Myelography can define the location(s) and the direction (ventral, dorsal, lateral) of the spinal cord compression. (11, 24) Myelography in stress positions of flexion or extension can be performed but offer significant risk of neurological deterioration. Linear traction myelography is a safer procedure and can distinguish a static from a dynamic lesion. Computed tomography (CT) myelography allows cross-sectional visualization of the spinal cord compression and determination of areas of spinal cord atrophy. (25) Magnetic resonance imaging (MRI) allows visualization of the spinal cord parenchyma, intervertebral disk, soft tissues and nerve roots, and the images can be obtained in any plane. (26, 27) Traction images (kinematic studies) can also be done with MRI. (26, 27) A study compared myelography and MRI in the diagnosis of CSMd in 18 Doberman pinschers.(26) The results showed that while myelography can identify the location of compressive lesions in the majority of dogs, it can also underestimate the severity of the lesion and not identify them precisely. The main advantage of MRI is its ability to visualize the spinal cord parenchyma, showing spinal cord signal changes. The study concluded that MRI was more accurate in predicting the site, severity and nature of spinal cord compression.

Treatment

The treatment of CSMd can be either surgical or medical (conservative).(1) Conservatively treated dogs should have restricted activity for at least 2 months in an attempt to minimize the dynamic component of spinal cord injury. Oral corticosteroids can also be used and many dogs respond well to corticosteroid therapy at least initially. Results of a recent study comparing medical and surgical treatment in 104 dogs with CSM indicate that medical treatment improves the neurological status of 54% of dogs, and stabilizes the condition in 27% of them. In contrast, surgical treatment improved the neurological status of 81% of dogs.(20)

Surgical treatment

Surgical treatment is generally assumed as the treatment of choice for CSMd. As most affected dogs have spinal cord compression, decompressing the spinal cord would, be in theory the definitive treatment. The decision to recommend surgical treatment should however, be based on several factors such as severity of neurological signs, presence of pain, type and severity of compressive lesion(s), response (or lack of response) to medical management, short- and long-term expectations of the owner, and presence of other concurrent neurological or orthopedic problems, or extra-neurologic diseases such as dilated cardiomyopathy that would affect the long-term outcome. The decision on the specific surgical technique can also be complicated. Twenty-three surgical techniques have been proposed to treat CSMd.(1, 28, 29) Surgical techniques for CSM can be grouped into 3 categories: direct decompressive techniques, indirect decompressive or vertebral distraction techniques, and motion preserving techniques.(1) Overall, the success rate of most surgeries is approximately 80%. This rate is superior to medical management; however, the median survival time of 76 dogs treated medically and surgically was found to be exactly the same (36 months) with both treatment methods. (20)

CERVICAL SPONDYLOTIC MYELOPATHY IN PEOPLE

Etiology

Cervical spondylotic myelopathy (CSMp) appears to have a multifactorial etiology. Factors that have been implicated in the development of CSMp include repeated occupational trauma, familiar predisposition, and Down syndrome.(30, 31) Smoking has been associated with intervertebral disc degeneration and it is a risk factor for CSMp.(30) A recent study (Patel et al, 2012), documented for the first time an inherited predisposition for CSMp.(31)

Pathogenesis

The mechanisms causing CSMp are essentially the same as those seen in dogs. People with CSM have a combination of static and dynamic factors. In humans, cervical stenosis is considered a fundamental condition for the development of clinical signs associated with CSMp, (32, 33) some authors even believe it may be the most important static factor.(17, 33) A narrow canal would lower the threshold at which the cumulative effects of various structures encroaching on the spinal cord cause signs of myelopathy.(33)

Spondylotic changes occur in the aging cervical spine secondary to intervertebral disc degeneration. Because of the disc degeneration, increased mechanical stresses occur at the vertebral endplates of adjacent vertebral bodies, which leads to subperiosteal bone formation and spondylosis.(30) Similar to CSMd , the concept of dynamic lesions, also called strain or stretch-associated injuries, seems very important in the development of clinical disease in humans.(14-16) A dynamic lesion is described as a lesion that worsens or improves upon different positions of the cervical spine, which is the observation made using flexion/extension myelographic views. The normal movement of the cervical spine causes significant changes in the relationship of the cervical structures. For instance, cervical extension in healthy humans caused an 11-16% reduction in the diameter of the vertebral canal due to infolding of the ligament flavum, annulus fibrosus, and posterior dura-mater.(34, 35) It also increased the spinal cord area by 9 to 17%. This is a manifestation of the Poison's effect, defined as an increase in cross-sectional area and a decrease in length, whereby the total volume remains the same.(36) Although no such studies have been performed in dogs, it is likely that the same principle applies. This would explain why cervical extension or dorsiflexion causes worsening of the spinal cord compression and clinical signs in CSM-affected dogs.(37) According to the Poison's effect, neck flexion generates the opposite effect in the spinal cord, stretching the spinal cord between C2 and T1 in up to 17.6% of its length in humans (while also reducing the cross-sectional area), with the maximal stretch occurring in the caudal cervical region.(38) With the spinal cord stretch, ventrally positioned space-occupying lesions, such as a protruded intervertebral disc, cause more severe spinal cord compression.

Clinical Signs and symptoms

The hallmark symptoms associated with CSMp are gait abnormalities and weakness or stiffness of the legs, which usually develop insidiously.(30) However, CSMp can present in various forms and the signs and symptoms have been classified in 5 different syndromes.(33) The lateral or radicular syndrome is characterized by pain and occasional gait abnormalities, and it is usually unilateral. The medial or myelopathic syndrome is characterized by gait changes without pain and it is often bilateral. Most patients have a combination of the radicular and myelopathic syndromes, also known as the combined form (third syndrome). The fourth syndrome is the vascular syndrome, which is the least common, and it is not associated with pain, but it can involve both upper and lower limbs. The fifth clinical presentation is the anterior syndrome, which is associated with upper extremity weakness only, without involvement of the lower limbs or pain. This is a rare form of CSMp that is associated with damage of the anterior (ventral) horns of the cervical spinal cord only.(33)

Imaging

Magnetic resonance imaging is the most commonly used test in the evaluation of human patients suspected of having CSMp because of its superior depiction of soft tissue anatomy.(30) Comparative studies using MRI and CT, CT-myelography or myelography have indicated that MRI is a superior diagnostic technique.(39-43) MRI studies in humans have also shown a correlation between the cross-sectional area of the spinal cord, and the severity of myelopathy.(44, 45) There is a consistent correlation between the degree of spinal cord compromise at the level of greatest compression and prognosis for recovery. In general, those patients with the greatest degree of spinal cord compression have a lesser chance of recovery after surgical decompression.(46)

Controversy exists as to whether spinal cord signal changes have prognostic significance in humans with CSMp. Some have proposed that hyperintensity in the T2-weighted images was associated with a poor outcome,(47, 48) whereas others found no association between hyperintensity and poor outcome.(49-51) More recently, it has been proposed that T2 hyperintensity reflects pathologically reversible changes, whereas the combination of T2 hyperintensity and T1 hypointensity suggests irreversible spinal cord changes and a poorer prognosis.(52-54) In addition to the combination of T2 hyperintensity and T1 hypointensity, the presence of spinal cord atrophy on pre-operative MRI is also associated with a poor outcome.(54)

Treatment and Natural History

Even though there have been 17 studies published on the natural history of CSM in people, the consensus is that strong evidence is still lacking in defining the natural history of CSMp.(55, 56) Studies indicate that the history of CSMp may be mixed, with many patients experiencing a slow, stepwise decline. Long periods of quiescence are not uncommon, however, and some patients may actually improve. In patients younger than 75 years of age with mild CSMp, it appears that non-operative management is associated with a stable clinical course over a period of at least 36 months. A recent study compared the outcome of medical and surgical treatment in 64 patients with a 10-year follow-up (the longest ever report in comparative studies).(56) All patients had mild or moderate forms of CSMp. No difference in outcome was seen between groups at the end of the 10 year period. Data on natural history of severe cases of CSMp is lacking. The consensus is that surgical treatment should be preferred over nonsurgical management for patients with severe and/or long-standing symptoms because the likelihood of improvement with nonsurgical treatment is low.(55)

Similar to the situation in dogs the optimum surgical treatment of CSMp has not been defined. Many surgical techniques with posterior and anterior approaches have been proposed. It appears that all techniques that effectively decompress the spinal cord are effective with regard to achieving short-term improvement in outcome.(54) These techniques include laminectomy, laminoplasty (all variants), laminectomy with arthrodesis, anterior decompression (anterior cervical diskectomy – ACD), ACD with fusion or instrumentation, corpectomy with plating, or cervical disc arthroplasty.(54)

CONCLUSION

Canine CSM and CSM in people show remarkable similarities. In both species the clinical signs typically develop insidiously and involve myelopathic and radicular signs. In both species, the fundamental pathogenesis also appears to be the same, a combination of static and dynamic factors causing spinal cord and nerve root compression. The controversies seen in the pathogenesis of the disease are also seen in people and dogs, such asymptomatic spinal cord compression.(10, 57-59) Data on outcome, primarily related to prospective studies correlating MRI findings and outcomes, is available for humans but it is lacking in dogs. The deterioration seen within 2 to 3 years after surgical treatment in dogs does not appear to occur in humans. It is important that the knowledge available in both species be shared, so that we can advance our understanding of the mechanisms leading to canine and human CSM.

REFERENCES

1. da Costa RC. **Vet Clin North Am Small Anim Pract** 2010;40(5)881-913. 2. Sharp NJ, et al. **J Invest Surg** 1989;2333. 3. Yarbrough CK, et al. **Adv Orthop** 2012;2012480643. 4. Palmer AC, et al. **Vet Rec** 1967;80(14)430-433. 5. Selcer RR, et al. **J Am Anim Hosp Assoc** 1975;11(2)175-179. 6. Mason TA. **Aust Vet J** 1977;53(9)440-445. 7. Olsson SE, et al. **Acta Vet Scand** 1982;23(1)65-78. 8. Jaggy A, et al. **J Am Anim Hosp Assoc** 1988;4453-460. 9. Burbidge HM. .**PhD thesis** 1999;121-135. 10. da Costa RC, et al. **Am J Vet Res** 2006;67(10)1601-1612. 11. Seim HB, et al. **J Am Anim Hosp Assoc** 1982;18241-251. 12. da Costa RC, et al. **Vet Radiol Ultrasound** 2012;53(1)64-70. 13. da Costa RC. .**Proc 25th Annu Meet ACVIM Forum** 2007;318-320. 14. Levine DN. **J Neurol Neurosurg Psychiatry** 1997;62(4)334-340. 15. Henderson FC, et al. **Neurosurgery** 2005;56(5)1101-13; discussion 1101-13. 16. Ichihara K, et al. **J Neurosurg** 2003;99(3 Suppl)278-285. 17. White AA3, et al. **Spine** 1988;13(7)856-860. 18. Panjabi MM, et al. Cervical Spine Kinematics and Clinical InstabilityThe Cervical Spine Research Society. The Cervical Spine4th 2005:55-78. 19. De Risio L, et al. **Vet Surg** 2002;31(5)418-427. 20. da Costa RC, et al. **J Am Vet Med Assoc** 2008;233(8)1284-1290. 21. Lewis DG. **J Small Anim Pract** 1989;30(12)657-665. 22. McKee WM, et al. **J Small Anim Pract** 1999;40(10)465-472. 23. da Costa RC, et al. **J Vet Int Med** 2009;23(3)740. 24. Sharp NJH, et al. **J Small Anim Pract** 1992;33(10)491-499. 25. Sharp NJH, et al. **Vet Radiol Ultrasound** 1995;36(2)100-108. 26. da Costa RC, et al. **Vet Radiol Ultrasound** 2006;47(6)523-531. 27. Lipsitz D, et al. **Vet Radiol Ultrasound** 2001;42(1)20-27. 28. De Decker S, et al. **Vet Surg** 2011;40(5)544-554. 29. Steffen F, et al. **Vet Surg** 2011;40(6)743-752. 30. Baron EM, et al. **Neurosurgery** 2007;60(1 Suppl 1)S35-41. 31. Patel AA, et al. **Spine (Phila Pa 1976)** 2012;37(1)26-29. 32. Yu YL, et al. **Neuroradiology** 1986;28(3)221-236. 33. Bernhardt M, et al. **J Bone Joint Surg Am** 1993;75(1)119-128. 34. Breig A, et al. **J Neurosurg** 1966;25(1)45-56. 35. Waltz TA. **Brain** 1967;90(2)395-404. 36. Panjabi MM, et al. Biomechanics of spondylotic cervical myelopathyDenaro V. Stenosis of the Cervical Spine1st Ed, 1991:43-50. 37. Seim HB. Diagnosis and treatment of cervical vertebral instability-malformation syndromes. Bonagura JD. Current veterinary therapy: small animal practice. 13th 2000:992-1000. 38. Muhle C, et al. **Spine** 2001;26(13)E287-93. 39. Brown BM, et al. **AJR Am J Roentgenol** 1988;151(6)1205-1212. 40. Larsson EM, et al. **Acta Radiol** 1989;30(3)233-239. 41. Modic MT, et al. **Radiology** 1986;161(3)753-759. 42. Breidahl WH, et al. **Australas Radiol** 1991;35(4)306-314. 43. Shafaie FF, et al. **Spine** 1999;24(17)1781-1785. 44. Fujiwara K, et al. **Spine** 1988;13(11)1212-1216. 45. Okada Y, et al. **Spine** 1994;19(20)2331-2335. 46. Bucciero A, et al. **J Neurosurg Sci** 1993;37(4)223-228. 47. Takahashi M, et al. **Radiology** 1989;173(1)219-224. 48. Mehalic TF, et al. **Neurosurgery** 1990;26(2)217-26 discussion 226-7. 49. Morio Y, et al. **Arch Orthop Trauma Surg** 1994;113(5)254-259. 50. Wada E, et al. **Spine** 1995;20(20)2226-2232. 51. Naderi S, et al. **Neurosurgery** 1998;43(1)43-9; discussion 49-50. 52. Morio Y, et al. **Spine** 2001;26(11)1238-1245. 53. Suri A, et al. **Spine J** 2003;3(1)33-45. 54. Fehlings MG, et al. **J Neurosurg Spine** 2009;11(2)97-100. 55. Matz PG, et al. **J Neurosurg Spine** 2009;11(2)104-111. 56. Kadanka Z, et al. **Eur Spine J** 2011;20(9)1533-1538. 57. De Decker S, et al. **Am J Vet Res** 2010;71(4)428-434. 58. Teresi LM, et al. **Radiology** 1987;164(1)83-88. 59. Boden SD, et al. **J Bone Joint Surg Am** 1990;72(8)1178-1184.

CRITICAL EVALUATION OF TREATMENT METHODS FOR CERVICAL SPONDYLOMYELOPATHY

Ronaldo C. da Costa, DMV, MSc, PhD, DACVIM (Neurology)
Columbus, Ohio

INTRODUCTION

The treatment of cervical spondylomyelopathy (CSM) is controversial. Few diseases in veterinary medicine have had so many surgical techniques proposed to treat it as CSM. To the author's knowledge at least 23 surgical techniques have been proposed to treat CSM and more continue to be put forward every year.(1-3) This clearly means one thing – we are treating a disease that we know very little about . This presentation discusses the pertinent issues related to treatment of CSM and presents the indications and outcomes of various surgical techniques.

NATURAL HISTORY

Ideally, the natural history of a disease should be known to properly decide on treatment recommendations. Unfortunately, the natural history of CSM has not been defined. It appcars that thc disease progresses slowly in many dogs with disc-associated CSM.(4) The same may be true for giant breed dogs with osseous compressions. Current evidence suggests that the progressive course of the disease is limited mostly to the lesion(s) seen at the time of initial diagnosis. Follow-up imaging studies of dogs with both disc- and osseous-associated CSM have not demonstrated the appearance of "domino" lesions or adjacent segment disease in dogs that had not been treated surgically. However, evidence in a large number of patients is lacking. It is possible that the development of "domino" lesions may be hastened by surgical treatment of CSM.

PATHOGENESIS

The pathogenesis of CSM is complex and still poorly understood. Many surgical techniques have been based on the premise of instability that has never been documented and it is unlikely to be present by the time a dog is presented with severe intervertebral disc degeneration or osteoarthritic changes.(5) As such, using instability to justify a surgical technique is unfounded. Based on current evidence it appears that depending on the surgical technique employed, instability may be present *after* surgery.(4)

CLASSIFICATION OF LESIONS AS STATIC OR DYNAMIC

Stress myelography, defined as the radiographic examination of the cervical spine in various positions (ventral flexion, dorsal extension, and linear traction), was proposed to differentiate static versus dynamic lesions.(6) Stress myelography, particularly traction, has been, for the last 30 years, considered essential to differentiate a dynamic from a static lesion.(6) This differentiation has since then been considered fundamental for surgical planning. It is interesting that such a widely used concept never had guidelines established for testing or interpretation.(7) Some indicate that a dynamic or traction-responsive lesion is one that improves with traction, while others consider a lesion dynamic or traction-responsive when the compression completely resolves .(8) The form of traction has never been standardized. Based on the current literature, the concept of dynamic and static lesions is highly subjective and dependant on personal opinion.(7, 8) Moreover, magnetic resonance imaging (MRI) has also revealed that almost any compressive lesion, whether it is dynamic or static, will improve with traction.(7) Additionally, the features seen on myelography do not always correlate with the imaging findings seen on MRI.(7)

TREATMENT METHODS

CONSERVATIVE TREATMENT

Traditionally, medical treatment for CSM has been considered a temporary measure to alleviate clinical signs. Without surgery, the disease was thought to be progressive, and euthanasia would have to be contemplated. The only evidence to support these statements came from a study of mostly Great Danes that essentially received no treatment, more than 30 years ago.(9) Medical management for CSM was recently revisited in two studies.(10, 11) A study compared the outcome of dogs treated medically and surgically and found that 54% of dogs treated medically improved and 27% were unchanged in a long-term follow-up.(10) Comparatively, surgical treatment led to improvement in 81% of dogs. This shows that, while surgery does offer the best changes of improvement for dogs with CSM, medical management is an acceptable alternative. The author also likes to start all dogs on medical management initially to evaluate the improvement obtained with it and to give owners the opportunity to decide on surgery. The response to medical management (exercise restriction and corticosteroids) can be used to indirectly assess the degree of reversible spinal cord lesions.

SURGICAL TREATMENT

Surgical techniques for CSM can be grouped into 3 broad categories: direct decompressive techniques, indirect decompressive or vertebral distraction techniques, and motion preserving techniques. Direct decompressive techniques reported include dorsal laminectomy, dorsal laminoplasty, ventral slot, inverted cone slot, and hemilaminectomy.(12-15, 15-18) Indirect decompressive techniques typically are grouped into the distraction-stabilization category, and have been reported using bone grafts of several types, pins (smooth, threaded) or screws and polymethyl methacrylate, interbody screws, washers, metallic spacers, metallic plates, plastic plates, k-wire spacer, Harrington rods, interbody polymethyl methacrylate plug, and fusion cage, all of these combined with either diskectomy, partial or complete ventral slots.(9, 19-37) Intervertebral disc fenestration has also been used, and more recently, motion preserving techniques, with disc arthroplasty (artificial disc replacement) have also been proposed.(38-40) With so many choices, the decision on which technique to choose may be difficult. The section below lists the indications and reported outcomes with various surgical techniques.

DIRECT DECOMPRESSIVE TECHNIQUES

Ventral Slot

Ventral slot is primarily indicated for single ventral "static" compressions. As previously discussed, the determination of a static or dynamic lesion is subjective, so it can be used for any ventral compression, as there is no data indicating that it is less successful or carries a worse outcome for "dynamic" lesions. Ventral slot offers adequate spinal cord decompression (if performed properly) and subsequent vertebral fusion is expected to occur at the slot site 8 to 12 weeks postoperatively.(13, 41) The reported long-term success rate of the ventral slot procedure for the treatment of CSM is 72%.(8, 12, 19)

Dorsal Laminectomy

Dorsal laminectomy is indicated for dorsal compressions associated with osteoarthritic changes of the articular facet joints, dorsal lamina malformation, or ligamentum flavum hypertrophy. It can be used to decompress one or multiple sites.(16, 42) When used in multiple sites it has been named continuous dorsal laminectomy.(15) Dorsal laminectomy has also been recommended for multiple ventral spinal cord compressions.(15) No direct comparisons of dorsal laminectomy with other surgical techniques have been published, so the choice of a dorsal laminectomy for ventral compressions is dependent on the surgeon's preference. A problem with the dorsal laminectomy technique is the high postoperative morbidity rate (70%) which can be challenging to manage in large and giant breed dogs.(16) Most dogs will improve over time (mean time to optimum recovery= 3.6 months) and the success rate for dorsal laminectomies ranges from 79 to 95%;(15, 16, 42) although 30% of dogs can have signs of recurrence postoperatively.(16)

Cervical Hemilaminectomy

A technique of lateral approach for cervical hemilaminectomy was used in the treatment of CSM in 5 dogs with either disc or osseous disease.(17) The authors reported no case of postoperative worsening of neurological status.(17) Although the number of treated dogs was small, hemilaminectomy should be considered a valuable alternative in dogs with lateralized compressions in which ventral or dorsal decompressions may not adequately decompress the spinal cord or nerve roots.

INDIRECT DECOMPRESSION - VERTEBRAL DISTRACTION TECHNIQUES

Various techniques have been reported to cause vertebral distraction. Vertebral distraction can be achieved via manual traction,(6) but it cannot be controlled, and as such cannot be recommended. It is suggested that a specific distraction instrument such as a Caspar distractor, or a modified Gelpi retractor be used when performing distraction techniques.(29)

Pins and Polymethyl methacrylate (PMMA)

This technique is mostly recommended for single ventral dynamic compressive lesions.(20) It can be used for two affected segments but the risk of failure increases. Either a partial or complete ventral slot is created. The metal implants used can be either two Steinmann pins, threaded pins or bone screws in each vertebra. Positive profile pins provide more rigid fixation than smooth pins.(43) The long-term success rate of this technique is 73%.(20) There are modifications of this technique using corticocancellous bone for distraction.(22) A recent modified version of this technique combined full thickness ventral slot with an U-shaped K-wire spacer and monocortical screws with PMMA for distraction.(34)

Screw-Bar-PMMA (Polymethyl methacrylate)

This is a novel technique recently proposed to avoid penetration of the vertebral canal or the transverse foramina. Cortical bone screws were inserted bilaterally into the transverse processes of each vertebrae. The screws penetrated the transcortices, and approximately 1 cm of the screw and head extended ventromedially. A non-threaded Steinmann pin was manually contoured to form a U-shape.(33) The contoured pin was wired to the screws heads using orthopedic wire. PMMA was placed to incorporate the screws, pins and wires.(33) Clinical results of this technique are not yet available.

Distraction using PMMA plug

This technique can be used for single or multiple ventral and dorsal compressions, either static or dynamic. The original report focused on ventral disc-associated dynamic lesions and the authors discussed that it would not be indicated for dorsal or ventral static lesions.(23) However, others have suggested that it can be used for up to three ventral sites and also be indicated to treat sites with dorsal compression by minimizing or eliminating the dynamic component of the spinal cord compression.(29, 44) A recent experimental study evaluated this technique in complete ventral slots in sheep and found that none of the animals studied developed bony fusion.(45) However, cancellous bone graft was not used in any of the animals studied. The incidence of bone fusion with the traditional technique is currently unknown. The reported long-term success with the PMMA plug technique is 82%.(23)

Locking Plate

Locking plates have been reported for the treatment of CSM.(3, 30-32) Screws of the locking plates are less likely to loosen than conventional screws and can be inserted monocortically with sustained stability, reducing the risk of vertebral canal penetration and spinal cord injury.(32) Locking plates were used with either partial ventral slots or discectomies. A variety of bone grafts was used in the reported cases ranging from cancellous autograft, cortical allograft with cancellous autograft, or cancellous bone block graft.(30-32) The success rate based on the follow-up information of 3 case series is 73%.(30-32) A significant disadvantage of this system is the cost of the plate and screws, which coupled with the modest success rate suggests that significant improvements are required to justify its routine use.

MOTION PRESERVING TECHNIQUES

Cervical disc arthroplasty

Disc arthroplasty is currently an area of intense investigation in humans with cervical spondylotic myelopathy. While no consensus exist in regards to the benefits of disc arthroplasty versus anterior (ventral) decompression with or without fusion in humans, disc arthroplasty appears to have some logical benefits in the treatment of CSM.(46, 47) Fusion or distraction may increase the risks of adjacent segment disease or "domino" syndrome by altering the biomechanics of the adjacent segments.(48, 49) Disc arthroplasty allows re-establishment of the normal disc height (width in dogs) with motion preservation of the affected intervertebral segment. Many different types of artificial discs are available in humans and several clinical trials are currently underway.(50) An artificial disc was recently introduced for treatment of CSM in dogs.(39, 40) The proposed system offers the advantage of allowing decompression of the spinal cord with a ventral slot prior to insertion of the artificial disc, therefore assuring direct decompression. The number of treated dogs to date is too limited to assess efficacy, but the initial experience warrants further investigation of disc arthroplasty in the treatment of CSM.

OUTCOME

Comparison of treatment methods for CSM is complicated due to the fact that the vast majority of the studies are retrospective, have limited numbers, varied follow-up times and different criteria for a successful outcome. In general surgical treatment for disc-associated CSM is usually successful, with approximately 80% (70% to 90%) of dogs improving after surgery. No surgical technique stands out as being clearly superior, even for dogs with disc-associated CSM. Intervertebral disc fenestration is not recommended because the reported success rate was only 33%. In contrast to older studies, new reports on medical management indicate an improvement rate of approximately 50% (45-54%). Considering the success rate of surgical and medical treatments for CSM, surgery more consistently leads to clinical improvement, and should almost always be considered in the treatment of dogs with CSM. Surgery, however, does not alter the long-term survival of dogs with CSM. The survival time of 76 dogs with CSM (33 dogs treated surgically and 43 dogs treated medically) was reported recently.(10) The median survival time of dogs with CSM was identical (36 months), regardless of whether the dog was treated medically or surgically. This finding indicates that CSM continues to progress independent of the method of treatment, and that the clinical deterioration seen months to years after treatment may not be due solely to failure of the surgery or development of adjacent segment disease, but also may occur secondary to other mechanisms, such as ischemia, apoptosis or other molecular changes within the spinal cord.

CONCLUSIONS

Newer surgical techniques are proposed continuously, but the criteria for patient evaluation and outcome assessment often do not allow meaningful comparisons among different surgical techniques. Objective inclusion criteria and valid methods of outcome assessment are needed to facilitate evaluation of treatments for CSM. Routine MRI investigation of dogs that deteriorate postoperatively is also important to address the mechanisms leading to late-onset postoperative deterioration.

All metallic implants, except those made of titanium, cause significant artifacts, precluding the use of MRI. Further understanding of the mechanisms causing CSM, along with prospective studies evaluating treatment methods, with long-term clinical and imaging follow-ups, could eventually lead to refinement of treatment strategies, and improve the outcome and survival times beyond those currently available for dogs with CSM.

REFERENCES

1. da Costa RC. **Vet Clin North Am Small Anim Pract** 2010;40(5)881-913. 2. De Decker S, et al. **Vet Surg** 2011;40(5)544-554. 3. Steffen F, et al. **Vet Surg** 2011;40(6)743-752. 4. da Costa RC, et al. **J Am Vet Med Assoc** 2007;231(2)243-250. 5. da Costa RC. **Proc 25th Annu Meet ACVIM Forum** 2007;318-320. 6. Seim HB, et al. **J Am Anim Hosp Assoc** 1982;18241-251. 7. da Costa RC, et al. **Vet Radiol Ultrasound** 2006;47(6)523-531. 8. Rusbridge C, et al. **The Journal of small animal practice** 1998;39(9)425-431. 9. Denny HR, et al. **J Small Anim Pract** 1977;18(2)117-132. 10. da Costa RC, et al. **J Am Vet Med Assoc** 2008;233(8)1284-1290. 11. De Decker S, et al. **J Small Anim Pract** 2009;50(3)136-142. 12. Chambers JN, et al. **J Am Anim Hosp Assoc** 1986;22(6)775-778. 13. Swaim SF. **J Am Vet Med Assoc** 1974;164(5)491-495. 14. Goring RL, et al. **J Am Anim Hosp Assoc** 1991;27(4)403-409. 15. Lyman R. **Proc 13th Kal Kan Symposium** 1989;13-16. 16. De Risio L, et al. **Vet Surg** 2002;31(5)418-427. 17. Rossmeisl JH,Jr, et al. **Vet Surg** 2005;34(5)436-444. 18. Jeffery ND. The 'wobbler' syndrome Jeffery ND. Handbook of small animal spinal surgery. 1995:169-186. 19. Bruecker KA, et al. **Vet Surg** 1989;18(3)197-203. 20. Bruecker KA, et al. **J Am Anim Hosp Assoc** 1989;25(6)677-683. 21. Bruecker KA, et al. **Vet Surg** 1987;16(1)84-85. 22. Ellison GW, et al. **J Am Vet Med Assoc** 1988;193(4)447-453. 23. Dixon BC, et al. **J Am Vet Med Assoc** 1996;208(1)61-68. 24. Queen JP, et al. **J Small Anim Pract** 1998;39(3)131-136. 25. Read RA, et al. **J Small Anim Pract** 1983;24(10)605-621. 26. McKee WM, et al. **J Small Anim Pract** 1999;40(10)465-472. 27. McKee WM, et al. **J Small Anim Pract** 1989;30(6)337-342. 28. McKee WM, et al. Cervical spondylopathy. Slatter DH. Textbook of small animal surgery. 3rd 2003:1180-1193. 29. Sharp NJH, et al. Cervical spondylomyelopathy. Sharp NJH and Wheeler SJ. Small Animal Spinal Disorders Diagnosis and Surgery. 2nd 2005:211-246. 30. Bergman RL, et al. **Vet Surg** 2008;37(6)530-536. 31. Trotter EJ. **Vet Surg** 2009;38(6)705-718. 32. Voss K, et al. **Vet Comp Orthop Traumatol** 2006;19(1)21-28. 33. Hicks DG, et al. **Am J Vet Res** 2009;70(6)719-726. 34. Shamir MH, et al. **Vet Surg** 2008;37(2)186-192. 35. Walker TL. Use of Harrington rods in caudal cervical spondylomyelopathy. Bojrab MJ. 1990:584-586. 36. Adrega Da Silva C, et al. **Vet Comp Orthop Traumatol** 2010;23(3)209-213. 37. Mason TA. **Vet Rec** 1979;104(7)142-145. 38. Lincoln JD, et al. **Vet Surg** 1985;14(3)240-246. 39. Adamo PF, et al. **Vet Surg** 2007;36(8)729-741. 40. Adamo PF, et al. **J Vet Int Med** 2009;23(3)710. 41. Gilpin GN. **J Am Vet Med Assoc** 1976;168(4)325-328. 42. Lipsitz D, et al. **Vet Radiol Ultrasound** 2001;42(1)20-27. 43. Koehler CL, et al. **American journal of veterinary research** 2005;66(4)678-687. 44. Galano H, et al. **J Vet Int Med** 2005;19(3)419. 45. Fransson BA, et al. **Vet Surg** 2007;36(5)449-457. 46. Heller JG, et al. **Spine (Phila Pa 1976)** 2009;34(2)101-107. 47. Bartels RH, et al. **Neurosurgery** 2010;66(6)1153-1160. 48. Rihn JA, et al. **Instr Course Lect** 2009;58747-756. 49. Hilibrand AS, et al. **J Bone Joint Surg Am** 1999;81(4)519-528. 50.Baaj AA, et al. **Neurosurg Focus** 2009;27(3)E10.

SESSION 35: THE NEUROMAP: A SIMPLE GUIDE TO LOCALIZING NEUROLOGICAL LESIONS – SPINAL CORD AND PERIPHERAL NERVOUS SYSTEM

Christine E. Thomson, BVSc(Hons), DACVIM(Neurology), DECVN, PhD
Palmerston North, New Zealand

The aim of the neurological examination is to determine whether neurological dysfunction is present and, if so, where the lesion causing the dysfunction is located in the nervous system. Once the lesion is localised, the possible causes can be listed and an appropriate diagnostic and subsequent treatment plan can be implemented. Whenever possible, we try to see if the signs can all be accounted for by a single lesion. If not, then the animal is considered to have a multifocal lesion, which makes certain diseases more likely. Localising neurological lesions can be challenging, particularly if it is not something you do frequently. The NeuroMap is a simple visual aid that has been designed as a ready reference guide to assist you in localising lesions. It is accompanied by a ready reference table. A key feature about the NeuroMap is that you don't need a degree in neuroanatomical-ese (terminology) to get a basic understanding of what could be happening in the nervous system of your patient.

The NeuroMap is based on mapping out the location of the main functions of the nervous system. You can read from the NeuroMap what neural functions may be disrupted with a lesion in a certain area and conversely, what neural functions will be normal.

Two key concepts underlie lesion localisation. 1) You need to note both the neural systems that are functioning normally as well as those that are dysfunctional (Figure 1). 2) It doesn't matter where a lesion is sited along the neural pathway (origin, middle or termination) the clinical signs of dysfunction will be similar. This is analogous to a battery (origin), wire (middle) and light bulb (termination). If there is a problem with any component, the light won't work. For example, an animal has a proprioceptive deficit and is standing on top of its paw (knuckling). The lesion could be in the sensory receptor (origin), the peripheral nerve or spinal cord pathway or brainstem (middle), or the sensory cortex of the forebrain (termination).

The neurological examination involves evaluating five neural functions – behaviour and arousal; ascending sensory systems (tactile, proprioception and nociception) motor function; cranial nerve function, and spinal reflexes. Spinal sensitivity (hyperpathia) may further localise a spinal cord lesion. Based on the results of the tests, the clinician determines whether the animal has neurological deficits and if so, which parts of the nervous system are functioning normally and which parts are functioning abnormally. Using the NeuroMap, the clinician can see where each of the neural functions being tested are located in the nervous system. Can a single lesion account for all the observed signs – both the normal and abnormal?

A number of functions are associated with most regions of the nervous system, either because neural pathways begin or end there, or are passing through it. If a lesion is in a particular region, then it could damage pathways in that region and cause signs of dysfunction. However if a neural function is not associated with that region, then it will not be affected. Identifying the neural pathways that are functioning normally indicates to the examiner that the lesion is probably NOT located in the region that those systems occupy.

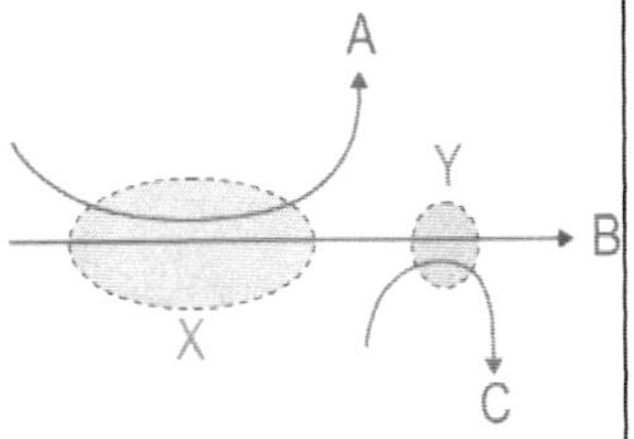

Figure 1. Normal AND abnormal function in the nervous system helps to localise a lesion. Consider the animal knuckling on its hindpaws – this indicates a conscious proprioception deficit (see later text) and can be represented by pathway B in Figure 1. The animal also has loss of the withdrawal reflex in that limb (pathway A), but the cutaneous trunci reflex (pathway C) is intact. The proprioceptive pathway (B) begins in the foot and travels up specific nerves to the spinal cord, along the spinal cord to the brainstem where it crosses sides and travels rostrally to the sensory cortex of the forebrain. The lesion could be anywhere along that pathway, but we know that A is also compromised and C is functioning normally. Thus the lesion must be at X and not Y. In this instance, X represents the lumbosacral spinal cord. Therefore that which is NORMAL (C), is just as important to identify as that which is ABNORMAL for localising the lesion.

Upper and lower motor neurons: UMN = upper motor neuron – you could also think of it as a 'central motor neuron' as the neuron is confined to the central nervous system (CNS = brain and spinal cord). LMN = lower motor neuron – you could also think of it as a 'peripheral motor neuron' as although the neuronal cell body is found in the CNS, the axon is largely in the periphery. LMNs supply striated muscle (e.g. limbs) and autonomic LMNs supply smooth muscle (e.g. the urinary bladder).

UMN disease damages UMNs (motor neurons in the brain or spinal cord), while LMN disease damages LMNs (motor neurons that are mainly in the peripheral spinal and cranial nerves.

Figure 2 and **Table 1.** The Neuro**RAT** – **R**eflexes, **A**trophy and **T**one.
To help remember how to differentiate between lesions affecting upper (central) motor neurons and lower (peripheral) motor neurons – think of the Neuro **RAT**.
R = Reflexes, **A** = Atrophy (of muscles), **T** = Tone. The table lists the changes that may be present caudal to the lesion.

Sign	UMN disease (damage to the UMNs)	LMN disease (damage to the LMNs)
Reflexes	Normal to increased	Decreased to absent
Atrophy	Disuse: Mild, generalised	Neurogenic: severe, specific muscles
Tone	Normal to increased	Decreased to absent

A spinal cord segment is a region of spinal cord to which is attached a pair of dorsal (sensory) nerve roots and a pair of ventral (motor) nerve roots. The dorsal and ventral roots fuse to form a spinal nerve; spinal nerves attach bilaterally to each spinal cord segment. These segmental spinal nerves form much of the peripheral nervous system (PNS) of the body and limbs; they may remain discrete (e.g. C3 spinal nerve), or spinal nerves may fuse in a plexus to form specific named nerves (e.g. brachial plexus giving rise to the radial nerve).

Functional regions of the spinal cord are based around where the limb innervation attaches. For the dog, this creates the regions: cervical (C1-5: neck), cervical intumescence (C6-T2: thoracic limbs), thoracolumbar (T3-L3: trunk), lumbosacral intumescence (L4-S3: pelvic limbs and viscera) and caudal segments (Cd1-5: tail). Because clinical signs of spinal cord disease are characteristic for each region, then we talk about animals having a region-specific localisation – e.g. thoracolumbar disease.

Reflexes: A reflex has three anatomical parts to it: input from a sensory receptor in the periphery, a central connection in the CNS (brain or spinal cord, for reflexes involving the cranial and spinal nerves, respectively) and a lower (peripheral) motor neuron that connects back to the muscle via a neuromuscular junction. Damage to any one of those components will cause reduction/loss of the reflex. The main reflexes used to assess limb innervation are the patellar and withdrawal reflexes in the pelvic limbs and the withdrawal reflex in the thoracic limbs. The perineal and the cutaneous trunci reflexes are also used to evaluate sacral and thoracolumbar spinal cord function, respectively.

UMN versus LMN disease: If an animal has a lesion at one of the intumescences (cervical or lumbosacral), where the limb innervation originates, then it is likely to compromise the lower (peripheral) motor neurons innervating that limb resulting in LMN signs: reduced **R**eflexes, **A**trophy of specific muscles within that limb and reduced muscle **T**one. These signs are quite localising and the clinician can tell which specific nerves, or which specific spinal cord segments, have been compromised. If an animal has a lesion cranial to the intumescence (e.g. thoracolumbar or cervical lesion), then the reflex arc and its LMN and the neuromuscular junction will be intact; the muscle will still be innervated. Thus, in the case of thoracolumbar disease, the pelvic limbs will have intact reflexes, and as the muscles still have their LMN innervation, they will have tone and any muscle atrophy will be mild due to disuse. As the upper (central) motor neuron has been compromised, the animal is described as having UMN signs to the pelvic limbs. Note: an animal will have UMN signs to the pelvic limbs if the lesion is located anywhere in the spinal cord cranial to L4. Thus UMN signs are not nearly as localising as LMN signs. If the lesion was in the cervical intumescence, the UMN supplying the pelvic limbs have been compromised, the LMN to the thoracic limbs will also be compromised. Thus in this example, the animal will have UMN signs to the pelvic limbs, but LMN signs to the thoracic limbs. A C1-C5 lesion will result in UMN signs to both the pelvic and thoracic limbs. Continence can also be affected (see below).

Increased muscle tone and reflexes: UMNs can facilitate or inhibit LMNs. Loss of facilitatory UMNs can result in paresis (decreased movement). Loss of inhibitory UMNs, can result in excess LMN activity and reflex function. As the LMNs supplying extensor (antigravity) muscles dominate those supplying flexor muscles, loss of inhibitory UMN may result in increased extensor tonus (spasticity).

Sensory function and proprioception: Also remember that it is not just motor neurons that will be compromised by a lesion in the spinal cord, but the sensory innervation will also be affected from the body caudal to the lesion. This includes sensory information about body position (proprioception), about the external environment (tactile, thermal and noxious stimuli) and the internal environment (e.g. viscera such as the fullness of the urinary bladder). There are two kinds of proprioception: 1) subconscious proprioception – the information that is projected to the ipsilateral cerebellum; 2) conscious proprioception – this information is projected to the contralateral forebrain for conscious awareness. Subconscious proprioception is particularly important for posture and locomotion; it's the sort of information that is used to keep the limbs located under the body's centre of gravity, both at rest and during movement. It originates in receptors in muscles and tendons. This information is projected to the cerebellum which then uses it to modulate motor function associated with postural and locomotory muscles. Subconscious proprioception will be compromised in an animal with cerebellar disease (affecting pathway termination). It may not know where its limbs are with respect to its centre of gravity, so it may stand base-wide or base-narrow, or cross its feet when ambulating. It will probably have ataxia (incoordination of gait). Conscious proprioception is important for awareness of movement, particularly movement that is voluntary and learned e.g. the kitten patting at a feather. Tactile input is a key component of this. An animal that has forebrain disease will still have good posture and locomotion (subconscious proprioception is intact), but the information from tactile receptors (e.g. from the paw) will not be received. Thus the animal may stand on top of its foot ('knuckling'). (Copies of subconscious proprioceptive information is sent to the forebrain, too (kinaesthesia) so the animal can be consciously aware of posture and locomotion.) Note: As both conscious and subconscious proprioceptive tracts travel up the spinal cord, spinal cord lesions often affect both types of proprioception; the parts of the body caudal to the lesion are often ataxic or the limbs have a stumbling gait, and the animal may stand on the top of its paw(s).

Nociception involves the transmission of noxious (tissue-damaging) stimuli by spinal nerves to the spinal cord where it stimulates local reflex function (e.g. withdrawal reflex) or is projected to the forebrain for conscious perception. It travels to the forebrain in many pathways distributed across the transverse area of the spinal cord. Thus to lose conscious perception of a noxious stimulus, requires that the spinal cord has been extensively damaged. In an animal with a transected thoracolumbar spinal cord, the reflex arc for the withdrawal reflex will still be present in the pelvic limbs, but the cranially directed nociceptive tracts will be interrupted. Thus when a noxious stimulus is applied to the hind foot, the animal will still pull its foot away, but it will not be able to consciously perceive the noxious stimulus. This implies severe spinal cord damage.

Innervation of the urinary bladder and anal sphincter: Sensory input from the bladder wall, urinary sphincter and perineum travels to the sacral spinal cord (S1-3). Cranially directed pathways convey information about the fullness of the bladder and rectum to the brain for conscious perception. The sacral spinal cord (S1-3) is the origin of LMNs supplying the bladder wall smooth muscle (parasympathetic innervation – this stimulates detrusor muscle contraction during voiding) and the striated muscle of the urinary and anal sphincters (contraction maintains continence). Sympathetic LMNs arise from the cranial lumbar spinal cord and stimulate contraction of smooth muscle at the neck of the bladder and relaxation of bladder wall muscle for urine storage. The LMNs innervating the bladder wall and the sphincters result in muscle tone and are necessary for reflex function. Thus lesions affecting the sacral spinal cord will result in a LMN signs with reduced detrusor muscle tone and perineal reflex, and poor tone in the anal sphincter. The bladder will be large, floppy, and because of poor tone in the striated sphincter, it will be easy to express and the animal will probably be dribbling urine. Similarly, the anal sphincter may be dilated and the animal will be faecally incontinent. If the lesion is cranial to the sacral spinal cord, then the LMNs supplying the bladder wall and the striated muscle of the urinary and anal sphincters will be intact. Thus the bladder will be full, turgid and difficult to express; the latter sign occurs because of loss of UMN inhibition to the LMNs innervating the urinary sphincter. Spinal cord disease often affects both urinary and faecal continence; uncommonly one elimination system only may be affected.

Using the NeuroMap: If you cover a region of the spinal cord in the NeuroMap (representing a lesion in it) you will be able to see which neurological functions will be compromised and which will still function normally. Conversely if you have an animal with particular signs, e.g. tetraparesis, knuckling on all four limbs, loss of thoracic limb reflexes, but with intact pelvic limb reflexes, you can identify on the NeuroMap where the wiring is for one of those functions, e.g. the thoracic reflexes, (C6-T2). What other functions are passing through this region? Yes, UMN to the pelvic limbs and sensory input from both the pelvic and thoracic limbs. That all fits. The pelvic limb reflexes are intact implying that their LMN are intact – yes, that is possible with a lesion in the C6-T2 region. Thus both the signs of dysfunction and signs of normal function have helped you localise the lesion.

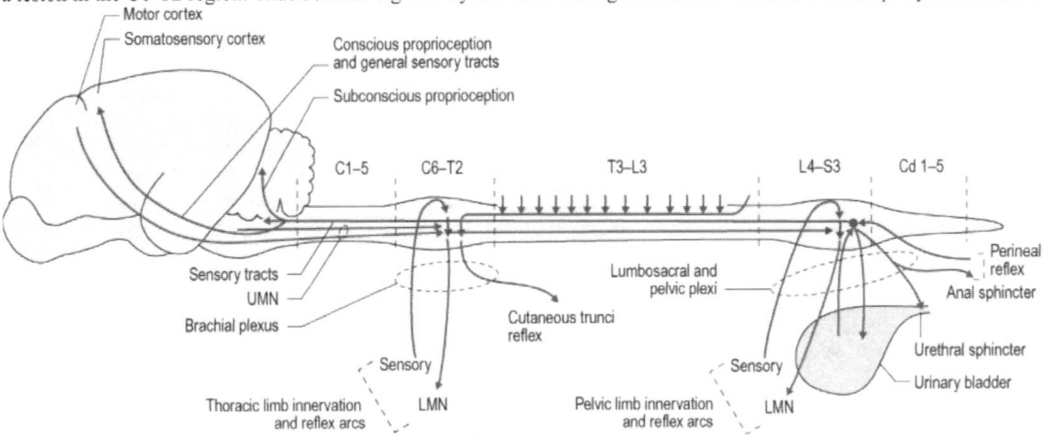

Figure 3. The NeuroMap for the spinal cord and spinal nerves

Table 2. Summary of effect of lesions in different areas of the spinal cord. TL = thoracic limb, PL = pelvic limb. a = UMN bladder – turgid, full, difficult to express. b = LMN bladder – flaccid, distended, easy to express, dribbling urine; c = animal will deposit faeces with good emptying of colon, but may not be aware that it is defaecating. d = may see some proprioceptive deficits or paresis with an S1-S3 lesion as S1 and S2 make variable contributions to the sciatic nerves. Note: whether or not all signs are present, depends on lesion severity.

Location of lesion	Loss of proprioception and sensory input	Presence of UMN signs	Presence of LMN signs to limbs Loss of other reflexes	Loss of urinary and faecal continence
Cervical C1-5	TL and PL both affected	TL and PL both affected	No LMN signs in TL and PL. Cutaneous trunci and perineal reflexes intact	UMN bladder[a] Faecal incontinence[c]
Cervical intumescence C6-T2	TL and PL both affected	PL only	TL, loss of cutaneous trunci reflex if lesion in C8-T2. Perineal reflex intact	UMN bladder Faecal incontinence[c]
Thoracolumbar T3-L3	PL only affected (TL normal)	PL only (TL normal)	No LMN signs in TL or PL Cutaneous trunci reflex lost caudal to lesion, perineal reflex intact	UMN bladder Faecal incontinence
Cranial lumbosacral L4-S1	PL only affected (TL normal)	No UMN signs to limbs	LMN signs in PL Cutaneous trunci reflex intact, perineal reflex intact	UMN bladder Faecal incontinence
Caudal lumbosacral S1-S3	TL normal, PL probably normal[d]	No UMN signs to limbs	LMN signs to pelvic viscera. Loss of perineal reflex	LMN bladder[b] LMN anal sphincter (dilated anus), faecal incontinence
Caudal nerves Cd1-5	Decreased tail sensation	No UMN signs to limbs	LMN signs to tail	Normal continence

REFERENCES

Figures and tables reprinted with permission from 'Veterinary Neuroanatomy: A Clinical Approach' by Christine Thomson and Caroline Hahn, Elsevier Health Sciences, May 2012, ISBN 9780702034824.

SESSION 36: THE NEUROMAP: A SIMPLE GUIDE TO LOCALIZING NEUROLOGICAL LESIONS – THE BRAIN AND CRANIAL NERVES

Christine E. Thomson, BVSc(Hons), DACVIM(Neurology), DECVN, PhD
Palmerston North, New Zealand

This Session follows on from Session 35, The NeuroMap: A Simple Guide to Localizing Neurological Lesions – Spinal Cord and spinal nerves. It use the same key concepts discussed in Session 35 as follows:

1) The NeuroMap, in which the five main neural functions tested in the neurological examination are mapped onto a diagram of the brain. These functions are behaviour and arousal; ascending sensory systems (tactile, proprioception and nociception); motor function (posture and gait); cranial nerve function, and spinal reflexes. Spinal sensitivity (hyperpathia) can further localise the lesion in spinal cord cases.
2) Using the neurological examination to identify both the neural functions that are NORMAL and those that are ABNORMAL (dysfunctional). See Figure 1, Session 35
3) Using the battery and light bulb analogy to remember that signs of dysfunction can be due to a lesion located anywhere along the neural pathway – at the origin, middle of the pathway or at the pathway termination.
4) Remembering the Neuro**RAT** (Session 35 fig 2 & Table 1) to help differentiate between lesions affecting upper (central) motor neurons (UMNs) and lower (peripheral) motor neurons (LMNs). See Session 35 re 'central' and 'peripheral' MNs
5) A reflex has three parts to it – input from a sensory receptor in the periphery, a central connection in the CNS (brain or spinal cord, for reflexes involving the cranial and spinal nerves, respectively) and a LMN that connects back to the muscle via a neuromuscular junction. Damage to any one of those components will cause reduction/loss of the reflex.

Localising the neurological lesions can be challenging, particularly if it is not something you do frequently. The NeuroMap is a simple, visual aid that can help with localizing lesions in the nervous system. It is accompanied by a ready reference table. A key feature about the NeuroMap is that you don't need a degree in neuroanatomical-ese (terminology) to get a basic understanding of what could be happening in the nervous system of your patient.

The NeuroMap is based on mapping out the location of the main functions of the nervous system. Thus it can be seen from the NeuroMap what neural functions may be disrupted with a lesion in a certain area and conversely, what neural functions will be normal.

In this session we'll draw a NeuroMap (see Figure 1) to help with localizing brain lesions and from that construct a ready reference table (Table 2) that can be used in conjunction with the BrainMap.

Functional brain divisions: Functionally, the brain can be divided into three main regions – the forebrain, brainstem (midbrain, pons and medulla oblongata) and the cerebellum. Because clinical signs of brain disease are characteristic for each region, then we talk about animals having a region-specific localisation – e.g. forebrain disease, or brainstem disease or cerebellar signs.

NEUROLOGICAL FUNCTIONS ASSESSED DURING THE NEUROLOGICAL EXAMINATION

Behaviour is primarily associated with the forebrain (limbic system), but its expression is dependent on other neural functions being normal.

Awareness and arousal are associated with the ascending reticular activating system (ARAS) of the brain stem and its diffuse projections into the cerebrum. Clinically, decreased consciousness can be due to focal lesions involving the brainstem or diffuse lesions of the cerebrum. One of the best ways to determine arousal levels is by observation. What is the animal's awareness like when it is left quietly in the environment? As soon as you start handling the animal it will be stimulated, so subtle changes in arousal might be missed. In veterinary medicine, five different levels of arousal can be described. Note: the commonly used term 'depressed' is a psychological term and is inappropriate:

1) Normal – bright, alert and responsive
2) Obtunded – dull, tends to fall asleep if left undisturbed, but can be aroused by non-noxious stimuli
3) Stuporous – somnolent; requires a noxious stimulus to arouse it
4) Comatose – unconscious; cannot be roused by even a noxious stimulus
5) Brain dead – no cerebrocortical electrical activity, no brainstem reflex function

Proprioception from the limbs and body travels cranially along the spinal cord. Subconscious proprioception projects to the ipsilateral cerebellum, whereas conscious proprioception projects to the contralateral forebrain (somatosensory cortex) (see session 35 notes). Proprioception for the head is from the vestibular apparatus in the inner ear.

Motor function: Upper (central) motor neurons in quadrupeds originate in the forebrain (motor cortex) and brainstem motor nuclei. UMNs influence lower (peripheral) motor neurons that exit the CNS in cranial and spinal nerves, from brainstem and spinal cord, respectively. The motor cortex of the forebrain functions primarily to direct voluntary, learned movements in the head, body, limbs and tail. The motor cortex influences LMNs in the brainstem that give rise to cranial nerves, and LMNs that exit the spinal cord as spinal nerves. The tracts passing into the spinal cord are called pyramidal UMNs as they pass through the structures in the caudal brainstem called the 'pyramids'. In quadrupeds, the brainstem UMN centres are important for posture and locomotion and semi-automatic activities such as chewing and swallowing. Brainstem UMN centres give rise to extrapyramidal UMNs (tracts do not pass through the pyramids) influencing LMNs in cranial and spinal nerves. Thus brainstem lesions can severely perturb gait (paresis) while forebrain lesions usually have minimal effect of gait. The cerebellum's job is to coordinate motor function to achieve posture, and the correct rate, range and force of movement ('metria'). It also coordinates the contraction/relaxation of agonist/antagonist muscles acting around the joint. To coordinate motor function, it has to know where the body parts are located in space, so it must constantly receive proprioceptive information from the limbs, body and head (subconscious proprioception). Thus cerebellar disease can result in altered posture, ataxia (incoordinated gait), hypermetria and tremor. As the cerebellum has a calming down (inhibitory) effect on motor function, cerebellar dysfunction can result in excess motor function; characteristically this causes spasticity.

Cranial nerves: Noting cranial nerve dysfunction can be quite localising. Specific cranial nerves are associated with each region of the brain, but clinically they are not associated with spinal cord function. Therefore an animal with spinal cord disease should have intact cranial nerve function (identifying normal function is just as important as identifying those neural functions that are abnormal in lesion localisation). Conversely in an animal with purely brain disease, the spinal reflexes should be intact. An animal that has multiple cranial nerve signs (e.g. cannot blink and has a droopy ear, has a head tilt, has reduced facial sensation and cannot swallow) probably has brainstem disease affecting cranial nerves VII, VIII, V, IX and X). Such a brainstem lesion will probably have compromised other functions in the brainstem e.g. mentation, cranially directed proprioceptive tracts and UMN centres. Testing the function of cranial nerves is outlined in table 1.

Reflexes: Both spinal and cranial nerve reflexes are tested. Spinal reflexes should be intact (present) in animals with brain lesions, as the wiring of their reflex arc is intact. Cranial nerve reflexes may be compromised if the cranial nerve, or the region of the CNS associated with the reflex has been damaged.

Table 1: testing the function of cranial nerves; a = afferent (input); e = efferent (output)

Head function	Innervation	Clinical testing	Dysfunction
Olfaction	Ia	Observation, odourant such as food	Dysosmia
Vision	IIa	Maze test Menace response – IIa & VIIe Pupillary light reflex – IIa & IIIe (parasympathetic)	Blindness, dilated pupil Note: mydriasis can also be due to primary ophthalmic disease or CN III dysfunction
Pupil size	IIa, IIIe parasympathetic, e sympathetic, e	Pupillary light reflex Pharmacological testing	Anisocoria, Mydriasis (↓ CN II (input) or CN III parasympathetic output) Miosis (↓ sympathetic innervation). (Note cerebellar lesions and primary ophthalmic lesions can also affect pupil size)
Eyeball position	VIIIa IIIe, IVe, VIe	Visual tracking of moving objects Eyeball position in different head positions Vestibulo-ocular reflex VIIIa & IIIe, IVe, VIe	Strabismus - static (LMN affecting cranial nerves III, IV or VI), - dynamic (vestibular)
Facial sensation	Va – all three branches (ophthalmic, maxillary, mandibular)	Tactile stimulation Va – different regions of the face – muzzle and ventral eyelid, dorsal eyelid, mandible Palpebral reflex – Va & VIIe (dorsal eyelid – ophthalmic br. ventral eyelid – maxillary br) Auriculo-palbebral reflex Va & VIIe (stimulate just in front of the external ear canal)	Facial hypoalgesia - ipsilateral with brainstem disease - contralateral with forebrain disease
Mastication	Ve – mandibular	Jaw tone Assess the bulk of the masticatory muscles e.g. temporalis and masseter muscles	Muscle atrophy Dropped jaw if bilateral
Facial expression	VIIe	Facial symmetry Palpebral reflex Va & VIIe Auriculo-palbebral reflexes Va & VIIe Movement of muzzle, external nares, eyelids, ears Symmetry of position of lip commissures when head is held up vertically.	Facial paresis/paralysis
Vestibular function	VIIIa	Head position Eyeball position (also involves III, IV, VI) Vestibulo-ocular reflex (VIIIa & IIIe, IVe, VIe) Gait and movement	Head tilt, circling, rolling Spontaneous nystagmus; abnormal vestibule-ocular reflex Strabismus Deranged body posture and ataxia
Pharyngeal function	IX & X, a & e	Gag reflex (IXa, Xa & IXe, Xe) Swallowing	Dysphagia Salivation
Laryngeal function	X & XI, a & e	Phonation Respiration	Dysphonia Respiratory stridor – laryngeal obstruction, Aspiration
Tongue	XIIe	Observation – LMN signs, usage Withdrawal from tactile stimulus	Atrophy, paresis/paralysis – unilateral or bilateral

The NeuroMap (Figure 1) depicts where the main neural functions tested in the neurological examination are located. If you cover a region of the brain on the NeuroMap, representing a lesion in it, you will be able to see which neurological functions will be compromised and which will still function normally. Conversely if you have an animal with particular signs, e.g. blindness in the right eye, knuckling on right forepaw and hind paw, relatively normal gait, and obtundation, you can identify on the NeuroMap where the wiring for one of the functions is, e.g. vision (forebrain). What other functions are passing through this region? Sensory input from both the pelvic and thoracic limbs and arousal are associated with the forebrain. That all fits. The right side of the body is affected. Both vision and conscious proprioception are received in the contralateral forebrain – thus the lesion must be in the left forebrain. As cranial nerves (III-XII) are functioning normally, and the animal has intact subconscious proprioception, reasonable gait (UMN from the brainstem) then the lesion is not in the brainstem or the cerebellum. Thus both the signs of dysfunction and signs of normal function have helped you localise the lesion.

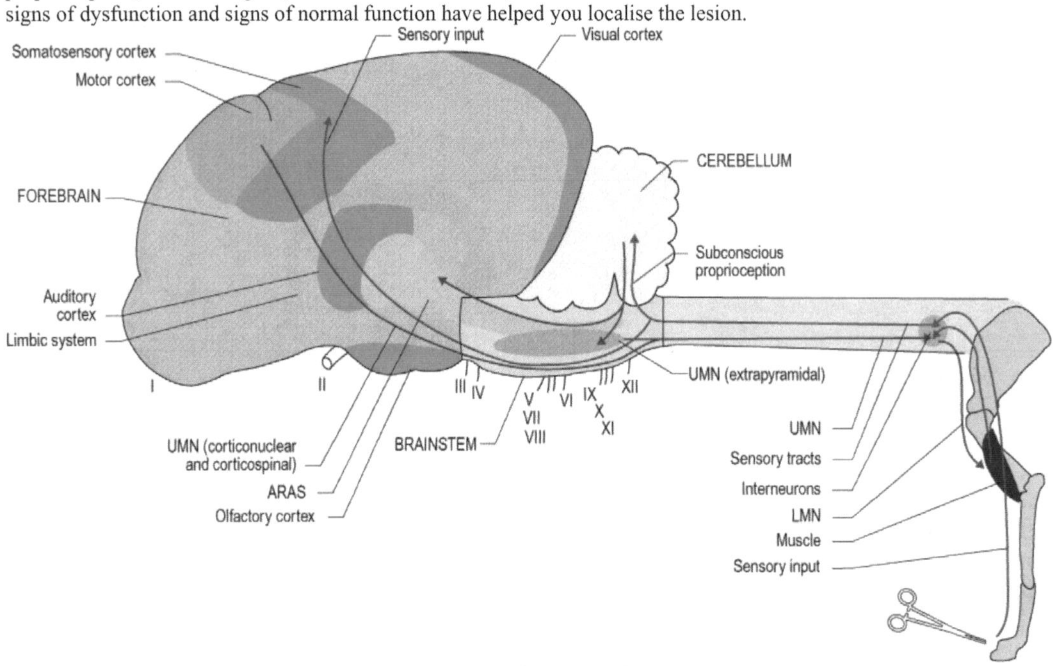

Figure 1: The NeuroMap of the Brain. Roman numerals refer to cranial nerves.

Table 2: Summary of neural signs that can occur with lesions in different parts of the brain. a=Part of CN XI (accessory nerve) arises from the cervical spinal cord, but it is hard to see signs of its dysfunction with lesions in this area of spinal cord.
Note: seizures will only occur with forebrain disease. b=important links between the cerebellum and vestibular system means that some vestibular signs may occur with lesions in certain parts of the cerebellum

Function	Forebrain	Brainstem	Cerebellum	Spinal cord
Proprioception a) Conscious b) Subconscious	 Yes, somatosensory cortex No	 Yes Yes, caudal brainstem	 No Yes	 Yes Yes
Nociception	Yes	Yes	No	Yes
Motor systems	Yes, important for planning motor function and voluntary movement Motor cortex (UMN), functions in voluntary motor activity, but has a minor role in gait. No LMNs.	Yes, major site of UMN nuclei in quadrupeds, important in posture and locomotion. Cranial nerves have LMNs innervating head and neck muscles	Yes, for motor coordination, but no UMNs (no weakness)	Yes UMN tracts LMNs innervating neck, body, limbs and tail
Behaviour, emotion and memory	Yes Limbic system	No	No	No
Arousal	Yes, the ARAS projects to widespread areas of forebrain	Yes (the ascending reticular activating system (ARAS)	No	No
Cranial nerves	CN 1, II	CN III-XII	CN VIII head and eyeball position[b]	No[a]
Spinal reflexes	Intact	Intact	Intact	Could be reduced/lost

REFERENCES

Figures and tables reprinted with permission from 'Veterinary Neuroanatomy: A Clinical Approach' by Christine Thomson and Caroline Hahn, Elsevier Health Sciences, May 2012, ISBN 9780702034824.

THE NEUROBIOLOGY OF MUSIC AND THE HEALING POWER OF THE BLUES

Charles H. Vite, DVM, PhD, DACVIM (Neurology)
Philadelphia, PA

"Musical notes and rhythm were first acquired by the male and female progenitors of mankind for the sake of charming the opposite sex" (Darwin 1871). Music is used to seduce others, to seduce ourselves, to create, to entertain, and to communicate. What is music? John Cage's composition of 1952 consists of four minutes and thirty-three seconds (hence the composition's title 4'33") of uninterrupted silence to be performed by "any instrument or combination of instruments". Is this music? Music appears universal, playing a significant role in all human cultures, even though it confers no obvious value to an organism (McDermott and Hauser, 2005). We eat to maintain ourselves, we have sex to maintain the species, we have music to... promote social cohesion in war and/or religion? to help pacify people? to serve a selected role in courtship?

"We are born with a musical wisdom and appetite" (Trevarthen 1999). From birth, children are attracted to music over spoken word (Nekata and Trehub 2004). Do animals have music? Birds, whales, and seals learn complex songs. Manual percussion, "drumming", occurs in African apes. Duets are sung by mated pairs of gibbons. However, tamarins prefer monkey chirps to screams but show no preference between consonant and dissonant musical sounds. Do animals then have music or "something else"? (Krumhansl 2005).

EFFECT OF MUSIC ON THE NERVOUS SYSTEM

The ability to perceive the emotional character of musical excerpts is generally consistent among listeners with less than a second required to elicit emotional judgments (Bigand et al 2005). Where does the emotional power of music come from? PET and fMRI data show the involvement of the neocortex and the limbic system in the emotional appraisal of music (Koelsch 2005). What is the biological basis of music? In 2008, an anthropologist underwent a year of singing lessons and a neuroscientist explored the impact on this training in his brain. Increased activity in Brodmann's area 22, 38, and 45 were found along with decreases in activity in other brain regions (Mithen and Parsons 2008). Schlaug et al (1995) and Hutchison et al (2003) found that the corpus callosum and cerebellum respectively are larger in male musicians than in non-musicians, with the intensity of musical training varying directly with the increase in size. Other studies have shown greater gray matter size of auditory cortex, inferior frontal gyrus, and superior parietal cortex in musicians compared to non-musicians, and a similar increase in myelination. These studies suggest that musical training directly influences brain anatomy.

NEUROLOGICAL DISORDERS AND MUSIC

Specific brain injury can result in selective loss of musical ability or selective loss in music recognition abilities, while global disease may fail to have an effect on musicianship (Stewart et al 2006; Alossa and Castelli 2008). Musical agnosia, or amusia, is a term used to designate acquired disorders of musical perception (receptive or sensorial amusia; amnesic amusia), performance (oral-expressive amusia, instrumental amusia, musical apraxia), and reading (musical alexia) or writing (musical agraphia) that is due to organic brain disease (Alossa and Castelli 2008). The composer Ravel developed apraxia, agraphia and alexia with eventual loss of musical creativity after a car accident; it was during this period that he wrote Bolero (Pearce JMS 2005). In addition to the study of patients with surgical excised tissue or strokes, certain hereditary disorders have also been associated abnormalities of musical perception. Williams syndrome is a neurogenetic developmental disorder characterized by aortic stenosis, elastin deficiency, hypercalcemia, scoliosis, and "pixie-like" facial features. Intellectual disability is also a feature. Patients are drawn to both listening to and performing music, experience higher levels of emotion while listening to music, and show increased musical memory. fMRI studies showed differences in temporal lobe activation in affected patients compared to control subjects (Levitan 2005).

Congenital amusia also occurs with those affected showing a deficit in the ability to discern different pitches and melodies, while being able to recognize normal lyrics and voices (Pearce JMS 2005). The prevalence of poor singing is approximately 10-15% of people. What are the phenotypes of poor singing? Some transpose pitch whey they imitate a melody; some transpose tempo suggesting that poor singing may be due to different mechanisms (Dalla Bella 2010). The use of fMRI and electrodiagnostics to evaluated differences between these populations will be discussed. A specific example using fMRI studies showed that the frontal inferior gyrus and posterior supratemporal regions are active in musicians compared to non-musicians. In contrast, tone-deaf people show abnormal morphometry of these regions. (Schlaug 2010). Similarly, the arcuate fasciculus was smaller in tone-deaf individuals compared to the control group (Schlaug 2010)

MUSIC THERAPY

Both making and listening to music may improve rehabilitation from both acute and chronic CNS injury through effects on attention, emotion, cognition, behavior, perception, and communication (Koelsch 2010). Music therapy has been used to treat brain injury, stroke, autism, pain, aphasia, dementia, anxiety, and depression. A specific example is the use of melodic intonation therapy in the treatment of stroke-associated aphasia. Patients unable to communicate verbally are able to communicate by singing words using just two pitches (Schlaug et al 2008). This training requires practice 1.5 hrs/day, 5 days a week, and >75 sessions in order to have significant effect. Diffusion tensor imaging of patients showed that melodic intonation therapy increased the size of the arcuate fasciculus and the MRI changes trended with improvement in communication ability (Schlaug et al 2010). Patients with stroke-associated motor dysfunction also showed improvement in fine and gross motor skills in response to music therapy (Altenmuller et al 2010). The clinical improvement was associated with movement-induced changes in EEG activity which were interpreted as an increased activity of brain motor regions. Depression may also be modulated by music therapy. fMRI and PET studies have shown that listening to music has effects on emotion processing centers in the brain similar to changes seen following antidepressant usage (Brandes et al 2009).

CONCLUSION

"Because no chimpanzee has ever spoken a sentence, ... scientists have been fascinated with the human ability for speech and pondered its origins. Consequently, much is known about the neurological substrates for language, and a robust literature is available for its hypothetical evolution. What has been forgotten, however, is that no chimpanzee ever played the violin" (Wallin NL et al 2010). In conclusion, we will consider what our animal population can teach us about music. My cockatoo dances with my sons every evening – what does it all mean?

REFERENCES

Alossa N and Castelli L. Eur Neurolo 2009; 61: 269; Altenmuller E, et al. Ann NY Acad Sci 2009; 1169:395; Bigand E, et al. Ann NY Acad Sci 2005; 1060: 427; Darwin C. The Descent of Man. 1871. Princeton University Press. Princeton N.J.; Hutchison S, et al. Cerebral Cortex 2003; 13: 943; Koelsch S. Ann NY Acad Sci 2005; 1060: 412; Koelsch S. Ann NY Acad Sci 2009; 1169: 374.; Krumhansl CL. Ann NY Acad Sci 2005; 1060: 1; Levitan DL. Ann NY Acad Sci 2005; 106-: 325; Mc Dermott J & Hauser MD. Nat Neurosci 2003; 6: 663; McDermott J & Hauser MD. Ann NY Acad Sci 2005; 1060: 6; Mithin SJ. The Singing Neandethals. 2006, Harvard University Press, Cambridge, MA; Mithen SJ & Parsons L. Cambridge Archaeolog J 2008; 18: 401; Nakata T & Trehub S. Inf Behav Dev 2004; 27: 455; Pearce JMS. Eur Neurol 2005; 54: 145; Schlaug G, et al. Neuropsychologia 1995; 33: 1047; Schlaug G, et al. Music Percept 2008; 25: 315; Schlaug G, et al. Ann NY Acad Sci 2009; 1169:385; Stewart L, et al. Brain 2006; 129: 2533; Trevarthen C. Musicae Scientiae Special Issue 1999; 155; Wallin NL et al. The Origins of Music.

ANALGESIA FOR NEUROLOGICAL PATIENTS

Phillip Lerche, BVSc, PhD, DACVA
Columbus, OH

INTRODUCTION

This lecture will:

- review the physiology of pain (physiologic pain, pathologic pain, neuropathic pain, neurotransmitters of the pain pathway)
- review pain terminology (hyperalgesia, allodynia, peripheral and central sensitization)
- review pain assessment in cats and dogs
- review the clinical pharmacology of the main classes of drugs used to provide analgesia in cats and dogs (opioid receptor agonists, sodium channel antagonists, prostaglandin inhibition, N-methyl-D-aspartate antagonists; $alpha_2$ adrenoceptor agonists)
- review the clinical pharmacology of drugs used as adjuncts to analgesia (anticonvulsants, antidepressants, tranquilizers, muscle relaxants, tramadol)
- provide strategies for managing the pain of acute and chronic pain in cats and dogs with surgical and non-surgical neurological diseases

PHYSIOLOGY OF PAIN

Pain in animals can be defined as an aversive sensory and emotional experience (a perception), which elicits protective motor actions, results in learned avoidance, and may modify species-specific traits of behavior, including social behavior. Pain is thus a complex experience, which, although exhibiting stereotypical physiologic and behavioral responses within a species (predator vs prey), is different for every individual. The experience of pain includes detection of tissue injury by the nervous system (*nociception*), conscious perception of pain, behavior modification, and variable degrees of sickness and suffering in response to pain.

Pain serves an important role in the interaction between the individual animal and its environment in that it signals the potential for tissue damage, and is thus protective in nature. This type of short-lived pain, where no or minimal tissue injury occurs, is referred to as *physiologic pain*. Pain that occurs following tissue injury is referred to as *pathologic pain*. Pathologic pain sensation can be also be experienced in the absence of a noxious stimulus (*spontaneous pain*), in response to a normally innocuous stimulus (*allodynia*), or as an exaggerated response to a noxious stimulus (*hyperalgesia*). Pain can also arise from nervous tissue itself (*neuropathic pain*).

Pathologic pain is somewhat arbitrarily divided based on duration into acute (hours) and chronic (days to years).

It is helpful to determine the mechanism, origin and severity of pain, as this may help in determining treatment options. The mechanism of pain can be due to inflammation, nerve injury, cancer, or idiopathic. Pain may be *visceral* (organ pain, e.g. peritonitis) or *somatic* (superficial e.g. laceration, or deep e.g. tibia fracture). Some diseases or surgeries may result in more than one of these types of pain.

PAIN PATHWAY

Nociception consists of *transduction* (transformation of noxious stimuli into electrical signals called action potentials), *transmission* (sensory impulses are conducted to the spinal cord), *modulation* (impulse amplification or suppression within the spinal cord), *projection* (impulses are transmitted to the brain), and *perception* (integration, processing and recognition of nociceptive information). In the absence of tissue damage, this process is somewhat 'hard-wired' or static and is characterized by a basic stimulus-response pattern. If the stimulus is persistent or tissue injury occurs this process becomes dynamic, potentially resulting in changes physiology and behavior.

Transduction of noxious stimuli (mechanical, chemical and thermal) is mediated by A-delta (myelinated) and C (unmyelinated) fibers peripherally. A-delta fibers mediate the fast onset sharp pain (*first pain*), whereas C fibers are responsible for longer lasting, burning pain (*second pain*) that follows. Visceral nociceptors are more sensitive to distension, volvulus, and ischemic injury. NSAIDs, opioids, local anesthetics and corticosteroids can inhibit transduction.

Transmission of painful stimuli from peripheral nociceptors to the spinal cord occurs via primary afferent sensory nerve fibers. The cell bodies of these afferents are located in the dorsal root ganglia of the spinal nerves, and sensory axons extend from there to synapse with neurons in the dorsal horn of the grey matter of the spinal cord. Local anesthetics and $alpha_2$ agonists can block transmission of nociceptive impulses.

Modulation of impulses occurs in the spinal cord. Impulses can be amplified or suppressed by excitatory or inhibitory interneurons. Sensory neurons in the dorsal horn are either wide-dynamic-range neurons that receive input from all sensory nerves and are sensitive to low- and high-threshold sensory input, or neurons that process high-threshold specific nociceptive information. Neurotransmitters that are important at this level of the nociceptive pathway include amino acids (glutamate, aspartate), prostaglandins (PGE_2), and peptides (substance P, neurokinin A). Local anesthetics, opioids, aplha$_2$ agonists, NSAIDs, N-methyl-D-aspartate receptor (NMDA) antagonists, tricyclic antidepressants and anticonvulsants all inhibit modulation of nociceptive information.

Projection of impulses from the dorsal horn to the brain occurs via nerve tracts. The spinothalamic tract ascends to the thalamus. The spinoreticular tract projects to the reticular formation and thalamus. The spinomesencephalic tract terminates in the periaqueductal gray matter, limbic system and hypothalamus of the midbrain, and the spinohypothalamic tract ascends to the forebrain and hypothalamus.

Perception is the integration and processing of nociceptive information in the brain to produce a response via descending tracts in the spinal cord. Responses transmitted from the brain to inhibitory neurons in the dorsal horn of the spinal cord can stimulate inhibitory neurons. Removal of this descending inhibitory activity, or *disinhibition*, can increase pain via decreased endogenous opioid release.

The areas of the brain involved in perception are the periaqueductal gray matter, the reticular activating system and the thalamus. Anesthetic agents, opioids, alpha$_2$ agonists, benzodiazepines and phenothiazines can all alter perception.

CONSEQUENCES OF UNTREATED PAIN

Pathologic pain has multiple consequences for the patient. These unwanted consequences include but are not limited to: physiologic adaptations, sickness syndrome, the development of chronic pain and dynamic changes to the nervous system, undesirable neuroendocrine responses, stress, discomfort, debilitation and suffering.

Tissue damage resulting in sustained noxious stimulation of the central nervous system (CNS) can dynamically and profoundly alter the function of neurons and pathways involved in pain. This alteration is known as *CNS plasticity* and can result in hypersensitivity in both acute and chronic pain. Both peripheral and central neurons show plasticity. Tissue trauma results in release of cellular mediators (ATP, H+, K+, prostaglandins, leukotrienes) from damaged cells and also attracts inflammatory cells, which in turn release additional mediators (serotonin, histamine, cytokines). These and other mediators (nerve growth factor, substance P) form a 'sensitizing soup' that increases the sensitivity (lowers the threshold) of peripheral nociceptors. Additionally, nociceptors that do not normally take part in transduction of pain, or *silent nociceptors*, now become sensitive to noxious input. Centrally, dorsal horn neurons that are stimulated by constant nociceptive input (*temporal summation*) become hyperexcitable and sensitive to low intensity stimuli. This is partially mediated via unblocking of NMDA-receptors and leads to central nervous system hypersensitivity, a phenomenon also referred to as 'wind-up'.

Pain also causes sympathetic stimulation leading to vasoconstriction, increased myocardial work and increased myocardial oxygen consumption. Skeletal muscle blood flow tends to increase whereas blood flow to the gastrointestinal and urinary tracts decreases.

Neurohumoral changes that occur in response to pain include but are not limited to release of ACTH, elevation in cortisol, norepinephrine and epinephrine, and a decrease in insulin. This results in a catabolic state and signifies a strong stress response. This stress response is important for immediate survival, but if left unchecked can lead to increased morbidity and mortality. Prolonged stress also leads to immunosuppression.

Distress is an extreme form of stress, and occurs when stressors negatively affect the animal's physiology and behavior, which includes pain. By definition an animal that is enduring pain is suffering. Additionally, painful patients may be reluctant or unable to eat with resultant wasting, and may also suffer from insomnia. Many of these symptoms are also seen with the *sickness response*, the body's reaction to inflammation that is mediated by the immune system. Additional symptoms of this syndrome include fever and hyperalgesia. This exaggerated pain response is mediated by pro-inflammatory cytokines which have been shown to alter nervous tissues at all stages involved in conveying pain information from the periphery to the brain.

ANALGESIC DRUG OPTIONS

No one single drug can eliminate all pain. Use of a combination of drugs with different mechanisms of action (multi-receptor analgesia) or target different points in the pain pathway (multi-modal analgesia) allows doses of each individual drug to be lowered, thus decreasing side effects. Side effects that are common to different classes of drugs (e.g. respiratory depression in dogs seen with opioids and alpha$_2$ agonists) may be cumulative, and thus close monitoring is recommended.

Opioid analgesics provide analgesia by occupying opioid receptors. The most profound analgesia is produced by agonism of the mu-opioid receptor. This is the basis for pain management of acute, severe pain, including peri-operative analgesia. Drugs that can agonize the mu-receptors fully include morphine, hydromorphone, methadone, and oxymorphone, all of which last from 3-4 hours in cats and dogs. Fentanyl, also a mu-agonist, has a short duration of action and is thus typically administered as a continuous rate infusion (CRI). This short duration of action is useful when repeat neurological examinations need to be performed, as temporarily stopping the CRI will cause the plasma concentration of fentanyl to fall rapidly. Side effects include effects on the CNS (dysphoria), GI (vomiting, increased defecation), cardiovascular (decreased heart rate at higher doses, histamine release with morphine), respiratory depression, urinary retention, thermoregulation (hyperthermia in cats). Side effects tend to be decreased when using buprenorphine (a partial mu-agonist with a duration of 6-8hrs) or butorphanol (a mu-antagonist, kappa agonist opioid with a duration of 1-2 hrs).

Corticosteroids and NSAIDs are anti-iflammatory though inhibition of prostaglandins and, as they act on the same pathway, should not be used concurrently. Side effects include GI ulceration, renal insufficiency and coagulopathy.

Local anesthetics (sodium channel blockers) can be administered as IV infusion in dogs (lidocaine 2mg/kg followed by 50 mcg/kg/min CRI, or as regional and local anesthetics (lidocaine, bupivacaine, ropivacaine. Side effects of excessive IV administration include anxiety, muscle fasciculations and seizures.

Ketamine at subanesthetic doses (10-20 mcg/kg/min) antagonizes NMDA receptors, which play a role in development of central sensitization. Ketamine is a sympathetic nervous system stimulant, and signs such as tachycardia or arrythmias may be seen even at low doses.

Dexmedetomidine can be used in so-called microdoses (i.e. doses much lower than those used for sedation or preanesthetic medication) 0.5-1 mcg/kg or 0.5-1 mcg/kg/hr, although sedation may occur which may make assessment of neurologic status challenging.

REFERENCES

1. CAROLYN M. MORTON, ET AL. 2005. AMERICAN JOURNAL OF VETERINARY RESEARCH 66(12):2154-2166. 2. L.R. WATKINS, S.F. MAIER. 2000. ANNUAL REVIEW OF PSYCHOLOGY, 51:29-57. 3. GAYNOR JS, MUIR WW. HANDBOOK OF VETERINARY PAIN MANAGEMENT 2ND ED. MOSBY, ST LOUIS, 2008. 4. POZZI A, ET AL. J AM VET MED ASSOC. 2006 JAN 1;228(1):53-60. 5. HANSEN B. VET CLIN NORTH AM SMALL ANIM PRACT. 2008 NOV;38(6):1353-63, VII.

COGNITIVE DYSFUNCTION SYNDROME: THE CANINE VERSION OF HUMAN ALZHEIMER'S DISEASE

C.W. Dewey, DVM, MS, DACVIM (Neurology), DACVS
Ithaca, NY

INTRODUCTION

Despite extensive research in human Alzheimer's disease (AD), the root cause(s) of this increasingly common disorder and effective treatment modalities remain a mystery. Canine cognitive dysfunction syndrome (CDS) is the dog analog of human AD and is a naturally occurring disease model for AD. Unfortunately, diagnostic and treatment options for CDS are even more of a challenge than for AD in people. The main purposes of this presentation are to discuss the clinical features of CDS as well as current and future treatment options for this disease.

PATHOPHYSIOLOGY

As with AD of people, the pathophysiology of CDS is unknown. In general, there are amyloid-dependent and amyloid-independent pathologic processes that act in concert to cause progressive brain deterioration. Recent evidence suggests that AD may begin as a vascular-associated phenomenon (oxidative damage to small blood vessels) with subsequent ischemia-activation of amyloid-processing enzymes. There are pathologic similarities between the brains of humans with AD and dogs with CDS. Cerebral vascular changes, meningeal thickening, gliosis, and ventricular dilatation occur in brains of both AD and CDS patients. More specifically, the progressive accumulation of a neurotoxic protein called beta-amyloid in the brain (in and around neurons) is a consistent feature in both AD and CDS. These accumulations coalesce to form plaques (neuritic plaques) and are most prominent in the frontal cerebral cortex and in the hippocampus in both human and veterinary disorders. In both disorders, the degree of beta-amyloid accumulation correlates with the extent of cognitive impairment. In addition to the accumulation of neurotoxic beta-amyloid (Aβ) protein in the aged canine brain, intraneuronal accumulation of a hyperphosphorylated microtubular-associated protein (tau protein) has also been demonstrated. Tau protein is the precursor to neurofibrillary tangles (NFT), another prominent histopathological feature of human AD. The absence of mature NFTs in the brains of dogs and cats with CDS has been argued as evidence against CDS of dogs being analogous to human AD. However, the absence of NFTs in dogs has a number of potential explanations. It is possible that dogs do not live long enough for the tau proteins to develop into NFTs as they do in people. While the amino acid sequence of Aβ protein is identical between humans and dogs, this is not the case for tau protein. The amino acid sequence of dogs differs from that in people; this different sequence may affect the ability of tau protein to form NFTs. Other structural abnormalities found in the aging canine brain that are similar to those in humans include cerebral atrophy, ventricular enlargement, blood vessel wall fibrosis and amyloid deposition (meningeal and parenchymal), microhemorrhages and infarcts, axonal degeneration with myelin loss, astroglial hypertrophy and hyperplasia, and intraneuronal accumulation of several substances (lipofuscin, polyglucosan bodies, and ubiquitin). The pathophysiology of CDS and AD is multifactorial and complex. There is evidence in both diseases that increased oxygen-free radical mediated cellular damage, decreased endogenous antioxidant defenses, inflammation (from various processes), decreased mitochondrial function, DNA damage, vascular compromise, and neurotransmitter imbalance are all interrelated processes that are involved in progressive cognitive impairment. Neurochemical changes that occur in the aging brain are believed to contribute to progressive cognitive impairment. An age-associated decline in the brain neurotransmitter levels of acetylcholine, dopamine, norepinephrine, and GABA have been documented in CDS and AD. Of these abnormalities, cholinergic dysfunction appears to have the highest and most consistent correlation with age-related cognitive impairment. Other neurochemical abnormalities identified in brains of CDS and AD patients include increased acetylcholinesterase levels (associated with cholinergic decline), increased monoamine oxidase B (catalyzes the breakdown of dopamine, with subsequent formation of free radicals), and elevated CSF levels of lactate, pyruvate and potassium.

CLINICAL FEATURES OF CDS

Cognitive dysfunction syndrome is recognized primarily in elderly dogs (more than 9 yrs), but should be suspected in animals 7 yrs or older that are demonstrating progressive cognitive impairment. Clinical signs of CDS are numerous and often nonspecific. They include inattentiveness, inactivity, aimless wandering (often pacing at night), demented behavior, disturbance of the sleep/wake cycle, urinary and/or fecal incontinence, difficulty navigating stairs, becoming lost in previously familiar environments, failure to recognize previously familiar people or animals, decreased interaction with family members, hearing loss, and excessive vocalization (often at night). Owners of CDS pets often describe their pets as acting "senile." Aging dogs appear to fall into three categories of cognitive function, similar to people. These categories are successful aging, mild cognitive impairment (MCI), and severe cognitive impairment (dementia). The third, most severe category is consistent with a diagnosis of AD. Standardized cognitive testing of older dogs (almost exclusively done in a research setting) has enabled a more objective determination of the extent of cognitive dysfunction in individuals, as well as the response to therapeutic interventions. In addition to the classic behavioral abnormalities indicative of CDS in dogs, the author occasionally encounters suspected CDS dogs with either transient central vestibular dysfunction or seizure activity of recent onset. Although not yet reported as a clinical feature of canine CDS, vestibulocerebellar dysfunction and seizures are reported as a potential consequence of AD in people.

Similar to AD of people, a diagnosis of CDS in a dog is based primarily on historical complaints indicative of progressive cognitive impairment. Before arriving at a presumptive diagnosis of CDS, the clinician should rule out other potential causes of cognitive dysfunction, such as metabolic disorders (e.g., hepatic encephalopathy) and structural brain disorders (e.g., brain tumor). In AD, MR imaging of the brain is usually performed as part of the diagnostic workup, and should ideally be part of the diagnostic plan for CDS patients. Brain imaging of AD patients can be normal, but may reveal brain atrophy, ventricular enlargement, and lesions in the medial temporal lobes of the cerebral cortex. Age-related changes appreciated on MR imaging of the brain in CDS patients are mainly reflective of brain atrophy and include ventricular enlargement, widened and well-demarcated cerebral sulci, and diffuse and scattered areas of T2-hyperintensity in periventricular white matter. Although these are consistent findings associated with the aging brain, they may be found in older patients without evidence of CDS. Recently, the thickness of the interthalamic adhesion as measured on transaxial T1 and T2-weighted MR images was found to be significantly smaller in dogs with CDS compared with dogs without CDS. An interthalamic adhesion thickness of 5.0 mm or less was found to be consistent with a diagnosis of CDS in dogs.

TREATMENT AND PROGNOSIS FOR CDS

There is no known cure for CDS. There are multiple proposed therapeutic approaches to CDS, with variable evidence of efficacy in improving cognitive function and/or delaying progression of cognitive decline. The use of oral L-deprenyl (selegiline), an irreversible inhibitor of MAOB, has been purported to improve cognitive function and slow progression of the disease in the majority of dogs with CDS. There is considerable variability in the degree of response achieved among patients, however. L-deprenyl is thought to exert its beneficial effects in the brain by restoring dopaminergic balance, as well as enhancing catecholamine levels, and decreasing levels of damaging free radical species. The dosage for dogs is 0.5—1.0 mg/kg every 24 hrs. Most patients will exhibit a positive response within the first month of therapy. Despite apparent positive responses of canine CDS patients to selegilene, there is some evidence that this drug does not have a significant effect on cognitive function in these patients, or in people with AD. The clinical efficacy studies supporting selegilene use in CDS are based primarily on owner response to questionnaires, rather than on standardized comparative cognitive testing procedures of treated and untreated patients. Since selegilene may produce non-specific low-level hyperactivity by increasing brain catecholamine levels, the "response" observed by owners may not truly be representative of improved cognitive ability. Selegilene is not considered an effective drug for human AD, due to variable responses and overall minimum improvement of cognitive function. The acetylcholinesterase inhibitor phenserine has exhibited efficacy in improving cognitive function in both dogs with CDS and humans with AD in clinical trials; to the author's knowledge, this drug is not yet commercially available for dogs. Behavioral changes in CDS patients may be alleviated with the use of GABA-ergic drugs, such as gabapentin or pregabalin. Because inflammatory changes have been identified in the brains of CDS dogs, the use of anti-inflammatory drugs (e.g., carprofen) has also been proposed. There is experimental evidence that certain naturally-occurring phytochemicals (e.g., curcumin, green tea catechins, resveratrol) may have therapeutic potential for AD and CDS via both anti-oxidant activity and decreasing brain amyloid burden. In a double-blinded, placebo controlled trial in CDS dogs, S-adenosylmethionine (SAMe) was found to be effective in improving clinical signs associated with mental decline. Finally, a large number of complimentary therapies have been suggested for the treatment of CDS, with the primary goals of calming the patient, reducing anxiety, and normalizing the sleep/wake cycle. These include melatonin, valerian root, dog-appeasing pheromone (DAP), phosphatidylserine, ginkgo biloba, DHA (an omega 3 fatty acid), and various antioxidants and mitochondrial co-factors. The evidence for efficacy of all these complimentary therapies is entirely anecodotal. There is convincing evidence that providing a diet fortified with antioxidants, mitochondrial cofactors, and essential fatty acids improves cognitive function and delays cognitive decline in dogs with CDS. This commercially available diet (Hills b/d) contains a mixture of fruits and vegetables, in addition to vitamins C and E, and mitochondrial cofactors (L-carnitine, DL-α-lipoic acid). Environmental enrichment, such as regular exercise and introduction of new toys, has also been demonstrated to improve cognitive function and delay cognitive decline in dogs with CDS. Progression of CDS appears to be more rapid in castrated versus intact male dogs, suggesting a potential role for hormone replacement therapy in this disease. The prognosis for CDS is guarded. Most affected patients are euthanized within 18—24 mo of onset of clinical signs, either due to progressive cognitive impairment or unassociated medical problems.

REFERENCES

[1]Yu C-H, et al. **J Comp Path** 2011; 145(1): 45-58. [2]Pimplikar SW, et al. **J Neuroscience** 2010; 30(45): 14946-14954. [3]Marchesi VT. **FASEB** 2011; 25:5-13. [4]Kim J, et al. **J Neurochem** 2010;112: 1415-1430.[5]Lee Y-J, et al. **Arch Pharm Res** 2010;33(10): 1539-1556.[6]Reme CA, et al. **Vet Ther** 2008;9:69-82.

HEREDITARY CEREBELLAR ATAXIA

Dennis O'Brien, DVM, PhD, DACVIM (Neurology)
Columbia, MO

The hereditary cerebellar ataxias are a heterogeneous group of diseases affecting a number of breeds. The severity of clinical signs runs the gamut from fairly benign phenotype compatible with a good quality of life to severe signs necessitating euthanasia. While the prevalence of any single syndrome is low, the number of breeds affected means that the conditions overall are not uncommon. The advent of modern molecular genetics has led to the identification of the mutations responsible for some of these diseases. Thus recognition of the disease is important to appropriately advise breeders of the steps needed to decrease the incidence of the disease in the breed.

This review will provide an overview of the development, anatomy and function of the cerebellum, review some of the syndromes which have been described, and discuss how the mutations which have been identified interfere with function.

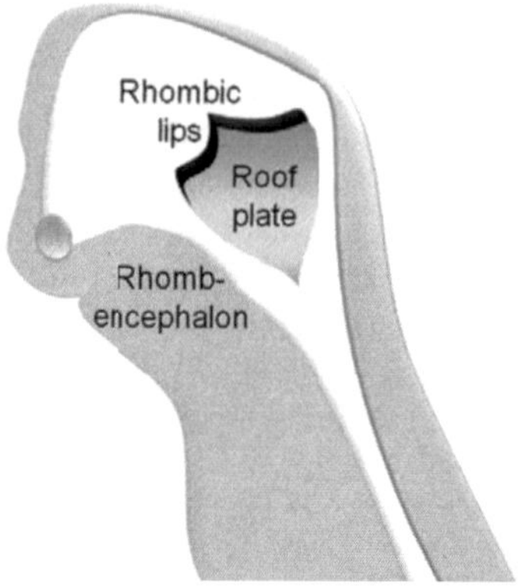

DEVELOPMENT & ANATOMY OF THE CEREBELLUM

The cerebellum develops from the rhombencephalon, so named because that portion of the neural tube which gives rise to the hindbrain is rhomboid in shape. While most of the neural tube is expanding in a medial-lateral direction, a portion of the rostral rhombencephalon rotates ~90° so that the cells form the rhombic lips which proliferate caudally & medially to form the cerebellum. As they do so, they cover the remnant of the open neural tube, the roof plate.

The progenitor cells adjacent to the nascent 4th ventricle first give rise to Bergman's radial glia which secret the signaling molecule called reelin. This directs the migration and development of the Purkinje cells from this same layer. The name comes from mutant mice missing the protein which have disordered cerebellar development and a "reeling" gait (cerebellar ataxia). A separate wave of migration of progenitor cells from the rhombic lips gives rise to the external granule cell layer on the surface. The signaling molecule sonic hedgehog is secreted by the Purkinje cells and induces the external granule cell neurons to differentiate into glutamatergic granule cells. Different signals from the Bergman's glia then direct migration of these cells past the Purkinje cells into the granule layer, leaving dendrites behind as parallel fibers.

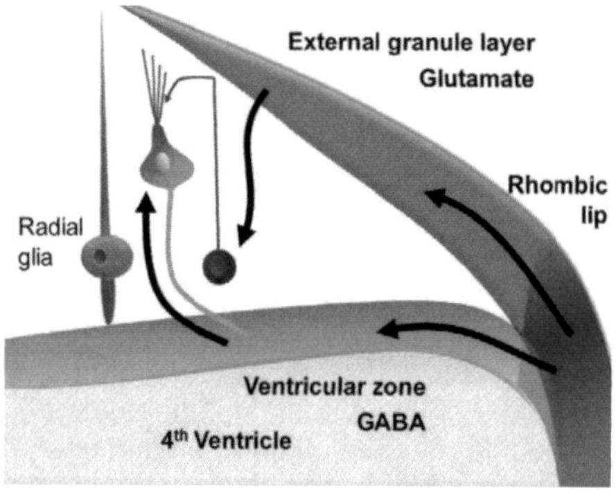

The function of the cerebellum is to utilize feedback from the vestibular system and from muscle and joint proprioceptors to fine-tune motor commands and produce coordinated movements. Together with the basal ganglia, the cerebellum is critical to motor learning. Motor commands are monitored through the middle cerebellar peduncle while sensory information comes through the caudal cerebellar peduncles. This information is integrated in the cerebellum through a network of synapses on the Purkinje cells, the sole efferent output from the cerebellar cortex. The Purkinje cell output projects forward through the rostral cerebellar peduncle to the motor systems to fine-tune the motor commands. Purkinje cells also project directly to the vestibular nuclei to regulate balance and eye movements.

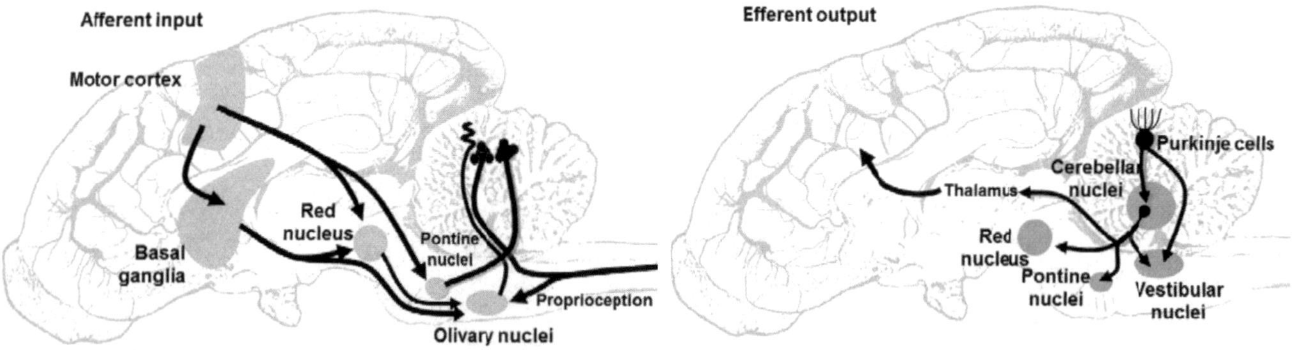

SIGNS OF CEREBELLAR DISESE

The signs of cerebellar disease reflect the loss of normal coordination of movement and balance. The classic clinical sign is cerebellar ataxia which is characterized by an exaggeration of movements (hypermetria). Typically this is most noticeable in the thoracic limbs, but in some cases, such as spinocerebellar ataxia in Jack Russell terriers, truncal & pelvic limb ataxia is also very prominent. The hypermetria will be particularly apparent when the animal is moving quickly or attempting to negotiate stairs. A course tremor of the head +/- the body may be seen when the animal fixes their attention on something. This is referred to as an intention or resting tremor and is most readily observed when the animal attempts to eat.

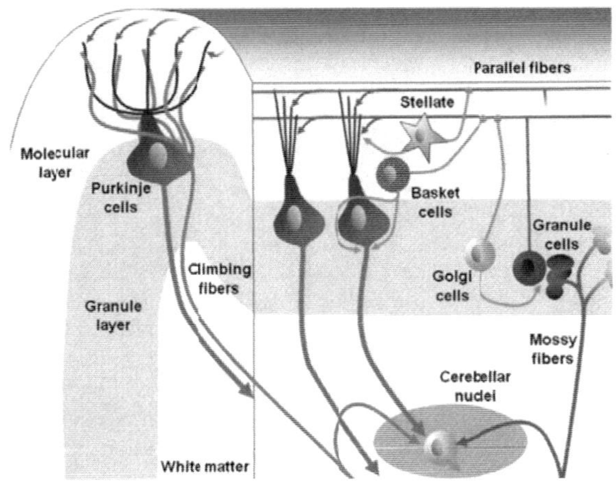

ell activity.
ltiple
ry nuclei
ith
ontrol

Strength is not affected in cerebellar disease, and though the animal may be unable to walk in severe cases, they can make strong voluntary movements. An animal that is recumbent from cerebellar disease may adopt a characteristic posture called decerebellate rigidity. The neck is dorsiflexed in opisthotonus and the thoracic limbs extended, while the hips are flexed.

Due to the intimate connections between the cerebellum and vestibular system, vestibular signs will also be present if the more ventral portions of the cerebellum are affected. Since hereditary diseases typically affect both sides equally, lateralized vestibular signs such as a head tilt or circling are not common. A vestibular type of ataxia with leaning and loss of balance may be noted. Sudden movements of the head may precipitate a vertigo episode. Sometimes this may result in dystonic posturing which the owner mistakes for a seizure. Nystagmus may be present at rest or readily elicited by dorsal flexion of the neck or by rolling the animal onto their back. The nystagmus will often be vertical but can vary. If the head is suddenly released after flexing the neck dorsally, an animal with cerebellar disease often cannot control the return of the head to a normal position adequately and will overshoot, sometimes banging their chin on the floor if not protected.

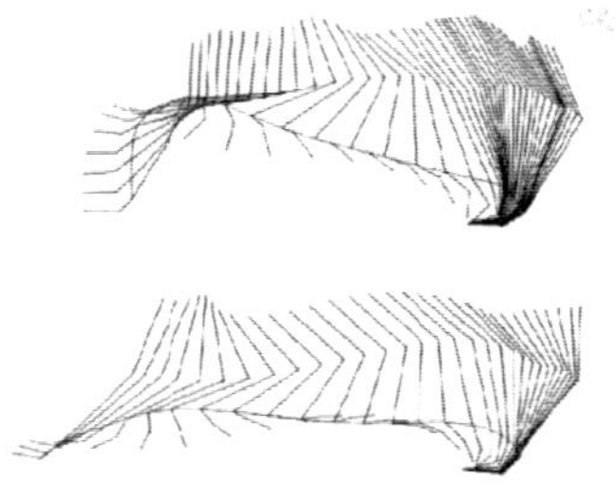

One unusual sign of cerebellar disease is a loss of the menace response. This is called a response because it is not a hard-wired reflex as one would expect, but rather a learned response. This is why the menace response is one of the last aspects of the neurologic exam to mature in a neonate. Because the cerebellum is critical for such motor learning, the response may never develop or subsequently be lost in cerebellar syndromes.

HEREDITARY CEREBELLAR DISEASES

A wide variety of hereditary cerebellar diseases have been described in a number of different breeds. A more complete listing of the syndromes can be found in neurology texts[1-3]. Some diffuse diseases of the nervous system such as leukodystrophies, neuroaxonal dystrophies, or lysosomal storage diseases will present initially with cerebellar signs. Inevitably, however, other signs such as blindness, weakness, personality changes, or seizures reveal the diffuse nature of these diseases. Here we will discuss some representative examples of conditions affecting primarily the cerebellum.

Congenital disease

Many congenital cerebellar diseases involve abnormal development of the cerebellum. Often this can be the result of an insult in utero such as hypoxia or infection, but a number of hereditary conditions in humans and rodents affect cerebellar development. The recognition of similar malformations in specific breeds suggests a role of genetics in domestic animals as well. At the most extreme, the cerebellum may fail completely to develop resulting in cerebellar hypoplasia. Failure of the final fusion of the developing cerebellum at the midline can result in varying degrees of aplasia of the vermis and cysts in the caudal fossa. The most well recognized syndrome of this type, the Dandy-Walker malformation, has been seen in several breeds[2].

Aside from gross malformations of the cerebellum, mutations in the genes for some of the molecules that signal the normal migration of neuronal precursors can lead to histologic derangements. Deletion of the gene for reelin in mice results in disruption of normal migration of the Purkinje cells. This results in a disorganized cerebellar cortex with mixing of the Purkinje cell and molecular layers. Jack Russell terriers with neonatal ataxia also show a disruption of the normal layered development of the cerebellum, but in this case the granule cells fail to develop normally[4]. Though it appears familial, the genetics of this granuloprival cerebellar ataxia has not been determined.

In neonatal ataxia in Coton de Tulears, the cerebellum looks normal on light microscopy, but the animals suffer from ataxia so severe that they are unable to stand or walk (astasia)[5]. In this case, the mutation is in the gene for the metabotropic glutamate receptor (MGluR1)[6] which is highly expressed in cerebellar Purkinje cells. Lack of normal MGluR1 expression in mice leads to a loss of normal pruning of redundant climbing fibers during development, but these changes are not visible on routine light microscopy. MGluR1 is a G-protein coupled receptor which activates the inositol tris phosphate (IP3)/diacyl glycerol (DAG) pathway. In mice and humans, mutations in several of the IP3 signaling pathway genes lead to cerebellar ataxia. The DAG half of this signaling pathway in the Purkinje cells down-regulates levels of the AMPA receptor, an ionotropic glutamate receptor. This down-regulation of AMPA receptors underlies long-term depression, one of the molecular bases of learning and memory. Thus without MGluR1 function, affected pups do not have normal motor learning and never develop the ability to walk. In mouse models, motor learning is quantified by the conditioned eye-blink response, the experimental equivalent of the menace response. Mice that lack MGluR1, like the affected pups, fail to develop the response.

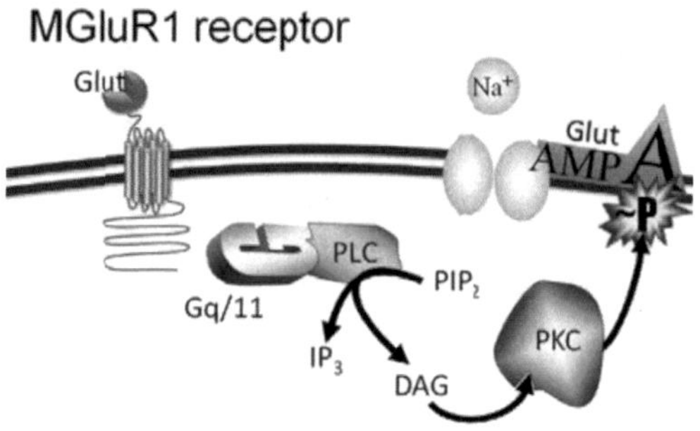

Delayed onset ataxia

In many of the hereditary ataxias reported in animals, there is a period of normal development followed by a progressive cerebellar ataxia. X-linked[7] and autosomal recessive forms have been described in animals though many dominant forms occur in humans. The onset of ataxia in dogs and cats can vary from weaning age to >10 years of age. The severity of the clinical signs can also vary from mild ataxia and intention tremors to marked ataxia and other motor signs. The nature of the histologic changes found also varies. In the simplest form, only Purkinje cells degenerate. In some cases, the degeneration can be regional producing the mildest signs. On histopathology, a normal dendritic tree of the basket cell interneurons can be found where the Purkinje cells should be. These "empty baskets", along with the period of normal motor function, indicates that the Purkinje cells develop and subsequently degenerate, a process sometimes referred to as abiotrophy. Typically an MRI will show cerebellar atrophy. Occasionally, there will be subsequent degeneration of other cells that communicate with Purkinje cells such as the granule cells and olivary nucleus.

So far, a mutation has been associated with only one Purkinje cell degeneration phenotype[8]. Arabian horses develop hereditary cerebellar ataxia at 1.5-4 months of ages. Linkage analysis mapped the locus to equine chromosome 2 and sequencing candidate genes within the locus identified a mutation in an area of overlap between two genes. The role of these genes in cerebellar function is not yet clear.

In several breeds, the degeneration is found in the spinocerebellar tracts rather than the cerebellum itself[9;10]. These are referred to as spinocerebellar ataxias, though in the human literature that term is often used in a broader sense to refer to ataxia caused by Purkinje cell loss as well. Breeds affected include the Jack Russell terrier where the ataxia is severe and often accompanied by myokymia and sometimes seizures, and the Brittany spaniel which have a very late onset of clinical signs.

Mutations in potassium or calcium channels in humans are associated with episodic cerebellar ataxia, presumably by altering membrane excitability. Episodic ataxia has been reported in a Bichon Frise[11], but it has not yet been shown to be familial.

Kerry Blue terriers and Chinese Crested dogs have an allelic condition characterized by an initial cerebellar ataxia at about 3 months of age[12;13]. At this stage of the disease, cerebellar atrophy is apparent on MRI and the only histologic change is Purkinje cell loss. By a year of age, however, more severe motor difficulties develop. Affected dogs have difficulty initiating movement and fall frequently. The development of these signs corresponds with degeneration of the caudate nucleus and substantia nigra, the former being readily apparent as increased T2 signal on MRI. Though the gene has been mapped[13], attempts to identify the causative mutation have thus far been unrewarding.

CONCLUSIONS

Progress is being made on identifying the mutations responsible for the hereditary ataxias in animals. This will provide breeders the tools to decrease the incidence of the disease. It can also identify the metabolic pathways important in ataxia of other causes. In humans for example, autoantibodies against the MGluR1 receptor have been identified in autoimmune cerebellar ataxia. Identify these pathways can also point toward therapeutic interventions to treat the signs of ataxia such as the use of the potassium channel blocker 4-aminopyridine (4AP) in treatment of episodic ataxia.

REFERENCES

[1]deLahunta A, Glass E. *Veterinary Neuroanatomy and Clinical Neurology*. 3rd ed. Philadelphia: WB Saunders, 2009.[2]Summers BA, Cummings JF, deLahunta A. *Veterinary Neuropathology*. St. Louis: Mosby, 1995.[c]Lorenz MD, Coates JR, Kent M. *Handbook of Veterinary Neurology*. 5th ed. St. Louis: Elsevier, 2011.[4]Coates JR, et al. *J VIM* 1996;10:176 [5]Coates JR, et al . *J VIM* 2002;16:680-689.[6]Zeng R *et al. J VIM* 25[2], 267-272. 2011. [7]O'Brien DP, et al. *Canine Practice* 1998;23:46.[8]Brault LS, et al. *Genomics* 2011;97:121-129.[9]Wessmann A, et al. *J VIM* 2004;18:515-521. [10]Higgins RJ, et al. *Acta Neuropathologica* 1998;96:97-101.[11]Hopkins AL, et al. *Journal of Small Animal Practice* 2010;51:444-446. [12]deLahunta A, Averill DR. *J VIM* 1976;168:1119-1124. [13]O'Brien DP, et al. *Journal of Heredity* 2005;96:727-34.

BIOMARKER DEVELOPMENT FOR CANINE DEGENERATIVE MYELOPATHY

Joan R. Coates, DVM, MS, DACVIM (Neurology)
Columbia, MO

INTRODUCTION

Amyotrophic lateral sclerosis, the most common adult motor neuron disease, is characterized by loss of motor neurons causing stiffness and slowing of muscle movements, difficulty speaking and swallowing, muscle atrophy and severe weakness. Patients typically die within 3-5 years of the onset of signs secondary to failure of respiratory muscles. The only currently FDA-approved drug for therapy in ALS is riluzole, an antiglutaminergic, which has shown marginal effects at slowing disease progression.[1-3] The pathologic hallmarks of ALS are muscle degeneration (*amyotrophy*), axonal degeneration and loss, extensive astrogliosis (*sclerosis*), intracellular protein aggregate formation and neuronal cell body death in upper and lower motor neurons in the spinal cord, brainstem and cerebral cortex. At disease onset, ALS patients typically present with only upper motor neuron (UMN) or lower motor neuron (LMN) signs that involve mainly the limbs (e.g., muscles innervated by spinal nerves) or the head and neck (bulbar) region (e.g. muscles innervated by cranial nerves). Thus, these forms of ALS are termed UMN onset, spinal onset or bulbar onset, respectively.[4] Virtually all human ALS patients eventually develop dysphagia, which can lead to aspiration and fatal complications. Some ALS patients also exhibit cognitive abnormalities that may precede or occur after the onset of symptoms, a pathologic continuum underlying this multisystem disorder.[5,6]

Canine DM is a spontaneously occurring, adult-onset, progressive disease that, like ALS, leads to paralysis and death.[7] The earliest clinical signs begin when the dog is 8 years or older. Asymmetric spastic weakness and general proprioceptive ataxia in the pelvic limbs progresses to paraplegia within 1 year from onset of signs. At disease onset, spinal reflexes are present, consistent with loss of UMN function. Dog owners often elect euthanasia when their dog becomes paraplegic. When euthanasia is delayed, LMN signs emerge as flaccid quadriplegia, widespread muscle atrophy, dysphagia, and inability to bark.[8,9] The pathologic features of DM that have been characterized to date are similar to those of ALS, including axonal degeneration with secondary demyelination and astroglial proliferation (sclerosis), most severe in the dorsal portion of the lateral funiculus and in the dorsal columns of the middle to lower thoracic region.[10] We have additionally documented denervation atrophy in muscle (amyotrophy), nerve fiber loss with axonal degeneration, and secondary myelin loss in myelinated fibers of peripheral nerves.[8] Thus, the distribution of lesions and clinical disease progression in canine DM are similar to that reported for the UMN-onset (predominant) ALS, which is characterized by UMN signs in the disease course that later progress to LMN signs.[8,11,12]

Another key link between DM and ALS is the genetic similarity. Familial ALS (fALS) accounts for 10% of all ALS cases; the rest appear as sporadic ALS (sALS).[13] Mutations in the superoxide dismutase (*SOD1*) gene, which encodes cytosolic Cu/Zn superoxide dismutase are most prevalent with the inherited form of ALS (fALS).[14-15] Although the etiology of the more common sALS is largely unknown, symptoms and pathology associated with *SOD1*-related ALS mirror those in sALS. Our group recently discovered that mutations in *SOD1* also underlie canine DM.[8] We first identified a *c.118G>A* transition in *SOD1* that predicts an E40K missense mutation in Pembroke Welsh Corgis (PWCs) with DM. In addition, we demonstrated that, as in ALS, cytoplasmic aggregates that bind anti-SOD1 antibodies were present in spinal cords of *SOD1:c.118A* homozygotes with DM. No similar SOD1 antigen-containing aggregates are found in spinal cords from *SOD1:c.118G* homozygotes.

While symptoms/clinical signs of ALS are dominated by motor dysfunction, other abnormalities occur outside the motor systems, in both ALS and canine DM. The cognitive and sensory dysfunctions for the ALS phenotype represent a pathologic continuum of a multi-system neurodegenerative disease.[4,16] In ALS patients, cognitive impairment and dementia can precede or occur after onset of symptoms.[5,6] Sensory abnormalities occur in10% to 23% of ALS patients, based on evidence of axonal loss in sensory nerves, abnormalities in somatosensory evoked potentials and diminished sense of vibration and touch-pressure.[17-20] Based on clinical signs and pathologic findings, DM is a multi-system degenerative disease that has many similarities to ALS. The general proprioceptive ataxia and dorsal column degeneration in DM parallels the sensory dysfunction in ALS. Characterization of cognitive dysfunction of DM-affected dogs has been limited but there is evidence of neurodegeneration in the brain.[21] Overall, ALS and DM share clinical signs of stiffness, weakness, muscle atrophy, progressive paralysis, and importantly, a common genetic mutation in the *SOD1* gene. These similarities are striking enough, and indicate that canine DM is a useful, spontaneous, large animal disease model for ALS.

BACKGROUND OF BIOMARKER DEVELOPMENT IN DISEASE

A *biomarker* is defined as any characteristic that is objectively measured and evaluated as an indicator of normal biological processes, pathogenic processes or pharmacogenomic processes to a therapeutic intervention.[22,23] Biomarkers are having an ever important role in the process of drug-development. Biomarkers incorporated into clinical trials can be subdivided into target biomarkers and efficacy biomarkers.[23] *Target biomarkers* bind to the target receptor and could be used to monitor receptor occupancy of drug by measuring displacement of the target marker. An *efficacy biomarker* must demonstrate a change in level of expected direction upon drug administration and must correlate with benefit in terms of clinical outcome. Thus, both should be incorporated into drug development process. Negative studies involving biomarkers directed toward specific targets would provide more efficient results for redirection of future studies.

A *surrogate end point* is defined as a biomarker that is intended to substitute for a known clinical end point, such as survival in the case of ALS. A surrogate end point is expected to predict benefit (or harm, or lack of benefit) based on epidemiologic, therapeutic, pathophysiologic or other scientific evidence.[22,24] Surrogate end points are frequently used to monitor disease progression and taken, as a measure of the effects of a specific treatment.[24] Surrogate end points also can help to identify underlying pathophysiology of disease. In clinical trials, surrogate end points can reduce patient numbers required, trial duration and study costs; thus, expediting drug development. Clinical end points, such as survival, have routinely been used; however this can be impractical owing to the relatively long time periods required for the end points to be achieved. Surrogate end points that can be reliably substituted for clinical end points will improve the efficiency of clinical trials. A *validated* surrogate marker must have established evidence of a drug-induced effect. Validation requires establishing sensitivity/specificity for the surrogate marker in a large cohort.[25] Stratification of ALS patients is key to design of biomarker studies and clinical trials since the heterogeneous patient population may mask potential candidate biomarker/drugs for subgroups of ALS.[23] In turn, biomarker/surrogate end point studies in canine DM may only translate to a subgroup of ALS patients.

CANINE DM AS A SPONTANEOUS DISEASE MODEL OF ALS

Therapeutic interventions that appeared promising when tested in rodent ALS models[26] have failed to ameliorate signs of ALS in people.[27-29] There are many potential explanations for "failure" of these mouse models, but the most prominent issues are (1) whether the *SOD1*-related ALS is an adequate disease model of the sporadic disease that represents most ALS patients and (2) whether the mouse is an adequate animal model for testing therapeutics for humans. Without an effective therapy in humans as a benchmark, it is difficult to know which of these issues are representative. We would argue that clinical similarities between *SOD1*-related ALS patients and sporadic ALS patients, and recent findings of SOD1 aggregates in pathology in sporadic ALS patients suggest that further study of SOD1 still is relevant to all ALS.[30]

Spontaneous canine models of diseases have played a key role in developing therapies for a number of disorders in people and other animals. Moreover, canine disease models such as DM offer a ready clinical population on which therapies can be evaluated in an environment that closely mimics human clinical trials. Therefore, development of the canine model could prove to be a major breakthrough in the assessment of therapeutic interventions for ALS. The pattern of clinical progression of canine DM is relatively homogeneous in comparison to the heterogeneity in human ALS. Study of a homogeneous disease pattern will facilitate longitudinal measures of disease progression and detection of therapeutic efficacy. Pathologic studies of ALS patients offer only a snapshot at the end-stage of disease. Because dogs are euthanized at varying disease stages based on the pet owner's decision, we can capture markers of disease at early and latter disease stages. If we can better follow the progression of the lower and upper motor neuron disease spectrum in DM-affected dogs, these dogs could serve as a valuable animal model for evaluating efficacy and safety of therapeutic interventions. Likewise, the proposed pathophysiologic processes of the *nature and chronology* of upper and lower motor neuron degenerations that underlie ALS need to be explored for canine DM.

When translating therapies from animals to people, sensitive and objective disease markers shared across species are necessary to diagnose and monitor disease progression so that therapeutic efficacy can be accurately predicted.[31] Developing functional measures for UMN and LMN disease components in canine DM and refining other clinical criteria in dynamic study of DM will provide sensitive and specific measures (markers) not only of disease progression but also of response to therapy that parallels a surrogate marker utilized in ALS patients and other ALS animal models. Such measures of disease progression will further validate canine DM as an ideal disease model for studying therapeutic interventions in human ALS.

Table 1 summarizes clinically-based platforms being used for biomarker discovery and development in ALS,[23,31-33] which can have potential for canine DM. In order to establish objective and effective UMN disease markers with use of neuroimaging is evolving from the current cross-sectional approach to longitudinal imaging studies and incorporation of functional studies. LMN disease markers has focused on surrogate markers using indirect techniques to assess motor unit loss but such studies such as motor unit number estimation lack a gold standard, a concern for reliability. A bioinformatics approach incorporating genomic studies will broaden analyses for biomarker studies in CSF and blood. Markers derived from protein or neuroimaging also have potential to identify patients in the earliest phase of disease when a treatment may be most effective.

REFERENCES

[1]Miller RG, et al. **Cochrane Database Syst Rev** 2007. [2]Lacomblez L, et al. **Neurology** 1996;47:S242. [3]Bensimon G, et al. **N Engl J Med** 1994;330:585. [4]Wijesekera LC and Leigh PN. **Orphanet J Rare Dis** 2009;4:1. [5]Strong MJ, et al. **Neurology** 1999;53:1665. [6]Wheaton MW, et al. **Neurology** 2007; 69:1411. [7]Averill DJ. **J Am Vet Med Assoc** 1973;162:1045. [8]Awano T, et al. **Proc Natl Acad Sci U S A** 2009:2794. [9]Coates JR, et al. **J Vet Intern Med** 2007;21(6):1323. [10]March PA, et al. **Vet Pathol** 2009; 241. [11]Coates JR and Wininger FA. **Vet Clin North Am Small Anim Pract** 2010;40:929. [12]Gordon PH, et al. **Neurology** 2009;72(22):1948. [13]Valdmanis PN, et al. **Curr Neurol Neurosci Rep** 2009;9:198. [14]Rosen DR, et al. **Nature** 1993;362:59. [15]Dion PA, et al. **Nat Rev Genet** 2009;10:769. [16]Leigh PN. Amyotrophic lateral sclerosis. In: Eisen AA, Shaw PJ editors. **Motor Neuron Disorders and Related Diseases**. Amsterdam: Elsevier, 2007, p. 249. [17]Georgesco M, et al. **Rev Neurol** 994;150:282. [18]Isaacs JD, et al. **J Neurol Neurosurg Psychiatry** 2007;78:750. [19]Pugdahl K, et al. **J Neurol Neurosurg Psychiatry** 2007;78:746. [20]Weis J, et al. **Neurology** 2011;76:2024. [21]Johnston PEJ, et al. **Vet Res** 2000;146:629-633. [22]Biomarkers Working Group. **Clin Pharmacol Therap** 2001;69:89. [23]Ganesalingam J and Bowser R. **Biomark Med** 2010;4:281. [24]Katz R. **J Am Soc Exp NeuroTherap** 2004;1:189. [25]Buyse M et al. **Clin Onc** 2010;7;309. [26]Gurney ME, et al. **Science** 1994;264:1772. [27]Turner BJ and Talbot K. **Progress Neurobiol** 2008;85(1):94. [28]Benatar M. Neurobiol Dis 2007;26(1):1. [29]DiBernardo AB and Cudkowicz ME. **Biochimica et Biophysica Acta** 2006;1762(11-12):1139. [30]Forsberg K, et al. Plos one 2010;5(7):1. [31] Turner MR, et al. **Lancet Neurol** 2009;8:94.[32]Pradat F, Dib M. **Mol Diag Ther** 2009;13:115. [33]Bowser R et al. **Exp Rev Mol** Diagn 2006;6:387.

Table 1. Disease measures in ALS patients correlated with disease

Category of Markers	Marker Description	Measurement Description	Additional Comments
Imaging Modalities	Measures of UMN dysfunction	Focuses on neuronal integrity, loss of motor neurons (corticospinal tracts)	Noninvasive; radiation-free mode; imaging biomarkers may change slowly over time – less sensitive biomarker; does not capture subcellular biological processes
MRI	Signal abnormalities	T2 W signal decreases during disease – accumulating corticospinal tract gliosis; longitudinal; easy to measure	Low sensitivity and limited specificity
Magnetic resonance spectroscopy	Detection of protein containing metabolites (N-acetylaspartate [NAA], NAAG, choline, creatine (Cr)	Decrease NAA:Cr suggests decrease in UMN integrity related to loss or dysfunction of motor neuron; Moderate correlation with disease severity	Lack of correlation with ALS-functional rating scale (FRS); questionable significance in rate of change in NAA and NAA:Cr; sensitivity is unknown
Diffusion tensor imaging (DTI)	Fractional anisotropy (FA) – evaluates movement of water; mean diffusivity reflects fiber integrity	Longitudinal measure: decrease of FA (water movement more directional – affected early in disease; elevation in mean diffusivity reflects chronicity	Lack of correlation with ALS-FRS; longitudinal measure may be limited to specific subgroups of ALS; DTI may detect changes pre-symptom
Other imaging modalities – voxel-based morphometry; arterial spin labeling; functional MRI, positron emission tomography			
Neurophysiological Measures	Measure of LMN dysfunction	Peripheral measures of motor units or muscle mass	Peripheral measures of motor unit or muscle mass may lack sensitivity for drug effects – more evaluation in clinical trials is needed to establish surrogate marker
Conventional EMG parameters	CMAP – indirect measure of number of innervated fibers; F-waves and H-reflex	CMAP – correlates well with muscle strength, correlation with MUNE; F-wave frequency is reduced in ALS,	CMAP does not enable detection of effects of collateral reinnervation, masks MU loss; Techniques are rapid and simple to perform; easily standardized; F-wave – correlation is poor with dynamic disease progression
Neurophysiologic index	Derived from CMAP, distal motor latency and F-wave frequency	Sensitive and reproducible; show larger change than clinical measures (e.g. ALS-FRS)	Objective, simple, does not require special methods
Motor unit number estimation (MUNE)	Provides a numerical value that reflects number of functioning motor units	Sensitive to clinical deterioration; useful for longitudinal study; stratify rapid vs. slow disease progression	Active denervation of ALS, absence of gold standard; operator dependence; correlates with CMAP
Electrical impedance myography	Current applied to measure electrical impedance	Muscle atrophy results in increased resistance	May be useful in monitoring disease progression or in modulation of muscle atrophy
Transcranial magnetic stimulation	Transcranial stimulation of cortical motor neurons; many types of measures	Prolonged central motor conduction time; Shorten cortical silent (inhibitory) period	May detect early disease alterations; lacks specificity; Cortical motor threshold, motor evoked potential amplitude, central motor conduction time; cortical silent period
Protein-based Measures	Markers in body fluids such as blood, CSF and urine	May be useful to monitor disease progression or as diagnostic biomarker	Large validation studies are required; main limitations in study are small numbers of patients; large patient cohort lacking
CSF markers *(correlated with disease duration)*	Cytokine, growth factors; CSF contains proteins and peptides released by cells of CNS	Changes may reflect any underlying neurologic disorder	Considered invasive; [growth factors – Glial-derived neurotrophic factor, VEGF, erythropoietin, VgF; cytokines – PG, MCP-1, substance P, transglutaminase]
Markers of axonal degeneration	neurofilament and Tau pNFH; pNFL	Increased pNFH higher in ALS patients , in UMN predominant disease; pNFL associated with rapid disease progression	Need to take into account heterogeneity of ALS, thus stratification and subgroup evaluations
Blood-based markers *(ALS is a systemic disease)*	cytokine, growth factor evaluations	variable correlations to ALS-FRS and disease progression rates	Ease of sample collection; [growth factors – TGF-B1, IGF; cytokines – fibronectin, hyaluronic acid IL-6, T cells]

161a

GENOME-WIDE ASSOCIATION ANALYSIS IDENTIFIES A MUTATION ASSOCIATED WITH ALASKAN HUSKY ENCEPHALOPATHY.

KM Vernau[1], JA Runstadler[2], EA Brown[1], JM Cameron[3], HJ Husan[2], RJ Higgins[1], C Ackerley[3], BK Sturges[1], PJ Dickinson[1], B Puschner[1], C Giulivi[1], GD Shelton[4], BH Robinson[3], DL Bannasch[1]. [1]University of California – Davis, Davis, CA, [2]University of Alaska Fairbanks, Fairbanks, Alaska, [3]Research Institute, The Hospital for Sick Children, Toronto, ON, Canada, [4]University of California-San Diego, La Jolla, CA.

Alaskan Husky Encephalopathy (AHE) is a fatal brain disorder, reported in 13 Alaskan Huskies. Dogs with AHE have an acute or chronic progressive waxing and waning history. They typically have multifocal central nervous system deficits including seizures, altered mentation, difficulty prehending food, absent menace response, hypermetria, proprioceptive positioning deficits, facial hypalgesia, ataxia and tetraparesis. Previous reports hypothesized that AHE was consistent with a mitochondrial encephalopathy.

We evaluated 10 Alaskan Huskies with AHE. The clinical, neuroradiological and neuropathological findings were similar to previous reports, except that 2 dogs have stable neurological deficits 1.5 and 3 years after diagnosis. Two dogs that had stable neurological deficits for 6 months and 2 years respectively after initial diagnosis deteriorated neurologically, and both dogs were euthanatized. One of these dogs had progression of lesions on MRI. Extensive diagnostic testing did not support a mitochondrial etiology for AHE. The purpose of this study was to search for a novel nuclear genetic mutation associated with AHE.

A whole genome scan was performed using the Illumina canine HD arrays with 8 affected and 20 controls dogs. A candidate gene with causal association was identified. All affected dogs were homozygous and none of the controls dogs were homozygous for this haplotype. 187 dogs from 51 breeds were genotyped and the mutant allele was not identified. 41 additional huskies including family members were tested and 16 of these dogs were carriers and the rest were wildtype.

161b

BIOLOGICAL IN-SITU TECHNIQUE FOR CONGENITAL VERTEBRAL ANOMALIES IN BRACHYCEPHALIC DOGS. Gorgi AA, Wininger FA, Fox DB, Essman SC, Sibigtroth CM, Coates JR.

Congenital vertebral anomalies are frequently present in many brachycephalic dog breeds. Although it is often an incidental finding, vertebral malformations could cause malalignment and abnormal angulations of the spine leading to severe neurological deficits. There is often instability associated with vertebral anomalies causing repetitive spinal cord injury. Treatment options often include medical management or surgical stabilization of vertebrae using implants with or without decompressive surgery. The hypothesis was that biological in situ fusion of abnormal vertebral segments in young dogs is an effective method to achieve physiologic stabilization of the vertebral column. This was achieved by a surgical technique utilizing autogenous and allogenic bone graft. To the best of our knowledge this technique has not been reported before in dogs. Our objectives were to minimize the progression of neurological signs by biologic stabilization, evaluate the long-term changes in vertebral deformity, and minimize the post-operative complications observed with more aggressive surgical techniques. Eight young brachycephalic dogs were included in this study. All dogs were non-ambulatory parapretic at presentation. All patients had a full neurological exam, spinal radiographs, MRI, and CT scan of the spine preformed. Recheck CT scan of the spine was performed 6-12 months post-operatively in all cases and most dogs had recheck CT scan about 8 weeks following surgery. All patients became ambulatory, had radiographic improvement of their vertebral deformity on follow-up evaluation, and had a minimum of 12 months follow-up. The clinical findings, pre and post-operative images, and surgical technique will be discussed during the presentation. We believe this is an effective treatment option for young dogs with congenital vertebral anomalies. This technique is associated with less morbidity, risks, and technical challenges than the other surgical vertebral stabilization techniques commonly used in these cases.

162a

PHARMACOKINETICS OF MIDAZOLAM AFTER INTRAVENOUS, RECTAL AND INTRAMUSCULAR ADMINISTRATION IN DOGS. M Schwartz, KR Muñana, JA Nettifee-Osborne, KM Messenger, MG Papich. North Carolina State University College of Veterinary Medicine, Raleigh, NC.

Prolonged epileptic seizures are frequent neurologic emergencies in dogs that require prompt medical intervention. Benzodiazepines are utilized as first-line drugs to treat seizures, and are most commonly administered rectally (PR) or intranasally when venous access is limited. The aim of this study was to compare the pharmacokinetics of midazolam when administered by the intravenous (IV), intramuscular (IM), and PR route. Six healthy dogs were randomly assigned to 3 treatment groups. Dogs in each group were administered 0.2 mg/kg midazolam IV, IM or PR in a crossover design. Whole blood samples were collected at baseline, and at regular intervals up to 480 minutes after administration for pharmacokinetic analysis. The study was repeated twice following a 72-hour washout, such that each dog received midazolam by the IV, IM and PR route. Plasma midazolam levels were measured by high-pressure liquid chromatography. Rectal administration resulted in erratic systemic availability with low to undetectable plasma concentrations. Geometric mean values and coefficient of variation (CV%) for midazolam peak plasma concentrations (C_{max}) were 0.80 µg/ml (39%) and 0.19 µg/ml (37%), following IV and IM administration, respectively. Time to peak concentration (T_{max}) after IM administration was 7.4 min (39%) with a bioavailability of 47% (34%). Findings suggest IM midazolam might be useful in treating seizures in dogs when venous access is unavailable, but higher doses may be required to account for bioavailability, while PR administration is likely to be of limited efficacy. Additional studies are needed to evaluate IM midazolam in the clinical setting.

162b

^{18}F-FLUORODEOXYGLUCOSE UPTAKE AND MAGNETIC RESONANCE IMAGING FINDINGS IN A CANINE MODEL OF ISCHEMIC STROKE USING A HIGH-RESOLUTION RESEARCH TOMOGRAPH-POSITRON EMISSION TOMOGRAPHY AND 7T-MR FUSION SYSTEM. BT Kang[1], JH Kang[1], G Kim[1], D Chang[1], MP Yang[1], HM Park[2]. 1. Chungbuk National University College of Veterinary Medicine, Cheongju, South Korea. 2. Konkuk University College of Veterinary Medicine, Seoul, South Korea.

Positron emission tomography (PET) is a useful imaging technique that can be used to monitor *in vivo* alterations in brain function. Magnetic resonance imaging (MRI) at 7 T provides high spatial and contrast resolution resulting in good visualization of canine brain structures. However, there have been no prior studies using PET and 7T-MR system in dogs with ischemic stroke. Thus the purpose of this study was to assess the ^{18}F-fluorodeoxyglucose (FDG) uptake and describe MR imaging findings at 7 T in the brain of ischemic dogs.

Ischemic stroke was induced by permanent middle cerebral artery occlusion using a silicone coated suture (3 beagle dogs) or autologous thrombus (2 beagle dogs). FDG-PET and T2-weighted MR imaging of the brain were performed on 3 dogs (suture model), and T2-weighted images were obtained on 2 dogs (thrombus model) at 24 hours after the stroke. Acquired PET and MR images were automatically co-registered by the image analysis software. The regions of interest (ROI) were manually drawn over the hyper- or hypointensity of the lesion on the T2-weighted images. A standard uptake value (SUV) and relative SUV ratio (rSUV = SUV of ROI in the ischemic hemisphere / SUV of the corresponding contralateral region) was calculated for each ROI. The MR examinations of suture model revealed unilateral, striatocapsular, and cerebral cortex lesions demonstrating hyper- or hypointensity. On thrombus model, hyperintensities were observed in the both hemispheres. Glucose hypometabolism was identified in the ischemic lesions of suture model. The mean SUV of ischemic lesions was 54% lower than that of corresponding areas of the contralateral hemispheres (rSUV: 0.54 ± 0.3). The metabolic changes on the FDG-PET corresponded well to the lesions seen on the MR images. FDG-PET may be a useful diagnostic tool for the evaluation of ischemic stroke in dogs when the metabolic data was combined with clinical and MR findings.

163a

DIFFUSION TENSOR IMAGING IN DOGS WITH CERVICAL SPONDYLOMYELOPATHY

R.C. da Costa, S. Choi, B. Dewey, S. Smith. College of Veterinary Medicine and College of Medicine, The Ohio State University, Columbus, OH; and Institute of Imaging Sciences, School of Medicine, Vanderbilt University, Nashville, TN.

Diffusion tensor imaging (DTI) is a relatively new magnetic resonance imaging technique. Anisotropic water diffusion in neural fibers such as the white matter of the spinal cord forms the basis for the utilization of DTI to track fiber pathways. Since DTI is sensitive to the underlying tissue microstructure, it provides a unique method of assessing the integrity of these neural fibers in the spinal cord. DTI has not been previously evaluated in dogs with cervical spondylomyelopathy (CSM). The goal of this study was to investigate the feasibility and usefulness of DTI in dogs with CSM having various degrees of spinal cord changes.

Fourteen dogs with clinical signs suggestive of CSM were prospectively studied. All dogs underwent MR imaging using a 3.0 T MR unit (Achieva, Philips, Cleveland, OH). In addition to a standard MRI protocol consisting of at least T1 and T2 weighted images in the sagittal and transverse planes, all dogs were studied using a DTI protocol specifically developed for the canine cervical spine. In short the protocol consisted of single-shot spin-echo echo planar imaging with TR of 24415 and TE of 42, 6 diffusion weighting directions, *b*-values: 0 and 500 s/mm2, slice thickness of 5 mm with no gap. Data was processed at Vanderbilt University using proprietary software to generate fractional anisotropy (FA) and mean diffusivity (MD) maps of the transverse sections. Three maps were acquired at each disc region from C3-4 to C7-T1. Sagittal and transverse T2 weighted MR images were evaluated and the 5 disc regions of all dogs classified into 3 categories: 1 = normal spinal cord (no compression), 2 = spinal cord compression with normal cord, and 3 = spinal cord compression with spinal cord signal changes. Comparisons of FA and MD maps were made among the 3 categories using a random-effects linear regression model.

Fourteen dogs were studied, 8 Doberman Pinschers, 5 Great Danes and 1 Bernese Mountain dog. Ten dogs had their main compression located at C6-7 and 4 dogs at C5-6. Eleven dogs had spinal cord signal changes. Out of all 70 spinal cord regions studied (5 per dog), 42 (60%) regions were graded as 1 (no compression), 14 (20%) graded as 2 (compression without cord changes), and 14 (20%) graded as 3 (compression with cord changes). The mean FA in regions graded as 1, 2 and 3 were 0.64, 0.66 and 0.57, respectively. No difference was seen between groups 1 and 2 (p=0.579), or 1 and 3 (p=0.282). The mean MD in regions graded as 1, 2 and 3 was 1.20, 1.16, and 1.44 (units: $10^{-3}mm^2/s$), respectively. The MD between groups 1 and 3 was significantly different (p=0.029), whereas between groups 1 and 2 it was not significant (p= 0.089).

Results indicate that DTI is a feasible technique in dogs with CSM. The significant differences seen in the MD maps, but not in the FA maps, suggests that the spinal cord white matter tracts are still functionally intact in compressive sites with or without spinal cord signal changes. The differences in MD maps may also indicate inflammatory changes in regions with signal changes.

163b

DIAGNOSTIC YIELD AND SIDE EFFECTS OF MINIMALLY INVASIVE BRAIN BIOPSIES IN 17 DOGS WITH ENCEPHALITIS. T. Flegel[1], A. Oevermann[2], K. Matiasek[3]. [1]Department of Small Animal Medicine, University of Leipzig, Germany, [2] Neurocenter, DCR-VPH, University of Bern, Switzerland, [3] Institute of Veterinary Pathology, Ludwig Maximilians University Munich, Germany.

Diagnosis of encephalitis is usually presumptive based on MR imaging and/or cerebrospinal fluid (CSF) analysis. Definitive diagnosis based on histopathology, however, is required for optimizing treatment strategies. Therefore, the objectives were to investigate diagnostic yield and side effects of minimally invasive brain biopsies (MIBB) in dogs with suspected encephalitis.

Dogs were included if MR images and CSF analysis were indicative for encephalitis and if one of the lesions was located in the forebrain since only forebrain lesions were biopsied. Two minimally invasive free-hand brain biopsies were taken from forebrain lesions through a 4 mm burr hole using a Sedan side-cutting needle with an outer diameter of 2.5 mm and a lateral opening of 10 mm. A third biopsy was taken if post biopsy MRI raised doubts that the lesion was correctly biopsied. Routine histopathological examination was performed. Side effects were assessed by evaluating post biopsy MRI (12/17) and by consecutive neurological examinations.

Seventeen dogs of the following breeds were included: Yorkshire terrier (n=4), Chihuahua (n=4), French bulldog (n=3), and one each of the following: Jack Russell terrier, Maltese, mixed breed dog, papillon, pug dog, and West Highland White terrier. The mean body weight was 6.1 kg (range: 2.1-14.0 kg).

The overall diagnostic yield with regard to a specific type of encephalitis was 82%. In an additional 12% encephalitis was evident, but a specific disease could not be determined, whereas one dog showed rather non-specific alterations in terms of astrocytosis, astrogliosis, microglial proliferation, and vasogenic edema, suggesting that just the periphery of the lesion was sampled. The following specific diagnoses were established: granulomatous meningoencephalitis (n=6), necrotizing leucoencephalitis (NLE, n=5), necrotizing meningoencephalitis (NME, n=2), canine distemper (n=1). In 3 dogs necropsy was available for comparison. The diagnosis of canine distemper encephalitis in a mixed breed dog as well as the diagnosis of NLE in a French bulldog were confirmed, whereas necropsy revealed NME in a Chihuahua, in which biopsy had shown inflammatory changes but was lacking the necrotizing component of NME.

Changes on post biopsy MRI did not require additional emergency intervention. There was no direct mortality caused by the biopsy procedure itself, but the indirect mortality was 5.9% since one dog died 2 days after the procedure due to suffocation because of a seizure during eating. The mean survival time of the other 16 dogs was 419 days (SD: 254; range: 6-1328 days). Morbidity was 35.3%. All neurological deficits caused by the biopsy procedure resolved within 3-14 days.

MIBB in suspected encephalitis leads to a specific diagnosis in the majority of dogs. Advantages of a definite diagnosis outweigh potential mortality and temporary neurological deficits.

164a

CONTROL OF URINATION IN PARAPLEGIC DOGS USING AN IMPLANTED VENTRAL ROOT STIMULATOR.

N Granger[1,2], S Mosse[3], D Chew[4], S Lacour[5], N Donaldson[3], M Craggs[6], J Fawcett[4], N Jeffery[1,7] [1] Dept. Veterinary Medicine, University of Cambridge, Cambridge, UK [2] Oncovet, Villeneuve d'Ascq, France [3] Implanted Devices Group, Dept. of Medical Physics, UCL, London, UK [4] Centre for Brain Repair, University of Cambridge, Cambridge, UK [5] Laboratory for Soft Bioelectronic Interfaces, EPFL, Lausanne, Switzerland [6] Division of Surgery and Interventional Science, UCL, London & Spinal Research Centre, RNOH, Stanmore, UK [7] Dept. Veterinary Clinical Sciences, College of Veterinary Medicine, Iowa State University, Ames, USA.

This study was conducted to determine the feasibility of providing bladder emptying control in paraplegic dogs using an implanted ventral root stimulator as a method to avoid the complications associated with other long-term urine voiding methods.

6 chronically paraplegic dogs with supra-sacral lesions were each implanted with a Brindley sacral anterior root stimulator (SARS), originally designed for use in human patients. Following a L6-S1 dorsal laminectomy, the L7 to S3 ventral roots were stimulated intra-operatively and the detrusor pressure (DP) simultaneously recorded. S2 stimulation produced the greatest increase in intra-bladder pressure in each case. Implanted stimulators consisted of a custom-made two-channel book electrode into which the S2 extra-dural roots were slotted. These were connected to a sub-cutaneous receiver that was activated by an external system controlling duration of on and off stimulation phases. The S2 dorsal roots were identified through anatomical and stimulation characteristics and sectioned intradurally to prevent reflex motor activation.

In 5 dogs, cystography showed DP above 70cmH_2O (the pressure needed to ensure good urine flow) during SARS activation immediately after surgery. All 6 dogs regained efficient bladder emptying characterized by rapid bladder emptying (<1 minute), minimal residual volume (<10% of emptied volume) and mean voided urine volume > 200mL/day by 1 month post-operatively. Long term follow-up has indicated persistent efficient bladder emptying and no urinary infections or other complications in all 6 dogs.

These data suggest that implanted Brindley SARS is a viable method to manage chronic incontinence caused by supra-sacral SCI in pet dogs.

164b

NEURONAL CELL CULTURES FOR INVESTIGATION OF THE NORMAL AND DISEASED NERVOUS SYSTEM. Thomson C[1], Griffiths I[2], Barnett S[3], McLaughlin M[3], Montague P[3], MacDonald J[1], Gilmour A[1], Clarke S[1], MacKay R[4], Mayhew I[1], Harding D[5], Kongara, K[1], Green R[6], Olson S[1]. [1] Institute of Veterinary, Animal and Biomedical Sciences, Massey University, New Zealand. [2] Glasgow, Scotland. [3]College of Medical, Veterinary and Life Sciences, University of Glasgow, Scotland. [4] Large Animal Medicine, University of Florida, Gainesville, USA. [5] Institute of Fundamental Sciences, Massey University, Palmerston North, New Zealand. [6] Graduate School of Biomedical Engineering, University of New South Wales, Australia

Cell cultures of the nervous system can be primary, using tissue harvested from the animal, or derived from cell lines. Primary cultures can be produced by culturing pieces of neural tissue (explants) or by dissociating the tissue to yield cultures of neurons, glial cells or mixed neuronal-glial cultures. In our lab we grow neuronal-glial cultures (from murine brain, spinal cord or spinal ganglia), neuronal stem cells, or cultures rich in astrocytes or oligodendocytes. We have also used cell lines such as PC12s, to assess interaction between nanomaterials and neurons, and other cell lines into which we transfect specific genes and assess their expression. We are currently working on developing a myelinating, synapsing model derived from murine forebrain.

Neurobiology and pathology studies: The mixed neuronal-glial cell cultures from the spinal cord were initially established using 120μm thick slices of fetal murine spinal cord, cultured as explants. This led to simpler procedure in which the fetal cord was dissociated. Cultures are rich in axons, myelin and synapses. The dissociated culture model has been used to investigate different aspects of neurobiology such as myelin proteins. We validated it as a model for investigating inflammatory disease by assessing the effects of cytokines on myelin. In current studies we are using it to evaluate the impact that astrocytes have on survival and myelination of axons. Astrocytes are highly variable both morphologically and functionally and their role in disease is a burgeoning area of research. Are they the Dr. Jekyll/Mr. Hyde cell of the nervous system?

Neurotoxicology studies: We have used mixed primary cultures of the CNS and spinal ganglia-Schwann cell cultures of the PNS to investigate whether the plant, *Hypochaeris radicata*, 'catsear' is the cause of Australian Stringhalt in horses. We have extracted the exudate from stressed and unstressed plants and assessed its effect on cultured neurons

Novel applications: we are using mixed cultures to assess biocompatibility of novel nanomaterials. Such work has application in neuroprosthetics.

Pros and cons of cell culture: Cell culture cannot mimic the complexity of the nervous system, but it can produce cost-effective, highly defined conditions in which to tease apart different aspects of normal biology, pathogenesis of disease and the effect of therapeutic agents.

STEREOTACTIC BRAIN BIOPSY: OUTCOMES, COMPLICATIONS AND RECOMMENDATIONS

Dr Peter. J. Dickinson. BVSc, PhD, DACVIM
Davis, CA

INTRODUCTION

Definitive diagnosis of intracranial lesions is based on histopathological examination, culture and occasionally additional analysis of tissue obtained from the lesion. In cases where surgical resection is not considered the optimal approach to treatment and diagnosis, minimally invasive biopsy techniques are considered the most appropriate way to obtain diagnostic data permitting a maximally informed approach to therapy. A variety of stereotactic approaches to brain biopsy have been described in dogs and cats, the majority of which have involved CT based systems, [1-6] however MRI based systems have also been reported recently, and are likely to become a mainstay of stereotactic biopsy in the future. [7] Both CT and MRI based systems have inherent advantages and disadvantages, and a comparison of the two modalities is not the specific subject of this review. Some data are available from relatively large case series looking at diagnostic yield and complications associated with stereotactic brain biopsy in dogs. [2,8,9] The purpose of this review is to summarize the major factors that influence successful performance and outcome following stereotactic brain biopsy, the majority of which are common to biopsy using both imaging modalities.

CASE SELECTION

Appropriate case selection has a major influence on morbidity and potential intra/post operative complications. Animals with severe neurological deficits and evidence of intracranial hypertension appear to have an increased risk of developing complications associated with both the procedure and anesthesia. Imaging data should be as current as possible to ensure appropriate targeting of rapidly progressing lesions, and as importantly, avoiding biopsy of lesions that have resolved in the period between initial imaging diagnosis and biopsy. MR based biopsy systems allow for detailed assessment of lesions at the time of biopsy, as part of the procedure. For CT based systems a decision must be made whether to re-acquire MR images prior to biopsy, particularly for poorly defined non contrast enhancing lesions. Biopsy of animals with coagulation disorders should be avoided if possible.

LESION LOCATION

Morbidity is generally lower for biopsies involving lesions involving the cranial fossa and cerebral hemispheres compared to the brainstem and cerebellum. Biopsy systems that allow for multiple trajectories such as center of arc systems allow for optimal trajectory planning to avoid ventricular structures, major vessels and sinuses.

PROCEDURAL COMPLICATIONS

Poor targeting

Poor targeting of lesions may be the result of many factors including operator error and incorrect input of target coordinates. A major influence on targeting accuracy is the length of the trajectory. It is essential that drill guides are seated firmly on the skull prior to drilling. Slippage of drills on acutely sloping bone surfaces is a common source of targeting error. Allowance must also be made for location of the biopsy port in relation to the needle tip.

Tissue collection

Most biopsy procedures are done with purpose designed side cutting brain biopsy needles. Standard human needles have a 10mm biopsy window. Availability of smaller veterinary designed needles may allow for more accurate collection of tissue, particularly with small lesions.

Prior to tissue collection, tissue is pulled into the biopsy chamber using negative pressure generated via a syringe. Excessive suction applied can result in normal tissue being pulled into the biopsy chamber, particularly at the periphery of lesions and if human needles are used. The appropriate amount of negative pressure varies depending on tumor type and operator experience is critical in obtaining high quality biopsy specimens.

Multiple biopsy samples increase the likelihood of obtaining diagnostic tissue, however there is an increased risk of complications. Some tumors , such as oligodendrogliomas will exfoliate relatively easily providing large volumes of tissue from a single sample. Intraoperative assessment of biopsy tissue is a critical step to ensure that diagnostic samples have been acquired. Rapid smear techniques using both alcohol fixed H&E and air died Wright's Giemsa stains provide complimentary information to determine whether diagnostic tissue has been collected for routine paraffin embedding, and to determine whether additional samples should be taken for culture or advanced diagnostic procedures such as PCR.

Inability to collect samples

Failure to obtain tissue from the biopsy needle may result from a highly fibrous tumor that is difficult to aspirate, or inability to generate adequate negative pressure at the biopsy port. Possible causes include compromise of the ventricular system or large cystic areas of the tumor resulting in aspiration of fluid. Rotation of the biopsy port, advancement or withdrawal of the port within the lesion (if possible), or as a last resort, using a different trajectory may be necessary to obtain a tissue sample.

Post biopsy procedures

Appropriate needle placement should be confirmed prior to biopsy, and imaging ideally should be done post biopsy to determine whether there is significant hemorrhage that may require further intervention.

MAJOR COMPLICATIONS

Overall reported complications rates are relatively low and generally less than 10%, however this is dramatically influenced by the animals clinical/neurological status and the lesion location.

Hemorrhage

Hemorrhage is one of the most common and potentially life threatening complications encountered during and after the biopsy procedure. Hemorrhage may be either intratumoral/intraparenchymal or subdural. Subdural hemorrhage usually results from inappropriate placement of the biopsy needle causing the dura to be pushed away from the skull resulting in tearing of blood vessels. The dura should always be penetrated with a sharp object (eg spinal needle) prior to placement of the blunt tipped biopsy needle. Intraparenchymal hemorrhage is often self limiting, however subdural bleeding may be extensive requiring surgical decompression in some cases.

Pneumocephalus

Presence of air within the brain parenchyma is a common finding post biopsy, presumably due to leakage down the needle outer cannula during biopsy collection. This does not appear to be a major cause of morbidity in most circumstances.

Infection

Post biopsy infection is uncommon in the author's experience, even when needle trajectories involve the frontal sinuses which requires drilling through both the inner and outer tables.

Tumor seeding

Although uncommon, biopsy tracts may act as pathways for tumor migration, and this should be considered when planning subsequent therapeutic plans such as radiation therapy.

Non diagnostic biopsy

Achieving a diagnosis following stereotactic biopsy is dependant on a multitude of factors. Appropriate biopsy procedure, with accurate targeting, choice of appropriate biopsy location, and collection of high quality samples is critical. Intraoperative assessment of tissue to determine biopsy quality and necessity for additional samples greatly increases the likelihood of a diagnosis. Inappropriate processing of samples can also decrease diagnostic yield. Most samples are small, and many, particularly from oligodendrogliomas, can be gelatinous in nature and easily lost during routine embedding. Agar embedding of small samples can ensure that valuable tissue is not lost following collection. Interpretation of both alcohol fixed and air dried smears, as well as small paraffin embedded sections requires specific skills and learning curves are steep.

In most cases where lesions are neoplastic, it is relatively easy to determine whether sufficient material has been collected based on intraoperative smears, and a specific diagnosis can be made (often from the smear preparation). Due to the small sample size tumor grading can be problematic and undergrading of tumors is a significant factor. Most diagnostic problems arise with inflammatory, infectious and degenerative processes where pathological findings can be non specific, and specific diagnoses are difficult based on limited tissue. Similar issues may arise with poor lesion targeting when peritumoral or necrotic tissue is collected. Intraoperative assessment of samples and collection of multiple samples from different areas can increase diagnostic yield in these cases.

REFERENCES

[1]Koblik P, et al. **Vet Rad Ultrasound** 1999;40(4):424-433. [2]Koblik P, et al. **Vet Rad Ultrasound** 1999;40(4):434-440. [3]Moissonnier P, et al. **Res Vet Sci** 2000;68:243-247. [4]Giroux A, et al. **Vet Rad Ultrasound** 2002;43(3):229-236. [5]Troxel M, et al. **Vet Rad Ultrasound** 2008;49(5):438-443. [6]Packer R, et al. **AJVR** 2011;72:109-121. [7]Chen A, et al. Vet Rad Ultrasound 2011;Epub. [8]Vernau K, et al. **Vet Pathol** 2001;38:47-57. [9]Moissonnier P, et al. JSAP 2002;43:115-123.

STRUCTURAL BRAIN PATHOLOGY IN CANINE PRIMARY EPILEPSY: DOES IT EXIST?

Marjorie E. Milne, BVSc (Hons), FANZCVS (Radiology), Garry Anderson, BAgSci, GradDipApplStats, MVSc, Kathleen E. Chow, BVSc (Hons), Dayle Tyrrell, BVSc MVSc, FANZCVS (Radiology), Terence J. O'Brien, MBBS, Melb, MD Melb, FRACP, Bradford Moffat BSc (Hons), MApp, Sci (Med. Phys), PhD, Sam N. Long, BVSc, PhD, DECVN

Melbourne, Victoria, Australia

INTRODUCTION

It is difficult to obtain an accurate estimate of the prevalence of primary ('idiopathic') canine epilepsy, but review of the literature indicates that dogs with 'primary' epilepsy represent approximately half of all accessions of seizing dogs to veterinary hospitals[1-3]. Review of the literature is further complicated by inconsistent case definition. In the past, the most common form of epilepsy was considered to be 'idiopathic' epilepsy, suspected in patients with generalized onset seizures, a young age of seizure onset (<5 years[3]), normal imaging findings and an absence of inter-ictal signs. Focal onset (or partial) seizures were considered uncommon, and indicative of underlying structural pathology. Careful questioning of the owner regarding the signs preceding generalized tonic clonic seizures, the use of visual recordings and EEG, have shown that focal onset seizures (with or without secondary generalization) are more common than previously thought. This highlights the difficulty in interpretation of some of the earlier case series for canine epilepsy, where diagnosis was often based on clinical presentation and inconsistent application of EEG or advanced brain imaging.

Treatment of dogs with 'primary' epilepsy has variable success. Medical management is most commonly used; although a cornucopia of anti-epileptic drugs are available, the most commonly used first-line anti-epileptic drugs remain phenobarbital, potassium bromide and benzodiazepines. Despite this, many dogs do not achieve satisfactory seizure control using medication and are regarded as having drug resistant epilepsy. Surgical options for treatment of dogs with drug resistant 'primary' epilepsy are currently limited. Transection of the corpus callosum[4] has been described, but is not widely in use.

In human epileptology, advances in brain Magnetic Resonance Imaging (MRI) techniques have led to the recognition of subtle structural pathologies in patients with focal onset epilepsies that would have previously been described as 'idiopathic' in etiology. The most common of these is hippocampal sclerosis in patients with temporal lobe epilepsy, but focal cortical dysplasia (FCD) and malformations of cortical development (MCD) are also recognized. Surgical resection of the epileptogenic lesion is widely practiced, and results in improved seizure control or elimination of seizures in many patients[5,6]. If similar epileptogenic lesions could be identified in dogs with drug resistant epilepsy, then surgical excision of the lesion may be an alternative to euthanasia of these patients.

In human epileptology, there is a move away from the ambiguous etiologic classification of 'idiopathic', 'symptomatic', and 'cryptogenic' epilepsies, with the International League Against Epilepsy (ILAE) proposing the classifications 'structural-metabolic', 'genetic', and 'unknown' epilepsies[7]. In veterinary medicine, a genetic basis for canine epilepsy has been identified in the miniature wire-haired Dachshund[8], and Belgian Shepherd[9]. If the new classification scheme is applied to the veterinary field, many canine epilepsies previously described as 'idiopathic' or 'cryptogenic' should now be regarded as epilepsies of unknown origin. This re-classification clearly acknowledges that little is known about the etiology of many canine epilepsies, and that structural-metabolic or genetic causes may be present but are not yet identified using currently employed diagnostic techniques.

MORPHOMETRIC ANALYSIS OF THE CANINE BRAIN

Our research group has been working on novel approaches to brain MRI scans in the search for subtle structural pathology in dogs with epilepsy of unknown origin.

Hippocampal Formation Volumetry

Using a 1.5 Tesla GE magnet, the epilepsy screening protocol includes a T1 weighted 3D FAST SPGR sequence, acquired in an oblique dorsal plane aligned perpendicular to the long axis of the hippocampal formation (HF). The HF is a complex three-dimensional structure consisting of the dentate gyrus, hippocampus, and subiculum. The HF is divided into the cranioventally located head, the body, and caudodorsally located tail. A technique for HF volumetry using a manual segmentation of the T1W 3D sequence was developed. The margins of the hippocampal head can be difficult to separate from the amygdala, and the tail can be difficult to separate from the thalamus. Delineation of boundaries is improved by reference to sagittal and transverse plane images. Manual segmentation had excellent intra-observer reproducibility with a Lin's Concordance Correlation coefficient of 0.97 (95%CI 0.94 to 0.99). Inter-observer repeatability was lower with a Lin's Concordance Correlation coefficient of 0.63 (95%CI 0.37 to 0.79), due to a tendency of the second observer to underestimate volume by not including all of the HF head.

The positive relationship between canine HF volume and body weight is accounted for by normalizing hippocampal volume to intracranial volume. Based on a sample population of 20 non-epileptic dogs without MRI identified structural brain disease, a lower 95% reference interval for normalized HF volume were established as 0.56 cm^3 (90%CI 0.52 to 0.60 cm^3) for the right HF, and 0.55 cm^3 (90%CI 0.52 to 0.58 cm^3) for the left HF.

Evaluation of epileptic dogs for hippocampal sclerosis

Hippocampal sclerosis (HS) is characterized histologically by cell loss and gliosis of the hippocampal formation, and is the most common pathology identified in human patients with drug resistant mesial temporal lobe epilepsy. Numerous case reports document the existence of hippocampal pathology in dead epileptic dogs, but a diagnosis of hippocampal sclerosis is currently a post-mortem one, and the significance of HS in canine epileptics is controversial[10,11]. In humans, antemortem diagnosis of HS utilizes the detection of hippocampal atrophy[12] using volumetric techniques or subjective detection of unilateral volume reduction, and detection of signal and structural changes within the hippocampal formation[13]. The MRI scans of 30 dogs with epilepsy of unknown cause were evaluated by two reviewers, using HF volumetry via the manual segmentation method and subjective analysis of the HF for structural and signal changes. The MRI scans were interspersed with MRI scans of non-epileptic dogs, and the reviewers were blinded to epilepsy status of each case. Reviewers evaluated the scans separately, and disparate findings were resolved by consensus. Using the previously established lower 95% reference interval for normalized HF volume, none of the 30 patients had a reduction in normalized HF volume. By subjective analysis, a mild reduction in unilateral HF volume at the level of the body was noted in 6/30 epileptic dogs and 5/18 non-epileptic dogs. A mild focal reduction in T1W signal intensity was noted in 3/30 epileptic dogs and 1/18 non-epileptic dogs. These mild changes to the HF did not appear to be useful in distinguishing between epileptic and non-epileptic groups.

Evaluation of epileptic dogs for Focal Cortical Dysplasia and other Malformations of Cortical Development

The MRI scans of the same 30 dogs with epilepsy of unknown cause were also scrutinized for evidence of focal cortical dysplasia or malformation of cortical development, using a detailed questionnaire. Focal cortical dysplasia may appear as focal thickening of cortical gray matter, poor gray-white matter differentiation, abnormal gyration, wide and deep sulci, and subcortical bands of gray matter traversing the white matter to connect cerebral cortex to the margin of the lateral ventricle. Malformations of cortical development included nodular heterotopia, band heterotopia, microlissencephaly, hemimegalencephaly, polymicrogyria and schizencephaly. By subjective evaluation, one of the epileptic dogs had mild widening of the left middle suprasylvian sulcus, but no further changes associated with Focal Cortical Dysplasia were identified. The remaining epileptic dogs did not have evidence of Focal Cortical Dysplasia or Malformations of Cortical Development.

THE FUTURE OF MORPHOMETRIC ANALYSIS OF THE CANINE BRAIN

The lack of evidence of structural brain disease in this cohort of epileptic dogs indicates that structural pathology is probably uncommon in dogs with epilepsy of unknown cause. However the limitations of this study must be kept in mind. Despite the use of a high resolution volumetric scan, at magnetic field strengths of 1.5 Tesla the inherent gray-white matter contrast of the canine brain is still poorer than that of humans, and the relatively smaller canine brain size reduces the available signal to noise ratio and spatial resolution. It is possible that high resolution scans performed at 3T may reveal abnormalities not visible at 1.5T. Restricting patient recruitment to those with drug resistant epilepsy and EEG confirmed focal onset epileptiform discharges will target evaluation to those patients who are more likely to have focal structural disease. An evaluation of human patients having an MRI for their first seizure found epileptogenic structural lesions in only 14% (38/263), while evaluation of patients having an MRI for drug resistant epilepsy identified lesions in 86% (97/113) of patients[14]. Finally, computer-assisted voxel-based morphometry would enable automatic segmentation of the hippocampal formation and other cortical structures, including quantification of cortical thickness and quantification of voxel intensity differences at the gray-white matter junction. These techniques are relatively new research tools in human neuroimaging, and are able to identify subtle cortical dysplasia on MRI scans that were previously thought to be 'normal'[15]. The development of a canine brain MRI atlas for image co-registration will further our knowledge of morphologic abnormalities in canine epileptic patients.

REFERENCES

[1]Zimmermann R, et al. **J Vet Intern Med** 2009; 23(5):970. [2]Pakozdy A, et al. **Acta Vet Hung** 2008; 56(4):471. [3]Podell M, et al. **J Am Vet Med Assoc** 1995; 206(11):1721. [4]Bagley RS, et al. **Vet Surg** 1995; 24(2):122. [5]Wiebe S, et al. **N Engl J Med** 2001; 345(5):311. [6]Fauser S, et al. **Brain** 2004; 127(Pt 11):2406. [7]Berg AT, et al. **Epilepsia** 2010; 51(4):676. [8]Lohi H, et al. **Science** 2005; 307(5706):81. [9]Oberbauer AM, et al. **BMC Genet** 2010; 11:35. [10]Buckmaster PS, et al. **J Vet Intern Med** 2002; 16(1):95. [11]Poncelet L, **Vet Rec** 2011; 169(14):359. [12]Cascino GD, **Magn Reson Imaging** 1995; 13(8):1133. [13]Jackson GD, **Magn Reson Imaging** 1995; 13(8):1081. [14]Vattipally VR, et al. **Neuroimaging Clin N Am** 2004; 14(3):349. [15]Bernasconi A, et al. **Nat Rev Neurol** 2011; 7(2):99.

ADVANCES IN THE PATHOGENESIS AND TREATMENT OF SPINAL CORD INJURY

Nicolas Granger, DVM, MRCVS, DECVN
Villeneuve d'Ascq, France

INTRODUCTION

Acute spinal cord injury (SCI) is primarily caused by *contusion*, and in many instances it is accompanied by persistent *compression*[1]. In dogs, intervertebral disc herniation is the main cause of SCI (accounting for 34% of the neurological cases in certain hospitals[2]) followed by road traffic accidents. The pathophysiology of acute SCI follows a biphasic pattern comprising primary and secondary pathological events that have been reviewed in detail elsewhere[1,3,4] and are briefly presented below.

PRIMARY AND SECONDARY DAMAGES

Primary damages correspond to physical disruption of cell membranes leading to ischemia and haemorrhage. Direct prevention of primary damages logically consists in surgical alignment and stabilisation of the spine and spinal cord (SC) decompression. There is strong pre-clinical evidence that early intervention is beneficial in SCI animal models[5]. In people, early intervention (i.e. between 24 and 72 hours, and ideally within 8 hours) is also recommended since several studies have detected neurological and functional benefits, without compromising the safety of the patient[5]. At the time of the surgery, physical interventions such as durotomy, CSF drainage and systemic hypothermia have been proposed, in order to prevent secondary damages. Durotomy at the time of SC decompression does not seem to add advantages[6]. CSF drainage is safe in human SCI[7] and might form a useful approach that probably needs further attention in dogs. Systemic hypothermia is gaining popularity for SCI treatment in people. There are pre-clinical data demonstrating behavioural benefits in rodent models[8], with the technical challenge to drop and maintain the body temperature between 30 and 34°C. In all cases, it is very important to mention that early surgical intervention must be accompanied by close monitoring of arterial blood pressure that should be maintained at all time with conventional therapies[4]. Secondary damages that further destroy the SC parenchyma split into: i/ acute changes[4] (vascular changes, ionic perturbances, excitotoxicity, free radical production and inflammation) that can be tackled by systemic administration of pharmacologic neuroprotective treatments[8]; ii/ and chronic changes[9] (resulting in cystic cavitations, the presence of CNS myelin debris containing inhibitory molecules, the glial scar, cut axons and demyelinated axons) that can be reversed by biological therapeutic interventions[10] (i.e. direct application of the treatment around or in the SC rather than systemic delivery) or cell therapies that will help to 'repair' or replace the damaged tissue.

MECHANISMS OF RECOVERY

Following the injury, three distinct mechanisms could in theory play a role in the recovery[11]: i/ compensation (i.e. adaptation to the new neurological status achieved by developing new movement strategies); ii/ plasticity of the SC below the lesion (i.e. reorganisation of spinal circuits under the influence of spared spinal tracts); iii/ repair. The recovery can be extensive in cases with incomplete SCI, via compensatory mechanisms and neural plasticity, which can be exploited by task-specific training. With complete SCI, repair mechanisms of the central nervous system are limited and axonal regrowth through the SC lesion inevitably fails. The recovery is slow, incomplete and mainly occurs within the first two to three months following the injury and tends to reach a plateau one year after injury in people[12] and around three months in dogs. A large number of therapies have been proposed, both in the acute and chronic phases[8,10,13] of injury, but currently, no effective treatment has emerged that can 'repair' the damaged SC. However, several therapies merit attention, evidenced by a growing number of clinical trials that have started both in dogs and people with SCI.

EARLY PHARMACOLOGIC NEUROPROTECTIVE INTERVENTION

Many compounds have been tested but it seems logical to focus here on drugs that can be administered intravenously and are already available, for example because they are used in other indications. As such, non-steroidal anti-inflammatory drugs may regain popularity, for example flufenamic acid. Indeed, a recent study has demonstrated the importance of a gene called *Trpm4* that is upregulated in the endothelium of SC blood vessels following SCI[14]. Inhibition of its metabolic pathway, which can be done with flufenamic acid, lead to reduction of haemorrhage and lesion size in a rodent model. Ibuprofen is another non-steroidal anti-inflammatory drug that may not only play an anti-inflammatory role but may also promote axonal sprouting[8]. Dosages and toxicity will need evaluation in dogs first. Riluzole, a sodium channel antagonist, and minocycline, a metalloproteinase inhibitor, have moved from the laboratory to phase I safety trials in humans[15]. Polyethylene glycol, that has the potential to seal damaged membranes and prevent ion influx, has been tested in dogs and is safe[16]. Further evaluation in a randomised and blinded fashion is ongoing[4], along with methylprednisolone sodium succinate. Finally, dimephosphon, a low-toxic organophosphorus compound, has been recently tested in dogs with experimentally induced SCI and shown to produce electrophysiological and histopathological benefits[17].

BIOLOGICAL THERAPEUTIC INTERVENTIONS

Three strategies aimed at promoting axonal growth have emerged and are likely to play an important role for SCI repair: i/ the degradation of inhibitory chondroitin sulfate proteoglycans (released by reactive astrocytes of the glial scar) with chondroitinase ABC(ChABC)[18,19]; ii/ the neutralization of CNS myelin debris (containing inhibitory molecules such as Nogo, myelin associated glycoprotein, semaphorin 4D and ephrin B3) primarily with anti-Nogo antibodies[20]; iii/ the inhibition of Rho activation[21]. The Rho-antagonist (trademarked as Cethrin®) and the antibody directed against Nogo (known as ATI355) have been tested in a phase I trial in humans for the first one[21] and is currently completing a phase I trial for the second one. ChABC has now been largely investigated in rodent models of SCI and is probably viewed by some neuroscientists as the most promising treatment for SCI[18,19].

CELL TRANSPLANTATION

Repairing the damaged SC can be achieved by providing a substrate or scaffold to the axons, most likely in the form of a cell transplant. A variety of cells (stem cells from the bone marrow, adipose tissue, umbilical cord blood cells, macrophages, oligodendrocyte progenitors, etc…) have been proposed for therapeutic intervention[13] and are currently being tested in 11 phase I trials throughout the world in people. However, no cell therapy so far has successfully completed a phase II trial, demonstrating the need for further proof of efficacy before translation to human patients. Amongst the various cell therapies that are currently developed for SCI repair, olfactory ensheathing cells (OECs) constitute a unique category of glial cells for which proof of efficacy in laboratory rodents is fairly well accepted[22]. Olfactory ensheathing cells are currently being tested in human patients in Poland in an open label non-randomised safety trial and previous trials, although on a limited number of patients, have demonstrated the safety of OEC transplantation[22]. An advantage of OECs is their differentiated state, as opposed to stem cells or embryonic cells, but wide range of beneficial properties when transplanted into the damaged SC. Indeed, OECs can: i/ promote and guide axonal growth through the lesion (glial scar); ii/ adopt a myelin ting phenotype similar to myelin ting Schwann cells; iii/ secrete a range of growth factors that can favour axonal growth. OECs are currently tested in pet dogs with chronic severe SCI[23]. Very recently in dogs, a report evaluated the potential of antilogous bone marrow stoma cells injected in the CSF[24]. Significant clinical improvement was found. However, only the motor function was evaluated and all dogs in this report remained incontinent.

COMBINATORIAL THERAPIES

The complexity of SCI is such that is it unlikely that a single treatment approach will repair the damaged SC. There is a clear need for a combinatorial therapy that will comprise early surgical intervention to prevent primary damages and early pharmacological intervention to limit the cascade of secondary injuries. For cases that will not recover, administration of a biological agent and/or a cellular transplant will be further necessary. Recently, the need for adjunctive rehabilitation to pharmacological treatments has been clearly demonstrated. For example, ChABC in combination with rehabilitation has been shown to promote functional recovery in acute SCI and has to be given in a narrow time window (around 4 weeks), at the time when rehabilitation usually starts (i.e. 2– 4 weeks after injury)[18]. These important findings highlight the facts that over the past decades, the medical advances that have had a direct impact on people living with SCI are primarily 'compensatory' therapies together with the management of complications such as urinary tract infections and should be associated with novel therapies. This is likely the same for companion animals. Firstly, physical rehabilitation forms the cornerstone of treatment for SCI patients[25]. Intense treadmill training with body weight support is given to encourage SC plasticity and is beneficial in patients with incomplete SCI[26]. Dogs can be trained in a similar way. Secondly, neuroprosthetic implants, by providing functional electrical stimulation of peripheral nerves, have had a tremendous impact on the patients with SCI. When placed on phoenix or sacral nerves, they can restore autonomous breathing (leaving the person free of a ventilator) or bladder emptying[27]. In dogs, neuroprosthetic is likely to play a role as well and sacral nerve root electrical stimulation has been recently tested in dogs with chronic SCI for bladder emptying.

REFERENCES

[1]Sekhon LH, et al. **Spine** 2001;26,S2-12.[2]Fluehmann G, et al. **J Small Anim Pract** 2006;47(10):582-7.[3]Rowland JW, et al. **Neurosurg Focus** 2008;25, E2. [4]Olby N, **Vet Clin North Am Small Anim Pract** 2010;40(5):791-807.[5]Furlan JC, et al. **J Neurotrauma** 2011;28(8):1371-99.[6]Loughin CA, et al. **Vet Comp Orthop Traumatol** 2005;18(3):141-6.[7]Kwon BK, et al. **J Neurosurg Spine** 2009;10(3):181-93.[8]Kwon BK, et al. **J Neurotrauma** 2011;28(8):1545-88.[9]Barnett SC, et al. **Nat Clin Pract Neurol** 2007;(3)152-61.[10]Kwon BK, et al. **J Neurotrauma** 2011;28(8):1589-610.[11] Curt A, et al. **J Neurotrauma** 2008;25(6):677-85.[12]Fawcett JM, et al. **Spinal Cord** 2007;45(3):190-205.[13]Tetzlaff W, et al. **J Neurotrauma** 2011;28(8):1611-82.[14]Gerzanich V, et al. **Nat Med** 2009;15(2):185-91.[15]Cadotte DW, et al. **Clin Orthop Relat Res** 2011;469(3):732-41.[16]Laverty PH, et al. **J Neurotrauma** 2004;21(12):1767-77.[17]Tumakaev RF, et al. **Eksp Klin Farmakol** 2010;73(5):41-4.[18]Wang D, et al. **J Neurosci** 2011;31(25):9332-44.[19]García-Alías G, et al. **Nat Neurosci** 2009;12(9):1145-51.[20]Maier IC, et al. **Philos Trans R Soc Lond B Biol Sci** 2006;361(1473):1611-34.[21]Fehlings MG, et al. **J Neurotrauma** 2011;28(5):787-96.[22]Su Z, et al. **Prog Neurobiol** 2010;92(4):517-532.[23]Jeffery ND, et al. **J Neurotrauma** 2005;22(11):1282-93.[24]Nishida H, et al. **Am J Vet Res** 2011;72(8):1118-23.[25]Wolfe DL, et al. in **Spinal Cord Injury Rehabilitation Evidence**, 010;3.1-3.44.[26]Dietz V, et al. **Brain Res Bull** 2009;78(1):I-VI.[27]Hamid S, et al. **Eur Spine J** 008;17(9)1256-69.

INTRAOPERATIVE ULTRASOUND FOR INTRACRANIAL SURGERY

A. Shores, DVM, MS, PhD, DACVIM (Neurology), M. Holland, DVM, MS, DACVR, J. Hudson, DVM, PhD, DACVR, E. Akin, DVM, M. Beasley, DVM, MS, N. Ortinau, DVM, J. Rich, DVM, E. Brinkman-Ferguson, DVM, DACVR, E. Baravik-Munsell, DVM, J. Haller, DVM, M. Thomas, DVM, MS, DACVR

Mississippi State, MS

Intraoperative imaging guidance is a rapidly advancing modality for veterinary intracranial surgery. We have utilized intraoperative ultrasound as our guidance modality in 25 patients over the course of four years. The purpose of this study was to present our initial experiences using intraoperative ultrasound imaging for intraaxial and extraaxial intracranial masses. This presentation includes details on the technique, instrumentation, intraoperative imaging characteristics, and its correlation with postoperative CT or MR imaging.

Twenty-five canine patients were presented to either the Mississippi State University Veterinary Neurology and Imaging (MSU-VNI) service or the Auburn University Small Animal Clinic (AUSAC) for neurologic disorders that were ultimately diagnosed as intracranial masses. Each of the 25 patients underwent a pre-operative evaluation consisting of routine blood and urine analyses and thoracic radiography. Some patients were evaluated with cardiac or abdominal radiography or both. Results of the pre-operative evaluations revealed no evidence of co-existing or metastatic neoplasia. Each of the patients underwent craniotomy or craniectomy procedures for removal / debulking of the intracranial mass using intraoperative ultrasound guided visualization of the mass. Mass removable was accomplished using an ultrasonic aspirator (CUSA Excel – MSU-VNI; CUSA NS200 – AUSAC). Twenty-four of the 25 patients survived surgery and 17 were eventually discharged to the owners' care. All 25 intracranial masses were characterized by microscopic evaluation, demonstrating the following distribution of tissue types: meningioma (13), glioma (10), nasal adenocarcinoma (1), cerebellar cyst (1). Twenty-two lesions were supratentorial and 3 were infratentorial.

In all patients, the intracranial mass was identified intraoperatively after craniectomy and durotomy using a 5 to 8 MHz probe enveloped in a sterile sleeve containing sterile ultrasound lubricant. The covered probe was placed directly on the cerebral, cerebellar, or brainstem tissue to visualize the mass after flooding the craniectomy site with sterile saline. The intracranial lesions were imaged in both sagittal and transverse planes. Most of each intracranial neoplasm appeared distinctly hyperechoic. The distinct imaging characteristics of the ventricles, falx cerebri, gray matter, and white matter were easily differentiated from the generally more heterogeneous hyperechoic intracranial tumors. Cystic areas of the tumors and the one cystic mass in this study were contrasted from surrounding structures by their hypoechogenicity. Size and shapes of the masses approximated the same parameters identified on advanced imaging (MRI, CT). Non-cystic portions of the meningiomas appeared most hyperechoic. The gliomas often were less heterogeneously hyperechoic than the meningiomas. Vascular supply to the intracranial tumors was detected by color Doppler and aided in directing the approach to the masses using the ultrasonic aspirator.

Correlating the preoperative MRI or CT images with the intraoperative ultrasound images was very valuable to the surgeons in gaining a perspective of the size of the mass, the proximity of ventricles, the eloquent areas of the brain, and the location of larger blood vessels. In addition, the serial images consistently gave the surgeons a definitive perspective on the completeness of the resection and helped differentiate grossly abnormal from grossly normal structures. Post-operative imaging correlated well with ultrasound imaging at the conclusion of the surgeries.

The authors conclude that the use of intraoperative ultrasound imaging is a valuable adjunct and correlates well with both preoperative and postoperative advanced imaging. Intraoperative ultrasound should be a standard imaging modality for intracranial procedures. Additional development of this technique, including the use of 3D ultrasound is warranted. Intraoperative ultrasonography is helpful in localizing and defining the margins of intracranial masses and accurately determines the extent of resection, as confirmed by postoperative MR or CT imaging.

REFERENCES

Gerganov V, et al, J Clin Neurosci, 2011, 18, 669. Unsgaard G, et al, Acta Neurochir 2005, 148, 235. Chen S, et al, Tz Chi Med J, 2004, 16, 85. Woydt M, et al, Acta Neurochir, 1996, 138, 1391. Hammoud M, et al, J Neurosurg, 1996, 84, 737.

217b

REPEATABILITY OF A MODIFIED DORSAL TECHNIQUE FOR ELECTRONIC VON FREY ANESTHESIOMETRY TO ASSESS SENSORY THRESHOLD IN SIX NORMAL DOGS. SA Moore, BF Hettlich. The Ohio State University, College of Veterinary Medicine, Columbus, OH.

Dogs are a powerful spontaneous model through which to evaluate interventions that improve sensory and motor outcomes after spinal cord injury (SCI). While much attention has been paid to assessing motor recovery for the purpose of clinical trials, little focus has been applied to quantitative sensory assessment in dogs. Here we examined the utility of a common sensory assessment used in rodent SCI research - electronic von Frey anesthesiometry (VFA). Six client-owned apparently normal dogs of various sizes, breeds and ages were evaluated. A modified von Frey technique was applied to the dorsolateral surface of the metatarsus, in the cutaneous autonomous zone for the Peroneal nerve. The amount of force required to elicit a behavioral response was termed the "sensory threshold" and was recorded for each dog. Mean sensory threshold values were 154.7 g $\pm$ 67.2 and 155.8 g $\pm$ 37.7 for the right and left pelvic limb, respectively. Because there was a positive correlation between sensory threshold and body weight ($r= 0.94$, $p= 0.005$), sensory threshold values were scaled to weight. This produced normal sensory threshold values of 12.9 g $\pm$ 7.2 and 12.5 g $\pm$ 5.2 for the right and left pelvic limbs. The same 6 dogs were evaluated on an additional occasion one week later, and sensory threshold values were equivalent to initial findings. This dorsal technique for VFA represents a viable means to quantify sensory threshold in normal dogs, and may represent an important additional outcome measure in clinical trials involving diseases affecting sensory threshold, such as SCI.

218a

VALIDATION OF NEUROPSYCHOLOGICAL TEST PROTOCOLS FOR COGNITIVE ASSESSMENT OF SENIOR DOGS AND CATS.

Milgram NW,[1] Yuanlong Pan[2] Landsberg GM[1,3] Araujo JA,[4] De Rivera C[5]

1. CanCog Technologies Inc., Toronto, ON 2, Nestlé Purina Research, Saint Louis, MO 3. North Toronto Animal Clinic, Thornhill, ON 4. Intervivo Solutions, Toronto, ON 5. Vivocore Inc., Toronto, ON

We have previously developed a series of neuropsychological tests for assessment of cognitive function in dogs, and validated these tests by demonstrating both age-dependent cognitive deficits and validation studies using diets, drugs and supplements. This presentation describes a a series of neuropsychological tests that we have developed for use in cats.

Subjects were cats ranging in age from 3-15 years. Cats were trained on a series of tests using both a t-maze and a "Feline Cognitive Test Apparatus" (FCTA), which consisted of a smaller version of the apparatus we've previously used in dogs. With cats, we obtained the following results:

Simple procedural learning did not show significant age sensitivity in cats. By contrast, cats, like dogs, showed age-dependent decline in reversal learning, which included positional reversal learning using a t-maze, and object discrimination reversal learning using the FCTA (Fischer LSD test; $p=0.035$), and in an egocentric discrimination reversal task (Analysis of variance; $p=0.039$). We also found significant age differences in an object discrimination learning task (Fisher LSD test; $p=0.04$), allocentric spatial learning (Landmark test; $p=0.04$) in acquisition of a spatial memory task (delayed-non-matching to position tasks; DNMP; $p=0.0002$). Cats, unlike dogs, were not able to learn a more complex three component DNMP; nor were they able to learn an object recognition task (Delayed-Non-Matching-To-Sample).

We have also shown that these tasks can be used to evaluate cognitive-modifying capabilities of therapeutics. Canine cognition has been positively impacted by dietary modification with antioxidants, MCTs, and a calcium channel modulators. We have also demonstrated beneficial effects of S-adenosylmethionine (Novifit) in both aged dogs and cats.

These findings provide validation for the use of a series of neuropsychological tests to assess cognitive function in dogs and cats and demonstrate that these tests can be used to assess therapeutics for treating CDS in both species.

218b

PROTEIN DISULPHIDE ISOMERASE COLOCALISATION WITH SUPEROXIDE DISMUTASE 1 IN CANINE DEGENERATIVE MYELOPATHY: EVIDENCE FOR ENDOPLASMIC RETICULUM STRESS. Long, S[1]. Chang, R[1]. Walker AK, Orian J[2], Atkin J[2]
[1]University of Melbourne Faculty of Veterinary Science, Melbourne, Australia; [2]La Trobe Institute for Molecular Science, LaTrobe University, Melbourne, Australia.

Canine Degenerative Myelopathy (DM) is a degenerative neurological condition which affects several dog breeds including German Shepherds, corgis, Chesapeake Bay retrievers, Boxers and Rhodesian Ridgebacks. Genome wide association mapping has revealed an autosomal recessive mutation within the Superoxide Dismutase 1 (SOD1) gene in affected dogs. Immunohistochemical staining has also identified SOD1-positive inclusions in the cytoplasm of motor neurons of dogs with DM. Mutations within the SOD1 gene and SOD1-positive inclusions in motor neurons are also found in familial cases of Amytrophic Lateral Sclerosis (ALS) in people, suggesting that DM is a potential large animal model of ALS. Recent studies in ALS have implicated endoplasmic reticulum (ER) stress and the unfolded protein response (UPR) in the pathogenesis of both sporadic and familial ALS cases. Studies have also shown that the enzyme protein disulphide isomerase (PDI) plays an important protective role in cells in ALS. Given the genetic similarity between familial SOD1-associated ALS and DM, we investigated whether ER stress and the UPR may also play a role in the development of canine DM.

Six dogs with DM and four age-matched control dogs were used for this study. Tissue taken from the thoracolumbar region of spinal cords examined post mortem were used for fluorescent immunohistochemcal staining of PDI and SOD1 and for western blotting to analyse the presence of the following proteins associated with ER stress and the UPR: BiP, PDI, CHOP, XBP1, and IRE1α. Quantitative densitometry was used to quantify the degree of expression of each protein.

Immunohistochemical staining revealed that 100% of SOD1-positive neurons were positive for PDI staining in affected dogs, and that PDI and SOD1 colocalised in a granular pattern similar to that reported in human ALS patients and SOD1 mouse models. Western blotting showed upregulation of all proteins: BiP, PDI, CHOP, XBP1, and IRE1α, with the degree of increase being in the order of two- to fourfold. This increase was statistically significant for CHOP, BiP and PDI. These results were similar to those reported in humans.

The colocalisation of PDI and SOD1 together with the upregulation of other proteins show that ER stress and the UPR are common to both human ALS (sporadic and familial) and canine DM. Therefore canine DM appears to share the same pathogenesis as ALS at a cellular level and thus provides a valuable spontaneous large animal model of the human disease.

219a

EVALUATION OF A POTENTIAL CANDIDATE GENE mGLuR4 FOR NECROTIZING MENINGOENCEPHALITIS IN DOGS. KM Vernau, EA Brown, PJ Dickinson, H Liu, RJ Higgins, BK Sturges, NC Pedersen, DL Bannasch. University of California – Davis, Davis, CA.

Glutamate is the major excitatory neurotransmitter in the CNS, which may also play a role in neuroinflammation. Glutamate levels are elevated in the brain of people with multiple sclerosis. Metabotropic glutamate receptors (mGluRs), are family of G-protein coupled receptors with 8 subtypes. Knockout mice without the mGluR4 subtype develop an earlier clinical onset and more severe form of experimental allergic encephalitis.

A whole genome scan was done using the Illumina canine HD arrays on 12 Pug Dogs with histopathologically confirmed necrotizing meningoencephalitis (NME) and 11 control Pug Dogs. An association was found across a large region on Cfa 12 at 5.28-8.83 Mb. The mGLuR4 gene is located within this interval at 6.34-6.45 Mb. We hypothesize that dogs with NME have a mutation in the GluR4 gene compared to control dogs.

DNA from 1 Pug Dog with NME, 1 Chihuahua with NME, 1 unaffected Pug Dog and an unaffected Dalmatian was examined for candidate mutations in the mGluR4 gene. Each of the 10 exons of mGluR4 was amplified and sequenced, including the start and stop codons including the regulatory regions, to evaluate the genomic DNA for any possible mutations. Semi-q RT-PCR was then performed on 12 Pug Dogs, 2 Chihuahuas, and 1 Maltese Terrier with histologically confirmed NME, and 7 healthy Pug Dogs, 5 Chihuahuas, 8 Maltese Terriers, 1 Dalmatian, 3 German Shepherd Dogs, 2 Labrador Retrievers, 1 German Shorthaired Pointer and 9 Yorkshire Terriers to evaluate the cDNA for differences in expression of mGLUR4 in the brain tissue of affected and unaffected dogs.

Homozygous mutations in the mGluR4 gene were found 11/12 affected Pug Dogs and none of the control Pug Dogs. However, homozygous mutations were found in control Maltese Terriers and Yorkshire Terriers, as well as in one affected Maltese Terrier, but none in affected Chihuahuas. Heterozygous mutations were found in all but one of the control breeds of dogs evaluated. No difference in RNA expression was noted between control and affected dogs. It is therefore unlikely that mutations in the mGluR4 gene are associated with NME in dogs.

219b

MATRIX METALLOPROTEINASE-2 AND -9 IN CANINE CENTRAL NERVOUS SYSTEM TUMORS CL Mariani. Comparative Neuroimmunology & Neurooncology Laboratory, North Carolina State University, Raleigh, NC.

Matrix metalloproteinase (MMP)-2 and -9 degrade type IV collagen and other extracellular matrix components and expression in brain tumors correlates with invasive behavior, higher tumor grade and recurrence in human patients. The purpose of this study was to investigate the expression of MMP-2 and -9 in canine central nervous system (CNS) tumors, as well as expression in cerebrospinal fluid (CSF) as a surrogate marker of tumor MMPs.

Tumor samples were from surgical biopsies or necropsy and protein was isolated and quantified. CSF samples were collected from patients during routine diagnostic evaluations. MMP expression was evaluated using gelatin zymography.

A total of 21 tissue (15 meningiomas, 4 gliomas, 2 other) and 55 CSF (21 meningiomas, 15 gliomas, 6 pituitary, 6 choroid plexus, 4 lymphoma, 3 other) samples have been evaluated. All tissue and CSF samples showed expression of latent MMP-2. Active MMP-2 was absent from CSF samples but was detected in 5/21 (24%) tissue samples. Latent and active MMP-9 was detected in 18 (33%) and 6 (11%) CSF samples and 17 (81%) and 5 (24%) tissue samples respectively. A higher percentage of choroid plexus tumors (83%) and round cell tumors (100%) expressed MMP-9 in the CSF compared to meningiomas (10%), gliomas (20%), and pituitary tumors (17%).

A subset of canine tumors express active forms of MMP-2 as well as active and latent MMP-9. CSF appears to be an insensitive surrogate marker of MMP activity within CNS tumors, but may be useful in predicting certain tumor types.

220a

SIGNALLING PROTEINS AND TOLL-LIKE RECEPTORS MAY INDUCE A TH-17 IMMUNE RESPONSE IN CANINE STEROID-RESPONSIVE MENINGITIS-ARTERITIS.

Maiolini[ab], M. Otten[c], R. Carlson[a] and A. Tipold[ab]

[a]Department of Small Animal Medicine and Surgery, University of Veterinary Medicine, Hannover, Germany; [b]Center for Systems Neuroscience, Hannover, Germany;

[c] Veterinary Practice Peter Gravert and Dr. Volker Otten, Gettorf, Germany.

Steroid-responsive Meningitis-Arteritis (SRMA) is a systemic inflammatory disease affecting mostly the cervical meninges and vessels. Clinical and laboratory findings include fever, a marked neutrophilic pleocytosis and leukocytosis. Increased production of acute-phase proteins and immunoglobulin A characterize the systemic immune response[1-4].

Several studies fuel the hypothesis that SRMA derives from a dysfunction of the immune system. It has been postulated that an overreacting Th2-mediated immune response induces high antibody production in patients with SRMA[5]. The exact pathomechanism of this disease is still not known. Recently, a new class of helper T cells, the Th17 cells, which play an active role in inflammatory response and in autoimmune diseases, were discovered[6-8]. The hypothesis that the aberrant immune response in SRMA might be associated with Th17 cells should be proven in the current study.

$CD4^+$ progenitors T cells differentiate towards Th17 subsets under the concomitant influence of IL-6 and TGF-β_1. Toll-like receptor (TLR) 4 and TLR9 can also influence the differentiation of the above mentioned Th17 lymphocytes. Therefore, we investigated the above mentioned cytokines and TLRs in dogs affected with SRMA and compared the values with data of dogs affected with different neurological and systemic diseases.

We found that in SRMA patients IL-6 was increased both intrathecally and systemically, while high levels of TGF-β_1 were found only intrathecally. Additionally, monocytes in SRMA were characterised by a relatively high expression of TLR4 and TLR9. These two TLRs activate dendritic cells to produce factors that together with IL-6 and TGF-β_1 drive $CD4^+$ progenitors to differentiate into Th17 cells[9,10].

We propose therefore, that the synergistic effect associated with simultaneous presence of both cytokines intrathecally and the increased expression of TLR4 and TLR9 may lead to a Th17-skewed immune response in SRMA maintaining an autoimmune reaction for several months.

1. Tipold A, Schatzberg SJ. J Small Anim Pract 2010;51:150-154.
2. Maiolini A, Carlson R, Schwartz M, et al. Vet J 2012;191:219-224.
3. Bathen-Noethen A, Carlson R, Menzel D, et al. J Vet Intern Med 2008;22:1149-1156.
4. Lowrie M, Penderis J, McLaughlin M, et al. J Vet Intern Med 2009;23:862-870.
5. Schwartz M, Puff C, Stein VM, et al. Vet J 2011;187:260-266.
6. Harrington LE, Hatton RD, Mangan PR, et al. Nat Immunol 2005;6:1123-1132.
7. Park H, Li Z, Yang XO, et al. Nat Immunol 2005;6:1133-1141.
8. Kolls JK, Linden A. Immunity 2006;24:677-688.
10. Montero Vega MT, de Andres Martin A. Allergol Immunopathol (Madr) 2009;37:252-263.

220b

THE MAGNETIC RESONANCE IMAGING APPEARANCE OF CANINE INTRA-AXIAL PRIMARY BRAIN TUMORS RELATES MORE STRONGLY TO TUMOR GRADE THAN TUMOR TYPE. RT Bentley[1], CP Ober[2], JR Ohlfest[3], KL Anderson[2], JF Naughton[1], DA Feeney[2], PD Constable[1], MG O'Sullivan[2], MA Miller[1], GE Pluhar[2]. 1. Purdue University School of Veterinary Medicine, West Lafayette, IN. 2. University of Minnesota College of Veterinary Medicine, Saint Paul, MN. 3. University of Minnesota Masonic Cancer Center, Minneapolis, MN.

The association between magnetic resonance imaging (MRI) features of canine intra-axial primary brain tumors and histopathological type and grade was determined.

Fifty six dogs with brain lesions were studied, including 34 cases of grade 2 to 4 tumors (17 astrocytomas, 14 oligodendrogliomas, 3 primitive neuroectodermal tumors) and 22 distractors. All 56 brain lesions had complete brain MRI and histopathological diagnosis. Five investigators blinded to the histopathologic diagnosis independently evaluated 22 radiological features according to a standardized grading system. Contingency tables were analyzed using Fishers exact test; $p<0.05$ was considered significant.

The severity of contrast enhancement was associated with increasing tumor grade, being more severe in the highest grade ($p=0.0046$), and less in grade 2 tumors ($p=0.001$). Grade 4 tumors more commonly exhibited a spherical-ovoid shape ($p=0.030$) and an isointense or hyperintense T1 signal ($p=0.038$). Gradient echo signal voids ($p=0.0007$), a single cyst ($p=0.0013$), and multiple intra-tumoral fluid accumulations ($p=0.046$) were more common in grade 3 and 4 tumors. Grade 2 tumors occurred more commonly in a diencephalic location ($p=0.023$), caused midline-shift or falcine herniation ($p=0.012$) and impinged on adjacent ventricles ($p=0.0017$). Astrocytomas more often had a caudal fossa location ($p=0.024$) than oligodendrogliomas, while oligodendrogliomas more commonly distorted adjacent ventricles ($p=0.0068$), caused caudal cerebellar herniation ($p=0.011$) and exhibited T1 hypointensity ($p=0.0064$); primitive neuroectodermal tumors were excluded from the analysis of tumor type due to low case numbers.

The MRI appearance of canine intra-axial primary brain tumors is significantly related to tumor grade, whereas differences associated with tumor type are less characteristic.

7a

ENHANCED MUTAGEN SENSITIVITY AS A PREDISPOSITION TO LYMPHOMA IN GOLDEN RETRIEVERS. DH Thamm,[1] SM Bailey,[1] DM Vail,[2] KK Grunerud,[1] BJ Rose.[1] 1. Colorado State University, Fort Collins, CO. 2. University of Wisconsin – Madison, Madison, WI.

Cancer is ultimately a disease of DNA. Deficiencies in repair of damaged DNA correlate with cancer susceptibility in some human cancers. We hypothesized that golden retrievers may have heritable defects in their ability to repair DNA damage that predispose them to the development of cancers such as lymphoma. In this pilot study we sought to determine whether golden retrievers with lymphoma exhibit enhanced sensitivity to DNA damaging agents compared with healthy golden retrievers or normal mixed-breed dogs.

Twenty golden retrievers presenting with untreated, histologically or cytologically confirmed lymphoma served as "cases" in this case-control study. Dogs with more than 1% circulating atypical lymphocytes were excluded. Twenty clinically normal, age-matched golden retrievers and 20 clinically normal, age-matched mixed-breed dogs served as controls. Peripheral blood mononuclear cells (PBMC) from each individual, were exposed to a DNA damaging agent, either ionizing radiation (gamma-rays) or bleomycin, and incubated for 1, 4 or 24 hours, then colcemid-arrested in metaphase for chromosome visualization. Prepared nuclei were dropped onto glass slides, Giemsa-stained, and evaluated by light microscopy for chromatid-type aberrations (gaps and breaks), informative biomarkers of defective DNA repair of DNA double-strand breaks (DSBs). The proportion of "DNA damage (mutagen) sensitive" dogs, defined as those dogs exhibiting deficient DNA repair as evidenced by chromatid-type aberrations in excess of the 75th percentile for the normal mixed-breed population, was assessed as a categorical variable and compared between groups using Fisher's exact test.

Golden retrievers with lymphoma demonstrated significantly higher frequencies of induced breaks than did either clinically normal mixed breed dogs or normal golden retrievers at multiple radiation doses and time points. A similar trend was observed in the bleomycin-treated samples. There were no significant differences between clinically normal golden retrievers and mixed-breed dogs under any conditions. There were no differences in the percent "DNA damage sensitive" between normal golden retrievers and normal mixed-breed dogs at any dose or time point; however, there was a significant difference between normal golden retrievers and golden retrievers with lymphoma in 3 of 6 radiation conditions, and the trend was preserved in almost all conditions. There were numerically more "DNA damage sensitive" golden retrievers with lymphoma in the bleomycin-sensitivity experiments as well, but this difference did not reach statistical significance.

In conclusion, the results of this study demonstrate that PBMC from golden retrievers with lymphoma experience higher levels of unrepaired DNA damage accumulation following mutagen exposure than do clinically healthy golden retrievers or clinically healthy mixed-breed dogs. These findings imply that heritable deficiencies in the ability to repair DNA damage, specifically double-strand breaks, may contribute to the development of lymphoma in golden retrievers.

7b

GS-9219/VDC-1101: EFFICACY OF A Q21 DAY ADMINISTRATION SCHEDULE IN CANINE LYMPHOMA. DH Thamm,[1] DM Vail,[2] ID Kurzman,[2] DB Tumas, N. Souza-Powers,[3] KE Burgess.[4] 1. Colorado State University, Fort Collins, CO. 2. University of Wisconsin – Madison, Madison, WI. 3. Gilead Sciences, Foster City, CA. 4. Tufts University, North Grafton, MA.

Non-Hodgkin's lymphoma (NHL) is one of the most commonly diagnosed cancers in both dogs and humans. While currently available therapies can induce remission in the majority of dogs and humans with NHL, drug-resistant relapse is common and there is a distinct need for novel agents with improved efficacy. The acyclic nucleotide phosphonate 9-(2-phosphonylmethoxyethyl)-guanine (PMEG) forms an active phosphorylated metabolite, PMEG diphosphate, in cells and causes cytotoxicity in dividing cells due to potent inhibition of the nuclear DNA polymerases α, δ, and ε, resulting in inhibition of DNA synthesis; however, the utility of PMEG as an anticancer agent is limited by its poor cellular permeability and nonspecific toxicity. VDC-1101 (GS-9219), a novel double prodrug of PMEG, was designed as a cytotoxic agent that preferentially targets lymphoid cells with significantly reduced systemic toxicity. We previously evaluated VDC-1101, administered on a variety of schedules, as monotherapy in 38 dogs with NHL. Objective responses were noted in 100% of chemotherapy-naïve dogs and 60% of refractory dogs, with a median remission duration (MRD) of 128 days. Given the ease of administration with a q21-day administration schedule, we sought to generate additional data regarding efficacy of this regimen.

21 dogs with NHL were treated with VDC-1101 as monotherapy at dosages ranging from 0.66 to 0.98 mg/kg free base, given as a 30-minute IV infusion once every 21 days. Dogs experiencing CR received 5 doses of VDC-1101, followed by monthly rechecks. Complete clinicopathological assessment and clinical assessment of remission and adverse effects (AEs) were performed every 21 days. Response was assessed according to RECIST criteria and AEs according to the VCOG-CTCAE v1.0.

7 dogs were treatment-naïve and 14 dogs had failed previous therapy. The overall response rate (ORR) was 76% (43% CR, 33% PR). The ORR was 100% in naïve dogs and 57% in pre-treated dogs. The MRD was 168 days (279 days for dogs experiencing CR). There was no significant difference in MRD between naïve and refractory patients (173 vs. 117 days). One dog was euthanized following hospitalization for grade 4 neutropenia, pneumonia and gastrointestinal (GI) signs. 2 dogs were euthanized in CR from presumed pulmonary fibrosis 117 and 173 days following treatment initiation. 2 dogs had asymptomatic grade 4 neutropenia, 2 dogs had transient grade 3 elevations in liver enzymes, and 1 dog each had grade 3 GI and cutaneous AEs.

In conclusion, VDC-1101 has substantial antitumor activity in dogs with both treatment-naïve and refractory NHL when administered on a q21 day schedule. AEs were generally mild to moderate and reversible, although additional characterization of the presumed idiosyncratic pulmonary toxicity is warranted, and concurrent corticosteroid treatment may mitigate this effect.

99a

THE SANARUS CASSI II^{TM} FREEZE CORE BIOPSY SYSTEM FOR BRAIN TUMOR BIOPSY: A PROSPECTIVE COMPARISON WITH SURGICAL TISSUE BIOPSY SECTIONS. DJ Marino, CA Loughin; Long Island Veterinary Specialists, Plainview, NY.

To describe an ultrasound assisted brain biopsy technique utilizing a freeze core needle harvest system (Cassi II, Sanarus, Pleasanton, CA), to report on the success of this technique in producing diagnostic quality samples in comparison to surgical biopsies, and to describe any associated complications associated with the technique.

Seventeen dogs and 3 cats with an MR imaging diagnosis of a brain mass which were treated with craniotomy and mass removal.

MRI neurolocalization, histopatholgy from the 10g Sanarus Cassi II^{TM} freeze core biopsy (CB) and from surgical biopsy (SB), post-operative complications, pre-and post-operative medical therapy were recored.

The seventeen dogs included various breeds. The 3 cats were domestic short hair breeds. The average age of the dogs was 9.4 years (range: 4-14 yrs), 15 years for cats.

The CB system was successful in producing biopsy samples that were of sufficient size and diagnostic quality to be evaluated histologically and were found to be in agreement with SB histopathology results in 95% (19/20) of the samples. The "stick-freeze" specimens, freezing artifact appears minimal and has not compromised interpretation. There were no intraoperative or post operative deaths or complications noted.

The CB system can be expected to safely yield diagnostic quality brain biopsy specimens without the need for stereotactic frame systems or re-imaging.

99b

SEVERE HYPERTHERMIA AFTER TRANSETHMOIDAL CUSA ASSISTED DIENCEPHALIC MASS REMOVAL IN 4 DOGS AND 2 CATS. DJ Marino, CA. Loughin, CW Dewey, LJ Marino. Long Island Veterinary Specialists,Plainviw, NY, Department of Clinical Sciences, College of Veterinary Medicine, Cornell University, Ithaca, NY.

Report clinical findings including severe hyperthermia after transethmoidal CUSA assisted diencephalic mass removal in 4 dogs and 2 cats.

Case records were reviewed and the following information was recorded: history, signalment, bloodwork (e.g., complete blood count, serum biochemistry profile), neurologic examination findings, MRI results, histopathology, and post-operative complications, pre-and post-operative medical therapy, and outcome. Neuroanatomic localization was confirmed with MRI in all patients.

A variety of tumor types were found in this small study group, an association with histopathology findings could not be assessed. Median time of the onset of hyperthermia was 3.5 hours (range: 1 to 6 hours) after extubation with a median high temperature at the onset of hyperthermia of 104.6° F, (range:103.2°F to 107°F). Median time of onset of the hypernatremia was 4.5 hours (range: 1 to 9 hours) after extubation with a median abnormal sodium level of 172 mmol/L (range: 168 mmol/L and 196 mmol/L). All patients died within 24 hours of extubation. The median time of death after hyperthermia was 10.5 hours (range: 6 to 13 hours). The median time of death after extubation was 13.5 hours (range: 11 to 15 hours)

Transethmoidal CUSA assisted diencephalic mass removal is associated with fatal hyperthermia. Both hyperthermia and subsequent hypernatremia are likely sequelae of hypothalamic damage.

9a

EVALUATION OF NBD PEPTIDE AS A NOVEL THERAPEUTIC IN CANINE LYMPHOMA.

G Habineza-Ndikuyeze[1], A Gaurnier-Hausser[1], M May[1], NJ Mason[1]. University of Pennsylvania's School of Veterinary Medicine, Philadelphia PA.

The NF-κB family of transcription factors plays a central role in the regulation of lymphocyte proliferation, differentiation and survival. NF-κB activation is normally highly regulated, however, aberrant, constitutive NF-κB activation has been identified in aggressive lymphoid malignancies in humans. Constitutive NF-κB activation contributes to lymphomagenesis and chemotherapeutic resistance by promoting the over-expression of NF-κB target genes that inhibit cellular apoptosis and promote cellular proliferation. The purpose of this study was to determine whether targeted inhibition of canonical NF-κB activation down regulates NF-κB target genes and promotes cellular apoptosis in dogs with large B cell lymphoma.

Using a combination of Electrophoretic Mobility Shift Assays, immunohistochemistry, immunoblots and qRT-PCR, malignant lymphoid tissue from dogs with large B cell lymphoma was analyzed for the presence of constitutive canonical NF-κB activation. Malignant canine B cells that showed evidence of aberrant constitutive activity were treated with the selective IKK inhibitor NEMO Binding Domain (NBD) peptide and evaluated for pathway inhibition by qRT-PCR. NBD peptide was administered to dogs with large B cell lymphoma in two separate clinical trials to evaluate its effects on the NF-κB pathway and the expression of canonical NF-κB target genes.

Constitutive NF-κB activity and increased NF-κB target gene expression was identified in the malignant lymph nodes of dogs with large B cell lymphoma. In vitro and in vivo administration of NBD peptide inhibited this activity and induced apoptosis of malignant lymphocytes. We conclude that selective inhibition of the IKK complex by NBD peptide can inhibit aberrant, constitutive NF-κB activity within canine malignant B cells and that this approach may represent a novel adjunct therapeutic agent for the treatment of dogs with primary or relapsed large B cell lymphoma.

9b

SAFETY OF METRONOMIC CYCLOPHOSPHAMIDE COMBINED WITH CARBOPLATIN FOR DOGS WITH CANCER. JH Burton, BJ Biller. Colorado State University Animal Cancer Center, Fort Collins. CO.

Maximum tolerated dose (MTD) chemotherapy provides limited survival benefit for most dogs with cancer and there is increasing evidence that it mobilizes circulating endothelial cells (CEC) which may contribute to tumor angiogenesis. Metronomic chemotherapy has been shown to inhibit tumor angiogenesis and blunt spikes in CEC that occur after MTD chemotherapy in rodent models of cancer. The objective of this study is to evaluate the safety of MTD carboplatin combined with escalating doses of metronomic cyclophosphamide and determine the impact on CEC in dogs with cancer.

Client owned dogs with gross or microscopic cancer were administered carboplatin at 300 mg/m^2 IV every three weeks for a total of four doses; oral cyclophosphamide was administered concurrently at a dose of 0, 12.5, 15, or 17.5 mg/m^2/day in cohorts of three dogs. A CBC and flow cytometric quantification of CEC were performed at days 0, 7, 14, and 21 for all dogs and additionally at days 42, 63, and 84 for dogs receiving metronomic cyclophosphamide.

Eleven of 12 planned dogs have been enrolled to date. The incidence and severity of neutropenia and thrombocytopenia was not significantly increased for dogs receiving the combination of carboplatin and metronomic cyclophosphamide at any dose as compared to dogs receiving carboplatin alone and no dogs experienced clinically significant gastrointestinal side effects. CEC were not significantly increased after MTD carboplatin alone, nor were CEC significantly decreased with concurrent administration of metronomic cyclophosphamide.

The combination of MTD carboplatin and metronomic cyclophosphamide appears to be well-tolerated in dogs; further study is required to determine if this combination therapy has improved efficacy over conventional dosing of carboplatin alone.

PET DOGS LEAD THE WAY AGAIN IN THE FIGHT TO STOP URINARY BLADDER CANCER

Deborah W. Knapp, DVM, DACVIM (Oncology)
West Lafayette, Indiana

OVERVIEW OF CANINE BLADDER CANCER

Although urinary bladder cancer comprises only 2% of all canine cancer, with the dog population in the United States exceeding 70 million, thousands of dogs will develop this cancer each year. Invasive transitional cell carcinoma (TCC) is the most common form of canine urinary bladder cancer.[1] Most TCCs are intermediate- to high-grade papillary infiltrative tumors. Other types of tumors can form in the bladder including squamous cell carcinoma, undifferentiated carcinoma, adenocarcinoma, rhabdomyosarcoma, lymphoma, hemangiosarcoma, fibroma, and other mesenchymal tumors.[1] TCC comprises more than 90% of all bladder cancer in dogs. The presentation will focus on the current and emerging strategies to manage canine TCC and the translational opportunities that could also impact humans with invasive bladder cancer. Naturally-occurring canine TCC bears remarkable resemblance to human invasive TCC, and positive results in canine studies are being translated into human studies.[2]

PRESENTATION, DIAGNOSIS, AND STAGING

Hematuria and stranguria are the most common clinical signs in dogs with TCC, but these signs are not specific for TCC and are more commonly associated with urinary tract infection. Similarly, although abnormal epithelial cells in urine, a positive urine bladder antigen test, and thickened wall or masses in the bladder and urethra raise the suspicion for TCC, other conditions such as other tumors, inflammatory conditions, and polypoid diseases can cause these same changes.[1] It is important to distinguish non-TCC conditions from TCC because the treatment and prognosis differ considerably and are dependent on the condition present. A diagnosis of TCC requires histopathologic confirmation. Tissue biopsies can be obtained by cystotomy, cystoscopy, and in some cases, by traumatic catheterization.[1,3,4] Cystoscopy provides the opportunity to visually inspect the urethra and bladder and to obtain biopsies in a non-invasive method. With the small size of cystoscopic biopsies, the operator must be diligent to collect sufficient tissue for diagnosis. Placing tissues in a specially designed histology cassette prior to processing helps prevent loss of small samples. In a recent report of 92 dogs, diagnostic samples were obtained by cystoscopy in 96% of female dogs and 65% of male dogs that ultimately had hisotopathologically-diagnosed TCC.[4] The more recent use of a wire basket designed to capture stones during cystoscopy allows collection of larger tissue samples, and is expected to increase the yield of diagnostic biopsy samples. Percutaneous biopsy methods can lead to tumor seeding and are best avoided.[1,5,6] In poorly differentiated carcinomas, immunohistochemistry for uroplakin III (UPIII) can be helpful in distinguishing TCC from other carcinomas. UPIII is expressed in >90% of canine TCC, and has been considered a specific marker for TCC.[7] Recently, UPIII expression has also been reported in canine prostate cancer,[8] although it is not known if the immunoreactive cells originated from the transitional epithelium of the prostatic ducts or elsewhere in the gland.

TCC is most often located in the trigone region of the bladder. It has also been reported to involve the urethra in 56% of dogs and the prostate in 29% of male dogs.[1] A thorough physical exam, including rectal exam, may reveal thickening of the urethra and bladder, enlarged lymph nodes, and sometimes a mass in the bladder or a distended bladder. A normal physical exam does not rule out TCC.

In dogs with confirmed or suspected TCC, overall health should be assessed (CBC, serum biochemistry profile, urinalysis, +/- urine culture) and the cancer staged (thoracic radiography, abdominal ultrasonography, and urinary tract imaging). Secondary urinary tract infections are common in dogs with TCC, and periodic urinalyses are indicated. To avoid the risk of seeding TCC through cystocentesis, urine may be collected by free catch or catheterization. If a catheter is to be passed, care must be taken to avoid penetrating the diseased bladder or urethral wall. Approximately 15-20% of dogs with TCC have metastasis at the time of diagnosis, and at least 50% develop distant metastasis prior to death. Common sites of metastases detected with thoracic radiography and abdominal ultrasonography include lymph nodes, liver, and lung. Less common sites include bone and other organs.[1] Urinary tract imaging is used to assess TCC location for potential surgical intervention and to map and measure TCC masses in order to subsequently determine response to medical therapy. Mapping TCC in the bladder, proximal urethra, and prostate can be accomplished by cystosonography, cystography, or computed tomography.[1,9,10] Regardless of the imaging technique used, to accurately track response to therapy, it is important to follow a consistent protocol from visit to visit for bladder distention and patient positioning. In addition, when using cystosonography to monitor therapy response, it is critical to have the same operator perform exams over multiple visits, and preferably using the same ultrasound machine. Preliminary information suggests that 3-dimensional ultrasound may improve the accuracy and repeatability of measurements of bladder masses (unpublished data, Naughton, Knapp).

TREATMENT

TCC-related death is typically due to urinary obstruction, metastases, or a combination of the two. The objectives of treatment are to eradicate the cancer when possible, and in cases when eradication is not possible, to control the cancer and provide as long a period of good quality life as possible. Cure of canine TCC is not common. Surgical removal of TCC in dogs is usually not possible because of the trigonal location, frequent urethral involvement of the cancer, in some cases, metastases. In addition, many dogs develop multifocal TCC in the bladder, consistent with the "field effect" proposed in humans in which the entire bladder lining is thought to undergo malignant change in response to carcinogens in the urine. In general, the indications for surgery in dogs with confirmed or suspected TCC include: (1) obtaining tissue for a diagnosis if tissue cannot be safely and adequately obtained nonsurgically, (2) removing the TCC within the bladder if lesions are away from the trigone, and (3) restoring urine flow. If surgery is performed, it is crucial to take measures to avoid seeding the cancer in the abdominal wall and other locations. TCC in the abdominal wall, once established, has responded very poorly to medical therapy.[6] Surgery and other "nonsurgical" procedures can be important in restoring or maintaining urine flow. Ureteral stents, when indicated, have traditionally been placed surgically, although less invasive techniques to place stents have recently been described.[11] Prepubic cystostomy tubes that bypass urethral obstruction are another option.[1] More recently, the placement of urethral stents has gained favor over cystotomy tubes in some cases because external tubes are avoided, and the pet owner does not have to drain the bladder.[12,13] The survival following stent placement can vary considerably from dog to dog. In preliminary findings from Purdue University, survival following stent placement in dogs ranged from a few days to a year.[13] Urethral stents can be placed nonsurgically with fluoroscopic guidance. Recently, a procedure has been described in which ultrasound-guided transurethral laser ablation was used to debulk TCC lesions, especially those causing urinary tract obstruction.[14] The risks versus the potential benefit of this approach requires further study. One of the challenges in laser resection, especially when done transurethrally, is the difficulty in judging the deeper aspects of the tumor and the risk of cutting too deep and perforating the urinary tract. The published results of radiation therapy in TCC have been disappointing in regards to side effects (urinary incontinence and cystitis with accompanying pollakiuria and stranguria). Further studies are pending.

Systemic medical therapy is the mainstay of TCC treatment in dogs, and usually consists of chemotherapy, cyclooxygenase (cox) inhibitors (nonselective cox inhibitors and cox-2 inhibitors), and the combination of these.[1,15-18] Although medical therapy is not usually curative, several different drugs can lead to remission or stable disease of TCC, and most therapies are well-tolerated. Although remission is obviously preferred, stable disease is considered a beneficial response to treatment when the quality of life is good. Resistance to one drug does not necessarily imply resistance to other drugs. Some of the best results are seen in dogs that sequentially receive multiple different treatment protocols over the course of their disease. The approach used in the Canine Bladder Cancer Clinic at the Purdue University Veterinary Teaching Hospital (PUVTH) is to obtain baseline measurements of the TCC masses, to initiate a starting treatment, to monitor the response (tumor size) to that treatment at 4-8 week intervals, and to continue that treatment as long as the TCC is controlled, side effects are acceptable, and quality of life is good. If cancer progression or unacceptable toxicity occurs, then a different treatment is instituted. Subsequent treatment changes are based on change in tumor size and stage, and treatment tolerability. By following this approach, TCC growth can be controlled in approximately 75% of dogs, the dogs' quality of life is usually very good, and survival times can extend well beyond a year. Although it could be tempting to simultaneously combine multiple chemotherapy agents in dogs with TCC, the benefit of this has not been determined, and the potential development of resistance to multiple drugs at the same time could limit the options for subsequent therapy.

The two broad types of medical therapy used in dogs with TCC are cox inhibitors and chemotherapy drugs. As a single agent, the nonselective cox inhibitor, piroxicam (0.3 mg/kg po daily), is a useful palliative treatment for dogs with TCC.[1] The quality of life in dogs that receive piroxicam is usually excellent. Responses in 62 dogs with TCC that received piroxicam as a single agent included 2 complete remission (CR, no evidence of cancer), 9 partial remission (PR, ≥50% decrease in tumor volume and no new tumor lesions), 35 stable disease (SD, < 50% change in tumor volume and no new tumor lesions), and 16 progressive disease (PD, ≥50% increase in tumor volume or new tumor masses). The survival time (median 195 days) compared favorably to that of 55 dogs in the Purdue Comparative Oncology Program Tumor Registry that were treated with cytoreductive surgery alone (median survival 109 days). Although most dogs tolerate piroxicam well, care must be taken to watch for gastrointestinal (GI) toxicity, particularly ulceration. If vomiting, melena, and anorexia occur, the drug must be withdrawn and supportive care provided as needed until the toxicity resolves. In these cases, it is usually considered safest to switch to a cox-2 inhibitor if further cox inhibitor treatment is indicated.

A cox-2 inhibitor, deracoxib (Dermaxx, Novartis), has been evaluated as a single agent at a dosage of 3 mg/kg PO daily in 26 dogs with TCC.[15] Tumor responses included 4 (17%) PR, 17 (71%) SD and 3 (12%) PD. The median survival following deracoxib and subsequent therapies was 323 days. Mild GI toxicity occurred in 20% of dogs, and 4% of dogs had renal or hepatic side effects. The 17% remission rate with deracoxib appears comparable to the remission rate with piroxicam, however, occasional CRs occur in dogs receiving piroxicam, whereas CR was not noted in dogs treated with deracoxib. In unpublished work, another cox-2 inhibitor, firocoxib (Previcox, Merial) has also had antitumor activity against canine TCC (unpublished data, Knapp and Henry). Results of a study of the effects of firocoxib as a single agent and the effects in enhancing the activity of cisplatin will be summarized in a research abstract at this Forum. It is not yet known if nonselective cox inhibitors and cox-2 inhibitors are equally effective in treating TCC. Cox-2 inhibitors, however, offer an advantage of less risk of gastrointestinal toxicity.

Several different chemotherapy protocols have been given to dogs with TCC. Two of the more frequently used protocols are mitoxantrone combined with piroxicam, and single agent vinblastine.[17,18] In a study of 55 dogs with TCC receiving mitoxantrone and piroxicam, 35% of dogs had remission with minimal toxicity, and the median survival was 291 days.[17] In a recent phase II study of vinblastine in 28 dogs with TCC, 36% of dogs had remission, and 50% of dogs had stable disease.[18] The majority of dogs had failed other therapies. A study of front line vinblastine is in progress. Other systemic therapies can also be given. Carboplatin combined with piroxicam induced remission in 38% of dogs with TCC, although toxicity was relatively common compared to other protocols.[16] The use of gemcitabine and piroxicam has also been reported in dogs with biopsy confirmed or cytologic evidence of TCC.[19]

Intravesical therapy is commonly used in humans with superficial TCC, and there has been interest in this approach to potentially treat the higher grade invasive TCC that occurs in dogs. A phase I clinical trial and pharmacokinetic (PK) study of intravesical mitomycin C (MMC) (1 hr dwell time /day, 2 consecutive days each month, escalating concentrations) was performed in dogs with TCC.[20] Tumor response was assessed in 12 of 13 dogs, and responses included 5 PR and 7 SD. Initially, the treatment appeared well tolerated, and the maximum tolerated dose based on local toxicity (bladder irritation of 1-2 days duration) was 700 μg/ml (1hr dwell time/day, 2 consecutive days/month). With further study, however, marked myelosuppression and severe GI upset occurred in 2 dogs, indicating the drug had been absorbed systemically. This occurred after the first treatment cycle in 1 dog and the fourth cycle in another dog. Although both dogs recovered with supportive care, the potential systemic exposure of the drug in other dogs is of serious concern. The amount of MMC being instilled in the bladder is great enough, that if a substantial proportion of the drug were to be systemically absorbed, it could be lethal. Therefore, intravesical therapy is not recommended for initial management in dogs with TCC, and the risks vs potential benefit must be carefully weighed if the treatment is to be considered at any point in the course of the disease.

New and emerging therapy approaches will be discussed including: (1) strategies to make more effective use of existing drugs such as metronomic therapy and demethylating agents, (2) emerging targeted therapies, and (3) innovative strategies to individualize therapy. A research report at the Forum will include discussion of folate-targeted therapy in dogs with TCC.

CAUSES AND PREVENTION OF TCC

Prevention of TCC would be far preferable to managing the cancer once it has formed. The etiology of canine bladder cancer is multifactorial. Risk factors include exposure to older generation flea control products and lawn chemicals, obesity, possibly cyclophosphamide exposure, female gender, and specific breeds.[1] The female:male ratio of dogs with TCC has been reported to range from 1.71:1 to 1.95:1.[1,21] Neutered dogs of both genders are more likely to develop TCC than intact dogs, although the mechanisms for this increased risk have not been determined.[1,21] There is a very strong breed associated risk for TCC in dogs.[1] Scottish terriers have an 18-20X increased risk. Shetland sheepdogs, West Highland white terriers, and beagles have a 3-5X increased risk.[1] Dogs in related breeds also appear to have higher risk of TCC. This breed-associated risk is being utilized to identify genetic variants that increase the risk of TCC in dogs, and potentially in humans (unpublished work, Ostrander and Knapp). Once genetic variants linked to TCC risk are identified in the dogs, these same variants will be assessed in humans. Chemical exposure has also been implicated in TCC risk in dogs. In a case control study of dogs of several breeds, an association between TCC and exposure to topical flea and tick dips was noted.[22] In the highest risk group (overweight female dogs), the risk of TCC was 28 times that of normal-weight male dogs not exposed to the insecticides.[22] The authors postulated that the "inert" ingredients (solvents and petroleum distillates), which accounted for more than 95% of the product, were possibly the carcinogens. Newer, spot-on type flea control products appear safer. In a case control study in Scottish terriers (STs), spot-on products containing fipronil were not associated with increased risk of TCC.[23] In this case control study in STs, exposure to lawn herbicides and pesticides was compared between STs with TCC ($n = 83$) and a control group of STs ($n = 83$, > 6 yrs of age, no history of urinary tract disease in the previous 2 years).[24] TCC risk was significantly higher in STs that had been exposed to lawn herbicides alone (Odds ratio [OR], 3.62; 95% confidence interval [CI] 1.17-11.19; $P < 0.03$) or herbicides and insecticides (OR, 7.19; 95% CI 2.15-24.07; $P < 0.001$) than in dogs not exposed.[24] An update on this area of research will be presented in a Research Report at this Forum.

A positive finding in the ST case control study was that dogs that ate vegetables at least 3 times a week, along with their normal diet, had a reduced risk of TCC (OR, 0.30; 95% CI 0.01-0.97; $P < 0.001$).[25] The specific type of vegetable with the most benefit could not be determined, but carrots, given as treats, were the most frequently fed vegetable in the study.

Pending results of prospective prevention studies, it would appear appropriate to limit dogs' exposure to lawn chemicals and older types of flea control products, and to feed dogs vegetables at least 3 times per week, especially in dogs in breeds with high risk for TCC. The owners of dogs in high-risk breeds should be informed of the TCC risk and encouraged to make note of urinary tract signs should they occur, and to pursue veterinary care in a timely fashion. Prospective studies to determine the value of TCC screening and early detection have not yet been reported.

SUMMARY

The outcome for dogs with TCC has improved substantially over the last several years. TCC has truly become a "treatable" cancer with at least 75% of dogs having a beneficial response to therapy. The potential outlook for dogs and humans who have TCC or who are at risk for developing TCC in the future, is even more exciting.

REFERENCES

[1]Knapp DW. Tumors of the urinary system. In: Withrow SJ, Vail DM, eds. Withrow and MacEwen's Small Animal Clinical Oncology. 4th ed. St. Louis, MO: Saunders, 2007;649. [2]Dhawan D, et al. Mol Cancer Ther 2010;9:1371. [3]Messer JS, et al. Clin Tech Small Anim Pract 2005;20:52. [4]Childress MO, et al. J Amer Vet Med Assoc, 2011;239:350. [5]Nyland TG, et al. Vet Radiol Ultrasound 2002;43:50. [6]Higuchi T, et al. J Amer Vet Med Assoc, in press, 2012. [7]Ramos-Vara JA, et al. Vet Pathol 2003;40:55. [8]Lai CL, et al. Prostate 2008;68:477. [9]Chun R, et al. J Vet Intern Med 1997; 11:279. [10]Hume C, et al. Vet Comp Oncol 2010;8:122. [11]Berent AC. J Vet Emerg Crit Care 2011;21:86. [12]Weisse C, et al. J Amer Vet Med Assoc 2006;229:226. [13]McMillan SK, et al. J Amer Vet Med Assoc 2012 in press. [14]Cerf DJ, et al. J Amer Vet Med Assoc 2012;240:51. [15]McMillan SK, et al. J Amer Vet Med Assoc 2011;239:1084. [16]Boria PA, et al. Vet Comp Oncol 2005;3:73. [17]Henry CJ, et al. Clin Cancer Res 2003;9:906. [18]Arnold EJ, et al. J Vet Intern Med 2011;25:1385. [19]Marconato L, et al. J Am Vet Med Assoc 2011;238:1004. [20]Abbo AH, et al. J Vet Intern Med 2010;24:1124. [21]Bryan JN, et al. Prostate 2007;67:1174. [22]Glickman LT, et al. J Toxicol Environ Health 1989;28:407. [23]Raghavan M, et al. J Am Vet Med Assoc 2004;225:389. [24]Glickman LT, et al. J Am Vet Med Assoc 2004;224:1290. [25]Raghavan M, et al. J Am Vet Med Assoc 2005;227:94.

IMRT, IGRT and SRT: Improving outcome for dogs with urogenital cancer

Dr. Susan M. LaRue
Fort Collins, Colorado

Canine urogenital tumors present a therapeutic challenge. Curative surgery is often not an option because of tumor location. Traditional fractionated radiation therapy is associated with severe acute intestinal toxicity. Chemotherapy and palliative radiation therapy can improve clinical signs, but duration of response is limited. Limited information is available on novel techniques such as local chemotherapeutic infusions or photodynamic therapy. Recent technological advances in the fields of radiation oncology and minimally invasive surgery have provided promising new options for the treatment and management of these cancers. This lecture discuss focus on the application of specific technologies, such as intensity modulated radiation therapy (IMRT), image guided radiation therapy (IGRT), dynamic adaptive radiation therapy (DART) and stereotactic radiation therapy (SRT) for treatment of tumors of the bladder, prostate and urethra. We will also discuss management of radiation/tumor effects with stenting of the ureter and urethra.

Tumors of the bladder, urethra and prostate

Tumors of the lower urinary tract make up 2% of all canine cancers, and it is speculated that with the increased availability of ultrasound and computed tomography that clinical diagnosis is becoming more common. Transitional cell carcinoma of the bladder is the most frequently encountered tumor histologic type. Key to the treatment of urinary tumors with radiation therapy is determining the location and extent of the tumor. The trigone of the bladder is the most common location although the entire bladder can be involved. Disease extension from the bladder to the urethra is seen in over 50% of patients while extension to the prostate occurs in almost 30% of male patients. Nodal (15%) and distant (14%) metastases can be present at initial presentation but is as high as 50% at the time of euthanasia. Therefore accurate staging is of great importance. Confirmed or suspected regional lymph nodes should be treated. Other tumor types observed in this region include squamous cell carcinoma, adenocarcinoma, undifferentiated carcinoma, rhabdomyosarcoma, lymphoma, hemangiosarcoma, fibroma and other sarcomas. Tumor histology does not impact radiation prescription because radiation dose is always limited by normal tissue tolerance. However histologic type is extremely important in selecting the most appropriate post radiation chemotherapy regimen.

Treatment with radiation therapy is technically challenging for a number of reasons. The bladder varies in size and location based not only on urine volume but on rectal content. (1) In addition, the target tissues move relative to the bony landmarks identifiable using standard port film techniques. Therefore when using a traditional radiation therapy technique such as three dimensional conformal radiation therapy (3D-CRT), the radiation treatment field necessary to include the tumor plus microscopic extension, (called the clinical target volume, or CTV) must be enlarged to take this variation into account.(2) This expanded radiation field is called the planning target volume (PTV). Adjacent normal tissue structures, such as the colon, rectum and intestines, are unnecessarily exposed to high doses of radiation, resulting in acute GI toxicity and consequential late effects. (3) This problem is exacerbated when disease has spread to regional nodes, further increasing the volume of tissue treated.

Three technological advances in radiation oncology have vastly improved the ability to safely and effectively treat urinary tract tumors. **Intensity Modulated Radiation Therapy (IMRT)** controls the intensity of each beam through motorized collimator leaves (MLC) that move during treatment, allowing the dose to be sculpted around important normal structures. This technique requires a specific treatment planning system that uses ***inverse planning.*** Inverse planning requires that the various tumor structures (Gross tumor volume [GTV], CTV and PTV) as well as critical normal tissue structures be identified. This is achieved by outlining (contouring) each structure of interest on each slice of the CT scan. Dose objectives for each structure are entered, and a sophisticated algorithm attempts to meet all objectives. The 2 major benefits associated with IMRT are: 1) Dose to adjacent normal tissue structures can be minimized, dramatically reducing acute effects and 2) Radiation dose to the tumor can be increased, improving the probability of tumor control. Both of these benefits were realized when IMRT became the standard of care for prostatic tumors in men.

Image guided radiation therapy (IGRT) is another key component in the treatment of urinary tract tumors in dogs. IGRT utilizes imaging systems built into the therapeutic accelerator to confirm the location of the target immediately prior to treatment. For tumors that are firmly associated with boney structures, such as brain tumors and nasal tumors, orthogonal radiographs taken at treatment can be compared, using intrinsic software, to the digitally reconstructed radiographs (DRRs) from the treatment plan. For bladder and prostate cancers, a more advanced imaging technique, such as a cone beam CT (CBCT) is required. Prior to treatment a CBCT is acquired and compared in 3 dimensions, to the CT used for the radiation treatment planning. IGRT allows the oncologist to confirm the location of the treatment target. The importance of this is that it allows a substantial decrease in the size of the PTV expansion, drastically decreasing the amount of normal tissues in the field.(1)

Dynamic Adaptive Radiation Therapy (DART) is a technique where the patient's CBCT is evaluated each day before treatment to determine size/location of bladder. Based on previous work by Nieset (2), variations in bladder size/location can be predicted. With DART, a "small", "medium" and "large" bladder plan is created. After the CBCT is obtained, the most appropriate treatment plan is selected.

Clinical Outcome

The results of a clinical study involving 21 client owned dogs with tumors of the lower urinary tract were encouraging. (personal correspondence, Michael Nolan). Dogs were treated using IGRT assisted IMRT, and in some patients DART. Mild to moderate GI toxicity was observed. Few patients had skin or urogenital toxicities. Late toxicities developed but were successfully managed using stents. The treatment was well tolerated and median survival was substantially better than any previous reports in the literature. Results of this study will be discussed.

Stereotactic Radiation Therapy (SRT) for Treatment of prostatic or urethral tumors

Stereotactic radiation therapy is a new technology that may have applications for tumors limited to the prostate or urethra. It is unlikely that this technique can be safely applied to the bladder. Bladders are not tolerant of large doses per fraction and because of variations in size/location of bladders, it is difficult to target with the precision that is required for SRT to be successful.

SRT involves the use of high doses per fraction but overcomes the radiobiological limitations with stereotactially verified positioning and treatment delivery techniques that leave a minimal volume of normal tissue in the high dose area. Stereotactic radiation therapy, by definition, requires: 1) a tumor for targeting (not microscopic disease); 2) treatment planning and administration that will provide a dramatic dose drop off between the tumor and the surrounding normal tissue structures; and 3) A method of stereotactically verifying patient positioning. The result is that normal late responding tissue structures are spared through dose avoidance rather than by administering small doses per fraction. The normal tissue structures still receive dose, and the dose per fraction is higher than for traditional radiation therapy. This technique is appealing because only 1-5 fractions of radiation are required and treatments are generally well tolerated by the patient. However, normal tissue tolerance for important regional late responding tissues such as rectum and colon, have not been well defined. An additional concern about this technology is that most canine tumors are transitional cell carcinomas. Unlike adenocarcinomas, these may not be restricted to the prostate.

Conclusions

The new renaissance in radiation oncology has introduced exciting technologies that may benefit canine and feline patients with urinary tract tumors. It is important that progress in this area be documented. Determining normal tissue tolerance of organs at risk is crucial to the development of optimal radiation therapy protocols. Minimally invasive procedures such as ureteral, urethral and rectal stents will become increasingly important as patients live longer and develop mild to moderate toxicities. As local tumor control improves, it is also critical that a chemotherapy protocols are carefully evaluated for efficacy.

REFERENCES

1. Nieset JR, Harmon JF, LaRue SM. Characterization of Canine Bladder Variations to Optimize Fractionated Radiation Therapy Protocols Using Cone-Beam Computed Tomography. Veterinary Radiology & Ultrasound 51(2), 235. 2009.
2. Nieset JR, Harmon JF, LaRue SM. Use of cone-beam computed tomography to characterize daily urinary bladder variations during fractionated radiotherapy for canine bladder cancer. Vet.Radiol.Ultrasound 2011;52:580-8. 3. Anderson CR, McNiel EA, Gillette EL, Powers BE, LaRue SM. Late complications of pelvic irradiation in 16 dogs. Vet.Radiol.Ultrasound 2002;43:187-92.

INTERVENTIONAL ONCOLOGY AND THE MANAGEMENT OF URINARY TRACT CANCER

William T. N. Culp, VMD, DACVS
Davis, CA

INTRODUCTION

Urinary tract neoplasia is a clinically important disease process that often presents a diagnostic and therapeutic challenge. Primary neoplasia of the kidney and ureter is rare, while bladder and urethral neoplasia are diagnosed with a much higher frequency. Transitional cell carcinoma (TCC), the most common tumor diagnosed in the bladder and urethra, often results in clinical signs such as pollakiuria, stranguria and hematuria[1]; additionally, prostatic carcinomas may also result in clinical signs related to the urinary tract.[2] This text will focus on minimally invasive procedures utilized in the treatment of urinary tract neoplasia (and tumors affecting the urinary tract) that are actively developing in veterinary medicine. Many of these therapies are in the early stages and long-term outcome data is not available. However, the evaluation of outcomes and procedures in human medicine may provide the opportunity for improved treatment options for veterinary patients.

The specialty of Interventional Oncology (IO) is based on the utilization of image-guided minimally invasive techniques for the diagnosis and treatment of neoplasia. The indications for IO in veterinary medicine are vast, and while many therapies are unproven, early results are encouraging. The use of IO techniques in the treatment of urinary tract neoplasia is particularly attractive. Current treatment options are often insufficient to immediately relieve obstructions of the urinary tract, and the severity of local disease is the cause of demise in many cases despite treatment with the usual modalities of surgery, chemotherapy and radiation therapy. The urinary tract IO techniques that are being utilized in veterinary medicine focus on two major areas: relief of urinary tract obstruction and treatment of the primary tumor.

INSTRUMENTATION

IR instrumentation and implants may be unique to the procedure that is being performed or can include basic equipment that can be found in every veterinary hospital. For vascular procedures, a standard needle or over-the needle catheter is necessary. These "vascular access needles" permit the blood vessel to be entered with minimal surrounding morbidity and allow for guidewires to be introduced into the blood vessel.

Guidewires have been employed for many uses in veterinary medicine. Guidewires come in a myriad of widths, lengths and configurations, and selecting the appropriate type can influence the success of a procedure. The most common guidewire utilized in our practice is the 0.035-inch hydrophilic guidewire with an angled tip. The tip of these guidewires is softer and more flexible than the remainder of the guidewire and thereby causes less trauma to structures that are accepting the guidewire. In some cases, a stiffer guidewire may be preferred or a guidewire with a steerable tip can be used.

Sheaths are often placed over guidewires into blood vessels or obstructed organs to protect tissue damage during exchanges of other instruments. Sheaths often have a sidearm that allows for injection of saline or contrast and are equipped with valves that prevent leakage of contrast or urine. Understanding the size of instrument/s that a sheath can accept will prevent missteps in the performance of procedures; having to replace a sheath that is too small for the necessary catheter or stent delivery system will delay the procedure and could potentially cause unnecessary tissue damage.

Catheters are generally categorized according to their tip (straight, pigtail, hook and angled) and the function/type (flush-type, pull-type, push-type) of the catheter. A full spectrum of catheters tips can be selected, and the interventional radiologist should have an appreciation of the anatomy that he or she is trying to access when choosing a catheter. Catheters are generally introduced into sheaths over guidewires; this allows for entry of a catheter with a curved tip into a lumen while preventing trauma.

Catheters, sheaths and guidewires have both vascular and non-vascular applications, and are often utilized in procedures that are performed through a natural orifice. Guidewires can be used as a guide for the placement of a marker catheter within the colon (such as with urethral stent placement) or as a means of accessing the bladder through the external genitalia. Sheaths are often placed in the distal urethra during stent placement or while performing cystourethrograms. Catheters can be used to administer contrast medium when a contrast study is being performed.

Stents are often integral for relieving urinary tract obstructions that occur secondary to neoplasia. They are constructed of various materials including metal, metal alloys or more pliable materials such as silicone and polyurethane. The type of material selected depends on the desired location of placement and function. Stents are often delivered on a delivery system, and metal or metal alloy stents are deployed through either self-expansion or balloon-expansion. The material and method by which stents are made and the location of placement determine the eventual expanded stent length. It is critical that an individual placing a stent has a thorough understanding of the device that he or she is placing as some stents foreshorten and some stents do not. Stents can be covered or uncovered depending on the desired function. Drug-eluting stents are covered stents that allow for local delivery of a medication as it elutes from the stent covering; some of the drugs that have been introduced into the covering of stents include anti-fibrotic agents and chemotherapeutic drugs.

URINARY TRACT INTERVENTIONAL ONCOLOGY APPLICATIONS

Urethral Stents

While not the predominating sign associated with bladder/urethral TCC and prostatic carcinoma, urethral obstruction can occur and may be life-threatening. Historically, treatment options for cases of urethral obstruction were limited to urethral catheterization, serial cystocentesis and tube cystostomy. While catheterization and cystocentesis may relieve the immediate problem of urine retention, these are likely ineffective long-term solutions. Tube cystostomy carries the challenge of continual monitoring and profound owner dedication.

Urethral stenting has emerged as an alternative to the previously mentioned options for relieving urethral obstructions and has been regularly reported in human medicine.[3-5] Urethral stenting has been advocated for multiple indications including the relief of strictures secondary to trauma, post-operative strictures, urethral detrusor-sphincter dyssynergia and urethral neoplasia. Outcomes are variable, but in general, urethral stents are well tolerated.

Stenting of the urethra is one of the most common IO technique performed in veterinary patients. The advantages of this technique are the minimally invasive nature and the potential for immediate relief of urethral obstruction. To perform this procedure, dogs are typically placed in lateral recumbency, and a guidewire is placed into the urethra and passed into the bladder. In male dogs, access is generally straightforward as guidewires can usually be passed easily into the penis; in female dogs, urethroscopic- or fluoroscopic-guidance may be necessary. Once a guidewire has been placed, a contrast cystourethrogram can be performed by utilizing a catheter or sheath that is placed over the guidewire and introduced into the distal urethra. The contrast study allows for localization of the obstruction and subsequent measurement of the normal and affected urethra so that an appropriately sized stent can be chosen.

One case series has evaluated the outcome in a cohort of dogs undergoing urethral stent placement.[6] In that series, eight dogs were diagnosed with prostatic carcinoma and four were diagnosed with bladder/urethra TCC. Both balloon-expandable and self-expanding stents were utilized. All dogs were able to urinate or be expressed after stent placement. Common complications in this group of dogs included continued stranguria and urinary incontinence. In one dog, a stent dislodged and migrated into the bladder.[6] Since that time, techniques for stent placement have evolved, and the procedure offers a life-saving and minimally invasive treatment options for many dogs.

Lower urinary tract neoplasia is rare in cats, and the use of urethral stents has only been described in two cats with malignant obstructions.[7-8] In both cats, the stents allowed for successful urination after placement.[7-8] In a female cat, placement of a urethral stent can be performed similar to a female dog. However, placement of a urethral stent in a male cat (that has not undergone a perineal urethrostomy) generally requires a small surgical approach to the bladder. In cats, stents are often placed for treatment of benign strictures, rather than malignant obstructions.

Ureteral Stents

Primary ureteral neoplasia is extremely rare. However, tumors originating in the bladder and urethra can expand and obstruct the flow of urine from the ureter into the bladder. Hydronephrosis and ureteral dilation can occur secondary to this obstruction and renal function can become compromised. Ureteral stenting emerged in human medicine as a means of relieving obstructions that occur from both urinary tract neoplasia and neoplasia causing extraluminal compression of the urinary tract. Several techniques have been described for the placement of ureteral stents in humans, and the majority are performed via a minimally-invasive approach (antegrade via access to the renal pelvis or retrograde with cystoscopic-guidance).[9-11]

Ureteral stent placement in dogs with malignant obstructions is often performed by utilizing an antegrade approach. Access into the ureter is obtained through ultrasound-guided placement of a needle or catheter into the renal pelvis. A guidewire is then passed antegrade into the ureter, through the obstruction into the bladder, and subsequently out of the patient via the urethra. Stents can also be passed retrograde with cystoscopic assistance; however, tumor obliteration of the ureterovesicular junction can make this difficult. With either technique, once guidewire access is obtained, stents can generally be passed over the guidewire and through the obstruction with the assistance of a sheath.

Recently, the use of ureteral stents to relieve malignant obstructions was described.[12] In this cohort of 12 dogs, a double-pigtail stent was placed with the goal of leaving the cranial pigtail in the renal pelvis and the caudal pigtail in the urinary bladder (beyond the obstruction). Improvement in the severity of hydronephrosis and ureteral dilation was noted in all cases that were evaluated by abdominal ultrasonography post-stent placement. Ureteral stent placement was determined to be safe and well tolerated in this cohort of dogs.[12]

Intra-arterial Chemotherapy

Intra-arterial chemotherapy involves the administration of chemotherapy into the artery or arteries directly supplying a tumor. This allows for the tumor to receive the first pass of chemotherapy as opposed to being exposed to a filtered amount of chemotherapy as is seen when chemotherapy is given intravenously. Advantages of intra-arterial chemotherapy include increased intra-tumoral concentration of a chemotherapeutic agent and potentially a decrease in chemotherapy-associated side effects.[13]

Intra-arterial chemotherapy is commonly employed in human medicine for hepatic, bladder, thyroid, lung and breast tumors. As some chemotherapeutic agents are considered radio-sensitizers, intra-arterial chemotherapy may also have improved efficacy when given in conjunction with radiation therapy.

The use of intra-arterial chemotherapy in veterinary medicine has been described.[14-17] One study evaluated the combined use of intra-arterial chemotherapy and radiation therapy in the treatment of urinary bladder carcinoma in two dogs.[14] In that study, access to the caudal aorta was obtained via the carotid artery, and intra-arterial chemotherapy was administered at the level of external iliac arteries. In those two dogs, side effects were minimal and an objective tumor response was determined via the use of double contrast cystograms. Direct comparisons were not made with a control group.

While larger scale studies on the use of intra-arterial chemotherapy for the treatment of bladder/urethral TCC and prostatic carcinoma are lacking, initial experience with the use of superselective catheterization and subsequent chemotherapy administration is hopeful. For superselective administration of chemotherapy, catheterization of the prostatic artery or caudal vesicular artery is attempted. Selecting a small order vascular branch such as these is advantageous as chemotherapy is being administered directly to the intra-tumoral blood supply and tissues unaffected by tumor are less likely to be exposed to high-dose chemotherapy. Theoretically, this results in greater tumor response and reduced systemic toxicity.

The IO techniques utilized in veterinary medicine are in the initial phases of implementation. Some of these procedures are able to palliate the clinical signs associated with neoplasia affecting the urinary tract, and the potential for utilizing IO techniques as a primary or neoadjuvant/adjuvant therapy exists. Further investigation is necessary to assess the long-term efficacy of these procedures, and additional advancements will likely be made.

REFERENCES

[1]Mutsaers AJ, et al. **J Vet Intern Med** 2003;17:136. [2]LeRoy BE, et al. **Vet J** 2009;180:149. [3]DeVocht TF, et al. **BJU International** 2003;91:627. [4]Denys P, et al. **J Urology** 2004;172:605. [5]Eisenberg L, et al. **J Urology** 2007;178:2456. [6]Weisse C, et al. **J Am Vet Med Assoc** 2006;229:226. [7]Newman RG, et al. **J Am Vet Med Assoc** 2009;234:236. [8]Christensen NI, et al. **Aust Vet J** 2010;88:478. [9]Barbalias GA, et al. **J Urology** 1997;158:54. [10]Ganatra AM, et al. **J Urology** 2005;174:2125. [11]Benson AD, et al. **J Urology** 2011;185:2217. [12]Berent AC, et al. **J Am Vet Med Assoc** 2011;238:1017. [13]Von Scheel JG, et al. **Arch Otorhinolaryngol** 1984;239:153. [14]McCaw DL, et al. **Vet Radiology** 1988;29:264. [15]Heidner GL, et al. **J Vet Intern Med** 1991;5:313. [16]Powers BE, et al. **Cancer** 1991;67:126. [17]Withrow SJ, et al. **Cancer** 1993;71:2484.

THE FUTURE OF PERSONALIZED MEDICINE FOR CANINE CANCER

Dr. Melissa C. Paoloni
Bethesda, MD

Novel approaches are needed to improve outcomes for cancer patients. In the last decade, advances in biological platforms and investigative tools has permitted the molecular characterization of cancer in a clinically relevant setting. Indeed, the field of personalized medicine (PMed) for cancer represents the realization such genomic, proteomic and epigenetic features, characterizations in the form of individualized cancer treatment that is no longer dependent on a histological description of a given patient cancer. [1,2,3,4] Goals of personalized medicine are to improve outcomes by revealing disease drivers or susceptibilities, potential toxicities, and resistance profiles unique to the patient. Although exciting, the applied clinical benefit of these time-consuming and costly techniques is yet unproven.

Molecular features of cancers have been the basis for selecting specific treatments for cancer for over 10 years. Such approaches have thus far been candidate based and include examples such as the use of imatinib (Gleevec) for bcr-abl translocations in AML, Her2Neu positive breast cancer patients with trustuzumab (Herceptin) and before this the use of tamoxifen in ER/PR positive breast cancer patients. [5,6] Such approaches represent some of the earliest forms of molecularly guided therapy. In its current form, PMed represents a large scale non-candidate based assessment of a given cancer rather than queries of a specific analyte. Such profiling is complex and does not necessarily lead to a single or defined intervention. Sophisticated mathematical models (algorithms) are needed to integrate these larger pools of molecular data and then model appropriate or reasonable therapeutic approaches. Examples of these non-candiate PMed studies have been reported. The Bisgrove trial, conducted by Van Hoff et al., treated 66 patients with refractory and metastatic cancers with regimens chosen through immunohistochemical and oligo expression profiling of each patient's tumor. [7] Progression free survival (PFS) improved compared to the immediately prior treatment approach in 27% of patients. [7] Although feasible and promising such PMed studies have paved the way for the controlled trials needed to definitively prove the therapeutic advantage of these approaches. The application of the PMed approach will also benefit from tissue collection approaches that have been optimized to yield molecular data, and from the rigorous assessment of the various PMed algorithms that convert molecular data to a suggested therapeutic intervention. Such optimization steps will be difficult to advance through the use of human clinical trials alone. Furthermore, it is not practical or scientifically reasonable to ask questions related to individualized therapy in clonally derived and expanded mouse models that are commonly used in drug development. Novel approaches to address questions in the PMed drug development path are necessary.

Comparative oncology is most often used to describe the study of cancer biology and therapy in pet animals with cancer. The heterogeneity and complexity of cancer in the our pet dog population and within cohorts of dogs with the same histological diagnoses is well-suited for modeling PMed. [8,9,10] The public availability of a progressively annotated canine genome and the advent of high throughput genomic techniques for the dog has enabled comparative oncology to describe canine cancer biology and define potential therapeutic targets in many of the same ways as human cancers. [11] Attributes of comparative models address many challenges of personalized medicine and provide an opportunity for progress. The field seeks to design adaptive trials incorporating individualized patient molecular profiling or target drug data iteratively. This allows investigators to modify stratification and determine if early trial termination is needed, so algorithms can be updated to include follow up data for on-going treatment arm assignments. Compressed progression times of veterinary oncology patients make this outcome data easier to compile and slot into these algorithms. It also permits phase II and phase III questions to be answered more expediently. Endpoints such as response rate (phase II), survival time, disease free interval or time to progression (phase III) can be simultaneously assessed within the same trial in veterinary cancer patients.

Personalized medicine success relies on combinational therapy. Since comparative modeling does not necessitate up-front treatment with standard of care regimens, novel and conventional agents can be tested together. Comparative oncology randomized control trials can be conducted in both the newly diagnosed and adjuvant setting. This will advance our understanding if primary tumor signatures or mutations predict best treatments for preventing metastases. Algorithms may define best "first," "second," "third" tier agents and comparative trials can combine them without restriction. Together with informatics and software advancements, comparative effectiveness studies between personalized medicine or biomarker algorithms can be designed, allowing for competing head-to-head trials. [12] Personalized medicine also focuses on wellness and disease prevention. Population studies in breeds predisposed to cancer may serve as excellent case-control cohorts for interventional studies on cancer prevention. As the field of personalized medicine grows, the clinical use of technologies such as expression arrays, exon or RNA sequencing, methylation, proteomics and target phospho-protein assays will increase. With this emphasis on personalized medicine, the pharmaceutical industry is also focused on the development of biomarkers and companion diagnostics for PMed approaches. [13] Comparative oncology models may help define tumor or surrogate biomarkers best suited for clinical development, especially if the most predictive represent a signature or pattern rather than a single molecule. Furthermore the opportunity to develop and test these novel assays in a relevant clinical setting will aid in translation to the human clinic.

To realize these advantages, a feasibility study was conducted through the Comparative Oncology Trials Consortium (COTC) to determine if the collection and processing of tumor samples from dogs with cancer, within a PMed framework, could be completed in a time period that would permit subsequent therapeutic studies of PMed in dogs with cancer. Tumor biopsies across multiple histologies and in cohorts of canine bladder transitional cell carcinoma (TCC), lymphoma and melanoma were collected and quality assurance/control measures applied to each step in the process of generating molecular data that would support a PMed derived therapeutic intervention. Results revealed high-quality, prospective tumor collections and large-scale target/drug identification studies in canine cancers are readily achievable. Data from this pilot study serves as rationale to now include the dog with cancer as a model to advance the field of PMed for human patients.

REFERENCES

1. Tian Q, Price ND, Hood L (2012) J Intern Med 271: 111-121. 2. Zhang L, Yang H, Xu J (2011) Curr Drug Metab 12: 455-459. 3. Xian W, McKeon F (2011) Genome Med 3: 22. 4. Stricker T, Catenacci DV, Seiwert TY (2011) Semin Oncol 38: 173-185. 5. Sawyers CL, et al. (2002) Blood 99: 3530-3539. 6. McKeage K, Perry CM (2002) Drugs 62: 209-243. 7. Von Hoff DD, et al. (2010) J Clin Oncol 28: 4877-4883. 8. Paoloni M, Khanna C (2008) Nat Rev Cancer 8: 147-156. 9. Vail DM, MacEwen EG (2000) Cancer Invest 18: 781-792. 10. Khanna C, et al. (2006) Nat Biotechnol 24: 1065-1066. 11. Lindblad-Toh K, et al. (2005) Nature 438: 803-819. 12. Ramsey SD, et al. (2011) Health Aff (Millwood) 30: 2259-2268. 13. Chiang A, Million RP (2011) Nat Rev Drug Discov 10: 895-896.

CACHEXIA IN FELINE CANCER PATIENTS

Erika L. Krick, VMD, DACVIM (Oncology)
Philadelphia, PA

Cachexia is a syndrome that occurs in several human and veterinary medical conditions. It is associated with involuntary weight loss, even in the face of adequate caloric intake, and loss of both fat and muscle mass occurs. Some of the more common diseases in humans that are associated with cachexia include AIDS, cancer, chronic obstructive pulmonary disease, renal failure, and heart failure.[1] The most common cancer types associated with cachexia in human medicine include gastrointestinal and colorectal cancers, lung cancer and pancreatic cancer.[2] It is estimated that > 1.3 million people in the United States experience cancer cachexia. Depending on the type of malignancy, up to 80% of cancer patients may exhibit signs of cachexia.[1,3] The clinical manifestations of cancer cachexia include involuntary weight loss, decreased muscle mass, reduced food intake and reduced performance status/quality of life scores.[4,5] In addition, cancer cachexia has significant physical and psychosocial implications, likely attributable to decreased muscle mass, malaise, and lethargy as well as anxiety regarding food intake and pressure from caregivers to increase food intake.[6-8] Patients with cancer cachexia have poor tolerance of and response to cancer treatment with a resulting shortened survival time.[5,9] It is estimated that 20-30% of cancer patients die from cancer cachexia.[4,10] The complex nature of the syndrome has made defining its clinical and metabolic components and developing effective treatments for it difficult.

Systemic inflammation, most commonly documented as elevations in pro-inflammatory cytokines such as TNFα, IFNγ, IL-1, IL-6 and C-reactive protein, is often associated with cancer cachexia.[4,5,10,11,12] These cytokines contribute to the cachectic state by increasing muscle loss, decreasing intestinal motility and gastric emptying leading to early satiety, and affecting production of hypothalamic neuropeptides, such as neuropeptide Y (NPY) and pro-opiomelanocortin (POMC), with the net result of decreased stimulation for food intake and increased protein catabolism.[11,13]

CACHEXIA IN VETERINARY PATIENTS

The clinical significance of body weight, body condition, and changes in body weight has been evaluated in some populations of veterinary patients, specifically canine and feline patients with heart disease and cancer. A study evaluating the relationship between body weight, body condition score (BCS), and survival for 108 dogs with congestive heart failure found a significant association between weight change after discharge from the hospital and survival, with dogs that gained weight having the longest survival time.[14] Dogs with the lowest and highest BCS had the shortest and longest survival, respectively, but the results were not statistically significant. A similar study evaluating 101 cats with congestive heart failure found that cats at the highest and lowest weights in the study population had significantly shorter survival times compared to cats in the mid-range weight group.[15] BCS was not associated with survival in either study, but because they were retrospective studies, BCS was performed by several different people and was not standardized. In contrast to dogs, weight changes after discharge from the hospital were not associated with survival in cats. Muscle mass score was not measured in either of these studies, likely due to their retrospective nature.

Among veterinary cancer patients, studies have suggested that cachexia is a more significant issue for cats compared to dogs. One study evaluating body weight and BCS of 100 dogs with cancer found that a minority of dogs had a BCS consistent with cachexia while 35% had some degree of muscle wasting (moderate to severe in 15%).[16] Interestingly, over half of the dogs had a BCS consistent with overweight or obese body condition. Dogs were evaluated at different time points during their treatment protocols (at the start of treatment, during treatment and in remission, during treatment and relapsed), and neither BCS nor body weight was associated with remission status. In contrast, almost half of 57 cats in a prospective study evaluating body weight, BCS, and fat and muscle mass in feline oncology patients had a BCS consistent with underweight or cachectic body condition.[17] In addition, the majority (91%) of cats had muscle loss at all of the 3 anatomical sites (temporal muscle, scapulae, hindlimbs) evaluated, and over half (60%) had fat loss at both of the anatomical sites (over the ribs and abdominal fat pad) evaluated. Both low body weight and BCS < 5/9 were associated with shorter survival times for all cats; only BCS was significantly associated with survival for the subset of cats with lymphoma. Cats in remission had a significantly higher mean BCS compared to cats that were not in remission, for both the entire study population and the subset of cats with lymphoma. On multivariate analysis, BCS and diagnosis of squamous cell carcinoma remained independent predictors of survival. All BCS assessments were performed by one individual in that study.

Changes in weight have also been shown to be associated with survival in cats with lymphoma. Cats that lost weight prior to being diagnosed with lymphoma had shorter survival times.[18] A more recent study examined the effect on survival of weight changes during treatment in 209 cats with lymphoma.[19] Cats were divided into weight change categories according to the percent weight change from baseline to the time point being assessed. If the time point weight was ≥ 5% less than the baseline weight, the cat was put in the 'lost weight' category, if the weight was ≥ 5% more than the baseline line weight, the cat was put in the 'gained weight' category, and if the weight was within 5% of baseline in either direction, the cat was put in the 'unchanged weight' group. Cats that lost ≥ 5% of weight compared to baseline 1 and 2 months into treatment had significantly shorter survival times compared to cats that had unchanged weight or gained weight. The same association was found for the subset of cats with large cell lymphoma, though it was only statistically significant at the 1 month time point. Overall, most of the cats lost weight over time, particularly the cats with large cell LSA. Interestingly, cats with lower baseline weights gained weight over time, while larger cats lost weight. The placement of an enteral feeding tube was more common in cats with small versus large cell LSA, and placement of a feeding tube was not associated with any change in weight.

Although these studies indicate that body weight and BCS can predict outcome in feline cancer patients, the role of cachexia as an independent prognostic factor remains unclear. For example, if patients with more aggressive tumors are at highest risk for cachexia, cachexia may be a marker of aggressive cancer, not necessarily a separate predictor of outcome. Future studies are needed to better clarify this issue.

COMPARATIVE ASPECTS

The feline patients described in the above studies seem to have clinical signs consistent with human cancer cachexia. For example, loss of both fat and muscle tissue has been observed in cats with cancer, and low body weight and BCS as well as weight loss are associated with shorter survival. In addition, human patients with gastrointestinal cancers are at higher risk for cancer cachexia, and gastrointestinal lymphoma is one of the most commonly diagnosed cancers in the cat. It is difficult to assess the role that cancer cachexia plays in chemotherapy tolerance in feline cancer patients due to the retrospective nature of most studies and the challenge in separating chemotherapy side effects from clinical signs of tumor progression, particularly in cats with gastrointestinal lymphoma. Although the clinical aspects are similar between cats and people, the metabolic abnormalities consistent with cancer cachexia in human patients have not yet been documented in cats. Elevation in inflammatory cytokines has been observed in feline cancer patients, but no distinction in these values between cats with and without cachexia has been made. Specifically, increases in serum α_1-acid glycoprotein have been documented in cats with cancer (both solid tumors and lymphoma) compared to healthy cats.[20,21] For cats with lymphoma, serum α_1-acid glycoprotein levels did not change significantly according to the patient's remission status, and the degree of elevation did not correlate with outcome.[21] Many studies can be undertaken to further understand the role of inflammatory cytokines in cats with cancer and healthy cats regarding body weight, body condition and muscle mass score. It is likely that cancer cachexia in cats is a multi-factorial syndrome as it is in humans, and as such, there are potentially multiple areas for therapeutic intervention. Our challenge is to identify these areas and develop a mechanism for organized use of therapeutic interventions for feline patients with both signs of and risk factors for cancer cachexia in order to maximize their effect.

POTENTIAL THERAPEUTIC TARGETS

The multi-factorial nature of cancer cachexia necessitates the exploration of multi-modality therapeutic interventions. For our feline patients, such potential interventions may include chemotherapy protocol alterations, decreasing inflammation or discomfort due to the cancer and/or cachexia syndrome, effective antiemetic control, appetite stimulation, and improving nutritional status by correcting clinically relevant deficiencies and offering calorie and/or protein dense diets.

Certainly, appropriate cancer treatment is an essential part of treatment for this condition, but this aspect is often challenging because of the poor physical condition and quality of life of patients with cancer cachexia. Selecting chemotherapy agents with a more tolerable toxicity profile, however, may allow fewer dose reductions, which could increase the patient's chance of responding. One such example is the substitution of vinblastine for vincristine. Vinblastine is associated with fewer gastrointestinal and neurologic side effects in people compared to vincristine and as such, may be better tolerated by cachectic feline cancer patients. A randomized clinical trial comparing the efficacy and toxicity of vincristine and vinblastine in a COP-based chemotherapy protocol was recently completed at the authors' institution, and the results will be presented as part of this lecture.[22]

Providing appropriate supportive care is particularly important for patients that are exhibiting clinical signs of cancer cachexia. Such examples may include alleviating pain or discomfort with anti-inflammatory medications (glucocorticoids, especially for lymphoma) and/or pain medications, ensuring adequate nausea control, and use of appetite stimulants. Ondansetron and maropitant are often effective in reducing tumor or chemotherapy related nausea, while cyproheptadine and mirtazapine are often used to stimulate appetite. It is important to attempt to effectively control nausea before instituting appetite stimulants. Cyproheptadine was recently shown to be effective for increasing weight gain in children with cancer cachexia.[23] Children in that study who did not respond to cyproheptadine received megestrol acetate, and most of those children gained weight as well. Megestrol acetate has been shown to increase appetite and produce weight gain in human patients with cancer cachexia in some studies, but a multi-modality treatment approach was most effective in one study.[24-27] There are several known risk factors associated with the use of megestrol acetate in cats.

Another potential factor related to cancer cachexia in cats is cobalamin deficiency. The majority of cats with small cell intestinal lymphoma are cobalamin deficient at diagnosis, but the prevalence among cats with large cell lymphoma remains unknown.[28] Cobalamin deficiency is associated with weight loss, poor appetite, and low BCS, so it would be helpful to determine if this could be a contributing factor to the cachexia-related signs seen in some cats with large cell lymphoma.[29,30] One major challenge in treating cachectic feline cancer patients is increasing their caloric intake. Often times these patients have a poor appetite due to a variety of causes, and even with supportive care it is difficult to improve their appetite enough to allow for weight gain. Encouraging clients to feed diets that are calorie and/or protein-rich may help, as patients are able to consume more calories from a smaller quantity of food, but to date there is no specific diet that has been formulated for cats with cancer.

REFERENCES

[1]Morley JE et al. **Am J Clin Nutr** 2006;83:735-743. [2]Dodson S et al. **Annu Rev Med** 2011;62:8.1-8.15. [3]von Haeling S et al. **J Cachexia Sarcopenia Muscle** 2010;1:1-5. [4]Fearon K et al. **The Lancet Oncol** 2011;12:489-495. [5]Fearon KC et al. **Am J Clin Nutr** 2006;83:1345-1350. [6]Tan BH et al. **Curr Opin Nutr Metab Care** 2008;11:400-407. [7]Hinsley R et al. **Int J Palliat Nurs** 2007;13:84-89. [8]Dahele M et al. **J Pain Symptom Manage** 2007;33:676-685. [9]Dewys WE et al. **Am J Med** 1980;69:491-497. [10]Tisdale MJ. **JNCI** 1997;89:1763-1773. [11]Stephens NA et al. **Curr Opin Support Palliat Care** 2008;2:267-274. [12]Krzstek-Korpacka M et al. **Clin Biochem** 2007;40:1353-1360. [13]Plata-Salaman CR. **Nutrition** 2000;16:1009-1012. [14]Finn E **J Vet Intern Med** 2010;24:1369-1374. [15]Slupe JL et al. **J Vet Intern Med** 2008;22:561-565. [16]Michel KE et al. **J Vet Intern Med** 2004;18:692-695. [17]Baez JL et al. **J Feline Med Surg** 2007;9:411-417. [18]Haddon AG et al. **J Vet Intern Med** 2008-22:153-157. [19]Krick EL et al. **J Feline Med Surg** 2011;13:976-983. [20]Selting KA et al. **J Vet Intern Med** 2000;14:503-506. [21]Correa SS et al. **J Am Anim Hosp Assoc** 2001; 37:153-158. [22]Krick E et al. **Vet Comp Oncol** 2011;9:e25. [23]Couluris et al. **J Pediatr Hematol Oncol** 2008;30:791-797. [24]Gebbia V et al. **Br J Cancer** 1996;73:1576-1580. [25]Lesniak et al. **Pol Arch Med Wewn** 2008;118:636-644. [26]Madeddu et al. **Clin Nutr** 2011; doi:10.1016/j.clnu.2011.10.005. [27]Mantovani G et al. **Oncologist** 2010;15:200-211. [28]Kiselow M et al. **J Am Vet Med Assoc** 2008;232:405-410. [29]Keesling PT et al, **J Anim Sci** 1975;41:317. [30]Reed N et al. **J Feline Med Surg** 2007;9:278-288.

COMPLEMENTARY AND ALTERNATIVE MEDICINE FOR VETERINARY CANCER PATIENTS: WHAT YOU AND YOUR CLIENTS SHOULD KNOW

Michael O. Childress, DVM, MS, DACVIM (Oncology)
West Lafayette, IN

WHAT IS "COMPLEMENTARY AND ALTERNATIVE MEDICINE"?

"Complementary and alternative medicine" (CAM) may encompass any of a number of herbal, nutraceutical, physical, or other treatments for veterinary cancer patients that have not been prescribed by a licensed doctor of veterinary medicine. Some authors make a distinction between *alternative medicine,* which is treatment intended to replace that recommended by a medical care provider, and *complementary medicine,* which is treatment given in addition to that prescribed by a medical care provider.[1,2] While alternative therapy is usually prescribed with the intent of directly treating a cancer, complementary therapy is often aimed at palliating clinical signs, such as pain or nausea, that are associated with cancer or cancer treatment.

PREVALENCE OF COMPLEMENTARY AND ALTERNATIVE CANCER THERAPIES

A recent meta-analysis reviewing data from over 65,000 human cancer patients estimated that 40% of patients employ some sort of CAM as part of their treatment,[3] although the use of CAM may exceed 90% in some human cancer patient populations.[4] Total estimated expenditure for all CAM therapies (used by both cancer patients and non-cancer patients) in the United States exceeds $30 billion per year.[5]

While the prevalence of CAM use in veterinary cancer patients is not extensively reported, a survey conducted at Colorado State University's Animal Cancer Clinic found that 76% of owners of dogs and cats with cancer used some form of CAM to treat their pets.[6] Nutritional supplements (including vitamins), special diets, and prayer were the most common CAM therapies used by survey respondents. It is noteworthy that 85% of respondents who admitted to using CAM for their pet had not discussed the use of such therapy with a veterinarian. This statistic suggests that many pet owners may feel uncomfortable discussing unconventional therapies with veterinarians or other members of the veterinary team.

ARE COMPLEMENTARY AND ALTERNATIVE CANCER THERAPIES HELPFUL?

CAM therapies encompass a vast array of medicinal, nutritional, physical, mental, or spiritual interventions. These include herbal products, nutraceuticals, megadose vitamins, special diets, massage therapy, acupuncture, prayer, and meditation. A number of different outcomes may be promised by proponents of these treatments, including cancer cure, cancer prevention, palliation of cancer- or treatment-related side effects, augmentation of immune system function, or enhancement of overall health and well-being. *For the most part, objective evidence to support claims of clinical anti-cancer efficacy associated with any of these therapies is lacking, particularly for those therapies that claim to cure cancer. This is an important fact to gently emphasize to owners who are considering pursuing a CAM therapy rather than an established cancer therapy, particularly if the established therapy has a high likelihood of therapeutic success.*

While few, if any, CAM therapies are likely to have clinically relevant anti-cancer efficacy, some therapies appear to be effective for palliating cancer- or treatment-related morbidities (such as pain or nausea), or improving the physical or psychological well-being of human cancer patients.[2] Here again though, there is a paucity of high-quality scientific evidence to support the efficacy of CAM therapies for palliative purposes.[7] Moreover, many of the therapies which do show promise, such as mind-body exercises (e.g. meditation, patient support groups), are poorly translatable to veterinary patients due to differences in animal and human cognition. However, a CAM therapy that may be useful in a palliative setting across species boundaries is acupuncture. Acupuncture has been shown to ameliorate acute vomiting following cancer chemotherapy and palliate cancer pain in a few well-designed randomized clinical trials involving human cancer patients,[8,9] although recent reviews[10,11] conclude that more study of this CAM modality is required in order to define its true therapeutic potential in the oncology clinic. Acupuncture has also shown some benefit in the management of non-cancer-related pain in canine studies.[12-14] *It should, however, be noted that acupuncture has not been critically evaluated for palliation of clinical signs related to cancer or cancer therapy in dogs or cats.*

ARE COMPLEMENTARY AND ALTERNATIVE CANCER THERAPIES HARMFUL?

Fortunately, most CAM therapies are unlikely to be harmful. However, some therapies may do significant harm if used inappropriately. *Any CAM therapy should be carefully scrutinized as to its potential for harm before its use is recommended or condoned.* In general, CAM therapies may be harmful in four ways:

1. They may be directly toxic or injurious to the patient
2. They may interact with another therapy in such a way as to be toxic or injurious to the patient
3. They may interact with another therapy in such a way as to make it less effective

4. Their use may prompt owners to delay seeking treatment by a more effective method, allowing the cancer to progress and rendering other treatments less likely to be effective

While there is little published evidence documenting harm done by unconventional therapy to veterinary cancer patients, some lessons may be drawn from the experience of human cancer patients. Knowledge of the molecular mechanisms of action of cancer therapies may also help to inform decisions as to which unconventional cancer therapies may be harmful. Some examples of harmful or potentially harmful CAM therapies include:

1. *Escharotics* – These are herbal medicinal preparations which often contain extract of the *bloodroot* plant (*Sanguinaria canadensis*).[15] They are usually manufactured as injectable or topical preparations. Escharotics are so named because of their ability to elicit necrosis with subsequent eschar formation following contact with living tissues. They are most commonly used for the treatment of cutaneous tumors. Products specifically marketed for companion animals include NeoplaseneX, XXTerra, and Cansema Black Salve Mix.[16-17] *Escharotics have the potential to cause significant tissue injury and their routine use in the treatment of animal cancer cannot be recommended at this time.*
2. *Herbal (botanical) medications* – Many antineoplastic drugs currently in use in humans and domestic animals are derived from plants or other natural sources. Examples include vincristine, vinblastine, paclitaxel, docetaxel, and etoposide. These drugs have been licensed for use by the Food and Drug Administration (FDA), and stringent regulations are in place to assure quality control in their manufacture. The efficacy and toxicity of these drugs are generally well established, as are their pharmacokinetic and pharmacodynamic profiles. In contrast, herbal medications available over the counter or over the Internet may have little quality control in their manufacture, and relatively little is known about their pharmacology.[18] Extracts from plants or other natural sources may contain a number of biologically active substances, some of which may have deleterious effects. Moreover, because their manufacture is not strictly regulated, there may be variable concentrations of active ingredients from one product to the next, and some products may contain harmful adulterants or excipients. Herbal medications may interact with enzymes critical for drug metabolism such as *cytochrome P450 enzymes* and *p-glycoprotein.*[18,19] Inhibition of these enzymes may alter the clearance of other drugs, including standard chemotherapy drugs, thereby augmenting their toxicity.
3. *Antioxidants* – Several anticancer therapies exert their effects through the generation of oxygen free radicals. Examples include radiation therapy and the chemotherapy drug doxorubicin. Antioxidants scavenge and detoxify free radicals, thereby rendering them harmless. While antioxidants are not likely to be harmful themselves, they may reduce the efficacy of therapies that kill cancer cells using free radicals. Therefore, many oncologists recommend that patients receiving chemotherapy or radiation therapy should refrain from taking antioxidants. Experimental data to support such a position are limited; however, a randomized clinical trial conducted by Bairati, et al. showed that α-tocopherol (Vitamin E) supplementation was associated with increased risk of tumor recurrence and second primary cancer formation, as well as increased risk of mortality due to all causes in a cohort of 540 human patients treated with radiation therapy for head and neck cancer.[20,21] Most studies fail to show a harmful effect of antioxidant supplementation in human cancer patients receiving chemotherapy, although these studies have been criticized for being inadequately designed and lacking in statistical power.[22,23]

DISCUSSING COMPLEMENTARY AND ALTERNATIVE THERAPY WITH PET OWNERS

The apparent popularity of CAM therapies among owners of pets with cancer underscores that veterinarians should develop a working knowledge of the benefits and detriments of CAM. Veterinarians should inquire as to whether owners are employing CAM to treat their pet's cancer while taking a complete medical history. When discussing CAM with pet owners, a dismissive or condemnatory attitude should be avoided. Such an approach may cause the owners embarrassment, may discourage more meaningful dialogue in the future, and may paradoxically result in greater willingness of owners to turn to unconventional therapy and eschew traditional medical treatment. The physician Franz Ingelfinger perhaps described this phenomenon best in stating that, "…an establishment indictment of a popular remedy is one of the best advertisements for that remedy."[24] An open and compassionate discussion with pet owners is likely to facilitate greater willingness of owners to divulge which unconventional therapies they use on their pets. This in turn may foster greater adoption of therapies that are potentially helpful and avoidance of those that are potentially harmful.

WHERE TO LOOK FOR MORE INFORMATION

The sheer number of CAM therapies marketed for the treatment of cancer in humans and companion animals is daunting. To compound this problem, the explosion of information available about these therapies on the Internet can make attempts to find scientifically reliable information on any single therapy an exercise in frustration. Much of the information on the Internet regarding unconventional therapies comes in the form of anecdotes, testimonials, or other unsubstantiated claims of efficacy. In contrast, there is a paucity of high quality information available to help pet owners and veterinarians decide which therapies are safe and/or potentially effective and which are ineffective or potentially harmful. Print and Internet resources providing scientifically sound information on CAM therapies for cancer are listed below in the references.[25-29]

REFERENCES

1. Metz JM. *Med Pediatr Oncol* 2000;34:20-26. 2. Ernst E, et al. *Eur J Cancer Care* 1999;35:1608-1613. 3. Horneber M, et al. *Integ Cancer Ther* 2011 Oct 21. [Epub ahead of print] 4. Bishop FL, et al. *Pediatrics* 2010;125:768-776. 5. Herman PM, et al. *BMC Complement Altern Med* 2005;5:11. 6. Lana SE, et al. *J Am Anim Hosp Assoc* 2006;42:361-365. 7. Bardia A, et al. *J Clin Oncol* 2006;34:5457-5464. 8. Alimi D, et al. *J Clin Oncol* 2003;21:4120-4126. 9. Shen J, et al. *JAMA* 2000284:2755-2761. 10. Ezzo J, et al. *Cochrane Database Syst Rev* 2006;2:CD002285. 11. Paley CA, et al. *Cochrane Database Syst Rev* 2011;1:CD007753. 12. Groppetti D, et al. *Br J Anaesth* 2011;107:612-618. 13. Joaquim JGF, et al. *J Am Vet Med Assoc* 2010;236:1225-1229. 14. Laim A, et al. *J Am Vet Med Assoc* 2009;234:1141-1146. 15. McDaniel S, et al. *Arch Dermatol* 2002;138:1593-1596. 16. Childress MO, et al. *J Am Vet Med Assoc* 2011;239:374-379. 17. White SD, et al. *Proceedings: 52nd Annual Convention of the American Association of Equine Practitioners.* San Antonio, TX:2006;p457-500. 18. Shord SS, et al. *Integr Cancer Ther* 2009;8:208-227. 19. Sparreboom A, et al. *J Clin Oncol* 2004;22:2489-2503. 20. Bairati I, et al. *J Natl Cancer Inst* 2005;97:481-488. 21.Bairati I, et al. *Int J Cancer* 2006;119:2221-2224. 22. Block KI, et al. *Cancer Treat Rev* 2007;33:407-418. 23. Lawenda BD, et al. *J Natl Cancer Inst* 2008;100:773-783. 24. Ingelfinger FJ. *N Engl J Med* 1977;296:1167-1168. 25. Fetrow CH, et al. *Professional's Handbook of Complementary and Alternative Medicines.* 3rd ed. Philadelphia: Lippincott Williams and Wilkins, 2003. 26. Robinson NG. Complementary and alternative medicine for patients with cancer. In: Withrow SJ and Vail DM, eds. *Small Animal Clinical Oncology.* 4th ed. St. Louis: Saunders-Elsevier, 2007;1126-1132. 27. National Center for Complementary and Alternative Medicine website. Available at: http://nccam.nih.gov/ 28. American Cancer Society website. Available at: http://www.cancer.org 29. National Cancer Institute website. Available at: http://www.cancer.gov/cancertopics/cam

8a

THYMIDINE KINASE 1 AND C-REACTIVE PROTEIN AS BIOMARKERS FOR SCREENING CLINICALLY HEALTHY DOGS FOR OCCULT DISEASE. [1]KA Selting, [2]CR Sharp, [3]R Ringold. [1]University of Missouri, College of Veterinary Medicine, Columbia, MO; [2]Tufts University, Cummings School of Veterinary Medicine, North Grafton, MA; [3]Veterinary Diagnostic Institute, Inc., Simi Valley, CA.

While most biomarkers of proliferation require fixed tissue, thymidine kinase (TK1) is soluble and can be measured in serum. TK1 can be elevated in a wide array of cancers, especially B-cell lymphoma and hemangiosarcoma. C-reactive protein (CRP) is a soluble biomarker of inflammatory conditions. Our purpose was to screen a large population of clinically healthy dogs, combining TK1 and CRP to create a Neoplasia Index (NI) to identify dogs for further diagnostic testing.

Serum was collected from clinically healthy dogs of at-risk breeds. TK1 was measured using a quantitative chemiluminescent assay.[a] CRP was measured using a canine-specific quantitative sandwich ELISA assay. High TK1 dictated cavity imaging whereas normal TK1 was followed for a minimum of 6 months by questionnaire. ROC analysis was used to determine optimal cutoff points to differentiate healthy from cancer-bearing dogs.

A total of 342 dogs were tested (157 goldens, 185 shepherds). Eleven dogs were diagnosed with cancer. Using the NI, the optimal cutoff of TK1 could be reduced from 6.0 U/L to 2.0 U/L, which increased sensitivity without loss of specificity (AUC=0.965 and 0.871 at 4 and 6 months respectively). Negative predictive value was 1.00 and 0.994 at 4 and 6 months, respectively. Abnormal results ranged from 0.4->100 for TK1; 0.2-46.6 for CRP; and 0.01-137 for NI. There were 16 deaths during the study: 3.4% of dogs with negative NI, 12.2% if positive and 50% if high positive.

Combining TK1 and CRP measurements distinguished between healthy dogs and those harboring occult neoplasia, justifying further testing.

[a]LIASON, DiaSorin, Stillwater, MN
[b]TECO Medical AG, Switzerland

8b

ONCOLOGIC OUTCOME AND PROGNOSTIC FACTORS IN 1134 DOGS WITH APPENDICULAR OSTEOSARCOMA TREATED AT A SINGLE INSTITUTION. L. Selmic[1], S. D. Ryan[1], N. P. Ehrhart[1], D. R. Worley[1], J. Eickhoff[2], S. J. Withrow[1]. [1]The Animal Cancer Center, Colorado State University, Fort Collins, CO. [2]University of Wisconsin, Madison, WI.

The purpose of the study was twofold, to report the outcome following treatment in a large population of dogs with appendicular osteosarcoma (OSA) and to evaluate for prognostic factors in the cohort of dogs that received amputation and chemotherapy.

Retrospective review of medical records at institution X identified 1134 dogs that received treatment for appendicular OSA between 1997 and 2010. Palliative-intent treatments were administered in 377 dogs (33.2%): analgesic medications alone (n=41, Median survival time (MST)=66 days), palliative radiation treatment ± bisphosphonates (n=177, MST=122 days), amputation only (n=147) and limb-spare only (n=12). Curative-intent treatments were administered in 757 dogs (66.8%): curative-intent radiation therapy and chemotherapy (n=44, MST=243 days), amputation and chemotherapy (n=488, MST=255 days) and limb-spare and chemotherapy (n=225, MST=389 days). Univariate analysis revealed that factors associated with a poorer survival included ALP > 142 IU/L ($p<0.03$), weight ≥ 40Kg ($p<0.006$), proximal vs. distal bone location ($p<0.005$) and humerus vs. all other sites ($p<0.001$). In the multivariate analysis, following adjustment for ALP, stage, weight and proximal vs. distal bone location, dogs with a femoral OSA location were found to have a significantly worse survival compared to dogs with radial OSA location ($p<0.03$).

Treatment outcomes in a large population of dogs with appendicular osteosarcoma are similar to those previously reported. Location may be a more important prognostic indicator than previously recognized.

100a

FOLATE - TARGETED THERAPY FOR CANINE URINARY BLADDER CANCER.

[1]D Dhawan, [1]R Rothenbuhler, [1]JF Naughton, [2]L Cheng, [1]JA Ramos-Vara, [1]PS Low, [3]CP Leamon, [3]N Parker, [3]PJ Klein, [3]I Vlahov, [1]L Fourez, [1]DW Knapp. [1]Purdue University, West Lafayette, IN; [2]Indiana University School of Medicine, Indianapolis, IN; [3]Endocyte, West Lafayette, IN.

More effective therapy is needed for canine invasive transitional cell carcinoma (TCC). Surgical removal of TCC in dogs is usually not possible. Although chemotherapy clearly prolongs survival, most dogs with TCC ultimately die from it. The purpose of this work was to define folate receptor (FR) expression and folate uptake in canine TCC; and to determine the maximum tolerated dose (MTD), antitumor activity, and toxicity of folate-targeted vinblastine in dogs with TCC.

FR expression in normal canine bladder and TCC was determined by immunohistochemistry (IHC) using the antibody PU17 which recognizes FRα and FRβ. Cellular localization, staining intensity, and % positive cells were recorded. Folate uptake was determined *ex vivo* by folate binding assay, and *in vivo* by nuclear scintigraphy with a ^{99m}Tc based FR targeted imaging agent. A dose escalation study of folate-targeted vinblastine (EC0905) was performed in dogs with biopsy confirmed, measurable TCC. EC0905 was administered IV once weekly (starting dose 0.2 mg/kg). Dose escalation was performed within and between dogs with at least 3 dogs in each dose group, and at least 6 dogs treated at the MTD. Toxicity was assessed by CBCs, serum chemistry profiles, urinalyses, physical exams, and owner observations. Toxicity was classified by VCOG criteria. The MTD was defined as the highest dose that resulted in 0 of 6 dogs having grade 4 toxicity and 0 or 1 of 6 dogs having grade 3 toxicity. Tumor response was assessed by physical exam, ultrasonography including detailed bladder mapping, and radiography.

FR expression was detected by IHC in 51 of 68 (75%) of primary tumors, in 7 of 12 (58%) nodal metastases, and in 10 of 21 (48%) of lung metastases, as well as in epithelial cells in 10 of 10 normal bladders. Folate binding to TCC tissues was detected in *ex vivo* studies. In 12 dogs undergoing nuclear scintigraphy, folate uptake in primary and metastatic lesions was observed, although radioactivity in urine hindered the detection of folate uptake in bladder masses in some cases. Distension of the bladder with saline, and individual patient positioning were important in the scintigraphy studies. The clinical trial of EC0905 included 10 dogs. The MTD was 0.25 mg/kg once weekly. No urologic toxicity was noted. Toxicity at doses above the MTD included gastrointestinal upset and myelosuppression. Even at the MTD, dose reduction became necessary after multiple cycles of treatment, and a more conservative starting dose would be 0.22 mg/kg once weekly. The tumor responses included partial remission ($\geq$50% reduction in tumor volume) in 5 dogs and stable disease (<50% change in tumor volume) in 5 dogs. The median length of treatment was 7 weeks (range 1-50 weeks).

In conclusion, FR expression was noted in the majority of TCC in dogs. Folate-targeted therapy had promising antitumor activity against TCC, and was generally well tolerated in dogs.

100b

DETECTION OF CHEMICALS IN THE URINE OF DOGS FOLLOWING HOME LAWN APPLICATION. DW Knapp, AR Diggs, WA Peer, BR Cooper, T Lee, PL Bonney, NW Glickman, LT Glickman, AS Murphy. Purdue University, West Lafayette, IN.

Transitional cell carcinoma (TCC) of the urinary bladder in dogs is most likely caused by a combination of heritable and environmental risk factors. Scottish terriers have an 18-20 fold increased risk of TCC compared to other dog breeds. TCC risk has been further increased 7 fold in Scottish terriers following exposure to lawn herbicides and pesticides. We hypothesize that these chemicals are eliminated through the urine following exposure, thus increasing TCC risk. The purpose of the present study was to measure concentrations of 3 commonly applied chemicals in the urine of dogs before and after application of the chemicals to lawns, by either the home owners or a commercial lawn company. Dogs living in homes with untreated lawns were also sampled.

Eligible pet dogs were those > 1 yr of age that lived in either a home with treated or untreated lawn in the West Lafayette, IN community. A drag sled method was used to collect dislodgeable residue from a 1.8 square foot area of grass prior to and at 24 and 48 hours after chemical application in treated lawns, and at one time point in untreated lawns. Free catch urine samples were collected from the dogs at these same time points. LC-MS was used to measure 2,4-dichlorophenoxyacetic acid (2,4-D), 4-chloro-2-methylphenoxypropionic acid (MCPP), and dithiopyr. Urine chemical concentrations were normalized to urine creatinine concentration. The length of time dogs spent on the lawn and on other grassy areas, history of the dog eating grass, and length of time the dog was restricted from the lawn following treatment were recorded.

Twenty-five treated and 8 untreated households were studied. One or more of the 3 chemicals was detected in urine of dogs in 20 of 25 households following treatment including 2,4-D in dogs from 7 households (range 113 – 1508 μg/g creatinine), MCPP in dogs from 15 households (range 0.6 – 506 ng/g creatinine), and dithiopyr in dogs in 4 households (range 0.05 to 2.3 ng/g creatinine). Following lawn treatment, chemical concentrations in the grass residues ranged from 0 – 109 $\mu g/m^2$ 2,4-D, 0 to 14 ng/m^2 MCPP, and 0 to 64 pg/m^2 dithiopyr, with concentrations still present at 48 hrs post treatment. The mean time dogs were restricted from the lawn after treatment was 5.8 $\pm$ 6.2 hrs. Chemicals were detected in dog urine prior to lawn treatment in 14 treated households, and in dog urine in 4 untreated households. Eight of these 18 dogs had access to other grassy areas. Chemicals were detected in grass residues prior to treatment in 20 treated households and in grass residues from 4 untreated households, possibly due to "drift" from adjacent areas.

The finding of lawn chemicals in the urine of 80% of dogs living in homes where the lawn was treated and in the urine of 50% of dogs living in homes where the lawns were untreated, suggests widespread chemical exposure of pet dogs to lawn chemicals. There is concern that some of these chemicals increase TCC risk. Therefore, exposure to these chemicals should be further investigated for the carcinogenic potential, including studies in dogs in TCC susceptible breeds.

TARGETED ANTIBODY THERAPIES; EXPECT MORE, PAY LESS

Nicola Mason, BVetMed, PhD, DACVIM
Philadelphia, PA

IMMUNOGLOBULINS – STRUCTURE AND FUNCTION

Antibodies are gamma globulin proteins that are produced by B cells and mediate target-specific neutralization and immunological attack against pathogens and neoplastic cells during innate and adaptive immune responses. Antibodies consist of two heavy and two light immunoglobulin (Ig) polypeptide chains, held together by intra-chain and inter-chain disulphide bonds (Fig. 1). The two antigen binding sites are each composed of variable portions of the heavy (VH) and light (VL) immunoglobulin chains that together determine the antigen specificity of the antibody. The fraction crystalizable (Fc) portion of the antibody mediates its immunological effects through interactions with cells and soluble factors such as complement. Five classes of antibodies exist IgM, IgG, IgA, IgD and IgE and each class is defined by their Fc region and their effector function. The IgG class is the major serum immunoglobulin and is further divided into subclasses IgG1, IgG2a and IgG2b, IgG3 and IgG4. IgG2a and IgG3 have significant cytotoxic capabilities through their ability to recruit cytotoxic cells such as NK cells and mediate Antibody Dependent Cell-mediated Cytotoxicity (ADCC). Additional mechanisms of cytotoxicity mediated by antibodies include Complement Dependent Cytotoxicity (CDC), induction of apoptosis following ligand binding, inhibition of cellular growth and sensitization to chemotherapy or radiation therapy. Antibody functions such as virus and cytokine together with the induction of apoptosis following ligand binding are mediated by antigen binding alone, without a requirement for the Fc portion of the immunoglobulin. The enormous diversity of antigen specificity within an individual's antibody repertoire is necessary to protect the individual from a wide variety of infectious agents and pathogens. This diversity is generated at the molecular level by random recombination of germline genes (Variable (V), Diversity (D) and Joining (J) genes) to generate billions of virtually unique nucleotide sequences that encode the antigen binding sites of the antibody. Three CDRs (Complementarity Determining Regions) are interspersed between 4 framework regions (FR) that are highly conserved between antibodies. It is the CDRs that are responsible for the antigen-specificity of the antibody molecule.

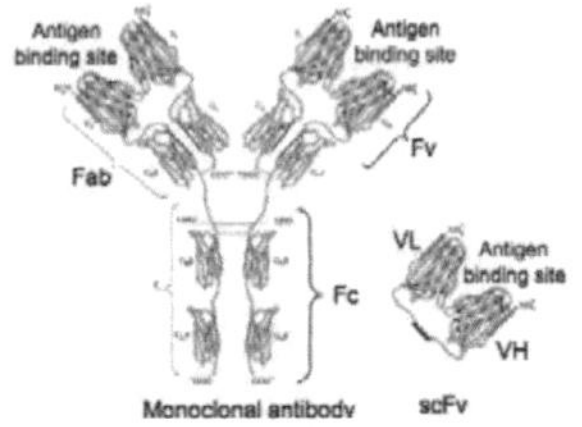

Fig. 1. Cartoon of MAb and single chain variable fragments (scFv)

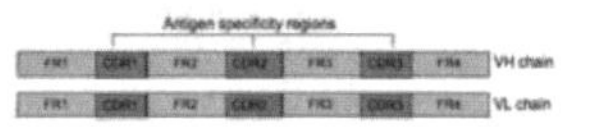

Fig. 2. Framework regions and Complementarity Determining Regions of VH and VL immunoglobulin chains

MONOCLONAL ANTIBODIES IN HUMAN MEDICINE

The pioneering work of Milstein and Kohler in the mid 70's resulted in the development of hybridoma technology and the ability to produce large quantities of monoclonal antibodies (MAbs) (antibodies that have specificity for one antigen only).[1] This led to an exponential rise in the generation of therapeutic and diagnostic monoclonal antibodies. Now there are 22 FDA approved MAbs in use in the human clinic and hundreds more are in various stages of development. About half of the approved MAbs recognize bona fide targets on lymphomas, leukemias and carcinomas (breast and colorectal) and their inclusion into standard of care chemotherapy regimes has revolutionized the treatment of these malignancies. Monoclonal antibodies such as rituximab (anti-CD20), bevacizumab (anti-Vascular Endothelial Growth Factor) and trastuzumab (anti-erb2) are shining examples of successful MAbs that specifically target tumors and their vasculature leading to prolonged progression free survival, overall survival and improved quality of life in human patients with B cell malignancies, colorectal adenocarcinoma and Her2 positive breast carcinoma respectively.[2-4] Other MAbs have been licensed for the treatment of auto-immune disease and transplant rejection through their ability to block the action of pro-inflammatory cytokines. Anti-viral MAbs and anti-platelet (GPIIb,IIIa) MAbs have also been licensed for use in patients with RSV and cardiovascular disease.

FROM MICE TO MEN

The first FDA approved MAb (Orthoclone; anti-CD3 for treatment of transplant rejection) was a murine (IgG2a) antibody. These xenogenic antibodies induced human immune responses leading to unwanted side-effects and reduced efficacy in repeat rounds of administration. As technology progressed, chimeric MAbs were genetically engineered where the less variable antibody framework regions of the original mouse antibody were replaced with human framework regions to reduce the in vivo generation of anti-MAb antibodies. Nowadays, fully human MAbs are being synthesized that have significantly less side effects following administration and less chance of being rejected by the patients' immune response.

THE SUCCESS OF MAB IN THE HUMAN CLINIC

The therapeutic success of MAbs such as rituximab and trastuzumab are well documented in the scientific literature. This success is underscored by reports of the global revenue generated by MAb sales. The world's MAb revenue for 2010 was \$44.6 billion and is expected to rise to \$62.3 billion by 2015 (www.visiongain.com). These astonishing figures emphasizing the ongoing success of monoclonal antibody therapy in the human field and cement their place at the forefront of immunotherapy for cancer, auto-immunity and transplantation.

MONOCLONAL ANTIBODIES IN THE VETERINARY CLINIC

Despite the well-documented success of MAbs in human medicine, MAbs have yet to make an impact in veterinary medicine. This is due to the fact that currently available MAbs 1) do not cross react with canine molecules (e.g. Rituximab an anti-

CD20 MAb that is effective in the treatment of many human B cell malignancies does not bind to canine CD20), 2) are human or human/murine chimeras and are therefore likely to be rejected when used repeatedly in dogs and 3) are prohibitively expensive (e.g. the anti-VEGF MAb (Avastin; bevacizumab) has an estimated cost per person of ~$80,000 per year). Only one MAb has been licensed by the USDA for use in the veterinary clinic. This MAb is CL/MAb 231, a murine IgG2a antibody that targets an unknown molecule on the surface of canine lymphoma cells. It exhibits is effect through Antibody Dependent Cell Mediated Cytotoxicity (ADCC) and Complement Dependent Cytotoxicity (CDC). While this MAb showed promise in initial investigations, it is no longer produced and the drug license has been returned to the Wistar Institute.

CANINE DERIVED ANTIBODY FRAGMENTS – BRINGING ANTIBODY TARGETED THERAPIES TO THE VETERINARY CLINIC

While MAbs have dominated the antigen-targeting armoury, the advent of recombinant antibody technology has led to the development of alternative therapeutic antibody formats that are now being investigated. Antibody fragments such as scFvs (single chain Fragment variable) and Fabs are being developed for therapeutic tumor targeting in man.[5,6] scFvs are the smallest antibody fragments that maintain the antigen specificity and binding affinity of the whole antibody and are composed of VH and VL chains joined by a flexible serine-glycine linker [7] (Fig.1). They are significantly smaller than MAbs (25kDa vs. 150kDa) and this endows them with excellent tumor penetrating properties making them highly attractive as tumor targeting agents. Recombinant technology advances have resulted in the ability to amplify and clone the variable regions of immunoglobulin heavy and light chains from peripheral blood lymphocyte cDNA and randomly recombine the products resulting in the generation of a scFv library that represents that individual's entire immunoglobulin repertoire.[8-10] The expression of scFv libraries on the surface of bacteriophage (a scFv phage display library) allows for the rapid screening of millions of scFvs and the selection of those scFvs that bind a particular antigen. Antigens bound by scFvs can be isolated and identified using standard biochemical techniques. In this way phage display has been used successfully to discover novel therapeutic targets and isolate antibody fragments that bind them.[11,12] One notable example of the use of scFv phage display libraries to identify and target tumor-associated antigens (TAA) was in a recent study using scFv phage display libraries derived from the peripheral lymphocytes of human cancer patients. Analysis of these libraries revealed an apparent breakdown of self-tolerance and generation of auto-antibodies to the $\alpha v\beta 3$ integrin, a heterodimer known to be associated with malignant metastases.[6] Furthermore, in that study the anti$\alpha v\beta 3$ scFv was isolated and used therapeutically in vivo to target metastatic cells, preventing hematogenous metastasis and inhibiting the growth of established metastatic foci in murine models of malignant mammary carcinoma.[6] As the fields of genomics, proteomics and metabolomics rapidly advance, novel targets in many different diseases are being identified and the diagnostic and therapeutic use of scFv technology is predicted to increase exponentially.

In an attempt to bring targeted antibody therapies to the veterinary arena, we have developed combinatorial antibody libraries from healthy dogs and canine cancer patients.[13] The recent sequencing of the canine genome has provided the genetic information required to generate diverse scFv libraries of canine origin. By designing PCR primers that recognize highly conserved sequences within VH and VL framework regions it has been possible to generate large libraries of canine VH and VL regions that have been joined together using simple molecular techniques into scFv libraries. These libraries are highly diverse and aim to represent the antigen-experienced immunological repertoire of the donor individual. The scFv libraries can be expressed on the surface of bacteriophage and can be screened in this format for binding to canine antigens of interest using simple immobilized antigen or whole cell panning techniques.

We have screened canine scFv libraries against immobilized antigens such as canine parvovirus capsid antigens and against VEGF. In the first instance we aimed to provide proof-of-principle that scFvs recognizing antigens that the individual had been exposed to could be isolated from scFv phage display libraries. Using CPV capsid antigen as a target, we isolated several different scFv that bind to CPV capsid antigen in phage ELISAs.[13] We next looked to see if we could isolate scFvs that recognize the self-antigen Vascular Endothelial Growth Factor (VEGF) from libraries generated from canine cancer patients. VEGF is expressed in many different canine tumor types including hemangiosarcoma (HSA), osteosarcoma, mammary carcinoma, malignant melanoma, nasal carcinoma, intracranial neoplasia, and mast cell tumors.[14-21] Furthermore, in dogs with HSA, osteosarcoma and malignant melanoma, serum levels of VEGF are increased and correlate with disease free interval and overall survival.[14,21,22] We hypothesized that dogs with HSA would generate antibody responses against VEGF and that canine VH and VL chains that bind to canine VEGF could be isolated from scFv phage display libraries. Using a simple immobilized panning approach we isolated several canine derived scFv that bound to canine VEGF. This suggests that as in human cancer patients, dogs with cancer may break peripheral tolerance to over-expressed self-antigens and libraries generated from these individuals may be rich in cancer specific targeting sequences. These libraries therefore may be useful not only in generating canine-specific targeting agents for use in canine cancer therapy but they may also enable identification of previously unknown tumor associated antigens that may be therapeutically relevant. Use of the scFv approach overcomes the previous obstacles to targeted antibody therapeutics in dogs by providing canine antibody fragments that recognize canine antigens. Furthermore, scFvs can be produced in bulk in *E. coli* obviating the need for expensive mammalian culture systems.

PRACTICAL USE OF SCFV TECHNOLOGY

While the small size of scFvs allow great tissue penetration in vivo, it also results in a relatively short serum half life as they are rapidly filtered through the glomerulus. While this property makes them exceptionally useful when conjugated to radionuclides for diagnostic imaging studies (on account of their low background signal), it is a drawback for their direct use as therapeutic targeting agents. Approaches to modify scFv to prolong their serum half-life include PEGylation of scFv fragments [23,24] and the generation of diabodies and mini bodies (scFv-CH3 dimers) that use the canine hinge and CH3 domain linked to the scFv. These approaches all increase targeting fragment size and have successfully prolonged serum $t_{1/2}$.[25,26] In their native form scFv do not contain an Fc region and are therefore unable to engage immune effector molecules. However, scFv molecules can be readily engineered to carry toxic moieties such as Pseudomonas exotoxin A that can then be targeted to the cell of interest. In some instances the VH and VL chain nucleotide sequences of selected, target specific scFv can be cloned into plasmids that contain the nucleotide sequences of the constant regions of the heavy and light chains leading to the production of monoclonal antibodies. This approach is being employed in our lab currently to generate canine MAbs based on scFv nucleotide sequences that could be used for diagnostic and therapeutic purposes.

Finally, to overcome the problems with short scFv half-life and the expense of generating more complex scFv formats such as diabodies and minibodies in mammalian expression systems, attention has turned to using scFv nucleotide sequences in gene therapy. In this approach viral vectors are used to transduce cells in vivo and direct the continued expression of scFvs. Initial reports of this approach are encouraging.[27,28] We are currently exploring the feasibility of this approach in dogs. Finally, given the recent success of adoptive immunotherapy using autologous T cells that have been genetically redirected using chimeric immunoreceptor constructs, efforts are now underway to generate similar constructs for use in adoptive immunotherapy strategies in the canine clinic. In this approach T cells are transduced with chimeric constructs that consist of a targeting scFv linked to a hinge region, transmembrane domain and then to co-stimluatory and CD3ζ domain. Genetically engineered T cells are then able to recognize antigen through a surrogate scFv (removing the need for MHC antigen-presentation) that when engaged delivers stimulatory signals to the T cell leading to antigen specific T cell proliferation, cytokine production and cytotoxic activity. This approach has shown recent and remarkable success in the human clinic.[29]

SUMMARY

Recombinant antibody technology has allowed the generation of MAbs, Fabs, and scFvs (and scFv derivatives) that have been employed for diagnostic and therapeutic purposes in the human clinic. The success of MAbs in human medicine paves the way for targeted antibody therapies to become a reality in veterinary medicine although many questions related to optimal format, production mechanisms and cost-of-production need to be answered. Until now, the exciting advances in the targeted therapy of cancer in human medicine have not been possible in dogs since monoclonal antibodies or scFvs *of canine origin* with known binding affinities to canine antigens have not been generated. The availability of canine-derived scFv phage display libraries represents a step forward in bringing antibody targeted therapies into the veterinary clinics but much work is still to be done to make this vision a reality.

REFERENCES

1. Kohler. Continuous cultures of fused cells secreting antibody of predefined specificity. *Nature* 256, 495-7 (1975). 2. Kabbinavar, F.F. et al. Combined analysis of efficacy: the addition of bevacizumab to fluorouracil/leucovorin improves survival for patients with metastatic colorectal cancer. *J Clin Oncol* 23, 3706-12 (2005). 3. Hiddemann, W. et al. Frontline therapy with rituximab added to the combination of cyclophosphamide, doxorubicin, vincristine, and prednisone (CHOP) significantly improves the outcome for patients with advanced-stage follicular lymphoma compared with therapy with CHOP alone: results of a prospective randomized study of the German Low-Grade Lymphoma Study Group. *Blood* 106, 3725-32 (2005). 4. Piccart-Gebhart, M.J. et al. Trastuzumab after adjuvant chemotherapy in HER2-positive breast cancer. *N Engl J Med* 353, 1659-72 (2005). 5. Begent, R.H. et al. Clinical evidence of efficient tumor targeting based on single-chain Fv antibody selected from a combinatorial library. *Nat Med* 2, 979-84 (1996). 6. Felding-Habermann, B. et al. Combinatorial antibody libraries from cancer patients yield ligand-mimetic Arg-Gly-Asp-containing immunoglobulins that inhibit breast cancer metastasis. *Proc Natl Acad Sci U S A* 101, 17210-5 (2004). 7. Glockshuber, R., et al. A comparison of strategies to stabilize immunoglobulin Fv-fragments. *Biochemistry* 29, 1362-7 (1990). 8. Huse, W.D. et al. Generation of a large combinatorial library of the immunoglobulin repertoire in phage lambda. *Science* 246, 1275-81 (1989). 9. McCafferty, J., et al. Phage antibodies: filamentous phage displaying antibody variable domains. *Nature* 348, 552-4 (1990). 10. Lerner, R.A., et al. On the use of combinatorial antibody libraries to clone the "fossil record" of an individual's immune response. *Proc Natl Acad Sci U S A* 88, 9705-6 (1991). 11. Topping, K.P., et al. Isolation of human colorectal tumour reactive antibodies using phage display technology. *Int J Oncol* 16, 187-95 (2000). 12. Ridgway, J.B. et al. Identification of a human anti-CD55 single-chain Fv by subtractive panning of a phage library using tumor and nontumor cell lines. *Cancer Res* 59, 2718-23 (1999). 13. Braganza, A. et al. Generation and validation of canine single chain variable fragment phage display libraries. *Vet Immunol Immunopathol* 139, 27-40 (2011). 14. Clifford, C.A. et al. Plasma vascular endothelial growth factor concentrations in healthy dogs and dogs with hemangiosarcoma. *J Vet Intern Med* 15, 131-5 (2001). 15. de Queiroz, G.F., et al. Vascular endothelial growth factor expression and microvascular density in soft tissue sarcomas in dogs. *J Vet Diagn Invest* 22, 105-8 (2010). 16. Kodama, A. et al. Establishment of canine hemangiosarcoma xenograft models expressing endothelial growth factors, their receptors, and angiogenesis-associated homeobox genes. *BMC Cancer* 9, 363 (2009). 17. Matiasek et al. Ki-67 and vascular endothelial growth factor expression in intracrainial meningiomas in dogs. *J Vet Intern Med* 23, 146-51 (2009). 18. Millanta, F., et al. Expression of vascular endothelial growth factor in canine inflammatory and non-inflammatory mammary carcinoma. *J Comp Pathol* 142, 36-42 (2010). 19. Rossmeisl, J.H., et al. Expression of vascular endothelial growth factor in tumors and plasma from dogs with primary intracranial neoplasms. *Am J Vet Res* 68, 1239-45 (2007). 20. Shiomitsu. Expression of epidermal growth factor receptor and vascular endothelial growth factor in malignant canine epithelial nasal tumors. *Vet Comp Oncol* 7, 106-14 (2009). 21. Taylor, K.H. et al. Expression of vascular endothelial growth factor in canine oral malignant melanoma. *Vet Comp Oncol* 5, 208-18 (2007). 22. Thamm, D. Serum vascular endothelial growth factor concentrations and post surgical outcome in dogs with osteosarcoma. *Vet Comp Oncol* 6, 126-32 (2008). 23. Chapman, A.P. PEGylated antibodies and antibody fragments for improved therapy: a review. *Adv Drug Deliv Rev* 54, 531-45 (2002). 24. Chapman, A.P. et al. Therapeutic antibody fragments with prolonged in vivo half-lives. *Nat Biotechnol* 17, 780-3 (1999). 25. Hu, S. et al. Minibody: A novel engineered anti-carcinoembryonic antigen antibody fragment (single-chain Fv-CH3) which exhibits rapid, high-level targeting of xenografts. *Cancer Res* 56, 3055-61 (1996). 26. Olafsen, T. et al. Characterization of engineered anti-p185HER-2 (scFv-CH3)2 antibody fragments (minibodies) for tumor targeting. *Protein Eng Des Sel* 17, 315-23 (2004). 27. Afanasieva. Single-chain antibody and its derivatives directed against vascular endothelial growth factor: application for antiangiogenic gene therapy. *Gene Ther* 10, 1850-9 (2003). 28. Frentzen. Anti-VEGF single-chain antibody GLAF-1 encoded by oncolytic vaccinia virus significantly enhances antitumor therapy. *Proc Natl Acad Sci U S A.* 106, 12915-20 (2009). 29. Kalos. T cells with chimeric antigen receptors have potent antitumor effects and can establish memory in patients with advanced leukemia. *Sci Transl Med* 3, 95ra73 (2011).

NOT YOUR FATHER'S HEMANGIOSARCOMA: INSIGHTS FROM GENETIC AND FUNCTIONAL STUDIES

Jaime F. Modiano, VMD, PhD and Erin B. Dickerson, PhD
St. Paul, MN

A HISTORICAL CONTEXT

Hemangiosarcoma is perhaps the most aggressive vascular tumor diagnosed in dogs. Although systematic surveys are not available to obtain accurate estimates of risk and incidence in companion animals, hemangiosarcoma appears to be relatively common in dogs, unlike it's counterpart in humans, commonly called angiosarcoma, which is vanishingly rare (<0.01% of human tumors). Reports of canine hemangiosarcoma are traceable to the early 1960's in the current medical literature. By the mid 1970's, Priester had recognized that hemangiosarcoma of the liver occurred 25 to 100 times more frequently in dogs than in humans,[1] and by the late 1970's, various investigators noted a breed predilection in German Shepherd Dogs in Europe.[2,3]

The breed predilection for hemangiosarcoma was confirmed and extended through various retrospective studies in the US in the 1980's and 1990's.[4-6] Seminal work from Priester and McKay published as an NCI monograph in 1980 provided the first reasonable estimate for the occurrence of hemangiosarcoma in domestic dogs in North America.[7] The data were obtained from the Veterinary Medical Data Program (VMDP), consisting of 15 veterinary schools in the U.S. and Canada between 1964 and 1977. Priester and McKay classified angiosarcomas in the category of hematopoietic neoplasms, reporting 165 cases. To put this number in context, the database included approximately 1.6 million records, and >41,500 had a recorded "tumor" event, although most "tumors" were benign. There were almost 10 times more lymphomas than splenic hemangiosarcomas in the VDMP, but about the same number of leukemias; splenic hemangiosarcomas accounted for ~0.3% of all benign and malignant tumors. Several of the common breeds we now associate with this disease, including Boxers and Golden Retrievers, were already overrepresented in the affected population of this systematic survey. Notably, Priester and McKay reported no significant differences in risk between male, female, or neutered animals. Arguably, the larger prevalence of sexually intact animals in that era might provide more accurate numbers to assess the potential effect of neutering on a dog's risk to develop this disease.

More recent estimates of risk and incidence of hemangiosarcoma in dogs can be obtained from health surveys privately commissioned by specific American Kennel Club National Breed Clubs.[8] These surveys are not normalized to a reference population and may contain considerable bias. Diagnoses, in a large proportion of cases, are unable to be directly confirmed by the investigators and sometimes by a veterinarian, so there is considerable variation in the numbers of responses obtained. These surveys are not meant to provide specific epidemiologic information per se, but rather to encourage fanciers of purebred dogs to support research in areas that are relevant to their breed of interest. Nevertheless, taken within the context of their limitations, breed health surveys can provide further insight into the epidemiology of hematopoietic neoplasms in dogs. These surveys have identified hemangiosarcoma as a highly prevalent disease in Golden Retrievers, Portuguese Water Dogs, Australian Shepherds, Salukis, and other popular, contemporary AKC breeds. In 2000, a review of spontaneously occurring tumors in companion animals by Vail and MacEwen estimated that hemangiosarcoma accounts for 7% of all malignancies in the dog.[9]

CLASSICAL PATHOLOGY

Hemangiosarcoma is classified as a vascular tumor, and specifically, as a tumor arising from malignant endothelial cells. Histologically, the tumors are cellular with moderate to extensive areas of hemorrhage and necrosis. They can be cavernous or solid, and the malignant cells can be highly pleomorphic with morphological features that are reminiscent of those seen in other sarcomas. The main distinction is that tumor cells are observed lining irregular vascular spaces (capillaries or sinusoids) filled with blood, and they express proteins commonly associated with endothelial differentiation.[10]

The histologic classification of these tumors has not been a source of controversy. While debate has raged on about other tumors regarding subclassification according to morphology or topology, "hemangiosarcoma" is a diagnostic monolith that describes tumors with diverse histologic properties associated with vascular structures in numerous organs.

NATURAL HISTORY

The common primary sites for hemangiosarcoma are the spleen, the right atrium of the heart, the liver, and the subcutis. The pattern of growth involves infiltration into normal tissues surrounding the tumor as well as distant metastasis. Yet, the disease is insidious; that is, it does not cause pain and the rate of growth in the early stages is relatively slow. Dogs harboring even large hemangiosarcomas may show no clinical signs or evidence that they have a life threatening disease. Generally, tumor-associated blood vessels are tortuous and malformed, and blood cells tend to pool in them and clot. The clots prevent blood and nutrients from reaching tumor cells, in turn causing these cells to die. This creates small ruptures in the tumor through which blood may escape into the abdomen, heart sac, chest, or subcutaneous space. Depending on the amount of blood lost, affected dogs may show constitutional signs, including lethargy and weakness, but these are transient and resolve as dogs reabsorb the blood components and make new blood cells. The clinical signs are recurrent, but they also are subtle enough to go unnoticed for some time. Microscopic, if not macroscopic metastasis almost certainly is present in most dogs at the time of diagnosis, and the eventual outcome for dogs with this disease often follows the rupture of a large or rapidly growing tumor, which results in acute, severe hemorrhage, collapse, shock, and death.

The response to treatment is unrewarding. Commonly cited median survival for dogs with splenic hemangiosarcoma (the most common form) treated with the standard of care ranges from 3 to 5 months,[11] with slightly better survival in dogs that had no gross metastasis at diagnosis and up to ~15-20% surviving >12 months. This is consistent with data from 112 cases seen at our hospital in a ~30-month period between 2009 and 2011.

THERAPY FAILURES: A CLUE TO ORIGIN OR AN ANATOMIC ANOMALY?

Few sarcomas show the metastatic potential seen in hemangiosarcoma. It also is not rare for sarcomas to be chemoresistant. Combined, the high rate of metastasis and chemoresistance places hemangiosarcoma alongside osteosarcoma as two prototypical canine sarcomas that are essentially incurable using conventional therapies. This has led to numerous controlled studies and anecdotal instances of experimental therapy for this disease. One study published in 1995 using liposome encapsulated muramyl tripeptide-phosphatidylethanolamine (L-MTP-PE) as adjuvant therapy for dogs with splenic hemangiosarcoma with no evidence of gross metastasis showed improvement in survival when compared to a historical control.[12] Development of this therapy then underwent a ~15 year hiatus due to unavailability of the drug. Recently, Takeda Pharmaceuticals resurrected this drug under the name Mifamurtide, which has received approval for use in children with osteosarcoma in Europe and rekindled interest for its potential to treat other tumors. Most other studies using combinations of cytotoxic drugs, metronomic dose schedules, autologous vaccines, and sundry immune-based and antiangiogenic therapies have not shown consistent improvement over the accepted standard of care. Several intrinsic properties of hemangiosarcoma could account for these results. One is potential widespread expression of drug efflux pumps. Recent data from our lab confirm that these cells express ABC transport proteins, although a diverse array of proteins is expressed by different tumors.[13] The hemodynamics within these tumors represent another obstacle. Blood flow is sludgy, creating widespread microscopic and macroscopic blood clots that may prevent even effective drugs from reaching their targets. The potential to use antiangiogenic agents to achieve vascular normalization and enhance chemosensitivity of these tumors has yet to be explored in detail. And finally, the microarchitecture of the tumors themselves might present significant obstacles.

This last possibility opens a segue to the question of what exactly is the cell of origin of hemangiosarcoma. Some clues should give pause to the assertion that this disease is simply a "tumor of endothelial cells." Unlike other tumors, hemangiosarcoma can occur in virtually any organ, albeit it tends to be associated with organs or regions that have complex vascular networks (spleen, heart, liver, lungs, brain, kidney, muscle, bone marrow, and skin). Visceral tumors show similar biological behavior, while cutaneous tumors are relatively indolent and can be curable. Unfortunately, these clinicopathological observations cannot answer the questions as to whether the tumors show site or substrate specificity and whether they originate from a single cell type. An apparent breed-specific predilection for different topological forms of the disease also has been noted,[14] suggesting tumors at different sites could have different etiology. On the other hand, these differences may be driven by other microenvironmental factors that are independent from the cell of origin.

Over the past decade, we have started to address these questions directly and systematically. First, we developed a resource of hemangiosarcoma-derived cell lines (>60 to date) with representation of anatomical and breed diversity (these cells can be made available through standard academic or industrial material transfer agreements managed by the University of Minnesota Office for Technology Commercialization). Next, we examined phenotypic and molecular commonalities among these cell lines, defining characteristics of cell surface antigen expression,[10,15] tumor suppressor gene inactivation,[16] and genome-wide gene expression profiles.[17,18] The sum of these data highlighted several findings that could change the paradigm for how this disease is thought to originate and progress. Two important aspects were the observation that hemangiosarcoma cells in isolation (cell culture) have phenotypic properties suggestive of bone marrow ontogeny and specifically of bone marrow derived (mesenchymal?) stem cells.[15] Alone, these data could not distinguish if the tumors themselves arise from a cell in the bone marrow that migrates to a vascular plexus, or if they originate from a bone fide stem cell. The gene expression profiling data offered some clues, however, as the recurrent signature of these genes was associated with enrichment of angiogenic and pro-inflammatory genes, with no evidence of tissue specificity.[17] These data gave rise to the questions that have occupied our labs for the past ~3-4 years.

ONE TUMOR IN MANY SITES, OR ONE SITE CREATING MANY TUMORS?

One way to test tissue or substrate specificity by the tumors directly, was to use a xenograft model developed originally by the Helfand group and subsequently adapted more widely by Kodama et al.[19,20] We developed a series of canine hemangiosarcoma cell lines expressing GFP and firefly luciferase, which allowed us to image tumor progression in vivo, as well as to detect xenograft cells within tumors and colonization at distant sites using molecular and immunohistochemical techniques. Collectively, the data show that canine hemangiosarcomas achieve systemic spread in murine hosts readily, and this is independent of the method of delivery (subcutaneous or intraperitoneal). In our experiments, hemangiosarcoma cells did not show site specificity for growth, but quite intriguingly, they appeared to respond to microenvironment cues and adopt various functions as part of this response. These functions included not only the potential to form anatomically distinct structures or tumors, but also the potential to direct other cells in their environment to do so. Concurrent studies assessed whether these cells conformed to the hierarchical ("cancer stem cell") model or to the stochastic model of cancer. Again, the data showed convincingly that hemangiosarcoma cell lines generally have a subpopulation that retains traits associated with cancer stem cells, including self-renewal, chemoresistance, and increased tumorigenicity in vivo. [13] Each tumor also appears to achieve these properties by independently disabling or enabling molecular networks associated with self-renewal and survival. A somewhat surprising result, however, was the observation that the hemangiosarcoma cancer stem cells also showed multipotency, which is not a common trait of cancer stem cells in other tumors. Not only did the cells maintain (or develop) properties generally associated with myeloid cells, which might have been predictable based on their previously established phenotypes, but they also were able to efficiently undergo adipogenic differentiation. Such results in vitro could be discounted as artifacts, but the data are supported by similar behavior of these cells in vivo, where they showed recurrent potential to form vascular, hematopoietic, and adipogenic tumors in mouse hosts. Finally, such multipotency can be re-created using gene expression profiling to subclassify canine hemangiosarcomas into tumors whose molecular properties uncover their distinct differentiation tendencies.

Much work remains to be completed. It is apparent that some or many properties of hemangiosarcoma are dependent on interactions between tumor cells and their local microenvironment. These interactions probably depend on the microanatomy of the niche and the capability of the cells to alter the niche by recruitment or reprogramming of local stromal cells. Local or systemic microangiopathy may precede the disease, but it also may be a consequence of the vascular disruption it creates. Platelets and platelet derived factors, inflammation, and hypoxia are probably key drivers of the disease. Yet, many critical questions remain. For example, are mesenchymal stem cells or pericytes (the presumed precursors of hemangiosarcoma cells) recruited from the bone marrow to a site of hypoxia and inflammation, or are resident pericytes responsible for the origin of the disease? Does loss of function in critical pathways alter their survival, leading to eventual tumorigenesis, or are mutations more likely associated with stromal interactions and reprogramming? And what relative component of therapy failure can we attribute to the make-up of the tumors, which includes an abundant component of non-malignant, tumor-associated stromal cells?

We are working to answer these questions through our ongoing studies, which include development of innovative "home-grown" treatment modalities,[21] as well as genetic manipulation of tumors to highlight potential interventions that can be used to attack their environmental niche. The game is afoot!

REFERENCES

1. W. A. Priester, *J Natl Cancer Inst* **57**, 451 (1976). 2. E. C. Appleby *et al.*, *Vet Rec* **102**, 449 (1978). 3. S. H. Dome *et al.*, *Vet Rec* **102**, 224 (1978). 4. C. Prymak *et al.*, *J Am Vet Med Assoc* **193**, 706 (1988). 5. N. O. Brown *et al.*, *J Am Vet Med Assoc* **186**, 56 (1985). 6. W. L. Spangler *et al.*, *J Am Vet Med Assoc* **200**, 829 (1992). 7. W. A. Priester *et al.*, *Natl Cancer Inst Monogr*, 1 (1980). 8. M. G. Ritt, in *Schalm's Veterinary Hematology*, (Wiley-Blackwell, Hoboken, NJ, 2010), pp. 427-32. 9. D. M. Vail *et al.*, *Cancer Invest* **18**, 781 (2000). 10. S. P. Fosmire *et al.*, *Lab Invest* **84**, 562 (2004). 11. C. A. Clifford *et al.*, *J Vet Intern Med* **14**, 479 (2000). 12. D. M. Vail *et al.*, *Clin Cancer Res* **1**, 1165 (1995). 13. B. H. Gorden *et al.*, submitted (2011). 14. N. J. Andersen *et al.*, in *Soft Tissue Tumors*, (InTech, 2011), pp. 149-174. 15. A. R. Lamerato-Kozicki *et al.*, *Exp Hematol* **34**, 870 (2006). 16. E. B. Dickerson *et al.*, *Vet Pathol* **42**, 618 (2005). 17. B. A. Tamburini *et al.*, *BMC Cancer* **10**, 619 (2010). 18. B. A. Tamburini *et al.*, *PLoS ONE* **4**, e5549 (2009). 19. N. Akhtar *et al.*, *Neoplasia* **6**, 106 (2004). 20. A. Kodama *et al.*, *BMC Cancer* **9**, 363 (2009). 21. J. T. Schappa *et al.*, submitted (2011).

VITAMIN D IN THE TREATMENT AND PATHOGENESIS OF CANINE MAST CELL TUMORS, OSTEOSARCOMA, AND CONGESTIVE HEART FAILURE

Ken Rassnick, DVM, DACVIM (Oncology)
Ithaca, NY

INTRODUCTION

Vitamin D deficiency increases the risk of neoplasia and heart failure in people. Vitamin D compounds have antitumor activity through induction of differentiation, antiproliferative effects, activation of apoptotic pathways, and inhibition of angiogenesis. In addition to anticancer effects as a single-agent, vitamin D compounds also potentiate the cytotoxicity of a diverse array of chemotherapy agents, steroids, and receptor tyrosine kinase inhibitors. This talk summarizes our work investigating the association between vitamin D status and canine tumors and heart failure as well as the effectiveness of calcitriol, the active from of vitamin D, for treatment.

CALCITRIOL INCREASES CISPLATIN CYTOTOXICITY AGAINST CANINE TUMOR CELLS[1]

Calcitriol (1,25-dihydroxyvitamin D_3), the principal biologically active form of vitamin D, has potent *in vitro* and *in vivo* antineoplastic activity. Calcitriol induces G1/G0 cell cycle arrest which might be mediated by increased expression of the cyclin-dependent kinase inhibitors $p21^{waf/kip1}$ and $p27^{kip1}$, decreased cyclin-dependent kinase-2 activity, and hyperphosphorylation of the retinoblastoma protein. In some cells, calcitriol induces apoptosis by down-regulating anti-apoptotic genes like Bcl-2. Calcitriol decreases expression of epidermal growth factor receptors, induces transforming growth factors 1 and/or 2, reduces insulin-like growth factor-1 signaling, and alters levels of hepatocyte growth factor. Inhibition of tumor invasion through decreased matrix metalloproteinase 2 and 9 activity, antiproliferative effects on tumor-derived endothelial cells, and inhibition of the prostaglandin pathway have also been reported.

In addition to anticancer activity as a single-agent, studies indicate calcitriol acts synergistically to potentiate the antitumor activity of numerous chemotherapeutics. The combination of calcitriol and cisplatin has been showed to be synergistic in a mouse squamous cell carcinoma model system. The increased cytotoxicity results from cisplatin-enhancement of calcitriol-induced apoptotic signaling through unregulated MEKK-1. Evidence also suggests calcitriol might enhance cisplatin-mediated cytotoxicity through inhibitory effects on DNA repair. In our work, we found that the combination of calcitriol and cisplatin had synergistic, antiproliferative activity against canine mammary carcinoma, mastocytoma and osteosarcoma cell lines. The concentrations of calcitriol and cisplatin required for 50% growth inhibition were 3 to 8-fold lower when the drugs were used in combination than when used individually, indicating this combination might have significant clinical utility in dogs.[1]

HIGH-DOSE INTRAVENOUS CALCITRIOL CAN BE GIVEN TO DOGS BUT IS ASSOCIATED WITH HYPERSENSITIVITY REACTIONS[1]

The antineoplastic effects of calcitriol are dose-related and high-dose pulse administration of calcitriol improves its therapeutic index. In a murine squamous cell carcinoma model, doses of calcitriol that resulted in C_{max} >10.0 ng/mL and AUC >40.0 h•ng/mL were effective at suppressing tumor growth. We conducted a phase I dose-escalation of IV calcitriol combined with cisplatin in tumor-bearing dogs. The maximum tolerated dose found in the study was 3.75-µg per kg calcitriol in combination with cisplatin (60-mg/m^2 plus diuresis). Calcitriol dosages >1.5 µg/kg were associated with serum levels ≥9.8 ng/mL and calcitriol dosages >1.0 µg/kg were associated with systemic exposure ≥45 h•ng/mL, indicating this combination might be clinically useful in the treatment of tumor-bearing dogs.[1]

Hypercalcemia is a well-described effect of high-dose calcitriol. As expected, in our phase I study, hypercalcemia was the dose-limited toxicity in dogs treated with the combination of calcitriol and cisplatin.[1] Hypersensitivity reactions (pruritis and urticaria) also occurred during many of the calcitriol infusions.[1] Hypersensitivity reactions were most likely due to polysorbate, the solvent in the IV calcitriol formulation.

INVESTIGATIONAL FORM OF ORAL CALCITRIOL HAS GOOD BIOAVAILABILITY IN DOGS[2]

Calcitriol is commercially available in 3 FDA-approved formulations: a parenteral formulation for intravenous administration as well as liquid and capsule formulations for oral administration. Administration of calcitriol by mouth would circumvent the polysorbate-associated hypersensitivity reactions seen with high-dose IV calcitriol, however this is not reasonable because the largest capsule size is 0.5-µg and administering the maximally tolerated 3.75-µg per kg dosage to an average-sized dog would require approximately 250 capsules. Also, unfavorable pharmacological conditions in the intestines associated with ingestion of such a large number of capsules might decrease calcitriol absorption. This has been shown to occur in people.

DN101 is a highly concentrated formulation of oral calcitriol (15- or 45-µg per capsule) designed specifically for cancer therapy. Studies in people show linear relationships between dosage and AUC and C_{max}; however substantial inter-individual variability was observed in people treated with high doses of DN101. We conducted a 2-way crossover study in with dogs randomly received a single 3.75-µg per kg dose of calcitriol either IV or oral (as DN101), followed by cisplatin (60-mg/m^2 plus diuresis). Three weeks later, the alternate form of calcitriol was given prior to another dose of cisplatin. Mean bioavailability as determined on the basis of AUC, was 71%. C_{max} was >10 ng/mL and AUC was >40 ng•h/mL (important murine PK parameters, as mentioned above) in most dogs treated with DN101, suggesting there would be clinical utility in the treatment of cancer-bearing dogs. Inter-individual variability in pharmacokinetic parameters was high in dogs treated with DN101; bioavailability ranged from 17% to 126%.[2] High inter-patient variability can lead to inconsistent clinical responses, so methods to minimize variation in systemic exposure are needed.

CANINE MAST CELL TUMOR EXPRESS VITAMIN D RECEPTOR (VDR), A TARGET FOR CALCITRIOL THERAPY[3]

Our work to evaluate *in vitro* synergism between calcitriol and cisplatin showed calcitriol had antiproliferative effects on C2 mastocytoma cells.[1] Effects of calcitriol are mediated primarily by binding to the vitamin D receptor (VDR), a member of the nuclear steroid hormone receptor superfamily. Murine bone marrow-derived mast cells, normal human cutaneous mast cells and at least one human mast cell leukemia cell line (HMC-1) constitutively express VDR. We performed western blot analysis and an immunohistochemical study of 69 canine mast cell tumor (MCT) specimens to determine whether canine MCTs express VDR, which would provide a molecular basis for our *in vitro* findings.[3] Western blot analysis in C2 mastocytoma cells resulted in a band with a molecular weight of approximately 52 kDa, consistent with VDR. Immunohistochemical results showed that most (97%) canine MCTs express VDR. VDR was also expressed by reactive canine mast cells (dermatitis) but the frequency of VDR expression and VDR score (frequency x intensity) was significantly greater in neoplastic mast cells. There was no difference in frequency of expression between different MCT grades, but the frequency of labeling in grade 3 MCTs was significantly higher than in reactive mast cells.[3] Collectively, our findings suggested that VDR is up-regulated in canine MCTs. Presence of this receptor within biopsy specimens suggests a novel target for therapy.

CALCITRIOL INCREASES MAST CELL TUMOR CHEMOTHERAPY AND RECEPTOR TYROSINE KINASE INHIBITOR CYTOTOXICITY *IN VITRO*[4]

As described above, we have shown *in vitro* synergism with calcitriol and cisplatin.[1] However, cisplatin is not a chemotherapeutic agent with routine use in the treatment of canine mast cell tumors (MCTs). Therefore, we evaluated the *in vitro* effects of calcitriol on C2 cells, a cell-line derived from a naturally occurring cutaneous MCT.[4] Calcitriol suppressed canine mastocytoma C2 cell viability in a dose-dependent manner. Also calcitriol exhibited synergistic, antiproliferative activity when used in combination with CCNU, vinblastine, imatinib or toceranib. The concentrations required for 50% growth inhibition were generally 2 to 6-fold lower when the drugs were used in combination than when used individually.[4] These results suggest combinations of calcitriol with conventional MCT chemotherapeutics or receptor tyrosine kinase inhibitor drugs might prove useful in the treatment of dogs with spontaneously occurring MCTs.

HIGH-DOSE ORAL CALCITRIOL INDUCES REMISSION IN DOGS WITH MAST CELL TUMORS[4]

A clinical trial of calcitriol in the treatment of canine mast cell tumors (MCTs) was a natural progression from our previous studies showing high-dose oral calcitriol can be given to dogs and has good bioavailabilty,[2] pharmacokinetic endpoints important in murine studies are achieved in dogs,[2] MCTs express vitamin D receptor (VDR),[3] and *in vitro*, calcitriol has cytotoxicity against C2 mastocytoma cells.[4] We treated 10 dogs with nonresectable MCTs with DN101, a concentrated oral calcitriol formulation specifically designed for cancer therapy. Four of 10 dogs achieved a remission (1 complete remission, 3 partial remissions). However, the clinical trial was terminated early due to a high frequency of adverse effects of oral calcitriol (4 dogs had hypercalcemia and 5 dogs had reversible azotemia). Together our results suggest calcitriol and calcitriol-combination therapies might have significant clinical utility in the treatment of canine MCTs but refinement of the calcitriol-dosing regimen must be done.

LABRADOR RETRIEVERS WITH MAST CELL TUMORS HAVE LOW SERUM VITAMIN D LEVELS[5]

Epidemiological data indicate that low serum vitamin D concentrations are associated with an increased risk of a variety of human tumors. Canine mast cell tumors (MCTs) express the vitamin D receptor[3] and vitamin D derivatives have *in vitro* and *in vivo* antitumor activity.[4] Taken together, these data suggest a possible role for vitamin D in the genesis and/or course of MCTs in dogs. We examined the association between vitamin D serum level and MCTs in Labrador Retrievers, a dog-breed predisposed to MCT development. To examine this association, we measured serum 25-hydroxyvitamin D_3

[25(OH)D] concentrations, the best indicator of an individual's vitamin D status. Mean 25(OH)D concentration in Labradors with MCTs was significantly lower than that of unaffected Labradors. The mean calculated dietary vitamin D intake per kg body weight in Labradors with MCTs was not statistically different from that of unaffected Labradors. These findings suggest low levels of 25(OH)D might be a risk factor for MCTs in Labrador Retrievers.[5]

VITAMIN D IN THE TREATMENT AND PATHOGENESIS OF CANINE OSTEOSARCOMA[6]

Calcitriol, the active form of vitamin D, has potent antitumor activity in different types of cancer and potentiates cytotoxicity of several chemotherapeutics. Evidence also suggests poor vitamin D status contributes to the pathogenesis of numerous cancers. Through immunohistochemistry and western blot analysis we have shown that the vitamin D receptor (VDR), a prerequisite for vitamin D therapy, is present in canine osteosarcoma (OSA). *In vitro*, calcitriol suppresses OSA 2.4 cell proliferation and the combination of calcitriol and cisplatin has synergistic, antiproliferative activity against OSA 2.4 cells.[1] A preliminary clinical trial of 15 dogs treated with calcitriol/cisplatin following amputation was undertaken as was investigation of the relationship between circulating serum 25-hydroxyvitamin D_3 [25(OH)D] concentration, the best indicator of an individual's vitamin D status, and risk of canine OSA.

DOGS IN CONGESTIVE HEART FAILURE HAVE LOW SERUM VITAMIN D LEVELS[7]

Vitamin D plays a pivotal role in cardiac function and there is increasing evidence that vitamin D deficiency contributes to the etiology and pathogenesis of congestive heart failure (CHF) in people. We examined the association between circulating 25-hydroxyvitamin D [25(OH)D], the hallmark of vitamin D status, and CHF in dogs. Mean 25(OH)D concentration in dogs with CHF was significantly lower than that of unaffected dogs. The mean calculated vitamin D dietary intake per kg body weight in dogs with CHF was actually significantly higher than that of unaffected dogs.

We also examined the impact of 25(OH)D status on clinical outcome in dogs with CHF. There was a significant effect of 25(OH)D level on time to another episode of CHF. These findings suggest low levels of 25(OH)D might be a risk factor for CHF and supplementation of vitamin D might be important in the treatment of some dogs with CHF.

CONCLUSION

In view of our substantial preclinical, clinical and epidemiological data described above supporting the potential role of vitamin D in canine cancers and congestive heart failure, careful studies to evaluate the impact of vitamin D supplementation and calcitriol (1,25-dihydroxyvitamin D_3) or other vitamin D analogues as approaches to disease prevention and treatment

REFERENCES

[1]Rassnick KM, et.al. **Cancer Chemo Pharm** 2008;62(5):881. [2]Rassnick KM, et.al. **Cancer Chemo Pharm** 2010;67:165. [3]Russel D, et.al. **J Comp Pathol** 2010;143:223. [4]Malone EK, et.al. Vet Compar Oncol 2010;8(3):209. [5]Wakshlag JJ, et.al. **Brit J of Nutr** 2011;106:60. [6]Rassnick KM, et.al. **Proceedings Vet Cancer Soc** 2010. [7]Kraus MS, et.al. **Proceeding Eur College Vet Intern Med - CA** 2011.

CANINE OSTEOSARCOMA- BENCH TO BEDSIDE MOLECULAR TARGETS AND THERAPEUTICS

Timothy M. Fan, DVM, PhD, DACVIM (Oncology, SAIM)
Urbana, IL

CANINE OSTEOSARCOMA

Osteosarcoma (OS) is a malignant tumor derived from primitive mesenchymal stem cells with the capacity to produce osteoid matrix. Consistent with its stem cell derivation, OS can adopt a spectrum of histomorphologic characteristics, permitting its categorization into chondroblastic, fibroblastic, osteoblastic, and telangiectatic subtypes. Clinically, OS develops predominantly in the metaphyseal regions of weight-bearing bones, most commonly the distal radius, proximal humerus, distal femur, proximal tibia, and distal tibia.[1,2] The incidence of OS is bimodal with the majority of cases arising in geriatric patients (greater than 7 years old) and a much smaller peak of OS observed in younger animals (2-3 years old).

Biologically, OS behaves as a high-grade sarcoma possessing aggressive osteolytic properties and the strong predilection for distant metastases. Typically, OS originates within the intramedullary cavity of metaphyseal bone. The growth and progression of appendicular OS result in effacement and erosion of the immediate bone microenvironment, including the marrow cavity and circumferential cortical and trabecular bone. In addition to localized skeletal perturbations, OS metastasizes to distant visceral organs, most commonly the pulmonary parenchyma via hematogenous dissemination. Effective treatment of appendicular OS necessitates the management of both local disease and distant metastases; however, even with the institution of multimodality therapy, the majority of dogs ultimately succumb to metastatic progression.

Despite the commonality of OS, a unifying etiopathogenesis for its development and progression remains incompletely elucidated. However, based upon molecular and preclinical investigations, OS genesis likely involves the complex interplay of aberrant molecular signals associated with environmental, bone metabolic and genetic factors. Based upon the complexity of OS etiopathogenesis, there is a clear and justified need to intensively investigate putative molecular pathways which participate in the development and progression of OS in dogs. The identification of drugable targets could lead to future therapeutic strategies that improve the long term management of OS.

MOLECULAR PATHOGENESIS

Unlike certain cancers which are characterized by recurrent chromosomal aberrations, the genetic derangements associated with OS tend to heterogeneous and chaotic. As such, the molecular origins of OS have been difficult to ascribe to specific intracellular signaling anomalies. Nonetheless, substantive evidence derived from experimental, preclinical, and comparative investigations have underscored the potential importance of key dysregulated intracellular signaling and cell survival pathways likely involved in OS development and progression. Many of these molecular pathways have been identified in not only immortalized cell lines, but also spontaneously arising OS samples; which further substantiates the importance of specific signaling pathways in naturally-occurring canine OS. It is the hope of cancer biologists and clinicians alike that the identification of drugable targets will promote the development and validation of novel treatment strategies for improving the long term outcome in dogs diagnosed with OS.

POTENTIAL DRUGABLE TARGETS

MET Proto-oncogene

The *MET* proto-oncogene encodes a tyrosine kinase receptor, which upon ligation with hepatocyte growth factor (HGF), mediates multiple cellular functions including cell scattering, motility, and proliferation. Excessive or dysregulated MET signaling in canine OS has been demonstrated to promote tumorigenic cellular characteristics including the augmentation of proliferation, migration, and scattering.[3-5] In addition to the observed *in vitro* effects of MET signaling in cell lines, a novel germline mutation which results in constitutive receptor phosphorylation and aberrant MET signaling has been identified primarily in the Rottweiler breed, and provides direct translational evidence for the participate of MET signaling in OS pathogenesis.[6] In addition to Rottweilers with germline *MET* proto-oncogene mutations, OS samples derived from diverse breeds co-express mRNA for both *MET* and *HGF*, suggesting the existence of a putative MET/HGF autocrine or paracrine feedback loop.[7] Primary OS with increased *MET* mRNA expression are associated with regional lymph node metastases; clinical findings which support the participation of MET signaling for metastases and cell invasion.

Growth Hormone and Insulin-like Growth Factor-1

The cellular effects of growth hormone (GH) are mediated through the hepatic production of insulin-like growth factor 1 (IGF-1). In osteoblasts, IGF-1 induces cell mitogenesis and protection from apoptosis, as well as promotes angiogenesis. Given the potent and pleiotropic effects of IGF-1 in bone cells, aberrant or excessive IGF-1 signaling likely participates in OS pathogenesis. The potential pro-tumorigenic effects of IGF-1 signaling in canine OS have been characterized through cell line studies.[8] The receptor for IGF-1 is expressed by canine OS cell lines and upon receptor ligation, IGF-1 promoted anchorage independent growth and invasion in some cell lines.[8] Complementing *in vitro* studies, the molecular targeting of the IGF-1signaling pathway has been assessed in dogs with spontaneously arising OS through the use of a long-acting somatostatin analogue (OncoLAR). Although significant reductions in circulating IGF-1 concentrations was achievable in dogs with OS administered OncoLAR, the extent of IGF-1 reductions (~43%) was probably not sufficient to attenuate the pro-tumorigenic effects of IGF-1 signaling, and hence no clinical benefit was detected in dogs treated with OncoLAR compared to placebo.[9]

***ErbB-2* Proto-oncogene**

The proto-oncogene *erbB-2* encodes human epidermal growth factor receptor 2 (HER-2), which is a tyrosine kinase receptor capable of promoting cell transformation and growth. The overexpression of HER2 protein as a result of gene amplification is a negative prognostic factor in mammary carcinoma in dogs; however less clarity exists for the role of HER2 overexpression in the pathogenesis of OS.[10,11] Given that HER2 is a drugable target in people (Herceptin), there remains potential interest in HER2 as a novel therapeutic target for canine OS. The overexpression of HER2 at the mRNA and protein levels have been demonstrated in canine OS cell lines, as well as spontaneously-arising OS tumor samples.[12] Based these investigative findings, HER2 antagonism in canine OS might remain a plausible strategy, however, additional studies are required to more thoroughly characterize the downstream pathways associated with HER2-associated signaling.

Mammalian Target of Rapamycin (mTOR)

The mammalian target of rapamycin (mTOR) pathway critically regulates cellular signals induced by growth factors, nutrients, energy, and stress for the purposes of regulating cell cycle progression and growth. Excessive or aberrant signaling through the mTOR pathway contributes to growth, survival, and chemotherapy resistance in multiple tumor types, and might participate in canine OS progression. Based upon work conducted in canine OS cell lines, the operative functionality of the mTOR pathway was supported by basal phosphorylation of mTOR and p70S6K, a downstream effector protein.[13] More importantly, phosphorylation of mTOR and could be inhibited by rapamycin; resulting in apoptosis and reduced growth of OS cells *in vitro*, suggesting that mTOR inhibition might serve as a therapeutic target. In a translational study conducted in dogs with OS, the feasibility of mTOR inhibition as a treatment strategy was supported in a phase I trial, which showed that biologically effective concentrations of rapamycin were safely obtainable in dogs, and that modulation of mTOR target proteins by rapamycin was achievable within the bone tumor microenvironment.[14]

Vascular endothelial growth factor (VEGF)

Vascular endothelial growth factor (VEGF) is a vascular endothelial cell mitogen and plays a central role in new blood vessel formation. Given that angiogenesis is a necessary step for tumor growth and metastases, strategies to antagonize VEGF-associated signaling holds great promise for the management of highly metastatic cancers such as canine OS. The receptor tyrosine kinase inhibitor, toceranib phosphate (Palladia™), has recently demonstrated preliminary anticancer activity across a broad range of tumor histologies, including metastatic pulmonary OS. Targets of toceranib include several members of the split-kinase family such as vascular endothelial growth factor receptor, platelet-derived growth factor receptor, and c-kit.[15] Based upon toceranib's ability to inhibit multiple signaling pathways, its potential activity was evaluated in 23 dogs with measurable pulmonary OS metastases and resulted in modest activity with a partial response rate of 4.3% (1/23 dogs) and stable disease rate of 43.5% (10/23 dogs).[16]

Survivin

Survivin is a small protein belonging to the inhibitor of apoptosis (IAP) family, and participates in the processes of cell division, as well as apoptosis inhibition. As a dimer, survivin inhibits both caspase-dependent and caspase-independent mediated apoptosis, and its expression can promote tumorigenesis. Given survivin's anti-apoptotic properties, its overexpression might provide a survival advantage to cancer cells and be associated with a negative prognosis. Survivin expression has been characterized in canine tumors including lymphoma and OS. For OS, survivin expression was detected in the majority of tissue samples (65/67), and expression intensity was associated with disease-free intervals.[17] As such, attenuating survivin protein expression via different strategies including siRNA technology or small molecule inhibitors remains an attractive novel strategy for treating dogs with OS, and additional studies are warranted.

REFERENCES

[1]Brodey RS et al. **J Am Vet Med Assoc** 1963;143:471-495. [2]Knecht CD et al. **J Am Vet Med Assoc** 1978;172:72-74. [3]Liao AT et al. **Vet Comp Oncol** 2007;5:177-196. [4]Liao AT et al. **Vet Comp Oncol** 2005;3:61-72. [5]MacEwen EG et al. **Clin Exp Metastasis** 2003;20:421-430. [6]Liao AT et al. **Anim Genet** 2006;37:248-252. [7]Fieten H et al. **Vet Pathol** 2009;46:869-877. [8]MacEwen EG et al. **J Cell Biochem** 2004;92:77-91. [9]Khanna C et al. **Clin Cancer Res** 2002;8:2406-2412. [10]Gorlick R et al. **J Clin Oncol** 1999;17:2781-2788. [11]Scotlandi K et al. **Eur J Cancer** 2005;41:1349-1361. [12]Flint AF et al. **Vet Pathol** 2004;41:291-296. [13]Gordon IK et al. **Am J Vet Res** 2008;69:1079-1084. [14]Paoloni MC et al. **PLoS One** 2010;5:e11013. [15]London CA et al. **Clin Cancer Res** 2003;9:2755-2768. [16]London CA et al. **Vet Comp Oncol** 2011. [17]Shoeneman JK et al. **Cancer Res** 2012;72:249-259.

TARGETED MONOCLONAL ANTIBODY THERAPY IN CANINE LYMPHOMA

Barb Biller, DVM, PhD, DACVIM, Rodney Page, DVM, DACVIM, Rhona Stein, PhD and Leah Mitchell, PhD
Fort Collins, CO

INTRODUCTION

Monoclonal antibodies are among the most effective novel cancer therapeutics for humans

Significant advances in the treatment of human cancers have been brought about by the use of therapeutic monoclonal antibodies (mAb) that selectively target malignant cells and stimulate an immune response that can potentially eliminate any residual neoplastic cells. In fact, mAb therapy represents one of the most successful new avenues of oncology research, resulting in products that have significantly improved the outcome for a number of malignancies including breast cancer, colorectal cancer, prostatic carcinoma and lymphoma[1,2]. One of the best examples is rituximab, an unconjugated or "naked" mAb directed against the B-cell surface antigen CD20 which is highly expressed by many human lymphomas. Unconjugated mAbs are designed to bind to specific targets on the surface of malignant cells and induce apoptosis of the targeted cell. Apoptosis often occurs through both direct and indirect effects; the indirect effects are typically mediated by the activation of secondary immune effector mechanisms such as antibody dependent cell-mediated cytotoxicity (ADCC) or the complement cascade. Rituximab was initially approved as a single agent therapy for patients with relapsed or refractory low-grade B-cell lymphoma, having an impressive overall response rate in this patient population of nearly 50%[1]. Rituximab's spectrum of efficacy has since been found to include high grade lymphomas and chronic lymphocytic leukemia; the mAb is now used in combination with chemotherapy as a first-line therapy for most human high grade lymphomas[3,4]. Other mAbs now used as first-line therapies include trastuzumab (Herceptin®), which targets cells overexpressing the receptor tyrosine kinase HER2/*neu*, in human patients with breast cancer and the anti-EGFR (epidermal growth factor receptor) antibody cetuximab, in people with colorectal carcinoma.

Despite the similarity between many human and canine tumors, therapeutic monoclonal antibodies are currently not available for canines. Although the immunoreactivity of human anti-CD20 antibodies (including rituximab) to canine malignant lymphocytes has recently been evaluated, no significant binding has been found[5]. A monoclonal antibody developed against canine lymphoma (mAb 231) was evaluated in mouse tumor models and in dogs many years ago, however this product is no longer available[6].

DEVELOPING A THERAPEUTIC MONCLONAL ANTIBODY FOR DOGS WITH LYMPHOMA

Targeting of MHC Class II expression on malignant B cells

With the goal of finding a mAb that targets malignant B cells in dogs, we initially investigated L243, a mouse anti-human HLA-DR IgG1 mAb that cross reacts with canine lymphoma cells[7]. The human leukocyte antigen (HLA)-DR is one of three polymorphic genes of the class II major histocompatibility complex (MHC II). Although HLA-DR is expressed on the surface of normal B cells and other immune cells such as dendritic cells, it is also very highly expressed (even more so than CD20) in hematopoietic malignancies such acute lymphoblastic leukemia, myelogenous leukemia and large B-cell lymphoma[8,9]. Similar to anti-CD20 antibodies, antibodies against HLA-DR are very potent apoptotic agents, inducing tumor cell death through direct and indirect mechanisms[9]. Hu1D10, the first anti-HLA-DR antibody to be evaluated in human clinical trials, was effective in treating relapsing and refractory indolent lymphomas[10,11]. However, administration of Hu1D10 was also associated with significant transfusion-related side effects including pan-lymphocyte depletion, neutropenia and gastrointestinal toxicity. Side effects were severe enough to warrant discontinuation of the clinical trial[10,11]. Because these side effects were primarily mediated through the antibody's secondary immunologic effector mechanisms, including ADCC and complement dependent cytotoxicity (CDC), newer antibodies that kill tumor cells independently of effector functions are now being investigated.

Based on this data we completed a pilot study investigating L243 in dogs with relapsed lymphoma[7]. *In vitro*, L243 was found to bind to malignant canine B-cells and had both apoptotic and anti-proliferative effects. Similar to the problems seen with administration of Hu1D10, however, L243 could be administered to dogs with lymphoma but the infusion period had to be progressively increased from 2 hours to 12 hours to minimize the severity of infusion-related side effects. Even when administered slowly, side effects were still prominent with 6 of 7 dogs experiencing at least grade 1 nausea, diarrhea and/or fever. Two dogs also had persistent grade 3 or 4 bone marrow toxicity. Only 2 of 7 dogs experienced temporary disease stabilization however many had previously failed multiple conventional chemotherapy protocols.

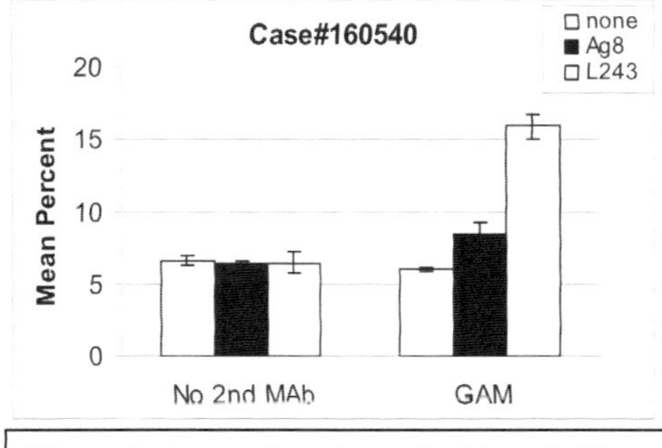

Figure 1: Apoptosis as determined by sub-G1 fraction of cells stained with propidium iodide and examined by flow cytometric assay in canine B lymphoma cells. Ag8 = control antibody, GAM = goat antimouse antibody.

Recently a modified version of L243, known as hL243γ4P (a humanized IgG4 mAb also called **IMMU-114**), was designed that retains the antigen-binding specificity and apoptotic properties of L243 for malignant B-cell but kills tumor cells independently of secondary immune effector mechanisms[8]. Therefore, in hopes of identifying a better tolerated mAb, we began preliminary investigations of this antibody in dogs with lymphoma. So far we have found that IMMU-114 binds *in vitro* to malignant lymphocytes obtained via fine needle aspiration from the majority of dogs with B cell lymphoma. In addition the mAb induces apoptosis in canine lymphoma cells when cross-linked with an anti-human IgG second antibody **(Figure 1)**. Evaluation of IMMU-114 in assays of ADCC and CDC performed on one canine lymphoma aspirate revealed that L243, but not IMMU-114, resulted in specific cell lysis of the lymphoma cells, thus suggesting a lack of immune effector functions (see **Figure 2)**. Consistent with this finding, administration of IMMU-114 to 2 normal dogs at 3 mg/kg i.v. over 90 minutes did not cause infusion reactions, however both dogs experienced a mild and transient lymphopenia[7].

Phase I study of IMMU-114 in dogs with B-cell lymphoma

We are now in the early stages of a larger phase I study of IMMU-114 in dogs with relapsed or naïve B cell lymphoma and a summary of the data will be presented in this talk. The goals of this study are to identify a safe and biologically active dosage of the mAb. Through a 3 x 3 dose escalation clinical trial we will first seek to identify the appropriate dosage of the mAb based on the toxicity profile and biologic endpoints such as induction of tumor cell apoptosis and lymph node measurements. Since the mAb is a human product, it will also be important to identify a "therapeutic window" for the antibody; this will be accomplished by monitoring the development of an anti-human IgG response by ELISA in the serum of treated dogs over time.

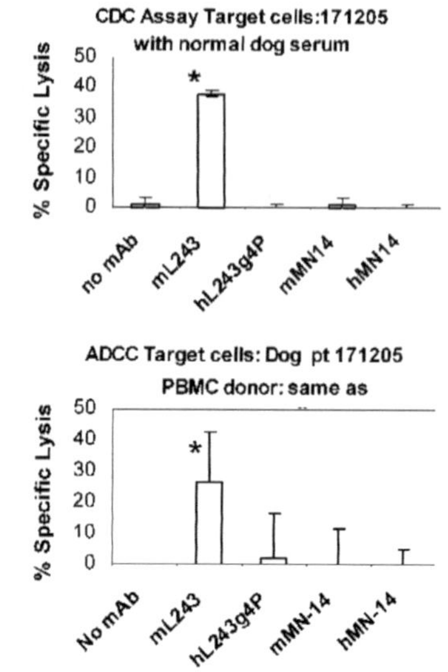

Fig. 2: Percent specific lysis in CDC and ADCC assays on FNA from a single dog with lymphoma. CDC assay was done using normal dog serum as a source of complement and ADCC was done using PBMCs from the same dog. Error bars, SD of 3 replicates, $P < 0.05$ relative to no mAb control. mMN14 and hMN14 are mouse and human antibody controls, respectively.

CONCLUSIONS AND FUTURE DIRECTIONS

The results presented here are early steps towards the goal of developing a therapeutic mAb for dogs with lymphoma. If the results of the phase I study with IMMU-114 are favorable, future clinical studies will be designed to assess the efficacy of multiple treatments with the mAb, both when used as a single-agent therapy and in combination with conventional chemotherapy protocols. The development of a "caninized" version of IMMU-114 is of significant interest to the antibody's manufacturer (Immunomedics, Inc.) given the large market potential for an effective product. Since MHC Class II antigens are expressed on the surface of other immune cells in dogs such as dendritic cells, investigation of the efficacy of IMMU-114 for canine malignant histiocytosis would also be warranted.

REFERENCES

1. Sharkey RM, et al. CA Cancer J Clin 2006;56:226. 2. Press O, et al. Hematology 2001;2:221. 3. Coiffer B. Clin Adv Hematol Oncol 2005;3:484. 4. Coiffer B. Clin Adv Hematol Oncol 2004;2:156. 5. Impellizeri J, et al. Vet J 2006;171:556. 6. Jeglum, K. Vet Clin North Am Small Anim Pract 1996;26:73. 7. Stein R, et al. Leuk Lymphoma 2011;52(2):273. 8. Stein R, et al. Blood 2006;108:2737. 9. Dechant M, et al. Sem Oncology 2003;30:465. 10. Brown K, et al. Clinical Lymphoma 2001;2:188. 11. Shi J, et al. Leuk Lymphoma 2002;43:1303.

CANINE LYMPHOMA – A CANCER GONE VIRAL?

Nicola Mason, BVetMed, PhD, DACVIM
Philadelphia, PA

INTRODUCTION

Lymphoma is the most common hematological malignancy in dogs and often follows a rapid and aggressive clinical course. Particular breeds of dogs are predisposed to developing Non-Hodgkin's Lymphoma and there is a clear breed-specific prevalence for lymphoma phenotype.[1] These findings suggest that risk factors for lymphoma development are inherited. As such, attention has focused on identifying genetic aberrations that may contribute to lymphomagenesis. In contrast, less attention has been paid to the possible role of environmental factors including infectious agents that may act alone or in concert with genetic factors, to trigger lymphomagenesis. Previous reports have suggested the possibility of a retroviral trigger for leukemia in dogs.[2] More recently serological responses against an Epstein Barr Virus (EBV)-like gammaherpesvirus (GHV) have been identified in healthy dogs suggesting that they may be infected with a GHV that has oncogenic potential. However, to date, no association has been made between exposure to/infection with an EBV-like GHV and lymphoma in dogs.[3,4]

EBV is the prototypical, double stranded DNA GHV that belongs to the subfamily lymphocryptovirus (LCV). One of the most prevalent human viral infections known, EBV infects up to 90% of the human population. Following acute infection, viral latency is established within the B cell compartment and in the majority of immune-competent human individuals chronic infection is asymptomatic.[5] However, some infected individuals develop EBV associated malignancies that reflect the restricted cellular tropism of EBV. These include the hematological malignancies (Burkitt's lymphoma (BL), Hodgkin's lymphoma (HL), Non-Hodgkin's Lymphoma (NHL) (including diffuse large B cell lymphoma (DLBCL) and NK/T cell lymphoma) and nasopharyngeal and gastric carcinoma.[6]

Dogs share a close phylogenetic relationship with humans and have co-habited with them for 15,000 years.[7,8] Given that genetic and environmental factors influence the initiation and progression of malignancy in humans, it is likely that similar genetic and environmental factors influence the development of cancer in the dog. Indeed, dogs develop spontaneous lymphoid malignancies including DLBCL and Burkitt-like lymphoma that share remarkable similarities in biological, behavioral, genetic and cytogenetic characteristics with their counterpart subtypes in humans.[9,10] For example, in human Burkitt's Lymphoma (BL) a chromosomal translocation places MYC under the control of the IgH promoter leading to its constitutive expression. The same chromosomal translocation leading to constitutive expression of MYC is present in dogs with spontaneous DLBCL and Burkitt-like lymphoma.[9,11] In addition, the canonical NF-κB pathway is constitutively active in the malignant lymphocytes of dogs with DLBCL, a finding that is shared with EBV-associated lymphoid malignancies in humans.[12] Indeed, c-myc and NF-κB are the two main transcription factors activated in the latency III phase of EBV infection and mediate the biological characteristics of EBV transformed cells.[13] Given these phylogenetic, environmental and genetic/cytogenetic similarities between dogs and humans and the B cell neoplasms they spontaneously develop we have begun to investigate the possibility that dogs and humans share a closely related oncogenic gammaherpesvirus.

Currently, there are no recognized GHV that infect dogs. Previous studies have revealed that following transfection with human CD21, the receptor for EBV, canine epithelial cells (Mardin Darby Canine Kidney (MDCK)) can be infected with EBV and sustain infection to establish stable EBV-infected canine clones.[14] In these cells, the viral genome exists as an episome and a latency type III gene transcription profile is expressed.[15] These results indicate that canine cells can support and maintain EBV infection. Furthermore, two reports suggest that healthy pet dogs can be naturally infected with an EBV-like herpesvirus.[3,4] These reports document serological responses against EBV-specific thymidine kinase, EBV encoded DNA binding protein, EBV-specific DNA polymerase and lymphoblastoid cell lines (LCLs) in healthy pet dogs which suggest the presence of an asymptomatic GHV carrier state, similar to that described in humans infected with EBV. Using In Situ Hybridization (ISH) on bone marrow aspirates and PCR on healthy lymphoid tissue respectively both reports showed evidence of viral RNA and DNA in a subset of healthy dogs.[3,4] However, efforts to identify viral sequences by PCR within 33 canine lymphoma samples failed to identify the presence of an EBV-like virus.[4] In that study, a wide variety of lymphoma samples were analyzed, including samples from skin (12), gut (5) and liver (2). A total of 8 malignant lymph nodes were evaluated by PCR and viral sequences were not identified.

In this lecture the previous and current evidence that suggests that dogs can be naturally infected with an oncogenic gammaherpesvirus will be discussed. Furthermore the hypothesis that infection may predispose to lymphomagenesis in a subset of dogs, either alone or in concert with genetic aberrations will also be discussed. Preliminary data that supports this hypothesis will also be presented.

Serological responses to EBV antigens in pet dogs

In humans, infection with EBV is accompanied by IgG antibody responses directed against the highly immunogenic, small Viral Capsid Antigens (VCA).[16] These IgG titers are extremely robust, minimally affected by immune suppression and persist for the lifetime of the individual.[17] As such, their identification is used routinely to confirm infection with EBV in humans (reviewed in [18]). Identification of antibodies that cross react with EBV VCAs is used to confirm infection with closely related LCVs that express homologs of EBV-VCAs particularly in the rhesus monkey and common marmoset.[19,20] Homologs of EBV-VCAs have also been identified in rhadinoviruses including Bovine Herpes Virus 4 (BHV4) and HHV8.[21,22]

Our studies have revealed that pet dogs have serological IgG responses that cross-react with specific EBV-VCAs. We found that dogs with and without lymphoma have detectable serological responses, suggesting that as with EBV infection in humans, infection with an EBV-like virus in many dogs is likely to be asymptomatic. Importantly as there are no *closely related* homologs of these VCAs in alphaherpesviruses (including the only known canine alphaherpesvirus CHV-1) or betaherpesviruses, these serological findings suggest that pet dogs have been naturally exposed to an EBV-like gammaherpesvirus. Interestingly dogs with the highest anti-viral titers also had spontaneous B cell lymphoma. Recent studies in humans have shown that patients with lymphoproliferative disorders including HL, NHL and PTLD have high anti-VCA IgG titers and viral load at the time of diagnosis.[23-25] In addition, retrospective studies have shown that high anti-VCA titers and viral loads may be predictive of lymphomagenesis in human patients although prospective, sero-epidemiological studies to confirm this are lacking.[26,27]

Identification of viral elements within canine lymphoid malignancies

In humans, detection of virus-specific sequences within lymphoid malignancies is used to indicate an association between viral infection and oncogenesis. We have employed ISH and IHC of canine B cell lymphomas together with PCR analysis to identify viral elements within canine lymphoma biopsy specimens. We found the presence of viral sequences with high homology to EBV present within the malignant lymph nodes of some dogs with DLBCL. We were also able to demonstrate the presence of Latent Membrane Protein-1 (LMP-1) a major oncogenic protein of EBV, in the malignant lymph nodes of some dogs with B cell lymphoma but not in normal healthy canine lymph node tissue. Interestingly, the principal mechanism by which LMP-1 promotes oncogenesis is through its ability to act as a CD40 mimic and drive constitutive NF-κB signaling within lymphocytes. This constitutive signaling leads to the over expression of genes that promote cellular proliferation (e.g. CyclinD1 and D2) and inhibit apoptosis (e.g. Bcl-2, c-FLIP, and XIAP). This pathway shown to be active in canine B cell lymphomas.[12]

In attempts to isolate the canine GHV, we used a previously published, cell-based culture method to grow canine B cells prior to analyzing them for the presence of intact viral particles by electron microscopy. Our results revealed the presence of viral like particles in 2 out of three cultures evaluated. Immunoelectron microscopy has yet to be performed on these sections to confirm the identity of these particles.

Taken together, these findings coupled with those of previous authors raise the intriguing possibility that an oncogenic gammaherpesvirus with high similarity to the highly prevalent EBV of humans may infect dogs and contribute to the development of lymphoma in a subset of individuals. Whether this virus

Future studies (Prospective longitudinal serological study in Golden Retrievers)

In humans, the presence of virus-specific sequences, in particular EBERs, within lymphoid malignancies together with high anti-VCA IgG titers and high EBV viral load at the time of lymphoma diagnosis suggest that viral infection/reactivation may contribute to lymphomagenesis.[28-30] Furthermore, in humans, genetic factors play an important role in determining susceptibility to virus-associated lymphoma. Given that the incidence of spontaneous lymphoma in the Golden Retriever dog is 1 in every 8 and that there is familial clustering associated with the development of lymphoma in this breed,[31] we now aim to look for correlations between infection and malignancy in pure-bred Golden Retriever dogs. We will perform a prospective, longitudinal serological study to evaluate the relationship between serological EBV antibody titers and spontaneous lymphomagenesis in age- and sex-matched Goldens. In addition, tumor tissue will be collected from enrolled Golden Retrievers to determine whether viral elements are present within their tumors. The database of tumor tissue generated from this study will allow us in the future to evaluate the presence of common genetic and cytogenetic abnormalities (e.g. c-myc translocation) in these dogs to determine if these tumors recapitulate the complexity of genetic and epigenetic factors present in human BL, HL or NHL. Moreover, these studies may lead to the identification of novel genetic factors that influence GHV-associated lymphomagenesis and could have a significant impact on our understanding of the role of an EBV-like virus in canine and human lymphomagenesis. Furthermore these studies may also lead to the identification of new targets for GHV-associated lymphoma therapies.

We aim to recruit approximately 500 dogs between 6 and 8 years of age (the average age of onset of clinical disease is 7 years.[32] Peripheral blood samples will be drawn from all dogs every 6 months from the time of enrollment to the end of the 2-year study period. End-point serum IgG titers against EBV-VCA will be determined by ELISA. At the time of lymphoma diagnosis, anti-VCA IgG titers will also be determined. Lymphomas that develop will be immunophenotyped and analyzed by histopathology. Samples will be analyzed for the presence of virus using ISH probes and qRT-PCR techniques that will be developed based on sequences derived from deep sequencing. We will monitor serological responses throughout standard of care L-CHOP chemotherapy to determine whether changes in antibody titers during therapy may reflect disease outcome. PBMCs will be collected at each time point and cryopreserved for future studies to determine viral load and to perform GWAS analysis that aims to shed light on genetic factors that may predispose to virus-associated lymphomagenesis. The pedigrees of all study dogs will be collected for determination of familial predispositions toward viral titer and lymphomagenesis. By studying the association between viral antibody titer and lymphoma development within a genetically restricted, highly predisposed breed, we aim to determine whether a causal relationship exists between natural EBV-like infection and lymphomagenesis in the dog.

If you wish to participate in this study by recruiting Golden Retriever dogs for serological analysis and tumor analysis please contact Dr. Nicola Mason at the University of Pennsylvania's School of Veterinary Medicine (Tel: 215 898 3996; E-mail: nmason@vet.upenn.edu).

REFERENCES

1. Modiano, J.F. et al. Distinct B-cell and T-cell lymphoproliferative disease prevalence among dog breeds indicates heritable risk. *Cancer Res* **65**, 5654-61 (2005). 2. Ghernati. Characterization of a canine long-term T cell line (DLC 01) established from a dog with Sézary syndrome and producing retroviral particles. *Leukemia* **13**, 1281-90 (1999). 3. Chiou, S.H., et al. Discovery of Epstein-Barr virus (EBV)-encoded RNA signal and EBV nuclear antigen leader protein DNA sequence in pet dogs. *J Gen Virol* **86**, 899-905 (2005). 4. Milman, G.,et al. Serological detection of Epstein-Barr virus infection in dogs and cats. *Vet Microbiol* **150**, 15-20 (2011). 5. Rickinson, A.B. & Moss, D.J. Human cytotoxic T lymphocyte responses to Epstein-Barr virus infection. *Annu Rev Immunol* **15**, 405-31 (1997). 6. Kutok, J.L. & Wang, F. Spectrum of Epstein-Barr virus-associated diseases. *Annu Rev Pathol* **1**, 375-404 (2006). 7. Lindblad-Toh, K. et al. Genome sequence, comparative analysis and haplotype structure of the domestic dog. *Nature* **438**, 803-19 (2005). 8. Vila, C. et al. Multiple and ancient origins of the domestic dog. *Science* **276**, 1687-9 (1997). 9. Breen, M. & Modiano, J.F. Evolutionarily conserved cytogenetic changes in hematological malignancies of dogs and humans--man and his best friend share more than companionship. *Chromosome Res* **16**, 145-54 (2008). 10. Vail, D.M. & MacEwen, E.G. Spontaneously occurring tumors of companion animals as models for human cancer. *Cancer Invest* **18**, 781-92 (2000). 11. Taub, R. et al. Translocation of the c-myc gene into the immunoglobulin heavy chain locus in human Burkitt lymphoma and murine plasmacytoma cells. *Proc Natl Acad Sci U S A* **79**, 7837-41 (1982). 12. Gaurnier-Hausser, A., et al. NEMO-Binding Domain Peptide Inhibits Constitutive NF-{kappa}B Activity and Reduces Tumor Burden in a Canine Model of Relapsed, Refractory Diffuse Large B-Cell Lymphoma. *Clin Cancer Res* **17**, 4661-71 (2011). 13. Faumont, N. et al. c-Myc and Rel/NF-kappaB are the two master transcriptional systems activated in the latency III program of Epstein-Barr virus-immortalized B cells. *J Virol* **83**, 5014-27 (2009). 14. Yates, J.L., et al. Stable replication of plasmids derived from Epstein-Barr virus in various mammalian cells. *Nature* **313**, 812-5 (1985). 15. Yang, L., et al. CD21-mediated entry and stable infection by Epstein-Barr virus in canine and rat cells. *J Virol* **74**, 10745-51 (2000). 16. van Grunsven, W.M., et al. Identification and molecular characterization of two diagnostically relevant marker proteins of the Epstein-Barr virus capsid antigen complex. *J Med Virol* **40**, 161-9 (1993). 17. Klutts, J.S., et al. Evidence-based approach for interpretation of Epstein-Barr virus serological patterns. *J Clin Microbiol* **47**, 3204-10 (2009). 18. Pattle, S.B. et al. The role of Epstein-Barr virus in cancer. *Expert Opin Biol Ther* **6**, 1193-205 (2006). 19. Moghaddam, A., et al. Infection of human B lymphocytes with lymphocryptoviruses related to Epstein-Barr virus. *J Virol* **72**, 3205-12 (1998). 20. Jiang, H., et al. Structural, functional, and genetic comparisons of Epstein-Barr virus nuclear antigen 3A, 3B, and 3C homologues encoded by the rhesus lymphocryptovirus. *J Virol* **74**, 5921-32 (2000). 21. Rao, P., et al. Cloning of the rhesus lymphocryptovirus viral capsid antigen and Epstein-Barr virus-encoded small RNA homologues and use in diagnosis of acute and persistent infections. *J Clin Microbiol* **38**, 3219-25 (2000). 22. Fogg, M.H.,et al. Reduced prevalence of Epstein-Barr virus-related lymphocryptovirus infection in sera from a new world primate. *J Virol* **79**, 10069-72 (2005). 23. Riddler, S.A., et al. Increased levels of circulating Epstein-Barr virus (EBV)-infected lymphocytes and decreased EBV nuclear antigen antibody responses are associated with the development of posttransplant lymphoproliferative disease in solid-organ transplant recipients. *Blood* **84**, 972-84 (1994). 24. Khan, G. et al. Phenotype and frequency of Epstein-Barr virus-infected cells in pretreatment blood samples from patients with Hodgkin lymphoma. *Br J Haematol* **129**, 511-9 (2005). 25. Fan, H. et al. Epstein-Barr viral load as a marker of lymphoma in AIDS patients. *J Med Virol* **75**, 59-69 (2005). 26. de-The, G. et al. Epidemiological evidence for causal relationship between Epstein-Barr virus and Burkitt's lymphoma from Ugandan prospective study. *Nature* **274**, 756-61 (1978). 27. Mueller, N. et al. Hodgkin's disease and Epstein-Barr virus. Altered antibody pattern before diagnosis. *N Engl J Med* **320**, 689-95 (1989). 28. Gulley, M.L. & Tang, W. Laboratory assays for Epstein-Barr virus-related disease. *J Mol Diagn* **10**, 279-92 (2008). 29. Wu, T.C. et al. Detection of EBV gene expression in Reed-Sternberg cells of Hodgkin's disease. *Int J Cancer* **46**, 801-4 (1990). 30. Weiss, L.M.,et al. Epstein-Barr virus and Hodgkin's disease. A correlative in situ hybridization and polymerase chain reaction study. *Am J Pathol* **139**, 1259-65 (1991). 31. Teske, E., de Vos, J.P., Egberink, H.F. & Vos, J.H. Clustering in canine malignant lymphoma. *Vet Q* **16**, 134-6 (1994). 32. Glickman. The Golden Retriever Club of America National Health Survey [cited 2000 Aug 15]. *Available from:* http://www.grca.org/healthsurvey.pdf. (2000).

CANINE LYMPHOMAS: RETHINKING A DIVERSE FAMILY OF NEOPLASMS

Michael O. Childress, DVM, MS, DACVIM (Oncology)
West Lafayette, IN

BIOLOGICAL DIVERSITY OF CANINE LYMPHOMAS

For years, canine lymphomas have been managed with little regard to the inherent heterogeneity of this group of neoplasms. Indeed, the term "lymphoma" (as opposed to "lymphomas") is often used to describe these cancers, with the implication that they are all essentially variations on the same theme. Accumulating evidence in humans, and more recently dogs, indicates that nothing could be farther from the truth. Lymphomas are possessed of stunning diversity, both in their ontogeny and in their biologic behavior. The World Health Organization - Revised European American Classification of Lymphoid Neoplasms (WHO-REAL) recognizes over 30 major subtypes of lymphoid neoplasia in humans, with many variants of these subtypes also recognized.[1] A similar WHO classification scheme for lymphoid neoplasia in domestic animals has also been recently proposed .[2,3]

The biological basis for this classification scheme is that each tumor catalogued therein represents a specific clinical disease entity. Each of these tumors originates from lymphocytes of a specific immunophenotype (B-cell, T-cell, or natural killer (NK) cell) that have arrested at a certain stage in their maturation process. Thus some lymphomas will be characterized by rapidly-dividing, immature cells ("large cell" or "lymphoblastic"), while others will be characterized by slowly-dividing, mature cells ("small cell" or "lymphocytic"). The former are often referred to as "high grade" lymphomas, while the latter are referred to as "low grade," although histopathologic grade cannot be assigned on the basis of cell morphology alone.[3,4] When classified according to the WHO scheme, the most common type of lymphoma in dogs is ***diffuse large B-cell lymphoma (DLBCL)***. This disease entity accounts for approximately 40-60% of all canine lymphomas,[3,5,6] and is analogous to the high-grade multicentric B-cell lymphoma most veterinarians probably imagine when they think of "lymphoma."

OPTIMAL DIAGNOSIS AND STAGING OF CANINE LYMPHOMAS

"The first and primary determinant of treatment is the histologic diagnosis." -Bruce Chabner, M.D., Director of Clinical Research, Massachusetts General Hospital Cancer Center

Despite significant advances in diagnostic pathology over the last two decades, cancer still remains a histologic diagnosis, made by observing the morphological and architectural relationships of neoplastic cells to one another and to surrounding normal cells. Whenever possible, lymphomas should be diagnosed by a tissue biopsy composed of adequate tissue with which to make an accurate diagnosis, and by which classification according to the WHO scheme can be made. For nodal lymphomas, an incisional wedge biopsy or surgical excision of a whole affected node are the preferred biopsy methods.

Standard hematoxylin and eosin histopathology should be complemented by immunohistochemical staining to assess immunophenotype. Most diagnostic pathology labs now offer immunohistochemical assays for detecting CD79a (a B-cell marker), CD3 (a T-cell marker), and potentially other immunophenotypic markers. Assessment of the tumor immunophenotype is essential for making a complete diagnosis of lymphoma, and is of prognostic relevance. B-cell lymphomas typically carry a more favorable prognosis than T-cell lymphomas,[7] although prognosis also varies within immunophenotypes depending upon the WHO histopathologic subtype.[5]

Veterinarians should resist the temptation to rely solely upon fine needle aspirate (FNA) cytology for the definitive diagnosis of lymphoma. Although it is less invasive and relatively less expensive than biopsy, FNA may not reliably distinguish the various WHO subtypes of lymphoma from one another. In additions, on some occasions, higher grade lymphomas may be found emerging from lower grade lymphomas in biopsy specimens. For instance, DLBCL may emerge from follicular lymphoma or marginal zone lymphoma; this may be very hard, if not impossible to interpret on FNA cytology. Moreover FNA may not be able to differentiate certain types of lymphoma from reactive lymphoid hyperplasia. FNA cytology is considered unhelpful for the diagnosis of lymphoma in humans,[8] and should be used only as a preliminary diagnostic tool rather than a confirmatory one.

In recent years, molecular diagnostics such as immunocytochemistry (ICC),[9] flow cytometry,[10,11] and PCR for antigen receptor gene rearrangement (PARR)[12] have become popular tools for diagnosing and staging canine lymphomas. ICC and flow cytometry are both used to detect CD79a, CD3, and other cell surface antigens in tissue samples. Diagnosis of lymphoma by ICC or flow cytometry relies on the fact that neoplastic samples will usually be composed of a monoclonal population of lymphocytes, either all B-cells or all T-cells (or rarely, all NK cells). To the contrary, lymphoid hyperplasia and other benign lymphoid pathologies will usually be composed of a mixture of B- and T-cells. However, both ICC and flow cytometry have the significant drawback in that they are performed on cytologic samples, and thus have the same limitations as FNA cytology – namely they have not been shown to reliably distinguish different WHO subtypes from one another. Moreover, both false positive and false negative results are possible, as some lymphomas may contain both B- and T-cells (e.g. T-cell rich B-cell lymphoma), and some benign pathologies may be characterized by monoclonal lymphocyte expansion (e.g. reactive lymphoid hyperplasia secondary to ehrlichiosis).[12] The author generally reserves ICC, flow cytometry, and PARR to assist in the diagnosis of lymphomas for which histopathology is not possible (e.g. in the rare patient that is too debilitated to undergo biopsy) or has yielded conflicting or confusing results.

Once a diagnosis of lymphoma is confirmed, staging tests are indicated to assess the extent of the cancer in the patient. The staging tests recommended for all canine lymphomas include a complete blood count (CBC), serum biochemistry profile, urinalysis, thoracic radiographs, abdominal ultrasound, and bone marrow aspirate or core biopsy. Although lymphoma may affect any visceral organ, common intrathoracic and intraabdominal sites of lymphoma are the liver, spleen, thymus gland, and internal lymph nodes. The CBC, serum biochemistry profile, and urinalysis primarily assess the general health status of the patient, particularly regarding renal, hepatic, and hematopoietic function, which will affect the patient's ability to metabolize, excrete, and tolerate chemotherapy. The CBC may also reveal circulating neoplastic lymphocytes, which is often indicative of bone marrow infiltration by the cancer, although this should be confirmed by bone marrow aspirate or biopsy. Some lymphomas may be more or less likely to involve multiple sites in the body. DLBCL, for instance, typically involves peripheral lymph nodes first, and rapidly spreads to liver, spleen, bone marrow, and potentially any other site. Other lymphomas, such as splenic marginal zone lymphoma and some low-grade nodal lymphomas often remain local or loco-regional cancers for a protracted period of time.[13] Thus, histopathological subtype correlates with clinical behavior of lymphomas, and this may in turn help to determine optimal staging and treatment protocols.

THERAPEUTIC IMPLICATIONS OF BIOLOGICAL DIVERSITY IN LYMPHOMAS

For many years, the "standard of care" treatment for most canine lymphomas has been doxorubicin-based combination chemotherapy, often referred to as CHOP chemotherapy. The acronym "CHOP" denotes the four drugs thought to have the greatest activity against lymphomas: **C**yclophosphamide, **H**ydroxydaunorubicin (doxorubicin), **O**ncovin (vincristine), and **P**rednisone. CHOP-based chemotherapy protocols induce complete remission (CR), that is, render the cancer clinically undetectable, in approximately 70-90% of dogs with lymphoma, and afford median survival times of approximately 1 year.[14-16] However, the studies from which these data derive focused mainly on dogs with multicentric forms of lymphoma, of which most were likely dogs with high-grade B-cell lymphomas such as DLBCL. Thus, these studies likely excluded a large number of dogs with other histologic subtypes of lymphoma. Nevertheless, CHOP is still considered by many to be the standard of care for most canine lymphomas.

More recent studies are beginning to question this wisdom. For instance, epitheliotropic T-cell lymphoma, a relatively indolent lymphoma of the skin and mucous membranes, seems to be more sensitive to the oral alkylating agent CCNU than to CHOP.[17,18] Multicentric T-cell lymphomas (such as peripheral T-cell lymphoma and anaplastic large T-cell lymphoma) may represent another group of lymphomas for which CHOP chemotherapy is not optimal. The median duration of first remission for T-cell lymphomas treated with CHOP has been historically disappointing, with reported lengths of approximately 3-5 months (as compared to 6-11 months for B-cell lymphomas).[19-21] A recent study[22] showed that the CR rate of T-cell lymphomas treated with a single dose of doxorubicin was only 17%, compared to 86% for dogs with B-cell lymphomas. These results are also very disappointing, considering that doxorubicin is considered by many to be the drug with the greatest single-agent activity against canine lymphomas. By comparison, another recent study[23] showed that 78% of dogs with T-cell lymphomas experienced complete remission when treated with MOPP (**M**echlorethamine, **O**ncovin, **P**rocarbazine, and **P**rednisone) and L-asparaginase. MOPP is traditionally used as a rescue therapy for dogs whose lymphoma is no longer responding to CHOP. These results suggest that the paradigm of CHOP chemotherapy as the gold standard for all lymphomas requires revision.

DIVERSITY WITHIN DIVERSITY – GENOMIC PROFILING AND LYMPHOMA SUBSETS

Veterinarians treating canine lymphoma patients have long recognized that some dogs seem to "defy the odds" and experience long-term survival in spite of an apparently aggressive cancer diagnosis. Others dogs, however, seem to respond poorly to therapy in the face of what would seemingly be a prognostically favorable lymphoma type. Tumor immunophenotype has historically been the characteristic most strongly associated with prognosis, and indeed this has spawned the pithy aphorism, "B is Better, T is Terrible." While this phrase does carry much truth, it certainly does not tell the whole story: some dogs with T-cell lymphomas experience long-term survival while some with B-cell lymphomas respond minimally to standard CHOP chemotherapy and die rapidly from their cancer. Histopathologic subtype likely explains some of this diversity in response – for instance, dogs with low-grade peripheral T-cell lymphomas (e.g. T-zone lymphoma) may be managed effectively with low-intensity chemotherapy (e.g. oral chlorambucil and prednisone) and do well in spite of what is commonly believed to be a prognostically unfavorable immunophenotype. This underscores the importance of an accurate histopathologic diagnosis of lymphoma subtype.

Recent human studies show that even within certain histopathologic subtypes of lymphoma, differences in gene expression patterns may segregate patients into prognostically favorable and unfavorable subsets. For example, two subsets of human DLBCL, designated "activated B-cell type" and "germinal center B-cell type" have recently been described.[24,25]

These subsets are characterized by different gene expression profiles and different survival rates. Approximately 60% of patients with germinal center B-cell DLBCL are alive at 5 years from diagnosis, as opposed to only 35% of patients with activated B-cell DLBCL.[25] A small-scale study in canine DLBCL was recently conducted, showing that many canine DLBCL have a gene expression profile similar to the human activated B-cell DLBCL.[26] This may serve in part to explain the relatively low cure rate of canine DLBCL (probably less than 5%) when compared its human counterpart (~50-60%).

In the future, it is likely that canine lymphomas will be treated based upon initial diagnosis using histopathology, followed by characterization of the molecular aberrations that have resulted in neoplastic transformation. This type of tumor characterization, based not only upon histomorphology and immunophenotype, but also according to the collective mutations that characterize an individual dog's cancer, may help to optimize therapy on an individual patient basis. The author eagerly looks forward to this forthcoming era of personalized cancer medicine for man's best friend.

REFERENCES

1. Swerdlow SH, et al, eds. *WHO classification of tumours of haematopoietic and lymphoid tissues.* Lyon, France, 2008, IARC Press. 2. Valli VE, et al. *Histological classification of hematopoietic tumors of domestic animals.* Washington, DC, 2002, Armed Forces Institute of Pathology and the World Health Organization. 3. Valli VE, et al. *Vet Pathol* 2011;48:198-211. 4. Valli VE, et al. *J Vet Diagn Invest* 2000;12:295-306. 5. Raskin RE, Fox LE. *Vet Clin Pathol* 2003;32:151 (abstract). 6. Vezzali E, et al. *Vet Comp Oncol* 2009;8:38-49. 7. Ruslander DA, et al. *In Vivo* 1997;11:169-172. 8. Hehn ST, et al. *J Clin Oncol* 2004;22:3046-3052. 9. Grindem CB. *Vet Clin North Am Small Anim Pract* 1996;26:1043-1064. 10. Comazzi S, et al. *Vet Immuno Immunopathol* 2006;112:243-252. 11. Lana S, et al. *J Vet Intern Med* 2006;20:1161-1165. 12. Burnett RC, et al. *Vet Pathol* 2003;40:32-41. 13. Valli VE, et al. *Vet Pathol* 2006;43:241-256. 14. Keller ET, et al. *J Vet Intern Med* 1993;7:289-295. 15. Zemann BI, et al. *J Vet Intern Med* 1998;12:465-470. 16. Garrett LD, et al. *J Vet Intern Med* 2002;16:704-709. 17. Risbon RE, et al. *J Vet Intern Med* 2006;20:1389-1397. 18. Williams LE, et al. *J Vet Intern Med* 2006;20:136-143. 19. Vail DM, et al. *Exp Hematol* 1996;24:807-815. 20. Chun R, et al. *J Vet Intern Med* 2000;14:120-124. 21. Rebhun RB, et al. *Vet Comp Oncol* 2010;9:38-44. 22. Beaver LM, et al. *J Am Vet Med Assoc* 2010;237:1052-1055. 23. Brodsky EM, et al. *J Vet Intern Med* 2009;23:578-584. 24. Alizadeh AA, et al. *Nature* 2000;403:503-511. 25. Rosenwald A, et al. *N Engl J Med* 2002;346:1937-1947. 26.Richards K, et al. *Proceedings:29th Annual Conference of the Veterinary Cancer Society.* Austin, TX:2009;p 70.

CANINE LYMPHOMA – IS THERE A STANDARD OF CARE?

Dennis B Bailey, DVM, DACVIM (Oncology)
Paramus, NJ

INTRODUCTION

Canine lymphoma is one of the most commonly diagnosed and treated cancers in veterinary oncology. Additionally, it is one of the most frequently published topics in veterinary oncology. Lymphoma was the subject matter for 25% of all oncology-related articles published in the *Journal of Veterinary Internal Medicine* from 1999 to 2007, four times more than any other tumor type.[1] Despite this, there are no well-established guidelines regarding initial diagnostic evaluation or treatment. The Veterinary Cooperative Oncology Group recently published a consensus document regarding response evaluation criteria for peripheral nodal lymphoma in dogs.[2] Establishing standards is important. It allows us to more accurately compare different studies, and ultimately to better care for our patients. This talk will review the literature regarding the clinical management of canine lymphoma patients, highlighting areas where there is apparent consensus as well as areas where there is discord.

REACHING A DEFINITIVE DIAGNOSIS

It is imperative to reach a definitive diagnosis before starting cancer treatment. Histopathology and cytology have been used most commonly. In human oncology, histopathology remains the gold standard. However, in a recent survey conducted at the 2009 Veterinary Cancer Society (VCS) Annual Conference, 96% of responding veterinarians recommended cytology of enlarged lymph nodes and/or other organs of suspected involvement, compared to only 28% who recommended histopathology.[3] Histopathology is required to accurately identify some other forms of canine lymphoma, such as the indolent lymphomas,[4] and immunohistochemistry remains one of the best ways to determine immunophenotype. However, the majority of lymphomas in dogs are diffuse and intermediate to large cell type, and for these forms cytology often is adequate for diagnosis. It is worth noting, though, that clinical pathologists often convey different levels of confidence in their diagnosis, and the level of confidence clinicians require before they take action varies. In an on-line survey of veterinarians who are members of the Veterinary Information Network (VIN), 51% of responders would initiate chemotherapy based on a cytologic diagnosis "consistent with lymphoma," 15% with a diagnosis of "probable lymphoma," 4% with a diagnosis "suggestive of lymphoma," and 2% with a diagnosis of "possible lymphoma."[5] When the same people were given the clinical scenario where their client is considering euthanasia if their dog has cancer, 63% would recommend euthanasia based on a cytologic diagnosis "consistent with lymphoma," 35% with a diagnosis of "probable lymphoma," 8% with a diagnosis "suggestive of lymphoma," and 2% with a diagnosis of "possible lymphoma."

Newer molecular tests, including flow cytometry and PARRs (PCR for antigen receptor rearrangement), are becoming are being used more commonly to help diagnose lymphoma, usually as an adjunct to cytology or histopathology. Flow cytometry uses antibodies to determine the immunophenotype of the lymphocytes in the sample collected. A single predominant immunophenotype or an aberrant immunophenotype supports the presence of a neoplastic population. Flow cytometry requires viable cells, and is most commonly run on whole blood (if neoplastic cells are seen in circulation) or an aspirate that has been placed in saline or another suitable liquid media. PARRs evaluates specific DNA sequences in the genes that encode the immunoglobulin and T-cell receptor genes. These genes undergo unique recombination events in B- and T-lymphocytes, respectively. The predominance of a single lymphocyte clone supports a diagnosis of cancer. PARRs does not require viable cells and can be run on a variety of samples, including blood, aspirated tissue in saline, and dried cytology slides (stained or unstained). However, PARRs yields a false negative result in 28% of dogs with lymphoma that has been confirmed on lymph node histopathology.[6]

INITIAL DIAGNOSTIC EVALUATION

Baseline staging diagnostics are recommended for cancer patients prior to initiating any therapy. The primary goals of these tests are to determine the clinical extent of the cancer and identify prognostic factors. In the broader context, they also are used to assess overall health and identify any unrelated diseases. The results of these tests are used to optimize the patient's treatment plan, guide patient monitoring, and assess prognosis.

While owner preference and finances can have a profound influence on the staging diagnostics that are performed, each clinician should have an "optimal" diagnostic plan that he or she recommends. In a recent survey conducted at the 2009 VCS Annual Conference, the most commonly recommended diagnostic tests were complete blood count (100%), chemistry panel (100%), urinalysis (85%), thoracic radiographs (84%), immunophenotyping (76%), bone marrow cytology (74% overall; 41% for all dogs, 33% only for dogs with abnormalities on their CBC), and echocardiogram for breeds considered to be at risk for underlying cardiac disease (69%).[3] Differences in staging diagnostic recommendations can result in stage migration, where more advanced diagnostic tests detect disease that otherwise would not have been identified.[7] The prognostic significance of a dog's numeric WHO stage has been inconsistent. Having said that, involvement of certain anatomic locations – such as the gastrointestinal tract, central nervous system, and skin – consistently carries a more guarded prognosis.[8-10] Additionally, the results of some diagnostic test results, such as immunophenotyping, do not impact stage but still have been strongly linked to prognosis.[11]

FIRST-LINE CHEMOTHERAPY TREATMENT

The most commonly used first-line chemotherapy protocols used to treat canine lymphoma are cyclic combination protocols that include L-asparaginase, vincristine, cyclophosphamide, doxorubicin, and prednisone. Specific protocol names include University of Wisconsin-Madison, COPLA, and VELCAP. These protocols and their many variations are often referred to by the generic name L-CHOP. Based on the 2009 VCS survey, when treating dogs with multicentric (stages III-V, substage a) high-grade B-cell lymphoma, 51% of responding veterinarians recommended an L-CHOP protocol, 30% recommend a CHOP protocol (omitting L-asparaginase), and 12% recommend a protocol that combines CHOP with one or more other chemotherapy drugs.[3] The omission of L-asparaginase by many responders likely reflects two retrospective studies demonstrating no difference in remission rate, remission duration, or overall survival with one dose of L-asparaginase is included at the start of a CHOP or COP (cyclophosphamide, vincristine, prednisone) based protocol compared to historical controls.[12,13] Both studies were retrospective studies resulting from a temporary drug shortage (i.e. L-asparaginase was omitted due to lack of availability, not due to clinician choice). No randomized or prospective studies have been performed to evaluate the benefit using L-asparaginase in combination with CHOP drugs for treating canine lymphoma.

First-line protocols used to treat dogs with multicentric (stages III-V, substage a) high-grade T-cell lymphoma were more varied. T-cell immunophenotype is a known negative prognostic factor. In a retrospective study of 24 dogs with T-cell lymphoma treated with CHOP, the initial response rate was high (88% complete remission, 8% partial remission, 4% no response). However, median remission duration was only 146 days, and median survival was 235 days.[11] In another case series evaluating L-asparaginase, mustargen, vincristine, procarbazine, and prednisone as a first-line protocol for dogs with T-cell lymphoma, median progression-free survival was 189 days, comparable to what has reported with CHOP-based protocols. However, 25% of dogs were still alive at 939 days.[14] Based on this information, veterinarians are more likely to try protocols other than standard L-CHOP or CHOP when treating T-cell lymphomas. In particular, they are more likely to include alkylating agents such as lomustine (CCNU), mustargen, and procarbazine in their first-line T-cell protocols.

Protocol length is impacted by efficacy, cost, and owner convenience. Published protocol lengths vary from 12-weeks to > 2 years. To date, there are no prospective, randomized studies to determine the impact of protocol length on remission duration or survival. However, one study comparing a 6-month L-CHOP protocol to an historical control where dog were treated for 2 years did not show any difference in disease-free interval or survival.[15] The results for protocols shorter than 6 months have been more variable. A 15-week L-CHOP protocol reported a median first remission duration of 140 days, and this was significantly shorter when compared to a historical control using an 18-month protocol.[16] A more dose-intense 12-week protocol reported a median first remission duration of 243 days, but 65% of the dogs in that study developed neutropenia and 65% experienced gastrointestinal toxicity.[17] Based on the 2009 VCS Survey, most veterinarians do not recommend using a protocol with a prolonged maintenance phase. Ten percent recommended using a protocol ≤16 weeks in length, 25% a protocol 17-23 weeks, 50% a protocol 24-28 weeks, 5% a protocol 29-52 weeks, and 10% a protocol >52 weeks.[3]

RESPONSE EVALUATION CRITERIA

When treating dogs with multicentric lymphoma, the response to therapy often is dramatic and easy to recognize. However, to make accurate comparisons across studies, standardized criteria are needed when assessing response. The Veterinary Cooperative Oncology Group has published a consensus document outlining response evaluation criteria for dogs with lymphoma where peripheral lymphadenopathy is the principal component of their disease presentation.[2] The criteria are adapted from the Response Evaluation Criteria for Solid Tumors (RECIST) criteria, currently considered to be the standard for assessing response in solid tumors in both human and veterinary oncology.

Target lesions are defined as affected peripheral lymph nodes with a longest-diameter measurement that is ≥20 mm at baseline. A minimum of 1 and a maximum of 5 target lesions should be selected. If there are >5 palpably enlarged lymph nodes, target lesions should be selected based on size (largest lymph nodes) and their suitability for accurate repeated measurements. All remaining measurable lesions are considered non-target lesions. Lesions only viewable with imaging (radiography or ultrasound), splenomegaly, hepatomegaly, bone marrow infiltrates, and effusions should *not* be used as target lesions, but they do qualify as non-target lesions.

For each target lesion, the longest diameter (LD) should be measured using calipers. Ideally, 2 people should measure each target lesion independently, and the mean should be calculated for each lesion. The mean LD measurements should be added together. This mean sum LD represents the baseline reference used to assess objective tumor response during treatment. For all non-target lesions, location, type, and number should be recorded. Relative size can be recorded where appropriate (e.g. splenomegaly), but an exact measurement of LD is not necessary.

Overall response is defined by objective assessment of target lesions, and subjective assessment of non-target lesions or any new lesions. For the target lesion, a complete response (CR) is defined as the disappearance of all evidence of disease. This can be a difficult assessment, though, since the size of a "normal" lymph node can be impacted substantially by breed size and body condition. A partial remission (PR) is defined as a reduction in mean sum LD by ≥30%. Stable disease (SD) is defined a decrease in mean sum LD by <30% or an increase by <20%. Progressive disease (PD) is defined by an increase in mean sum LD by ≥20%. Additionally, given the difficulty of accurately measuring small (< 10 mm) lesions, at least one target lesion must be > 15 mm and have increased by at least 5 mm to confirm progression. For all non-target lesions, responses are classified as CR when there is complete resolution of all non-target lesions, PD if there is unequivocal progression, and nonCR/nonPR for everything else. The appearance of any new lesions constitutes PD. The responses in target and non-target lesions are then combined to create an overall response (Table 1).

While there are specific objective criteria to characterize the extent of responses to treatment, more important is the duration of that response. The Veterinary Cooperative Oncology Group recommends using progression-free survival. Progression-free survival is measured from the first day of treatment until either the first date the criteria for progressive disease are met or the date of death from any cause. The maximum response attained (i.e. smallest tumor burden) is used as the reference criterion for assessing progressive disease, but the extent of that response (CR, PR, SD) does not matter when assessing progression-free survival. This is done for two reasons. First, as discussed above, the definition of a "normal" lymph node is subjective, making it difficult to accurately

Table 1: Criteria for assessing overall response in dogs with peripheral lymph node lymphoma.[2]

Overall Response	Target Lesions	Non-Target Lesions	New Lesions
Complete Response	CR	CR or None	No
Partial Response	CR	Non-CR / Non-PR	No
	PR	Non-PD	No
Stable Disease	SD	Non-PD	No
Progressive Disease	PD	CR, Non-CR / Non-PD, PD, or None	Yes or No
	CR, PR, SD or PD	PD	Yes or No
	CR, PR, SD, or PD	CR, Non-CR / Non-PD, PD, or None	Yes

distinguish a CR from a good PR. Second, since a cure is rarely attained regardless of the extent of the objective response, documenting progression is one of the most repeatable and widely applicable response measures available.

CONCLUSIONS

It is not practical to insist that every dog with lymphoma undergo the same initial diagnostic evaluation and treatment. Individual choices will be impacted by veterinarian experience, availability of chemotherapy drugs and equipment, patient overall health status, owner goals and preferences, and finances. However, it is very reasonable to work to establish a consensus outlining what we as a profession consider ideal regarding initial diagnostic evaluation, initial treatment, and assessment of response to therapy. We then should strive to recommend and adhere to these guidelines as much as reasonably possible. More importantly, we should insist that investigators use these guidelines when evaluating new treatment options for canine lymphoma. All dogs in a clinical study should undergo the recommended initial diagnostic evaluation and be assessed using the VCOG-proposed response evaluation criteria. We also should work towards more randomized, controlled studies, using the proposed initial treatment protocol as the standard control. This will allow for better comparisons across studies, and greatly accelerate our understanding of how to provide the best possible care for dogs with lymphoma.

REFERENCES

1. Sahora A, et al. **J Vet Intern Med** 2010;24:51. 2. Vail DM, et al. **Vet Comp Oncol** 2009;8:28. 3. Regan RC, et al. **Vet Comp Oncol** in press. 4. Valli VE, et al. Vet Pathol 2006;43:241. 5. Christopher MM, et al. **J Vet Intern Med** 2010;24:496. 6. Lana SE, et al. **J Vet Intern Med** 2006;20:329. 7. Flory AB, et al. **J Vet Intern Med** 2007;21:1042. 8. Rassnick KM, et al. **J Vet Intern Med** 2009;23:317. 9. Snyder JM, et al. **J Vet Intern Med** 2008;22:172. 10. Fontaine J, et al. **Vet Comp Oncol** 2009;7:1. 11. Rebhun RB, et al. **Vet Comp Oncol** 2011;9:38. 12. MacDonald VS, et al. **J Vet Intern Med** 2005;19:732. 13. Jeffries AB, et al. **J Am Anim Hosp Assoc** 2005;41:221. 14. Brodsky EM, et al. **J Vet Intern Med** 2009;23;578. 15. Garrett LD. **J Vet Intern Med** 2002;16:704. 17. Moore AS, et al. **J Vet Intern Med** 2001;15:348. 18. Simon D, et al. **J Vet Intern Med** 2006;20:94.

CANINE CUTANEOUS MAST CELL TUMOR PROFILE/PANEL: WHAT DOES IT MEAN? SHOULD I ORDER IT?

Debra A Kamstock, DVM, PhD, DACVP
Fort Collins, CO

INTRODUCTION

This lecture will review the parameters evaluated within a canine cutaneous mast cell tumor profile/panel and the rationale for performing each of these tests based on peer reviewed published studies. Specific parameters discussed will include proliferation indices (mitotic index, Ki67, AgNORs, PCNA) KIT expression, and *c-kit* mutation status.

The mast cell tumor profile/panel is a battery of tests, predominantly molecular, performed in addition to routine microscopic evaluation and grading. This set of tests is based on the premise of furnishing the clinician with additional prognostic as well as predictive information to help guide decisions in determining the optimal therapeutic plan.

PROLIFERATION INDICES

Proliferation indices reflect the growth fraction of a tumor and have been shown to carry prognostic relevance in many mast cell tumor studies.[2-12,14-16,19] Mitotic index, Ki67 expression, AgNOR counts, and PCNA expression are all proliferation indices; each representing a different aspect of cellular proliferation.

MITOTIC INDEX (MI)

What is it? Mitotic index is the number of cells in mitosis (M phase) within an indicated number of high-power (400x) fields (hpf). In an attempt to standardize reporting in veterinary oncologic pathology, it has been recommended that mitotic index be reported as the total number of mitotic figures (MFs) per *ten* high-power fields.[1] This is typically the standard means of reporting MI; however, variation amongst pathologists does exist. Additionally, there is variability across individual microscopes and objectives, which inherently affects the field of view. Thus, even with standardization of reporting, interobserver variability cannot be completely eliminated.

Why does it matter? A number of canine cutaneous mast cell tumor studies have demonstrated MI carries prognostic relevance with regard to biologic activity as well as overall survival.[2-7] Moreover, it has been demonstrated that the prognostic significance of MI may be independent of tumor grade.[2] As such, while MI is taken into account when determining tumor grade, it is often reported as an independent variable in a mast cell tumor profile/panel.

How is it evaluated? Mitotic index is determined by the pathologist via light microscopy on a hematoxylin and eosin (H&E) stained section. Mitotic figures, which are visible via routine light microscopy, are counted in ten high-power (400x) fields from regions rich with neoplastic cells (e.g. paucicellular areas such as regions with hemorrhage, necrosis, or abundant stroma are avoided) and with high mitotic activity as determined from a low power scan. The mitotic index is reported as X# MFs/10hpf.

Ki67

What is it? Ki67 is a nuclear protein encoded by the *MKI67* gene. It is necessary for cellular proliferation and is involved in ribosomal RNA transcription. Ki67 is expressed during all active phases of the cell cycle (G_1, S, G_2 & Mitosis); therefore, Ki67 is an excellent marker for cellular proliferation (tumor growth fraction). The Ki67 labeling index has been correlated with biologic behavior/clinical course in many human cancers and in some veterinary cancers, including canine cutaneous mast cell tumors.[6, 8-12] **Why does it matter?** In canine cutaneous mast cell tumors, the labeling index or score has been shown to be of prognostic value both independently and in combination with other markers.[6, 8-12] It has been demonstrated to correlate with increased incidence of local recurrence, increased rate to local recurrence, increased rate to distant metastases, and decreased overall survival.[6, 8-12]

How is it evaluated? Ki67 expression is determined via immunohistochemistry (IHC) using an anti-Ki67 antibody. Ki67 is a nuclear protein; therefore, when expressed, results in a nuclear stain. The relative number of Ki67-positive cells is used to determine the proliferation index or the number of cells actively involved in the cell cycle. Webster et. al. more specifically determined a significant Ki67 index > 23 cells/grid area. This index was determined by counting the number of immunopositive cells present in a 10x10 mm grid area using a 1cm^2 10x10 grid reticle at 400x magnification within highly proliferative areas. The number of immunopositive cells per grid area was then evaluated over 5 high-power fields and subsequently averaged to obtain the growth fraction or Ki67 index.[8]

ARGYROPHILIC NUCLEOLAR ORGANIZING REGIONS (AgNORs)

What is it? AgNORs are argyrophilic (silver-staining) nucleolar organizing regions. Interphase AgNORs are structural-functional units of the nucleolus in which all the components necessary for ribosomal synthesis are located. Two agyrophilic proteins, nucleolin and nucleophosmin, are involved with rRNA transcription and processing of the interphase

AgNORs. These proteins permit the stainability with silver methods and subsequent AgNOR visualization via light microscopy. The number of AgNORs is strictly related to rRNA transcriptional activity and, in continuously proliferating cells, to the rapidity of cellular proliferation. The faster the rate of cellular proliferation, the greater the AgNOR quantity. As such, AgNOR analysis is unique to other proliferation indices in that it does not just identify cells in proliferation but, more specifically, provides information on the rate of proliferation.

Why does it matter? AgNOR quantification has proven to be a significant independent prognostic factor in a number of human cancers and, in select human tumors, a predictive factor as well.[13] Similarly, AgNOR analysis has been determined to hold prognostic relevance in a number veterinary cancers, including canine cutaneous mast cell tumors.[6-8,10,12-18] More specifically, AgNOR analysis has been demonstrated to be prognostic for overall survival, local recurrence, disease free interval, and, in one study in combination with Ki67, for local recurrence and metastatic disease.[6]

How is it evaluated? AgNORs are identified as discrete black nucleolar foci when tissue is stained with a histochemical sliver-staining method. The total number of AgNORs is counted in a said number of cells at 1000x magnification after which the average number of AgNORs/cell is calculated.

PROLIFERATING CELL NUCLEAR ANTIGEN (PCNA)

What is it? PCNA or proliferating cell nuclear antigen is a subunit of DNA polymerase delta. It acts as a homotrimer and helps increase the processivity of leading strand synthesis during DNA replication. It is involved in a number of additional nuclear functions including DNA repair, specifically post-replication repair (PRR). It is expressed during late G1 and S-phase with maximal expression in S-phase.

Why does it matter? PCNA, like Ki67, is a nuclear protein expressed during cellular proliferation and thus can be exploited as a biomarker to reflect the growth fraction of a tumor. In contrast however, PCNA is only expressed in S-phase (and to a lesser extent in G_1) while Ki67 is expressed during all active phases (G_1, S, G_2 & Mitosis) of cellular proliferation. Regardless, PCNA has been explored in a number of canine cutaneous mast cell tumor studies to determine its potential prognostic relevance.[7-10,16,19] Most studies reveal PCNA is an inferior prognostic marker when compared to Ki67 and AgNORs. However, rare studies conclude prognostic significance and one school of thought suggests a combination of all proliferation indices is superior to individual indices; therefore, PCNA expression may still be performed in the MCT panel (and is thus addressed here).[7,8,19] **How is it evaluated?** PCNA expression is determined via IHC using an anti-PCNA antibody. Like Ki67, PCNA is a nuclear protein; therefore, when expressed, results in a nuclear stain. Similarly, the relative number of PCNA-positive cells is used to determine the proliferation index or the number of cells actively involved in cellular proliferation.

KIT (CD117)

KIT/CD117 is a transmembrane receptor tyrosine kinase (RTK), encoded for by the proto-oncogene *c-kit*. It is expressed on mast cells amongst other cells including hematopoietic progenitor cells, melanocytes, and interstitial cells of Cajal. KIT is activated by the ligand stem cell factor (SCF) also known as mast cell growth factor. This activation results in receptor dimerization leading to phosphorylation of KIT's cytoplasmic domain and subsequent initiation of downstream signaling cascades ultimately impacting cellular survival, differentiation, proliferation, and various cellular functions. If KIT activation becomes dysregulated such that there is ligand-independent hyperactive kinase activity then this ultimately results in an unregulated growth stimulus and potential tumor formation.

KIT EXPRESSION PATTERN / LOCALIZATION

What is it? The expression pattern or localization of KIT refers to its cellular location. As mentioned, KIT is a transmembrane protein and thus, in normal mast cells, is typically associated with and expressed on the cell surface/cell membrane with minimal presence in the cytoplasm. In neoplastic mast cells, especially those that are less well-differentiated, it has been recognized that KIT localization is often in the cytoplasm.[20,21] This cytoplasmic location is often referred to as aberrant localization. **Why does it matter?** The recognition of this variation in cellular localization of KIT between normal and variably differentiated neoplastic mast cells led to further studies in an attempt to correlate the KIT expression pattern with clinical outcome.[21] It was determined that these various expression patterns / localization of KIT correlated with biologic behavior such that increased cytoplasmic KIT was significantly associated with decreased overall survival and an increased rate of local recurrence.[21]

How is it evaluated? The KIT expression pattern is determined via IHC using an anti-KIT antibody. Three expression patterns have been characterized: (1) Type I expression characterized by membranous staining, (2) Type II expression characterized by stippled cytoplasmic or focal perinuclear staining, and (3) Type III expression characterized by diffuse cytoplasmic staining.[21] Type II and III expression patterns have been proven to carry prognostic significance as indicated above.

C-kit MUTATION STATUS

What is it? *C-kit* is the proto-oncogene which encodes for the receptor tyrosine kinase KIT. The mutation status refers to the presence or absence of one or more mutations in the *c-kit* gene. Mutations in *c-kit* can lead to ligand-independent constitutively activated KIT which can subsequently play a role in tumorigenesis. Investigations in canine cutaneous mast cell tumors led to the identification of mutations in the *c-kit* gene; most notably of exon 11 within the juxtamembrane domain in the form of internal tandem duplications (ITDs) and less often, deletions.[22-27] Mutations in exon 8 and 9 have also been identified but appear to occur at a lesser frequency.[26]

Why does it matter? Studies performed to correlate the presence of *c-kit* mutations with clinical outcome revealed that canine MCTs with *c-kit* mutations were significantly associated with an increased incidence of recurrent disease and decreased overall survival.[28] The same study also determined a significant association between *c-kit* mutations and aberrant protein localization.[28] In addition to prognostic relevance, identification of a *c-kit* mutation may also serve as a predictive factor with regard to response to therapy with tyrosine kinase inhibitors (TKIs).[29,30] Many veterinary studies are ongoing; however, in human medicine, tumor-associated *c-kit* mutations, especially as seen in gastrointestinal stromal tumors (GISTs) indicate the use of TKIs as the presence of a mutation is predictive of a positive therapeutic response.[31]

How is it evaluated? The *c-kit* mutation status is determined via polymerase chain reaction (PCR) and DNA amplification utilizing probes specific for the respective mutation regions. The assay can be performed on unstained formalin fixed paraffin embedded tissue sections or on fresh tissue aspirates or impression smears which may mitigate the potential of false negatives as the tissue has not been altered by formalin fixation.

CONCLUSIONS

In conclusion, parameters evaluated in the mast cell tumor profile/panel are done so with the intent of providing the clinician and owner with additional prognostic and possibly predictive information to help guide decisions in determining the optimal therapeutic plan for a given patient. Based on research and taking into consideration all parameters (e.g. tumor grade, stage, location, excisional status, etc.) this battery of tests may be most beneficial when attempting to determine the optimal therapeutic plan for an intermediate grade mast cell tumor as these seem to present the greatest variability in biological behavior based on routine microscopic evaluation alone. It should also be recognized that while multiple parameters are often run simultaneously, selection of individual tests could also be pursued.

REFERENCES

[1]Kamstock, DA, et al. **Vet Pathol** 2011;48(1):19. [2]Romansik, EM, et al. **Vet Pathol** 2007;44(3): 335. [3]Elston, LB, et al. **Vet Pathol** 2009;46(2):362. [4]Patnaik AK, et al. **Vet Pathol** 1984;21:469. [5]Kiupel, M, et al. **Vet Pathol** 2011;48(1):147. [6]Thompson, JJ, et al. **Vet Pathol** 2011;48(1):169. [7]Simoes JP, et al. **Vet Pathol** 1994;31:637. [8]Webster, JD, et al. **Vet Pathol** 2007;44(3):298. [9]Abadie JJ, et al. **J Am Vet Med Assoc** 1999;215:1629. [10]Scase, TJ, et al. **J Vet Intern Med** 2006;20(1):151. [11]Maglennon, GA, et al. **Vet Comp Oncol** 2008;6(4):268. [12]Webster, J. D., et al. **BMC Vet Res** 2008;4:32. [13]Pich, A, et al. **Micron** 2000;**31**(2):133. [14]Kravis LD, et al. **J Am Vet Med Assoc** 1996; 209:1418. [15]Bostock, DE, et al. **Br J Cancer** 1989;59(6):915. [16]Ettinger, SN, et al. **J Am Vet Med Assoc** 2006;228(7):1053. [17]Kiupel, M, et al. **J Comp Pathol** 1998;119(4):407. [18]Preziosi, R, et al. **Res Vet Sci** 2002; 73(1):53 [19]Seguin, B., et al. **J Vet Intern Med** 2006;20(4):933. [20]Reguera, MJ, et al. **Am J Dermatopathol** 2000;22(1):49. [21]Webster, JD, et al. **Vet Pathol** 2004;41(4):371. [22]London, CA, et al. **Exp Hematol** 1999;27(4):689. [23]Ma, YB, et al. **J Invest Dermatol** 1999;112(2):165. [24]Zemke, D, et al. **Vet Pathol** 2002;**39**(5):529. [25]Downing, S., et al. **Am J Vet Res** 2002;63(12):1718. [26]Letard, S., et al. **Mol Cancer Res** 2008;6(7):1137. [27]Webster, JD, et al. **BMC Cancer** 2006;6: 85. [28]Webster, JD, et al. **Neoplasia** 006; 8(2):104. [29]Isotani, M, et al. **J Vet Intern Med** 2008;22(4):985. [30]London, CA, et al. **Clin Cancer Res** 2009;**15**(11):3856. [31]Vetto, JT. **Onco Targets Ther** 2009;2:151.

TREATMENT RECOMMENDATIONS FOR CANINE MAST CELL TUMORS

Dennis B. Bailey, DVM, DACVIM (Oncology)
Paramus, NJ

INTRODUCTION

Canine mast cell tumors (MCT) are one of the most common tumors in veterinary oncology. They are also one of the most frustrating to treat. The biologic behavior of these tumors varies widely. Many have a benign behavior and are cured with surgery alone. However, some mast cell tumors are malignant. They can be very locally invasive, they can metastasize, and they can be fatal. Therefore, when treating canine MCTs it is important to be appropriately aggressive, ensuring that patients are neither subjected to unnecessary treatments nor deprived of needed therapy. There are several prognostic factors used to guide therapy decisions, including a newly proposed grading scheme and prognostic panel. This talk will discuss how these and other prognostic factors should be used to help guide treatment decisions.

DIAGNOSTIC EVALUATION FOR DOGS WITH MAST CELL TUMORS

Staging Diagnostics

Initial staging diagnostics should include CBC, chemistry panel, and urinalysis. When MCTs metastasize, regional lymph nodes are affected most commonly, followed by liver and spleen, and occasionally the bone marrow. The regional lymph node should be assessed either with cytology or preferably histopathology (by surgically removing the lymph node at the same time as the primary tumor). Abdominal ultrasound is advised prior to surgery. Aspirates of the liver, spleen, or other intra-abdominal organs should be performed only if they appear abnormal on imaging; routine aspiration of organs that image normally is not recommended. The author recommends thoracic radiographs, particularly in geriatric patients, as part of a pre-anesthetic screen. However, pulmonary metastasis is rare. Bone marrow cytology usually is reserved for patients with high grade tumors (especially if aggressive local therapy, such as radiation therapy, is going to be pursued) or patients with disseminated disease. Buffy coat profiles are neither sensitive nor specific, and are not recommended.

Mast Cell Tumor Histopathology and Tumor Grade

The grading scheme first proposed by Patnaik divided MCTs into three groups based on cellular morphology, mitotic index, and depth of invasion.[1] The grading scheme strongly correlated with survival after surgery alone: 93% of dogs with grade I tumors were alive 1500 days after surgery, compared to 44% of dogs with grade II tumors and 6% of dogs with grade III tumors. However, this grading system has lost much of its clinical utility. In the original study published by Patnaik of 83 MCTs, 30 (36%) were classified as grade I, 36 (43%) as grade II, and 20 (17%) as grade III. Currently the vast majority of canine MCTs are reported as grade II.[2] Additionally, grading criteria have been modified and can vary significantly among pathologists. In a study where 10 pathologists evaluated and graded the same 60 MCTs, there were only 4 tumors where all pathologists agreed on grade and 6 tumors that received grades of I, II, and III by different pathologists.[3]. Recently, a new 2-tier grading system was proposed.[2] Using these new guidelines, MCT are considered high-grade if they meet any of the following criteria: ≥7 mitoses in 10 high-powered fields, ≥3 multinucleated cells (3 or more nuclei) in 10 hpf, ≥3 bizarre nuclei in 10 hpf, karyomegaly. With this new grading system, dogs with low-grade MCT treated with surgery alone had a median survival of >2 years, compared to 4 months for high-grade MCT. It is essential that clinical veterinarians familiarize themselves with the components of these MCT grading systems to ensure that the microscopic description is in agreement with the grade assigned to the tumor.

Mast Cell Tumor Prognostic Panels

MCT prognostic panels are becoming more widely available. The parameters assessed in these panels include Ki-67 and proliferating cell nuclear antigen (PCNA) immunostaining, argyrophilic nucleolar organizing region (AgNOR) histochemical staining, Kit (CD117) immunostaining, and Kit PCR to screen for common mutations in the juxtamembrane domain. High Ki-67 and AGNOR scores have been associated with local tumor recurrence, MCT metastasis, and MCT-associated death.[4-6] The prognostic significance of PCNA is less clear. One study showed that dogs that died from their MCT had higher PCNA counts than dogs that did not, but there was great overlap between groups.[6] Other studies have not demonstrated any prognostic significant for PCNA counts.[4,5] Immunostaining for the receptor tyrosine kinase (RTK) Kit can identify abnormal localization of the protein within the cytoplasm instead of the cell membrane. Increased localization of Kit in the cytoplasm is correlated with higher Patnaik tumor grade and higher Ki-67 score,[7] although the impact on clinical outcome has not been reported. PCR can be used to screen for internal tandem duplications (ITD) in the negative regulatory juxtamembrane domain (most commonly in exons 11 or 12). These mutations result in constitutive activation of Kit in the absence of ligand binding, and are associated with increased risk of local recurrence, increased risk of metastasis, and higher tumor proliferation indices.[4,8]

There are a few important caveats with prognostic panels. First, not every lab offers all of the tests listed above. It is important to know exactly what is included before ordering a prognostic panel. Second, proliferation indices (Ki-67, PCNA, and AgNOR) can be scored using a variety of systems that can yield substantially different cut-off values that define a favorable or unfavorable prognosis. It is important to know the cut-off values for the specific lab you are using, and you should ensure that those values have been validated. Third, it can easily take 1-3 weeks to get the full prognostic panel results. For many MCT, this additional time will not impact prognosis, but for advanced stage or rapidly growing tumors the results might not be timely enough.

Recommendations regarding when to order a MCT prognostic panel are still being refined. While we are still learning, it is not wrong to recommend ordering a full panel for all MCT. However, as we gain more experience, particularly with cases where there is discordance within the panel or between the panel and histologic grade, this recommendation might change. Others currently are recommending the MCT panel primarily for all low-grade MCT in an effort to identify those occasional tumors that have favorable features on routine histopathology but a more aggressive phenotype. At a minimum, the panel should be considered for patients where there is disagreement among other prognostic factors (e.g. a very small, well-circumscribed dermal tumor that on histopathology has no abnormal morphologic features but a high mitotic index).

TREATMENT OPTIONS

Surgery

For dogs with no evidence of metastasis, wide surgical excision is the treatment of choice whenever possible. Historically, recommendations have been made to remove MCTs with lateral margins of 3 cm and a deep margin of 1 fascial plane. However, >90% of low grade MCT (grades I and II using the earlier Patnaik system) are completely excised with a lateral margin of 2 cm and a deep margin of 1 fascial plane.[9,10] For larger or more invasive MCTs, though, wider surgical margins should still be attempted. Similarly, for tumors with substantial peritumoral edema secondary to mast cell degranulation, it can be difficult to visually define the extent of the tumor at surgery. A wider margin is also recommended for these tumors to help ensure complete excision.

If a MCT is classified as low grade, has a favorable prognostic panel, and is completely excised with clean margins, no further treatment is indicated. Continued monitoring is always recommended, though. The author recommends a physical examination 1, 2, 4, 6, 9, and 12 months after surgery, and then every 6 months thereafter. If a low-grade MCT is incompletely excised, historically adjuvant radiation therapy has been recommended to improve local control (see below). However, radiation probably is not necessary for all dogs in this category. Reported local recurrence rates are only 18-30%.[11,12] Since some of the parameters in the MCT prognostic panel have been linked to local recurrence and/or metastasis, perhaps these results can be used to decide whether adjuvant radiation is indicated or whether careful monitoring might be appropriate.

For MCT that are high grade and/or that have unfavorable prognostic panel results, surgery is still indicated if the disease is localized. However, due to the increased risk for metastasis, adjuvant chemotherapy is recommended (see below).

Radiation Therapy

Radiation therapy is most effective when combined with surgery. Definitive protocols typically consist of a total dose of 48-57 Gy delivered over 15-19 fractions on a Monday-through-Friday basis. For dogs with low grade (Patnaik grade I or II) MCT that are incompletely excised and then treated with adjuvant radiation therapy, long-term local control is achieved in >90%.[13,14] Again, though, since surgery alone provides long-term local control in >70% of dogs, not every dog with an incompletely excised low grade MCT needs to undergo radiation therapy. Hopefully the newer grading system and prognostic panel will help us to better recognize which patients will benefit from this additional therapy.

For dogs with incompletely-excised high grade (Patnaik grade III) MCT, adjuvant radiation therapy is still a reasonable option as long as there is no gross evidence of distant metastasis. However, even after adequate local control is attained, about half will ultimately go on to develop metastasis.[15,16] Perhaps the MCT prognostic panel will help identify which dogs are more likely to develop metastasis. For those with a more favorable panel, a combination of radiation therapy and chemotherapy might be best. In contrast, for those with an unfavorable panel and high risk for metastasis, perhaps the improved local control afforded by radiation therapy would not impact long-term prognosis, and chemotherapy alone would be a more reasonable option.

Radiation therapy also can be used as a primary modality for tumors that are too large or invasive for surgery to be possible. In one study, local disease was controlled for a median for 12 months.[17] In the author's experience, though, the duration of local disease control varies greatly from one patient to the next, and for large, infiltrative tumors it often is substantially shorter than the reported median. Additionally, there is increased risk for degranulation reactions when gross disease is irradiated.

Chemotherapy

Systemic chemotherapy is indicated as a primary treatment modality for dogs with gross metastatic disease. It also is recommended as an adjuvant therapy, after adequate local/regional control has been attained, for dogs considered to be at high risk for metastasis and MCT-related mortality. However, the definition of "high risk" is still being refined. All oncologists would agree that dogs with high grade MCT should be treated with adjuvant chemotherapy. Using the new 2-tier system, when dogs with high grade MCT were treated with surgery alone, median survival was 4 months and 90% of dogs died due to their MCT.[2] Most would also agree that dogs with unfavorable prognostic panel results also should be considered for adjuvant chemotherapy. However, questions still to be answered include what to do when the prognostic variables are discordant (i.e. some indicate a favorable prognosis, and others an unfavorable one), and what to do when the tumor grade and prognostic panel disagree (i.e. a low-grade tumor with unfavorable panel, or a high-grade tumor with a favorable panel). MCT associated with the oral cavity, lips, and muzzle are associated with higher rates of regional lymph node metastasis and MCT-associated mortality.[18,19] Adjuvant chemotherapy should be considered for dogs with tumors in these locations. In contrast, inguinal and perineal MCT are not inherently more aggressive.[20,21] Adjuvant chemotherapy should not be recommended automatically for tumors in these locations unless warranted by other prognostic factors. The prognostic significance of regional lymph node involvement is uncertain. When 19 dogs with MCT of various grades (predominantly Patnaik grade II) were treated with surgery and radiation to attain adequate regional control, and no adjuvant chemotherapy, disease progression was identified in 6 (31%) and median progression-free survival was 1240 days.[22] Adjuvant chemotherapy should be considered for dogs with regional lymph node involvement, but if relatively few mast cells are seen on histopathology and all other prognostic variables are favorable, monitoring might be reasonable as well.

Once a decision is made to pursue chemotherapy, the next decision is what drug or combination of drugs to use. Several conventional chemotherapy drugs have been evaluated as single agents and in various combinations (**Table 1**). Additionally, over the past few years the tyrosine kinase inhibitors (TKI) toceranib phosphate (Palladia) and masitinib (Kinavet) have become available. These two classes of drugs work very differently. Conventional chemotherapy drugs typically interfere with some aspect of cell metabolism that is critical for progression through the cell cycle. In contrast, TKIs inhibit receptor tyrosine kinases, cell surface receptors that are important for initiating signal transduction cascades that ultimately trigger cell division. As mentioned above, up to 30% of canine MCT have mutations in the RTK Kit that result in constitutive activation. Moreover, these mutations are associated with increased risk of local tumor recurrence, increased risk of metastasis, and higher tumor proliferation indices. Kit likely represents the primary target of Palladia and Kinavet, however tumor responses are still seen in tumors that do not have the most common genetic mutations.[23,24] There are several possible explanations for this: these MCT might have other Kit mutations for which there are no commercially available screening tests, wild-type Kit might still play an important role in MCT progression, or other RTKs targeted by Palladia and Kinavet might also play a role in MCT progression.

For dogs with tumors where Kit mutation status is known, the author currently recommends using TKI drugs as first-line therapy for Kit mutation positive tumors because they have the highest overall response rate. However, for Kit mutation negative tumors, starting with conventional chemotherapy drugs is recommended since they have higher reported overall response rates. (It is worth noting, though, that the impact of Kit mutation status on response rates with conventional drugs has not been evaluated.) For dogs with measurable disease, response to therapy can easily be assessed. However, when treating dogs in the adjuvant setting, response to treatment cannot be easily assessed unless progressive disease is identified. Some oncologists are therefore recommending that dogs in this latter group be treated sequentially with conventional drugs first followed by a TKI. Protocols using conventional drugs and TKIs concurrently are not recommended until safe dosages are established. One study demonstrated unacceptable hematologic toxicity when vinblastine and Palladia were administered concurrently.[25]

Table 1: Efficacy of chemotherapy drugs used to treat canine mast cell tumors (measurable disease)

Protocol	Number Treated	Complete Response (%)	Partial Response (%)	Overall Response Rate (%)	Median Remission Duration	Comments
Prednisone	25	4	16	20	Not given	Response assessed 28 days after starting therapy (Ref 26)
Vinblastine (2.0 mg/m^2 q1 wk)	25	0	12	12	77 d	(Ref 27)
Vinblastine (3.5 mg/m^2 q2 wk)	26	4	23	27	CR = 63 d PR = 28 d	(Ref 27)
CCNU (lomustine)	19	5	37	42	CR = 440 d PR = 77 d	(Ref 28)
Hydroxyurea	46	4	24	28	CR = 352 d PR = 46 d	(Ref 29)
Vinblastine & Prednisone	15	33	14	47	154 d	(Ref 30)
CCNU, Vinblastine & Prednisone	17	30	35	65	CR = 141 d PR = 66 d	(Ref 31)
Cyclophosphamide, Vinblastine, & Prednisone	11	46	18	64	Not Given	Progression-Free survival was 74 days. (Ref 32)
Toceranib Phosphate (Palladia)	145	15 (assessed at 42 d)	28 (assessed at 42 d)	43 (assessed at 42 d)	84 d	Response rate 60% for Kit ITD+ tumors, 31% for Kit ITD- tumors (Ref 23)
Masitinib (Kinavet)	202	11 (assessed at 168 d)	5 (assessed at 168 d)	16 (assessed at 168 d)	Time to Progression 118 days	(Ref 24)

REFERENCES

1. Patnaik AK, et al. **Vet Pathol** 1984;21:469. 2. Kiupel M, et al. **Vet Pathol** 2011;48:147. 3. Northrup NC, et al. **J Vet Diagn Invest** 2005;17:245. 4. Webster JD **Vet Pathol** 2007;44:298. 5. Scase TJ, et al. **J Vet Intern Med** 2006;20:151. 6. Abadie JJ, et al. **J Am Vet Med Assoc** 1999;215:1629. 7. Gil da Costa RM, et al. **BMC Vet Res** 2007;3:19. 8. London CA, et al. **Clin Cancer Res** 2009;15:3856. 9. Simpson AM, et al. **J Am Vet Med Assoc** 2004;224:236. 10. Fulcher RP, et al. **J Am Vet Med Assoc** 2006;228:210. 11. Michels GM, et al. **J Am Anim Hosp Assoc** 2002;38:458. 12. Misdorp W. **Vet Q** 1987;9:279. 13. Poirier VJ, et al. **J Am Anim Hosp Assoc** 2006;42:430. 14. Frimberger AE, et al. **J Am Anim Hosp Assoc** 1997;33:320. 15. Hume CT, et al. **J Am Anim Hosp Assoc** 2011;47:37. 16. Hahn KA, et al. **J Am Vet Med Assoc** 2004;224:79. 17. LaDue T, et al. **Vet Radiol Ultrasound** 1998;39:57. 18. Gieger TL, et al. **J Vet Intern Med** 2003;17:687. 19. Hillman LA, et al. **J Am Vet med Assoc** 2010;237:936. 20. Cahalane AK, et al. **J Am Vet Med Assoc** 2004;225:401. 21. Sfiligoi G, et al. **J Am Vet Med Assoc** 2005;226;1368. 22. Chaffin K, et al. **Vet Radiol Ultrasound** 2002;43:392. 23. London CA, et al. **Clin Cancer Res** 2009;15:3856. 24. Hahn KA, et al. **J Vet Intern Med** 2008;22:1301. 25. Robat C, et al. **Vet Comp Oncol** 2011; Epub ahead of print. 26. McCaw DL, et al. **J Vet Intern Med** 1994;8:406. 27. Rassnick KM, et al. **J Vet Intern Med** 2008;22:1390. 28. Rassnick KM, et al. **J Vet Intern Med** 1999;13:601. 29. Rassnick KM, et al. **Vet Comp Oncol** 2010;8:103. 30. Thamm DH, et al. **J Vet Intern Med** 1999;13:491. 31. Rassnick KM, et al. **Vet Comp Oncol** 2010;8:138. 32. Camps-Palau MA, et al. **Vet Comp Oncol** 2007;5:156.

MAST CELL TUMOR CASES AND CONTROVERSIES

Laura D. Garrett, DVM, DACVIM (Oncology)
Urbana, IL

INTRODUCTION

Mast cell tumors (MCT) are the most common malignant skin cancer in dogs, and significant variability exists in their biological behavior. Most of the tumors are cured with appropriate local therapy, but a subset shows very malignant behavior and the potential to spread to lymph nodes, liver, spleen, and other areas and to thus become a systemic cancer. Because of this variable behavior, it is difficult to predict how an individual tumor is going to behave, and thus difficult to know if staging tests are needed and what treatments to offer in many cases. In addition to controversies over systemic disease, controversies also exist over what treatment is needed to best attain local control of these tumors.

PROGNOSTIC FACTORS FOR SYSTEMICALLY AGGRESSIVE BEHAVIOR

HISTORY AND PHYSICAL EXAMINATION

Some factors that can be obtained from a history and physical examination that are considered to carry a more guarded prognosis in dogs with MCTs include recent, rapid tumor growth; certain locations (mucocutaneous, muzzle, pinnae, inguinal, nail bed); and fixed, ulcerated tumors.

Mucocutaneous Location

In limited published cases eyelid margin MCT appeared to have relatively benign behavior and were effectively treated with local therapy, although one dog was reported to have regional lymph node metastasis.[1-3] MCT of the conjunctiva may be of concern only locally, without reported metastasis in three dogs.[4, 5] In contrast, two case reports of MCT of the lip showed aggressive behavior and local metastatic disease at diagnosis, with survival time of 6 months or less.[6, 7] Five dogs were reported with MCT of the tongue, two presented with lymph node and/or systemic metastasis, with two of the remaining three having post-operative local recurrence leading to euthanasia.[8] In a paper evaluating chemotherapy for high risk MCT patients, 11 dogs with mucous membrane MCT (vulva, prepuce, conjunctiva, oral cavity) had significantly shorter median survival times (MST) than 50 dogs with MCT of haired skin.[9] A recent paper of 32 dogs with 33 conjunctival MCT treated with surgery showed prolonged survival times, with only 2 dogs have local recurrence despite incomplete margins in 25 cases, and no dogs dying of mast cell related disease.[10]

Muzzle/Perioral/Oral Location

Two studies have confirmed a more aggressive behavior for muzzle MCT, with one showing similar behavior for perioral and oral cases as well.[11, 12] The rate of documented metastasis to local (mandibular) lymph nodes was similar in both papers, at 58% and 59%, compared to <10% rate for other cutaneous sites. The median survival times in the studies were 30 and 52 months, which shows that despite a more aggressive tumor behavior, dogs with these tumors may have prolonged survival times. In both studies, dogs with lymph node metastasis had significantly shorter survival times than dogs without nodal metastasis, with a median of 14 months in one paper, and an unreported median (less than 20 months) in the other.[11, 12]

CASE 1: Ginger, a 6 year old FS mixed breed dog with a nasal planum MCT.

Inguinal Location

Historically, the "back half" of the dog was considered to carry a worse prognosis, and while this is generally not supported, there are still concerns about MCT located in the inguinal areas, especially involving the prepuce or scrotum. One study reported on dogs with MCT in the perineal and/or inguinal region treated aggressively, most receiving tri-modality therapy with surgery, radiation and chemotherapy. The median survival was 37 months, and the authors concluded that with appropriate therapy dogs with perineal or inguinal MCT can do well.[13] It is important to note that these dogs received a more aggressive treatment protocol than do the majority of dogs with MCT at other cutaneous sites. A later study compared the outcome of dogs with inguinal or perineal MCT to dogs with MCT in other cutaneous locations; these dogs received a range of therapies. While there was no statistically identified difference between the locations, preputial or scrotal MCT, included in the inguinal category, were also analyzed separately from other inguinal or perineal tumors. The disease-free interval (DFI) for these 12 dogs was significantly shorter (4.2 months) than for the 84 dogs with tumors in other cutaneous locations (33.9 months).[14] Dogs with inguinal or perineal tumors were also significantly more likely to have received chemotherapy, thus potentially biasing the results. The authors stated that the power of the study may not have been adequate to identify a difference between the locations. Overall, many oncologists still feel that scrotal and preputial MCT can have a more malignant biological behavior.

CASE 2: Digger, a 3 year old M Vizsla with preputial MCT.

Subcutaneous Location

Recently, several papers reported information regarding MCT in the subcutaneous (SQ) location.[15-17] As the grading system (see below) was developed only with cutaneous tumors, it cannot necessarily be applied to SQ tumors with prognostic accuracy. Survival times for the SQ tumors was found to be prolonged, with the majority of dogs in both papers (53 dogs and 306 dogs) not succumbing to mast cell disease.[15, 16] Out of the 306 dogs, only 8% had local recurrence, and only 4% had metastasis.[16] Risk factors identified for local recurrence and metastasis included mitotic index, infiltrative vs circumscribed histologic pattern, the presence of multinucleation, and age at diagnosis as well as Ki67, Ki67+AgNOR, and KIT cellular localization pattern.[16, 17]

HISTOLOGIC INFORMATION

Grade

A histologic grading scheme (I to III) was developed for classification of MCTs affecting haired skin and is still one of the most accepted prognostic indicators of canine MCT behavior. However, histologic assessment is prone to operator subjectivity, giving rise to extremely variable grading results for the same tumor among different pathologists.[18, 19] In one study with10 pathologists evaluating the same 60 MCT, there was agreement on grade by all pathologists for only 4 tumors, and 6 of the MCT had all 3 grades assigned. Differences in the references used to grade the MCT was the suspected reason for the disparity, and despite 6 different references being used in this study group, at least 80% of the pathologists agreed on the grade of 50% of the tumors.[19] In a follow-up study involving the same tumors and pathologists, the Patnaik grading scheme was used by all. While mean agreement improved to 62.1%, only 16/60 tumors received the same grade by all, and 3/60 still received all three grades.[18] The subjectivity and variability between pathologists bring into question the heavy reliance placed on the grade of a MCT to predict its behavior. In addition, most MCTs are designated grade II and most of these are cured with surgery. However, a subset of grade II tumors show aggressive behavior, and the question of how best to identify this subset of high risk cases is a matter of great debate.

Recently, a two-tier (high vs low grade) grading scheme was proposed in a study of 95 dogs with MCT treated with surgical excision alone.[20] High grade tumors had one of the following criteria: MI (see below) of ≥ 7, ≥ 3 multinucleated cells or cells with bizarre nuclei in 10 hpf, or karyomegaly. Areas of highest mitoses or anisokaryosis were evaluated. Median survival time for high grade (10 dogs) was 3.6 months, vs median not reached (>2 years) for low grade (85 dogs). Further studies validating this proposed system are warranted; currently the author's university pathology department provides the standard grade, a MI, and the two-tier grade for every sample.

Case 3: Sadie, approx 11 yr old (rescue) FS mixed breed, 2 cm SQ (deep dermis) mass – Grade II but "high grade"

Other Assessments on Biopsies

As grade is so variable and subjective, many other prognostic factors, including DNA aneuploidy, c-kit–staining pattern, presence of c-kit mutations, microvessel density, Ki67, PCNA, and mitotic index (MI), have been evaluated in an attempt to better predict the behavior of canine MCTs and to pick out the "bad" grade II tumors. Two groups have evaluated the MI (total number of mitotic figures counted in 10 high-power fields – areas with the highest mitoses are the ones to be counted) and found that it is predictive of survival time, even within the grade II tumor category. The groups did identify different values for the index, with the first paper showing a median survival of > 70 months for an index score ≤ 5 and a survival < 2 months for a score > 5.[21] The second group wrote a letter to the editor in response to the first paper and confirmed that high scores have very short survival times, with index scores broken down into three groups: MI=0, median survival time not reached ; MI=1 to 7, median survival time = 18 months; and MI > 7, median survival = 3 months.[22]

STAGING INFORMATION

In general, staging tests for asymptomatic dogs with cutaneous MCT are extremely low yield. The test most often positive is regional lymph node aspiration. Cytologic assessment of the locoregional LN is important, even if the node is not enlarged. In the study of dogs with muzzle MCT, 4/11 lymph nodes with metastases were of normal size.[11] Lymph node involvement has been associated with a worse prognosis in a number of studies.[12, 23-25] Fine needle aspirate (FNA) is ideally performed prior to excision of the primary MCT, as surgical treatment of the tumor can produce confusing LN results due to local inflammation. Due to the fact that mast cells are a normal feature in LNs, some MCT cases will not have a definitive answer on LN cytology. Criteria for LN involvement has been proposed[23] and morphometrical analysis seemed useful in a preliminary study.[24]

Case 4: Jo, a 10 year old FS lab with MCT over metatarsus and + popliteal LN

Abdominal ultrasound (US) also shows low positive results. I strongly recommend US of the abdomen to evaluate the sublumbar LNs if the MCT is on the back half of the patient. The benefit of fine needle aspiration and cytology of ultrasonographically normal-appearing liver and spleen for staging of dogs with mast cell tumors (MCT) is a topic of ongoing debate. Mast cells can be found in moderate numbers in normal canine liver and spleen, and while increased numbers were found in splenic aspirates in 51 MCT-bearing dogs with normal spleens on US compared to 32 unaffected dogs, the splenic cytology did not correlate with systemic behavior.[27] In another study, 52 dogs with MCT underwent ultrasound evaluation and FNA with cytology of their liver and spleen.[28] Cytologic criteria of MCT infiltration included clustering of well-differentiated mast cells, large numbers of well-differentiated mast cells, or mast cells with atypical morphology (pleomorphic and poorly granulated). The dogs were separated into 2 groups; those without MCT infiltration into either organ (n=42) and those with infiltration into one or both organs (n=10); survival time between these groups was significantly different at 733 vs 34 days, respectively (p<.0001). This dramatic difference in survival time supports that the cytologic findings were consistent with systemic MCT. Survival time based on US appearance alone in the dogs was not significant, suggesting that cytologic evaluation of the spleen and liver may be indicated for complete staging of dogs with MCT regardless of the ultrasonographic findings.[28] More recently, 19 dogs with aggressive grade II or III MCT treated with vinblastine/CCNU chemotherapy were evaluated to determine the specificity and sensitivity of ultrasound findings in determination of

MCT infiltration of the liver and/or spleen, using cytologic assessment to determine infiltration.[29] Seven dogs had MCT infiltration, and the sensitivity of US for detection of infiltration was 43% for the spleen and 0% for the liver. Dogs with cytologic determination of infiltration of the spleen or liver had significantly shorter survival than dogs without infiltration (100 vs. 291days and 100 vs 276 days, respectively, p<.0001). This clinical survival thus corroborating the cytologic finding of systemic disease. Due to the poor sensitivity of ultrasound in detection of cytologically confirmed organ infiltration and the decreased survival associated with identified infiltration, FNA and cytology of the spleen and liver is recommended for staging of dogs with MCT thought to be at high risk for metastasis. [29]

Case 5: Ginger, an 8 year old FS mix with muzzle MCT; splenic aspirate

LOCAL CONTROL

Surgical Margins

While surgery is the mainstay of therapy for MCT, and is curative in a large percentage of cases, there is still much controversy over surgical margins. Removing the tumor with gross margins of 2-3 cm lateral margins and a deep margin one tissue plane beyond the tumor has been recommended. Twenty-three MCT in 21 dogs were evaluated for completeness of excision at 1, 2, and 3 cm from the tumor edge; 100% of tumors had complete excision at the 2 cm margin mark. Only grade 1 (3) and grade 2 (20) MCTs were in the study, so it is not clear if 2 cm margins will be adequate for grade 3 tumors.[30] In a subsequent study, 91% (21/23) of grade 1 and 2 tumors were completely excised with 2 cm margins and had no local recurrence for >538 days.[31] Thus, 2 cm margins provide good results for grade 1 and 2 MCT with decreased morbidity. More recently, a report on 100 dogs with 115completely resected MCT showed no recurrence or metastasis in grade 1 and 2 tumors when excised with lateral margins ≥ 10 mm and deep margins ≥ 4 mm. The margin around the gross tumor was assessed on the histologic sample, so some shrinking would have occurred, but it does appear that even 2 cm margins may be more than is needed for successful removal of these tumors.[32]

Histologic Margins

Another controversy rages over what to do about clean vs dirty margins. MCT will recur both with clean and dirty margins, and the % of such recurrence varies widely between papers. With recurrence defined as a new mass within 2 cm of previous incision developing up to 1 year post-operatively, two papers stated 19-37.5% recurrence with clean margins.[33, 34] Other papers site lower recurrence with clean margins, from 0-11%.[32, 35-37] With incomplete resection, one paper found a 63.6% recurrence rate.[34] This is in contrast to a number of papers which report a much lower recurrence rate for dirty margins, generally ranging from 12-30%.[16, 35, 38-41] In one of these papers, only 23% of 28 dogs with incompletely resected grade II tumors experienced local regrowth. The combination of Ki67 and PCNA scores were prognostic for local recurrence, and dogs with such recurrence had significantly decreased survival times.[39] Thus, in cases where further local therapy is being discussed, these proliferation indices may be beneficial. Also, SQ MCT recurrence was associated with a number of histologic markers (see above). With such low recurrence rates for "incomplete" resections, more aggressive surgery or radiation therapy to treat the "residual" disease becomes a controversial topic and one for further discussion with pet owners.

Case 6: Fred, a 6 year old MN Pitbull with incompletely resected MCT of the flank, high grade

SYSTEMIC THERAPY

Additional therapy (chemotherapy, TKI therapy) may be considered in dogs with poor prognostic indicators (e.g., grade III, high MI, poor location) after excision of the MCT. Dogs with nonexcisable tumors may also be considered for chemotherapy or TKI therapy; both of these therapies show better success against smaller tumors, but certainly can be beneficial in cases with bulky tumors. The role for TKIs in microscopic residual disease (poor prognostic cases) is still being investigated.

REFERENCES

[1] Northrup NC, et al., **JAHAA** 2004; 40(4): 309.[2] Roberts SM, et al., **JAVMA** 1986; 189(10): 1355.[3] Jacobi S, et al., **Vet Ophth** 2008; 11(6): 395.[4] Barsotti G, et al., **Vet Ophth** 2007; 10(1): 60.[5] Johnson BW, et al., **JAHAA** 1988; 24: 439[6] Herman LH, et al., **JAVMA** 1967; 151(10): 1322.[7] Patnaik AK, et al., **Vet Path** 1982; 19(6): 608.[8] Beck ER, et al., **JAHAA** 1986; 22(4): 525.[9] Thamm DH, et al., **J Vet Med Sci** 2006; 68(6): 581[10] Fife M, et al., **Vet Ophth**2011; 14(3): 153.[11] Gieger TL, et al., **JVIM** 2003; 17(5): 687.[12] Hillman LA, et al., **JAVMA** 2010; 237(8): 936.[13] Cahalane AK, et al., **JAVMA** 2004; 225(3): 401.[14] Sfiligoi G, et al., **JAVMA** 2005; 226(8): 1368. [15] Newman SJ, et al., **J Comp Path** 2007; 136(4): 231.[16] Thompson JJ, et al., **Vet Path** 2011; 48(1): 156.[17] Thompson JJ, et al., **Vet Path** 2011; 48(1): 169.[18] Northrup NC, et al., **J Vet Diagn Invest** 2005; 17(6): 561.[19] Northrup NC, et al., **J Vet Diagn Invest** 2005; 17(3): 245.[20] Kiupel M, et al., **Vet Path** 2011; 48(1): 147.[21] Romansik EM, et al., **Vet Path** 2007; 44(3): 335.[22] Elston L, et al., **Vet Path** 2009.[23] Krick EL, et al., **Vet Comp Oncol** 2009; 7(2): 130.[24] Marconato L, et al., **Vet Comp Oncol** 2008; 6(3): 162.[25] Murphy S, et al., **Vet Rec** 2006; 158(9): 287.[26] Endicott MM, et al., **Vet Comp Oncol** 2007; 5(1): 31.[27] Finora K, et al., **Vet Comp Oncol** 2006; 4(3): 178.[28] Stefanello D, et al., **JVIM** 2009; 23(5): 1051.[29] Book AP, et al., **Vet Rad US** 2011; 52(5): 548.[30] Simpson AM, et al., **JAVMA** 2004; 224(2): 236.[31] Fulcher RP, et al., **JAVMA** 2006; 228(2): 210.[32] Schultheiss PC, et al., **JAVMA** 2011; 238(11): 1464. [33] Jaffe MH, et al., **JSAP** 2000; 41(1): 7.[34] Grier RL, et al., **JSAP** 1995; 36(9): 385.[35] Michels GM, et al., **JAHAA** 2002; 38(5): 458. [36] Seguin B, et al., **JAVMA** 2001; 218(7): 1120.[37] Weisse C, et al., **JAHAA** 2002; 38(1): 71.[38] Misdorp W, **Vet Q** 1987; 9(3): 279. [39] Seguin B, et al., **JVIM** 2006; 20(4): 933.[40] Brocks BA, et al., **Vet Surg** 2008; 37(5): 472.[41] Murphy S, et al., **Vet Rec** 2004; 154(24): 743.

SIRS, SEPSIS, MODS, MOF AND MORE

Claire R. Sharp, BSc, BVMS, MS, DACVECC
North Grafton, MA

"Animals that cannot sense endotoxin may die if they are infected by Gram-negative bacteria. Animals that sense endotoxin and respond too vigorously, as in the cases of sepsis and septic shock, may also die, victims of their own inflammatory reactions." Munford R.S. 2005.

The systemic inflammatory response syndrome (SIRS) and sepsis are surprisingly common and associated with high morbidity and mortality in dogs and cats. Similarly, severe sepsis is the leading cause of death in people in noncoronary ICUs. Despite significant advances in our understanding of the pathophysiology of SIRS and sepsis, identifying sepsis early is still a challenge. Current evidence suggests that early detection and evidence-based treatment strategies dramatically reduce morbidity and mortality in people with sepsis; the same is likely to be true in dogs and cats. With a greater understanding of the pathophysiology of SIRS and sepsis, and the sequelae of these conditions, notably MODS, you can positively impact your patients' outcome through early detection of sepsis and its complications, and prompt, appropriate intervention.

SIRS and Sepsis - The spectrum of disease

The systemic inflammatory response syndrome (SIRS) occurs due to an imbalance between pro- and anti-inflammatory mediators on a systemic scale, in response to an insult. SIRS can be the result of an infectious insult (sepsis) or a non-infectious insult such as trauma, neoplasia, and various sterile inflammatory diseases such as pancreatitis. Frustratingly noninfectious SIRS and sepsis can be clinically indistinguishable. Why? Their pathophysiology is almost identical, with the exception only of the initial insult. In sepsis, pathogen associated molecular patterns (PAMPs) expressed by the pathogen, stimulate pattern recognition receptors (PRRs) in the host. PAMPs include motifs such as lipopolysaccharide (LPS or endotoxin) on gram negative bacteria or peptidoglycan of gram positive bacteria. Virus, fungal organisms and parasites also have PAMPs that stimulate receptors of the host's innate immune system. Stimulation of PRRs on immune cells (monocytes, macrophages and neutrophils) drives an intracellular signaling cascade that ultimately results in the upregulation of nuclear transcription of a variety of inflammatory mediators (more on this below).

In non-infectious SIRS tissue damage results in the expression or release of damage associated molecular patterns (DAMPs) on/from host cells (eg. heat shock proteins (HSPs), HMGB-1 and other cellular components such as ATP and DNA). These DAMPs stimulate also stimulate PRRs stimulating the same intracellular signaling cascade. The resultant inflammatory cascade is indistinguishable from that induced by a pathogen / PAMPs.

Remember however, that inflammation is a normal response to infection (as well as non-infectious insults). Inflammation, characterized by the 5 cardinal signs of heat, pain, redness, swelling and loss of function, is protective when localized to the area of the initial insult. However SIRS results when inflammation is no longer contained, and rather affects the whole body. This is lack of containment is likely multifactorial in its origin; contributory factors include patient genetics, appropriateness of treatment, exuberance of the host response, imbalance between pro- and anti-inflammatory responses and the presence of a potential 2nd hit. When inflammation becomes systemic, then it has the potential to adversely affect the host.

SIRS is a clinical diagnosis, made on the basis of the TPR (temperature, pulse and respiratory rates), and white blood cell count (WBCC, from your complete blood count). While specific definitions are published there is no consensus in veterinary medicine, and what denotes SIRS varies between species. Essentially SIRS is considered present if the animal fulfills 2 or 3 out of 4 of the following SIRS criteria: 1) Abnormal temperature (either fever or hypothermia – the latter is more common in cats). 2) Abnormal heart rate (either tachycardia or bradycardia – the latter is more common in cats. 3) Tachypnea; and/or 4) A change in the leukon (leukocytosis, leucopenia or a significant left shift.... Depending on the definition >3% or >5% or >10% bands).

Identification of these clinical criteria in a patient should then prompt a search for the cause / underlying disease; either non-infectious or infectious (and aggressive treatment). In order to diagnose sepsis, a patient must fulfill SIRS criteria and have a confirmed (or highly suspected) infection. Infection can be diagnosed on cytology, histopathology or culture. In some cases the source of inflammation is immediately obvious, while in others it is not. The preliminary search for infection may include point-of-care infectious disease testing (FIV/FeLV in cats, 4Dx in dogs), thoracic radiographs (looking for evidence of pneumonia, especially if lower respiratory signs, or metastatic neoplasia), abdominal ultrasound (looking for free fluid – septic peritonitis, hepatobiliary sepsis etc.); and/or urine sediment examination.

Severe sepsis refers to the presence of sepsis with one or more organ dysfunctions. While there is not a consensus definition for what denotes an organ dysfunction in veterinary medicine, things to consider include:

- Renal dysfunction (rise in [Creatinine] > 0.5mg/dL, in the absence of pre-renal or post-renal causes)
- Cardiovascular dysfunction (septic shock vs. myocardial dysfunction)

- Respiratory dysfunction (requirement of supplemental oxygen or mechanical ventilation, ALI/ARDS)
- Hepatic dysfunction (Tbili >0.5mg/dL)
- Coagulation dysfunction, DIC (thrombocytopenia, prolonged PT, PTT or ACT, abnormal TEG)
- Gastrointestinal dysfunction (vomiting, regurgitation, ileus, constipation, diarrhea)
- Endothelial dysfunction (vascular leak); and
- Laminitis (horses).

The presence of two or more of these organ dysfunctions is referred to as multiple organ dysfunction syndrome (MODS); this can occur due to either sepsis or non-infectious causes. Septic shock refers to hypotension non-responsive to intravascular volume expansion (i.e. vasopressor dependent). Not surprisingly, mortality increases dramatically as the patient progresses from sepsis to severe sepsis to septic shock. What constitutes organ failure vs. organ dysfunction is also poorly defined; perhaps the most obvious is acute kidney failure that we may recognize with the onset of oligoanuria.

The Pathophysiology of SIRS and Sepsis

The innate immune response represents the first line of defense in the ability of a host to recognize and respond appropriately to microbial invasion and control infection; and thus is key to the pathophysiology of sepsis. Innate immunity is not specific to the invading pathogen and does not generate immunology memory. The foundation of the innate immune response is the innate ability to recognize microbial products. Various microbial components, known as PAMPs, are able to active the innate immune system including LPS, peptidoglycan, lipotechoic acid, lipoprotein and bacterial CpG DNA. Nonclonal, cellular, PRRs of host macrophages, monocytes and neutrophils recognize these conserved PAMPs and induce a cellular response.[1] This response can include the release of proinflammatory cytokines, phagocytosis or expression of co-stimulatory molecules by antigen presenting cells (APCs), the latter which facilitates interaction with the adaptive immune response.

LPS, the principal component of the cell wall of gram negative bacteria, is a prime example of a unique and highly conserved microbial surface molecule that engages with the innate immune system of the mammalian host via PRRs on a wide range of host cells.[2] Humans and other mammals have evolved exquisite sensitivity to low concentrations of LPS to combat infection.[3] Monocytes and macrophages orchestrate the innate immune response to LPS by expressing a wide variety of inflammatory mediators. Induction of an innate immune response and the resultant expression of pro-inflammatory mediators is normally tightly controlled and localized to sites of microbial invasion.[4] When an animal senses endotoxin, the usual outcome is successful antimicrobial defense.[4] However, in the cases of sepsis, septic shock and MODS, an exaggerated and systemic pro-inflammatory response results in substantial patient morbidity and mortality.[3,6]

The interaction of bacterial PAMPS with components of the innate immune system occurs initially extracellularly, before engaging with cell surface toll-like receptors (TLRs) and initiating an intracellular signaling cascade.[7] During gram negative sepsis, LPS is elaborated into circulation where it binds the acute phase plasma protein LPS binding protein (LPSBP or LBP) facilitating its delivery to the GPI-anchored cell surface receptor CD14 on cells of the immune system.[8] While alternative mechanisms for LPS transfer to CD14 may exist, this process occurs much more efficiently in the presence of LBP.[9] Interestingly, some clinical studies indicate that endotoxin is found in the systemic circulation of the majority of human patients with septic shock and this endotoxemia is largely independent of the nature of the infecting microorganism.[7] CD14 is a myeloid marker antigen that exists in 2 forms. Membrane bound CD 14 (mCD14) is expressed on the surface of myeloid cells and anchored via a glycosylphosphatidylinositol (GPI) tail.[3] Soluble CD14 (sCD14) it is thought to mediate LPS activation of CD14-negative cells, such as endothelial and epithelial cells. [7,10] mCD14 is considered a PRR; in addition to binding LBP-opsonized LPS, it also binds a variety of other microbial constituents, such as peptidoglycan, lipotechoic acid, and fungal antigens.[7] Given that mCD14 lacks an intracellular domain, a coreceptor that spans the cell membrane is necessary to effect intracellular signaling.[7,11]

In the case of LPS, this transmembrane signaling receptor is primarily TLR4.[11,12] The TLRs are a family of type 1 transmembrane receptors expressed on a large number of immune cells that act as PRRs thereby playing a vital role in the activation of the innate immune response to microbial pathogens.[11,12] Specifically, leucine-rich repeats in the extracellular domain of TLRs are enable pathogen recognition.[3] Each TLR recognizes a different repertoire of PAMPS, such that collectively, the 10 currently recognized human TLRs are able to detect most infectious organisms including bacteria, viruses, fungi and protozoa.[13] While it is generally accepted that TLR4 is the predominant LPS receptor, there are a couple of exceptions to this rule that have been demonstrated thus far in vitro. Nonenteric LPS preparations derived from both *Leptospira interrogans* and *Porphyromonas gingivalis* have been shown to signal vial TLR2.[14,15] In addition to recognizing gram negative bacteria and the lipid A portion of bacterial LPS, other identified ligands for TLR4 include other microbes and their products (BCG cell walls, bacterial glycolipids, M*yobacterium tuberculosis,* fusion (F) protein of the respiratory syncytial virus, *Cryptococcus neoformans*, *Aspergillus fumigatus* hyphae), other endogenous macromolecules (heat shock protein 60, fibrinogen, heparin sulfate and saturated fatty acids) and synthetic ligands .[12,13]

MD2, an accessory protein, is thought to directly bind LPS, independent of CD14 or LPSBP, enhancing LPS sensitivity when coexpressed with TLR4.[11,12] MD2 likely functions in multimeric complexes; either binding to LPS directly before interacting with TLR4, or inducing stabilization of TLR-4 dimers in receptor comlplexes.[16,17] In vitro studies using immunostaining have demonstrated that when macrophages become activated by LPS, mCD14, MD2 and TLR4 cluster on the cell surface in close proximity to one another, forming "raft"-like arrays.[18]

Following ligand binding, TLR4 homodimerizes, facilitating binding to specific adaptor molecules. Adaptor proteins, which all contain a Toll/IL-1R (TIR) domain, provide a structural platform to which various kinases, particularly tyrosine kinases and mitogen activated protein kinases (MAPK), and substrates are recruited.[3,7,12] This initiates a signal cascade that culminates in the activation of specific DNA binding proteins that activate transcription. Adaptor proteins involved in TLR4 signaling include myeloid differentiation factor (MyD88); MyD88-adaptor-like (Mal) or TIR-associated protein (TIRAP); toll-receptor-associated activator of interferon, also called TIR domain-containing adaptor molecule; and toll-receptor associated molecule or TIR-domain-containing adapter molecule-2.[13] It is evident that TLR4 has both a MyD88-dependent and –independent pathway. MyD88-dependent signaling involves IRAK and TRAF6. TAK1-TAB2 and MEKK1-ECSIT complexes phosphorylate Iκ, whose subsequent degradation permits nuclear translocation of NF-κB/Rel complexes, such as p50/p65. While the MyD88-dependent pathway is the most thoroughly studied of the two broad signaling pathways, 75% of genes activated through TLR4 in macrophages are actually activated through MyD88-independent pathways. In the MyD88-independent pathway, through the adapter TRIF, TLR4 signals through similar mechanisms as TLR3, resulting in the production of IFNβ. Three main transcription factors are activated through these pathways by TLR4; nuclear factor (NF)-κB, activator protein-1 (AP-1) and interferon response factor-3 (IRF-3).[13]

Gene products of these transcription factors include cytokines [e.g. tumor necrosis factor-alpha (TNF-α), interleukin 1 beta (IL-1β), IL-6, IL-10, chemokines (e.g. CXCL-8 or IL-8), nitric oxide, products of the complement and arachidonic acid cascades and reactive oxygen species.[19] These (predominantly) pro-inflammatory mediators drive the development of inflammation, which systemically results in the manifestations we attribute to sepsis and ultimately MODS. Some of these are discussed below.

TNF-α, IL-1β, and IL-6 are potent, prototypic mediators of systemic inflammation, produced primarily by activated monocytes and macrophages. When given systemically, independent of PAMPs or DAMPs, these cytokines induce systemic inflammation. These pro-inflammatory cytokines are pleotropic and redundant such that their functions will be considered collectively for these purposes. These cytokines promotes endothelial cell permeability, neutrophil release from bone marrow and expression of endothelial leukocyte adhesion molecules. TNF in particular is directly cytotoxic to some cells by induction of apoptosis. These cytokines also stimulate the adaptive immune response, stimulating differentiation of B cells to produce immunoglobulins and potentiating the activity of Th cells, thereby promoting both humoral and cellular immune responses. These cytokines also influence the hypothalamus inducing fever, increase hepatic production of acute phase proteins (such as CRP and SAA), trigger muscle catabolism, peripheral lipolysis and hepatic lipogenesis. Additionally these early pro-inflammatory cytokines not only provide positive feedback for their own production creating a cytokine storm, but also increase the expression of a wide variety of other pro-inflammatory mediators (e.g. NO and cyclooxygenase 2 of the arachidonic acid cascade), chemokines (e.g. CXCL-8) and adhesion molecules.[20,21] IL-6 in particular may also be involved in mediating gastrointestinal barrier dysfunction in shock.

IL-8 or CXCL-8 is a key chemokine whose primary function is in the recruitment and activation of neutrophils.

Interleukin 10 is a potent anti-inflammatory cytokine that down regulates the inflammatory response through various mechanisms.[22] IL-10 acts to inhibit synthesis of many pro-inflammatory mediators (IFN-γ, IL-3, IL-2 and TNF) as well as enhancing production of IL-1RA by monocytes. IL-10 also blocks IFN-γ, binding at a key step in its signal transduction pathway.[22] IL-10 is expressed and secreted by a variety of cell types; predominantly T helper 2 (Th2) cells but also monocytes/macrophages and dendritic cells; expression is regulated differently by these different cells types and occurs in response to an activation stimulus.[22,24]

Nitric oxide is a ubiquitous biological mediator that plays an important role in the pathophysiology of sepsis. Microbial products and inflammatory mediators stimulate exuberant production of NO via upregulation of inducible nitric oxide synthase (iNOS) in many cells, including macrophages, hepatocytes, and endothelial cells.[8] Excessive production of NO by iNOS is thought to be a key mediator of the myocardial depression, hypotension and vascular hyporeactivity associated with sepsis, severe sepsis and septic shock.[8,25] Nitric oxide also has profound immunomodulatory functions in the face of systemic infection, thought its effects on leukocyte adhesion and activation, preventing normal leukocyte-endothelial interactions and the neutrophil oxidative burst via direct effects on NADPH oxidase.[8,26,27] In addition NO contributes to the hemodynamic derangement in sepsis via inhibition of platelet aggregation and formation of platelet plugs.[8]

Systemic inflammation also leads to systemic activation of the coagulation system.[24] PAMPs and pro-inflammatory cytokines activate platelets, promote thrombogenicity of endothelial surfaces, induce intravascular expression of tissue factor (TF) and consume circulating anticoagulants (tissue factor pathway inhibitor [TFPI], antithrombin [AT] and activated protein

C [aPC]). Additionally, endotoxin and pro-inflammatory cytokines down regulate fibrinolysis, resulting in a procoagulant state that may culminate in DIC and the formation of microvascular thrombi resulting in microcirculatory derangements, tissue ischemia and MOD. Activation of coagulation also potentiates systemic inflammation. During sepsis, anticoagulants, pro-coagulants, such as thrombin, have potent pro-inflammatory properties. Thrombin promotes inflammation via leukocyte activation and induction of pro-inflammatory cytokines, further contributing to the self-perpetuating nature of the systemic inflammatory reaction.

MODS is a multifactorial phenomena that occurs secondary to both noninfectious SIRS, and likely more commonly, sepsis. Some of the specific organ dysfunctions are adequately explained by the aforementioned pathophysiology. For example coagulation dysfunction / DIC is explained by the concurrent activation of coagulation, with an initial prothrombotic phase, followed by a consumptive coagulopathy. Cardiovascular dysfunction / septic shock is at least in part explained by the actions of NO and the resultant vasodilatory shock. Additionally relative adrenal insufficiency can contribute to vascular hyporeactivity, a phenomenon that is corticosteroid responsive. Respiratory dysfunction is explained by leukocyte infiltration and fluid leak into alveoli, as well as impaired blood flow in the pulmonary circulation associated with thrombosis; leading to profound diffusion barrier and V/Q mismatch. Interestingly other organ dysfunctions are less well understood. For example acute kidney injury and acute hepatic dysfunction do not appear to be associated with significant inflammation in these organs. In fact histopathology of these organs in patient that succumb to sepsis may be normal; in these cases cytopathic hypoxia is often implicated in organ dysfunction. That being said histopathologic evidence of DIC (both microvascular hemorrhage and thrombosis) is common; while necrosis and apoptosis are less commonly identified but can occur. The mechanisms of these organ dysfunctions are an area of active research.

REFERENCES

1. Medzhitov R, et al. *N Engl J Med* 2000;343:338. 2. Hodgson JC. *J Comp Pathol* 2006;135:157. 3. Guha M et al. *Cell Signal* 2001;13:85. 4. Munford RS. *J Endotoxin Res* 2005;11:69. 6. Angus D et al. *Crit Care Med* 2001;29:1303. 7. Opal SM, Gluck T. *Crit Care Med* 2003;31:S57. 8. Kirkeboen KA, Strand OA. *Acta Anaesthesiol Scand* 1999;43:275. 9. Fenton MJ, Golenbock DT. *J Leukoc Biol* 1998;64:25. 10. Medzhitov R et al. *Nature* 1997;388:394. 11. Cristofaro P, Opal SM. *Drugs* 2006;66:15. 12. Lien E, Ingalls RR. *Crit Care Med* 2002;30:S1. 13. Leaver SK et al. *Crit Care Med* 2007;35:1404. 14. Werts C et al. *Nat Immunol* 2001;2:346. 15. Hirschfeld M et al. *Infect Immun* 2001;69:1477. 16. Visintin A et al. *Proc Natl Acad Sci U S A* 2001;98:12156. 17. Viriyakosol S et al. *J Biol Chem* 2001;276:38044. 18. Pfeiffer A et al. *Eur J Immunol* 2001;31:3153. 19. Cohen J. *Nature* 2002;420:885. 20. Goodman RB et al. *Cytokine Growth Factor Rev* 2003;14:523. 21. Dinarello CA. *Crit Care Med* 2005;33:S460. 22. Gosain A, Gamelli RL. *J Burn Care Rehabil* 2005;26:7. 24. Scumpia PO, Moldawer LL. *Crit Care Med* 2005;33:S468. 25. Abraham E, Singer M. *Crit Care Med* 2007. 26. Clancy RM et al. *J Clin Invest* 1992;90:1116. 27. Kubes P et al. *Proc Natl Acad Sci U S A* 1991;88:4651.

GI MICROFLORA AND THE HYGIENE HYPOTHESIS

James G. Fox, DVM, DACLAM
Cambridge, MA

ORIGIN OF THE HYGIENE HYPOTHESIS

During the 1980s and 1990s Strachen and Matricardi noted in separate studies that having a number of older siblings strongly correlated with a reduced risk of hay fever [**1**, **2**]. These observations were consistent with the premise that protective postnatal infections, transmitted within the sibling cohort, might be lost with the adoption of modern hygiene. These findings were recapitulated with studies illustrating that children living in a farming environment were protected against allergies. These earlier studies were supported by a recent large cohort study in Europe documenting that children raised in a farm environment were indeed protected from developing hay fever and atopy [**3**]. Subsequent epidemiological studies and experiments conducted in the laboratory suggested that allergic conditions were ameliorated in individuals or animals exposed to helminths, atypical mycobacteria, and intestinal microorganisms.

Although these studies emphasized the apparent protection due to various microbial exposures at early childhood, there was a lack of mechanistic underpinnings to prove or disprove the hygiene hypothesis. Renewed interest in this field has been prompted during the last several years with several exciting publications that have begun to dissect probable biological mechanisms that support the original studies published by Strachen & Matricardi.

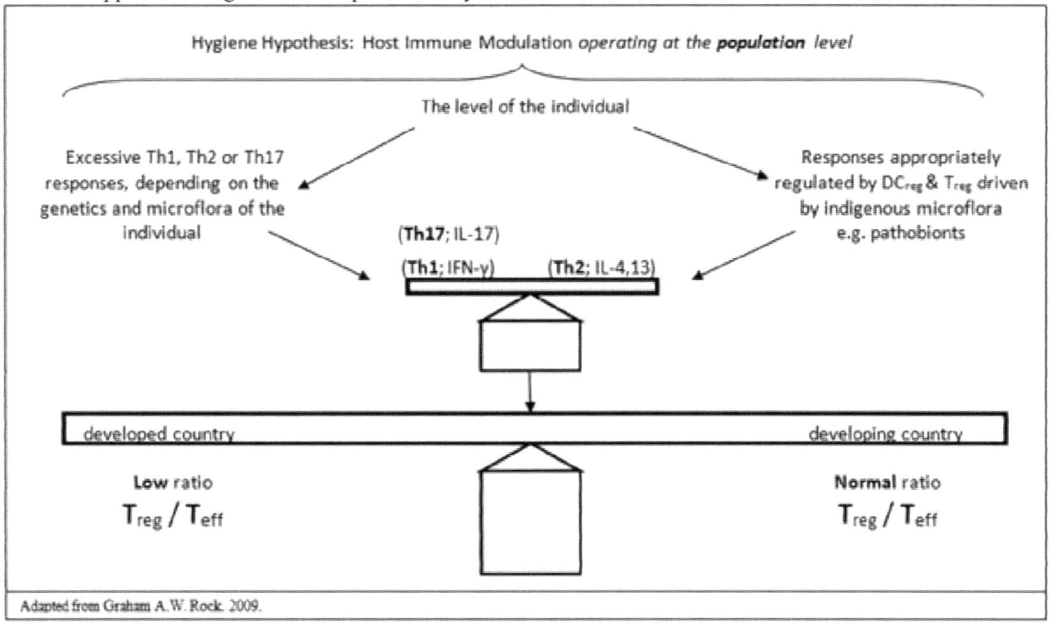

Host Immune Modulation

HELMINTHS AFFECT HOST IMMUNE RESPONSES

Helminths infect virtually every mammal living in a natural environment, and these parasites indeed are a focus of disease interest in veterinary medicine; many believe that parasites have coevolved with their hosts, providing a strong rationale why helminths have a restricted host range. It is estimated that over 1 billion humans are infected with a variety of different helminth species, many of which can chronically infect individuals, or can frequently reinfect individuals living in endemic areas with high parasite burdens. Importantly many of these parasites are sensitive to freezing or their intermediate hosts require tropical climates to survive. Transmission of these parasites commonly occurs through stages of the parasite being shed into the environment via host excretions.

Individuals infected with helminths have a Th2 and IL10 skewed immune response directed against the particular parasite. Children residing in Cameroon had regulatory IL10 and TGFβ responses to mitogens in vitro that directly correlated to burden of infection with *Ascaris lumbricoides* and *Trichuris truchiura* [**4**]. Compelling experimental evidence of helminths' ability to modulate immune response and affect IBD is illustrated by a multitude of mouse studies. Helminth infections in mice also protect against Type 1 diabetes, asthma, Graves disease, and multiple sclerosis.

Developed Industrialized Countries Have Limited Parasite Burdens

With the advent of public health measures and modern sanitation practices, parasitic infections during the 20th Century have dramatically decreased in industrialized populations, particularly after World War II and the introduction of antihelminthic drug treatments.

Increasing Prevalence of IBD

Prior to the 20th Century, IBD was seldom diagnosed. For example in London prior to the 1880s, ulcerative colitis had not been diagnosed and between 1884-1909, only sporadic cases were observed. However, during the last 50 years of the 20th Century, IBD has been increasing dramatically in populations residing in developed countries. Crohn's disease wasn't recognized until the 1930s and along with ulcerative colitis, was increasingly recognized in the United States and Europe. Studies published in the early 1990s suggested that being raised in rural parts of southern U. S. protected individuals from IBD and likewise blue collar employees working in dirty environments had lower incidence of IBD [**5**]. While IBD continues to be prevalent in the US and Europe, it is less frequently seen in Eastern Europe, Asia, Africa, and South America [**6**]. As countries develop socioeconomically, in these regions IBD incidences increase. In Japan and South Korea the prevalence of IBD is dramatically increasing. These statistics argue that environmental conditions influence susceptibility to development of IBD, for example dramatic changes in dietary habits; increase in smoking however appears to protect individuals from developing ulcerative colitis. These persuasive arguments supported the hypothesis that the disappearance of helminths in populations in developed countries increased the risk of developing IBD.

TREATMENT OF IBD IN HUMANS WITH THE SWINE PARASITE *TRICHURIS SUIS*

Trichuris suis, related to the human whipworm, *T. trichuria,* is currently being used in clinical trials to treat patients with IBD. In humans, *T. suis* doesn't complete its life cycle and is therefore clinically safe to administer. In a small open-label study, 2500 *T. suis* were administered to 7 patients with IBD. All of the treated patients had clinical improvement [7] In another trial, 29 Crohn's patients were treated with *T. suis* ova every 3 weeks; 24 weeks post treatment 79% had clinical improvement [**8**]. A followup randomized trial for treatment of UC with *T. suis* also demonstrated a protective effect of patients receiving *T. suis* ova [**9**]. Is there a clinical condition in dogs and cats with IBD that may benefit from this approach?

THE AFRICAN ENIGMA

The epidemiologic observation of the 'African Enigma' was coined when populations with a high prevalence of *H. pylori* infection but a low incidence of gastric cancer was postulated to be due, in part, to the high prevalence of helminths in this population. Indeed, experimental infection of mice with a nematode, *H. polygyrus*, known to induce pronounced Th2 and regulatory T cells that effectively suppress Th1 and Th17 responses, attenuated the severity of Helicobacter spp.- induced gastritis and premalignant lesions of gastric cancer [**10**]. Studies conducted by our laboratory in Colombia, South America suggest that parasite burden in a population of low risk of gastric cancer despite a >95% infection rate of *H. pylori*, may be explained by the protective effect provided by a high intestinal parasite burden. In the same country another population with a high risk of *H. pylori* associated gastric cancer have a low burden of intestinal helminths [**11,12**].

INDIGENOUS MICROBES PROTECT AGAINST OR PROMOTE CHRONIC INFLAMMATION

Several epidemiological studies have reported an inverse relationship between *Helicobacter pylori* infection and asthma [**13**, **14**]. Experimentally, *H. pylori* infection in a mouse model prevents allergic asthma through the induction of regulatory T cells. *H. pylori* infection, particularly in neonatal mice efficiently protected mice from airway hyperresponsiveness, goblet cell metaplasia and pulmonary inflammation [**15**]. A recent study expanded on these observations and presented data that several specific glycolipids from *H. pylori* induced a population of regulatory T cells that inhibit asthma in a mouse model, possibly through a (T?) regulatory T cell mechanism [**16**]. These findings support the notion that loss of the host's ancestral indigenous microflora, rather than a decline in childhood illnesses, accounts for the increased incidence of asthma in developed countries [**17**].

Animal models have also illustrated that extragastric microbes can impact diseases of the stomach. Concurrent *H. bilis* infection of the lower bowel significantly attenuated gastric inflammation due to *H. pylori* infection [**18**]. The same investigators subsequently have demonstrated that these immune modulatory effects are microbe species specific. Preinoculation with the intestinal *Helicobacter muridarum*, similar to that demonstrated with *H. bilis*, attenuated gastric injury induced by *H. pylori* [**19**]. However, preinfection with *H. hepaticus* reciprocally increased the severity of *H. pylori* gastric disease these effects were postulated to be mediated by differential activation of IL7; levels of IL7 were increased by *H. hepaticus,* but were reduced by *H. muridarum*, implying that the intestinal microbiota serve as a biological rheostat positioned to either amplify or dampen microbially induced inflammation in the stomach [**20**].

CONCLUSION

Dogs and cats are historically known to be infected with a myriad of intestinal parasites. These parasites have deleterious effects on the host and development of effective antihelminthics have systematically and effectively eliminated many of these parasites from their respective hosts. Similarly, progressive loss of indigenous microbial populations from pet animals may have also occurred while their environments experienced increased hygiene and widespread use of antimicrobial agents. Are the apparent increases in atopy, allergies, and autoimmune diseases in dogs and cats a reflection, in part due to the profound immune-modulatory effects of the dynamic shift in their microbial colonization of the gut? Studies being conducted in humans and mouse models suggest that similar studies in dogs and cats should be conducted.

REFERENCES

[1]Strachan DP. **Brit Med Journal** 1989; 299:1259. [2]Matricardi PM, et al **J Allergy Clin Immunol** 1998; 101:439. [3] Ege MJ, et al. **NEJM** 2011;364:701. [4]Turner JD. **J Med Micro** 2008;197:1204. [5] Loftus EV. **Gastroenterology** 2004;126:1504. [6] Molodecky NA, et al. **Gastroenterology** 2012;142:46. [7]Summers RW, et al. **Am J Gastro** 2003;98(9):2034. [8]Summers RW, et al. **Gut** 2005. 54:87. [9] Summers RW, et al. **Gastroenterology** 2005;128:825. [10] Fox JG, et al. **Nat Med** 2000;6:536. [11]Whary, et al. **Cancer Epidem Biomar Prev** 2005;14:1464. [12] Ek, et al. **Helicobacter** 2012. [13]Chen Y, Blaser MJ. **J Infect Dis** 2008;198:553. [14]Reibman J, PlosOne, 2008 3;12:e4060. [15]Arnold IC, et al. **J Clin Invest** 2011;121:3088. [16]Chang YJ, et al. **J Clin Invest:**2011;121:57. [17]Blaser MJ, Falkow S. Nat Rev Microbiol 2009;7:887. [18]Lemke LB, et al. **Infect Immun** 2009;77:2147. [19]Ge Z, et al. **Inf Immun** 2011;79:3861. [20]Peek RM, Kuipers EJ. **Gut** 2012;61:1.

DNA METHYLATION: UNDERSTANDING DOGS TO UNDERSTAND HUMANS

Jeffrey N. Bryan, DVM, MS PhD, DACVIM (Oncology)
Kansas City, Missouri

INTRODUCTION

First reported in 1949, methylcytosine in DNA sequences has only relatively recently received attention in the development, progression, diagnosis, and management of human cancers [1]. Methylcytosine is created post-replication by the addition of a methyl group to the 5 position of the cytosine ring when a cytosine is followed by a guanine (CpG) in sequence. The addition is made by DNA methyltransferases (DNMTs) using the methyl donor s-adenosyl methionine. Across mammalian genomes, CpG dinucleotides are relatively rare events. Sporadically, CpG may be found in high density in constructs called CpG islands that are frequently associated with the control regions of genes. Hypermethylation of these islands was originally thought to result in persistent silencing of the associated gene. New work has demonstrated, however, that chromatin marks on histones are responsible for the active silencing of the gene, which can often be expressed even when methylated. These hypermethylated genes will revert to a silenced state in relatively short order when the signal for the histone changes recedes.[2] Recently, a third concept of CpGs has emerged. The CpGs residing in the lower density regions near CpG islands are considered to reside on the "shores" of those islands and appear to play a significant role in cell development and stem cell function. The first reported examination of DNA methylation in canine cancer was in 2003 when lymphomas were shown to be characterized by the same global loss of DNA methylation seen in human lymphomas.[3] Subsequently, the often hypermethylated *FHIT* gene was studied in canine lymphoma cell lines and found to be aberrantly expressed, but hypermethylation could not be documented to be the cause.[4] Conversely, the *DLC1* gene was shown to be significantly hypermethylated in canine lymphoma patient samples, but no effect on the expression of the gene could be identified.[5] Put most simply, epigenetic changes like DNA methylation are the mechanism by which a liver cell remains a liver cell and a skin cell remains a skin cell, yet they share identical DNA within the same body. Certain sets of genes in each cell are mostly permanently turned on or off to limit the set of proteins that can be expressed. These mechanisms become disrupted in the course of carcinogenesis, changing the gene sets available to cancer cells and resulting in significant genetic and epigenetic plasticity.

Cancer has long been known to be a genetic disease. However, recent reports have identified at least as many genes (sometimes hundreds) that are dysregulated by gain or loss of DNA methylation as are mutated, deleted, translocated, or amplified.[6] In fact, some diseases, such as prostate cancer, are more frequently defined by characteristic hypermethylation in genes than by characteristic mutation. The degree to which epigenetic derangement characterizes some cancers is startlingly underscored by a series of experiments in which mouse melanoma nuclei were reprogrammed by nuclear transplantation into an oocyte and implantation for development or chimeric growth.[7] Cells with transplanted melanoma nuclei were able to participate as embryonic stem cells in growth and organ differentiation. Adult chimeras had a high rate of melanoma formation, but were otherwise normal, functional mice. These experiments demonstrate that epigenetic reprogramming of embryogenesis abrogates many of the changes that characterize cancer, raising the possibility that therapeutic induction of such reprogramming could lead to apoptotic death of the cells in patients with cancer.

Thousands of papers (15,500 reports in Pubmed on 1/29/12) have now been published on DNA methylation in cancer. Significant progress has been made in understanding the role of DNA methylation in development, stem cell generation and reprogramming, and in gene expression in individual cancer types. However, these understandings have not yet translated into strong biomarker development for identification of risk, early diagnosis, or prognosis in many cancers. Nor have huge strides been made in the abrogation of cancer risk for individuals with known environmental and lifestyle risk factors. Because of the relatively shorter lifespan of dogs, the shared environment as human companions, and the relatively identifiable breed risk of many cancers, the comparative evaluation of DNA methylation changes in dog cancers would seem a fruitful pursuit to understand the interaction of environment, genes, epigenetic responses to environmental stresses, and the development of cancers. DNA methylation changes could be identified and studies as markers of risk, early diagnosis, prediction of therapeutic response, prognosis, and testing of novel therapeutic interventions.

EPIGENETIC PROGENITOR ORIGIN OF CANCER

Recently, a model of carcinogenesis has been proposed that has been called the epigenetic progenitor origin of cancer.[8] Unlike the clonal genetic model of cancer development, this model suggests that progenitor cell populations in tissues are affected by environmental interactions that lead to an expanded pool of epigenetically altered progenitor cells with altered differentiation potential and polarity. On this background of expanded progenitor cells a gatekeeper genetic mutation occurs characteristic of the particular type of tumor. This then becomes the primed source of a clonal cancer. Following this is a period of epigenetic and genetic plasticity that results in expansion of the clone and development of the hallmarks of cancer.

DNA METHYLATION MARKS AS BIOMARKERS

Because DNA methylation patterns are altered early in the process of carcinogenesis, it is likely that DNA methylation marks could serve as biomarkers of disease risk. In fact, recent work has demonstrated that differentially methylated regions of CpG island shores, CpG islands, and even intergenic regions reliably distinguish normal from neoplastic colon tissue.[9] Further, benign polyps had an intermediate frequency of methylation changes between normal and neoplastic epithelium. These same marks are reliably altered in several other types of cancer in a similar fashion to colon cancer and are enriched in regions differentially methylated during stem cell reprogramming during induced pluripotent stem cell formation, supporting the concept that it is a progenitor cell population that is initially affected in cancers.[9] If such regions are reliably affected, it stands to reason that their detection in tissue might serve as a harbinger of increased risk of full-blown carcinogenesis. Because of the difficulty and expense of longitudinal studies in human subjects, identification and validation of methylated DNA risk biomarkers in dogs could streamline the development of such assays in humans.

BIOMARKERS OF DISEASE

Methylated DNA in tumor-associated genes can be detected in tissue samples as well as remote media such as plasma and urine. Gene-specific methylation has been identified to be strongly associated with large number of tumor types. *GSTP1* in prostate cancer. ADAM23 in breast cancer is frequently methylated and associated with a more aggressive, more highly metastatic form of the disease.[10] The company Exact Sciences has released a test of four methylation markers with a reported 100% sensitivity to detect colon cancer.[10] Methylated genes can be assayed in tissue biopsies to guide therapy decisions. When the drug metabolizing gene *MGMT* is methylated and silenced in glioblastomas and other brain tumors, human patients have a better response to temozolamide.[11;12] When unmethylated, the gene product is expressed which results in removal of DNA adducts and loss of drug efficacy. Expansion of biomarker development in naturally occurring canine tumors will accelerate the application of such assays to human disease.

EPIGENETICALLY ACTIVE THERAPY

DNA methylation and histone modifications can be modified with drug therapy, causing the re-expression of silenced genes and resulting in cell death and tumor regression *in vitro* and in mouse models of disease. However, these drugs have translated incompletely to the clinic. Currently, Vidaza (5-azacytidine) and Decitabine (5-aza,2-deoxycitidine) (DNMT inhibitors) have been found to be useful in treating myelodysplastic syndrome in humans, resulting in transfusion independence.[10] Histone deacetylase (HDAC) inhibitors such as Vorinostat (SAHA) have found application in a number of tumors with SAHA being approved for epitheliotropic lymphoma in humans. Clinical trials of DNMT inhibitors as well as HDAC inhibitors have been combined with conventional chemotherapy with demonstrated tolerability and better efficacy compared to chemotherapy alone.[10] Given the delicate balance of gene control across many cell types within the body, it is likely that generalized inhibition of DNMTs or HDACs is not desirable, even for individuals with cancer, and particularly for a protracted course of therapy. Brief courses of inhibition to sensitize to chemotherapy are attractive for better management of disease. Ideally, inhibition of these molecules will be targeted selectively to cancers, avoiding these effects on the rest of the somatic cells. Animal models of naturally occurring cancers could model such targeted therapies and be used to identify potentially optimal drug combinations.

IDENTIFICATION OF DNA METHYLATION

A multitude of techniques exist to query chromatin for epigenetic marks related to cancer.[13] Individual genes can be studied using bisulfite techniques to identify methylation changes that alter gene expression or serve as a mark typical of a particular histology. DNA methylation arrays exist for humans and have been designed for dogs that allow elucidation of DNA methylation patterns by isolating a pool of fragmented DNA enriched for methylcytosine-containing regions. Hybridization with a DNA methylation array identifies the genomic regions containing significant methylcytosine compared to normal tissue of the same origin. Next generation sequencing techniques now allow a pool of fragmented DNA enriched for methylcytosine to be sequenced. This DNA can be in native form, identifying gene regions with higher methylcytosine content, or can be bisulfite sequenced to determine the methylcytosine patterns at a single-base level. The advent of the sequencing of the canine genome provides the bioinformatic framework to identify the sequences of desired genes or to locate reads from large-scale sequencing experiments. The output from these experiments can then be used to compare methylation patterns across species, potentially identifying the unifying changes that are most at the root of these diseases. These more recent, high throughput, high output methods yield massive quantities of data. Such output almost invariably yields statistically significant results. The challenge of this field has been identifying clinically significant results. The comparative evaluation of similar cancers across species promises to identify the subset of information that represents the underlying malfunctions in gene regulation leading to cancer. Identifying such marks could lead to preventive interventions in high risk groups, earlier cancer diagnoses, markers of prognostic and therapeutic import, and sensitive markers of residual disease. Companion animals offer highly relevant models of naturally occurring cancers that can be evaluated at much less expense in a much shorter time-frame than humans with similar disease. Exploiting these models could lead to significant progress in understanding these important mechanisms of disease.

REFERENCES

1.Hotchkiss RD. Colloques intern.centre natl.recherche sci.Unites biol.douees contin.genet.(Paris) 1949;8:57-65. 2. Raynal NJ, et al. Cancer Res. 2012 epub ahead of print. 3. Pelham JT, et al. Res.Vet.Sci. 2003;74:101-104. 4. Hiraoka H, et al. J.Vet.Med.Sci. 2009;71:769-777. 5. Bryan JN, et al. BMC.Genet. 2009;10:73. 6. Jones PA, Baylin SB. Cell 2007;128:683-692. 7. Hochedlinger K, et al. Genes Dev. 2004;18:1875-1885. 8. Feinberg AP, et al. Nat.Rev.Genet. 2006;7:21-33. 9. Hansen KD, et al. Nat.Genet. 2011;43:768-775. 10. Costa FF. Cancer Manag.Res. 2010;2:255-65.:255-265. 11. Hegi ME, et al. N.Engl.J.Med. 2005;352:997-1003. 12. Donson AM, et al. Pediatr.Blood Cancer 2007;48:403-407. 13. Bryan JN, et al. Cancer Therapy 2008;6:137-148.

TO CLOT OR NOT TO CLOT: DISORDERED HEMOSTASIS AND WHAT TO DO ABOUT IT

Kari V. Lunsford, DVM, MS, DACVIM
Mississippi State, MS

NORMAL HEMOSTASIS:

The mechanisms involved in normal hemostasis are complex, intricately interrelated and dynamic processes by which both vascular integrity and the fluidity of blood are maintained. These hemostatic systems remain in a delicate balance serving as the body's natural defense against blood loss. If this balance is disrupted, with a shift toward clot formation, unwanted pathological and potentially life threatening thrombosis may occur. In order to fully understand and react to thromboembolic complications in our patients, it is necessary first to review the major players involved in normal hemostatic processes.

Thrombus vs. Clot

There is an important distinction to be made between a thrombus and a clot. A thrombus forms only within the vasculature and only during life. The thrombus is a solid coagulated mass composed of platelets, fibrin and entrapped cellular elements which forms in either an artery or a vein. This is in contrast to a blood clot which may form outside the vascular space (hematoma), outside the body and/or after death.

Components

The normal vascular endothelium is often given less attention than it should in discussions of hemostasis. The primary purpose, after all, of the hemostatic system is to maintain the integrity of the vascular endothelium. Both the endothelial cells themselves and the subendothelial constituents of the vascular wall play major roles in normal hemostasis by interactions with platelets and coagulation proteins to initiate a cascade of events which ultimately results in the generation of insoluble fibrin and thrombus formation.

THROMBUS ORIGIN

Arterial thrombosis

Thrombosis and subsequent thromboembolism are recognized as major contributors to the morbidity and mortality associated many diseases in human medicine and as such is widely studied. In people, arterial thrombosis involving the coronary circulation is the most common cause of myocardial infarction and thrombotic occlusion of the cerebral arteries is a leading cause of stroke. Other significant sequelae to arterial thrombosis include ischemic injury and subsequent gangrene involving the extremities and renal disease and hypertension associated with occlusion of the renal arteries.[1]

In humans arterial thrombosis is most commonly triggered by rupture of atherosclerotic plaques exposing the lipid rich central core to circulating blood. This initially leads to predominately platelet activation but ultimately the coagulant proteins are activated as well. These thrombi are typically found to extend into the damaged vessel wall as well as into the vessel lumen. In the case of partial occlusion, increased shear stresses resulting from reduced blood flow cause further activation and aggregation of platelets and fibrin deposition leading to complete occlusion over time if left untreated.[2]

Cardiogenic embolism, a leading cause of ischemic stroke in people, results primarily from blood stasis secondary to alterations in intra-cardiac blood flow. However, emboli may also result from intra-cardiac thrombus formation secondary to disruption in the endocardial surface related to underlying cardiac disease. Arterial embolism occurs when a portion of non or partially occlusive arterial thrombus breaks off and occludes a smaller more distal artery.

Cardiogenic embolism is the more frequent form of arterial thromboembolic disease in veterinary patients. In cats arterial thromboembolism occurs when emboli break away from thrombi that have formed in the left atrium or ventricle secondary to cardiomyopathy.[3,4] In dogs arterial thromboemboli have been reported associated with gastric dilation and volvulus, neoplasia and hyperadrenocorticism. [5-7]

Venous thrombosis

Venous thrombosis in people, particularly of the large proximal leg veins often leads to life threatening pulmonary embolism. Venous thrombosis is also associated with postphlebitic syndrome. Postphlebitic syndrome or chronic venous insufficiency results from functional insufficiency of the venous valvular system resulting in increased venous pressure, bidirectional flow of venous blood secondary to incompetent valves and inadequate cardiac venous return. Secondary changes resulting from this venous valvular insufficiency include increased venous pressures, further dilation and distention of the venous system, spread of valvular insufficiency and perpetuation of the thrombophilic state.

In contrast to the composition of arterial thrombi, in which platelet aggregates predominate, the thrombi that form in the venous circulation are composed primarily of fibrin and red blood cells. Venous thrombosis occurs in both human and veterinary patients, when pro-coagulant stimuli overwhelm the counter-regulatory mechanisms that normally ensure hemostatic balance. This is often in the form of inappropriate or excessive activation of the coagulation system, or "hypercoagulability." The potential for pathologic thrombus formation is heightened when this "hypercoagulability" occurs in the face of other abnormalities such as vascular injury or blood stasis.

In human patients vascular trauma as occurs with surgery has been shown to initiate excessive coagulation and venous thrombus formation in the veins, and indwelling central venous catheters may also be a source of vessel injury initiating pathologic thrombus formation in the upper extremities of people. Human venous thrombi most frequently develop either in the muscular veins of the calf or in the deep valve cusp pockets of the deeper calf veins under low-flow conditions.[8] Due to the avascular nature of these valve cusps, they depend on local luminal blood flow to maintain oxygenation. When venous stasis occurs, local hypoxemia develops and the expression of tissue factor is induced in the valve cusps. This does not tend to be the case in veterinary medicine. Venous thrombosis tends to occur when hypercoaguable conditions combine with localized alterations in blood flow and disruptions in endothelial integrity. Thrombi then form at these sites of local pathology or body wide as in the case of DIC when endothelial alterations occur system wide.

Inflammation

Inflammation also plays a role in the activation of the coagulation process. Tissue damage, surgical procedures and medical conditions all may generate inflammatory cytokines that can contribute to a pro-coagulant state. These cytokines activate endothelial cells causing them to express adhesion molecules to bind circulating leukocytes. When monocytes bind, they synthesize tissue factor and express receptors for both factor X and fibrinogen thus promoting coagulation on their surfaces.[9] Activated neutrophils generate free radicals and hydrolytic enzymes which perpetuate endothelial damage thus exposing more tissue factor, initiating platelet activation and enhancing thrombus formation.

EVIDENCE AND RECOMMENDATIONS

Thromboprophylactic Therapy

The etiology of thrombus origin may be important when determining appropriate therapy, whether thrombolytic or prophylactic, as the components of the clot vary with origin. Thrombi that form secondary to blood stasis tend to have a greater proportion of fibrin and red blood cells, while those that form secondary to endothelial disruptions tend to have a larger proportion of platelets. With this in mind, it would seem that therapies to prevent thrombi originating in the arterial vasculature (formed under high shear stress conditions and mediated by platelet endothelial interactions) should target platelet activity, while the prevention of venous thrombosis (formed under low shear conditions and mediated by coagulant proteins) should involve interruption of the coagulation cascade. Unfortunately, the trigger or initiating event for thrombus formation is not always clear, particularly in the case of dogs predisposed to thrombus formation by conditions such as immune mediated hemolytic anemia (IMHA).

Current Evidence

There are few studies evaluating the efficacy of these prophylactic therapies in dogs with IMHA and the evidence that does exist seems to be relatively equally split between advocating anti-platelet therapies such as aspirin and anti-coagulant therapies such as heparin. While it would seem that therapies geared toward preventing venous thromboembolism would be most effective if they targeted coagulation proteins rather than platelets, one study[10] demonstrated increased survival in dogs with IMHA that received low dose aspirin. Ongoing studies,[11] however are suggesting that dose adjusted heparin may be the treatment of choice for the prevention of PTE in dogs with IMHA. It may be that both therapies are beneficial in this setting, as the etiology of thrombus formation in IMHA is likely due to multiple factors and triggering events involving both platelets and activation coagulant proteins.

Even when the origin of thrombus formation is well understood as is the case of venous thromboembolism (VTE) in people undergoing orthopedic surgeries, contradictory therapeutic recommendations are made and a consensus is difficult to reach in light of discordant study results. These differences center, in part, around the use of aspirin for the prevention of venous thromboembolism. As is the case with canine IMHA, some human studies do demonstrate efficacy for aspirin in the prevention of venous thromboembolism. While there are certainly differences in collecting data and evaluating evidence from study to study and these differences may to some degree influence results, these discordant conclusions may also suggest that we do not yet have a clear understanding of the role of the platelet in venous thrombus formation and that the platelet may play a larger role than was previously thought and this role for the platelet may be particularly significant in inflammatory conditions such as IMHA.

Recommendations

Until further evidence is gathered in the form of prospective clinical trials evaluating therapeutic efficacy, it may be prudent to base therapeutic decisions on what is known about the pathogenesis of thrombus formation. When pathogenesis is uncertain, combination therapies may be of benefit.

REFERENCES

[1] Weitz JI, et.al. **Chest** 2004;126:265S. [2] Fuster, V, et.al. **N Engl J Med** 1992;326:242. [3] Smith SA, et.al. **Vet Clin N Am-Small**. 2004;34(5):1245. [4] Pion P. **Vet Clin N Am-Small** 1988; 18(1):79. [5] Carter WO. **JAVMA** 1990; 196(11):1829. [6] Santamarina G ,et.al. **JVIM** 2003;17 (6): 917. [7] Boswood A, et.al. **JSAP** 2000; 41(3):109. [8] Lee A, et.al. **Circulation** 2001;104:74. [9] Johnson LR. Editorial: **JVIM** 2002; 16:501. [10] Weinkle TK, et.al. **JAVMA** 2005; 226(11):1869. [11] Helmond SE, et.al. **JVIM** 2010; 24(3):597.

ANTIPLATELET AND ANTICOAGULANT THERAPY: WHAT DO WE KNOW? WHERE ARE WE GOING?

Marilyn Dunn, DMV, MVSc, DACVIM
St-Hyacinthe, Québec, Canada.

ANTIPLATELETS
CYCLOOXYGENASE INHIBITORS
Aspirin : What do we know?

Aspirin is the most common antiplatelet drug used in veterinary medicine. It acts through irreversible acetylation of the platelet cyclooxygenase site leading to decreased thromboxane A_2 synthesis. Thromboxane promotes platelet activation and aggregation by signalling through a G-protein-coupled receptor. 70% of dogs have a dysfunction of this protein explaining variability in platelet response to thromboxane ('aspirine resistant')[1]. The effects of aspirin are permanent and last for the lifespan of the platelet (7-10 days). The antiprostaglandin effect of aspirine may decrease prothrombotic inflammation in some disease states (IMHA).

Dogs

A study of aspirin in healthy dogs administered at 0.5mg/kg and 10mg/kg SID showed anti-platelet effects on TEG-PM however decreased platelet aggregation only occurred at the 10mg/kg dose.[2] In another study, healthy dogs given aspirin 0.5mg/kg SID and 5mg/kg BID showed decreased platelet aggregation.[1,3] Prolonged PFA closure times were shown in healthy dogs receiving aspirin 8mg/kg SID.[4] In a retrospective study of IMHA in dogs, improved survival was attributed to low dose (0.5 mg/kg PO) aspirin administration.[5] Doses of 1 mg/kg/day failed to decrease thromboxane urine metabolites, suggesting inadequate *in vivo* anti-platelet effects of this dose in healthy dogs.

Cats

Aspirin has been commonly recommended for cats at risk of thromboembolism (TE) at a dose of 81 mg/cat PO q72h , but efficacy is lacking. A retrospective study of long-term therapy in cats with previous arterial TE showed that aspirin (5 mg/cat q72h) was associated with similar or lower rates of clot recurrence when compared to other thromboprophylactic therapies, and adverse effects were rare.[6] No controlled studies have evaluated low dose aspirin in cats.

Aspirin: Where are we going?

The ideal prevention dosing protocols to prevent thrombosis in dogs and cats remains unknown. Further investigations will most likely involve aspirin being compared to other antiplatelet agents.

ADP RECEPTOR ANTAGONISTS

Thienopyridines irreversibly inhibit the binding of ADP to specific platelet receptors (P2Y12). Receptor blockade impairs platelet release reaction and ADP-mediated activation of GPIIb/IIIa, thereby reducing primary and secondary aggregation response. These drugs must be metabolized by hepatic cytochrome p450 as only the metabolites are active.

Clopidogrel : What do we know?
Cats

Clopidogrel, given to cats at doses of 18.75 to 75 mg PO q24 h was well tolerated and resulted in significant antiplatelet effects.[7] Clopidogrel may counteract the vasoactive properties of serotonin in aortic thromboembolism and may improve collateral vessel development. In vitro addition of Clopidogrel to TPA did not increase thrombolysis.[8] FAT CAT study: ongoing: comparison of aspirin (81mg/cat q72hrs) and clopidogrel (18.75 mg/cat /day) in prevention of recurrence of arterial TE.

Dogs

Healthy dogs receiving Clopidogrel showed decreased platelet aggregation at doses of 0.5 and 1mg/kg/day. Despite previous reports mentioning delayed onset of effect (3-5 days), an effect was seen within 90-180 minutes following administration. In dogs receiving Clopidogrel for 5 days, decreased platelet aggregation was documented 8 days after stopping therapy, indicating a prolonged antiplatelet effect.[9] A prospective study in dogs with IMHA receiving low dose aspirin and/or Clopidogrel showed no difference in survival rates.[10]

Clopidogrel: Where are we going?

Dosing interval must still be defined. A loading dose, as had previously been reported does not appear necessary in dogs. Some variability in response among individuals, possible polymorphism of the P2Y12 receptor? Monitoring may therefore be important (TEG platelet mapping, aggregometry, Plateletworks).[11]

Ticlopidine; What do we know?

Ticlopidine, a thienopyridine, administered to healthy cats caused decreased platelet aggregation however also resulted in dose-dependent anorexia and vomiting.[12] Experimental studies in dogs have demonstrated antiplatelet activity and resulted in less vascular lesions in dogs experimentally infected with heartworm.[13]

Ticlopidine: Where are we going?

Nowhere in cats. Potential therapeutic benefits in dogs.

Prasugrel has been shown to be a more potent platelet inhibitor in dogs than Clopidogrel.[14]

Gp IIb/IIIa RECEPTOR ANTAGONISTS

What do we know?

Activation of GPIIb/IIIa is the final common pathway of platelet aggregation. Drugs capable of inhibiting fibrinogen binding to GPIIb/IIIa are therefore potent antiplatelet agents. Three intravenous GPIIb/IIIa antagonist drugs have been developed for clinical use in people: monoclonal antibodies that bind to the receptor (abciximab), and competitive mimetics of fibrinogen (eptifibatide, tirofiban). In a study of cats with induced arterial injury, abciximab combined with aspirin reduced thrombus formation compared with aspirin alone.[15] Administration of eptifibatide to cats caused circulatory failure and death, and is therefore contraindicated. Although dogs have been used in pharmacokinetic and pharmacologic studies of GPIIb/IIIa antagonists, clinical studies have not been performed.

Where are we going?

Lack of clinical data. IV route of administration limits its use.

ANTICOAGULANTS

Warfarin: What do we know?

Warfarin is a vitamin-K antagonist that alters the synthesis of vitamin-K dependent clotting proteins (Factors II, VII, IX, X) and the anticoagulants, proteins C and S. Warfarin interferes with hepatic reductase activity leading to impaired post-translational carboxylation. Warfarin's anticoagulant activity is delayed (4-5 days) as the newly synthesized inactive clotting proteins gradually replace their functional counterparts. Therapy is monitored based on aPT, with a target prolongation to 1.5 times baseline and INR. An INR target range of 2 to 3 is considered optimal. Bleeding complications commonly (20%) occur.

Dogs

Administration of 6mg/day for 2 days following vascular surgery achieved prolongation of aPT and INRs within therapeutic range.[16]

Cats

A pharmacodynamic/kinetic study of warfarin showed great interindividual variations with the average dose needed to obtain a therapeutic range varying from 0.061-0.088mg/kg/day.[17] Close monitoring for dosage adjustment is essential because warfarin's anticoagulant effect is highly variable from one patient to another (enterohepatic recirculation).

Warfarin: Where are we going?

No recent studies. Unpredictable pharmacokinetics and bleeding complications have caused it fall out of favor.

Heparin: What do we know?

Unfractionated heparin (UH) a heterogeneous mixture with 1/3 of molecules possessing a pentasaccharide sequence that binds to antithrombin (AT). Thrombin inactivation requires formation of a ternary complex involving heparin, AT, and thrombin. UH has been the anticoagulant of choice for thromboprophylaxis in human and veterinary medicine. Despite its widespread use, UH has a complex pharmacokinetic profile that produces an unpredictable anticoagulant effect. UH binds to numerous plasma proteins and to endothelial cells, macrophages, and platelets. Due to its extensive binding to macrophages and endothelial cells, low-dose subcutaneous UH has poor bioavailability.

Dogs

Empiric UH therapy varies widely, encompassing SC doses of 50 to 500 IU/kg *q* 6-12 hr. Pharmacokinetic studies in healthy dogs have shown that doses of 250 to 500 IU/kg result in target anti-Xa activities, however bleeding occurred in dogs treated with 500 IU/kg SQ q8 hr.[18] UH therapy is monitored by measurement of aPTT, with dose adjustments to prolong values to 1.5 to 2.5 times baseline. Preliminary reports in dogs with IMHA indicate that high doses (> 300 IU/kg SQ q6hr) are required to attain target anti-Xa levels.[19] A randomized prospective study compared the survival of dogs with IMHA receiving either a standard dose of UH (150U/kg SC q 6hr) or an adjusted dose based on monitoring of anti-Xa activity. A possible survival advantage with adjusted dosing was found.[20]

Cats

Healthy cats showed maximal prolongations of aPTT when given SC injections of UH 100U/kg (1-2 hrs post) and 400U/kg (3-4 hrs post).[21]

Heparin Where are we going?

Surprisingly few clinical studies. Utility in acute thrombosis as a constant-rate-infusion (short duration of effect, antidote). Future studies compare heparin to other anticoagulants.

Low molecular weight heparins (LMWH)

LMWH are produced by depolymerization of heparin. LMWHs bind to AT and accelerate its inhibition of Factor Xa. LMWHs bind poorly to plasma proteins and cells and undergo first order renal clearance. While anti-Xa activity values are useful for defining UH and LMWH pharmacokinetics and comparing different dosage regimens, its *in vivo* anticoagulant effects are complex and not limited to Xa inhibition. A meta-analysis of human trials found that once daily dosing of LMWHs was as effective in thromboprevention as q12h dosing although once daily dosing resulted in 12 to 14 hour periods with a sub-therapeutic anti-Xa level.[22] In people, peak anti-Xa activity appears to correlate more closely to efficacy, however the relationship between target anti-Xa activity and clinical outcome remains to be fully defined.

LMWH: What do we know?

Dogs

The pharmacokinetics (PKs) of 2 LMWH's have been investigated using anti-Xa activity to monitor anticoagulant effect. In dogs, dalteparin has a half-life of 123 minutes, a volume of distribution of 50-70mls/kg BW and an absolute bioavailability of 104% when administered subcutaneously.[23] Enoxaparin, given to dogs at 0.8 mg/kg q6h, produced sustained anti-Xa activity in the range of 0.5-2.0 U/ml.[24] Dalteparin given to dogs at 150 U/kg q8h resulted in anti-Xa activity in the range 0.2 –1.0 U/mL.[25] A number of other LMWHs have been used experimentally in dogs (tinzaparin, nadroparin) and appear to have properties that are similar. In a study of 6 ill dogs receiving dalteparin at 150 U/kg q 12h, anti-Xa activity peaked predictably 3 hours following administration and fell below therapeutic range 12 hours after the dose.[26] Mestre et al reported a thrombosis-protective effect in dogs with an anti-Xa activity of 0.55 U/ml.[27] A study of ill dogs considered at risk for TE received either high dose heparin, low dose heparin or dalteparin, variable Xa levels were obtained and the high dose dogs suffered bleeding complications. [28]

Cats

PK studies of dalteparin and enoxaparin in healthy cats revealed rapid absorption and elimination with peak effects at 2-3 hours post administration. The predicted doses required to maintain anti-Xa activity $\geq$ 0.5 U/mL were 150 IU/kg q4h for dalteparin and 1.5 mg/kg q6h for enoxaparin.[29] A retrospective study of 57 cats with cardiac disease treated with 100 IU/kg dalteparin SID or BID showed that it was well tolerated and infrequently caused clinical bleeding.[30] This study was not designed to test dalteparin's efficacy, however, and anti-Xa activity was not measured. A study in cats with experimentally induced thrombosis showed that administration of enoxaparin at a dose of 1mg/kg SC q12 hrs decreased thrombus formation as compared to controls and that thrombosis inhibition did not correlate with Xa inhibition.

LMWH: Where are we going?

Identify monitoring methods that correlate to clininal anticoagulation (TEG, thrombin generation…) Clinical trials to optimize LMWH therapy for specific canine and feline thrombotic disorders and defined clinical outcome measures.

Dirext Xa inhibitors: What do we know?

Rivaroxaban is a novel, direct factor Xa inhibitor used in people for the prevention and treatment of thromboembolic disorders at a fixed-dose and without the need for laboratory monitoring. Its *per os* administration makes it an attractive anticoagulant for clinical use. It has been shown in vitro to exert Xa inhibition and resultant decreased thrombin generation in both cats and dogs.[30,31] We currently have a study evaluating the pharmacokinetics/dynamics of rivaroxaban in healthy dogs.
A pharmacokinetic study of fondaparinux in cats suggested a dose of 0.2mg/kg SC q12hrs inhibits Xa activity sufficiently to be clinically therapeutic.[32] Fondaparinux must be administered parenterally thus making it a less attractive.

Other anticoagulants are being evaluated experimentally in dogs and cats (direct thrombin inhibitors, recombinant activated protein C, tissue factor pathway inhibitor).

REFERENCES

1.Sharpe et al. AJVR 2010;71:1294-1304, 2.Puglia et al. IVECCS abstract 2010, 3. Brainard et al. Am J Vet Res 2007:68:251-257, 4.Nielsen et al. Vet Clin Path 2007:36;267-273, 5.Weinkle et al. J Am Vet Med Assoc 2005:226:1869, 6. (Smith SA, *et al. J Vet Intern Med* 2003;17(1):73, 7. Hogan et al. J Am Vet Med Assoc 2004;225:1406, 8. Hogan et al. J Vet Res 2004;65:715-719, 9. Brainard et al. Am J Vet Res 2010;71:822-830, 10. Brainard et al. Am J Vet Res 2010;71:822-830, 11. Hamel-Jolette et al. Can J Vet Res 2008;73;73-76, 12. Hogan et al. Am J Vet Res 2004;64:327-332, 13. Boudreaux *et al. Am J Vet Res* 1991;52(12):2000, 14. Niitsu et al. Eur J Pharmacol 2008;579:276-282, 15. Bright et al. Vet Ther 2003(4):35. 16. Monnet et al. Am J Vet Res 2000;61:48-50, 17. Smith et al. J Vet Pharmacol Ther 2000;23:339-344, 18. Mischke et al. Am J Vet Res 2001;62:1887, 19. Breuhl et al. J Vet intern Med 2005;19:399, 20. Helmond et al. J Vet Intern Med 2010;24:597-605, 21. Mischke et al. abstr ECVIM 2010, 22. Dolovich et al. Arch Int Med 2000;160:181, 23. Grebe et al., Berl Muench Tieraerztl Wochenschr 2000;113:103, 24. Lunsford et al. Proceed ACVIM 2005, 25. Mischke et al. Am J Vet Res 2001;62:1887, 26. Charland et al. Méd Vét Qué 2006;35:86, 27. Mestre et al. Thromb Res 1985:38:389-399, 28. Scott et al. IVECCS 2009;74-80, 29. Alwood et al. J Vet Intern Med 2007;21(3):378-387, 30. Smith et al. J Am Vet Med Assoc 2004;225:237, 30. Brainard et al. Abstr ACVIM 2011, 31. Conversy et al. abstr ACVIM 2012, 32. Hogan. Proceedings ACVIM 2010.

MONITORING PLATELET FUNCTION DURING ANTI-PLATELET THERAPY

John M. Thomason, DVM, DACVIM
Starkville, MS

INTRODUCTION

Normal blood hemostasis is a complicated process that involves the vascular endothelium, numerous coagulation factors, and several interacting cells. When exposed to injured or damaged endothelium, either due to normal cell turnover or vessel trauma, platelets undergo several functional responses, including platelet adhesion and aggregation, release of pro-coagulant substances, and surface expression of procoagulant receptors.[1] The formation of a platelet plug during primary hemostasis is the first step in coagulation, and plays a major role in preventing blood loss. Inadequate platelet number and function will increase the risk of hemorrhage; while continual platelet activation, especially in hypercoagulable states, may lead to inappropriate thrombus formation. The risk of thromboembolism has been identified as a significant complication in many commonly encountered diseases, and can be a devastating complication of otherwise treatable conditions. In veterinary medicine, the diseases/conditions that are associated with an increased risk of thromboembolic complications include immune-mediated hemolytic anemia (IMHA), protein-losing nephropathies, hyperadrenocorticism, sepsis, disseminated intravascular coagulation, and systemic inflammatory response syndrome.[2,3] Because of the devastating complications associated with a thromboembolic event, preventative therapy is an essential component of the treatment of hypercoagulable patients. Unfortunately, many of the medications used in veterinary medicine that inhibit coagulation factors, such as heparins, can only be given by injection or are cost prohibitive for long term administration, making anti-platelet therapy the most commonly administered preventative therapy.

PLATELET PHYSIOLOGY

Platelet activation and aggregation is dependent on several pathways that ultimately lead to the formation of a platelet plug. Activation of the platelet surface glycoprotein integrin αIIbβ3 (formerly called the GPIIb/IIIa complex) by thrombin, one of the most potent platelet activators, enhances the interaction between platelets and fibrinogen and stabilizes the platelet plug. However, the direct stimulus of the αIIbβ3 complex may not reach the threshold needed for platelet activation, leading to the need for secondary/amplification pathways that aid in platelet plug development. One of these amplification pathways involves thromboxane A_2 production via the cyclooxygenase (COX) enzyme pathway. Thromboxane A_2 is a prostaglandin that contributes to hemostasis by causing local vasoconstriction as well as promoting platelet aggregation. A second amplification pathway involves the release of adenosine diphosphate (ADP) from dense granules, which in turn activates G protein-coupled receptors (such as P2Y12) located on the platelet surface. Both of these amplification pathways lead to the stimulation of platelet surface αIIbβ3 receptors and binding of fibrinogen to form a platelet plug. These pathways of platelet activation have become targets for anti-platelet medications that inhibit platelet function.

ANTI-PLATELET MEDICATIONS

Aspirin

Aspirin has been used for many years for its anti-inflammatory and analgesia properties, but is also the most commonly used anti-platelet medication. By inhibiting the COX enzyme, aspirin prevents the synthesis of thromboxane A_2 through blockage of the arachidonic acid conversion pathway, and permanently inhibits platelet function.[4] An anti-inflammatory or high dose of aspirin (10 mg/kg twice daily) consistently inhibits platelet function by decreasing thromboxane production.[5] Unfortunately, high dose aspirin can result in gastrointestinal and renal side effects, making this dose undesirable for preventative therapy. Lower doses of aspirin (0.5-1 mg/kg twice daily) can be administered in an attempt to maintain the anti-platelet effects of the medication without the associated side effects. Unlike the higher, anti-inflammatory dose, a low dose of aspirin will inhibit platelet thromboxane synthesis and function without permanently inhibiting COX function of the vascular endothelium, allowing prostacyclin synthesis and vasodilation to continue.[6] Unfortunately, however, low dose aspirin does not consistently inhibit platelet function in all dogs. We recently demonstrated that about one-third of aspirin treated dogs respond to low dose aspirin as expected, with prolongation in PFA-100® closure times, while the remaining dogs either do not respond or inconsistently respond to the low dose.[7]

Thienopyridines

Clopidogrel and ticlodipine are thienopyridines that have platelet inhibitory activities, and function independent of arachidonic acid metabolism. These medications require hepatic metabolism by hepatic cytochrome p450 enzymes to form metabolites that irreversibly block the platelet ADP receptor P2Y12. Both drugs are capable of inhibiting platelet aggregation induced by collagen and thrombin, although as the concentration of these agonists increase, the inhibitory effects are overwhelmed. The majority of the drugs' anti-platelet activity is therefore due to the inhibition of ADP-induced aggregation.[4] Although veterinary experience with these medications is minimal, the use of clopidogrel to prevent thromboembolic complications in IMHA patients maybe as effective as low dose aspirin.[8]

Integrin αIIbβ3 (GPIIb/IIIa) inhibitors

Even though aspirin and thienopyridines are effective at inhibiting platelet function, the effects are only partial because these drugs do not inhibit other platelet agonists (i.e. thrombin) or the final common pathway, the platelet-fibrinogen interaction. Therefore, there is great interest in developing therapies that antagonize the final common pathway by inhibiting the glycoprotein integrin αIIbβ3 (GPIIb/IIIa) complex.[9] This glycoprotein has become the target of several anti-platelet drugs that utilize either monoclonal antibodies against the receptor (abciximab), natural receptor inhibitors found in snake venom (eptifibatide), or synthetic non-peptide receptor antagonists (tirofiban) to inhibit platelet function.[9] With these drugs, when about 80% of targeted receptors have been blocked, platelet aggregation is almost completely inhibited but bleeding times are only mildly prolonged.[10]

EVALUATING PLATELET FUNCTION

Evaluation of platelet function was originally used to screen patients for bleeding disorders and defects in primary hemostasis. These diagnostic methodologies were soon also used to predict which patients would bleed during medical and surgical procedures. More recently, methods of evaluating platelet function has focused on assessing the effectiveness of anti-platelet therapies. With insufficient inhibition of platelet function, patients could remain at risk for thromboembolism, while excessive platelet inhibition could lead to excessive hemorrhage. Since some patients do not adequately respond to anti-platelet therapy (for example, patients with aspirin resistance), evaluating platelet function during therapy is essential for maintaining adequate platelet inhibition. Monitoring platelet function during therapy allows for more accurate adjustments in drug dosing.

There are multiple tests that evaluate platelet function, but deciding which test is the most appropriate requires an understanding of platelet physiology, the pharmacology of anti-platelet therapy, and clinical practicability. There is not a single ideal test that provides all the information necessary to clinically monitor platelet function. Multiple diagnostic tests may be necessary to determine if platelet function has been inhibited or if the patient has not appropriately responded to anti-platelet therapy. Understanding the mechanism of action of anti-platelet therapies plays an important role in deciding which test of platelet function is the most appropriate. One test of platelet function may be appropriate to monitor aspirin-induced platelet dysfunction, while the same test may not provide any valuable information about clopidogrel-induced platelet dysfunction. The clinical practicality of these tests is a major contributing factor in decisions regarding how to best evaluate response to anti-platelet therapy. The ideal test of platelet function should be easy to use and reproducible, and should directly measure platelet function. Some instruments are expensive, and require advanced training, while other instruments are point-of-care and simple to use. Unfortunately, there is not a consensus on which instrument provides the best clinical assessment of platelet function.

BLOOD SAMPLE COLLECTION

Platelets are inherently prone to aggregation, and can become activated by a multitude of agonists and stimulates, including collection technique and the length and temperature of storage. Maintaining high sample quality, with minimal platelet activation, is important during analysis because previously activated platelets will provide an inaccurate assessment of platelet function. To minimize artificial activation, it is recommended to perform a non-traumatic venipuncture with a steady blood flow using at least as 21 gauge needle, ensure the appropriate ratio between the anticoagulant and blood, maintain all blood tubes at room temperature, and perform immediate analysis (ideally within 2 hours).

TESTS OF PLATELET FUNCTION

Historically, the most readily available clinical assessment of platelet function has been the buccal mucosal bleeding time (BMBT) and the toe nail clip. These techniques require making either a small incision on the buccal mucosa or clipping a toe nail and recording how long it takes the bleeding to stop. Unfortunately, these methods have several limitations. Similar to other tests of platelet function, the results can be adversely affected by thrombocytopenia. Also, both of these procedures can create a mild amount of discomfort for the patient, and may require sedation or even anesthesia, especially if the patient is aggressive or fractious. Even though these tests can be performed in most veterinary clinics, and can provide a rough assessment of platelet function, these techniques do not provide objective results, and are too insensitive to monitor drug induced platelet dysfunction.

Platelet aggregometry has been utilized for many years to evaluate platelet function, and has long been considered the gold standard for evaluating platelet function. Aggregometry uses agonists to activate platelets in either platelet rich plasma (turbidometric or light transmission aggregometry) or whole blood (impedance aggregometry), and assesses the ability of platelets to aggregate. With turbidometric aggregometry, platelet function is evaluated by detecting the change in light transmission through platelet rich plasma; the greater the amount of platelet aggregation, the greater amount of light transmission. Impedance aggregometry, which typically uses whole blood instead of platelet rich plasma, evaluates platelet response to several agonists by measuring the electrical resistance in blood that is generated by platelet aggregation. Turbidometric aggregometry is more sensitive than impedance aggregometry to subtle changes in platelet function. Although both turbidometric and impedance aggregometry are capable of detecting drug-associated platelet dysfunction, these methodologies are not routinely used to monitor and adjust anti-platelet therapy. Platelet aggregometry, especially turbidometric aggregometry, is expensive and requires specialized equipment and training, thus making it difficult to utilize in a clinical setting.

The Siemens Platelet Function Analyzer-100 (PFA-100®) is an instrument that evaluates platelet function, but is more clinically useful than aggregometry. This instrument is an *in vitro* platelet function analyzer that measures the time needed to form a platelet plug. Platelet agonists such as collagen, epinephrine and ADP are used to activate platelets under high shear conditions. The instrument measures the time, in seconds, needed to activate platelets and form a plug that occludes an aperture designed to replicate damaged vascular endothelium. The PFA-100® utilizes two types of cartridges with two platelet agonists per cartridge (collagen/epinephrine and collagen/ADP) to evaluate platelet function. The PFA-100® collagen/epinephrine cartridge is used to detect drug-induced platelet dysfunction and the collagen/ADP cartridge is used to evaluate congenital and acquired platelet function defects.[11] Prolongation of the PFA-100® closure time as measured with the collagen/epinephrine cartridge has become established as a sensitive indicator of aspirin-induced platelet dysfunction in both humans and dogs.[12] However, the PFA-100® is insensitive at detecting the inhibitory effects of ADP antagonists. Unfortunately, there are several other limitations with the PFA-100®. According to the manufacturer, a blood sample with less than 150,000/μl platelets does not provide an accurate assessment of platelet function when compared to samples with platelet numbers within normal reference ranges. We have recently established that references ranges can be generated with the collagen/ADP cartridges in dogs with mild thrombocytopenia. However, with great variation in closure times, reference ranges could not be established in samples with a marked thrombocytopenia and with the collagen/epinephrine cartridge. Another limitation of the PFA-100® is that a decreased hematocrit may artificially prolong the closure time. However, when compared to the buccal mucosa bleeding time and the toe nail clip, which are subjective, difficult to standardize, and prone to error if not performed correctly, the PFA-100® provides a better objective clinical assessment of platelet function.

Other whole blood assays that assess platelet function by evaluating platelet aggregation include the VerifyNow® and PlateletWorks® instruments. The VerifyNow® instrument binds fibrinogen-coated beads to activated platelets and measures the light transmitted during platelet aggregation. However, since canine platelets do not consistently respond to the agonists typically used with instrument (arachidonic acid and thrombin receptor activation peptide), this instrument may not provide the most consistent assessment of platelet function in dogs.[13] The second instrument, the PlateletWorks®, indirectly evaluates platelet function by measuring platelet counts before and after activation by several different agonists, and reporting the results as percent aggregation and percent inhibition. The PlateletWorks® assay may be an effective instrument to assess the effects of clopidogrel on feline platelets.[14] However, since there is minimal experience with these instruments in dogs, the clinical effectiveness of these instruments is unknown.

Several instruments claim to evaluate platelet function under low shear conditions by detecting viscoelastic changes in whole blood. These devices use a viscoelastometric system to detect mechanical changes that occur within a clot, and provide information on coagulation, fibrin gel formation, and clot retraction. Instead of simply detecting the time taken to form fibrin strains and a blood clot, these instruments also measure the strength of the clot, the time to maximum clot strength, and fibrinolysis. The three technologies that are most commonly used in veterinary medicine include thrombelastography (TEG), thrombelastometry (ROTEM), and the Sonoclot® coagulation and platelet function analyzer. Unfortunately, these technologies are designed to focus on the formation of a fibrin clot, and although platelets are involved in clot formation and clot retraction, these instruments typically do not provide a sensitive assessment of platelet function and drug-induced platelet dysfunction.

Flow cytometry detects fluorescence from labeled monoclonal antibodies that are bound to surface proteins to identify and quantitate the expression of a particular protein on a cell surface. Unlike aggregometry, where platelet activation is necessary, flow cytometry can evaluate surface protein expression both before and after platelet activation, and thereby monitor the efficacy of anti-platelet therapy. Some of the surface markers that detect platelet activation include P-selectin, phosphatidylserine, and fibrinogen bound to glycoprotein IIb-IIIa.[15,16] Flow cytometry techniques have been developed to identify unactivated and activated platelets as well as microparticles released from activated platelets.[17,18] Even though flow cytometry does not evaluate the platelet's ability to form a platelet plug, understanding which proteins are expressed during clinical diseases and anti-platelet therapy helps determine the activation status of platelets. It was long believed that COX-1 was the only COX isoform expressed in circulating platelets. However, we recently described a flow cytometry technique that identified and quantified the expression of both COX-1 and COX-2 in circulating canine platelets.[5] We have also demonstrated that high and low doses of aspirin are capable of altering the expression of both COX isoforms,.[5,7] potentially leading to a better understanding of how aspirin influences platelet function. Flow cytometry, however, is a technically challenging procedure that is usually limited to specialized research laboratories.

Thromboxane concentration is a reliable indicator of COX function, both in hypercoagulable states and during aspirin therapy. Urinary concentrations of 11-dehydro-thromboxane B_2 (the most abundant urinary metabolite of thromboxane) and 2,3-dinor thromboxane B_2 are increased in human diseases that are associated with a hypercoagulable state.[19] Elevated urinary concentrations of 11-dehydro-thromboxane B_2 have also been identified in dogs with gastric dilatation-volvulus.[19] Measurement of urinary 11-dehydro-thromboxane B_2 has been utilized in previous studies as an indicator of aspirin-induced COX inhibition.[19-21] Levels of thromboxane metabolites in plasma or serum can be artifactually increased due to platelet activation during sample collection and processing. The use of urine to measure thromboxane metabolites avoids the problem of *in vitro* platelet activation associated with blood collection, and is therefore considered to be a more reliable indicator of platelet thromboxane synthesis.[20,22] Urinary assessment of 11-dehydro-thromboxane B_2 and/or 2,3-dinor thromboxane B_2 are useful indicators of *in vivo* thromboxane production, however, the 2,3-dinor thromboxane metabolite is a more sensitive indicator of aspirin induced platelet dysfunction.[19,21,23] In humans, greater than 95% reduction in thromboxane synthesis is required for aspirin to have a sufficient clinical effect on platelet function.[24] Unfortunately, assessment of thromboxane metabolites currently requires access to specialized equipment that is not available in practice, limiting the effectiveness of these assays in a clinical setting. Also, since neither thienopyridines nor αIIbβ3 inhibitors inhibit the COX enzyme, these tests are only useful for assessing the effects of aspirin and other non-steroidal anti-inflammatory drugs on platelet function.

REFERENCES

1. McMichael M. *J Vet Emerg Crit Care* 2005;15(1):1. 2. Goggs R, et al. *J Vet Emerg Crit Care* 2009;19:30. 3. Wells R, et al. *J Vet Intern Med* 2009;23:295. 4. Patrono C, et al. *Chest* 2001;119:39S. 5. Thomason J, et al. *J Vet Intern Med* 2011;25:1106. 6. Rackear D, et al. *J Am Anim Hosp Assoc* 1988;24(1):23. 7. Dudley A, et al. 29th ACVIM Forum (abstract) 2011. 8. Mellet A, et al. *J Vet Intern Med* 2011;25:71. 9. Rosove M. *Best Pract Res Clin Haematol* 2004;17(1):65. 10. Scarborough R, et al. *Circulation* 1999;100:437. 11. Kundu S, et al. *Semin Thromb Hemost* 1995;21:106. 12. Nielsen L, et al. *Vet Clin Pathol* 2007;36:267. 13. Christopherson P, et al. *Vet Clin Small Anim* 2012;42(1):173. 14. Hamel-Jolette, et al. *Can J Vet Res* 2009;73(1):73. 15. Moritz A, et al. *Vet Clin Pathol* 2003;32(1):6. 16. Moritz A, et al. *Am J Vet Res* 2005;66:325. 17. Weiss DJ, et al. *J Vet Intern Med* 2006;20(3):682. 18. Weiss D. *Vet Clin Pathol* 2002;31(2):72. 19. Baltzer W, et al. *Am J Vet Res* 2006;67(1):78. 20. Perneby C, et al. *Thromb Res* 1999;96:427. 21. McConnell J, et al. *Neurosci Lett* 2001;313:88. 22. Catella F, et al. *Proc Natl Acad Sci USA* 1986;83:5861. 23. Hoh C, et al. *Am J Vet Res* 2011;72:1038. 24. Weber A, et al. *Herz* 2008;33(4):287.

USE AND MISUSE OF DIAGNOSTIC PANELS IN CANINE VECTOR BORNE DISEASE TESTING.

Linda Kidd DVM, PhD, DACVIM
Pomona, CA

INTRODUCTION

Vector borne disease is an important cause of morbidity and mortality in dogs. Clinical findings caused by infection include those commonly associated with idiopathic immune mediated disease. Interestingly, like vector borne disease, the incidence of idiopathic immune mediated disease has been shown to be seasonal in some studies.[1,2] It has been hypothesized that many patients with immune mediated disease suffer from undiagnosed underlying infection with vector borne and other infectious agents.[3,4] Inappropriately declaring that vector borne disease has been ruled out in a patient with immune mediated disease because of a "negative tick panel" may significantly impact morbidity and mortality.

"Tick panels" and "vector borne disease panels" offered by diagnostic laboratories or in house test kits give clinicians the ability to test peripheral blood for multiple agents using serology, polymerase chain reaction (PCR) or a combination of the two. Choosing a panel requires the clinician to consider which organisms to test for, and which methodology (PCR or serology) to employ. Knowing the epidemiology and clinical findings that are most commonly associated with each organism help determine which agents should be included in a diagnostic panel. The sensitivity of serology as compared to PCR varies with characteristics of the host, the assays, and pathophysiologic characteristics of the organism. Therefore, using either PCR or serology exclusively may overlook the presence of infection. Knowing when clinical signs occur, when and if organisms circulate in peripheral blood, and whether they circulate in high or low number helps determine whether serology and/or PCR, or acute and convalescent serologic testing are most appropriate for an individual patient.

WHICH ORGANISMS SHOULD BE INCLUDED IN A PANEL?

Knowing epidemiology and common clinical findings helps determine which organisms should be included in a panel for a particular ill patient. This consideration may be more important for serologic testing, as most serologic panels are not as comprehensive as widely available PCR panels. It is important to be sure that agents that are commonly associated with the clinical findings for a particular patient are included in diagnostic testing. For example, in a dog with IMHA, a negative result for a "tick panel" that tests for antibodies to *Borrielia burgdorferi*, *Rickettsia rickettsii* and *Ehrlichia canis* would not effectively rule out vector borne disease as a cause for IMHA, as *E. canis* is the only one of these three organisms that has been associated with IMHA in dogs.[5] *Babesia*, *Anaplasma*, *Bartonella* and haemotropic *Mycoplasma* species would also be important to include in diagnostic testing for a dog with IMHA.[6,7,8,9,10] Furthermore, serologic crossreactivity among different species within a genus does not always occur. It is important to be sure that all the species of a genus that are likely in a particular clinical scenario are included in a serologic panel. For example, a Staffordshire terrier with hemolytic anemia due to infection with *B. gibsoni* may test negative for antibodies to *B. canis*. Knowing that *B. gibsoni* is a likely cause of hemolytic anemia in this breed, helps guide the clinician to use a panel that includes this agent. Consideration of geographic locale is also important. For example, it would be theoretically more important for a clinician practicing in Wisconsin to include testing for *A. phagocytophilum* for a thrombocytopenic dog with no travel history than a clinician practicing in South Dakota. Summaries of common clinical signs and geographic distributions of infectious agents are available.[11] Comprehensive testing can also be beneficial, because the geographic distribution of vector borne agents is expanding, as are the clinical descriptions of diseases associated with these agents. Restricting diagnostic testing to a few organisms may overlook infection in some patients.

WHICH METHODOLOGY SHOULD I USE, PCR OR SEROLOGY?

Most serologic tests for vector borne disease agents indirectly document infection by detecting antibodies. Whole organisms or select antigens are used as targets for antibodies in the assays. An important limitation of serology is that for most assays, a positive result for a single sample indicates exposure but not necessarily active infection for some agents. This makes it difficult to determine if clinical findings can be attributed to infection, particularly in endemic areas. Despite this limitation, serology has been the mainstay of diagnostic testing for vector borne disease for decades due to technical difficulties associated with isolating and culturing these organisms from patient samples. PCR detects DNA of organisms themselves without the need for culture. Recently, advances in molecular biology have made PCR panels easy to perform and widely available. PCR makes it possible to directly test for the presence of multiple organisms in peripheral blood and other samples with high sensitivity and specificity, assuming appropriate laboratory controls are in place. Despite these advances in molecular biology, serology still has an important role in the diagnosis of vector borne disease. Consideration of the pathophysiology of the organism and the strengths and limitations of the assays in the context of the clinical presentation can help the clinician determine which methodology is most appropriate and whether combination testing is indicated. Specifically, considering whether the organism is likely to be circulating in blood at the time of presentation, and the significance of a positive or negative test result can help determine whether PCR or serology should be used or if additional testing is warranted.

Is the organism likely to be present in the blood sample at the time of presentation?

In order for PCR to be positive, the organism has to be present in the sample. Many current PCR assays are very sensitive and can reliably detect the presence of only one organism in a sample. Therefore, if organisms are circulating in blood in high numbers at the time they cause clinical signs, it is very likely that a PCR test will detect the organism. However, clinicians should keep in mind that the aliquot of blood used to extract DNA for PCR is small (eg. 200uL) and the volume of extracted DNA used in the assay is even smaller (eg. 5uL). Therefore the organism's DNA might not be present in a test sample even if the organism is present in peripheral blood in high numbers, and the PCR assay is exquisitely sensitive. For example, dogs infected *A. phagocytophilum* exhibit clinical signs acutely after infection. *Anaplasma phagocytophilum* infected granulocytes circulate in peripheral blood and can be present in relatively high numbers during this acute phase. Therefore PCR testing of peripheral blood is sensitive.[12] However, if the organism does not happen to be present in a particular aliquot used for the assay, or the patient has received medication such as doxycycline that decreases circulating copy number, PCR can be "falsely" negative. Additional PCR testing (if doxycycline has not been given) of the acute sample or acute and convalescent serologic testing can confirm active infection if the initial PCR is negative.

Documenting seroconversion is particularly useful for diagnosing acute infection with agents that circulate in peripheral blood in low copy number. In this scenario, PCR is inherently insensitive due to the pathophysiology of the organism. For example, *R. rickettsii* causes acute illness in dogs, and these endotheliotropic organisms circulate in low copy number in peripheral blood. Therefore, PCR is not as sensitive as is demonstrating a four-fold change in titer.[13]

Serologic testing can also help diagnose infection with agents that cause chronic infection. Interestingly, clinical signs of Lyme borreliosis do not occur in experimentally infected dogs until 2-6 months after tick exposure. Organisms do not circulate in peripheral blood in high numbers.[14] Consequently, serologic testing rather than PCR is routinely used to document exposure to *B. burgdorferi* using peripheral blood samples. *Ehrlichia canis* can cause clinical signs during acute or chronic infection in dogs. This organism may circulate in blood in low copy number and/or intermittently during chronic infection. Similarly, *Bartonella* species also may circulate only intermittently or in low copy number during chronic infection.[8] Serology can indicate exposure if PCR is negative and thereby facilitate diagnosis.[8,15] Alternatively, to document infection with *Bartonella*, sterilely collected samples of whole blood can be grown in enrichment media prior to PCR to increase sensitivity.[8,16] In summary, for organisms that circulate in low copy number or intermittently at the time clinical signs occur, serology may facilitate diagnosis.

What is the significance of a positive result in my patient?

Assuming appropriate laboratory controls are in place, a positive PCR result indicates active infection or the presence of circulating dead organisms. Response to therapy will help determine if presenting clinical signs can be attributed to infection or if coinfection or other underlying disease should be considered. PCR assays can be positive at the level of genus or species, depending on the specific DNA sequence that the primers target. If sequencing is not performed to verify the infecting species, it is possible, albeit unlikely, that a species is actually different from the reported species. This may have implications with regard to treatment and prognosis, among other factors. Considering whether or not infection with the specific agent makes sense given the signalment, clinical signs and geographic locale is helpful. If the results don't make sense for a patient, consulting the laboratory and further testing, including DNA sequencing can be considered.

When considering the significance of a positive titer on a single sample it is again useful to consider the pathophysiology of the organism in the context of the individual patient because a positive serologic test may indicate exposure and not necessarily active infection. For example, *R. rickettsii* and *A. phagocytophilum* primarily cause acute rather than chronic disease in dogs. However, previous infection may result in long lived antibody titers.[12,17] If serologic testing is positive for these agents and clinical signs have been present for many months, previous exposure, but not active infection is likely. In this case co-infection with other vector borne agents, the presence of cross reacting antibodies, or the presence of other underlying disease should be considered. Additional testing of banked or newly obtained samples using PCR and/or serology may be useful in such cases.

Serologic crossreactivity among species within a genus and between genera also affects the interpretation of a positive serologic test. As is the case with PCR, considering whether or not infection with the specific agent makes sense given the signalment, clinical findings and geographic locale is helpful. For example, there is extensive serologic crossreactivity among spotted fever group (SFG) *Rickettsia*. In general the species of infecting *Rickettsia* is been presumed based on geographic locale, so if a dog from the South Eastern US seroconverts to *R. rickettsii,* Rocky Mountain spotted fever (RMSF) becomes the presumptive diagnosis. However, other species of SFG *Rickettsia* that infect people are present in the US and it is also likely that these organisms infect dogs and induce disease.[18,19] Furthermore, exposure to nonpathogenic SFG *Rickettsia*, some of which are common endosymbionts in ticks, may be a common cause of positive titers, particularly low and persistent titers, to *R. rickettsii* in dogs.[20] Serologic cross reactivity with SFG *Rickettsia* also appears to occur in dogs infected with *Bartonella henselae*.[21] Therefore a single positive titer to *R. rickettsii* may not represent infection with that organism. PCR testing on banked acute EDTA samples can facilitate species verification and acute PCR and convalescent serology can help confirm or rule out the presence of active infection in seropositive patients.

When interpreting results of positive serologic assays in individual patients, clinicians should also keep in mind that infection with novel species and expanding geographic distribution of known species occurs, as this phenomenon is responsible for many outbreaks of emerging infectious disease. For example, the geographic distribution of RMSF in the United States has increased beyond the distribution of its primary tick vectors, *Dermacentor variabilis* and *D. andersonii*. *Rhipicephalus sanguineous,* a ubiquitously distributed tick, recently caused an outbreak of RMSF with associated mortality in a non-endemic area of Arizona.[22] Retrospectively, it was shown that infection existed in the dog population before the fatal outbreak in people.[23] This illustrates how a positive titer in a nonendemic region does not necessarily represent a false positive result and may represent disease emergence.

What is the significance of a negative result in my patient?

The pathophysiology of an infecting agent can also affect the overall sensitivity of PCR testing. As alluded to previously, PCR may have a high sensitivity in that it can reliably detect one copy of target DNA in a sample (absolute sensitivity), but a low sensitivity with regard to determining whether an organism is infecting a patient (clinical sensitivity) if organisms normally circulate only transiently during a particular phase of infection or in low numbers in peripheral blood. Thus even the most sensitive PCR may not indicate that an animal is infected with a particular agent, and the presence of infection cannot be definitively ruled out based on a negative result. Therefore, repeating PCR or performing serologic testing is necessary to rule out infection in some cases. PCR may also be negative because the infecting species is not targeted by primers. Primers that target genus rather than species facilitate testing for a broad range of species in an individual patient.

Serologic testing on a single sample may also be negative in the face of infection if the organism causes clinical signs in the acute phase of disease, prior to the time when antibodies are formed. For example, serology might not indicate exposure in dogs acutely infected with agents like *R. rickettsii*, *E. canis*, *Babesia* species and *A. phagocytophilum,* because clinical signs can sometimes manifest prior to seroconversion.[12,20,15,24] Convalescent testing documenting seroconversion or PCR on acute samples may confirm active infection in these acutely infected patients.

Surprisingly, and unfortunately, serologic testing can also be negative during chronic infection. For some organisms, such as *Babesia* and *Bartonella*, antibodies are not detectable in some chronically infected patients even after seroconversion should have already occurred.[8,25] PCR is necessary to document infection in these cases.

Sensitivity can also differ among different types of serologic assays testing for the same organism. For example, some dogs actively infected with *E. canis* test seropositive using IFA and seronegative using ELISA. This may be related to differences in the nature of antigens used in the respective assays.[26]

Finally, the presence of some organisms cannot be tested for using serology because they can't be grown in culture, they have never been isolated, or a commercial assay is not available. Examples would include haemotropic *Mycoplasmas*, *E. ewingii* and *Babesia conradae*. PCR or microscopic examination of blood smears is needed to document infection with these agents.

THE IMPORTANCE OF BANKING SAMPLES.

In order to enable appropriate testing using PCR and serology, both serum and EDTA samples can be obtained at initial presentation. Based on the presenting scenario, a decision can be made to determine if testing using serology or PCR or both is most appropriate initially. If initial PCR or serologic testing is negative, and both testing methodologies were not employed, banked samples can be accessed to test using alternative methodology. Convalescent serologic testing can be considered in order to confirm active infection, particularly when there is an acute history of illness. Repeating PCR on initial or additional samples may also be advantageous in some cases.

SUMMARY

Consideration of the epidemiology, associated clinical findings, and pathophysiology of vector borne disease agents is helpful in determining which organisms to include in testing and which methodologies should be employed for a given presenting scenario. Combination testing using PCR and/or acute and convalescent serologic testing may be necessary to accurately diagnose infection with vector borne agents in some patients and should be considered. Most importantly, clinicians should remember that a "negative panel" using either serology or PCR, does not necessarily rule out the presence of underlying vector borne disease in a patient with clinical signs of immune mediated disease.

REFERENCES

1. Klag et al JAVMA 1993 783. 2. Weinkle et al JAVMA 2005 1869. 3. Breitschwerdt EB, personal communication. 4. Greene CT, personal communication. 5. Frank et al Am J Vet Res 2010 1443. 6. Adachi et al J Vet Med Sci 1992 1081. 7. Carli et al Vet Parasit 2009 51. 8. Breitschwerdt J Vet Emerg Crit Care Soc 2010 8. 9. Mazepa J Am Anim Hosp Assn 2010 405. 10. Messick Vet Clin North Amer 2003 1453. 11. Greene CT Inf ectious diseases of the dog and cat. 2012. 12. Carrade J Vet Inter Med 2009 1129. 13. Kidd Vet Micro 2008 294. 14. Greene CT et al. Infectious diseases of the dog and cat 2012 455. 15. Harrus et al. Infectious diseases of the dog and catt 2012 227. 16. http://www.galaxydx.com. 17. Breitschwerdt Am J Vet Res 1990 1312. 18. Beeler et al Am J Trop Med Hyg 2011 244. 19. Nicholson Trends Parasit 2010 205. 20. Breitschwerdt Am J Vet Res 1987 1436. 21. Solano-Gallego Vet Res 2004 585. 22. Demma Ann NY Acad Sci 2006 342. 23. Nicholson Ann NY Acad Sci 2006 519. 24. Fukumoto et al J Vet Med Sci 2001 977. 25. Birkenheuer J Am Anim Hosp Assn. 1999 125. 26. Chandrashekar Am J Vet Res 2010 1443.

POLYPRENYL IMMUNOSTIMULANT: ADJUNCT FOR TREATMENT OF FELINE RHINOTRACHEITIS AND FELINE INFECTIOUS PERITONITIS

Alfred M. Legendre, DVM, MS DACVIM
Knoxville, TN

INTRODUCTION

Polyprenyl Immunostimulant is a plant based biological agent that is believed to modify the immune response through its action on Toll-like receptors. The agent directs the immune response toward a cell mediated immune response by up regulating Th-1 cytokines. Cell mediated cytotoxicity is important to the destruction of virus infected cells thereby reducing the viral load. Preliminary studies supported benefit in cats with respiratory conditions in a shelter environment. Further studies have been done to validate the use of polyprenyl immunostimulants in cats.

SAFETY STUDIES

Safety studies were done dosing over 350 normal cats in 10 states to determine the safety of the polyprenyl immunostimulants and to meet the US Department of Agriculture (USDA) safety requirements. Many of the cats were kittens of a similar age to the kittens in the rhinotracheitis study. Other than some cats not liking the taste of the agent no significant adverse events were noted. Cats have been dosed with a 10X dose of Polyprenyl Immunostimulant with no adverse effects and mice have been given 100X doses without adverse effects.

FELINE HERPES VIRUS INFECTIONS- RHINOTRACHEITIS

Specific pathogen free kittens 6-10 weeks of age were challenged with a pathogenic strain of feline rhinotracheitis obtained from the USDA. The virus challenge was given intranasally and in the eyes. This is the same strain of virus used to test commercial vaccines used to protect against rhinotracheitis. Ten kittens received placebo and 10 received Polyprenyl Immunostimulants orally twice daily. The kittens were evaluated daily for nasal discharge, nasal obstruction, conjunctivitis, dehydration, weight loss and salivation. A point system was established to quantify the severity of clinical signs. The observers were blinded to the treatment and placebo group. The kittens were evaluated once a day for seven days before challenge and for 21 days after virus challenge. There was a statistically significant difference in the two groups for conjunctivitis and rhinitis. The polyprenyl treated group had a shorter duration of clinical signs and less severe clinical signs.

FELINE INFECTIOUS PERITONITIS

Initial studies supported a benefit to using Polyprenyl Immunostimulant in the treatment of the "dry" form of Feline Infectious Peritonitis (FIP). No benefit was noted in treatment of the effusive or "wet" form of FIP. With the support of the Winn Feline Foundation a study was initiated to assess the benefit of treatment of "dry" form FIP cats with the Polyprenyl. An article was published in 2009 on preliminary findings in 3 cats with dry form FIP. One of these initial study cats, is still alive more than 5 years after diagnosis. The diagnosis of FIP in that cat had been made by biopsy of a granulomatous lesion in the abdomen that had areas of coronavirus antigen detected by immunofluorescence. The article generated many calls and emails from owners asking that their cats be included in the study. We worked with the primary care clinicians and the veterinary specialist on qualifying the cats to enter into the study.

The diagnosis of "dry" form FIP can be challenging. Criteria for inclusion included an adequate history and diagnostic work-up that should have identified other diseases that could mimic FIP. Definitive cases included histology or histopathology that identified a pyogranulomatous reaction in a tissue without evidence of fungal or other organisms in addition to the other criteria specified. Probably cases of FIP did not have cytology or histopathology but had a prolonged history of lethargy, fever, weight loss and unresponsiveness to antibiotic therapy. Some cats had clinical signs of uveitis, others had neurologic signs, a number has lymphadenitis of mesenteric lymph nodes detected on ultrasound, pyogranulomatous involment of the intestinal tract at the ileocolic junction was noted in some cats. Hyperglobulinemia and hypoalbuminemia were criteria for inclusion as was a moderate to high coronavirus antibody titer. Each case was evaluated to verify that the work up was extensive enough to have identified other conditions. All cats entered into the study were started on polyprenyl immunostimulants orally three times a week. The cats were evaluated monthly. There was not a placebo group in this study. In a disease generally recognized as 100% fatal, it was not considered ethical to have a placebo group.

Over 100 cats with definitive and probable "dry" form FIP were enrolled in the study. Over 50 cats were eventually in the study group. Cats that were lost to follow up, cats that died before receiving the polyprenyl and cats that did not survive 7 days after starting the treatment were excluded. It was not considered realistic to include cats dying before 7 days because an immune modulator would not be expected to rapidly reverse an advanced disease process. We will do our final tabulation of the data when the last cat entered into the study has died or has survived for a year. At this time, it appears that about 25% of cats in the study have survived for 6 months or longer. There is little information about survival times for cats with "dry" form FIP that were untreated or treated with steroids and antibiotics or given feline interferon.

It appears that Polyprenyl Immunostimulant improves the survival times in cats with "dry" form FIP. It appears to improve well being of the cats and improves appetite. This is not the final answer to treatment of FIP but perhaps a step in the right direction. An antiviral treatment in conjunction with Polyprenyl Immunostimulant should improve clinical outcome.

REFERENCES

1. Legendre AM, Bartges JW: Effect of Polyprenyl Immunostimulant on the survival times of three cats with the dry form of feline infectious peritonitis. Journal of Feline Medicine and Surgery (2009) 11, 624-626.

HISTOPLASMOSIS: CURRENT DIAGNOSTIC AND TREATMENT STRATEGIES

Audrey K. Cook, BVM&S, MRCVS, DACVIM, DECVIM-CA
College Station, TX

INTRODUCTION

Histoplasma capsulatum is a soil-borne dimorphic fungus, found in temperate and sub-tropical areas. Within the USA, cases are most often reported around the Mississippi, Ohio and Missouri Rivers. Infection usually occurs via inhalation of the microconidia, and is followed by hematogenous and lymphatic spread.[1] The gastrointestinal (GI) system may be the portal of entry in dogs. Many mammalian species are susceptible hosts and clinical disease is reported in both dogs and cats. Any organ system can be affected, resulting in a granulomatous inflammatory reaction. Common sites of infection in dogs are the lungs, GI tract, lymph nodes, liver, spleen, eyes and bone marrow. The distribution in cats is slightly different, as GI involvement is rare in this species. Affected animals are often young, without evidence of underlying immunodeficiency. This finding contrasts with human medicine, in which the majority of patients have AIDS or other reasons for a compromised immune response.

The incubation period is relatively short, at 2-3 weeks. Many infections are likely transient, clear spontaneously and are essentially sub-clinical. In patients who fail to self-clear, clinical signs may be non-specific, but include anorexia, depression and weight loss. Patients with pulmonary involvement may present with overt dyspnea or (less commonly) persistent cough. Dogs with GI involvement may have chronic diarrhea, often with hematochezia. Physical examination may reveal pallor, fever and poor body condition, Lymphadenopathy or organomegaly may be noted. Ocular involvement should be suspected if uveitis or chorioretinitis is evident.

Clinicopathological findings are similarly non-specific, and may include non-regenerative anemia, hypoalbuminemia and hyperglobulinemia. Liver enzymes may be increased in patient with hepatic involvement. Hypercalcemia may occur but is relatively uncommon in both species.

The classic presentation for an affected patient is therefore a young to middle aged dog or cat, in an endemic area, with a non-specific febrile disease. The index of suspicion is increased by the presence of an interstitial pulmonary pattern on thoracic radiographs, organomegaly, lymphadenopathy or uveitis. GI histoplasmosis should be considered in any dog with chronic diarrhea, particularly if the history suggests large bowel involvement.

DIAGNOSIS: HISTORICAL APPROACH

Establishing a diagnosis of histoplasmosis traditionally relies upon demonstration of the organism in circulating monocytes or within phagocytic cells in affected organs.[1] Cytological identification of the organisms is fairly straight forward, as ovoid or round yeast-like structures, 2-5 µm in diameter, with a central, spherical lightly basophilic body surrounded by a clear halo, may be seen with standard modified Wright stains. In contrast, histopathological identification requires special stains, such as periodic acid-Schiff or Gomori's silver. Organisms may be missed with routine H and E stains. Although organisms may be noted within macrophages on a blood smear examination, the collection of appropriate cytological or histological samples may require sedation or anesthesia, and carry a variable degree of risk (including hemorrhage, iatrogenic infection, pneumothorax, etc.).

Serological tests are of limited use in the diagnosis of histoplasmosis in both dogs and cats. False negatives are a particular issue, with several studies documenting seropositivity in <50% of affected animals.[2-5] This may reflect slow seroconversion; studies in human patients suggest that 2-6 weeks are needed for a measurable antibody response.[6] Conversely, prior exposure may result in false positive serological results in cats and dogs with clinical disease due to other disorders.

Fungal culture may be performed, but isolation and identification of the organism can take up to 2 weeks. In addition, the organism is pathogenic in the cultured state, and therefore poses some risk to laboratory personnel.

ANTIGEN TESTING

In patients with active infection, *Histoplasma* antigens are released into affected tissues and enter body fluids including plasma, urine, and cerebrospinal fluid. In the last decade, detection of cell wall galactomannan antigen in blood, urine or other body fluids has been used routinely to identify disseminated histoplasmosis in human patients.[7] Antigen assays have been shown to be a reliable diagnostic modality in this species, with good sensitivity and specificity. In addition, these tests may be used to monitor response to therapy or indicate the onset of relapse in human patients. The currently available antigen assay uses an enzyme-linked immunoassay method, based on a murine-derived monoclonal antibody (MAb) against *Histoplasma capsulatum.* In short, a fixed quantity of the monoclonal antibody is incubated with the test fluid (which may be urine, cerebrospinal fluid, serum, bronchoalveolar lavage fluid). If antigens are present, these will bind with the MAb and antigen:antibody complexes will form proportionate to the amount of antigen in the sample. The MAb-sample is then incubated with a known concentration of *H. capsulatum* antigen on an ELISA plate. The MAb-antigen complexes will be washed away, whilst the free MAb can be determined using standard ELISA methodology. Results are then compared to a standard curve based on known quantities of *H. capsulatum* antigen.

In human patients, antigen assays appear to be most useful in individuals with acute disseminated disease, and less sensitive in those with chronic localized infections. Cross-reactivity has been reported in human patients with a variety of other disseminated fungal infections, including blastomycosis, paracoccidioidomycosis, penicilliosis marneffei, and less frequently, aspergillosis and coccidioidomycosis.[7] In the United States, blastomycosis is the most common cause of cross reactivity, and approaches 100% in dogs and humans.[8,9] In areas endemic to both infections, other diagnostic methods would be needed to differentiate the two.

There is very little published data regarding the use of the Histoplasma antigen tests in dogs and cats. This author has evaluated antigen testing in cats (JFMS: manuscript in press) and found the sensitivity to be 94.4% (using a gold standard of cytological/histopathological identification of the organism). Determination of specificity was difficult, as none of the antigen positive cats underwent a necropsy examination. Based on this case series, we felt that urine testing may be superior to serum testing, although recent data in human patients has shown that pre-treatment of serum with ethylene diamine tetraacetic acid at 100°C improves sensitivity for detection of antigenemia. It may be prudent to test both urine and serum to maximize sensitivity.

TREATMENT

Long-term antifungal therapy is the mainstay of treatment for affected patients, along with supportive care. As fungal cultures are rarely performed on veterinary patient with histoplasmosis, susceptibility information is not available and empirical treatment is necessary. It is also helpful to remember that numerous factors will influence patient outcome, including host immune status, drug susceptibility, location of infection, time between onset of signs and initiation of therapy, and effective administration of the most appropriate dose of the most potent and safe antifungal drug. Few of these variables are under the control of the clinician. In addition, many antifungal drugs have marked variability in blood concentrations due to inconsistent absorption, metabolism, elimination or interaction with concomitant medications.

Good responses have been reported with itraconazole, at 5mg/kg q12 hrs); some sources suggest 10 mg/kg q12-24 hrs.[1,3] Itraconazole is available as a branded product (Sporanox®, Janssen) and various generics. Although occasional reports in the human literature and anecdotal comments in the veterinary community suggest that generic products may be less efficacious, larger studies indicate that FDA approved generic formulations can be safely selected.[10,11] There is a distinct difference however between an FDA approved generic agent and a product compounded from bulk ingredient. Without FDA oversight, drug content and bioavailability cannot be assured, and compounded products should be used with extreme caution. One recent study in penguins indicated that peak serum concentration and area under the curve were significantly lower for compounded itraconazole when compared to a generic medication.[12]

Bioavailability of itraconazole depends upon the formulation, and is substantially higher for the liquid product containing cyclodextrin (i.e, Sporanox). This has a 30% greater AUC than the branded capsule when the same dose is administered.[13] GI uptake is also impacted by gastric events. The liquid is best given on an empty stomach about 1 hour before eating, and it not significantly impacted by gastric pH. In contrast, the bioavailability of the capsule is substantially enhanced by administration with a fatty meal, and is reported to be around 55% in healthy human volunteers taking the brand product. Absorption of the capsule form is decreased by concurrent administration of a gastric acid suppressor, but tends to be more unpredictable even under ideal circumstances than the cyclodextrin liquid formulation.

Numerous drug interactions are reported with itraconazole, due to its inhibition of CYP3A4. These may result in increased plasma concentration of the other medication, or a change in the pharmokinetics of the itraconazole. Increased plasma concentrations of itraconazole can occur with macrolide antibiotics (such as erythromycin), whereas decreased levels are expected with concurrent use of CYP3A4 inducers, such as anticonvulsants.[13]

Fluconazole (2.5-10 mg/kg q12-24 hrs) may be a more appropriate choice in patients with ocular or neurological involvement, due to superior penetration of affected organs. Although a direct comparison between the efficacy of itraconazole versus fluconazole for companion animals with histoplasmosis has not been reported, similar overall outcomes were described for the two agents in dogs with blastomycocis.[14] Treatment was longer for the fluconazole group, but overall cost was substantially less. However, in human patients with disseminated histoplasmosis and AIDS, treatment outcomes with fluconazole were poorer than for itraconazole.[15] Unlike the other triazole agents which are cleared by the liver, fluconazole undergoes renal excretion. The incidence of hepatotoxicity seems to be fairly similar for these two agents. Human studies have suggested that toxicity is often related to serum concentrations and may be mitigated by dose reduction.[16]

In recent years, new triazole drugs have become available, namely voricoanzole and posaconazole. Voriconazole (Vfend®, Pfizer) is a broad-spectrum antifungal agent, with enhanced efficacy against *Aspergillus* spp. It has excellent oral bioavailability, with 99% reported in human volunteers.[13] Despite this, plasma levels are still unpredictable, and appear to be impacted by age and hepatic status. Posaconazole (Noxafil®, Schering-Plough) is the most recently approved triazole, and has enhanced efficacy against filamentous fungi. At the present time, based on evidence in the human field there is little reason to reach for either agent in a dog or cat with histoplasmosis, unless standard treatment is unsuccessful and resistance is suspected.[15]

All of the triazoles (with the possible exception of fluconazole) are known to have substantial inter- and intra-individual variability of concentration in the blood, and therapeutic drug monitoring is routinely performed in human patients.[13] Indeed, efficacy is clearly impacted by blood concentrations in this species, and the cost of monitoring is quickly offset by improved outcomes. Itraconazole has a relatively long half-life in people (>1 day), so it takes about 2 weeks days to reach steady state. As a general guide, doubling the dose results in approximately a 3x increase in plasma concentrations.

The pharmacology diagnostic laboratory at Auburn University College of Veterinary Medicine now offers an assay for itraconazole using an HPLC method. The fee as of Jan 2012 for this service is $75.00. Either serum (no separator tubes) or plasma can be used, and information regarding diagnosis, dose and concurrent medications should be provided. Sample timing is not critical. Based on human recommendations, random concentrations of at least 1.0 µg/ml are probably appropriate.[15]

Amphotericin B may be helpful in severely affected patients, at 0.5 mg/kg IV every 48 hrs until a total dose of 8-10 mg/kg (dog) or 4-8 mg/kg (cat) is reached. Amphotericin B is a macrolide polyene drug with a broad spectrum of anti-fungal activity. It appears to be primarily fungistatic, but may be cidal against some organisms depending on drug concentration. Significant nephrotoxicity has been reported following administration of this agent, although the more expensive lipid based formulations appear to be less toxic with improved tissue penetration.[17] There has been some controversy in the human literature about the concurrent use of amphotericin B with azole drugs, as the former binds to ergosterol within the fungal cell membrane causing leakage of cellular contents, whilst the latter inhibit the synthesis of ergosterol.

Terbinafine (Lamisil®, Novartis) is commonly used in human patients, particularly for toe-nail fungus. This is a squaline epoxidase inhibitor and may work synergistically with the azole drugs. It is generally well tolerated by canine patients and positive results have been reported in dogs with *Malassezia* dermatitis and one dog with pythiosis.[18,19] Doses reported ranged from 5.5 mg/kg BID to 30 mg/kg SID. The author is not aware of any published reports describing the use of this agent in veterinary patients with histoplasmosis, although one study describing the pharmacokinetics in normal dogs suggests that it would be helpful in these cases.[20]

Caspofungin (Cancidas®, Merck) is an echinocandin antifungal agent. This is a new class of anti-fungal agent, and works on the fungal cell wall by inhibiting glucan synthesis (this contrasts to many other antifungals which target the cell membrane). This homopolysaccharide is a major component of the cell wall of many pathogenic fungi, but is absent from mammalian tissues. It plays a key role in osmotic stability and cell growth /division. Caspofungin may be synergistic with both amphotericin B and the triazole drugs and does not impact the cytochrome P450 enzyme system.[21] It is licensed in people for invasive and refractory aspergillosis and candidiasis. There are conflicting reports regarding the efficacy of this agent in murine models of histoplasmosis, and the author is not aware of any studies in companion animals.

MONITORING

Traditionally, clinical status and resolution of visible lesions (such as pulmonary nodules) have been used to guide therapy. Treatment generally lasts >6 months, and azole therapy administered for at least 1 month beyond apparent 'cure'. In human patients, diminishing antigenemia / antigenuria are known to correlate with response to treatment, and antigen levels are expected to fall below the detection threshold when therapy is complete.[9] Although we do not have similar published data for dogs and cats, personal experience suggests that serial antigen testing may be a reliable way to document resolution. Serial urine antigen assay results were available for a small group of cats (n=7) diagnosed with histoplasmosis at TAMU (unpublished data). All seven cats were treated with oral itraconazole (initial dose: 9.4-10mg/kg/day), +/- terbinafine (n=2; initial dose: 10mg/kg/day), +/- amphotericin B (n=1). The urine antigen concentration had decreased in all seven patients when first rechecked at 5-17 weeks post diagnosis, and was negative in all cases within 6 to 20 weeks from initial diagnosis. Measurement of a baseline antigen assay is advisable, even in patients with cytological confirmation of disease, as repeated testing provides an objective indication of patient response. In general, azoles are often administered for 2-3 weeks before substantial clinical response is noted, so a drop in antigen concentration after 4-8 weeks of therapy would be a positive prognostic indicator.

OUTCOME

The overall prognosis for dogs and cats with histoplasmosis is fair. However, prolonged therapy is necessary and proactive supportive care (such as feeding tube placement) is important. Clients must be carefully counseled about the expected cost of treatment and be prepared for a slow recovery and the possibility of relapse.

REFERENCES

[1]Bromel C, et al. **Clin Tech Small Anim Pract** 2005;20:227. [2]Wolf AM. **J Vet Intern Med** 1987;1:158. [3]Hodges RD, et al. **J Vet Intern Med** 1994;8:409. [4]Davies C, et al. **J Am Anim Hosp Assoc** 1996;32:380. [5]Johnson LR, et al. **J Am Anim Hosp Assoc** 2004;40:165. [6]Wheat LJ. **Trends Microbiol** 2003;11:488. [7]Wheat LJ, et al. **Am J Med** 1989;87:396. [8]Spector D, et al. **J Vet Intern Med** 2008;22:839. [9]Connolly PA, et al. **Clin Vacc Immunol** 2007;14:1587. [10]Pasqualotto AC, et al. **Int J Antimicrob Agents** 2007;30(1):93. [11]Veronin MA, **J Pharm Health Serv Res** 2011;2(3):135. [12] Smith JA, et al. **J Zoo Wildl Med** 2010;41(3):487. [13]Andes D, et al., **Antimicrob Agents Chemother** 2009;53(1):24. [14]Mazepa ASW, et al. **J Vet Intern Med** 2011;25(3):440. [15]Wheat LJ, et al. **Clin Inf Dis** 2007;45:807. [16]Lestner JM, et al. **Clin Inf Dis** 2009;49(6):928. [17]Johnson PC, et al. **Ann Intern Med** 2002;137(2):105. [18]Guillot J, et al. **Vet Derm** 2003;14(3):153. [19]Hummel J, et al. **Med Myc** 2011;49(5):539. [20]Sakai MR, et al. **Vet Derm** 2011;22(6):528. [21]Letscher-Bru V, et al. **J Amtimicrob Chemother** 2003;51(3):513.

CANINE PARVOVIRUS TYPE: DOES IT IMPACT PREVENTION, DIAGNOSIS OR PROGNOSIS?

Julia K. Veir, DVM, PhD, DACVIM (SAIM)
Fort Collins, CO

ETIOLOIGY AND PATHOGENESIS

Parvoviruses are single stranded, non-enveloped, DNA viruses. Due to their lack of an envelope, they are extremely environmentally hardy and resistant to disinfection. Until the late 1970's, feline panleukopenia was the major disease associated with parvoviral infection in small animal medicine and canine parvovirus-1 (CPV-1) was a relatively non-pathogenic virus most commonly associated with illness in neonatal dogs. However, after the emergence of canine parvovirus-2 (CPV-2) from a carnivore parvovirus (likely cats, mink, or other closely related hosts) the infectious enteritis associated with CPV-2 infection became a significant disease in dogs. Canine parvovirus-2 evolved in a relatively short time into two variants, CPV-2a and CPV-2b, both of which are geographically widespread and associated with similar clinical signs and significant pathogenicity. It should be noted that CPV-2b is capable of infecting and causing disease in the cat, therefore, cross-species exposure of ill animals should be avoided. A new variant, CPV-2c, was first reported in Italy in 2000, with subsequent reports of detection in the United States starting in 2005. This newer variant is closely related to CPV-2a and CPV-2b and therefore, should in theory be similar in pathogenicity and virulence. Anecdotally, there are reports of failure of current vaccines to adequately protect from disease associated with CPV-2c; however, most published reports suggest there is adequate protection.

DIAGNOSIS

Diagnosis is based on presence of the appropriate clinical syndrome and ideally, demonstration of recent parvoviral infection.

Demonstration of Recent Parvoviral Infection

The most clinically useful method of detection of virus in the ill animal is via point of care enzyme linked immunosorbent assays (ELISA) (IDEXX, Pfizer, Synbiotics). These assays detect viral antigen in feces or rectal swabs and results and are relatively sensitive and specific, however, both false negatives and false positives can occur. False negatives can occur with infection of any CPV-2 variant secondary to relatively low concentrations of viral particles due to either decreased shedding in the later stages of the disease or due to dilutional effects from voluminous diarrhea. The impact of infection with CPV-2c on point of care ELISA results is controversial at this time. There are reports of CPV-2c causing clinical disease but negative point of care ELISA results, however, other studies have demonstrated similar results using such assays for CPV-2c as for CPV-2a and -2b (Decaro, 2010). Preliminary work in our laboratory has identified three animals in a series of 49 dogs with a clinical diagnosis in which CPV was not detected in feces using point of care ELISAs but CPV was detected using quantitative PCR. All three of these isolates were later determined to be the CPV-2c variant (Veir, unpublished data). However, as a complete history and clinical illness were not available for all of these animals, the same reasons for any false negative may be the cause. False positives have been reported in association with recent vaccination with modified live parvoviral diseases. One manufacturer of point of care assays has an unpublished study demonstrating lack of reaction after vaccination in a group of 60 beagles vaccinated with modified lived (MLV) parvoviral vaccines. In a study performed by our laboratory of 12 puppies vaccinated with a MLV parvoviral vaccine, feces from one puppy produced a positive result on a point of care ELISA five days after vaccination (Burton, 2008).

With the advent of PCR "panels" available from major diagnostic laboratories, PCR may become more commonly used in the diagnosis of CPV enteritis. Unfortunately, qualitative PCR using whole blood was frequently positive in the above study of 12 puppies as soon as two days after vaccination and continued until the study endpoint at 14 days post-vaccination. Our laboratory has attempted to decrease the frequency of false positives caused by recent vaccination via PCR using quantitative, instead of qualitative, PCR assays. Using samples from the same 12 puppies and an additional eight naturally infected puppies positive for CPV using a point of care ELISA, a significantly higher viral load was detected via quantitative PCR in the naturally infected group (Veir, 2009). While not yet commercially available, qPCR may prove useful in differentiating naturally infected and recently vaccinated animals. Virus isolation, hemaglutination inhibition, and electron microscopy can all be used to demonstrate recent infection, however, due to the turnaround time, are often reserved for confirmation of disease or in group health situations.

TREATMENT

Supportive and Symptomatic Care

<u>FLUID THERAPY</u> Maintenance of hydration is at the core of supportive care as severe ongoing fluid losses due to vomiting and diarrhea are common in CPV enteritis.

Intravenous crystalloids (lactated Ringer's solution (LRS), Normosol-R, Plasmalyte) are the replacement fluid of choice for the overwhelming majority of cases. At presentation, many animals are severely dehydrated; therefore, absorption of fluids from the subcutaneous space will be impaired. Intravenous access improves care tremendously but if peripheral vasoconstriction prevents placement of an intravenous catheter, an intraosseous catheter will allow for fluid replacement until peripheral vasoconstriction is corrected and intravenous access is restored. Once dehydration is corrected, attention must be paid to electrolyte balance and glucose levels. Hypokalemia is common and at minimum, maintenance fluids should be supplemented to a level of 20 mEq/L. If hypokalemia is present or develops in the face of fluid therapy, additional supplementation based on those tables available in veterinary texts is recommended. Some authors suggest supplementing all maintenance fluids with 2.5%-5% dextrose regardless of levels, however, in the author's hospital, dextrose is not commonly added to maintenance fluids unless glucose levels start to trend toward the lower end of the reference intervals. For ease of fluid rate adjustment, a maintenance bag of fluids with electrolyte supplementation and if being used, dextrose supplementation, is run at a rate of 2-4 ml/kg/hour to account for maintenance needs and insensible losses. A second bag of a plain, non-supplemented crystalloid fluids is then used to replace sensible losses (most commonly gastrointestinal losses, estimated by nursing staff on an hourly basis, more frequently if animal is critically ill) over the following hour.

ONCOTIC SUPPORT The fluids lost in CPV enteritis can be quite high in protein, therefore, hypoproteinemia can develop in affected animals. The author administers colloidal support if the dog develops peripheral edema or evidence of third spacing, total solids decrease below 3.5 mg/dl or albumin approaches 1.5 mg/dl. Two types of colloids are available for administration. Naturally occurring colloids include fresh frozen plasma, whole blood, human albumin, and canine albumin. Traditionally, fresh frozen plasma has been espoused as the treatment of choice because of the possibility of additional benefits such as coagulation factors and CPV specific antibodies. However, in the absence of a coagulopathy, fresh frozen plasma may not be the best choice for colloidal support. Plasma has a relatively low oncotic pressure and may actually be associated with an increase in pro-inflammatory cytokines. Even in plasma taken from a recently recovered animal, anti-CPV antibody concentrations are minimal and large volumes are required to bring albumin into a more acceptable range. Concentrated sources of albumin, either human or canine, are now commercially available. In theory, 25% human albumin or canine albumin provides natural colloidal support without the volume required for fresh frozen plasma. Expense can be a significant issue and to date, no studies have been published evaluating efficacy of albumin in CPV enteritis as compared to the more affordable alternative, synthetic colloids. If significant anemia develops, whole blood transfusions or packed red blood cell transfusions can be used, but provide little oncotic support.

The most commonly used synthetic colloids in veterinary medicine are hetastarch and dextran 70. Both provide a much greater oncotic pull per unit volume than the naturally occurring colloids while remaining relatively affordable and with a long shelf life. For these reasons, synthetic colloids are the treatment of choice of the author if colloidal support alone is needed. Due to dilution of coagulation factors, excessive administration of any synthetic colloids can be associated with coagulopathies. Judicious use (maintenance rate of 20 ml/kg/day) will avoid this hypocoaguable state but periodic monitoring of coagulation parameters is recommended if prolonged use is needed.

NUTRITION Enteral nutrition is associated with improved mucosal integrity and repair and starvation is associated with small intestinal mucosal atrophy. Early enteral nutrition starting 12 hours after admission, regardless of status, was associated with improved weight gain and clinical signs when compared to the more traditional approach of withholding food until 12 hours after cessation of vomiting (Mohr, 2003). In the study referenced, the authors used nasoesophageal feeding of a balanced liquid canine diet, though syringe feedings of a low residue, easily digestible diet can be attempted as well.

ANTIEMETICS As noted, many of the complications of CPV enteritis are at least in part associated with vomiting and anorexia. Maropitant (Cerenia, Pfizer) is a recently released centrally acting anti-emetic that acts via neurokin-1 (NK-1) receptor antagonism. In a placebo controlled clinical trial in dogs with CPV enteritis, administration of maropitant (1 mg/kg q24h, IV) as compared to ondansetron (0.5 mg/kg q8h, IV) was associated with weight gain while in hospital compared to weight loss in the ondansetron group (Sullivan, unpublished data). It should be noted that the maropitant was administered daily in dogs as young as nine weeks of age and no adverse effects seen despite no days off. Metoclopramide, a centrally and peripherally acting anti-emetic, has also been used (2 mg/kg/day constant rate infusion). Metoclopramide can increase intestinal motility which in theory could increase the risk of intussusception. For this reason, maropitant is the anti-emetic of choice in our hospital.

ANTIBIOTICS Gastrointestinal mucosal barrier disturbances leading to increased bacterial translocation as well as the leukopenia common in many cases leads to an increased risk of sepsis. Mildly affected animals without leukopenia may not need antibiotic therapy. The author uses cefoxitin (22 mg/kg q 8h, IV) for moderately affected animals and for animals that are felt to need broader gram negative coverage, a combination of a penicillin (ampicillin, 22 mg q8h, IV) and a fluoroquinolone (enrofloxacin, 5 mg/kg q24h, IV) is used. Cartilage abnormalities have been associated with use of enrofloxacin in young large breed dogs and clients are warned of this possibility. An alternative to enrofloxacin is an aminoglycoside (gentamicin, 6 mg/kg q24h, IV) but avoidance of nephrotoxicity requires animals to be volume replete which is difficult to guarantee early in the disease.

MISCELLANEOUS TREATMENTS The use of equine endotoxin antiserum in animals affected by CPV enteritis to reduce the effects of gram negative bacterial toxins has been reported with mixed results. At this time, because of the lack of proven efficacy and the possibility of ill effects, the use of the endotoxin antiserum in CPV enteritis is not supported in the author's hospital. Recombinant human granulocyte colony stimulating factor (rhG-CSF) was evaluated for use in puppies with CPV enteritis but has not been shown to improve outcome (Rewerts, 1998). Empiric administration of anthelminics, proton pump or H2 blockers, pain medications, and gastrointestinal protectants may be used in animals with CPV enteritis at the discretion of the practitioner. There are no case controlled studies supporting or discrediting the use of these therapies.

Antiviral Treatments

IMMUNOTHERAPY Passive immunotherapy using convalescent plasma collected from recently recovered dogs has been anecdotally reported to improved clinical outcome. However, in a recent blinded, placebo controlled study of naturally infected dogs in our hospital, the administration of an average of 2.2 ml/kg IV once on admission did not significantly improve any measured parameters, including survival, cost or duration of hospitalization, severity of clinical signs, or viral load. The dose was empirically chosen but closely corresponds to that which is recommended for passive immunotherapy in the human population. Recombinant feline interferon-omega (Virbagen, Virbac) showed promise in two small placebo controlled trials. Mortality and clinical signs were decreased in the treatment group (deMari, 2003). The product is not commercially available in the United States and can be cost prohibitive when available. It is therefore not commonly used, however, if the agent becomes more widely available it may warrant further investigation.

ANTIVIRAL MEDICATIONS Oseltamivir (Tamiflu, Roche) is a neuraminidase inhibitor commonly used for treatment of influenza infections due to the heavy reliance of influenza viruses on neuraminidases. It has been proposed as a treatment in animals infected with CPV but not yet showing significant clinical signs. In a published placebo controlled study (Savigny, 2010), an improvement in weight gain and maintenance of white blood cell counts was noted in the treatment group as compared to the control group, however, no significant difference was noted in survival or clinical signs. Based on this data, it is unclear the role that oseltamivir may play in treatment of CPV enteritis, however, it should be noted that oseltamivir resistance is on the rise in humans with influenza infections. For this reason, use of the drug in veterinary patients without a clearly documented advantage must be carefully considered.

PROGNOSIS

Preliminary data in our laboratory of an investigation of 49 naturally infected dogs suggests a slightly prolonged time to recovery as measured by days of hospitalization when only infecting subtype was taken into account. However, as this was a retrospective study, confounding variables such as variations in treatment, co-morbid conditions, or financial constraints were not evaluated and so this data must be considered in light of these limitations. To this author's knowledge, no other published data considering effect of CPV subtype on prognosis is available at the time of this writing.

PREVENTION

Sporadic reports of infection with CPV-2c in well vaccinated adult dogs have been published (Decaro, 2009). An additional review of infection in vaccinated dogs did detect CPV-2c in vaccinated dogs with clinical signs, but at a similar rate to detection of CPV-2b (Kapil, 2007) and a thorough investigation of concurrent disease or confounding factors was not possible. Experimental challenge with CPV-2c after standard immunization with currently available commercial vaccines did not produce significant clinical signs in a small group of dogs (Schultz 2007). At this time, it seems that it is unlikely that widespread vaccine failure will result from the spread of CPV-2c in the canine population, however, further data is needed.

REFERENCES

Burton JH, Veir JK, Morris AK, et al: Detection of canine parvovirus DNA from blood and antigen in feces collected from healthy puppies after administration of modified live vaccine, *J Vet Intern Med* 22:703, 2008. Decaro N et al: Severe parvovirus in a 12-year-old dog that had been repeatedly vaccinated *Vet Rec* 164, 593-595, 2009. Decaro N, Desario C, Beall, MJ, et al: Detection of canine parvovirus type 2c by a commercially available in-house rapid test, *Vet J* 184:373-375, 2010. deMari K, Maynard L, Eun HM, et al: Treatment of canine parvoviral enteritis with interferon omega in a placebo-controlled field trial, *Vet Record* 152:105-108, 2003. Kapil S et al: Canine Parvovirus Types 2c and 2b Circulating in North American Dogs in 2006 and 2007, *J Clin Microbiol* 45(12): 4044–4047, 2007. Mohr AJ et al: Effect of early enteral nutrition on intestinal permeability, intestinal protein loss, and outcome in dogs with severe parvoviral enteritis, *J Vet Intern Med* 17:791-798, 2003. Rewerts JM, McCaw DL, Cohn LA, et al: Recombinant human granulocyte colony-stimulating factor for treatment of puppies with neutropenia secondary to canine parvovirus infection, *J Am Vet Med* 213:991, 1998. Savigny MR, Macintire DK: Use of oseltamivir in the treatment of canine parvoviral enteritis, *J Vet Emerg Crit Car* 20:132-142, 2010. Schultz RD, Larson LJ. Current canine parvovirus type 2 (CPV-2) vaccines provide excellent immunity to all genotypes of CPV-2 (eg CPV-2a, 2b, and 2c). 88th Conf Res Workers in Anim Dis 2007, p. 113. Sullivan, unpublished data, 2011. Veir JK, Duffy AL, Dow SW, et al: Comparison of quantitative PCR and conventional endpoint PCR for amplification of parvovirus DNA in blood from naturally infected and recently vaccinated dogs, *J Vet Intern Med* 23:769, 2009.

ANTIBIOTICS AND ANTIMICROBIAL RESISTANCE

Joseph M. Blondeau, MSc, PhD, RSM(CCM), SM(AAM), SM(ASCP), FCCP
Saskatoon, Saskatchewan, Canada

INTRODUCTION

Antibiotics inhibit or kill bacteria when interfering with biological processes that are necessary for organism survival and growth.[1] Antibiotics can be broadly classified into those which are bacteriostatic and those which are bactericidal.[2] The differentiation between the bacteriostatic versus bactericidal antibiotic is based on an in vitro measurement in the laboratory that determines the reduction in viable bacterial cells in the presence of the drug and depending on the magnitude of the reduction determines whether or not a bacteriostatic versus bacteriocidal effect is seen. While it is tempting to think that a bactericidal antibiotic may be better than the bacteriostatic antibiotic, evidence from the human infectious diseases literature has failed to consistently show substantial differences in clinical outcome in patients treated with either static or cidal drugs. There is certainly evidence indicating a more rapid eradication of bacteria in the presence of cidal agents but clinical outcome has remained comparable between cidal versus static drugs and this may be a flaw in study design.

Antimicrobial resistance can be divided into intrinsic versus acquired resistance.[3] For intrinsic resistance, an organism is resistant to the antibiotic because it either lacks the target that is necessary for the drug or the target is inaccessible to the drug. The net result is that the organism is not inhibited in the presence of the drug. Acquired resistance implies that the organisms were initially susceptible to the drugs but through one of four major mechanisms, organisms acquired resistance and is no longer inhibited at clinically achievable or sustainable drug concentrations. There are essentially four major mechanisms by which an organism may become resistant to an antimicrobial agent:

1. Enzymatic activation is a mechanism whereby the organism produces an enzyme. When the enzyme interacts with the antimicrobial molecule, it results in a structural modification to the drug thereby rendering the drug biologically inactive against that organism. The best-known example of enzymatic inactivation is with the beta-lactamase enzymes which hydrolyze the beta-lactam ring of beta-lactam antibiotics such as penicillins. A more recent global concern is the evolution of extended spectrum beta-lactamase producing (ESBL) Gram-negative bacilli. These organisms are resistant extended spectrum cephalosporins including our third-generation agents (i.e. ceftriaxone, cefotaxime).
2. Decreased uptake in or altered membrane permeability are mechanisms which affect the ability of the drug to diffuse into the bacterial cell. The net result is a decreased accumulation of drug within the bacterial cell. When the intracellular concentration is insufficient to inhibit or kill the organism, then the organism is rendered resistant.
3. Efflux defines a mechanism whereby the drug diffuses into the bacterial cell and is then effluxed (pumped) out of the bacterial cell. Efflux pumps are responsible for pumping the antibiotic back out of the bacterial cell. The net result of efflux is a decreased accumulation or concentration of antibiotic within the bacterial cell. When the intracellular concentration is insufficient to inhibit or kill the organism, then the organism is rendered resistant. Macrolide antibiotics are one drug class where efflux is a recognized mechanism of resistance.
4. Alteration to target proteins. For this mechanism of resistance, the target protein to which the antimicrobial must bind in order to exert its biological activity has been altered. This may occur as a result of a mutation in the gene that encodes for that protein. The net result is that more antimicrobial agent is required to inhibit or kill the organism from a practical point of view, antimicrobial resistance can be defined based on a breakpoint value but in reality, resistance occurs when the amount of drug required to inhibit or kill the organism exceeds the amount that can safely be achieved in the body by conventional dosing. Drug-resistant organisms are associated with clinical failure. Macrolide and fluoroquinolone antibiotics are affected by altered target proteins as a mechanism of resistance.
5. Co-or cross resistance is defined as resistance to multiple antimicrobial agents from within similar classes of drugs or different classes of drugs. Some multi-drug-resistant pathogens have also been given the designation of superbugs and such organisms include vancomycin resistant enterococcus, methicillin resistant *Staphylococcus aureus* and perhaps now extended spectrum beta-lactamase producing gram-negative bacilli. The unfortunate hallmark of superbugs and other multi-drug-resistant organisms is there pan resistance to many different antimicrobial agents. Such bugs can become therapeutic dilemmas as a number of antimicrobial agents available for therapy is limited and in some instances, consideration for combination therapy to be essential.

Antimicrobial resistance is well defined for a number of antimicrobial agents against a number of companion animal pathogens including *Escherichia coli*, *Staphylococcus pseudintermedius* and *Pseudomonas aeruginosa*.[4] Geographical variation in resistance rates is also likely problematic for companion animals and this parallels what has been seen with human pathogens. How should such data influence an antimicrobial choice to treat infection? Does inadvertently choosing an antimicrobial agent to which the pathogen is resistant automatically imply clinical failure? Similarly, does choosing an antimicrobial agent to which the pathogen is susceptible automatically imply clinical success? Is there a role for combination therapy – where and when?

In this presentation, an overview of antimicrobial resistance will be presented along with a brief summary of the molecular mechanisms of resistance. Additionally, an overview of the current prevalence of antimicrobial resistance of companion animal pathogens in North America will be reviewed. There will also be an emphasis on some new strategies for drug use including an explanation of the mutant prevention concentration (MPC) which is a novel measurement to determine the likelihood for antimicrobial agents to select for antimicrobial resistance bacteria at clinically relevant drug concentrations.[5,6] Optimization of therapy will be discussed.

REFERENCES

[1]Blondeau JM. **Clin Therapeu** 1999;21:3-40. [2]Blondeau JM. **Exp Opin Pharmacother** 2002;3(8):1131-1151. [3]Blondeau JM. **J Infect Dis Pharmacother** 2001;4(Suppl.2):1-28. [4]Blondeau JM. STAT: Steps to Antimicrobial Therapy: Strategies to optimize therapy and minimize resistance when treating bacterial infections. Introducing the mutant prevention concentration (MPC) and includes special chapter on Exotics (small mammals, birds and reptiles). Canada - Hensall ON; USA - Port Huron, MI: North American Compendiums, 2009. [5]Blondeau JM. **Vet Dermatol** 2009;20:383-396. [6]Blondeau JM, *et al.* **J Chemother** 2004;16(Suppl. 3):1-19.

CHOOSING ANTIMICROBIAL AGENTS IN A WORLD OF ANTIMICROBIAL RESISTANCE

Joseph M. Blondeau, MSc, PhD, RSM(CCM), SM(AAM), SM(ASCP), FCCP
Saskatoon, Saskatchewan, Canada

INTRODUCTION

The vast majority of antimicrobial agents are prescribed empirically. The dictionary.com definition of empiric is "a member of an ancient sect of physicians who based their practice on experience alone disregarding all theoretical and philosophic considerations". Another definition would be "without the knowledge of". In medical microbiology and infectious diseases, empiric therapy is for an infectious process in which an antimicrobial agent is prescribed. Antibiotic decisions are based on a syndromic approach. For example, a number of different organisms may be associated with respiratory tract infections. An antibiotic prescribed for treatment of a respiratory tract infection needs to have a spectrum of activity to include the most probable causes. A broad spectrum antimicrobial agent does not imply the drug is optimal against all potential pathogens.

Penicillin was discovered in the late 1920's and approved for clinical use by the Federal Drug Administration in the mid 1940's. Within 10 years of the introduction of penicillin into clinical practice, 70-80% of *Staphlyococcus aureus* strains were resistant to penicillin as the result of a penicillinase enzyme that hydrolyzed the beta-lactam ring of the molecule. This early warning about the potential for antimicrobial resistance seemed to largely go unnoticed or underappreciated as the subsequent history of antimicrobial development, with numerous broad spectrum and potent compounds, has been paralleled by increasing antimicrobial resistance. Such data might suggest 2 things: 1) antimicrobial resistance is inevitable and it is only a matter of time or 2) we simply have not figured out how to optimally use these agents to affect a satisfactory clinical cure while minimizing the probability that resistance will occur. Perhaps there is also a third option and that is that optimization of therapy will delay the onset of resistance but ultimately, resistance is inevitable.

Acquired antimicrobial resistance refers to organisms that were initially susceptible to an antimicrobial agent but subsequently became resistant as a result of one or multiple mechanisms. Mechanisms of resistance include enzymatic inactivation, decreased uptake/altered membrane permeability, efflux and altered target proteins. A single bacterium my simultaneously contain multiple resistance mechanisms thereby conferring multi-drug resistant phenotypes.

PK/PD

Shorter Durations of Therapy: In human medicine, prolonged antimicrobial therapy was recommended for many infectious diseases. Such lengths of therapy could extend from 7-10, 10-14 and even as long as 21 days of therapy for conditions that can now be treated with shorter durations of therapy. For conditions such as osteomyelitis longer durations of therapy (6 weeks) are still recommended out of necessity. Shorter durations of therapy likely impact of reducing resistance selection as evidence shows and increased risk for resistance with longer durations of therapy! Shorter durations of therapy are now recommended (based on good clinical evidence) in humans for acute uncomplicated urinary tract infections, acute maxillary sinusitis, acute exacerbations of chronic bronchitis and for community acquired pneumonia.[1-5]

Antimicrobial agents can be differentiated based on being concentration dependent versus time dependent agents.[6] For concentration dependent agents, the maximum serum drug concentration to MIC ratio (C_{max}/MIC) or area under the drug concentration curve to MIC value (AUC/MIC) are important PK/PD parameters. A C_{max}/MIC ratio of 8-12 and an AUC/MIC ratio of >100 has been shown to be associated with a favourable clinical response and minimization of resistance. For time dependent agents the time the serum drug concentrations exceed the MIC over the duration of the dose (40-50%) was shown to be associated with a favourable clinical response. How often are such parameters considered when an antimicrobial agent is chosen and is such information readily available and in a useable format to be considered when prescribing decisions are being made?

TARGETED ORGANISMS THERAPY

Target organism therapy implies that the infecting pathogen is known and that antimicrobial therapy is targeted specifically towards that organism and this allows the choice of an agent that could optimize therapy. Optimization of therapy includes a satisfactory clinical outcome while at same time minimizing the likelihood for resistance selection during therapy. As previously stated, antimicrobial agents are generally prescribed in a syndromic approach where an agent is selected to cover all of the potential and most common pathogens associated with a specific infection. For target organism therapy, this could imply that once the infecting pathogen is known that a reassessment of the antimicrobial choice be made and a decision to alter ongoing therapy made. There are numerous practical issues associated with this approach. One reluctance is a unwillingness on behalf of the clinician to change therapy in a patient showing a favourable clinical response. As well, changing an antimicrobial agent to which a patient is tolerating and new or additional side effects may be seen may also hinder a decision to change antimicrobial therapy. Finally, costs and inconvenience are also practical considerations especially if a full course of antimicrobial therapy has already been purchased for the initially prescribed drug. Having said all that, suboptimal therapy may still yield a favourable clinical response but at a consequence of an increased likelihood for resistance selection during therapy.

NOVEL *IN VITRO* MEASUREMENTS PREDICTING RESISTANCE

The Mutant Prevention Concentration (MPC) was described in 1999 as a novel *in vitro* measurement based on testing high density bacterial populations against antimicrobial agents.[7] The basis for this approach is that in high density populations, spontaneous resistant cells may be present and in the presence of drug therapy, suboptimal treatment may allow resistant bacterial cells to survive in the presence of the drug. High density bacterial populations are present in a number of acute infections. Determining the MPC drug concentration provides a target drug concentration to prevent resistant selection.[8-10] A big question relates to the likelihood that MPC drug targets can be obtained with conventional dosing of agents used in companion animal infections.

During this presentation, data will be shown commenting on issues pertaining to antimicrobial use, antimicrobial resistance and newly identified strategies to differentiate antimicrobial agents based on their likelihood for resistance selection against key veterinary pathogens.[11] A review of current therapy options for key companion animal infections and various antimicrobial agents will be reviewed as will data on specific companion animal pathogens (*E. coli, S. pseudintermedius, P. aeruginosa*) considering the potential for resistance selection/prevention with various drugs. A number of the practical considerations for antimicrobial use will also be discussed. It does appear that our past thinking on antimicrobial use needs to be changed. One limitation is availability of peer reviewed data that can provide us with additional guidance.

REFERENCES

[1]Blondeau JM. STAT: Steps to Antimicrobial Therapy: A Guide to Management of Infectious Diseases. 3rd edition ed. Canada-Hensell, ON: USA-Port Huron, MI: North American Compendium Inc., 2011. [2]Hooton TM, *et al.* **Clin Infect Dis** 2010;50:625-663. [3]Klossek JM, *et al.* **Int J Clin Practice** 2005;59(2):230-238. [4]Balter MS, *et al.* **Can Respir J** 2003;10(Suppl. B):3B-32B. [5]Mandell LA, *et al.* **Clin Infect Dis** 007;44(Suppl. 2):S27-S72. [6]Schentag JJ, *et al.* **Clin Infect Dis** 2001;32(Suppl.1):S39-46. [7]Dong Y, *et al.* **Antimicrob Agents Chemother** 1999;43:1756-1758. [8]Blondeau JM, *et al.* **J Chemother** 2004;16(Suppl. 3):1-19. [9]Hesje C, *et al.* **Exp Rev Respir Med** 2007;1(1):7-16. [10]Blondeau JM. **Vet Dermatol** 2009;20:383-396. [11]Blondeau JM. STAT: Steps to Antimicrobial Therapy: Strategies to optimize therapy and minimize resistance when treating bacterial infections. Introducing the mutant prevention concentration (MPC) and includes special chapter on Exotics (small mammals, birds and reptiles). Canada - Hensall ON; USA - Port Huron, MI: North American Compendiums, 2009.

71a

ON THE TRACE OF NEW BIOMARKERS: DIFFERENTIAL EXPRESSION OF CIRCULATING MICRORNAS IN LEAN, OBESE AND DIABETIC CATS. S. Fleischhacker[1], S. Bauersachs[2], K. Weber [1]. 1. Clinic of Small Animal Medicine, LMU University, Munich, Germany. 2. Laboratory for Functional Genome Analysis, LMU University, Munich, Germany.

MicroRNAs are small single-stranded RNA molecules that regulate gene expression by suppressing specific mRNAs. The purpose of our study was to apply both microRNA-array and qPCR techniques on small volumes of serum samples from healthy lean, obese and diabetic feline patients to test whether microRNA-based diagnostics are feasible in a clinical setting.

Blood samples were taken from healthy lean cats (LEAN), healthy obese cats (OB), diabetic cats (DB) and diabetic cats in remission (DR) (n=8 each). Total RNA was extracted from 400 µl of serum. The samples from eight LEAN and eight DB cats were labeled and hybridized on microRNA microarrays representing 1136 distinct mouse microRNA targets. To confirm the results, specific assays for four targets were tested on samples of LEAN, OB, DB and DR cats. RNA was reverse transcribed and amplified using the miScript SYBR Green PCR Kit on a 7500 Real Time PCR System.

244 distinct microRNAs were detected on the microarrays, and 11 were found to be differentially expressed in DB compared to LEAN cats. Using specific microRNA assays, miR-122, miR-193b and miR-483* were found to be up-regulated 100-fold, 20-fold and 6-fold in DB compared to LEAN cats. OB and DR cats were found to have a similar expression pattern as the LEAN group.

Small blood samples of feline patients yield sufficient material to detect microRNA serum profiles on high density mouse microarrays. Differential expression could be confirmed using qPCR. We consider microRNA serum profiles to have the potential to detect disease development in small animals.

71b

TEST DEVELOPMENT FOR PRECLINICAL DETECTION OF HYPOADRENOCORTICISM (HOAC) IN DOGS. M Rick[1], TA Mueller[2], K Meek[1,2]. 1. Pathobiology and Diagnostic Investigation. 2. Microbiology and Molecular Genetics. Michigan State University, East Lansing, MI.

Approximately 70-90% of HoAC cases in human medicine result from immune-mediated adrenalitis. Although clinical signs are not present until ~90% of the adrenal cortex is destroyed, several anti-adrenal auto-antibodies (ABs) can be identified in serum. In humans, presence of these ABs constitutes the primary criteria for an early diagnosis of adrenalitis. Such a diagnostic tool is not available for dogs, but it would have the potential to be life- and money-saving by enabling an early diagnosis. HoAC is a significant problem in dogs with a prevalence of 0.6 to 3.1 per 1000. Currently, more than one-third go unrecognized until there is an, often fatal, acute adrenal crisis. Preliminary attempts, using a commercially available radioimmunoassay (RIA) for the detection of human 21-hydroxylase (21-OH) ABs (Kronus®), and an in-house RIA, inconsistently suggested the presence of 21-OH ABs in some samples of affected dogs.

We therefore hypothesized that, as in human HoAC, anti-21-OH AB production occurs in canine HoAC, with AB production preceding clinical disease. The specific aims were to determine whether anti-adrenal auto-ABs are present in dogs with HoAC and to establish a diagnostic test to detect canine anti-adrenal auto-ABs. To prove our hypothesis, we expressed and purified the canine 21-OH. The purified protein was then used to create positive controls for assay development and for assay development itself.

His- and GST-tagged canine 21-OH fusion proteins were expressed in *E. coli* using standard techniques. Proteins were purified using electro-elution out of SDS-PAGE gel, and fast-protein liquid chromatography. Presence of 21-OH was confirmed by comparing the actual protein size with the calculated on SDS-PAGE, by Western Blot (WB) using anti-GST or anti-His as the primary ABs, and by Matrix-assisted laser desorption/ionization. Two rabbits and two dogs were then immunized with the purified 21-OH-GST. The sera from these animals were used as negative (pre-immunization) and positive (post-immunization) controls in immunohistochemistry (IHC), and, using the 21-OH-His as antigen, in WB, and enzyme-linked immunosorbent assay (ELISA).

No specific labeling for 21-OH could be identified using IHC. With WB, samples from the dogs and rabbits post immunization showed immunoreactivity, but not before. With ELISA, the optical density in the two dogs was significantly increased post immunization. Then, a small array of sera from dogs with naturally occurring primary HoAC was tested. The data obtained with WB, as well as ELISA, showed that approximately 30% of dogs with naturally occurring primary HoAC had measurable ABs against 21-OH, which proved part of our hypothesis, that as in human HoAC, anti-21-OH AB production occurs in dogs with HoAC. Ongoing development and evaluation of the ELISA will help answer the question whether auto-AB production precedes clinical disease, therefore implying a role in the pathogenesis of HoAC, and will enable epidemiologic and genetic studies on breed, sex, and age based predispositions of canine HoAC.

72a

CHRONIC PANCREATITIS IN THE ENGLISH COCKER SPANIEL SHOWS A PREDOMINANCE OF $IgG4^+$ PLASMA CELLS IN SECTIONS OF PANCREAS AND KIDNEY. PJ Watson[1], F Constantino-Casas[1], CJ Saul[2] ; MJ Day[3]. [1]Department of Veterinary Medicine, University of Cambridge, UK; [2]Chestergates Referral Hospital, Chester, UK; [3]School of Veterinary Sciences, University of Bristol, UK.

Chronic pancreatitis (CP) in the English Cocker Spaniel (ECS) has been described as a histologically distinctive disease which is duct destructive and shows similarities to human autoimmune CP. Human autoimmune CP is characterised by a predominance of $IgG4^+$ plasma cells in the pancreas and is associated with $IgG4^+$ disease in other organs including liver, kidney and salivary and tear ducts, prompting the new term 'IgG4-related sclerosing disease'. It was hypothesised that CP in ECS would be predominantly associated with $IgG4^+$ plasma cells, whereas CP in other breeds would be involve a mixed population of plasma cells. It was also hypothesised that ECS would show multi-organ $IgG4^+$ disease.

Sections of pancreas from 5 ECS with confirmed CP; 4 dogs of other breeds with confirmed CP and 4 dogs without pancreatitis (one ECS and 3 other breeds) were available for immunolabelling together with sections of kidney with glomerulonephritis and interstitial nephritis from one of the ECSs with CP. Canine lymph node and duodenum acted as positive controls. Sections were immunolabelled with a rabbit anti-human polyclonal IgG at 1:5000 dilution and a panel of 4 previously validated mouse monoclonal antibodies specific for canine IgG subclasses at a 1:400 dilution.

The canine IgG1 specific antibody was ineffective in formalin fixed tissue, as in previous studies, but the polyclonal IgG and the monoclonals specific for canine IgG2, IgG3 and IgG4 worked well. Two ECSs had $IgG4^+$ CP with multifocal dense accumulations of plasma cells which were predominantly $IgG4^+$ with 20-50 cells per high power field (hpf). Two ECSs had end-stage CP with marked fibrosis and very few plasma cells. One ECS had periductular plasma cell infiltrates with lesser numbers of $IgG4^+$ plasma cells (6-10 per hpf) and a similar number of $IgG2^+$ plasma cells. This dog also had renal disease with $IgG4^+$ nephritis with 40-70 $IgG4^+$ plasma cells per hpf and very few other IgG subtypes. The normal pancreas sections had very few plasma cells of any subtypes, occasionally seen individually and predominantly in blood vessels. Other breeds with CP had mild to moderate plasma cell infiltrations in inflammatory foci which were polyclonal with a predominance of $IgG2^+$ cells.

Two ECS showed histological findings typical of $IgG4^+$ CP in man. One dog had less convincing pancreatic lesions, but renal histology typical of $IgG4^+$ sclerosing disease in man. More investigations are indicated to investigate whether this is the only cause of CP in ECS, its prevalence and treatment.

72b

BABESIA CONRADAE INFECTION IN COYOTE HUNTING DOGS INFECTED WITH MULTIPLE BLOOD-BORNE PATHOGENS. JD Dear, SD Owens, A Biondo, BB Chomel, M Marcondes, JE Foley, JE Sykes. Department of Medicine & Epidemiology, School of Veterinary Medicine, University of California, Davis, California.

Babesia conradae is a small canine intra-erythrocytic piroplasm. Although ticks are suspected vectors, transmission routes remain unknown. We identified *B. conradae* infection using PCR in dogs from a kennel of 20 sighthounds with a history of coyote fights. The study's goal was to examine risk factors for, and laboratory abnormalities of infection. Blood was collected from each dog and from 6 non-sighthound dogs within the same colony for CBCs, biochemistry panels, conventional and real-time PCR assays (*Babesia* spp., hemoplasmas, *Ehrlichia canis*, *Bartonella* spp., *Anaplasma* spp., and *Rickettsia* spp), ELISA serology (IDEXX 4Dx SNAP test), and IFA serology and culture for *Bartonella* spp. The association between *B. conradae* infection and coyote fighting, age, genetic background and laboratory abnormalities was investigated.

The prevalence of *B. conradae*, *Mycoplasma haemocanis* and *Mycoplasma haematoparvum* infection in the sighthounds was 9/20, 10/20, and 5/20, respectively. The seroprevalence of *Bartonella* spp. infection was 12/20. *E. canis* and a *Hepatozoon felis*-like organism were detected in 1 dog each. For the non-sighthounds, the prevalence of *B. conradae*, *M. haemocanis*, and *M. haematoparvum* infection was 0/6, 3/6 and 4/6, respectively. Infection with *B. conradae* was significantly associated with coyote fighting; lower hematocrits, leukocyte counts, MCHCs, platelet counts and serum albumin; and higher MCVs, mean platelet volumes, and serum globulin. The results of this study suggest that as for *B. gibsoni*, aggressive interactions with other dogs play a role in *B. conradae* transmission. B *conradae* infection may cause thrombocytopenia, leukopenia and anemia along with chronic antigenic stimulation.

73a

MESENCHYMAL STEM CELLS IN EXPERIMENTAL FELINE ASTHMA: A PILOT STUDY

CR Reinero[1], CH Chang[1], T Webb[2], C Davis[2], H Liu[1], J Quimby[2], S Dow[2]. [1]Comparative Internal Medicine Laboratory, College of Veterinary Medicine, University of Missouri, Columbia MO; [2]Center for Immune and Regenerative Medicine, Colorado State University, Fort Collins, CO.

Feline allergic asthma is associated with airway eosinophilia, hyperresponsiveness and remodeling. Mesenchymal stem cells have emerged as novel immunotherapeutic options for a variety of inflammatory lung diseases. Their beneficial role is likely mediated through paracrine effects which modulate the immune system. We hypothesized that intravenous infusion of allogeneic mesenchymal stem cells (MSCs) would dampen airway inflammation in experimentally asthmatic cats.

Asthma was induced in 5 cats using Bermuda grass allergen (BGA) at the University of Missouri (MU). Four cats were treated with 3 i.v. infusions of allogeneic MSCs (days 0, 14 and 28); one cat received iv placebo. The first two infusions used 2X10E6 and 4X10E6 MSCs respectively which were shipped frozen and thawed immediately prior to use; while the 3rd infusion used 4.7X10E6 cells which were shipped live. Evaluation of bronchoalveolar lavage fluid (BALF) eosinophilia and blood IL-10 production, T regulatory cells (Tregs), and T and B cell phenotype were performed prior to MSC treatment and at day 47, with BGA aerosol challenging preceding collections by 24 hours.

No adverse reactions were observed with MSC treatment. A modest reduction in airway eosinophilia between days 0 and 47 in cats treated with MSC compared to the control cat (MSC group mean±SEM BALF eosinophils 75±2% and 32±5%, placebo 61% and 57%) was found. IL-10 concentrations were higher in MSC-treated versus placebo treated cats at day 47 by ELISPOT (268±33 vs 205 IL-10 producing cells) and by ELISA after stimulation with 1μg/ml LPS (0.293±.051ng/ml vs <lower limit of detection of 0.125ng/ml). Flow cytometric evaluation of whole blood Tregs, T cells, and B cells did not reveal significant differences between treatment groups: (Tregs 3.4±0.8% vs 3.0%, T cells 63±3% vs 66%, and B cells 28±2% vs 34%).

Our pilot data in asthmatic cats show that repeated infusions of allogeneic MSCs were well-tolerated and altered some aspects of the asthmatic immune response in cats in a sustained fashion. Future studies will focus on determining optimal dosing and preparation strategies for MSC administration to ameliorate asthma in this feline model. Additionally, studies that include airway function testing and larger numbers of animals are likely to provide additional insight.

73b

EVALUATION OF THE ASSOCIATION BENAZEPRIL/RENAL DIET TO CONTROL RENAL PROTEINURIA IN NON-AZOTEMIC DOGS.

O.Cortadellas[1], J. Talavera[2], Mª J. Fernández del Palacio[2]. [1]Clinica Veterinaria Germanías. Gandía -Valencia [2] Departamento de Medicina y Cirugía Animal. Universidad de Murcia, Spain.

The possible role of persistent proteinuria in promoting progressive renal injury has encouraged different therapeutic recommendations to reduce proteinuria in patients with renal disease. Angiotensin-converting enzyme inhibitors have demonstrated its efficacy controlling proteinuria in dogs with chronic kidney disease. Renal diets (RD) are moderately protein restricted and have beneficial effects in azotemic dogs. However, its role in proteinuric non-azotemic (PNAz) dogs has not been well documented. This study investigates if the combination of benazepril (Be)/RD improves proteinuria and nutritional status in PNAz dogs compared to Be/maintenance diet (MD).

Twenty-two non-azotemic dogs with renal proteinuria (UPC > 1, twice consecutively in 2 weeks once other causes of proteinuria were excluded) were included in the study. Dogs were randomly assigned to the Be/MD (5.5 g protein/100 Kcal EM) group or to the Be/RD (3.7 g protein/100 Kcal EM) group during 60 days. Dogs (n=5) with serum albumin < 2 g/dl also received aspirin (1 mg/kg, twice a day). A physical examination, blood pressure measurement, CBC, biochemistry panel, complete urinalysis and UPC were performed at day 0 and day 60. Changes in UPC were considered significant according to the criteria proposed by Nabity et al, (JVIM 2007). Data were statistically compared (Mann-Whitney and Wilcoxon signed-rank test, $p<0.05$).

At day 0 there were not statistically significant differences between study groups for age, sex, weight, albumin, serum creatinine and urea concentrations, systolic blood pressure and UPC. Benazepril dose (mg/kg) did not differ between groups (Be/MD: 0.55±0.25; Be/RD: 0.52±0.21; p=0.569). During the study period, UPC did not significantly decrease in the Be/MD group. Median (range) at day 0 was 3.53(1.83-7.35) vs 2.87(0.14-7.35) at day 60 (p = 0.534). Individually, 4/11 dogs had a significant reduction of UPC, 5 had no significant changes and 2 had a significant UPC increment. A significant decrease (p = 0.026) in UPC was present in the Be/RD group: day 0= 3.07(1.03-8.50); day 60= 1.25(0.15-10.47). UPC significantly decreased in 7 dogs, remained stable in 3 and increased in 1 dog. Weight and serum albumin concentration did not change significantly in either groups, but there was a trend for positive increment in weight in Be/MD group (p=0.067) and for serum albumin in Be/RD group (0.058).

The results of this study suggest that administration of Be/RD improves proteinuria control in PNAz compared to Be/MD, without inducing significant malnutrition.

74a

MINIMALLY-INVASIVE REMOVAL OF UROLITHS IN FEMALE CATS. H Albasan[1,2], JP Lulich[1], CA Osborne[1], F Alkan[3]. 1. University of Minnesota, College of Veterinary Medicine, Minnesota Urolith Center, St. Paul, MN. 2. Ankara University, Faculty of Veterinary Medicine, Department of Internal Medicine, Ankara, Turkey. 3. Selcuk University, Faculty of Veterinary Medicine, Department of Surgery, Konya, Turkey.

In this study, effects of minimally-invasive removal of urocystoliths by lithotripsy and/or basket retrieval from female cats were evaluated.

Lithotripsy was performed during anesthesia with a holmium-YAG laser via cystoscopy. Basket retrieval and voiding urohydropropulsion were used to remove small uroliths and urolith fragments. Post procedural contrast cystography was used to assess the efficacy and safety.

With owner preference and consent, a total of 17 cats with naturally occurring urocystoliths and/or urethroliths presented to Veterinary Teaching Hospital at the University of Minnesota were included in this study. The mean age of cats was 7.4 years.

Urolith removal was complete in 94% of cats (16 of 17). Incomplete urolith removal was detected in 1 case in which uroliths were blasted successfully but, stone basket retrieval was unsuccessful because of urethral constriction. Laser procedure was only applied to uroliths in 7 of 17 cats; uroliths in remaining 10 cats were small enough to pass through the urethra via urolith-retrieving basket. The average time of stone removal procedure in 16 cases was 38 minutes; however, the average time of stone removal procedure was increased to 54 minutes when only laser lithotripsy cases were considered in 6 cats.

Results of this study suggested that laser lithotripsy with basket retrieval was an effective alternative to surgical removal of urocystoliths and urethroliths in female cats.

74b

ASSESSMENT OF ALLOGENEIC CRYOPRESERVED MESENCHYMAL STEM CELLS FOR THE TREATMENT OF FELINE CHRONIC KIDNEY DISEASE. JM Quimby, TL Webb, SW Dow. Center for Immune and Regenerative Medicine, Department of Clinical Sciences, Colorado State University, Fort Collins, CO.

Feline chronic kidney disease (CKD) is characterized by chronic tubulointerstitial nephritis, and inflammation contributes to the progression of renal fibrosis. Mesenchymal stem cells (MSC) have demonstrated anti-inflammatory and anti-fibrotic effects in rodent CKD models. The purpose of this study was to evaluate the use of allogeneic cryopreserved MSC for treatment of feline CKD.

MSC were isolated from the adipose of SPF research cats, expanded in culture, cryopreserved in liquid nitrogen and thawed for use as cats were enrolled. Six CKD cats (group A) received 2×10^6 MSC i.v. and 5 CKD cats (group B) received 4×10^6 MSC i.v. at 2, 4, and 6 weeks. CBC, chemistry, urinalysis were performed at weeks 0,2,4,6, and 8. Group B cats and two control CKD cats had iohexol glomerular filtration (GFR) performed at zero and 8 weeks.

Group A cats received 3 doses of MSC without apparent adverse effects. However, Group B cats experienced transient adverse effects, including vomiting (2 cats), increased respiratory rate (4 cats). Serum creatinine concentrations decreased significantly ($p=0.01$) in group A cats, but was unchanged in group B cats. Group B cats had an overall trend towards improvement in GFR ($p = 0.056$), with individual increases of (75%, 10%, 61%, 7%, 6%) compared to untreated CKD control cats (7%, -18%).

Administration of MSC may improve renal function in CKD cats, but cryopreserved cells may be associated with unacceptable adverse effects related to MSC infusion. Future studies will examine the effectiveness of administration of non-preserved MSC for CKD.

NOT "JUST" TRACHEAL COLLAPSE: PATHOPHYSIOLOGY AND MANAGEMENT

Eleanor C. Hawkins, DVM, DACVIM (SAIM)
Raleigh, NC

TRACHEAL COLLAPSE vs TRACHEOBRONCHOMALACIA

Tracheal collapse in dogs has been thought of as a congenital cartilage defect of small breeds that results in a honking cough, and that sometimes progresses to signs of upper airway obstruction. Several studies have demonstrated ultrastructural differences in the tracheal cartilage of toy breed dogs with collapsed tracheas, compared to those with normal tracheas, such as areas within the rings that are hypocellular or contain fibrocartilage or fibrous tissues, and decreased chondroitin sulfate and calcium.[1-3] In many of these dogs, signs do not develop until later in life. One plausible explanation for the delay of signs is that they are initiated by an "acute on chronic" event. In this scenario, an exacerbating problem develops in an affected dog which results in increased respiratory efforts, airway inflammation, and/or cough. Exacerbating problems could include infectious tracheobronchitis, heart enlargement or failure, or parasitic disease, perhaps with contributions from obesity, exposure to tobacco smoke, or poor oral health. Changes in intrathoracic and airway pressures during increased respiratory efforts or cough contribute to narrowing of the trachea and stretching of the dorsal ligament. With severe collapse, fluttering or physical trauma to the mucosa may further stimulate cough. Inflammation also contributes to an ongoing cycle of cough and of collapse. Collagenases and proteases released by inflammatory cells may weaken the structure of the airways. Damage to the tracheal epithelium and changes in mucus composition and secretion impair airway clearance. Previously tolerable irritants and organisms may perpetuate inflammation and cough.

We now know that many dogs with tracheal collapse also have bronchomalacia, with narrowing of principal and/or lobar bronchi,[4] that not all dogs with tracheal collapse are toy breeds,[5,6] and that some dogs have bronchial collapse without tracheal collapse.[6,7] In human medicine, the broader term tracheobronchomalacia (TBM) is used, and is further classified as primary (congenital) or secondary (acquired).[8,9] Some propose a distinction between atrophy of the longitudinal muscle and elastic fibers of the dorsal tracheal membrane, "excessive dynamic airway collapse" (EDAC), and true softening of the cartilage (TBM).[9] Both of these conditions can occur in the same individual, or either independently.[9] Although in the human literature TBM is at times discussed separately from tracheomalacia, often the term TBM is used to describe malacia of either or both trachea and bronchi.

Further complicating the diagnosis and management of dogs with TBM is that this syndrome not only shares similar presenting signs as those of chronic airway inflammation (e.g., idiopathic chronic bronchitis, eosinophilic bronchopneumopathy, bacterial bronchitis, parasitic disease), TBM can be seen concurrently with these diseases. In some dogs, it is likely that the chronic airway inflammation actually causes the TBM through inflammation and cough, perhaps with contributing factors related to the genetic make-up of an individual dog related to cartilage structure or balance of pro- and anti-inflammatory mediators. It is possible that the TBM contributes to airway inflammation, as described above with respect to congenital disease. In people, chronic airway inflammation is associated with secondary TBM, and it may be present in as many as 50% of people evaluated for COPD.[10]

EXTRA-THORACIC vs INTRA-THORACIC SIGNS

TBM, then, can affect be primary or secondary, and can affect the trachea and/or bronchi. Equally important, from a clinical perspective, is whether collapse occurs predominantly in the cervical trachea or thoracic inlet, or whether it occurs in the intra-thoracic airways. Dogs with predominantly extra-thoracic tracheal collapse can present with signs of upper airway obstruction, including respiratory distress most pronounced on inspiration and audible stertorous sounds. If respiratory distress occurs in dogs with intra-thoracic airway collapse, it tends to be more pronounced on expiration and an audible, loud, wheeze/cough can often be heard.

It is possible that a relationship exists between primary and secondary TBM and abnormalities in breathing, with primary TBM (i.e. congenital tracheal collapse) more often resulting in extra-thoracic obstruction and secondary TBM more often resulting in intra-thoracic obstruction. This conjecture is partially supported by a study of tidal breathing flow-volume loops in toy or small breed dogs with tracheal collapse and <u>no</u> evidence of other respiratory disease, in which abnormalities were seen predominately in the inspiratory limb.[11] Dogs with tracheal collapse had decreases in IF75, IF50, IF25, and IF12.5, lower inspiratory flow values relative to expiratory flow values at all time points (PEF/PIF, EF75:IF75; EF50:IF50; EF25:IF25; EF12.5:IF12.5), and increased inspiratory time relative to expiratory time (TE:TI; TI:TTOL).[11] In a study of 18 dogs with bronchomalacia, but no tracheal collapse, inflammation was identified on bronchoalveolar lavage and bronchial biopsy, and the presenting cough was described as mild and wheezing, often occurring at any time of day, including at rest.[7]

PREVALENCE

The true prevalence of TBM in dogs is not known. Studies often originate from referral institutions and may over-represent dogs with poorly responsive signs, and making the diagnosis is difficult. In a report of bronchoscopies performed on 58 dogs, half had some form of airway collapse.[4] Bronchial collapse was reported in 35 of 40 (87.5%) of brachycephalic dogs undergoing bronchoscopy.[12] We reported findings from 115 dogs with chronic cough, of which 59 (51%) had tracheobronchomalacia.[6] In addition, 31 of 32 (97%) toy breed dogs had TBM among their diagnoses.[6] It is also difficult to be certain how frequently airway inflammation is present in dogs with TBM. Factors that interfere with accurate determination include bias in case selection, anti-inflammatory or anti-microbial therapy, and method of specimen collection (e.g. bronchoalveolar lavage vs bronchial brushing).[13]

PROBLEMS IN DIAGNOSING

The first problem in diagnosis TBM in dogs is that the degree of airway collapse in healthy patients is not known. It should be safe to assume that at rest, during tidal breathing, healthy dogs do not have evidence of tracheal collapse (i.e. without forced expiration or cough), although this assumption may not be true for lower airways, based on personal experience and the high reported occurrence of bronchial collapse in brachycephalic dogs.[12] A degree of collapse occurs in healthy people during forced exhalation or cough. The term "dyanamic airway collapse" (DAC) describes this change. Historically, a decrease in the airway cross-sectional area of <50% during forced exhalation or cough was considered to be normal, and this figure has been used in the veterinary literature. Recent studies have shown that healthy people can have collapse of up to 70%, and that some individuals can experience total collapse of specific lower airways.[14] The definition of TBM in people will likely be modified to describe collapse of >70%.[14]

The second problem in the diagnosis of TBM is that the degree of collapse is dynamic, and related to the pressures within and external to the airway (intra-thoracic pressure). Malacic airways within the thorax can be of normal caliber during tidal breathing, but then collapse completely during forced exhalation or cough. It is possible, in some dogs, to induce forced exhalation or cough during bronchoscopy or fluoroscopy, but breathing effort information is rarely provided in the literature. In other patients, it simply may not be possible to induce vigorous efforts. Even when strong breathing or cough can be induced, the intensity of these efforts is difficult to quantitate or reproduce, preventing even consistent follow-up of the same patient. Because of the difficulty in controlling degree of effort, it is also difficult to determine whether one diagnostic modality is "better" than another. Bronchoscopy may be most sensitive if a light plane of anesthesia allows for the induction of cough. However, if such a cough cannot be induced during the procedure, fluoroscopic evaluation during cough may be more sensitive. Maximizing collapse of extra-thoracic airways is even more difficult.

CERVICAL LUNG LOBE HERNIATION: ANOTHER TWIST

Review of the radiology reports from 121 dogs undergoing tracheal fluoroscopy for evaluation of cough at our institution showed that 85 (77%) had evidence of cervical lung lobe herniation during cough or forced exhalation. (Nafe LA, et. al., manuscript in progress) Of the 85 dogs with herniation, 82 (96%) had fluoroscopic evidence of TBM. Interestingly, 39% of dogs with cervical lung lobe herniation had extra-thoracic kinking of the trachea during cough. It is likely that kinking of the trachea results in some trauma to the airways, and may be another factor in the progression of cough in some dogs.

MEDICAL MANAGEMENT AND RESPONSE

The above sections reviewed the complexity of patients traditionally considered as having tracheal collapse. TBM can be primary or secondary, clinical signs may reflect extra-thoracic or intra-thoracic large airway collapse, degree and extent of airway collapse varies, and an accurate diagnosis of TBM can be difficult. Further, concurrent or complicating respiratory disease can also be present. As such, it is unlikely that one approach to the treatment of dogs with TBM will be uniformly successful. Ideally, a thorough diagnostic evaluation is performed and treatment is tailored to the individual patient and modified based on their specific response.

A positive clinical ramification of the cycle of collapse, cough and inflammation is that breaking this cycle will result in relief from clinical signs in the majority of dogs. Medical management provides adequate relief for most animals. In a study of 100 dogs by White, et. al., medical therapy resulted in resolution of signs for at least 1 year in 71% of cases.[15] Medical treatment modalities have the greatest likelihood of success in those patients with cough or with intra-thoracic collapse. Whenever possible, the exacerbating event is identified and treated. If presenting signs are acute and severe, subsequent therapy may need to be aggressive and then gradually lessened once cough and inflammation are back in control. Treatment targets beyond specific complicating disease include: decreasing inflammation and small airway obstruction through environmental management, bronchodilators, and corticosteroids; addressing factors that lead to increased respiratory efforts, such as obesity, hyperadrenocorticism, elongated palate, or stenotic nares; avoiding situations that induce cough, such as pressure from collars or intensely exciting activities; and cough suppression, usually with narcotic anti-tussives. The use of stanzolol to improve tracheal integrity has also been reported. Tracheal stenting remains a valuable option when medical therapy fails to control clinical signs adequately.

Treat All Exacerbating or Concurrent Respiratory Tract or Cardiac Disease

Potential exacerbating events that can be treated directly include bacterial infection, brachycephalic airway syndrome, and left atrial enlargement or congestive heart failure.

Decrease Inflammation and Small Airway Obstruction

Environmental factors: Irritants such as smoke (from tobacco or fireplace) and perfumed products are avoided. Motivated clients can take steps to improve the air quality in their home, such as cleaning or eliminating carpet, furniture, and window treatments; cleaning the furnace and regularly replacing air filters; and using an air cleaner. The American Lung Association has a useful website with non-proprietary recommendations for improving indoor air quality (www.lung.org).

Bronchodilators: Bronchodilators are indicated particularly in dogs with clinical signs resulting from intra-thoracic TBM. If effective, decreased resistance to air flow within smaller airways should result in increased intra-luminal pressures within the larger airways, countering the tendency for collapse. It is possible that relaxation of smooth muscle could decrease airway rigidity and worsen the tendency for collapse; and, although I have not observed deterioration in any of my patients, the response in each dog should be assessed. Theophylline has potential benefits (unproven in dogs) beyond bronchodilation, including anti-inflammatory effects (which may be synergistic with those of steroids), improvement in mucociliary function, decrease in fatigue of respiratory muscles, and inhibition of release of mast cell mediators of inflammation. Sympathomimetic drugs are preferred by some clinicians.

Corticosteroids: Corticosteroids are often effective in controlling cough in dogs with TBM. Potential negative effects in patients with TBM include an increased susceptibility to infection; a tendency toward obesity, hepatomegaly, and muscle weakness that may adversely affect ventilation; and pulmonary thromboembolism. Prednisone is initially given at a dose of 0.5 to 1.0 mg/kg every 12 hours, with a positive response expected within 1 week. If a positive response is seen, the initial dose is continued for 2-3 weeks before tapering of dose is initiated. If no response is seen to an adequate dose, the drug is discontinued. Dogs that require relatively high dosages of prednisone, have unacceptable adverse effects, or have conditions for which glucocorticoids are relatively contraindicated (such as diabetes mellitus), may benefit from local treatment with fluticasone propionate via metered dose inhaler.

Minimizing Breathing Efforts and Avoiding Situations That Induce Cough

Increased breathing efforts necessitated by obstructed upper airways in dogs with brachycephalic airway syndrome can worsen signs of TBM. Although brachycephalic airway syndrome is generally associated with clinical signs due to increased inspiratory efforts, the narrowed airways can presumably also impede normal expiration resulting in the need to create relatively greater intra-thoracic airway pressures during expiration. ***Surgical correction of stenotic nares and elongated soft palate*** can potentially benefit in some dogs with TBM, whether extra- or intra-thoracic. Patients that are overweight and/or unfit can benefit from ***weight loss and exercise.*** Exercise should be tailored to the dog's current fitness level and degree of pulmonary dysfunction to avoid causing excessive respiratory efforts or even death. Contributions from other respiratory disorders, cardiac disease, or systemic disease are treated. ***Harnesses*** should be used instead of collars, and excessive excitement should be avoided. ***Sedatives,*** such as phenobarbital, or ***anxiolytics,*** are prescribed for some animals, and these can be administered before known stressful events.

Cough Suppression

Cough suppressants are often necessary for successful management of tracheal collapse. They may be needed for a short-course (i.e. 1-3 weeks) to break the cough cycle during an exacerbation of signs. As the disease progresses, chronic administration may be necessary. A narcotic cough suppressant, such as butorphanol (0.5 mg/kg q6-12h, orally) or hydrocodone (0.25 mg/kg q6-12h, orally) is generally required. Limiting administration to times of the day when coughing is most severe may preserve some beneficial effect of cough. For dogs with severe cough, hydrocodone may provide the greatest relief and the dosage can be increased to effect while monitoring for side-effects.

Stanzolol

A novel approach to treating TBM has been reported by Adamama-Moraitou, et. al. using stanzolol to improve tracheal wall strength.[16] Possible mechanisms include enhanced protein or collagen synthesis, increase of chondroitin sulfate content, an increase in lean body mass, and decreased inflammation.[16] Dogs with tracheal collapse, but not bronchitis, were treated with 0.3 mg/kg stanzolol divided BID for 2 months then tapered for 15 days. Dogs in the stanzolol group had improved clinical signs by some measures after 30 days, and improvement in grade of collapse was seen on tracheoscopy at 75 days.

ROLE OF STENTS

Tracheal stenting should be considered for animals that are no longer responsive to medical management, usually because of large airway obstruction. The introduction of intraluminal stents has made the option of tracheal stenting much more practical.[17] The most commonly used stents are self-expanding and made of nickel-titanium alloys. In experienced hands, they are simple to place during a short period of anesthesia using fluoroscopic or bronchoscopic guidance. There is minimal morbidity associated with stent placement, and response is immediate and often dramatic. However, clinical signs (particularly cough) may not completely resolve, collapse of airways beyond the trachea and concurrent conditions are not directly addressed (often resulting in the continued need for medical management), and complications such as granuloma formation and stent fracture can occur, limiting their long-term success.

There also may still be a role for external, cervical stenting using prosthetic rings. A recent report by Becker, et. al. found that median survival was not influenced by the presence of intra-thoracic airway collapse in dogs treated with extra-luminal cervical rings.[18] Median survival time was >2500 days for cervical tracheal collapse alone and 1500 days for cervical and intrathoracic tracheal collapse.[18] However, complications included laryngeal paralysis requiring tie-back immediately postoperatively in 7 of 33 dogs (21%), laryngeal collapse in 8 dogs (including 5 of those requiring the tie-back), and respiratory distress requiring tracheostomy in 1 dog. Three dogs (9%) died within 1 week of surgery.[18]

New research in people is showing promise for a technique of central airway stabilization using mesh on the posterior aspect of the trachea and bronchi in patients with severe TBM and COPD. In selected patients, improvement has been demonstrated in pulmonary function tests and scores related to dyspnea and quality of life.[10]

CONCLUSIONS

Treatment of dogs with "tracheal collapse" can be greatly improved by considering the wide-range of presentations of TBM, recommending thorough diagnostic evaluation, and tailoring treatment to the specific abnormalities and response of each patient. On a larger scale, there is potential for significant strides to be made in the management of dogs with TBM as this condition is better characterized.

REFERENCES

[1]Dallman JM, et. al. Am J Vet Res 1988;49:2117. [2]Dallman JM, et. al. Am J Vet Res 1985;46:2110. [3]Done SH, et. al. J Small Anim Pract 1976;17:783. [4]Johnson LR, et. al. J Vet Intern Med 2010;24:298. [5]Johnson LR, et. al. J Am Vet Med Assoc 2001;219:1247. [6]Hawkins EC, et. al. J Vet Intern Med 2010;24:825. [7]Adamama-Moraitou KK, et. al. Vet J 2010; 191:261. [8]Carden KA, et. al. Chest 2005;127:984. [9]Murgu SD, et. al. Respirology 2006;11:388. [10]Ernst A, et al. Chest 2011;140:1162. [11]Pardali B, et. al. J Vet Intern Med 2010;24:832. [12]DeLorenzi, D, et al. J Am Vet Med Assoc 2009;235:835. [13]Hawkins EC, et. al. Am J Vet Res 2006;67:160. [14]Ridge CA, et al. J Thorac Imaging 2011;26:278. [15]White RAS, et al. J Small Anim Pract 1994;35:191. [16]Adamama-Moraitou, KK, et. al. Intl J of Immunopathol Pharmacol 2011;24:111. [17]Sura PA, et al. J Am Vet Med Assoc 2008;232:228. [18]Becker WM, et al. Vet Surg 2012;41;(epub).

FELINE UPPER RESPIRATORY ASPERGILLOSIS: HOW DIFFERENT IS IT FROM CANINE SINONASAL ASPERGILLOSIS?

Vanessa R. Barrs, BVSc(hons), MVetClinStud, FANZCVSc (Feline Medicine) GradCertEd (Higher Ed)
Sydney, NSW, Australia

INTRODUCTION

Canine sinonasal aspergillosis (SNA) was first reported in 1897 shortly after the first reports of SNA in humans in 1895[1]. Reports of a similar disease in cats did not appear until the 1980s[2-4]. Upper respiratory tract aspergillosis (URTA) is an emerging feline infection with over two-thirds of the approximately 50 reported cases being published in the last 5 years[5-12]. Key differences between canine and feline disease include the infecting species, clinical presentation and response to therapy. In dogs the major form of disease (>99% of cases) is SNA; a non-invasive mycosis confined to the sinonasal cavity, while in cats sino-orbital aspergillosis (SOA) is the more common form (63% of cases)[2-3, 5-7, 10-15]. SOA is a deeply invasive mycosis that originates in the sino-nasal cavity and extends to involve orbital and other paranasal tissues.

ETIOLOGY

Controversies in fungal nomenclature and identification (ID)

More than 250 species have been ascribed to the Genus *Aspergillus,* which is subdivided into 8 subgenera; *Aspergillus, Fumigati, Circumdati, Candidi, Terrei, Nidulantes, Warcupi* and *Ornati.* Each subgenus comprises from 1 to 4 sections[16].
The most common isolates from canine and feline URTA are from the subgenus *Fumigati* section *Fumigati,* also known as the *Aspergillus fumigatus* complex[12,17-18,28]. There are occasional reports of isolates from the subgenus *Circumdati (A. flavus* and *A. niger)*[8,19-21] causing SNA in both cats and dogs and from the subgenus *Nidulantes* (*A. nidulans*)[21] causing SNA in dogs.

The *A. fumigatus* complex contains asexual members (anamorphs, mitotic phase), many of which also have sexual forms (teleomorphs, meiotic phase). There has been controversy around fungal taxonomy because teleomorphs in this complex have been assigned to a different genus – *Neosartorya.* Classical nomenclature required that an organism with alternate names be labeled with that of its sexual phase. This created confusion for organisms such as *A. fumigatus* where the teleomorph (*Neosartorya fumigata*) was only recently discovered [22]. Fortunately, in sweeping reforms to the *International Code of Nomenclature for algae, fungi and plants* a "one-fungus, one-name" principle was adopted in 2011 [23]. The revised nomenclature code will be published in 2013. In short, it is likely that the oldest generic name (eg. *A. fumigatus*), irrespective of whether it is typified by a species name with a teleomorphic or an anamorphic type, will be used.

It is now well established that members of the *Aspergillus fumigatus* complex cannot be reliably identified solely on the basis of phenotypic features. Some, termed *A. fumigatus*-mimetic species, have very similar anamorph colony morphology to *A. fumigatus.* This is exemplified by a case of feline SOA where the fungal pathogen was identified phenotypically as *A. fumigatus* but on comparative sequence-based analyses of the Internal Transcribed Spacer (ITS) and beta-tubulin regions was correctly identified as *A. udagawae (Neosartorya udagawae)*[5]. Mis-identification of *A. fumigatus* in human patients with invasive aspergillosis (IA) has important implications for treatment and prognosis. In a review of 86 isolates from human patients with IA previously identified as *A. fumigatus*, 12 isolates were subsequently identified as *A. udagawae* based on beta-tubulin and rodlet A sequences[24]. In this subset of patients the median duration of illness was 7 times longer and disease was refractory to standard therapy. Similarly, a distinctive form of IA characterised by chronicity, propensity to spread across anatomical planes and reduced susceptibility to antimycotic drugs was initially attributed to infection by *A. fumigatus*. The molecular identity was *A. viridinutans.*[25] *A. fumigatus*-mimetic species have higher *in vitro* minimum inhibitory concentrations for amphotericin-B and triazole antifungal drugs (eg. voriconazole) than *A. fumigatus*[26].

Advances in ID of fungal pathogens in feline and canine URTA

Molecular ID of isolates causing feline and canine URTA has been reported since 2007[5, 27-28]. Isolates from 23 Australian cats were identified using PCR and sequencing of the ITS region. *A. fumigatus* was identified as the cause of SNA in 4 of 6 cats[12]. All 17 isolates from cats with SOA were identified as *A. fumigatus*-mimetic species from the *Aspergillus fumigatus* complex. Individual species determination was not possible using ITS sequence analysis alone. In a subsequent study of these and other isolates at the CBS KNAW Fungal Biodiversity Centre in Utrecht, sequence comparisons of the beta-tubulin, calmodulin and rodlet A regions identified *A. lentulus, A. thermomutatus* (*N. pseudofischeri)* and a novel species of *Aspergillus* from cats with SNA. The majority of isolates from cats with SOA were identified as the same novel species of *Aspergillus*. *A. udagawae* was identified as a cause of feline SOA in a single case report from Japan[4]. While *A. fumigatus* was implicated as the etiological agent in previous reports of feline SOA, ID was based on morphologic criteria alone[7,10,15]. Current evidence suggests that *A. fumigatus* is the most common cause of SNA in cats and that SOA is only caused by *A. fumigatus*-mimetic species. Analysis of more cases is required to confirm this.

Phenotypic ID of a large number of isolates from dogs with SNA has implicated *A. fumigatus* as the etiologic agent in most cases[17-18]. However, molecular ID has been performed on a small number of isolates only[12,28]. In 14 dogs with clinically confirmed SNA, DNA of *Penicillium* or *Aspergillus* spp. was detected in nasal mucosal biopsies using a genus-specific real-time PCR assay. Species specific PCRs to detect DNA of *A. fumigatus, A.niger, A. terreus* or *A. flavus* were positive for *A. fumigatus* in 7 dogs and negative in the other 7 dogs. It is possible that these isolates may have been *Penicillium* spp. or another *Aspergillus* spp. Alternatively there may have been insufficient *A. fumigatus* DNA present to be positive with the specific PCR. In another study, ITS sequencing of archival tissues from the sino-nasal cavity of 7 dogs with SNA identified *A. fumigatus* in all 7 cases[12]. Further molecular studies are necessary to determine the frequency of involvement of individual species within the *Aspergillus fumigatus* complex and from subgenera other than *Fumigati.*

PATHOGENESIS

Like SNA in dogs, infection in cats starts in the sinonasal cavity. Direct extension of infection into contiguous structures including the orbit and paranasal tissues occurs in SOA. This is evidenced by progression of SNA to SOA[13], sinonasal cavity involvement on imaging or at necropsy in cases of SOA[7,11-12], and by isolation of the same novel species of *Aspergillus* that causes SOA in a case of SNA . Extension of infection from the sinonasal cavity is usually through the orbital lamina, situated between the orbit and frontal sinus. Given that only *A. fumigatus*-mimetic species have been isolated from cats with SOA, infecting fungal species may be a major determinant of progression. A major difference between SOA in cats and SNA in dogs is the invasiveness of the mycosis. In canine SNA fungal hyphae do not invade the nasal mucosa and are located in adjacent superficial necrotic plaques[29]. By contrast, SOA in cats is an invasive mycosis. Fungal hyphae invade the respiratory epithelium and form granulomas within the orbit and paranasal tissues.

EPIDEMIOLOGY

In both cats and dogs URTA typically occurs in young to middle-aged animals. Of 49 reported feline cases the mean age at diagnosis was 6.5 years with a range from 1.5 to 13 years[2-15,19-20,34]. In case-series of canine SNA, mean age at diagnosis was 5 to 6 years[17-18,30-33] Cats with SNA tend to be older at diagnosis (mean/median 8 y) than those with SOA (mean 6 y/median 8 y). No sex-predisposition has been identified for feline URTA. Of 47 reported feline cases where sex was specified, there were 26 males and 21 females. Males appear to be over-represented in canine SNA. In combined reports of 377 cases there were 245 males and 132 females with a male to female ratio of 1.9:1. A striking difference between feline and canine URTA is the facial conformation of affected animals. As more cases of feline URTA are reported an over-representation of brachycephalic breeds has become apparent. Whilst nearly half of all cases (24) were domestic crossbred cats (22 domestic shorthair, 2 domestic longhair), 40% (21) were of brachycephalic conformation, mostly Persian (10) or Himalayan (6) breeds. By contrast dolicocephalic and mesaticephalic breeds of dogs are predisposed to SNA - infection in brachycephalic breeds is rare. In most case-series comparisons with hospital populations to calculate Odd's ratios were not performed, thus overt breed predispositions have not been confirmed. However, breeds repeatedly identified include Rottweilers, Labrador Retrievers, Golden Retrievers and German Shepherd dogs[17,31-33] In the largest study of feline URTA 4 of 6 cats with SNA and 5 of 17 cats with SOA were brachycephalic, but the difference was not statistically significant[12]. Interestingly, when all published cases are combined, overall 11 of 18 (61%) cats with SNA were brachycephalic while 10 of 31 (33%) cases of SOA were brachycephalic.

The reasons why brachycephalic cats and dolicocephalic or mesaticephalic dogs are predisposed to URTA remain elusive. In both dogs and cats the upper respiratory tract is the major site of localised disease. By contrast, in humans, localised infections are most common in the lower respiratory tract, particularly in immunocompromised patients. There is no evidence of an association between retrovirus infection and URT aspergillosis, with only one FeLV positive case reported[4]. Most dogs with canine SNA are systemically healthy. When URTA does occur in humans, risk factors include decreased sinus aeration and drainage of respiratory secretions secondary to infection, polyps and allergic rhinosinusitis. Reduced drainage of URT secretions due to brachycephalic conformation could be a risk factor in cats. However, since brachycephalic dogs are under-represented for SNA, it is likely that additional risk factors are present in cats. These could include heritable defects in mucosal immunity, previous viral URT infection and previous antibiotic treatment favouring fungal colonisation[34].

CLINICAL PRESENTATION

In dogs the triad of muzzle pain, profuse mucopurulent to haemorrhagic chronic nasal discharge and depigmentation, crusting or ulceration of one or both nares is highly suggestive of SNA. Presenting signs in cats with SNA are more subtle and include a history of sneezing and unilateral or bilateral serous to mucopurulent nasal discharge.[4,8,9,12,20,34] Intermittent epistaxis occurs in 40% of cases. Occasionally, a discharging sinus or soft-tissue mass is identified overlying the frontal sinus or the nasal bone as a result of bony lysis and fungal proliferation[12,19]. Stertor is variably present due to excessive nasal secretions or a caudal nasal cavity/nasopharyngeal fungal granuloma. Nasal depigmentation or ulceration has not been documented in cats with SNA. Cats with SOA are usually presented for investigation of unilateral exophthalmos. Clinical signs are referable to invasive expansion of a fungal granuloma in the ventromedial orbit. In addition to exophthalmos, these include dorsolateral deviation of the globe, conjunctival hyperemia, third eyelid prolapse and exposure keratitis. An oral mass or ulcer in the ipsilateral pterygopalatine fossa behind the last molar tooth is usually present. Extension of infection outside the sinonasal cavity may cause facial distortion, including swelling of the nasal bridge, periorbital tissues and soft tissues adjacent the maxilla. At the time of presentation, nasal signs are absent in 40% of cases but the medical history will reveal sneezing or nasal discharge in the previous 6 months in almost all cases.[10,12,15] Neurological signs have been reported in 15% of cases including blindness, circling, facial muscle fasciculation and hyperesthesia[7,10,12,15]

DIAGNOSIS

Definitive diagnosis of feline URTA requires cytological or histological ID of fungal hyphae in affected tissue and ID of the fungal pathogen. Similar to canine SNA, a definitive diagnosis may require various combinations of diagnostic tests including serology, computed tomography, endoscopy, cytology, histology and fungal culture. Serum anti-*Aspergillus* antibodies can be detected by numerous methods including counter-immunoelectrophoresis (CIE), agar gel immunodiffusion (AGID) or ELISA. These tests have been applied to individual cases of feline URTA only, thus the sensitivity and specificity of antibody detection for diagnosis of feline aspergillosis is currently not known. Five of 10 reported cases (9 SNA, 1 SOA) tested seropositive [4,8,12,20,34]. By contrast, two recent studies evaluated serology for diagnosis of canine SNA[17,35]. In one study AGID was compared with fungal culture of nasal biopsies[17], and in the other the AGID and ELISA were compared[35]. For both methods a purified aspergillin preparation composed of extracts of *A. fumigatus*, A. *niger* and *A. flavus* was used. In one study fungal culture was more sensitive (81%) than serologic testing (67%) while specificity was high for both fungal culture (100%) and serology (98%) [17]. In the second study ELISA had higher sensitivity (88%) than AGID (76%) and specificity was high for both methods (ELISA 97%, AGID 100%). These studies demonstrate that seropositivity for *Aspergillus* spp. is highly suggestive of SNA in dogs but that negative test results do not rule out aspergillosis.

Serum galactomannan (GM) measurement for diagnosis of feline URTA was recently evaluated using the Platelia™ ELISA[36]. GM was measured in healthy cats (n=44) including juvenile (n=31) and adult cats (n=13) as well as in cats with URTA (n=13; 6 SNA, 7 SOA), non-fungal upper respiratory tract disease (n=15) or being treated with β-lactam antibiotics (n=14). The overall sensitivity and specificity was 24% and 78% respectively, for a cut-off optical density index (ODI) of 1.5. High numbers of false positive results were identified in juvenile cats and in cats receiving β-lactam antibiotics. Using the same ELISA, GM was measured in dogs with SNA[35]. Using a cut-off ODI of 0.5, 24% of dogs with SNA tested positive, as did 11% of dogs with nasal tumours, 9% of dogs with non-fungal rhinitis and 24% of healthy dogs. Overall sensitivity was 24% and specificity was 82%. Currently available antigen tests cannot be recommended in the evaluation of feline or canine patients with URTA.

TREATMENT

In general, the prognosis for treatment of feline and canine SNA is good, but treatment can be challenging. Since canine SNA is a noninvasive mycosis, topical therapy is considered more effective than systemic therapy because of direct contact with fungal plaques[29]. However, multiple applications are often required to achieve clinical cure. In a multicentre, retrospective study of 81 dogs with SNA treated with topical clotrimazole or enilconazole using catheters placed intranasally and/or via sinus trephination, a single treatment was successful in 47%[32]. Techniques reported to improve efficacy of topical antifungals include endoscopic/sinuscopic debridement of sinonasal fungal plaques prior to therapy[37], endoscopic guidance of infusion catheters into the caudal frontal sinus[37] and use of depot preparations of 1% clotrimazole or1% bifonazole cream[35]. Information on treatment of feline SNA is limited. Of 14 feline cases with available follow-up, signs resolved in 11 (78%)[4,8,12,20,34] The most common successful treatment regimes were systemic antifungal therapy alone (triazole +/- amphotericin-B) (n=5), systemic triazole therapy combined with topical intranasal clotrimazole (n=2) or topical therapy alone (single 1% intranasal clotrimazole infusions) (n=2). Similar to canine SNA, the importance of debridement of fungal plaques is illustrated by resolution of signs in one case[34] after rhinoscopy and sinonasal cavity lavage and in another after sinusotomy/rhinotomy and instillation of iodoform paste[4]. Also, debridement of gross fungal plaques in cats with SNA treated with systemic antifungal therapy could have contributed to treatment success[12]. Presence or absence of orbital involvement in cats with URTA has important prognostic significance. Few cases of SOA have been treated successfully despite aggressive therapy including orbital exenteration and use of newer generation fungicidal triazoles (posaconazole or voriconazole), liposomal amphotericin-B and the echinocandin micafungin[5,7,10-12,15]. Fluconazole should not be used for treating feline URTA because most species in the *Aspergillus fumigatus* complex are resistant *in vitro*[12,26]. Posaconazole or itraconazole are recommended for first-line therapy. The pharmacology of posaconazole has not been determined in cats but it is well tolerated after oral administration[12]. Although isolates show *in vitro* susceptibility to voriconazole, this drug should not be used initially due to the reported high frequency of adverse neurological events [9-10,12]. Where disease is confined to the nasal cavity topical treatment is also recommended. For SOA use of triazole antifungals in combination with amphotericin-B may improve outcomes. Molecular ID of fungal pathogens causing feline URTA will enable development of optimal treatment protocols.

REFERENCES

[1] Black L, et al. **Vet Rec** 1973;92;447. [2]Peiffer R, et al. **J Am Vet Med Assoc** 1980; 176:449. [3]Wilkinson G, et al. **J Small Anim Pract** 1982; 23:127. [4]Goodall S, et al. J Small Anim Pract 1984;25:627. [5]Kano R, et al. Mycoses 2008; 51:360. [6]Karnik K, et al. **Vet Radiol Ultrasound** 2009;50:65. [7]Barachetti L, et al. **Vet Ophthalmol** 2009;12:176. [8]Furrow E, et al. **J Am Vet Med Assoc** 2009;235:1188. [9]Quimby J, et al.**J Vet Int Med** 2010;24:647. [10]Smith L, et al. **Vet Ophthalmol** 2010:13:190. [11]Giordano C, et al. **J Fel Med Surg** 2010; 12:714. [12]Barrs V, et al. **Vet J** 2012:191:58. [13]Halenda RM, et al. **Vet Radiol Ultrasoun** 1998;38:208. [14]Hamilton H, et al. **J Am Anim Hosp Assoc** 2000;36:343. [15]McLellan G, et al. **J Am Anim Hosp Assoc** 2006;42:302. [16]Samson R, et al. In: Aspergillus: molecular biology and genomics, Caister Academic Press 2010:19. [17]Pomrantz J, et al. **J Am Vet Med Assoc** 2007;230:1319. [18]Pomrantz J, et al. **J Am Vet Med Assoc** 2010:236:757. [19]Malik R, et al. **J Fel Med Surg** 2004;6:383. [20]Whitney B, et al. **J Feline Med Surg** 2005;7:53. [21]Sharp N. In, Greene C (Ed) **Infectious Diseases of the Dog and Cat**, 2nd Ed WB Saunders 1998:404. [22]O'Gorman CM et al. **Nature** 2009;457:471. [23]Miller J, et al. **PhytoKeys** 2011:5:1. [24]Sugui J, et al. **J Clin Microbiol** 2010;48:220. [25]Vinh D, **Emerg Infect Dis** 2009;15:1292. [26]Alcazar-Fuoli L, et al. **Antimicrob Agents Chem** 2008;52:1244. [27]Barrs V, et al. **J Vet Intern Med** 2007; 21:579 (abstr). [28]Peeters D, et al. **Vet Micrbiol** 2008;128:194. [29]Peeters D, et al. **J Comp Path** 2005;132:283. [30]Johnson L, et al. **J AmVet Med Assoc** 2006;228:738. [31]Billen F, et al. **Can Vet J** 2010:51:164. [32]Sharman M, et al. **J Sm An Pract** 2010:51:423. [33]Mathews K, et al **J Am Vet Med Assoc** 1998; 213: 501. [34]Tomsa K, et al. **J Am Vet Med Assoc** 2003; 222:1380. [35]Billen F, et al. **Vet Microbiol** 2009;133:358. [36]Whitney J, et al. **ANZCVsc Science Proceedings** 2011;9. [37]Zonderland J, et al **J Vet Med Assoc** 2002;221:1421.

PULMONARY DISEASE FROM CANINE- AND FELINE-ASSOCIATED INTESTINAL HELMINTHS

Dwight D. Bowman, MS, PhD
Ithaca, NY

INTRODUCTION

The common intestinal helminth parasites of dogs and cats also compromise the lungs of the infected host. It is often forgotten that hookworms and roundworms cause lung lesions as they make their way to the site where the adults are found. Also, these lesions in the lungs can be significant and interfere with the diagnosis of other helminth infections of the lungs, specifically, *Dirofilaria immitis* and *Aelurostrongylus abstrusus*. The worms that cause lung and gastrointestinal disease in dogs are *Toxocara canis*, *Ancylostoma canium*, and *Ancylostoma braziliense*; the ones in cats are *Toxocara cati*, *Ancylostoma tubaeforme*, *Ancylostoma braziliense*, and *Toxocara canis*. It may be that *Toxocara cati* causes disease in dogs, but this has not been examined. In parts of the world where Angiostrongylus vasorum occurs, there has also not been any significant effort to determine to what extent the migration of *T. canis* larvae through the lungs might interfere with the diagnosis of the condition or perhaps worsen the outcome of infection. Similarly, although lung disease is a regular part of a *T. canis* infection, there has been no real effort to determine what effects a concomitant infection might have on heartworms.

PARASITE PREVALENCE

The prevalence of these parasites in the non-cared-for-pet remains very high. Nationally in an ongoing survey of shelter animals that consists of several thousand canine and fecal samples from around the nation, 34% of dogs had eggs of *Ancylostoma*, 13% of dogs had eggs of *Toxocara*, and for cats 9% had eggs of *Ancylostoma* and 22% had *Toxocara* (Dr. B.L. Blagburn, Auburn University, personal communication). For cared-for-pets fairly high rates of infection can be seen on the national prevalence maps at the Companion Animal Parasite Council website www.capcvet.org where 6.1% of 780,000 feline samples were present for *Toxocara* eggs as were 2.9% of the 4 million canine samples. Similarly, 3.6% of the canine samples were positive for *Ancylostoma* eggs; only 0.2% of the feline samples were positive for *Ancylostoma* eggs.

THE INSULT

The gastrointestinal hookworms and roundworms migrate to the lungs as a routine part of the life cycle. In the case of *T. canis* and T. cati, the larvae undergo a liver-lung migration after infective eggs are ingested and then make their way via the respiratory escalator to the larynx where they are then coughed up and swallowed. With both these parasites, some percentages of the larvae also get carried via the bloodstream to other tissues such as the muscle and brain where they enter a hypobiotic state. The hookworms, *A. caninum*, *A. tubaeforme*, and *A. braziliense* that penetrate the skin of their host make their way to the intestine also via the lungs where they are carried in the venous circulation. Again the hookworm larvae make their way up the respirator tree before returning via the pharynx to the intestinal tract. As with roundworms, some of the hookworm larvae are also shifted into the various tissues of the body where they undergo hyperbiosis.

As the larvae make their way through the lungs they cause significant disease. What is often not realized is the effect that *Toxocara cati* has on the lungs of the infected cat. The majority of work done in this regard was published only in the form of a thesis submitted in 1969 to the University of Connecticut (Swerczek, 1969). In this work, it was presented in great detail that the experimental infection of cats with *T. cati* induced severe lesions, medial hypertrophy of the pulmonary vessels, in the lungs. It was concluded that these studies showed "that the lesion is produced earlier and is much more extensive than in *A. abstrusus* infection. The earliest change noted in the arteries was mild intimal proliferation 2 weeks post infection, and medial hyperplasia was observed as early as 3 weeks and was severe by 6 weeks postinfection." It was felt that the observed lesions seen in random-sourced cats was probably most commonly caused by *T. cati*. Also, the lesions did not clear in cats that had been infected, and it was considered that the induced lesions were irreversible. Also noted in these cats, like with *A. abstrusus* infections, was the occurrence of peribronchial mucus gland hyperplasia. A single egg inoculum caused the lesion, and repeated reinfection simply caused them to worsen and to persist for longer periods.

Swerczek also reported on the disease induced in cats after their experimental infection with the eggs of the canine roundworm, *Toxocara canis*. About this infection, he stated that "the focal pneumonia associated with the migration of *T. canis* larvae was much more severe than that seen in *T. cati* infection." and "Muscular hyperplasia of the bronchioles and terminal alveolar ducts was also seen and was more severe in cats with evidence of sever ascarid infection." A cat was found naturally infected with larvae identified as those of *T. canis* with very large granulomas in the kidneys that also had significant lesions is the lungs similar to those described by Swerczek in his thesis.[1] The lesions in the cat were similarly reproduced in cats that were experimentally infected with *Toxocara canis*, and these lesions also included severe lung disease with medial hypertrophy of the pulmonary vessels.[2]

Unfortunately, there is much less known about the effects of the canine or feline roundworm on the lungs of dogs. Also, there is almost nothing known about the effects of hookworms on the canine and feline lung. How much such lesions lead to the pathology associated with canine heartworm disease is difficult to ascertain. It is known that in areas where people are routinely infected with hookworms that there are lesions that appear in the lungs each season when the worms again become active or when people again undertake activities that put them at risk. Also, it is known that when people go to an area such as Florida or Thailand and develop cutaneous larva migrans that this is also accompanied by associated pulmonary lesions. There is every reason that similar disease will also appear in cats and dogs.

PREVENTION

Overall, based on the high prevalence of *T. cati* seen in cats, the knowledge that this parasite causes lung lesions, and the fact that infections with *T. cati* and *T. canis* can occur all season long, is an argument for keeping cats on preventive therapy. When coupled with the amount of lung disease associated with heartworms and lungworm, the three together argue very strongly for year-round heartworm and broad spectrum parasite control in cats.

Unfortunately, it is fairly unknown to what extent the different canine and feline preventives would minimize the effects of these repeated infections and insults to the canine lung. It is not expected that many of the drugs in the doses and formulations used would be stopping the migration of the larvae, since they do not prevent the transplacental infection of puppies with larvae from the mother in the case of *T. canis* or the transmammary infection of puppies with *A. caninum*. However, there is reason to suspect that some of these molecules in the different formulations that exist may minimize the effects on the dog or cat to which they are administered. However, it would seem that the continued use of products in dogs and cats would minimize the numbers of animals seeing the stages that induce the lesions.

It needs to be remembered that the use of various heartworm and internal nematode preventives on a seasonal or year-round basis will mask the repeated infection of a host with hookworms and roundworms. These drugs are designed and labeled to kill the adult worms, not the migrating larvae. Thus, although the fecal of the animal may therefore remain free of hookworm and roundworm eggs, it will not necessarily be free of regular infection through the ingestion of roundworm eggs or the skin penetration of hookworm larvae.

CONCLUSIONS

Roundworms cause significant medial hypertrophy of the pulmonary vessels; this is similar to what occurs in cats with heartworms and *Aelurostrongylus* infections, and these lesions do not appear to be associated with clinical signs of pulmonary hypertension. Thus, in looking at lung disease in cats, *T. cati* needs to be considered as one of the rule-ins or rule-outs when such histological changes are noted in surveys of feline lungs showing similar lesions. It appears that unlike with *A. abstrusus* in cats and heartworms in dogs and cats that the radiographic changes associated with *T. canis* and *T. cati* will not be associated with enlarged pulmonary arteries in the caudal lung lobes on radiographic evaluation.[3]

The various heartworm and internal parasite preventives that we are routinely using in pets are markedly reducing the prevalence of gastrointestinal parasites compared to pets that do not receive regularly veterinary care. However, dogs and cats continue to be at risk of contracting these parasites from the environment and of developing disease due to these forms that may not be associated any longer with stages that can be diagnosed on routine fecal examinations. Work needs to progress to determine the actual magnitude of the disease that is induced and the means by which such disease can be presented. In people such infections and pulmonary signs are regularly associated with marked eosinophilia, but dogs and cats appear to respond to these infections with only minimal peripheral eosinophilias as compared to humans. Of course, this might be different in dogs that are adult and naïve to the infections or which have had regular abbreviated infections through the routine use of preventive products.

REFERENCES

[1]Parsons JC, et al. **J Comp Pathol** 1988;99:343. [2]Parsons JC, et al. **Int J Parasitol** 1989; 19:479. [3]Dillon AR. **Bayer Cutting Edge Symposium/NAVC** 2011,
http://www.bayeranimalhealthsymposium.com/dillon.shtml

REVIEW OF PARASITIC DISEASES ASSOCIATED WITH THE NEONATAL DOG AND CAT

Dwight D. Bowman, MS, PhD
Ithaca, NY

INTRODUCTION

There are parasites that routinely are part of the neonatal package of disease agents in puppies and kittens. These include helminths, protozoa, and arthropods. Although many of these agents cause gastrointestinal disease, other organ systems, e.g., hepatic, renal, pulmonary, nervous, etc., can be compromised in these young animals. The reason why these parasites are of such significance in these young animals is due to many reasons that include the small size of the host, the poorly developed immune response of the young, and the fact that the parasites have life cycles specifically derived for the purpose of infecting the prenatal, neonatal, or perinatal host.

MODES OF INFECTION

Puppies and kittens are often infected by ingesting eggs and cysts of parasites from the environment into which they are born. The ingestion of these environmentally contaminating stages is the major source by which young animals become infected and develop neonatal coccidiosis and toxascaris due to the roundworm *Toxascaris leonina*. Puppies can also develop massive infections with roundworms, *Toxocara canis*, hookworms, *Ancylostoma caninum*, and the threadworm, *Strongyloides stercoralis*, from the environment, often serving as their own source of the stages that are infecting them and ultimately inducing the development of the hyperinfections with these parasites that can prove fatal. This is also what can occur within the environment where the fairly immobile young are perfect prey for fleas, *Ctenocephalides felis* and *C. canis*, and lice that can rapidly bloom to very high numbers. However, many of the agents causing significant disease in neonates have life cycles that permit the direct infection of the prenatal or neonatal animal for the purpose of infecting the animal at a stage when its immune system is minimally protective. Two parasites are deadly when the kitten or puppy is infected transplacentally, i.e., in cases of prenatally acquired toxoplasmosis and neosporosis. *T. canis* is always a threat to puppies because of the ability of the larvae to leave their hypobiotic state and migrate into the liver and lungs of the developing canine fetus. Some parasites have developed the means of infecting the next host via transmission in colostrum of milk; this is the mode of infection by which dogs are infected as puppies with *A. caninum* and *S. stercoralis*, and both of these infections can have fatal outcomes. Kittens are susceptible to several infections from their mothers via milk if the queen is infected during lactation. This is the case with two trematodes, the intestinal flukes of the genus *Alaria* and the lung fluke *Paragonimus kellicotti*; kittens; this is also what occurs when kittens are nursing on a queen that has recently ingested the eggs of the feline roundworm, *Toxocara cati*.[1] Two canine parasites are probably transmitted from the mother to the neonatal puppy through saliva as she nuzzles her young; one of these parasites is the nasal mite *Pneumonyssoides caninum* and the other is the lungworm *Filaroides hirthi* – both of which have developed the means of regularly infecting puppies soon after birth. Dogs will also shed the larvae of *F. hirthi* in the feces, and these larvae are directly infective to puppies. Similarly, both dogs and cats can be shedding oocysts of *Cryptosporidium* and the cysts of *Giardia* in their feces, and these stages will be directly infective to the offspring. Two other parasites infect puppies and kittens not uncommonly, but in a less direct fashion. First, the tapeworm *Dipylidium caninum* can build up to massive numbers in neonatal puppies and kittens due when there is an overabundance of fleas and flea larvae and mom is already infected. Second, the maggots of different *Cuterebra* species can be brought home to the puppies or kittens by a mother cat or dog that has been hunting around rabbit or rodent burrows and has the fresh maggots on her fur that can then be transferred to the kitten or puppy where they can enter a body opening and begin to mature in the neonatal animal.

MAJOR CLINICAL SIGNS

The parasites described above that make up the neonatal package often produce signs that are fairly distinctive and recognizable. The problem is that when present in large numbers, many of these infections can be life threatening. Veterinarians practicing in a given area or with certain housing situations, shelters, breeding kennels, catteries, often learn to recognize the common signs of the agents that are most likely to affect their specific population. However, even the best practitioner can be caught off guard when the disease presents atypically or when it presents for the first time. It also must be remembered that very typically puppies and kittens will be infected with several different agents making it harder to classify the lesions that are observed.

Dysentery or Diarrhea with Melena or Blood

This is the typical presentation with coccidiosis or prenatal or neonatal hookworm disease. The puppies (and with coccidiosis, to some extent kittens) will lose large quantities of blood into the stool from epithelial erosion in the case of coccidia and from the feeding activities of the developing stages and adults of hookworms. Sign of such infections will be associated with marked anemia and dehydration. The puppy may be depressed with a respiratory rate around 30, a heart rate near 200, with a low temperature (unless a secondary bacterial infection), with mucous membranes that are tacky and pale, and a hematocrit of 10% or lower. Death due to hookworms can occur as early as 10 days after the animal is born.

Anemia

In the case of toxocariasis and flea induced anemia, there is likely to be no associated diarrhea. Both situations have a potential fatal outcome. Puppies and kittens have been exsanguinated by fleas in just about every environment. In the case of toxocariasis, the anemia is often a medical emergency caused by intestinal perforation in puppies 4 to 8 weeks of age. Flea infestations can kill animals at a much younger age if the number of fleas in the area where the puppies or kittens is housed has high enough levels of fleas and developing stages present.

Diarrhea

Cryptosporidium, *Giardia*, and *S. stercoralis* will often cause diarrhea without anemia. Simple, uncomplicated giardiasis will present with no blood. Similarly, severe cryptosporidiosis typically presents without the appearance of blood in the stool. Strongyloidiasis can sometimes be associated with bloody stools or melena, but this would usually be in cases of massive autoinfection with erosion of the small intestinal epithelium.

Impaction

The parasites most likely associated with impaction in the neonate are infections with *Toxocara canis* in puppies, *Toxascaris leonina* in puppies or kittens, and perhaps most commonly before the development of the highly efficacious modern flea products with Dipylidium caninum. These parasites can be present in massive numbers in very young animals, and death can occur anywhere after 3 weeks of age. The very short life cycle of *D. caninum* makes this tapeworm a risk to puppies and kittens if they are being cared for in an area with flea developmental stages present.

Respiratory signs

Infections with *F. hirthi* are going to present usually as a simple dry cough. In heavy infections, a diffuse pattern will appear on radiographs of the lung. Massive hyperinfection and death has occurred when dogs with existing infections have been immunosuppressed for extended periods with prednisolone. *P. caninum* that lives in the nasal sinuses is reported to cause reverse sneezing in infected dogs. *Alaria* and *P. kellicotti* only rarely probably ever cause neonatal disease in dogs and kittens, but the larvae of *Alaria* are liable to migrate to the lungs before returning to the intestinal tract, and very young kittens in Michigan have been found with patent *P. kellicotti* infections which is the basis for the belief that they were infected via the transmammary route. Toxoplasmosis in prenatally infected dogs and cats can present as pneumonia associated with tachyzoites developing within the lung tissue.

Neurologic disease

Toxoplasmosis and neosporosis can both cause severe neurologic disease in the prenatally infected dog. Prenatally acquired neosporosis very typically presents with the classic signs of flaccid hindlimb paresis. Puppies and kittens infected prenatally can also be born with severe cerebral deficits due to tachyzoite damage. When the maggots of *Cuterebra* migrate into the neurologic tissue, the lesions will depend on whether the maggots have entered the brain, brain stem, or spinal cord.

Systemic disease

Toxocara can cause significant disease that appears unrelated to the gastrointestinal location of the adult worms because sometimes young adults and mature worms will migrate into ectopic sites. Thus, these worms can cause biliary obstruction and pancreatitis with associated death by migrating into the pancreatic duct. The maggots of *Cuterebra*, although typically making their way successfully back to the subcutaneous warble in the infected dog or cats after there several week sojourn in the viscera, will occasionally get lost and besides causing neurologic disease can be expected to turn up in just about any atypical location.

DIAGNOSIS

The diagnosis of the intestinal agents is done by routine fecal examinations in association with clinical signs. However, after that it gets more complicated. Emergency presentations with anemia may be diagnosed from the fleas and flea dirt present, but if due to perforation by an ascarid, it will probably require exploratory surgery. The classic neosporosis presentation can be diagnosed by the clinical presentation, but after this it will require significant clinical work up, perhaps with a neurologist and multiple serological tests. Most infestations with the nasal mite will be overlooked, and the infections with *F. hirthi* are liable to be missed in most cases unless the disease becomes severe. *Cryptosporidium* and *Giardia* are fairly easy to diagnose, but this then leads to the continual question as to what next?[2]

TREATMENT

Many of these diseases can be prevented by routine parasite control in the mother and the offspring, beginning animals on year-round internal and ectoparasite control as soon after birth as feasible. In areas with lots of fleas where *D. caninum* is still present, a product containing praziquantel might be a useful addition to puppy and feline healthcare. Coccidiosis remains a disease where there is still not a good regular treatment in the US.[3]

CONCLUSIONS

Year-round parasite control of the mother and routine treatment of puppies and kittens as soon after birth as possible cannot prevent all of these infections, but they can significantly reduce the severity of the infections and the chance of complications.

REFERENCES

[1]Coati N, et al., **Parasitol Res** 2004, 92;142. [2]Montoya A, et al. **Parasitol Res** 2008; 103:1141. [3]Schimmel A, et al., **Parasitol Res** 2011; 109:s1.

LYME DISEASE: WHICH TEST IS BEST?

Meryl P. Littman, VMD, DACVIM
Philadelphia, PA

INTRODUCTION

When a dog is "positive (+) for Lyme disease" does it matter which test was used? Does the dog have Lyme disease or does being "Lyme +" only mean the dog has antibodies to one or more *Borrelia burgdorferi* antigens? Proof of natural exposure, having compatible clinical signs, ruling out common differential diagnoses, and response to therapy may suggest Lyme disease, but true documentation of disease and causation may be difficult. Doxycycline may treat many tick-borne diseases that mimic Lyme disease and even non-infectious diseases may respond to its anti-inflammatory and anti-arthritic properties.[1] In some endemic areas, 70-90% of asymptomatic dogs were found to be seropositive,[1] so Lyme+ status found in a sick dog may be just a coincidence. Although none of the following tests prove Lyme disease as cause of illness nor predict future illness, some do offer helpful information to clinicians.

TESTS FOR LYME ANTIGENS

In experimental studies, the sites of *Ixodes* tick bites are controlled and recorded. Skin biopsies from those sites can be cultured for *Borrelia burgdorferi* (Bb) organisms or tested by polymerase chain reaction (PCR) using primers to detect Bb organismal DNA or plasmidial DNA. Even a year after experimental exposure, tests show viable organisms (by culture) or Bb DNA (by PCR) in the skin biopsies of previous tick bite sites in untreated dogs and in perhaps 10% or more of treated dogs.[1] But in field cases, Bb cultures and PCR tests are rarely positive even when testing targeted areas (joint fluid, synovia, or kidney) since signs are caused by immune-mediated responses in predisposed individuals.[1] Lyme arthritis in experimental Beagle puppies (adults do not show illness) occurs first in the leg closest to where ticks attached, thus Bb travels to joints by tissue migration and not hematogenously.[1] Testing a dog's blood sample for Lyme antigens by culture or PCR is not fruitful and Lyme+ dogs can be used as blood donors.

TESTS FOR LYME ANTIBODIES

Since we cannot easily find Lyme antigens, we seek to find proof of exposure by testing patient sera for Lyme-specific antibodies. There are a variety of tests now available but not all of the information they give is helpful. Here are some questions:

1. Does the test differentiate between vaccinal and natural exposure antibodies? Are there any other cross-reactions?
2. Will the test give a positive result at the time of presentation with illness, or do we need to do paired titers (acute and convalescent)? Answer: The experimental model of Lyme arthritis (there is no model for Lyme nephritis) shows a long incubation (2-5 months) before clinical signs occur, so seroconversion for all of the following antibody tests will have occurred by the time of presentation.[1] Paired tests are not necessary for Lyme disease, but may be indicated for other infectious diseases mimicking Lyme disease, eg, Anaplasmosis, Ehrlichiosis, Rocky Mountain Spotted Fever, Leptospirosis.
3. Does the test result indicate acute or chronic infection? Does it matter how long the dog has been infected? Some tests are positive 1-2 weeks earlier than others, however since the incubation is long, the dog will probably not present ill yet. It is debatable if it matters whether the dog was exposed in the last few months or whether it was exposed years ago. We may only be able to tell when it was last exposed, not when it was first exposed, and this information may not be clinically helpful.
4. Does the test give a qualitative result (positive or negative) or a quantitative result? Is a quantitative result helpful?
5. Does the test become negative or wane if the dog is successfully treated? How long does it take to see a change?
6. Does the test result prove causation of current illness or predict illness in the future? Unfortunately none of the tests currently available prove causation of current illness nor predict illness whether illness is likely to occur in the future. There appears to be no correlation between seropositivity or heighth of titer and illness; the vast majority (95%) of seropositive dogs in experimental models and field cases have high quantitative levels and remain non-clinical and non-proteinuric.[1,2]
7. Does the test include information about other infectious diseases? Is it a point-of-care test?

Indirect Fluorescent Antibody (IFA, whole cell), Enzyme-Linked Immunosorbent Assay (ELISA, whole cell)

Historically, the first test developed in the 1980's as an assay for antibodies directed against Bb was a whole cell IFA test (the second was the whole cell ELISA). The tests use antigens derived from cultured Bb, which express the same antigens as Bb does in the tick (including ospA), and not necessarily the combination of antigens expressed by Bb in the host during natural infection, but there is enough overlap of expression so that positive results are obtained. These tests may pick up cross-reacting antibodies to other *Borrelia* spp. and possibly to other spirochetes. The tests do not differentiate Bb antibodies directed against vaccinal antigens from natural exposure antigens since vaccinal antigens are also derived from cultured Bb. Titer results are quantitated but the heighth of the titers do not correlate with illness or its severity, nor predict illness, and many asymptomatic dogs have very high titers.[1,2] Early seroprevalence studies showed that less than 5% of seropositive dogs had limb or joint disease signs over a 20 month observation period,[1] similar to the number of seronegative dogs that showed those signs. Whole cell ELISA antibody titers did not decline as much as C6 antibodies (see below) after treatment.[2,3]

IgM versus IgG Lyme Titers

These antibodies are found by IFA or ELISA methods, using anti-canine IgM or IgG antibodies. IgM antibodies are the first antibodies to respond to infection, while IgG antibodies are produced later after seroconversion. The clinical information obtained is not usually helpful in canine practice, because there is no acute phase of illness in the dog, compared with humans, and dogs do not present sick with Lyme disease before IgG antibodies are present. In the experimental model of Lyme disease, in which infected ticks were placed on puppies which were observed/tested over a year, clinical signs were not seen until 2-5 months after exposure, well after they had seroconverted and had IgG antibodies present.[1] By 2-5 months after exposure, IgM may be present or declining, and IgG antibodies would be present. It would be unusual to find a sick dog that is Lyme IgM+ and IgG negative. If a dog presented with clinical signs at that time, it may be a coincidental finding and the clinical signs may be due to something else, e.g., Anaplasmosis. Another confounder is that since Bb manifests antigenic variation in an untreated carrier dog (whether it is symptomatic or not), there are new antigens presenting to the host over time, and thus new IgM peaks will respond to those new antigens, so finding a dog with IgM+/IgG+ status does not necessarily help to answer the question of whether the dog was exposed for the first time recently or if perhaps the dog is a chronic carrier, exposed long ago, and now either recently exposed again or perhaps showing new IgM production for new antigens expressed as the Bb goes through its antigenic repertoire, as the organisms evade the host's immune system.

Western blot

The Western blot (WB) test uses cultured Bb antigens separated by their molecular weights, to which serum antibodies (if present) bind and can be marked, giving a pattern of dark bands (precipitations). The WB was the first test to help differentiate between vaccinal antibodies and natural exposure antibodies, because different antigens are preferentially expressed by Bb in the tick vs. in the host. Bacterin vaccines are made from killed organisms that are grown in culture, which express the antigens that Bb generally express when living in a tick. Many bands are seen in bacterin vaccinated dogs, including an ospA band (see ospA below). Non-exposed dogs vaccinated with a recombinant ospA subunit vaccine would have just a solitary band at the ospA level (p31). In comparison, dogs with natural exposure antibodies are less likely to show a band at p31. During tick feeding, changes in gene expression cause some expected surface protein changes, for instance, an ospC band (p23) at early infection, and an ospF band (p28) during later infection. In the chronic carrier, random and spontaneous mutations and recombinations in expressed and silent cassettes in Bb gene families allow for many more antigens to be expressed and new bands will appear.[4] Interpretation of the WB banding pattern is not always clear-cut especially in late or chronic infections.[4] Chronic carriers may even show bands seen with vaccination with bacterin, since Bb shows the host its antigenic repertoire over time, including antigens normally expressed in ticks, or *in vitro*, or in bacterins. For instance, 17% of non-vaccinated dogs showed ospA bands.[4,5] In about 10% of non-vaccinated human patients (usually HLA-DR4+ haplotype), chronic antibiotic-resistant Lyme arthritis occurs, correlated with very high anti-ospA antibody levels and ospA-sensitized T cells, as arthritis worsens.[6] Thus, ospA can be seen in the infected host and are not always vaccinal antibody (see ospA below). More study needs to be done to determine if certain bands are associated with serious disease, eg, Lyme nephritis.

C6 ELISA[7,8]

The C6 ELISA test detects antibodies directed against a synthesized protein (C6) which mimics a conserved portion (IR6) of the VlsE antigen (variable major protein-like sequence, expressed) which is only expressed by Bb when it is in the host and not within the tick or culture, and therefore not present in any Lyme vaccines. The C6 ELISA is a specific and sensitive test for natural exposure antibodies. Reagents are not species-specific so the test has been used to study seroprevalence in cats and horses. The C6 antibodies are detected 3-5 weeks after natural exposure in dogs (before clinical signs are seen in the experimental model), remain high in the carrier state (whether sick or not), and wane by 6 months after treatment.[1,2] The in-house SNAP kits give qualitative results (negative or positive) for heartworm antigen and antibodies to Lyme C6 antigen, Ehrlichia canis/chaffeensis (SNAP-3Dx), plus Anaplasma phagocytophilum/platus (SNAP-4Dx), and plus Ehrlichia ewingii in the new SNAP-4Dx Plus Test (IDEXX). The Lyme Quant C6 (IDEXX) gives a quantitative result, which can be compared over time, but the heighth of the Quant C6 value does not predict illness, since many healthy asymptomatic, non-proteinuric dogs have very high values.[2] Decisions about treatment with antibiotics may not necessarily rely on the Quant C6 results,[9] but if a dog is treated, comparisons between pre- and 6 month post-treatment levels may be useful, because the Quant C6 has been shown to wane after treatment with antibiotics[1,2] while the whole-cell ELISA antibodies test do not wane as much.[2,3] In experimental dogs that were treated early (several months after exposure), post-treatment Quant C6 often dropped into the range considered negative (<30).[10] Field cases may not be treated as early but ≥ 50% decline is usually seen, probably indicating decreased antigenic load, decreased circulating immune complexes, and possible clearance, although Quant C6 may not always drop into the negative range.[2] If the decline is not great, it could be due to lack of clearance, re-exposure, or immune memory. Especially in chronic cases and when the pre-value is not very high, the decline may not be appreciable despite response to therapy.[1] Perhaps due to chronicity, immune memory may allow for a long-term positive low level of antibodies. Comparing pre- and 6 month post-treatment Quant C6 may be helpful to check for a decline, and to get a new baseline for future comparisons should signs recur; e.g., if increased again, retreatment may be justified, and if not increased over the new baseline, other differentials should be entertained. The qualitative C6 on the SNAP tests is very sensitive and may remain positive after treatment for many years.

OspA Antibody

OspA is the outer surface protein on Bb that is usually expressed when Bb is living in the tick. This lipoprotein acts to anchor Bb to the tick's midgut. During tick feeding, the expression of ospA is down-regulated, allowing the organism to unhook from the midgut and enter into the tissues of the host that the tick is feeding on, thus transmitted Bb organisms have less or no ospA on them. The transmission of Bb during tick feeding was found to occur at 52-53 hours in the mouse model, indicating the time it takes for decreased expression of ospA to occur.[4] OspA is expressed when Bb is cultured *in vitro*, and so is present in killed bacterins which come from Bb cultures. OspA is also present in subunit recombinant ospA vaccines, which work by increasing anti-ospA antibody in the vaccinated dog, so that during tick feeding, anti-ospA antibodies will enter the tick's midgut and kill Bb organisms there, thwarting transmission to a vaccinated dog. OspA antibodies rise and fall soon after vaccination; it is unknown what a protective titer level is.

We used to think that ospA was not expressed in the host, but perhaps some Bb entering into the host may still be expressing some ospA, so that ospA may be present in the host very early after exposure. During chronic infections, when Bb is going through its antigenic repertoire to avoid the host immune system, ospA can be expressed again on Bb while in the host (see Western blot above). OspA is pro-inflammatory and anti-ospA antibodies may be associated with immune-mediated disease in humans.[6]

OspC Antibody

OspC is the outer surface protein on Bb that is up-regulated and expressed more and more on Bb as the tick is feeding, so that it is one of the antigens that is first seen by the host's immune system after transmission of Bb into the host. OspC antibodies are detected at the p23 band (see Western blot above). OspC antibodies rise at 2-3 weeks after exposure, which is earlier than C6 or ospF antibodies are detected (see below), and wane 3-5 months after exposure,[11] which is still probably before many dogs become ill, since canine Lyme disease has such a long incubation. Since the dog does not present ill at this time, it is not necessarily helpful to know if OspC antibodies are present. Since experimentally, puppies did not show signs of Lyme arthritis until 2-5 months after exposure, a dog coming to the clinician sick with Lyme disease would probably have waning or negative ospC antibodies at the time of presentation, if it only had been exposed that once. If a dog has re-exposure, ospC may rise again, and so be found not necessarily in relation to when Lyme disease was first contracted. Since the experimental models only use a single time of tick exposure, many of the findings concerning Lyme tests are simplified, and not as complicated as in field cases which are likely to have multiple exposures.

OspF Antibody

OspF is the outer surface protein that is expressed on Bb in the host later in the course of infection.[4] Anti-ospF antibodies generally rise 6-8 weeks after exposure and remain high during the carrier state.[11] They would probably be detected by the time clinical signs are apparent, since the experimental model showed no acute phase of illness and signs of Lyme disease would not be expected until 2-5 months post-exposure. As with other quantitative tests, the heighth of the titer does not predict illness. More study is needed to see if ospF antibodies wane after treatment. In most clinical situations, by the time a dog presents ill with Lyme disease, I suspect that ospC antibodies may be absent (unless there are multiple exposures) and ospF antibodies would be present.

Multiplex Testing (Cornell University, Animal Health Diagnostic Center)[11]

This test includes a Western blot and quantitation of antibodies directed against ospA (usually considered due to vaccine but possibly not), ospC (as a marker for recent natural exposure), and ospF (as a marker for chronic infection, carrier status, or natural exposure obtained at least 6-8 weeks ago). See above for the information that is obtained from each of these tests.

AccuPlex4 (Antech Diagnostics)

This test may pick up antibodies earlier than other available tests for Lyme, Ehrlichia, Anaplasma, and for the antigen of heartworm. For Lyme, a variety of proprietary antigens are used as "bait" to bind antibodies, and an algorithm is used to compare the magnitude and relationships of response to 5 markers, including those usually found in vaccinates vs. natural exposure, and in early vs. late Bb infections. Since it is debatable whether specific heighth of titers is clinically helpful in diagnosing or managing Lyme disease, the results are not given in quantitative terms.

Abaxis ELISA Quantitative Lyme Test (Abaxis Veterinary Reference Laboratories)

This new test measures a combination of antibodies to synthetic (proprietary) peptides which mimic certain antigens, including C6, which are not found in vaccinates or non-exposed dogs. The test gives quantitative results and its sensitivity and specificity are described as comparable to the Quant C6. The antibody level rises rapidly after exposure and declines rapidly after successful treatment. When above a defined threshold, active disease is presumed rather than vaccination or exposure, according to announcements. I wonder though, if similar to C6, if immune memory could cause persistent low levels after treatment of chronic infection. A qualitative screening in-house test is being developed.

Lastly and perhaps most importantly:

For all dogs (sick or not) that are found to be positive for natural exposure antibodies against Bb, it is recommended to screen and monitor them for proteinuria, since Lyme nephritis is a rare but most serious form of canine Lyme disease.[1,12] Screening for co-infections may also be indicated, since being Lyme+ is a marker for tick and wildlife exposure. Havine a Lyme+ dog is a sentinel and a time to educate owners about tick control, tick-borne diseases, and public health concerns.

REFERENCES

[1]Littman MP, et. al. **J Vet Intern Med** 2006;20(2):422. [2]Levy SA, et al. **Clin & Vacc Immunol** 2008;15(1):115. [3]Straubinger RK, et al. **J Infect Dis** 2000;181:1069. [4]Littman MP. **Vet Clin N Am (Sm An)** 2003;33(4):827. [5]Guerra MA, et al. **J Clin Microbiol** 2000;38(7):2628. [6]Littman MP. **ACVIM Proc** 2000:629. [7]http://www.idexx.com/animalhealth/testkits/4dx/faq.jsp [8]http://www.idexx.com/view/xhtml/en_us/smallanimal/reference-laboratories/testmenu/innovative-tests/quant-c6.jsf [9]Littman MP. **Clin Brief** 2011;9:13. [10]Philipp MR, et al. **J Inf Dis** 2001;184:870.
[11]http://ahdc.vet.cornell.edu/docs/Lyme_Disease_Multiplex_Testing_for_Dogs.pdf [12]Littman MP. **Vet Clin N Am (Sm An)** 2011:41(1):31.

ADVANCES IN THE PREVENTION & MANAGEMENT OF *BORRELIA BURGDORFERI* INFECTIONS IN DOGS

Richard E. Goldstein, DVM, DACVIM, DECVIM-CA
New York, NY

INTRODUCTION

Lyme disease and its causative tick-borne agent *Borrelia burgdorferi* have become a extremely common in many areas of the United States and Europe over the last 20 years in both humans and dogs (1). Anecdotally, it appears that 2009 stood out in terms of the emergence of high concentrations of Lyme-positive dogs in areas that previously were thought to have only occasional cases. Despite the large number of clinical and nonclinical Lyme cases seen in small animal practice in Lyme-endemic areas, the veterinary practitioner faces many unanswered questions regarding this disease. These include even the most basic question— how to treat a clinical case of Lyme disease or even how to recognize infection with *Borrelia burgdorferi* in a sick dog? Additional questions most practitioners struggle with, perhaps even more often, are if, when, and then how to monitor nonclinical dogs for evidence of *Borrelia burgdorferi* infection? And perhaps most important of them all is how to prevent Lyme disease? How good is tick control? Should we or should we not vaccinate? Who should we vaccinate and do vaccines prevent or promote the most severe manifestation of Lyme disease— Lyme nephritis? After a brief overview of *Borrelia burgdorferi* pathogenesis necessary for the understanding of these issues, this discussion will revolve around the ways that veterinarians can answer these questions best in their practice.

CLINICAL SIGNS OF LYME DISEASE

The lack or apparent lack of clinical signs in most dogs with active Lyme infection makes both the diagnosis and the study of this disease very difficult. In dogs, overt clinical signs are observed only in approximately 5% to 10% of infected cases (2). These signs tend to occur 2 to 5 months after the infection and include lameness (mono- or polyarthritis, **Figure 1**), lymphadenopathy, lethargy, and fever (3). Unlike in humans with acute infections, skin lesions are uncommon in dogs. The classic Lyme signs of arthritis, pain, fever, and lethargy typically resolve within approximately 3 days, in some cases only with antibiotic therapy (1). Some questions remain regarding more serious, less common syndromes that have been associated with Lyme infection in dogs including renal disease (Lyme nephritis), cardiac disease (myocarditis), and neurologic disease. Another question yet to be answered is whether some dogs get the devastating chronic recurrent disease seen in some infected humans.

DIAGNOSIS

Bacterial Culture or PCR Identification

This is very difficult in the case of *Borrelia* due to the small number of infecting organisms and the complicated techniques involved in their successful culturing.

Serology

Clinically we must rely on serology in conjunction with clinical signs to diagnose this disease. There are currently three types of serologic tests commercially available: nonspecific or kinetic enzyme-linked immunosorbent assay (ELISA), or KELA; C6 antibody testing; and Western blot.

Nonspecific or Kinetic ELISA (KELA)

This is a very sensitive test aimed at identifying any antibodies produced against *Borrelia* whole cell antigen. It does NOT differentiate between antibodies produced in reaction to Lyme infection versus Lyme vaccination and will be positive in both instances. Since we can usually not be sure of infection status and many times of vaccination status, a positive nonspecific ELISA should ideally be followed up with an additional test that would confirm infection, such as a Western blot (4) or a C6 antibody test (5).

C6 Antibody Testing

There are currently two commercially available tests for canine antibodies against the Lyme C6 peptide. This peptide is expressed only during infection; therefore, these tests are meant to be positive only in the event of natural exposure and negative in naive dogs or dogs vaccinated for Lyme (6). These tests include the in-house SNAP® 3Dx® or SNAP® 4Dx® Tests and the quantitative C6 antibody test available through IDEXX.

The SNAP 4Dx is an excellent test for screening dogs for Lyme infection. This can be done as part of a screening program for nonclinical dogs or when Lyme disease is suspected. A positive result is indicative of active Lyme infection (7).

The quantitative C6 antibody test checks for similar antibodies as the 4Dx but in a quantitative fashion. We still have a lot to learn about the value of knowing the quantitative C6 titer in all cases of Lyme infection, but we do know that the quantitative C6 titer does drop with therapy (8) and we have shown that the titer also correlated very well with circulating anti-Lyme immune complexes (9) and is likely in the future to be a useful tool in treatment decisions of nonclinical dogs.

Western Blot

This technique involves a blot smeared with Lyme antigens located in known locations. This is a relatively expensive and labor-intensive type of test and its interpretation requires expertise, and therefore this will never be in-house technology. As we have learned more about the 4Dx and quantitative C6 assays, they have taken over some of the role of the Western blot as a confirmatory assay. At this time I would recommend using the Western blot only in dogs where the vaccinal status is important to the veterinarian.

LYME PREVENTION

Tick Removal

It takes time for an infected tick to transmit *Borrelia* to a dog; this typically can happen no sooner than 48 hours from the beginning of the blood meal (1). Therefore, daily tick removal especially immediately after possible tick exposure is beneficial. People should handle possibly infected ticks with care though, since there is the potential for transmission to the person from the tick. Tick control is crucially important and is an indispensable part of any Lyme prevention strategy. A recent study demonstrated that tick control with fipronil was successful in preventing Lyme infection in dogs when exposed to infected ticks compared with nontreated control dogs (10).

Should We Vaccinate??

Whenever contemplating the use of a vaccine one needs to weigh the pros and cons. These are the considerations that, in my opinion, should be evaluated in such cases:

• How common is the pathogen? How common is the disease? How severe is the disease? How treatable is the disease? How good is the natural immunity? How expensive, reliable, and safe is the vaccine? Is there a zoonotic potential?

For canine Lyme disease vaccines:

• Common? The disease is very common. Even though most dogs are asymptomatic, the infection rate is so high in the Northeastern United States that it is a very common disease. Ten percent of infected dogs showing clinical signs out of 50% to 75% of the dogs in some areas is a lot!.

• Severity? Lyme disease is not that bad in most cases. What we do not know is the prevalence and significance of more severe syndromes such as Lyme nephritis and chronic recurrent disease.

• Treatable? Lyme disease is relatively treatable in most cases in terms of eliminating clinical signs. But in many dogs, based on experimental studies, we may never get rid of the bacteria even with antibiotic therapy (11). So perhaps Lyme diseases is not that treatable. And then of course there is Lyme nephritis, which at best seems very hard to treat.

• Natural immunity? Not very good, because the immune response appears to dampen over time. This is likely a result of the bacteria "hiding" themselves from the immune system, mainly in tissue such as synovial membranes, and down-regulating their immunogenic surface proteins (12).

• How expensive, reliable, and safe is the vaccine? Three kinds of vaccines currently are commercially available. All appear to be relatively inexpensive. Two are whole cell bacterin based and the other a recombinant outer surface protein A (OspA) single antigen vaccine. There is very little published safety data. All three vaccines are approved for dogs and so had to go through safety testing. All three types of vaccine are likely to be effective. How do these vaccines work? One mechanism of action common to all three vaccines is via production of anti-OspA antibodies. These antibodies produced in the dog in response to vaccinal OspA antigen work in the gut of the tick to bind the bacteria during the blood meal (13), thus sterilizing the gut of the tick and preventing the transmission of the bacteria into the dog.

• Safety of Lyme vaccines? There are different aspects to this question: Regarding acute vaccine reactions, one could assume that all the dogs in the vaccine studies seemed fine so the vaccine must be safe. Obviously that is not a definitive answer. The vaccine studies are too small to pick up uncommon reactions, are too short in duration, are invariably conducted by the vaccine manufacturers, and many of these studies are never published. While safety studies that are much larger are conducted on every vaccine, these are typically looking for acute reactions and most often not published. The more pressing question is: Do vaccines contribute to clinical signs associated with Lyme disease or to the syndrome known as Lyme nephritis? There is a large amount of new and exciting data to try to answer this question, and it is a good question because we now know that Lyme nephritis is not caused by an inflammatory response to renal invasion of *Borrelia burgdorferi* organisms (13). Lyme nephritis is a condition associated with an accumulation of immune complexes in the kidney. We have shown that all Lyme-positive dogs have a measurable level of Lyme-specific circulating immune complexes in their blood (9). Recently we have also been able to purify renal immune complexes from kidneys of dogs that died of Lyme nephritis and to identify Lyme-specific antibodies present within those immune complexes. Studies are still ongoing regarding possible co-infections or other antigen–antibody complexes in those kidneys. Other evidence for the lack of a key role for vaccine-induced antibodies in Lyme nephritis is the fact that, despite looking very closely, we have yet to identify a dog with Lyme nephritis with solely vaccinal antibodies within the renal immune complexes. Vaccinal antibodies in the immune complexes are also fairly rare in dogs that have been infected as well as vaccinated against Lyme disease.

To answer the question directly, though, we have just completed a large study at Cornell assessing the possible effects of vaccination with either a whole cell bacterin or recombinant OspA vaccines on immune complex formation in dogs. Lyme-negative as well as Lyme-positive dogs were included in different groups in the study. Briefly, the study suggests that there is very little cause for concern when vaccinating Lyme-negative dogs. When vaccinating Lyme-positive dogs, there may be cause for some concern, especially if the C6 antibody titer is very high and possibly if the whole cell bacterin vaccine is being used. Alternative protocols involving treatment prior to vaccination and the use of recombinant vaccines in Lyme-positive dogs are currently being studied. and are summarized below.

TREATMENT

I believe there is little question about treating Lyme-positive dogs with clinical signs. What should we use? There is definitely no proven benefit in vaccinating a dog with clinical Lyme disease from a therapeutic standpoint and that is **not** recommended. That leaves antibiotics: amoxicillin and doxycycline both work well and should be used for 4 to 6 weeks. Doxycycline will also likely be better at treating additional organisms that may have co-infected our patient along with Lyme disease and is therefore the therapy of choice (14).

The more difficult question is what about the nonclinical majority, that is, Lyme-positive dogs with nor clinical signs? Why not treat them all? Although it is hard to look the concerned owner of a young Lyme-positive golden retriever in the eye and refuse to treat a possibly fatal disease, theoretically we should approach the treatment question from a risk/cost-benefit standpoint. At this time I am not sure we have enough information to make an informed choice in this matter.

What do we know?

• Every Lyme-positive dog should be tested for proteinuria and if proteinuric (urine protein-creatinine ratio [UPC] > 0.5–1), these dogs should be treated as clinical cases. They should be given antibiotics and should not be vaccinated, at least not until after being treated.
• Lyme-positive dogs may remain infected with Lyme at low numbers with or without treatment (11).
• There is good evidence to show that treatment tends to lower Lyme titers faster than they would decline without treatment (especially C6 antibody titers)(8). This is also true for circulating immune complexes (15) and therefore may be true for bacterial load as well. Based on experimental and a growing number of field studies data, IDEXX Laboratories is currently recommending treatment above a certain C6 quantitative titer and then rechecking titers 6 months after the initial positive test (or 5 months after the treatment has ended) later to evaluate the success of the treatment.. The more we gain experience with this approach, and as more controlled field studies are published, the better this looks. We hope to have a more definitive answer in the near future for this pivotal question.

LYME NEPHRITIS

Current Recommendations for Screening and Treatment of Dogs with Suspected Lyme Nephritis

These recommendations are based on limited experience and theory, and not yet on strong clinical data (14):.

• Monitor dogs in endemic areas for Lyme infection.
• Screen all positive dogs for signs of proteinuria or microalbuminuria.
• Screen all dogs that present with proteinuria or microalbuminuria for Lyme.
• Consider treating any dogs positive for both Lyme and proteinuria or microalbuminuria with 4 to 6 weeks of doxycycline.
• If proteinuria persists or worsens (based on urine protein/creatinine ratio):
  ○ Continue doxycycline
  ○ Consider a low-protein diet and an angiotensin-converting enzyme (ACE) inhibitor
  ○ Consider renal biopsy
  ○ If the renal biopsy is consistent with immune-mediated glomerulonephritis, consider immunosuppression with drugs such as mycophenolate, azathioprine, chlorambucil, cyclophosphamide or cyclosporine in consultation with a veterinary internist.

REFERENCES

1.Greene CE, Straubinger RK. Borreliosis. In: Greene CE (ed): Infectious Diseases of the Dog and Cat, 3rd ed. Philadelphia: WB Saunders Co, 2006, pp 417-435.2.Levy SA, Magnarelli LA. Relationship between development of antibodies to Borrelia burgdorferi in dogs and the subsequent development of limb/joint borreliosis. J Am Vet Med Assoc. 1992;200:344–347. 3.Appel MJG, Allen S, Jacobson RH, et al. Experimental Lyme disease in dogs produces arthritis and persistent infection. J Infect Dis. 1993;167:651-664.4.Lindenmayer J, Weber M, Bryant J, et al. Comparison of indirect immunofluorescent-antibody assay, enzyme-linked immunosorbent assay, and western immunoblot for the diagnosis of Lyme disease in dogs. J Clin Microbiol. 1990;28:92-96.5.Levy S, O'Connor TP, Hanscom JL, et al. Utility of an in-office C6 ELISA test kit for determination of infection status of dogs naturally exposed to Borrelia burgdorferi. Vet Therap. 2002;3:308-315.6.Liang FT, Jacobson RH, Straubinger RK, et al. Characterization of a Borrelia burgdorferi VlsE invariable region useful in canine Lyme disease serodiagnosis by enzyme-linked immunosorbent assay. J Clin Microbiol. 2000;38:4160-4166.7.Bacon RF, Biggerstaff BJ, Schriefer ME, et al. Serodiagnosis of Lyme disease by kinetic enzyme-linked immunosorbent assay using recombinant VlsE1 or peptide antigens of Borrelia burgdorferi compared with 2-tiered testing using whole-cell lysates. J Infect Dis. 2003;187:1187-1199.8.Levy SA, O'Connor TP, Hanscom JL, et al. Quantitative measurement of C6 antibody following treatment of Borrelia burgdorferi antibody-positive nonclinical dogs. Clin Vaccine Immunol. 2007;Nov 14 [Epub ahead of print].9.Goldstein RE, Atwater DZ. Serologic and circulating immune complex analysis in dogs naturally infected with Borrelia burgdorferi (abstract). ACVIM Forum. 2006.10.Jacobson R, McCall, J, Hunter J, et al. The ability of fipronil to prevent transmission of Borrelia burgdorferi, the causative agent of Lyme disease to dogs. J Appl Res Vet Med. 2004;2:1.11.Straubinger RK, Straubinger AF, Summers BA, et al. Status of Borrelia burgdorferi infection after antibiotic treatment and the effects of corticosteroids:An experimental study. J Infect Dis. 2000;181:1069-1081.12.Straubinger RK. PCR-based quantification of Borrelia burgdorferi organisms in canine tissues over a 500-day postinfection period. J Clin Microbiol. 2000;38(6):2191-2199.13.Hutton TA, Goldstein RE, Njaa BL, et al. The search for Borrelia burgdorferi in kidneys of dogs with suspected "Lyme nephritis." J Vet Intern Med. 2008;22(4):860-865. 14.Littman MP, Goldstein RE, Labato MA, et al. ACVIM small animal consensus statement on Lyme disease in dogs: diagnosis, treatment, and prevention. J Vet Intern Med. 2006;20(2):422-434.15.Goldstein RE, Atwater DZ. Serology and circulating immune complexes in dogs naturally infected with Borrelia burgdorferi before and after doxycycline therapy (abstract). ACVIM Forum. 2006.

109a

MISOPROSTOL PROMOTES GASTRIC BARRIER DEFENSE AGAINST NON-NSAID MEDIATED ACID-INDUCED INJURY. T. L. Hill, D. Lascelles, A. T. Blikslager

Department of Clinical Sciences, North Carolina State University College of Veterinary Medicine, Raleigh, NC.

The prevalence of gastric ulceration in dogs is largely unknown but considered to be comparable in causes, scope, and mortality to that of humans. Prevention of gastric injury is a major focus of the pharmaceutical industry. Misoprostol, a synthetic PGE_1 analog, has historically been used for the treatment of NSAID-induced gastric injury. A role for misoprostol in prevention of acid-mediated injury alone is less well examined. We used an *ex vivo* Ussing chamber model system to examine the protective effect of misoprostol on acid-induced injury to canine gastric mucosa.

Gastric antrum was collected immediately post-mortem from 6 random-source dogs scheduled for euthanasia by a local animal control facility. Antral mucosa devoid of the seromuscular layers was maintained in Ussing chambers. After a 30-minute equilibration, acid Ringer's solution (pH 1.2) was applied to the mucosa for 45 minutes. Misoprostol (10^{-4}M) was added to the mucosa and serosal bathing reservoirs commensurate with mucosal exposure to acid. Effect on gastric barrier function was assessed by calculating the transepithelial electrical resistance (TER) and transmucosal flux of ^{3}H-mannitol. Tissue samples were collected from each treatment group, stained with hematoxylin and eosin, and examined via light microscopy by a blinded investigator.

Mucosal application of acid Ringer's solution significantly decreased TER and increased flux of ^{3}H-mannitol across canine gastric mucosa. Both of these measures of barrier function were significantly ameliorated by the presence of misoprostol. Acid injury was associated with partial disruption of the superficial gastric epithelium that was also lessened in tissues treated concurrently with misoprostol. These findings suggest that misoprostol may promote gastric barrier function against a wider variety of acid-induced injuries in addition to its role in preventing NSAID induced ulceration. The Ussing chamber model represents a powerful tool with which the specific mechanisms of PGE_1 action in promoting barrier defense can be further investigated.

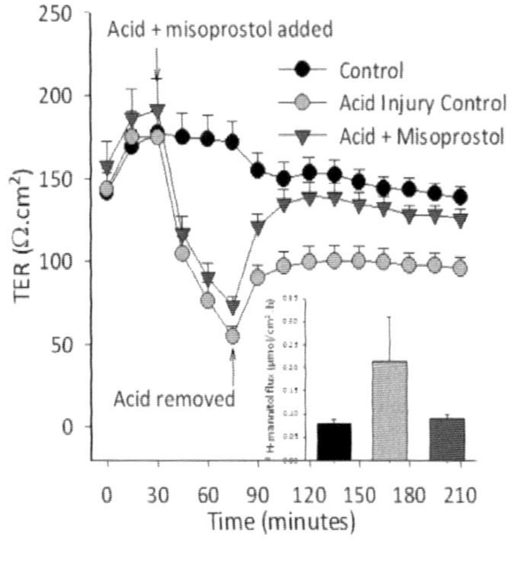

109b

ANTIMICROBIAL ACTIVITY OF CEFOVECIN (CONVENIA®) AGAINST BORRELIA BURGDORFERI AND ITS IMPACT ON EARLY LYME BORRELIOSIS IN DOGS.

J Johnson[1], B Wagner[2,3], D Garcia-Tapia[1], N Honsberger[1], H Freer[2,3], A Rollins[2,3], V King[1], C Earnhart[4], R Marconi[4] and P Meeus[1] [1] Pfizer Animal Health, Kalamazoo, MI, [2]Department of Population Medicine and Diagnostic Sciences and [3]Animal Health Diagnostic Center, College of Veterinary Medicine, Cornell University, Ithaca, NY ; [4] Virginia Commonwealth University, Richmond, VA.

Lyme disease is caused by the tick-borne spirochete *Borrelia burgdorferi* (*Bb*), a bacterium infecting mammals and birds. Clinical signs are well characterized in dogs and include fever, acute arthritis, lameness, and nephritis in some cases. However, most exposed dogs remain asymptomatic with only an estimated 5 % developing the clinical syndrome called Lyme disease. Because of this and because an optimal therapy has not been established, due to lack of suitable laboratory models and strain information, treatment strategies remain a topic of discussion. Therapies, as currently recommended, are intensive and of long duration, and therefore conducive to non-compliance.

In vitro MIC data suggested that cefovecin (Convenia®, Pfizer Animal Health), a novel, extended-spectrum semisynthetic cephalosporin, would be efficacious against *Bb* in dogs at the recommended dose of 8 mg/kg and 14 day dosing interval. We therefore performed two laboratory studies, using dogs infected with wild-caught *Ixodes scaplularis* ticks, to evaluate the efficacy of cefovecin against *Bb* strains of different origin and to compare its effectiveness with the recommended treatment protocols of doxycycline at 10 mg/kg per os SID and amoxicillin at 20 mg/kg per os TID for 1 month. Treatment outcomes monitored were: (1) the present or absence of spirochetes in skin and joints using both culture and PCR, (2) antibody titers using the SNAP 4Dx® and the Quant® C6 ELISA from Idexx, as well as a novel sensitive fluorescent bead multiplex assay which allows for the simultaneous detection of immune responses to several antigens, (3) clinical signs associated with *Borrelia* infection including lameness, ataxia and fever, and (4) histological evaluation of synovial joint, bladder and skin tissues for inflammatory and other changes compatible with Lyme disease, as described.

In both studies dogs treated with cefovecin had a marked reduction in circulating antibodies against *B. burgdorferi*, had PCR and culture-negative skin and joint samples post-treatment, and had reduced cutaneous and articular inflammation when compared to untreated controls. The variability of the parameters and the limited number of dogs used (8 per group) did not allow for statistical hypothesis testing, but the 2 dose cefovecin treatment appears at least as good as the currently recommended standard of care of sustained doxycycline and amoxicillin administration for one month.

110a

IDENTIFICATION OF THE GENE FOR BURMESE AUTOSOMAL RECESSIVE HYPOKALEMIC POLYMYOPATHY BY GENOME-WIDE ASSOCIATION STUDY.

LA. Lyons[1], B. Gandolfi[1], R. Malik[2], B. Jones[3], E-M, Prinzenberg[4], G. Erhardt[4], TJ. Gruffydd-Jones[5]. 1. Population Health & Reproduction, School of Veterinary Medicine, University of California – Davis, Davis, CA. 2. Centre for Veterinary Education, University of Sydney, Sydney, NSW, Australia 3. Centre for Companion Animal Health, Institute of Veterinary, Animal & Bio-medical Sciences, Massey University Palmerston North, New Zealand 4. Institute of Animal Breeding & Genetics, Justus Liebig University, Giessen, Germany 5. The Feline Centre, School of Veterinary Science, University of Bristol, Langford, Bristol.

Hypokalemia is the presence of a low serum potassium concentration, which can occur as a primary problem (known as *hypokalaemic periodic paralysis* in human patients, and *hypokalaemic polymyopathy* in cats) or a secondary disorder. The classic presentation of hypokalemia is a skeletal muscle weakness that is frequently episodic in nature, and either generalized, or somewhat localized to the cervical muscles. Sometimes muscle weakness is incessant, rather than episodic. Low serum potassium [K^+] is accompanied by markedly elevated serum creatine kinase (CK) activity (50 – 97,000 IU/L). Clinical signs typically become evident between 2 to 6 months-of-age and may be induced by stress or exercise. Hypokalemia is well documented in Burmese on three continents, but NOT in Burmese in North America. Past and present breed prevalence and pedigree analysis has strongly suggested hypokalemia in non-USA Burmese to be an autosomal recessive condition, with high penetrance. Previous genetic studies of *CACNL1A3 and SCN4A* of the cat has not indicated causative mutations. The domestic cat research community has developed a Feline 60K DNA array produced by Illumina, Inc. and financially supported by Hill's Pet Food, Inc. A case – control study was performed using 30 Burmese cats with hypokalemia and 30 control Burmese cats without hypokalemia. The association analysis was performed using PLINK and identified a region on cat chromosome E1. Within 1.2 Mb are two candidate genes that affect potassium regulation. Both genes were directly sequenced and a nonsense mutation producing a stop mutation was identified. All cases were homozygous for the mutation, some controls were carriers, but most were wildtype. A Devon Rex and domestic shorthair cat with hypokalemia were also homozygous for the mutation. Mutation frequency is being evaluated. Overall, the identification of the Burmese hypokalemia mutation will allow the development of a genetic test. The genetic test could be used in random bred cats to help decipher a heritable, primary cause of hypokalemia from a secondary cause and the test could be used to eradicate the disease in Burmese. Once this disease is eradicated or can be monitored via genetic testing, more outcrossing could be performed to increase the genetic health of worldwide Burmese populations.

Also, the test would prevent the spread of the disease to other breed populations that use Burmese, such as Bombay, Singapura, Burmilla, Tonkinese, and Australian Mist. Finally, the identified gene could always be considered for other species that have potassium channel anomalies.

110b

EFFECTS OF ENTEROCOCCUS FAECIUM SF68 ON STRESS DIARRHEA. A. M. Gore, A. Reynolds. Nestle Purina PetCare Research, St. Louis MO.

Prior research has demonstrated that the probiotic Enterococcus Faecium SF68 (SF68) provides several benefits that support GI health: stimulation of fecal IgA and enhanced response to vaccines in puppies, and reduction in diarrhea in kittens and in sheltered cats (Benyacoub 2003; Czarnecki-Maulden 2007; Bybee 2011). In humans, a double-blind clinical trial documented efficacy for this probiotic against antibiotic-induced diarrhea (Wunderlich 1998). The study presented here was undertaken to determine the effect of SF68 on acute stress diarrhea in dogs. Stress diarrhea is a common problem among competitive canine athletes, including Alaskan sled dogs. This real-world situation provided a reliable model to study stress-induced diarrhea in dogs.

A placebo-controlled, double-blinded study was designed and owners/handlers of active sled dog teams in the Fairbanks, Alaska, area were solicited to participate. Upon the occurrence of diarrhea in a dog, the handlers were provided with either SF68 or a placebo, based on random allotment, which they administered by adding it on top of each dog's daily meal for 7 days. Both investigators and handlers were blinded as to which treatment was in each packet. Handlers monitored the dogs' diarrhea by recording fecal scores daily, using a 1 – 3 scale (1 = liquid stool; 2 = soft stool; 3 = normal stool; diarrhea = 1 or 2). The data were analyzed to assess severity and duration of diarrhea. A Poisson regression and Kaplan-Meier survival curve analysis were used to determine the effect of treatment on duration of diarrhea, and longitudinal analysis was used to further evaluate the speed of resolution.

Twenty-six adult Alaskan sled dogs completed the study. Longitudinal analysis showed that diarrhea improved more quickly in dogs given the SF-68 compared to the placebo group ($p < 0.05$), and dogs in the SF68 group had less diarrhea ($p < 0.05$) from day 3 onward. Dogs given the SF-68 supplement had significantly fewer days with diarrhea, improving, on average, 2 days sooner than the placebo group. The number of days that the dogs experienced the most severe form – watery diarrhea -- also was significantly less on SF68 than on placebo. By day 4, 92% of the dogs in the SF68 group had normal stools, with 100% resolution by day 5 in this group. Resolution occurred more slowly in the placebo group, and never reached 100% within the 7-day study.

This study documents that Enterococcus Faecium SF68 contributes to rapid resolution of stress-induced diarrhea, with improvement occurring within 5 days.

111a

RESPONSE OF CATS CHRONICALLY INFECTED WITH *CYTAUXZOON FELIS* TO DIMINAZENE DIACETURATEC.

LA Cohn[1], KM Lewis[1], JR Dodam[1], HMarr[2], AJ Birkenheuer[2]. 1 – University of Missouri – College of Veterinary Medicine, Columbia, MO 2 – North Carolina State University College of Veterinary Medicine, Raleigh, NC.

Cytauxzoon felis is a hematoprotozoan parasite of cats that causes substantial morbidity and mortality, but cats that survive acute infection may remain persistently parasitemic and may serve as a pathogen source for vector ticks. For many hematologic infections, parasite numbers in circulation vary markedly over periods of time as brief as hours, and often parasite numbers vary with time of day. If this were true for *C. felis*, it could impact a veterinarian's ability to identify chronically infected carrier cats.

Parasite burden was estimated in the peripheral blood of five apparently healthy cats with naturally acquired chronic *C. felis* infections. Peripheral blood was obtained twice in a 24 hour period, 12 hours apart, from each cat via jugular venipuncture on 4 separate occasions. In total, 20 samples were obtained at approximately 9 am and 20 samples obtained at approximately 9 pm into EDTA containing Vacutainer tubes. Blood smears were prepared immediately, and the reminder of the blood was used to perform CBC in routine fashion and for real time PCR for *C. felis*. Each slide was stained with Wright-Giemsa and examined thoroughly for piroplasms, and the number of piroplasms per 1500 RBC was noted. Cycle threshold (Ct) was also used as a semi-quantitative measure of parasite burden in peripheral blood. Ct was compared between times of day using a two way repeated measures ANOVA (two factor repetition).

Piroplasms were observed at low numbers on all blood smears at all time points, and all blood samples were PCR positive for *C. felis*. While at least a single RBC containing piroplasm was seen on every slide, the number of piroplasm infected RBCs per 1,500 RBC ranged from 0 to 4, with no difference between samples obtained in the morning versus the evening. The mean Ct for all cats on samples obtained in the morning was 24.1 ± 1.2 and the mean in the evening was 24.7 ± 2.0. No significant difference was detected in the Ct between samples obtained in morning versus the evening ($p = 0.107$). However, parasite burden was observed to vary from day to day in a given cat ($p = 0.041$). This suggests that timing of blood collection to detect the carrier state is unimportant despite some variability in parasite burden from day to day. PCR is a more sensitive method for detection of the carrier state, but if blood smear analysis is used, the entire smear must be carefully evaluated as piroplasms often occur in extremely low numbers.

111b

EFFECTS OF THE PROBIOTIC VSL#3 ON MICROBIOLOGICAL, HISTOLOGICAL, AND IMMUNOMODULATORY PARAMETERS IN DOGS WITH INFLAMMATORY BOWEL DISEASE. G Rossi[1], G Pengo[2], M Caldin[3], JM Steiner[4], and JS Suchodolski[4]. 1. School of Veterinary Medical Sciences, University of Camerino, Camerino, Italy. 2. Clinic "St. Antonio", Cremona, Italy. 3. San Marco Laboratories, Padova, Italy. 4. Gastrointestinal Laboratory, Texas A&M University, College Station, TX, USA.

Similar to human inflammatory bowel disease (IBD), three main factors are considered fundamental in the pathogenesis of canine idiopathic IBD: the interactions between the mucosal immune system, host genetic susceptibility, and environmental factors (e.g., microbiota, nutrition). While probiotics are used frequently in small animal practice, there are few published studies regarding their clinical effect in dogs with IBD. The aim of the study was to investigate clinical, microbiological, histological, and immunomodulatory effects of a commercial multi-strain probiotic (VSL#3, VSL Pharmaceuticals, Inc.; containing 4 strains of lactobacilli, 3 strains of bifidobacteria, and 1 strain of Streptococcus). We hypothesized that VSL#3 would modulate enteric bacteria leading to induction and maintenance of remission in dogs with IBD.

Ten privately owned dogs of different breeds and ages with a long-time diagnosis of IBD were selected on the basis of recurrence of clinical signs and absence of any treatment for a month. Dogs received on average 450 billion lyophilized bacteria daily for 60 days. Fecal and plasma samples were collected, and dogs were assessed clinically, endoscopically, and histologically at baseline (T0) and 30 days after the end of treatment (T1; 90 days after T0). Plasma citrulline concentrations were measured by mass spectrometry. The fecal microbiota was analyzed by 16S rRNA gene 454-pyrosequencing and qPCR assays, and compared to data obtained from ten healthy pet dogs. Histology scores, the expression of TGF-β, CD3, and Fox-p3 positive cells, and the clinical activity index (CIBDAI) were compared at T0 and T1 using paired t-tests or Wilcoxon matched pairs tests, where appropriate. Statistical significance was set at $p<0.05$.

After treatment with VSL#3, significant decreases were observed for CIBDAI ($p=0.006$), histology scores ($p<0.001$), and infiltration with CD3+ T-lymphocytes ($p=0.007$). In contrast, plasma citrulline concentrations ($p=0.013$), TGF-β ($p=0.002$) and Fox-P3 ($p=0.002$) positive cells increased after treatment. Significant differences (Unifrac distance metric; $p<0.05$) in microbiota composition were observed between the healthy dogs and T0 of the IBD dogs. Selected microbial groups such as *Facealibacterium* ($p<0.05$) increased after treatment with VSL#3, but the overall microbiota composition was not significantly different between T0 and T1.

While dogs with IBD showed clinical improvement after VSL#3 treatment, their microbiota was still altered when compared to healthy dogs. Immunomorphological parameters suggest a potential anti-inflammatory effect of VSL#3 associated with a reduction of mucosal infiltration, which was observed 30 days after the end of the treatment. In humans, VSL#3 has already strong proven clinical efficacy in IBD and this is the first study conducted in dogs.

112a

ANTI-ALPHA ENOLASE ANTIBODIES IN THE SERUM OF CATS WITH AND WITHOUT AZOTEMIA STRATIFIED BY VACCINATION STATUS. MR Lappin[1], C Sonius[1], JC Whittemore[2], JR Hawley[1], L Schoenberg[3]. 1. Colorado State University, Fort Collins, CO, 2. University of Tennessee, Knoxville, TN, 3. Just Cats Hospital, Cleveland, OH.

In humans, spontaneously occurring antibodies against the autologous glycolytic pathway enzyme alpha enolase (AE) are associated with immune-mediated nephritis. Anti-AE antibodies in cats can occur spontaneously but also can be induced by administration of feline vaccines grown on cell lines. In a pilot study, we detected an association between anti-AE antibodies and azotemia in vaccinated privately-owned cats. However, in that study samples were not available from age-matched cats that had not been vaccinated within the last 5 years. The purpose of this study was to assess for associations amongst anti-AE antibodies and azotemia cats stratified by vaccine types administered over the previous 5 years.

Clinically stable privately owned cats $\geq$ 10 years of age, with and without azotemia (creatinine >2 mg/dl), and with an available vaccine history for $\geq$ 5 years were recruited. Sera were assayed for creatinine concentrations and anti-AE antibodies. Results from the anti-AE antibody assay were evaluated using ANOVA, including age, group (azotemic versus non-azotemic) and the 5 year vaccination history (vaccines grown on cell lines; vaccines not grown on cell lines; no vaccines) as covariates. Additionally, anti-AE antibody assay results were categorized as positive or negative and evaluated using a generalized linear model appropriate for binomial outcomes. This model also included age and the different groups as covariates.

Median ages were 15 years (range: 10-21) and 12 years (range: 10-17) for cats with (n=79) and without azotemia (n=73), respectively. Results stratified by vaccine status are shown in Table 1.

Table 1	G1	G2	G3	All AZO	G4	G5	G6	All no AZO
# cats	24	24	31	79	20	25	28	73
% anti AE +	37.5	33.3	48.4	40.6	25	52	50	43.8
Mean anti-AE AB	69.4	63.9	71.9	68.7	53.6	83.8	77.8	73.3

G1/G4 = cell line vaccines; G2/G5 = intranasal or recombinant vaccines only; G3/G6 = no vaccines; G1/G2/G3 = azotemia (AZO); G4/G5/G6 = no azotemia.

Anti-AE antibodies were detected in cats with and without recent vaccine administration and cats in each vaccine group. Significant differences amongst individual and combination group results were not detected. However, there is a numerical trend for increased proportions of anti-AE antibody positive cats and antibody levels in the azotemic cat group administered cell line vaccines (G1) compared to non-azotemic G4 cats existed as previously reported.

Anti-AE antibodies occur spontaneously in cats like in humans and do not consistently correlate to the 5 year vaccine status or azotemia. Research assessing renal enolase expression is needed.

112b

DETERMINATION OF THE PEPTIDE-BINDING MOTIF OF A PREVALENT CANINE MHC CLASS Ia ALLELE. P Ross[1], AS Buntzman[2], JC Holmes[1], BG Vincent[3], GS Gojanovich[1], KR Miller[3], JA Frelinger[2], EC Collins[3], PR Hess[1]. 1. North Carolina State University College of Veterinary Medicine, Raleigh, NC. 2. University of Arizona School of Medicine, Tucson, AZ. 3. University of North Carolina at Chapel Hill School of Medicine, Chapel Hill, NC.

CD8+ T-cell responses are triggered by T-cell receptor recognition of short peptides presented by MHC class Ia molecules on the target cell surface. Remarkably, of the large number of peptides that potentially can be derived from a particular cytosolic protein, <3% are presented. To a great extent, this limit results from the preference of any given class I allele for binding specific amino acids at peptide anchor positions, which defines the motif, and gives rise to immunodominance, where cytotoxic T-cell responses are predictably focused on the same few peptides in all individuals sharing the allele. While this phenomenon makes the study of CD8+ T cells in populations feasible, the extensive polymorphisms of class I genes require identification of prevalent alleles, which are well-established in humans, but not in the dog. Our objectives were to 1) find a frequently shared allele at the canine class Ia locus, *DLA-88*; and 2) determine the binding motif. By sequence-based genotyping within breeds, where MHC diversity is restricted, we found that ~40% of unrelated Golden Retrievers (*n*=47) carry the **50801* allele. To discover the relevant motif, recombinant *DLA-88*50801* was immunoprecipitated from lysates of a transfected canine cell line, and bound peptides were acid-eluted, isolated by size exclusion and affinity purification, and analyzed by tandem mass spectrometry. Thirty-six sequences were obtained. A motif was determined by sequence alignment and *in silico* modeling, and validated by successfully folding *DLA-88*50801*with a synthetic motif+ peptide. This study fulfills some important prerequisites for studying CD8+ T cells in dogs.

THE DIFFICULT DIABETIC: ACROMEGALY, CUSHING'S, AND OTHER CAUSES OF INSULIN RESISTANCE

Mark E. Peterson, DVM, DACVIM
New York, NY

Diabetes mellitus is one of the most common endocrine disorders in dogs and cats. In both cats and dogs with diabetes, apparent resistance to the action of insulin can develop during treatment. Clinical insulin resistance should be suspected in any dog and cat in which marked clinical signs (polyuria and polydipsia) and hyperglycemia persists throughout the day despite insulin doses of > 1.5 U/kg per injection.[1,2]

A logical stepwise approach should be used to document insulin resistance and rule out recognized causes of insulin resistance in diabetic dogs and cats receiving large daily doses of insulin. Prior to performing an in-depth and expensive workup for insulin resistance, one should first eliminate the simple or obvious causes for poor diabetic control associated with client management problems, insulin injection site problems and problems with insulin activity.[2]

DIFFERENTIAL DIAGNOSIS OF INSULIN-RESISTANT DIABETES IN DOGS AND CATS

Insulin resistance may occur as a result of any process that alters the availability of insulin in the circulation or any condition that alters insulin action at the target cell level. Factors that alter availability of insulin include decreased absorption of injected insulin, development of anti-insulin antibodies, and increased rates of insulin degradation. Conditions leading to changes in insulin action at the cellular level include infection, inflammation, obesity, secretion of counter-regulatory hormones (cortisol, growth hormone, thyroid hormone), and various hormonal or metabolic factors.

In dogs and cats, the most common causes of insulin resistance include inflammation or bacterial infections (UTI, dental disease); pancreatitis; hyperlipidemia; or renal, hepatic, or cardiac insufficiency. In addition, severe, chronic hyperglycemia itself can create a state of insulin resistance, both in the diabetic dog or cat. Insulin-resistant diabetes that develops secondary to a concurrent endocrinopathy (e.g., hyperadrenocorticism, diestrus, acromegaly, or thyroid disease) is also common in both dogs and cat (Table 1).

Table 1. Causes of Insulin Resistance

Immunologic insulin resistance (i.e., insulin-binding antibodies)
Infection/inflammation (e.g., dental disease, urinary tract infection, pancreatitis)
Ketoacidosis or concurrent disease
Poor subcutaneous absorption of insulin
Severe post-prandial hyperglycemia
Severe obesity
Drugs (e.g., glucocorticoids and progestational agents)
Endocrine disorders
- Diestrus and progesterone excess (bitch)
- Acromegaly (dog, cat)
- Cushing's syndrome (dog, cat)
- Hyperthyroidism (cat)
- Hypothyroidism (dog)
- Hyperlipidemia (dog)

Hyperadrenocorticism in Dogs

Alterations of glucose metabolism occur frequently in dogs with hyperadrenocorticism (Cushing's syndrome). About 40-50% of dogs have fasting hyperglycemia, whereas the prevalence of overt diabetes mellitus in dogs with Cushing's disease is about 10%.[1-3] Most dogs with Cushing's syndrome do not become diabetic because increased insulin secretion compensates for the insulin resistance and maintains plasma glucose concentrations near normal. However, under conditions of limited insulin reserve or pronounced insulin resistance, progressive hyperglycemia, overt diabetes mellitus and ketoacidosis will develop. Because of the marked insulin resistance commonly associated with hyperadrenocorticism, diabetic dogs with Cushing's syndrome are also resistant to the glucose-lowering effects of exogenous insulin treatment and most will require high daily insulin doses to control clinical signs of hyperglycemia.

Identifying dogs that have concurrent hyperadrenocorticism and diabetes can be very challenging. In most dogs, diabetes is diagnosed first whereas Cushing's syndrome is suspected later, as higher and higher insulin doses are needed to maintain adequate glycemic control. Unfortunately, it can be difficult to diagnose Cushing's syndrome in dogs with overt diabetes mellitus on clinical grounds alone, because many clinical signs (i.e., polyuria, polydipsia, polyphagia, and hepatomegaly) are common to both diabetes and hyperadrenocorticism. Underlying hyperadrenocorticism should be suspected in any diabetic dog that has or develops hair loss, abdominal distention, calcinosis cutis, or other clinical signs suggestive of glucocorticoid excess. In many diabetic dogs with untreated hyperadrenocorticism, resistance to the glucose-lowering effects of exogenous insulin therapy also develops; concurrent hyperadrenocorticism should be excluded in any diabetic dog that has sustained hyperglycemia and glycosuria despite high daily insulin doses (> 1.5 units/kg/dose).[1-3]

Specific laboratory tests (ACTH stimulation, dexamethasone suppression) must be used to definitively confirm hyperadrenocorticism and to differentiate pituitary-dependent adrenal hyperplasia from adrenal tumors. Caution must be used when interpreting test results, however, because the stress associated with poorly controlled diabetes mellitus alone commonly alter the test results to produce false-positive results. Dogs with diabetes should not be evaluated for hyperadrenocorticism with these testing procedures until azotemia and ketoacidosis have resolved and diabetic control with insulin has stabilized for a few weeks.

If only slightly abnormal test results suggestive of hyperadrenocorticism are obtained, it is usually wise to withhold treatment for hyperadrenocorticism and reconfirm the abnormal test results 1 to 3 months later, in order to prevent unnecessary and potentially dangerous treatments.

In dogs with concurrent hyperadrenocorticism and diabetes, insulin treatment is most important, and the diabetes stabilized prior to treatment of hyperadrenocorticism. Because of cortisol induced insulin resistance, many diabetic dogs with untreated hyperadrenocorticism will require high daily doses of insulin to control severe hyperglycemia and prevent ketoacidosis. Correction of hyperadrenocorticism with surgical adrenalectomy, mitotane, or trilostane removes the cause of insulin resistance (i.e., cortisol excess) and reduces the daily insulin requirements in these dogs and makes the diabetes easier to regulate.

Hyperadrenocorticism in Cats

Naturally occurring Cushing's syndrome is uncommon in cats. Like the dog, approximately 85% of the cats with the naturally-occurring disease have pituitary-dependent hyperadrenocorticism (PDH), whereas the remaining cats have a unilateral adrenal tumor. Compared with dogs, cats with hyperadrenocorticism are highly predisposed to developing hyperglycemia and overt diabetes. Approximately 80-90% of reported cats with hyperadrenocorticism have concurrent diabetes mellitus.[4] As in diabetic dogs with untreated hyperadrenocorticism and cortisol-induced insulin resistance, many cats will require relatively high daily insulin doses to control severe hyperglycemia and glucosuria. However, the degree of insulin resistance (i.e., dose of insulin required to control diabetic state) is not as marked as observed in most cats with acromegaly (see below). Some diabetic cats with hyperadrenocorticism respond nicely to "routine" doses of insulin (2-4 U twice daily) whereas some cats never develop diabetes at all.

Hyperadrenocorticism should be suspected in any cat that has insulin-resistant or poorly controlled diabetes mellitus, especially if other characteristic signs of Cushing's syndrome (especially hair loss or pendulous abdomen) are also present.[4] In almost all of these cats, the diabetes is diagnosed first and Cushing's syndrome is suspected later, as truncal hair loss and pendulous abdomen develop, together with difficult-to-control diabetes.

The definitive diagnosis of hyperadrenocorticism should be established by use of ACTH stimulation and dexamethasone suppression tests, abdominal ultrasonography, and endogenous plasma ACTH concentrations. Pituitary imaging (MR or CT), if available, is a useful means to identify a mass in the region of the pituitary or adrenal gland. These findings also help to differentiate pituitary-dependent hyperadrenocorticism from cortisol-secreting adrenocortical tumor.

Definitive treatment of spontaneous hyperadrenocorticism in cats may be difficult. Use of mitotane (Lysodren) is not recommended in cats because of their sensitivity to chlorinated hydrocarbons. Recently, medical control with trilostane has been reported, and this promises to become the medial treatment of choice.[4] Other treatments that can be considered include unilateral adrenalectomy in cats with hyperfunctioning adrenocortical tumors; bilateral adrenalectomy, followed by mineralocorticoid and glucocorticoid replacement therapy in cats with PDH; and external pituitary radiation therapy for cats with PDH.

Acromegaly in Dogs (Progesterone Excess)

Insulin resistance secondary to diestrus should always be considered in any newly diagnosed or poorly controlled, intact female dog. This is true regardless of when the last known estrus occurred. Development of insulin resistance during diestrus can be a direct result of the high circulating progesterone concentration.[5,6] However, circulating progesterone in the dog can stimulate the release of growth hormone (GH) from the mammary gland, and GH excess appears to be primary reason for the insulin resistance associated with estrus in this species. Attempts to suppress estrus by administration of long-term progestogens (such as medroxyprogesterone acetate) may also lead to GH excess and signs of acromegaly in dogs.[5,6]

Growth hormone, especially in dogs and cats, displays powerful diabetogenic activity and appears to provoke hyperglycemia mainly by inducing peripheral insulin resistance. As progesterone secretion from the corpora lutea increases, so does the plasma growth hormone concentration from the dog's mammary tissue. Insulin resistance can quickly become marked, and life-threatening diabetic ketoacidosis can develop. Bitches with transient DM secondary to diestrus have a high likelihood of developing permanent insulin-dependent DM during the next estrus phase.

Attempts to closely regulate a dog's diabetic condition while it is under the influence of hyperprogesteronism can be ineffective and involve extensive time and expense. Therefore, the recommended treatment for dog with diabetes secondary to diestrus is ovariohysterectomy. The overt diabetes sometimes resolves once progesterone and GH levels return to normal, whether through ovariohysterectomy, spontaneous regression of the corpus lutea, or withdrawal of progestogen therapy. The high GH concentrations will normalize rapidly following ovariohysterectomy and more slowly following withdrawal of progestogen drugs.[5,6] The insulin requirement for GH-induced diabetes mellitus will also decline, but the reversibility of the diabetes depends on the insulin reserve of the pancreatic beta islet cells. Responsiveness to insulin can begin to improve within hours after ovariohysterectomy and is often significantly better within 7 days.

Acromegaly in Cats

Chronic hypersecretion of GH results in acromegaly, a disease characterized by overgrowth of connective tissue, bone, and viscera. As mentioned previously, canine acromegaly is most often caused by endogenous or exogenous progestogens that induce GH overproduction. In contrast, progestogens in cats do not stimulate GH hypersecretion. Instead, the predominant cause of acromegaly in cats, as in humans, is a GH-secreting tumor of the pituitary gland.[7-10]

Acromegaly occurs in older cats (90% are male), and the associated conformation alterations, including an increase in body size, prognathia inferior, and weight gain despite poorly regulated diabetes, have an insidious onset and are often not noticed by owners (or veterinarians). Because of this insidious onset, severe insulin resistance and clinical diabetes are usually the earliest and most recognized clinical manifestations of acromegaly. Insulin resistance can be quite severe, many times requiring more than 10-20 U of insulin twice daily.[7-10]

Diagnosis of acromegaly in cats requires documentation of high serum concentrations of GH or insulin-like growth factor (IGF-1). Growth hormone assays are not readily available in the United States, but IGF-1can be measured by most commercial veterinary laboratories. However, diagnosis is often based on identification of conformational changes in a cat with insulin-resistant diabetes, persistent increase in body weight despite diabetes, and evidence of a pituitary mass on CT or MR imaging.

Definitive treatment of acromegaly in cats may be difficult. Surgical excision of the GH-secreting tumors has not been evaluated in cats with acromegaly. At this time, external pituitary radiation offers the best chance for control of the disease, but availability of such radiation therapy is limited. However, the short-term prognosis for cats with acromegaly appears to be relatively good. Severe insulin-resistant diabetes mellitus can generally be satisfactorily controlled using large doses of insulin in divided daily doses. The insulin requirement for GH-induced diabetes mellitus will also decline, but the reversibility of the diabetes depends on the insulin reserve of the pancreatic beta islet cells.

Hyperthyroidism in Cats

Insulin resistance has been noted to occur with feline hyperthyroidism. The exact mechanism is unknown. Suggested mechanisms in humans include decreased insulin synthesis and secretion, impaired insulin-receptor binding affinity, postreceptor defects, a disproportionate increase in nonfunctional proinsulin by the beta cells, and increased insulin clearance.

Thyroid hormone administration in normal cats has been reported to result in decreased glucose clearance and endogenous insulin resistance, and similar alterations have been reported in cats with naturally occurring hyperthyroidism.[11,12] Cats have been shown to have a persistent decrease in glucose clearance 6 months after treatment with radioiodine. Therefore, the clinical effect of treating hyperthyroidism with diabetic regulation is not always consistent. However, resolution of diabetes can occur after treatment in some cats, especially if diabetes is of recent onset. Even if diabetes persists, daily insulin requirements decrease in most diabetic cats after resolution of their hyperthyroidism.

MANAGMENT OF INSULIN-RESISTANT DIABETES IN DOGS AND CATS

Once insulin resistance is documented, one should initiate an investigation to identify the underlying cause. Correction of the underlying disorder will often lead to resolution of insulin-resistance and allow for better regulation of the diabetes mellitus.

If the cause of the insulin resistance cannot be identified or definitive treatment of the underlying disorder is not possible (e.g., removal of a GH-secreting pituitary tumor responsible for acromegaly), one can generally overcome the resistance to the insulin and control marked hyperglycemia by administering large and divided doses of insulin.[2-5] It is important to realize that in any cat or dog with severe insulin resistance, no matter what the cause, at least twice-daily administration of insulin is essential. In diabetic dogs with insulin resistance, use of an intermediate-acting insulin (e.g., NPH or Lente insulin) is generally preferred over a longer-acting insulin (such as PZI or glargine) because the intermediate-acting insulins have a more rapid onset of action, are more rapidly absorbed into the bloodstream, and therefore have a more potent glucose-lowering effect than do the longer-acting insulin preparations. In many dogs and cats with severe insulin resistance, use of a combination of short- and longer-acting insulin administered as a 1:2 ratio (e.g., one-third of total dose given as Regular insulin and two-thirds administered as NPH, PZI, glargine) may be required to control severe hyperglycemia. The added short-acting insulin helps to "overcome" severe insulin resistance, and minimize postprandial hyperglycemia; this allows the longer-acting insulin to better maintain acceptable glycemia until time of the next meal.[13-15]

In summary, there are many potential causes of insulin resistance in a clinical setting. Many of these causes can be determined by a thorough history and physical examination, if one is aware of those concomitant diseases or drug therapies that can result in insulin resistance. There are several diagnostic tests that can be performed to determine whether underlying metabolic or endocrine disorders are contributing to insulin resistance. In spite of attempts to rule out causes of insulin resistance, a specific cause may not always be apparent and large and divided doses of insulin may be needed to control diabetes in some dogs and cats.

REFERENCES

[1]Eigenmann JE, et al. *Vet Clin North Am Small Anim Pract* 1981;14:837. [2]Peterson ME. *Vet Clin North Am Small Anim Pract* 1995;25: 691. [3]Peterson ME, et al. *J Am Vet Med Assoc* 1981;178:66. [4]Peterson ME. *Manual of Canine and Feline Endocrinology* 2004; 205. [5]Eigenmann JE. *Vet Clin North Am Small Anim Pract* 1984;14:827. [6]Peterson ME. *Manual of Canine and Feline Endocrinology* 2004;187. [7]Peterson ME, et al. *J Vet Intern Med* 1990;4:192. [8]Berg RI, et al. *J Vet Intern Med* 2007;21:892. [9]Niessen SJ, et al. *J Vet Intern Med* 2007;21:899. [10]Peterson ME. *J Vet Intern Med* 2007;21:889. [11]Hoenig M, et al. *J Endocrinol* 1989;121:249. [12]Hoenig M, et al. *Res Vet Sci* 1992;53:338. [13]Bohannon NJV. *Postgrad Med* 2003;113:39. [14]Gil GV, et al. *Difficult Diabetes*. 2001. [15] Randeva HS, et al. *Insulin Resistance: Insulin Action and Its Disturbances in Disease*. 2005;535.

GALLBLADDER MUCOCELES: THE ENDOCRINE CONNECTION

Audrey Cook, BVM&S, MRCVS, DACVIM (SAIM), DECVIM-CA
College Station, TX

INTRODUCTION

In the last decade, the gallbladder (GB) mucocele (MC) has become one of the most common reasons for biliary tract surgery in dogs. The etiology of this disorder is unclear; genetic factors appear to play a substantial role, but recent reports have suggested that concurrent endocrinopathies, primarily hyperadrenocorticism (HAC), diabetes mellitus (DM) and hypothyroidism (HypoT) may predispose patients to MC formation. It is possible that various underlying pathologies result in the same end point, or that a combination of abnormalities is required for a MC to arise. This lecture will review the current body of knowledge on the canine GB MC and discuss the impact of endocrinopathies in this disorder.

BACKGROUND

Bile is produced continuously by hepatocytes and is collected in the hepatic canaliculi. From the canaliculi, it drains into interlobular ducts, which merge to form hepatic ducts; these subsequently join the cystic duct to form the common bile duct (CBD). Although some bile flows directly into the small intestine from the liver, most of it is temporarily stored in the GB prior to release into the duodenum.

The bile produced by the hepatocytes is composed of water, bile acids, bilirubin, cholesterol, phospholipids and electrolytes. Bile acids are synthesized from cholesterol, and represent the quantitatively most important pathway for removal of cholesterol from the body. The synthesis of a full complement of bile acids requires 17 enzymes. The expression of selected enzymes in the pathway is tightly regulated by nuclear hormone receptors and other transcription factors, which ensure a constant supply of bile acids in an ever changing metabolic environment. Glucocorticoids, thyroid hormone and insulin are all known to impact bile acid production.

During storage in the GB, bile undergoes substantial modification, with the removal of sodium, chloride and water, and the addition of hydrogen ions. After the ingestion of food, a bicarbonate-rich mucin is secreted by the GB mucosal epithelial cells, which mixes with the stored hepatic bile. Numerous gastrointestinal (GI) hormones play a role in GB function. Both secretin and vasoactive intestinal polypeptide (VIP) trigger the secretion of alkaline mucin, whilst cholecystokinin (CCK) triggers GB contraction. CCK is secreted by the enterocytes in response to the presence of fat in the proximal small intestine. Other influences (either stimulatory or inhibitory) on GB contractility include substance P, neuropeptide Y, somatostatin, VIP, and the autonomic nervous system. In healthy patients, these influences appear to be fairly small.

A MC is defined as an abnormally distended GB containing a buildup of luminal mucus. Grossly, the GB appears enlarged (unless it is ruptured) and contains gelatinous material: this is essentially a congealed mixture of mucus and inspissated bile. At present, it is unclear what drives MC formation: is it a reflection of impaired GB motility, is it generated by a primary dysfunction of the GB epithelium, or is it due to abnormal bile constituents? Histopathological examination of the GB wall predictably reveals cystic mucinous hyperplasia with variable associated inflammatory infiltrates, but this may be a response to abnormal GB contents rather than a primary event.

Ultrasonographically, a MC can be distinguished from biliary sludge by its immobility (in contrast, sludge will be gravity dependent) and the classic stellate /starfish /kiwi appearance. Sludge is more uniformly hyperechoic, and lacks the striated appearance seen with a MC. Sludge is also referred to as inspissated bile, and is a mobile mixture of precipitated cholesterol crystals, bile pigments, bile salts, and mucin. In human medicine, inspissated bile is considered abnormal, and is associated with cholestasis, choleliths, cholecystitis, and biliary infection. Having said that, one longitudinal study following human patients suggested that sludge may essentially come and go over time.[1] Although some metabolic changes have been associated with the formation of biliary sludge, abnormal GB motility is thought to play a key role in people, and sludge is widely regarded as the starting point of gallstone formation. At the present time, the significance of biliary sludge in canine patients is unknown, and it is often seen in clinically normal geriatrics.[2] One recent study has suggested however that GB emptying is compromised in dogs with either sludge or a MC.[3] Whether dysmotility is a cause or consequence of abnormal GB contents is unclear. Although it seems logical that biliary sludge would precede the development of a MC, this hypothesis has not been proven.

GENETIC INFLUENCES

It is clear that strong breed predispositions exist for GB MC formation, with Shetland sheepdogs at particularly high risk.[4] A recent genetic analysis indicates that an insertion mutation in ABCB4 is associated with GB disease in Shelties and other breeds, with the aberrant gene present in 14/15 Shelties with a MC, versus only 1/21 clinically healthy dogs of the same breed.[5] This genetic mutation results in compromised secretion of phosphatidylcholine (PC) across the hepatocyte canalicular membrane. This agent reduces bile salt cytotoxicity and protects the biliary epithelium from damage. Defects in PC secretion are associated with numerous biliary disorders in people, so it is interesting that a similar defect may play a role in canine GB disease. This finding also suggests that the classic cystic mucinous hyperplasia noted histopathologically in affected GBs may simply reflect the response of the biliary epithelium to a primary bile abnormality, rather than indicate a primary mucosal disorder.

ENDOCRINE INFLUENCES

In the last 20 years, about 250 canine GB MC cases have been reported in the peer reviewed veterinary literature.[6-11] In addition to identifying clear breed predispositions, these studies have routinely reported concurrent endocrinopathies, including HAC, DM and HypoT. It is unclear however if a common process stands behind these associations, e.g., hypercholesterolemia or other dyslipidemias, or if different pathologies play a role, e.g., GB dysmotility associated with HypoT, versus changes in bile acid production secondary to HAC.

Hyperadrenocorticism

Iatrogenic or spontaneous HAC has been frequently reported in dogs with GB MC. In one recent study, the odds ratio for a dog with a MC having HAC was over 29.[12] The reason for the link between HAC and MC formation is unclear, although recent studies conducted by Kook *et al* in Zurich have shed some light on the effect of sustained exogenous cortisol excess on biliary function in beagles.[13-15] In these studies, six dogs received approximately 8 mg/kg of hydrocortisone PO q12hrs (equivalent to approximately 2 mg/kg of prednisone) for 12 weeks; 6 dogs served as controls.

Based on the assumption that microprecipitates likely form within clear bile as a transitional stage to the formation of a classic MC, the biliary tract of these dogs was examined every month during glucocorticoid administration and for 3 months after steroids were discontinued. Although GB sludge appeared in all the treated dogs by day 56, it was also noted in 3/6 control dogs, and no significant differences were reported between the two groups. In addition to examining the GB ultasonographically, bile sample were collected periodically for measurement of bilirubin, cholesterol and calcium concentrations (all of these compounds are implicated in the formation of sludge in human patients). Somewhat surprisingly, concentrations all three constituents decreased during the study period. Interestingly, bile pH was significantly less acidic in the treated dogs compared to the controls.

In dogs with spontaneous MC, the incidence of infection is unclear, with reported rates ranging from 9-43%. Common isolates include *Escherichia coli, Enterobacter, Enterococcus* and *Clostridium* spp. These results are based on standard culture methods, and may be impacted by prior administration of antimicrobials. In patients with confirmed biliary infection, it is impossible to determine if this plays a role in MC generation or is simply secondary to bile stasis. If bacterial infection does in fact precipitate MC development, dogs with HAC may be vulnerable to ascending infection due to the immunosuppressive effects of cortisol excess. This hypothesis was tested in the Zurich study, in which bile cultures were performed at 28 day intervals.[14] In this small population, three months of high-dose hydrocortisone administration did not result in significant bactibilia in the treated dogs, at least based on cytological examination of GB contents and traditional culture methods.

One of the more interesting findings in the Kook study was the impact of glucocorticoid administration on bile acid profiles. The dogs receiving hydrocortisone had significant and reversible, increases in the concentration of unconjugated bile acids and a decrease in total taurine-conjugated bile acids within the GB.[15] Glycine conjugation was not affected by steroid administration. Conjugation decreases the hydrophobic properties of bile, thereby mitigating injury to the biliary epithelium. It seems reasonable to deduce that a shift to a more caustic bile acid profile may result in irritation of the GB epithelium, with subsequent mucinous hyperplasia and eventually MC formation. If this is the case, this process would suggest that in both Shelties and dogs with HAC, the caustic properties of the bile itself appear to trigger mucinous hyperplasia as a protective response by the biliary epithelium.

In addition to their effect on bile acid biochemistry, glucocorticoids may also substantially increase their synthesis. Early studies using rodent tissue culture models indicate that bile acid synthesis is markedly increased by both dexamethasone and endogenous glucocorticoids. This has not been directly investigated in dogs, but it seems likely that a similar effect occurs.

Hypothyroidism

In the human field, HypoT has long been associated with biliary tract disease, and it has been clearly established that patients with subnormal thyroid function are predisposed to cholelithiasis. Although these patients are often hypercholesterolemic, it seems that stone formation is not solely due to changes in bile composition, but also reflects compromised GB emptying. Decreased bile flow has also been implicated. Thyroxine has a pro-relaxing effect on the sphincter of Oddi, and hypothyroidism therefore predisposes patients to biliary stasis.[16] Interestingly, MC are very uncommon in human patients, whereas gallstones are rare in dogs. However, if hypoT does impact bile flow and GB emptying in dogs, it may be hypothesized that prolonged retention of bile in the GB results in progressive modification/concentration of the contents, resulting in epithelial irritation, secondary mucinous hyperplasia and a MC.

Two dogs with GB MC and spontaneous HypoT were described by Walter in 2008.[17] Both dogs were treated with ursodiol and levothyroxine, +/- SamE, omega 3 fatty acids, famotidine and fenbendazole. The MC in both dogs completely resolved within 2-3 months. In the Mesich study, HypoT was strongly associated with development of MC, but concern was expressed about the reliability of the diagnosis in some of these cases.[12] Evaluating thyroid status in clinically ill dogs can be problematic, as acute or severe illness may impact both total T4 and free T4 concentrations.[18] An increased TSH concentration would be strongly supportive of spontaneous hypoT, but a low value would not exclude the diagnosis.

Diabetes mellitus

Concurrent DM has been reported in several dogs with GB MC. In people, DM is strongly associated with GB disease, particularly cholelithiasis. Additional risk factors for gallstone formation in this population include hypertriglyceridemia and hypercholesterolemia. Compromised GB motility has been implicated, and emptying following CCK challenge in human diabetics maybe impaired, particularly in those with evidence of autonomic neuropathy. It has been shown that endogenous CCK levels following dietary challenge maybe substantially elevated in some diabetics, along with a diminished number of receptors on the GB wall.[19] Similar studies have not been reported in dogs, but it seems likely that similar processes may occur in this species.

SUMMARY

Although it appears that an inherited hepatic synthetic disorder may be the driving force behind formation of a MC in certain dog breeds, endocrine influences appear to play a key role in many other patients. It seems unlikely that the GB itself is the root of the problem; rather, changes in bile composition or emptying most likely trigger a reactive mucinous hyperplasia within the GB which then leads to MC formation. In non-Sheltie breeds, it seems wise to screen for concurrent endocrinopathies, although the diagnosis of both HAC and HypoT can be challenging in clinically ill patients. Several anecdotal reports suggest that administration of ursodeoxycholic acid (ursodiol) may be beneficial in these patients, due to its hydrophilic and choleretic properties. In patients with hypercholesterolemia or increased serum triglyceride concentrations, dietary fat restriction should also be considered.

REFERENCES

[1]Janowitz P, et al. **Hepatology** 1994;20(2):291. [2]Bromel C, et al. **Vet Radiol Ultrasound** 1998;39(3):206. [3]Tsukagoshi T, et al. **Vet Radiol Ultrasound** 2011;DOI:10.1111/j.1740-8261.2011.01868.x. [4] Aguirre AL, et al. **J Am Vet Med Assoc** 2007;231(1):79. [5]Mealey KL, et al. **Comp Hepatol** 2010;9:6. [6]Pike FS, et al. **J Am Vet Med Assoc** 2004;224(10):1615. [7]Besso JG, et al. **Vet Radiol Ultrasound** 2000;41(3):261. [8]Newell SM, et al. **J Am Anim Hosp Assoc** 1995;31(6):467. [9]Worley DR, et al. **J Am Vet Med Assoc** 2004;225(9):1418. [10]Amsellam PM, et al. **J Am Vet Med Assoc** 2006;119(9):1451. [11]Crews LJ, et al. **J Am Vet Med Assoc** 2009;234(3):359. [12]Mesich MLL, et al. **J Small Anim Prac** 2009;50(12):630. [13]Kook PH, et al. **The Vet J** 2012 (in press). [14]Kook PJ, et al. **J Vet Inter Med** 2010;24(1):224. [15]Kook PH, et al. **Am J Vet Res** 2011;72(12):1607. [16] Laukkarinen J, et al. **J Clin Endo Met** 2007:92(11):4260. [17]Walter RW, et al. **J Am Vet Med Assoc** 2008;232(11):1688. [18]Schoeman JP, et al. **Vet Parasit** 2007;148(2):75. [19]Fraquelli M, et al. **Dig Liver Dis** 2003;35(suppl 3):12.

UNDERSTANDING ACTH STIMULATION TESTS

Patty Lathan, VMD, MS, DACVIM
Mississippi State, MS

INTRODUCTION

Adrenocorticotropic hormone (ACTH) stimulation tests are used for the diagnosis of hypoadrenocorticism and hyperadrenocorticism, and for the monitoring of treatment of hyperadrenocorticism. Although initial protocols for the test were modified from human medicine, research over the past two decades has assessed the efficacy of different formulations, doses, and routes of delivery.

BACKGROUND

ACTH is a 39-amino acid peptide hormone synthesized by the pituitary gland. It is secreted by the pituitary following stimulation by corticotropin releasing hormone (CRH) released from the hypothalamus. ACTH travels through the blood to bind to receptors in the adrenal cortex, where it stimulates the synthesis of cortisol and other adrenocortical hormones.

The goal of an ACTH stimulation test is to achieve maximum stimulation of the adrenal cortex, as measured by adrenocortical hormone levels. In most situations, cortisol is the target hormone, although aldosterone and other hormones are sometimes measured. Maximal stimulation requires the administration of a supra-physiologic dose of ACTH. When testing for hypoadrenocorticism (Addison's disease), the clinician hopes to stimulate the adrenal gland so much that, if no cortisol is released, she can definitely determine that the patient's adrenal gland is non-functional. When testing for hyperadrenocorticism (Cushing's syndrome), the clinician hopes to demonstrative excessive cortisol production from a hyperplastic adrenal cortex. Treatment for hyperadrenocorticism is aimed at preventing the adrenal cortex from producing enough cortisol to cause clinical signs, while maintaining enough adrenocortical reserve that the patient does not go into crisis. Thus, ACTH stimulation tests are used to help keep the adrenocortical reserve within a relatively tight range.

The ACTH stimulation test is the only test that can be used to definitively diagnose naturally-occurring and iatrogenic hypoadrenocorticism, and to monitor treatment of hyperadrenocorticism with mitotane or trilostane. However, other tests are available for the diagnosis of hyperadrenocorticism. Although the ACTH stimulation test is more specific than the low dose dexamethasone suppression (LDDS) test and the urine cortisol creatinine ratio, it is much less sensitive. I prefer to use the LDDS test in most cases of suspected hyperadrenocorticism. The ACTH stimulation test is reserved for patients in which a false positive seems likely, such as those with a concurrent illness such as diabetes mellitus.

PROTOCOLS IN DOGS

Two forms of ACTH have been used for the stimulation test. One form is a synthetic analogue of ACTH (cosyntropin or tetracosactrin), which consists of the first 24 amino acids of ACTH. This portion of the molecule appears to maintain the full biologic activity of ACTH. The other formulation of ACTH is corticotropin, purified ACTH that is generally obtained from porcine pituitary glands. Cosyntropin is available as a commercial product in the US. Corticotropin, however, is currently only readily available as a compounded veterinary product. Although an FDA-approved corticotropin product is available for human use in the US, it is prohibitively expensive for veterinary use.

In the 1970s, multiple protocols utilizing the intramuscular administration of corticotropin were recommended by various authors[1], and the efficacy of intramuscular cosyntropin was proven.[2] An early study demonstrated that the intramuscular administration of both porcine corticotropin (2.2 IU/kg of a commercially available formulation) and synthetic cosyntropin (0.25 mg/dog) resulted in statistically similar adrenocortical stimulation in healthy dogs, and that the test could be performed with either preparation.[3] Later, intravenous administration of cosyntropin was shown to be as effective as intramuscular.[4] Corticotropin is not given IV due to concerns about immunoreactivity.

The rising cost of cosyntropin, and decreased accessibility of a commercial corticotropin preparation, prompted the search for less expensive testing protocols. Compounding pharmacies started selling their own corticotropin formulations, and researchers evaluated the use of lower doses of cosyntropin. Unfortunately, one small study suggests that the use of compounded corticotropin, compared to cosyntropin, may yield more borderline results when testing for hyperadrenocorticism.[5] However, several studies have documented that the dose of 5 μg/kg of synthetic cosyntropin produces maximal adrenocortical stimulation in dogs, and can be used to diagnose hyperadrenocorticism and hypoadrenocorticism.[6,7,8,9] Additionally, this low dose is just as effective whether given intramuscularly or intravascularly, for the diagnosis and monitoring of hyperadrenocorticism.[10] Intramuscular administration of cosyntropin is not recommended when testing for Addison's disease, as the effect of decreased perfusion on test results is unknown. The use of compounded corticotropin is not recommended for the diagnosis of hypoadrenocorticism, either, due to concerns about inconsistent results and the need for intramuscular injection.

Another study by financially-minded investigators showed that leftover reconstituted cosyntropin can be stored at -20 C for up to six months without losing activity.[11] It must be kept in a plastic container (such as a syringe), since ACTH binds to glass. I prefer to store the solution in 0.2 mL aliquots to minimize waste.

INTERACTIONS WITH THE TEST

Administration of glucocorticoids prior to the ACTH stimulation test can alter the test in two different ways. Several synthetic glucocorticoids, including prednisolone and methylprednisolone, actually cross-react with the assay used to measure cortisol. This could result in a falsely-elevated cortisol concentration. Dexamethasone does not cross-react with the assay, and can be given prior to completion of testing, as is often done in cases of Addisonian crisis.

The other way in which glucocorticoids can interfere with the test is by suppression of the hypothalamic-pituitary-adrenal axis (HPAA). Administration of glucocorticoids (even topical, particularly in small dogs) inhibits the release of ACTH. As the adrenal cortex is deprived of ACTH, it atrophies. The longer the steroids are given, the more exaggerated the response. Ideally, no short-acting glucocorticoids should be given within a month of testing. Longer periods are needed for longer-acting glucocorticoids.

REFERENCES

[1]Feldman EC, et.al. **Can Fel Endo Repro** 1987; 161. [2]Feldman EC, et al. **JAAHA** 1977; 13:494. [3]Feldman EC, et. al. **AJVR** 1982; 43(3):522. [4]Hansen BL, et. al, **JAAHA** 1994; 30(1):38. [5]Hill K, et. Al. **Vet Med** 2004; 99(2):134. [6]Watson ADJ. **AVJ** 1998; 76(4):255. [7]Frank LA, et. Al. **JSAP** 2000; 41:308. [8]Kerl ME, et. Al. **JAVMA** 1999; 214:1497. [9]Lathan PL, et.al. **JVIM** 2008;22(4): 1070. [10]Behrend EN, et. Al. **JAVMA** 2006;229(4):528. [11]Frank LA, et. al. **JAVMA** 1998; 212(10):1569.

ADVANCED CONSIDERATIONS OF ESOPHAGEAL DYSFUNCTION

Michael Willard, DVM, DACVIM
College Station, TX

ESOPHAGITIS

Esophagitis is relatively common. The difficult arises from the fact that esophagitis can present with clinical signs that are subtle or seem more like vomiting instead of regurgitating. Mild esophagitis may only cause minor signs (mild regurgitation of mucus and phlegm) while severe esophagitis can cause so much pain that patients refuse to swallow water or even saliva. It is important to diagnose esophagitis promplty because delayed diagnosis can have serious repercussions. Substantial inflammation of the esophageal mucosa causes muscular weakness by interrupting the reflex arcs within the esophagus and/or between the esophagus and the brain.[1] Patients can have minor esophageal distention but major signs. Barium esophagrams can have relatively minor changes and not reflect the severity of the esophagitis. Esophagoscopy typically shows an edematous, reddened, bleeding esophageal mucosa, ± structure formation, making it the diagnostic method of choice. In rare cases, there may be more subtle changes with thickening and discoloration (especially at the lower esophageal sphincter of cats).

There is a wide range of causes of esophagitis. Anesthetic procedures may result in gastric acid pooling in the esophagus for relatively long periods of time.[2] However, gastroesophageal reflux from any cause can be responsible.[3,4] Hiatal hernias occasionally are responsible for such reflux. Rare animals ingest caustic substances (e.g., lye), and cats will like caustic disinfectants off their fur. Tetracyclines, NSAIDs, ciprofloxacin and clindamycin have substantial potential to cause esophagitis. Pills and capsules are notorious for lodging in the esophagus of cats[5], and doxycycline is a recognized cause of esophageal stricture in cats. Esophagitis may also be secondary to any cause of protracted vomiting. Parvovirus enteritis may cause such intense vomiting that esophagitis results. If a vomiting animal has the character of its vomitus change and suggest regurgitation, consider the possibility that esophagitis has occurred secondary to the persistent vomiting. Gastroesophageal reflux may be potentiated by or caused by esophagitis (which may be caused by reflux in the first place). Thus, there may be a positive feedback loop which can be hard to break (i.e., esophagitis causes more reflux which causes more esophagitis which causes more reflux which causes ...). Rarely there can be spontaneous inflammation, as seen with eosinophilic esophagitis in dogs.[6] Brachycephalic dogs seem to have an increased incidence of gastroesophageal reflux, esophagitis and perhaps hiatal hernia.[7]

Minimize gastroesophageal reflux by keeping the stomach as empty as possible by using prokinetics such as metoclopramide or, preferably cisapride. Mosapride should become available soon. Cisapride is more effective than metoclopramide. Metoclopramide can be given by injection; a useful fact in animals that are regurgitating profusely. Gastric acid secretion should be minimized and preferably abolished. H-2 receptor antagonists (e.g., cimetidine, ranitidine, famotidine) are competitive inhibitors. Omeprazole, lanosprazole and pantoprazole are non-competitive inhibitors of gastric acid secretion. Therefore, these drugs are clearly more effective. You can try to achieve greater efficacy with the H-2 receptor antagonists by doubling or tripling their dose. Sucralfate is of uncertain value in patients with esophagitis. Unless there is some gastric acid reflux into the esophagus (which you are desperately trying to stop in the first place), it is doubtful that the sucralfate is of much use. If you use it, it should be administered as a slurry.

A combination of omeprazole and cisapride seems to be particularly effective. Placing a PEG tube has advantages in patients with severe disease.

CICATRIX

Cicatrix (i.e., benign scarring) may occur after severe esophagitis from any cause (including foreign objects). It is particularly easy to miss this problem on a barium swallow if only liquid barium is used. If radiographs using liquid barium are nonrevealing, repeat the study using barium mixed with food. Endoscopy is sensitive for finding such lesions; however, you must consider the size of the patient as you evaluate the esophageal lumen. A partial stricture will be very obvious in a 5 kg dog or cat but may not be apparent in a 45 kg dog. Balloon-dilatation or bougienage are both effective. Surgical resection should be a last ditch resort and only used if esophageal ballooning[8] or bougienage[9] has failed despite repeated dilatations. Endoscopic administration of intralesional steroids may help minimize reformation of the stricture.[10] Typically we inject 1-2 ml of triamcinolone at the site of the stricture either before or after ballooning. Special needles are used to perform injections through the endoscope without damaging the biopsy/aspiration channel. Another technique used in difficult strictures is to make 3-4 equidistant cuts into the stricture using an electrocautery device (i.e., either a snare or a knife) prior to ballooning. This helps the stricture to "break" open at multiple spots with the idea that there will be 3 or 4 smaller, less deep lacerations at the stricture site instead of one major, deep laceration which is more likely to restricture. However, you should not attempt to use electrocautery through an endoscope unless you have some training less you cause too much trauma to the tissues or destroy your endoscopic equipment. Another technique is to "paint" the site where the stricture was broken down with Mitomycin (NOT mithromycin C, there is a difference).[11] A 5 mg bottle is reconstituted and soaked up into a gauze sponge. Then this sponge is endoscopically placed on the site where the stricture was broken up for 5 min. Then it is flushed off with 60 ml of water. For particularly difficult cases, stents may be used.[12] Depending upon the type of stent, these must sometimes be sutured in place. Stents custom made for dogs may be obtained from the Infiniti corporation (http://www.infinitimedical.com/p_stents.html). The device that allows you to suture the stent in place by means of a flexible endoscope is made by Pare Surgical Inc (http://www.infinitimedical.com/p_stents.html).

ESOPHAGEAL FOREIGN OBJECTS

Esophageal foreign objects may consiste of bones, rawhide treats, food, dental chew toys, toys, balls, rocks, wood, etc. They usually lodge at the thoracic inlet, base of the heart, or lower esophageal sphincter. A history of a patient that begins to regurgitate (as opposed to vomit) acutely is very suggestive of acquired esophageal obstruction due to a foreign object. These patients may continue to drink water, but they typically refuse solid food because the food bolus cannot pass by a partial esophageal obstruction and causes pain whenever it tries to. Realization that the patient is regurgitating (as opposed to vomiting) should be a "red flag". Foreign bodies can erode and perforate the esophagus much quicker than they would stomach or intestines.

Plain radiographs are the most helpful diagnostic. Bones are a common cause of obstruction, and plain films that are made with proper technique and then carefully evaluated are diagnostic in most cases. Poultry bones are not as radiodense as the patient's bones, which means that excellent radiographic technique is required to see them. Foreign bodies in the esophagus can perfectly mimic pulmonary or mediastinal masses. If there is poor contrast in the region of the esophagus, pleural effusion or pneumothorax, one must consider esophageal perforation and mediastinitis. If plain films are not diagnostic, barium provides excellent contrast. Iodide contrast is safer if there is an unsuspected perforation. Esophageal perforation may occur at variable times after ingestion of a foreign object. If a foreign object is tightly lodged in the esophagus, it can cause ischemia and perforation in 2-3 days. The prognosis for animals with esophageal perforation and severe mediastinitis is guarded to poor, depending upon their condition at the time of diagnosis.

Endoscopy is the preferred method of removing foreign objects. Fluoroscopic[13] and surgical[14] techniques can be effective if the operator is well trained. Rigid endoscopes allow much more control of the foreign object and are preferred to flexible scopes for removal of these foreign objects. It is especially useful to be able to pull the object into a rigid endoscope and then withdraw it and the scope as a unit, thus protecting the esophagus. The main disadvantage of rigid endoscopes is that they are often not long enough in larger dogs. If a large object or a bone cannot be easily dislodged, you can use rigid equipment to "chew" it up and hopefully dislodge it. If that fails, passing a large Foley catheter behind the foreign object and inflating the balloon often helps; it distends the esophagus (thus freeing the foreign object) and then is used to pull the object out. If you cannot pull a foreign object out of the esophagus, you can try to push it into the stomach. However, do not push bones or other foreign objects into the stomach unless you are sure that it is smooth on the aborad side and will not further damage the mucosa. Finally, be careful if you insufflate the esophagus lest you rupture a weakened area in the mucosa and/or cause a fatal tension pneumothorax.

Most fish hooks can often be successfully removed endoscopically. Fish hooks have usually penetrated the mucosa (and sometimes the muscular tunics); you will often have to use rigid equipment to carefully force the tip of the hook back out of the mucosa.

After removing the foreign object, retake plain chest radiographs to be sure that a pneumothorax (which would indicate a perforation) is not present. Antibiotics are indicated if there is substantial esophageal mucosal ulceration (and especially if you remove a fish hook which had been used with various baits that can harbor anaerobic bacteria). Depending upon the amount of damage, corticosteroids may be used to try to prevent cicatrix formation; however, it is not clear that they are effective. Rarely there may be severe hemorrhage.[15]

HIATAL HERNIAS

Shar Pei's seem to have a relatively high incidence of hiatal hernias.[16] They can be difficult to diagnose. Sometimes seen on plain radiographs and simple barium contrast radiographs, the more occult cases sometimes need endoscopy for diagnosis. Sometimes one must manually put pressure on the abdomen during film exposure to try to push the stomach through the hernia and into the chest so that it can be diagnosed radiographically. The big question is whether the hiatal hernia is causing a problem or is an "innocent bystander". In particular, if you have an older dog or cat (i.e., > 1-2 years old) that just started having clinical signs, you should strongly consider that the hiatal hernia is a fortuitous finding that is not responsible for the clinical signs.

ESOPHAGEAL WEAKNESS

A common misconception is that all dogs with significant esophageal disease will have regurgitation and obvious changes on plain radiographs. Plain radiographs are often diagnostic, but they miss some cases. Barium contrast radiographs/fluoroscopy are the most definitive means of detecting this problem, but sometimes they can also miss the diagnosis. Segmental esophageal weakness (especially in the cervical region) can easily be missed by the novice diagnostician. It is important to include the cervical region when doing radiographs or fluoroscopy to look for esophageal dysfunction. Some animals have partial esophageal dysfunction that is sufficient to cause major clinical signs but is not seen on plain radiographs or even contrast radiographs unless fluoroscopy is used. Positioning of the patient is important when performing barium contrast radiographs.[17]

Acquired weakness

Acquired esophageal weakness is usuallyeasy to distinguish from obstruction, especially when a barium contrast radiograph is performed. The severity of radiographic lesions does not always correlate well with clinical severity. Acquired esophageal weakness is typically difficult to resolve because it is hard to find the underlying cause. Myopathy, neuropathy, myasthenia gravis, dermatomyositis, dysautonomia, esophagitis, Addison's disease, *Spirocerca lupi*, tick paralysis, central nervous system disease, or infiltrative non-obstructive esophageal tumors are possible causes. Generalized myopathies and neuropathies often affect the esophagus because it is composed of striated muscle in the dog. Signs of lower motor neuron disease in these patients are sometimes seen and can include loss of muscle mass, weakness, an inability to bark, or a change in the quality of the bark. Some clients report that their animal has laryngitis, which may seem likely because these pets typically have repeated respiratory infections due to aspiration pneumonia. Treatment of the myopathy or neuropathy should resolve the problem, but symptomatic therapy for the esophageal dilatation is indicated.

Localized myasthenia in the dog is a syndrome in which the esophagus is the only muscle which is obviously weak. Up to 25-30% of dogs with acquired esophageal weakness have this syndrome. Third degree heart block may also be seen in some patients with megaesophagus due to myasthenia. This is diagnosed in dogs with esophageal weakness by detecting serum antibodies to acetylcholine receptors. The antibodies are relatively stable and require little special handling other than refrigeration. If myasthenia is strongly suspected but the titer is negative, it can be valuable to repeat the titer as they sometimes seroconvert later. Myasthenia gravis will sometimes spontaneously resolve. Treatment for myasthenia gravis that does not spontaneously resolve may include anti-acetylcholinesterase drugs and/or cytotoxic agents. Azathioprine and mycophenolate seem to be effective drugs for this purpose. In general, we try to avoid steroids as they seem to be associated with more problems. In really severe cases, we can place a percutaneous gastrostomy tube to support the patient and lessen aspiration while waiting for the drugs to have an effect. However, a gastrostomy tube will not prevent all aspiration as the dog is still swallowing saliva which can be regurgitated and aspirated.

Hypoadrenocorticism may be responsible for causing esophageal weakness, even when the serum electrolytes are normal. This is especially true in standard sized, black poodles, but it can occur in any breed. Treatment for hypoadrenocorticism is steroids, which can make the esophagus start functioning again. However, if your diagnosis is wrong and you give steroids because you suspect the dog may have hypoadrenocorticism, all you are doing is making aspiration pneumonia and subsequent death that much more likely.

Idiopathic weakness

Idiopathic megaesophagus (i.e., either congenital megaesophagus or acquired megaesophagus for which a cause cannot be found) can only be treated with symptomatic therapy, which usually consists of feeding the animal 3-4 meals of gruel from an elevated platform and making the pet remain in the near vertical position from 5-10 minutes after eating. Near-vertical means just that. It is useless for the dog to just lift its head up while eating; it should be standing on its back legs. If necessary, use a portable ladder or put the dog in a large trash can to help it remain vertical during this time. Google "Bailey chair" on the internet; this will give you some ideas of how to practically accomplish this. However, this time-honored treatment does not always work. Some animals with idiopathic esophageal weakness are controlled as well (or better) if they are fed free-choice dry food from an elevated platform. Some can even be fed from the floor. Free-choice feeding encourages the pet to eat small amounts of food throughout the day, thus avoiding intermittent large meals which are more likely to be retained and further dilate the esophagus. If there is any esophageal motility remaining, the dry food may be easier for the esophagus to propel then gruel. It is difficult to predict which feeding regime will work best for a particular patient, and both of these feeding regimes may need to be tried. While most dogs with idiopathic megaesophagus die from aspiration, there are enough of them that respond well that it is very much worth trying. A reasonable percentage of dogs with idiopathic, congenital megaesophagus will spontaneously improve and have normal or near normal function.[18] You cannot predict response to therapy or spontaneous remission; all you can do is support the patient and see what happens.

REFERENCES

[1]Sellon RK, et al. Esophagitis and esophageal strictures. **Vet Clin N Am** 2003;33:945. [2]Wilson DV, et al. Postanesthetic esophageal dysfunction in 13 dogs. **J Am Anim Hosp Assoc** 2004;40:455. [3]Gualtieri M, et al. Reflux esophagitis in three cats associated with metaplastic columnar esophageal epithelium. **J Am Anim Hosp Assoc** 2006;42:65. [4]Han E, et al. Feline esophagitis secondary to gastroesophageal reflux disease: clinical signs and radiographic, endoscopic, and histopathologic findings. **J Am Anim Hosp Assoc** 2003;39:161. [5]Westfall DS, et al. Evaluation of esophageal transit of tablets and capsules in 30 cats. **J Vet Int Med** 2001;15:467. [6]Mazzei MJ, et al. Eosinophilic esophagitis in a dog. **J Am Vet Med Assoc** 2009;235:61. [7]Poncet C, et al. Prevalence of gastrointestinal tract lesions in 73 brachycephalic dogs with upper respiratory syndrome. **J Small Anim Pract** 2005;46:273. [8]Leib MS, et al. Endoscopic balloon dilation of benign esophageal strictures in dogs and cats. **J Vet Int Med** 2001;15:547. [9]Bissett SA, et al. Risk factors and outcome of bougienage for treatment of benign esophageal strictures in dogs and cats: 28 cases (1995-2004). **J Am Vet Med Assoc** 2009;235:844. [10]Altintas E, et al. Intralesional steroid injection in benign esophageal strictures resistant to bougie dilation. **J Gastroenterol Hepatol** 2004;19:1388. [11]Gillespie MG, et al. Role of mitomycin in upper digestive tract stricture. **Head and Neck** 2007;29: 12. [12]Fiorini A, et al. Self-expandable metal coil stents in the treatment of benign esophageal strictures refractory to conventional therapy: a case series. **Gastrointest Endoscop** 2000;52:259. [13]Moore AH. Removal of oesophageal foreign bodies in dogs: use of the fluoroscopic method and outcome. **J Small Anim Pract** 2001;42:227. [14]Sale C, et al. Results of transthoracic esophagotomy retrieval of esophageal foreign body obstructions in dogs: 14 cases (2000-2004). **J Am Anim Hosp Assoc** 2006;42:450. [15]Cohn LA, et al. Fatal hemothorax following management of an esophageal foreign body. **J Am Anim Hosp Assoc** 2003;39:251. [16]Guiot L, et al. Hiatal hernia in the dog: a clinical report of four Chinese shar peis, **J Am Anim Hosp Assoc** 2008;44:335. [17]Bonadio CM, et al. Effects of body positioning on swallowing on esophageal transit in healthy dogs. **J Vet Int Med** 2009;23:801. [18]Bexfield N, et al. Esophageal dysmotility in young dogs. **J Vet Int Med** 2006;20:1314.

PATHOPHYSIOLOGY OF HEPATIC ENCEPHALOPATHY IN COMPANION ANIMALS: NEW INFORMATION

Dr. Richard J. Mellanby, BSc, BVMS, PhD, DSAM, DECVIM-CA, MRCVS
Edinburgh, UK

BACKGROUND

Hepatic encephalopathy (HE) is a common cause of morbidity and potentially mortality in dogs with liver disease, most notably in dogs with congenital portosystemic shunts (cPSS).[1] Dogs with cPSS usually have a single, extra-hepatic vessel which links the portal vein and the caudal vena cava, although a range of vascular abnormalities have been described.[2] Dogs with a cPSS can present with signs that are similar to HE in humans ranging from mild lethargy to ataxia, disorientation and coma.[3,4] Dogs can be screened with good sensitivity and specificity for cPSS by measurement of serum bile acids and ammonia[5] and the abnormal vessel can be definitively identified in most cases by abdominal ultrasonography, computed tomography or magnetic resonance imaging.[6-8] In the small proportion of equivocal cases, the disease can be definitively diagnosed by portovenography during laparotomy .[2] Dogs with cPSS are initially typically teated with lactulose and antibiotics.[9] Once their condition has been stabilised, they normally undergo a staged ligation of the shunting vessel after which the prognosis is reasonably good.[10,11] Minimally invasive surgical treatments have recently emerged as a novel therapeutic strategy for dogs with cPSS.[12,13] For dogs which are not candidates for surgery, extended medical management is normally undertaken. Crucially, a significant proportion of dogs with cPSS have a poor response to treatment and there is a clear need to develop better therapies for these dogs.[14-16] As a significant proportion of dogs with a cPSS which respond poorly to therapy have refractory HE, it is hoped that a better understanding of the pathophysiology of HE will open up novel therapeutic options.

PATHOGENESIS OF CANINE HE

Hyperammonemia is commonly cited as the most important mediator of HE in dogs[3,17] yet there is a paucity of data on how ammonia correlates to the severity of HE and how concentrations alter following either successful, or unsuccessful, medical management of HE. The lack of studies on this topic is particularly surprising given that it has recently being demonstrated in dogs with urea cycle enzyme deficiencies that hyperammonemia alone is insufficient to cause HE.[18] Other studies have also suggested a role for gastrointestinal-derived endogenous benzodiazepines,[19] hypercortisolism,[20] and altered tryptophan metabolism[21] in the development of HE in dogs with spontaneous liver disorders. There is a lack of consensus on the role of inflammation in the development of HE in dogs.[22,23] A recent study of the brain magnetic resonance imaging characteristics found that the majority of dogs with cPSS had hyperintense focal areas in the lentiform nuclei on T1-weighted images that did not enhance after intravenous gadolinium; leading to the authors speculating that this abnormality is likely to be due to manganese deposition.[24]

PATHOGENESIS OF HUMAN HE

In contrast to canine HE, there has been a concerted effort in recent years to better understanding the pathophysiology of HE in humans. Hepatic encephalopathy in humans has been estimated to occur in approximately 30-45% of patients with cirrhosis and 10-50% of patients with transjugular intrahepatic portosystemic shunt, while minimal hepatic encephalopathy has been estimated to affect approximately 20-60% of patients with liver disease.[25] The estimated prevalence of chronic liver disease and cirrhosis in the United States is approximately 5.5 million cases and in 2003 there were over 40 000 patients hospitalized for a primary diagnosis of hepatic encephalopathy, resulting in total charges of approximately $932 million.[25] In addition, trends over the past 10 years suggest that the burden of HE is increasing, as indicated by increases in hospital admissions and higher charges per stay.[25] Furthermore, because of the inconsistencies in coding for HE, it is likely that the currently available prevalence and cost data under-estimate the true burden of HE.[25] Because the impairment in mental, physical and social activities due to HE translates into very substantial social and economic costs, there has been a considerable research drive to better understand the underlying pathophysiology of HE in humans.[25-28] From this work a number of metabolic derangements have emerged as key factors in driving HE.

Ammonia

Whilst ammonia has long been incriminated in the pathogenesis of HE in humans,[29] reports of an absolute correlation between blood ammonia levels and severity of HE are not consistent;[30] indeed, it is well recognised that some patients with severe HE have normal ammonia concentrations.[31] Furthermore, two controlled trials of ammonia administration to patients with cirrhosis did not document development of HE.[32,33] Consequently, a wide range of other studies have been undertaken examining additional factors which might act either alone, or in synergy with ammonia, to initiate and perpetuate clinical signs of HE in humans.

Inflammation

Additional research has focussed on understanding the role of other contributing factors which may allow hyperammonemia, in the setting of advanced liver disease, to cause HE in a high proportion of patients. Several lines of investigation have implicated a role for both inflammation and hyperammonemia in the development of HE.[29] Recent studies have found a strong positive correlation between circulating levels of TNFα and severity of HE in patients with cirrhosis.[34,35] In addition, induction of hyperammonemia by administration of an amino acid solution to cirrhotic patients without overt HE resulted in the development of HE only in the presence of a systemic inflammatory response.[36] These findings suggest that inflammation, particularly mediated by TNFα, plays a crucial role in the neuropsychological effects of hyperammonaemia in patients with liver disease. Thus, the interaction between ammonia and inflammation/ infection appears to be crucial in the development of HE although the mechanism is unclear.[29]

Manganese

Other studies have suggested that manganese is involved in the development of HE. Although manganese is an essential trace element, manganese toxicity is a well described syndrome with profound effects on the central nervous system and in particular the basal ganglia where astrocytes are a site of early dysfunction and damage.[37] Patients with a wide range of liver pathologies have been shown to exhibit increased tissue concentrations of manganese.[38-40] Since manganese is excreted by the hepatobiliary route, this elevation is believed to be due to decreased biliary excretion and to increased systemic availability due to portal-systemic shunting associated with chronic liver disease.[37] Manganese neurotoxicity appears to involve activation of microglia and/or astrocytes, leading to damage of nearby neurons.[41] Manganese may also potentiate the production of proinflammatory cytokines by microglia.[42] Interestingly, ammonia and manganese have been shown to have additive effects on the induction of free radicals in cultured astrocytes, further implicating manganese in the pathogenesis of HE [43].

Altered permeability of gastrointestinal tract

Patients with liver disease often have increased intestinal permeability which is a possible contributory factor in the development of HE, through altered increased portal uptake of inflammatory mediators, bacteria, antigens and toxins.[44,45]

NOVEL DEVELOPMENTS IN UNDERSTANDING OF PATHOPHYSIOLOGY OF HE IN DOGS

Ammonia

We have recently re-examined the relationship between ammonia and presence of HE in dogs with a cPSS. In a large series of over 150 dogs we have demonstrated that plasma ammonia concentrations were significantly higher in dogs with a cPSS which had HE compared to dogs with a cPSS which were asymptomatic. However, several dogs in this case series had profound clinical signs of HE but had normal or only mildly increased plasma ammonia concentrations implying that factors other than hyperammonemia were important in driving clinical signs of HE in some cases.

Manganese

Given the well established neurotoxic effects of manganese, we have recently examined manganese metabolism in dogs with a cPSS, dogs with non hepatic illnesses and healthy control dogs. Importantly, we have demonstrated that dogs with cPSS have significantly increased concentrations of whole blood manganese.[46]

Inflammation

We have also begun to examine the relationship between inflammation and presence of HE in dogs with a cPSS. We have demonstrated that dogs with a cPSS and HE have significantly higher serum CRP concentrations than asymptomatic dogs with a cPSS.

Collectively, our recent studies have demonstrated that dogs with a cPSS and HE have many of the same metabolic disturbances that humans with HE exhibit. Additional studies examining plasma amino acid and cytokine profiles in dogs with cPSS which either have HE or are asymptomatic are ongoing.

REFERENCES

1. Tobias KM, Rohrbach BW. Association of breed with the diagnosis of congenital portosystemic shunts in dogs: 2,400 cases (1980-2002). J Am Vet Med Assoc 2003;223:1636-1639. 2. White RN, Macdonald NJ, Burton CA. Use of intraoperative mesenteric portovenography in congenital portosystemic shunt surgery. Vet Radiol Ultrasound 2003;44:514-521. 3. Maddison JE. Hepatic encephalopathy. Current concepts of the pathogenesis. J Vet Intern Med 1992;6:341-353.4. Winkler JT, Bohling MW, Tillson DM, et al. Portosystemic shunts: diagnosis, prognosis, and treatment of 64 cases (1993-2001). J Am Anim Hosp Assoc 2003;39:169-185. 5. Gerritzen-Bruning MJ, van den Ingh TS, Rothuizen J. Diagnostic value of fasting plasma ammonia and bile acid concentrations in the identification of portosystemic shunting in dogs. J Vet Intern Med 2006;20:13-19. 6. d'Anjou MA, Penninck D, Cornejo L, et al. Ultrasonographic diagnosis of portosystemic shunting in dogs and cats. Vet Radiol Ultrasound 2004;45:424-437. 7. Bruehschwein A, Foltin I, Flatz K, et al. Contrast-enhanced magnetic resonance angiography for diagnosis of portosystemic shunts in 10 dogs. Vet Radiol Ultrasound 51:116-121. 8. Stieger SM, Zwingenberger A, Pollard RE, et al. Hepatic volume estimation using quantitative computed tomography in dogs with portosystemic shunts. Vet Radiol Ultrasound 2007;48:409-413. 9. Mas A. Hepatic encephalopathy: from pathophysiology to treatment. Digestion 2006;73 Suppl 1:86-93. 10. Sereda CW, Adin CA. Methods of gradual vascular occlusion and their applications in treatment of congenital portosystemic shunts in dogs: a review. Vet Surg 2005;34:83-91. 11. Mehl ML, Kyles AE, Hardie EM, et al. Evaluation of ameroid ring constrictors for treatment for single extrahepatic portosystemic shunts in dogs: 168 cases (1995-2001). J Am Vet Med Assoc 2005;226:2020-2030. 12. Bussadori R, Bussadori C, Millan L, et al. Transvenous coil embolisation for the treatment of single congenital portosystemic shunts in six dogs. Vet J 2008;176:221-226. 13. Weisse C, Mondschein JI, Itkin M, et al. Use of a percutaneous atrial septal occluder device for complete acute occlusion of an intrahepatic portosystemic shunt in a dog. J Am Vet Med Assoc 2005;227:249-252, 236. 14. Watson PJ, Herrtage ME. Medical management of congenital portosystemic shunts in 27 dogs--a retrospective study. J Small Anim Pract 1998;39:62-68. 15. White RN, Burton CA, McEvoy FJ. Surgical

treatment of intrahepatic portosystemic shunts in 45 dogs. Vet Rec 1998;142:358-365. 16. Greenhalgh SN, Dunning MD, McKinley TJ, et al. Comparison of survival after surgical or medical treatment in dogs with a congenital portosystemic shunt. J Am Vet Med Assoc 236:1215-1220. 17. Taboada J, Dimski DS. Hepatic encephalopathy: clinical signs, pathogenesis, and treatment. Vet Clin North Am Small Anim Pract 1995;25:337-355. 18. Zandvliet MM, Rothuizen J. Transient hyperammonemia due to urea cycle enzyme deficiency in Irish wolfhounds. J Vet Intern Med 2007;21:215-218. 19. Aronson LR, Gacad RC, Kaminsky-Russ K, et al. Endogenous benzodiazepine activity in the peripheral and portal blood of dogs with congenital portosystemic shunts. Vet Surg 1997;26:189-194. 20. Sterczer A, Meyer HP, Van Sluijs FJ, et al. Fast resolution of hypercortisolism in dogs with portosystemic encephalopathy after surgical shunt closure. Res Vet Sci 1999;66:63-67. 21. Holt DE, Washabau RJ, Djali S, et al. Cerebrospinal fluid glutamine, tryptophan, and tryptophan metabolite concentrations in dogs with portosystemic shunts. Am J Vet Res 2002;63:1167-1171. 22. Howe LM, Boothe DM, Boothe HW. Endotoxemia associated with experimentally induced multiple portosystemic shunts in dogs. Am J Vet Res 1997;58:83-88. 23. Peterson SL, Koblik PD, Whiting PG, et al. Endotoxin concentrations measured by a chromogenic assay in portal and peripheral venous blood in ten dogs with portosystemic shunts. J Vet Intern Med 1991;5:71-74. 24. Torisu S, Washizu M, Hasegawa D, et al. Brain magnetic resonance imaging characteristics in dogs and cats with congenital portosystemic shunts. Vet Radiol Ultrasound 2005;46:447-451. 25. Poordad FF. Review article: the burden of hepatic encephalopathy. Aliment Pharmacol Ther 2007;25 Suppl 1:3-9. 26. Schomerus H, Hamster W. Quality of life in cirrhotics with minimal hepatic encephalopathy. Metab Brain Dis 2001;16:37-41. 27. Wein C, Koch H, Popp B, et al. Minimal hepatic encephalopathy impairs fitness to drive. Hepatology 2004;39:739-745. 28. Groeneweg M, Quero JC, De Bruijn I, et al. Subclinical hepatic encephalopathy impairs daily functioning. Hepatology 1998;28:45-49. 29. Shawcross D, Jalan R. The pathophysiologic basis of hepatic encephalopathy: central role for ammonia and inflammation. Cell Mol Life Sci 2005;62:2295-2304. 30. Odeh M. Relationship between circulating levels of tumour necrosis factor-alpha and Child-Pugh scores in liver cirrhosis. Int J Clin Pract 2007;61:519; author reply 520. 31. Ong JP, Aggarwal A, Krieger D, et al. Correlation between ammonia levels and the severity of hepatic encephalopathy. Am J Med 2003;114:188-193. 32. Eichler M, Bessman SP. A double-blind study of the effect of ammonium infusion on psychological functioning in cirrhotic patients. J Nerv Ment Dis 1962;134:539-542. 33. Cohn R, Castell DO. The effect of acute hyperammoniemia on the electroencephalogram. J Lab Clin Med 1966;68:195-205. 34. Odeh M, Sabo E, Srugo I, et al. Serum levels of tumor necrosis factor-alpha correlate with severity of hepatic encephalopathy due to chronic liver failure. Liver Int 2004;24:110-116. 35. Odeh M, Sabo E, Srugo I, et al. Relationship between tumor necrosis factor-alpha and ammonia in patients with hepatic encephalopathy due to chronic liver failure. Ann Med 2005;37:603-612. 36. Shawcross DL, Davies NA, Williams R, et al. Systemic inflammatory response exacerbates the neuropsychological effects of induced hyperammonemia in cirrhosis. J Hepatol 2004;40:247-254. 37. Normandin L, Hazell AS. Manganese neurotoxicity: an update of pathophysiologic mechanisms. Metab Brain Dis 2002;17:375-387. 38. Mizoguchi N, Nishimura Y, Ono H, et al. Manganese elevations in blood of children with congenital portosystemic shunts. Eur J Pediatr 2001;160:247-250. 39. Ikeda S, Sera Y, Yoshida M, et al. Manganese deposits in patients with biliary atresia after hepatic porto-enterostomy. J Pediatr Surg 2000;35:450-453. 40. Versieck J, Barbier F, Speecke A, et al. Manganese, copper, and zinc concentrations in serum and packed blood cells during acute hepatitis, chronic hepatitis, and posthepatitic cirrhosis. Clin Chem 1974;20:1141-1145. 41. Chang JY, Liu LZ. Manganese potentiates nitric oxide production by microglia. Brain Res Mol Brain Res 1999;68:22-28. 42. Filipov NM, Seegal RF, Lawrence DA. Manganese potentiates in vitro production of proinflammatory cytokines and nitric oxide by microglia through a nuclear factor kappa B-dependent mechanism. Toxicol Sci 2005;84:139-148. 43. Jayakumar AR, Rama Rao KV, Kalaiselvi P, et al. Combined effects of ammonia and manganese on astrocytes in culture. Neurochem Res 2004;29:2051-2056. 44. Ramachandran A, Balasubramanian KA. Intestinal dysfunction in liver cirrhosis: Its role in spontaneous bacterial peritonitis. J Gastroenterol Hepatol 2001;16:607-612. 45. Fujii T, Seki T, Maruoka M, et al. Lactulose-L-rhamnose intestinal permeability test in patients with liver cirrhosis. Hepatol Res 2001;19:158-169. 46. Gow AG, Marques AI, Yool DA, et al. Whole blood manganese concentrations in dogs with congenital portosystemic shunts. J Vet Intern Med 2010;24:90-96.

UNDERSTANDING ADAPTIVE IMMUNITY IN CANINE IBD: CYTOKINES AND MUCOSAL PROTECTION

Silke Schmitz, DrMedVet, DECVIM
London, United Kingdom

INTRODUCTION

Inflammatory bowel disease (IBD) in humans and dogs is a chronic inflammatory condition of the gut with a complex pathogenesis. It is now commonly accepted that – on the genetic background of a susceptible individual – the breakdown of "oral tolerance" towards luminal bacteria through both innate and adaptive immune mechanisms is involved in the development and maintenance of these diseases. Even though it is believed that the intestinal microflora plays a crucial role in the initiation of inflammation, the exact sequence of events in disease development and progression remains unknown. Much progress has been made in the last decade to understand possible mechanisms. This knowledge has mostly been derived from rodent animal models of IBD. Unfortunately, little is known about the similarity of pathomechanisms of canine IBD to experimental forms of IBD in mice and we are only now starting to understand some of the characteristics of the canine disease. The picture becomes even more complicated when considering that the two forms of human IBD (Crohn's disease [CD] and ulcerative colitis [UC]) have distinct characteristics on a clinical, histological and also molecular level; whereas canine IBD seems to be much more diverse and can be difficult to distinguish from food-responsive or antibiotic-responsive forms of chronic enteropathy.

This manuscript/talk thus wants to focus on knowledge about the pathogenesis of human IBD and rodent models and will try to elucidate both possible similarities as well as discrepancies between these species/models and the canine disease. The phenomena of mucosal protection/ tolerance and especially possible T helper (Th) subsets involved is discussed in detail and new research from our group on canine IBD is presented.

ORAL TOLERANCE AND IMMUNE EXCLUSION – THE STEADY STATE IN A HEALTHY INDIVIDUAL

The gastrointestinal (GI) tract constitutes an important interface between host and environment (especially the trillions of intestinal microbes) and, as such, has the dual role of excluding pathogens whilst remaining tolerant towards most food antigens and the commensal non-pathogenic microflora.

The first line of defense is the exclusion of any source of bacterial and food antigenic stimulation. This **"immune exclusion"** is achieved both by the production of a viscous mucus layer from goblet cells in the colon, the production of antimicrobial peptides (see below) and the production of secretory antibodies (mostly IgA) from B-cell blasts and plasma cells, thereby counteracting microbial colonization and mucosal penetration [1-3].

The harmless antigens that are in contact with intestinal epithelial cells in a healthy individual will elicit **"oral tolerance"**, which is commonly defined as the induction of both local and systemic non-responsiveness to nonpathogenic antigens and plays a vital physiological role in preventing hyper-reactive immune responses[4]. The initial response to both tolerogenic and immunogenic forms of orally administered antigen is the same: both induce transient T cell activation and proliferation that is observed primarily in the gut-associated lymphoid tissue (GALT)[5-7]. The subsequent response – ie. tolerance vs. inflammation – seems to depend on several mechanisms:

1) The presence of low levels of co-stimulation: Soluble antigens (as opposed to cell-bound antigens) are presented by antigen-presenting cells (APCs) without any co-stimulatory signals (ie. without MHC class molecules) and bind to the high-affinity CD80/86 ligand CTLA-4 on T-cells instead[8]. CTLA-4 then mediates non-responsiveness, and suppresses the activation of T cells.

2) The presence of "tolerogenic" dendritic cells (DCs). These DCs of the lamina propria (LP) seem to be of a more "immature" phenotype and are able to sample luminal antigens without inducing their own maturation[9]. Immature, gut-derived DCs then constitutively migrate from the mucosal epithelium to the mesenteric lymph nodes, and induce tolerance[10]. It has been postulated that mucosal DCs which fail to mature and do not secrete Interleukin (IL)-12 (the main pro-inflammatory T-cell stimulatory factor) may be central to the generation of regulatory T cells (Tregs) in the GALT[11]. Furthermore, a novel population of DCs has been described that secrete anti-inflammatory IL-10 when stimulated with microbe-associated molecular patterns (MAMPS) like lipopolysaccharides (LPS)[12].

3) The development of Tregs: either "natural" $CD4^+CD25^+FoxP3^+$ Tregs[13] bearing cell-surface TGFβ which inhibits naïve T cell maturation and proliferation of Th effector cells[14] or the influence of so called Tr1 regulatory cells which produce vast amounts of IL-10[15].

Basal anti-inflammatory COX-2 dependent prostaglandin (PGE2) production by stromal cells in the lamina propria of the intestine is another major contributor to the anti-inflammatory response of the GALT[16, 17].

THE ROLE OF T HELPER CELLS IN THE DEVELOPMENT OF IBD

From animal models of IBD and research in humans, there is no doubt that the induction and persistence of inflammatory T cell populations in inflamed tissue is one of the main characteristics of IBD. In humans, the ratio of $CD4^+$ to $CD8^+$ T cells in the

LP is 2:1[18], whereas in dogs they were reported to be 3:2[19]. Increased numbers of $CD4^+$ cells in the LP of dogs with IBD (as well as antibiotic-responsive diarrhea, but not with food-responsive diarrhea) have been reported[20]. Lamina propria (LP) $CD4^+$ T cells reacting to the enteric flora are the effector cells causing experimental IBD in many rodent models.[21] Once activated, $CD4^+$ T cells can differentiate into functionally distinct subsets. **Th1 cells** preferentially produce Interferon gamma (IFNy) and IL-2 and evoke cell-mediated immunity and phagocytosis. **Th2 cells** preferentially secrete IL-4, IL-5 and IL-13 and evoke strong antibody responses/ eosinophil recruitment. An additional subset identified are **Th17 cells** (producing IL-17A, IL-22 and IL-23), which have been shown to be important for recruitment, activation and migration of neutrophils[22] as well as being implicated in many autoimmune diseases[23]. On the contrary, **Tregs** produce IL-10 and TGFβ and are important in immune regulation (see above). In humans, it is now widely accepted that CD is a condition driven by a mixed Th1/Th17 response, characterized by the production of IL-17A and – to a greater extent - IFNγ, whereas UC is a "Th2-like" disease with overproduction of IL-4 in the acute and IL-5 plus IL-13 in the chronic stages of disease. These cytokines were only recently discovered to be mainly produced by natural killer T (NKT) cells in the mucosa[24]. Attempts to classify canine IBD as a disease with a certain bias towards a Th cell lineage have been disappointing so far[25-27]. This is complicated by the paucity of available canine-specific antibodies that could be used to detect cytokines on the protein level. Instead, most research so far has assessed the expression of cytokines on the mRNA level, where results were inconclusive. In one study, mRNA for Th1-related cytokines (IL-2, IL-12p40 and TNFα), as well as Th2 cytokines (IL-5) and Treg cytokines (TGFβ) were up-regulated[25] in the mucosa of dogs with IBD compared to controls; another study could not find any differences in expression of IFNy or TNFα in dogs with chronic diarrhea compared to controls[26], whereas lymphoplasmacytic colitis was associated with increased IL-2 and TNFα expression (but not IL-12p34, IL-12p40, IL/4, IL/6 or IFNγ) and was thus deemed to be a Th1 cell driven disease[28]. More recent investigations did not necessarily address involvement of specific cytokines or Th cells in IBD pathogenesis, but reported a reduction of both inflammatory cytokines (IFNy, IL-12p35, IL-5, IL-13, IL-17C, IL-27) and anti-inflammatory cytokines (IL-10, TGF) after treatment[29]. In a recent study, our group has investigated the role of Th17-related cytokines in canine IBD[27]. Surprisingly, expression of IL-17A was significantly lower in duodenal tissue from German Shepherd dogs (GSDs) with IBD ($n = 18$) and dogs of other breeds with IBD ($n = 33$) compared to 15 healthy controls ($p = 0.014$), but no difference in the expression of IL-22 ($p = 0.839$), IFNγ ($p = 0.359$), IL-10 ($p = 0.085$) or TGFβ ($p = 0.551$) across groups was detected. Thus it seems that Th17 or Th1 cell up-regulation does not occur in canine IBD. Research is ongoing to assess if a specific Th cell type is contributing to the pathogenesis of canine IBD.

THE ROLE OF EPITHELIAL CELLS AND PATTERN-RECOGNITION RECEPTORS IN THE DEVELOPMENT OF IBD

Intestinal epithelial cells (IECs) not only have important barrier, absorption and transport function, but they also play an important role in innate and adaptive immunity including management of the luminal microbial ecology. The inducible excretion of regulatory molecules by IECs is consistent with the notion that these cells may play a role in initiating and regulating physiologic and pathological mucosal responses, associated with the development of IBD. Moreover, human IECs may be an important target of the dysregulated immune response due to the effects of barrier disruptive cytokines (IFNy, IL-14, IL-13, TNFα). IECs are the first host cells to detect and subsequently interact with luminal antigens and could therefore be regarded as crucial cells in discriminating between friend and foe, self and non-self or pathogen and commensal. All four principal intestinal epithelial cells (absorptive enterocytes, Paneth cells, goblet cells, enteroendocrine cells), as well as subepithelial myofibroblasts, and various professional immune cells (monocytes/macrophages, dendritic cells can initiate and orchestrate the local immune response to microorganisms by using pattern recognition receptors (PRRs), as for example Toll-like receptors (TLRs), to recognize characteristic structural features of bacteria, such as LPS, peptidoglycans or bacterial DNA[30]. The chronic inflammatory state that characterizes IBD is thought to be a result of an exaggerated host defense reaction to the commensal flora. Altered expression of TLRs and other PRRs hence plays a crucial role in the development of the disease. Primary IECs from normal human mucosa constitutively express TLR3 (recognizing double-stranded RNA, poly I:C) and TLR5 (recognizing flagellin) at high levels, while TLR2 (recognizing lipopeptides) and TLR4 (recognizing LPS) are only detectable at low levels[30]. In active CD, TLR3 is significantly down-regulated in IECs, whereas TLR4 is strongly up-regulated in both UC and CD[30]. TLR2 and TLR5 expression usually remain unchanged. How exactly bacterial components stimulate these types of receptors i*n vivo* in order to initiate, maintain or limit disease is currently unknown, but individual genetic susceptibilities/ mutations in PRRs are likely to play a role (as exemplified by the mutations of the NOD-2 gene that was identified as a risk factor for human CD[31]). Regarding the situation in dogs, only a few studies have addressed the expression of TLRs so far. In one study, TLRs 2,4 and 9 were up-regulated in the duodenum of dogs with steroid-responsive disease, but not in food-responsive disease[32]. In another, only TLR2 (but not TLR4) was up-regulated in IBD, and there was a mild correlation between TLR2 expression and disease severity[33]. These findings seem to be complicated by evidence of breed variations. One study looked at TLR expression in GSDs with IBD compared to healthy Greyhound controls and found a different picture: here, TLR2 and TLR4 expression were not different between groups, whereas TLR4 expression was increased and TLR5 expression decreased in all segments of the gut[34]. Also, only now do we start to understand the genetic background that might predispose dogs to the development of IBD. For example, our group has recently shown that polymorphisms in the TLR4 and TLR5 genes have been found to be significantly associated with IBD in GSDs[35]. Furthermore, one of those polymorphisms in TLR5 could be shown to be associated with an increased risk of development of IBD in 38 different breeds[36].

The role of IECs in the adaptive immunity of the gut is less well understood, but there is mounting evidence that in humans, IECs are an important source of various chemokines and may therefore play a role in the recruitment of neutrophils and T cells to the epithelial layer in IBD, which may initiate and/or promote intestinal inflammation. IECs are capable of producing a wide range of pro-inflammatory cytokines, such as IL-1β and TNFα, both implicated in the pathogenesis of human IBD. On the other hand, they are also able to produce TGFβ and IL-10, and it has been shown that signaling through CD1d can induce IL-10 production by IECs and therefore dampen epithelial pro-inflammatory signals. These findings suggest a central role for IECs in the complex network of pro- and counter- regulatory cytokines in IBD.

THE ROLE OF THE MUCUS LAYER AND ANTIMICROBIAL PEPTIDES

Mucus forms a gel layer covering the GI tract, acting as a semi-permeable barrier between lumen and epithelium. Goblet cells synthesize secretory and epithelial-bound mucins[35] and trefoil factor peptides (TFF), which determine the thickness and properties of mucus and aid in mucosal healing after injury[36]. In human patients with IBD, alterations in mucins have been described involving genetic mutations in mucin genes, changes in mucin mRNA and protein levels, degree of glycosylation, sulfation and degradation of mucins [37]. Rodent models for IBD have demonstrated how changes in mucins, in particular in MUC2, imposed by immunological or microbial factors, may contribute to the development and/or perpetuation of chronic IBD[38]. Trefoil factor expression has been shown to be up-regulated in human IBD both in inflamed tissues[39] and in serum[40], which may represent a non-specific repair function of the intestinal cells. To the author's knowledge, no studies investigating mucin production in canine IBD are available; however, in a pilot study in our lab we could show that both canine TFF1 and 3 were significantly up-regulated in duodenal and colonic tissues from dogs with IBD compared to controls (unpublished data).

IECs express various antimicrobial peptides such as defensins and cryptidins, which are part of the innate immunity and contribute to local intestinal host defense. In the human gut, there are at least 9 different defensins, most of them expressed by Paneth cells (specialized epithelial cells localized at the bottom of the epithelial crypts). The β-defensin HBD-1 is constitutively expressed in human colonic tissues while HBD-2 is induced in IBD (mostly in UC)[41, 42]. Defensins might thus control bacterial invasion in IBD, although the functional significance of this remains to be established. To the author's knowledge no studies evaluating defensin expression in canine IBD are currently available.

THE ROLE OF THE INFLAMMASOME

The inflammasome is an intracellular multi-protein complex consisting of NOD-like receptors (NLRs), caspases and adaptor proteins that initiates the inflammatory cascade during infections, resulting in the production of IL-1β and IL-18. Recently, NLRs (NLRP1, NLRP3, NLRC4 and NLRP6) have emerged as important regulators of intestinal homeostasis in humans. In genome-wide association studies, single nucleotide polymorphisms (SNPs) in the NLRP3 gene have been linked to the susceptibility to CD in people[43]. Long before the inflammasome itself was characterized, IL-1β and IL-18 were considered causally related to human IBD[44, 45]. In *in vitro* studies the addition of IL-1β to epithelial cells lead to increased intestinal permeability by altering tight junction function[46]. Rodent models confirmed the importance of IL-1β, as a soluble IL-1 receptor antagonist could be used to suppress immune-complex colitis, whereas administration of IL-1 receptor antibodies increased inflammation and mortality[47]. Evidence for protection from colitis through anti-IL-18 antibodies was provided later, where a dramatic attenuation of TNBS-induced colitis was seen in mice[48]. Additionally, *Nlrp3* knockout mice have been shown to be more prone to colitis[49]. In human CD, alongside other cytokines, IL-1β mRNA[50] and IL-18[45] are up-regulated in the diseased mucosa. Currently, there is no information available about the potential role of the inflammasomes in canine IBD. However, in a pilot study in our lab we could demonstrate that canine NLRP3 is expressed in the duodenum of about 50% of IBD cases but in none of the healthy control dogs examined (unpublished data). Investigating IL-1β or IL-18 expression in canine IBD seems to be an attractive target for further studies.

CONCLUSION

Although phenotypically similar, the pathogenesis of canine IBD is possibly quite distinct to the situation in people. Thus conclusions drawn from human or experimental rodent models might not necessarily apply to the dog. There is accumulating evidence that genetic analysis might bring some insight into predispositions and pathogenesis of the disease in dogs. Also, it seems likely that disturbances in all parts of the complex defense and tolerance mechanisms that operate in the GI tract – from the composition of the microflora, the mucus layer, function of epithelial cells, signaling of innate and adaptive immune receptors to dendritic cells and lymphocytes – can potentially lead to disturbances in the delicate balance maintaining the steady state.

Consequently, a whole variety of abnormalities might lead to the same – or at least very similar – phenotype, just as it seems to be the case in humans. Much more research needs to be done to elucidate potential mechanisms leading to the inflammatory state in canine IBD.

REFERENCES

1. Brandtzaeg P, et. al., **Immunol Today** 1999;20(3):141. 2. Brandtzaeg P, et. al., **Immunol Rev** 1999;171:45. 3. Brandtzaeg P, et. al., **Gastroenterology** 1989;97(6):1562. 4. Mowat AM, **Curr Opin Gastroenterol** 1999;15(6):546. 5. Sun J, et. al., **J Immunol** 1999;162(10):5868. 6. Blanas E, et. al., **J Immunol** 2000;164(6):2890. 7. Shi HN, et. al., **J Immunol** 2000;165(11):6174. 8. Fowler S, et. al., **Eur J Immunol** 2002;32(10):2997. 9. Rescigno M, et. al., **Nat Immunol** 2001;2(4):361. 10. Sallusto F, et. al., **J Exp Med** 1999;189(4):611. 11. Roncarolo MG, et. al., **J Exp Med** 2001;193(2):F5. 12. Wakkach A, et. al., **Immunity** 2003;18(5):605. 13. Shevach EM, **J Exp Med** 2001;193(11):F41. 14. Chen CH, et. al., **J Exp Med** 2003;197(12):1689. 15. Groux H, et. al., **Nature** 1997;389(6652):737. 16. Newberry RD, et. al., **J Immunol** 2001;166(7):4465. 17. Newberry RD, et. al., **Nat Med** 1999;5(8):900. 18. Lefrancois L, in Handbok of Mucosal Immunology, et.al, Ed. (Academic Press, San Diego, 1994), 287-298. 19. Elwood CM, et. al., **Vet Immunol Immunopathol** 1997;58(3-4):195. 20. German AJ, et. al., **J Vet Intern Med** 2001;15(1):14. 21. Elson CO, et. al., **Gastroenterology** 1995;109(4):1344. 22. Lin L, et. al., **PLoS Pathog** 2009;5(12):e1000703. 23. Ghoreschi K, et. al., **Trends Immunol** 2011;32(9):395. 24. Strober W, et. al., **Gastroenterology** 2011;140(6):1756. 25. German AJ, et. al., **Dig Dis Sci** 2000;45(1):7. 26. De Majo M, et. al., **Vet Res Commun** 2008;32 Suppl 1:S275. 27. Schmitz S, et. al., **Vet Immunol Immunopathol** 2012;doi:10.1016/j.vetimm.2012.01.013:(in press). 28. Ridyard AE, et. al., **Vet Immunol Immunopathol** 2002;86(3-4):205. 29. Lynch A, et. al., ECVIM, (2011). 30. Cario E, et. al., **Infect Immun** 2000;68(12):7010. 31. Ogura Y, et. al., **Nature** 2001;411(6837):603. 32. Burgener IA, et. al., **J Vet Intern Med** 2008;22(3):553. 33. McMahon LA, et. al., **Vet Immunol Immunopathol** 2010;135(1-2):158. 34. Allenspach K, et. al., **Vet Microbiol** 2010;146(3-4):326. 35. Kim YS, et. al., **Curr Gastroenterol Rep** 2010;12(5):319. 36. Kalabis J, et. al., **J Biol Chem** 2006;281(10):6434. 37. Einerhand AW, et. al., **Eur J Gastroenterol Hepatol** 2002;14(7):757. 38. Lu P, et. al., **Inflamm Bowel Dis** 2011;17(10):2047. 39. Shaoul R, et. al., **J Pediatr Gastroenterol Nutr** 2004;38(5):488. 40. Vestergaard EM, et. al., **Scand J Clin Lab Invest** 2004;64(2):146. 41. Wehkamp J, et. al., **Eur J Gastroenterol Hepatol** 2002;14(7):745. 42. Wehkamp J, et. al., **Dig Dis Sci** 2002;47(6):1349. 43. Villani AC, et. al., **Nat Genet** 2009;41(1):71. 44. Mahida YR, et. al., **Gut** 1989;30(6):835. 45. Pizarro TT, et. al., **J Immunol** 1999;162(11):6829. 46. Al-Sadi RM, et. al., **J Immunol** 2007;178(7):4641. 47. Ferretti M, et. al., **J Clin Invest** 1994;94(1):449. 48. Siegmund B, et. al., **Proc Natl Acad Sci U S A** 2001;98(23):13249. 49. Zaki MH, et. al., **Immunity** 2010;32(3):379. 50. Sawa Y, et. al., **Int J Mol Med** 2003;11(2):175.

ADVANCED ENDOSCOPY TECHNIQUES IN SMALL ANIMAL GASTROENTEROLOGY: WHAT ARE THE POTENTIALS AND PITFALLS?

Dr. Caroline Mansfield
Melbourne, Australia

INTRODUCTION

Endoscopy in its earliest forms began in the 19th century, but did not become widely used in veterinary medicine until the 1970's. Flexible endoscopes are now widely used in veterinary medicine for assessment of mucosal disease in hollow structures such as the gastrointestinal, respiratory and urinary tracts. The two types of flexible endoscopes in regular use are fiberoptic and video endoscopes. With fiberoptic endoscopes, image transmission and illumination occurs via two bundles of fiberoptic fibres, whose point of origin at the tip of the endoscope precisely matches that at the eyepiece. In contrast, video endoscopes rely upon a microelectronic charge coupled device (CCD) chip at the tip of the endoscope that senses the image for subsequent electronic transfer for processing and direct display on a video monitor. Video endoscopes, particularly those incorporating high-definition technologies, have vastly improved the quality of image resolution and capture in veterinary medicine. Technical capacities for miniaturization of the CCD chip have limited the smallest diameter video-endoscope able to be manufactured.

However conventional endoscopy is frequently limited as gross morphological findings may only be detectable late in disease, may not correlate with histopathological evaluation or lesions may be subtle and easily missed. In addition, even utilizing upper and lower gastrointestinal endoscopy, the entire gastrointestinal tract is not able to be evaluated. Adjunctive application of topical dyes to highlight abnormalities as with chromoendoscopy or technology such as narrow band imaging and confocal endomicroscopy additionally enhance identification as well as classification of subtle lesions; whilst capsule endoscopy and push enteroscopy both provide a means for evaluation of a far greater length of the gastrointestinal tract than was previously achievable.

LIGHT SOURCE AND IMAGE RECONSTRUCTION TECHNIQUES

In video white-light endoscopy (WLE), xenon white-light is preferably used for illumination as it provides greater contrast and visibility than halogen light sources. The illumination is transferred to the tip of the endoscope via fiberoptic bundles. These fibers are randomly arranged bundles, as specific and orderly transfer of light is less important compared with the coherent transfer required for image reconstruction in fiberoptic endoscopes. Depth of penetration of photons within the mucosa depends upon their wavelength. Superficial penetration occurs with the blue band, intermediate for green and in-depth for the red band. Differing absorption and reflection of these wavelengths then provides the overall endoscopy image. Hemoglobin is primarily responsible for absorption of blue and green wavelengths which explains the overall red appearance of the mucosa, and why vasculature appears red with WLE. Image transfer in WLE occurs when the CCD chip transfers the full range of white-light for reconstitution of the natural color in a single step, or the tissue is alternately illuminated by each of the three bandwidths of light by a rotating interference broadband filter interposed with standard xenon white-light. Each of the three overlapping bands are then separately transferred for image reconstruction.

Chromoendoscopy

This is a wide-field of view, WLE technique that involves topical application of stains or dyes by use of a spray-catheter or lavage. The various stains and dyes used may be absorptive, contrasting or react with particular cellular components. Absorptive agents aid identification of epithelial cell changes due to preferential absorption or diffusion of light. A variety of topical dyes have been evaluated including Lugol's solution, methylene blue, toludine blue and crystal violet. Lugol's solution (1.5 – 3%) is an iodine based stain which shows affinity for glycogen, particularly in non-keratinized squamous cells. In people this makes it particularly useful for highlighting regions of dysplasia, neoplasia or inflammation within the esophagus, such as with Barrett's esophagus, esophageal carcinoma or esophagitis. Cells within these regions tend to be glycogen depleted and therefore stain weakly or not at all.[1-3] Similarly, failure of cellular uptake or heterogenous staining with methylene blue (0.1 – 0.5% solution) within the upper intestinal tract or colon indicates metaplastic, neoplastic or inflammatory change. Concern over the potential of methylene blue to cause oxidative damage to cellular DNA following exposure to white-light has led to its declining popularity.[2,4] Toluidine blue (1% solution) stains cellular nucleic acids (both DNA and RNA) with an increased affinity for premalignant and malignant cells partially due to an increased mitotic rate.[3] Therefore, rather than a lack of staining in abnormal tissue as with Lugol's and methylene blue, deep blue staining is seen within regions of dysplasia or neoplasia. However false positive results may also be seen with inflammatory tissue.[2] Cresyl violet (0.05 – 0.1%) also stains nucleic acids and can be used in a similar fashion, but has technical issues as discussed below.[2]

These absorptive stains may be used alone, but have also been utilized in conjunction with contrast stains such as indigo carmine (0.1 – 0.8% solution). Compared with absorptive stains, contrast agents are not actively taken up by cells and instead tend to seep into and pool within mucosal crevices thereby highlighting surface topography and mucosal irregularities.[2,4] Reactive stains produce a chemical reaction with specific cellular components resulting in a color change similar to a pH indicator. Congo red (0.3-0.5% solution) for example undergoes a color change from red to blue or black in the presence of acid (pH <3) and has been used to specifically highlight regions of acid-secretion within the gastric mucosa. Acetic acid (1.5 – 3%), another reactive stain, reversibly denatures cytoplasmic proteins. Application to the surface mucosa increases cellular opacity which masks the underlying capillary network, whitening the epithelial surface. This effect is prolonged in normal tissue, whilst dysplastic or neoplastic tissue takes on a reddish hue due to congestion of the associated capillaries which then enhances mucosal architecture within these abnormal regions.[2,3]

Although chromoendoscopy is relatively inexpensive and easy to perform, gastric and intestinal surface application can be time consuming and cumbersome, regardless of the particular stain used. Surface washing with mucolytic agents and prolonged dwell times are sometimes required for adequate stain distribution or cellular uptake, with subsequent vigorous washing required to remove excessive dye. Failure at any of these steps can result in variable and patchy staining increasing the false positive and negative rate.

Narrow Band Imaging and Virtual Chromoendoscopy

Narrow band imaging (NBI) has been developed to overcome the technical shortcomings of chromoendoscopy. Exploitation of differential penetration, absorption and reflection of each of the red/green/blue (R/G/B) wavelengths of light forms the basis for NBI.[5,6] As with standard WLE, optical interference filters are interposed with the xenon white-light used for illumination. However, in NBI these filters restrict incident light much more than standard broadband filters and not only clearly separate each of the R/G/B wavelengths, but also restrict tissue exposure to only green (540nm) and blue (415nm) wavelengths. This results in a significant difference in the average depth of penetration of each band within the mucosa.[5-7] However, in order to recreate the image on the video monitor, reassignment of these two wavelengths to each of the three input channels (R/G/B) must occur. This is achieved by assigning the blue wavelength (415nm) to both the green and blue channels, whilst the green wavelength (540nm) is allocated to the red channel.[7] Image transfer via the CCD chip occurs as for standard WLE with each of these bands transferred either in a single step, or sequentially.

Selective absorption and reflection of these narrower bands improves definition of the surface mucosa in the reconstructed image. Mucosal micro-vasculature is enhanced due to selective absorption of blue wavelength light via hemoglobin resulting in reproduction of the superficial capillary network in a brownish color. The green wavelength penetrates more deeply, is outside the absorption band for hemoglobin and therefore provides detail of deeper, larger vessels.[5-7] The final image represents the superficial and deeper detail superimposed as a single image and this enhances the definition of flat lesions, and improves evaluation of the microvasculature.

In comparison to NBI, which relies upon optical filtration of the light source, virtual chromoendoscopy units use computed spectral estimation algorithms to post-process and optimize the captured WLE image. Using this method, reflected photons of set wavelengths can be isolated, and assigned to each of the R/G/B bands in order to provide optimal contrast, producing a similar image to that with NBI, but with more choice in the combination of wavelength settings. The most common use of this modality in people is to observe flat polypoid type lesions that are not as easily visible using WLE.

CONFOCAL ENDOMICROSCOPY

Confocal endomicroscopy (CEM) encompasses confocal microscopy technology into endoscopy in one of two ways. In one type, the confocal bundle is passed through the channel of a working scope. In the prototype confocal endomicroscopes that our group is using, the confocal microscopes are incorporated into a flexible video endoscope and rigid endoscope. The flexible confocal endomicroscope utilizes a 488 nm laser, and is capable of 7 µm axial resolution, with image assessment possible up to a depth of 250 µm. Within the gastrointestinal tract, CEM can be utilized to assess the mucosa when the endoscope is in contact with the surface to give a real-time, virtual histopathologic image. This histologically equivalent image is possible because light that reaches the image detector is filtered to a large degree such that only that light which exactly matches the focal plane within the tissue is detected. This eliminates light scatter which would otherwise contribute to 'noise' and produce foreground and background blur, limiting depth of view.[5] Imaging via CEM is achieved in gastroduodenoscopy in conjunction with the use of fluorophores (contrast agents) and adjusting focal-plane depth as required to assess deeper structures.[6-9]

The most commonly utilized and studied fluorophores in human and laboratory studies include fluorescein, acriflavine and cresyl violet.[7-10] The latter has been evaluated to a far lesser extent that the former two, but has the added potential of allowing for wide-field of view evaluation via chromoendoscopy and identifying areas of interest that can then be assessed more closely using CEM. The images subsequently obtained are adequate to assess cellular detail; however precipitation of cresyl violet on contact with the mucosal surface can potentially complicate interpretation by resulting in patchy distribution which may create artefactual lesions on wide-field assessment. In addition , this precipitant may adhere to the light fiber on the tip of the endoscope, and the wide-field view is then blocked.

Acriflavine (0.05% solution) is a topical antiseptic agent that complexes with intranuclear DNA and polyanionic biopolymers thus highlighting nuclear features.[11] To a lesser extent cytoplasmic components are stained. In some studies acriflavine has also been used to highlight bacterial populations – either via active uptake of acriflavine by specific bacterial populations, or via removal of luminal contents, *ex vivo* staining and re-introduction of this specifically stained content. In doing so, the distribution of bacteria, and potentially bacterial translocation of organisms in disease can be observed. In our preliminary study acriflavine provided the best cellular detail compared to fluorescein or cresyl violet in the stomach.

Fluorescein is the most commonly utilized fluorophore in studies detailing CEM. In comparison to cresyl violet and acriflavine, this is administered intravenously. Fluorescein highlights vascular detail and provides cytoplasmic detail without specifically highlighting nuclear contents. Variable fluorescence can occur due to changes in cellular pH and this can be used to identify regions of disease where affected cells often have a more acidic cytoplasm and therefore a lower degree of fluorescence. We have determined that an IV dose of 15 mg/kg is safe in dogs (minimum body weight 12 kg), with minimal to no side effects apart from temporary discoloration of skin and urine and highlights the cellular structures well.

Clinically in people, gastrointestinal CEM has been used for a wide range of different disease assessments. Much of the work performed thus far has been in its use to assess transformation of dysplastic disease to neoplasia in Barret's esophagus, gastritis and ulcerative colitis.[5] In these circumstances, and particularly using topical fluorophores such as acriflavine, nuclear abnormalities can be detected, but in addition vascular and crypt patterns that correlate with specific disease states have been described. CEM has also been used to evaluate the edges of lesions resected via endoscopy to detect residual disease, evaluate the number of epithelial gaps or the degree of epithelial shedding in inflammatory disease in response to various mucosal challenges, assess numbers of goblet cells and to evaluate the distribution of bacterial populations to identify translocation.[1-7] Another advantage of this methodology is the ability to target biopsy sampling, and to gain a real-time appreciation of the villi morphology. Currently, the biggest disadvantage of this technology remains the expense and availability of the equipment. In addition, obtaining diagnostic quality images and interpreting them requires considerable training and experience.

VIDEO ENDOSCOPIC CAPSULES

Video capsules are miniaturized chips contained in a capsule that takes multiple images as it transits the intestinal tract, transmitted to a wireless receiver attached to the body. Lead position allows for the approximation of anatomical location during reading of the studies. Video capsules are currently recommended as the most appropriate diagnostic modality in people with a positive occult fecal blood test that have had normal standard WLE examination of the upper and lower gastrointestinal tract.[12] Capsules can have cameras at both ends (designed for examination of the esophagus), or more standardly, in one end along with a light source (light-emitting diodes) and 8-hour battery. They capture seven images per second, resulting in over 50,000 images in a typical study. The short battery life may be problematic in some dogs that could have longer gastric emptying than this, resulting in evaluation of the stomach mucosa only. In our experience, pre-treatment with metoclopramide did not result in faster gastric transit of the capsule. Fasting for 12-16 hours is recommended, and treatment with a bowel preparation (PEG) may enhance visibility. Once the capsule has passed the stomach, a small meal can be given along with gentle exercise to promote intestinal transit prior to the end of the battery life.

Contraindications include strictures and intestinal obstruction, and they are not recommended in people with cardiac pacemakers. Anecdotally, some gastroenterologists have reported that the site of capsule obstruction is often determined to be a previously unsuspected neoplastic process following surgical removal of the capsule. The minimum size of dogs that can safely ingest these capsules without risk of obstruction is unknown, but the procedure is used in children following what is called the 'Jelly bean test'. In veterinary medicine, potential applications include motility studies,[13] assessing for causes of occult bleeding and non-invasive assessment of luminal pathology. However, the expense (approx $450 USD per capsule), time and training required for interpreting the images make this a modality that is unlikely to become widespread. Additionally, any lesion identified still requires histologic evaluation, along with specific therapy to result in clinical benefit

DEEP (PUSH) ENTEROSCOPY

The inability to reach all parts of the small intestine has led to the development of techniques to reach the distal small bowel, otherwise called deep or push entersocopy. The three techniques use overtubes to facilitate progression of the endoscope, with the overtubes being spiral tips, single or double balloons. In people, both antegrade and retrograde approaches are used, but in dogs it is typically antegrade approach that is most common. Bowel preparation is not necessary for the antegrade approach, but the dog should be fasted for at least 16 hours. To date, this procedure has been described in medium sized dogs, but not in cats.[14-15] With the double balloon technique, generally two people are required to operate the system, whereas with experience the spiral overtube and single balloon techniques can be performed by a single operator. For double balloon, the endoscope with the overtube (polyurethane with a latex balloon) is passed normally into the pylorus. The endoscope is pushed at least 30 cm in to the duodenum and the endoscope balloon is inflated to anchor the endoscope. The overtube is advanced distally and its balloon inflated, then both the overtube and endoscope are 'pulled' towards the mouth together, pleating the small bowel. The endoscope balloon is then deflated and the endoscope advanced distally (with the overtube balloon still anchored). This 'push-pull' cycle is repeated until the length of the small bowel has been assessed. Successful visualization of the entire small bowel of dogs has been reported.[14-15]

These procedures are quite long, and therefore in dogs it should be carefully considered before they are used. The most common indication in people is when there is melena but no lesion has been identified on standard gastroduodenoscopy or video endoscopic capsule imaging. Isolated jejunal lymphocytic-plasmacytic enteritis has been identified using this technique,[16] and it could be considered an alternative to exploratory laparotomy in some cases. Training guidelines for medical gastroenterology suggests that expert endoscopists need to perform 5-10 of these procedures before they gain adequate skill,[15] and that due to the lack of landmarks in the small intestine, fluoroscopy should be used initially to guide placement during training. As in human gastroenterology, it may also be preferable for carbon dioxide insufflation as opposed to air to reduce discomfort following the procedure. There are also therapeutic models of this equipment that allow for hemostasis, mucosal ablation, polyp resection and stricture dilation in the distal small intestine.

REFERENCES

[1]Marcon NE et al.. **Atlas of Endomicroscopy**. 2008:1 [2]Amateau S et al., **Current Opin Gastroenterol** 2010;26:445 [3]Curvers WL et al, **Best Pract Res Clin Gastroenterol** 2008;22:687 [4]Hurlstone DP et al., **Gut** 2004;53:284 [5]Kiesslich R et al, **Atlas of Endomicroscopy**. 2008 [6]De Palma GD. **World J Gastroenterol** 2009;15:5770 [7]Dunbar K & Canto M. **Current Opin Gastroenterol** 2008;24:631 [8]Dunbar K & Canto M. **Tech Gastro Endo** 2010;12:90 [9]Kiesslich R, et al. **Gastroint End Clinics N Am** 2005;15:715 [10]Goetz M, et al. **Gastro Endo** 2009;70:959 [11]Tiffe HW et al. **Histochem** 1977;53:63 [12]Schnoll-Sussman & Kelsey, **Successful Training in Gastro Endo**; 2011 pp 97 [13]Boillat et al., **Am J Vet Res**, 2010 71: 898[14]Latorre et al, **Vet Rec** 2007; 161:587 [15]Ayala et al., **J Am Anim Hosp Assoc** 2011; 47:262 [16]Okolo & Buscaglia, **Successful Training in Gastro Endo**; 2011 pp 109

ENHANCING APPETITE IN THE FELINE PATIENT

Jessica Quimby, DVM, DACVIM
Fort Collins, CO

INTRODUCTION

Anorexia is one of the most common clinical signs of disease in cats presenting to the veterinary hospital. Unfortunately it is non-specific and a thorough medical workup is required to determine the underlying cause. Nutritional support is critical during the period of time that the primary problem is being sought and addressed. Additionally, several chronic diseases, such as chronic kidney disease and neoplasia, are common in cats and may result in inadequate caloric intake. Thus management of long term hyporexia is important. Poor body condition is associated with decreased prognosis in several species and has a negative effect on immune function, wound healing and strength.[1] Most importantly owners are distressed when their pet does not eat and it has a negative effect on owner perception of quality of life.

THE UNIQUE FELINE

Cats have a higher requirement for protein and amino acids than other species. When nutrition is inadequate, energy is derived from mobilization of amino acids from muscles stores as opposed to fat.[1] Elderly cats are also unique in comparison to other species as they have stable to increased level of metabolism, as opposed to decreased metabolism.[2] A reduced ability to digest protein and fat has been documented in elderly cats. These combined factors make the utilization of high quality, easily digestible food product critical for these patients.[3] Cats are also particularly sensitive to changes in environmental factors, such as the timing and location of feeding, as well as the food type; smell and "mouth feel" may also play into their willingness to eat.[4]

PATHOPYSIOLOGY

A meal can be thought of as having three parts; initiation, maintenance and termination. Different stimuli are responsible for each phase. Initiation of a meal depends heavily upon environmental stimuli such as olfactory, visual, tactile and auditory cues.[4] Termination of a meal occurs as a result of negative feedback mechanisms such as gastric distention or presence of nutrients in the intestine. Suppression of appetite may occur as a result of psychological, environmental or physical factors such as decreased olfaction, nausea, pain, fever, ileus, vomiting that interfere with this process.[1,4] Circulating inflammatory mediators such as TNF-a, IL-1, and IL-6 may play a role in suppression of appetite.[1] As vomiting is a common manifestation of disease in cats, and nausea is likely involved even if vomiting does not occur, examining the pathophysiology of vomiting centers and receptors may help in determining the etiology of the inappetence and appropriate medical management.

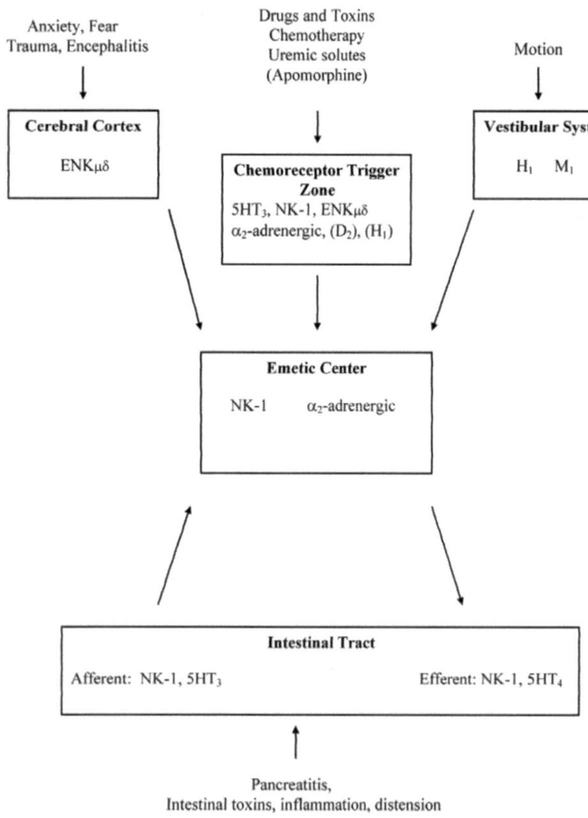

LEARNED FOOD AVERSION

Care should be taken to select the appropriate patients for appetite enhancement as learned food aversion is thought by most to be prevalent in cats.[4] Learned food aversion occurs when the patient associates nausea, pain or other physical manifestations of disease with the act of eating or the sight or scent of food. Even after the underlying illness is resolved, this aversion may remain. Therefore it is critical that cats that are overtly nauseous – drooling, gagging, turning away from food - particularly in hospital or in acute illness are not forced to eat lest food aversion be created.[4]

If cats are too nauseous or critical to even consider oral feeding, or have not responded to appetite encouragement after 3-5 days, placement of an enteral feeding tube should be considered. Nasoesphageal, esophageal or gastrotomy tube can be chosen depending on the type and duration of feeding desired.[1] Parenteral feeding should be considered in cats that cannot tolerate enteral feeding. Additionally many clinicians feel that prescription diets (i.e. renal) should not be fed in hospital during a crisis lest an aversion be created to the diet desired for long term management. The best candidates for pharmacological enhancement of appetite are cats leaving the hospital with their acute crisis resolved, and cats with chronic disease in the home environment.

THERAPIES

Medical management of underlying disease

Whenever possible it is obviously ideal to address the underlying disease condition. However in some chronic diseases, i.e. kidney disease, this is not possible and therefore medical management of complications of chronic disease are ideal. Dehydration and anemia can both potentially play a role in inappetence and therefore should be addressed when applicable.

Environmental factors and food choices

As cats are particularly sensitive to environmental factors, suggestions for appetite enhancement include: quiet environment with exclusive access without interference from bothersome household members, novel food type in cases of possible aversion, similar food type in case of food exclusivity, warming the food particularly if olfaction is an issue, alternatively chilling the food if aromas appear to result in nausea, social interaction while eating, feeding small frequent meals as premature satiety is associated with many disease states.

Pain management

For both acute and chronic medical conditions, appropriate management of pain is critical to improving appetite. Opioids are perhaps most commonly utilized. Buprenorphine has a significantly longer half-life and may be useful for in home palliation of chronic pain. Butorphanol may also have some visceral analgesic properties. Opioids should be used with caution in patients with ileus. Maropitant may also have some visceral analgesic properties.[5]

Anti-nausea/anti-emetic medications

Assessment of nausea in the feline patient is of course subjective, but if suspected based on the underlying condition, treatment is merited both to prevent learned food aversion and to improve appetite in general. Ondansetron and dolasetron have been used in human renal failure patients and appear to have potent anti-nausea and anti-emetic properties.[6,7] Maropitant is an anti-emetic, but is thought to also have anti-nausea properties and clinical trials for its use in kidney disease cats are currently underway. Mirtazapine demonstrates anti-nausea properties in addition to its appetite stimulating properties as it acts at the 5HT3 receptor similarly to ondansetron.[8] All four of these drugs work centrally as well as peripherally on receptors. Many are available as an injection which may be a welcome alternative to owners. Metoclopramide is thought to be a weak anti-emetic in cats do to a lack of receptors in the CRTZ, but may be useful as a promotility drug.[9]

Appetite stimulants

Cyproheptadine has been used for some time as an appetite stimulant and has anecdotal efficacy in many patients, however its efficacy has never been scientifically evaluated. Twice daily administration is necessary in many cases and this can prove a challenge for owners, particularly longterm. Mirtazapine has become more commonly used and recent exploration of its pharmacodynamics and pharmacokinetics have provided information for more effective use in cats.[10,11] Pharmacodynamic studies have illustrated that it can be a potent appetite stimulant, but higher doses are more commonly associated with side effects (hyperexcitability, vocalization, tremors). Smaller, more frequent doses are recommended. The half-life is short enough that it could be administered daily in normal cats. Renal disease delays clearance and in these patients, every other day administration is recommended.[11] Owner should be aware mirtazapine and cyproheptadine cannot not be administered concurrently, cyproheptadine is in fact used as an antidote for serotonin effects of mirtazapine overdose. The ghrelin receptor agonist, capromorelin, which mimics the activity of the "hunger hormone" ghrelin in stimulating appetite, is currently being assessed for use in cats which may add additional options for appetite stimulation in chronically hyporexic patients.

Drug	Receptor	Location of action	Dosage
Butorphanol	ENK	Cerebral cortex and CRTZ	0.2 – 0.4 mg/kg SQ, IV q 6-8 hrs
Buprenorphine	*mu*	Various	0.005 – 0.01 mg/kg SQ, buccal mucosa
Metoclopramide	D2 (weak)	CRTZ (weak)	0.2 – 0.4 mg/kg SQ, IV q 6-8hrs
Ondansetron	5HT3	CRTZ and GI afferent	0.2 – 1.0 mg/kg IV, SQ, PO q 8 –12 hrs
Dolasetron	5HT3	CRTZ and GI afferent	0.6 – 1.0 mg/kg IV, SQ, PO q 24hrs
Maropitant	NK-1	Emetic center, CRTZ, GI	1 mg/kg SQ, PO q 24hrs
Mirtazapine	5HT3	CRTZ and GI afferent	15mg tablet: 1/8 q 24 hrs in normal cats, q 48 hrs in kidney disease.
Cyproheptadine	H1	Various	2-4 mg per cat q 12-24 hrs

REFERENCES

[1]Chan D. **J Fel Med Surg** 2009;11:925. [2]LaFlamme DP. **Vet Clin Small Anim** 2005;35:713. [3]Sparkes AH. **Top Companion Animal Med** 2011; 26:37. [4]Michel KE. **J Fel Med Surg** 2001;3:3. [5]Boscan et al. **Am J Vet Res** 2011; 72:1576. [6]Santos LCP et al. **Vet Anaesth Analg** 2011;38:320. [7]Ljutic D et al. **Kidney Blood Press Res** 2002;25:61. [8]Reichelmann RP et al. **Am J Hospice and Palliative Med** 2010; 27:106. [9]Trepanier L. **J Fel Med Surg** 2010;12:225. [10]Quimby JM. **J Vet Pharmacol Therap** 2010;34:388. [11]Quimby JM. **J Vet Intern Med** 2011;25:985.

NEW DEVELOPMENTS IN THE MANAGEMENT OF FELINE DIABETES

Jacquie Rand, BVSc, DVSc, MACVS, DACVIM
Brisbane, Australia

The primary goal of management of a newly diagnosed diabetic cat has changed from just controlling clinical signs and avoiding clinical hypoglycaemia, to one where the primary goal is to achieve diabetic remission, and in some cats, a cure for diabetes. Preliminary data show that about 25-30% of cats that achieve remission, have no evidence of glucose intolerance on a glucose tolerance test, meaning essentially that they are cured. The remaining 70% have various levels of impaired glucose intolerance, and hence should be classified as pre-diabetic.

With appropriate therapy, remission is achievable in approximately 90% of newly diagnosed diabetic cats, that is, cats are able to maintain euglycemia without the need for insulin therapy. However, remission rates are substantially lower in cats with long-standing diabetes where hyperglycemia has been inadequately controlled because of inappropriate insulin type or dose adjustments, or because underlying disease such as acromegaly precludes achieving excellent glycemic control. Typically, long-term, poorly controlled cats referred for management of their diabetes, have remission rates of 30% or less, because there are inadequate numbers of functioning beta cells remaining to maintain euglycemia without exogenous insulin therapy.

Whatever the underlying cause of diabetes in an individual cat, chronic hyperglycemia causes glucotoxic and lipotoxic damage to beta cells which initially is reversible but later irreversible. Therefore, therapy needs to be aimed at rigorously controlling blood glucose and should be instituted as soon as possible after diagnosis. Early institution of rigorous glycemic control results in remission rates of > 80%, whereas waiting 6 months or longer to institute the same rigouous control results in substantially lower remission rates, which are in the order of 30%. The key factors in achieving high remission rates in newly-diagnosed diabetic cats are firstly, use of long-acting insulin, secondly, feeding a low carbohydrate diet and thirdly, frequent monitoring, ideally home monitoring, together with appropriate adjustment of insulin dose. For those cats not achieving remission, the goals are to resolve clinical signs and avoid hypoglycemia.

Insulin therapy

The long-acting insulins glargine and detemir remain the preferred treatment for diabetes mellitus in cats to maximize the probability of remission. Their effectiveness and safety is enhanced when combined with a low-carbohydrate diet. Achieving good glycemic control and hence remission with intermediate-acting potent insulins such as NPH and lente is often difficult, and increases the risk of clinical hypoglycemia. Published data suggests that the long-acting insulins glargine and detemir provide better glycemic control and reduced risk of clinical hypoglycemia when given twice daily and combined with a low-carbohydrate diet. More importantly, their use results in a significantly higher probability for remission.

Cats presented with diabetic ketoacidosis can be treated with subcutaneous glargine during stabilisation. When used intramuscularly or intravenously, glargine acts like regular insulin and can be used for initial stabilisation in place of regular insulin.

Glargine can be safely instituted at 0.5 IU/kg q12h and serial blood glucose curves should be obtained daily for 3 days either in hospital or at home. When evaluating the blood glucose curve using glargine, it is often more useful to assess pre-insulin glucose concentration rather than the nadir glucose. We have found it often takes 3--5 days for a good glucose-lowering effect to be seen in the glucose curves, possibly because of the long duration of action and carry-over effect of glargine. Almost all cats will need to have their initial dose reduced within 2 weeks and many will achieve remission within 4-6 weeks.

In contrast to glargine, detemir is a newer synthetic insulin analogue with long duration of action through modification of the insulin molecule via addition of an acylated fatty acid chain. This modification facilitates reversible binding to plasma proteins, particularly albumin, from where it is released slowly into plasma. The modification also prolongs self-association in the injection depot, which prolongs absorption from subcutaneous tissue at the injection site and contributes to the long duration of action.

Detemir results in similar remission rates and time to remission as glargine. The same dosing prococol as for glargine can be used. Although, the median maximum dose reported from one study (1.75 IU/cat q12h) was approximately 30% less than with glargine (2.5 IU/cat q12h) for a similarly designed study, this might be because the cats on detemir were older and had higher rates of concurrent renal disease, which might have decreased the renal clearance of detemir. This is unlike the situation in dogs, where the dose of detemir is approximately ¼ of that for glargine. The advantages of detemir is that it can be diluted with a special diluting medium supplied by the manufacturer (although it appears to be difficult to obtain from the manufacturer in USA) and that the duration of action might be longer and more predictable than with glargine. The advantage of glargine is that when administered intramuscularly or intravenously, it has a similar action profile to regular insulin and can be used in the treatment of cats with diabetic ketoacidosis.

Monitoring and adjusting insulin dose when using glargine or detemir should be based on a number of parameters, including pre-insulin and nadir glucose concentration, water intake, urine glucose concentration and clinical assessment as shown in Table 1. Cats treated with glargine or detemir should have a negative, 1+ or 2+ urine glucose (scale 0--4+) and a value of 3+ or 4+ probably indicates that a dose increase is required. Owners should be encouraged to do home glucose monitoring to facilitate tight glycemic control. Meters designed for use with feline blood are advantageous in achieving remission because the accuracy is greater than meters designed for human blood which tend to read lower than the blood glucose actual value. Owners performing home monitoring still need to return frequently for veterinary visits, ideally once a week, until remission is achieved, so the home log of glucose concentrations and insulin dose can be reviewed and dose adjusted as required.

New developments in management of human diabetes include GLP-1 agonists such as exenatide, which now exists in a once-a-week form, however, no clinical trial data have yet been reported from diabetic cats. This drug stimulates insulin secretion but only in the presence of glucose, so is not associated with severe clinical hypoglycemia in the way that insulin therapy is. In addition it stimulates beta cell regeneration. Another new development is a once-a-week insulin for human use, which potentially could provide a once-a-day or once-every-other-day therapy for cats.

Feeding: Low carbohydrate, restricted phosphorus diet

Diets low in carbohydrate reduce postprandial hyperglycemia and insulin concentrations in healthy cats and are associated with increased remission rates in diabetic cats. Data from diabetic cats indicate that low-carbohydrate diets result in better clinical control, reduced insulin requirements and increased rates of diabetic remission One study demonstrated a significantly higher remission rate using a diet with approximately 12% of energy from carbohydrate (Hills m/d) compared to 26% carbohydrate (Hills w/d); the protein contents were similar. In the first few days of treatment, if cats refuse these low-carbohydrate diets, they should be offered any palatable food. However, a commercial low-carbohydrate diet should be used in diabetic cats, unless contraindicated by other disease. When required, dietary management of renal disease using a restricted-protein diet should take precedence over dietary management of diabetes; acarbose can be used to reduce glucose absorption from the gastrointestinal tract. However, acarbose works well only in cats that consume their food within a short time after it is offered and is much less effective in cats which eat multiple small meals during the day, which is the feeding pattern more often observed in cats with IRIS stage 3 or 4 renal disease. Acarbose substantially decreases postprandial blood glucose in cats fed a high carbohydrate diet, but has no real benefit in cats eating a low carbohydrate diet.

Obesity in cats markedly reduces insulin sensitivity, and hence obese diabetic cats should be fed restricted energy so they lose 1--2% bodyweight per week. Weight loss improves insulin sensitivity, and may reduce insulin requirements. Although many diabetic cats achieve remission without substantial weight loss, achieving an ideal bodyweight is critical for maintenance of long-term remission. Due to the decreased postprandial hyperglycemia with a low-carbohydrate diet, diets with less than 20% of energy from carbohydrate should be used for obese diabetic cats during the energy-restriction phase. In contrast to feline diabetic diets, most feline weight loss diets are low-fat, high-carbohydrate diets and are not recommended for newly diagnosed diabetic cats being managed to achieve remission.

There is increasing data now available linking renal disease to diabetes in cats. Therefore, given the frequency of renal disease in diabetic cats (17% in one study in 5-10 year old cats), attention should be paid to the phosphate content of the diet. Many grocery-line low-carbohydrate--high-protein feline maintenance diets are high in phosphate. Cats in stage 1 or 2 renal failure which do not require protein or phosphate restriction should be placed on a low-carbohydrate diet formulated for feline diabetics that is low in phosphate; high-phosphate, ultra low-carbohydrate diets are best avoided.

In conclusion, the principal goal of therapy has changed from controlling clinical signs to that of aiming for remission, and in some cases, cure of their diabetes. Key factors in achieving these goals are the use of the long-acting insulins glargine and detemir, low carbohydrate, restricted-phosphorous diets, and close monitoring, ideally with home monitoring, and appropriate dose adjustments. New therapies are emerging from human studies and they may have a future role in improving management of feline diabetes and reducing frequency of therapy.

REFERENCES

Roomp P K, **Rand** JS (2009). Evaluation of Detemir In Diabetic Cats Managed With a Protocol For Intensive Blood Glucose Control, *Journal of Veterinary Internal Medicine* vol 23 (3), p. 697.Roomp, **Rand** JS (2009). Intensive Blood Glucose Control is Safe and Effective in Diabetic Cats Using Home Monitoring and Treatment with Glargine Journal of Feline Medicine and Surgery;11(8), pp.668-682.Marshall RD, **Rand** JS, Gunew MN, Menrath VH (2010). Glargine administered intramuscularly is effective for treatment of feline diabetic ketoacidosis (2010). *Journal of Veterinary Internal Medicine* vol 24 (3) p 686-687

FELINE CHRONIC PAIN: ASSESSMENT AND MANAGEMENT IN COMPLEX MEDICAL CASES

Lisa Moses, VMD, DACVIM
Boston, MA

INTRODUCTION

As our profession has grown more skillful in managing chronic disease, our patients are living longer than many of us could have imagined when we started practicing medicine. Unfortunately, we have only recently started to focus on the chronic pain and disability that accompanies chronic disease and extended survival times. Many assumptions and obstacles have shaped our approach to this problem. When surveyed, veterinarians often cite the perceived difficulties of chronic pain assessment as the biggest obstacle to treatment. Underlying that may be difficulty believing that chronic pain exists in many of our patients, since it is hard to "see". Perhaps even more pervasive is the assumption that pain accompanying chronic disease will automatically and successfully resolve when the primary disease process is controlled. These barriers will be explored in this discussion. Additionally, treatment strategies and outcome measures specific to cats will be detailed and case examples presented during the lecture. The goal of this session is to focus attention on the under recognized problem of chronic pain states that accompany common medical problems.

CHRONIC PAIN PATHOPHYSIOLOGY

Fundamental to a discussion of chronic pain is an understanding of nervous system physiology involved in normal pain processing and how this becomes pathological in chronic pain states. Pain itself is a multidimensional experience, defined by the International Association for the Study of Pain (IASP) as an unpleasant sensory and emotional experience (perception) associated with actual or potential tissue damage or is described in terms of such damage[1]. Pain is the sum total perception of the sensory, affective/emotional ("suffering") and functional components. Being a subjective perception distinguishes pain from other clinical signs that can usually be externally and concretely measured. Nociceptive pain (transient pain without tissue damage) and pathological pain (pain with tissue damage) are well understood physiological processes[2]. These types of acute pain are physically linked to noxious stimuli, generally serve a protective function, and resolve when tissue damage heals.

Chronic pain is a maladaptive pathological process in which pain may persist after tissue healing. There are several important implications to this: pain can be present when no structural lesion exists and persistent or severe, inadequately treated, pathological pain can be precursors to chronic pain states. This is because the nervous system actually changes in response to pain stimulation (neuroplasticity).[3]

Chronic pain is characterized by hyperalgesia (enhanced perception of pain in a local, often expanding area) and allodynia (pain from a stimulus that does not normally provoke pain). [1, 2] Chronic pain is often episodic, can be spontaneous, and subject to waxing and waning severity. [1,2,4] These features challenge our typical approach to assessment since this type of pain may not be elicited on physical examination. The pathological changes that occur in pain signaling and processing appear to be closely conserved in many different mammalian species studied including dogs, cats, rodents and humans.

WHY TREAT PAIN ACCOMPANYING CHRONIC DISEASE?

Our best defense against debilitating chronic pain states is prompt and aggressive treatment of acute pain, specifically aimed at reducing peripheral and central sensitization.[5] Chronic pain is a disease of the central nervous system that depends upon three main mechanisms: hypersensitization of the spinal cord, activation of nociceptive pathways projecting to the medulla and midbrain, and activation of descending pain facilatory pathways.[2] To reverse or change the reorganization in the nervous system requires specific therapeutic interventions separate from treatment of the original disease process. Animals that develop chronic pain states will not have spontaneous resolution of their pain when their primary disease process is improved or resolved. We are lacking methods to determine what percentage of feline patients who have chronic disease will develop persistent pain states. The numbers vary in human patients and are heavily influenced by genetic factors and past history of pain, besides the original disease process.

Treating chronic pain accompanying chronic disease is important for more basic reasons of good medical practice. Addressing pain associated with disease is obviously part of our therapeutic goals. Perhaps treating pain would improve response in the many patients whose clinical signs are only incompletely resolved with treatment of their primary disease. In particular, visceral cavity pain is implicated in poor appetite, ileus, nausea, and lethargy.

Arguably, perceived discomfort is the reason many owners seek veterinary care. As clinicians we intuitively understand that many of the chronic diseases we manage cause pain or discomfort at least some of the time. Our clients may be more focused than us on the idea that the animals feel discomfort daily while treatment is beginning or progressing. It seems unlikely that our patients understand our hope that treatment will result in alleviation of pain sometime in the future. Ethically we should consider whether this is acceptable or not.

OVERCOMING OBSTACLES OF ASSESSMENT

Several research groups have been involved in an effort to produce validated pain scales for use in dogs, but these do not yet exist for cats. Multidimensional validated scales exist to assess canine post-operative pain, chronic osteoarthritis pain and health related quality of life in cancer patients. The validated scales for chronic pain are based on owner observed changes in behavior and functional movements.

Dr. Sheilah Robertson of Florida and Dr. Jacqueline Reid of Glasgow are currently collaborating on the development of a validated pain scale for cats. Dr. Robertson, in particular, is involved in identifying feline specific pain behaviors. Work done on cats with osteoarthritis demonstrated that owners are able to distinguish changes in behavior associated with reduced pain in cats given analgesics compared to those given placebo.[6, 7] Until this work is completed, the use of client/patient specific outcome measures is a useful approach. This method of assessment depends upon careful history taking in order to identify behavior changes that have occurred since pain began and the return of more "normal" behaviors with therapy. For example, cats with osteoarthritis may stop jumping on the bed, but may resume with treatment of their pain. The author uses an intake form that asks owners to comment on functional mobility and social behavior (i.e. grooming, jumping, sleeping position and location, stretching, playing, etc.). The client is asked to repeat filling out this form at regular intervals during therapy.

Another useful approach is having owners keep a pain diary. This can be in the form of a standardized questionnaire or a few questions specifically tailored to the patient. It is important to limit it to a few, mostly y/n questions, and ask owners to record entries about the same time every day to capture trends. A pain diary is particularly useful for clients who have trouble assessing their cat's overall progress (or lack thereof) or whose medical problems are sufficiently complex to confuse pain assessment.

Given that pain is, by definition, subjective and until interspecies communication is perfected, the most useful method of chronic pain assessment consists of clinical suspicion and response to treatment. We are often not comfortable with this approach since it counters our usual diagnostic process. Since the consequences of untreated chronic pain are so significant for our patient's quality of life and the treatments are generally low risk, our discomfort does not dismiss our responsibility to address chronic pain.

COMMON DISEASES ASSOCIATED WITH CHRONIC PAIN

Almost all of the chronic diseases diagnosed in cats may be associated with enough discomfort to cause observable changes in behavior. Recognition of associated pain depends upon asking careful questions of a caretaker about the daily routine and being open to the idea that some clinical signs of disease may be direct pain behaviors. An easy place to start is in cases where appropriate therapy has not resulted in expected resolution of clinical signs. For cats, this often means persistent poor appetite, sleeping more than usual/lethargy and decreased grooming behavior.

While many chronic diseases are easily recognized as manifesting as pain states (i.e. osteoarthritis, stomatitis), others may be overlooked. Diseases of visceral abdominal organs cause episodic or constant discomfort, but clinicians typically assess abdominal pain only with palpation. Palpation has a low sensitivity for detecting chronic visceral pain, in part because of poor localization and the variable intensity of this type of pain. Cancer pain is multifactoral in origin; there are usually multiple, changing mechanisms of pain generation in an individual as cancer progresses. Inflammatory mammary carcinoma is a typical example of this. Besides the obvious inflammatory pain in the tumor, neuropathic pain develops as nerves are impacted and hypersensitivity and/or allodynia develops in surrounding, seemingly normal tissue. Many clinicians rarely consider treating pain as a separate problem in cases of visceral abdominal disease or even cancer until the evidence for pain is unequivocal. Since cats are very capable of hiding signs of pain, under treatment of chronic pain is likely more prevalent than it is in dogs. The challenge given here is to consider treating pain as a specific clinical problem, worthy of being addressed early in the progression of disease, before you have glaring evidence of pain.

Common disease processes to consider chronic pain in:

- Lower urinary tract disease
- Chronic pancreatitis
- IBD/chronic hepatitis
- Chronic constipation/megacolon
- Degenerative neurological disorders (e.g. diabetic polyneuropathy)
- Chronic ocular disease
- Neoplasia (e.g. fibrosarcoma, GI LSA, oral and nasal SCC)
- Periodontal disease, stomatitis

PERSISTENT POST-SURGICAL AND POST-TRAUMA PAIN

Persistent post-surgical, therapeutic, or traumatic pain are clinical problems that have traditionally been difficult to manage. Examples of these cases are post-onchyectomy neuropathy and persistent pain after fibrosarcoma excision or radiation therapy. Persistent post-surgical pain syndromes can happen after any type of surgery, but are most often associated with nerve trauma (either due to the disease process or surgery or both). These cases often do not have demonstrable structural lesions and pain may not be elicited on palpation. Owners will sometimes report odd behavior, like chewing fur off a localized area unrelated to the original problem or spontaneous vocalization and running. This is usually due to neuropathic pain syndromes where the pain may be spontaneous and episodic. These cases may show signs of opioid hyperalgesia and usually respond better to treatment with anti-convulsants, anti-depressants, acupuncture, and anti-inflammatory medications. They generally require layered therapy and response to treatment may take several months, but can be very successful.

FELINE MUSCULOSKELETAL PAIN AND DEGENERATIVE JOINT DISEASE

Degenerative joint disease (DJD) including osteoarthritis (OA) of synovial appendicular joints deserves particular attention when considering chronic pain in cats. Radiographic evidence of DJD is exceedingly common in cats. The reported prevalence ranges from 90% to 22% in various studies.[8] In recent years multiple researchers and authors have written about the mismatch between radiographic prevalence of feline OA and lameness.[6-9] Investigation into this central question has led to multiple conclusions: Lameness does not appear to be the most common or reliable sign of DJD in cats, radiographic evidence of DJD is not necessarily associated with pain in cats, but maybe most importantly, radiographically normal joints may actually be painful. Joint thickening, effusion, and crepitus correlate well with radiographic DJD in contrast to pain on manipulation of joints.[9] Lascelles reminds us of an important point; radiographs should be used to confirm our suspicions of DJD, but not to guide treatment.

DJD appears to have other features in cats that distinguish it from the disease in dogs. Cats form less new bone and have less radiographic pathology with DJD than dogs do. Clinical features are also different. Changes in ability to jump up or down, changes in the height of jumps, decreased overall movement, avoiding contact with others and irritability when handled are better indicators of DJD associated pain in cats than lameness. Improvements in all these signs and return to using the litter box, grooming, stair climbing, claw sharpening and playing are indicators of treatment success.

Treatment of DJD in cats is very rewarding in spite of the complexities of NSAID therapy in this species. In 2010, The International Society of Feline Medicine and the American Association of Feline Practitioners published excellent, comprehensive guidelines on the use of long term NSAID therapy in cats.[10] The guidelines summarize recent work on the pharmacology of NSAIDs in cats and make specific recommendations on use with concurrent disease and medications. Down load these guidelines for free at www.isfm.net/toolbox/info_sheets/NSAIDs_guidelines.pdf .

The author uses combinations of Adequan® subcutaneously, acupuncture, anti-convulsants (i.e. gabapentin) and environmental modification for treatment of feline DJD. NSAIDs are incorporated into therapeutic plans as required, even in cases with other diseases typical of elderly cats.

GENERAL THERAPEUTIC STRATEGIES FOR FELINE CHRONIC PAIN

When chronic pain is addressed as a problem worthy of designated intervention, it can be approached like any other medical problem. Although the issue of chronic pain may be more often brought to surgeons, internists are uniquely qualified to manage it. These cases often require manipulation of multiple therapies and monitoring of the impact of therapy on organ function. This is the case even in the straightforward case of feline DJD without co-morbidities. The ultimate goal of this session is to convince our colleagues that treatment of chronic pain is possible with current knowledge and will improve outcome for our patients with chronic diseases.

Key Points:

- Response to treatment may be the only "test" we have for chronic pain; treat for *expected* pain, not confirmed pain
- Multimodal therapy is usually required to successfully treat chronic pain
- Non-pharmaceutical therapy is particularly useful for chronic pain and reduces adverse drug events
- Co-morbidities and concomitant drug therapy does not justify withholding pain management
- Ensure that all potential discomfort causing clinical problems are addressed by the pain management plan
- Use nutraceuticals and disease-modifying OA agents (DMOAD) rationally and carefully
- Use adjunctive pain medications to reduce reliance on medications more likely to cause adverse drug events
- Design client and/or patient specific outcome measures and give clients strategies to assess progress
- Explicitly discuss therapeutic goals and educate clients about reasonable expectations of chronic pain management
- Update pain management plans regularly, but no more often than every 3-4 weeks
- If therapy becomes ineffective, look for new problems
- Collect baseline CBC, chemistry profile and urinalysis before beginning drug therapy and at regular intervals after.
- Any patient receiving drug therapy that becomes sick should be examined and have lab work evaluated

SUMMARY

Chronic pain states in cats frequently coexist with, or are caused by, complex medical conditions that we regularly treat. Diagnosis of chronic pain can most often be inferred from owner's observations about reductions in function mobility, deviations from typical behaviors and changes in social interactions. Even when a diagnosis of chronic pain is uncertain, trials of appropriate analgesia may assist in confirming the clinical suspicion. Treatment of pain that may accompany chronic illness can improve overall response to treatment and improve clinical signs that linger. Besides chronic pain caused by visceral organ diseases, diseases that manifest primarily as pain producing or disabling are under diagnosed in cats. Degenerative joint disease is particularly common, produces different radiographic and clinical signs compared to dogs and is frequently left untreated due to fears of drug side effects. Multimodal therapy reduces risk of adverse events and improves outcome and can often greatly improve quality of life.

REFERENCES

[1] Mersky H, et al.Classifications Of Chronic Pain, 1994, IASP Press. [2] Vanderah TW. **Med Clin N Am** 2007;91:1. [3]Grubb T. **Top Comp Anim Med** 2010;25(1):45. [4]Roberston S, et al **J Feline Med Surg** 2010;12:188. [5]Pozzi A, et al. **JAVMA** 2006;228(1):53. [6]Lascelles et al. **JVIM** 2007;21:410. Bennett D et al. **J Feline Med Surg** 2009;11:997. [8]Kerwin SC. **Top Comp Anim Med** 2010;25(1):218. [9]Lascelles D, et al. **J Feline Med Surg** 2010;12:200. [10]Sparks A, et al. **J Feline Med Surg** 2010;**12:**521.

FIGURE OUT FELINE PANCREATITIS

M. A. Forman, DVM, DACVIM
Stamford, CT

INTRODUCTION

Pancreatitis in cats has been recognized as a relatively common disorder causing significant morbidity and infrequently, mortality.[1-35] The frequency of the ante mortem diagnosis of pancreatitis in cats has increased with advances in pancreatic serology and imaging.[8,15,16] This is despite non-localizing clinical signs of inappetence, lethargy, vomiting, weight loss and less commonly, diarrhea and physical examination findings of dehydration, hypothermia, and icterus. Additionally biochemical abnormalities of increased liver enzyme activities, hyperbilirubinemia, hypercholesterolemia, hypoglycemia, ionized hypocalcemia, and hypoalbuminemia are non-specific to the pancreas.[8,15,16,28,33] Multiple generations of serologic tests for pancreatitis have been developed and evaluated with the sensitivity, negative predicative value, specificity and positive predicative value established for a majority of these tests (see Table 1). Multiple pancreatic imaging techniques have also been evaluated (see table 1) including advanced imaging of the pancreas with computed tomography and endosonography. This lecture will review the new generations of feline pancreatic-specific serology including patient side testing, advances in imaging including transabdominal ultrasound imaging of the normal pancreas and serial ultrasounds in cats with pancreatitis. Lastly, a classification scheme of six categories (see Table 2) will be proposed as a tool in the diagnosing the likelihood of feline pancreatitis.

Pancreatic Serology

The next generation of pancreatic serology testing, following the serum Trypsin-Like Immunoreactivity (TLI), was 2 similar but technically different feline pancreas-specific lipase assays. The feline pancreatic lipase immunoreactivity assay (fPLI) is a polyclonal sandwich radioimmunoassay designed and previously (not currently in use) only performed at the Gastrointestinal Laboratory, Texas A&M University, College Station, TX with a sample size of 100 µL. This assay sensitivity and specificity was established utilizing histopathology (n=29 (21 ill & 8 healthy cats)) as the gold standard and for a measure of pancreatitis severity.[36,37] The reference interval for fPLI was 2.0–6.8 µg/L, with concentrations >10 µg/L considered consistent with pancreatitis. The assay was both sensitive in cats with moderate to severe pancreatitis (100%), and specific in the healthy cats (100%), however a lower sensitivity (54%) was noted in cats with mild pancreatitis and poor specificity (33%) in ill cats without pancreatitis.[15]

IDEXX Laboratories, in conjunction with the Gastrointestinal Laboratory, Texas A&M University, College Station, TX then developed a sandwich enzyme-linked immunosorbent assay (ELISA) to measure feline pancreas-specific lipase (Spec fPL® Test). This assay utilizes monoclonal antibodies (in contrast to polyclonal antibodies in the fPLI assay) to bind a single specific binding site. The assay smaller sample size is 25 µL (compared to 100 µL for fPLI) and is performed at the Gastrointestinal Laboratory, Texas A&M University and IDEXX Laboratories. A correlation study with 786 stored serum samples with known fPLI (range 2.0–128 µg/L) revealed a high correlation with Spec fPL ($R^2 = 0.938$). This assay sensitivity and specificity was established utilizing a pancreatitis likelihood classification scheme (n=182 (141 ill & 41 healthy cats) as a gold standard incorporating pancreatic cytohistopathologic evidence in select case (similar number of ill cats but not percentage to the fPLI study) (cytology (n = 5), histopathology (n = 17)).[38] The reference interval for Spec fPL was 0.7–3.5 µg/L, with concentrations >5.4 µg/L considered consistent with pancreatitis. Using the Spec fPL concentration of 5.4 µg/L as the diagnostic cut off and excluding cats with indeterminate pancreatitis, the sensitivity, specificity (and their respective 95% confidence intervals) was 79.4 (63.2 – 89.7) and 79.7 (67.7 – 88.0). The likelihood of a positive test was 3.9 and a negative test, 0.3.

Following the development of Spec fPL, a patient side (in hospital use) feline pancreas-specific lipase (SNAP fPL® Test) was developed and released. The SNAP fPL is a serum based, semi-quantitative, monoclonal antibody test that is reported by IDEXX laboratories to be optimized to match the performance of the Spec fPL test. Two result options are displayed, normal or abnormal. A normal result correlates directly with the normal reference interval for the Spec fPL (0.7–3.5 µg/L), however an abnormal result correlates with both the grey zone (3.6 – 5.3 µg/L) and the interval consider consistent with pancreatitis (>5.4 µg/L). The grey zone was included to permit the SNAP fPL to be utilized as a screening test (higher negative predicative value, a patient tested negative is unlikely to have the disease), however at the expense of a lower positive predicative value.

With the broader utilization of pancreatic serologic testing, questions have been raised on the significance of elevated concentrations concurrently with a potentially more significant disorder (as with an inflammatory bowel disease (IBD)) and persistent increases despite resolution of clinical signs attributed to pancreatitis. The former is explained by the high frequency of reported concurrent disorders (up to 2/3rd) in cats with pancreatitis including IBD, hepatic lipidosis, cholangitis, diabetes mellitus[14,15,19,22,39] however the clinical challenge is avoiding a diagnostic delay for a disorder that is ultimately causing the patient's clinical signs (as with an obstructive intestinal foreign). Ultimately additional research is necessary to make strong clinical recommendations in this situation, however the author recommends screening for concurrent disorders or complications of pancreatitis when cats are being treated for pancreatitis and fail to improve. The explanation for latter is one part of recently completed study (data being evaluated at the time of this writing) involving serial serologic (Spec fPL) and ultrasonographic monitoring of cats with pancreatitis. This will be discussed during the lecture portion and is potentially explained by subclinical elevated pancreatic concentrations associated with chronic pancreatitis.

Pancreatic Imaging

Unlike serology, transabdominal ultrasound is considered challenging and operator, equipment and patient compliance dependent; however it permits detection of non-pancreatic concurrent disorders, screening for causes of pancreatitis and ultrasound guided fine needle aspiration cytology of the pancreas and peripancreatic fluid accumulations.[1,2,40-47] For the normal feline pancreas, the majority of the pancreas can be visualized with smooth pancreatic borders, relatively isoechoic pancreas parenchyma compared to liver with no pancreatomegaly, pancreatic duct enlargement, peripancreatic fluid accumulation, pancreatic nodules or cysts, or enlarged peripancreatic lymph nodes. Pancreatomegaly is defined as left limb larger than 9.5 mm in thickness, body greater than 8.5 mm in thickness or the right limb larger than 6 mm in thickness. The pancreatic duct is considered enlarged if greater than 2.5 mm in diameter. Ultrasonographic changes associated with pancreatitis include pancreatomegaly, hypoechoic pancreatic parenchyma, hyperechoic peripancreatic fat / mesentery, dilated pancreatic or bile duct (s), dilation of the gall bladder, thickened gastric wall and corrugated and thickened duodenal wall. [1,2,15,40-47]

The minimum requirements for the ultrasonographic diagnosis of pancreatitis in cats has not been determined, and it is possible for cats with pancreatitis to have no ultrasonographic abnormalities. Certain abnormalities often associated with pancreatitis such as enlargement of pancreatic duct and pancreatic parenchyma hyperechogenicity may be potentially seen in aged cats without pancreatitis.[42,43,46,48-52] Additionally, certain abnormalities, pancreatic pseudocysts, nodules or duct dilation may suggest recent, but currently inactive pancreatitis while a concurrent non-pancreatic disorder is causing the current clinical signs.[53] Wide and conflicting sensitivities for the diagnosis of pancreatitis by abdominal ultrasound have reported from 20-35% to 62% with a specificity of 73% based on limited numbers of symptomatic cats without pancreatitis.[15,28,32,42,45,47] Advances in ultrasonographic equipment, ultrasonographer training and study methodology may have contributed to wide range of sensitivities.

In contrast to feline pancreatic ultrasound and the human experience with advanced imaging of the pancreas, improvements in diagnosing pancreatitis have not yet been realized with standard pancreatic computed tomography or endosonography. Two studies[15,16] utilizing computed tomography with IV contrast failed to identify differences in cross section diameter, contours, enhancement, and pre- and postcontrast density between normal and ill cats, however a recent case report[18], performed following a feasibility study[54] suggested a benefit of concurrent utilizing radiolabeled leukocytes to aid in identifying pancreatitis. Pancreatic margins and parenchyma was better visualized with endosonography, compared to transabdominal ultrasound, but failed to provide a diagnostic benefit.[55]

In a manner similar to pancreatic serologic testing, questions have been raised on the significance and/or presence of persistent ultrasound pancreatic abnormalities following an episode of pancreatitis. This was evaluated in recently completed study (data being evaluated at the time of this writing) involving serial serologic (Spec fPL) and ultrasonographic monitoring of cats with pancreatitis. This will be discussed during the lecture portion.

Table 1: Reported Diagnostic Utility of Non-Invasive Tests in Feline Pancreatitis[15,16,28,32]

TEST	Sensitivity	Negative Predicative Value	Specificity	Positive Predicative Value
Amylase, Lipase	Not Useful	Not Useful	Not Useful	Not Useful
Abd. Ultrasound*	24-67*%	57%	73%	80%
Endosonography	Not Useful	Not Useful	Not Useful	Not Useful
Computed tomography	20%	Not Useful	Not Useful	Not Useful
TLI*	28%	41%	75%	71%
PLI*	67%	62%	91%	92%
Spec fPL^	79.4%	87%	79.7%	69%

*N=29 (21 ill & 8 healthy cats), ^N=182 (141 ill & 41 healthy cats)

Classification scheme in the diagnosing the likelihood of feline pancreatitis

Due to the inherent challenges in diagnosing pancreatitis in cats we proposed a classification scheme during the evaluation of the Spec fPL of clinically defined groups for pancreatitis in cats incorporating the degree of abnormalities in clinical signs, laboratory evidence, ultrasound imaging, cytology and/or histopathology. For the purpose of that study, Spec fPL was not included in the laboratory evidence portion however; we would suggest it is appropriate in future studies to include this parameter in the classification scheme. Likelihood classification schemes have been utilized in other difficult to diagnosis disorders including feline infectious peritonitis and hypertension.[56-58]

The ability to definitively include and exclude pancreatitis in cats, dogs and humans is challenging.[15,59,60] Histopathology, the gold standard for other gastrointestinal disorders including inflammatory bowel disease[61], has limits since pancreatitis can be multifocal and focally severe and therefore, pancreatic inflammation can be inadvertently not detected without serial sections of the pancreas[4], which is contraindicated in live clinical cats. Additionally the degree and type of pancreatic inflammation which correlates with the clinical signs of pancreatitis have not been definitively established.[15] Intestinal and hepatic inflammation commonly noted concurrently with pancreatitis and termed 'Triaditis', can cause similar clinical, physical examination and biochemical and hematologic abnormalities and confuse the diagnosis of pancreatitis.

A classification scheme of six categories (see Table 2) was developed by which to classify each cat as to the likelihood that the cat had pancreatitis. Within each category several criteria were evaluated including: cytohistopathologic (if available), ultrasound, and clinicopathologic findings and the presence or absence of another convincing disease. Histopathologic evidence of pancreatitis or lack of pancreatitis was weighted most heavily and a cat could not be classified as definitely having pancreatitis unless there was histopathologic evidence of the disease. A cat was categorized as definitely having pancreatitis if acute inflammation and/or necrosis was noted to any degree in any section of the pancreas. The absence of the above lesions was evidence that that cat did not have pancreatitis although a cat could be classified as definitely not have pancreatitis even if histopathology of the pancreas was not available for evaluation.

Convincing ultrasonographic evidence of pancreatitis needed to be present for a cat to be classified as probably having pancreatitis (in the absence of histopathologic abnormalities) and a normal pancreatic ultrasound was necessary for a cat to be classified as definitely not having pancreatitis. Cats with weak or no evidence of pancreatitis could be classified as possibly having, possibly not having or probably not having pancreatitis. Clinicopathologic evidence of pancreatitis included the presence of compatible clinical signs as well as laboratory evidence of disease. Clinical signs considered consistent with pancreatitis included: lethargy, inappetence or anorexia, abdominal pain, weight loss, dehydration, vomiting and/or diarrhea. Laboratory findings of pancreatitis include: increased liver enzymes, hyperbilirubinemia, hypoalbuminemia, hypocalcemia, hypokalemia, prerenal azotemia, hypercholesterolemia and/or hyper or hypoglycemia.

In a subset of cats, the absence or presence of another convincing disease to explain clinical signs was used to categorize the cats as possibly or possibly not having pancreatitis respectively. In these cats there was a clinical suspicion of disease but no histopathology was available, ultrasound was normal or not convincing and laboratory results were normal, not convincing or compatible with another disease process. In the absence of normal pancreatic histology being available, the presence of another convincing disease was also used in conjunction with a normal pancreatic ultrasound to classify a cat as definitely not having pancreatitis.

Table 2: Feline pancreatitis classification scheme

Likelihood		**Definition**
Pancreatitis	Strong (definite)	Clinicopathologic and cytohistopathologic evidence of pancreatitis independent of ultrasonographic findings.
	Moderate (probable)	Compatible clinicopathologic with convincing ultrasonographic evidence of pancreatitis and no available cytohistopathologic evidence.
	Weak (possible)	Compatible clinical signs with weak or no ultrasonographic or laboratory evidence and no available cytohistopathologic evidence. A clinical suspicion of pancreatitis without alternative disease.
Not Pancreatitis	Weak (possible not)	Compatible clinical signs with weak or no ultrasonographic or laboratory evidence and no available cytohistopathologic evidence. A clinical suspicion of pancreatitis is present; however a convincing alternative disease is diagnosed.
	Moderate (probably not)	Weak or inconsistent clinicopathologic and/or ultrasonographic evidence of pancreatitis and no available histopathologic evidence.
	Strong (definitely not)	Histopathologic evidence of the absence of pancreatitis; or normal ultrasound of pancreas with alternative disease, not known to be associated with pancreatitis.

CONCLUSION

The expanded utilization of pancreatic serology and imaging is changing the dogma of the challenging nature of antemortem
diagnosis of feline pancreatitis.[15,23,32] The next diagnostic challenge with feline pancreatitis is accurate interpretation and correlation to the pet's clinical signs of elevated pancreatic serologic concentrations and abnormal pancreatic ultrasonographic findings.

REFERENCES

1. Bailiff NL, et al. J Am Anim Hosp Assoc 2004;40:69-74. 2. Bennett PF, et al.. J Am Anim Hosp Assoc 2001;37:466-473. 3. Bruner JM, et al. J Am Vet Med Assoc 1997;210:1757-1760. 4. De Cock HE, et al. Vet Pathol 2007;44:39-49. 5. Owens JM , et al. J Am Anim Hosp Assoc 1975;11:83-89. 6. Akol KG, et al. J Vet Intern Med 1993;7:205-209. 7. Arendt T. Dig Dis Sci 1993;38:39-44. 8. Allen HS, et al. Can J Vet Res 2006;70:313-316. 9. Antonucci F, et al. Scand J Gastroenterol Suppl 1980;62:53-57. 10. Borgstrom A, et al. Z Physiol Chem 1980;361:625-631. 11. Ercan MT, et al. Nucl Med Commun 1993;14:798-804. 12. Fernandez-del Castillo C, et al. Pancreas 1992;7:263-270. 13. Ferreri JA, et al. J Am Vet Med Assoc 2003;223:469-474. 14. Forcada Y, et al. J Feline Med Surg 2008;10:480-487. 2008 Jul 2018. 15. Forman MA, et al. J Vet Intern Med 2004;18:807-815. 16. Gerhardt A, et al. J Vet Intern Med 2001;15:329-333. 17. Grumbein SL. Feline Practice 1981;11:23-26. 18. Head LL, et al. Vet Radiol Ultrasound 2005;46:263-266. 19. Hill RC, et al. J Vet Intern Med 1993;7:25-33. 20. Hines BL, et al. J Am Anim Hosp Assoc 1996;32:147-152. 21. Karanjia ND, Widdison AL, Jehanli A, et al. Pancreas 1993;8:189-195. 22. Kimmel SE, Washabau RJ, Drobatz KJ. J Am Vet Med Assoc 2001;219:1105-1109. 23. Kitchell BE, et al. Am J Vet Res 1986;47:1170-1173. 24. Lapointe JM, et al. Vet Pathol 2000;37:282-284. 25. Mansfield CS, et al. J Fel Med Surg 2001;3:117-124. 26. Mansfield CS, et al. J Fel Med Surg 2001;3:125-132. 27. Pedersen NC, et al. Vet Microbiol 2000;73:281-300. 28. Rademacher N, et al. J Vet Intern Med 2008;22:1310-1316. 2008 Sep 1316. 29. Rothenbacher H, et al. J Am Vet Med Assoc 1963;143:1099-1102. 30. Ryan CP, et al. Feline Practice 1981;11:31-34. 31. Sakai M, et al. J Vet Med Sci 2006;68:1331-1333. 32. Saunders HM, et al. J Am Vet Med Assoc 2002;221:1724-1730. 33. Swift NC, et al. J Am Vet Med Assoc 2000;217:37-42. 34. Strombeck DR, et al. Am J Vet Res 1981;42:1966-1970. 35. Suter PF, et al. J Am Vet Rad Soc 1969;10:4-11. 36. Steiner JM. et al. Comp Biochem Physiol B Biochem Mol Biol 2003;134:151-159. 37. Steiner JM, et al. Can J Vet Res 2004;68:309-314. 38. Forman MA, et al. Manuscript in preparation 2012. 39. Hirsch VM, et al. J Am Vet Med Assoc 1983;182:1223-1226. 40. Fahie MA, et al. J Am Anim Hosp Assoc 1995;31:478-482. 41. Garvey MS, et al. Vet Clin North Am Small Anim Pract 1984;14:1231-1246. 42. Hecht S, et al. Clin Tech Small Anim Pract 2007;22:115-121. 43. Hecht S, et al. Vet Radiol Ultrasound 2007;48:45-50. 44. Hittmair KM, et al. Vet Radiol Ultrasound 2001;42:149-155. 45. Lamb CR. Vet Clin North Am Small Anim Pract 1999;29:307-342. 46. Leveille R, et al. J Vet Intern Med 1996;10:296-299. 47. Moon ML, et al. Vet Radiol Ultrasound 2003;44:199-203. 48. Baron ML, et al. Vet 2010;51:86-89. 49. Etue SM, et al. Vet Radiol Ultrasound 2001;42:330-336. 50. Hecht S, et al. Vet Radiol Ultrasound 2006;47:287-294. 51. Larson MM, et al. Vet Radiol Ultrasound 2005;46:238-242. 52. Wall M, et al. J Am Anim Hosp Assoc 2001;37:49-53. 53. VanEnkevort BA, et al. J Vet Intern Med 1999;13:309-313. 54. Head LL, et al. Vet Radiol Ultrasound 2003;44:420-428. 55. Schweighauser A, et al. J Feline Med Surg 2009;11:492-498. 2008 Dec 2018. 56. Paltrinieri S, et al. J Vet Diagn Invest 2007;19:266-272. 57. Shanmugam R. J 34:161-177. 58. Addie DDJ, O. In: Green CE, ed. Infectious Diseases of The Dog and Cat, Second ed. 1998:63. 59. Steiner JM, et al. Vet Ther 2008;9:263-273. 60. Steer ML, et al S. N Engl J Med 1995;332:1482-1490. 61. Day MJ, et al. Journal of Comparative Pathology 2008;138:S1-S43.

SEPSIS BIOMARKERS

Claire R. Sharp BSc, BVMS, MS, DACVECC
North Grafton, MA

Sepsis, defined as the systemic inflammatory response to infection, is a serious problem in dogs and cats that is associated with high morbidity and mortality.[1-3] Unfortunately, the diagnosis of sepsis in veterinary patients is challenging and delays in diagnosis and thus initiation of therapy may to contribute to the high mortality rates. In recent years, the importance of establishing biomarkers for the diagnosis and staging of sepsis has become evident.

A biomarker, or biological marker, is defined as "a characteristic that is objectively measured and evaluated as an indicator of normal biological processes, pathogenic processes, or pharmacologic responses to a therapeutic intervention".[4] Another perhaps more simple definition is a "quantifiable measurement(s) of biological homeostasis that define(s) what is 'normal', therefore providing a frame of reference for predicting or detecting what is 'abnormal'.[5] Historically biomarkers were primarily physiologic indicators; in the case of sepsis these would be the SIRS criteria (i.e. fever or hypothermia, tachycardia or bradycardia, tachypnea and changes in the leukon). More recently however the term biomarker has been used to refer to endogenously produced biochemical compounds that can be used to diagnose, stage, prognosticate and/or monitor response to treatment. These molecular biomarkers have biophysical properties, which allow their measurements in biological samples such as whole blood, plasma, serum, cerebrospinal fluid, bronchoalveolar lavage fluid, pleural fluid, peritoneal fluid, and tissue. Molecular biomarkers also include nucleic acid-based biomarkers such as gene mutations or polymorphisms; and quantitative analysis of the gene expression of peptides, proteins, lipids metabolites, and other small molecules. The fields of genomics, proteomics, lipidomics and metabolomics are predominantly beyond the scope of this discussion but are certainly emerging areas of study in relation to sepsis.

Some biomarkers may help guide patient management.[6-10] The utility of biomarkers lies in their ability to provide timely clinically relevant information, beyond that which is readily available from routine physiologic data and clinical examination.[11] Identifying biomarkers that could enable clinicians to make a rapid diagnosis of sepsis and subsequently initiate early therapy, particularly antibiotic therapy[12], may benefit patients with sepsis. In addition to their role as a diagnostic tool it hoped that sepsis biomarkers would allow for temporal categorization of systemic inflammation such that tailored therapeutic interventions can be implemented. Biomarkers are also being investigated to facilitate the identification of potential therapeutic targets for immunomodulation and predict patient response to novel treatment strategies. In addition, biomarkers that correlate with disease severity and predict morbidity and mortality are under investigation as there are useful for patient / family / client education and patient stratification in clinical trials, especially in human medicine.[7, 9, 10, 13]

The performance of a biomarker can be evaluated by variety of measures including its sensitivity, specificity, positive and negative predictive values and likelihood ratios. A likelihood ratio reflects the ratio of the measured true positives (sensitivity) to the measured false positives (1-specificity). A likelihood ratio of less than one indicates that the test performs better than random chance; the usefulness of the test increase as the likelihood ratio diverges from one. Likelihood ratios at differing values of the test can be plotted graphically to produce a receive operating characteristics curve, such that superior performance is reflected in a higher area under the curve.[11] Evaluating biomarker performance is important to aid in determination of their clinical utility.

Potential biomarkers are identified by several approaches. Firstly "a biomarker may be selected on the basis of a biologically compelling association with a disease state or a candidate therapeutic intervention".[11] For example circulating endotoxin would be a logical biomarker for Gram-negative infection, and may identify a patient that would benefit from treatment with an endotoxin-neutralizing agent. Secondarily biomarkers may be identified by apparent association with a disease in the absence of a biologically plausible link. For example procalcitonin (PCT) was first observed to be increased in patients with infection, and decrease in response to antibiotic therapy.[14] More recently the search for biomarkers has employed techniques such as microarrays to detect mRNA transcripts (i.e. genomics), and mass spectrometry to detect peptides (i.e. proteomics), in biological samples. These are somewhat unbiased approaches that facilitate screening of the entire genome for candidate biomarkers by identifying parameters which are differentially expressed in the population of interest.[15]

Determining the most ideal biomarkers for the diagnosis, staging, and prognostication of sepsis is complex. There are many potential biomarkers for sepsis; more than 100 distinct molecules have been proposed[16]. This discussion will review biomarkers that have received some attention in the veterinary literature such as clinical parameters, endotoxin (lipopolysaccharide) activity (EA), cytokines (TNF, IL-1β, IL-6, IL-10) and chemokines (CXCL-8), vasoactive mediators (NO, AVP, copeptin), coagulation parameters, acute phase proteins (CRP), and natriuretic peptides (ANP, BNP and CNP). It is important to be cognizant of the inconsistencies in the literature, including discordant conclusions regarding biomarker relevance in certain settings. Just some of the sources of variability in the measurement of biomarkers in sepsis include variable assay methodology (e.g. bioassay vs. immunoassay [ELISA, bead, RAI]), reference standards, site of sampling (i.e.

cellular (membrane, cytoplasmic, nuclear) vs. regional (e.g. BALF) vs. systemic (blood) (e.g. WB, plasma, leukocytes, leukocyte subsets)), requirements for activation (e.g. protein C, TGF-B); and the presence of soluble receptors, circulating inhibitors, and carrier proteins which may interfere with circulating concentration of the biomarker of interest.

Clinical parameters

Rectal temperature, heart rate, respiratory rate and leukocyte count have been used as markers for SIRS in dogs. In one study of 500 dogs with a variety of illnesses, not just sepsis, presenting to a veterinary teaching hospital, mortality was significantly associated with the number SIRS criteria fulfilled.[17] While these parameters are often used as part of the clinical definition of sepsis, it is important to note that these criteria are not necessarily specific for sepsis (i.e. infection). The sensitivity and specificity of SIRS criteria for the diagnosis of sepsis is variable based on the study and the cut off criteria used but ranges from a sensitivity of 77-97% and specificity of 64-77% when 2 of the 4 criteria were met.[18] Physical examination and basic blood work parameters are also used as biomarkers for MODS. For example an increase in creatinine of > 0.5mg/dL, in the absence of pre-renal or post-renal causes is diagnostic for acute kidney injury in the face of sepsis.

Endotoxin Activity (EA)

It would seem logical that diagnosis of sepsis could be expedited by rapid detection of circulating pathogen associated molecular patterns (PAMPs). Measurement of EA in people at the time of ICU admission has a high negative predictive value (NPV) for the diagnosis of Gram-negative infection and sepsis; such that it may have a role in the exclusion of sepsis (Se = 85.3%, Sp = 44%, NPV 98.6%).[19] Unfortunately however EA does not seem to be useful in human medicine in predicting disease severity and assessing the risk of development of severe sepsis / organ dysfunction; thus it cannot be used to aid in resource allocation or guide therapy. A small number of veterinary studies have evaluated EA; two in dogs and one in horses with colic.

The first of the canine studies evaluated 22 healthy dogs, 11 sick dogs without evidence of SIRS, and 8 dogs with suspected sepsis (defined by the authors as SIRS with toxic neutrophils on blood smear evaluation).[20] They found that EA was significantly higher in dogs with sepsis compared to those without SIRS and healthy dogs. The authors thus concluded that the EAA is useful for endotoxin measurement in dogs and proposed a clinical cut off value of 0.346, since only one of the healthy dogs had an EA above that value. The second study evaluated the ability of the EAA to identify infection and predict outcome in 100 dogs consecutively admitted to a university teaching hospital ICU.[21] In contrast to the first study, EAA did not correlate with the presence of infection (p>0.1) but was useful was part of a multivariate prognostic model to predict survival. Clearly further studies are warranted regarding the use of EA as a biomarker for sepsis in dogs.

Lipopolysaccharide Binding Protein

LBP is constitutively expressed in serum and concentrations increase as part of the acute-phase response.[22] LBP has been shown to predict severity of ALI and mortality in people with severe sepsis,[23] have a high sensitivity and specificity for diagnosing Gram-negative bacteremia in people with febrile neutropenia, and is a better marker of sepsis than IL-6 and CRP in critically ill human neonates over 2 days old.[24] This is in contrast to a study evaluating the use of LBP as a sepsis biomarker in the setting of a human surgical ICU, which concluded that the correlation of LBP with disease severity and outcome is weak compared with outer potential biomarkers, such that the use of LBP as a biomarker in the surgical ICU population is not warranted.[25] While the molecular expression and tissue distribution of canine LBP has been investigated,[26] LBP has not been evaluated as a sepsis biomarker in small animal veterinary patients to the knowledge of the author. Serum LBP has been evaluated in horses with colic (n=765), but did not correlate with outcome, underlying disease process, or portion of the gastrointestinal tract affected.[27]

Tumor Necrosis Factor (TNF)

While TNF is not necessarily a useful diagnostic biomarker for sepsis in people, it has been shown promise prognostically; with higher TNF concentrations correlating with mortality and persistent elevations being associated with a worse prognosis in a variety of human studies.[28-30] A variety of works have investigated the potential role of TNF As a biomarker in veterinary patients with sepsis. Plasma TNF activity has been demonstrated in hospitalized dogs with canine parvoviral (CPV) enteritis with a significant association between increasing TNF activity and morality in this population.[31] In addition TNF-α has been evaluated as a potential biomarker for dogs with septic peritonitis. In 23 dogs with septic peritonitis, TNF (measured by ELISA) was not detectable in the serum of 17 dogs on day 1; whereas TNF was detected in the abdominal fluid of 16 dogs. Plasma TNF bioactivity at hospital presentation in dogs with sepsis, non-infectious SIRS and healthy controls has been evaluated by investigators at the University of Missouri, Comparative Internal Medicine Laboratory. We found that significantly more dogs in the sepsis or SIRS groups had measurable TNF bioactivity in their plasma compared to healthy controls, but there was no difference between the sepsis and SIRS groups.[32] Additionally, plasma TNF bioactivity was not significantly correlated with mortality.

Interleukin 6 (IL-6)

In humans with intra-abdominal sepsis there is a direct correlation between plasma IL-6 and the risk of death with plasma IL-6 concentrations alone predicting outcome with >80% accuracy.[33] Persistent production of IL-6 seems to be more important than initial or peak plasma concentrations in regards to patient outcome. IL-6 has documented diagnostic superiority to procalcitonin (PCT) in human neonates with sepsis but not adults. Plasma IL-6 concentration has also recently been found to be predictive of outcome in canine SIRS and sepsis and thus may be a valuable biomarker for assessing critically ill dogs.[34] In our aforementioned study significantly more dogs in the sepsis or SIRS groups had detectable plasma IL-6 compared to healthy controls, but there was no difference between sepsis and SIRS groups. Additionally, plasma IL-6 concentration was not significantly correlated with mortality.[32] We have also evaluated IL-6 in cats. Cats with sepsis were more likely to have detectable concentrations of IL-6 than cats with SIRS or healthy cats, and IL-6 concentrations significantly correlated with mortality in the sepsis group.

Nitric Oxide

Since NO is a highly reactive free-radical molecule with a half-life of only a few seconds it cannot be easily measured, however, NO degrades to the stable end products nitrate and nitrite.[35] Plasma nitrate and nitrite are readily measured and have been used extensively as indicators of NO production.[35-38] A prospective study by Jones et al. found that dogs with critical illness have significantly higher nitrate and nitrite concentrations than non-critically ill and healthy dogs and that nitrate/nitrite are correlated with not only disease severity but also survival.[36] We have evaluated NO in dogs with sepsis, and found that dogs with sepsis had significantly higher plasma NO concentrations than healthy control dogs; however there was no difference between septic dogs and dogs with NSIRS. Nor was there a difference in plasma NO concentrations between dogs that suffered in-hospital mortality and those that survived to discharge in either the sepsis or SIRS groups.[39] This is in contrast to the conclusion of Jones et al that critically ill dogs with the highest NO concentrations had greater mortality.

Procalcitonin (PCT)

Procalcitonin is perhaps the single most useful diagnostic and prognostic biomarker for bacterial sepsis in adult human patients. Additionally PCT has been useful to guide antibiotic therapy (both commencement, cessation and de-escalation of antibiotics).[40, 41] PCT is the precursor peptide of the hormone calcitonin; it is typically expressed in the C cells of the thyroid gland, but also secreted by non-thyroidal tissues. In human patients and experimental models it appears that circulating concentrations of PCT begin to increase within 2-6hrs of induction of bacteremia and/or endotoxemia, with a circulating half-life of approximately 24 hours.[42-44] PCT concentration will also be transiently increased with trauma, such as surgery; but return to baseline within 24 hours in the absence of infection or sepsis. Another interesting feature of PCT as a biomarker is that it is useful in the setting of chemotherapy induced sepsis, since it is a tissue-based rather than leukocyte-based biomarker.

Unfortunately the ELISA for PCT in humans cannot be used for dogs due to differences in the carboxy terminus of the PCT molecule, such that the only study evaluating PCT in dogs investigated PCT mRNA expression by qRT-PCR. [45] This study demonstrated greater PCT mRNA in sick dogs (n=56; categorized into groups with infectious, inflammatory/non-infectious, neoplastic and other 'diseases) compared to healthy controls (n=11) ($p<0.001$), but no there was no difference in PCT mRNA expression based on the primary disease process.[45] Interestingly in five dogs from each group, PCT mRNA expression was determined at two time points; and in each dog the changes in its expression were in agreement with the clinical evaluation of patient improvement or deterioration, suggestive of a possible prognostic role.[45] Investigators are actively working to develop an assay for PCT in dogs.

Coagulation Biomarkers

While coagulation parameters are not particularly useful diagnostic biomarkers for sepsis, alterations in coagulation are directly correlated to disease severity and mortality in human patients with sepsis.[22] Hypercoagulability is also a common finding in dogs with naturally occurring sepsis as evidenced by elevations in prothrombin time (PT), partial thromboplastin time (PTT), D-dimer and fibrinogen degradation products (FDP) concentrations, and depletion of AT and aPC. [20] It appears that aPC may act as a useful prognostic indicator in dogs with sepsis. Evaluation of hemostasis in dogs with CPV enteritis also documented decreased AT activity and hypercoagulability in affected dogs.[23] Von Willebrand factor antigen concentration has been evaluated as a marker of endothelial activation and is an independent predictor of survival in humans with sepsis. In a study of 14 dogs with sepsis and 11 control dogs, vWF antigen concentration was significantly higher in the sepsis group compared to control ($p<0.005$) but there was no difference between the survivors and non-survivors in the sepsis group.[46]

Natriuretic Peptides (NP)

The NP system consists of three structurally related members; atrial natriuretic peptide (ANP), brain or B-type natriuretic peptide (BNP) and C-type natriuretic peptide (CNP). ANP and BNP are made primarily in the heart, whereas CNP is produced predominantly by vascular endothelial cells, but also other organs including the brain, kidneys, gastrointestinal tract, and bone. NPs are synthesized as prepropeptides; the signal peptide is removed by signal peptidase. The propeptide is then cleaved by propeptide convertase to produce an inactive N-terminal (NT) peptide and a biologically active C-terminal mature peptide. In general, the NT-peptide has greater plasma half-life and stability compared to the biologically active c-terminal peptide and thus is a preferred assay target.

ANP and BNP have been extensively evaluated for their potential role as biomarkers, both for sepsis and a variety of other diseases, most notably cardiovascular disease. Mid-regional (MR) pro-ANP has shown the most promise as a sepsis biomarker in people, due to its longer half life than ANP, with particular utility for individual risk assessment in septic patients with admission pro-ANP levels significantly higher in survival than nonsurviving patients ($p=0.001$).[47] In this setting pro-ANP performed better than PCT and CRP and had a similar AUC to the APACHE II score which is a well established tool for prognostication and stratification in people.[47] BNP has also been shown to be a useful diagnostic and prognostic biomarker in human sepsis.[48] ANP has not been evaluated in dogs with sepsis. In pilot study there was no significant difference in serum NT-pBNP concentrations between dogs with sepsis and healthy controls. We have also investigated NT-pro-CNP as a biomarker for sepsis in dogs. We found that dogs with sepsis ($n=11$) had significantly greater serum NT-pCNP concentrations than dogs with SIRS ($n=11$) or healthy controls ($n=49$) ($p<0.001$).[49] Using a cut-off value of 10.2pmol/L, the sensitivity, specificity, PPV and NPV of serum NT-pCNP for differentiating sepsis from SIRS in dogs was 82%, 94%, 90% and 89% respectively.[49]

C-reactive protein (CRP)

CRP is an acute phase protein that has been evaluated extensively as a biomarker both in human and veterinary medicine. A CRP assay has been validated for use in dogs[50] and increased CRP has been found in dogs associated with surgery, immune mediated disease,[51, 52] pancreatitis,[53] gastrointestinal disease and a variety of infectious insults including sepsis.[54, 55] CRP concentrations appear to be higher in critically ill dogs compared with controls.[56] The change in CRP over time has also been suggested to predict survival.[57] CRP appears to be less useful in cats, with one study demonstrating that feline CRP did not change significantly between clinically normal cats and hospitalized cats. Other acute phase proteins, such as serum amyloid A were more useful biomarkers of the acute phase response in cats in that study.[58]

REFERENCES

[1]Brady C et al. *J Am Vet Med Assoc*. 2000; **217:**531.[2]Costello MF et al. *J Am Vet Med Assoc*. 2004; **225**(6): 897.[3]Sergeeff JS et al. *J Vet Intern Med*. 2004; **18**(3): 295.[4]Group BDW. *Clin Pharmacol Ther*. 2001; **69**: 89.[5]Dalton WS, Friend SH. *Science*. 2006; **312**(5777): 1165.[6]Bienvenu J et al. *Clin Chem Lab Med*. 2000; **38**(4): 267.[7]de Werra I et al. *Crit Care Med*. 1997; **25**(4): 607.[9]Friedland JS et al. *Crit Care Med*. 1996; **24**(11): 1775.[10]Bozza FA et al. *Crit Care*. 2007; **11**(2): R49.[11]Marshall JC, Reinhart K. *Crit Care Med*. 2009; **37**(7): 2290.[12]Garnacho-Montero J et al. *Crit Care*. 2006; **10**(4): R111.[13]Martin C et al. *Crit Care Med*. 1997; **25**(11): 1811.[14]Assicot M et al. *Clin Infect Dis*. 2005; **41 Suppl 7**: S427.[16]Marshall JC et al. *Crit Care Med*. 2003; **31**(5): 1560.[17]Okano S et al. *Vet Rec*. 2002; **150**: 245.[18]Hauptmann J et al. *Vet Surg*. 1997; **26**: 393.[19]Marshall JC et al. *J Infect Dis*. 2004; **190**(3): 527.[20]Kjelgaard-Hansen M et al. *Comp Immunol Microbiol Infect Dis*. 2008; **31**(6): 477.[21]Dickinson AE et al.. IVECCS; 2009; pp. 685.[22]Tobias et al. Participation. *Am J Respir Cell Mol Biol*. 1992; **7**(3): 239.[23]Villar J et al. *PLoS One*. 2009; **4**(8): e6818.[24]Pavcnik-Arnol M. *Intensive Care Med*. 2004; **30**(7): 1454.[25]Sakr Y et al. *Crit Care Med*. 2008; **36**(7): 2014.[26]Usui M et al. *Biochim Biophys Acta*. 1998; **1397**(2): 202.[27]Vandenplas ML et al. *Am J Vet Res*. 2005; **66**(9): 1509.[28]Martinez M et al. *Crit Care Med*. 1999; **27**: 1303.[29]Oberhoffer M et al. *Clin Chem Lab Med*. 1999; **37**(3): 363.[30]Gogos CA et al. *J Infect Dis*. 2000; **181**(1): 176.[31]Otto C et al. *JVIM*. 1997; **11**: 65.[32]DeClue AE et al. *JVIM*. 2012.[33]Patel RT et al. *Br J Surg*. 1994; **81**(9): 1306.[34]Rau S et al. *Vet Clin Pathol*. 2007; **36**(3): 253.[35]deLaforcade A et al. *JVIM*. 2003; **17**: 315.[36]Jones J et al. *JVECC*. 1999; **9**(4): 195.[37]Doughty LA et al. *Crit Care Med*. 1996; **24**(7): 1137.[38]Wong HR et al. *Arch Dis Child*. 1996; **74**(6): 482.[39]Osterbur K et al. Vet Record; 2011: 544.[40]Christ-Crain M et al. *Lancet*. 2004; **363**(9409): 600.[41]Stocker M et al. *Neonatology*. 2009; **97**(2): 165.[42]Meisner M et al. *Shock*. 2003; **19**(2): 187.[43]Dandona P et al. *J Clin Endocrinol Metab*. 1994; **79**(6): 1605.[44]Nijsten MW et al. *Crit Care Med*. 2000; **28**(2): 458.[45]Kuzi S et al. *J Vet Diagn Invest*. 2008; **20**: 629.[46]Rogers CL, Rozanski EA. *J Vet Intern Med*. 2010;24:229.[47]Morgenthaler NG et al. *Crit Care*. 2005; **9**(1): R37.[48]Chen Y, Li C. *Am J Emerg Med*. 2009; **27**(6): 701.[49]DeClue AE et al. *JVIM* 2011;238:890.[50]Kjelgaard-Hansen M et al. *J Small Anim Pract*. 2008; **49**(6): 282.[51]Griebsch C et al. *Vet Clin Pathol*. 2009.[52]Lowrie M et al. *J Vet Intern Med*. 2009; **23**(4): 862.[53]Mansfield CS et al. *J Am Vet Med Assoc*. 2008; **233**(6): 936.[54]Dabrowski R et al. *Theriogenology*. 2009; **72**(4): 471.[55]Nakamura M et al. *J Vet Med Sci*. 2008; **70**(2): 127.[56]Chan DL. *J Vet Intern Med*. 2009; **23**(3): 559.[57]Gebhardt C et al. *JVECC* 2009; **19**(5): 450.[58]Kajikawa T et al. *Vet Immunol Immunopathol*. 1999; **68**(1): 91. [59]Shapiro NI et al. *Crit Care Med*. 2009; **37**(1): 96.

TOP 10 TIPS FOR SURVIVING SEPSIS

Claire R. Sharp, BSc, BVMS (Hons), MS, DACVECC
North Grafton, MA

Sepsis is a potentially devastating disease syndrome, associated with potentially high morbidity and mortality in both human and small animal veterinary medicine. There is growing evidence based medicine regarding best practice for management of people with sepsis and that outcome can be improved by implementing treatment recommendations.[1] The tips presented here are based on recommendations from the Surviving Sepsis Campaign[1] and personal experience.

1. Have a high index of suspicion for sepsis. Sepsis is recognized / diagnosed on the basis of fulfillment of criteria for the systemic inflammatory response syndrome (SIRS) in the presence of a suspected or confirmed infection. While there is no consensus definition of sepsis in small animal patients as there is in people, SIRS is essentially diagnosed on the basis of meeting 2 or 3 out of 4 of the following clinical criteria; notably an abnormal temperature, heart rate, respiratory rate or a change in the leukon (leukocytosis, leucopenia, or significant left shift).[2, 3] Identification of SIRS criteria in any dog or cat should encourage a proactive diagnostic approach to rule in or out infection. In some cases the source of infection is immediately obvious, while in others it is not. In addition to a thorough physical examination, the preliminary search for infection may include point-of-care (POC) infectious disease testing (FIV/FeLV in cats, 4Dx in dogs), thoracic radiographs (looking for evidence of pneumonia, especially if lower respiratory signs), abdominal ultrasound (looking for free fluid – septic peritonitis, hepatobiliary sepsis etc.), urine sediment examination and/or evaluation of any abnormal fluid accumulations. Infection can be definitively diagnosed on the basis of cytology (intracellular bacteria), culture or less commonly histopathology. In the future, biomarkers (biochemical markers) may play a role in identifying sepsis. I also use the word sepsis in the ER; so if I have a dog with pneumonia + fever + leukocytosis or UTI + fever + tachycardia I say that it is septic, rather than just referring to its diagnosis as pneumonia or UTI. I believe that this helps keep us vigilant for potential adverse sequelae of sepsis.

2. Obtain samples for culture prior to (but don't delay) initiation of antibiotics. This may include abdominal fluid, airway lavage fluid, joint fluid, pleural fluid, urine and/or blood, depending on the case. Even if you are not sure if you are going to submit samples, it is best to obtain them upfront. If you are submitting, submit promptly so as to ensure results ASAP.

3. Give appropriate IV antibiotics early. Perhaps the single most important factor in stabilizing the septic patient (following early recognition of sepsis) is early appropriate use of parenteral antibiotics. This cannot be overemphasized, especially in the emergency room setting. Evidence in people with severe sepsis and septic shock suggests that every hour delay on administration of appropriate antibiotics increases mortality quite significantly.[4, 5] It is not just about giving an antibiotic, rather you need to give an appropriate antibiotic.[6] This means understanding the origin of sepsis and the likely pathogens involved. There is evidence to suggest that using a combination of antimicrobials to optimize both gram negative and gram positive coverage is ideal.[7] It is also vital to be aware of local resistance patterns, to ensure that the empiric antibiotics that you choose are likely to be effective against the suspected pathogen. For example we face a high burden of fluoroquinolone resistant *E.coli* in dogs with community (and hospital) acquired septic peritonitis in our area; thus we use amikacin (if normal renal function) or cefotaxime (if azotemic) as the empiric antibiotic for gram negative coverage in our ER. Broad spectrum is then achieved with the addition of clindamycin. For animals that develop hospital acquired sepsis empiric antibiotic choice is usually based on the assumption of a multidrug resistant (MDR) organisms. Of biggest concern are methicillin resistant Staphylococcus species, resistant *E.coli* and resistant Enterococcus species.

4. Identify and treat shock early and aggressively. This is the early goal directed therapy (EGDT) part of the 6-hour sepsis bundle. Measure serum lactate immediately (as part of your emergency minimum data base). Identify clinical signs of shock; these may include depressed level of consciousness, abnormal heart rate, abnormal CRT, weak pulses and hypotension. Remember that cats in shock look different to dogs. If signs of shock are present resuscitate to end-points. Start with an isotonic crystalloid; these are your mainstay. Consider colloids (weigh up their pros and cons). Blood products may be required in special circumstances. Avoid over hydration. Avoid fluid overload especially in cats. Consider the potential for underlying cardiac and renal disease, which may reduce fluid tolerance. Remember that fluids can contribute to a dilutional coagulopathy. It is also vital to consider that shock in sepsis is not just hypovolemic shock; but rather these animals usually have microcirculatory disturbances, maldistribution of blood flow and potentially poor myocardial function. Don't be afraid to start vasopressor agents following restoration of adequate intravascular volume (you should be more afraid of a recumbent patient with a reliable MAP < 60mmHg and signs of end-organ compromise such as no urine!). When you start pressors be brave; dopamine is an acceptable first line agent but norepinephrine may be required in septic shock. Role of vasopressin (AVP) deficiency.

5. Instrument early. Severe sepsis and septic shock require multimodal pharmacologic intervention and management of organ dysfunctions is challenging and intensive. For the hemodynamically unstable patient you will need multiple sites of vascular (usually venous) access, so why wait. Get lines in early, before the edema starts, since it only gets progressively more challenging. It is also imperative to be prepared if going into the OR. We have a rule of at least 2 sites of vascular access prior to entering the OR. It is easier to place too many IV catheters (obviously within reason) and pull them when no longer needed. Many times, one of your lines will be an aseptically placed sampling catheter since you will be rechecked blood work frequently. Arterial catheterization is recommended if giving vasopressors for septic shock or during anesthesia. Additionally placing a urinary catheter can be very helpful in septic patients with acute kidney injury (AKI) or chronic renal insufficiency/failure so as to closely monitor urine output and help guide fluid balance. It is also imperative to de-instrument appropriately. Once you patient has 'turned the corner' remove any unused IVCs; we do this as early as the day after source control surgery.

6. Ensure early source control. Following identification of sepsis, sample collection, administration of IV antibiotics and resuscitation from shock, the next step of management of sepsis is source control. Without effective source control you are unlikely to have treatment success in situations of an enclosed infection (e.g. septic peritonitis, septic arthritis, pyothorax etc.). This too must be pursued as soon as is reasonably possible following initial stabilization of the patient. This often means emergency surgery; the septic focus cannot wait! Sepsis is not a situation where you can really afford to wait to take to the OR. Since source control often requires surgery (e.g. exploratory celiotomy or thoracotomy), provision of analgesia becomes an important part of management, so consider your options here in regards to efficacy of analgesics, effects on the immune system, and side effects.

7. Don't trust a post-op GI surgery. Dehiscence following gastrotomy, enterotomy, or intestinal resection and anastomosis are some of the more common causes of hospital acquired sepsis (R&A > enterotomy > gastrotomy). Early detection of dehiscence is vital to maximize chances of successful management, however sometimes this is hard. We don't exactly know the best indicator of dehiscence but worrisome signs in the post-operative period include new onset vomiting, ongoing anorexia or new anorexia when they had been eating, development of a fever when they had been afebrile and a dramatic increase in drain output when there had been relatively constant or decreasing drain output (from closed suction abdominal drains). Obviously identifying septic peritonitis requires identification of septic (intracellular bacteria) suppurative abdominal fluid and then dehiscence is confirmed on re-explore. Remember that risk factors for dehiscence (albeit inconsistently reported) include hypoalbuminemia, per-operative peritonitis, intestinal foreign body, hypotension and corticosteroid administration.[8]

8. Be vigilant for hospital acquired sepsis. In addition to hospital acquired abdominal sepsis secondary to dehiscence, other sources of HA sepsis include catheter associated blood stream infections, urinary tract infections, incisional infections and pneumonia (including ventilator associated pneumonia). Sources of vascular access should checked daily for evidence of infection. Additionally they should be inspected / interrogated if a hospitalized pet develops a new onset fever. Central venous catheters must be placed with strict aseptic technique. They should be bandaged and the bandages changed daily. Unused catheters should be removed. Aseptic technique is also very important when handling TPN. The use of indwelling urinary catheters should be avoided; particularly the convenience u-cath. If urinary catheters are necessary they should be placed as aseptically as possible and kept clean.[8] There is evidence that regularly cleaning the external part of the catheter and flushing the prepuce or vaginal vault with chlorhexidine reduces the risk of infection. The risk of incisional infections is minimized by having incisions covered with sterile adhesive dressings while the animal is in the ICU. If an incisional infection develops, the site should be explored in an OR, and you should be prepared to go into the body cavity (for abdominal and thoracic surgeries). Ultrasound can also be used to guide fluid collection for cytology and assess deep tissue involvement.[9] Sometimes the skin is just the tip of the iceberg.

9. Don't underestimate the 2nd hit. Our septic patients live on the edge; while they may present to you more or less stable, it is likely that their body is only just coping with the systemic inflammatory response it is experiencing. Anything that we do to place further physiologic stress on the animal has the potential to result in destabilization. Perhaps the most common 2nd hit scenario is general anesthesia and the potential for associated vasodilation, hypotension, hypothermia and hemorrhage. While we generally can't avoid the second hit, minimizing its severity (i.e. minimizing hypotension, maintaining body temperature and transfusing aggressively in the event of significant intraoperative blood loss) is imperative. Additionally monitoring for the development of complications in the early post-operative period is critical.

10. Be prepared for multiple organ dysfunction syndrome (MODS). While there is not a consensus definition for what denotes an organ dysfunction in veterinary medicine, things to consider include the list below modified from Kenney et al.[10]:

- Renal dysfunction (rise in [Creatinine] > 0.5mg/dL, in the absence of pre-renal or post-renal causes)
- Cardiovascular dysfunction (shock, myocardial dysfunction)
- Respiratory dysfunction (requirement of supplemental oxygen or mechanical ventilation)
- Hepatic dysfunction (Tbili >0.5mg/dL)
- Coagulation dysfunction, DIC (thrombocytopenia, prolonged PT, PTT or ACT, abnormal TEG)
- Gastrointestinal dysfunction (vomiting, regurgitation, ileus, constipation, diarrhea); and
- Endothelial dysfunction (vascular leak).

If organ dysfunction is evident then early aggressive intervention is warranted. For example if creatinine is rising and/or urine output is inappropriate, give a fluid challenge to assess if there is a pre-renal component. If not then you need to rapidly progress through pharmacologic options (furosemide, fenoldopam etc.) and consider renal replacement therapy in the presence of AKI. Similarly if you have a septic patient that develops respiratory distress, supplement oxygen; if you are worried about opening the oxygen cage door, then the animal probably needs mechanical ventilation. Also if your patient is deteriorating rapidly in multiple organ failure, then it is vital to be upfront with the owners. These patients have a grave prognosis and euthanasia should be discussed.

REFERENCES

[1]Levy MM et al. *Crit Care Med.* 2010; **38**(2): 367.[2]Brady CA et al. *J Am Vet Med Assoc.* 2000; **217**(4): 531.[3]de Laforcade AM et al. *J Vet Intern Med.* 2003; **17**(5): 674.[4]Kumar A et al. *J Infect Dis.* 2006; **193**(2): 251.[5]Zubert S et al. *Crit Care Med.* 2010; **38**(4): 1211.[6]Kumar A et al. *Chest.* 2009; **136**(5): 1237.[7]Kumar A et al. *Crit Care Med.* 2010; **38**(9): 1773.[8]Ralphs SC et al. *J Am Vet Med Assoc.* 2003; **223**(1): 73.[9]Trout NJ et al. *J Am Vet Med Assoc.* 1994; **205**(11): 1565.[10]Kenney EM et al. *J Am Vet Med Assoc.* 2010; **236**(1): 83.

SEPSIS: WHAT THE FUTURE HOLDS?

Claire R. Sharp BSc, BVMS, MS, DACVECC
North Grafton, MA

Sepsis and its sequelae are a common cause of death in critically ill human and veterinary patients. Despite advances in our understanding of this disease, mortality remains high. Current therapy still relies largely on intensive supportive care and attempts at eliminating the underlying microbial etiology (source control and early appropriate antibiotics). The future of sepsis management is worth discussing in the context of biomarkers, the Surviving Sepsis Campaign, care bundles and novel therapies. Sepsis biomarkers and the surviving sepsis campaign were discussed in previous lectures.

CARE BUNDLES

One of the biggest changes in the management of sepsis in human medicine in the last decade has been the introduction of care bundles, a type of protocolized medicine. Several studies and a recent meta-analysis have shown that bundled care can improve survival in severe sepsis. A care bundle is a set of interventions that when used together, significantly improve patient outcomes. Generally these bundles involve multidisciplinary teams working together to deliver the best possible care, supported by evidence-based research and practices.

We have investigated an "abdominal sepsis protocol" in our hospital. Key elements of this protocol include rapid identification of abdominal sepsis and early therapeutic intervention. Rapid identification is based on consistent history/clinical signs, laboratory testing and imaging. An index of suspicion of abdominal sepsis is encouraged if two or more of the following are present: vomiting, fever, abdominal pain, abdominal effusion (on screening AUS performed in the ER), and hypoglycemia. The next step is doing POC blood testing; a NOVA, PCV/TS and blood smear, as well as evaluation of the abdominal fluid. We perform both cytology and a blood-fluid glucose and lactate differential. If abdominal sepsis is confirmed then the protocol dictates that the primary clinician starts broad spectrum antibiotics (amikacin and clindamycin or cefotaxime and clindamycin), fluid resuscitates the patient to end-points and notifies a senior clinician (3rd year resident or faculty member). The protocol also involves preparation for surgery, notifying the anesthesia and surgery teams and post-operative care, including daily checking of the culture until results are finalized. Implementation of this protocol over a 6 month period reduced the average time to administration of antibiotics and improved survival.

Clearly there is a role for care bundles in both human and veterinary medicine. The key is getting your team on board with a protocol. This has been a struggle in human medicine too, where compliance in some hospitals is poor.

NOVEL THERAPIES

In human medicine, novel interventions that have shown to improve mortality, at least in certain populations of septic patients include early goal directed therapy (EGDT) in the emergency room and ICU, activated protein C in severe sepsis, and ARDSNet recommendations for mechanical ventilation. Tight glycemic control may also be beneficial, although studies are conflicting. Many novel therapies have shown huge promise in experimental models of sepsis and failed to translate into the clinical realm. Unfortunately thus far there is no magic bullet for sepsis in human medicine, and unlikely to be one any time soon in veterinary medicine. Even activated protein C which was shown to improve mortality in early studies in people, may not be as beneficial as originally thought, and is antigenic in dogs and thus cannot be used. I will discuss a variety of other potentially promising treatments.

Eritoran (E5564), a specific, TLR4 antagonist, is currently one of the more promising sepsis drugs in human clinical trials. Eritoran is a synthetic antagonist of the lipid A portion of LPS that functions at the level of the TLR4 receptor itself. A double-blind, placebo-controlled trial demonstrated that Eritoran blocked some of the toxic effects of endotoxin infusion in healthy human volunteers, ameliorating the fever, tachycardia, leukocytosis and rise in pro-inflammatory cytokine concentrations.[1] A trend towards improved mortality was demonstrated in a phase II trial in septic humans and phase III clinical trials are currently underway.[2]

Polymyxin (PM) is a cyclic cationic polypeptide antibiotic that binds LPS, preventing it from activating inflammatory cells. Administration of polymyxin may be an efficacious, affordable and safe means of inhibiting Gram negative bacteria-induced inflammation in veterinary patients. In a placebo controlled clinical trial, dogs with naturally occurring parvoviral sepsis treated with polymyxin E (12,500 IU/kg, IM q 12h) had significantly improved hemodynamic status and lower plasma TNF concentrations than the control group. Potential dose-dependent side effects of PM include: respiratory depression, nephrotoxicosis, cardiovascular compromise, and neurotoxicosis. However, the dose needed to achieve anti-endotoxic effects is low, making these side effects less likely. Polymyxin B is routinely used in the treatment of Gram-negative sepsis in horses with anecdotal reports of good success, and experimental evidence of efficacy. PMB has also been shown to decrease TNF production in response to experimental endotoxemia in cats.[3] Although our experience in dogs and cats is limited, further study is warranted.

Intravenous immunoglobulin (IVIG) is an immunoglobulin preparation from the pooled plasma of human donors. It contains a broad repertoire of antibodies but is fractionated to contain predominantly IgG (at least 90%), with a normal subclass distribution; as well as small quantities of IgA, IgM, IgD and IgE, and trace amounts of soluble CD4, CD8 and human leukocyte antigen (HLA) molecules, cytokines and intact Fc (fragment, crystallizable) molecules. Following infusion in human patients, the half-life averages 21to 25 days, with ~55% distributing extravascularly. The half life in dogs is relatively short at 7 to 9 days.[4] The mechanisms of IVIG are complex and reflect the numerous functions of circulating immunoglobulins including Fc receptor blockade, attenuation of complement and cytokine mediated damage, and modulation of B- and T-lymphocyte activation and effector functions. In experimental studies it has been shown that polyvalent Igs can provide broad cross-reactivity against a number of potential antigens (both known and unknown). Proposed mechanisms of action of polyvalent immunoglobulins in sepsis include their ability to improve opsonization of infecting microbes facilitating phagocytosis and killing, prevention of nonspecific complement activation, protection against antibiotic induced liberation of LPS; and neutralization of LPS and a wide variety of other superantigens.

There is ongoing debate about the efficacy of polyvalent immunoglobulins in preparations such as IVIG, as adjunctive therapy for sepsis or septic shock. A recent meta-analysis evaluating the use of IVIG in sepsis included 27 trials with a total of 2,202 human patients .[5] This review concluded that polyvalent Igs exert a significant positive effect on survival in sepsis and septic shock (relative risk of death of 0.79) with a strong trend in favor of an Ig preparation enriched with IgA and IgM (IgGAM; reducing mortality by 34% in adults and 50% in neonates) compared with preparations containing only IgG.[5] This is likely since the IgGAM preparations contain higher titers of antibodies to a greater number of bacterial pathogens, and supply more opsonins.[6, 7] The use of human IVIG has not been reported in septic veterinary patients.

CONCLUSIONS

Hopefully the future of sepsis and SIRS management in veterinary medicine will involve ongoing, well designed basic research studies, animal model work and clinical trials in order to better understand these syndromes in our patients and improve outcome.

REFERENCES

[1]Lynn M et al. *J Infect Dis.* 2003; **187**(4): 631.[2]Barochia A, et al. *Expert Opin Drug Metab Toxicol.* 2011; **7**(4): 479.[3]Sharp CR et al. *J Feline Med Surg.* 2010; **12**(4): 278.[4]Scott-Moncrieff JC, Reagan WJ. *Semin Vet Med Surg (Small Anim).* 1997; **12**(3): 178.[5]Kreymann KG et al. *Crit Care Med.* 2007; **35**(12): 2677.[6]Trautmann M et al. *Clin Exp Immunol.* 1998; **111**(1): 81.[7]Garbett ND et al. *Clin Exp Immunol.* 1989; **76**(1): 8.

LIVER SAMPLING METHODS AND CONSIDERATIONS: IS LAPAROSCOPY REALLY BEST

Keith Richter, DVM, DACVIM
San Diego, CA

INTRODUCTION

Hepatobiliary diseases can be challenging to diagnose with clinical signs and laboratory abnormalities alone, and often require hepatic tissue sampling for cytologic or histopathologic evaluation. Recent advances in imaging technology, the use of multiple imaging modalities, and newer biopsy methods have resulted in improvement in the veterinary clinician's ability to safely procure hepatic tissue for evaluation. There are several means of obtaining hepatic samples including fine needle aspiration (FNA), ultrasound-guided biopsy, laparoscopy, and laparotomy. All techniques have both advantages and disadvantages, which should be carefully considered before choosing the appropriate sampling method.

PRE-BIOPSY CONSIDERATIONS

Persistent elevation of serum hepatic enzymes, elevated serum bile acids, focal or diffuse hepatic masses or echotextural changes, or the need to assess response to therapy are all indications for hepatic sampling. The primary concern for any hepatic sampling technique is the chance for hemorrhage, although it is unclear what the best predictor of hemorrhage is in patients about to undergo hepatic sampling. Ewe studied 200 human patients and evaluated bleeding laparoscopically, and found no correlation with any *in vitro* coagulation test and "liver bleeding time".[1] These findings have been subsequently observed in other studies.[2,3] Studies have shown that clotting times assessing proteins induced by vitamin K antagonism (PIVKA) are more sensitive in detecting coagulopathies in patients with hepatic disease.[4,5] However, in a pilot study performed at the authors' hospital, hepatic bleeding times assessed via laparoscopy did not correlate with PIVKA time. Thus it appears that peripheral blood coagulation indices are generally unreliable guidelines regarding the risk of hemorrhage following liver biopsy. In most cases of significant hemorrhage, technical errors such as damaging a large vessel are the cause rather than persistent oozing from a needle biopsy site. In one study in normal dogs, biopsies taken from the left lateral hepatic lobe using a biopsy punch, biopsy needle, ligature method, laparoscopic biopsy forceps, and ultrasonically activated scalpel resulted in minimal hemorrhage (<2ml).[6] However, this did not assess the risk of hemorrhage in dogs or cats with hepatic diseases.

FINE NEEDLE ASPIRATES

Fine needle aspiration is typically performed with ultrasound guidance and involves obtaining a small amount of hepatic tissue for cytologic analysis. Using a 22 gauge, 1.5 inch needle without a syringe, the needle is rapidly agitated in and out of the hepatic tissue and simultaneously twisted to obtain a sample. This method relies on capillary action rather than suction to acquire tissue within the needle, which results in less hemodilution. The sample is then expelled onto a glass slide for cytologic examination. Often three to five separate attempts are made to increase the sample size and diversity. Ultrasound imaging allows the sampling of focal lesions or diffuse abnormalities. Aspiration sites must be chosen based on accessibility and limited risk for damaging local structures including the gallbladder and large vessels. The clinician must decide whether the relative risk of sampling such lesions is the best decision, or whether alternative methods would be more appropriate. Sedation or anesthesia may be necessary depending on the individual patient's temperament.

Fine needle aspiration is a quick and inexpensive method of hepatic sampling that requires little to no sedation and carries a low risk for hemorrhage. Multiple sites can be easily sampled and the procedure can usually be performed on an outpatient basis. The primary disadvantage of FNA is the questionable accuracy. The sample size often limits the number of available cells to obtain an accurate diagnosis, and hemodilution makes it difficult to assess whether inflammatory cells were present in the liver or peripheral blood. There are several important elements used to interpret pathologic information including lobular architecture, presence and location of inflammation within a lobule, presence and severity of fibrosis, metal accumulation, vascular abnormalities, and lobule heterogeneity. These criteria cannot be accurately examined using a cytologic sample obtained using FNA. Several studies have compared FNA cytology to biopsy with histopathology. In one study with a total of 34 cases, there was good correlation in 35% of cases, partial correlation in 35% cases, and no correlation in 30% of cases.[7] Poor correlation was found with a variety of histologic changes including vacuolar change, lipidosis, cholestasis, inflammation, and neoplasia. In a similar study with 97 cases, complete agreement between FNA and histopathology was seen in only 30% of cases of dogs; 25% agreement with inflammation, 14% agreement with neoplasia (mainly carcinoma), and 64% agreement with vacuolar hepatopathy.[8] In cats, there was overall agreement in 51% of cases; 27% agreement with inflammation, 33% agreement with neoplasia (lymphoma), and 64% agreement with vacuolar hepatopathy.[8] Although vacuolar hepatopathy was the most sensitive diagnosis, it was also the most common misdiagnosis using cytology. In another study, the best correlation between hepatic cytology and biopsy was seen with lipidosis, lymphoma, and carcinoma whereas the worst performance was seen with inflammatory and fibrotic disorders.[9]

In summary, given its clear limitations, FNA is best used as an adjunctive diagnostic modality in conjunction with other techniques or clinical findings, and certainly does not replace histopathology. The clinician must be aware of its inherent inaccuracy before undertaking FNA and relying on the cytologic findings.

ULTRASOUND-GUIDED BIOPSIES

Ultrasound-guided hepatic biopsy uses a cutting-type needle as a sampling tool. Automated needles are preferred and should be either completely automated or semi-automated. Completely automated needles thrust the inner obturator followed by the outer cutting sheath into the liver in a fraction of a second. The needles can be operated with one hand while the other hand operates an ultrasound transducer to allow precise placement of the biopsy instrument. There is minimal displacement of the liver, a shorter intra-parenchymal phase, and a more reliable yield of tissue. Using the rapid cutting action, the hepatic tissue tends to be less fragmented. Semi-automated needles require manual placement of the internal obturator into the liver, followed by an automatic thrusting of the outer sheath by a spring-loaded mechanism. These needles have the additional advantage of control over the final needle position, since the tip of the needle can be precisely localized before the outer cutting sheath is deployed. The authors generally use a 16-gauge needle for ultrasound-guided hepatic biopsy.

Local anesthesia and minimal sedation or general anesthesia is used depending on the patient. Care must be taken prior to taking samples to ensure that vessels and other organs are not within the path of the needle. For diffuse lesions, the transducer is typically placed caudal and to the left of the xiphoid, and aimed at the left medial or lateral lobes. In patients with a small liver, it may be difficult to adequately visualize the needle without gastric gas interference. Placing these animals in a 45-degree right lateral oblique position can reduce this interference. If the animal is under general anesthesia, an assistant can compress a re-breathing bag to hold the animal in deep inspiration which serves to move the diaphragm and liver caudally to improve visualization. The area is surgically prepped, and a stab incision is made in the skin at the desired needle insertion site. The ultrasound transducer is covered with sterile wrap and sterile lube is used to enhance skin contact. While one hand maneuvers the transducer, the other hand advances the needle into the liver under direct ultrasound visualization. The image should be optimized to maximize the chance of recognizing the needle within the liver. To allow distinction of the needle from other echogenic structures, the needle can be gently moved in and out with minimal movements (attempting to move the liver within the abdominal cavity rather than the needle within the liver). Occasionally the needle cannot be seen, and indirect evidence of organ penetration must be used such as movement of the liver or visualization of movement at the liver border. The needle is then directed so the trajectory will avoid other structures when it is fired. The needle is fired, and then removed. For most cases, four to five samples are obtained, and are submitted for aerobic/anaerobic culture, histopathologic evaluation, and metal (copper, zinc, and iron) quantification. Careful examination for post-biopsy hemorrhage is then performed. External digital pressure may be used to help control hemorrhage in smaller patients; an abdominal compression wrap is usually ineffective for this.

This method requires minimal sedation in many patients, has a low to moderate cost, and provides adequate samples for histopathology. In one study, 96 percutaneous transabdominal hepatic needle biopsy samples were obtained with no adverse consequences noted.[10] However, this study was performed in normal dogs, and in the authors' experience this sampling method has the highest risk of any sampling method for hemorrhage. Additional disadvantages include the need for sedation or anesthesia in some patients, difficulty of imaging small livers, difficulty of obtaining liver tissue in patients with fibrosis, and samples that carry a questionable representation of the underlying hepatic pathology. This questionable accuracy is in most part due to potential for sampling error. This method still results in a relatively small sample size, possible fragmentation of fibrous tissue, and may not enable sampling of abnormalities located in other lobes (the left medial or lateral lobes are generally sampled due to their ease of imaging). In one study, percutaneous hepatic sampling using core biopsies resulted in 92% diagnostic quality samples, however these were not compared to large wedge biopsy to assess the accuracy of this method.[11] In another study, the diagnostic accuracy of the Tru Cut needle biopsy was compared to the gold standard of surgical wedge biopsy of the liver in 124 patients.[12] The overall discordance between the two methods was 53% in dogs and 50% of cats, with >60% discrepancy occurring with chronic hepatitis or cirrhosis, cholangitis/cholangiohepatitis, portosystemic vascular anomalies, microvascular dysplasia, fibrosis, and miscellaneous disorders. These are the disorders that are the most commonly seen among dogs and cats with hepatobiliary disease. The greatest accuracy was with neoplasia (80% concordance). In this study, needle biopsies commonly had <4 portal triads. Use of a 14-gauge needle versus 18-gauge may reduce this discordance as it raises the number of portal triads sampled from an approximate mean of four to seven. However the larger needle size carries a higher risk of complications, including hemorrhage. One study in human beings concluded that at least 6 to 8 portal triads are necessary for proper evaluation of hepatic biopsy specimens.[13] Another study in normal dogs showed that the use of a 16-gauge biopsy needle resulted in inconsistent procurement of adequate tissue samples (42% of samples had <6 portal triads, 17% were fragmented to the extent that they were unacceptable for evaluation).[14]

In summary, ultrasound-guided hepatic biopsy is relatively easy to perform, but has questionable accuracy. The accuracy may be increased by using a larger gauge needle, but this carries a greater risk of post-biopsy hemorrhage. If the patient is suspected of having inflammatory disease, vascular abnormalities, significant fibrosis, or is at risk for hemorrhage, laparoscopy or laparotomy should be considered.

LAPAROSCOPIC BIOPSIES

Laparoscopy is performed under general anesthesia with the patient in dorsal recumbency tilted 45 degrees to the left. A Veress needle is placed at the level of the umbilicus into the peritoneal space through a stab incision using a #11 blade. After ensuring no obstruction and negative pressure using the infusion and aspiration of saline, the abdomen is insufflated with carbon dioxide gas and maintained at a pressure of approximately 12 mm Hg. A scope port (cannula) is then placed 4 cm right lateral to the Veress needle. The Veress needle is then removed and replaced with an instrument port. Hepatic sampling is achieved using a "spoon" or oval cup biopsy forceps. Multiple samples are obtained under direct visualization, and samples are submitted for aerobic/anaerobic culture, histopathologic evaluation, and metal (copper, zinc, and iron) quantification. Following procurement of all the biopsy specimens, the sites are inspected for hemorrhage. The abdomen is then decompressed, and lidocaine and bupivacaine are infused into the peritoneal cavity through either port. Both the instrument and scope ports are removed, and the port site incisions are closed using either a cruciate or simple interrupted pattern in the body wall, subcutaneous tissue, and skin.

This technique enables gross evaluation of the entire liver, extra-hepatic biliary system, and surrounding structures while obtaining multiple large specimens of liver. The ability to obtain multiple samples decreases the risk for sampling artifact in cases of regional diversity within the liver. Additionally, by directly visualizing the hepatic parenchyma, the clinician can correlate the histopathologic findings and clinical data with the gross appearance of the liver to render the most accurate diagnosis. In one study, "spoon" or oval cup forceps used for hepatic biopsy in normal dogs resulted in consistent diagnostic samples with a mean of 16 to18 portal triads per biopsy specimen.[14] The authors concluded that biopsy samples obtained using this instrument yielded accurate representation of the hepatic parenchyma. Another study in clinical patients revealed diagnostic quality biopsy specimens in all 80 dogs that underwent laparoscopic liver biopsy.[15] They were all considered to contain a sufficient number of portal triads to obtain a histologic diagnosis. The number of portal triads in each sample ranged from 8 to 13, and the mean total number of portal triads submitted per patient was 39. However, disagreements in histologic diagnoses were identified in 7 of 49 (14%) of the dogs for which biopsy samples from multiple sites were available for review.[15] This suggests that multiple samples should always be obtained, ideally from different lobes. Laparoscopy also enables the visualization of smaller masses and irregularities that may not be evident with ultrasonographic imaging.

There is generally minimal bleeding during this procedure, even in patients with *in vitro* coagulopathies. Using a "spoon" or oval cup forceps typically results in a marked decrease in the amount of hemorrhage when compared to needle biopsies. Any hemorrhage can be directly visualized for adequate clot formation. If hemorrhage persists, direct pressure using a blunt probe for five minutes can be used. If the site continues to bleed, electrocautery can be applied to the biopsy site or a topical hemostatic agent (Gel Foam®) can be placed directly on the biopsy site using laparoscopic forceps. In one study, 76 of 80 dogs (95%) survived to hospital discharge after undergoing laparoscopic liver biopsy.[15] Three (4%) of dogs required conversion to laparotomy (none of which were due to hemorrhage). Another 3 dogs (4%) required a blood transfusion, all of which were anemic prior to the biopsy. In this study, the complication rate was too low to identify risk factors that could be used to predict the development of complications.[15] In the author's experience, conversion to laparotomy or significant hemorrhage requiring blood transfusion is extremely rare. The main disadvantages of laparoscopy are the need for general anesthesia, increased cost, and the need for technical expertise and equipment. There is also the potential for inadvertent trauma to adjacent organs during trocar/cannula placement.

SURGICAL BIOPSIES

Wedge biopsy via laparotomy is another potential method for obtaining hepatic biopsies. If a random liver biopsy is needed and a section of liver is protruding, a guillotine suture can be used. A pre-formed encircling ligature of 4-0 monofilament absorbable suture material is placed around the protruding section of liver. The ligature is then tightened until it has crushed the hepatic parenchyma. After completing several throws in the knot, the sample is excised 1-2 mm distal to the ligature using Metzenbaum scissors or a scalpel blade. If a specific area of liver is needed, a sample can be obtained using the transfixation method or a biopsy punch. The transfixation method entails placing a ligature through the liver lobe approximately 8-10 mm from its edge. The ligature is tightened to crush through the hepatic parenchyma of half the desired biopsy specimen. An additional throw is made at a right angle to the first ligature, and this throw is tightened to crush the parenchyma of the second half of the specimen. The sample is removed 1-2 mm distal to the crushed area using a scalpel blade or Metzenbaum scissors. If the desired area does not lie near the edge of a liver lobe, a 6 mm biopsy punch can be utilized. The biopsy punch should be inserted into the liver ensuring not to penetrate the opposite surface. If the biopsy site is close to the hilus, extra caution must be used so that no more than half of the thickness of the liver is penetrated. The biopsy sample is removed from the liver using scissors. Hemorrhage can be controlled by filling the defect with a topical hemostatic agent (Gel Foam®) and applying digital pressure for 3 to 5 minutes, or by suturing the hepatic capsule with fine absorbable monofilament suture in a cruciate pattern. Akin to laparoscopy, this method has similar advantages and disadvantages as listed above. It is more invasive, but it allows easier biopsy of other abdominal organs (such as intestine and mesenteric lymph node) and the ability to perform therapeutic maneuvers (such as hepatic mass removal or biliary diversion).

REFERENCES

1. Ewe K: **Dig Disc Sci** 1981; 26:388. 2. McVay PA, et al: **Am J Clin Pathol** 1990; 94:747. 3. Bigge LA, et al: **J Am Anim Hosp Assoc** 2001; 37:228. 4. Center SA, et al: **J Vet Intern Med** 2000; 14:292. 5. Center S A, et al: **J Vet Intern Med** 1998; 12:214. 6. Vasanjee SC, et al: **Vet Surg** 2006; 35:86. 7. Fondacaro JV, et al: **Abst. Proceedings of the 17th ACVIM Forum** 1999; 719. 8. Wang, KY, et al: **JAVMA** 2004; 224:75. 9. Roth L: **Vet Clin Pathol** 2001; 30:35. 10. Hitt ME, et al: **Am J Vet Res** 1992; 53:785. 11. Barr F: **J Small Anim Pract** 1995; 36:105. 12. Cole TL, et al: **JAVMA** 2002; 220:1483. 13. Bravo AA, et al: **N Engl J Med** 2001; 344:495. 14. Vasanjee SC, et al: **Vet Surg** 2006; 35:86. 15. Petre SL, et al: **J Am Vet Med Assoc** 2012; 240:181.

DOGMA AND DILEMMAS IN DIC

Stephanie A. Smith, DVM, MS, DACVIM
Champaign, IL

The majority of critically ill patients have some degree of ongoing activation of hemostatic mechanisms. Coagulation activation is always present to some degree with inflammation due to the profound crosstalk between inflammatory and hemostatic pathways. Activation of coagulation systems is an appropriate physiologic response to injury that prevents blood loss and initiates pathways necessary for debris removal and eventual healing. The point at which coagulation activation crosses from being locally confined and protective to being systemically distributed and destructive is rarely clear, and not easily identified clinically. Coagulation activation in the critically ill is a continuum of appropriate hemostatic responses that may gradually progress to loss of inhibitory mechanisms, microvascular thrombosis, and eventually defective coagulation with inability to prevent hemorrhage.

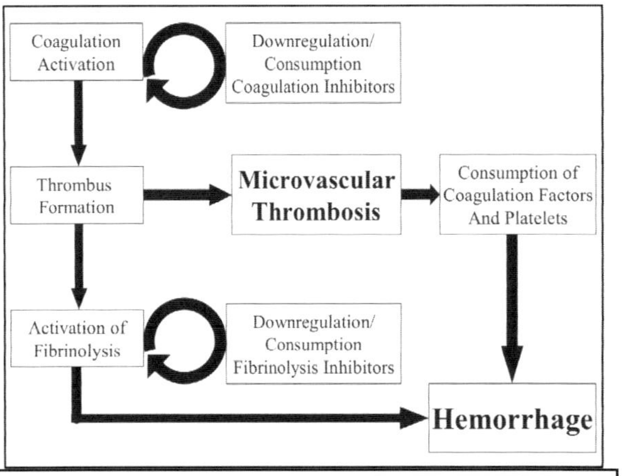

LABORATORY DIAGNOSIS OF DIC

There are no widely agreed upon criteria for diagnosis of DIC in either veterinary or human patients. In theory, DIC should be associated with evidence of activation of coagulation (thrombin production) and fibrinolysis, with inhibitor consumption and end-organ damage. Traditionally, veterinary clinicians have diagnosed DIC when any 3 of the common coagulation tests (PT, APTT, platelet count, FDP's, fibrinogen) are abnormal. However, this "3 abnormal" criterion is not specific for DIC, as these conditions may be met in patients with coagulopathy due to anticoagulant rodenticide exposure, hepatic disease, and combination disorders. Recently, new diagnostic tests have been developed in human medicine that are more specific indicators of consumption as the cause of the coagulopathy. One study evaluated the utility of routinely available coagulation tests in identification of DIC in a human ICU population. While the routine screening tests of platelet count, PT, APTT, and TT were highly sensitive, their specificity was low. A decrease in fibrinogen level had 100% specificity, but only 22% sensitivity. Elevated FDP's and elevated D-dimer were highly sensitive (100%, 91% respectively) with reasonable sensitivity (67%, 68% respectively). When these two tests were combined as a screening tool, the combination had a sensitivity of 91%, specificity of 94%, and correctly identified (efficiency) 95% of patients as positive or negative for DIC. Note that the combination of PT+APTT+FDP (which is more routinely available in veterinary medicine) was nearly as efficient at 86%. Although decreased AT alone was not highly specific, it was very sensitive. However, a variety of other studies have indicated that the *degree* of AT deficiency has valuable prognostic significance in critical illness, suggesting that the usefulness of this test is in assessing severity of disease rather than identification of whether or not DIC is present. Note that although schistocytosis is associated with microangiopathic disease, identification of schistocytes is relatively uncommon in both humans and dogs (10%) with DIC.

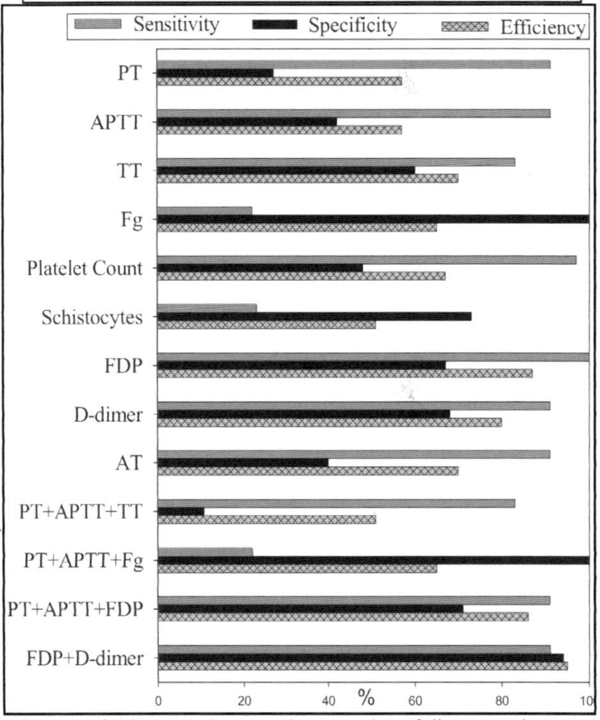

The Scientific Subcommittee on DIC of the International Society of Thrombosis and Hemostasis (ISTH) proposed algorithms for the diagnosis of DIC. Verification of the prognostic utility of these scoring systems is ongoing. Similar scoring systems have been proposed for veterinary patients, but not yet widely adopted.

NON-OVERT VS OVERT DIC

Unfortunately, there are no widely accepted criteria for use in the veterinary patient that categorize coagulation activation as non-overt DIC or overt DIC.

In general, non-overt DIC is the sub-clinical (and rarely identified) phase where coagulation is being activated, generating thrombin and causing fibrin to be formed, leading to intravascular thrombus formation. Non-overt DIC should be associated with laboratory evidence of thrombin formation (increased TAT complexes), fibrin formation (FDP's), fibrinolysis (FDP's and D-dimers) and inhibitor consumption (decreased AT and PC). Note that global tests for coagulation (PT and APTT) are not prolonged until factor levels have dropped to approximately 30%. Consequently, in non-overt DIC these tests, as well as the platelet count, will be within reference range or only mildly abnormal. Serial evaluations may indicate rising clot times and falling platelet counts. Once coagulation activation has tipped from non-overt to overt DIC, consumption of coagulation factors has progressed to the point that PT and APTT are prolonged, platelet counts are well below normal, and fibrinogen will drop (or TT as a surrogate for fibrinogen will rise). In animals, overt DIC is the form of consumption coagulopathy most commonly identified. At this point of the process DIC is often associated with inappropriate hemorrhage.

MANAGEMENT OF DIC AS A COMPLICATION

Unfortunately, there is little agreement regarding appropriate treatment for DIC. The single exception to this lack of consensus is that in all cases the underlying disorder needs to be controlled or eliminated. A variety of non-specific supportive measures should be implemented to limit ongoing endothelial and end-organ damage and the resulting continued activation of coagulation. Tissue oxygenation should be directly addressed where possible with appropriate measures to maintain perfusion. Therapy should include volume replacement, oncotic support to help maintain intravascular volume and limit tissue edema, and pressor support as indicated. Even with adequate perfusion, oxygen delivery may be limited with decreased oxygen carrying capacity and/or oxygen saturation. Anemia, and hypoxemia should be addressed whenever possible. Lastly, acid base imbalances should be corrected when present.

Triggers for platelet transfusion in humans	
Condition	if count:
Any patient	<5,000/µl
Sepsis	<20,000/µl
Active bleeding	<50,000/µl
Prior to invasive procedure	<50,000/µl

SPECIFIC THERAPIES DIRECTED AT COAGULATION/FIBRINOLYSIS BALANCE

Since DIC is a complication secondary to many different disease processes, clinical trials evaluating specific therapies for DIC are difficult to conduct due to the lack of consistency in the patient population. Some therapies have been critically evaluated in human patients by limiting the underlying disease process (primarily to severe sepsis). No specific DIC therapy has ever been critically evaluated in a controlled, outcome based study in veterinary patients.

The complex pathophysiology of DIC results in loss of pro- and anti-coagulant proteins as well as pro- and anti-fibrinolytic proteins. FFP contains all plasma proteins in normal balance. Regardless of the imbalance, administration of FFP should theoretically shift every protein level in a normal direction. Experience with use of FFP in DIC is largely empirical. However consensus guidelines developed by a variety of human medical expert panels recommend the use of FFP in DIC. There is some disagreement about the phase of DIC where FFP should be administered, but the most often quoted recommendations are:

FFP Indicated	FFP Not-indicated
With active or impending hemorrhage Prior to invasive procedures	In the absence of hemorrhage With chronic, compensated DIC without hemorrhage For the normalization of laboratory tests only

Ideally, in humans FFP should be administered to maintain antithrombin levels at >70% normal. One study of FFP administration to critically ill dogs (mean baseline AT of 69%) indicated that administration of 10-15 ml/kg FFP q12h was helpful in maintaining AT levels ay baseline values, but ineffective at increasing AT levels. Unfortunately, the ability to transfuse adequate plasma volume in order to have significant impact on factor levels may be limited by volume and/or cost. Due to the variable half-lives of coagulation factors, repeated FFP administration may result in severe factor imbalances.

Platelet transfusion may also be indicated in DIC, although specific platelet count triggers have not been identified for veterinary patients. Neither FFP nor stored blood contain functional platelets in relevant quantities. Fresh whole blood will contain functional platelets as well as plasma proteins, but is not appropriate for the non-anemic patient, and carries the risks of transfusion reactions, infectious disease transmission, and volume overload. Platelet concentrates are the safest and most effective means of delivering platelets. However, production of this product is time and resource intensive, and may be limited to large blood banking centers.

Heparin has the potential to be useful during the phase of DIC where coagulation activation and fibrin deposition predominate. However, because there are no clear criteria for identifying this phase of the process, it is difficult to definitively identify which patients at which times would most benefit from this therapy. Controlled clinical trials are lacking in both human and veterinary medicine. The issue in human medicine is further compounded by the high prevalence of deep venous thrombosis (DVT) in the critically ill. Human patients at risk for development of overt DIC often are receiving heparin for prevention of DVT, making it difficult to evaluate the impact of heparin on the clinical course of DIC. Furthermore, veterinary patients rarely are definitively diagnosed with DIC during the early phase where heparin has the greatest potential to be of benefit. Heparin therapy is indicated in human patients with DIC where a thrombotic disorder predominates in DIC (eg. purpura fulminans secondary to meningococcemia) and when the underlying disease process is associated with a high risk of DVT. Neither of these conditions occurs with any frequency in veterinary patients, except for as a complication of immune-mediated hemolytic anemia. The bioavailability of unfractionated heparin is reduced in the presence of acute phase proteins. Consequently, use of low molecular weight heparin may be preferred. Heparin therapy is contraindicated in humans with central nervous system injury, fulminant hepatic failure, and once coagulation factor consumption has progressed to overt bleeding. Use of heparin is also probably inappropriate in our patients with these conditions.

Recommendations for unfractionated heparin therapy in dogs and cats with DIC have been widely made, but never substantiated. Dosage recommendations vary enormously, from 10 U/kg to 300 U/kg. There is also no clear consensus on how frequently and by what route heparin should be administered.

SUMMARY

Although great progress has been made in understanding the pathophysiology of coagulation activation in critical illness, these advances have yet to be translated into better survival in veterinary patients. Improvement in patient outcome is limited by both lack of consistency in definitions of DIC and consensus regarding appropriate diagnostics. Management guidelines are primarily based on personal experience, dogma, and extrapolation from human medicine. Scientific validation of specific therapies is lacking and desperately needed.

REFERENCES

Bateman SW et al. J Am Vet Med Assoc 1999; 215(6): 798-804. Bateman SW et al.. J Am Vet Med Assoc 1999; 215(6):805-810. Bick RL.. Clin Appl Thromb Hemost 2002; 8(1):1-31. Dempfle CE. Blood Rev 2002; 16 Suppl 1:S23-8.:S23-S28. Levi M et al. Blood Rev 2002; 16(4):217-223. Stokol T et al.. Am J Vet Res 2000; 61(4):393-398. Taylor FB et al. Hematol Rep 2002; 1(1):34-40. Taylor FB et al. 2001; 86(5):1327-1330. Toh CH, et al. Hematol 2003; 8(2):65-71. Wada H et al. Am J Hematol 2003; 74(1):17-22. Yu M et al. Crit Care Med 2000; 28(6):1777-1780. Rozanski EA et al. J Vet Emerg Crit Care 2001; 11(1):15-21. Toh CH et al. Hematology 2003; 8(2):65-71.

MAKING SENSE OF THE NONSENSE –
DIAGNOSIS AND TREATMENT OF ENTEROPATHOGENIC BACTERIAL INFECTIONS

Stanley L. Marks, BVSc, PhD, DACVIM (SAIM, Oncology), DACVN
Davis, CA

The clinical significance of isolation of enteropathogenic bacteria causing diarrhea in dogs and cats is clouded by the existence of many of these organisms as normal constituents of the indigenous intestinal flora. The primary enteropathogenic bacteria most commonly incriminated in canine and feline diarrhea includes *Clostridium perfringens*, *Clostridium difficile, Campylobacter* spp., *Salmonella* spp., and *Escherichia coli* associated with granulomatous colitis. Veterinarians are faced with a quandary when attempting to diagnose small animals with suspected bacterial-associated diarrhea because the isolation rates for putative bacterial enteropathogens are often similar in diarrheic and nondiarrheic animals, and because the incidence of bacterial-associated diarrhea is extremely variable. The indications for performing fecal enteric panels are poorly defined, resulting in indiscriminate testing, and misinterpretation of results. The results of fecal cultures may take up to 72 hours to complete, and many animals are administered antibiotics indiscriminately before results of fecal cultures or toxin assays are available. In addition, the resolution of clinical signs of diarrhea is often wrongly equated with eradication of the "incriminating" pathogen. Fecal cultures and toxin analysis should be reserved for dogs and cats with an acute onset of bloody diarrhea (with or without concurrent evidence of sepsis), and in diarrhea outbreaks occurring in multiple animals in a densely populated shelter or kennel environment. In addition, screening for *C. difficile*, *Campylobacter* spp., or *Salmonella* spp. is indicated when zoonotic implications are present.

A fecal bacteriologic panel was evaluated in 260 dogs with diarrhea and yielded 28 (10.8%) diagnostic results ($650/positive test result).[1] In addition, among these 28 positives, many may have been false positives, because causality was not established. This impressive cost derives from the relative insensitivity of tests for the most likely pathogens and the poor selection of specimens being cultured. The advent of real-time PCR panels for dogs and cats with diarrhea has provided a new paradigm for the rapid and sensitive detection of toxin genes or organisms associated with disease. Interpretation of these panels can be problematic however because virtually all of these bacterial organisms have been frequently isolated from the feces of clinically healthy dogs and cats.

CLOSTRIDIUM PERFRINGENS

Clostridium perfringens is an anaerobic, spore-forming, gram-positive bacillus that has been associated with outbreaks of acute, often severe diarrhea in humans, horses, dogs, and cats. The elaboration of 4 major toxins (α, β, ι, and ε) is the basis for typing the organism into 5 toxigenic phenotypes, A-E. Each type may also express a subset of at least 10 other established toxins, including *C. perfringens* enterotoxin (CPE), a well-characterized virulence factor whose production is co-regulated with sporulation. ***The true role of C. perfringens in enterocolitis in dogs and cats is unclear***. A study evaluating 843 *C. perfringens* isolates collected from 103 dogs revealed that all isolates tested were type A, with 15% harboring the enterotoxin gene (*cpe*).[2] Enterotoxigenic *C. perfringens* type A has been associated with human food poisoning and sporadic diarrhea, canine acute and chronic, large and small bowel diarrhea, and an acute hemorrhagic diarrheal syndrome (AHDS).[1-3] Although several studies have shown an association between the immunodetection of CPE in fecal specimens and canine diarrhea, the pathogenesis of *C. perfringens*-associated diarrhea in the dog is not fully understood, as CPE is also detected in up to 14% of nondiarrheic dogs.[2] The role of *C. perfringens* in cats is equally unclear because CPE was only detected in 9/219 cats with diarrhea (4.1%) and in 1 of 54 cats without diarrhea (2%) in a recent study.[4]

DIAGNOSIS

There is currently no gold standard for the diagnosis of canine *C. perfringens*-associated diarrhea, and clinical signs are extremely variable. ***The optimal diagnostic approach utilizes a combination of fecal CPE immunodetection* via ELISA *and molecular techniques for detection of cpe (enterotoxin gene). Caution should be heeded upon relying on the outdated and insensitive test of evaluating stained fecal smears for C. perfringens endospores.***

Clinical Signs

There are no pathognomonic clinical signs indicative of *C. perfringens*-associated diarrhea in the dog and cat, and the spectrum of disease attributed to this organism varies greatly. *Clostridium perfringens*-associated diarrhea can manifest with small intestinal, large intestinal, or diffuse clinical signs.[1] *C. perfringens*-associated diarrhea should also be considered as a potential cause of acute hemorrhagic diarrhea in dogs.[1]

Culture

Because *C. perfringens* is a normal commensal of the intestinal microflora, the mere culture of an isolate from the stool is of little clinical significance.[2,5] In addition, isolation rates for *C. perfringens* are similar in diarrheic and nondiarrheic dogs.[2] Culture may be useful in procuring isolates for the application of molecular techniques like PCR for detection of specific toxin genes or molecular typing of isolated strains to establish clonality in suspected outbreaks.

Fecal Endospores

Because sporulation is co-regulated with enterotoxin production, fecal endospore counting of Wright or Gram stained fecal smears ($\geq$ 3 spores per high power field) has been suggested as a tool for diagnosing enterotoxigenic *C. perfringens*-associated disease.[6] Several studies have reported ***no association*** between fecal endospore counts and the presence of diarrhea, or between spore counts and the detection of CPE in fecal specimens.[2,5,7] ***The diagnostic value of finding increased endospores in a fecal smear from a diarrheic dog is virtually zero.***

Fecal Enterotoxin (CPE) Immunodetection

Detection of CPE in fecal specimens is the most widely used diagnostic tool for *C. perfringens* in both human beings and animals. There is only one commercially available ELISA kit (TECHLAB Inc., Blacksburg, VA) for detection of CPE in fecal specimens; however, the performance characteristics of this assay have not been validated in the dog or cat to date. Sensitivities of immunodetection methods are extremely important when one considers that disease associated with CPE may be dependent on the concentration of CPE present within the intestinal lumen. This phenomenon is underscored by the finding that up to 14% of healthy dogs have detectable concentrations of CPE utilizing the qualitative TECHLAB ELISA.[2] In addition, time delays in sample processing can result in false positive or false negative results (unpublished observation).

Molecular Techniques

PCR detection of enterotoxigenic *C. perfringens* can be performed on fecal specimens or isolates collected via culture. Non-diarrheic dogs are far less likely to be positive for both CPE and the enterotoxin gene (*cpe*) (4%), compared to diarrheic dogs (28%).[2] Toxin DNA and RNA quantification for the α-toxin gene and *cpe* has been shown to be significantly higher in diarrheic dogs and cats compared to non-diarrheic animals (unpublished observation).

THERAPY

There is little objective information defining when and how to treat dogs and cats with Clostridial-associated diarrhea, and mild cases of diarrhea can be self-limiting without the need for antimicrobial therapy. Animals with acute disease (hemorrhagic gastroenteritis) merit appropriate antimicrobial therapy, although antibiotics are commonly administered even in cases of chronic diarrhea. ***Optimal antibiotics for the treatment of canine C. perfringens-associated diarrhea include ampicillin, metronidazole, and tylosin.***[8] A study documented a high rate (21%) of *in vitro* resistance to tetracycline in dogs.[9] Preliminary studies in the author's laboratory have documented a high incidence (96%) of isolates that carry a gene encoding an efflux pump for tetracyclines (the *tetA*(P) gene). These findings are extremely important in light of the fact that tetracycline has been touted as a drug of choice for the treatment of dogs with suspected *C. perfringens*-associated diarrhea or small intestinal bacterial overgrowth, and subinhibitory concentration have actually been recommended as a long term therapeutic approach.[6] Furthermore, these results raise speculation of the potential for canine commensal *C. perfringens* to serve as a reservoir capable of both interspecies and intraspecies transfer of antibiotic resistance genes.

CLOSTRIDIUM DIFFICILE

Clostridium difficile is a gram-positive, anaerobic spore-forming bacillus, and is the major cause of antibiotic-associated pseudomembranous colitis in people. *Clostridium difficile* has also been associated with diarrhea and enterocolitis in foals and adult horses, as well as diarrhea in dogs and cats. *C. difficile* infection (CDI) is less common in cats, and a recent study by the author documented an incidence of 5% in diarrheic cats based on detection of toxin A via ELISA. To date, three toxins produced by *C. difficile* have been described: toxin A (an enterotoxin), toxin B (a cytotoxin), and CDT (an ADP-ribosyltransferase). Diseases associated with *C. difficile* have primarily been attributed to the activity of toxins A and B, and strains have historically been thought to produce both toxins (toxigenic isolates), or neither (nontoxigenic). ***There are increasing reports of variant strains isolated from human, canine, and equine clinical cases of CDI that produce only toxin A or toxin B.***[10] The role that *C. difficile* plays in the dog and cat is not as well documented as it is in humans and horses, and most of the literature focusing on *C. difficile* in dogs and cats is based on isolation of the organism which ranges from 0% to 40% in apparently healthy and diarrheic animals.[2,11]

DIAGNOSIS

Clinical signs

Clinical signs that have been associated with canine CDI range from asymptomatic carriage to a potentially fatal acute hemorrhagic diarrheal syndrome.[2] As with *C. perfringens*, there does not appear to be a specific anatomic localization of clinical signs, and recent studies have shown that dogs and cats with suspected CDI commonly have signs of small and large intestinal diarrhea as well as diffuse disease characterized by involvement of both the small and large intestine concurrently.[2]

Detection of the Organism via Culture

Fecal culture of *C. difficile* from diarrheic specimens should be attempted, despite similar isolation rates between nondiarrheic and diarrheic dogs (0 to 40%).[2] ***The main potential role of fecal culture in diagnosis is to rule out CDI, because a negative result from a properly collected and processed fecal specimen has a high negative predictive value.*** Fecal culture may also be useful for procuring strains for detection of toxin genes and typing.

Detection of the Organism via ELISA

Antigen ELISA can be useful because it is easy to perform, rapid, and may be highly sensitive. This type of test detects "common antigen" (glutamate dehydrogenase [GDH] that is predominantly present in toxigenic and non-toxigenic *C. difficile* strains and a few uncommon *Clostridium* species. This test has the same limitations as culture in terms of detection of non-toxigenic strains and colonization, although ***a negative result for GDH can rule out infection with C. difficile*** if the test is highly sensitive.

Fecal toxin detection

The current gold standard assay is the cell culture cytotoxicity assay, which detects toxin B activity; however, this assay is expensive and requires up to 48 hours for conformation of a negative result.[12] Routine diagnosis of *C. difficile*-associated diarrhea has been made based on positive fecal assays for toxins A or B. There are currently several commercially available immunoassays for detection of *C. difficile* toxin A or detection of both toxins A and B in fecal specimens; however, the performance characteristics of the human-based immunoassays is unacceptably low in dogs.[13] Because of increasing reports of toxin A-negative, toxin B-positive strains isolated from clinical cases, the use of ELISA kits that detect both toxins is gaining preference. However, none of the commercial ELISA kits currently available have been validated in dogs or cats and there is no standard assay utilized by veterinary diagnostic labs.

Molecular Techniques

The detection rates of toxigenic *C. difficile* strains following culture is similar in diarrheic and nondiarrheic dogs reducing the diagnostic value of this study in dogs. Direct PCR on stool to detect toxin B genes is increasingly used in human hospitals because of the high sensitivity of the test when used in combination with appropriate clinical case definition.

Combination Testing

Currently, ***the use of a combination of toxin testing by ELISA and organism detection (culture, antigen ELISA or real-time PCR) is recommended for the diagnosis of CDI in dogs and cats.*** The chosen ELISA should detect both toxin-A and toxin-B, because toxin-A-negative, toxin-B-positive strains have been documented to in canine isolates. Positive toxin detection by ELISA and concurrent detection of organism is presumptive for a diagnosis of CDI. Detection of toxin but failure to identify *C. difficile* should be interpreted with caution, particularly considering the relatively high sensitivity of organism and antigen detection methods compared to toxin ELISA. In such cases, it is plausible that the toxin ELISA result is false positive, rendering a diagnosis of *possible* CDI. ***Antigen or culture positive but toxin negative results are difficult to interpret because of the marginal sensitivity of available toxin tests. Such results should be considered to indicate a possible diagnosis of CDI in an animal with diarrhea and no other identifiable cause of diarrhea but such testing cannot be considered definitive.***

THERAPY

Metronidazole is the drug of choice for dogs and cats with suspected CDI. Although metronidazole-resistant *C. difficile* isolates obtained from foals and adult horses have been reported, a study evaluating the *in vitro* susceptibilities of 70 canine *C. difficile* isolates did not demonstrate any evidence of resistance to metronidazole.[9] The second drug of choice in human patients, and occasionally in horses, is vancomycin; however, it is used only in cases of nonresponsive CDI or when metronidazole-resistant strains have been demonstrated.

ZOONOTIC IMPLICATIONS

The risk of zoonotic transmission currently is unclear. Transmission of *C. difficile* from animals to humans has not been documented. However, because *C. difficile* is an important human pathogen and the strains of *C. difficile* that infect dogs often are indistinguishable from those found in people with CDI, it is prudent to consider *C. difficile* as potentially zoonotic.

CAMPYLOBACTER SPP.

Campylobacter spp. are small (0.2 to 0.5 µm x 0.5 to 5 µm), microaerophilic, gram-negative, c urved rod-shaped bacteria. *Campylobacter* species that have been implicated in canine enteric disease include *C. jejuni*, *C. coli*, *C. helveticus*, and *C. upsaliensis*. It has recently been shown that some selective media can have an inhibitory effect on a number of *Campylobacter* spp, resulting in more sensitive species, such as *C. upsaliensis* or other catalase-negative or weakly-positive species being missed.[14] Fecal shedding of *C. jejuni* is significantly greater in dogs < 6 months old and during the summer and autumn.[15] The higher prevalence of infection in pups versus adult dogs may reflect increased exposure of young animals to fecal excrement and confinement to a limited space. In addition, the unexposed immune system of the pups may increase the susceptibility to intestinal colonization by *C. jejuni*. Other enteric pathogens, such as parvovirus, *Giardia*, or *Salmonella* may play a synergistic role. ***The isolation of Campylobacter spp. from a diarrheic animal does not necessarily implicate Campylobacter as a cause of the diarrhea.*** Several authors have concluded that Campylobacters are not a primary pathogen in dogs because of similar isolation rates between nondiarrheic (0 to 49%) and diarrheic animals (0 to 74%).[16,17] Additional studies are warranted to investigate host factors and co-factors that are necessary to induce clinical disease in animals infected with *Campylobacter* spp.

DIAGNOSIS

Clinical signs

Clinical signs include mucus-laden or watery diarrhea, with or without blood and leukocytes, partial anorexia, occasional vomiting, and slight fever of 3 to 7 days duration.[18] Severity of the disease is dependent on the number of organisms ingested by the host, as well as previous exposure and development of protective antibody. Concurrent infections with other enteric pathogens such as parvovirus, *Giardia*, or *Salmonella*, may play a synergistic role and worsen clinical signs.[18]

Stained fecal smears

Campylobacter-like organisms (CLO's) can be identified by examining stained smears (Gram stain or Romanovsky-type stain) of fresh feces from the patient. The organism's characteristic morphology (slender, curved rods with an "S" shape or sea gull-shaped appearance) allows it to be identified relatively easily. The major limitation of direct examinations is that the procedure fails to differentiate between *Campylobacter* spp., or between related organisms including *Helicobacter* spp., and *Anaerobiospirillium* spp. In addition, ***identification of CLO's is not sufficient to warrant a diagnosis of Campylobacter-associated diarrhea, as many healthy dogs and cats can harbor CLO's in their intestinal tract.***

Fecal culture

For optimal recovery of *Campylobacter* spp., feces or fecal swabs should be fresh or placed immediately into anaerobic transport medium before refrigeration at 4°C. For isolation, the use of a formulated selective medium containing antimicrobial agents (e.g., Campy-CVA containing cefoperazone, vancomycin, and amphotericin B) gives better recovery than other direct-plating selective media.[19] Microaerophilic incubation conditions should be maintained, and the plates should be incubated at 37 C or at 42 C, when isolation of *C. jejuni* and *C. coli* from feces is attempted. Suspect colonies should be gram-stained and sub-cultured to 5% SBA. Biochemical tests can then be performed to speciate Campylobacter-like organisms isolated, although this method has limitations and is being complimented with multiplex PCR assays. The use of selective medium containing cefaperazone should be used when attempting to isolate *C. upsaliensis*, as the organism is more resistant to cefoperazone than to cephalothin. Characterization of *Campylobacter* infections or mixed infections with *Helicobacter* and *Campylobacter* spp. is best accomplished utilizing molecular structure-based diagnostics, employing genus- and species-specific PCR, RFLP analysis, and 16S rRNA sequence analysis.

Molecular techniques

Routinely used phenotypic tests often result in unreliable identifications, requiring additional testing such as PCR to identify and differentiate *Campylobacter* spp. These assays include direct sequencing of the 16S rDNA and comparison with databases such as GenBank, DNA hybridization with probes specific for different species, PCR amplification of specific regions of 16S rDNA or the *lpx*A gene, and amplified fragment length polymorphism.[20] These tests have been examined for their ability to differentiate a variety of *Campylobacter* species such as *C. coli*, *C. jejuni*, *C. lari*, and *C. upsaliensis*.[20] Recently, a cpn60-based direct fecal real-time PCR assay has been described that can detect 14 different *Campylobacter* species.[21]

THERAPY

The majority of cases are uncomplicated, self-limiting and will resolve with supportive therapy alone. Because isolation of Campylobacter does not necessarily imply causation of clinical signs, treatment may not be warranted and may further disrupt the intestinal microflora. However, in immunocompromised or febrile animals, or in animals with evidence of hemorrhagic diarrhea, antimicrobial treatment may be indicated, particularly when *C. jejuni* is isolated.

Macrolides or fluoroquinolones are most commonly used to treat Campylobacter infections, although fluoroquinolones should be avoided in young animals due to their possible adverse effects on cartilage and increasing documentation of resistance. Antimicrobial drug resistance has been reported to both of these drug classes, but routine antimicrobial sensitivity testing is rarely performed in veterinary diagnostic laboratories because it is difficult and time consuming. In 1 study that examined the MIC values of Campylobacter isolates from dogs and cats, some resistance to enrofloxacin and ciprofloxacin was observed.[22] Erythromycin, 10-15 mg/kg PO q8h or azithromycin, 5-10 mg/kg PO q24h can be given for 5-21 days as treatment. Azithromycin is better tolerated, but to the authors' knowledge, no published studies regarding efficacy of azithromycin for treatment of campylobacteriosis in dogs or its comparison to other macrolides or fluoroquinolones are available. In general, treated dogs will have a 50-73% response to treatment; 50% of cats will respond.[23] It is not known whether treatment failures reflect resistance to the antimicrobial drug used or an incorrect diagnosis of *Campylobacter*-associated diarrhea.

SALMONELLA SPP.

The salmonellae are Gram-negative, motile, non-spore-forming facultative anaerobic bacilli that belong to the family *Enterobacteriaceae*. The genus *Salmonella* consists of only 2 species, *Salmonella enterica* and *Salmonella bongori*. *S. enterica* is divided into 6 subspecies: *S. enterica* subsp. *enterica, S. enterica* subsp. *salamae*, *S. enterica* subsp. *arizonae*, *S. enterica* subsp. *diarizonae*, *S. enterica* subsp. *houtenae* and *S. enterica* subsp. *indica*.[24] Salmonellae are ubiquitous organisms that can infect or be isolated from a variety of mammals, birds, reptiles and insects. The prevalence of *Salmonella* in healthy dogs and cats can be similar to that observed in diarrheic animals.[25]

DIAGNOSIS

Clinical Signs

Salmonellosis primarily is an acute disease although it should be suspected as a cause in any acute or chronic gastrointestinal illness in dogs and cats. The clinical signs are highly variable. Acute episodes of illness are thought to occur 3-5 days after exposure but clinical signs also have been demonstrated to occur after only 12 hours. Fever, malaise, anorexia followed by vomiting, abdominal pain and diarrhea are common. The diarrhea frequently is watery or mucoid and can be bloody in severe cases. The severity of infection in dogs and cats varies with the individual animal. Most dogs that shed *Salmonella* have no clinical abnormalities, although some dogs may manifest clinical signs of sepsis.[26]

Culture

Salmonella are facultative anaerobes that grow readily at 37°C. There are a large variety of commercial media available including MacConkey agar, XLD agar, and Brilliant Green agar, and the unlimited combinations of pre-enrichment, selective enrichment (selenite F broth, tetrathionate broth, Gram-negative broth) and selective culture media make it practically impossible to determine which combination of methods is most likely to result in the successful isolation of *Salmonella* from feces. Most diagnostic laboratories use a combination of selective enrichment broth followed by subculture to 1 or more selective agar plates and will go on to identify presumptive *Salmonella* colonies using biochemical techniques. Isolates identified as *Salmonella* can be further discriminated by serotyping that typically is performed by specialized reference laboratories.

Molecular Techniques

PCR assays vary and can detect *Salmonella* in a variety of different matrices, from water to human stool samples.[27] Despite the availability of rapid, cost effective real-time PCR methods that have been developed within the last few years, few clinical diagnostic laboratories have fully embraced this technology and there have been no multi-center validations for the use of PCR to detect *Salmonella* in dog or cat feces. It is recommended that PCR after overnight enrichment in a non-selective broth be adopted as the gold standard, and that all positive PCR samples be cultured using selective enrichment to isolate and identify the infecting organism.[28] Assays that have been validated in the literature still must be verified as "fit for purpose" in the laboratory in which they will be used. However, this is not a laborious process and it is imperative that these steps be taken in an effort to better standardize the laboratory diagnosis of *Salmonella*. Given that the sensitivity of conventional culture is much lower than that of PCR, multiple cultures should be performed if relying on this method alone. When culturing equine feces it has been recommended that at least 5 serial samples be cultured to increase sensitivity (estimated at around 55%).[29] If the specificity of the test is estimated at 99% and the sensitivity of any culture method for dog and cat feces is greater than 45%, 3 consecutive negative cultures allows one to be only 90% confident that the sample is truly negative and 6 cultures would be required to be 99% confident (Helen Aceto, personal communication).

The traditional diagnosis of salmonellosis is made based on isolation of the organism in conjunction with clinical signs and assessment of potential risk factors such as hospitalization, age, environmental exposure, and antibiotic administration. However, isolation of *Salmonella* is not necessarily indicative of involvement in disease, as similar isolation rates can be detected in healthy nondiarrheic animals. Hematological abnormalities are variable, and include a nonregenerative anemia, lymphopenia, thrombocytopenia, and neutropenia with a left shift. Toxic neutrophils are found in animals with systemic disease and endotoxemia, findings similar to those documented with canine parvovirus.

Therapy

The treatment of Salmonella in diarrheic dogs and cats is a controversial topic, with many microbiologists and infectious disease specialists arguing that antibiotics are not advocated for treating animals with uncomplicated *Salmonella* gastroenteritis. Instead, ***antibiotics should be reserved for animals with concurrent signs of systemic infection or a history of immunosuppresion.*** Fecal shedding can be prolonged with the injudicious administration of antibiotics, and infecting organisms may acquire transferable (plasmid-mediated) resistance.[30] For septic dogs and cats, antibiotics reported to be effective against *Salmonella* include fluoroquinolones, chloramphenicol, trimethoprim-sulfonamide, and amoxicillin. Determination of a susceptibility profile is recommended for selection of optimal antimicrobials.

PUBLIC HEALTH CONSIDERATIONS

Both *Salmonella* and *Campylobacter* spp. have zoonotic implications, and the importance of isolating either organism in diarrheic dogs and cats allows one to minimize risk of transmission to people by implementing appropriate strategies in the household. These strategies include the restriction of high-risk contact, the prompt cleaning and disinfection of "accidents" in the house, and the rigorous practice of hand hygiene. Infected animals should not be walked in public places for at least one month beyond the clinical resolution of diarrhea.

OPTIMIZING THE COLLECTION, HANDLING AND SHIPPING OF FECAL SAMPLES

Proper collection and preservation of feces frequently are neglected but important requirements for the isolation of putative enteropathogens. Approximately 2-3 grams of fresh feces should be collected into a clean, sealed and leak-proof cup or sterile container and transported to the laboratory as soon as possible to maximize survival of *Salmonella* and *Campylobacter* spp. Specimens should be processed within 2 hours after collection. If the laboratory is on-site, no transport medium is required. Transport medium such as Cary-Blair or Amies gel should be used for specimens that cannot be cultured within 2 hours of collection. Rectal swabs are suboptimal for bacterial isolation given the limited volume of feces obtained. If rectal swabs are used, the specimen should be collected with a sterile swab, placed in Amies transport medium, and transported to the laboratory as soon as possible. Specimens should be kept cool at 4-10°C, but not frozen. Fecal specimens submitted for ELISA testing should not be placed in transport media.

CONCLUSION

The lack of well-scrutinized practice guidelines for veterinarians that provide objective recommendations for implementing fecal bacterial testing combined with the clinical documentation of enteropathogenic bacteria in diarrheic and healthy animals has resulted in indiscriminate testing and misinterpretation of results. This problem has been compounded by the lack of validated immunoassays for dogs and cats, and the acceptance of molecular-based testing as the "holy grail" in our diagnostic armamentarium. Molecular-based testing should be carefully integrated with conventional testing while recognizing the benefits and limitations of each modality. Veterinarians should be cognizant of the fact that ***most bacterial enteropathogens are associated with self-limiting diarrhea***, and the injudicious administration of antimicrobials could be more harmful than beneficial. Supportive therapy and appropriate hygiene control should be considered in all animals with suspected or confirmed bacterial-associated diarrhea (with the exception of *E. coli* associated with granulomatous colitis in which antimicrobial therapy is warranted), and ***antimicrobials should only be administered to animals manifesting systemic signs of illness***.

REFERENCES

[1]Cave NJ, et al. **J Am Vet Med Assoc** 2002;221:52-59. [2]Marks SL, et al. **J Vet Intern Med** 2002;16:533-540. [3]Meer RR, et al. **Rev Environ Contam Toxicol** 1997;150:75-94. [4]Queen EV, et al. **J Vet Intern Med** 2012;26:54-60. [5]Marks SL, et al. **J Am Vet Med Assoc** 1999;214: 357-360. [6]Twedt DC: *Clostridium perfringens*-associated diarrhea in dogs, in DeNovo R (ed), *Proceedings of the Eleventh Annual Veterinary Medical Forum. American College of Veterinary Internal Medicine.*, Madison, WI, Omnipress, 1993, 121-125. [7]Weese JS, et al. **J Vet Intern Med** 2001;15: 374-378. [8]Greene CE: Enteric Bacterial Infections, in Greene CE (ed), *Infectious Diseases of the Dog and Cat,* Philadelphia, WB Saunders, 1998, 243-245 [9]Marks SL, et al. **Vet Microbiol** 2003;94: 39-45. [10]Barbut F, et al **J Clin Microbiol** 2002;40:2079-2083. [11]Riley TV, et al. **Epidemiol Infect** 1991;107:659-665. [12]Landry ML, et al. **J Clin Microbiol** 2001;39:1855-1858. [13]Chouicha N, et al. **J Vet Diagn Invest** 2006;18(2):182-8. [14]Steinhauserova I, et al. **Lett Appl Microbiol** 2000;31:209-212. [15]Torre E, et al. **Am J Vet Res** 1993;54:260-262. [16]Hosie BD, et al. **Vet Rec** 1979;105:80. [17]Malik R, et al. **Aust Vet Practit** 1989;19:16-18. [18]Fox JG, et al. **J Am Vet Med Assoc** 1983;183:1430-143. [19]Aspinall ST, et al. **J Appl Bacteriol** 1996;80:645-650. [20]Engvall EO, et al. **J Appl Microbiol** 2002;92:47-54. [21]Chaban B, et al. **BMC Microbiology** 2010;10:73-79. [22]Rossi M, et al. **Vet Microbiol** 2008;129:304-314. [23]Fleming MP. **Vet Rec** 1983;113:372-374. [24]Guibourdenche M, et al. **Res Micro** 2010;161:26-29. [25]Mackel, D. C, et al. J **Infect Dis** 1952;91:15-18. [26]Allison, C. J. **Vet Rec** 1969;85:564. [27]Ellingson JL, et al. **Mol Cell Probes** 2004;18:51-57. [28]Ward MP, et al. **J Vet Diagn Invest** 2005;17:118-23. [29]van Duijkeren E, et al. **Vet Q** 1995;17:63-6. [30]Arroyo L, et al. **J Vet Int Med** 2004;18:387-388.

MITOTANE VS. TRILOSTANE: WHAT'S THE EVIDENCE?

Ellen N. Behrend, VMD, PhD, DACVIM, Joezy Griffin Professor
Auburn, AL

Note: to convert cortisol in nmol/L to µg/dl, divide the concentration in nmol/L by 27.6.

HOW TO USE TRILOSTANE

A synthetic steroid analogue that inhibits the adrenal enzyme 3β-hydroxysteroid dehydrogenase, trilostane suppresses production of progesterone and its end-products including cortisol and aldosterone. Additional enzymes, e.g. 11β-hydroxylase and 11β-hydroxysteroid dehydrogenase, may also be affected[1]. Trilostane (Vetoryl®) is FDA-approved for treatment of canine hyperadrenocorticism (HAC). Below is a discussion of latest information and outstanding questions.

Question #1: dose

The original European package insert recommended a starting dose of 2-10 mg/kg once daily. The package insert in the U.S. has a starting dose of 2.2-6.7 mg/kg daily. A current anecdotal recommendation, the one used by the author, is to start at 2 mg/kg daily, divided if possible. Regardless of where therapy begins, dosage adjustments, either up or down, will be required during the course of treatment in most dogs. Authors of one study noted that in most dogs an initial sensitivity to the drug existed followed by a need for a dose increase. After time, the dose required often hit a plateau[2].

Despite recommendations in the package insert for once-daily dosing, trilostane may begin to lose effectiveness at 8-10 hrs post-pill[3-5]. Indeed, trilostane has been administered twice-daily with good efficacy[5-8]. Whether once- or twice-daily administration is better is unclear (see question #3 below). A recent paper even reported using TID dosing.[8]

Trilostane absorption from the GI is increased by food. Thus, to be consistent, a patient should always be either fasted or fed when the trilostane is administered. Giving the pill with food may help to decrease the dose required for disease control, but this has not been evaluated.

Question #2: timing of dose adjustments

Even though the initial recheck should be approximately 10-14 days after starting therapy and veterinarians used to increase the dose then if ideal cortisol concentrations had not been achieved, lately some specialists have been recommending that if any improvement has been noted at the first recheck, "hold the course" until the 30 day recheck. The basal and ACTH-stimulated cortisol concentrations usually decrease until the 4-wk recheck even if the dog remains on the same dose. The dose is only raised at the day 14 assessment if no improvement has been seen, clinical signs are still striking and/or the post-ACTH cortisol concentration is markedly above ideal. The single caveat is that if the cortisol concentrations are below ideal at any time, the trilostane dose should be lowered.

Question #3: timing of monitoring

The optimal post-ACTH serum cortisol concentration remains to be determined. The goal for post-ACTH cortisol concentration has varied from <250 nmol/L[9] to 30-70 nmol/L[10] to 25-125 nmol/L[2]. The manufacturer recommends approximately 40-150 nmol/L. A recent publication provides evidence of the appropriateness of this target[8].

The ideal timing of post-pill sampling also needs to be elucidated. Post-ACTH cortisol may vary with the interval between dosing and testing[3]. Doing an ACTH stimulation test 3-8 hr[2] and at 4-6 hrs post-pill[11] have both been advocated. The package insert recommends a 4-6 hr interval. For BID therapy, although one study recommended starting the test 8-12 hrs post-pill[7], such timing has not been critically evaluated. This author uses the manufacturer's recommendations and thinks it quite important to keep the interval constant for each patient from test to test.

An early paper suggested that monitoring the urine cortisol:creatinine ratio (UCCR) before a trilostane dose could indicate duration of action[2]. However, 2 additional studies yielded conflicting results. In 22 dogs with HAC, a morning UCCR pre-dose was lower in dogs with good control compared to those with poor control (14.8 vs. 47.5); however, overlap existed between the 2 groups[5]. On the other hand, in 18 PDH dogs, the UCCR was measured in samples collected before or 6 hr after trilostane administration. Samples at neither time were significantly different between dogs with adequate vs. inadequate control.[12]

Interestingly, in one study, of 6 dogs that had UCCRs within the reference range "most of the times", 3 developed hypocortisolism and in 3 it was suspected based on clinical signs but was not confirmed. The hypocortisolism was documented or suspected between 16 weeks and 21 months after the trilostane dose was judged to be optimal[12].

Question #4: efficacy

Overall, trilostane appears to be highly effective in suppressing cortisol secretion and controlling clinical signs in the majority of patients[2, 5-7, 9, 10, 13]. As with mitotane, many clinical signs of HAC typically quickly resolve with disease control, but certain ones, e.g. dermatological abnormalities, can take up to 3 months. Other abnormalities such as calcinosis cutis or pseudomyotonia may not fully resolve. A small proportion of dogs with PDH are not well controlled with trilostane[2, 10, 14].

Special consideration should be given to twice-daily dosing in dogs in which breaks in control of the HAC could be detrimental, e.g. dogs with concurrent diabetes mellitus or with pulmonary thromboembolism secondary to HAC. To this author, one of the biggest concerns with using trilostane, especially if once daily, is that the duration of control of cortisol secretion varies between dogs and is not easily determined. Although clinical signs may be in remission, we do not know how much control of cortisol secretion is needed to improve or prevent the serious complications of HAC. Adequate control of HAC with mitotane significantly decreases blood pressure and proteinuria[15]. Control of hypertension and proteinuria has not yet been assessed in dogs on trilostane. The effect of control by either drug on development of complications such as recurrent infection or thromboembolic disease has never been evaluated. In a small retrospective study of 8 diabetic dogs, insulin requirements and fructosamine concentrations were not consistently reduced during trilostane treatment for HAC[16].

Question #5: safety

Reported adverse effects of trilostane for the most part are relatively mild, including lethargy and vomiting, but fatality has occurred[2, 10, 14]. Although some studies found relatively low incidence of side effects, one non-peer-reviewed report states mild, self-limiting side effects such as diarrhea, vomiting and lethargy occur in 63% of treated dogs[11]. Compared to mitotane, trilostane has less effects on suppression of aldosterone concentrations regardless of level of control of PDH[17].

As with mitotane therapy, excess adrenal gland suppression can occur and warrants discontinuing trilostane temporarily and lowering of the dose. Although, in theory, the effects of trilostane as an enzyme inhibitor should be rapidly reversible (e.g. within a couple days), suppression can last weeks to years[2, 7, 18, 19]. With oversuppression, recommendations from many sources are to discontinue the medication for a few days and then begin again. I would prefer to perform an ACTH stimulation test to document return of function. One dog developed hypocortisolism after only 3 trilostane doses, and it lasted for at least one year [18]!

Adrenal necrosis can occur secondary to trilostane administration[20], potentially due to the elevations in ACTH concentration that occur with treatment[21]. The hypoadrenocorticism reported after complete adrenocortical necrosis in one dog lasted for at least 3 months, but likely would be permanent. How often acute iatrogenic hypoadrenocorticism will occur with trilostane is unknown, but is likely more common than originally believed. In one study, 4 of 6 dogs with PDH and 1 of 1 with AT treated with trilostane had some degree of adrenal necrosis at necropsy. In 2 dogs, the damage was severe enough to potentially cause hypoadrenocorticism. Both dogs had received mitotane therapy before trilostane but had been on trilostane for 15 and 22 months[22], so the contribution of each drug is unclear. Adrenal rupture, possibly secondary to adrenal necrosis, may have occurred (Vetoryl® package insert).

Final recommendations

Given all the information, my current recommendations for using trilostane are:

1. Start as close to the bottom end of the dosing interval as possible: 2 mg/kg daily or 1 mg/kg BID. I prefer BID in general to get as much control as possible throughout the day. Definitely start BID in diabetics and in others in which elevated cortisol concentrations during the day may be detrimental, e.g. high risk for pulmonary thromboembolism.
2. Make sure the patient is always fed or always fasted when the pill is given. Giving with food may help decrease the dose.
3. Be cautious with dose increases!
4. Do the ACTH stimulation test 4-6 hrs post-pill. Keep the timing and feeding consistent with every test.
5. If cortisol concentrations are too low, perform an ACTH stimulation test before reinitiating trilostane therapy. Suppression can be prolonged!

WHEN TO USE MITOTANE VS. TRILOSTANE

There are multiple ways that trilostane and mitotane can be compared to make a decision as to which to use.

General efficacy for pituitary-dependent hyperadrenocorticism (PDH)

As stated above, overall, trilostane appears to be highly effective in suppressing cortisol secretion and controlling clinical signs in the majority of patients [2, 5-7, 9, 10, 13]. In 184 dogs with PDH treated with mitotane for a mean of 2 years, the response was judged as excellent in 83%, fair in 16% and poor in 0.6%[23]. Survival with trilostane therapy is at least as good as with mitotane. In one study, median survival on trilostane given SID was 662 days (8-1971 d) while survival on mitotane was 708 days (33-1399 d)[13]. In dogs with PDH given trilostane BID, median survival was 930 days[7].

General efficacy for adrenal tumors

Surgical removal of cortisol-secreting adrenal tumors (AT) is the recommended treatment, but circumstances exist where surgery is neither possible nor desired. Trilostane has been used to treat dogs with AT[24-27]. For trilostane, not enough information is available to ascertain whether the treatment protocol or efficacy varies if treating PDH versus AT. One dog with an AT did receive a maximum dose of 17.2 mg/kg[26], higher than typical for PDH.

In a recent paper, mitotane and trilostane were compared in a total of 37 dogs with AT. The aim of both drugs was control of cortisol secretion as for PDH, not ablation of adrenal tumor tissue as has been recommended by some. The median survival was not significantly different between the drugs (277 d overall). The only factor that affected survival was presence of metastases[27].

Of 32 dogs with an AT treated with mitotane with the goal of tumor ablation, 66%, 28% and 6% were judged by their owners to have a good to excellent, fair response and poor response, respectively. Mean survival time was approximately 16 months with a reported range of 20 days to 5.1 years[28].

In dogs with AT, mitotane could theoretically be preferred as it is truly chemotherapeutic in this instance, killing primary neoplastic cells and, perhaps, metastatic ones as well. Trilostane controls tumoral secretion, not growth. In fact, in dogs treated with trilostane, adrenal gland size increases[22, 29]. If the goal is to prepare a patient for adrenalectomy, trilostane may be of choice.

Efficacy for control of more serious complications of hyperadrenocorticism (HAC)

To reiterate, one of the biggest concerns with using trilostane to this author, especially if once daily, is that the duration of control of cortisol secretion varies between dogs and is not easily determined. When serious complications of HAC are present or the patient is diabetic, twice-daily therapy should be considered. The effect of control by either drug on development of complications such as recurrent infection or thromboembolic disease has never been evaluated. Mitotane does decrease proteinuria and hypertension with adequate control of PDH; whether trilostane does as well has never been investigated.

Adverse effects

Trilostane has more adverse effects than originally believed (see above). Adverse effects of mitotane are generally gastrointestinal or neurological. One or more adverse effects occur in approximately 25% of dogs with PDH during loading and include weakness, vomiting, anorexia, diarrhea and/or ataxia[23]. Approximately 33% of dogs on maintenance mitotane therapy for PDH develop adverse effects, typically shortly after initiation of the maintenance dosage or during relapse when daily therapy is reinstituted. Adverse effects occur in approximately 60% of dogs with AT treated with mitotane.

Approximately 33% of dogs will have a serum cortisol concentration less than ideal (e.g. post-ACTH cortisol concentration <30 nmol/L) after induction with mitotane for PDH. In most dogs, serum cortisol concentrations will rise into the ideal range within 2 to 6 wks, but up to 18 mths may be needed[23]. Complete mineralocorticoid and glucocorticoid deficiency is seen in approximately 6% of dogs anywhere from 1 mth to years after initiation of maintenance therapy and is usually permanent[23].

Contraindications

The trilostane package insert states that primary hepatic disease and renal insufficiency are contraindications. In Plumb's Veterinary Drug Handbook, it states that dogs with preexisting renal or hepatic disease should receive mitotane with caution and with more intense monitoring. A single case report of mitotane-induced hepatotoxicity exists[30]. Hyperbilirubinemia has been documented in 3 dogs receiving trilostane for PDH[9], but whether the trilostane was the cause was not established. The true risk of using either drug in the presence of liver or renal issues remains unknown. Similarly, whether one drug is a better choice has not been established.

Ease

During the maintenance phase, mitotane is typically given 2-3 times per week. While some owners enjoy not having to give pills daily, others find it hard to remember to give the pill if it is not given every day.

Induction with mitotane seems to be scarier than initiation of trilostane therapy. Recommendations for mitotane induction by some authors include intense monitoring – personally calling the owners every day starting day 3. The impression exists that it can be easier to use trilostane if the owners are not very observant or hard to reach by phone or if the dog does not have obvious polyphagia. I am not sure the comfort level with trilostane is deserved; cortisol decreases with both drugs over approximately equal time spans.

Final recommendations

Clearly, which drug to use is not a clear-cut decision. I suggest the following:

1. Talk to the owner describing both drugs, protocols, etc. so they make an informed decision.
2. If the dog has an AT and surgery is not an option, consider mitotane as first-line therapy.
3. If the dog has an AT and therapy is to be initiated to prepare for adrenalectomy, consider trilostane.
4. If the dog is a diabetic or has a serious complication of HAC, consider mitotane as first-line therapy.

REFERENCES

[1] Sieber-Ruckstuhl NS, et al. *Domest Anim Endo* 2006;31:63-75; [2] Braddock JA, et al. *Aust Vet J* 2003;81:600-607; [3] Witt A, Neiger R. *Vet Rec* 2004;154:399-400; [4] Bell R, et al. *Vet Rec* 2006;159:277-281; [5] Vaughn MA, et al. *J Am Vet Med Assoc* 2008;232:1321-1328; [6] Clemente M, et al. *Vet Rec* 2007;161:805-809; [7] Alenza DP, et al. *J Am Anim Hosp Assoc* 2006;42:269-276; [8]Feldman EC. *J Am Vet Med Assoc* 2011;238:1441-1451; [9] Neiger R, et al. *Vet Rec* 2002;150:799-804; [10] Ruckstuhl NS et al. *Am J Vet Res* 2002;63:506-512; [11] Neiger R. *Proc 22nd Ann Vet Med Forum* 2004;699-701; [12] Galac S, et al. *J Vet Int Med* 2009;1214-1219; [13] Barker EN, et al. *J Vet Int Med* 2005;19:810-815; [14] Neiger R, Ramsey IK. *Proc 20th Ann Amer Coll Vet Int Med Forum* 2002;544-546; [15] Ortega TM, et al. *J Am Vet Med Assoc* 1996;209:1724-1729; [16] McLauchlan G, et al. *J Small Anim Pract* 2010;51:642-648; [17] Reid LE, et al. *J Vet Int Med* 2011;25:682; [18] Ramsey IK, et al. *Aust Vet J* 2008;86:491-495; [19] Galac S, et al. *Vet J* 2010;183:75-80; [20] Chapman PS, et al. *J Small Anim Pract* 2004;45:307-310; [21] Burkhardt WA, et al. *Domest Anim Endocrinol* 2011;40:155-164; [22] Reusch CE, et al. *Vet Rec* 2007;160:219-224; [23] Kintzer PP, Peterson ME. *J Vet Int Med* 1991;5:182-190; [24] Machida T, et al. *J Vet Med Sci* 2007;70:317-320; [25] Benchekroun G, et al. *Vet Rec* 2008;163:190-192; [26]Eastwood JM, et al. *J Small Anim Pract* 2003;44:126-131; [27] Helm JR, et al. *J Vet Int Med* 2011;25:251-260; [28] Kintzer PP, Peterson ME. *J Am Vet Med Assoc* 1994;225:54-61; [29] Mantis P, et al. *Vet Radiol Ultrasound* 2003;44:682-685; [30] Webb CB, Twedt DC. *J Am Anim Hosp Assoc* 2006;42:298-301.

TWO ARE NOT BETTER: FELINE CHRONIC KIDNEY DISEASE AND HEART DISEASE

Ashley B. Saunders, DVM, DACVIM (CARDIOLOGY)
College Station, TX

INTRODUCTION

Chronic kidney disease (CKD) and heart disease are both common in cats, often with an increased incidence in the aging population. Both are typically irreversible and progressive with highly variable survival rates. The interactions between the heart and kidneys are commonly referred to as the "cardiorenal syndrome." The pathophysiology is complex and involves reduced cardiac output, renal hypoperfusion, activation of the renin angiotensin aldosterone system (RAAS) and sympathetic nervous system and is often complicated by systemic hypertension, electrolyte abnormalities, anemia, concurrent diseases (diabetes mellitus, hyperthyroidism) and polypharmacy. Poor perfusion to the kidneys activates the RAAS system resulting in sodium and water retention and increased circulating blood volume contributing to overload of the cardiovascular system. What is good for one system is counterproductive for the other. Is it possible to strike a balance?

SYSTEMIC HYPERTENSION

The International Renal Interest Society (IRIS) treatment recommendations for CKD include discontinuing nephrotoxic medications, identifying and treating pre-renal abnormalities, and measuring urine protein to creatinine ratio (UP/C) and blood pressure.[1] Systolic blood pressure > 180mmHg increases the risk of end organ damage to the eyes, brain, kidneys, and cardiovascular system. Hypertension is most common in aging cats and is associated with hyperthyroidism and CKD. In one study, systemic hypertension was documented in 19.4% of cats at the initial diagnosis of CKD.[2] Azotemia and left ventricular hypertrophy have been reported in up to 62% and 48%, respectively, of cats with systemic hypertension.[3-5] Persistent elevations in systemic blood pressure and/or evidence of end organ damage warrants therapy with amlodipine 0.625 to 1.25 mg PO once daily, a dihydropyridine calcium channel blocker. Combined therapy with an angiotensin converting enzyme (ACE) inhibitor is recommended in cases where blood pressure remains above 160mmHg or UP/C remains above 0.4.[1] ACE inhibition results in dilation of the renal efferent arterioles but the effects are too weak to significantly reduce blood pressure alone.[6,7] Benazepril therapy has been show to reduce proteinuria, a marker of survival, in cats with CKD[8-10] A marked increase in creatinine following initiation of an ACE inhibitor would suggest an adverse drug effect.[1] Therefore monitoring is important.

HEART FAILURE

Medical therapy in preclinical heart disease is not well defined in cats. ACE inhibitors, which block profibrotic effects of angiotensin II and aldosterone, have been evaluated in multiple studies with mixed results while they have proven favorable in delaying the progression of CKD.[6] Enalapril did not adversely affect blood pressure or creatinine in a small retrospective evaluation of cats with preclinical and clinical hypertrophic cardiomyopathy.[11] In a study of 102 cats with hypertrophic cardiomyopathy, azotemia was present in 58.8%.[12] Therapy of congestive heart failure typically consists of furosemide and an ACE inhibitor to reduce fluid accumulation. Important complications that can be encountered while managing heart failure and contribute to reduced renal perfusion includes systemic hypotension and dehydration/volume contraction associated with diuretic therapy. The use of inodilator therapy (pimobendan) has been reported in two retrospective evaluations of heart disease in cats, one focused predominately on heart failure and one on systolic dysfunction in heart failure.[13,14] Whether or not it has effects on renal perfusion has yet to be investigated but it's use did not significantly affect creatinine concentrations in these two studies. It is important to remember several issues inherent in managing a cat with CKD and heart failure. Cats receiving diuretic therapy have dilute urine and can have pre-renal azotemia without a characteristic urine specific gravity making it important to use surrogates for assessing hydration. Fluid therapy for CKD can easily overload the cardiovascular system.

ASSESSMENT AND THERAPEUTIC TIPS

Things to consider when making recommendations:

1. Can the owner give medications (multiple, oral)?
2. Which medications interact and what are the adverse effects?
3. What is causing the cat's clinical signs – heart disease or renal disease?
4. Blood work and thoracic radiographs go hand in hand in heart failure. If azotemia is present, what do the lungs look like? If congestion is present (pleural effusion, pulmonary edema), what do the kidneys look like?
5. Is blood pressure accurate and repeatable (a serious challenge in some cats)?
6. Be careful with fluids.
7. Encourage drinking and eating.

N-terminal pro-B-type natriuretic peptide (NTproBNP) is a diagnostic biomarker shown to be elevated in systemic hypertension[15] and heart failure.[16] While NTproBNP concentrations can be elevated in dogs with azotemia, concentrations were not reported to be significantly elevated in cats with mild to moderate normotensive CKD.[15,17]

Polypharmacy can include the use of NSAIDS, a potential cause of acutely worsening renal disease when receiving an ACE inhibitor.[18] Efferent arteriole vasodilation induced with an ACE inhibitor combined with NSAID induced cyclooxygenase inhibition creates vasoconstriction of the afferent arteriole which can result in a reduction in glomerular capillary pressure and glomerular filtration rate. Changes in renal function can result in drug toxicity by affecting the clearance of medications.

Recheck evaluations typically consist of at least blood pressure, thoracic radiographs and serum biochemistry panel including electrolytes. Although an echocardiogram is require to diagnose the type of cardiac disease present, once the diagnosis has been established, echocardiography is often reserved for assessing patients that are not responding to therapy and to evaluate for pericardial effusion, ventricular dysfunction and atrial thrombus formation. The goal is to maintain appetite, body condition, address the clinical signs, and provide a good quality of life (ie. If the cat has dyspnea and congestive heart failure, diuretic therapy is indicated along with continued monitoring of the kidney disease. Mild azotemia, in the absence of clinical signs of CRD, may be acceptable and well tolerated.)

This discussion will elaborate on the challenges of addressing opposing goals when managing combined CKD and heart disease in cats using case examples.

REFERENCES

[1]Iris-kidney.com [2]Syme HM, et al. J Am Vet Med Assoc 2002;12:1799. [3]Chetboul V, et al. J Vet Intern Med 2003;17:89. [4]Henik, et al. J Am Anim Hosp Assoc 2004;40:359. [5]Sampedrano C, et al J Vet Intern Med 2006;20:1106. [6]Lefebvre HP, et al. Curr Pharmaceut Design 2007;13:1347. [7]Steele JL, et al. Vet Ther 2002;3:157. [8]Chakrabarti S, et al. J Vet Intern Med 2012;26:Epub. [9]Syme HM, et al. J Vet Intern Med 2006;20:528. [10]Mizutani H, et al. J Vet Intern Med 2006;20:1074-1079. [11]Rush J, et al. J Am Anim Hosp Assoc 1998;34:38. [12]Gouni V, et al. J Vet Cardiol 2006;10:117. [13]MacGregor J, et al. J Vet Cardiol 2011;13:251. [14]Gordon S, et al. J Am Vet Med Assoc 2012;In Press. [15]Lalor SM, et al. J Vet Cardiol 2009;11:S71. [16]Fox PR, et al. J Vet Cardiol 2009;11:S51. [17]Schmidt MK, et al. J Vet Cardiol 2009;11:S81. [18]Lefebvre HP, et al. J Vet Pharmacol Therap. 2004;27:265.

WHEN TWO ARE NOT BETTER: FELINE CHRONIC KIDNEY DISEASE AND NEPHROURETEROLITHIASIS

Joe Bartges, DVM, PhD, DACVIM, DACVN
Knoxville, TN

Upper urinary tract urolithiasis (nephroureterolithiasis) is being recognized more frequently with a 10-fold increase in occurrence during a 20-year interval.[1] There appears to be an association of nephroureteroliths in cats with acute or chronic renal failure.[2] Many patients are asymptomatic and nephroureteroliths are discovered when lower urinary tract uroliths are creating clinical signs or when abdominal imaging is performed for other reasons. Nephroureteroliths can lead to obstruction to urine flow; deterioration of renal function; serve as a nidus for bacterial urinary tract infection; or cause hematuria, or pain. Clinical signs, when present, may include systemic signs associated with renal failure or pyelonephritis, vague signs (abdominal pain, arched back, vomiting, or anorexia) or hematuria. Sometimes animals present for an abrupt change in behavior or an acute onset of abdominal pain or vomiting when an acute ureteral obstruction occurs; this may be mistaken for intervertebral disk disease. Small, multiple uroliths are common and usually found in both kidneys and/or ureters. Small ureteroliths appear to migrate over time and can move retrograde in the ureter.[3] Approximately 15% of cats with upper urinary tract uroliths have lower urinary tract uroliths.[2] When lower tract uroliths occur in association with upper tract uroliths, it is unknown whether they formed in the lower urinary tract or whether they passed through the ureter and into the urinary bladder. A common scenario is to find upper tract uroliths in a cat with chronic renal failure. When there is unilateral nephrolithiasis, often they occur in the larger and more functional kidney.

ETIOLOGY AND PATHOPHYSIOLOGY

Calcium oxalate accounts for most (>70%, and probably > 90%) nephroureteroliths.[1] Struvite, particularly infection-induced, and other mineral types do occur. Risk factors for calcium oxalate nephroureteroliths are the same as for calcium oxalate urocystourethroliths.[4] Breeds considered at higher risk for forming calcium oxalate uroliths include long-haired and Siamese cats. They occur with approximately equal frequency in males and females in cats. Nephroureteroliths usually occur in middle-aged or older cats, but have been described in animals < 6 years. Approximately 1 in 3 cats have hypercalcemia, usually idiopathic in nature even in the face of chronic kidney disease. In these cats, serum total and blood ionized calcium concentrations are increased, but parathyroid hormone concentration is low. Struvite uroliths occur either due to a bacterial infection with a urease-producing microbial organism (infection-induced struvite) or without an infection (sterile struvite). Although sterile struvite is the most common mineral found in feline uroliths, it appears to occur rarely in nephroureteroliths. Usually nephroureteroliths occurring in cats are infection-induced struvite. *Staphylococcus spp* is the most common urease-producing microbial organism associated with infection-induced struvite uroliths; however, other urease-producing organisms include *Proteus spp, Enterococcus spp, Streptococcus spp,* and uncommonly *Klebsiella spp*. Occasionally, other minerals such as urate and cystine form uroliths in the upper urinary tract.

THERAPY

Not all nephroureteroliths require treatment especially by surgery. In one study, presence of nephroliths without ureteral obstruction was not associated with progression of chronic kidney disease.[2] If uroliths are composed of struvite or urate, medical dissolution can be attempted.[4] Indications for treatment of nephroureteroliths include obstruction to urine flow, increase in urolith size and/or number despite appropriate therapy, if uroliths are composed of a mineral amenable to medical dissolution therapy, recurrent or persistent urinary tract infections, compromise of renal function due to uroliths, recurrent or persistent clinical signs associated with renal pain (e.g. vomiting or abdominal pain), or severe hematuria. Many cats with nephroureteroliths already have renal compromise. Removal of nephroureteroliths in cats with chronic kidney disease, especially if ureteroliths, improves survival; however, there is an associated morbidity and mortality associated with surgery.

Medical management[5]

At present, there is no means to medically dissolve calcium oxalate uroliths. Infection-induced struvite uroliths can be dissolved by feeding a low magnesium and phosphorous diet that induces aciduria and administering an appropriate antimicrobial agent. Average dissolution time for infection-induced struvite uroliths is 8 weeks; however, nephroliths often take longer to dissolve. Sterile struvite uroliths can be dissolved by feeding a low magnesium and phosphorous diet that induces a diuresis; administering an antimicrobial agent is unnecessary. Average dissolution time is 2-4 weeks; however, nephroliths often take longer to dissolve. Urate uroliths can be dissolved by feeding an ultra-low protein diet that induces alkaluria and administering allopurinol. This protocol is effective in approximately 40% of dogs with urate uroliths; however, is not effective in dogs with portovascular anomalies. In most cases, uroliths dissolve in 4 weeks. Dissolution time is variable and often several months. Because urate uroliths are radiolucent typically, monitoring dissolution requires ultrasonography, CT, or contrast excretory urography Allopurinol has not been evaluated in cats.

In human beings, alpha blocking agents have been shown to promote movement of ureteroliths into the bladder (called medical expulsion therapy) especially if the uroliths are in the distal 1/3 of the ureter; however, the effectiveness in cats has not been evaluated.[6] Glucagon has also been used to facilitate passage of ureteroliths after fragmentation by lithotripsy; however, was not effective in cats with ureteroliths.[7] Alpha-blocking agents (prazosin: 0.5 mg PO q12h) and anti-inflammatory drugs (prednisone: 0.5 mg/kg PO q12h, dexamethasone: 0.05 m/kg IV q12h) may facilitate passage with fluid therapy. In one study, ureteroliths passed in 9 out of 52 cats treated with IV fluids and diuretics.[8] Amitriptyline (0.5 – 2.0 mg/kg PO q24h) has been shown to relax ureteral smooth muscle and facilitate expulsion of ureteroliths in humans.[9] Spontaneous passage of ureteroliths has been shown to occur. Ureteroliths have also been reported to move retrograde up to 4 cm in the ureter even back into the renal pelvis resulting in improvement of urine flow and azotemia.[3]

Dialysis[10-12]

Dialysis is the process of removing solutes and/or water from one solution (plasma) to another solution (dialysate) by an osmotic gradient through a semi-permeable membrane. The 3 types of dialysis are peritoneal dialysis, hemodialysis, and enteric dialysis using an oral sorbent. Acute renal failure is the prevailing indication for dialysis in cats, although hemodialysis may be used to manage chronic kidney disease.[13] Conditions where dialysis is indicated include: acute renal failure, pre-surgical treatment of uremia (e.g. uroabdomen or renal transplantation), electrolyte abnormalities unresponsive to medical therapy (e.g. hypernatremia and hyperkalemia), metabolic and acid-base disturbances (e.g. hepatic encephalopathy and severe metabolic acidosis), severe fluid overload, and toxicities where the toxin is dialyzable (e.g. ethylene glycol, acetaminophen or salicylate toxicity, barbiturate toxicity, and grape/raisin toxicity). Dialysis involves the use of one type of semi-permeable membrane that allows the passage of small molecules while preventing the transfer and loss of large molecules. In peritoneal dialysis, the peritoneal membrane, interstitial tissue, and capillary endothelium act as the membrane. In hemodialysis, blood is circulated outside of the body through a dialyzer.

Surgical management[14]

Surgical removal is not necessary for all upper urinary tract uroliths. Indications for surgery are similar to those listed previously. Surgery is not indicated for relatively small, non-obstructive upper urinary tract uroliths. Small upper urinary uroliths may pass into the urinary bladder. It is unknown what a safe time period to allow for ureteroliths passing without resulting in irreversible renal damage. Renal function does not recover n dogs with unilateral ureteral ligation for 40 days. If a cat is ill due to ureteral obstruction, it is not probably appropriate to wait for the ureteroliths to pass or to medically manage them for a prolonged period of time if renal function is to be preserved. If a urinary tract infection is present, removal of the upper urinary tract urolith may be necessary to control the infection; however, it is not possible to determine prior to surgery if the uroliths are the source of infection. If bilateral upper urinary tract surgery is required, the procedures should be staged with an approximate separation of 4 weeks in order to re-evaluate renal function and to allow recovery from the first procedure. In general, the side with the most renal function should be operated first in order to preserve as much renal function as possible. Nephroliths can be removed either nephrotomy or by pyelolithotomy; however with either technique, it is not always possible to retrieve all uroliths. Nephrotomy should be avoided, if possible, because it causes renal damage and loss of nephrons. Pyelolithotomy is preferred because renal parenchymal tissue is not incised and so loss of renal function is less likely to occur when compared with nephrotomy; however, pyelolithotomy cannot be performed unless the renal pelvis is dilated. Often it requires microsurgical technique Ureterotomy is technically difficult to perform because of the diameter of the ureter, especially in cats and stricture formation often occurs Nephrectomy and ureterectomy are indicated for a severely hydronephrotic, non-functional kidney.

Comparison of medical vs surgical management

In a retrospective analysis of medically versus surgically managed patients with ureteroliths, surgically managed patients tended to do better over a longer time. There was a recurrence rate of 40%.[8]

Lithotripsy[15]

Extracorporeal shock wave lithotripsy (ESWL) is a standard of care for many human patients with upper urinary tract uroliths, and has been performed successfully in cats.[15] Struvite is most easily fragmented, followed by calcium oxalate, then urate, then cystine. When upper urinary tract uroliths fragment, the small fragments pass into the urinary bladder over several weeks; their passage is not instantaneous. Animals that have undergone ESWL should have periodic examinations including serum biochemical analysis, urinalysis, and radiographs to monitor for successful passage of fragments and to detect any compromise of renal function due to the lithotripsy procedure. Acceptable candidates are those with nephroliths smaller than 2-3 cm in their greatest dimension or ureteroliths. With bilateral nephroliths, both kidneys are treated at the same time, unless there is concern about compromising renal function further. Feline upper urinary tract uroliths appear to be more difficult to fragment with ESWL than in dogs, and renal function is more likely to be compromised. Successful fragmentation has been reported to occur in < 20-25% of feline cases.[15,16] Although renal function was normal in 4 healthy cats undergoing lithotripsy,[17] lithotripsy of clinical cases of nephroureterolithiasis in cats suggests that many cats, particularly those with pre-existing renal disease, experience renal function compromise or worsening of their renal failure. While lithotripsy is considered safer and less invasive than surgical removal of upper urinary tract uroliths, there are risks. Abdominal pain, hemorrhage, and bruising of the kidneys occur, and hematuria may be observed immediately after the procedure. More significant hemorrhage within or around the kidney may occur in some cases. Residual stone fragments often take several weeks to move from the kidney into the urinary bladder. Transient or permanent ureteral obstruction can occur. If permanent and progressive ureteral obstruction occurs, it requires re-treatment by lithotripsy or surgical intervention. Uncommon complications include pancreatitis, bowel irritation, hemolysis, and systemic hypertension.

Percutaneous nephrolithotomy

A rigid scope can be inserted through the renal parenchyma into the dilated renal pelvis. The nephrolith is visualized and either retrieved or fragmented using laser lithotripsy and fragments retrieved.

Ureteral stent

In patients where nephroureteroliths cannot be managed surgically, urinary diversion may be accomplished by placing a ureteral stent.[18] A double pig-tailed stent is placed surgically, fluoroscopically, or via cystoscopy. One of the pig-tails is placed so that it is within the dilated renal pelvis and the other pig-tail is placed so that it is within the urinary bladder. The body of the stent connects the 2 pigtails and provides diversion of urine flow around the obstructive ureteroliths

Ureteral ileal graft

Recently, resection of an obstructed ureter and replacement with a vascularized ileal graft was described in a cat with ureteral obstruction and progressive renal azotemia.

Renal transplantation[19]

A retrospective study of 19 cats with renal failure associated with calcium oxalate urolithiasis has been published There were 13 females and 7 males All cats were azotemic and 17 were anemic Hypercalcemia was present in 7 cats Mean duration of survival in all cats was 605 days 8 cats were alive 282-1,005 days (median = 1,305 days) 11 cats died 2-1,197 days (median = 300 days) 5 cats formed uroliths in their allograft kidney 2 were hypercalcemic.[20]

MANAGEMENT STRATEGIES

Unfortunately, there are no standard protocols or large long term studies that guide management of nephroureterolithiasis in cats; however, recommendations can be made.

- Most feline nephroureteroliths are composed of calcium oxalate; however, other minerals exist that may be amenable to medical dissolution
- Cats with chronic kidney disease should undergo periodic abdominal radiography particularly if there is an abrupt decline in renal function because of the association of chronic kidney disease with nephroureterolithiasis
- The presence of nephroliths does not necessitate aggressive management. Appropriate preventative therapy and monitoring may be all that is required. Nephroliths are not associated with progression of chronic kidney disease; however, ureteroliths are associated with decreased survival time.
- Because often nephroureteroliths are associated with chronic kidney disease and it is unknown how long they have been present, aggressive therapy may not result in improvement in azotemia
- Cats that present with chronic kidney disease and uremia associated with ureteral obstruction should be managed medically first for 2 to 4 days in order to see if ureteroliths will be expelled or to stabilize the patient for intervention. This may involve medical expulsion therapy using anti-inflammatory drugs and ureteral smooth muscle relaxants (amitriptyline or alpha-blockers) or dialysis
- In cats with ureteroliths, surgical management by ureterectomy and neoureterocystostomy is an effective means if the ureter is dilated and there is at least 1 cm of proximal ureter available.
- In cats with ureteroliths where surgical intervention is not possible, ureteral stenting is a reasonable option

REFERENCES

1.Lekcharoensuk C, Osborne CA, Lulich JP, et al. Trends in the frequency of calcium oxalate uroliths in the upper urinary tract of cats. J Am Anim Hosp Assoc 2005;41:39-46. 2.Ross SJ, Osborne CA, Lekcharoensuk C, et al. A case-control study of the effects of nephrolithiasis in cats with chronic kidney disease. J Am Vet Med Assoc 2007;230:1854-1859. 3.Dalby AM, Adams LG, Salisbury SK, et al. Spontaneous retrograde movement of ureteroliths in two dogs and five cats. J Am Vet Med Assoc 2006;229:1118-1121. 4.Lulich JP, Osborne CA, Albasan H. Canine and feline urolithiasis: diagnosis, treatment, and prevention In: Bartges J,Polzin DJ, eds. Nephrology and Urology of Small Animals. West Sussex UK: Wiley-Blackwell, 2011;687-706. 5.Bartges JW. Diseases of the urinary bladder In: Birchard SJ,Sherding RG, eds. Saunders Manual of Small Animal Practice. 2nd ed. Philadelphia: WB Saunders, 2000;943-957. 6.Singh SK, Agarwal MM, Sharma S. Medical therapy for calculus disease. BJU Int 2011;107:356-368. 7.Forman MA, Francey T, Fischer JR, et al. Use of glucagon in the management of acute ureteral obstruction in 25 cats. J Vet Intern Med 2001;18:390-391. 8.Kyles AE, Hardie EM, Wooden BG, et al. Management and outcome of cats with ureteral calculi: 153 cases (1984-2002). J Am Vet Med Assoc 2005;226:937-944. 9.Achar E, Achar RA, Paiva TB, et al. Amitriptyline eliminates calculi through urinary tract smooth muscle relaxation. Kidney Int 2003;64:1356-1364. 10.Langston C. Hemodialysis In: Bartges J,Polzin DJ, eds. Nephrology and Urology of Small Animals. West Sussex UK: Wiley-Blackwell, 2011;255-285. 11.Acierno MJ. Continuous renal replacement therapy In: Bartges J,Polzin DJ, eds. Nephrology and Urology of Small Animals. West Sussex UK: Wiley-Blackwell, 2011;286-292. 12.Labato MA. Peritoneal dialysis In: Bartges J,Polzin DJ, eds. Nephrology and Urology of Small Animals. West Sussex UK: Wiley-Blackwell, 2011;293-305. 13.Cooper RL, Labato MA. Peritoneal dialysis in cats with acute kidney injury: 22 cases (2001-2006). J Vet Intern Med 2011;25:14-19.

14.Tobias K. Renal and ureteral surgery In: Bartges J,Polzin DJ, eds. Nephrology and Urology of Small Animals. West Sussex UK: Wiley-Blackwell, 2011;596-616. 15.Adams LG, Goldman CK. Extracorporea shock wave lithotripsy In: Bartges J,Polzin DJ, eds. Nephrology and Urology of Small Animals. West Sussex UK: Wiley-Blackwell, 2011;340-348. 16.Lane IF. Lithotripsy: an update on urologic applications in small animals. Vet Clin North Am Small Anim Pract 2004;34:1011-1025, vii. 17.Gonzales A, Labato M, Solano M, et al. Evaluation of the safety of extracorporeal shock-wave lithotripsy in cats. ACVIM Forum 2002;810. 18.Weisse C, Berent A, Todd K, et al. Evaluation of palliative stenting for management of malignant urethral obstructions in dogs. J Am Vet Med Assoc 2006;229:226-234. 19.Adin CA. Renal transplantation In: Bartges J,Polzin DJ, eds. Nephrology and Urology of Small Animals. West Sussex UK: Wiley-Blackwell, 2011;306-328. 20.Aronson LR, Kyles AE, Preston A, et al. Renal transplantation in cats with calcium oxalate urolithiasis: 19 cases (1997-2004). J Am Vet Med Assoc 2006;228:743-749.

TWO ARE NOT BETTER: FELINE CHRONIC KIDNEY DISEASE AND DIABETES MELLITUS

Claudia A. Kirk, DVM, PhD, DACVN, DACVIM
Knoxville, TN

INTRODUCTION

Chronic kidney disease (CKD) and diabetes mellitus (DM) are common disorders of aged cats. Logically, cats affected concurrently with both disorders occur relatively frequently in this age group. Clinical signs of both disorders include polyuria/polyphagia, weight loss, lethargy, poor coat and weakness. A key clinical feature differentiating disorders is the presence of a marked increase in appetite in cats with DM compared to a gradual decrease in food intake common in CKD. In later disease states or with diabetic complications such as Diabetic Ketoacidosis (DKA)or Hyperglycemic Hyperosmolar Syndrome (HHS), appetite may no longer be a differentiating clinical sign. Fortunately, both disorders are readily identified by biochemical changes identified in standard diagnostics: CBC, biochemical profile and urinalysis. Due to similarities in clinical signs and complex metabolic derangements, the management of cats with concurrent CKD and DM can pose unique considerations and outcomes that are less favorable than that of each individual disorder.

Risk factors and Prevalence

The common risk factors for development DM include advancing age, obesity, indoor environment, inactivity, male gender, neutered reproductive status and breed (Burmese).[1,2] Risk factors for CKD in cats include advancing age and a reduced body condition score.[2]. Age alone is not the only shared risk between the disorders as metabolic derangements increases the risk of developing the other disease concurrently. Complications of DM include the development of diabetic nephropathy while CKD contributes to insulin resistance and the risk of DM.

The prevalence of CKD in all cats has been reported at 2.2% of cats in the US[3] and approximately 33% in cats over 12 years.[4] Diabetes mellitus has been reported at 0.5% or 1 in 200 in domestic cat breeds and up to 2 % (i.e. 1 in 50) in the Burmese cats.[5] While studies evaluating the rate of reciprocal occurrence of disease are not available, Goossens[6] reported 13% of diabetic cats were concurrently affected with CKD. Necropsy results from cats in this same study reported 25% (13 of 53) cats had evidence of glomeruloscerosis pointing to undiagnosed CKD in several cats.

Diabetic nephropathy takes 10-20 years to develop in humans. While diabetic nephropathy is reported as an uncommon complication in cats with DM, these histopathological changes suggest diabetic nephropathy may be overlooked. Conversely, 2 studies report an increased risk of cats with renal disorders developing DM. Burmese cats treated for CKD have a greater risk for DM[7]as did cats undergoing renal transplant.[8]

METABOLIC INFLUENCE ON COMORBIDITIES

Metabolic alterations and comorbid conditions of CKD and DM have significant overlap. Common shared complications include: negative energy balance with weight loss, dehydration, hypertension, metabolic acidosis, electrolyte imbalance, proteinuria, and oxidative stress and urinary tract infections. Other factors are specific to CKD (e.g. Azotemia, phosphate retention, and hyperkalemia) or DM (ketosis, hyperglycemia, glycosuria, DKA, HSS) and can negatively influence outcome or complicate treatment plans if ignored.

Water balance

Loss of renal concentrating ability in CKD and osmotic diuresis in DM result in profound polyuria and polydipsia in both disorders. Maintaining adequate hydration status is critical in the face of profound water loss. Chronic dehydration reduces renal perfusion and leads to activation of renin-angiotension system and can contribute to intrarenal or systemic hypertension and progressive renal damage. Dehydration and poor renal perfusion increases the risk of DKA and HHS in diabetic cats subsequent to the retention of metabolic acids and/or glucose.

Stage I-II CKD may not be readily recognized in cats with DM since loss of concentrating ability may go unnoticed. Cats with DM typically have urine specific gravity levels between 1.025-1.035 despite profound polydipsia. The osmotic effect of urine glucose at 2% (4+) can increase the USG reading by 0.001and mask renal concentrating defects and overt isosthenuria.[9]

Energy Balance

Weight loss is common in cats with DM and CKD. Decreased food intake in response to hyporexia, nausea and altered taste is responsible for weight loss in CKD. Cats with DM have cellular energy deficits due to low cellular glucose entry and activation of metabolic pathways to support energy production to the cell (i.e. lipolysis, glycogenolyis and gluconeogenesis). Lean and fat tissues stores are utilized to meet these energy demands. Endogenous proteins and amino acid turnover results in lean mass loss, generation of metabolic acids, and increased urea production. In cats with CKD, negative energy balance contributes to azotemia, metabolic acidosis, weakness, and lethargy via increased protein turnover. In the diabetic cat, amino acids contribute substrate for hepatic gluconeogenesis to increase blood glucose. With negative energy balance, the cat with DM mobilizes fat stores in response to low insulin activity. Beta- oxidation of fatty acids increases ketones production as a byproduct of lipid metabolism. Left unchecked, negative energy balance can further complicate both CKD and DM by contributing to lean mass loss, azotemia, metabolic acidosis and ketosis.

Acidosis

As noted above, metabolic acidosis and ketosis are common in both CKD and DM. Cats with concurrent DM and CKD are doubly challenges as increased levels of metabolic acids are produced in both conditions. The ability to excrete hydrogen ions and resorb bicarbonate is lost as kidney function declines resulting in retention of nonvolatile acids. Cats respond to acidosis with increased renal ammoniagenesis which can contribute to renal damage via complement activation.[10] A vicious cycle of renal damage and progressive azotemia results subsequent to increased renal ammonia levels. With acidosis the risk of hypokalemia increases as potassium ions move to the extra cellular fluid and are excreted in the urine. Acidosis is a major contributor to progressive renal disease in cats and occurs in up to 15% of cats with Stage III CKD and up to 53% of cats with Stage IV disease.[10] In the cat with DM, uncompensated acidosis with a blood pH < 7.35 and ketosis contributes to the clinical decompensation and DKA[11].

Electrolyte alterations

Electrolyte abnormalities are common in cats with CKD and DM. Acid-base disturbances, hyperglycemia, and osmotic diuresis contribute to the whole body deficit of potassium, magnesium, and phosphorus. As CKD severity increases, phosphorus retention counterbalances phosphorus loss typical of DM. Serum electrolyte levels may not reflect whole body stores upon initial presentation. Following rehydration, correction of acidosis and administration of insulin, redistribution of these electrolytes into the cell can result serious decreases in serum levels. Such deficits contribute to muscle weakness (i.e. hypokalemic and hypomagnesemic myopathy), arrhythmias (secondary to hypomagnesemia), hemolysis (due to hypophosphatemia), anorexia (due to hypokalemia) and contribution to ongoing renal damage (i.e. hypokalemic nephropathy). As CKD progresses to failure retention of potassium, phosphorus may occur.

Proteinuria

Proteinuria is a common complication in cats with CKD[10] or DM[9]. Because bacterial urinary tract infections are common with both diseases, urinalysis and culture should be performed to rule out such infections. In cats without a urinary tract infection, urine protein:creatinine ratios greater than 0.4 are considered elevated.[10] Diabetic nephropathy, hypertension and primary glomerular damage are potential factors in proteinuria. Control of hypertension should be implemented using angiotensin converting enzyme inhibitors such as benazepril. However, reduction of dietary protein and fatty acid supplementation is often recommended. The later strategy may at odds with nutritional strategies for DM.

Insulin resistance

The goal of managing cats with DM is to correct metabolic abnormalities and regulate blood glucose levels. A number of cats will active remission reduction of insulin resistance via weight reduction and reversal of glucose toxicity. Most cats with DM will require exogenous insulin along with dietary therapy. Diabetic cats with CKD tend to be a greater challenge to regulate with insulin therapy due to mild to moderate insulin resistance associated with the CKD.[6,9] Reduced insulin sensitivity has been noted in cats with CKD.[6,8,9] Decreased renal function may result in decreased renal insulin clearance, prolonged duration of insulin effect and risk for hypoglycemia in cats receiving exogenous insulin. Insulin resistance, poor glucose regulation and risk for hypoglycemia have been reported in both people and cats with DM and CKD.[9] Furthermore, evaluating response to insulin therapy is challenging in the cat with DM and CKD. Improvement in PU and PD is an important clinical outcome supporting adequate glucose control. With CKD, PU/PD may not improve significantly in well controlled cats with DM. Home glucose monitoring becomes important in assessing these patients as weight loss may reflect CKD and not glucose control in DM .

Diabetic Nephropathy

Diabetic Nephropathy is a serious complication in people with DM. Without renal biopsy, the etiology of CKD is not always evident. The prevalence CKD with DM has been reported between 13-19% from clinical studies[6,12] but 25% on histopathology.[6] Diabetic nephropathy results from glomerular damage leading to glomerularsclerosis. The mechanism is thought to be caused by chronic intraglomerular hypertension and renal hyperperfusion secondary to chronic glycosuria. With increased glomerular pressures, messangial expansion and altered glomerular blood flow contribute to glomerularsclerosis.[9] The hallmark of diabetic nephropathy is proteinuria and progressive renal damage leading to azotemia and CKD. Prevention and control occurs by providing excellent glucose control. Thus, the challenge, as CKD contributes to insulin resistance and poor glucose regulation, poor diabetic control contributes to ongoing renal damage and progressive CKD. Outcome studies in cats with DM found cats with CKD had mediocre to poor glucose regulation[6]. Survival analysis indicated median survival cats with good to mediocre control was 26 months compared to a median survival time of 17 months in cats with mediocre to poor control.[6]

TREATMENT CHALLENGES

Determining the priority for therapeutic selections is often a challenge with comorbid conditions. Detailed description of DM and CKD management is beyond the scope of this paper and the reader is referred to reviews specific to this area. Diabetes management targets glucose control through diet and exogenous long-acting insulin (e.g. PZI, glargine or detimer) in most cases. Control of CKD relies on dietary therapy and other treatment bases on severity and clinical signs include phosphate binders, potassium supplementation, bicarbonate therapy, H2 blockers, antiemetics, appetite stimulants, SQ fluids, and ACE inhibitors. While there is little contraindication to most treatment strategies between CKD and DM, a significant differences in ideal nutritional composition of foods for CKD and DM do exist.

Diet

Current recommendations for nutritional management of cats with DM suggest foods high in protein, low in carbohydrates and moderate in fat. Protein levels in therapeutic foods designed for diabetic management vary from 40% to57% dry matter. Because phosphorus is a component of animal proteins, phosphorus levels are 0.7% to 1.0% dry matter are common. Carbohydrate recommendations vary but general acceptance of levels approximately 15% or less of metabolic energy have been suggested.[5] Limited information is available about optimal carbohydrate source. A singles study evaluated corn and sorghum compared to rice and found overall glucose levels greater in rice suggesting rice has a higher glycemic index compared to corn or sorghum.[5] Other nutritional factors such as enhanced vitamins and mineral compensate for urinary losses associated with high volume output. Antioxidant supplementation is considered valuable in reducing oxidative stress that contributes to several diabetic complications, including diabetic nephropathy. Fiber enhancement is included in many diets and has been shown to delay glucose absorption across the gut, depending on fiber amount and structural properties.

Nutritional management of renal disease has been shown to increase survival in cats by nearly 2.5 years.[13] Nutrition is the mainstay of therapy for cats in Stage II - III CKD to help prevent progression of renal disease and critical to late stages (IV and V) to ameliorate clinical signs associated with uremia. Key nutritional factors include protein restriction, high energy density with moderate to increased carbohydrates and fats, phosphate restriction, omega-3 fatty acid supplementation, potassium supplementation, dietary buffers, sodium control and fiber enhancement. Similar to diet composition for DM management, enhanced levels of vitamins, minerals and antioxidants support increased losses in the urine help squelch oxidants that contribute to ongoing renal damage.[14] Soluble fiber supplements may provide some benefit by increasing urea loss in the feces when fermentation by gut microbes acts as a nitrogen sink following microbial biomass expansion. The therapeutic conflict between treatments lies in the nutrition difference between diets designed for DM ant that of CKD. Protein, carbohydrate, fat and phosphate levels targets are opposite for each disorder each disorder.

Elevated protein levels are beneficial in cats with DM in maintaining lean mass during weight loss and serving as an energy source in replacement for carbohydrate reductions. As cats are obligate carnivores, the increased levels appeal to those want a "wild-type" diet composition. Protein also helps to increase satiety and may benefit by supporting weigh loss. Carbohydrate restriction avoids postprandial blood glucose increases helping to control hyperglycemia, stabilize insulin dose and improve glucose control. Cats fed restricted carbohydrate foods have a greater rate of diabetic remission than more traditional foods for DM with high fiber and moderate carbohydrates.[15] Control of dietary fat helps reduce the energy density of the diet and limit the role of dietary fat in lipotoxicity. Fiber benefits in helping delay nutrient absorption across the intestine to modulate postprandial glucose uptake and improve satiety. More traditional high fiber therapeutic foods are still appropriate as they control postprandial glucose uptake across the gut and lowers caloric density of the food. High fiber therapeutic foods are very helpful in achieving weight loss by enhancing satiety and cutting calories while still providing a larger food portion.

High protein foods contribute to azotemia, increased metabolic acid production, renal ammoniagenesis, and phosphorus intake in cats with CRD. As such, the high protein low carbohydrate foods designed for cats with DM are contraindicated in cats with CRD. The elevated fat and carbohydrate in renal therapeutic foods maintain energy balance with a smaller food volume thereby helping to reduce nausea, anorexia and food aversions that come with CKD. As protein levels are reduced in diets designed for CKD, carbohydrate levels must naturally increase. Carbohydrate levels in the form of rice are highly digestible, palatable and support energy balance in cats with CKD.

Generally, CKD has taken precedence in priority of treatment considerations with insulin dose adjusted as needed to achieve glucose regulation. This may not always be the best choice if control of progression of renal damage and diabetic nephropathy is dependent on glucose regulation. It is important to have good understanding of the nutritional profile of therapeutic foods to avoid complications that may arise in feeding nutrient levels contraindicated for that disease.

CONCLUSIONS

Diabetes and CKD are common comorbid conditions in aged cats. It is important to carefully screen and monitor for the occurrence of each disease as metabolic changes increase the risk for development of comorbidity in each disorder. Multiple factors contribute to more difficult treatment of cats with both CKD and DM. Insulin resistance from CKD and diabetic nephropathy in DM contribute to poor control of both conditions. While most therapeutic modalities are compatible in treating DM and CKD, nutritional strategies are often in conflict. Typically renal disease will take precedence and DM control can be accomplished by adjusting insulin dose.

REFERENCES

[1]Scarlett J, et. al. J Am Vet Med Assoc 1998; 212:1725, [2]Lund E, et. al. J Am Vet Med Assoc. 1999: 214:1336, [3]Kirk C, et. al. Proceed Waltham Internl Symp, abst. 2001; [4]Lulich J, et. al. Compen Contin Educ 1992; 14: 127, [5] Rand J, et. al. JN 2004; 134:2072, [6]Goossens M. et al. J Vet Intern Med 1998;12:, [7]Lederer R, et. al. J Vet Intern Med 2003; 17:433, [8] Case, et. al. J Am Vet Med Assoc 2007;15:230, [9] Feldman E and Nelson R eds., Canine and Feline Endocrinology and Reproduction 2004 Elsevier; 539, [10]Elliot D, in Consultations in Feline Internal Medicine 2010, August J ed , Elsevier; 137.[11] Hess R, in Consultations in Feline Internal Medicine, 2010, August J ed , Elsevier; 297. [12]Goossens M, et al. J Vet Intern Med abst 1995;9:181, [13]Ross S, et. al. J Am Vet Med Assoc 2006; 229:949, [14]Yu S, et. al, Vet Res Commun 2006; 30(4):403, [15] Bennet N, et. al. J Feline Med Surg 2006; 8 (2):73.

TWO ARE NOT BETTER: FELINE CHRONIC KIDNEY DISEASE, INFLAMMATORY BOWEL AND LIVER DISEASE

Jacqueline C. Whittemore, DVM, PhD, DACVIM
Knoxville, TN

CHRONIC KIDNEY DISEASE

Chronic kidney disease (CKD) occurs commonly, affecting up to 20% of cats in some populations. The prevalence of CKD increases with increasing age. Due to the increased use of geriatric wellness testing, many patients are diagnosed with CKD prior to development of clinical signs of disease. Clinical signs, when present, may be vague or insidious in onset. Common abnormalities include alterations in behavior, lethargy, hyporexia, weight loss, vomiting and polyuria/polydipsia (PUPD). Physical examination may be unremarkable. Alternately, abnormalities that may be identified include dehydration, weight loss, poor body condition score, a poor or ill-kempt haircoat, pale mucous membranes, halitosis, abnormal renal palpation, and retinal hemorrhages or separation. In cases with significant PUPD, a large bladder may also be palpable. The most common histopathological finding in cats with CKD is lymphocytic-plasmacytic interstitial nephritis.

Staging of CKD

Chronic kidney disease should be staged using the International Renal Interest Society (IRIS) guidelines to facilitate optimal patient care and medical communication. The initial stage is assigned based on the patient's creatinine concentration (Stage 1: <1.6 mg/dL, 2: 1.6-2.8 mg/dL, 3: 2.9-5.0 mg/dL, and 4: >5 mg/dL). This determination should be made in the fasted and appropriately hydrated patient over a period of days to weeks. In cases that are presented while ill or in uremic crisis, staging should be postponed until both the patient and the creatinine concentration have been stabilized.

Because of their impact on ongoing nephron loss and long-term patient survival, patients should be substaged based on the presence or absence of proteinuria and systemic hypertension. The urine protein:creatinine ratio (UPC) is assessed after elimination of pre- or postrenal causes of protein loss. Patients are categorized as nonproteinuric (NP: UPC <0.2), borderline proteinuric (BP: UPC 0.2-0.4), or proteinuric (P: >0.4). Indirect determination of blood pressure should be performed by an experienced technician or veterinarian in a calm setting. Patients are then substaged as no/minimal (0: systolic <150 mmHg), low (1: systolic 150-159 mmHg), moderate (2: systolic 160-179 mmHg) or high (3: systolic ≥180 mmHg) risk. Patients with complications of hypertension are denoted with a 'c'. Finally, if a substage is altered through treatment, the current substage is used but with the modifier (T) added to denote that the categorization is the result of treatment. Examples of the use of the IRIS staging system in daily practice may be found on the IRIS website (http://www.iris-kidney.com).

Diagnostic testing

Initial diagnostic testing for a patient with CKD includes indirect determination of arterial blood pressure, complete blood count, serum or plasma biochemical profile including electrolytes, feline leukemia virus and immunodeficiency virus screening, serum thyroid hormone measurement, urinalysis, and quantitation of renal proteinuria. Urine culture, abdominal radiographs, and abdominal ultrasound are also indicated to identify clinically relevant or potentially correctable sequelae (bacterial cystitis, ascending pyelonephritis, urolithiasis, etc), to assess disease severity, and to identify concurrent non-renal issues. Evaluation of parathyroid hormone and ionized calcium concentrations may be indicated on a case-by-base basis. Although patients are at risk of B-vitamin deficiencies due to PUPD, hyporexia, vomiting and diarrhea, testing is not generally indicated because deficiencies generally correct with dietary change to a prescription renal diet.

Therapeutic considerations

The goals for management of a patient with CKD include maintenance of a normovolemic state and appropriate body condition score (BCS), control of factors that perpetuate renal damage and nephron loss (hyperphosphatemia, hypokalemia, proteinuria, hypertension, and urinary tract infection), and management of factors that impact patient quality of life (uremia, acidemia, hypergastrinemia, gastritis, and anemia).

The hallmark therapy for management of IRIS Stage 2 or greater CKD is transition to a prescription renal diet. Such diets have a relatively high caloric density due to increased fat content (to maintain body mass), provision of highly digestible protein in restricted quantities (to maintain a neutral or positive nitrogen balance without excessive generation of nitrogenous wastes), phosphorus restriction (to limit hyperphosphatemia and secondary hyperparathyroidism), potassium citrate supplementation (to correct intracellular hypokalemia and improve acid-base balance), omega fatty acid supplementation (to decrease inflammation), and some diets also feature sodium restriction (to mitigate systemic hypertension). Supplemental fluid therapy may be provided by feeding canned food diets, strategic use of circulating water bowls and other attractions, and administration of subcutaneous fluids. Placement of feeding tubes may be indicated in some patients to allow ongoing replacement of fluid deficits, ensure adherence to a renal diet and maintenance of a neutral or positive energy balance, and administer medications without degradation of the human-animal bond.

Additional therapies may include use of angiotensin-converting enzyme inhibitors or calcium channel blockers (to control proteinuria and hypertension), H2 receptor blockers like famotidine (to mitigate negative effects of hypergastrinemia), phosphate binders (to reduce intestinal phosphorus absorption and limit secondary renal hyperparathyroidism), bioavailable oral iron supplementation with darbopoietin or erythropoietin (to correct non-regenerative anemia due to erythropoietin deficiency), B vitamins (to correct deficiency and promote appetite), and appetite stimulants like mirtazapine (to assist in maintenance of a neutral or positive energy balance and decrease nausea and emesis). Though calcitriol administration improves survival in dogs with CKD and renal hyperparathyroidism, available research does not support improved outcome in cats.[1] Probiotics containing urease-producing bacteria may be utilized to trap nitrogenous products in the gastrointestinal lumen, though documented efficacy in cats with naturally-occurring CKD is lacking.

FELINE CHRONIC ENTEROPATHIES AND LIVER DISEASE

Chronic enteropathies and inflammatory liver disease occur commonly in cats. Chronic enteropathies (CE) are distinguished by the presence of chronic (>3 weeks) gastrointestinal signs that may be relapsing and remitting, elimination of other causes of clinical signs, and failure to respond to symptomatic therapies, including parasiticides and gastrointestinal protectants. Common clinical signs include lethargy and change in activity level; weight loss accompanied by a decreased, normal, or increased appetite; episodic to persistent vomiting of bile, phlegm, or food; and small, large, or mixed bowel diarrhea. Physical examination may be normal. Alternately, clinical dehydration, weight loss, poor body condition score, a poor or ill-kempt haircoat, or thickened intestinal loops may be identified.

Feline inflammatory liver disease is currently subcategorized into neutrophilic and lymphocytic cholangitis. Neutrophilic cholangitis is believed to result from ascending bacterial infection of the biliary tree. Concurrent inflammatory bowel disease (IBD) and pancreatitis have been documented in one-third of cats with neutrophilic cholangitis.[2] It is postulated that IBD alters biliary flow and function of the Spincter of Oddi, allowing bacterial reflux into the biliary tree. Although it has been postulated that lymphocytic cholangitis results from chronic neutrophilic cholangitis, repeated studies have failed to identify a bacterial association[3] and the disease is currently believed to result from immune dysregulation. Cats with neutrophilic cholangitis typically present for acute onset of malaise, vomiting, weight loss, hyporexia, and fever. Clinical signs are generally more indolent in cats with lymphocytic cholangitis. Based on comparison of clinical and necropsy data, cats with neutrophilic cholangitis appear more likely to die from their disease.[2]

Diagnostic testing

Because CE are often diagnosed based on exclusion of other disease processes and response to therapeutic trials, diagnostic testing is often extensive. Initial testing includes indirect determination of arterial blood pressure, complete blood count, serum or plasma biochemical profile including electrolytes, feline leukemia virus and immunodeficiency virus screening, serum thyroid hormone measurement, urinalysis, screening for fecal parasites (direct smear, zinc sulfate, and sugar flotation, with or without testing for enteropathogens), and quantitation of pancreatic lipase immunoreactivity (PLI) concentrations. Common abnormalities on lab work include hyperproteinemia consistent with free water deficit, hypophosphatemia, increased liver enzyme activities, and increased PLI concentrations. Though uncommon in patients with CE alone, anemia is prevalent in cats with inflammatory liver diease.[2] Thiamin, cobalamin, and folate deficiencies occur commonly due to hyporexia, vomiting, intestinal malabsorption, and altered utilization due to liver dysfunction. Thiamin deficiency is particularly problematic because of its negative impact on appetite. Cobalamin deficiency has been identified in roughly 20% of cats with CE[4] and up to 80% of cats with well-differentiated alimentary lymphosarcoma,[5] also known as small cell lymphoma. Correction of hypocobalaminemia has been associated with significant improvements in body weight, clinical signs and biochemical abnormalities.[6] Abdominal radiographs and abdominal ultrasound are indicated to evaluate for evidence of gastrointestinal neoplasia, assess the likelihood of concurrent extra-gastrointestinal disorders, and identify clinically relevant sequelae. In cases with clinical signs suggestive of neutrophilic cholangitis, liver aspirates and cholecystocentesis should be performed to obtain samples for Gram stain and bacterial culture.

After completion diagnostic testing and stabilization of urgent diseases like neutrophilic cholangitis, therapeutic trials are performed to better characterize CE. Generally, patients are first treated with one or more paraciticides (such as fenbendazole) to rule out occult gastrointestinal parasitism. If clinical signs do not abate, an antibiotic trial (metronidazole, amoxicillin with clavulanate, or tylosin) may be performed to rule out antibiotic responsive disease. Finally, one or more diet trials (novel protein source, hydrolyzed protein, and highly digestible gastrointestinal diets) are generally performed to rule out food responsive disease (FRD) prior to collection of gastrointestinal biopsies. Depending on the study, up to 50% of cases with CE may be diagnosed with FRD; almost half of these cases may remain in remission after rechallenge with their primary diets. Although cats with FRD are typically younger than those with IBD and more commonly have lower intestinal signs, significant overlap is recognized. It is also noteworthy that histopathologic inflammation commonly occurs with FRD. For these reasons, performance of endoscopy prior to food trial will result in miscategorization of patients. In spite of widespread adoption of histopathologic standards for diagnosis, distinguishing severe IBD from small cell lymphoma remains challenging. Increased performance of ileoscopy and immunophenotypic analysis during histologic evaluation may improve diagnostic accuracy and prognostication, though both IBD and small cell lymphoma are commonly managed using prednisone and chlorambucil.

Therapeutic considerations

The goals for management of a patient with CE include dietary therapy to eliminate or mitigate clinical signs, maintenance of an appropriate BCS and normovolemic state, correction of factors that may impact response to therapy (hypocobalaminemia, secondary infections, etc). Immunomodulation is administered as necessary based on disease subcategorization and response to other therapies. Multiple studies have documented that histopathologic changes persist in spite of clinical resolution of FRD or IBD. The goal is therefore long-term maintenance with dietary therapy alone, not resolution of histologic abnormalities. Cats with FRD typically achieve clinical resolution of signs within 10-14 days of initiation of dietary change. In cases that fail dietary therapy alone and are diagnosed with IBD, prednisone or prednisolone is initially administered at 2 mg/kg/d, in concert with administration of a new hypoallergenic diet; immunosuppression is tapered over 8-10 weeks then discontinued. Between 60-80% of patients with CE can be anticipated to respond to dietary and medical management. Patients with refractory IBD or concurrent inflammatory liver disease, however, may require long-term steroid administration or treatment with other immunosuppressive agents, like chlorambucil or cyclosporine. Cats with neutrophilic cholangitis are initially treated with aggressive supportive care and antibiotic therapy, adjusted based on the results of bacterial cultures. After stabilization, further work-up for and treatment of underlying CE is indicated.

Patients with hypocobalaminemia should receive parenteral supplementation weekly until the deficit is corrected; this typically requires 4-6 weeks of supplementation. Some cats with severe disease may require ongoing supplementation. Additional therapies that may be considered include gastrointestinal protectants, appetite stimulants like mirtazapine (to assist in maintenance of a neutral or positive energy balance and decrease nausea and emesis), supplementation of omega-3 fatty acids (for anti-inflammatory effect), and administration of prebiotics in cases with large bowel signs (to increase bacterial production of polyunsaturated fatty acids, a preferred energy substrate for colonocytes). Because dysregulation of mucosal immune function has been postulated to play a significant role in FRD and IBD, probiotics – postulated to reset normal immune tolerance – are conceptually attractive. Controlled trials demonstrating efficacy remain lacking, however.

THE UNFORTUNATE INTERSECTION OF DISEASES

Because CKD and CE, with or without liver and pancreatic involvement, commonly occur in cats, only the extremely fortunate of clinicians will not eventually be faced with balancing the two disease processes. Each disease may have an insidious onset and result in clinical signs of malaise, ill-thrift, dehydration, weight loss, poor appetite, vomiting, and diarrhea. It may be difficult to determine clinically whether a cat's clinical signs reflect only CKD or are due to a combination of CKD with CE. Because different disease mechanisms underlie development of these clinical signs – and some signs may be self-perpetuating or worsen the severity of other signs – a patient with both diseases may have more severe clinical signs than would be present due to either disease alone. For example, dehydration due to obligate polyuria and increased intestinal losses may worsen uremia and acidemia, increasing nausea and malaise, hyporexia and vomiting, and negative energy and nitrogen balances. Diarrhea may increase the risk of urinary tract infection.

Diagnostic testing

Accurate staging of CKD in a patient with active CE with or without inflammatory hepatic disease may be difficult. Normalization of the cat's water balance may be challenging and require immunosuppression. Proteinuria may reflect glomerular or tubular dysfunction, but altered glomerular permselectivity due to inflammation may also play a role. Accurate assessment of phosphorus and calcium balance may be difficult because circulating phosphorus generally increases with development of renal secondary hyperparathyroidism as CKD progresses, while CE is associated with hypophosphatemia. Serum phosphorus concentration should be reevaluated after control of CE is achieved to assess the need for phosphate binders in patients with initially normal phosphorus concentrations. There is also particular risk of marked B-vitamin deficiencies in cats with concurrent CKD and CE or liver disease. For these reasons, assessment of B-vitamin concentrations is particularly important. Because immunosuppression may need to be continued indefinitely, increased surveillance for occult urinary tract infection is also indicated.

Therapeutic considerations

Successful long-term management for CKD and CE hinges on use of special dietary therapy, and no commercially-prepared diet currently exists that is ideal for management of both. Preparation of a home-cooked diet is currently the only way to achieve optimal balance of the nutritional requirements for both diseases, but this is untenable for many clients. Novel protein source and hydrolyzed diets contain substantially greater protein concentrations than diets formulated for the management of CKD. Based on studies comparing long-term outcome in CKD patients fed maintenance versus renal diets, it would be reasonable to anticipate earlier onset of uremic crises in cats with IRIS stage 2 or greater CKD that are fed non-renal diets. In the author's experience, successful clinical control of most patients with CKD and CE (due to either IBD or FRD) generally requires concurrent use of a renal diet and long-term immunosuppression.

Although immunosuppression for CE is generally performed using glucocorticoids, their use may worsen PUPD – resulting in difficulty maintaining patient hydration, increased risk of urinary tract infection, and overall poor quality of life. Glucocorticoids may also increase renal protein loss, complicating assessment and management of proteinuria. Therefore, alternate immunosuppressives may be preferable in the patient with concurrent CKD. Because of extensive first pass metabolism, budesonide may offer a balance between effective local glucocorticoid effect for CE and inflammatory liver disease and decreased systemic effects. Chlorambucil and cyclosporine have no glucocorticoid effects but may cause hyporexia and vomiting. Sequential trials may be necessary to determine the ideal immunomodulator for any given patient. Once clinical remission is achieved, the patient may be weaned to the lowest effective dose required to prevent clinical signs and maintain quality of life. Hypocobalaminemia can be corrected in most patients with CE and inflammatory liver disease through parenteral supplementation over 6-10 weeks, but ongoing loss of B-vitamins should be anticipated in at least a subset of patients with concurrent CKD. Chronic surveillance and supplementation may be necessary to ensure patients do not become recurrently hypocobalaminemic. In patients with concurrent CKD and inflammatory liver disease, it may be difficult to discern whether anemia reflects chronic disease or erythropoietin deficiency. When treating for erythropoietin deficiency, achieving adequate iron supplementation may be challenging in patients with multiple causes for nausea.

Appetite stimulants, such as mirtazapine, may be particularly helpful in mitigating malaise, hyporexia, nausea and emesis in cats with concurrent CKD and CE. In one placebo-controlled trial, a dose of 1.88 mg/cat was effective in stimulating appetite while minimizing side effects such as excessive vocalization, hyperactivity, and muscle tremors.[7] Based on a follow-up single-dose pharmacokinetic study in geriatric cats with and without CKD, a 48-hour dosing interval should be used in cats with CKD to avoid drug accumulation.[8] A non-significant increase in maximal mirtazapine concentration was noted in cats with CKD, which may indicate increased intestinal absorption. Because it is unknown how CE and inflammatory liver disease impact intestinal absorption and metabolism of mirtazapine, care should be taken to monitor for signs of drug accumulation in cats with these concurrent diseases. Probiotic administration may be considered in an effort to increase nitrogen trapping in the intestinal lumen while modulating mucosal immune function. Use of a quality product containing high concentrations of a variety of bacterial strains is recommended. Breaking of manufacturer capsules has been associated with lack of clinical efficacy and is not recommended.[9]

REFERENCES

[1]Hostutler RA, et.al. **JVIM** 2006;20:1307. [2]Clark JEC, et al. **JFMS** 2011;13:570. [3]Warren A, et al. **Vet Path** 2010; 48:627. [4]Jergens A, et al. **JVIM** 2010;24:1027. [5]Kiselow MA, et al. **JAVMA** 2008;232:405. [6]Ruaux CG, et al. **JVIM** 2005;19:155. [7]Quimby J, et al. **J Vet Pharmacol Ther** 2011;34:388. [8]Quimby J, **JVIM** 2011;25:985. [9]Rishniw M, et al. **JFMS** 2011;13:405.

GENERAL QUALITY ASSURANCE: BASIC DEFINITIONS AND CONCEPTS

Kendal E. Harr, DVM, MS, DACVP
Seattle, Washington

COPYRIGHT STATEMENT AND ACKNOWLEDGEMENT

The following abstract is partially excerpted from draft ASVCP Allowable Total Error Guidelines which are under review and will be finalized in 2012. The author acknowledges the collective knowledgebase and time commitments of contributors to this document.

INTRODUCTION

Quality assurance (QA) is the overall system for assuring the quality of laboratory test results. QA includes monitoring and assessment of all laboratory systems and procedures, with the objective of identifying problems, making corrections, and continuously improving the quality of testing service such that pre-analytical, analytical, and post analytical error are minimized.[1] Internal or in-clinic quality control (QC) validation and reference interval generation are part of the instrument/method validation and verification process. (See ASVCP POCA, General QA guidelines, and reference intervals found here http://www.asvcp.org/pubs/qas/index.cfm) Once instruments/methods are judged to be suitable for routine testing, QA ensures that ongoing instrument performance is stable and that the errors inherent to the instrument/methods do not exceed levels that would invalidate the interpretation of test results. Important QA tools include regular quality control (QC) procedures and participation in an external quality assurance (proficiency testing) program.

The concept of quality requirements is the foundation for quality planning. Quality requirements can help guide interpretation of laboratory test results because they provide perspective about variability of results within an acceptable interval and potential significance of abnormal findings.[2] A hierarchy of quality requirements has been proposed[3], and the most stringent quality requirements are based on clinical outcomes and clinical decision thresholds. Quality requirements may also be based on data about biologic variation of an analyte, analytical performance criteria (e.g., as mandated by CLIA for human medicine[4]), expert opinion and/or technological state-of-the-art. [3]

BIAS (inaccuracy)

Total systematic error which includes constant and proportional bias. Presence of bias may be due to multiple factors. It is the difference between the measured result and the concentration of a known standard. Generally used to describe the inaccuracy of a method relative to a comparative method in a method comparison experiment. The term bias has a specific meaning in the statistical t-test and in difference plot analysis, where bias (expressed in analyte units) equals the difference between the mean values of the two methods being compared or the average of all the differences between the paired sample values, i.e. the intercept on a Bland Altman graph. Bias may also be expressed as a percentage according to the formula bias(%) = mean target – mean measured/meantarget x 100.

IMPRECISION (random error)

Lack of repeatability or reproducibility of the same result; represented by the standard deviation (in units of the test) or coefficient of variation (percent). Standard Deviation (SD) is a measure of variability or diversity and shows how much variation or dispersion there is from the average (mean or expected value). A small SD indicates that the data points tend to be very close to the mean, whereas large SD indicates that the data points are spread out over a wide range of values. SD is the square root of a dataset's variance. Coefficient of Variation (CV) is standard deviation divided by the mean and expressed as a percentage. CV for purposes of determination of calculated TE should be determined from precision studies using quality control materials at various levels of analytes, known standards and/or patient samples.

ALLOWABLE TOTAL ERROR

Westgard was the first to introduce the concept of total error in 1974.[5] Analytical imprecision (reproducibility of the result) and bias (systematic error) were combined into a single measure of the uncertainty of a test result. The ideal situation is to have highly accurate and precise measurement, i.e. low bias and low CV or SD, respectively. Westgard originally used TE = bias(%) + 1.65CV, but a coefficient (z value) as high as 6 has been used by some authors for method validation studies.[2]

The ASVCP guideline defines total allowable error as **bias(%) + 2CV** which is consistent with CLIA recommendations. (CLIA '88 Proficiency Testing Limits, U.S. Federal Register). If units of the test are used, then the equation**, bias (expressed in units of the test) + 2SD**, is used to calculate allowable total error. **Observed or calculated total error (TEobs)** is the sum of random error (imprecision) and systematic error (bias can be calculated from instrument performance data according to the formula), as defined in this guideline, 2CV + bias(%) or 2SD + bias (quantity of analyte). The calculated TE is specific for a single instrument/method. **Allowable total error (TEa)** or desirable total error is an analytical quality requirement that sets a limit for both the imprecision (random error) and inaccuracy (systematic error or bias) that are tolerable in a single measurement or single test result to insure clinical usefulness. Tables for allowable total error are found at the end of this document.

Instrument performance evaluation procedure and calculations

The following procedure may be performed when, a new instrument is being considered for purchase, a new instrument is evaluated to ensure that it performs according to manufacturer's claims, routine internal quality control of an instrument - conducted to ensure adequate ongoing performance, evaluating performance during an external quality assurance (proficiency) program.

Three methods that may be used:

1. Comparison with peer group means in an external QA program participation. This typically must be done using an external quality assessment program that is employed to help insure quality laboratory results. While some external quality assurance programs use assayed materials in human medicine, typically in veterinary medicine, unassayed materials are used and there is reliance on the peer group mean. Peer group is defined by same instrument and/or method as that upon which the result is obtained. An external quality assessment using comparison to a peer group is dependent on the numbers of instruments included in the peer group as well as other laboratories' maintenance of equipment and quality control.
2. Comparison with target values provided by manufacturers of assayed quality control materials.

An assayed **quality control material (known standard)** may be repeatedly measured for 5 days to determine mean, bias, SD and coefficient of variation. In this situation, the mean of the results should be compared to the mean of the assayed mean to determine bias. These data can then be used to calculate total error (TEobs) of the analyte. The assayed QCM should be specific for the equipment and methods being evaluated; the instrument manufacturer should be consulted if there is any doubt regarding QCM suitability. Please consult the manufacturer to insure that it is appropriate for the equipment and methods.

3. Based on comparison with known gold standards for various analytes (standards provided by external regulating or governmental organizations and agencies)

As external quality control programs currently in existence are method specific and methods used by in-clinic laboratories are frequently not represented, option 1 is often not available to the in-clinic laboratory. The following steps are designed for quality assurance assessment for the in-clinic laboratory but may also be used by reference laboratories. All steps should be carried out by appropriately trained personnel who are knowledgeable regarding the analyzer's operation and the facility's quality assessment program. Calculations can easily be carried out using commercially available software programs. Calculations should be performed for each analyte and each QCM.

Measure each QCM daily for a minimum of five days.[6] Using these data, for each QCM and each analyte, calculate: Mean (average), Standard deviation (SD), and Coefficient of variation (CV). This value represents between-day (interassay) imprecision of the analyzer. The mean, SD, and CV of the analyzer derived from these QC data are referred to as the calculated or observed SD, and CV.

Calculate the analyzer's measured bias using the measured mean and the QCM manufacturer's reported mean for the assayed control material (using same instrument and/or method as that used by the analyzer) according to the formula. QCM manufacturer's reported means are commonly found in the QCM package insert, categorized according to the instrument and method producing the assayed values.) Measured bias may be a positive or a negative number, depending upon whether the analyzer's results are lower or higher than the manufacturer's reported mean. If bias is a negative number (e.g., – 5.0%), then the *absolute number* (5.0%) should be used below.

Calculate the analyzer's observed total error (TE_{obs}), using measured CV and measured bias. Compare measured TE_{obs} to TEa. If $TE_{obs} < TEa$ (or very close to it), then the quality requirement is met and instrument is considered suitable for measurement of that analyte. If $TE_{obs} > TEa$, then several options exist.

TEobs Interpretation and Assessment of External Quality Assessment Results

Calculated total error (TEobs) for all analytes determined on in-house or reference lab equipment should be compared to the ASVCP total allowable error guidelines found in Table 1. If calculated total allowable error (TEobs) is greater than that which is acceptable (TEa), attempts should be made to identify and correct causes of imprecision (high CV) and inaccuracy (high bias).

If these sources of error cannot be corrected or if problems occur repeatedly, the manufacturer of the instrument and/or a boarded clinical pathologist with expertise in QA should be called upon for further assessment. Further assessment may include attempts to improve performance capability by analyzer adjustments, operator training, replacement of reagent with new reagent or different manufacturer product, or, potentially, analyzer replacement. Alternately, the initial quality requirements may be relaxed. This is not recommended but is possible only if potential additional error can be tolerated in diagnostic judgment. This requires education of ALL clinicians using the analyzer regarding amended total allowable error of the analyte(s) in question. Any changes outside of the recommended TEa in this document must be justified and documented in a laboratory handbook. This should be done only upon consultation with a boarded veterinary clinical pathologist.

Three to five boarded clinicians (ACVIM or ECVIM with various specialties) gave opinions upon clinically desired TEa at low, mid, and high analyte concentrations and activities, except for troponin where the opinion of a single cardiologist was used. Total allowable error was calculated directly from reference equipment used by QALS members using the equation 2CV + bias% = TEa% to insure that TEa was possible. NCR=Not Clinically Relevant. *CAP* College of American Pathologists Participant Summary, April 2004. *CLIA* - CLIA '88 Proficiency Testing Limits, U.S. Federal Register. *BV* -Spanish Society of Clinical Chemistry and Molecular Pathology (SEQC) table of Desirable Quality Specifications based on Biological Variation, 2004 Update. For details, visit www.westgard.com/guest26.htm (accessed 27 Sept 2011) and http://www.dgrhoads.com/db2004/ae2004.php (accessed 19 November

2011) *CFX* "Canadian Fixed Limits", The College of Physicians and Surgeons of Saskatchewan *RCPA* Royal College of Pathologists of Australasia and the Australasian Clinical Biochemist Association Quality Assurance Program

Analyte	Low Analyte Values	Within RI	High Values	CLIA Value
Albumin	15%	15%	15%	10%
Alkaline Phosphatase	NCR	25% (20% desirable)	25% (20% desirable)	30%
Alanine Amino Transferase	NCR	20%	20%	20%
Ammonia	NCR	20%	20%	Not found
Amylase	NCR	20%	20%	30%
Aspartate Amino Transferase	NCR	25%	25%	20%
Bicarbonate	20% (15% desirable)	20% (15% desirable)	20% (15% desirable)	10% (RCPA) to 20% (CAP)
Bile Acids[1,2,3]	20%	20%	20%	None found
Cholesterol	20%	20%	20%	10%
Chloride	5%	5%	5%	5%
Creatine Kinase	NCR	30%	30%	30%
Creatinine[10]	20%	20%	20%	15%
Gamma Glutamyl Transferase	NCR	20%	20%	15% (RCPA) to 30% (CFX)
Glutamate Dehydrogenase	NCR	30%	25%, >90IU 20%	None found
Glucose	10%	20%	20%	6% Low, 10% High
Iron	30% (15% desired)	30%	30%	20%
Potassium[10]	5%	5%	5%	0.5mmol/L
Lactate[8,9]	NCR	40%	40%	10 (RCPA) to 30% (CFX)
LDH	NCR	20%	20%	20%
Magnesium[4,5,6]	15% desirable, 20% acceptable	15% desirable, 20% acceptable	15% desirable, 20% acceptable	25%
Sodium[10]	5%	5%	5%	4mmol/L
Phosphorus[11,12]	20%	15%	15%	10-23% (CAP)
Sorbitol Dehydrogenase	NCR	20%	20%	None found
Total Bilirubin	NCR	30% (25% desirable)	30% (25% desirable)	0.4 mg/dl, 20%
Total Calcium[5,11,12]	10%	10%	10%	2% (BV) to 8% (CFX)
Total Protein	10%	10%	10%	10%
Triglyceride	NCR	25%	25%	25%
Troponin[7]	NCR	70%	70%	20%CV maximal with around 50% TEa if calculated
Urea[10]	15%	12%	12%	2mg/dl, 9%
Uric acid[13]	10%	10%	10%	17%

Table 1. Total Allowable Error as defined in the ASVCP draft guidelines.

REFERENCES

[1]Reed, SM et al. Equine Internal Medicine. 1998. WB Saunders, London. [2]Hoffman, WE, et al. 1987.. Am J Vet Res 48: 1343. [3]Mendle, VE. 1988. Am J Vet Res 49: 572 [4]Khanna C, et al.. 1998. JVIM 12:304-309. [5]Martin LG. Hypercalcemia and Hypermagnesemia. In Advances in Fluid and Electrolyte Disorders/Vet Clinics of North America, May 1998. 28: 565-585. [6]Cortes YE, Moses L. Compendium. 2007;420-427. [7]Porciello, F., et al. Austral Vet J, 86: 390–394. 2008. [8]Moore JN, et al. Equine Vet J 1976;8:49-54. [9]Lagutchik MS, et al. JVECCS 1998;8:117-127. [10]de Papp E, et al. JAVMA 1999; 215(1):49-52. [11]Savage , C. Vet Clin Equine 2008;24:387–40. [12]Toribio, RE. Vet Clin Equine 2011;27:129–147 [13]Harr KE. Vet Clin Pathol. 2002;31(3):140-151.

In Text: [1] **http://www.westgard.com/assuring-quality-through-total-quality-management.htm** (Accessed January 30, 2012) [2]Westgard JO. **Clin Lab Manage Rev** 1996 Jul-Aug;10(4):377-81, 384-93, 396-403. [3]Kenny, D. **Scan J of Clin Lab Invest** 1999;59:585. [4] http://www.dgrhoads.com/db2004/ae2004.php (Accessed January 20,2012) [5]Westgard JO, et al. **Clin Chem** 1974;20:825-833. [6]Rishniw, M., et al. **Vet Clin Pathol** 2012; doi: 1939-165X.2011.00386.x

ALTERNATIVE OPTIONS FOR MANAGING URINARY TRACT DISEASE IN THE DOG AND CAT

Mary Bowles, DVM, DACVIM
Stillwater, OK

INTRODUCTION

Although standard, mainstream, accepted therapy should always play a prominent role in treating patient problems, alternative therapeutic approaches can be valuable as adjunctive treatment options. In addition to increasing the arsenal of therapeutic agents available, the use of alternative treatments have the potential for reducing the toxicity and expense of more commonly recommended protocols. Chronic or recurring urinary tract disease may benefit from alternative treatment options as discussed below.

CIPROFLOXACIN

Fluroquinolones, especially ciprofloxacin, have been advocated for use in the treatment of transitional cell carcinoma in cats and dogs. Studies have been published which suggest that ciprofloxacin and other fluroquinolones have antineoplastic activity against human transitional cell carcinoma cell lines *in vitro*.[1,2,3] Study results indicate that ciprofloxacin may have the ability to inhibit cell growth and induce apoptotic cell death in transitional cell carcinoma of the urinary bladder, making this antibiotic potentially useful as a preventive and therapeutic agent for TCC. One *in vitro* study indicated that the initiation of apoptosis in ciprofloxacin-treated TCC cells of human origin was related to disruption of mitochondrial function.[4] To the speaker's knowledge no human, canine, or feline *in vivo* study results have been published in regard to the efficacy of ciprofloxacin (or other fluroquinolones) as a therapeutic agent for TCC. However, a clinical trial by the UCDavis oncology group is being conducted to determine if intratumoral concentrations of enrofloxacin and its metabolite ciprofloxacin reach high enough levels in a variety of tumors to potentially be efficacious in antineoplastic protocols. Although fluroquinolones such as ciprofloxacin cannot currently be routinely recommended as an effective antineoplastic agent for TCC, it may be a desirable antibiotic choice for canine and feline TCC patients with secondary urinary tract infection.

PROBIOTICS AND PREBIOTICS

Probiotics are live microorganisms, which when administered in adequate amounts, confer a health benefit on the host. Prebiotics are defined as non-living, nondigestible food ingredients such as oligosaccharides which beneficially affect the host by selectively stimulating the growth of and/or activating the metabolism of one or a limited number of health promoting bacteria. When a probiotic is combined with a prebiotic for the purpose of producing a synergistic effect, the mixture is referred to as a synbiotic. Probiotics with or without prebiotics are usually administered orally and appear to have the most potential for use in dogs and cats as nutritional aids in the treatment and prevention of diarrhea. However, other organ systems besides the gastrointestinal tract may also benefit from the use of such supplements.

One synbiotic product (Azodyl™, Vetoquinol) has been marketed as a preparation that will lower the concentration of nitrogenous waste products (especially, BUN and creatinine) in the blood of canine and feline patients with chronic kidney disease (CKD) through enteric metabolism of uremic solutes by the probiotic organisms. This action, according to the marketing company, helps prevent further kidney damage and minimizes the clinical and biochemical consequences of declining renal function. Azodyl® is administered to patients as an enteric-coated capsule which contains *Enterococcus thermophilus*, *Lactobacillus acidophilus*, and *Bifidobacterium longum* as probiotic organisms and oligosaccharides as prebiotic agents. Although reduction in azotemia has been reported in anecdotal, uncontrolled clinical trials in dogs and cats with CKD and in a limited, controlled human clinical study[5,6], the studies on which the company's claims are primarily based were performed in nephrectomized rat and minipig models.[7,8] An additional concern raised about efficacy of Azodyl lies in the fact that this product can be difficult to administer to cats in its capsule form. Consequently, the contents of the capsule are frequently sprinkled on the feline patient's food for administration, raising the question of whether the probiotic agent will be sufficiently active to be effective following GI transit. Results of a study were recently published which evaluated the effect of Azodyl in CKD cats when sprinkled on the food or administered as a slurry rather than being administered in capsule form.[9] No reduction in azotemia was detected during the 2-month clinical trial period, but it remains to be determined if this failure was related to lack of efficacy of the synbiotic itself or to the manner in which it was administered. As of this writing, although Azodyl would not be considered harmful to canine and feline patients with CKD, its efficacy in regard to the manufacturer's claims is questionable.

Probiotics have also been investigated in regard to their potential for decreasing the recurrence of urinary tract infection (UTI). One recent controlled study in women suggested that use of intravaginal suppositories containing *Lactobacillus* reduced the incidence of UTIs in subjects previously identified as experiencing repeated UTI episodes.[10] Oral probiotics may also have the ability to repopulate the vaginal area with beneficial organisms that reduce the incidence of UTI through immunomodulation and pathogen displacement.[11] Larger trials in humans as well as controlled studies in canine and feline patients are needed to further define the potential role of probiotics in the prevention of recurring UTIs.

It has also been suggested that probiotics may aid in the prevention of oxalate urolith formation. Studies in humans and rats have shown a decrease in urine oxalate concentration with colonization of the intestinal tract with *Oxalobacter formigenes*, a microorganism that facilitates degradation of oxalate in the gut and leads to decreased excretion of oxalate in the urine.[12,13] The potential for use of this organism as a means of reducing oxalate urolith formation in dogs and cats has not been adequately investigated. Furthermore, the microbe which most consistently degrades oxalate in the gut, *Oxalobacter formigenes*, is not contained in the common probiotic preparations used in the dog and cat.

RHUBARB

Recently the product Rubenal™ by Vetoquinol has been marketed for use in slowing the progression of chronic kidney disease in dogs and cats. The product contains a medicinal rhubarb, Rheum officinale, which reportedly acts as an antifibrotic, anti-inflammatory agent and, in various lab animal species, has decreased the severity of experimental glomerulosclerosis and related proteinuria.[14] These actions have the potential to slow down the progression of kidney disease and result in a better quality of life. A study was recently completed which examined the effects of Rebenal versus benazepril in CKD cats.[15] Neither of the agents administered produced objective improvement in the status of the cats although many owners of patients receiving medicinal rhubarb reported subjectively that the cats had an improved sense of well being. The Rubenal pill administered is large and may be difficult to give to some cats. Few side effects have been observed but include vomiting and lethargy. Administering Rubenal to patients with a history of calcium oxalate urolithiasis should be avoided due to the potential for increased oxalate excretion in the urine with rhubarb consumption.

CRANBERRY SUPPLEMENT

Cranberry supplement may help prevent recurring bacterial UTI, especially infection caused by E. coli. It probably does not acidify the urine sufficiently to be effective in that way, but is thought to contain proanthocyanidins which disrupt the iofilm that helps protect certain bacteria and allow them to attach to the uroepithelium. Studies in humans have supported efficacy of cranberry supplementation in altering biofilm bacterial load and contributing to prevention of UTIs.16,17 A recent canine study indicated that cranberry extract decreased adhesion of E. coli to uroepithelium.18

Cranberry supplementation is best used concurrently with appropriate antibiotic therapy and/or immediately after antibiotic therapy has eradicated infection. Cranberry juice can be used in both dogs and cats, but it is bitter tasting and it is difficult to get a pet to consume enough to accomplish the desired effect. UroMAXX™ by Animal Nutritional Products has cranberry extract which is combined with other nutriceuticals for urinary tract health and is an easy formulation for administering one to two times daily (I would recommend twice daily dosing). In addition, many other over-the-counter cranberry extract preparations can be purchased. Dosing is variable, but is often recommended at approximately 10 – 20 mg/kg every 12 – 24 hours mixed with food.

Conflicting evidence exists in regard to the effect of cranberry ingestion on the formation of oxalate uroliths. Consequently, cranberry supplementation should probably be avoided, at present, in patients suffering from or having a history of oxalate uroliths since such supplements may contribute to the excretion of oxalate and favor oxalate formation due to its acidifying qualities

METHENAMINE

For prevention of recurring UTI, especially with E. coli, the use of an urinary antiseptic can be considered (methenamine hippurate- 500 mg/dog PO q12 hrs or 250 mg/cat PO q12 hrs) once the UTI is appropriately treated with antibiotics and under control.19 Meta-analysis of human controlled trials have indicated that methenamine hippurate may be a method of safe and effective short-term prophylaxis in UTI patients provided that individuals do not have a neuropathic bladder or upper urinary tract abnormalities.20 However, published, controlled studies examining methenamine's efficacy in dogs and cats are currently lacking.

Methenamine is not thought to be as effective with urease-producing organisms like some strains of Staph, Proteus, Enterobacter, and Pseudomonas which increase urine pH; consequently, a urinary acidifier may be required to be given concurrently to increase efficacy. The compound methenamine hippurate has hippuric acid added to help acidify the urine to aid \ in the hydrolysis of methenamine to formaldehyde and ammonia. A urine pH of < 5.5 is optimum for break down of methenamine. Both formaldehyde and hippuric acid have nonspecific antibacterial activity against various gram positive and gram negative organisms. Methenamine may also have antimicrobial activity against fungal organisms. It is inadvisable to administer methenamine to patients that are pregnant, lactating, dehydrated, or experiencing renal or hepatic insufficiency. Adverse effects resulting from methenamine can be more prevalent in the cat and are primarily related to nausea, anorexia and vomiting although dysuria may be produced occasionally in response to high urine formaldehyde concentrations.21

OMEGA-THREE FATTY ACIDS

Fatty acid supplements have become popular as therapeutic and preventative agents over the last several years for a variety of disorders, including renal disease, cancer, osteoarthritis, hypertriglyceridemia/hypercholesterolemia, dermatitis, urogenital infections, and cardiac disease. Although many of the claims of efficacy are unproven, the potential for use as anti-inflammatory agents seems to be supported both in theory and in the clinical setting. The fatty acid supplements most commonly used are omega-3 and omega-6 fatty acids. Omega-3 fatty acid supplements include eicosapentanoic acid (EPA) and docosahexaenoic acid (DHA), which are derived from oils of fish such as salmon, trout and menhaden and from flaxseed. Omega-6 fatty acid supplements include linoleic acid (LA) and gamma-linolenic acid (GLA), which are derived from oils of seeds such as evening primrose, black currant, and borage.

Fatty acid anti-inflammatory effect comes about through the mechanism of cell injury and the subsequent inflammatory cascade that ultimately involves the lipooxygenase and cyclooxygenase enzyme pathways. When cell membranes are injured, phospholipids in the membranes are broken down enzymatically, producing the omega-3 fatty acid , eicosapentaenoic acid, and the omega-6 fatty acid, arachidonic acid. Metabolism of eicosapentaenoic acid results in the formation of products called eicosanoids that are non-inflammatory, non-immunosuppressive, and anticoagulatory (inhibit platelet aggregation and clotting). Conversely, metabolism of arachidonic acid produces eicosanoids that are primarily pro-inflammatory, immunosuppressive, and pro-coagulatory (cause platelet aggregation and clotting). An exception to the production of inflammatory eicosanoids from arachidonic acid is the formation of the anti-inflammatory agent prostaglandin E_1. In addition, the omega-6 fatty acid di-homo-gamma-linolenic acid (DGLA) produces eicosanoids that are less inflammatory than those produced from arachidonic acid. Consequently, dietary supplementation of omega-3 fatty acids and the omega- 6 fatty acid DGLA would be expected to result in increased production of non-inflammatory eicosanoids and decrease in inflammation from cell injury.

Since fish oils can have an objectionable odor and occasionally cause GI upset, flaxseed oil is sometimes used as a source of omega-3 fatty acids. Flaxseed oil does contain some omega-6 fatty acids, but is also a source of the omega-3 fatty acid alpha- linolenic acid (ALA), which can be converted to EPA and DHA. However, dogs and cats are not thought to have good ability to convert ALA to more actively non-inflammatory components, limiting its usefulness as a dietary source of beneficial omega-3 fatty acid.[22,23]

As stated previously, supplementation of omega-3 fatty acids increases the production of non-inflammatory eicosanoids, including prostaglandins, which can be beneficial in reducing inflammation in renal disease as well as improving blood flow to the kidney through vasodilation. In addition, such supplementation can decrease the production of inflammatory eicosanoids. Published study data related to CKD has been most supportive of the benefits of omega-3 fatty acids in dogs although some data exists suggesting potential benefits in cats with CKD fed diets supplemented with omega-3 fatty acids.[24,25,26] Studies in dogs with CKD have indicated that those dogs receiving omega-3 fatty acid supplementation had improved renal function (reduction of proteinuria and prevention of glomerular hypertension) and fewer histopathologic lesions (glomerulosclerosis, renal interstitial cellular infiltrates) in the kidneys. The benefits to CKD cats from omega-3 fatty acids have not been well established although studies have indicated that potential benefits may include improvement in glomerular filtration rate and survival time as well as reduction in mean blood pressure values.[26,27] It is also possible that both canine and feline CKD patients may derive a beneficial synergistic effect from combining antioxidants with omega-3 fattyacids.[28]

A wide range of doses (50 – 300 mg/kg/day) can be found for omega-3 fatty acid supplementation, especially, EPA and DHA. In general, starting out at the lower end of the dosage range and, depending upon the patient response, gradually increasing the dose is the standard recommendation for many patients. The national research council (NRC) has established a safe upper limit for daily EPA and DHA administration for dogs based on metabolic weight (370 mg/kg$^{0.75}$ or 2,080 mg for a 10-kg dog).[29] A recommended dose of EPA and DHA for dogs with kidney disease has been published based on metabolic weight (140 mg/kg$^{0.75}$/day or 790 mg/day for a 10-kg dog).[23] An NRC upper safe limit has not been established for an EPA and DHA dose in the cat. Due to the lack of reliable information concerning safe feline doses the general rule of starting at the lower end of the dosage scale for omega-3 fatty acids is particularly important. Although doses of omega-3 fatty acids have been emphasized in this discussion, omega-6 fatty acids may be beneficial in combination with omega-3 fatty acids for renal disease. However an ideal ratio of these components for supplements has not been determined.

Fish oil supplements are usually in the form of capsules or liquids. Fish oils can oxidize and become rancid, prompting manufacturers to add Vitamin E or to extract oxygen from fish oil capsules in order to prevent rancidity. At higher doses, fish oils may decrease platelet aggregation and prolong bleeding time, increase fibrinolysis, and may reduce von Willebrand factor. astrointestinal signs such as nausea, gas formation, and diarrhea may be associated with fish oil supplementation. Administering these supplements with food and increasing doses gradually may help limit or prevent gastrointestinal side effects. An additional concern in cats is evidence that higher doses of omega-3 fatty acids may be associated with immunosuppression,[30] providing additional support to the previously made recommendation of using lower end dosages in the cat.

In addition to supplements, many prescription diets for pets have omega-3 fatty acids incorporated into the formulation, including canine and feline renal diets. The amount and exact types of omega-3 fatty acids are not always easily determined from the pet food label, but their presence should be one of the considerations when contemplating fatty acid supplementation.

REFERENCES

[1] Ebisuno S, et al. *Cancer* 1997; 80:2263. [2] Aranha O, et al. *Clin Cancer Res* 2000;6:891. [3] Kamat AM, et al. *Urology* 2004; 63:457. [4]Aranha O, et al. *J Urol* 2002; 167:1288. [5] Palmquist R. *J Amer Holist Vet Med Assoc* 2006; Jan-Mar:23. [6]. Ranganathan N,et al. *Curr Med Res Opin* 2009; 25:1919. [7]Ranganathan N, et al. *Scientific World Journal* 2005; 5:652. [8] Ranganathan N, et al. 3rd World Congress of Nephrology poster presentation, 2005, Singapore. [9]Rishniw M, et al. *J Feline Med Surg* 2011; 13:405. [10] Stapleton AE, et al. *Clin Infect Dis* 2011; 52:1212. [11] Reid G,et al. *Curr Opin Clin Nutr Metab Care* 2009; 12:583. [12] Hatch M, et al. *Kidney Int.* 2006; 69: 691. [13] Sidhu H, et al. *J Urol* 2001; 166:1487. [14] Zhang G, et al. *Nephrol Dial Transplant* 1996;11:186. [15] Hanzlicek AS, et al. 2011 *ACVIM Forum Proceedings* abstract. [16] Reid, G, et al. *Spinal Cord* 2001; 39:26. [17]Masson P, et al. *Infect Dis Clin North Am* 2009: 23:355. [18] Smee N, et al. 2011 *ACVIM Forum Proceedings* abstract. [19].Pressler B,et al. In *Veterinary Internal Medicine*, 7th ed. 2010:2036. [20] Lee BSB, et al. Cochrane Library, 14 APR 2010. http://onlinelibrary.wiley.com/doi/10.1002/14651858.CD003265.pub2/full [21].Plumb DC. In *Plumb's Veterinary Drug Handbook*, 7th ed. 2011. [22] Dunbar BL, et al. *J. Nutr* 2002;132:1701S. [23] Bauer JE. *JAVMA* 2011; 239:1441. [24]Brown SA, et al. *J Clin Lab Med* 1998; 131:447. [25] Brown SA, et al. *J Lab Clin Med* 2000; 135:275. [26] Plantinga EA,et al. *Vet Rec* 2005; 157:185. [27] Brown SA. *Compend Contin Edu Pract Vet* 1999; 21: 11S. [28] Brown SA. *Vet Clin North Am Small Animal Pract* 2008; 38:157. [29] National Research Council. In: *Nutrient Requirements of Dogs and Cats.* Washington,D.C:National Academy Press, 2006; 359. [30] Chew BP, et al. In *Recent Advances in Canine and Feline Nutrition. Vol. III. Iams Nutrition Symposium Proceedings.* 2000; 55.

OBESITY AND STRESS: RISK FACTORS FOR FELINE LOWER URINARY TRACT DISEASE?

Jodi L. Westropp, DVM, PhD, DACVIM
Jennifer A. Larsen, DVM, PhD, DACVN
Davis, CA

INTRODUCTION

Feline lower urinary tract disease (FLUTD) is an inclusive term used to describe any disorder affecting the urinary bladder or urethra of cats. Signs include variable combinations of pollakiuria, stranguria, periuria, dysuria and hematuria, and have been reported to occur in 1.3-1.7% of cats examined at private veterinary practices in the United States.[1] Indoor-housed cats have been reported in several studies to be at risk for LUTD.[2-3] The reasons for this are unclear and are likely multifactorial; indoor-housed cats may have excess body weight and decreased activity, which may contribute to clinical signs.

The LUTD reported in clinical cases is not always fully characterized, and lower urinary tract signs (LUTS) in cats can occur due to cystic calculi, bacterial urinary tract infections, neoplasia, or idiopathic causes. In 2/3 of cats that present with LUTS no definitive diagnosis can be made;[4] therefore this syndrome is called feline idiopathic (or interstitial) cystitis (FIC). The two most commonly diagnosed feline LUTD reported in the literature are urolithiasis and FIC (obstructive or non-obstructive). Bacterial urinary tract infections are not commonly reported in younger, otherwise healthy cats when evaluating data from referral institutions.[4]

RISK FACTORS FOR FELINE UROLITHIASIS

Unfortunately, there is a lack of published studies in cats to document specific risk factors for urolithiasis. There are likely intrinsic and extrinsic factors that predispose certain cats to form stones, including breed, age, and environment.[5-6] Struvite stones were reported significantly more often in younger cats compared to calcium oxalate (CaOx) uroliths.[5] Stress related events and feeding management have not been evaluated in cats with stone disease, but some hypothesize these factors may contribute to stone formation in some cats.

Indoor-housed cats are at risk for LUTS (and possibly urolithiasis) as are neutered males and spayed females.[6] While the exact link remains elusive, indoor-housing as well as neutering are associated with a sedentary lifestyle and obesity.[9] More than a third of adult cats are overweight or obese, and one study showed that overweight and obese cats had a higher prevalence of urinary disease (defined as acute cystitis, feline urological syndrome, obstruction, infection, and urolithiasis) compared to those reported to be normal or underweight.[10] Additionally, there are reports of associations between obesity and urolithiasis in humans[11] where the prevalence of CaOx and uric acid stones has been reported to increase with increasing body size.[12] In fact, the association between increased body weight, body mass index, waist circumference and kidney stones has been well documented in humans.[11] Finally, there are studies suggesting that stressful life events can result in more clinical events of urinary stone disease in humans.[13] Studies investigating body mass index, body weight, and environmental stressors and urinary stone disease are warranted in feline patients to see what, if any, relationships exist.

RISK FACTORS FOR FIC

In one study evaluating obstructive FIC, the proportion of indoor-outdoor cats was significantly lower in affected cats compared to the control group, and the proportion of cats consuming only dry food was higher.[14] Therefore, like urolithiasis, indoor housing has been reported to be a risk factor as well as excessive body weight and decreased activity.

Studies have been conducted to investigate why stress may increase clinical signs in cats predisposed to FIC. Alterations in the neuroendocrine systems have been reported;[15-16] a decreased functional sensitivity of the alpha-2 adrenoceptors (α-2 AR) has also been documented in cats with FIC.[17] In addition to sensory and efferent neural abnormalities, the stress response system can be altered by a variety of cytokines binding to cell surface receptors and may even be associated with stressors that occur in utero. Studies in mice have suggested that events experienced during development may affect visceral sensory systems and result in chronic idiopathic disorders.[18]

A recent study of healthy cats and cats with FIC found that environmental stressors, also resulted in increased number of sickness behaviors (e.g. vomiting, lethargy, anorexia) in cats with FIC when the results were controlled for other factors.[19] Cats with FIC have a variable combination of comorbid disorders such as behavioral, endocrine, cardiovascular and gastrointestinal problems,[2-3,20] and it is imperative that the complete medical history, including signs not related to the LUT, be evaluated in these cats. Detailed environmental histories are also important, as this may need to be addressed as part of the therapeutic approach for management of the disease.

TREATMENT PROTOCOLS

While urolithiasis and FIC are not the same disease and the therapeutic approach for the acute presentation is quite different, the long-term management of these two diseases actually has a very similar foundation. In each case, it is important to obtain a thorough environmental history, medical history, and diet history. The clinician must also gain an understanding of the owner's attachment to the cat and their willingness to adhere to the suggestions provided, since stress management is a critical part of treatment. A body condition score and accurate weight should be obtained at the initial visit. For cats with urolithiasis, once the stone has been removed, a dietary change may be indicated. For cats with FIC, after treating the initial bout of clinical signs with analgesics (e.g. buprenorphine), a diet with increased moisture concentration is often recommend. The specific diet chosen for either disease needs to account for other comorbid conditions that may be present such as obesity, or kidney disease.

Increasing water intake

For both diseases, a diet high in moisture is indicated, assuming the owner is willing to feed it and the cat is willing to eat it. Increasing moisture intake is the cornerstone of therapy for urolithiasis in human[21] and veterinary medicine. A high moisture diet is recommended for cats with a history of stones to decrease the urine concentration of mineral precursors and to promote more frequent micturition secondary to increased urine volume. Likewise, increasing the water intake of cats with FIC may help improve clinical signs by encouraging frequent micturition and reducing urine specific gravity, therefore reducing the accumulation of sloughed cellular material and mucus. This approach may be clinically efficacious, as LUTS recurred in only 11% of affected FIC cats during one year of feeding a canned diet, while recurrence occurred in 39% of cats fed the dry version of the same formulation.[22] However, while this suggests that increased moisture may be important, dry and canned versions of the same formula differ not only in moisture content but in many other ways, and the exact reasons for this effect remain to be determined.

Increasing water intake can be achieved most easily by feeding a canned diet. If the cat is not accustomed to consuming canned food, repeated, gradual introductions may be necessary to encourage acceptance. Warming the food and ensuring that fresh servings are offered at least twice daily may help increase consumption of a canned diet. Rather than complete withdrawal of the current diet, the new canned food should be offered in a separate container next to the cat's regular diet. If the owners or the cat are not willing to accept a new canned food, an alternative is to add water to dry kibble to achieve higher moisture content. To attain 80-85% moisture using this method, adding 4-5 cups of water for every cup of dry food is necessary and may not be practical. Urine specific gravity should be periodically monitored until the adequacy of the current strategy is assessed. Urine specific gravity should be evaluated from samples obtained at home to gain a better understanding of owner compliance and water intake in the home environment; typically non-absorbent litter is necessary to accomplish the collection of urine. This will also prevent multiple, potentially stressful visits to the veterinarian.

Choosing an appropriate diet

There are a variety of diets that are specifically formulated for FLUTD, and there have been a number of studies investigating the efficacy of these products. While product testing studies provide useful information, there remains a lack of data on the effects of any single strategy or of modifying specific dietary variables. For any individual case, the specific diet chosen will depend not only on the specific clinical manifestation of the particular disease process, but also on the chronicity and severity of the disease as well as other conditions the cat may have. For example, if the cat with FIC or urolithiasis is overweight, high moisture, veterinary therapeutic diets specifically formulated for weight loss are indicated. For urolithiasis management, the diet should be evaluated for the minerals of interest (e.g. calcium, oxalate, magnesium) in addition to the target urine pH, depending on what type of stone was removed from the cat.

Veterinary therapeutic diets formulated for active weight loss are generally less energy dense than other diets. The calories are diluted out on a volume basis with fiber (may also provide satiety effect), water (canned diets), and/or air (puffed and/or irregularly/large shaped kibble). Of course, the latter strategy does not promote satiety and is aimed at altering the owner's tendency to overfeed by encouraging a perception of increased volume of food. Weight loss diets are not necessarily low in fat, and they vary widely in nutritional profiles, ingredients, and fiber levels, so there are appropriate options for a wide range of patients. An important characteristic is a higher concentration of protein and other essential nutrients relative to calories, so that when the amount of energy fed is reduced, the risk of deficiencies is lower. Since maintenance diets are formulated with the assumption of average intake of energy, the profile of these diets is such that essential nutrients are provided in concentrations that may be adequate only when individual intake is average. As such, using a maintenance diet for weight loss carries the risk of deficiency since these modifications have not been incorporated, and the diet will be fed for an extended period during the weight loss program.

Both over-the-counter and veterinary therapeutic diets marketed as "high protein/low carbohydrate" are typically high in fat and energy dense, so the volume amount fed is typically very low and might not be acceptable to the owner or the cat. Conversely, diets with low energy density achieved with high fiber should be used carefully in extremely obese cats; a gradual transition and ensuring adequate water intake is necessary to help reduce the risk of constipation. For dietary recommendations in cats with multiple problems like obesity and FIC or urolithiasis, consultation with a specialist in veterinary nutrition is advised.

Patient assessment and achieving weight loss

Body condition scoring (BCS) systems are an invaluable tool in the assessment of all patients. In obese cats, the BCS should be used to monitor the patient during weight loss. This provides a subjective yet quantitative way to estimate the amount of excess adipose present and the ideal body weight. Using the 9-point system (which has been validated using more objective measures of adiposity), each point is 10-15% of body weight.[23] For cats, an ideal BCS is 5 on a scale of 1 to 9. Obesity is defined, as in people, as weighing about 25-30% or more over ideal (equivalent to 7 or more on a 9 point scale). Assessment for potential concurrent problems related to obesity should not be overlooked, including associated problems such as diabetes mellitus and hyperlipidemia.

When instituting a weight loss plan, the goal rate of loss is 1% of body weight per week; however, any lesser weight loss that is achieved should be encouraged and celebrated. Weight loss rates approaching or exceeding 2% of body weight may risk disproportionate loss of lean body mass as well as begging or food seeking behaviors. Additionally, faster rates of loss in cats with concurrent problems may confound assessment of ongoing or unstable diseases. It is important to prepare owners for a weight loss plan that is a long-term commitment; when a cat with a BCS of 8/9 is losing at the goal rate of 1% of initial body weight per week, the ideal body condition will be achieved no sooner than 36 weeks, and in reality it often takes much longer. Focusing the owner on the process (ie the rate of loss) is often more successful than focusing on a specific ideal body weight that may not be realized for many months. As a general rule, in multi- cat households or in cats that are free fed, the owner should be instructed on how to change to multiple meal feedings, separate the cats, or consider "fat-cat houses" with cutout silhouettes for feeding the various cats in the household separately. Owners should be encouraged to use a gram scale, which is the most accurate way to measure both canned and dry food.

Establishing the amount of food to feed when beginning a weight loss plan isn't necessarily straightforward. Ideally, the current intake should be calculated. If the current intake cannot be determined due to ad lib feeding, multi-cat households, access to unmonitored feed sources, variable treating, or insufficient/questionable diet history, calculate RER (resting energy requirement; RER = 70 * (BW in kg)^3/4). Start the weight loss plan by feeding a total of 80% of the current intake (if known), or the RER. Monitoring and adjustments will be necessary. Evidence suggests that exercise improves the success rate of weight loss programs.[24] If possible, encouraging increased activity will facilitate more efficient weight loss) and, once achieved, help ensure maintenance of a lean body condition. In addition, increasing activity with play activities (chasing toys or laser pointers) may help manage stress and boredom.

Environmental modification and stress reduction

There are few studies on the clinical effects of stress management protocols on the incidence and recurrence rate of LUTD in cats. One study evaluated the use of a tailored, well-designed multimodal environmental modification protocol in cats with LUTS and reported an improvement in many stress-related behaviors in addition to reduced LUTS.[20] More extensive, controlled studies with larger numbers of cats are warranted to fully evaluate the usefulness of customized plans to decrease stress levels and optimize the indoor environment for cats. Once an environmental and dietary history has been evaluated, such a plan may be an effective adjunctive treatment to LUTD and can help the owner to reach the desired goals for their cat. When making modifications to the home management practices for affected cats, it is advisable to avoid changing more than one thing in the house at a time (including diet) so the cat and the owner are not overwhelmed. Issues such as food and water, dietary needs, litter boxes, social contact and conflicts, space and activity should be addressed and will be discussed in lecture.

MONITORING

It is important when managing chronic diseases in cats like urolithiasis, FIC, obesity, or a combination of these that a treatment program, including dietary changes, environmental adjustments, and monthly follow up (either by phone or in the hospital) are done to ensure success.[25] Often these monitoring programs can be maintained with the help of a dedicated nurse in the practice. The client should be contacted 3-5 days after beginning the program and small adjustments made at that time. Continued follow up consultations are necessary to adequately evaluate the cat's weight, body condition, and other changes in behavior. The time frame for these rechecks will depend on the client's needs, and what progress has or has not been made. Specific goals should be outlined and positive reinforcement for the owner when goals are met is helpful. If obesity is being simultaneously managed, photographs can easily be taken and logged into the record for future comparisons and encouragement.

CONCLUSION

LUTD in cats can be frustrating to manage. Stress and obesity may play a role in both urolithiasis and FIC. Collection and evaluation of comprehensive historical information is necessary to effectively address multiple complex issues related to environment, diet, and other management practices. Addressing these issues is often warranted in order to achieve successful outcomes in many cases.

REFERENCES

1. LUND EM, ET AL. J AM VET MED ASSOC. 1999;214(9):1336. 2. BUFFINGTON CA, ET AL. J AM VET MED ASSOC. 2006;228(5):722. 3. BUFFINGTON CA. J AM VET MED ASSOC. 2002;220(7):994. 4. BUFFINGTON CA, ET AL. J AM VET MED ASSOC. 1997;210(1):46. 5. CANNON AB, ET AL. J AM VET MED ASSOC. 2007;231(4):570. 6. APPEL SL, ET AL. CAN VET J. 2010;51(5):493. 7. FETTMAN MJ, ET AL. RES VET SCI. 1997;62(2):131. 8. KANCHUK ML, ET AL. J NUTR. 2003 JUN;133(6):1866. 9. SCARLETT JM, ET AL. INT J OBES RELAT METAB DISORD. 1994;18 SUPPL1:S22. 10. LUND EM, ET AL. INTERN J APPL RES VET MED. 2005;3(2):88. 11. TAYLOR EN, ET AL. J AM MED ASSOC. 2005;293(4):455. 12. CHOU YH, ET AL. UROL RES. 2011;39(4):283. 13. NAJEM GR,ET AL. INT J EPIDEMIOL. 1997;26(5):1017. 14. SEGEV G, ET AL. J FELINE MED SURG. 2011;13(2):101. 15. WESTROPP JL, ET AL. J UROL. 2003;170(6):2494. 16. WESTROPP JL, ET AL. AM J VET RES. 2006;67(4):731. 17. WESTROPP JL, AM J VET RES. 2007;68(2):203. 18. HOLMES A, ET AL. NEUROSCI BIOBEHAV REV. 2005;29(8):1335. 19. STELLA JL, ET AL. J AM VET MED ASSOC. 2011;238(1):67. 20. BUFFINGTON CA, ET AL. J FELINE MED SURG. 2006;8(4):261. 21. TISELIUS HG. BJU INT. 2003;91(8):758. 22. MARKWELL PJ, ET AL. J AM VET MED ASSOC. 999;214(3):361. 23. LAFLAMME D. FELINE PRACT. 1997;25:13. 24. ROUDEBUSH P, ET AL. J AM VET MED ASSOC. 2008;233(5):717. 25. YAISSLE JE, ET AL. J AM VET MED ASSOC. 2004;224(12): 1932-5.

FELINE URETHRAL OBSTRUCTION: TRADITIONAL AND ALTERNATIVE APPROACHES TO TREATMENT

Edward Cooper, VMD, MS, DACVECC
Columbus, OH

PATHOGENESIS OF OBSTRUCTION

It has long been held that urethral obstruction (UO) is caused by the presence of a physical obstruction occluding the urethral lumen. While mechanical obstruction of the urethral lumen certainly plays a roles in some cats, calculi or plugs are not identified in approximately 50% of cases.[1,2] It has been speculated that cats with functional UO may in fact develop urethral spasm and mucosal edema secondary to unrecognized feline idiopathic cystitis (FIC) The pathogenesis of FIC is still unclear, as attempts to identify a causative bacterial or viral (e.g. calicivirus) agents have been unsuccessful. Instead, neurohumoral alterations suggests that there is an imbalance between the sympathetic and hypothalamic-pituitary-adrenal axis responses to stressful situations in affected cats.[3] This imbalance results in impaired blood flow to the bladder wall and local release of inflammatory mediators, resulting in mucosal edema, smooth muscle spasm, and lower urinary tract pain. Pain, in turn, may result in or exacerbate urethral smooth muscle dysfunction and urethral mucosal inflammation. These conditions, either independently or in conjunction with a physical obstruction, ultimately may lead to urethral obstruction. As such, addressing this functional component of obstruction could have important implications for therapeutic management.

PATHOPHYSIOLOGY OF OBSTRUCTION

Continued accumulation of urine in cats with complete urethral obstruction results in in pressure necrosis and mucosal injury. Pressure within the urinary bladder is transmitted up the ureters to the kidneys, and back through the entire nephron to Bowman's capsule. When hydrostatic pressure within Bowman's capsule exceeds glomerular filtration pressure, GFR is reduced to zero. Within 24-48 of obstruction the kidney's excretory ability ceases, and azotemia, acidemia and electrolyte disturbances result in the typical clinical picture associated with UO.

As described elsewhere, severe uremia in obstructed cats leads to rapid-onset, nausea, vomiting, anorexia, and obtundation. Hypovolemia may occur secondary to decreased intake of water and increased loss through the gastrointestinal tract. Severe hyperkalemia is the most life-threatening complication of UO because of its effects on the cardiovascular system. Hyperkalemia increases resting membrane potential without affecting threshold potential. The resultant overstimulation of Na^+ channels causes them to deactivate with a net reduction in Na^+ conductance, diminished rate and magnitude of cardiac myocyte depolarization, and progressive bradycardia. The increase in myocyte membrane resting potential has an opposite effect on K^+ channels, and results in a marked increase in conductance (and a more rapid and dramatic repolarization). These changes ultimately result in cessation of all cardiac electrical activity, and asystole.

Hyperkalemia also results in vasodilation and venous pooling of blood. Severe metabolic acidosis induces protein denaturation, enzymatic dysfunction, and catecholamine hyposensitivity. Although these changes should in theory result in hypotension and cardiovascular collapse, mean arterial blood pressure of cats with naturally-occurring UO at time of diagnosis (as measured by oscillometric methods) is not typically decreased, even in patients with severe metabolic derangements (BUN/Creat as high as 352 mg/dL and 27.6 mg/dL, pH as low as 6.97).[4] Instead, cats with more severe metabolic derangements tend to be normotensive whereas less severely affected patients tend to be hypertensive (though this difference did not reach significance in the one study reporting a possible association). The authors suggested that several factors (such as pain, stress, etc.) may have served to offset any tendency toward hypotension.

DIAGNOSIS

Diagnosis of feline urethral obstruction is typically not challenging, often being made based on history and initial physical exam. As such, initial assessment/diagnostics are geared toward assessment of severity of metabolic derangement. Physical exam findings can be predictive of the severity of hyperkalemia:, concurrent bradycardia (HR < 140) and hypothermia (T < 96.6°F) are 98% predictive of serum K^+ concentration >8 mmol/L in cats with UO (although this subset of "sick" blocked cats makes up only 10-20% of UO cases).[5] Point-of-care diagnostic testing should be performed (ECG, PCV/TP, blood gas analysis, chemistry profile, etc.) to further gauge severity metabolic derangement, guide initial treatment, and to determine later response to therapy.

INITIAL STABILIZATION OF THE "SICK" BLOCKED CAT

Fluid therapy should be started immediately to support vascular volume and dilute serum potassium concentration, even if bladder decompression cannot be performed immediately. Normal saline (0.9% NaCl) has been suggested historically to be preferable to alternative, K^+-containing fluids, because of the greater dilutional effect on serum K^+. However, 0.9% NaCl may exacerbate metabolic acidosis. Conversely, balanced electrolyte solutions are alkalinizing, but contain small amounts of K^+. A prospective clinical study comparing 0.9% NaCl and a balanced electrolyte solution (Normosol R®) showed no difference in outcome (patient survival or length of stay) or in reduction of serum K^+ concentration, although acid-base abnormalities corrected more rapidly in the Normosol R®-treated group.[6] A more recent study using a model of UO found similar results comparing 0.9% NaCl and lactated Ringer's solution.[7] It is therefore likely that use of any commercially-available replacement fluid can be used, provided that an appropriate volume is administered to meet the individual patient's needs.

Significant bradycardia and hyperkalemia-associated ECG changes warrant immediate intervention. Calcium gluconate offsets the electrophysiological effects of hyperkalemia by raising threshold potential and reestablishing a more normal gradient with resting potential. Administration of calcium can then be followed by insulin, dextrose, and/or sodium bicarbonate to promote intracellular shift of K^+.

Cystocentesis can also be a part of initial stabilization by allowing immediate relief of pressure within the urinary tract and an increase in glomerular filtration rate. In addition, only minimal sedation is usually necessary to perform cystocentesis, which may be of benefit in patients with more significant metabolic derangement. Relieving back-pressure against the obstruction (whether due to stone, plug or spasm) may ease passage of a urinary catheter. Although there is theoretical concern that cystocentesis in cats with UO could result in tear or rupture of a distended and friable bladder, a prospective study of UO cats at our institution,[8] failed to identify abdominal effusion (as assessed by ultrasound) following cystocentesis performed at time of diagnosis.

URETHRAL CATHETERIZATION

Generally speaking, passage of a urinary catheter to relieve physical obstruction is considered to be an essential in the management of UO. In order to minimize risk of urethral trauma, heavy sedation and analgesia, or general anesthesia, are recommended. Injectable (ketamine [5-10 mg/kg] and diazepam or midazolam [0.25-0.5 mg/kg], buprenorphine [0.01-0.02 mg/kg] and acepromazine [0.03-0.05 mg/kg], +/- propofol [1-4 mg/kg, to effect]) or inhalant anesthesia may be used in the stable patient. Diazepam, in combination with buprenorphine or hydromorphone (0.05-0.2 mg/kg), is oftentimes sufficient in unstable cats. Epidural analgesia can also be utilized in an effort minimize the cardiovascular effects of most anesthetics. A recent case series has demonstrated that coccygeal epidural (as opposed to lumbosacral) may be safe and effective for these patients.[9] Vocalizing or movement during catheterization attempts, reflecting insufficient sedation, is likely to be associated with significant urethral spasm and an increased risk of urethral trauma.

Open-ended catheters should be used to initially relieve urethral obstruction. A mixture of saline and sterile lubricant (3:1 ratio) as the flush solution may decrease likelihood of urethral damage or irritation and secondary urethrospasm. Another helpful technique is to pull the prepuce caudally once the catheter is seated within the penile urethra. This will straighten out the urethra, thus easing passage of the catheter and reducing urethral trauma. Urethral or vesicular instillation of local anesthetic agents (such as lidocaine) does not appear to be of benefit. Once the obstruction is relieved the bladder can be emptied and flushed.

ADDITIONAL DIAGNOSTICS

Careful selection of additional diagnostics to be performed can be very important in UO, especially in cases with limited financial resources. A complete blood count rarely yields useful information beyond that provided by PCV/TP. Urinalysis may reveal crystalluria, along with hematuria and pyuria. Interestingly, a recent study found that cats with UO are significantly more likely to have these changes than patients with non-obstructive FIC.[10] However, this information is unlikely to affect acute case management. The prevalence of bacterial cystitis is very low in cats with lower urinary tract disease, and culture of urine is unlikely to be useful. However, several recent studies from Europe have challenged this paradigm, reporting infection rates of 12-44%.in cats with lower urinary tract disease.[11-13] As there is potential to introduce a UTI during catheterization, it is recommended that urine culture and sensitivity be performed at the time of catheter removal, though the overall incidence of catheter-associated infections in cats with UO is unknown. Given the significant potential for contamination during catheter removal, the practice of submitting the catheter tip for culture should be avoided. It is also prudent to perform abdominal radiographs to assess for cystic/urethral calculi, being certain to include the entire lower urinary tract.

POST-OBSTRUCTIVE CARE

Maintaining fluid balance is an extremely important component of post-obstructive care. This is especially true in patients that develop a post-obstructive diuresis (POD), which may last for several days and could result in massive urine production. This diuresis is thought to occur secondary to a combination of blood accumulation of osmotically active substances, renal papillary pressure necrosis, medullary washout, and/or antidiuretic hormone resistance. There is also evidence that accumulation of kinins, which have both diuretic and natriuretic effects, and atrial natriuretic peptide may play a role in POD. In a recent study POD occurred in 46% of male cats with naturally-occurring UO.[14] Cats with acidemia (pH <7.35) were significantly more likely to develop POD, and no association with magnitude of azotemia or hyperkalemia at time of diagnosis. Administration of large volumes of intravenous fluids may in turn perpetuate dieresis, and gradual reduction may be needed for restoration of renal concentrating ability. Conversely, there is also potential concern for inadequate urine production (i.e. <1ml/kg/hr) after obstruction is relieved. Obstructive uropathy (i.e. intrinsic renal damage secondary to post-renal obstruction) may result in oliguria, but anecdotal and personal experience suggests that this is uncommon in cats with UO.

Cystitis, urethritis, urethral catheterization, and pain may each contribute to re-obstruction secondary to urethrospasm. Buprenorphine (0.01-0.02 mg/kg) offers sufficient pain control in most cats, and has the benefit of oral administration. If buprenorphine is not sufficient, this author favors use of fentanyl constant rate infusion (2-4 mcg/kg/hr) and tends to avoid hydromorphone because of the potential to cause excitability and hyperthermia. Acepromazine (0.05 mg/kg IV/IM or 0.5 mg/kg PO) may minimize patient stress and agitation, but is appropriate only in metabolically stable patients. The non-specific α-antagonist effects of acepromazine may also promote urethral relaxation, thus decreasing likelihood of re-obstruction, as it has been shown to decrease maximal urethral closure pressure.[15]

Post-obstructive care (along with monitoring, serial blood work, supportive care etc.) should be continued until the urinary catheter is removed and spontaneous urination is observed. Administration of antibiotics to hospitalized cats is not recommended, both because of the previously mentioned low prevalence of bacterial urinary tract infections, and because antibiotics do not prevent catheterization-associated infection.

ALTERNATIVE MANAGEMENT PROTOCOLS

Treatment options discussed above are oftentimes limited by owner financial constraints. As previously discussed, there is likely a significant component of functional obstruction (secondary to urethral edema and spasm) as a physical one (plug or stone) associated with UO. A recent study has demonstrated that pharmacological manipulation (analgesia and sedation), a low-stress environment, and intermittent cystocentesis may result in spontaneous urination without the need for catheterization.[16] This approach was offered in lieu of euthanasia when traditional treatment for UO was prohibited by financial constraints. Fifteen cats were treated using this protocol, with successful outcome (resumption of spontaneous urination and hospital discharge) in 11 cases (73%). Major complications included the development of uroabdomen or hemoabdomen, with patients having severe azotemia likely being at greater risk. Cats that were successfully discharged had continued treatment at home as outlined below, and there did not appear to be a greater risk of immediate or long-term re-obstruction when compared historically to traditional management. These results suggest that this protocol could serve as an alternative to euthanasia, though no direct comparison to traditional management has been made.

Amitriptyline, a tricyclic antidepressant that also has some urethral relaxant effects, may be considered in some cats with urethral obstruction, 20 cats with UO treated with amitriptyline, rather than urethral catheterization, had a 100% rate of spontaneous resolution of UO and azotemia within 72 hours in one study.[17] However, this study did not indicate how the hospitalized patients were managed until spontaneous urination occurred. There also does not appear to have been any further investigation into this therapeutic option or evidence to support its use since that study was performed.

In some cases, financial limitations may preclude any hospitalization. For cats presenting in the earlier stages of obstruction that are not yet significantly ill, it may be possible (as a last resort) to provide care on an outpatient basis by providing sedation and analgesia (e.g. acepromazine and buprenorphine) and bladder decompression through either passage of a urinary catheter or cystocentesis. There is no peer-reviewed evidence to support the merits of either of these approaches, nor is there information regarding the likelihood of success or recurrence.

AT-HOME CARE

Depending on the underlying cause, and the study reporting, there is an approximately 25-40% incidence of recurrence with UO.[2,12] As such, regardless of how cats are treated while hospitalized, continued at-home care is recommended to decrease the likelihood of immediate or delayed re-obstruction. Continued analgesia and sedation (e.g. acepromazine and buprenorphine) for 5-7 days after discharge anecdotally may decrease the likelihood of UO recurrence. If there is persistent straining during urination following urethral catheter removal, prazosin (0.25-0.5 mg/cat q12-24) may provide some urethral relaxation, improve comfort, and also decrease rate of UO recurrence. Antibiotics should only be dispensed based on results of urine culture taken at the time of catheter removal. However, one recommendations which has been shown to help decrease the risk of FIC (and therefore, presumptively also may reduce the prevalence of re-obstruction) is increasing water intake (via use of wet food, flavored water, running-water bowl, etc). Cats consuming less water have an increased risk of FIC compared to a control population,[10] and cats with UO are more likely to consume a dry food only diet when compared to a control population.[12] Given the potential association between UO and FIC, and the role that environmental stress can play in the latter, providing environmental enrichment[18] may also offer some benefit, though this has not been directly assessed.

REFERENCES:

[1]Barsanti JA, et al. J Vet In Med 1996;10(1):34. [2]Gerber B, et. Al. **J Fel Med Surg** 2008;10:16. [3]Westropp JL, et al. **Vet Clin North Am Small Anim Prac** 2004;34:1043. [4]Malouin A, et. al. **J Vet Emerg Crit Care** 2007;17(1):15. [5]Lee JA, et. al. **J Vet Emerg Crit Care** 2006;16(2):104. [6]Drobatz KJ, et. al. **J Vet Emerg Crit Care** 2008;18(4):355. [7]Cunha MG, et. al. **Am J Vet Res** 2010;71(7):840. [8]Cooper ES, et al. Under review. [9]O'Hearn AK, et. al. **J Vet Emerg Crit Care** 2011;21(1):50. [10]Defauw PAM, et. al. **J Fel Med Surg** 2011;13:967. [11]Eggertsdottir AV, et. al. **J Fel Med Surg** 2007;9:458. [12]Segev G, et. al. **J Fel Med Surg** 2011;13:101. [13]Saevik BK, et. al. **J Fel Med Surg** 2011;13:410. [14]Francis BJ, et. al. **J Fel Med Surg** 2010;12:606. [15]Marks SL, et. al. **Am J Vet Res** 1996;57(10):1497. Cooper ES, et. al. **J Am Vet Med Assoc** 2010; 237(11):1261. [17]Achar E, et. al. **Kidney Int.** 2003;64:1356. [18]Buffington CAT, et. Al. **J Fel Med Surg** 2006;8(4):261.

MANAGING ACUTE RHABDOMYOLYSIS IN HORSES

Erica C. McKenzie BSc, BVMS, PhD. DACVIM, DACVSMR
Corvallis, OR

INTRODUCTION

Acute rhabdomyolysis in horses arises from a variety of etiologies, and reflects severe muscle necrosis variably accompanied by systemic consequences including pain, inflammation, deranged acid base status, electrolyte anomalies, dehydration, shock, renal dysfunction and recumbency. Severely affected horses can present a challenging medical situation, and evaluation should include attempts to identify the probable underlying cause of disease to ensure that thorough and appropriate treatment is provided.

Causes of acute rhabdomyolysis

Acute rhabdomyolysis may arise as a result of exertional and non-exertional phenomena. In regard to exertional rhabdomyolysis, it is now clear that many affected horses have hereditary muscular defects provoking their disease. Specific forms of hereditary exertional rhabdomyolysis that have been identified include polysaccharide storage myopathy (PSSM) and recurrent exertional rhabdomyolysis (RER). Polysaccharide storage myopathy in Quarter horses, Paints and Appaloosas is usually associated with heterozygosity for a mutation of the glycogen synthase 1 gene (*GYS1*).[1] Horses possessing this mutation are classified as having 'Type 1 PSSM' while horses with biopsy findings consistent with PSSM (heavy staining intensity with Periodic Acid Schiff stain, and amylase resistant polysaccharide inclusions) that lack the *GYS1* mutation are classified as having 'Type 2 PSSM', which is believed to be a common phenomenon in Warmblood horses. Horses that are homozygous for the *GYS1* mutation tend to be at greater risk of acute exertional rhabdomyolysis. Furthermore, a small percentage of Quarter horses that concurrently possess the *GYS1* mutation in addition to the malignant hyperthermia related mutation of the ryanodine receptor gene (*RYR1*) are at risk of severe clinical disease.[2]

RER reflects a calcium regulation disorder of Thoroughbreds and the causative mutation remains unknown.[3] Other breeds at risk of acute exertional rhabdomyolysis include polo ponies, endurance horses, and Standardbred racehorses. The cause of disease in these breeds has not been clearly elucidated, however, evidence suggests that polo ponies suffer episodes of rhabdomyolysis early in the season as a result of intense efforts above their level of conditioning, while Standardbreds may have a disorder analogous to that of Thoroughbreds.[4,5] Although unpublished research recently performed by the author found that Arabian endurance horses that developed exertional rhabdomyolysis during racing did not possess the *GYS1* mutation, further research is required to fully investigate etiology of disease in this breed. Some endurance horses develop severe sporadic myopathy as a component of the exhausted horse syndrome, which has been reported to occur in other intense equine athletic activities, and likely reflects the cumulative damage evoked by a combination of hyperthermia, electrolyte imbalance, dehydration, substrate depletion and over-exertion.[6,7]

Multiple non-exertional causes of rhabdomyolysis can be associated with severe acute disease, including nutritional myodegeneration, atypical myopathy, traumatic myopathy (anesthesia or injuries), malignant hyperthermia, and immune mediated phenomena associated with respiratory viruses and *Streptococcus equi* infection. Toxic rhabdomyolysis can occur due to exposure to ionophores, plant toxins, or in a more recent report, a compounded pharmaceutical.[8]

Clinical signs

Clinical signs of acute exertional rhabdomyolysis develop during or soon after exercise, and often consist of inappropriate sweating, tachypnea, stiffness or non-specific lameness, and reluctance to continue. Occasional horses display signs consistent with colic, such as pawing and stretching out. Endurance horses may display inappropriate sweating and tachycardia during exercise, or delayed heart rate recovery, lethargy and tachypnea at check points. Discolored urine may be observed if significant myoglobinuria occurs. However, a proportion of horses participating in different athletic events may develop profound derangements of serum creatine kinase (CK > 100,000 U/L) consistent with severe rhabdomyolysis but with few clinical indications beyond discolored urine. Hence the clinical appearance of acute exertional rhabdomyolysis in some horses correlates poorly with serum CK, and mild clinical signs do not preclude the possibility of substantial rhabdomyolysis.

Clinical signs associated with non-exertional rhabdomyolysis are somewhat dependent on the underlying cause. Foals with nutritional myodegeneration may display stiffness, weakness, abnormal head position and dysphagia due to lingual or pharyngeal dysfunction. Myoglobinuria and signs of cardiac disease (murmurs, tachycardia, sudden death) may also occur. Adult horses with nutritional myodegneration can display similar clinical signs in addition to ptyalism, masseter muscle swelling or pain, and abnormal jaw motion associated with masseter myositis.[9] Immune mediated causes of acute rhabdomyolysis are currently categorized as immune mediated myositis, acute streptococcal myositis, and infarctive hemorrhagic purpura.[10] Immune mediated myositis has been reported in multiple breeds ages (ranging from several months to years). Many affected horses have a history of recent respiratory infection, and initially display stiffness and lethargy followed by rapid and profound gluteal and epaxial muscle atrophy. In horses with active *S. equi* infection, severe acute generalized myositis may occur on rare occasions, typically in young adult Quarter horses. Affected horses become stiff and reluctant to move, and develop firm, swollen and painful muscles particularly in the epaxial and gluteal regions. Many become recumbent and uncontrollably painful, necessitating euthanasia.[10] In addition mature horses with recent exposure to *S. equi* may develop a highly fatal syndrome of infarctive hemorrhagic purpura. Clinical signs include stiffness, limb edema and petechiation of mucus membranes consistent with typical purpura, but affected horses also develop focal intramuscular swellings and/or signs of severe colic related to extensive infarction of the skeletal muscles and gastrointestinal tract.[10]

Atypical myopathy is most prevalent in European countries but is a disease of growing concern in the USA. The clinical signs of atypical myopathy are quite similar to acute nutritional myodegeneration or ionophore toxicosis, with sudden onset of severe rhabdomyolysis unassociated with exertion in a previously healthy horse. Atypical myopathy tends to affect horses less than three years of age grazing poor quality pasture, often in conjunction with an adverse weather event. Affected horses frequently become recumbent and myoglobinuric, and die within 72 hours. Surprisingly, they often present with a fairly bright demeanor and good appetite in contrast to horses with nutritional myodegeneration or ionophore toxicity. The cause atypical myopathy is hypothesized to be acquired multiple acyl-CoA dehydrogenase deficiency resulting in catastrophic disturbance of aerobic metabolism in the muscle cell, though the causative agent involved in this acquired phenomenon has yet to be identified.[11]

Anesthetic myopathy may affect focal areas of muscle or more extensive muscle groups, and may result from prolonged anesthesia combined with inappropriate limb positioning and/or inadequate padding, potentially exacerbated by hypotension and hypoxemia. Horses in lateral recumbency may develop myopathy in the dependent triceps, masseter, biceps femoris and vastus lateralis muscles. Horses positioned in dorsal recumbency may develop myopathy of the longissimus and gluteal muscles, and the adductor muscles of heavy horses in particular are also at risk.[12] Clinical signs consistent with anesthetic myopathy include prolonged recovery often with multiple violent attempts to stand. Violent fasciculations, sweating and reduced weight bearing on affected limbs may be evident once the horse is standing, and swelling of involved muscle groups may be evident visually and on palpation. Signs consistent with classic malignant hyperthermia (hyperthermia, tachycardia, sweating, myoglobinuria and rigidity in response to halothane or succinylcholine) have been reported in a variety of equine breeds, and an associated mutation of the *RYR1*gene has been reported in more than ten Quarter horses displaying this phenomenon.[13]

Diagnosis

Assessment of horses with suspected acute rhabdomyolysis should commence with collection of a thorough history to elucidate any associated causes, including unaccustomed exertion, exertion after an unusual rest period, dietary factors, recent respiratory infection, trauma, anesthesia or possible toxicoses. Physical examination should include thorough visual inspection and palpation of all muscle groups for asymmetry, pain, tightness, swelling or atrophy, including the masseter and neck muscles.

Serum biochemistry analysis including measurement of serum CK and aspartate transaminase (AST) provides invaluable information regarding the severity and time course of a rhabdomyolysis episode, and assists identification of possible complications including azotemia, hypoalbuminemia and electrolyte derangements. Acute severe rhabdomyolysis is often associated with profound serum CK increases (> 100,000 U/L) in addition to hyponatremia, hypochloremia, hypocalcemia, hyperphosphatemia and hyperkalemia with concurrent metabolic acidosis. Normal CK and AST activities in a horse that is reluctant to move, recumbent, or producing discolored urine should prompt consideration of other diagnoses including lameness, colic, laminitis, pleuropneumonia, tetanus, aorto-iliac thrombosis, intravascular hemolysis, bilirubinuria, or neurological disease. Horses suffering from exhaustion syndrome may have some of the described electrolyte derangements, but more commonly display hypokalemia and metabolic alkalosis reflecting profound sweat losses. A CBC can provide useful evidence of infection and inflammation, particularly in horses with acute streptococcal myositis or infarctive purpura hemorrhagica in which prominent neutrophilia and hyperfibrinogenemia may be apparent. Blood gas analysis provides more rapid information regarding electrolyte derangements, in addition to allowing complete assessment of acid base status. Additional samples that should be considered include whole blood for measurement of selenium concentration or glutathione peroxidase activity, and plasma for measurement of vitamin E concentrations if nutritional myodegeneration is suspected. Serum protein M titer is often increased in horses with infarctive hemorrhagic purpura, and whole blood or hair samples can be submitted for genetic testing in horses with possible PSSM or malignant hyperthermia. A urine sample (preferably free catch) should be obtained to evaluate specific gravity as a potential indicator of renal function, and to determine the presence and severity of myoglobinuria. Dipstick analysis can be performed, while respecting the fact that the orthotoluidine pads on many dipsticks are relatively insensitive for hemoglobin and myoglobin. A positive result that might indicate myoglobin should therefore be differentiated from hemoglobin or hematuria by investigation of urine cytology, serum coloration, serum CK and CBC.

Imaging (ultrasound or thermography) of musculature is most useful in diseases provoking asymmetric or focal changes, such as traumatic myopathies, infarctive hemorrhagic purpura or acute streptococcal rhabdomyolysis. Imaging may help confirm the presence and extent of diseased musculature or assist in selection of a biopsy site. Echocardiography should be considered in nutritional myodegeneration, suspected monensin toxicity or atypical myopathy since myocardial damage is present in a significant proportion of affected horses with these diseases. Abdominal ultrasound and abdominocentesis should be strongly considered in horses that display abdominal pain in conjunction with suspected infarctive purpura, and demonstration of findings consistent with gastrointestinal infarction in these cases is associated with an extremely guarded prognosis.

Muscle biopsy can be useful to confirm a suspected differential diagnosis before instituting potentially detrimental therapies (such as corticosteroids) or to distinguish muscle diseases with similar clinical presentations. Techniques for biopsy have been extensively described, however, emphasis should be placed on collecting a sample from an obviously affected muscle group versus choosing a biopsy site that may be more technically convenient. In general the semimembranosus muscle is an appropriate choice for horses with exertional rhabdomyolysis or generalized acute non-exertional rhabdomyolysis, whereas biopsy of the epaxial and gluteal musculature is usually diagnostically superior for immune mediated phenomena (which can be readily diagnosed and distinguished on muscle histopathology). Samples should preferably be submitted as fresh muscle shipped overnight to a laboratory that can process frozen sections to ensure a full complement of diagnostic staining can be done if necessary. Given the high frequency of the *GYS1* mutation in Quarter horses, Belgian drafts, Appaloosas and Paints; genetic testing of hair or blood is a convenient alternative to biopsy, however histopathologic changes consistent with PSSM may not be appreciable in affected horses under two years of age.

Treating and Managing Acute Rhabdomyolysis

The objectives of care in acute rhabdomyolysis include preventing further injury, maintaining mobility and circulation to muscle tissue, relieving anxiety, mania and pain, correcting fluid, acid base and electrolyte derangements, and preventing renal compromise.

Horses suspected of having acute exertional or non-exertional rhabdomyolysis should be immediately restricted from exertion (including walking or trailering). If a horse must exert itself to relocate for transport and treatment, it should not be pushed beyond a slow hand walk with frequent rest to avoid further stress and muscle tension. Horses requiring trailering should be left loose with as much room as possible and the trailer bedded with straw or shavings in case of recumbency during transport. Ramp trailers allow easier movement on or off the trailer whether standing or recumbent. Horses arriving in recumbency should have an IV catheter placed for rapid diagnostics (blood gas) and bolus fluid administration (5 to 10 L of isotonic crystalloid) prior to moving to reduce the potentially detrimental impact of sedation or IV anesthesia that may be required to allow transport into the clinic. This will also facilitate rapid identification of potential contraindications for anesthesia such as hyperkalemia ($K^+ > 5.5$ mEq/L) or profound metabolic acidosis (pH < 7.1). Recumbent horses should be moved to a large, deeply bedded stall with an overhead hoist to facilitate slinging. They should have their feet and distal limbs wrapped, a head protector applied and lubricant placed in their eyes. Laterally recumbent horses should be regularly rotated to relieve dependent muscles, and where appropriate, slinging can be attempted.

Horses with gross myoglobinuria should receive balanced polyionic fluids via nasogastric or IV administration until urine is clear and/or dehydration and any underlying azotemia has resolved. Horses with significant metabolic acidosis (pH < 7.2, bicarbonate < 15 mEq/L) may be candidates for IV bicarbonate supplementation, since aciduria can potentially exacerbate myoglobinuric nephrotoxicity. Hypertonic saline can be added to isotonic crystalloid fluids (at 50 to 100 mls/L) in horses with significant hyponatremia or hypochloremia. Immediate administration of large volumes of hypertonic saline (1 to 2 liters) to horses with acute rhabdomyolysis should probably be avoided to prevent exacerbation of interstitial fluid deficits and systemic acidosis. Horses with possible nutritional myodegeneration should receive an IM injection of a selenium product (1 to 5 mg/ml concentration) at a dose of 2.5 to 3 mg per 45 kg of body weight. Daily supplementation with vitamin E (600-1000 mg/day PO) should be performed until results of analyses are known. Parenteral or nasogastric nutrition and provision of a slurry diet is often necessary for days to weeks in horses with significant masseter myositis. Dantrolene can be useful in these cases to prevent ongoing necrosis and to relieve trismus via potent muscle relaxation effects. Valium may also promote muscle relaxation though patients should be monitored carefully for ataxia.

After terminating anesthesia, active cooling is critical in the management of malignant hyperthermia, and also for hyperthermic exhausted horses, and can be achieved via provision of cold intravenous and nasogastric fluids, cold water hosing and enemas, application of cold packs over major superficial vessels, and cold air convection. Although intravenous dantrolene preparations constitute an invaluable treatment for malignant hyperthermia, the cost is usually seriously prohibitive in horses.

Treatment of horses with acute rhabdomyolysis or infarctive purpura associated with *S. equi* infection is challenging and usually unsuccessful. However, if attempted, antibiotic therapy with penicillin ± rifampin is usually indicated in addition to corticosteroids to suppress inflammatory and immune responses. The prognosis for immune mediated myositis is much better and these cases usually respond very well to administration of dexamethasone, prednisolone or triamcinolone, with gradual decline of serum CK and AST and return of muscle mass. Horses should be monitored for relapses with subsequent respiratory infections.

Diuretics are indicated in horses that display little or no urine production hours after commencing high rate fluid therapy. Diuretics should not be administered to dehydrated horses or horses that are not receiving any fluid therapy. Furosemide can be administered as periodic boluses (0.5-1 mg/kg IV or IM q 2 to 12 hours) however, continuous rate infusion (0.12 mg/kg IV loading dose then 0.12 mg/kg/hr for 8 hours) tends to be more effective. If oliguria persists for 24 hours despite aggressive fluid and furosemide administration, additional diuretics (glucose, mannitol) or introduction of a dopamine infusion is indicated.

Analgesia can be extremely challenging in severe anesthetic myopathy, streptococcal rhabdomyolysis, and infarctive purpura, and affected horses may be dangerously violent. Horses that have received adequate volumes of IV fluids and which do not have azotemia and gross myoglobinuria can be treated with full dose NSAID's (phenylbutazone or flunixin). Analgesia can be expanded by the addition of opioids including butorphanol or fentanyl. Parenteral morphine should be avoided due to its negative impact on gastrointestinal activity; however it is a very useful analgesic for epidural administration if severe myopathy of one or both hind limbs is present. Constant rate infusions of lidocaine or even detomidine may be required to control particularly painful horses. Acepromazine can provide sedation with minimal ataxia, and may also improve muscular perfusion via vasodilatory effects.

Daily monitoring of serum CK can help guide treatment continuing therapy. Dantrolene is likely a useful addition to any situation with known or anticipated ongoing muscle necrosis, such as post anesthetic myopathy, nutritional myodegeneration and atypical myopathy. Dantrolene prevents calcium release from the sarcoplasmic reticulum of the muscle cell, resulting in muscle relaxation and inhibition of necrosis. Contrary to previous accounts, dantrolene is not well absorbed in horses that have been fasted for more than 4 hours.[14] It can be administered 2 to 4 times daily at a dose of 4 to 6 mg/kg PO. Although recent research suggests that dantrolene is efficacious for preventing anesthetic myopathy, evidence from a recent clinical trial suggests that it may cause hyperkalemia and cardiac arrhythmias in some horses (in a study of non-fasted horses pre-medicated with 6 mg/kg via nasogastric tube 60 minutes before general anesthesia). Hence dantrolene administration should also be avoided in hyperkalemic horses or in those with a possibility of hyperkalemic periodic paralysis. Adjunct therapies that may benefit acute rhabdomyolysis include anti-oxidant preparations (DMSO, vitamin C), therapeutic ultrasound, massage, acupuncture, and hyperbaric oxygen therapy. In addition, horses that are willing to eat should be fed a low starch, high fiber diet and monitored for signs of laminitis.

REFERENCES

[1]McCue M, et al. **J Vet Intern Med** 2008;22:1228; [2]McCue M, et al. **Neuromuscul Disord** 2009;19:37; [3]Ward TL, et al. **Am J Vet Res** 2000;61(3):242; [4]McGowan CM, et al. **Vet Rec** 2002;150(17):535; [5]Isgren CM, et al. **PLoS ONE** 2010;5(7): e11594. doi:10.1371/journal.pone.0011594; [6]Fielding CL, et al. **J Vet Emerg Crit Care (San Antonio)** 2009;19(5):473; [7]Macleay JM. **Equine Internal Medicine** Reed SM, et al (eds) 2010;3:488; [8]Thompson JA, et al. **J Am Vet Med Assoc** 2011;239(6):842; [9] Pearson EG, et al. **Vet Rec** 2005;156(20):642; [10]Valberg S. **Large Animal Internal Medicine** Smith B (ed) 2009;4:1388; [11]van der Kolk JH, et al. **Mol Genet Metab** 2010;101(2-3):289; [12]Dodman NH, et al. **J Am Vet Med Assoc** 1988; 193(1):83; [13]Aleman M, et al. **J Vet Intern Med** 2009;23(2):329. [14]McKenzie EC, et al. **Equine Vet J** 2010;42 Suppl 38:613.

CYTOLOGY OF NEOPLASTIC DISEASES AND ITS CORRELATION WITH HISTOPATHOLOGY

Rose E. Raskin, DVM, PhD, DACVP
West Lafayette, IN

INTRODUCTION

Does cytology correlate well with its tissue counterpart and how sensitive can it be to detect neoplasia? These are common questions especially for oncologists. It is well recognized that cytology is excellent for identification of infectious agents as well as for descriptions of individual cell morphology. When a mass is noticed or neoplasia suspected should cytologic biopsy be chosen over excisional or incisional biopsy? This choice has benefits related to cost-conscious efforts, ease of procedure, speed of diagnosis, as well as reduced anesthetic risk. Clinicians should have a sense about the sensitivity, specificity, and diagnostic accuracy for the cytology procedure intended for the suspected neoplasm. Several studies [1-5] have examined these issues in an evidence-based manner using STARD (Standards for Reporting of Diagnostic Accuracy) to fully disclose the study parameters.[6] In most cases, the gold-standard for diagnosis is histology alone but also consideration is made for clinical outcome where possible. This presentation will review the known correlations relative to several selected neoplastic diseases.

Considerations for Evaluation of Correlation Studies

- Type of study (retrospective or prospective), terminology or classification used (diagnosis that is general or specific), level of certainty, experience of pathologist (cytology or histology), blinded or biased to clinical signs, whether substantiated with advanced diagnostic testing for either cytology or histopathology
- Uniformity of tissue sampled relative to tissue architecture related to the degree of exfoliation (limitations of cytology with variability in sites and degree of tissue injury e.g., focal lesion, necrosis, fibrosis, hemorrhage)
- Cytologic method performed e.g., fine needle aspiration, core aspiration, impression smear, washes, scraping, discharges
- Inclusion or exclusion of inadequate cellular specimens in the study and correlation statistics
- Availability and collection of tissue for architectural diagnoses e.g., incisional, excisional, cutting-needle or wedge, necropsy
- Histologic diagnosis dependent on architectural pattern and immunochemistry for specific classification but grouped in broad categories cytology e.g., perivascular wall tumors, cutaneous basilar cell neoplasms
- Use of modifiers for cytologic interpretation provides questionable diagnosis especially for retrospective studies[7]

CUTANEOUS AND SUBCUTANEOUS MASSES

A retrospective study of 292 palpable cutaneous and subcutaneous masses collected by fine needle aspiration from 242 dogs and 50 cats from 1999-2003 were compared with histopathology.[1] Of those, 49 were excluded due to low or poor cellularity. Histologic diagnoses for the 49 excluded cases were chronic inflammation (n = 15), sarcoma (n = 9), sebaceous adenoma (n= 5), hemangioma (n = 5), carcinoma (n = 3), benign follicular tumor (n = 2), histiocytoma (n = 2), epidermal cyst (n= 2), papilloma (n = 2), and 1 each of fibroma, chondroma, lipoma, and mast cell tumor. The cytologist was aware of the signalment data, site, and gross morphology of the lesion. Histopathologic sections were examined by an anatomic pathologist who was unaware of the cytologic diagnosis. Cytologic samples were obtained by FNA and histopathologic samples were collected by surgical biopsy or at necropsy. The agreement between the diagnoses of neoplasia versus non-neoplasia was evaluated using histopathology as the gold standard. Overall, the cytologic diagnosis was in agreement with the histopathologic diagnosis in 90.9% (221/243) of cases. A cytologic diagnosis of neoplasia was obtained in 176 cases (175 true positives and 1 false positive). Sixty-seven cytologic samples were classified as non-neoplastic (46 true negatives, 21 false negatives). For diagnosing neoplasia, cytology had a sensitivity of 89.3%, a specificity of 97.9%, a positive predictive value of 99.4%, and a negative predictive value of 68.7%. The false negative diagnosis of neoplasia in a cytologic sample was due to inflammation with a lack of neoplastic cells (n = 13), or the presence mainly of keratin or necrotic debris (n = 5). Half of those missed were mesenchymal neoplasms. Round cell tumors were diagnosed by cytology in 52 cases, but in 10 cases, cytomorphology could not determine the specific cell type or whether malignant. The histologic diagnosis in these 10 cases was lymphoma (n = 6), cutaneous histiocytoma (n = 3), and cutaneous extramedullary plasmacytoma (n = 1).

The authors concluded that a major limitation of diagnostic cytology is the ability to obtain a satisfactory specimen (16.8% of cases in this study). While screening and resampling could be attempted, some lesions, particularly mesenchymal neoplasms exfoliate poorly. Despite the limitations, overall agreement between cytology and histopathology to determine neoplasia versus non-neoplasia for cutaneous and subcutaneous masses was > 90% of cases.

BONE

A retrospective study was conducted to compare the agreement between cytology and histopathology to distinguish between neoplastic, inflammation, and non-neoplastic proliferative canine bone lesions.[2] A standardized format for reporting diagnostic accuracy (STARD) was followed.[6] A computerized search of canine medical records at the University of Minnesota Veterinary Medical Center identified 65 cases evaluated between 2002 through 2006 by incisional biopsy (IB) and/or excisional biopsy (EB) but 13 were excluded relative to the low cellularity on cytology sufficient for a diagnosis. Cytology vs IB and cytology vs EB were compared pairwise for agreement. The correlation in disease process between cytology and IB was 71%, and for EB was 71% but these two histologic procedures only had a 54% concurrence for the 13 cases that had both procedures. The study brings up the issue of differences between reference methods used as the gold standard in that a needle tissue biopsy may be less accurate compared with the wedge biopsy. When both histologic methods were combined and compared against cytology, there was an overall agreement of 73%. Strong agreement (92%) occurred between cytology and histology relative to the diagnosis of neoplasia. However, there was poor agreement (27%) for the combined inflammatory and non-neoplastic proliferative conditions. Cytology cellularity significantly affected rates of correlation. Cytologic diagnoses from highly cellular samples are more likely to correlate with histopathology than those from less cellular samples.

In a very recent study, the authors evaluated the diagnostic accuracy of an uncommon method of aspiration cytology relative to the standard fine needle aspiration and histopathology in 27 dogs with lytic and /or proliferative bone lesions.[8] The novel method used a 16 gauge bone marrow core aspiration (CA) needle (Jamshidi) in comparison to a 22 gauge standard injection needle (FNA). Alkaline phosphatase (ALP) staining, a known method to recognize osteoblasts producing the bone enzyme, assisted in the cytologic diagnosis of osteosarcoma (OSA). ALP staining was 100% sensitive for the diagnosis of OSA. Histopathologic evaluation followed immediately after CA collection or within two weeks from the amputated limb. Histopathology revealed twenty cases as OSA, two as non-OSA sarcoma, and 5 as reactive bone. OSA was accurately diagnosed in 85% and 95% of FNA and CA, respectively. The higher accuracy of this study[8] to detect neoplasia compared with the former study[6] likely involved the improved cytologic identification of OSA and the focused comparison to one type of neoplasia. Both FNA and CA techniques were identical, 5/5 for the diagnosis of reactive bone. CA using a bone marrow core biopsy needle allowed for penetration of cortical bone and provided a more representative aspirate cytology using a larger bore needle than FNA; however, there was no significant difference in diagnostic accuracy between techniques due to the small case numbers. Needle size appeared to have little influence on sample quality as cellularity was similar in most cases between the two collection methods.

Another study evaluated diagnostic accuracy of ultrasound-guided fine needle aspiration in agreement with histopathology.[9] Aspirations from 36 aggressive appendicular bone lesions having cortical lysis or periosteal proliferation were evaluated by cytology and histology. A 20 gauge needle attached to a 12 ml syringe was used to aspirate the involved site. If cortical destruction was present, the needle penetrated the cortex into the medullary cavity before applying negative pressure. If the cortex was intact, the needle penetrated the soft tissue area adjacent to the cortex having the greatest periosteal reaction. Cytologic diagnosis of sarcoma lesions were further examined following alkaline phosphatase staining. A diagnostic sample with sufficient cellularity was obtained in 32 of the 36 cases. The four non-diagnostic samples were identified as osteosarcoma. Of the 32 diagnostic samples, cytology indicated sarcoma, with a sensitivity of 97% and a specificity of 100%; however this may be skewed by exclusion of the non-diagnostic samples. When a diagnosis of sarcoma was made on cytology, alkaline phosphatase staining suggested osteosarcoma, sensitivity was 100%, specificity 67%, positive predictive value 96%, and negative predictive value 100%.

GASTROINTESTINAL TRACT

The level of agreement between fine-needle aspirates and impression smears of gastrointestinal tract tumors in dogs and cats relative to a reference method of histopathology was evaluated in a retrospective case series.[10] Animals involved 38 dogs and 44 cats with histologically confirmed gastrointestinal tract tumors. Fourteen animals (3 cats and 11 dogs) had gastric tumors. Histologic diagnoses included lymphoma (n =6, including all 3 cats), adenocarcinoma (2), carcinoma (2), histiocytic sarcoma (1), leiomyoma (1), leiomyosarcoma (1), and mast cell tumor (1). Sixty-nine animals (41 cats and 28 dogs had intestinal tumors. Histologic diagnoses for the 41 cats included lymphoma (n =26, adenocarcinoma (6), leiomyosarcoma (4), mast cell tumor (2), carcinoma (1), histiocytic sarcoma (1), and liposarcoma (1). Histologic diagnoses for the 28 dogs included lymphoma (n = 9), adenocarcinoma (7), leiomyosarcoma (6), carcinoma (5), and 1 neuroendocrine tumor (gastrinoma). Cytologic examination of fine-needle aspirates (n = 67) or impression smears (31) were compared with the histologic diagnosis, and extent of agreement was classified as complete, partial, none, or undetermined. Complete agreement was defined as agreement in regard to both cell lineage and cell type. Partial agreement was defined as agreement in regard to cell lineage, but a lack of agreement in regard to cell type. No agreement was defined as a lack of agreement in regard to cell lineage or pathologic process. Extent of agreement was classified as undetermined if the cytologic specimen was unsatisfactory due to hypocellularity, hemodilution, or necrosis.

For 48 of the 67 (72%) fine-needle aspirates, there was complete or partial agreement between the cytologic and histologic diagnoses. For 12 (18%) aspirates, the extent of agreement could not be determined because the cytologic specimen was considered unsatisfactory. For 29 of the 31 (94%) impression smears, there was complete agreement between the cytologic and histologic diagnoses, and for 2 (6%), there was partial agreement. None of the impression smears were considered unsatisfactory. A higher proportion of samples with complete, complete or partial agreement were significantly higher for impression smears than for fine-needle aspirates. In this study there was moderate agreement between results of cytologic examination of fine-needle aspirates from dogs and cats with gastrointestinal tract neoplasia and the definitive histologic diagnosis. The agreement between results of cytologic examination of impression smears and the histologic diagnosis appeared to be higher demonstrating that histologic correlation may be influenced significantly by the method of cytologic collection for gastrointestinal tumors. It is not surprising as the impression smears were collected during surgery or necropsy where better visualization of lesions is possible compared with ultrasound-guided percutaneous aspiration.

LIVER

To improve diagnostic accuracy, a study evaluated using immunochemistry for Ki-67, a proliferation marker, on liver cytologic specimens[11]. Cytologic results from the 30 dogs with hepatic disease were compared with histologic interpretation with and without immunochemistry. Of these cases, nine were identified with malignant neoplasia (2 lymphoma, 4 hepatocellular carcinoma, 2 metastatic carcinoma, and 1 cholangiocarcinoma). Cytologic evaluation alone identified all lymphoma cases, all metastatic carcinomas, the one case of cholangiocarcinoma, and 2 of 4 hepatocellular carcinomas. The two not identified as malignant were considered to be benign on cytology. No cases of hepatic degeneration or inflammation were falsely interpreted by cytology to be neoplasia. Without Ki-67 immunochemistry, the agreement of cytologic evaluation to histopathology was 78%; however, this sensitivity increased to 100% using Ki-67 immunochemistry. Despite the low sample population, there is good evidence that previously low diagnostic accuracy of cytology for liver malignant neoplasia may be markedly improved with ancillary diagnostic tests.

LYMPHOID (SPLEEN)

A recent study was conducted to evaluate the diagnostic accuracy between ultrasound-guided fine-needle aspirates of splenic lesions compared to the reference test of histologic evaluation.[12] This retrospective study involved splenic conditions that were benign, neoplastic, or inflammatory from 29 dogs and 3 cats. Of the 32 records meeting the inclusion criteria of both cytologic and histologic evaluation, one case was excluded due to insufficient cytologic material which was diagnosed from a surgical biopsy as a hemorrhage cyst with reactive lymphoid hyperplasia. Cytologic samples were obtained by fine-needle aspiration, and histologic specimens were obtained via necropsy, surgical biopsy, or ultrasound-guided cutting-needle biopsy. The cytologic diagnosis corresponded to the histologic diagnosis in 19 of 31 (61.3%) cases. When broken down by histologic diagnosis, 11 of 14 (78.5%) animals with non-neoplastic conditions (reactive lymphoid hyperplasia or extramedullary hematopoiesis) and 8 of 15 (53%) neoplastic conditions agreed between cytologic and histologic evaluations. Lack of agreement occurred in 5 of 31 (16.1%). Histologic evaluation of tissue architecture was required to distinguish between reactive and neoplastic conditions in 7 of 31 (22.6%) cases as the results were equivocal. The case selection in this study was too small to significantly evaluate the agreement within specific splenic neoplasms, reactive lymphoid hyperplasia, extramedullary hematopoiesis, or inflammatory cases. It may be concluded that the moderate agreement between cytology and histology in these cases may reflect the need to evaluate tissue architecture.

MAMMARY TISSUE

One prospective study evaluated multiple (>4 per tumor) fine-needle aspirate samples on 50 mammary tumors from dogs.[5] Cytologic and histopathologic examinations were compared for agreement using the histologic diagnosis as the reference method. Seven of 50 (14%) samples were inadequate for diagnosis or inconclusive owing to too few intact cells, excessive blood contamination, thick smears, or preparation artifacts that compromised cell morphology. Of these inadequate samples, one was malignant and 6 were benign by histopathologic examination. The cytologic diagnosis agreed with the histologic diagnosis for benign and malignant tumors in 25/26 (96%) and 15/17 (88%) cases, respectively and overall diagnostic accuracy was 93%. For malignant mammary tumors, cytologic examination had a sensitivity and specificity of 88% and 96% and positive and negative predictive values of 93% and 93%, respectively.

Most recently, a larger study[13] evaluated 90 canine mammary neoplasms to determine diagnostic accuracy of fine-needle aspiration cytology compared with histopathology. Three aspirations were performed on each mammary gland mass using a 22-gauge needle attached to a 5-ml syringe before the mammary glands were surgically excised and submitted for histopathologic examination. Twenty-five (27.7%) of 90 samples were classified as inadequate for cytologic diagnosis related to paucicellularity (n = 12), significant blood contamination (n = 6), inflammatory cells (n = 5), and indistinguishable morphology (n = 2). Histopathologic diagnosis of the inadequate samples were complex adenocarcinoma (n = 13), carcinosarcoma (n = 8), osteosarcoma (n = 2), simple adenocarcinoma (n = 1) and lipoma (n = 1). The diagnostic accuracy, sensitivity and specificity of cytologic examination for diagnosing malignancy were 96.5%, 96.2% and 100%, respectively. The author summarized two studies since 2007 in addition to that mentioned in the previous paragraph and found diagnostic accuracy for canine mammary tumors ranged from 88.5-92.9%. However, an earlier study with 25% diagnostic accuracy used one aspirate biopsy per tumor and a smaller (3ml) syringe than in other studies (5-10ml). It appears cytologic-histologic correlation may be influenced by a number of factors which can improve the diagnostic accuracy.

NASAL TISSUE

A retrospective study to determine the diagnostic accuracy of brush cytology in differentiating non-neoplastic and neoplastic diseases in dogs with chronic intranasal disease involved samples collected over a 12-year period.[3] Histologic diagnosis, follow-up clinical information, or both were used as the gold standard. Dogs free of disease or with no progression of disease at 1 year were considered negative for neoplasia. Samples of nasal brushings from 138 dogs were evaluated and indices of brush cytology agreement in detecting neoplasia included sensitivity, specificity, positive and negative likelihood ratios, and diagnostic odds ratio. Overall diagnostic accuracy was 86.2%. Of 62 cases of neoplastic disease, true-positive and false-negative diagnoses using cytology was 44 (71.0%) and 18 (29.0%) cases, respectively. False-negative diagnoses of neoplasia were not attributed to low cellularity, but to the presence of inflammatory cells that influenced the significance of the neoplastic cells. Brush cytology had a sensitivity of 0.71, specificity of 0.99, positive likelihood ratio of 53.94, negative likelihood ratio of 0.29, and diagnostic odds ratio of 188.33. As brush cytology tends to collect a deeper cell population than surface cytology, diagnostic accuracy is good for chronic intranasal lesions in dogs in differentiating between neoplastic and non-neoplastic conditions.

NEOPLASTIC EFFUSION

The diagnostic accuracy for the diagnosis of ovarian carcinoma in dogs was recently evaluated relative to the cytologic features of effusions.[4] After determination of the cytologic features associated with ovarian carcinoma (OC) e.g., papillary epithelial structures, a blinded study was conducted to identify these features within 82 pleural, pericardial, and abdominal effusions resulting from OC (n = 7), other neoplasms (n = 40), and non-neoplastic disorders (n = 35). Two clinical pathologists with variable length of experiences evaluated the samples and their diagnostic accuracy was evaluated separately. Effusions associated with OC contained blood and had moderate to high cellularity, with neoplastic cells arranged in a prominent tight papillary pattern with minimal intercellular spaces; however this arrangement is not unique to OC and could be found in metastatic carcinomas (mammary, prostatic), but not with mesotheliomas. However, reactive mesothelium from idiopathic pericarditis can be problematic as they produce papillary clusters. For the 2 clinical pathologists, sensitivity ranged from 86% to 100% and specificity from 57% to 97%. Overall accuracies in distinguishing OC from other effusions were 98.8% and 93.9%.

PROSTATE

Cytologic evaluation of canine prostatic disease was compared with histopathologic evaluation.[14] Cytologic and histopathologic specimens of prostate or paraprostatic tissue from 25 adult dogs were retrospectively evaluated. Cytologic samples were obtained by ultrasound-guided fine needle aspiration (US-FNA), prostatic massage, or direct impression smears or aspirates of tissue at surgery. Histopathologic sections were processed from tissue collected by percutaneous ultrasound-guided cutting needle biopsy or incisional/excisional biopsy at surgery or necropsy. One cytology case suggested neoplasia while the surgical biopsy missed it that which was later discovered on necropsy. One cytology case missed the neoplasia identifying only the presence of benign prostatic hyperplasia which was present. Accuracy for neoplasia was 92%, sensitivity 87.5%, specificity 94.1%, positive predictive value 87.5%, and negative predictive value 94.1%. In conclusion, the study suggests there is strong agreement between cytologic and histologic diagnosis for prostatic neoplasia.

REFERENCES

[1]Ghisleni et al. **Vet Clin Pathol** 2006;35:24. [2]Berzina et al. **Vet Clin Pathol** 2008;37:332. [3]Caniatti et al. **Vet Clin Pathol** 2012;41:133. [4]Bertazzolo et al. **Vet Clin Pathol** 201241;127. [5]Simon et al **Vet Clin Pathol** 2009; 38:521. [6]Bossuyt et al. **Vet Clin Pathol** 2007;36:8.[7]Christopher et al. **J Vet Intern Med** 2010;24:496. [8]Neihaus et al. **J Am Anim Hosp Assoc** 2011;47:317. [9]Britt et al. **Journal of Small Animal Practice** 2007;48:145. [10]Bonfanti et al. **J Am Vet Med Assoc** 2006;229:1130. [11]Neumann et al. **Vet Clin Pathol** 2005;34:132. [12] Ballegeer et al. **J Am Vet Med Assoc** 2007;230:690. [13]Sontas et al. **Reprod Dom Anim** 2012;47:125. [14]Powe et al. **Vet Clin Pathol** 2004;33:150.

Cytology in the Diagnosis of Inflammation and Infectious Disease

Jim Meinkoth, DVM, PhD, DAVCP
Stillwater, OK

INTRODUCTION

Inflammatory lesions are commonly encountered in small animal practice. These may present as masses, ulcers, plaques or draining tracts. Inflammatory cells are often easier for the practitioner to identify than the various neoplastic cells, because of a familiarity with the basic cells types from examination of peripheral blood. Therefore, inflammatory lesions can often be diagnosed with confidence by examining slides in-clinic. If cells other than inflammatory cells are present, or if infectious organisms are suspected but not confidently recognized, slides can be sent off for evaluation by a clinical pathologist. However, it is important to recognize that any of the basic inflammatory cells may have a variety of different appearances in cytologic preparations because of aging changes or because of proliferation at the site of inflammation. Sometimes, inflammatory cells, particularly reactive macrophages, may have features that can mimic those of neoplasia and could potentially lead to misdiagnoses. It is important to be able to recognize neutrophils (degenerate vs. non-degenerate), various forms of macrophages, lymphocytes, plasma cells, eosinophils and mast cells. If a lesion is comprised predominantly/exclusively of inflammatory cells, you should (1) try to characterize the type of inflammation as this may suggest a certain underlying process or list of potential processes and (2) examine for infectious organisms. Most infectious organisms stain well with cytologic stains and can be easily recognized with practice. One of the strengths of cytology is rapid identification of infectious organisms.

CYTOLOGIC PATTERNS OF INFLAMMATION

NEUTROPHILIC (ACUTE) INFLAMMATION

Preparations in which greater than ~85% of the cells are neutrophils are referred to as neutrophilic, suppurative or purulent. Non-degenerate (normal) neutrophils are those with tightly clumped, basophilic nuclear chromatin. A few to many of the neutrophils may be hypersegmented or pyknotic (nuclear chromatin broken into tight, round spheres). These are aging changes that may occur *in vivo* or *in vitro* and generally indicated a relatively non-toxic environment. Degenerate neutrophils are those that have lost their ability to control water homeostasis and are undergoing hydropic degeneration. As water diffuses into the cell and into the nucleus via nuclear pores, it causes the nucleus to swell, fill more of the cytoplasm, and stain homogeneously eosinophilic. These often resemble band neutrophils in peripheral blood. The presence of degenerative neutrophils suggests a bacterial cause because these changes are due primarily to toxins produced by bacteria (e.g., endotoxin). However, the absence of degenerative neutrophils does not rule out a bacterial cause because bacteria may be present in low numbers and some bacteria produce low levels of toxin. Therefore, when degenerative neutrophils are present your suspicion of a bacterial infection is increased. Although all cell types are subjected to the same toxin, degenerative change is evaluated in neutrophils only. Degenerative changes can be induced by autolysis, therefore, samples made from dead animals or samples collected and held for a long time before smears are made may show degenerative as well as pyknotic changes.

The most common infectious cause of neutrophilic inflammation is bacterial infection but other organisms (e.g., Sporothricosis) and many noninfectious disorders (e.g., necrotic areas in tumors, immune mediated disorders, trauma, chemical irritants, etc) can also cause neutrophilic inflammation. If no organisms are found, culture may still be informative as organisms can be present in numbers low enough that they cannot be seen on direct smears. This is especially true if the animal has been on antibiotics. Often, the important finding is that the lesion is strictly inflammatory and not neoplastic. Even if the cause of the inflammation is not obvious, this can be an important finding.

PYOGRANULOMATOUS (CHRONIC ACTIVE) INFLAMMATION

Preparations with an inflammatory population that contains both neutrophils and a prominent fraction of macrophages (~15 to 50% macrophages) are termed pyogranulomatous. Macrophages are larger than neutrophils. They tend to have a round nucleus and abundant cytoplasm that is often vacuolated. Macrophages phagocytize large structures and may contain fungal organisms, foreign material, and other cells/cell debris. When macrophages are activated they become larger in size, so some size variation is normal in these cells. Usually the difference in size results from differences in the amount of cytoplasm present rather than nuclear size differences. Macrophages can also become *binucleate* or *multinucleated* and may form large, *inflammatory giant cells*. Sometimes, macrophages with have abundant, evenly basophilic cytoplasm without the typical vacuolation and phagocytized cell debris often noted with these cells. Such macrophages are referred to as 'epithelioid' macrophages as they may resemble epithelial cells. These changes must not be confused with malignant changes seen in tissue cells. Caution is warranted evaluating criteria of malignancy in cells when there is evidence of inflammation, because macrophages can be easily confused with neoplastic cells by inexperienced cytologists.

Pyogranulomatous inflammation suggests a cause other than "routine" bacterial infection. Fungal infections (e.g., Blastomycosis), higher bacteria (e.g., *Actinomyces spp./ Nocardia spp.*), protozoa, and some chronic, noninfectious disorders (e.g., foreign bodies, necrosis) are common causes of pyogranulomatous inflammation.

GRANULOMATOUS (CHRONIC) INFLAMMATION

These are preparations in which the majority of the cells present are macrophages. Multinucleated inflammatory giant cells, reactive fibroblasts, and lymphocytes may be present also. Granulomatous inflammation can be caused by many of the same things as pyogranulomatous inflammation. Cryptococcosis, mycobacterial lesions and steatitis (panniculitis) are some lesions that can often result in lesions have a marked predominance of macrophages with few if any other cells.

EOSINOPHILIC INFLAMMATION

Eosinophilic inflammation or an eosinophilic component to inflammation is indicated in preparations in which large proportions of the cells are eosinophils (greater than 10 to 20% eosinophils). The other cell types are often an admixture of neutrophils, macrophages, mast cells, and lymphocytes. High numbers of eosinophils occur with immune/allergic reactions, eosinophilic granuloma complex in cats, parasitic disorders (e.g., lungworms, heartworm disease), certain fungal infections (e.g., phycomycosis, rarely cryptococcosis), and as a paraneoplastic response to certain neoplasia (e.g., mast cell tumors, lymphoma). Allergic (or 'hypersensitivity' responses) may occur in response to some common bacterial infections or insect bites and result in an eosinophilic component. Occasionally, eosinophil granules may either stain poorly or may stain a tan or muddy-brown (rather than the expected orange color) in tissue preparations making them somewhat more difficult to recognize. However, presence of distinct granules allows for their identification. Also, neutrophils occasionally have a fine eosinophilic stippling in thick exudates and should not be confused with eosinophils. Neutrophil granules are not as large and distinct as eosinophil granules, even if they do stain more prominently than normal. Eosinophil nuclei are somewhat larger than neutrophils and often less segmented, often having a bi-lobed appearance.

LYMPHOCYTIC OR LYMPHOCYTIC/PLASMACYTIC INFLAMMATION

These preparations contain a large proportion of small lymphocytes and mature plasma cells with variable numbers, but generally few, large lymphocytes. These lesions must be differentiated from lymphoma which typically consists of a predominance of medium to large lymphocytes. Small lymphocytes are smaller than neutrophils and consist mostly of nucleus with only a small rim of cytoplasm visible only a small portion of the circumference of the nucleus. Mature plasma cells are B lymphocytes that have been stimulated to produce antibodies. They are larger than lymphocytes because they have more abundant, deep blue cytoplasm. The nucleus is usually about the same size as that of small lymphocytes and is eccentric. There is usually a clear zone in the cytoplasm next to the nuclear ('golgi area' or 'paranuclear clear zone'). By contrast large lymphocytes ('immature lymphocytes' or 'lymphoblasts') are often larger than neutrophils with enlarged nuclei that may have prominent nucleoli.

Cutaneous lesions yielding lymphocytic/plasmacytic inflammation occurs with some injection site reactions, arthropod bites, and some foreign body reactions. This type of inflammation is also commonly seen in feline stomatitis/gingivitis, inflammatory bowel disease and in lymphocytic-plasmacytic hepatitis/cholangiohepatitis.

BACTERIA

With the routine blood stains (Wrights, Diff-Quik) almost all bacteria, whether Gram-positive and Gram-negative, stain blue to purple. *Mycobacterium sp.* are an exception. These organisms have a lipid cell wall and therefore do not stain, but appear as non-staining small rods. These are relatively uncommon, but are occasionally encountered and should be considered in certain clinical conditions. These are usually associated with a granulomatous inflammatory response.

Intracellular bacteria indicate a bacterial infection (primary or secondary) while extracellular bacteria may represent a bacterial infection or contamination. Bacterial infections often have bacteria intracellularly and extracellularly while contamination has only extracellular bacteria. Gram Stains can be used but it is very difficult to get reproducible, accurate staining of bacteria when they are in exudative effusions. Pathogenic bacterial cocci in veterinary medicine are typically Gram-positive (*Staphylococcus spp. and Streptococcus spp.*). Small bacterial rods are typically gram-negative, and all pathogenic small, bipolar rods are Gram-negative. Larger bacterial rods may represent *Bacillus spp.* or *Clostridium spp.* These are often found as postmortem overgrowth in samples collected from animals at necropsy. Elongated, filamentous rods are usually *Actinomyces* or *Nocardia spp.* They are characterized by long, slender (filamentous) strands that stain pale blue, and have intermittent, small, pink or blue areas. It is important to differentiate true filamentous bacteria (one long organism) from chains of bacterial rods. In the latter, clear areas can usually be seen between the individual organisms. Contaminants from the oral cavity like *Simonsiella spp.* are considered nonpathogenic. *Simonsiella* organisms are found extracellularly and often adhered to superficial squamous epithelial cells. They are bacteria that divide lengthwise and line up in parallel rows giving the appearance of a single, large bacterium. Their presence indicates oropharyngeal contamination. They are most commonly found in respiratory washes with oropharyngeal contamination, but may also be found in cutaneous samples if the patient has been licking at the lesion

COMMON PATHOGENIC FUNGI

Fungal infections are common in certain geographic regions. Fungal organisms are first differentiated as to whether they are yeast (systemic, dimorphic fungi) or hyphating in tissue. The yeast can generally be specifically identified by their appearance. Hyphating fungi must be cultured to be identified, although a guess can be made by morphology and location of the lesion (i.e., hyphae from the nasal cavity of a German Shepherd dog is likely *Aspergillus spp.*). Yeast are easily differentiated based on size and morphology.

YEAST

Sporothrix schenckii are round to oval to fusiform (cigar-shaped) and are about 3-9 um long and 1-3 um wide. They stain pale to medium blue with a slightly eccentric pink or purple nucleus. They are small enough that more than one can be present in a neutrophil and many can be found in a macrophage. They must be differentiated from *Histoplasma spp.* organisms which are about the same size. Histoplasma organisms are always round and do not have elongated ('cigar body'forms). Sporothrix are most commonly found in draining tracts of cats. Touch preps will have extremely high numbers of organisms are found. In dogs and horses, organisms are in very low numbers and difficult to see. Often diagnosis in these species requires a clinical suspicion of the disease and culture or histopathology using special fungal stains. Sporothrix is zoonotic and many veterinarians have found this out the hard way after being bitten or scratched by an infected cat.

Histoplasma spp. are round to slightly oval, but are not fusiform. They are about 2-4 u in diameter, stain pale to medium blue, and contain an eccentric pink to purple staining nucleus that is often crescent shaped. There is usually a thin clear halo around the yeast. They can be found in cutaneous lesions in both cats (more common) and dogs. They are more commonly found in other sites. Dogs typically have GI lesions and organisms are found in rectal scrapings. In cats they are often found in transtracheal washes, oral lesions. In either species, disseminated disease can occur with organisms found in bone marrow, blood or internal organs.

Blastomyces spp. are blue, spherical, 8-20 um in diameter, and thick walled. Occasional organisms showing broad-based budding may be found. These organisms are about the size of a neutrophil or bigger, so are easily differentiated from the small yeast. Blastomycosis is most commonly seen in dogs. It is commonly found in cutaneous lesions (multiple), lymph node aspirates and washes of the respiratory system (TTW, BAL). Sometimes it is seen in aspirates from the anterior chamber of the eye.

Cryptococcus spp. are extremely variable in size but are generally 4-15 u in diameter without their capsule and 8-40 um in diameter with their capsule (thick mucoid). The organism stains pink to blue-purple and may be slightly granular. The capsule is usually clear and homogenous. Also, nonencapsulated forms of Cryptococcus may be found. These appear similar but have a thin clear capsule. Occasional organisms showing narrow-based budding may be found. Most typically found as a cutaneous lesion in cats, they may also be seen in dogs. In disseminated disease, organisms may be found in lymph nodes.

Coccidioides spp are most common in the southwestern US. The organisms range from 10 to greater than 100 u in diameter. They appear as blue or clear, double contoured spheres with finely granular protoplasm and will often appear folded or crumpled. Round endospores, 2-5 um in diameter may be seen in some of the larger organisms. The organisms can be seen in cutaneous lesions, commonly invade bone resulting in osteomyelitis and can also be found in the respiratory system (TTW, BAL).

FUNGAL HYPHAE

Fungi that form hyphae require culture for specific identification. Some hyphating fungi do not stain well with the hematologic-type stains and are seen only as negative images surrounded by inflammatory cells.

SELECTED NONINFECTIOUS INFLAMMATORY LESIONS:

CYSTIC LESIONS

Fluid from a variety of cystic lesions (salivary mucocele, seroma, hygroma) is commonly evaluated cytologically. Most of the different cystic lesions cannot be differentiated via cytology. Whether you are dealing with a seroma, pressure point hygroma, etc is determined clinically, as the cytologic findings of all of these lesions are similar. Cytologically, it is important to look for evidence of (1) infection/acute inflammation and (2) tumor cells (to identify cystic neoplasia – common in mammary glands). If neither of these is found, the sample is often read out as 'benign cystic fluid' and it is up to the clinician to determine the specific lesion. Most cystic fluids contain numerous macrophages, although some cysts are of low cellularity. The macrophages may be relatively small, unstimulated macrophages resembling blood monocytes or large, actively phagocytic macrophages with abundant vacuolated cytoplasm. It is important to note that macrophages may divide in tissues and thus few to moderate numbers of mitoses may be seen and this should not be considered evidence of neoplasia.

Neutrophils may be present in variable numbers resulting from inflammation induced by the pressure on surrounding tissue. Intracystic hemorrhage occasionally occurs and is recognized by the presence of RBC breakdown products (hemosiderin, hematoidin). Fluid from a salivary mucocele can often be recognized by the presence of amorphous blue mucus "tufts or clouds or pillows" scattered amongst the cells. If fluid is obtained when aspirating a mass that is a suspected tumor, the fluid may be submitted for cytology but it is important to also collect additional aspirates if a solid tissue component remains following removal of the fluid. Often tumor cells do not exfoliate into the cystic portions of tumors and finding from solid tissue aspirate will be markedly different than from the fluid itself.

Injection site reactions

Reactions at the sites of injections are common lesions sampled in small animal practice. These may be neutrophilic, pyogranulomatous, granulomatous, or lymphocytic. Occasionally amorphous, homogeneous, brightly eosinophilic material may be seen extracellularly and within macrophages with vaccine reactions. It is common to find spindle cells that represent reactive fibroblasts in these lesions. This is of concern in cats, where injections site sarcomas occur. If suspicious spindle cells are seen, the lesions should either be carefully monitored to ensure it regresses in a short period of time or removed and submitted for histopathology.

EOSINOPHILIC GRANULOMA COMPLEX LESIONS

These are often suspected based on their clinical appearance. Cytology is done to confirm the nature of the lesion. If the lesion is thin and plaque-like, scrapings may be the best method of collection. Scrapings or aspirates usually contain high numbers of eosinophils. Often, especially in scrapings, numerous free eosinophil granules are present in the background of the smear. Fibroblasts are present in variable numbers.

STEATITIS (FAT NECROSIS, PANNICULITIS)

This is a fairly common cutaneous lesion. It may present as a single lesion, where it usually is the result of prior tissue trauma. Sometimes, it presents as multiple nodules across wide areas of skin. It has been associated with vitamin E/selenium deficiency and pancreatitis, but often is idiopathic. Aspirates from areas of inflamed fat typically consist of inflammatory cells and high numbers of variably sized lipid vacuoles. Granulomatous (macrophagic) inflammation is most common, although pyogranulomatous and purulent inflammation may be seen. The macrophages often contain numerous small, clear, lipid vacuoles within their cytoplasm. Large, multinucleated macrophages are commonly seen and must not be confused with neoplastic cells.

Case Studies in Infectious Disease Cytology

Jim Meinkoth, DVM, PhD, DAVCP
Stillwater, OK

INTRODUCTION

Identification of infectious organisms is a powerful use of cytology. This session will present cases, some common and some unusual, in which cytologic examination resulted in a specific etiologic diagnosis. The features of the infectious organisms as well as hints that might raise the clinical index of suspicion will be highlighted.

POST SURGICAL PERITONEAL EFFUSION IN A DOG

Case Presentation

An 11-year-old, intact male, Poodle was presented at Oklahoma State University, Center for Veterinary Health Sciences (OSU-CVHS) with a history of vomiting, abdominal distention and elevated biochemical liver values. Abdominal ultrasound revealed a moderate amount of abdominal fluid, widespread vascular mineralization, and a markedly thickened and irregular gallbladder wall. Cytologic examination of abdominal fluid revealed marked pyogranulomatous inflammation with abundant golden-to-dark green pigment, consistent with bile, present extracellularly and within macrophages. Cytologic interpretation was bile peritonitis. No infectious organisms were identified, and subsequent aerobic and anerobic bacterial cultures were negative.

The patient was started on intravenous ampicillin/sulbactam and enrofloxacin. Exploratory abdominal laparotomy revealed a ruptured gall bladder and a cholecystectomy and duodenotomy were performed. Following surgery, the effusion persisted despite antibiotic therapy and the patient's condition began to deteriorate.

Cytologic Findings

Abdominal fluid collected 6 days post-surgery cytologic exam revealed marked pyogranulomatous inflammation with numerous fungal yeasts present extracellularly as well as within neutrophils and macrophages. The organisms were round-to-oval, basophilic, ~3 – 8 um in diameter and had a clear, circumferential halo. Few organisms showed narrow-based budding. Rare elongated yeast or pseudohyphae were noted phagocytized by macrophages. No bacteria were seen.

Cytologic Interpretation and Clinical Outcome:

The findings were interpreted as septic peritonitis with fungal yeast morphologically compatible with *Candida spp.* A second exploratory with laparotomy was performed and revealed an abscess at the gallbladder fossa which was debrided and omentalized. The patient subsequently developed anuric renal failure and was euthanized 2 days after the second surgery. While septicemia was suspected, blood cultures were not performed and necropsy was declined by the owner.

Culture of the second abdominal fluid sample subsequently revealed the presence of *E. coli* and 2 morphologically distinct fungal colonies consistent with *Candida spp.* Colonies were identified as *Candida albicans* and *Candida glabrata,* respectively. No antifungal treatment was initiated in this patient. Histopathology of the gallbladder showed severe, diffuse, chronic-active, necrosuppurative cholecystitis with fibroplasia.

Discussion

Candida are commensals normally inhabiting the alimentary tract as well as other tissues. Overgrowth and colonization of fungal organisms may occur and subsequent to colonization, these opportunistic pathogens can spread to the peritoneum when there is disruption of the mucosal barrier resulting from conditions such as surgery, perforation, or neoplasia.

Candida peritonitis is relatively common in humans and is associated with significant morbidity and mortality compared to sterile or bacterial peritonitis.[1,2] Recognized risk factors include peritoneal dialysis, hollow viscous perforation, previous abdominal surgery (+/- anastomotic leakage), trauma, neoplasia, pancreatitis, previous antibiotic administration, and immunosuppressive therapy. *Candida albicans* peritonitis has been reported in dogs only twice.[3,4] We have identified several cases of dogs with peritonitis complicated by *Candida* spp over the last few years. They presented with duodenal perforation due to NSAID use, intestinal resection and anastamosis following post spay dehiscence, intestinal foreign body removal, cholecystectomy due to gall bladder rupture and dehiscence of intestinal biopsies following exploratory laparotomy. In all cases, initial diagnosis of fungal peritonitis was made via cytologic examination of peritoneal effusions which revealed marked suppurative to pyogranulomatous inflammation with numerous 3-8μm oval, deeply basophilic yeast organisms with thin clear capsules noted within macrophages or neutrophils and extracellularly. Additionally, Candida spp. may form hyphae and pseudohyphae. *Candida* peritonitis should be considered in patients with peritoneal contamination of the gastrointestinal or biliary contents.

CRUSTY SKIN LESIONS IN A KITTEN:

Case Presentation

A young kitten was presented by a Good Samaritan who had found the animal by the side of the road while bicycling. They reported that they took the kitten back to their house approximately 10 miles carrying it inside their bicycle shirt. Upon physical exam the kitten was in poor body conditions and had several alopecic, dry, crusty, dermal lesion. Skin scrapings of the lesions were collected and submitted for cytologic examination.

Cytologic Findings

The smears were highly cellular and contained a mixture of mature squamous epithelial cells and inflammatory cells with abundant cellular debris and scattered hair shaft. The inflammatory cells with predominantly neutrophils with lower numbers of macrophages. Numerous organisms were present, predominantly extracellularly and often in dense clusters. These are small round to rectangular ("brick shaped") basophilic structures approximately 2-4 um in diameter and were surrounded by a thin clear halo. Similar structures were seen to be elongated from hyphae-like structures. Rarely, organisms were seen associated with hair shafts.

Cytologic Interpretation and Clinical Outcome:

The findings were interpreted as neutrophilic inflammation with dermatophytosis. The owner subsequently developed classic ringworm lesions on his chest where he had carried the kitten inside his shirt.

Discussion

Dermatophytosis is often suspected by clinical presentation and classically confirmed by culture. Dermatophytes stain well with routine cytologic stains (Diff-Quik, Wright's) and have a distinctive morphology allowing rapid identification, although speciation is not possible via cytology alone.[5] Skin scraping is the most effective way of collecting samples. Organisms are often abundant in the lesions and the most difficult part of diagnosing ringworm via cytology is getting cells and organisms from scrapings to stick to the slides and not wash off in the stain, since these are usually dry lesions. An effective method is to scrape sufficiently deep to produce exudation of a little blood/serum from the site. This serum will help to adhere the cells to the slide. Serum may be collected onto the scalpel blade from one such scraping to help adhere more superficial cells collected at other sites. Scrapings that are adequate should contain high numbers of squamous epithelial cells, not only superficial, anucleate cells but more basal cells. Often lesions will have dense populations of inflammatory cells, but not always. Characteristic organisms are significant even in the absence of inflammatory cells. Organisms may be seen adhered to or within hair shafts if adequate numbers remain adhered to the slides. However, more often free organisms are found in the background of the slides not associated with hair shafts.

CEREBROSPINAL FLUID FROM A SEIZURING DOG

Case Presentation

A five-year-old, spayed female, Shetland sheepdog mixed-breed dog presented to Oklahoma State University-Boren Veterinary Medical Teaching Hospital with a history of two seizures, progressive hind limb ataxia, and polyuria/polydipsia. The first seizure occurred 3 years prior and the patient made a complete recovery. Two and a half months prior to presentation a hypermature cataract with capsular rupture was diagnosed in the right eye and a Neomycin/Polymyxin B/Dexamethasone ophthalmic solution was prescribed. Approximately one month later, polydipsia was noted. A second seizure occurred one month prior to presentation. The patient remained lethargic after the second seizure and developed hind limb ataxia one week before examination. Vaccinations and monthly heartworm preventatives were current.

Initial physical examination findings included normal temperature (100.8° F), heart rate, pulse, hydration status, mucous membranes and capillary refill time. A neurologic examination demonstrated conscious proprioceptive deficits, normo- to hyper-reflexia in all limbs, ataxia when wheel-barrowing and no pain on palpation of the spinal column. An orthopedic examination showed good range of motion in all limbs. Ophthalmic examination on day two revealed aqueous flare, diffuse posterior synechiae and a hypermature cataract with neovascularization across the lens of the right eye, culminating in a diagnosis of chronic anterior uveitis and hypermature cataract. No abnormalities were found in the left eye. Results of a CBC, chemistry profile and T4 were unremarkable. Urinalysis was within normal limits, with exception of hyposthenuria (1.006). A computed tomography scan of the brain performed on day 2 showed no abnormalities. Cerebrospinal fluid (CSF) was collected via atlanto-occipital cisternal puncture and submitted for cytologic examination.

Cytologic Findings

Fluid analysis revealed a marked pleocytosis (total nucleated cell count 1,106/µL); due to insufficient quantity, CSF protein was not measured. On a cytocentrifuged preparation stained with Wright-Giemsa a marked eosinophilic pleocytosis (41% eosinophils, 39% large mononuclear cells, 11% large granular lymphocytes, and 9% non-degenerate neutrophils) was evident. In neutrophils and macrophages rare, 6-8 micron, oval to kidney-bean shaped organisms with basophilic, granular cytoplasm that obscured nuclei and were surrounded by thin non-staining cell walls were seen. Although no endosporulation was seen, rare empty casings were noted.

Cytologic Interpretation and Clinical Outcome:

The cytologic interpretation was eosinophilic pleocytosis with probable *Prototheca* organisms. Based on the severity of the patient's clinical signs and the grave prognosis associated with protothecosis in dogs, the client requested euthanasia two days after presentation. Gross necropsy examination showed no obvious gross changes other than pulmonary congestion. Histologic examination revealed mononuclear inflammation and intralesional algal organisms in the cerebral leptomeninges, the ependymal lining of the lateral ventricles with extension into the sub-ependymal neuropil, the choroid of the right eye, peri-ocular fascia of the left eye, the myocardium, and the renal tubules and interstitium. Although the gastrointestinal tract was examined for a potential primary site of infection, no inflammation or *Prototheca* organisms were found. Culture of CSF and a sub-dural/olfactory bulb swab yielded colonies typical of *Prototheca* species. Intestinal culture produced marked numbers of normal flora, but *Prototheca* was not isolated. A portion of the 18S rDNA from the cultured organisms was amplified by PCR using universal eukaryote primers. Sequencing of the amplicons identified the organisms as *Prototheca zopfii*.

Discussion

Prototheca species are saprophytic, achlorophyllous algae closely related to *Chlorella* species. *Prototheca* is ubiquitous in the environment world-wide; however, clinically significant disease is relatively rare. The two pathogenic species, *Prototheca zopfii* and *P. wickerhamii* cause disease in various species. Gastrointestinal signs are the most common manifestation of disseminated disease in dogs followed by ocular and neurologic manifestations. The prognosis in dogs is guarded to poor. This case was unusual because the chief complaints were hind limb ataxia and seizures with no clinical involvement of the gastrointestinal tract. Although rare cases of protothecosis manifesting as solely neurologic disease have been previously reported, the majority present with hemorrhagic colitis.[6]

Diagnosis of protothecosis is usually accomplished by culture and/or cytologic examination and visualization of organisms in rectal scrapings, lymph node aspirates, vitreous fluid or, sometimes, urine sediment. On cytologic examination *Prototheca* organisms vary in size depending on the species and stage of development and range from 1.3 to 13.4 microns in diameter and from 1.3 to 16.1 microns in length. They have thin non-staining cells walls, basophilic granular cytoplasm and small centrally placed nuclei and reproduce by endosporulation.

An eosinophilic pleocytosis in CSF in association with protothecosis has been published previously.[6,7] Thus, an eosinophilic pleocytosis should raise the index of suspicion for protothecosis. Diligent cytologic examination is required due to the scarcity of organisms, and lack of identification on cytology does not preclude isolation via culture.[6] Other causes of a marked eosinophilic pleocytosis (>20% eosinophils) in dogs include idiopathic eosinophilic meningitis, cryptococcosis, neosporosis, and migrating helminths.[8]

REFERENCES

1. Miles R, et al *Kidney International* 2009; 76: 622. 2. Prakash A et al *Indian J Gastroenterol.* 2008;27(3):107. 3. Rogers CL, et al. *J Vet Emerg Crit Care* 2009; 19(2):193. 4. Ong KC, et al. *J Vet emerg Crit Care* 2010; 20(1):143. 5. Caruso K, *Vet Clin Path* 2002;31(1),13. 6. Stenner VJ, et al. *Med Mycol.* 2007; 45:249. 7.Gupta A, et al. *Vet Clin Pathol.* 2011; 40(1):105. 8.Windsor RC, et al. *J Vet Intern Med.* 2009; 23:275

QUALITY ASSURANCE FOR POINT-OF-CARE ANALYZERS IN VETERINARY MEDICINE: WHAT EVERY USER SHOULD KNOW

Bente Flatland, DVM, DACVIM, DACVP
Knoxville, TN

COPYRIGHT STATEMENT AND ACKNOWLEDGMENT

Information and recommendations made herein are adapted from a draft guideline for point-of-care testing in veterinary medicine, currently being written by the American Society for Veterinary Clinical Pathology (ASVCP) Quality Assurance and Laboratory Standards (QALS) Committee, and are used with permission of the ASVCP. This draft guideline represents the work of multiple hard-working individuals, and the author acknowledges these contributors. Once finalized, the ASVCP POCT guideline document will be published on-line at www.asvcp.org ("Publications" link).

INTRODUCTION

Point-of-care testing (POCT), also known as "bedside" or "near-patient" testing, is performed using various instruments (point-of-care analyzers, POCA) of varying complexity. In human medicine in the United States, POCT is subject to federal regulatory oversight governed by the Clinical Laboratory Improvement Amendments of 1988 (CLIA) and administered by the Centers for Medicare and Medicaid Services (CMS).[1] CLIA classifies laboratory tests according to complexity as "moderate", "high", and "waived".[1,2] Waived tests (non-instrumental testing systems and small, hand-held devices) are not subject to regulatory oversight.[1,2] Years ago, POCT fell primarily in the "waived" category. Technological advances and increasing device sophistication mean that, today, POCA can also be found in the other categories. At this time, veterinary laboratory medicine (in any setting) is not governed by regulatory oversight, which means that veterinarians ideally should demonstrate a commitment to quality assurance (QA) and quality control (QC) from within the profession.[3]

Concern about the quality of veterinary in-practice testing has been expressed by veterinarians themselves in published literature[4,5]; however, little, if any, concise and practical guidance is available to veterinary practitioners on this topic. Veterinary guidelines and textbooks are aimed at laboratory professionals and complex laboratory equipment as found in reference laboratories. Teaching about laboratory QA/QC in veterinary curricula is generally scant, leaving new graduates with little training in how to establish, evaluate, and maintain the quality of in-practice laboratory testing. Acknowledging this void, the Quality Assurance and Laboratory Standards (QALS) Committee of the American Society for Veterinary Clinical Pathology (ASVCP) formed a POCT subcommittee in 2009 charged with writing guidelines for POCT in veterinary medicine. Our goal is to finalize these recommendations in 2012. In a related development, the National Association of Veterinary Technicians of America (NAVTA) recently approved a clinical pathology specialty for veterinary technicians.[6] This development should facilitate training of veterinary technicians with expertise in laboratory quality management, and ASVCP is actively reaching out to NAVTA so that our respective organizations may collaborate on this effort.

FORMALIZING POCT

Veterinary hospitals of all sizes having POCA should establish a formalized approach to POCT that includes a written quality plan and/or manual. Academic veterinary medical teaching hospitals and large specialty practices should consider forming a POCT committee or working group that oversees POCT policies and acquisition, maintenance, and quality management. Suggestions regarding composition, responsibilities, and function of POCT committees and working groups are available[7-9], but generally all major POCT stakeholders should be represented in such a body (e.g., clinicians, technicians, medical records, billing, and information technology). The individual veterinary facility should decide whether any POCT committee's role is primarily as an advisory body or whether it also has a policing and enforcement role.

QA, QC, QUALITY PLAN, QUALITY MANUAL, AND DOCUMENT CONTROL

In the clinical laboratory, *quality assurance* (QA) refers to procedures that minimize all types of error (pre-analytical, analytical, and post-analytical). *Quality control* (QC) refers to specific steps taken to minimize analytical error. A *quality plan* is a concise, written statement summarizing the philosophy, goals, and framework upon which a facility's quality system is based.[10] A quality plan may be part of a more comprehensive *quality manual* that also contains detailed policies, standard operating procedures (SOPs), and forms relevant to all aspects (operational, analytical, and reporting) of running the in-practice laboratory. It is important that policies and SOPs are updated regularly and maintained according to a *document control policy* which ensures that only current, approved document copies are in circulation. Veterinary practices having in-house laboratory equipment are encouraged to use such documents.

EQUIPMENT OPERATOR TRAINING AND ASSESSMENT

Adequate equipment operator training is an essential component of QA.[3] The hospital manager should ensure that all personnel performing POCT are properly trained, and provision should be made for both initial training and continuing education. *Examination audits* (competency assessments) should be carried out by the manager (or other qualified individual) to document personnel competence; audits should follow initial training and may be performed periodically thereafter at the manager's discretion. Maintenance of operator training logs or other training records is also recommended.[11]

EQUIPMENT MAINTENANCE

Manufacturer's recommendations for maintenance of equipment should be followed and documented. Periodic software updates should be performed as needed/recommended by the manufacturer. An instrument performance log should be maintained to document any problems with the POCA or its results, any trouble-shooting performed, and any corrective action(s) taken as a result. Efficacy of corrective actions should be confirmed.[11]

QUALITY MANAGEMENT OPTIONS FOR VETERINARY PRACTICES SEEKING TO MAINTAIN POCA

Quality management options include (1) use of non-statistical QA procedures, (2) comparability testing, (3) participation in an external quality assurance program (EQA program, a.k.a. proficiency testing), and (4) use of quality control materials (QCM) with analysis of resulting "control data". These options are not mutually exclusive, and multiple strategies are ideally employed.

Non-statistical QA procedures

Non-statistical quality assurance procedures address all aspects of laboratory testing. Such safeguards involve many "common sense" procedures already in place in many well-run veterinary practices. Examples of non-statistical QA procedures include (but are not limited to) use of written policies, SOPs, and forms; documentation of personnel training; analysis of properly obtained and handled samples; use of non-expired, properly stored reagents and QCM; documentation of instrument maintenance and repairs; correlation of laboratory results with other clinical findings; monitoring of trends in patient data; use of repeat and medical review criteria; use of appropriate reference intervals; and ensuring accurate transference of laboratory results into patient medical records.

Comparability testing

Comparability testing refers to analysis of a patient sample by two or more laboratories to ascertain agreement of results. E.g., confirming a high calcium concentration from an in-house analyzer by sending an aliquot of sample to a reference laboratory. A limitation of this recommendation for POCA operators has been knowing what constitutes "acceptable" agreement between results. Forthcoming guidelines from the ASVCP should help.

EQA programs (proficiency testing) and comparability testing

In an EQA program, multiple laboratories (typically on a quarterly basis) measure aliquots of the same sample and send results to the program administrator. The program administrator then compiles these data and reports results back to the participants together with relevant statistics. The value of an EQA program is directly proportional to the number of participants, and participating laboratories or practices should ensure that their results are compared to an appropriate peer group using the same or similar analytical methods. Marked deviation of an individual laboratory's or practice's results from the peer group mean should prompt trouble-shooting to determine the underlying cause. Ideally, a veterinary-specific EQA program should be used. A limitation of this recommendation for POCA operators is that EQA programs currently available to veterinarians are aimed at reference laboratories using more complex equipment. EQA options aimed at veterinary in-clinic laboratories are needed.

Use of quality control materials (QCM) ("running controls")

The best way to determine whether a laboratory instrument is performing adequately is to assay material having known analyte concentrations.[5] Use of QCM is encouraged and is the *only* way to confirm proper function of the entire POCA system (instrument, reagents, and operator). POCA may have internal, built-in, electronic (or other) QC functions. These are valuable; however, further studies are needed to assess adequacy of these as sole QC strategies. At this time, internal instrument QC functions should not be considered a substitute for using QCM or for non-statistical QA procedures.

A *control run* refers to one analysis (measurement) of a QCM. *Control data* refers to laboratory results generated by analysis of QCM. *Control limits* are cut-off points or thresholds outside which control data are considered unacceptable ("out-of-control"). Control data should be recorded and archived.[12] POCA software programs may allow storage and analysis of control data, either locally by the operator and/or distantly by the instrument manufacturer.

QCM are available through biochemistry supply companies worldwide and may be designated as "assayed" or "non-assayed" materials.[13] Use of assayed QCM is recommended. For assayed QCMs, the manufacturer reports a mean analyte concentration (frequently with a range and standard deviation) for each analyte in the QCM package, together with relevant analytical methods used.[13] Veterinarians should ideally purchase QCM having a mean determined by the same analytical method(s) as used by the POCA in question.

How often should controls be run?

QC should be performed frequently enough that QCM are used cost-effectively and that POCA analytical error can be reliably detected. Daily QC (i.e., analyzing QCM each day that the POCA is to be used for patient samples) is an ideal recommendation that may not be suitable for all clinics, depending upon the POCA analyzer type, clinic operating budget, and caseload. Each individual veterinary clinic using POCA must determine frequency of their QC.

What should you do with control data?

Control data analysis can take several forms. The simplest analysis is to determine whether results fall within the manufacturer's reported ranges for the QCM. If yes, adequate POCA performance is assumed. If no, and/or if data are accompanied by instrument error flags, results of the control run should be rejected and trouble-shooting commence. A limitation of using QCM manufacturer ranges as the control limits is that these do not take the individual POCA's analytical performance into account; as a result, there is a possibility that clinically significant analytical error could be missed.

A more rigorous approach is to graph control data using *control charts*, which plot analyte concentration on the y-axis and time on the x-axis (e.g., Levy-Jennings control charts). Traditionally, control limits used with control charts are based on a 1_{2s} control rule; that is, any one (*1*) result falling outside ± 2 standard deviations (*2s*) from the mean is rejected as unacceptable ("out-of-control"). Control charts are useful for spotting individual unacceptable results and for spotting "drifts" in results over time. The 1_{2s} control rule has good probability for detecting analytical error (probability of error detection, Ped); however, a limitation of 1_{2s} is that it may cause false rejection of a small percentage of acceptable control runs (resulting in increased expense if QCM are re-analyzed to verify an unacceptable result).[14] Generic application of 1_{2s} as the only control rule is no longer recommended when statistical QC using validated rules is possible.[14,15]

Many potential control rules exist. The most rigorous analysis of control data (often referred to as "statistical QC") uses validated control rules tailored to a specific instrument's analytical performance and a pre-determined quality requirement. A limitation of this approach in the past has been that the veterinary community lacked consensus about quality requirements for commonly measured analytes in the major domestic species; forthcoming ASVCP guidelines should help. All clinically used laboratory instruments have some degree of inherent analytical error. If statistical QC is desired, an *instrument performance study* is needed to measure a given instrument's *imprecision* (representing random analytical error) and *bias* (representing systematic analytical error). Values for imprecision and bias, in turn, can be used to calculate a given instrument's *observed total error* (TE_{obs}). TE_{obs} can be compared to a predetermined *quality requirement* (such as *allowable total error*, or TEa) to determine whether instrument performance is acceptable and robust enough to permit statistical QC.

QC validation is the process of choosing appropriate control rules for a given instrument and requires information about that instrument's imprecision and bias in additional to a quality requirement (such as TEa). It is recommended that selected control rules have a probability of analytical error detection (Ped) ≥ 85 to 90% and a probability of false rejection of control data (Pfr) ≤ 5%.[15,16] Statistical QC using validated rules is only possible (i.e., will only provide optimal Ped and Pfr) if instrument analytical performance is adequately robust (as determined by measured imprecision and bias and the chosen quality requirement).

Based on a recent study, it is possible to monitor analytical performance of biochemistry POCA in veterinary medicine using a simple control rule called 1_{3s} and one or two QCM.[16] The 1_{3s} rule states that any one data point (*1*) falling outside ± 3 standard deviations (*3s*) from the mean is rejected as unacceptable ("out-of-control"). In instruments with adequately robust analytical performance, using the 1_{3s} rule and one QCM or two QCM (of differing analyte concentrations), Ped of ≥ 85% and ≥ 90% were achievable, respectively. Achievable Pfr was 0%.[16] Presumably, this rule may be used with similar success in other POCA types (e.g., hematology analyzers); however, this remains to be investigated.

REFERENCE INTERVALS

Instrument manufacturers commonly provide reference intervals to aid laboratory data interpretation. Ideally, such reference intervals should be validated if there is any doubt regarding suitability of the manufacturer's reference interval for the patient population being tested at the POCT site. For details concerning establishment and validation of reference intervals in veterinary medicine, see the ASVCP reference interval guideline document.[17]

SUMMARY

A formalized and comprehensive approach to the quality management of POCA is recommended in all veterinary settings. Facilities should document laboratory quality procedures by means of a written quality plan or manual that includes policies, SOPs, and forms. Logs and records should be maintained and archived. Equipment operator training and assessment should be documented and ongoing. Non-statistical QC procedures should be routinely used. QCM and interpretation of control data should be used where applicable to monitor POCA performance and ensure optimal quality of patient laboratory data. Reference intervals for patient data interpretation should ideally be validated.

REFERENCES

[1]Ehrmeyer SS. Regulatory issues regarding point-of-care testing. In: Price CP, St. John A, Kricka LL, eds. *Point-of-Care Testing: Needs, Opportunity, and Innovation*, 3rd ed. Washington, DC:AACC Press;2010:237-249. [2]Categorization of tests. Available at http://www.cms.gov/CLIA/10_Categorization_of_Tests.asp#TopOfPage. Accessed January 11, 2012. [3]Flatland B, Freeman KP, Friedrichs KR, Vap LM, Getzy KM, Evans EW, Harr KE. ASVCP quality assurance guidelines: control of general analytical factors in veterinary laboratories. *Veterinary Clinical Pathology* 2010; 39:264-277. [4]Mitzner BT. It's time for in-house quality assurance (editorial). JAAHA 2002; 38:12-13. [5]Freeman KP, Evans EW, Lester S. Quality control for in-hospital veterinary laboratory testing (commentary). *Journal of the American Veterinary Medical Association* 1999; 215: 928-929. [6]NAVTA announces 11th veterinary technician specialty. Available at https://www.navta.net/press/navta-announces-11th-veterinary-technician-specialty. Accessed January 11, 2012. [7]Kost GJ (ed.) *Principles and Practice of Point-of-Care Testing*. Philadelphia, PA: Lippincott Williams & Wilkins, 2002. [8]Price CP, St. John A, Hicks JM (eds.) *Point of Care Testing*, 2nd edition. Washington, DC: AACC Press, 2004. [9]Medicines and Healthcare Products Regulatory Agency. Device Bulletin Management and use of IVD point of care test devices. DB2010(02). London, England: Medicines and Healthcare Products Regulatory Agency; February 2010. [10]Sacchini F, Freeman KP. Quality documentation challenges for veterinary clinical pathology laboratories. *Journal of Veterinary Diagnostic Investigation* 2008; 20: 266-273. [11]American Society for Veterinary Clinical Pathology Quality Assurance Guidelines. Available at www.asvcp.org. Accessed January 11, 2012. [12]Barry PL. QC – The records. In Westgard JO, etc. *Basic QC Practices*. 3rd ed. Madison, WI: Westgard QC, Inc; 2010:123-127. [13]Quam EF. QC – the materials. In Westgard JO, ed. *Basic QC Practices*. 3rd ed. Madison, WI: Westgard QC, Inc; 2010:161-169. [14]Westgard JO. QC – The Westgard Rules. In Westgard JO, editor. *Basic QC Practices*, 3rd edition. Madison, WI: Westgard QC, Inc.; 2010:73-86. [15]Westgard JO. QC – The chances of rejection. In: Westgard JO, editor. Basic QC Practices, 3rd edition. Madison: Westgard QC; 2010. p. 215-230. [16]Rishniw M, Pion P, Maher T. The quality of veterinary in-clinic and reference laboratory biochemistry testing. *Veterinary Clinical Pathology*; in press. [17]American Society for Veterinary Clinical Pathology Guidelines for the Determination of Reference Intervals in Veterinary Species and Other Related Topics. Available at www.asvcp.org. Accessed January 11, 2012.

INTERPRETING VALUES OUTSIDE REFERENCE INTERVALS: IS IT THE PATIENT OR THE MACHINE?

Kendal E. Harr, DVM, MS, DACVP
Seattle, Washington

COPYRIGHT STATEMENT AND ACKNOWLEDGEMENT

The following abstract is partially excerpted from ASVCP Reference Interval Guidelines finalized 2011 which will be published as a special report in the Veterinary Clinical Pathology Journal in 2012. This guideline was put together by a core team of laboratorians with consultation from biostatisticians and editorial from the ASVCP Quality Assurance and Laboratory Standards Committee. Additionally, the ASVCP General Quality Assurance Guidelines were recently revised and finalized in 2009 with consultation from over 50 specialists and provides at least a portion of the preanalytical component. Both guidelines may be found here: http://www.asvcp.org/pubs/qas/index.cfm (Accessed January 30, 2012) The author acknowledges the collective knowledgebase and time commitments of these contributors.

INTRODUCTION

When interpreting medical data, there are several factors which must be considered to ensure correct diagnosis. While many clinicians focus on the physiologic or pathologic mechanism that may be causing the "abnormal" result(s), there are several "non-animal" factors which may contribute to results outside of reference intervals. Within the past couple of years, the ASVCP Quality Assurance and Laboratory Standards committee has finalized several guidelines to address reference intervals, preanalytical, analytical, and postanalytical error in both point of care as well as reference laboratories. The following discussion will review the recent ASVCP guidelines with which your laboratories and in-clinic analyzers should be in accordance, so that you can depend on your results. Commonly observed discrepancies will be discussed so potential pitfalls can be avoided.

REFERENCE INTERVALS

Reference intervals are important and asking the correct questions can change understanding and use of your report. Are your reference intervals correct for your patient? Are they method, species, and population specific? Where did they come from and were they generated according to current ASVCP guidelines? Is partitioning required to improve acuity of diagnosis? You need to ask your laboratory for this information if it is not apparent to you on your record. According to current guidelines, if there is any deviation from normal generation in statistical or population analysis, this must be directly annotated on the report. If RI generation has been in accordance with guidelines, this should be available to the clinician in a written report (RI study summary document). This information may include, but is not limited to: Reference population demographics and number of reference subjects sampled, subject preparation and time or season of collection, if relevant, sample type and handling, confidence intervals around the reference limits.[1] If you are unaware of the specific of the reference intervals you are currently using, please ask! The numbers that you use likely on a daily basis for diagnosis are typed in by a technician from data generated by one or a handful of hopefully qualified quality assurance specialists (a.k.a. human beings who make mistakes).

When was the last time that the reference interval was reviewed and revalidated? Current recommendations are revision every 3-5 years. Even at reputable institutions this can be problematic. Recently at UC Davis, it was realized that bovine hematology reference intervals that were used and essentially disregarded by clinicians as most results were "out of bounds" were last evaluated in the 1950s.[2] Amazing what written archives sitting on a slightly older emeritus professor's shelf found by an exuberant veterinary student can elucidate; not likely to happen in the digital age.

History

Reference intervals are a relatively recent concept that was originally proposed in 1969.[3] Remarkably, this cornerstone in medical diagnosis is less than 50 years old! In the 1970s, while the International Federation of Clinical Chemistry pursued official guidelines and manuscripts, significant theoretical discussion and publication on reference intervals took place with numerous philosophies and mathematical approaches suggested and tested. In this soup of ideas, one can find a publication on almost any approach to the establishment of reference intervals. Although many of these approaches are not currently used, articles can be cited to justify outdated statistical evaluation, creating significant difficulty in review of manuscripts submitted for publication and confusion for some authors. As an associate editor of the Journal of Zoo and Wildlife Medicine, I have found this to be especially true in other journals focused on a practitioner audience. Official recommendations have existed in the human field since the 1980s when the Expert Panel on the Theory of Reference Values (EPTRV) authored a 6-part series on the production of reference values, which was adopted by several professional organizations including the Clinical and Laboratory Standards Institute (CLSI).[4-9] In 2008, the Clinical Laboratory Standards Institute finalized a 3rd revision of guidelines, promoting transference and easing the strict number of required reference individuals (previously n = 120) by use of the "robust" method.[10] A summary of the guidelines for human populations has been published.[11] These guidelines are relevant to veterinary medical diagnostics, but do not address species differences, the inclusion of low numbers of reference individuals (<80) that frequently occurs in veterinary medicine, and reference populations that are often afflicted with undiagnosed subclinical disease. Therefore, the ASVCP consensus veterinary guidelines on the establishment of reference intervals have been developed and are available with free and open access. Interestingly, both human and veterinary guidelines specifically state that reference intervals should be population- and method-specific and that reference intervals from the literature should not be used as they may result in misdiagnosis.[10]

Current Use

Practicing in a medical field with seemingly limitless species, logistic reality dictates that reference intervals cannot be created specific to all patients. It is the responsibility of veterinary diagnostic laboratories to serve as many species as possible, but the likelihood of having reference intervals for tapirs is small. Reference intervals from the literature are regularly used by veterinary practitioners; therefore, appropriate description of the details of reference interval generation in a publication has been addressed in the veterinary guidelines. A full description of the health and population characteristics as well as the preanalytical and analytical methods is mandatory to avoid misdiagnosis. Until very recently, most articles in the veterinary literature lacked adequate descriptions of populations and analytical methods to allow appropriate comparison and use. Additionally, in the human field established consensus reference intervals are partitioned to age and gender and occasionally to race. Partitioned reference intervals are not typically provided by veterinary diagnostic laboratories, but more publications are addressing the effect of physiologic variables on histogram dynamics. As one moves forward through time and looks through articles published in the past few years in *Veterinary Clinical Pathology*, we see continual improvement in complete descriptions of population characteristics, health status of animals, preanalytical methods, and analysis, as well as reports of larger numbers of animals with consistent statistics applied. We hope to continue this trend with our collaborations with JVIM, JAVMA, JZWM, JWD, and other journals.
Additionally, computer support programs have been developed and are being updated so that statistical analysis may be performed according to reference interval guidelines. The purpose is to streamline the process of correct generation of reference intervals in veterinary diagnostics. Scientists at the French veterinary college in Toulouse have generated freeware that can be downloaded here http://www.biostat.envt.fr/spip/spip.php?article63 (accessed January 30, 2012) Reference Value Advisor is a set of macroinstructions for Excel that compute reference intervals using the standard and robust methods with and without generalized Box-Cox data transformation. Please see the website for further information and citations.

PREANALYTICAL ERROR

Preanalytical error is any potential interferent compound, change in the sample due to environment, storage and transfer prior to actual analysis on the machine. Preanalytical error may result in minor change or result in results that are not interpretable. Samples with moderate to severe hemolysis should in general should not be used for interpretation. Please consult your laboratory directly for which analytes may still be used (e.g. PCV) in these situations. While this may make diagnosis of certain disease categories such as IMHA difficult, it is better to have no data than to have bad data and misdiagnosis. Better to rely on physical exam and clinical assessment in these situations.

Coagulation

>90% of error associated with coagulation analysis in human medicine is attributed to preanalytical error.[12-15] For this reason, a flow chart for correct sampling procedure for venipuncture has been finalized and is available here http://www.asvcp.org/pubs/pdf/CoagSampGuide.pdf Whole blood should be collected in trisodium citrate anticoagulant in a 9:1 ratio and thoroughly mixed. This typically is accomplished by filling to the indicated mark on the appropriate blood tube. Specimens that do not conform to this dilution should be rejected.[16] Citrate volume may need to be adjusted for samples from very anemic and polycythemic animals.[17] For tests requiring plasma, the citrated tube is centrifuged and chilled to 2-8Â°C. Plasma should be separated from blood cells and transferred to a plastic tube (not glass).[18,19] Specimen stability at room or refrigerated temperature (2-8°C) is 4 hours and 24 hours for APTT and PT, respectively. If testing is not performed within these time intervals, samples should be frozen at -20°C.[20] Fresh citrated whole blood used for platelet function or other coagulation analyses should be ideally kept for <1 hour. (Giger, per communication). If samples are mailed into a laboratory for testing versus direct transport, plasma should be placed in a plastic tube then frozen, then be packed on ice and shipped to arrive frozen within 24 hours.

Hematology

Blood films made in the clinic should not be refrigerated and should be protected from condensation and freezing during transport to the laboratory. Anticoagulated samples for hematology that are found to have macroclots that can be found upon visual inspection will produce variably erroneous results. The clinic should be contacted either in writing or by phone and informed that the sample will produce erroneous results. Because the varying degree of inaccuracy cannot be predicted, clotted samples are unsuitable for analysis and it is not recommended that these samples be analyzed. If samples of questionable or substandard quality are analyzed, any procedures and possible inaccuracies should be documented in writing by the laboratory. Additionally, any possibly inaccurate results should have easily seen comments on the report to the clinician that clearly state those values may be inaccurate and misleading.

My recommendation is to send in known clotted samples to your laboratory as a cross check. You WANT to get a phone call telling you that your sample is clotted. If your lab is either missing or not reporting clotted samples, they are not functioning within guidelines and are likely providing you with inaccurate results a percentage of the time.

Urinalysis

Identification of the urine collection method is important when interpreting the presence and concentration of potential contaminants including blood and bacteria. The submitter should clearly state the method by which the urine was obtained, such as free flow (midstream, early, or late), catheterization, cystocentesis, or from the floor or metabolism cage. Clear specimen containers can be used to facilitate gross examination if urine will be examined within 30 minutes. However, if urinalysis will be delayed, urine should be protected from exposure to UV light to prevent degradation of urine constituents (eg, bilirubin). Lids should be secure to prevent evaporation and/or volatilization of urine constituents (eg, ketones).

Urinalysis

Optimally, urine should be examined within 30 minutes of collection. If immediate examination is not possible, urine should be stored at refrigerated temperatures to minimize changes in urine physical and chemical make up and to inhibit bacterial growth. Strict recommendations for duration of refrigerated storage cannot be made, because this depends on specific urine components.[21] Storage for a maximum of 24 hours in the refrigerator is generally recommended (Osborne cautiously suggests 6-8 hours), but urine may be stable for shorter or longer periods depending on its initial make up. Chemical constituents that are particularly unstable include bilirubin and glucose, and pH if bacteria are present.[21,22] Stability of formed elements depends on urine pH and concentration. Crystals may form in vitro during storage at either room temperature or under refrigeration.[23,24] If crystalluria is a clinical concern, freshly collected urine should be examined immediately. Refrigerated samples should be brought to room temperature prior to analysis. Because urinalysis results may be affected by storage duration and temperature, the time the urine was collected, the time it arrived in the laboratory, and method of storage should be recorded. Alternative methods of preservation are available for stabilization of urine chemistry, inhibition of bacterial growth, and preservation of formed elements. Manufacturer's claims should be followed regarding intended use of particular preservative and duration of storage.

Quantitative microbiologic culture techniques are recommended for determining presence of significant bacteriuria. Urine specimens collected by cystocentesis are preferred, but specimens properly collected by catheterization and free-catch are acceptable if quantitative culturing methods are employed. Urine should be submitted for the microbiological culture prior to urinalysis procedure to avoid contamination of the specimen. Alternatively, a sterile aliquot can be set aside for possible microbiological culture subsequent to urinalysis procedure. Refrigerated urine samples are acceptable for microbiological culture for at least 6 hours and often up to 24 hours. Refrigeration of urine specimens for 24 hours may result in false negative culture results.[25] If bacteriostatic transport media is used, urine samples do not need to be refrigerated.

A thorough review of all preanalytical issues cannot be provided in this space, please consult the QAS guidelines, http://www.asvcp.org/pubs/qas/newQas/index.cfm as well as manufacturer's recommendations regarding specific methods employed.

Table 1. Procedural steps for de novo determination of RI for new methods or new populations.
(Page 27 of ASVCP reference interval guideline, references are to said guideline.)

1. Perform literature search for information about analytes to be measured (preliminary investigation).
2. Define reference population and establish selection, inclusion and exclusion criteria (Section 1 and Table 1).
3. Develop questionnaire to be completed by examining clinician, owner/caretaker or both in order to determine if reference individual fits the selection or partitioning criteria (Section 2).
4. Determine number of reference individuals available or the number required to establish reference intervals with desired level of certainty (as reflected by 90% CI around the reference limits) (Section 3).
5. Select reference individuals, preferably by direct methods (Section 4).
6. Collect and handle reference samples in standardized manner (Section 5).
7. Analyze reference samples using well-controlled methods (Section 6 and 7).
8. Prepare histogram (Section 8).
9. Identify outliers (Section 9). This may require prior transformation to appropriately apply outlier detection methods and may need to be repeated after initial outliers are eliminated.
10. Determine distribution of reference data (Gaussian or non-Gaussian) (Section 10). If using parametric methods, transform data if it is not Gaussian and retest distribution. Transformation may improve the performance of the robust method. Nonparametric methods do not require any particular distribution.
11. Calculate upper and lower reference limits using an appropriate statistical method based on distribution of data and number of samples (Section 11 and Table 2). Calculate confidence intervals for the upper and lower reference limits.
12. Determine the need for partitioning only if there are sufficient numbers of reference samples or there is evidence for clinical importance (Section 12).
13. Document all previous steps for a comprehensive reference interval summary report (Section 13)

REFERENCES

[1]Plebani M. **Clin Chim Acta** 2004;346(1):25-35. [2]George, J. W., et al. **Vet Clin Pathol** 2010; 39: 138–148. [3]Grasbeck R, Saris NE. **Scand J Clin Lab Invest** 1969;26(Suppl 110):62-63. [4]Solberg HE. **Clin Chim Acta**. 1987;165:111-118. [5]PetitClerc C, Solberg HE. **Clin Chim Acta**. 1987;170:S1-S11. [6]Solberg HE. **Clin Chim Acta**. 1988;177:S1-S12. [7]Solberg HE, Stamm D. **J Automat Chem** 1991;13:231-234. [8]Solberg HE. **Clin Chim Acta** 1987;170:S13-S32. [9]Dybkær R. **Clin Chim Acta** 1987;170:S33-S42. [10] Zakowski J. Clinical and Laboratory Standards Institute (CLSI). **CLSI document C28-A3.** 2008;Vol. 28(3). [11]Geffré A et al.**Vet Clin Pathol** 2009;38:288-298. [12]Lippi, et al., **Blood Coagula Fibrinolysis** 2006;17:1-7. [13]Valenstein PN, et al. **Clin Lab Med** 2004;24:979–996. [14] Bonini P, et al. **Clin Chem** 2002;48:691–698. [15]Dale JC, Novis DA. **Arch Pathol Lab Med** 2002;126:416–419. [16] Adcock D, et al. **Blood Coagul Fibrinolysis** 1998; 9:463–470. [17]Stockham SL, Scott MA. **Fundamentals of Veterinary Clinical Pathology**, 2nd ed.2008.p.277. [18]Fiebig EW, et al. **Am J Clin Pathol** 2005;124:902–909. [19]Kratz A, et al. **Arch Pathol Lab Med** 2006;130:39–44. [20]Adcock DM, et al. **Am J Clin Pathol** 1998;9:595–599 [21] Rabinovitch A, et al. Urinalysis –approved guideline, 3rd ed. **Clinical and Laboratory Standards Institute document GP16-A3, 2009**. [22]Osborne CA, Stevens JB. **Urinalysis: A clinical guide to compassionate patient care.** 1999:17-24, 51-62, and 125-131. [23]Albasan H, et al. **J Amer Vet Med Assoc** 2003;222:176-179. [24]Sturgess, CP, et al. **J Fel Med Surg** 2001;3:81-85. [25]Padilla J, et al. **J Amer Vet Med Assoc** 1981;178(10):1077-81.

QUALITY ASSURANCE IN ACTION: EXAMPLES FROM THE FIELD

Bente Flatland, DVM, DACVIM, DACVP
Knoxville, TN

CASE ONE

Presented by Dr. Kendal Harr.

CASE TWO

Blood from an adult dog. CBC obtained using an Advia 120 (Siemens Medical Solutions Diagnostics, Tarrytown, NY, USA)

Raw data output from the analyzer:

Analyte	Units	Result	Reference Interval
RBC	$\times 10^6/\mu L$	8.44	5.60-8.70
Hct	%	56	41-60
Hgb	g/dL	19.2	14.7-21.6
MCV	fL	66	62-74
MCHC	g/dL	34.2 **L**	34.5-36.3
MCH	pg	22.7	22.0-26.2
RDW	%	11.9	11.4-13.6
WBC	$\times 10^3/\mu L$	8.02	5.10-14.00
% Neut	%	52.4	
% Lymph	%	27.3	
% Mono	%	20	
% Eo	%	0.3	
% Baso	%	0.1	
# Neut	$\times 10^3/\mu L$	4.20	2.65-9.8
# Lymph	$\times 10^3/\mu L$	2.20	1.10-4.60
# Mono	$\times 10^3/\mu L$	1.60 **H**	0.65-0.85
# Eo	$\times 10^3/\mu L$	0.02	0.0-0.85
# Baso	$\times 10^3/\mu L$	0.00	0.0-0.1
Plt	$\times 10^3/\mu L$	249	147-423
MPV	fL	9.5	7.8-12.4

Medical technologist's manual WBC differential count:

Neut	46%	$3.69 \times 10^3/\mu L$	
Bands	0%	$0.00 \times 10^3/\mu L$	
Lymph	24%	$1.92 \times 10^3/\mu L$	
Monos	6%	$0.48 \times 10^3/\mu L$	**L**
Eo	21%	$1.68 \times 10^3/\mu L$	**H**
Baso	3%	$0.24 \times 10^3/\mu L$	**H**

Question:

1. What is a potential explanation for the discrepancy between the automated and manual WBC differential counts?

CASE THREE

Blood from an adult dog. CBC obtained using an Advia 120 (Siemens Medical Solutions Diagnostics, Tarrytown, NY, USA) and a Scil ABC (Scil Animal Care Company, Gurnee, IL, USA).

Raw data output from the analyzers:

		Advia 120		Scil ABC	
Analyte	**Units**	**Result**	**Reference Interval**	**Result**	**Reference Interval**
RBC	$\times 10^6/\mu L$	2.72 **L**	5.60-8.70	3.15	5.50-8.50
Hct	%	20.2 **L**	41-60	23.7 **L**	37.0-55.0
Hgb	g/dL	8.1 **L**	14.7-21.6	10.5 **L**	12.0-18.0
MCV	fL	74.3 **H**	62-74	75 **H**	60-72
MCHC	g/dL	40.3 **H**	34.5-36.3	44.4 **H**	34.0-38.0
MCH	pg	29.9 **H**	22.0-26.2	33.3 **H**	19.5-24.5
RDW	%	16.1 **H**	11.4-13.6	7.6 **L**	12.0-16.0
WBC	$\times 10^3/\mu L$	547.00 **H**	5.10-14.00	62.9 **H**	6.0-17.0
% Neut	%	5.2		91.4*	
% Lymph	%	66.3		6.1	
% Mono	%	28.6		2.5	
% Eo	%	0.0		Error	

% Baso	%	39.0		NR	
# Neut	$\times 10^3/\mu L$	28.17 **H**	2.65-9.8	57.6* **H**	3.5-12.0*
# Lymph	$\times 10^3/\mu L$	362.3 **H**	1.10-4.60	3.8	1.2-4.5
# Mono	$\times 10^3/\mu L$	156.3 **H**	0.65-0.85	1.5 **H**	0.3-1.0
# Eo	$\times 10^3/\mu L$	0.11	0.0-0.85	Error	
# Baso	$\times 10^3/\mu L$	213.2 **H**	0.0-0.1	NR	
Plt	$\times 10^3/\mu L$	255	147-423	224	200-500
MPV	fL	13.8 **H**	7.8-12.4	7.6	6.1-10.1

*Granulocyte count

Questions:
1. What concerns do you have about these data?
2. What are your differential diagnoses for the abnormal findings?
3. What steps should be taken to confirm your most likely differential diagnosis?

CASE FOUR

Blood from an adult dog. CBC obtained using an Idexx Lasercyte (Idexx Laboratories, Inc., Westbrook, ME, USA).

Raw data output from the analyzer:

Analyte	Units	Result	Reference Interval
RBC	$\times 10^6/\mu L$	7.43	5.50-8.50
Hct	%	53.8	37.0-55.0
Hgb	g/dL	19.7 **H***	12.0-18.0
MCV	fL	72.4	60.0-77.0
MCHC	g/dL	36.6*	30.0-37.5
MCH	pg	26.5*	18.5-30.0
RDW	%	15.2	14.7-17.9
WBC	$\times 10^3/\mu L$	0.34 **L***	5.50-16.90
% Neut	%	39.3	
% Lymph	%	51.1	
% Mono	%	5.2	
% Eo	%	2.2	
% Baso	%	2.2	
# Neut	$\times 10^3/\mu L$	0.13 **L** *	2.00-12.00
# Lymph	$\times 10^3/\mu L$	0.17 **L** *	0.50-4.90
# Mono	$\times 10^3/\mu L$	0.02 **L** *	0.30-2.00
# Eo	$\times 10^3/\mu L$	0.01 **L** *	0.10-1.49
# Baso	$\times 10^3/\mu L$	0.01*	0.00-0.10
Plt	$\times 10^3/\mu L$	209	175-500
MPV	fL	12.8	
*Flags	Temperature out of range (TI), WBC abnormal distribution (WD), HGB channel error (HC)		

Questions:
1. What concerns do you have about these data, if any?
2. What would you do next?

REFERENCES

1.Iazbik MC, Couto CG. Vet Clin Pathol 2005;34:140-143.2.Giori L, et al. Vet Clin Pathol 2011;40:475-483.3.Zaldívar-Lopez S, et al. Vet Clin Pathol 2011;40:414-425.4.Lilliehöök I, Tvedten HW. Vet Clin Pathol 2011; 40:450-458.5.Papasouliotis K, Cue S, Crawford E, Pinches M, Dumont M, Burley K. Vet Clin Pathol 2006; 35:295-302.6.Becker M, Moritz A, Giger U. Vet Clin Pathol 2008; 37:373-384.7.Welles EG, Hall AS, Carpenter DM. Vet Clin Pathol 2009; 38:20-29.

HOW TO CHOOSE THE RIGHT DIALYSIS MODALITY

Cathy Langston, DVM, DACVIM (SAIM)
New York, NY

TERMINOLOGY

There are a variety of dialysis modalities that can be used for renal replacement therapy. Extracorporeal therapies are those that take blood out of the body for purification. Peritoneal dialysis, in which the purification takes place inside the body, is considered intracorporeal therapy. The extracorporeal therapies have conventionally been divided into intermittent or continuous modalities, but the lines between those have become blurred recently. Chronic hemodialysis, for people with end-stage kidney disease, has traditionally been intermittent treatments of 3 to 4 hours duration 3 days a week. Whether this is an appropriate dosing scheme is debatable, but because the vast majority of veterinary dialysis is provided for acute kidney injury, recommendations for treating CKD in people lack validity in this setting.

Intermittent hemodialysis (IHD) is usually provided with a dialysis delivery machine that produces the dialysate from purified water. The dialysate is not sterile, but modern machines have several filtration steps that can render the dialysate ultrapure, meaning almost free of bacterial contamination and with very low concentrations of endotoxins. The dialysate flow rate can be varied, but typically in the range of 300-800 ml/min (18-48 L/hr). Blood flow rates are maximized (up to 500 ml/min). This very fast dialysate flow rate compared to the blood flow rate provides highly efficient diffusive clearance of small solutes such as urea, creatinine, and potassium.

Continuous renal replacement therapy (CRRT) is typically performed with a different type of machine using pre-packaged sterile fluids for dialysate. The traditional schedule for CRRT is an intention to treat 24 hours a day until the patient recovers or is stable enough to transition to IHD. This may be a few days up to a few weeks. In practice, disruptions in treatment (i.e., because of filter clotting, scheduled filter changes, for procedures such as imaging, etc.) limit treatment to something less than 24 hours a day. Dialysate flow rates may vary from 0 to 8 L/hr (0-133 ml/min). CRRT typically relies more heavily on convective clearance compared to diffusive clearance. Convective solute clearance removes solutes dissolved in fluid that is extracted from the patient (using hydrostatic forces with extracorporeal therapies and osmotic forces with peritoneal dialysis). To avoid volume depletion, replacement fluid is administered intravenously to the patient. Although not approved by the FDA, many units use dialysate and replacement fluids interchangeably, as both are sterile. Replacement fluid rates may range from 0 to 8 L/hr. The decrease in efficiency is overcome by a much longer treatment time. CRRT typically uses slower blood flow rates (up to 200 ml/min) than IHD.

CRRT can be further divided into modes. Continuous venovenous hemofiltration (CVVH) involves solely convective clearance via a combination of fluid removal by ultrafiltration and fluid replacement, without any dialysate for diffusive clearance. CVVHD involves diffusive clearance via dialysate, with minimal contribution from convective clearance from the fluid removed by ultrafiltration to control overhydration. CVVHDF involves a combination of replacement fluid and dialysate, in a proportion determined by the operator. Theoretically, CVVH improves middle molecule clearance compared to CVVHD. CVVH may be associated with more filter clotting compared to CVVHD.

"Hybrid" therapies are variations on a theme. Sustained low-efficiency dialysis (SLED) uses an intermittent hemodialysis machine, slow blood flow rates, and long treatment times (e.g., 6-12 hours). Sustained low-efficiency daily dialysis with filtration (SLEDD-f) adds in a component of convective clearance. Extended daily dialysis (EDD) involves intermittent treatments (usually 6-8 hours) 6 to 7 days a week. Intermittent CRRT is, as the name suggests, treatment provided with a machine designed for CRRT but used for a shorter treatment duration. Other novel machine types are available, such as the Genius system, which generates a fixed amount of dialysate (75 L) from purified water for use during a single treatment. Another machine, the NxStage machine, uses pre-packaged sterile fluids, and is marketed for use either in a CRRT mode or for daily dialysis. As new models of the most popular IHD and CRRT machines are introduced, they are becoming more versatile, in that the ranges of dialysate flow rate, replacement fluid rate, and blood flow rate are expanded, blurring the distinction between types of machines. A suggested name for these methods of use is prolonged intermittent renal replacement therapy (PIRRT), which removes the connotation of type of machine used.

CHOICE OF MODALITY

In the absence of data, strong and animated debate about the virtues and perceived superiority of one modality over the other was a popular topic in nephrology. Informal surveys showed that nephrologists, who are most familiar with IHD for their CKD patients, tend to choose IHD for their AKI patients. Criticalists tend to choose CRRT. Because of a perception that CRRT maintains hemodynamic stability better than IHD, there tends to be a bias towards sicker patients preferentially being started on CRRT, which has confounded retrospective data review.

One possible criterium for choosing a modality would be the survival rate. In a prospective randomized multi-center study, Mehta et al found worse mortality rates in the CVVH arm (65.5%) compared to the IHD arm (47.6%, $p = 0.02$). However, despite randomization, the CVVH patients had higher APACHE III scores at entry. In a single center study by

Augustine et al, there was no difference in mortality (67.6% in CVVHD vs 70.0% in IHD). These patients were equal in illness severity score. In a study by Uehlinger et al, again there was no difference in mortality (47% in CVVHDF vs 51% in IHD). In both the Hemodiafe and SHARF studies, survival rates were similar in CVVH vs IHD arms. A meta-analysis by Bagshaw et al found no difference in the overall survival. In all of these studies, length of hospitalization was similar between groups (mean, 21-42 days). All the studies mentioned so far compared CRRT to IHD prescriptions in which treatment duration was less than 5 hours and patients were treated 3 to 7 days a week. Marshall et al retrospectively evaluated mortality in 3 hospitals who converted from CRRT to PIRRT. They found no difference in the observed mortality with PIRRT. In a randomized prospective single-center study, Abe et al compared CVVHDF to SLEDD-f and found better survival in the SLEDD-f group.

Renal recovery rates also tend to be similar between CRRT and IHD groups. In the studies mentioned above, renal recovery rates ranged from 12.5-50% in the CRRT arm, and from 10-42% in the IHD arm. One thing to point out is that these are recovery rates for all patients. Because death is a competing outcome, studies that report renal recovery in only the survivors may have much higher recovery rates. In Jacka et al, the renal recovery rate of 87% in the CRRT arm compared to 36% recovery in the IHD arm was significantly different. However, more patients died in the CRRT arm, and when considering all patients, the magnitude of difference lessens, in that 32% of all CRRT patients recovered renal function, compared to 18% of IHD patients. Abe et al found a higher renal recovery rate in the SLEDD-f group (60% in CVVHDF vs 80% in SLEDD-f, $p < 0.05$).

Based on these data, survival, length of stay, and renal recovery rates are similar between CRRT and IHD. CRRT is preferred over IHD in hemodynamically unstable patients because of the perception that CRRT is superior in that setting. Several prospective studies specifically enrolled critically ill patients in a randomized manner to test that hypothesis. In studies by Uehlinger et al. Vinsonneau et al, and Augustine et al, there was no difference in mean arterial pressure between groups. Augustine et al found a decrease in MAP during treatment compared to pre-treatment in the IHD group but not in the CVVHD group. However, the average decrease was only 2.6 mmHg. Kielstein et al found no difference in hemodynamic parameters (mean arterial pressure, heart rate, cardiac output, and systemic vascular resistance) between critically ill ventilated patients treated with CRRT compared to PIRRT.

Patients with AKI and absolute or relative oliguria may develop substantial fluid accumulation over the course of resuscitation and treatment before starting renal replacement therapy. CRRT may be chosen as the therapy in those cases, to allow for more gradual but sustained fluid removal. Because fluid is removed from the vascular space, and the vascular volume then refills from the interstitial space, a rate of fluid removal that exceeds the refill rate can cause intravascular volume depletion leading to hypotensive episodes. In the study by Augustine et al, patients on IHD had a net gain of fluid on day 2 (a non-dialysis day), and although they had a negative balance on day 3, the CRRT patients had a more negative net balance on day 3. This is despite a greater decrease in urine output in the CRRT group in that study, indicating that CRRT was more effective at fluid removal compared to IHD. Bouchard et al found a negative net balance at 10 days in the CRRT group, while the IHD group sustained a positive fluid balance by day 10. Improvement in fluid control may better protect cerebral perfusion in patients with fulminant hepatic failure, acute brain injury, or cerebral edema, and in that subset, CRRT may be superior to IHD.

CVVH vs CVVHD vs CVVHDF

CRRT can provide clearance via convection, diffusive, or a combination of the two. CVVH is a purely convective clearance mode. With CVVH, a large volume of fluid (i.e., 20-35 ml/kg/hr) is removed from the patient to clear the solutes dissolved in that fluid. The fluid used to replace that volume is a balanced polyionic solution. The replacement fluid can be delivered to the extracorporeal blood circuit before the blood reaches the filter. This dilutes the blood entering the filter and decreases the efficiency of solute removal. Alternately, the replacement fluid can be added after the filter, maintaining maximum efficiency of clearance. In that set-up, the large amount of plasma water removal as the blood passes through the filter hemoconcentrates the blood and increases the risk of clotting in the filter. With CVVHD, clearance is predominantly diffusive, and filter clotting may be less severe than with CVVH (in pre-dilution or post-dilution configurations). Diffusive clearance is excellent for small molecules (< 100 molecular weight), but clearance decreases as molecular weight increases, with almost no molecules over 1000 MW being cleared. Convective clearance can remove solutes up to 10,000 MW. Because cytokines and other inflammatory mediators fall in the middle molecule size range, theoretically convective clearance may be superior to diffusive. In practice, however, this does not translate into better survival or other outcome measures.

Peritoneal Dialysis

Peritoneal dialysis (PD) may seem like an attractive choice because it does not require a specialized dialysis machine. Phu et al compared PD to CVVH in patients with infectious causes of AKI (primarily malaria) and found the PD patients were 5 times more likely to die. The CVVH group had faster resolution of azotemia and acidosis and those patients were on renal replacement therapy for a shorter period of time. The cost of CVVH was half that of PD. However, in a study by Gabriel et al comparing PD to daily IHD, no difference was found in solute control, mortality, or renal recovery, and the IHD patients required longer courses of treatment than the PD patients.

NON-MEDICAL ISSUES IMPACTING CHOICE OF THERAPY

One of the main criticisms of CRRT is that it is labor intensive and therefore costly. CRRT roughly costs 2.5 times as much as IHD (Forni and Hilton, Manns et al, Rauf et al). The main costs of CRRT include the fluids, which are expensive to produce and ship, and the allocation of ICU nursing services. The main costs of IHD are nursing costs, which are determined by the frequency and duration of treatment. The cost of providing renal replacement therapy is highly variable based on the institution, and the cost is only a part of the total cost of care for the AKI patient.

The abilities of the nursing staff are also a factor to consider in deciding about modality. If the nurses are comfortable with providing CRRT, that may be a great choice, but if the nurses are not adequately trained and lack sufficient clinical experience to have confidence, CRRT may be a poor choice for the unit and the patient. In addition to the level of training, the amount of support plays a role in deciding about modality. Small, low-volume units may be understaffed to provide continual care. Using PIRRT may alleviate some of the staffing issues by shifting treatment times to first and second shift time slots and avoiding the need for an overnight shift.

CONCLUSION

There is no right choice for RRT modality. From a medical perspective, no therapy has proven to be superior in survival, renal recovery, stability, or efficacy for people, and veterinary experience is far too limited at this stage to provide data. In human hospitals, the choice of therapy appears to be made based on availability and physician preference, and this is likely going to be the case in veterinary medicine. In a unit that can provide all of the types of therapy, each has advantages and disadvantages for an individual case that may play a role in choosing. For units that offer a limited number of therapies, developing a team that is proficient at providing that type of therapy is likely the best strategy for optimizing patient outcomes.

REFERENCES

Abe et al, Artificial Organs 2010; 34:331. Augustine et al, Am J Kid Disease 2004; 44:1000. Bagshaw et al, Crit Care Med 2008; 36:610. Bouchard et al, Kidney Int 2009; 76:422. Farese et al, Artificial Organs 2009; 33:634. Forni and Hilton, N Engl J Med 1997; 336:1303. Gabriel et al, Kidney Int 2008; 73:S87. Jacka et al, Can J Anesth 2005; 52:327. James et al, Sem Dial 2011; 24:215. Kielstein et al, Am J Kidney Dis 2004; 43:342. Lins et al, Nephrol Dial Transplant 2009; 24:512. Manns et al, Crit Care Med 2003; 31:449. Marshall et al, Nephrol Dial Transplant 2011; 26:2169. Mehta et al, Kidney Int 2001; 60:1154. Phu et al, N Engl J Med 2002; 347:895. Rauf et al, J Crit Care Med 2008; 23:195. Uehlinger et al, Nephrol Dial Transplant 2005; 20:1630. Vinsonneau et al, Lancet 2006; 368:379.

TRANSFUSION THERAPY IN IMMUNE-MEDIATED HEMATOLOGIC DISEASE

Ann E. Hohenhaus, DVM, DACVIM (Oncology & SAIM)
New York City, NY

INTRODUCTION

Immune mediated hematologic disease encompasses several disorders. The 5th edition of Schalm's Veterinary Hematology, Section on Immunohematology lists immune mediated hemolytic anemia (IMHA), thrombocytopenia (ITP) and neutropenia as the primary immune hematologic diseases. Systemic lupus erythematosus is also included in the section, because it may include some or all of these disorders. Evan's syndrome (concurrent IMHA and ITP) is not specifically discussed nor are the recently described cases of concurrent ITP and immune-mediated neutropenia.[14] Immune-mediated hematologic diseases occur in both dogs and cats, but are more frequently reported in dogs and for that reason, this manuscript will focus on dogs.

When transfusion is considered in immune-mediated hematologic disorders, administration of red blood cells most commonly comes to mind; however, transfusion of fresh frozen plasma for the treatment of disseminated intravascular coagulation, transfusion of human intravenous immunoglobulin as an immunosuppressive agent and transfusion of platelets via fresh whole blood, platelet concentrate or lyophilized platelets are important therapeutic interventions in immune-mediated hematologic diseases. Infusion of hemoglobin based oxygen carriers also plays a role in the management of immune-mediated hematologic disorders. At this time, transfusion plays a no role in the treatment of immune mediated neutropenia in veterinary patients. Although transfusion of a variety of blood products is used in the treatment of immune mediated disease, this manuscript will focus on red blood cells, fresh frozen plasma and platelets.

VETERINARY CURRENT STATE OF THE ART

Table 1. Red blood cell transfusions in canine IMHA*

Author	Year	Number of dogs	Number transfused	%	Comments
2	2002	72	40	55	Products not specified.
16	2005	151	106	70	Higher acute mortality. Longterm survival not correlated with transfusion. Products not specified.
10	2006	149 A/P	98	66	N transfusion Hct = 18% Transfusion Hct = 12%
15	2009	28	26	93	All pRBC. Median 23 ml/kg. No difference in transfusion requirement IV Ig and placebo
3	2010	128	79	62	Transfusion prior to admission = greater decrease in lactate
7	2010	110	78	71	Transfusion did not change time to discharge
11	2011	73 P only	56	76	A/P and P equal transfusion requirement
*Ten years 2002-2011; >25 cases included in study; transfusion data reported					

Table 2. Red blood cell transfusions in canine ITP

Author	Year	Number of dogs	Number transfused	%	Comments
13	2002	24	18	75	pRBC 10 treated with prednisone pRBC 8 treated with prednisone-vincristine Median pRBC transfusions = 1
12	2008	30	5	17	pRBCs
1	2009	18	15	83	pRBC 14.7 ml/kg IVIg group pRBC 21.2 ml/kg Placebo group
9	2011	73	26	36	Product not specified. FWB or pRBCs Median transfusions = 2

Table 3. Fresh frozen plasma transfusions in canine IMHA*

Author	Year	Number of dogs	Number transfused	%	Comments
2	2002	72			18/31 DIC. Products not specified
16	2005	151	106		Survival not correlated with transfusion. Products not specified.
10	2006	149 A/P	0?	0?	All transfused with "blood"
15	2009	28	0	0	DIC not reported
3	2010	128	0?	0?	All transfused with "blood"
7	2011	110	0?		
11	2011	73 P only	0?	0?	All transfused with "blood"
*Ten years 2002-2011; >25 cases included in study; transfusion data reported					

Table 4. Fresh frozen plasma transfusions in canine ITP

Author	Year	Number of cases	Number transfused	%	Comments
4	1985	26	0	0	
13	2002	24	0	0	
12	2008	30	0	0	
1	2009	18	2	11	Indication not specified
9	2011	73	2	3	Prevent dilutional coagulopathy

Table 5. Platelet transfusions in canine ITP

Author	Year	Number of cases	Number transfused	%	Comments
5	1990	15	6 Platelet rich plasma	4	6/8 dogs with single episode of ITP 0/7 dogs with relapsing ITP Limited benefit
Fishkin	unpub	7	7 Cryopreserved platelets	100	Cryopreserved platelet pilot study 4/7 red blood cell transfusion Benefit not apparent

Table 6. Fresh whole blood transfusions in canine ITP

Author	Year	Number of cases	Number transfused	%	Comments
4	1985	26	6	23	6 FWB
13	2002	24	0	0	
12	2008	30	11	37	11 FWB 93% survived >14 d
1	2009	18	6	33	6 FWB
9	2011	73	26	36	Product not specified. FWB or pRBCs Median transfusions = 2

CURRENT STATE OF TRANSFUSION THERAPY IN HUMAN IMMUNE-MEDIATED DISEASE

The incidence of IMHA in humans is 0.8/100,000, making it a rare disease. The most common form of autoimmune hemolytic anemia in humans results from the presence of warm autoantibodies which interfere with antibody screening and compatibility testing. Transfusion is recommended in cases of brisk hemolysis, inadequate response to therapy and worsening anemia. Consensus does not exist as to the best possible method to identify the optimal unit of red blood cells for transfusion. Removal of the autoantibody can sometimes be accomplished by adsorbing it using autologous or allogeneic red blood cells. Another strategy is to determine an extended red blood cell phenotype and transfuse cells matching the patient's phenotype. Leukoreduced red blood cells are recommended to decrease the occurrence of febrile non-hemolytic transfusion reactions which mimic acute hemolytic transfusion reactions.[6]

Immune mediated thrombocytopenia is more common in humans than IMHA. The 2011 American Society of Hematology Clinical evidence based guideline update on ITP identified a need for additional studies of regarding the management of serious bleeding. Evidence for platelet transfusion is limited to case reports of cases with life threatening hemorrhage and platelet transfusions administered q 30-480 minutes in conjunction with a CRI of IV Ig.[8]

POTENTIAL MARKERS OF TRANSFUSION NEED IN DOGS

Like all medical therapies, transfusion carries both a benefit and a risk. The decision to transfuse any blood product should be made after a risk benefit analysis. Most small animal clinicians would agree coagulation testing consistent with disseminated intravascular coagulation is a strong indication for FFP transfusion in any immune mediated hematologic disorder; and although not proven, treatment with FFP appears to have clinical benefit in controlling life threatening hemorrhage. The clinical benefit of platelet or FWB transfusion in the face of immune destruction of platelets is less clear and some authors have reported little clinical benefit. Possibly the most difficult transfusion decision surrounds the use of red blood cell transfusions in IMHA because of the inherent difficulties in identifying compatible red blood cells and the concern for adding "fuel" to the hemolytic fire.

Biomarkers of red blood cell transfusion need

A recent retrospective study of dogs with IMHA evaluated lactate as an outcome measure and as an assessment of response to transfusion. The authors demonstrated a correlation of elevated lactate and known prognostic indicators for IMHA and a 100% survival of dogs whose lactate fell below 2.0 mmol/L within 6 hours of admission. Lactate also correlated with admission hematocrit. A greater decrease in lactate was associated with blood transfusion within the first 6 hours of admission and the authors suggest a target lactate below 2.0 mmol/L may have a beneficial impact on outcome. Packed red blood cell transfusions will clearly be a component of this strategy.

A recent prospective study of dogs with IMHA, hospitalized dogs and clinically normal dogs evaluated cardiac troponin I as a measure of myocyte damage. Dogs with IMHA had a greater level of myocyte damage than the other 2 groups. Multiple factors likely contributed to the increase in myocardial damage: myocarditis, myocardial necrosis and hypoxia due to anemia; however there was no difference in the troponin level between dogs with moderate and marked anemia. The utility of cardiac troponin as a surrogate marker of transfusion has potential utility but requires additional investigation.

RECOMMENDATIONS FUTURE INVESTIGATIONS

This analysis of the current status of transfusion in canine immune mediate hematologic disorders has identified deficiencies in the reporting of transfusions in these disorders. All transfusion data should be reported clearly and completely. A complete report of transfusion would ideally include: pretransfusion criteria be set forth prospectively, the product transfused completely described and response to transfusion reported using an objective measure such as increase in hematocrit, or platelet count or changes in coagulation parameters.

Administration of intravenous immunoglobulin for immune mediated disorders is currently an area requiring further investigation. Intravenous Ig has shown efficacy in treatment of ITP, but is costly. Until a randomized trial comparing IV Ig to vincristine treatment is completed, its true clinical benefit will remain speculative. Fresh whole blood has been commonly administered to increase platelet count in dogs with ITP, but control of hemorrhage and platelet increment has not been reported.

An adequately powered study to evaluate clinical benefit of IV Ig in IMHA is lacking.

REFERENCES

[1]Bianco D, J Vet Intern Med 2009;23:1071.[2]Carr AP, et al. J Vet Intern Med 2002;16:504.[3]Holahan ML, et al. Vet Emerg Crit Care 2010; 20:413[4]Jackson ML, et al. Can Vet J 1985;26: 245.[5]Jans HE, et al. J Vet Intern Med 1990;4: 4.[6]King KE, et al. Sem Hematol 2005;42:131.[7]McAlees TJ. Aust Vet J 2010;88:25.[8]Neunert C, et al. Blood 2011;117:4190.[9]O'Marra SK, et al. J Am Vet Med Assoc 2001;238.[10]Piek CJ. J Vet Intern Med 2008;22:366.[11]Peik CJ, et al. BMC Vet Res 2011;7:15.[12]Putsche JC, et al. J Am Anim Hosp Assoc 2008;44:250.[13]Rozanski EA, et al. J Am Vet Med Assoc 1990;220: 447.[14]Vargo CL, et al. Taylor Can Vet J 2007;48:1159.[15]Whelan MF, et al. J Vet Emerg Crit Care. 2009;19:158.[16]Weinkle TK, et al. J Am Vet Med Assoc 2005;226:1869.

THROMBOPROPHYLAXIS IN IMMUNE-MEDIATED HEMOLYTIC ANEMIA

Sarah Helmond, BVSc(Hons), DACVIM
Livermore, CA

Over the past 40 years, IMHA has been reported to have mortality rates of anywhere between 20 & 86%. Survival in reports from the past 6 years has improved but papers as recent as 2011 are still showing mortality rates of 30-40%. First reported in 1989[1] PTE is a significant cause of mortality, occurring in up to 80% of dogs who die of IMHA.[2]

PREVALENCE OF HYPERCOAGULABILITY

Several studies document hypercoagulability in IMHA, both on admission and during hospitalization. A 1998 study showed increased FDPs, decreased platelets, coagulation factors and AT in dogs with IMHA.[3] In 2001 a prospective study found evidence of DIC in 13/20 dogs.[4] A 2002 review identified or suspected DIC in 40% of 72 dogs with IMHA, with thromboemboli found in 20/25 dogs who had necropsies.[2] Thromboelastography® (TEG) has been used to investigate hypercoagulability in IMHA. A retrospective report of 39 dogs with IMHA using TEG, found 33/39 to be hypercoagulable.[5] The 6 dogs with normal traces all died. However because the study was retrospective, with variable immunosuppressive and thromboprophylactic therapies, conclusions regarding outcomes cannot be made. A prospective study of 11 dogs with IMHA found that they were all hypercoagulable on admission, prior to any therapy.[6] Similarly, a prospective study of 30 dogs with IMHA used serial TEG monitoring over the first 5 days finding that they were all hypercoagulable.[7] Again, a higher MA was associated with survival. All dogs had high C-reative protein, which was positively correlated with fibrinogen, and remained high at day 5, while antithrombin *decreased* over 5 days (36% of dogs had subnormal AT at day 5). D-dimers *increased* over the 5 days. While this paper showed TEG results prior to any therapy, it is difficult to assess the findings of the serial monitoring due to the heterogeneity of therapies used in the study. It is interesting to see a persistence in risk of thrombotic disease during hospitalization, whether this is because of treatments given or independent changes in disease progression.

CAUSE OF HYPERCOAGULABILITY

The key to effective prevention is understanding the underlying cause. Hemolysis as a trigger mechanism for DIC and consumption-mediated coagulopathy in the dog has been documented by experimental studies.[3] Several studies in humans have demonstrated critical links between inflammation and hypercoagulability. Specifically, inflammatory mediators activate platelets and monocytes, causing them to become procoagulant by releasing phosphatidylserine-rich membrane-derived microparticles (MPs) and inducing tissue factor (TF) expression.[8–12] MPs promote thrombin generation through provision of a vast phosphatidylserine surface, which binds coagulation factors and amplifies their activity. TF is the main initiator of coagulation *in vivo,*[12] and its induced expression on monocytes activates coagulation, eventually forming a fibrin thrombus. IMHA is intrinsically an inflammatory disease. Inflammation and necrosis are evident in multiple organs in necropsy and affected dogs have increased serum levels of inflammatory mediators, e.g. C-reactive protein (CRP).[13]

Recently, a study showed increased expression of tissue factor in dogs with IMHA.[14] There are other studies currently in progress investigating expression of tissue factor and levels of circulating platelet-derived microparticles in dogs with IMHA.

The treatments used in IMHA can cause hypercoagulability. Catheterization and blood transfusions have been proposed to increase coagulability. Administration of hIVIgG to healthy dogs was associated with increases in fibrin/fibrinogen degradation, thrombin-antithrombin complexes and C-reactive protein concentrations.[15] Steroids have been shown to be procoagulant.[16,17] Recently a study showed prednisolone at 1 & 4 mg/kg/day resulted in hypercoagulable TEG traces, but variable effects on thombin generation when using Thrombin Generation Analysis.[18]

Cyclosporin has been shown to cause hypercoagulabily in humans. A prospective study evaluated P-selectin and COX expression via flow cytometry, platelet aggregation via PFA-100® analysis, and clot formation via SonoClot® viscoelastometry in healthy dogs after a week of cyclosporin (10mg/kg BID and 5mg/kg SID).[19] They did not show a general increase in platelet reactivity, although they did find a variety of responses in individual dogs, suggesting "individual dog-to-dog variable responsiveness to cyclosporine may explain variations in drug pharmacodynamics, platelet thromboxane receptor function, thromboxane-mediated degranulation, and P-selectin expression". The study did not investigate the effects of chronic therapy.

IS ROUTINE THROMBOPROPHYLAXIS REQUIRED?

Routine thromboprophylaxis in IMHA has been reported since 2001[4] and has become standard of care in many institutions. While a variety of therapies have been reported, a study in 2005[20] led to the widespread use of 'ultralowdose' aspirin. Recently a study suggested that anticoagulation was not routinely required.[21] This paper showed 1 year mortality rates of 36% for dogs treated with prednisolone alone, and 31% for dogs treated with both Azathiaprine and prednisolone, however there were a large number of dogs lost to follow up.

Nevertheless, some of the highest reported survival rates were when using routine thromboprophylaxis: 82% survival at 30 days, (69% at 1 year) with low-dose aspirin and azathiaprine (retrospective study of 151 dogs)[20]; 83% 90-day survival with clopidogrel or low-dose aspirin therapy (a prospective study of 24 dogs)[22]; 180 day survival of 87% with individually-adjusted unfractionated heparin (prospective study of 15 dogs);[23] 75% survival using enoxaparin (prospective study of 8 dogs)[24] and 80% 30-day survival using low dose aspirin (prospective study of 30 dogs)[7]

There appears to be a spectrum of severity of IMHA reported in the literature. This may explain the wide variety in mortality rates and response to or need for anticoagulation. Use of a scoring system such as the "CHAOS" score may provide a criteria for determining severity of illness in dogs with IMHA and help stratify studies designed to examine outcome.[25]

PLATELET INHIBITORS

Aspirin

Aspirin has been used in several studies without any obvious detrimental effect, although questions remain regarding the safety of a combining COX inhibitor with prednisolone in dogs with potentially poor perfusion to the gastrointestinal tract. Several studies over the past 3 years have assessed the effect of aspirin on platelet function in normal dogs at different doses, with conflicting results.

- Shearer et al assessed doses of 0.5, 1.0, 2.0 and 10.0 mg/kg/d in healthy dogs.[26] They found *no statistically significant effect on platelet function at 0.5mg/kg/d* with either optical aggregometry or the PFA-100®. ADP-induced aggregometry detected inhibition by day 7 for 1.0 and 2.0 mg/kg/d aspirin doses. No effect on platelet aggregation was detected with 10 mg/kg/d aspirin. The PFA-100® detected significant inhibition of function for all aspirin doses, except the 0.5mg/kg/d dose. Some dogs had consistently low or high plasma metabolite levels regardless of the dose administered, suggesting variability in individual absorption.
- Sharpe et al used whole blood impedence aggregometry and flow cytometry to assess platelet function at the 0.5mg/kg/day dose.[27] Aspirin *significantly delayed platelet aggregation onset* with ADP at 20 µmol/L, *attenuated maximum aggregation* with various concentrations of ADP and collagen, and *slowed aggregation rate* with the highest ADP and collagen concentrations, Depending on the parameter tested, *up to 30% of dogs failed to have an aspirin effect.* Thrombin stimulation significantly increased P-selectin expression in platelets and platelet-leukocyte aggregates, but *aspirin did not alter basal or thrombin-stimulated P-selectin expression.*
- Puglia et al compared the effect of low dose aspirin measured by TEG Platelet Mapping™ with whole blood impedance aggregometry in healthy dogs.[28] Aspirin was administered for 7 days at 0.5, 2, and 10 mg/kg, PO, q24h. There was a significant difference from baseline for mean TEG-PM percent aggregation with the 0.5 mg/kg and 10 mg/kg doses. *A significant difference was seen with aggregometry for only the 10 mg/kg dose*, and only with collagen or arachadonic acid (not ADP). This suggested that platelet mapping was more sensitive or overestimated the aspirin effect.
- Thomason et al investigated the effect of aspirin (10mg/kg) on platelet COX expression.[29] They demonstrated that canine platelets express both COX-1 and COX-2. Aspirin caused increased COX-1 expression despite impairment of platelet function. COX-2 expression tended to decrease in response to aspirin, but varied markedly between individual dogs.
- Similarly, Dudley et al investigate the effect of 1mg/kg/day aspirin, using PFA-100®, flow cytometry for COX expression and ELISA for 11-dehydro-thromboxane B2 (11-dTXB2).[30] There was an increase in COX 1 & 2 expression, and a significant decrease in mean urine 11-dTXB2:creatinine. Although thromboxane levels decreased in most dogs, platelet function was consistently affected in only 36% of dogs.
- Urinary Thromboxane metabolites were also investigated by Hoh et al with conflicting results.[31] They found that Urinary 11-dTXB2 levels are low and not impacted by either 1 or 10mg/kg doses of aspirin, concluding that this metabolite did not accurately reflect platelet thromboxane metabolism in dogs. Urinary 2,3-dTXB2 levels significantly decreased after a single 10mg/kg dose however, 1mg/kg did not consistently impact urinary 2,3-dTXB2 levels, suggesting that this dose is insufficient to reliably inhibit canine platelets.

Three studies have come closer to demonstrating the effect of aspirin in dogs with IMHA.

- A prospective study used TEG to follow 20 dogs with IMHA, treated with prednisolone, azathiaprine and either aspirin.[32] 80% of the dogs were hypercoagulable. There was a mortality rate of 50%, with all dogs that appeared normocoagulable dying. Since there was no comparison group, the effect of aspirin is unknown.
- A prospective, placebo controlled randomized study with healthy dogs investigated the effects 0.5mg/kg aspirin and 2mg/kg prednisolone on TEG parameters.[33] They found that Prednisone causes hypercoagulability in healthy dogs based on increased G, MA, and CI, and that the addition of ultra-low dose aspirin to prednisone has no effect on the parameters measured in this study.
- A similar study found that Prednisone administered at approximately 2 mg/kg/d, PO, for 7 days with or without concurrently administered aspirin (0.5mg/kg/day) increased clot strength and decreased clot lysis in healthy dogs.[34]

Clopidogrel

Clopidogrel irreversibly and selectively inhibits ADP-dependent platelet aggregation. It is superior at prevetion of arterial thrombotic disease in humans when compared to aspirin, with less side effects. The first study that showed pharmacokinetics in dogs showed a delayed effect due to p450 enzyme metabolism to the active drug.[35] A loading dose of 10mg/kg provided maximal platelet inhibition in 90 minutes, thereafter 2-4mg/kg/day provided adequate platelet inhibition, resolving by day 7 after discontinuation. Platelet inhibition increased with repeated dosing. A study of 0.5mg/kg & 1mg/kg, using TEG Platelet Mapping™ showed that by 3 hours after the initial dose, MA had decreased to a mean of 20% of baseline, regardless of dose.[36] Impedance aggregometry showed a rapid decrease in ADP-induced platelet aggregation by 2 hours.

Shearer et al investigated a dose of 2mg/kg and 4mg/kg, demonstrating significant inhibition of platelet function using optical aggregometry and the PFA-100 by day 3.[26] This study used the same methodology to investigate aspirin, finding clopidogrel to be superior.

The use of Clopidogrel in IMHA has become more popular, athough so far there is little evidence to support its use over aspirin. A prospective, positive-controlled, unmasked clinical trial with 24 dogs randomized in 3 treatment groups to receive either 0.5mg/kg/day aspirin, or Clopidogrel (10 mg/kg PO on day 1, followed by 2–3mg/kg q24h) or both failed to show any difference in survival or thromboembolic events, however this may have been affected by low numbers and high survival (6/8, 7/8 and 7/8 respectively).[22]

Given that it is the metabolite of Clopidogrel that provides the platelet inhibition, its effect can be increased by rifampin and decreased by cimetidine. In humans, interaction with omprazole have shown to decrease its effects, but this has not been demostrated in dogs. In addition, in humans, different genotypes have different metabolism of Clopidogrel, leading to some people being 'resistent' to its effect.

Other platelet function inhibitors such as ticlopidine, prasugrel, and ticagrelor have been investigated in dogs (as a model for humans). The inhibition of platelet aggregation by prasugrel was about 10 and 300 times more potent than clopidogrel and ticlopidine, respectively.[37] Ticagrelor is the first reversibly binding oral P2Y12 receptor antagonist, and has been shown in a canine coronary thrombosis model to be more effective than clopidogrel at preventing re-occlusion.[38]

ANTICOAGULANTS:

Unfractionated Heparin

UFH use in IMHA has been reported for several years. A retrospective study showed very high mortality rates using low-dose unfractionated heparin.[20] Similarly a prospective study comparing low fixed-dose UFH (150u/kg q6-8h) to individually-adjusted showed very high mortality in the low fixed-dose UFH group.[23] Sub-therapeutic dosing of heparin is a risk factor for thrombosis in humans, however effective dosing and monitoring of UFH is difficult in the dog. It is likely that fixed dose heparin protocols are inadequate due to differences in protein-binding, consumption and excretion of the heparin molecules. Based on extrapolation from human medicine, heparin therapy is often monitored in dogs using the aPTT with the dose adjusted to achieve an aPTT prolongation of 1.5 to 2.5 fold. aPTT may not be a consistent or reliable method of monitoring heparin therapy in dogs, especially those with inflammatory diseases. The efficacy of heparin therapy may be improved by monitoring plasma heparin concentration, however, the main drawback of anti-Xa monitoring is the cost and the lack of availability to most clinicians.

- Breuhl et al monitored UFH in dogs with IMHA using both APTT and anti-Xa monitoring, using a dose of 300u/kg q 6h. They found that only 8 of the 18 dogs attained anti-Xa activity greater than 0.35U/mL (therapeutic range considered to be 0.3-0.7U/mL).[39]
- Individually-adjusted heparin therapy, using anti-Xa monitoring to adjust the UFH dose so as to maintain a therapeutic plasma level, resulted in 87% 180-day survival in a pilot study. Doses between 150 and 566U/kg q6-8h were required.[23]

TEG® has been proposed as a way of monitoring UFH use. In a study of healthy dogs, TEG®, aPTT and anti-Xa activity monitoring were performed after unfractionated heparin dosing (200 U/kg q8h SQ x 3 days)[40] Statistically significant prolongation of the R time and a decrease in angle and MA was seen in all dogs by hour 3. Only 1 dog had a prolonged aPTT. Anti-Xa activity levels increased on Day 4 but not on Day 1. They concluded that the extensive prolongation of the R time indicates that TEG® may be too sensitive.

Sonoclot® was recently evaluated for use in monitoring UFH use in normal dogs.[41] Using a dose of 300u/kg q8 they evaluated Sonoclot parameters, APTT and anti-Xa, finding good correlation between anti-Xa & Sonoclot® (*R = 0.648-0.687*), although to a lesser degree than aPTT *(R = 0.745)*. APTT has been shown to be problematic in monitoring UFH in IMHA, it remains to be seen whether Sonoclot® is superior.

Low molecular weight heparin

In humans, LMWH are popular due to their greater ease of use and predictability. In humans their excretion follows 1st order kinetics (renal excretion) making plasma levels more predictable. They are also less likely to bind to proteins, leading to greater bioavailability. Lastly their slower excretion (in humans) makes once daily dosing feasible. These qualities are all desirable in dogs, and LMWH has been used in IMHA studies recently. It is important to note that findings for enoxaparin cannot be extrapolated to dalteparin due to the differences between the molecules.

- A prospective study compared the effects of 3 heparin protocols on anti-factor Xa activity (AXa) in 18 dogs at risk for venous thrombosis.[42] Only 3 dogs had IMHA. The LDH group received a constant rate infusion (CRI) of UFH at 300 U/kg/d, the HDH group received a bolus of 100 U/kg of UFH IV, then a CRI of 900 U/kg/day, and the DP group received 100 U/kg Dalteparin SC at 0, 12, and 24 hours. They were treated for only 24 hours. Neither DP nor LDH increased anti-Xa to to the therapeutic range. HDH increased anti-Xa to this range in 2 of 6 dogs, but had unpredictable effects on aPTT and resulted in hemorrhage in 4 dogs.
- A retrospective study of 8 dogs with IMHA treated with Enoxaparin showed high survival.[24] All dogs survived to discharge, two dogs were euthanized at day 28 and 87, one due to suspect thromboembolic complications. Enoxaparin therapy was tapered until discontinuation at home in 7 dogs within 6 to 21 days after discharge and discontinued abruptly in one dog.

Pharmacokinetic studies in healthy dogs suggest a dose of 0.8mg/kg Enoxaparin q 6h to sustain therapeutic anti-Xa activity (0.5-2U/mL), but the therapeutic relevance of this target anti-Xa activity remains controversial.[43] This point was emphasized in a study that used a venous stasis model in *cats* where plasma anti-Xa activity was not significantly correlated with thrombus formation.[44] Measurement of *in vitro* heparin concentration may not adequately reflect degree of *in vivo* anticoagulation. In addition, APTT is unaffected by LMWH due to its lack of inhibition of factor IIa. A recent study of 24 healthy dogs proposed that Calibrated Automated Thrombogram (CAT) might provide a new approach for monitoring LMWH therapy.[45] They measured Endogenous Thrombin

Potential for 24 hours after a single dose of 50, 100 or 150 U/kg of dalteparin, finding significant time and dose-dependent decreases in ETP for up to 6 hours depending on dose. Further studies investigating this modality as a potential 'point of care' test are ongoing.

Synthetic Xa inhibitors are new drugs that are promising due to their homogeneity and hence predictability compared to natural heparins but have not been studied in dogs.

WHICH DRUG IS BEST?

Very few studies have compared antithrombotic therapies in IMHA. A retrospective study showed similar efficacy between low dose aspirin and IA UFH in 33 dogs with IMHA, however small sample size and retrospective nature of the study might have influenced the findings.[(46)] Currently Ontario Veterinary College is conducting a blinded clinical trial comparing enoxaparin at 0.8 mg/kg SC q6h to aspirin (1mg/kg/day) and clopidogrel. At the University of Minnesota a randomized blinded clinical trial is underway comparing individually adjusted UFH and 0.5mg/kg/day aspirin.

DURATION OF THERAPY:

The appropriate length of anticoagulation treatment in IMHA is unknown, with no published studies investigating this question. Several studies have shown death from thrombotic disease several weeks after diagnosis, indicating that continuation of antithrombotic therapy after discharge is recommended. Gradually weaning heparin is recommended due to reports of 'rebound hypercoagulability' in both humans and dogs[(40)] although a follow up study did not show this after abrupt discontinuation of 3 days of 300u/kg q8h UFH[(47)]

LIMITATIONS OF TEG

Thromboelastography has been used in many of the above mentioned studies, but the conclusions of those studies must take into account the limitations of TEG. TEG is affected by anemia[(48)], and hemolysis.[(49)] There are also differing methodologies regarding activation: Kaolin, TF, or recalcification of citrated whole blood. One study[(33)] found significant day-to-day variability in some TEG parameters in normal dogs, and sampling method can significantly alter TEG results. Importantly, many of the pharmacokinetic studies using TEG use healthy rather than hypercoagulable dogs. The question remains what do we aim for if we are treating a hypercoagulable dog - normal or hypocoagulable traces? The finding by several TEG studies in IMHA dogs that those with relatively hypocoagulable tracings have a worse outcome (most likely due to severe consumptive coagulopathy) makes using TEG to monitor effective anticoagulant therapy difficult.

CONCLUSION

Well powered, prospective controlled studies, are required to provide definitive information regarding appropriate anticoagulant use in IMHA. Until then, evidence supports the use of thromboprophylaxis. Further studies investigating the cause of the hypercoagulability may provide a clearer target for antithrombotic therapy. A 0.5mg/kg/day dose of aspirin may be inadequate in some dogs, clopidogrel may have a more beneficial effect, although this is yet to be shown in a large study. Heparin, both UFH and LMWH may be of greater effect in hospitalized animals, however it appears that a high dose is important, requiring monitoring. Newer drugs show promise, but as yet there is inadequate data to promote their use in IMHA. Hopefully some of the studies on the horizon will help answer questions regarding the most appropriate therapy for dogs with IMHA.

REFERENCES

[1]. Klein MK et al **J. Am. Vet. Med. Assoc**. 1989 Jul 15;195(2):246–50. [2]. Carr AP et al **J. Vet. Intern. Med.** 2002 Oct;16(5):504–9. [3]. Mischke R. **Dtsch. Tierarztl. Wochenschr.** 1998 Jan;105(1):13–6. [4]Scott-Moncrieff JC et al **J Am Anim Hosp Assoc**. 2001 Jun;37(3):220–7. [5]Sinnott VB et al **J Vet Emerg Crit Care**. 2009 Oct;19(5):484–8. [6]Fenty RK et al **J Am Vet Med Assoc**. 2011 Feb 15;238(4):463–7. [7]. Goggs R et al **Vet. J**. 2011 [8]. Joop K et al **Thromb. Haemost.** 2001 May;85(5):810–20. [9] Ogura H et al **J Trauma**. 2001 May;50(5):801–9. [10]Soriano AO et al **Crit. Care Med.** 2005 Nov;33(11):2540–6. [11]Cermak J et al **Blood**1993 Jul 15;82(2):513–20. [12]Eilertsen K-E et al **Blood Coagul. Fibrinolysis**. 2004 Oct;15(7):521–38. [13]Griebsch C et al **Vet Clin Pathol**. 2009 Dec;38(4):421–5. [14]Piek CJ et al **Vet. Immunol. Immunopathol**. 2011 Dec 15;144(3-4):346–54. [15]Tsuchiya R et al **J Vet Intern Med.** 2009 Nov 1;23(6):1164–9. [16]. de Kruif MD et al **The Journal of Immunology.** 2007 Feb 1;178(3):1845–51. [17]. Bottles KD et al **Blood Coagul. Fibrinolysis**. 1993 Jun;4(3):405–14. [18]. Rose LJ et al **Vet Clin Pathol**. 2011 Dec;40(4):426–34.[19]. Thomason J et al **J Vet Intern Med.** 2010;24(3):681 (Abstract). [20]. Weinkle TK et al **J. Am. Vet. Med. Assoc**. 2005 Jun 1;226(11):1869–80. [21]. Piek CJ et al **BMC Vet. Res.** 2011;7:15. [22]. Mellett AM et al **J. Vet. Intern. Med**. 2011 Feb;25(1):71–5. [23]. Helmond SE et al **J. Vet. Intern. Med**. 2010 Jun;24(3):597–605. [24]. Bianco, D et al. (abstract) **Proceedings of ECVIM Congress**, Sevilla, Spain: 2011 [25]. Whelan MF, et al. **J Vet Intern Med.** 2006;20(3):715 (Abstract). [26]. Shearer L et al **J Vet Intern Med**. 2009;23(3):745. [27]. Sharpe KS et al **Am. J. Vet. Res**. 2010 Nov;71(11):1294–304. [28]. Puglia GD et al (abstract) **Proceedings International Veterinary Emergency and Critical Care Symposium** 2010. [29]. Thomason J et al **J. Vet Intern Med** 24(3):681 (abstract). [30]. Dudley A et al **J. Vet Intern Med.** 2011;25(3):698 (abstract). [31]. Hoh C, et al, **Proceedings of ACVIM Congress, SAIM Research Focus**, 2010. [32]. Flint SK et al **J Vet Intern Med.** 2010;24(3):681 (abstract). [33]. O'Kell A et al **J. Vet Intern Med**. 2011;25(3):698. [34]. Flint SK et alAm. **J. Vet. Res**. 2011 Oct;72(10):1325–32. [35]. Goodwin J et al **J Vet Intern Med**. 2007;21(3):609 (abstract). [36]. Brainard BM et al **J Vet Intern Med**. 2008;22(3):739 (abstract). [37]. Niitsu Y et al **Eur. J. Pharmacol.** 2008 Jan 28;579(1-3):276–82. [38]. Wang K et al **Thromb. Haemost.** 2010 Sep;104(3):609–17. [39]. Breuhl EL et al. **J Am Anim Hosp Assoc.** 2009 Jun;45(3):125–33. [40]. Pittman JR, et al **J Vet Emerg Crit Care** (San Antonio). 2010 Apr 1;20(2):216–23. [41]. Babski DM et al (abstract) **Proceedings IVECCS Symposium** 2010. [42]. Scott KC et al **J Vet Emerg Crit Care** (San Antonio). 2009 Feb;19(1):74–80. [43]. Lunsford KV et al **J Am Anim Hosp Assoc**. 2009 Dec;45(6):261–7. [44]. Van De Wiele CMet al. **J. Vet. Intern. Med**. 2010 Feb;24(1):185–91. [45]. Gara-Boivin C et al **J Vet Intern Med.** 2011;25(3):698 (abstract). [46]. Orcutt E et al **J Vet Intern Med**. 2009;23(3):693 (Abstract). [47]. Pittman JR et al (abstract) **Proceedings IVECCS Symposium** 2010. [48]. Ruttmann et al **Anaesth Intensive Care.** 2003 Feb;31(1):40–3. [49]. Bauer et al **Vet Clin Pathol.** 2010 Jun;39(2):180–9.

IMMUNOMODULATION: A REVIEW WITH A FOCUS ON T-CELLS

J. Corey Orava, DVM
Manchester, VT

INTRODUCTION

The normal immune system is dynamic and balanced. Numerous disease states arise when any part of the system does not function properly. Immunomodulation encompasses all therapeutic interventions aimed at modifying the immune response for a therapeutic intent. These therapies have one of two possible goals: either increase the immune response or decrease the immune response. This review will briefly examine both aspects with a focus on the biology of T-lymphocytes and strategies to modify their behavior and numbers.

Increasing the Immune Response

Two main reasons exist for desiring a more active immune response: a functional immune system is present but is inadequate for a specific condition, or the immune system is pathologically decreased (i.e. immunodeficient). The most utilized form of immunomodulation in veterinary medicine involves the use of vaccines. Another therapy with the intent of increasing the immune system involves immunomodulation to fight cancer and certain recalcitrant infections (e.g. leishmaniasis and tuberculosis).[1]

Decreasing the Immune Response

When an immune response is excessive it is termed a hypersensitivity reaction. These hypersensitivity reactions have been divided into four categories based on the underlying mechanisms.[2]

Type I – immediate hypersensitivity mediated by IgE (e.g. food allergy, allergy to drugs, anaphylaxis)
Type II – antibody dependent (IgG or IgM) cytotoxicity (e.g. immune-mediated hemolytic anemia/thrombocytopenia)
Type III – immune complexes deposited in tissues (e.g. glomerulonephritis, immune-mediated polyarthritis)
Type IV – delayed (T-cell dependent) hypersensitivity (e.g. contact dermatitis, type I diabetes)

In some of the aforementioned hypersensitivity reactions the causative agent is external (i.e. allergies). In such cases, the simplest solution is to remove the inciting antigen. Alas, in many instances this is easier said than done. When exposure cannot be avoided, anti-inflammatory and/or anti-histamine drugs are usually employed. These drugs mask symptoms rather than address the underlying cause.

The corollary of hypersensitivity reactions caused by an extraneous antigen is when the antigen is intrinsic – i.e. it is part of the 'self.' In these diseases, termed autoimmune, pathology occurs due to a defect of self tolerance. This situation can be more complicated because an autoimmune disease can develop secondarily to an allergy. Early approaches to treating autoimmunity focused on non-specific immunosuppression. As all medical professionals are aware such indiscriminant immunosuppression, especially when implemented over the long-term, carries significant side effects.

Newer therapies seek to be curative (i.e. long-term effect with minimal adverse consequences) rather than symptomatic. In order for a cure to a hypersensitivity or autoimmune reaction to occur, the immune system must be reprogrammed to not attack the inciting antigen. Simplified, this entails elimination of all offending T-cells and reseeding the body with antigen presenting cells (APCs) that express the antigens concerned. The term used to describe such reprogramming is *induced tolerance.*[3]

TOLERANCE

Tolerance is a general term meaning that the immune system does not attack an antigen. Tolerance is categorized according to the location where it occurs. Central tolerance occurs in the thymus while peripheral tolerance occurs elsewhere in the body.[4]

Central Tolerance

Early T cells, called thymocytes, produced in the bone marrow enter the cortex of the thymus. These cells are double positive (DP) for CD4 and CD8 and are pre-programmed to die unless they receive a signal. That signal is provided by binding with a peptide-MHC complex (the MHC can be either a Class I or II molecule). If the molecule is Class I then the T cell retains CD8 and downregulates CD4, thus becoming $CD8^+CD4^-$ or more simply a $CD8^+$ T cell. If the molecule is Class II then the opposite occurs with the end result being a CD4+ T cell. The peptides exposed to these selected T cells represent self antigens. As part of the selection process the T cell receptor (TCR) molecule rearranges so that future exposure to such antigens does not elicit a response. This type of cell selection has been termed, positive selection. After positive selection the cells are directed to the medulla of the thymus.

The majority of peptide-MHC complexes leading to positive selection are relatively weak. When a cell signal is more intense (due to a strong peptide) the DP cells are not rescued but rather instructed to undergo a TCR-mediated apoptosis. This is termed negative selection or clonal deletion. This process occurs predominantly in the medulla of the thymus, while positive selection is a feature of the cortex. In order for the effective removal of self-reactive thymocytes, the thymus must have a complete representation of tissue-specific self antigens. A review by Kyewski and Derbinski[5] noted that genes expressed by medullary thymic endothelial cells (mTECs) represent most, if not all tissues of the body. Tissue representation even crosses gender boundaries – e.g. female mTECs express genes specific for male reproductive organs and vice versa. The vast majority (>95%) of DP cells that enter the thymus do not survive these combined selection processes.[3,4,6,7]

Peripheral Tolerance

Since not all self-reactive T-cells are removed by the thymus, other mechanisms exist in the periphery of the body to prevent autoimmune disorders. The immune system employs multiple strategies with the simplest being minimizing exposure of potentially-reactive T cells to various organs. This is accomplished by the way naïve T cells circulate through the body – from blood to secondary lymphoid organs, to efferent lymph ,then back to blood.[7] Three other strategies exist for peripheral T-cell tolerance: anergy, deletion and immune suppression. [8]

Anergy occurs when a T-cell encounters its appropriate peptide under specific circumstances and thereafter enters a state of unresponsiveness. One such circumstance occurs when the peptide-containing cell lacks appropriate co-stimulatory molecules (e.g. CD80 and CD86). The T-cell does not become activated, persists in the body and thereafter will not become activated even if it meets the appropriate peptide with the necessary co-stimulatory molecules.[8,9]

Deletion is the apoptotic death of T-cells after repeat exposure to an appropriate antigen. This process is called antigen induced cell death (AICD) and likely has several pathways, but the best understood involves signaling of CD95 (Fas). The importance of this process is made evident by mice and humans with defective CD95 reliably developing various autoimmune diseases.[8,10,11]

The last mechanism of peripheral tolerance and perhaps the most clinically relevant involves specialized T-cells that have the ability to suppress other T-cells. These cells were originally called T-suppressors but have been renamed regulatory T-cells, or Tregs. Numerous cell types qualify as Tregs, including type 3 T-helpers (Th3), type 1 regulatory T cells (Tr1), and most importantly $FOXP3^{+}CD4^{+}CD25^{+}$ T cells. Tregs can either develop naturally in the thymus ($FOXP3^{+}CD4^{+}CD25^{+}$ T cells) or they can be created in the periphery by the activation of non-regulatory T-cells (Th3, Tr1). The former group is referred to as nTregs (natural) while the later as iTregs (induced).[8,12]

Tregs

Research on regulatory T-cells has developed so rapidly that several conferences have been held to discuss just this topic. The importance of Tregs is best illustrated in vivo when there is a deficiency or defect in the Tregs or when there is an excess of Tregs. When the thymus is depleted of nTregs, multiple organ-specific autoimmune disorders develop.[13] Conversely, an excess of nTegs is associated with cancer. In the dog, Biller et. al.[14] reported the percentage of $FOXP3^{+}CD4^{+}CD25^{+}$ T-cells in normal dog's blood and lymph nodes to be 4.3% and 9.8% respectively. The values in dogs with cancer were 7.5% and 17.1% respectively.

It should come as no surprise that much of the research being done on Tregs involves enhancing their activity for autoimmune disease and graft rejections but also limiting their deleterious effects in tumor development. Many issues need to be overcome before this therapy reaches the bedside, most notable is establishing safety. There are currently only two clinical trials in humans involving the administration of Tregs (www.clinicaltrials.gov, trial # NCT01446484, NCT01210664). Both of which are in phase I, the primary goal of which is to demonstrate safety.

Mesenchymal Stem Cells

Mesenchymal stem cells were originally believed to work predominantly through differentiation. Within the past 10 years there has been a paradigm shift in our understanding of how MSCs behave in vitro and in vivo. Strong evidence now suggests that the major mechanisms of action of MSCs involve regulating apoptosis, fibrosis and immune responses. The immunomodulatory capabilities of MSCs have been extensively studied and many mechanisms have been discovered. These include: induction of T-cell anergy, reduction of T-cell proliferation, reduction of pro-inflammatory cytokines (e.g. Il-1, IFN-γ) and the upregulation of Tregs. In addition, MSCs do not express the co-stimulatory molecules CD80 and CD86 and are thus immunoprivileged, minimizing reactions when used allogeneically.[15,16,17,18]

Unlike the two FDA trials with Tregs, over 200 trials with MSCs have been registered, several of which, including some demonstrating efficacy, have already been completed (www.clinicaltrials.gov, trials # NCT00992485, NCT01274975, NCT00395200, NCT00294112, NCT00475410).

CONCLUSION

The immune system is arguably the most complex component of the mammalian body. Our understanding of its mechanisms of action has grown tremendously in the past decade. Still, many pieces of the picture are missing. It is well established that the normal immune system continually produces effector cells along with regulatory cells that keep the former in check. When the balance is disrupted, disease ensues. Correcting any discrepancies of the immune system has historically involved a non-specific approach. This frequently leads to undesirable side effects. New cell-based modalities are in development to provide better outcomes with few side effects.

REFERENCES

1. Gea-Banacloche J, **Immunomodulation**, in Principles of Molecular Medicine Humana Press2006: 1304p 2. Tizard I, Veterinary Immunology 8th ed Saunders 2008: 592p 3. Sprent J, et.al., **Phil Trans R Soc Lond** 2001;356(1409):609 4. Sakaguchi S, et.al., **Cell** 2008;133(5):775 5. Kyewski B, et.al., **Nat Rev Immunol** 2004;4(9):688 6. Griesemer A, et.al., **Transplant** 2010;90(5):465 7. Mueller D, **Nat Rev Immunol** 2010;11(1):21 8. Romagnani S, **Intern Emerg Med** 2006;1(3):187 9. Powell J, **Clin Immunol** 2006;120(3):239 10. Green D, et.al., **Immunol Rev** 2003;193:70 11. Zhang J, et.al., **Cellular Mol Immunol** 2004;1(3):186 12. Garden O, et.al., **Int Immunopharm** 2011;11(5):576 13. Buckner J, **Nat Rev Immunol** 2010;10(12):849 14. Biller B, et.al., **Vet Immun Immunopath** 2007;116(1-2):69 15. Singer N, et.al., **Annu Rev Pathol Mech Dis** 2011;6:457 16. Ben-Ami E, et.al., **Autoimmun Rev** 2011;10(7):410 17. Gonzalez-Rey E, et.al., **Ann Rheum Dis** 2010;69(1):241 18. Ge W, et.al., **Transplant** 2010;90(12):1312

CLOSTRIDIAL DISEASES IN LARGE ANIMALS AND HORSES

J. Scott Weese, DVM, DVSc, DACVIM
Guelph, Ontario, Canada

INTRODUCTION

Clostridium is an extremely diverse Genus of Gram positive to Gram variable sporeforming rods. Clostridia can be found in virtually every individual, human or animal, and they play an important role in both health and disease. The differences between different clostridial species and their properties are striking. Some can produce some of the most potent toxins ever identified (*C. botulinum*), some can cause life-threatening myositis (*C. perfringens, C. septicum*) and some can cause severe systemic illness (*C. piliforme, C. novyi*). The greatest overall role of clostridia, however, is likely as a cause of enteric disease. Yet, at the same time, many members of this Genus are common commensals that can be found in large numbers in the intestinal tract.

In defense of clostridia

Clostridium has been referred to as a 'much-maligned' Genus, with good reason. The Genus consists of over 300 different species, with very few recognized as enteropathogens. In fact, clostridia are common components of the intestinal microbiome of large animals and probably play a critical role in gut and systemic health.

But...

While clostridia may be 'much-maligned', some of the negative perceptions about this Genus are warranted. Despite the fact that only a small number of clostridia are recognized enteropathogens, these species can account for significant morbidity and mortality in large animals. The two most important species involved in enteric disease of large animals are *C. difficile* and *C. perfringens*, but undoubtedly others can play a role in some situations.

CLOSTRIDIA AS ENTEROPATHOGENS IN LARGE ANIMALS

Clostridium difficile

Clostridium difficile is a critically important cause of diarrhea in some species but commonly found in healthy individuals of many different species. In humans, it is the leading cause of antimicrobial-associated and hospital-associated diarrhea, and an increasing cause of disease in low-risk people in the community. In animals, the relevance is highly variable.

The pathogenesis of disease is poorly understood but relates in large part to the production of toxins in the intestinal tract. Most toxigenic (and therefore clinically relevant) strains of *C. difficile* produce toxin A (an enterotoxin) and toxin B (a cytotoxin). A smaller percentage of strains produce a binary toxin (CDT), the role of which is currently unclear. Factors that trigger development of disease versus colonization are not well understood and likely relate in large part to the complex interaction between the bacterium, the endogenous intestinal microbiome and the host.

Horses

In horses, *C. difficile* is the most commonly diagnosed cause of colitis in some regions.[1-3] It can cause disease in horses of any age, including foals less than 24 hours of age. Mortality rates can be high, although a true understanding of the mortality of *C. difficile* infection (CDI) in horses is hampered by the limited number of studies that have been performed and the focus on referral populations (which presumably consists of more serious cases). Regardless, CDI can range from mild and self-limiting to rapidly fatal. As opposed to the situation in humans, equine CDI is typically community-associated[3,4] and hospital transmission seems to be uncommon. Hospital-associated outbreaks have been reported,[5] but these seem to be quite rare, and outbreaks appear to be of greatest relevance on farms, particularly involving young foals on breeding farms. While most often a cause of enterocolitis, *C. difficile* has also been implicated as a cause of duodenitis/proximal jejunitis in horses.[6] This has only been shown in one study involving one region, and broader study is needed to determine whether this is repeatable.

While an important cause of disease, *C. difficile* can also be found in a small percentage of non-diarrheic horses, particularly young foals and individuals treated with antimicrobials.[1,3,7-9] Longitudinal study indicates healthy horses may transiently shed *C. difficile* and the cumulative prevalence can be high, suggesting that horses are frequently exposed to *C. difficile* yet colonization is transient.[10] Indeed, given the ability of *C. difficile* spores to persist in the environment, it is possible that exposure to *C. difficile* from the environment, other horses, food and water is a regular (if not daily) event.

Pigs

In pigs, *C. difficile* is an important cause of neonatal diarrhea in some regions, yet rare to non-existent in others. As opposed to other species, CDI in pigs is essentially restricted to 1-7d old piglets, where it can cause severe colonic disease with high morbidity and mortality.[11-15] Yet, *C. difficile* shedding is very common in healthy young piglets, with prevalences above 90% in some groups of young (>1 wk old) piglets and cumulative prevalence rates nearing 100%, even in the absence of recognized CDI on a farm.[16,17] Clearly, CDI is driven by more than simple exposure to *C. difficile*, something that is true in horses as well but perhaps most dramatic in piglets, and reasons for some individuals developing disease while others remain unaffected are as of yet unclear. Shedding rates tend to decrease greatly over time and low prevalences (<10%) are present by slaughter age.[11,17-19]

Ruminants

In contrast, the role of *C. difficile* in disease of ruminants is somewhat enigmatic. While associations between *C. difficile* and diarrhea has been reported in calves,[20,21] the true role of this organism in bovine neonatal diarrhea is still unclear and experimental attempts to produce disease have failed.[22] Colonization rates of healthy calves can be high in the absence of disease,[20,21,23] as in some other species, and the reason for this is equally perplexing in ruminants. There is no suggestion that *C. difficile* is a cause of enteric disease in older cattle, despite the fact that the organism can be found in a small percentage of adult cattle.[24] Shedding rates in feedlot cattle tend to be quite low.[24]

Types

The types of *C. difficile* that are found in large animals raise many questions about the epidemiology of this organism. In food animals, one strain tends to predominate. Depending on the typing method that is used, this strain is referred to as ribotype 078, North American pulsotype 7/8 and toxinotype V.[17,25,26] This is an interesting strain that is uncommon in humans, but appears to be an increasing cause of human CDI internationally, particularly in community-associated cases.[27-29] It can also be found in retail meat, and concern has been expressed that it might be an emerging zoonotic and foodborne pathogen.[30-32]

The situation is somewhat different in horses. Ribotype 078 has been found in horses[26,33] and recent evidence suggests this strain may be on the rise in some regions, however a variety of other strains are also found, typically strains that are also identified in humans with CDI.[26,34-36]

Clostridium perfringens

Clostridium perfringens is rather ubiquitous organism that can cause disease of various body systems, but also be found in a large percentage of healthy individuals in some animal species. Diagnosis of *C. perfringens* enteritis can be quite challenging in some species, hampering our understanding the true role of this organism in disease, but it has been called the most important clostridial enteropathogen in animals.[37] *Clostridium perfringens* isolates are classified into 5 major types, A through E, based on the production of 4 major toxins (alpha, beta, epsilon, iota). Additionally, isolates may possess a range of other toxin genes. Most information is available for two of these additional toxins, enterotoxin (CPE) and beta-2 toxin, and these may be the most important virulence factors in some animal species. Particularly in ruminants, *C. perfringens* can be associated with disease of a wide range of body systems as well as overwhelming systemic disease, however enteric manifestations are outlined below.

Horses

As opposed to many other species, where *C. perfringens* can be found in a large percentage of healthy individuals, *C. perfringens* is uncommonly identified in healthy horses, with the exception of young foals and broodmares, where shedding rates can be quite high.[10,38-43] While this would seem to facilitate diagnosis, with the presence of *C. perfringens* in a diarrheic individual indicating a role in disease, it is not this straightforward because of the ability of *C. perfringens* to proliferate following disruption of the intestinal microflora. Thus, finding large numbers of *C. perfringens* in feces of a diarrheic horse does not differentiate cause from effect. A lack of good diagnostic tests, namely tests to detect a variety of *C. perfringens* toxins, further complicates diagnostic testing.

Despite these challenges, it is hard to dispute that *C. perfringens* is certainly a cause of enterocolitis in horses. Most convincing information revolves around strains producing CPE and beta-2 toxin,[2,3,39,44] toxins that are not restricted to specific Types, however these can also be found in healthy horses. Type C (and rarely Type D) have also been implicated as a cause of serious disease, mainly in foals.[45-48] The role of Type A strains that only produce alpha toxin (not additional toxins such as CPE or beta-2 toxin) is questionable.

Disease can occur in horses of any age, including neonatal foals. Limited information is available regarding risk factors for disease. As with *C. difficile,* severity of disease may range from mild and self-limiting to rapidly fatal hemorrhagic colitis.

Pigs

Clostridium perfringens types A and C are recognized causes of necrotizing enterocolitis in piglets.[49] CPE and beta-2 toxin also appear to be of relevance.[50] A large percentage of healthy pigs, particularly young individuals, can harbour *C. perfringens* in the absence of disease.[12]

Ruminants

Clostridium perfringens is an important enteropathogen in ruminants, particularly young animals, but it can also be found in a large percentage (>50%) of young healthy individuals,[51] complicating diagnosis. Type A *C. perfringens* has been associated with ruminal and abomasal tympany, abomasitis and abomasal ulceration in young calves,[52] and (with or without beta-2 toxin) suggested as a cause of hemorrhagic bowel syndrome in adult cattle, a severe syndrome that typically affects the jejunum.[53-55] Various case reports have implicated Type A *C. perfringens* in a range of enteric disease in ruminants of different ages, typically through isolation of large numbers of the organism from feces or intestinal contents,[56] however the baseline prevalence and ability of this organism to overgrow in response to other enteric diseases clouds the diagnosis.

Type B is a cause of lamb dysentery, a severe disease of neonatal lambs that predominantly occurs in Britain and South Africa. Type C causes a similar severe syndrome in calves and lambs. Type D causes enterotoxemia in young sheep and cattle, but the gastrointestinal component is not typically clear because of the rapid onset of severe systemic disease.

The roles of CPE and beta-2 toxin are less clear. In addition to a potential role of hemorrhagic bowel syndrome of adult cattle, beta-2 toxin has been suggested as a cause of necrotic enteritis in young calves.[57]

Other clostridia

It is likely that other clostridia are enteropathogens, particularly given the large percentage of suspected infectious enteric disease cases that remain idiopathic. While more often a cause of wound infection, *C. septicum* can cause enteric disease (braxy) in sheep and calves.[58] The disease is typified by abomasitis and varying degrees of systemic compromise is young lambs and calves, and has also been suggested as a cause of disease in foals.[59] *Clostridium sordellii* is a widespread bacterium that is periodically implicated in severe disease in various animal species although its true role is unclear.[60]

Determining a role for other clostridia is complicated by our poor understanding of the range of clostridia that may be present in the intestinal tract, our limited understanding of the intestinal microflora, difficulties isolating many clostridia and problems differentiating overgrowth of species in response to disease from disease caused by overgrowth. Diagnosis of other clostridia as causes of enteric disease is difficult. General anaerobic culture could be performed, but many clostridia are not readily isolated using standard methods. Further, it is difficult to impossible to ascribe disease to a certain bacterial species that is present in a fecal sample in the absence of organized research studies because of the relatively poor overall knowledge of the equine clostridial microflora. Presence of a specific *Clostridium* at high numbers in a diarrheic fecal sample might raise suspicion, but it can certainly not indicate infection. This was highlighted by a report of experimental lincomycin-induced colitis, where *Clostridium cadaveris* was initially implicated as the cause;[61] something that was subsequently refuted.[62]

REFERENCES

[1] Baverud V, et al. ***Equine Vet J*** 1997;29:279-284. [2] Donaldson MT, et al. ***J Am Vet Med Assoc*** 1999;215:358-361. [3] Weese JS, et al. ***Equine Vet J*** 2001;33:403-409. [4] Weese J, et al. ***Equine Vet J*** 2006;38:185-188. [5] Madewell BR, et al. ***J Vet Diagn Invest*** 1995;7:343-346. [6] Arroyo L, et al. ***J Med Microbiol*** 2006;55:605-608. [7] Baverud V, et al. ***Equine Vet J*** 2003;35:465-471. [8] Medina-Torres CE, et al. ***Vet Microbiol*** 2011. [9] Avbersek J, et al. ***Anaerobe*** 2009;15:252-255. [10]Schoster A, et al. *ACVIM Forum* . Denver, CO, 2011. [11] Alvarez-Perez S, et al. ***Vet Microbiol*** 2009. [12] Baker AA, et al. ***Appl Environ Microbiol*** 2010;76:2961-2967. [13] Waters EH, et al. ***J Vet Diagn Invest*** 1998;10:104-108. [14] Yaeger M, et al. ***J Vet Diagn Invest*** 2002;14:281-287. [15] Yaeger MJ, et al. ***J Vet Diagn Invest*** 2007;19:52-59. [16] Norman K, et al. ***Anaerobe*** 2009. [17] Weese JS, et al. ***Anaerobe*** 2010;16:501-504. [18] Weese JS, et al. ***Zoonoses and Public Health*** 2011;In press. [19] Hoffer E, et al. ***J Food Prot*** 2010;73:973-975. [20] Hammitt, et al. ***Vet Microbiol*** 2007. [21] Rodriguez-Palacios A, et al. ***Emerging Infect Dis*** 2006;12:1730-1736. [22] Rodriguez-Palacios A, et al. ***Vet Microbiol*** 2007;124:166-172. [23] Costa MC, et al. ***Veterinary Microbiology*** 2011. [24] Rodriguez-Palacios A, et al. ***Appl Environ Microbiol*** 2011:1-28. [25] Debast SB, et al. ***Environ Microbiol*** 2009;11:505-511. [26] Keel K, et al. ***J Clin Microbiol*** 2007;45:1963-1964. [27] Goorhuis A, et al. ***Clin Infect Dis*** 2008;47:1162-1170. [28] Rupnik M, et al. ***J Clin Microbiol*** 2008;46:2146. [29] Jhung M, et al. ***Emerging Infect Dis*** 2008;14:1039-1045. [30] Songer JG, et al. ***Emerging Infect Dis*** 2009;15:819-821. [31] Metcalf D, et al. ***Can Vet J*** 2010;51:873-876. [32] Rupnik M. ***Clin Microbiol Infect*** 2007;13:457-459. [33] Medina-Torres CE, et al. ***Veterinary Microbiology*** 2011. [34] Arroyo L, et al. ***J Med Microbiol*** 2005;54:163-166. [35] Arroyo LG, et al. ***Vet Microbiol*** 2007;120:179-183. [36] Songer JG, et al. ***J Vet Diagn Invest*** 2009;21:377-380. [37] Songer J, et al. ***J Vet Diagn Invest*** 2005;17:528-536. [38] Browning GF, et al. ***Equine Veterinary Journal*** 1991;23:405-409. [39] Herholz C, et al. ***J Clin Microbiol*** 1999;37:358-361. [40] Kanoe M, et al. ***Microbios*** 1990;64:153-158. [41] Netherwood T, et al. ***Epidemiol Infect*** 1996;117:375-383. [42] Tillotson K, et al. ***J Am Vet Med Assoc*** 2002;220:342-348. [43] Waggett BE, et al. ***Equine Vet J*** 2010;42:494-499. [44] Garmory HS, et al. ***Epidemiol Infect*** 2000;124:61-67. [45] Diab SS, et al. ***Vet Pathol*** 2011. [46] Drolet R, et al. ***Can Vet J*** 1990;31:449-450. [47] Howard-Martin M, et al. ***Journal of the American Veterinary Medical Association*** 1986;189:564-565. [48] Pearson EG, et al. ***Journal of the American Veterinary Medical Association*** 1986;188:1309-1310. [49] Jäggi M, et al. ***Schweiz Arch Tierheilkd*** 2009;151:369-375. [50] Wollschläger N, et al. ***Schweiz Arch Tierheilkd*** 2009;151:377-382. [51] Bartels CJM, et al. ***Preventive veterinary medicine*** 2010;93:162-169. [52] Van Kruiningen HJ, et al. ***Can Vet J*** 2009;50:857-860. [53] Ceci L, et al. ***J Vet Med A Physiol Pathol Clin Med*** 2006;53:518-523. [54] Dennison AC, et al. ***Journal of the American Veterinary Medical Association*** 2005;227:132-138. [55] Dennison AC, et al. ***Journal of the American Veterinary Medical Association*** 2002;221:686-689. [56] Van Immerseel F, et al. ***J Comp Pathol*** 2010;143:289-293. [57] Morris WE, et al. ***J Vet Diagn Invest*** 2011;23:254-259. [58] Songer J. ***Clin Microbiol Rev*** 1996;9:216-234. [59] Jones SL, et al. ***Cornell Vet*** 1993;83:143-151. [60] Hibbs CM, et al. ***Vet Med Small Anim Clin*** 1977;72:256-258. [61] Prescott JF, et al. ***Equine Vet J*** 1988;20:417-420. [62] Staempfli HR, et al. ***Can J Vet Res*** 1992;56:233-236.

CRITICAL ASSESSMENT OF CLOSTRIDIAL DISEASES IN DOGS AND CATS

Stanley L. Marks, BVSc, PhD, DACVIM (SAIM & Oncology), DACVN
Davis, CA

Infectious diarrhea in dogs and cats is one of the most common maladies facing small animal veterinarians today, but the role of enteropathogenic bacteria is poorly understood. *Clostridium perfringens* and *C. difficile* are two of the most commonly incriminated enteropathogenic bacteria in canine diarrhea;[1,2] however, the clinical documentation of Clostridial-associated diarrhea in dogs and cats is clouded by the presence of these organisms existing as normal constituents of the indigenous intestinal microflora. The challenge to diagnose Clostridial-associated diarrhea in dogs is further perpetuated by the similar isolation rates of *C. perfringens* and *C. difficile* in diarrheic and nondiarrheic dogs.[3] This problem is compounded by the lack of validation of commercial toxin assays commonly utilized in dogs and cats. The indications for performing fecal enteric panels on diarrheic dogs and cats are poorly defined, resulting in indiscriminate testing, and misinterpretation of results. The results of fecal cultures may take up to 72 hours to complete, and many dogs are administered antibiotics indiscriminately before results of fecal cultures or toxin assays are available. In addition, the resolution of clinical signs of diarrhea is often wrongly equated with eradication of the "incriminating" pathogen.

CLOSTRIDIUM PERFRINGENS

Clostridium perfringens is an anaerobic, spore-forming, gram-positive bacillus that has been associated with outbreaks of acute, often severe diarrhea in humans, horses, dogs, and cats. The elaboration of 4 major toxins (α, β, ι, and ε) is the basis for typing the organism into 5 toxigenic phenotypes, A-E. Each type may also express a subset of at least 10 other established toxins, including *C. perfringens* enterotoxin (CPE), a well-characterized virulence factor whose production is co-regulated with sporulation. A study evaluating 843 *C. perfringens* isolates collected from 103 dogs revealed that all isolates tested were type A, with 15% harboring the enterotoxin gene.[3] Enterotoxigenic *C. perfringens* type A has been associated with human food poisoning and sporadic diarrhea, canine acute and chronic, large and small bowel diarrhea, and an acute hemorrhagic diarrheal syndrome (AHDS).[1,3,4] Although several studies have shown an association between the immunodetection of CPE in fecal specimens and canine diarrhea, the pathogenesis of *C. perfringens*-associated diarrhea in the dog is not fully understood, as CPE is also detected in up to 14% of nondiarrheic dogs.[3] In addition, a number of other virulence factors such as the beta2 (β2) toxin may also play a role in diarrhea. *C. perfringens* β2 toxin has been associated with necrotic enteritis in piglets and typhlocolitis in horses.[5,6] The role of β2-toxigenic *C. perfringens* in dogs is less well understood as only a single study evaluating 24 isolates collected from diarrheic dogs has been published to date. In that study, isolates were positive for the enterotoxin gene (33%), the β2-toxin gene (33%), or both (17%); however, interpretation of the results of that study is difficult due to the small number of dogs and lack of a nondiarrheic control population.[7] The role of *C. perfringens* in cats is equally unclear because CPE was only detected in 9/219 cats with diarrhea (4.1%) and in 1 of 54 cats without diarrhea (2%) in a recent study.[8]

PATHOGENESIS

Canine *C. perfringens*–associated diarrhea has been attributed to *C. perfringens* enterotoxin (CPE), which has been shown to induce fluid accumulation and diarrhea in a dog model when administered orally or directly in the intestinal lumen.[9] The role of CPE in the development of diarrhea is clouded because CPE is detected in up to 34% of diarrheic dogs, and in 5 to 14% of nondiarrheic dogs.[3] In contrast to humans in whom *C. perfringens*-associated diarrhea is usually a result of ingestion of enterotoxigenic isolates, *C. perfringens*-associated diarrhea in dogs appears more likely to be secondary to disruption of the intestinal microenvironment, enabling proliferation or sporulation of commensal enterotoxigenic *C. perfringens*. The enterotoxin is a 35-kDa protein encoded by the *cpe* gene, whose expression is co-regulated with sporulation of the organism. *C. perfringens* strains that carry a chromosomal *cpe* have primarily been associated with human food-borne disease, whereas strains with a plasmid *cpe* gene have been associated with human non-food-borne diseases and animal diseases, including canine diarrhea.[10] Non-foodborne diseases associated with CPE are thought to involve commensal enterotoxigenic strains that are somehow triggered to undergo massive sporulation. The trigger may be one of several different factors including changes in diet, antibiotic administration, or co-infection with another intestinal pathogen. Once released into the intestinal lumen, CPE interacts with specific epithelial tight junction proteins forming a small protein complex of ~90 kDa, where it then becomes trapped on the membrane surface. The small CPE complex then interacts with additional host proteins, forming several larger complexes, including a ~155 kDa complex and a ~200 kDa complex that contains the tight junction protein, occludin.[11] Recent studies have suggested the ~155 kDa complex is responsible for the cytotoxic and histopathological damage which provides CPE access to occludin, causing alterations in tight junction structure and function, leading to paracellular permeability changes that contribute to diarrhea.[11]

DIAGNOSIS

There is currently no gold standard for the diagnosis of canine *C. perfringens*-associated diarrhea, and clinical signs are extremely variable. The optimal diagnostic approach utilizes a combination of fecal CPE immunodetection and molecular techniques, and caution should be heeded on relying upon the outdated and insensitive test of evaluating stained fecal smears for *C. perfringens* endospores.

Clinical Signs

There are no pathognomonic clinical signs indicative of *C. perfringens*-associated diarrhea in the dog, and the spectrum of disease attributed to this organism varies greatly. *Clostridium perfringens*-associated diarrhea can manifest with small intestinal, large intestinal, or diffuse clinical signs.[1] *C. perfringens*-associated diarrhea should also be considered as a potential cause of acute hemorrhagic diarrhea in dogs.[1]

Culture

Because *C. perfringens* is a normal commensal of the intestinal microflora, the mere culture of an isolate from the stool is of little clinical significance.[3,12] In addition, isolation rates for *C. perfringens* are similar in diarrheic and nondiarrheic dogs.[3] Culture may be useful in procuring isolates for the application of molecular techniques like PCR for detection of specific toxin genes or molecular typing of isolated strains to establish clonality in suspected outbreaks.

Fecal Endospores

Because sporulation is co-regulated with enterotoxin production, fecal endospore counting of Wright or Gram stained fecal smears ($\geq$ 3 spores per high power field) has been suggested as a tool for diagnosing enterotoxigenic *C. perfringens*-associated disease.[13] Several studies have reported no association between fecal endospore counts and the presence of diarrhea, or between spore counts and the detection of CPE in fecal specimens.[3,12,14] The diagnostic value of finding increased endospores in a fecal smear from a diarrheic dog is virtually zero.

Fecal Enterotoxin Immunodetection

Detection of CPE in fecal specimens is the most widely used diagnostic tool for *C. perfringens* in both human beings and animals. There is only one commercially available ELISA kit (TECHLAB Inc., Blacksburg, VA) for detection of CPE in fecal specimens; however, the performance characteristics of this assay have not been validated in the dog or cat to date. Sensitivities of immunodetection methods are extremely important when one considers that disease associated with CPE may be dependent on the concentration of CPE present within the intestinal lumen. This phenomenon is underscored by the finding that up to 14% of healthy dogs have detectable concentrations of CPE utilizing the qualitative TECHLAB ELISA.[3] In addition, time delays in sample processing can result in false positive or false negative results (unpublished observation).

Molecular Techniques

PCR detection of enterotoxigenic *C. perfringens* can be performed on fecal specimens or isolates collected via culture. Non-diarrheic dogs are far less likely to be positive for both CPE and the enterotoxin gene (*cpe*) (4%), compared to diarrheic dogs (28%).[3] Toxin DNA and RNA quantification for the α-toxin gene and *cpe* (enterotoxin gene) has been shown to be significantly higher in diarrheic dogs and cats compared to non-diarrheic animals (unpublished observation).

THERAPY

There is little objective information defining when and how to treat dogs and cats with Clostridial-associated diarrhea, and mild cases of diarrhea can be self-limiting without the need for antimicrobial therapy. Animals with acute disease (hemorrhagic gastroenteritis) merit appropriate antimicrobial therapy, although antibiotics are commonly administered even in cases of chronic diarrhea. Optimal antibiotics for the treatment of canine *C. perfringens*-associated diarrhea include ampicillin, metronidazole, and tylosin.[2] Recent evidence has shown a high rate (21%) of *in vitro* resistance to tetracycline.[15] Preliminary studies in the author's laboratory have documented a high incidence (96%) of isolates that carry a gene encoding an efflux pump for tetracyclines (the *tetA*(P) gene). These findings are extremely important in light of the fact that tetracycline has been touted as a drug of choice for the treatment of dogs with suspected *C. perfringens*-associated diarrhea or small intestinal bacterial overgrowth, and subinhibitory concentration have actually been recommended as a long term therapeutic approach.[13] Furthermore, these results raise speculation of the potential for canine commensal *C. perfringens* to serve as a reservoir capable of both interspecies and intraspecies transfer of antibiotic resistance genes.

SUMMARY

Clostridium perfringens-associated diarrhea in dogs and cats appears to be a complex disease with a poorly understood pathogenesis. It is still uncertain whether or not *C. perfringens* is a primary cause of diarrhea, or a secondary phenomenon in dogs and cats. In addition, much of the historical and anecdotal information regarding *C. perfringens* in dogs and cats needs to be revised. Current research has shed light on the following aspects of *C. perfringens*-associated diarrhea in dogs and cats:

- *C. perfringens*-associated diarrhea is not limited to clinical signs of colitis.
- Fecal endospore counting is not a reliable test for establishing *C. perfringens*-associated diarrhea.
- Caution should be heeded when interpreting results of CPE immunodetection assays, particularly the RPLA, which has been shown to produce discordant results when compared with other immunodetection methods.
- Isolation of nonenterotoxigenic strains from a diarrheic specimen does not preclude involvement in disease as other virulence factors may be involved.
- The use of tetracyclines for management of *C. perfringens*-associated diarrhea is discouraged.

- The optimum diagnostic approach for *C. perfringens*-associated diarrhea is the use of an ELISA for CPE in conjunction with PCR for detection of enterotoxigenic strains.

CLOSTRIDIUM DIFFICILE

Clostridium difficile is a gram-positive, anaerobic spore-forming bacillus, and is the major cause of antibiotic-associated pseudomembranous colitis in people. *Clostridium difficile* has also been associated with diarrhea and enterocolitis in foals and adult horses, as well as diarrhea in dogs and cats. *C. difficile* infection (CDI) is less common in cats, and a recent study by the author documented an incidence of 5% in diarrheic cats. To date, three toxins produced by *C. difficile* have been described: toxin A (an enterotoxin), toxin B (a cytotoxin), and CDT (an ADP-ribosyltransferase). Diseases associated with *C. difficile* have primarily been attributed to the activity of toxins A and B, and strains have historically been thought to produce both toxins (toxigenic isolates), or neither (nontoxigenic). There are increasing reports of variant strains isolated from human, canine, and equine clinical cases of CDI that produce only toxin A or toxin B.[16] The role that *C. difficile* plays in the dog and cat is not as well documented as it is in humans and horses, and most of the literature focusing on *C. difficile* in dogs and cats is based on isolation of the organism which ranges from 0% to 40% in apparently healthy and diarrheic animals.[3,17]

PATHOGENESIS

Antibiotic administration, especially in hospital environments, is the most commonly reported predisposing factor for the development of CDI in people and horses, although there have been reports of CDI in the absence of antibiotic administration.[18] In contrast, administration of antibiotics does not appear to predispose dogs to CDI.[3,14]

Toxins A and B

Toxins A and B are the principal virulence factors thought to be involved in the pathogenesis of CDI. The genes encoding these two toxins have been sequenced, and together with three accessory genes, form an approximately 19.6-kb pathogenicity locus, which is found only in toxigenic strains. The mechanism of action for both toxins involves inactivation of Rho proteins by glycosylation, thereby causing the depolymerization of actin filaments, disruption of the cytoskeleton, cell rounding, and cell death.[19] Toxin A has been shown to induce histological damage with hemorrhagic necrosis of the mucosa as well as hemorrhagic secretion in small intestinal loops of rabbits, hampsters, and mice. In contrast, no tissue damage or fluid secretion is seen in toxin B treated intestinal loops, although it is cytotoxic for *in vitro* cell lines. However, several studies suggest a possible synergistic effect for toxins A and B, where mucosal damage caused by toxin A enables cytotoxic activity of toxin B.[20] The traditional view that pathogenic strains of *C. difficile* always produce both toxins A and B is now being questioned due to increasing reports of toxin A negative, toxin B positive strains isolated from clinical *C. difficile* cases.

ADP-ribosyltransferase (CDT)

The *C. difficile* ADP-ribosyltransferase, is a binary toxin consisting of two independently coded protein components; a binding component (CDTb), and an enzymatic component (CDTa), which catalyzes the ADP-ribosylation of monomeric actin, inducing alterations in the cytoskeleton. To date, all *C. difficile* strains that harbor the CDTa and CDTb genes, also contain some part of the pathogenicity locus containing the toxin A and B genes. The incidence of binary-positive strains isolated from clinical cases has been reported between 6 and 13% for human strains, and 24% for equine strains.[21] To date, CDT-positive *C. difficile* strains have not been found in any canine isolates.

DIAGNOSIS

Clinical signs

Clinical signs that have been associated with canine CDI range from asymptomatic carriage to a potentially fatal acute hemorrhagic diarrheal syndrome.[3] As with *C. perfringens*, there does not appear to be a specific anatomic localization of clinical signs, and recent studies have shown that dogs and cats with suspected CDI commonly have signs of small and large intestinal diarrhea as well as diffuse disease characterized by involvement of both the small and large intestine concurrently.[3]

Culture

Fecal culture of *C. difficile* from diarrheic specimens should be attempted, despite similar isolation rates between nondiarrheic and diarrheic dogs (0 to 40%).[3] The main potential role of fecal culture in diagnosis is to rule out CDI, because a negative result from a properly collected and processed fecal specimen has a high negative predictive value. Fecal culture may also be useful for procuring strains for detection of toxin genes and typing.

Fecal toxin detection

The current gold standard assay is the cell culture cytotoxicity assay, which detects toxin B activity; however, this assay is expensive and requires up to 48 hours for conformation of a negative result.[22] Routine diagnosis of *C. difficile*-associated diarrhea has been made based on positive fecal assays for toxins A or B. There are currently several commercially available immunoassays for detection of *C. difficile* toxin A or detection of both toxins A and B in fecal specimens; however, the performance characteristics of the human-based immunoassays is unacceptably low in dogs.[23] Because of increasing reports of toxin A-negative, toxin B-positive strains isolated from clinical cases, the use of ELISA kits that detect both toxins is gaining preference. However, none of the commercial ELISA kits currently available have been validated in dogs or cats and there is no standard assay utilized by veterinary diagnostic labs.

Molecular Techniques

The detection rates of toxigenic *C. difficile* strains following culture is similar in diarrheic and nondiarrheic dogs reducing the diagnostic value of this study in dogs. Direct PCR on stool to detect toxin B genes is increasingly used in human hospitals because of the high sensitivity of the test when used in combination with appropriate clinical case definition.

THERAPY

Metronidazole is the drug of choice for dogs and cats with suspected CDI. Although metronidazole-resistant *C. difficile* isolates obtained from foals and adult horses have been reported, a study evaluating the *in vitro* susceptibilities of 70 canine *C. difficile* isolates did not demonstrate any evidence of resistance to metronidazole.[15] The second drug of choice in human patients, and occasionally in horses, is vancomycin; however, it is used only in cases of nonresponsive CDI or when metronidazole-resistant strains have been demonstrated.

ZOONOTIC IMPLICATIONS

The risk of zoonotic transmission currently is unclear. Transmission of *C. difficile* from animals to humans has not been documented. However, because *C. difficile* is an important human pathogen and the strains of *C. difficile* that infect dogs often are indistinguishable from those found in people with CDI, it is prudent to consider *C. difficile* as potentially zoonotic.

SUMMARY

C. difficile is a significant enteropathogen in human beings and horses where infection is most commonly associated with disruption of the normal microflora followed by colonization by toxigenic strains. The role that *C. difficile* plays in dogs and cats is not well defined, and there are only a few studies evaluating the presence of toxins in diarrheic and nondiarrheic animals. There is even less information regarding CDI in cats, with only a single report of disease in 2 cats in a household.[24] Whether this indicates a lower susceptibility to disease, less frequent exposure or underreporting because of less frequent testing is unclear. Colonization rates of 0-21% of cats in the general population have been reported[25] although colonization rates can be higher (9-38%) in cats in veterinary hospitals.[26] Current research has shed light on the following aspects of CDI in dogs and cats:

- *C. difficile*-associated diarrhea is not limited to clinical signs of colitis.
- The lack of a standardized diagnostic assay makes interpretation of positive fecal toxin assays difficult. A combination of detection of common antigen (GDH) and toxin detection is currently recommended.
- The implementation of PCR for detection of toxin A or B genes is of limited diagnostic value.

REFERENCES

[1]Cave NJ, et al. **J Am Vet Med Assoc** 2002;221:52-59. [2]Greene CE: Enteric Bacterial Infections, in Greene CE (ed), *Infectious Diseases of the Dog and Cat*, Philadelphia, WB Saunders, 1998, 243-245. [3]Marks SL, et al. **J Vet Intern Med** 2002;16:533-540. [4]Meer RR, et al. **Rev Environ Contam Toxicol** 1997;150:75-94. [5]Garmory H.S, et al: **Epidemiology and Infection** 2000;124:61-67. [6]Herholz C, et al. **J Clin Microbiol** 1999;37:358-361. [7]Thiede S, et al. **Vet Rec** 2001;149:273-274. [8]Queen EV, et al. **J Vet Intern Med** 2012;26:54-60. [9]Bartlett ML, et al. **Appl Microbiol** 1972;23:193-197. [10]Miyamoto K, et al. **Infect Immun** 2002;70:4261-4272. [11]McClane BA. **Toxicon** 2001;39:1781-1791. [12]Marks SL, et al. **J Am Vet Med Assoc** 1999;214: 357-360. [13]Twedt DC: *Clostridium perfringens*-associated diarrhea in dogs, in DeNovo R (ed), *Proceedings of the Eleventh Annual Veterinary Medical Forum. American College of Veterinary Internal Medicine.*, Madison, WI, Omnipress, 1993, 121-125. [14]Weese JS, et al. **J Vet Intern Med** 2001;15: 374-378. [15]Marks SL, et al. **Vet Microbiol** 2003;94: 39-45. [16]Barbut F, et al **J Clin Microbiol** 2002;40:2079-2083. [17]Riley TV, et al. **Epidemiol Infect** 1991;107:659-665. [18]Ho M, et al. **Clin Infect Dis** 1996;23:S102-S106. [19]Just I, et al. **J Biol Chem** 1995;270:13932-13936. [20]Mitchell TJ, et al. **Gut** 1986;27:78-85. [21]Braun M, et al. **FEMS Microbiol Letters** 2000;184: 29-33. [22]Landry ML, et al. **J Clin Microbiol** 2001;39: 1855-1858. [23]Chouicha N, et al. **Journal of Veterinary Diagnostic Investigation** 2006;18(2):182-8. [24]Weese JS, et al. **J Am Vet Med Assoc** 2001;218:1436-1439. [25]al Saif N, et al. **J Med Microbiol** 1996;45:133-137. [26]Madewell BR, et al. **J Vet Diagn Invest** 1999;11:50-54.

ISCAID ANTIMICROBIAL GUIDELINES WORKING GROUP RECOMMENDATIONS: RESPIRATORY INFECTIONS

Michael R. Lappin, DVM, PhD, DACVIM
Fort Collins, Colorado

The Antimicrobial Guidelines Working Group of the International Society for Companion Animal Infectious Diseases (ISCAID) is comprised of a group of individuals with a diverse range of expertise in infectious diseases. The Working Group was formed because it was believed there are significant limitations in objective, published information relative to the treatment of infectious canine and feline respiratory diseases. The Working Group has drafted guidelines for treatment of bacterial causes of canine and feline respiratory tract diseases based on available data along with expert opinion while considering principles of infectious diseases, antimicrobial therapy, antimicrobial resistance, pharmacology, and internal medicine. Similar guidelines used in human medicine were evaluated as examples with consideration given to the major differences between species.[1,2]

The Working Group believes that the antimicrobial use guidelines for treatment of bacterial respiratory tract infections in dogs and cats to be presented at the ACVIM Forum should be interpreted as general recommendations that are reasonable and appropriate for the majority of cases. The potential for variability between cases should be acknowledged and the guidelines should not be considered standards of care that must be followed in all circumstances. Guidelines such as these should be considered an aid to decision-making for a majority of cases with the potential that different or additional approaches may be required in the minority. While the guidelines were designed to be used internationally, the Working Group recognizes that regional differences in antimicrobial resistance rates, antimicrobial availability, and prescribing regulations exist.

The Working Group believes that respiratory tract disease is commonly associated with primary or secondary bacterial infections in dogs and cats and that antimicrobial therapy is indicated in some cases. However, it is difficult to distinguish between bacterial colonization and infection in many patients as bacterial can be cultured from many parts of the respiratory tract. There is a relative lack of controlled treatment studies for most infectious respiratory tract disease syndromes in dogs and cats and further research in the field is needed. The Working Group suggests that veterinary clinicians obtain an infectious disease consultation on cases from which multiple antimicrobial agent resistant organisms are cultured.

The Working Group has divided the guidelines into the following categories:

Treatment Recommendations for Cats with Suspected Acute Bacterial Upper Respiratory Infections (URI);
Treatment Recommendations for Cats with Suspected Chronic Bacterial URI;
Treatment Recommendations for Dogs with Uncomplicated Bacterial Causes of Canine Infectious Respiratory Disease Complex;
Treatment Recommendations for Dogs or Cats with Suspected Bacterial Bronchitis;
Treatment Recommendations for Dogs or Cats with Suspected Bacterial Pneumonia; and
Treatment Recommendations for Dogs or Cats with Pyothorax.

For each section of the document, a summary of the primary organisms involved, the recommended diagnostic evaluation, the recommended treatments, and suggestions for therapeutic monitoring will be provided.

REFERENCES

[1]Lionel A, at al. Clin Inf Dis 2007; 44:S27–72.; [2]Bradley JS, et al. Clinical Infectious Diseases 2011; 53:617-630.

FELINE HAEMOPLASMOSIS - A ONE MEDICINE APPROACH TO PATHOGENICITY, DIAGNOSIS & TREATMENT

Séverine Tasker, BSc, BVSc(Hons), PhD, DSAM, DECVIM-CA, PGCertHE, MRCVS
Bristol, UK

INTRODUCTION

The haemotropic mycoplasmas (haemoplasmas) are small bacteria that parasitise red blood cells and can induce haemolysis, causing anaemia. They were formerly classified as Rickettsial organisms (when they were named *Haemobartonella* spp.) but sequencing and resulting phylogenetic analysis showed that they were actually mycoplasmal in nature. Reclassification and renaming of these organisms then occurred. A resumé of currently recognised feline haemoplasma species is shown in Table 1.

Table 1: Feline haemoplasma species, their prevalence and pathogenicity

Species Name	Reported Prevalences	Pathogenicity
Mycoplasma haemofelis	0.4 – 46.6%	Acute infection often results in haemolytic anaemia
'*Candidatus* Mycoplasma haemominutum'	8.1 – 46.7%	Acute infection can induce a drop in erythrocyte parameters but not usually severe enough to cause anaemia unless cat has concurrent disease or is immunocompromised e.g. retrovirus infection
'*Candidatus* Mycoplasma turicensis'	0.4 – 26%	

PATHOGENICITY

Different species....different strains

Mycoplasma haemofelis is the most pathogenic of the feline haemoplasma species. Acute infection often results in severe haemolytic anaemia although in some cases only mild anaemia results. Chronic infection is not usually associated with significant anaemia. Cats do not need to be immunocompromised or splenectomised to succumb to clinical disease with *M. haemofelis*. In the author's experience, young cats may be more likely to develop severe clinical disease compared to older cats. Epidemiological studies have, however, only variably demonstrated associations between anaemia and *M. haemofelis* infection. This may be because these studies usually include chronically *M. haemofelis*-infected asymptomatic cats. Persistent autoagglutination or positive Coombs' testing, indicating the presence of erythrocyte-bound antibodies, have been demonstrated in anaemic cats with acute *M. haemofelis* infection. One detailed study (Tasker, *et al* 2009b) showed that erythrocyte-bound antibodies reactive at 4°C (cold reactive antibodies; both IgM and IgG) appeared a few days earlier than those reactive at 37°C (warm reactive antibodies; primarily IgG). However, in most cats, these antibodies appeared only *after* the development of anaemia had started. The absence of erythrocyte-bound antibodies at the onset of development of anaemia could reflect a problem with the sensitivity of their detection or that erythrocyte-bound antibodies appear as a result of haemoplasma-induced haemolysis rather than initiating it. In line with the latter, we have found that these antibodies disappear with antibiotic and supportive treatment alone, without the need for specific glucocorticoid treatment.

Although '*Candidatus* Mycoplasma haemominutum' infection can cause a *drop* in red blood cell parameters, anaemia is not usually induced except in cats with concurrent problems e.g. FeLV infection. '*Candidatus* M. haemominutum' has also been associated with the development of myeloproliferative disease in cats with FeLV infection in one experimental study (George, *et al* 2002). Although concurrent problems are usually present in '*Candidatus* M. haemominutum' infected cats that develop anaemia, cases of so-called primary '*Candidatus* M. haemominutum' anaemia (i.e. without any apparent concurrent disease or infection present) have also been reported, so infection with this species cannot be ruled out as a cause of anaemia in an individual case. However conflicting data do exist, as one recent US study actually documented that '*Candidatus* M. haemominutum' infected cats were less likely to be anaemic than non-'*Candidatus* M. haemominutum' infected cats (Sykes, *et al* 2007).

'*Candidatus* Mycoplasma turicensis' infection has resulted in anaemia or a small drop in red blood cell parameters in some experimental studies, but generally anaemia is uncommon following infection. Concurrent disease and immunosuppression are both thought to be involved in the pathogenesis of '*Candidatus* M. turicensis' disease, in a similar way to the pathogenesis described for '*Candidatus* M. haemominutum'. Determining the pathogenicity of '*Candidatus* M. turicensis' in naturally infected cats has been difficult in epidemiological studies as cats are often co-infected with other haemoplasma species, confounding disease associations.

It is possible that different strains of each of the feline haemoplasma species reported exist, and that these vary in pathogenicity. This might explain some of the conflicting data reported in different studies. However, other factors such as the underlying health status of the cat are also likely to play an important role in the outcome of haemoplasma infection.

Transmission may influence pathogenicity

The natural route of transmission of haemoplasma infection between cats in the field has not yet been determined but it is possible that transmission may affect the pathogenicity of infection, although this has not been specifically investigated. The cat flea, *Ctenocephalides felis*, has been implicated in transmission, but only very transient *M. haemofelis* infection has been reported in cats experimentally infected via the haematophagous activity of fleas, and clinical and haematological signs of *M. haemofelis* infection were not induced in the recipient cat. However, haemoplasma DNA has been found in fleas and ticks collected from both pet and stray cats. The clustered geographical distribution of infection in some studies supports the role of an arthropod vector in haemoplasma transmission. Recent studies in Switzerland have found that subcutaneous inoculation of '*Candidatus* M. turicensis'-containing blood resulted in infection transmission, whereas the same inoculation method using '*Candidatus* M. turicensis'-containing saliva, did not. This suggests that haemoplasma transmission by social contact (saliva via mutual grooming etc.) is less likely than transmission by aggressive interaction (blood transmission during a cat bite incident) (Museux, *et al* 2009). Vertical transmission has not been definitively shown using molecular methods with feline haemoplasma infection but has been suggested for other haemoplasma species (Hornok, *et al* 2011). Blood transfusion is another potential route of transmission, and blood donors should be screened for haemoplasma infection (Tasker 2010).

DIAGNOSIS

Culture......if only!

Haemoplasmas are currently unculturable *in vitro* despite numerous attempts in our and other laboratories. Recently a number of haemoplasmas have been subjected to whole genome sequencing, including work performed by our group in sequencing two well characterised feline haemoplasma isolates; *M. haemofelis* strain Langford 1 (Barker, *et al* 2011) and '*Candidatus* M. haemominutum' strain Birmingham 1. These data have highlighted the limited metabolic capabilities of these important pathogens (glucose is their only energy source), which likely contribute to the haemoplasmas' current uncultivatable status. Such knowledge of haemoplasma metabolic capabilities will allow us to direct future *in vitro* cultivation attempts.

Cytology is not reliable

Cytology of blood smears may show haemoplasmas on the surface of erythrocytes but this is known to be very insensitive for diagnosis, and cytology cannot easily differentiate between haemoplasma species. The untrained eye may also fail to distinguish stain precipitate and Howell-Jolly bodies from true haemoplasma organisms.

Polymerase chain reaction (PCR) assays are usually great

PCR is now the diagnostic method of choice for haemoplasma infection. When properly designed and executed, PCR is far more sensitive than cytology. Specificity is also a valuable feature, with diagnostic PCRs being available for each of the feline haemoplasma species. The feline haemoplasma PCR assays described so far are all based on the 16S rRNA gene. Both non-quantitative conventional PCR (cPCR) and real-time quantitative PCR (qPCR) assays exist, but qPCRs are increasingly used. These allow quantification of haemoplasma DNA in the sample being analysed (usually a defined volume of blood, which is subjected to PCR) so that we can monitor the course of haemoplasma infection and evaluate any response to treatment e.g. a decrease in the level of haemoplasma DNA in the blood following institution of effective antibiotic treatment. Real-time quantitative PCRs have enabled us to describe the *in vivo* kinetics of experimental haemoplasma infection too. Cats experimentally infected with haemoplasmas initially show a rapid increase in copy number with peak numbers typically being reached around 2 to 4 weeks after infection, although *M. haemofelis* copy numbers can fluctuate greatly, especially in the first few weeks post-infection. Some (up to 1/3) of *M. haemofelis*-infected cats continue to show very large fluctuations in *M. haemofelis* copy number for several months following experimental infection; this should be considered when interpreting qPCR results because if such fluctuations occur in naturally infected cats, changes in organism number cannot always be deemed to be a consequence of antibiotic treatment or host immune response. In contrast, '*Candidatus* M. haemominutum'- and '*Candidatus* M. turicensis'-infected cats show very little fluctuation in copy number over time. The reasons for the marked fluctuations in blood *M. haemofelis* copy number over time has been a focus of our recent research. We have not been able to show any evidence of sequestration of *M. haemofelis* organisms in tissues such as the spleen and liver at times of cyclical low copy number, so this is unlikely to be the cause (Tasker, *et al* 2009a). The marked increase in *M. haemofelis* copy number seen in the blood immediately after initial infection confirms that rapid multiplication of organisms is possible in infected cats, so this may well be responsible for the marked rapid increases in *M. haemofelis* copy number seen during *M. haemofelis* cycling. The rapid decreases in *M. haemofelis* copy number could then arise due to rapid effective clearance from the blood as a result of the host's immune reponse. Antigenic variation may mediate such fluctuations. Indeed analysis has shown that a very large portion (62%) of the *M. haemofelis* genome encodes a set of uncharacterized hypothetical proteins arranged in series of paralogous repeats; these could mediate antigenic variation through differing expression of haemoplasma surface proteins over time (we have confirmed *in vivo* expression of some of these proteins), thus enabling *M. haemofelis* to evade the host's immune response (Barker, *et al* 2011).

Serology may become available in the future

The development of haemoplasma protein-based serological assays has been limited by our inability to culture haemoplasmas *in vitro* preventing the easy acquisition of adequate amounts of haemoplasma proteins for use in such assays. Studies in our laboratory have evaluated the feline serological response to haemoplasma infection using antigen preperations obtained both from purified haemoplasma organisms collected at peak parasitaemia and recombinant haemoplasma proteins expressed from haemoplasma genes identified by shotgun cloning. Additionally an ELISA has been developed based on recombinant *M. haemofelis* DnaK (Barker, *et al* 2010), and this has been used to screen the sera from experimentally infected cats. Experimentally infected cats became seropositive following infection, with a greater antibody response recorded in those cats inoculated with *M. haemofelis*, compared to '*Candidatus* M. haemominutum' and '*Candidatus* M. turicensis'. This could be due to the humoral immune response being directed against conserved, haemoplasma clade-specific, and/or species-specific epitopes on *M. haemofelis* DnaK, or a measure of the degree to which the immune response to DnaK is triggered by the infecting haemoplasma species due to the severity of disease. Antibody levels were maximal in the early (~2-4 weeks) post-infection period, suggesting that antibody levels may help differentiate acute from chronic *M.*

haemofelis infection. Colleagues in Switzerland have also reported the successful use of a recombinant *M. haemofelis* DnaK ELISA to detect serological responses to haemoplasma infection (Wolf-Jackel, *et al* 2010). Further work is required to determine the specificity of these assays before they can be used successfully in naturally infected cats.

TREATMENT

Antibiotics

A number of antibiotics, notably tetracycylines (primarily doxycycline) and fluoroquinolones (e.g. marbofloxacin, pradofloxacin), have been shown to be effective in reducing haemoplasma copy numbers in experimental studies. In most studies these antibiotics have also improved clinical signs and haematological abnormalities. The majority of studies have been done using *M. haemofelis*. Doxycycline (10 mg/kg daily PO) is often used to treat haemoplasmosis, and longer treatment courses (up to 6 weeks) are recommended by some to increase the chance of eliminating infection, although no antibiotic treatment regime has yet been reported that consistently and predictably clears haemoplasma infection. One study (Dowers, *et al* 2009) suggested that pradofloxacin (at two doses; both the standard 5 mg/kg daily PO, as well as a higher dose of 10 mg/kg daily PO) may be more effective at clearing *M. haemofelis* than doxycycline. Another found that '*Candidatus* M. haemominutum' infection only temporarily responded to marbofloxacin (2 mg/kg daily PO), with no evidence for clearance of infection seen (Tasker, *et al* 2006). Variability in response to treatment in different studies may thus arise due to differences between haemoplasma species, isolates, host factors and/or route of administration.

Corticosteroids

Corticosteroids have been recommended as adjunct treatment for haemoplasmosis, to treat any immune-mediated component of anaemia, although their efficacy has not yet been proven. In our experience, clinically ill cats, including those that are Coombs' positive, respond to antibiotic treatment and supportive care alone without the need for corticosteroids. Indeed immunosuppressive doses of corticosteroids have been used experimentally to exacerbate haemoplasma infection, so their routine use is not advised.

Supportive care

Supportive care is also required for acute haemoplasmosis treatment. This should include correction of dehydration with fluid therapy, and blood transfusion or treatment with an oxygen carrying haemoglobin compound, if available, if the anaemia is severe.

ONE MEDICINE – ONE HEALTH:

Human haemoplasma species or animal haemoplasmas in humans?

PCR methods using generic haemoplasma PCR assays are now used to reliably investigate human haemoplasma infections, although limited human epidemiological studies have failed to detect significant infections (Tasker, *et al* 2010, Willi, *et al* 2009). However studies describing human haemoplasma infections have been reported in China (Hu, *et al* 2009, Yang, *et al* 2007), but these descriptions have not described clinical disease, PCR methodology or infecting species, making interpretation difficult. Others have described the presence of existing veterinary haemoplasma species DNA in humans; *Mycoplasma suis* (Yuan, *et al* 2007, Yuan, *et al* 2009), *M. haemofelis* and/or *Mycoplasma haemocanis* (Kallick 2007, Santos, *et al* 2008) and *Mycoplasma ovis* (Sykes, *et al* 2010). However any association between haemoplasma infection and clinical signs in these reports is difficult to fully ascertain due to either lack of clinical information provided or concurrent *Bartonella henselae* infection.

We have recently published (Steer, *et al* 2011) a human case report documenting acute haemolysis and pyrexia in association with infection with a novel haemoplasma species. The patient responded clinically to doxycycline, but moxifloxacin was subsequently added to aid infection clearance. qPCR documented a fall in haemoplasma copy number during treatment, with eventual negative qPCR results obtained. The origin of the haemoplasma species identified in this patient is not known but it is possible that it represents a zoonotic infection or that the novel haemoplasma is a species whose primary host is humans.

Haemoplasma species may thus impact human and animal health and future studies should evaluate zoonotic potential and transmission of these emerging pathogens.

REFERENCES

Barker, E.N., et al. (2011) **Vet Res**, 42, 83. Barker, E.N., et al. (2010) **Clin Vaccine Immunol**, 17, 1926. Dowers, K.L., et al. (2009) **Am J Vet Res**, 70, 105. George, J.W., et al. (2002) **Am J Vet Res**, 63, 1172. Hornok, S., et al. (2011) **Vet Microbiol**, 152, 411. Hu, Z., et al. (2009) **Emerg Infect Dis**, 15, 1139. Kallick, C.A. (2007) Sphingomonas Research Partners, L.P. (Lemont, IL, US), United States. Patent. Museux, K., et al. (2009) **Vet Res**, 40, 45. Santos, A.P., et al. (2008) **Emerg Infect Dis**, 14, 1922. Steer, J.A., et al. (2011) **Clin Infect Dis**, 53, e147. Sykes, J.E., et al. (2007) **J Vet Intern Med**, 21, 685. Sykes, J.E., et al. (2010) **J Clin Microbiol**, 48, 3782. Tasker, S. (2010) **J Feline Med Surg**, 12, 369. Tasker, S., et al. (2006) **Microbes Infect**, 8, 653. Tasker, S., et al. (2009a) **Microbial Pathogenesis**, 47, 334. Tasker, S., et al. (2010) **J Med Microbiol**, 59, 1285. Tasker, S., et al. (2009b) **Vet Microbiol**, 139, 323. Willi, B., et al. (2009) **J Clin Microbiol**, 47, 4049. Wolf-Jackel, G.A., et al. (2010) **Clin Vaccine Immunol**, 17, 1917. Yang, A., et al. (2007) **Rev Med Microbiol**, 18, 1. Yuan, C., et al. (2007) **Ann Microbiol**, 57, 467. Yuan, C.L., et al. (2009) **Am J Vet Res**, 70, 890.

ANTIGEN TESTING FOR DEEP MYCOSES

Dr. Jane Sykes, BVSc (Hons), PhD, DACVIM
Davis, CA
Lawrence J. Wheat, MD; Emily J. Kirsch; Michelle M. Durkin, MA; and Patricia A. Connelly, MA
Indianapolis, Indiana

INTRODUCTION

Major progress has been made in diagnostic assays for fungal disease in recent years. Cytology, histopathology and culture are useful for diagnosis of deep mycoses, but sometimes organisms are undetectable with cytology and histopathology, and culture carries the risk of laboratory-acquired infections for laboratory personnel. A history of antifungal drug use may render these assays falsely negative. In addition, these techniques may require invasive specimen collection techniques that are risky to the patient. In recent years, there has been an increased availability of antigen assays for diagnosis of deep mycotic infections. These were initially developed for diagnosis of mycoses in human patients, but their use has now been extended to companion animal species. Although assays for *Cryptococcus* spp. polysaccharide antigen have been used for accurate diagnosis of cryptococcosis in dogs and cats for several decades, serologic assays for other mycoses have been limited to antibody tests. Antibody-detection assays are most useful for diagnosis of coccidioidomycosis and nasal aspergillosis, provided they are performed and interpreted correctly by the laboratory that performs the assays. The purpose of this article is to update the reader on the results of recent studies that have evaluated the use of antigen assays for diagnosis of deep mycoses in dogs and cats. The use of the assays should be considered in dogs or cats with signs of antibiotic-nonresponsive fever; nodular, cavitary, or lobar pulmonary lesions; osteolytic bone lesions; ocular inflammation such as uveitis or chorioretinitis; diskospondylitis; lymphadenomegaly; or cutaneous nodules. However, an understanding of the sensitivity and specificity of these assays is required for proper interpretation of the results.

Cryptococcus neoformans and *Cryptococcus gattii*

Cryptococcus spp. possess a large, branching polysaccharide capsule that is composed of glucuronoxylomannan [GXM] and galactoxylomannans [GalXM] and is a key virulence factor. Latex agglutination assays that detect cryptococcal capsular polysaccharide antigens have been routinely used to detect cryptococcal capsular antigen in serum, urine, or CSF of human patients. The use of these assays was first reported for the diagnosis of systemic cryptococcosis in companion animals in 1990.[1] Two studies evaluated their sensitivity and specificities in cats, each using a different assay.[1,2] The reported sensitivities were 95 and 100%. Both studies reported specificities of 100%, although specificity of different assays in the human literature has ranged from 93 to 100%.[3] We recently showed that the sensitivity of the CALAS assay (Meridian Diagnostic Inc, Cincinnati, OH) in 53 cats with cryptococcosis was 96%.[4] The 2 cats that tested negative had disease localized to the nasal cavity and eye, respectively. In dogs, the sensitivity appears to be lower than that in cats; only 16 of 19 dogs with cryptococcosis at the University of California, Veterinary Medical Teaching Hospital had positive titers, and titers in dogs were significantly lower in dogs than in cats.[4] The dogs with negative titers had disseminated infections, and in 2 of the dogs, CSF was positive for cryptococcal antigen. We also detected low positive titers (≤ 200) in 4 cats and a single dog that did not appear to have cryptococcosis based on the results of further diagnostic work-up, which suggested that the specificity of these assays is not 100%.[4]

Coccidioides spp.

Diagnosis of coccidioidomycosis in dogs is usually based upon the detection of immunoglobulin G (IgG) anti-*Coccidioides* antibodies by agar gel immunodiffusion (AGID). Thus, an antibody titer to *Coccidioides* spp. ≥ 1:16 in a dog with suggestive clinical abnormalities is strongly suggestive of active coccidioidomycosis, and false negative test results are rare. The sensitivity of antibody assays for coccidioidomycosis may vary between laboratories. Low positive titers have been documented in some healthy dogs that reside in endemic areas. In a prospective, longitudinal study from Arizona, only five of 28 dogs (18%) with positive tests for *Coccidioides* antibody were judged to have clinical disease.[5] Furthermore, titers overlapped in dogs with clinical coccidioidomycosis and those without clinical disease, supporting the need for additional tests, such as cytology, histopathology and/or culture to establish the diagnosis of coccidioidomycosis in some dogs.[5] In human patients, detection of antigen in urine and serum has been reported to complement the results of serologic testing for antibodies and histopathology in human patients with coccidioidomycosis.[6,7] These assays are especially useful for markedly immunocompromised humans, such as those with HIV-AIDS, which may have false negative antibody test results. *Coccidioides* antigenuria was detected in 71% of human patients with moderately severe or severe coccidioidomycosis.[6] Specificity was 99% in humans without fungal infection, but cross-reactions were noted in 10% of those with other endemic mycoses. However, the sensitivity of a *Coccidioides* antigen assay when applied to serum or urine of 60 dogs from Arizona with consistent clinical signs of coccidioidomycosis and positive antibody titers of ≥ 1:16 was only 20%.[8] Positive assay results occurred in 3/45 dogs with positive antigen tests for *Histoplasma* and 1/26 dogs with positive antigen tests for *Blastomyces*, which suggested that serological cross-reactivity may occur with *Blastomyces* and *Histoplasma* antigens, as documented in humans. Seventy-five dogs with other diseases, including 13 dogs with systemic mold infections tested negative. Thus at least this study suggested that this antigen assay is not suitable for diagnosis of coccidioidomycosis in dogs, primarily because of low sensitivity.

Blastomyces dermatitidis

Antibody assays have been shown to have low sensitivity for diagnosis of blastomycosis in dogs.[9,10] In 46 dogs with confirmed blastomycosis, the sensitivity of antibody detection by AGID was only 17.4% and an enzyme immunoassay had a sensitivity of 76.1%.[9] In other studies, the sensitivity of AGID was 50% or less for diagnosis of pulmonary blastomycosis.[10,11] In addition, false positive test results may occur as a result of exposure in endemic areas, although the specificity of the AGID has not been well studied. Assays for *Blastomyces* cell wall galactomannan antigen have now replaced antibody assays for diagnosis of canine blastomycosis. The performance of the MiraVista antigen assay (MiraVista Diagnostics, Indianapolis, IN) was evaluated in 46 dogs with confirmed blastomycosis. When urine was used, the sensitivity was 93.5%, whereas when serum was assayed sensitivity was 87%.[9] Only 1 of 43 control dogs without blastomycosis had a positive test result. Considerable cross-reactivity occurs with *Histoplasma* galactomannan antigen, and so false positive test results might be expected in dogs with histoplasmosis. In human patients, 96% of humans with histoplasmosis were positive for *Blastomyces* antigen. It is possible that cross-reactivity with other endemic mycoses may also occur. Antigen concentration decreased with treatment. In another study, radiographic resolution of pulmonary lesions occurred before antigen assay results became negative. It is not yet clear whether there is an advantage to continuation of treatment until antigenuria becomes undetectable.

Histoplasma capsulatum

As with *Blastomyces dermatitidis*, antibody assays for *Histoplasma capsulatum* have had poor diagnostic sensitivity,[12] and false positives reportedly occur in recovered animals, although the specificity of antibody assays is not well documented in the literature. A urine antigen assay for *H. capsulatum* antigen (MiraVista Diagnostics, Indianapolis, IN) is widely used in human patients for diagnosis of histoplasmosis. Recently, the use of this assay has been applied for the diagnosis of histoplasmosis in cats. The urine of 17/18 cats with confirmed histoplasmosis was positive, whereas none of 26 cats with alternate diagnoses tested positive.[13] Additional studies that evaluate larger numbers of cats and dogs with confirmed histoplasmosis are required to establish the sensitivity and specificity of this assay, but initial results are promising.

Aspergillus spp.

The performance of serological assays for diagnosis of nasal aspergillosis in dogs differs considerable that that for diagnosis of systemic aspergillosis. Serology for detection of *Aspergillus* spp. antibodies was initially reported as non-specific for diagnosis of nasal or systemic aspergillosis in dogs.[14] However, studies from the University of California, Davis that evaluated the performance of an AGID assay for diagnosis of nasal aspergillosis showed that it had a low sensitivity (67%) but high specificity (98%) for this purpose.[15] Thus, a positive antibody titer to *Aspergillus* spp. in a dog suspected to have nasal aspergillosis is strongly suggestive of the disease, but a negative titer does not rule out the diagnosis. In contrast, results of the AGID assay in dogs with systemic aspergillosis showed a sensitivity of only 20%.[16] Because systemic aspergillosis is likely to occur in immunocompromised patients, the use of antigen, rather than antibody assays, represents a logical choice for diagnosis. Serum galactomannan antigen assays have been used for diagnosis of invasive aspergillosis in humans. In humans, this assay is sensitive and specific, with an overall sensitivity that has ranged from 40% to 71% and specificity from 53% to 89%.[17-21] We recently evaluated the sensitivity and specificity of this assay in dogs with and without pulmonary or disseminated aspergillosis. A total of 11 of 13 dogs with systemic aspergillosis tested positive. All dogs that had disseminated disease were positive, whereas the 2 dogs with negative test results had localized pulmonary infections. As occurs in humans, some dogs with other invasive mold infections (such as *Paecilomyces* and *Penicillium* spp.) tested positive. High-level false positive results also occurred in dogs treated with Plasmalyte infusions, but specificity appeared to be high in dogs treated with penicillin-based antimicrobial drugs, which have been associated with false positives in humans. Positive results in dogs without disseminated mold infections were rare, and low in magnitude. Both serum and urine appeared to be useful specimens, although the use of urine may increase specificity. This assay appears insensitive for diagnosis of nasal aspergillosis.

CONCLUSIONS

In dogs with suspected deep mycoses, there is evidence that antigen assays may be useful for diagnosis in dogs and cats, with the possible exception of coccidioidomycosis. For assays that detect galactomannan, cross-reactivity with other deep mycoses can occur, so positive results may indicate only that a fungal infection is present. Assays for *Histoplasma* cross-react with *Blastomyces* galactomannan antigen (and vice versa), and assays for *Aspergillus* galactomannan antigen primarily cross-react with the antigen of other molds. Excluding this phenomenon, these assays appear to be highly specific. Cross-reactivity may not be of concern to clinicians given that currently similar treatments are indicated for these infections. However, an attempt to make a specific diagnosis of the organism involved with cytology, histopathology or culture is still recommended if feasible without risk to the patient, because specific diagnosis is important for an understanding of the epidemiology of these pathogens, and might have implications for antifungal drug susceptibility and prognosis. In addition, the sensitivity and specificity of these assays should be evaluated with larger numbers of animals, including those with other mycoses.

REFERENCES

[1]Medleau L, et al. **J Am Vet Med Assoc** 1990;196:1470. [2]Malik R, et al. **Aust Vet J** 1996;74:358. [3]Tanner DC, et al. **J Clin Microbiol** 1994;32:1680. [4]Trivedi SR, et al. **J Am Vet Med Assoc** 2011;239:357. [5]Shubitz LE, et al. **J Am Vet Med Assoc** 2005;226:1846. [6]Durkin M, et al. **Clin Infect Dis** 2008;47:e69. [7]Durkin M, et al. **Clin Vaccine Immunol** 2009;16:1453. [8]Kirsch EJ, et al. **Clin Vaccine Immunol** 2012;In press. [9]Spector D, et al. **J Vet Intern Med** 2008;22:839. [10]Crews LJ, et al. **J Am Vet Med Assoc** 2008;232:222. [11]Klein BS, et al. **Am J Vet Res** 2000;61:554. [12]Hodges RD, et al. **J Vet Intern Med** 1994;8:409. [13]Cook AK, et al. **J Vet Intern Med** 2011;25:632. [14]Harvey CE, O'Brien JA. In: Kirk RW, ed. CVT VIII. Philadelphia, PA: WB Saunders Co;1983:236. [15]Pomrantz JS, et al. **J Am Vet Med Assoc** 2007;230:1319. [16]Schultz RM, et al. **J Vet Intern Med** 2008;22:851. [17]Salonen J, et al. **Scand J Infect Dis** 2000;32:485. [18]Klont RR, et al. **Clin Infect Dis** 2004;39:1467. [19]Scotter JM, et al. **Pathol** 2005;37:246-253. [20]Pfeiffer CD, et al. **Clin Infect Dis** 2006;42:1417-1427. [21]Wheat LJ, Walsh TJ. **Eur J Clin Microbiol Infect Dis** 2008;27:245.

TREATMENT OF FUNGAL INFECTIONS: THE WHICH, WHY AND HOW OF ANTIFUNGAL DRUG THERAPY

Dr. Jane Sykes, BVSc(Hons), PhD, DACVIM
Davis, CA

INTRODUCTION

The use of antifungal drugs in human and veterinary medicine has increased in recent years. In dogs and cats, the use of potent immunosuppressive drugs such as cyclosporine for treatment of immune-mediated disease may have led to an increased prevalence of fungal infections. There may also have been increased recognition of the importance of fungal infections in dogs and cats, and the cost of many antifungal drugs has decreased. Most (if not all) antifungal drugs used in dogs and cats are not approved in these species. Therefore, veterinarians use human drugs off-label, with indications and dosing protocols often extrapolated from the human use. This article reviews the classification, mechanisms of action, spectrum of activity, resistance mechanisms, tissue penetration, clinical use, and adverse effects of the major systemic antifungal drugs used to treat dogs and cats.

Compared to antibacterial treatment, antifungal treatment is often more prolonged. Fungal organisms grow much more slowly, and the drugs used (primarily azoles) are fungistatic, not fungicidal. Therefore, long-term treatment is needed to inhibit fungal growth and allow for the animal's immune system, which is often compromised, to eradicate the infection. Treatment of animals with deep mycoses with antifungal drugs must often be continued for months, and in some animals it may be weeks before any clinical improvement is evident. Drug interactions are common with antifungal agents.

Cure with antifungal drug treatment is most likely to be achieved in dogs with localized histoplasmosis, blastomycosis, and coccidioidomycosis, or in dogs with mold infections that have occurred secondary to immunosuppressive drug treatment and the immunosuppression can be reversed. Some cats with cryptococcosis may also be cured, but many require life-long antifungal drug treatment. The prognosis is poorest for dogs with disseminated mycoses that are predisposed breeds or that lack a clear underlying cause for their disease. The inflammatory response that follows organism lysis with treatment may lead to significant worsening of clinical signs in the first few days of treatment. Use of NSAIDs or, when the brain is involved, a short period of anti-inflammatory doses of glucocorticoids, may be required until the signs begin to resolve.

AZOLE ANTIFUNGALS

Azole antifungal drugs inhibit sterol 14α-demethylase, a cytochrome P450-dependent fungal enzyme involved in synthesis of ergosterol from lanosterol. The result is the accumulation of 14α-methylsterols, which disrupt the fungal cell membrane (Figure 1). The use of ketoconazole has largely been replaced by fluconazole or itraconazole, because of a higher rate of adverse effects such as hepatotoxicity with ketoconazole. However, because of its low cost, ketoconazole continues to be used in veterinary medicine when the cost of other antifungal drugs is prohibitive to the client, and it remains efficacious for treatment of *Malassezia* spp. dermatitis and feline nasal and cutaneous cryptococcosis. Some fungal infections respond to treatment with single agent fluconazole or itraconazole. The cost of fluconazole has reduced considerably, but itraconazole remains relatively expensive. Itraconazole is often more effective than fluconazole, especially when bone lesions are present. Also, molds such as *Aspergillus* have intrinsic resistance to fluconazole, so fluconazole should not be used to treat mold infections. Fluconazole penetrates the brain and the urinary tract, whereas itraconazole may not. Some compounded formulations of fluconazole and itraconazole may lack efficacy. When treatment failure occurs, serum drug levels can be performed to ensure adequate drug concentrations are present. Itraconazole levels should be performed 2 weeks after initiating treatment, because itraconazole can take this long to reach maximum serum concentrations. Increasingly, it is apparent that resistance to one azole does not always imply resistance to other azole antifungal drugs.

Newer azole antifungal drugs include voriconazole and posaconazole. Voriconazole is second-generation triazole that is derived from fluconazole. In humans, it is currently the drug of choice for treatment of invasive aspergillosis.[1] It is also used to treat other serious and refractory mold infections. It is at least as active as itraconazole against veterinary isolates of *Cryptococcus* spp., *Candida* spp. and *Aspergillus fumigatus*.[2] Unlike fluconazole, active drug does not enter the urinary tract. It has good penetration of the CNS and has been used with some success to treat systemic mold infections in dogs.[3] If therapeutic drug monitoring can be performed, the recommended trough concentration target is between 1 and 5 mg/L, and peak concentrations < 6 mg/L).[1,4,5] Because of its high cost, adverse effects have not been well documented in dogs, but the author has observed decreased appetite, increased serum liver enzyme activities and uncommonly CNS signs, including ataxia and staring. Tachypnea and marked pyrexia after IV infusion was observed in one dog. Cats are highly sensitive to adverse effects of voriconazole, and develop inappetence, CNS and ocular signs that have included ataxia, pelvic limb paresis, mydriasis, apparent blindness, decreased pupillary light responses, and a decreased menace response.[6] Cardiac arrhythmias and hypokalemia were also noted in cats treated with voriconazole. A safe dose of voriconazole has not been established for cats. Posaconazole is an itraconazole analog that has demonstrated good efficacy for treatment of several refractory deep mycoses in animals, including *Aspergillus* spp. and *Mucor* spp. infections in cats and invasive aspergillosis in dogs.[3,7-9] It has the broadest spectrum of activity of all the azoles. Its spectrum of activity is similar to that of voriconazole, with the addition of zygomycetes.[10] Posaconazole is available as an oral suspension. Like itraconazole, in humans its absorption is promoted by the concurrent presence of fatty food. Therapeutic drug monitoring is recommended in human patients, with peak concentrations > 1.48 mg/L.[4,5] Most administered posaconazole is eliminated in the feces. Data on CSF concentrations of posaconazole are not yet available. Adverse effects appear to be less common in cats when compared with voriconazole, but information is limited.[7-9,11]

AMPHOTERICIN B

For animals with disseminated disease, treatment with amphotericin B (AMB) should be considered. Amphotericin B has antifungal activity against all of the important small animal fungal pathogens. Resistance to AMB is rare, although high MIC values have been observed for some molds including some isolates of *Aspergillus terreus*, *Paecilomyces* spp., as well as *Sporothrix schenckii*. AMB irreversibly binds sterols in fungal cell membranes, forming pores with subsequent leakage of ions. AMB is virtually unabsorbed from the gastrointestinal tract, so it is formulated for IV infusion (Fungizone®, AMB-D) as a complex with the bile salt deoxycholate. Addition of electrolyte to the solution causes it to aggregate, so it is administered in D5W. Penetration of the CSF and vitreous humor is poor, and so drugs such as fluconazole or voriconazole may be needed in animals with dissemination to the CNS or eye. The major adverse effect of AMB-D is nephrotoxicity, which is dose-dependent and transient when detected early and the drug is discontinued. Renal tubular acidosis, severe nephrogenic diabetes insipidus, hypokalemia and hypomagnesemia can occur with treatment. Rarely, aggressive intravenous fluid therapy and potassium supplementation and may be required. Loading with sodium before the infusion decreases nephrotoxicity.[12] Slow administration in a large volume of fluid also decreases nephrotoxicity. Concurrent use of other nephrotoxic drugs, such as aminoglycosides, should be avoided. In humans, AMB-D can also cause fever, chills, headache, nausea, vomiting, and rarely cytopenias and anaphylaxis.[12] Fever, inappetence and vomiting also appear to occur in some dogs treated with AMB-D, and rarely nephrogenic diabetes insipidus. Phlebitis can occur at the IV infusion site. Non-steroidal anti-inflammatory drugs can be administered to decrease pyrexia during treatment.

Lipid Formulations of AMB

Lipid formulations of AMB are associated with a considerable reduction in nephrotoxicity compared with AMB-D. Decreased nephrotoxicity results from a reduced rate of transfer of AMB to mammalian cell membranes and increased drug clearance from the blood by the mononuclear phagocyte system. This allows administration of a higher dose of the drug, sometimes with improved treatment efficacy. Sodium loading is usually not necessary, and the drug can be administered over a shorter period of time. Although nephrotoxicity is reduced, renal function should still be monitored. Most experience in veterinary medicine has been reported with Abelcet®, which is a mixture of AMB, dimyristoylphosphatidylcholine and dimyristoylphosphatidylglycerol that forms ribbon-like sheets. Abelcet® has been used in small animal patients to treat cryptococcal meningitis, disseminated coccidioidomycosis, blastomycosis, histoplasmosis, and pythiosis.

FLUCYTOSINE (5-FLUOROCYTOSINE)

Flucytosine has activity only against *Cryptococcus neoformans* and *Candida* spp. These fungi deaminate flucytosine to 5-fluorouracil, a potent antimetabolite. Drug resistance arising during therapy is very common, and so flucytosine must ALWAYS be used in combination with other drugs. Penetration of the CSF is excellent and high urinary concentrations can be achieved. The most common adverse effects are myelosuppression and gastrointestinal upset. The drug should be avoided when renal failure is present. The CBC should be monitored during treatment. Administration of flucytosine should be avoided in dogs, as development of a severe (but reversible) drug eruption within 2-3 weeks of starting treatment is common. Unfortunately, flucytosine can be unaffordable for many owners of pets with cryptococcosis or candidiasis.

PNEUMOCANDINS/ECHINOCANDINS

The echinocandins inhibit formation of beta(1,3)-D-glucans in the fungal cell wall. The prototype drug is *caspofungin acetate*. Other drugs in this class are micafungin and anidulafungin. Caspofungin is given once daily as a slow IV infusion. Caspofungin is effective against resistant *Candida albicans*, and it also has efficacy against *Aspergillus*. It is ineffective against *Cryptococcus*. Cost is slightly less than for lipid formulations of AMB, but nevertheless the need for daily IV infusion together with the cost means it is rarely used in veterinary medicine at this time. In addition, the optimum dose for dogs and cats and its adverse effects in these species are not known. Caspofungin was used successfully to induce remission in a dog with disseminated aspergillosis (1 mg/kg IV in 250 mL 0.9% NaCl over 1 hour, q24h).[3]

TERBINAFINE

Terbinafine (Lamasil®) inhibits fungal squalene epoxidase, blocking fungal ergosterol synthesis. Terbinafine is well absorbed, although there is a high hepatic first pass effect in humans. It accumulates in skin, nails and fat. In veterinary medicine, it has been most commonly used to treat dermatophytosis, and it is well tolerated. Its efficacy for invasive fungal infections has not been well investigated, although there are a few published and some anecdotal reports of its use to successfully treat some deep mycoses, especially when used in combination with other drugs.

Table 1. Suggested drug dosages for treatment of deep mycoses in dogs and cats

Drug	Dose Range	Frequency	Route
Ketoconazole	5 – 10 mg/kg (cats and dogs)	q12h	PO
Itraconazole	5 mg/kg (cats and dogs)	q12-24h	PO
Fluconazole	Cats: 50 mg/cat Dogs: 5 – 10 mg/kg	q12-24h q12h	PO
Voriconazole	Dogs: 4 – 5 mg/kg	q12h	PO
Posaconazole	Cats: 5 mg/kg Dogs: 5 – 10 mg/kg	q24h q12-24h	PO
Deoxycholate amphotericin B	Dogs: 0.5 mg/kg Cats: 0.25 mg/kg	Mon-Wed-Fri until a total dose of 4-6 mg/kg or azotemia occurs*	IV over 1 to 6 hours in D5W. Total fluid volume administered should be based on body weight and ability to tolerate a fluid load.
Abelcet® (amphotericin B lipid complex)	Dogs: 3 mg/kg Cats: 1 mg/kg	Mon-Wed-Fri for 9-12 treatments*	IV; dilute in D5W to 1 mg/kg; give over 1-2 h
Flucytosine	Cats: 25 – 50 mg/kg	q6–8h	PO
Terbinafine	30 – 40 mg/kg (cats and dogs)	q24h	PO

* Check BUN and creatinine prior to each administration. Discontinue if azotemia occurs.

REFERENCES

[1]Lat A, Thompson GR, 3rd. **Infect Drug Resist** 2011;4:43-53. [2]Okabayashi K, et al. **Nihon Ishinkin Gakkai Zasshi** 2009;50:91-94. [3]Schultz RM, et al. **J Vet Intern Med** 2008;22:851-859. [4]Hope WW, et al. **Curr Opin Infect Dis** 2008;21:580-586. [5]Goodwin ML, Drew RH. **J Antimicrob Chemother** 2008;61:17-25. [6]Quimby JM, et al. **J Vet Intern Med** 2010;24:647-649. [7]Evans N, et al. **Med Mycol** 2011;49:194-197. [8]Wray JD, et al. **J Feline Med Surg** 2008;10:523-527. [9]McLellan GJ, et al. **J Am Anim Hosp Assoc** 2006;42:302-307. [10]Pound MW, et al. **Med Mycol** 2011;49:561-580. [11]Krockenberger MB, et al. **J Feline Med Surg** 2010;12:231-236. [12]Laniado-Laborin R, Cabrales-Vargas MN. **Rev Iberoam Micol** 2009;26:223-227.

REGRESSIVE AND PROGRESSIVE FELV INFECTIONS: CLINICAL RELEVANCE

Katrin Hartmann, Prof., Dr. med. vet., Dr. habil., DECVIM
Munich, Germany

INTRODUCTION

Feline leukemia virus (FeLV) is one of the most important infectious agents in cats. For a long time, FeLV was considered to account for most disease-related deaths, and to be responsible for more clinical syndromes than any other single agent in cats. However, in recent years prevalence and consequently importance of FeLV have been decreasing. Recently, new diagnostic tools, including very sensitive PCR methods provide new data that question the traditional understanding of FeLV pathogenesis.[1] Cats believed to be immune to FeLV infection were found to remain provirus-positive. Antigen-negative, provirus-positive cats are FeLV carriers and, following reactivation, may act as an infection source. The clinical relevance of antigen-negative, provirus-positive cats is not yet clear. The provirus is integrated into the cat's genome and it is unlikely to be cleared over time. These cats do not shed virus, but reactivation with reoccurring virus shedding is possible.[2] A new classification has been proposed, in which the stages of FeLV infection are described as (1) abortive infection (comparable to the former "regressor cats"), (2) regressive infection (comparable to the former "transient viremia" followed by "latent infection"), (3) progressive infection (comparable to the former "persistent viremia"), and (4) focal or atypical infection (Table 1).[3,4,5,6]

(1) Abortive infection. After initial infection, virus replicates in the local lymphoid tissue in the oropharyngeal area. In some immunocompetent cats (formerly called "regressor cats)", viral replication may be stopped by an effective humoral and cell-mediated immune response; these cats never become viremic. They have high levels of neutralizing antibody, but neither FeLV antigen nor viral RNA or proviral DNA can be detected in the blood at any stage. Abortive infection is caused by low dose exposure to FeLV.[7] It is currently unknown, how often this situation truly occurs in nature, because studies using very sensitive PCR methods have found that in many of the formerly considered "regressor cats" virus actually can be retrieved later on, and it appears likely that no cat or only very few cats can completely clear FeLV infection from all cells.

(2) Regressive infection is accompanied by an effective immune response, and virus replication and viremia are contained prior to or shortly after bone marrow infection. After initial infection, replicating FeLV spreads systemically within mononuclear cells (lymphocytes and monocytes). Cats have positive results on tests that detect free antigen in plasma (e.g., ELISA) and shed virus. In cats with regressive infection, this viremia is terminated within weeks or months (formerly therefore called "transient viremia"). In some cats, viremia may persist longer than 3 weeks. After about 3 weeks of viremia, bone marrow cells become involved and affected hematopoietic precursor cells produce infected granulocytes and platelets that circulate in the body. Even if bone marrow becomes infected, a certain percentage of cats can clear viremia but cannot eliminate the virus from the body even if they terminate viremia because the information for virus replication (its proviral DNA) is present in bone marrow stem cells. This stage has been called "latent infection" (now a stage of regressive infection). Although proviral DNA remains, no virus is actively produced, and cats with regressive infection have negative results from routine tests that detect FeLV antigen. The molecular basis of latency is the integration of a copy of the viral genome (provirus) into cellular chromosomal DNA. During cell division, proviral DNA is replicated and the information given to the daughter cells. Thus, complete cell lineages may contain FeLV proviral DNA. However, proviral DNA is not translated into proteins, and no infectious virus particles are produced. Therefore, regressively infected cats do not shed FeLV and are not infectious to others. Sensitive PCR methods may detect provirus in the blood of cats with regressive infection that are antigen-negative. In a study in Switzerland it was shown that in addition to 7% antigen-positive, provirus-positive cats, 10% of the cat population that were negative for antigen were positive for proviral DNA in blood.[8] Regressive infection can be reactivated because the information for producing complete viral particles is present and can potentially be reused when antibody production decreases (e.g., after immune suppression).

In cats with **(3) progressive infection**, FeLV infection is not contained early in the infection, and extensive virus replication occurs, first in the lymphoid tissues and then in the bone marrow and in mucosal and glandular epithelial tissues. These cats will remain persistently viremic and infectious to other cats for the remainder of their life. This condition was called "persistent viremia" and is now classified as progressive infection. These cats develop FeLV-associated diseases, and most of them die within a few years. Regressive and progressive infections can be distinguished by repeated testing for viral antigen in peripheral blood; initially both are accompanied by persistent FeLV proviral DNA in blood but later are associated with different FeLV loads.[3,9]

(4) Focal infections or atypical infections have been reported up to 10% of experimentally infected cats. Focal infections or atypical infections may also rarely be observed in natural infections, consisting of a persistent atypical local viral replication (e.g., in mammary glands, bladder, eyes). This can lead to intermittent or low-grade production of antigen, and therefore, these cats may have weakly positive or discordant results in antigen tests, or positive and negative results may alternate.

Table 1. Outcome of FeLV infection[6]

Outcome of FeLV infection	FeLV p27 antigen in blood	Virus blood culture	Viral RNA in blood	Viral DNA in blood	Viral tissue culture	Viral shedding	FeLV-associated disease
Progressive	Positive	Positive	Positive	Positive	Positive	Positive	Likely
Regressive	Negative	Negative	Negative	Positive	Negative	Negative	Unlikely
Abortive	Negative	Negative	Negative	Negative	Negative	Negative	Unlikely
Focal	Negative	Negative	Not tested	Not tested	Positive	Variable	Unlikely

CLINICAL SIGNS IN PROGRESSIVE INFECTION

Progressive FeLV infection can cause variable clinical signs. Although the virus was named after the contagious tumor that first garnered its attention, most infected cats are presented not for tumors but for anemia or immunosuppression. Clinical signs associated with FeLV infection can be classified as tumors, hematologic disorders, immunosuppression, immune-mediated diseases, and other syndromes (including neuropathy, reproductive disorders, and fading kitten syndrome).[10]

FeLV is a major oncogene that causes different **tumors** in cats, most commonly lymphoma and leukemia, less commonly other hematopoietic tumors and rarely other malignancies (including neuroblastoma, osteochondroma, and others). There is a decrease in prevalence of FeLV infection in cats with lymphoma that also indicates a shift in tumor causation in recent years. Whereas 59% of all cats with lymphoma or leukemia were FeLV antigen-positive in one German study from 1980 to 1995, only 20% of the cats were FeLV antigen-positive in the years 1996 to 1999 in the same institution. The most important mechanism by which FeLV causes malignancy is by insertion of the FeLV genome into the cellular genome near a cellular oncogene (most commonly *myc*), resulting in activation and over-expression of that gene. These effects lead to uncontrolled proliferation of that cell (clone). A malignancy results in absence of an appropriate immune response. FeLV may also incorporate the oncogene to form a recombinant virus (e.g., FeLV-B, FeSV) containing cellular oncogene sequences that are then rearranged and activated. When they enter a new cell, these recombinant viruses are oncogenic. In a study of 119 cats with lymphomas, transduction or insertion of the *myc* locus had occurred in 38 cats (32%).[11] Thus, FeLV-induced neoplasms are caused, at least in part, by somatically acquired insertional mutagenesis in which the integrated provirus may activate a proto-oncogene or disrupt a tumor suppressor gene. A recent study suggests that the *U3-LTR* region of FeLV transactivates cancer-related signalling pathways through production of a non-coding 104 base RNA transcript that activates *NF kappaB*.[12] Common integration sites for FeLV associated with lymphoma development have been identified in six loci: *c-myc*, *flvi-1*, *flvi-2* (contains *bmi-1*), *fit-1*, *pim-1*, and *flit-1*. Oncogenic association of the loci includes that *c-myc* is known as a proto-oncogene, *bmi-1* and *pim-1* have been recognized as *myc*-collaborators, *fit-1* appears to be closely linked to *myb*, and *flit-1* insertion was shown to be associated with over-expression of cellular genes, e.g., *activin-A receptor type II-like 1* (*ACVRL1*).[13] *Flit-1* seems to have an important role in the development of thymic lymphomas and appears to represent a novel FeLV proviral common integration domain that may influence lymphomagenesis as insertional mutagenesis. Among 35 FeLV-related tumors, 5/25 thymic lymphomas demonstrated proviral insertion within *flit-1* locus, whereas 0/4 alimentary, 5/5 multicentric lymphomas, and 1/1 T-lymphoid leukemia examined had rearrangement in this region. Expression of *ACVRL1* mRNA was detected the two thymic lymphomas with *flit-1* rearrangement, whereas normal thymuses and 7 lymphoid tumors without *flit-1* rearrangement had no detectable *ACVRL1* mRNA expression.[14]

Hematopoietic disorders, particularly cytopenias caused by bone marrow suppression, are a common finding in cats infected with FeLV. Hematologic disorders described in association with FeLV include anemia (non-regenerative or regenerative), persistent, transient, or cyclic neutropenia, platelet abnormalities (thrombocytopenia and platelet function abnormalities), aplastic anemia (pancytopenia), and panleukopenia-like syndrome. Anemia is a major non-neoplastic complication that occurs in a majority of symptomatic FeLV-infected cats.[15] Anemia in FeLV-infected cats may have various causes. Approximately 10% of FeLV-associated anemias are regenerative, most FeLV-associated anemias, however, are non-regenerative and caused by the bone marrow suppressive effect of the virus resulting from primary infection of hematopoietic stem cells and infection of stroma cells that constitute the supporting environment for hematopoietic cells. *In vitro* exposure of normal feline bone marrow to some strains of FeLV causes suppression of erythrogenesis. In addition to the direct effect of the virus on erythropoiesis, other factors can cause non-regenerative anemia in FeLV-infected cats (e.g., anemia of chronic disease promoted by high concentration of cytokines). FeLV infection may cause decreased platelet counts. It also may be responsible for platelet function deficits, and the lifespan of platelets is shortened in some FeLV-infected cats. Thrombocytopenia (resulting in bleeding disorders) may occur secondary to decreased platelet production from FeLV-induced bone marrow suppression or leukemic infiltration. Platelets harbor FeLV proteins as a result of infection, and megakaryocytes are frequent targets of progressive FeLV infection. Immune-mediated thrombocytopenia, which rarely occurs as a single disease entity in cats, often accompanies autoimmune haemolytic anemia in cats with underlying FeLV infection. FeLV infection also may cause decreased neutrophil or lymphocyte counts. Neutropenia is common in FeLV-infected cats and generally occurs alone or in conjunction with other cytopenias. In some cases, myeloid hypoplasia of all granulocytic stages is observed, suggesting direct cytopathic infection on neutrophil precursors by FeLV. In some neutropenic FeLV-infected cats, an arrest in bone marrow maturation may occur at the myelocyte and metamyelocyte stages. It has been hypothesized that an immune-mediated mechanism is responsible in cases in which neutrophil counts recover with glucocorticoid treatment ("glucocorticoid-responsive neutropenia"). Feline panleukopenia-like syndrome, also known as FeLV-associated enteritis or myeloblastopenia, consists of severe leukopenia with enteritis and destruction of intestinal crypt epithelium that mimics feline panleukopenia caused by feline panleukopenia virus (FPV) infection. However, FPV antigen has been demonstrated by immunofluorescence staining in intestinal sections of cats that died from this syndrome after being experimentally infected with FeLV.[16] FPV was also demonstrated by electron microscopy despite negative FPV antigen tests. It appears that this syndrome may actually not be caused by FeLV itself, as previously thought, but by co-infection with FPV. The syndromes also has been referred to as FeLV-associated enteritis in cats with progressive FeLV infection because the clinical signs observed are usually gastrointestinal, including hemorrhagic diarrhea, vomiting, oral ulceration or gingivitis, anorexia, and weight loss.[17,18] It is still unclear whether all theses syndromes have the same origin and are simply caused by co-infection with FPV (and even modified life FPV vaccines have been discussed) or if they are caused by FeLV itself.[16] Hematopoietic neoplasia ("myeloproliferative disorders"), including leukemia, may also cause bone marrow suppression syndromes by crowding out of bone marrow cells. Myelodysplastic syndrome, characterized by peripheral blood cytopenias and dysplastic changes in the bone marrow, is a pre-stage of acute myeloic leukemia. It was found that changes of the *LTR* region of the FeLV gene (presence of 3 tandem direct 47-bp repeats in the upstream region of the enhancer) are strongly associated with the induction of myelodysplastic syndrome.[19] Myelofibrosis, another cause of bone marrow suppression, is a condition characterized by abnormal proliferation of fibroblasts resulting from chronic stimulation of the bone marrow, such as chronic bone marrow activity from hyperplastic or neoplastic regeneration caused by FeLV. In severe cases, the entire endosteum within the medullary cavity can be obliterated.

Diseases secondary to **immunosuppression** account for a large portion of the syndromes seen in FeLV-infected cats. Clinical signs in these cats are characterized by the secondary infections that commonly occur in immunosuppressed FeLV-infected cats. In addition to immunosuppression, FeLV-infected cats can develop **immune-mediated diseases** caused by an overactive or dysregulated immune response to the virus. Immune-mediated diseases described in FeLV-infected cats include autoimmune haemolytic anemia, glomerulonephritis, uveitis with immune complex deposition in iris and ciliary body, and polyarthritis.

Other syndromes directly caused by FeLV infection include FeLV-associated neuropathy, reproductive disorders, and fading kitten syndrome. In FeLV-associated neuropathy, direct neurotoxic effects of FeLV have been discussed as pathogenetic mechanisms. FeLV envelope glycoproteins may be able to produce increased intracellular free calcium leading to neuronal death. A polypeptide of the FeLV envelope was found to cause dose-dependent neurotoxicity associated with alterations in intracellular calcium ion concentration, neuronal survival, and neurite outgrowth. FeLV-infected queens can transmit the virus transplacentally. **Reproductive failure** in form of fetal resorption, abortion, and neonatal death is common if *in utero* FeLV infection occurs. Kittens born to infected queens may become exposed to FeLV transplacentally, but heavy exposure also occurs at birth and throughout the nursing period. Some kittens become immune, but most become progressively infected and die at an early age of the so-called **fading kitten syndrome**, characterized by failure to nurse, dehydration, hypothermia, thymic atrophy, and death within the first two weeks of life.

CLINICAL SIGNS IN REGRESSIVE INFECTION

Most cats with regressive infections are healthy because for the majority of pathogenic mechanisms by which FeLV causes clinical signs, active virus replication is necessary. This is not the case in regressive FeLV infections, in which the virus is harbored in a "dormant" and non-productive form. However, besides the potential risk of reactivation, regressive infections may have some impact on a cat's health. Regressive infection may explain how myelosuppression or hematopoietic malignancy could be FeLV-related in FeLV antigen-negative cats. FeLV provirus can be inserted at many different sites in the host's genome, carrying potent regulatory signals. In the development of **myelosuppressive disorders** or tumors, integrated FeLV provirus may interrupt or inactivate cellular genes in the infected cells, or regulatory features of viral DNA may alter expression of neighboring genes. In addition, because bone marrow microenvironment cells (e.g., myelomonocytic progenitor cells and stromal fibroblasts) provide a reservoir of regressive FeLV infections, it seems possible that the integrated provirus may alter cellular functions and contribute to the pathogenesis of myelosuppressive disorders. Additionally, FeLV provirus may cause bone marrow disorders by inducing the expression of antigens on the cell surface, resulting in an immune-mediated destruction of the cell. In a recent study, 2/37 cats with non-regenerative cytopenias and negative FeLV antigen test were positive by bone marrow PCR suggestion that regressive FeLV infection indeed can cause myelosuppression.[20]

FeLV proviral DNA was detected in **lymphomas** of cats that tested negative for FeLV antigen, also suggesting that the virus may be associated with a larger proportion of lymphomas than previously thought. FeLV has been shown to incorporate cellular genes; several such transducted genes that are also present in regressively infected cells have been implicated in viral oncogenesis.[21,22,23] Some studies showed involvement of FeLV provirus in lymphomas of FeLV antigen-negative cats.[24,25] FeLV antigen-negative cats from FeLV cluster households had a 40-fold higher rate of development of FeLV-negative lymphoma than did those from the general population. FeLV-negative lymphomas have also occurred in laboratory cats known to have been infected previously with FeLV.[23] It is still unclear, how common regressive FeLV infection is responsible for FeLV-associated tumors in the field as study results have been controversial. Proviral DNA was detected in formalin-fixed, paraffin-embedded tumor tissue in 7/11 FeLV-negative cats with lymphoma.[24] However, other groups found evidence of provirus in only 1/22[22] and in 0/50 FeLV antigen-negative lymphomas.[26]

REFERENCES

[1]Torres AN, et al. **Vet Immunol Immunopathol** 2008;123:81. [2]Cattori V, et al. **Mol Cell Probes** 2006;20:172. [3]Torres AN, et al. **Virology** 2005;332:272. [4]Hofmann-Lehmann R, et al. **Vaccine** 2007;25:5531. [5]Hofmann-Lehmann R, et al. **Vet Immunol Immunopathol** 2008;123:119. [6]Levy J, et al. **J Feline Med Surg** 2008;10:300. [7]Major A, et al. **Vet Res** 2010;41:17. [8]Hofmann-Lehmann R, et al. **J Gen Virol** 2001;82:1589. [9]Pepin AC, et al. **Virus Res** 2007;127:9. [10]Lutz H, et al. **J Feline Med Surg** 2009;11:565. [11]Tsatsanis C, et al. **J Virol** 1994;68:8296. [12]Forman LW, et al. **FEBS Lett** 2009;58:1386. [13]Fujino Y, et al. **Vet Immunol Immunopathol** 2008;123:138. [14]Fujino Y, et al. **Virology** 2009;386:16. [15]Gleich S, et al. **J Vet Intern Med** 2009;23,552. [16]Lutz H, et al. **Vet Immunol Immunopathol** 1995;46:21. [17]Kipar A, et al. **Vet Pathol** 2000;37:129. [18]Kipar A, et al. **Vet Pathol** 2001;38:359. [19]Hisasue M, et al. **Int J Cancer** 2009;124:1133. [20]Stuetzer B, et al. **J Vet Intern Med** 2010;24:192. [21]Rezanka LJ, et al. **Cancer Invest** 1992;10:371. [22]Sheets RL, et al. **J Virol** 1993;67:3118. [23]Rohn JL, et al. **J Virol** 1994;68:2458. [24]Jackson ML, et al. **Can J Vet Res** 1993;57:269. [25]George JW, et al. **Am J Vet Res** 2001;63:1172. [26]Stuetzer B, et al. **J Feline Med Surg** 2011;13:81.

IS THERE AN EPIDEMIC OF FELINE BARTONELLOSIS? DIAGNOSIS, TREATMENT, PREVENTION, AND ZOONOTIC ASPECTS

Michael R. Lappin, DVM, PhD, DACVIM
Fort Collins, Colorado

Cats have proven by culture or DNA amplification to be infected by *Bartonella henselae*, *B. clarridgeiae*, *B. koehlerae*, *B. quintana* and *B. bovis*.[1,2] Cats are the main reservoir hosts for *B. henselae*, *B. clarridgeiae* and *B. koehlerae*. *Bartonella* spp. have both intra-endothelial and intra-erythrocytic phases of infection.[3,4] Based on results of serologic, culture, or polymerase chain reaction (PCR) assay studies, cats are commonly exposed to or infected by *Bartonella* spp. around the world.[1,2,5-8]

Bartonella henselae is the most common cause of Cat Scratch Disease as well as bacillary angiomatosis, peliosis hepatis, and bacterial endocarditis. Recently, *B. henselae* has been documented as a cause of chronic disease syndromes like fever, headaches, polyarthritis, and chronic fatigue in immunocompetent veterinary health care providers or *Bartonella* researchers.[9] Most medical doctors may not recognize this differential and should be informed if you are exhibiting these problems. The use of BAPGM culture media with PCR assay has been shown to be one of the most sensitive ways to prove *Bartonella* spp. bacteremia.[10] This assay is available at one commercial laboratory (www.galaxydx.com).

Bartonella henselae, *B. clarridgeiae*, *B. koehlerae* are transmitted amongst cats by *Ctenocephalides felis* and so prevalence is greatest in cats from regions where fleas are common. In a recent study in the United States, we collected fleas from cats and attempted to amplify *Bartonella* spp. DNA from flea digests as well as the blood of the cat.[7] The prevalence rates for *B. henselae* in cats and their fleas were 34.8% and 22.8%, respectively. The prevalence rates for *B. clarridgeiae* in cats and their fleas were 20.7% and 19.6%, respectively. Results are similar in other studies performed around the world including recent studies completed in England and Australia.[5] In Scotland, we showed the *Bartonella* spp. seroprevalence and DNA prevalence rates to be 15.3% and 5.8%, respectively.[6] In healthy cats in shelters, we amplified DNA of a *Bartonella* spp. from 56.9% of healthy cats exposed to fleas and from 100% of the fleas collected from the cats.[8]

Bartonella henselae survives in flea feces for days after being passed by infected *C. felis*.[11] Infected flea feces are likely to contaminate cat claws during grooming and then *Bartonella* are inoculated into the human when scratched. It is also possible that open wounds are contaminated with infected flea feces. Thus, administration of flea control products, avoiding bites and scratches, and thorough cleansing of wounds or areas contaminated with flea feces is indicated to potentially decrease risk of acquiring bartonellosis.[2,12]

We recently completed a study in which SPF cats were exposed naturally to *C. felis* allowed to feed on cats with *B. henselae* administered IV.[13] Some cats were administered imidocloprid-moxidectin monthly and other cats were maintained as untreated controls. At the end of the study, all cats in the untreated control group become infected but none of the cats administered imidocloprid-moxidectin. Fever requiring treatment was detected in 2 of 6 untreated cats which apparently responded quickly to administration of enrofloxacin administration. At this time only imidacloprid-moxidectin has been shown to block transmission of *Bartonella* spp. amongst cats. Using this model, we have recently shown that *C. felis* can transmit *B. henselae* to puppies (Unpublished data, 2012).

Most cats with serological evidence of exposure to a *Bartonella* spp., a *Bartonella* spp. cultured from blood, or *Bartonella* spp. DNA amplified from blood by PCR assay are clinically normal. However, *Bartonella* spp. infection of cats has also been associated directly or indirectly with a variety of clinical manifestations like fever, lethargy, lymphadenopathy, uveitis, gingivitis, endocarditis, myocarditis, hyperglobulinemia, osteomyelitis, cutaneous vasculitis, and neurological diseases.[2,13-21] How often cats become ill from *Bartonella* spp. infections is unknown and more information is needed. For example, the association of *B. henselae* infection to uveitis in a cat was first made in an individual case with uveitis that ultimately responded to doxycycline therapy.[14] We subsequently found *Bartonella* antibody production and DNA in the aqueous humor of cats previously presumed to have idiopathic uveitis. A series of clinical cases of feline ocular disease that were responsive to antibiotic therapy was recently reported.[15] Thus, it appears likely that *Bartonella* spp. causes ocular disease in some cats. However, it can be difficult to determine which cats have been exposed and which cats are diseased.[2,20] For example, in recent studies, the prevalence rates for *Bartonella* spp. antibodies in feline sera were not significantly different for cats with or without seizures, cats with or without stomatitis, cats with or without anemia, or with or without increased feline pancreatitic lipase immunoreactivity.[22-27] It is also still also still unclear as to why some cats develop *Bartonella* associated illness and others do not. For example, we failed to induce *Toxoplasma gondii* or *Bartonella* spp. uveitis when we inoculated *Bartonella* IV into cats with chronic toxoplasmosis.[28]

Blood culture, PCR assay on blood, and serologic testing can be used to assess individual cats for *Bartonella* infection. Cats that are culture-negative or PCR-negative and antibody-negative and cats that are culture-negative or PCR-negative and antibody-positive are probably not a source of flea, cat, or human infection. However, bacteremia can be intermittent and false-negative culture or PCR results can occur, limiting the predictive value of a single battery of tests. With PCR, false positive results can occur and positive results do not necessarily indicate that the organism is alive. While serologic testing can be used to determine whether an individual cat has been exposed, both seropositive and seronegative cats can be bacteremic, limiting the diagnostic utility of serologic testing. Thus, testing healthy, client-owned cats for *Bartonella* spp. infection is not currently recommended in the United States.[12] Testing should be reserved for cats with suspected clinical bartonellosis.

In one study, testing for *Bartonella* spp. IgM alone had no greater positive predictive value when compared to testing for IgG.[29] The combination of serology and PCR assay or culture is likely to give the best predictive values and is available in some laboratories like Antech Diagnostics, North Carolina State University, Galaxy Diagnostics, and Colorado State University (www.dlab.colostate.edu/).[16] Some cats can have low level bacteremia and specialized media may be required to grow the organism from some cats as mentioned for humans. The combination of culture and PCR may be required to diagnose infection.[10] If the results of *Bartonella* tests are negative in a clinically ill cat, the organism is not likely the cause of the clinical syndrome unless the infection was peracute and serological testing was used as the diagnostic test. If the results of *Bartonella* tests are positive, the agent remains on the differential list, but other causes of the clinical syndrome must also be excluded.

In a recent study in our laboratory, we showed that *Bartonella* is not killed by CPDA-1 solution.[30] Because of these findings and the fact that some cats become ill after IV inoculation with the organism, Colorado State University recommends only using *Bartonella* spp. PCR or culture negative and *Bartonella* spp. seronegative cats as blood donors. I believe this data supports the Colorado State University recommendation that all cats used as blood donors be housed indoors when possible and be maintained on imidacloprid containing flea control products year round.

In experimental studies, administration of doxycycline, tetracycline, erythromycin, amoxicillin-clavulanate, or enrofloxacin can limit bacteremia but does not cure infection in all cats.[13,31-33] To date, use of antibiotics in healthy cats has not been shown to lessen the risk of cat scratch disease.[12] Thus in the United States, treatment is generally recommended for clinically ill cats.[2] If clinical bartonellosis is suspected, the AAFP Panel Report recommends doxycycline at 10 mg/kg, PO, daily for 7 days as the initial therapeutic trial. In the United States, I have my doxycycline prescriptions formulated into a flavored suspension to avoid esophageal strictures. Using the drug twice daily is also acceptable and may increase the chance of eliminating bacteremia. If a positive response is achieved, continue treatment for 2 weeks past clinical resolution of disease or for a minimum of 28 days. If a poor response is achieved by day 7 or doxycycline is not tolerated and I still believe bartonellosis is a valid differential diagnosis, I consider fluoroquinolones as second choices. Enrofloxacin at 5 mg/kg, PO, daily has been used most frequently to date.[13] Recently, *B. henselae* isolates from people or cats were shown to rapidly develop resistance to azithromycin and so this drug should not be used to treat feline bartonellosis.[33] In my experience, *Bartonella* spp. positive cats that have failed to respond after administration of 2 different drugs with presumed anti-*Bartonella* activity generally have another cause of the clinical syndrome. There is no clinical utility in rechecking *Bartonella* serological test or PCR test results in cats.

To lessen the likelihood of acquiring a *Bartonella* spp. infection from a cat, the following are adaptations of what is recommended to HIV-infected people and other cat owners by the Centers for Disease Control and the American Association of Feline Practitioners.[2,12]

- Flea control should be initiated and maintained year-round.
- If a family member is immune deficient and a new cat is to be acquired, adopt a healthy cat > 1 year of age and free of fleas.
- Immune deficient individuals should avoid contact with cats of unknown health status.
- Declawing of cats is generally not required but claws should be trimmed regularly.
- Bites and scratches should be avoided (including rough play with cats).
- Cat-associated wounds should be washed promptly and thoroughly with soap and water and medical advice sought.
- While *Bartonella* spp. have not been shown to be transmitted by direct contact with saliva, cats should not be allowed to lick open wounds.
- Keep cats indoors to minimize hunting and exposure to fleas and other possible vectors.

REFERENCES

[1]Breitschwerdt EB, et al. **J Vet Emerg Crit Care (San Antonio)**. 2010;20:8-30.; [2]Brunt J, et al. **J Feline Med Surg**. 2006;8:213-226.; [3]Kordick DL, et al. **J Clin Microbiol** 1995;33:1655-1656.; [4]Billeter SA, et al. **Vet Microbiol.** 2011 Oct 20. [Epub ahead of print].; [5]Barrs VR, et al. **Aust Vet J** 2010;88:160-165.; [6]Bennett AD, et al. **J Feline Med Surg.** 2011;13:553-557.; [7]Lappin MR, et al. **J Fel Med Surg** 2006;8:85-90.; [8]Lappin MR, et al. **Vet Dermatol**. 2009;20:509-514.; [9]Breitschwerdt EB, et al. **Emerg Inf Dis** 2007;13:938-941.; [10]Duncan AW, et al. **J Microbiol Methods**. 2007;69:273-281.; [11]Finkelstein JL, et al. **J Med Entomol** 2002;39:915-919.; [12]Kaplan JE, et al. ***Recommendations and Reports.*** **MMWR** 2009;58(RR04);1-198.; [13]Bradbury CA, et al. **J Am Vet Med Assoc.** 2010;236:869-873.; [14]Lappin MR et al. **J Am Vet Med Assoc** 1999;214;1205-1207.; [15]Ketring KL et al. **J Am Anim Hosp Assoc** 2004;40:6-12.; [16]Lappin MR, et al. **J Fel Med Surg** 2009;11:141-148. ; [17]Sykes JE, et al. **J Feline Med Surg**. 2010;12:631-636.; [18]Nakamura RK, et al. **J Vet Cardiol**. 2011;13:277-281.; [19]Cozzani E, et al. **Clin Exp Dermatol**. 2011 [Epub ahead of print]; [20]Stiles J. **Vet Ophthalmol**. 2011;14:S9-14.; [21]Varanat M, et al. **Vet Pathol**. 2011 Apr 13. [Epub ahead of print].; [22]Dowers KL, et al. **J Vet Int Med** 2005;19:471.; [23]Pearce L, et al. **J Fel Med Surg** 2006;8:315-320.; [24]Ishak AM, et al. **J Feline Med Surg.** 2007;9:1-7. [25]Bayliss DB, et al. **J Feline Med Surg**. 2009;11:663-637. ; [26]Quimby JM, et al. J Feline Med Surg. 2008;10:66-72.; [27]Belgard S, et al. **Berl Munch Tierarztl Wochenschr**. 2010;123:369-376.; [28]Powell CC, et al. **J Fel Med Surg** 2002;4:107-110.; [29]Ficociello J, et al. **J Vet Intern Med.** 2011;25:1264-1269.; [30]Bradbury CA, et al. **ACVIM 2010** (abstract).; [31]Kordick DL, et al. **Antimicrobial Agents Chem** 1997;41:2448-2455.; [32]Greene CE, et al. **J Clin Microbiol** 1996;34:1682-1685.; [33]Biswas S, et al. **J Clin Microbiol.** 2010;48:617-618.

BREED SPECIFIC HEPATOPATHIES: ENGLISH SPRINGER SPANIEL, DALMATION AND STANDARD POODLE

Penny J. Watson, MA, VetMD, CertVR, DSAM, DECVIM
Cambridge UK
David Twedt, DVM, DACVIM
Fort Collins, CO

INTRODUCTION

Liver disease is very common in dogs with a reported prevalence of canine chronic hepatitis in old dogs at post mortem in one UK study of 12% [1] and a larger number of dogs being affected if reactive and vacuolar hepatopathies are included. The current prevalence of acute and chronic hepatitis in the USA is unknown. Canine hepatitis can have a number of causes: some cases are 'reactive' secondary to disease in other organs or part of a multi-organ systemic infectious or inflammatory process. Many cases are confined to the liver where they are often idiopathic, although a number can be caused by copper storage disorders or other defined diseases. Hepatitis can be recognized in any breed of dog or crossbreeds but several studies have demonstrated increased prevalences in certain breeds. Some breed specific hepatopathies have been recognized in several studies in dogs from different countries at different times (eg copper storage disease in Bedlington terriers; chronic hepatitis in English cocker spaniels and Doberman Pinschers) whereas other breed specific hepatopathies seem to have a more limited geographical and/or temporal association – such as the Dalmatian and English Springer spaniel diseases discussed here. Chronic hepatitis in these breeds has only recently been reported and seems to predominate in Dalmatians in the USA and English springer spaniels in the UK. The reported differences in the genetic make-up of Pedigree dogs in the UK and USA, consistent with a 'founder effect' in USA dogs, might provide some explanation for these differences [2]. It is important to ensure that any reported increased prevalence of disease in a breed is real and backed up with strong data comparing the diseased group with a good control group, to avoid assigning a falsely increased breed prevalence due to bias. The degree of inbreeding in pedigree dogs makes it very likely that breed specific hepatopathies have the same cause within one breed, although probably different causes between breeds. Once a breed-related disease has been confirmed, it should provide excellent opportunities to work to understand the causes of the disease in that breed and thus allow prevention and effective treatment.

ENGLISH SPRINGER SPANIELS

Prevalence of disease in the UK and other countries

Dr Watson was approached by the health co-ordinators for the English Springer Spaniel (ESS) breed clubs in the UK in 2005 with a request to investigate a suspected emerging problem of hepatitis in the breed. The breed society thought the disease was much more common in show lines than working lines. Initial skepticism on the part of the author was replaced with interest as it became apparent that the breed society were right. A retrospective study that reviewed 551 cases of histologically confirmed chronic hepatitis in dogs in the UK between 2001 and 2008 compared with a control population of dogs registered by a UK microchip company (175,442 dogs in 2001 and 311,085 dogs in 2008) found overrepresentation of ESS, together with some previously reported breeds (Labrador Retrievers, American and English cocker spaniels, Dalmatians and Doberman pinschers) and also some novel breeds (Cairn terriers, Great Danes, and Samoyeds) [3]. It is interesting to note that this study identified an increased prevalence in Dalmatians in the UK for the first time, although no further work has been undertaken on the pathogenesis in this breed in the UK. ESS, Dalmatians and Doberman pinschers with CH in that study were significantly younger than Cairn terriers, English cocker spaniels and Labrador retrievers with CH. The mean age of all dogs with chronic hepatitis was 8 years, whereas ESS had a mean age of 5 years (range 1 year 2 months to 11 years). Female ESS were over-represented with a 3:1 ratio female:male. The UK Dalmatians showed an even more marked female predominance. This study therefore showed that ESS had an increased prevalence of CH in the UK over this recent time period. To the authors' knowledge hepatitis has not yet appeared with increased prevalence in ESS in the USA. However, anecdotally, there is also an increased prevalence of hepatitis in ESS in Norway. Ellen Skancke and Wenche Farstad at the Norwegian School of Veterinary Science had reports of hepatitis in ESS pre-dating the UK reports in the 1990s and the disease is reportedly more common in Norway than Sweden. Interestingly, the breed society in the UK confirmed that Norwegian ESS show dogs were more closely related to UK than North American dogs, whereas Swedish ESS have strong North American founder effects. There have also been sporadic reports of hepatitis in ESS in Australia, another country which has used largely UK blood lines. Initial pedigree analysis of affected UK dogs failed to find a common ancestor within 6 generations. All this evidence argues for a disease with some genetic involvement which emerged relatively recently, probably appearing in UK dogs longer ago than the 6 generation pedigree but after the founder dogs went to North America.

Clinical appearance

A recently published study of chronic hepatitis in ESS documented the clinical and histological signs[4]. This study described 68 ESS diagnosed with chronic hepatitis in the UK between 2006 and 2010 for which the authors had access to formalin fixed liver tissue. The female predominance was confirmed with 48 females and 20 males. Twenty of the females were entire and seven of these showed onset of clinical signs within 2 months of estrus. The median age of affected dogs was 3 years 7 months (range 7 months to 8 years 5 months) confirming that this form of chronic hepatitis appears to occur in younger dogs than usual. Clinical signs tend to be severe with only five dogs being asymptomatic at presentation and most dogs showing lethargy and anorexia (which is often the clinical sign which leads to presentation to the primary vet). Half of the dogs were icteric at presentation (a clinical sign which is not so common in chronic hepatitis in other breeds) and also half showed vomiting. Ascites was clinically obvious in 17 dogs at presentation and 24 dogs were hyperthermic.

Clinical investigation

All dogs had blood tests and abdominal ultrasound examinations and 33 dogs had radiographs of the abdomen. All dogs had elevations of alkaline phosphatase and alanine aminotransferase which were usually marked. Bilirubin was increased in 69% of cases. A neutrophilia was common. These findings made biliary tract disease such as extrahepatic biliary tract obstruction or biliary tract infection a definite differential diagnosis for these cases so abdominal imaging was certainly indicated. Sixty two dogs had a change in hepatic echogenicity on ultrasound. The biliary tract and gall bladder appeared normal in all dogs but 49 dogs had an obvious reduction in liver size. Ascites was found on ultrasound in 27 dogs (- 10 more cases than appreciated clinically). Liver biopsies were taken in all cases either by ultrasound guidance (n=35) or at laparoscopy (n= 8) or at laparotomy (n=22) and all biopsies were examined with a variety of stains including rhodanine for copper. Moderate to severe chronic hepatitis was diagnosed in all dogs with a lymphoplasmcytic infiltrate which was multifocal. Many dogs also had a neutrophilic inflammatory infiltrate and hepatocyte necrosis and apoptosis was seen in all cases. All dogs had evidence of fibrosis and this was bridging in 38 dogs, of which 22 were classified as cirrhosis. Rhodanine staining was positive in 21 dogs but multifocal and only scored as grade 1 in 19 dogs and 2 in two dogs suggesting it was secondary and not primary.

Potential causes, treatment and outcome

The cause of the disease in ESS is currently unknown. The histological appearance would be consistent with viral, autoimmune or toxic causes. Copper storage disease has been ruled out. PCRs for canine adenovirus-1; canine herpesvirus; canine parvovirus and pathogenic *Leptospira* species were negative. Potential viral causes are still being actively investigated and DLA analysis and genome-wide association studies are being undertaken. A limited retrospective study failed to find any beneficial effect of steroid therapy in affected dogs and anecdotally some dogs seem to have deteriorated after steroid therapy, although a properly controlled prospective trial of immunosuppressive therapy would be warranted to rule in or out autoimmune disease. Currently, treatment of affected dogs is supportive only and the prognosis is generally poor. Affected ESS seem to have a shorter life expectancy than many other breeds with chronic hepatitis: median survival time in the study was 189 days. Thirty-eight dogs died within 3 months, although 12 dogs survived more than a year from diagnosis.

DALMATIONS

Copper-associated liver disease has been described in Dalmatians with several reports in the literature suggesting it is likely a familial disease. The disease tends in some ways to share similarities to the copper-associated liver disease in the Bedlington terrier. Dalmatians are reported to have hepatic copper concentrations as high as 9000μg/g liver dry weight (LDW) paralleling that often found in the Bedlington. It differs from Bedlingtons in disease of Dalmatians is generally characterized as occurring most often younger dogs, having a rather acute onset and rapid progression with generally a poor prognosis. Dogs have high serum activities of alanine aminotransferase and aspartate aminotransferase, severe necrosis of zone 3 of the liver lobule with the surviving macrophages and hepatocytes containing excessive copper.

We reported on a series of ten affected Dalmatians with copper associated liver disease seen at Colorado State University over a four-year span.[1] These dogs tended to be young at the time of presentation with a mean age of five and a half years (range of 2-9 years). There was no obvious sex predilection (4 males, 6 females). Two of the dogs were related (mother and son). In another report of an affected dog stated two of the littermates died from a similar type of liver disease. Nine of the 10 Dalmatians presented for acute gastrointestinal symptoms, including anorexia, vomiting, or diarrhea. One dog presented 2 weeks after whelping. Most of the dogs were reported to be previously healthy and three were icteric at presentation. In all cases, the initial chemistry panel revealed marked increases in ALT and AST enzyme activity with lesser elevations in ALP enzyme activity. In the 10 cases the relative increase in ALT activity was much greater (average 10.8 times above normal) than the relative increase in ALP activity (average 5.5 times above normal) suggesting predominant hepatocellular, rather than cholestatic, liver disease. In all of the cases, hepatic copper concentrations were found to be abnormal. The mean hepatic copper concentration for nine Dalmatians was 3197 μg/g LDW, (range = 754-8390 μg/g). In five of these nine dogs, hepatic copper concentration exceeded 2000 μg/g. The greatest accumulation of copper (8390 μg/g) was seen in the youngest dog. Mean hepatic copper concentration for dogs less than 6 years old (n=4) was 4371 μg/g, whereas mean copper concentration for dogs greater than 6 years old (n=5) was 2256 μg/g. In two case reports of single Dalmatian dogs, both less than 2 years old had hepatic copper concentration of 9424 μg/g LDW[2] and 7664 μg/g LDW.[3]

Necroinflammatory changes are the notable histologic finding. Piecemeal necrosis and bridging fibrosis were commonly observed and two cases had micronodular cirrhosis. Generally, the inflammatory infiltrate was either primarily lymphocytic or neutrophilic which may have simply reflected the stage of the disease process. Morphologic features of cholestasis were generally not the predominate finding. Hepatic copper based on histochemical staining was most often in zone 3 however it was not possible to define an exact zonal distribution because of the extensive hepatic necrosis, fibrosis, and disrupted architecture on histopathology in the majority of the cases examined. Three of 4 dogs having urinary evaluation had both glycosuria and proteinuria suggested a concurrent fanconi-like syndrome. It is unknown if these dogs had abnormal renal copper concentrations affecting the kidneys as well.

The therapy is directed at copper chelation using penicillamine or trientine, a low copper diet and antioxidant and other general liver support. Because of the disease often has an acute onset and rapid course makes effective chelation therapy difficult. Eight of the 10 Dalmatians in our report were euthanized, one dog died, and one dog was managed long term with chelation therapy. The average time from initial presentation to death our series was 80 days.

STANDARD POODLES

We were contacted by a member of the Poodle Club of America (PCA) reporting that Standard Poodles (SP) have a chronic hepatitis (CH) that was thought to be a breed-associated condition. The PCA has also listed CH under the major health concerns of that breed and there are numerous anecdotal reports of certain lines of SP having a high incidence of chronic liver disease. In reviewing current publications relating to canine CH the SP is mentioned in many of those reports. We have recently received funding from the PCA to attempt to characterize the disease both clinically and histologically and to then begin investigation into the etiology as well as to collect DNA from normal and affected SPs for subsequent studies.

We have currently treated a number of SP with CH and are actively collecting DNA, histopathology, clinical information and pedigrees of affected dogs. What we know at this point is that CH appears to be a breed specific disease of SP. Affected dogs can be as young as 3 or 4 years of age at time of presentation however most dogs present with evidence of liver disease and or elevated liver enzymes around 6-9 years of age. Female dogs are overrepresented. Histology shows a necroinflammatory changes in predominately zone 1 but the inflammation can extend further into the lobule with bridging fibrosis and ultimately in some cirrhosis. Histochemical staining for copper using rhodanine as well as hepatic copper quantitation shows normal or modest copper accumulation that is thought to be likely secondary to cholestasis rather than a primary copper associated hepatopathy. Our findings are different than the reports of a lobular dissecting hepatitis that had been described in three related young SP.[4] Lobular dissecting hepatitis is a rare condition of unknown etiology.

The etiology of SP CH is yet unknown. Because SP in general tend to have a high incidence of immune disorders we speculate that immune mechanisms could possibly play a role in the etiology of CH. This is supported by the fact that we have treated a small number of SP having CH with immunosuppressive agents and observing some improvement. Preliminary DNA studies investigating MHC class II expression in SP with CH has been unrewarding. MHC class II expression is sometimes observed in immune disease but it was not identified as yet in SP. We are currently soliciting liver biopsies of SP suspected of having CH and have funding to cover comprehensive histological and hepatic metal evaluation for those cases. For more information I can be contacted at: twedt@colostate.edu.

REFERENCES

English Springer Spaniels: [1]Watson JP, et al **JSAP** 2010; 51:566-572, [2]Angles JM, et al **Tissue Antigens.** 2005;66(3):173-84, [3]Bexfield N, et al **Vet J.** 2012 Jan 4th epub ahead of print, [4]Bexfield N, et al **Vet Rec** 2011; 169: 415-419.

Dalmatians and Standard Poodles: [1]Webb CB, et al **JVIM** 2002;16:665-668, [2]P. Napier, **Canadian Veterinary Journal** 1996; 37:45, [3]L.J. Noaker, et al **JAMA** 1999;214:1502–1506, [4]Jensen AL, et al **Zentralbl Veterinarmed A**. 1991;38:194-197.

BREED RELATED HEPATOPATHIES: LABRADOR RETRIEVER

Cynthia R.L. Webster, DVM, DACVIM
North Grafton, MA

INTRODUCTION

The Labrador retriever is the most popular breed registered in the US, Canada and the UK. Labradors are also one of the most popular assistance dogs, as well as one of the most widely used dogs in search and rescue and customs and border protection. When you consider all this and the fact that they make great hunting dogs and phenomenal family pets, it almost seems like Labradors have evolved to be the perfect breed of dog. Any disease which compromises the health of this breed affects the veterinary profession worldwide. Over the last decade it has become increasingly apparent that Labradors have a breed predisposition for chronic hepatitis (CH) that progresses to cirrhosis and eventually to death from hepatic failure.

Chronic hepatitis (CH) in the dog was originally reported in 1975 when a manuscript describing the copper associated CH in Bedlington terriers was published. This was followed in the 1980's with reports of CH in Doberman pinschers, Cocker spaniels West Highland white terriers and Standard poodles. A copper associated CH in the Dalmatian was described in the late 1990's. Most recently a retrospective study of breed associated CH in the UK added the English Springer spaniel, Samoyed, Cairn and Great Dane to the list of dogs with an increased incidence of CH. Early case series of CH from Canada in the 1980's, failed to see a Labrador breed predilection.[7] The first report of an increased incidence of CH in Labradors was from a retrospective study of CH in 1991 from Sweden.[1] In 2006 (Europe)[11] and 2007 and 2009[18,19], (US) the first descriptive studies of Labradors with CH were published. Since that time publications dealing with the role of diet and genetics in the disease have been appeared in the literature.

CLINICAL PRESENTATION

Before beginning a comprehensive comparison of the available literature on Labrador hepatopathy it is important to note the inclusion criteria for the studies that have been done. In 2/5 of the largest case series, inclusion criteria included the presence of positive copper staining on hepatic biopsy as well as histological signs of CH.[11,19] In these studies the authors note that somewhere between 30-40% of the archival cases of Labrador CH fell into this copper associated category.[11,15] In the other 3 retrospective studies the inclusion criteria did not include the presence of positive copper staining.[3,14,18] Collectively, these publications suggest that there may be 2 populations of Labrador that get CH one that is marked by the presence of significant copper stores and one that is not. Whether these populations share other common etiopathogenic factors remains to be determined.

Signalment

A female predominance (appr 70%) has been reported in 2 case series (one from the US and one from Europe)[11,19] in which copper associated disease was reported and in one study of all Labradors with CH (UK).[3] In 2 additional studies both from the US, there were equal numbers of males and females.[14,18] A female predisposition occurs in CH in Dobermans and Springer spaniels. Median age at diagnosis was 6.4-9.3 yrs (range, 2-15) which is comparable to that reported for CH in the general canine population (5.5-7.7, range 0.3 -17).

Clinical Signs and Physical Examination

Some Labradors are asymptomatic at the time of presentation (with only increased serum liver enzymes) or present for another disease and liver disease is found serendipitously on biochemical screening. It appears that clinical signs develop late in the course of the disease process, similar to CH in other breeds. Signs, when they occur, are most often non-specific. The most commonly reported signs are gastrointestinal and include decreased appetite or anorexia (43-100%) and vomiting (38-53%). Other clinical signs include weight loss (20-25%), diarrhea and PU/PD (about 8%). Only one study reported on the presence of concurrent diseases in detail.[18] In that study concurrent conditions were present in many dogs including endocrine disease (diabetes mellitus 1/24, hypothyroidism 2/24), allergic skin disease (10/24), splenic masses (2/24) and one each with megaesophagus, SARDS, a necrotic intra-abdominal lipoma and gastric dilation. In another study 2/15 dogs were post-whelping. No dogs were on concurrent hepatotoxic drugs (this was an exclusion criteria) and all were apparently eating commercial dog food, although details on diet history in all the studies is incomplete. Findings on physical examination included icterus (9/24, 5/15), ascites (2/24, 2/15), hyperthermia (3/24), abdominal pain (2/24). Overall most dogs were in good body condition with a mean BCS in one study of 6/9 (range 2-8) and reports of only 2/15 dogs in another with a poor BCS.

Clinicopathological Findings and Diagnostic Imaging.

Clinical pathology was only reported in 2 studies.[11,18] All dogs had increases in at least one serum liver enzyme. Serum ALT (8-10 x the upper limit of normal -ULN) was typically higher than serum ALP (3.4X ULN). Serum AST was consistently increased, but to lesser extent than ALT. GGT was variably increased in about 1/3 of dogs. Hypoalbuminemia was present in about a quarter of the dogs. Hyperbilirubinemia was noted in about 50% of the dogs in one study while both studies reported about 40% of the dogs had increased in aPTT and 20-40% increases in PT. Fibrinogen was decreased in 5/5 and 6/9 dogs in which it was evaluated. No bleeding or thrombotic complications were reported at the time of presentation. However, in a recent study of canine portal vein thrombosis 6/31(20%) of the dogs were Labradors with CH.[16] Most dogs had exposure to another predisposing cause for thrombosis such as concurrent therapy with corticosteroids, recent surgery or an immune mediated blood dyscrasia. The PCV was normal in most dogs or mildly decreased and some dogs had a mild neutrophilic leukocytosis. In one study 5/24 dogs had thrombocytopenia. Notable abnormalities on the urinalysis were bilirubinuria in 7/14 dogs and glucosuria in 4/14. Three of the 4 dogs with glucosuria had normal blood glucose. The glucosuria was transient and could have represented the transient Fanconi's syndrome reported in dogs with copper storage disease.[2,8]

Imaging finding were reported for 33/39 dogs in 2 reports. The findings were nonspecific and typical of those seen with CH in dogs. Hepatomegaly was seen in 2 dogs and microhepatica in 12. The liver was hyperechoic in 6, hypoechoic in 5, inhomogenous in 8 and nodular in 9. Ascites was noted in 4 dogs. In 7 dogs the liver was normal on ultrasound examination.

DIAGNOSIS

Hepatic biopsy was obtained in all dogs. In a study looking at all dogs regardless of copper status[18], there was a mild to moderate infiltrate of lymphocytes and macrophages which was most prominent in the peri-portal area with about ½ of dogs having a significant influx of neutrophils as well. Twenty one of 24 dogs had increased fibrosis, 6 dogs had bridging fibrosis and 4 cirrhosis. Two studies reported on the histopathology in dogs with copper associated CH.[11,19] In the first study hepatic histopathlogy showed varying degrees of necrosis and apoptosis, centrilobular mononuclear inflammation regeneration and fibrosis. Fibrosis with architectural distortion was present in 9/15 with 4 dogs having cirrhosis.[11] In the second study multifocal to coalescing centrilobular hepatitis characterized by a macrophages and smaller numbers of neutrophils and mononuclear inflammatory cells was described. They also noted the presence of multifocal pigment granulomas. Eight dogs had bridging fibrosis and 6 dogs nodule formation.[19] In summary, histopathology in affected Labradors was consistent with an inflammatory hepatopathy that was often associated with fibrosis and in some cases cirrhosis. Those dogs with copper associated disease appeared to have lesions centered on the centrilobular area an observation which has been reported in copper associated disease in other breeds.[22]

Normal values for hepatic copper in the Labrador were reported as 163-345 PPM[19] and 120-304 PPM[11] and are in line with previous reports in of normal hepatic copper in mixed breed dogs (< 400 PPM).[22,23] Histochemical staining of hepatic biopsy material for copper was positive in all dogs in the 2 studies in which dogs were selected for copper associated disease (inclusion criteria). In these studies copper was concentrated in centrilobular and midzonal hepatocytes and within Kupffer cells.[11,19] In the US study in which all Labradors were evaluated,[18] copper staining was positive in 15/17 dogs and did not appear to have a centrilobular orientation. Staining was mild in 5 dogs and moderate to severe in 10. In a study from the UK in which all dogs with CH were evaluated, no Labradors had positive hepatic copper staining.[3] Only one of the retrospective studies reported quantitative hepatic copper analysis. In this study of copper associated CH in Labradors, values were available from 12 dogs. Ten dogs had hepatic copper greater than 2000 PPM (2375PPM to 3369PPM), one had 1490 PPM and one 1990 PPM.[19] Hepatic copper concentration did not correlate with age suggesting that there may not be progressive copper accumulation in the liver as is the case in Bedlingtons and humans with Wilsons disease.[14]

The results of hepatic copper evaluation in Labradors in the initial reports suggested that there are 2 groups of Labradors with CH, one that accumulates copper preferentially in the centrilobular area and a group that does not. To complicate the matter, however, later studies examining heritability and dietary influences in Labradors with CH have reported dogs with high hepatic copper levels that also clearly have hepatic copper overload on histochemistry, but do not have CH. Hepatic copper overload in the absence of CH has also been reported in some Bedlingtons and West highland white terriers.[22,23] The reasons for the lack of liver pathology in these individuals is unknown. One explanation might be the presence of different genetic mutations in copper metabolism proteins. Over 300 mutations in the ATP7B gene occur in people with Wilson's disease and the mutations are associated with considerable phenotypic diversity in disease presentation and treatment outcome. Alternatively the discrepancy may reflect the need for an additional environmental factor (toxin, infection, immune aberration) to trigger the inflammatory process.

Currently there are no noninvasive methods to replace hepatic copper determination for diagnosis of copper overload. Dogs do not get the low ceruloplasmin or Kayser-Fleisher corneal rings used to diagnosis Wilson's disease in people and one study failed to show that copper levels in hair or nail clippings correlated with hepatic copper concentrations.[4]

OUTCOME

In a retrospective on copper associated CH, 9 dogs died or were euthanized due to hepatic disease from 1-50 days (mean 10.4 days), none of which received copper chelation therapy. It was noted in the manuscript that 5 dogs treated with penicillamine had resolution of clinical signs (although one dog also received prednisone). Long term follow-up for the additional dog was not given. In the US study of all Labrador with CH, median survival was 374 days (range,1-2645 days). Eleven dogs died within 2 months and 13 dogs survived for over a year after diagnosis. Of the 14 dogs that died, 10 died from complications related to hepatic disease. Dogs that died within 2 months were more likely to have a prolonged PT and thrombocytopenia while dogs with anorexia, serum hypoglobulinemia or prolonged aPTT had shorter overall survival. No significant associations were found with overall survival and histopathologic grade, degree of serum liver enzyme elevation, hyperbilirubinemia, serum albumin or blood glucose. There was a tendency for histopathologic stage (degree of fibrosis) to be negatively correlated with survival time. A clinical score was generated in this study patterned after the Child-Pugh score in humans. It incorporated information on prognostic factors from the current study as well as previous studies on canine CH. The final model included the presence of clinical signs associated with hepatic failure (ascites, HE, anorexia, PU/PD) and biochemical evidence of decreased hepatic synthesis (decreases in serum globulin and albumin and increases in bilirubin and aPTT). The clinical score assigned to the Labradors using this model was significantly correlated with over-all survival time and histological stage. With validation in a larger sample of dogs of varying breeds from different geographical areas, this clinical scoring system may prove useful as a non-invasive tool to evaluate response to therapy or predict prognosis in population of animals. It would decrease reliance on hepatic biopsy which is fraught with sampling and interpretative error.

ETIOLOGY

Putative causes of CH in dogs include hepatotoxins, drugs, infectious agents, immune aberrations and defects in copper metabolism, although admittedly most cases end up being idiopathic. Bedlington terriers get the prototypical copper metabolism defect which is due to a mutation in the COMMD1 gene.[5] The exact function of the COMMD1 gene is not known although in the liver it is clearly evolved in copper handling. Other breeds in which CH is associated with hepatic copper accumulation include the Dalmatian, West Highland white terrier and Doberman, although no genetic basis for copper overload in these breeds has been reported. Some preliminary investigation into a genetic basis for the Labrador copper associated hepatitis has been completed. In this study some evidence of heritability was found in a small group of Labradors in the Netherlands.[9] The Netherlands group has also reported the preliminary results of a screen for 14 candidate copper related genes (including ATP7B and COMMD1) in 60 Labrador retrievers.[12] The phenotype for the trait was defined was defined as an elevated hepatic copper concentration above 600 PPM with localization of copper in zone 3 of the liver lobule. The study concluded that all the candidate genes were not involved in hepatic copper accumulation in Labrador retrievers if it is assumed that a single disease gene exists.

Two studies have investigated a possible link between diet and Labrador CH. The NRC minimal daily copper requirement for the dog is 0.06 mg/kg/day(2.9 mg/kg dry weight of food) while the AAFCO recommendations are for 7.3 mg/kg dry matter. Commercial dog foods are relatively high in copper content when compared with this recommendation, ranging from 12 to 25 mg/kg dry matter. This would suggest that dogs eating commercial foods are subjected to dietary copper excess. The first study done in the Netherlands suggested that hepatic copper content could be manipulated by dietary copper concentration.[10] In this study they took Labradors with copper associated CH most of which had high hepatic copper values (around 800 PPM with a range of 70 to 2810) and placed them either on a copper restricted diet alone (4.8 PPM/kg dry matter which delivered about 1.2 mg of copper/day) or on the restricted diet plus zinc gluconate (10 mg/kg BID PO). Hepatic copper was evaluated 8 and 16 months later. Hepatic copper decreased in both groups to a comparable extent from a mean of 770 PPM to 236 to 320 PPM and 720 to 328 to 340 PPM in the diet and diet and zinc treated groups, respectively. There was no change in qualitative hepatic copper staining scores, histological indices of inflammation or serum liver enzyme activity (which was in the normal range before and after the treatment periods). These findings indicate that hepatic copper concentration in Labradors is sensitive to dietary manipulation.

A second diet related study was prompted by the researchers observation that normal canine hepatic copper concentrations reported in the literature have increased over time from 7 PPM in 1929 to 80 PPM in 1956 to a value of 200-400 PPM since the late 1970's.[14] The current normal value for canine hepatic copper is 10 times that seen in other mammals (10-30 PPM). Coincident with the progressive increase in hepatic copper concentration was a switch by the pet food industry in 1996 from using cupric or cuprous oxide as dietary supplements to copper sulfate or copper chelates which are absorbed more efficiently. The authors designed a study to evaluate the effect of this dietary change on hepatic copper. They looked retrospectively at 4 groups of Labradors: dogs from 1980-1997 with and without CH and dogs from 1998-2011 with and without CH.[14] Copper concentrations were significantly higher in dogs with CH compared to those without, 614 PPM (104-4234) to 299 PPM (93 -3810), respectively. Hepatic copper was also significantly higher in the later time period in both normal 177 PPM (93-453) vs 753 PPM (101-3810) and CH dogs 346 PPM (104-3300) vs 1115 (191-4234). These results suggest Labradors accumulate copper in their liver in relation to the amount in the diet. This finding does not eliminate the possibility of a genetic mutation in copper metabolism in this breed as the inability to handle high copper loads may be genetically based. It does suggest however along with Netherland study that copper overload in the Labrador may respond to dietary management.

Increased sensitivity to dietary copper levels is seen in other animal models of hepatic copper imbalance. Rats and mice with mutations in Wilson's gene, ATP7B, demonstrate unique sensitivity to copper overload when placed on high copper intake. Mice with a hepatic specific knockout of the Bedlington gene, COMMD1 gene, accumulate 20 times more hepatic copper than normal mice when placed on a high copper diet. No dietary studies have been done with Bedlington's but it is known that most dog foods contain excess copper so a role for diet in hepatic copper overload in the Bedlington's cannot be ruled out. There is also a large animal model of increased sensitivity to dietary copper. The North Ronaldsay sheep is a primitive breed which has adapted to a living in a copper impoverished environment (<5 PPM) and develop hepatic disease due to copper overload when put in a copper replete environment (11 PPM). The reason for this unique susceptibility is unknown although a role for COMMD1 mutation has been ruled out.

TREATMENT

Retrospective studies have convincingly documented that Labradors get a slowly progressive CH that eventually results in the development of cirrhosis, hepatic failure and death. Evidence suggests that there can be a long period in which the disease may remain clinically silent, but severe liver pathology may be developing. Thus increases in serum transaminases should not be ignored in Labradors. Instead these abnormalities should be followed in series and if progressive or persistent, should prompt acquisition of a hepatic biopsy for evaluation of copper concentration. Biopsy material should be submitted for histopathology with H&E, Trichrome or Sirius red staining (for fibrosis) and rhodanine or rubeanic acid (copper staining). A wedge biopsy is best although large gauge (16 or 14 guage) needle biopsies may suffice. Needle core hepatic copper values may be less than those in seen in wedge samples, but likely will reflect copper overload.[13] Since cirrhotic nodules may have less copper than adjacent inflammatory tissue, quantitative copper values may go down in late stage disease or be discordant with results of qualitative assessment. It should be remembered that copper quantification can be done accurately off of de-parafinized tissue.[13]

When does hepatic copper accumulation in the Labrador warrant therapeutic intervention? The answer to this question is not straight forward because the hepatic copper level associated with hepatotoxicity is not known Earlier literature suggested that 2000 PPM[22] was the level above which hepatic toxicity occurred, while newer studies have lowered this to 1000-1500 PPM.[14 22,23] The true answer to this questions awaits further study. For now we can only offer guidelines for treatment based on our current level of understanding of the Labrador CH. This author suggests that treatment approach be dictated by a combination of factors: the results of both qualitative and quantitative hepatic copper evaluation, histopathological evaluation of grade (inflammation, necrosis, apoptosis) and stage (fibrosis) of hepatic disease, the presence of clinical signs, the degree of elevation in serum liver enzymes particularly the aminotransferases and exclusion of other causes of liver disease (toxic, infectious, immune). Exclusion of concurrent disease is particularly important since hepatic copper accumulation can predipose to liver injury from other agents, but may not itself cause the damage.

If qualitative copper staining is primarily centrilobular or if it is severe (4-5+) and not just associated with areas of inflammation and/or the quantitative hepatic copper is greater than 1500 PPM, then the dog should get copper lowering therapy with either pencillamine or zinc and a low copper diet should be initiated. Concurrent treatment with anti-oxidants such as SAMe, vitamin E or ursodeoxycholate should be started. If on re-check, the dog is asymptomatic and liver enzymes are decreasing the current treatment should be continued until liver enzymes are normal. Ideally discontinuation or tapering of copper lowering medication should be based on hepatic biopsy in 6 to 8 month to document adequate de-coppering. Whether or not the de-coppering agent should be stopped in dogs whose hepatic copper levels have returned to normal levels is uncertain. In Bedlington's and Wilson's disease patients chelation therapy is life-long, although the dose can be safely reduced when hepatic copper levels are normal. In Labradors livers de-coppered with penicillamine, have been shown to re-accumulate copper when the drug is stopped and the dog is on a normal commercial dog food.[10] Whether a low copper diet would be adequate to prevent re-accumulation of copper is unknown.

If hepatic copper scores are 1-3 + and not centrilobular, but instead associated with areas of inflammation (particularly if it is peri-portal) and quantitative analysis is less than 1000 PPM, copper lowering drugs are not indicated. If the hepatic biopsy in these patients does not show high activity (severe inflammation, interface hepatitis) and the dog does not have a high serum ALT > 4-5 X ULN, then initiating treatment with hepatoprotective agents is indicated and activity monitored with periodic assessment of serum liver enzymes. If the hepatic copper is greater than 400 PPM then a low copper diet is started as well. If the hepatitis is of high grade on histopathology or serum transaminases are >4-5 times the ULN then a trial of immunosuppressive therapy with corticosteroids or cyclosporine is indicated. The author cautions against the use of corticosteroids in dogs with late stage disease as it may pre-dispose to thrombotic events and/or the development of gastrointestinal bleeding.

Another consideration in treating Labradors with CH with copper lowering therapy is the fact that hepatic copper deficiency can exacerbate hepatic injury. Moreover, copper deficiency has been reported in Belington's on long term chelation with penicillamine. Thus it may be best to monitor copper levels especially in dogs on both copper lowering drugs and a low copper diet.

REFERENCES

1Andersson M, et. al. **J Small Ani Pract** 1991;32: 1-5. **2** Appleman EH, et. al. **J Vet Intern Med**. 2008;22:1038-42. **3** Bexfield NH, et. al. Vet J. 2012 (In Press). **4** Bode P, et. al.Anal Bioanal Chem. 2008;390:1653-8. **5** Favier RP et. al. Vet Q. 2011;31:49-60. **6** Fuentealba I et. al. **J Comp Pathol.** 1989;100:1-11. **7** Fuentealba C, et al. **Can Vet J**. 1997;38:365-73. **8** Hill TL et. al. **J Vet Intern Med**. 2008;22:219-22. **9** Hoffman G, et . al. Anim Gen 2008;39:454. **10** Hoffmann G, et.al. J Vet Intern Med. 2009;23(5):957-63.11 Hoffmann G, et. al. **J Vet Intern Med.** 2006 ;20:856-61 **12**, House JV, et.al. Proc Brit Small Ani Vet Ass 2008 pp508. **13** Johnston AN, et. al. **Am J Vet Res**. 2009;70:1502-11. **14** Johnston AN, et. al.**J Am Vet Med Assoc** 2012 (in Press). **15** Poldervaart JH, et.al. **J Vet Intern Med**. 2009;23:72-80. **16. Respess M et al J Vet Intern Med 2012 (In press).** 17 Seguin MA, et. al. **J Am Vet Med Assoc**. 2001 15;218:1593-7, 1580 **18** Shih JL, et. al. **J Vet Intern Med**. 2007;21:33-9. **19** Smedley R, et. al. **Vet Pathol**. 2009;46:484-90. **20** Spee B, et. al. **J Vet Intern Med**. 2006;20:1085-92. **21** Sriramachari S, et. al. **Indian J Med Res**. 2008 ;128:93-6m. **22** Thornburg LP. **J Vet Diagn Invest** 2000;12:101-10. **23** Thornburg LP, et. al. **Vet Pathol**. 1990;27:81-88 **24** Vonk WI, et. al. **PLoS One**. 2011;6:e29183. **25** van den Ingh TS, et. al.**Vet Rec**. 2007;161:728.

Breed Specific Hepatopathies: Scottish Terriers & Maltese Dogs

S.A. Center, DVM, DACVIM
Ithaca, NY

Scottish Terrier Degenerative Vacuolar Hepatopathy

In dogs, increased serum alkaline phosphatase (ALP) activity reflects necroinflammatory, neoplastic, or cholestatic disorders involving hepatic biliary or pancreatic ductal systems, canalicular cholestasis (i.e. altered organic anion pumps as in sepsis), or simply induction phenomena (endogenous or exogenous glucocorticoids, or acute phase response). Vacuolar hepatopathy (VH), a relatively common canine syndrome, accompanies many systemic illnesses (inflammatory, neoplastic, infectious, primary and secondary hepatic disorders). VH is linked with increased hepatic ALP gene transcription, enhanced production of the corticosteroid isoenzyme, and abnormally increased serum ALP activity (3-fold or much greater). Hepatocytes in VH accumulate excessive cytosolic glycogen leading to cell distention and increased fragility. While some clinicians consider this a "benign" syndrome, in some cases it unfortunately leads to diffuse hepatic remodeling (degenerative VH), a nodular cirrhotic appearing liver, and intrasinusoidal and splanchnic hypertension leading to acquired portosystemic shunts, ascites, and hepatic insufficiency. While any dog breed or mixed breed dog can be affected with degenerative VH, among most severely affected dogs are Scottish Terriers (ST). A pioneering study demonstrated the relationship between high serum ALP and excess serum adrenocortical steroid hormone concentrations in apparently healthy ST, and correlation between older age and androstenedione concentrations after ACTH stimulation. For clinicians that only occasionally examine Scottish Terriers, recognition of syndromic features usually leads to a presumptive diagnosis of either typical or atypical hyperadrenocorticism. Conventional adrenal-modulatory treatment with Lysodren has led to acute deaths or rapid onset of hypoadrenocorticism while use of Trilostane has made little difference in enzymology, hepatic features or clinical signs, and sometimes has ill effects.

To more fully understand the clinical spectrum of this syndrome we retrospectively investigated liver biopsies from ST (1985 to 2012) available in the Anatomic Pathology archives of Cornell University. Tissues were evaluated with observers naive to signalment or clinical information and cases were segregated into two groups: Group 1= degenerative VH and Group 2= normal liver, necroinflammatory hepatic lesions, or congenital vascular malformations. Thereafter, signalment, historical and physical examination findings, clinicopathologic data, liver size (abdominal palpation, survey lateral radiograph, ultrasonographic [US] imaging), and US features (parenchymal echotexture, detection of parenchymal nodules, nodule dimensions and echotexture, adrenal dimensions, detection of adrenal nodules, and the presence of ascites or APSS or other abdominal abnormalities) were collated.

Of 97 biopsies, ~70% had a marked VH, including a small subset of dogs with copper-associated hepatopathy (~10%). A concerning discovery was that ~30% of dogs with chronic degenerative VH also had histologically confirmed hepatocellular carcinoma (single or multiple). Additionally, other dogs displayed what appear to be precursor foci of cellular atypia (adenomatous hyperplasia). These lesions are being further characterized using immunohistochemical assessments of cell proliferation. Most liver biopsies were pursued because of concern over persistently and/or progressively (mths to yrs) increased ALP activity associated with ultrasonographic features suggesting diffuse liver disease transforming hepatic architecture. There is no significant association between age and ALP activity in this dataset; high ALP activity was usually recognized by 3-4 years and there was no significant difference in fold increase among age categories. Examination of sequential data for individual dogs also did not consistently demonstrate progressive ALP increase. Most dogs demonstrated a normal to large liver size and hyperechoic hepatic parenchyma contrasting against hypoechoic nodules ranging in size form 3-15mm. Some chronically affected dogs (followed clinically > 4 years) demonstrated the same US pattern but additionally had a nodular liver surface contour and normal to small liver size. Approximately 23% of dogs had adrenals judged as large or had asymmetric nodular lesions on US assessment; unilateral and bilateral adrenomegaly was recognized in some ndividuals. LDDST (n=8), urine cortisol:creatinine (home collected urine, n=2), ACTH cortisol response (n=31), ACTH stimulated adrenal steroid panel (UT Endocrinology Laboratory, n=17) were inconsistent in implicating a specific hormonal aberration among dogs and were not convincing for simple overproduction of cortisol. Concentrations of $\geq$ 1 adrenocortical hormone exceeding reference values were detected in every dog studied, similar to findings previously published initially describing the syndrome. Adrenal sex hormone profiles demonstrated increased androstenedione and 17-hydroprogesterone concentrations in most but not all dogs.

High ALP activity and degenerative VH in Scottish Terriers likely reflects a genetic defect affecting steroidogenesis, in some ways resembling 21-hydroxylase deficiency adrenal hyperplasia in humans. Overproduction of androgenic hormones may also explain the long recognized prostatomegaly common in neutered males of this breed. Likewise, appearance of HCCA in these dogs also may be linked with chronic increased androgens, as recognized in other species. The molecular phenotype of this trait has not been characterized at the time these notes were prepared. Adrenal tests implicate classic hyperadrenocorticism in relatively few dogs whereas sex hormone profiles pre and post ACTH commonly demonstrate $\geq$ abnormally increased steroidogenic hormone consistent with atypical hyperadrenocorticism. Nevertheless, clinical features of hyperadrenocorticism are inconsistent and often lacking. Treatment with Lysodren has led to iatrogenic illness, hypoadrenocorticism, or death and treatment with Trilostane has either failed to effect clinical or clinicopathologic benefit or has resulted in lethargy, inappetence, diarrhea (blood tinged), and vomiting. While adrenal modulation with these drugs usually fails to attenuate serum ALP activity, it does blunt cortisol concentrations sometimes to a symptomatic level. Use of ketoconazole for adrenal modulation has not been explored in many ST with degenerative VH; provision of physiologic glucocorticoid support is recommended if low dose treatment is considered.

Approximately 25% of ST with a degenerative VH progressively develop severe hepatic remodeling and slow progression to hepatic insufficiency associated with jaundice and bleeding tendencies (terminal stage). A subset develop abdominal effusion reflecting portal hypertension. Insidious progression toward lobular collapse and intrahepatic sinusoidal hypertension may culminate in hepatic failure as early as 6 years but is more commonly observed in elderly dogs ($\geq$10 years). While some dogs with this lesion may live a good quality life for up to 16 yrs, why some dogs develop a progressive degenerative lesion is not understood (may reflect trait penetrance, environment, concurrent disease status and many other factors).

Signalment and clinicopathologic parameters in dogs with ST VH associated with HCCA compared to those lacking HCCA have no predictive value (age, serum enzyme activity, others). Furthermore, dogs with copper associated hepatotoxicity also cannot be definitively deduced using clinicopathologic assessments (ALT is not significantly higher or ALP significantly lower compared to ST degenerative VH-likely because all ST are predisposed to VH). Thus liver biopsy is required.

We recommend US imaging for middle aged ST with degenerative VH and progressively increasing serum ALP activity in an effort to monitor for HCCA (also urinary bladder TCCA surveillance) but warn against simplistic intervention with adrenal modulatory agents conventionally used for classic hyperadrenocorticism. Dietary protein modification and restriction has no role in chronic management of these dogs unless signs of hepatic insufficiency manifest. Affected dogs appear to tolerate glucocorticoids in modest dosages to manage inflammatory conditions (i.e. inflammatory bowel disease, chronic atopy) without exaggerated adverse effects. Since the primary hepatic lesion is degenerative with gradual hepatocyte dropout leading to parenchymal collapse and altered sinusoidal perfusion, antioxidants may provide benefit (speculative). The disorder is not highly fibrotic and is not typically associated with accumulation of transition metals. Thus, targeted intervention for these phenomena are not recommended. Increased total serum bile acids (TSBA) are a late syndrome manifestation. Ursodeoxycholate at that stage may be advantageous. Further mechanistic investigations of ST degenerative VH are necessary to formulate safe appropriate interventions for dogs with rapid syndrome progression.

Maltese Dog Zone 3 Hepatopathy

Maltese have high prevalence of congenital hepatic vascular malformations, with hepatoportal microvascular dysplasia (MVD) vastly outnumbering occurrence of congenital portosystemic vascular anomalies (PSVA). A multi-breed association mapping study of the PSVA/MVD trait in small dog breeds at Cornell University has included Maltese dogs and confirms that this trait can have a prevalence as high as 60%-90% within kindreds. Total serum bile acid (TSBA) concentrations are conventionally used to economically phenotype the PSVA/MVD trait in young dogs. However, clinicians and Maltese dog enthusiasts have been confused by an early publication that proposed TSBA in Maltese was confounded by interfering analytes. Genetic mapping work and liver biopsy from Maltese with increased TSBA concentrations strongly refute that supposition. We recognize an inflammatory Zone-3 (centrilobular) hepatic lesion in some dogs with the PSVA/MVD trait and this lesion associates with increased serum ALT activity and is strongly represented in Maltese. The histologic lesion ranges in severity, may be progressive leading to cirrhosis, and is seemingly associated with concurrent inflammatory bowel lesions (splanchnic circulation of inflammatory cytokines and inflammatory cells to the liver).

To more fully understand the clinical spectrum of this lesion and to objectively determine its prevalence among Maltese with suspected liver disease we retrospectively investigated clinical records of Maltese dogs presented for suspected liver disease (n=175) and liver biopsies from these dogs (n=136, submitted to the Anatomic Pathology Diagnostic Service at Cornell University [1980 to 2011]). Tissues were evaluated with observers naive to signalment or clinical information with cases segregated into three groups: Group 1= inflammatory Zone 3 lesion; Group 2= absence of inflammatory zone 3 lesion; Group 3= cirrhosis. Biopsies were evaluated and numerically scored to indicate the presence or absence of features diagnostic for portal hypoperfusion (i.e. increased arteriole profiles in portal triads, close approximation of portal triads and hepatic venules representing lobular atrophy rather than lobular collapse, small or inconspicuous portal veins, small hepatocytes reflecting deprivation of hepatotrophic substances, increased numbers of binucleate hepatocytes, prominent throttling smooth muscle with increased sagittal sections of hepatic venules, orphan arterioles randomly encountered within hepatic parenchyma) and the presence or absence of inflammatory infiltrates and lipogranulomas. Thereafter, signalment, historical and physical examination findings, clinicopathologic data, and results of US hepatic imaging (parenchymal echotexture, detection of nodules, assessment of gallbladder size, comparative dimensions of hepatic venule, portal vein, hepatic artery within hepatic parenchyma, presence PSVA or APSS, ascites, or other abdominal abnormalities) were collated.

Of 136 biopsies, >90% demonstrate features of portal hypoperfusion (PSVA=22%, MVD without PSVA ~60%). Congenital hypoperfusion could not be evaluated in most dogs with cirrhosis. Every dog with the portal hypoperfusion had high TSBA. In most cases, the presence of an increased serum ALT in conjunction and high TSBA concentrations initiated collection of a liver biopsy.

Inflammatory infiltrates (usually lymphoplasmacytic +/- eosinophils) adjacent to hepatic venules (zone 3) were identified in >80% of dogs. In ~50% of these dogs, lipogranulomas (foamy macrophage aggregates) also were adjacent to hepatic venules and often partially or completely obscured hepatic venule profiles. However, changes were variable among and within biopsy sections in an individual dog. Severe lesions correlated with intrahepatic sinusoidal hypertension and APSS confirmed by imaging studies or grossly at surgery or necropsy. In these, connective tissue was confirmed to obscure/occlude the lumen of hepatic venules (a so called venoocclusive lesion). While cirrhosis was confirmed in ~6% of biopsy samples, ascites was detected clinically in ~ 23% of dogs (n=175). Ascites was associated with US detected hepatic parenchymal nodules and nodular surface contour consistent with parenchymal remodeling. Diminution of hepatic venules was often notable on US color-flow during vascular interrogation. Randomly located lipogranulomas were identified in >35% of biopsy samples from dogs with portal hypoperfusion (PSVA/MVD lesion) Inflammatory infiltrates involving zone 1 (portal triad) were uncommon (<24% of biopsies), a notable finding considering that zone 1 is the classic location of canine chronic necroinflammatory hepatitis not involving copper hepatotoxicity. Thus, the most common form of inflammatory liver disease in Maltese dogs associates with increased TSBA concentrations and zone 3 perivenular inflammatory infiltrates +/- lipogranulomas. In dogs with cirrhosis, the zone of initial disease could not be ascertained due to extensive parenchymal extinction and regenerative nodule formation. Five dogs had copper associated hepatotoxicity (Rhodanine staining and quantified tissue copper ≥ 1,200 ppm). Increased iron in Kupffer cells, macrophages, and lipogranulomas was present in ~65% of biopsies. Stored iron increases risk of transition metal amplified oxidative injury. In a subset of dogs with simultaneous gastric and small intestinal biopsies, >90% had IBD lesions and concurrent Zone 3 inflammation +/- hepatic venule associated lipogranulomas.

Treatment of dogs with this Zone 3 hepatopathy has focused on management of associated IBD using hypoallergenic diets (home cooked nutritionist formulated or commercially available), metronidazole (7.5 mg/kg PO BID, low dose due to portal hypoperfusion lesion and observed risk of neurotoxicity with conventional dosages), and glucocorticoids avoiding use of those with mineralocorticoid attributes (dexamethasone used every 3rd day, antinflammatory dose). Vitamin E (10 IU/kg/day) and SAMe (20 mg/kg/day) are recommended for antioxidant and antifibrotic benefit. Because a subset of severely affected dogs have developed thromboembolic lesions of hepatic venules and/or portal veins, minidose aspirin (0.5 mg/kg PO SID to BID) has also been initially prescribed in dogs with severe lesions until inflammation abates (assumed based on decline in liver enzyme activity with normal synthetic markers). Inflammation also has been successfully managed using budesonide rather than a systemic glucocorticoid. In all cases treated with glucocorticoids, dosage is titrated to response (enzymes, clinical features) to minimize undesirable effects. In dogs requiring continued immunomodulation or displaying unacceptable glucocorticoid effects, azathioprine or cyclosporine have been combined with or have replaced glucocorticoid treatment. Ascites (18% clinical cases) has often been successfully managed with sodium restriction and combined furosemide and spironolactone administration. A loading dose of spironolactone (2-4 mg/kg, PO) is given once and followed by 1-2 mg/kg PO BID. Furosemide is dosed at 1 mg/kg PO SID to BID. Diuretic treatment has often been suspended when ascites is in remission.

REFERENCES

Zimmerman K, et al:Hyperphosphatasemia and concurrent adrenal gland dysfunction in apparently healthy Scottish Terriers. J Am Vet Med Assoc 2010;71:178-186. Tisdall PLC, et al: Post-prandial serum bile acid concentrations and ammonia tolerance in Maltse dogs with and without hepatic vascular anomalies. Aust Vet J 1995:72:121-126

NEUROENDOCRINE TUMORS IN HUMANS

Lowell Anthony, MD, FACP
Lexington, KY

Background and Epidemiology

Neuroendocrine tumors (NETs) comprise a heterogeneous group of neoplasms that arise from neuroendocrine cells that are widely dispersed throughout the body but concentrated in certain areas such as the pituitary, thyroid, lung, pancreas, gastrointestinal tract and the adrenal medulla. These tumors can be benign and multi-focal such as insulinomas or highly malignant and metastatic such as small cell lung cancer. The more specific term "well-differentiated neuroendocrine tumor" is preferred over the more general term "carcinoid" to more accurately describe biologic behavior of a specific primary site rather than a term linked to an indolent disease process characterized by serotonin over-secretion.

The incidence of human NETs is 5 cases per 100,000 of the U.S. population which represents an increase from 1 case per 100,000 thirty years ago. This increase is most likely due to multiple factors that include better reporting, greater awareness and education as treatment options have increased and more screening procedures performed. NETs constitute the second most prevalent (35 cases per 100,000) gastrointestinal neoplasm because of their indolent nature and interventions that can effectively "turn back the clock" in metastatic disease patients. The majority of well-differentiated neuroendocrine carcinomas (NECs) or about two-thirds, arise from the distal small bowel and proximal large intestine. The bronchopulmonary system (BPS) accounts for roughly one-fourth of the well-differentiated and intermediate-grade tumors that are referred to as carcinoid and atypical carcinoid, respectively. It is estimated that approximately one fourth of patients present with metastatic disease at the time of initial diagnosis.

NECs are divided into low (well-differentiated), intermediate (moderately well-differentiated) and high grade (poorly-differentiated) categories according to their mitotic or proliferative index. Patients with low- or intermediate-grade NETs can be further divided into either functional (secretory) or nonfunctional (nonsecretory) categories depending upon whether a hormonally recognizable clinical syndrome is present. Functional NETs include those that over-secrete amines and/or peptides such as well-differentiated and intermediate-grade tumors such as those arising from the lung, thymus, stomach, pancreas, ovary, small and large bowel. Specific functional tumor types include carcinoid syndrome, insulinomas, gastrinomas (Zollinger-Ellison syndrome), VIPomas (Verner-Morrison syndrome, WDHA [Watery Diarrhea, Hypokalemia and Achlorhydria syndrome], pancreatic cholera), glucagonomas ("Sweet" syndrome), somatostatinomas and other more rare syndromes such as ghrelinomas, GRFomas (growth hormone releasing factor), serotonin-producing islet cell (pancreatic carcinoid), LHomas (lutinizing hormone), reninomas, IGF-2-omas, erythropoeitinomas, neurotensinomas, MSHomas (melanocyte stimulating hormone), parathyrinomas, PTH-rPomas (parathyroid hormone-related peptide), calcitoninomas and ACTHomas.

Nonfunctional NETs (NF-NETs) can arise from the lung, thymus, skin, stomach, pancreas, appendix, ovary, kidney, small and large intestine. Those with a pancreatic origin are sometimes referred to as PPomas as pancreatic polypeptide levels are frequently elevated but no clinical syndrome is present. These NF-NETs have similar histologic characteristics to the functional subgroup. Elevated hormonal levels of chromogranin A, pancreastatin, human chorionic gonadotropin subunits (HCG), neuron specific enolase (NSE), neurokinin A (NKA or Substance K) and pancreatic polypeptide may be present and assist with NET diagnosis.

NET patients and particularly NF-NETs are commonly misdiagnosed and/or diagnosed late in their disease course as their symptoms are nonspecific, insidious in onset and present for years. Presenting symptoms include those attributable to a specific clinical syndrome and/or tumor-related symptoms such as involuntary weight loss, pain, jaundice, night sweats, nausea or emesis. All functional / syndromic NETs patients are symptomatic but not all symptomatic NETs patients are functional /syndromic.

NETs of unknown primary site are well described and are usually of intestinal origin. Patients may present with liver only disease and no apparent primary detectable on scans or endoscopy. As intestinal primary sites may be small and submucosal, regional lymphatic metastases are suggestive of a 'midgut' primary. These tumors may be functional or nonfunctional.

A paraneoplastic complication of intestinal NETs is predominantly right-sided heart failure or "carcinoid heart disease". Tricuspid valvular insufficiency and pulmonic stenosis are the hallmarks of this complication. NETs are in the differential diagnosis of anyone presenting with right-sided congestive heart failure. The scarring of the valves and the retraction of the papillary muscles is related to prolonged exposure to high serotonin levels and other possible mediators such as TGF-β and FGF. The incidence of carcinoid heart disease has dramatically decreased from 45-50% in the pre-somatostatin era to about 5% presently. This correlates with the overall improvement in survival of patients with metastatic disease as the presence of carcinoid heart disease predicts a survival of less than 3 years.

Imaging Studies

CT (Figure 1) and MRI offer sensitive detection and tumors are best visualized using either a multiphasic contrasted CT or MRI as these tumors are highly vascular. Endoscopic ultrasound can be useful in the evaluation of esophageal, gastric, insulinomas, small pancreatic tumors and rectal primaries. Nuclear medicine studies such as the OctreoScan (Figure 2) visualizes approximately 90% of well-differentiated NETs through the expression of the somatostatin receptor subtype 2 (sst2). Meta-iodobenzylguanidine (MIBG) scanning detects the presence of the catecholamine transporter protein that is present in nearly all pheochromocytomas but only about two-thirds of intestinal well-differentiated NETs and one-third of bronchial and pancreatic NETs. Scanning with I-123 MIBG has therapeutic implications as those tumors that exhibit significant uptake, therapy with I-131 MIBG is an option.

Tumor histology predicts the likelihood of which nuclear medicine scans will be useful. Well-differentiated tumors generally exhibit sst2 and the In-111 pentetreotide scan would be the initial nuclear medicine scan to order. High-grade histology predicts FDG-PET positivity with SUVs > 2.5. Intermediate-grade tumors may exhibit both In-111 pentetreotide and FDG-PET positivity with the latter showing lower SUVs than in the high-grade tumors. PET scanning using Ga-68 linked to somatostatin analogs improves sensitivity when compared to traditional somatostatin receptor scintigraphy but is considered investigational in the U.S.

Treatment

As NETs comprise a heterogeneous group of neoplasms, definitive therapy is dependent upon primary site, tumor grade and stage. Surgical extirpation for localized NET disease is the only curative modality. For unresectable low- and intermediate-grade tumors, debulking is commonly performed to not only improve symptoms and tumor control but to decrease disease-related complications such as pancreatic insufficiency from chronic pancreatic duct obstruction and intestinal ischemia arising from obstruction of mesenteric vessels by lymphatic metastases.

Local hepatic therapies (hepatic embolization or chemoembolization, radioembolization, radiofrequency ablation) may be combined with systemic therapy on an individual basis. For patients with 30-60% liver involvement by tumor, hepatic debulking prior to progression is suggested. Prescribing local therapy for patients with > 60% liver involvement is based on individual factors and comorbidities.

Medical therapy for advanced NETs is considered palliative with symptom and tumor control comprising the major goals. Anti-hormonal (somatostatin analogs), cytotoxic and targeted agents are options based on the tumor's proliferative index. Systemic first-line therapies for well-differentiated tumor with hormonal syndrome include the somatostatin analogs (octreotide, lanreotide); Intermediate-grade tumor without hormonal syndrome, cytotoxic agents (temozolamide, 5-FU, streptozocin, and doxorubicin) are prescribed. For intermediate-grade tumors with hormonal syndrome, therapy with somatostatin analog combined with cytotoxic agents (temozolamide, 5-FU, streptozocin and doxorubicin) may be used. For high grade tumors, cytotoxic agents in combination such as etoposide and cisplatin are commonly used.

Survival Outcomes

The overall 5-year survival for localized and metastatic midgut well-differentiated NEC ranges between 52-77%. The size of the primary and depth of invasion predicts metastases to local-regional and metastatic sites. Approximately 75% of primary tumors exceeding 1 cm in size metastasize vs 31.5% for those between 6 to10 mm and only 16% for those tumors less than 6 mm. The overall 5-year survival for non-metastatic midgut well-differentiated NEC is approximately 90%. Poorly differentiated NEC, regardless of primary site, has the shortest survivals with localized disease patients having a median 18 month survival while metastatic disease patients' survival would be approximately 12 months.

Figure 1. CT coronal image showing (see arrows) a large primary pancreatic NET mass and liver metastases.

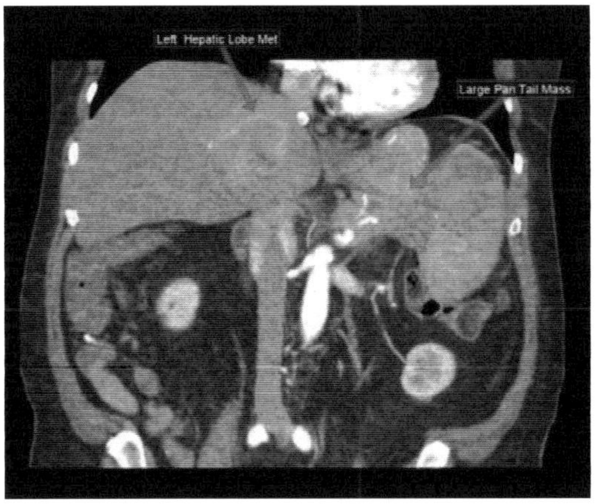

Figure 2. OctreoScan with 24 hour planar anterior and posterior images of another pancreatic NET patient with a somatostatin receptor type 2 (sst2) positive primary and liver metastases.

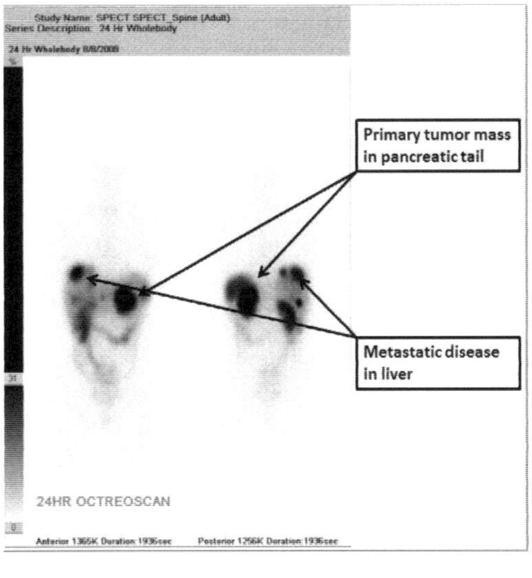

REFERENCES

1. Ardill JE, et al. Ann Clin Biochem 2008;45(Pt 6):539-559. 2. Arnold, et al. Best Prac & Res Clinl Gastro 2005;19(4):491-505. 3. Bomanji JB, et al. Clin Onc (Royal College of Radiologists) 2003 Jun 1915;193-198. 4. Boudreaux JP, et al. Ann Surg 2005 Jun 241;839-845. 5. Kelekis NL, et al. European Radiology 1997 2007;875-886. 6. Kennedy AS, et al. Am J Clin Onc 2008 Jun 1931;271-279. 7. Klimstra DS, et al. Am J Surg Path 2010;34(3):300-313. 8. Kvols LK, et al. NEJM 1986;315(11):663-666. 9. Moertel C, et al. Cancer 1961 1914;901-912. 10. Norton JA, et al. Surgery 2003;134(6):1057-1063. 11. O'Connor DT, et al. NEJM 1986;314(18):1145-1151. 12. Pelage JP, et al. Abdominal Imaging 1999;24(3):240-245. 13. Rubin J, et al. JCO 1999 Feb 1917;600-606. 14. Ruszniewski P, et al. Digestion 2000;62 Suppl 1:79-83. 15. Yao JC, et al. JCO 2008 Jun 20 1926;3063-3072.

NUCLEAR IMAGING IN VETERINARY ENDOCRINOLOGY

Gregory B. Daniel, DVM, MS, DACVR
Blacksburg, VA

INTRODUCTION

Scintigraphic studies are used in veterinary medicine to image thyroid, parathyroid, adrenal and pancreatic diseases. By far, the most common of these studies is thyroid scintigraphy. Thyroid scintigraphy has been described in a variety of domestic species including feline, canine, equine and avian. Feline scintigraphy has been the most extensively studied. Parathyroid scintigraphy showed promise after an initial case report but further work was disappointing in terms of sensitivity and specificity. Adrenal scintigraphy has been reported but is limited to a case report on a pheochromoctyoma. Somatostatin receptor scintigraphy has also been reported to image a variety of tumors such as paraganglioma, and insulinomas. This presentation will review these applications of scintigraphy to image endocrine disease in dogs and cats.

THYROID SCINTIGRAPHY

Thyroid scintigraphy is used to evaluate the functional status of the feline thyroid gland (hyperthyroidism, euthyroidism, or hypothyroidism). It is also used to differentiate between unilateral and bilateral thyroid disease which especially relevant for surgical treatment of feline hyperthyroidism. It is used to detect and localize hyperfunctioning ectopic thyroidal tissue or thyroid metastatic lesions.

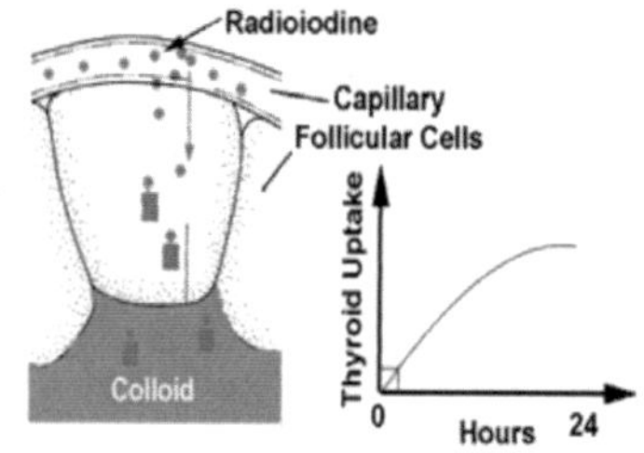

Radionuclides used for Imaging

Radioiodine

Iodine is required for synthesis of thyroid hormone. All isotopes of iodine are metabolized identically; therefore, the radioisotopes (^{131}I and ^{123}I) can be substituted for stable iodine (^{127}I) and used for scintigraphic imaging. The first step of iodide uptake by the thyroid follicular cells is the trapping by the sodium-iodide symporter (NIS). Oxidised radioiodine is then combined with the amino acid tyrosine to form a radioactive thyroglobulin molecule in a process called organification. The thyroid gland can store this iodine-laden hormone in the thyroid follicles in the form of thyroglobulin or released it into the blood.

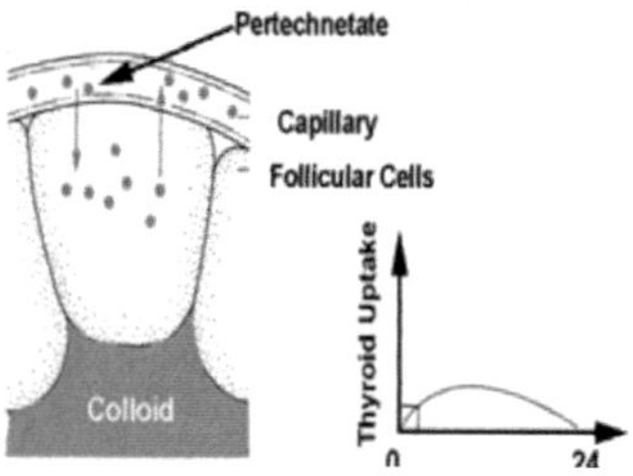

Uptake of iodine is regulated by TSH, which influences the rate of iodine trapping primarily through NIS expression. Specific effects of TSH on the thyroid gland include increased proteolysis of thyroglobulin, increased iodination of tyrosine, increased size and secretory activity of thyroid cells, increased number of thyroid cells and increased activity of the iodine pump. The salivary glands, gastric mucosa, and choroid plexus of the brain also trap circulating radioiodine.

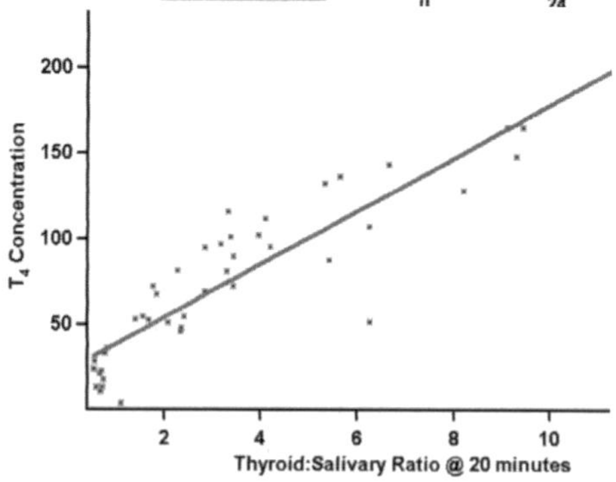

Sodium ^{99m}Tc-Pertechnetate

^{99m}Technetium is a transitional metal that imitates the actions of the halogens. The biologically important halogens are iodide and chloride. ^{99m}Technetium in the form of pertechnetate ($^{99m}TcO_4^-$) is trapped and concentrated by the follicular cells of the thyroid gland. Pertechnetate does not undergo organification therefore it is not incorporated into active hormone. The thyroid will initially take up pertechnetate just as radioiodine but radioactivity peaks earlier, then pertechnetate is cleared from the thyroid.

Normal Feline Thyroid Scan

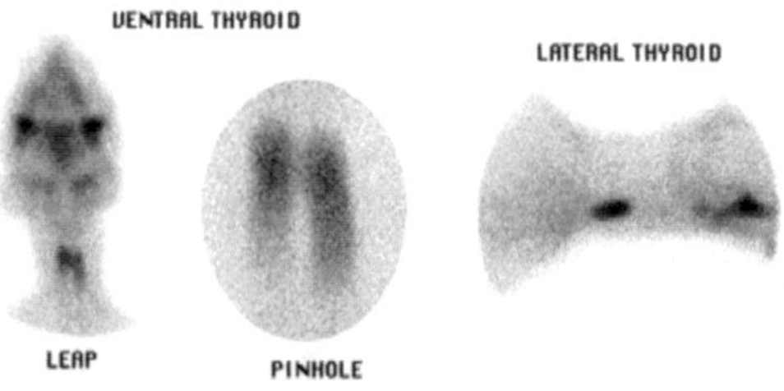

Normal Canine Thyroid Scan

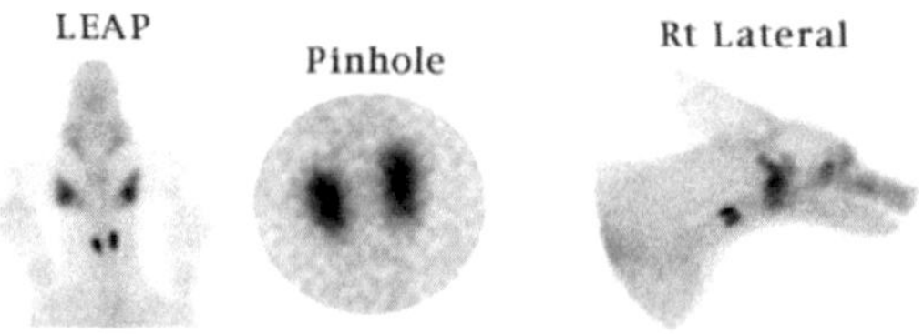

Principles of Interpretation

Thyroid uptake can be used to determine the metabolic function of the thyroid in cats. Thyroid uptake has been positively correlated to thyroid hormone levels (r=0.93)[1].

Thyroid uptake can be quantified by determining % dose uptake, thyroid:salivary ratio or thyroid:background ratio. Normal values for thyroid:salivary ratio and % dose uptake are shown below in Tables 1 & 2.

Table 1. Normal Feline

Reference	**Thyroid: Salivary Ratio @ 20 min Mean ± SD**	**Range**	**% Dose Uptake**
[2]	0.87 ± 0.13	0.60 - 1.03	NR
[3]	0.81 ± 0.05	0.73 - 0.90	0.21 ± 0.06
[4]	Mean not given	0.51 - 1.65	0.38 ± 0.09
[5]	0.94 ± 0.12	0.74 - 1.06	0.38 ± 0.09

Table 1. Normal Canine

Reference	Thyroid: Salivary Ratio @ 20 min Mean ± SD	Range	% Dose Uptake
[6]	1.12 ± 0.13	0.9 – 2.2	0.15 ± 0.15

Dogs with thyroid carcinoma may exhibit either increased, decreased or a mixed pattern of radionuclide uptake but the tumor usually does not produce functional thyroid hormone. Unlike the cat, increase uptake in the canine thyroid does not correlate with T_4. Since most dogs with thyroid carcinoma have nonfunctional lesions in terms of hormone production, the contralateral thyroid lobe radionuclide uptake will not be suppressed. In the minority of cases in which the canine thyroid tumor is producing excess hormone, there will be markedly reduced or absent radionuclide uptake by contralateral thyroid lobe due to the TSH suppression through the pituitary axis.

PARATHYROID SCINTIGRAPHY

Parathyroid scintigraphy is used for the pre-operative localization of hyperfunctioning parathyroid tissue (parathyroid adenomas and parathyroid hyperplasia) in patients with primary hyperparathyroidism.

Radionuclides used for Imaging

Because of the small size of the parathyroid glands, diagnosis is difficult with most imaging modalities. Hyperfunctioning parathyroid tissue can be imagined using nuclear medicine techniques.. Parathyroid scintigraphy has been attempted using a variety of radionuclides or radiopharmaceuticals $^{99m}TcO_4$/^{201}Thallum subtraction, ^{123}I/^{201}Thallum subtraction, ^{99m}Tc-sestamibi and ^{99m}Tc-tetrofosmin. ^{99m}Tc-sestamibi rapidly accumulates in thyroid and hyperfunctional parathyroid tissue. ^{99m}Tc-sestamibi is the most commonly used in people and its use has been reported in dogs.

^{99m}Tc-sestamibi

^{99m}Tc-sestambi is a lipophilic cationic molecule that accumulates in metabolically active cells with high mitochondrial concentration. The net mitochondrial concentration is a function of both plasma and mitochondrial membrane potentials. ^{99m}Tc-sestambi will concentrate within myocardium, liver, kidney, muscle, and various malignant tumors. ^{99m}Tc-sestamibi will initially concentrate in normal thyroid tissue and in hyperfunctioning parathyroid tissue such as parathyroid adenoma and hyperplastic parathyroid tissue. The mechanism by which ^{99m}Tc-sestamibi accumulates in hyperfunctional parathyroid tissue is not completely understood. Parathyroid oxyphil cells contain abundant mitochondria, prompting speculation that these cells are responsible for uptake and retention of ^{99m}Tc-sestamibi. Lesion detectability correlates with parathyroid gland size, with larger parathyroid glands being more readily detected than smaller lesions.

Principles of Interpretation

^{99m}Tc-sestamibi clears from normally functioning thyroid tissue, but persists in hyperfunctioning parathyroid tissue. Using this principle of differential washout, images are acquired 10-15 minutes after administration of ^{99m}Tc-sestamibi and repeated 2-3 hours later. Hyperfunctioning parathyroid tissue is seen as focal areas of increased radionuclide uptake during the delayed phase. The initial phase images helps with anatomic localization of the abnormal parathyroid tissue.

In people, this technique has good sensitivity (80-90%) for detection of parathyroid adenomas but the sensitivity is lower for detection of hyperplastic glands. In some dogs, thyroid uptake of ^{99m}Tc-sestamibi is evident by 20 minutes after intravenous administration and mostly gone by 4 hours. However, in a large percentage of dogs, the thyroid and parathyroid glands are never or only faintly visualized.

^{99m}Tc-sestamibi parathyroid scintigraphy has been reported to have localized a parathyroid adenoma in a dog[7], but overall this technique appears to have poor sensitivity and specificity. In a later study of 15 hypercalcemic dogs, sensitivity of dual-phase ^{99m}Tc-sestamibi scintigraphy for detecting hyperfunctioning parathyroid tissue was 11% and specificity was 50%, with an overall accuracy was 27%[8]. The sensitivity for detecting parathyroid adenomas was 25%, and for hyperplastic glands was 0%.

ADRENAL SCINTIGRAPHY

Adrenal scintigraphy is used to evaluate adrenal masses or tumors of the adrenal medulla. It is also an effective method for whole body imaging for localization of extra-adrenal pheochromocytoma

Radionuclides used for Imaging

The University of Michigan first reported the detection of a pheochromocytoma using ^{131}I-metaiodobenzyl guanidine (MIBG) in early 1980's. Initially, MIBG availability was limited to research applications and facilities with radiochemistry labs. MIBG is now commercially available with ^{123}I radiolabel from GE Healthcare (AdreView™). MIBG has a molecular structure similar to norepinephrine and is taken up and stored in chromaffin cells and therefore useful for imaging normal and abnormal sympathetic adrenergic tissue. There is one case report of using ^{123}I-MIBG to image an adrenal tumor in a dog.[9] Tumors arising from tissues of neural crest origin that can be imaged with MIBG including neuroblastomas, carcinoid, medullary thyroid carcinoma and paraganglionomas.

NEUROPEPTIDE SCINTIGRAPHY

Somatostatin (Octreotide) receptor scintigraphy is used to image of tumors arising from neuroendocrine cells. Neuroendocrine tumors that can be imaged by this technique include: insulinomas[10-12], paraganglionomas[13], medullary thyroid carcinoma, pheochromoctomas, neuroblastoma, and functional pituitary adenomas or adenocarcinomas.

Radionuclides used for Imaging

Neuroendocrine cells secrete somatostatin. There is a high density of somatostatin receptors in a variety of neuroendocrine and non neuroendocrine tumors. Somatostatin has a very short biological half-life (less than two minutes), so radiolabeling the somatostatin molecule is impractical. Octreotide is an eight-chain amino acid peptide that was shown to be a somatostatin analog. Octreoscan® (^{111}In-DTPA-D-Phe1-octreotide) is commercially available for imaging tumors that express the somatostatin receptor. SPECT imaging would be preferred to planar imaging to increase sensitivity of detection.

REFERENCES

1. Daniel, G.B., et al. Quantitative thyroid scintigraphy as a predictor of serum thyroxin concentration in normal and hyperthyroid cats. *Veterinary Radiology & Ultrasound*, 2002. 43(4): p. 374-382. 2. Beck, K.A., et al. The normal feline thyroid - technetium pertechnetate imaging and determination of thyroid to salivary-gland radioactivity ratios in 10 normal cats. *Veterinary Radiology*, 1985. 26(1): p. 35-38. 3. Nieckarz, J.A. et al. , The effect of methimazole on thyroid uptake of pertechnetate and radioiodine in normal cats. *Veterinary Radiology & Ultrasound*, 2001. 42(5): p. 448-457. 4. Henrikson, T.D., et al. . Thyroid to salivary ratios determined by technetium-99(m) pertechnetate imaging in thirty-two euthyroid cats. *Veterinary Radiology & Ultrasound*, 2005. 46(6): p. 521-523. 5. Lee, W.R., et al. The effects of iohexol administration on technetium thyroid scintigraphy in normal cats. *Veterinary Radiology & Ultrasound*, 2010. 51(2): p. 182-185. 6. Adams, W.H., et al. Quantitative Tc-99m-pertechnetate thyroid scintigraphy in normal beagles. *Veterinary Radiology & Ultrasound*, 1997. 38(4): p. 323-328. 7. Matwichuk, C.L., et al., Use of technetium Tc 99m sestamibi for detection of a parathyroid adenoma in a dog with primary hyperparathyroidism. *Journal American Veterinary Medical Association*, 1996. 209(10): p. 1733-&. 8. Matwichuk, C.L., et al. Double-phase parathyroid scintigraphy in dogs using technetium-99m-sestamibi. *Veterinary Radiology & Ultrasound*, 2000. 41(5): p. 461-469. 9. Berry, C.R., et al., Use Of I-123 Metaiodobenzylguanidine Scintigraphy For The Diagnosis Of A Pheochromocytoma In A Dog. *Veterinary Radiology & Ultrasound*, 1993. 34(1): p. 52-55. 10. Garden, O.A., et al., Somatostatin receptor imaging in vivo by planar scintigraphy facilitates the diagnosis of canine insulinomas. *Journal Of Veterinary Internal Medicine*, 2005. 19(2): p. 168-176. 11. Lester, N.V., et al., Scintigraphic diagnosis of insulinoma in a dog. *Veterinary Radiology & Ultrasound*, 1999. 40(2): p. 174-178. 12. Robben, J.H., et al. Comparison of ultrasonography, computed tomography, and single-photon emission computed tomography for the detection and localization of canine insulinoma. *Journal Of Veterinary Internal Medicine*, 2005. 19(1): p. 15-22. 13. Buchanan, J.W., et al., Left atrial paraganglioma in a dog: Echocardiography, surgery, and scintigraphy. *Journal Of Veterinary Internal Medicine*, 1998. 12(2): p. 109-115.

DIAGNOSIS AND TREATMENT OF CANINE PHEOCHROMOCYTOMA

Claudia Reusch, Prof. Dr., DECVIM
Zurich, Switzerland

INTRODUCTION

The cells of the adrenal medulla are called pheochromocytes or chromaffin cells. They can be regarded as modified postganglionic sympathetic neurons lacking axons. Innervation is by preganglionic fibres of the sympathetic nervous system which induces the release of catecholamines into the bloodstream. Some extra-adrenal chromaffin tissue is also present adjacent to the aorta, in the carotid bodies, viscera and within sympathetic ganglia. Pheochromocytomas (PHEO) are rare catecholamine-producing neuroendocrine tumors which arise from the adrenal medulla in the vast majority of cases. Occasionally tumors derive from extra-adrenal chromaffin tissue, and are referred to as paragangliomas.

CATECHOLAMINE SYNTHESIS, STORAGE, SECRETION AND METABOLISM

Catecholamines include epinephrine, norepinephrine and dopamine, which are synthesized from tyrosine by a process of hydroxylation and decarboxylation. Catecholamines are stored within the chromaffin cells in cytoplasmatic vesicles, together with various other substances, e.g. chromogranin-A, somatostatin, enkephalins, synapophysin, vasoactive intestinal polypeptide, ACTH, CRH.

All of the epinephrine in the circulation derives from the adrenal medulla, whereas circulating norepinephrine mostly comes from postganglionic sympathetic neurons and only to a small extent from the adrenal medulla. Secretion of catecholamines is part of the activation of the sympathetic nervous system. Examples for stimuli are exercise, perceived danger, surgery, hypovolemia, hypotension and hypoglycemia. Plasma half life of catecholamines is very short (1 – 3 minutes). They are metabolized into the inactive compounds normetanephrine, metanephrine and vanillylmandelic acid. They may also be inactivated in the liver by conjugation with sulfate or glucuronide. Excretion is via urine.

In the normal adrenal medulla of dogs, cats and humans, 60 – 80% of the catecholamine content is epinephrine, which is therefore the major catecholamine. In human PHEO, norepinephrine is the predominant catecholamine secreted and in some tumors it may be the only catecholamine produced. On rare occasions tumors may secrete only epinephrine. The secretory patterns of canine and feline PHEO have not yet been thoroughly investigated, according to preliminary results the situation in dogs may be similar to humans.

CATECHOLAMINE ACTIONS

Catecholamines bind to two types of receptors: α- and β-adrenergic receptors, which are further subdivided into subtypes α_1, α_2, β_1, β_2 and β_3. The α_1 receptor subtype mediates vascular and smooth muscle contraction and stimulation of the receptor causes vasoconstriction and increased blood pressure. Activation of α_2 receptors inhibits norepinephrine release, suppresses sympathetic outflow and decreases blood pressure. Stimulation of β_1 receptors results in positive inotropic and chronotropic effects on the heart. β_2 stimulation causes bronchodilatation and vasodilatation, and β_3 receptors regulate energy expenditure and lipolysis.

Norepinephrine and epinephrine both interact with α- and β-receptors. However, compared with norepinephrine, epinephrine has a higher affinity for adrenergic receptors in general and β_2 receptors in particular and is more potent.

PREVALENCE AND BIOLOGICAL BEHAVIOUR

PHEO is considered rare in dogs and even more uncommon in cats. However, due to the diagnostic difficulties quite a few may be overlooked and therefore prevalence may be higher as generally assumed. Most tumors are unilateral, occasionally both adrenal glands are affected. PHEO may coexist with glucocorticoid producing adrenocortical tumors, ACTH-producing pituitary tumors or other endocrine tumors and as such be part of the multiple endocrine neoplasia syndrome (MEN). PHEO are potentially malignant in dogs and local invasion, distant metastasis, luminal compression of vessels and tumor rupture is possible. Tumor size is extremely variable and ranges between a few millitmeters and more than 10 cm in diameter at the time of diagnosis.

SIGNALEMENT AND CLINICAL PRESENTATION

The disease may occur at any age, however it is most commonly seen in older dogs ($\geq$ 8 years). There is no apparent sex and breed predisposition.

The clinical presentation of animals with PHEO is highly variable and ranges from complete absence of signs to dramatic and life-threatening signs; the latter is usually the result of a hypertensive crisis. Clinical signs depend on the type of catecholamine produced by the tumor and the amount and frequency of catecholamine release into the circulation. Clinical signs are usually episodic; they may occur several times per day or may only recur after weeks or months. The most common owner complaints are weakness and episodic collapse. Other signs are: anorexia, weight loss, lethargy, anxiety, tachypnoe, panting, tachycardia, arrhythmias, acute blindness due to retina detachment and/or ocular hemorrhage, pacing, muscle tremor, seizures, pu/pd, vomiting and diarrhoea. Large tumors may cause abdominal distension, ascites and hind-limb edemy. Intra-abdominal or retroperitoneal hemorrhage may occur.

Hypertension is one of the hallmarks of the disease. However, hypertension is not pathognomonic for PHEO and is also frequently found in hypercortisolism.

Ninety to 95% of human patients with PHEO have hypertension. Blood pressure patterns vary: about 50% of patients have sustained hypertension with marked fluctuations and severe episodic blood pressure peaks, approximately 25% are normotensive between hypertensive episodes and the remaining patients have stable and sustained hypertension. Hypertension varies from mild to severe and the systolic rise during a hypertensive paroxysm may be up to 200 mm Hg. Interestingly, episodic hypotension or even syncope may occur in patients with tumors that secrete mainly epinephrine.

Current information on blood pressure values in canine PHEO is limited to less than 50 dogs, approximately half of which had hypertension. Similar to the situation in human medicine, the increase in blood pressure in dogs may range from mild to severe; the maximum systolic pressure reported was 325 mmHg.[1]

CLINICAL PATHOLOGY

There are no consistent abnormalities on complete blood count, serum biochemistry and urinalysis. Various findings are possible, none of which would raise suspicion of PHEO: mild nonregenerative anemia, stress leukogram, mild thrombocytopenia or thrombocytosis, increased alkaline phosphatase, increased alanine aminotransferase, azotemia, hypoalbuminemia, mild hyperglycemia, hypercholesterolemia, proteinuria, hypo- or isosthenuria. Laboratory results may also be completely normal.

DIAGNOSTIC IMAGING

Since clinical signs are often vague and unspecific, PHEO is often only considered after an adrenal mass has been detected by ultrasonography. PHEO range in size between a few millimeters to more than 10 cm in size. Masses may be hypoechoic, isoechoic or hyperechoic compared to renal cortex or may have mixed echogenicity. They may also contain anechoic, non-far enhancing areas which represent foci of necrosis and hemorrhage. Mineralisation seems to be extremely rare in PHEO, in contrast to cortisol-producing masses. Tumors are usually unilateral, in these cases the non-involved adrenal gland has normal shape and size. However, exceptions may occur in cases of bilateral PHEO, co-existence with other types of tumors (e.g. cortisol-producing) or with PDH. Ultrasonography is helpful to detect invasion into surrounding vessels. According to currently available studies no pattern of echogenicity or architecture is specific for PHEO.[2]

In humans CT and MRI have a sensitivity of approximately 90% and a specificity of 70-80% for the diagnosis of PHEO being ≥ 5 mm in diameter. Because of the limited specificity of CT and MRI, the use of advanced functional imaging techniques is recommended. The latter include scinthigraphy with (123)I-Metaiodobenzylguanidine (MIBG) and positron emission tomography (PET) using agents such as (18)F-Fluorodopamine (F-FDA) and (18)F-3,4,-dihydroxyphenylalanin (DOPA).[3] Szinthigraphy and PET take advantage that the radiopharmaceuticals target the catecholamine synthesis, storage and secretion pathways of chromaffin cells. The use of these techniques have so far only been described in a very small number of dogs with PHEO and no information is available with regard to sensitivity and specificity.

MEASUREMENT OF CATECHOLAMINES AND METANEPHRINES

In humans diagnosis of PHEO is mainly based on biochemical detection of excessive amounts of catecholamines (epinephrine, norepinephrine) and their metabolites (metanephrine, normetanephrine). There is currently some controversy on whether blood or urine testing should be used preferentially. The parameters can be measured either by HPLC with electrochemical or mass spectrometric detection and by immunoassays such as RIA and ELISA. However, there is indication that immunoassays may underestimate levels compared with HPLC-based methods, potentially leading to diagnostic error.[4]

Measurement of urinary catecholamines and metanephrines (by HPLC with electrochemical detection) was recently introduced into veterinary medicine. So far, we have performed three consecutive studies. Since 24 h urine sampling is impractical in dogs the parameters were measured in spot urine samples and concentration was expressed as ratio to urine creatinine concentration. In the first study the influence of stress was investigated by taking urine samples at several occasions (prior to, during and after a hospital visit). Samples taken during the hospital visit revealed higher catecholamine/metanephrine levels in healthy client-owned dogs but not in staff-owned dogs familiar with the environment, pointing to the influence of stress.[5]

In the second study urinary catecholamine/metanephrine to creatinine ratios were measured in dogs with histologically confirmed PHEO. Samples of all dogs with PHEO collected in the hospital revealed higher ratios than the control groups, suggesting that sampling during initial evaluation might nevertheless be practicable. This was considered relevant as diagnosis would not be delayed due to the necessity of sending animals home with the advice to collect a urine sample. Additionally, urine sampling and processing are subject to certain conditions such as acidification and light protection, which may be more reliably done by a vet in the hospital than by an owner at home. Dogs with PHEO had significantly higher urinary epinephrine, norepinephrine and normetanephrine to creatinine ratios than controls, the least overlap was seen with regard to the urinary normetanephrine to creatinine ratio. The normetanephrine to creatinine ratio therefore was considered to be the most promising parameter.[6]

The finding is similar to the situation in human medicine and may be explained as follows. The metabolism of epinephrine and norepinephrine into the inactive compounds metanephrine and normetanephrine occurs in the same cells. Since sympathetic nerves do not contain the responsible enzyme COMT metanephrine and normetanephrine are derived exclusively from non-neuronal sources i.e. from the adrenal medulla. PHEO also express COMT and therefore produce metanephrine and normetanephrine, which are secreted directly into the circulation. The higher sensitivity of metanephrines over catecholamines is attributable to their continuous production and secretion, whereas catecholamines are often released intermittently. Because norepinephrine is usually the most abundant catecholamine in PHEO its metabolite normetanephrine has the highest concentration and sensitivity.

Based on our results we propose that the situation is similar in canine PHEO. In the third study urinary catecholamine/metanephrine to creatinine ratios were measured in dogs with hyperadrenocorticism (HAC) and compared to the results in dogs with PHEO. HAC is one of the most important differential diagnosis for PHEO. Both diseases might have similar clinical signs such as weakness, tachypnea, panting, pu/pd, hypertension-associated bleeding as well as similar abnormalities detected by ultrasonography. Urinary epinephrine, norepinephrine and metanephrine to creatinine ratios did not differ between dogs with HAC and dogs with PHEO, whereas the norepinephrine to creatinine ratio was significantly higher. Using a cut-off of 4 times the highest normetanephrine to creatinine ratio measured in healthy dogs there was no overlap between HAC and PHEO.[7] In human medicine a similar approach is taken by using cut-off between 2 and 4 times the upper limit of normal.

TREATMENT

Adrenalectomy is the treatment of choice; however, there is a high risk of hypertensive and hypotensive crises, cardiac arrhythmias and hemorrhage. Surgery should only be performed by a team of experienced surgeons and an anesthetist. In humans, α-adrenergic blockade (usually phenoxybenzamine) is used for at least one week prior to adrenalectomy. The aim is to reverse vasoconstriction and hypovolemia and control fluctuations of blood pressure and heart rate during anesthesia. The recommended starting dose of phenoxybenzamine in dogs is 0.25 mg/kg BID, which is gradually increased every few days until signs of hypotension or adverse drug reaction occur or the maximum dose of 2.5 mg/kg BID is attained.

A recent study showed that although there was no difference in intra- and postoperative hypo- and hypertensive episodes and arrhythmias, dogs treated with phenoxybenzamine had a significantly decreased mortality compared with untreated dogs.[8] In cases in which adrenalectomy is not an option, phenoxybenzamine should be used on a long-term basis to control blood pressure. Additional β-adrenergic therapy may be necessary in patients with severe tachycardia, but it should not be given without prior α-blockade in order to avoid severe hypertension.

REFERENCES

[1]Reusch CE, et.al. **Vet Clin North Am Small Anim Pract** 2010;40(2):335. [2]Besso JG, et.al. **Vet Radiol Ultrasound** 1997;38(6):448. [3]Timmers HJ, et.al. **J Clin Endocrinol Metab** 2009;94(12):4757. [4]Singer J, et.al. **Nat Rev Endocrinol** 2011;7:749. [5]Kook PH, et.al. **J Vet Intern Med** 2007;21:388. [6]Kook PH, et.al. **Vet Rec** 2010;166:169. [7]Quante S, et.al. **J Vet Intern Med** 2010;24:1093. [8]Herrera MA, et.al. **J Vet Intern Med** 2008;22(6):1333.

CREATININE: REASSESSMENT OF AN OLD BIOMARKER

David J. Polzin, DVM, PhD, DACVIM
St. Paul, Minnesota

It is broadly accepted that measurement of glomerular filtration rate (GFR) is the most useful tool currently available for recognizing and staging kidney disease. However, measurement of GFR using clearance techniques, whether traditional clearances requiring measured, timed urine collections or single-injection disappearance methods, is generally viewed as too cumbersome for screening and staging in practice settings. Despite an extensive search for biomarkers to identify the presence and severity of kidney disease, serum creatinine concentration (SCr) remains the standard surrogate for GFR. In addition to recognition of kidney disease, classification schemes designed to guide diagnostics, treatment and prognosis of chronic kidney disease have been developed using SCr as a staging tool. The IRIS CKD staging system is used for this purpose in dogs and cats while the K/DOQI clinical practice guidelines for chronic kidney disease have been developed for use in humans with chronic kidney disease. While serum creatinine values are used for the IRIS CKD Staging system, the K/DOQI clinical practice guidelines for humans uses an estimated GFR value (eGFR) for classification purposes. The "eGFR" is based SCr that is adjusted to take into account gender, race, age, and other clinical parameters. Similar corrections are as yet not available for dogs and cats.

Despite its near universal application, SCr has several limitations to its use as a measure of kidney function. It is a relatively insensitive estimate of glomerular filtration rate until a substantial reduction in overall kidney function has already occurred. It requires about a 75% reduction in GFR before serum creatinine values consistently exceed the upper limit of normal for many laboratories. The insensitivity of serum creatinine results from at least two important factors. The first is the innate relationship between GFR and serum creatinine and the second is that factors other than intrinsic renal disease can alter SCr, including its endogenous production in muscle. In dogs, size and sex may influence the true normal range.

The relationship between serum creatinine and GFR is such that every time the GFR declines by half, the SCr doubles. For example, if a dog has a baseline creatinine of 0.5 mg/dl and its GFR declines by 50%, the serum creatinine only increases to 1.0 mg/dl, still well within the normal range. If a further reduction of 50% in GFR occurs (to 25% of the original GFR), the creatinine will rise to 2.0 mg/dl where it might begin to be recognized as elevated. A further decline of 50% in GFR (to 12.5% of normal) and the creatinine increases to 4.0 mg/dl. As can be seen by this example, the slope of the rise in SCr becomes much more acute after a substantial loss of GFR has already occurred, but the slope is flat over most of the range of GFR. It follows that the wider the normal range is for SCr, the less sensitive it is for intrinsic kidney disease. However, if the range becomes too narrow, serum creatinine will lose specificity because other factors may influence the creatinine value including body muscle mass, breed of dog, and other factors. This conundrum is intrinsic to the nature of serum creatinine and limits its utility in diagnosis of early CKD.

The search for biomarkers of kidney disease is largely driven by the desire to detect kidney disease at an earlier stage. Early recognition of either CKD or acute kidney injury (AKI) is deemed important because of the perceived potential value of early detection leading to earlier therapeutic and renoprotective interventions that will prevent or delay progression to the more advanced stages of kidney disease. While SCrs do rise as GFR declines, even in its earliest stages, the decline is small and typically "hidden" in the normal range and minor variations in determination of SCr.

Several studies have indicated that serum creatinine values are reasonably consistent from reference laboratory to reference laboratory. However, despite the consistency between labs, interpretation of normal ranges may interfere with detection of kidney disease. As a general rule, the patient's SCr is compared to the upper limit of the reference range provided by the evaluating reference laboratory. While the specific values obtained at different laboratories are typically consistent with one another, interpretation of the actual value may vary greatly from lab to lab because of differences in published reference ranges. Standards for determination of reference ranges are lacking in most veterinary labs and this presents a serious limitation on the use of SCr for the purpose of detecting early (mild) kidney disease or injury.

While a single measure of SCr may correlate poorly with GFR, serial evaluation of SCrs within the same individual can be quite sensitive in detecting progressive changes in GFR, so long as the actual measurement of serum creatinine is highly reliable and repeatable. As a consequence, serially evaluating SCrs at regular intervals in patients suspected as being at risk for CKD may markedly increase the sensitivity of this test. In general, this approach to utilizing the patient's own "normal" SCr as a standard against which subsequent determinations of serum creatinine values can be compared will require: 1) confirmation of the reliability and repeatability of serum creatinine concretions provided by the reference laboratory (using a measure of serum creatinine measurement that minimizes/eliminates the measurement of non-creatinine chromagens), 2) collection of serum samples for analysis only when the patient has been fasted and is confirmed to be well hydrated, and 3) concurrent determinations of body weight, body condition score and urine specific gravity. Ideally an initial baseline serum creatinine would be collected when the patient is between 2-3 years of age (full grown and adult weight and muscling) and reconfirmed 4 to 6 months later. Subsequent determinations of serum creatinine (collected under the same conditions and with the same supplemental data) would be compared back to the patient's baseline values considering changes in body weight and body condition scores. Regular follow-up determinations of SCrs would be ideal for detection of early kidney disease. Should an increase in SCr be detected, it should be reconfirmed by follow-up measurements and the patient should be evaluated for evidence of prerenal or postrenal causes for an increase in SCr. Urinalysis results may be helpful in interpreting changes in creatinine; however, early detection of a pattern of rising SCr may precede loss of adequate urine concentrating ability. Further serial determinations of SCr may be necessary to confirm the pattern as indicative of kidney disease.

REFERENCES

National Kidney Foundation: Am J Kidney Dis: 2006:39(2 Suppl 1):S1; IRIS: IRIS Staging system 2006.[http://www.iris-kidney.com]; Peake M and Whiting M: Clin Biochem Rev: 2006:27:173; Ulleberg et al: 2011:53:25; Braun JP et al: 2003:32:163; Boozer L et al: 2002:JVIM 16:354.

NEW BIOMARKERS OF KIDNEY DISEASE: AN UPDATE

Mary B. Nabity, DVM, PhD, DACVP
College Station, TX

INTRODUCTION

The search for better non-invasive biomarkers of kidney disease is an active area of research, and several promising urine and serum biomarkers have been investigated in acute and chronic renal diseases. Although many urinary proteins indicating renal damage and/or dysfunction have been variably studied in dogs with renal disease, only a few have been studied sufficiently to begin to glean a sense of how they might be useful in detecting and monitoring chronic kidney disease (CKD) in patients. This session provides an update on some of these biomarkers with regard to their clinical utility based on several recent studies.

URINARY BIOMARKERS OF TUBULAR DAMAGE/DYSFUNCTION

Proteins of tubular origin can be present in the urine due to release by damaged tubular cells or by impaired absorption of proteins normally present in the glomerular filtrate. These proteins have great potential as early biomarkers of tubulointerstitial (TI) damage. While their sensitivity in acute renal disease has been documented, their ability to identify TI damage earlier than conventional tests (e.g., serum creatinine (sCr)) in dogs with CKD needs further evaluation. In particular, their specificity for TI damage must be verified. While associated structural changes have been observed histologically in dogs with elevated urinary biomarkers,[1,2] some studies call into question the influence of proteinuria due to glomerular damage on these tubular proteins.[2,3] For instance, in dogs with progressive CKD due to X-linked hereditary nephropathy (XLHN), several urinary biomarkers were increased as compared with normal dogs well before an increase in sCr, but typically only after development of proteinuria.[2] Factors such as in vitro hematuria, pyuria, and bacteriuria did not appear to have a significant effect on urinary N-acetyl B-D-glucosaminidase (NAG) and retinol binding protein (RBP);[4] however, the influence of in vivo inflammation and hemorrhage have not been adequately assessed for any of the urine biomarkers. Three tubular proteins evaluated in multiple recent studies are discussed in more detail below.

N-acetyl B-D-glucosaminidase (NAG)

NAG is a lysosomal enzyme present in proximal renal tubular cells, and it is released from the cells secondary to tubular damage. Therefore, it is generally accepted as a biomarker of tubular injury, which has been supported by the increase seen secondary to administration of nephrotoxic drugs.[5] Analytically it is an attractive protein because it can be measured using an enzymatic assay and is therefore easily adapted to different species, unlike many of the immunoassays that require a species-specific antibody. Because of this, it is one of the most studied of the non-albumin urinary proteins in renal disease. In normal dogs, activity of this enzyme is generally low,[1,2,6-8] although leukocyte contamination and semen contamination can result in increased activity. While there is some overlap with that seen in normal dogs, activity is typically higher in dogs with renal disease.[2,7] Activity was increased in dogs with pyometra-associated renal disease, and it significantly decreased, typically to a level comparable with normal dogs, after ovariohysterectomy.[1,8] While there was moderate correlation with severity of glomerular and tubular damage in dogs with progressive CKD due to XLHN, correlation with proteinuria was higher.[2] In addition, urinary NAG activity was compared with severity of glomerular and tubular damage based on renal biopsy in ~100 dogs with various renal diseases, typically manifested as protein-losing nephropathy (PLN). NAG activity demonstrated moderate correlation with severity of glomerular damage but poor correlation with severity of TI damage (Nabity, unpublished observations). Therefore, whether urinary NAG is more likely present due to glomerular or tubular damage in dogs with PLN is questionable, and further investigation of its urinary origin needs to be performed. Even if present due to ongoing tubular damage, preliminary results suggest that NAG activity does not appear to be a useful indicator of the severity of tubular damage in dogs.

A similar finding is likely in cats, although no study relating renal biopsy findings and NAG activity is available. However, NAG was correlated with proteinuria but was not associated with sCr in geriatric cats,[9] and it was not a significant factor in predicting the development of azotemia in cats.[10]

Retinol binding protein (RBP)

RBP is a low molecular weight (LMW) protein that, when circulating freely, will readily pass through the glomerular filtration barrier but is essentially completely reabsorbed by the proximal tubules in health. As tubular function decreases, reabsorption of RBP also decreases, resulting in increased urinary excretion. While it is therefore considered a marker of tubular dysfunction, glomerular proteinuria can interfere with absorption of LMW proteins. RBP is increased in dogs with proteinuric renal disease as compared with normal dogs, and it is also typically increased in dogs with pyometra-associated proteinuria, with resolution after treatment of the pyometra.[1,2,7,8,11,12] A study of dogs with XLHN showed that RBP correlated well with sCr and GFR, and it correlated more strongly with histologic indicators of renal injury severity than did sCr and GFR.[2] In addition, RBP was found to be markedly increased in the urine of XLHN dogs with proteinuria and early azotemia as compared with when those dogs were proteinuric but not azotemic.[13] These studies show promise in the utility of RBP for the early diagnosis and monitoring of TI dysfunction. However, Raila et al evaluated several subsets of dogs with various combinations of proteinuria, azotemia, and reduced creatinine clearance. This study found that urinary RBP was of no diagnostic value for the early detection of reduced GFR in nonazotemic dogs, and it demonstrated a moderate correlation with UPC, which could suggest a greater influence of proteinuria than tubular damage on its urinary presence.[3] Therefore, further evaluation of RBP is required before its clinical utility in dogs can be adequately assessed.

Urinary RBP has also been shown to be elevated in cats with CKD and with hyperthyroidism.[14,15] However, pretreatment concentrations did not predict which cats would develop decreased GFR after treatment for hyperthyroidism.[16] Additional studies assessing its clinical utility in feline renal diseases are lacking.

Cauxin

Cauxin is a unique urinary protein in cats produced by the proximal tubule that is present in particularly high concentrations in intact males.[17] Because it can comprise a major proportion of the protein present in normal cats, removal may help to enhance the accuracy of urine dipsticks in detecting abnormal proteinuria.[18] When evaluated in cats with varying severity of TI damage, cauxin was shown to decrease in both the kidney and urine as severity of TI injury increased.[19] In another study, urinary cauxin:creatinine did not differ between cats with varying degrees of azotemia.[20] Interestingly, a high ratio was predictive of geriatric cats developing azotemia, but a large overlap in the ratio was observed with cats that didn't develop azotemia. Therefore, its usefulness in predicting azotemia in an individual cat may be limited.[20]

URINARY BIOMARKERS OF GLOMERULAR DAMAGE/DYSFUNCTION

The predominant urinary protein present with glomerular damage is albumin. However, large proteins such as intact immunoglobulins may also be present, and it is thought that the greater the compromise to the filtration barrier, the larger the proteins that can pass through. Because so little protein is present in the urine of normal animals, even mild glomerular damage can often easily be detected. The most studied non-albumin glomerular protein is immunoglobulin G (IgG). IgG is a high molecular weight protein (~160 kDa) that is normally present in very small amounts in normal canine urine.[1,2,7,8] Like NAG and RBP, it is increased in the urine of dogs with CKD and pyometra-associated proteinuria, and it typically decreases to normal concentrations with treatment of pyometra.[1,2,7,8] In a study of dogs with progressive CKD due to XLHN, IgG did not correlate any better with the severity of glomerular damage than UPC.[2] However, in ~90 dogs with various protein-losing nephropathies it did correlate much better than UPC with severity of glomerular damage (Nabity, unpublished observations). It may therefore be useful as a non-invasive indicator of the extent of glomerular injury. Interestingly, despite its much larger size (~900 kDa), urinary IgM did not correlate as well as IgG with severity of glomerular damage in these dogs with PLN (Nabity, unpublished observations).

ENDOGENOUS MARKERS OF GFR

In addition to urinary markers of renal damage, the search for endogenous serum/plasma markers of GFR that can perform better than sCr and BUN continues. One promising molecule is symmetric dimethylarginine (SDMA), which originates from hydrolysis of methylated proteins. This molecule is increased in people with CKD,[21] and preliminary studies in dogs show that it correlates well with sCr but that it might have a tighter reference interval in normal dogs as compared with sCr (IDEXX, unpublished observations). Therefore, it might be of more use than conventional tests in indicating a decreased GFR based on a single blood sample. However, additional studies are in progress to further evaluate this molecule.

REFERENCES

[1]Maddens B, et al. **J Vet Intern Med** 2011; 25(5):1075. [2]Nabity MB, et al **J Vet Intern Med** 2012; in press. [3]Raila J, et al **Am J Vet Res** 2010; 71(11):1387. [4]Smets P, et al. **J Vet Diagn Invest** 2010; 22(6):896. [5]Grauer GF, et al. **J Vet Intern Med** 1995;9(5):324. [6]Brunker JD, et al **Am J Vet Res** 2009; 70(2):297. [7]Smets P, et al. **J Vet Intern Med** 2010; 24(1):65. [8]Maddens B, et al. **J Vet Intern Med** 2010; 24(6):1263. [9]Jepson RE, et al. **Am J Vet Res** 2010; 71(2):241. [10]Jepson RE, et al. **J Vet Intern Med** 2009; 23(4):806. [11]Maddens B, et al **Vet Immunol Immunopathol** 2010; 134(3-4):259. [12]Vinge L, et al. **Nephrol Dial Transplant** 2010; 25(8):2458. [13]Nabity MB, et al. **Vet Clin Pathol** 2011; 40(2):222. [14]van Hoek I, et al. **J Immunol Methods** 2008; 329(1-2):208. [15]van Hoek I, et al. **J Vet Intern Med** 2009; 23(5):1031. [16]van Hoek I, et al. **Domest Anim Endocrinol** 2009; 36(1):45. [17]Miyazaki M, et al. **Comp Biochem Physiol B Biochem Mol Biol** 2006; 145(3-4):270. [18]Miyazaki M, et al. **Vet J** 2011; 189(3):312. [19]Miyazaki M, et al. **Res Vet Sci** 2007; 82(1):76. [20]Jepson RE, et al. **Am J Vet Res** 2010; 71(8):982. [21]Schwedhelm et al. **Nat Rev Nephrol** 2011; 7(5):275.

APPLICATION OF NOVEL BIOMARKERS IN DOGS WITH EXPERIMENTAL ACUTE KIDNEY INJURY

Carrie A. Palm, DVM, DACVIM
Davis, CA

Acute kidney injury (AKI) represents a spectrum of disease severity, ranging from injury that is clinically non-detectable to severe damage resulting in fulminate acute renal failure (ARF). ARF is the most severe stage of AKI and is associated with the highest morbidity and mortality in both humans and animals. Mortality rates often exceed 60%, in large part because of the inability to recognize renal injury soon after an initial insult.[1] Creatinine is a marker of renal function, but is insensitive for early detection of renal injury as elevations in creatinine outside of the reference range do not occur until there is loss of approximately 75% of renal function. In addition, factors such as breed, age, body condition, fluid volume status and laboratory variation also contribute to the insensitivity of creatinine for early AKI diagnosis. Due to the limitations of utilizing serum creatinine as a maker for early renal injury, a staging system for AKI has recently been adopted from human medicine for use in veterinary medicine.[2] This staging system allows for categorization of patients into 5 stages of AKI, ranging from non-azotemic AKI to severe AKI and ARF. These categories are based on absolute levels of azotemia, as well as changes in azotemia over a specified period of time, stressing the need to identify renal injury early in its course.[2] Early diagnosis of renal injury may allow for therapeutic interventions that could lead to better prognoses. In addition, recognition of renal injury before significant elevations in creatinine occur, may give a better understanding of variables causing injury that are not currently obvious.

Because serum creatinine is not ideal for early diagnosis of AKI, there is an interest in evaluating renal biomarkers to assess for evidence of early kidney damage. Biomarkers are natural substances that can be detected in urine or blood that may allow for early diagnosis of AKI, either individually or in combination. An ideal marker of AKI should be easily detectable, should be sensitive and specific, have rapid laboratory or point-of-care turnaround time and of course, should have significant and earlier elevations as compared to creatinine. Many candidate markers are being evaluated in human medicine, including Neutrophil-Gelatinase Associated Lipocalin (NGAL), Kidney Injury Molecule-1 (KIM-1), N-acetyl-b-d-glucosaminidase (NAG) and cystatin-C. None of these have been widely investigated in veterinary medicine although studies are published evaluating their utility in animal models of AKI as well as investigations in human medicine.[4] These markers, and many others, hold promise for allowing early diagnosis of AKI and may provide an opportunity for veterinarians to treat AKI in its earliest stages before high mortality occurs.

NGAL is a 25kDa protein that can be measured in both blood and urine. It has been shown to be one of the earliest and most robustly elevated markers in AKI.[4] In human medicine, cardiac surgery is commonly associated with AKI. In one study, a large number of children under going elective cardiac surgery developed AKI (as defined by an increase in serum creatinine by 50%) within 2-3 days post-procedure. Significant elevations in NGAL occurred within 2 hours of surgery, several days prior to the changes in creatinine.[5] This supports that NGAL may be of benefit to detect early declines in renal function. A canine specific ELISA NGAL kit is currently available for evaluation in the research setting (Abbott Laboratories, Abbott Park, IL).

KIM-1 is a transmembrane protein that is highly over-expressed in proximal tubule cells after ischemic or nephrotoxic renal injury in animal models.[6] It has been reported that urinary KIM-1 can be used to distinguish ischemic AKI from prerenal azotemia and from chronic kidney disease.[6] Both a human and rat ELISA, as well as a point-of-care (Rena-strip) KIM-1 tests (BioAssay Works®, LLC, Ijamsville, MD,) are available, but these tests have not been validated in the dog.[6-8] Both the KIM-1 ELISA and Rena-strip have been shown to be sensitive and specific markers of ARF in humans and rats. In addition, the KIM-1 Rena-strip also has the advantage of being a point-of-care test, which allows for a rapid turnaround time.

Because these two urinary markers seem promising in human and animal models, we have started investigations to evaluate their utility in dogs. The first part of the study evaluated if these specific markers could be detected in the urine of dogs. The second part determined if these markers differentiated AKI from other types of urinary tract diseases. Finally, biomarkers were evaluated in 5 dogs with experimentally induced AKI in the laboratory setting, to determine if these markers had significant increases sooner than serum creatinine. Daily blood and urine samples were collected from dogs before and during experimental induction of kidney injury (as defined by a 50% increase in creatinine values from baseline). In addition, blood and urine samples were obtained during renal recovery, to allow for comparison of biomarker concentrations through all stages of AKI.

Evaluating urinary biomarkers may lead to a better understanding of the progression to ARF and will determine if there is a more sensitive and specific test to diagnose AKI before it develops into a life-threatening condition. Early AKI recognition may allow for early intervention and a reduction in the mortality that may occur with this disease. In addition, the availability of biomarkers that allow for early recognition of AKI may give a better understanding of processes that are causing significant renal injury.

REFERENCES:

1Parikh CR. **Crit Care Med** 2008; 36 (4Suppl):159. 2Cowgill LD et al. **In: Bartges J, Polzin D. Nephrology and urology of small animals. 1st ed**; 2011:472. 3Devarajan P. **Scand J Clin Lab Invest** 2008;Suppl 241:89. 4Cruz DN et al. **Seminars in Dialysis** 2011;24(2):124. 5Mishra J et al. **Lancet** 2005;365:1231. 6Won H et al. **Kidney International** 2002;62:237. 7Vaidya V et al. **Kidney International advance online publication** 2009;April. 8Ozer J et al. **Nature Biotechnology** 2010;28(5):486.

APPLICATION OF NOVEL BIOMARKERS IN NATURALLY OCCURRING ACUTE KIDNEY INJURY

Gilad Segev, DVM, DECVIM-CA
Jerusalem, Israel

INTRODUCTION

Acute kidney injury (AKI) is characterized by an abrupt and sustained decrease in the glomerular filtration rate (GFR). It is a common disorder in companion animals and humans, and is associated with high treatment costs as well as high morbidity and mortality. Four phases are currently recognized in AKI: initiation, progression, maintenance, and recovery. Using the common clinicopathologic markers (e.g., serum creatinine), the disease is characteristically recognized only in the maintenance phase, when clinical signs are overt.

Despite advances in the management of AKI, including the introduction of renal replacement therapies, the mortality rate among human and animal patients remains unacceptably high. Over the past 50 years, mortality rates of human patients with AKI in intensive care units have remained as high as 70%.[1] One of the speculated reasons for the high mortality is the late recognition of the disease and consequently the narrow window of opportunity for therapy. Therefore, there is a need to recognize the disease early, before overt renal failure is evident, and when therapeutic intervention could potentially be more effective. The need for early diagnosis is further emphasized in veterinary medicine, because renal replacement therapies are not readily available.

Despite the diagnostic advancements made in other medical fields (e.g., the use of biomarkers in cardiology), serum creatinine concentration is still being used as a marker for AKI, despite its multiple shortcomings. Substantial changes in GFR may be associated with relatively small changes in serum creatinine concentration within the first 24–48 hours following initiation of AKI, thus preventing an accurate estimate of the timing of injury and an accurate assessment of the severity of dysfunction. Mild increases in serum creatinine concentration, as little as 0.5 mg/dL, have been associated with increased mortality.[2] Moreover, a transient rise of serum creatinine (for 1–3 days) was associated with increased odds ratio for in-hospital mortality,[3] and with the need for chronic dialysis over the ensuing 3 years.[4]

In the recent years research in nephrology has been growing in an attempt to identify sensitive and specific biomarkers of AKI in human patients. Data regarding the utilization of such biomarkers in veterinary medicine is still scarce, and further research is warranted. Each of the investigated biomarkers exhibits advantages and weaknesses in diagnosing AKI. It is thus most likely that a future array of biomarkers will be used, not only for early diagnosis of AKI, but also in determining its etiology and prognosis.

Such biomarkers will prove useful if patients at risk for AKI will be tested before overt renal failure is evident. Some of the potential biomarkers of AKI assessed in human medicine have been; anecdotally evaluated in veterinary medicine.

POTENTIAL BIOMARKERS

Gamma-glutamyl transpeptidase (GGT)

This enzyme is present in proximal renal epithelial cells, and its urine activity can be measured. Due to its instability, samples must be analyzed shortly after urine collection, thus limiting its clinical utility. In one study, 24 hour urine GGT activity was measured in dogs with experimentally induced AKI using gentamicin. Increased urine GGT activity was recorded as early as on day 1, whereas serum creatinine concentration increased only 7 days after AKI was initiated,[5] suggesting that urine GGT activity is a highly sensitive and early marker of AKI.

N-acetyl-β-glucosaminidase (NAG)

NAG is a proximal tubule lysosomal enzyme. Increased NAG levels have been reported in human patients with nephrotoxicity,[6] in delayed renal allograft function,[7] and following cardiopulmonary bypass procedures.[8] Increased NAG activity has also been shown to precede increases in serum creatinine concentration by 12 hours to 4 days,[9] allowing early therapeutic intervention. NAG activity is also associated with the severity of the injury, and increased activity is associated with the need for renal replacement therapy as well as with mortality.[10] However, NAG activity may increase in several other non-renal disorders (e.g., rheumatoid arthritis and hyperthyroidism), limiting its specificity, and thus false positive predictions may occur.

α1 and β2-microglobulins

β2-microglobulin is an 11.8-kD protein, globally expressed on nucleated cell surfaces. It is typically filtered in the glomerulus, and is later reabsorbed and catabolized almost entirely by renal proximal tubular cells. Increase in its urine concentration has been shown to be an early marker of tubular injury in several settings, including nephrotoxicity, cardiac surgery,[11] and renal transplantation.[12] Increases in β2-microglobulin has preceded increases in serum creatinine concentration by 4–5 days.[13] Its use is limited by its instability in urine. α1-microglobulin is a 30-kD protein, synthesized by the liver. Its free form is readily filtered by the glomerulus and reabsorbed by renal proximal tubular cells. α1-microglobulin is stable in urine, and is thus considered a more practical marker for proximal tubule dysfunction. Unfortunately, its serum concentration is affected by several non-renal causes (e.g., liver disease), therefore limiting its specificity.[14]

Retinol Binding Protein (RBP)

RBP is a 21-kDa protein, synthesized by the liver, and is freely filtered by the glomerulus. It is subsequently reabsorbed and catabolized by renal proximal tubular cells. Increased urinary RBP concentration has been found to be a highly sensitive indicator of renal tubular dysfunction and to precede increases in urinary NAG activity.[15] For example, in infants, following birth asphyxia, increased urinary RBP concentration was predictive of AKI.[16] Similar to other renal biomarkers, serum RBP levels are influenced by other factors, and thus false negative predictions occur.[13] Urinary RBPs were assessed in dogs with pyometra as markers of proximal tubular function and found to be significantly increased compared to healthy controls. In this study, 68% of 17 non-azotemic dogs presented increased urinary concentrations of renal biomarkers, indicating that dogs with pyometra do sustain AKI, although this is often undetected using routine markers.[17]

Cystatin-C

This 13-kD protein is a cysteine protease inhibitor that has been studied relatively extensively as a marker of human AKI. Unlike serum creatinine concentration, it is unaffected by sex, age, and muscle mass, and is thus considered a more specific renal function marker. Due to its low molecular weight cystatin-C is freely filtered by the glomerulus. It is subsequently reabsorbed and catabolized, but not secreted by renal tubules. Urinary and serum cystatin-C concentrations are sensitive markers of AKI, superior to serum creatinine concentration,[18,19] and their levels were significantly associated with the prognostis.[20]

Kidney Injury Molecule-1 (KIM-1)

This is a type-I cell membrane glycoprotein, which is substantially up-regulated during AKI. Increases in its urine concentration is considered a sensitive marker of AKI.[21,22] Urinary KIM-1 was found to increase within 12 hours after an ischemic renal insult, prior to detection of cylinduria or increase in serum creatinine concentration.[22] Urinary KIM -1 concentration was also found to be an outcome predictor in 202 human patients with established AKI.[10]

Neutrophil Gelatinase-Associated Lipocalin (NGAL)

This 25-kD protein is up-regulated above 10-fold within the first few hours post ischemic renal injury. Increased urinary NGAL concentration was also recorded as an early biomarker of ischemic renal injury,[23] and can be detected in mice urine as early as three hours after cisplatin administration.[24] In pediatric patients undergoing cardiopulmonary bypass, increases in urinary NGAL concentration preceded increases of serum creatinine by 1–3 days.[25] Serum NGAL concentration has been shown to increase in a number of inflammatory and infectious conditions. In such cases, NGAL is filtered through the glomerulus and its urinary concentration increases, thereby decreasing its specificity for renal damage.

Interleukin-18 (IL-18)

This cytokine levels increase in a variety of inflammatory conditions. Nonetheless, it is constitutively expressed by renal distal tubular epithelium and its urinary levels increase in AKI patients. IL-18 concentration has been shown to increase in patients with delayed graft function, compared to normal patients with other urinary disorders, such as prerenal azotemia, urinary tract infection, chronic kidney disease, and nephrotic syndrome.[26] In critically-ill human adults with acute respiratory distress syndrome, increased urinary IL-18 was an early marker of AKI, preceding increases of serum creatinine by 1–2 days, and urinary IL-18 concentration was significantly associated with the prognosis.[27]

The lecture will present data regarding the use of routine and novel biomarkers in dogs with naturally-occurring AKI.

REFERENCES

1. Waikar SS, et al. Clin J Am Soc Nephrol 2008;3:844. 2. Chertow GM, et al. J Am Soc Nephrol 2005;16:3365. 3. Uchino S, et al. Nephrol Dial Transplant 2010;25:1833. 4. Wald R, et al. Jama 2009;302:1179. 5. Greco DS, et al. Am J Vet Res 1985;46:2332. 6. Emeigh Hart SG. J Pharmacol Toxicol Methods 2005;52:30. 7. Ikenaga H, et al. Clin Sci (Lond) 1993;84:469. 8. Ascione R, et al. Ann Thorac Surg 1999;68:493. 9. Westhuyzen J, et al. Nephrol Dial Transplant 2003;18:543. 10. Liangos O, et al. J Am Soc Nephrol 2007;18:904. 11. Dehne MG, et al. Anaesthesist 1995;44:545. 12. Schaub S, et al. Am J Transplant 2005;5:729. 13. Tolkoff-Rubin NE, et al. Clin Lab Med 1988;8:507. 14. Penders J, et al. Clin Chim Acta 2004;346:107. 15. Bernard AM, et al. Clin Chem 1987;33:775. 16. Roberts DS, et al. Arch Dis Child 1990;65:1021. 17. Maddens B, et al. J Vet Intern Med 2011;25:1075. 18. Conti M, et al. Clin Chem Lab Med 2006;44:288. 19. Uchida K, et al. Clin Chim Acta 2002;323:121. 20. Herget-Rosenthal S, et al. Clin Nephrol 2005;64:41. 21. Vaidya VS, et al. Am J Physiol Renal Physiol 2006;290:F517. 22. Han WK, et al. Kidney Int 2002;62:237. 23. Mishra J, et al. J Am Soc Nephrol 2003;14:2534. 24. Mishra J, et al.. Am J Nephrol 2004;24:307. 25. Wagener G, et al. Anesthesiology 2006;105:485. 26. Parikh CR, et al. Am J Kidney Dis 2004;43:405. 27. Parikh CR, et al. J Am Soc Nephrol 2005;16:3046.

NOVEL TREATMENT STRATEGIES FOR UNCOMPLICATED UTI's IN DOGS

Jodi L. Westropp, DVM, PhD, DACVIM, Davis, CA
Dennis Chew, DVM, DACVIM, Columbus, OH
Jane E. Sykes, BVSc, PhD, DACVIM, Davis, CA

INTRODUCTION

The clinical signs of a bacterial cystitis can include stranguria, pollakiuria, inappropriate urinations, dysuria and hematuria. Occult urinary tract infections (UTIs) in dogs have also been reported.[1, 2] Bacterial UTIs occur in approximately 14% of dogs in their lifetime,[3] with a variable age of onset. Spayed female dogs have been reported to be at increased risk for a UTI, which is likely due to anatomic differences.[4] An uncomplicated UTI is defined as a UTI without structural, metabolic, or functional urinary abnormalities, as well as the absence of other comorbidities that could favor a more serious adverse outcome.

Uncomplicated, community- acquired UTIs in dogs generally occur in healthy dogs and the most common urinary pathogen has been reported to be strains of *E. coli*, accounting for up approximately 45% of all pathogens isolated in one study.[5] When added together, *Staphylococcus, Proteus, Klebsiella and Enterococcus* spp. account for approximately an additional 40% of all isolates examined in that same study. With regard to *E. coli* UTIs in dogs and humans, the phylogenetic group of *E. coli,* B2, was the most common subtype identified.[6]

In order to help prevent *E. coli* and other pathogens from entering the urinary tract, several built in host "defense mechanisms" are present. Micturition itself and frequent and complete voiding can help remove bacteria. In addition to anatomic structures and urine voiding, the mucosal surface of the urinary tract has intrinsic mucosal antimicrobial properties and the glycosominoglycan layer can also act as a protective mechanism. High urine osmolality and high concentrations of urea can also inhibit bacterial growth. While dilute (<1.018) urine may predispose an animal to bacterial infection, it is likely the disease that produces the dilute urine allows the infections to happen. In dogs, submitting a urine culture based on a low urine specific gravity (<1.013) does not appear to be cost-effective in the absence of active urine sediment or high clinical suspicion for UTI.[7]

DIAGNOSTICS

Quantitative urine culture is the gold standard for diagnosis of bacterial urinary tract infections. The drugs chosen for susceptibility testing by individual laboratories are variable. Likewise, the method of susceptibility testing used is subject to laboratory preference and has direct implications on how results are reported. The clinician should consult with his or her microbiologist to determine whether the susceptibilities reported are based on concentrations of the antimicrobials in the serum or the urine. Current *in vitro* susceptibility testing is based on utilizing bacterial inocula of 10^5 cfu/mL. Choosing an antimicrobial that exceeds the MIC for the infecting bacteria by at least fourfold is usually recommended. If the average *urine concentration* is four times the MIC value or higher, it is likely that the drug will be at least 90% effective in eradicating the offending organism in an otherwise healthy patient. Attainable urine concentrations of some antimicrobials can achieve up to 100 times that attained in serum. Consequently, sterility within the urinary tract can be achieved at times when it would otherwise appear that these drugs would fail to do so at concentrations achieved in the plasma.

While a urine culture is the gold standard for diagnosis of a UTI, it can be costly, and in practice culture kits have been marketed for companion animal use. We have recently evaluated one of these kits, the Uricult® Vet paddles which appear to be a useful screening tool for identification of bacterial growth. However, inaccuracies occurred when paddles were used for organism identification, particularly when multiple uropathogens were present. When growth occurs, paddles or urine should be submitted to a commercial microbiology laboratory for identification and antimicrobial susceptibility testing. Furthermore, according to the International Society for Companion Animal Infectious Diseases (ISCAID), "bacterial isolation should only be attempted in clinics with appropriate laboratory facilities, proper equipment, proper biosafety level 2 (BSL-2) containment and waste management, and adequately trained individuals".[8] Quantitative culture should be performed, which may be difficult using the Uricult vet paddles. Finally, if the clinician chooses to evaluate urine cultures using plates or paddles, they should contact their laboratory to determine if those items will be accepted for identification confirmation and MIC testing.

TREATMENT OF UNCOMPLICATED UTI

For empiric therapy, the Antimicrobial Use Working Group of the ISCAID recommends urinary antibacterial drugs that are likely to be effective against more than 90% of the urinary isolates when this information is available. In general, ISCAID recommends initial therapy for uncomplicated UTI with amoxicillin (11–15 mg/kg PO Q 8 H) or trimethoprim–sulfonamide (15 mg/kg PO Q 12 H); amoxicillin–clavulanic acid is not recommended for initial treatment in these cases due to lack of evidence for the need for clavulanic acid in addition to amoxicillin.[8]

By convention, uncomplicated UTI has routinely been treated for 7-14 days. In human medicine, numerous studies exist evaluating the efficacy of short duration antimicrobial therapy for uncomplicated UTI in women. In human patients, studies evaluating the efficacy of 1 and 3 day treatment regimens for uncomplicated UTIs has been described.[9, 10] Few studies exist that have evaluated the efficacy of any dosing regimen of antimicrobial therapy for canine UTI. No studies of short duration oral antimicrobial protocols for treatment of naturally occurring canine UTI have been published, but older studies evaluating a one-day and 3-day treatment regimen in an induced *E. coli* canine cystitis model has been reported .[11] One-day treatment using either injectable amikacin or oral trimethoprim-sulfadiazine were not effective in eradicating uropathogens in eight dogs (4 female, 4 male). The 3-day treatment protocol using trimethoprim-sulfadiazine resulted in bacteriological sterile urine 14 days after treatment in 4 of 4 female dogs but in none of 4 male dogs.

Studies have shown cefovecin to have good *in vitro* activity against all major aerobic and anaerobic bacterial pathogens associated with skin, urinary tract and periodontal disease from isolates submitted from dogs in parts of the United States and Europe,[12 13]but *in vivo* UTI studies are limited. In one clinical trial cefovecin was administered subcutaneously (8mg/kg) to 61 dogs with an uncomplicated UTI and the authors reported it to be effective in 88% (59/67) of the dogs they evaluated.[13] In this study, it appears that urine cultures were evaluated 21 days after the antibiotic was administered, however urinary pathogens isolated from free catch samples were also evaluated in that study, and the sample collection method at each visit was not standardized. They compared the dogs receiving cefovecin to dogs that received cephalexin (15mg/kg) orally twice daily for 14 days. While 57.4% of the dogs receiving cephalexin had negative urine cultures, only 75% of these dogs had urinary pathogens susceptible to cephalexin. The clinical cure rate was 82.4% for dogs receiving cefovicin and 70% for dogs receiving cephalexin.

We have recently evaluated the use of a high dose short duration (HDSD) enrofloxacin protocol for canine UTI. In this study, dogs with uncomplicated UTI were randomly assigned to two groups. In group one, the dogs received 20mg/kg of enrofloxacin once daily for 3 days and group 2 dogs received 13.75mg/kg of amoxicillin-clavulanic acid twice daily for 14 days. Urine cultures and presence or absence of clinical signs were evaluated 7 days after the cessation of the antimicrobial treatment. At interim analysis, we have found a microbiologic cure rate of 83% and 78% for group 1 and 2, respectively. All dogs had resolution of clinical signs such as stranguria, pollakiruia and/or hematuria except for one dog.

CONCLUSION

Treatment for uncomplicated canine UTI should be based on urine cultures and susceptibility testing. If empiric therapy is utilized, "first line" antimicrobials have been recommended. If the animal fails to respond to the selected drug, it is imperative the urine is cultured for quantitative analysis and susceptibility testing. Short duration protocols for treatment of uncomplicated UTI are underway and may be considered. Additional research should also be undertaken to determine if a shortened protocol will positively impact owner compliance and evaluate what, if any, effect these protocols may have on antimicrobial resistance patterns in fecal flora or in organisms not initially eradicated from the urinary tract.

REFERENCES

1.McGuire NC, Schulman R, Ridgway MD, Bollero G. Detection of occult urinary tract infections in dogs with diabetes mellitus. J Am Anim Hosp Assoc. 2002; **38**(6): 541-4.2. Litster A, Moss S, Platell J, Trott DJ. Occult bacterial lower urinary tract infections in cats-urinalysis and culture findings. Vet Microbiol. 2009; **136**(1-2): 130-4.3.Ling GV. Therapeutic strategies involving antimicrobial treatment of the canine urinary tract. J Am Vet Med Assoc. 1984; **185**(10): 1162-4.4.Seguin MA, Vaden SL, Altier C, Stone E, Levine JF. Persistent urinary tract infections and reinfections in 100 dogs (1989-1999). J Vet Intern Med. 2003; **17**(5): 622-31.5.Ling GV, Norris CR, Franti CE, Eisele PH, Johnson DL, Ruby AL, et al. Interrelations of organism prevalence, specimen collection method, and host age, sex, and breed among 8,354 canine urinary tract infections (1969-1995). J Vet Intern Med. 2001; **15**(4): 341-7.6.Johnson JR, Kaster N, Kuskowski MA, Ling GV. Identification of rovirulence traits in Escherichia coli by comparison of urinary and rectal E. coli isolates from dogs with urinary tract infection. J Clin Microbiol. 2003; **41**(1): 337-45. 7.Tivapasi MT, Hodges J, Byrne BA, Christopher MM. Diagnostic utility and cost-effectiveness of reflex bacterial culture for the detection of urinary tract infection in dogs with low urine specific gravity. Veterinary clinical pathology / American Society for Veterinary Clinical Pathology. 2009; **38**(3): 337-42.8.Weese JS, Blondeau JM, Boothe D, Breitschwerdt EB, Guardabassi L, Hillier A, et al. Antimicrobial use guidelines for treatment of urinary tract disease in dogs and cats: antimicrobial guidelines working group of the international society for companion animal infectious diseases. Vet Med Int. 2011; **2011**: 1-9.9.Bleidorn J, Gagyor I, Kochen MM, Wegscheider K, Hummers-Pradier E. Symptomatic treatment (ibuprofen) or antibiotics (ciprofloxacin) for uncomplicated urinary tract infection?--results of a andomized controlled pilot trial. BMC Med. 2010; **8**: 30.10.Arav-Boger R, Leibovici L, Danon YL. Urinary tract infections with low and high colony counts in young women. Spontaneous remission and single-dose vs multiple-day treatment. Arch Intern Med. 1994; **154**(3): 300-4.11.Rogers KS, Lees GE, Simpson RB. Effects of single-dose and three-day trimethoprim-sulfadiazine and amikacin treatment of induced Escherichia coli urinary tract infections in dogs. Am J Vet Res. 1988; **49**(3): 345-9.12. Stegemann MR, Passmore CA, Sherington J, Lindeman CJ, Papp G, Weigel DJ, et al. Antimicrobial activity and spectrum of cefovecin, a new extended-spectrum cephalosporin, against pathogens collected from dogs and cats in Europe and North America. Antimicrob Agents Chemother. 2006; **50**(7): 2286-92. 13.Passmore CA, Sherington J, Stegemann MR. Efficacy and safety of cefovecin (Convenia) for the treatment of urinary tract infections in dogs. J Small Anim Pract. 2007; **48**(3): 139-44.

MANAGEMENT OF COMPLICATED UTI

Dennis J. Chew, DVM, DACVIM
Columbus, OH
Jodi L. Westropp, DVM, PhD, DACVIM
Davis, CA

INTRODUCTION

The likelihood for success or failure in long-term eradication of bacterial UTI involves interactions between the uropathogen, predisposing factors within the patient (functional, anatomical/structural, metabolic), and decisions made by the attending clinician and the owner regarding diagnostic evaluation and therapy. A complicated UTI is defined as an infection that is not readily eliminated with a single course of antibacterial treatment or one in which the urinary tract does not remain sterile for long periods after initial eradication of the UTI. In these instances, there is often an identifiable underlying anatomical, metabolic, or urinary functional problem. Less commonly the UTI is considered complicated because of inherent resistant patterns of the uropathogen (e.g. multiple drug resistance), the organism has special virulence factors, and when multiple organisms are isolated. Recurrent UTI occurs following treatment with or without clinical signs – clinical signs can be surprisingly infrequent in dogs with recurrent UTI. Recurrent UTI is caused by either reinfection with a new microorganism after a period of sterility within the urinary tract, or failure to completely clear the initial organisms from the urinary tract (persistent or relapsing infection)[1]. Similar to dogs with uncomplicated UTI, *E. coli* was the primary pathogen reported in two studies; two or more bacterial isolates were identified in dogs in one of those studies. [1,2]. Uropathogens were resistant to achievable serum concentrations of all commonly prescribed oral antibiotics in 30% of dogs with recurrent UTI in one study. Potentially predisposing factors for recurrent UTI were identified in about 70% of dogs in the same study[1].

REINFECTION

In patients with reinfection (with or without clinical signs), the urinary tract is initially sterilized following antimicrobial treatment but a new infection occurs within weeks to months after treatment has been stopped. A different bacterial isolate was found in 50% of dogs with recurrent UTI in one study[1.] It is easy to identify a reinfection when the quantitative urine culture reveals a different bacterial genus and species than that originally isolated. It can be difficult to differentiate with certainty when the same genus and species of microorganism is isolated after a period of sterility especially with E. coli infections. Strains of E. coli cannot be differentiated based purely on the antibiogram result; two organisms with the same antibiogram may have vastly different virulence properties[3,4]. A new infection could occur due to the same genus and species following re-inoculation from the animal's own flora from feces or from the distal urethra and genital tracts. Contamination with the same organism from the environment could also occur at this later date. Multiple new infections suggest that the animal's host defense mechanisms are not operating optimally. Though the urinary tract is easily sterilized in these instances, underlying predispositions are often present that allow new infection to ascend and become established again in the urinary tissues. Routine oral antibacterial treatment (see Uncomplicated UTI in previous session) is often quickly effective in the eradication of organisms responsible for reinfection. Neither big-gun antibiotics nor long treatment courses of treatment are needed for most recurrent UTI due to reinfections.

RELAPSING OR PERSISTENT INFECTION

Relapsing UTI (with or without clinical signs) is caused by the identical organism initially isolated. It could be due to the persistence of an organism that was never completely eradicated, a finding that occurred in 42% of dogs with recurrent UTI in one study[1]. Relapse suggests the organism is deep-seated in the tissues (inaccessible to the antibacterial, as in the kidney, prostate, or polyps), that sub-therapeutic concentrations of the antibacterial were achieved in the urine and/or urinary tissues, or that the organism is resistant to the chosen antimicrobial. Classically, positive quantitative urine culture results are identified soon after discontinuation of treatment, usually within days to one week. Results from urine culture reveal no growth while the patient is receiving urinary antimicrobial treatment (in-vivo urine susceptibility testing or so called "therapeutic culture"). The ability to sterilize the urine does not guarantee eradication of the organism from the urinary tissues. Persistent UTI is a variant of relapsing infection in which bacterial cultures remain positive with the same organism during antimicrobial treatment. In this instance, the organism has not been eradicated, even transiently despite administration of a urinary antimicrobial agent. Persistent implies severe abrogation of local host defenses or that the organism is highly resistant to the administered antimicrobial drug.

No studies have been reported for the proper dosing schedule or duration of treatment for dogs or cats with relapsing or persistent infections. Long-term therapy with an appropriate antimicrobial for 30 to 60 days is often prescribed for those with initial failure to sterilize the urinary tract. An antibiotic that achieves high tissues concentrations (e.g. fluoroquinoloines, trimethoprim or ormetoprim potentiated sulfas) is advised. Antibiotic sensitivity testing (preferably with MIC) should be used to choose an antibacterial based on levels achieved in the serum rather than that of urine since relapsing UTI is often associated with deep tissue infection rather than more superficial mucosal infection. An increased dose or frequency of drug administration should be considered for drugs in which this is an option and appropriate. This approach may be necessary in patients with sequestration of uropathogens in locations that do not allow easy access to antibacterial agents (kidney, prostate, polypoid cystitis, proliferative urethritis, pods of intracellular bacterial communities). An expanded panel of drugs for susceptibility testing can be requested when susceptibility testing reveals resistance to many of the commonly employed oral urinary antibacterials. Furthermore, dogs with recurrent non-clinical UTIs may not always require treatment. The clinician must decide if the lower urinary tract infection may ascend and cause clinical pyelonephritis. Judicious restraint in the use of antimicrobials may help prevent the emergence of greater bacterial resistance.

DIAGNOSTIC EVALUATION OF PATIENTS WITH RECURRENT UTI

Although antibacterial treatment is pivotal in sterilizing the urine and tissues of the urinary tract, identification and correction of predisposing anatomic/structural, functional, or metabolic problems is essential in order to allow initial sterilization of the urine and to prevent new infections from developing. Anatomic defects were the most commonly identified predisposition for recurrent UTI followed by abnormal micturition and altered uroepithelium in one study of dogs[1]. Predisposing factors facilitate ascent of bacteria and development of an initial UTI and also can make it more difficult to eradicate UTI once established in the urinary tissues. A search for anatomical or structural defects such as a recessed vulva, ectopic ureter(s), urachal diverticulum, polypoid cystitis, proliferative urethritis, urolithiasis, foreign bodies, prostatic disease, and neoplasia within the urethra or bladder should be undertaken. Infection in the kidney is inherently more difficult to eradicate than in the lower urinary tract, especially when chronically infected or in the face of partial obstruction. Micturition problems such as urinary incontinence or urine retention should be identified and corrected when possible. Those with metabolic abnormalities such as diabetes mellitus, hyperadrenocorticism, hyperthyroidism, CKD may more readily acquire a new UTI or fail to easily resolve an existing UTI, as is similarly true for patients receiving corticosteroid therapy or other immunosuppressive drugs. It can be difficult to impossible to achieve long-term urinary tract sterility in patients that continue to receive immunosuppressive drug treatment.

Imaging studies that include survey abdominal radiographs, contrast cystourethrogram, and abdominal ultrasound are an important part of the evaluation of patients with recurrent lower UTI in order to rule out structural and anatomical problems. Excretory urography and renal ultrasonography may be needed to evaluate the kidneys more completely, especially to help exclude an element of obstruction within the upper urinary tract. Cystoscopy should be considered if the initial imaging tests do not document structural or anatomical abnormalities. Even when no obvious lesions are observed during cystoscopy, we recommend submitting a biopsy of the bladder mucosa for both histopathology and culture to evaluate the patient for deep-seated infections. Positive bacterial growth is sometimes isolated from the bladder mucosa and uroliths of dogs when prior urine culture has been negative[5]. The possibility that intracellular bacterial communities (so called "pods") could live within a protective layer of biofilm inside uroepithleial cells is intriguing, as this could explain relapsing UTI in some patients. This environment would impair penetration of antibiotics to the site of the infection, so these pathogens could survive and then re-inoculate the urine once antibiotics have been discontinued. Evidence in support of this concept exists in mouse models using *E. coli* as the uropathogen[6].

ADJUNCT AND PROPHYLACTIC THERAPY

Cranberry juice or extract may be effective in reducing the frequency of recurrent UTI due to new infections in some humans following initial eradication of UTI with antimicrobial drugs. This beneficial effect has been attributed to impairment of bacterial adherence to uroepithelium by proanthocyanidins found in cranberries; this effect may be more specifically directed against attachment of P-fimbriated *E coli* to uroepithelium. Adherence to uroepithelium is an initial first step in colonization and further growth resulting in the establishment of UTI. Anti-adherence properties against uropathogenic *E coli* were demonstrated in two studies using urine from dogs treated with cranberry extract. Both of these studies used surrogates for attachment of uropathogenic *E. coli* to uroepithelium in clinical patients.[7,8] Studies demonstrating the ability of cranberry extract to reduce the frequency of recurrent UTI in dogs have yet to be published. Moreover, the degree to which uropathogenic *E coli* contribute to the overall incidence of UTI in dogs is unknown, and the effect of cranberry therapy on pathogens lacking P fimbriae (eg, *Staphylococcus*, *Enterococcus*, or *Proteus* spp.) is also unknown.

Prophylactic treatment for dogs and cats with recurrent clinical infections have been employed in veterinary medicine for decades, but no studies have been reported on the effectiveness of these protocols or on the effect of this treatment on resistance patterns of uropathogens or fecal organisms. They should not be used in those in which the recurrent UTI is known to be relapsing. Although they are sometimes employed, the effectiveness of daily or every-other-day versus pulse dosing protocols for the use of urinary antimicrobials in dogs and cats with recurrent UTI has not been specifically reported.

CONCLUSION

Successful treatment of complicated UTI requires evaluation of quantitative urine culture with susceptibility before (diagnostic), during (therapeutic), and after (surveillance) therapy. Correction of underlying predisposing factors increases the likelihood for successful long-term eradication of organisms from the patient's urine and urinary tissues. Minimally, urine should be cultured following collection by cystocentesis before treatment is started and then 5 to 7 days after treatment has been discontinued. Treatment is further informed when quantitative urine culture is performed 3 to 5 days after treatment has been started to rule out a persistent UTI and then again 3 to 5 days before treatment is stopped to exclude the possibility of superinfection with a new organism. Urine cultures are recommended at 1, 3 and 6 months to detect the presence of UTI due to reinfection (or possibly late relapse) regardless of clinical signs.

REFERENCES

[1]Seguin MA, et.al. **J Vet Intern Med** 2003;17:622-31. [2]Norris CR, et.al. **J Am Anim Hosp Assoc** 2000;36:484-92. [3]Drazenovich N, et, al. **J Vet Intern Med** 2004;18:301-6. [4]Freitag T, et.al. **J Vet Intern Med** 2006;20:245-9. [5]Gatoria IS, et.al. **J Small Anim Pract** 2006;47:727-32. [6]Anderson GG, et.al. **Science** 2003;301:105-7. [7]Howell AB, et.al. **J Vet Intern Med** 24:678, 2010. [8]Smee N, et. Al. **J Vet Intern Med** 25:716-717, 2011.

IAD AND RAO – WHAT HAVE WE LEARNED IN THE PAST TEN YEARS?

Melissa R. Mazan, DVM, DACVIM
North Grafton, MA

INTRODUCTION

Over the past decade, our understanding of inflammatory airway disease and recurrent airway obstruction in horses has burgeoned. The research areas can be categorized, loosely, as molecular biology, genetics, etiology and etiopathology, phenotype, mucus, physiology, diagnosis, the role of the immune system/allergies, and treatment. Environmentally induced respiratory diseases are also being investigated in some ruminants, particularly in bovine respiratory disease complex and in respiratory diseases of goats.

THE CYTOKINE DEBATE

There has been considerable attention paid to mRNA/cytokine evidence for the etiopathology of IAD and heaves. Whether the Th2 system or the Th1 system, or a combination, is at work in either of these diseases, remains up for debate, although the picture is somewhat clearer now for horses with mast-cell mediated IAD (see below). [1]Part of the dilemma may be that different tissues and cells (bronchial epithelial, all cells in BALF v. lymphocytes, macrophages, or neutrophils) and varying PCR techniques as well as disease models have been used. [2], [3,4], [5], [6], [7], [6], [8], [9]. Moreover, the issue is further clouded by lack of consistency in timepoints that have been examined. [2], [10]. Further work has been done in trying to understand the role of transcription factors in initiating these cytokine responses, especially the roles of AP-1 and CREB. [11]

PHENOTYPE

Equally important has been the development of more in-depth phenotype profiles for horses with IAD. Importantly, several workers have confirmed that IAD is more than a BAL finding, in that it results in abnormal blood gases and poor performance. [12], [13], [14] The importance of mucus in the trachea as a signifier of lower airway disease has been brought up, with multiple studies showing that increased mucus impairs performance in racehorses [15] and sport horses. [13]Physical findings with respect to airway mechanics [16]and reactivity, as well as cytokine differences between horses with mast cell-mediated disease v. neutrophil-mediated disease has clarified the identification of mast cell horses as having a Th2 cytokine profile as well as predictable airway reactivity. [17,18], [19]. Interestingly, the clinical suspicion that lower airway inflammation can result in upper airway dysfunctions has been demonstrated. [20] Considerable work has also enhanced our understanding of the RAO phenotype – notably, that severe RAO can be associated with systemic disease [21]and surfactant dysfunction. [22] Moreover, airway obstruction, smooth muscle hyperplasia, and elevated levels of BALF myeloperoxidase were shown to be chronically present even in horses ostensibly in remission. [12], [23], [24]

DIAGNOSIS

Improvements have been made in our ability to diagnose these diseases, including the utility of owner-questionaires for diagnosis of RAO, the repeatability of new PFT modalities such as open plethysmography, and the lack of sensitivity for old favorites such as radiographs for IAD. [25], [26], [27]

ETIOLOGY

We still don't know exactly what causes some horses to develop RAO and not others. Environmental contributions to RAO and IAD have been closely investigated, with stable dust, and various components including beta-glucans, endotoxin, and silicates being involved in BALF neutrophilia and airway dysfunction. [28], [29] Importantly, multiple studies have shown that stable dust and its components cause respiratory abnormalities in control horses as well, however, only RAO horses are sufficiently affected to result in severe clinical disease. Likewise, known asthma triggers such as cold air cause airways dysfunction and inflammation in normal horses, not just RAO. [30]This finding may help to explain, however, the hypothesis that IAD and RAO closely related diseases that display a spectrum of clinical signs. (Viel) The potential influence of this environment on human airways has also been explored. [31], [32], [33] The ever-intriguing question of whether viral disease plays a role in the development of lower airway inflammation was again brought up when it was found that EHV-2 virus was more commonly found in the trachea of horses with IAD. [34] The question of the innate immune system and the role of a truly allergy-mediated mechanism in horses with RAO has also been explored, with a relationship between genetics, disease, and allergen-specific IgGa and IgE being seen in one study, [35], however, in vitro and intradermal allergy tests do not definitively support an IgE mediated mechanism. [36], [37], [38], [39]The role of anti-oxidants in RAO and IAD is likewise not completely clear. Acute organic dust exposure caused significant antioxidant depletion in the equine trachea in RAO, [40] and exercise alone in unaffected racehorses caused increases in markers of oxidative stress. [41] Further studies showed in contrast that the redox system was upregulated during exposure to dusty straw and hay to combat oxidative stress. [42] An important contribution to the field of RAO research has been in genetics, with a mixed inheritance pattern [43] at least in certain breeds, and certain chromosomal patterns and genes being associated with RAO. [44], [45] Other investigators are beginning to probe whether RAO can be considered a systemic disease. During exacerbation of a form of RAO, endothelin is elevated in the systemic circulation ;[46] enhanced platelet activation during exacerbation of RAO can also be viewed as evidence of systemic disease. [47] Cardiovascular disease may also be an end result of RAO. [48]

TREATMENT

Despite significant new information about non-septic inflammatory lower airway diseases in the horse, we have gained little in terms of treatment. It has been well-established that environmental management is critical to the management of RAO [49], and that improvements in PFTs may not be associated with improvements in BALF inflammatory cells or inflammatory molecules in BALF. [50] Certain treatments, such as sound wave therapy, phosphodiesterase inhibitors, and single acupuncture treatments have been shown to be ineffective. [51], [52] , [53] Other potential treatments, such as MAPK p38 inhibitors, neurokinin-2 receptors, and manipulation of CpG motifs, remain to be further studied. [54], [55], [46]

ENVIRONMENT AND FOOD ANIMAL RESPIRATORY DISEASE

Bovine respiratory disease complex is considered the most costly disease of beef cattle in North America. [56] The effect of dust on development of BRDC was explored in the 1980s, and investigators at that time concluded that particles between 2-3.3 um had an impact on development of BRDC. [57] Unfortunately, little data has been forthcoming since that time except in goats. Several studies have shown that aerosolization of sterile dust has a negative impact on respiratory health in goats, with a decrease in the ability to clear fungal spores from the lung, [58] as well as increasing blood lymphocyte counts and rectal temperatures .[59] It appears that this is an area that would profit from further research.

REFERENCES

1. Hughes KJ et al **Veterinary immunology and immunopathology** 2011;140:82-89. 2. Ainsworth DM et al **Veterinary immunology and immunopathology** 2003;96:83-91. 3. Cordeau ME, et al **Veterinary immunology and immunopathology** 2004;97:87-96. 4. Horohov DW et al **Veterinary immunology and immunopathology** 2005;108:237-245. 5. Riihimaki M et al **Veterinary immunology and immunopathology** 2008;122:256-264. 6. Kleiber C et al **Veterinary immunology and immunopathology** 2005;104:91-97. 7. Joubert P et al **Veterinary immunology and immunopathology** 2011;142:236-242. 8. Laan TT et al **Journal of veterinary internal medicine / American College of Veterinary Internal Medicine** 2006;20:167-174. 9. Berndt A et al **American journal of physiology** 2007;292:L936-943. 10. Giguere S et al **Veterinary immunology and immunopathology** 1999;67:1-15. 11. Couetil LL et al **Veterinary immunology and immunopathology** 2006;113:11-20. 12. Couetil LL et al **American journal of veterinary research** 2001;62:538-546. 13. Widmer A et al **Veterinary journal** 2009;182:430-435. 14. Evans DL et al **Research in veterinary science** 2011;90:510-515. 15. Holcombe SJ et al **Equine veterinary journal** 2006;38:300-304. 16. Richard EA, et al **Equine veterinary journal** 2009;41:384-389. 17. Lavoie JP et al **Journal of veterinary internal medicine / American College of Veterinary Internal Medicine** 2011;25:322-329. 18. Bedenice D et al **Journal of veterinary internal medicine / American College of Veterinary Internal Medicine** 2008;22:1022-1028. 19. Beekman L et al **Journal of veterinary internal medicine / American College of Veterinary Internal Medicine** 2012;26:153-161. 20. Van Erck E. **Equine veterinary journal Supplement** 2011;43 Suppl 40:18-24. 21. Gehlen H et al **Journal of veterinary internal medicine / American College of Veterinary Internal Medicine** 2008;22:1014-1021. 22. Christmann U et al **Journal of veterinary internal medicine / American College of Veterinary Internal Medicine** 2008;22:1452-1455. 23. Herszberg B et al. **The Journal of allergy and clinical immunology** 2006;118:382-388. 24. Art T, et al **Canadian journal of veterinary research = Revue canadienne de recherche veterinaire** 2006;70:291-296. 25. Gerber V et al **Equine veterinary journal** 2011;43:153-158. 26. Nolen-Walston RD et al **Journal of veterinary internal medicine / American College of Veterinary Internal Medicine** 2009;23:631-635. 27. Mazan MR et al **Equine veterinary journal** 2005;37:541-545. 28. Pirie RS et al **Clinical and experimental allergy : journal of the British Society for Allergy and Clinical Immunology** 2003;33:676-683. 29. Beeler-Marfisi J et al **American journal of veterinary research** 2010;71:682-689. 30. Davis MS et al **Equine veterinary journal Supplement** 2006:535-539. 31. Mazan MR et al **Occupational medicine** 2009;59:220-225. 32. Riihimaki M et al **Canadian journal of veterinary research = Revue canadienne de recherche veterinaire** 2008;72:432-439. 33. Berndt A et al **Veterinary journal** 2010;183:54-57. 34. Fortier G et al **Veterinary microbiology** 2009;139:34-41. 35. Scharrenberg A et al **Animal genetics** 2010;41 Suppl 2:131-137. 36. Tahon L et al **Veterinary immunology and immunopathology** 2009;127:85-93. 37. Wong DM et al **American journal of veterinary research** 2005;66:1348-1355. 38. Kunzle F et al **Journal of veterinary medicine A, Physiology, pathology, clinical medicine** 2007;54:40-47. 39. Moran G et al **Veterinary research communications** 2010;34:719-726. 40. Deaton CM et al **Equine veterinary journal** 2006;38:417-422. 41. Michelotto PV, Jr., et al **Journal of veterinary internal medicine / American College of Veterinary Internal Medicine** 2010;24:414-419. 42. Tan RH et al **American journal of veterinary research** 2010;71:1500-1507. 43. Gerber V et al **Journal of veterinary internal medicine / American College of Veterinary Internal Medicine** 2009;23:626-630. 44. Shakhsi-Niaei M et al **Animal genetics** 2010;41:559-560. 45. Swinburne JE et al **Mammalian genome : official journal of the International Mammalian Genome Society** 2009;20:504-515. 46. Costa LR et al **Journal of veterinary internal medicine / American College of Veterinary Internal Medicine** 2009;23:1239-1246. 47. Dunkel B et al **Veterinary immunology and immunopathology** 2009;131:25-32. 48. Johansson AM et al **Journal of veterinary internal medicine / American College of Veterinary Internal Medicine** 2007;21:302-307. 49. Couetil LL et al **American journal of veterinary research** 2005;66:1665-1674. 50. Couetil LL et al **Journal of veterinary internal medicine / American College of Veterinary Internal Medicine** 2006;20:399-406.51. Wilson DV et al **Equine veterinary journal** 2004;36:489-494. 52. Goncarovs KO et al **Journal of veterinary internal medicine / American College of Veterinary Internal Medicine** 2010;24:1503-1508. 53. Lavoie JP et al **Journal of veterinary internal medicine / American College of Veterinary Internal Medicine** 2006;20:175-181. 54. Klier J et al **Veterinary immunology and immunopathology** 2011;144:79-87. 55. Lavoie JP et al **Equine veterinary journal** 2008;40:577-583. 56. Taylor JD et al **The Canadian veterinary journal La revue veterinaire canadienne** 2010;51:1095-1102. 57. MacVean DW et al **American journal of veterinary research** 1986;47:2676-2682. 58. Purdy CW et al **American journal of veterinary research** 2008;69:501-511. 59. Chirase NK et al **Journal of animal science** 2004;82:1219-1226.

PULMONARY REGENERATIVE THERAPY: FROM NATURE TO REGENERATIVE NURTURE

Melissa R. Mazan, DVM, DACVIM and Julia Paxson, DVM, PhD, DACVIM
North Grafton, MA

INTRODUCTION

The job of the respiratory system is daunting. By the time the organism is born, the lung has to be ready for continuous exposure to an extremely toxic substance (oxygen), airborne pathogens, continuously eliminate another toxic substance from the body (carbon dioxide), detoxify those substances, and clean and humidify the air that delivers all of this. Meanwhile, it has to do all this while being only a few cell layers thick, and it must do its job repetitively for the life of the organism, never taking a break. To add a layer of difficulty and complexity to all this, the respiratory system must develop at least 40 different cell types to accomplish the task.

GENERAL STRUCTURE OF THE LUNG

Large Structure: Although lobes vary among species, the basic building blocks of the lung are the same. A pair of primary bronchi lead to the right and left lobes. In humans the left lobe is divided into two lobes, and the right into three. The domestic animals, except notably for the horse, have 4 subdivisions on the right – the cranial, middle, caudal, and accessory, and the left has two, the cranial and caudal, with the cranial lobe being divided into cranial and caudal portions. Ruminants and pigs are also different in that the right cranial lobe emanates from the tracheal bronchus, and this right cranial lobe is further divided into cranial and caudal parts. The horse, curiously, has lungs with far less recognizable divisions than others – but the bronchi divide so as to want to give the horse a cranial, middle, caudal, and accessory lobe on each side.

Branching airways: After the trachea has divided into the two mainstem (right and left) bronchi, and in cattle and ruminants, the tracheal bronchus, secondary bronchi then branch throughout the lungs, eventually becoming bronchioles, which are the final conducting airways of the lungs which will eventually lead into the respiratory portion of the lungs.

Respiratory portion of lungs: there is some variation among species – for instance, rodents do not have respiratory bronchioles, whereas horses have short respiratory bronchioles, and they are present in carnivores and primates. Either terminal bronchioles in rodents or respiratory bronchioles in most other species, lead to alveolar ducts, which then lead to the clusters of alveoli that are called alveolar sacs, wherein the epithelium is thinly apposed to the endothelium. This is where the primary job of the lung – gas exchange – is performed.

CELLS OF THE LUNG

Conducting portion of the lung: The conducting airways, from the trachea to the beginning of the respiratory bronchioles, have, depending on the species, columnar or pseudostratisfied epithelium, containing ciliated cells that function to propel mucus upward to be expectorated, goblet cells to produce mucus, basal cells which are relatively undifferentiated and serve as progenitor cells for other airway cells, Clara cells, and neuroendocrine cells.

Respiratory portion: The respiratory bronchioles are composed of the same epithelium as the conducting portion, although ciliated cells are less common in the distal airways, and alveoli, which are composed of squamous epithelium – type I pneumocytes – and cuboidal type II pneumocytes.

Cells of mesenchymal origin: the lung also contains airway smooth muscle which encircles bronchioles, endothelium, vascular smooth muscle cells, pericytes, fibroblasts, lipofibroblasts, myofibroblasts, and mesothelial (pleural cells).

Immune system: Critically important to the health of the lung, as well, are the alveolar macrophages, interstitial macrophages, and in some species, including horses, pigs, ruminants, and cats, pulmonary intravascular macrophages.

REVIEW OF DEVELOPMENT

In order for the lung to develop properly, there must be coordination among the ventral foregut endoderm, which will go on to form the lung epithelium, and the lateral plate mesoderm that will go on to form the basic architecture of the lung, including the connective tissue, the vasculature, the cartilage, the lymphatics, and the pleura.

Developmental stages of the lung have been described in rodents and humans; these stages, while not described for all species, provide a useful framework for understanding lung development.

Embryonic phase (Embryonic day 9-11.5 in mouse, 3-7 weeks in human): At this time, the ventral foregut is recognized as the eventual site of the lungs. Signaling interactions (see below) between mesenchyme and endoderm are essential to ensure that the dorsal portion will become the esophagus, and the ventral portion the trachea. This phase ends with the first bilateral lung buds.

Pseudoglandular phase (E11.5-16.5 mouse, 5-17 weeks in human): This stage is characterized by branching morphogenesis as well as the beginning of vasculogenesis (vessels arising *de novo*) and angiogenesis (vessels arising from pre-existent structures, e.g. the aortic sac). The lung tubules begin to narrow from proximal to distal, the epithelium of the conducting airways begins to differentiate, and the mesenchyme is also beginning to differentiate into fibroblasts, myoblasts, and chondrocytes.

Canalicular phase (E16.5 – 17.5 in mouse, 16-26 weeks in human): At this point, the branching morphogenesis in the human is almost complete. The airways become thinner and narrower, and the respiratory part of the lung begins its big push for development. Alveolar type I and type II cells start to develop.

Sacular phase (E17.5-postnatal day 5 in mouse, 24-38 weeks in human): The alveolar ducts and sacs proliferate, and the stroma becomes thinner and thinner, so that alveolar epithelium and capillaries are in close apposition. The primitive septae are very thick, and each has a double capillary network. Very importantly, ATII cells become functional, producing surfactant.

*Alveolar phase (postnatal day 5-28 in mouse, 38 weeks-maturity in human):*secondary septae form in the alveoli, the alveoli are considerably thinner, and each subsequent alveolus has a single capillary network.

TRANSCRIPTIONAL CONTROL OF LUNG DEVELOPMENT

There are hundreds of genes that are expressed in the process of a lung developing from the primordial endoderm and mesoderm to the eventual creation of a gas exchange organ. There are well over a hundred important transcription factors. This paper will focus on only a very few of the most important of these genes at the major developmental stages. The references here mark review papers that discuss the topic in depth.

Embryonic: All are intimately familiar with NK2 homeobox 1 transcription factor (Nkx2-1), otherwise known as thyroid transcription factor (Ttf-1) very well indeed, for without it, our lungs would never have even started to develop. It is Nkx2-1 that is expressed in the endoderm (which will become the epithelium of the lung) that determines that the ventral portion of the ventral foregut will become the trachea, and not the esophagus. (Rawlins) At this stage, there is also critical cross-talk signaling going on between the surrounding mesenchyme and the endoderm, including various BMPs, FGFs, and Wnts. (Morrissey and Hogan) The developing lung is very plastic and very undifferentiated at this stage.

Pseudoglandular: During this phase, there is an absolute requirement for the interactions between FGF10 and its receptor, FgfR-2. FGF10 is produced by the surrounding mesoderm, and FgfR2 is produced by the lung buds of the endoderm. Without these growth factors, lung buds will not appear, and branching morphogenesis will not happen. In addition, it appears that BMPs, Wnts, Shh, and retinoic acid signaling are all also necessary for epithelial/mesenchymal interactions that underlie appropriate branching morphogenesis.

Canalicular: During this period, transcriptional control of differentiation of airway cells is most important. Basal cells are marked by p63, Clara cells Ttf-1 and Foxa1/a2, and ciliated cells Foxj1 and Sox 17. Interestingly, the concentration of these factors differs proximally to distally in the airways. (Maeda)

Sacular: At this stage, the endoderm of the distal lung is starting to differentiate into ATI and ATII cells. Retinoic acid signaling becomes very important – in a Goldilocks way. It cannot be too much or too little – it must be just right. (Morrissey) Glucocorticoid receptors, not surprisingly, are important in perinatal lung maturation. The difficult part is that GCs may enhance maturation of surfactant production and help regulate fluid homeostasis in the lung (Maeda) , but on the other hand, may inhibit secondary septae formation (Morrissey and Hogan) thus contributing to an abnormally simplified lung. Ttf-1, FOXA2, NFATC3, among others, also are important to perinatal lung maturation.

Alveolar: Production of elastin and other matrix molecules are integral to formation of secondary septae, and these processes rely on retinoic acid signaling. It is likely that mesenchymal stem cells are integral to this process.

Postnatal: interestingly, proper regulation of GATA-6 and TTF-1 appear to be necessary to postnatal alveolarization as well.

REGENERATIVE THERAPIES FOR THE LUNG

Knowing how nature makes a lung, is it possible for us to assist in regeneration of portions of a lung, or in some utopian world, an entire lung? The main approaches to this problem are 1) stimulating or delivering endogenous stem cells, 2) use of exogenous stem cells and 3) whole-lung engineering.

What are the endogenous stem cells of the lung? In embryonic development, the cells at the distal tips of branching airways have been characterized as multipotent progenitor cells that self-renew during development and can give rise to multiple different cell types. These cells have a different gene expression and faster cell cycle kinetics than more proximal cells (Rawlins) and lineage tracing experiments using one of these genes (Id2) as a knock-in show that these cells indeed differentiate to all bronchiolar and alveolar cell types. (Rawlins) What is far less clear is the identity and role of stem cells in the adult mammal. Warburton's group has described 5 putative niches and their accompanying stem/progenitor cells in the lung – 1) airway submucosal glands that can reconstitute surface tracheal epithelia, 2) Keratin-14 –expressing basal cells that can repopulate denuded airway epithelium, 3) variant Clara cells that are located within neuroendocrine bodies, and 5) BASC, variably known also as variant Clara cells at the bronchoalveolar duct junctions, which can potentially repopulate the distal airway and alveolus, and type II alveolar cells which can regenerate ATI cells after injury, and mesenchymal stem cells. (Garcia) The mesenchymal lung stem cell has only recently been the focus of much interest, with the realization that both in development and in repair, the interaction between epithelium and mesenchyme is essential to proper growth. (McQualter) Our lung has explored the role of LMSCs in healing elastase-induced injury. Recently, a multipotent lung stem cell in the human lung was described – further work is necessary to further understand the significance of this finding. (Kajstura)

Exogenous stem cells: Both hematopoietic and non-lung-derived mesenchymal stem cells have been used in lung injury with the aim of modulating inflammation and perhaps stimulating endogenous lung stem cells. Work is currently ongoing to determine if induced pluripotent stem cells and embryonic stem cells will either be able to develop into appropriate lung cells or stimulate regeneration or ameliorate inflammation in the injured lung. (Garcia)

Whole lung regeneration: Whole lung regeneration in many ways is looked on as the holy grail. Again, because of the recognition of the complexity and necessity of the mesenchyme scaffold of the lung in development, the common approach is to find a scaffold that can be repopulated with both epithelial cells and cells of mesenchymal origin. The lung can be stripped of its cells (de-cellularized), and seeded with lung. This has been done successfully with a tracheobronchial transplant in a human, but much work remains to be done for a whole lung or lobule. (Macchiarini)

REFERENCES

Cardoso WV et al. *Development* 133, 1611-1624, 2006 Rawlins E. *Developmental Dynamics* 240; 463-476, 2011 Morrissey EE et al. *Developmental Cell* ;18, 2010 Maeda Y et al. *Physiol Review;*87, 219-244, 2007 Hong KU, et al. *Am J Respir Cell Mol Biol*, 2001, 24:671-681 Giangreco A, et al. *Am J Pathol* 2002, 161:173-182 Kim CF, et al. *Cell* 2005;121, 823-835 Garcia O, et al. *British Medical Bulletin 2012, 1-15* McQualter JL et al. *Proc Natl Acad Sci USA* 2010; 107, 1414-1419 Hoffman AM et al. *Stem Cells Development*, 20; 1779-1792 Kajstura J et al, *N Eng J Med* 2011;364:1795-806 Macchiarini P, et al. *Regenerative Medicine* 2011;6:14-15

EVIDENCE-BASED ANTIMICROBIAL TREATMENT STRATEGIES FOR INFECTIONS IN LLAMAS AND ALPACAS

Daniela Bedenice, Dr. med vet., DACVIM, DACVECC. Grafton, MA

Bacterial infections requiring extended antimicrobial therapy (such as pneumonia, peritonitis, sepsis, uterine and dental infection) [1,2] are a significant cause of morbidity and mortality in camelids. Traditionally, plasma drug concentrations in relation to the minimum inhibitory concentration (MIC) for the bacterial pathogen of interest have been used to predict potential efficacy of antimicrobial agents. For respiratory tract infections, pulmonary epithelial lining fluid (ELF) and alveolar macrophages (AM) have been advocated as important sites of infection with common extracellular and intracellular respiratory pathogens, respectively. Drug penetration and drug concentrations in lung compartments should therefore assist in the selection of appropriate antibacterial therapy and design of dosing regimens [3]. In order to reach the alveolar ELF and lung interstitium antimicrobials must cross the blood-bronchus barrier. Factors that affect penetration include drug protein binding, the degree of lipophilicity, administration of the drug in a non-ionized form, pH at the site of infection as well as the degree of pulmonary inflammation. Additionally, the pharmacokinetic-pharmacodynamic parameters that correlate with the efficacy of specific antimicrobial agents include (1) the percentage of time during which unbound (free) drug concentrations remain above the MIC of an infecting micro-organism for time dependent drug; (2) the ratio of the maximum unbound drug concentration to the MIC of an infecting micro-organism in the case of concentration dependent antimicrobials (3) the ratio of the unbound drug's area under the curve (AUC) during a 24-hour time period to the MIC of an infecting micro-organism [in the case of antimicrobials such as fluoroquinolones, aminoglycosides, azithromycin, clarithromycin, doxycycline and vancomycin] [3].

To date only seven antimicrobials have been evaluated in the peer-reviewed literature and subsequently proven to have favorable pharmacokinetics for therapeutic use in camelids [4-9]. For example, bacterial isolates obtained from both llamas and alpacas have demonstrated favorable sensitivity patterns to the antimicrobial florfenicol [10]. Florfenicol is a broad-spectrum antibiotic with penetration into most body tissues including internal organs, lung, skeletal muscle, milk, synovial fluid and to a lesser extent aqueous humor and spinal fluid. However, recent data suggests that the subcutaneous dosing of 40mg/kg florfenicol in the form of **Nuflor®** in alpacas may be inadequate for some bacterial pathogens unless multiple doses are administered. On the other hand, there is evidence of adverse effects caused by Nuflor® after repeated 40mg/kg SC doses. Holmes et al investigated the pharmacokinetics of florfenicol (Nuflor®; 300 mg/mL; Schering-Plough Corp., Kenilworth, NJ, USA) in eight healthy adult alpacas, following single dose injection of 20mg/kg IM (intramuscular) and 40mg/kg SC (subcutaneous) using a randomized, cross-over design and 28-day washout period. Subsequently, 40mg/kg Nuflor® (florfenicol) was administered SC every other day for 10 doses to evaluate long-term effects. Maximum plasma florfenicol concentrations (C_{max}, measured via high-performance-liquid-chromatography) were achieved rapidly, leading to a higher C_{max} of 4.31+/-3.03 μg/ml following administration of 20mg/kg IM than 40mg/kg SC (C_{max}: 1.95+/-0.94 μg/ml). Multiple SC dosing at 48hr intervals achieved a C_{max} of 4.48+/-1.28 μg/ml at steady state. The area under the curve and terminal elimination half-lives were 51.83+/-11.72μg/ml.h and 17.59+/-11.69 hours after single 20mg/kg IM administration, as well as 99.78+/-23.58μg/ml.h and 99.67+/-59.89 hours following 40mg/kg injection of Nuflor® SC, respectively. Florfenicol decreased the following hematological parameters after repeated administration between weeks 0 and 3: total protein (6.38 vs. 5.61 g/dL, P<0.0001), globulin (2.76 vs. 2.16 g/dL, P<0.0003), albumin (3.61 vs. 3.48 g/dL, P=0.0038), white blood cell count (11.89 vs. 9.66 x10^3/μL, P<0.044), and hematocrit (27.25 vs. 24.88%, P<0.0349). Significant clinical illness was observed in one alpaca.[9]

The florfenicol product **Nuflor Gold®** (300 mg/mL; Schering-Plough) was more recently introduced to the US market, in addition to standard Nufor®. This new formulation is claimed to show better absorption and leads to higher plasma drug concentrations in cattle following single 40mg/kg SC dosing (Nuflor Gold® C_{max}: 5.93 mcg/mL; Nuflor® C_{max}: 4.69 mcg/mL).[11] Similarly, a 2011 study in camelids suggests that the maximum plasma florfenicol concentration is 3-4 times higher in alpacas receiving a single dose of 40mg/kg Nuflor Gold® SC vs. 40mg/kg Nuflor® SC. In alpacas, plasma levels remained above 1 mcg/mL for 24 hours (mean concentration of 1.94 +/- 1.5 mcg/mL at 18 hrs) following the administration of Nuflor Gold®, thus showing superior pharmacokinetics for use in alpacas in comparison to conventional Nuflor®.

The disposition of five therapeutic antimicrobial agents was studied in llamas following intravenous bolus administration by Christensen et al.[5] Six llamas were each given **ampicillin, tobramycin, trimethoprim / sulfamethoxazole, enrofloxacin** and **ceftiofur** at a dose of 12 mg/kg, 1 mg/kg, 3 mg/kg / 15 mg/kg, 5 mg/kg, and 2.2 mg/kg of body weight, respectively, with a wash out period of at least 3 days between treatments. Plasma concentrations of these antimicrobial agents were determined by reverse phase HPLC over 12 hours following IV bolus dosing. Based on the disposition parameters determined, and assumptions of likely effective minimum inhibitory concentrations (MIC), the dose and dosing interval for the following antimicrobial agents was suggested as follows: 6 mg/kg every 12 h for ampicillin; 4 mg/kg once a day or 0.75 mg/kg every 8 h for tobramycin; 5 mg/kg every 12 h for enrofloxacin; and 2.2 mg/kg every 12 h for ceftiofur sodium for llamas. Intrapulmonary penetration studies have been conducted with β-lactam agents since the early 1990s, specifically focusing on the use of penicillins and cephalosporins in humans and animals. However, the percentages of ELF penetration differs substantially both based on cephalosporin type and animal species.

The pharmacokinetics and bioavailability of **trimethoprim-sulfamethoxazole** (TMP-SMX) were also studied in six healthy male-castrate alpacas after intravenous (IV) or oral (PO) drug administration of 15 mg/kg TMP-SMX using a crossover design with a 2-week washout period. After 90 days one group (n = 3) was given a PO dose of 30 mg/kg TMP-SMX and the other group (n = 3) was given a PO dose of 60 mg/kg TMP-SMX. After IV administration of 15 mg/kg of TMP-SMX the mean initial plasma concentration (C0) was 10.75 +/- 2.12 microg/mL for trimethoprim (TMP) and 158.3 +/- 189.3 microg/mL for sulfamethoxazole (SMX). Elimination half-lives were 0.74 +/- 0.1 h for TMP and 2.2 +/- 0.6 h for SMX. The mean residence times were 1.45 +/- 0.72 h for

TMP and 2.8 +/- 0.6 h for SMX. The areas under the respective concentration vs. time curves (AUC) were 2.49 +/- 1.62 microg h/mL for TMP and 124 +/- 60 microg h/mL for SMX. Total clearance for TMP was 21.63 +/- 9.85 and 1.90 +/- 0.77 mL/min kg for SMX. The volume of distribution at steady state was 2.32 +/- 1.15 L/kg for TMP and 0.35 +/- 0.09 L/kg for SMX. After intragastric administration of 15, 30 and 60 mg/kg the peak concentration (C_{max}) of SMX were 1.9 +/- 0.8, 2.6 +/- 0.4 and 2.8 +/- 0.7 microg/mL, respectively. The AUC was 9.1 +/- 5, 25.9 +/- 3.3 and 39.1 +/- 4.1 microg h/mL, respectively. Based upon these AUC values and correcting for dose, the respective bioavailabilities were 7.7, 10.5 and 7.94%. Trimethoprim was not detected in plasma after intragastric administration. These data demonstrate that therapeutic concentrations of TMP-SMX are not achieved after PO administration to alpacas.[12]

The disposition of **ceftiofur** sodium was studied in llamas following intramuscular (IM) administration and in alpacas following intravenous and IM administration. Eleven adult llamas were given ceftiofur sodium by IM injection. Each animal received either a standard dose of 2.2 mg/kg or an allometrically scaled dose ranging from 2.62 to 2.99 mg/kg in a crossover design. Ten adult alpacas were given ceftiofur sodium by IV and IM injections. Each animal received a standard dosage of 1 mg/kg or an allometrically scaled dose ranging from 1.27 to 1.44 mg/kg IV, and 1.31-1.51 mg/kg IM. Blood samples were collected at 0, 0.25, 0.5, 1, 2, 4, 8, 12, 24, 36, 48, and 72 h after administration of ceftiofur. Pharmacokinetic parameters of ceftiofur in llamas and alpacas were similar following IM dosing at both dose levels. Overall, ceftiofur pharmacokinetics in llamas and alpacas were similar, and comparable to reported parameters for sheep and goats.[7]

Both single and multiple dose trials to determine the pharmacokinetics of **gentamicin** sulfate have been performed in clinically normal, adult male llamas. Nineteen llamas were given gentamicin (5 mg/kg of body weight, IV) as a single bolus, and serum gentamicin concentration was monitored over the next 48 hours. Two months later, 10 llamas were given gentamicin (2.5 mg/kg) IV for the first day, then IM every 8 hours for 7 days. The 5 mg/kg IV kinetic study revealed $t_{1/2}$ alpha of 14.5 +/- 5.06 minutes and $t_{1/2}$ beta of 166 +/- 20.5 minutes. Similarly, the 2.5 mg/kg IV kinetic study revealed $t_{1/2}$ alpha of 17.7 +/- 6.59 minutes and $t_{1/2}$ beta of 165 +/- 40.3 minutes. Peak serum gentamicin concentration averaged 10.1 micrograms/ml in the multiple-dose trial, and trough concentration averaged 1.5 micrograms/ml. Dose effects were not observed for gentamicin clearance, volume of distribution, or half-lives.[4] However, it has been shown that once daily dosing of a higher aminoglycoside concentration is associated with less renal toxicity than multiple lower daily doses, based on studies in humans and horses. In this context, it was identified that renal tubular uptake (and thus toxicity) of aminoglycoside is most dependent of the duration of drug exposure. Overall, gentamicin pharmacokinetic variables in llamas resemble those in other ruminant species. Aminoglycosides are known for poor lung penetration and varied concentrations in the lung tissue depending on the anatomical site that is sampled. In general, studies in humans have shown that aminoglycosides concentrations in ELF are significantly lower than plasma concentrations during the first 1.5 hours after administration. However, the concentrations in ELF become similar to those in plasma by 2 or 3 hours after administration [3].

Gandorf et al determined the plasma concentrations of **enrofloxacin** and the active metabolite ciprofloxacin after PO, SC and IV administration in alpacas. A crossover design was used for administration of 3 single-dose treatments of enrofloxacin to alpacas, which was followed by an observational 14-day multiple-dose regimen. Single-dose treatments consisted of IV and SC administration of injectable enrofloxacin (5 mg/kg) and PO administration of enrofloxacin tablets (10 mg/kg) dissolved in grain to form a slurry. Plasma enrofloxacin concentrations were measured by use of high-performance liquid chromatography. The multiple-dose regimen consisted of feeding a mixture of crushed and moistened enrofloxacin tablets mixed with grain. Behavior, appetite, and fecal quality were monitored throughout the 14-day treatment regimen and for 71 additional days following treatment. Mean half-life following IV, SC and PO administration was 11.2, 8.7, and 16.1 hours, respectively. For SC and PO administration, mean total systemic availability was 90.18% and 29.31%, respectively; mean maximum plasma concentration was 3.79 and 1.81 microg/mL, respectively; and area under the curve (AUC) was 50.05 and 33.97 (microg x h)/mL, respectively. The SC or PO administration of a single dose of enrofloxacin yielded a ratio for AUC to minimum inhibitory concentration >100 for many grampositive and gram-negative bacterial pathogens common to camelids. The administration of enrofloxacin (5 mg/kg, SC or 10 mg/kg, PO) may be thus be appropriate for antimicrobial treatment of adult alpacas, although the pharmacokinetic data following SC use are superior.

A study in dogs has shown that mean concentrations of enrofloxacin in ELF were greater than in plasma (ELF to plasma ratio of 13.5). Enrofloxacin concentrations in AM and ELF largely exceeded the MICs of the major bacterial pathogens and surpassed (by about two times) the breakpoint MIC of that drug. These results suggest that sufficient antimicrobial activity is present in AM and ELF of dogs following oral administration of 5 mg/kg enrofloxacin to be effective in the treatment of lower respiratory tract infections involving susceptible organisms[13].

REFERENCES:

[1] Niehaus AJ, Anderson DE. **J Am Vet Med Assoc** 2007;231:284-289. [2] Smith J. **Vet Clin North Am: Food Animal Practice** 1989;5:101-143. [3] Rodvold KA, Yoo L, George JM. **Clinical pharmacokinetics** 2011;50:689-704. [4] Lackey MN, Belknap EB, Greco DS, et al. **Am J Vet Res** 1996;57:1193-1199. [5] Christensen JM, Smith BB, Murdane SB, et al. **J Vet Pharmacol Ther** 1996;19:431-438. [6] Junkins K, Boothe DM, Jensen J, et al. **J Zoo Wildl Med** 2003;34:9-15. [7] Drew ML, Johnson L, Pugh D, et al. **J Vet Pharmacol Ther** 2004;27:13-20. [8] Gandolf AR, Papich MG, Bringardner AB, et al. **Am J Vet Res** 2005;66:767-771. [9] Holmes K, Bedenice D, Papich MG. **Vet Pharmacol Ther** 2011; Epub ahead of print. [10] Anderson D. **International Camelid Health Care Conference** 2009. [11] Schering-Plough-Animal-Health. **Freedom of information summary**. Original new animal drug application, 2008. [12] Chakwenya J, Lakritz J, Tyler J, et al. **J Vet Pharmacol Ther** 2002;25:321-327. [13] Hawkins EC, Boothe DM, Guinn A, et al. **Journal of veterinary pharmacology and therapeutics** 1998;21:18-23.

HYPERCOAGULABILITY IN PULMONARY DISEASE: WHERE'S THE EVIDENCE

Elizabeth Rozanski, DVM, DACVIM (SAIM), DACVECC
North Grafton, MA

Pulmonary disease is a common in small animals. Most approaches to pulmonary disease classically focus on localization and/or management. The purpose of this talk at the Forum is to discuss the potential role of inflammation and subsequent development of hypercoagulability in magnifying and perpetuating lung injury. The aspect of pulmonary medicine is particularly appealing, as a variety of clotting tests have been introduced to clinical medicine practice and long-term anticoagulant therapy has become more practical in dogs and cats.

Laboratory diagnosis of hypercoagulability- Traditionally, hypercoagulability has been difficult to detect with routine laboratory based methods for evaluating coagulation. While increased clot formation and subsequent lysis can be detected by documenting increased circulating degradation products of cross-linked fibrin (D-dimers), this is only an indirect suggestion of hypercoagulability. Thrombin-antithrombin complexes may also be measured. Thromboelastography (TEG) however, is a simple, point-of-care test which has recently become widely commercially available and with particular utility for detection of hypercoagulability. Thromboelastography produces a standardized tracing, the components of which represent the different stages and quality of clot formation and lysis. The reaction time (R time, minutes) represents the rate of initial fibrin formation; the clotting time (K time, minutes), represents the time until a fixed level of clot firmness is reached; the angle (α, degrees), represents the rate of clot growth; the maximal amplitude (MA, mm), measures the maximal strength of the developed clot; and the shear elastic modulus strength (SEMS or G, dynes/cm^2), is a parametric measure of clot firmness expressed in metric units and calculated from MA [G = (5000 x MA)/(100-MA)]. The A_{60} value is the amplitude of the TEG tracing 60 minutes after the MA has been reached and is useful in measuring clot retraction or lysis. A hypercoagulable TEG tracing is represented by a short R time, short K time, steep angle, and a wide MA.

Local clot formation may also be detected on histopathology, which may confirm the presence of microvascular thrombosis, or larger clots may be seen on ultrasound or CT scanning. This scan shows a pulmonary thrombus in a dog with metastatic thymoma. No clinical signs of respiratory impairment were present. The thrombi are highlighted by the arrows.

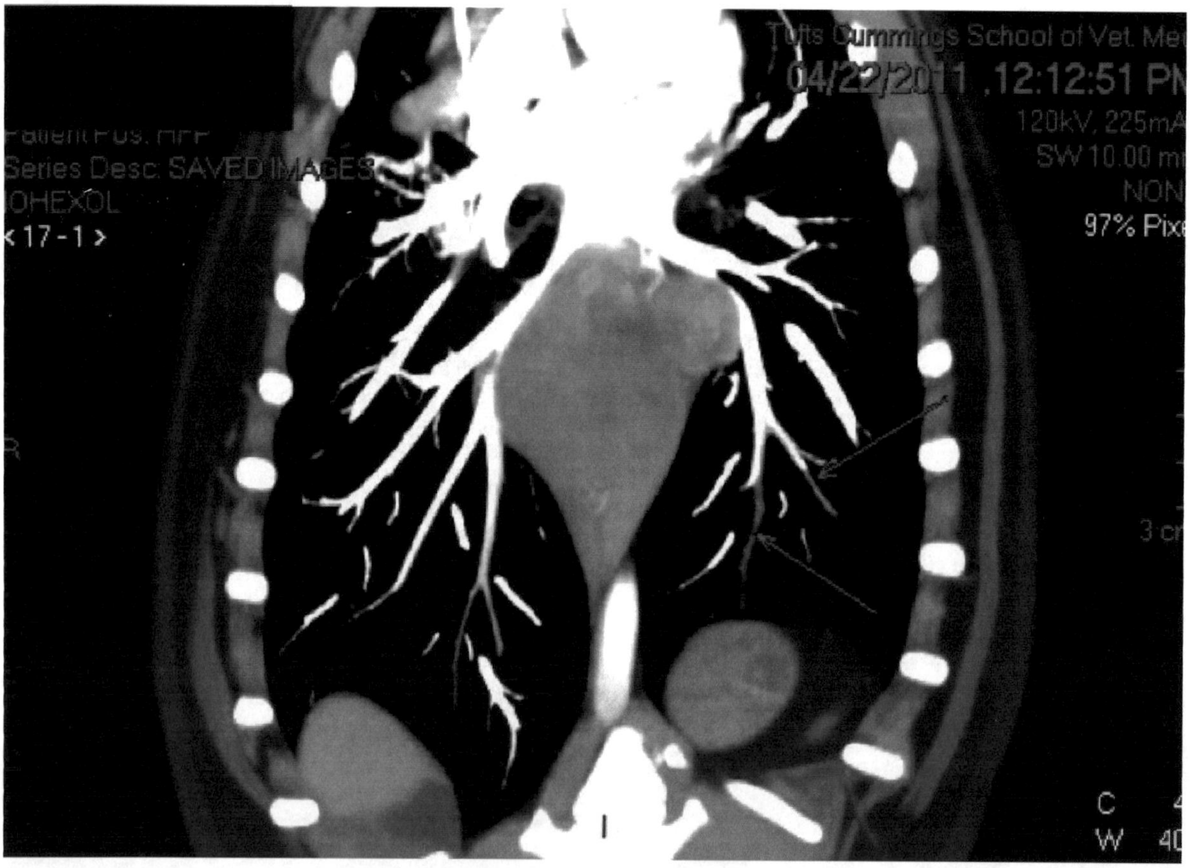

The major pulmonary diseases that warrant consideration for hypercoagulability include acute lung injury/acute respiratory distress syndrome (ALI/ARDS), chronic obstructive pulmonary disease, pulmonary fibrosis, and possibly hypoxemia alone. Lung disease may result in hypercoagulability through the presence of inflammation, vascular stasis, altered blood viscosity, and endothelial activation. Sleep apnea in people may also be associated with a pro-inflammatory condition that may result in hypercoagulability. Finally, and of clinical importance, pulmonary disease is often treated with glucocorticoids, which are well-recognized to promote a hypercoaguable state.

ARDS/ALI is a complication of critical illness associated with the development of an inflammatory non-cardiogenic (permeability) pulmonary edema. Histopathology of non-survivors demonstrates diffuse alveolar damage, with associated microvascular thrombi and the formation of excessive intra-alveolar fibrin. An imbalance between coagulation and fibrinolysis appears for favor the formation of thrombi, both systemically and within the lungs. Some researchers have implemented tissue factor (TF) pathways in the pathogenesis of ARDS. Excessive TF promotes inflammation and coagulation and will promote fibrin deposition. Additionally, altered fibrinolysis may be a key player. Recommendations for anticoagulation in ARDS/ALI exist, and include treatment with heparin, activated Protein C (when accompanying severe sepsis) and plasminogen activators. Canine cases of ARDS are less commonly recognized, but alveolar fibrin deposition is a feature as well. The criticalist should recognize that inflammation and coagulation are closely linked, and attempts to attenuate inflammation may affect hemostasis.

COPD is a chronic inflammatory lung disease in people, and similar equivalents are appreciated in dogs and cats. As an inflammatory disease, COPD is considered pro-coagulant, and fibrin deposition is a feature in some cases of severe airway disease. Additionally, anti-inflammatory doses of prednisone are the mainstay of therapy, and these promote coagulation. Severe COPD may also result in chronic hypoxemia, and this may be pro-coagulant in of itself.

De Boer and colleagues in a review article this January discussed that asthma is associated with a procoagulant state specific to the bronchoalveolar space, and that this is worsened by impaired local activities of the anticoagulant protein C system and fibrinolysis. Recently, protease activated receptors have been implicated as the molecular link between coagulation and allergic inflammation in asthma which represents and exciting opportunity to intervene and potentially prevent the worsening of some severe asthmatics. .

Pulmonary fibrosis, while considered most likely to represent an alteration in epithelial repair has also been association with the development in intravascular and alveolar thrombi. One study (Markart et al) evalatuated treating IPF human patients with inhaled heparin, and sound that it was safe and well-tolerated. The role of adjuvant inhaled heparin in IPF patients remains to be determined.

Pulmonary hypertension represents a unique from of pulmonary dysfunction and may occur secondary to pulmonary thromboembolism (PTE). Additionally, and of clinical relevance, initially non-thrombotic pulmonary hypertension may ultimately be associated with the development of PTE due to vascular stasis and endothelial activation and some physicians advocate the addition of anticoagulants to any person with PHTH.

Sleep apnea - Obstructive sleep apnea (OSA) is a common syndrome in people characterized by episodes of total and/or partial collapse of the upper airways alternating with normal breathing during sleep. Obstructive sleep apnea is associated with chronic intermittent hypoxia, oxygen de-saturation, snoring, sleep fragmentation, and arousal from sleep. Excessive daytime sleepiness is associated with poor quality of life as well as a multitude of health problems including the development of cardiovascular (CV) and thromboembolic (TE) disorders (coronary artery disease, hypertension, cardiac arrhythmias and stroke). While CV and TE disease in people with OSA is undoubtedly multi-factorial, hypercoagulability is a key etiologic factor.

The English bulldog has been used as a naturally occurring model of OSA, with documented chronic intermittent hypoxia, episodes of oxygen de-saturation, frequent snoring, excessive daytime sleepiness, and exercise intolerance associated with abnormal upper airway anatomy and disordered respirations. Additionally, brachycephalic dogs have a higher risk of cardiovascular disease and shorter life span than dogs with normal facial conformation. Given the species similarities between people and dogs with regards to OSA, it may be that some of the early onset CV disease and other complication are associated with a hypercoagulable state.

In people with OSA, hypercoagulability can be documented by TEG and increased D-dimer activity, and successful treatment results in normalization of clotting times. Specifically, Guardiola determined that people with OSA had significantly shortened R times, and von Kanel reported patients with OSA have elevated plasma fibrinogen levels, exaggerated platelet activity, and reduced fibrinolytic capacity. In addition, Bokinsky reported increased nocturnal platelet activation and aggregation in patients with OSA, and Shitrit documented increased D-dimers in people with OSA, with greatest abnormalities documented in people with the most severe hypoxemia. Hypercoagulability (as evidenced by shortened R and K times, but not widened MA) has also been demonstrated in a rodent model of OSA.

Recommendations and clinical significance

Pulmonary disease and/or hypoxemia is considered inflammatory. Inflammatory disease and coagulation are closely linked, and perpetuation of inflammation is likely to result in altered coagulation. Evidence from human and experimental animal studies support the development of a hypercoagulable state with pulmonary disease, and this is parodoxically worsened by steroids. Diagnosis of alteration in coagulation may be based upon systemic activation of the clotting cascade or by more specialized testing such as TEG. Histopahthology of lung tissue may show fibrin deposition, and advanced diagnostic imaging may reveal thrombi. The question if it is warranted to anticoagulant dogs with chronic pulmonary disease or ARDS is unknown, but may be considered.

Figure 2. TEG from a dog with severe chronic inflammatory airway disease and pulmonary hypertension. This is consistent with marked systemic hypercoagulability. The normal trace is given in the dotted graph.

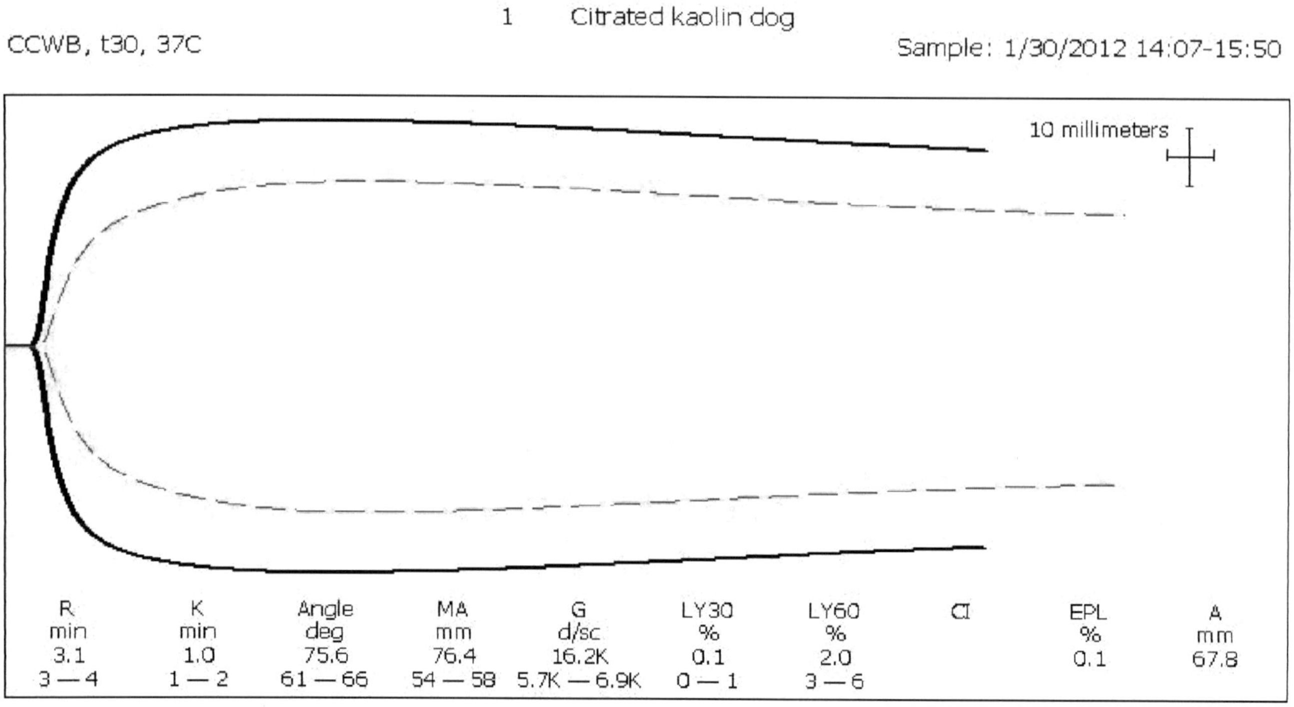

REFERENCES

1) Markart et al Safety and tolerability of inhaled heparin in idiopathic pulmonary fibrosis. J Aerosol Med Pulm Drug Deliv. 2010 Jun;23(3):161-72.2) Sebag SC et al. Therapeutic Modulation of Coagulation and Fibrinolysis in Acute Lung Injury and the Acute Respiratory Distress Syndrome. Curr Pharm Bioechnol 2011 (epub) 3) Hofstra JJ et al. The role of bronchoalveolar hemostasis in the pathogenesis of acute lung injury. Semin Thromb Hemost 2008, 34*5) 475-484. 4)Donahue SM, Otto CM. Thromboelastography: a tool for measuring hypercoagulability, hypocoagulability, and fibrinolysis. *J Vet Emer Crit Care.* 15;2005:9-16 5.)Budhiraja R, Parthasarathy S, and Quan SF. Endothelial dysfunction in obstructive sleep apnea. *J Clin Sleep Med.* 3;2007: 409-156.) Hendricks JC, Kline LR, Kovalski RJ et al. The English bulldog: a natural model of sleep-disordered breathing. *J Appl Physiol.* 63; 1987:1344-50 7) Othman M, Gordon SP, Iscoe S. Repeated inspiratory occlusions in anesthetized rats acutely increase blood coagulability as assessed by thromboelastography. *Resp Physiol & Neuro.* 171;2010:61-66 8) Guardiola JJ, Matheson PJ, Clavijo LC et al. Hypercoagulability in patients with obstructive sleep apnea. *Sleep Med.* 2;2001:517-23 9) Sanner BM, Konermann M, Tepel M et al. Platelet function in patients with obstructive sleep apnoea syndrome. *Eur Respir J.* 16;2000:648-52. 10) de Boer at al. Asthma and Coagulation, Blood Jan 2012 (epub).

EXPANDING THE DIAGNOSTIC APPROACH FOR FELINE ALLERGIC ASTHMA: FUNCTIONAL AND IMMUNOLOGIC TESTING

Carol Reinero, DVM, DACVIM, PhD
Columbia, MO

OVERVIEW

Feline asthma is a common diagnosis made in cats with respiratory signs. There are multiple differentials for this lower airway disorder including chronic bronchitis, *Aleurostrongylus abstrussus*, heartworm associated respiratory disease (HARD) and ascariasis (i.e., phase when migrating through the lung). Although they have overlapping clinical and radiologic features, they are all distinct disorders with different pathophysiologic bases of disease and requiring different treatments. These differentials need to be discriminated with further testing; however, currently available diagnostics have limitations. Traditionally, a combination of an appropriate history and physical examination findings (e.g., cough, wheeze, and/or expiratory respiratory distress), thoracic radiographic features (bronchial or bronchointerstitial pattern), and bronchoalveolar lavage fluid analysis demonstrating airway eosinophilia has been used to support a diagnosis of feline asthma. Ancillary diagnostics (fecal examination, heartworm serology) may or may not provide definitive proof of an underlying parasitic infection.

In humans with asthma, pulmonary function testing is used to discriminate fixed and non-fixed airway obstruction, assess airway hyperresponsiveness (AHR) and determine if there is a reduction of airflow limitation in response to a bronchodilator. Differences in pulmonary mechanics can be appreciated with different types of airway pathologies. Additional information, reviewed in this lecture, is being gathered on pulmonary function testing in cats and may ultimately prove to be beneficial. Another key feature unique to feline allergic asthma and different from chronic bronchitis and the other parasitic diseases is its immunopathogenesis. There is ample evidence that asthma in cats is due to a type I hypersensitivity reaction to inhaled allergens. Understanding more about how aeroallergens are taken up and processed by mucosal dendritic cells for presentation and activation of T helper 2 cells which orchestrate the allergic inflammatory process may open up doors for improved immunodiagnostics specific for asthma.

PULMONARY MECHANICS

In human asthmatics, pulmonary function testing helps confirm the presence of obstructive airway disease. Spirometry is employed to measure airflow and volumes in and out of the lung and requires a respiratory maneuver dependent on patient cooperation and effort as they are instructed to inhale and exhale forcibly through a mouthpiece. Airflow may be impeded by bronchoconstriction/AHR, edematous or inflamed airway walls with a narrow caliber, or intralumenal airway secretions in asthma. Spirometry is beneficial in diagnosis, stratification of disease severity, determination of response to therapy and prognosis. Key features of asthma include AHR and reversibility (at least in part) of airflow limitation by a bronchodilator.

Given that cats are unable to respond to commands to forcibly exhale through a mouthpiece, other strategies for assessing lung function have been evaluated. Non- to minimally-invasive methods of lung function include tidal breathing flow-volume loops using a tight fitting facemask, barometric whole body plethysmography (BWBP), and forced expiratory flow-volume curves using a thoracic compression technique. Invasive methods of pulmonary mechanics are the most reliable measures of airway resistance and other lung function parameters but require anesthesia. One of the newer techniques to acquire this data is using ventilators capable of measuring airflow and pressure [1].

To avoid anesthesia, over the last several years there have been multiple studies using BWBP for airway diseases in cats. However, the index of airflow limitation measured by this technique called enhanced pause (Penh) is likely not a good measure of airway resistance [2,3]. This is underscored by some discrepant results in studies using one technique versus the other. For example, when inhaled N-acetylcysteine was assessed by BWBP, no changes in Penh were noted [4]. When the same inhalant drug was delivered using ventilator-acquired pulmonary mechanics, increases in airway resistance (on average an increase of 150% over baseline) were noted in every cat [1] However, the non-invasive nature of BWBP, which does provide an outcome measure that may relate to clinical improvement, likely still has merit as long as it is interpreted with caution.

Ventilator-acquired pulmonary mechanics can be used if the ventilator has a sensitive pneumotachograph capable of measuring flows and pressures; resistance can subsequently be calculated and displayed in real-time on the monitor screen. Resistance values, when used in tandem with bronchoprovocation trials (see below) may help determine AHR by traditional calculations of dose-response curves—i.e., determination of the effective concentration of bronchoprovocant leading to a 200% increase in baseline resistance, the EC200Raw. Additionally, certain ventilators are capable of breathhold maneuvers to allow determination of end inspiratory plateau pressures (Pplat) and positive end expiratory pressures (PEEP). The measurement of Pplat-PEEP reflects lung resistive pressures, an inverse correlate of respiratory system compliance.

Increases in Pplat-PEEP may indicate airway constriction, edema or inflammation. Interestingly, there is evidence in experimentally asthmatic cats that measurement of Pplat-PEEP may be able to detect changes in lung mechanics when the "gold standard" EC200Raw is not able to[5].

For airway diseases like asthma, assessment of AHR is important and can be applied in tandem with different methods of pulmonary mechanics measurement. Bronchoprovocation can be performed using direct (e.g., methacholine, carbachol, histamine) or indirect (mannitol, AMP) agents. Direct bronchoprovocants cause airway constriction by stimulating airway smooth muscle cells themselves. Indirect bronchoprovocants including adenosine monophosphate (AMP) and mannitol cause airway constriction by acting on inflammatory cells or activated airway neural pathways. With this method, pre-existing airway disease must be present for indirect bronchoprovocants to have an effect. It is thought that indirect bronchoprovocation is more important in assessing AHR induced by inflammation (thus separating healthy from asthmatic individuals) and measuring the effectiveness of anti- inflammatory treatment. A study using BWBP with AMP has been performed in healthy research cats without airway disease, in pet cats with chronic bronchitis or asthma [6]. None of the nine healthy cats reacted to AMP, whereas 2/6 cats with chronic bronchitis and 7/9 cats with asthma reached the endpoint of the study suggesting AMP induces airflow limitation in cats with pre-existing inflammation.

IMMUNOLOGIC BASIS FOR ALLERGIC ASTHMA

In susceptible individuals, when an allergen is inhaled into the airways, it can be picked up by local antigen presenting cells (usually dendritic cells) for presentation to the immune system. Antigen presenting cells process antigen, so that a portion can be presented on the surface of the cell in conjunction with the major histocompatibility complex (MCH) II. Local naïve CD4+ T helper 0 (Th0) cells have a T cell receptor that can recognize the allergen in conjunction with MHC II. With appropriate costimulation, these cells are subsequently directed to become Th2 cells, which produce a variety of cytokines including IL-4, IL-5, IL-10 and IL-13 among others. Thus, the Th2 cell is considered the major cell which orchestrates the subsequent inflammatory response. The cytokines produced by Th2 cells can induce the maturation, differentiation and survival of eosinophils, and stimulate B cells to produce antibodies of the IgE class. IgE antibodies that are secreted become bound to mast cells just below the surface of the airways. Upon reexposure to the allergen, two IgE antibodies are crosslinked, send an intracellular message and trigger mast cell degranulation. A variety of preformed mediators within mast cell granules are immediately released causing bronchoconstriction, increases in vascular permeability and inflammatory cell influx. Other eicosanoids and cytokines are elaborated by the mast cells, additional cells infiltrate into the airways and a second wave of bronchoconstriction ("late phase response") can occur hours to a couple days later. An emerging cell of importance is the T regulatory cell (Treg) which can be produced during an immune response to help regulate inflammation. It is postulated that T regs might be deficient in numbers or function in asthmatics, leading to uncontrolled airway inflammatory responses.

IMMUNODIAGNOSTICS

An allergen-driven strongly polarized Th2 response will lead to downstream activation of allergic pathways in asthmatic individuals. Detecting these immunologic alterations in affected individuals may be a way to assist in the diagnosis of asthma and discriminate it from other conditions. Most of the research investigating the immune response in asthma has been performed in an experimental model of feline allergic asthma which mimics the disease in pet cats. Broadly, the immunodiagnostics can be broken down into four categories: lymphocyte phenotype identification, lymphocyte proliferation assays, cytokine profiles, and serology (IgE).

In an experimental model of feline asthma, cats are sensitized to and challenged with a clinically relevant allergen—house dust mite or Bermuda grass to replicate an asthmatic phenotype. This model has enabled development of molecular "tools" with which to understand more about the immunopathogenesis of the disease and assess how the immune response is altered with therapeutic intervention. Lymphocyte phenotype including pan B, pan T, and selective T cell markers (CD4 and CD8) can be used on peripheral blood leukocytes or BALF cells using flow cytometry to provide information about how these adaptive immune cell profiles are different in asthmatics. Previously, increases in peripheral blood B cells (which may be important in generation of IgE antibodies) have been found to be increased compared with control cats [7]. Similarly, flow cytometry can be used to identify Tregs by using antibodies that recognize CD4, CD25 and foxP3. Tregs have been shown to increase with allergen-specific immunotherapy, a treatment which is intended to restore immune tolerance to what should be a benign antigen [8]. Activated lymphocytes are triggered to proliferate and in allergic asthma, do so in an allergen-specific fashion. Lymphocyte proliferation assays can be performed using ELISA technology (relative crude) or flow cytometry; the latter has the advantage of being able to identify the phenotype of the proliferating cell. Experimentally asthmatic cats have lymphocytes which readily proliferate with ex vivo incubation with specific allergen; hypoproliferation can be induced with successful therapy[8]. Allergen-specific CD4+Th cells produce a variety of cytokines which can be measured by ELISA, PCR or ELISPOT. The ratio of Th2 (eg, interleukin 4) to Th1 (eg, interferon gamma) cytokines has been used in humans to gage how strongly polarized the immune response is in allergic asthma. A decrease in IL-4:IFN-g ratios suggest attenuation of the allergic response [9]. As allergen-specific IgE is believed to be of prominent importance in driving clinical signs of allergic asthma (by triggering mast cell degranulation), identification of this subclass of antibody has diagnostic importance. Both allergen-specific IgE and total IgE can be measured in serum from cats[10,11]; intradermal skin testing can also be performed, although is less commonly performed in pet cats for a variety of reasons. Serum allergen-specific IgE results appear to be highly dependent on the laboratory performing the assay, with highly unreliable results coming from one laboratory tested[12].

CONCLUSION

Allergic asthma has a distinct immunopathophysiology from other lower airway inflammatory disorders. Although some of the clinicopathologic features overlap those seen in chronic bronchitis and parasitic disorders (e.g, clinical signs, radiographic evidence of a bronchial or bronchointerstitial pattern), an effort should be made to discriminate these disorders. This is important both so that specific treatment targeting the underlying or perpetuating trigger (if applicable) can be initiated, but also so that an appropriate prognosis can be given. Developing better pulmonary function testing and immunodiagnostics may help with this goal.

REFERENCES

1. Reinero C, et al. *J Fel Med Surg* 2011;13:69-73. 2. Bates J, et al. *Am J Respir Cell Mol Biol* 2004;31:373-374. 3. Kirschvink K. *Vet J* 2008;176:125-126. 4. Shibly S, et al. *27th Symposium of the Veterinary Comparative Respiratory Society, Plymouth, MA* 2009. 5.Lee-Fowler T, et al. *Int Arch Allergy Immunol* in press. 6. Hirt R, et al. *Vet J* 2011;187:54-59. 7. Norris Reinero C, et al. *Int Arch Allergy Immunol* 2004;135:117-131. 8. Reinero C, et al. *Vet J* in press. 9. Lee-Fowler T, et al. *Vet Immunol Immunopathol* 2009;129:49-56. 10. Delgado C, et al. *J Fel Med Surg* 2010;12:991-994. 11. Reinero C. *Vet Immunol Immunopath* 2009;132:13-20. 12. Lee-Fowler T, et al. *Vet Immunol Immunopath* 2009;132:46-52.

THE RECOVER INITIATIVE - WHAT THE EVIDENCE TELLS US ABOUT CPR IN SMALL ANIMAL AND COMPARATIVE RESEARCH

Manuel Boller, Dr. med. vet., MTR, DACVECC
Philadelphia, PA

INTRODUCTION

Today, almost every dog and cat that experiences a cardiopulmonary arrest in the hospital (other than during anesthesia) does not survive to hospital discharge; survival is approximately 20 % in humans that experience in-hospital cardiac arrest.[1,2,3] Despite many differences between humans and dogs or cats, this discrepancy suggests that cardiac arrest could be survivable for a considerably higher proportion of veterinary patients. Many barriers to more successful management of cardiopulmonary arrest exist, including financial limitations to provide costly post-cardiac arrest care in the face of uncertain outcome. A comprehensive treatment strategy, that reaches from prevention/preparedness measures, to basic and advanced life support, to post-cardiac arrest care are necessary to be successful.[4,5,6] Recommendations for a resuscitation strategy were previously deducted from guidelines for humans, i.e., the American Heart Association guidelines for acute cardiac care, or were based on expert opinion.[7,8] The findings of a recent survey published in the Journal of Veterinary Emergency and Critical Care (JVECC) suggest considerable variation in CPR technique among veterinary practitioners, and variable compliance with peer-reviewed published recommendations.[9] It is likely that the lack of true evidence-based, consensus CPR guidelines for veterinarians (and therefore the lack of standardized veterinary CPR training) are major contributors to this variation. More importantly, this variability in approach may also compromise patient outcomes. In human medicine, the advent of a widely employed, cohesive educational strategy in CPR was only possible when consensus was reached on optimal CPR technique. Furthermore, systematic evaluation of the evidence on human CPR unveiled major knowledge gaps and allowed prioritization of collaborative research efforts to fill these gaps.[10] Thus, in February of 2011, the Reassessment Campaign On Veterinary Resuscitation (RECOVER) was established to strategically assess and reassess the scientific evidence on how to best conduct CPR in domestic animals, to summarize these findings in a consensus statement on veterinary CPR, and to generate evidence-based consensus guidelines for clinical management of cardiopulmonary arrest in veterinary medicine. It is the goal of RECOVER to generate the first evidence-based consensus guidelines for veterinary CPR, to disseminate these findings, to aid in the development of teaching tools to implement these guidelines, and finally to facilitate development of research instruments to fill critical knowledge gaps in veterinary CPR. In sum, the goal was to significantly advance the field of veterinary CPR. This lecture will describe what and who the RECOVER initiative is, what methodologies of evidence evaluation were employed, how human and basic research weighs into clinical guideline generation and finally will provide examples from the new RECOVER CPR guidelines.

RECOVER AS AN ORGANIZATION

RECOVER, co-chaired by Dr. Dan Fletcher and Dr. Manuel Boller, brought together experts in veterinary resuscitation science with professional contributors from other disciplines, and employed a process similar to that used to generate the AHA guidelines for CPR.[11] The 2011 initiative was overseen by an Advisory Committee including the ACVECC President, leadership from VECCS, EVECCS, and AVECCT, the co-chair of the International Liaison Committee on Resuscitation (ILCOR, the organization responsible for the scientific evaluation used by AHA to generate the 2010 CPR Guidelines), and other selected experts from various specialties and environments. ACVECC and VECCS both endorsed and financially supported RECOVER during the process. But what made RECOVER truly unique was the group of over 85 volunteering specialists at its core. Importantly, 52% of participating specialists were working in a private specialty practice setting. These specialists were assigned to one of 5 different CPR domains: (1) Preparedness/Prevention; (2) Basic Life Support; (3) Advanced Life Support; (4) Monitoring; (5) Post-cardiac Arrest Care. Participating specialist conducted in-depth, standardized evaluations of the literature to answer the most pressing questions in veterinary CPR. Each of these individuals invested an average of 40 hours in answering their clinical CPR question by completing a so-called worksheet.

WORKSHEET PROCESS

The scientific data evaluation process was closely modeled after the International Liaison Committee on Resuscitation (ILCOR) C2005 and C2010 processes (http://www.ilcor.org).[12] The knowledge that ILCOR generated was then used by the regional resuscitation groups (American Heart Association in the US, European Resuscitation Council in Europe, etc) to generated the respective regional human CPR Guidelines (e.g. AHA CPR guidelines) that best meet the requirements of their populations. Each worksheet was based on a specific question of clinical importance and was worded in a standardized PICO format (PICO format: Patient/population, Intervention, Comparison, Outcome). The generic format for questions related to therapeutic interventions was: "In (P)atients, does the (I)ntervention, when compared with (C)omparator, improve (O)utcome?".

Guidelines on how to conduct a standardized literature search were provided to the worksheet authors and each author developed a search strategy to identify evidence related to the PICO question. Out of all hits, relevant articles were selected. General inclusion criteria for these articles were: English language, original research (no reviews), and peer review. Internet based collaboration (Basecamp®) and reference management (Mendeley®) tools were used to allow file and information exchange. Each relevant study was rated using the four criteria outlined in Table 1.

Table 1. Evidence rating criteria used in the RECOVER initiative

A. Direction of support

Supporting	Neutral	Opposing
Study outcome supports the PICO question	Study outcome is neutral to the PICO question	Study outcome is opposing the PICO question

B. Level of Evidence

LOE1	LOE2	LOE3	LOE4	LOE5	LOE6
Randomized controlled trial in the target species	Clinical studies in the target species using concurrent controls without randomization	Experimental (laboratory) studies in the target species	Retrospective studies in the target species using controls	Clinical studies without controls (case series)	Studies not directly related to the specific patient/population (e.g. different patient population, different species, (e.g. humans), mechanical models etc.)

C. Quality of Evidence

Good	Fair	Poor
Met most or all of the quality criteria for the type of study	Met some of the quality criteria for the type of study	Met few of the quality criteria for the type of study

D. Outcomes Assessed

A	B	C	D	E
Return of Spontaneous Circulation (ROSC)	Survival of event	Survival to hospital discharge	Intact neurologic survival	Other endpoint

For a given PICO question, once all relevant publications were categorized as above, these categorizations were arranged in a grid, to provide a visual representation of the overall quantity of evidence in support, neutral or opposing the question. Quality criteria assigned to each article then allowed assessing the overall strength of the evidence in support, neutral or opposing the question. This process was used to generate conclusions and treatment recommendations regarding the question. Finally, short summaries of each of the papers reviewed were written and the information relevant to the PICO question was emphasized. All worksheets can be viewed at the RECOVER website (http://www.acvecc-recover.org).

CLINICAL GUIDELINES

The reviewed and accepted worksheets served as the basis to generate draft veterinary clinical CPR guidelines by the RECOVER co-chairs and were reviewed in conjunction with the RECOVER domain chairs and the worksheet authors. All the drafted guidelines (along with the worksheets) were uploaded to a RECOVER website (http://www.acvecc-recover.org) and were made available to the ACVECC and VECCS membership as part of the consensus process. The website was set up to allow blog-like commenting and discussion from any interested registered individual. The most controversial recommendations were discussed in a dedicated three-hour session at the International Veterinary Emergency and Critical Care Symposium in September of 2011 that was attended by approximately 200 interested individuals. The overall strength of recommendation was graded according to the benefit-risk ratio of a recommendation (class of recommendation), and the level of evidence in support of that recommendation (level of recommendation) (Table 2). The wording used in the guidelines (i.e. should; reasonable; may be considered) is specific for the class of recommendation.

Table 2. Grading System for Clinical Recommendations in RECOVER CPR Guidelines

A. Class of Recommendation

I	IIa	IIb	III
Benefit >>> Risk (SHOULD be performed)	Benefit >> Risk (REASONABLE to perform)	Benefit >= Risk (MAY BE CONSIDERED)	Risk > Benefit (NOT to be performed since it is not helpful and may be harmful).

B. Level of Recommendation

A	B	C
(Multiple Populations) Multiple high quality and/or high level of evidence studies	(Limited Populations) Few to no high quality and/or high level of evidence studies	(Very Limited Populations) Consensus opinion, expert opinion, standard of care.

After a period of additional web-based comment, open feedback, and editing that ended in February 2012, seven distinct manuscripts containing materials and methods, summaries of the evidence for each domain (including knowledge gaps and the final guidelines) will be published in JVECC in 2012. Examples of the new guidelines will be highlighted during this lecture. The RECOVER committee believes this process has the potential to have a profound impact on veterinary emergency and critical care medicine, and will pave the way for future consensus processes evaluating other issues of importance in veterinary medicine.

FUTURE DIRECTIONS

The methodology that was used for the RECOVER initiative enabled a very efficient, systematic review of a large body of literature, primarily due to the contributions of a large number of volunteer worksheet authors. Furthermore, more than half of the worksheet authors were practicing in private specialty hospitals and the remainder held academic positions, indicating that this initiative was effective in reaching out to and including those that otherwise face obstacles in conducting clinical research. These worksheet authors contributed considerable time to the project (mean± STD; 32±17 hours), which amounted to more than 2000 hours total, and yet 94% of the worksheet authors indicated that they would like to participate in future initiatives. Thus it is a realistic possibility that the RECOVER initiative on an in itself or as a model for other initiatives may prove to be a sustainable effort, leading not only to reevaluation of the CPR evidence in 5 years or so, but even to extend the scope to other issues relevant to resuscitation. Next to the evidence evaluation and grading of clinical CPR recommendations, RECOVER will also support the use of this information for teaching initiatives, including web based applications. The preparedness domain of the guidelines also contains recommendations in regards to the frequency of training for CPR providers.

Next to the support of educational elements, the RECOVER roadmap also includes development of consensus on how to report clinical CPR case material for both dogs and cats. This effort is underway with support of ILCOR and others, and will be the foundation for the construction of a veterinary CPR registry that allows capture of a large number of cases at different participating institutions. With this infrastructure in place, it is the hope to be able to address some of the knowledge gaps that are currently limiting our understanding on how to best perform CPR in dogs and cats by conducting observational studies and interventional trials. This all with the overarching goal in mind of improving outcomes from cardiac arrest for dogs, cats and humans alike in the years to come.

REFERENCES

[1]Hofmeister EH, et al. **JAVMA** 2009;235:50. [2]Nadkarni VM, et al. **JAMA** 2006;295:50. [3]Peberdy MA, et al. **Resuscitation** 2003;58:297. [4]Iwami T, et al. **Circulation** 2009;119:728. [5]Lund-Kordahl I, et al. **Resuscitation** 2010;81:422. [6]Boller M, et al. **JAVMA** 2012;240(5):1. [7]Field JM, et al. **Circulation**. 2010;122(18 suppl 3):S640. [8]Cole SG, et al. **J Vet Emerg Crit Care** 2002;12(4):261. [9]Boller M, et al. **J Vet Emerg Crit Car** 2010;20(6):558. [10]Gazmuri RJ, et al. **Resuscitation** 2007 ;75 :400. [11]Sayre MR, et al. **Circulation** 2010;122(18 Suppl 3):S657. [12]Morley PT, et al. **Circulation** 2010;122(16 suppl 2):S283.

THE RECOVER INITIATIVE –
CONSENSUS ON BASIC AND ADVANCED LIFE SUPPORT DURING CPR

Daniel J. Fletcher, PhD, DVM, DACVECC
Ithaca, NY

Veterinary protocols for cardiopulmonary resuscitation (CPR) have been largely adapted from those promoted by the American Heart Association for people. The Reassessment Campaign on Veterinary Resuscitation (RECOVER) recently completed a systematic review of the literature relevant to veterinary CPR and developed the first evidence-based, consensus CPR guidelines for small animals (http://www.acvecc-recover.org). This web site contains thorough worksheets documenting the literature supporting the recommendations contained in this document. A summary of the consensus guidelines is presented here.

BASIC LIFE SUPPORT (BLS)

Basic Life Support (BLS) should be initiated as quickly as possible following diagnosis of CPA using the Circulation, Airway, Breathing (CAB) concept. Circulation should be addressed first, as ventilation will be ineffective if there is no cardiac output, and evidence suggests that outcome worsens as delay to the initiation of chest compressions increases.

Circulation: Chest Compressions

Patients with CPA have no forward blood flow out of the heart and no delivery of oxygen to the tissues. An immediate consequence is the exhaustion of cellular energy stores, cell depolarization and thus loss of organ function. This quickly results in increasing severity of ischemic organ injury and sets the stage for escalating reperfusion injury upon reinstitution of tissue blood flow. The initial goals of chest compressions are to provide (1) pulmonary blood flow for oxygen uptake and CO_2 elimination, and (2) tissue perfusion for oxygen delivery to restore cellular metabolic activity. Experimental evidence suggests that even well-executed external chest compressions produce at best 30% of normal cardiac output, making proper technique critical. Chest compressions should be started as soon as possible after diagnosis or suspicion of CPA. Delay in the start of high quality chest compressions reduces the likelihood of return of spontaneous circulation (ROSC).

Chest compressions should be done with the dog or cat in lateral recumbency with a compression depth of 1/3-1/2 the width of the chest at a rate of 100-120 compressions per minute regardless of size or species. Use of aids to ensure correct compression rate, such as a metronome or a song with the correct tempo (e.g., "Staying Alive") is recommended. Leaning on the chest between compressions must be avoided to allow full elastic recoil. Chest compressions should be delivered without interruption in cycles of 2 minutes, and a new compressor should take over after each cycle to reduce the effect of rescuer fatigue. Any interruption in compressions should be as short as possible, as it takes approximately 60 seconds of continuous chest compressions before coronary perfusion pressure (CPP) reaches its maximum. CPP in turn is a critical determinant of myocardial blood flow and the likelihood of ROSC.

The physiology of blood flow generation is fundamentally different during CPR compared to spontaneous circulation. Two distinct theories exist to explain how chest compressions lead to systemic blood flow. The cardiac pump theory is based on the concept that the left and right ventricles are directly compressed, increasing the pressure in the ventricles, opening the pulmonic and aortic valves and providing blood flow to the lungs and the tissues respectively. Recoil of the chest between compressions due to the elastic properties of the rib cage creates negative pressure within the chest, improving filling of the ventricles before the next compression. The thoracic pump theory is based on the concept that external chest compressions raise overall intrathoracic pressure, forcing blood from intrathoracic vessels into the systemic circulation, with the heart acting as a passive conduit.

Given the chest wall stiffness in medium and large dogs, blood flow generated by the thoracic pump mechanism likely predominates in these patients. Therefore, it is recommended that the chest be compressed over the highest point on the lateral thoracic wall with the patient in lateral recumbency (i.e., the widest part of the chest). In contrast, in very keel chested dogs (e.g., Doberman Pinschers, sight hounds), it is reasonable to do chest compressions directly over the heart as the cardiac pump mechanism likely predominates. In markedly barrel chested dogs (e.g., English Bulldogs), compressions over the sternum with the patient in dorsal recumbency may be more effective in eliciting the thoracic pump mechanism than lateral chest compressions. In these and other large dogs with low chest compliance, considerable compression force is necessary for CPR to be effective. The compressor should maintain locked elbows with one hand on top of the other, and the shoulders should be directly above the hands. This allows compressions to be done using the core muscles rather than the biceps and triceps, reducing fatigue and maintaining optimal compression force. If the patient is on a table and the elbows cannot be locked, a stool should be used or the patient should be placed on the floor.

Most cats and small dogs tend to have higher thoracic compliance and narrower chests than larger dogs, making the cardiac pump mechanism achievable in these patients; therefore, chest compressions should be done directly over the heart.

Compressions may be performed using the same two-handed technique as described above for large dogs, or may be done using a single-handed technique where the compressing hand is wrapped around the sternum and compressions are achieved from both sides of the chest by squeezing. Circumferential compressions of the chest using both hands may also be considered.

Airway and Breathing: Ventilation

If an endotracheal tube and laryngoscope are available, the patient should be intubated as soon as possible. Both dogs and cats can be intubated in lateral recumbency, so chest compressions should continue during endotracheal tube placement. If an endotracheal tube is not readily available, mouth to snout ventilation will provide improved oxygenation and CO_2 removal. The patient's mouth should be held closed firmly with one hand. The neck is extended to align the snout with the spine, opening the airway as completely as possible. The rescuer makes a seal over the patient's nares with his/her mouth and blows firmly into the nares to inflate the chest. The chest should be visually inspected during the procedure and the breath continued until a normal chest excursion is accomplished. An inspiratory time of approximately 1 second should be targeted.

In non-intubated patients ventilated using the mouth to snout technique, ventilation cannot be performed simultaneously with chest compressions. Therefore, 30 chest compressions should be delivered, immediately followed by two breaths. Alternating compressions and ventilations should be continued for 2-minute cycles, and the rescuers rotated every cycle to prevent fatigue. Chest compressions and ventilations should be performed simultaneously in intubated patients because the inflated cuff of the endotracheal tube allows alveolar ventilation during a chest compression and interruptions in chest compressions are minimized. Intubated patients should be ventilated at a rate of 10 breaths per minute with an inspiratory time of approximately 1 second. If a spirometer is available, a tidal volume of approximately 10 ml/kg should be targeted. This low minute ventilation is adequate during CPR since pulmonary blood flow is reduced. Care should be taken not to hyperventilate the patient, as low arterial CO_2 tension leads to cerebral vasoconstriction, decreasing oxygen delivery to the brain.

ADVANCED LIFE SUPPORT (ALS)

Once BLS procedures have been implemented, the CPR team should initiate Advanced Life Support (ALS), which includes monitoring, drug therapy, and electrical defibrillation. Drug therapy is preferably administered by the intravenous or intraosseus route. Therefore, placement of a peripheral or central intravenous or intraosseous catheter is recommended, but should not interfere with continuation of BLS.

Monitoring

Many commonly employed monitoring devices are of limited use during CPR due to their susceptibility to motion artifact and the likelihood that decreased perfusion will compromise accurate readings. Low yield monitoring devices include pulse oximeter and indirect blood pressure monitors, including Doppler and oscillometric devices. The two most useful monitoring devices during CPR are the electrocardiogram (ECG) and end tidal CO_2 monitor ($ETCO_2$).

Electrocardiogram (ECG)

Although the ECG is highly susceptible to motion artifact and is of limited use during ongoing chest compressions, an accurate rhythm diagnosis is essential to guide drug and defibrillation therapy. The goal of ECG monitoring during CPR is to diagnose which of the three most common arrest rhythms are present: (1) asystole, (2) pulseless electrical activity (PEA), or (3) ventricular fibrillation (VF). The ECG should be quickly evaluated while compressors are being rotated between 2-minute cycles of CPR, the rhythm diagnosis should be called out to the group by the team leader, and differing opinions on the diagnosis should be solicited. Discussion about the rhythm diagnosis should not prevent rapid resumption of chest compressions.

End-Tidal Carbon Dioxide ($ETCO_2$)

$ETCO_2$ data can be used in multiple ways during CPR, and regardless of the technology used is highly resistant to motion artifact. The presence of measureable CO_2 by $ETCO_2$ monitoring is supportive of (but not definitive for) correct placement of the endotracheal (ET) tube. Because $ETCO_2$ is proportional to pulmonary blood flow, it can also be used as a measure of chest compression efficacy under conditions of constant quality of ventilation. Upon return of spontaneous circulation (ROSC), $ETCO_2$ dramatically increases due to the rapid increase in circulation, and therefore is a valuable early indicator of ROSC during CPR.

Drug Therapy

Depending on the arrest rhythm, the use of vasopressors, parasympatholytics, and/or anti-arrhythmics may be indicated in dogs and cats with CPA. In addition, in some cases the use of reversal agents, intravenous fluids, and alkalinizing drugs may be indicated. Strict adherence to evidence-based CPR algorithms is recommended to increase the quality of CPR, the likelihood of ROSC and the chance of survival to hospital discharge.

Vasopressors

Regardless of the arrest rhythm vasopressors are recommended to increase peripheral vasoconstriction. Because cardiac output is low even during optimal external chest compressions, shunting of blood away from the periphery and towards the core (e.g., the heart, lungs, and brain) is essential to maintain perfusion to these vital organs. Epinephrine causes peripheral vasoconstriction via stimulation of α_1 receptors. It is a nonspecific catecholamine that also acts on β_1 and β_2 receptors, but the α_1 effects have been shown to be the most beneficial during CPR. Initially low doses (0.01 mg/kg IV/IO every other cycle of CPR) are recommended, but after prolonged CPR, a higher dose (0.1 mg/kg IV/IO every other cycle of CPR) may be considered. Epinephrine may also be administered via ET tube (0.02 mg/kg low dose; 0.2 mg/kg high dose) by feeding a long catheter through the ET tube and diluting the epinephrine 1:1 with isotonic saline.

Vasopressin is an alternative vasopressor that exerts its vasoconstrictive effects via activation of peripheral V1 receptors. It may be used interchangeably with epinephrine during CPR at a dose of 0.8 U/kg IV/IO every other cycle of CPR. Potential benefits of vasopressin include continued efficacy in acidic environments in which α_1 receptors may become unresponsive to epinephrine and lack of β_1 effects (positive inotropy and positive chronotropy), which may cause increased myocardial oxygen consumption and worsened myocardial ischemia upon ROSC. Vasopressin may be administered via ET tube using the technique described above.

Anticholinergics

Atropine is an anti-cholinergic, parasympatholytic drug that has been extensively studied in CPR. Although only a few studies have shown a beneficial effect, there is limited evidence of a detrimental effect, and atropine at a dose of 0.04 mg/kg IV/IO may be considered during CPR in dogs and cats, and is reasonable in all dogs and cats with asystole or PEA associated with increased vagal tone. Atropine may also be administered via ET tube (0.08 mg/kg).

Anti-arrhythmics

Although non-perfusing VF/ventricular tachycardia (VT) should be treated as early as possible with electrical defibrillation, patients with VF refractory to defibrillation may benefit from treatment with amiodarone at a dose of 2.5-5 mg/kg IV/IO. This drug has been associated with anaphylactic reactions and hypotension in dogs, so patients should be closely monitored for signs of peripheral vasodilation, wheals, or hives once ROSC is achieved. Treatment with diphenhydramine (2 mg/kg IM) and/or anti-inflammatory corticosteroids (0.1 mg/kg dexamethasone sodium phosphate IV) is warranted should these signs be noted.

If amiodarone is not available, patients with VF refractory to electrical defibrillation may benefit from lidocaine 2 mg/kg slow IV/IO push. Although this drug has been shown to increase the defibrillation threshold in dogs in one study, benefit was evident in others.

Reversal Agents

Although specific evidence of efficacy is not available, the use of reversal agents in dogs and cats in which reversible anesthetic/analgesic drugs were recently administered may be considered. Naloxone (0.01 mg/kg IV/IO) may be used to reverse opioids, flumazenil (0.01 mg/kg IV/IO) for benzodiazepines, and atipamezole (0.1 mg/kg IV/IO) or yohimbine (0.1 mg/kg IV/IO) for α_2 agonists.

Intravenous (IV) Fluids

The routine use of IV fluids in euvolemic or hypervolemic patients are not recommended during CPR, but is reasonable in patients with documented or suspected hypovolemia. In euvolemic or hypervolemic patients, fluids administered IV serve solely to increase right atrial pressure, which results in decreased perfusion of the brain and heart and should be avoided. However, in hypovolemic patients, IV fluids will help to restore adequate circulating volume, and will increase the efficacy of chest compressions and improve perfusion.

Corticosteroids

The routine use of high-dose corticosteroids during CPR in dogs and cats is not recommended. Although one retrospective study showed an association between administration of corticosteroids and increased rate of ROSC in dogs and cats, the type and dose of steroids administered were highly variable and the study design did not allow determination of a cause and effect relationship (Hofmeister et al, 2009). Other studies have shown no benefit or harm from the use of steroids during CPR. Non-CPR studies have demonstrated that single high doses of corticosteroids in dogs frequently lead to gastrointestinal ulceration and bleeding, which could also cause other ill effects such as bacterial translocation. In addition, corticosteroids suppress the immune system and reduce prostaglandin production in the kidney, a primary mechanism employed by the kidney to maintain perfusion in the face of hypotension. Because the documented risks of high-dose steroids far outweigh the potential benefit shown in one study, the use of steroids is not recommended in patients with CPA.

Alkalinization Therapy

In patients with prolonged CPA of greater than 10-15 minutes, alkalinization therapy with administration of sodium bicarbonate (1 mEq/kg, once, diluted IV) may be considered. Prolonged CPA commonly leads to severe acidemia resulting from both metabolic acidosis, due to lactate and uremic acids, and venous respiratory acidosis due to inadequate peripheral perfusion and accumulation of CO_2. This acidemia can cause severe vasodilation and inhibition of normal enzymatic and metabolic activity. Because these issues may be rapidly resolved once ROSC is achieved, bicarbonate therapy should be reserved for patients with prolonged CPA and with documented severe acidemia ($pH < 7.0$) that are not hypoventilating.

Electrical Defibrillation

Early electrical defibrillation in patients with ventricular fibrillation (VF) has been associated with increased ROSC and survival to discharge in numerous studies, and is superior to anti-arrhythmic medical therapy. The goal of defibrillation is to stop the ventricular myocardial cells from contracting by driving them all simultaneously into a refractory period, allowing the pacemakers to take over and drive coordinated contractions of the heart. If the duration of VF is known or suspected to be of duration of 4 minutes or less, chest compressions should be continued until the defibrillator is charged and the patient should then be defibrillated immediately. If the duration of VF is known or suspected to be more than 4 minutes, one full cycle of CPR should be done before defibrillating to allow the myocardial cells to generate enough energy substrate to restore a normal membrane potential, thereby increasing the likelihood of success.

Defibrillators may be either monophasic (delivering a current in one direction across the paddles) or biphasic (delivering a current in one direction, the reversing and delivering a current in the opposite direction). The use of biphasic defibrillators is recommended over monophasic defibrillators because a lower current (and hence less myocardial injury) is required to successfully defibrillate the patient. For monophasic defibrillators, an initial dose of 4-6 J/kg should be used, while biphasic defibrillation should start at 2-4 J/kg. The second dose may be increased by 50%, but subsequent doses should not be further increased.

After defibrillation, chest compressions should be resumed immediately and a full 2-minute cycle of CPR administered before reassessing the ECG and determining if the patient is still in VF and should be defibrillated again. Brief assessment of the ECG immediately after defibrillation to determine if a perfusing rhythm has resulted is reasonable, but should minimally delay resumption of chest compressions.

OPEN CHEST CPR

Direct cardiac massage during open chest CPR generates superior cardiac output compared to closed chest CPR, but is more invasive, expensive, and requires surgical closure and more intensive care if ROSC is achieved. Indications for open chest CPR include pleural space disease, pericardial disease, flail chest or other significant thoracic wall defects. Direct cardiac massage should also be done in patients undergoing thoracic or abdominal surgery that develop CPA under anesthesia during the surgery. In the case of abdominal surgery, an incision can be made in the diaphragm and cardiac massage performed from an abdominal approach. Finally, in very large, broad chested dogs, closed chest CPR is unlikely to be effective, and open chest CPR should be initiated as soon as possible if the owner has consented.

REFERENCES

Boller, M. et al. **JVECC** 2010; 20(6):558. Cole, S.G., et al. **JVECC** 2002; 12(4):261. Cole, S.G., et al. **JVECC** 2003; 13(1):13. Hofmeister, E.H. et al. JAVMA 2009; 235(1):50. Neumar, R.W. et al. **Circulation** 2010; 122(18_suppl_3):S729. Plunkett, S.J. et al. **JVIM** 2008; 22(1):9. Travers, A. H. et al. **Circulation**; 122(18_suppl_3): S676.

THE RECOVER INITIATIVE –
CONSENSUS ON PREPAREDNESS, TRAINING, AND POST-ARREST CARE

Daniel J. Fletcher, PhD, DVM, DACVECC
Ithaca, NY

PREPAREDNESS

Delays in initiation of CPR for patients with cardiopulmonary arrest (CPA) consistently result in poor outcomes, making it crucial that veterinary practices are well prepared for early recognition of CPA. Studies in human medicine have shown that a combination of didactic CPR training and opportunities to practice psychomotor skills is more effective than either technique alone. Training is recommended for all veterinary personnel who may be called upon to assist in a crisis, including both technicians and veterinarians. Until structured training programs are available in veterinary medicine, practice owners must devise accessible training programs for their staff. Refresher training and drills at least every 6 months have been shown to improve performance in human medicine, and structured assessment and feedback after training maximizes effectiveness.

A fully stocked crash cart routinely audited for content should be available. Because the incidence of CPA in many practices is low, systems should be in place to ensure regular restocking of the crash cart. CPR algorithm charts and emergency drug dosing charts improve adherence to guidelines and individual performance during CPR (See examples in Figures 1 and 2). They should be available in a central location in he practice and reviewed with any staff that may be called upon to acquire drugs for Advanced Life Support efforts.

Following all CPR attempts, debriefing sessions during which team performance is discussed and critically evaluated should be held. The team leader should facilitate the discussion, but all team members should be encouraged to participate and offer their insights.

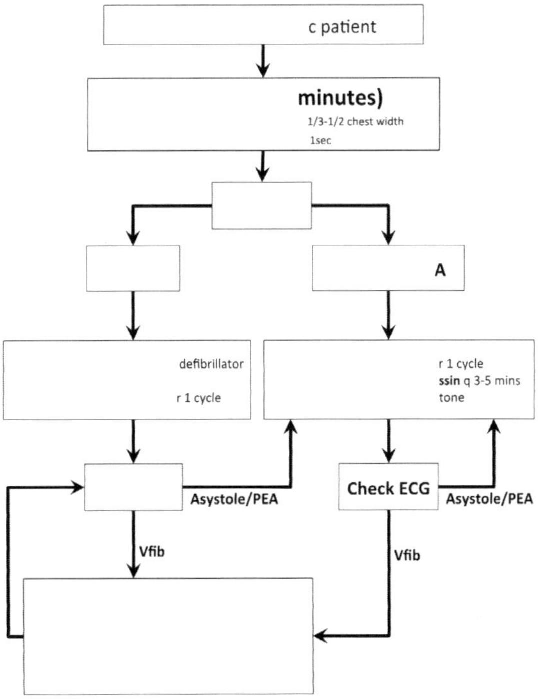

		Weight (kg)	2.5	5	10	15	20	25	30	35	40	45	50
		Weight (lb)	5	10	20	30	40	50	60	70	80	90	100
	DRUG	DOSE	ml	ml	ml	ml	ml	ml	ml	ml	ml	ml	ml
Arrest	Epi Low (1:1000)	0.01 mg/kg	0.03	0.05	0.1	0.15	0.2	0.25	0.3	035	0.4	0.45	0.5
	Epi High (1:1000	0.1 mg/kg	0.25	0.5	1	1.5	2	2.5	3	3.5	4	4.5	5
	Vasopressin (20 U/ml)	0.8 U/kg	0.1	0.2	0.4	0.6	0.8	1	1.2	1.4	1.6	1.8	2
	Atropine (0.54 mg/ml)	0.05 mg/kg	0.25	0.5	1	1.5	2	2.5	3	3.5	4	4.5	5
Anti-Arrhyth	Amiodarone (50 mg/ml)	5 mg/kg	0.25	0.5	1	1.5	2	2.5	3	3.5	4	4.5	5
	Lidocaine (20 mg/ml)	2-8 mg/kg	0.25	0.5	1	1.5	2	2.5	3	3.5	4	4.5	5
Reversal	Naloxone (0.4 mg/ml)	0.04 mg/kg	0.25	0.5	1	1.5	2	2.5	3	3.5	4	4.5	5
	Flumazenil (0.1 mg/ml)	0.01 mg/kg	0.25	0.5	1	1.5	2	2.5	3	3.5	4	4.5	5
	Atipamezole (5 mg/ml)	50 ug/kg	0.03	0.05	0.1	0.15	0.2	0.25	0.3	035	0.4	0.45	0.5
Defib	External Defib (J)	2-10 J/kg	20	30	50	100	200	200	200	300	300	300	360
	Internal Defib (J)	0.2-1 J/kg	2	3	5	10	20	20	20	30	30	30	50

DIAGNOSIS OF CARDIOPULMONARY ARREST (CPA)

CPA is an important differential diagnosis for any acutely unresponsive patient. It is a clinical diagnosis based on the presence of unconsciousness, lack of breathing and absence of a palpable pulse. Regardless of the clinician's index of suspicion for CPA in an individual patient, a rapid assessment focused on ruling out CPA should be undertaken immediately in any unresponsive patient. A standardized approach based on evaluation of Airway, Breathing, and Circulation (ABC) will quickly identify the condition and allow immediate intervention should the diagnosis be made. Because the benefits of starting CPR immediately in a patient with CPA far outweigh the risks of performing CPR on an unresponsive patient not in CPA, the clinician should not delay starting CPR in any patient in which there is a suspicion of CPA.

If CPA cannot be definitively ruled out, CPR should be initiated immediately rather than further diagnostic assessment. This is important as (1) several studies in human medicine have shown that pulse palpation is an insensitive test for diagnosis of CPA, and this may also be the case in dogs and cats, and (2) a large body of literature supports the notion that even short delays in initiating CPR in pulseless patients reduce the likelihood of successful resuscitation. Therefore, the ABC assessment should take no more than 10-15 seconds to complete.

POST-CARDIAC ARREST CARE

Patient outcome is determined by the events that led to cardiopulmonary arrest and the duration of the cardiac arrest, but also by the processes unfolding during and after reperfusion. Two-thirds of human cardiac arrest victims that achieve ROSC die during the post-resuscitation phase. In veterinary medicine, only 16% of dogs and cats initially successfully resuscitated were found to survive to hospital discharge in one study (Hofmeister et al., 2009).

Abnormalities in the post-resuscitation phase result from a combination of anoxic brain injury, post-ischemic myocardial dysfunction, the systemic response to ischemia and reperfusion and the persistent precipitating pathology (e.g. underlying disease processes). Consequently, the clinical abnormalities in these post-cardiac arrest cases are highly variable and therapy should be aimed at alleviating the resulting clinical signs.

Early hemodynamic optimization, similar to algorithms described for severe sepsis and septic shock, has been shown to be effective in human CPA survivors, and should be considered in hemodynamically unstable small animals after cardiac arrest. Central venous oxygen saturation of at least 70% and/or normalization of lactate levels should be used as global perfusion metrics and endpoints for resuscitation. Central venous pressure monitoring is also a useful guide for fluid therapy, as left ventricular dysfunction may increase the risk for hydrostatic pulmonary edema, while ischemia-reperfusion injury can lead to high permeability pulmonary edema. Cerebral autoregulation of blood flow may be impaired and adequate cerebral blood flow may depend on a sufficiently high cerebral perfusion pressure, requiring a mean arterial pressure of 80 mm Hg or higher.

Mild therapeutic hypothermia (core temperature of 32-34°C [89.6° to 93.2°F]) for 12-24 hours after ROSC improves neurological outcome after CPA in humans and other species, including dogs, that remain comatose 1-2 hours after resuscitation. While therapeutic hypothermia may not be feasible in veterinary patients, rapid active rewarming post-arrest is probably detrimental, and hyperthermia is certainly harmful. It is therefore important to monitor body temperature after CPA. Active rewarming should be avoided and passive rewarming should be limited to at most 0.5°C per hour. Patients should also be carefully monitored for seizures and treated aggressively with diazepam (0.5 mg/kg IV/IO) if they occur.

Short term mechanical ventilation to assure optimal arterial oxygen tension (80 to 100 mm Hg) and CO_2 levels (35-40 mm Hg) and to prevent respiratory arrest in the comatose post-cardiac arrest patient is optimal if available, but is not required for patients that are ventilating sufficiently. In all cases, adequacy of ventilation should be monitored using end-tidal or venous/arterial blood gas analysis. Ample experimental evidence substantiates the detrimental effects of oxidative injury after ischemia-reperfusion, particularly when reperfusion occurs under hyperoxic conditions. Titration of oxygen to a normoxic level (e.g., targeting an SpO_2 of 94 to 96%) is a reasonable strategy to reduce oxidative injury while maintaining adequate tissue oxygenation. Additional supportive patient care is determined by the patients' underlying disease processes and by concomitant organ dysfunctions such as ileus and renal failure.

PROGNOSIS

There is limited data in the veterinary literature regarding prognosis for patients receiving CPR after cardiopulmonary arrest. Although overall survival rates have been reported to be quite low, the underlying cause of the arrest may contribute significantly. Patients that experience CPA as a consequence of severe, untreatable or progressive chronic diseases are less likely to experience good outcomes. However, peri-anesthetic CPA carries a better prognosis for survival to discharge (as high as 47% in one recent retrospective veterinary study), and aggressive, prolonged CPR attempts in these cases are reasonable. Adherence to these evidence-based CPR guidelines may help improve the likelihood of survival in these cases.

REFERENCES

Boller, M. et al. **JVECC** 2010; 20(6):558. Cole, S.G., et al. **JVECC** 2002; 12(4):261. Cole, S.G., et al. **JVECC** 2003; 13(1):13. Hofmeister, E.H. et al. JAVMA 2009; 235(1):50. Neumar, R.W. et al. **Circulation** 2010; 122(18_suppl_3):S729. Plunkett, S.J. et al. **JVIM** 2008; 22(1):9. Travers, A. H. et al. **Circulation**; 122(18_suppl_3): S676.

PATHOPHYSIOLOGY OF ISCHEMIA AND REPERFUSION –WHAT MATTERS DURING ARREST AND CPR

Manuel Boller, Dr. med. vet., MTR, DACVECC
Philadelphia, PA

INTRODUCTION

The majority of cardiac arrest patients that initially achieve return of spontaneous circulation (ROSC) will not survive to hospital discharge. This makes cardiac arrest the most profound and deadly form of (global) ischemia and reperfusion, in both veterinary and human medicine. The epidemiology of the disease in humans is as follows. In the United States alone, an estimated 300'000 people experience out-of-hospital cardiac arrest. In 150'000 of these (50%), resuscitation is attempted; 40'000 of these (26.7%) achieve return of spontaneous circulation (ROSC) with admittance to ICU ensues. Of these, 13'000 patients (32.5%; 4.3% of total arrests) live to hospital discharge, and the majority of those survivors (10'000, 77%) achieve good neurological outcome.[1]

This means that between 60% and 70% of sudden cardiac arrest (SCA) victims originally resuscitated go on to die. Similarly, 70% of the people suffering from in-hospital cardiac arrest will succumb to their disease despite being initially successfully resuscitated. There is no epidemiological information available for OHCA in veterinary medicine, but likely the numbers may be even worse. For in-hospital cardiac arrest, individual veterinary centers report survival to discharge rates of 2 to 10% for dogs and cats, despite initial ROSC rates of 35 to 45%. Patients die during a lethal post-cardiac arrest syndrome that is characterized by four major components: anoxic brain injury, multiorgan failure, cardiogenic shock, and the sequela of preexisting diseases.

The severity of these conditions varies with the duration of ischemia or untreated cardiac arrest.[2] Medical intuition dictates that the duration of cardiac arrest is of crucial importance to the severity of post-resuscitation abnormalities. Becker and Weisfeldt proposed different phases of ventricular fibrillation-associated SCA depending on its duration.[3] In an early phase of cardiac arrest (0 to approximately 4 minutes), patients can be resuscitated by defibrillation only. In a second, later phase (after approximately 4-10 minutes), successful resuscitation depends upon provision of flow by administration of chest compressions, as the myocardium is depleted off high energy substrates such as ATP. In this environment defibrillation will lead to asystole without replenishing energy stores first. After prolonged cardiac arrest (> 10 minutes), a metabolic phase is entered in which interventions that address oxidative stress, inflammatory sequela, endothelial injury, apoptosis and energy generation are necessary for these patients to survive.

The many deaths sustained after ROSC has widened the attention from CPR itself to include the post-resuscitation phase, in the search of optimizing therapy to allow a higher proportion of patients to survive. In this lecture, three select post-cardiac arrest care strategies are selected, titrated reoxygenation, goal-directed hemodynamic optimization and mild therapeutic hypothermia. The pathophysiologic reasoning for these interventions and benchtop and clinical evidence for their efficacy is provided.

OXIDATIVE INJURY AFTER CARDIAC ARREST

The debate is ongoing on whether injury during ischemia is more substantial than during reperfusion or vise versa. Nevertheless there is little doubt that injury is sustained during ischemia itself. Rapid decrease in oxygen delivery occurs with onset of cardiac arrest and tissue PO_2 decreases rapidly dependent on tissue metabolic activity and oxygen stores available. Glycolysis is not a sustainable source of energy-rich phosphates in the majority of cell types. In the brain, ATP stores were found to be completely depleted within 4 minutes of complete cerebral ischemia in dogs, while this required 20 minutes in the intestines and 40 minutes in the myocardium.[4-6] With energy depletion depolarization of cell membranes ensues (=ischemic depolarization). Any kind of cardiac electrical activity as detected by ECG indicates that global ischemic myocardial depolarization has not yet taken place, or, during reperfusion, has been restored once again. Ischemic depolarization will also lead to influx of large amount of sodium, chloride and calcium into the cytoplasm, leading to cell swelling and membrane disruption. Cytosolic calcium concentration is thereby of particular interest. It is the result of calcium influx, efflux, binding and sequestration. Efflux is executed by an ATP-driven transporter and via electrogenic $3Na^+/Ca^{2+}$ exchanger, both of which are impaired during ischemia. Protein binding is reduced due to increased intracellular H^+. And last, energy depleted mitochondria and endoplasmatic reticulum (ER) will lose their ability to sequester cytosolic calcium. The sum of these processes, and the circumstance that the extracellular calcium concentration is 10'000 times higher than the cytosolic concentration, lead to a steep rise in intracellular calcium concentration. It is current thinking that the combined influences of increased exposure to calcium, oxidative stress and energy depletion will lead to mitochondrial permeability transition. This will lead to release of cytochrome c into the cytoplasm where it forms together with other proteins a large complex (apoptosome) that then activates a cascade of proteolytic enzymes (caspases) that ultimately initiate DNA fragmentation and dismantling of the affected cells.[7] In an asphyxial rabbit cardiac arrest model, administration of the mitochondrial permeability transition inhibitor cyclosporine A at the time of reperfusion was effective in reducing myocardial reperfusion injury, cardiovascular dysfunction and multiple organ failure.[8]

The rapid decrease in tissue PO_2 is accompanied by an increase in PCO_2 and H^+ concentration during cardiopulmonary arrest and CPR. The myocardial PCO_2 increase can reach beyond 200 mm Hg after a few minutes of ventricular fibrillation and is mostly due to buffering of acids rather than CO_2 production or lack of CO_2 removal.[9] Hypercarbia reduces defibrillation success, myocardial contractile function and may impair survival from cardiac arrest.[10, 11] Rapid normalization of oxygen, carbon dioxide, and pH is the implied goal of CPR, and reoxygenation following prolonged global ischemia is an essential goal for saving lives from cardiac arrest. However clinical approaches to optimally reoxygenate and restore function while minimizing reperfusion injury remain controversial, as both hyperoxia and hypoxia are considered injurious.

The pathobiology of ROS during ischemia and reperfusion in the context of cardiac arrest has been reviewed in greater detail by Becker and Neumar.[12, 13] Free radicals, oxygen free radicals and reactive oxygen species are generally used as interchangeable terms. The radical superoxide (O_2^-) results when oxygen accepts a free electron. Approximately 5% of all oxygen reduced during oxidative phosphorylation in the mitochondria is not directly leading to the end product of water, but rather superoxide radicals first. The free electrons are quasi leaking from the electron transfer chain. An additional source of superoxide radicals is the NADPH oxidase that is upregulated during ischemia and reperfusion. During decreased intracellular pH as occurs in the context of ischemia, superoxide is protonated to $^{\cdot}HO_2$, which is a more reactive form and capable of lipid peroxidation and cell membrane injury. However, the endogenous antioxidant enzyme superoxide dismutase (SOD) will accelerate the transformation of $O^{\cdot-}$ into hydrogen peroxide (H_2O_2). H_2O_2 is of limited toxicity and is rapidly scavenged by catalase or glutathione peroxidase (GSH) to form H_2O. This system has remarkable capacity for averting oxidative injury under physiologic conditions, but exhaustion of the system (e.g. increasing amount of oxidized GSH) ultimately occurs and can be used to quantify oxidative stress. If the ROS load increases drastically, such as in ischemia and reperfusion, the catalase/GSH is overwhelmed, H_2O_2 accumulates and becomes increasingly more likely to generate more reactive products, namely hydroxyl radicals ($^{\cdot}OH$). Free metal ions and particularly iron that is more readily available during the low pH condition of ischemia, will accelerate that process (Fenton reaction). Highly reactive hydroxyl radicals will then lead to cell membrane damage, lipid peroxidation, DNA damage and protein alterations. Moreover, superoxide can also react with nitric oxide (NO), which is similarly upregulated during ischemia and reperfusion, to form the short lived peroxynitrite ($ONOO^-$), which is in its own right a potent inducer of cell death. The mitochondrial origin of the above ROS generation during ischemia was demonstrated by blocking specific sites along the mitochondrial electron transfer chain that inhibited ROS generation. However, as already noted above, limitation of inspiratory oxygen concentration during the first 60 minutes of reperfusion did not only lead to improved functional and morphological outcomes, but also limited brain lipid peroxidation and protein nitration.

An extensive body of literature has demonstrated that arterial hyperoxemia during the first 60 minutes after ROSC following cardiac arrest increases oxidative brain injury, increases neurodegeneration and worsens functional neurologic outcome, but also negatively affects overall survival. Most recently, Kilgannon et al. demonstrated an independent association between hyperoxia (PaO_2 > 300 mm Hg) during the first day of ICU admission after ROSC and in-hospital mortality.[14] Using the same population, the same investigators later showed that a PaO_2 increase of 100 mm Hg increased the risk of in-hospital mortality by 24%, or with other words, a dose-dependent association between super normal oxygen administration and non-survival.[15] In an experimental study involving dogs resuscitated from 10 minutes of cardiac arrest, post-resuscitation titration of oxygen administration to a SpO_2 of 94-96% compared to a consistent FiO_2 of 1.0, lead to significant improvement in neurological outcome and a reduction in neuronal degeneration in the particularly vulnerable hippocampal CA1 region.[16] Numerous other studies using canine models of cardiac arrest with varying severity of injuries and endpoints corroborate these findings.[12] Another piece of supporting evidence for the occurrence of oxidative stress after cardiac arrest originates from reports documenting the generation of large amount of reactive oxygen species after CPR. [17-19] In these studies the elaboration and liberation of ROS is peaking early (15 to 60 minutes) after reperfusion (ROSC), whereby the maximum ROS generation increases with duration of ischemia/cardiac arrest.

The clinical recommendation by the American Heart Association (2010) for administration of oxygen after successful resuscitation from cardiac arrest is that a titrated oxygenation strategy guided by arterial blood gas and/or pulse oximetry (e.g. >94% and ≤98%) is reasonable to consider.[20]

Next to other tissues, such as brain, heart, liver, lung and kidney, endothelial generation of ROS during ischemia and reperfusion has been demonstrated.[21] Besides oxidative alterations of lipids, proteins and DNA, ROS also lead to activation of transcription factors such as NF-κB. Rapid and sustained activation will enhance expression of inflammatory cytokines/chemokines, thereby building a connection between oxidative stress and inflammatory response. Cytokine activated endothelial cells in turn express intercellular adhesion molecule-1, E- and P-selectin and modulate leukocyte recruitment, adhesion and transgression. Microvascular dysfunction is another important consequence.

POST-CARDIAC ARREST DISEASE: A SEPSIS-LIKE SYNDROME

The return of spontaneous circulation after the global ischemic event of cardiopulmonary arrest (CPA) leads to a profound whole-body ischemia-reperfusion syndrome that has been recognized as 'post-resuscitation disease' for more than three decades.[22] It resembles severe sepsis in many aspects, with respect to its effects on inflammation, coagulation, and the endothelium.[23-36] Gando et al. showed in cardiac arrest patients that the endothelial activation and injury markers soluble vascular cell adhesion molecule (sVCAM-1), soluble intercellular adhesion molecule (sICAM-1), soluble E-selectin (sE-selectin), soluble thrombomodulin (sTM) and neutrophila elastase, a marker of neutrophil activation, became elevated 6 to 24 hours after return of spontaneous circulation (ROSC).[28] The same group also demonstrated by quantifying tissue factor (TF) and tissue factor pathway inhibitor (TFPI) concentration in patient serum that activation of the extrinsic coagulation pathway occurs early in the post-resuscitation phase, if not already during CPR itself.[37] Bottiger et al. showed the occurrence of neutrophil activation and neutrophil-endothelial interaction during the first 24 hours after ROSC.[38] Geppert et al. found that von Willebrand factor (vWF) antigen and sICAM concentrations were elevated after cardiac arrest and that the degree of their elevation may contain prognostic significance when used in combination with other metrics.[44] They also showed in a prospective case-control study in people that two thirds of ROSC patients fulfilled the systemic inflammatory response syndrome (SIRS) criteria the day after CPR.[39] Adrie et al., in response to their observation of neutrophil and endothelial activation paired with high concentrations of circulating cytokines (TNF-α, IL-6, IL-9, IL-10) in the post-arrest phase, coined the term "sepsis-like-syndrome" to describe this post-resuscitation characteristic in their pivotal paper in 2002.[40] In severe sepsis, endothelial cells become "activated" and their surface shifts from an anticoagulant to a procoagulant and proadhesive substrate, providing fertile ground for reduction of capillary blood flow by obstruction.[41, 42] Furthermore, the endothelium loses its barrier function, leading to extravasation of intravascular fluid, hemoconcentration and blood flow stasis. This effect is augmented by reduced blood cell deformability.[43] Tissue edema may ensue in the same context. The considerable cross-talk that can occur between the inflammatory and coagulation systems is exemplified by the syndrome of disseminated intravascular coagulation (DIC) in which systemic activation of coagulation leads to microvascular thrombosis and simultaneous consumption of coagulation factors and platelets.[44] The global dimensions of cardiac arrest and resuscitation thereof will further explain the systemic nature of the post-resuscitation disease, including the occurrence of multi-organ failure. In veterinary medicine, these changes are often further accentuated by the influence of preexisting, precipitating disease, rendering the post-resuscitation population highly heterogenic and challenging to treat. Patient heterogeneity in disease severity and response to treatment can be responded to with a hemodynamic optimization strategy[45] comparable to that described by Rivers et al.[46] for early goal-directed treatment of sepsis and septic shock. This is herein recommended for the treatment of small animal patients with hemodynamic instability during the post-resuscitation phase. Global hemodynamic and perfusion variables should be employed as endpoints, with IV fluid administration, vasoactive and positive inotropic drugs, and blood products used as needed and titrated to reach optimized mean arterial pressure, central venous pressure, urine output, hemoglobin concentration, and venous oxygen saturation.

POST-RESUSCITATION DISEASE AND MILD THERAPEUTIC HYPOTHERMIA

In addition to titrated reoxygenation and hemodynamic optimization, mild therapeutic hypothermia (MTH), that is the lowering of the patient's core temperature to 32 - 34°C, has been added to the resuscitation algorithm in an increasing number of human medical centers for those patients that do not wake up shortly after ROSC. The wide-spread application of MTH this indication is in response to the successful use of post-arrest therapeutic hypothermia in two landmark trials.[47, 48] The multitude of mechanisms by which hypothermia exerts its effects, renders this intervention difficult to understand but indicates that it is in essence a combination therapy that employs several pathways, the overall effect of which is protection from ischemia-reperfusion injury. Nonetheless, the protective effect of MTH even when administered hours after ROSC proofs that some of the injury is evolving after reperfusion and is amenable to treatment. While MTH is typically administered systemically and impacts all organ systems including the heart, effects on the brain are of greatest interest and best examined. If hypothermia is present at onset of ischemia the time until ischemic depolarization is delayed thereby markedly reducing injury sustained by ischemia at the first place.[49] Next to the reduced rate of neuronal energy depletion MTH may also attenuate intracellular acidosis associated with ischemia.[50, 51] The formation of glutamate and other neurotransmitter that are increased in ischemia and reperfusion and lead to injurious accumulation of intracellular calcium is attenuated in the presence of MTH.[52] This generally leads to reduced excitotoxicity during reperfusion.[53] ROS play a key role in post-cardiac arrest brain injury (and probably systemic injury). The neuronal elaboration of hydroxyl radicals and hydrogen peroxide after onset of reperfusion was markedly reduced with mild to moderate hypothermia during or after ischemia.[54,55] Anti-inflammatory effects of MTH may also be protective given the pro-inflammatory nature of ischemia and reperfusion injury. ICAM-1 expression, neutrophil rolling and endothelium transgression and IL-1β, IL-8 expression and NFκB activation were all reduced in various models of ischemic brain injury compared to normothermic controls.[56-58] Some studies also indicate the hypothermia attenuates cytochrome c release after ischemia reperfusion and reduces markers of apoptosis, such as DNA fragmentation.[59-61] While these and many more studies provide support for potential protective mechanisms of hypothermia, an equally large and diverse body of research supports the notion that hyperthermia worsens the neurological sequela of ischemia and reperfusion injury. Thus it is crucial to prevent hyperthermia/pyrexia, although the clinical data is less strong to support that notion. The findings in clinical human research studies though are largely in unison in regards to the benefit of MTH and it is now considered standard of care. However, clinical implementation issues persist. Clinical MTH is performed by technologies such as surface cooling blankets or endovascular cooling catheters, but can also be achieved with ice-cold intravenous infusions or application of ice packs. Accomplishing MTH as soon as possible after reperfusion (ROSC) is likely important and is more readily accomplished in small animal patients than in humans. The optimal duration of mild therapeutic hypothermia is not clear and may well depend on severity of injury. The current AHA recommendation is 12 to 24 hours in humans.[20]

Whether MTH is a feasible and effective strategy to use in comatose veterinary patients after CPR needs to be examined, but the science strongly supports its use for such patients.

REFERENCES

[1]Nichol G, et al. **JAMA** 2008;300(12):1423. [2]Neumar RW, et al. **Circulation** 2008;118(23):2452. [3]Weisfeldt ML et al. **JAMA** 2002;288(23):3035. [4]Nishijima MK, et al. **AJP - Heart Circ Physiol** 1989;257(6):H1860. [5]Blum H, et al. **Ann Surg** 1986;204(1):83. [6]Jennings RB, et al. **Am J Pathol** 1978;92(1):187. [7]Halestrap AP, et al. **Biochimie** 2002;84(2-3):153. [8]Cour M, et al. **Eur Heart J** 2011;32(2):226. [9]Johnson BA, et al. **J Appl Physiol** 1995;78(4):1579. [10]Kette F, et al. **Crit Care Med** 1993;21(6):901. [11]Idris AH, et al. **Chest** 1995;108(2):522. [12]Neumar RW. **Curr Opin Crit Care** 2011;17(3):236. [13]Becker LB. **Cardiovasc Res** 2004;61(3):461. [14]Kilgannon JH, et al. **JAMA** 2010;303(21):2165. [15]Kilgannon JH, et al. **Circulation** 2011;23(23):2717. [16]Balan IS, et al. **Stroke** 2006;37(12):3008. [17]Idris AH, et al. **Crit Care Med** 2005;33(9):2043. [18]Basu S, et al. **FEBS Lett** 2000;470(1):1. [19]Basu S, et al. **Free Radic Res** 2003;37(3):251. [20]Peberdy MA, et al. **Circulation** 2010;122(18 Suppl 3):S768. [21]Adams JA. **Crit Care Med** 2006;34(12 Suppl):S458. [22]Negovsky VA. **Resuscitation** 1972;1(1):1. [23]Negovsky VA. **Crit Care Med** 1988;16(10):942. [24]Adrie C, et al. **Curr Opin Crit Care** 2004;10(3):208. [25]Adrie C, et al. **J Am Coll Cardiol** 2005;46(1):21. [26]Bottiger BW, et al. **Curr Opin Crit Care** 2001;7(3):176. [27]Fries M, et al. **Crit Care Med** 2003;59(1):105. [28]Gando S, et al. **Intensive Care Med** 2000;26(1):38. [29]Gando S, et al. **Intensive Care Med** 1999;25(6):588. [30]Geppert A, et al. **Crit Care Med** 2003;31(3):805. [31]Hekimian G, et al. **Shock** 2004;22(2):116. [32]Ito T, et al. **Resuscitation** 2004;62(1):55. [33]Kern KB. **Cardiol Clin** 2002;20(1):89. [34]Kern KB, et al. **J Am Coll Cardiol** 1996;28(1):232. [35]Mussack T, et al. **World J Surg** 2001;25(5):539-43. [36]Oppert M, et al. **Intensive Care Med** 1999;25(12):1386. [37]Gando S, et al. **Thromb Res** 1999;96(2):107. [38]Bottiger BW, et al. **Crit Care Med** 2002;30(11):2473. [39]Geppert A, et al. **Crit Care Med** 2000;28(7):2360. [40]Adrie C, et al. **Circulation** 2002;106(5):562. [41]Vincent JL, et al. **Crit Care Clin** 2005;21(3):469-477. [42]Hoesel LM, et al. In: Aird W editor. **Endothelial Biomedicine**. 1st ed. New York, NY:Cambridge University Press; 2007, pp. 1294. [43]Piagnerelli M, et al. **Intensive Care Med** 2003;29(7):1052. [44]Levi M, et al. **Circulation** 2004;109(22):2698. [45]Neumar RW, et al. **Circulation** 2010;122(18 Suppl 3):S729. [46]Rivers E, et al. **N Engl J Med** 2001;345(19):1368. [47]Bernard SA, et al. **N Engl J Med** 2002; 346:557. [48]Hypothermia after Cardiac Arrest Study Group. **N Engl J Med** 2002; 346:549. [49]Takeda Y, et al. **Crit Care Med** 2003;31:255. [50]Laptook AR, et al. **Stroke** 1995;26:1240. [51]Laptook AR, et al. **Pediatr Res** 1995;38:919. [52]Graf R, et al. **Stroke** 1992;23:150. [53]Globus MY et al. **J Cereb Blood Flow Metab** 1991;11:S10. [54]Globus MY et al. **J Neurochem** 1995;65:1250. [55]Lei B, et al. **Neurosci Lett** 1997 ;222 :91. [56]Inamasu J, et al. **Neurol Res.** 2001; 23: 105-111. [57]Sutcliffe IT, et al. **J Cereb Blood Flow Metab** 2001;21:1310. [58]Goss JR, et al. **J Neurotrauma** 1995;12:159. [59]Zhang Z, et al. **Mol Brain Res** 2001 ;95 :75-85. [60]Niwa M, et al. **Brain Res** 1998;794:338. [61]Yenari MA, et al. **J Cereb Blood Flow Metab** 2002;22:29.

PLATELET ACTIVATION IN CRITICAL ILLNESS

Armelle de Laforcade, DVM, DACVECC
North Grafton, MA

Recognition of hemostatic disorders associated with critical illness in veterinary medicine is on the rise. In recent years, thromboelastography has been used to document hypercoagulable changes in diseases including parvoviral enteritis, protein losing nephropathy and enteropathy, immune mediated hemolytic anemia, and in sick dogs admitted to an intensive care unit. The relationship between inflammation and coagulation is well described, and microvascular thrombosis is accepted as a contributing factor to the development of organ failure in critically ill people. With better documentation of hypercoagulability in sick dogs and in dogs with systemic inflammation, it seems likely that microvascular thrombosis also influences the morbidity and mortality of critically ill dogs.

Cytokine mediated tissue factor expression triggering the extrinsic pathway of coagulation is thought to contribute to a hypercoagulable state in states of systemic inflammation. As a powerful modulator of the immune response, thrombosis and fibrinolysis, the platelet's role in potentiating hypercoagulable states must not be overlooked. Altered platelet number and function have been documented in association with human critical illness, and thrombocytopenia is accepted as a poor prognostic marker and a predictor of mortality in this patient population. In particular, studies suggest enhanced platelet activation and aggregation in diseases such as sepsis, acute lung injury and ARDS, and in diabetes. These and other studies have led some to believe that anti-platelet therapies may be useful in reducing organ dysfunction in hospitalized critically ill people. To date, several small studies have suggested a survival benefit to such anti-platelet therapy but studies on a larger scale are needed.

In animals, increased platelet activation has been described in dogs with inflammatory diseases, neoplasia, immune-mediated hemolytic anemia, nephrotic syndrome, and in dogs with mitral valve prolapse. In a study of 20 dogs with systemic inflammation, flow cytometry documented greatly increased platelet-associated P-selectin expression in 8/11 dogs with SIRS and in 4/9 dogs with sepsis compared to normal dogs, suggesting a potential pro-thrombotic state in response to inflammatory stimuli. Along these lines, another study investigating platelet reactivity in research dogs infused with interleukin-6 showed both increased platelet number and increased platelet reactivity in response to this cytokine. Two studies of dogs with immune hemolytic anemia have used flow cytometry to identify circulating activated platelets, with one study finding a correlation between platelet activation and severe thrombocytopenia. In a recent study evaluating platelets activity in dogs hospitalized in our intensive care unit, flow cytometry, platelet aggregometry and thromboelastography were used to investigate platelet activity in sick dogs hospitalized in an intensive care unit. Preliminary data suggest that sick dogs were more likely to show changes consistent with hypercoagulability on thromboelastography. Sick dogs had increased platelet aggregation and ADP-mediated platelet activation compared to control dogs. None of these findings correlated with severity of illness scores or the development of multiple organ failure.

Given the increased awareness of hypercoagulable states in veterinary medicine and the interrelationship between inflammation, platelet activation and coagulation, the use of anticoagulants in sick dogs is on the rise. In many of the studies referenced above, thrombosis was identified in a proportion of sick dogs, in some cases despite anticoagulation. Veterinary studies investigating various modes of anticoagulation are sparse, and large multi-center studies are needed to truly establish evidence based guidelines for routine anticoagulant and anti-platelet therapy. To date, the correlation between hypercoagulability as documented by thromboelastography or other means and microvascular fibrin deposition has not been confirmed in dogs, and the true need to prescribe expensive anticoagulant or antiplatelet therapy remains in question. In a study of critically ill foals, fibrin deposition was documented in the liver, kidney and spleens of non-survivors. Given the diverse and potent role of the platelet in disease states, and the probable role of increased platelet activation in the development of microvascular thrombosis and organ failure, comprehensive evaluation of hemostasis should be performed in critically ill dogs.

The most common anti-platelet agents used in veterinary medicine include aspirin and clopidogrel. In addition to their effects on platelets, both drugs may also impact the inflammatory response. Aspirin exerts its antiplatelet effect by blocking cyclooxygenase mediated thromboxane production. Aspirin is inexpensive and widely used in veterinary medicine to treat hypercoagulable states. Some studies in people have suggested a potential anti-inflammatory role for aspirin, though results in general have been mixed. Clopidogrel is an irreversible inhibitor of the platelet ADP P2Y12 receptor, blocking the effect of ADP on platelet activation. There is stronger evidence supporting an anti-inflammatory role for clopidogrel in people with inflammatory condition. Interestingly when given to endotoxin treated mice, clopidogrel reduced platelet activation, reduced platelet mediated activation of neutrophils, and reduced the upregulation of pro-inflammatory cytokines.

Recognition of hypercoagulable states requires a sound understanding of the link between inflammation and coagulation, and a raised index of suspicion as to when to suspect a hemostatic imbalance in sick dogs. In particular, the potential for the platelet to contribute to hypercoagulable states should not be overlooked. Studies documenting hypercoagulability in animals with naturally occurring disease states are on the rise, and correlation between these studies and those documenting gross and microvascular thrombosis are needed to determine the degree to which anticoagulation should be prioritized.

REFERENCES

1. THOMAS JS, ROGERS KS. PLATELET AGGREGATION AND ADENOSINE TRIPHOSPHATE SECRETION IN DOGS WITH UNTREATED, MULTICENTRIC LYMPHOMA. J VET INTERN MED 1999;13:319-322. 2. WINNING J, NEUMANN J, KOHL M, ET AL. ANTIPLATELET DRUGS AND OUTCOME IN MIXED ADMISSIONS TO AN INTENSIVE CARE UNIT. CRIT CARE MED 2010;38(1):32-37. 3. KATZ J, KOLAPPA K, BECKER R, ET AL. BEYOND THROMBOSIS: THE VERSATILE PLATELET IN CRITICAL ILLNESS. CHEST 2011;139(3):658-668. 4. WEISS D, BRAZELL J. DETECTION OF ACTIVATED PLATELETS IN DOGS WITH PRIMARY IMMUNE-MEDIATED HEMOLYTIC ANEMIA. J VET INTERN MED 2006;20:682-686. 5. PENG J, FRIESE P, GEORGE J, ET AL. ALTERATION OF PLATELET FUNCTION IN DOGS MEDIATED BY INTERLEUKIN-6. BLOOD 1994;83(2):398-403

6. OLSEN L, KRISTENSEN AT, HAGGSTROM, ET AL. INCREASED PLATELET AGGREGATION RESPONSE IN CAVALIER KING CHARLES SPANIELS WITH MITRAL VALVE PROLAPSE. J VET INTERN MED 2001;15:209-216. 7. WAGG C, BOYSEN S, BEDARD C. THROMBOELASTOGRAPHY IN DOGS ADMITTED TO AN INTENSIVE CARE UNIT. J VET CLIN PATH 2009;38(4):453-461 8. MORTIZ A, WALCHECK BK, WEIS DJ. EVALUATION OF FLOW CYTOMETRIC AND AUTOMATED METHODS FOR DETECTION OF ACTIVATED PLATELETS IN DOGS WITH INFLAMMATORY DISEASE. AM J VET RES 2005;66(2):325-329.

LUNG FUNCTION AND VENTILATION IN THE CRITICALLY ILL PATIENT

Elizabeth Rozanski, DVM, DACVIM (SAIM), DACVECC
North Grafton, MA

Overview

Adequate ventilation is clearly crucial to successful patient outcome. When managing critically ill patients, respiratory failure is a principal cause of morbidity and death. Pulmonary disease may be loosely divided into several categories, including 1) presenting complaint 2) hospital acquired 3) neurological or neuromuscular failure 4) upper airway and 5) cats. Cats may of course be affected by #1-#4, but it is wise to recognize that the feline response to critical illness is different from the dog. Pleural space disease (air/fluid/hernia) may be responsible for distress but may be addressed via thoracocentesis or operative interventions. From a therapeutic and prognostic standpoint, it is prudent to consider etiology and therapy separately and focus specifically on action steps that will result in better patient outcome. As respiratory distress may be immediately life threatening, some guidelines may be helpful to consider to improve patient outcome.

Guidelines

1) Early broad-spectrum antibiotic therapy is warranted with any suspicion of sepsis. While a culture is ideal, untreated pneumonia may be promptly fatal. Pneumonia is common in critically ill dogs, particularly following anesthesia. In large referral hospitals, MDR strains may predominate, and coverage for anticipated pathogens is warranted. Cats are uncommonly affected with pneumonia; although it is possible. In the last few years, there have been several reports of H1NI infection in pet cats; thus questioning of possible exposure to human viral illness in cats with unclear interstitial to alveolar infiltrates.
2) Echocardiography is very useful in determining volume status or volume overload; pulmonary hypertension is more common that potentially appreciated. Hypertrophic cardiomyopathy is common in cats; murmur intensity (or even presence) does not correlate well to severity of heart disease.
3) Pulmonary thromboembolism may be a player in some canine respiratory distress; diagnosis is challenging. The D-dimer is less helpful than anticipated. Echocardiography is occasionally helpful, while CT angiography is considered the gold standard, but requires anesthesia and access to a CT scanner. While reports of cats with PTE exist, almost invariably there were no clinical signs that would make the clinician suspect PTE.
4) Acute lung injury/Acute respiratory distress syndrome (ALI/ARDS) is an uncommon but possible cause respiratory distress in hospitalized dogs with underlying critical illness. ARDS is a secondary lesion, not a primary disease. It is debatable if cats get ARDS; they may have a form of permeability (non-cardiogenic) edema.
5) Steroids are beneficial in some lung diseases and upper airway fixed and dynamic obstructions, and in patients with chronic lung disease exacerbations may be associated with infection

Therapy for respiratory distress is directed at the underlying cause. However, decreasing hypoxemia and work of breathing will improve patient outcome. Adequate oxygen delivery to tissue beds is dependent on cardiac output and oxygen concentration of the arterial blood. Major options for improving oxygenation include delivery of supplemental oxygen, and maintaining sternal recumbency, particularly in deep-chested or obese dogs. Supplemental oxygen may be delivered by several methods; all of which are familiar to the ACVIM Forum attendees and will not be further covered. One caveat is that it is essential to recognize that panting may negate most benefit from nasal oxygen. For example, if a dog is panting, and receiving 3 liters/min of supplemental oxygen, with an inspiratory flow rate of 1L/sec, with a 50 msec inspiratory time, the FiO2 would be 25%, while if the same dog on 3 liters/min that was not panting (eg. inspiratory flow rate of 400ml/sec and inspiratory time 1 second) the FiO2 would be 31%.

Body position is an often overlooked aspect of hypoxemia. Two studies (MacMillan, Rozanski) have confirmed alterations in healthy and diseased dogs with lateral body positions. Other potential therapies for hypoxemia include heliox (a mixture of helium and oxygen) which may be useful in upper airway obstruction and potentially in lower airway disease due to lower viscosity. In human medicine, non-invasive ventilation is the main-stay, including continuous positive airway pressure (CPAP) or a non-rebreathing valve. However, evaluation of face mask ventilation in animals is limited by their lack of cooperation, and experimental studies have supported a similar degree of sedation is required for ventilation with a face mask as via intubation. Heavy sedation without intubation in critically ill dogs may increase the risk of pneumonia.

POSITIVE PRESSURE VENTILATION

Positive pressure ventilation is life-saving in some cases, and it is essential consider providing this level of supportive care in high-end referral and academic practices. Ventilation should be considered when a patient has evidence of ventilatory failure, respiratory failure, post-operative support of some thoracic procedures, impending sense of respiratory arrest and in some cases following CPR (See RECOVER proceedings in other sessions here at the Forum). The major requirements for successful ventilation include at least 1 clinician and 1 technician with enthusiasm for providing ventilation, a ventilator, and some space. There are a multitude of high quality human critical care ventilators that may be used in dogs and cats. There is not one specific model that is associated with better outcome although some of the new modes maybe be preferable in a specific patient. (or more "fun" for the clinician) Older human ventilators are commonly retired from human hospitals, and may be available at very low costs.
[eg. EBAY 1/28/12- ventilator search included the workhorse, BEAR 1, for $99.88 USD.]

Critical care ventilators have several options that might be of interest. When discussion ventilation, it is wise to recall the relationship between volume, pressure, flow and timing. "Healthy" lungs rarely require any specific type of ventilation. Diseased lungs tend to have low compliance (stiffer lungs), which may be impacted by the mode chosen. Ventilation may be volume-limited or pressure-limited, with little or no clinical difference in outcome. In the states, the east of the Mississippi River tends be more a volume-limited region, while west of the Mississippi tends to be more pressure limited, although this reflects regional training differences. Volume limited means that the ventilator will deliver a set volume, no matter what the pressure required to do that, this may lead to barotrauma if the lung compliance falls. Pressure limited ventilation means that the tidal volume is limited to a specific inspiratory pressure (eg. 15 cmH20), this has the advantage or preventing barotrauma, but if the lungs are stiffer, then hypoventilation (with subsequent hypercarbia) may develop. Ventilation may also be delivered using different modes, the most common being assist/control where either at a set rate (eg 15 breaths/minute) the ventilator delivers a breath, or when the ventilator senses negative flow at a set rate (FLOW trigger) ; this mode may lead to hyperventilation in dogs that pant. Another major mode is Synchronous intermittent mandatory ventilation (SIMV); this mode delivers breaths at a pre-set rate; any additional voluntary efforts by the patient are not supported. There is no mode that does not permit spontaneous efforts. Positive end expiratory pressure (PEEP) is used to help prevent de-recruitment and collapse of the alveoli. It is prudent to recall that the normal negative inspiratory pressure supports cardiac filling, and excessive PEEP in a hemodynamically unstable patient will worsen cardiac output. Pressure support is available on some ventilators, and in this mode when the breath is initiated by a patient, the ventilator completes the breath by supporting to a specific inspiratory pressure (eg. 8 cmH20 above PEEP).

Almost all dogs and cats require anesthetic depth deep enough to require intubation in order to permit PPV. In dogs and cats with anticipated long-term ventilation, tracheostomy maybe considered to limit the degree of sedation required. There are no ideal sedation/anesthetic agents for ventilation. Pentobarbital was popular for years, but is no longer readily available. It should be recalled that propofol is not a particularly safe drug, and maybe quite cardiovascularly depressing. We typically used combination of opioids (fentanyl), low doses of dexdormitor, and propofol. Diazepam or midazolam may be added, and ketamine may be useful. For weaning (see later paragraph), transition to a propofol CRI may help with more rapid waking. In all cases, long term infusions may require a longer time frame to metabolize, and over time alterations in dosing may be required. In rare cases, atracurium may be used to prevent "bucking" the ventilator (eg. patient –ventilator dyssynchrony) but be careful to combine with adequate analgesics.

In general, the major advantage of ventilation (positive pressure ventilation- PPV) is decreasing work of breathing, improving oxygenation and promoting CO2 elimination. Arterial hypercarbia, or a CO2 of > 50-60 mmHg will be associated with increasing acidosis, and this will be associated with mental depression, and vasodilation. Initial steps in addressing hypercarbia should include decreasing any respiratory depressants (high narcotic rates), and placing in sternal body position. Most cases of spontaneous hypoventilation are "not the lungs' fault", meaning that these reflect CNS disorders, high cervical lesions and rare toxicity (drug, tick and snake bite). These are wonderful cases to ventilate, as the underlying disease resolves, so does the need for PPV. No specific recommendation for ventilation exist for these patients, but guidelines include a tidal volume adequate to maintain normal CO2 (usually 8-10 ml/kg; 10-15 breaths per minute), with no or limited supplemental oxygen and no or little (2-3 cmH20) PEEP.

When considering hypoxemia, lung protective approach should be considered, with the idea to limit barotrauma and worsening lung injury. Low tidal volume ventilation is advised in people, although it has not be explored in canine patients with ARDS. One study in beagles confirmed that low tidal volume ventilation is well tolerated in healthy dogs. (Oura et al, JVECC 2012). In people, ARDSNET has created guidelines for ventilating people, and supports the use of higher PEEP, and limiting plateau pressures. (www.ARDSNET.gov) . While initiatively is common to believe that pulse oximetry should be maintained as close to 100% as possible, in people it has been recognized that a SPO2 of> 88% is adequate and higher values may be associated with perpetuation of lung injury due to barotrauma. End-tidal CO2 is typically continuously monitored during ventilation. One important caveat is that with low tidal volume ventilation and/or higher respiratory rates there may be a large gradient between the measured arterial CO2 and the end-tidal CO2. This is due to increase dead space, which decreases the amount of end-tidal CO2, not a "malfunction" of the ETCO2 monitoring.

Major complications of ventilation include progression of the underlying disease. Similar to dialysis in AKI, ventilation does not specifically treat lung injury, but supports the lung during the time frame required for healing to occur. Pneumothorax is common, and should be suspected in the hypoxemic patient treated with higher tidal volumes that acutely desaturates. Ventilator –associated lung injury is possible, and airway colonization (and possibly infection) is possible. Other complications include "down" patient issues, including urinary tract infection, peripheral edema including tongue and facial swelling, corneal ulceration and gi ulceration.

Weaning from ventilation should be considered when the patient's condition has improved. Practically in veterinary medicine, most ventilator survivors have relatively short clinical courses, although long-term ventilation has been described. Weaning may be attempted if the patient is stable on <50 % FiO2, < 5-6 cmH20 of PEEP and importantly with relatively low PIP (eg, < 15 cm H20 or so). If a high PIP is required to maintain adequate oxygenation, then spontaneous ventilation will be inadequate and the patient will quickly tire and fail weaning.

REFERENCES

MacMillan et al. J Vet Emerg Crit Care (San Antonio). 2009 Dec;19(6):564-70; Rozanski et al. CJVR Can J Vet Res. 2010 Jan;74(1):34-9. Hopper K et alJ Am Vet Med Assoc. 2007 Jan 1;230(1):64-75. Oura et al. JVECC 2012 (in press)

RE-EVALUATION OF FEEDING STRATEGIES IN CRITICALLY ILL PATIENTS

Daniel L. Chan, DVM, DACVECC, DACVN, FHEA, MRCVS
North Mymms, Hertfordshire, UK

INTRODUCTION

For many years, the major controversy surrounding critical care nutrition was in fact "Do critical care patients actually need nutrition?" In years past (and perhaps even today) provision of nutrition to critically ill patients typically received a very low priority. This problem was later described as "in-hospital starvation" and is most common in the care of elderly human patients. When the effects of malnutrition on patient morbidity and mortality were realised, a reactive movement ensued in critical care and many patients were fed quite aggressively, leading to the term "hyperalimentation," which we have now recognised as causing its own set of problems. Much like other aspects of critical care, our paradigm of critical care nutrition constantly changes; previously held assumptions become less relevant and new research uncovers novel strategies.

Nutritional support is now being considered essential for the recovery of post-operative, critically ill, and injured human patients. Whilst there is convincing evidence of the deleterious effects of malnutrition in people, the optimal nutritional strategies for critically ill and post-operative animals remain controversial and largely are unknown. A recent study aimed to evaluate the impact of nutrition support (both enteral and parenteral) in hospitalized dogs and cats (not necessarily "critically ill") and concluded that a positive relationship on outcome was associated with instituting nutritional support and that outcome was further improved when nutritional support approached or exceeded nutritional targets (ie, meeting resting energy requirements (RER).[1] The study was retrospective in nature, was non-randomised and suffered from other design flaws, making conclusions difficult to discern. Despite the lack of definitive answers it must be emphasised that recommendations for nutritional support of critically ill animals are based on current understanding of the metabolic response to injury and the limited clinical information available. This should not discourage the implementation of nutritional support for critically ill or injured animals. In fact, with proper patient selection, sound nutritional planning, and careful monitoring, nutritional support can be an integral part in the successful recovery of many critically ill animals.

METABOLIC CHANGES ASSOCIATED WITH ILLNESS

One of the major metabolic alterations associated with critical illness involves body protein catabolism, in which protein turnover rates may become markedly increased. In people, protein-malnutrition is associated with poorer outcomes. Whereas healthy animals primarily lose fat when deprived of sufficient calories (simple starvation), sick or traumatised patients catabolise lean body mass when they are not provided with sufficient calories (stressed starvation). During the initial stages of fasting in the healthy state, glycogen stores are used as the primary source of energy. Within days, a metabolic shift occurs towards the preferential use of store fat depots, sparing catabolic effects on lean muscle tissue. In diseased states, the inflammatory response triggers alterations in cytokines and hormone concentrations and rapidly shifts metabolism towards a catabolic state. Glycogen stores are quickly depleted, especially in strict carnivores such as the cat, and this leads to an early mobilisation of amino acids from muscle stores. As cats undergo continuous gluconeogenesis, the mobilisation of amino acids from muscles is more pronounced than that observed in other species. With continued lack of food intake, the predominant energy source is derived from accelerated proteolysis (muscle breakdown), which in itself is an energy-consuming process. Muscle catabolism that occurs during stress provides the liver with gluconeogenic precursors and other amino acids for glucose and acute-phase protein production. The resultant negative nitrogen balance or net protein loss has been documented in critically ill dogs and cats. One study estimated that 73% of hospitalised dogs (including post-operative patients) evaluated in four different veterinary referral centres were in a negative energy balance.[2] However, with proper surveillance and nutritional protocols in place, adequate nutritional support can be provided in veterinary ICU settings.[3]

The consequences of continued lean body mass losses include negative effects on wound healing, immune function, strength (both skeletal and respiratory muscle strength), and ultimately on overall prognosis. In the context of post-operative patients, this could lead to greater risk of surgical wound dehiscence and post-operative infections. Due to the metabolic alterations associated with critical illness, and in part due to an inability or reluctance of many critically ill and post-operative dogs and cats to take in sufficient calories, this patient population is at increased risk for rapid development of malnutrition. Given the serious sequelae of malnutrition, preservation or reversal of deteriorating nutritional status via nutritional support is paramount. Nutritional support is therefore aimed at minimising the impact of malnutrition and enhancing rate of recovery.

TIMING OF NUTRITIONAL SUPPORT

In dogs, a period as short as 3 days of anorexia has been documented to produce metabolic changes consistent with those seen associated with starvation in people. However, these dogs would not necessarily exhibit any easily detectable abnormalities on clinical assessment suggestive of being malnourished. Dogs with overt signs suggestive of malnutrition usually have a more protracted period (usually weeks to months) of disease progression. In cats detectable impairment of immune function can be demonstrated in healthy cats subjected to acute starvation by day 4, and so recommendations to institute some form of nutritional support in any ill cat with inadequate food intake for more than 3 days have been made. In both dogs and cats, there is some consensus that there is an urgent need to implement a nutritional intervention (e.g., place feeding tube) when a dog or cat has not eaten for more than 5 days. Currently, the major recommendation for support of the critically ill animal is to initiate early enteral feeding as soon as it is feasible. Some veterinary studies have implemented enteral nutrition as early as within 10 hours of admission.[4] Although it is difficult to assess whether these measures actually influence outcome, one thing that has been established is that many patients previously assumed to require "bowel rest," actually tolerated early enteral feeding. The most notable studies demonstrating this tolerance to feeding evaluated puppies with parvoenteritis,[4] cats with acute pancreatitis[5] and most recently, dogs with severe and acute pancreatitis.[6] It is therefore reasonable to pursue enteral feeding even in cases believed to have significant gastrointestinal dysfunction.

PARENTERAL NUTRITION: BAD INTERVENTION OR JUST BAD TIMING?

The optimal timing of implementing parenteral nutrition (PN) is very controversial at the moment, even in people. Recent publications specifically evaluating timing of initiating PN support (eg, early enteral followed by early PN vs early enteral followed by delayed PN) have suggested that early initiation (ie, within 48 h of admission to ICU) was associated with higher morbidity and mortality.[7] Other studies have corroborated these findings and there is a growing consensus against initiating PN until day 8 of ICU admission.[8,9] However, there should be considerable caution in extrapolating these results to companion animals. First, the differences were rather modest (6% increase in likelihood of being discharged alive, P= 0.04) and the clinical relevance of such modest differences needs to be considered.[7] A perhaps more important distinction in the provision of nutritional support, is the optimal calorie target. There is ample evidence in both the human and veterinary literature demonstrating the deleterious effects of over-feeding in both morbidity and perhaps mortality. For these reasons, recent veterinary recommendations have centred on simply targeting RER. Although there are similar recommendations in human ICUs, an analysis of the caloric target in the recent studies,[7,8] which demonstrated longer hospitalisation times, longer ICU stays, shows that patients who did worse were essentially "aggressively fed" and likely overfed compared to. Patients that received enteral feeding only for first 8 days, basically received hypocaloric nutrition (previously shown to be beneficial) and only received PN if they failed to consume 50% of targeted calories after day 8.[7] This was in contrast to the PN group which were fed much greater number of calories (within 48 hours) and therefore it is reasonable to question whether complications weren't simply a reflection of problems with overfeeding rather than related to the timing of PN implementation. Future studies implementing an iso-caloric component to these trials is needed to confirm or refute initial findindgs. In animals, where a combination of enteral and parenteral nutrition was associated with better outcome, the total target was less than RER.[10] It is therefore possible that positive outcomes can be achieved with supplemental PN, as long as the total caloric target is set very conservatively at RER.

VETERINARY RECOMMENDATIONS

A major goal of nutritional support is to minimise metabolic derangements and the catabolism of lean body tissue. During hospitalisation, recovery of normal body weight is not the top priority as this should occur when once the animal is discharged from the hospital and complete their recovery at home recovering from a state of critical illness their disease. There is much that remains unclear regarding the nutritional requirements of critically ill animals in general. In certain circumstances assumptions are made that nutritional requirements in animals are similar to people afflicted with similar diseases. However, it is important to recognise that there may be significant species and disease differences that make direct comparisons or extrapolations less applicable. Experimental data suggests dramatic changes in energy requirements in animals with thermal burns, however, there are virtually no clinical data to support this notion. In experimental models, dogs with thermal burns experienced increased energy requirements, accelerated gluconeogenesis, glucose oxidation, lipolysis and increased amino acid oxidation. In the absence of definitive data to suggest otherwise, current recommendations are to start nutritional support as soon as it is deemed safe and initially target RER but to reassess the patient continually as energy requirements may exceed 2 x RER. The goal of nutritional support is to optimise protein synthesis and preserve lean body mass. Feeding at least 6-7 g protein per 100 kcal (25-35% of total energy) may be necessary. It is unknown whether nutrients such as glutamine and arginine would provide extra benefits in this patient population.

INNOVATIONS IN VETERINARY NUTRITIONAL SUPPORT

As alluded to earlier, for many years conventional therapy actually ignored nutritional needs of critically ill patients. As more and more evidence illustrated the consequences of malnutrition there was a gradual change to ensure that all patients received adequate nutrition. Typical delays in starting nutrition decreased from weeks to 10 days, and now debates involve how many hours should nutrition be delayed. The concept of early enteral feeding is becoming more widely practiced and there are some evaluations of its utility in managing dogs with pancreatitis, hemorrhagic gastroenteritis. Results support the notion that employing early enteral nutrition is achievable in many patients and therefore, this should be pursued in virtually every patient. Another innovation is the use of fluoroscopic guidance to place naso-enteric catheter as a means for providing nutritional support.[11] With greater adoption of this technique in practice means that many more animals which will receive appropriate nutritional care. Another recent assumption, which has been challenged is the perception that critically ill animals need to be fed via continuous rate infusions rather than receive boluses of food into their feeding tubes. A recent trial found no difference in tolerance or ability to reach nutritional targets by feeding animals using boluses rather than continuous rate infusions.[12]

Overall effect on outcome is unfortunately beyond these small trials. However, an important point about how early to intervene is that any patient receiving nutritional support should be cardiovascularly stable and hydration, electrolyte and acid-base disturbances should be addressed before nutritional intervention. Also, commencing feeding after tube placement should really be held until at least the animal has recovered from anaesthesia. Recumbent animals receiving feedings are at risk of aspiration. Patients with compromised gastrointestinal motility (e.g., anaesthetised patients, patients on opioids analgesics, patients with ileus) are also at risk for complications and should be monitored closely.

FUTURE DIRECTIONS

While most can agree that nutritional support is important, and that early intervention is better that delayed, the question remains, how early to intervene? Is it days or hours? The most aggressive approach is to place a feeding tube as soon as possible and start feedings within hours. Is this necessary? Studies in canine patients with parvoviral enteritis, severe acute pancreatitis and haemorrhagic gastroenteritis support the premise that early nutritional intervention is at least well tolerated and produces few complications. Other areas requiring further investigation include, what role does PN have in veterinary critical care? Should it be delayed similarly to what has been recently recommended in people (ie, not until day 8)? Does the composition of enteral diet or PN solution alter patient outcome? The area of immune-nutrition has not been properly evaluated in veterinary medicine and perhaps future work in this aspect may yield useful and interesting results.

SUMMARY

Nutritional support continues to be a controversial topic in critical care. From the problem of under-feeding (ie, in-hospital starvation) to grossly overfeeding (ie, hyperalimentation), these may be important areas requiring further clarification. The timing of intervention, the optimal caloric goal (eg, RER vs less than RER vs RER x illness factor) should be clarified. Previous assumptions about feeding the critically ill animals also seemed to be been successfully challenged and the way we approach some critically ill animals has dramatically changed. Newer techniques, newly developed nutrient components will also enhance the way we support patients. These innovations along with new research findings illustrate how dynamic the field of critical care nutrition is, and highlight the need to continually reassess and question our knowledgebase.

REFERENCES

1. Brunetto MA, et al. J Vet Emerg Crit Care 2010;20(2):224; 2. Remillard RL, et al. Vet Ther 2001;2(4):301; 3. Michel KE, et al. J Vet Emerg Crit Care 2006; 16(Suppl):S2; 4. Mohr AJ, et al. J Vet Intern Med 2003;17(6):791; 5 Klaus JA, et al. J Vet Emerg Crit Care 2009;19(4):327; 6. Mansfield CS, et al. J Vet Intern Med 2011;25:419; 7. Casaer MP, et al. N Engl J Med 2011;365(6):506; 8. Kutsogiannis J, et al. Crit Care Med 2011;39(12):2691; 9. Cahill NE, et al. JPEN J Parenter Enteral Nutr 2011;35(2):160; 10. Chan DL, et al. J Vet Intern Med 2002; 16(4):440; 11. Beal MW, et al. J Vet Emerg Crit Care 2011;21(2):151; 12. Campbell JA, et al. J Vet Emerg Crit Care 2010;20(2):232

DIAGNOSIS AND NURSING CARE OF TETANUS AND BOTULISM IN HORSES

Allison Stewart, BVSc(hons), MS, DACVIM (LAIM), DACVECC
Auburn, AL

TETANUS (LOCKJAW)

INTRODUCTION

Tetanus is caused by toxins secreted by *Clostridium tetani*, a common soil inhabitant, especially soil heavily contaminated by fecal material. Under anaerobic conditions (such as necrotic tissue and pus) spores germinate into the vegetative form and produce toxins:

a. Tetanolysin - increases tissue necrosis
b. Tetanospasmin - binds to nerves where it inhibits inhibitory interneurons in the ventral horn of the spinal cord.
c. Nonspasmogenic toxin - causes over-stimulation of the sympathetic nervous system.

CLINICAL SIGNS

Incubation period: 2-4 weeks

1. Stiff gait - initially
2. Sawhorse stance
3. Prolapse of 3rd eyelid
4. Flared nostrils, lips retracted, erect ears
5. Signs are exaggerated with stimuli
6. Respiratory paralysis in terminal stages

DIAGNOSIS

By clinical signs, rarely confirmed by culture.

TREATMENT

1. Non-stimulating environment - dark stall, ears plugged with mineral oil soaked cotton.
2. Sedation - acepromazine at 4-6 hour intervals; chloral hydrate; diazepam; sodium pentobarbital
3. Muscle relaxants in combination with sedatives: glycerol guaiacolate (guaifenesin) IV drip to effect
- methocarbamol - 10-20 mg/kg TID
4. Slings or deep bedding
5. Traditional penicillin therapy has been replaced by **metronidazole** (IV at 15 mg/kg q 6 hours or orally at 20 mg/kg q 6 hrs or 25 mg/kg q 8 hrs) to treat the wound infection. A hoof abscess should be well debrided and can be flushed with dilute hydrogen peroxide and packed with crushed metronidazole tablets made into a paste.
6. Antitoxin **5,000 to 50,000 IU** sub Q; infiltrate the wound with antitoxin. In humans intrathecal antitoxin is also administered
7. Parenteral nutrition and intravenous balanced electrolytes for nutritional support; or feed and water by NG tube.

PROGNOSIS

Is good if:

1. The horse can still drink
2. The horse remains standing (80% mortality rate if recumbant)
3. The horse survives more than one week after onset of signs

PREVENTION

Although horses are traditionally vaccinated with tetanus toxoid each year, protection probably is longer lasting. An antitoxin can be given to horses to provide very rapid, but short-lasting immunity to the toxin produced by *Clostridium tetani*. At least 1500 units of antitoxin can be given to horses to confer immediate protection against the disease. Tetanus prophylaxis recommended for a wounded horse that has never been vaccinated for tetanus or whose vaccination history is unknown, or has not been vaccinated for tetanus within a year, is 1500 units of tetanus antitoxin and simultaneous vaccination with tetanus toxoid. It has been argued that tetanus antitoxin should not be given to horses unnecessarily because an acute, highly fatal liver disease (Theiler's disease) occasionally occurs in horses given a biologic prepared from horse serum. However, the incidence of Theiler's disease has greatly decreased in recent years. Because Theiler's disease has not been reported in horses under 2 years of age, tetanus antitoxin is safe to use in foals, and should be given to foals whose dams were not vaccinated for tetanus late in gestation.

BOTULISM

INTRODUCTION

Botulism is caused by a toxin produced by *C. botulinum* - 8 types of toxin: A, B, Ca, Cb, D, E, F, G have been recognized. Each type is different in its geographic distribution and species susceptibility. Type B organisms are common in the soil of the mid-atlantic states and Kentucky.

3 routes of infection

1. **Ingestion of performed toxin** - usually occurs in adults. Associated with forage contaminated with a carcass or noncarion-associated botulism (**forage poisoning**). Usually associated with Type C or D. Most commonly occurs when silage is fed.
2. **Wound botulism** - at least 1 reported case - secondary to castration and secondary to an IM injection. Can occur with any deep anaerobic wound.
3. **Toxicoinfectuous** - occurs in foals (shaker foal syndrome) - Spores are ingested; they germinate and toxin is produced and absorbed from the GI tract. Originally it was thought to be associated with gut lesions such as gastric ulcers, although this has never been proven and is unlikely.

PATHOGENESIS

Preformed toxin or toxin produced in the GI tract or toxin from wounds binds at the neuromuscular junction. It causes paralysis of the cholingeric nerve fibers by **blocking the release of acetylcholine**. Signs occur after ingestion of preformed toxin, and appear in 3-7 days.

CLINICAL SIGNS

1. Afebrile progressive muscular paralysis.
2. Restlessness, incoordination ataxia.
3. Inability to rise; shaking as they rise (hence the term "shaker" foal).
4. Sternal recumbency with head at the flanks.
5. Dysphagia is often the first sign.
6. Decreased tongue and tail tone

DIAGNOSIS

Usually by clinical signs; also by demonstration of toxin in feces, serum or tissue - since horses are so sensitive, toxin is not usually demonstrated in the serum or tissue. Circumstantial or a tentative diagnosis is reached by demonstrating toxin and the organism in feed. Electrodiagnostic studies may be of value in diagnosis.

TREATMENT

1. Neutralization of circulating toxin with specific multivalent antiserum. The antitoxin has no effect on the toxin attached to nerve cells. Before antitoxin was available, mortality was >90%; with antiserum mortality rate is <30%. Antitoxin can be gotten from New Bolton Center which covers all serotypes (610-444-5800, ext. 2383). Antiserum to either type B or type C can be purchased from a number of other sources at a much cheaper price.
2. Restriction of movement
3. Assisted ventilation and/or oxygen therapy
4. Antibiotics (for secondary complication - like aspiration pneumonia). **Avoid** ones that potentiate neuromuscular weakness - aminoglycosides, tetracycline, and procaine penicillin. If you are treating simple pressure sores maybe use trimethoprim sulfonamides. If you have any serious secondary infections (ie aspiration pneumonia) use intravenous potassium penicillin G, enrofloxacin and metranidazole . Antibiotics are ineffective in eradicating the organism from the intestinal tract.
5. Cathartics - avoid magnesium products
6. Parasympathomimetic drugs may provide temporary improvement but have been shown to increase mortality.
7. Nursing care - catheterization of the bladder, evacuate rectum/enemas, fluid therapy, alimentation, lubricate corneas, flip every 4 hours, keep clean and dry. Nursing care is intensive, but essential for a good result.

PREVENTION

C. botulinum type B toxoid - (Bot-Tox: Neogen Corp) - to prevent shaker foals, vaccinate the mare 3 times at monthly intervals; last dose 2-3 weeks before foaling. During subsequent pregnancies a single injection during the last month is recommended. For protection of adults, yearly vaccination after initial series of 3 doses; however, there is no cross protection between Type B and other toxins.

REFERENCES

Mayhew J. Large Animal Neurology 2nd Ed 2008 340. Reed et al. Equine Internal Medicine 3rd Ed 2010 637. Smith BP Large Animal Internal Medicine 4th Ed 2009 1089, 1584, 69. Galey FD. Vet Clin North Am Equine Pract 2001(3):579

RABIES IN LARGE ANIMALS; TRANSMISSION, DIAGNOSIS, PREVENTION AND OCCUPATIONAL SAFETY

Allison Stewart, BVSc (hons), MS, DACVIM (LAIM), DACVECC
Auburn, AL

ETIOLOGY

Highly neurotropic, invariably fatal RNA virus of genus *Lyssavirus*; family *Rhabdoviridae*. At least 6 genotypes, with different major and minor antigenic variants, with resulting differences in host range and pathogenicity. "Fixed" strains have been serially propagated through laboratory animals with a resultant attenuation of virulence for other species, but increased pathogenicity in the species that they have been passaged. "Street" strains are obtained from natural infections from warm-blooded animals.

EPIDEMIOLOGY

Australia, New Zealand, Britain, Hawaii and Scandinavia are rabies free. Switzerland was declared rabies free in 1998 after 20 years of oral wildlife vaccination. In Central and South America and Malaysia, rabies is epizootic (tropical climate with large population of infected feral mammals). The vampire bat often transmits the paralytic form resulting in economic losses due to cyclic outbreaks in cattle. Annual losses exceed 50,000 cattle. Europe: sylvatic form from red fox; endemic in Yugoslavia and Turkey, with progressive spread throughout the continent. Africa: wildlife hosts include wild dogs, jackals and mongoose. Canada: fox movement south from the arctic and northerly movement of the striped skunk from North Dakota. Endemic in southern Ontario.

USA: During 2010 there were 6,153 reported cases of rabies, down from 6,690 in 2009 and 7,067 in 1999. More than 92% of cases in 2011 were in wild animals: raccoons (36.5%), skunks (23.5%), bats (23.2), foxes (7%), whereas 8% were in domestic species; cats (4.9%; 303 cases), cattle (1.1%; 71 cases mostly from Virginia, Pennsylvania, South Dakota, Texas, Kansas, North Carolina and New York), dogs (1.1%; 69 cases), horses and mules (37 cases). Similar to other species, equine cases have been gradually declining (zero in 2006, 63 in 2003 and 65 in 1999).[1] The majority of equine cases were from non-vaccinated animals, although reportedly there have been some cases of equine rabies in horses vaccinated greater than 12 months previously. In 2010 there were only 6 rabid sheep and goats and 1 pig, which was down from previous years.[2]

Infections in horses and livestock tend to coincide with epizootic in the wildlife reservoirs.[3] There is an increased incidence in late summer when wildlife populations have peaked. Foxes, cotton rats and coyotes are extremely susceptible; cats, rabbits and cattle are highly susceptible; sheep, goats and dogs are moderately susceptible and opossums seem to be relatively resistant. Distinct virus variants can be identified by monoclonal antibody panels or by genetic analysis. There are also some rabies-like viruses such as the Australian bat lyssavirus, European bat lyssavirus I and II, Makola virus, the Lagos bat virus and Duvenhage virus. Only some of these are cross-protected by the conventional rabies vaccines.

TRANSMISSION

Bites from infected animals, or contamination of a skin wound with fresh contaminated saliva. Bites from rabid animals are only infective if the virus has migrated down to the sensory nerve terminals in the mouth. Virus can be present in the milk of infected animals, with transmammary and transplacental infection only very rarely seen. The virus is fragile; susceptible to desiccation and most common disinfectants. At 54° C the virus is inactivated in less than one hour, but it can survive for years at sub-freezing temperatures. Inhalation of virus from infected bat caves is possible, though inter-bat spread is primarily from bites. Human and laboratory animals have developed fatal infections after respiratory exposure to viral aerosols. Ingestion of recently deceased infected carrion (commonly bats) is considered a method of natural spread to carnivores. Certain rabies strains have low virulence in skunks and bats leading to chronic asymptotic infections with these species acting as reservoirs over successive years. In subclinical bats, the virus can reside in fatty tissues without causing neuronal invasion. Due to their migration, bats are important in the geographic dissemination of the virus. Outbreaks in domestic animals often occur when wildlife numbers increase, with peak incidences corresponding to the large-scale movement of wildlife at mating times or when young are forced from their parent's territory.

PATHOGENESIS

After subcutaneous or intramuscular inoculation, there is replication in these tissues. The neurotropic virion then binds to the acetylcholine receptors (and other receptors), specifically via their G glycoprotein (this is the target of neutralizing antibodies). The virions then enter the cells by endocytosis and after uncoating, release their nucleocapsid. There is intracytoplasmic replication, with the newly budded virions released into the intracellular spaces. The rabies proteins are highly immunogenic, but as the virus is not highly cytopathic, there is no cell lysis during replication or maturation, thus little antigen is presented to the immune system. No humoral or cell-mediated immune response is detected in the movement phase from the bite site to the CNS. The virus spreads centripetally up neuronal axons to the CNS usually via the spinal cord. There is no hematogenous spread and thus no viremia.

The virus replicates extensively in the limbic system affecting the cortical control of behavior with the manifestation of furious rabies. If there is replication in the neocortex then the dumb form of rabies is seen. From the CNS the virons move centrifugally in the peripheral nerves (at 3mm/hr) to all highly innervated body organs. In the salivary gland there is budding from the plasma membranes of the mucous cells resulting in high concentrations within the saliva.

ZOONOSIS

From 1980 to 1995 there have been 28 human cases in the USA. From 2011 -2010 there have been 21 cases of human rabies, with 2 cases in 2010 and 4 in 2009.[2] In 2004 there were 4 iatrogenic human cases which were transplant recipients of a liver, 2 kidneys and an artery from another previously undiagnosed rabies case. There has only ever been one survivor- an 8 year-old girl who was put into a medically induced coma.[2] In the last decade, 20 of the 22 cases of rabies in humans in the USA, were determined by genetic analysis, to be associated with bat variants of the rabies virus. Only one of these cases had a definite history of an animal bite. It is likely that bat bites were either unnoticed or ignored and subsequently forgotten.[4] Infection rates are much higher in developing countries with 91% of human infections secondary to dog bites, 2% from cats, 3% from bats, 2% from foxes, other wild animals <1% and 2% from other domestic animals.[5]

CLINICAL FEATURES

Encounters between domestic animals and rabid wild animals are rarely witnessed. The puncture-like wounds are very difficult to find. Incubation period: is variable and up to 9 months,[3] generally horses show clinical signs after 2 to 9 weeks, [3,6,7]

The virus will cause multifocal dysfunction of peripheral nerve rootlets, basal ganglia, or behavioral centers in the rhomencephalon. • Furious rabies: extreme hyperexcitability, fear or rage. • Dumb rabies: depression • Paralytic rabies: flaccid tetraparesis or paraparesis • Sudden death. EARLY RABIES CAN LOOK LIKE ANYTHING! [3,6-15]

Early clinical signs are often non-specific with depression, anorexia and mild ataxia. Later there may be repetitive twitching, hyperasthesia, hypermetria, and conscious and unconscious proprioceptive deficits may be seen. Horses often present for colic or lameness. There may be superlibedo, with mounting of other animals or innate objects or regional pruritis due to paraesthesia. Pruritis may lead to intense rubbing or biting, causing loss of hair or wool or self-mutilation. Some cases show belligerousness and run frantically through fences, charging in response to auditory or tactile stimuli. They may bite the handler or attack themselves. Periods of violence may be intermittent, and interspersed between periods of profound depression or normalcy. Spinal reflexes may be normal, increased, decreased or absent. The paralytic form is more common in cattle, with an unexplained ataxia or shifting leg lameness that is rapidly followed by paraparesis or paraplegia. Spinal reflexes and muscular tone may be decreased or absent. Most horses become recumbent in 5-7 days, but may continue to eat and drink normally, until anorexia and encephalitic signs develop with coma, grand mal seizures and death. The furious form may result in aggressiveness and indiscriminant attack. In cattle there may also be bloat, tenesmus, ptyalism, pollakiuria, hypersexuality, paraphimosis, flaccidity of tail and anus, bellowing, silent bellowing, recumbency, coma or convulsions. The dumb form results in depression, inappetance and fever (>39.4° C or 103° F). There is head dropping, flaccid facial musculature, ptosis, profuse ptyalism, pharyngeal paralysis, yawning, nibbling lip motions, tenesmus, paraphimosis and marked ataxia. There may also be flaccidity of the bladder, tail, anus and tongue, and muscle tremors. There may be strabismus, nystagmus, blindness, circling, head-pressing, head tilt, abnormal vocalization and odontoprisis. Prior to death the animal may be comatose or seizure. Rabies is rapidly progressive and uniformly fatal with death from cardiorespiratory failure usually within 10 days. In horses rabies can be visually indistinguishable from togaviral encephalitis, lekoencephalomalacia, hepatoencephalopathy, EHV-1, EPM, space-occupying masses, and meningitis. Most horses die within 5 days of showing clinical signs,[3,6] but up to 14 days has been observed.[13] The incubation period, clinical signs, and survival time are related to the strain pathogenicity, the dose obtained, the immune status of the host and the proximity of the bite wound to the CNS.

CLINICAL PATHOLOGY

Virus has been detected in the blood, CSF, urine and milk of some species therefore all laboratory samples should be labeled appropriately and handled with latex gloves. CSF may be normal, or show pleocytosis (5 -30 mononuclear cell/UL, occasionally neutrophils or rarely eosinophils) and a normal to moderately increased protein (60-200mg/dl). Xanthochromia is rare, but more likely to be seen early in the course of the disease.[3]

SEROLOGY

Serum and CSF neutralization tests and fluorescent antibody tests on salivary gland, saliva, cornea, skin, tactile hair follicles and CSF have been used as antimortum tests, but false negatives and positives make interpretation difficult.[1,3,16] Positive results in the face of clinical signs should be considered highly likely, but not definitive. PCR can be used with primers that amplify the genomic RNA and viral mRNA sequences from antimortem and postportem samples.[17]

Virus neutralization tests are available, but the presence of antibodies is not diagnostic. Immune status can be determined with passive hemagglutination, complement fixation, indirect FAT or radioimmunoassay, but will not assist in diagnosis. ELISA can also be used to measure rabies specific antibody in sera.

NECROPSY

Definitive diagnosis is from cryostatically sectioned fluorescent antibody stained impression smears of hypocampus, medulla oblongata, gasserian ganglia or cerebellum. This takes two hours and has a 98-99% reliability.[16] Monoclonal antibodies specific for epitopes of rabies virus envelope proteins are also highly specific. H&E histopathology may show a non-supportive encephalitis, ganglioneuritis with neuronal necrosis and Negri bodies, (intraneuronal eosinophilic cytoplasmic inclusions of viral ribonucleoprotein, in the hippocampus and purkinje cells of the cerebellum of ruminants and some horses) are highly suggestive, but is no longer considered pathognomonic.[1] They are not seen in every case, especially if the animal was euthanized in the early clinical course of the disease.[3,16] Approximately 30-53% of FAT confirmed rabid horses were **negative for Negri bodies**, being more likely to be seen in horses that had clinical signs for longer than 4 days prior to death.[3]

A hemorrhagic poliomalacia was recorded in the spinal cord of some horses. Overall, considering the impressiveness of clinical signs, there is often remarkably little evidence of cellular injury, and histopathology is often unimpressive.[3,18] There may be cerebral edema, meningeal congestion, focal areas of hemorrhage, perivascular cuffing by mononuclear cells, gliosis, glial nodules, neuronophagia and neuronal degeneration. These changes may be most severe in the dorsal root ganglia. Electron microscopy and FAT of specimens from animals and humans that have died from the disease reveal that almost all neurons in the CNS are infected with the virus.[17]

There may also be thymic atrophy, absent germinal centers to lymph nodes and changes in the spleen. An immunoperoxidase test can be used on formulin-fixed tissue sections. Intracerebral inoculation of weanling mice with brain homogenates of suspected cases followed by fluorescent antibody testing (FAT) of brain sections after the development of neurologic signs (usually 11-12 days, range of 4-18 days, with a negative test only given after 21 days) is the gold standard test. A neuroblastoma cell line tissue culture infection test may be used to replace the mouse inoculation test, as it allows staining within 4 days.

An antigen-capture immunodiagnostic test using salivary gland samples captures the rabies antigen onto a solid phase using purified polyclonal antibodies. The Se & Sp are similar to the FAT, but the test is rarely used in the USA.

TREATMENT

As rabies is a reportable disease, management of a rabies suspect must be undertaken in consultation with state or federal health authorities. Any horse bitten by a wild animal should be considered potentially exposed to rabies whether or not the wild animal is available for testing.

Immediate irrigation of the wound[19] with 20% soft soap solution or Zephiran is the best method of prevention of establishment of infection. Others suggest iodine or quaternary ammonium disinfectants.[20]

In accordance with the Compendium of Animal Rabies Control, fully vaccinated animals, that were vaccinated more than 30 days before the bite should be revaccinated immediately and observed for 45 days. An exposed but unvaccinated animal should be euthanized immediately, or kept under quarantine for 6 months. As herbivore to herbivore transmission is rare, it is not necessary to restrict the rest of the herd if only one animal had been exposed.

PROTECTION OF VETERINARY PERSONAL

If rabies is a differential diagnosis the animal should only be handled by personal who have had appropriate rabies pre-exposure vaccination. A list of "in-contact" and potential "in-contact" individuals including owners should be kept on a clip-board close to the stall. Any person who handles the animal or biologic samples from the animal should sign the list. Laboratory personal should be prewarned that rabies is a differential prior to submitting any samples (blood, CSF) for diagnostic testing. "Rabies suspect" should be written on all containers and submission paperwork.

When examining any horse that has rabies as a possible differential gloves, face shields/masks and eye goggles should be used during all examinations. For any personal performing decapitation and brain removal protective Tyvek coveralls and a mist mask rated for biohazards (>N95 rating) should be used.

If a positive rabies test is confirmed then all interested parties should be immediately notified including the state veterinarian. Exposed individuals should consult their physician and local and state health authorities regarding post exposure treatment. For human exposure the WHO recommends vaccination postexposure on days 0, 3, 7, 14 and 30.

SAMPLE SUBMISSION

If a rabies-exposed horse develops neurologic signs it should be euthanized and the head removed and shipped in situ (well sealed and cooled) for histology and FAT. Freezing or fixation will interfere with FAT and should not be performed.

Post-exposure vaccination of animals is unlikely to be useful in animals as death occurs prior to the development of useful immunity. Euthanasia of suspect animals already exposed to humans should be avoided until clinical signs are shown, as progression of the disease within the CNS is required to obtain a diagnosis. Remember: rabies is a notifiable disease.

CONTROL

Pre-exposure vaccination of dogs and cats, ferrets, horses and valuable livestock is recommended and required prior to interstate transport.[21] Rabies is an excellent immunogen and vaccines induce a strong serologic response within 45 days after a single dose. Many vaccines are now available for use every 3 years, however no vaccine has been approved to provide protection for horses and cattle for greater than one year, therefore annual vaccination is still recommended.[23] Any FDA approved vaccines can be used for revaccination, even if the product is not the same as previously administered. Pre-exposure immunization of veterinarians or high-risk groups using inactivated (killed) virus produced in human diploid cells.[22] The Compendium of Animal Rabies Control published annually by the National Assoc. of State Public Health Veterinarians Inc [23] details pre-and post exposure management. http://www.cdc.gov/MMWR/preview/mmwrhtml/rr6006a1.htm In livestock a primary vaccination is given with a second dose after 1 month followed by boosters given annually using an inactivated cell culture vaccine. This results in neutralizing antibodies after one month, which last for 12 months. AAEP guidelines on vaccination are available at Available at http://www.aaep.org/rabies.htm. For foals from unvaccinated dams give the primary dose at 3 months of age, a second dose at 12 months of age, then annually. For foals born to vaccinated dams give the first dose at 6 months, second dose at 7 months, boost again at 12 months, then annually. Vaccination of pregnant mares can be performed 4-6 weeks before foaling. The rabies vaccine is considered highly protective, but breaks in vaccine protection have been reported in several species,[24] and in two young horses with a positive post mortem rabies diagnosis that reportedly had had their primary vaccination. A two dose primary vaccination series for foals is therefore recommended by some.[3] Antibody levels can be determined using the rapid fluorescence focus inhibition test (RFFIT) available through the Kansas State Veterinary Diagnostic Laboratory, www.vet.k-state.edu/depts/dmp/service/rabies/guideline.htm.

Due to zooinotic implications rabies should never be excluded from the differential list even in vaccinated horses showing undifferentiated, rapidly progressive neurologic deterioration.[7]

PROTOCOL POST EXPOSURE

Livestock exposed to a rabid animal that have an up-to-date vaccination record should be vaccinated immediately and observed for 45 days. Unvaccinated livestock should be euthanized or kept under close observation for 6 months. A report describes that a post-exposure protocol with immediate vaccination, with a booster prior to release from quarantine with a 90-day strict isolation was 99.7% successful in unvaccinated cattle, horses, sheep, goats and pigs.[24] Research in other countries describes that after natural infection of cattle and horses the use of four injections of a modified-live vaccine were protective. Post inoculation of experimentally infected sheep had no effect on incubation period, clinical symptoms or mortality rate, and in this study antibody titers in the vaccinated animals had no predictive value for protection. Postexposure vaccination and concurrent treatment with human antirabies immunoglobulins improved survival. However, no USDA-licensed biologics for postexposure prophylaxis of previously unvaccinated domestic animals exist, however immune globulin may be considered for very valuable animals/zoological specimens.

VACCINATION OF WILDLIFE.

Reduction of the wildlife population by trapping or poisoning is unsuccessful in reducing enzootic rabies. Vaccinia recombinants expressing the rabies G glycoprotein induce protective immunity from production of neutralizing antibodies, T helper cells and cytotoxic T cells. The products in the oral vaccine are self-replicating and thus unintentional human exposure must be avoided. Mass oral vaccination using vaccine-baits is an effective and feasible and acceptable control method. Over 100 million doses of vaccine-laden bait has been distributed over 6-million square kilometers of Europe.[25] Such programs have also been effective in Texas and Canada and are presently being undertaken in Europe. After using this oral immunization program between 1989 and 1994, the prevalence of rabies was reduced to 20% the 1989 level. This was also successful in Ontario. It took 7-17 days for 80% of foxes to eat their first bait after an aerial bait distribution. During 2000 1.85 million doses of the V-RG vaccine were distributed over Texas, and no cases of the dog-coyote variant of the rabies virus have since been reported.[2]

A vaccinia recombinant rabies vaccine confers a minimum of 12 months protection to fox cubs and 18 months to adults, and is stable in the field for 1 month. Field trials resulted in a dramatic reduction of rabies prevalence in Belgium, and campaigns are currently underway in several European countries.

Skunks are refractory to many oral rabies vaccines, but a vaccinia virus recombinant expressing the rabies virus glycoprotein has been tested with good results, and is also useful in wild swine.

A modified live vaccine is effective in dogs and cats. These are not approved for use in livestock and may produce fatal allergic or viral encephalitis in large animals. Titer persistence is also greater in horses with the killed vaccine compared to the modified live vaccine. A modified live vaccine used overseas protects adult cattle for 4 years, and for 2 years when used in calves less than 2 months.

REFERENCES

[1]Radostits O, et al. *Veterinary Medicine:* 9th ed. 2000;1201. [2]Blanton JD, et al. *J Am Vet Med Assoc* 2010;239(6):773.tp://www.cdc.gov/rabies/location/usa/surveillance/index.html [3]Green SL, et al. *J Am Vet Med Assoc* 1992;200:1133. [4]Noah D, et al. *Am J Public Health* 1996;86:1149. [5]Fishbein DB. *Infect Dis Clin North Am* 1991;5:53. [6]Green SL. *Vet Clin North Am Equine Pract* 1993;9:337. [7]Green SL. *Vet Clin North Am Equine Pract* 1997;13:1. [8]Hahn D. *Vet Med Small Animal Clin* 1982;9:1409. [9]Joyce J, et al. *Compend Contin Educ Pract Vet* 1981;2:S56. [10]Marler RJ, et al. *J Am Vet Med Assoc* 1979;175:293. [11]Meyer EE, et al. *J Am Vet Med Assoc* 1986;188:629. [12]Smith LL, et al. *Can Vet J* 1972;13:193. [13]Siger S, et al. *Equine Pract* 1989;1:6. [14]Striegel P, et al. *Mod Vet Pract* 1983;12:983. [15]West GP. *Vet Clin North America Equine* 1985;69:415. [19] *J Ark Med Soc* 1998;95:84. [20]Schroeder WG. *J Am Vet Med Assoc* 1969;155:1842. [21]Sedmak P, et al. *Compend Contin Educ Pract Vet* 1984;1:49. [22]Schneider L. *Devel Biol Stand* 1995;84:49. [23]*J Am Vet Med Assoc* 2011;239:5:609. [24]Clark KA, et al. *J Am Vet Med Assoc* 1996;208:1827. [25]Stohr K, et al. *Vet Rec* 1996;139:32.

EVALUATION OF THE NEUROLOGIC PATIENT

Stephanie Gilliam, RVT, BAS, CCRP
Columbia, MO

INTRODUCTION

The neurologic examination is a skill that takes practice, but one that every technician can be proficient in. It is vital in localizing the lesion and therefore confirming a diagnosis in a neurologic patient. The goals of this presentation are to: 1) understand basic neuroanatomy, 2) become proficient in obtaining a complete neurologic history, 3) understand how to perform a neurologic examination and 4) understand the basics of localizing a neurologic lesion.

The neurologic examination consists of 5 parts: observation, postural reaction testing, cranial nerve examination, spinal reflexes and pain perception. The goal of the neurologic examination is to be able to effectively localize the lesion to a certain area of the nervous system (in some cases it may be multifocal or diffuse).

NEUROANATOMY

The nervous system is divided into two divisions, the upper and lower motor neurons. The upper motor neuron system (UMN) is composed of the brain and spinal cord (CNS). These neurons are responsible for voluntary motor function and modulating the activity of the lower motor neurons. UMN lesions produce a set of clinical signs caudal to the level of the injury. The classic clinical signs of an UMN lesion are increased muscle tone, normal to hyper reflexes and no significant muscle atrophy.[2]

The lower motor neuron (LMN) system is composed of the peripheral nerves that connect the central nervous system with muscles and glands. The lower motor neurons are located in the brain stem and all spinal cord segments. Each spinal cord segment has a pair of spinal nerves originating from it. The axons extending from these segments form the spinal and cranial nerves and consists of a dorsal (sensory) and ventral (motor) root. The lower motor neuron unit consists of these cell bodies, axons, the neuromuscular junction and the muscle itself. The classic clinical signs associated with a LMN lesion are loss of muscle tone, decreased to absent reflexes and severe muscle atrophy. [2]

NEUROLOGIC HISTORY

Obtaining a complete history from the owner is essential and can provide evidence that a neurologic problem is present. The history should begin with the patient's signalment; species, breed, age and sex. Many diseases are more or less likely to occur in certain groups or breeds of animals. Young animals are more likely to have congenital disorders while older animals are more likely to have neoplastic conditions. The onset of clinical signs is the next important part of the history. It is important to distinguish if the signs are peracute (minutes-hours), acute (day-week) or chronic (week-months). A timeline is the next part of the history. It can sometimes be exact or may be difficult to determine, but it is important to distinguish if the progression of the disease has been slow, fast, static, improving or waxing and waning. Animals may also have episodic disorders in which they are normal between signs.

Some clinical presentations can be difficult to classify, such as syncope versus a seizure. If an animal is presenting with such signs it is important to first obtain a complete description of the episodes. Did the animal urinate or defecate during the episode? The duration of the episode is also important. A typical seizure is about 1-2 minutes long while a vestibular episode could last much longer. It is also important to determine whether or not the animal was conscious during the episode. Did the animal respond to the owner during the episode? Was there a post ictal phase or was the animal normal right after the episode occurred? These are the questions that can help distinguish if the animal truly experienced a seizure or if it was something else such as syncope or a vestibular episode.

NEUROLOGIC EXAMINATION

The equipment needed for a neurologic examination is a bright light source such as a transilluminator, a reflex hammer (pleximeter), a hemostat and cotton balls. The observation portion of the exam can be done while taking the history from the owner. The dog should be allowed to move freely in the exam room or in an open area. The dog's mental status, posture and gait can be evaluated at this time. The animal's mental status can be documented as alert, depressed, stuporous or comatose. Behavioral changes such as aggression can also be documented at this time. When evaluating a patient's posture it is important to evaluate the whole animal beginning with the head, the trunk and finally the limbs. A common abnormality with the head is a tilt suggestive of vestibular dysfunction or a turn suggestive of forebrain disease. Abnormal posture of the trunk can consist of scoliosis (lateral deviation), lordosis (swayback) or kyphosis (dorsal deviation). Abnormal posture of the limbs includes increased or decreased muscle tone and improper positioning.[2]

Gait should be evaluated by having an assistant walk, trot and turn the animal in order to observe. It may also be helpful to watch the animal walk freely in an enclosed area. Abnormalities of gait may include paresis (weakness) or ataxia (uncoordinated gait). Paresis essentially means weakness while paralysis suggests loss of all voluntary motor function. The prefix para- means the hind limbs are involved. Tetra- means all four limbs are involved and hemi- means the limbs of one side are involved. For example, if a dog is paraparetic it is weak in the hind limbs. There are 3 types of ataxia; proprioceptive, vestibular and cerebellar. Signs of proprioceptive ataxia include an inability to know where the feet are in space. These animals will cross over and knuckle their feet when walking. Vestibular ataxia can cause circling, a head tilt and torticollis. Animals with cerebellar ataxia will have a dysmetric gait, which is characterized by long movements (overreaching) and overflexion of the joints. This can look like a "goose-stepping" gait.[2]

The responses that maintain an animal's upright position are the postural reactions. Abnormalities of one of these reactions do not provide localizing information because lesions in one of several areas of the nervous system could affect this reaction. However, postural reactions are an important part of the neurologic examination because they can detect minimal deficits. The hopping reaction is evaluated by lifting the three limbs that you are not testing and pushing the animal laterally to observe their ability to hop on each limb. Equal responses should be seen on all limbs. General proprioception (GP) can confirm that a neurologic disease is present. The animal is given support under the chest or between the pelvic limbs and the paw is turned over so that the dorsal surface of the paw is on the ground. The animal should immediately return the paw to the normal position. CP deficits indicate a neurologic lesion and are not caused by orthopedic disease.[2]

The cranial nerve examination is the next part of the neurologic exam. An abnormality of a cranial nerve indicates a problem in a localized area that is not provided by postural reactions. There are 12 cranial nerves. Cranial nerve (CN) I is the olfactory nerve, which is responsible for the conscious perception of smell. It may be inferred from the history or assessed by watching the animal walk around the room to see if they are sniffing. Irritating substances such as ammonia should not be used to assess CN I because they can also stimulate the trigeminal nerve in the nasal mucosa. CN II is the optic nerve and is responsible for sensory of vision and the pupillary light reflex (PLR). Vision can be assessed by watching the animal navigate around a room. It may also be assessed by using the menace response. To perform the menace response the examiner makes a threatening gesture with a hand or finger at one eye being careful not to touch whiskers or push air into the face. The normal response is a blink. The PLR is induced by shining a light in each eye and observing pupillary constriction in both eyes (testing both direct and consensual PLR).

CN III (oculomotor nerve) is responsible for pupillary constriction and also innervates the dorsal, medial and ventral recti, the ventral oblique and the levator palpebrae muscle of the upper eyelid. This nerve is assessed by the PLR, observing for abnormal eye movements (nystagmus) and by looking for a ptosis (drooping upper eyelid). A physiologic nystagmus, or doll's eye, can be seen by moving the head laterally. The eyes should track towards the direction you are moving the head. This is a normal nystagmus. CN IV (trochlear nerve) is responsible for motor function to the dorsal oblique muscle of the eye and is difficult to assess. A lesion of this nerve may cause a lateral rotation of the eye.[2]

CN V (trigeminal nerve) is responsible for motor to the muscles of mastication and sensory to the face. There are 3 branches of the trigeminal nerve; mandibular, ophthalmic and maxillary. The mandibular nerve innervates the temporal and masseter muscles. A bilateral lesion could cause a dropped jaw. A unilateral lesion could cause decreased jaw tone. The palpebral reflex tests 2 branches of the trigeminal nerve. The medial canthus tests the ophthalmic branch and the lateral canthus tests the maxillary branch. Stimulation of the nasal mucosa will also test the ophthalmic and maxillary. The mandibular branch can be assessed by touching or pinching the jaw.[2]

CN VI (abducent nerve) innervates the retractor bulbi and the lateral rectus muscles. Lesions of this nerve can cause a medial strabismus and an inability to retract the globe. It is assessed using eye movements and the palpebral reflex. CN VII (facial nerve) is responsible for motor to the muscles of facial expression, sensory for taste and the inner surface of the pinna. A lesion can cause the lips, eyelids or ears to droop and will usually be asymmetric. CN VIII (vestibulocochlear nerve) has 2 branches; the cochlear branch and the vestibular branch. The cochlear branch is responsible for hearing and may be tested using a loud noise or by performing a BAER (brainstem auditory-evoked response). The vestibular branch is responsible for orientation of the head and may produce many clinical signs if a lesion is present. Unilateral disease can cause ataxia, nystagmus and a head tilt towards the side of the lesion.[2]

CN IX (glossopharyngeal nerve), CN X (vagus nerve) and CN XI (accessory nerve) are commonly considered together. The glossopharyngeal nerve innervates the pharynx along with the vagus nerve. It also supplies some motor fibers to the salivary glands and sensory to part of the tongue. The vagus nerve innervates the larynx, pharynx and the palate. These nerves can be assessed by eliciting a gag reflex by inserting a tongue depressor into the pharynx. The spinal root of the accessory nerve supplies the motor pathway to the trapezius muscle and can be assessed by palpating for atrophy. CN XII (hypoglossal nerve) provides motor to the tongue and is assessed by observing the animal extend its tongue and looking for atrophy.[2]

Spinal cord reflexes test the integrity of specific peripheral nerves and spinal cord segments. Decreased or absent reflexes indicate a LMN lesion. Normal to increased reflexes indicate a UMN lesion. Reflexes should be done while the animal is in lateral recumbency. Muscle tone and range of motion can also be assessed at this time. The forelimb reflexes include the biceps, triceps, extensor carpi radialis and the flexor (withdrawal) reflex. The biceps reflex tests the musculocutaneous nerve, which originates from C6-8. To do this reflex the examiner puts a finger on the tendon and taps it with the pleximeter. The response will be either a slight flexion of the elbow or movement of the skin that overlies the biceps brachii.. The triceps reflex tests the radial nerve, which originates from C7-T1 and is done in the same fashion as the biceps. The response with the triceps should be an extension of the elbow or a contraction of the muscle. It is an unreliable reflex, therefore a decreased or absent response should not be considered abnormal. The extensor carpi radialis reflex tests the radial nerve as well (C7-T1). To do this reflex the carpi radialis muscle is tapped with a pleximeter and extension of the carpus is observed. The flexor reflex is also known as the withdrawal. Sensory input for the withdrawal is through the radial, median and ulnar nerves. The motor output is through the axillary, musculocutaneous, median and ulnar nerves (C6-T2). To do this reflex the toe is pinched in order to elicit flexion of the shoulder, elbow and carpus.[1]

The pelvic limb reflexes include the patellar, gastrocnemius, cranial tibial and the flexor reflex. The patellar reflex tests the femoral nerve, which originates from L4-6. To do this reflex the limb is supported under the femur and the patellar tendon is tapped with the pleximeter in order to elicit extension of the stifle. The gastrocnemius reflex tests the tibial nerve. The tibial nerve is one branch of the sciatic nerve, which originates from L6-S1. In order to test this reflex the hock must be flexed in order to put some tension on the tendon. The tendon of the gastrocnemius muscle is then tapped with the pleximeter and extension of the hock or contraction of the caudal thigh muscles is observed. The cranial tibial reflex tests the peroneal nerve, which is the other branch of the sciatic nerve (L6-S1). The belly of the cranial tibial muscle is tapped with the pleximeter and flexion of the hock is observed. The flexor reflex (withdrawal) of the pelvic limb tests the sciatic nerve and is performed in the same fashion as for the thoracic limb. Flexion of the entire limb (hip, stifle and hock) is observed.[1] The flexor reflex is a reflex and therefore does not require involvement of the brain. An animal that has no perception of pain can still have an intact flexor reflex. The perineal reflex is tested by light stimulation of the perineum. The response observed will be contraction of the anal sphincter and flexion of the tail. This reflex tests the pudendal nerve which originates from S1-3. The cutaneous trunci (panniculus) response is done by pinching the skin on each side of the vertebrae between T2-L4 and a twitch of the skin is observed. The cutaneous trunci is innervated by the lateral thoracic nerve that originates from C8-T1.[2]

The last portion of the examination is the sensory examination. These tests should be done at the end of the examination so that you do not loose cooperation from the patient at the beginning. This includes testing for any areas of hyperesthesia as well as ensuring that the animal has a conscious perception to noxious stimuli. Hyperesthesia is by definition an increased sensitivity of nociceptive fibers to normal stimulation.[2] To test for spinal hyperesthesia the examiner should begin caudally around L7 and press on each spinous process firmly moving cranially. The severity of the stimulus is then increased gradually from touch to deep palpation. If there is a painful area you will observe a behavioral reaction. The order of loss of function with a neurologic lesion is conscious proprioception, motor function, superficial pain, and lastly deep pain. Therefore, if the animal does not have motor function it is necessary to check for pain sensation of the feet. To test for superficial pain sensation the skin in between the toes is pinched with a hemostat. A behavioral response such as aggression or vocalization should be observed. A withdrawal is not a behavioral response. To test for deep pain sensation a hemostat is used to squeeze a digit. Again, a behavioral response should be observed.

LESION LOCALIZATION

After completing the neurologic examination an attempt is made to localize the lesion to one anatomic site of the nervous system. The nervous system is divided into five levels of the spinal cord and five sections of the brain (or peripheral vestibular). If it is not possible to localize to one region then it may be a multifocal or diffuse process. The levels of the spinal cord are divided as follows: C1-C5, C6-T2, T3-L3 and L4-S2. The brain is divided as follows: cerebrum, diencephalon, brainstem, vestibular system and the cerebellum.

When localizing a spinal cord lesion the first step is to determine if the signs present are consistent with a UMN or LMN lesion. Classic signs of a LMN lesion are decreased or absent reflexes, loss of muscle tone and muscle atrophy. Classic signs of a UMN lesion are normal to increased reflexes, no significant atrophy and increased muscle tone. Paresis or paralysis is a primary finding with both UMN and LMN disease.[2] Paresis of all four limbs indicates a lesion cranial to T3. The spinal reflexes and cranial nerves are what help to localize further. If cranial nerves are all within normal limits the brain is excluded. Decreased reflexes indicate LMN disease; therefore the lesion would be C6-T2. If the reflexes were normal the lesion would be localized to C1-C5. If the animal is paraparetic (hind limb weakness) with normal forelimbs and a normal cranial nerve examination we know the lesion is caudal to T3. Decreased hindlimb reflexes would indicate L4-S2 while normal to increased hindlimb reflexes would indicate T3-L3.

Lesions that are localized to the brain are further localized to a certain area of the brain according to the clinical signs that are present. Lesions of the cerebrum will usually cause behavioral changes, seizures, vision loss with intact PLR, contralateral decreased facial sensation, contralateral hemiparesis and postural reaction deficits. Lesions of the diencephalon (thalamus or hypothalamus) may cause UMN signs in all limbs or a hemiparesis and postural reaction deficits. It may also cause circling, CN II, III IV and VI deficits depending on the extent of the lesion. The most common signs of a lesion here are related to the function of the hypothalamus. Brainstem lesions will also cause UMN signs in all limbs or on one side (hemiparesis). Cranial nerve deficits are more extensive with brainstem lesions than with other areas. Mental status may also be altered with a brainstem lesion. Signs of vestibular disease include falling, rolling, head tilt, circling, strabismus, nystagmus and ataxia. A vestibular lesion can be further localized as being central or peripheral. Distinguishing between these two is very important because of the differences in treatment and prognosis. A central lesion will cause postural reaction deficits while peripheral disease will not. Central disease may also cause an altered mental status or CN V and VII deficits.[2]

REFERENCES

[1]Chrisman, C. et al. ***Neurology for the Sm An Pract.***2003, Teton NewMedia. [2] Lorenz, M.D. et al. ***Handbook of Vet Neurology (5th ed.).*** 2011, Elsevier Saunders.

DEGENERATIVE MYELOPATHY DIAGNOSIS AND MANAGEMENT: THE TECHNICIAN'S ROLE

Stephanie Gilliam, RVT, BAS, CCRP
Columbia, MO

INTRODUCTION

Degenerative myelopathy (DM) is a progressive disease of the spinal cord that affects many dog breeds. There are not any evidence based medical approaches for the treatment of DM; therefore veterinary technicians can play an important role in the management of these cases.

PATHOGENESIS

DM was first described in the German Shepherd Dog, and has historically been considered a disease that results in upper motor neuron dysfunction of the pelvic limbs.[1] It has recently been realized that DM affected dogs also develop lower motor neuron signs later in the progression of the disease.[2] Dogs with upper motor neuron disease exhibit a spastic gate or paralysis; while dogs with lower motor neuron disease exhibit a flaccid paralysis.[3] DM is now classified as a degenerative disease that affects the axons of the spinal cord and peripheral nerves.

SIGNALMENT

Age at onset of clinical signs is usually older than 8 years with a mean of 9 years in large breeds.[1] A study of affected Pembroke Welsh Corgis reported a mean age at onset of 11 years.[4] There is no sex predilection. The disease has been histopathologically confirmed in many breeds with some of the most common being the German Shepherd Dog, Pembroke Welsh Corgi, Rhodesian ridgeback, Chesapeake Bay retriever and the boxer.[5]

CLINICAL SIGNS

The key clinical signs of DM are progressive asymmetric ataxia (incoordination), spastic paraparesis (hind end weakness), and lack of spinal hyperesthesia (pain on palpation of the spine). These signs will progress to paraplegia and eventually will involve the thoracic limbs as well. The disease course to paraplegia can vary but most large breed dogs will progress to nonambulatory paraparesis within 6 to 9 months after onset of signs.[5] Owners of large breed dogs often elect to euthanize when their dog is no longer able to support weight in the hind limbs. Small breeds are more easily cared for by the owner for a longer period of time. The median disease duration in the Pembroke Welsh Corgi was 19 months.[4]

The earliest clinical signs are asymmetric general proprioceptive ataxia and mild paraparesis. Asymmetry of signs is commonly reported but as the disease progresses the signs become symmetric. The paraparesis will progress to paralysis and ascend to affect the thoracic limbs. Flaccid tetraplegia occurs in the late stage. Muscle atrophy becomes evident in the pelvic limbs as the signs progress. Atrophy of the appendicular muscles occurs in the late stage. Difficulty swallowing, inability to bark and urinary and fecal incontinence can also occur in the late stage of the disease.[2,4]

DIAGNOSIS

A presumptive diagnosis is based on clinical signs along with a series of neurodiagnostic procedures to exclude other disease that can mimic DM. The neurologic examination is critical in order to develop a diagnostic approach. The lack of spinal hyperesthesia is a key sign that distinguishes DM from other compressive myelopathies.[6] Cerebrospinal fluid (CSF) analysis is done to rule out inflammatory disorders, electrodiagnostic procedures can be done to rule out other peripheral nerve and muscle disorders, and myelography, CT and/or MRI of the spinal cord can be done to rule out a compressive disease such as intervertebral disk disease or neoplasia.[3] If no signs of other disease are found on these tests a presumptive diagnosis of DM is made. A definitive diagnosis can only be made through postmortem histopathology.[1]

A mutation in the superoxide dismutase 1 (*SOD1)* was discovered as the underlying cause for most forms of DM.[2] Mutations of this gene in humans are an underlying cause for some forms of amyotrophic lateral sclerosis (ALS- Lou Gehrig's disease). A DNA test for the *SOD1* mutation is now commercially available. Dogs that test homozygous for this mutation are at risk for developing DM and will contribute one chromosome with the mutation to their offspring. However, some dogs that test homozygous never develop clinical signs, which suggest an age-related incomplete penetrance. Dogs that test heterozygous are considered carriers and are less likely to develop clinical signs but could pass on a chromosome to their offspring. Dogs that test homozygous normal are unlikely to develop DM and will pass on a protective normal allele to their offspring.[5]

MANAGEMENT STRATEGIES

Pharmacotherapies for the treatment of DM have been empiric with a lack of evidence-based medical approaches for determining efficacy. Aminocaproic acid has been promoted for long-term management; however, a recent study evaluated a combined therapy of aminocaproic acid and N-acetylcysteine with vitamins B, C and E, found no beneficial effects.[7] Kathmann *et al.* reported data from 22 affected dogs that received varying degrees of physical rehabilitation.[8] Dogs that received intensive physical rehabilitation had longer survival times (mean of 255 days) compared to dogs that received moderate (mean of 130 days) or no physical rehabilitation (mean of 55 days). The rehabilitation regimen consisted of active and passive exercises. These results suggest that survival time was positively associated with the physical rehabilitation performed but other studies are needed to measure the efficacy of physical rehabilitation in DM-affected dogs.

The goals of a physical rehabilitation program for an affected dog should be to maintain joint range of motion, reduce spasticity, and to retard disuse muscle atrophy and loss of neuromuscular function. When developing and beginning a rehabilitation program it is important to avoid fatigue. If the dog shows persistent weakness or morning fatigue after exercise on the previous day the therapist should readjust the exercise program in order to prevent further muscle damage. Nonstrenuous activity is recommended in ALS patients and has been shown to increase quality of life. Activities should include periods of activity with rest in between.[9] Ideally a rehabilitation plan will be implemented at the first signs of DM.

Active exercises are recommended as long as some voluntary movements remain intact. This includes standing and weight shifting exercises, as well as sit to stand exercises. Balance exercises such as walking on an uneven surface (air mattress or cushion) and other proprioceptive exercises such as weaving through obstacles or walking over obstacles (cavaletti poles) may also be performed. These exercises should generally be started slowly, beginning with about 5 minutes once a day. The frequency and duration of exercises can then be adjusted accordingly for each individual patient.[10]

Hydrotherapy is a commonly used modality in canine rehabilitation. An underwater treadmill can facilitate active movements while supporting the dog's body weight through buoyancy effects. For example, a dog bears 91% of its body weight with the water at the level of the hock. Increasing the water level to the stifle the dog bears 85%, and to the level of the hip the dog bears 38% of its body weight. The resistance of the water also provides resistance, which allows for an increase in muscle strength. Hydrostatic pressure of the water has been shown to reduce edema and swelling. Walking or swimming in water has also been shown to improve general circulation.[11] The water temperature should be kept warm because of the beneficial effects of heat when applied to tissues. Heat provides an increase in elasticity and blood flow of the tissue.[10]

Passive range of motion exercises are indicated to help prevent joint contracture, maintain mobility of soft tissue, enhance circulation and improve synovial fluid production. To perform this exercise the patient should be relaxed in either a lying or standing position. The therapist will gently flex the limb to the point of resistance then extend in the same manner. The limb is also manipulated in a "bicycling" manner to simulate a normal gate pattern. A repetition of 15-20 times is recommended at least 3 times daily.[10]

As a dog begins to lose proprioception protective boots to protect the paws can be recommended to the owner. The protective footwear can vary but should be properly fitted and water resistant. As the dog becomes more paraparetic sling support is needed. There are many commercially available slings. It should be properly fitted in order to prevent chaffing. A wheel cart should be considered for dogs that become nonambulatory. Wheel carts provide independence for a DM affected dog but the owner needs to be aware of monitoring for fatigue. It is also important to ensure that the cart fits appropriately.[6]

The secondary consequences of DM are decubitus ulcers, pneumonia, urinary tract infections and contracture. The veterinary technician has an important role in being aware of these disease complications. If the dog is recumbent and unable to get in to a sternal position rotation is performed every 4 hours. The recumbent dog will need to be kept on soft padded bedding and the skin over bony prominences should be assessed daily for redness or irritation. The bedding should be absorbent in order to wick away moisture. This is critical to prevent fecal or urine scalding.[6]

As the disease progresses to paraplegia the patient will likely lose the ability to voluntarily urinate. The technician will need to teach the owner how to manually express the bladder by applying gentle pressure to the abdomen. This will need to be done 2-3 times daily. If manual expression is not possible, intermittent catheterization should be performed. A urinalysis will need to be performed routinely to evaluate for the presence of a urinary tract infection.[10]

REFERENCES

[1]Averill,DR. **JAVMA** 1973;162(12):1045-51. [2]Awano,T. et al. **Proc Natl Acad Sci USA** 2009;106:2794-5. [3]Coates, JR. et al. **JVIM** 2007;21:1323-1331. [4]Lorenz,MD. et al. ***Handbook of Vet Neuro*** 2011, Elsevier Saunders. [5]Coates,JR. et al. **Vet Clin Small Anim** 2010;40:929-950. [6]Bonagura,JD. et al. ***Kirk's Current Vet Therapy XV*** 2011, Elsevier Science. [7]Polizopoulou,ZS. et al. **Acta Vet Hung** 2008;56(3):293-301. [8]Kathmann,I et al. **JVIM** 2006;20:927-932. [9]Umphred,D. ***Neuro Rehab*** 2007, Mosby Elsevier. [10]Lindley,S. et al. ***BSAVA Man of Can and Fel Rehab*** 2010, BSAVA. [11]Millis,DL. et al. ***Can Rehab and Phys Ther*** 2004, Elsevier Saunders.

ADRENAL INSUFFICIENCY IN CRITICAL ILLNESS: IS IT JUST ADDISON'S?

David Liss, RVT, VTS (ECC)
Los Angeles, CA

INTRODUCTION

Critical illness-related corticosteroid insufficiency (CIRCI), previously known as relative adrenal insufficiency, is a syndrome of relative, as opposed to absolute adrenal dysfunction observed in critically ill patients. Initially observed in septic patients, CIRCI has been documented in many other forms of critical illness, primarily in human medicine, such as pancreatitis and trauma. Recognition, diagnosis, and treatment are essential to affect survival in this relatively small patient population.

The hypothalamic-pituitary-adrenal axis is a complex system of negative feedback loops and hormones controlling the release and restraint of cortisol. As cortisol receptors are present in almost every cell in the body it seems apparent that tight regulation of this hormone is in order. The hypothalamus responds initially to insults from the body (such as stress, hypoglycemia, etc) and release coricotropin releasing hormone (CRH). CRH travels from the hypothalamus to the anterior pituitary where it's agonist affects stimulate the release of adrenocorticotrophic hormone, or ACTH. Once ACTH is released it traverses the body, via the bloodstream, to the adrenal glands. There its agonist activity stimulates the release of free cortisol into systemic circulation. Once adequate cortisol levels are established ACTH, and subsequently CRH, levels dwindle as they are inhibited by the presence of cortisol.

BACKGROUND

During times of physiologic stress the hypothalamic-pituitary-adrenal, HPA, axis increases plasma concentrations of adrenocorticotrophic hormone, ACTH, and subsequently raises plasma cortisol levels. Initially documented in critically ill human patients, CIRCI appears to be a relative, or reversible, state of adrenal dysfunction whereas the body's ability to release cortisol during a stressful period, such as a septic insult, is blunted. The adrenal cortex is intact and functioning, as evidenced by a basal cortisol level and cortisol production post-ACTH challenge, but this level is insufficient to maintain homeostasis. In contrast to absolute adrenal failure, Addison's disease, long-term glucocorticoid and mineralocorticoid replacement is not required in these patients, after they have recovered from their critical state. Cortisol is a highly important and complex molecule with effects on the cardiac, renal, hematologic, vascular, and metabolic systems. Many body processes rely on cortisol for effective function. Cortisol is involved with nutrient metabolism, maintenance of cell membranes, maintenance of vascular tone, catecholamine activity, cardiac contractility, and immune system and endothelial cell function.

CIRCI is thought to originate from underlying systemic inflammation and cytokine release. Pro-inflammatory cytokines, such as interleukin-1 (IL-1), IL-6, and tumor necrosis factor alpha (TNF-a) potentially suppress the HPA axis causing this transient disturbance in cortisol release and activity.

PRESENTATION AND CLINICAL SIGNS

Adrenal deficiency (relative or absolute) could, theoretically, present with any clinical signs that an "Addisonian" crisis patient might present with: collapse, hypotension, hypovolemia, dehydration, bradycardia, hyperkalemia, hyponatremia. In addition, in cases of sole glucocorticoid deficiency, waxing and waning signs such as vomiting, diarrhea, lethargy, anorexia, may also be present. However, with regards to relative adrenal dysfunction, as is the case in CIRCI, these clinical signs are usually masked in the face of critical illness. The most reproducible clinical manifestations of CIRCI is refractory hypotension in the face of severe illness; despite volume-loading and vasopressor therapy. A patient with persistent hypotension, despite initial interventions such as fluids and a vasopressor, should be considered suspicious for the CIRCI complex.

The incidence of CIRCI in critically ill humans varies greatly from study to study. This is due, in part, to the lack of standardization of diagnosing/identifying adrenal dysfunction. Yet, these incidences range from 30-50% with varied population sizes (Cohen and Vankatesh, 2010). A 2007 veterinary study identified an incidence of 48% of critically ill dogs (Burkitt et al, 2007).

DIAGNOSIS

Diagnosis of adrenal dysfunction in critical illness is hotly debated in human medicine. Several methods of clinicopathologic identification of relative adrenal dysfunction in the critically ill exist in human medicine and these include: random cortisol screening, corticotropin testing, free cortisol estimation, metapyrone testing, and glucocorticoid trial and response assessment. The typical method involved in diagnosing CIRCI in veterinary patients involves using a standard synthetic ACTH stimulation test. It is not the basal or post-stimulation cortisol values that are explicitly important, rather the difference between these two values, termed the delta cortisol (Δ cortisol). When this value is <9 mg/dL, this is considered, in addition to clinical signs, to point towards CIRCI. It has also been suggested that a narrowed criteria of a Δ cortisol of ≤ 3 mg/dL is highly suggestive of CIRCI.

Standard ACTH testing (protocol below) was compared to low-dose ACTH administration in critically ill patients. Using a 0.5ug/kg (vs. standard 5ug/kg) dose of ACTH proved to identify adrenal insufficiency in this population. This regimen identified several dogs with adrenal dysfunction not previously identified with the standard protocol. Thus, using an ACTH dose of 0.5ug/kg would be recommended in the diagnosis of patients with CIRCI.

Additional testing modalities to identify transient adrenal dysfunction in the critically ill include using a cortisol assay aimed at providing a true cortisol level- not a pre and post-stimulated value. As in many adrenal testing protocols, measuring a free-cortisol level would be ideal, as a total value can encompass free hormone with hormone that is protein-bound.

Criteria for establishing CIRCI in a critically ill veterinary patient:

- Patients with a normal/elevated basal cortisol level whose post-stimulation cortisol does not increase greater than 5% of the baseline.
- Patients with a delta cortisol ≤ 3 mg/dL

ACTH Stimulation Protocol:

- Draw 2mL pre-sample
- Administer 0.5ug/kg of synthetic ACTH (Corsynotropin)
- Obtain a 1 hr. post-ACTH sample

TREATMENT

There are several studies documenting increased survival rates in critically ill vasopressor-dependent septic patients who receive steroids after bouts of refractory hypotension. Steroid doses used in this case should be at physiologic or replacement doses. These are subsequently less than those used for anti-inflammatory or immunosuppressive effects. Hydrocortisone is used in human medicine most frequently. The dose is typically 200-300 mg for an average adult per day. This equates to a 3-4mg/kg/day dose. Using relative potencies to determine doses of other steroids yields equivalent doses of prednisone at 0.7-1mg/kg/day and dexamethasone at 0.1-0.4mg/kg/day (see Figure 1). This daily dose is typically divided into 4 doses given every 6 hours.

Figure 1

Steroid	Dose per day	Dose every 6 hours
Hydrocortisone	3-4 mg/kg/day	0.75-1 mg/kg
Prednisone	0.7-1 mg/kg/day	0.18-0.4 mg/kg
Dexamethasone	0.1-0.4 mg/kg/day	0.025-0.1 mg/kg

It seems plausible to also administer a "loading" dose of a steroid and administer a CRI for the remainder of the daily amount.

CASE DISCUSSION

- **Signalment:** Mabel 12 year old, spayed female wheaten terrier 15.8kg
- **Presenting Complaint**: Acute weakness, two episodes of vomiting, one episode of diarrhea
- **Initial Assessment:** Mabel was tachycardic, hypothermic, and hypotensive (Doppler measured 30mmHg on right pelvic limb), with no Doppler blood pressure obtainable on the left pelvic limb. She had pale/cold mucous membranes, and a capillary refill time of 3 seconds. She had a 2/6 heart murmur of unknown etiology. Her extremities were also cool. She lacked left femoral arterial pulses.
- **Interventions:** Mabel presented in early decompensated shock. A mixed acid-base disorder consisting of a metabolic acidosis and a respiratory acidosis was diagnosed. Brief ultrasound showed no free abdominal or thoracic fluid. Initial packed cell volume was 43% and the total solids was 7.6 g/dL. After a 40mL/kg bolus of a balanced isotonic crystalloid (Normosol-R®), the repeat Doppler blood pressure obtained was 60mmHg and the packed cell volume was 24% with a total solids of 4.0 g/dL. A 5mL/kg artificial colloid (Hetastarch) bolus was also given. Coagulation times revealed a prothrombin time of 16 seconds (normal is 12-17) and an activated partial thromboplastin (aPTT) time of 121 seconds (normal is 70-102). The delayed aPTT time may be a dilutional effect and/or poorly understood effects of synthetic colloids. After initial stabilization, plans were made to admit her to the

ICU. Blood pressures continued to temporarily improve with boluses of crystalloids and colloids, but did not stabilize for several hours. Her blood pressure finally stabilized 8 hours after admission with a Doppler systolic holding at 90 mmHg. Broad spectrum antimicrobials were started to treat a possible septic focus and gastrointestinal (GI) translocation. A central line was placed in a saphenous vein for administration of additional fluids and repeated blood draws. Packed red blood cell and fresh frozen plasma transfusions were started for anemia and a potential dilutional coagulopathy after resuscitation with multiple blood volumes of crystalloids and colloids. Hematochezia was noted and gastrointestinal blood loss was added to her problem list. Pain control was instituted using buprenorphine (0.015mg/kg) administered intravenously. Gastrointestinal protectants and anti-emetics were also added to address vomiting and gastrointestinal hemorrhage. Enoxaparin, a low-molecular weight heparin, was started to prevent further clot formation.

- **Presence of criteria for CIRCI:** By day 4, Mabel had been tentatively diagnosed with Acute Pancreatic Necrosis and SIRS. Her initial blood pressure reading of the day revealed hypotension (64 mmHg via Doppler). Her CVP was 6 cm H20 so her preload/volume status was judged to be adequate. In lieu of this information Dopamine was started at 5ug/kg/min. Because she had some previous bouts of fluid unresponsive hypotension the attending clinician sent out an ACTH stimulation test the evening prior. Results revealed an improper response to ACTH stimulation: resting cortisol level was 2.2mg/dL (ref range: 1.0-5.0 mg/dL) and post-ACTH cortisol level of only 4.2 mg/dL (ref range; 8-17 mg/dL). She did not fit the criteria of primary hypoadrenocoriticism (Addison's disease) because her resting cortisol level was within the reference range. However, her delta cortisol was 2 (4.2 – 2.2) revealing an inadequate physiologic response to adrenal stimulation. Based on meeting the criteria of a delta cortisol of < 3 and the presence of refractory hypotension despite fluid and pressor therapy Mabel was started on hydrocortisone (Solu-Medrol®) at 2mg/kg/day divided into 4 doses. A dose of 0.5mg/kg was given initially and within 15 minutes her blood pressure increased to 75mmHg. It began to slowly climb and held at around 90-100mmHg for a few hours until Mabel unfortunately arrested; most likely due to a cerebral thromboembolic event.

CONCLUSION

Adrenal dysfunction in human patients remains somewhat controversial- with several camps opposed and several touting the "real-ness" of this syndrome. As this syndrome has been identified in veterinary patients, and the prognostic importance of correcting hypo-adrenal related hypotension in critically ill patients it remains plausible to incorporate CIRCI into the "critical care syndromes" such as: DIC, MODS, ARDS, etc. Veterinary technicians can assist with the intervention and treatment of patients with CIRCI by being diligent about monitoring blood pressure and understanding vasopressor pharmacology, knowing proper testing protocols for adrenal axis testing, and being ready to prepare and administer steroid CRI's.

REFERENCES

Burkitt, Jamie. Relative Adrenal Insufficiency. In: *Small Animal Critical Care Medicine*. Eds. Silverstein and Hopper. St. Louis. Mo. Elsevier. 2009, Burkitt, et al. Relative Adrenal Insufficiency in Dogs with Sepsis. *Journal of Veterinary Internal Medicine*. 2007; 21: 226-231, Durkan et al. Suspected relative adrenal insufficiency in a critically ill cat. *Journal of Veterinary Emergency and Critical Care*. 2007; 17 (2): 197-201, Peyton, Jamie L. and Burkitt, Jamie M. Critical illness-related corticosteroid insufficiency in a dog with septic shock. Journal *of Veterinary Emergency and Critical Care Medicine*. 2009. 19 (3): 262-268. Martin, Linda G. and Groman, Reid P. Relative adrenal insufficiency in critical illness. *Journal of Veterinary Emergency and Critical Care*. 2004; 14 (3): 149-157. Walker, et al. Incidence and Outcomes of Critical Illness-Related Corticosteroid Insufficiency in Trauma Patients. *The American Surgeon*. 2011; 77 (5): 579-585. Cohen, J and Venkatesh, B. Relative adrenal insufficiency in the intensive care population; background and critical appraisal of the evidence. In Anesthesia Intensive Care. 2010; 38 (3): 425-436. Behrend & Martin. Diagnosis and Management of Adrenal Dysfunction. Conference proceedings IVECCS 2011.

SPLEEN PHYSIOLOGY AND THE TREATMENT OF COMMON DISEASES

Amy N. Breton, CVT, VTS (ECC)
Waltham, MA

ANATOMY

The spleen is the largest of the lymphatic organs. It is located in the upper left region of the abdomen and is attached to the stomach via small blood vessels. The spleen is much like a large lymph node, except that instead of containing lymph fluid it contains blood and is divided into chambers. Each chamber is like an individual lymph node in the sense that it helps to filter the blood through the spleen. The chambers of the spleen are lined with *red pulp* and *white pulp*. The *white pulp* consists mainly of T and B lymphocytes. The *red pulp* contains arterial capillaries and small venous vessels.

PHYSIOLOGY

The functions of the spleen include hematopoiesis, circulation, destruction and preservation of red blood cells and immunologic function. Because it is directly connected to the circulation of blood, it responds faster than other lymph nodes to blood borne antigens. The spleen stores between 10-20% of the total blood volume. It also produces lymphocytes, most of the monocytes and plays an important role in producing antibodies. All blood, at some point, gets filtered through the spleen. Despite its importance, there are no tests that are specific to splenic function.

SPLENIC COMPLICATIONS

There are two kinds of insults to the spleen that a veterinary technician may have to deal with: biological or traumatic. Biological insults are more common and are something that most veterinary technicians will come across at some point in their career. There are three kinds of biological insults: splenic masses, splenic torsion and splenomegaly. A traumatic event to the spleen is less common. Traumatic events to the spleen have the potential to produce more life threatening situations. There are two types of traumatic events: blunt or penetrating.

BIOLOGICAL INSULTS

Splenic Masses

Splenic masses are common in the canine. Certain canine breeds have a pre-disposition to getting splenic masses (German Shepherds, Golden Retrievers) however no genetic link has been proven yet. One study at Ohio State University indicated that of 85 dogs that required a splenectomy, 20% were found to have non-neoplastic tumors in nature versus 50% that were neoplastic. The other 30% of the dogs had their spleens removed for non-tumor related causes.

Most of the time, clients describe very vague symptoms. Lethargy, weakness, vomiting and anorexia are the most common complaints. Many times clients will simply state "my dog is just not right." Dogs may have a splenic mass for months and never show symptoms until the spleen ruptures. Within hours, the dog's disposition may go from active and playful to lethargic. It is important to remember to never dismiss a client simply because the symptoms described do not seem serious enough.

Diagnostic workup should start with a complete physical exam. Most tumors of the spleen are very vascular and when ruptured they start to bleed. Sometimes the bleeding cannot stop or is significant in quantity. These animals will typically present in very critical debilitated states. Owners may report they were fine in the morning and found them collapsed later in the day. Physical findings may include symptoms of shock: tachycardia, pale mucous membranes, ataxia, tachypnea and poor pulses. The patient may be hyper- or hypothermic. Occasionally the mass is large enough to palpate in the abdomen. If the mass has not ruptured, physical findings may be few to none. Slight pain in the abdomen and a distended abdomen may be all the physical exam findings.

Depending on how the patient presents, the clinic team may need to quickly work together to stabilize the patient. Intravenous fluids and blood products may need to be given quickly in order for further diagnostics to continue.

Bloodwork should be run and analyzed. A complete blood count (CBC) may show a regenerative anemia. Thrombocytopenia may occur secondary to the mass or from disseminated intravascular coagulation (DIC). The chemistry profile may show hypercalcemia, hyperglobulinemia, hemoglobinemia and hyperbilirubinemia. In cats, an eosinophilia may occur. If the splenic mass has ruptured you will typically see a low PCV and sometimes a low total protein.

Radiographs should be taken. Remember that these animals may be painful in the abdomen. Putting pressure on the abdomen to lift the pet onto the radiology table should be avoided. Generally large tumors are evident on radiographs, but sometimes they are not. A more effective tool is the ultrasound. Most skilled ultrasongraphers can readily detect splenic masses. An ultrasound-guided biopsy can be done to help determine the nature of the mass.

Common neoplastic splenic tumors include lymphosarcoma (LSA), fibrosarcoma (FSA) and hemangiosarcoma (HSA), of which the latter is most common in the canine. Overall prognosis for HSA is poor. By the time the client brings in their pet, roughly 50% of patients will have metastatic disease, meaning that the cancer has already spread. The median survival with chemotherapy is around 6-9 months with most cancerous splenic tumors. Many veterinarians will give owners the option of euthanizing their pet on the surgery table if it appears that other organs are affected. Overall, if a patient presents with a splenic mass, the prognosis is poor.

The most common non-neoplastic tumor is a hematoma in dogs, while in cats benign tumors are quite rare. Underlying disease like splenic lymphoma should be ruled out as a cause of the hematoma. More often than not, no underlying causes are present in a dog that has a splenic hematoma. The problem is simply that the hematoma has either grown too large and is impeding the normal function of the GI tract, or it has burst and is causing a hemoabdomen.

Splenic Torsion

A splenic torsion very rarely occurs by itself. Typically it is associated with gastric dilatation-volvulus (GDV) in dogs. Since the spleen is attached to the stomach, when the stomach twists the spleen may also twist. If the spleen is acutely twisted, the dog may present in shock and cardiovascular collapse simply because the blood flows to and from the spleen was stopped so dramatically. A twist will cause an increase in the spleen size, just like the stomach increases in size when it becomes twisted. Depending on the force of the twist, it may also cause the spleen to rupture. This situation can become life threatening very quickly. If the splenic torsion occurs because of a GDV the surgeon will decide whether the spleen is viable and whether to save it or remove it completely.

How to treat a splenic torsion that occurred by itself is still controversial. Some veterinarians will untwist the spleen and decide whether the spleen was damaged enough to warrant removal. If the spleen is still able to function normally, untwisting it will cause it to return to normal size within minutes. Other veterinarians will always remove the spleen. There is no way to secure the spleen permanently, so it is possible for the spleen to twist again if it is not removed.

Splenomegaly

Splenomegaly is simply defined as the enlargement of the spleen. Cats seem to be more prone to splenomegaly than dogs. A variety of diseases can cause the spleen to become enlarged. Any tumor, neoplastic in nature or not, can cause the spleen to enlarge. Certainly, splenic torsion can also cause the spleen to enlarge in size. There are four categories of splenomegaly: Inflammatory, Hyperplastic, Congestive and Infiltrative. When examining a patient, it is important to note that you may not always be able to palpate the spleen, but that does not mean it is not enlarged. Some spleens that can be palpated may not actually be enlarged.

- *Inflammatory splenomegaly* is one of the most common reasons for an increased spleen size. Inflammatory splenomegaly typically occurs when the spleen becomes inflamed from a disease process. Both viral and bacterial infections may lead to either generalized or localized splenitis. Diseases such as toxoplasmosis, canine hepatitis, leishmaniasis, feline infectious peritonitis and brucellosis may all cause the spleen to enlarge Bacterial infections such as sepsis, pyometra and salmonella have also been known to cause splenitis.

- *Hyperplastic splenomegaly* occurs less frequently. The spleen commonly reacts to stimulation that occurs with the hyperplasia that comes with the destruction of red blood cells. Hyperplasia is also common with infectious and autoimmune diseases such as IMHA and Heinz Body hemolysis.

- *Congestive splenomegaly* occurs when the portal vein or splenic vein (the main veins of the spleen) becomes congested (blocked). Certain drugs, like barbiturates, can increase blood pooling and therefore cause a slight "back up" of blood to the spleen. In these patients you may see a decrease in PCV. This is because the blood cannot filter properly. Right sided congestive heart failure, obstruction of the caudal vena cava (such as with heartworm disease) and splenic torsion can all cause congestive splenomegaly.

- *Infiltrative splenomegaly* is also one of the most common causes of splenomegaly. In human medicine infiltrative splenomegaly is a separate category from cellular causes, such as tumors of the spleen. In veterinary medicine there is no differentiation between the two, so tumors are included as a cause of infiltrative splenomegaly. Infiltrative splenomegaly means that something has invaded the spleen causing it to become enlarged. Besides actual masses, leukemias, malignant lymphoma and multiple myeloma are all causes of infiltrative splenomegaly.

The typical splenomegaly patient will present with non-descriptive signs. Owner may describe weight loss, inappetance, vomiting, diarrhea, polyuria/polydipsia and fever. Most of these symptoms are present due to the underlying cause of the splenomegaly. Again, it is important to never dismiss an owner simply because they think their pet is "just not acting right." In human patients with non-symptomatic splenomegaly a certain level of discomfort was still noted in the abdomen region. Most commonly, the patient described a "dragging pain" and "pressure" in the area of the spleen. In human patients, the pressure on the colon from the increase size in the spleen can lead to irregular bowel movements which may include diarrhea.

TRAUMATIC INSULTS

At some point, most veterinary technicians will encounter a patient who has had a splenic injury due to a traumatic event. It has been said that the spleen is second behind the liver as the most injured organ in the abdomen. Cats are less likely to have a splenic trauma because their rib cage protects almost the entire spleen. Cats also are more agile during a fall or when thrown into the air. There are two types of splenic trauma: blunt or penetrating trauma.

Blunt trauma is when the pet was impacted at either tremendous speed, weight or a combination of both. This can occur for many reasons including if the pet was hit by a car, fell out of a building or was hit with an object intentionally. Each pet will present to the clinic differently. Due to the release of adrenaline, the pet may seem "normal" to the client. It is important to urge the client to still come in to be seen by the veterinarian. Sometimes an internal injury to the spleen may not present itself for hours.

When the patient arrives, it is important to do a complete physical exam. Signs that splenic trauma may have occurred are: pale mucous membranes, pain in the abdomen, broken ribs, distended abdomen, tachycardia, hypotension and heart arrhythmias. The blunt trauma patient may hide its symptoms. Even though the owner says their pet is "fine" it is important to have a complete work up for each trauma patient.

When the spleen undergoes a traumatic event it can go through a splenic contraction, in which the spleen will enlarge and release a large quantity of red blood cells. This event can falsely elevate the PCV shortly after the traumatic event. It is therefore important to obtain a baseline PCV and total protein when the patient arrives so that it can be used as a comparison later if the patient stays for monitoring.

Radiographs are an important tool in diagnosing trauma to the spleen. They can show any obvious fluid in the abdomen. An ultrasound is a better diagnostic tool for determining whether the spleen is intact or ruptured and whether there is any free fluid in the abdomen. An abdominocentesis may help to prove whether there is blood in the abdomen or not. The presence of blood in the abdomen may mean the spleen has ruptured.

Similar to human medicine, patients that have a slow bleed are not necessarily taken to surgery right away. An abdominal wrap may be placed to help apply pressure to the bleeding area. However, if the patient is not hemodynamically stable, a total or partial splenectomy to remove the bleeding spleen is still the treatment of choice.

Penetrating traumatic events to the spleen are not as common. When they do occur, however, owners typically respond very quickly because the pet has an obvious wound. Gun shot, sticks impaled through the abdomen or an animal attack on a pet are all examples of penetrating traumas that can occur to the spleen. The pet usually presents bleeding, sometimes quite excessively. It is important to apply pressure or a bandage to the area to help control the bleeding if possible. If the spleen suffered a direct insult because of a penetrating object, it is important that the patient is stabilized as much as possible and taken to surgery for a splenectomy.

THE SPLENECTOMY

As previously discussed, the spleen is a very important organ, but it can also be removed with minimal complications to the body. Many of the disease processes described above may end with the patient getting a splenectomy. It is up to the veterinarian to decide whether the patient would benefit from surgery or not. The goal of the splenectomy is to stop the bleeding that is occurring and therefore to stop of the detrimental effects that the bleeding is causing on the body. Geriatric patients or patients that are immunocompromised are at the greatest risk to complications from surgery. One of the biggest risks during a splenectomy is bleeding either intraoperatively or postoperatively. To minimize the risk, patients ideally should be stabilized prior to surgery. If the cause of the patient's shock is a bleeding spleen that, despite efforts, continues to bleed, then surgery may be necessary and emergent. It may not be possible to stabilize the patient before surgery. If the patient has a PCV less than 20% or is clinical for anemia, a blood transfusion prior to surgery is usually necessary. Many patients undergoing a splenectomy require blood transfusion intra-op as well as pre-op.

Anesthetic considerations include placing the patient on oxygen only for 15 minutes pre and post-surgery. Typically these patients are anemic, and increasing the oxygen level in their blood stream may help with any complications during induction or recovery. Acepromazine should be avoided in these patients because it can significantly decrease blood pressure. Typically these patients have compromised vascular systems already.

Most notably during and after surgery, splenectomized patients can have heart arrhythmias. This is not a common side effect in humans, but is a common occurrence in small animals. Any trauma to the spleen can cause heart arrhythmias. Commonly, these patients will have ventricular premature contractions (VPC). It is known that patients that have ruptured splenic masses have a higher incident of heart arrhythmias including ventricular tachycardia. Current thinking as to why arrhythmias occur includes:

1) Myocardial Depressant Factor: This is a theory that is thought to be the primary cause of the arrhythmias. The thought is that the spleen releases factors that decrease the function of the myocardium of the heart.
2) Sympathetic Overstimulation: This may be the cause when dealing with acute trauma or GDV. Since splenic masses are slow growing, the shock to the sympathetic nervous system is probably limited, but certainly the removal of the spleen can set off the process leading to cardiac arrhythmias.

Most cardiac arrhythmias do not lead to severe complications. Many arrhythmias, if left untreated, will eventually resolve on their own. Treatment usually occurs if the patient is experiencing clinical symptoms or the heart rate remains high for a long period of time.

Overall there are few reported complications that occur when the spleen is removed. It is known that patients that have had their spleen removed may be more susceptible to infection. The functions of the spleen are taken over by the liver and lymphatic system. As for what specific system takes over what job, that still remains a mystery. It is only known that the body does compensate for the removal of the organ without any interruption to the rest of the system.

In human patients, a splenectomy is considered a last resort. Partial splenic resections and managing the spleen rather than removal are becoming increasing popular in both human and veterinary medicine. Laparoscopic surgery to remove the spleen in people is becoming increasingly popular. Since new information is being gathered about the spleen all the time, it seems evident that veterinary medicine may opt for alternative treatments rather than a splenectomy.

REFERENCES

1. Ettinger S: Textbook of Veterinary Internal Medicine, 1989., 2. Bowdler A: The Complete Spleen. Structure, Function, and Clinical Disorders, 2002., 3. Nelson R, Couto C: Small Animal Internal Medicine, Mosby, 1998., 4. Ettinger S, Feldman E: Textbook of Veterinary Internal Medicine (fifth edition) W.B. Saunders Company, 2000., 5. Fossum T, Hedlund C, Hulse D, Johnson A, Seim H, Willard M, Carroll G: Small Animal Surgery, Mosby, Inc, 1997., 6. Tilley L, Smith F: The 5-Minute Veterinary Consult Canine and Feline (third edition) , Lippincott Williams and Wilkins, 2004., 7.Hole J: Essentials of Human Anatomy and Physiology (fifth edition), Wm. C. Brown Publishers, 1995

MEDICAL MANAGEMENT OF FELINE CHRONIC KIDNEY DISEASE

Jessica Quimby, DVM, DACVIM
Fort Collins, CO

INTRODUCTION

Chronic kidney disease (CKD) remains a leading cause of illness and death in cats in the United States, with as many as 15-30% of cats affected. Clinical signs include polyuria, polydipsia, decreased appetite, weight loss and vomiting. Generally at the time that routine laboratory tests reveal the presence of kidney dysfunction, approximately 75% of renal function is already gone. As renal function declines, azotemia, hypokalemia, hyperphosphatemia, anemia and dehydration progressively worsen. This is a progressive disease, and currently no treatment short of renal transplantation has been shown to reverse or halt declining renal function for any significant period of time. Renal transplant is not widely available and is too expensive for many clients. Additionally not all CKD cats are acceptable candidates for transplant. Medical management therefore is the mainstay of treatment and can help patients cope with metabolic complications of the disease and improve quality of life.[1,2]

IRIS STAGING

The International Renal Interest Society has developed a staging system for chronic renal disease. www.iris-kidney.com With this system, cats are grouped into four stages (IRIS I-IV) based on serum creatinine measurements assessed on at least two occasions in a stable patient. Substaging is performed based on the measurement of systemic blood pressure and proteinuria, both of which are a concern as they are thought to hasten disease progression.

THERAPIES

Nutrition and Diet

Several recent studies[3-5] have documented the therapeutic value of specially formulated diet in the management of CKD, including decreased incidence of uremic crisis and increased survival. These diets typically contain restricted amounts of high quality protein, adequate non-protein calories, and are restricted in phosphorus. The failure of the patient to eat the diet negates the benefit of dietary management, and therefore a key therapeutic target for these patients is the maintenance of appetite and food intake. Cats with CKD often suffer from poor appetite. Management of long term hyporexia is important. Poor body condition is associated with decreased prognosis in several species and has a negative effect on immune function, wound healing and strength.[6] Most importantly it also has a negative effect on owner perception of quality of life.

Anti-nausea Therapy

Cats with CKD likely suffer from nausea, vomiting and inappetence as a result of uremia – a buildup of toxins in blood– that affects the chemoreceptor trigger zone in brain. In addition, uremia may have effects on the intestinal tract that lead to further unwillingness to eat. Several anti-nausea therapies have become recently available. These include maropitant, ondansetron and dolasetron. These drugs work at nausea center in brain as well as in gut and can be given as an injection. Ondansteron has been documented to be helpful in human patients suffering from uremia.[7] Mirtazapine demonstrates anti-nausea properties in addition to its appetite stimulating properties as it acts at the 5HT3 receptor similarly to ondansetron.[8] Limiting gastric acidity with the use of H_2 blockers such as famotidine anecdotally appears to palliate inappetence in some patients, however both the degree of hyperacidity present in CKD and the efficacy of these medications remains unproven.

Appetite Stimulant Therapy

Cyproheptadine has been used for some time as an appetite stimulant and has anecdotal efficacy in many patients, however its efficacy has never been scientifically evaluated. Twice daily administration is necessary in many cases and this can prove a challenge for owners, particularly longterm. Mirtazapine has become more commonly used and recent exploration of its pharmacodynamics and pharmacokinetics have provided information for more effective use in cats.[9,10] Pharmacodynamic studies have illustrated that it can be a potent appetite stimulant, but higher doses are more commonly associated with side effects (hyperexcitability, vocalization, tremors). Smaller, more frequent doses are recommended. The half-life is short enough that it could be administered daily in normal cats. Renal disease delays clearance and in these patients, every other day administration is recommended. [10]Owners should be aware mirtazapine and cyproheptadine cannot not be administered concurrently; cyproheptadine is in fact used as an antidote for serotonin effects of mirtazapine overdose. The ghrelin receptor agonist, capromorelin, which mimics the activity of the "hunger hormone" ghrelin in stimulating appetite, is currently being assessed for use in cats which may add additional options for appetite stimulation in chronically hyporexic patients.

Systemic Hypertension

When a small number of functional nephrons are left in the kidney, those that remain are subjected to hyperfiltration as they are forced to accommodate the same amount of blood. This phenomenon leads to intraglomerular hypertension, which is then exacerbated by systemic hypertension. Systemic hypertension appears to be common in cats with CKD, but the exact relationship is unknown. Elderly cats, particularly those with renal impairment should be routinely screened. Systemic hypertension can have other deleterious effects such as retinal hemorrhage and detachment, neurologic and cardiac impairment. Amlodipine is documented to be an effective treatment for hypertension in the cat and is the drug of choice.[2]

Proteinuria

Intraglomerular hypertension as well as diseases of the glomerular apparatus can lead to proteinuria. The renal tubular epithelial cells are responsible for processing this protein and this increased their metabolic load and oxidative stress. Proteinuria is associated with poorer prognosis in several species. Several studies looked at efficacy of ACE inhibitors (benazepril) for treatment of proteinuria in cats and it was documented that benazepril decreases proteinuria in cats with CKD, but results were only suggestive of long-term benefit, not conclusive. Nonetheless it is recommended that cats with urine protein creatinine ratio > 0.5 on two repeated measurements should be treated.[2]

Hydration

As the kidney becomes disease, the ability to concentrate the urine is lost. Adequately maintaining hydration by giving SQ fluids anecdotally appears to substantially help quality of life, improve appetite and activity. It can be a very helpful tool for owners in management of disease. Feeding canned food instead of dry, or adding water to food is another way to potentially increased water consumption. Paying special attention to water sources in the house – fresh, accessible, water fountains etc. is also key.

Anemia

The kidney is responsible for producing the hormone erythropoietin which stimulates the bone marrow to produce red blood cells. As kidney disease progresses, the hormone levels decrease and anemia can result. Chronic low-grade gastrointestinal hemorrhage may also contribute to anemia and iron deficiency may be noted. The hormone can be supplemented with artificial products such as recombinant human erythropoietin (Epogen) or Darbepoetin alpha (Aranesp). Epogen is given as a SQ injection and is effective in treating anemia secondary to CKD. However as it is a foreign protein, approximately 1/3 of cats will develop antibodies against both the foreign erythropoietin and their own, and potentially become transfusion dependent. For this reason, the product is not recommended until PCV falls below 20%. Darbepoetin is a longer-acting form of erythropoietin and is thought to have less association with anti-epo antibodies. It is currently the product of choice, but expense can limit its use for some owners.[2] A feline-specific erythropoietin is also currently under development and clinical trials may soon be enrolling participants.

Hypokalemia

Inadequate dietary intake, increased urinary loss and activation of the renin-angiotensin-aldosterone system are all thought to contribute to hypokalemia. Hypokalemia is associated with development and worsening of CKD in humans and appears to exacerbate damage to tubular epithelial cells. Serum potassium levels are not representative of systemic tissue potassium levels and cats with low normal serum potassium may actually be systemically depleted. Potassium is vital for normal muscle function, GI motility and some clinicians recommend supplementation even when serum potassium is in the low normal range[11], with a goal of maintaining serum potassium levels above 4 mg/dL.

Hyperphosphatemia

The kidneys are the main route of phosphorus excretion and as kidney function declines, hyperphosphatemia develops. Hyperphosphatemia is detrimental because it contributes to renal secondary hyperparathyroidism, tissue mineralization and progression of CKD. Thus controlling phosphorus intake through diet and phosphate binders is important part of CKD management.[12] If a renal diet is initiated and phosphorus is still elevated after 4-6 weeks, then a phosphate binder is recommended with the goal of keeping serum phosphorus in the low normal range.[2] These medications actually bind the phosphorus in the food and make it less bioavailable so it is critical that the medication is given with each meal. Several different phosphate binders are available including aluminum hydroxide, calcium carbonate, lanthanum carbonate, however the main problem with administration is palatability.

Calcitriol

Calcitriol, the active form of Vitamin D, is produced in the kidney. This production is decreased in CKD and hyperphosphatemia further inhibits its synthesis. Calcitriol is an important inhibitor of parathyroid hormone and low levels leads to renal secondary hyperparathyroidism which has multiple deleterious consequences. In dogs and people, use of calcitriol has been shown to be beneficial; a study performed in cats was not supportive. Nonetheless, controversy remains and many clinicians recommend supplementation. Careful monitoring of calcium and PTH levels is required as hypercalcemia can result from over-supplementation.[2]

Anti-oxidant Therapy.

Several processes in the diseased kidney potentially result in oxidative stress including hypoxia, proteinuria, and inflammation. Oxidative stress documented in CKD cats[13] and potentially contributes to the progression of disease. Vitamins E, C and B-carotene shown to decrease oxidative stress in CKD cats and a clinical trial examining Vitamin E supplementation is currently underway.

Fatty Acid Therapy

Omega- 3 Fatty acids are anti-inflammatory and the underlying process in CKD is inflammatory in nature, supplementation is thought to be beneficial. Several studies have demonstrated evidence of benefit in human CKD patients and rodent models demonstrate decreased oxidative stress, inflammation and scarring. Many diets are now supplemented with fatty acids.

REFERENCES

[1]Roudebush P et al. **J Feline Med Surg** 2009; 11:95. [2]Polzin DJ. **Vet Clin North Am Small Anim Pract** 2011;41:15. [3]Elliott J et al. **J Small Anim Pract** 2000;41:235. [4]Ross SJ et al. **J Am Vet Med Assoc** 2006;229:949. [5]Plantinga EA. **Vet Rec** 2005;157:185. [6]Chan D. **J Fel Med Surg** 2009;11:925. [7]Ljutic D et al. **Kidney Blood Press Res** 2002;25:61. **J Vet Intern Med** 2011;25:985. [8]Reichelmann RP et al. **Am J Hospice and Palliative Med** 2010; 27:106. [9]Quimby JM. **J Vet Pharmacol Therap** 2010;34:388. [10]Quimby JM. **J Vet Intern Med** 2011;25:985. [11]Theisen SK et al. **J Vet Intern Med** 1997;11:212. [12]Kidder A and Chew D. **J Feline Med Surg** 2009;11:913. [13]Keegan RF and Webb C. **J Vet Intern Med** 2010;24:514.

PRACTICAL GUIDE TO DIAGNOSIS AND PREVENTION OF CANINE LEPTOSPIROSIS

Chad M. Johannes, DVM, DACVIM (SAIM)
Raleigh, NC

INTRODUCTION

Canine leptospirosis, historically considered most prevalent in rural/farm dogs, is becoming recognized as an emerging infectious disease more frequently diagnosed in dogs located in urban/suburban areas.[1-4] While enhanced awareness amongst veterinary staff likely plays a role in the increased diagnoses, contact with common backyard wildlife (i.e., raccoons, skunks) may result in increased exposure to the *Leptospira* serovars that commonly infect dogs.[5,6] Recognition of the common sources of infection, presenting clinical signs and diagnostic considerations are key to understanding the best mechanisms of disease prevention. As an important member of the veterinary health care team having frequent communication time with dog owners, veterinary technicians play a vital role in client education and awareness regarding the latest information on canine leptospirosis.

EXPOSURE POTENTIAL IN DOGS

Leptospira spp. are thin, flexible, spiral bacteria that thrive in stagnant or slow-moving warm water and may remain viable for several months in moist soil. Increased season incidence of diagnosis is noted in summer and late fall, likely related to increased exposure to water sources harboring the organism. While >200 serovars of *Leptospira interrogans sensu lato* have been identified worldwide, the most common serovars infecting dogs in the US include Icterohaemorrhagiae, Canicola, Grippotyphosa, Pomona and Bratislava. Canicola is most adapted to the canine host with the dog serving as the primary reservoir for this serovar. Common reservoir hosts (carry, shed the organism) for the remaining serovars recognized in dogs include the rat, raccoon, skunk, opossum, mouse, squirrel, cow, and pig.

Transmission from the reservoir host to dogs can occur directly (contact with infected urine, bite wounds) or indirectly (contaminated water sources, soil, food, bedding).[7] Mucous membranes or abraded skin are common points of entrance with incubation period within the infected dog ranging from 2 to 30 days (7 days most common). Serologic studies indicate that 17-20% of pet dogs in various regions of the US (WA, MI) had exposure to one or more of the common *Leptospira* serovars, regardless of breed.[8,9]

CLINICAL PRESENTATION

The severity of clinical disease experienced by an infected dog typically depends on the infecting *Leptospira* serovar, number of organisms, and host age and susceptibility (i.e., immune status, vaccination history). Many infected dogs will have an acute presentation of sudden illness. Multiple body systems can be affected including renal, hepatic, vascular (vasculitis, coagulopathies), pulmonary, ocular and muscular (myopathies). The result is a myriad of potential non-specific clinical signs ranging from fever, vomiting, diarrhea, hematochezia, hematemesis, dehydration, and tachypnea, to epistaxis, icterus, petechiae, and pain on muscular or renal palpation. Although less commonly noted, more protracted presentation may occur. Clinical signs in this more chronic disease process may include fever, inappetance, weight loss, polydipsia, polyuria, and progressive deterioration in renal or hepatic function.[7]

DIAGNOSIS IN PRACTICE

Obtaining an accurate and detailed history, along with clinical signs and physical examination changes noted at presentation, remain vital components of appropriately considering leptospirosis as a potential diagnosis. Changes noted on initial diagnostics (CBC, serum biochemistry profile, urinalysis, abdominal ultrasound) may increase the index of suspicion of leptospirosis, especially in cases of acute renal failure or hepatopathy.

The microscopic agglutination test (MAT) remains the standard serologic test for the diagnosis of leptospirosis. Single timepoint titers ≥1:800 along with corresponding clinical signs is supportive of the diagnosis of leptospirosis. In acute cases, initial titers (first 7 to 10 days) may be negative. In these cases where the initial titer is negative but leptospirosis is suspected, it is important to manage the case as a possible leptospirosis case and follow-up with repeat titers 2 to 4 weeks later with a 4-fold change in values being supportive of the diagnosis.[7] Recent studies looking at *Leptospira* titers from clinical patients evaluated at multiple diagnostic laboratories has revealed frequent discordant results regarding the specific infecting *Leptospira* serovar reported. Findings indicate that while the MAT is a valuable diagnostic tool for the diagnosis of canine leptospirosis, it is not a reliable predictor of the infecting serovar.[10]

Polymerase chain reaction (PCR) testing of urine is another valuable clinical tool for the diagnosis of leptospirosis which compliments and is frequently used in conjunction with the MAT. PCR is a very sensitive, less specific test for *Leptospira* organisms and does not identify the infecting serovar. This test can be helpful early in the course of infection to identify infected dogs that have not yet had time to produce antibodies (seroconversion) and thus are negative on MAT. Additional utility resides in confirming active infection in dogs with a positive MAT that also have a history of vaccination against *Leptospira* spp. This test can also be quite helpful in detecting infection in dogs with chronic renal or hepatic disease. It is important to recognize that false positives are possible and that a negative result does not exclude the diagnosis of leptospirosis; results should be interpreted in light of clinical signs.[11-13]

ZOONOTIC POTENTIAL

While the annual reported incidence of leptospirosis amongst people in the US is relatively low, the zoonotic potential exists. Human infection typically occurs via recreational exposure (i.e., swimming in contaminated water) which often results in small regional outbreaks of the disease or via occupational exposure (i.e., people who work outdoors or with animals). Infected dogs often shed *Leptospira* organisms in the urine, serving as a possible source of infection to people. Veterinary personnel and caregivers in frequent, close contact with urine or urine-contaminated bedding from infected dogs would seem to be at increased risk for exposure. One recent report indicated that 2.5% of 511 veterinarians tested demonstrated evidence of previous leptospiral exposure based on seriology. [14]

The duration of organism shedding by infected dogs is unclear, as is the optimal treatment time to diminish shedding. In general, the risk of exposure via urine from an infected dog is believed to be small within 24 hours of starting appropriate antimicrobial therapy but caution in handling urine is recommended for 3 to 5 days. The urine PCR test can be considered to help further verify the lack of urine shedding in an infected dog undergoing treatment.

Knowledge regarding the transmission of *Leptospira* spp. is an important component to preventing infection amongst veterinary personnel in close contact with infected dogs. Any dog suspected of potentially being infected with *Leptospira* spp. based on history, clinical signs and/or diagnostic findings should be treated as having leptospirosis until determined otherwise. Preventative measures include wearing appropriate protective attire (gloves, body gown, goggles) when handling suspected dogs or their urine/bedding, washing hands after handling patients and preventing splashing or aerosolization of urine when cleaning cages/kennels.

PREVENTION

Vaccination

Immunization via *Leptospira* bacterin administration has been effective in reducing the prevalence and severity of canine leptospirosis. Vacciniation with bacterin products has been shown to protect a group of dogs against renal colonization and urine shedding, as well as against symptomatology and leptospiremia.[7,15] Multiple newer bacterins containing the 4 most common serovars infecting dogs in the US (Icterohaemorrhagiae, Canicola, Grippotyphosa, Pomona) are now available: Vanguard®L4 (Pfizer), RECOMBITEK® 4 Lepto (Merial), Novibac® Lepto 4 (Merck). The bacterins appear to provide protection for 12 months.[13,16]

Assessing which dogs are at risk will vary from region to region. Annual vaccination with 4-serovar vaccines is recommended for at-risk dogs, regardless of breed. In regions where infection occurs in urban, backyard dogs, all dogs may be at risk and 4-serovar bacterin may be considered part of a core vaccination protocol. In other locations, only dogs that contact wildlife, swim, hunt or roam on farmland may be at risk.[13]

When considering whether or not the *Leptospira* bacterin should be a component of a practice's core vaccination protocol, important factors to consider include: 1) Regional prevalence; 2) Significant morbidity/mortality for infected, non-vaccinated dogs; 3) Zoonotic disease (risk to owners' families and hospital staff); 4) Availability of safe and effective bacterins which protect against the 4 most common serovars currently diagnosed in US dogs.

Environment

Regardless of a dog's geographic location (urban, suburban or rural) and vaccination status, steps can be taken to reduce the risk of exposure to *Leptospira* spp. One key element is reducing rodent/wildlife infestations and access to kennels/bedding, food storage sites and water sources. Areas with high risk of transmission (standing water in dog's normal habitat, farm ponds, etc) should be avoided. Especially important in a veterinary clinic setting is the isolation of infected dogs or dogs suspected of being infected with leptospirosis based on history and clinical signs.

SUMMARY

Canine leptospirosis is increasingly recognized clinically in wide and varying geography. Education of veterinary hospital staff as to the exposure risks for dogs, clinical signs, diagnostic approach and preventative measures plays an important role in disease recognition and prevention through client education. Understanding which dogs are at risk for exposure and implementing appropriate vaccination strategies can help play a key role in protecting at-risk dogs.

REFERENCES

[1]Moore GE, et. al. **Emerg Infect Dis** 2006;12:501; [2]Ward MP, et. al. **J Am Vet Med Assoc** 2002; 220(1):53; [3]Raghavan R, et. al. **Prev Vet Med** 2011;101:241; [4] Ward MP, et. al. **J Am Vet Med Assoc** 2004; 225(1):72; [5]Richardson DJ, et. al. **Vector-Borne Zoonot** 2003;3(4):187; [6]Mitchell MA, et. al. **J Wildlife Dis** 1999;35(2):347; [7]Greene, **Infectious Diseases of the Dog & Cat**, 2006; [8]Daniels MA, et. al. **Zoonoses Public Hlth** 2008;55:436; [9]Stokes JE, et. al. **J Am Vet Med Assoc** 2007;230(11):1657; [10]Miller MD, et. al. **J Vet Intern Med** 2011;25:426; [11]Harkin KR, et. al. **J Am Vet Med Assoc** 2003;222(9):1224; [12]Harkin KR, et. al. **J Am Vet Med Assoc** 2003;222(9):1230; [13]Sykes JE, et. al. **J Vet Intern Med** 2011;25:1; [14]Whitney EA, et. al. **J Am Vet Med Assoc** 2009;234(7):938; [15]Schreiber P, **Vet Microbiol** 2005;108(1-2):113; [16]Klaasen HL, et. al., **Vet Microbiol** 2003;95(1-2):121.

ORAL TUMORS IN DOGS AND CATS

Tracy Gieger, DVM, DACVIM, DACVR
Baton Rouge, LA

INTRODUCTION

Oral tumors are common in dogs and cats and their incidence increases with age. Presenting complaints can include a visible mass detected by the owner, bloody or increased saliva, pawing at the face, loose teeth, gingival proliferation or bleeding, facial deformity, weight loss, gagging, dropping food out of the mouth, and less commonly, enlarged mandibular lymph nodes or salivary glands or exophthalmos. Different histologic tumor types have similar appearances (with the exception of black tumors, which are almost always melanomas). A thorough oral exam should be part of every physical exam; this includes lips, cheeks, tongue/sublingual region, salivary glands, palate, teeth, and tonsils. Oral cavity tumors may also extend into the pharynx, nasal cavity, and retro-orbital space. In fractious animals, sedation may be required for a thorough oral exam. In addition, oral examination is recommended on any anesthetized/intubated patient.

STAGING FOR ORAL TUMORS

If an oral tumor is detected, staging should include 3D measurement of the tumor with calipers and drawing/recording the tumor on a body map or oral cavity exam form. Fine needle aspiration (FNA) and cytology can be performed on oral tumors; however, secondary inflammation and sometimes infection can be present, possibly causing a false negative for the diagnosis of a tumor (i.e., cytology does not show tumor cells since inflammation is present). In addition, copious bleeding may occur after FNA due to the vascular nature of the oral mucosa. Regional lymph nodes should always be aspirated if they are accessible, even if they are normal in size. Bloodwork should be performed to examine for evidence of systemic disease prior to anesthesia or therapy for oral tumors. All cats with oral tumors should be FeLV/FIV tested. Thoracic radiographs (3 view) should be considered in any patient with a malignant oral tumor. Abdominal imaging (specifically ultrasound) should be considered in older patients prior to aggressive local therapy for oral tumors to stage them for evidence of systemic disease (other tumors or metastasis from their oral tumor, especially for malignant melanomas).

Skull and/or dental quality radiographs can be useful to examine the underlying bone for lysis, and are more useful for mandibular tumors than maxillary tumors. Advanced imaging techniques such as computed tomography (CT) or magnetic resonance imaging (MRI) can be useful to detect bony lysis, to determine the extent of the tumor, and to examine all of the regional lymph nodes for evidence of neoplasia. CT is useful for radiation therapy treatment planning, and both CT and MRI are useful in surgical planning.

Ideally, biopsy procedures should be planned after imaging tests are completed to avoid disruption of fascial planes and creation of hematomas or other surgical wounds that could confound visualization of the entire tumor. Biopsies may be incisional or excisional, depending on the location, blood supply, and size of the tumor. Smaller, pedunculated tumors may be best biopsied via excisional biopsy while larger masses are biopsied with incisional biopsy techniques such as wedge, punch, or tru-cut biopsy. Obtaining a diagnosis and other biopsy-related information (i.e., vascular invasion, grade, etc.) prior to contemplating a definitive treatment plan is helpful, especially for tumors where complete surgical excision is not possible.

In general, the prognosis is best for rostrally-located tumors, tumors <2 cm, and tumors without bony lysis and/or regional lymph node involvement. As with any tumor type, early diagnosis and appropriate therapy provides the best chance for long-term tumor control.

SYMPTOMATIC CARE FOR PATIENTS WITH ORAL TUMORS

Secondary inflammation and infection are commonly present in patients with oral cavity tumors due to continued disruption of the normal oral mucosa/flora by the tumor. Antibiotic therapy is almost always warranted, and can be challenging in hyporexic/anorexic animals and/or patients that are difficult to pill. Antibiotic choices include clindamycin and amoxicillin-clavulanic acid. In cats or dogs that cannot be medicated, subcutaneous administration of antibiotics can be helpful. Long-acting antibiotics such as cefovecin (Convenia®) may be useful in such patients as well. Anti-inflammatory medications such as NSAIDs or steroids (not to be given concurrently) are usually indicated in patients with oral tumors to alleviate inflammation/discomfort. Other pain medications including tramadol and other opiates may be indicated in some patients. Oral rinses may help to keep the mouth moist and decrease food and saliva buildup around a tumor.

Feeding tube placement may be needed for nutritional support and to give medications during RT or post-operatively. Feeding tubes are typically temporary while the surgical site heals or while the radiation therapy is helping to shrink the tumor. Esophagostomy tubes are most commonly used in these patients.

CANINE ORAL TUMORS

The most common malignant oral tumors in dogs are melanoma (OMM), squamous cell carcinoma (SCC), and fibrosarcoma (FSA). Dental ligament tumors such as canine acanthomatous ameloblastoma are locally invasive into bone but do not metastasize. Papillomas (caused by a papovavirus) are seen in young dogs and are contagious via direct contact/play biting behavior. Other tumors including multilobular tumors of bone (MLO), plasma cell tumors, mast cell tumors, granular cell tumors, transmissible venereal tumors, osteosarcoma, and other sarcomas occur in the oral cavity. Occasionally, cutaneous/epitheliotrophic lymphoma will involve the oral cavity, and may appear as oral plaques, severe gingivitis, or ulcerated lesions.

Oral malignant melanoma (OMM) has a high potential for local invasiveness/bone destruction as well as regional lymph node and distant metastasis. Melanomas on the lip are usually less malignant in behavior than those inside of the oral cavity. Breeds with a large amount of oral pigment such as Cocker spaniels, Chow chows, Rottweilers, and Golden retrievers are predisposed. Melanomas may be black or non-pigmented (termed amelanotic melanoma). Aspiration of a pigmented OMM or metastatic lymph node yields fluid that is black. Complete systemic staging (as described above) is recommended for dogs with OMM, and if staging is negative for metastasis, aggressive local therapy (surgery or radiation therapy) should be considered. Since OMM is likely to metastasize, options such as the Merial® melanoma vaccine or possibly chemotherapy should be considered as an adjunct treatment. Many dogs with non-resectable melanomas can be managed with radiation therapy (RT); about 75% of dogs with gross disease will have a response to RT, but most ultimately die of metastasis within 6-9 months. If dogs with OMM present with visible metastasis, the prognosis is poor since chemotherapy drugs are minimally effective in treating gross metastatic disease. Chemotherapy drugs that may have efficacy against melanoma include carboplatin, dacarbazine, lomustine, and melphalan.

Oral squamous cell carcinoma (SCC) in dogs typically appears as an erosive or proliferative mass, and sometimes as a fistula such as an oronasal fistula. This tumor is locally invasive and sometimes metastasizes to the regional lymph nodes (usually mandibular LN). If staging for metastasis is negative, aggressive local therapy is warranted. Surgical resection of the primary tumor is often possible with partial mandibulectomy or maxillectomy. If surgery is not an option due to size or location of the tumor, RT is usually helpful to control the local disease. Non-steroidal drugs may also be helpful in symptomatic therapy (pain, ulceration) and possibly anti-tumor effects for SCC. If complete removal with surgery is achieved, long-term tumor control is possible since distant metastasis is uncommon. Follow-up chemotherapy is rarely warranted and has not been shown to be definitively effective in treating or preventing metastatic disease. In most studies of dogs with SCC treated with combinations of surgery and/or radiation therapy, median survival times are approximately 1-1.5 years.

Fibrosarcoma of the oral cavity is most often termed, "histologically low grade, biologically high grade" due to its typically locally very invasive nature but bland histologic appearance and lack of metastasis. These tumors are most commonly seen on the maxillae of Golden retrievers but can be seen in other breeds and in other locations within the oral cavity. They are often mistaken for "benign" swellings or trauma to the face, and cytology often produces false negative results. Often, incisional biopsies reveal only inflammatory cells/fibrous connective tissue. Unfortunately, bone destruction by the tumor is common, so cross-sectional imaging is warranted in every case. It is common for these tumors to invade into the nasal cavity, orbit, and sometimes brain. If surgery is feasible, excision should be performed by a surgeon experienced in reconstructive surgery. A recent study showed that with aggressive local resection, 88% of dogs were alive 1 year after surgery. Radiation therapy is useful to slow the progression of non-resectable tumors, but cure is unlikely; RT typically controls tumor growth for 6-9 months before progression. Chemotherapy may be useful in some cases for palliation, but rarely provides long-term control.

Tumors of odontogenic origin were recently classified by their clinical and histologic appearance; the most common forms include canine acanthomatous ameloblastoma (CAA), peripheral odontogenic fibroma, and focal fibrous hyperplasia (thought to result from chronic oral inflammation rather than neoplastic origin). CAA is the most common form, and occurs most commonly in the rostral mandible. It does not metastasize (carcinomas are rare), but it frequently invades the underlying bone. If this tumor is incompletely excised ("shaved off" at the gumline), underlying bone destruction continues. Aggressive local therapy such as mandibulectomy or maxillectomy is warranted for cases where bone lysis is present secondary to the tumor. Imaging with dental radiographs (mandibular) or CT scan (maxillary) is indicated to determine the extent/presence of bone lysis. If tumors are non-resectable, RT is indicated for local control. Definitive radiation therapy schemes that deliver 40-50 Gy are warranted, and long-term local control after RT is possible.

Lymphoma may occur in the oral cavity and may appear as severe gingivitis/gingival hyperplasia or masses, nodules, or oral plaques. Any animal with oral cavity lymphoma should be completely staged for systemic involvement, and if lesions are confined to the oral cavity, RT and/or chemotherapy is frequently helpful to manage these lesions. The long-term prognosis for a dog with oral cavity lymphoma is usually poor, and significant pain due to the lesions is common.

Tongue tumors in dogs are relatively uncommon. Melanoma, SCC, plasma cell tumors, and granular cell tumors are seen most commonly. Partial glossectomies are useful in many dogs with tongue tumors, and clean surgical margins can be achieved if 1 cm of tissue around the gross tumor can be resected. The prognosis is better for completely excised tumors, smaller tumors, and benign histologies. Radiation therapy may be useful for non-resectable tumors.

Tonsillar tumors are rare in dogs, and SCC and melanoma are most common. Regional lymph node metastasis is more common than with SCC of intraoral locations. Surgical resection is usually impossible due to the fixed location/invasiveness of these tumors. Tonsillar tumors are usually treated with palliative courses of RT to help with local pain control and alleviate clinical signs, but long-term tumor control is rarely possible. Survival times >6 months are uncommon unless complete surgical margins can be achieved.

Salivary gland tumors in dogs are also rare, and carcinomas occur most commonly. Many dogs with salivary gland tumors are presented for examination because the owner notices a firm swelling on the neck related to the tumor and regional lymph node involvement (frequently firm/fixed primary tumor and mandibular/retropharyngeal LN). Treatment typically involves RT and possibly chemotherapy because surgical excision is rarely possible. Treatment is usually palliative, but long-term control is possible in some cases. In one study of dogs and cats with salivary gland tumors treated with surgery followed by radiation therapy, median survival times of >500 days were reported.

FELINE ORAL TUMORS

Oral tumors make up 10% of all feline tumors, and they are frequently associated with significant morbidity because they are not detected until late in the course of the disease. Typically, older cats (>15 years) are affected, and SCC makes up the great majority of feline oral tumors. Other tumors are uncommon and include lymphoma, fibrosarcoma, melanoma, and mast cell tumors. Benign tumors that affect the oral cavity include ameloblastoma, eosinophilic granuloma complex, and nasopharyngeal polyps. Tonsillar and salivary gland tumors are rare in cats.

Oral squamous cell carcinoma (SCC) makes up the majority of feline oral tumors. Cats that live in smoking households have a 2x increase in development of this cancer. Epidemiologic studies have also implicated use of flea collars and a canned food diet as increased risk factors for SCC development. These tumors are usually sublingual or mandibular in location, but can be maxillary or pharyngeal. In some cats, a palpably thickened mandible without evidence of a soft tissue tumor is seen. Bone lysis is common, and dental radiographs should be performed in any cat with loose teeth or gingival proliferation. Most SCC cannot be surgically resected due to their location. Since distant metastasis is uncommon, however, if surgery is possible, it is often the cat's best chance for long term survival. In a study of cats treated with mandibulectomy for SCC, 40% of tumors recurred after surgery, but 43% of cats were still alive at 1 and 2 years post-operatively. In general, cats tolerated mandibulectomies, but long-term effects such as mandibular drift, ptyalism, and dysphagia occurred in 76% of cats. For sublingual and other non-resectable tumors, RT often helps to decrease inflammation and improve pain control. The long-term prognosis for this tumor is poor, and most cats are euthanized shortly after diagnosis. If cats with SCC are not eating or grooming, serious discussion with the owner about quality of life should occur. For cats that are relatively healthy besides the tumor, RT is usually the best option; the median survival time for cats with SCC treated with RT is approximately 4 months. Chemotherapy is rarely helpful in the management of this tumor.

SUMMARY

Oral tumors are most easily treated when they are small and well defined. Client education is essential for early detection, and veterinary personnel should demonstrate to owners show to brush their pet's teeth and examine the oral cavity. Frequent dental cleanings and oral cavity exams under anesthesia should be a part of the pet's routine health care. Once tumors are detected, proper staging and therapy will provide the pet's best chances for long term survival and quality of life.

REFERENCES

Fiani F, Verstraete FJ, Kass PH, et al. Clinicopathologic characterization of odontogenic tumors and focal fibrous hyperplasia in dogs. J Am Vet Med Assoc 2011; 238: 495-500. Theon AP, Rodrguez C, Madewell BR. Analysis of prognostic factors and patterns of failure in dogs with malignant oral tumors treated with megavoltage radiation. J Am Vet Med Assoc 1997; 210: 778–784. LaDue-Miller T, Price GS, Page RL, Thrall DE. Radiotherapy of canine non-tonsillar squamous cell carcinoma. Vet Radiol Ultrasound 1996;37:74–77. Theon AP, Rodriguez C, Griffey S, Madewell BR. Analysis of prognostic factors and patterns of failure in dogs with periodontal tumors treated with megavoltage irradiation. J Am Vet Med Assoc 1997; 210: 785–788. Syrcle JA, Bonczynski JJ, Monnette S, et al. Retrospective evaluation of lingual tumors in dogs. J Am Anim Hosp Assoc 2008; 44(6): 308-19. Mas A, Blackwood L, Cripps P, et al. Canine tonsillar squamous cell carcinoma. J Sm Anim Pract 2011; 52(7): 359-64.

ANTIEMESIS THERAPEUTIC CONSIDERATIONS FOR THE ONCOLOGY PATIENT

Chad M. Johannes, DVM, DACVIM (SAIM)
Raleigh, NC

INTRODUCTION

Chemotherapy has played a critical role in improving outcomes and increasing survival times for a number of types of cancer in recent decades, both in human and veterinary oncology. As more chemotherapeutic agents have become available and more intensive protocols have developed, one of the most common challenges encountered from an adverse effect aspect is chemotherapy-induced nausea and vomiting (CINV). Significantly impacting health-related quality of life in cancer patients, it is important to understand the incidence and pathophysiology of CINV and therapeutic interventions that can be made to reduce its effect on patient care and clinical outcomes.

IMPACT OF CINV IN HUMAN CANCER PATIENTS

As many as 80% of human cancer patients receiving chemotherapy experience some level of adverse effects, the most common and dreaded of which is CINV. The impact CINV can have on human cancer patients is profound: adversely affecting quality of life and daily activities, including work productivity; giving rise to medical complications (poor nutrition, dehydration, electrolyte imbalances); causing treatment delays or refusal to pursue additional therapies; increasing emergency room/hospital visits and subsequent cost of medical care.[1-3]

In spite of the availability of more effective antiemetics (newer 5-HT_3 antagonists, NK_1 antagonists) in recent years, control of CINV in human cancer patients remains a challenge. The role of healthcare practitioners is vital to improving the control of CINV. Studies indicate that underestimating the incidence, failing to understand the pathophysiology and ineffectively implementing antiemetic treatment guidelines contributes to the ongoing clinical burden of CINV.[1,4,5]

Nearly two-thirds of human patients receiving chemotherapy experienced one or more episodes of CINV in one prospective study. Delayed CINV (greater than 24 hours after administration) occurred nearly twice as often as acute CINV (within 24 hours of administration).[5] Also of note is that nausea occurred twice as often as vomiting. Patients indicated that nausea was more distressing than vomiting and had a greater impact on their quality of life.[5,6] This point has particular significance for veterinary health professionals where evaluation of nausea in patients is much more subjective but nonetheless important to monitor.

In human oncology, chemotherapeutic agents and radiation therapy locations/procedures have been categorized as to their emetic risk (high, moderate, low, minimal) and CINV treatment guidelines are published to serve as recommendations for effectively controlling signs in patients being treated.[7] Although studies have found that patients receiving antiemetic therapy according to these guidelines experience less frequent delayed CINV compared to those receiving inadequate therapy, findings also indicate that the guidelines are frequently not followed. Several studies have identified that healthcare providers greatly overestimate the control of delayed nausea and emesis following both highly emetogenic and moderately emetogenic chemotherapeutics and that patients do not always adequately report symptoms.[4,5,8,9]

CINV PATHOPHYSIOLOGY IN HUMAN ONCOLOGY

CINV is typically classified as acute or delayed. Acute CINV occurs within 24 hours of chemotherapy administration whereas delayed CINV is defined as nausea or vomiting occurring 24-120 hours after chemotherapy. The distinction between acute and delayed CINV implies more than timing. Physiologic differences in the pathways involved account for acute versus delayed CINV. Acute CINV appears to be primarily mediated by serotonin (5-HT_3) pathways while delayed CINV is primarily substance P (NK_1) pathway mediated. This difference becomes important when preventative therapies are considered. Breakthrough CINV is nausea and vomiting that occurs in spite of antiemetic therapy and requires the use of rescue (additional antiemetic) treatment.[10-12]

Initiation of emesis (vomiting) requires activation of the vomiting center within the brainstem which has three main regional components (area postrema, nucleus tractus solitarius, dorsal vagal complex) that integrate the emetic responses. In general, activation of the vomiting center can occur as the result of stimuli from many sources including drugs (i.e., chemotherapeutics), gastrointestinal input/irritation/stretching, motion, smells, sights or emotions.

CINV in particular is believed to result from the presence of chemotherapeutic agents or their metabolites within the blood stream and/or the cerebrospinal fluid acting on the chemoreceptor trigger zone (CRTZ) in the area postrema (lies outside the blood-brain barrier). Signals from the area postrema are relayed to the nucleus tractus solitarius and the emetic reflex is induced. Chemotherapeutic drugs may also cause the release of serotonin and substance P from the gastric mucosa (enterochromaffin cells) which triggers signals via vagal sensory fibers to the nucleus tractus solitarius, initiating the emetic reflex via another route. In addition to the substance P (NK_1) and serotonin (5-HT_3) pathways, cannabinoid and dopamine

(D_2) pathways are believed to be involved in CINV. Alternative pathways involved in nausea and vomiting do not appear to be activated in CINV, including: acetylcholine/muscarinic (M), histamine (H_1), endorphin and gamma-aminobutyric acid pathways.[1,12]

These findings have practical implications as antiemetic therapeutic options for CINV are considered. Antiemetic therapy is administered prophylactically to minimize the clinical impact of CINV and the treatment utilized depends on the chemotherapy regimens employed. Patient-related risk factors and prior response to treatment regimens are also carefully considered.[7,12] It has been well-documented that patients experiencing acute CINV have a higher risk of developing delayed CINV compared to those who are completely protected during the acute phase. This recognized carry-over effect from the acute to the delayed phase is an important consideration and emphasizes the importance of preemptive emetic control strategies.[12]

PRACITAL CONSIDERATIONS FOR CLINICAL MANAGEMENT OF CINV

Neurokinin (NK) Receptor Antagonists

Substance P (a tachykinin) has been recognized as a key component in the emetic reflex by stimulation of NK_1 receptors in the emetic center, CRTZ and abdomen/gastrointestinal tract. Development of effective NK_1 receptor antagonists which block the binding of Substance P to NK_1 receptors provided a significant breakthrough in the control of CINV in human oncology. As a result, this class of drug is part of the recommended protocol for patients receiving chemotherapeutics characterized by a high emetic risk.[7]

Maropitant (Cerenia™, Pfizer) was approved by the FDA for the prevention and treatment of acute vomiting in dogs in 2007. Two separate studies utilizing maropitant for the prevention of CINV in dogs receiving the highly emetogenic chemotherapy agent cisplatin showed that dogs treated with maropitant experienced significantly fewer subsequent emetic events compared to placebo-treated dogs.[13,14] When maropitant was administered prior to cisplatin treatment in a prevention regime, 94.9% did not vomit compared with only 4.9% of placebo-treated dogs.[13]

A recent report from Tufts University revealed that maropitant, when administered orally for 5 days following doxorubicin chemotherapy, was effective in preventing delayed vomiting induced by doxorubicin. Interestingly, dogs receiving maropitant also experienced less frequent and severe diarrhea than placebo-treated dogs. The authors indicated that prophylactic treatment with maropitant might improve quality of life and decrease the potential need for dose reductions in certain dogs receiving doxorubicin therapy.[15] Additional studies in dogs indicate that maropitant may be beneficial in reducing visceral pain[16,17] which may be another mechanism by which it helps reduce the effects of CINV. Although not approved for use in cats, published data on the safety and efficacy of maropitant as an antiemetic in cats is available.[18]

5HT3 Antagonists

Selective 5-HT_3 antagonists block the binding of serotonin to receptors in the gut and brain (nucleus tractus solitarius).[12] Newer generation drugs of this category (i.e., palonosetron) are recommended in the antiemetic protocols for human oncology patients receiving highly and moderately emetogenic chemotherapeutic drugs as well as for many patients at risk for radiation-induced nausea and vomiting.[7] While not approved for use in dogs and cats, ondansetron and dolasetron are utilized fairly commonly as antiemetic agents in dogs and cats[19] due to their generic availability, relatively low cost and empirical efficacy.

Metoclopramide

An antagonist of dopamine D_2 receptors (CRTZ) at lower dosages, metoclopramide may inhibit 5-HT_3 receptors in the gut and CRTZ at higher dosages. While human cancer patients receiving the high-dose metoclopramide experienced less CINV subsequent to cisplatin therapy, extrapyramidal reactions (involuntary movements of face, tongue extremeties; trismus; tremors) were experienced by a number of patients. Diphenhydramine was often administered to alleviate these signs and studies showed that addition of dexamethasone improved the antiemetic effect compared to metoclopramide alone. As a result, the three-drug combination of high-dose metoclopramide, a corticosteroid and diphenhydramine was considered the standard antiemetic regimen for human chemotherapy patients until the development of 5-HT_3 and NK_1 antagonists.[12] Based on the findings that the dopaminergic D_2 pathway plays little role in the development of CINV and the side effects observed with high-dose therapy, metoclopramide is typically only utilized in combination therapy for patients experiencing nausea or vomiting in spite of optimal prophylaxis.[7]

Other agents

Butorphanol was commonly used as part of standard antiemetic protocols in dogs receiving cisplatin prior to the availability of 5-HT_3 and NK_1 antagonists.[20] A number of studies have evaluated the use of corticosteroids as antiemetic agents in human oncology patients. Studies have indicated that, as single agents, corticosteroids have an antiemetic effect in patients receiving moderately emetogenic chemotherapies. Additionally, it has been shown that corticosteroids increase the effect of other antiemetics (i.e., low-dose and high-dose metoclopramide, 5-HT_3 antagonists, NK_1 antagonists).[12] As a result, dexamethasone is commonly utilized as a component of therapy for CINV associated in human cancer patients being treated with high, moderate and low emetogenic chemotherapy agents as well as radiation patients at high or moderate risk of nausea or vomiting.[7] The potential role of corticosteroids such as dexamethasone as antiemetic agents in dogs and cats is not well documented and this category of drug is not routinely included in CINV prophylactic protocols for veterinary patients.

SUMMARY

While specific antiemetic treatment guidelines have yet to be established for veterinary patients receiving chemotherapy, preemptive control of CINV is important for veterinary patients as well. A growing body of literature supports this assertion. Prevention and control of CINV can profoundly impact the quality of life of patients and potentially reduce treatment delays or dosing reductions. Veterinary technicians play a vital role in recognizing CINV and educating clients in order to affect the best outcomes.

REFERENCES

[1]Hawkins R, et. al. **Clin J Oncol Nurs** 2009;13(1):54; [2]Hamadani M, et. al. **J Oncol Pharm Pract** 2007;13(2):69; [3]Shih YCT, et. al. **Cancer** 2007;110(3):678, [4]Grunberg SM, et. al. **Cancer** 2004;100(10):2261; [5]Ihbe-Heffinger A, et. al. **Ann Oncol** 2004;15(3):526; [6]Bloechl-Daum B, et. al. **J Clin Oncol** 2006;24(27):4472; [7]Basch E, et. al. **J Clin Oncol** 2011;29:4189; [8]Erazo Valle A, et. al. **Curr Med Res Opin** 2006;22:2403; [9]Majem M, et. al. **Support Care Cancer** 2011;19:1983; [10]Jordan K, et. al. **Oncologist** 2007;12(9):1143; [11]Schwartzberg L. **J Support Oncol** 2006;4(2, Suppl. 1):3; [12]Herrstedt J. **Nat Clin Pract Oncol** 2008;5(1):32; [13]Vail DM, et. al. **Vet Comp Oncol** 2007;5(1):38; [14]de la Puente-Redondo VA, et. al. **Am J Vet Res** 2007;68(1):48; [15]Rau SE, et al. **J Vet Intern Med** 2010;24:1452; [16]Boscan P, et. al. **Am J Vet Res** 2011;72(12):1576; [17]Alvillar BM, et. al. **Vet Anaesth Analg** 2011;online doi:10.111/j.1467-2995.2011.00670.x; [18]Hickman MA, et. al. **J Vet Pharmacol Therap** 2008;31:220; [19]Ogilvie GK. **J Am Anim Hosp Assoc** 2000;36:481; [20]Moore AS, et. al. **J Am Vet Med Assoc** 1994;205(3):441.

POLYUNSATURATED FATTY ACIDS AND CANCER: FRONTIERS FOR CURE AND CONTROL

Gregory K. Ogilvie, DVM, DACVIM (SAIM, Oncology), DECVIM-CA (Oncology)
Carlsbad CA

Cancer is a major cause of morbidity and mortality in dogs, cats and women worldwide. The ultimate goal for the treatment of malignancies is to eliminate all evidence of cancer resulting in a cure. With high grade malignancies, this is usually accomplished with a combination of surgery, chemotherapy, radiation, and more recently, molecular approaches of cancer therapy. Despite decades of intense effort and billions of dollars of expenditure, cures for high grade cancers have remained elusive. Fortunately, new knowledge about new and older therapeutic agents such as the n-3 polyunsaturated fatty acid (PUFAs), docosahexaenic acid (DHA) and eicosapentaenoic acid (EPA) have resulted in improved quality and quantity of life. Many referring veterinarians around the world have noted that the medical, surgical and radiation oncologists are treating with n-3 PUFA's such as docosahexaenoic acid to enhance the effect of chemotherapy and radiation therapy as an adjuvant therapy to prevent cancer progression and metastasis. These same caring veterinarians have often asked:

- Should all cancer patients be treated with PUFA's such as DHA?
- Should PUFA's such as DHA be administered to cancer patients to enhance the efficacy of chemotherapy and radiation therapy?
- Should PUFA's such as DHA be prescribed to reduce the toxicity of radiation therapy?
- Should antioxidants such as vitamin E be administered to cancer patients receiving DHA or EPA?

The use of fatty acids such as docosahexaehoic acid is designed to enhance disease free interval, survival and quality of life after surgery by reducing the rate of cancer development or incidence. This concept, known as 'cancer prevention by delay' has recently been recognized and is an important mechanism behind the successes of several therapeutic agents including (Lippman and Hong, 2002):

For the last decade, laboratories in Japan, France, Colorado and now at Angel Care Cancer Center in California has been involved in the search for dietary lipids associated with a delay in cancer relapse. Dietary lipids such as DHA appear to influence the growth of many types of cancer including breast and prostatic cancer. (Franceschi et al., 1995; Braga et al., 1997, Fay et al., 1997, Thompson et al., 1996). From a cohort of women treated for localized presentations of breast cancer, the group in France used adipose tissue sampled during surgery as a biomarker of past dietary intake of polyunsaturated fatty acids (Bougnoux et al., 1994). They found elevated n-3 fatty acids, especially DHA to be associated with a higher metastasis-free survival, suggesting that these fatty acids could potentially delay metastasis by decreasing tumor growth or development. Furthermore, using a case control approach comparing the fatty acid composition of adipose breast tissue obtained at the time of the surgical removal of either malignant or benign breast tumors, the French group found both alpha linoleic acid and docosahexaenoic acid (DHA) to be positively associated with a decreased risk of having breast cancer (Maillard et al., 2002). They also explored the role of n-3 PUFA on mammary tumor growth using the experimental system of NMU-induced mammary tumors in rats. Because fatty acids are substrates for lipid peroxidation processes, the group in France studied the effects of n-3 fatty acids on tumor growth in interaction with anti- or pro-oxidant compounds. They found that dietary n-3 fatty acids in the form of fish oil that contains docosahexaenoic acid inhibited tumor development. Furthermore, this tumor growth inhibition was most evident in the absence of the antioxidant vitamin E. Inhibition of tumor growth was even greater when the n-3 fatty acids were given in the presence of pro-oxidants (Cognault et al., 2000). Such effects were not found when the lipid diet was low in fatty acids. These data suggested that oxidized n-3 fatty acids have an inhibiting role on tumor growth and emphasize the importance of the interaction of anti- and pro-oxidant compounds with n-3 fatty acids. There is a growing body of data based on our work as well as work does by others that suggests the effect of n-3 fatty acids such as DHA seem to involve several steps of tumor formation. N-3 fatty acids such as DHA and EPA:

i. Inhibit tumor vessel formation (angiogenesis)
ii. Inhibit cell proliferation in several epithelial cell lines.
iii. Enhance the rate of tumor cell death
iv. Induce lipid peroxidation which enhances the efficacy of radiation and chemotherapy induced cancer cell death; this effect is diminished or reduced dramatically with vitamin E.
v. Suppress the expression of cyclooxygenqse-2 in tumors therefore decreasing cancer cell proliferation.
vi. Suppression of NF kappa B activation and bcl-2 expression thus allowing apoptosis of cancer cells

PUFAs and Chemotherapy

Dietary lipids such as DHA have been suggested to modify the sensitivity of tumors to ROS-generating anticancer drugs which is the reason it is incorporated into the treatment regimen of some patient's world wide. For example, when dogs with lymphoma were treated with doxorubicin chemotherapy and a diet supplemented with n-3 fatty acids in the form of fish oils, that there was a direct correlation between the level of docosahexaenoic acid in the blood and improved disease free interval (Ogilvie et al, 2000). Another study using the same randomized study design was used to assess at the efficacy of n-3 fatty acids in combination with doxorubicin chemotherapy to improve the disease free interval in dogs with hemangiosarcoma, a highly metastatic, rapidly fatal malignancy. There was a statistically significant positive correlation between the n-3 fatty acids and disease free interval (Richardson et al, in press). A similar approach was used in rats bearing autochthonous, NMU-induced mammary tumors. We found that dietary supplementation with fish oil or DHA increased the sensitivity of mammary tumors to anthracyclines, compared with dietary supplementation with saturated fatty acids (de Poncheville et al., 2000). Since DHA is the most polyunsaturated of the polyunsaturated fatty acids, lipoperoxydation is suspected to be a likely molecular mechanism implied in the enhancement of the response of the cancer cells to cytotoxic drugs. Addition of vitamin E to diet provided to mammary tumors bearing rats abolished the enhancing effect of DHA on tumor sensitivity to anthracyclines. (de Poncheville et al, 2000). In all studies done to date, there has been no clinically significant toxicity other than transient gastrointestinal distress linked to the dietary change (Ogilvie et al, 2000, McNiel et al, 1999, Swaim et al, 1989). Therefore, based on the safety and efficacy profile of n-3 fatty acids , it seems reasonable to further define the efficacy of n-3 fatty acids, especially docosahexaenoic acid for the treatment of spontaneously occurring cancer in dogs with the intent to provide evidence for its use in randomized human clinical trials.

PUFAs and Radiation Therapy

Radiation therapy is currently the most effective treatment for many localized malignancies including those within the nasal cavity. Research is underway at Angel Care Cancer Center to identify methods to maximize efficacy of radiation while minimizing the adverse effects associated with radiation therapy. Among the agents being evaluated to minimize the normal tissue damage are long chain polyunsaturated fatty acids of the n-3 series such as docosahexaenoic acid and eicosapentaenoic acid, which readily are incorporated into cell membranes and ameliorate inflammation and carbohydrate dyshomeostasis. In a study involving 12 dogs with histologically confirmed malignant carcinomas of the nasal cavity were randomized to receive isocaloric amounts of a diet supplemented with menhaden fish oil including DHA (experimental diet) or an otherwise identical diet supplemented with corn oil (control diet). Megavoltage radiation was delivered to all dogs. The data in that study suggested that feeding a diet supplemented with fish oil and arginine is associated with decreased concentrations of inflammatory mediators involved with radiation damage in skin and mucosa, and improved performance scores in dogs with malignant nasal tumors (Hansen et al, in press 2003).

Tumor sensitisation to radiation by polyunsaturated fatty acids has been investigated. Vartak et al. (1997) (1998) studied the in vitro response of a chemically-induced rat malignant astrocytoma cell line to radiation after the cell culture medium was supplemented with gamma-linoleic acid (GLA) or long chain n-3 PUFA (EPA or DHA), and found that n-3 PUFAs enhanced radiation-induced cell cytotoxicity. Colas et al (2003) documented enhanced radiosensitivity of rat autochthonous mammary tumors by dietary docosahexaenoic acid.

Whether dietary n-3 fatty acids can lead to enhanced sensitivity of tumor tissue in the absence of a similar increase in the radiosensitivity of non-tumor tissue remains a critical issue. Several studies have suggested that PUFAs do not sensitize normal tissues to radiation. For example, since ionizing radiation generate ROS; we initiated a study to determine whether dietary DHA might sensitize mammary tumors to irradiation using a model where mammary tumors were induced by N-methylnitrosourea (NMU) in Sprague-Dawley rats. In the study, we showed that dietary DHA sensitized mammary tumors to radiation. The addition of vitamin E inhibited the beneficial effect of DHA, suggesting that this effect might be mediated by oxidative damage to the peroxidizable lipids (Colas et al, 2003).

REFERENCES

Appel, MJ, Van Garderen-Hoetmer, AM. and Woutersen, RA. Effects of dietary conjugated linoleic acid on pancreatic carcinogenesis in rats and hamsters. Cancer Res 54, 2113-2120, 1994. Aro A, Mannisto S, Salminen I, Ovaskainen ML, Kataja V, Uusitupa M. Inverse association between dietary and serum conjugated linoleic acid and risk of breast cancer in postmenopausal women. Nutr Cancer 2000;38(2):151-7 Banni, S and Martin, JC. Conjugated linoleic acid and metabolites. In: Trans fatty acids in human nutrition,. JL. Sébédio (ed.), Dundee: The oily press, 1998, pp. 261-302. Bougnoux, P. (1999) n-3 polyunsaturated fatty acids and cancer. Curr. Opin. Clin. Nutr. Metab. Care, **2**, 121–126 Bougnoux, P., Koscielny, S., Chajes, V., Descamps, P., Couet, C. & Calais, G. (1994) Alpha-linolenic acid content of adipose breast tissue: a host determinant of the risk of early metastasis in breast cancer. Br. J. Cancer, **70**, 330–334 Bougnoux P, Germain E, Chajes V, Hubert B, Lhuillery C, Le Floch O, Body G, Calais G. Cytotoxic drugs efficacy correlates with adipose tissue docosahexaenoic acid level in locally advanced breast carcinoma. Br J Cancer 1999b, 79 :1765-1769 Braga, C., La Vecchia, C., Negri, E., Franceschi, S. & Parpinel, M. (1997) Intake of selected foods and nutrients and breast cancer risk: an age- and menopause-specific analysis. Nutr. Cancer, **28**, 258–263Chajès, V., Sattler, W., Stranzl, A. & Kostner, G.M. (1995) Influence of n-3 fatty acids on the growth of human breast cancer cells in vitro: relationship to peroxides and vitamin E. Breast Cancer Res. Treat., **34**, 199–212 Cognault, S., Jourdan, M.L., Germain, E., Pitavy, R., Morel, E., Durand, G., Bougnoux, P. & Lhuillery, C. (2000) Effect of an alpha-linolenic acid-rich diet on rat mammary tumor growth depends on the dietary oxidative status. Nutr. Cancer, **36**, 33 Colas S, Paon L, Denis F, Prat M, Louisot P, Hoinard C, Le Floch O, Ogilvie G, Bougnoux P. Enhanced radiosensitivity of rat autochthonous mammary tumors by dietary docosahexaenoic acid. Int J Cancer. 2004;109(3):449-54.Connolly JM, Gilhooly EM, Rose DP. Effects of reduced dietary linoleic acid intake, alone or combined with an algal source of docosahexaenoic acid, on MDA-MB-231 breast cancer cell growth and apoptosis in nude mice. Nutr Cancer 1999;35(1):44-9De Stefani, E., Deneo-Pellegrini, H., Mendilaharsu, M. & Ronco, A. (1998) Essential fatty acids and breast cancer: a case-control study in Uruguay. Int. J. Cancer, 76, 491–494Fay, M.P., Freedman, L.S., Clifford, C.K. & Midthune, D.N. (1997) Effect of different types and amounts of fat on the development of mammary tumors in rodents: a review. Cancer Res., **57**, 3979–3988 Franceschi, S., Favero, A., La Vecchia, C., Negri, E., Dal Maso, L., Salvini, S., Decarli, A. & Giacosa A. (1995) Influence of food groups and food diversity on breast cancer risk in Italy. Int. J. Cancer, **63**, 785–789Franceschi, S., Favero, A., Decarli, A., Negri, E., La Vecchia, C., Ferraroni, M., Russo, A., Salvini, S., Amadori, D., Conti, E., et al. (1996) Intake of macronutrients and risk of breast cancer. Lancet, **347**, 1351–1356 Germain E, Lavandier F, Chajes V, Schubnel V, Bonnet P, Lhuillery C, Bougnoux P. .Dietary n-3 polyunsaturated fatty acids and oxidants increase rat mammary tumor sensitivity to epirubicin without change in cardiac toxicity. Lipids 1999;34 Suppl:S203 Additional References Available Upon Request

SCIENCE BEHIND THE PLACEBO EFFECT AND THE MIND BODY CONNECTION IN VETERINARY ONCOLOGY

Gregory K. Ogilvie, DVM, DACVIM (SAIM, Oncology), DECVIM-CA (Oncology)
Carlsbad, CA

The function of the body, the immune system and psychosocial state are all interconnected and influenced by each other. Essentially, the mind, body, soul and immune system can modify the function or state of each other, and all are influenced by the interaction with the environment. That environment includes the presence of other people or animals. This complex interaction is called the mind body connection sometimes referred to as the field of psychoneuroimmunology. Unfortunately, veterinarians, physicians and other healers often under estimate and appreciate the value of this complex relationship in influencing the art and science or healing. This area of healing was essentially unknown 20 years ago, but it now a highly active specialty of healing, research and interest. The value of the mind body connection to veterinarians and people with animals is extraordinary as our interaction with animals provides healing and disease on both sides of the relationship. The purpose of this brief document is to introduce the reader to the vast field of the mind body connection from one veterinarian's perspective.

Bilateral Health Benefits of Animals

The relationship between animals and man has long been celebrated for the richness, joy, delight and even frustration that this association has brought human beings for centuries. Veterinarians bring direct health benefits to animals, but this healing extends beyond medication and surgery. Indeed, our presence in the life of animals has been shown to result in positive and negative effects on their health and well being. Similarly, animals bring health and healing to people around them. Called the human animal bond, family-pet-veterinary bond or just the animal human relationship has cataloged quantitative health benefits of animals in our lives. These include but are not limited to:

- Lower blood pressure
- Decreased stress
- More positive outcome and a better attitude
- Decreased hospitalization and increased longevity of the elderly
- Decreased duration of hospitalization of the elderly
- Improved social interaction with others among children or incarcerated individuals
- Decreased loneliness, depression, anger
- Improved survival among patients with coronary artery disease
- Enhanced recovery of those who have mental illness by providing empathy and "therapy"; providing connections that can assist in redeveloping social avenues; serving as "family" in the absence of or in addition to human family members; and supporting self-efficacy and strengthening a sense of empowerment.

The Science behind the Mind Body Connection

The mechanisms of this amazing relationship between mind, body, immune system and soul is intricate and complex to say the least. The immune system has been shown to be innervated by nerves confirming that the brain has input on the mind. Similarly, the mind initiates or modifies systematic change in cholinergic, serotonergic, dopaminergic, and other neurochemical pathways and opioid, benzodiazepine, and other neural receptor sites that mediate and modulate various subjective, cognitive, and affective states. But that is not all. Neuropeptides centered in the limbic system of the brain that control emotions formation influence lymphocytes and in spleen monocytes creating a linkage with the immune system and the brain confirming a bidirectional communication between mind and body, brain, and immune system. Therefore, the mood/mental state is linked with cellular defense and repair mechanisms

One might find this superficial discussion of the mechanism interesting but may ask how this applies to the life of a veterinarian or caregiver. As suggested above, factors that cause stress such as fear, depression, feelings of loss of control and apprehension are known to alter mental state, our immune system, and also our bodily function. For example, it has been shown that 3 episodes of depression double the risk of developing cancer. Thus, this documented relationship gives veterinarians reason to manage stress within their lives, hospitals and within the homes of clients who have animals that are being presented for care. Stressful states alters speech, ability to process information and finally, social contact. People and animals often go into an altered mental state secondary to stress.

There is little evidence to doubt that people who care for their pets, relatives or friends with cancer are also at risk of undergoing mental, physical, neurohumoral and immunologic changes. This can result in changes that range from depression to overt evidence that resembles post traumatic stress syndrome. The mechanism of altered mental states and health is complex but often involves altered hormonal levels with widespread implications. The normal diurnal rhythm of the body is critical to enhance and maintain health and wellness. Stress creates an imbalance in the diurnal rhythm of many hormones. Normally, cortisol is 4 times higher in the morning than in the evening however depressed or stressed people have elevated levels throughout the day. In chronic stress, cortisol levels remain elevated. This is a maladaptive stress. The cortisol levels in human breast cancer patients have been reported to be abnormally high throughout the day. Patients with relatively abnormal levels tended to die sooner than those that had more normal levels.

One consequence of stress is depression that can influence health and wellness. This was highlighted in a study in which 125 women with metastatic breast cancer completed an evaluation for depression symptom at baseline and then were randomly assigned to a treatment group or to a control group that received educational materials. Median survival time was 53.6 months for women with decreasing scores of depression over 1 year and 25.1 months for women with increasing evidence of depression. Thus, decreasing depression symptoms over the first year were associated with longer subsequent survival for women with metastatic breast cancer.

Interestingly enough, people who experience fear and deal with it, have cortisol levels tend to be more normal and this coincides with more normal mental and physical health. Support to allow veterinary health care team members and clients alike to express emotion and to allow people to face feelings directly may be healing. There is some benefit expressing one's fear and worry in a group. Thus, a welcoming veterinary health care team, a comfortable waiting room full of people who are going through the same experiences, or the presence of group therapy by a skilled facilitator may be very powerful in allowing people to deal with their emotions appropriately as well as well as to reduce negative health consequences to others and animals around them. A randomized trial of breast cancer patients had group discussions. This helped psychologically and made them able to be more in control of their emotions. The women with cancer were much more in control of their emotions. It reduced their PTSD. Psychotherapy was shown in five studies to improve survival, reduced stress and pain. Helping people manage their fear has a positive impact on their minds and bodies and improves survival and quality of life.

REFERENCES

1.Wisdom JP, Saedi GA, Green CA. Another breed of "service" animals: STARS study findings about pet ownership and recovery from serious mental illness. Am J Orthopsychiatry. 2009 Jul;79(3):430-436. 2.Barker SB, Wolen AR. The benefits of human-companion animal interaction: a review. J Vet Med Educ. 2008 Winter;35(4):487-495. 3.Knight S, Edwards V. In the company of wolves: the physical, social, and psychological benefits of dog ownership. J Aging Health. 2008 Jun;20(4):437-455. 4.Friedmann E, Thomas SA. Pet ownership, social support, and one-year survival after acute myocardial infarction in the Cardiac Arrhythmia Suppression Trial (CAST). Am J Cardiol. 1995 Dec 15;76(17):1213-7. 5.Sephton SE, Dhabhar FS, Keuroghlian AS, Giese-Davis J, McEwen BS, Ionan AC, Spiegel D. Depression, cortisol, and suppressed cell-mediated immunity in metastatic breast cancer. Brain Behav Immun. 2009 Nov;23(8):1148-55. 6.Spiegel D. Mind matters in cancer survival. JAMA. 2011 Feb 2;305(5):502-503. 7.Innominato PF, Palesh O, Dhabhar FS, Lévi F, Spiegel D. Regulation of circadian rhythms and hypothalamic-pituitary-adrenal axis: an overlooked interaction in cancer. Lancet Oncol. 2010 Sep;11(9):816-817. 8.Palesh O, Butler LD, Koopman C, Giese-Davis J, Carlson R, Spiegel D. Stress history and breast cancer recurrence. J Psychosom Res. 2007 Sep;63(3):233-9.

T17 FELINE HYPERTHYROIDISM: NEW THERAPY (IT'S IN THE BOWL!)

Vicky Ograin, MBA, RVT
Topeka, KS

INTRODUCTION:

Hyperthyroidism is now recognized as the most common endocrinopathy of older cats. Despite worldwide occurrence, the pathogenesis of feline hyperthyroidism remains unclear. Traditional methods of managing feline hyperthyroidism include thyroidectomy, antithyroid medications, and radioactive iodine. Surgery and radioactive iodine are designed to provide permanent solutions, oral antithyroid drugs are used to control hyperthyroidism and must be given daily to achieve and maintain their effect. All three modes of therapy are effective, but none are without risks. Recent studies document that a fourth option now exists for hyperthyroid cats. Feeding a low-iodine food decreases thyroid hormone concentrations and alleviates clinical signs of hyperthyroidism in cats. Management of hyperthyroidism is now as safe and easy as feeding your cat.

DIAGNOSIS:

Hyperthyroidism is generally considered a disease of middle-aged to older cats, with 12 to 13 years being the average age of onset. Less than 5% of cats are younger than 10 years of age at the time of diagnosis.[1] Functional adenomatous hyperplasia of one (30%) or, more commonly, both (70%) thyroid lobes is the most common abnormality associated with feline hyperthyroidism.[1] Typical clinical signs include weight loss with polyphagia, tachycardia, polyuria and polydipsia, nervousness or agitation, alopecia or unkempt coat, increased vocalization, heat intolerance and less often, vomiting, anorexia and lethargy. Most cats also have a palpable thyroid nodule.[2] Diagnosis is generally based on the presence of one or more typical clinical signs and elevated serum thyroxine (T_4) concentrations. The diagnosis of hyperthyroidism should not be excluded on the basis of a single normal serum T_4, especially in a cat with typical clinical signs and a palpable thyroid nodule. In these cases rule out concurrent nonthyroidal illness and repeat the total serum T_4 or evaluate simultaneous free T_4 and TSH measurements.[1]

RISK FACTORS:

Despite the increasing worldwide prevalence, the pathogenesis of feline hyperthyroidism remains a mystery. A number of epidemiologic studies have attempted to elucidate causes for the apparent increased incidence of hyperthyroidism over the last 20 years[3-5] Although increased awareness, access to thyroid hormone assays and an increasingly aging cat population contribute to increased recognition these factors may not completely explain the documented increase in prevalence of hyperthyroidism.[3] It is likely that the underlying cause is multifactorial, with both genetic predispositions and environmental factors contributing to the manifestation of disease. Investigators have postulated that immunological, infectious, nutritional (eg, iodine), environmental (eg, toxins, goitrogens), and genetic factors may play a role in the development of hyperthyroidism in cats[6-9]

TRADITIONAL THERAPY:

Traditional methods of managing feline hyperthyroidism include thyroidectomy, antithyroid medications, and radioactive iodine. Each therapy has advantages and disadvantages.

ANTITHYROID MEDICATIONS:

The thioureylene antithyroid drugs (methimazole, carbimazole and propylthiouracil) are the most common medical management of both human and feline hyperthyroid patients because of their reliable ability to inhibit the synthesis of thyroid hormones and thereby lower serum thyroid hormone concentrations. These drugs do not affect the trapping of inorganic iodide by, or the release of preformed hormones from the thyroid gland. They are widely recommended to stabilize hyperthyroid cats prior to surgery and are the only drugs that can be used chronically for management of hyperthyroidism.[1]Chronic administration of antithyriod medications has the advantage of being readily available, requiring no special facilities, and being reasonably inexpensive, at least initially. Almost all cats are potential candidates unless thyroid carcinoma is suspected. Anesthesia and hospitalization are avoided. On the other hand, daily administration of medication is required. Carbimazole and methimazole can also be formulated in a pluronic lecithin organogel (PLO) for transdermal application with reasonable efficacy. However there are no transdermal products approved for use in hyperthyroid cats. Custom formulation increases expense of therapy and stability of the product is not guaranteed. Methimazole and carbimazole have no effect on the underlying pathologic lesion in hyperthyroid cats so after months to years of treatment, thyroid nodules may continue to enlarge. This hyperplasia commonly results in the need to progressively increase the daily dose of antithyroid drugs. Indeed some cats may progress such that they no longer respond to antithyroid drugs and require an alternate form of therapy[1] While many cats have been successfully managed long term on oral therapy the risk of adverse reactions should not be overlooked. Most adverse reactions associated with methimazole or carbimazole occur within the first 3 months of therapy. Reported reactions include vomiting with or without anorexia and depression, self-induced excoriations of the head and neck, and mild to serious hemotologic complications.

SURGICAL THYROIDECTOMY:

Surgical thyroidectomy is a relatively simple and curative procedure. However, to decrease the cardiac and metabolic complications associated with anesthetizing hyperthyroid cats, prior medical management and control of the thyrotoxicosis are required. Once euthyroidism has been achieved, the main surgical considerations are whether to perform a unilateral or bilateral thyroidectomy, the type of technique to use, and the potential postoperative complications.[1] Hypoparathyroidism resulting in hypocalcaemia is the most significant potential postoperative complication. This generally arises only if the parathyroid glands are injured, devascularized, or inadvertently removed during a bilateral thyroidectomy. Mild hypocalcaemia develops in most cats post bilateral thyroidectomy, but treatment is not necessary unless clinical signs develop. Fortunately hypoparathyroidism is rarely permanent, and recovery of parathyroid function may occur days to months after surgery.

RADIOACTIVE IODINE:

Radioiodine treatment is often considered the best option for many cats since it has the potential to cure the disease with a single treatment, is not dependent on the location of the hyperfunctional thyroid tissue, no general anesthesia is required and side effects are minimal. Markedly elevated pretreatment T_4 levels, increasing size of the goiter and the severity of clinical thyrotoxicosis as well as preexisting kidney disease all potentially adversely affect the response to this therapy.[1] The main limitation to widespread use of radioactive iodine is the requirement for special licensing and the isolation of the cat for variable periods after treatment This can range from several days to several weeks depending on state or local radiation regulations and, particularly, the dose.[1]

NUTRITIONAL MANAGEMENT:

Excessive production of thyroid hormone is the hallmark of feline hyperthyroidism. Production of thyroid hormone requires uptake by the thyroid gland of sufficient amounts of iodine which is provided by dietary intake.[1] It is not surprising that many metabolic pathways are upregulated in the thyroid glands of hyperthyroid cats. Indeed recent genomic studies have shown that the metabolic pathways responsible for transporting iodine into the thyroid cells and into the colloid for inclusion into thyroid hormones are considerably upregulated in hyperthyroid cats.[10] This observation led to the question; if iodine transport systems are upregulated and dietary iodine is necessary for the production of thyroid hormones, what happens when dietary iodine is restricted in hyperthyroid cats? A new therapeutic food, Hill's Prescription Diet® y/d® Feline, the only food for feline hyperthyroidism, was developed based on the hypothesis that feline hyperthyroidism can be managed nutritionally by limiting the amount of dietary iodine available for production of thyroid hormones. Three studies have documented the safety and efficacy of Hill's Prescription Diet® y/d®Feline as the sole management in cats with naturally occurring hyperthyroidism.[11-13] The results of these studies support that a therapeutic food with dietary iodine levels at or below 0.32 ppm dry matter basis (DMB) provides an effective and safe management for cats with naturally occurring hyperthyroidism. Serum total thyroxine concentrations returned to the normal range within 8 to 12 weeks of initiating nutritional management in cats fed foods with ≤ 0.32 ppm iodine DMB. All hyperthyroid cats maintained on foods with ≤ 0.32 ppm iodine DMB as the sole source of nutrition remained euthyroid. In all of these studies biochemical features of renal function remained stable and no other biochemical abnormalities were observed. Feline y/d should not be fed to kittens or pregnant queens, they require higher levels of iodine than found in feline y/d.

NEWLY DIAGNOSED PATIENTS:

For a patient newly diagnosed or with recurrence of hyperthyroidism post thyroidectomy without other complications (hypertension, cardiac disease) Hill's Prescription Diet® y/d®Feline is appropriate as the sole management. A laboratory workup, including T4 and serum chemistry should be done before starting the food. Most cats require 7 days to transition from their current food to y/d Feline. Palatability and acceptance of this food are excellent. It is crucial that owners understand that success of this method depends on y/d Feline being the sole source of nutrition for their cat to the exclusion of treats or human foods. The first recheck evaluation should be scheduled 4 weeks after beginning the transition to y/d Feline or 3 weeks after y/d Feline becomes the sole food and should include physical exam, T4, serum chemistries and UA. Clinical improvement including weight gain, improved haircoat and decreased tachycardia/cardiac murmur are often noted by the first evaluation. By 8 weeks more clinical improvement should occur and 85% of cats will be euthyroid. By 12 weeks all cats should have normal T_4 values. If values remain abnormal poor adherence to dietary therapy should be suspected and a thorough history should be taken.

TRANSITIONING FROM ANTI-THYROID MEDICATIONS:

For cats with normal T_4, gradually decreasing the antithyroid drug dose will provide some control of T_4 while the cat is being transitioned to y/d Feline. Decrease daily medication dose by 50%, discontinue drugs after exclusively eating y/d Feline for 1-2 weeks.

GUIDELINES FOR MANAGING CATS WITH PERSISTENT HYPERTHYRODISM:

To date all hyperthyroid cats managed with Hill's Prescription Diet® y/d®Feline as the sole source of nutrition have become and remained euthyroid as long as the cat had no access to other sources of dietary iodine. Persistently elevated T_4 should raise concerns of poor adherence to dietary recommendations. Maintaining y/d Feline as the sole source of nutrition can be challenging for some owners. Discovering the source of dietary iodine intake can be a challenge for the veterinary health care team. In keeping with the principles of client centered communications, owner should be questioned in an open-ended, non-threatening manner. Sources of dietary iodine that may alter the response to this therapy include treats, flavored or compounded medications, access to 'people food', consumption of wild caught prey, and access to other pet foods.

ADVERSE EFFECTS OF NUTRITIONAL MANAGEMENT:

Based on the studies completed to date, no adverse effects of nutritional therapy have been noted. In all cats treated to date biochemical features of renal function have remained stable and no other biochemical abnormalities have been observed. There have been no reports of hypothyroidism or progression of kidney disease. Cats with IRIS stage 1 kidney disease have been managed with y/d Feline without progression of the disease. One hundred percent of hyperthyroid cats whose sole source of nutrition is y/d Feline have become euthyroid. Hypothyroidism has not been reported in cats managed exclusively with y/d Feline.

CONCLUSION:

Hyperthyroidism is common older cats. While the pathogenesis of feline hyperthyroidism remains unclear a variety of treatment options are available. Traditional methods of managing feline hyperthyroidism include thyroidectomy, antithyroid medications, and radioactive iodine. All three modes of therapy are effective but none are without risk of side effects. Hill's Prescription Diet® y/d Feline® is the first therapeutic food specifically formulated to manage hyperthyroidism in cats. Studies document that when y/d Feline is the sole source of nutrition 100% of hyperthyroid cats return to and remain euthyroid. Management of hyperthyroidism is now as safe and easy as feeding your cat.

REFERENCES:

1. Mooney CT. Hyperthyroidism In: Ettinger SJ, et.al.; Textbook of Veterinary Internal Medicine: Diseases of the Dog and Cat. Seventh Edition ed. St. Louis, MO: Saunders 2010;1761-1779. 2.Thoday KL, et. al.; Historical, clinical and laboratory features of 126 hyperthyroid cats. The Veterinary record 1992;131:257-264. 3. Edinboro CH, et al.; Epidemiologic study of relationships between consumption of commercial canned food and risk of hyperthyroidism in cats. J Am Vet Med Assoc 2004;224:879-886.

4. Olczak J, et al.; Multivariate analysis of risk factors for feline hyperthyroidism in New Zealand. N Z Vet J 2005;53:53-58. 5. Scarlett JM, et. al.; Feline hyperthyroidism: a descriptive and case control study Prev Vet Med 1988;6:295-309. 6. Scarlett JM. Epidemiology of thyroid diseases of dogs and cats. The Veterinary clinics of North America Small animal practice 1994;24:477-486. 7 Gerber H, et al.; Etiopathology of feline toxic nodular goiter. The Veterinary clinics of North America Small animal practice 1994;24:541-565. 8.Taylor JA, et al.; Perspectives on the diagnosis of feline hyperthyroidism. The Canadian veterinary journal La revue veterinaire canadienne 1989;30:477-481. 9. Gunn-Moore D. Feline endocrinopathies. The Veterinary clinics of North America Small animal practice 2005;35:171-210, vii. 10. Gao X, et al.; Gene Expression Analysis of Feline Thyroid Tissue and Blood from Cats with Evidence of Mild or Marked Hyperthyroidism Reveals Potential Molecular Causes of the Disease and Identifies Future Routes for Intervention (abstract). Journal of Veterinary Internal Medicine 2011;25:678. 11. Melendez LM, et. al.; Titration of dietary iodine for maintaining normal serum thyroxine concentrations in hyperthyroid cats (abstract) Journal of Veterinary Internal Medicine 2011;25:683. 12. Melendez LM, et al.; Titration of dietary iodine for reducing serum yhyroxine concentrations in newly diagnosed hyperthyroid cats (abstract). Journal of Veterinary Internal Medicine 2011;25:683. 13. Yu S, et al.; Controlled level of dietary iodine normalizes serum total thyroxine in cats with naturally occurring hyperthyroidism (abstract). Journal of Veterinary Internal Medicine 2011;25.

CATS AND CARBOHYDRATES: WHAT SHOULD FLUFFY EAT ?

Kara M. Burns, MS, MEd, LVT
President Academy of Veterinary Nutrition Technicians
Wamego, KS

The role of cats of in our lives has changed dramatically over the past 20 years. Data from the American Pet Products Manufacturers Association (APPA) 2007-2008 National Pet Owners Survey reports that there are approximately 83.3 million cats owned in the United States (U.S.) with 34% of U.S. households owning at least one cat and 56% of those cat owners having 2 or more cats! It is well accepted that proper nutrition and care throughout a cat's life maximizes health, longevity and quality of life. One of the first questions many cat owners ask of their veterinary healthcare team is 'What should I feed my cat'? There continues to be considerable discussion on the role of carbohydrates in the pathogenesis of feline diseases. This presentation reviews the nutritional requirements of cats and explores the role of carbohydrates in feline nutrition particularly related to obesity and diabetes mellitus, thus providing veterinary technicians with information to answer cat owner nutritional questions.

NUTRITIONAL REQUIREMENTS OF CATS

Cats are carnivores with unique anatomic, physiologic, metabolic and behavioral adaptations that support eating foods higher in protein and lower in carbohydrates as compared to dogs.[1,2] Cats are well adapted hunters with increased visual acuity, upright ears, sensitive facial whiskers, tactile hairs, retractable claws and conical, sharply pointed teeth. Cats have low liver glucokinase activity which limits the ability to metabolize large amounts of simple carbohydrates. Cats also demonstrate decreased intestinal and pancreatic amylase activity, the enzymes responsible for the digestion and uptake of starches and sugars. These adaptations support consumption of natural prey, such as small rodents which are estimated to contain around 55% protein, 45% fat and 1-2% carbohydrate on a dry matter basis. Despite these specific adaptations cats efficiently use carbohydrates and carbohydrates provide a good source of energy. Dry foods containing 40% or more dietary carbohydrates with an average digestibility of 85% are well tolerated by normal cats.

CARBOHYDRATES AND CAT FOODS

The primary purpose for adding carbohydrates to pet foods is to supply energy. During periods of high energy needs such as growth, gestation and lactation, carbohydrates become conditionally essential. In queens, carbohydrates spare protein necessary to sustain blood glucose concentrations and provide a substrate for lactose during milk production. In addition to providing a source of energy, carbohydrates provide structural integrity for dry foods, essentially holding the kibble together and preventing crumbling. This is a very practical consideration in manufacturing quality pet foods which meet the aesthetic 'needs' of pet owners.

There are some interesting facts about cat owner's choice of food and method of feeding their cats. The manufacture of commercial pet foods began in the late 1950s and top sellers at that time were canned (moist) fish varieties for cats. Over the past several decades there has been a shift toward purchase of dry foods. Results from a study evaluating feeding habits of cat owners revealed that most owners feed dry food alone or dry food mixed with moist food. Additionally, of those cat owners that feed dry food, more than 50% feed free choice or keep the cat's bowl full at all times. Owners that feed canned food are more likely to feed their cats 1 or 2 meals daily. A consequence of free choice feeding is excessive caloric intake which leads to obesity.

It's also important to consider the evolution of cat's roles in our lives. In the past most cats lived outside, were free-roaming and relied primarily on hunting prey or tossed out table scraps to sustain them. Today cats have a more home-centered life, live indoors, are neutered, more sedentary and eat a commercial cat food. The veterinary healthcare team must consider these changing factors when evaluating the health and wellness of today's pet cats.

OBESITY IN CATS

The number of cats that are overweight or obese has reached epidemic proportions in the United States. It's estimated that 35% of adult cats are overweight or obese and the percentage of obese cats increases with age. There are many risk factors for obesity in cats including genetics, gender, neuter status, age, activity level, food and type of food fed and method of feeding. Mixed breed cats are generally more obese than purebred cats. Male cats are predisposed to being overweight and neutering further increases the risk of obesity. Cats that are spayed or neutered have decreased metabolic rates and increased food intake, body weight, body conditions score and amount of body fat. One study demonstrated that energy intake increased by 50% within 12 weeks post-gonadectomy and in a separate study of neutered cats, those that were fed free choice became overweight, however, when fed a controlled amount of the same food those cats successfully lost weight. The risk of obesity also increases with age, reduced activity, free choice feeding and increased caloric intake.[3]

Veterinary technicians and all members of the healthcare team need to remember and educate clients on the many health conditions associated with obesity in cats. These include diabetes mellitus, cancer, skin diseases, dental disease and lower urinary tract problems. Obese cats are also more difficult to manage in terms of sample collection (blood, urine) and catheter placement and may be more prone to treatment complications including difficulty intubating, respiratory distress, and slower recovery time and delayed wound healing. It is widely believed that obesity affects quality of life and leads to reduced life expectancy.

Currently available evidence does not support a direct cause-and-effect relationship between increased carbohydrate intake and obesity in cats. Rather increased dietary carbohydrates is an indirect cause due to free-choice feeding of dry foods, and thus pet owners must be educated on the consequences of feeding free choice.

DIABETES MELLITUS IN CATS

More than 80% of cats are thought to have type II Diabetes Mellitus (DM), characterized by peripheral insulin resistance and dysfunctional β cells. Clinical managements of cats with type II DM includes oral hypoglycemic drugs or exogenous insulin combined with nutritional management with the goal of achieving optimal glycemic control and weight control.[4]

Risk factors associated with DM include increasing age, male gender, Burmese greed, being neutered, decreased physical activity and obesity. Most of these risk factors are also risk factors for being overweight or obese. There have been several published studies that have evaluated the role of environmental and nutritional factors in the pathogenesis of DM. These studies have demonstrated that 1) indoor confinement and low physical activity were associated with DM 2) body weight plays a role in insulin sensitivity and glucose effectiveness and 3) gonadectomy stimulates food intake resulting in weight gain. Currently available evidence does not support a direct cause-and-effect relationship between increased carbohydrate intake and DM in cats.[4]

Nutritional management, including maintenance of an ideal body condition, is important in the successful treatment of cats with DM. The goals of nutritional management are to 1) blunt postprandial hyperglycemia 2) control body weight 3) support nutrient needs 4) improve insulin sensitivity 5) minimize diabetic complications and 6) coordinate peak nutrient uptake with insulin activity. Feeding increased fiber which contributes to weight control and impacts carbohydrate absorption from the gastrointestinal tract has been associated with improved glycemic control in cats with DM. Low carbohydrate foods may be effective for maintaining glycemic control in cats with DM, however may not be ideal for all diabetic cats. Proteins, amino acids and fat stimulate pancreatic secretions and use of moderate-protein, low-fat, and high-carbohydrate foods have been recommended for cats with acute or active pancreatitis. High –protein foods should be avoided in cats with liver or kidney disease. Current evidence supports that foods ranging from 5-26% of calories as carbohydrates are associated with diabetic remission and improved glycemic control.

As healthcare team members we need to increase our efforts to educate pet owners about the importance of disease prevention in their cats. It is important for cats to maintain ideal body condition throughout their life and this requires constant communication with cat owners about the risk factors for obesity and DM, tips for maintaining a healthy weight in their cats and discussions about food choice and feeding methods.

SUMMARY

Proper nutrition and care throughout a cat's life maximizes health, longevity and quality of life. Foods high in carbohydrate content have been implicated in contributing to the cause of obesity and DM in cats. Cats are carnivores with specific adaptations to utilize foods high in protein; however, cats also efficiently use carbohydrates. Obesity and DM are two diseases that commonly affect cats and share similar risk factors. Currently available evidence does not support a direct cause-and-effect relationship between increased carbohydrate intake and obesity or DM in cats. It is likely that a combination of factors is involved and that increased carbohydrate intake is an indirect cause due to free-choice feeding of dry foods and associated weight gain or obesity. One of the most important discussions healthcare team members can have with cat owners is the importance of maintaining ideal body condition in their cats.

REFERENCES

1. Kirk CA, Debraekeleer J, Armstrong PJ. Normal cats, Small Animal Clinical Nutrition, 4th ed., Walsworth Publishing Company, Missourri, 2000;291-347. 2. Zoran DL. The Carnivore Connection to Nutrition in Cats. J Am Vet Med Assoc 2002;221:1559-1567. 3. Laflamme DP. Understanding and Managing Obesity in Dogs and Cats. Vet Clin Small An Pract 2006;36:1283-1295. 4. Kirk CA. Feline Diabetes Mellitus: Low Carbohydrates Versus High Fiber? Vet Clin Small An Pract 2006;36:1297-1306.

ENHANCING APPETITE IN THE FELINE PATIENT

Jessica Quimby, DVM, DACVIM
Fort Collins, CO

INTRODUCTION

Anorexia is one of the most common clinical signs of disease in cats presenting to the veterinary hospital. Unfortunately it is non-specific and a thorough medical workup is required to determine the underlying cause. Nutritional support is critical during the period of time that the primary problem is being sought and addressed. Additionally, several chronic diseases, such as chronic kidney disease and neoplasia, are common in cats and may result in inadequate caloric intake. Thus management of long term hyporexia is important. Poor body condition is associated with decreased prognosis in several species and has a negative effect on immune function, wound healing and strength.[1] Most importantly owners are distressed when their pet does not eat and it has a negative effect on owner perception of quality of life.

THE UNIQUE FELINE

Cats have a higher requirement for protein and amino acids than other species. When nutrition is inadequate, energy is derived from mobilization of amino acids from muscles stores as opposed to fat.[1] Elderly cats are also unique in comparison to other species as they have stable to increased level of metabolism, as opposed to decreased metabolism.[2] A reduced ability to digest protein and fat has been documented in elderly cats. These combined factors make the utilization of high quality, easily digestible food product critical for these patients.[3] Cats are also particularly sensitive to changes in environmental factors, such as the timing and location of feeding, as well as the food type; smell and "mouth feel" may also play into their willingness to eat.[4]

PATHOPYSIOLOGY

A meal can be thought of as having three parts; initiation, maintenance and termination. Different stimuli are responsible for each phase. Initiation of a meal depends heavily upon environmental stimuli such as olfactory, visual, tactile and auditory cues.[4] Termination of a meal occurs as a result of negative feedback mechanisms such as gastric distention or presence of nutrients in the intestine. Suppression of appetite may occur as a result of psychological, environmental or physical factors such as decreased olfaction, nausea, pain, fever, ileus, vomiting that interfere with this process.[1,4] Circulating inflammatory mediators such as TNF-a, IL-1, and IL-6 may play a role in suppression of appetite.[1] As vomiting is a common manifestation of disease in cats, and nausea is likely involved even if vomiting does not occur, examining the pathophysiology of vomiting centers and receptors may help in determining the etiology of the inappetence and appropriate medical management.

LEARNED FOOD AVERSION

Care should be taken to select the appropriate patients for appetite enhancement as learned food aversion is thought by most to be prevalent in cats.[4] Learned food aversion occurs when the patient associates nausea, pain or other physical manifestations of disease with the act of eating or the sight or scent of food. Even after the underlying illness is resolved, this aversion may remain. Therefore it is critical that cats that are overtly nauseous – drooling, gagging, turning away from food - particularly in hospital or in acute illness are not forced to eat lest food aversion be created.[4]

If cats are too nauseous or critical to even consider oral feeding, or have not responded to appetite encouragement after 3-5 days, placement of an enteral feeding tube should be considered. Nasoesphageal, esophageal or gastrotomy tube can be chosen depending on the type and duration of feeding desired.[1] Parenteral feeding should be considered in cats that cannot tolerate enteral feeding. Additionally many clinicians feel that prescription diets (i.e. renal) should not be fed in hospital during a crisis lest an aversion be created to the diet desired for long term management. The best candidates for pharmacological enhancement of appetite are cats leaving the hospital with their acute crisis resolved, and cats with chronic disease in the home environment.

THERAPIES

Medical management of underlying disease

Whenever possible it is obviously ideal to address the underlying disease condition. However in some chronic diseases, i.e. kidney disease, this is not possible and therefore medical management of complications of chronic disease are ideal. Dehydration and anemia can both potentially play a role in inappetence and therefore should be addressed when applicable.

Environmental factors and food choices

As cats are particularly sensitive to environmental factors, suggestions for appetite enhancement include: quiet environment with exclusive access without interference from bothersome household members, novel food type in cases of possible aversion, similar food type in case of food exclusivity, warming the food particularly if olfaction is an issue, alternatively chilling the food if aromas appear to result in nausea, social interaction while eating, feeding small frequent meals as premature satiety is associated with many disease states.

Pain management

For both acute and chronic medical conditions, appropriate management of pain is critical to improving appetite. Opioids are perhaps most commonly utilized. Buprenorphine has a significantly longer half-life and may be useful for in home palliation of chronic pain. Butorphanol may also have some visceral analgesic properties. Opioids should be used with caution in patients with ileus. Maropitant may also have some visceral analgesic properties.[5]

Anti-nausea/anti-emetic medications

Assessment of nausea in the feline patient is of course subjective, but if suspected based on the underlying condition, treatment is merited both to prevent learned food aversion and to improve appetite in general. Ondansetron and dolasetron have been used in human renal failure patients and appear to have potent anti-nausea and anti-emetic properties.[6,7] Maropitant is an anti-emetic, but is thought to also have anti-nausea properties and clinical trials for its use in kidney disease cats are currently underway. Mirtazapine demonstrates anti-nausea properties in addition to its appetite stimulating properties as it acts at the 5HT3 receptor similarly to ondansetron.[8] All four of these drugs work centrally as well as peripherally on receptors. Many are available as an injection which may be a welcome alternative to owners. Metoclopramide is thought to be a weak anti-emetic in cats do to a lack of receptors in the CRTZ, but may be useful as a promotility drug.[9]

Appetite stimulants

Cyproheptadine has been used for some time as an appetite stimulant and has anecdotal efficacy in many patients, however its efficacy has never been scientifically evaluated. Twice daily administration is necessary in many cases and this can prove a challenge for owners, particularly longterm. Mirtazapine has become more commonly used and recent exploration of its pharmacodynamics and pharmacokinetics have provided information for more effective use in cats.[10,11] Pharmacodynamic studies have illustrated that it can be a potent appetite stimulant, but higher doses are more commonly associated with side effects (hyperexcitability, vocalization, tremors). Smaller, more frequent doses are recommended. The half-life is short enough that it could be administered daily in normal cats. Renal disease delays clearance and in these patients, every other day administration is recommended.[11] Owner should be aware mirtazapine and cyproheptadine cannot not be administered concurrently, cyproheptadine is in fact used as an antidote for serotonin effects of mirtazapine overdose. The ghrelin receptor agonist, capromorelin, which mimics the activity of the "hunger hormone" ghrelin in stimulating appetite, is currently being assessed for use in cats which may add additional options for appetite stimulation in chronically hyporexic patients.

REFERENCES

[1]Chan D. **J Fel Med Surg** 2009;11:925. [2]LaFlamme DP. **Vet Clin Small Anim** 2005;35:713. [3]Sparkes AH. **Top Companion Animal Med** 2011; 26:37.
[4]Michel KE. **J Fel Med Surg** 2001;3:3. [5]Boscan et al. **Am J Vet Res** 2011; 72:1576. [6]Santos LCP et al. **Vet Anaesth Analg** 2011;38:320. [7]Ljutic D et al. **Kidney Blood Press Res** 2002;25:61. [8]Reichelmann RP et al. **Am J Hospice and Palliative Med** 2010; 27:106. [9]Trepanier L. **J Fel Med Surg** 2010;12:225. [10]Quimby JM. **J Vet Pharmacol Therap** 2010;34:388. [11]Quimby JM. **J Vet Intern Med** 2011;25:985.

ALTERNATIVE AND RAW FOOD DIETS

Kara M. Burns, MS, MEd, LVT
President, Academy of Veterinary Nutrition Technicians
Wamego, KS

Each year in the US there are 76 million reported cases of food borne illness in people. Although the majority are self limiting 323,000 people are hospitalized and ~ 5,200 deaths are attributed to food borne illnesses each year. These figures are important because homemade and raw diets for pets are made with ingredients from the same food supply chain. In veterinary medicine few topics are associated with such strong opinions as the feeding of raw meat to dogs and cats. Both proponents and opponents are firmly entrenched in their views. Unfortunately, much of the information on both sides is based on anecdotes, theoretical concerns and personal opinion. While objective information is increasing, this is still a field based largely on passion and opinion. This is an attempt to present a summary of the objective data for the nutritional adequacy, safety and benefits of home prepared wellness and therapeutic diets and raw food diets.

Some health-conscious pet owners prefer preparing foods for their pets daily. Home preparation provides owners with a sense of involvement and allows them to select for or against specific ingredients. To maintain the nutritional adequacy and avoid contamination, owners must truly be committed to the long-term maintenance of proper hygiene and preparation methods. The meat and eggs produced for human consumption and used to prepare homemade pet foods are often contaminated with microbes [1,2,3] Research indicates that many people are careless about cross-contamination during food preparation.[4] The potential for food-borne illness in pets is reduced if owners use high-quality ingredients that have been stored properly, heat the foods to temperatures sufficient to destroy pathogens, and prepare only amounts that are readily consumed. Nutritional adequacy of home prepared diets is also of concern. Veterinary healthcare team members will inevitably work with pet owners that prefer alternative diets for their pets, and thus should be familiar with the objective data surrounding the nutritional adequacy, safety and benefits of home prepared wellness and therapeutic diets and raw food diets.

RAW MEAT DIETS

Raw meat feeding was popularized by Dr. Ian Billinghurst after the publication of his two books; *Give Your Dog A Bone* and *The BARF Diet*.[5,6] The latter book introduced the term "BARF", "Biologically Appropriate Raw Food' or "Bone And Raw Food", which has become widespread terminology. These books, which lack any scientific evidence, are based on the author's belief that commercial pet food is contributing to serious health problems in pets.

Many proponents have made very strong claims, including 'Dogs fed cooked and processed food and no bones will always develop a weakened immune system and poor dental health".[7] Some have reported that raw diets are useful for the prevention or control of a wide range of conditions including chronic digestive, allergic and metabolic disease, bladder stones, feline lower urinary tract disease, intermittent vomiting or diarrhea, seborrhea sicca and recurrent ear infections.[7] More specific claims of the 'disappearance' of many dental problems, many skin problems, ear infections, anal sac problems, arthritis, incontinence, keratoconjunctivitis sicca, diabetes and reproductive problems, and improvement of behavioral problems have also been made.[8] An interesting claim is that 'As a general rule, any genetic fault that your pet may have inherited will have minimum chance of expression when the BARF diet is fed.' [8] There is frequent discussion about presumed negative effects of cooking, including destruction of vitamins, enzymes, 'antiageing factors', reducing protein value and availability, and production of 'foreign foods'. [7]

Proponents tend to dismiss infectious disease concerns, with statements such as '*Salmonella* and *Campylobacter* are of absolutely no consequence to a healthy dog'. [7] While Billinghurst claims that there are numerous feeding trials that show dogs fed raw foods are much healthier than dogs fed cooked foods, there is no objective evidence suggesting a beneficial health effect of feeding raw foods.

Nutritional adequacy, foreign body (i.e. bone) ingestion and infectious diseases are all cited as concerns regarding feeding raw foods. Infectious disease risks have received the most attention and is the area where more published research is currently available. As with any raw meat, there is the potential for contamination of raw meat-based pet foods. *Salmonella* spp has gathered the greatest attention as a possible risk; however *Campylobacter, Clostridium difficile, C. perfringens, C. botulinum, E. coli, Yersinia enterocolitica, Listeria monocytogenes* and enterotoxigenic *Staphylococcus aureus* are also of concern. [8,9,10] Recent studies also suggest that the potential exists for foodborne disease in pets eating raw meat but also disease in humans who are in contact with raw meat or the pet. Additionally pets fed raw meat are capable of contaminating the environment.

Bacterial contamination can be present in meat prior to slaughter in sick animals, however contamination occurs more commonly during slaughter, evisceration, processing and packaging. Ground meat, with its higher surface area to volume ratio and the need for additional processing, is more likely to be contaminated. The term 'human grade' has no regulatory definition and its use has been prohibited. However, since a significant portion of food sold for human consumption may be contaminated with a variety of pathogens, the use of 'human grade' meat does not indicate an absence of contamination. Once purchased, improper storage can result in bacterial growth. Some pathogens can grow at room temperature, while others can produce enterotoxins that cause disease.

In general, storage guidelines for raw meat are intended to prevent bacterial *growth*, not kill contaminants. Improper storage may create an environment which favors increases in bacterial numbers. As a result clinically irrelevant levels may become levels that can cause disease. Many organisms grow quickly at room temperature and sub-optimal refrigeration temperature highlighting the need for proper storage guidelines and concerns about growth in food residues in food bowls. [11]

Ingestion of pathogens is the first step to developing disease. Many pathogens of concern, such as *Salmonella* spp, are dose dependent and large numbers are necessary to cause disease in most hosts, particularly if the host is otherwise healthy. However, some pathogens such as enterohemorrhagic *E. coli* can cause disease at very low doses and some individuals may be more susceptible.

EVIDENCE BASED CLINICAL NUTRITION

Studies documenting nutritional inadequacies

A 1992 survey of 200 homemade diets for both dogs and cats recommended by veterinarians for diagnosis and / or management of adverse food reactions found the majority to be nutritionally inadequate.[12] Another study compared 85 published home-cooked recipes for dogs and cats to AAFCO (American Association of Feed Control Officials) standards and found 86% inadequate in at least one nutrient, 55% deficient in protein and perhaps most importantly 77% of those diets deficient in taurine. [13]

Studies documenting contamination of raw foods

One study reported the isolation of a variety of serovars of *Salmonella* spp from 45% of raw meat samples.[14] A more sensitive DNA probe detected *Salmonella* in 66% of samples. *Salmonella* spp were found in 56% of samples of meat fed to Greyhounds in another study.[15] A small study reported isolation of *Salmonella* spp from 8/10 (80%) homemade raw meat diets.[12] A study of commercial frozen raw meat diets in Ontario reported isolation of coliforms from all diets ranging from 3.5 x 103 to 9.4 X 106 CFU/g, *Salmonella* spp and *Clostridium perfringens* from 20% each, *C. difficile* from 4%, but no *E. coli* O157 or *Campylobacter* spp.[11] A larger Canadian study reported isolation of *Salmonella* spp from 22% of commercial frozen diets, spore forming bacteria from 100% and *C. difficile* from 8%. In this study, chicken diets were 4 times more likely to contain *Salmonella* spp than other types.[16] *E. coli* O157:H7 is a significant human pathogen with a very low infective dose, and it is has been identified in raw dog food.[17]

Studies documenting fecal shedding of pathogens

Joffe and Schlesinger 2002 reported isolation of *Salmonella* spp from 30% dogs fed raw chicken, but 1/10 dogs fed commercial diets.[12] Murphy reports that feeding raw meat or raw bones to dogs was associated with shedding of resistant *E. coli*. [18] A recent study evaluated *Salmonella* colonization of dogs fed commercial raw dog food that was identified as naturally contaminated with *Salmonella* spp. Seven/16 (44%) dogs fed *a single meal* of contaminated raw meat shed *Salmonella* compared to 0/12 dogs fed *Salmonella*-free raw meat. No abnormal clinical signs developed in dogs that were shedding *Salmonella*, but they shed *Salmonella* for up to 11 days. [19] Preliminary results of a prospective study of pathogen acquisition and shedding by dogs involved in therapeutic visitation programs has identified a high incidence of *Salmonella* shedding in raw meat-fed dogs, and a cluster of infected dogs that were fed meat from the same butcher. This group is of additional concern because of contact with hospitalized humans, who must be considered at higher risk of infection. [13]

Studies documenting clinical infection

In one report, Salmonella spp were isolated from feces and food samples of Greyhounds with diarrhea, leading the authors to conclude that the diet is the primary vehicle for Salmonella infection in this breed.[16] Fatal Salmonella Newport gastroenteritis and septicemia was reported in two cats fed a home-prepared raw-meat diet. An indistinguishable isolate was found in the diet, strongly suggesting it was the source.[20]

Studies documenting environmental contamination

A recent study of cleaning and disinfection practices for experimentally-inoculated food bowls reported Salmonella could be isolated from food bowls experimentally inoculated with a residue of Salmonella-containing raw meat for the entire 2 week study period. Further, this study reported that common cleaning and disinfection practices were inadequate for complete elimination of Salmonella from bowls.[21]

Studies documenting human infection

The potential public health risks associated with handling and feeding raw diets may be of greater importance than disease in animals. One study has reported an outbreak of salmonellosis in people associated with dried-beef containing pet treats, indicating the potential for human disease from *Salmonella*-contaminated pet products.[22]

RECOMMENDATIONS

There is no objective information indicating a health benefit of feeding raw diets. There is evidence to support concerns about raw diets with respect to nutritional adequacy, pathogen contamination, fecal shedding and environmental contamination. Veterinarians that recommend and / or sell raw meat diets should be consider their legal liability if either pet or human disease occurs and proper counseling of risks and infection control measures has not been performed and documented.

For those who wish to feed raw foods, careful consideration should be given to the individual circumstances. Raw food may be contraindicated in households where the implications of infection are particularly serious and situations where clinical infection is more likely to occur. In humans, infants, the elderly and immunocompromised (disease or treatment) individuals are classified into these groups. The Centers for Disease Control and Prevention (CDC) guidelines for preventing opportunistic infections among high risk groups such as hematopoietic stem cell transplant recipients and HIV infected persons state that eggs, poultry and meat products fed to pets must be well cooked.[23,24] Similarly, feeding of raw meat is probably inappropriate in situations where animals in the household are likely at higher risk (i.e. breeding operations, immunocompromised pets).

Veterinarians and veterinary technicians should be responsible for carefully explaining household infection control practices to clients. Standard guidelines for handling raw meat and prevention of foodborne infections are available.[25] Hand hygiene is the most important general infection control practice. Remind owners that hands should be thoroughly cleaned after handling raw meat. Raw meat should be stored so it will not come into contact with other food items. All items that contact the food should be disinfected after use. Cutting boards used for raw meat should **not** be used for anything else. Many bacteria can grow quickly at room temperature, so advise that raw meat should not be thawed at room temperature or allowed to sit in bowls. Bowls should be cleaned thoroughly, shortly after feeding. Any remaining food should be removed; the bowl should be scrubbed to remove debris and then disinfected. High risk individuals should not have contact with food bowls. Water bowls should be treated with similar caution. Feces should be handled with care. Fecal contamination of the environment should be cleaned promptly, with hands washed thoroughly after contact with feces.

REFERENCES

1. Notermans S, Gallhoff G, Zwietering MH, et al. (1995) The HACCP concept: specification of criteria using quantitative risk assessment. *Food Microbiology*; 12:81-90. 2. Fenlon DR, Wilson J, Donachie W (1996) The incidence and level of *Listeria monocytogenes* contamination of food sources at primary production and initial processing. *Journal of Applied Bacteriology*; 81:641-650. 3. Remillard RL. (2005) Advising clients who feed raw diets to pets. *North American Veterinary Conference Clinician's Brief*; 3:29-30. 4. Patil SR, Cates S, Morales R. (2005) Consumer food safety knowledge, practices, and demographic differences: findings from a meta-analysis. *Journal of Food Protection*; 68:1884-1894 5. Billinghurst I (1993) *Give Your Dog A Bone*. 6. Billinghurst I (2001) *The BARF Diet*. 7. Stogdale L, Diehl G. (2003) In Support of Bones and Raw Food Diets. *Can Vet* J; 44:783. 8. LeJeune JT, Hancock DD. Public health concerns associated with feeding raw meat diets to dogs. J Am Vet Med Assoc 2001 Nov 1;219(9):1222-1225. 9. Weese JS, Rousseau J, Arroyo L. Bacteriological evaluation of commercial canine and feline raw diets. Can Vet J 2005; Jun;46(6):513-516. 10. Joffe DJ, Schlesinger DP. Preliminary assessment of the risk of Salmonella infection in dogs fed raw chicken diets. Can Vet J 2002 Jun;43(6):441-442. 11. Mann JE, Smith L, Brashears MM. Validation of time and temperature values as critical limits for Salmonella and background flora growth during the production of fresh ground and boneless pork products. J Food Prot. 2004 Jul; 67(7):1389-1393. 12. Roudebush P, Cowell CS. Results of a hypoallergenic diet survey of veterinarians in North America with a nutritional evaluation of homemade diet prescriptions. *Vet Dermatol* 1992; 3: 23-28. 13. Lauten S, Smith T, Kirk C, Bartges J, Adams A. Computer Analysis of Nutrient Sufficiency of Published Home-Cooked Diets for Dogs and Cats. ABSTRACT ACVIM Forum Baltimore, MD 2005 14. Chengappa MM, Staats J, Oberst RD, Gabbert NH, McVey S. Prevalence of Salmonella in raw meat used in diets of racing greyhounds. J Vet Diagn Invest. 1993 Jul; 5(3):372-7. 15. Stone GG, Chengappa MM, Oberst RD, Gabbert NH, McVey S, Hennessy KJ, Muenzenberger M, Staats J. Application of polymerase chain reaction for the correlation of Salmonella serovars recovered from greyhound feces with their diet. J Vet Diagn Invest. 1993 Jul; 5(3):378-85. 16. Finley R. MSc Thesis, University of Guelph, 2005. 17. Freeman LM, Michel KE Evaluation of raw food diets for dogs. J Am Vet Med Assoc. 2001 Mar 1;218(5):705-709 Erratum in: J Am Vet Med Assoc 2001 May 15;218(10):1582. 18. Murphy CM. MSc Thesis, University of Guelph, 2004 19. Finley R, Ribble C, Aramini J, Vandermeer M, Popa M, Litman M, Reid-Smith R. The risk of salmonellae shedding by dogs fed Salmonella-contaminated commercial raw food diets. Can Vet J. 2007 Jan; 8(1):69-75. 20. Stiver SL, Frazier KS, Mauel MJ, Styer EL. Septicemic salmonellosis in two cats fed a raw- meat diet. J Am Anim Hosp Assoc. 2003 Nov-Dec; 39(6):538-42. 21. Weese JS, Rousseau J. Survival of Salmonella Copenhagen in food bowls following contamination with experimentally inoculated raw meat: effects of time, cleaning, and disinfection. Can Vet J. 2006 Sep; 47(9):887-9. 22. Pitout JD, Reisbig MD, Mulvey M, et al. Association between handling of pet treats and infection with Salmonella enterica serotype newport expressing the AmpC beta-lactamase, CMY-2. J Clin Microbiol. 2003 Oct;41(10):4578-82. 23. Centers for Disease Control and Prevention (CDC), MMWR 2000;49(RR10):1. 24. CDC. http://wonder.cdc.gov/wonder/prevguid/m0038328/m0038328.asp 25. CDC, http://www.cdc.gov/ncidod/dbmd/diseaseinfo/foodborneinfections_g.htm#consumersprotect 25. CDC, http://www.fightbac.org/main.cfm. 24th Annual ACVIM - 471 - 2006, Louisville, KY

CHEMOTHERAPY ADMINISTRATION-KEEPING YOU SAFE!

Kim Albin, LVT, VTS (Oncology)
DeWitt, Michigan

INTRODUCTION

Working in a veterinary healthcare environment has its hazards but nothing can be more hazardous than the handling of cytotoxic drugs. The National Institute for Occupational Safety and Health (NIOSH) has recommended that all veterinary institutions that handle cytotoxic drugs institute a program that provides appropriate protective measures for all workers who may be exposed to these drugs. (1)

Exposure

The Center for Disease Control (CDC) routinely supplies a list of drugs they consider hazardous. (2) The CDC describes hazardous drugs as being those that have specific health effects and a high toxicity at low doses. Exposure to these drugs happens throughout as the drugs travels; from handling vials, administration, disposal, touching of contaminated surfaces, or cleaning of waste and animals. (3) The most likely exposure comes from skin absorption, inhalation and ingestion. Neelesticks and other sharps injuries also pose a risk for the veterinary care provider. (4) Many of these drugs can be excreted within feces and urine for 72 hours or more, and some oral medications in emesis. (5)

Handling Hazards

NIOSH has provided some safe handling policies and procedures for veterinary workers. These include: Hazardous drugs only prepared and administered by designated personnel in designated areas only. Posting signage to warn all of the hazardous drugs being handled in the area. Warn pregnant, breastfeeding and other members of reproductive age of the health risks especially during the first trimester. Documentation of worker training and understanding of all procedures. (1) Since DNA changes are shown to occur with exposure to cytotoxic drugs, care should be taken to inform all pregnant, and nursing mothers of the risk to their unborn child and infant. It is recommended that these mothers sign an informed consent of their acknowledgment of hazards and risks associated with them.

All hazardous drugs should be handled at all times with personal protection equipment (PPE) which includes a double layer of chemotherapy gloves, non-permeable gown with cuffs, N95 mask, face splash protection and shoe covers. Storing of these drugs should be separate from all other drugs; where no food or drink is allowed, and with spill kits within close vicinity. (1) OSHA requires that all employees be fit tested for respirator masks. (6)

When dealing with preparation of cytotoxic drugs, great care must be taken. NIOSH has very specific guidelines. Preferably a negative pressure clean room for all handling, storage, and administration of drugs. No eating, drinking, chewing of gum or application of cosmetics in the drug preparation area. Use of PPE throughout entire procedure. Spill kits and under pads used. Use of a containment device, preferably a 100% vented biological safety cabinet. Horizontal laminar flow chemotherapy hoods do not protect the worker. Use of a closed-system drug transfer device. Properly clean all surfaces that may have come in contact with drugs. There is not universal cleaner but the steps of: Ph4 acid soap, water, Ph10 acid soap, water, isopropyl alcohol should provide drug removal for most drugs. Always wash hands with soap and water after each gloving. (1)

Administration and Waste

Administration of these drugs is recommended within the clean room. (1) If this is not possible, then each animal should be administered drug in their own dedicated cage/kennel with it's own drainage area. (1) Always use PPE and proper technique. Always prime IV tubing with saline prior to attaching closed-system with drug. Never disengage IV tubing once it has been attached. Leave the IV line closed and intact for disposal. All waste is to go into an appropriate chemotherapy waste receptacle. Always discard all PPE after each use. Do not keep gowns and masks for future use. Cytotoxic drugs will remain on the gown. Always wash hands, and clean area of administration with above cleaning steps.

Spills

One person trained for dealing with hazardous drugs should always handle spills. Always take the time to apply all layers of PPE. All spill materials go into a chemotherapy waste receptacle.

Medical Monitoring

It is recommended that all institutions conduct general health and reproductive questionnaires to their employees. This should be done once a year along with a physical examination, CBC, reticulocyte count, WBC and hormone levels. Any exposure or health changes should prompt an examination of the employee and additional blood work. (1)

Final Thoughts

Safety can be achieved through diligence and knowledge. Knowledge is power, and we can all be powerful if we take the time and care.

REFERENCES

1. Workplace Solutions from NIOSH. Publication No. 2010-150 2. NIOSH List of Antineoplastic and Other Hazardous Drugs in Healthcare Settings 2010. Publication No. 2010-167 3. Meijster T, Fransman W, Veldhof R, Kromhout H (2006). Exposure to antineoplastic drugs outside the hospital environment. Ann Occup Hyg 50: 657-644. 4. NIOSH Preventing worker deaths and injuries when handling mannitol. Publication No. 2007-124. 5. Cass Y, Musgrave CF (1992). Guidelines for the safe handling of excreta contaminated by cytotoxic agents. AM J Hosp Pharm 49: 1957-1958. 6. OSHA 29 Code of Federal Regulations 1910.134

ELECTROCHEMOTHERAPY: THE WAVE OF THE FUTURE!

Chelsea D. Tripp, DVM, MS, DACVIM (Oncology)
Seattle, WA

DEFINITIONS

Electropulsation – *the direct delivery of electric pulses to cells* [1]

Electropermeabilization/Electroporation/Electrochemotherapy – *using exogenous substances (chemotherapy, gene transfer, DNA, etc) to enter cells after electropulsation* [1]

INTRODUCTION

Electropulsation has received considerable attention in the last 15 years as an emerging way of delivering therapies to different tumor types. [2] In this discussion, we will focus on its use with chemotherapy. Gene transfer and vaccines are also currently being investigated but are not part of this discussion.

When a cell is exposed to sufficiently large electric field, its plasma membrane is changed so that molecules that normally cross the membrane only in minute amounts can more readily pass through the barrier. [3] This phenomenon was first described in 1972. [4] Almost a decade later, gene transfer was achieved through this method of electropulsation. [5] This technique is now widely used as a method for introducing foreign genes into cells in vivo, electrofusing cells to tissues, increasing the uptake of certain drugs by cells and infusing cells with monoclonal antibodies. [6]

Electrochemotherapy can be used as a single modality treatment for gross disease or as an adjuvant treatment in humans and our pet species. It is an anti-cancer treatment that couples local administration of chemotherapy drugs to the delivery of trains of electric pulses having a square or biphasic waveform. The application of these pulses allow for alterations in the cell membrane through an electrically mediated reorganization of the plasma membrane such that the chemotherapy drug and flow down its concentration gradient. Ultimately this leads to cancer apoptotic death. [7]

Electropulsation can be used in all kinds of isolated cells as well as in tissues. The electric field to which the target cells are exposed to has to be of sufficient strength and duration. The magnitude of the electric field to be used depends on a variety of factors; the cell type, size, orientation, density, pulse duration and number of pulses. The selection of the pulse parameters is influenced by the size and type of molecule that is to be internalized. Future strategies are aimed at the treatment of cancer based on direct effects of pulsed electric fields on cells, independent of the induced influx of material. This would include tumor ablation but irreversible electropulsation by the application of intense nano-second-duration pulses which could induce direct apoptosis and destruction of the tumor vasculature. [8]

Chemotherapy is widely used for a broad range of cancers. In many instances, the response rates are low. This is because the drug cannot reach its intracellular site of action. The cell membrane is often the barrier that reduces the effectiveness by restricting drug access. The two drugs of choice are bleomycin and cisplatin; as they both are relatively large molecules and poorly permeable on their own. These drugs are then permeabilized and driven across the membrane by simple diffusion. The structural alterations that happen to the membrane are able to reseal through metabolic processes.

It has also been further demonstrated that endothelial cells are sensitive to bleomycin and cisplatin delivered by electropulsation. The cytotoxic effect is not necessarily limited to the cancer cells. Stromal cells, including endothelial cells of blood vessels can also be disrupted leading to endothelial cell death, disruption of tumor blood flow, and consequent death of the tumor cells surrounding the vessels. This may help explain the vascular-disrupting effect of electrochemotherapy.

Histologic evaluation of the tissues after treatment reveals a large detectable inflammatory reaction around and inside the treated regions. Macrophages and lymphocytes are usually present in large numbers as well as eosinophilic infiltration is possible within the first few days of treatment. Appearance of the tumor cells changes into enlarged cells with pyknosis and cytoplasmic vacuolization, indicative of massive necrosis. [7]

Preclinical trials and trials in humans have demonstrated excellent anti-tumor effectiveness of electrochemotherapy on various tumor types, minimal toxicity and safety of the procedure. Veterinary studies have demonstrated that electrochemotherapy is very effective in the treatment of cutaneous and subcutaneous tumors of different histologic subtypes in dogs, cats and horses. [1] [9] [10] [11] [12]

The process does require heavy sedation but is overall well tolerated and many tumors can be treated in 1-3 treatment sessions. Some tumors may require additional future treatments to maintain remission. The advantages of electrochemotherapy are that it is easy to perform on an outpatient basis, low cost and requires only one or two sessions of heavy sedation to short anesthesia versus standard full course radiation therapy.

Types of tumors to treat:

The following are examples of tumor types which are amenable to electrochemotherapy. In general, tumors need to be superficial or easy to access the tumor bed to be successful with treatment:

- Fibrosarcoma in the rostral mouth, eyelid, foot pad or distal limb
- Melanoma in the mouth, eyelid or foot pad. Digit should be resected first then treated
- Squamous cell carcinoma in mouth, eyelid, ear, nasal planum or footpad. Digit should be resected first then treated
- Incompletely resected soft tissue sarcomas where aggressive surgical treatment would necessitate amputation
- Feline vaccine-associated fibrosarcoma after surgery
- Localized cutaneous lymphoma in dogs or cats
- Low to intermediate incompletely resected low grade mast cell tumors on distal limb, trunk and head
- Perianal & rectal tumors (when can be externalized)
- Operation site margins where there is a significant risk of remaining tumor tissue
- Sarcoids and squamous cell carcinomas in horses
- Superficial tumors on exotics (small mammals, birds & zoo animals)

REFERENCES

1. Cemzar M., et al. **JVIM** 2008; 22, 826-831. 2. Spugnini EP, et al. **J Exp Clin Cancer Res** 2005; 24,245-54. 3. Miklavcic D, et al. **J Membrane Biol** 2010; 236,1-2. 4. Neumann E, et al. **J Membr Biol** 1972; 10, 279-290. 5. Neumann E, et all. **EMBO J** 1982; 1, 841-845. 6. Heller R, et al.,**Cancer,** July 1, 1998; 83, 148-157. 7. Spugnini EP, et al. **Journal of Trans Med** 2007; 5 8. Nuccitelli R, et al. **Biochem Biophys Res Commun** 2006; 343, 351-360. 9. Spugnini. EP, et al. **In Vivo** 2007; 21, 819-822. 10. Spugnini EP, et al.. **JVIM** 2011; 25,407-411. 11. Spugnini EP, et al. **Cancer Chemother Pharmacol** 2007; 59, 375-381. 12. Mir LM, et al. **British J of Cancer** 1997; 76(12), 1617-1622

THE ROLES FOR RADIATION IN CANCER THERAPY

Laura D. Garrett, DVM, DACVIM (Oncology)
Urbana, IL

MECHANISM OF ACTION OF RADIATION THERAPY

Radiation causes cell death or sterilization via ionization and thus damage to cellular DNA. Photon particles can directly ionize and thus damage cellular DNA. Two-thirds of radiation-induced damage is indirect, via the ionization of water molecules into free radicals that then damage the DNA. Interaction of these free radicals with oxygen leads to greater damage.

Factors affecting radiosensitivity

Both normal and tumor tissue are susceptible to DNA damage from radiation. Tumor related factors that affect response to ionizing radiation include doubling time/growth fraction, blood supply (oxygenation), size, location and histologic type. Timing and dose of radiation delivered depends upon four factors:

Repair:

Both normal and tumor cells can repair themselves after exposure to radiation. Slowly responding tissues have a greater capacity for repair than rapidly responding tissues (such as tumors).

Repopulation:

This refers to the replacement of cells killed by radiation by division of cells not damaged by the radiation. Both tumor and normal cells may be stimulated to undergo mitosis to aid in replacement of cells that have been killed. The longer the radiation treatment course, the more repopulation may inhibit tumor control.

Redistribution:

Radiation is most effective at killing cells in M and G2 phases of the cell cycle. Cells that are in more resistant phases (S, G1) during exposure to radiation will eventually redistribute into M or G2 and become sensitive to radiation damage. This is one reason why multiple doses of radiation are necessary (or chronic exposure, in the case of brachytherapy).

Reoxygenation:

Hypoxic cells are resistant to radiation therapy (see above MOA). As sensitive, oxygenated tumor cells die off, the hypoxic cells become oxygenated and thus become sensitive to radiation therapy. The fact that oxygenated tumor cells are more sensitive to RT is important for treatment planning. Large, bulky tumor masses do not respond as well to RT as microscopic (post-surgical) tumors do.

SIDE EFFECTS

While all proliferating cells are inherently sensitive to radiation therapy, tumor cells are less able to repair themselves following exposure to ionizing radiation. This difference in repair ability provides a therapeutic window for radiation treatment; however, it is expected that normal cells within the radiation therapy field will have side effects. The severity of the effects on normal tissue depends somewhat on the type of radiation used and on the tissue irradiated. These effects can be broken into acute and chronic categories.

Acute side effects

Acute effects are seen in rapidly dividing tissues, most often epithelial. Fraction size and overall treatment time influences acute side effects development. These effects arise towards the end of RT, and may actually get worse after the completion of therapy (like a sunburn gets worse for the next day or two post sun-exposure). Reactions include moist desquamation, oral mucositis, keratoconjunctivitis sicca (KCS) and alopecia. Acute effects are not dose-limiting; in other words, the severity of the acute effect will not help determine the total dose of radiation. Acute effects are temporary, self-limiting and do not affect quality of life in the long run. Management of moist desquamation includes gentle cleansing of the wound (vigorous cleaning will wipe away healing epithelium). Oral mucositis may inhibit eating; oral rinses with (cool) black tea or viscous lidocaine solutions may be helpful. Rarely, in extreme cases, placement of a feeding tube may be necessary until the oral cavity has healed (2 to 4 weeks after completion of therapy)

Chronic (late) side effects

Late effects can occur in epithelial or mesenchymal tissues. Late effects are influenced by the dose per fraction but not the overall treatment time. Timing for these effects is 6 months to years after completion of RT. These adverse effects range from permanent hair loss to KCS and cataract formation to bone or CNS necrosis. Because these problems can affect long term quality of life and even be life threatening, radiation therapy protocols are designed to minimize these late effects.

RADIATION THERAPY SOURCES

Ionizing radiation can be delivered by various sources; the most well investigated and frequently used in small animal patients is external beam radiation, or teletherapy. Cobalt and linear accelerators are the radiation machines used most often in veterinary medicine. High-precision radiotherapy machines that use computer-controlled linear accelerators to deliver intensity-modulated radiation therapy (IMRT) are becoming available for our veterinary patients, as are TomoTherapy units, which deliver IMRT via a linear accelerator and concurrent helical CT scanning.[1, 2] Osteosarcoma is one tumor type that has been treated with some success with IMRT.[3]

GOALS FOR RADIATION THERAPY

The goal of radiation therapy may be either curative or palliative. Curative, or definitive, therapy involves many small frequent doses (fractions) of radiation to the affected area. Tumors may be treated in this manner to address local microscopic disease that exists beyond the surgical field (either post-operatively, "adjuvant therapy", or pre-operatively, "neo-adjuvant therapy") or for certain macroscopic diseases such as brain and nasal tumors. Palliative therapy involves only a few large fractions administered less frequently. The goal of palliation is to make the patient more comfortable, not to put the tumor into remission. One tumor, however, in the gross disease setting responds better to the large fractions than to small frequent ones; that tumor is canine oral melanoma (80% response rate, 60% complete remission). Thus, for melanoma the large, infrequent fractionation is considered a locally "definitive therapy", not "palliative".

DEFINITIVE THERAPY

Microscopic disease

In dogs, radiation is often curative for microscopic disease remaining post-operatively from low to moderate grade soft tissue sarcomas (90% 2 year control rate, 75% 5 year survival) and mast cell tumors (85-95% 2 year control rate). The time to treat these cases is immediately after the histopathology comes back with "dirty margins", not after multiple recurrences and surgeries. In cats, unfortunately, vaccine-associated sarcomas with residual microscopic disease do not respond as well to radiation as soft tissue sarcomas do in dogs. Studies have shown that radiation extends the tumor free interval in these cats, but it is not likely to be curative as the tumors may recur both in and out of the radiation field. Residual microscopic disease for carcinomas can also be effectively addressed with radiation therapy, including such sites as perianal, anal sac, ear canal, salivary, and cutaneous. Round cell tumors such as lymphoma, plasma cell tumors, and transmissible venereal tumors are very radiosensitive and can be locally cured with lower doses of radiation than sarcomas or carcinomas. In some instances, these tumors respond to the large dose, infrequent fractionation scheme similar to melanomas.

Macroscopic disease

For macroscopic disease, radiation is the treatment of choice in only a few instances.

Nasal tumors:

The standard of care for nasal tumors is external beam radiation as single agent therapy. Surgical treatment alone does not increase survival times and may in fact decrease them. Limited studies have shown that surgery combined with radiation for nasal tumors may be of benefit in some cases; further study is needed. At this time, single agent radiation is the standard of care, but although radiation is given in a "curative" manner, few cures occur – the goal is improvement in clinical signs and better quality of life. A recent study of 139 dogs with nasal carcinoma that were NOT treated showed a median survival of only 95 days. With radiation therapy, survival times range from a median of 8-14 months, depending on the study.[4] Roughly 50% of dogs with nasal carcinomas or sarcomas are alive at 1 year with roughly 25% alive at 2 years. Cats have similar survival times with radiation therapy for nasal carcinomas. Nasal lymphoma in the cat responds very well to radiation, with median survival times of 2-3 years, also depending on the study.[5, 6]

Brain tumors

Brain tumors in the dog have been treated with a range of therapies, including symptomatic treatment, chemotherapy, surgery, and radiation. Dogs that receive radiation (either as single agent therapy or with other treatments) have increased survival times over those that do not, with medians ranging from 160-700 days (versus <1 month without radiation). Studies that evaluated single therapy radiation for non-surgical brain tumors in dogs revealed median survival times of 250-700 days, with the longer time from a more recent study.[7] This improvement may be due to earlier detection and treatment of tumors, as well as improvement in radiation treatment planning and delivery. Additionally, neurosurgery is being performed with greater frequency in our veterinary patients, especially with meningiomas, which tend to be located superficially in the brain. A study in dogs with meningiomas showed a median survival of 7 months with surgery alone and 16.5 months when radiation was used post-operatively.[8] In another series of cases treated with radiation post-operatively, the 2 year progression-free survival was 68%.[9] Cats with meningiomas have extended survival times with surgery alone; radiation does not appear to be needed in this patient population.[10, 11]

Pituitary tumors

Pituitary tumors in the dog and cat may also be helped with radiation therapy. In dogs with macroadenomas treated with radiation, the MST was not reached, vs. 359 days in the untreated group. Also, the 1, 2, and 3 year estimated survival was 93, 87, and 55% in the treated group, vs. 45, 32, and 25% in the untreated dogs. With radiation therapy, 1/3 of the dogs with hyperadrenocorticism had their endocrine problem resolve, and about 2/3 of the dogs with neurologic signs showed improvement of those signs.[12] Cats with pituitary tumors can also respond well to radiation, with average survival times of 15-17 months (most cats died of non-tumor related illness). The majority of the cats with endocrine disorders secondary to their pituitary tumor (e.g. Cushing's disease, acromegaly) showed improvement in their endocrine status.[13, 14]

PALLIATIVE THERAPY

Palliative radiation can be used in an attempt to increase patient comfort with virtually any neoplasm.[15] The tumor type most often treated in this manner is appendicular osteosarcoma in the dog. The precise fractionation scheme varies depending on the institution, but typically these dogs receive radiation given once weekly for a total of 3 or 4 doses.[16] Two doses of radiation given over two consecutive days, which may be easier for owners who have to travel, seems to be equally effective.[17] Pain relief occurs for two reasons: rapid relief (within 24 hours) may be seen due to decreased pain mediator production by normal bone cells, with a slower onset in relief due to some re-ossification of the tumor site. Improvement in clinical signs is seen in 70-90% of patients, with a median duration of 2.5-4 months. Some patients may be retreated with a similar protocol if pain recurs locally. Bisphosphonate treatment, particularly zoledronate, to improve palliation provided by radiation is being investigated and appears to hold exciting potential; studies are ongoing.[18]

REFERENCES

[1]Hunley DW, et al., **Can Vet J** 2010; 51(3): 293. [2]Lawrence JA, et al., **Vet Radiol Ultrasound** 2010; 51(5): 561. [3]Coomer A, et al., **Vet Comp Oncol** 2009; 7(1): 15. [4]Yoon JH, et al., **Res Vet Sci** 2008; 84(1): 140. [5]Sfiligoi G, et al., **Vet Radiol Ultrasound** 2007; 48(4): 388. [6]Haney SM, et al., **J Vet Intern Med** 2009; 23(2): 287. [7]Bley CR, et al., **J Vet Intern Med** 2005; 19(6): 849. Axlund TW, et al., **J Am Vet Med Assoc** 2002; 221(11): 1597. [9]Theon AP, et al., **J Am Vet Med Assoc** 2000; 216(5): 701. [10]Gallagher JG, et al., **Journal of the American Veterinary Medical Association** 1993; 203(10): 1437. [11]Gordon LE, et al., **Vet Surg** 1994; 23(2): 94. [12]Kent MS, et al., **J Vet Intern Med** 2007; 21(5): 1027. [13]Mayer MN, et al., **J Vet Intern Med** 2006; 20(5): 1151. [14]Sellon RK, et al., **J Vet Intern Med** 2009; 23(5): 1038. [15]Gieger T, et al., **J Am Anim Hosp Assoc** 2008; 44(3): 116. [16]Green EM, et al., **J Am Anim Hosp Assoc** 2002; 38(5): 445. [17]Knapp-Hoch HM, et al., **J Am Anim Hosp Assoc** 2009; 45(1): 24. [18]Fan TM, et al., **J Vet Intern Med** 2008; 22(2): 380.

CONTINUOUS RATE INFUSION: DRUGS AND CALCULATIONS

Meri F. Hall, RVT, LATG, LVT, CVT, VTS (SAIM)
Athens, PA

There are many medications that are infused at a constant rate. We refer to them as CRI's. There are many formularies available for CR's. Many CRI's are used for pain management, cardiac support and administration of insulin.

This method administers a precise calculated amount at a constant rate. The efficacy of the drug and the total daily dose may be significantly reduced. For medications that are ultra short in duration, a loading dose (LD) is commonly administered before beginning a CRI.

Most CRI doses are recorded as micrograms per kilogram body weight per minute. The calculation of the drug dose must be performed with the utmost of accuracy. For the most accurate administraton, the use of a syringe pump is recommended since the drug is not diluted prior to use. If a syringe pump is not available, the drug may be combined with a compatible diluent. This is commonly 0.9% NaCl, although other intravenous fluids may be used.

If a syringe pump is not used then the following calculations must be used. The length of time a fluid bag will last at the desired rate of infusion is determined by dividing the volume of fluid by the rate.

$$Volume\ in\ fluid\ bag\ in\ ml \times {}^{per\ hour}/_{fluid\ administration\ rate} = length\ of\ infusion\ in\ hours$$

For some medications the dose is by the minute, so the number of hours of the fluid in the bag will last by 60 to determine the number of minutes the bag will last.

$$Length\ of\ infusion\ in\ hours \times {}^{60\ minutes}/_{hr} = length\ of\ infusion\ in\ minutes$$

If the dose is by the day, divide the number of hours the bag will last by 24 to determine the number of days in each bag.

$$Length\ of\ infusion\ in\ hours \times {}^{day}/_{24\ hours} = length\ of\ infusion\ in\ days$$

Most drugs are dosed in mg or µg per kg body weight.

$$wt\ in\ kg \times {}^{mg}/_{kg} = dose\ in\ mg$$

If the drug dose is mg/kg/hr, multiply the dose by the hours each bag will last. The result is the amount of the drug in mg to add to the bag.

$$dose\ in\ mg \times length\ of\ infusion = {}^{mg}/_{liter}$$

For drugs dosed by µg/kg/minute, multiply the dose in µg by the number of minutes the bag will last. Divide this number by 1000 to convert to milligrams (since most medications have concentrations of mg/ml). The result is the number of µg to add to the bag.

$$dose\ in\ \mu g \times length\ of\ infusion\ in\ minutes \div {}^{mg}/_{1000\ \mu g} = dose\ in\ mg$$

For the most accurate dosing, the amount of the drug to be added should be removed from the base fluid.

BULLYING AND LATERAL VIOLENCE AMONGST NURSES AND TECHS

Lisa Allison, BS, LVT, CVT, VTS (Oncology)
Heber City, UT
Joshua Estrin, MS, LCSW, CMFSW
Plantation, FL

The behavior goes by many names: lateral or horizontal violence, incivility, sabotage, bullying, "nurses eating their young", vertical aggression. Regardless of what we choose to call it, the behavior is on the rise within the workplace and creates not only an unpleasant environment but a dangerous one.

What do these behaviors look like? Lateral violence may be overt and easy to spot. It can be name calling, bickering, criticism, gossip, yelling, finger pointing (blame), or negative facial expressions. Covert forms are more difficult to notice: unfair assignments, refusing to help a coworker or work with certain coworkers, sabotage, exclusion, ostracism, or lying. These examples are by no means exhaustive. In general, the acts "are a consistent pattern of behavior designed to control, diminish or devalue a peer (or group) which creates a risk to health or safety".

Are nurses particularly susceptible to lateral violence versus other professions? And if so, why? While no comparative studies are to be found, there is a strong belief this may be very prevalent within the profession.. "One of the real challenges is that most nurses are being worked off their feet. Lateral violence may be due to nurses displacing stress and aggression on one another." Furthermore, when nurse are viewed as a group in the entire context of a hospital's hierarchy we may see yet another reason for the inter-group conflict. "Nurses generally don't have sufficient control over their work environment and have a high degree of accountability coupled with a low degree of autonomy. When nurses don't have control but must be accountable, you can see where they might not be happy with one another."

How do we prevent it? We must first admit lateral violence exists. It is not "life as a nurse" nor a rite of passage for new graduates or team members. A "zero tolerance policy related to disruptive behavior, including a professional code of conduct and educational and behavioral interventions to assist nurses in addressing disruptive behavior" should be implemented within the workplace. Prevention and promotion of a culture of safety are of primary importance. Policies that swiftly and definitively address transgressions must be enacted, while at the same time support is provided to any individual harmed by the interaction. Bullying creates a dangerous work environment for our patients and ourselves. Failure to communicate effectively can cause life threatening errors. By tolerating or participating in these behaviors we risk our patients' safety and therefore violate our ethical code.

A survey was conducted within the nursing profession during the Spring of 2011. 955 members responded via the internet or paper hardcopy. The majority of those participating were women (879/50). However. it was interesting to note that men experienced higher incidences of lateral violence than did their female counterparts. Age and level of education did not appear to significantly affect the incidence of experience or witnessing of these behaviors. The lecture will look more deeply into these findings and discuss the conclusions as well as work to create a template for the workplace environment to foster awareness and create preventative strategies.

REFERENCES

Farrell G.**Journal of Advanced Nursing** 2005; 35(1):26-33, Pugh A **Cross Currents** 2006, Center for American Nursing 2008, Dumont C. Nursing 2012;42(1)44-49.

TOP TEN MISTAKES MADE WHEN ADMINISTERING MEDICATIONS TO ANIMALS

Mark G. Papich, DVM, MS, DACVCP
Raleigh, North Carolina

INTRODUCTION

In small animal therapeutics there are many legends, out-dated treatments, misconceptions, misleading advertising, and myths that persist. Data is now available through pharmacologic studies to dispel many of these old thoughts and concepts. For the instances when there is not specific data to prove or disprove the value or harm of a drug treatment, we can make some educated predictions based on our evaluation and interpretation of experimental data, even when it has been derived from human or laboratory animal studies. In addition to dispelling misconceptions, there are other mistakes made when prescribing medications to animals that should be addressed and corrected. Some of these mistakes and misconceptions will be discussed in this presentation.

This presentation will focus on drug treatments. Time will not allow an exploration of one of the largest areas for which there is controversy, and perhaps unproven treatments: alternative and complimentary medical therapies, holistic treatments, and the use of dietary supplements ("nutriceuticals").

#1: NOT COMMUNICATING DRUG PRESCRIBING INSTRUCTIONS TO OWNERS

Veterinarians and their staff often assume that pet owners understand the jargon and shorthand used when prescribing medications. To some pet owners, "give medication three times a day" may mean 8:00am, 12:00 noon, and 3:00pm. Instructions to give the medication "every 8 hours" is more accurate and ensures that the dose interval will produce the best therapeutic outcome. Likewise, abbreviations such as "sid", "bid", and "tid" are often misunderstood. A dose interval in hours in plain language is more easily understood and will avoid errors. The abbreviation "sid" is not an accepted pharmacy abbreviation and has been confused with 5/d (five times a day) or other intervals on a hand- written prescripton. Many pet owners do not understand "per os" and it is much better to simply say "by mouth". (Owners have been known to insert medication tablets in the rectum when given the instructions "per os".)

Proper storage of medications (eg, insulin and other injectable products) is important to the drug's activity. Do not assume that pet owners understand how to give medications, when to give them (with food, without food, in the morning, in the evening), and techniques for administering medications. Many cat owners have walked out of the veterinary hospital with a prescription for their cat for oral capsules wondering how they are to manage this feat. A few minutes rehearsing administration of the medication with the pet owner before they leave the hospital will ensure that the pet is receiving the medication properly.

#2: ASSUME THAT COMPOUNDED DRUGS EQUIVALENT TO PROPRIETARY DRUGS

Compounding is the practice of modifying a dose form in order to facilitate administration to a patient. Compounding can be a therapeutic necessity because we lack the availability of medications to treat the variety of diseases in our patients. For example, human dose formulations must be modified to administer to a dog, cat or exotic animal. Sometimes compounding is performed to ease administration, mask a bitter taste, or accommodate an animal's small size. Compounding is allowed by federal law, provided that it is performed on a patient by patient basis and other restrictions are met. The FDA Center for Veterinary Medicine provides guidance on what is legally allowed on their web site. Individual states may have pharmacy legislation that further restricts some forms of compounding.

Compounded formulations may provide an equivalent therapeutic response to a proprietary formulation in many instances. However, there may be problems with solubility, stability, and potency for some drugs (Papich, 2005). The inactive ingredients and excipients added to drug formulations are done so to ensure the stability of the drug, provide an optimum chemical environment, pH, or increase the ease of packaging or handling. Adding other chemicals, flavorings, vehicles, or interfering with protective coatings of tablets in the course of compounding may interfere with the stability of the drug, decreasing its potency, oral absorption, and efficacy. The most common interaction is that from a change in pH. When veterinarians compound formulations in their own practices, or write an order for this to be done in a pharmacy, they should be cognizant of the potential interactions and alterations that may compromise the stability and potency of the active ingredient. Drugs known to be compromised when compounded for animals include omeprazole, fluoroquinolone antimicrobials, diazepam, and antifungal drugs (eg, itraconazole). Itraconazole, for example is unstable and insoluble. Compounded formulations do not produce equivalent blood concentrations as the proprietary formulation of Sporanox. Fluoroquinolone are subject to chelation with cations (eg, Fe^{+3}, Ca^{++}, Al^{+3}, Mg^{+3}) that may be added as supplement. The chelation will inhibit oral absorption.

#3: YOU DON'T PAY ATTENTION TO OBVIOUS SIGNS OF DRUG INACTIVATION

There may be clues that compounding has affected drug quality, purity, or potency. Oxidation is often visible through a color change (color change to pink or amber for example). Loss of solubility may be observed through precipitation. Some drugs are prone to hydrolysis from moisture. A rule-of-thumb for veterinarians is that if a drug is packaged in blister packs or moisture proof barrier, it is probably subject to loss of stability and potency if mixed with aqueous vehicles. If compounded formulations of solid dose forms show cracking or "caking", or swelling, the formulation has probably accumulated moisture and may have lost potency. Another rule-of-thumb is that if the original packaging of a drug is in a light-protected or amber container it is probably prone to inactivation by light. Vitamins, cardiovascular drugs, and phenothiazines are labile to oxidation from light during compounding. Also, as a general rule, if an antibiotic is available in a powder that must be reconstituted in a vial or oral dispensing bottle prior to administration, it should not be mixed with other drugs.

#4 ASSUME THAT DRUGS ARE ABSORBED THROUGH THE SKIN

In reality, only a small number of drugs are absorbed through the skin. But, to meet the growing demands for more transdermal medications for pets, compounding pharmacies have prepared existing drugs (both human and veterinary drugs), into transdermal formulations. Most drugs cannot be absorbed across the skin without some enhancement. Their lipophilicity or solubility characteristics otherwise prevent the drug from penetrating the skin. Therefore, veterinary compounding pharmacies have mixed drugs with "penetration enhancers" (PE) to facilitate transdermal absorption. The most popular of the PE used by veterinary pharmacists is pleuronic lecithin organogel (PLO), which is lecithin (derived from eggs or soybeans) mixed with isopropyl palmitate and a poloxamer (Pluronic). The ingredients in PLO act as surfactants, emulsifiers, and solubilizing agents. There is little data available to suggest that drugs applied in a PLO vehicle are actually absorbed systemically in animals. Most published reports of transdermal application of drugs to cats showed that absorption was incomplete, nonexistent, or highly inconsistent among cats. Drugs examined so far have included glipizide, dexamethasone, buspirone, amitriptyline, fentanyl, morphine, fluoxetine, and diltiazem. While in a pharmacokinetic investigation, methimazole was shown to be poorly absorbed (Hoffman, et al 2002), clinical investigations provided evidence of efficacy with repeated transdermal applications (Sartor et al, 2004; Hoffman et al 2001). Other drugs formulated in a PLO which have been poorly absorbed are morphine, fentanyl, and enrofloxacin.

Drug absorption through the skin is actually more challenging than expected. The barriers of stratum corneum, hair, and first-pass metabolism limit this route. In addition, some drugs may not be soluble or compatible in a transdermal vehicle. Many drugs are not potent enough to be delivered in the small volume (often 0.1 mL) required for practical transdermal therapy.

#5 ASSUME IT IS UNSAFE TO ADMINISTER DRUGS TO PEDIATRIC PATIENTS

Actually, for many drugs the dose may need to be *increased* in pediatric patients compared to adult animals. Neonates have a higher percentage of body water and decreased drug protein binding compared to adults. Total body water in a neonate may be as high as 80-85%, compared to 55 to 60% in an adult. The most significant change in body composition is in the relative volume of extracellular water (ECF), which may be twice as high in neonates compared to adults. This translates to a larger volume of distribution (V_D) and greater dilution for water-soluble drugs. Examples of antibiotics affected include aminoglycosides, penicillins, and cephalosporins.

Both renal and hepatic drug clearance can be high in young animals. Kidney function reaches adult values rapidly in most animals. Usually, by 4 weeks of age, most animal's kidneys can excrete drugs at near-adult capabilities. Hepatic clearance of drugs also matures quickly. There is a 4 to 6 week period of development of the function of microsomal drug metabolizing enzymes following birth. At one month of age the metabolizing capabilities begin to show rapid increase in development and by six weeks of age they are nearly at adult levels. After the age of approximately 6 weeks, there is no rationale for altering drug doses in animals to account for hepatic clearance.

#6 ASSUME IT IS UNSAFE TO ADMINISTER DRUGS TO GERIATRIC PATIENTS

Pets seem to be living longer, and veterinarians are treating more age-related problems in pets. Unfortunately, there are few studies published in the veterinary literature to document changes in drug disposition in geriatric animals compared to younger adults, and practically nothing available to guide veterinarians on drug administration to older animals compared to younger adults. Recommendations published in clinical reviews are based on extrapolations from studies in people, or the *assumption* that drug disposition probably is different in aged animals. Since organ function declines with age, we presume that drug clearance is affected.

Older pets may have other concurrent diseases that may complicate drug therapy. They are also more likely to be receiving other drugs chronically for treatment of a disease compared to an older pet. These drugs can increase the likelihood of adverse reactions or produce a drug interaction. However, old age alone is not a reason to reduce the drug dose, alter the dosing interval, or withhold potentially helpful medications. If there is accompanying organ failure (liver, kidney) dose adjustments may be necessary if these organs are used to clear the drug from the body. If multiple other drugs are used, adjustments may be needed to avoid an interaction. However, *age alone* is not a good reason to adjust the regimen.

#7 FAILURE TO TREAT PAIN AGGRESSIVELY

Some veterinarians are reluctant to administer analgesic drugs because they are (1) not familiar with the drug's pharmacology and use, (2) not willing to use medications controlled by the DEA, (3) concerned that the analgesic will mask other clinical signs that should be monitored, or (4) concerned about adverse effects. In reality, none of these are good reasons for not providing patients with adequate analgesic treatment. The medications available such as the approved nonsteroidal anti-inflammatory drugs (NSAIDs) are safe in most patients as long as prescribing recommendations are followed. The most serious problem is gastrointestinal ulceration and perforation, which is actually rare when these medications are used according to manufacturer's recommendations. Acute kidney disease, another concern, can be avoided of the patient's renal function is monitored and they are not dehydrated. Opiates, another important group of analgesic drugs are extremely safe. They have high margins of safety and can be reversed if problems arise. They can be administered with a wide range of other medications without risk of interactions. There are *very few* contraindications to using an opiate. Instead, many veterinarians rely on unproven treatment such as tramadol, transdermal medications, topical treatments, and some alternative therapies.

#8 BELIEVING EVERYTHING YOU HEAR ABOUT NSAIDS

The nonsteroidal anti-inflammatory drugs (NSAIDs) all have as their most common adverse effect, gastrointestinal (GI) toxicity. Common reactions are vomiting, diarrhea, and nausea. The more severe reactions are GI ulcers and perforation and acute kidney disease. Starting in the late 1990s, we became convinced through developments of new drugs for people that the COX-2 inhibitors ("Coxibs"), drugs specific for the COX-2 isoenzyme, may be safer than traditional therapies. Then, in late 2004 studies revealed – again from human medicine – that these drugs may be unsafe. Two of them (rofecoxib and valdecoxib) later were withdrawn from the market because of concerns for adverse cardiovascular events.

There are now several veterinary-approved COX-2 inhibitors. There is no convincing published data to indicate that these drugs are safer, or more effective than NSAIDs that are less selective for the COX-2 enzyme. Most studies are performed using a "non-inferiority" analysis, which indicated that they were simply *noninferior* to other medications. In veterinary medicine, it is difficult to compare efficacy and gastrointestinal safety among drugs because the numbers of patients enrolled in clinical studies are relatively small. Most comparisons have come from evaluations from limited clinical trials used for drug registration. These trials are available from the drug sponsor and, in general, have shown that the incidence of common GI signs such as vomiting, nausea, and diarrhea are not much different among the NSAIDs.

#9 BELIEVING THAT YOU CAN PERFORM ANTIBIOTIC SUSCEPTIBILITY EASILY IN YOUR HOSPITAL LABORATORY

Reliable antimicrobial susceptibility can only be performed using approved clinical standards for testing. The most widely accepted standard for susceptibility testing is from the Clinical Laboratory Standards Institute (CLSI, 2008). The CLSI sets standards and strict procedural details for culturing bacterial isolates, testing, and interpreting results. The procedural details are provided in an extensive document (CLSI 2008) and it is unlikely that most veterinary hospital laboratories can perform these tests reliably unless the personnel are capable of adhering to these standards. Instead, a reliable laboratory that adheres to CLSI standards is recommended.

#10 ASSUMING THAT ADMINISTRATION OF ANTIBIOTICS WILL CAUSE DRUG RESISTANCE PROBLEMS

The effect of antibiotic administration to small animals on drug resistance requires additional on-going investigation and surveillance. It is inaccurate to suggest that administration of antibiotics *causes* bacterial resistance. However, antibiotic administration – if not active enough to eliminate resistant isolates – can select for these resistant strains, which can multiply and flourish. Inadequate antibiotic treatment consisting of doses too low, infrequent administrations, or selection of a poorly active drug, is probably the most common reason for emergence of drug resistance. Because inadequate treatment may produce resistance, a strategy should be advocated that employs the use of highly active drugs, administered using appropriate regimens to attain pharmacokinetic-pharmacodynamic (PK-PD) targets, for as short of a duration as possible (Amyes et al, 2007). This suggests that in companion animals, veterinarians should not be reluctant to administer highly active drugs, provided appropriate regimens are used for only as long of a duration as necessary. When possible, therapy then can be *de-escalated* to less active, more narrow spectrum drugs, based on culture and susceptibility results and clinical response. If administered rationally, and at correct dosages, the administration of an antibiotic can eradicate the infection and actually decrease resistance. Resistance emerges both through horizontal spread – through transfer of genetic elements carrying genes for resistance – as well as mutations arising during treatment through selection of resistant strains.

REFERENCES

Amyes SGB, et al. J Antimicrob Chemother 59: 825-826, 2007. CLSI: Approved Standard—Third Edition. CLSI document M31-A3. Wayne, PA: Clinical and Laboratory Standards Institute; 2008. Hoffman G, et al. J Vet Intern Med. 2001;15:299. Hoffman SB, et al. J Vet Pharmacol Ther. 2002;25:189-193. Papich MG. AAPS Journal. 2005; 07(02): E281-E287. Sartor LL, et al. J Vet Intern Med. 2004;18:651-655. Syakalima M, et al. *J Vet Med Sci.* 1997;59:905-909.

ECG BASICS: TIPS TO GET THE BEST OUT OF IT

Romain Pariaut, DVM, DACVIM, DECVIM-CA (Cardiology)
Baton Rouge, LA

INTRODUCTION

Electrocardiogram artifacts have two potentially undesirable effects: They may make it difficult to interpret the recording and they may lead to incorrect diagnosis.

WHAT IS AN ELECTROCARDIOGRAM?

The electrocardiogram is usually described by the acronyms ECG or EKG (Danish word). It was developed by William Einthoven in 1903. The ECG records the electrical activity that is generated by the heart from electrodes attached to the skin of the patient.

The heart is a pump that circulates the blood. The heart is made of two top chambers, the left and right atria that receive blood from the veins and push the blood into the bottom chambers, the left and right ventricles. The left and right ventricles eject blood into the arteries. The sequential organized contraction of the cardiac chambers is possible because of specialized cells generating an electrical current that is transmitted to the cardiac muscle and "tell the cardiac muscle" when and how fast to contract.

The ECG indicates if the electrical activity, and therefore the contraction of the heart (mechanical activity), is normal or not. If the electrical current does not propagate properly, the mechanical contraction of the heart is altered. If no electrical impulse is generated, the heart doesn't contract, the blood circulation stops and the animal collapses if the delay before the next electrical impulse is too long. If an electrical impulse occurs to soon, the heart contracts sooner than expected which can also result in poor perfusion of the body.

The ECG is made of 5 waves: P, Q, R, S, T. The P wave reflects the electrical activity that propagates across the left and right atrium and triggers their contraction. The Q, R and S waves reflect the impulse traveling across the left and right ventricles. This is known as the QRS complex. The T wave results from the cells of the ventricles going back to a resting state. It always follows the QRS complex.

The ECG machine is made of electrodes connected to a recorder. Nowadays, electronic systems can provide detailed analysis of the data and an interpretation of the tracing. You should never only rely on the computer-generated description of the ECG, as it is commonly erroneous when the rhythm is complex. If the tracing is sent to a company for remote interpretation of the ECG by a cardiologist, you should concentrate on recording the best quality ECG, as the quality of the interpretation will depend on the quality of the recording.

LEAD PLACEMENT AND SET UP

The patient should be placed on its right side, as normal values in dogs and cats have been established with animals kept in this position. If the patient is restrained in a different position, the ECG tracing morphology may be slightly different.

Electrodes are attached to the front and back legs. An electrode is attached to the left front leg (Black), the right front leg (White), the left back leg (Red) and the right back leg (Green) [Snow (white) and grass (green) are always on the ground (closes to table)]. The green electrode grounds the patient to the ECG machine to decrease interference. The ECG records the electrical activity between 2 electrodes, which is known as a lead. When the ECG is recorded from electrodes attached to the limbs, the leads are called limb leads. Recording the electrical activity between the right and left arms displays an ECG tracing called lead I. Recording between the right arm and the left leg displays an ECG tracing called lead II. Recording between the left arm and the left leg displays an ECG tracing called lead III. The advantage of recording multiple leads rather than a single lead, allows the assessment of the propagation of the electrical impulse from different directions, which increases the likelihood that important information about the electrical activity of the heart is not missed. The ECG machine is able to display additional leads calculated from the information obtained from the 3 electrodes. They are known as aVR, aVF, AVL.

The ECG tracing is displayed on a calibrated paper. The horizontal axis is the speed and can usually be selected at 12.5 mm/sec, 25 mm/sec, 50 mm/sec. 50 mm/sec spreads out the ECG and makes it easier to read recordings from patients with a fast heart rate. The vertical axis is the amplitude and can be selected at 5 mm/mV, 10 mm/mV or 20 mm/mV. On the calibrated paper, each little box is 1 mm. Speed and amplitude are displayed on the corner of the ECG and should always be checked before interpreting an ECG.

ELIMINATING ECG ARTIFACTS

Interpretation of an ECG starts with a good quality recording. There are many factors that can alter the quality of the recording:

Poor contact between the skin and the electrodes

Alligator clips or electrode patches can be used. Patches require clipping an area of the skin. Alligator clips are usually used for short-term recordings. Patches are preferred for long-term recordings. Proper contact of the electrodes with the skin is obtained by wetting the skin with alcohol. It is important to only wet the area around the electrode and not a large surface because it could create an electrical bridge and interference. Coupling gel (not ultrasound gel that is used for ultrasonography and is water-based) should be preferred. If an electrode does not have good connection with the skin, the electrode with poor contact can be identified by determining which leads display the artifact. For example: if there is poor contact with the red electrode, lead II and III will display the artifact but lead I won't. Poor electrode contact usually causes rapid, large and erratic trace deflections. It can also cause the baseline to wander.

Electrical interference from the network and appliances

Because the ECG records electrical current, it will not only record the activity of the heart but also any current that may be in close proximity. It is therefore important to record ECGs away from any equipment that may generate an electrical field. Moreover, it is common that a system that is not grounded adequately will cause artifacts, known as 60-cycles (60 Hz is the frequency of the AC current in the USA). Metallic tables should be covered with a rubber mat or a fleece to avoid contact between the electrodes and the table. Electrical interference usually result in a wide homogeneous baseline.

Electrode motion

Motion is the most common cause for artifacts. The patient should be adequately restrained, and mildly sedated if necessary (Butorphanol 0.2 mg/kg, Diazepam or Midazolam 0.2 mg/kg IM or IV). Breathing artifacts, caused by the motion of the chest, are usually avoided by attaching the electrodes more distally on the legs, below the elbows and the knees. Contact between 2 electrodes should be avoided by holding the electrodes still or placing a towel between 2 opposite electrodes. Also, legs should lie flat not touching one another.

ECG machines are equipped with filters. It is usually recommended to set the Filter at 150 Hz. The filter can be turned on to remove electrical noise. For example, a low pass filter at 40 Hz will remove all the electrical noise that is above 40 Hz (2400 / min) but also alter the shape of the ECG complexes: P waves may disappear if the filter is on. It is important to assess the ECG tracing with and without the filter.

BASIC ECG INTERPRETATION

First, always record multiple leads. Some systems display one lead at the time, others display the 6 limb leads simultaneously.

Always, start by looking at lead II. Lead II shows the electrical activity traveling from the right arm towards the left leg. This is the main direction of the electrical current of the heart in most animals.

Identify the waves. The QRS complex is usually the largest wave and the easiest to identify. Because the QRS results from the current traveling in the ventricles and initiates the mechanical contraction of the ventricles, it is always present if the animal is alive. Life is not possible if the ventricles don't contract, i.e. if there are no QRS complexes. It is important to note that the QRS complex doesn't tell us if the contraction of the cardiac muscle is strong or not. In most animals, there will be a P wave in front of the QRS complex. This indicates that current traveled first in the left and right atrium, which are the top chambers of the heart, and then propagated towards the ventricles. However, the left and right atria have a limited contribution to cardiac contraction. On occasion, there are no P waves, indicating that the left and right atria don't contract, but the animal is alive and may behave normally. The QRS is always followed by a T wave, because the tissue of the ventricles has to go back to a resting state. If the T wave is not visible in lead II, it is important to look at other leads, i.e. look from different directions. It is possible that the current generating the T wave was not traveling towards lead II.

There are different techniques to measure the heart rate. Usually, it is not important to know the exact heart rate between two beats but instead to have an idea of the average heart rate. The "pen-method" is easy to measure an average heart rate. At a speed of 50 mm/sec, a regular pen (15 cm -150 mm) represents 3 seconds (50 + 50 +50 = 150 = 1 + 1 +1 seconds). Place the pen at the beginning of a QRS complex and count the number of QRS under the pen, i.e. during 3 seconds. By multiplying by 20 (20 X 3 seconds = 60 seconds = 1 minute) you obtain the heart rate for 1 minute. At a speed of 25 mm/sec, a regular pen represents 6 seconds (25 +25 +25 + 25 +25+ 25 = 150 = 1 +1 +1 +1 +1 +1 seconds). The number of QRS complexes under the pen is the number of beats during 6 seconds. Multiply by 10 to have the heart rate over 1 minute.

Abnormalities usually result from extra-beats (which cause palpitations in people) or missing beats. Extra-beats can originate from the atrium and the current propagating within the heart produces a tracing on the ECG that is very similar to a normal PQRST tracing. However, those extra-beats will come much sooner that you expect. They are known as atrial premature contractions (APCs). Extra-beats can originate from the ventricles. They will be displayed as wide and bizarre QRS complexes that may be inappropriately identified as artifacts. Missing beats will result in a long pause on the tracing, i.e. a flat line. A pause longer than 8 seconds will cause the animal to collapse (syncope).

In conclusion, the quality of the recording is critical to allow a correct interpretation of the ECG.

REFERENCES

Kittleson MD, et al. Electrocardiography. In: **Small Animal Cardiovascular Medicine** 1998:72-94. Harvey AM, et al. B. **J Vet Intern Med** 2005;19(4):533-566. Coleman MG, et al. **Am J Vet Res** 2005;66(2):233-237.

EMERGENCY MANAGEMENT OF ARRHYTHMIAS

Deborah M. Fine, DVM, MS, DACVIM (Cardiology)
Columbia, MO

INTRODUCTION

Arrhythmias requiring urgent therapy may result from a variety of conditions including underlying cardiac disease, trauma, anesthesia, electrolyte abnormalities, and hypoxia. Of these, cardiac disease is the common cause of life-threatening arrhythmias. Determining the specific nature of the underlying cardiac disease ultimately determines the likelihood of successful short and long-term management of an arrhythmia.[1]

The first decision that must be made in arrhythmia management is deciding if an arrhythmia requires therapy at all. Important considerations in this decision include:

1. Is the arrhythmia hemodynamically significant (i.e. causing clinical signs)?
   - If the answer to this question appears to be "no", then emergency therapy is not indicated. The exception to this occurs if the next consideration appears likely.
2. Does the arrhythmia appear likely to degenerate into a less stable rhythm?
3. Can a correctable underlying cause for the arrhythmia be identified?
   - If this is the case, then successfully treating the underlying abnormality is likely to result in resolution of the arrhythmia.
4. What are the risks of beginning therapy?
   - All anti-arrhythmic therapies have the potential to be pro-arrhythmogenic. Many of them also have profound effects on cardiovascular function. Therefore the potential benefit associated with therapy must outweigh the potential risks.

BRADYARRHYTHMIAS

Drug therapy and complications of anesthesia are the most common non-cardiac causes of pathologically slow heart rates. Cardiovascular depression can result from sedatives, analgesics, anesthetic agents, and hypothermia. Appropriate therapy includes reversal agents (e.g. naloxone for opioids), decreasing the rate or dose of anesthetic administration, administration of atropine, and warming the patient. Less common causes of bradycardia include head trauma/brain tumors, and malignant systemic hypertension. These abnormalities ultimately mediate their effects by increasing vagal tone; however the underlying mechanisms are quite different.

The most common bradyarrhythmias requiring emergency therapy are:

1. **Severe second degree atrioventricular heart block**
   - ECG features: intermittent and frequent failure of the P wave to be conducted to the ventricles. Therefore many P waves are not followed by a QRS complex. This results in the ventricular rate being slower than the atrial rate.
   - Etiology: can be the result of increased vagal tone, cardiomyopathy, or idiopathic degeneration of the AV node.

2. **Third degree atrioventricular heart block**
   - ECG features: There is complete dissociation between P waves and QRS complexes. No P waves are conducted to the ventricles. The ventricular depolarization rate (30 to 60 bpm) is much slower than the atrial depolarization rate.
   - Etiology: most commonly the result of idiopathic degeneration of the AV node, or cardiomyopathy. Other less common causes include cardiac glycoside toxicity (e.g. digoxin, Lilly of the Valley, oleander, bufo toads), infectious myocarditis (e.g. Ehrlichia, Bartonella), and infiltrative cardiac neoplasia.

3. **Sick sinus syndrome**
   - ECG features: prolonged periods of asystole (no P waves or QRS complexes), often alternating with periods of tachycardia. This arrhythmia is often intermittent and the patient may have a completely normal rhythm between episodes.
   - Etiology: most commonly idiopathic and most commonly seen in female miniature schnauzers.

4. **Atrial standstill**
   - ECG features: complete absence of P waves preceding any of the QRS complexes.
   - Etiology: hyperkalemia (e.g. urethral blockage, Addisonian crisis), or idiopathic atrial cardiomyopathy (seen most often in English springer spaniels).

In the absence of a treatable underlying cause such as hyperkalemia, most symptomatic bradycardias require implantation of a permanent transvenous or transdiaphragmatic pacemaker for alleviation of clinical signs. Medical therapy with atropine, theophylline, or propantheline may improve symptoms; however improvement is usually temporary.[2]

5. **Asystole**
   - ECG features: flat-line in multiple leads (must rule-out artifact caused by lead detachment)
   - Etiology: occurs when the heart's electrical system fails to generate a depolarization. It may occur as a severe vagal reflex during ocular manipulation, or with vomition. More commonly it occurs due to severe myocardial hypoxia subsequent to ventricular fibrillation, and commonly occurs after unsuccessful attempts at defibrillation.
   - Treatment: atropine, epinephrine, and vasopressin; but the outcome is often fatal regardless of therapy. Asystole does not respond to emergency pacemaker therapy.

TACHYARRHYTHMIAS

Most clinically significant tachycardias occur in the context of underlying heart disease. Therefore, long-term prognosis for patients presenting with symptomatic tachycardia is usually guarded. Nonetheless, therapy can be rewarding and may result in marked improvement in clinical signs, even if only temporarily. There is no absolute number which defines a hemodynamically significant tachycardia. Holter evaluation of dogs shows that the sinus node can briefly depolarize at rates of 250 bpm during periods of extreme excitement. Blood pressure measurement can help assess the hemodynamic consequences of a tachyarrhythmia. Tachyarrhythmias requiring medical therapy are much more common in dogs then in cats.

The most common tachyarrhythmias requiring emergency therapy are:

1. **Atrial or AV nodal tachycardia**
   - ECG features: persistent or intermittent bursts of supraventricular tachycardia ("supraventricular" refers to normal, narrow appearing QRS complexes, indicating the depolarization is being conducted via the normal conduction system). P waves may be absent, abnormal in appearance, or buried in the preceding T wave.
   - Etiology: may occur subsequent to cardiac disease causing atrial enlargement. Atrial enlargement and fibrosis serve as a substrate for abnormal pacemaker activity. This is more common in older animals. In younger animals, this form of tachycardia is often the result of accessory conduction tissue. These bypass tracts allow abnormally fast conduction between the atria and the ventricles.
   - Treatment: calcium channel blockers (diltiazem, verapamil) are the first choice if systolic dysfunction is present. Beta blockers (esmolol, atenolol) will decrease systolic function and should be used very cautiously in patients with myocardial dysfunction. Digoxin is rarely used for emergency control of supraventricular tachyarrhythmias, but rapid digitalization is sometimes necessary for patients with systolic dysfunction in which a calcium channel blocker or beta blocker is initially unsafe.

2. **Ventricular Tachycardia**
   - ECG features: persistent or intermittent bursts of wide QRS complexes with no visible P waves present. The depolarization originates in the ventricular myocardium, and therefore is not propagated through the normal conduction system. As a result the QRS complexes are bizarre in shape and prolonged in duration. The T wave is large and usually of opposite polarity to the QRS complex and a flat ST segment is absent. Sustained ventricular tachycardia may degenerate into ventricular fibrillation if not corrected.
   - Etiology: usually occurs in the presence of severe underlying heart disease. Ventricular tachycardia can also be induced by marked metabolic disturbances, toxicities, and trauma, but these abnormalities are more likely to cause an accelerated idioventricular rhythm (see below).
   - Treatment: intravenous lidocaine is the treatment of choice; if patients are non-responsive then IV procainamide can be tried.

3. **Ventricular Fibrillation**
   - ECG features: rapid, irregular undulations of the ECG which may be coarse or fine. There is a complete absence of recognizable P waves or QRS complexes. Hemodynamically this is disastrous; death ensues rapidly if not corrected.
   - Etiology: severe end-stage cardiac disease or other abnormality which results in profound myocardial hypoxia.
   - Treatment: electrical defibrillation is the only reliable therapy. The probability of successful defibrillation declines by 50% for every 3 – 5 minute delay. Drug therapy can be used to reduce the energy required to electrically

defibrillate the heart. Intravenous amiodarone, bretylium, and magnesium have been used in animals that are refractory to repeated attempts at electrical defibrillation.

4. **Atrial fibrillation**
   - This is the most common tachyarrhythmia seen in veterinary medicine; however it rarely requires emergency therapy.
   - ECG features: supraventricular QRS complexes (narrow), irregularly irregular R to R intervals, absence of P waves, +/- fine baseline fibrillation waves ("f waves")
   - Etiology: severe atrial enlargement and myocardial fibrosis are the necessary underlying substrates. Increased circulating catecholamines also increase action potential propagation and decrease the refractory period allowing perpetuation of the arrhythmia.
   - Treatment: most patients will present with heart rates that are elevated, but usually not to the same extent as the arrhythmias listed above. Therefore, in most instances initiating oral therapy will be adequate for controlling clinical signs. However, in dogs with atrial fibrillation and dilated cardiomyopathy, rapid intravenous digitalization may help to control signs of heart failure in the acute setting.

Accelerated idioventricular rhythm

This is the most common ventricular arrhythmia seen in veterinary medicine and is often mistaken for ventricular tachycardia. Consequently patients presenting with this arrhythmia are often inappropriately treated with lidocaine. Accelerated idioventricular rhythms have wide bizarre biphasic ventricular complexes; however the rate is usually less then 180 bpm. The patients usually do NOT have underlying cardiac disease. Rather, accelerated idioventricular rhythms are secondary to a vast array of non-cardiac abnormalities including neoplasia, gastric dilation-volvulus, trauma, sepsis, autoimmune disease, and end-stage organ failure. This arrhythmia is not hemodynamically significant, but is important because it is a marker of an underlying abnormality. Anti-arrhythmic therapy is rarely necessary, and rarely successful. Successful treatment of the underlying abnormality will result in resolution of this arrhythmia.

Ventricular tachycardia is distinguished from an accelerated idioventricular rhythm by examination of the ECG features

Ventricular tachycardia shows the following features:

1. Rate is significantly higher than the underlying sinus rhythm (usually > 220 bpm)
2. Onset is always premature; the R-R interval of the last normal sinus complex to the first abnormal ventricular complex is shortened compared to the normal sinus R-R intervals (e.g. It starts with a ventricular premature complex)
3. Associated with significant underlying cardiac disease
4. Usually responds to intravenous lidocaine with conversion to a sinus rhythm

Accelerated Idioventricular rhythms show the following features:

1. Rate is only slightly higher than the underlying sinus rhythm (typically about 10% faster)
2. Onset is variable; it can start with a premature complex, or it can start after a pause. It can even start at the same time that a sinus complex is depolarizing the ventricle, in which case a fusion beat will result (a fusion beat is intermediate in appearance between a sinus origin complex and a ventricular origin complex)
3. Associated with underlying systemic disease
4. Resolves with resolution of the underlying disease. Almost never requires antiarrhythmic therapy as it is not hemodynamically significant of itself. Does not typically respond to lidocaine.

EMERGENCY DRUG FORMULARY

All hospitalized patients should have emergency drug doses calculated on admission.[3] The dose sheet should be kept in an easily accessible location (i.e. chart or cage door). It is impossible to memorize the dose of all emergency drugs that may potentially be used, and looking up drug dosages during an emergency is extremely inefficient. There are a number of drug calculators available on the web that only requires entering the patient's species and weight, and the doses are automatically calculated.

REFERENCES

[1]Moise NS. Textbook of canine and feline cardiology 1999:331,386. [2]Estrada AH. J Vet Cardiol. Dec;11(2):79. [3]Laste NL. Vet Clin North Am 2001:31(6):1231.

CARDIAC PACEMAKERS IN SMALL ANIMALS

H. Edward Durham Jr., CVT, LATG
Columbia, MO

INTRODUCTION

Cardiac pacemaker implantation is increasingly available for treatment of symptomatic bradyarrhythmia in small animals. The goals of this presentation are to: 1) familiarize the veterinary technician with the indications for cardiac pacing, 2) introduce the types of equipment and systems available, 3) review implantation of a pacemaker and 4) discuss proper handling techniques and care for patients that have received cardiac pacemakers. Although veterinary cardiologists perform most pacemaker implantations, increasingly veterinary technicians in private practice will see dogs with a pacemaker since these patients return to their primary veterinarian for veterinary care unrelated to the pacemaker. Awareness of appropriate care for pacemaker patients can alleviate potential problems with the pacemaker system.

Pacemaker implantation is more commonly performed in dogs than cats since dogs are more apt to develop bradyarrhythmias and have a slower ventricular escape rate. It is this slow heart rate that leads to clinical signs.[1] The first reported use of cardiac pacing in veterinary medicine was in a horse in 1967.[2] The first use of a cardiac pacemaker in a dog was just a year later.

INDICATIONS

The most common indications for cardiac pacing in veterinary medicine are symptomatic bradycardias.[1] The predominant bradyarrhythmias that necessitate pacing are 3rd Degree atrioventricular block (AVB), 2nd Degree AVB, Sick Sinus Syndrome (SSS), and Persistent Atrial Standstill. Emergency pacemaker implantation may be required for 3rd degree AVB, which can result in sudden death. Familiarity with the normal electrocardiogram (ECG) will aid the technician in identifying these arrhythmias.

Third degree AVB is a complete disassociation of atrial to ventricular impulses. It is characterized clinically by profound bradycardia, weakness, and/or syncope. Electrocardiographically, 3rd degree AVB appears as non-conducted P waves with a junctional or ventricular escape rhythm. The patient's hemodynamic status is maintained by the ventricular escape rhythm. Heart rates between 30 and 60 beats per minute are not uncommon in dogs with 3rd Degree AVB[1], although escape rates can be faster in cats (100-110 bpm).

Sick Sinus Syndrome is a cessation of impulses from the sinoatrial node causing asystole for a period of time sufficient to cause syncope. This arrhythmia is typically unresponsive to atropine or related drugs. The asystole can be interrupted by another sinus impulse or an escape impulse that "restarts" the heart and the patient recovers from the syncope. This condition is not generally considered an emergency. High-grade 2nd degree AVB is also not usually considered an emergency, but is an arrhythmia that should be promptly addressed. It is defined as a normal to prolonged P to R interval with many non-conducted P waves. High-grade second-degree AVB can progress to third degree AVB and is therefore an indication for permanent cardiac pacemaker implantation.[1] Persistent atrial standstill is rare and associated with atrioventricular muscle disease and a lack of atrial contractions. This differs from hyperkalemia-associated atrial standstill which can be treated medically.

EQUIPMENT

The pacemaker system consists of two main components: the pulse generator and the pacing lead. The pulse generator delivers the electrical impulse through an electrode in the pacing lead to depolarize the heart causing it to contract. Either the ventricle alone (single chamber pacing) or the ventricle and the atrium (dual chamber pacing) can be paced. In recent years dual chamber pacing has become more common. The pulse generator is powered by a lithium iodide battery that has a life expectancy of 8-10 years when new. The pulse generators used in veterinary medicine are usually human units that have exceeded their shelf life for human implantation. These pulse generators have 4-6 years of battery life when acquired for veterinary use, which meets the needs of most veterinary patients.

There are two types of generators available, unipolar and bipolar; each uses a lead of corresponding polarity. A bipolar system uses a lead with two electrodes at the tip. Current is delivered through one electrode and returns to the generator via the other electrode and the pacing lead. Unipolar systems use the patient's body as the return conduction system. The bipolar system is preferred because it is less likely to cause muscle fasciculation as the impulse returns to the pulse generator.

There are also two basic lead placement configurations available for cardiac pacing: endocardial and epicardial. Epicardial pacing systems consist of a generator with a pacing lead that is attached to the outer surface (epicardium) of the left ventricle. This system is typically implanted using an abdominal incision with a transdiaphragmatic approach to the pericardium. The lead is literally screwed into the epicardium of the left ventricular apex with a corkscrew-type tip on the lead. The lead is then brought through the diaphragm and attached to the generator, which is generally placed within the muscle layers of the abdominal wall, or in the subcutaneous tissue outside of the abdominal cavity. Epicardial pacing is most commonly used in very small patients (e.g. cats), when vascular access is a problem, when there is another reason for a celiotomy (e.g. abdominal tumor), or when training or equipment for transvenous pacing is lacking. Epicardial leads are typically unipolar, but newer bipolar leads are becoming available.

Endocardial transvenous pacing is less invasive than epicardial pacing since it does not require the opening of a body cavity. Endocardial leads can be either unipolar or bipolar. In the endocardial transvenous pacing system, the pacing lead is passed via the jugular vein (transvenous) through the right atrium into the apex of the right ventricle. Either jugular vein may be used. This procedure must be done with fluoroscopic guidance to insure proper placement of the lead. Once placed, the lead is either actively screwed into the endocardium with a corkscrew-like tip or is held in place passively by grappling hook-type projections or *tines* that snag the trabeculae of the right ventricle. These systems hold the lead tip in place until scar tissue surrounds the lead tip, more securely anchoring it. The jugular vein is ligated distal to the lead insertion point and the lead is secured with the proximal portion of the vessel (ligation of the jugular vein causes no problems in dogs due to their excellent collateral vascularization). The remaining distal portion of the lead is then tunneled under the superficial muscles of the neck and connected to the pulse generator. The lead is secured to the pulse generator with setscrews. These tiny screws are essential for establishing a secure the attachment **and** electrical contact between the lead and the generator. The pulse generator is sutured into a pocket under the superficial muscles on the dorsolateral aspect of the neck or near the scapula. The endocardial lead can also be attached to temporary pulse generator while the generator pocket is dissected and the permanent system connected. Currently, the endocardial pacing system is most commonly used due to equipment availability, ease of placement, lower morbidity, and decreased anesthetic complication rate compared to epicardial pacing.[3]

The pacemaker batteries, though relatively low voltage (about 3 volts), are capable of producing sufficient energy, measured in volts and is known as *amplitude,* to depolarize the heart. This voltage delivered as a "pulse" of brief duration, measured in milliseconds, known as *pulse width*. The *current* is the amplitude multiplied by the pulse width. The lead and heart tissue constitute the resistance (measured in ohms) that must be overcome to stimulate the heart.

The relationship (Ohm's Law) is: **current = voltage / resistance**

Current needs to be controlled such that sufficient energy is delivered to depolarize the heart, but battery strength is not wasted. The pacemaker discharge rate (i.e., the heart rate) also must be controlled. These and other pacemaker settings can be manipulated with a programmer that controls the pulse generator non-invasively with a magnetic field guided by a computer. A programmer is available for all pacemakers, although they are not interchangeable. Each pacemaker manufacturer also makes a programmer for their systems. Modern programmers can program a variety of models from one manufacturer. The available parameters that are programmable vary with the model of pulse generator, but all pacemakers allow control of the pulse (heart) rate, voltage amplitude, pulse width, and level of sensing native heartbeats. Newer generators are highly programmable allowing for programming of sleep modes, variable rates, automatic threshold detection, and arrhythmia recording along with other parameters.

Pacing modes are identified by a 4 letter code system. This system designates the first letter to be the chamber paced, the second letter is the chamber sensed, and the third letter indicates the pulse generator function when a native beat is sensed, the final letter indicates if a pacemaker increases pace rate in response to activity. The abbreviations used are: V= ventricle, A= atria, O= nothing, I= inhibit, R = rate responsive. The most basic pacing mode is VOO; meaning ventricle paced, no chamber senses and no pacemaker response other than pacing the ventricle. VOO mode paces but does not allow for any native activity. Most systems are programmed to fire only when the patient's rate drops below a preset value (known as VVI or demand mode). In VVI mode the designation indicates that the ventricle is paced, the ventricle sensed, and the pulse generator is inhibited from firing when a native beat is detected. This inhibition keeps the pacemaker from firing at inappropriate times which could occur with VOO pacing and may initiate ventricular tachycardia. Other pacing modes such as dual chamber pacing to achieve a more physiologic contraction are now available. Pacing modes and variables can be changed non-invasively using the pacemaker programmer.

PRE-IMPLANTATION EVALUATION

In the case of emergency pacing the suspected diagnosis is confirmed with an ECG; radiographs are performed to determine the anesthetic risk and identify concurrent disease. The presence of concurrent heart disease that could alter the long-term prognosis is also evaluated by echocardiogram. In the situation of a non-emergency pacing, time permits a full workup, including blood serum chemistries, complete blood count, urinalysis, thoracic radiographs, ECG, echocardiogram and non-invasive blood pressure. In some cases a 24 hour ambulatory ECG (Holter Monitor) may be required to confirm the diagnosis. Some clients will also elect to have an abdominal ultrasound performed to rule out previously detected neoplasia before investing is a pacemaker. These data will be used to rule out other causes of the arrhythmia that may obviate the need for cardiac pacing and give indication of prognosis.

When acquiring blood for analysis; ***venipuncture should not be performed using either jugular vein prior to referring a patient for pacemaker implantation!*** A hematoma around the jugular may preclude using that vein for the pacing lead. The lateral saphenous vein is preferred for blood collection so the cephalic veins can be preserved for IV catheter placement.

Patients with signs of congestive heart failure should be stabilized appropriately with medications such as furosemide, nitroglycerin and sometimes with or without temporary pacing. In some cases, a constant rate infusion of isoproterenol or dopamine may be administered to increase heart rate of escape rhythms and augment cardiac contractility. **Lidocaine, digoxin and β-blockers, or calcium channel blockers are contraindicated.** These drugs can worsen the bradycardia, decrease contractility or suppress the ventricular escape beats that are sustaining the patient.

Patients with severe bradycardia or escape activity that changes rate erratically may require temporary transvenous pacing with a special pacing catheter to reduce the risk of sudden death from anesthesia. A combined anesthetic protocol of heavy intravenous sedation, neuromuscular blockade, intubation and ventilation, and local anesthesia can be used to gain vascular access and place a pacing lead for this function. An external temporary pulse generator is then attached using sterile connectors. This allows control of the heart rate during the procedure and allows safe administration of inhalation anesthesia. An alternative to temporary catheter pacing is transthoracic pacing. In transthoracic pacing, two wide electrodes are adhered to each side of the shaved thorax of the patient through which just enough current is applied to depolarize the heart with an external pacing system.

POSTOPERATIVE CARE

Transvenous endocardial pacing is the primary method of pacing today. The surgical site is small and post-operative pain is minimal compared with epicardial pacing. Post-implantation care involves analgesia, sedation, and continuous electrocardiographic monitoring. Antibiotics are given peri-operativly and post-operatively. Opiate sedation/ analgesia also keep the patient quiet after surgery to avoid dislodging the lead thereby necessitating surgical repositioning. A sterile bandage is placed around the neck to protect the surgical site and reduce seroma formation. Radiographs are taken in the right lateral and dorsoventral views the following day to confirm the lead placement. Avoid positioning the patient on its back during the first two weeks following implantation as this may dislodge the lead. After implantation **neck leads or collars should NEVER be used on a dog with a transvenous pacemaker system**; they should always wear a harness. An ECG is used to verify pacemaker function and rate. Pulse generator parameters are programmed to a base setting and the patient is released from the hospital 1-3 days after implantation. At discharge, the client is instructed how to take the patient's heart rate and asked to check it daily until the next visit. If the patient's heart rate goes below the minimum pacing rate, the client is instructed to call the clinic immediately.

The clients are asked to return for suture removal 10 to 14 days from discharge. At this visit an ECG and thoracic radiographs are performed. The pulse generator is also evaluated with the programmer. As the endocardium develops a reaction around the pacemaker lead tip where it contacts the heart, increased resistance to the pacing impulse can develop. The minimum current for the pulse generator needed to "capture" the ventricle, causing it to depolarize and contract is determined. To verify capture, a continuous ECG is performed while programming the pacemaker while the current is gradually reduced until capture is lost. When this "threshold" is determined, the pacemaker is programmed at an output setting of two times the threshold to assure capture in the event that resistance increases. Typically, this allows a sufficient margin of safety.

A similar follow up visit is scheduled about 3 months post implant at which time radiographs and pulse generator evaluation are performed. Unless complications arise, then the client is asked to see their local veterinarian for biannual radiographs and to visit the cardiologist annually for pacemaker system evaluation. At all follow-up visits an ECG, echocardiogram and pacemaker programming are repeated and any other tests that may be indicated.

COMPLICATIONS

Complications of pacemakers have been well documented [1, 4] and can range from mild to serious. The most common complications seen are seroma, or hematoma around the pulse generator, dislodgment of the pacing lead from the heart, congestive heart failure from concurrent heart disease or subsequent myocardial failure, tissue reaction sufficient to prevent pacing capture (known as exit block), and malfunctions of the pacing system. Infections occur infrequently, but can occur months after the implantation from hematogenous spread. System malfunctions include lead fracture or dislodgement, battery depletion, or damage to the pulse generator circuitry from electrical impulses (i.e. electrocautery or defibrillation).

CAVEATS FOR OTHER VETERINARY CARE

- Keep conspicuous records of pacemaker details, including rate settings, in the patient's chart along with the pacemaker system's location in the patient.
- When the patient is in the hospital label cage ID with "Pacemaker" sign.
- **ALWAYS** use a harness for pacemaker patients. **NEVER** put leashes or collars around the patient's neck
- **NEVER** attempt venipuncture in the vein that carries the pacing lead.
- Avoid placing the patient on its back or over extending the neck.

Although the stable pacemaker patient can resume a substantially normal life, there are details of patient care that must be remembered. In epicardial pacing, the pacemaker generator is usually located in the abdomen or in between abdominal muscle layers. It must be remembered where the pacemaker is located during palpation so as not to mistake it for a mass. For endocardial transvenous systems the generator and leads are more superficial and greater care must be exercised. During the physical exam the generator is usually palpable and should not be moved, rotated, or mistaken for a mass. Likewise the lead in the jugular vein is easy to feel. In obtaining blood for laboratory analysis, it is crucial to know where the pacemaker and lead are located. ***One should NEVER attempt venipuncture in the jugular vein that carries the pacing lead.*** Striking the lead with a hypodermic needle could cause damage to the insulation around the lead or the actual metallic core of the lead resulting in immediate malfunction of the system and potential life threatening complications for the patient.

The veterinary technician should become acquainted with the electrocardiogram of pacemaker patients.[5] Patients with pacemakers have a distinctive ECG pattern that shows a "pacing spike" and typically demonstrate wide QRS complexes. Knowing the pacemaker rate setting can also alert the technician to depletion of battery life in a pulse generator. Pulse generators will automatically reduce their rate when the battery weakens, indicating it is time to replace the pacemaker. Modern pacemakers are not susceptible to interference from microwave ovens, garage door openers or pet identification microchip readers. Patients with pacemakers should remain clear of strong magnetic fields such as magnetic resonance imaging or children playing with magnets. Patients that have pacemakers implanted cannot wear electric shock collars like those used with invisible fencing systems or have electrocautery during other surgeries.

REFERENCES

[1]Moïse, N. ***Textbook of Can and Fel Cardio*** 1999, WB Saunders pp400. [2]Taylor, D. et al. **JAVMA** 1967; 151(9):pp1172. [3]Bonagura, J. et al. **JAVMA** 1983; 182: pp149. [4]Kittleson, M. ***Small Ani Cardiovasc Med*** 1999, Mosby pp525. [5] Durham, H.E. **Vet Tech** 2000, 21(11): pp642

EFFECTIVE LABORATORY MEDICINE IN EQUINE PRACTICE

Richard Hepburn, BVSc, MS, CertEM(IntMed), DACVIM(LAIM), MRCVS
Gloucestershire, UK

INTRODUCTION

Hematology and biochemistry are the most commonly used components of clinical pathology in equine practice. Automated cell counters and chemistry analyzers are increasingly affordable, and when paired with effective cytology techniques provide a huge amount of information regarding the presence of inflammation and pathologic processes in different locations. When establishing a laboratory, an individual technician should have responsibility for sample processing, and report to a suitably experienced clinician. Automated cell counters are easy to use and extremely reliable, as are dry chemistry analyzers. Wet chemistry processing is more cost effective but analyzers can be temperamental. Even though many analyzers self calibrate, more regular quality control is essential[1]. Informally this can involve sending a proportion of samples to an external laboratory for comparison. Many equipment suppliers, universities and external laboratories offer more formal programs. All quality control results should be recorded and equipment calibrated as necessary. A medicine specialist should check cytology slides; air-dried slides can also be sent away for comparison. The use of appropriate normal ranges is vital, which should reflect the population of horses seen. Ideally they should be derived from a representative sample of clinically normal horses, if this is not possible then the use of published reference ranges is acceptable. Separate ranges for adults[2] and neonates[3] are required.

HEMATOLOGY

A complete blood count covers assessment of red and white blood cells, and platelet numbers. Effective sampling and handling routines are vital. Excitement and recent exercise cause splenic contraction, releasing stored red blood cells (RBC) into the circulation, adding up to 15% packed cell volume (PCV) points. A physiologic neutrophilia can also occur as neutrophils detach from the vascular endothelium. Red blood cells sediment rapidly, so adequate mixing is required before measurement.

Hemogram[4,5]

A manual centrifugal PCV is the most accurate (±1%) way of measuring the hematocrit, and should be performed in all cases. Automated cell counters will measure the main RBC indices: hemoglobin concentration [Hb], red blood cell count (RBCC), mean corpuscular volume (MCV), mean corpuscular hemoglobin (MCH) and hemoglobin concentration (MCHC), and red cell distribution width (RDW); and are most accurate if they have algorithms designed for equine blood. A peripheral blood smear (Diff-Quik stained) should be used to assess erythrocyte morphology, this is also a great way of improving cytology skills..

The two main RBC abnormalities are erythrocytosis and anemia: Erythrocytosis describes increased PCV, RBCC and [Hb] above the normal range. Absolute erythrocytosis, due to increased bone marrow erythrogenesis, is rare. Relative erythrocytosis reflects increased circulating RBCs due to splenic contraction, hemoconcentration (dehydration) or a combination of both. Anaemia is seen as a reduced PCV. Three basic causes are: blood loss; hemolysis (assess PCV÷ [Hb], if <1/3 suggests intravascular hemolysis); and non-regeneration. Bone marrow RBC regeneration should start within 3-4d and be complete within 21d (depending upon severity of the anemia). Increased mean cell volume (MCV) is often the only sign of a regenerative response; reticulocytosis and polychromasia (seen on blood smear) are rare. Non-regeneration is associated with either anemia of chronic inflammation (rare) or iron deficiency. Iron status can be assessed by measuring serum iron concentration $[Fe]_s$ and total iron binding capacity (TIBC). With iron deficiency $[Fe]_s$ is reduced and TIBC is elevated, with anemia of chronic inflammation $[Fe]_s$ is low, but TIBC is normal.

Leukogram[6,7]

The presence of inflammation should be identified by assessment of total and differential white blood cell counts. The use of prepackaged diluent manual counting systems (*Unopette, Becton-Dickinson & Co*) is not recommended as the inherent error is high (20%), and the technique is time consuming. Automated counters have a low error (5%), although this can increase if cell counts are very high or very low. Ideally the differential count should be compared to a Diff-Quik smear, which also allows identification of any toxic changes.

An increase in the total circulating number of leukocytes above the normal range is termed leukocytosis, a reduction is leucopenia. In equine medicine neutrophils are pathophysiologically most reactive. They have a half life ($t_{1/2}$) of 6-14h, migrate into tissues within 2h of injury and once there will die. Neutropenia (reduced neutrophil count) reflects increased movement of cells out of the blood and into tissues. The speed of migration relative to the release from bone marrow dictates the severity of neutropenia. The proportion of band neutrophils and presence of cytoplasmic toxicity should be assessed manually. A degenerative left shift occurs when band neutrophils outnumber mature ones. Toxic changes (cytoplasmic vacuolation, basophilic stippling, inclusion bodies) are suggestive of marked inflammation. Neutrophilia (increased neutrophil count) is caused by increased production and release of neutrophils from bone marrow, and occurs approximately 24-48h after inflammation starts, and reduces as the inflammation becomes localized or inactive.

Lymphocytes are produced in bone marrow and lymphoid tissue. Unlike neutrophils, lymphocytes recirculate from tissues to blood and have a life span of several years. B-cells transform into plasma cells and are responsible for antibody production. The peripheral lymphocyte count reflects the balance between cells entering and leaving the blood, rather than just the rate of their production. Lymphocyte counts tend to be quite stable although two changes can occur: Mild lymphopenia is common in sick horses, with recirculating cells trapped in lymph nodes. Lymphocytosis can occur with chronic antigenic stimulation in bacterial, viral, and Rickettsial infection. Lymphocytic leukaemia is rare in horses.

Monocytosis mainly occurs with monocyte specific conditions (equine monocytic *Ehrlichiosis*), although a reflex monocytosis can occur after systemic inflammation. Monocytopaenia is hard to identify as the reference range often includes zero. Eosinopenia is very hard to identify as the normal reference range also includes zero. Eosinophilia is associated with parasitism, but is uncommon. Basophil changes are difficult to interpret as so few of these are present normally. Automated counters will often interpret toxic neutrophils (with basophilic stippling) as basophils, reporting an erroneous basophilia. Blood smear cytology will reveal the truth.

Platelets[8]

Platelets are small anuclear discs of cytoplasm that are a vital component of hemostasis. Their life span is approximately 5-9 days and they cannot multiply. The spleen contains 30-40% of all circulating platelets, so thrombocytosis (increased platelet count) can occur with splenic contraction due to excitement. Automated counting is typically performed on an EDTA sample, although aggregation can occur with this anticoagulant in the horse, leading to an incorrect diagnosis of thrombocytopenia (reduced platelet count, usually below 100,000/uL). Platelet clumping will also be seen at the leading edge of the blood smear. Thrombocytopenia should be confirmed with a sample in citrate anticoagulant; any case with petechial hemorrhage should have this sample taken as a routine. Spontaneous hemorrhage doesn't usually occur until the platelet count falls below 25,000/uL.

ROUTINE BIOCHEMISTRY

Biochemical profiling aims to identify specific pathophysiologic processes and localize disease to particular organs. It can be split into blood proteins, enzyme activities and clinical chemistry levels. Dry chemistry automated analyzers typically have different tests grouped together as profiles, although some machines allow individual tests to be selected. In order to limit costs if very high results are expected (such as with rhabdomyolysis) dilution prior to first analysis may be required. Wet chemistry analyzers allow much more freedom when selecting which variables will be run, and many will perform dilutions automatically.

Blood Protein Analysis[9,10]

Total plasma protein is measured by refractometer, albumin and globulin concentration are assessed automatically. Hyperproteinemia occurs with dehydration, and both albumin and globulin should be equally raised (panhyperproteinemia), an uneven increase suggests albumin loss (e.g. diarrhea). Hyperglobulinemia is rare. Hypoproteinemia initially occurs due to hypoalbuminemia, as albumin is a smaller molecule than the globulins. Hypoalbuminemia reflects one or more of three pathophysiologic processes: decreased production (malnutrition, hepatic disease); sequestration (edema formation); or most commonly increased loss (protein losing enteropathy). The $t_{1/2}$ of albumin is long, so decreased production takes a long time to become clinically apparent.

The acute phase proteins fibrinogen and serum amyloid A are useful indicators of inflammation in the horse. Fibrinogen is an acute phase protein that is produced by the liver and increases approximately 24-48h after initiation of an inflammatory process however the magnitude of hyperfibrinogenemia correlates poorly with the severity of the inflammation, and it persists once the inflammation has resolved. Serum amyloid A increases more rapidly than fibrinogen and to a greater degree, but importantly it also reduces rapidly and consistently once inflammation has resolved, making it a useful indicator of resolution of inflammation.

Enzyme Activities[11,12,13,14]

Increased enzyme activity allows localization of disease to different organs. When deciding how frequently to monitor treatment response (seen by reduced enzyme activity) it is useful to appreciate the $t_{1/2}$ of individual enzymes. In cases of exertional rhabdomyolysis dilution of serum samples is often required before measurement (1:5 to 1:10).

SDH – sorbitol dehydrogenase – liver specific (hepatocytes), $t_{1/2} \approx 12h$

γ-GT – gamma glutamyl transferase – liver specific (periportal cells and hepatocytes), $t_{1/2} \approx 3d$

ALP – alkaline phosphatase – predominantly liver (biliary epithelial cells, hepatocytes), $t_{1/2} \approx 3d$

CK – creatinine kinase – muscle specific, predominantly skeletal muscle (cardiac form short $t_{1/2}$ and small amount), $t_{1/2} \approx 2h$

AST – aspartate aminotransferase – non-specific, released from liver, muscle, RBCs, $t_{1/2} \approx 12h$

LDH – lactate dehydrogenase – non-specific as many tissue specific isoenzymes, $t_{1/2} \approx 10h$

In cases of liver disease GGT, ALP and bile acids are the tests of greatest diagnostic utility, with ALP reduction also being a good indicator of response to therapy. In cases of muscle disease the differing kinetics of CK and AST can be used to assess the timing of disease. Increased CK activity occurs within 4-6 hours of the onset of muscle disease, and peaks at 6-12 hours. Persistently high CK activity indicates ongoing muscle damage. Unexpectedly high CK activity found <2h after the onset of a single bout of muscle disease suggests that recent subclinical muscle damage has occurred. AST activity increases more slowly and due to its long $t_{1/2}$ will persist after the muscle damage has resolved.

Clinical Chemistries[9,11,15,16]

Bilirubin, glucose, creatinine, blood urea-nitrogen, serum triglycerides, bile acids and blood lactate are commonly measured. Bilirubin is carried in the blood bound to albumin in an unconjugated form, is taken up by the liver and converted to a conjugated form. A moderate unconjugated hyperbilirubinemia occurs in fasted and anorexic horses (<100mg/L). Larger increases occur with hemolysis and with significant liver disease (both conjugated and unconjugated). Blood glucose levels are tightly regulated, hypoglycaemia is commonest in sick neonates. Hyperglycemia can occur with excitement and endocrine dysfunction. Accurate blood glucose measurement requires the use of Flu/Ox vacutainers. Creatinine is predominantly cleared by glomerular filtration and so is the most accurate indicator of renal function. Increased creatinine (azotemia) can be classified as pre-renal (dehydration), renal (renal disease) or post-renal (bladder rupture). Blood urea-nitrogen is an inaccurate measure of renal function in horses due to its extensive enterohepatic recycling. Serum triglycerides can be elevated during states of negative energy balance in diseases such as enterocolitis, where hypertriglyceridemia is associated with prolonged inappetence. Lactate is produced by anaerobic metabolism during tissue hypoxia (sustained heavy exercise, bowel ischemia), very high blood lactate concentrations can indicate a poor prognosis in colic cases. Bile acids undergo enterohepatic recirculation, consistent increases occur with biliary obstruction and are strongly predictive of significant liver disease.

CYTOLOGY[17,18,19,20]

Examination of tracheal wash and bronchoalveolar lavage fluids, joint and body cavity aspirates is easily performed. Measurement of total cell count is required in all samples except airway fluids and should be performed automatically. The differential count should be done manually as many samples will be too degenerate for reliable automated reading. This can either be performed as a smear, or as a cytocentrifuge slide. As disparities can exist between the two methods, individual practices should develop a standard method of sample processing. If samples are very dilute then centrifugation (<300G), and resuspension in a smaller diluent volume can be helpful. Diff-Quik staining is sufficient for the majority of equine cytology samples, although if a lot of airway samples are expected Toluidine blue (30 mins stain time) is required to identifying mast cells, which will not show up on Diff-Quik. Ideally 2-400 cells should be counted, using a consistent approach, to give a representative differential. A gram stain should be performed on any sample where sepsis is suspected. In cases of suspected synovial sepsis where only a small sample volume is obtained, a wet drop smear should be prepared. A mean of 0-2 leukocytes per high power field (x400) corresponds to a total leukocyte count of $<1.3\times10^9$/L.[18] Total protein measurement by refractometer is appropriate for joint and body cavity fluids. Measurement of creatinine in abdominal fluid is useful in cases of suspected bladder rupture, as is lactate in cases of suspected bowel ischemia.

In any sample where abnormal or unexpected cells are seen then an unstained, air dried slide should be prepared and sent along with a sample of fluid to an external laboratory for second opinion. This is particularly important in cases of suspected abdominal neoplasia, as inflamed equine mesothelial cells are extremely reactive and so can have a very variable appearance.

REFERENCES

[1]Mahaffey E **Vet Lab Med Clin Path 4th ed** 2003:331, [2]Parry B **Current Therapy in Equine Medicine 6** 2009:956, [3]Axon J & Palmer J **Vet Clin Equine** 2008;24(2):357, [4]Lording PM, **Vet Clin Equine** 2008;24(2):225, [5]Brockus C & Andreasen C **Vet Lab Med Clin Path 4th ed** 2003:3, [6]Carrick J and Begg A, **Vet Clin Equine** 2008;24(2):239, [7]Latimer K & Prasse K **Vet Lab Med Clin Path 4th ed** 2003:46, [8]Topper M & Welles E **Vet Lab Med Clin Path 4th ed** 2003:99, [9]Evans E & Duncan J **Vet Lab Med Clin Path 4th ed** 2003:162, [10]Crisman et al **Vet Clin Equine** 2008;24(2):285, [11]Bain P **Vet Lab Med Clin Path 4th ed** 2003:193, [12]Bender H **Vet Lab Med Clin Path 4th ed** 2003:260, [13]Durham A et al **Equine Vet J** 2003;35(6):542, [14]Durham A et al **Equine Vet J** 2003;35(6):554, [15]Tennent-Brown B **Compend Contin Educ Vet** 2011;33(12):E1, [16]Dunkel B et al **Equine Vet J** 2003;35(6):590, [17]Couetil L et al **J Vet Intern Med** 2007;21(2):356, [18]Hoffman A **Vet Clin Equine** 2008;24(2):423, [19]Steel C **Vet Clin Equine** 2008;24(2):437, [20]Rakich P & Latimer K **Vet Lab Med Clin Path 4th ed** 2003:304

COMPARATIVE ANATOMY AND PHYSIOLOGY OF THE LARGE ANIMAL FEMALE REPRODUCTIVE TRACT

Shirley I. Sandoval, BAS, LVT-VTS (LAIM)
Pullman, WA

INTRODUCTION

Reproduction is an important component of large animal practice. However, the majority of instruction in veterinary technology programs is aimed at small animal practice. With the major concern of pet overpopulation, there is more emphasis put on education of the surgical altering of dogs and cats rather than the production aspects of the large animal species. Therefore, it is the objective of this seminar to provide information of a comparative nature to establish a good foundation for reproductive work in the large animal setting.

COMPARATIVE ANATOMY OF THE LARGE ANIMAL FEMALE REPRODUCTIVE TRACT

The major female reproductive anatomical structures are the ovaries, oviducts, uterus, cervix, vagina and vulva[1].

Ovaries

Supported in the abdominal cavity by the broad ligament, the ovaries are paired structures located bilaterally near each kidney. The ovarian structure is made up of the cortex and medulla. The cortex contains a finite number of ova; the total number varies by species (i.e. approximately 36,000 in the mare, 120,000 in the cow[2]).

The ovarian structure in most species is made up of a central medulla surrounded by the cortex. In these females, follicles can develop anywhere on the cortex surface. In the mare, these structures are reversed with the cortex surrounded by the medulla. Follicles in the mare develop from the cortex and the resulting ova ovulate through a specialized structure called the ovulation fossa. The equine ovary is reniform in shape with the convex surface (dorsal) attached to the broad ligament and the concave (ventral) surface housing the ovulation fossa.

Oviduct

Providing the pathway for ovum from the ovary to the uterus, the oviduct is also suspended by the broad ligament in the abdomen. The oviduct is divided into the infundibulum, ampulla, and isthmus. Ciliated cells line portions of the oviduct. The cilia, along with muscle contractions, assist in the movement of the ovum to the ampulla. Conversely, the spermatozoa are also assisted with migration to the ampulla where fertilization occurs. The fertilized ova are housed in the oviduct until the resulting zygote is ready to enter the uterus through the uterotubal junction (UTJ). In the mare, only fertilized ova pass through the UTJ[2].

Uterus

The uterus in all domestic large animals is bicornuate, or heart shaped. It is a dynamic tubular structure consisting of two horns and a body. As with the ovaries and oviduct, the broad ligament supports the uterus in the abdomen. It is connected cranially to the oviduct by the UTJ and caudally to the vagina via the cervix. Depending upon the species, there is a variation in the level of development of the uterine horns and body[1]. Mares have a large uterine body and poorly developed horns, while sows have well developed horns and a short body. The level of development in the ruminant uterine body and horns varies in structure between the mare and sow; but in the camelid, the left horn is slightly larger than the right[2]. The uterus is made up of three tissue layers: the perimetrium, myometrium and endometrium. The sow and mare exhibit many endometrial folds in the uterus, while ruminants have small non-glandular areas called caruncles. The uterus is a dynamic organ that changes between being edematous, toned or flaccid under the influence of fluctuating hormones[2]. It also greatly increases in size when gravid and involutes quickly post-partum.

Cervix

The cervix is a tubular structure that connects the uterus and vagina. It varies in size, shape and structure depending upon the species. In most large animals, the cervix consists of a number of cervical rings. In the mare, however, there are longitudinal folds that transition into the uterus to endometrial folds[2]. As with the uterus, the cervix is also dynamic and changes its appearance under hormonal influences.

Vagina

The vagina contains the caudal aspect of the cervix, the vestibule, the urethral orifice and in the maiden, the hymen.

Vulva

The outer most structure to the female reproductive tract is the vulva. It is made up of a pair of labia (vulvular lips). These structures should remain closed with the exception of urination and parturition. Elongation of the vulva occurs just prior to parturition as the perineal region becomes relaxed.

COMPARATIVE LARGE ANIMAL REPRODUCTIVE PHYSIOLOGY OF THE FEMALE

Cyclicity

The estrous cycles are classified by frequency. Polyestrus females cycle year around and include the bovine, camelid and porcine. Seasonal polyestrus females cycle for specific periods throughout the year and are further subdivided into long-day breeders

(Mares) and short-day breeders (ewes and does). With seasonal polyestrus breeders, initiation of cyclicity is influenced by the photo period, temperature and location. The length of these intermittent breeding periods also varies by species and breed[1].

Within the estrous cycle, there are two main phases. The follicular phase, where the dominant hormone produced is estrogen; and the luteal phase, with progesterone being the dominant hormone. However, seasonal polyestrus animals also have anestrus periods where the ovaries do not have follicular growth or luteal tissue[1].

Ovulation

Follicular growth is regulated via positive and negative hormonal feedback mechanisms. The decrease of progesterone (P4) causes an up regulation in the secretion of gonadotrophin releasing hormone (GnRH), which in turn initiates the release of follicle stimulating hormone (FSH) and luteinizing hormone (LH). As the follicles grow under the influence of FSH, they secrete estradiol. The larger the follicle, the more estradiol they produce. As estradiol levels near their peak, inhibin is released to stem the secretion of FSH. Once the estradiol threshold is obtained, there is a surge of LH which causes ovulation in those species which spontaneously ovulate (bovine, caprine, equine, ovine & porcine). Conversely in camelids, induction of ovulation also requires copulation and the presence of the chemical "ovulation inducing factor" (OIF) from the seminal plasma[1]. If induced ovulators are not bred, the dominant follicles regress and a new follicular wave begins.

Progesterone

Post ovulation, the granulosa and theca cells from the ovulation site join to form luteal tissue in the structure of a corpus luteum (CL) which secretes P4. The uterus requires sufficient levels of P4 to prepare for embryo attachment and maintain the conceptus during pregnancy[1]. Progesterone is produced by the CL to maintain the pregnancy in the camelid, caprine, and porcine species throughout the pregnancy; however in the bovine, equine, and ovine the placenta takes over the production of P4 at 200, 70, & 50 days respectively[2].

Fertilization

Fertilization occurs in the oviduct and the cellular division begins. Depending upon the species, the stage at which the developing embryo enters the uterus varies. Once in the uterus, the blastocyst hatches and in the mare remains round; but in the cow, ewe, camelid and sow it becomes filamentous[1].

Maternal Recognition

Upon reaching the uterus, the embryo must alert the dam to its presence and protect the CL from prostaglandin secretion and subsequent CL luteolysis. This is termed maternal recognition of pregnancy[1]. A chemical is produced by the blastocyst of the bovine and ovine; bovine Interferon τ (bIFN-τ) and ovine Interferon τ (oIFN-τ) respectively[1]. In the mare, the embryo migrates throughout the uterine body and horns multiple times between day 12-14 post ovulation as a signal to the dam of its existence[1]. For maternal recognition in the sow, there must be at least two embryos in each horn. In addition, the sow will produce PGF2α, but the embryos also produce significant levels of estradiol (E_2) which shifts the direction of the prostaglandin secretion into the uterine lumen[1].

Placentation

The embryos of large animals require some form of placental attachment to the uterine endometrium to facilitate metabolic functions for the duration of gestation. The type of placenta varies between species and is classified by the chorionic villi. Diffuse placentas are found in the camelid, equine and porcine species; in addition the mare and camelid placenta is further delineated as microcotyledonary. Unique to the mare, between 35-60 days, the placenta develops "endometrial cups" which secrete equine chorionic gonadotrophin (eCG). These cups subsequently stop secreting and slough off into the uterine lumen[1].

The placentas of ruminants are classified as cotyledonary. Unlike the diffuse placenta, cotyledonary placentas are attached to the uterus via the caruncles. The combination of fetal cotyledons and maternal caruncles is known as the placentome. Interestingly, the placentomes in the sheep are concaved, while in the cow they are convex[1]. Also located in the ruminant placenta are binucleated giant cells. They secrete pregnancy specific protein B (PSPB)[1]. This protein can be detected in the circulating blood of ruminants using an ELISA test[3]. Sampling must occur no sooner than 22 days for sheep and 30 days for cattle and goats[3].

REFERENCES

[1]Senger, P. *Pathways to Pregnancy and Parturition, 2nd ed.* 1999. [2]Youngquist, R. et.al. *Current Therapy in Large Animal Theriogenology, 2nd ed.* 2007. [3]Bio Check Veterinary diagnostics and technologies, 2009 http://biochecklabs.com/?s=&p=2 accessed 9/25/11

UNDERSTANDING COMMON DISORDERS OF SMALL RUMINANTS

Thomas Passler, DVM, PhD, DACVIM
Auburn, AL

INTRODUCTION

Small ruminants enjoy a growing popularity in North America and are kept for a variety of purposes including production of meat, milk, and fiber, brush control, showing and competition in livestock shows, and as pet animals. More than 50% of goat farms in the US have fewer than 10 animals, and almost all US goat farms (99.7%) are considered small-scale operations with fewer than 500 animals.[1] Over 67% of goat operations have been in existence for less than 10 years[1] resulting in varying levels of knowledge among small ruminant owners. Many of the common diseases of small ruminants are caused by an underlying component of poor management, and hence can often be prevented by appropriate care, including housing, monitoring, and feeding. As compared to cattle, the requirements for energy availability and quality of forages are greater for small ruminants and this is reflected in high selectivity when grazing or browsing. Sheep are grazers and select higher quality portions of available roughage. In contrast, goats perform poorly on monoculture pastures, and if given a choice, prefer a diet of approximately 80% browse and 20% grasses.[2] As is true for many illnesses, proper nutrition and management can prevent the most common diseases of small ruminants, including obstructive urolithiasis, intestinal parasites, and pregnancy toxemia. Prevention is much easier and rewarding than medical or surgical treatment.

UROLITHIASIS

Etiology and Pathophysiology

Although the formation of uroliths occurs in both sexes, clinical signs are almost exclusively observed in male ruminants. The anatomy of the male urinary tract may lead to accumulation of calculi and subsequent obstruction. The most common sites of urethral obstruction in small ruminants are the urethral process followed by the sigmoid flexure. Urolithiasis should be considered a metabolic disease, and different risk factors underlie its development. The mineral composition of the calculi is dependent on geographic region and diet. Small ruminants on high-grain diets such as show animals, pets, and finishing lambs most commonly develop struvite uroliths (magnesium ammonium phosphate) or apatite uroliths (calcium phosphate).[3] Diets containing a large portion of legumes predispose to calcium carbonate uroliths, and silicate uroliths occur in animals foraging on silicaceous plants as present in the western United States.[3] Castration before puberty is considered a risk factor for the development of urolithiasis. Wethers castrated at 5 months of age have significantly larger penile and urethral diameters as compared to those castrated at either 2 weeks or 3 months.[4] Castration of small ruminants anticipated to be at risk for urolithiasis should therefore be performed as late as possible.

Diet and water consumption are important risk factors for the development of urolithiasis. Diets high in cereal grains contain large amounts of phosphorus, and their consumption provides much additional phosphorus that has to be excreted by the kidneys. In ruminants, another route for phosphorus excretion is through salivary secretion with subsequent fecal elimination; however, high grain diets (especially in pelleted forms) elicit little cud-chewing activity and salivation, adding to the need for renal excretion.[5] Reduced water intake and resulting urine concentration are also important in the pathophysiology of urolithiasis. The disease is more common in the hot and cold months, when water intake is inadequate to facilitate high renal excretion. A change of water sources to less palatable water (e.g. chlorinated city water) can promote decreased water intake with resulting urine concentration and urolithiasis.

Clinical Signs

Affected animals show signs of abdominal pain and tread, swish their tails and grind their teeth.[5] Depression, lethargy, and recumbency are frequently observed. Tachycardia, tachypnea, and rumen stasis are common. Because a mild rumen bloat is frequently present and other clinical signs are similar, grain overload and rumen acidosis should be considered as differential diagnoses. Although urine dribbling may still be present, many affected animals are anuric. Evidence of gritty precipitates on the preputial hairs suggests the etiology. On ultrasound, a distended bladder, thickened bladder wall, and abdominal fluid may be observed.[5] In small ruminants, the bladder is more common to rupture, but as in cattle, urethral rupture may also be observed. On blood sample analysis, a stress leukogram, increases in BUN and creatinine, hyponatremia, hypochloremia, and acidemia may be present.[6] The serum potassium concentration is frequently low to normal as result of inappetence but can also be elevated.[6]

Therapy and Prevention

Both, surgical and medical management is possible and should be considered according to financial constrains, chronicity of the illness, and severity of the obstruction. In acute cases in which urine dribbling is still observed, medical management may be sufficient, but animals must be hospitalized for close observation and the possibility for timely intervention. Medical therapy consists of fluids to correct dehydration and electrolyte imbalances, anti-inflammatory drugs (flunixin meglumine at 1.1 to 2.2 mg/kg IV), and urine acidification (ammonium chloride, 200 – 300 mg/kg PO). Amputation of the urethral process is frequently necessary, when urine dribbling is absent or inadequate. To extend the penis, sedation should be given (acepromazine hydrochloride, 0.05 to 0.1 mg/kg IV or IM or diazepam 0.1 mg/kg IV).[3] The urethral process should be dissected at an acute angle using a scalpel blade. Scissor may crimp the urethral stump, decreasing the diameter of the new opening.[7] Although amputation of the urethral process provides immediate relief, the long-term prognosis for continued urination is guarded, especially if management changes are not implemented.

Another treatment option for struvite urolithiasis is the use of Walpole's solution (pH 4.5) that is instilled into the bladder and dissolves struvite uroliths and crystals that form only at alkaline ph. Under general anesthesia an 18 gauge 4 inch needle is inserted into the bladder under ultrasound-guidance. Once the bladder is emptied, 30-50 mL of Walpole's solution is instilled and again removed. This can be repeated 3-4 times until an additional 30-50 mL of Walpole's solution is instilled and left in bladder. Urine flow may resume in 24-36 hours with normal voiding occurring in 3-5 days. Some animals may require a second cystocentesis in 2-3 days.[8]

Surgical options include the perineal urethrostomy, tube cystotomy, or bladder marsupialization.[7] Each option can provide effective urine flow and bladder emptying, but the perineal urethrostomy should be considered a salvage procedure, as urethral strictures develop commonly. Tube cystotomy and bladder marsupialization procedures require several days of post-surgical after care, and because urine scalding occurs frequently with bladder marsupializations owners should be aware of cost and complications with each procedure.

When an animal presents with clinical urolithiasis, the remaining herd members should be considered at risk, and management changes implemented. In the prevention of struvite uroliths, dietary management is most important. The reduction of grain for male ruminants or complete elimination of grain from pet goat diets is essential. Increases in roughage to grain ratios, feeding of loose rather than pelleted feeds, and correction of the calcium to phosphorus ratio to 2:1 are useful. In feedlot animals or when owners chose not to reduce the grain intake, quantity and quality of water should be assessed. The addition of 5% NaCl to diets increases thirst and water intake. Ammonium chloride supplementation (0.5% - 1.0% dry matter) can be useful but reduces the palatability of feedstuffs and animals may sort out loose salts.

INTESTINAL PARASITES

Infections with parasites are common in small ruminants, but their impact on health largely depend on management. *Haemonchus contortus* is the most important nematode in the majority of small ruminant operations in the United States, but is found worldwide with in increasing distribution.[9] Other economically important nematodes include *Teladorsagia circumcincta, Cooperia curticei, Trichostrongylus* spp.[10] In addition, coccidiosis can be a major problem in young animals, and especially recently weaned lambs and kids in close housing conditions.[10]

Adult *Haemonchus contortus* attach to the mucosa of abomasum and cranial small intestine and consume blood. In contrast, *Trichostrongylus* spp. and other nematode parasites live in the small intestine and do not consume blood. Hence, blood loss and anemia are clinical features of only *H. contortus* infections. Other clinical signs of parasitism include unthriftyness, rough hair coat, submandibular edema (hypoproteinemia), and death. The development of nematodes from eggs to adult worms depends on environmental moisture and temperatures, and is only a few days in warm, moist climates. In the southern United States, parasite infections are a problem throughout the year. During periods of drought or cold weather, *H. contortus* undergoes hypobiosis, and large numbers of eggs are shed when weather conditions become more suitable for parasite development. Similarly, coccida must sporulate outside of the host, and sporulation is influenced by environmental conditions. Moderated temperatures and moisture enhance sporulation, and both unsporulated and sporulated oocyst can survive for extended periods of time in the environment.[10]

Some differences exist in the environmental risk factors for intestinal parasite infestation. Recently weaned lambs and kids are at greatest risk for coccidiosis when congregated on small pastures that result in excessive accumulation of oocysts. Clinical coccidiosis is rare when animals have ample excess to pasture and browse. Additional risk factors include stress of transport, weaning and feed changes.[10] Overstocking is among the largest risk factors for parasitism.[11] When close housing is unavoidable, routine removal of feces and feeding animals above the ground from elevated troughs and hay-racks is essential. Goats and sheep on permanent pastures without access to browse are at risk for nematode infestations. As nematode larvae climb onto blades of grass up to a height of 2-4 inches, over-grazing is detrimental.[11] This may be especially true for meat goats that, if given the choice, are browsers, and, in general, have less immunity against intestinal nematodes. These animals should have abundant access to woody browse and will prefer young growth of bushes and trees over pastures grasses. Within a herd, 30-35% of animals habor the majority of parasites.[10] Identification and removal of high shedders from herds, and implementation of rotational grazing programs will reduce the nematode burden of all herd members.

The high prevalence of resistance to dewormers must shift the focus of parasite control away from deworming to alternative management options, which include:

1) Improvement of housing and pasture
    (a) Provision of browse and high quality feeds
    (b) Implementation of rotational grazing
    (c) Feeding in elevated feed bunks and hay racks
2) Assessment of parasite burden in herd by:
    (a) Routine fecal flotation
    (b) Egg-count quantification by McMaster's technique
    (c) Assessment of anemia with FAMACHA program
3) Identification of anthelmintic resistance
    (a) McMaster on fecal samples before and 7-10 days after deworming
        - Deworming must result in >90 % reduction in egg count
    (b) DrenchRite assay
        - Cultures parasites *in vitro*, similar to bacterial sensitivity testing
4) Removal of highly infested shedders and genetic selection of small ruminants inherently resistant to parasites
5) Appropriate deworming of only those animals that need it
6) Biosecurity to keep animals with resistant parasites off the farm

7) Alternative control methods:
   a) Copper oxide particles
   b) Mixing livestock species
   c) High tannin-containing forages: e.g. *Sericea lespedeza*
   d) Nematode trapping fungi

PREGNANCY TOXEMIA

Etiology and Pathophysiology

Pregnancy toxemia is among the most common metabolic disorder of sheep and goats, and is the result of a negative energy balance. Various risk factors exist but a decreasing plane of nutrition and resulting hypoglycemia and hyperketonemia are central to the pathophysiology. Pregnancy toxemia typically occurs in the last trimester (especially last month) of pregnancy. At that time, the energy demand of the growing fetus may exceed the energy availability. Ewes and does carrying multiple fetuses are at increased risk, as females carrying twins require 180%, and those with triplets 240% more energy than females carrying singletons.[12] Large or multiple fetuses inhibit the abdominal space, further limiting the amount of food intake. Primary pregnancy toxemia occurs when the plane of nutrition is decreased in late pregnancy, but other stressors and risk factors exist. Over-condition at the end of pregnancy is a risk factor, because excessive amounts of fatty acids are released from storage and reach the liver. The liver is unable to process the large amount of fat and the glucose production is further inhibited. Any stressors, including inclement weather, illness, and handling can tip the balance between subclinical and clinical negative energy balance. The physiological response to negative energy balance is the mobilization of fatty acids for energy production through the Krebs cycle, but at the same time, a high fetal demand for glucose exists. A conflict between gluconeogenesis and energy production develops as liver stores for oxaloacetate are diminished. As free fatty acids can only be incompletely removed from the liver, and energy demands persist, ketone bodies are produced. Although ketone bodies provide energy to many tissues, they further decrease appetite and glucose production.

Clinical signs

Clinical signs are those of a hypoglycemic encephalopathy. Affected animals are separated from the herd, depressed or somnolent. Tremors, opisthotonos, stumbling and falling can be observed in later stages. Signs of abdominal discomfort, including grinding of teeth may be seen. Affected animals appear blind, walk into objects, and stand in abnormal postures, such as star-gazing. Intermittent convulsions can be observed and are followed by more normal periods.[12] Blood glucose concentrations are often low early in the disease but increase upon death of fetuses. Elevated ketone body concentrations in blood or urine are suggestive of the etiology.

Therapy and prevention

Depending on the stage of the disease, financial constraints, and whether dam or offspring are the primary interest, medical and/or surgical therapies are possible. The goal of therapy must be the discontinuation of the fetal energy demand, and a caesarean section is often the best choice for therapy. For surgical candidates, appropriate fluid therapy and glucose replacement should be chosen. The termination of pregnancy by injection of prostaglandins and/or steroids is suggested when the value of the patient does not allow surgery,[13] but may not induce parturition quickly enough. Administration of 50% dextrose (100 to 250 ml) temporally restores blood glucose concentrations and decreases the glucagon: insulin ratio, which inhibits further fat mobilization. Preferably, 5% dextrose solutions should be given as continued-rate infusions. Further supportive care includes the administration of rumen transfaunate from a healthy donor, high quality feeds, and oral propylene glycol (60 ml BID) as a ruminal propionate precursor.

The separation of dams into high and low risk groups aids in prevention as appropriate feeding plans can be designed for each group. To prevent secondary pregnancy toxemia, stressors such as transport or regrouping of animals should be avoided in late pregnancy. At-risk animals may be drenched with propylene glycol and/or receive ionophores which increase the availability of ruminal propionate for gluconeogenesis. Ionophores are beneficial for prevention but not treatment, because ruminal adaption to these antibiotics takes 3-5 days and results in temporarily decreased feed intake.

REFERENCES

[1]USDA APHIS NAHMS, **http://www.aphis.usda.gov/animal_health/nahms/smallscale/downloads/Small-scale_goat.**pdf, 2011; [accessed Jan 25, 2012]; [2]Rankins DL, et al. in: **Sheep and Goat Medicine**, Pugh DG, et al. 2012:18. [3]Jones M, et al. in **Sheep and Goat Medicine**, Pugh DG, et al. 2012:325. [4]Bani Ismail ZA, et al. **Bulg J Vet Med** 2007;10:29. [5]Van Metre DC in: **Large Animal Internal Medicine**, BP Smith. 2009:949. [6]George JW, et al. **J Am Vet Med Assoc** 2007;230:101. [7]Wolfe DF, et al., **Large Animal Urogenital Surgery**. 1998. [8]Janke JJ, et al. **J Am Vet Med Assoc** 2009;234:249. [9]Waller PJ, et al. **Vet Parasitol** 2004;122:207. [10]Miller JE, et al. in **Sheep and Goat Medicine**, Pugh DG, et al. 2012:106. [11]Fleming SA, et al. **J Vet Intern Med** 2006;20:435. [12]Edmondson MA, et al. in **Sheep and Goat Medicine**, Pugh DG, et al. 2012:150. [13]Brozos C, et al. **Vet Clin North Am Food Anim Pract** 2011;27:105.

EQUINE REHABILITATION

Anna Firshman, BVSc, PhD, DACVIM, DACVSMR
Saint Paul, MN

Equine Sports Medicine is a relatively new discipline, concerned with the diagnosis and treatment of medical and surgical problems of equine athletes. The discipline of equine sports medicine naturally encompasses the field of equine rehabilitation, which is the treatment of injury or illness, to decrease pain and restore form and function. Much of the practice of equine rehabilitation to date is based on clinical experiences and/or extrapolation from human studies, without formal research to validate specific therapeutic interventions, specifically in equine patients. In the last couple of years however, there have been significant advances in the research and evaluation of some rehabilitation treatments in horses, with the result that a body of scientific literature is steadily growing on the subject.[1,2,3,4,5,6,7] Recently, the American College of Veterinary Sports Medicine and Rehabilitation (ACVSMR) was formed to advance the art and science of veterinary sports medicine by promoting expertise in the structural, physiological, medical and surgical needs of athletic animals, and the restoration of normal form and function after injury or illness (www.ACVSMR.org). The aim of this new college is to strengthen the field of veterinary sports medicine, by encouraging research and training, in order to develop more effective prevention, management, and treatment of disorders affecting animal athletes.

Physical therapy is a well-established and integral component of human health care and the benefits of its procedures and treatments are supported by a solid base of basic and clinical research. Human physical therapy can trace its roots back to ancient Greek and Chinese medicine, but in recent times owes much of its development to its application in the treatment of disorders affecting victims of war, industrial accidents and disease outbreaks (such as polio epidemics).[8,9] The American Physical Therapy Association defines physical therapy as the diagnosis and management of movement dysfunction and enhancement of physical and functional abilities; the restoration, maintenance, and promotion of optimal physical function, wellness and fitness, and quality of life as it relates to movement and health; and the prevention of the onset, symptoms, and progression of impairments, functional limitations, and disabilities that may result from diseases, disorders, conditions, or injuries. It is important to understand that the practice of physical therapy is not simply the application of passive modalities such as therapeutic ultrasound, heat, cold, massage, and joint mobilization. Although these are commonly used tools in physical therapy, the physical therapist is specifically trained to treat underlying neurophysiologic and pathophysiologic contributors to pain and movement dysfunction, caused by a specific disorder. The type of treatment chosen will concentrate on alleviating sensory, neuromotor, and mechanical dysfunction. Techniques used involve pain modulation, manual therapy, proprioceptive and motor retraining, therapeutic exercise, and functional and sport-specific activities.[8]

Today some veterinary schools are teaching the principles of techniques used in equine rehabilitation to veterinary students, with the result that more equine veterinarians are and will be incorporating these principles into their management of equine sport related injuries. The AVMA and AAEP define veterinary physical therapy as the use of noninvasive techniques (such as massage, stretching, laser, electrical stimulation, magnetic, ultrasound, rehabilitative exercises, hydrotherapy, heat, and cold) for the rehabilitation of injuries. In addition, both the AVMA and AAEP state that veterinary physical therapy must be performed under a referral of a veterinarian after a veterinary diagnosis. Both these organizations assert that veterinary physical therapy should be performed by a licensed veterinarian, or where in accordance with state practice acts, by a licensed, certified, or registered veterinary or animal health technician educated in veterinary physical therapy, or a licensed physical therapist educated in nonhuman animal anatomy and physiology. It has become customary in recent years to call this type of therapy "rehabilitation" rather than "physical therapy", in order to avoid misrepresentation of the term "physical therapy" by non-physical therapists. These regulations emphasize the importance of collaboration of the veterinary physical therapist, veterinary rehabilitation technician and veterinarian. It is essential that, prior to implementation of a rehabilitation plan, a veterinary examination and diagnosis be made, to identify conditions in which particular physical therapy interventions may be contraindicated. If possible the veterinarian should determine the pathological diagnosis, treatment (which may include analgesics, non-steroidal anti-inflammatory drug (NSAIDS), casts and bandages, and surgical interventions), and the integration of physical therapy. Good communication between the various professions and owner is paramount to ensure the best outcome for the horse.

Common candidates for rehabilitation therapy include equine patients with sport related injuries, neurologic or musculoskeletal disorders, or those recovering from a surgical procedure. There are numerous conditions that may be primarily cared for by an internal medicine service that can benefit from rehabilitation therapy. Examples include musculoskeletal disorders and neurological disorders in foals, adult horses with neurologic disorders, muscle disorders, or simply those patients who are confined to a stall due to their requirements for continuous treatments (e.g. continuous rate infusions). The detrimental effects of stall rest in a horse have not been studied in the same detail as the effect of bed rest in humans; however it is well known that the absence of mechanical loading within the musculoskeletal system induces maladaptive responses in bones, muscles, joints, and ligaments.[8] It is also generally accepted that biological tissues have a great capacity for adaptive change, and one of the most potent factors leading to adaptive responses is chronic mechanical loading or stress.[8] These principles are a stark contrast to the age old recommendations for prolonged stall rest in injured horses. Thus, controlled loading of injured tissues based on scientific principles is one of the underlying concepts behind many rehabilitation exercises used in horses.[6] There are many examples of such exercises that can be carried out on an equine patient while on stall rest and these can be used to improve balance, core stability and strength. Many of these exercises are colloquially known as "carrot stretches". For further descriptions of exercises that can be utilized in the rehabilitation of horses, the reader is directed to other sources.[10]

In addition to therapeutic exercises, there are a variety of physical modalities that can be utilized in a rehabilitation plan in order to optimize tissue healing and minimize the effects of disuse or immobilization. These rehabilitation modalities include cold therapy, superficial heat therapy, therapeutic ultrasound, electrotherapy (such as TENS and NMES) and hydrotherapy (such as swimming and underwater treadmill exercise). The decision to use a particular modality should be based on stage of healing, effect of the modality, and any contraindication for its use. For complete descriptions of theses modalities the reader is directed to other sources.[1]

Equine rehabilitation is a new and exciting field. While the veterinarian in charge of the patient should be involved in the assessment and decisions about rehabilitation treatments, there are many opportunities for the involvement of veterinary technicians in the rehabilitation team. Equine rehabilitation practices often have a high throughput of cases and well trained veterinary technicians can help enormously in maintaining high standards of care. There are many skills possessed by veterinary technicians that can be incorporated into a rehabilitation service including basic physical assessments and history taking, client education and performing rehabilitation exercises and modalities.

REFERENCES

[1]McGowan C et al, Animal Physiotherapy: Assessment, Treatment and Rehabilitation of Animals, Blackwell Publishing 2007. [2]Clayton HM, et al, Evaluation of biomechanical effects of four stimulation devices placed on the hind feet of trotting horses. Am J Vet Res. 2011;72(11):1489-95. [3]Stubbs NC. Osseous spinal pathology and epaxial muscle ultrasonography in Thoroughbred racehorses. Equine Vet J. 2010;42 Suppl 38:654-661. [4]Clayton HM. Dynamic mobilisations in cervical flexion: Effects on intervertebral angulations. Equine Vet J. 2010;42 Suppl 38:688-694. [5]Clayton HM. Swing phase kinematic and kinetic response to weighting the hind pasterns. Equine Vet J. 2011;43(2):210-215. [6]Stubbs NC. Dynamic mobilisation exercises increase cross sectional area of musculus multifidus. Equine Vet J. 2011;43(5):522-529. [7]Clayton HM. Hindlimb response to tactile stimulation of the pastern and coronet. Equine Vet J. 2010;42(3):227-233. [8]Paulekas R, Haussler KK. Principles and Practice of Therapeutic Exercise for Horses. J Eq Vet Sci 2009;29(12):870-893. [9]Pagliarulo, MA. Introduction to Physical Therapy 4th ed. 2007 Elsevier Mosby, St. Louis, MO. [10]Stubbs NC. Activate your horse's core. 2008 Sport Horse Publications, Lansing, MI.

EQUINE CARDIOLOGY: PHYSIOLOGY, ACQUIRED AND CONGENITAL DISEASE

H. Edward Durham, Jr., CVT, LATG
Columbia, MO

INTRODUCTION

Horses have one of the most interesting and amazing cardiovascular systems. Much of human interaction with horses and other equines is a result of athletic pursuits and their amazing ability to perform work. The basic physiologic and pathophysiologic principles are broadly similar between mammalian species, yet breed and species differences do exist.

EXERCISE PHYSIOLOGY

Exercise physiology is an important component of equine cardiology. Assessing the athletic ability of horses is an area of ongoing research. Many cardiovascular indices have been evaluated or are being actively investigated, but no single parameter has been found that will accurately predict a horse's future performance.

Perhaps one of the simplest indices is the evaluation of heart rate (HR). It is well accepted that the better an athlete's conditioning the lower their HR at a given level of exercise. In fact the relationship of HR to work effort is linear at sub-maximal exercise in humans and horses. The healthy horse has a resting HR of 24-48 bpm, but at peak exercise the HR may reach 240-260 bpm. One method that standardizes evaluation of HR at submaximal performance is known as V_{140} and V_{200} or the horse's velocity at a HR of 140 or 200. Simply stated, this test asks the question, how fast is the horse running when its HR reaches 140 or 200 bpm? By running a horse on a treadmill and attaching an ambulatory electrocardiograph (ECG) monitor, one can record the velocity of the treadmill when the HR reaches the target. A better performer should have a higher V_{140} or V_{200}; however this test may be better at comparing a horse to itself during training than predicting performance between two horses.

Another method of assessing cardiovascular performance is to look at the maximal uptake of oxygen (O_2) possible for an individual athlete. This index is called VO_2 max and also has a linear relationship with HR, which makes the use of HR monitoring useful since it is so closely linked with O_2 demand. When prolonged sub maximal work is undertaken the HR tends to plateau and stabilize until a higher level of work is performed, although some cardiac drift may occur, especially in dehydrated animals (as seen with prolonged exercise). This observation seems to correlate well with VO_2 max. Interestingly, stroke volume reaches its maximum at 40% of VO_2 max and further increases of cardiac output result from increases in HR. Cardiac output, or the volume of blood pumped by the heart in one minute increases from about 20 liters at rest to nearly 300 liters during peak exercise.

Post exercise HR has also been used to evaluate fitness. Following exercise the HR falls quite quickly to a steady-state higher than the resting HR for a short period then gradually decreases to normal. Horses with a higher post exercise heart rate tend not to perform as well as those with lower post exercise HR. Research demonstrated that endurance horses that had a resting HR of greater than 65-70 after 30 minutes were much more likely to not finish or to finish poorly and horses with a HR 60 bpm after 30 minutes (HR >50 bpm not uncommon post exercise) had less evidence of dehydration during a rest period echocardiogram[1]. Rapid HR recovery (post exercise) implies better cardiovascular health.

Several studies have investigated the use of ECG changes to predict performance. The Heart Score is the mean of the QRS durations (ms) in the frontal plane bipolar leads and was shown to be correlated with Thoroughbred earnings in one study[2]. Additional research has not supported those findings. Another method evaluated changes in T wave morphology as a sign of myocarditis in horses with poor race performance. Again further research has brought doubt on T wave morphology as a clinical tool. At this time, the ECG is not thought to give much information on performance other than ruling out clinical arrhythmias that might reduce performance. Echocardiographic indices of heart size have been related to performance in which better performing horses also showed slightly larger cardiac size by echocardiography in some but not all studies. However, the predictive value may be limited and be best suited for equine endurance athletes[3].

PHYSICAL EXAMINATION

Histories of exercise intolerance, poor performance, and identification of arrhythmias and/or murmurs are the most common indication for a cardiology work up. Some horses present with ventral subcutaneous edema, fever of unknown origin or rarely, coughing. Many murmurs are detected during routine veterinary care or pre-purchase examinations. When a significant cardiac condition is suspected a complete physical examination should be performed, with attention to those aspects of the examination specific to the cardiovascular system. Before exciting the patient with palpations, the examiner should observe the horse's respiratory effort and rate. One should observe the patient's jugular veins for distension or pulsation. Pulsations that extend higher than the lower third of the neck are abnormal, implying elevated right atrial pressure and tricuspid regurgitation (TR). Care must be given in this assessment not to confuse carotid pulses under the jugular vein as jugular pulsations. This false positive finding is especially common in thin horses. Arterial pulses can be felt in the facial artery under the proximal third of the jawbone. Hypokinetic pulses are associated with conditions that cause hypotension. Hyperkinetic pulses can be detected with significant aortic regurgitation (AR). The facial artery can also be used to palpate pulses during auscultation to evaluate for pulse deficits. One should also palpate both sides of the thorax for the presence of palpable vibrations or "thrills" associated with murmurs and for the apical impulse. Auscultation typically begins over the mitral valve area, which is located about the 5th-6th intercostal space just caudal to the left elbow. The S_1 sound and murmurs of mitral regurgitation (MR) are best heard in this area. Auscultation proceeds cranially and dorsally where the semilunar valves are evaluated. Note the HR and rhythm for arrhythmias. Foals have higher HR than adults; around 80-120 bpm that gradually slows as they age. Auscultation of the right hemithorax should also be performed. Murmurs associated with TR and ventricular septal defects (VSD) are best heard on the right side.

During auscultation it is normal to hear two, three or four sounds of the cardiac cycle; the first and second heart sounds (lub & dub) typically heard in all mammals and two additional sounds (S_3 and S_4) associated with ventricular filling are usually appreciated in horses. The third heart sound, S_3 is associated with early ventricular filling; the S_4 sound also called the "atrial contraction sound" denotes atrial systole.

Murmurs indicate turbulent blood flow in the heart. A number of murmurs may be detected in horses. Two of the most common murmurs are functional murmurs and MR from valvular degeneration. Functional murmurs are usually systolic murmurs which occur in a structurally normal heart. One study reported functional murmurs occurring in 66% of horses[1]. Another common cause of systolic murmurs, especially in young horses, is the VSD. This murmur generally radiates towards the right sternum and can be quite loud. Diastolic murmurs of AR, which have been described as a "dive bomber" sound, are also heard in the clinically normal animal and are not uncommon. One can sometimes hear a systolic/diastolic combination murmur or "to/fro" murmur. This unusual murmur can be heard in healthy animals with functional murmurs or in aged horses with valvular degeneration. It is distinguished from the continuous murmur by a short break in the sound that occurs when the heart changes from systole to diastole. Continuous murmurs of patent ductus arteriosus can be heard during examination of young foals less than a week old but are extremely rare in horses older than that. Also, pre-systolic S_4 and S_3 filling murmurs may also be heard in clinically normal animals.

One important clinical challenge is to distinguish MR from functional murmurs. They can partly be differentiated by their timing. The murmur of MR is typically holosystolic (lasting throughout the duration of systole), whereas functional murmurs tend to be restricted to early systolic to mid-systolic periods. Functional murmurs are best heard over the heart base and MR murmurs best over the left apical impulse. Functional murmurs tend to change with exercise. After exercise the murmur may increase or decrease in intensity. Functional murmurs typically have a crescendo or decrescendo pitch in contrast to the MR sound of a plateau shaped murmur that changes little in intensity. Echocardiography is the best diagnostic tool to identify the cause of a murmur and definitively differentiate the two.

A summary of major equine murmurs by their timing reveals the systolic murmurs to include: functional murmur, MR, TR, and the VSD. The murmurs of diastole include: presystolic S_4 or S_3 filling, and AR. There is always the potential for a to/fro murmur as well. Diseases of the pericardium can produce rubbing, or knocking sounds. These sounds can occur at any time during the cardiac cycle when the heart is in motion.

DIAGNOSTICS

Electrocardiography is a valuable tool for assessing equine cardiac rate and rhythm and diagnosing arrhythmias. Due to the conduction system of the horse, assessment of Mean Electrical Axis and inferences of chamber enlargement are less useful. Horses have a different cardiac conduction system compared with dogs. In the dog, the Purkinje fibers extend along the endocardium and the impulse is carried to the epicardium via cell to cell depolarization. In horses, the conduction fibers extend to the epicardium and endocardium allowing for nearly simultaneous depolarization of the entire myocardium at once. Consequently, it becomes unnecessary to perform a recording in the frontal plane to create the hexaxial lead system common in small animals. The standard method of recording the ECG in horses is from the base-apex lead. With the electrocardiograph set to print lead I or II the base-apex lead is created by placing a negative electrode at the right jugular furrow (typically white) and a positive electrode (black or red) at the left precordial apical impulse with a ground electrode attached in an unobtrusive place. A typical equine ECG recorded in the fashion demonstrates a positive "P" wave, negative "QRS" complex and positive "T" wave for which normal values are published[1].

Echocardiography can be used for assessment of cardiac chamber size and systolic function. By imaging from both sides of the horse, the entire heart can usually be visualized except in very large draught horses. Precise chamber measurements can be taken, systolic function assessed, and murmurs can be located using color flow Doppler. The experienced echo-sonographer can calculate pressure differences based on the velocity of blood movement through the heart. Cardiac masses and pericardial effusion can be identified. An echocardiogram can be useful for horses with AF about to undergo quinidine therapy to assess the overall heart function before it is suppressed by therapy.

ARRHYTHMIAS

Arrhythmia detection and characterization represent an important part of the examination as arrhythmias are commonly associated with poor performance. Most arrhythmias detected in horses are physiological and manifest no clinical signs. Perhaps the most common is physiological 2nd degree atrioventricular block. This arrhythmia is related to strong parasympathetic vagal tone seen in horses and abates with exercise. Impulses from the sinus node are blocked at the atrioventricular node when the horse is calm and sympathetic tone is low. When stimulated and sympathetic tone increases, it overrides the vagal tone and the block disappears. Other arrhythmias of no clinical significance are transient sinus arrhythmia that may be seen following exercise, sinoatrial block, wandering pacemaker, and 1st degree atrioventricular block. Equine patients presenting with gastrointestinal disease may show premature atrial complexes on an ECG that have been detected during auscultation.

Atrial fibrillation (AF) is an abnormal cardiac rhythm that can occur in horses with normal hearts, and is of clinical importance because it is a leading cause of non-lameness performance reduction. Atrial fibrillation may sometimes occur transiently during or after exercise. If AF is sustained it can decrease cardiac performance leading to a reduction in cardiac output and therefore overall performance. Atrial fibrillation in horses is characterized by an undulating baseline with no clear P waves. Although the rhythm is irregular, the rate is typically within normal limits due to the strong vagal system. The QRS complexes have a normal morphology.

Atrial Fibrillation occurs in ostensibly normal myocardial tissue in many horses. In this case, treatments that convert the arrhythmia to a sinus rhythm are preferred. If underlying cardiac disease is the cause of AF then treatment to control the ventricular rate is preferred. In most cases, AF is treated with oral quinidine sulphate administered via nasogastric tube. Before treatment begins a standard base-apex ECG should be obtained. The QRS complex duration should be measured and recorded before initiating therapy. If the QRS duration increases by more than 25% during treatment, quinidine therapy should be discontinued. Quinidine has a narrow limited therapeutic range of only 2-5 μg/ml with several side effects including hypotension, tachycardias, alterations of conduction, CHF, sudden death, colic, diarrhea, ataxia, laminitis, airway obstruction, urticaria and/or wheals and depression. Close observation is an essential part of quinidine therapy. Continuous ECG monitoring is ideal to watch for prolongation of the QRS complex, ventricular arrhythmias or supraventricular tachycardias.

Quinidine is usually administered every 2 hours for the first four doses and then blood levels are checked. Dosing is reduced to approximately every 6 hours until conversion occurs. Blood quinidine levels should be checked as needed. If after two days the AF persists, additional treatment with digoxin can be implemented. This protocol requires extreme care since quinidine and digoxin compete for the same receptor site and can predispose the patient to digoxin toxicity. Toxic side effects of digoxin overlap those of quinidine a great deal so the assault to the gastrointestinal system and the cardiovascular system is especially severe. An alternative protocol calls for the cessation of treatment for 24 hours after the first two days of therapy. Some horses will convert during this drug holiday; if the AF persists then quinidine can be re-initiated as before.

Recently, new antiarrhythmics such as flecainide and amiodarone have been evaluated for converting AF to a sinoatrial rhythm. Research is ongoing, with amiodarone holding promise; however flecainide may be limited to use in acute AF. Perhaps the boldest therapy is transvenous electrocardioversion. During this cardiac catheterization procedure, special electrode-tipped catheters are advanced into the right heart and an electrical current applied to defibrillate the atrial tissue returning the patient to sinus rhythm. Electrocardioversion requires special equipment and general anesthesia, so it is offered only in limited facilities.

Ventricular arrhythmias and 3rd degree or complete AVB are also seen in horses. Ventricular arrhythmias can be treated with lidocaine or in the case of refractory ventricular tachycardia, magnesium sulphate. Horses with complete AVB must receive a permanent pacemaker, but will never return to athletic activity.

CONGENITAL HEART DISEASE

Ventricular septal defects (VSD) are the most commonly reported congenital defect in horses. These developmental defects can range from life threateningly severe to incidental. Murmurs of VSD are typically ≥ grade 3 systolic right sided murmurs that radiates towards the sternum. Many VSDs are small and restrictive, meaning the diameter is small enough that most flow through the defect is inconsequential. It is possible to find a restrictive VSD as an incidental finding during echocardiogram in an otherwise normal horse. These defects are in the membranous region of the septum just under the aortic valve. If the defect is large, the shunt volume can over circulate the lungs and the left heart and can ultimately lead to left sided congestive heart failure (CHF). Alternatively, the aortic valve apparatus may be deformed and serious AR may result leading to CHF. A less common VSD can be seen in the infundibular region near the pulmonic valve. This defect is interesting in that the resulting murmur sounds like pulmonic stenosis (PS). However since PS and aortic stenosis are extremely rare in horses (unless they are seen as part of a more complex set congenital defects) it can be easily misdiagnosed by auscultation alone. Even with echocardiography the infundibular VSD can be difficult to diagnose. Complex congenital defects are most often seen in foals and are usually life-threatening, causing early death. Echocardiography is extremely helpful in evaluating congenital cardiac disease.

ACQUIRED HEART DISEASE

As horses age they may develop a chronic myxomatous valvular degeneration that primarily affects the aortic and mitral valves. Valvular degeneration can lead to incompetency, murmurs and, if regurgitation is severe enough, CHF. Mitral regurgitation from valvular degeneration is one of the most common causes of murmurs in horses. It is also a common cause of heart failure and can lead to pulmonary rupture and sudden death. Echocardiograms reveal thick valve leaflets that may prolapse into the left atrium. Color flow Doppler demonstrates large color jets crossing into the left atrium during systole creating a systolic murmur. Degeneration that affects the aortic valve may cause a diastolic murmur. As the aortic valve degenerates and more leakage occurs, the volume overload from blood moving back from the aorta eventually causes the ventricle to fail.

Dilated cardiomyopathy is virtually the only cardiomyopathy seen in horses and has been associated with CHF. Echocardiographically it has the same hallmarks as seen in dogs; a dilated left ventricle with poor contractility. The prognosis is poor for this condition and therapy for CHF is the only course.

Inflammatory conditions such as endocarditis and pericarditis are also seen in horses and both can present with a fever. The diagnosis of both of these conditions is aided by echocardiography. Pericardial effusion can be seen readily by ultrasound which can then be used to guide placement of cannula for fluid removal. Vegetative lesions on the valve due to endocarditis may be apparent by echocardiography, thereby supporting the clinical diagnosis and helping to assess the damage.

Lastly, toxicities such as ionophore toxicity and White snakeroot poisoning can lead to sudden death in horses from ventricular arrhythmias. Ionophore toxicity is typically introduced from cross contamination of cattle feed into the equine food source. Horses affected with ionophore toxicity need care for several months after exposure as the full effects can be delayed. White snakeroot toxicity (*Eupatorium rugosum*) can occur in any grazing stock in the Midwest, and can also lead to heart failure in horses.

Treatment of heart failure in horses is typically limited to the use of furosemide, digoxin and angiotensin converting enzyme inhibitors.

REFERENCES

[1] Marr C, 1999 **Cardiology of the Horse,** Saunders [2] Steele **Austrail Medical Pub** 1963 [3] Young LE, Wood JLN, **ACVIM Proceed** 2003

HYPERTROPHIC CARDIOMYOPATHY SCREENING IN CATS

Charlotte Pace, BA (Hons), RVN, VTS (Cardiology)
Hatfield, UK

INTRODUCTION

Screening for hypertrophic cardiomyopathy (HCM) is relevant because of the prevalence of the disease in the cat population. Recent studies have reported that up to 15% of apparently healthy cats may be affected (1, 2). HCM can affect cats of any age, sex, pedigree and non-pedigree breeds. Prognosis varies with the severity of the disease, but if a cat is affected, the consequences can be very serious. HCM can result in pleural effusion, pulmonary oedema, aortic thromboembolism (ATE) or sudden death. It is probable that sudden death is under-reported. For breeders and owners, even focusing clinical trials to target specific groups of cats, screening could be very useful. It is hoped that information gained could provide more informed guidelines as to when and how to treat affected cats.

DEFINTION

HCM is a primary (idiopathic) cardiomyopathy, and it is believed to have a genetic cause in most cases. It is a disease characterised by a thickening (hypertrophy) of the left ventricle. Left ventricular hypertrophy causes stiffness of the heart muscle meaning that the ventricle cannot relax as it should. The increased thickness of the heart muscle may also exceed the myocardial blood supply, causing localized areas of ischaemia. This further exacerbates stiffness and inefficiency of the ventricle. Some parts of the ventricle get thicker than others, and if the increased thickness occurs on the interventricular septum, it can cause an obstruction to blood flow out of the left ventricle into the aorta (sometimes referred to as hypertrophic 'obstructive' cardiomyopathy or systolic anterior motion of the mitral valve leaflet). This compromises cardiac output further, because the ventricle has to counter increased resistance. With either form of HCM, the left atrium may increase in size to try and accommodate the increased pressure in the ventricle. Sometimes this can lead to left sided congestive heart failure. The pattern of hypertrophy can be variable and this may be related to prognosis.

DIAGNOSIS

Diagnosis of HCM is difficult. Physical examination may or may not be helpful. Cats with the obstructive form of the disease may have a heart murmur, but those without outflow tract obstruction may not have a murmur at all. In a recent retrospective study of cats presenting to a cardiology referral centre, 71% of cats had heart murmurs (3), although this number is probably biased because cats are more likely to be referred for investigation of a heart murmur. The same study showed that the presence of a gallop rhythm and/or arrhythmia were more likely to present with congestive heart failure. This suggests these physical exam findings are either unreliable as an indicator of the presence of HCM, or are too late for screening. Echocardiography therefore, is the preferred method for diagnosing HCM.

SCREENING

There are multiple reasons why screening cats with HCM is useful. Breeders are often very aware of the potential risks and prevalence of the disease. Existing data, such as the work done on Maine Coons and Ragdolls (4) has made breeders very keen to screen before breeding. Cat owners sometimes also actively request screening of a sibling or relative, if they have had a cat affected, or die from HCM. Veterinarians often wish for a cat to be screened before intravenous fluid therapy or for suitability for anaesthesia.

PROBLEMS

Echocardiography is regarded as the 'gold standard' for diagnosing HCM. Rather surprisingly however, there seems to be little consensus on exact measurement criteria. This is particularly relevant for breeders, when a misplaced calliper can mean the difference between diagnosing a cat as 'affected' or 'healthy'. Understandably, many technicians and veterinarians find this particularly challenging. The differences occur because some cardiologists will use M-Mode measurements to define left ventricular hypertrophy, whereas others will use 2-D echocardiography (2). There is also disagreement on what is classified as 'thick'. Upper limit references can vary between 5mm and 6mm. The difference with this criteria makes assessment very difficult, and so standardisation of echocardiographic measurement is required. Current recommendations include (5): -

- Specify technique used (2D or M-Mode)
- Measurement technique should be specified to describe
  - Maximal wall thickness or mean of several measurements
  - Maximal focal thickness or wall thickness affecting at least 50% of wall segment
  - Specification of the number and precise planes used for 2D
  - Specification of whether or not endocardial surfaces are included in measurement
  - Avoidance of false tendons
  - Record, quantify and report any dynamic outflow tract obstruction
  - Specify if any papillary muscle hypertrophy

Other screening options are oral (cheek) swabs for known genetic mutations. This is useful for testing one of the known mutations (such as those identified in Ragdolls and Maine Coons (4)), but many cats can be gene positive and phenotype negative. Also, there may be other genetic mutations, so it is possible that some cats are gene negative but phenotype positive. Biomarkers such as NT-ProBNP have been shown to be able to distinguish healthy cats from affected cats and from cats with congestive heart failure (6, 7).

FUTURE

Due to the potential seriousness of HCM, there is a general desire to know more about the disease in cats. To be able to predict with some accuracy those cats that are at risk of developing CHF and thromboembolic disease would be invaluable. If more information became available, it could also mean that clinical trials could be targeted at specific groups of cats. This might then be able to guide veterinarians treat cats with HCM appropriately.

REFERENCES:

1. Paige CF, et al. **JAVMA** 2009; 234:1398. 2. Wagner T, et al. **J Vet Cardiol** 2010; 12:171. 3. Payne J, et al. **J Small Anim Pract** 2010; 51:540. 4. Meurs KM, et al. **Genomics** 2007; 90:261 5. Luis Fuentes V, **ACVIM proceedings** 201; 148-149. 6. Connolly D, et al. **J Vet Cardiol** 2009; 11; S41-S50. 7. Connolly D, et al. **J Vet Cardiol** 2009; 11; S51-S61

DIAGNOSIS AND TREATMENT OF CHRONIC MITRAL VALVE DEGENERATION IN DOGS

Charlotte Pace, BA (Hons), RVN, VTS (Cardiology)
Hatfield, UK

INTRODUCTION

Chronic mitral valve degeneration is the most common acquired cardiac disease in dogs. Popularity of the breeds that are predisposed to mitral valve disease and the relationship between mitral regurgitation and morbidity and mortality linked to congestive heart failure (CHF) (1) make diagnosis and treatment an everyday concern for many veterinary practitioners. Management of chronic valve disease varies depending upon the stage of heart disease present. The recent ACVIM consensus (2) provided guidelines for classification and treatment. Notably there were some areas where consensus was not reached. Technicians should be aware of the classification systems, as this can guide diagnostic tests and treatment.

CLASSIFICATION SYSTEMS

Classification systems are beneficial because they can provide guidance regarding treatment and monitoring. They are also helpful in establishing severity of the disease process. There are different classification systems available, but perhaps the most useful is the 2001 American College of Cardiology/American Heart Association (3), which was adapted for the ACVIM consensus statement on canine chronic valvular disease (2).

STAGE A: PATIENTS AT RISK

Patients in this category are predisposed to developing heart disease, such as small-medium size terrier breeds (4). At this stage, patients do not have any structural cardiac abnormalities.

Diagnosis

At risk patients should have regular physical examination checks, which should include auscultation. Annual vaccination checks are ideal to monitor for development of a heart murmur.

Treatment

No treatment is recommended at this stage.

STAGE B: STRUCTURAL HEART DISEASE, NO SIGNS OF HEART FAILURE

Patients in this category have had a left apical systolic heart murmur detected, but show no signs of heart failure.

Diagnosis

Physical examination should be routine, and include recording the intensity and location of the murmur. Thoracic radiography is highly recommended at this stage. It will provide baseline comparison for disease progression for the individual patient, which can be more sensitive than vertebral heart scoring (1). Blood pressure measurement is recommended to rule out any other underlying diseases. Echocardiography is also suggested to confirm diagnosis and take baseline measurements for mitral regurgitation velocity, left atrial size and systolic function. From these diagnostic tests, this category is then sub-divided into two groups: -

- Stage B1: Patients with no evidence of cardiac remodeling on radiographs or echocardiography;
- Stage B2: Patients with evidence of cardiac remodeling on radiographs or echocardiography.

Treatment

Stage B1: No treatment is recommended at this stage.
Stage B2: Treatment at this stage is controversial. Recent literature has produced conflicting results regarding the use of angiotensin converting enzyme inhibitors (ACEi) at this stage (5, 6). One study that was examining the use of beta-blockers at this stage was terminated. There is currently a study in progress investigating the use of pimobendan at this stage (7).

STAGE C: PREVIOUS OR PRESENT CLINICAL SIGNS OF HEART FAILURE

This category includes patients that have had, or are experiencing clinical signs of heart failure as a result of chronic valvular disease. This is a wide spectrum ranging from management of the outpatient to the patient presenting with acute heart failure.

Diagnosis

Due to the wide difference in presentation of this class, diagnostic tests and timing will vary. Physical examination and monitoring respiratory rate and patterns are crucial. The outpatient may require radiographs to assess for signs of pulmonary congestion and/or assess any other underlying disease process. Echocardiography is also useful to confirm diagnosis, if diagnosis has not already been made. Blood sampling is also recommended, either for baseline results for renal parameters, or to exclude any other disease, such as pre-existing renal disease, or to monitor response to medical therapy. Blood pressure monitoring is also indicated. The acute heart failure patient will need stabilization before diagnostic tests are performed.

Treatment

Outpatient: Treatment goals are to remove abnormal fluid retention, neurohormonal blockade and improve hemodynamic function. Owners should be encouraged to monitor respiratory rate and effort, demeanour, exercise tolerance, and appetite.

Acute heart failure patient: Treatment goals for the emergency patient include increasing oxygenation, removal of effusions if appropriate, diuresis, optimizing hemodynamic function, and supporting arterial pressures if necessary. Monitoring and assessment of the patient will guide further treatment.

STAGE D: REFRACTORY HEART FAILURE

This category includes patients that have become unresponsive to standard CHF therapy.

Diagnosis

These patients will already have been diagnosed with CHF due to chronic valvular degeneration. Monitoring and titrating therapy is the main aim at this stage. Recording respiratory rate and effort, monitoring blood pressure and measuring renal and electrolyte parameters are important.

Treatment

Treatment goals at this stage are to eliminate abnormal fluid retention using a combination of diuretics (sequential nephron blockade). Thoracocentesis and/or abdominocentesis might also be necessary. Mechanical ventilation might be recommended to allow time for therapy to act. Arterial vasodilation using sodium nitroprusside, hydralazine, or amlodipine, might be suitable for some patients requiring afterload reduction. Cachexia is also a concern at this stage. Omega-3 fatty acid supplementation has been shown to have a beneficial effect on dogs with cachexia (8).

REFERENCES:

1.Lord PF et al, 2011, **JVIM**, 25:1312-1319 2. Atkins et al, 2009, **JVIM**, 23:1142–1150. 3. American College of Cardiology/American Heart Association 2001, Oyama M, In **BSAVA Manual of Canine and Feline Cardiorespiratory Medicine** 112-120. 4. Swenson L et al, 1996 **JAVMA**, 15; 208 (12): 2009-12. 5. Kvart C et al, 2002, **JVIM**; 16:80-88. 6. Atkins CE et al, 2007, **JAVMA**; 231; 7; 1061-1069. 7. EPIC study, Boswood A et al, 2011 **Vet Record** 168(8): 222. 8. Freeman LM et al, 1999, **JVIM**, 12, 440-448

TREATING THE FINAL STAGE OF HEARTWORM DISEASE: CAVAL SYNDROME

Bonnie Heatwole, CVT
Gainesville, FL

INTRODUCTION

Although Caval Syndrome is considered uncommon, it is a serious and potentially fatal disease. Patients diagnosed with this stage of Heartworm Disease can die from cardiogenic shock, metabolic acidosis, disseminated intravascular coagulation, multiple organ failure and/or anemia within 12 to 48 hrs of presentation. Dogs that have developed heavy burdens of worms are prone to acutely developing caval syndrome. A mass of worms can become entangled in the right atrium and tricuspid valve causing an obstruction of blood flow, intra- vascular hemolysis and right sided heart failure. Patients diagnosed with caval syndrome require surgical removal of the worms.

Echocardiography is the diagnostic test of choice to confirm the stage of heartworm disease and to access the worm burden. Worms can be visualized in the right atrium and in the posterior vena cava. Worms have also been found in the hepatic veins and can induce paradoxal septal wall motion which is caused by tricuspid regurgitation. Right ventricular dilation and hypertrophy will be noted as well as a small, volume depleted left ventricle.

ANESTHETIC PROTOCOL

Due to cardiovascular depression and high mortality rates, avoidance of volatile anesthesia is desirable. At the University of Florida, Small Animal Hospital, our current anesthesia protocol calls for total intravenous anesthesia (TIVA) using injectable drugs that are short acting and that have minimal effects on myocardial depression and blood pressure. We also pre-treat patients with dexamethasone and diphenhydramine to avoid a serious anaphylactic reaction from the systemic release of antigens during worm removal and possibly worm crushing.

The following is our recommended anesthetic protocol when performing surgical worm removal for caval syndrome:

Pretreatment:

Dexamethethasone- 0.14mg/kg IV
Diphenhydramine- 2mg/kg IM
Atropine- 0.02mg/kg IV
Heparin- 500units/kg SQ
Lidocaine Local Block
Oxygen

Induction:

Etomidate- 1mg/kg IV bolus

Maintenance:

Propofol CRI- 50 – 300mcg/kg/min
Fentanyl CRI- 0.5mcg/kg/min
100% oxygen

RETRIEVAL DEVICES

There are a number of devices on the market that may be utilized for adult heartworm removal. At the University of Florida, our preferred method is the use of the Ishihara retrieval forceps. The benefit of these retrieval forceps is that they are flexible and steerable. Additionally, they have flat jaws that do not completely close, thus significantly reducing the chance of worm crushing leading to dangerous complications. However, these retrieval forceps are no longer manufactured and this has led us to investigate other possible retrieval devices. A basket retrieval device used with a check flow introducer passed into the right atrium has also been used by our group. This was successful and inexpensive but required removing the entire check flow catheter with each passage of the system. A few other devices which may be used are endoscopic retrieval devices such as the Ensnare retrieval device, a rubber tipped grasping forcep or a soft wire basket retrieval device. The latter two are manufactured by Olympus. The soft wire basket device is the preferred tool used by cardiologists at both Louisiana State and Texas A&M Universities. Another tool we have attempted to use is a brush like retrieval device using a red rubber catheter and fine suture tied to the end of the tip to make a brush; however for that particular patient, a Chihuahua, the 5 Fr red rubber catheter was too big to pass through the vessel and a 3French red rubber catheter was too flimsy.

CASE REPORT

Peanut is a 4yr old 15.1kg castrated male mixed breed canine who presented for two day duration of anorexia and lethargy. Physical exam findings consisted of a temperature of 99.0, respiratory rate of 24 breaths per minute, a heart rate of 130 beats per minute, capillary refill time of < 2sec, pale pink mucus membranes, with abdominal distention and harsh lung sounds. Base line diagnostics were performed by the emergency service and consisted of a HW snap test which was a strong positive and a CBC revealing a regenerative anemia with a PCV of 28%. On the chemistry panel, Peanuts' BUN and glucose were moderately increased at 29mg/dl and 162mg/dl respectively. A blood type and cross match for blood transfusions were done at this time as well.

Thoracic radiographs demonstrated right sided heart enlargement with tortuous and enlarged pulmonary arteries. Echocardiography confirmed the diagnosis of caval syndrome with the presence of worms in the right atrium and entangled in the tricuspid valve.

SET UP, PREP AND SURGERY

For this case, the extraction was performed in the cardiac catheterization laboratory with a transesophageal echocardiographic ultrasound unit also available for assistance in imaging during the worm extraction. Peanut was prepared for surgery by clipping and sterilely prepping the right side of his neck from just below the jaw to the thoracic inlet. Surgical supplies consisted of a minor pack, Ishihara forceps, a #10 surgical blade, sterile gauze, 2.0 silk, sterile gloves, gown, cap & mask, sterile scrub, heparin flush, and a sterile drape.

Depending on how stable the patient is upon presentation, sedation may not be required and jugular access can be gained using only a local anesthetic. Peanut was pre-medicated with 1.5mg dexamethasone (2mg/ml) IV, 12mg diphenhydramine IM (50mg/ml) and 0.7mg glycopyrolate IM (0.2mg/ml). The anesthetic protocol for Peanut was used to achieve conscious sedation using 12mg etomidate (2mg/ml) IV and 75mcg remifentanyl (50mcg/ml) IV. Peanut was intubated with 100% oxygen and maintained with a CRI of dopamine (400mg/10ml) 5mg/kg/min, midazolam (5mg/ml) 2.5mg IV, remifentanyl 5mcg/kg/IV.

A cutdown was performed over the right jugular vein and using blunt dissection, the jugular vein was isolated. A venotomy incision was made and with fluoroscopic guidance the Ishihara forceps were advanced into the right heart. Several passes were made with the forceps using fluoroscopic and transesophageal echocardiographic guidance. During the procedure, a transfusion of packed red blood cells and plasmalyte was started. The procedure was considered to be complete when several passes with the forceps failed to retrieve any more worms and no worms were visualized in the right heart with echocardiography. A total of 68 worms were removed from Peanuts heart! The jugular vein was ligated with 2.0 silk and the subcutaneous tissue was closed with 3.0 PDS. A subcuticular pattern using 3.0 monocryl was used to close the skin.

RECOVERY, DISCHARGE AND POST-OP CARE

From prepping to extubation, the procedure took 1 hour and 10 minutes. Peanut was transferred to the ICU for overnight care and monitoring. Two hours post infusion a repeat PCV/TP was taken and results were 36% and 6.0mg/dl respectively. His recovery was uneventful, vital signs remained within normal limits and Peanut was discharged the following day. Medications sent home to be given prior to adulticide therapy consisted of prednisolone to minimize the inflammatory response and doxycycline to suppress systemic infection/inflammation from the Wolbachia bacteria that can be released from dead worm fragments.

Prednisolone- 10mg PO q/12hr

Doxycycline 75mg PO q/12hr

Exercise restriction- Strictly controlled leash walks only, no strenuous activity such as running, jumping or excessive excitement. Peanut was started on a HW preventative immediately to eliminate any microfilaria.

A standard split dosing of melarsomine (Immticide®) 2 to 4 weeks after Peanut fully recovered was administered. A repeat blood chemistry and complete blood count was performed before starting the melarsomine treatment to ensure that liver and kidney functions were within normal limits.

CONCLUSION

Even though we now have a number of drugs available for heartworm prevention, we still see several dogs every year with infestations. It is extremely important to be familiar with all the stages of this disease and to act quickly when presented with a patient in caval syndrome. In the Louisiana delta, which has the in highest incidence of heartworm infections in the US, most practitioners do not have access to a million dollar cardiac catheterization lab and many do not even have an ultrasound machine. However, these veterinarians have been doing heartworm removals for years. For dogs diagnosed with Caval Syndrome, surgical removal is the only option to prevent them from dying, which is why despite the risk, the poor condition of the patient and the lack of tools or techniques, these doctors do whatever they can to save their patient. The procedure itsself does not take long, does not require heavy sedation or general anesthesia, can be performed with a number of different retrieval devices and can be done in either a surgery suite or treatment room.

REFERENCES

Etienne C, et al.. **Clinical Veterinary Advisor** 2010; 477. Ware W, **Cardiovascular Disease in Small Animal Medicine** 2007; (24): 362. American Heartworm Society 2-15-2010, **Canine Heartworm Guidelines, Surgical Extraction of Adult Worms**, 8 & 9. **Heartworm Infestation and Disease**, 252. Rush J, et al. **A practical Approach to Radiographic Diagnosis of Heart Disease in Small Animals**, 30. Haskins S, **Journal of Veterinary Emergency and Critical Care** 2006; 16(4) 300.

ADVANCED ENDOSCOPIC TECHNIQUES

Dr. Caroline Mansfield
Melbourne, Australia

INTRODUCTION

Endoscopy in its earliest forms began in the 19th century using a tin tube and a candle, but did not become widely used in veterinary medicine until the 1970's. Flexible endoscopes are now widely used in veterinary medicine for assessment of mucosal disease in hollow structures such as the gastrointestinal, respiratory and urinary tracts. The two types of flexible endoscopes in regular use are fiberoptic and video endoscopes. With fiberoptic endoscopes, image transmission and illumination occurs via two bundles of fiberoptic fibres, whose point of origin at the tip of the endoscope precisely matches that at the eyepiece. In contrast, video endoscopes rely upon a microelectronic charge coupled device (CCD) chip at the tip of the endoscope that senses the image for subsequent electronic transfer for processing and direct display on a video monitor. Now with high definition video screens and advanced image capture systems the quality of images available has improved considerably.

However conventional endoscopy is frequently limited as gross morphological findings may only be detectable late in disease, may not correlate with histopathologic evaluation or lesions may be subtle and easily missed. In addition, even utilizing upper and lower gastrointestinal endoscopy, the entire gastrointestinal tract is not able to be evaluated. Adjunctive application of topical dyes to highlight abnormalities as with chromoendoscopy or technology such as narrow band imaging and confocal endomicroscopy additionally enhance identification as well as classification of subtle lesions; whilst capsule endoscopy and push enteroscopy both provide a means for evaluation of a far greater length of the gastrointestinal tract than was previously achievable. Another recent advancement is the improved ability to perform therapeutic procedures during endoscopy.

LIGHT SOURCE AND IMAGE RECONSTRUCTION TECHNIQUES

In video white-light endoscopy (WLE), xenon white-light is preferably used for illumination as it provides greater contrast and visibility than halogen light sources. The illumination is transferred to the tip of the endoscope via fiberoptic bundles. These fibers are randomly arranged bundles, as specific and orderly transfer of light is less important compared with the coherent transfer required for image reconstruction in fiberoptic endoscopes. Depth of penetration of photons within the mucosa depends upon their wavelength. Superficial penetration occurs with the blue band, intermediate for green and in-depth for the red band. Differing absorption and reflection of these wavelengths then provides the overall endoscopy image. Hemoglobin is primarily responsible for absorption of blue and green wavelengths which explains the overall red appearance of the mucosa, and why vasculature appears red with WLE. Image transfer in WLE occurs when the CCD chip transfers the full range of white-light for reconstitution of the natural color in a single step, or the tissue is alternately illuminated by each of the three bandwidths of light by a rotating interference broadband filter interposed with standard xenon white-light. Each of the three overlapping bands are then separately transferred for image reconstruction.

Chromoendoscopy

This is a wide-field of view, WLE technique that involves topical application of stains or dyes by use of a spray-catheter or lavage. We have found that performing this in the stomach requires use of an endoscopic spray catheter and large volume force on the dye to be injected in order to coat a sufficient area. If only a small area is under evaluation alternatively the dye can be 'dripped' onto the area of interest. The presence of food and the use of flushing to clear the field of vision interfere with the visibility of this process. Surface washing with mucolytic agents and prolonged dwell times are sometimes required for adequate stain distribution or cellular uptake, with subsequent vigorous washing required to remove excessive dye. Failure at any of these steps can result in variable and patchy staining. Dyes that have been evaluated include Lugol's solution, methylene blue, toludine blue and cresyl violet. Concern over the potential of methylene blue to cause oxidative damage to cellular DNA following exposure to white-light has led to its declining popularity.[1,2] The staining of pre-malignant cells with these dyes is one of the theoretical benefits of this technique. However, false positive results may also be seen with positively staining inflammatory tissue.[1] Cresyl violet (0.05 – 0.1%) also stains nucleic acids and can be used in a similar fashion, but it can precipitate in the tip of the endoscope, requiring the endoscope to be withdrawn and cleaned before being reinserted.

Narrow Band Imaging and Virtual Chromoendoscopy

Narrow band imaging (NBI) has been developed to overcome the technical shortcomings of chromoendoscopy. Exploitation of differential penetration, absorption and reflection of each of the red/green/blue (R/G/B) wavelengths of light forms the basis for NBI.[3,4] As with standard WLE, optical interference filters are interposed with the xenon white-light used for illumination. However, in NBI these filters restrict incident light much more than standard broadband filters and not only clearly separate each of the R/G/B wavelengths, but also restrict tissue exposure to only green (540nm) and blue (415nm) wavelengths. This results in a significant difference in the average depth of penetration of each band within the mucosa.[3-5] However, in order to recreate the image on the video monitor, reassignment of these two wavelengths to each of the three input channels (R/G/B) must occur. This is achieved by assigning the blue wavelength (415nm) to both the green and blue channels, whilst the green wavelength (540nm) is allocated to the red channel.[7] Image transfer via the CCD chip occurs as for standard WLE with each of these bands transferred either in a single step, or sequentially.

Selective absorption and reflection of these narrower bands improves definition of the surface mucosa in the reconstructed image. Mucosal micro-vasculature is enhanced due to selective absorption of blue wavelength light via hemoglobin resulting in reproduction of the superficial capillary network in a brownish color. The green wavelength penetrates more deeply, is outside the absorption band for hemoglobin and therefore provides detail of deeper, larger vessels.[3-5] The final image represents the superficial and deeper detail superimposed as a single image and this enhances the definition of flat lesions, and improves evaluation of the microvasculature.

In comparison to NBI, which relies upon optical filtration of the light source, virtual chromoendoscopy units use computed spectral estimation algorithms to post-process and optimize the captured WLE image. Using this method, reflected photons of set wavelengths can be isolated, and assigned to each of the R/G/B bands in order to provide optimal contrast, producing a similar image to that with NBI, but with more choice in the combination of wavelength settings. The most common use of this modality in people is to observe flat polypoid type lesions that are not as easily visible using WLE.

MUCOSAL ABLATION AND HEMOSTASIS

The use of electricity in vascular hemostasis relies on the principles that electricity always takes the path of least resistance, always seeks ground and there must be a complete circuit for electricity to flow. Monopolar circuits use the patient's body to complete the circuit, and examples include needle-knifes and polypectomy snares. Bipolar/multipolar modes do not require grounding as the circuit is completed between two points on the active electrode. Cutting during endoscopy is achieved when tissue is quickly heated to over 100°C, with short electrical arcs (sparks) starting at around 200V. It is desirable for units to reach the spark mode quickly to acheive more effective cutting. Endoscopic cutting instruments are composed of thin wires and don't directly touch the tissue, as they are surrounded by a steam barrier of vaporized intracellular fluid. Coagulation techniques can be direct (using low voltage but with pressure on the area) or indirect where the current is transmitted by electrical arcs under high voltage, over a larger area than cutting modes to halt superficial and diffuse tissue bleeding. Argon plasma coagulation (APC) is a new modality that is a monpolar non-contact device where the current is delivered through ionized argon gas. This technique is useful for control of superficial, diffuse bleeding without causing avulsion of tissue. Technology can now combine cutting and coagulation techniques in one machine. Extensive flushing is often required with these techniques, so addition of an endoscopic flush pump is optimal as the flushing through the endoscope channel may not be sufficient.

Endoscopic mucosal resection (EMR) involves injection of saline into the submucosa. An electric snare can then be used to remove mucosa in the desired area moving from the lateral edges towards the middle. Endoscopic submucosal dissection (ESD) takes this a step further and allows for en bloc removal of lesions. Both of these procedures require combined cautery/coagulation units, double channel endoscopes, an endoscopic flush pump, polypectomy snares, and extensive experience in routine endoscopy and endoscopic interventional techniques.

CONFOCAL ENDOMICROSCOPY

Confocal endomicroscopy (CEM) is a new (and not readily available) technology that encompasses confocal microscopy technology into endoscopy in one of two ways. In one type, the confocal bundle is passed through the channel of a working scope. In the prototype confocal endomicroscope that our group is using, the confocal microscope is incorporated into a flexible video endoscope and rigid endoscope. The flexible confocal endomicroscope utilizes a 488 nm laser, and is capable of 7 μm axial resolution, with image assessment possible up to a depth of 250 μm. Within the gastrointestinal tract, CEM can be utilized to assess the mucosa when the endoscope is in contact with the surface to give a real-time, virtual histopathologic image. This histologically equivalent image is possible because light that reaches the image detector is filtered to a large degree such that only that light which exactly matches the focal plane within the tissue is detected. This eliminates light scatter which would otherwise contribute to 'noise' and produce foreground and background blur, limiting depth of view.[3] Imaging via CEM is achieved in gastroduodenoscopy in conjunction with the use of fluorophores (contrast agents) and adjusting focal-plane depth as required to assess deeper structures.

VIDEO ENDOSCOPIC CAPSULES

Video capsules are miniaturized chips contained in a capsule that obtains multiple images as it transits the intestinal tract, transmitted to a wireless receiver attached to the body. Lead position on the body allows for the approximation of anatomical location during reading of the studies. Video capsules are currently recommended as the most appropriate diagnostic modality in people with an occult fecal blood test that have had normal standard WLE examination of the upper and lower gastrointestinal tract.[6] Capsules can have cameras at both ends (designed for examination of the esophagus), or more standardly, in one end along with a light source and 8-hour battery. The short battery life may be problematic in some dogs that could have longer gastric emptying than this, resulting in evaluation of the stomach mucosa only. In our experience, pre-treatment with metoclopramide does not result in faster gastric transit of the capsule. Fasting for 12-16 hours prior to administration of the capsule is recommended, and treatment with a bowel preparation (PEG) may enhance visibility. Once the capsule has passed the stomach, a small meal can be given along with gentle exercise to promote intestinal transit prior to the end of the battery life.

Contraindications for capsule administration include strictures and known intestinal obstructions and they are not recommended in people with cardiac pacemakers. Anecdotally, some gastroenterologists have reported that the site of capsule obstruction is often determined to be a previously unsuspected neoplastic process following surgical removal of the capsule. The minimum size of dogs that can safely ingest these capsules without risk of obstruction is unknown, but the procedure is used in children following what is called the 'Jelly bean test', where children are sequentially fed larger lollies up to the size of a jelly bean to ensure they can pass it easily. In veterinary medicine, potential applications include motility studies,[7] assessing for causes of occult bleeding and non-invasive assessment of luminal pathology. However, the expense (approx $450 USD per capsule), time and training required for interpreting the images make this a modality that is unlikely to become widespread. Additionally, any lesion identified still requires histologic evaluation, along with specific therapy to result in clinical benefit.

DEEP (PUSH) ENTEROSCOPY

The inability to reach all parts of the small intestine has led to the development of techniques to reach the distal small bowel, otherwise called deep or push enteroscopy. The three techniques use overtubes to facilitate progression of the endoscope, with the overtubes being spiral tips, single or double balloons. In people, both antegrade and retrograde approaches are used, but in dogs it is typically antegrade approach that is most common. Bowel preparation is not necessary for the antegrade approach, but the dog should be fasted for at least 16 hours. To date, this procedure has been described in medium sized dogs, but not in cats.[8-9]

Spiral overtubes are 118 cm long with a 48Fr diameter, with a 21-cm raised helical element at the distal end. Use of this modality usually requires two people- one to manipulate the endoscope and keep the lumen in view, whilst the other manipulates the overtube. The overtube needs to be prepared prior to the procedure by injecting a specific lubricant to coat the entire inner surface of the overtube before inserting the endoscope. The endoscope is then inserted into the esophagus (it may require clockwise rotation to advance through the upper esophageal sphincter) and into the duodenum with minimal gastric dilation. It is then necessary to a 'Cantero maneuver', where there is simultaneous clockwise rotation and overtube withdrawal. Then, engagement is achieved by gentle forwards pressure and clockwise rotation of the overtube until the desired point is reached. Withdrawal requires and anti-clockwise rotation.

The single balloon splinting tube is made of silicon and is 140-cm long with an inner hydrophilic surface, connected to a balloon at the distal end, with an external controller for inflation and deflation. This procedure can be performed by a single operator. Once in the jejunum, the endoscope is held in place by angluating the tip of the endoscope and suctioning the mucosa onto the endoscope. The overtube is then advanced to the tip of the scope and the balloon inflated. The entire scope (including overtube) is then withdrawn orally to allow the small intestines to pleat the intestine, and the procedure is repeated until the small intestine has been assessed.

With the double balloon technique, generally two people are required to operate the system, whereas with experience the spiral overtube and single balloon techniques can be performed by a single operator. For double balloon enteroscopy, the endoscope with the overtube (polyurethane with a latex balloon) is passed normally into the pylorus. The endoscope is pushed at least 30 cm in to the duodenum and the endoscope balloon is inflated to anchor the endoscope. The overtube is advanced distally and its balloon inflated, then both the overtube and endoscope are 'pulled' towards the mouth together, pleating the small bowel. The endoscope balloon is then deflated and the endoscope advanced distally (with the overtube balloon still anchored). This 'push-pull' cycle is repeated until the length of the small bowel has been assessed. Successful visualization of the entire small bowel of dogs has been reported.[8-9]

These procedures are quite long, and therefore in dogs it should be carefully considered when they should be used. The most common indication for this procedure in people is when there is melena but no lesion has been identified on standard gastroduodenoscopy or video endocapsule. Training guidelines for medical gastroenterology suggests that expert endoscopists need to perform 5-10 of these procedures before they gain adequate skill,[10] and that due to the lack of anatomical landmarks in the small intestine, fluoroscopy should be used initially to guide localization. As in human gastroenterology, it may also be preferable for carbon dioxide insufflation as opposed to air to reduce discomfort following the procedure. There are also therapeutic models of this equipment that allow for hemostasis, mucosal ablation, polyp resection and stricture dilation in the distal small intestine.

CONCLUSION

The technical ability to perform a wide range of diagnostic and therapeutic procedures in human medicine is expanding rapidly. The application of these procedures to veterinary medicine is growing, but is limited by cost, availability and training required in a number of those modalities.

REFERENCES

[1]Amateau S et al., **Current Opin Gastroenterol** 2010;26:445 [2]Hurlstone DP et al., **Gut** 2004;53:284 [3]Kiesslich R et al, **Atlas of Endomicroscopy**. 2008 [4]De Palma GD. **World J Gastroenterol** 2009;15:5770 [5]Dunbar K & Canto M. **Current Opin Gastroenterol** 2008;24:631 [6]Schnoll-Sussman & Kelsey, **Successful Training in Gastro Endo**; 2011 pp 97 [7]Boillat et al., **Am J Vet Res**, 2010 71: 898 [8]Latorre et al, **Vet Rec** 2007; 161:587 [9]Ayala et al., **J Am Anim Hosp Assoc** 2011; 47:262 [14]Okolo & Buscaglia, **Successful Training in Gastro Endo**; 2011 pp 109

MANAGEMENT OF METHICILLIN-RESISTANT STAPHYLOCOCCAL INFECTIONS

J. Scott Weese, DVM, DVSc, DACVIM
Guelph, Ontario, Canada

INTRODUCTION

Despite being inherently susceptible to virtually every known antimicrobial class, staphylococci are notorious antimicrobial resistant pathogens. The ability of staphylococci to become resistant was repeatedly demonstrated as the introduction of new drugs was typically followed (sometimes rapidly) by emergence of resistance. Included in this pattern was resistance to methicillin, which emerged not long after the use of methicillin began in 1959. While resistance to methicillin itself is not a major concern, methicillin-resistance is caused by production of an altered penicillin binding protein that has a poor affinity for beta-lactam antimicrobials and confers resistance not just to methicillin, but to virtually all beta-lactams (penicillins, cephalosporins, carbapenems). Methicillin-resistant staphylococci are also often resistant to a wide range of other antimicrobials, severely limiting treatment options. While was first noted as a problem in humans, methicillin-resistance has emerged as a pressing issue in companion animals, and there have been striking recent changes in the epidemiology of methicillin-resistant staphylococcal infection and colonization in dogs and cats.

Staphylococcal species: what's in a name?

It is critical to remember that *all staph are not created alike*. The clinical relevance, prevalence of colonization, incidence of disease and potential for interspecies transmission vary greatly between different staphylococcal species. Inadequate identification practices, variable and sometimes improper susceptibility testing methodology, and variable reporting of results can lead to confusion and possibly suboptimal patient care.

There are two main groups of staphylococci; coagulase positive and coagulase negative. Coagulase positive species (e.g. *S. aureus, S. pseudintermedius, S. schleiferi coagulans*) are the most virulent and of greatest clinical relevance. Differentiation of these, or at a minimum, differentiation of *S. aureus* from other coagulase positive staphylococci is needed for both interpretation of zoonotic risks and the use of proper antimicrobial susceptibility testing methodology (since there are different standards for *S. aureus* versus other staphylococci). Coagulase negative staphylococci tend to be of limited virulence and after often considered together as a single group, although there is some evidence of variation in virulence and clinical relevance. Knowing the type of *Staphylococcus* that is involved, therefore, is critical for proper management.

Methicillin-resistant *S. pseudintermedius* (MRSP)

Staphylococcus pseudintermedius is the most clinically important *Staphylococcus* in dogs and a common commensal that can be found in or on a large percentage, if not all, healthy dogs, and a smaller percentage of cats.[1-5] As with methicillin-susceptible *S. pseudintermedius*, MRSP can be found in or on healthy dogs and cats,[2,4,6-8] and it appears that the rate of colonization is increasing in many regions. MRSP appears to have emerged and disseminated internationally in companion animals at a truly amazing rate, with rapid development of a very high level of drug resistance. MRSP infections are being identified virtually everywhere that people are looking, and the increase in incidence of disease, while not objectively reported, seems to be dramatic, particularly among dogs with pyoderma. Most infections are community-associated, particularly involving dogs with pyoderma and otitis, but hospital-associated transmission does occur and is of particularly concern among surgical patients.

MRSP is an opportunistic pathogen and colonization does not necessarily lead to disease. Indeed, it is likely that the vast majority of colonized animals never develop a clinical infection. The risk of infection in MRSP carriers has not been reported, but it is reasonable to assume that MRSP carriers are at some increased risk of MRSP infection, at least in certain situations. MRSP carriage is probably of limited concern to individual healthy dogs in the general population, but of greater relevance in dogs with comorbidities that increase the likelihood of any opportunistic infection (e.g. atopy) and those that undergo surgery. Additionally, MRSP presumably poses a risk to veterinary facilities as a source of hospital-associated transmission.

There is no indication that MRSP infections are inherently more serious than infections caused by methicillin-susceptible strains, however they could ultimately be associated with increased morbidity and mortality because of failure of initial empirical antimicrobial therapy and limited treatment options

Methicillin-resistant *S. aureus* (MRSA)

MRSA tends to receive a higher profile than MRSP because of its huge impact in human medicine, however it is a much less common cause of infection in dogs and cats than MRSP. Regardless, attention must be paid to this organism because of the potential for serious and difficult-to-treat infections, and the greater zoonotic implications.

As with MRSP, MRSA causes opportunistic infections and can be found in a small percentage of healthy dogs. It is assumed that the emergence of MRSA in companion animals is directly related to MRSA in humans. As MRSA expanded in humans in the general population in the 1990s and 2000s, exposure of pets was inevitable. Humans are thought to be the ultimate source of infection for a large percentage, if not the vast majority, of infections in pets, and changes in the epidemiology of MRSA in humans can be reflected in changes in trends in pets. Unlike in human medicine, most MRSA cases in animals are not associated with veterinary hospitals, but hospital outbreaks can occur. High MRSA carriage rates can also be found in specific dog populations such as households where another pet has an MRSA infection,[9] or during outbreaks in breeding or rescue kennels.[10,11] Being owned by a human healthcare worker and participation in hospital visitation programs have been identified as risk factors for MRSA colonization in dogs, and are logical based on the increased likelihood of exposure to colonized people.[12,13] Contact with children has also been identified as a risk factor.[13] While these, and potentially other, risk factors should be considered, MRSA can be identified in any animal and absence of known risk factors should not lead to excluding MRSA from consideration.

Methicillin-resistant *S. schleiferi* (MRSS)

Staphylococcus schleiferi consists of two subspecies, the coagulase positive *S. schleiferi* subsp *coagulans* and coagulase negative *S. schleiferi* subsp *schleiferi*. These are less common causes of infection compared with *S. pseudintermedius* and *S. aureus*, but a large percentage of diagnostic laboratories do not attempt to different *S. schleiferi coagulans* from *S. pseudintermedius* or *S. schleifer schleiferi* from other coagulase negative species. so there are limitations in understanding of the role of these species in disease. *S. schleiferi coagulans* is most commonly implicated in pyoderma and otitis externa in dogs, but other opportunistic infections such as urinary tract infection and pneumonia have been reported.[2,14-19] Less in known about *S. schleiferi schleiferi*, but has been implicated as a cause of pyoderma and otitis,[15,17,18] and there is increasing concern that it may be more virulent than other coagulase negative staphylococci.

Methicillin-resistant coagulase negative staphylococci (MR-CoNS)

MR-CoNS are common commensals and generally of limited virulence. MR-CoNS infections may be overdiagnosed because MR-CoNS can be isolated as contaminants from various superficial body sites. In human medicine, MR-CoNS are primarily a concern in hospitalized individuals.[20] The situation may be similar in dogs can cats, with most CoNS being of minimal pathogenicity and mainly of relevance in highly compromised patients. While CoNS are generally considered as a group, it is possible that there are differences between CoNS species. This mainly relates to three organisms; *S. schleiferi* subspecies *schleiferi, S. epidermidis* and *S. felis*. *Staphylococcus schleiferi schleiferi* and *S. epidermidis* have been implicated as a cause of pyoderma and otitis,[15,17,18,21] and may be inherently more pathogenic in skin and ear infections than other CoNS. A primary role of *S. felis* in urinary tract infections in cats has also been suggested.[22] These species may, therefore, be more relevant and potential causes of community-associated disease. As with other staphylococci, methicillin-resistant strains are inherently no more pathogenic than methicillin-susceptible strains, and the implications of colonization with MR-CoNS are typically inconsequential.

TREATMENT

Treatment of methicillin-resistant staphylococcal infections follows the same principles as treatment of susceptible infections, and the simple presence of MRSA does not mean that longer or more intensive treatment is indicated. Methicillin-resistance is an antimicrobial resistance determinant, not a true virulence factor, so MRS infections should be no more virulent than those caused by susceptible infections *if the infection is identified and proper treatment is provided.*

Some superficial infections may be treated topically with antibacterials or antiseptics such chlorhexidine, silver sulfadiazine, fusidic acid, mupirocin, povidone iodine or other topical antimicrobials. Topical chlorhexidine can be highly effective for superficial pyoderma caused by MRSP and with the high prevalence MRSP in some areas, it is becoming a treatment of choice. Miconazole, while an antifungal, has excellent anti-staphylococcal properties and is another topical option. Honey can also be effective, particularly with large wound infections.

In most situations, systemic antimicrobial therapy will be required, alone or in conjunction with local therapy. Drug choices should be based on *in vitro* antimicrobial susceptibility testing in combination with other patient and drug factors (e.g. drug penetration, concentration at the site of infection, activity at the site of infection, drug interactions, etc.) The level of resistance possessed by MRSA isolates can be quite variable, ranging from resistance to only beta-lactams to resistance to most antimicrobials. Identification of an MRSA infection does not mean that the latest generation or most potent antimicrobials are required, particularly in non-life-threatening infections. Fluoroquinolones tend not to be good options for treatment because resistance can emerge very quickly and because *in vitro* susceptibility does not always correlate well with clinical response. Inducible clindamycin resistance is common with MRSA (but rare with MRSP).[23] This phenomenon is characterized by *in vitro* susceptibility to clindamycin but resistance that is induced *in vivo* and with specific testing. Inducible resistance is a concern with erythromycin-resistant/clindamycin-susceptible isolates if testing for inducible resistance has not been performed. Extended susceptibility testing may need to be requested from diagnostic laboratories if there are no viable options among the antimicrobials that are routinely tested. Ideally, drugs that are important for treatment of critically ill humans should be avoided.

PUBLIC HEALTH ISSUES:

MRSA

Zoonotic transmission is of particular concern with MRSA, although neither the true role of companion animals in human infection nor the occupational risks to veterinary personnel are well understood. Overall, it is likely that pets play at best a very limited role in human MRSA infection, however zoonotic infections certainly cannot be dismissed and in some situations an infected or colonized animal could pose a reasonable risk to humans. The greatest concerns probably involve handling pets with clinical MRSA infections (especially infections of external sites) and individuals that have contact with large numbers of different animals (i.e. veterinary personnel).

Colonization of people in contact with infected or colonized dogs and cats (often with demonstration of the same strain in humans and animals) has been widely reported.[12,24-29] Pets have been implicated as sources of infection for humans in households,[26,28] however conclusions are often overstated since finding MRSA in both people and pets does not identify direction of transmission, and human-pet transmission is most likely in many situations. Zoonotic transmission can and does occur, however, and care must be taken to reduce the risk of this in veterinary clinics and households.

Veterinary personnel are at high risk for MRSA colonization, as have been shown in various studies.[30-33] The health impacts of this are unclear, although there are limited reports of MRSA infection in veterinary personnel that have worked with infected or colonization animals,[34,35] along with numerous anecdotal reports. Further, the high rates of colonization could increase the risk of infection in some situations such as if colonized veterinary personnel undergo surgery or are admitted to hospital.

MRSP

Despite the fact that exposure to *S. pseudintermedius* must be very common in pet owners (given the high prevalence of colonization in pets), human *S. pseudintermedius* infections are rarely reported and colonization rates in veterinary personnel and pet owners are rather low (1-4%),[4,36] suggesting poor host adaptation to humans. However, since *S. pseudintermedius* (including MRSP) infections can occur in humans,[37-39] efforts are required to reduce this risk.

Other staphylococci

While minimally investigated, it is currently thought that the risk of zoonotic transmission of other staphylococci from pets to people is very low.

INFECTION CONTROL IN VETERINARY HOSPITALS

While there are few published reports of MRSA transmission in veterinary hospitals, anecdotal information abounds and MRSA was the 2nd most commonly reported cause of outbreaks in a survey of veterinary teaching hospitals.[31] Additionally, MRSP appears to be an increasingly common hospital-associated pathogen, particularly as a cause of surgical site infections. Therefore, efforts to control MR-staphylococcal transmission are required from both patient and public health perspectives.

Challenges encountered in MR-staphylococcal control include the presence of MR-staphylococcal colonization in animals at admission, colonization of clinically normal veterinary personnel and lack of objective data regarding control measures. Information from human medicine, where abundant research has been performed, can provide some information but the many differences hamper direct extrapolation of human data. In general, MR-staphylococcal control in veterinary hospitals revolves around good general infection control practices because of the large cohort of 'silently' infectious patients (and personnel). Standard hygiene practices (especially hand hygiene), routine cleaning and disinfection practices, cohorting of risk groups, isolation of infected or colonized animals, judicious antimicrobial use, passive surveillance (monitoring culture results and disease incidence rates) and ensuring prompt culture of infected sites form the backbone of veterinary hospital staphylococcal control.

Active screening for MRSA is widely used in humans but is rare (and probably rarely indicated) in companion animals. Active surveillance can identify colonized animals to allow for prompt isolation and help understand aspects of MR-staphylococcal transmission in facilities. However, cost and logistical challenges are often prohibitive, and surveillance may be best reserved in most facilities as a research and/or outbreak investigation tool. However, as MRSP rates rise, particularly among surgical patients, screening of high risk groups such as dogs undergoing orthopedic surgery may be indicated, particularly if easy and effective interventions (e.g. changes in peri-operative prophylaxis) can be implemented to reduce the likelihood of infection.

Very rarely is MRSA screening of personnel indicated or useful. Facilities should assume that one or more of their personnel are colonized and structure routine infection control practices to reduce the implications of colonized personnel, since logistical, privacy, legal and efficacy challenges preclude the use of personnel screening as an effective and practical tool.

REFERENCES

[1] Fazakerley J, et al. ***Vet Dermatol*** 2009;20:179-184. [2] Griffeth GC, et al. ***Vet Dermatol*** 2008;19:142-149. [3] Hartmann FA, et al. ***Vet Microbiol*** 2005;108:119-131. [4] Hanselman BA, et al. ***Can Vet J*** 2009;50:954-958. [5] Saijonmaa-Koulumies LE, et al. ***Vet Dermatol*** 2002;13:123-130. [6] Epstein CR, et al. ***Infect Genet Evol*** 2009;9:283-285. [7] Hanselman BA, et al. ***Vet Microbiol*** 2007;126:277-281. [8] Vengust M, et al. ***Lett Appl Microbiol*** 2006;43:602-606. [9] Faires MC, et al. ***J Am Vet Med Assoc*** 2009;235:540-543. [10] Loeffler A, et al. ***Vet Microbiol*** 2009. [11] Floras A, et al. ***Vet Rec*** 2010;166:826-827. [12] Boost M, et al. ***Clin Microbiol Infect*** 2007;13:731-733. [13] Lefebvre SL, et al. ***J Am Vet Med Assoc*** 2009;234:1404-1417. [14] Bes M, et al. ***Vet Rec*** 2002;150:487-488. [15] Frank L, et al. ***J Am Vet Med Assoc*** 2003;222:451-454. [16] Igimi S, et al. ***Int J Syst Bacteriol*** 1990;40:409-411. [17] May ER, et al. ***J Am Vet Med Assoc*** 2005;227:928-931. [18] Morris D, et al. ***Vet Dermatol*** 2006;17:332-337. [19] Rich M, et al. ***Vet Rec*** 2007;161:107. [20]Archer GL, et al. *Principles and practice of infectious diseases*. 2005;2352-2360. [21] Penna B, et al. ***Vet Dermatol*** 2010;21:292-296. [22] Litster A, et al. ***Vet Microbiol*** 2007;121:182-188. [23] Faires M, et al. ***Vet Microbiol*** 2009. [24] Weese J, et al. ***Vet Microbiol*** 2006;115:148-155. [25] Leonard F, et al. ***Vet Rec*** 2006;158:155-159. [26] Manian FA. ***Clin Infect Dis*** 2003;36:e26-28. [27] van Duijkeren E, et al. ***J Clin Microbiol*** 2005;43:6209-6211. [28] Sing A, et al. ***N Engl J Med*** 2008;358:1200-1201. [29] Vitale C, et al. ***Emerging Infect Dis*** 2006;12:1998-2000. [30] Anderson ME, et al. ***Vet Microbiol*** 2007;129:410-417. [31] Burstiner LC, et al. ***Vet Surg*** 2010;39:150-157. [32] Hanselman B, et al. ***Emerg Infect Dis*** 2006;12:1933-1938. [33] Wulf M, et al. ***Emerging Infect Dis*** 2006;12:1939-1941. [34] Weese J, et al. ***Emerging Infect Dis*** 2005;11:430-435. [35] Weese J, et al. ***Vet Microbiol*** 2006;114:160-164. [36] Talan DA, et al. ***J Clin Microbiol*** 1989;27:2393. [37] Sasaki T, et al. ***J Clin Microbiol*** 2007;45:1118-1125. [38] van Duijkeren E, et al. ***Vet Microbiol*** 2008;128:213-215. [39] Stegmann R, et al. ***J Antimicrob Chemother*** 2010;65:2047-2048. [40] Frank LA, et al. ***Vet Dermatol*** 2009;20:496-501. [41] Stegmann R, et al. ***J Antimicrob Chemother*** 2010:1-2. [42] Van Hoovels L, et al. ***J Clin Microbiol*** 2006;44:4609-4612. [43] Pottumarthy S, et al. ***J Clin Microbiol*** 2004;42:5881-5884.

RESPIRATORY MECHANICS & MONITORING

Bea Biddinger, LVT, VTS (ECC)
E. Lansing, MI

FUNDAMENTAL RESPIRATORY MECHIANICS

Basic lung function is designed to exchange oxygen and carbon dioxide. In order to transfer oxygen from atmospheric air to the blood stream three functions must be in place: ventilation, diffusion, and perfusion. Ventilation is the process of air moving into and out of the lungs. Diffusion is the movement of gases between the alveoli and the blood in the capillaries. And perfusion occurs when the cardiovascular system pumps blood throughout the lungs. Where these three things meet are the interface of each tiny alveolus that is wrapped with a capillary bed consisting of both pulmonary artery (carrying oxygen deficient blood) and pulmonary vein (carrying newly oxygen rich blood to the heart). It is the surface area of millions of alveoli that allow for the movement of carbon dioxide out of the blood and oxygen into the blood.

PHYSICAL EXAM AND PATIENT HANDLING

Normal breathing is woefully underappreciated and completely overlooked until something goes wrong with the process of respiration. A normal breath should appear natural, free, and easy. Inspiration is normally no longer than one second with moderate expansion of the chest and abdomen occurring simultaneously. Exhalation is normally 2-3 seconds and should not be forced in character. Respiratory distress/dyspnea, as defined, is observable respiratory difficulty or physically labored breathing; clinically evident by an inability to ventilate and/or oxygenate adequately. Prolonged inspirations may indicate upper airway disease while prolonged exhalations may indicate lower airway disease. Anxiety ridden or exaggerated breathing efforts may indicate obstructed ventilation and hyperventilation respectively. Physical manifestations of respiratory distress can vary widely from tachypnea and loud panting to slow stridorous whistling attempts to breathe. Cats may demonstrate dyspnea by open mouth panting. All of these presentations demand fast action to deliver supplemental oxygen while addressing the underlying cause of the dyspnea.

Gentle handling should be employed in order to decrease anxiety in patients presenting with respiratory distress. Patients should be stabilized first with oxygen. Handling should correspond to the urgency of the situation. If a patient is in distress upon presentation, but is still calm, handling that patient should proceed slowly and gently to avoid increasing anxiety. However if a patient is in extreme distress and in danger of respiratory failure, then the physical exam and diagnostics, such as radiography, should not take precedence before stabilizing comfortable breathing. That may mean a rapid procedure such as a chest tap be done within minutes of presentation in order to improve ventilation.

Oxygen should be delivered in the least stressful manner possible. Methods include; flow by oxygen, oxygen mask, oxygen hood (or e-collar with loose baggy hood), nasal cannulation, oxygen cage, intratracheal catheter, or endotracheal tube. The delivery mode should correlate to the patient's need and should not cause any further anxiety to the patient. Physical exam of the patient can now take place as tolerated by the patient and will include auscultation of upper and lower airways and all lung fields. Cats may require a hands-off approach with oxygen being supplied into an oxygen cage and the physical exam waiting for 5-10 minutes. If the cat's dyspnea is not decreasing some degree of intervention is needed to assess the patient. Appropriate oxygen therapy mode varies depending on the severity of the patient's condition. A patient should never die from respiratory distress. To provide oxygen in times of distress may require intervention such as sedation or anesthetic induction followed by intubation and positive pressure ventilation. These measures would allow for time to diagnose disease and implement therapeutic measures.

RESPIRATORY MONITORING METHODS

It should always be remembered to watch for trends in respiratory monitoring. Evaluation of a patient's respiratory status should not be determined from one reading alone of either a pulse oximeter or blood gas analysis. Careful observation of a patients comfort level , thorough auscultation, and serial monitoring will provide a more complete evaluation of respiratory function.

Pulse Oximetry

Pulse oximetry measures the percentage of hemoglobin saturated with oxygen (SpO_2; units=%) within the blood, and it is generally the first objective method of assessing a patient's respiratory status. Pulse oximetry works by spectrophotometry and measures two forms of hemoglobin that circulate in arterial blood; those forms of hemoglobin are oxyhemoglobin (saturated) and deoxyhemoglobin (unsaturated). Carboxyhemoglobin (bound to carbon monoxide), and methemoglobin (oxidized) are two additional forms of hemoglobin that would be measured using a cooximeter. As pulse oximetry does not differentiate carboxyhemoglobin and methemoglobin readings are not considered reliable in patients with smoke inhalation or acetaminophen toxicosis. Pulse oximetry probes transmit wavelengths of red to infrared light waves across vascular beds of tissue and read saturated or unsaturated hemoglobin. The amount of light that is reflected or passed through is evaluated during pulsatile flow.

Limitations to pulse oximetry are considerable; tissue perfusion (poor cardiac output, vasoconstriction, decreased body temperature), icterus, skin pigmentation or thickness, dry mucous membranes, patient movement, and ambient light all affect readings. Moist mucous membrane areas (tongue, buccal membrane) or thin, light colored, skin areas (point of hock, ear, inter digits) work the best. In order to be considered reliable, a pulse oximetry reading should demonstrate an accurate heart rate and stable, repeating, waveform oscillations. Some units will have a signal strength indicator as a bar graph or indice number. Probe types vary from clip on (gentle clamp) types that transmit light through tissues like the ear, lip, or tongue to reflectance probes that lay against tissue beds such as the inner rectum, prepuce, or under tail base. Be aware that once the probe has been in place for several minutes vascular beds may become compressed and readings may decrease. A normal SpO_2 reading for patients breathing room air should be at least 96%. Measurements of <92% SpO_2 require supplemental oxygen delivery and/or ventilatory assistance.

Arterial Blood Gas Analysis

While pulse oximetry is generally the first tool used to assess a patient; arterial blood gas analysis is considered the gold standard in evaluating a patient's ability to oxygenate and ventilate. It measures the lung's ability to diffuse oxygen into arterial capillary blood (oxygenation) while extracting CO_2 from venous capillaries (ventilatory efficiency). Because blood gas analysis also measures $PaCO_2$ (partial pressure of carbon dioxide in arterial blood) it can be used to evaluate for evidence of hypoventilation whereas pulse oximetry only addresses oxygenation status.

Blood gas analysis works by measuring the partial pressure of oxygen (PaO_2; units=mmHg) and carbon dioxide ($PaCO_2$; units=mmHg) dissolved in the plasma. It does this by using electrodes to compare a known electrolyte solution to the blood sample across a semi-permeable membrane (permeable to CO_2 and O_2). Normal PaO_2 is 80-100mmHg for patients breathing room air. Hypoxemia is defined as PaO_2<80mmHg; with values <60mmHg PaO_2 requiring oxygen supplementation and/or ventilatory assistance. Normal canine $PaCO_2$ is 35-45mmHg; feline normal may be slightly lower. Hypercarbia is defined as $PaCO_2$ levels >45mmHg. Analyzing blood gases is becoming much more practical as bedside units are becoming more affordable and easy to use.

Obtaining arterial gas samples does require practice and causes some stress to the patient. Measures should be taken to minimize stress because anxiety and struggling may alter results (hyperventilation). Potential sites for arterial sampling are the dorsal pedal, femoral, and auricular arteries. After sampling, any site used must be held off for 10 minutes or bandaged to control hemorrhage and to preserve the artery for future sampling. Using the dorsal pedal artery will require less manipulation of the patient's position as compared to puncturing the femoral artery, and the metatarsal site is also easily bandaged after use. Anatomical proximity of the femoral artery to the femoral vein makes the possibility of obtaining a venous or mixed venous/arterial sample greater when puncturing the femoral artery. Placement of an arterial catheter will facilitate repeated sampling and serial monitoring. Obtaining arterial gas samples from cats requires extensive experience and may require an anesthetized patient.

RELATIONSHIP BETWEEN OXYGEN SATURATION AND PaO_2

In order to compare the values received from pulse oximetry to arterial blood gas analysis a technician needs to understand the oxyhemoglobin dissociation curve. Pulse oximetry measures the percentage hemoglobin saturated with oxygen while PaO_2 represents the partial pressure of oxygen gas dissolved in arterial blood. These two measurements relate to each other in a sigmoid relationship called the oxyhemoglobin dissociation curve. The relatively flat area of the curve represents a small change of approximately 10% saturation while reflecting a potential drop in PaO_2 of up to 40mmHg. What this means to patients is a smaller change in pulse oximetry percent that would correlate to a significant decrease in PaO_2. In example, a pulse oximetry reading of 90% roughly equals a PaO_2 of 60mmHg, which would qualify a patient as hypoxemic. When analyzing an arterial blood gas, a current patient body temperature should be programmed into the analyzer due to the fact that hypothermia shifts the oxyhemoglobin dissociation curve to the left. Hypothermia (left shift of dissociation curve) results in a decrease in oxygen release from hemoglobin and therefore decreased amounts of oxygen available for tissue pickup. A decrease in pH and $PaCO_2$ will also occur with hypothermia. The reverse is true of changes due to hyperthermia.

EVALUATION OF RESPIRATORY FUNCTION

It is important to remember to distinguish between evaluating a patient's oxygen saturation and evaluating a patient's ventilatory function. Each value (pulse oximetry and arterial blood gas analysis) is helpful and they relate but not interchangeably. It's a bit like comparing apples to oranges. Just imagine your teenage child's drivers training instructor giving you feedback that they are driving a car well. And compare that to feedback that they are driving the car well in heavy traffic during a time crunch. Which feedback gives you more information? In order to thoroughly evaluate ventilator function it is best to perform the calculation known as the A-a gradient.

Calculating A-a gradients

Calculation of the alveolar (PAO_2)-arterial (PaO_2) oxygen gradient, commonly referred to as the A-a gradient, gives an estimate of the effectiveness of gas transfer while accounting for the extent of ventilation. A-a gradients are based on theory that the majority of oxygen that enters the lungs should pass into arterial circulation. As lung function decreases the oxygen gradient (difference) between the alveoli and the arterial capillaries increases.

The first step in evaluating the respiratory components of a patient's sample is to verify if a blood sample taken was indeed an arterial sample. This is especially important because ventilatory function cannot be evaluated from venous blood. If the hemoglobin saturation (SpO_2) reading of the blood gas analysis is >85% the sample is most often assumed arterial. Saturations less than 75% are most likely venous samples and values between 75-85 are a gray zone and can depend on the severity of pulmonary disease. Certain circumstances may account for very low saturations. A mixed arterial/venous sample, high altitude, or significant lung disease may be responsible. Once the blood sample has been confirmed as arterial the next step is taken to quantify respiratory function. Calculating A-a gradients is not accurate when a patient is receiving supplemental oxygen.

Calculating an A-a gradient is as follows: PA02 = (barometric pressure – water vapor pressure) x FI02 – 1.2 x PaC02
The A-a gradient then equals PA02 – Pa02. Partial pressure of alveolar oxygen (PA02) is calculated using these factors; barometric pressure at sea level is 760mmHg; water vapor pressure is 47mmHg; FI02 is fractional inspired oxygen concentration (21%); 1.2 is the respiratory quotient (value ranges from .8 – 1.2). PaC02 and Pa02 (partial pressure of arterial gases) are obtained from the arterial blood gas sample.

Another approach to calculating A-a gradient is a more simplified equation that can only be used when at sea level. At sea level, water vapor pressure (roughly 50) is subtracted from barometric pressure (760) then multiplied by inspired room air oxygen concentration (21%). This gives a PA02 value of 150. Equation form this would look like (760-50) x .21 = 150. This simplified formula provides a constant value of PA02 as long as a patient is breathing room air at sea level. Putting in the respiratory quotient and the values from the arterial blood gas gives a simplified A-a gradient calculation of: A-a gradient = [150 – PaC02(1.1)] – Pa02

At sea level, patients with normal lung function breathing room air (21% oxygen) will have an A-a gradient ranging from 0-10. Values of 10-20 may be considered mild pulmonary dysfunction and values > 20 would be considered moderate pulmonary dysfunction. Patients exhibiting A-a gradients of >30 would require careful assessment of ventilation to determine if mechanical ventilation is required. Again, it must be said that A-a gradients should only be evaluated for patients breathing room air and a reasonably accurate barometric pressure should be determined.

Pa02/FI02 ratio

Pa02/FI02 ratio is a calculation often used for patients receiving increased inspired oxygen concentrations. A general rule of thumb states the Pa02 should equal roughly 5x the inspired oxygen concentration. Pa02/FI02 ratio would be helpful for assessing oxygenation in patients maintained on a ventilator or anesthetized patients that are receiving known concentrations of inspired oxygen. In normal lung function Pa02/FI02 ratio is nearly 500mmHg and values below 200 indicate severe venous admixture. Calculating Pa02/FI02 ratio of patients receiving nasal or mask oxygen is not practical since the inspired oxygen concentration would be difficult to determine. This usually warrants taking the patient off oxygen for 5-10 minutes in order to calculate an A-a gradient on room air. Serial monitoring of A-a gradients on room air will give the clinician objective information on a patients progress.

REFERENCES

[1]Davis HD, Vet ICU Book 2002; 258. [2]Haskins SC, IVECCS Proceedings 2005; 159. [3]Martin L, All You Really Need To Know To Interpret Arterial Blood Gases 1999; xxi, 14, 48.

PLEURAL SPACE DISEASE; AND CHEST TAPS AND TUBES

Bea Biddinger, LVT, VTS (ECC)
E. Lansing, MI

INTRODUCTION

Pleural space disease is a common cause of respiratory distress in emergent and critical patients. Air, fluid, exudates, chyle, blood, or herniated abdominal organs may be present in the chest cavity. When any space occupying condition occurs, the lungs cannot expand normally and pulmonary function may fail. Patients must be stabilized first with oxygen if in distress. And for patients with rapid respiratory compromise a thoracocentesis (chest tap) should be performed before diagnostics such as radiographs.

CHEST TRAUMA AND WOUNDS

Chest wounds must be investigated to ensure the pleural lining of the thorax is intact. "Sucking chest wounds" are penetrating wounds to the pleura, and the name is derived from the sound of air sucking in upon inspiration. A bandage of sterile lubricating gel and sterile gauze should be placed over the area to restore a seal. Palpation of the entire chest area is necessary to find any small penetrating injuries such as bite wounds or gunshot. The presence of subcutaneous emphysema indicates a penetrating injury, skin wound, or tracheal damage. Injuries to the trachea such as crushing, laceration, avulsion, or a migrating esophageal or tracheal foreign body may not exhibit obvious external trauma. However, air leakage at the site can dissect through tissue planes causing varying degrees of subcutaneous emphysema, pneumomediastinum, or pneumothorax. Wounds are cleaned and evaluated for tissue trauma, dead space, or infection. Surgical exploration of the chest is often recommended. Tracheal wounds should be cleaned and lightly bandaged with sterile materials to minimize accumulation of subcutaneous emphysema. Intrathoracic tracheal injury would require strict cage rest and possibly surgical correction. Radiographs of the chest are taken when the patient is stable and aerobic and anaerobic cultures taken of any penetrating wound.

PNEUMOTHORAX

Blunt trauma causing rib fractures and flail chest often account for pneumothorax and hemothorax by lacerating lung pleura and vessels. Rib fractures are often indiscernible until radiographs are taken. Flail chest can be observed due to the defect it produces, and is defined as a segment of two or more adjacent ribs fractured in both dorsal and ventral locations. Segments of flail chest will move freely and paradoxically with respiration. In addition to potentially lacerating lung tissue or vessels, flail chest also negatively affects ventilatory efficiency by decreasing the amount of negative pressure (tidal volume) generated during inspiration. Likewise respiratory function is decreased due to areas of pulmonary contusion in direct proximity with rib fractures and flail segments. Rib fractures generally do not require direct treatment but patients do benefit from treatment of pain. Flail chest may require external splinting or surgical fixation to limit movement of the fractured segment.

Pneumothorax is one of the most common results of chest trauma and is quite common in patients exhibiting fractures. Pneumothorax is an accumulation of free air located outside the lungs but within the chest cavity; and can be classified as open when accompanied by a flesh wound that communicates to the chest or closed when the leak of air occurs from within the thorax. Closed pneumothorax can be relatively minor and self-limiting when the amount of air leakage is small. Tension pneumothorax is a term used to describe severe forms of closed pneumothorax where a defect in the lung leaks air during inspiration and then closes during the rest of the respiratory cycle. A one-way valve is formed letting air into the pleural space and trapping it there. Large accumulations of air inhibit lung expansion and venous return to the heart. Alveolar collapse is progressive and will result in V/Q mismatch, intrapulmonary shunting, and lung lobe atelectasis. Treatment of pneumothorax is removal of air by chest tap or chest tube placement if air repeatedly or rapidly accumulates. Radiographs, oxygen therapy, and pain relief are all typical during diagnosis and treatment.

HEMOTHORAX

Accumulation of blood in the chest may occur as a result of trauma, coagulopathy, or neoplastic conditions. Removal of the accumulated blood is not necessary unless respiratory compromise is present. Treatment of hemothorax is supportive in nature and may include fluids and blood products if the patient is hypotensive. Diagnosis and treatment of underlying disease would take priority before tapping the chest of a patient with a concurrent coagulopathy in all cases except where a chest tap is immediately necessary to alleviate respiratory distress.

DIAPHRAGMATIC HERNIA

A diaphragmatic hernia may be present from birth (congenital defect) or may be result of traumatic injury. The blunt force of trauma causes a rent in the diaphragm and abdominal organs slide into the pleural space. The organs most commonly involved are liver, spleen, omentum, and stomach. Diagnosis is generally made upon radiographs although gut sounds may be discernable on auscultation. Treatment would be supportive in nature until surgical intervention is practical.

PLEURAL EFFUSIONS

Pleural effusion can result from increased vascular permeability, increased hydrostatic pressure, decreased oncotic pressure, lymphatic obstruction, infection, or hemorrhage from trauma, coagulopathy, or neoplasia. Specific examples of disease processes can include: congestive heart failure, cardiomyopathy, neoplasia, chylothorax (idiopathic or lymphatic obstruction), sepsis, feline infectious peritonitis, pancreatitis, hypoalbuminemia, lymphangiectasia, lymphoma, or lung lobe torsion. Evaluation of fluid sampled during chest tap will likely yield information as to the cause of the effusion. Samples of the fluid should be placed into EDTA tubes for cytologic evaluation, red top clot tubes for biochemical analysis, sterile tubes or swabs for culture, and possibly test specific tubes for PCR (polymerase chain reaction) used to help diagnose feline infectious peritionitis.

Many clinicians are familiar with the very traditional classification scheme of pleural fluids being categorized as transudate, modified transudate, or exudates which describe fluids by color, total protein value, and total nucleated cell count. However, due to overlap of some values, those categories have more recently been modified and expanded to assist in determining etiology which separates fluids into transudates (either protein rich or poor), exudates (either septic or non-septic), or effusions caused by (either ruptured vessel/viscus or cell exfoliation). A diagnosis will likely be made by gross examination of the fluid, laboratory examination of the fluid, imaging of the chest, and of course patient history and physical findings.

Transudates are either protein rich or protein poor (<2.5 g/dl). Using a refractometer to measure is a quick and easy scanning tool. Examples of associated disorders, respectively, would be congestive heart failure and hypoalbuminemia (GI or renal system loss of albumin or decreased albumin production by liver). Exudates are turbid in appearance and usually have very high nucleated cell counts. These fluids are classified as either septic or non-septic. Examples of associated disorders, respectively, include penetrating wounds or migrating foreign bodies and pacreatitis or feline infectious peritonitis. Effusions caused by a ruptured vessel are most commonly caused by hemorrhage (trauma) but can be chylous as well when lymphatic drainage is impaired. And finally neoplasias will cause effusions by cell exfoliation.

PERICARDIAL EFFUSION

Pericardial effusion is an accumulation of fluid (usually blood) within the pericardial sac. The most common causes are neoplasia, idiopathic effusion, congestive heart failure, coagulopathy, and infection. Pericardial effusion is life threatening due to the restrictive nature it has on the heart. Decreased cardiac output and systemic venous congestion result from cardiac tamponade. Thoracic ultrasound to view the heart is a very quick and non-stressful method of diagnosing the presence of pericardial effusion. Pericardiocentesis is palliative treatment that will restore cardiac output.

CHEST TAP EQUIPMENT

For cats, a butterfly catheter 19 or 21 gauge does a very nice job with a three way stopcock and syringe attached. For an obese cat, a 20 gauge 1½ inch needle may be needed. For smaller dogs (<15 kgs) equipment can vary from butterfly catheters, 20 gauge 1½ inch needles, or 20 gauge over the needle peripheral catheters. For larger dogs 16 or 18 gauge catheters are often used. Thoracic drainage catheters work well for all but the smallest of dogs and will be discussed separately. Pros and cons exist for any instrument chosen but all are generally safe if used in an appropriate manner. Butterfly catheters and 20 gauge needles have the sharp end that will remain in the chest during evacuation. To prevent lung laceration, the tip of the needle should be angled closer to the chest wall. As greater volumes are removed, angle the needle parallel with the internal chest wall. Peripheral catheters do not risk lung laceration once the stylet is removed; **however** these catheters kink very easily between layers of skin, intercostal muscle, and pleura making drainage impossible. Typically this means several more attempts which is why the author does not recommend using over the needle catheters. Regardless of instrument chosen to perform the tap, an extension set with 3-way stopcock facilitates rapid evacuation and minimizes trauma caused by movement.

Chest Tap Procedure

Clip a 4" square area over the 7-10th intercostal space. Air will be removed easiest from the dorsal third of the chest with the patient in sternal or lateral recumbency. Ventral taps will remove the most fluid with the patient in sternal recumbency or standing. An initial scrub of the skin is performed. Local anesthesia of lidocaine or bupivicaine in the skin and intercostal muscles can be helpful. Another complete scrub of the skin is performed. Sterile gloves are worn to guard against infection. Identify the 7-8th or 8-9th intercostal space. Grip close to the tip to ensure the amount of penetration into the chest does not exceed what is necessary. Puncture into the chest on the cranial aspect of a rib. If rib is struck, the needle is "walked" off the cranial aspect of the rib in order to avoid the intercostal arteries, veins, and nerves that run along the caudal curvature. Once inside the pleural space, the extension set is connected as quickly as possible or a sterile finger plugs the needle to decrease the chance of air entering. An assistant holds the syringe and aspirates as the needle is advanced in small increments. The assistant indicates whether each aspiration is negative (receives negative pressure) or yields air or fluid. With the extension set, 3-way stopcock,and syringe system, air or fluid evacuated from the chest can be quickly expelled by closing the stopcock to the patient and pushing it out the open port. The 3-way is then turned off to the open port and evacuation is resumed.

THORACIC DRAINAGE CATHETERS

For evacuation of large volumes, drainage catheters perform superiorly whereas other peripheral catheters or needles may drain only small volumes and need to be repeated on both sides of the chest. Placement of a drainage catheter is usually sufficient to completely drain the entire chest. This is a big advantage for the patient to receive only one needle puncture. For cats a drainage catheter is fairly large and would likely only be used for instances when it would be left as an indwelling chest tube. A drainage catheter of 8fr. size would be appropriate in those indwelling situations. Small canine patients would also benefit from the smaller 8fr. size catheter. A 10fr. is appropriate for medium and large dogs (weighing >15kg). Disadvantages of the thoracic drainage catheter are price and size.

Drainage Catheter Placement Procedure

As a rule, most every patient should receive sedation, analgesia, and local block. Clip a large area of the chest (rib spaces 4-13). An initial scrub of the skin is performed. Infiltrate skin and muscles with local anesthetic. Sterile gloves are worn and a drape is placed over the site. A full thickness stab incision through the skin is made with a number 11 or 15 scalpel blade. Intercostal rib spaces 7-8, 8-9, or 9-10 are appropriate. Before beginning, make certain the catheter is not adhered to the stylet by twisting it. Hold the catheter with the non-dominant hand wrapped firmly around the catheter approximately 1 ½" from the trocar tip. This will ensure the chest is not penetrated too far. The dominant hand is used on the driving end, and the catheter is held perpendicular to the chest. Once the thoracic cavity is entered the catheter is aimed toward the patient's opposite elbow. Next the stylet is quickly unscrewed and the catheter is advanced off the stylet into the chest. These catheters have multiple perforations that must be placed inside the chest very quickly in order to avoid a pneumothorax. In general, more length of the catheter is fed into the chest initially than may be needed. The catheter can be backed out later if placed too far inside, but due to contamination risk, advancing the catheter further into the chest later is never allowed. Radiographs would be indicated if the catheter will be indwelling. The end of a drainage catheter has a female connector to fit an extension set, 3way stopcock, and syringe. Once aspiration of the chest is finished, the catheter is removed with a quick, continuous, pull while the skin is pinched at the insertion site. Fingertip pressure is maintained for several minutes, and a light bandage using a triple antibiotic ointment applied. If preferred, a skin staple can be placed at the insertion site.

CHEST TUBES

A size 10 or 12fr. (Argyle®) chest tube is appropriate for most cats and small dogs while medium and large dogs need a 20 or 28fr. tube. Connecting adapters, suction tubing, and tube occluding clamps need to be ready before placement. Consideration needs to be given whether the patient will be placed on continuous underwater drainage. A three way stopcock may be necessary within the connections in order to infiltrate the pleural space with local anesthetic or flush the chest with sterile saline. Only three way stopcocks with leur lock fittings should be used. A chest tube clamp can be purchased or made from a straight hemostat with lengths of red rubber catheters cut to fit over each jaw. Placement of an indwelling chest tube is indicated to remove recurring air, large recurring volumes of fluid, or to lavage the chest. A commonly accepted guideline recommends chest tube placement if three chest taps are needed within 24 hours or if a pneumothorax is very rapidly recurring.

Chest Tube Placement Procedure

With the anesthetized patient in lateral recumbency, clip and prep an area from the scapula to well past the last rib closely approaching the spine and the sternum. An assistant pulls the patient's skin cranial ventral as far as possible and holds it until placement of the tube into the chest. When the stretched skin is released, it will provide a tunneled space guarding against air communication. Intercostal space 9-10 is identified and a skin incision is made in the dorsal third of the chest. Before placement, the chest tube is measured from the last rib to the elbow to approximate the length of tube to be placed inside the chest. There are two common methods of placement. Blunt dissection is performed with hemostats through the subcutaneous fat and the intercostal muscles followed by intertion of the tube. Or there is the trocar method where the end of the tube is held with the non-dominant hand approximately 1 ½ -2" from the tip while the dominant hand is on driving end of the tube. The tube is directed perpendicularly and is driven with a sharp strike of base of palm until the pleural space is popped through. Regardless of method choice the tube is then aimed toward the opposite (down) elbow. The tube and stylet is advanced together only until it has entered the thorax and then the tube is quickly advanced into the chest while the stylet is held in place. Once the chest is penetrated the anesthetist must provide controlled ventilation to avoid a pneumothorax. The stylet is removed and the chest can be evacuated using sterile connections. A purse string suture is placed in the skin, and a finger trap suture is used to hold the tube in place. Once the chest tube is secured, a light sterile wrap is applied. Radiographs are obtained while the patient is still anesthetized to confirm correct placement of the tube. Connections should be placed to utilize either intermittent or continuous suctioning, and secured to insure that all seals are air tight. Suture, wire suture, tape, or thin strips of adhesive dressing can be used to secure various connections. Sterile gauze and triple antibiotic ointment at the insertion site provide the first layer of bandage. Kling and Elastikon work nicely for a secure bandage. Always remember to tape a loop of the tubing to the bandage so the patient cannot pull directly on the chest tube.

REFERENCES

[1]Dempsey SM, et al. JAAHA 2011;47(1):1. [2]Macintire DK, et al. Manual SA ECCM 2005; 128. [3]Jutkowitz LA. MSU Vet Med student course packet.

BRAIN TUMORS IN SMALL ANIMALS: DIAGNOSIS AND MANAGEMENT

R. Timothy Bentley, BVSc, DACVIM
West Lafayette, IN

INTRODUCTION

Brain tumors in small animals are relatively common, occurring at an age-standardized incidence rate of 20 cases per 100,000 dogs per year[1]. In dogs, meningiomas and gliomas are most common[2], whereas in cats meningiomas predominate[3,4]. Other primary brain tumors include choroid plexus tumors, primary CNS lymphoma or histiocytic sarcoma, and primitive neuroectodermal tumors [2,3].

Pituitary tumors are most often included with secondary brain tumors[5]. Secondary brain tumors include hemangiosarcoma (especially in dogs), secondary CNS lymphoma and intravascular lymphoma, carcinomas, nasal tumors, secondary histiocytic sarcomas, and melanomas[5,6].

Occasionally, the simultaneous occurrence of a meningioma with a glioma, a choroid plexus tumor or another meningioma in the same dog is reported[2,7–10], and while cats sporadically present with a meningioma concurrent to lymphoma or another secondary brain tumor, it is slightly more common for cats to present with 2 to 4 or even more concurrent meningiomas[4,11].

The outcome of dogs and cats treated for brain tumors with supportive therapy only has not been closely studied. In 45 dogs with brain tumors receiving either no treatment or supportive therapy, a median survival of only 0.2 months was reported[12].

Table 1. Epidemiological Factors – Small Animal Primary Brain Tumors.

Tumor	Age	Breeds	Percentage of Primary Brain Tumors	Gender Predisposition
Meningioma	9.6-11.1yr with 40% increased risk per year compared to other brain tumors (Dogs) [2,10,13] 10.3-12.2yr (Cats) [4,14]	Golden Retriever, other dolichocephalic breeds and Boxers[2,10] Domestic Shorthairs[4,14,15]	45% (Dogs)[2] 59-73% (Cats)[3,4]	Female dogs and male cats proposed but inconsistently reported[2,4,10,15,16]
Glioma	8-9yr (Dogs)[2] 9-13yr (Cats)[4]	Brachycephalic Dogs, especially Boxers and Boston Terriers[2] Domestic Shorthairs[4]	36% (Dogs)[2] 8% (Cats)[4]	Female dogs[17,18] and male cats[4]
Choroid Plexus Tumor in Dogs	5yr (papilloma)[19] 7yr (carcinoma)[19]	Golden Retrievers[19]	7%[2]	Males and females equal[19]
Primary CNS Lymphoma	7.4yr (Dogs)[2] 10.5yr (Cats)[4]	Domestic Shorthairs[4]	4% (Dogs)[2] 5% (Cats)[4]	Slightly more female cats[4]

INTRACRANIAL MENINGIOMA

Pathology

Meningiomas arise from the arachnoid cap cells. The majority of canine meningiomas are benign (grade I) or atypical (grade II), but malignant meningiomas (grade III) are occasionally reported[10].The meningothelial and transitional histopathological subtypes are most common, and subtype may be related to outcome[10,13].

Presentation and Diagnostic Features

Slower progression and normal mental status at the time of presentation are more common in dogs with meningioma than with other tumor types[2,12]. There is a predisposition to the forebrain, specifically the fronto-olfactory area, and seizures are the most repeatable clinical sign; the inter-ictal neurological examination may be normal early in the disease course[2,8,10,13]. The mean time from seizures and/or behavior change to developing deficits is 90 days[20]. Other forebrain deficits reported include change in personality, circling, and deficits in the menace response and postural responses[8,21]. Around 20% of canine meningiomas affect the caudal fossa[10], causing generalized ataxia or vestibular syndrome, paresis and cranial nerve deficits[2,13]

Most feline meningiomas occur in various locations superficial to the cerebrum or within the falx cerebri[4,14], with about 20% within the third ventricle (tela choroidea)[4]. Associated clinical signs include altered consciousness (26%), seizures (25%) and circling (23%)[4]. Paresis of 1 to 4 legs has been reported in 83% of cats[15]. Cats may present with only non-specific clinical signs such as lethargy and anorexia, and meningiomas have been reported to be detected incidentally (23%)[4].

When viewed with magnetic resonance imaging (MRI), meningiomas are typically severely and homogenously enhancing masses with a clear extra-axial origin and distinct margins[3,10]. Most are iso- to hyperintense on T2-weighted images, T1 iso- to hypo-intense, and have at least some degree of peri-tumoral edema[3,10] in both dogs and cats. A "dural tail" has been reported in the majority of cases, a finding which is most common with meningioma but is not pathognomic[3,10,22,23]. In dogs, supratentorial meningiomas are more likely to have diffuse edema and intra-tumoral fluid accumulations, while skull base meningiomas may be plaque-like[10]. While only 2 of 33 feline meningiomas were reported to be cystic, meningiomas were the only type of extra-axial tumor to show a cystic structure in this study[4]. Hyperostosis is noted in the adjacent skull bone in 49-73% of cats[3,15], and there can even be skull erosion[22]; hyperostosis is seen with 0-23% of dogs[10,24]. In dogs, histiocytic sarcoma can have an indistinguishable MRI appearance but is less common[25].

A moderately to severely elevated CSF protein with mild mixed or neutrophilic pleocytosis is common in dogs and cats[2,4].

Treatment and Prognosis

Dogs with meningiomas have a longer survival than with other brain tumors[12]. Median survival times with conventional surgery have been 5-7 months[26-28]. Median survival times with advanced surgical equipment, such as a cavitating ultrasonographic aspirator or endoscopic assistance, have been 42 and 70 months respectively[13,29]. Cats treated surgically had a much longer median survival time than cats treated with radiation or chemotherapy[4]. While surgery in cats is considered to be the optimal treatment, with reported median survivals of 22 or 23 months, it is notable that perioperative mortality approaches 20%[4,15]. Chemotherapy with hydroxyurea, post-operatively[13] or alone[30], has been reported in dogs but not well studied.

Radiation therapy has been better studied in dogs. The median survival for radiation therapy of extra-axial, non-pituitary masses without specific histological diagnosis was 11.5 months or 19 months[31,32]. Post-operative radiation following conventional surgery significantly improves the long-term outcome, with median survivals of 14, 16.5 or 63 months[28,31,33].

It is worth noting that while aggressive therapies are associated with relatively good results in both species, outcome is extremely variable, with reports including individual patients being lost during attempted treatment, and other individuals achieving permanent remission. The time to presentation (0 to 30 months), and the time from the initial clinical signs to necropsy (0 to 13.5 months) also indicate variable progression[2,20]. The prognosis is therefore extremely variable and hard to predict for individual patients.

INTRACRANIAL GLIOMA

Pathology

The majority of canine gliomas are astrocytomas or oligodendrogliomas, but undifferentiated gliomas, mixed oligoastrocytomas and mixed astrocytoma-ependymoma are occasionally reported[2,34]. Astrocytomas may be slightly more prevalent than oligodendrogliomas[2]. Astrocytomas may be reported as either grade II or III astrocytomas or as glioblastoma (grade IV astrocytoma), while oligodendrogliomas occur as oligodendrogliomas (grade II) or anaplastic oligodendrogliomas (grade III)[18]. Classification of small animal gliomas into different subtypes has not been universalized, but a system is available for humans[35].

Presentation and Diagnostic Features

The majority of both oligodendrogliomas and astrocytomas affect the forebrain, and clinical signs such as seizures, vision loss and mentation change are common; brainstem deficits such as vestibular syndrome are also possible[2,18].

The MRI appearance of gliomas is much more variable than it is for meningiomas. Although usually intra-axial, occasional astrocytomas of apparent extra-axial origin[2], diffuse leptomeningeal astrocytomas[36], and oligodendrogliomas contacting the meninges[18] have been reported. Excepting contact with the meninges, no significant differences were found between astrocytomas and oligodendrogliomas; both showed variable degrees of signal uniformity, contrast enhancement or rim enhancement, hemorrhage and cystic areas, but both were typically T1-hypointense, T2-hyperintense and showed perilesional T2 hyperintensity[18]. Contrast enhancement was associated with increasing tumor grade[18]. Gliomas can appear similar to cerebrovascular accidents in dogs[17].

In dogs, a moderately elevated CSF protein with a variable degree of mononuclear or mixed pleocytosis is typical[2].

Treatment and Prognosis

The median survival with radiation therapy for canine intra-axial masses without histological diagnosis has been reported to be 26 months[32]. In another report of canine intra-axial masses, the median survival following surgery ± radiation therapy was 10 months, compared with 9 months for radiation therapy alone[31].

In 3 cats treated for cerebral astrocytoma with anticonvulsants, effective seizure control was achieved for 1 to 3 years[37], however other limited reports of cats with gliomas indicate a rapid progression to euthanasia[38,39].

CHOROID PLEXUS TUMORS AND EPENDYMOMAS

Pathology

Tumors of the lateral, third and fourth ventricles include tumors arising from the choroid plexus, such as choroid plexus papillomas (CPP) or choroid plexus carcinomas (CPC), and ependymoma, which originates from the lining of the ventricles.

Presentation and Diagnostic Features

The frequency of canine choroid plexus tumors increases moving caudally, with around half being in the 4th ventricle[2,19]. As a result, seizures are less common than with other primary brain tumors, while deficits such as vestibular syndrome, severe mentation change and neck pain are more common[2,19]. In cats, ependymomas, occur within the lateral or third ventricles[4].

The CSF protein is apparently always elevated in dogs, in the range of 27-380mg/dL, with average concentrations of 80 or 153mg/dL[2,19]. Higher CSF protein was significantly associated with CPCs rather than CPPs, and a CSF protein above 80mg/dL was only observed with CPCs[19]. While malignant cells are typically not observed[2], re-evaluation of banked CSF lead to the detection of neoplastic choroid epithelium in 7 of 15 dogs with CPCs but in no CPP cases[19]. Mild mixed or mononuclear pleocytosis is typical[2,19].

Choroid plexus tumors are typically intensely enhancing on MRI[19]. CPPs are either globular or papilliform, whereas CPCs are nearly always globular[19]. They are largely T2-hyperintense, but unlike most other primary brain tumors, there is a roughly even distribution between hypo-, iso- and hyper-intensity on T1-weighted imaging[19]. Ventriculomegaly, which in relation to the mass may be either rostral or both rostral and caudal, is present in about 3 out of 4 cases[19]. Detection of apparent metastasis to other sites within the CNS was seen in 35% of CPCs but no cases of CPP[19].

Treatment and Prognosis

Little information is available regarding the outcome of choroid plexus tumors. The choroid plexus is extremely vascular, making surgery a challenging prospect; one dog dying of intracranial hemorrhage the day after surgery has been briefly reported[12].

PRIMARY CNS LYMPHOMA

Pathology

Primary CNS lymphoma is less common than secondary lymphoma in cats and dogs[2,4,5], and cats are rarely FeLV positive[4].

Presentation and Diagnostic Features

As this disease can either focally or diffusely affect single or multiple regions of the brain, the presentation is variable[4,5]. Occasionally, the optic chiasm is affected, causing acute blindness[40].

Lymphoblasts were only reported in the CSF of 1 of 6 cats with intracranial lymphoma, and 1 of 5 dogs with primary CNS lymphoma[4,5]. Although potentially undetectable[3], lymphoma variably causes single or multifocal lesions that are either intra-axial or extra-axial, and was the only neoplasia to cause diffuse cerebral or diffuse brainstem disease in cats[4,5]. Lymphoma was also the only other tumor affecting the meninges besides meningioma in cats[4], and the only other tumor to cause a dural tail sign[3]. While location(s) varies significantly, lesions were repeatably T2 iso- to hyper-intense, T1 iso- to hypo-intense and markedly contrast enhancing in cats[3].

Treatment and Prognosis

The median survival with corticosteroids was 21 days in cats[4], and one dog receiving prednisone, vincristine and doxorubicin survived 2 months[40], but chemotherapy and radiation therapy have not been well studied.

REFERENCES

1. Dobson J, et al. *Journal of Small Animal Practice*. 2002;43:240-246. 2. Snyder JM, et al. *Journal of Veterinary Internal Medicine*. 2006;20(3):669-675. 3. Troxel MT, et al. *Journal of Veterinary Internal Medicine*. 2004;18(2):176-89. 4. Troxel MT, et al. *Journal of Veterinary Internal Medicine*. 2003;17(6):850-9. 5. Snyder JM, et al. *Journal of Veterinary Internal Medicine*. 2008;22(1):172-7. 6. Kent M, et al. *Veterinary Radiology & Ultrasound*. 2001;42(6):504-10. 7. Espino L, et al. *Acta Veterinaria Hungarica*. 2009;57(3):389-97. 8. McDonnell JJ, et al. *Journal of the American Animal Hospital Association*. 2007;43(4):201-8. 9. Stacy BA, et al. *Journal of Veterinary Internal Medicine*. 2003;17(3):357-9. 10. Sturges BK, et al. *Journal of Veterinary Internal Medicine*. 2008;22(3):586-595. 11. Forterre F, et al. *Journal of Feline Medicine and Surgery*. 2007;9(1):36-43. 12. Heidner GL, et al. *Journal of Veterinary Internal Medicine*. 1991;5(4):219-26. 13. Greco J, et al. *Journal of the American Veterinary Medical Association*. 2006;229(3):394-400. 14. Tomek A, et al. *Journal of Feline Medicine and Surgery*. 2006;8(4):243-53. 15. Gordon L, et al. *Veterinary Surgery*. 1994;23(2):94-100. 16. Mandrioli L, et al. *Journal of Veterinary Science*. 2007;8(2):111-5. 17. Cervera V, et al. *Veterinary Radiology & Ultrasound*. 2011;52(1):33-40. 18. Young BD, et al. *Veterinary Radiology & Ultrasound*. 2011;52(2):132-41. 19. Westworth D, et al. *Journal of Veterinary Internal Medicine*. 2008;22(5):1157-1165. 20. Foster ES, et al. *Journal of Veterinary Internal Medicine*. 1988;2(2):71-4. 21. Bagley R, et al. *Clinical Techniques in Small Animal Practice*. 1998;13(3):179-84. 22. Gutierrez-Quintana R, et al. *Journal of Feline Medicine and Surgery*. 2011;13(4):296-9. 23. Graham JP, et al. *Veterinary Radiology & Ultrasound*. 1998;39(4):297-302. 24. Hathcock JT. *Veterinary Radiology*. 2000;37(4):257-63. 25. Tamura S, et al. *Veterinary Radiology & Ultrasound*. 2009;50(2):178-181. 26. Niebauer G, et al. *Journal of the American Veterinary Medical Association*. 1991;198(1):89-95. 27. Glass E, et al. *Journal of the American Animal Hospital Association*. 2000;36(1):43-50. 28. Axlund TW, et al. *Journal of the American Veterinary Medical Association*. 2002;221(11):1597-600. 29. Klopp LS, et al. *Journal of Veterinary Internal Medicine*. 2009;23(1):108-115. 30. Tamura S, et al. *The Journal of Veterinary Medical Science*. 2007;69(12):1313-5. 31. Brearley MJ, et al. *Journal of Veterinary Internal Medicine*. 1999;13:408-412. 32. Bley CR, et al. *Journal of Veterinary Internal Medicine*. 2005;19(6):849-54. 33. Platt SR, et al. *Journal of Veterinary Internal Medicine*. 2006;20(3):663-8. 34. Hayes KC, et al. *Veterinary Pathology*. 1969;6(2):94-116. 35. Louis DN, et al. *Acta Neuropathologica*. 2007;114(2):97-109. 36. Kraft S, et al. *Journal of Computer Assisted Tomography*. 1990;14(4):555-60. 37. Sarfaty D, et al. *Journal of the American Veterinary Medical Association*. 1987;191(8):976-8. 38. Dickinson PJ, et al. *Veterinary Pathology*. 2000;37(2):160-167. 39. Sant'Ana FJF, et al. *Veterinary Pathology*. 2002;39(6):759-761. 40. Davidson M, et al. *Journal of the American Veterinary Medical Association*. 1991;199(6):755-8.

THERAPEUTIC LASERS IN VETERINARY MEDICINE

Kristin Kirkby, DVM, MS, CCRT, DACVS
Kirkland, WA

LASER PHYSICS

Lasers were first used surgically to cut or seal tissue. It is only recently that therapeutic (also known as cold laser or low-level) lasers have gained wide recognition in veterinary medicine. Lasers, along with the sun, ordinary light bulbs, x-ray machines, and microwave ovens, emit electromagnetic radiation. Energy from these sources travels at the speed of light in packets known as photons. Photons travel in waves, and the type of radiation is distinguished by its wavelength. Electromagnetic radiation is typically described as a spectrum from very short (gamma rays, 1000-1 fm) to long (radio waves, 1000-1m) wavelengths. Laser radiation falls within the center of this spectrum, with wavelengths in the visible light (400-800 nm) and infrared (1000-0.8 um) range.

Laser is an acronym for "light amplification by stimulated emission of radiation". In this process, electromagnetic energy is harnessed into an intense, coherent, monochromatic beam of light. The properties of monochromaticity (all waves are same length/ color) and coherence (all waves in phase) are characteristics of all lasers.

A laser is made using a material (gas, liquid, solid) that when stimulated by an external energy source such, such as electricity, will release photons of a single color or wavelength. For example, a helium-neon (HeNe) laser will emit only wavelengths of approximately 633 nm, which fall within the visible light spectrum and produce the color red. Gallium arsenide (GaAs) lasers emit wavelengths of 904 nm, and are thus invisible within the infrared portion of the spectrum. Wavelength is inversely proportional to energy. This can be appreciated by recognizing differences between harmless radio waves (long wavelength, low energy) and gamma rays that are used in cancer radiation therapy (short wavelength, very high energy). Wavelengths between 300-400 nm represent UVA and UVB rays. These rays possess dangerous ionizing properties that cause sunburn. The wavelength of therapeutic lasers all fall between 400 and 980 nm, with the previously mentioned HeNe and GaAs lasers being the most frequently studied and employed clinically.

Therapeutic lasers are commonly referred to as "low level lasers" or "cold lasers", in contrast to high-powered lasers that are used to cut tissue. True low level lasers have a power output less than 500 miliwatts (mW) and cannot cut tissue. Power output is significant, because a laser with a higher wattage will reach the desired dose more quickly. Amongst low level lasers, power output (watts, W) varies greatly, anywhere from 3.5mW to 500 mW. Recently, therapeutic lasers with power output greater than 500 mW have been introduced to the veterinary field. While these lasers are able to treat a large surface area more quickly, they produce discernable heat and have the capacity cause considerable tissue heating and risk burning skin.

There are several other parameters that can differ between lasers. Laser output can be continuous or pulsed. When a laser is pulsed, the power output will reach a peak and return to zero at varying frequencies (Hz), or duty cycles. Therefore, the amount of power delivered will be the average power output.

Power density, or intensity, is the amount of power concentrated on a given area and is measured in W/cm^2. If the laser beam is spread over a larger area (larger spot size), the amount of energy at each point becomes less, compared to concentrating the energy at a single, small point. This will be influenced by the beam diameter (mm) and spot size (cm^2) that are specific to each laser. Some lasers allow for adjustment of beam diameter and spot size.

Lasers are divided into four classes (I, II, IIIa, IIIb, and IV). A common misconception is that these classes distinguish the efficacy or quality of the laser. Rather, laser class is determined by the ability to cause eye injury and is based on power output. Class I-IIIa lasers, including supermarket scanners, laser pointers and remote controls, are considered safe. Class IIIb lasers pose a risk of eye injury, and eye protection is recommended. Any laser with greater than 500 mW of power falls into class IV, which is considered to be an acute hazard to the skin and eyes from direct and scattered radiation.

Dosage (also known as energy density or fluence) appears to be the most important laser parameter in clinical application. Lasers that are available commercially and marketed for medical or veterinary use will likely come with recommended doses for various conditions pre-programmed into the unit or listed in the instruction manual. While the optimal treatment dose has not been established for any condition, generally recommended doses fall between 2-10 J/cm^2.

Dosage is measured in J/cm^2, and can be calculated using the following formula when treating superficial targets:

$$\text{Dose} = \frac{P \times t}{A}$$

P= laser's output power (W)
t= treatment time (seconds)
A= area treated (cm^2)
1 J= 1 W/s

The depth of laser penetration depends on the wavelength of the laser, with longer wavelengths resulting in deeper penetration. A GaAs laser (904 nm) can reach tissue depths of 3-5 cm, while a HeNe (633 nm) will have more superficial penetration, near 1cm. Photons emitted by a laser interact with tissue in four ways: reflection, transmission, scatter and absorption. Factors that may influence the depth of penetration include hair coat, skin coloration, and tissue composition. Much of the radiation energy may be absorbed by hair, dark pigmented skin, and highly vascular tissue such as muscle. However, unlike sound waves, light can penetrate bone. Reflection of light off of the tissue surface can decrease absorption, thus, when targeting deep tissue, the laser should be held in contact with the skin.

PHOTOBIOMODULATION

Photobiomodulation is the application of light in order to modify a biologic process. Depending on the wavelength, power and other factors, light can cause beneficial or harmful effects in cells and tissue. Therapeutic lasers can be used to stimulate favorable effects in tissue, including enhanced wound healing and modulation of pain. The exact mechanism of tissue healing has yet to be fully elucidated; but, there is growing evidence to support lasers influencing each phase of wound healing. Laser energy is absorbed by chromophores, such as cytochrome *c* oxidase, in the mitochondria. This results in increased cellular metabolism and ATP production and stimulation of DNA and RNA formation. These events are followed by an increased expression of growth factors, cytokines, and genes related to cell proliferation and migration.

The precise role of laser therapy in the inflammatory phase is debatable. Some studies have shown enhancement of the inflammatory response and increased production of growth factors such as TGF-b_1. Immune system up-regulation has also been demonstrated. Others have demonstrated acceleration of the inflammatory phase, with rapid progression into the proliferative phase of healing. Furthermore, laser therapy has been shown to reduce inflammatory mediators, such as COX-2 and PGE, and decrease inflammatory cells and edema.

An increase in vascular activation (hyperemia) has been seen within the first 36 hours following laser treatment to open wounds. This may be due to a local, sub-sensory increase in temperature, which can influence cell membranes and ion exchange and, ultimately, vascular tone.

While laser therapy does appear to influence the inflammatory phase, it is clear from a large number of studies that the proliferative phase of wound healing is greatly enhanced by low-level lasers. Research conducted on cell cultures has found that various wavelengths and doses are effective at increasing fibroblast proliferation, but HeNe laser (633 nm) at a dose of 5 J/cm^2 appears to stimulate the greatest amount of cell proliferation and migration.

The final phase of wound healing is maturation or remodeling of scar tissue. Lasers influence this phase by enhancing the organization of collagen fibers within wounds. Furthermore, there are reports of using laser therapy (particularly 585 nm) to treat hypertrophic scars in humans.

Lasers are used routinely in veterinary rehabilitation to relieve pain. The mechanism of pain relief is thought to be due to several mechanisms, including increased secretion of seretonin, increased release of endogenous opiates, and blockage of afferent C fiber depolarization.

Based on the aforementioned properties of laser, the indications for use in veterinary medicine are numerous, but primarily include enhancement of cutaneous wound and tissue healing and amelioration of acute and chronic pain. Because laser is recognized to enhance neovascularization, irradiation of tumors or wounds that may contain cancer cells is contraindicated. Furthermore, lasers pose a known risk to the eye, so irradiation of or near the eye should not be performed. Additional contraindications include irradiation over a pregnant uterus and open growth plates.

Finally, it is important to note that a therapeutic window exists for photobiostimulation. At sub-therapeutic doses, cells will not be stimulated and no reactions will occur; at extremely high doses, detrimental effects can be seen. Interestingly, it is believed that low-level lasers, at any dose, have minimal effect on normal, uninjured tissue. The modulatory effects may also be wavelength specific and vary with other laser parameters such as polarization or pulse frequency.

CONCLUSION

In conclusion, therapeutic lasers are gaining acceptance in veterinary medicine as an adjunctive tool in wound healing and pain management. Therefore, it is important for practitioners to increase their knowledge in this area in order to make informed recommendations for such therapy. While a plethora of anecdotal reports, case reports and experimental evidence exist in support of therapeutic lasers, large clinical studies are currently lacking. It is anticipated that clinical veterinary studies will emerge in the near future that enhance our understanding of the usefulness of this modality.

REFERENCES

1. Corazza AV, et al: Photomed Laser Surg 25:102-106, 2007. 2. Bjordal JM, et al: Aust J Physiol 49:107-116, 2003. 3. Mendez TM, et al:. J Clin Laser Med Surg 22: 19-25, 2004. 4. Woodruff LD, et al: Photomed Laser Surg 22: 241-247, 2004. 5. Yu W, Naim JO, et al: Lasers Surg Med 15: 281-289, 1994. 6.Montesinos M. LASER. J Eur Med Laser Assn 1(3): 2-7, 1988. 7. Wakabayashi H, et al: Lasers Surg Med 13: 605-610, 1993. 8. Lucroy MD, et al: Vet Surg 28: 292-295, 1999. 9. Wong-Riley MT, et al: J Biol Chem 280: 4761-4771, 2005. 10. Fulop AM, et al. Photomed Laser Surg. 27(5):695-702. Additional references available upon request.

PATIENT CARE OF NON-AMBULATORY DOGS WITH SPINAL CORD INJURY

Jessica Weber, BS, LVT, VTS (SAIM, Neurology)
Seattle, WA

Treatment of the non-ambulatory dog can be time-intensive and frustrating. It can also be very rewarding to help a dog that arrived unable to walk regain that function. Many of the treatment options, and the outcome of those treatments, are dependent on diligent patient care. While there are many reasons a dog may be unable to walk, such as weakness due to metabolic, cardio-pulmonary and orthopedic disease, this presentation focuses on care of dogs with spinal cord injury. Topics include bedding choice, positioning, urinary considerations, rehabilitation and rest.

Neurologic disease leading to the inability to walk can usually be classified as "paresis" or "plegia." **Paresis** is the inability of one or more limbs to support weight against gravity. The limb can be consciously moved, though this function may be much decreased, and conscious pain perception in that limb is present. **Plegia** is the absence of movement in one or more limbs. Conscious pain perception may or may not be present. Though the animal cannot consciously move the limb, reflexive movements, such as the withdrawal and myotatic reflexes, may still be present. Paresis or plegia are usually combined with descriptions of which limbs are involved. **Para** involves the rear limbs, such as a paraplegic Dachshund. This is indicative of mid to low spinal injury, e.g. T3-L3 or L4-S2 spinal cord injuries. In a paraparetic/plegic animal the front limbs should be normal. **Tetra** involves all the limbs, for example a tetraparetic animal is weak in all four limbs. This is usually due to a cervical spinal cord or brain injury. **Mono** involves one limb. A monoplegic animal cannot move one limb. This can be due to disease in the central or peripheral nervous system, such as a brachial plexus avulsion. **Hemi** involves two ipsilateral – or same-sided – limbs; therefore a hemiparetic animal could be weak only on the left side due to a brain tumor in the right forebrain.

The examples above are only a small subset of possible differentials as paresis and plegia are caused by a number of different disease processes or injuries. The most common injury is spinal cord trauma due to intervertebral disk disease (IVDD). Vertebral fracture, fibrocartilageous embolism (FCE), high-velocity disk protrusion or trauma to the brain or peripheral nerves may also lead to a non-ambulatory dog. Paresis can be due to generalized weakness due to metabolic disease, like diabetes or hypothyroidism, or systemic peripheral nervous system disorders such as Myasthenia Gravis. Central nervous system disease such as neoplasia, vascular incidents, inflammatory or infectious disease can also lead to paresis.

Once a dog is unable to walk normally various complications can occur. In the case of tetraparetic animals, pulmonary atelectasis – collapsed lung – or pneumonia are the most pressing concerns. Lying on one side too long can lead to lung disease and decubital ulcers. Animals are also more susceptible to aspiration pneumonia when recumbent. Damage to the bladder from distention and infection can occur in animals that are unable to urinate normally. Limbs that are not receiving neurologic stimulus may atrophy quickly and joints that are immobile may have a decreased range of motion. Once neurologic function returns the animal may be too weak or painful to move the affected limb. Treatment of a down dog takes a large commitment of time and patience from technical staff as well as owners. Significant improvement may not be seen for weeks, leading to compassion fatigue in caretakers, as well as lethargy and reduced motivation in the patient.

POSITIONING AND BEDDING CHOICE

Providing plenty of soft bedding is the first step to help our recumbent patients and avoid complications. The bedding should be changed or refreshed often to prevent decubital ulcers, also known as bedsores or pressure sores, formed when skin is deprived of blood flow due to outside pressure. Soft, fluffy bedding – such as egg crate foam or an orthopedic mattress – can reduce pressure on bony points like elbows or hips. Any soiled or wet bedding should be removed as soon as possible. Urine scald caused by lying in soiled bedding can increase the chance of decubital ulcers and can lead to skin infections. Soiled animals should be cleaned and dried thoroughly. Touch-up cleaning can be accomplished with baby wipes.

In addition to soft bedding, frequent position changes are also necessary to prevent ulcers. Dogs that have motor function and are reasonably aware will usually move to a new position as needed, allowing for adequate blood flow to tissues compressed against the bedding. Dogs that are profoundly paretic to plegic or mentally inappropriate may not realize that an area is becoming sore. These types of animals needs to be moved at regular intervals. Position changes should occur every four to six hours, as long as the dog is not suffering from an unstable fracture or other disease that might worsen with movement. After spinal surgery, position changes should be done gently. Caution should be used when moving painful or anxious patients to prevent bites or flailing. Common resting positions include full lateral recumbency on the right or left side, partial recumbency with the front limbs in a sternal position and the back limbs in lateral recumbency or a full sternal position in which all limbs are placed underneath the dog in a sphinx-like pose. When changing positions on an animal in partial recumbency it is also important to move the front limbs to slightly different positions, not just the back limbs. Tetraplegic, stuperous or comatose patients have additional considerations for positioning and bedding. If possible these animals should be kept sternal, or rotated into sternal position at least four times daily to reduce atelectasis. The head should be elevated approximately 20 to 30 degrees, especially in stuperous or comatose patients, to reduce the chance of aspiration pneumonia and to reduce intracranial pressure.[1]

URINARY COSIDERATIONS

The next concern is usually urination. Can the dog urinate voluntarily, or is he or she leaking urine? Urinary complications of non-ambulatory dogs include urinary tract infections, pyelonephritis (kidney infection), damage to the bladder including atonic bladder, skin infections from urine scald and in extremely rare cases bladder rupture.

In order to anticipate these complications it is important to understand the neurologic control of urination, i.e. micturition. When bladder function is compromised due to neurologic disease the dysfunction is characterized as upper or lower motor neuron in origin. Upper motor neuron (**UMN**) bladder dysfunction is typical of T3-L3 or thoracolumbar spinal lesions, such as a paralyzed dachshund, but is also possible in cervical spine and brain lesions as well. UMN bladder dysfunction leads to increased bladder and urethral tone.[1] This increased tone may make it difficult for the dog to start urination and to fully empty the bladder. Bladder expression or catheterization may be needed to prevent damage to the bladder, such as atony wherein the bladder wall muscles are stretched for long periods and then are unable to contract properly. Dogs with UMN bladders may appear to urinate large amounts due to overflow, but may not be able to empty the bladder fully on their own. It is important to note if the dog is able to consciously urinate, e.g. the dog is positioning to urinate or moving the tail when urinating. If they are not consciously urinating, regular bladder palpation or visualization via ultrasound is helpful to evaluate how well the bladder is emptying or if it is becoming overly full. Bladder expression is performed every six to eight hours until the dog is able to urinate on its own. Forceful bladder expression in these dogs could lead to bladder rupture, so care must be taken not to palpate too aggressively.

Lower motor neuron (**LMN**) bladder dysfunction is typical of L4-S2 or lumbosacral spinal lesions, such as cauda equina syndrome in the German Shepherd Dog or peripheral neuropathies involving the nerve roots. LMN dysfunction leads to decreased bladder and urethral tone with a flaccid and easily expressed bladder. This decreased tone leads to urinary incontinence.

Bladder expression is often recommended in cases of both UMN and LMN disease. For UMN bladders, expression of the bladder is often required to prevent bladder wall damage due to over-extension, as these dogs may not be able to start or complete urination on their own. For LMN bladders, expression of the bladder can be helpful to decrease episodes of incontinence. Bladder expression is not without complications. As previously discussed, in cases of UMN disease, the urethral tone is often increased, sometimes to the point that moderate pressure on the bladder wall is unable to overcome the urethral tone and allow for urination. This may lead to bladder wall damage or rupture if too much force is used. It can also be a source of anxiety or discomfort in patients, especially those recovering from spinal surgery. In these cases pharmaceutical agents, like diazepam or prazosin, can be used to facilitate expression, or a urinary catheter can be placed.

There are several common techniques for bladder expression. Choosing the correct technique depends on the situation and practice. When first practicing bladder expression it is often easier to learn hand placement and the feel of the bladder and surrounding structures on an anesthetized animal. Check with the veterinarian first, but any animal being prepared for surgery – other than bladder surgery – may be a candidate for bladder expression practice. Once successful with anesthetized patients, start practicing on awake patients. Bladder expression can be performed in lateral recumbency or in a standing position. Place your hands on either side of the caudal abdomen. The bladder usually rests approximately halfway between the last rib and the pelvis but can sometimes be located more caudally. Carefully palpate the abdomen moving hands medially, dorsally and caudally until the bladder is felt. The bladder should feel like a water balloon. Once the bladder is isolated, use steady even pressure until urine is expressed. Continue the steady even pressure until the bladder becomes too small to isolate. If the expression requires more than moderate pressure to produce urine stop and talk to a veterinarian.

Learning to correctly express a bladder takes patience, time and practice. Owners of paralyzed dogs are often required to learn a skill within days that medical professionals may take weeks of practice to do so with ease. If possible several lessons and trial bladder expression attempts should be practiced before the owner goes home with a dog needing bladder expression. Additional photos, videos and support groups can be found at http://www.dodgerslist.com.

Pharmaceutical therapy can be used to aid urination in neurologically compromised patients.[1] These therapies may aid in management of the neurologic bladder but should not replace regular nursing care and monitoring.

Bethanacol increases bladder wall contraction and can be useful in both UMN and LMN urinary dysfunction. It does not reduce urethral tone, though, so dogs with UMN disease may still not be able to urinate as the urethral tone is increased.

Phenoxybenzamine decreases urethral sphincter tone, so is helpful for UMN lesions. It does take a few days to take effect though and can cause hypotension and syncope as it is an α antagonist.

Prazosin, like phenoxybenzamine, decreases urethral sphincter tone. It is quicker acting, needing only 30 to 90 minutes to be effective and can also cause hypotension and syncope due to α antagonist activity.

Diazepam given intravenously or orally relaxes the external sphincter tone, as well as reducing anxiety in dogs requiring frequent bladder expression. Diazepam is a skeletal muscle relaxant.

Phenylpropanolamine (PPA) increases internal urethral sphincter tone and can be useful for incontinence due to LMN lesions.

Urinary catheterization may be recommended in cases where there is trauma to the abdomen and in recumbent or critical patients. The ability to measure and characterize urine production can be useful to monitor kidney function, hydration and look for urinary tract infection. Catheterization is also useful in the care of large recumbent dogs to maintain cleanliness of the dog and the bedding. Depending on veterinarian preference, long-term indwelling or intermittent catheterization may be used.

There are significant complications associated with urinary catheterization. Up to 50% of dogs with indwelling urinary catheters will contract a urinary tract infection after four or more days of catheterization.[1] Utmost care must be taken to place the catheter aseptically. After placement the catheter should also be maintained and cleaned regularly to help prevent ascending infection. Damage to the urethra and/or bladder can occur with aggressive or improper placement of the catheter. Self-trauma by the patient or accidents such as occluding or pulling on the line of a closed collection set may require emergency care.

Unlike urination, **defecation** usually proceeds without assistance in animals with neurologic disease. Owners are often concerned with the possibility of constipation, as this is a common problem with humans who are sedentary or on opiate medications.

Typically dogs will continue to defecate normally despite neurologic disease, though the normal defecation schedule noted by the owner may differ; usually less frequent defecation. Constipation may be caused by opiates in some dogs or by changes in diet and stress in a hospital setting. Conversely, diarrhea may occur with other medications, such as steroids, as well as with changes in diet. Dogs with lower motor neuron disease can have urinary and fecal incontinence. In these dogs and dogs that are para or tetraplegic, regular expression of feces can help with general cleanliness.[1]

REHABILITATION

The next problem to address is muscle and nerve degeneration from disuse, which can cause a slower recovery. Rehabilitation exercises help to hasten recovery in neurologically compromised patients.[2] Through a combination of modalities a patient's mobility and muscle strength can be maintained and increased while pain is decreased. These aspects work together to increase client satisfaction and patient attitude; both of which are essential components to successful treatment.

The first of these modalities is range of motion or **ROM** exercises. ROM exercises can be either passive or active activities. With passive ROM, the patient's limb is moved through a series of positions. With active ROM, the patient's muscles assist or resist the therapist's movements. Active ROM can be difficult to obtain with veterinary patients, since we cannot ask them to tense or relax certain muscles. The benefits of passive ROM exercises include increased joint mobility, circulation, lymph drainage and muscle lengthening as well as decreased tissue adhesions. With active ROM exercises all of the previous benefits are obtained as well as decreased muscle atrophy – a common effect of disuse.[2]

Exercises are started 24 hours after surgery as long as the dog's pain is well controlled. In cases of severe trauma, exercises should not be started until the animal is stable. For cases of confirmed FCE or high velocity disk protrusion, exercises are started soon after diagnostics are completed as dogs with this diagnosis are generally not painful and rehabilitation exercises have been shown to significantly improve patient outcome. The passive ROM exercises consist of moving the affected limb(s) as if the dog is walking or stretching. Each joint from the hip/shoulder down to the toes is brought through its normal range of motion several times. The leg is also moved in a circular pattern as if the dog is riding a bicycle to simulate a walking movement. The sessions last five to fifteen minutes on each affected limb, and are repeated every six to eight hours. Use care when flexing or extending traumatized, geriatric or sensory deprived limbs, as the movements may be painful or cause additional trauma.

Massage can also be used in conjunction with the ROM exercises. Some of the benefits of massage are similar to ROM exercises; e.g. increased circulation and lymph drainage, decreased tissue adhesions and decreased pain. Massage therapy can also relax the patient and loosen stiff or sore muscles allowing for a greater range of motion.[2] Common massage techniques in dogs include stroking, effleurage and compression. Stroking often begins a massage or therapy session by softly running a hand over the dog from neck to tail and down the legs. This technique is soothing and relaxing. Effleurage begins from a distal area and using medium pressure moves the tissues proximally while gliding the hand over skin and fur. This technique helps to encourage lymph drainage towards the body. Compression of tense or stiff muscles can help to decrease tone. Once a stiff or tense muscle is found and isolated, it is then compressed with mild to medium pressure with one or two fingers. The finger is turned over the muscle, without gliding over the skin, and then released. This technique is thought to help circulation by restricting flow then releasing pressure. Massage is contraindicated in dogs suffering from shock, fever or over an area of acute inflammation.

Supported walking allows dogs with limited mobility the chance to stand and walk. A variety of pre-made slings and appliances are readily available. For smaller dogs a simple sling can be made with a towel under the abdomen. For larger dogs, or dogs that are more permanently plegic, wheelchairs are available for rent or to own. Supported walking or standing decreases limb edema from prolonged recumbency. The upright stance also decreases atelectasis and aspiration pneumonia. Dogs are often more encouraged to move their legs freely when supported and may be more apt to urinate on their own. Once dogs are able to bear some weight on their own, supported walking is an active ROM exercise. The use of slings or carts should be monitored closely. If a dog is unable to move a limb or replace a foot properly, a bootie should be used to decrease trauma to the top of the foot and nails. Repeated dragging of a limb or foot can lead to ulcerations of the foot and premature wearing of the nail. In cases of unstable lesions, fractures and luxations, the lesion may be worsened by sling use. Any appliance may become distressing to a dog not used to needing help; wheelchairs and slings should be introduced slowly and any discomfort or anxiety addressed immediately.

STRICT REST

The final step to caring for a down dog is strict rest. In cases of IVDD and spinal surgery the best treatment is rest. By keeping the dog as quiet as possible, the spinal cord has the best and quickest chance of recovery. This rest period is generally at least two to four weeks plus two additional weeks once the dog is recovered. The strict rest period is best achieved with cage confinement so that the dog is unable to engage in walking, running, playing or walking on stairs, especially when unsupervised. This confinement can be achieved with the use of airline crates, playpens or small rooms – such as a bathroom. Exercise should be limited to the rehabilitation and massage sessions, with short, on-leash walks for urination and defecation. During the rest period, the walks should be no longer than five minutes. Small dogs should be carried outside and set down near the area they will most likely urinate. Large dogs should be slowly sling-walked outside. If the dog is unable to walk with sling support it should be carried outside using a cart or gurney. While in the crate or confinement area, entertainment such as Kong™ or other food-based puzzle toys can be used. Balls and other chew-type toys can be used provided the dog is not overly excited by the toy and at risk of reinjuring itself by playing. Dogs that are not crate trained can be sedated if they are unable to overcome crate-induced anxiety.

REFERENCES

1. A Practical Guide to Canine and Feline Neurology, Curtis Dewey DVM, 2nd Edition, 2008; 2. Canine Physical Therapy and Rehabilitation, Darryl Millis MS DVM, David Levine PhD PT, Robert Taylor MS DVM, 2004

ECHOCARDIOGRAPHY: ROOM SET UP, RESTRAINT AND SEDATION FOR THE DOG AND CAT

Anne Myers RVT, VTS (Cardiology)
Raleigh, NC

INTRODUCTION

With more veterinary cardiologists, radiologists and internal medicine specialists working in small animal practice and the increasing availability of traveling sonographers, many veterinary technicians may find themselves assisting with a cardiac ultrasound or **echocardiogram** on a dog or cat with confirmed heart disease or a newly diagnosed heart murmur. While chest radiographs can reveal the overall size of the heart from the outside, an echocardiogram (echo) uses sound waves to generate a real time two-dimensional image inside the atria, ventricles, and great vessels to aid in diagnosis and treatment. The room used to perform echoes should be a **quiet zone**, especially when handling a nervous dog or cat. Ultrasound supplies, adjustable lighting, the echo machine and a functional table to lay the patient on are needed. In addition, effective restraint and positioning of the animal by support staff is necessary to obtain diagnostic images in a timely manner. This will sometimes require the administration of intravenous (IV) or intramuscular (IM) sedatives to an overly anxious or aggressive patient. Some will be asymptomatic with mild heart disease, while others could be hemodynamically unstable or in heart failure. Sedation should be administered cautiously, using low doses of agents with the least adverse effects on cardiac and respiratory function. At NCSU, a low dose of the tranquilizer **Acepromazine** is used and often combined with the opiates **Buprenorphine** or **Butorphanol**. Before proceeding with sedation, or if it is not an option, there are some distraction techniques that can be tried to avoid it. The goal is to have a stable animal cooperative enough to place a transducer on the thorax for ten to fifteen minutes of imaging.

ECHO SUPPLIES

Make sure the echo sonographer has everything needed within reach. This will include a set of clippers, a spray bottle of alcohol to part the hair-coat over the chest and plenty of ultrasound gel. Alcohol breaks down skin oils present in the dog and cat and allows for good contact of the gel. Gel is placed on the active end of the echo probe or **transducer** to enhance image quality by eliminating air that causes the sound waves to refract or scatter. In addition, a tap cart on wheels stocked with supplies should be nearby in case large volume pericardial, pleural or peritoneal effusion is discovered and needs to be removed.

ECHO ROOM LIGHTING

A dimly lit room is optimal for viewing the black, white, and grays of echo video clips and stills. A combination of fluorescent overhead lights which can be turned off and recessed lighting in the ceiling that can be dimmed during the study seems to work best. A small lamp placed on a counter in the room and turned away from the echo screen will also provide dim lighting.

ECHO TABLE DESIGN

The echo table should be constructed of sturdy material such as fiberglass to support a heavy patient. A soft, non-slip pad placed on top can provide comfort for an older or frail patient. A hydraulic lift to raise and lower the table via a foot pedal allows for easier lifting and dismounting of large dogs. It also ensures a comfortable standing position for staff holding the patient during the study. A cut out section should be present in the middle of the echo table or at either end. This allows the operator to hold the transducer underneath the patient to obtain images from the down side lateral thorax. The table should be designed to slant downwards at a slight angle towards the echo sonographer to allow the heart to fall toward the transducer.

ECHO MACHINE

The echo machine used for canine and feline studies should employ phased array transducers with varying degrees of penetration to create a fan shaped beam for adequate two dimensional images of the heart. The ability to perform motion-modes, color flow, and spectral Doppler studies is paramount to an accurate diagnosis by the veterinarian. Care and maintenance of ultrasound equipment is important. A cleaning with a mild soap solution should be compatible with most systems. Keep air filters, track balls, and control panels free of animal hair, dander and ultrasound gel. If air intake filters become clogged with hair, the machine can over heat and may shut down unexpectedly. The transducers must be handled carefully and never dropped as they contain piezoelectric crystals which send out and receive sound waves to generate images. Archiving patient studies onto storage media may also be the responsibility of veterinary staff. Use an ultrasound company with good customer support.

RESTRAINING THE DOG OR CAT FOR AN ECHO

Most studies begin with the patient held in right lateral recumbence by one or two staff members to obtain the right parasternal long axis and short axis views of the heart. Placing the patient on its side allows the heart to lay closer to the chest wall and compresses the lungs to decrease air artifact. The person restraining the front end of the patient pulls the right down leg forward to facilitate placement of the transducer onto the chest wall at the level of the heart. The top leg is allowed to hang loosely down. If only one person is available to hold a large dog, an elongated sand bag or **"sand snake"** can be placed over the back end. Shaking or panting in dogs results in excessive chest motioninterfering with image quality and can be minimized by either laying a hand gently but firmly on the "up" chest wall or by closing the mouth for a few seconds during image acquisition. Giant breed dogs may have to stand for echoes as the table may not be large or strong enough to support their weight. The staff member standing by the patient will have to pull the right or left foreleg forward and away from the chest for the echo sonographer to image the heart. Most echo views should be obtained with the patient attached to a corresponding electrocardiogram (ECG) tracing at the bottom or top of the screen to correlate stills and clips with either systole or diastole. Observation and treatment of a tachy or bradyarrhythmia at the beginning of the echo may need to be documented with a separate diagnostic ECG machine before switching to left apical views in left lateral recumbence.

PATIENT DEMEANOR AND THE SEVERITY OF HEART DISEASE

An echocardiogram is a non-invasive procedure and generally well-tolerated. Some patients that are challenging to work with in an exam room with their owners may be cooperative when taken for imaging. That said, the fear or anxiety experienced by some companion animals separated from their owners can result in shivering, shaking, excessive panting, or aggressive behavior such as biting, striking, or growling. Tachycardia caused by increased sympathetic tone in heart failure patients can become worse with stress. Older patients may be painful due to arthritis. Puppies and kittens can become frightened and struggle. Other patients may have trouble breathing due to congestive heart failure and be in need of chest radiographs or a brief initial standing echo until stabilized with oxygen, diuretics and anxiolytic medications. Complications obtaining echo views can occur and may include exacerbation of cardiac arrhythmias or formation of pulmonary edema. Some patients will respond to distraction while others may need to be muzzled or sedated. If intravenous sedation is to be administered, using a hind leg vein such as the medial saphenous in the cat and the lateral saphenous in the dog is recommended.

DISTRACTION TECHNIQUES

Tactics used with uncooperative patients to avoid sedation for an echo include offering small amounts of food such as baby food, peanut butter, or dog and cat treats; having a third person by the head petting and talking to the animal; or having the owner present in the echo room during the exam.

ECHO SEDATIVES USED AT NCSU

1. **Acepromazine** is an effective tranquilizer that does not provide analgesia. Low doses are recommended for animals with cardiac disease because of its adrenaline blocking effects on vascular smooth muscle resulting in vasodilatation. Lowering arterial blood pressure normally does notaffect a patient with adequate systolic function; however, if cardiac output is already poor due to conditions such as dilated cardiomyopathy or pericardial effusion, acepromazine can cause hypotension and collapse. Acepromazine may also lower the seizure threshold in some canine breeds. At NCSU, it is used alone or combined with Buprenorphine or Butorphanol. The pharmacy dispenses it in a dilute form of 1milligram (mg) per milliliter (ml), allowing small doses to be given accurately.
2. **Buprenorphine** is a partial opiate agonist that blocks pain at the mu receptor in the central nervous system and is used primarily as an injectable long lasting analgesic in the dog and cat. It is compatible with acepromazine and can also be applied to the buccal mucosa. It is considered safe to give to patients with cardiac disease and is used frequently at NCSU. It comes in 1 ml vials at the strength of 0.3mg/ml.
3. **Butorphanol** is a partial opiate kappa agonist and antagonist at the mu receptor and is used as an analgesic for treating mild to moderate pain. It is compatible with acepromazine and comes in 1mg and 2 mg per ml vials and at 10mg per ml in a 10 ml bottle.
4. **Ketamine** inhibits NMDA or memory receptors in the CNS and is considered a dissociative anesthetic agent. It increases sympathetic tone, so effects on the cardiovascular system may include tachycardia and hypertension at higher doses. Ketamine is not recommendedin cats with hypertrophic cardiomyopathy. At NCSU, ketamine is used in low doses on fractious cats that are not adequately sedated after receiving acepromazine combined with an opiate. Ketamine comes in 10 ml vials at 100mg/ml. Many cats experience pain when injected IM; administering the agent slowly with a small gauge needle may alleviate this. Eyes will remain open after administration, so don't forget to lubricate them.

COMMONLY USED ECHO SEDATION PROTOCOLS AT NCSU

Dog

Acepromazine 0.01-0.03 mg/kg used alone or mixed with Buprenorphine 0.0075mg/kg IV or IM

Acepromazine 0.01-0.03 mg/kg mixed with Butorphanol 0.1-0.2 mg/kg IV or IM

Cat

Acepromazine 0.01-0.02 mg/kg used alone or mixed with Buprenorphine 0.01 mg/kg IV or IM

Acepromazine 0.01-0.02 mg/kg mixed with Butorphanol 0.2 mg/kg IV or IM

CONTRAINDICATED SEDATIVE

Dexmedetomidine is an alpha2 agonist that depresses sympathetic tone, reducing contractility of the heart muscle and causing vasoconstriction of peripheral arteries resulting in bradycardia and transient hypertension. It will adversely affect systolic function measurements taken by echo and should not be used for this imaging procedure. Even more importantly, its negative inotropic effects make it an unwise choice for sedating dogs and cats with heart disease.

THE FRACTIOUS FELINE

Fractious cats can be especially challenging to restrain for an echo. The best advice is to stay clear of the "danger zone" of front claws and face. Work slowly and quietly from the side and handle with thick towels placed over the shoulder blades. One strategy used to pull the front down leg forward for an echo,is to loop a long strip of tape around it with a large tab at the end. This keeps the restrainer's right hand far away from the front end. Face muzzles may be difficult to place on an unsedated cat. If sedation is needed; a protocol employed at NCSU is to administer **Acepromazine 0.02mg/kg mixed with Butorphanol at 0.2 mg/kg IM. If more sedative is needed, Ketamine is administered at 5-10 mg IV or IM.**

CONCLUSION

In North Carolina, the number of boarded veterinary cardiologists has increased from four in 1998 to eight in 2012. Cardiology Residents in training at the NCSU Veterinary Health Complex have grown from one to three DVMs. While this increase may not be typical for all areas of the country, it could be predictive of a future trend. As the number of these specialists rise, there is going to be more need for knowledgeable and trained staff to help them perform echocardiograms on dogs and cats.

REFERENCES

1Abbo LA, et al, **Pharmacokinetics of Burprenorphine Following Intravenous and Oral Transmucosal Administration in Dogs** Veterinary Therapeutics 2001; Vol. 9, No 2 2Bonagura, JD, Kirk RW, **Kirk's Current Veterinary Therapy XII Small Animal Practice** 1995; 773-779 3Fields EL, **Ultrasound: What You Need to Know** 2011; 16th Annual NCVC 4Fox P, et al, **Textbook of Canine and Feline Cardiology** 1999; 853-8655Hopfensperger M, **Humane Handling of Dogs to Minimize Fear and Maximize Safety in the Veterinary Clinic Setting** 2011; 16th Annual NCVC: 6Plumb DC, **Plumb's Veterinary Drug Handbook Fifth Edition** 2004; 2: 98: 10: 399: 439: 480. 7Sherman BL, **Managing Fear Responses in Dogs** 2011; 16th Annual NCVC 8Simpson W, **Beyond the "Scruff & Stretch": Low Stress Handling to Make Your Life Easier Working with Cats** 2011; 16th Annual NCVC: 9Stepien RL, et al, **Cardiorespiratory effects of Acepromazine Maleate and Buprenorphine Hydrochloride in Clinically Normal Dogs** American Jour of Vet Research 1995; Vol. 56, Issue: 78-84

HOW CAN VETERINARY REHABILITATION HELP MY PATIENT?

Kristin A. Kirkby, DVM, MS, CCRT, DACVS
Kirkland, WA

The American Association of Rehabilitation Veterinarians defines veterinary physical rehabilitation as "the treatment of injury or illness to decrease pain and restore function." This philosophy is not dissimilar to that of neurologic surgery or internal medicine which strive to return a patient to normal function and health. With these common goals in mind, it is clear that a synergistic relationship should exist between the specialists and rehabilitation veterinarian such is the standard of care in human medicine and surgery.

The overall aim of veterinary rehabilitation is to facilitate tissue healing, restore muscle strength, increase endurance and proprioception, prevent further injury, and re-condition the animal to its prior activity level. The first step in this process is to establish a **thorough** diagnosis. A complete orthopedic and neurologic examination should be performed. Particular attention should be paid to spinal alignment and muscular tightness or trigger points. It is common for animals that have an injury isolated to one limb to develop compensatory changes in other limbs or their back. For example, many dogs with paresis or paralysis will develop muscle tightness and strains in the shoulders and neck. Likewise, dogs with osteoarthritis of the hips will develop tightness and fasciculations in the lumbar epaxial muscles. During the physical examination measurements should be taken that can be utilized to objectively evaluate the outcome of therapy. These include thigh (or brachial) circumference, goniometric measurement of joint range of motion, body weight and weight-to-height ratio, and assessment of mobility, balance and functional limitations.

Once an assessment of the animal's condition is made, a rehabilitation plan is developed and individually tailored to the patient. It is tempting to develop "protocols" for certain conditions; however, a better method of practice is to develop guidelines for progression through rehab. That is, rather than changing or instituting new treatment at certain time points, functional or physiologic goals should be set as criteria for an animal to progress from one phase of treatment to the next. The rehabilitation veterinarian must work with the surgeon in assessing the progression of healing. "Red flags" or indications of complications with the surgery should be brought to the attention of the surgeon immediately.

Physical rehabilitation encompasses a whole-body approach to patient assessment and treatment. Most commonly musculoskeletal and neurologic disorders are addressed through the use of manual therapy, exercise therapy, and various physical modalities. Many rehabilitation practitioners also include other complementary therapies, such as acupuncture and chiropractic, into their practice. Nutritional support is also a very essential component of physical rehabilitation.

Rehabilitation following surgery

The processes of healing and recovery are similar regardless of the type of surgery performed. The following is a very basic overview of phases of healing and the potential rehabilitation interventions available as therapy progresses. With regard to the neurological surgery patient, crucial aspects of the recovery process include nursing care, including bladder expression and recumbancy care. Ideally, with implementation of a rehabilitation program aimed at restoring motor and patient mobility, the need for intensive nursing care will be minimized.

Phase 1: Inflammatory phase

Inflammation begins at the time of surgery and is a normal and essential part of the healing process.The inflammation phase include pain and swelling. Goals of rehabilitation during inflammation are to minimize pain and swelling. Excessive swelling can perpetuate pain and impair wound healing. Pain should be addressed using a multi-modal approach that may include steroids or NSAIDs, opiates, local anesthetics, and epidural analgesia. Modalities recommended to decrease swelling and inflammation include cyotherapy, low level laser therapy, TENS and acupuncture. Laser therapy can also be applied to the wound to accelerate healing. The inflammatory phase *typically* last 3 days; however, clinical signs of inflammation (heat, pain, redness, swelling, loss of function) should be used to gauge the end of this phase.

Phase 2: Weight bearing, range of motion, mobility

Early, controlled weight bearing should be encouraged through the use of assistive devices and hydrotherapy in order to minimize muscle atrophy, maintain joint health, and encourage motor pathway restoration. Prior to the patient being able to consistently bear weight, joint compressions and mobilizations should be performed to stimulate proprioceptors within the joint. Neuro-muscular electrical stimulation (E-stim) can be used to stimulate muscle contraction. Passive range of motion (ROM) should also be employed at this stage. All joints in the effected limbs should be placed through range of motion exercises. Once the animal is able to ambulate with minimal support, active ROM can be instituted. That is, rather than the therapist (or owner) placing the joint through ROM, the animal is led through exercises which require them to flex and/or extend their joints. Examples included crawling, sit-to-stand, down-to-stand.

Phase 3: Strengthening, functional acitivites and proprioception

Once the animal is consistently ambulating, the aim of rehabilitation is to strengthen the musculature of the effected limbs and core and regain normal proprioception and balance. Controlled therapeutic exercises including leash walks, underwater treadmill therapy, balance ball, and cavalettis are the keys to increasing strength. Activities of daily living should be incorporated into the exercise program including stairs, destination jumping, and maintaining a squat position.

Phase 4: Endurance

This phase will overlap with strengthening. As an animal becomes stronger and more confident in ambulation, the length of leash walks should be gradually increased. Underwater treadmill sessions can also be intensified. The degree of endurance training will differ based on the expected activity level of the dog following therapy; however, some degree of endurance training or conditioning should be implemented in the rehabilitation of all dogs following surgery.

Remember than the above phases are purely guidelines for rehabilitating an animal following neurologic surgery. It is essential that programs be tailored to each individual animal and that the animal is re-assessed regularly. Compensatory changes in should be managed if diagnosed. The duration of rehabilitation will vary from patient to patient, but can be expected to last 2-6 months.

THERAPEUTIC MODALITIES

Veterinary rehabilitation relies on establishing a thorough assessment of the patient and prescribing an individualized treatment plan. The foundation for rehabilitation is therapeutic exerciseis to restore muscle strength, endurance and proprioception. Therapeutic modalities are tools that deliver a physical agent to the body for therapeutic purposes, or in other words, a transfer of energy to the body for a therapeutic purpose. Different types of energy can be utilized, including thermal (ice and heat), electromagnetic (laser), sound (shock wave and ultrasound), electrical (E-stim), and mechanical (massage or compression). These tools can be combined with therapeutic exercises to facilitate healing and restoration of normal function. The basic principles of some of the most common therapeutic modalities are discussed.

Cryotherapy (ice)

Cryotherapy is the application of low temperatures to remove heat from the body and is most commonly performed using an ice pack. Cryotherapy is indicated to reduce inflammation and/or pain. The mechanism of reducing inflammation is through localized vasoconstriction and decreased histamine and protease release. This leads to a reduction in interstitial fluid and decreased swelling. Pain is temporarily alleviated through decreased production of prostaglandins and decreased nerve conduction velocity. Typical recommendations for cryotherapy include:

- Placing a moist towel over the treatment area, placing the ice pack on top of the towel, and wrapping the limb with an Ace bandage or dry towel
- Treatment for 15-20 minutes, every 4-6 hours post acute inflammation

Thermotherapy (heat)

Thermotherapy is the application of heat to the body in order to soothe pain or stimulate blood flow to an area; it should not be used during the acute inflammatory phase. The mechanism of action of heat for pain reliefis not completely understood, but the soothing action is thought to break the pain-spasm-pain cycle that commonly accompanies musculoskeletal injuries. Topical heat packs can be used, however the maximum depth of penetration is only 1 cm. The skin should be protected from direct contact with the heat pack and treatment times should range between 15-20 minutes. It is essential that the animal's skin be monitored frequently and if any doubt whether the pack is too warm it should be removed.

Low level laser therapy

Laser therapy is the application of light energy (photons) to stimulate tissue. Low level lasers, or cold lasers, stimulate cellular metabolism, fibroblast proliferation, collagen deposition and enhance angiogenesis, but are not capable of cutting tissue. Indications for laser therapy include enhancement of wound healing, increasing circulation, and pain relief. Studies in lab animals have also shown therapeutic lasers to enhance the regeneration of nerves and decrease injury following spinal cord trauma. Several lasers are currently marketed for veterinary use and differ based on their power and wavelength, which influence the depth of penetration and time required for treatment, and their ability to cause injury to the eye.

Neuromuscular electrical stimulation

Neuromuscular electrical stimulation is the application of electrical energy to the body. Typical E-stim that is used for rehabilitation applies energy in pulses that are capable of causing muscular contractions. Indications for E-stim include: prevention of muscle atrophy, muscle re-education, muscle strengthening, increasing range of motion, and decreasing edema. Another form of electrical stimulation is transcutaneous electrical neuromuscular stimulations (TENS), which utilizes continuous electrical energy in order to provide a counter-irritant effect for pain management and muscular relaxation.

Therapeutic ultrasound

Therapeutic ultrasound is typically used for deep (3-5 cm) heating of tissues and facilitation of tissue healing. Most of the benefits of ultrasound can be accomplished with low level laser therapy, which is more commonly utilized in veterinary rehabilitation. Therapeutic ultrasound can be used to deliver topical medication into tissue (phonophoresis).

Hydrotherapy/ underwater treadmill

Water offers several beneficial properties that make hydrotherapy and the underwater treadmill very useful modalities in veterinary rehabilitation. Depending on the height of the water, the buoyancy of the water will provide varying levels of support for animals that are either weak or painful. The hydrostatic pressure of water exerts a force on the body that can help decrease limb edema, particularly when combined with movement under the water. The cohesion and turbulence of the water provide a resistant force when walking through the water. This property is beneficial for increasing range of motion of joints as the animal typically needs increased joint flexion to break the surface of the water when walking. Additionally, it is more challenging to walk through water leading to increased muscle strength and endurance following regular therapy in an underwater treadmill. Water temperature in most underwater treadmills or pools is kept around 85 degrees F, which helps increase blood flow to the limbs and provides a soothing environment.

The importance of a complete diagnosis, assessment, therapeutic plan, and regular reassessment, cannot be over-emphasized. Veterinarians or the physical therapist (who must work directly with a veterinarian in most states) that have attained certification in canine rehabilitation have spent hundreds of hours in continuing education to learn the intricacies of rehabilitation, and patients with neurologic and orthopedic conditions that would benefit from rehabilitation should ideally be referred to these professionals.

REFERENCES

1. Millis DL, et al. Canine Rehabilitation and Physical Therapy. Saunders, St. Louis, MO, 2004. 2. Marsolais GS, et al. J Am Vet Med Assoc 2002; 220: 1325-1330. 3. Levine D, et al. Canine Rehabilitation and Physical Therapy. Saunders, St. Louis, MO, 2004.
4. Gandini G, et al. J Small An Pract 44(2):76-80, 2003. 5. Byrnes KR, et al. Lasers Surg Med. 36: 171 – 185, 2005. 6. Chow RT, et al. Lancet. 374 (9705), 2009: pp 1897 – 1908. 7. Rochkind S, et al. Lasers Surg Med. 28(3): pp 216 – 219, 2001.

**Would suggest a listing of the credentialed rehabilitation schools such as:

Accredited University of Tennessee Certificate Program in Canine Rehabilitation:
http://www.utc.edu/Academic/CanineRehabilitation/

Canine Rehabilitation Institute, Florida
http://www.caninerehabilitation.com

Integrated Touch Therapy, Inc.
http://www.integratedtouchtherapy.com/index.asp

Healing Oasis
http://www.thehealingoasis.com

INFECTIOUS AND IMMUNE MEDIATED ARTHROPATHIES

Colleen Ruderman, RVT, VTS (SAIM)
Atlanta, Ga

IMMUNE MEDIATED ARTHROPATHY

Immune mediated polyarthropathy (IMPA), also known as idiopathic non-erosive non-infectious arthritis, is the most common immune mediated joint disease in dogs. It is less common in cats, but has been reported. IMPA is one of the leading causes of cases that present with fever of unknown origin. It can affect single joints however the polyarticular form is most common. IMPA is divided into four types. Type I is idiopathic, or primary immune related, type II is associated with infection, type III is associated with gastrointestinal or hepatic disease, and type IV is associated with neoplastic causes.

Type I is idiopathic and the most common form of IMPA and causes a non-deforming arthropathy. As with many immune-mediated diseases it occurs most often in purebred dogs between 1-6 years old. Clinical symptoms include a cyclic fever of unknown origin, anorexia, lameness, lymphadenopathy, and generalized stiffness. Swelling and heat can be noted in the distal joints and some may exhibit a palpable joint effusion palpable on flexion/extension of the joint. The most common laboratory abnormalities with type 1 IMPA include a leukocytosis with neutrophilia. Radiographs are often unremarkable with the exception of soft tissue swelling around the affected joints.

The diagnostic test of choice is the collection of synovial fluid via arthrocentesis. Heavy sedation or complete anesthesia is recommended as movement can affect clean sample collection. The patient is placed in lateral recumbency and the joints to be sampled are clipped and surgically prepared. The most common joints sampled are the carpus, tarsus, and stifle; however, samples can also be obtained from the elbow, shoulder, and hip. If only a single joint is affected, the sample should be obtained from that joint; however, in most cases multiple joints are sampled. Normal synovial fluid should be clear to slightly yellow and can be collected in very small quantities (<0.5mls) depending on the size of the animal. Large amounts of synovial fluid can be considered an abnormal finding. Gross evaluation of abnormal fluid can show a turbid, dark, or blood tinged fluid sample, however it should be noted that blood contamination is the most common complication with arthrocentesis. If a large enough sample is obtained it can be placed in EDTA and submitted for fluid analysis to obtain a cell count. Normal fluid has very low cellularity (<10% cells), abnormal fluid most often contains high numbers of neutrophils and can contain blood cells. If only small quantities of synovial fluid are obtained slides can be prepared for cytology. Synovial fluid cells can degenerate quickly and slides should be made immediately. Samples should also be submitted for bacterial culture and sensitivity. When multiple joints are sampled the fluid can be pooled for culture.

Treatment includes immune suppression and supportive care with pain management. Most often prednisone is used for its quick onset of action (2-7 days) at doses ranging from 1-4mg/kg. Treatment with prednisone can range from 3-6 months with a slow taper. Shorter treatment times are often associated with relapse. Approximately 1/3 of cases typically respond with prednisone alone but second line treatment is often needed if remission is not achieved within 2 weeks. Other immune suppressive medications include cyclosporine, azathioprine, and leflunomide. Second line treatment can also be used to allow the patient to be tapered off prednisone earlier than 3-6 months as side effects can significantly affect the quality of life of the patient and the client. True remission is considered a clean joint tap (not often performed) and resolution of clinical sings. Analgesics are often needed until remission is achieved.

Canine rheumatoid arthritis (RA) is a type I non-infectious erosive inflammatory joint disease that affects small, toy breeds ranging from 8 months to 8 years of age. The clinical symptoms include shifting leg lameness, swelling, joint laxity, and most often affect the carpal/tarsal joints, but can affect any joint including vertebral articulations. Other symptoms include fever, lethargy, anorexia, and lymphadenopathy. Radiographic abnormalities typically show loss of bone density at the joint, soft tissue swelling, luxations/subluxations, and narrowing or widening of the joint spaces. There is no specific rheumatoid factor testing in dogs due to lack of sensitivity; however antinuclear antibodies (ANA) can occasionally be detected in canine RA. Arthrocentesis will yield an inflammatory fluid analysis with elevated white blood cell count. The treatment for canine rheumatoid arthritis consists of supportive care and life-long immune suppressive doses of corticosteroids and other second line immune suppressive medications. Azathioprine is most often used over cyclosporine due to the long term nature of treatment. Most often a combination of multiple immune suppressive medications is needed for resolution of the clinical symptoms of canine RA. Leflunomide is showing come promise as well but can be cost prohibitive. These patients require close monitoring of liver/kidney values and urinalysis to monitor for sub-clinical urinary tract infections.

Systemic Lupus Erythematosus (SLE) is an immune mediated disorder that is not common and most often affects purebreds (canine and feline), but can affect mixed breeds as well. Clinical signs are associated with multiple immune mediated diseases including IMPA. The diagnosis of SLE is typically obtained by performing an antinuclear acid (ANA) test and exclusion of other underlying causes. The treatment for SLE related polyarthritis is single or combination immune suppressive therapy. Synovial fluid can have cells with phagocytosed nuclear chromatin in SLE.

Similar to immune mediated hemolytic anemia, some medications can incite an immune response that affects the joints of dogs. These medications include sulfonamides, penicillin, erythromycin, lincomycin, and cephalosporins. Drug related antibody-antigen complexes are deposited throughout the body in microvasculature and cause immune response. The clinical symptoms are similar to IMPA, but can also include hemorrhagic rashes and acute vasculitis. Symptoms will typically present themselves 10-21 days following the start of the medication. Treatment with general supportive care and discontinuation of implicated medication most often yields a quick resolution; however, some patients will require some immune suppressive medications.

INFECTIOUS ARTHROPATHY

Bacterial arthritis can be blood born (hematogenous), or local. Common causes for blood borne bacterial arthritis include infections of the genitourinary tract, skin, oral and respiratory systems. Most often blood born bacterial arthritis is seen with puppy strangles. Septicemia with progression into the joint cavity as a result of pyelonephritis, cystitis, and prostatitis has also been noted. Bacterial arthritis can affect proximal joints as well as intervertebral discs. Local causes for bacterial arthritis are most often a result of surgical contamination, non-sterile arthrocentesis, and penetrating wounds. Common organisms for bacterial arthritis include *Streptococcus canis*, *Methicillin-resistant Staphylococcus aureus* (MRSA), *Pasteurella* sp., *Brucella* sp., and *Salmonella* sp.

Clinical symptoms of bacterial arthritis include weight bearing and non-weight bearing lameness, fever, swelling and redness of the joint and surrounding tissue, non-healing wounds, and draining tracts. Imaging results that are often noted with bacterial arthritis are thick synovial membranes, distended joint capsule, displacement of fiscal planes, widening of the joint space, destruction of the synovial membrane, and osteomyelitis. The diagnosis of bacterial arthritis is made by collection and culture of the synovial fluid. Concurrent cultures of blood and urine can also be performed. Treatment of bacterial arthritis is comprised of general supportive care and antibiotic therapy based on culture results. If blood and/or urine cultures are positive for bacterial growth it is often considered the primary cause for the bacterial arthritis. Local infusions of antibiotics can be used in addition to systemic bactericidal medications.

L-form bacterial arthritis is a disease that affects cats. L-form bacteria are cell wall deficient and are similar to Mycoplasma. L-form bacteria cause subcutaneous draining wounds and are most often attributed to bite wounds. For this reason they are often located in the distal portion of the limbs and invade the joint cavity via the blood stream. Culture of synovial fluid in these cases are often negative, however treatment with tetracycline antibiotics is often successful. Mycoplasma arthritis is rare and typically associated with immune compromise and localization of mycoplasma to other areas (most commonly respiratory).

Viral arthritis is not common in small animals; however, there are reports of a limping syndrome in cats with concurrent calici virus infectionor following the use of live feline calicivirus vaccination. Effusive forms of feline infectious peritonitis have also been associated with potential viral arthritis in cats. Live vaccination has also been shown to cause limping in Weimaraners. Distemper virus can also affect the joints of dogs and may possibly lead to rheumatoid conditions later in life.

Fungal arthritis is also rare in small animals and is most often secondary to systemic fungal disease such as coccidioides, blastomycosis, histoplasmosis, aspergillosis, and cryptococcus. Fungal culture on synovial fluid is the diagnostic test of choice in cases of systemic fungal infections with acute arthritic symptoms; however identification of fungal organisms on cytology can also be performed. In the case of fungal arthritis the infection most often starts in the lungs with 85% of patients showing lesions on thoracic radiographs. The infection then becomes systemic via lymphatic and vascular routes. Treatment is comprised of supportive care combined with antifungal medications such as, amphotericin, fluconazole, and itraconazole.

Rickettsial and ehrlichial arthritis are forms of vector borne bacterial infections that can have joint related symptoms. The most common rickettsial disease is Rocky Mountain Spotted Fever (RMSF) which is caused by *R. rickettsii* an intracellular bacterium transmitted by *Dermacentor sp.* (deer ticks). RMSF can cause severe vasculitis and associated polyarthritis. The clinical symptoms are often joint, muscle, and neurologic pain, as well as a stiff gait caused by generalized inflammation, lymphadenopathy and fever. Ehrlichial infections are transmitted by the *Rhipicephalus sp.* (brown dog tick) and cause a non-erosive polyarthritis. Different types of ehrlichia are associated with more severe forms of arthritis (*E. ewingi, E. equi*) and are considered to be the granulocytic forms. Cross reactivity in diagnostic testing will not differentiate between various types of ehrlichial disease. The clinical symptoms seen with Ehrlichia are similar to RMSF and include muscle stiffness, lameness, polyarthritis, lymphadenopathy, fever, and stilted gait. Canine granulocytotropic anaplasmosis (*A. phagocytophilum*) was a previous member of the Ehrlichia genus and is also an obligate intracellular organism that causes musculoskeletal pain, stiffness, and lameness. It is transmitted by the *Ixodes* sp. tick and is diagnostically and clinically similar to Ehrlichiosis. Diagnostic testing for rickettsial and ehrlichial arthritis is needed to distinguish it from other forms of polyarthritis. Clinical history is needed to determine exposure to ticks as well as general laboratory abnormalities such as thrombocytopenia. The granulocytic forms of ehrlichia can be seen on blood film evaluation; however PCR, ELISA, and titers are the diagnostic tests of choice. The treatment for rickettsial and ehrlichial infections include tetracycline antibiotics therapy, supportive care, analgesia, vector control, and client education.

Lyme disease, or Lyme borelliosis, is associated with a spirochete bacterial arthritis. *Borellia burgodoferi* is a spirochete bacterium that is transmitted by ticks. The vector associated with this disease is *Ixodes* sp. The musculoskeletal symptoms of Lyme disease are most often joint related in the limb closest to the tick bite. Acute polyarthritis and shifting leg lameness are the most common clinical signs and can become chronic intermittent. Other symptoms include joint swelling and effusion, fever, and lymphadenopathy. General laboratory findings are often normal; however, other forms of Lyme disease can result in abnormal biochemical and CBC findings. Radiographs of the limbs will typically reveal soft tissue swelling and non-erosive arthritis. Several diagnostic tests exist for Lyme disease and include ELISA, Western Blot, and IFA testing. Titers can also be performed; however, they are affected by the vaccination status of the animal. In addition PCR testing from skin close to the affected joint and the synovial fluid can be diagnostic. Most often suppurative inflammation is noted on joint fluid cytology. Cultures of the synovial fluid are typically not performed due to the difficulty in isolating the bacteria from joint fluid. The treatment for arthritis associated with Lyme disease is multifocal. Non steroidal anti-inflammatory medications in addition to opioids are typically used to manage pain. Third generation cephalosporins, erythromycin, doxycycline, and ampicillin/amoxicillin are used to treat the spirochete bacteria itself. In addition to direct treatment vector control, vaccination in endemic areas and client education are important aspects of the treatment of Lyme disease. Anti-inflammatory doses of glucocorticoids may also be used; however, it is often contraindicated due to their immunosuppressive properties.

REFERENCES

1. Infectious Diseases of the Dog and Cat; Greene, C, 2006; Ed. 3; pp533-665, 1-34, 417-419, 203-242, 2. Textbook of Veterinary Internal Medicine, Ettinger, 2000; Ed 5(2), pp 1875-1885, 3. Diagnostic Cytology and Hematology of the Dog and Cat, Cowell et. al.; 1999; pp 105-108, 4. Laboratory Procedures for Veterinary Technicians; Hendrix et. al.; 2007, Ed 5, pp322-325, 5. Clements, D et.al; Type 1 immune-mediated polyarthritis in dogs; JAVMA; Vol 224, No. 8, pp1323-1327, 6. Efficacy of leflunomide for treatment of immune-mediated polyarthritis in dogs, Colopy et. al.; JAVMA, 2010; 236(3)

LIVER: PHYSIOLOGY & DISEASE

Amy N. Breton, CVT, VTS (ECC)
Waltham, MA

LIVER FUNCTIONS

All products of digestion are filtered through the liver making the liver's functions numerous and very important. The liver is responsible for excretion of copper, enhancing the absorption of fatty acids, fat soluble vitamins, carbohydrate metabolism, storage of glycogen, globulin synthesization, coagulation, detoxification and thermoregulation.

INFECTIOUS DISEASES

Infectious Canine Hepatitis causes primary liver failure. Infectious hepatitisis a viral disease of canines that is transmitted through the oro-nasal route and works by initially attacking the Kupffer cells and damaging the hepatocytes. Signs include fever, anorexia, hepatomegaly, disseminated intravascular coagulation (DIC), jaundice, and abdominal pain with or without effusion. There are several tests specific for infectious canine hepatitis: IgM, IgG and ELISA. On a serum chemistry profile the liver enzymes ALT, AST, ALP and GGT are usually are elevated. Canine hepatitis can also cause hypoglycemia, hypoalbunemia, orbilirubinuria. It's important to note that infectious canine hepatitis is different than chronic hepatitis. Chronic hepatitis is not a disease and is merely a description of the liver being inflamed.

Leptospirosis is a primary cause of liver failure. It is a bacterial disease caused by one or more of several known serovars. Most leptospirosis serovars affect the liver. Dogs are the primary hosts and typically pass the disease from urine into mucous membranes or an open wound. Leptospirosis is a zoonotic disease. Gloves should be used when handling animals suspected to have leptospirosis. The disease targets the hepatocytes in the liver. Leptospirosis is rarely seen in cats. Symptoms usually are seen within 4-12 days after infection. Clinical signs include vomiting, fever, jaundice, anorexia, polydipsia, polyuria, melena, and hematuria. In the northeast the most common leptospirosis serovars are the *L. pomona* and *L. grippotyphosa*. There are specific tests for all known serovars. Initial bloodwork usually reveals anemia, thrombocytopenia, high liver enzymes, and azotemia.

Histoplasmosis is a fungal infection which causes secondary liver dysfunction. The animal becomes infected by directly inhaling or ingesting the fungus. It is not contagious from one animal to another. Dogs and cats under 4 years of age appear to mostly represented. The respiratory system is the primary organ that is affected. In about 1/3rd of cats affected an enlarged liver is noted. Signs include dyspnea, abnormal lung sounds in about 50% of cats (less than 50% of dogs), fever, lethargy and severe diarrhea. In order for the diagnosis to be definitive, cytology must be performed on the liver. A complete blood count (CBC) typically reveals nonregenerative anemia and thrombocytopenia. A chemistry profile may show increase total bilirubin, AST, ALT and ALKP.

Coccidiomycosis is another systemic fungal infection which originates in the lungs. Coccidiomycosis has also been called "San Joaquin Valley fever" or "valley fever." Like histoplasmosis, it is not contagious and is usually transmitted by inhalation of the fungus. Periods of rain cause the fungus to rise to the top of the soil and become airborne in the wind. Coccidiomycosis can spread to almost any organ, including the liver. More than 50% of cats have cutaneous lesions on the liver which may also include abscesses. Signs include dyspnea, abnormal lung sounds, fever, weight loss, bone lesions, and central nervous system signs. Similarly to histoplasmosis a CBC most commonly reveals a nonregenerative anemia. A serum chemistry profile may show an increase in total bilirubin, AST, ALT and ALKP.

Babesiosis is caused by *Babesia*, an intracellular protozoa, which is more common in the southern United States, Africa and Asia. With an increase of importing of dogs from the south to other areas of the country, it is important to remember to obtain a complete history on the animal. Typical signs include lethargy, pale mucous membranes, fever, and weight loss. Icterus is more common in dogs than in cats. Due to the severe anemia which is common with babesia, liver enzymes typically become elevated. Thrombocytopenia can occur along with the life threatening anemia.

Cytauxzoonosis is caused by the *Cytauxzoon felis* protozoan and usually only affects cats in the southcentral and southeastern United States. The protozoan specifically attacks the vascular system of the kidneys, brain, lungs, liver and spleen. Most of the time owners will notice that their cats are depressed, not eating, and have pale mucous membranes. Cats usually are anemic, icteric and severely ill when they present. Within 5-6 days after infection, you will typically see a decrease in the packed red blood cell volume. There is a 99.9% mortality rate associated with this disease. Liver enzymes are typically elevated.

Liver Flukes (*Platynosum concinnum*) occur mainly in cats in tropical areas, specifically Florida and Hawaii. Few infected cats exhibit signs. Those that do are usually jaundiced, vomiting, and anorexic. The adult liver fluke reside in the bile duct and gallbladder. The eggs of the liver fluke hatch in a snail and the sporocysts are then ingested by a second intermediate host, most commonly the anole lizard. The cat then eats the lizard and becomes infected. The fluke's eggs are passed back into the environment through a cat's feces which are then ingested by a snail to start the cycle over again. Typically you can diagnosis liver flukes by seeing the eggs in the fecal sediment. The liver enzymes ALT, AST, ALP and total bilirubin are most commonly elevated. A urine sediment will show bilirubinuria.

METABOLIC DISEASES

Hepatic Lipidosis occurs secondary to another disease process or can be an idiopathic disease. It is defined as the intracellular accumulation of lipid within the liver. Anorexia is usually the predisposing factor causing hepatic lipidosis. In dogs hepatic lipidosis is usually an incidental finding, while in cats hepatic lipidosis is the most common liver disease. When a feline does not eat or their dietary protein is restricted the apoproteins (needed for transporting lipids out of the liver) decrease. As a result lipids start to accumulate in the liver and eventually cause "fatty liver" syndrome also known as hepatic lipidosis. Obese cats have an increase risk for developing this disease when they go through a period of anorexia. Typically owners will describe anorexia, jaundice and lethargy. Most commonly on blood work you'll see increases in ALT, ALP, AST and bilirubin. Ultrasound helps to diagnose hepatic lipidosis.

Liver Shunt, also known as a portosystemic shunt (PSS) is an abnormal blood vessel that bridges the portal and systemic circulation allowing the blood to bypass the liver. Animals with PSS usually have stunted growth, intermittent vomiting and diarrhea and may also have polyuria and polydipsia. Animals may also exhibit head pressing, stupor or dementia. Cats may become aggressive. Approximately 25-50% of animals with PSS may experience seizures. Bloodwork may show mild elevations in ALT, ALP and ASP. Ammonia levels are commonly elevated in animals with PSS. The most common blood test used for helping to diagnose PSS is the bile acid test. An elevation in the postprandial test indicates the liver's failure to eliminate bile acid. Ultrasonography and radiography can also be used to help diagnosis a shunt, but the most accurate test is a portogram (a contrast angiography). This test is very time consuming, expensive and requires the patient to undergo anesthesia. If a shunt is suspected most veterinarians will recommend an abdominal exploratory. If the surgeon is unable to locate the PSS, then a portogram may be considered.

Primary Hepatic Neoplasia is not very common in dogs or cats (roughly only 0.8%-2.3%). Metastases from neoplasia on other organs are much more common to the liver. In dogs roughly 7%-36% of dogs that had cancer primarily affecting other organs also had metastasis to the liver. Owners typically report that their pets are lethargic, anorexic or are experiencing vomiting and/or diarrhea. On physical only 18-30% of dogs will appear jaundice. Most animals that have primary liver neoplasia will also have hepatomegaly. Ultrasound is more accurate at detecting tumors of the liver than radiographs. Bloodwork usually reveals increases in liver enzymes: ALP, ALT, AST, and bilirubin. In dogs, anemia is one of the most common findings on a CBC with primary hepatic neoplasia.

Copper Storage Disease is associated with canines. Copper is the third most abundant trace element in the body with the liver having the highest concentrations. Copper plays a role in the body's metabolism. Excessive copper is excreted in bile. Copper can accumulate in the liver because of a metabolic defect. The excessive copper can cause hepatocellular damage. Copper accumulates in the body due to an abnormal metabolism process. The three ways that copper can accumulate is from abnormal absorption (though this is rare), defective excretion mechanism or cholestasis, which stops the normal excretion process. Certain breeds such as Bedlington Terriers, West Highland White Terriers, Doberman Pinchers and Skye Terriers are predisposed to copper accumulation. Signs include those associated with progressive liver failure (vomiting, diarrhea, lethargy, anorexia, jaundice, ascites, anemia). Anemia can result from the oxidative damage of oxygen free radical to the red blood cells caused by unbound copper. Bloodwork usually supports chronic liver failure. Biopsies of the liver will help to diagnosis a copper storage disease.

Pancreatitis is a disease that can be very frustrating because it often has an unknown etiology. Canine pancreatitis can be divided into two different forms: chronic and acute. Typically it is the acute form of pancreatitis that causes liver failure. The chronic form typically does not involve the liver. Roughly 26% of dogs were found to be icteric when they had an acute fatal pancreatitis. Cholestasis can occur secondarily from pancreatitis. This is because of the pathway bile follows (see chart 1.2).A partial or complete obstruction in any part of the pathway can affect the organs involved in the pathway, such as the liver. Animals with pancreatitis can exhibit anorexia, vomiting and signs current with severe abdominal pain such as walking with a hunched over appearance or hiding in cats. In severe cases the animal may present in hypovolemic shock. A CBC usually reveals a leukocytosis and neutrophilia with a left shift. A serum chemistry panel may show azotemia and an increase in liver enzymes and total bilirubin. Radiographs are usually non-diagnostic. Ultrasound is usually the diagnostic instrument of choice along with sending out pancreas specific enzyme tests like a PLI.

TOXINS/TRAUMA

Trauma can be blunt or penetrating. Certainly the liver can be damaged if a penetrating wound, like a gun shot, directly damages the liver. Blunt traumas may not be so obvious. Typically patients present to a veterinary clinic in shock. Because of adrenaline, underlying injury to the liver may be masked. One study of 600 dogs involved in a motor vehicle accident showed that the liver was the primary organ affected. Blunt trauma to the liver typically shows an increase in ALT. Generally this resolves in 4-10 days post injury. It's important to monitor the animal for a couple of days to ensure the liver enzymes are continuing to decrease.

Toxins are another type of insult to the liver. There are different kinds of toxins; the two main types being biological and synthetic. Examples of biologic toxins would be fungi and plants. Examples of synthetic toxins would be pharmaceuticals. Toxins can cause both acute and chronic liver failure. There are hundreds of toxins that can cause liver failure as the liver is the primary organ for detoxification. Many pharmaceuticals have the potential to cause chronic liver failure. The larger the dose and the longer the exposure to any medication, the more risk to the liver. Most toxins enter the liver directly through the portal system. See table 1-1 for some of the most common synthetic and biologic toxins.

Malignant Hyperthermia, also known as heat stroke, can affect almost every system in the body. It's important to remember that any heat stoke patient can experience liver failure. Because of the high risk of systemic inflammatory response syndrome (SIRS) and DIC, the liver can become affected. Heat stroke can cause fatal liver injury. Typically you will see elevations in AST, ALT and total bilirubin indicating liver damage.

RECOVERING THE LIVER

It is important to obtain a diagnosis when dealing with liver disease. Without one, the treatment given will be general and not specific to the disease. The overall treatment goal for any liver disease is to provide supportive care. Throughout treatment it is important to manage the complications that can occur with liver failure. Some complications which can occur include: gastrointestinal ulceration, anemia, infection, ascites, edema and coagulopathy.

As a veterinary technician it is important to monitor all systems of the animal (cardiovascular, respiratory, neurological). Veterinary technicians likely spend more time with the patient and therefore can notify the veterinarian of early signs of edema, anemia, or infection. Doing a complete physical exam each time you are with the animal is important. By forgetting to check your patient's mucous membrane color you may not notice that your patient's pink mucous membranes are now pale. The jaundice color of your patient can change on an hourly basis and it is important to notify your veterinarian of such changes.

Aggressive fluid therapy is one of the most important treatments when dealing with liver disease. Fluid therapy helps with hepatic circulation and also helps prevent complications such as disseminated intravascular coagulation, shock or renal failure. Sometimes 0.45% NaCl will be used to help prevent sodium retention and portal hypertension. Colloids, such as hetastarch, may be used to help restore oncotic pressure. Electrolytes should be monitored and adjustments to fluids should be made accordingly. Since infection with liver disease is always a complication, antibiotics may be used as prevention.

Nutrition plays an important role in recovering a patient from liver disease. This is particularly true in cats where hepatic lipidosis can occur from anorexia alone. During the initial treatment stages animals may be fasted. This is because many times they present nauseas or vomiting. However, as long term care continues, for many of these patients, nutrition must be considered. Nutrition is important for any animal to recover. Placement of a feeding tube should be considered if the animal is unwilling to eat. This allows for more controlled and effective feedings, rather than syringe force-feedings. Veterinary technicians play an important role because they often can encourage patients to eat by offering them a variety of foods or spending extra time hand feeding them. Some liver diseases require special diets, so it is important that the veterinarian be consulted before any feedings are initiated.

As complications arise from liver disease, certain treatments may be necessary. For example, in severe liver disease cerebral edema may develop. Mannitol and furosemide may be used to help reduce the edema. Notifying the veterinarian of changes in pupil size (anisocoria) or neurological status can make the difference in saving the animal's life. A veterinary technician may be the first one to notice petechiae on the patient emphasizinga complete physical exam should be done every time a patient is observed. Fresh frozen plasma may be given if the animal starts to become coagulopathic.

Some patients recover quickly from liver disease. Within 24 hours an animal can go from being very jaundice and lethargic to eating and drinking with no yellow coloration anywhere present on the body. However, some animals experience severe complications from liver disease. It is important to remember that even if the animal is not suffering from primary liver disease, they can still experience severe complications simply because their liver is affected.

REFERENCES

1. Dyce K, Sack W, Wensing C: *Textbook of Veterinary Anatomy*. Philadelphia, PA, W.B.Saunders Company, 1987, 2. Hole J: *Essentials of Human Anatomy and Physiology* (fifth edition). Dubuque, Iowa, Wm. C. Brown Publishers, 1995., 3. Aspinall V: *Essentials of Veterinary Anatomy and Physiology*. London, Elsevier, 2005. 4. Dimski D: *The Veterinary Clinics of North America*: Small Animal Practice: Liver Disease. Philadelphia, W.B.Saunders Company, 1995., 5. Tilley L, Smith F: *The 5-Minute Veterinary Consult: Canine and Feline* (second edition). Philadelphia, Lippincott Williams and Wilkins, 2000., 6. Ettinger S, Feldman E: *Textbook of Veterinary Internal Medicine* (fifth edition). Philadelphia, W.B.Saunders Company, 2000., 7. Shell L: Hepatic Lipidosis Paper Found at www.vin.com, March 26, 2006., 8. Armstrong P: *Feline Hepatic Lipidosis* (V161), Western Veterinary Conference 2005, 9. Kolata RJ, *Johnston DE: Motor Vehicle Accidents in Urban Dogs*: A Study of 600 cases, J. AM Vet Med Association 167:938, 1975., 10. Wingfield W, Raffe M: *The Veterinary ICU Book*. Jackson Hole, Wyoming, Teton NewMedia, 2002., 11. Osweiler G: *The National Veterinary Medical Series: Toxicology*. Philadelphia, Williams & Wilkins, 1996., 12. Guyton A, Hall J: *Textbook of Medical Physiology*. Philadelphia, W.B. Saunders Company, 1996., 13. Aspinall V, O'Reilly M: *Introduction to Veterinary Anatomy and Physiology*. Edinburgh, Butterworth Heinemann, 2004.

Table 1.1

Synthetic Drugs Associated with Liver Toxicity[10, 11]
Acetaminophen
Aspirin
Carprofen (Rimadyl)
Diazepam (cats)
Griseofulvin (cats)
Mebendazole (dogs)
Phenobarbital
Phenols (found in cleaners)
Phosphorus (found in rodenticide)
Thiacetarsamide (dogs)
Vitamin A

Biological Agents Associated with Liver Toxicity[10, 11]
Amanita mushrooms (dogs)
Blue-green algae
Cycad Seeds
Hornet Sting
Sago Palm

Chart 1.2 Bile Duct Pathway

Bile is first secreted by the hepatocytes in the liver and then follows the pathway.

Bile canaliculi (small tubes that collect bile secreted by the hepatocytes)
↓ drain into
Canals of Hering (small bile ducts)
↓ drain into
Bile ductules (small channels that connect the interlober ductules of the right and left ducts)
↓ drain into
Left and right hepatic ducts
↓ merge together to create
Common hepatic duct
↓ bile finally exits the liver and joins
Cystic duct
↓ the common hepatic duct and cystic duct merge together to create the
Common bile duct merges into
↓ Pancreatic duct
↓ the union of the common bile duct and pancreatic duct form theampulla of Vater
↓ bile then finally enters
Duodenum

Author Index

A

B

Diagnostic Utility of Immunocytochemistry and Cytochemisxtry for Clinical Cytology

Barrs, Vanessa, BVSc (hons), MVetClinStud, MACVSc (Small Animal), FACVSc (Feline), GradCertEd (Higher Ed)

Feline UR Aspergillosis - How DIfferent Is It from Canine Sinosal Aspergillosis?

Bartges, Joseph, DVM, PhD, DACVIM, DACVN

Two Are Not Better: Feline Chronic Kidney Disease and Nephrolithiasis

Behrend, Ellen, VMD, PhD, DACVIM, Joezy Griffin Professor

Milotane vs. Trilostane: What's the Evidence?

Bentley, R., DACVIM (Neurology)

The Magnetic Resonance Imaging Appearance of Canine Intra-Axial Primary Brain Tumors Relates More Strongly to Tumor Grade than Tumor Type

Brain Tumors in Small Animals: Diagnosis and Management

Bianco, Domenico, DACVIM

Human Intravenous Immunoglobulin: Current Indications in Smal Animal Medicine

Biddinger, Bea, LVT, VTS (ECC)

Respiratory Mechanics and Monitoring

Pleural Space Disease: Chest Taps and Tubes

Biller, Barbara, DVM, PhD, DACVIM

Targeted Monoclonal Antibody Therapy in Canine Lymphoma

Blondeau, Joseph, MSc, PhD, RSM(CCM), SM(AAM), SM(ASCP), FCCP

SOTA: Antibiotics and Antibiotic Resistance: A Practical Discussion

SOTA: Choosing Antimicrobial Agents in a World Antimicrobial Resistance

Boller, Manuel, Dr. med. vet., MTR, DACVECC

The RECOVER Initiative: What the Evidence Tells Us About CPR in Small Animal and Comparative Research

Pathophysiology of Ischemia and Reperfusion: What Matters During Arrest and CPR

Bonagura, John, DVM, DACVIM

Assessment of Systolic Function

Boothe, Dawn, DVM, PhD, DACVIM, DACVCP

Antimicrobial Use Guideliens: Treatment of UTI

Boswood, Adrian, MA, VetMB, MRCVS, DVC

Pimobendan Improves Survival in Doberman Pinscher's with Preclinical DCM: The Results of Protect a Double-Blind, Placebo-Controlled, Randomized Clinical Trial

Bowles, Mary, DVM, DACVIM

Alternative Options for Managing Urinary Tract Disease in the Dog and Cat

Bowman, Dwight, MS, PhD

Pulmonary Disease from Canine- and Feline-Associated Intestinal Helminths

Review of Parasitic Diseases Associated with the Neonatal Dog and Cat

Boyle, Ashley, DACVIM

Breton, Amy, CVT, VTS (ECC)
Spleen Physiology and the Treatment of Common Diseases
Liver: Physiology and Disease

Bryan, Jeffrey, DVM, MS, PhD, DACVIM (Oncology)
DNA Methylation: Understanding Dogs to Understand Humans

Buchanan, James, DVM, MMedSci, DACVIM
Veterinary Cardiology History

Buczinski, Sebastien, DACVIM
The Diagnostic Criteria Used in Bovine Bacterial Endocarditis: A Meta-Analysis Of 460 Published Cases from 1973 to 2011
Resident Training: Clinical Experiences Outside of the Teaching Hospital

Burns, Kara, MS, MEd, LVT
Cat and Carbohydrates: What Should Fluffy Eat?
Alternative and Raw Diets

Burton, Jenna, DACVIM
Safety of Metronomic Cyclophosphamide Combined with Carboplatin for Dogs with Cancer

Byers, Stacey, DVM, MS, DACVIM (LAIM)
Study of Metabolism in Pregnant and Postpartem Ovine

C

Cave, Nick, BVSc, MVSc, PhD, DACVN
The "Human Factor" to Feeding Pets
Dietary Isoflavones and their Biological Effects

Cebra, Christopher, VMD, MA, MS, DACVIM
Coccidiosis in New World Camelids
Camelid DIsease of Public Significance

Center, Sharon, DVM, DACVIM
Breed Specific Hepatopathies: Scottish Terrier and Maltese
Diagnostic and Managerial Pearls for Selected Hepatobiliary Disorders

Cerda-Gonzalez, Sofia, DVM, DACVIM (Neurology)
Surgical Management of Congenital Brain and Cervical Spine Disorders in Dogs
Cognitive Dysfunction Syndrome: The Canine Version of Human Alzheimer's Disease

Chaffin, Morgan, DVM, MS, DACVIM
Thoractomy for Persistent Pleural Abscesses Associated with Equine Pleuropneumonia

Chamorro, Manuel, DACVIM
Antibody Titers in Calves Fed Colostrum or a Replacer

Chan, Daniel, DVM, DACVECC, DACVN, FHEA, MRCVS

Intralipid Therapy: From Nutrition to Antidote

Chew, Dennis, DVM, DACVIM
Management of Complicated UTI

Childress, Michael, DACVIM
Complementary and Alternative Medicine for Veterinary Cancer Patients: What You and Your Clients Should Know
Canine Lymphomas - Rethinking a Diverse Family of Neoplasms

Coates, Joan, DVM, MS, DACVIM (Neurology)
Biomarker Development for Canine Degenerativ Myelopathy

Coetzee, Johann (Hans), BVSc, Cert CHP, PhD, MRCVS, DACVCP
Update on the Diagnosis and Treatment of Bovine Anaplasmosis

Cohn, Leah, DVM, MS, PhD, DACVIM (SAIM)
High Dose Diamenazine Diaceturate Does Not Eliminte Parasitemia in Cats with Chronic Cytauxzoon Felis Infection

Cook, Audrey, BVM&S, MRCVS, DACVIM (SAIM), DECVIM-CA
Histoplasmosis: Current Dignostic and Treatment Strategies
Gallbladder Mucoceles: The Endocrine Connection

Cooper, Edward, VMD, MS, DACVECC
Microcirculatory Imagine: A New Frontier in Hemodynamic Monitoring
Feline Urethral Obstruction: Traditional and Alternative Approaches to Treatment

Corcoran, Brendan, MVB, DPharm, PhD, MRCVS
Tissue Engineered Canine Mitral Valves: Value in Valve Disease Research

Cortadellas, Oscar
Evaluation of the Association Benazepril/Renal Diet to Control Renal Proteinuria in Non-Azotemic Dogs

Craig, Thomas, DVM, MS, PhD
SOTA: Anthelmintic Resistance in Small Ruminants
SOTA: Control of Helminth Parasites in Cow Calf Operations

Culp, William, VMD, DACVS
Interventional Oncology and the Management of Urinary Tract Cancer

d

da Costa, Ronaldo, DMV, MSc, PhD, DACVIM (Neurology)
Comparison of Cervical Spondylotic Myelopathy in People and Cervical Spondylomyelopathy in Dogs
Critical Evaluation of Treatment Methods for Cervical Spondylomyelopathy

da Costa, Ronaldo, DMV, MSc, PhD, DACVIM (Neurology)
Diffusion Tensor Imaging in Dogs with Cervical Spondylomyelopathy

Nuclear Imaging in Veterinary Endocrinology

De Laforcade, Armelle, DVM, DACVECC
Platelet Activation in Critical Illness

Dear, Jonathan, DACVIM
Babesia Conradae Infection in Coyote Hunting Dogs Infected with Multiple Blood-Borne Pathogens

Dickinson, Peter, BVSc, PhD, DACVIM (Neurology)
Stereotactic Brain Biopsy: Outcomes, Complications and Recommendations
UCD Brain Tumor Trial

Dore, Elizabeth, DACVIM
Evaluation of Environmental Samples to Predict Mycobacterium Avium Subsp. Paratuberculosis Prevalence in Quebec Dairy Herds

Dow, Steven, DVM, PhD, DACVIM (SAIM)
T-Cell Targets for Immune Modulation
Mesenchymal Stem Cells for Immunotherapy: Nature's Alternative to Steroids

Dow, Steven, DVM, PhD, DACVIM (SAIM)
Monocytes Rediscovered: Their Critical Role in Cancer and Infection

Dunkel, Bettina, DVM, PhD, DACVIM, DACVECC, MRCVS
The Role of Platelets in Equine Inflammation and Immunity
Equine Interstitial Lung Diseases: Clinical and Radiographic Assessment

Dunn, Marilyn, DMV, MVSc, DACVIM (SAIM)
Antiplatelet and Anticoagulant Therapy: What Do We Know? Where Are We Going?

Durando, Mary, DVM, PhD, DACVIM (LAIM)
Comparison of Endocardiographic Measurements in Elite and Non-Elite Arabian Horses

Durham, Andy, DECEIM, DRCVS
Is Metformin Useful in EMS Cases?

E

Edward Durham, H., CVT, LATG
Cardiac Pacemakers in Small Animal Cardiology
Equine Cardiology: Physiology, Acquired and Congenital Disease

Elfenbein, Johanna, DACVIM (LAIM)
Salmonella Typhimurium Genetics During Intestinal Inflammation

Estrin, Joshua, MS, LCSW, CMFSW
Bullying and Lateral Violence Amongst Nurses and Techs

F

Canine Osteosarcoma - Bench to Bedside Molecular Targets and Therapeutics

Fecteau, Gilles, DVM, DACVIM (LAIM)
The Best Papers of 2011

Fecteau, Marie-Eve, DACVIM
Novel Strategies for the Prevention of Johnne's Disease in Cattle

Fine, Deborah, DVM, MS, DACVIM (Cardiology)
Hemodynamic & Biochemical Alterations in Canine Lymphoma
Emergency Management of Arrythmias

Finno, Carrie, BSc, DVM, DACVIM (LAIM)
Genetic and Nutritional Aspects of Equine Neuroaxonal Dystrophy

Firshman, Anna, BVSc, PhD, DACVIM, DACVSMR
Equine Rehabilitation

Flatland, Bente, DVM, DACVIM (SAIM), DACVP
Quality Assurance for Point-of-Care Analyzers in Veterinary Medicine: What Every User Should Know
Quality Assurance in Action: Examples from the Field

Flegel, Thomas, DACVIM
Minimally Invasive Brain Biopsy in Canine Encephalitis

Fletcher, Daniel, PhD, DVM, DACVECC
The RECOVER Initiative - Consensus on Basic and Advanced Life SUpport During CPR
The RECOVER Initiative - Consensus on Preparedness, Training and Post-Arrest Care

Forman, Marnin, DACVIM
Figure out Feline Pancreatitis

Foster, Derek, DVM, MS, DACVIM (LAIM)
Enterocyte Response to Cryptosporidium Parvum

Fox, James, DVM, DACLAM
GI Microflora and the Hygiene Hypothesis

G

Garden, Oliver, DACVIM
All Creatures, Great and Small: A Review of Regulatory T-Cells in Mice, Humans, Dogs and Other Domestic Species

Garrett, Laura, DVM, DACVIM (Oncology)
Mast Cell Tumor Cases and Controversies
Oncology Reviews: Current Controversies in Canine Mast Cell Tumors

Garrett, Laura, DVM, DACVIM (Oncology)
The Roles for Radiation in Cancer Therapy

Dietary Impact on Phenotypic Measurements

Equine Metabolic Syndrome - New Insights on Pathophysiology

Geor, Raymond, BVSc, PhD, DACVIM (LAIM)
Equine Metabolic Syndrome - Nutritional Management

Gieger, Tracy, DVM, DACVIM, DACVR
Diagnosis and Management of Oral Tumors in Dogs and Cats

Giguere, Steeve, DVM, PhD, DACVIM (LAIM)
Emergence of Widespread Macrolide and Rifampin Resistance in Rhodococcus Equi Isolates from a Horse Breeding Farm

Gilliam, Stephanie, RVT, BAS, CCRP
Evaluation of the Neurologic Patient

Degenerative Myelopathy Diagnosis and Management: The Technician's Role

Goehring, Lutz, DVM, MS, PhD, DACVIM (LAIM)
Anti-Inflammatory Drugs and EHV-1 Infection

Goldstein, Richard, DVM, DACVIM (SAIM), DECVIM-CA
New Developments in Quantitative Testing and Vaccine Safety Canine Lyme Disease

Gorgi, Alireza, DACVIM
Biological In-Situ Fusion Technique for Congenital Vertebral Anomalies in Brachycephalic Dogs

Gow, Adam George, DECVIM, DRCVS (Small Animal Medicine)
Pathophysiology of Hepatic Encephalopathy in Companion Animals: New Information

Granger, Nicolas, DVM, MRCVS, DECVN
Urination Control by Sacral Root Stimulator

Advances in the Pathogenesis and Treatment of Spinal Cord Injury

Gregory, Clare, DVM, DACVS
Non-Glucorticoid Immunomodulators

H

Haggstrom, Jens, DVM, PhD, DECVIM-CA (Cardiology)
Molecular Genetic Research in Dogs from a Veterinary Cardiologist's Perspective

Haggstrom, Jens, DVM
Controversies in Mitral Valve Disease in Dogs

Hall, Meri, RVT, LVT, CVT, RLATG, VTS (SAIM)
Continuous Rate Infusion: Drugs and Calculations

Harr, Kendal, DVM, MS, DACVP
General Quality Assurance: Basic Definitions and Concepts

Interpreting Values Outside Reference Intervals: Is It the Patient or the Machine

Hart, Kelsey, DVM, DACVIM (LAIM)

Hartmann, Katrin, Prof., Dr. med. vet., Dr. habil., DECVIM
Regressive and Progressive FeLV Infections: Clinical Relevance

Hawkins, Eleanor, DVM, DACVIM (SAIM)
Not "Just" Tracheal Collapse: Pathophysiology and Management

Heatwole, Bonnie, CVT
Treating the Final Stage of Heartworm Disease: Caval Syndrome

Helmond, Sarah, DACVIM
Thromboprophylaxis in Immune-Mediated Hemolytic Anemia

Hepburn, Richard, BVSc MS(Hons) CertEM(IntMed) DACVIM (LAIM) MRCVS RCVS MRCVS
Equine Glandular Ulceration: Epidemiology and Pathophysiology
Equine Glandular Ulcertain: Comparative Aspects and Management

Hepburn, Richard, BVSc, MS, CertEM(IntMed), DACVIM(LAIM), MRCVS
Effective Laboratory Medicine in Equine Practice

Hess, Paul, DACVIM (Oncology & SAIM)
Determination of the Peptide-Binding Motif of a Prevalent Canine MHC Class la Allele

Hewson, Joanne, DVM, PhD, DACVIM (LAIM)
Diagnosing BRD: Have we Gotten any Better Over the Years?

Hill, Tracy, DVM, DACVIM (SAIM), DECVIM-CA, DECVCP
Misoprostol Promote Gastric Barrier Defense Against Non-NSAID Mediated Acid-Induced Injury

Hinchcliff, Kenneth, BVSc, PhD, DACVIM (LAIM)
Physiological Responses of the Horse to Exercise - Causes of Poor Performance

Hodgkiss-Geere, Hannah, BVM&S, MSc, MRCVS
Canine Adult Stem Cells: Their Biology and Therapeutic Potential in Heart DIsease

Hohenhaus, Ann, DVM, DACVIM (Oncology & Internal Medicine)
Transfusion Therapy in Immune-Mediated Hematologic Disease

J

Johannes, Chad, DVM, DACVIM (SAIM)
Practical Guide to Diagnosis and Prevention of Canine Leptospirosis
Antiemesis Therapeutic Considerations for the Oncology Patient

K

Kamstock, Debra, DVM, PhD, DACVP
Canine Cutaneous Mast Cell Tumor Profile: What Does It Mean?

Kang, Byeong-Teck, DVM, PhD
18F-Fluorodeoxyglucose Uptake and Magnetic Resonane Imaging Findings in a Canine Model of

7T-MR Fusion System

Katja Hoglund
BP, HR and Catecholamines in Genotyped Healthy Dogs in a Clinical Situation

Kay, Nancy, DVM, DACVIM (SAIM)
Transforming Your Clients' Internet Research into a Win-Win Situation
E-mail Communication with Your Clients and Colleagues: The Good, the Bad, and the Ugly

Keene, Bruce, DVM, MSc, DACVIM (Cardiology)
Efficacy of Bay 41-9202 (Bisoprolol Oral Solution) for the Treatment of Chronic Valvular Heart Disease (CVHD) in Dogs.

Kellihan, Heidi, DVM, DACVIM (Cardiology)
Pulmonary Hypertension: Physiology and Diagnosis
Clinical Application of Pulmonary Hypertension Diagnosis and Treatment

Kent, Michael, DACVIM
Stereotactic Radiosurgery for Brain Tumors: Clinical Trial Update

Kidd, Linda, DACVIM
Use and Misuse of Diagnostic Panels in Canine Vector-Borne Disease Testing

Kirk, Claudia, DVM, PhD, DACVIM (SAIM), DACVN
Two Are Not Better: Feline Chronic Kidney Disease and Diabetes Mellitus

Kirkby, Kristin, DVM, MS, CCRT, DACVS
Therapeutic Lasers in Veterinary Medicine
How Can Veterinary Rehabilitation Help my Patient?

Kittleson, Mark, DVM, PhD, DACVIM (Cardiology)
Feline Hypertrophic Cardiomyopathy: Screens and Genes

Knapp, Deborah, DVM, DACVIM (Oncology)
Pet Dogs Lead the Way in the Fight to Stop Urinary Bladder Cancer
Folate-Targeted Therapy for Canine Urinary Bladder

Knapp, Deborah, DVM, DACVIM (Oncology)
Detection of Chemicals in the Urine of Dogs Following Home Lawn Application

Kocan, Katherine, PhD
Bovine Anaplasmosis: An Overview of Current Challenges

Kraft, Susan, DVM, PhD, DACVR, DACVR-RO
Advanced MRI for Cancer Diagnosis, Staging and Evaluation of Treatment Response

Krick, Erika, VMD, DACVIM (Oncology)
Cachexia in Feline Cancer Patients

L

Lacombe, Veronique, DVM, PhD, DACVIM (LAIM), DECEIM

Lacorazza, Daniel, PhD

SOTA: Mouse Models to Study Normal and Malignant Hematopoiesis

SOTA: Differentiation of Effector and Memory T Cells upon Immune Response

Landsberg, Gary, BSc, DVM, DACVB, DECVBM-CA

Validation of Neuropsychological Test Protocols for Cognitive Assessment of Senior Dogs and Cats

Langston, Catherine, DVM, DACVIM (SAIM)

Controversies in RRT: Is IHD, CRRT, or PD Better?

Lappin, Michael, DVM, PhD, DACVIM (SAIM)

ISCAID Antimicrobial Guidelines Working Group Recommendations: Respiratory Infections

Anti-Alpha Anolase Antibodies in the Serum of Cats With and Without Azotemia Stratified by Vaccination Status

Lappin, Michael, DVM, PhD, DACVIM (SAIM)

Is There an Epidemic of Feline Bartonellosis? Diagnosis, Treatment, Prevention, and Zoonotic Aspects

Larsen, Jennifer, DVM, PhD, DACVN

Obesity and Stress: Risk Factors for Feline Lower Urinary Tract Disease?

LaRue, Susan, DVM, MS, PhD

New Radiation Oncology Techniques for Treatment of Urogenital Tumors

Lathan, Patty, VMD, MS, DACVIM (SAIM)

Understanding ACTH Stimulation Tests

Le Bobinnec, Gerard, DVM, DECVIM (Cardiology)

Auscultation of Murmurs in the Dog and Cat: How Color Flow Mapping IMproves Our Understanding

Utility of Flecainide for the Treatment of Ventricular Arrhythmias

LeBlanc, Amy, DVM, DACVIM (Oncology)

PET Imaging in Veterinary Medicine: Oncology and Beyond

Legendre, Alfred, DVM, MS, DACVIM (SAIM & Oncology)

Polyprenyl Immunostimulant: Adjunct for Treatment of Feline Rhinotracheitis and Feline Infectious Peritonitis

Leguillette, Renaud, DMV, MSc, PhD, DACVIM (LAIM)

Upper and Lower Airway Inflammation in Horses: Endoscopic Scoring, Diagnosis and Treatment

Effects of Hyperbarbic Oxygen Therapy in Horses

Lerche, Phillip, BVSc, PhD, DACVA

Analgesia for Neurological Patients

Lester, Guy, BSc (Hons) BVMS PhD DACVIM (LAIM)

Weight Gain in Hospitalized Foals

Liss, David, RVT, VTS (ECC)

Adrenal Insufficiency in Critical Illness: Is It JUST Addison's?

Lyme Disease: WhichTest is Best?

Ljungvall, Ingrid, DVM
Controversies in Mitral Valve Disease in Dogs

Lloyd, David, PhD, BVetMed, FRCVS, DECVD
Antimicrobial Therapy of Superficial Bacterial Folliculitis in Dogs

Long, Sam, BVSc, PhD, DECVN
Protein Disulphide Isomerase Colocalisation with Superoxide Dismutase 1 in Canine Degenerative Myelopathy: Evidence for Endoplasmic Reticulum Stress

Lunn, Paul, BVSc, MS, PhD, MRCVS, DACVIM (LAIM)
Havemeyer Highlights: Focus on EHV-1 Workshop

Lunsford, Kari, DVM, DACVIM (SAIM)
Disordered Hemostasis and What to do About it

Lyons, Leslie, PhD
Identification of the Gene for Burmese Autosomal Recessive Hypokalemic Polymyopathy by Genome-Wide Association Study

M

MacGregor, John, DVM, DACVIM (Cardiology)
Cardiologists' Treatment of Canine Tachyarrhythmias
Rational Use of Pimobendan: What Uses are Evidence-Supported?

Maiolini, Arianna
Signalling Proteins and Toll-like Receptors May Induce a Th-17 Immune Response in Canine Steroid-Responsive Meningitis-Arteritis

Mansfield, Caroline, BSc, BVMS, MACVSc, MVM, DECVIM-CA
Advanced Endoscopy Techniques in SA Gastroenterology: What Are the Potentials and Pitfalls?
Advanced Endoscopic Techniques

Mariani, Christopher, DACVIM (Neurology)
Matrix Metalloproteinase-2 and -9 in Canine Central Nervous System Tumors

Marik, Paul, MD, FCP(SA), FRCP(C), FCCP, FCCM, FACP
SOTA: Critical Illness Related Corticosteroid Insufficiency; Diagnosis and Management
SOTA: The HPA Axis and the Stress Response

Marino, Dominic, DVM, DACVS, DACCT, CCRP
Freeze Core Brain Biopsy Comparison with Surgical Biopsy
Fatal Hyperthermia After Diencephalic Mass Removal

Marks, Stanley, BVSc, PhD, DACVIM (SAIM & Oncology), DACVN
Critical Assessment of Clostridial Diseases in Dogs and Cats
Making Sense of the Nonsense - Diagnosis and Treatment of Enteropathogenic Bacterial Infections

Mason, Nicola, DVM, PhD, DACVIM (SAIM)

Targeted Antibody Therapies: Expect More, Pay Less

Mason, Nicola, DVM, PhD, DACVIM (SAIM)
Canine Lymphoma - A Cancer Gone Viral?

Mayo, Christie
Epidemiology of Blue Tongue Virus Infection Among California Ruminants

Mazan, Melissa, DVM, DACVIM (LAIM)
IAD and RAO - What have we Learned in the Last Ten Years?
Pulmonary Regenerative Therapy: How Can We Make a New Lung?

McCue, Molly, DACVIM
Biochemistry, Morphmettics and Breed Comparisons

McFarlane, Dianne, DVM, PhD, DACVIM (LAIM)
Comparison of Plasma ACTH Concentration, Plasma ?-MSH Concentration, and Overnight Dexamelthasone Suppression Test for Diagnosis of PPID

McKenzie, Erica, BSc, BVMS, PhD, DACVIM (LAIM)
Muscular Causes of Poor Performance in Athletic Horses

McKenzie, Erica, BSc, BVMS, PhD. DACVIM (LAIM)
Evaluating and Managing the Neurologic or Recumbent Camelid
Managing Acute Rhabdomyolysis in Horses

McVey, David, DVM, PhD, DACVM
Arthropod Borne Food Animal Diseases

McVey, Shawn, MA, MSW
Money & Medicine
Hiring Winers: Emotional Intelligence-Based Hiring

McVey, Shawn, MA, MSW
Managing Change and Growth: Key Performance Indicators
Managing in an Economic Downturn

Meeus, Patrick, DVM, PhD, DEVPC, DACVM (Parasitology)
Antimicrobial Activity of Cefovecin (Convenia®) Against Borrelia Burgdorferi and its Impact on Early Lyme Borreliosis in Dogs

Meinkoth, James, DVM, PhD, DAVCP
Cytology in the Diagnosis of Inflammation and Infectious Diseases
Case Studies in Infectious Disease Cytology

Meurs, Kathryn, DVM, PhD
Interpretation and Use of Genetic Testing for Cardiac Disease

Milne, Marjorie, BVSc(Hons), FANZCVS(Radiology)
Structural Brain Pathology in Canine Primary Epilepsy: Does It Exist?

Mitchell, Mark, DVM, MS, PhD, DECZM (Herpetology)
Making a Contriubtion to Evidence-Based Medicine: How to Design a Study
Making a Contribution to Evidence-Based Medicine: How to Analyze a Study

Insights from Genetic and Functional Studies in Hemangiosarcoma

Moore, Sarah, DVM, DACVIM (Neurology)
Repeatability of a modified dorsal technique for Electronic Von Frey Anesthesiometry to Assess Sensory Threshold in Six Normal Dogs.

Moses, Lisa, VMD, DACVIM
Feline Chronic Pain:Assessment and Management in Complex Medical Cases

Munana, Karen, DACVIM (Neurology)
Pharmacokinetics of Midazolam After Intravenous, Rectal and Intramuscular Administration in Dogs

Muzzi, Ruthnea, MV, Msc, Dcs, Pós-Doutora
Longitudinal Strain and Strain Rate by Two-Dimensional Speckle Tracking in Felines
Vp on Diastolic Function in Dogs

Myers, Anne, RVT, VTS (Cardiology)
Effective Restraint and Sedation Techniques for Performing Echocardiograms on the Dog and Cat

N

Nabity, Mary, DVM, PhD, DACVP
New Biomarkers of Kidney Disease: An Update

Navarre, Christine, DVM, MS, DACVIM (LAIM)
Disaster Preparedness for Food Animal Practitioners and Producers

Nichols, Sylvain, DMV, MS, DACVS
What's New in Gastrointestinal Surgery?
Complicated or Challenging Umbilical Surgery in Calves

O

O'Brien, Dennis, DVM, PhD, DACVIM (Neurology)
Hereditary Cerebellar Ataxia

O'Grady, Michael, DACVIM (Cardiology)
Pathophysiology of Dilated Cardiomyopathy

O'Sullivan, Lynne, DACVIM
Treatment of Dilated Cardiomyopathy

Ogilvie, Gregory, DVM, DACVIM (SAIM & Oncology), DECVIM-CA
Polyunsaturated Fatty Acids and Cancer: Frontiers for Cure and Control
Science Behind the Placebo Effect and the Mind-Body Connection in Veterinary Oncology

Ograin, Vicky, MBA, RVT
Feline Hyperthyroidism: New Therapy (It's in the Bowl!)

Myxomatous Mitral Valve Disease in Dogs

Orava, J., DVM
Modulating Immune Responses in Disease States

Orton, E., DVM, PhD, DACVS
Transcatheter Mitral Valve Implantation (TMVI) for Dogs

Orvalho, Joao, DACVIM
Real-Time Three-Dimensional Echocardiography: From Diagnosis to Intervention

Oyama, Mark, DVM, DACVIM (Cardiology)
Predicting First Onset of Congestive Heart Failure in Dogs with Mitral Valve Disease
Diuretic Resistance in Dogs Treated for Heart Failure

Oyama, Mark, DVM, DACVIM (Cardiology)
Cardiology Reviews: Diagnosis, Prognosis and Risk Stratification of Heart Disease Using NT-proBNP Assay

P

Pace, Charlotte, BA (Hons), RVN, VTS (Cardiology)
Hypertrophic Cardiomyopathy Screening in Cats
Diagnosis and Treatment of Chronic Mitral Valve Degeneration in Dogs

Palm, Carrie, DACVIM (SAIM)
Application of Novel Biomarkers in Dogs with Experimental Actute Kidney Injury

Paoloni, Melissa, DVM, DACVIM (Oncology)
The Future of Personalized Medicine for Canine Cancer

Papich, Mark, DVM, MS, DACVCP
The Challenges of Empirical Antimicrobial Therapy: A Focus on UTI and Pyoderma
Unique Pharmacokinetics in Camelids

Papich, Mark, DVM, MS, DACVCP
Top Ten Mistakes Made When Administering Medications to Animals

Pariaut, Romain, DVM, DACVIM (Cardiology), DECVIM-CA (Cardiology)
ECG Basics: Tips to Get the Best Out of It

Passler, Thomas, DACVIM
BVD Virus in Wildlife and Domestic Species Other than Cattle

Passler, Thomas, DVM, PhD, DACVIM
Understanding Common Disorders of Small Ruminants

Payne, Jessie, PhD
Prognostic Indicators in Feline Hypertrophgic Cardiomyopathy

Peek, Simon, BVSc, PhD, DACVIM
Idiopathic Hemorrhagic Pericardial Effusion

The Problem Diabetic: Acromegaly, Cushing's, and Other Causes of Insulin Resistance

Platt, Simon, BVM&S, MRCVS, DACVIM (Neurology), DECVN
UGA Brain Tumor Trial

Podell, Michael, MSc, DVM, DACVIM (Neurology)
Neurology Reviews: Epilepsy 101 for Internists
Advanced Treatment Options for Epilepsy

Polzin, David, DVM, PhD, DACVIM (SAIM)
Creatinine: Reassessment of an Old Biomarket

Potter, Tim, BVetMed, PhD, MRCVS
PK-PD Integration and Modeling for Antimicrobials Used in the Management of BRD

Prado, Maria, MV PhD DACVIM (LAIM)
Mastitis, Biofilms, and Bioluminescent Streptococcus Uberis

Q

Quimby, Jessica, DVM, DACVIM (SAIM)
Assessment of Allogeneic Cryopreserved Mesenchymal Stem Cells for the Treatment of Feline Chronic Kidney Disease
Enhancing Appetite in the Feline Patient

Quimby, Jessica, DVM, DACVIM (SAIM)
Medical Management of Feline Chronic Kidney Disease

Quimby, Jessica, RVT, VTS (Cardiology)
Enhancing Appetite in the Feline Patient

R

Ralston, Sarah, VMD, PhD, DACVN
Metabonomic Studies: Potential for Identification and Prevention of OCD in Foals
Forage-Based Total Mixed Rations: An Economic Alternative?

Rand, Jacqueline, DACVIM
New Development in Management of Feline Diabetes

Raskin, Rose, DVM, PhD, DACVP
Cytology of Neoplastic Diseases and Its Correlation with Histopatholody

Rassnick, Kenneth, DVM, DACVIM (Oncology)
Vitamin D in the Treatment and Pathogenesis of Canine Mast Cell Tumors, Osteosarcoma and Congestive Heart Failure

Reinero, Carol, DVM, PhD, DACVIM (SAIM)
Expanding the Diagnostic Approach for Feline Allergic Asthma: Functional and Immunologic Testing

Reusch, Claudia, Prof. Dr., DECVIM
Diagnosis and Treatment of Canine Pheochromocytoma

Reuss, Sarah, VMd, DACVIM (LAIM)
Evaluation of Employment Opportunities, Career Satisfaction, and Financial Compensation of Recently-Boarded ACVIM -LAIM Diplomates

Reynolds, Arleigh, DACVN
Effects of Enterococcus Faecium SF58 on Stress Diarrhea

Richter, Keith, DVM, DACVIM
Liver Sampling Methods and Considerations: Is Laparoscopy Best?

Rick, Markus, Med Vet, PhD
Test Development for Preclinical Detection of Hypoadrenocorticism (HOAC) in Dogs

Riley, Christopher, DACVS
Mare-Associated Predictors of Passive Antibodies in Foals

Rishniw, Mark, BVSc, MS, PhD, DACVIM (Cardiology & SAIM)
Cardiology Mythbusters: Dispelling Diagnostic Dogma in Veterinary Cardiology

Rossi, Giacomo, DVM, PhD
Effects of the Probiotic VSL#3 on Microbiological, Histological, and Immunomodulatory Parameters in Dogs with Inflammatory Bowel Disease

Roussel, Allen, DVM, MS, DACVIM (LAIM)
The Other Best Papers of 2011

Rozanski, Elizabeth, DVM, DACVIM (SAIM), DACVECC
Hypercoagulability in Pulmonary Disease; Where is the Evidence
Lung Function and Ventilation in the Critically Ill Patient

Ruderman, Colleen, RVT, VTS (SAIM)
Infectious and Immune-Mediated Arthropathies

S

Sandoval, Shirley, BAS, LVT-VTS (LAIM)
Comparative Anatomy and Physiology of the Large Animal Female Reproductive Tract

Saunders, Ashley, DVM, DACVIM (Cardiology)
Two Are Not Better: Feline Chronic Kidney Disease and Heart Disease

Scansen, Brian, DVM, MS, DACVIM (Cardiology)
Echocardiographic Assessment of RV Function in 150 Dogs with Pulmonic Stenosis

Schmitz, Silke, DrMedVet, DECVIM
Understanding Adaptive Immunity in Canine IBD: Cytokines and Mucosal Protection

Schott, Harold, DVM, PhD, DACVIM (LAIM)
Efficacy of Prascent for Treatment of PPID

Probiotics: Influencing the Immune Response "Naturally"

Methicillin-Resistant Staphylococcus Aureus (MRSA) in Horses

Scott Weese, J., DVM, DVSc, DACVIM (LAIM)

Clostridial Diseases in Large Animals and Horses

Management of Methicillin-Resistant Staphylococcal Infections

Scott, Morgan, DVM, PhD

Issues in Antimicrobial Resistance

Segev, Gilad, DVM, DECVIM-CA

Applicaiton of Novel Biomarkers in Naturally Occurring Acute Kidney Injury

Selmic, Laura, BVetMed (Hons), MRCVS, DACVS

Oncologic Outcome and Prognostic Factors in 1134 Dogs with Appendicular Osteosarcoma Treated at a Single Institution

Selting, Kim, DACVIM

Thymidine Kinase 1 and C-Reactive Protein as Biomarkers for Screening Clinically Healthy Dogs for Occult DIsease

Sharp, Claire, BSc, BVMS, MS, DACVECC

SIRS, Sepsis, MODS, MOF and More

Sepsis Biomarkers

Sharp, Claire, BSc, BVMS, MS, DACVECC

Top Ten Tips for Surviving Sepsis

Sepsis: What the Future Holds

Shearer, Jan, DVM, MS

Welfare Issues in Beef and Dairy Cattle (Beef Cattle Emphasis)

Proper Euthanasia Methods

Shores, Andy, DVM, MS, PhD, DACVIM(Neurology)

Intraoperative Ultrasound for Intracranial Surgery

Sleeper, Margaret, VMD, DACVIM (Cardiology)

Therapeutic Gene Delivery: How Close is It to Prime Time?

Smith, Stephanie, DVM, MS, DACVIM (SAIM)

Dogma and Dilemmas in DIC

Stewart, Allison, BVSc (Hons), MS, DACVIM (LAIM), DACVECC

Diagnosis and Nursing Care of Tetanus and Botulism in Horses

Rabies in Large Animals: Transmission, Diagnosis, Prevention and Occupational Safety

Sykes, Jane, BVSc (Hons), PhD, DACVIM (SAIM)

Antigen Testing of Deep Mycoses

Treatment of Fungal Infections: The Which, Why and How of Antifungal Drug Therapy

T

Endotoxemia as a Predisposition for Laminitis

Tasker, Severine, BSc, BVSc(Hons), PhD, DSAM, DECVIM-CA, PGCertHE, MRCVS
Feline Haemoplasmosis - a One Medicine Approach to Pathogenicity, Diagnosis and Treatment

Taylor, Sandra, DVM, PhD, DACVIM (LAIM)
Factors Associated with Survival in 50 Horses with Potomac Horse Fever

Thamm, Douglas, VMD, DACVIM (Oncology)
Mutagen Sensitivity in Golden Retriever Lymphoma
GS-9219/VDC-1101 Efficacy in Canine Lymphoma

Thomason, John, DACVIM
Monitoring Platelet Function during Anti-Platelet Therapy

Thomson, Christine, DACVIM
The NeuroMap: A Simple Guide to Localizing Neurological Lesions - Spinal Cord and Peripheral Nervous System
The NeuroMap: A Simple Guide to Localizing Neurological Lesions - the Brain Map

Thomson, Christine, DACVIM
Neuronal Cell Cultures for the Investigation of the Normal and Diseased Nervous System

Tinkler, Stacy, DACVIM
Sodium Bicarbonate Administration Exacerbates Post Exercise Hypokalemia in Standardbreds Undergoing a Simulated Race Protocol

Trepanier, Lauren, DVM, PhD, DACVIM, DACVCP
Turning a Clinical Question Into a Testable Hypothesis
Fellowship Opportunities and Funding Sources for Internists

Tripp, Chelsea, DVM, MS, DACVIM (Oncology)
Electrochemotherapy: The Wave of the Future

Twedt, David, DVM, DACVIM (SAIM)
Breed Specific Hepatophathies: Dalmation and English Springer Spaniel

V

Valberg, Stephanie, DVM PhD DACVIM (LAIM), ACVSMR
Movement Disorders in Horses: Shivers, Stringhalt and Other Conundrums

van Eps, Andrew, BVSc, PhD, MACVSc, DACVIM (LAIM)
Digital Hypothermia Applied After the Onset of Lameness Prevents Lamellar Failure in the Oligofructose Laminitis Model

van Erck, Emmanuelle, DVM, PhD, DECEIM
Field Exercise Testing of Horses - Clinmical Applications
Cardio-Respiratory Causes of Poor Performance

Van Metre, David, DVM, DACVIM (LAIM)
Zoonotic Diseases: Case Scenarios

Canine Parvovirus Type: Does it Impact Prevention, Diagnosis or Prognosis?

Vengust, Modest, DVM, DACVIM (LAIM)
Starling Forces in the Lung of Horses

Vernau, Karen, DACVIM (Neurology)
Genome-Wide Association Analysis Identifies a Mutation Associated with Alaskan Husky Encephalopathy
Evaluation of a Potential Candidate Gene mGLuR4 for Necrotizing Meningoencephalitis in Dogs

Vite, Charles, DVM, PhD, DACVIM (Neurology)
The Neurobiology of Music and the Healing Power of the Blues

Vollmar, Andrea, Dr.med.vet., DECVIM (Cardiology)
Pimobendan Vs. Benazepril Vs. Metildigoxin in Irish Wolfhounds with Preclimical DCM or AF

W

Walz, Paul, DVM, MS, PhD, DACVIM (LAIM)
Efficacy of Vaccinating Cattle Against Bovine Viral Diarrhea Virus

Watson, Penny, DECVIM, DRCVS
Breed Specific Hepatophathies: Dalmation and English Springer Spaniel
Chronic Pancreatitis in the English Cocker Spaniel Shows a Predominance of IgG4+ Plasma Cells

Weber, Jessica, BS, LVT, VTS (SAIM, Neurology)
Patient Care for Non-Ambulatory Dogs with Spinal Cord Injury

Weber, Karin, DMV
On the Trace of New Biomarkers: Differential Expresson of Circulating MicroRNAs in Lean, Obese and Diabetic Cats

Webster, Cynthia, DVM, DACVIM
Breed Specific Hepatopathies: Labrador Retriever
Drug Induced Liver Injury in Dogs

Weinstein, Peter, DVM, MBA
Systematizing the Money Talk to Clients
How Transparent Comunication can Help the Bottom Line

Westropp, Jodi, DVM, PhD, DACVIM
Novel Treatment Strategies for Uncomplicated UTIs in Dogs
Obesity and Stress: Risk Factors for Feline Lower Urinary Tract Disease?

Wheat, Joseph, MD
Antigen Testing of Deep Mycoses

Whittemore, Jacqueline, DVM, DACVIM (SAIM), PhD
Two Are Not Better: Feline Chronic Kidney Disease, Inflammatory Bowel and Liver Disease

Willard, Michael, DVM, DACVIM (SAIM)
Advanced Considerations of Esophageal Dysfunction

Anthelminthic Resistance and Application of FAMACHA in Camelids

Witonsky, Sharon, DACVIM (LAIM)
Immunity to Sarcocystis Neurona

Institute of Physics
Dirac House, Temple Back
Bristol BS1 6BE UK

CURRAN ASSOCIATES INC
proceedings.com

References

[1] Fong P M, Keil D C, Does M D, Gore J C 2001 Polymer gels for magnetic resonance imaging of radiation dose distributions at normal room atmosphere Phys. Med. Biol. 46 3105-13

[2] Maryanski M J, Gore J C, Kennan R P, Schulz R J 1993 NMR Relaxation enhancement in gels polymerized and cross-linked by ionizing radaiton: a new approach to 3D dosimetry by MRI Magn. Res. Imaging 11 253-8

[3] Senden R J, De Jean P, McAuley K B, Schreiner L J 2006 Polymer gel dosimeters with reduced toxicity: a preliminary investigation of the NMR and optical dose response using different monomers Phys. Med. Biol. 51 3301-14

[4] Pappas E, Maris T, Angelopoulos A, Paparigopoulou M, Sakelliou L, Saidilos P, Voyiatzi L, Vlachos L 1999 A new polymer gel for magnetic resonance imaging (MRI) radiation dosimetry Phys. Med. Biol. 44 2677–84

[5] Gore J C, Kang Y S, Schulz R J 1984 Measurement of radiation-dose distributions by Nuclear Magnetic-Resonance (NMR) Imaging Phys. Med. Biol. 29 1189-97

[6] Adamovics J, Maryanski M J 2006 Characterisation of PRESAGE^{TM}: A new 3-D radiochromic solid polymer dosemeter for ionising radiation Radiation Protection Dosimetry 120 107-12

[7] DOSGEL 2006 J. Phys. Conf. Ser., 56

[8] Fuxman A M, McAuley K B, Schreiner L J 2003 Modeling of free-radical crosslinking copolymerization of acrylamide and N,N_-methylenebis(acrylamide) for radiation dosimetry Macromol. Theory Simul. 12 647–62

[9] McAuley K B 2004 The chemistry and physics of polyacrylamide gel dosimeters: why they do and don't work J. Phys.: Conf. Ser. 3 29-33

[10] McAuley K B 2006 Fundamentals of polymer gel dosimeters J. Phys.: Conf. Ser. 56 35-44

[11] Durmaz S, Okay O 2000 Phase separation during the formation of poly(acrylamide) hydrogels Polymer 41 5729–35

[12] Maryanski M, Audet C, Gore J C 1997 Effects of crosslinking and temperature on the dose response of a BANG polymer gel dosimeter Phys. Med. Biol. 42 303-11

[13] Babic S, Schreiner L J 2006 An NMR relaxometry and gravimetric study of gelatin-free aqueous polyacrylamide dosimeters, Phys. Med. Biol. 51 4171–87

[14] Warman J M, Luthjens L H, de Haas M P 2008 In-situ radiation dosimetry based on radio-fluorogenic copolymerization, proc. DOSGEL 2008, 303-7

[15] De Deene Y, Vergote K, Claeys C, De Wagter C 2006 The fundamental radiation properties of normoxic polymer gel dosimeters : a comparison between a methacrylic acid based gel and acrylamide based gels Phys. Med. Biol. 51 653-73

[16] Lepage M, Whittaker A K, Rintoul L, Back S A J, Baldock C 2001 The relationship between radiation-induced chemical processes and transverse relaxation times in polymer gel dosimeters Phys. Med. Biol. 46 1061-74

[17] Fuxman A M, McAuley K B and Schreiner L J 2005 Modelling of polyacrylamide gel dosimeters with spatially non-uniform radiation dose distributions Chem. Eng. Sci. 60 1277–93

[18] Karlsson A, Gustavsson H, Månsson S, McAuley K B, Bäck S A J 2007 Dose integration characteristics in normoxic polymer gel dosimetry investigated using sequential beam irradiation Phys Med. Biol. 52 4697-706

[19] Katchalsky A, Blauer G 1951 Kinetics of methacrylic acid polymerization in aqueous solution Trans. Faraday Society 47 1360-70

[20] Eliassaf J, Silberberg A, Katchalsky A 1955 Negative Thixotropy of Aqueous Solutions of Polymethacrylic Acid Nature 4493 1119

[21] Moussaid A, Schosseler F, Munch J P, Candau S J 1993 Structure of polyacrylic acid and polymethacrylic acid solutions: a small angle neutron scattering study J. Phys. II France 3

573-594

[22] Hayashi S-I, Munenori Y, Usui S, Haneda K, Takahiro T 2008 The role of gelatin in a methacrylic acid based dosimeter, Proc. DOSGEL 2008, 5th International Conference on Radiotherapy Gel Dosimetery 94-96

[23] Salomons G J, Park Y S, McAuley K B, Schreiner L J 2002 Temperature increases associated with polymerization of irradiated PAG dosimeters Phys. Med. Biol. 47 1435–1448

[24] Bamford CH and Shiiki Z 1968 Free-radical template polymerization Polymer 9 595-8

[25] Polowinski S 2002 Template polymerisation and co-polymerisation Prog. Polym. Sci. 27 537-77

[26] Fernandes J P, Pastorello B F, de Aurojo D B, Baffa O 2008 Formaldehyde increases MAGIC dosimeter melting point and sensitivity Phys. Med. Biol. 53 N53-N58

[27] Koeva V I, Csaszar E S, Senden R J, McAuley K B and Schreiner L J 2008 Polymer gel dosimeters with increased solubility: a preliminary investigation of the NMR and optical dose-response using different crosslinkers and co-solvents Macromol. Symp. 261 157-66

[28] De Deene Y, Pittomvils G, Visalatchi S 2007 The influence of cooling rate on the accuracy of normoxic polymer gel dosimeters. Phys. Med. Biol. 52 2719–28

[29] Papadakis E, Maris T G, Zacharopoulou F, Pappas E, Zacharakis G, Damilakis J 2007 An evaluation of the dosimetric performance characteristics of n-vinylpyrrolidone-based polymer gels Phys. Med. Biol. 52 5069–83

[30] Jirasek A, Hilts M, Shaw C, Baxter P 2006 Investigation of tetrakis hydroxymethyl phosphonium chloride as an antioxidant for use in x-ray computed tomography polyacrylamide gel dosimetry Phys. Med. Biol. 51 1891

[31] Djabourov M, Leblond J, Papon P 1988 Gelation of aqueous gelatin solutions. I. structural investigation J. Phys. France 49 319-32

[32] Djabourov M, Leblond J 1987 Thermally reversible gelation of the gelatin-water system Am. Chem. Soc. Symp. Ser. 350 211-23

[33] Maquet J, Théveneau H, Djabourov M, Leblond J, Papon P 1986 State of water in gelatin solutions and gels: An 1H NMR investigation Polymer 27 1103-10

[34] Djabourov M, Maquet J, Theveneau H, Leblond J, Papon P 1985 Kinetics of gelation of aqueous gelatin solutions Br. Polym. J. 17 169-74

[35] Rees D A 1969 Structure, conformation, and mechanism in the formation of polysaccharide gels and networks Adv.Carbohydrate Chem.Biochem. 24 267-332

[36] Eagland D, Pilling G, Wheeler R G 1975 Studies of the collagen fold formation and gelation in solutions of a monodisperse alpha gelatin Discuss.Faraday Soc. 57 181-200

[37] De Deene Y, Hanselaer P, De Wagter C, Achten E, De Neve W 2000 An investigation of the chemical stability of a monomer/polymer gel dosimeter Phys. Med. Biol. 45 859-78

[38] Normand V, Muller S, Ravey J, Parker A 2000 Gelation kinetics of gelatin: A master curve and network modeling. Macromolecules 33 1063-71

[39] Lepage M, Whittaker A K, Rintoul L, Bäck S Å J, Baldock C 2001 The relationship between chemical processes and transverse relaxation times in polymer gel dosimeters Phys. Med. Biol. 46 1061-74

[40] Lepage M, Gore J C 2004 Contrast mechanisms in magnetic resonance imaging J. Phys.: Conf. Ser. 3 78-86

[41] Maryanski M J, Zastavker Y Z, Gore J C 1996 Radiation dose distributions in three dimensions from tomographic optical density scanning of polymer gels: II Optical properties of the BANG polymer gel Phys. Med. Biol. 41 2705-17

[42] Olding T, Holmes O, Schreiner L J 2008 Development of a quality assurances scattering phantom for cone beam optical CT Prelim. Proc. 5th Int. Conf. Radiother. Gel Dosim., Crete, Greece 169-73

[43] Lopatiuk-Tirpak O, Langen K M, Meeks S L, Kupelian P A, Zeidan O A, Maryanski M J

2008 Performance evaluation of an improved optical computed tomography polymer gel dosimeter system for 3D dose verification of static and dynamic phantom deliveries Med. Phys. 35 3847-59

[44] Hilts M, Jirasek A, Duzenli C 2005 Technical considerations for implementation of x-ray CT polymer gel dosimetry Phys Med Biol. 50 1727 – 45

[45] Koeva V I, McAuley K B, Jirasek A, Schreiner L J. 2008 Preliminary investigation of the NMR, optical and x-ray CT dose-response of polymer gel dosimeters with cosolvents and increased crosslinker levels. Prelim. Proc. 5th Int. Conf. Radiother. Gel Dosim., Crete, Greece.

[46] Baldock C, Lepage M, Bäck S Å J, Murry P J, Porter D, Kron T 2001 Dose resolution in radiotherapy polymer gel dosimetry: Effect of echo spacing in MRI pulse sequence Phys. Med. Biol. 46 449-60.

Formaldehyde increases MAGIC gel dosimeter melting point and sensitivity

Juliana P. Fernandes, Bruno F. Pastorello, Draulio B. de Araújo, Oswaldo Baffa

Departamento de Física e Matemática, FFCLRP, Universidade de São Paulo, Av. Bandeirantes 3900, 14040-901 – Ribeirão Preto, SP, Brazil.

juliana@biomag.usp.br

Abstract. Polymeric gel dosimeters are being used to verify three-dimensional (3D) dose distributions of different types of radiotherapy treatments, especially the most complexes ones. An important factor that can limit the wider use of this kind of dosimeter is temperature, as gel melting can destroy 3D information. This work shows that the addition of formaldehyde to the gel preparation increases the melting point, allowing its use in warmer environments, including up to body temperature. An addition of 3% in mass of the formaldehyde solution to a MAGIC type gel dosimeter increased its melting point from 25 to 69°C. Also important were a 10.5% increase in gel sensitivity and an expressive decrease in relaxation rate R2 uncertainty that added to the agreement of the depth dose distribution measurement with the expected dosimetric data of the LINAC show the potential applicability of this new dosimeter for 3D dose verification.

1. Introduction

Gel dosimeters are a new type of dosimeters for the evaluation of complex dose distribution in radiation therapy, with the unique characteristic of allowing three dimensional dose visualization [1]. The normoxic polymeric [2] gels are gaining special attention because they are easier to be manufactured compared to the gels initially proposed [3].

With the advances in treatment techniques that use ionizing radiation and the need of verifying these procedures before the treatment of the patients, gel dosimeters are really a promising alternative dosimetric system. However, there are several factors that may disturb their ability to precisely measure dose distributions, one of them is temperature [4], since at relatively low temperatures, around 25ºC, the gel, in its current formulation, can melt and 3D dose information is lost.

Madsen [5] used formaldehyde to increase the melting point of certain gels used in ultrasound phantoms simulations with good results. Studies to increase the gelatin melting temperature aiming the use of gelatin as drug capsules for oral intake were also performed by other authors [6,7,8,9]. These studies show that the presence of formaldehyde increases cross-linking, leading to an increase of melting temperature. Thus, motivated by these previous studies, we herein study the influence of formaldehyde in MAGIC gels as an alternative to increase its melting point and the influence of formaldehyde addition in the gel dosimetric response.

Journal of Physics: Conference Series **164** (2009) 012004 doi:10.1088/1742-6596/164/1/012004

2. Materials and Methods

2.1. Gel manufacturing

Gelatin (bovine skin, 250 Bloom, Gelita®) was added to water at room temperature and then heated to 45°C, where the mixture was kept until the gelatin had completely melted, after approximately 30 minutes. The heater was then turned off and the mixture was cooled to 35°C, when ascorbic acid (Vetec®), copper sulfate (Vetec®) and the formaldehyde (Merck®), as a water solution with 37% minimum, stabilized with 10% methanol, were added. All reagents were used without additional purification. After approximately 5 minutes the methacrylic acid (Acros®) was finally added. The solution was stirred continuously during the entire mixing procedure. The ideal concentration of formaldehyde to increase the melting point was studied. Table 1 shows the gel compositions analyzed.

Table 1 – Gel composition for a 50 ml volume with varying amounts of formaldehyde to assess its effect on gel's melting point. In all samples 4.1g of gelatin bovine skin, 250 Bloom; 17.6 mg of ascorbic acid; 2.95g of methacrylic acid and 1 mg of copper sulfate were used. The formaldehyde solution used contains a minimum of 37% and was stabilized in 10% methanol.

Sample	Ultra Pure Deionized Water (ml)	Formaldehyde Solution (ml)	Formaldehyde Solution (% mass)	Melting Point (°C)
1	42.00	---	---	25.0
2	41.50	0.50	1	24.5
3	41.25	0.75	1.5	26.5
4	41.00	1.00	2	27.0
5	40.75	1.25	2.5	29.0
6	40.50	1.50	3	69.0

2.1.1. *Melting point study.* To study the melting point, each sample (table 1) was poured in a 50ml beaker. All gels were stored in a refrigerator, at a temperature of 10ºC, for one day before use. The samples were heated and their temperatures were monitored with a digital thermometer until completely melting of the gel, we considered the melting point as the temperature in which the gel lost its gel appearance and became a viscous solution.

2.1.2. *Sensitivity study: MAGIC x MAGIC with formaldehyde.* To compare the sensitivity of MAGIC gels two sets of gels were prepared at the same time and with the same conditions, except for the addition of formaldehyde to one of the samples (same composition as sample 6, table 1) and the other was left without it (same composition as sample 1, table 1). For this study, the gels were poured into a 5ml cylindrical blood collection tube with low vacuum and sealed for further irradiation and imaging.

2.1.3. *Gelatin Bloom study.* As an alternative to increase the melting point of gels, Mcjury [10] have shown that using gelatin with higher strength, or Bloom, would increase the melting point of the gel. Hence, such influence was studied over MAGIC gels with formaldehyde. For this study, we prepared two samples of 50ml of gel (same composition as sample 6, table 1) using gelatin of 300 and 250 bloom. We also poured the gel into the 5ml cylindrical vacuum tubes.

2.1.3. *Depth Dose Measurement.* To measure the depth dose distribution we prepared 100 ml of gel (same proportions of sample 6, table 1) and poured the gel into a test-tube.

2.2. Irradiation

The gel phantoms for the sensitivity and the bloom study were irradiated with cobalt 60, in a Gammatron-Siemens unit. We used single beans with its central axis parallel to the diameter of the cylindrical tubes inserted in a plastic slab that assured the build-up region at the front surface of the gel. Doses of 1, 2, 3, 4, 5 and 10 Gy were delivered and one tube was not irradiated, as a reference. For the depth dose measurements a 6 MV photon beam delivered by a Varian 600CD linear

Journal of Physics: Conference Series **164** (2009) 012004 doi:10.1088/1742-6596/164/1/012004

accelerator was used. Each test-tube was irradiated separately in a water phantom which provided lateral scattering, with radiation incident normal to the bottom of the test-tube, following the work of Oldham *et al* [11]. Both equipments are routinely checked for its dose rate according to the 398 IAEA protocol, yielding an uncertainty smaller than 3%.

2.3. MRI Acquisition

MR images were acquired after the thermal equilibrium of the gel tubes with the MR scanner room temperature, 1 day after the irradiation, to allow enough time for reaction completion and uniform thermal equilibrium.

MRI images to evaluate R2 were acquired using a 1.5 T scanner (Siemens, Magneton Vision) with a head coil and single spin echo sequences with echo times of 22, 60 and 120 ms, a repetition time of 3000 ms and a matrix size of 128 × 256 pixels. The slice thickness was 5mm and the FOV was 240 mm. The transverse relaxation rate R2 was calculated using a specific program developed by our group [12] in MatLabR_ 6.5 (Mathworks Inc). The transverse relaxation rate R2 was evaluated on a pixel wise basis of a selected region of interest (ROI) and a histogram was used to assess its distribution. The ROI was selected on the entire gel area and a Gaussian distribution was assumed with its peak value representing the mean R2 value; its standard deviation was also calculated.

3. Results and Discussions

The presence of formaldehyde in the gel preparation increased the gel melting point as indicated in table 1. For gel preparation with 3% in volume concentration of the formaldehyde solution (sample 6) the melting point was 69°C, which is high enough to allow convenient gel manipulation in all clinical environments, without the need of a thermal protection during its transportation to prevent the loss of the dose distribution in the volume irradiated.

Comparing the radiation dose–response curves of MAGIC gel with and without formaldehyde (figure 1 - a), we can see that the gel containing formaldehyde presents a higher sensitivity and smaller errors bars for low and high dose points - this seems to indicate that the presence of this compound yield a more uniform gel. Formaldehyde increases gel melting point by increasing the cross-linking reactions in gelatin molecules [13,14] possibly requiring a higher thermal energy to break down the chemical bonds.

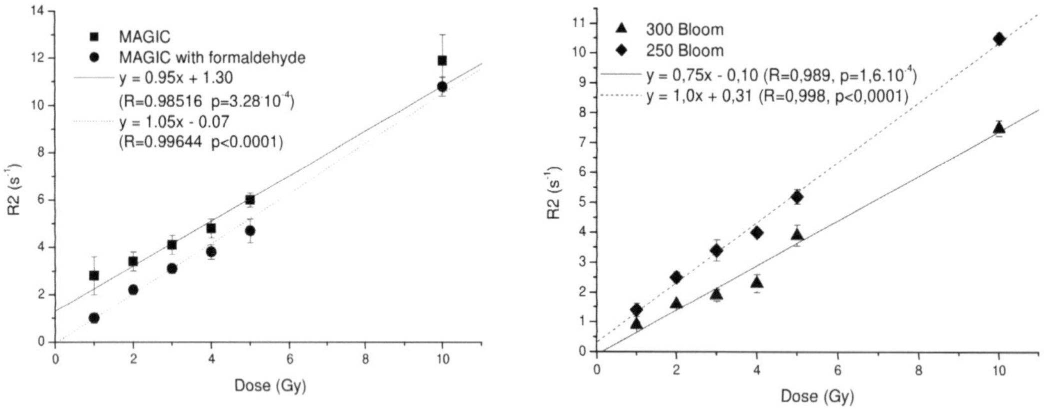

(a) (b)

Figure 1 – (a) Dose response curves for the two MAGIC gels, with (sample 6) and without formaldehyde (sample 1). (b) Influence of the gelatin Bloom in the sensitivity of the MAGIC gel dosimeters. It can be seen that the higher the Bloom the smaller is the sensitivity to radiation. For both figures, the lines connecting the experimental points represent a linear fit of the data and the error bars represent the standard deviation of the mean values in the selected ROI.

The increase in the gelatin bloom for increasing the melting point is not appropriate for our goals, since a decrease in the gels' sensitivity was also observed (figure 1 - b).

The test-tube image used to access the depth dose distribution and the R2 distribution are shown in figure 2 – (a) and (b). Thc plot of the depth dose normalized to its maximum value is in great concordance with the expected values according to the dosimetric data of the used LINAC (figure 2 – c), showing the potential of the MAGIC gel with formaldehyde to measure dose distributions.

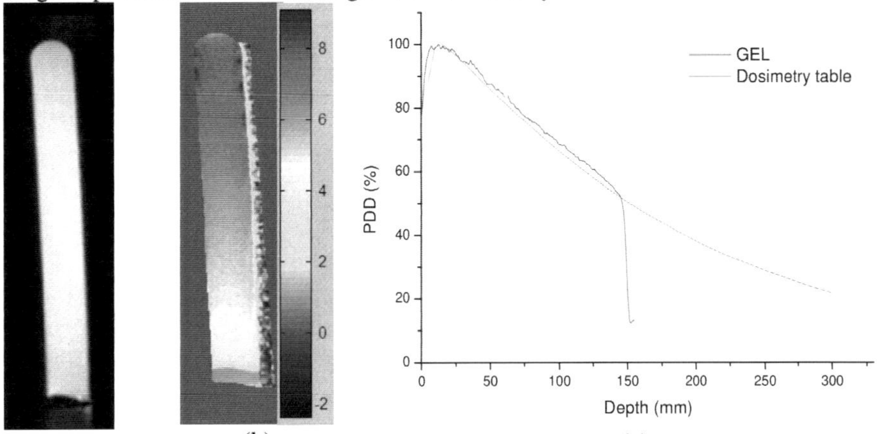

(a) (b) (c)

Figure 2 – (a) MRI image of the test-tube (b) R2 distribution in the MRI image (c) Depth dose distribution normalized to the maximum value compared to the dosimetry table.

4. Conclusions

The addition of formaldehyde to MAGIC type gel dosimeters increases the melting point, up to 69°C, and the dosimeter sensitivity by about 10.5% when compared to standard MAGIC gels. It also decreased the uncertainty in R2. The depth dose measurement in accordance to the dosimetric data confirms the potential of the new dosimeter to measure dose distributions. We can also say that the new formulation of this gel is more reliable for 3D dose distribution measurements and is easier to handle.

References

[1] De Deene Y, De Wagter C, B. Duyse V, Derycke S, De Neve W and Achten E. "Three-dimensional dosimetry using polymer gel and magnetic resonance imaging applied to the verification of conformal radiation therapy in head-and-neck cancer". Radiotherapy and Oncology 48 (1998) 283–291.

[2] Fong P M, Keil D C, Does M D and Gore J C. "Polymer gels for magnetic resonance imaging of radiation dose distributions at normal room atmosphere". Phys. Med. Biol. 46 (2001) 3105–3113.

[3] Maryanski M J, Schulz R J, Ibbott G S, Gatenby J C, Xie J, Horton D and Gore J C. "Magnetic resonance imaging of radiation dose distributions using a polymergel Dosimeter". Phys. Med. Biol. 39 (1994) 1437-55.

[4] De Deene Y, Vergote K, Claeys C and DeWagter C. "The fundamental radiation properties of normoxic polymer gel dosimeters: a comparison between a methacrylic acid based gel and acrylamide based gels". Phys. Med. Biol. 51 (2006) 653–673.

[5] Madsen EL, Zagzebski JA and Frank GR. "Oil-ingelatin Dispersions for use as Ultrasonically Tissue-Mimicking Materials". Ultrasound in Medicine and Biology 8 (1982) 277-287.

[6] Lefebvre DR, Han J, Lipari JM, Long MA, McSwain RL and Wells HC. "Dielectric Analysis for In-Situ Monitoring of Gelatin Renaturation and Crosslinking". Journal of Applied Polymer Science 101 (2006) 2765–2775.
[7] Gold TB, Buice RG, Lodder RA and Digenis GA. "Determination of extent of formaldehyde-induced crosslinking in hard gelatin capsules by near-infrared spectrophotometry". Parmaceutical Research 14 (1997) 1046-50.
[8] Ofner III CM, Zhang Y, Jobeck VC, Bowman BJ. "Crosslinking Studies in Gelatin Capsules Treated with Formaldehyde and in Capsules Exposed to Elevated Temperature and Humidity". Journal of Pharmaceutical Sciences 90 (2001) 79-88.
[9] Tengroth C, Gasslander U, Andersson FO and Jacobsson SP. "Cross-Linking of Gelatin Capsules with Formaldehyde and Other Aldehydes: An FTIR Spectroscopy Study". Pharmaceutical Development and Technology 10 (2005) 405–412.
[10] Mcjury M, Oldham M, Cosgrove VP, Murphy PS, Doran S, Leach MO and Webb S. "Radiation dosimetry using polymer gels: methods and applications". Br. J. Radiol. 73 (2000) 919–29.
[11] Oldham M, McJury M, Baustert I B, Webby S and Leach MO. "Improving calibration accuracy in gel dosimetry". Phys. Med. Biol. 43 (1998) 2709–2720.
[12] Carneiro AAO, Fernandes JP, de Araújo DB, Elias JJr, Martinelli ALC, Covas DT, Zago MA, Ângulo IL, St. Pierre TG and Baffa O. "Liver Iron Concentration Evaluated by Two Magnetic Methods: Magnetic Resonance Imaging and Magnetic Susceptometry". Magnetic Resonance in Medicine 54 (2005) 122–128.
[13] Hall TJ, Bilgen M, Insana MF and Krouskop TA. "Phantom materials for elastography". IEEE Trans. Ultrason. Ferroelectr. Freq. Control 44 (1997) 1355–65.
[14] Rice JR, Milbrandt RH, Madsen EL, Frank GR, Boote EJ and Blechinger JC. "Anthropomorphic 1H MRS head phantom". Med. Phys. 25 (1998) 1145–56.

Journal of Physics: Conference Series **164** (2009) 012005 doi:10.1088/1742-6596/164/1/012005

Characterization of a normoxic polyacrylamide gel using MRI and optical CT

Malcolm Heard[1], Geoffrey Ibbott[1], David Followill[1], R. Allen White[2], Edward Jackson[3], Mohammad Salehpour[1]

[1]Radiation Physics, University of Texas M.D. Anderson Cancer Center, 1515 Holcombe Blvd, Houston, Texas 77030
[2]Biostatics and Applied Math, University of Texas M.D. Anderson Cancer Center, 1515 Holcombe Blvd, Houston, Texas 77030
[3]Imaging Physics, University of Texas M.D. Anderson Cancer Center, 1515 Holcombe Blvd, Houston, Texas 77030

mheard@mdanderson.org

1. Introduction

The field of gel dosimetry grew out of the observation that changes in certain gel formulations in response to radiation could be demonstrated through magnetic resonance imaging (MRI) [1]. MRI continues to be the most widely used method of evaluating gels. The major disadvantages of MRI are the difficulties experienced in gaining access to an imager and the expertise necessary to achieve the accurate quantitative imaging required for 3D dosimetry. Since 1993, the availability of MR scanners has significantly increased. Presently, several radiotherapy departments are investigating the use of MRI for treatment planning [2]. The increased availability of MRI should increase the access for gel imaging. The 3D dosimetry community, through DOSGEL meetings and the discussion forum, has helped to inform users on the optimal techniques for the imaging of polymer gels with MRI.

Gore et al. [3] developed an optical CT system to provide an alternative to MRI for imaging polymer gels. The idea was to provide a method for imaging the gels that would be easily accessible and low-cost. A major limitation of optical imaging has been the amount of time required to perform the imaging. The development of new optical scanners and upgrades to the designs of already existing models has significantly reduced the imaging time.

Both imaging modalities have been used in studies that characterized polymer gel formulations and in studies demonstrating the ability of polymer gels to be used as a clinical tool. Despite the growth in both areas, few studies have been done using both imaging modalities. The purpose of this study was to characterize a polymer gel dosimeters using both MRI and optical CT and to compare the characteristics of the two imaging modalities.

2. Materials and Methods

For this study, a PAGAT gel containing 4.5% *N,N'*–methylene-bis-acrylamide (bis), 4.5% acrylamide (AA), 5% gelatine, 5 mM tetrakis (hydroxymethyl) phosphonium chloride (THPC), 0.01 mM hydroquinone (HQ) and 86% H_2O was manufactured according to procedures developed by Venning et al. [4]. A slight change was made to the manufacturing procedure as the gels were not made in normal atmospheric conditions. Instead, all gels were made in an AtmosBag™ (Sigma-Aldrich,

Milwaukee, WI) in a nitrogen environment. The gels were mixed and poured in PET jars, 5 cm in diameter and 5 cm in height, and sealed. The gels were immediately placed in zip-top bags and sealed before removal to normal atmospheric conditions. This change to the manufacturing procedure prevented oxygen contamination through the polypropylene cap of the PET jar. The gels were stored for 24 hours before irradiation.

Four different batches of PAGAT gel were made for this study. Three batches were used to determine the dose response, inter-batch reproducibility, and dose resolution. Each dosimeter was placed in a water tank and irradiated using a 20 x 20 cm^2 pair of parallel-opposed 6 MV beams. Each batch of dosimeters was irradiated to doses from 1 Gy to 10 Gy. The fourth batch of gel was used to evaluate the spatial stability of the gel and the intra-batch reproducibility. Each gel was placed in a water tank and irradiated using a half-blocked 5 x 10 cm^2 field from a 6 MV accelerator to create a steep dose gradient. Three dosimeters were irradiated to 2 Gy at a depth of 5 cm, three dosimeters were irradiated to 5 Gy at a depth of 5 cm, and three dosimeters were irradiated to 8.25 Gy at a depth of 5 cm. Radiochromic film was also irradiated using the same beam configuration in a phantom of water-equivalent plastic. Dose levels of 2 Gy, 5 Gy, and 8.25 Gy were delivered to the film at a depth of 5 cm.

Immediately following irradiation, all gels were placed in the MRI scanning room to equilibrate to room temperature overnight. Imaging was performed 24 hours after irradiation using a Bruker 4.7 T research MRI scanner. T_2 weighted imaging was performed using 64 equidistant spin-echoes with a TE of 9 ms, TR of 6500 ms, slice thickness of 2 mm, matrix size of 64 x 64, FOV of 80 mm in both the frequency and phase encoding directions with one acquisition. The base images were transferred to a personal computer and fitted to a mono-exponential decay curve using MATLAB™ (The Math Works, Inc.). Additional software was written in MATLAB™ to compute R_2 maps and perform additional analysis.

Immediately after MR imaging of the gels, they were transferred to the optical CT room. The fast-scanning platform of the OCT-OPUS™ laser CT scanner (MGS Research Inc., Madison, CT) was used. An in-plane resolution of 1 x 1 mm^2 was used.

3. Results and Discussion

The dose response of three different batches of PAGAT gel is displayed in Figures 1 and 3. The dose response using optical CT could only be determined up to 7 Gy as dosimeters irradiated to higher doses were too optically dense to be imaged. The response of the dosimeters at each dose level was compared to determine the inter-batch reproducibility. The differences in R_2 were as large as 4.6% at 5 Gy and the differences in net OD were as large as 20% at 7 Gy. Figures 2 and 4 show the dose resolution using a 95% confidence limit. The best dose resolution was achieved when optical CT was used because of the large Δ net OD/Δ Dose.

Figures 5-8 are profiles from gels irradiated with a half-blocked field. The profiles were normalized at 15 mm from the central axis. Figures 5 and 6 show excellent agreement between film and PAGAT gel. No dose overshoots were observed near the steep dose gradient. Figure 7 and 8 show disagreement between film and gel in the low dose region. These differences are attributed to the performance of the optical CT system when measuring low OD values. The uncertainty in the determination of the OD can be as large 5.5%. In addition, the gels with low OD values are affected by a ring artifact that resulted in the "wave" structure seen in the 2 Gy profile. Figure 9 displays the maximum percent difference between dosimeters within the same batch irradiated to the same dose. Difference as large as 3.7% were observed for gels imaged with optical CT and differences as large as 7.3% were observed in gels imaged with MRI. The data suggest that these dosimeters should be used as relative dosimeters since there can be significant differences in the response of gels within the same batch.

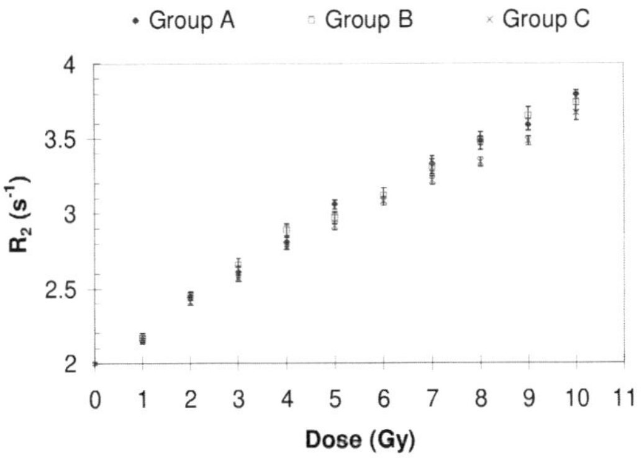

Figure 1. Dose response of 3 batches of PAGAT gel imaged with MRI. Error bars represent one standard deviation.

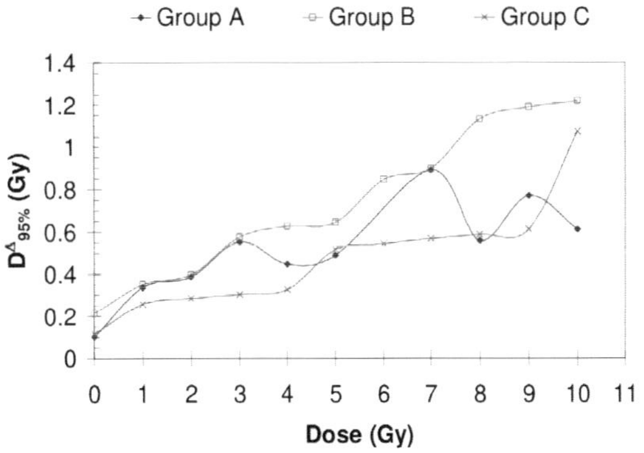

Figure 2. Dose resolution of 3 batches of PAGAT gel imaged with MRI.

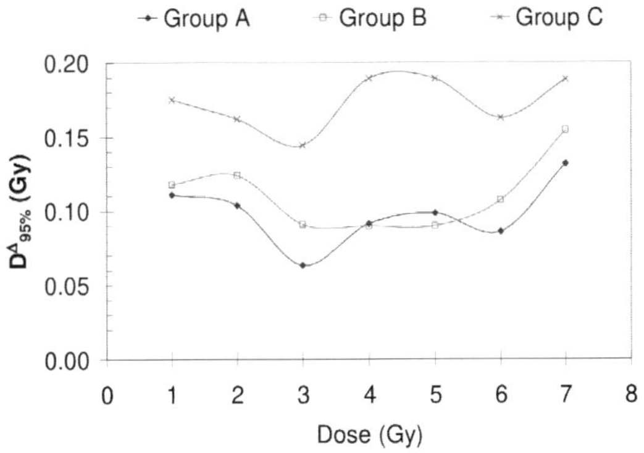

Figure 3. Dose response of 3 batches of PAGAT gel imaged with optical CT. Error bars represent one standard deviation.

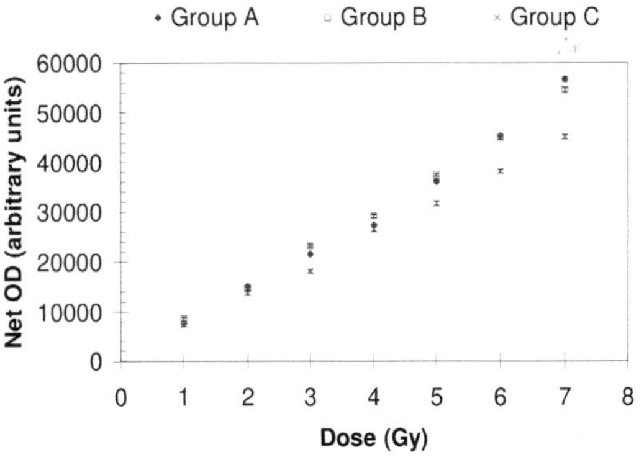

Figure 4. Dose resolution of 3 batches of PAGAT gel imaged with optical CT.

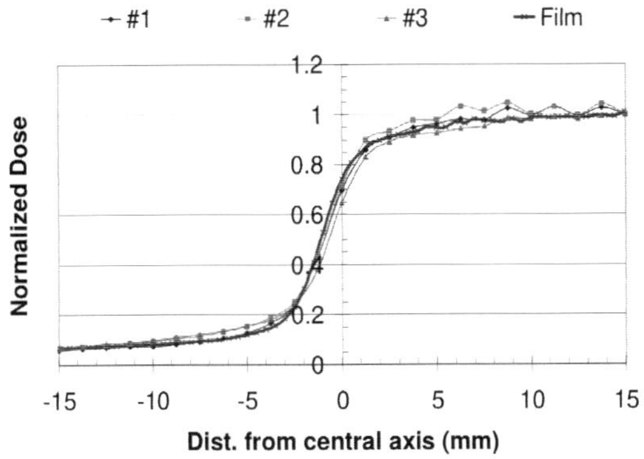

Figure 5. Profile through gels irradiated to 8.25 Gy with a half-blocked field and imaged with MRI.

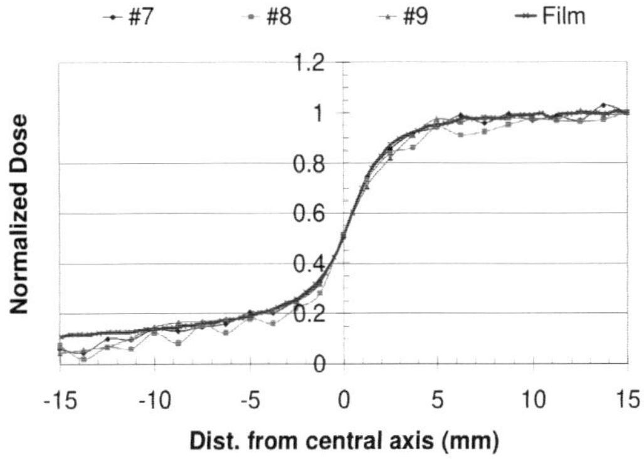

Figure 6. Profile through gels irradiated to 2 Gy with a half-blocked field and imaged with MRI.

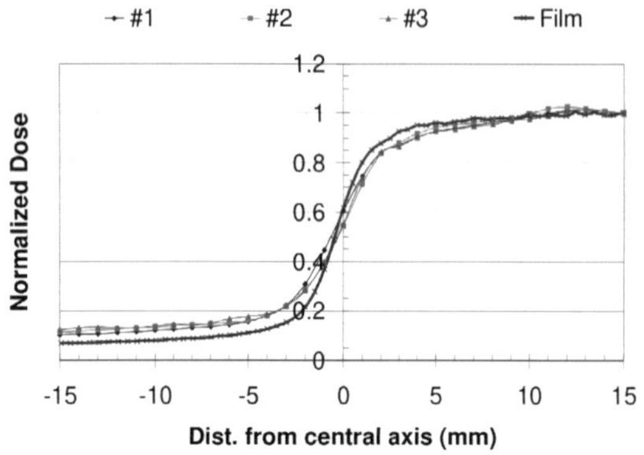

Figure 7. Profile through gels irradiated to 8.25 Gy with a half-blocked field and imaged with optical CT.

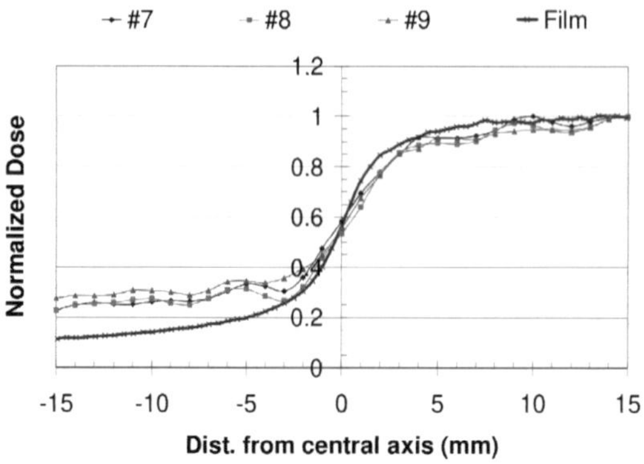

Figure 8. Profile through gels irradiated to 2 Gy with a half-blocked field and imaged with optical CT.

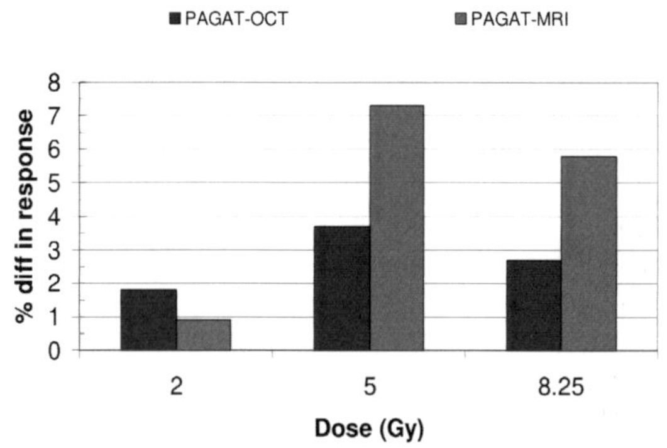

Figure 9. Intra-batch reproducibility of PAGAT gel.

References

[1] Maryanski, M. J., Gore, J. C., Kennan, R. P., and Schulz, R. J., 1993. NMR relaxation enhancement in gels polymerized and cross-linked by ionizing radiation: a new approach to 3D dosimetry by MRI. Magnetic Resonance Imaging **11** 253-258.

[2] Chen, L., Price Jr., R. A., Nguyen, T. B., Wang, L., Li, J. S., Qin, L., Ding, M., Palacio, E., Ma, C. M., and Pollack, A., 2004. Dosimetric evaluation of MRI-based treatment planning for prostate cancer. Phys. Med. Biol. **49**, 5157–5170.

[3] Gore, J. C., Ranade, M., Maryanski, M. J., and Schulz, R. J., 1996. Radiation dose distributions in three dimensions from tomographic optical density scanning of polymer gels: I. Development of an optical scanner. Phys. Med. Biol. **41** 2695-2704.

[4] Venning, A. J., Hill, B., Brindha, S., Healey, B. J., and Baldock, C., 2005. Investigation of the PAGAT polymer gel dosimeter using magnetic resonance imaging. Phys. Med. Biol. **50** 3875-3888.

Evaluation of the radiation-sensitizer/protector and/or antioxidant efficiencies using Fricke and PAG dosimeters

Ridthee Meesat, Jean-Paul Jay-Gerin, Abdelouahed Khalil, Martin Lepage

Département de médecine nucléaire et de radiobiologie, Faculté de médecine et des sciences de la santé, Université de Sherbrooke, Sherbrooke (Québec) J1H 5N4, Canada

Martin.Lepage@USherbrooke.ca

Abstract. In this study, our aim is to assess the potential of Fricke and polyacrylamide gel (PAG) dosimeters to quantitatively evaluate the efficiency of potential radiation sensitizers/protectors and antioxidants. These compounds are of importance in radiotherapy as well as in disease prevention and promotion of health. The basic principle of the Fricke dosimeter is the radiation-induced oxidation of Fe^{2+} to Fe^{3+} in an aerated aqueous 0.4 *M* H_2SO_4. The production of ferric ions is most sensitive to the radical species produced in the radiolysis of water. Using this method, we observed that cystamine (one of the best of the known radioprotectors) can prevent oxydation of Fe^{2+} from reactive radiolysis species. However, one obvious disadvantage of the Fricke dosimeter is that it operates under highly acidic conditions (pH 0.46), which may degrade biological compounds. In contrast, the pH of the polyacrylamide gel (PAG) dosimeter is almost neutral, such that degradation of compounds is less probable. A change in R_2-dose sensitivity was observed in the presence of radiosensitizers/radioprotectors and antioxidants. The protective effect of Trolox (a well-known antioxidant) and thiourea (a radioprotector) was readily observed using the PAG dosimeter. Incorporation of iodinated radiation sensitizers such as NaI and an iodine contrast agent led to a quantifiable sensitizer enhancement ratio. These studies suggest that the Fricke and the PAG dosimeters have the potential to evaluate the efficiency of radiation sensitizers/protectors and antioxidants.

1. Introduction

Radiation sensitizers and radiation protectors play an important role in clinical radiotherapy [1,2]. It is therefore important to understand the mechanisms of the action of such compounds in order to help better control their biological effects. For example, a chemical compound designed at protecting DNA from radiation damage needs not only to reach its cellular target location but also retain its radioprotecting activity to produce an effective response. Using clonogenic assays, the radiation survival of cells can be measured, but this only provides information on the addition of these two properties. If a compound proves to be less effective than anticipated, is it because it is in fact not a good radioprotector, or is it that it did not reach its target location? This study aims at characterizing compounds at the molecular or chemical level, such that this information can aid in the design and the optimization of compounds that will be therapeutically useful *in vivo*.

Radiosensitizers are intended to enhance tumor cell killing while having little effect on normal tissues. Enhancement of the radiation damage by radiosensitizers (also called "dose enhancers")

originates from the direct action of radiation on the sensitizing compounds. This mechanism of radiosensitization can generate a variety of toxic effects: production of secondary electrons and Auger electron cascades [3], X-ray fluorescence [3], high-LET α-particles and charged fragments [4] as well as free-radical species [5]. These products can damage critical cellular targets such as DNA, RNA, proteins or lipids, increasing the extent of initial radiation injury and thus improving tumor killing. Inversely chemical radioprotectors act by a decreasing radiosensitivity, especially of normal tissues [6]. This may be associated to the capacity of the radioprotector to scavenge free radicals produced by the ionizing radiation (antioxidant activity).

In this study, we have developed a method, fast and easy to apply, to assess the efficiencies of radiation sensitizers and protectors (antioxidants), based on the use of the Fricke and polyacrylamide gel dosimeters.

1.1. The Fricke dosimeter

The Fricke dosimeter has found extensive use because it gives absolute results with a high accuracy. The basic principle of this chemical dosimeter is the radiation-induced oxidation of Fe^{2+} to Fe^{3+} in an aerated aqueous H_2SO_4 solution. The production of ferric ions is most sensitive to the radical species produced in the radiolysis of water. The presence of any other compounds during irradiation may affect the availability of those free radicals and will therefore change the radiation chemical yield of Fe^{3+} [$G(Fe^{3+})$]. Normally, under low-linear-energy-transfer (LET) irradiation conditions, $G(Fe^{3+})$ is equal to 15.6 ± 0.3 molecules/100 eV. In the presence of a radioprotector (or an antioxidant), $G(Fe^{3+})$ is expected to be lower than 15.6, while in the case of a radiosensitizer, it is expected to be higher than 15.6. Using this method, Jayson and Wilbraham [7] have studied the protective action of cystamine, a well-known biological radiation-protective compound. Their results showed that cystamine can prevent oxidation of Fe^{2+}, as observed by a lower yield of Fe^{3+}. In this case, cystamine and Fe^{2+} competed in the reaction with oxidizing radicals. The Fricke dosimeter was used more recently by Herold *et al.* [8] who measured a dose enhancement from gold microspheres in solutions irradiated with 200 kVp X-rays and ^{137}Cs γ-rays. Together, these studies suggest that the Fricke dosimeter could be a useful tool to evaluate the radiation protective (antioxidant) and radiosensitizing efficiency of water-soluble compounds.

1.2. The polyacrylamide gel (PAG) dosimeter

The PAG dosimeter is a three-dimensional radiation dosimeter that is composed of monomers such as acrylamide (AA) and N,N'-methylene-bis-acrylamide (BIS) dissolved in an aqueous gelatin matrix [9]. The principle of the dosimeter is based on the polymerization of the monomers initiated by radical species produced by the radiolysis of water [10]. In the evaluation of polymer gel dosimeters by magnetic resonance imaging (MRI), the spin-spin (or transverse) relaxation time (T_2) of protons in the gel following irradiation can be determined in two or three dimensions. The quasi-linear relationship between the inverse of T_2 (R_2) and the absorbed dose, at low doses, is called the "R_2-dose sensitivity" [11]. More radical species are formed when a radiosensitizer (or "dose enhancer") is added to the gel dosimeter and this should increase the extent of polymerization for a given dose. Under these conditions, the R_2-dose sensitivity will be increased. In a recent study, Boudou *et al.* [12] have used this method to measure the dose enhancement in polymer gels doped with iodine (a high-Z element). Unlike radiosensitizers, radioprotectors and antioxidants can scavenge radiation-induced water free radicals, such that their presence is expected to lead to a decrease of the R_2-dose sensitivity.

2. Experimental procedure

2.1. Fricke dosimeter

The chemicals used in the Fricke dosimeter solution are ferrous ammonium sulfate (as the hexahydrated salt) (99.99 %, Sigma Aldrich), sulfuric acid (98%, Sigma Aldrich), and double-distilled water. The solutions were prepared from 0.342 g of $Fe(NH_4)_2(SO_4)_2.6H_2O$ dissolved in 0.4 *M* H_2SO_4.

The concentration of ferrous ions in the solution was 1 m*M* and the density of the solution was 1.024 g/cm^3. Special attention was given to cleaning glassware according to the standard procedures used in radiation chemistry.

The samples were irradiated up to 80 Gy at 25°C by ^{60}Co γ-rays (1.25 MeV) in a Gammacell 220 (Atomic Energy of Canada Limited). The dose rate was 3.44 Gy/min.

The concentration of ferric ions was calculated from the increase in optical density (absorbance) measured by UV spectrophotometry at 304 nm using a value of 2205 M^{-1} cm^{-1} for the molar extinction coefficient of Fe^{3+}. $G(Fe^{3+})$ was obtained from the gradient of the linear curve between the Fe^{3+} concentration and the absorbed dose.

2.2. PAG dosimeter

The dosimeter gels were prepared with acrylamide (AA) and N,N'-methylene-bis-acrylamide (BIS, 99+%, electrophoresis grade, Aldrich), gelatin (~300 bloom, Aldrich), and water (de-ionized). Gelatin was added to water at room temperature and left to soak for 10 min and subsequently heated and maintained at 45°C. The monomers (AA and BIS) were successively added and magnetically stirred for typically 15 min until complete dissolution. The gels were prepared under a controlled N_2 atmosphere inside a glove box. The solution was poured into glass vials having a Teflon-lined screw-top cap [10,11].

The samples were irradiated up to 50 Gy at 25°C in the ^{60}Co Gammacell 220 (see above) or with a 150 kVp X-ray therapy system (Therapax HF150T) whose dose rate was 0.86 Gy/min.

The T_2 relaxation times of water protons in the gel samples were determined using a head coil in a Siemens Sonata 1.5 T MRI scanner. The sample vials were imaged with a single slice in the coronal (horizontal) plane at 21°C using a multi-echo spin-echo pulse sequence. The R_2 (R_2 = $1/T_2$)-dose sensitivity was obtained for gels containing various concentrations of selected compounds.

3. Results and discussion

3.1. Fricke dosimeter: Effect of cystamine (a radioprotector/antioxidant)

Figure 1 shows a marked decrease of $G(Fe^{3+})$ upon addition of cystamine in the presence or in the absence of oxygen in the Fricke dosimeter. Our results clearly confirm the earlier data of Jayson and Wilbraham [7] showing the radiation protective property of this compound.

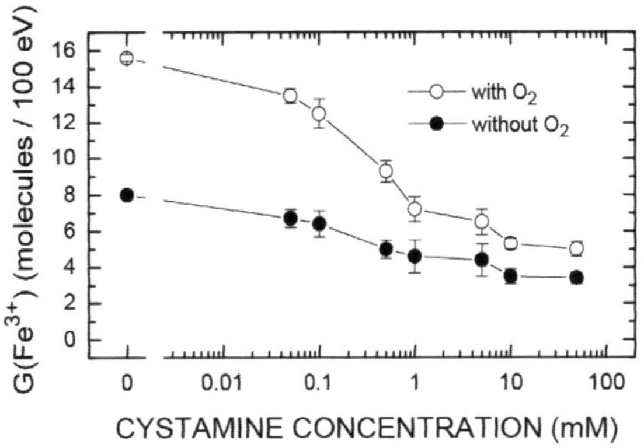

Figure 1. Plot of $G(Fe^{3+})$ as a function of the concentration of added cystamine in the presence and in the absence of oxygen.

3.2. PAG dosimeter: Effect of Trolox (a radioprotector/antioxidant)

Trolox is a water-soluble tocopherol (vitamin E) analogue, well-known for its antioxidant properties. The data obtained with the PAG dosimeter irradiated with ^{60}Co γ-rays in the presence of various

Journal of Physics: Conference Series **164** (2009) 012006 doi:10.1088/1742-6596/164/1/012006

concentrations of Trolox show a clear decrease in the R_2-dose sensitivity, which varies from 0.31 to 0.20 (± 0.01) s^{-1} Gy^{-1} in the dose range ~1-5 Gy (see Fig. 2). Similar results were observed with thiourea, another radiation-protective compound (results not shown). These results provide convincing evidence of the reliability of our method of protection evaluation.

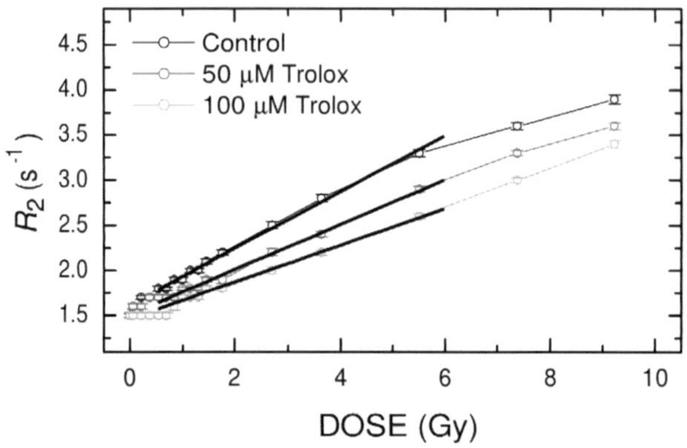

Figure 2. Plot of R_2 against absorbed dose for two different concentrations of Trolox.

3.3. PAG dosimeter: Effect of NaI and an iodinated contrast agent (ICA) (two radiosensitizers)

Several studies have reported the radiosensitization properties of iodine and some iodinated compounds for clinical use, such as iododeoxyuridine (IUdR) [13]. We first monitored the effect of iodine by adding sodium iodide (NaI) in various concentrations to the dosimeter gel and irradiating the system with X-rays from the Therapax HF150T machine. The average energy of photons was ~40 keV. Typically, at an absorbed dose of 5 Gy, the results shown in Fig. 3a indicate that the R_2-dose sensitivity decreases (taking the control line in the absence of NaI as a reference) at the lowest NaI

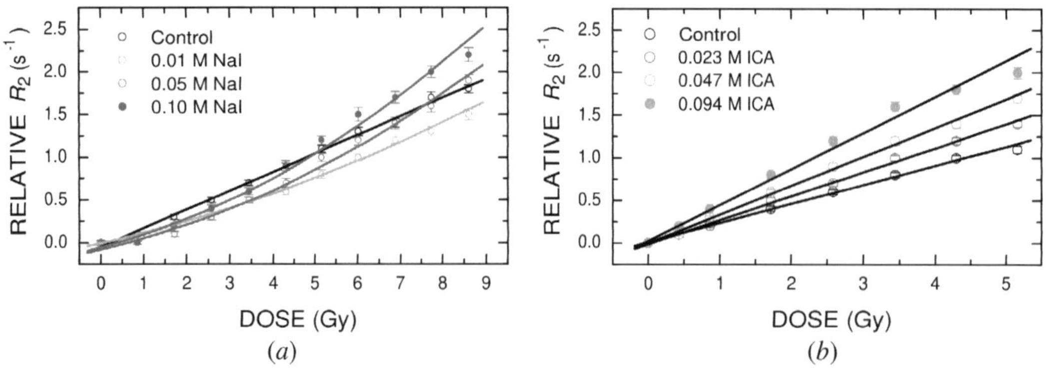

Figure 3. Plot of relative R2 against absorbed dose for three different concentrations of (*a*) NaI and (*b*) iodinated contrast agent (ICA).

concentration considered (0.01 *M*) and then increases at our highest concentrations (0.05 and 0.1 *M*). This peculiar behavior can be explained as follows. At low enough concentrations, NaI exhibits protective properties. Indeed, in neutral water, NaI is dissociated into Na^+ and I^- such that I^- can rapidly scavenge (via electron-transfer-type reactions) the radiolytically formed OH radicals [14] thus preventing the initiation of the polymerization reaction. However, as the NaI concentration increases, the direct action of the radiation on iodine becomes progressively dominant and radiosensitizing effects can be observed. This interpretation was confirmed by using iothalamate meglumine

(Conray®30), an ICA, where the iodine atom is covalently bound to the molecule and is therefore not present as I^- in the gel dosimeter. Figure 3b shows a clear increase of the R_2-dose sensitivity from 0.23 s^{-1} Gy^{-1} in the absence of ICA to 0.42 s^{-1} Gy^{-1} at an ICA concentration of 0.094 *M*. At a given radiation dose, this sensitizing property of ICA increases as the drug concentration increases, as expected.

4. Conclusion

We conclude from these experiments that the Fricke and the PAG dosimeters can be used to evaluate quantitatively the efficiency of radiation-sensitizers/protectors and antioxidants. Work is currently in progress to confirm the present results in biochemical and biological systems so that the reliability of our method can be firmly established.

References

[1] P. Wardman, Chemical radiosensitizers for use in radiotherapy. *Clin. Oncol.* **19**, 397-417 (2007).
[2] C. K. K. Nair, D. K. Parida and T. Nomura, Radioprotectors in radiotherapy. *J. Radiat. Res.* **42**, 21-37 (2001).
[3] H. H. Ertl, L. E. Feinendegen and H. J. Heiniger, Iodine-125, a tracer in cell biology: Physical properties and biological aspects. *Phys. Med. Biol.* **15**, 447-456 (1970).
[4] R. F. Barth, J. A. Coderre, M. G. H. Vicente and T. E. Blue, Boron neutron capture therapy of cancer: Current status and future prospects. *Clin. Cancer Res.* **11**, 3987-4002 (2005).
[5] P. Wardman, The mechanism of radiosensitization by electron-affinic compounds. *Radiat. Phys. Chem.* **30**, 423-432 (1987).
[6] C. F. Dunne-Daly, Principles of radiotherapy and radiobiology. *Semin. Oncol. Nurs.* **15**, 250-259 (1999).
[7] G. G. Jayson and A. C. Wilbraham, The utilisation of the Fricke dosimeter for evaluating the biological radiation-protective potential of water-soluble organic compounds. *Chem. Commun.* 641-642 (1968).
[8] D. M. Herold, I. J. Das, C. C. Stobbe, R. V. Iyer and J. D. Chapman, Gold microspheres: A selective technique for producing biologically effective dose enhancement. *Int. J. Radiat. Biol.* **76**, 1357-1364 (2000).
[9] M. J. Maryanski, J. C. Gore, R. P. Kennan and R. J. Schulz, NMR relaxation enhancement in gels polymerized and cross-linked by ionizing radiation: A new approach to 3D dosimetry by MRI. *Magn. Reson. Imaging* **11**, 253-258 (1993).
[10] M. Lepage, A. K. Whittaker, L. Rintoul, S. Å. J. Bäck and C. Baldock, The relationship between radiation-induced chemical processes and transverse relaxation times in polymer gel dosimeters. *Phys. Med. Biol.* **46**, 1061-1074 (2001).
[11] M. Lepage, P. M. Jayasakera, S. Å. J. Bäck and C. Baldock, Dose resolution optimization of polymer gel dosimeters using different monomers. *Phys. Med. Biol.* **46**, 2665-2680 (2001).
[12] C. Boudou, I. Troprès, J. Rousseau, L. Lamalle, J. F. Adam, F. Estève and H. Elleaume, Polymer gel dosimetry for synchrotron stereotactic radiotherapy and iodine dose-enhancement measurements. *Phys. Med. Biol.* **52**, 4881-4892 (2007).
[13] R. Nath, P. Bongiorni and S. Rockwell, Iododeoxyuridine radiosensitization by low- and high-energy photons for brachytherapy dose rates. *Radiat. Res.* **124**, 249-258 (1990).
[14] M. Quintiliani, Physico-chemical basis of radiosensitization by iodine compounds. *Radiat. Phys. Chem.* **30**, 409-422 (1987).

A procedure to mathematically amend possible thickness disuniformities in gel-layer dosimetry

Mauro Carrara[1], Grazia Gambarini[2], Giacomo Bartesaghi[2], Carlo Fallai[3], Anna Negri[2]

[1]Medical Physics Unit, Fondazione IRCCS "Istituto Nazionale Tumori", Milan, Italy

[2] Physics Department of the Università degli Studi and INFN, Milan, Italy

[3]Radiotherapy Unit, Fondazione IRCCS "Istituto Nazionale Tumori", Milan, Italy

Email: Grazia.Gambarini@mi.infn.it

Abstract. The continuous development of conformal radiotherapies requires a corresponding improvement of dosimetric techniques. Fricke gel-layer dosimetry, coupled with a suitably developed software, has proven to be a reliable technique for 3D absorbed dose distribution verification. The method is based on the evaluation of the light transmitted by gel-layers due to the fact that th optical density difference between irradiated and non irraddiated dosimeters is proportional to the absorbed dose. Measurement are performed by means of a planar illumination source and a computer-controlled CCD camera. Recently, adoption of gel layers with increased superficial dimensions has evidenced a possible thickness disuniformity of layers, thus introducing a new source of measurement inaccuracy. In this work, a method to mathematically amend this possible source of error is proposed. Dose profiles along the central axis of phantoms irradiated with X-ray therapy and Boron Neutron Capture Therapy fields, obtained with and without the application of the proposed correction procedure, were compared with ionization chamber measurements and Monte Carlo simulations, respectively. The obtained results show a good reliability of the proposed procedure.

1. Introduction

Gel dosimetry has proven to be a valid dose distribution verification system for various radiotherapy applications [1]. Different kinds of dosimeters have been proposed by the scientific community, and different kinds of imaging techniques have been developed [2]. In the last years, our group has significantly improved the Fricke gel-layer dosimetry technique by defining their best chemical composition [3], by optimizing the image acquisition assessment [4] and by developing a dedicated software for image analysis [5]. These improvements allowed to adopt Fricke gel-layers for both X-ray radiotherapy [6] and BNCT [7] dose distribution measurements. A particular attention was dedicated to detect possible systematic or random sources of errors which might determine inaccuracies in the evaluation of the delivered dose distributions. Algorithms for artifacts amendment due to eventual small defects in the gel structures or inhomogeneous illumination of the sample were proposed [8]. Recently, adoption of gel layers with increased superficial dimensions has shown that the layers thickness might not be uniform, thus introducing a further source of inaccuracy. In this work, a method to mathematically correct this source of error is proposed.

2. Materials and methods

2.1. Fricke gel-layer dosimeters and experimental assessment

The adopted tissue-equivalent gel dosimeters were radiochromic Fricke gels in form of layers having a thickness of 3mm and a 11x11cm^2 surface. Their standard composition was: Porcine skin in the amount of 3% of the final weight, ferrous sulphate solution [1mM $Fe(NH_4)_2(SO_4)_2 \cdot 6H_2O$], sulphuric acid [25mM H_2SO_4] and Xylenol-Orange [0.165mM $C_{31}H_{27}N_2Na_5O_{13}S$]. The experimental assessment consisted of a planar illumination source, a CCD camera for the acquisition of optical transmission images of the gel layers and a computer for the CCD control and for the storage and elaboration of the acquired images. Further details regarding the instrumentation may be found elsewhere [3].

The method for attaining images of dose distribution is based mainly on the following steps: i) grey-level images of the transmitted light at about 585nm are detected for each dosimeter before ($GL_{585,b}$) and after ($GL_{585,a}$) irradiation; ii) the difference in optical density $\Delta(OD)$, which is linearly correlated to the dose D until saturation, is evaluated for each pixel with coordinates (m,n) applying the function

$$\Delta(OD)(\mathrm{m,n}) = \log\,[GL_{585,b}(\mathrm{m,n}) \,/\, GL_{585,a}(\mathrm{m,n})] \qquad (1)$$

iii) knowing the linearity factor between D and $\Delta(OD)$, superficial matrices of dose distribution are finally obtained.

A dedicated software was developed with the programming environment MATLAB® (The MathWorks Inc, Natick MA, USA) to automatically perform steps i) to iii) and to interactively determine dose profiles, surfaces as well as isodose curves [4].

2.2. Thickness disuniformity correction of gel-layers

In this paper, a method to enhance the accuracy of the previously described procedure is proposed. The method is based on the acquisition of a further light-transmittance image of the non irradiated gel-layer with an optical band-pass filter centered around 430nm ($GL_{430,b}$). In fact, assuming that before irradiation ferrous ions inside the dosimeters are homogeneously concentrated, possible differences in gray level values of the image are due only to a thickness inhomogeneity of the layer. The additional information can be adopted to develop normalization matrices N_b (before irradiation) and N_a (after irradiation) which can be applied for amending the images acquired at 585nm. Funcion (1) becomes:

$$\Delta(OD_{cor})(\mathrm{m,n}) = \log\,[GL_{585,b}(\mathrm{m,n}) \cdot N_b(\mathrm{m,n}) \,/\, GL_{585,a}(\mathrm{m,n}) \cdot N_a(\mathrm{m,n})] \qquad (2)$$

The obtained algorithm (2) was integrated in the previously developed software for the achievement of planar and volumetric dose distributions. Figure 1 illustrates the image processing chain of a single gel layer applying algorithms (1) and (2). Superficial representations of matrices $GL_{585,b}$(m,n), $GL_{585,a}$(m,n), N_b(m,n), N_a(m,n), $GL_{585,b}$(m,n) $\cdot$ N_b(m,n), $GL_{585,a}$(m,n) $\cdot$ N_a(m,n), $\Delta(OD)$(m,n) and $\Delta(OD_{cor})$(m,n) are shown.

2.3. Depth dose measurements

Experiments were performed in order to check the reliability of the proposed method. Depth dose profiles in a cubic polystyrene phantom (30 cm side) were detected at the center of a 10x10cm^2 X-beam field (Varian Clinac 2100c, 18 MV). Dose profiles obtained applying equations (1) and (2) were then compared with measurements achieved by means of a cylindrical ionization chamber (Scanditronix RK, 0.12cc).

Moreover, a series of dose measurements were carried out placing the gel dosimeters both in a standard water phantom and in a cylindrical 16cm wide and 14cm high tissue-equivalent phantom, and irradiating them in a BNCT thermal column reactor. Some of the adopted dosimeters were doped with

40ppm of ^{10}B and therefore permitted to obtain boron dose. Results were compared with Monte Carlo calculations of thermal neutron flux performed with MCNP5 [9].

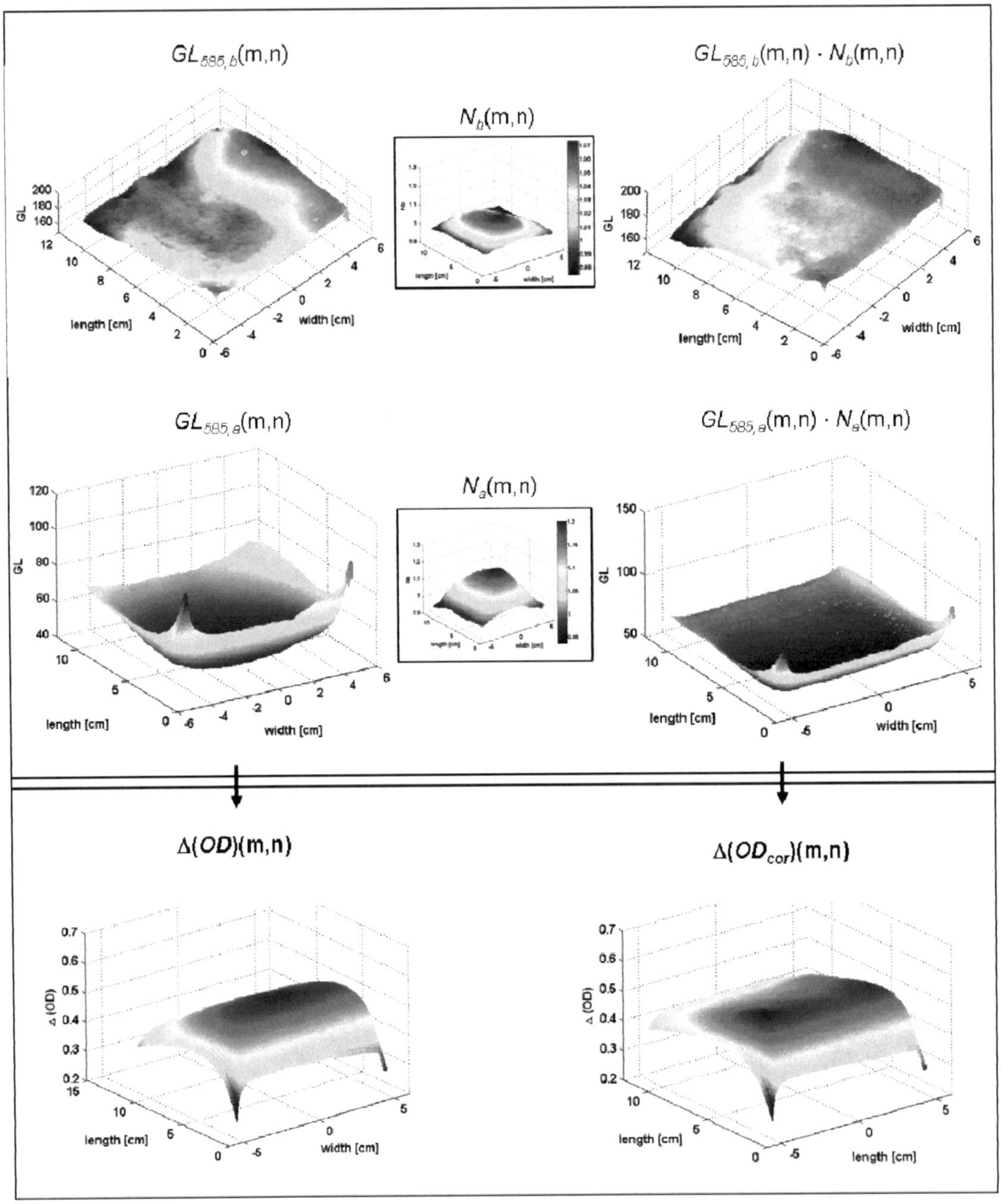

Figure 1: Partial and final results of the image processing chain on a single gel layer applying algorithms (1) (left column) and applying the correction algorithm (2) (right column). The normalization matrices N_b (before irradiation) and N_a (after irradiation) are the smaller ones, represented in the central column.

Journal of Physics: Conference Series **164** (2009) 012007 doi:10.1088/1742-6596/164/1/012007

3. Results

The dose profiles measured by means of two different Fricke gel-layers exposed to an X-ray radiotherapy field, processed with and without thickness disuniformities amendment, are reported in figure 2. Ionization chambers measurements are shown for comparative aims.

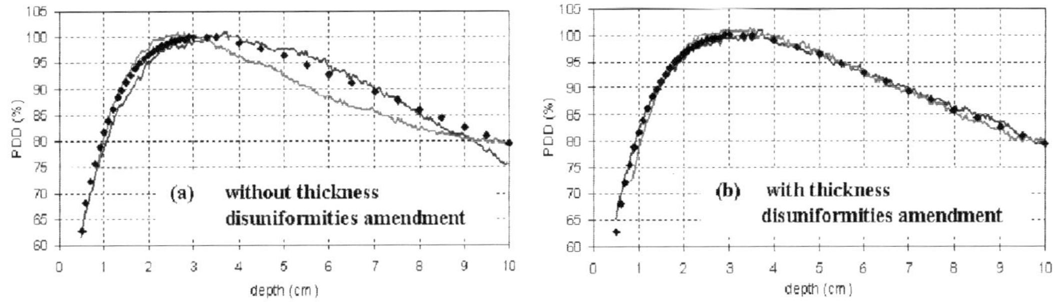

Figure 2: Percentage depth dose (PDD) on-axis profiles obtained with 2 Fricke-gel layers exposed to a 18 MV X-ray radiotherapic beam (Varian Clinac 2100c) (a) without thickness disuniformities amendment and (b) with thickness disuniformities amendment. Profiles were compared to PDDs obtained by means of a ionization chamber.

The dose profiles measured in two different experimental set-ups with two different Fricke gel-layers exposed to an epithermal neutron beam for BNCT, processed with and without thickness disuniformities amendment, are reported in figure 3. Monte Carlo simulations are shown for comparative aims.

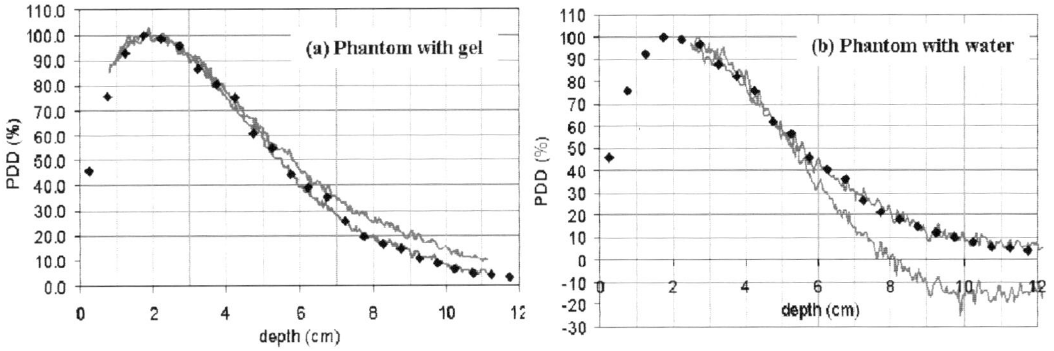

Figure 3: Percentage Depth Dose (PDD) profiles for thermal neutrons measured in (a) a water phantom and in (b) a gelatin phantom exposed to an epithermal neutron beam for BNCT. In green and red are reported data without thickness disuniformities amendment and with thickness disuniformities amendment, respectively. Black dots represent data obtained by means of Monte Carlo simulations.

4. Conclusion

The reported results show that the proposed procedure for the mathematical amendment of possible thickness disuniformities of gel-layer dosimeters is reliable. When larger dosimeters are involved, the thickness disuniformity effect cannot be neglected. In those cases, the procedure is necessary for the correct achievement of the delivered dose distributions. After appropriate optimization, the procedure is ideally applicable not only to Fricke gel-layers, but to any kind of gel dosimeter in form of layer.

Acknowledgments
The work was partially supported by the National Institute of Nuclear Physics (INFN), Italy.

References

[1] G. S. Ibbott 2006 Clinical Applications of Gel Dosimeters *J Phys. Conf. Series* **56**, 108-131

[2] C. Baldock 2006 Historical overview of the development of gel dosimetry: a personal perspective. *J Phys. Conf. Series* **56**, 14-22

[3] G. Gambarini, C. Birattari, M. Mariani, R. Marchesini, L. Pirola., P. Prestini, M. Sella, S. Tomatis 2004 Study of light transmittance from layers of Fricke-xylenol-orange-gel dosimeters. *Nucl. Instr. and Meth. B* **213**, 321-324

[4] M. Carrara, G. Gambarini, S. Tomatis 2005 Experimental Imaging and 3-D Rendering of Absorbed Dose by Means of Piled-up Dosimetric Sheets. *IEEE Nuclear Science Symposium Conference Record* 870-873 (ISBN 0-7803-9221-3)

[5] G. Gambarini, M.Carrara, M.Valente 2006 3D-Reconstruction of absorbed dose obtained from gel-dosimeter-layers. *9th ICATPP Conference on Astroparticle, Particle, Space Physics, Detectors and Medical Physics Applications*, 705-709, ed. World Scientific (ISBN 981-256-798-4)

[6] S. Tomatis, M. Carrara, G. Gambarini, R. Marchesini, M. Valente 2007 Gel-layer dosimetry for dose verification in intensity modulated radiation therapy. *Nucl. Instr. Meth. A* **580**, 506-509

[7] G. Gambarini, R.L. Moss, M. Mariani, M. Carrara, G.G. Daquino, V.A. Nievaart, M. Valente, E. Vanossi 2007 Gel Dosimeters as Useful Dose and Thermal-Fluence Detectors in Boron Neutron Capture Therapy (BNCT*). Radiat Eff Def S* **162**, 777-783

[8] M. Carrara, G. Gambarini, S. Tomatis, M. Valente 2007 Dose distribution measurements by means of gel-layer dosimeters. Evaluation of algorithms for ayrtifacts amendment. *Nucl. Instr. Meth. A* **579**, 334-338

[9] MCNP – A General Monte Carlo N-Particle Code, Version 5, X5 Monte Carlo Team, Los Alamos National Laboratory, 2003

Journal of Physics: Conference Series **164** (2009) 012008 doi:10.1088/1742-6596/164/1/012008

Aspects of radiation beam quality and their effect on the dose response of polymer gels: photons, electrons and fast neutrons

Andreas Berg[1], Christian Bayreder[1], Dietmar Georg[2], Achim Bankamp[3], Gerd Wolber[3]

[1]Dept. for Medical Physics, Medical University of Vienna, Vienna, Austria
[2] Div. Medical Radiation Physics, Dept. of Radiotherapy University Hospital Vienna, Vienna, Austria
[3] Dept. for Medical Radiation Physics, German Cancer Research Centre, Heidelberg, Germany

andreas.berg@meduniwien.ac.at

Abstract Polymer gels are generally assumed to exhibit no significant dependence of the dose response on the energy or type of irradiation for clinically used beam qualities. Based on reports on differences in dose response for low energy photons and particle beams with high linear energy transfer (LET) we here investigate the dose response and energy dependence for a normoxic methacrylic acid polymer gel (MAGAT) for X-rays (100 kV), high energy photon beams (E = 1.2 MeV (^{60}Co), 6 MV and 15 MV) and for three different electron energies (4, 12 and 20 MeV). Due to the possible impact also the sensitivity of the dose response to the dose rate is reported. A reduction in polymer gel relaxation rate has been observed for proton and carbon beams due to the high Linear Energy Transfer (LET) of these types of radiations. We here report on the dose response of an acryl-amide polymer gel (PAG) in a fast neutron field along with collimation as proposed for Boron neutron capture therapy (BNCT).

1. Introduction

Ideal dosimeters are expected to exhibit a dose response independent of the type (electromagnetic waves vs. particles) and beam energy. The measurement result should not be influenced by the way and circumstances of application for instance time between dosimeter preparation and irradiation, temperature during scanning or dose rate. Polymer gels are generally assumed to exhibit no sensitivity to the photon energy in the range of clinical used beams except for very low energy (E≤60 keV) Brachytherapy [1]. Mariansky et al [2] compared the dose response for 6 MV photons with 15 MeV electrons for BANG-2™ polymer gels, without finding significant energy dependence. The same result was found comparing 300 kV x-rays with ^{60}Co photons, 6 MV and 8 MV photons [3]. Further investigations on BANG™ and PAG type polymer gels confirmed the independence of the dose response from beam quality [4]. However, a small increase of the relaxation rate with photon energy was observed in a normoxic methacrylic acid polymer gel dosimeter (nMAG) for a photon beam of 6 and 25 MV [4] in the high dose saturation region. In the first part of our study we systematically investigate the dose response for a similar type of polymer gel using **T**etrakis-**H**ydroxy-methyl-**P**hosphonium-**C**hloride (THPC) with regard to energy dependence for X-rays (100 kV), for higher photon energies (E =1.2 MeV (^{60}Co), 6 MV and 15 MV) and for three electron energies (4, 12 and 20 MeV).

Due to tissue equivalence, simple 3D-evaluation and high spatial resolution polymer gels are also candidates for the newly developing Hadron beam therapy based mainly on protons and

Journal of Physics: Conference Series **164** (2009) 012008 doi:10.1088/1742-6596/164/1/012008

carbon ions. The first polymer gel investigations on such ion particle beams revealed sensitivity to the biologically relevant Linear Energy Transfer [6, 7]. Polymer gel dosimetry has also been investigated for its potential use for Boron neutron capture dosimetry [3]. An increase in dose response with Boron content could be detected in polymer gels. However no data on the dose response to neutrons of a PAG dosimeter was offered but proposed to be useful. In the second part of our study we present data on the dose response of a PAG-type polymer gel in a fast neutron field and compare the sensitivity to that of Co60-photons.

2 Materials We investigated two separate types of polymer gels based on:
a) Bis-Acryl-Amide using Nitrogen flushing through the Gel: **P**oly-**A**crylamide **G**el **(PAG)** to remove oxygen, which suppresses polymerization. This type of polymer gel is used for comparing the dose response in a field of fast neutrons to that of a photon field (^{60}Co). The polymer gel was manufactured at the German Cancer Research Center (DKFZ/Heidelberg) from water, gelatin: 6% w/w, monomer: acryl amide (3%) cross-linker: BIS (3%) using nitrogen flushing to reduce oxygen to a concentration below 0.003 mg/l (conf. table 1). We used BAREX^{TM} made small elliptical containers (MGS Research Inc., Madison, USA) for reasons of water equivalence and oxygen barrier characteristics.
b) Methacrylic **A**cid **G**el **A**nd **T**etrakis-hydroxy-methyl-phosphonium-chloride **(**THPC) as oxygen scavenger (**MAGAT**). This type of polymer gel is manufactured in our laboratory at Medical Univ. of Vienna [7], [8]. THPC is used due to its superior oxygen scavenging capabilities.

Components	PAG-gel	MAGAT-gel
Deionized water	88%(w/w)	87% (w/w), 870 g
Gelatin	6%(w/w)	8% (w/w), 80 g
Methacrylic acid		5% (w/w), 50 g
THPC		0,3316 g (2mM)
Acrylamide	3%(w/w)	
N,N`-methylen-bis-acrylamide	3%(w/w)	

Table 1 Chemical composition of the two types of polymer gels. MAGAT is used for comparing the dose response of photons and electrons in a wide energy range. PAG represents the dosimeter material for investigating differences between fast neutrons and ^{60}Co-photons.

3. Methods and results

3.1 Aspects of radiation quality: photon and electron energy [8]

The dose response and energy dependence for a normoxic MAGAT polymer gel was investigated for X-rays (100 kV), for higher photon energies (E = 1.2 MeV (^{60}Co), 6 MV and 15 MV) and for three electron energies (4, 12 and 20 MeV). All samples were irradiated with absolute dose levels D = 0, 1, 2, 4, 6, 8 and 12 Gy within about 30 hours. The MR-evaluation was completed within the next 2 days. The dose rates are determined by the specific radiation units available. DR $\approx$ 0,5 Gy/min were applied for X-rays and the ^{60}Co photon field. The linear accelerators offered 5 Gy/min for the higher electron and photon energies (4-20 MeV).

3.1.1 *X-ray irradiation* Irradiation with an energy of 100 kV was performed by a Gulmay D3300 orthovoltage therapy unit (Gulmay Medical Ltd., Chertsey, UK) using a tungsten target. The inherent filtration consists of 3 mm Be and the quality of the irradiation beam is determined by 2.8 mm HVL (half value layer) aluminium. A waterproof thimble Farmer-type chamber (PTW, M-30006) connected to a UNIDOS therapy dosimeter (PTW) was used for absolute dosimetry. Depth dose measurements were performed using a waterproof Roos^{TM} chamber (PTW, M-34001). The irradiation was adjusted such that the indicated absolute dose levels were applied to a depth of 6 cm under water level and about 2 cm distant from the bottom in the gel container.

3.1.2 *^{60}Co irradiation* The irradiations were performed in a 40x40x39 cm^3 water phantom using a ^{60}Co source (780 Elite Theratronics, Canada, source surface distance = 80 cm) in a 15 x 15 cm^2

field. The bottom of the gel container coincides with the water surface. The polymer gel response was evaluated in about 5 mm distance at the depth dose maximum. Absolute dose levels were verified using a calibrated Farmer-type chamber (0.6 cm^3, PTW type 30013, Germany). The field uniformity across the centre of 10 cm of the field is better than 2%.

3.1.3 *High energy photon and electron irradiation* MAGAT polymer gels were irradiated in a water phantom using a linear accelerator (Varian C linac 2100C/D). The containers were placed at a water depth, such that the dose maximum appears in the centre of the vial. Absolute dosimetry in the photon field (field size: 10x10cm^2; SSD:100cm) was performed with a calibrated ionization chamber following the IAEA 398 protocol. A Markus chamber (Roos^{TM}) served for electron dosimetry. Two photon energies (6, 15MV) and three electron energies (4, 12, 20MeV) were investigated.

3.1.4 *MR-evaluation* Parameter selective T2 imaging with a head coil was performed on two subsets of the polymer gels comprising about 20 samples each. We used a multiple spin echo sequence (CPMG) with equidistant echoes (TE=20, 40...400 ms, TR = 10.5 s, FOV = 18 x 18 cm^2, Mtx: 128 x 128, 23 slices, slice thickness = 1 mm, NEX = 4).

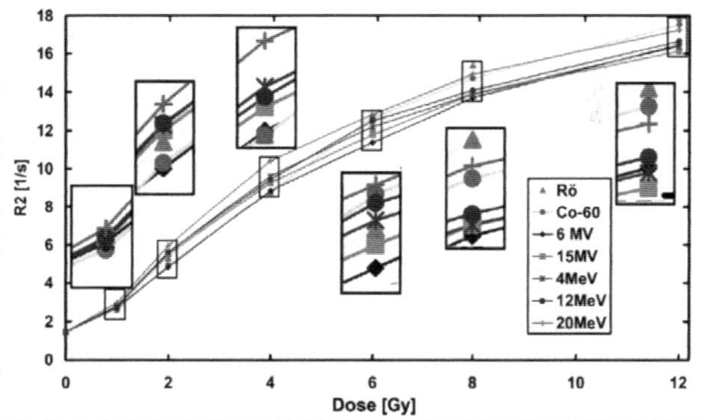

Fig. 1 MR-scanned dose-response (MAGAT-dosimeter) for several photon (100 kV-15 MV) and electron energies (4,12 and 20 MeV). The larger boxes indicate a 3 times enlargement of the small boxes next.

3.1.5 *Results* Dose-R2-response-curves for X-rays (100 kV), ^{60}Co, 6 and 15 MV photons, 4, 12 and 20 MeV electrons are shown in Fig. 1. The sensitivities calculated on the basis of a linear regression analysis in between D = 0 and D = 6 Gy are plotted in fig. 2. No significant differences are observed for high energy photons and electrons provided by a linear accelerator (LINAC); only a tendency to higher sensitivities at higher beam energy is observed.

On the first view the MAGAT polymer gel appears to exhibit significant higher sensitivities for the low energy ^{60}Co- and X-ray photons. However these beam qualities also differ with regard to dose rate from high energy photon and electron beams provided by a LINAC radiation. We therefore investigated also the dose rate dependence for a separate batch of MAGAT gels (fig. 3). The data on the dose response clearly indicates a suppression of the relaxation rate (R2) at increasing dose rates for the MAGAT-type gel. The reduction in dose sensitivity with higher dose rates DR >2 Gy/min) is significant (fig. 3b). Thus, the observed lower sensitivities for high-energy LINAC-photons and electrons (DR≈ 5 Gy/min) with reference to ^{60}Co-photons and X-ray (DR≈ 0.5 Gy/min) can be explained by a reduction in the R2-response at high dose rate for the MAGAT polymer gel. For the dose range investigated, the independence of the dose response on the photon energy is in agreement with investigations on a nMAG polymer gel [4] comparing the energy response of LINAC photon fields at 6 and 25 MeV max. energy. Similar to our results a slight tendency to higher dose rates at high energy was observed in this

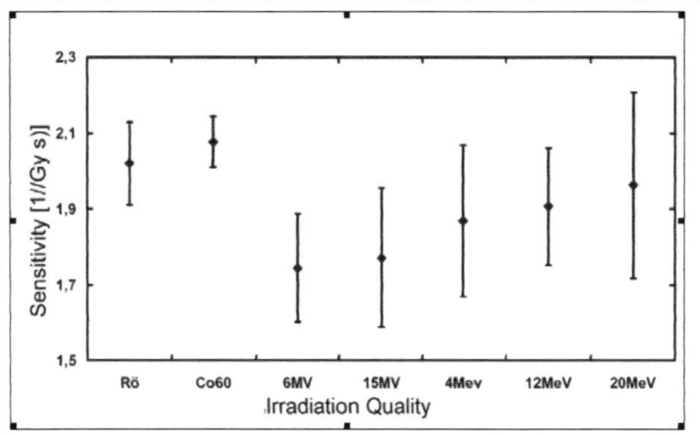

Fig. 2 Sensitivity as determined from linear regression analysis (1 Gy < D < 6 Gy on the dose response data of fig. 1 for different beam qualities. Error bars indicate a 95% confidence interval.

Journal of Physics: Conference Series **164** (2009) 012008 doi:10.1088/1742-6596/164/1/012008

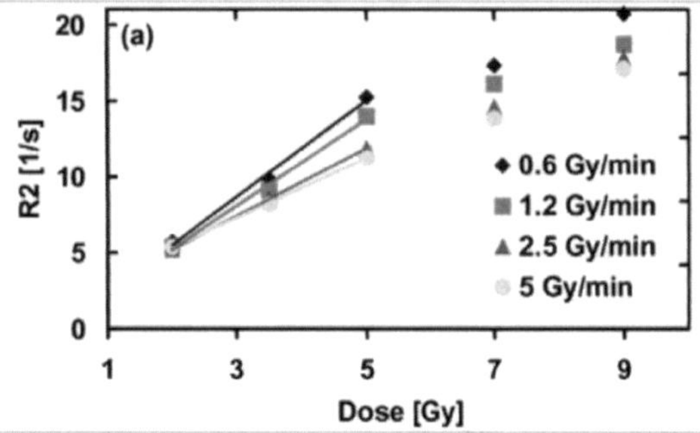

Fig. 3a Impact of the dose rate [Gy/min] on the sensitivity of a MAGAT polymer gel.

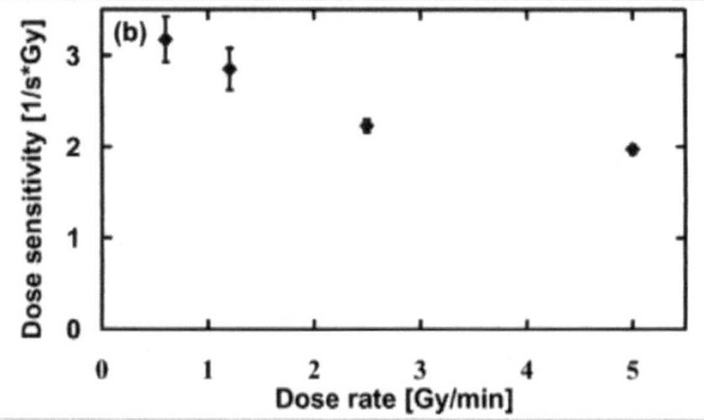

Fig. 3b Dose sensitivity as obtained from linear regression analysis in the dose region between 2 and 5 Gy (fig. 3a) for different dose rates. Error bars indicate a 95% confidence interval.

study. A significant energy dependence for various BANG™ gels, a VIPAR and a MAGIC gel was shown for photon energies smaller than 60 kV [1].

3.2 Aspects of radiation quality: neutron-fields [9]

3.2.1 *Materials*: Polymer gels (PAG). The polymer gel was manufactured at the German Cancer Research Centre (DKFZ/Heidelberg) from water, gelatin: 6% w/w, monomer: acryl amide (3%) cross-linker: BIS (3%) using nitrogen flushing to reduce oxygen to a concentration below 0.003 mg/l. We used BAREX™ made small elliptical containers (MGS Research, Inc. Madison, CT 06443 USA) for reasons of water equivalence and oxygen barrier characteristics.

3.2.2 *Methods: radiation beam characteristics, reference dosimetry and MR-evaluation* A radiation field (13x13 cm^2) of fast neutrons (n) was generated at compact cyclotron (Scanditronic MC32 NI, 32 MeV protons) at DKFZ/Germany irradiating a Beryllium target with 16 MeV protons. The generated neutrons exhibit a broad energy spectrum with highest energy of 16 MeV and medium energy of 6-7 MeV (fast neutrons). Reference dosimetry on the neutron field for calibration of the monitor units was performed using a Boron ionization chamber in the same experimental arrangement as prepared for the polymer gel irradiations. The γ-contribution to the overall dose amounts to about 5%.

Magnetic resonance micro-imaging at high signal-to-noise ratio was performed on a 3T human MR scanner using sensitive birdcage-type resonators (20 echoes, ΔTE=20ms, CPMG, slice thickness: 1 mm; FOV: 5x5 cm^2, Mtx: 128x128). A histogram analysis over ROIs (≈300 pixel) offered mean and standard deviation for error estimation.

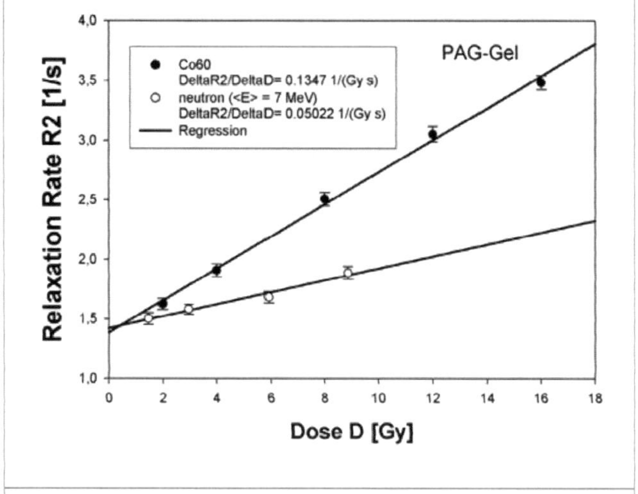

Fig. 4 Dose response of PAG-type gel for Co60-photons (E ≈1,2 MeV) and a field of fast neutrons (<E>≈6 MeV). The different sensitivities ΔR2/ΔD of the gel determined from linear regression analysis are indicated.

3.2.3 *Photon versus neutron dose response: results* The dose response for ^{60}Co-photons may be compared to the data on the field of fast neutrons (fig. 4). The sensitivity of the acryl amide polymer to fast neutrons gel is significantly reduced with reference to ^{60}Co-photons. Similarly to the procedure for determining the radiobiological efficiency of TLDs an efficiency factor can be determined as ratio of the relative sensitivities [5] to the fast neutron field and ^{60}Co-photons: $\varepsilon = \dfrac{\Delta R2 / \Delta D_{neuton}}{\Delta R2 / \Delta D_{Co60-photon}} = 0.37$. The reduced efficiency of the polymer gel in the neutron field is most likely related to the high linear energy transfer (LET) in neutron fields, determined by the low energy backscattered protons. The energy dissipation of neutrons is controlled by the scattering of neutrons by hydrogen. Its contribution to the KERMA of neutron fields in biological soft tissue amounts to about 90% [10] for the medium neutron energy of 6 MeV. A reduced sensitivity of polymer gels to high LET-irradiation was also reported for carbon-ions [5] and protons [6]. We interpret the reduced sensitivity for high-LET radiation by the increased presence of radicals, which results in termination of the polymerization process by two radical carrying partners. The length of polymer chains is reduced and the mobility of water molecules and multimers with reference to low LET-radiation increased.

References

[1] Pantelis E, Karlis A K, Kozicki M, Papagiannis P, Sakelliou L and Rosiak J M 2004 Polymer gel water equivalence and relative energy response with emphasis on low photon energy dosimetry in brachytherapy Phys. Med. Biol. **49** 3495–3514

[2] Maryanski M J, Ibbott G S, Eastman P, Schulz R J and Gore J C 1996 Radiation therapy dosimetry using magnetic resonance imaging of polymer gels Med. Phys. **23** 699-7

[3] Farajollahi A R 1999 An investigation into the applications of polymer gel dosimetry in radiotherapy Ph.D. theses Med. Phys. **26** 493

[4] De Deene Y 2004 Essential characteristics of polymer gel dosimeters J. Phys. Conf. Ser. **3** 34-57

[5] Ramm U, Weber U, Bock M et al. 2000 Three-dimensional BANG gel dosimetry in conformal carbon ion radiotherapy Phys. Med. Biol. **45** N95-102

[6] Gustavsson H, Bäck S, Medin J et al., 2004 Linear energy transfer dependence of a normoxic polymer gel dosimeter investigated using proton beam absorbed dose measurements Phys. Med. Biol. **49** 3847-55

[7] Bayreder C, Georg D, Moser E, Berg A 2006 Basic investigations on the performance of a normoxic polymer gel with tetrakis-hydroxy-methyl-phosphonium chloride as an oxygen scavenger: Precision, accuracy, stability, and dose rate dependence Med. Phys. **33** 2506-13

[8] Bayreder C submitted may 2008 High resolution resonance based dosimetry using normoxic polymer gels PhD Thesis, Medical University of Vienna, Vienna, Austria

[9] [9] Berg A, Wolber G, Bankamp A, Moser E 2003 MR-basierte Polymergel-Dosimetrie: Neutronen- und Photonen-Antwort Tagungsband (ISBN 3-925218-77-7) der 34. Jahrestagung der Deutschen Gesellschaft für Medizinische Physik (DGMP) Heidelberg/Germany 104-105

[10] Reich H (ed.) 1990 Dosimetrie ionisierender Strahlung B.G. Teubner, Stuttgart p.85

Volume-dependent internal temperature increase within polymer gel dosimeters during irradiation

Mahbod Sedaghat[1], Vincent Hubert-Tremblay[2], Luc Tremblay[1], Rachel Bujold [2], and Martin Lepage[1]

[1]Centre d'imagerie moléculaire de Sherbrooke, Département de médecine nucléaire et de radiobiologie, Université de Sherbrooke, Sherbrooke, QC, Canada and [2]Département de radio-oncologie, Centre hospitalier universitaire de Sherbrooke, Sherbrooke, QC, Canada

mahbod.sedaghat@Usherbrooke.ca

Abstract. The increase in temperature within polymer gel dosimeters due to exothermal polymerization reactions was measured during irradiation inside two normoxic polymer gels (MAGAT and PAGAT) with different volumes (from 18 ml to 257 ml) at a dose rate of 3.0 and 0.5 Gy.min^{-1}. The same samples were evaluated by MRI the day after irradiation. The temperature rise is clearly a function of the irradiated gel volume and for the MAGAT gel dosimeter it is almost twice that of the PAGAT dosimeter. Herein, we report on the experimental set-up, preliminary temperature measurements and the corresponding MRI results.

1. Introduction

In performing dose measurements with polymer gels a chain of procedures from fabrication of the gel, filling the appropriate phantoms, transporting, positioning, irradiating, calibration and finally scanning is to be followed, each of which requires careful attention to the details that may impact on the gel response. Accordingly numerous authors have reported the effect and severity of different factors that may alter the final dose determination. Among these is the temperature, which is known to affect most dosimetric measurements. However, polymer gel dosimeters are unique in that not only the ambient temperature affects their results [1], but the exothermal polymerization reaction [2] causes the temperature of the gel to increase during irradiation. Although this temperature rise has been recognized and its possible effects were previously hypothesized [2,3], no experimental attempt has been made to fully evaluate whether or not the gel response as determined by MRI is altered by these internal temperature changes.

The magnitude of the temperature rise is a function of the number of moles of reactions occurring per unit time [2], therefore it depends on the nature and concentration of the monomers, on the irradiated volume of gel, on the dose rate as well as on the rate of heat loss associated with the material and thickness of the container.

It was suggested that if the temperature increase in gel dosimeters affects the dynamics of polymerization reactions or the final polymer morphology, then the test tube calibration method might not always provide accurate results since the small volume of gel in the test tubes would not experience the same temperature changes as larger volumes of a phantom of interest [2,3].

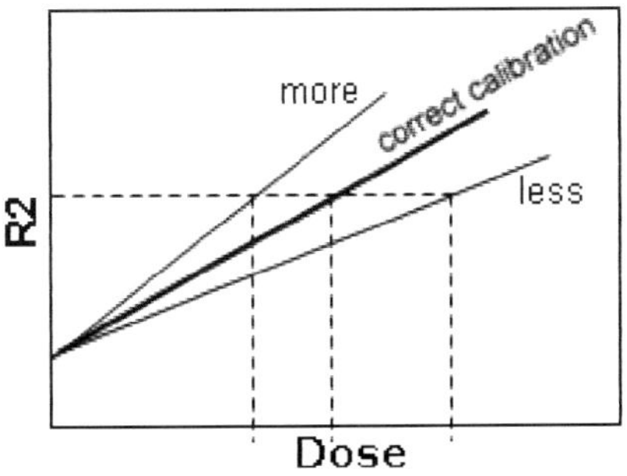

Figure1. Less polymerization in test tubes results in dose overestimation and vice versa.

In fact several groups have observed dose deviations between calibration vials and larger phantoms originating from the same batch of gel [4]. Different potential causes were hypothesized for this phenomenon [3,4] and some discrepancies are difficult to reproduce [5,6].

Assuming that less polymer is formed in small calibration test tubes, the dose in the phantom of interest would be overestimated by that calibration and *vice versa*. This is depicted in Figure 1 where three presumed calibration curves are drawn, showing the correlation between the spin-spin relaxation rates (R_2) and absorbed dose.

The objective of this work is to experimentally investigate the extent of temperature increase in different polymer gel dosimeters for different doses, dose rates and volumes of gel as well as its possible impact on the gel dose-response. Our preliminary temperature measurements are presented here.

2. Materials and Methods

Eight MRI-compatible optical temperature sensors (NeOptix Inc, Quebec, Canada, accuracy ±1°C, resolution 0.1°C) were used to measure the temperature inside the phantoms. The standard probes were protected by a thin Teflon jacket. We found this jacket released oxygen in the polymer gel dosimeter such that areas where R2 remained constant after irradiation were detected around every probe (Fig. 2a). At our request, the company developed a new polyimide sheathing that does not cause oxygen contamination of the dosimeter gel (Fig. 2b). The outer diameter of the polyimide-coated optical fibers was 0.7 mm. Four cylindrical gel containers were constructed from Delrin® with an identical wall thickness of 2.5 cm and height of 15 cm. The outer diameters ranged in an increasing order from 6.0, to 7.5, 8.5 and 10 cm; the corresponding volumes were 18, 57, 150 and 275 ml respectively. The thickness of the base of the containers was 4 mm. Four holes were drilled at the same location in the walls of each container, allowing temperature probes to be inserted into the gel at different positions (Fig. 2c). Two normoxic gel formulations were studied, namely the polyacrylamide based gel (PAGAT) and the methacrylic acid based gel (MAGAT) dosimeters. Tetrakis (hydroxymethyl) phosphonium chloride (THPC) was used in both dosimeter gel formulations. The optimized formulations and fabrication procedures were reported [6,7]. The probes were inserted into desired positions, the phantoms were filled with the gel and placed in a water tank in a conventional refrigerator for about 5 hours for the gel to set. The water tank was then transferred to the radiotherapy department to acclimatize with the room temperature for 24 hours. In some cases pre-irradiation MR images were obtained to verify the homogeneity of the gel before irradiation in the four phantoms. The phantoms were previously CT-scanned and the CT images were transferred into a Pinnacle™ Treatment Planning System with which the dose delivery has been optimized for a 10 x 10 cm^2 field with 6 MV photons such that identical depths in all four phantoms received identical doses. A dose of 12 Gy to D_{max} in water was delivered with high (300 MU/min) and low (50 MU/min) dose rates for the first series of experiments reported here [Siemens Oncor™ Linear Accelerator (Siemens Medical Solutions AG, Germany)]. Each phantom was transferred from the first water tank into a second water tank with an identical stable water temperature, it was irradiated then transferred back to

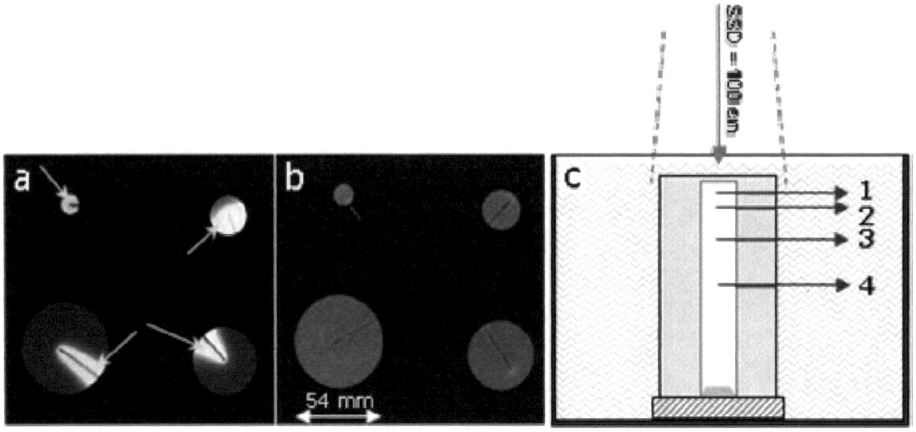

Figure 2. T_2-weighted MR images of four phantoms containing optical temperature sensors. (a) Sensors with a Teflon jacket through which oxygen could diffuse resulted in the inhibition of the polymerization of the gel around the probes in the phantoms (green arrows). (b) The Teflon jacket was replaced by polyimide sheathing which has a much lower permeability to oxygen. The MRI slices shown correspond to position 3 in the experimental set-up depicted in (c).

the first water tank. The temperature measurement continued for a few hours then the phantoms were taken to the MRI room and scanned 24 hours later with a 1.5 T clinical scanner (Siemens Sonata, Erlangen, Germany) according to a previously optimized multi-spin echo protocol. The probes were not removed from the gel to avoid distortion to the surrounding gel.

3. Results and Discussion

Figure 3 shows the temperature recorded by the probe in position 3 as a function of time (as shown in Fig. 2c) for high (a-d) and low (e-h) dose rates and the two gel dosimeters. The dose at position 3 of all four phantoms was 10 Gy in these experiments. The temperature increase in position 1 was always lower than in position 2 and 3 even if the dose at this point was higher. This is tentatively rationalized by the geometry of the phantomes where the base is thinner than the side walls. The maximum temperature rise for MAGAT as can be seen on (h) is up to two times larger than for the PAGAT dosimeter. The temperature rise for the PAGAT dosimeter was slightly higher at the higher dose rate for all volumes of gel. For the MAGAT dosimeter, on the other hand, a lower temperature increase was observed in the smallest container with the low dose rate (compare a and e) and a higher temperature increase was observed in the largest phantom with low dose rate (d and h).

For high dose rate irradiation of PAGAT gel dosimeter our preliminary MRI measurements reveal no significant difference between profiles of R_2 along the central axis of the three larger phantoms, but the smallest phantom (18 ml) shows higher R_2 values of the order of 6% along the depth. This suggests that termination reactions are favored in the larger containers by the heat that is produced in the gel dosimeter through polymerization.

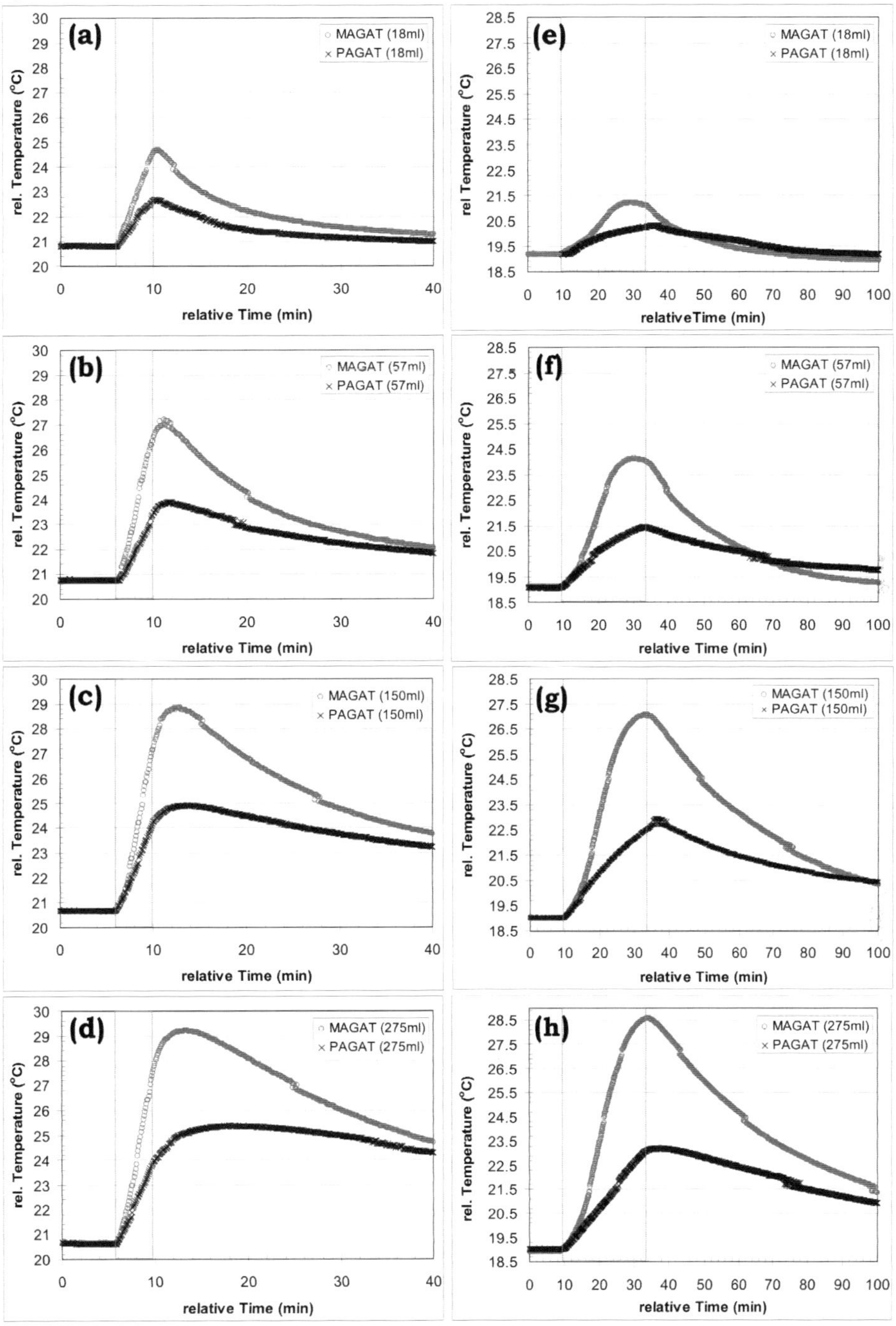

Figure 3. Temperature differentials in position 3 as shown in figure (2c) for different volumes of MAGAT (red circles) and PAGAT (blue crosses) gels irradiated with 6 MV photons at two dose rates. The dose at this position was 10 Gy. Graphs a-d, refer to a dose rate of 300 MU/min and graphs e-h depict temperature changes for a dose rate of 50 MU/min. Start and end of irradiations are also shown by vertical lines. The phantoms were in water before, during and after irradiation.

The results are qualitatively similar to those we have previously reported [8]. We have yet not observed reproducible discrepancies between the four phantoms for other studies with MAGAT and low dose rate irradiations. As the rate of heat release may be considered proportional to the rate of polymerization, these studies may provide additional information on the nature of polymerization reactions in polymer gel dosimeters.

References

[1] Y. De Deene, Essential characteristics of polymer gel dosimeters. *J. Phys.: Conf. Ser.* 3, 34–57 (2004).
[2] G. J. Salomons, Y. S. Park , K. B. McAuley, L. J. Schreiner, Temperature increases associated with polymerization of irradiated PAG dosimeters. *Phys. Med. Biol.* **47**, 1435–1448 (2002).
[3] V. P. Cosgrove, P. S. Murphy, M. McJury, E. J. Adams, A. P. Warrington, M. O. Leach, S. Webb, The reproducibility of polyacrylamide gel dosimetry applied to stereotactic conformal radiotherapy. *Phys. Med. Biol.* **45,** 1195–1210 (2000).
[4] Y. De Deene, On the accuracy and precision of gel dosimetry. *J. Phys.: Conf. Ser.* **56**, 72–85 (2006).
[5] D. A. Low, J. F. Dempsey, R. Venkatesan, S. Mutic, J. Markman, E. M. Haacke, J. A. Purdy, Evaluation of polymer gels and MRI as a 3-D dosimeter for intensity-modulated radiation therapy. *Med. Phys.* **26**, 1542-1551 (1999).
[6] A. J. Venning, B. Hill, S. Brindha, B. J. Healy, C. Baldock, Investigation of the PAGAT polymer gel dosimeter using magnetic resonance imaging. *Phys. Med. Biol.* **50**, 3875–3888 (2005).
[7] C. Hurley, A. Venning, C. Baldock, A study of a normoxic polymer gel dosimeter comprising methacrylic acid, gelatin and tetrakis (hydroxymethyl) phosphonium chloride (MAGAT). *Appl. Rad. Isotop.* **63**, 443–456 (2005).
[8] E.-M. Dumas, G. Leclerc, M. Lepage, Effect of container size on the accuracy of polymer gel dosimetry. *J. Phys.: Conf. Ser.* **56**, 239–241 (2006).

Journal of Physics: Conference Series **164** (2009) 012010 doi:10.1088/1742-6596/164/1/012010

Reproducibility study of normoxic polyacrylamide gel (nPAG) dosimeters

Giuliana Russo[2], Jan Vandecasteele[1], Tom Vercauteren[1], Salvatore De Pasquale[2], Yves De Deene[1]

[1]Research group QMRI, Dept. for experimental cancerology, nuclear medicine and radiotherapy, Ghent university hospital, De Pintelaan 185, 9000 Gent, Belgium
[2]Dept. of Physics "E.R.Caianiello", University of Salerno, Italy

E-mail: russo@sa.infn.it

Abstract. In this study, the overall accuracy of normoxic polyacrylamide gel (nPAG) dosimeters is considered. Different badges of nPAG are fabricated, poured in containers of glass and Barex™ and irradiated with a 6MV square photon beam. The polymer gel dosimeters were read out using MRI. The overall reproducibility and accuracy of nPAG gel dosimeters was determined by comparison with depth-dose profiles acquired with a Pin Point ionization chamber. Additionally, the effect of the container wall on the depth-dose profile and the effect of temperature changes before and after irradiation on the R2-dose response have been investigated. The average standard deviation and maximum deviation between 8 gel-measured depth-dose profiles and a depth-dose profile measured with an ionization chamber amounted to 0.543 Gy (2.5%) and 2.579 Gy (11.7%) respectively.

1. Introduction

Polymer gel dosimeters are radiation sensitive chemical dosimeters that were first proposed in 1950 by Day. In the 90s, gel dosimetry has been considered for the use of radiation dose verification in radiotherapy. Polymer gel dosimeters contain vinyl monomers dispersed in an aqueous gel matrix rendering these systems suitable for 3D dose verification. Early polymer gel dosimeters were highly sensitive to oxygen. Oxygen inhibits the radiation induced polymerization, as oxygen is as a free radical scavenger. In the first generation of polymer gel dosimeters oxygen was removed by flushing the gel with inert gasses such as nitrogen and argon. In recent polymer gel dosimeters the effect of oxygen is compensated by the addition of an antioxidant. These polymer gels can be fabricated under normal atmospheric conditions and are therefore called 'normoxic'. Studies have been performed to investigate the dosimetric properties (such as tissue equivalence, temperature dependence, energy dependence and dose-rate dependence) of different polymer gel dosimeters.

Until now polymer gel dosimeters have not been widely used in the clinic. To gain confidence in the applicability of a specific polymer gel dosimeter in the radiotherapy clinic, an overall reliability study is necessary. Such a study should not be restricted to one step in the dosimetry process (fabrication, irradiation, read out and calibration) but should encompass all steps. This study aims at providing an overall figure of reliability of an nPAG dosimeter by comparing a gel measured depth-dose curve with a corresponding depth-dose curve obtained with an ionization chamber ('golden standard'). Several volumetric gel dosimeters from different batches of nPAG gel are irradiated with a square field. Each experiment is accompanied by a proper calibration using a series of calibration vials irradiated in a water phantom at reference depth. Barex™ and glass have been used as cast material for the actual gel dosimeters.

2. Materials and methods

2.1. Preparation

All gel dosimeters studied in this work contained 8 % (w/w) gelatin (300 bloom), 3% (w/w) acrylamide, 3% (w/w) N, N'-methylene-bisacrylamide and 3 mM tetrakis-hydroxymethyl-phosphonium-di-sulphate (THPS).

Gelatin is added in half of the amount of de-ionized water and is heated to 50°C until a clear solution is obtained. The monomers are added in the other half of the amount of water and the solution is heated to 60°C. After cooling down the monomers and the gelatin solution to 40°C, both solutions are mixed. The antioxidant is added when the solution is cooled down to 35°C. Subsequently the Barex™ box shaped (10 cm x 10 cm x 20 cm) and/or cylindrical (diameter = 10 cm, height = 35 cm) glass volumetric phantoms and sixteen glass test tubes (10 mm OD, 150 mm length, 1 mm thick wall) are filled with gel. After filling, all phantoms are placed in a large water basin at 35°C in order to obtain a uniform and equal cooling process in both calibration vials and volumetric gel phantom.

2.2. Irradiation

All gel phantoms are irradiated with 6MV photon beams delivered by a clinical liner accelerator (Elekta SL-18). The gel filled test tubes that serve as calibration phantoms are placed in a small water phantom at reference depth (5 cm) and are irradiated with a square beam of 10 cm-by-10 cm (SSD = 95 cm). The delivered dose in the calibration vials ranged from 0 Gy to 30 Gy in steps of 2 Gy. The volumetric phantoms were irradiated with 2000 Monitor Units (MU), with a field size of 5cm x 5cm and at an SSD of 95cm.

2.3. MRI measurements

The volumetric gel phantom and test tubes, corresponding to the same batch were scanned together in a clinical MR scanner (Siemens, Magnetom Symphony 1.5T). A multi-echo sequence with 32 echoes and CPMG phase encoding scheme was used for the evaluation of irradiated polymer gel dosimeters. The imaging parameters are TR = 6 s, TE = 40 ms, field-of-view=220 mm for the Barex™ phantom and 240 mm for the glass phantom, matrix size=256 x 256 and slice thickness=10 mm. To eliminate errors in the determination of the R2 measurements caused by temperature changes, the irradiated gel was acclimatized in the MR scanner room prior to scanning for 24 hours. The homogeneity in the R2 maps was investigated by scanning a non-irradiated ('blank') volumetric gel with the same imaging parameters. The homogeneity was within 0.30%.

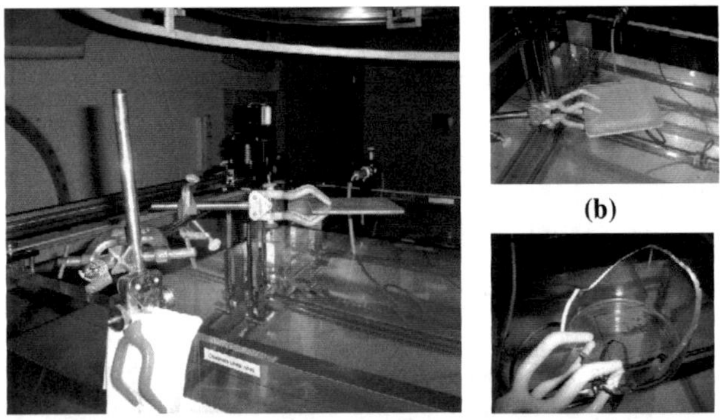

(a) (b) (c)

Figure 1. Set-up for measuring the influence of the phantom wall material on the depth dose profile (a). A close up of the Barex™ (b) and the glass sheet (c) are shown. The source-to-surface distance of both sheets was 80 cm.

2.4. Ionization chamber measurements

To investigate the effect of the phantom wall material on the shape of the depth dose profile, dose measurements with a PinPoint ionization chamber were performed. The ionization chamber was mounted on a stepper motor driven stage inside an automatic water phantom and was connected to electrometers. The source-to-surface distance was 95 cm. The initial position of the ionization chamber was at the water surface and along the beam axis. The phantom with ionization chamber was irradiated with a 5 cm-by-5 cm 6 MV photon beam corresponding to the same radiation configuration as in the gel experiment. A depth dose curve was recorded by driving the ionization chamber in the beam axis direction in the water bath with the beam on. To investigate the effect of the walls of the gel containers, a Barex™ and a glass sheet were placed between the collimator and the water phantom respectively (figure 1).

2.5. Combined ionization chamber / gel dosimetry experiment

A Barex™ phantom with two inserts for an ionization chamber was constructed (figure 2). The two Perspex inserts were constructed at 5 cm and 10 cm depth on opposite sides of the phantom so that the top of the ionization chamber was in the centre of the beam.

(a) (b)

Figure 2. The Barex™ phantom with two inserts (a) for simultaneous dose measurement with an ionization chamber during irradiation (b).

2.6. Temperature investigation

To investigate the effect of temperature variations in the gel after fabrication [1], three series of test tubes filled with the same batch of gel, were studied. Initially, all gel samples were stored in a water container (20 l) with the water temperature at 35 °C, the same temperature of the gel after fabrication. Temperature probes were inserted in one sample of each series. The temperature history of each of the gel samples is shown in figure 6a. The temperature of the first series of gel samples was increased to 25 °C for 4 hours before the irradiation (blue line in figure 6). The temperature of the second series was increased to 26 °C after irradiation and before the MRI acquisition (green line in figure 3). The third series was cooled normally (pink line). All gel samples were scanned together at 22 °C.

3. Results and discussion

3.1. Gel dosimeter calibration

In figure 3 the R2-dose response curve of different batches of nPAG dosimeters is shown.

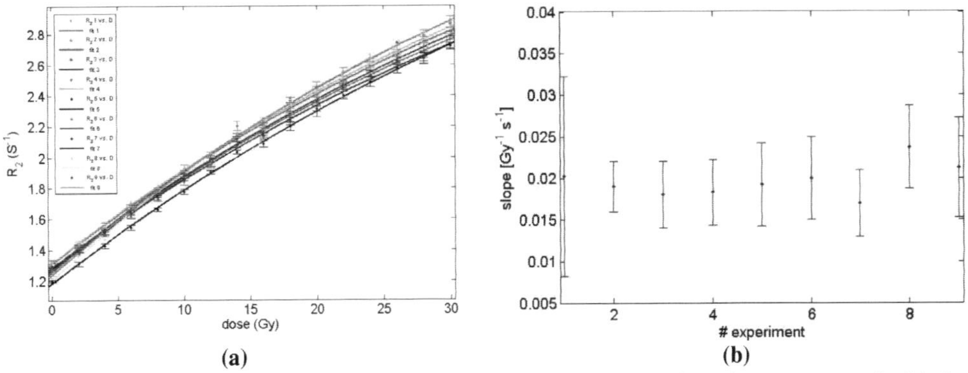

(a) (b)

Figure 3. R2 versus dose calibration plots for different gel batches (a). The colours correspond with the experiments discussed further. The slope of the R2-dose for the different experiments is shown in (b).

From figures 3a and b it can be seen that the slope of the R2-dose response of the different calibration curves does not differ significantly.

3.2. Depth dose: comparison with ionization chamber

The depth dose profile for each gel experiment is provided in figure 4a and b for gels contained in a Barex™ and glass container respectively. The black line in each of the figures corresponds to an ionization chamber measurement.

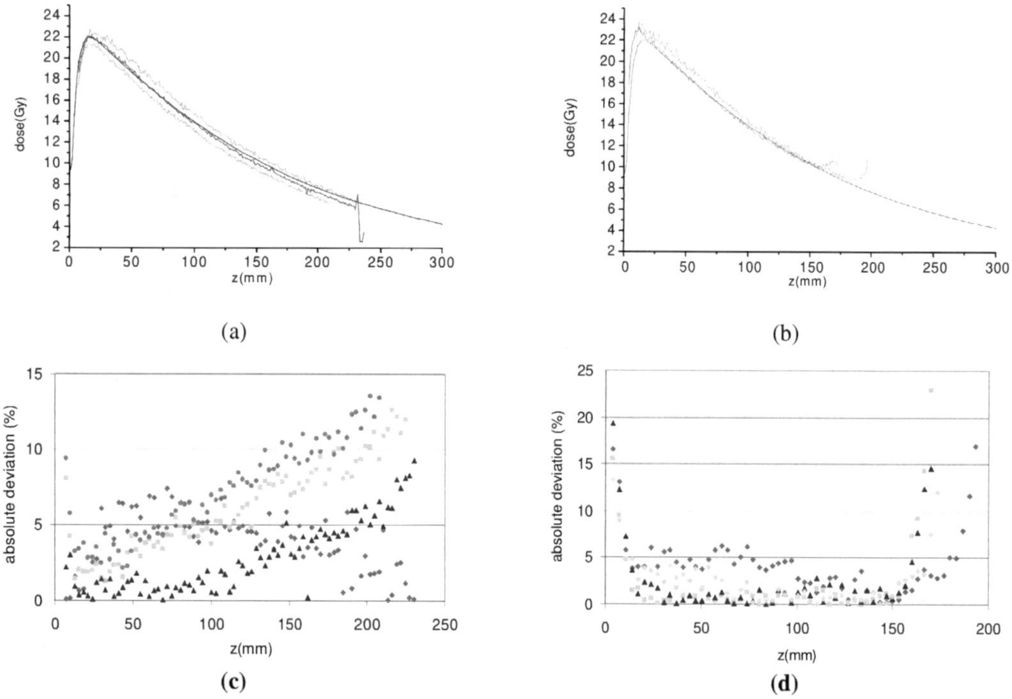

Figure 4. Comparison of gel measured (coloured lines) and ionization chamber measured (black line) depth dose curves for gels contained in a glass (a) and Barex™ container respectively. Corresponding percentage deviations of the dose calculated in the glass phantoms (c) and Barex™ phantoms (d) are also shown.

From figure 4a and b it can be seen that the depth dose response is different for the Barex™ and glass phantoms. The increase in measured dose at the top and bottom of the Barex™ phantom (figure 4b and d) may be attributed to oxygen contamination. The average standard deviation and maximum deviation of all gel-measured depth-dose profiles and a depth-dose profile measured with an ionization chamber amounted to 0.543 Gy (2.5% relative to the dose at isocenter) and 2.579 Gy (11.7% relative to the dose at isocenter) respectively in a region between 10 mm and 150 mm in the phantom.

3.3. Combined ionization chamber / gel dosimetry experiment

Average dose values at 5 cm and 10 cm depth in the phantom obtained with pin-point ionization chamber and gel dosimetry are listed in table 1.

Table 1. Average dose values recorded at 5 cm and 10 cm depth in the phantom using a Pin Point ionization chamber and gel dosimetry in the Barex™ and glass phantoms. The number between square brackets indicates the number of experiments. The values between round brackets indicate the standard deviation as calculated from the different experiments.

Dose value	Pin Point chamber [#6]	Barex™ phantom [#4]	Glass phantom [#7]
5 cm	18.18 (± 0.15 Gy)	18.40 (± 0.72 Gy)	18.00 (± 0.66 Gy)
10 cm	13.35 (± 0.10 Gy)	13.47 (± 0.61 Gy)	13.13 (± 0.64 Gy)

The average dose values obtained with the nPAG gel in the Barex^{TM} and glass phantoms correspond with the dose values measured with the ionization chamber to within 1.6 % at both depths. However larger deviations are seen amongst the different gel experiments.

3.4. Radiological influence of the container
From independent depth dose measurements with an ionization chamber no significant influence from the container material could be detected (figure 5).

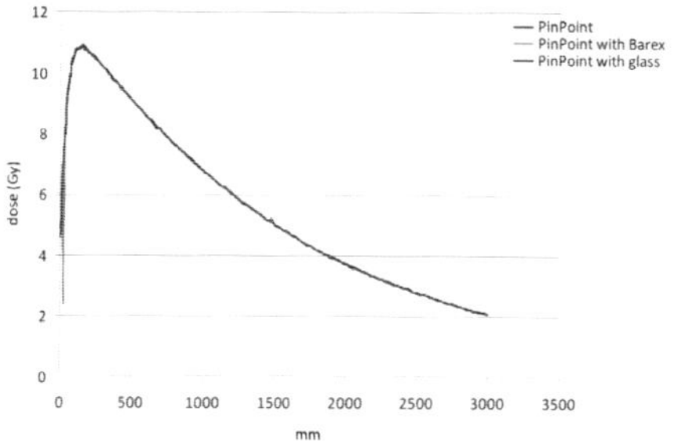

Figure 5. Depth dose profiles recorded with an ionization chamber with different cast materials in place.

3.5. Temperature investigation
The effect of temperature variations before and after irradiation of the nPAG gel dosimeter on the R_2-dose response in the nPAG gel dosimeter has been investigated. R2-dose calibrations curves are shown in figure 6b. No significant effect of dose-R2 variations are found in the dose range of 0 Gy to 20 Gy. A slightly lower response is found at doses higher than 20 Gy in the gel samples that were heated to 26 °C after irradiation.

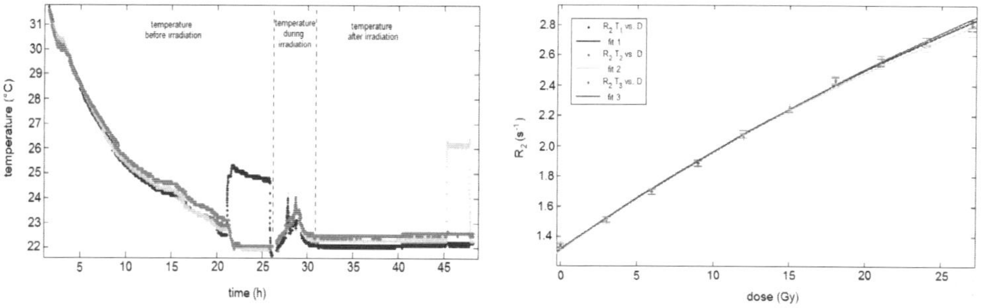

(a) (b)

Figure 6. Temperature history of three different sets of gel samples (a) and corresponding (same colors) R2-dose response curve (b).

4. Conclusions

The average standard deviation between depth dose profiles obtained with nPAG gel dosimeters and a depth dose profile recorded with an ionization chamber in the region between 10 mm and 150 mm depth in the phantom was found to amount 0.543 Gy (2.5%). The maximum dose difference amounted to 2.579 Gy (11.7%). A systematic dose deviation with ionization chamber measurements which increased with depth was found for gel dosimeters poured in a glass container. These variations could not be explained by a photon energy shift caused by the glass. Depth dose profiles in Barex^{TM} phantoms are susceptible to large variations near the container wall which may be caused by oxygen contamination. Temperature variations in the nPAG gel dosimeter before and after irradiation were not found to have a significant effect on the R2-dose response.

Acknowledgements
The authors thank Mr. Bart Truyens for constructing the Barex^{TM} phantoms.

References

[1] De Deene Y, Pittomvils G, Visalatchi S, The influence of cooling rate on the accuracy of normoxic polymer gel dosimeters *Phys. Med. Biol.* **52** (2007) 2719-2728

Radiochromic Gel Dosimeter (FXG) Chemical Yield Determination for Dose Measurements Standardization

M.A. Bero[*] and M. Zahili

National Radiation Metrology Laboratory (NRML), Protection and Safety Department, Atomic Energy Commission, Damascus P.O.Box 6091, Syria.

mbero@aec.org.sy

Abstract. Different gel dosimetry techniques have been proposed as methods for three dimensional dose distribution measurements. Gel dosimeters were proposed for radiotherapy dosimetry at which the calibration of any dose-measuring technique is the main requirement for its proper practical application. The standardisation of the technique requires the establishment of a method for the calibration of each type of these gel materials. The aim of this study is to determine the chemical yield for the radiochromic "Ferrous-sulphate and Xelenol-orange Gelatin" (FXG) gel. Using standard absorbed dose to water protocol, the radiation field is calibrated first; then a standard Fricke solution dosimeter is calibrated. The obtained chemical yield for Fricke is $G(Fe^{3+}) = 1.68\times10^{-6}$ [mol. J^{-1}]. Finally, FXG dosimeter was calibrated against the standard Fricke solution before calculating the system chemical yield and comparing its value with Fricke system. The results show that □.G (FXG) = 6093.1 [m^2. J^{-1}] = 19.5×□.G (Fricke). Compared with the Fricke system, the FXG dosimeter has much higher sensitivity. Additionally, the overall accuracy and reproducibility of the gel detector was also comparable to those reported for Fricke solution. However, the FXG dosimeter properties could be improved further through the establishment of a defined manufacturing procedures and calibration method; as well as precise selection of sensitive materials and radiation effects evaluation techniques.

1. Introduction

A dosimetric system that can precisely measure radiation dose variations in three dimensions with high spatial resolution is highly demanded for various applications. An obvious example is radiotherapy dosimetry, where assessing and measuring the absorbed dose accurately is essential to achieve a successful treatment. In practice, due to the complexity of the treatment procedures and the use of sophisticated irradiation facilities, most of the required information is obtained using complex computational methods. Such techniques are usually time consuming and need extremely accurate calibration. The theoretical calculations and computer simulations are also unlikely to have absolute accuracy, and they are susceptible to human errors and unexpected problems. Therefore, in all 3-D dosimetry investigations the prime objective is to find practical systems that are able to measure the dose distribution in three dimensions with sub-millimetre resolution in a non-invasive, non-destructive and accurate way. Such measurements would be valuable on their own, and they might also be used to check whether the expected dose distributions for computationally controlled photon beams are

[*] Corresponding Author: e-mail: scientific@aec.org.sy, (M. A. Bero), mbero@aec.org.sy, Fax: 00963116112289

achieved in practically. One of the main aims of treatment planning in radiotherapy is that the absorbed dose distribution inside the target volume must be within closely specified limits in order to ensure complete destruction of malignant cells. It is also important to achieve very steep dose gradients are achieved at the boundaries of the irradiated region in order to protect all healthy surrounding tissues. Accurate absorbed dose measurement is an essential factor to attain these objectives. The previously presented dosimetry approach aims at obtaining a complete 3-D dose distribution within materials that are tissue-equivalent, sensitive to ionizing radiation, and they can form part of a phantom to imitate the tumor volume in size, shape and/or composition. The method utilizes high-resolution tomographic imaging methods capable of mapping complex dose distributions such as those applied in conformal radiotherapy, using suitable materials in which radiation effects can be visualised and quantitatively measured.

In summary, radiation dose measurement is an integral part of all radiation treatment procedures. An accurate dosimetry protocol helps not only ensure a high probability of short term success in radiotherapy only, but also helps to improve the health outcome for people undergoing this kind of treatment. As a matter of fact, the objective of radiation therapy is always to deliver a sufficient dose to the tumor while minimizing both the short and long-term risks. Carefully planned radiotherapy procedures must always consider the twin objectives of delivering adequate radiation energy to the targeted tumor, and at the same time minimizing the exposure of the surrounding normal tissues.

2. Experimental Methods

Absorbed dose measurements following standard protocol [1] is performed in order to define the radiation field. A standard ionization chamber is always mounted at the center of the beam to monitor all delivered doses. Fricke dosimeter was prepared and used according to standard procedure [2]. Then FXG is prepared according to a defined method described in [3] and calibrated against the Fricke system.

2.1. Irradiation

A Theratron-Elite-80 ^{60}Co radiotherapy machine with 10×10 cm^2 collimated square field at source to surface distance (SSD) of 100 cm is used for irradiation. The dose rate of 2.1 Gy/min at the gel samples position 5 cm depth in water phantom was applied to obtain a dose response to γ-rays for both Fricke and FXG dosimeter. Samples were used in each irradiation and the delivered doses are increased up to 10 Gy, figure 1. Irradiation is usually made within 12 hours after the preparation of the sensitive materials, and the optical absorption measurements are recorded 30 minutes after irradiation. Standard ion chamber (0.6 cc) is mounted at the center of the beam to give immediate dose readings.

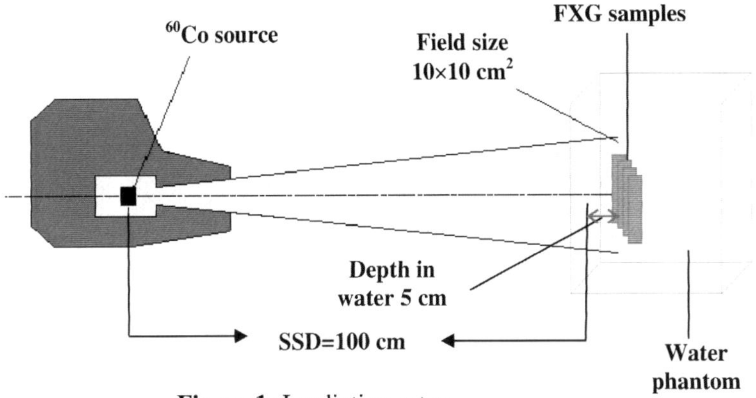

Figure 1: Irradiation setup.

2.2. Optical spectroscopic measurements

A powerful double-beam UV-visible spectrophotometer SPECORD 210 (analyticjena AG, Konard-Zuse-Str., 1-07745, Jena, Germany) is used in order to perform the optical density measurements (OD), in 1 cm path length in the gel samples. Samples placed in cuvette containers are evaluated before and after irradiation. Wavelengths scan mode is used first to obtain full absorption spectrum in the visible region. Then, a suitable wavelength for FXG dosimeter is selected to make quantitative measurements. The selected wavelength λ=565 nm corresponds to the maximum changes induced by ionizing radiation interactions at 10 Gy absorbed dose level.

3. Results and Discussion

A dosimeter is said to be absolute if it can be used to measure the absorbed dose deposited in it without requiring calibration in a known radiation field. The Fricke solution is perceived as being capable of absolute dose measurements in some cases when strict requirements are met. It is usually employed as a relative dosimeter because its response depends upon the conversion coefficient called the chemical yield G(Fe^{3+}) that has to be accurately determined in national standards laboratories. The established chemical yield value for the conventional standard Fricke solution does increase by adding organic substances to the system. Getting higher chemical yield makes the system standardization hard to achieve. Gel dosimeters usually contain additional organic materials to initiate a chain reaction that leads to a higher chemical yield [4]. Similarly, FXG contains large quantities of organic materials, mainly gelatin plus the metal ion indicator which complex chain reaction that lead to higher chemical yield.

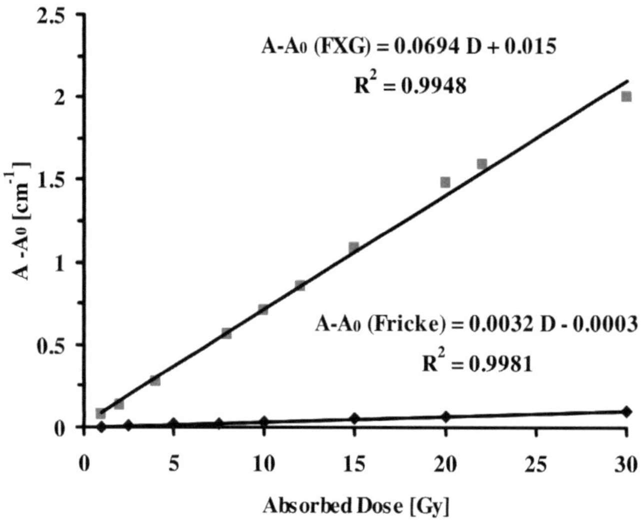

Figure 2: A comparison between the dose response of Fricke and FXG dosimeters.

Therefore, we propose a procedure for the calibration of FXG gel dosimeter against the standard Fricke system since both systems are similar in terms of the initial chemical reaction as a result for the radiation interactions. FXG response compared to that of the Fricke dosimeter in the dose range of 0-30 Gy is plotted in figure 2.

Chemical yield G is the basic quantity that can be used to calibrate FXG against the Fricke solution. Accurate determination of G is difficult; however, it is possible to take the product "ε.G" instead of the individual values for molar linear absorption coefficient and the chemical yield, ε and G, as it was recommended for the standard Fricke dosimeter [5]. Doing so help avoiding large systematic errors found in the measurements of "ε".

Journal of Physics: Conference Series **164** (2009) 012011 doi:10.1088/1742-6596/164/1/012011

$$\varepsilon \cdot G = \frac{\Delta A}{D\ .\rho..d}$$

Where, ΔA is the measured changes in the optical absorbance, *D* the given dose ρ is the sample's density and *d* is the optical path length which is normally 1 cm. The evaluation of the ε.G give a good estimate of the relative sensitivity of the radiochromic FXG gel dosimeter, which can be used as a calibration procedure if it is compared with the standard Fricke system, table 1.

Table 1: Fricke and FXG dosimeters are compared and the FXG relative sensitivity was calculated.

Dosimeter Property	Fricke Dosimeter	FXG dosimeter
ΔA	0.0032	0.0694
ρ	1.024×10^{-3} kg.m^{-3}	1.139×10^{-3} kg.m^{-3}
εG	312.5 m^{2}.J^{-1}	6093.1 m^{2}.J^{-1}
Relative Sensitivity	1	19.5

The system stability can be divided into pre-irradiation stability and post-irradiation stability. First of all, the dosimeter properties should be stable before being put to use. Storage conditions including temperature, light, humidity, etc., as well as the purity of the system materials may cause some changes with time and may affect the dosimeter reading in the absence of radiation. Ferrous ions in the FXG oxidize slowly and as a chemical system suffer the same sort of pre-irradiation changes as the Fricke system. This self-oxidation is reported to be proportional to the square of the ferrous ion concentration [6]. Self-oxidation process cannot be avoided because dissolved oxygen exists in the system is also needed for the system to function properly. However, the effect can be minimized by cutting in half the amount of ferrous ions in the system, which reduces the natural oxidation process by two folds. In all cases, however, leaving one sample from the same preparation batch un-irradiated, as a control, provides a correction for this effect and, hence, dose measurements should always be normalized by the control reading.

The dosimeter reading stability, after it has been exposed to radiation, is an important quality in any dosimetry system and especially the one that provide integrated dose measurements. Most systems show some sort of unavoidable changes. It is advantageous to make these changes as small and as reproducible as possible, therefore the post irradiation effects can be estimated and suitable corrections are applied during the standardization of the techniques. It was found that keeping the irradiated samples in a cool dark place minimizes these post irradiation effects.

The concept of reproducibility describes the effects of system properties fluctuations, the ambient conditions, and the nature of the measured radiation fields. The reproducibility can be expressed by calculating the random errors and can be estimated from the data obtained in repeated measurements. Usually, these random changes are expressed in terms of standard deviation. The FXG system consists of a number of chemicals each having its own effect on the system performance, plus the effects of environmental conditions during the preparation, irradiation, evaluation and storage. All mentioned causes contribute to the noticed variations in the system response between different dosimeter batches. However, for a set of samples prepared at the same time from the same materials stock and under similar ambient conditions, the system gives adequate reproducibility and satisfactory results. The main sources of errors can be traced to firstly, errors occurring during the preparation, and to the type of used chemicals. Secondly some errors might originate during the irradiation process and dose measurements calibration. Thirdly, standard errors originate from various equipment and tools employed or the methodology being followed; human errors are also expected at the stage of data acquisition and analysis. Finally, the dose calculation, using FXG depends on the value of the chemical yield factor which is not precisely defined and therefore leads to unavoidable calibration errors. Experimentally, the errors of a particular preparation are estimated by measuring a set of un-irradiated samples together with another set, from the same preparation, irradiated to 10 grays in a ^{60}Co gamma radiation field.

Journal of Physics: Conference Series **164** (2009) 012011 doi:10.1088/1742-6596/164/1/012011

Table 2: FXG precision in particular set of samples taken from the same batch.

Description	A cm^{-1} at 565 nm	σ %
Un-irradiated samples	0.157	3.181
Un-irradiated samples measured after 24 hours delay	0.160	3.125
Sample irradiated to 10 Gy	0.372	3.433

Table 2, shows the mean values and the percentage standard deviation in the reading of ten samples from the same batch. Here the readings are expressed in terms of the optical density at the measuring wavelength 565 nm, for un-irradiated FXG samples and another set of FXG samples exposed to 10 Gy.

4. Conclusions

It is essential to calibrate the gel system before using it in practice for radiotherapy dosimetry. The method described above offers the possibility to calibrate the gel dosimeter against standard system of similar type like the standard Fricke solution. Using standard procedure to prepare and use the gel detector will minimize errors and provide higher accuracy and reproducibility for the system.

Acknowledgments

The authors would like to thank Prof. I. Othman, AECS general director, for his support and interest in this work. Thanks also to Dr. M.S. Al-Masri for useful discussions of the project.

References

[1] INTERNATIONAL ATOMIC ENERGY AGENCY, 2000, Absorbed Dose Determination in External Beam Radiotherapy: An international Code of Practice for Dosimetry Based Standards of Absorbed Dose to Water, Technical Reports Series **No. 398**, IAEA, Vienna.

[2] ASTM, American Society for Testing and Materials, 1995, Standard Practice for using the Fricke Reference Standard Dosimetry System, **ASTM-E 1026**, Annual Book of ASTM Standards, Vol.12.02, ASTM, Philadelphia, PA, USA.

[3] Bero MA, Gilboy WB and Glover PM, 2001, Radiochromic gel dosemeter for three-dimensional dosimetry, *Rad. Phys. Chem*, **61**, 433-5.

[4] Day MJ, 1990, Radiation dosimetry using nuclear magnetic resonance: an introductory review, *Phys. Med. Biol.* **35**, 1605-9.

[5] International Commission on Units and Measurements, 1984, Radiation Dosimetry: Electron Beams with Energies Between 1 and 50 MeV, (International Commission on Radiation Units and Measurements, **Report No. 35**), ICRU, Bethesda, MD,.USA.

[6] Fricke H and Hart EJ, 1966, Chemical Dosimetry in *Radiation Dosimetry*, **Volume II**, *ed.*, Attix FH and Roesch WC, Academic Press, New York, pp 167.

The examination of fundamental characteristics of polymer gel detectors in the irradiation of 210MeV protons

T. Tominaga[1], M. Yoshioka[1], S. Usui[1], S. Hayashi[1], K. Haneda[1], D. Suga[2], K. Katahira[2], and Y. Tsunei[2]

[1]Department of Clinical Radiology, Hiroshima International University, Higashi-Hiroshima, Japan

[2] Hyogo Ion Beam Medical Center, Tatsuno, Japan

t-tomi@hs.hirokoku-u.ac.jp

Abstract. This paper is intended as a study of the radiological characteristics of the polymer gel detector in the irradiation to 210MeV protons. The depth dose distributions in the gels were examined with regard of its dose and dose rate dependences. The results indicate the energy dependence of dose-response relations due to LET effect.

1. Introduction

The purpose of our study is to examine the possibility of a polymer gel detector's utilization for the treatment plans in particle therapy and their 3D dose verifications. Since a polymer gel detectors has been introduced as new domestic tool for 3D dosimetry, their radiological characteristics in the irradiation of photon beam have been well investigated so far (e.g. reviewed in [1]). With regard to the irradiation of high linear energy transfer (LET) particles to polymer gels, a few studies have been reported [2-4]. In these studies, strong suppression of the gel's response at the Bragg peak has been observed. And it has been indicated that the suppression was related to the increase in LET at the Bragg peak region [5].

In the study of biological effects of ionizing radiation, track simulations for charged particles by Monte Carlo method have been developed including chemical reactions among water radicals, moreover the yield of hydroxyl radical induced by radiation has been estimated by empirical model of deterministic kinetics calculations [6].

The present paper deals with the idea of this model in order to interpret our experimental results on the dose response of the gel in the irradiation of 210MeV protons.

2. Materials and methods

We prepared MAGAT type gel for this study. The gel consists of water, gelatin (10%w/w), methacrylic acid (5%w/w) and tetrakis hydroxymethyl phosphonium chloride (THPC 2mM). The compounded gelatin solution was divided into cylindrical PMMA containers of 3cm diameter and 40cm length for the measurements of spread-out depth dose in the gel. All samples were irradiated in a fixed horizontal 210MeV proton beam at the facility of the Hyogo Ion Beam Medical Center. Precede the irradiation to the samples, the ionization chamber was irradiated to determine the prescribed irradiation dose at the center of the spread-out Bragg peak (SOBP) in a PE phantom, and also spread-out depth dose was measured in water. In irradiations, cylindrical containers were placed along the beam line including full proton's ionization range. Firstly various doses from 1 to 20Gy were delivered to eight containers to obtain the response - dose relations, and examine the consistency in the percentage depth dose distribution (PDD) on the irradiated dose independently. The dose rate of 4Gy/min typically applied in treatments was adopted in these irradiations. Secondary the dose rate was varied from 0.4 to 4Gy/min and delivered 4Gy to five containers to examine the dose rate effect on the spread-out depth dose and SOBP in the gel. MR images of samples were recorded by 1.5T PHILIPS Gyroscan Intera scanner with a multiple spin-echo sequence at TE_1=20ms, TE_2=100ms and TR=10s, and the transverse relaxation rate R_2 images were analyzed.

3. Results and discussions

Figure 1 shows the depth R_2 distributions in the gels that were irradiated spread-out depth dose of protons with various amount of dose. In the analysis, the image data of pre-irradiation were subtracted from the R_2 data in advance to eliminate an effect of no uniform background in the gel.

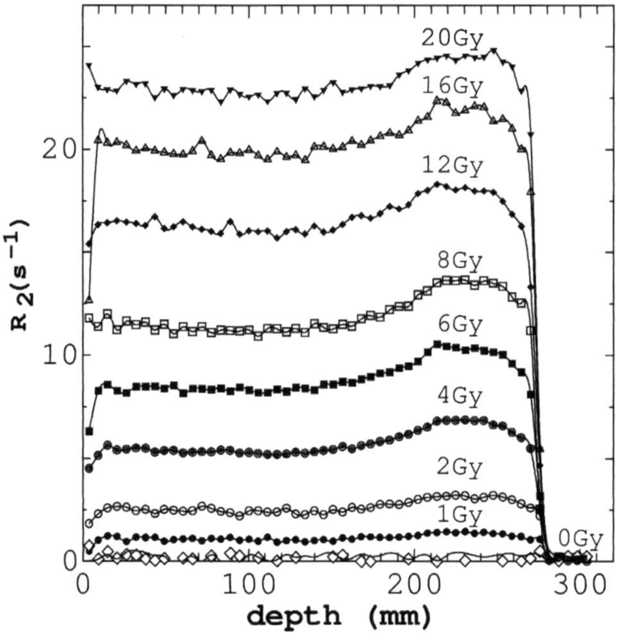

Figure 1. The depth R_2 distributions of 210MeV proton beam in the gel.

Eight cylindrical containers of MAGAT gels from the same production batch were irradiated in various doses as 1, 2, 4, 6, 8, 12, 16, 20Gy.

The horizontal axis presents the depth from the top of container where incident beam in. No constant R_2 background in each sample were subtracted from R_2 data in advance, but still ambiguity in R_2 are left near the entrance of beam due to the effect of the edge of container.

These depth-R_2 curves are supposedly correspond to the depth dose distributions, and yet R_2-dose relations are required to convert each other. The dose response curve of R_2 for MAGAT polymer gel irradiated 210MeV protons is presented in Figure 2. It must be noted that the R_2 in this figure are the values at the center of SOBP where the prescribed dose has been calibrated by ionization chamber. The data are fitted with the summational function of exponential as

$$R_2=R_{2max}(1-\exp(-A(\text{dose}-d_0)))+B\exp(-C\,\text{dose}) \qquad (1)$$

where R_{2max} is the saturation relaxation rate. Second exponential function and d_0 parameter represent the "induction" effect, which is generally seen as gently slope of response curve at low dose range [7].

In Figure 3 the PDD are graphed for the gels irradiated 4, 12, 20Gy, which were derived from the depth-R_2 relations and R_2-dose relations of the expression (1). In the comparison with corresponding PDD measured by ionization chamber, the suppression at the Bragg peak and the excess at the incident side of beam are recognized in the response of gels. This suppression is consistent with previously reported results by other groups [2], while the excess like our results has not been reported previously. And what has to be noticed is that the excess is more remarkable as much as dose increases, despite PDD should not depend on the irradiated dose, theoretically.

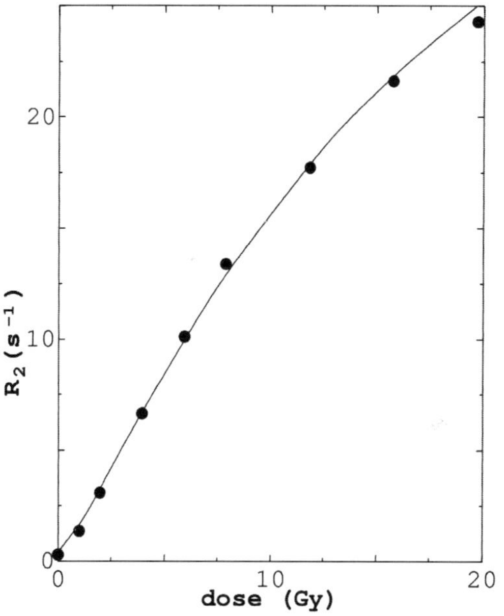

Figure 2. R_2 versus dose for the gel in the irradiation of 210MeV protons.

R_2 in this figure correspond to the value at SOBP of each depth R_2 distribution in Figure 1.

The data are fitted with the following summational function of exponential.

$$R_2 = 37.5(1\text{-}\exp(-0.0575(\text{dose}-0.55))) + 1.625\exp(-1.15\text{dose})$$

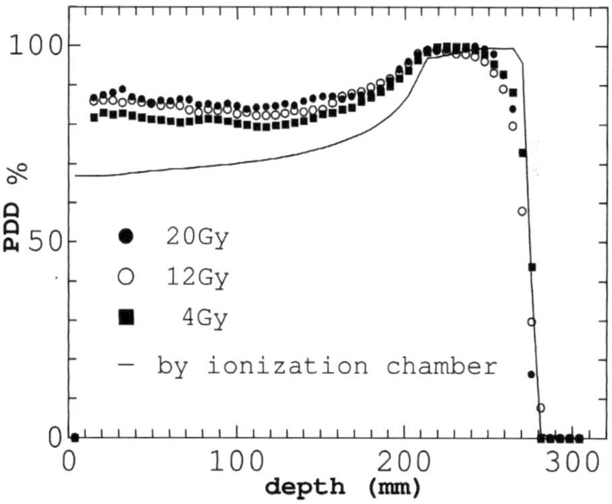

Figure 3. The percentage depth dose (PDD) of 210MeV proton beam in the gel.

Dose at each depth were converted from R_2 based on R_2-dose relation in Figure 2, and PDD were calculated as dose/dose$_{max}$.

PDD measured in water by ionization chamber were normalized at the end of our data on the horizontal axis.

These results can be explained by the depth dependence of the R_2-dose relation in the gel. The R_2-dose relation at the various depths on the PDD representing in Figure 4 are considered. The absorbed dose at each depth was estimated from the PDD measured by ionization chamber, and applied it to the depth-R_2 relations in Figure1. The R_2-dose relations at each depth appear in Figure 5.
The data at each depth are fitted with a summation of exponential function as eq.(1) with same parameters as the expression in Figure 2, except R_{2max}. The value of R_{2max} decreases with increasing the depth.

The diffusion model for a "spur"[6], which is sought of a cluster of radicals, may interpret the depth dependence, in other words "the energy dependence", of the R_2-dose relation in the gel. Though this empirical model still leaves room for improvements, the concept of this idea is reasonable and acceptable phenomenologically. In this model, a spur is initially produced along the paths of an incident particle and secondary electrons, and propagates into a medium like a diffusion of elementary wave. The radicals inside the spur interact with other radicals in the nearby spur at meeting point of both spurs. As the distance between successive initial spurs should depend on a local energy loss of an incident proton, the yields of radical also depend on the energy along its trajectory. With the energy loss advances, the interactions of radicals increase abruptly owing to decrease of distance between the spurs. In consequence the maximum number of radical and R_{2max} must be suppressed gradually as the incident proton energy decreases.

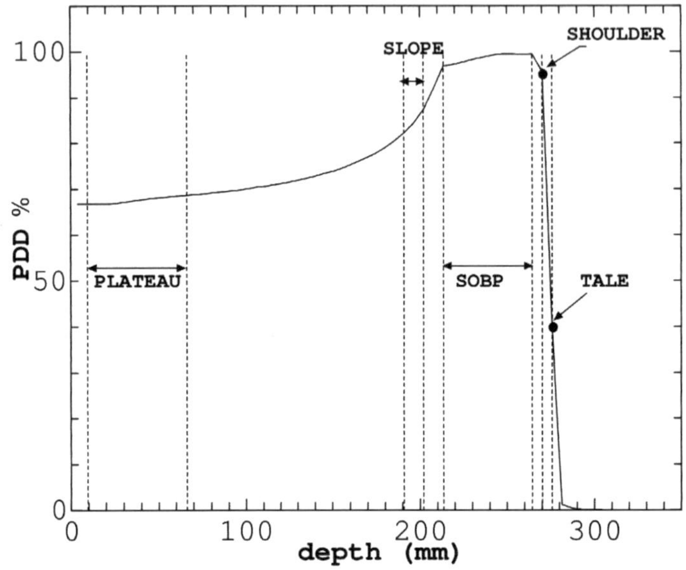

Figure 4. On the basis of SOBP measured by ionization chamber, the examinable depths are selected.

Average PDD at each depth is as follows:

PLATEAU: 68±1%,

SLOPE : 85±3%,

SOBP:99±1%,

SHOULDER: 96%,

TALE: 40%.

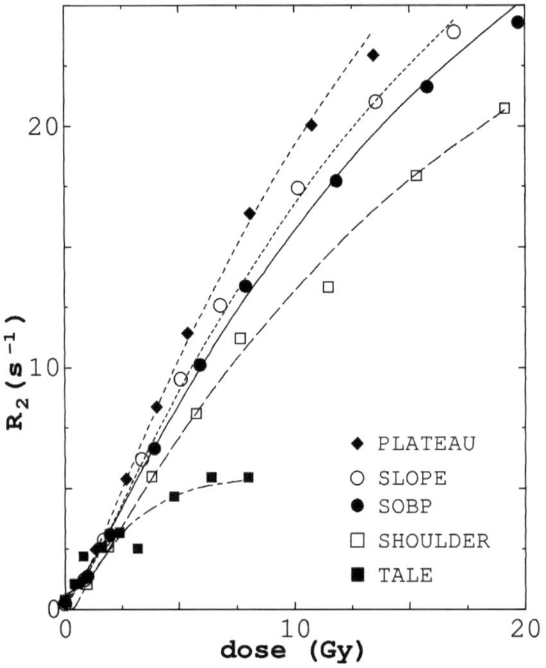

Figure 5. R_2 versus dose in the irradiation with 210MeV protons at various depths of PDD.

Each data are fitted with the following summational function of exponential,

$$R_2 = R_{2max}\,(1-\exp(-0.0575(\text{dose}-0.55))) + 1.625\exp(-1.15\text{dose})$$

where R_{2max} represents 45.75[s^{-1}], 40.0[s^{-1}], 37.5[s^{-1}], 31.5[s^{-1}], 5.5[s^{-1}] for PLATEAU, SLOPE, SOBP, SHOULDER, and TALE respectively.

Another result, which may support this model, is presented next. Figure 6 shows the seriese of SOBP in the gels with increasing dose rate. From the figure it is clear that absolute values of R_2 strongly depend on dose rate such that R_2 increase with decreasing dose rate.

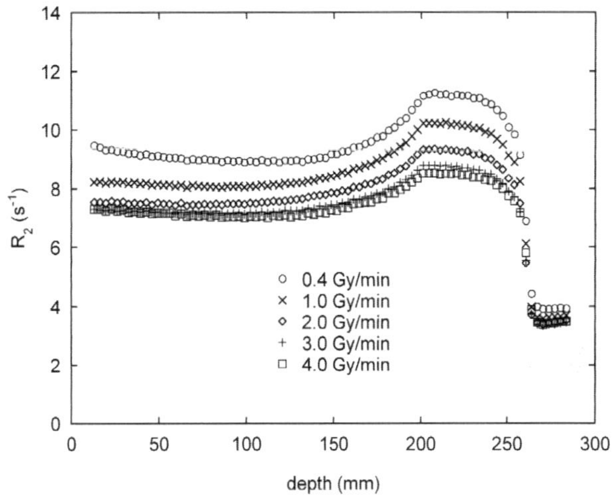

Figure 6. The depth R_2 distribution for the various dose rates of protons irradiating 4Gy.

The absolute value of R_2 decreases with increasing dose rate duc to quenching the spur with other spurs from nearby protons, while the shapes of SOBP and the suppression at SHOULDER are consistent with each other without dose rate dependence.

It is inferred from the fact that the quenching would occur along even single particle trajectory.

This fact of the dose rate dependence of R_2 can be interpreted as that the decrease of distance between nearby protons with increasing dose rate would increase the spur's interaction probability, accordingly the polymerization rate decrease.

These results lead us to the remark on the dose estimation from R_2 in the irradiation of protons. That is, the energy and density of incident proton at each location should be considered in deducing a dose distribution from R_2 distribution because gel's sensitivity depends on them.

4. Conclusion

In this work, we concerned with the dose response of MAGAT gels in the irradiation of protons. The empirical diffusion model for a spur interprets the depth R_2 distributions in the gel detector. What the polymer gel detector observes is just a distribution of radicals, and it must be quite complicated process required to deduce a dose distribution, especially in the irradiation of charged particles. It is concluded that the polymer gel detector would not be useful in dose measurements for particle therapy so much, but it can provide us a significant information as the distribution of radical induced by irradiations, which other dosimeter can not offer.

A further direction of this study will be to examine more detail on the characteristics of the dose response to clarify the relation between the radical distributions and dose distributions.

References

[1] De Deene Y, Pittomvils G, Visalatchi S. *Phys. Med. Biol.* 2007; **52** 2719-28
[2] Heufelder J, Stiefel S, Pfaender M et al. *Med. Phys.* 2003; **30** 1235-40
[3] Jirasek A, Duzenli C. *Med. Phys.* 2002; **29** 569-77
[4] Ramm U, Weber U, Bock M, et al. *Phys. Med. Biol.* 2007; **52** 2719-28
[5] Gustavsson H, Back S A, Medin J, et al. *Phys. Med. Biol.* 2004; **49** 3847-55
[6] Yamaguchi Y, Uchihori Y, Yasuda N, et al. *J. Radiat. Res.* 2005; **46** 333-41
[7] Hayashi S, Yoshioka M, Usui S, et al. *submitted to Phys. Med. Biol.*

Journal of Physics: Conference Series **164** (2009) 012013 doi:10.1088/1742-6596/164/1/012013

A new polymer gel dosimeter composed of methacrylic acid, agarose gel and THPC with gelatin

M Yoshioka[1,2], S Hayashi[1], S Usui[1], K Haneda[1], H Numasaki[2], T Teshima[2] and T Tominaga[1]

[1]Department of Clinical Radiology, Faculty of Health Sciences, Hiroshima International University, Higashi-Hiroshima, Hiroshima, 739-2695, Japan.

[2]Department of Medical Physics & Engineering, Osaka University Graduate School of Medicine, Suita, Osaka, 565-0871, Japan

m-yoshi@hs.hirokoku-u.ac.jp

Abstract. In this paper, a new type of methacrylic acid based gel dosimeter is presented. This gel contains both agarose and gelatin with deferent roles respectively. The agarose conducts itself as a gelling agent, while the gelatin relates to the graft reaction of methacrylic acid. This new type of gel excels in the long-term stability of R_2 after irradiation. The characteristics of this gel were studied by the measurements of R_2 with MRI and the direct measurements of temperature in the gel during the irradiation.

1. Introduction

The normoxic methacrylic acid (MA) based polymer gel dosimeter (MAG system) has been developed in 2001 (Fong *et al* 2001), and is considered as one of the most standard gel dosimeters so far.

The MAG gel has superior in the sensitivity and resolution to dose, however it has still unstable factors relevant to the dose responses due to the temperature dependence of gelatin's viscosity. To overcome the difficulties, it is one of solutions to use other gelling agent such as agarose. On the other hand, it has been reported recently that gelatin was the indispensable component to the polymerization reaction as a "trigger" (Hayashi *et al* 2008).

In this paper, we demonstrate the characteristics of a new type of MAG system, which contains gelatin as only a "trigger", but not as a matrix. In the point of using gelatin to initiate the polymerizations and using agarose as a gelling agent, our gel is essentially different from the conventional MAG system.

2. Materials and methods

2.1. Gel preparation

Table 1 shows composition of gels. In the table, A0.3, G1.0 and M5.0 have same compositions, but just their expressions are different. We used agarose (Agarose S for electrophoresis, Nippon Gene), gelatin (swine skin, 300 Bloom, Sigma Aldrich), MA (99%, wako), tetrakis hydroxymethyl phosphonium chloride (THPC) (80% solution in water, Sigma Aldrich), and distilled water, as component elements of gels.

The procedure of fabricating 500 g gel (0.3% agarose, 1.0% gelatin, 5.0% MA) is shown as following.

- Add 5.0 g gelatin and 1.5 g agarose to 420 g water in a beaker, and wrap it.
- Heat it up with stirring until the solution just starts bumping.
- Cool it down to 40°C in a water bath.
- Add 25.0 g MA, and stir it for 15 minutes.
- Add 180 μl THPC solution dissolved in 48.5 g water, and leave it for 1.5 minutes with stirring.
- Divide it into small containers (approximately 50 ml glass vial).
- Store the divided samples in an incubator at 23°C (same temperature as MRI room) for one day.

Table 1. Composition of polymer gel dosimeters. All values (except THPC) represent weight percentages. In the expression of the sample, "A", "G" and "M" stand for the gel remarked on the concentration of "agarose", "gelatin", and "MA", respectively. And the variations in the concentration are shown in bold.

	Concentration		
Chemical	A0.2, A0.3, A0.4	G0.5, G1.0, G2.0, G3.0	M2.0, M3.0, M5.0, M7.0
Water	93.6 – 93.8%	91.7 – 94.2%	91.7 – 96.7%
Agarose	**0.2 – 0.4%**	0.3%	0.3%
Gelatin	1%	**0.5 – 3%**	1%
MA	5%	5%	**2 – 7%**
THPC	2 mM	2 mM	2 mM

2.2. Gel irradiation

The gel samples were irradiated to 6 MV photon beam of a medical linear accelerator (EXL12SP, Mitsubishi) in a water phantom at 100 cm SSD, the top surface of vials was set at the same level of water surface. The samples were irradiated 0 - 15 Gy for the study of dose-R_2 response relations.

2.3. MR imaging

All images were acquired with 0.3 T open MRI system (AIRIS-II comfort, Hitachi) using a multiple spin-echo sequence. All of dose-R_2 response curves were acquired after one day post-irradiation. The long-term stability of R_2 in A0.3 was examined by the acquisitions at 3 - 240 hours post-irradiation.

2.4. Temperature measurements

The probe of a digital thermorecorder (TR-52, T&D) was inserted into the gel. The variations in the temperature of the gel owing to exothermic reactions were recorded every two seconds from the start to after the irradiation.

3. Results and discussion

Figure 1 shows dose-R_2 response curves of all gels in this experiment. All of gels have the non-liner dose-R_2 response curves. In Figure 1(a), each dose-R_2 curve is constant in its shape, while the absolute value of R_2 increases as the concentration of agarose increases. And the maximum variation in the gel's temperature is almost same irrespective of agarose concentration, as Table 2 shows. These results show that the agarose would not participate in the polymerization reaction, but just conduct itself as a gelling agent.

On the other hand, with increasing gelatin concentrations, the gradients of dose-R_2 response curves increase at the high dose region [Figure 1(b)] and the gels' maximum variations in temperature increase [Table 2]. With increasing MA concentrations, the gradients of dose-R_2 response increase slightly at the low dose region [Figure 1(c)] and the gels' maximum variations in temperature also increased [Table 2].

Above experimental results can be explained as follows. In the case of low gelatin concentration, the graft reaction of MA on gelatin occurs to some extent at the low dose region. As a result, the relatively large variation in R_2 is found (the reaction might cause larger structural change of gelatin molecule). After the consumption of gelatin, the MA propagation reaction is then predominant at the high dose region. The resulting MA polymers cause small variation in R_2. On the other hand, with increasing gelatin concentration, the reaction between gelatin and MA is still predominant at the higher dose region and steep gradients of the dose-R_2 response are obtained. The excessive amounts of MA over gelatin in M5.0 might be consumed in the propagation reaction. As a result, the maximum variations in temperature of M7.0 would be larger than that of M5.0.

Figure 2 shows a long-term stability of R_2 in A0.3 after irradiation. The R_2 of this gel is much less dependent on the post-irradiation time than that of the standard MAG system (De Deene *et al* 2006). The reason of this stability is interpreted that the agarose in this gel would not participate in the polymerization reaction. It is the advantage of agarose as a gelling agent.

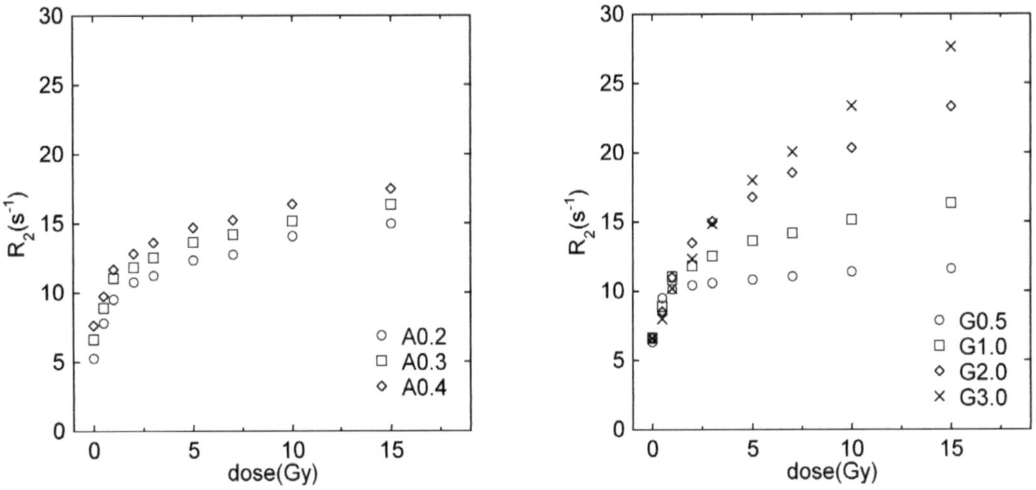

(a) (b)

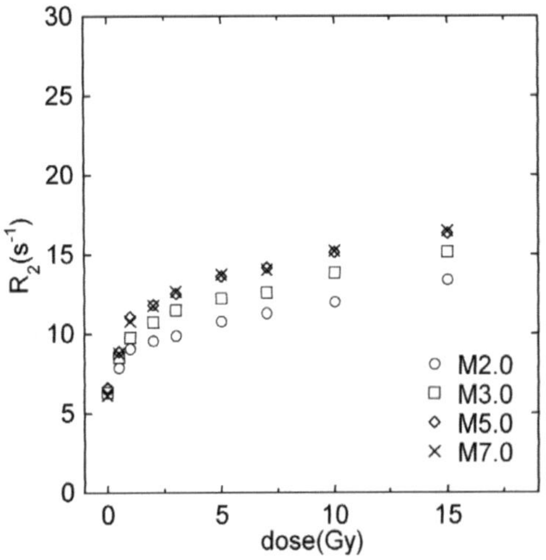

(c)

Figure 1. Dose-R_2 response curves for A0.2 - A0.4 (a), G0.5 - G 3.0 (b) and M2.0 - M7.0 (c).

Table 2. Maximum variations in temperature with 5 Gy irradiaiton.

Maximum variations in temperature (°C)										
A0.2	A0.3	A0.4	G0.5	G1.0	G2.0	G3.0	M2.0	M3.0	M5.0	M7.0
2.1	2.2	1.9	1.6	2.2	2.5	2.6	0.7	1.1	2.2	2.7

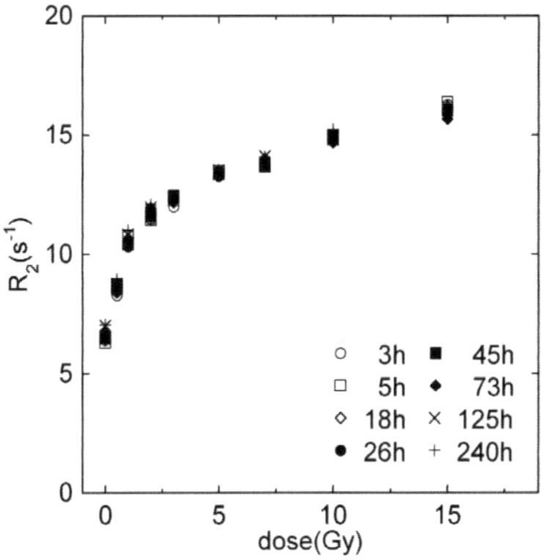

Figure 2. The long-term stability of R_2 in A0.3.

4. Conclusion

In this study, we investigated the characteristics of a new MAG type polymer gel dosimeter using agarose as a gelling agent. The dose-R_2 response is comparable with that of the conventional MAG system and the thermal stability is superior. Our gel appears to be a promising dosimeter because it minimizes both the structure relaxation of gelatin at storing and the melting of the gel by the exothermic polymerization reaction. Further studies on developing of the more stable dosimeters using combination of gelatin and the other gelling agents are undergoing in our group.

References

[1] Fong PM, Keil DC, Does MD, Gore JC 2001 Polymer gels for magnetic resonance imaging of radiation dose distributions at normal room atmosphere. *Phys. Med. Biol.* **46** 3105-13

[2] S Hayashi, M Yoshioka, S Usui et al. *to be appeared in this conference*

[3] De Deene Y, Vergote K, Claeys C, De Wagter C 2006 The fundamental radiation properties of normoxic polymer gel dosimeters: a comparison between a methacrylic acid based gel and acrylamide based gels. *Phys. Med. Biol.* **51** 653-73

Journal of Physics: Conference Series **164** (2009) 012014 doi:10.1088/1742-6596/164/1/012014

Internal Calibration of gel dosimeters: A feasibility study

J. V. Trapp, T. Kairn, S. Crowe, A. Fielding

School of Physical and Chemical Sciences

Queensland University of Technology

GPO Box 2434, Brisbane 4001, Australia

j.trapp@qut.edu.au

Abstract. In this work we test the feasibility of a new calibration method for gel dosimetry. We examine, through Monte Carlo modelling, whether the inclusion of an organic plastic scintillator system at key points within the gel phantom would perturb the dose map. Such a system would remove the requirement for a separate calibration gel, removing many sources of uncertainty.

1. Introduction

For gel dosimeters to be clinically useful the acquired dose information must be reliable, accurate and precise. There are several sources of uncertainty in absorbed dose measurements taken with gel dosimeters, broadly summarized as chemical, radiation, scanning [1], and mathematical contributions [2]. Much of the uncertainty arises during the calibration process, and there is often a tradeoff between benefits and drawbacks with different calibration methods.

To use a gel dosimeter one would normally produce a large enough volume so that a 'phantom gel' and 'calibration gel' are taken from the same manufacturing batch, therefore ensuring initial chemical consistency. The calibration gel is irradiated to a range of doses such that a sufficient number of data points are available to determine a calibration function of 'signal vs dose' (the signal being R2, optical attenuation, density etc, depending on imaging modality, and noting that the calibration gel is imaged under the same conditions as the phantom gel). The calibration function is then used to convert images of the phantom gel to dose.

Several methods of handling the calibration gel have been presented in the literature. These methods include dividing the calibration gel into several small vials which are irradiated to varying doses [3], irradiating a single large flask of calibration gel with several small beams [4, 5] or with electron depth-dose information [6] and irradiation along the axis of gel-filled test tubes using linac depth-dose data in water [7].

Some calibration methods eliminate the need for a calibration gel by obtaining data points directly from the phantom gel, therefore eliminating some setup and positional uncertainties. These methods include comparison of selected regions of interest in the phantom gel with planning system predictions [8] and equalization of dose-area histogram data from an image slice of the gel to that of a film placed at an equivalent depth [9].

There is significant potential for introducing errors into the irradiation process when separately irradiating the phantom gel and the calibration gel. Taylor *et al* [10, 11] examined several variations of the above methods and found that errors in the dose to the calibration phantom tend to be within 1% for the 'small vials' and 'large flask' method, and about 2% for the 'test tube' method. In addition to uncertainties in the delivered dose, because the calibration gel is separate to the phantom gel additional errors may be introduced due to different temperature histories, irradiation conditions and imaging conditions [1]. For the methods which do not utilize a separate calibration gel, there is the possibility of errors due to region-of-interest selection, planning system inaccuracies, scattered radiation contributions and image slice selection.

The ideal solution to reduce calibration uncertainty would be to place reference dosimeters within the phantom gel at key locations to obtain the calibration points. The reading obtained from the dosimeter could then be accurately applied to the post-irradiation images, thus eliminating the sources of error described above. Unfortunately, placing dosimeters such as ionization chambers and thermoluminscent dosimeters inside the phantom gel would perturb the radiation field, thus altering the very dose information that was intended to be measured. One alternative would be to utilize a gel equivalent dosimeter for this task, and as such, organic plastic scintillators with a density of 1.032 g/cm^3 could prove suitable. In this work we present a Monte Carlo feasibility study for the use of organic plastic scintillators to provide calibration data points from directly from within the phantom gel.

2. Methods and Materials

Monte Carlo simulations were performed using the EGSnrc/BEAMnrc software [12]. The setup simulated was a 20 × 20 × 20 cm volume of polyacrylamide gel (PAG) [13] embedded with scintillators of volume 2 × 2 × 0.25 cm. A pre-commissioned model [14] of an Elekta Precise linac was used with a source to surface distance of 90 cm and field size of 10 × 10 cm. Voxel size was 0.25 × 0.25 × 0.25 cm. The nominal photon energy was 6 MeV.

Figure 1 shows the geometry of the scintillator/gel setup. Three scintillators were positioned on the beam central axis at depths of 1.5 cm, 10 cm, and 18.5 cm. In practical use, organic plastic scintillators are coated with a reflective paint and opaque plastic tape [15] and so the simulations were repeated to include scintillator only, and scintillator with reflecting paint and light blocking tape. The atomic compositions of the various materials are given in Table 1 [10]. For the simulations where paint and tape were used the materials were approximated to be proportionally 'mixed' together with the gel in the voxels immediately surrounding the scintillators having a paint layer of 100 μm thickness and tape layer of 500 μm thickness. Additionally, two simulations of a 'gel only' phantom were obtained with different random number seeds for comparison of results.

Figure 1 – Layout of phantom geometry

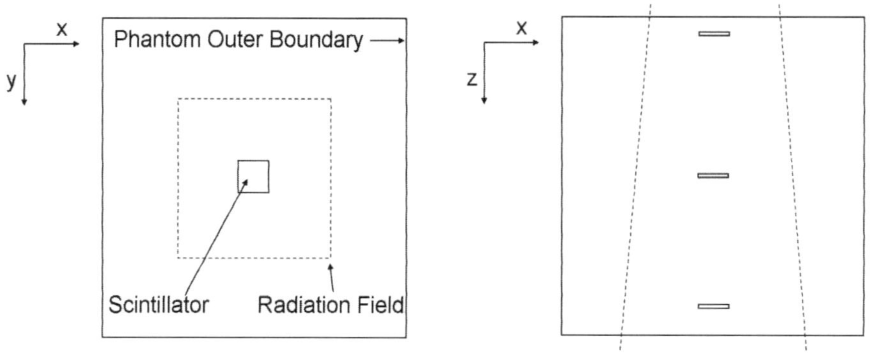

Table 1 - Different simulations performed

Setup	Simulation	Chi-square (compared to Simulation 1)
1	PAG only	
2	PAG with three scintillators on central axis as shown in Figure 1	0.000070
3	Same as Setup 2, however voxels surrounding the scintillator have paint and tape contributions	0.000069

Table 2 - Composition of Materials

Material	Density (g/cm^3)	Atomic Composition – Fraction by Weight				
		H	C	N	O	Ti
PAG	1.02	0.107	0.047	0.017	0.829	
Scintillator	1.032	0.085	0.915			
Paint, tape and gel combination	1.008	0.112	0.227	0.021	0.634	0.006

3. Results

Figure 2 shows the central axis depth dose information for the various simulations. The depth dose is normalized to the dose in a 'PAG only' phantom at 10 cm depth. Included in Table 1 is a chi-square comparison between each simulation and the 'PAG only' phantom. From the figure and the chi-square results, it can be seen that the inclusion of an organic plastic scintillator (with or without the paint and tape) in the gel has little overall perturbation of the dose in the gel. This is also confirmed by Bland and Altman testing [16] (not shown).

Figure 2 – depth dose data for the systems examined

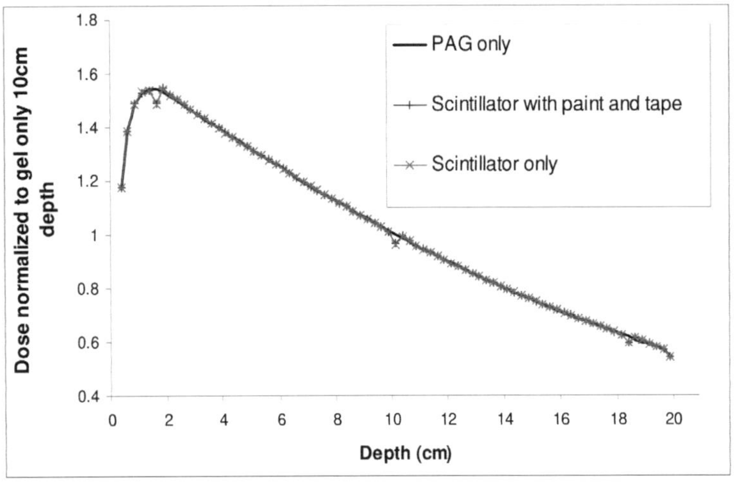

Figure 3 shows the data from Figure 2 in the region close to d_{max}. Figure 4 is a plot of the percentage difference in dose with depth between the 'PAG only' simulation and the simulation with gel, paint and tape. From the figures it can be clearly seen that there is a decrease in dose of approximately 3.52 % at the location of the scintillators as compared to the 'PAG only' simulation, followed by a slight increase dose of approximately 1% in the gel for approximately 1 cm downstream of the scintillator. The decrease in dose at the scintillator can be overcome through careful calibration of each scintillator for dose-in-water. The increased dose in the 1 cm following the location of the scintillator is thought to be due to increased electron scatter from the scintillator; however this requires further investigation. Nevertheless, the initial solution would be to ensure that the 1 cm depth of gel immediately following the scintillator is not be used in analysis of dose maps, and that scintillators not be placed within 1 cm of key regions of the phantom gel.

Figure 3 – Depth dose information close to d_{max}

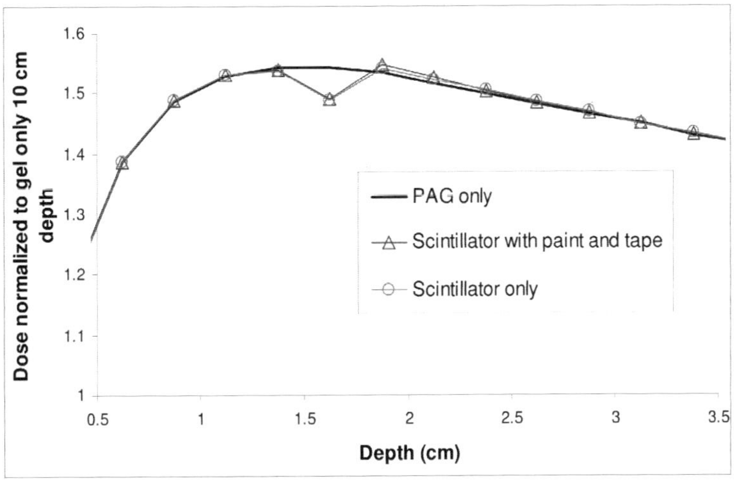

Figure 4 – Percentage difference in depth dose between Simulations 1 and 3

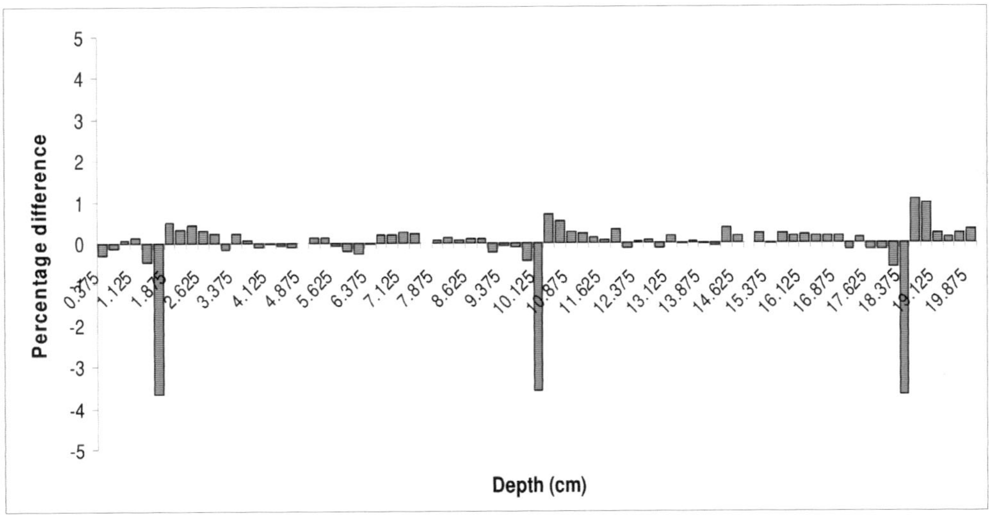

4. Discussion

This work has shown that, in principle, the direct measurement of dose within the phantom gel can be achieved through the use of well calibrated organic plastic scintillators. This work is ongoing however initial results indicate that the scintillators should be located such that they are further than 1 cm 'upstream' from any key locations within the phantom gel.

5. Acknowledgements

This work was supported by an Australian NHMRC project grant. Computational resources and services used in this work were provided by the HPC and Research Support Unit, Queensland University of Technology, Brisbane, Australia.

References

[1] De Deene, Y., *On the accuracy and precision of gel dosimetry.* J. Phys. Conf. Ser., 2006. **56**: p. 72-85.

[2] Trapp, J.V., G. Michael, P.M. Evans, C. Baldock, M.O. Leach, and S. Webb, *Dose resolution in gel dosimetry: Effect of uncertainty in the calibration function.* Phys. Med. Biol., 2004. **49**: p. N139-N146.

[3] Appleby, A., E.A. Christman, and A. Leghrouz, *Imaging of spatial radiation dose distribution in agarose gels using magnetic resonance.* Med. Phys., 1987. **14**(3): p. 382-384.

[4] Maryanski, M.J., J.C. Gore, R.P. Kennan, and R.J. Schulz, *NMR relaxation enhancement in gels polymerized and cross-linked by ionizing radiation: A new approach to 3D dosimetry by MRI.* Magn. Reson. Imaging, 1993. **11**: p. 253-258.

[5] Oldham, M., I.C. Baustert, C. Lord, T.A.D. Smith, M. McJury, A.P. Warrington, M.O. Leach, and S. Webb, *An investigation into the dosimetry of a nine-field tomotherapy irradiation using BANG-gel dosimetry.* Phys. Med. Biol., 1998. **43**: p. 1113-1132.

[6] Trapp, J.V., M. Partridge, V.N. Hansen, P. Childs, J. Bedford, A.P. Warrington, M.O. Leach, and S. Webb, *The use of gel dosimetry for verification of electron and photon treatment plans in carcinoma of the scalp.* Phys. Med. Biol., 2004. **49**: p. 1625-1635.

[7] Oldham, M., M. McJury, I.C. Baustert, S. Webb, and M.O. Leach, *Improving calibration accuracy in gel dosimetry.* Phys. Med. Biol., 1998. **43**: p. 2709-2720.

[8] Love, P.A., P.M. Evans, M.O. Leach, and S. Webb, *Polymer gel measurement of dose homogeneity in the breast: comparing MLC intensity modulation with standard wedged delivery.* Phys. Med. Biol., 2003. **48**: p. 1065-1074.

[9] Trapp, J.V., A.P. Warrington, M. Partridge, A. Philps, M.O. Leach, and S. Webb, *Three dimensional measurement of radiation dose in grid therapy.* Phys. Med. Biol., 2004. **49**: p. N317-N323.

[10] Taylor, M.L., R. Franich, P.N. Johnston, M. Miller, and J.V. Trapp, *Systematic variations in polymer gel dosimeter calibration due to container influence and deviations from water equivalence.* Phys. Med. Biol., 2007. **52**: p. 3991-4005.

[11] Taylor, M.L., R.D. Franich, J.V. Trapp, and P.N. Johnston, *A comparative study of the effect of calibration conditions on the water equivalence of a range of gel dosimeters*, in *2007 IEEE Nuclear Science Symposium Conference Record*, B. Yu, Editor. 2007, IEEE: Piscataway, NJ, USA. p. 3170-3175.

[12] Rogers, D.W.O., B.A. Faddegon, G.X. Ding, C.M. Ma, and J. We, *BEAM: A Monte Carlo code to simulate radiotherapy treatment units.* Med. Phys., 1995. **22**(5): p. 503-524.

[13] Baldock, C., R.P. Burford, N. Billingham, G.S. Wagner, S. Patval, R.D. Badawi, and S.F. Keevil, *Experimental Procedure for the Manufacture and Calibration of Polyacrylamide Gel (PAG) for Magnetic Resonance Imaging (MRI) Radiation Dosimetry.* Phys. Med. Biol, 1998. **43**: p. 695-702.

[14] Kairn, T., D. Cassidy, P.M. Sandford, and A.L. Fielding, *Radiotherapy treatment verification using radiological thickness measured with an amorphous silicon electronic portal imaging device: Monte Carlo simulation and experiment.* Phys. Med. Biol., 2008. **In Press**.

[15] Williams, K., N. Robinson, J.V. Trapp, T. Ackerly, R. Das, P. Kemp, and M. Geso, *A portable organic plastic scintillator dosimetry system for low energy X-rays: A feasibility study using an Intraoperative X-ray unit as the radiation source.* Journal of Medical Physics, 2007. **32**(2): p. 73-76.

[16] Bland, J.M. and D.G. Altman, *Statistical methods for assessing agreement between two methods of clinical measurement.* Lancet, 1986. **1**(8476): p. 307-310.

Dosimetric performances of optically detected Fricke-agarose-Xylenol Orange gel

A. M. Luciani[ab], A.Palma[ab], S. Grande[ab], P. Sordi[a], L. Guidoni[ab], V. Viti[ab], E. Di Castro[c], C. De Felice[c], A. Lo Bosco[c]

[a]Dipartimento Tecnologie e Salute, Istituto Superiore di Sanità and [b]INFN Gruppo Collegato Sanità, Viale Regina Elena 299, 00161, Roma (Italy)

[c]Istituto di Radiologia, Fisica Sanitaria Policlinico Umberto I, Università "La Sapienza" Roma

Corresponding author's e-mail: guidoni@iss.it

Abstract. A first prototype of optical scanner based on a CCD camera and the use of Fricke-agarose-XO has allowed us to reconstruct 3D dose maps, monitoring doses in the range 0.5-3 Gy. Measurements performed with laboratory tests and with clinical beams would indicate that the uncertainty involved in dose mapping are compatible with requirements of a 3D dosimetry for the verifications of treatment plans.

1. Introduction

In the field of gel dosimetry, we are examining Fricke-agarose-Xylenol Orange (XO) gels for obtaining 3D dose maps. This kind of gels have the advantage of a simple and inexpensive preparation. Addition of the dye XO complexing the oxidized metal ions in Fricke-gel systems provides the opportunity to measure the dosimeter absorbance in the range of visible light, enhancing both sensitivity and specificity of the optical analysis. The main drawback of the system, that is ion diffusion after irradiation can be overcome by the use of optical techniques particularly by utilizing a CCD for detection, due to very quick acquisitions of 2D projections. The presence of XO in the gel further reduces ferric ion diffusion.

A first prototype of optical scanner based on a CCD and the use of Fricke-agarose-XO has allowed us to reconstruct 3D dose maps, monitoring doses in the range 0.5-3 Gy. Laboratory tests and measurements with clinical beams were then performed.

2. Preparation and measurements details.

Optical readings were made at measurements of the radio-chromic Fricke-agarose-Xylenol Orange gel. Optical differences ΔOD of samples before and after irradiation were measured. For this purpose, a CCD based optical scanner was used. Details on the dosimeter preparation and on optical measurements are here reported.

2.1. Choice of the wavelength

To exploit the formation of a coloured complex between the dye Xylenol Orange (XO) and ferric ions, it was necessary to take into consideration characteristics of the complex and optimum wavelength readings, that were the subject of many studies. In particular detailed studies [1] have shown that, for determination of unknown amounts of ferric ions, it is not convenient to read the solution at the peak absorbance, near 543 nm, because in that spectral region a relevant contribution of free dye is also present. A correction for this contribution is possible, but strongly dependent on the extinction coefficient of both components, that may change due to differences in pH and to the impurities often present in the dye, depending on dye source. It is therefore suggested to read the Optical Density (OD) at 560 nm, in order to minimize the contribution of free XO. At this frequency, the absorption of the free dye is very low. According to the study reported in [1], the estimated error at 560 nm is 1.6% in a solution of 25 mM H_2SO_4 , 25 mM Fe^{+3}, 125 mM XO. The light box of our scanner was set for emitting monochromatic light at 567 nm. The choice of 567 nm further decreases this error, with negligible effect on gel sensitivity.

2.2. Binning of data

To increase signal to noise, we used the binning procedure. Binning is a process of combining the charges in adjacent pixels on the CCD, prior to readout, which effectively increases the size of the pixel. The net result is a charge that is the sum of the charges of the binned pixels which yields an improved signal-to-noise ratio. Binning quantities are user-set in any aspect-for example 2 x 2, 4 x 4, 1 x 100, etc. We have considered 3x3, 6x6 and 11x11 pixels.

2.3. Dosimeter composition (2D Projections)

For accurate measurements, the absorbance of the complex must be independent by the relative concentration of XO. This condition is satisfied when concentration of XO is at least three times that of ferric ion. According to reference [1], a ratio of 5 gives safer margins.

2D measurements were then performed, by examining 2D projections, in order to control the performance as far as composition and stability are concerned for the dosimeter gel. Sucrose was included in gel formulation and the resulting dosimeters were compared. The presence of sucrose, keeping the agarose molecules apart, is intended to avoid the formation of lumps, allowing to obtain a

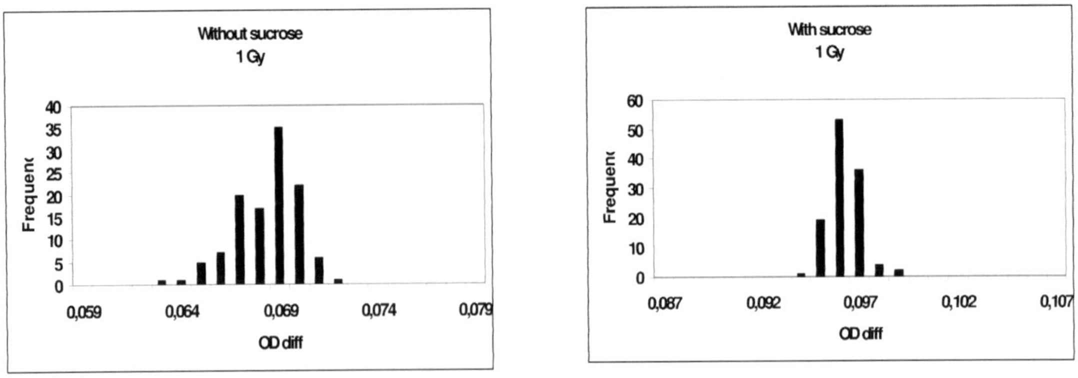

Figure 1. Distributions of the optical difference values (ODdiff) for a 2D projection (binning 6x6 pixel) for irradiated gels without and with addition of sucrose for a homogeneous irradiation with 1 Gy.

very homogeneous gel with a consequent small uncertainty of pixel values. Figure 1 shows the distribution of ΔOD values for a 6x6 pixel of gel without (left) and with (right) sucrose in a region of interest (ROI) of homogeneously irradiates gels. In Table 1, comparison of the performances of gels with (Gel II) and without sucrose (gel I) is shown for two irradiation doses and by binning data with different pixel sizes. Mean values and percent standard deviation are reported.

From the present results we may infer:

- percent standard deviation is always lower in gel containing sucrose; this should be a consequence of a more uniform gel preparation.
- sensitivity is higher for gel containing sucrose.

Table 1. ΔOD mean values ΔOD_m and percent standard deviation from gel without and with sucrose, homogenously irradiated with a ^{137}Cs source at two different doses and binning data with different numbers of pixels.

	No sucrose			**Sucrose**	
Binning	**ΔOD_m**	**% st dev**		**ΔOD_m**	**% st dev**
			1 Gy		
3x3	0.068	3.07		0.095	1.3
6x6	0.068	2.43		0.096	1
11x11	0.068	2.05		0.095	1.1
			3 Gy		
3x3	0.18	1.41		0.29	1.2
6x6	0.18	1.21		0.09	1.1
11x11	0.18	1.14		0.29	1.2

From data reported in Table 1, it is possible also to observe the reduction in the percent standard deviation by increasing numbers of pixels for data binning.

2.4. Differences in gel preparations

Performance of different gel preparations with the same composition was then studied. Gel stability as a function of time was checked showing that the optical density of the gel samples was stable from 24 hr up to 32 hr after the gel preparation for both control and irradiated samples keeping the gel in the temperature range 22-25 °C

Comparison between four samples deriving from preparations of different days were then compared, by taking measurements from two samples from the same preparation (a and b). Parameters obtained by linear fitting the dose response curve with the equation ΔOD=m*Dose, in the range 0-4 Gy, are shown in Table 2. These data show that the same preparation had very close parameters, while larger differences were observed in different days. Consequently, this kind of dosimeter can be used only for relative dosimetry.

Table 2. Parameters of linear fitting of dose response curves for samples from the same preparation (cuvette a and b) and from different preparations (marked with the number of the experiment).

N. Exp	**m_a**	**SE_a**	**R^2_a**	**m_b**	**SE_b**	**R^2_b**	**$\frac{(m_a - m_b)}{\sqrt{(SE_a)^2 + (SE_b)^2}}$**
1	0.096	0.001	0.9997	0.097	0.001	0.9996	0.9
2	0.091	0.004	0.9998	0.091	0.003	0.9994	0.03
3	0.097	0.001	0.99	0.099	0.001	0.9998	1.6
4	0.102	0.001	0.9991	0.1091	0.0005	0.9986	0.1

3. 3D dose maps

The gel was irradiated in a cylinder 0.7 cm (r) x 10 cm (h). 180 2D projections were obtained by rotating the sample and then 3D dose maps were obtained by means of an *ad hoc* software based on a back projection algorithm.

Results from two irradiation sections are here presented. The first one was relative to a laboratory test performed by irradiating the sample in a ^{60}Co cell where a dose gradient was obtained by shielding part of the sample during irradiation. The second one was relative to an irradiation with clinical beams of a Linear Accelerator (LA) of 6 MV.

3.1. Laboratory tests

Uncertainty of pixel values is the main source of uncertainty for dose measurements. To evaluate this uncertainty, the gel containing cylinder was irradiated in a ^{60}Co cell with a dose of 10 Gy and by shielding part of the gel. TLD dosimetry was used for calibration. With this arrangement, dose was homogeneous within approximately 2 % in the axial direction, that is perpendicular to the longitudinal axis where the dose gradient was obtained. Figure 2 shows the percent standard deviation of the optical density difference in 10 mm raws, with a spatial resolution of 0.51 cm^3, as a function of the dose. It is possible to observe that, with this spatial resolution, percent standard deviation exceeded 2 % only in the lowest dose range.

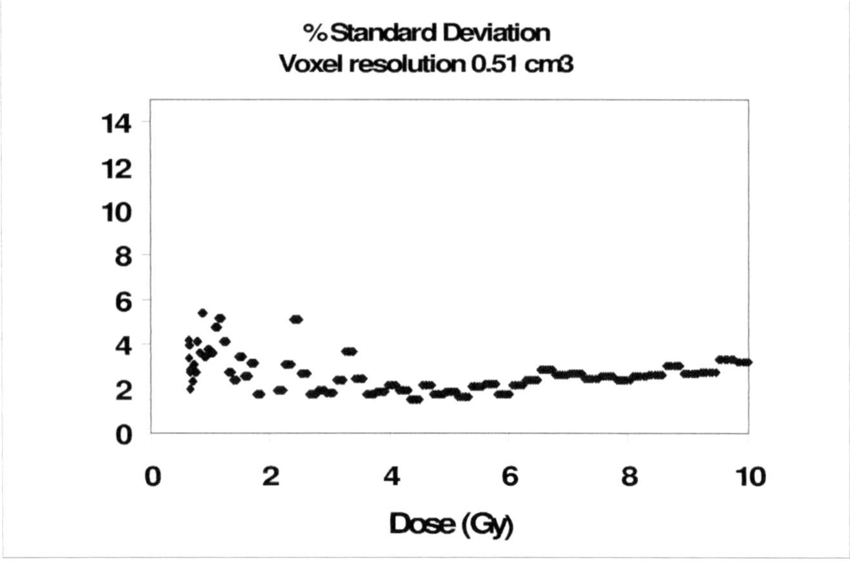

Figure 2- Percent standard deviation of the optical density in a 10 mm raw with a spatial resolution for voxels of 0.51 cm^3, as a function of the dose.

3.2. Dose maps with clinical beams

Irradiation of the gel containing vessel set in a polystyrene phantom was performed with a LA at 6 MV, as shown in Figure 3a. The irradiation field was 10x10, with 3 Gy to the isocenter. Doses were measured with a ionization chamber I.C. (0.13 cc) to give the profile in water phantom. Its accuracy was 1.9%. Absolute dose measurement was performed in polystyrene phantom with a I.C. of 0.6 cc. Figure 3b shows one longitudinal section from the reconstructed 3D map obtained from the 2D projections through the software for the 3D reconstruction.

a) c)

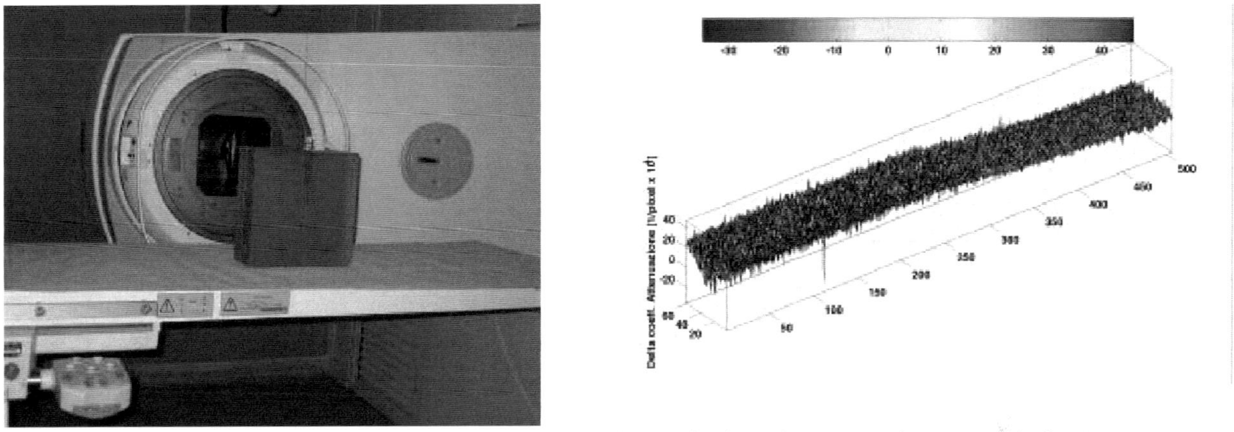

Figure 3. a) Geometry of irradiation of the Fricke-gel dosimeter in the polystyrene phantom with the 6 MV LA; b) a longitudinal section from the reconstructed 3D map.

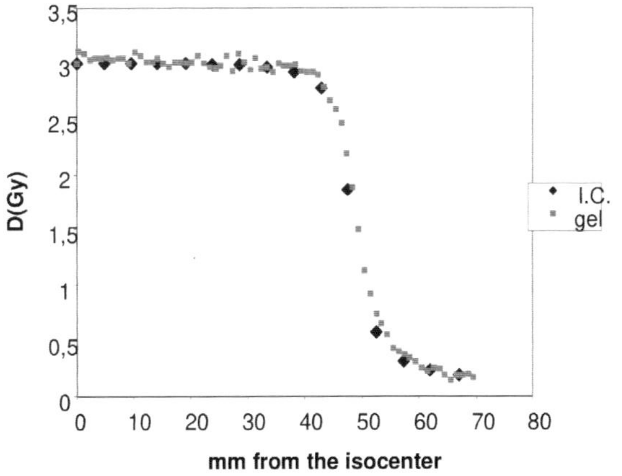

Figure 4. Comparison between dose values obtained with the I.C. and the gel (longitudinal column) with a spatial resolution $1 mm^3$.

Comparison between relative doses measured with I.C. and that measured with Fricke-gel is shown in Figure 4. These data show that in the plateau region with a $1 mm^3$ sptial resolution the standard deviation in this gel dosimeter was 0,045822, that is 1.5%.

4. Conclusions

Measurements performed with laboratory tests and with clinical beams would indicate that the uncertainty involved in dose mapping are compatible with requirements for a 3D dosimetry for the verifications of treatment plans.

Acknowledgments - The paper has been performerd in the framework of the "European Integrated Project MAESTRO" (Methods and Advanced Equipment for Simulation and Treatment in Radio-Oncology, N° LSHC-CT-2004-503564, 1st May 2004 - 30th April 2009).

References

[1] Gay C, Collins J, Gebicki M. Determination of iron in solutions with the Ferric-Xylenol Orange complex. Anal. Biochem. 273, 143-148 (1999).

Journal of Physics: Conference Series **164** (2009) 012016 doi:10.1088/1742-6596/164/1/012016

A dosimetric evaluation of tissue equivalent phantom prepared using 270 Bloom gelatin for absorbed dose imaging in Gamma Knife radiosurgery

C C Cavinato[1], O Rodrigues Jr[1], J H Cervantes[2], S R Rabbani[2] and L L Campos[1]

[1] Radiation Metrology Centre, Instituto de Pesquisas Energéticas e Nucleares – CNEN/SP, Sao Paulo, Brazil

[2] Physics Institute, University of Sao Paulo, Sao Paulo, Brazil

E-mail: ccavinato@ipen.br

Abstract. Tissue equivalent gel phantoms have been widely studied in radiation therapy for both relative and reference dosimetry. A Fricke xylenol gel (FXG) spherical phantom was evaluated by means of magnetic resonance image method (MRI) to measure absorbed dose distribution resulted from gamma knife irradiation. The FXG phantom was prepared using 270 Bloom gelatin. The gelatin is a tissue equivalent material, of easy preparation, can be used to mold phantoms into different shapes and volumes, is commercially available and inexpensive. The results show that the Fricke gel phantom prepared with 270 Bloom gelatin satisfy the requirements to be used for the quality control in stereotactic radiosurgery using Gamma Knife technique and may constitute one more option of dosimeter in radiation therapy applications.

1. Introduction

Among the high-precision techniques used in stereotactic radiosurgery for treatment of intracranial lesions, which they use narrow ionizing radiation beams [1], the Gamma Knife is the most precise modality currently available [2]. This technique allows obtaining a local energy release in tumors saving the surrounding health tissues. Therefore there is the necessity of getting the three-dimensional (3-D) images of in-phantom absorbed dose, achieving high precision and the most reliable resolution [3].

Gel dosimetry is a technique for dose imaging based on water-equivalent gels. Phantoms can easily be obtained and used as dosimeter to obtain 3-D dose distribution. Different kinds of gel dosimeters have been developed, the two main investigated groups are the aqueous gels, the Fricke gel dosimeters [4,5,6,7,8,9] and the based on radiation-induced polymerization and cross-linking of acrylic monomers, the polymer gels [3,10,11,12,13]. Imaging techniques such us spectrophotometry and MRI scanning, have been proposed [14,15,16,17]. Compared with other imaging methods, magnetic resonance imaging (MRI) presents superior properties for planning, guiding, monitoring and controlling therapeutic interventions, including minimally invasive procedures [16,17,18]. Applied in radiation therapy, the main advantages of MRI are the high spatial resolution and the direct acquisition of multiplanar and 3-D imaging data.

The Fricke gel dosimeter has been studied for determine both dose and dose distribution with accuracy and precision [12,13] in radiation treatments. Oxidation of ferrous ions (Fe^{+2}) to ferric ions (Fe^{+3}) occurs when the dosimeter is exposed the ionizing radiation [14] with the oxidation yield proportional to the absorbed dose and the absorbed dose distributions can be imaged by MRI. The studied Fricke gel solutions have been prepared using 300 Bloom gelatin [8,14,19,20] that is imported and very expensive in Brazil.

The purpose of this work is to make the dosimetric evaluation of a FXG spherical phantom prepared using 270 Bloom gelatin to measure absorbed dose distribution resulted from gamma knife irradiation.

2. Materials and methods

2.1. Spherical phantom preparation

The spherical glass housing (1 mm thick wall, 12.5 mm diameter and 1.0 L volume) was filled with Fricke xylenol gel solution (FXG) in order to simulate the human brain. The FXG solution was prepared with 1 mM ferrous ammonium sulphate [$Fe(NH_4)_2(SO_4)_2 \cdot 6H_2O$], 1 mM sodium chloride (NaCl), 0.1 mM xylenol orange dye ($C_{31}H_{28}N_2Na_4O_{13}S$), 50 mM sulphuric acid (H_2SO_4), tri-distilled water [4] and 5% by weight 270 Bloom porcine gelatin.

2.2. Phantom irradiation and calibration

The phantom positioned at the headframe was irradiated to a prescription dose of 40 Gy and 50% isodose surface using a Leksell Gamma Knife system model B.

For dose response calibration, gel samples were irradiated at predefined ^{60}Co gamma dose levels between 1 and 90 Gy.

2.3 MR Imaging

The evaluation technique employed was the Magnetic Resonance Imaging (MRI) using the PHILIPS GYROSCAN S15/ACS scanner.

MR scans of the calibration gel samples were used to establish a dose response curve as a function of the calculated relaxation rates R_1. This curve is the reference curve for dose readings in the phantom.

Before and after gamma knife irradiation the phantom MR images were acquired, the T_1 weighted multi-slice of the phantom were also obtained and, using localized spectroscopy, T_1 was measured at the irradiated volume and far from it in order to correlate the pixel intensity with the irradiation dose. The standard Z orientation slices were used in image acquisition (transaxial section image).

A MATLAB program was created to calculate the pixel T_1 values and reconstruct the dose in 3D to the patient from the measured gamma knife images.

3. Results and discussions

The calibration curves, R_1 and NMR signal intensity as a function of the dose are presented in figure 1. The 3D phantom representation is present in figure 2. A 33 MRI slices (zSlices in figure 2) were processed in Matlab program. The dose colormap in figures 2, 3, 4 and 5 were reported as percentage of the prescription dose (% PD). Selected MRI slices are presented in figure 3.

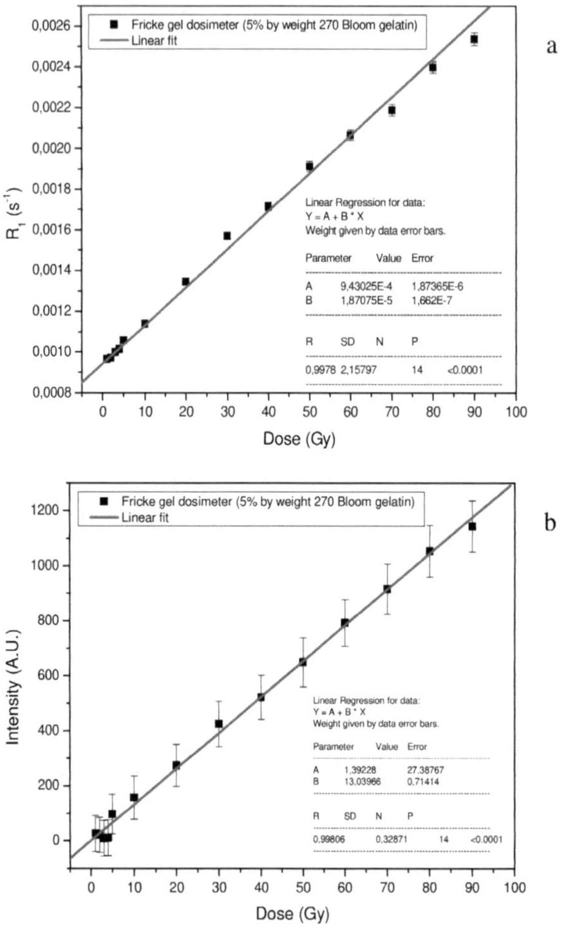

Figure 1. Dose-response curves calculated using R_1 (a) and NMR signal intensity (b).

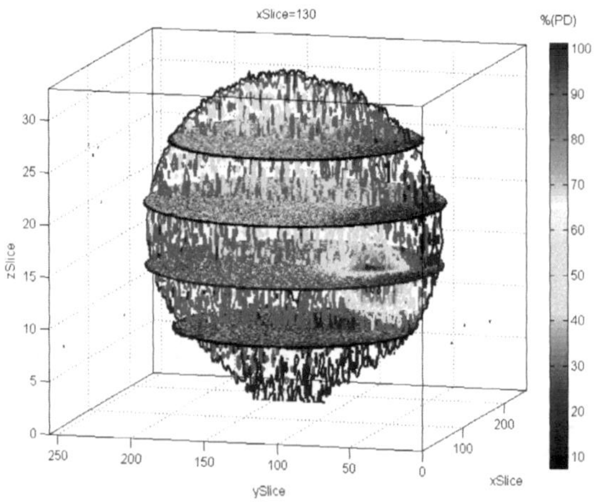

Figure 2. 3D phantom representation showing MRI slices orientation.

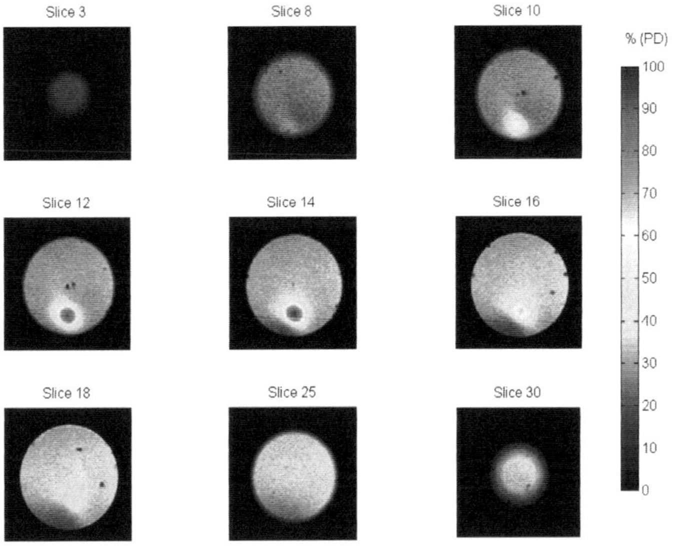

Figure 3. Selected MRI zSlices.

A reconstructed three-dimensional phantom image is presented in figure 4.

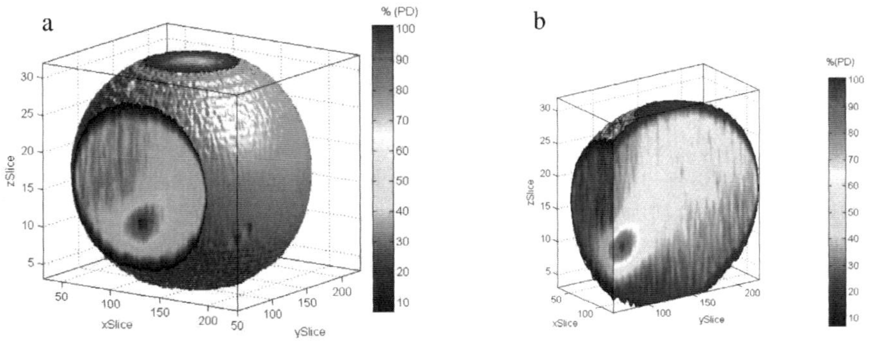

Figure 4. A reconstructed three-dimensional phantom image: cut in ySlice = 50 to highlight the centre of the irradiated volume (a); cut in ySlice = 60 and xSlice = 120 showing irradiation cone projection (b).

4. Conclusions

The obtained results show that the Fricke gel phantom prepared with 270 Bloom gelatin satisfy the requirements to be used for the quality control in stereotactic radiosurgery using Gamma Knife technique and may constitute one more option of dosimeter in radiation therapy applications. The phantom prepared with 270 Bloom porcine gelatin presents appropriate consistence and integrity (i.e., does not occurs formation of fungus) at room temperature (~ 25° C) during the time between irradiation and evaluation. The 270 Bloom porcine gelatin is of easy preparation at laboratory, inexpensive and commercially available. The dose delivered to any point in the phantom can be measured, the dose distribution can be mapped in three-dimensional spaces, in any slice for any orientation.

References

[1] Tozer-Loft S M, Walton L, Forster D M C and Kemeny A A 1999 An improved technique for comparing Gamma Knife dose-volume distributions in stereotactic radiosurgery *Phys. Med. Biol.* **44** 1905-19

[2] Ma L, Chin L S, Shepard D, Amin P and Slawson R 2000 The effect of user-defined variables on dosimetry consistency in Gamma Knife planning *Phys. Med. Biol.* **45** N43-7

[3] Mariani M, Vanossi E, Gambarini G, Carrara M and Valente M 2007 Preliminary results from a polymer gel dosimeter for absorbed dose imaging in radiotherapy *Radiat. Phys. Chem.* **76** 1507-10

[4] Olsson L E, Petersson S, Ahlgren L and Mattsson S 1989 Ferrous sulphate gels for determination of absorbed dose distributions using MRI technique: basic studies *Phys. Med. Biol.* **34** (1) 43-52

[5] Podgorsak M B and Schreiner L J 1992 Nuclear magnetic relaxation characterization of irradiated Fricke solution *Med. Phys.* **19** (1) 87-95

[6] Healy B J, Zahmatkesh M H, Nitschke K N and Baldock C 2003 Effect of saccharide additives on response of ferrous-agarose-xylenol orange radiotherapy gel dosimeters *Med. Phys.* **30** (9) 2282-91

[7] Bero M A and Kharita M H 2004 Effects of ambient temperature on the FXG radiochromic gels used for 3-D dosimetry *Journ. Phys. CS* **3** 236-39

[8] Schreiner L J 2004 Review of Fricke gel dosimeters *Journ. Phys. CS* **3** 9-21

[9] Bero M A 2007 Dosimetric properties of a radiochromic gel detector for diagnostic X-rays *Nucl. Instr. and Meth. in Phys. Res. A* **580** 186-9

[10] Karaiskos P, Petrokokkinos L, Tatsis E, Angelopoulos A, Baras P, Kozicki M, Papagiannis P, Rosiak J M, Sakelliou L, Sandilos P and Vlachos L 2005 Dose verification of single shot gamma knife applications using VIPAR polymer gel and MRI *Phys. Med. Biol.* **50** 1235-50

[11] Papagiannis P, Karaiskos P, Kozicki M, Rosiak J M, Sakelliou L, Sandilos P, Seimenis I and Torrens M 2005 Three-dimensional dose verification of the clinical application of gamma knife stereotactic radiosurgery using polymer gel and MRI *Phys. Med. Biol.* **50** 1979-90

[12] Novotny Jr J, Spevacek V, Dvorak P, Hrbacek J, Novotny J, Tlachacova D, Schmitt M, Vymazal J, Tintera J and Cechak T 2004 Application of polymer-gel dosimetry in stereotactic radiosurgery *Journ. Phys. CS* **3** 288-92

[13] Mayanski M J, Schulz R J, Ibbott G S, Gatenby J C, Xie J, Horton D and Gore J C 1994 Magnetic resonance imaging of radiation dose distributions using a polymer-gel dosimeter *Phys. Med. Biol.* **39** 1437-55

[14] Bero M A, Gilboy W B and Glover P M 2001 Radiochromic gel dosemeter for three-dimensional dosimetry *Radiat. Phys. Chem.* **61** 433-35

[15] Gore J C, Kang Y S and Schulz R J 1984 Measurement of radiation dose distributions by nuclear magnetic resonance (NMR) imaging *Phys. Med. Biol.* **29** (10) 1189-97

[16] Hendee W R 2002 New imaging techniques *Oncologic Imaging* **2** ed D.G. Bragg, P. Rubin and H. Hricak (W. B. Saunders, Philadelphia)

[17] Chu W C 2001 Radiation dosimetry using Fricke-infused gels and magnetic resonance imaging *Proc. Natl. Sci. Counc. ROC(B)* **25** (1) 1-11

[18] Jolesz F A and Blumenfeld S M 1994 Interventional use of magnetic resonance imaging *Magn. Reson. Q.* **10** (2) 85-96

[19] Audet C and Schreiner L J 1997 Multiple-site fast exchange model for spin-lattice relaxation in the Fricke-gelatin dosimeter *Med. Phys.* **24** (2) 201-9

[20] Bero M A, Gilboy W B, Glover P M and El-masri H M 2000 Tissue-equivalent gel for non-invasive spatial radiation dose measurements *Nucl. Instr. and Meth. in Phys. Res. B* **166-167** 820-5

Preliminary investigation of the NMR, optical and x-ray CT dose-response of polymer gel dosimeters with cosolvents and increased crosslinker levels

V.I. Koeva[1], K. B. McAuley[1], A.Jirasek[2], L.J. Schreiner[3, 4]

[1] Chemical Engineering Dept., Queen's University, Kingston, Canada, K7L 3N6
[2] Dept. of Physics and Astronomy, University of Victoria, Victoria, Canada, V8W 6V5
[3] Cancer Centre of Southeastern Ontario, Kingston, Canada, K7L 5P9
[4] Depts. of Oncology and Physics, Queen's University, Kingston, Canada, K7L 3N6

Kim.McAuley@chee.queensu.ca

Abstract. Three co-solvents (glycerol, N-propanol and isopropanol) have been investigated for increasing the solubility of N,N'-methylene-bisacrylamide (Bis) crosslinker in polymer gel dosimeter recipes. Using isopropanol, the crosslinker solubility increased from approximately from 3 to 10 % by weight, enabling the manufacture of polymer gel dosimeters with higher levels of crosslinking than was previously possible. The new dosimeter recipes can be imaged effectively using Nuclear Magnetic Resonance (NMR), optical and x-ray Computed Tomography (CT) techniques.

1. Introduction

Recently, Senden *et al.* [1] investigated the dose-response behaviour of new gel dosimeter recipes in which Acrylamide (Aam) is replaced by N-isopropylacrylamide (NIPAM), which is less toxic. As summarized by Senden *et al.* [1] all polymer gel dosimeter recipes (except methacrylic-acid-based dosimeters) use Bis as the crosslinker. Limited water solubility and low crosslinking efficiency are the main concerns with the use of Bis in gel dosimeters. Low crosslinking efficiency arises due to cyclization, which occurs due to the formation of a favorable seven-membered ring [2] so that as much as 80% of the pendant vinyl groups from the Bis are consumed [3,4]. In our research group, several new polymer gel dosimeters were prepared and irradiated using Aam or NIPAM and the following potential crosslinkers: poly(ethylene glycol) diacrylate, di(ethylene glycol) divinyl ether, tri(ethylene glycol) divinyl ether, ethoxylated trimethylolpropane triacrylate, 1,3,5-triacryloylhexahydro-1,3,5-triazine, N,N'-diallyltartardiamide, calcium diacrylate and N,N'-ethylenebis(acrylamide) [5] in the hope that higher crosslink densities could be obtained. Unfortunately, none of the alternative crosslinkers investigated is a suitable replacement for Bis in gel dosimeter systems. The main problems encountered were low reactivities, limited water solubility and the presence of free-radical inhibitors that are added to liquid crosslinkers to prevent polymerization during shipping and handling. Since a good replacement for Bis could not be found, we have directed our current investigations toward increasing the level of crosslinking by adding co-solvents to the recipe.

The aim of our current research is to develop improved NIPAM-based polymer-gel-dosimeter recipes that will produce higher dose sensitivities. Three different cosolvents (glycerol, isopropanol and N-propanol) are tested to determine whether they can increase the solubility of Bis. Additional Bis in the recipe should result in higher levels of crosslinking and in more sensitive and accurate dosimeters. Higher dose sensitivity is particularly important if X-ray CT will be used as an imaging modality [6].

2. Materials and Methods

The polymer gel dosimeters were manufactured in a fume hood under normal atmospheric conditions using standard experimental procedures [7]. Gelatin (300 Bloom Type A) was allowed to swell in 80% of de-ionized water and cosolvent for 10 minutes at room temperature before heating to 50 °C. A water bath was used to ensure uniform temperature around the flask. While stirring continuously, the crosslinker was dissolved at 40 °C, requiring typically ~5 minutes. After the gelatin-crosslinker mixture was cooled to ~32 °C, the monomer (NIPAM) was added. A solution of the antioxidant Tetrakis (hydroxymethyl) phosphonium chloride (THPC) was prepared using the remaining 20% of the water, and added to the solution. The resulting solution was transferred into small plastic vials, which were filled to a height of 2 cm and then capped. The caps were wrapped with sealing film to prevent additional oxygen from entering the samples. The samples were wrapped with aluminium foil to prevent photo-polymerization, and placed in a refrigerator to set. The dosimeters were irradiated up to 40 Gy at room temperature using a T780 Cobalt-60 tomotherapy benchtop (MDS Nordion, Kanata, Canada), approximately 24 h post-manufacture using a 10x10 cm^2 field directed perpendicular to their length. NMR relaxation rates (R_2) were determined between three hours and three days after irradiation, so that the time evolution polymerization reactions could be studied. The uniformly irradiated gel dosimeters were characterized using an Oxford Instruments MARAN 20/35 benchtop NMR spectrometer operating at 20 MHz. An Ultrospec 1000 UV/Visible Spectrophotometer was used to quantify the fraction of 633 nm light that could pass through the samples.

3. Results and Discussion

The use of three cosolvents, glycerol, isopropanol and N-propanol, was investigated in an attempt to increase the amount of crosslinker that can be dissolved in the aqueous gel solution. Cosolvents can be mixed with the main solvent to adjust the solubility properties of solutes. For example, isopropanol-water mixtures have been used to enhance solubility during the production of biopolymers [8]. In the current work, glycerol, isopropanol and N-propanol were selected as potential cosolvents because they are miscible with water and non-volatile, and their atomic compositions indicate that they will not interfere with the tissue-equivalence of the dosimeter [9]. Note that these cosolvents will not alter the pH of the solution significantly, which may be important because pH can influence R_2 [10]. The effects of the three cosolvents on the solubility of Bis in aqueous solutions were tested using various ratios of water and cosolvent. These experiments were performed at 50 °C to be consistent with the dissolution temperature used during the gel manufacture. Note that the solubility of Bis at room temperature is lower than at 50 °C. As shown in Table 1, glycerol is an effective cosolvent, which increases the solubility of Bis from 3 wt % to 5 wt%, but isopropanol and N-propanol are even more effective, increasing the solubility of Bis to 10 and 17 wt %, respectively. Since glycerol is a less-effective cosolvent, and is viscous and difficult to use, only isopropanol and N-propanol were further investigated as cosolvents in dosimeter recipes.

Several gel dosimeters were produced using 30 wt % N-propanol or isopropanol in the recipes. The mass of water was reduced to accommodate the cosolvent. In each case, the dosimeter with the new composition was compared to a standard dosimeter (6 %T, 50 %C, with NIPAM monomer, Bis crosslinker and no cosolvent). Figures 1a and 1b compare the NMR and optical dose responses, respectively, for the standard dosimeter and the same dosimeter made with isopropanol. The NMR dose-response results are similar, indicating that the presence of the cosolvent had very little influence on the polymerization. Figure 1b shows that the isopropanol had a mild influence on the optical properties of the gel, possibly because the cosolvent influences the size and number of polymer particles that are formed.

Table 1. Solubility of Bis in aqueous solution at 50 °C with different cosolvents.

Water : cosolvent ratio (by weight)	Solubility of bisacrylamide (wt%) in Glycerol [9]	Solubility of bisacrylamide (wt%) in Isopropanol	Solubility of bisacrylamide (wt%) in N-propanol
10:0	3	3	3
9:1	4.5	7	7
8:2	5	8	10,5
7:3	5	10	14
6:4	5	10	17

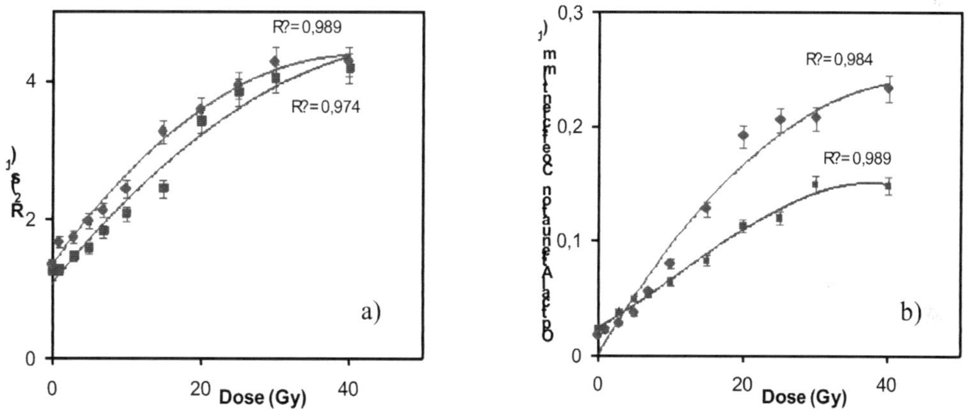

Figure 1. NMR and optical dose responses of 6% T, 50 %C NIPAM/Bis dosimeters containing 5 mMol THPC. Results are shown for a dosimeter containing 30 wt% isopropanol (♦) and no cosolvent (■). Curves were added to guide the eye.

Figure 2 compares the R_2 and optical responses of a standard 6 %T 50% C dosimeter with dose responses of 10% T 50 %C dosimeters prepared using isopropanol or n-propanol cosolvents. As expected, the 10%T dosimeters with cosolvents produced larger dose sensitivities than the standard dosimeter, because more Bis was dissolved and available for crosslinking. The optical responses for the dosimeters produced using the two cosolvents were very similar. However, Figure 2a shows that the R_2 dose sensitivity of the dosimeter containing N-propanol was lower than that for the dosimeter containing isopropanol. This trend was reproducible using several batches of gels (results not shown). We are unsure why isopropanol produces a different R_2 response than N-propanol. Perhaps the difference is related to the fact that isopropanol is a chain-transfer agent in NIPAM polymerization [11] that may have an important influence on the molecular weight of the polymer that is produced.

Figure 3 shows the reproducibility of NMR responses of 10 %T 50%C dosimeters prepared using 30% isopropanol. The responses in Figure 3a indicate that considerable polymerization took place between 3 h and 20 h post irradiation, but that little additional polymerization occurred after 20 h. Figure 3 b shows results from three replicate batches 20 h after irradiation, indicating that the results are reproducible. Figure 4a shows three 10% T 50% C dosimeters produced with 0, 10 and 30 wt% isopropanol, respectively. The dosimeters made with 10 and 30 wt% of isopropanol gave very similar dose responses, which suggests that using 10 wt% of isopropanol is enough to dissolve all of the available Bis.

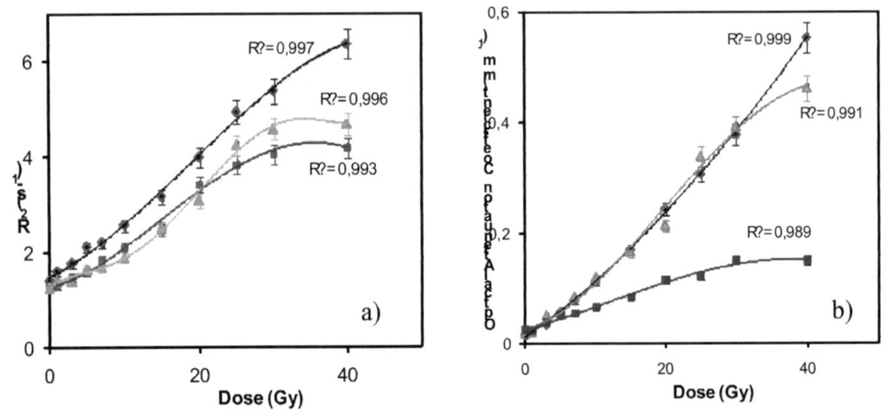

Figure 2. NMR and optical dose responses of three dosimeters: 10% T, 50 %C, 30 wt% isopropanol (♦),10% T, 50 %C, 30 wt% N-propanol (▲) and 6% T, 50 %C, no cosolvent (■). All dosimeters were prepared using 5 mM THPC.

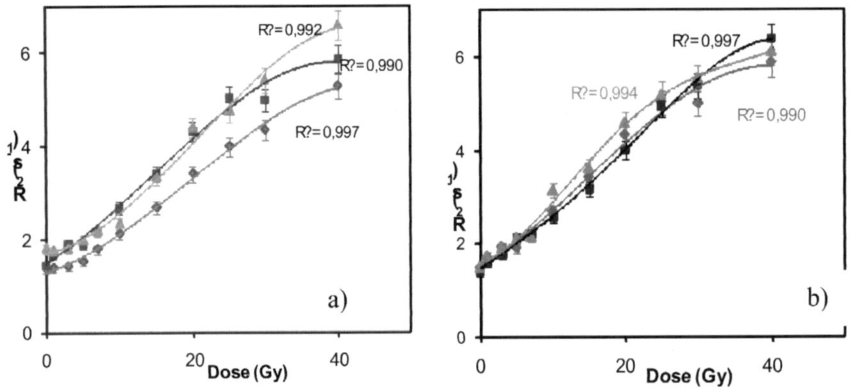

Figure 3. NMR response of 10% T, 50 %C NIPAM/Bis dosimeters containing: (a) 30 wt% isopropanol 3 hours (♦), 20 hours (■) and 42 hours (▲) after irradiation; (b) 30 wt% isopropanol for three replicate batches 20 hours after irradiation.

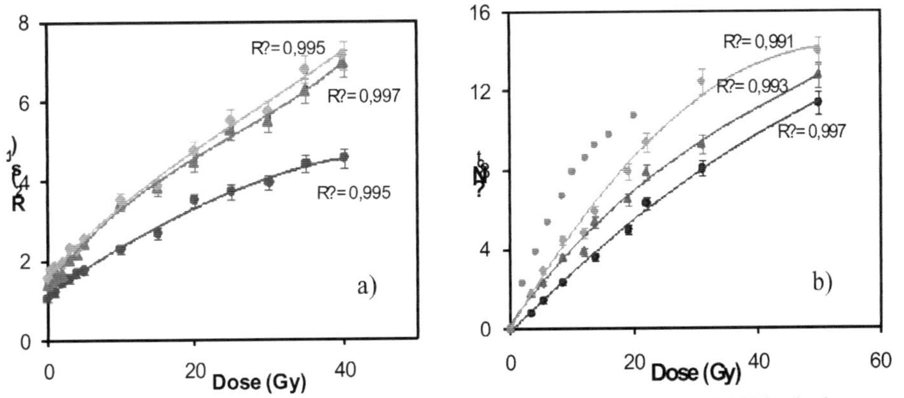

Figure 4. NMR and ΔN_{CT} dose responses of 10% T, 50% C NIPAM/Bis dosimeters containing 0 (●), 10 (▲) and 30 (♦) wt% of isopropanol. In Figure 4b an anoxic PAG dosimeter (6% T 50% C) (●) is included for comparison.

We also tested the dosimeters containing isopropanol using X-ray CT read-out, which relies on CT contrast due to density changes occurring in irradiated polymer gels. Hilts *et al* [12] have shown that the dose values and high dose gradients of the derived profiles match very well using CT and MRI methods [12,13] and that the X-ray CT response is relatively insensitive to the gel temperature during imaging [12]. Figure 4b shows that 10%T 50 %C polymer gels containing isopropanol give better response than those without co-solvent, but are still less sensitive than traditional (anoxic) 6% T 50 %C polyacrylamide gels.

4. Conclusions

The solubility of Bis crosslinker can be increased substantially by adding either N-propanol or isopropanol as a cosolvent. Dosimeters produced using isopropanol cosolvent and up to 5 wt% Bis by weight were shown to produce dose response curves (using NMR, optical and X-ray CT read-out methods) with significantly higher dose sensitivity than standard NIPAM-based polymer gel dosimeters containing 3 wt% Bis. Unfortunately, the dose sensitivity of the new normoxic 10 %T 50 %C dosimeters is not as high as the dose sensitivity for traditional PAG dosimeters produced using anoxic conditions. Further recipe optimization will be required to determine whether cosolvents can help to produce polymer gel dosimeters that are sensitive and accurate enough for routine X-ray CT read-out.

5. Acknowledgements

Financial support was provided by the Natural Science and Engineering Research Council of Canada (NSERC) and the Canada Institutes for Heath Research (CIHR).

6. References

[1] Senden R J, DeJean P, McAuley K B and Schreiner L J (2006) Phys. Med. Biol. 51, 3301-3314
[2] Gopalan, A., P. Venuvanalingam, S. P. Manickam, K. Venkatarao, and N.R. Subbaratnam (1982), *Eur. Polym. J.* **18**, 531-4.
[3] Tobita, H., and A. E. Hamielec (1990), *Polymer* **31**, 1546-52.
[4] Naghash, H. J., and O. Okay (1996), *J. Appl. Polym.Sci.* **60**, 971-9.
[5] Koeva V.I., E.S. Csaszar, R.J. Senden, K.B. McAuley, L.J. Schreiner,(2008) *Macromol. Symp.* **261**,157–166.
[6] Jirasek A., P. Baxter, M. Hilts (2007) Med. Phys. **34 (6)**, 1934-1943

[7] De Deene, Y., C. Hurley, A. Venning, K. Vergote, M. Mather, B. J. Healy, and C. Baldock (2002b), *Phys. Med. Biol.* **47**, 3441-63.
[8] D.Martino, D.Estenoz, DL Reyna, and J. Warner (2006), 2nd Mercosur Congress on Chemical Engineering *(Rio De Janeiro, Brazil).*
[9] De Deene, Y., K. Vergote, C. Claeys, and C. De Wagter (2006), *Phys. Med. Biol.* **51**, 653-73.
[10] Kennan, R.P., K.A. Richardson, J. Zhong, M.J. Maryanski, and J. C. Gore. (1996). *Journal of Magnetic Resonance, Series B*. **110**: 267-277
[11] Chang Y., Mumick P.S., Worldwide K.C. (2001) *United States Patent*, Patent # US 6 268 449 B1
[12] Hilts, M., C. Audet, C. Duzenli, and A. Jirasek (2000), *Phys. Med. Biol.* **44**, 2559-71.
[13] Babic, S., and L. J. Schreiner (2006), *Phys. Med. Biol.* **51**, 4171–4187

A study on the characteristic of normoxic polymer gel dosimeter according to its composition

Sam Ju Cho[1], Sangwook Lim[2], Suk Lee[3], Sang Hoon Lee[4], Hyun Do Huh[5], Kwang Hwan Cho[4], Jisun Jang[4], Jin Ho Choi[6], Dong Oh Shin[7], Soo Il Kwon[4], Hyong Geun Yun[1]

[1]Department of Radiation Oncology, Donguk University, Seoul 100-715, [2]Deprtment of Radiation Oncology, Kosin University College of Medicine, Busan 602-702, [3]Department of Radiation Oncology, College of Medicine, Korea University, Seoul 136-701, [4]Department of Medical Physics, Kyonggi University, Suwon 443-760, [5]Department of Radiation Oncology, College of Medicine, Inha University, Incheon 400-711, [6]Department of Radiation Oncology, Gachon University of Medicine and Science, Inchoen 406-799, [7]Department of Radiation Oncology, Kyung Hee University, Suwon 130-701

atilla@paran.com

Abstract. In this study, to find the optimal composition of the gel as therapeutic radiation, the amounts of methacrylic acid and gelatin were varied. The polymer gel with various compositions were evaluated for its sensitivity, reproducibility, and accuracy. As the concentration of the gelatine is high, the threshold R2 value increases and the dose response was decreases. As the concentration of the methacrylic acid is high, both the threshold R2 value and the dose response were decrease. As both concentrations of the gelatine and the methacrylic acid is high, the sensitivity to the dose was increases within some range. It was found that the polymer gel composed in this study can be optimized for measuring the therapeutic radiation.

1. Introduction

The polymer gel can be specified as its composition [1-5]. In this study, the polymer gel was composed in a laboratory with focusing for therapeutic radiation. To find the optimal composition of the gel as therapeutic radiation, the amounts of methacrylic acid and gelatin were varied. The polymer gel with various compositions were evaluated for its sensitivity, reproducibility, and accuracy.

2. Materials and methods

The polymer gels were grouped into nine as varying ingredients ratio to see the dose response and sensitivity (Table 1). The polymer gel composed under normal oxygen condition was bottled in the glass vials with 1.6 cm diameter and 6 cm height. The glass vials were full-filled with the polymer gel to remove the air bubbles to minimize the oxygen effect.

The vials were placed at the reference position in the water phantom (WP 34, Scanditronix-Wellhofer, Germany) and were irradiated with 6 MV therapeutic radiations to deliver 0, 1, 3, 5, 7, 9, 12, 15, and 20 Gy respectively.

Table 1. Nine groups with various composition of gel.

	Gelatin (wt%)	MAA (wt%)	Water(wt%)	THPC (mM)	HQ (mM)
Group 1	6	6	88	10	0.05
Group 2	6	8	86	10	0.05
Group 3	6	10	84	10	0.05
Group 4	8	6	86	10	0.05
Group 5	8	8	84	10	0.05
Group 6	8	10	82	10	0.05
Group 7	10	6	84	10	0.05
Group 8	10	8	82	10	0.05
Group 9	10	10	80	10	0.05

In order to see the reproducibility and the accuracy of the gel, 8% gelatin, ten vials of the normoxic polymer gel were composed with 10 mM THPC, and 0.05 mM HQ. The five vials were irradiated 2 Gy and the rest of five vials were irradiated 4 Gy. The deviations of T2 times were analyzed with MR and their reproducibility was evaluated.

Calibration curve was acquired by delivering 0, 2, 4, and 6 Gy to each vial. The calibration curve was applied to the irradiated ten vials previously used for evaluation of reproducibility and dose accuracy was evaluated.

In this study the 1.5T Magneton Avanto MR scanner (Siemens Co., USA) was used with conditions of multi-spin echo sequence, head coil, 256×256 of matrix, 5 mm of slice thickness, and 180×180 mm^2 of FOV.

3. Results and discussion

It was found that as the concentration of the gelatine is high, the threshold R2 value increases and the dose response was decreases. As the concentration of the methacrylic acid is high, both the threshold R2 value and the dose response were decrease (Figure 1). As both concentrations of the gelatine and the methacrylic acid is high, the sensitivity to the dose was increases within some range (Figure 2).

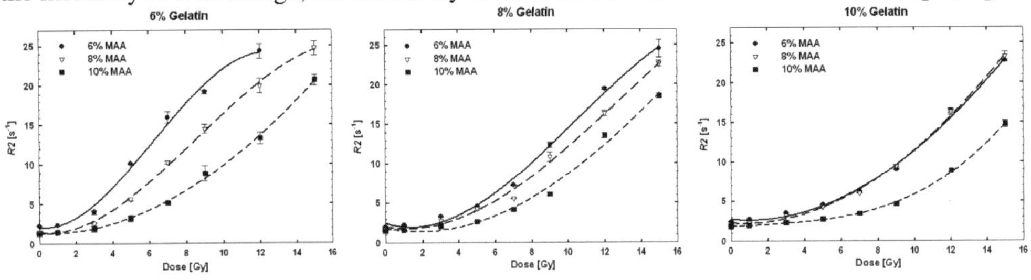

Figure 1. The R2-dose response curve (a) 6% gelatin, (b) 8% gelatin and (c) 10% gelatin with 6, 8 and 10% MAA respectively.

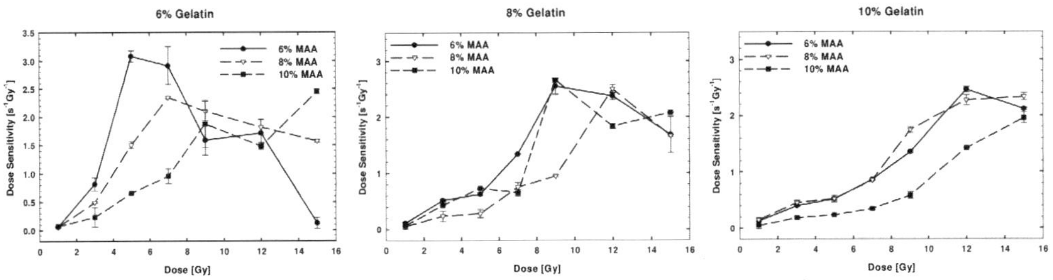

Figure 2. The dose sensitivity curve (a) 6% gelatin, (b) 8% gelatin and (c) 10% gelatin with 6, 8 and 10% MAA respectively.

The dose reproducibility was 3.39% and 2.21% for 2 Gy and 4Gy respectively, and the accuracy 0.34% and 1.13% for 2 Gy and 4 Gy respectively (Table 2).

Table 2. Reproducibility and accuracy for 2 Gy and 4 Gy.

	Reproducibility	Accuracy
2 Gy (%)	3.39	0.34
4 Gy (%)	2.21	1.13

4. Conclusion

It was respected that the polymer gel dosimeter can be optimized for its use, since the characteristics of the polymer gel can be varied according to its concentrations in this study.

References

[1] Maryanski MJ, Gore JC, Kennan RP, *et* al. NMR relaxation enhancement in gels polymerized and cross-linked by ionizing radiation: A new approach to 3D dosimetry by MRI. Magn Reson Imaging 11;253-258:1993

[2] Maryanski MJ, Ibbott GS, Eastman P. et al. Radiation therapy dosimetry using magnetic resonance imaging of polymer gels. Med Phys 23;699-705:1996

[3] Fong PM, Keil DC, Mark DD, et al. Polymer gels for magnetic resonance imaging of radiation dose distribution at normal room atmosphere. Phys Med Biol 46;3105-3113:2001

[4] Cho SJ, Shin D, Huh HD, et al. Development of an novel normoxic polymer gel dosimeter(TENOMAG), Jouranl of the Korean Physical Society 52;1798-1805:2007

[5] Deene YD, Hurley C, Venning A, *et* al: A basic study of some normoxic polymer gel dosimeters. Phys Med Biol 47;3441-3463:2002

Journal of Physics: Conference Series **164** (2009) 012019 doi:10.1088/1742-6596/164/1/012019

The role of gelatin in a methacrylic acid based gel dosimeter

Shin-ichiro Hayashi[1], Munenori Yoshioka[1,2], Shuji Usui[1], Kiyofumi Haneda[1] and Takahiro Tominaga[1]

[1]Department of Clinical Radiology, Faculty of Health Sciences, Hiroshima International University, Higashi-Hiroshima, Hiroshima, 739-2695, Japan

[2]Department of Medical Physics & Engineering, Osaka University Graduate School of Medicine, Suita, Osaka, 565-0871, Japan

rin@hs.hirokoku-u.ac.jp

Abstract. In this work, the dose response of a methacrylic acid based gel dosimeter containing different concentrations of gelatin is examined to clarify the role of gelatin in the polymerization process. Direct temperature monitoring of the gel throughout the irradiation process is adopted as the method. In addition, to investigate the function of gelatin on polymerization, another experiments are performed after replacing gelatin with some acids. From the results, it is concluded that gelatin is an indispensable component in the polymerization reaction which acts as a "trigger", and that the peptide structure in gelatin also assumes an important role.

1. Introduction

With rapid advances being made in radiotherapy treatment, the three-dimensional (3D) polymer gel dosimeter has been studied as a possible dosimeter for measuring the complex 3D dose distribution directly.

In a polymer gel dosimeter, gelatin is generally used as a matrix to coagulate the solution and to prevent diffusion of the created polymer. The reaction between gelatin matrix and monomers has been investigated in several previous studies [e.g., 1-5]. However, the mechanism of action of the polymer gel dosimeter and the influence of various characteristics of the polymer gel dosimeter (such as stability and susceptibility) have not yet been fully clarified.

In this work, the dose response of a methacrylic acid based gel dosimeter containing different concentrations of gelatin is examined to clarify the role of gelatin in the polymerization process. Since polymerization in the polymer gel dosimeter is an exothermic reaction, direct temperature monitoring of the gel throughout the irradiation process was adopted as the method for studying the gel's mechanism.

In addition, to investigate the function of gelatin on polymerization, another experiments were performed after replacing gelatin with some acids.

2. Materials and methods

MAGAT type gels were prepared for this study. As shown in the table, nine different concentrations of gelatin were compounded with an equal volume of water and equal concentrations of methacrylic acid and THPC. The only sample of 120 g/L gelatin gelatinized at room temperature (22°C). In addition to the gels described above, we prepared another type of solution, which included amino acid (glycine, alanine and proline) instead of gelatin, to examine the effect of gelatin structure and residues. The composition of these solutions was almost the same as that shown in the table, except that 7% w/w amino acid was added instead of gelatin.

The irradiations were performed using the 6 MV photon beam of a medical linear accelerator (MITSUBISHI Co. EXL-12SP). The three beams delivered 5, 10, and 20 Gy in each kind of sample with a constant dose rate of 2 Gy/min. The temperature measurements were performed using a digital thermometer. The probe of a thermometer was vertically inserted into the center part of the cylindrical glass vial (4 cm diameter and 7 cm height) through a hole in the cap. The temperature was measured consecutively for one hour including the period of irradiation.

After temperature measurement, the opacity of the gel dosimeter was estimated by using pictures of the samples. The irradiated vial was put in front of a light table with an optical step tablet and a picture was taken with a digital still camera. The mean pixel value (from 0 to 255) around the center of the vial was obtained roughly by using the suitable ROI.

Table Methacrylic acid based polymer gel compositions prepared in this study.
The concentrations per volume were used for all MAGAT type gels, instead of a weight ratio.

Chemical	Concentration
water	
MA	0.7 M
gelatin	0, 2, 3, 5, 10, 20, 40, 60, 120 g/L
THPC	2 mM

3. Results and Discussion

Figure 1 shows one of the temperature variations as a function of time during and after irradiation. The case of 40 g/L of gelatin concentration is presented as an example, because a same tendency was also observed for the other concentrations. As the figure shows, a rapid rise in temperature due to radiation-induced polymerization is observed at the beginning of irradiation. As the dose increases, the variation becomes more obvious.

The maximum values in the temperature variation are plotted as a function of the concentration of gelatin in Figure 2. In the figure, the value increases rapidly and reaches its maximum around 20-40 g/L of gelatin irrespective of irradiation dose. It is remarkable that no variation in temperature was observed in the gelatin free solution (0 g/L) despite an increase in temperature was seen even in a very low gelatin concentration solution, such as 2 g/L (about 0.2% w/w). This result suggests that gelatin is an indispensable component of the polymerization reaction in the MAGAT type dosimeter.

The opacity of the irradiated gel was obtained from the pixel values of the images. As shown in Figure 3, the values decrease as the dose increases in the region of gelatin concentration generally used (about 40–120 g/L). However, it was found that the tendency turned at concentrations of less than 10 g/L. This result suggests that the mechanism of polymerization in a low concentration of gelatin is different from that in a higher concentration.

Moreover, in order to investigate the role of gelatin in polymerization, another temperature measurements were performed with amino acids (glycine, alanine and proline) instead of gelatin. Figure 4 shows the results with different solutions containing each amino acid. In the figure, increases in temperature were observed in all three samples containing an amino acid after 20 Gy of irradiation. Although the change in opacity was not observed in these samples, the viscosity of the solutions increased after irradiation. In a measurement using acetic acid and propionic acid, which do not have amino groups, no responses were observed. The results indicate that polymerization can occur in solutions without gelatin but with an amino acid. Therefore, it is suggested that the -NH-CH-CO- skeleton, which is a common structure in gelatin and amino acids, contributes to the polymerization reaction rather than the higher order structure of gelatin or any other particular amino acid residues.

4. Conclusion

In this work, we investigated gelatin's additional role in the polymerization reaction in a methacrylic acid based gel. From the results of temperature monitoring of samples during the irradiation process, it is concluded that gelatin is an indispensable component in the polymerization reaction which acts as a "trigger", and that the peptide structure in gelatin also assumes an important role. Various amino acids and their derivatives are expected to carry out a similar role.

On the basis of our results, the unprecedented usage of gelatin or amino acids as a "trigger" is required for gel composition. A further direction of this study will be to examine the characteristics of such a system [6], and clarify the mechanism of polymer gel, with the aim of developing it to a high performance polymer gel dosimeter.

References

[1] De Deene Y, Vergote K, Claeys C and De Wagter C 2006 The fundamental radiation properties of normoxic polymer gel dosimeters: a comparison between a methacrylic acid based gel and acrylamide based gels *Phys. Med. Biol.* **51** 653-73.

[2] Karlsson A, Gustavsson H, Månsson S, McAuley K B and Bäck S A J 2007 Dose integration characteristics in normoxic polymer gel dosimetry investigated using sequential beam irradiation *Phys. Med. Biol.* **52** 4697-4706

[3] Lepage M, Whittaker A K, Rintoul L, Bäck S A J and Baldock C 2001a The relationship between radiation-induced chemical processes and transverse relaxation times in polymer gel dosimeters *Phys. Med. Biol.* **46** 1061-74

[4] Lepage M, Whittaker A K, Rintoul L, Bäck S A J and Baldock C 2001b Modelling of post-irradiation events in polymer gel dosimeters *Phys. Med. Biol.* **46** 2827-39

[5] McAuley K B 2006 Fundamentals of polymer gel dosimeters *J. Phys.: Conf. Ser.* **56** 35-44

[6] Yoshioka M et al. *To be appeared in this conference.*

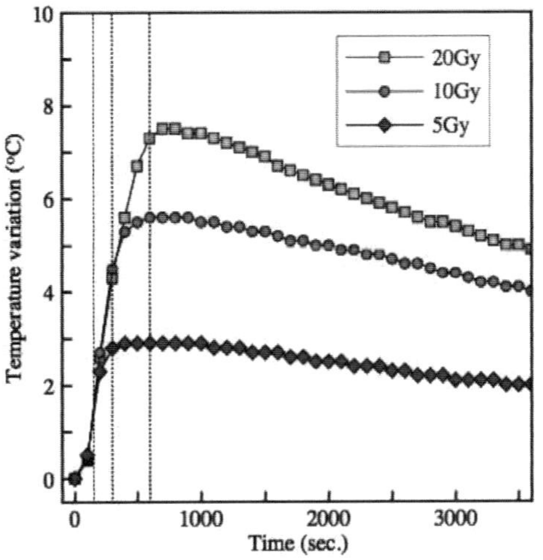

Figure 1 The temperature variations with time and different doses (5, 10 and 20 Gy) for the MAGAT type gel with 40 g/L of gelatin. The dotted lines show the end time of irradiation, 150 s, 300 s and 600 s from the origin.

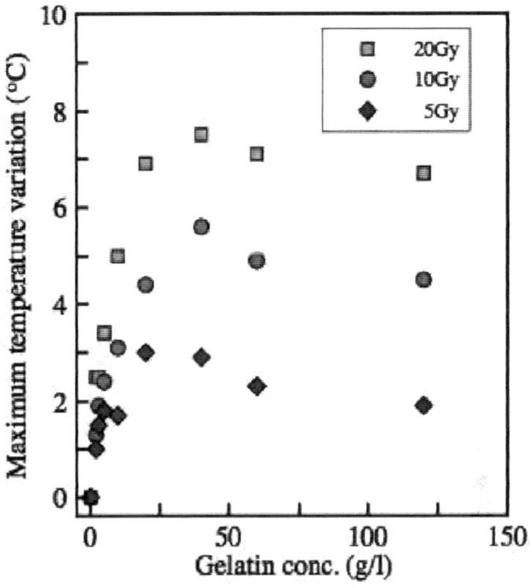

Figure 2 The changes of maximum temperature variations with different gelatin concentrations and doses.

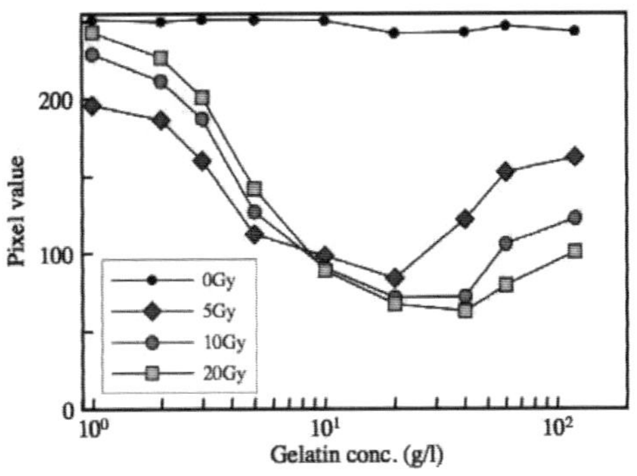

Figure 3 The changes of opacity with different gelatin concentrations and doses.

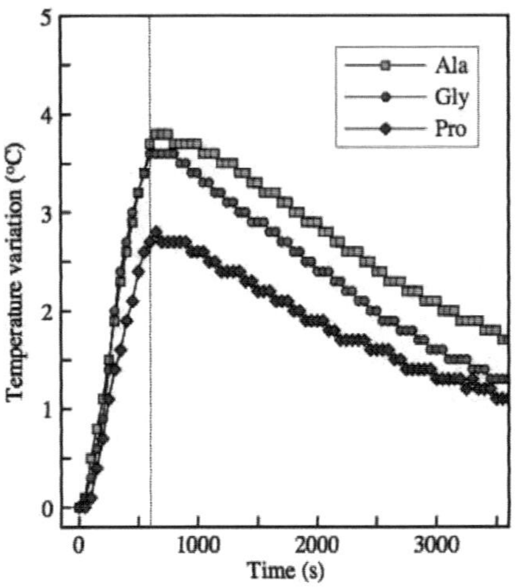

Figure 4 The temperature variations in the amino acid solutions irradiated with a dose of 20 Gy. The vertical dotted line indicates the end time of irradiation, 600 s from the origin.

The history and principles of optical computed tomography for scanning 3-D radiation dosimeters: 2008 update

Simon J Doran

CRUK Clinical Magnetic Resonance Research Group, Institute of Cancer Research, Sutton, Surrey, UK

Department of Physics, University of Surrey

Simon.Doran@icr.ac.uk

Abstract. The current status of optical CT for 3-D radiation dosimetry is reviewed. The technique is first placed in its historical context, pointing out the relationship with other methods of optical imaging and showing how optical-CT has emerged independently in several different fields and under different names. The theoretical background of the method is described briefly and this provides the foundation for an explanation of the different types of scanner. The relative advantages and disadvantages of instruments based on scanned lasers and pixelated (area) detectors are presented. The latest generation of "fast laser scanners" is described and the review is concluded with a discussion of the different radiation-sensitive materials used as samples in optical CT.

1. Introduction

It is now more than a decade since the first optical computed tomography (CT) data for a polymer gel radiation detector were published [1, 2]. Since that date, there have been considerable advances in both 3-D imaging techniques and dosimeter formulation. Developments have continued apace since the last reviews of the area in 2006 [3, 4] and this paper is an updated version of [3]. Whilst it is not possible to discuss fully all the latest contributions in the limited space available, the reference list has been updated with the intention that it can provide a definitive record of the activity in the area.

Of course, the roots of 3-D imaging via optical CT may be traced back much further than 1996. The first part of this review will place the current work in its historical context and also explain how it is related to other methods of optical imaging. An expanded theoretical description of the principles of optical CT will then be given, which will allow the reader to appreciate the differences between the various types of scanner currently being developed and also to give an idea of the different types of data processing that are employed. The article will discuss in detail the relative merits of scanners based round pixelated (area) detectors, compared with those using a single laser beam, and will compare the latest crop of so-called "fast laser scanners". Central to our ability to image dose distributions has been the development of novel materials whose optical properties change in response to radiation, and a survey of these will be given. Finally, I will speculate briefly on the future of the technique.

Journal of Physics: Conference Series **164** (2009) 012020 doi:10.1088/1742-6596/164/1/012020

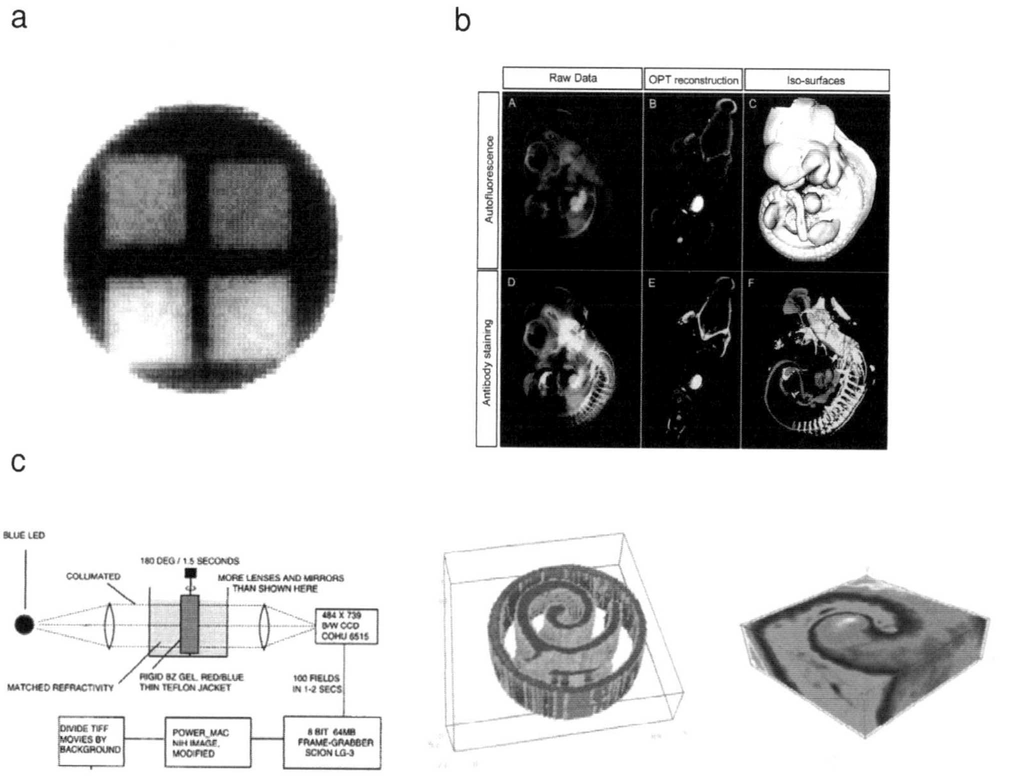

Figure 1: Examples of work from the three different fields in which optical CT emerged independently: (a) gel dosimetry, reproduced from Fig. 5 of; [1] (b) 3-D microscopy, reproduced from Fig. 3 of [7]; (c) visualization of chemical waves, composite of Figs. 1, 4 and 5 of [6].

2. General overview

Although 3-D radiation dosimetry using magnetic resonance imaging (MRI) was already well established by the time of Gore and Maryanski's seminal 1996 paper [1], the idea of detecting radiation changes optically has, in fact, a much longer history than the use of MRI. Chemical dosimeters date back at least as far as 1927 [5] and, at the outset , used UV spectroscopy as the readout method. Two factors contributed to the relatively late appearance of optical CT on the scene for gel dosimetry. Firstly, whilst MRI (like X-ray CT) was developed primarily for medical applications and then applied as a mature imaging modality to 3-D radiation dosimetry, there was no parallel medical development of optical CT, because the human body is opaque at visible wavelengths. Secondly, it has been only since the late 1990's that the technology needed for such measurements has been available. The consumer imaging boom of the past decade has given the impetus for the production of high quality, cheap CCD's; high performance photodiode detectors and receiver electronics have also become readily accessible; and sophisticated image reconstruction algorithms have been imported from other branches of medical imaging. It may be argued that as soon as the conditions were right, the technique was "discovered".

This is borne out by the fact that optical CT emerged independently in at least three completely different fields. Contemporaneously with Gore and Maryanski's invention, Winfree presented [6] what appears to have been the first CCD-based optical-CT scanner, in an elegant experiment to investigate self-organising chemical structures. In 2002, Sharpe [7] published the first optical micro-CT images, and patented his approach, which he termed optical projection tomography (OPT), apparently unaware of the prior developments. More recently, optical CT has been re-introduced with the name optical transillumination tomography in studies of tissue-engineered blood vessels [8-11].

Optical CT is part of a family of different techniques of optical imaging. Each is applicable to only a certain range of samples and has different advantages and disadvantages.

Confocal microscopy [12], optical coherence tomography (OCT) [13] and diffuse optical tomography (DOT) [14] are techniques that are all commonly used to image biological tissues. Laser scanned confocal microscopy is capable of producing high-resolution (1μm in-plane and 5μm axially) images at video frame rates via a process known as optical sectioning, in which only the region of the object close to the focal plane of the lens system is imaged efficiently. Other 3-D microscopy techniques are reviewed by Steltzer [15], but not discussed further here. OCT is an interferometric technique that combines back-scattered light from tissue with a reference beam. A succession of high-resolution (potentially sub-micron) real-time, 2-D images may be reconstructed from different depths using methodology that is analogous to ultrasound pulse-echo imaging. Both confocal microscopy and OCT are limited (as used conventionally) to a maximum penetration depth of the order of a few mm by the highly scattering nature of tissue samples. However, given that 3-D radiation dosimeters scatter relatively weakly in comparison with tissue, it is probable that variants of OCT could be used successfully for dosimetry. DOT is used for larger (up to tens of cm) samples that are highly scattering. Using sophisticated mathematical techniques for solving what is a very complex inverse problem, both optical absorption and scattering coefficients may be reconstructed, though with a spatial resolution only of order 1 cm.

The configuration used for our parallel-beam optical CT scanner [16-19] has much in common with schlieren and shadowgraphy methods [20], which find application in areas as diverse as the optics of flames and particle velocimetry. The distinction of parallel-beam optical-CT from these techniques is that, while they aim specifically to image inhomogeneities in refractive index, we would rather avoid them.

3. Principles of computed tomography

3.1. Beer's Law

The starting point for our discussion is *Beer's Law*, which describes how light and X-rays are attenuated as they pass through a medium[1]. In a uniform substance of linear attenuation coefficient μ, the light intensity, as measured by a detector placed at depth d is given by

$$I(d) = I_0(d)\exp(-\mu d), \tag{1}$$

[1] Note in passing that the applicability of Beer's Law is something that needs to be formally demonstrated for each chemical system and is not a given [21]. To our knowledge the limits of applicability in 3-D dosimeters have not yet been investigated, but one may anticipate that, for the range that can be probed via CT with current technology (up to 14-bit ADC, i.e., integrated optical density up to ~4), then Beer's Law will hold for absorbing media. However, for systems such as polymer gel dosimeters, where contrast is obtained via scattering, then Beer's Law will "break down" much earlier.

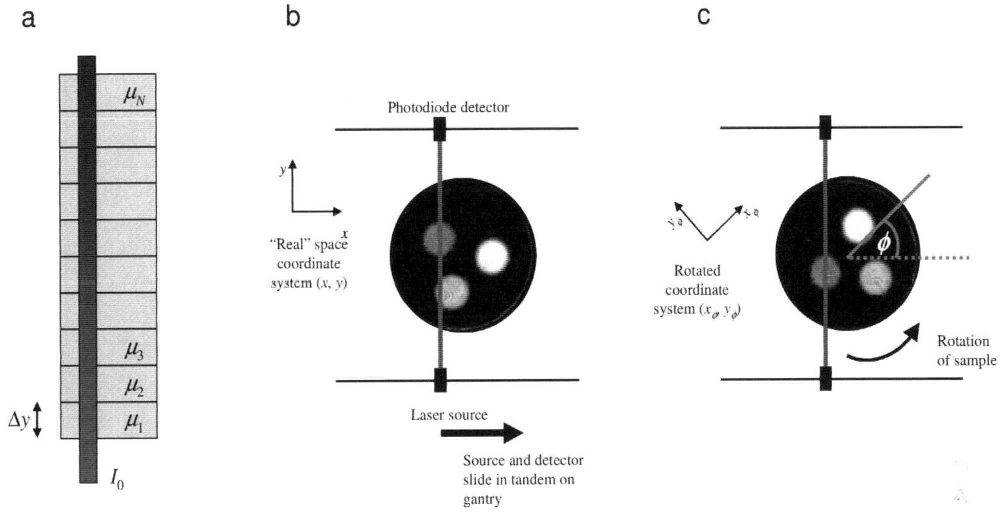

Figure 2: (a) Graphical representation of Beer's law; (b) Schematic diagram of a laser optical-CT scanner: a laser is projected through the sample and the attenuated beam is detected by a photodiode; the source and detector move in tandem along a gantry; (c) The sample rotates, allowing projection data to be acquired at different angles.

where I_0 is the intensity measured at depth zero. Suppose we now consider a set of N blocks of different material, each of width Δy, as shown in Figure 1(a). The X-ray intensity measured at the exit of the set of blocks is

$$I(N\Delta y) = I_0 \exp(-\mu_1 \Delta y) \cdot \exp(-\mu_2 \Delta y) \cdot \exp(-\mu_3 \Delta y) \ldots = I_0 \exp\left\{-\sum_{i=1}^{n} \mu_i \Delta y\right\}. \quad (2)$$

For the limit $\Delta y \to 0$, $N \to \infty$, this becomes

$$I = I_0 \exp\left\{-\int_{\text{across sample}} \mu(y)\, dy\right\}. \quad (3)$$

3.2. The Radon transform and sinograms

As shown in Figure 2(b), the laser-scanned CT apparatus (often described as a first-generation system) consists of a source and detector, placed on either side of the object to be imaged. These slide along in tandem. Consider the intensity of the attenuated laser beam received by the detector when the source-detector assembly is at position x:

$$I(x) = I_0 \exp\left\{-\int \mu(x, y)\, dy\right\}, \quad (4)$$

where μ(x,y) is now the 2-D distribution of optical attenuation coefficient. μ is related to the optical density (OD)[2] by the relation μ = OD ln 10. In a medical X-ray scanner, it would be normal for the the source-detector track to rotate around the sample. However, a major design simplification where the object to be scanned is inanimate is to make the sample rotate instead. (This makes no difference to the principles involved, but the mathematical formalism below must be modified in a minor way from that seen in many textbooks by adding a negative sign to the rotation angle.)

As the sample rotates by a positive angle ϕ in Figure 1(c), the function describing the attenuation coefficient changes to μ'(x, y) by the following transformation of co-ordinate system:

$$\mu'(x,y) = \mu(x',y'), \text{ where } \begin{pmatrix} x' \\ y' \end{pmatrix} = R_{-\phi} \begin{pmatrix} x \\ y \end{pmatrix}, \quad (5)$$

where $R_{-\phi}$ is the standard 2-D matrix for a rotation through angle $-\phi$, such that

$$\begin{aligned} x' &= x\cos\phi + y\sin\phi \\ y' &= -x\sin\phi + y\cos\phi \end{aligned} . \quad (6)$$

To take a concrete example, suppose the sample has rotated through +30°. A point now at $x = 1$, $y = 0$ and thus contributing to the $I_{30°}$ profile at $x = 1$ used to be at (√3/2, –0.5) at the start of the experiment.

The "profile" or "projection signal" when the sample has rotated through angle ϕ is

$$I_\phi(x) = I_0 \exp\left\{-\int \mu'(x,y)\,dy\right\} = I_0 \exp\left\{-\int \mu(x',y')\,dy\right\}. \quad (7)$$

We define the *Radon transform* — often the word "projection" is used interchangeably — as

$$P_\phi(x) = \int_{\text{sample}} \mu(x',y')\,dy = -\ln\left(\frac{I_\phi(x)}{I_0}\right) \quad (8)$$

and create a new "space", called *Radon space*, in much the same way as one defines the reciprocal Fourier domain in MRI[3]. As shown in Fig. 3, Radon space has two dimensions, x and ϕ, and, at the general point (x , ϕ), we "store" the result of the projection $P_\phi(x)$. Taking lots of projections at a complete range of x and ϕ "fills" Radon space with data in much the same way that one fills Fourier space with 2-D MRI data, and these may be presented as a *sinogram*.

---

[2] Note that for an optical component such as a film or neutral density filter, we often talk about the optical density of the component as a whole and define OD as $-\log_{10} T$, where T is the transmittance I / I_0. In that case, OD has no units. However, in this article, we use the definition found at http://en.wikipedia.org/wiki/Optical_density: OD = $-(\log_{10} T)$ / *sample length*, in which case OD *does* have units and these may be expressed in terms of cm^{-1} or mm^{-1} as appropriate. Note that this makes sense, because OD then becomes an *intensive property* of the sample (http://en.wikipedia.org/wiki/Intensive_and_extensive_properties), that is susceptible of being imaged on a voxel-by-voxel basis. This is exactly the same distinction as that between absorbed dose and total absorbed energy.

[3] Strictly speaking, Radon space is not exactly like Fourier space in the sense that a Fourier transform may be exactly inverted to return to the original data — the data and its transform are simply two different *representations* of the same information. The back-projection reconstruction technique does not allow us to get back *exactly* to the original attenuation coefficient distribution.

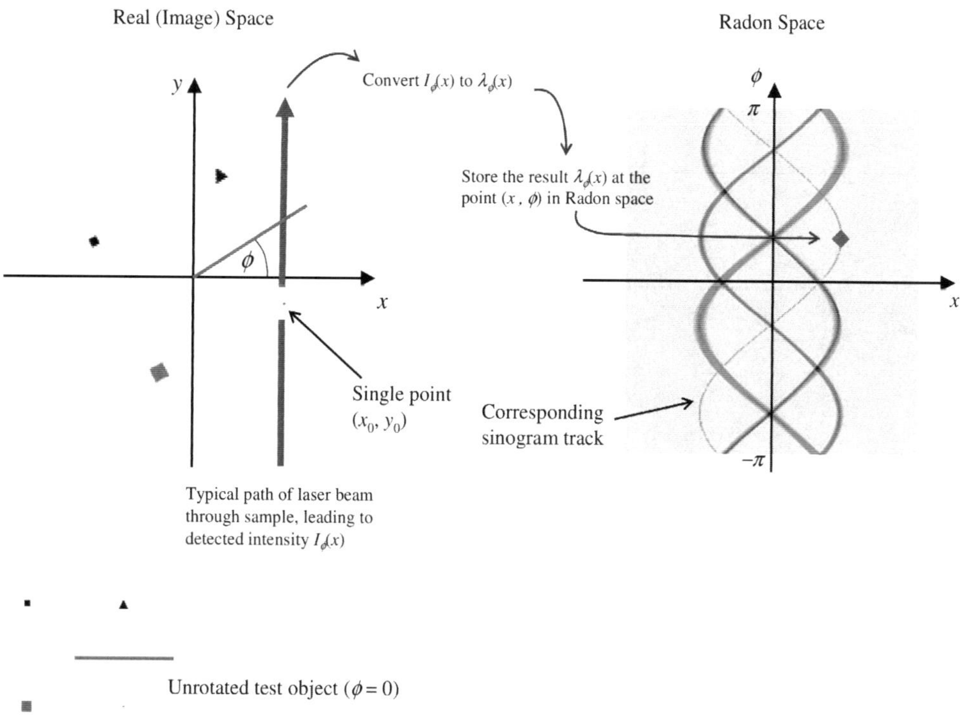

Figure 3: (a) Relationship between real space and Radon space. The presentation of data in Radon space in 2-D image form is often called a sinogram. Note how the square feature in the object leads to a sinogram track with substructure and this may be interpreted by saying that the square is made up of many points, each producing a sinogram track of its own, and these latter are overlaid.

Consider what the sinogram looks like for a sample consisting of a *single point* in real (image) space. For a given sample rotation angle ϕ, all positions x on the profile lead to $\lambda_\phi(x) = 0$, except the one coinciding with the point *to* which (x_0, y_0) has been rotated. Thus, all points in the Radon space corresponding to the single-point object are zero, except along the track

$$x = x_0 \cos\phi - y_0 \sin\phi = R\cos(\phi + \phi_0) \qquad (9)$$

where $R = (x_0^2 + y_0^2)^{1/2}$ and $\phi_0 = \tan^{-1}(y_0 / x_0)$. If we have a composite object, then the filled Radon space is simply the sum of all the individual points making up the object (i.e., multiple sinusoids, with different values of R and ϕ_0). See Figure 3 for an illustration of this.

3.3. Reconstruction via backprojection

The aim of optical CT is to obtain the optical attenuation of the sample (which is related to the absorbed dose) as a function of spatial position by acquiring a set of projections. In other words, given the complete set of projection data $P_\phi(x)$, i.e., a full Radon space, we wish to reconstruct the function $\mu(x,y)$. This is performed by a process known as *back-projection* (see Figure 4), which works as follows:

1. Consider one row of the sinogram, corresponding to angle ϕ. Note how in Figure 4, the value of the Radon transform $P_\phi(x)$ is represented by the grey level of the pixel. When we look at a single row (i.e., a 1-D set of data), we can draw this as a graph.

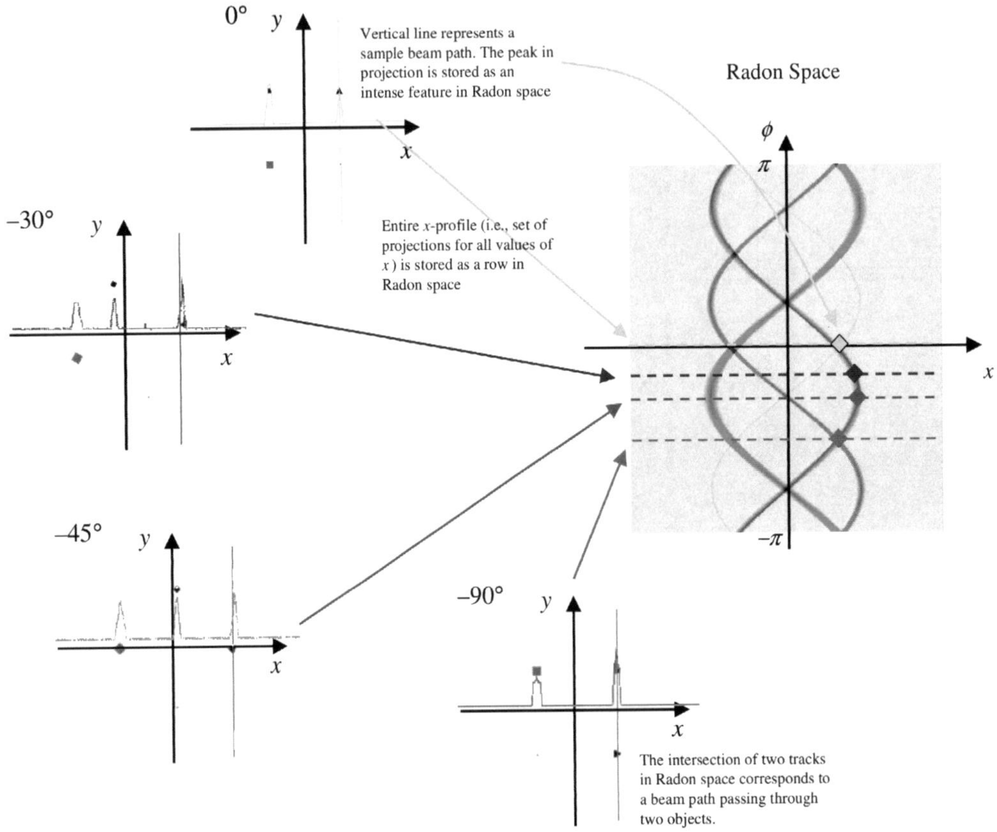

Figure 4: Illustration of the relationship between acquired profiles and sinogram tracks in Radon space

2. Place the sinogram row an angle ϕ in real space. Then "smear it out" evenly all the way along the perpendicular direction. This is called *back-projecting* the data.

3. Repeat steps 1 and 2 for all the lines in the sinogram — see Figure 4. Where the back-projections overlap, the signal adds constructively to give high-intensity image regions. Try out the excellent "Reconstruction Demo" in the medical section of the IDL distribution (http://www.ittvis.com/idl/).

3.4. Mathematical description of filtered back-projection
This is not quite the whole story. It turns out that the image that is produced by this method is blurred. To get the right representation of the object, we need an additional mathematical "trick" called filtering. In order to understand what filtering is mathematically and how the complete filtered backprojection technique works, we must first see the relationship between the Fourier transform of the image and those of its projections.

For the simple case of an un-rotated object, the Fourier transform of the projection is:

Journal of Physics: Conference Series **164** (2009) 012020 doi:10.1088/1742-6596/164/1/012020

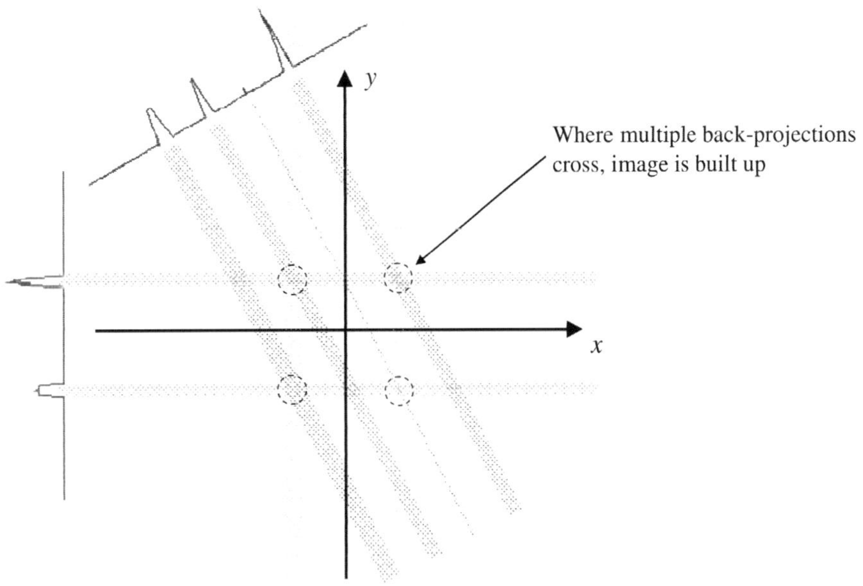

Figure 5: Back-projection of sinogram rows to form an image. The high-intensity areas of the image correspond to the crossing points of all three back-projections of the profiles.

$$\tilde{P}_0(k_x) = \int_{-\infty}^{\infty} P_0(x)\exp(-ik_x x)dx = \int_{-\infty}^{\infty}\int_{-\infty}^{\infty} \mu(x,y)dy \cdot \exp(-ik_x x)dx \tag{10}$$

If there were an extra factor $\exp(-ik_y y)$ after the integral sign, then we would have exactly the expression for the 2-D Fourier transform of the image. Instead, we can imagine that there is an extra multiplying factor of 1 "hidden" in this equation. Furthermore, we know that $1 = \exp(0) = \exp(-ik_y.0)$ whatever k_y might be. Hence, we can deduce the result that

$$\tilde{P}_0(k_x) = \int_{-\infty}^{\infty}\int_{-\infty}^{\infty} \mu(x,y)\exp(-i[k_x x + k_y y])dxdy = \tilde{\mu}(k_x, k_y = 0) \tag{11}$$

What this means is that the 1-D Fourier transform of the projection at zero degrees is simply the central line of the 2-D Fourier transform of the image. This idea will be very familiar to any MRI scientists, for whom it is a well known fact that the Fourier transform of the zero phase-encode step of an MR scan gives the projection of the sample in the phase-encoding direction.

Since the co-ordinate system can be chosen arbitrarily, it follows intuitively — and can easily be proved formally — that an equivalent result applies for projections taken at any angle. The 2-D Fourier transform of μ, evaluated along the line at angle ϕ in k-space is the same as the 1-D Fourier transform of the projection at angle ϕ.

This important result is known as the Fourier slice theorem and it is the basis of backprojection reconstruction. Let us now write the desired final image in terms of its Fourier transform:

$$\mu(x,y)=\frac{1}{2\pi}\int_{-\infty}^{\infty}\int_{-\infty}^{\infty}\tilde{\mu}(k_x,k_y)\cdot\exp(+i[k_x x+k_y y])\,dk_x\,dk_y \tag{12}$$

It is very useful to write the expression on the right hand side of the equation in terms of a polar rather than Cartesian representation of k-space. Thus $k_x = k\cos\phi$ and $k_y = k\sin\phi$, yielding

$$\mu(x,y)=\frac{1}{2\pi}\int_{0}^{2\pi}\int_{0}^{\infty}\tilde{\mu}(k,\phi)\cdot\exp(+ik[x\cos\phi+y\sin\phi])\,k\,dk\,d\phi \tag{13}$$

We note that the expression $\tilde{\mu}(k,\phi)$ is the same as $\tilde{P}_\phi(k)$, by the Fourier slice theorem, and that the expression inside the square brackets above is exactly what we wrote down earlier for x'. So we end up with the following expression directly linking the measured projections and the image that we are trying to reconstruct:

$$\mu(x,y)=\frac{1}{2\pi}\int_{0}^{2\pi}\left\{\int_{0}^{\infty}\tilde{P}_\phi(k)\cdot\exp(+ikx')\,k\,dk\right\}d\phi \tag{14}$$

Notice that there is an x on the LHS and an x' on the right of Eq. (14). Going back to the explicit expression for x' in terms of x and y and looking at exactly which point on the profile contributes to the reconstructed μ at position (x,y), it should be apparent that this is the mathematical formulation of the back-projection operation itself. In this "continuous version" of the formula, to which our acquired data are a discrete approximation, the final image, μ, is made up from a differential contribution (i.e., weighted by $d\phi$) of the projection from each angle between zero and 2π.

Again, we see that the contents of the integral in the braces are almost a Fourier transform. Without the k, the result of this integral would simply be the original projection P_ϕ. The k term modifies or "filters" the Fourier transform by giving more weight to the higher spatial frequencies. Our final formula for reconstructing images is thus:

$$\mu(x,y)=\frac{1}{2\pi}\int_{0}^{2\pi}Q_\phi(x')\,d\phi, \tag{15}$$

where Q_ϕ is the *filtered projection*, which can be obtained from the measured data.

In order to arrive at a practical reconstruction algorithm, it is important to take account of a number of other considerations that are beyond the scope of this review. More detailed descriptions of the back-projection process may be found in [22] and [23]. Finally, it is worth pointing out that filtered back-projection is not the only method of image reconstruction. A family of methods based on the *algebraic reconstruction technique* (ART) is also available. These are potentially more flexible, allowing one to cater for various deficiencies in the raw projection data and different models for how the projections are formed[4], but they are more time-consuming to run. See Ch. 7 of [22] for a general description and [8] for an application in optical CT.

[4] Examples of the type of problems that might be solved in this way are reconstruction from a limited number of projections [8]; dealing with refractive index inhomogeneities in the sample; and accounting for the fact that projections from a CCD scanner are not correctly modeled by line integrals of the form shown in Eq. (4), but, rather, are the result of absorption over a "double cone", as described in [17].

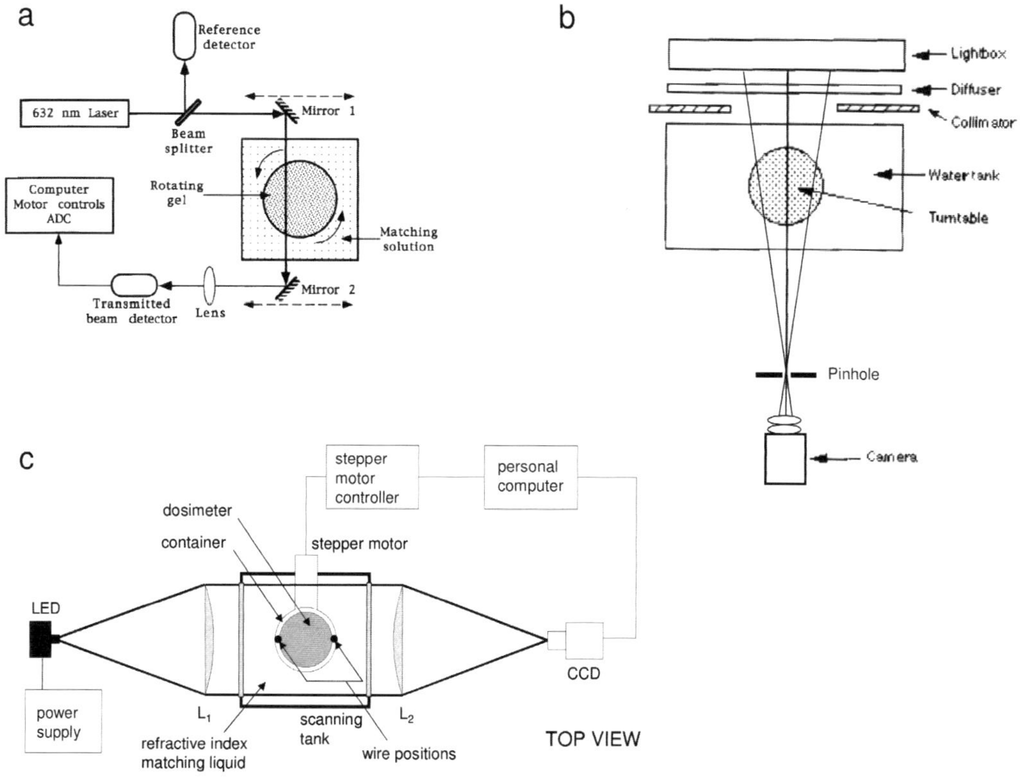

Figure 6: Optical-CT scanning geometries in current use: (a) first-generation laser configuration (diagram reproduced from [1]); (b) cone-beam CCD configuration (diagram modified from [67]); (c) parallel-beam CCD scanner (diagram reproduced from [17]).

4. Optical CT in 3-D radiation dosimetry

4.1. First generation laser scanners

The first-generation scanner geometry illustrated in Fig. 2(b) was the one employed by Gore *et al.* [1]. Fig. 6(a) gives a more detailed schematic of their apparatus, which developed into what was marketed under the name OCTOPUS by MGS Inc., for a number of years the only commercial optical-CT scanner. At approximately the same time, Tarte *et al.* were scanning gel sections using a single laser beam [24] and, in the years that have followed, a number of other groups have published work of increasing sophistication using first-generation scanners [25-52].

Excellent reviews by Jordan [53-55], presented at previous DOSGEL conferences, have discussed in outline a number of the optical problems to be addressed when designing such systems. These include: minimisation of interference effects and stray light; scatter from optical components and the radiochromic gels themselves [56]; reflection; dynamic range; wavelength selection; wall corrections [57]; the plasma discharge from lasers; temperature changes; and the characterisation of detectors. Further detailed work on characterising a custom-built optical-CT laser system was performed by Oldham et al. [36, 39].

First generation scanners tend to produce high quality images and, in the absence of a systematic comparison of all the different scanner types on standard samples, the OCTOPUS scanner has, on occasion, been referred to as a "gold standard" [52].

However, a major disadvantage of first-generation systems is their slow scanning speed. In order to obtain a profile, the laser beam and photodiode must be stepped mechanically in parallel across the sample. A typical performance figure for the original OCTOPUS system is found in [35], where slices of 128 × 128 pixels were acquired at a rate of 12 minutes per slice. To obtain multiple slices, the sample was raised or lowered mechanically an appropriate distance. True-3D scans, with isotropic high resolution and a large field-of-view in the slice direction are not feasible using this methodology and even images with fairly modest numbers of slices take many hours. More recently, improved imaging times of order 5 minutes per slice have been demonstrated using an improved version of OCTOPUS scanner [48], but there is a limit as to how far this technology can be pushed and a full 3-D scan still took of order 16 hours.

4.2. Scanners based on pixelated (area) detectors

4.2.1. Parallel- and cone-beam geometries. A faster alternative is the use of scanners based on CCD or CMOS area detectors. These chips are now extremely widespread because of their use in digital cameras. Whereas laser systems acquire data in a point-by-point fashion, imaging detectors allow us to obtain a complete 2-D projection in one go — see Figure 2 of [53]. Each 2-D projection gives the data required for creating a row in the sinogram for *every* slice in a 3-D reconstruction. A modern scientific CCD camera will have a matrix size of typically 1000 × 1000 pixels and so it is easy to see that speed gains of more than two orders of magnitude might result if 3-D data are needed. (This is of course a big "if", since large numbers of slices and an ultra-high spatial resolution are not always required.) In practice, the speed when using a CCD-based system is often limited by the data-throughput rate. Particular bottlenecks can occur in transferring data off the CCD chip to the camera and then in the transfer process out of the camera to the host computer.

Two sub-classes of CCD scanner have so far been presented in the gel dosimetry literature: parallel-beam and cone-beam.

Our group has developed scanners based on the parallel-beam geometry [16-19, 58, 59], as illustrated in Fig. 6(c). Although conceived entirely independently, these are similar to the early design of Winfree [6], which pre-dates them. A key element of the parallel-beam design is the use of *telecentric* optics and this can be achieved using either a combination of a large converging lens close to the matching tank and a standard camera lens [17] or a specialist, but more expensive, telecentric lens [52]. The parallel-beam scanner created by Oldham *et al.* is a dual-purpose instrument, capable of imaging both in absorption and emission modes and it can be applied equally to the areas of radiotherapy dosimetry and of biological microscopy [60-63], as popularized by Sharpe [7, 64-66].

The cone-beam geometry (Fig. 6(b)) was first introduced by Wolodzko *et al.* [67] and later pursued further by Jordan *et al.* [68, 69]. A commercial scanner is now marketed by Modus Medical Devices Inc. under the trade name Vista™. Since its release in 2004, a number of preliminary evaluations of this product have been made [70-74].

To the best of this author's knowledge, there are at present no comparative data in the literature to determine whether either of the two geometries is more advantageous. However, several practical comparisons may be drawn. (i) The cone-beam design is generally cheaper and is easily scaleable [69]. It has the advantage of being able to scan larger objects without recourse to expensive optical

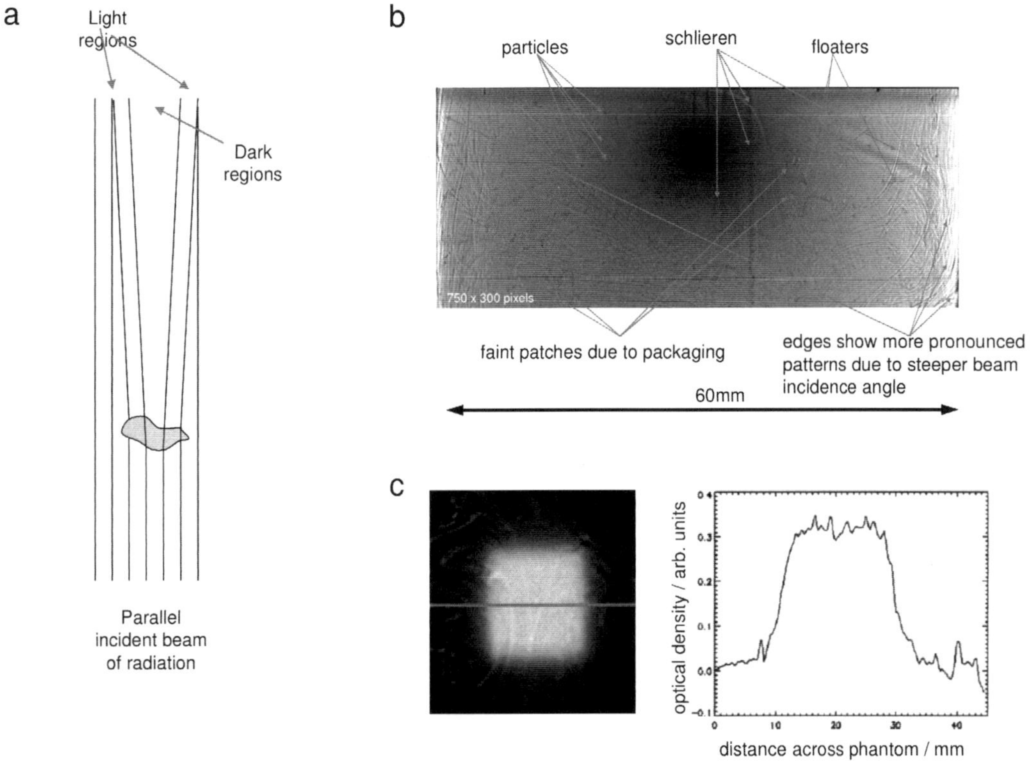

Figure 7: (a) Illustration of the principle behind the generation of schlieren effects; (b) an extreme example of schlieren and other effects that degrade the quality of projection images. The dark region in the centre is a brachytherapy irradiation; (c) an extreme example of the problems schliere can cause in the final reconstructed image and profile.

components. However, whilst large optics are expensive, they still represent only a relatively small fraction of the typical purchase price of a scanner. (ii) The cone-beam scanner is more compact. (iii) The parallel-beam scanner uses less light power (single LED vs. 2-D array of LED's behind a diffuser), suggesting that it may be easier to guard against stray light. (iv) The original parallel-beam scanner [18] was unsuitable for imaging scattering dosimeters, but the most recent design [17] has a more controllable and, potentially, better scatter rejection than the cone-beam design.

It should also be noted that a considerable body of work has also been performed using so-called "stacked-gel" detectors, together with 2-D CCD imaging, rather than a full 3-D readout [75-84]

4.2.2. Particular considerations for the use of pixelated detectors

(i) A general disadvantage of scanners based on pixelated detectors together with a wide beam is the possible introduction of artefacts by refractive index inhomogeneities (schlieren). Such variations in refractive index result, typically, from incomplete mixing of the various components of either the dosimeter itself or the matching liquid, or from local temperature gradients, which lead to convection currents. If these inhomogeneities are present in the dosimeter, as it cools during the manufacturing process, then they are "frozen in". By contrast, the effects in the matching liquid can be minimised by leaving the liquid for some time after inserting the dosimeter, so that settling occurs.

As light rays pass through the schliere, they are deviated and the effect on the projections is two-fold. Firstly, in some cases, this causes several rays to overlap, leading to bright regions of the image, whilst the regions away from which the light has been refracted are dark. This is precisely the mechanism exploited in shadowgraphy [20]. Fig. 7(a) illustrates the phenomenon schematically, whilst Fig. 7(b) is an extreme example that we observed. The consequence for the final image is difficult to predict *a priori*, and manifests itself as a structured, but complex artifact, as in Fig. 7(c).

The second effect is more subtle: light rays no longer travel in straight lines through the sample and this means that Eq. (4), on which the whole imaging experiment is predicated, is invalid. The only work to have been performed in investigating the implications of this source of systematic error appears to be that of Oldham and Kim [39], who observed no deviation in the measured positions for their needle phantom in the presence of radiation-induced refractive index inhomogeneities. However, the importance or otherwise of this effect for quantitative dose images has yet to be assessed and, since Oldham and Kim's work used a laser scanner, further work is needed to clarify the effects for area scanners.

(ii) *Ring artefacts* are generated when a feature not associated with the dosimeter sample is present in the same place in all projections. A typical cause might be a bubble or scratch on the wall of the tank containing the matching liquid. Whilst such artefacts have been studied both in the context of laser scanning [39] and CCD scanning [18], an additional problem for CCD scanners, particularly those using low-end chips is a non-uniform pixel response to incident light and/or isolated "dead" pixels. Additional calibration or other steps must be taken. In our case, purchase of a high-quality CCD (the ORCA BT-1024G from Hamamatsu) has largely eliminated this problem, although cleaning of the scanning tank must be very thorough. Recently, an algorithm has been developed in the context of X-ray CT [85] to ameliorate any residual image artefacts.

(iii) *Calibration* for a possible non-linear response of the CCD chip is necessary to ensure that the correct ratio I / I_0 is calculated in Eq. [8]. Some CCD's have a non-linear response. It is also necessary to subtract the value of the CCD "dark current" from pixel values in both the light-field and dosimeter images [86]. An allowance needs to be made for the transmission properties of the lens systems and the quality of the beam. This may be done conveniently by acquiring a "light-field" image prior to insertion of the sample. Alternatively, a complete dataset may be acquired from the dosimeter prior to irradiation. This has two important advantages: firstly, the reconstructed image contains only the changes induced by the radiation and no background; and secondly, effects of residual refractive index mismatches at the walls are greatly reduced, leading to a much larger useable volume of dosimeter.

(iv) *Dynamic range*: A potential concern for the design of any optical CT system to be used for accurate metrology is the range of optical densities that may successfully be reconstructed and the artefacts that will be generated if the integrated optical density across the sample exceeds this. In many cases, the dynamic range of the photodiode detector used in laser scanning systems is superior to that of a CCD, which might be as low as 8-bit. Given the presence of the CCD dark-current, it may well be difficult to image successfully optical densities greater than 2 (i.e., $I / I_0 = 0.01$). This question is discussed in depth in [86]. The ORCA BT-1024G has a dynamic range of approximately 20000:1 and has a 14-bit ADC and thus allow us to overcome these problems.

(v) *Image signal-to-noise ratio and accuracy*: When describing the performance of an imaging CCD, manufacturers often quote the *full-well capacity* (i.e., how many electrons a pixel on the chip can hold) and the *RMS readout noise* in electron units. The dynamic range of the camera is defined as the ratio of these two numbers. For our camera, these figures are 80,000 and 4, giving a dynamic range of 20,000 : 1. However, it is important to realise that this is not the same as the image signal-to-noise ratio (SNR). In an excellent didactic article [87], Healey and Kondepudy describe the roles of shot

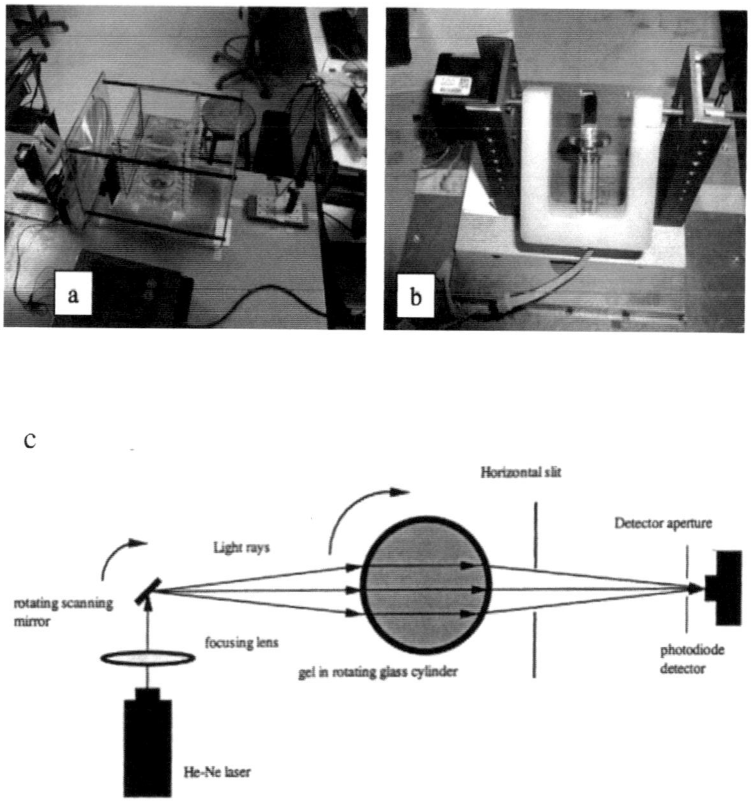

Figure 8: Examples of previous "fast-laser" scanner designs: (a) and (b) are two views of the scanner of Conklin et al., reproduced from. (c) schematic of the initial design of Maryanski

noise (i.e., photon counting statistics), dark current, readout noise and "scene variation" (i.e., genuine random changes in the variable being imaged) in determining the SNR performance of a machine vision system.

(vi) *Type of sensor*: Currently, CCD's are superior to CMOS detectors in terms of noise performance and dark current. However, future developments of pixelated detectors are likely to favour CMOS devices, which will be quicker and cheaper, with the capability of detecting greater OD dynamic range and better resolution. CMOS is a standard lithographic technique for microchips, so adding more logic for each pixel is easy. This in turn enables dual mode operation (linear or logarithmic). Our focus should be on the logarithmic mode, since it should allow us to obtain a more uniform OD resolution across the full OD dynamic range. Extending the dynamic range of CCD's is possible by programming multiple acquisitions with different exposure times [86], but this is in many cases time consuming and runs the risk of images being corrupted by *blooming* [88].

4.3. Fast laser scanners

For some while, research has been ongoing to try and create a novel design of scanner that combines the advantages of a single-beam laser scanner with the speed of a CCD scanner. A first step was taken by Maryanski [89], with a design the contained two novel features. Firstly, the laser was reflected by a rotating mirror, which allowed the beam to be scanned very rapidly across the sample in a single dimension. Secondly, there was no matching tank and the sample was used to refract the beam in order to obtain a parallel set of beams through the sample. Both of these features are present in the latest scanner from MGS, the IQ-SCAN. Van Doorn's group [90] and Conklin *et al.* [91] have also demonstrated the feasibility of a rotating mirror approach. Two of these scanners are illustrated in Figure 8. Perhaps the most sophisticated fast laser scanner built so far was introduced by Krstajic and Doran [92] and makes use of ideas drawn from confocal microscopy. Instead of using simple rotating mirrors, this instrument (described in detail in [92]and characterized further in [93]) manipulates the laser beam using a pair of galvanometer controlled mirrors. This allows the beam to be scanned in 2-D, with the potential for control of the beam using adaptive optics to reduce errors. It is well known in the microscopy and OCT communities that this type of arrangement can be used to output images at video frame rates, leading to the potential (not yet realized in 3-D dosimetry) of 3-D scans in under a minute.

5. Dosimeter materials

As described earlier, chemical dosimetry was initially based around so-called Fricke solution, which absorbs in the UV (304 nm). Pioneering work investigating how to shift this absorption into the visible range was performed by Gupta, who authored some 20 papers on the ferrous xylenol-orange system (e.g., [94]). More recently, since the advent of optical CT in 3-D radiation dosimetry, a large number of other authors have investigated different properties of ferrous xylenol-orange gels (FXG) [58, 84, 95-104]. FXG is easy to manufacture and its performance is not hindered by the presence of atmospheric oxygen (in fact, this positively helps). However, its major disadvantage is the mobility post-irradiation of the ferric ion complex. Diffusion causes an unacceptable blurring of the dose pattern within a few hours of irradiation and this has led many groups to move away from Fricke gels to other chemical systems. Nevertheless, Babic *et al.* have demonstrated that, once the problems involved are well understood and controlled, then excellent results may be obtained [46, 105].

Polymer gels were developed to overcome the diffusion problem. Initially developed for MRI-based dosimetry, their potential in optical CT studies was obvious from early on [2]. Whilst contrast in FXG is developed by means of *absorption* — the orange gel turns purple on exposure; see [18] for a typical spectrum — polymer gels attenuate light by *scattering* and hence change from clear to opaque white. Several polymer gel formulations have been studied or developed specifically for optical CT [106, 107] and a variety of different applications have been pursued [35, 36, 39, 48, 49, 80, 89, 108-111]. A disadvantage of some polymer gels is their sensitivity to atmospheric oxygen.

Without doubt, one of the most exciting developments during recent years has been the introduction [112] of PRESAGE™, a novel transparent plastic dosimeter. The material is rigid and easily machineable. It is insensitive to oxygen; moreover, as it is not a gel, PRESAGE™ does not dry out. Hence, one can remove the container and, with it, two optical interaces. This makes modelling of the optical path through the dosimeter [18] much simpler. PRESAGE™ is stable during a prolonged irradiation period, and has good post-irradiation storage properties. It has a linear response at low energies and over an exceptionally wide range of doses [113]. Not surprisingly, the last four years have seen a flurry of interest in this dosimeter [16, 17, 42, 44, 47, 50-52, 59, 92, 93, 114-123].

6. Future perspectives and conclusions

Optical CT in 3-D radiation dosimetry has come a long way over the last 12 years. As pointed out in the introduction, the competing readout systems, MRI and X-ray CT, were already mature imaging modalities when first applied to dosimetry. Nevertheless, the capability of optical CT instruments is now approaching and, in areas such as image resolution and acquisition speed, exceeding that of MRI. Naturally, there is less experience with the technique and a number of the most promising dosimeter materials are only a few years old. Their formulations are still changing, with key properties not yet measured. Thus, it would probably be fair to say that the state-of-the-art in optical dosimetry is where MRI was 3–5 years ago. We still have to undertake much more painstaking verification work, investigation of anomalous results and the pursuit of absolute dosimetry, topics which have characterised the last two decades of work with MRI and continue to do so.

Future work needs to address one issue above all. The optical CT workflow needs to be optimized. Imaging needs to become more routine and less of a specialised "research" activity. The process needs to be "less messy", no longer requiring users to spending time worrying about issues such as refractive index matching of their samples. There needs to be better quality assurance built into the scanners. File formats need to be standardised. Finally, with many different designs of scanner available, there need to be more test objects and widely agreed standards for how to compare systems.

Two years ago, I wrote the following confident statement: *Optical CT scanners cost between one and two orders of magnitude less than an MRI system and this should ensure that the community of researchers will grow rapidly during the next five years.* This hope has certainly been fulfilled by this year's crop of DOSGEL abstracts. At DOSGEL 2008, there were 21 abstracts with an optical theme, presented by 7 different groups. This year, the programme features 27 such submissions, from 15 groups, with an increased diversity of applications.

I ended the previous review with the words: *It is possibly premature to sound the death knell of MRI, but we are catching up fast!* A recent, preliminary analysis of the published scientific literature for the period 2006 – 2008 yielded 78 full papers on gel dosimetry. Of these 34 used MRI as the readout modality and 34 used optical methods. Other methods, including X-ray CT and ultrasound, accounted for the other 10.

Optical CT has now arrived!

References

[1] Gore, J.C., M. Ranade, and M.J. Maryanski, *Radiation dose distributions in three dimensions from tomographic optical density scanning of polymer gels: I. Development of an optical scanner.* Physics in Medicine and Biology, 1996. **41**(12): p. 2695-2704.

[2] Maryanski, M.J., Y.Z. Zastavker, and J.C. Gore, *Radiation dose distributions in three dimensions from tomographic optical density scanning of polymer gels: II. Optical properties of the BANG polymer gel.* Physics in Medicine and Biology, 1996. **41**(12): p. 2705-2717.

[3] Doran, S. and N. Krstajic, *The history and principles of optical computed tomography for scanning 3-D radiation dosimeters.* J Phys. Conf. Ser., 2006. **56**: p. 45-57.

[4] Oldham, M. *Optical CT scanning of polymer gels.* in *DOSGEL 2006*. 2006. Sherbrooke.

[5] Fricke, H. and S. Morse, *The chemical action or roentgen rays on dilute ferrosulphate as a measure of dose.* Am J. Roent. Rad. Ther., 1927. **18**: p. 430-432.

[6] Winfree, A.T., et al., *Quantitative optical tomography of chemical waves and their organizing centers.* Chaos, 1996. **6**(4): p. 617-626.

[7] Sharpe, J., et al., *Optical Projection Tomography as a Tool for 3D Microscopy and Gene Expression Studies.* Science, 2002. **296**(19 april): p. 541-545.

[8] Haidekker, M.A., *Optical transillumination tomography with tolerance against refraction mismatch.* Computer Methods and Programs in Biomedicine, 2005. **80**(3): p. 225-235.

[9] Yao, G. and M.A. Haidekker, *Transillumination optical tomography of tissue-engineered blood vessels: a Monte Carlo simulation.* Applied Optics, 2005. **44**(20): p. 4265-4271.

[10] Haidekker, M.A., *A hands-on model-computed tomography scanner for teaching biomedical imaging principles.* International Journal of Engineering Education, 2005. **21**(2): p. 327-334.
[11] Gladish, J.C., et al., *Optical transillumination tomography for imaging of tissue-engineered blood vessels.* Annals of Biomedical Engineering, 2005. **33**(3): p. 323-327.
[12] Webb, R.H., *Confocal Optical Microscopy.* Reports on Progress in Physics, 1996. **59**: p. 427-471.
[13] Tomlins, P.H. and R. Wang, *Theory, developments and applications of optical coherence tomography.* Journal of Physics D: Applied Physics, 2005. **38**: p. 2519-2535.
[14] Gibson, A.P., J.C. Hebden, and S.A. Arridge, *Recent advances in diffuse optical imaging.* Physics in Medicine and Biology, 2005. **50**(4): p. R1-R43.
[15] Stelzer, E.H.K., *Three dimensional light microscopy*, in *Handbook of computer vision and applications*, B. Jahne, H. Haussecker, and P. Geissler, Editors. 1999, Academic Press: London, UK. p. 541-562.
[16] Krstajic, N. and S.J. Doran, *Characterization of a parallel-beam CCD optical-CT apparatus for 3D radiation dosimetry.* Physics in Medicine and Biology, 2007. **52**(13): p. 3693-3713.
[17] Krstajic, N. and S.J. Doran, *Focusing optics of a parallel beam CCD optical tomography apparatus for 3D radiation dosimetry.* Physics in Medicine and Biology, 2006. **51**: p. 2055-2075.
[18] Doran, S.J., et al., *A CCD-based optical CT scanner for high-resolution 3D imaging of radiation dose distributions: equipment specifications, optical simulations and preliminary results.* Physics in Medicine and Biology, 2001. **46**(12): p. 3191-3213.
[19] Bero, M.A., et al. *Faster optical tomography with parallel-beam white light for three-dimensional dosimetry.* in *DOSGEL '99, 1st International Workshop on Radiation Therapy Gel Dosimetry.* 1999. Kentucky.
[20] Settles, G.S., *Schlieren and Shadowgraphy Techniques, Visualizing Phenomena in Transparent Media.* 2001, Berlin: Springer Verlag.
[21] Dick, V.P., *Applicability limits of Beer's law for dispersion media with a high concentration of particles.* Applied Optics, 1998. **37**(21): p. 4998-5004.
[22] Kak, A.C. and M. Slaney, *Principles of computerized tomographic imaging.* 2001: SIAM.
[23] Hsieh, *Computed Tomography: Principles, Design Artifacts and Recent Advances.* 1 ed. 2003: SPIE Press. 387.
[24] Tarte, B.J., P.A. Jardine, and T. van Doom, *Laser-scanned agarose gel sections for radiation field mapping.* Int. J. Radiat. Oncol. Biol. Phys., 1996. **36**: p. 175-179.
[25] Knisely, J.P.S., et al., *3D dosimetry for complex stereotactic radiosurgery using a tomographic optical density scanner and bang polymer gels.* International Journal of Radiation Oncology Biology Physics, 1997. **39**(2): p. 214-214.
[26] Kelly, R.G., K.J. Jordan, and J.J. Battista, *Optical CT reconstruction of 3D dose distributions using the ferrous-benzoic-xylenol (FBX) gel dosimeter.* Medical Physics, 1998. **25**(9): p. 1741-1750.
[27] Maryanski, M.J., et al., *High-resolution 3D dosimetry for endovascular brachytherapy using optical laser CT microimaging of BANG polymer gels.* Medical Physics, 1998. **25**: p. A107.
[28] Oldham, M., et al., *High resolution gel-dosimetry by optical-CT and MR scanning.* Medical Physics, 2001. **28**(7): p. 1436-1445.
[29] Liang, R., C. Wuu, and M. Maryanski, *Accuracy of high-resolution optical CT scanning system for polymer gel dosimetry as applied to intravascular brachytherapy.* Medical Physics, 2002. **29**(6): p. 1206.
[30] Oldham, M., et al., *Achieving and utilising the 3rd dimension in optical gel-dosimetry.* Medical Physics, 2002. **29**(6): p. 1370.
[31] Oldham, M., *Light transport in optical-CT gel-dosimetry: Optical Monte-Carlo modeling.* Medical Physics, 2002. **29**(6): p. 1370-1370.

[32] Wuu, C., et al., *Implementation of 3D dose verification for IMRT using polymer gel dosimeters and laser-beam optical CT scanner.* Medical Physics, 2002. **29**(6): p. 1365-1366.
[33] Witten, M., et al., *Radiation dose distributions using a polymer gel dosimeter: A comparison of analyses using magnetic resonance imaging and laser optical scanning.* Medical Physics, 2002. **29**(6): p. 1207.
[34] Xu, Y., C. Wuu, and M. Maryanski, *An algorithm for optimized CT scanning of gel dosimeters.* Medical Physics, 2002. **29**(6): p. 1205-1206.
[35] Islam, K.T.S., et al., *Initial evaluation of commercial optical CT-based 3D gel dosimeter.* Medical Physics, 2003. **30**(8): p. 2159-2168.
[36] Oldham, M., et al., *Optical-CT gel-dosimetry I: basic investigations.* Medical Physics, 2003. **30**(4): p. 623-634.
[37] Xu, Y., C.S. Wuu, and M.J. Maryanski, *Performance of laser CT scanning of polymer gels as a tool for three dimensional dose verification.* Medical Physics, 2003. **30**(6): p. 1426-1426.
[38] Xu, Y., C.S. Wuu, and M.J. Maryanski, *Determining optimal gel sensitivity in optical CT scanning of gel dosimeters.* Medical Physics, 2003. **30**(8): p. 2257-2263.
[39] Oldham, M. and L. Kim, *Optical-CT gel-dosimetry II: Optical artifacts and geometrical distortion.* Medical Physics, 2004. **31**(5): p. 1093-1104.
[40] Xu, Y., C. Wuu, and M. Maryanski, *Sensitivity calibration procedures in optical-CT scanning of polymer gel dosimeters.* Medical Physics, 2004. **31**(6): p. 1833.
[41] Wuu, C., et al., *Verification of whole-body dosimetry in an IMRT treatment planning system.* Medical Physics, 2005. **32**(6): p. 1989-1989.
[42] Guo, P.Y., J. Adamovics, and M. Oldham. *Simple 3D validation experiments for PRESAGE(TM)/optical-CT dosimetry.* in *DOSGEL 2006, fourth international conference on radiotherapy gel dosimetry.* 2006. Sherbrooke.
[43] Oldham, M., et al. *Towards four dimensional (4D) dosimetry for radiation-therapy.* in *DOSGEL 2006, fourth international conference on radiotherapy gel dosimetry.* 2006. Sherbrooke.
[44] Oldham, M., et al. *IMRT verification using a radiochromic/optical-CT dosimetry system.* in *DOSGEL 2006, fourth international conference on radiotherapy gel dosimetry.* 2006. Sherbrooke.
[45] Ravindran B, P., et al. *Investigation of an in-house laser CT scanner for reading normoxic gel.* in *DOSGEL 2006, fourth international conference on radiotherapy gel dosimetry.* 2006. Sherbrooke.
[46] Babic, S., J. Battista, and K. Jordan, *Three-dimensional dose verification for intensity-modulated radiation therapy in the radiological physics centre head-and-neck phantom using optical computed tomography scans of ferrous xylenol-orange gel dosimeters.* International Journal of Radiation Oncology Biology Physics, 2008. **70**(4): p. 1281-1291.
[47] Oldham, M., et al., *An investigation of the accuracy of an IMRT dose distribution using two- and three-dimensional dosimetry techniques.* Medical Physics, 2008. **35**(5): p. 2072-2080.
[48] Lopatiuk-Tirpak, O., et al., *Performance evaluation of an improved optical computed tomography polymer gel dosimeter system for 3D dose verification of static and dynamic phantom deliveries.* Medical Physics, 2008. **35**(9): p. 3847-3859.
[49] Wuu, C.S. and Y. Xu, *Three-dimensional dose verification for intensity modulated radiation therapy using optical CT based polymer gel dosimetry.* Medical Physics, 2006. **33**(5): p. 1412-1419.
[50] Guo, P.Y., J. Adamovics, and M. Oldham, *A practical three-dimensional dosimetry system for radiation therapy.* Medical Physics, 2006. **33**(10): p. 3962-3972.
[51] Adamovics, J. and M.J. Maryanski, *Characterisation of PRESAGE (TM): A new 3-D radiochromic solid polymer dosemeter for ionising radiation.* Radiation Protection Dosimetry, 2006. **120**(1-4): p. 107-112.

[52] Sakhalkar, H.S. and M. Oldham, *Fast, high-resolution 3D dosimetry utilizing a novel optical-CT scanner incorporating tertiary telecentric collimation.* Medical Physics, 2008. **35**: p. 101-111.

[53] Jordan, K.J. *Developmental issues for optical CT and gel dosimetry.* in *DOSGEL '99, 1st International Workshop on Radiation Therapy Gel Dosimety.* 1999. Lexington.

[54] Jordan, K.J., *Fundamentals of optical measurements in gel dosimetry.* Proc. of DOSGEL 2001 2nd Int. Conf. on Radiotherapy Gel Dosimetry, 2001: p. 66-72.

[55] Jordan, K.J., *Advances in optical CT scanning for gel dosimetry.* Journal of Physics: Conference Series, 2004. **3**: p. 115-122.

[56] Jordan, K.J., L. Coelho, and K. Chu. *Reduction of optical scatter and diffusion in radiochromic gel containing glycerol.* in *World Congress on Medical Physics and Biomedical Engineering.* 2000. Chicago.

[57] Jordan, K.J., G. Garvey, and J.J. Battista. *Measuring complete optical CT datasets using thin-walled vessels for 3D gel dosimetry.* in *Proc. World Congress on Medical Physics and Biomedical Engineering.* 2000. Chicago.

[58] Bero, M.A., et al., *Three-dimensional radiation dose measurements with Ferrous Benzoic Acid Xylenol Orange in Gelatin gel and optical absorption tomography.* Nuclear Instruments and Methods in Physics Research A, 1999. **422**: p. 617-620.

[59] Doran, S.J., et al. *Optical CT scanning of PRESAGE(TM) polyurethane samples with a CCD-based readout system.* in *DOSGEL 2004, third international conference on radiotherapy gel dosimetry.* 2004. Ghent.

[60] Oldham, M., et al., *Three-dimensional imaging of xenograft tumors using optical computed and emission tomography.* Medical Physics, 2006. **33**(9): p. 3193-3202.

[61] Sakhalkar, H.S., et al., *Functional imaging in bulk tissue specimens using optical emission tomography: fluorescence preservation during optical clearing.* Physics in Medicine and Biology, 2007. **52**(8): p. 2035-2054.

[62] Oldham, M., et al., *Three-dimensional imaging of whole rodent organs using optical computed and emission tomography.* Journal of Biomedical Optics, 2007. **12**(1).

[63] Oldham, M., et al., *Optical clearing of unsectioned specimens for three-dimensional imaging via optical transmission and emission tomography.* J Biomed Opt, 2008. **13**(2): p. 021113.

[64] Sharpe, J., *Optical Projection Tomography.* Annual Review of Biomedical Engineering, 2004. **6**: p. 209-228.

[65] Walls, J.R., et al., *Correction of artefacts in optical projection tomography.* Physics in Medicine and Biology, 2005. **50**: p. 4645-4665.

[66] Walls, J.R., et al., *Resolution improvement in emission optical projection tomography.* Physics in Medicine and Biology, 2007. **52**(10): p. 2775-2790.

[67] Wolodzko, J.G., C. Marsden, and A. Appleby, *CCD imaging for optical tomography of gel radiation dosimeters.* Medical Physics, 1999. **26**(11): p. 2508-2513.

[68] Jordan, K.J., et al. *Optical cone beam tomography with low pressure sodium light.* in *DOSGEL 2001, Second International Conference on Radiotherapy Gel Dosimetry.* 2001. Brisbane.

[69] Jordan, K. and J. Battista. *Small, medium and large optical cone beam CT.* in *DOSGEL 2006, fourth international conference on radiotherapy gel dosimetry.* 2006. Sherbrooke.

[70] Bosi, S.G., P. Naseri, and C. Baldock. *Initial investigation of a novel phantom for simulating optical scattering of dosimetric gels.* in *DOSGEL 2006, fourth international conference on radiotherapy gel dosimetry.* 2006. Sherbrooke.

[71] De Jean, P., et al. *Initial experience with a commercial cone beam Optical CT unit for polymer gel dosimetry I: Optical dosimetry issues.* in *DOSGEL 2006, fourth international conference on radiotherapy gel dosimetry.* 2006. Sherbrooke.

[72] De Jean, P., et al. *Initial experience with a commercial cone beam Optical CT unit for polymer gel dosimetry I: Clinical potential.* in *DOSGEL 2006, fourth international conference on radiotherapy gel dosimetry*. 2006. Sherbrooke.
[73] Jordan, K. and J. Battista. *Linearity and image uniformity of the Vista™ optical cone beam scanner.* in *DOSGEL 2006, fourth international conference on radiotherapy gel dosimetry*. 2006. Sherbrooke.
[74] Sarabipour, S., et al. *A preliminary study of the measurement of slice-width dose profiles (SWDP) on diagnostic x-ray CT scanners using PAGAT polymer gel dosimeters with optical CT read-out.* in *DOSGEL 2006, fourth international conference on radiotherapy gel dosimetry*. 2006. Sherbrooke.
[75] Vanossi, E., et al., *Study of polymer gel for dose imaging in radiotherapy.* Radiation Measurements, 2008. **43**(2-6): p. 442-445.
[76] Valente, M., et al., *Gel dosimetry measurements and Monte Carlo modeling for external radiotherapy photon beams comparison with a treatment planning system dose distribution.* Nuclear Instruments & Methods in Physics Research Section a-Accelerators Spectrometers Detectors and Associated Equipment, 2007. **580**(1): p. 497-501.
[77] Tomatis, S., et al., *Gel-layer dosimetry for dose verification in intensity-modulated radiation therapy.* Nuclear Instruments & Methods in Physics Research Section a-Accelerators Spectrometers Detectors and Associated Equipment, 2007. **580**(1): p. 506-509.
[78] Mariani, M., et al., *Preliminary results from a polymer gel dosimeter for absorbed dose imaging in radiotheraphy.* Radiation Physics and Chemistry, 2007. **76**(8-9): p. 1507-1510.
[79] Gambarini, G., et al., *Gel dosimeters as useful dose and thermal-fluence detectors in boron neutron capture therapy.* Radiation Effects and Defects in Solids, 2007. **162**(10-12): p. 777-783.
[80] Gambarini, G., et al., *Optical analysis of gel dosimeters: Comparison of Fricke and normoxic polymer gels.* Nuclear Instruments & Methods in Physics Research Section B-Beam Interactions with Materials and Atoms, 2007. **263**(1): p. 191-195.
[81] Gambarini, G., et al., *Dose distributions in phantoms irradiated in thermal columns of two different nuclear reactors.* Radiation Protection Dosimetry, 2007. **126**(1-4): p. 640-644.
[82] Castellano, G., et al., *An optimized Monte Carlo (PENELOPE) code for the characterization of gel-layer detectors in radiotherapy.* Nuclear Instruments & Methods in Physics Research Section a-Accelerators Spectrometers Detectors and Associated Equipment, 2007. **580**(1): p. 502-505.
[83] Carrara, M., et al., *Dose distribution measurements by means of gel-layer dosimeters. Evaluation of algorithms for artifacts amendment.* Nuclear Instruments & Methods in Physics Research Section a-Accelerators Spectrometers Detectors and Associated Equipment, 2007. **579**(1): p. 334-338.
[84] Gambarini, G., et al., *Dose imaging with gel-dosemeter layers: Optical analysis and dedicated software.* Radiation Protection Dosimetry, 2006. **120**(1-4): p. 144-147.
[85] Sijbers, J. and A. Postnovz, *Reduction of ring artefacts in high resolution micro-CT reconstructions.* Physics in Medicine and Biology, 2004. **49**(14): p. N247-N253.
[86] Krstajic, N. and S.J. Doran. *Characterization of a parallel beam CCD optical-CT apparatus for 3D radiation dosimetry.* in *DOSGEL 2006, fourth international conference on radiotherapy gel dosimetry*. 2006. Sherbrooke.
[87] Healey, G.E. and R. Kondepudy, *Radiometric Ccd Camera Calibration and Noise Estimation.* Ieee Transactions on Pattern Analysis and Machine Intelligence, 1994. **16**(3): p. 267-276.
[88] Al-Nowais, S. and S.J. Doran. *CCD-based optical CT scanning of highly attenuating phantoms.* in *DOSGEL 2008*. 2008. Hersonissos, Crete.
[89] Wuu, C.S., et al., *Dosimetry study of Re-188 liquid balloon for intravascular brachytherapy using polymer gel dosimeters and laser-beam optical CT scanner.* Medical Physics, 2003. **30**(2): p. 132-137.

[90] Van Doorn, T., et al., *A fast, high spatial resolution optical tomographic scanner for measurement of absorption in gel dosimetry.* Australasian Physical & Engineering Sciences in Medicine, 2005. **28**(2): p. 76-85.

[91] Conklin, J., et al. *Fast laser optical CT scanner with rotating mirror and Fresnel lenses.* in *DOSGEL 2006, fourth international conference on radiotherapy gel dosimetry.* 2006. Sherbrooke.

[92] Krstajic, N. and S.J. Doran, *Fast laser scanning optical-CT apparatus for 3D radiation dosimetry.* Physics in Medicine and Biology, 2007. **52**(11): p. N257-N263.

[93] Krstajic, N. and S. Doran. *Initial characterization of fast laser scanning optical CT apparatus for 3-D dosimetry.* in *DOSGEL 2008.* 2008. Hersonissos, Crete.

[94] Gupta, B.L., U.R. Kini, and R.M. Bhat, *Sensitivity of Ferrous Sulfate Benzoic Acid Xylenol Orange Dosimeter (Fbx System) to B-10(N, Alpha)Li-7 Recoils.* International Journal of Applied Radiation and Isotopes, 1976. **27**(1): p. 31-34.

[95] Chu, K.C., et al. *A novel Fricke dosimeter using PVA cryogel.* in *DOSGEL '99, first international workshop on radiotherapy gel dosimetry.* 1999. Lexington, KY.

[96] Bero, M.A., et al. *A coloured gel system for optical three-dimensional dosimetry.* in *DOSGEL '99, first international workshop on radiotherapy gel dosimetry.* 1999. Lexington, KY.

[97] Bero, M.A., W.B. Gilboy, and P.M. Glover, *Radiochromic gel dosemeter for three-dimensional dosimetry.* Radiation Physics and Chemistry, 2001. **61**(3-6): p. 433-435.

[98] Bero, M.A., W.B. Gilboy, and S.J. Doran. *Investigation into the radiochromic (FXG) gel dosemeter: stability and uncertainty in optical measurements.* in *DOSGEL 2001, second international conference on radiotherapy gel dosimetry.* 2001. Brisbane.

[99] Brindha, S., et al. *PVA-FX gel for three dimensional dose distribution measurement: A preliminary study on dose response.* in *DOSGEL 2001, second international conference on radiotherapy gel dosimetry.* 2001. Lexington, KY.

[100] Chu, K.C. and K.J. Jordan. *Manufacturing PVA-Fricke gels.* in *DOSGEL 2001, second international conference on radiotherapy gel dosimetry.* 2001. Lexington, KY.

[101] Hill, B., et al. *Manufacturing method and absorbed dose measurements of ferrous sulphate polyvinyl alcohol gel.* in *DOSGEL 2001, second international conference on radiotherapy gel dosimetry.* 2001. Brisbane.

[102] Jordan, K.J., et al. *Refrigeration temperature gradients generate dose response artifacts in Fricke gelatin dosimeters.* in *DOSGEL 2001, second international conference on radiotherapy gel dosimetry.* 2001. Brisbane.

[103] Moreira, M.V. *FXG mass attenuation coefficient evaluation for radiotherapy routine.* in *DOSGEL 2004, third international conference on radiotherapy gel dosimetry.* 2004. Ghent.

[104] Healy, B. *Characterization of the ferrous-xylenol orange-gelatin (FXG) gel dosimeter.* in *DOSGEL 2004, third international conference on radiotherapy gel dosimetry.* 2004. Ghent.

[105] Babic, S., J. Battista, and K. Jordan, *An apparent threshold dose response in ferrous xylenol-orange gel dosimeters when scanned with a yellow light source.* Physics in Medicine and Biology, 2008. **53**(6): p. 1637-1650.

[106] Maryanski, M.J., Y. Xu, and C.S. Wuu, *Dose sensitivity modification in polymer gel dosimeters improves the accuracy of optical CT scanning for gel dosimetry.* Medical Physics, 2003. **30**(6): p. 1478-1478.

[107] Maryanski, M., *New polymer gel dosimeter with high refractive index for artifact-free optical CT scanning.* Medical Physics, 2005. **32**(6): p. 1998-1998.

[108] Alwan, R., et al., *Calibration of laser tomography as a new optical diagnostic tool applied to dosimetric polymer gels.* Nuclear Instruments & Methods in Physics Research Section B-Beam Interactions with Materials and Atoms, 2008. **266**(5): p. 834-840.

[109] Koeva, V.I., et al., *Polymer gel dosimeters with increased solubility: A preliminary investigation of the NMR and optical dose-response using different crosslinkers and co-solvents.* Macromolecular Symposia, 2008. **261**: p. 157-166.

[110] Bosi, S., et al., *Initial investigation of a novel light-scattering gel phantom for evaluation of optical CT scanners for radiotherapy gel dosimetry.* Physics in Medicine and Biology, 2007. **52**(10): p. 2893-2903.

[111] Senden, R.J., et al., *Polymer gel dosimeters with reduced toxicity: a preliminary investigation of the NMR and optical dose-response using different monomers.* Physics in Medicine and Biology, 2006. **51**(14): p. 3301-3314.

[112] Adamovics, J. and M.J. Maryanski, *New 3D radiochromic solid polymer dosimeter from leuco dyes and a transparent polymeric matrix.* Medical Physics, 2003. **30**(6): p. 1349-1349.

[113] Al-Nowais, S., et al. *An attempt to determine the saturation dose for PRESAGE(TM).* in *DOSGEL 2008*. 2008. Hersonissos, Crete.

[114] Adamovics, J. and M.J. Maryanski, *A new approach to radiochromic three-dimensional dosimetry-polyurethane.* Journal of Physics: Conference Series, 2004. **3**: p. 172-175.

[115] Adamovics, J. and M. Maryanski, *OCT scanning properties of PRESAGE - A 3D radiochromic solid polymer dosimeter.* Medical Physics, 2004. **31**(6): p. 1906-1906.

[116] Doran, S., et al. *True-3D scans using PRESAGE™ and Optical-CT: A case study in proton therapy.* in *DOSGEL 2006, fourth international conference on radiotherapy gel dosimetry.* 2006. Sherbrooke.

[117] Guo, P.Y., J.A. Adamovics, and M. Oldham, *Characterization of a new radiochromic three-dimensional dosimeter.* Medical Physics, 2006. **33**(5): p. 1338-1345.

[118] Wai, P., et al. *Dosimetry of the Amersham 6711 Oncoseed™ using PRESAGE™ and Optical CT.* in *DOSGEL 2006, fourth international conference on radiotherapy gel dosimetry.* 2006. Sherbrooke.

[119] Adamovics, J., K. Jordan, and J. Dietrich. *PRESAGE(TM) - Development and optimization studies of a 3D radiochromic plastic dosimeter - Part 1.* in *DOSGEL 2006, fourth international conference on radiotherapy gel dosimetry.* 2006. Sherbrooke.

[120] Adamovics, J., et al. *PRESAGE(TM) - Development and Optimization Studies of a 3D Radiochromic Plastic Dosimeter - Part 2.* in *DOSGEL 2006, fourth international conference on radiotherapy gel dosimetry.* 2006. Sherbrooke.

[121] Heard, M., J. Adamovics, and G. Ibbott. *Measurement of a 200 MeV proton beam using a polyurethane dosimeter.* in *DOSGEL 2006, fourth international conference on radiotherapy gel dosimetry.* 2006. Sherbrooke.

[122] Manzoor, A., et al., *An exploration of new formulations for PRESAGETM 3D dosimetry.* Medical Physics, 2006. **33**(6): p. 2069-2069.

[123] Sakhalkar, H., et al., *Characterization of a novel, fast, in-house, CCD-based optical-CT scanner for 3D dosimetry in radiochromic dosimeters.* Medical Physics, 2007. **34**(6): p. 2610-2610.

Journal of Physics: Conference Series **164** (2009) 012021 doi:10.1088/1742-6596/164/1/012021

Light scattering artefacts in a funnel phantom using optical CT

Stephen G. Bosi[1,2], Saxby Brown[2,3], Sarvenaz Sarabipour[4], Yves De Deene[5] and Clive Baldock[2]

[1]Dept of Radiation Oncology, Prince of Wales Hospital, Sydney;
[2]Institute of Medical Physics, School of Physics, University of Sydney;
[3]Dept of Radiation Oncology, Royal North Shore Hospital, Sydney;
[4]Dept of Nuclear Medicine & Radiobiology, University of Sherbrooke, Canada;
[5]Radiotherapy Department, Ghent University Hospital, Belgium

Email: c.baldock@physics.usyd.edu.au

Abstract A gelatin phantom containing a funnel-shaped region of high opacity (or optical density OD) was used to examine light-scattering induced artefacts in a cone-beam optical CT scanner used for gel dosimetry. To correctly simulate polymer gel dosimeters, the opacity was introduced by adding a colloidal scatterer to the gelatin. In line profiles of OD taken from 3-D reconstructions of the funnel, those profiles with a long pathlength through high OD regions exhibited a "dishing" artefact, while those of short pathlength exhibited the opposite effect - "doming". These phenomena are accounted for by a model that includes the effect of stray, scattered light.

1. Introduction

Dosimetric polymer gels (such as PAGAT [1]) undergo a change in opacity (measured as optical density) when exposed to radiation due to the formation of cross-linked polymer chains. Optical CT [2,3] can be used to evaluate exposed dosimetric polymer gels. These cross-linked regions result in local microscopic variations in refractive index which act as scattering centres, causing incident light to scatter, resulting in extinction of light from the incident beam. For small to moderate abundances of these polymerised regions, the degree of extinction obeys the Beer-Lambert law:

$$T = \exp(-x\,\Delta) \tag{1}$$

where T is optical transmittance, x is optical pathlength and Δ is optical density (OD or extinction coefficient). OD is proportional to concentration of scattering centres and hence radiation dose over a useful range of dose [3].

Polymer gel dosimeters have the advantage over the radiochromic Fricke gels [4] of being more stable over time. However, unlike (optically absorbing) Fricke gels, the mechanism of extinction in polymer gels is scattering [5] which produces stray light that can result in artefacts in dose distributions deduced from reconstructions of optical CT scans.

In an earlier paper [5], a finger phantom was developed to investigate the effects of scatter on optical density measurements in polymer gels (and hence dose) using optical CT. In the phantom, the commercial antiseptic Dettol™ (Reckitt Benckiser) was added to porcine gelatine to produce "opalescence" or

"Tyndall scattering" typical of a polymerised gel dosimeter. Antiseptic concentration was varied to simulate regions of differing dose.

In that paper, it was found that because of stray light, optical CT would underestimate larger ODs. Also, reconstructions of regions with OD known to be uniform, exhibited a "dishing" or "cupping" artefact; a plot of OD versus position dips lower in the central part of a region compared with its periphery (e.g. Figure 1 right). As dishing can also result from beam hardening, testing this effect in a simulated phantom in which radiation plays no role, allowed the contribution of stray light to this effect to be analysed in isolation.

The severity of both the underestimation of OD and the dishing artefact increases with increasing OD. The aim of the following research was to examine the effects of increasing pathlength and background OD on these two stray light induced artefacts in an optical CT scanner. By developing practical models of the behaviour of stray light in systems like this, it becomes possible to correct for artefacts in dosimeters of this nature but some aspects can also apply to imaging modalities such as X-ray CT which are subject to similar artefacts.

2. Experimental procedure and results

2.1. Scanner

A cone beam optical CT scanner [6] (Vista™ Optical CT Scanner by Modus Medical Devices Inc.) was employed in this study. The scanner (described in detail elsewhere [5]) consists of a diffuse light source illuminating a dosimeter/phantom in a PETE cylindrical screw-top vessel mounted in an aquarium filled with water to match the gel refractive index. A CCD camera takes snapshots of the transmitted projections of the jar which is then rotated incrementally between projections. From these, a 3D map of OD is reconstructed using the Feldkamp back-projection algorithm [7].

2.2. Phantom

Two gels of differing OD were produced by adding differing concentrations of Dettol™ to porcine gelatine. The PETE vessel was filled with the lower OD gel which was allowed to set with a glass funnel centrally mounted in it. Upon removal a funnel-shaped cavity was left, which was then back-filled with the higher OD gel (Figure 1 left). Another PETE vessel was filled entirely with the lower OD gel, to act as a baseline reference (analogous to an unexposed dosimeter) for the scanner. Nominally, the OD within the funnel region of the phantom should be uniform, although it is possible that there is some variation due to uncontrollable inhomogeneities in the rate of setting of the gel. Unlike the finger phantom [5], Dettol™ was also deliberately added to the baseline reference gel to increase the effect of stray light on the process of baseline subtraction.

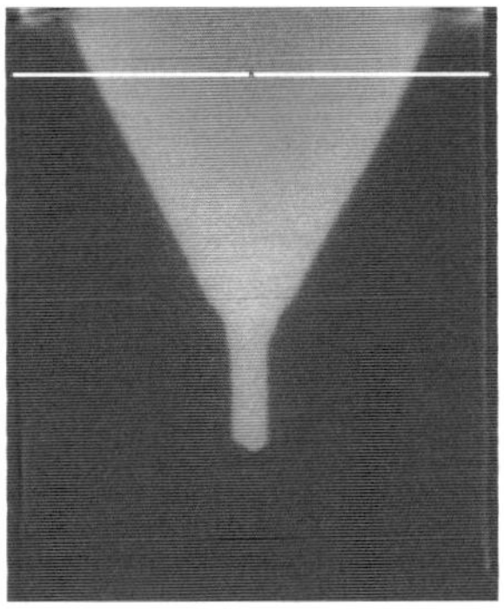

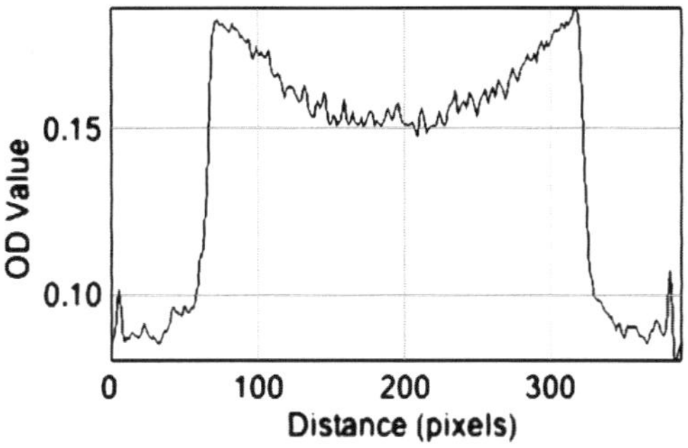

Figure 1: Reconstructed slice of the Funnel Phantom and profile plot of OD along the line selection.

2.3. Reconstruction

The reconstruction (Figure 1 left) was analysed by generating plots of OD profiles (Figure 1 right) using ImageJ (open source, the National Institute of Health USA).

The gels were designed to have uniform OD within each of the the two separate contiguous regions, so any appreciable deviation from uniformity evident in the OD profiles is an artefact. OD profiles were plotted for selections taken through different cross-sections of the funnel, so as to examine the effect of increasing pathlength of the light through the higher OD region. Some examples are plotted in Figure 2. The width of each funnel cross-section (and hence maximum pathlength of light through it) is evident in Figure 2 from the horizontal separation of the shoulders of each plot.

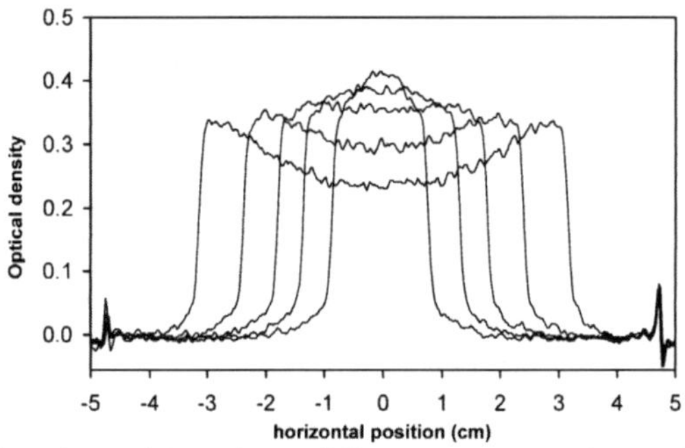

Figure 2: Profile plots of OD along different cross-sections of the funnel.

As expected and in line with findings for the finger phantom, the wider the cross-section (the longer the pathlength) the more severe the dishing and the more severe the underestimate of OD. However, another effect is evident. For the narrowest cross-sections, the peak exhibits an inverted dishing or "doming". In the finger phantom [5], the cause of the dishing was shown to be stray light entering the scanner camera from in front of the more opaque region, thus underestimating the OD of this region. Dishing occurs because for regions of longest pathlength, the true transmittance is lower, so the stray light represents a larger fraction of the apparent transmittance and so the underestimate of OD through the *centre* of a region is more severe there. However, explaining doming requires a more complex model than was used to explain the finger phantom results.

2.4. Modelling Data

Modelling was applied to the raw projection images taken by the scanner camera rather than the reconstructed profiles. The light received by the camera was assumed to consist of two components - directly transmitted light (assumed to obey the Beer-Lambert law [Eqn 1]) and stray light, scattering back into the camera indirectly. The intensity of the stray light was assumed to be roughly uniform across contiguous regions of the images in the camera snapshots. This is a simplified but quite a good assumption [5] except in the penumbra region near interfaces. The stray light could either enter the detector directly from scatter within the medium in front of the more opaque region (Figure 3 right), or by reflection off the opaque region itself (Figure 3 left). These two contributions were treated separately in the model. In the model, the baseline reference phantom was treated as a single region, while the funnel phantom was divided into 4 regions: the funnel itself, the background gel on either side of the phantom as seen by the camera, the background gel behind the funnel and the background gel in front of the funnel. The reference phantom and the two background regions were modelled to have two light components:

direct transmittance and a back-scatter component (Figure 3 right). The funnel was assumed to have an extra component - front-face illumination (Figure 3 left).

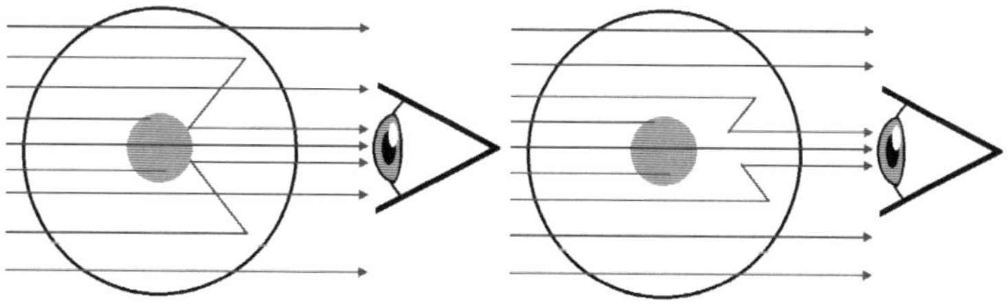

Figure 3: Front-face illumination (left) and backscatter from the intervening medium (right).

Line intensity profiles through the funnel region of the raw projection images were plotted. These were divided by the intensities of line profiles through the baseline reference phantom, to yield an (apparent) transmittance profile plot. The negative logarithm of these plots yielded apparent OD plots. Model apparent OD plots were also generated for each experimental profile and the parameters (true OD and stray light intensity) for each region fitted using a least squares regression. The experimental plots fitted closely, except in the penumbra region near the boundary between background and funnel (see two examples in Figure 4). In this region, the intensity of the stray light tapers off gradually because of shadowing by the high OD funnel region.

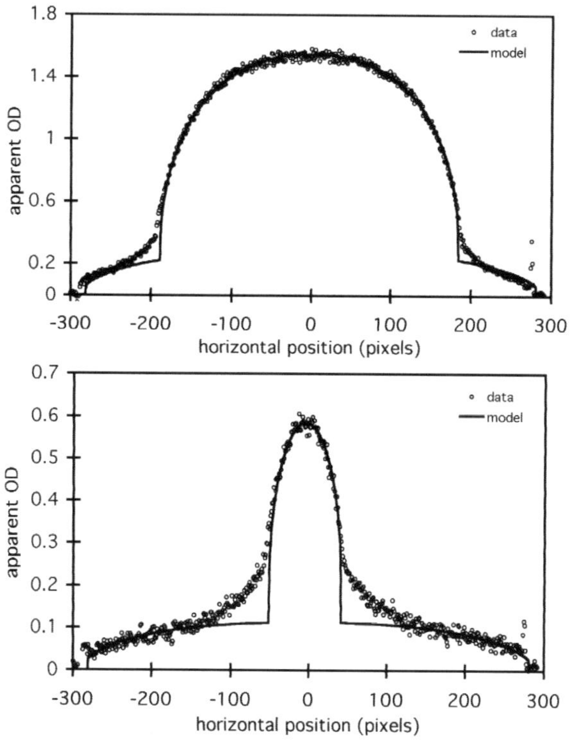

Figure 4: Examples of model OD curves fitted to experimental OD profiles taken from raw projection images.

3. Discussion

The simple assumption of uniform stray light in each region of the image in the model accurately fits the optical behaviour except in the penumbra. To fit this region requires a more sophisticated treatment of how stray light behaves near the interface between regions of strongly differing OD. Perhaps a Monte Carlo treatment of this region could provide this.

Doming is explained by the fact that in the funnel phantom, the background gel outside the funnel and the baseline reference phantom were made with an OD higher than was used for the background and baseline reference in the finger phantom [5] and so the effect of stray light in these two regions cannot be ignored. The result is that OD in the baseline and background gels, is *also* underestimated. Moreover, there is a hidden dishing artefact in the baseline reference data which when divided out of the phantom data has the effect of increasing the apparent OD in the middle of the funnel which partially compensates for dishing in the funnel. For small cross-sections of the funnel where the dishing is weak, dividing by the dished baseline results in doming.

4. References

[1] Venning A J, Hill B, Brindha S, Healy B J, Baldock C 2005 Investigation of the PAGAT polymer gel dosimeter using magnetic resonance imaging *Phys. Med. Biol.* **50** 3875-3888

[2] Oldham M, Siewerdsen J H, Shetty A, Jaffray D A 2001 High resolution gel-dosimetry by optical-CT and MR scanning *Med. Phys.* **28** 1436-1445

[3] Gore J C, Ranade M, Maryanski M J, Schulz R J 1996 Radiation dose distributions in three dimensions from tomographic optical density scanning of polymer gels: I. Development of an optical scanner *Phys. Med. Biol.* **41** 2695-2704

[4] Baldock C, Harris P J, Piercy A R, Healy B 2001 Experimental determination of the diffusion coefficient in two-dimensions in ferrous sulphate gels using the finite element method *Australas. Phys. Eng. Sci. Med.* **24** 19-30

Appleby A, Leghrouz A 1991 Imaging of radiation dose by visible color development in ferrous–agarose–xylenol orange gels *Med. Phys.* **18** 309-312

[5] Bosi S, Naseri P, Puran A, Davies J, Baldock C 2007 Initial investigation of a novel light-scattering gel phantom for evaluation of optical CT scanners for radiotherapy gel dosimetry *Phys. Med. Biol.* **52** 2893-2903

[6] Miller J, Adamovics J, Dietrich J 2005 Cone Beam Optical CT Scanner for 3D Dosimetry *Med. Phys.* **32** 2138

[7] Feldkamp L A, Davis L C, Kress J W 1984 Practical cone–beam reconstruction *J. Opt. Soc. Am.* A **1** 612-619

Initial characterization of fast laser scanning optical CT apparatus for 3-D dosimetry

Nikola Krstajić[1], Simon Doran[2]

[1] Kroto Innovation Centre, Nanoscience and Technology Building, Broad Lane, North Campus, S3 7HQ, University of Sheffield, Sheffield, UK

[2] CRUK Clinical Magnetic Resonance Research Group Institute of Cancer Research, 15 Cotswold Rd, Belmont, Sutton Surrey SM2 5NG, UK

N.Krstajic@physics.org and Simon.Doran@icr.ac.uk

Abstract. Laser scanning optical computed tomography (optical CT) is potentially a better technique for analyzing 3-D dosimeters than CCD based dosimetry due to the fact that narrow cone of light passing through the sample is not affected by the refractive index inhomogeneities of surrounding structures. We have previously demonstrated fast laser scanning of 3-D dosimeters using an optical arrangement borrowed from confocal microscopy. The main rationale for this design is that it guarantees good telecentricity at high scanning speeds. Here we demonstrate a number of advantages of this arrangement as compared to broadbeam optical CT based around CCD detectors. Additionally, we provide initial characterization of the apparatus in terms of signal-to-noise and dynamic range.

1. Introduction

Optical computed tomography (optical CT) is potentially the cheapest avenue for measuring dose distributions in 3-D [1-5]. Alternatives include Magnetic Resonance Imaging (MRI) [6], X ray CT and ultrasound [7]. All modalities have been discussed in detail in review articles in previous DOSGEL conferences [8-10]. We present here an initial effort in characterizing the fast laser scanning optical CT apparatus first described in [11].

2. Methods

The fast laser scanning setup is shown in Figure 1 (reproduced from [11]). Figure 1(a) shows the diagram of the setup while Figure 1(b) shows the relay optics consisting of two galvanometer mirrors and two paraboloidal mirrors. The arrangement ensures fast raster scanning from a *stationary* pivot point on the galvo mirror 2. The high quality test target (HQTT) (Edmund Optics, Barrington, NJ, US, part NT54-803) is used to calculate the modulation transfer function (MTF). MTF is calculated from a line profile of the transmission image of the HQTT in air.

A signal-to-noise ratio study was performed applying the same methodology as in [12]. The only difference is that here, the scans done were performed over 400 views rather than 1000 views. The voxel size in the reconstruction domain was set to 1 mm and 0.5 mm. Note that MTF was measured from distortion-corrected projections, while the uniformity study used raw images. Comparisons are

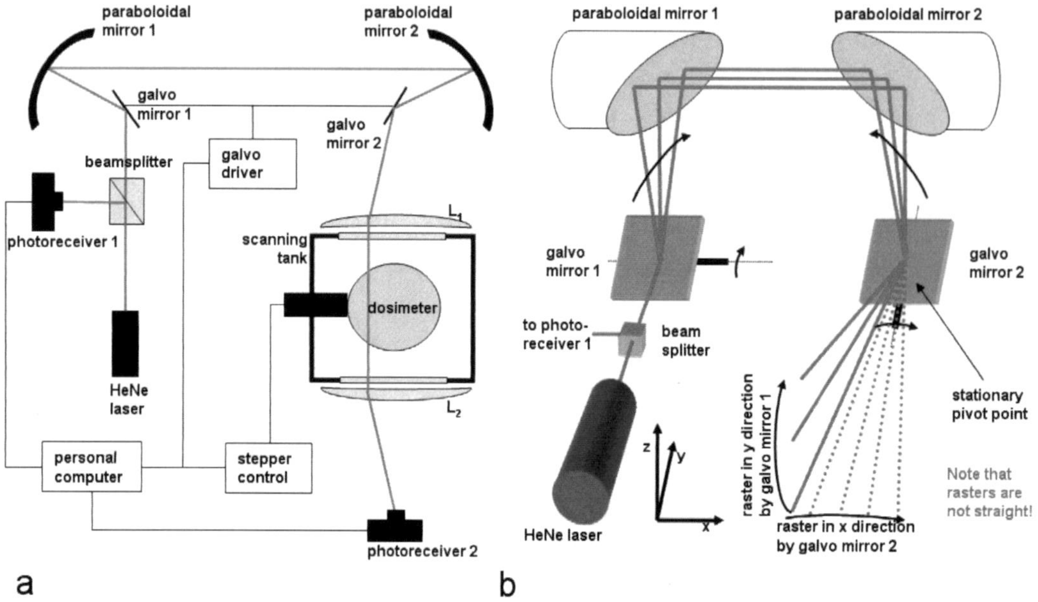

Figure 1: (a) Top-view diagram of the fast laser scanning apparatus, (b) relay optics ensures a stationary pivot point of the laser beam going into L_1. Reproduced from [11].

made for MTF versus voxel size and SNR versus voxel size. Linearity in absorbance measurements and projection signal-to-noise ratio (SNR) were shown previously [11].

Lastly, we present a scan of a PRESAGE^{TM} (Heuris Pharma, NJ, USA) sample irradiated with a proton beam. The sample was irradiated four times, using an unmodulated beam, collimated to a 12mm diameter circular cross-section. The beam was applied for different durations (of the order of tens of seconds), such that estimated doses of 1, 2, 16 and 20 Gy were applied at the Bragg peak.

3. Results

Figures 2(a) and 2(b) show modulation transfer functions for the system in two experiments. Figure 2(a) compares the MTF for a 512 × 512 image with the HQTT at or near the focal plane of L_1 (i) with the same configuration but with the HQTT further from or nearer to L_1; and (ii) with the CCD-based scanner operating at a similar field-of-view and nominal pixel resolution. The MTF for the laser-based apparatus is better than the corresponding broadbeam apparatus up to 2 lp/mm. After that, the MTF drops sharply to zero. This is a clear indication of the finite size of the laser focal spot, which blurs out any spatial frequencies above a certain threshold. It might have been expected that (for the planar HQTT, at least) the nominal laser spot size should not have been a limitation, since in the focal plane of L_1, the beam diameter should be extremely small. Although this needs further investigation, a possible reason is that the HQTT was not placed *exactly* in the focal plane. The numerical apertures (NA) and pixelation of both systems are similar and therefore both should be capable of similar spatial resolution. Figure 2(b) shows the expected sharp decline in MTF for reduced image resolution (i.e. increased voxel size). It should be stressed that one has the possibility of using a laser beam with smaller diameter [17] should this prove necessary.

The uniformity study results are shown in Figures 2(c) and 2(d). In Figure 2(c) the reconstruction SNR for 1mm voxel is better than 90:1 from absorbance range 0.2 to 1.6. Note that for most of the absorbance range (0.5 to 1.6) it is better than 200:1. Therefore the instrument provides a unique

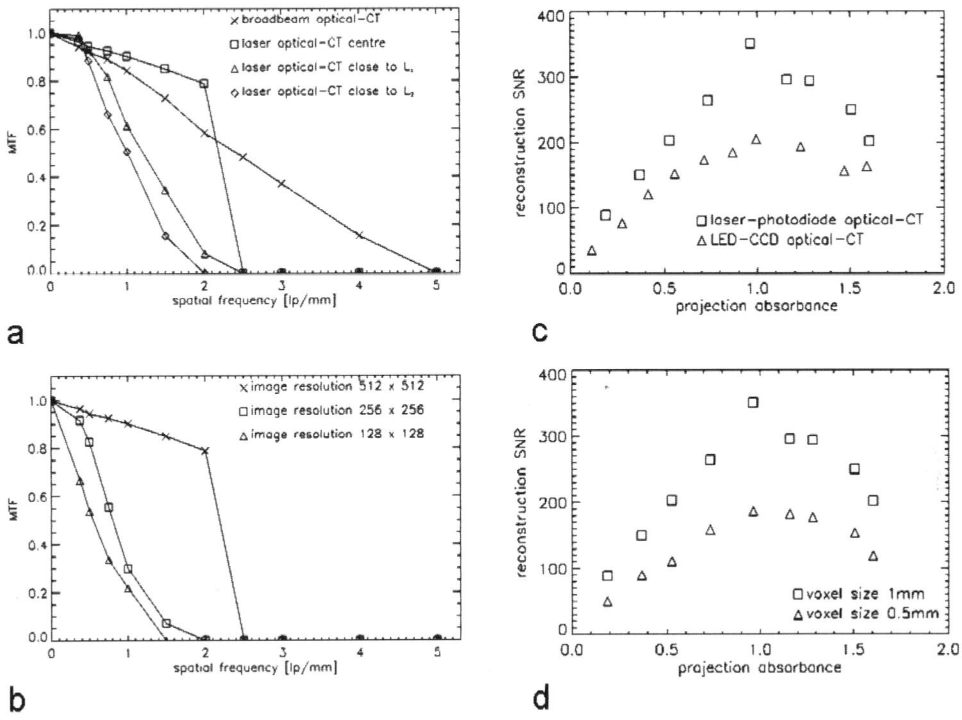

Figure 2: (a) MTF of laser optical CT in focus (centre) and out of focus (close to L_1 and L_2); (b) MTF of fast laser scanner images with different matrix sizes and constant FOV; (c) SNR study results for laser based optical-CT and comparison with results from CCD based optical CT [12].

potential for speed, high quality readout and good dose resolution. Figure 2(d) shows a fall in SNR with reduced voxel size. However, SNR for 0.5mm voxel in laser scanning arrangement is similar to 1mm voxel in broad beam arrangement.

The scan of PRESAGE sample is shown in Figure 3. To illustrate the reconstruction process we show the projection as read by photodiodes (Figures 3(a) and 3(b)), the absorbance of that projection which is obtained after calibration using the HQTT [12] (Figures 3(c) and 3(d)) and finally, the reconstruction slice (Figures 3(e) and 3(f)). Notice the high SNR of all the profiles. The profile across the slice (Figure 3(f)) shows that the irradiated circles are not uniform. We are currently investigating whether this is a scanner artifact or whether the dose deposited really had this form.

4. Discussion

Fast laser scanning is an established field [13, 14] and most of the innovations presented here have been borrowed from a confocal microscopy scanning arrangement [15]. The acquisition speed relies on fast galvanometer driven mirrors and appropriate photoreceivers. This has not been exploited to the full in optical CT. For example, the fastest galvanometer mirrors (driven at mechanical resonance, so called resonant galvanometer) can perform up to 10kHz line scanning, which is equivalent to video rate imaging. This suggests that the potential for development of the scanner is great.

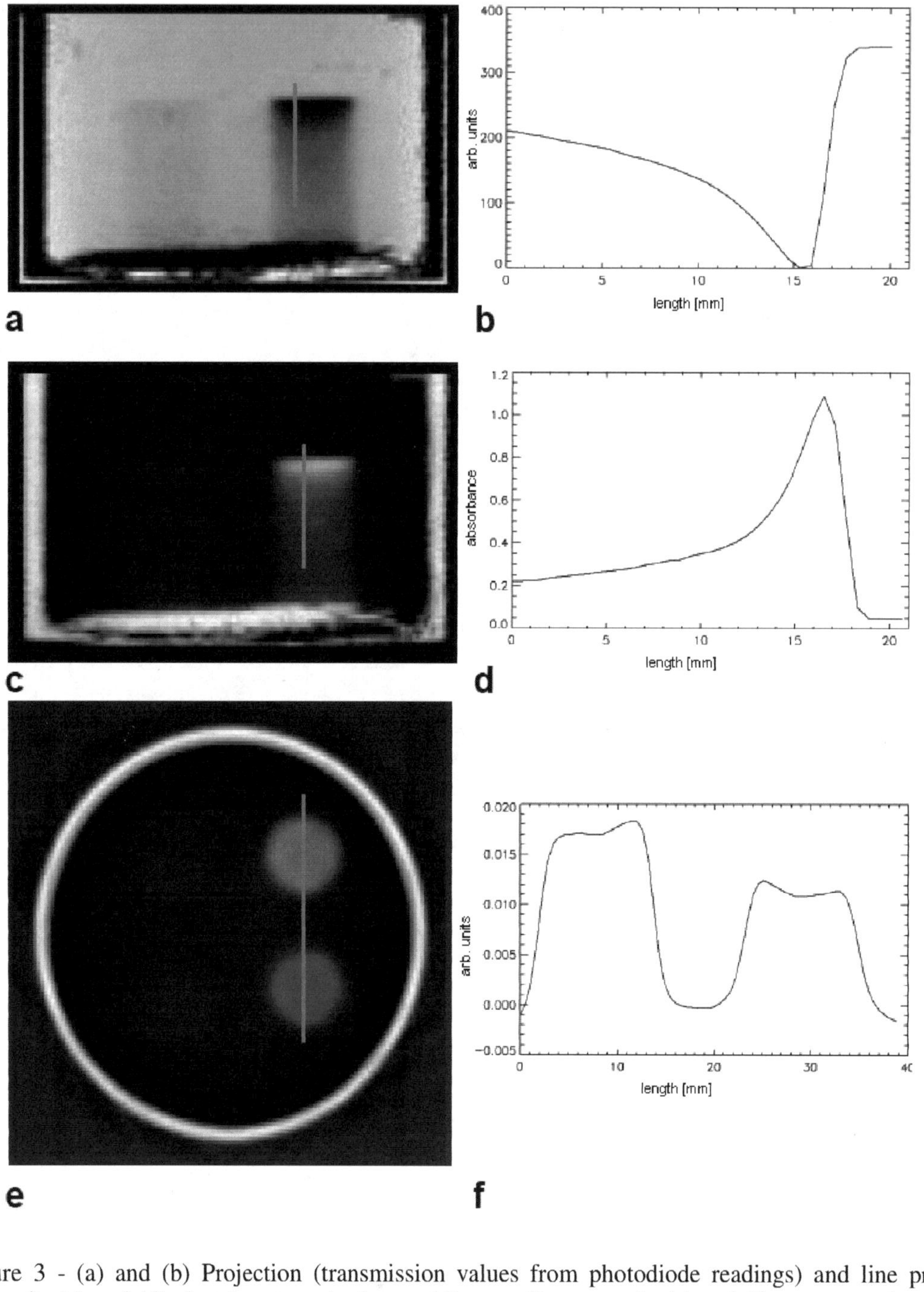

Figure 3 - (a) and (b) Projection (transmission values from photodiode readings) and line profile across it, (c) and (d) absorbance projection and line profile across it, (e) and (f) reconstruction slice and line profile across it.

It has been shown that the positioning of the sample plays an important role in determining the MTF of the laser system. The results with the HQTT placed away from the focal plane of L_1 are worse than the one done in focus. However, again, one can exploit existing advances in extended depth-of-field techniques which are directly applicable to our arrangement [16].

5. Conclusion

Exhaustive characterization has yet to be done and we present an initial insight into performance of fast laser scanning apparatus. The results are very promising, but need further verification and comparison with other dosimetry techniques such as radiochromic film and ionization chambers.

Acknowledgments

Work presented here was financed by the EPSRC Doctoral Training Grant. We would like to thank John Adamovics for providing PRESAGE™ samples, BLASER charity for additional funds provided for equipment, John-William Brown for help with the optomechanical arrangements of the apparatus.

References

[1] Doran, S. and N. Krstajic. *The history and principles of optical computed tomography for scanning 3-D radiation dosimeters. Proceedings of DOSGEL 2006*. 2006. Sherbrooke, Canada.editors: M. Lepage, A. Jirasek, and L.J. Schreiner

[2] Jordan, K.J., *Fundamentals of optical measurements in gel dosimetry.* Proc. of DOSGEL 2001 2nd Int. Conf. on Radiotherapy Gel Dosimetry, 2001: p. 66-72.

[3] Maryanski, M.J., Y.Z. Zastavker, and J.C. Gore, *Radiation dose distributions in three dimensions from tomographic optical density scanning of polymer gels: II. Optical properties of the BANG polymer gel.* Physics in Medicine and Biology, 1996. **41**(12): p. 2705-2717.

[4] Oldham, M. *Optical CT scanning of polymer gels. DOSGEL 2006*. 2006. Sherbrooke.editors:

[5] Gore, J.C., M. Ranade, and M.J. Maryanski, *Radiation dose distributions in three dimensions from tomographic optical density scanning of polymer gels: I. Development of an optical scanner.* Physics in Medicine and Biology, 1996. **41**(12): p. 2695-2704.

[6] De Deene, Y., *Fundamentals of MRI measurements in gel dosimetry.* Journal of Physics: Conference Series, 2004. **3**: p. 87-114.

[7] Baldock, C., *X-ray computer tomography, ultrasound and vibrational spectroscopic evaluation techniques of polymer gel dosimeters.* Journal of Physics: Conference Series, 2004. **3**: p. 136-142.

[8] *DOSGEL 2004. Third International Conference on Radiotherapy Gel Dosimetry*. 2004. Ghent.editors: Y. De Deene and C. Baldock

[9] *DOSGEL 2001. Second International Conference on Radiotherapy Gel Dosimetry*. 2001. Brisbane.editors: Y. De Deene and C. Baldock

[10] *DOSGEL 2006. Fourth International Conference on Radiotherapy Gel Dosimetry*. 2006. Sherbrooke.editors: M. Lepage, A. Jirasek, and L.J. Schreiner

[11] Krstajic, N. and S.J. Doran, *Fast laser scanning optical-CT apparatus for 3D radiation dosimetry.* Physics in Medicine and Biology, 2007. **52**: p. N257-N263.

[12] Krstajic, N. and S.J. Doran, *Characterization of a parallel beam CCD optical-CT apparatus for 3D radiation dosimetry.* Physics in Medicine and Biology, 2007. **52**: p. 3693-3713.

[13] Beiser, L., *Fundamental architecture of optical scanning systems.* Applied Optics, 1995. **34**(31): p. 7307-7317.

[14] Marshall, G.F., *Applied optics and optical engineering volume VI - Chapter 6 Scanning devices and systems.* Applied optics and optical engineering, ed. R. Kingslake and B.J. Thomson. Vol. 6. 1980.

[15] Amos, W.B., *Achromatic scanning system.* 1991, Medical Research Council, Cambridge, UK

[16] Marks, D.L., et al., *Three dimensional tomography using a cubic-phase plate extended depth of field system.* Optics Letters, 1999. **24**(4): p. 253-255.

[17] Wuu C S et al., *Dosimetry study of Re-188 liquid balloon for intravascular brachytherapy using polymer gel dosimeters and laser-beam optical CT scanner*, Med. Phys. **30**(2), 132-137

Journal of Physics: Conference Series **164** (2009) 012023 doi:10.1088/1742-6596/164/1/012023

CCD-based optical CT scanning of highly attenuating phantoms

Shamsa Al-Nowais[1] and Simon J Doran[2,1]

[1]Department of Physics, University of Surrey, Guildford, UK
[2]CRUK Clinical MR Research Group, Institute of Cancer Research, Sutton, UK

Corresponding author e-mail: Simon.Doran@icr.ac.uk

Abstract. The introduction of optical computed tomography (optical-CT) offers economic and easy to use 3-D optical readout for gel dosimeters. However, previous authors have noted some challenges regarding the accuracy of such imaging techniques at high values of optical density. In this paper, we take a closer look at the 'cupping' artefact evident in both light-scattering polymer systems and highly light absorbing phantoms using our CCD-based optical scanner. In addition, a technique is implemented whereby the maximum measurable optical absorbance is extended to correct for any errors that may have occurred in the estimated value of the dark current or ambient light reaching the detector. The results indicate that for absorbance values up to 2.0, the optical scanner results have good accuracy, whereas this is not the case at high absorbance values for reasons yet to be explained.

1. Introduction

As has been described in previous DOSGEL conferences [1, 2], optical CT is an attractive imaging readout modality for 3-D dosimetry experiments, because of its speed, spatial resolution and low cost. However, the scanners currently available are the result of only a limited development effort by a small number of academic teams and this is in marked contrast to the 30 years of major investment by multinational companies that is the case with Magnetic Resonance Imaging, the main competing modality.

One result of this is that a number of artefacts may still be present in optical CT images and, while careful image acquisition can usually eliminate these, their existence may reduce confidence in the technique. In this study, we examine further the "cupping" artefact described by De Jean *et al.* at the 2006 DOSGEL meeting [3]. Figure 1 of their paper demonstrates the effect with data acquired on the commercial Vista cone-beam scanner. The authors comment:

As expected from the multiple scattering conditions in polymer systems, gels irradiated to uniform but high doses with significant opacity exhibit cupping in CT numbers as one crosses the phantom. Gels that are irradiated with distributions that preserve transparency throughout most of the phantom... show well behaved dose response.

The suggestion of De Jean *et al.* is that the effect is caused by multiple scattering, and thus represents, effectively, a breakdown in Beer's Law (an implicit assumption of the CT method). However, this is not the only possible explanation for this type of observation. In this paper, we set out to reproduce the effect and discuss other potential causes.

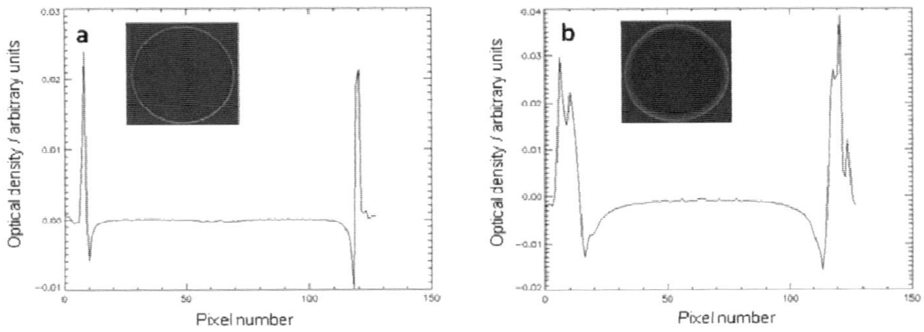

Figure 1: Results of image acquisitions to demonstrate the effect of wall artefacts. Profile across sample with image inset for (a) Teflon® FEP and (b) PET containers.

2. Materials and methods

A uniform and stable water phantom was created, whose optical attenuation was varied either by the addition of an absorbing blue dye (food colouring), or a scattering agent (Dettol ™) [4]. Two containers with different refractive indices were investigated. The first was clear plastic jar of approximately 7 cm diameter made of polyethylene terephthalate (PET, n=1.575), the material used for the Vista scanner. The second container was formed from a transparent, thermoplastic sleeve made of Teflon® FEP (DuPont, n=1.341) approximately 7.9 cm in diameter. The bottom of the sleeve was sealed to an aluminium base plate using a thin layer of a warm gelatin mixture and subsequently cooled in a refrigerator. Both containers were filled with deionised water (Milli-DI® systems, Millipore), then inserted in the scanning tank and surrounded by deionised water. Optical CT images were acquired using the method and apparatus described in [5]. For the second series of scans, the Teflon container was filled with deionised, degassed water, to which varying concentrations of blue dye or Dettol™ were added, and optical CT images acquired.

As will be shown in the following section, these images demonstrated cupping artefacts. Since the effect appeared with *both* scattering and absorbing phantoms, we eliminated multiple scattering as a cause of the cupping in our case (although this may have been the origin in the case of De Jean *et al.*). It was hypothesised that this effect was caused by the low detected signal at the centre of the phantom, which is only slightly above the level of the dark-current image. As shown in [6], the dynamic range of the CCD should have been sufficient to measure successfully the expected absorbance range encountered. However, any errors in the estimated value of the dark current, or ambient light reaching the detector not accounted for, would cause corresponding errors in the measured absorbance values.

In order to discover whether this was the cause of the discrepancy, the dynamic range extension method of Krstajić and Doran ([6]) was employed. Briefly, this consists of the following procedure.

(a) Acquire one set of projection images (P_1) with the normal light intensity level, chosen such that the largest signal value is close to the detector maximum, but no saturation occurs. (For highly absorbing phantoms, the minimum transmitted light intensity will be very close to the detectable threshold — this is the problem that we are setting out to solve.)

(b) Acquire a second set of images (P_2) with increased light intensity. The low absorption regions of the projection will now be saturated but *bona fide* data will now be detected in the centre.

(c) Construct composite projections (P_{comp}) pixel-by-pixel, using data from P_1, wherever the signal is above a pre-determined threshold (related to the dark current) and suitably scaled data from P_2 where the data fall below this threshold. Scaling may be accomplished either by adjusting the light intensity such that certain regions of the projection contain valid data for both P_1 and P_2 or by acquiring an additional light-field image with a calibrated neutral density filter in the beam path.

(d) Reconstruct projections P_{comp} as normal.

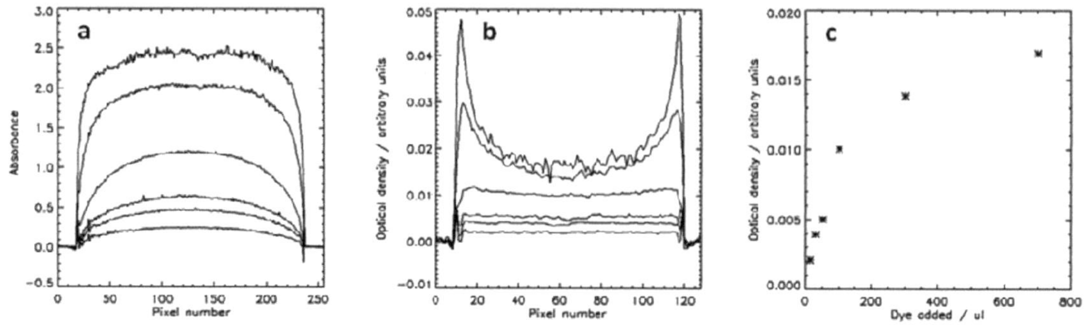

Figure 2: (a) absorbance profiles across sinogram data of uniform cylinder with gradually increasing dye concentration; (b) profiles across reconstructed optical density images; (c) relationship between quantity of dye added and reconstructed image intensity at the centre of the profile — we would expect this to be linear.

If the cause of the problem was indeed an erroneous estimation of the ambient light or the dark image, then projections P_2, which are obtained with a much higher light level than P_1, should be less affected by the artefact.

3. Results and discussion

Figure 1 demonstrates the importance of using a container whose refractive index is close to that of its contents and surroundings. Wall effects in the reconstructed slices in Figure 1(a) for PET are significantly worse than those of Figure 1(b) for Teflon FEP. In the former case, the field-of-view over which reliable quantitative measurements can be made is reduced to less than 50% of the phantom diameter, compared with around 80% for Teflon FEP. Although in a practical measurement on a dosimeter, a pre-irradiation scan of the phantom [7] will correct for much of the error, it is still better to minimise the potential problems by good data acquisition practice and match the refractive index of the container to that of the gel dosimeter.

Figure 2 shows the results of increasing the concentration of the attenuating additive. In each case, profiles have been corrected by dividing by the zero-concentration scan. It is seen that, as the concentration increases, the shape of the absorbance profile changes from a smooth arc of form

$$A(x) = \mu y = 2\mu(r^2 - x^2)^{1/2} \qquad [1]$$

into a profile with much steeper sides and a flatter top (Figure 2(a)). This, in turn, leads to an image, whose edges are much brighter than expected and whose middle is depressed (Figure 2(b)). If one plots the image intensity at the centre, one sees that the expected linear relationship between dye concentration and reconstructed optical density breaks down (Figure 2(c)).

In implementing the extended dynamic range method, we expected to encounter the problem of *blooming* of the CCD and Figure 3 shows graphically how serious this can be. When the CCD receives a light signal that is too high, i.e., too many photons are incident on a given pixel on the chip, the signal output saturates and the charge-accumulating well "overflows". Photo-electrons are detected in adjacent wells (pixels). Although the method described above *expects* the saturation, the corruption of data in the pixels surrounding the saturated region is a problem. Furthermore, in extreme cases the output amplification stage of the camera may be overloaded causing complete breakdown and zero recorded signal, as illustrated in Figure 3(b). The solution is straightforward for this experiment, with its simple cylindrical geometry — in the second scan we simply masked the regions in which signal overflowed — but this would not be possible in the general case and this would limit severely the

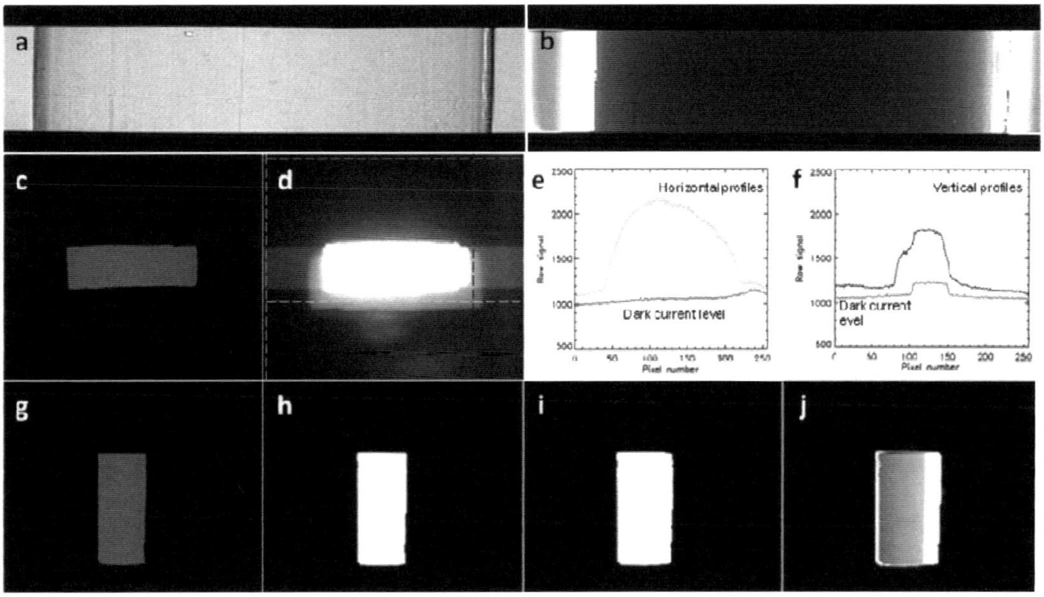

Figure 3: Blooming — a cautionary tale. When attempting to implement the extended absorbance measurement method, it is vital to take account of the blooming effects that occur when the CCD is saturated. (a) image of the phantom at normal light level with water and no dye inside; (b) high intensity light field and blue-dye filled phantom — values of attenuation are as expected in the sample centre, but incorrect at both edges; (c) masked region of CCD at normal light level; (d) same region with high intensity; (e) horizontal profiles through sample at locations shown in (d); (f) vertical profiles; (g)–(j) images of masked region at gradually increasing light levels — note how the portion of the image containing a signal gets wider and then exhibits the same behavior as seen in (b).

application of this method in the general case. (This issue is not of course a problem with the fast laser scanner [8]).

Figure 4(a) analyses the results of an imaging experiment with normal light intensity level for a highly absorbing phantom. Owing to the quality of the CCD, low noise data are obtained even at high values of absorbance. However, the systematic error causing the wings in Figure 2(b) is evident. Data are expected to satisfy Eq. [1] (smooth green line), but this is patently not the case. Initially, we hypothesised that the cause of this was an incorrect dark image level used in the correction of the data. In order to obtain our values for absorbance we use the following equation [6]:

$$A = -\log \frac{S - D}{S_0 - D} = -\log_{10} \frac{\text{Raw data} - \text{Dark image}}{\text{Correction data} - \text{Dark image}} \qquad [2]$$

It can be shown that if D is in error by an amount □, then the corresponding error in absorbance A is □ / $(S\text{–}D).\log_{10}e$. If $S>>D$, then this poses little problem. However, for highly attenuating samples, S approaches D (i.e., the contribution made to the signal by the light we are actually trying to measure becomes small). Under these circumstances, any slight error in the estimation of D (or equivalently, failing to take account of changes in ambient light) becomes very significant. The blue (middle) line in Figure 4(a) demonstrates how, by adding an appropriate (~10%) constant to the value of D in Eq. [2] before performing the data processing, we "improve" the fit dramatically. The high light field experiment would confirm this was the correct thing to do, because if the hypothesis were true, the new signal at the centre of the phantom should be much less sensitive to inaccuracies in the estimation of D. — i.e., a good fit without including a □ term, since D is inherently much greater than S.

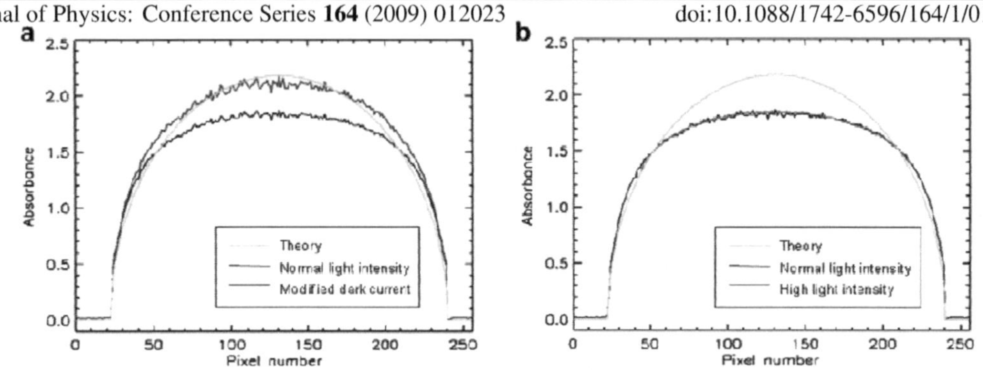

Figure 4: (a) Imaging experiment performed with normal light intensity, showing raw data (black line, bottom), theoretical data (green line, top) and data that would be obtained if we assumed there to be an error in the background light level/dark current image and corrected this. (b) Data obtained from high light intensity image. The deviation from the theoretical value remains unexplained.

An independent measurement of the optical density value of the neutral density filter was performed by filling the scanning tank with de-ionized water and increasing the light intensity until the detected signal was just below saturation level. A single projection was taken, then a reflective neutral density filter (Edmund Optics NT43-815) of nominal absorption 1.0 and tolerance ±5% was placed in front of the LED, and another projection taken. The experimental optical density 0.95165 (SD=0.005, SEM=5×10^{-5}) was used to correct the data in Figure 4(b) and it should be noted that there are *no adjustable parameters* in the scaling between the overlying red and black lines. The increase in light intensity leads to data that have higher SNR, as expected [6], but demonstrates categorically that our hypothesis concerning the cause of our incorrect measurement of high absorbance values is incorrect.

4. Conclusions

We have reproduced an artefact very similar to that demonstrated in [3] using different apparatus. Analysis has shown that the incorrect image profile is not specific to a scattering sample but occurs in other cases where high absorbance values are incorrectly measured by the scanner. The hypothesis that the cause of the effect is an incorrect measurement of the dark current or unaccounted for ambient light has been proved to be incorrect. Further investigations are ongoing.

References

[1] Krstajic, N and Doran, S J. The history and principles of optical computed tomography for scanning 3-D radiation dosimeters. *J. Phys Conf. Ser.*, **56**, 45–57 (2006)

[2] Oldham, M. 3D dosimetry by Optical-CT scanning. *J. Phys Conf. Ser.*, **56**, 58–71 (2006)

[3] DeJean P., S.R., McAuley K., Rogers M., Schreiner L J. Initial experience with a commercial cone beam optical CT unit for polymer gel dosimetry II: Clinical potential. *J. Phys Conf. Ser.*, **56**, 183–186 (2006)

[4] Bosi, S *et al.*, Initial investigation of a novel light-scattering gel phantom for evaluation of optical CT scanners for radiotherapy gel dosimetry. *Phys. Med. Biol.*, **52**(10) 2893-2903 (2007)

[5] Krstajic, N and Doran S J. Focusing optics of a parallel beam CCD optical tomography apparatus for 3D radiation gel dosimetry. *Phys. Med. Biol.*, **51**(8) 2055-2075 (2006).

[6] Krstajic, N and Doran S J, Characterization of a parallel-beam CCD optical-CT apparatus for 3D radiation dosimetry. *Phys. Med. Biol.*, **52**(13) 3693-3713 (2007)

[7] Doran, S J *et al.*, A CCD-based optical CT scanner for high-resolution 3D imaging of radiation dose distributions: equipment specifications, optical simulations and preliminary results. *Phys. Med. Biol.*, **46**(12) 3191-3213 (2001)

[8] Krstajic, N and Doran S J, Fast laser scanning optical-CT apparatus for 3D radiation dosimetry. *Phys. Med. Biol.*, **52**(11) N257-N263 (2007).

Optimization of a fast optical CT scanner for nPAG gel dosimetry

Jan Vandecasteele[1] and Yves De Deene[1,2]

[1]Research group QMRI, Department for experimental cancer research, nuclear medicine and radiotherpy (ECNURAD), Ghent University Hospital, De Pintelaan 185, 9000 Gent, Belgium

[2]Medical image and signal processing (MEDISIP), Ghent University Hospital, De Pintelaan 185, 9000 Gent, Belgium

E-mail: Jan.Vandecasteele@ugent.be

Abstract. A fast laser scanning optical CT scanner was constructed and optimized at the Ghent university. The first images acquired were contaminated with several imaging artifacts. The origins of the artifacts were investigated. Performance characteristics of different components were measured such as the laser spot size, light attenuation by the lenses and the dynamic range of the photo-detector. The need for a differential measurement using a second photo-detector was investigated. Post processing strategies to compensate for hardware related errors were developed.
Drift of the laser and of the detector was negligible. Incorrectly refractive index matching was dealt with by developing an automated matching process. When scratches on the water bath and phantom container are present, these pose a post processing challenge to eliminate the resulting artifacts from the reconstructed images Secondary laser spots due to multiple reflections need to be further investigated. The time delay in the control of the galvanometer and detector was dealt with using black strips that serve as markers of the projection position. Still some residual ringing artifacts are present. Several small volumetric test phantoms were constructed to obtain an overall picture of the accuracy.

1. Introduction

In optical CT scanning, optical transmission projections are acquired at different projection angles. Four different optical scanner types have been proposed in the past [1]. The first generation optical CT scanner was made of a laser beam and a photo-detector that moved simultaneously compared to the volumetric gel phantom to measure one projection. After each projection, the volumetric gel phantom was rotated and a new projection was measured. The second type of optical CT scanner was based on the principle of a cone beam shaped diffuse light source that illuminates the volumetric gel phantom and transmission images were acquired using a CCD camera. Again after each projection the volumetric gel phantom was rotated. The third kind of optical CT scanner used a parallel incident light beam generated by a point light source and a large plano-convex lens. After passing through the volumetric gel phantom a second plano-convex lens was used to focus the light beam onto a CCD camera. The last generation optical CT scanner, also called the fast laser scanning configuration uses a scanning laser beam and a photo-detector.

The optical CT scanner build at the Ghent University was based on this last type of scanner using a galvanometer mirror system to sweep the laser beam towards a plano-convex lens. At this point the laser beam is refracted so that the scanning beam is perpendicular to the rotation axis of the volumetric gel phantom. The first images acquired were contaminated with several imaging artifacts. The origins of those artifacts were investigated. First, blank measurements (without phantom) were performed to isolate scanner-related artifacts. Drift of the laser, drift of the detector, incorrectly matching of the refractive index, scratches on water bath and phantom container, secondary laser spots due to multiple reflections and the time delay in the control of the galvanometer and detector were investigated.

Performance characteristics of different components were also measured such as the laser spot size, light attenuation by the lenses and the dynamic range of the photo-detector. The need for a differential measurement using a second photo-detector was investigated. Post-processing strategies to compensate for hardware related errors were developed. Finally, several small volumetric test phantoms were constructed to obtain an overall picture of the accuracy.

2. Materials and methods

2.1. Optical CT scanner

An optical CT scanner (OPTOSCAN) (figure 1) was constructed in house. We use a 2.0 mW, 632.8 nm HeNe-laser (JDS Uniphase, Model 1122p, CA,USA) with a beam diameter of 0.63 mm ($1/e^2$ points). Before reaching a galvanometer mirror system (shortened: galvanomirror), the laser beam is narrowed by two pinhole collimators to sharpen the laser beam size. The galvanomirror directs the laser beam to a plano-convex lens (Melles-Griot, Albuquerque, USA) which is positioned at focus distance (450mm) from the galvanomirror. The laser beam leaving the plano-convex lens travels in parallel lines through the gel dosimeter until it passes another plano-convex lens which guides the laser beam towards the photodetector.

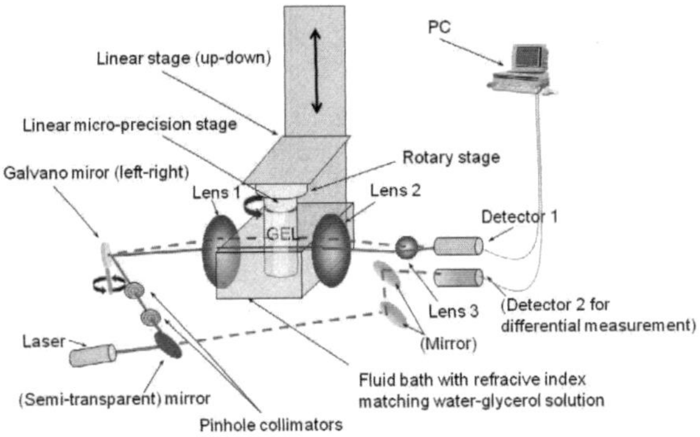

Figure 1. Schematic basic design of the in house optical laser scanner. The laser beam travels towards a rotating galvanomirror. Before reaching this galvanomirror the laser beam is guided by two pinhole collimators. The galvanomirror directs the laser beam to a plano-convex lens. From this part on the laser beam travels in parallel lines through the gel dosimeter until it passes another plano-convex lens which guides the laser beam towards the photodetector. In a second configuration, the mirror next to the laser was replaced by a semi-transparent mirror whereby the transparent part of the laser beam travels towards a second detector to perform a differential measurement instead.

The photodetector is a large area balanced photoreceiver (Thorlabs, model 2307, Munich, Germany). In between the lenses a large fluid bath (32.0 x 32.0 x 25.0 cm^3) is placed. Two black strips are attached to the walls of the fluid. These strips are used to correctly align all measured projections per angular increment. The refractive index matching procedure is performed automatically based on a dry and wet off axis refraction measurement and using an automated pump system with a refractive index matching fluid (water-glycerol solution).

In a second hardware configuration, a semi-transparent mirror (pellicle beam splitter, BP145B1, Thorlabs) was used to divide the laser beam into two beams. One fraction (45%) is guided towards the galvanomirror, the second fraction (55%) is guided towards a second detector eye. With this second detector eye, it was possible to make a differential measurement of the laser beam to compensate for possible drift of the laser.

Matlab (The MathWorks, Natick, USA) code was developed in house to control the optical scanner, construct sinograms and optical density images.

2.2. Post-processing

Before the actual measurement, a blank projection (no phantom present) and a dark current are recorded. During the measurements, data is recorded and stored in a text file. The file also contains all the spatial information of the measured data points. The first post-processing step involves aligning all the measured projections in the sinogram. This is performed by searching the two black reference strips on the wall of the fluid bath in each projection. Each projection is then correctly aligned according to the positions of the black strips in the sinogram. This alignment step is also performed for the blank projection. The optical density OD' is calculated from the transmission signal intensity using equation 1

$$OD' = \log_{10}\left(\frac{(SB-SD)}{(S(x,\theta,z)-SD)}\right) \tag{1}$$

were SB is the mean of a blank projection (no phantom present), SD is the dark current of the photodetector and S(x,θ,z) represents the signal at position x, angle θ and vertical slice position z.

3D rendered image reconstructions were composed using Image J (Image J 1.38x, Bethesda, USA).

2.3. Performance measurement of different building blocks

The laser stability was evaluated by measuring the laser intensity during 30 min. The laser spot was placed at a fixed position in the centre of the lenses. Detector stability was evaluated by measuring the dark current during 30 min. The dark current was measured by covering the detector eye so that no light could enter the detector. The laser spot size was measured in between the lenses. The photodetector was mounted inside the same linear micro-precision stage where the phantom is normally mounted. Before the detector eye, a circular disc with an aperture of 50 μm was placed. In discrete steps of 0.125 mm the detector was moved perpendicular to the laser beam manually using the linear micro-precision stage. For each positional increment, a measurement was carried out and averaged over 100 data points.

Attenuation of the lenses was evaluated by measuring the laser intensity averaged over 1 minute with and without the lenses. A measurement of the laser intensity was performed placing the laser spot in the centre and off centre (approximately 1 cm from the edge of the lens). Finally a measurement of the laser intensity was performed with the third lens positioned in front of the detector.

Influence of the fluid bath was evaluated by acquiring 20 blank projections with the fluid bath empty. The influence of the volumetric phantom container was evaluated by acquiring 20 projections, with and without phantom container which was filled with refractive index matching fluid.

2.4. Volumetric test phantom fabrication

Three volumetric phantoms were used in this study (R: 2.6 cm, h: 6.7 cm). Two gel phantoms were fabricated with gelatine gel from porcine skin (7% (w/w)) (bloom 300, Sigma-Aldrich, Bornem, Belgium). In the first phantom needles were positioned in a spiral configuration. In the second phantom, an air space was created in the shape of a funnel, which was subsequently filled with a chemical induced polymerised nPAG gel (Gelatin (6% (w/w), acrylamide (Aam) (3%, (w/w)), N,N'-methylene-bis-acrylamide (Bis) (3% (w/w)), 0.02 g ammonium persulphate and 0.2 ml N,N,N',N'-Tetramethyl-ethylene-diamine (TEMED) all obtained from Sigma-Aldrich. A third large volumetric test phantom was constructed (R: 4.75 cm, h: 15 cm) which was filled with the same refractive index matching fluid (glycerol-water) as the fluid bath.

3. Results and discussion

3.1. Laser and detector stability

The photodetector drift was found negligible over a period of 30 min. The mean value was -8.89 10^{-4}± 4.14 10^{-4} V. The discretisation levels of the detector can be clearly seen in figure 2a.

The laser stability was recorded over a period of 30 min after switching on the laser. The change in detected laser intensity is shown in figure 2b. Variations were significant in the first 10 minutes after start-up but saturate after this period (figure 3b). The mean value of the detected laser intensity after 15 minutes was 1.58 ± 0.01 V. From these measurements, it was concluded that before starting a measurement the laser should be turned on for at least 15 minutes. The low frequency variations in the laser intensity have not been explained.

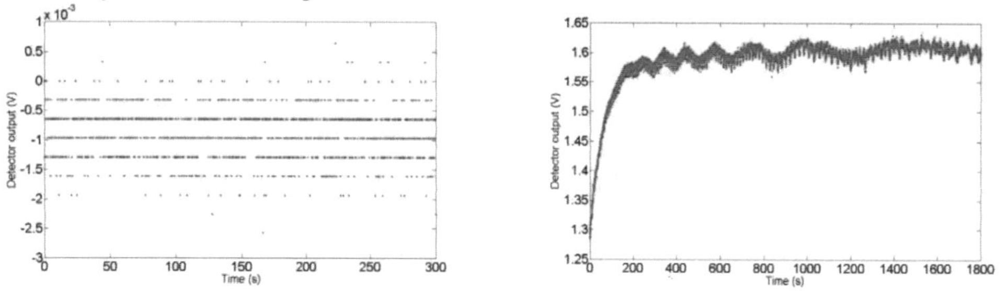

(a) (b)

Figure 2: the detector signal over a course of 1800s (a) and the laser intensity signal over a course of 1800s (b).

3.2. Laser spot size

The measured intensity profile across the laser spot is show in figure 3. A Gaussian function was fitted to the profile and a full-width-at-half-maximum of 1.27 mm was measured.

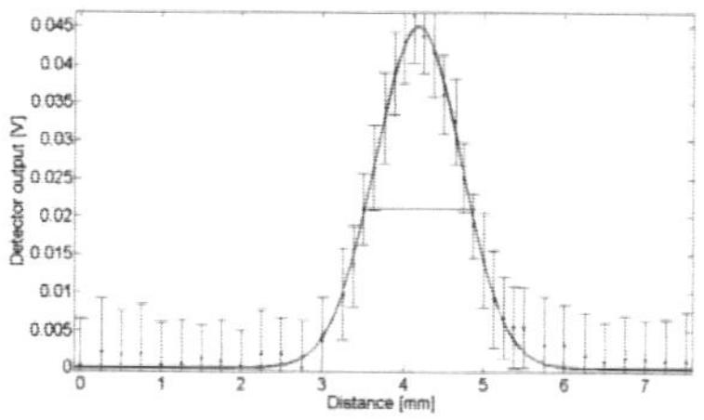

Figure 3. The red line represents the Gaussian fit of the data measured using the pinhole aperture disc in front of the detector. The detector was moved in discrete steps along the width of the laser beam.

3.3. Attenuation of the lenses

The attenuation of the laser intensity by the lenses was evaluated. A 9.43% attenuation of the laser beam by lenses 1 and 2 was detected (figure 1) with the laser beam travelling through the centre of the lenses. When the laser beam travels 5.4cm off centre through the lenses, the attenuation amounted to 12.52%. With lens 3 in place, the attenuation of the whole lens configuration measures 16.81% for the laser beam travelling through the centre and 16.65% for the laser beam travelling off centre. The reason for the difference between attenuation in the centre and off centre is probably due to a larger portion of multiple reflections reflected away from the detector when the laser beam enters the lens by an angle.

3.4. Differential measurement

The need for a differential measurement was investigated. The exiting laser beam was split in two fractions by a semi transparent mirror (transmission: 55%, reflection: 45%). The reflected fraction travelled towards the galvanomirror. The transmitted fraction travelled towards the second eye of the detector. A differential measurement would correct for possible laser drift. To see if a differential measurement performed better, measurements were compared with and without the second measurement. It was seen that the use of a differential measurement resulted in only an offset in the intensity data. It was concluded that a differential measurement is not required. This was also consolidated from the detector an laser stability measurements.

3.5. Refractive index matching fluid

To match the refractive index of the water-glycerol solution with the refractive index of the gel dosimeter an automated procedure has been developed. The principle is based on a dry and wet off-axis refraction measurement of the gel dosimeter. Based on the determined refractive index of the gel and the fluid, the required glycerol concentration is determined. The concentration dependence of the refractive index was also determined experimentally. Using recorded data on the concentration glycerol in the fluid, the difference in concentration can be adjusted using an automated pump and mixing system (figure 4).

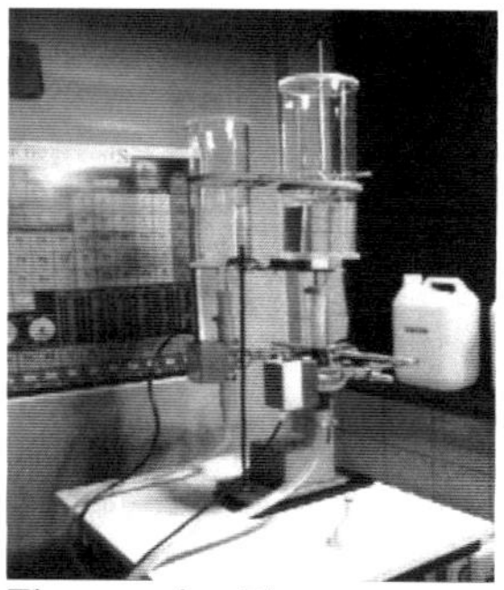

Figure 4. The automated pump and mixing system.

3.6. Influence of the water basin and volumetric phantom container

Projections of in the empty fluid bath were recorded to investigate scratches on the glass wall (figure 5). When the refractive index matching fluid was pumped in the fluid bath the variations in the data signal became significantly smaller. This may be explained by filling of the scratches by the refractive index matching fluid. Refractions of the laser beam will not be enhanced as much when passing from glass to matching fluid as from glass to air. Turbulences and little dust particles in the water could also induce variations in the signal. It was found advantageous to wait at least 1 hour after filling the fluid bath before measuring.

Projections with a blank volumetric phantom, *i.e.* a cylindrical phantom filled with the refractive index matching fluid, were measured to investigate the influence of the container on the signal. It was found that scratches on the wall of the volumetric phantom induce artefacts in the projections (figure 5).

For measurements of small volumetric phantoms, a small fluid bath (14.2x7.5x5.5 cm³) was used to minimize the amount of refractive index matching fluid.

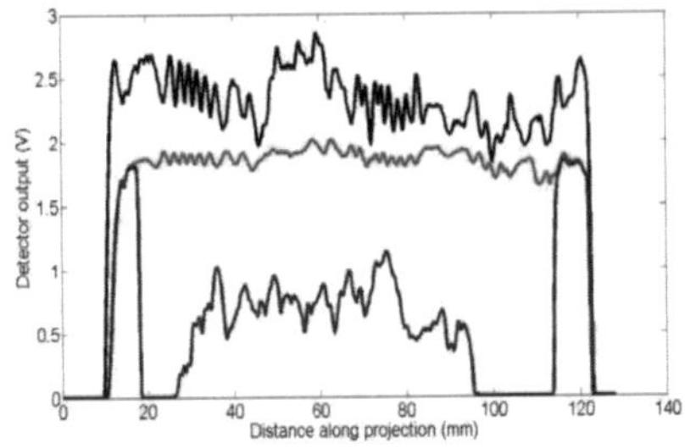

Figure 5. Showing 3 averaged projection: black line is a projection taken with no fluid in the bath, red line is a projection taken with fluid in the bath and the blue line is a projection taken with fluid in the bath and a blank volumetric phantom.

3.7. Post processing

Synchronization problems between the input of the projection data and the output of the galvanomirror were detected . This resulted in shifted projection in the sinogram which caused severe artifacts. Also when OD' calculations were performed with misaligned data projection versus bank projection, a magnification of the ringing artifacts became apparent. To compensate for these errors a Matlab (The MathWorks) script was used to detect signal voids in the top and bottom of the sinogram from black strips on the wall of the fluid bath (figure 6).

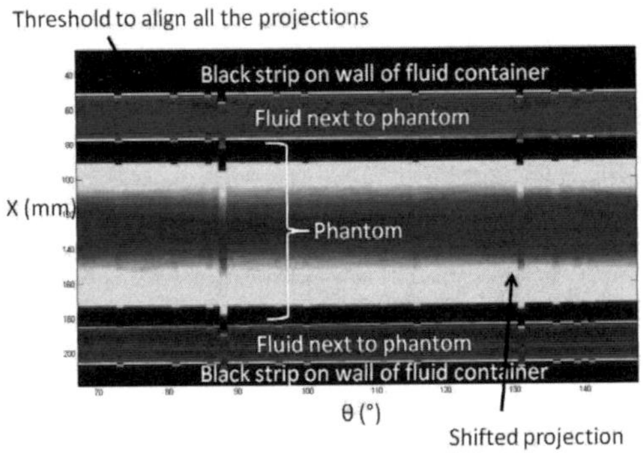

Figure 6. Sinogram to demonstrated shifted projection and the location of the black strips on the wall of the fluid bath.

Severe misalignments were dealt with using the black strips. However, some sub-pixel shifts between acquired projections and blank measurements could still result in minor artifacts. Further optimisation is needed to minimize further ringing artifacts.

3.8. Test QA volumetric phantoms

The first 3D reconstructing images of QA volumetric phantoms were very promising (figure 7).

Figure 7. Reconstructed 3D images of test volumetric phantoms.

4. Conclusion

A fast laser scanning optical CT scanner with a fully automatic refractive index matching system refractive was constructed and optimized at the Ghent University. The first images acquired were contaminated with several imaging artifacts. The origins of artifacts were investigated. Drift of the laser and the detector was found negligible after 15 minutes warm up of the laser. When scratches on the water bath and phantom container are present, these pose a post processing challenge to eliminate the resulting artifacts from the reconstructed images. Secondary laser spots due to multiple reflections on the lens system needs to be further investigated. The time delay in the control of the galvanometer and detector were dealt with using black strips that serve as markers of the projection position. Still some residual ringing artifacts are present. Several small volumetric test phantoms were constructed to obtain an overall picture of the accuracy. Further optimisation is required to obtain a better accuracy and precision.

References

[1] Oldham M 2004 Optical-CT scanning of polymer gels *J. Phys.: Conf. Ser.* **3** 122-135

A prototype fan-beam optical CT scanner for polymer gel dosimetry

A. Jirasek[1], D. Rudko[1,2], and D. Wells[2]

1. Dept. of Physics and Astronomy, University of Victoria, Victoria, BC, CANADA

2. BC Cancer Agency, Vancouver Island Centre, Victoria, BC, CANADA

jirasek@uvic.ca

Abstract. A prototype optical CT scanner has been developed for scanning polymer gel dosimeters used in 3D radiation therapy dose verification. The scanner is based on a 3rd generation, fan-beam CT design and employs a 543 nm laser fanned through a 60° line-generating lens. A series of 5 detector arrays or 64 elements each collects light on the distal side of the water tank. Pre-detector collimators are employed to reject light scattered away from incident ray trajectory. Results of scanner characterisation studies, as well as preliminary data acquired on irradiated polymer gels, are discussed.

1. Introduction

Optical computed tomography (OptCT) has been employed by a number of groups in the imaging of polymer gel dosimeters for radiation dosimetry. Two basic OptCT designs have emerged in recent years. The first is based on a 1st generation translate-rotate CT design, employing, typically a single detector and laser beam, scanned in parallel across the dosimeter. While potentially a very accurate method for imaging dosimeters, the systems can be slow [1]. The second category of designs employ a cone beam of light and, typically, a CCD detector to simultaneously image multiple "slices" of the dosimeter [2]. While offering rapid imaging times, scatter rejection and collimation are still outstanding issues in these systems.

This study reports on the use of a 3rd generation CT design to image polymer gel dosimeters. The system utilizes a fan beam of 543 nm laser light. Light emerging from the distal end of the scanning tank is collected using an array of 320 photodetectors arranged concentric with the focal point of the fan beam. Custom manufactured scatter-rejection collimators are positioned in front of each detector element. The system thus offers possibilities for rapid imaging (~1 min/slice) and scatter rejection. The OptCT scanner is still in a prototype phase and currently ring artefacts in the resultant images limit the overall dose accuracy attainable with the system.

2. Methods

2.1. Instrument Design

A schematic and image of the OptCT scanner are shown in figure 1. A 2mW, 0.8 mm beam diameter, 543 nm laser (Research Electro-Optics, Boulder CO) is incident on a 60° line generating lens (Edmunds Optics, Barrington, NJ). Fanned laser light is incident through a semi-circular matching medium tank, designed such that light is incident normal to the tank faces. The matching medium consists of water and glycerol for index matching to gel refractive index, as well as blue dye used for attenuation matching. Polymer gels are rotated and translated utilizing computed controlled rotation and translation stages (Newport, Irvine, CA). Gels are housed in 1L cylindrical containers (Modus Medical, London, ON). Laser light is collected on the distal side of the tank using five concentrically arranged photodetector arrays (Hammamatsu, Japan), with each detector array housing 64 photodetector elements of 0.7 x 0.8 mm. Detector pitch is 0.8 mm. A collimator array, designed such that each collimator element is concentric with the fan beam focal spot, is positioned in front of the detector array. A custom-built microcontroller interfaces detector output and synchronization to a PC. Data collection and reconstruction is performed using Matlab (The Mathworks, Natick, MA).

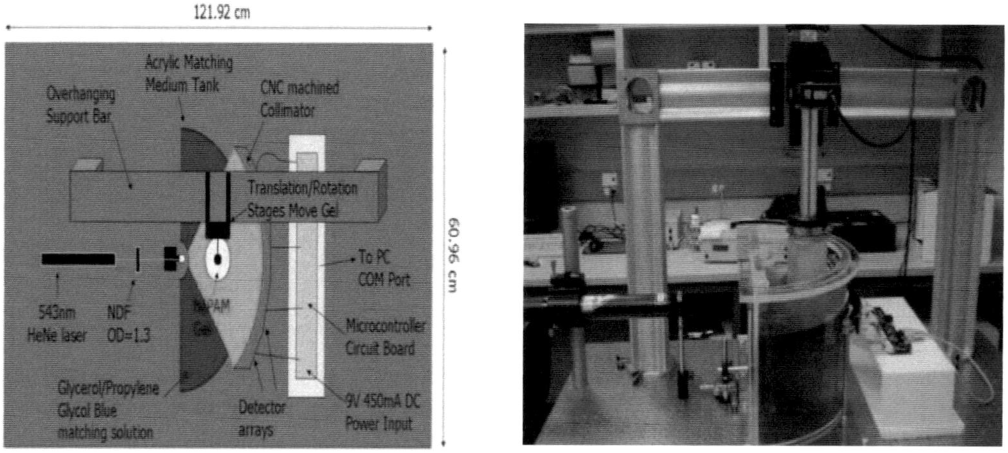

(a) (b)

Figure 1: (a) Schematic and (b) image of OptCT scanner.

2.2. Instrument characterisation and data post processing

A series of instrument characterization tests were performed. The uniformity of detector intensities when subject to the light fan beam was measured. System temporal stability, slice acquisition time, system noise, spatial resolution, and spatial distortion were characterized.

Image reconstruction was performed using the filtered backprojection algorithm built in to Matlab. Further postprocessing was performed to optimize image quality. Specifically, image filtering in the image domain, the FFT domain (FFT of image), and filtering of the sinogram (pre-reconstruction) were all tested using a range of image filters. Window sizes in the image filters were also tested.

2.3. Polymer gel imaging

Two cases utilizing NIPAM polymer gel (3% NIPAM, 3% bis-acrylamide, 5% gelatin, 5mM THPC) irradiated with a 6MV photon beam were used to test the imaging capabilities of the system. In the first case, a square 3x3 cm^2 field (maximum dose ~5 Gy) was used to irradiate polymer gel along the length of the 1L cylinder container. In the second case a 7 field IMRT irradiation was performed on a 1L cylindrical gel phantom (6 Gy at isocentre). Isodose overlays, profile plots, and dose difference maps between planned and imaged dose distributions were used to test the ability of the system to image irradiated polymer gel dosimeters.

3. Results and Discussion

3.1. Instrument characterisation

Figure 2a illustrates the uniformity of detector intensities when a fan beam is passed through the tank full of matching medium solution and with no polymer gel in place. Figure 2b illustrates laser/detector temporal stability. Typically, stability is within 0.8% of the mean detector intensity.

Table 1 shows the effects of index of refraction mismatch on image distortion (magnification). With optimized index matching solution, magnification distortions are reduced to within 0.01 cm.

Spatial resolution was determined by utilizing profiles through a needle phantom. The needle line function was deconvolved from the image and the 30% value of the MTF of the resultant profile was used as a determination of the system spatial resolution. Using this method, a spatial resolution of 0.25 mm/pixel was obtained, matching well the grid size of the reconstructed image (0.22 mm/pixel).

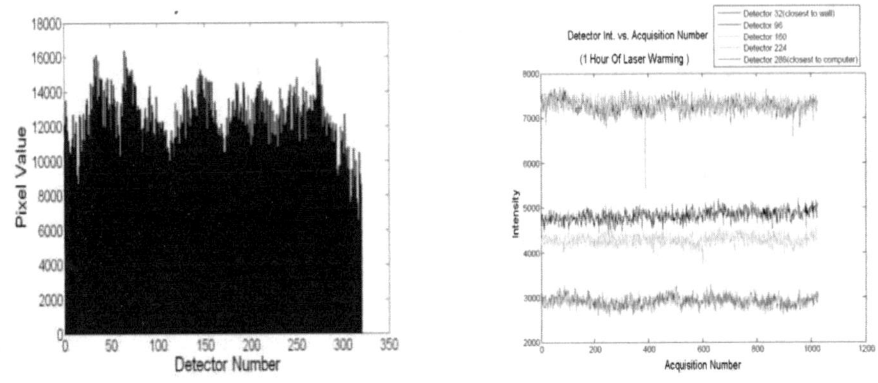

(a) (b)

Figure 2: (a) Signal intensity variation over the full detector array. (b) Laser/detector temporal stability as a function of acquisition number (i.e. time).

Table 1: Effects of index of refraction mismatch on image distortion (magnification). True physical distance was measured utilizing x-ray CT imaging.

	Gel Cylinder Diameter(cm)	Needle Separation(cm)
True Physical Distance	9.64±0.01	3.00±0.01
Image Distance(n=1.33)	9.51±0.01	2.94±0.01
Image Distance(n=1.35)	9.64±0.01	3.00±0.01

Figure 3a illustrates the effects of multiple averages per angle on the resultant standard error on the mean detector signal. As illustrated in figure 3b, the decreasing standard error as a function of increasing the number of averages comes at the expense of increased integration time. Figure 3b was calculated using 1° per angular increment. It is anticipated that large improvements to the results in figure 3b can be realized with improved data transfer speeds between the microcontroller and PC.

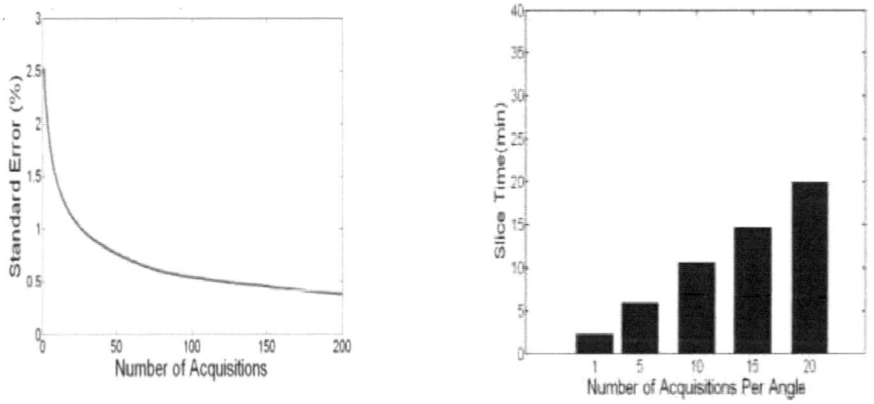

(a) (b)

Figure 3: (a) Standard error of mean pixel intensity as a function of the number of acquisition averages. (b) Slice time as a function of the number of acquisition averages.

3.2. Polymer gel imaging

Figure 4 illustrates a OptCT images of the 3x3cm^2 square field irradiated gel. Shown are results for a raw image (figure 4a), a Gaussian filtered image (figure 4b) and a sinogram median filtered image (figure 4c). Figure 4d illustrates plan contours overlayed over the OptCT sinogram-filtered image and shows the ability of the system to accurately reconstruct the overall position and shape of the irradiation. Figure 4e illustrates a dose difference map between the OptCT image (sinogram filtered) and treatment plan. Figure 4f illustrates profile plots through OptCT images and the treatment plan. Figures 4 shows that, in general, filtering in the sinogram domain (i.e. prior to image reconstruction) offers the greatest improvements in image quality. Still, ring artefacts can be observed in the images and remain an outstanding issue for future work.

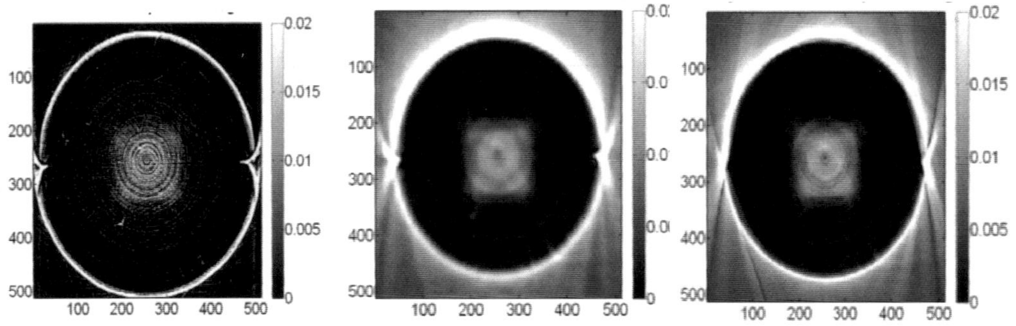

(a)
(b) (c)

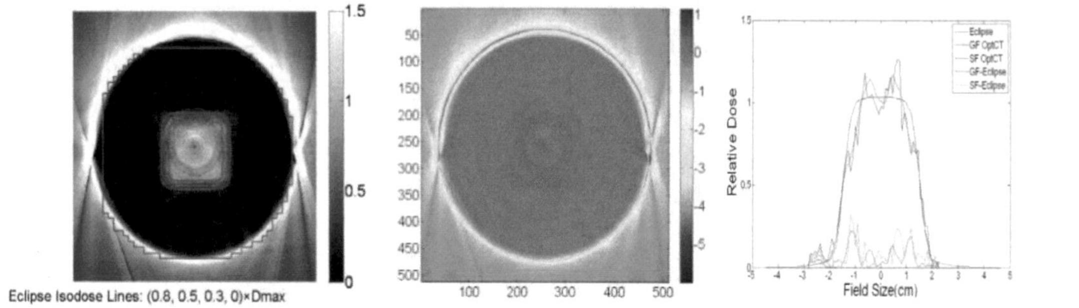

(d) (e) (f)

Figure 4: (a) Raw OptCT image of 3x3 cm^2 irradiated polymer gel, (b) Gaussian filtered image, (c) sinogram median filtered image, (d) plan contour overlay, (e) dose difference map between OptCT image and treatment plan, (f) profile plot through OptCT images and plan.

Figure 5 illustrates the results for OptCT imaging of polymer gel irradiated with a 7 field IMRT plan. Figure 5a illustrates the raw image. Figure 5b illustrates the treatment plan. Figure 5c illustrates the sinogram median filtered image, while figure 5d illustrates the planning contour overlay. The dose difference map between the sinogram filtered OptCT image and the treatment plan is shown in figure 5e. Figure 5f shows representative profile plots through the gel cylinder. While the general shape of the dose distribution is well captured in the OptCT images, ring artefacts still complicate more detailed quantitative analysis.

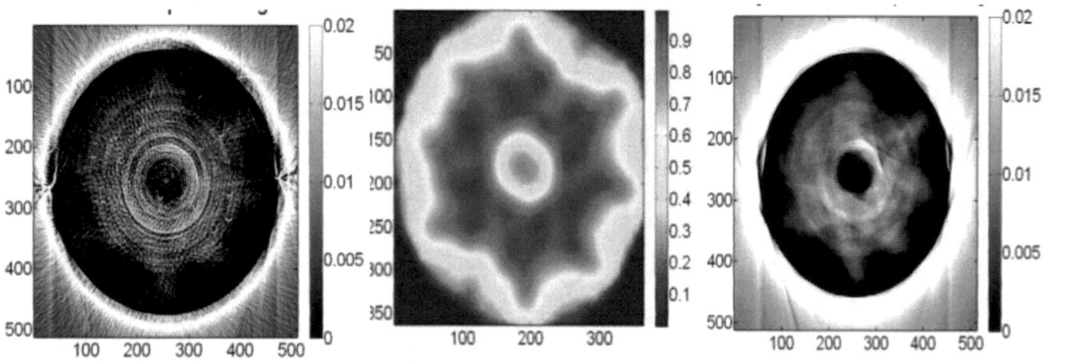

(a) (b) (c)

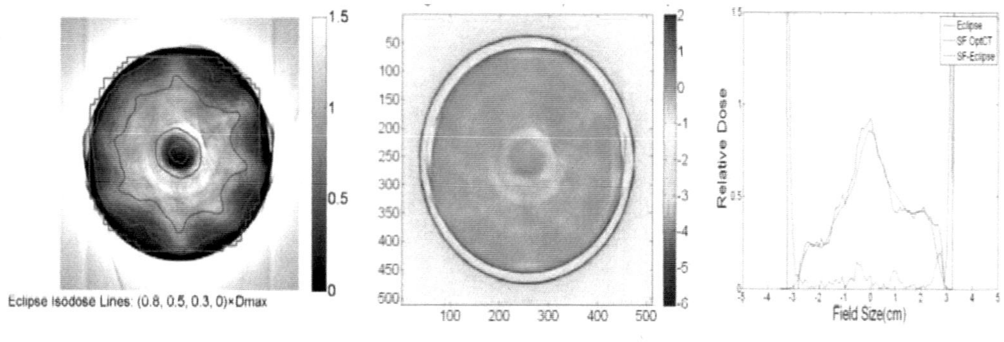

(d) (e) (f)

Figure 5: (a) Raw OptCT image of 7-field IMRT irradiated polymer gel, (b) treatment plan, (c) sinogram median filtered image, (d) plan contour overlay on OptCT image, (e) dose difference map between treatment plan and sinogram filtered image, and (f) representative profile plots through images.

4. Conclusions

A prototype fan beam optical CT scanner based on a 3rd generation CT design has been developed. The system utilizes a semi-circular design, pre-detector collimation, and photodetector array to acquire projection data from irradiated polymer gel dosimeters. While in the prototype stage, the system can reconstruct dose distributions delivered to the polymer gel. However, ring artefacts remain an outstanding issue for future work. Suppression of ring artefacts will enable further quantitative analysis on OptCT images of irradiated polymer gel.

Acknowledgments

Funding from the Canadian Foundation for Innovation, the BC Knowledge Development Fund, the Natural Sciences and Engineering Research Council of Canada, and the BC Cancer Agency is acknowledged.

References

[1] M. Oldham, H. Sakhalkar, P. Guo, and J. Adamovics. An investigation of the accuracy of an IMRT dose distribution using two- and three-dimensional dosimetry techniques. *Med. Phys.*, **35**(5):9, 2008.
[2] N. Krsta jic and S. Doran. Characterization of a parallel-beam CCD optical-CT apparatus for 3D radiation dosimetry. *Phys. Med. Biol*, **52**:3693–713, 2007.

Journal of Physics: Conference Series **164** (2009) 012026 doi:10.1088/1742-6596/164/1/012026

Optical evaluation of Fricke xylenol orange gel by light scattered at 90 degrees

J. Svoboda[1,2], R. Alwan[1], F. Guermeur[1], L. Makovicka[1], Y. Bailly[1], V. Spěváček[2], T. Čechák[2], E. Martin[3]

[1]IRMA/ENISYS/FEMTO-ST UMR 6174 CNRS, Université de Franche-Comté, Pôle Universitaire, BP 71427, 25211 Montbéliard Cedex, France

[2] Dept. of dosimetry and application of ionizing radiation, FNSPE, Czech Technical University in Prague, Břehová 7, Praha 1, 115 19, Czech Republic

[3] Department of radiotherapy, CH Belfort-Montbéliard, 25209 Montbéliard, France

svobojir@kmlinux.fjfi.cvut.cz

Abstract. This communication presents optical method for evaluation of Fricke xylenol gel (FXG) using light scattered at 90 degrees to initial direction. Although Fricke gel is predominantly absorbing, gelatine matrix scatters enough light which could be collected and related to dose delivered to gel. Initials experiments were oriented to determination applicability of this approach.

1. Introduction

This research is focused to development of an optical method based on measuring of a light primary scattered on gelatine and then attenuated by absorption. The basic is to established relations between scattered light intensity and doses. Moreover used apparatuses are very simple and thanks to utilization of the rectangular containers instead of cylindrical this technique avoids problems related to reflections.

2. Materials and Methods

The experimental setup consists of laser, piezoelectric apparatus providing light thin light sheet, focusing lens and support with sample. Scattered light is collected by a 12 bits CCD camera (fig 1). As a dosimeter a Fricke xylenol orange gel [1] was used for all experiments. Once the raw image is de-noised and smoothed analyse [2] begin. Target is to find attenuation coefficient either for whole cell or for small region of pixels. Next step consists in comparison of attenuation at different point of a one sample. Last one lies in measuring on heterogeneous sample and reconstruction of attenuation distribution.

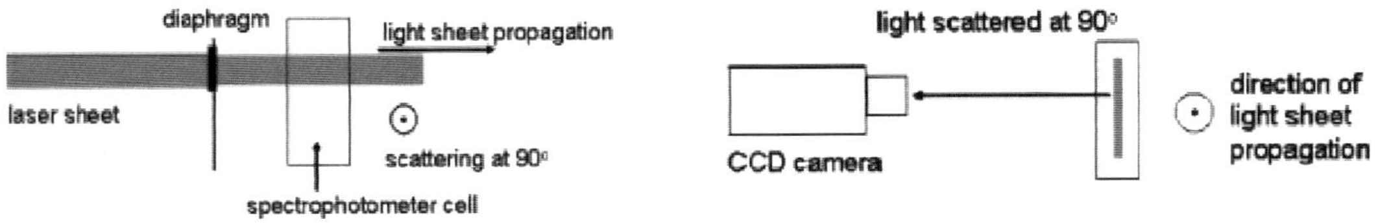

Figure 1. Schema of experimental setup

2.1. Spectrometric cells

In this first experiment a set of 7 irradiated spectrometric cells (0, 1, 2, 4, 6, 9 and 12 Gy) was subsequently illuminated by laser sheet passing through the centre of a cell (fig. 2). Theses wavelengths were used 476, 488, 514 and 633 nm. For reduction of noise every image was taken as an average of 320 frames. After that, images were processed using Matlab software. Then the intensity of images was examined in order to verify exponential decrease $I = I_0 exp(-ax)$ and its parameter a where I is observed image's intensity I_0 is initial intensity and x is distance in pixels. The objective of this measurement was to determine the response of our systems.

Figure 2. Example of typical acquired image

2.2. Uniform phantom

Glass 5x5x8 cm glass container filled with nonirradiated gel was placed on the support so laser sheet passed through just behind wall closer to camera, then container stepped by 1 mm toward the camera until laser reached remoter wall. In this way 50 images were acquired which allowed to complete a 3D data set. Aim of this measurement was examination of decreases of images' intensity in both directions – parallel and perpendicular to propagation of a laser sheet. For this purpose attenuation in parallel direction was directly estimated from the images and also attenuation in smaller region (100x100 pixels) was calculated and compared. In case of perpendicular direction corresponding data in different images were plotted in graph and fitted.

2.3. Phantom with artificial cavities

For this phantom same glass container as in previous experiment were used. Three artificial cavities, two circular with diameter of 1 and 1,3 cm and one rectangular 1x3,5 cm were filled with FXG coloured by addition of known amount of Fe^{3+} (0,4 and 0,6 mM). By now only homogenous samples were measured so the experiment was used to test capability to distinguish difference among coloured gels and reconstruction of the attenuation distribution. To assign appropriate value of a in layer between two consecutive positions of the laser sheet (514 nm), it is necessary to make correction for precedent layers. Initially the a for small regions (50x50 pixels) in first image (closest to CCD camera) is calculated. Next image intensity of the second layer is modified in order to compensate a lost of signal due to passage of the first layer. Then modified image is used to find a for the second image. In this way a for all laser sheet position is determined.

3. Results and Discussion

3.1. Spectrometric cells

Images' intensities for all cells were plotted in graph (fig. 3). Parameter *a* was found to be in range from 3,266.10^{-5} pixel^{-1} for nonirradiated gel to 7,59.10^{-5} pixel^{-1} for gel irradiated to 12 Gy as shown in next graph (fig. 4). Note that *a* is an experimental value and an exact relation between the *a* and the coefficient of extinction has to be established yet. However *a* seems to be related to dose and thus it could be used for dose measurement.

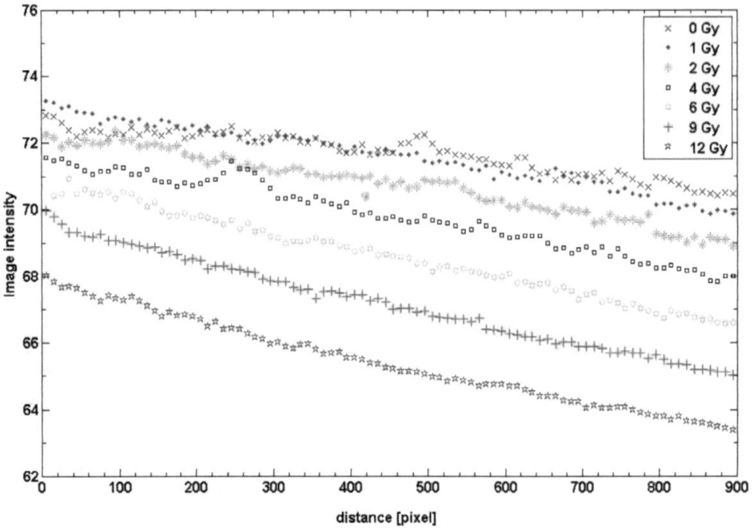

Figure 3. Intensity of images for irradiated gel. Red laser (633 nm) passes through cells from the left to the right.

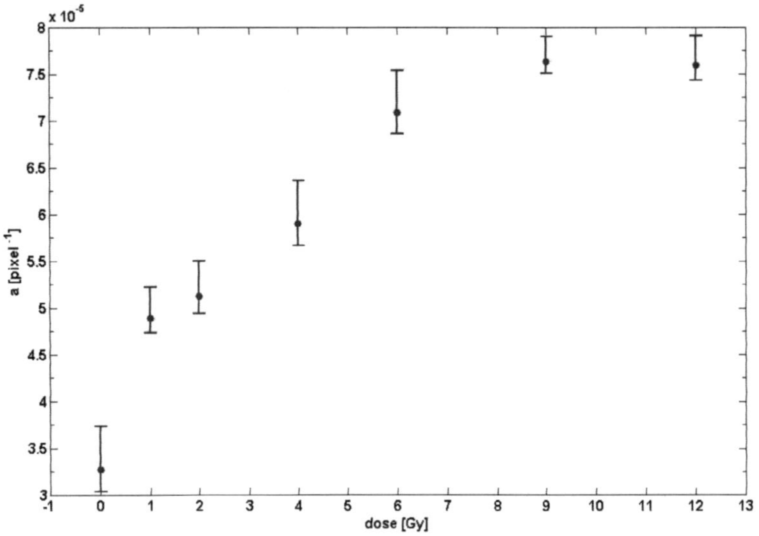

Figure 4. Parameter *a* as a function of the dose measured at 633 nm.

3.2. Uniform phantom

Parameter *a* for planes parallel to the laser sheet was measured to be 2,302±0,367.10^{-4} pixel^{-1}. For perpendiculars planes intensity of different images with same x coordinates were used (Fig. 5) and *a* was found to be 2,592±0,576.10^{-4} pixel^{-1}. Additionally *a* was calculated for smalls region of the image

(100x100 pixel) and the result is $2{,}359 \pm 0{,}995.10^{-4}$ pixel^{-1}. All three values are in agreement although the standards deviations are important.

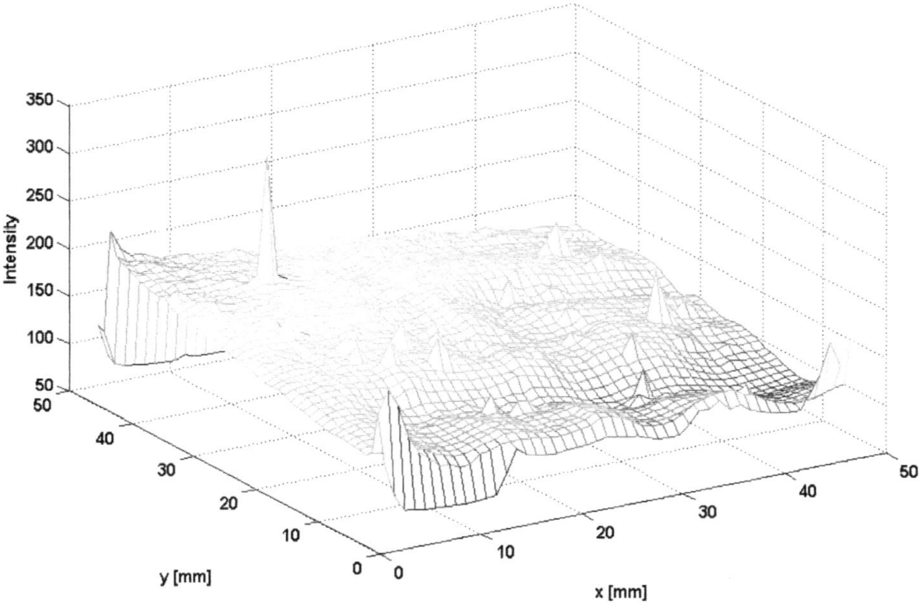

Figure 5. This image illustrates signal's diminution in gel measured for 5x5 cm phantom. The laser sheet (514 nm) is propagated along x axis. The entering surface is that with coordinates [0, y] while CCD camera is placed next to surface [x, 50].

3.3. Phantom with artificial cavities

The result of the last experiment is presented at the following image (Fig. 6). One pixel in original image corresponds to ~ 0,05 mm, however spatial resolution in reconstructed image is about 5 times worse as larger region (50x50 pixel) had to be used because of noise . Coefficient *a* was calculated $2{,}417 \pm 0{,}305.10^{-4}$ pixel^{-1} for clear gel, $3{,}406 \pm 0{,}571.10^{-4}$ pixel^{-1} (cavity 2), $5{,}079 \pm 0{,}468.10^{-4}$ pixel^{-1} (cavity 1) and $5{,}127 \pm 0{,}275.10^{-4}$ pixel^{-1} (cavity 3). These values agree with previous experiment.

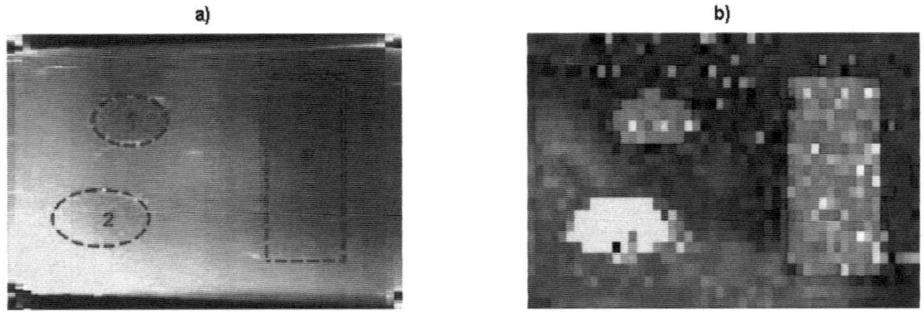

Figure 6. Phantom with cavities – a) Composed image of acquired data, each column represents one position of the laser sheet which was moved with 1 mm step across the container. Areas 1, 2 and 3 correspond to coloured gel (1 and 3 - 0,6 mM while 2 - 0,4 mM of Fe^{3+}). b) Estimated coefficient *a*.

4. Conclusion

This method seems to be usable for 3D dosimetry. Next step consists of amelioration of images processing, spatial resolution and using more appropriated wavelength corresponding to the absorption maximum of irradiated gel and thus increase sensibility.

5. Acknowledgement

The authors acknowledge the financial support of the LCC (Ligue Contre le Cancer), Région Franche-Comté and the CAPM (Communauté d'Agglomération du Pays de Montbéliard).

References

[1] M.A. Bero, W.B. Gilboy and P.M. Glover, 2001, Radiochromic gel dosemeter for threedimensional dosimetry *Radiat. Phys. Chem.,* 166, 433–5.

[2] R. Alwan, F. Guermeur, Y. Bailly, L. Simonin, J. Svoboda, L. Makovicka, E. Martin, 2008, Calibration of laser tomography as a new optical diagnostic tool applied to dosimetric polymer gels, *Nucl. Instrum. Methods B*, 266, 834-840.

Journal of Physics: Conference Series **164** (2009) 012027 doi:10.1088/1742-6596/164/1/012027

Three-dimensional radiation dosimetry with optical projection tomography

AE Papadakis[1], G Zacharakis[2], J Ripoll[2], F Zacharopoulou[1], TG Maris[1], J Damilakis[1]

1. Medical Physics Department, University Hospital of Heraklion, University of Crete, Heraklion, POBox 1352, 71110, Crete, Greece.
2. Institute of Electronic Structure and Laser – Foundation of Research and Technology Hellas, P.O. Box 1527, 71110 Heraklion Crete, Greece

Email: apapad@iesl.forth.gr

Abstract. Here we present initial results on the study of the dosimetric properties of a normoxic N-Vinylpyrrolidone based polymer gel dosimeter using a newly developed optical projection tomography (OPT) system. The system employs a sensitive CCD camera, a rotation stage allowing full 360 deg rotation and a wide field homogeneous light source which transilluminates the gel sample. This setup is capable of producing high resolution images of the light attenuation map inside the sample's volume in three dimensions (3D). The experimental procedure involves the capturing of two-dimensional projection images of the polymer sample in rotation steps of 1 deg. Data analysis is performed by back projecting the photons using an inverse Radon transform resulting in the reconstruction of 3D images of the attenuation coefficient. The resulting 3D attenuation map of the irradiated volume can be directly correlated to the radiation dose imparted. The OPT system may operate in 'dry' conditions without the use of any refractive index matching fluids. The sensitivity and dynamic range offered by the methodology covers the range of radiotherapy doses in modern clinical practice.

1. Introduction

Optical projection tomography (OPT) is a new imaging technique, which has been successfully used in genetic research for three-dimensional imaging of small animal models.[1],[2],[3],[4],[5] OPT may be considered the optical equivalent of x-ray CT scanning, provided that the sample under investigation is transparent and the photons are traveling through the sample without any significant absorption or scattering. Three dimensional radiation dosimeters including polymer gels and radiochromic dosimeters fulfill the above requirement. Being originally translucent, their optical properties are appropriately modified in the presence of ionizing radiation. Several optical instrumentations have been developed and evaluated to determine the imparted dose through the quantification of the changes in the optical properties of the irradiated dosimeters.[6],[7],[8],[9] Guo et al,[10] demonstrated accurate 3D dosimetry using the PRESAGE™/optical-CT dosimetry system incorporating a commercial, laser-based optical-CT dose-readout scanner (OCTAPUS™, MGS Research Inc.). More recently, Sakhalkar et al,[11] in an attempt to reduce scanning time, demonstrated the application of an optical CT scanner, which incorporates a uniform-broad beam source, customized image forming optics and a charged coupled device (CCD) camera. Despite the advances in the research and development of the 3D radiation dosimetry techniques much effort is still placed towards the development of practical methodologies which could be conveniently applied in the routine radiation therapy practice.

In this study we present initial results on the readout of a normoxic N-Vinylpyrrolidone based polymer gel dosimeter with a newly developed optical projection tomography (OPT) system.

2. Materials and Methods

2.1 Gel Preperation

A new normoxic N-Vinylpyrrolidone based polymer gel dosimeter with tetrakis (hydroxymethyl) phosphonium chloride (VIPET) was employed in this study. The VIPET[12] polymer gels were composed of 4% wf VIPE, 4% wf bis, 7% wf gelatin, and 5mM THPC. The manufacturing procedure was performed in normal atmospheric conditions under a fume hood as follows: the gelatin was added to the double distilled deionized water at room temperature (25 ^{0}C) and allowed to dissolve. The solution was then heated to 50 ^{0}C and once the temperature was stabilized the bis was added. Heating was achieved through a hot-plate and stirring unit. When the solution became transparent, about 20 min later, the mixture was cooled down to approximately 32 ^{0}C and VIPE was added. Once the constituents were completely dissolved after about 5 min, THPC (Sigma-Aldrich) was added in the solution. Three minutes after the addition of THPC the gel was transferred to cylindrical glass vials. The vials were tightly sealed with their caps and parafilm. The gels were allowed to solidify overnight before irradiation at room temperature.

2.2 Irradiation

Five cylindrical glass vials (22 ml) containing the VIPET gel material were irradiated to five radiation doses up to 32 Gy, using a 6 MV LINAC x-ray beam. A parallel opposed dual beam arrangement was applied to produce a uniform dose distribution throughout the entire gel sample. The above samples provided the dose calibration data. Another cylindrical glass tube (150 ml) containing the VIPET gel was irradiated coaxially with the same x-ray radiation beam in a 2 x 2 cm^2 arrangement. That beam delivered 8 Gy at the depth of the maximum dose.

2.3 Optical projection tomography apparatus

A schematic drawing of the OPT apparatus is shown in figure 1. A broadband light field illumination is provided by a LED array, which is consisted of 30 diodes. The gel sample is positioned on a motorized rotation stage (8MR190-2, Standa, Lithuania) the movement of which is controlled from a personal computer via a microcontroller. The specimen is illuminated through a light diffuser which is placed in between the LED array and the gel sample. The sample is rotated stepwise through a 360^{0} revolution with an image projection captured at each step. The rotational step was 1 deg with a total of 360 images acquired in a complete specimen revolution. Typical exposure times ranged from 200 ms to 1000 ms per projection and the total imaging time ranged from 4 min to 8 min. Capturing of the projection images is performed with a monochrome 12 bit, 640 x 480 pixel, CCD camera (Sony, model: ICX-414AL) through a zoom lens. The camera uses the IEEE-1394 data transferring protocol and it is capable of acquiring images at a rate of up to 86 frames/s.

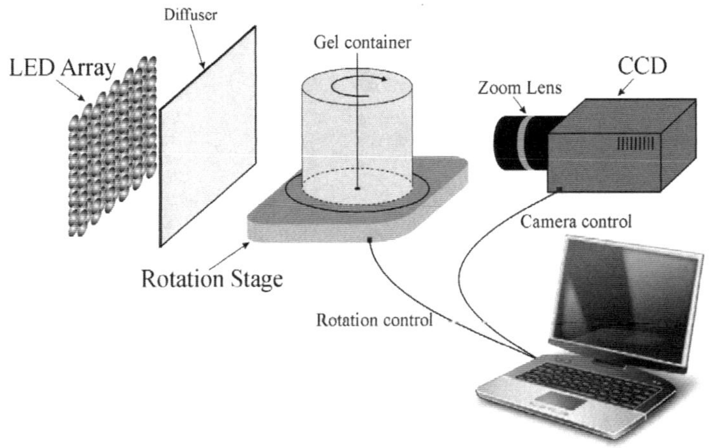

Figure 1. A schematic diagram of the optical projection tomography apparatus.

The iris diaphragm of the zoom lens is set such that the depth of focus spans the full diameter of the gel container. In-focus data are thus recorded from all parts of the gel volume for every image projection. Specially developed software (in-house LabView code) is employed for the control of the camera and the rotation stage as well as the 3D reconstruction of the acquired projection data. Image projections are transformed according to Beer's law to derive sums of attenuation coefficients. The CCD is oriented such that the rows of CCD pixels are aligned perpendicular to the rotational axis. The temporal sequence from a row of pixels forms a sinogram that reconstructs the corresponding slice using standard convolution filtered back-projection with the Hamming filter. Prior to the 3D reconstruction the software determined automatically the position of the specimen's rotational axis. This was done by reconstructing a series of images with differently assumed positions of the rotational axis, and calculating the variance of each reconstruction. The reconstructed image with the maximum variance corresponds to the least offset between the true and the assumed position of the rotational axis and is thus the least blurred.[13],[14]

The reconstruction uniformity and spatial resolution of the OPT system were evaluated. In particular, an unirradiated gel sample was scanned to evaluate the imaging reconstruction uniformity and a thin optical fiber 360 μm in diameter was scanned to evaluate the imaging resolution.

3. Results and Discussion

Figure 2 illustrates the OD-dose response curve obtained at 24 h post-irradiation using the OD data obtained from the gel vials irradiated to doses between 0 and 32 Gy. The linear regression analysis showed that OD and dose are linearly related in the range from 0 to ~ 20 Gy (OD = 0.017 x D – 0.0023, $r^2 = 0.999$).

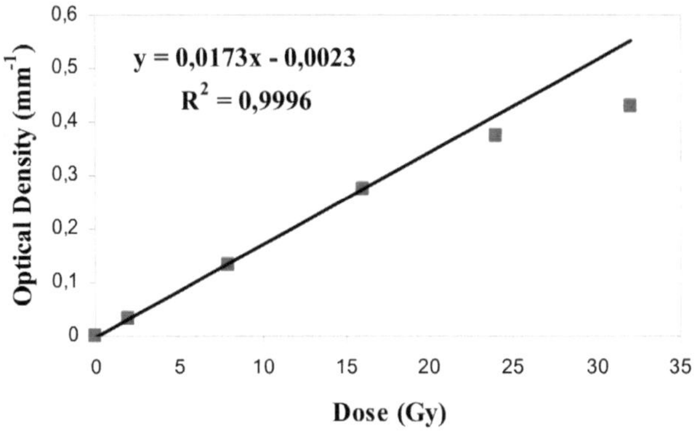

Figure 2. Optical density (OD) vs. dose calibration data. A linear response is preserved for doses up to approximately 20 Gy.

Figure 3(a) shows a projection image of the unirradiated gel sample. The dark regions along the boundaries between the gel and the cylindrical glass wall are attributed to the differences in the refractive index among the gel material, glass and air. Figure 3(b) shows a reconstructed axial slice along the red dashed line of (a), and Figure 3(c) illustrates a thick line profile curve along the area enclosed within the white frame of (b). The profile curve shows that the unirradiated gel material is reconstructed in 'bowl' shaped image slices. The difference in the reconstructed values between the center and the periphery regions of the homogenous gel is up to 40%. This difference was attributed to the fact that the gel vial is surrounded by air without being immersed to refractive index matching fluid.

Figure 4(a) shows a projection image of the irradiated gel sample and figure 4(b) shows a reconstructed axial slice along the red dashed line of (a). Figure 4(c) illustrates a thick line profile curve along the area enclosed within the white frame of (b). The reconstructed profile values of figure 4(c) were normalized against the corresponding values of the unirradiated gel (figure 3c) to compensate for the 'bowl' effect. The derived values may be converted to optical density and subsequently to dose values based on the OD-dose calibration equation given in Figure 2.

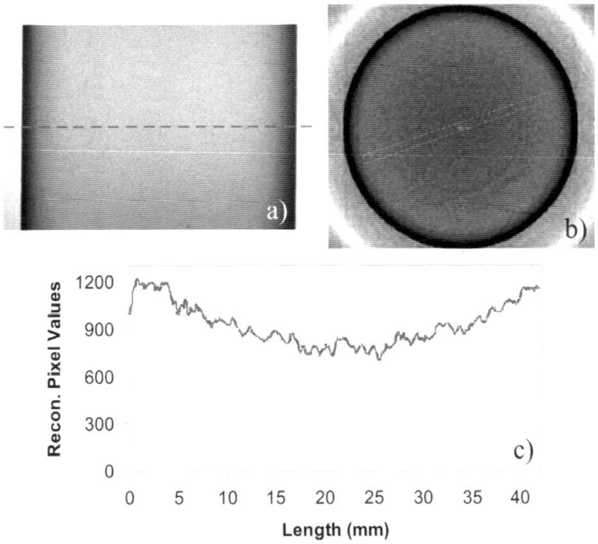

Figure 3. (a) A projection image of the unirradiated gel, (b) reconstructed axial slice at the level of the red dashed line, and (c) thick-line profile curve of the reconstructed values along the area within the white frame in (b).

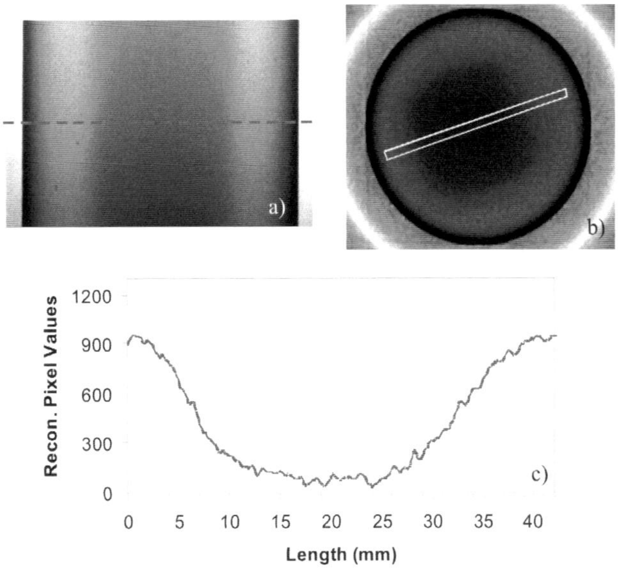

Figure 4. (a) A projection image of the 2 x 2 cm^2 6 MV beam irradiated gel, (b) a reconstructed axial slice at the level of the red dashed line, (c) a thick-line profile curve of the reconstructed values along the area within the white frame in (b).

At the zoom level setting used in this experiment the reconstructed image slices are approximately 1 mm thick. The full width at half maximum of the reconstructed fiber was 860 μm. The calculated MTF is greater than 10 % down to approximately 1.1 mm^{-1}, which is better than the 2 mm^{-1} limit prescribed by the ICRU for radiation therapy dosimetry. Better z-axis and x-y spatial resolution is anticipated at a higher zoom level.

4. Conclusion

The OPT apparatus presented in this study may operate without the need of cumbersome refractive index matching liquid baths. This allows the readout process to be performed in 'dry' conditions with the gel container surrounded only by air. Further studies are needed to evaluate the performance of the OPT system presented in this study. These studies need to evaluate various parameters such as the geometric accuracy and distortion, linearity, signal to noise and contrast to noise ratios, and the presence of image artifacts. The initial findings presented in this study indicate that 3D dosimeters might be potentially scanned without being immersed in a refractive index matching fluid facilitating their convenient use for the verification of routine radiation treatment plans.

References

[1] Sharpe J. Ahlgren U, Perry P, Hill B, Ross A, Hecksher-Sorenser J, Baldock R, Davidson D, "Optical projection tomography as a tool for 3D microscopy and gene expression studies," Science, 296, 541-545, 2002.

[2] Kerwin J, Scott M, Sharpe J, Puelles L, Robson SC, Martínez-de-la-Torre M, Ferran JL, Feng G, Baldock R, Strachan T, Davidson D, Lindsay S., "3-dimensional modeling of early human brain development using optical projection tomography, BMC Neurosci., 5, 27, 2004.

[3] Lasser T, Soubret A, Ripoll J, Ntziachristos V, "Surface reconstruction for free-space 360^0 fluorescence molecular tomography and the effects of animal motion," IEEE Trans Medical Imaging, 27, 188, 2008.

[4] Deliolanis N, Lasser T, Hyde D, Soubret A, Ripoll J, Ntziachristos V, "Free space fluorescence molecular tomography utilizing 360^0 geometry projections," Optics Letters, 32, 382, 2007.

[5] Sarma S, Kerwin J, Puelles L, Scott M, Strachan T, Feng G, Sharpe J, Davidson D, Baldock R, Lindsay S., "3D modelling, gene expression mapping and post-mapping image analysis in the developing human brain. Brain Res Bull., 66:449-53, 2005.

[6] Krstajic N, Doran S, "Characterization of a parrallel beam CCD optical-CT apparatus for 3D radiation dosimetry," Phys. Med. Biol. **52**, 3693-3713, 2007.

[7] Krstajic N, Doran S, "Fast laser scanning optical-CT apparatus for 3D radiation dosimetry," Phys. Med. Biol. **52**, N257-N263, 2007.

[8] Oldham M, Siewerdsen J, Kumar S, Wong J, Jaffray D, "Optical-CT gel dosimetry: Basic investigations," Med Phys 30, 623, 2003.

[9] Oldham M, Siewerdsen J, Shetty A, Jaffray D, "High resolution gel-dosimetry by optical-CT and MR scanning," Med Phys 28, 1436, 2001.

[10] Guo P, Adamovics J, Oldham M, "A practical three-dimensional dosimetry system for radiation therapy," Med. Phys. 33, 10, 2006.

[11] Sakhalkar H, Oldam M, "Fast, high-resolution 3D dosimetry utilizing a novel optical-CT scanner incorporating tertiary telecentric collimation," Med. Phys. **35**(1), 101-110, 2008.

[12] Papadakis A, Maris T, Zacharopoulou F, Pappas E, Zacharakis G, Damilakis J, "An evaluation of the dosimetric performance characteristics of N-vinylpyrrolidone based polymer gels," Phys. Med. Biol. **52**, 5069-5083, 2007.

[13] Walls J, Sled J, Sharpe J, Henkelman M, "Correction of artifacts in optical projection tomography," Phys. Med. Biol. 50, 4645-4665, 2005.

[14] Groen A, Young T, Lighthart G, A comparison of difference focus functions for use in autofocus algorithms, Cytometry, **6**, 81-91, 1995.

Scatter measurements for optical cone-beam computed tomography

Kevin Jordan and Jerry Battista

London Regional Cancer Program, London Health Sciences Centre and Department of Medical Biophysics, University of Western Ontario, London, ON, Canada

kevin.jordan@lhsc.on.ca

Abstract. Both x-ray and optical cone-beam CT (CBCT) scanners are limited by scattered photons that contaminate the primary images. Transparent radiochromic gels and plastics are anticipated to produce less scattering than polymerization gels and hence result in reconstructed dose distributions with greater contrast and dynamic range. Four methods of scatter measurement were investigated to quantify scatter fractions using an in-house optical CBCT scanner. The methods consisted of generating either "coin" shadows or primary beamlets in transmission images. The four methods generated similar results of 10% scatter fractions in gel transmission images for this particular scanner configuration with both leuco crystal violet and carbon black micelle gels. The scatter contributions were distributed as 5% from micelle gel, 2% from Teflon wall, 2% from refractive index matching liquid and 1% other sources. Applying a uniform background subtraction of 2% of open field intensity resulted in 0.2% difference between laser CT and cone-beam CT reconstructed attenuation coefficients for a black micelle gel finger phantom.

Introduction

Several geometries of optical computed tomography (CT) scanners have been developed for three-dimensional (3D) quantitative measurement of optical attenuation in radiation sensitive materials. Single-ray, single-detector scanners are efficient at rejecting scattered light, while with cone-beam optical CT (CBCT) low angle forward scattered light can contaminate signals on nearby CCD detector pixels. Polymerization gels form white, translucent images and initial evaluations of scatter have been reported for commercial laser CT[1-3] and CBCT[4,5] scanners. Scattered light contamination of transmission data is expected to be much less for transparent radiochromic gels and plastics relative to polymerization gels. For radiochromic ferrous xylenol orange gel, slightly lower contrast than expected was observed in a study comparing treatment plan calculation, EBT radiochromic film, laser and CBCT of gel for the same intensity modulated radiotherapy treatment[6]. An improved scatter correction would have resulted in contrast equivalence for the laser CT and CBCT reconstructions. Scattered radiation has been extensively investigated in order to improve contrast in images. With the introduction of x-ray CT scanners interest in evaluating sources of scatter radiation and its impact on quantitative imaging was renewed[7,8]. Currently work to quantify scatter in cone beam x-ray CT is being reported[9]. Several approaches for measurement of scatter with x-ray CBCT have been reported and methods to remove scatter prior to reconstruction are promising[10]. In this study, several methods of measuring scatter are compared for an in-house CBCT scanner based on the work of Wolodzko et al[11] . The objective is to examine different approaches to quantifying the scatter fraction (scatter/total) in optical CBCT. The quantitative scatter measures will provide important information for the design of scanners and scatter correction algorithms.

Methods and materials

A schematic of the in-house CBCT scanner is shown in Figure 1. The scanner components include: a diffuse light source, a refractive-index-matching-liquid-filled aquarium, gel in a 96 mm diameter PFA Teflon cylinder mounted on a rotational platform, and a CCD camera (Apogee Alta 47+, Tamron zoom lens). The lens (F11) was adjusted to provide a depth of focus that spanned the front and back of the gel cylinder. The distance from the rotation axis to lens aperture was 85 cm and the light diffuser face was 75 cm from the axis. The first scatter experiment involved placing a black disk, diameter 25 mm, at the face of the light diffuser, Figure 1a. The pixels from the central third of the shadow were compared to pixels outside but near the shadow. The second method consisted of drilling two 3 x 3 array of 10 mm holes into two black metal sheets. The pair of sheets were separated by 10 cm and formed an array of divergent, collimated beamlets, Figure 1b. Images with and without the collimator in place were compared. The third approach involved printing onto a transparent plastic sheet an array of 2 mm diameter dots on a 6 mm spacing and then placing the sheet at the entrance face of the aquarium, Figure 1c. This approach generated an array of shadows in the transmission image. The last approach consisted of drilling an array of 3 mm diameter holes with a 10 mm spacing into a metal black sheet and placing the sheet at the entrance face of the aquarium, Figure 1d. A pair of images with and without each of these sheets (a-d) were taken for each scatter geometry: air, aquarium with water, aquarium with refractive index matching liquid, aquarium with water and Teflon cylinder and finally gel in Teflon cylinder and refractive index matching liquid. The purpose of this sequence of measurements is to quantify the relative scatter contributions of the components placed in the beam path.

For CBCT scanning a series of 400 transmission images were acquired, in 12 minutes, as the gel is rotated through 360 degrees. Both a reference scan (pre-irradiation) and a data scan (post-irradiation) are acquired. Corrections for readout noise and scattered light and rebinning of divergent rays were applied before reconstruction using filtered back projection (Matlab function iradon). In this work the four methods of measuring scattered light were compared. A 3D CT scan of a black micelle gel finger phantom was also acquired, see Figure 2. The fingers had outer diameters of 25.4, 12.6 and 4.8 mm respectively. In this case the reference scan consisted of a similar jar filled only with colourless micelle gel. The clear, colourless, polyethlylene terephthalate jars (Modus Medical Devices, London Ontario) and were 9.3 cm diameter. A horizontal slice near the scanner's optical axis was examined for this study. Comparisons with the equivalent slice scanned with an in-house laser CT scanner were conducted. The laser scanner results are assumed to be "scatter-free" and without spectral hardening artefacts.

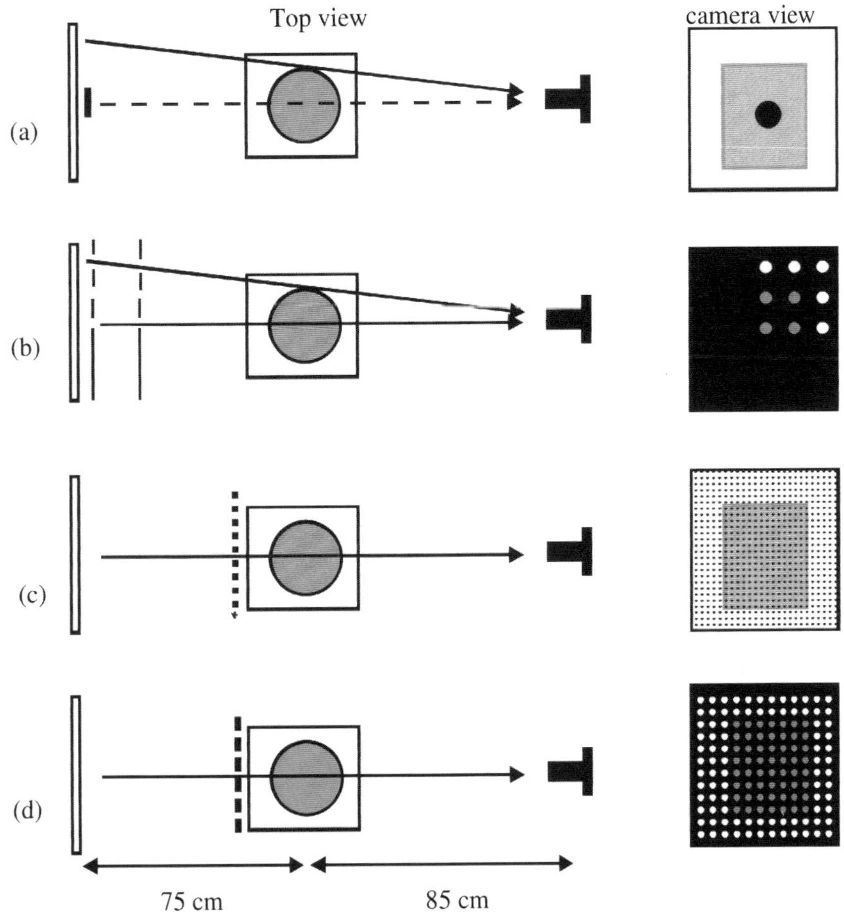

Figure 1. Left: Top view schematic of optical CBCT scanner (left to right; light diffuser, scatter sheet, aquarium, gel and camera). Right: transmission images for measuring scatter fraction.

Results and discussion

The results with the collimated beamlets, (geometry 1b) were examined fully initially. The "in air" measurement revealed the beamlets (primary) intensities were 99% of the open field values. The water-filled aquarium yielded the beamlet value to 99% and the Teflon cylinder walls further reduced the primary to 96%. The refractive index matching liquid, 8% propylene glycol in water reduced the primary to 93%. Finally the system with gel, (4 % gelatin, 4 mM Triton X-100 and 1 mM leuco crystal violet) reduced the primary to 88% along the cylinder diameter. Hence a scatter fraction of 12% was observed under these conditions.

Comparison of the primary fraction for the four scatter conditions revealed similar results within 2%, demonstrating near equivalence for this particular set of dimensions. Blocking large areas of the light field had little impact on the local primary measurements, demonstrating that low angle scatter is the dominant scatter component. Also, measurements indicated a near uniform background intensity as well. The magnitude of this background varied from 0.5% (shadow between central holes, geometry 1d) to 5% (outside open light field). Reconstructions were performed with different values of uniform background subtracted from all of the images prior to reconstruction. The attenuation coefficient was within 1% for each of the finger sizes, but 3.5% low relative to laser CT value if no uniform background subtracted. However, using a constant 2% uniform background subtraction the values with this correction were 0.3776 cm^{-1} (std 0.0095) for laser CT and 0.3783 cm^{-1} (std 0.0060) for CBCT.

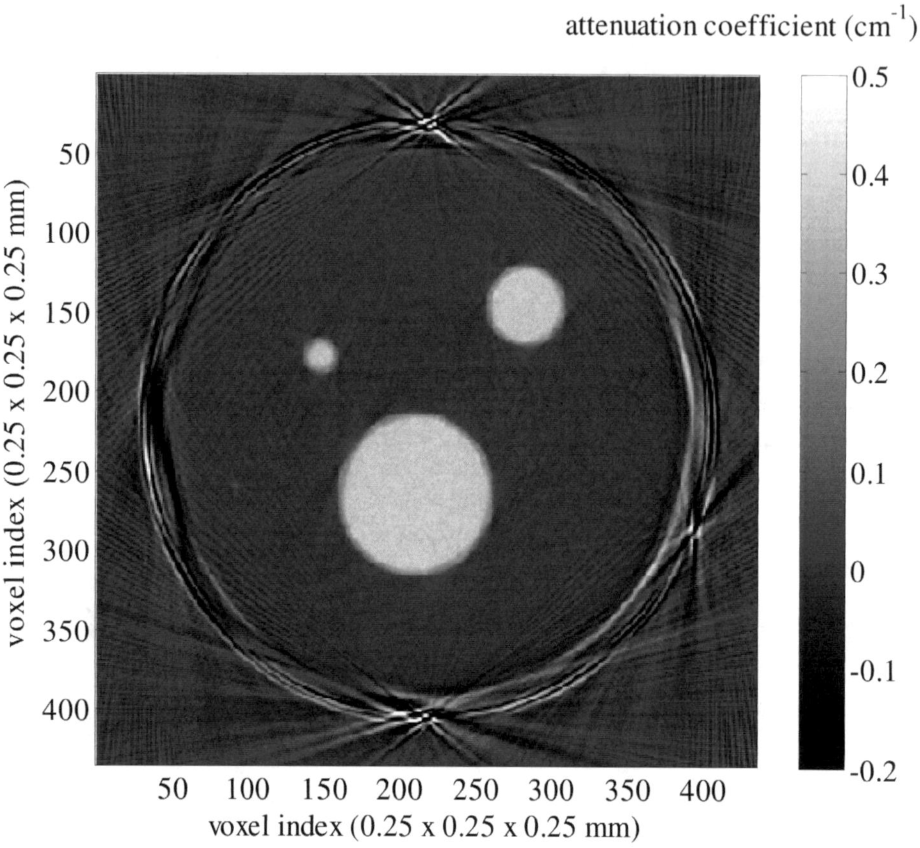

Figure 2. Reconstructed transverse slice of black micelle gel finger phantom. Note, CT streaking artefacts due to the two seams of plastic jar at top and bottom of image.

Conclusions

Several sources contribute to the scatter component of the transmission images. For this particular gel and scanner geometry, 90% of the image intensity was due to primary rays (and very small angle scatter). Placing a single black metal sheet with an array of 3 mm holes at the entrance face of aquarium(Figure 1d) resulted in scatter measurements which agreed with the more cumbersome, geometry specific, divergent array pair (Figure 1b). The closest agreement between laser CT and CBCT reconstructed attenuation coefficient occurred when subtracting a uniform background of approximately 2% of the open field from each image prior to taking ratios. This simple scatter correction resulted in reconstructed attenuation coefficients that agreed with laser CT results to within 1% for the same gel finger phantom set. Future work to determine a more accurate method to measure this uniform background correction, (possibly regional) will allow CBCT and laser CT comparisons to be performed with increased precision.

Acknowledgements: This work was supported by the London Regional Cancer Program Endowment Fund, Varian Research Fund, previous grant from NCIC and the Ontario Research and Development Challenge fund (ORDLF – OCITS project).
Disclosure: The authors have a licensing with Modus Medical Devices concerning optical cone beam CT scanning.

References

1. Maryanski M J, Zastavker Y Z and Gore J C 1996 Radiation dose distributions in three dimensions from tomographic optical density scanning of polymer gels: II. Optical properties of the BANG polymer gel *Phys. Med. Biol.* 41 2705-17

2. Xu Y, Wuu C S and Maryanski M J 2004 Performance of a commercial optical CT scanner and polymer gel dosimeter for 3-D dose verification *Med. Phys.* 31 3024-33

3. Oldham M 2004 Optical-CT scanning of polymer gels *J Phys: Conf Series* 3:122-35

4. De Jean P, Senden R, McAuley, Rogers M and Schreiner L J 2006 Initial experience with a commercial cone beam optical CT unit for polymer gel dosimetry I: Optical dosimetry issues J Phys: Conf Series 56 179-82

5. Bosi S, Pourandokht N, Puran A, Davies J and Baldock C, 2007 Initial investigation of a novel light-scattering gel phantom for evaluation of optical CT scanners for radiotherapy gel dosimetry, *Phys. Med. Biol.* **52** 2893-2903

6. Babic S, Battista J and Jordan K 2008 Three-dimensional dose verification for intensity-modulated radiation therapy in the Radiological Physics Centre head and neck phantom using optical computed tomography scans of ferrous xylenol orange dosimeters *Int. J. Rad. Oncol. Biol. Phys.* 70 1281-1291

7. Joseph P M and Spital R D 1982 The effects of scatter in x-ray computed tomography *Med. Phys.* 9 64-72

8. Johns PC and Yaffe M, 1982 Scattered radiation in fan beam imaging systems, Med Phys 9:231-39

9. Kyriakou Y, Meyer M and Kalender W A 2008 Technical note: comparing coherent and incoherent scatter effects for cone-beam CT, Phys *Med Biol* 53 N175-85

10. Rinkel J, Gerfault L, Esteve F and Dinten J-M 2007 A new method for x-ray scatter correction: first assessment on a cone-beam CT experimental setup, Phys *Med Biol* 52 4633-52

11. J. G. Wolodzko, C, Marsden and A. Appleby 1999 CCD imaging for optical tomography of gel radiation dosimeters, *Med. Phys.* 26 2508-13

Journal of Physics: Conference Series **164** (2009) 012029 doi:10.1088/1742-6596/164/1/012029

Optical CT scanning of cross-linked radiochromic gel without cylinder wall

Kevin Jordan

London Regional Cancer Program, London Health Sciences Centre and Department of Medical Biophysics, University of Western Ontario, London, ON, Canada

kevin.jordan@lhsc.on.ca

Abstract. Genipin cross-linked gelatin hydrogels with 0.2 M sulphuric acid are radiochromic and have sufficient sensitivity for investigating doses less than 50 Gray. Because of the gelatine cross-links, these gels have sufficient strength to allow removal from the vessel container in which they were cast. Placing the gels in the same liquid that was used for preparing the gel allows the radiochemistry to the same throughout the gels and provides physical support. In this buoyancy neutral environment the gel has the same shape and the preparation vessel. This allows optical CT scanning without wall artefacts due to reflection, refraction and optical activity. A gel was irradiated to dose of 25 Gray with a 10 MV photon beam of 20 x 20 mm cross section. Full 3D optical CT scanning was performed with a Vista 10 optical cone-beam CT scanner. Central beam axis profiles and depth dose agree with diode and parallel plate ion chamber measurements. These results demonstrate that genipin cross-linked gel can be used for accurate 3D dosimetry, including surface dose measurements.

1. Introduction

Optical CT scanning of radiation sensitive materials is complicated by the need for vessel walls to provide mechanical support for hydrogels and chemical isolation from the environment. For example, polymer gels such and BANG and PAG require isolation from oxygen and ferrous xylenol orange gelatine gels (FX) require a uniform oxygen concentration. Removing the vessel wall from an FX gel and immersing the gel in a FX solution permits wall-less scanning, but this approach can only be used at low temperatures, and dose response is not uniform due to the dissolved oxygen gradient near the gel surface. Also, additional materials for refractive index matching of the surrounding liquid can diffuse into the gel and modify the dose response near the gel surface. Finally, the gelatin itself will slowly dissolve into the surrounding liquid.

Genipin cross-links many types of polymers to form hydrogels. Gelatin is one example. The resulting cross-linked gels are blue in colour and can have much higher melting points than analogous gelatin gels that rely on hydrogen bonding. By lowering the pH of these cross-linked, genipin gels, transparent, radiochromic materials can be formed that have sufficient rigidity and dose sensitivity for radiotherapy dosimetry[(1)]. These materials have several desirable features for 3D radiation dosimetry. In particular, stable dose images both in space and time are formed immediately following irradiation. The gelatin cross-linking also prevents the gel from dissolving into the external liquid.

Removal of the vessel wall eliminates some refraction, reflection and optical activity artefacts[(2-4)]. True refractive index matching also allows the optical CT scanning of complex 3D shapes that can be cast into any shaper with hydrogels. Currently, the only material that has been practical to optically

CT scan, without walls, has been the radiochromic plastic, Presage[5] . Having a material with uniform dose response throughout its entire volume (i.e. up to surfaces) and uniform optical properties would dramatically increase the range of dosimetry problems that could be accurately studied in hydrogels.

2. Materials and methods

Genipin was purchased from the WACO Chemical Company of Japan. Porcine gelatin, 300 bloom and sulphuric acid (ACS reagent grade) were purchased from Sigma-Aldrich. Two percent by mass gelatin was dissolved in water at 43°C in a water bath. Genipin, $2x10^{-5}$ by mass was added. Over 18 hours of continual stirring, the solution changed from colourless to blue. Sulphuric acid was added to give a concentration of 0.2 M in the final gel. Other than a slight dilution of colour intensity, no spectral change was observed as the acid concentration increased. Beakers were coated with a thin film of petroleum 'jelly' grease to act as a mould release agent and then filled with gel. Samples were stored at 4°C for two days prior to scanning. Gels were released by adding 0.2 M sulphuric acid, with a syringe and tube, between the gel and beaker wall. The beaker was turned nearly upside down and tapped on the bench top. The intact gel slid out and was placed into a modified Vista (Modus Medical Devices, London, Canada) PETE jar half-filled with 0.2 M sulphuric acid. The bottom of the jar had been cut off and a jar lid inserted and taped in place. As a means of holding the gel from moving two small pins were inserted through the jar bottom. A nylon strap fixed as an inverted arch gently pressed the gel to the base plate and pins. The strap also prevented wobbling of the gel top during CT rotation. Pre and post irradiation optical CT scans were collected with the Vista10 (second generation optical cone-beam CT (CBCT) scanner from Modus Medical Devices) at 590 nm. A scan time of 8 minutes was required for a full 3D scan of 512 images over 360 degrees. Reconstruction of a 256^3 array required an additional 12 minutes at 0.5 x 0.5 x 0.5 mm voxel size. The gel was irradiated in an anterior to posterior geometry with a 10 MV photon beam, source to gel top distance of 100 cm and a beam size of 20 x 20 mm at the surface. The liquid in the jar was adjusted to provide a level surface with the edge of the gel and approximately 1 mm of coverage at the gel centre. The nylon restraining arch also slightly depressed the centre of the gel top, estimated to be less that 1 mm of curvature. A dose of approximately 25 Gray was delivered to the gel, at 25 mm depth. The post irradiation scan began 10 minutes after the irradiation was completed. All measurements were performed at a temperature of 21°C.

3. Results and discussion

Figure 1 is a transmission image recorded by the Vista10 CBCT scanner. The gel appears blue and transparent with white light illumination but has low transmission for yellow light. The gel had a density slightly greater than that of water and was essentially buoyancy neutral in water.

Figure 2 is a reconstructed slice at a depth of 20 mm below the top of the gel. The cross section of a 20 by 20 mm photon beam is evident. Note that the gel has moved laterally between the pre and post scans due to handling. The gel had sufficient strength to remain intact when removed from the casting beaker and transferred by hand into the modified Vista jar for scanning. However, the movement associated with handling between pre and post scan caused the small pins through the base to cut the gel and allow it to settle approximately 1 mm from the original position. In order to prevent this shifting from occurring in future scans, a base ring mount will be designed to provide a larger surface area to restrict the gel position. Without the top pressure to hold gel onto the base pins the gel would have floated out of position with the slightest disturbance. Ideally the gel edge should be not detectable in the reconstruction if there is no absorbed dose.

Figure 3, compares profiles at a 20 mm depth for the gel and a diode detector scanned in water. The sharp peak at +20 mm is the edge of the gel that shifted between pre and post scans. The features at -23 mm and -3 mm are reconstruction artefacts that can be seen as rings in the slice presented and spheres in the 3D data set. This artefact occurs near the scanner's optic axis.

Figure 4 shows the percent depth dose from merged data sets recorded with a parallel-plate ion chamber near surface and diode detector beyond 20 mm. For comparison, normalized attenuation coefficients from the full 3D reconstructed dose distribution are also plotted. A depth offset of 1 mm was applied to the gel data in order to improve agreement. The amount of water covering the gel during irradiation had been visually estimated to be near 1 mm. However, no accurate depth reading was obtained for this feasibility study. The agreement along the entire curve demonstrates the gel dose response linearity, water equivalence and refractive index matching. Since all are required for good depth-dose matching. Without refractive index matching, the gel could not have been scanned to within 2 mm of the surface. Future scans where the top of the gel in only in contact with water will dramatically reduce artefacts in reconstruction very near the surface.

4. Conclusions

Cross-linked genipin gelatin hydrogels were produced with adequate dose sensitivity and mechanical rigidity for optical CT scanning in water with the vessel wall removed. The gel was minimally restrained within a water bath that provided uniformity for irradiation, refractive index matching for scanning and neutral buoyancy for support of the gel. Excellent agreement between depth dose measured with parallel-plate ion chamber and gel near beam entrance surface demonstrate that near surface dose measurements can be obtained with wall-less optical cone beam CT scanning of this gel.

Acknowledgements: This work was supported by the London Regional Cancer Program Endowment Fund and the Ontario Research and Development Challenge fund (ORDCF-OCITS project).
Disclosure: K Jordan has a licensing agreement with Modus Medical Devices concerning optical cone-beam CT scanning.

References

[1] Jordan K, June 2008 Zero diffusion radiochromic genipin-gelatin dosimeter for three-dimensional radiation dosimetry, *Conference Proceedings COMP annual meeting*,

[2] Jordan K 1999 Developmental Issues for optical CT and gel dosimetry, *Dosgel'99 Conference proceedings*

[3] Jordan K 2004 Advances in optical CT scanning for gel dosimetry, *J. Phys.: Conference Series* **3** 115-21

[4] S. J. Doran, K. K. Koerkamp, M. A, Bero, P. Jenneson, E. J. Morton and W. B. Gilboy 2001 A CCD-based optical CT scanner for high-resolution 3D imaging of radiation dose distributions: equipment specifications, optical simulations and preliminary results, *Phys. Med. Biol.* **46** 3191-3213

[5] Adamovics J, Maryanski MJ, 2006 Characterization of Presage: A new 3-D radiochromic solid dosimeter for ionising radiation, *Radiat Prot Dosimetry* **120** 107-12

Figure 1. Transmission image of genipin gel, without containment vessel, immersed in a 0.2 M sulfuric acid solution. The gel was contained inside a plastic jar for optical CT scanning.

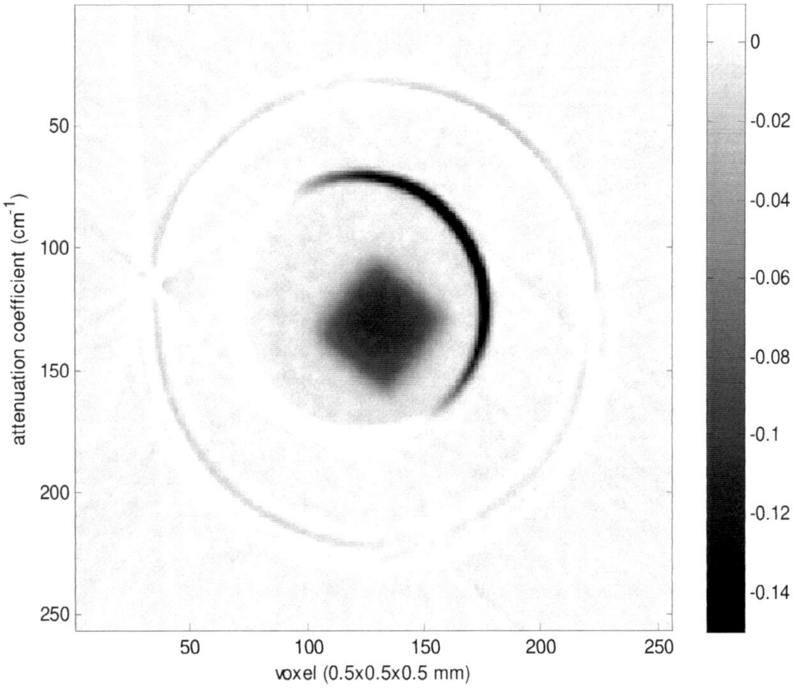

Figure 2. Reconstructed image of change in attenuation coefficient at 590nm, transverse plane at 20 mm depth due to 25 Gy of absorbed dose from 10 MV photon beam. Note the bright and dark crescents at edge of gel are due to lateral movement of gel between pre and post-irradiation scans.

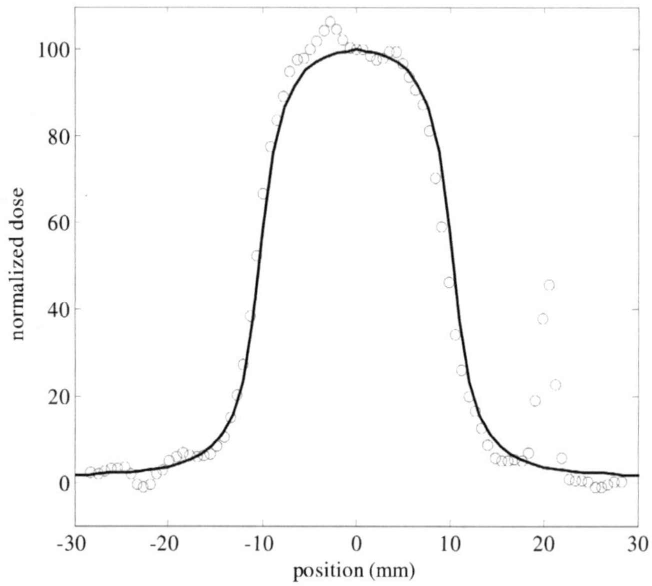

Figure 3. Normalized dose profiles at 25 mm depth (diode scan in water (solid line), gel (o)).

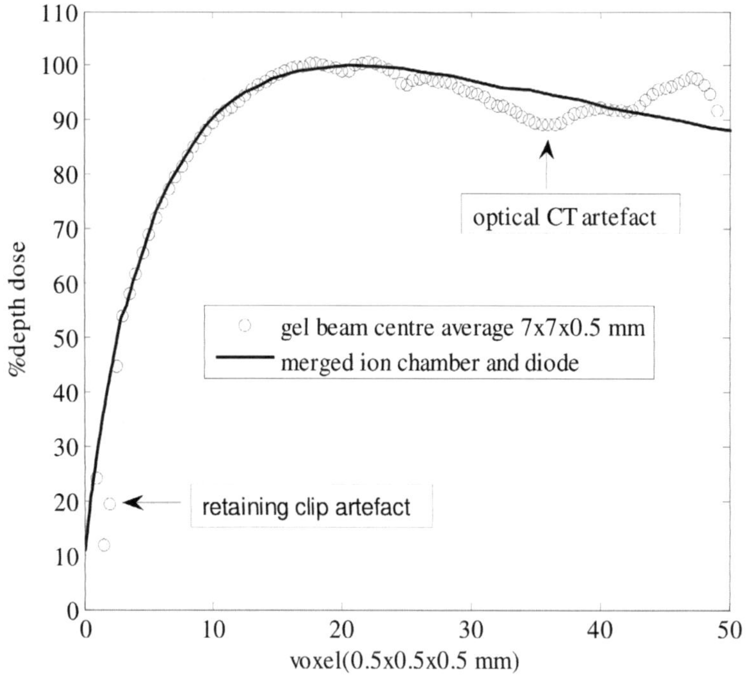

Figure 4. Central axis normalized depth doses (merged parallel-plate ion chamber and diode measurement (solid line), gel (o)).

Development of a Quality Assurance Scattering Phantom for Cone Beam Optical CT

Tim Olding[1], Oliver Holmes[1], L. John Schreiner[2,3]

[1]Department of Physics, Queen's University

[2]Medical Physics Department, Cancer Centre of Southeastern Ontario

[3]Departments of Physics and Oncology, Queen's University, Kingston, ON, Canada

Tim.Olding@krcc.on.ca

Abstract. Standard water-based acrylic emulsion scattering solutions have been developed for use in optical scattering gel phantoms. These solutions display good uniformity and long term stability, and can be index matched to polymer gels through the addition of propylene glycol or molten gelatin. Preliminary quality assurance (QA) performed on the Vista cone beam optical computed tomography (OptCT) scanner (Modus Medical, London, ON) using these standard scattering solutions provides valuable information toward the establishment of protocols for scanner warm-up and dosimeter/bath temperature control. A characterization of Vista scanner response over a range of scattering concentrations indicates that improved dosimeter performance may be realized through lowering the %T concentration of the polymer gel into the linear response range of the scanner A gel molding process has been developed for these scattering solutions, with promising results obtained for early stage scattering gel phantoms.

1. Introduction

An important step in the qualification of a new optical CT scanner is to separate the evaluation of optical measurement integrity and performance limitations from the uncertainties in dose delivery and gel dosimetry. Independent optical absorption [1,2] and optical scattering [3] phantoms have previously been developed toward this end. A scattering phantom is particularly useful in the characterization of OptCT scanner intended for use in polymer gel dosimetry. In this paper we describe the characterization of novel calibration scattering solutions, initial scattering gel phantom development, and an associated performance evaluation of the VISTA cone beam optical CT scanner (Modus Medical, London, Canada) for use in polymer gel dosimetry.

2. Materials and Methods

A low cost water-based acrylic emulsion has been identified that has a comparable average particle size (~350 nm) to that that of polymer gel dosimeters [1], a reasonable index of refraction (~1.5) and which exhibits good solution stability over time. This emulsion (Duramax B-1000, Rohm & Haas) is easily diluted in propylene glycol-water mixtures and maintains uniformity of colloidal distribution in heated gelatin solutions. Acrylic scattering solutions incorporating 12% propylene glycol by weight for refraction matching were prepared in 1 L polyethylene terephthalate (PET) jars (9.2 cm in diameter) and evaluated using the VISTA cone beam OptCT scanner. Reference and data scans were taken using 633 nm LED illumination and a 1024x768 pixel CCD camera (410 projections over 360°). Reconstructions were completed via Feldkamp backprojection with a Hamming filter to a 0.5 mm voxel size. Portions of the scattering solutions were separated and also measured using an Ultrospec 1000 UV-VIS spectrophotometer (Amersham Pharmacia, Uppsala, Sweden).

Heated acrylic scattering solutions containing 5% gelatin by weight and varied amounts of propylene glycol for refraction matching were cast into both gelatin and machined acrylic molds forming usefully shaped and sized scattering phantoms for evaluation of scanner performance capabilities and for comparison to similar structures in polymer gel dosimeters.

N-isopropylacrylamide (NIPAM) based polymer gels were prepared in the manner according to Senden *et al.* [4], with 50 %C and total monomer and crosslinker (%T) in the range of 4-6 wt%. Gels were poured into 1 L polyethylene terephthalate (PET) bottles for imaging and irradiated using a T780 Cobalt-60 tomotherapy benchtop (MDS Nordion, Kanata, Canada) approximately 24 hours post-manufacture using 1x1cm^2 pencil beams.

3. Results and Discussion

Varied concentration acrylic scattering solutions containing 12 wt.% propylene glycol (PG) show good long term stability and uniformity over a period of the seven months to date. Linear regression fits to mean attenuation coefficient against concentration of scattering solution at 0 months and seven months agree within error (Figure 1). These acrylic scattering solutions can then be used as standards for scanner QA and to optimize the dosimetry of polymer gels against scanner performance.

The compatibility of the acrylic emulsions with heated gelatin solutions allows for the development of scattering gel phantoms that can be used for rapid calibration of the scanner. A five finger scattering phantom with 0 wt.% propylene glycol-based scattering solutions in the scattering regions (Figure 1 bottom left) provides a single scan assessment of VISTA scanner contrast linearity that could be used in daily scanner QA. Two additional effects are worthy of mention. First, an obvious difference is observed between the mean attenuation coefficient of the 0 wt.% and 12 wt.% scattering solutions at equivalent scattering concentrations. This is primarily sourced from the addition of propylene glycol, which increases the refractive index of the solution relative to that of the scattering particles, leading

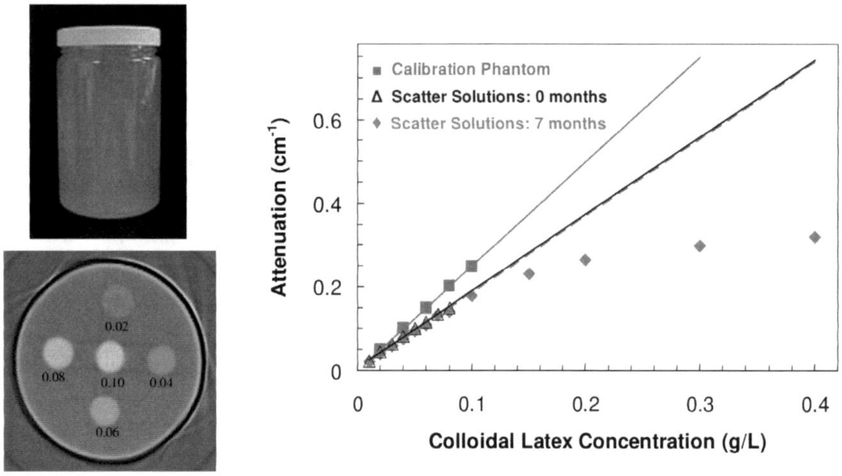

Figure 1: OptCT-derived mean attenuation coefficients of standard acrylic scattering solutions (top left) measured at 0 months (right, in blue) and 7 months (right, in pink) agree within error. A simple gel phantom (bottom left) with several scattering fingers derived from acrylic scattering solutions (0% wt. propylene glycol) is used to generate a scanner calibration curve (right, in green). Error bars are smaller than symbol size.

to reduced scatter and hence, a reduction in reported mean attenuation coefficient [5]. Second, the effect of multiple scattering is clearly observed for the larger scattering volume 1 L PET jars in the departure from linearity of mean attenuation coefficient at higher scattering concentrations.

Examples of the use of standard scattering solutions for scanner QA are illustrated in Figures 2 and 3. A time trial involving hourly scans of a 0.1 g/L standard scattering solution reveals an instability in the mean attenuation value calculated from the reconstructed images over 4 hours of scan time (Figure 2). To help determine the source of the instability, the mean intensity value was observed over time within the scattering (jar) region of the projection bitmaps. The reported mean intensity value is nearly identical when both the light source and panel are allowed to warm up as when just the light source is turned on for 3 hours while the camera is blocked (CB) from light exposure. This suggests that the camera is the primary source of the time dependent drift in reported intensity value. The camera response stabilizes roughly within the first 2 hours of camera warm up, with an overall apparent reduction up to 3% reported in average attenuation value at steady state. A second experiment in which the light source is switched off for 15 minutes then turned on again (Figure 3 left) illustrates the importance of even a small power interruption on scanner operation. Temporal effects can be minimized by setting protocols requiring a minimum warm up time of 2-3 hours for the scanner and ensuring that both the camera and light source remain on between reference and data scans.

A second example of the use of standard scattering solutions for procedural QA is shown in Figure 3. In this experiment, a 0.1 g/L scattering solution was cooled to 10°C before scanning. It was found that the reconstructed mean OptCT attenuation coefficient was elevated by more than 3% compared to a room temperature scan of the same scattering solution. This effect is likely due the increasing difference between the refraction index of the solvent and of the scattering particles with decreasing temperature. An evaluation of the mean intensity value within the jar region from the projection bitmaps (as in Figure 2) shows the increase in intensity as the jar warms up (Figure 3 right).

The information gained from this type of scattering phantom trial is important for the establishment of scanner quality assurance protocols. Acrylic emulsion based scattering solutions and scattering gel phantoms provides a calibration of OptCT scanner performance that is independent of dose or spatial uncertainties that could arise with the use of irradiated polymer gels for QA.

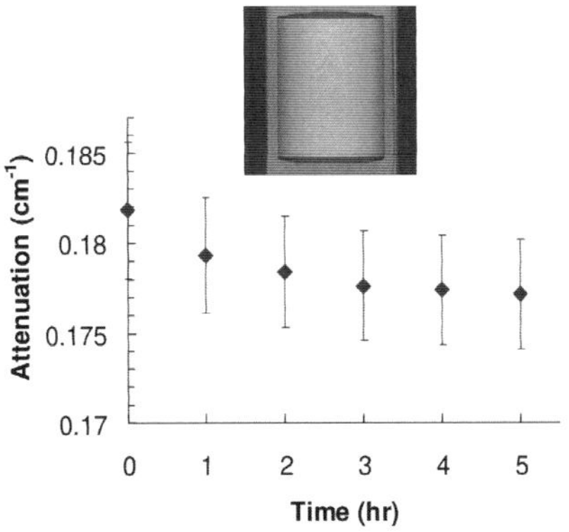

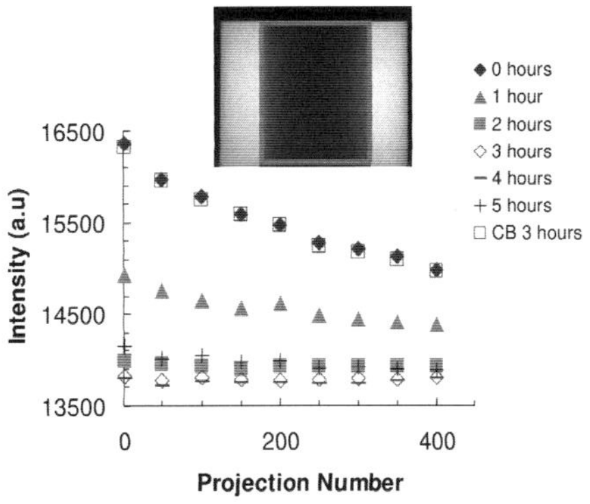

Figure 2: The mean reconstructed attenuation coefficient is determined from reconstructed scans over time of a 0.1 g/L standard scattering solution(left), with selected region of interest (ROI) shown in yellow. The mean intensity value was observed within the jar region on the projection bitmaps over a period of 5 hours (right, ROI in red) for simultaneous power up of camera and light source, and for the situation where the camera was blocked during the first 3 hours after light source power up (CB 3 hours, right).

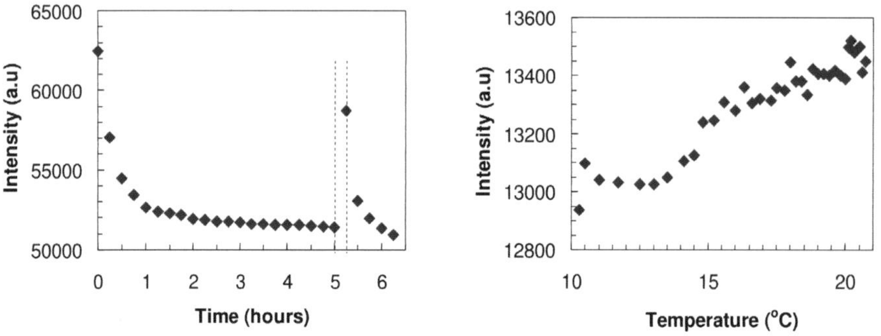

Figure 3: A 15 minute power off of the OptCT scanner light source causes a significant change in camera pixel response (left), observed from the mean intensity value. Re-stabilization of the camera response occurs within 30 minutes. A temperature trial shows an increase in mean intensity value from projection data with increased temperature (right). The mean intensity value ROIs were the same as the one described in Figure 2.

As stated previously, the effects of multiple scattering are clearly observed in scattering solution OptCT data (Figure 1). This highlights the need for a scatter correction strategy to increase the range of the cone beam OptCT scanner for polymer gel dosimetry. However, a simple adjustment of gel chemistry to lower the sensitivity into the linear performance range of the scanner significantly improves the dosimetry for present use (Figure 4). A gamma map of 500 cGy intersecting pencil beam treatment plan delivered to a standard 6%T dosimeter indicates regions of significant failure in the dosimetry for 3% 3mm gamma criteria. From the data obtained in scattering gel phantom QA we have independently determined that these regions correspond to operation outside the linear performance range of the OptCT scanner. The dosimetry is improved by changing the polymer gel preparation to 4%T to restrict dosimeter behavior in a well-behaved region. Then 3D readout shows that actual dose delivery passes the 3% 3mm criteria for the 4%T dosimeter. This approach unfortunately does involve a loss of dose sensitivity through the %T reduction. Scatter correction strategies explored further in an accompanying paper in these proceedings may allow polymer gel sensitivity to be raised back to the level of the standard 6%T dosimeter.

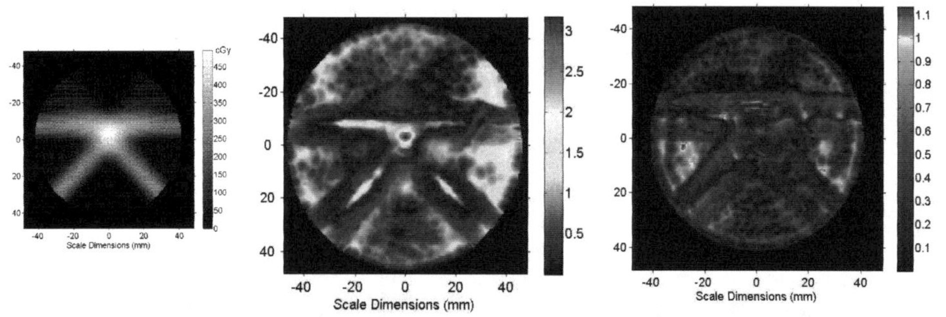

Figure 4: Gamma maps of 500 cGy intersecting pencil beam treatment plan (left) delivered to a standard 6%T dosimeter (centre) and a 4%T dosimeter (right).

Gelatin molding techniques can be employed in combination with a cross-linker to set gelatin shapes such that they cannot be re-melted. Unique scattering shapes (Figure 5) can be realized that have a background scatter concentration representing the base scattering level of the polymer gel and an elevated scatter concentration representing dose delivery to that volume. The phantom in Figure 5 roughly mimics a 500 cGy delivery to a 4%T polymer gel dosimeter with a "background" scatter concentration of 0.025 g/L and a "dose" scatter concentration of 0.10 g/L. The ability to produce such phantoms is anticipated to be a powerful tool in the assessment of OptCT performance capabilities and the establishment of quality assurance protocols.

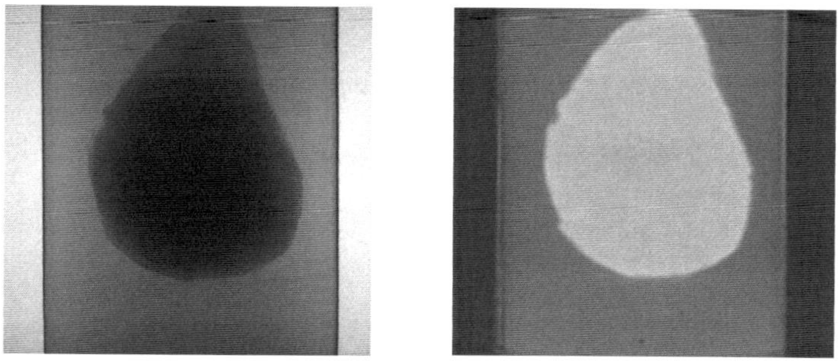

Figure 5: Raw projection (left) and OptCT reconstructed images (right) of a prototype crosslinker-stabilized gel phantom mimicking a 500 cGy delivery to a 4%T polymer gel dosimeter.

4. Conclusions

Independent scattering solutions with good uniformity and long term solution stability have been developed to validate the integrity of the optical CT measurements and investigate different means of multiple scatter reduction in cone beam OptCT. These scattering solutions are compatible with heated gelatin solutions for forming rigid scattering shapes that are suitable for incorporation into a gel phantom for cone beam OptCT quality assurance.

5. Acknowledgements

Research funding has been provided by the Cancer Centre of Southeastern Ontario, the Ontario Research and Development Fund and the Canadian Institutes of Health Research (CIHR).

6. References

[1] Oldham M, Siewardsen J H, Kumar S, Wong J and Jaffray D A 2003 Optical-CT gel dosimetry I: basic investigations *Med. Phys.* **30**, 623-634

[2] Guo P, Adamovics J and Oldham M 2006 Simple 3D validation experiments for PRESAGE^{TM}/optical-CT dosimetry *J. Phys.: Conference Ser.* **56**, 187-190

[3] Bosi S G, Naseri P, Puran A, Davies J and Baldock C 2007 Initial investigation of a novel light-scattering gel phantom for evaluation of optical CT scanners for radiotherapy gel dosimetry *Phys. Med. Biol.* **52**, 2893-2903

[4] Senden R J, DeJean P, MacAuley K and Schreiner L J 2006 Polymer gel dosimeters with reduced toxicity : a preliminary investigation of the NMR and optical dose-response using different monomers *Phys. Med. Biol.* **51**, 3301-3314

[5] Franklin J and Wang Z Y, 2002 Refractive Index Matching: A general method for enhancing the optical clarity of a hydrogel matrix *Chem. Mater.* **14**, 4487-4489.

Scatter Corrections for Cone Beam Optical CT

Tim Olding[1], Oliver Holmes[1], L. John Schreiner[2,3]

[1]Department of Physics, Queen's University

[2]Medical Physics Department, Cancer Centre of Southeastern Ontario

[3]Departments of Physics and Oncology, Queen's University, Kingston, ON, Canada

Tim.Olding@krcc.on.ca

Abstract. Cone beam optical computed tomography (OptCT) employing the VISTA scanner (Modus Medical, London, ON) has been shown to have significant promise for fast, three dimensional imaging of polymer gel dosimeters. One distinct challenge with this approach arises from the combination of the cone beam geometry, a diffuse light source, and the scattering polymer gel media, which all contribute scatter signal that perturbs the accuracy of the scanner. Beam stop array (BSA), beam pass array (BPA) and anti-scatter polarizer correction methodologies have been employed to remove scatter signal from OptCT data. These approaches are investigated through the use of well-characterized phantom scattering solutions and irradiated polymer gel dosimeters. BSA corrected scatter solutions show good agreement in attenuation coefficient with the optically absorbing dye solutions, with considerable reduction of scatter-induced cupping artifact at high scattering concentrations. The application of BSA scatter corrections to a polymer gel dosimeter lead to an overall improvement in the number of pixel satisfying the (3%, 3mm) gamma value criteria from 7.8% to 0.15%.

1. Introduction

The effects of multiple scatter on the measurement integrity of optical computed tomography (OptCT) for polymer gel dosimetry is an important issue that remains to be addressed. From preliminary work in characterizing the VISTA OptCT scanner [1,2] it appears that optical scatter effects are not unlike in-scatter effects seen in x-ray CT [3]. Different pre-processing and post-processing schemes have been employed in cone beam x-ray CT with varying degrees of success [4] One strategy employs anti-scatter grids to reject X-ray scatter signal prior to image acquisition [5,6]. Another approach is the beam stop method, where an array of opaque disks is added at the beam input side of the scanner [7,8]. Scatter data is then acquired and interpolated to form scatter maps that are subtracted from the image projections prior to tomographic reconstruction. This method has been used in the calculation of stray light values for optical absorption in patent blue violet dye solutions [9], but has yet to be applied to optically scattering media. In this work, we describe the application of anti-scatter polarizers, a beam stop array (BSA) and a beam pass array (BPA) toward the removal of scatter signal from OptCT data. Scatter corrections for standard well-characterized scattering solutions (described in an accompanying paper in these proceedings) and polymer gel dosimeters are investigated.

2. Materials and Methods

Standard scattering solutions based on Duramax B-1000 colloidal latex binder and incorporating 12 wt.% propylene glycol have been prepared in 1 L polyethylene terephthalate (PET) jars. The solutions were evaluated using the VISTA cone beam OptCT scanner and with: (a) a 4mm diameter dot BSA inserted at the front face of the aquarium (see Figure 1), (b) polarizing films placed at the front and back faces of the scanner aquarium, or (c) a 2mm diameter pinhole BPA inserted at the front face of the aquarium. Reference and data scans were taken using 633 nm LED illumination and a 1024x768 pixel CCD camera (410 projections over 360°, 15 minutes per scan). Image reconstruction was completed using Feldkamp backprojection with a Hamming filter, to a 0.5 mm voxel size.

N-isopropylacrylamide (NIPAM) based polymer gels were prepared according to Senden *et al.*[10] (50 %C, 4%T). Gels were poured into 1 L polyethylene teraphthalate (PET) bottles for imaging and irradiated using a T780 Cobalt-60 tomotherapy benchtop (MDS Nordion, Kanata, Canada) approximately 24 hours post-manufacture using 1x1cm^2 pencil beams. OptCT imaging of the polymer gel dosimeters was completed using the BSA method.

3. Results and Discussion

Both anti-scatter polarizers and BSA corrections yield an improvement in the overall linearity of the VISTA OptCT scanner through rejection of multiple scatter from the signal (Figure 1). However, the anti-scatter polarizer method did not remove the cupping artifact observed at higher solution concentrations despite the appearance of linearity. This method also introduces additional optical artifacts in the projection images that compromise the measurement integrity in some locations.

In comparison, the BSA method extends the range of scanner measurement integrity with significant cupping artifact reduction (Figure 2). As the beam stop array reduces the overall primary

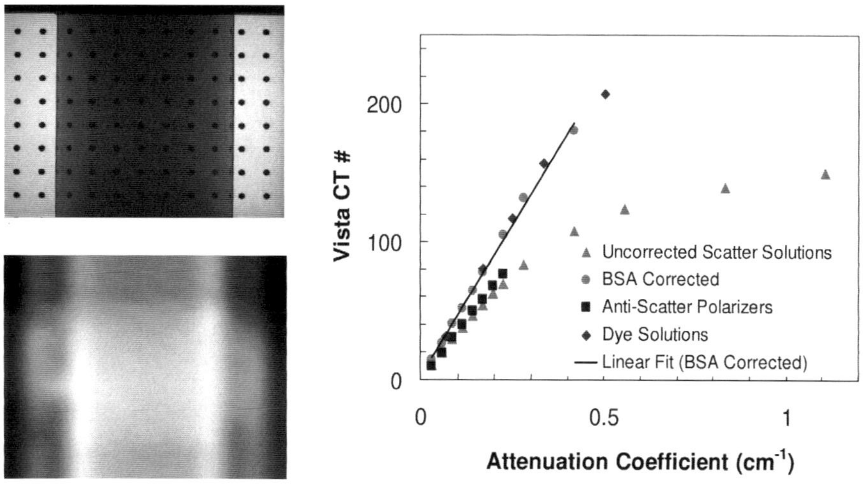

Figure 1: A 4mm diameter dot BSA was inserted at the front of the aquarium (top left) and average scattered light values determined at the centre of the blockers. These values were used to develop interpolated scatter maps (bottom left) which are subtracted from the reference and data projections prior to reconstruction.. Mean reconstructed VISTA CT numbers (right) are determined from a 80 mm diameter, 80 mm high cylindrical region of interest (ROI) along the central axis of the PET jar. Attenuation coefficients are determined from spectrophotometer measurement of the same solution. Error bars are smaller than symbol size.

signal, this is not an exact correction strategy (see Figure 3). However, the corrected scatter solutions show good agreement with the optically absorbing dye solutions in the relationship between attenuation coefficient and VISTA CT number. A finer array dot spacing than the one used in this experiment (16 mm spacing between 4 mm diameter dots) with smaller dots would provide a more

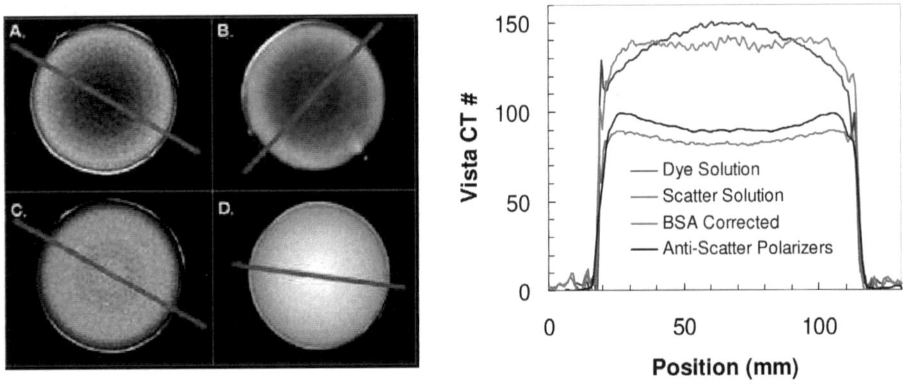

Figure 2: The application of a BSA scatter correction to a uniform 0.1 g/L phantom scattering solution in a 9 cm diameter PET jar results in a significant reduction (left, C) in the observed cupping artifact (left, A) and overall increased average attenuation value (right). Polarizer corrected data (left, B), shows an increased average attenuation value, but without significant cupping artifact reduction. Highly absorbing uniform patent blue violet dye solutions display a bowed attenuation profile (left, D) consistent with results of previous work by Jordan *et al.* [9].

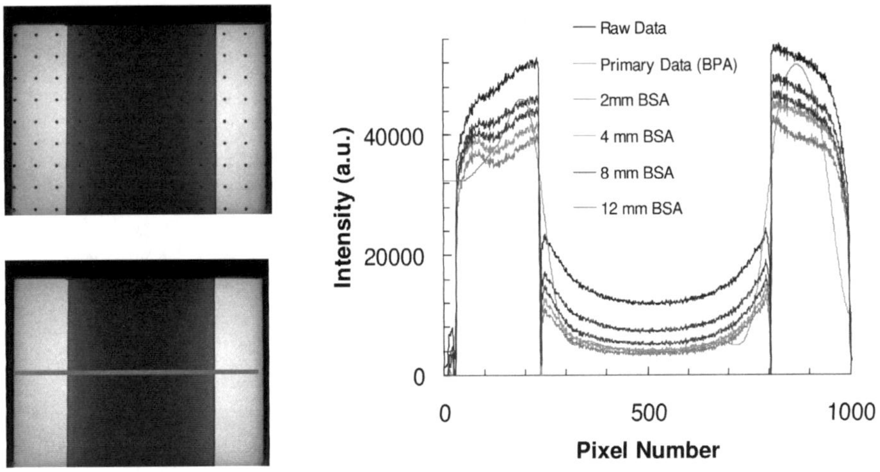

Figure 3: Different sized diameter dot beam stop arrays such as the preferred 2 mm array (top left) were inserted at the front of the aquarium and average scattered light values determined at the centre of the blockers. These values were used to develop interpolated scatter maps (as in Figure 1) which are subtracted from the reference and data projections prior to reconstruction. A reduction in scatter signal estimation is observed with increasing blocker dot size (right) along a line in the projection image (bottom left). This result is from increased area blocking of the light source.

accurate scatter map correction near the jar edges and at the central axis, with the additional benefit of reduced primary signal loss from the beam stop array (Figure 3). The 4 mm BSA blocks close to 5% of the light source intensity, whereas an equivalent 2 mm BSA would block only 1%.

Application of the BSA scatter correction method to polymer gels irradiated with a pencil beam calibration pattern leads to considerable improvement in the dosimetry (Figure 4). The middle third of the gel dosimeter is particularly susceptible to scatter artifacts. In a comparison of the calculated reference and OptCT measured dose distributions, 2D gamma values greater than 1 may be observed in 7.8% of the pencil beam calibration pattern pixels within this region. This can be reduced to less than 0.2% pixel failure with a BSA scatter correction, even in gels irradiated with highly scattering pencil beams.

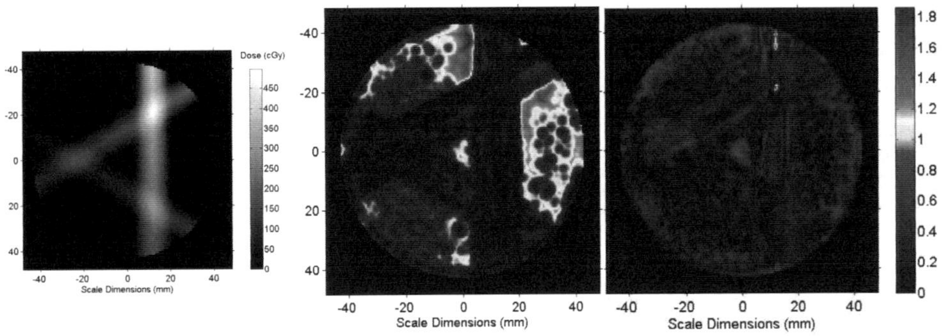

Figure 4: A 4%T polymer gel dosimeter irradiated with a 500 cGy intersecting pencil beam pattern (left) shows significant pixel failure (7.8%) for 3%, 3mm 2D gamma criteria (centre). A BSA scatter correction applied to the OptCT data reduces pixel failure to 0.15% (right).

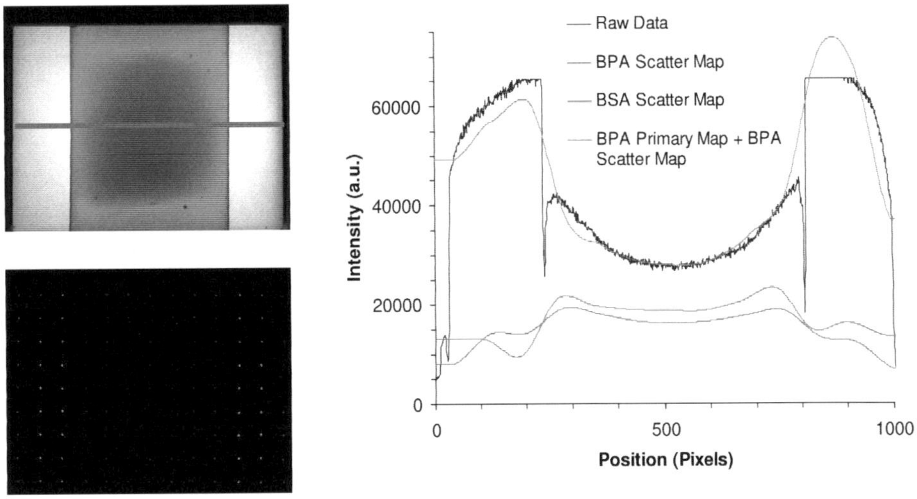

Figure 5: Evaluation of the light intensity along a line in the projection image of a 4%T polymer gel dosimeter irradiated with a 400 cGy test prostate plan (OptCT projection, top left). The BPA method (bottom left) yields a closer approximation (right) of the primary to scatter signal than the BSA method.

OptCT projection data from polymer gels irradiated with a larger scatter volume test prostrate plan is shown in Figure 5. As previously stated, the BSA method underestimates the scatter signal due to blocking of primary signal from the light source. An alternative method of scatter correction utilizes a beam pass array, which essentially consists of a grid of 2 mm diameter pinholes. In this method, the primary OptCT signal is estimated from the pinhole points, then subtracted from the raw OptCT projection (primary + scatter) to obtain the scatter signal at the pinhole point. The scatter values at the pinhole points are then interpolated using a bicubic spline to form a scatter map as in Figure 1. While the BPA method does not suffer from inaccuracy due to blocking of primary signal, it is significantly perturbed by refraction effects and minor imperfections in the gel dosimeters at different projection angles, leading to streak artifacts in the reconstructed image. Further results are forthcoming on the development of adaptive BSA and BPA scatter maps correcting individual projections prior to image reconstruction.

4. Conclusions

Beam stop array and beam pass array corrections have exciting potential for multiple scatter corrections of cone beam OptCT in polymer gel dosimetry. While BSA scatter corrections are limited in cone beam x-ray CT by the realities of increased x-ray dose and acquisition time, the benefits to cone beam OptCT are well worth the time afforded to such corrections. Irregularities in the reference scan (i.e. the base level of the dosimeter) are well-addressed by such corrections and adaptive scatter maps correcting individual projection images prior to reconstruction have promise to provide further correction to artifacts introduced in highly scattering irradiated dosimeters.

5. Acknowledgements

Research funding has been provided by the Ontario Research and Development Fund with the Cancer Centre of Southeastern Ontario (CCSEO), and the Canadian Institutes of Health Research (CIHR).

6. References

[1] DeJean P, Senden R, MacAuley K, Rogers M, and Schreiner L J 2001 Initial experience with a commercial cone beam optical CT unit for polymer gel dosimetry I & II *J. Phys.: Conference Ser.* **56** 179-186

[2] Sarabipour S, Bosi S, Hill B and Baldock C 2006 A preliminary study of the measurement of slice-width dose profiles (SWDP) on diagnostic x-ray CT scanners using PAGAT polymer gel dosimeters with optical CT read-out *J. Phys.: Conference Ser.* **56** 280-282

[3] Joseph P M and Spital R D 1982 The effects of scatter in x-ray computed tomography *Med.Phys.* **9** 464-472

[4] Rinkel J, Gerfault L, Esteve F and Dinten J-M 2007 A new method for x-ray scatter correction: first assessment on a cone-beam CT experimental setup *Phys. Med. Biol.* **52** 4633-4652

[5] Endo et al. 2001 Effect of scattered radiation on image noise in cone beam CT *Med. Phys.* **28** 469-474

[6] Siewardsen J H, Moseley D J, Bakhtiar B, Richard S and Jaffray D A 2004 The influence of antiscatter grids on soft-tissue detectability in cone-beam computed tomography with flat-panel detectors *Med. Phys.* **31**, 3506-3520

[7] Siewardsen J H and Jaffray D A 2001 Cone-beam computed tomography with a flat-panel imager: magnitude and effects of x-ray scatter *Med. Phys.* **28** 220-231

[8] Ning R and Tang X 2004 X-ray scatter correction algorithm for cone beam CT imaging *Med. Phys.* **31** 1195-1202

[9] Jordan K and Battista J 2006 Linearity and image uniformity of the Vista^{TM} optical cone beam scanner *J. Phys.: Conference Ser.* **56** 217-220

[10] Senden R J, DeJean P, MacAuley K and Schreiner L J 2006 Polymer gel dosimeters with reduced toxicity : a preliminary investigation of the NMR and optical dose-response using different monomers *Phys. Med. Biol.* **51**, 3301-3314

Journal of Physics: Conference Series **164** (2009) 012032 doi:10.1088/1742-6596/164/1/012032

NMR mechanisms in gel dosimetry

L.J. Schreiner

Departments of Oncology and Physics, Queen's University, and Cancer of Centre Southeastern Ontario at KGH, Kingston, Canada, K7L 5P9

john.schreiner@krcc.on.ca

Abstract. Nuclear magnetic resonance was critical to the development of gel dosimetry, as it established the potential for three dimensional dosimetry with chemical dosimeter systems through magnetic resonance imaging [1]. In the last two decades MRI has served as the gold standard for imaging, while NMR relaxometry has played an important role in the development and understanding of the behaviour of new gel dosimetry systems. Therefore, an appreciation of the relaxation mechanisms determining the NMR behaviour of irradiated gel dosimeters is important for a full comprehension of a considerable component of the literature on gel dosimetry. A number of excellent papers have presented this important theory, this brief review will highlight some of the salient points made previously [1-5].

The spin relaxation of gel dosimeters (which determines the dose dependence in most conventional MR imaging) is determined principally by the protons on water molecules in the system. These water protons exist in different environments, or groups (see Figure 1): on bulk water, on water hydrating the chemical species that are being modified under irradiation, and on water hydrating the gel matrix used to spatially stabilize the dosimeter (e.g., gelatin, agarose, etc). The spin relaxation depends on the inherent relaxation rate of each spin group, that is, on the relaxation rate which would be observed for the specific group if it were isolated. Also, the different water environments are not isolated from each other, and the observed relaxation rate also depends on the rate of exchange of magnetization between the groups, and on the fraction of protons in each group. In fact, the water exchanges quickly between the environments, so that relaxation is in what is usually termed *the fast exchange regime*. In the limit of fast exchange, the relaxation of the water protons is well characterized by a single exponential and hence by a single apparent relaxation rate. In irradiated gel dosimeters this observed rate is a function of the absorbed dose.

In this review I will first develop the fast exchange model for the spin lattice relaxation Fricke gel dosimeter system, as this is conceptually the simpler system. Fundamental concepts such as relaxivity (which specifies the ability of some species to enhance the relaxation of water protons) and chemical yield will be presented. The linear dose relationship for Fricke gel dosimeters, and the reduced dose sensitivity of Fricke dosimeters containing chelators, will be explained. The model will then be extended to the spin-spin relaxation of polymer gel systems and the main differences discussed. The reasons for the enhanced dose sensitivy and dynamic range for spin-spin (R2) rather than spin-lattice (R1) relaxation will be presented.

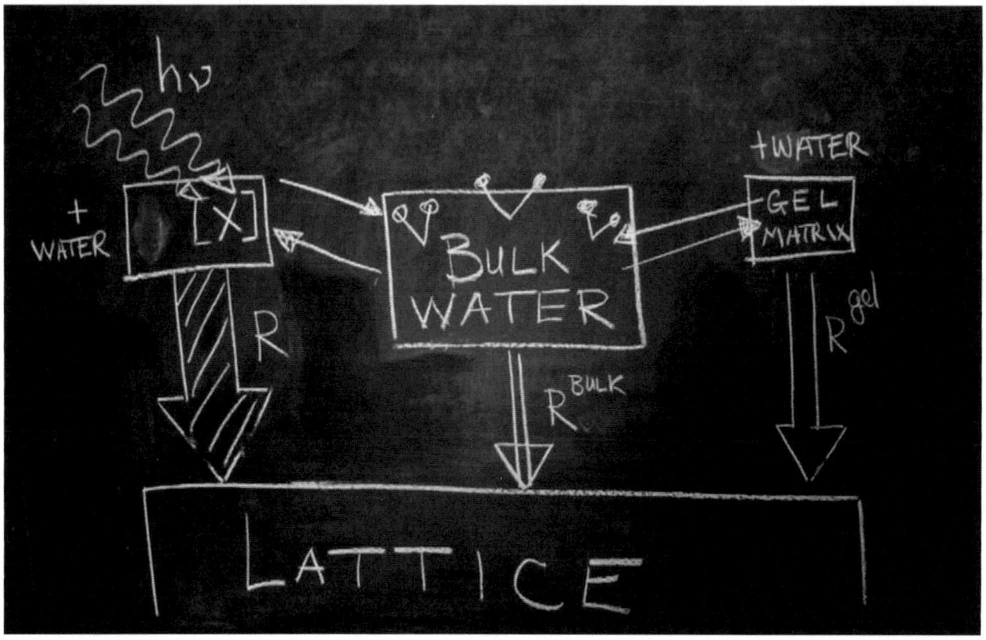

Figure 1: A cartoon sketching the water environments and exchange and relaxation paths for the spin-lattice of an irradiated gel dosimeter containing some species X having a concentration dependence on absorbed dose. As the concentration [X] increases, the amount of water hydrating the species also increases. The final observed relaxation rate is a weighted sum of the individual groups' inherent rates. Because bulk water is typically a very inefficient relaxing proton group (i.e., has a long relaxation time and hence small relaxation rate) the relaxation of the hydration water of the radiation mediated species will dominate the observed relaxation. Hence the dosimeter will exhibit a dose dependence.

Reference List

Note: I have intentionally not referenced the articles compiled in the Proceedings of the previous four International Workshops on Radiation Therapy Gel Dosimetry

[1] Gore J C, Kang Y S and Schulz R J 1984 Measurement of radiation dose distributions by nuclear magnetic resonance (NMR) imaging. *Phys Med Biol* **29** 1189-97

[2] Audet C and Schreiner L J 1997 Multiple-site fast exchange model for spin-lattice relaxation in the Fricke-gelatin dosimeter. *Med Phys* **24** 201-9

[3] Babic S and Schreiner L J 2006 An NMR relaxometry and gravimetric study of gelatin-free aqueous polyacrylamide dosimeters. *Phys Med Biol* **51** 4171-87

[4] Lepage M, McMahon K, Galloway G J, De Deene Y, Back S A and Baldock C 2002 Magnetization transfer imaging for polymer gel dosimetry. *Phys Med Biol* **47** 1881-90

[5] Lepage M, Whittaker A K, Rintoul L and Baldock C 2001 ^{13}C-NMR, ^{1}H-NMR, and FT-Raman Study of Radiation-Induced Modifications in Radiation Dosimetry Polymer Gels. *J Appl Polym Sci* **79** 1582-71

Journal of Physics: Conference Series **164** (2009) 012033 doi:10.1088/1742-6596/164/1/012033

Review of quantitative MRI principles for gel dosimetry

Yves De Deene[1,2]

[1]Research group QMRI, Department for experimental cancer research, nuclear medicine and radiotherapy (ECNURAD), Ghent University Hospital, De Pintelaan 185, 9000 Gent, Belgium

[2]Medical Image and Signal Processing (MEDISIP), Ghent University Hospital, De Pintelaan 185, 9000 Gent, Belgium

E-mail: Yves.DeDeene@UGent.be

Abstract. The radiation dose distribution absorbed by polymer gel dosimeters can be read out by several methods such as magnetic resonance imaging (MRI), optical CT, X-ray CT and ultrasound. MRI was the first method that was explored to read out polymer gel dosimeters. Although MRI was soon recognised as a promising technique, limited access to MRI scanners and the often (wrongly perceived) complexity in optimizing the imaging protocol has led to a search for other imaging modalities. In this paper we hope to unveil the mysticism of magnetic resonance imaging. Firstly, the basic principles of magnetic resonance image acquisition will be explained. Also, quantitative properties to describe the image quality are defined. Secondly, some sequences for quantitative T1 and T2 imaging will be analysed and specific issues concerning optimization and accuracy will be highlighted. Thirdly, we provide the reader with some easy guidelines and tools to investigate the accuracy of quantitative imaging sequences. It should also be noted that many of the parameters that describe the accuracy of the imaging technique also apply to other imaging modalities.

1. Introduction

In the early days of gel dosimetry, magnetic resonance imaging (MRI) was suggested as thé method to read-out the gel dosimeters. The use of MRI as a non-destructive measurement method of a dosimeter gel was first proposed in 1984 by Gore *et al* [1] who showed that ferrous sulfate chemical dosimeters initially developed in 1927 [2] could be probed by nuclear magnetic relaxometry and hence by magnetic resonance imaging (MRI) [1]. Gore *et al* investigated the nuclear magnetic resonance (NMR) relaxation properties of irradiated Fricke or ferrous sulfate dosimetry solutions showing that radiation-induced changes, in which ferrous (Fe^{2+}) ions are converted to ferric (Fe^{3+}) ions, could be quantified using MRI and subsequently showed that Fricke dosimetry solutions dispersed throughout a gel matrix could be used to obtain three-dimensional (3D) spatial information using MRI. It was subsequently shown that irradiated Fricke-type gel dosimeters did not retain a spatially stable dose distribution due to ion diffusion within the irradiated dosimeters [3]. Fricke solutions with various gelling agents such as gelatin, agarose, sephadex and polyvinyl alcohol (PVA) were investigated. Chelating agents to reduce diffusion in Fricke gels, such as xylenol orange (XO), had only limited success [4] and diffusion remained a significant problem in the advancement of gel dosimetry. Different models have been described that explain the mechanism of how the relaxation rates are affected by the paramagnetic substances [1, 5-7]. As explained elsewhere in these proceedings the spin-lattice relaxation rate (R1 = 1/T1) and the spin-spin relaxation rate (R2 = 1/T2) in Fricke gels is altered significantly upon irradiation. As R1 of an unirradiated Fricke gel dosimeter is small as compared to the R2 of an unirradiated Fricke gel dosimeter, the dynamic range of the Fricke gel dosimeter in relative terms is higher for R1 than R2. For this reason, R1 mapping is preferred to R2 mapping for Fricke gel dosimeters. Also R1 maps (at least in the early days) can be obtained with a shorter acquisition time than R2 maps which is of crucial importance in avoiding diffusion related blurring of the dose distribution.

Polymer systems for the use of radiation dosimetry were first proposed as early as 1954, where Alexander *et al* discussed the effects of ionizing radiation on polymethylmethacrylate [8]. In 1992, Kennan *et al* reported on NMR longitudinal relaxation studies performed on an irradiated aqueous solution of N,N'-methylene-bis-acrylamide (Bis) and agarose, which showed that the relaxation rates increased with absorbed dose [9]. Polymer gel dosimeters are based on the conversion of co-monomers to polymer aggregates upon irradiation. This reaction alters the mobility of surrounding water molecules which also results in a change in R1 and R2 [10]. The dose-response of R2 in gelatin based polymer gel dosimeters however is more pronounced than of R1. To explain the effect that the radiation-induced polymerization has on the R2 relaxation rate, a model of fast exchange [11] is adopted [12–14]. It is shown in later studies that not only the relaxation rate can be used as an imaging parameter but also other MR contrasts such as magnetization transfer [15–17] and chemical shift [18]. For a further explanation of these mechanisms, the reader is also referred to other papers in these proceedings. Different imaging sequences can be used to acquire quantitative images. These imaging sequences may differ in performance in terms of accuracy, precision and speed. It will be shown that for a specific imaging sequence, these three properties are interconnected.

The target figure of accuracy that is aimed in gel dosimetry for high-precision radiotherapy is about 3-5% of the maximum dose in regions of homogeneous dose and a spatial error of less than about 2-3 mm in regions of high-dose gradients. This figure of accuracy encompasses the overall dosimetry experiment. The problem in evaluating the overall accuracy of the dose maps obtained with gel dosimetry is that there is no "golden dosimetric standard" to compare with. The most reasonable strategy is to compare doses obtained with gel dosimetry with doses obtained by the most reliable dosimetry techniques that apply to a certain spatial dimension. For example, dose profiles of a single field (photons and electrons) can be compared with dose profiles obtained with an ionization chamber or diamond detector [19–20]. In two dimensions, gel dosimetry can be compared with film dosimetry [19, 21–22]. Dose distributions obtained with gel dosimetry have been compared with those calculated with treatment planning software [21, 23–29]. Errors that compromise the accuracy may occur at different stages of the dosimetry procedure [30].

At the stage of imaging the dosimeter, several imaging artefacts may cause errors in the final dose map. These errors may be classified in dose inaccuracies or in deformations of the dose maps. Studies of these different artefacts have resulted in different compensation strategies. The verification of the treatment plan can be seen as the major purpose of gel dosimetry in radiotherapy quality assurance. Besides the possibility of systematic errors, the dose maps will also contain stochastic noise. To minimize the stochastic noise in the images, the imaging sequence parameters should be optimized.

2. Basic principles of magnetic resonance imaging

2.1. Basic components of an MR scanner and basic principle

In this paragraph, the basic components and principles behind MRI are explained very briefly for the reader that is completely unfamiliar with MRI. For a more comprehensive treatise of the basic principles of MRI, the reader is referred to more specialized literature [31-32]. There are also some interesting online textbook available [33-34].

The MR scanner is composed of three major components:

(1) A strong magnet which is mostly a cryogenic magnet that produces a homogeneous strong main magnetic field. The magnetic field strength in clinical scanners ranges typically from 0.3 T to 7 T. The signal-to-noise ratio (SNR) increases with the magnetic field strength.

(2) Magnetic field gradient coils that are able to produce a magnetic field which is also aligned in the same direction as the main magnetic field but which varies in field strength along a certain direction. Three gradient coil configurations produce magnetic field gradients in one of the three orthogonal directions (x, y, z). Typical magnetic field gradients in modern clinical scanners range in strength from 10 mT/m to 40 mT/m.

(3) A radiofrequency (RF) coil (and corresponding electric circuitry) which is able to produce and/or receive an electromagnetic wave. There exist transmit and receive coils. Transmit coils produce an electromagnetic wave that excites the nuclei of the scanned subject while receive coils capture the signal produced by the precessing magnetic dipole moments (MDMs) of the excited nuclei. In many text books on nuclear magnetic resonance this is the place where the concept of 'spin' is introduced. The concept 'spin' is actually a quantum mechanical property. For the sake of simplicity, in this paper, we will keep to a classical mechanics (electromagnetic) approach. The reader is referred to other textbooks for a more comprehensive quantum mechanical explanation. It can be shown that the behaviour of the magnetic dipole moments in a classical mechanics approach is equivalent to the quantum expectation value of the spin magnetic moment in the quantum mechanical approach.

Apart from these three components, the MR scanner typically also contains shim coils in which a DC current can be applied in order to homogenize the static magnetic field. The scanner is also surrounded by a 'cage of Faraday' to eliminate interference of the electromagnetic waves with waves exterior to the scanner such as from radio-transmitters and other electronic instruments.

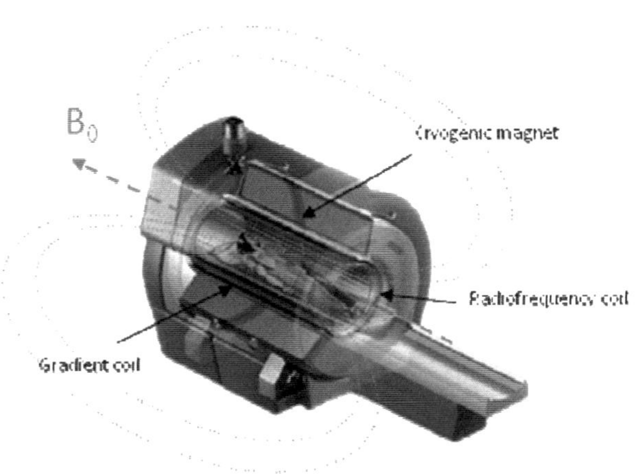

Figure 1. Major components of a clinical MRI scanner

In the absence of a magnetic field, the MDMs will be aligned in random directions. When the patient is slid into the magnet of the MR scanner, the MDMs will precess around the direction of the main magnetic field. Due to magnetic dipole interactions with the surroundings, the spins will align with the main magnetic field. This process is called 'spin-lattice' or 'T1 relaxation' (figure 2a). In human tissue or polymer gels this occurs in a time span of typically 0.1 s to 3 s. In order to get a signal from the scanned subject, the MDMs are excited from their equilibrium state (figure 2c). The condition for excitation is that the frequency of the exciting RF wave (f_{RF}) has a frequency which corresponds to the Larmor frequency given by

$$f_{RF} = \frac{\gamma}{2\pi} B \tag{1}$$

with γ the gyromagnetic ratio which is specific for the kind of nucleus (for hydrogen protons γ = 267.54 $s^{-1}.T^{-1}$) and B the magnetic field at the nucleus. The angle over which the MDM is flipped towards the plane transverse to the magnetic field (the transverse plane) is dependent on the length and intensity (amplitude) of the RF pulse. After excitation the MDMs will precess which results in a MR signal induced in a receive coil (figure 2d).

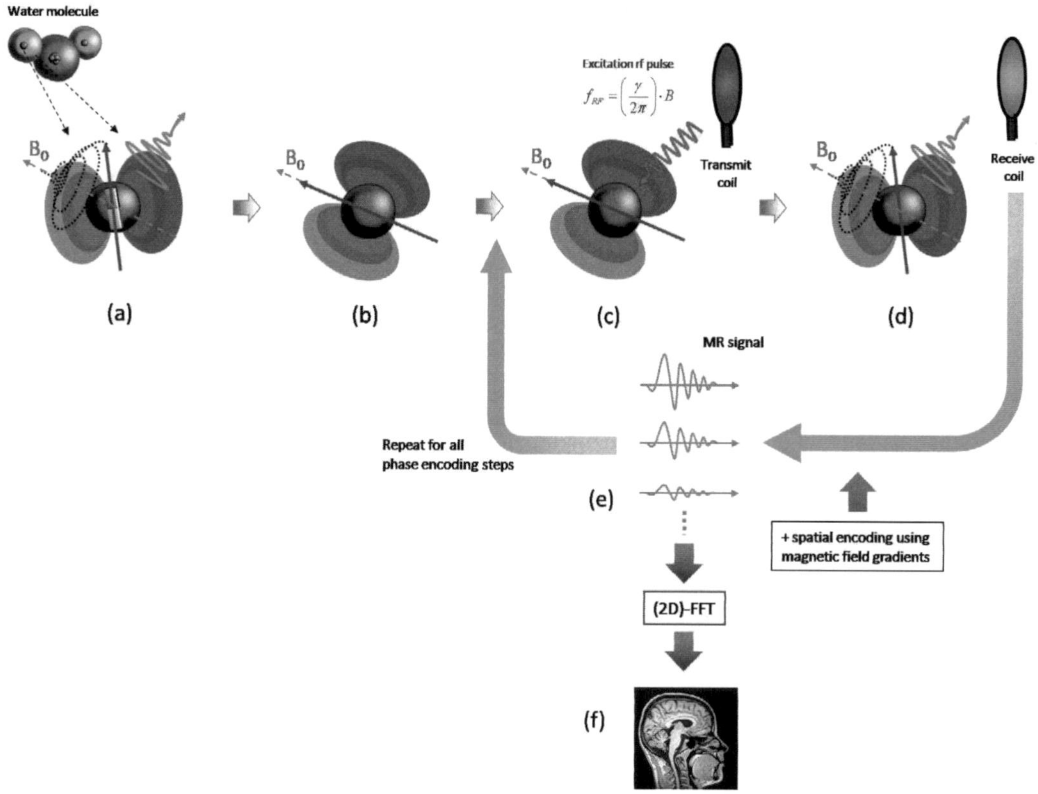

Figure 2. Basic principle of MR signal formation involving several steps: **(a)** When hydrogen protons are placed in a magnetic field (B_0), their magnetic dipole moments will start to precess. Due to (lossy) interactions with (magnetic fields of) surrounding nuclei and electrons (the lattice), the magnetic dipole moments will rapidly (0.1 s – 3 s) align with the main magnetic field **(b)**. In this equilibrium state no signal is received. Subsequently, a transmit coil will send an RF pulse to the sample. If the RF pulse has the Larmor frequency, the nuclei will be excited and the magnetic dipole moments are brought out of equilibrium so that they will precess around the main magnetic field **(c)**. This precession results in an electromagnetic wave that induces a current in the receive coil **(d)**. This signal is then digitized and stored. The sequence of excitation and signal reception **(c-d)** can be repeated while different magnetic field gradients are applied to encode the MR signal spatially **(e)**. After several encoded MR signals are recorded, a (2D/3D) Fourier transform is applied to reconstruct the MR image **(f)**.

It should be noted that the MR signal decays with a time constant much faster than T1. This (T2*) decay is attributed to the presence of local magnetic field inhomogeneities within each voxel. These magnetic field inhomogeneities lead to phase dispersion in the precessing MDMs as the tiny variations in field strength result in slightly different Larmor frequencies and thus accumulating dephasing.

Sometimes the same coil is used for transmitting and receiving. To be able to extract spatial information from the received MR signal, the signal is encoded using magnetic field gradients. By applying a magnetic field gradient, the Larmor frequency is made space dependent (i.e. B in equation 1 becomes a function of spatial coordinates (and time) in other words B(x,y,z,t)). The sequence of excitation and reception (figure 2c-d) is repeated with different phase encoding gradients. The recorded signals are digitized and stored in a matrix also called 'k-space' (figure 2e). Applying a fast (2D or 3D) Fourier transform (FFT) on k-space yields the MR image(s) (figure 2f).

2.2. Spatial encoding: building blocks

In an MR sequence three different spatial encoding steps can be recognized:

(1) Slice selection: When a gradient is applied in combination with an RF excitation pulse, only one slab of MDMs will be excited in the scanned subject.
(2) Phase encoding: When a gradient is applied after the MDMs have been excited but before signal reception, the MDMs will precess at a different frequency depending on their position in the gradient direction during the time this gradient is on. This leads to different phases in the phase encoding direction after a phase encoding gradient is applied.
(3) Frequency encoding: Upon signal reception a frequency encoding gradient is switched on so that the MDMs will precess at different frequencies depending on their position in the gradient.

In conventional imaging sequences, the direction of the three encoding gradients is taken orthogonal. Spatial encoding can be easily explained by using an analogy in acoustics.

2.2.1. Slice selection

The principle of slice selection is illustrated in figure 3.

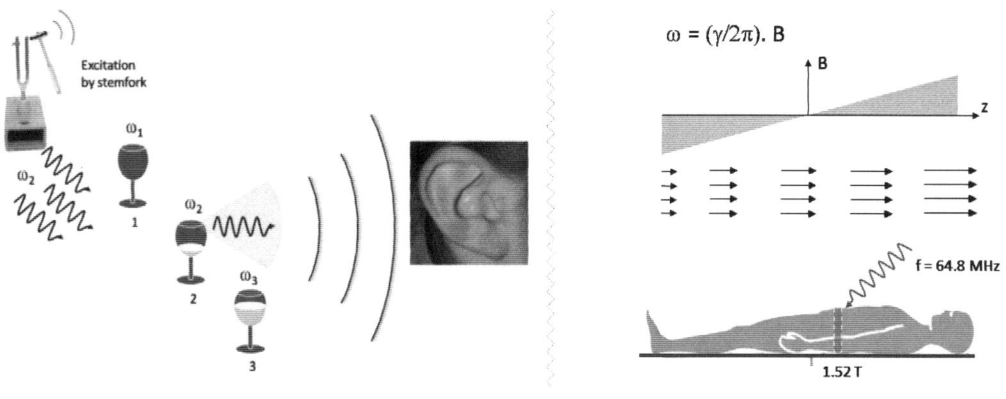

(a) **(b)**

Figure 3. Principle of slice selection illustrated in analogy with an acoustics experiment of resonating wine glasses (a) and in MRI (b).

An analogy with acoustics is provided in figure 3a. Suppose three glasses are filled with different amounts of wine (Glass 1 is empty and glass 3 is filled). When an acoustic wave with a certain wavelength is created with a stem fork only the glass of which the mechanical resonance frequency corresponds to the frequency of the acoustic wave will vibrate. When the acoustic wave produced with the stem fork is abruptly stopped, the vibrating glass will still produce an acoustic wave with the same frequency.

Similarly in MRI (figure 3b), when a magnetic field gradient (slice selection gradient) is applied and a radiofrequency electromagnetic wave (RF pulse) is produced by a radiofrequency coil (antenna), only those MDMs of which the magnetic field and thus the Larmor frequency (equation 1) corresponds with the frequency of the electromagnetic wave will be excited. If the RF pulse has the correct pulse length and duration (energy), the MDMs can be flipped over an angle of 90 degrees in the transverse plane. As mentioned before, this will produce an alternating electromagnetic field (MR signal) which can be received using another (or the same) antenna. Thus the MR signal comes only from a slab of the scanned object.

2.2.2. *Frequency encoding*

The principle of frequency encoding is illustrated in figure 4.

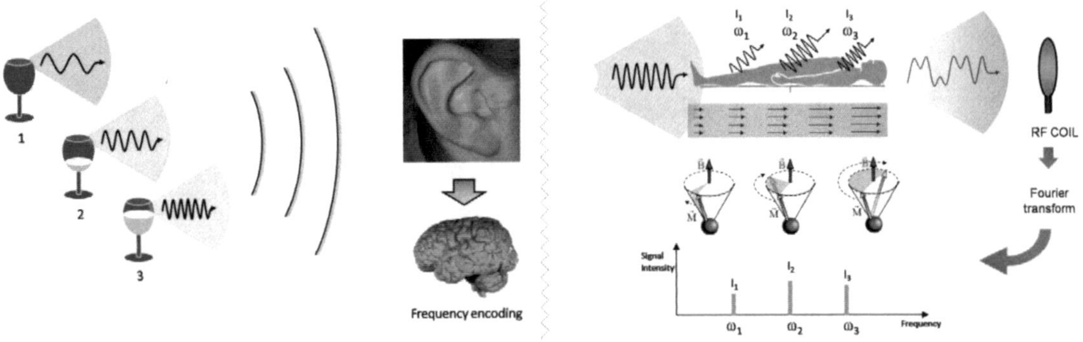

Figure 4. Principle of frequency encoding selection illustrated in analogy with an acoustics experiment of resonating wine glasses (a) and in MRI (b).

As an analogy, again three wine glasses are considered (figure 4a). Suppose that initially the wine glasses are empty and that they are excited by an acoustic wave (having a frequency corresponding to the resonance frequency of an empty glass). The three glasses thus vibrate at the same frequency. Suppose now that while the wine glasses are still vibrating, they are filled to different levels of wine. As a result the vibration frequency will change. An observer will hear an acoustic wave composed of different frequencies. If this experiment would be repeated with different amounts of wine glasses (e.g. 6 empty, 2 half filled and 5 completely filled glasses) a well-practiced observer would be able to determine how many wine glasses of each filling level are present just by hearing the produced sound.

Similarly in MRI, when MDMs are precessing in the transverse plane after an excitation RF pulse, they will start to precess at different rates when a gradient is applied. The precession rate (frequency) for each MDM corresponds with its location in the magnetic field gradient as is described by the Larmor equation (equation 1). Performing a Fourier transform on the acquired electromagnetic MR signal will give a spectrum of which the frequencies correspond with unique spatial positions. The amplitude of the frequency components corresponds with the MDM density at that specific position.

2.2.3. Phase encoding

Phase encoding is very similar to frequency encoding. If a gradient is applied during a certain period of time and with a certain amplitude, the MDMs will gain a phase difference with respect to each other depending on the gradient strength, the duration of the gradient and their position in that gradient. If this procedure is repeated for different gradient strengths, the different frequency components (and thus the spatial position) can be extracted.

2.3. Spatial encoding: mathematical formalism

The total MR signal received by the receiver coil is the sum of all signals induced in the RF coil by the excited MDMs in the selected slice. The signal originating from one infinitesimal small voxel of MDMs has an amplitude and phase. The amplitude is determined by the concentration of MDMs within the voxel (the density), the attenuation due to relaxation and a scaling factor which is determined by the receptivity of the coil. The phase is determined by the history of the magnetic field at the location of the voxel and is caused by the precession of the MDMs at that location. Mathematically the magnetization in each voxel m_{xy} with coordinates (x,y,z) at time t can be written as

$$m_{xy}(x,y,z,t) = m_0 \cdot \rho'(x,y,z,t) \cdot e^{j\Gamma(x,y,z,t)} = m_x(x,y,z,t) + j \cdot m_y(x,y,z,t) \qquad (2)$$

The MR signal $m_{xy}(x,y,z,t)$ in the voxel is thus a complex signal containing a real (m_x) and imaginary (m_y) part. The density ρ' is the density of MDMs also taking into account the relaxation. The phase Γ(x,y,z,t) acquired during a certain time period t in a magnetic field B(τ) is

$$\begin{aligned} \Gamma(x,y,z,t) &= -\gamma \int_0^t B(\tau)\, d\tau \\ &= -\gamma \int_0^t B_0(\tau)\, d\tau - \gamma \cdot x \int_0^t G_x(\tau)\, d\tau - \gamma \cdot y \int_0^t G_y(\tau)\, d\tau - \gamma \cdot z \int_0^t G_z(\tau)\, d\tau \end{aligned} \qquad (3)$$

The total received signal is the sum of all the precessing moments and will also be complex:

$$M(t) = \iiint_{slice} m_{xy}(x,y,z,t) \cdot dx\, dy\, dz = M_x(t) + jM_y(t) \qquad (4)$$

The complex MR signal can be detected using a quadrature detector.

Using the down-regulated MR signal (undone from the carrier frequency), the first term on the left side from equation 3 accounting for the free precession in the main magnetic field vanishes leaving only the gradient terms. If we assume that slice selection is performed the gradient induced phase dispersion occurs only in the x and y direction.

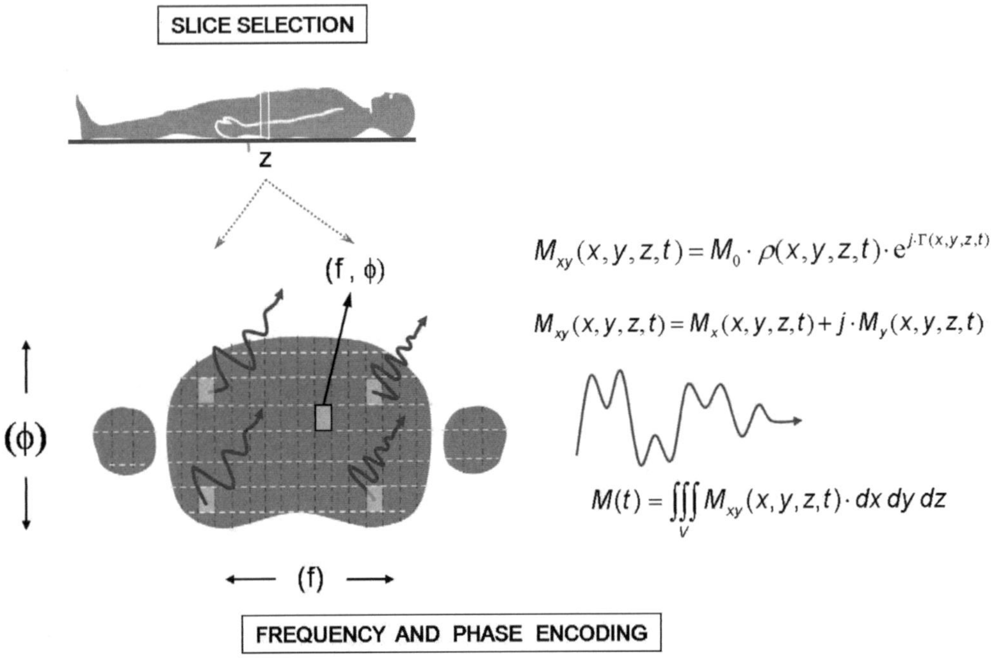

Figure 5. Spatial encoding in MRI. After slice selection, the slab of MDMs is further encoded by phase and frequency encoding gradients. The MR signal is the sum of all the signals produced by precessing MDMs in the excited slab.

Assuming slice selection in the z-direction, the MR signal can be written as:

$$S(t) = S_0 \cdot \iiint_{slice} \rho'(x,y,z=z_0) \cdot e^{-j(k_x(t)x + k_y(t)y)}.dx\,dy\,dz \tag{5}$$

with:

$$k_i = \gamma \cdot \int_0^t G_i(\tau)\,d\tau \qquad i = x, y$$

Equation 5 has the shape of a Fourier transform. The inverse Fourier transform gives the 'density'of the protons in image space (i.e. the image).

$$\rho(x,y,z=z_0) = C \cdot \iiint_{slice} S(k_x(t),k_y(t)) \cdot e^{j(k_x x + k_y y)}.dk_x\,dk_y \tag{6}$$

Journal of Physics: Conference Series **164** (2009) 012033 doi:10.1088/1742-6596/164/1/012033

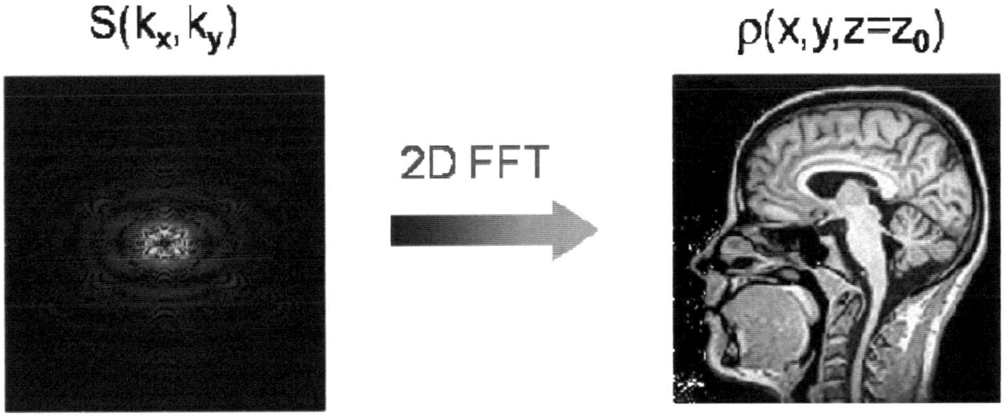

Figure 6. The MR image is the two-dimensional Fourier transform from the k-space which is a matrix containing the time signal data.

2.4. The spin-echo sequence

The spin-echo sequence, discovered by Erwin Hahn in 1950, is a basic MR sequences on which many of the current imaging sequences are based. The spin-echo sequence is drawn in figure 7. The first line shows the radiofrequency pulses while the second line shows the MR signal.

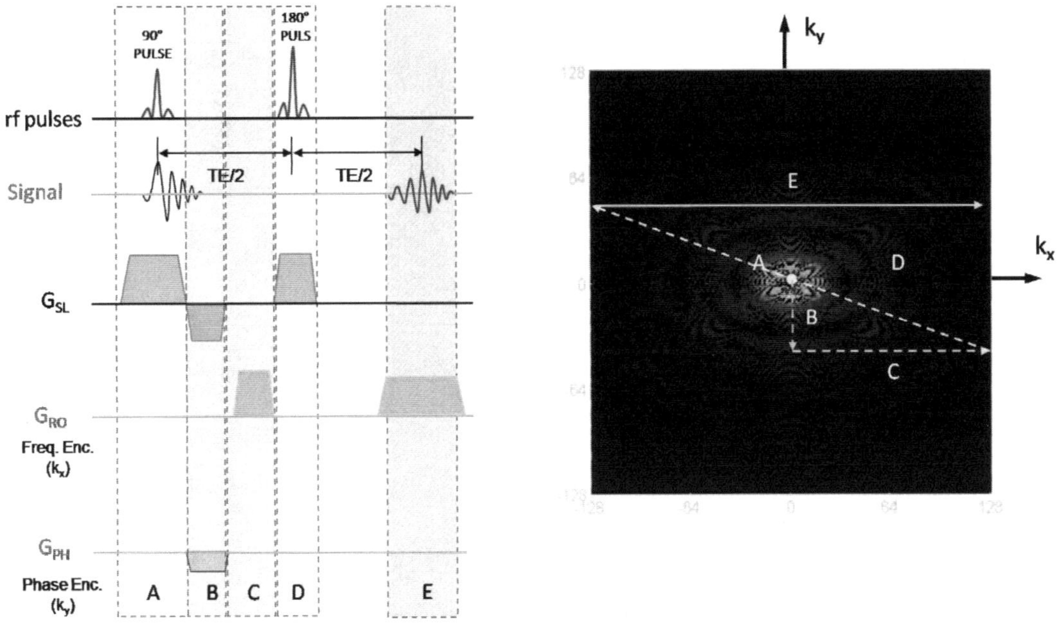

Figure 7. Spin-echo sequence and filling of the k-space matrix. The phase encoding is in the y-direction while the frequency encoding is in the x-direction. The letters A-E in k-space correspond with the events in the sequence indicated by A-E.

In the absence of imaging gradients, after excitation (90° pulse), a free induction decay (FID) signal is produced. The signal of the FID decays with a time constant T2* (= 1/R2*). An important component in this decay process is related to phase dispersion by magnetic field inhomogeneities that originate from the applied static magnetic field at one hand and microscopic field inhomogeneities caused by the inhomogeneous microstructure at the other hand. After a time TE/2 (when the FID signal has decayed) a 180° pulse is applied. As we have seen previously this is achieved by an RF pulse with a double amplitude or duration. After a time TE/2 after the 180° pulse a spin-echo signal appears. After the 180° pulse the transverse component of the MDMs will be flipped in the opposite direction. The MDMs that were dephasing at a faster rate (due to local magnetic field homogeneities) before the 180° pulse will rephrase at the same (faster) rate after the 180° pulse. The MDMs will be in phase after the same time as was given for the rephrasing (see also figure 8).

As explained in section 2.3, the k-space (figure 7, right) is filled by the digitized MR signal for different phase encoding steps. The filling of k-space can be easily understood by following the different steps (A-E) in figure 7. As discussed previously, the k-space has coordinates (k_x, k_y) which correspond with the second integral expression in equation 5. Upon excitation of the MDMs, the longitudinal magnetization is flipped in the transverse plane, placing us in the centre of the k-space (A) with coordinates (k_x, k_y) = (0, 0) (A). The (negative) phase gradient in B moves our pointer in the (negative) direction of k_y (phase encoding direction). A positive gradient in the frequency encoding direction moves the pointer along path C to the right edge of k-space. The 180° RF pulse moves the pointer to a position which is point-mirrored with respect to the centre (D) at the left edge of k-space. Upon acquisition while a frequency encoding gradient is applied, one k-space line is recorded along E. This sequence is repeated after a time TR (the repetition time) but with slightly different phase encoding gradients so that successive k-space lines (E) are recorded. The same logic can be followed to understand how k-space is filled with other sequences. This logic can also be extended to 3D imaging sequences.

3. Quantitative imaging sequences for measuring T1, T2 and MT

In principle, it is possible to use any kind of image in which the image contrast is monotonically related to the absorbed dose for dose mapping. Using calibration vials the image contrast can be related to the absorbed dose. Theoretically, it is possible to use a T1-weighted, T2-weighted or MT-weighted image. However, due to the inhomogeneity of the radiofrequency field, the image uniformity in these images is so poor that in practice a very bad accuracy is obtained. A far better performance in terms of accuracy is reached by using quantitative images such as T1-maps, T2-maps and MT-maps. In this section, different quantitative imaging sequences are discussed.

3.1. Quantitative R1 imaging sequences

3.1.1. The spin-echo sequence

The general shape of the spin-echo sequence is given in figure 8. The repetitive unit consists of an excitation RF pulse with flip angle α and a 180° refocusing pulse at TE/2. The repetition time (i.e. the time between two successive excitation RF pulses α) is TR. In the example given below a Carr-Purcell approach is taken where both the excitation pulse α and the 180° refocusing pulse are given along the y-axis. In the literature on pulse sequence design, this is often written as α_y-180_y. Note that any other scheme (e.g. α_x-180_y) would give similar results.

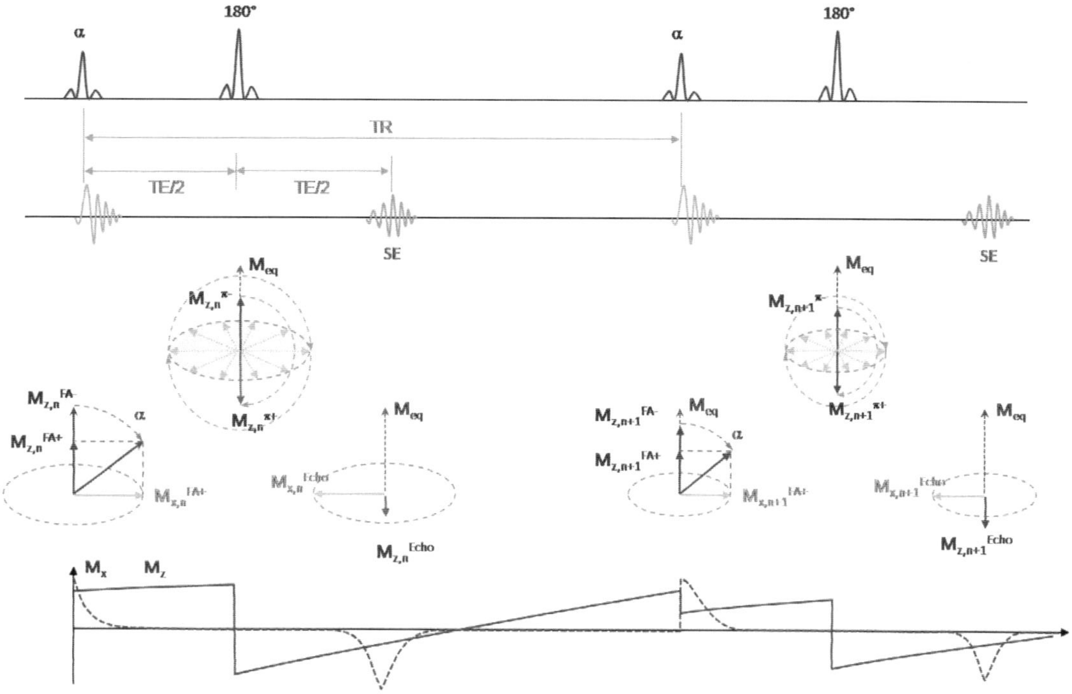

Figure 8. RF pulse scheme and signal formation of a spin-echo sequence with magnetization history of the longitudinal component (M_z) and the transverse component (M_x).

The signal in a spin-echo sequence is weighted by T1 (R1 = 1/T1) through the choice of the repetition time TR and by T2 (R2 = 1/T2) through the choice of TE. Thus by acquiring several spin-echo images recorded with different repetition times TR, the spin-lattice relaxation rate R1 can be fitted. The signal of the spin-echo sequence acquired with an excitation pulse with flip angle α in the steady state is given by

$$S = S_{eq} \cdot \frac{\left(1 - 2 \cdot \exp\left(-R1 \cdot \left(TR - \frac{TE}{2}\right)\right) + \exp\left(-R1 \cdot TR\right)\right)}{\left(1 + \cos\alpha \cdot \exp\left(-R1 \cdot TR\right)\right)} \cdot \exp\left(-R2 \cdot TE\right) \cdot \sin\alpha \tag{7}$$

with S_{eq} the equilibrium signal corresponding with full excitation and full signal recovery (i.e. the signal obtained when $\alpha = 90°$ and $TR \rightarrow \infty$).

In the special case that $\alpha = 90°$ ($\cos\alpha = 0$, $\sin\alpha = 1$) equation 3 reduces to

$$S = S_{eq} \cdot \left(1 - 2 \cdot \exp\left(-R1 \cdot \left(TR - \frac{TE}{2}\right)\right) + \exp\left(-R1 \cdot TR\right)\right) \cdot \exp\left(-R2 \cdot TE\right) \tag{8}$$

The parameters R1 and R2 can be obtained from equations 3-5 can be fitted for various TR and TE using a χ^2 minimization fitting such as a Levenberg-Marquardt algorithm. It is advisable to treat the flip angle α as an unknown variable to account for imperfect excitation.

In the special case that the echo time TE is much smaller than the repetition time TR (and $\alpha = 90°$), equation 4 can be further simplified to

$$S = S_{eq} \cdot \left(1 - \exp\left(-R1 \cdot TR\right)\right) \cdot \exp\left(-R2 \cdot TE\right) \tag{9}$$

This function can be used to fit only for the spin-lattice relaxation time R1.

In practice, several base images are recorded with differing repetition times (and echo times). The signal intensity in each pixel position of the set of base images is fitted against the functions 3, 4 or 5. The fitted parameters R1 and R2 for each pixel of the base images is plotted in a new image, respectively the parametric R1 and R2 map. A background filter on the base images is also often applied before fitting. This background filter makes that only those pixels with a value larger than a certain threshold value in the base images are taken into account for fitting. The pixels smaller than the threshold are considered as noise and are set to zero in the parametric maps.

3.1.2. The saturation recovery sequence (SRGE)

The shape of a saturation recovery sequence is given in figure 9. The saturation recovery sequence is a special case of the FLASH sequence (Fast Low Angle Shot) but with the flip angle of the excitation pulses equal to 90°. This way, all longitudinal magnetization that is rebuild during the repetition time TR is flipped in the transverse plane and gives way to signal.

The signal can thus be written as

$$S = S_{eq} \cdot (1 - \exp(-R1 \cdot TR)) \cdot \exp(-R2^{*} \cdot TE) \tag{10}$$

Using the saturation recovery sequence, several base images are recorded with different repetition times TR. The spin-lattice relaxation rate R1 can be obtained by fitting the signal intensity for different TR on a pixel-by-pixel basis in the base images as described before. Spoiler gradients are applied after each gradient echo recording to remove phase coherence in the transverse magnetization.

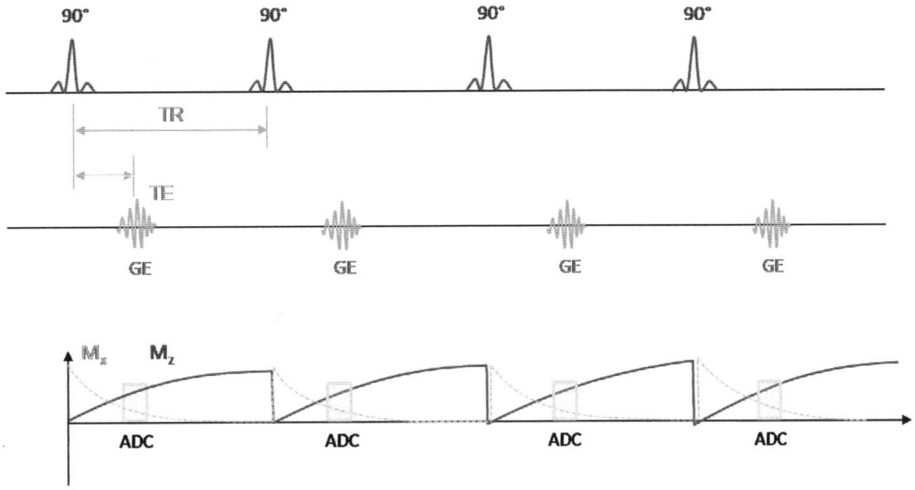

Figure 9. RF pulse scheme and signal formation of a saturation recovery sequence with magnetization history of the longitudinal component (M_z) and the transverse component (M_x).

Spoiled gradient recalled echo imaging (SPGRE) in which the 90° pulses are replaced by α-pulses can be found in a technique called DESPOT in which base images are recorded with various flip angles α [37].

3.1.3. The inversion recovery sequence (IRGE)

With the inversion recovery sequence (IRGE), a high T1-weighted contrast can be obtained.

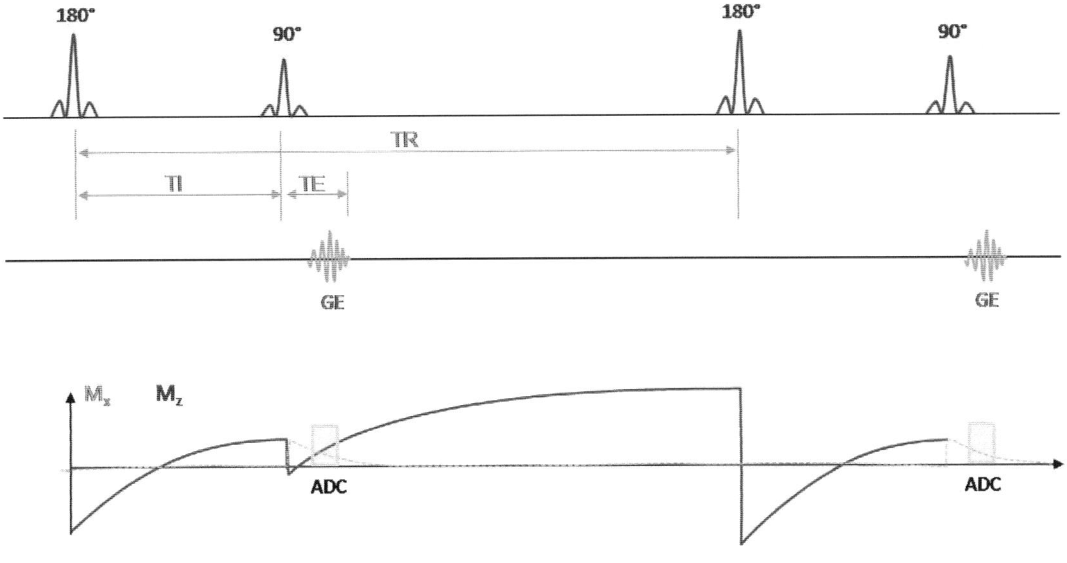

Figure 10. Inversion recovery sequence with magnetization history of the longitudinal component (M_z) and the transverse component (M_x).

The first RF pulse in the IRGE sequence is a 180° inversion pulse. This inversion pulse, inverts the longitudinal magnetization (figure 10). The magnetization is restored during a time TI (the inversion time) by T1 relaxation. After a time TI, the magnetization is flipped in the transverse plane by a 90° RF pulse.

The z-component of the magnetization recovers exponentially according to:

$$M_z = M_0 \cdot (1-(1-\cos\theta_{inv}).\exp(-R1\cdot t)+\exp(-R1\cdot TR)) \tag{11}$$

with θ_{inv} the flip angle of the inversion pulse (in figure 10, θ_{inv} = 180°).

After the 90° pulse, the signal thus becomes (in the case that TE << TR):

$$S = S_{eq}\cdot(1-(1-\cos\theta_{inv})\cdot\exp(-R1\cdot TI)+\exp(-R1\cdot TR))\cdot\exp(-R2^{*}\cdot TE) \tag{12}$$

Note that the signal in equation 11 can be negative. Amplitude images are strict positive. To improve fitting it may therefore be more appropriate to use the complex signal.

It should be mentioned that in reality there are no exact 90° and 180° slice selective pulses (figure 11). Instead the flip angle varies within the excited slice. The flip angle as a function of the distance perpendicular to the slice is called the slice profile. It should be mentioned that in all of the above equations a perfect slice is assumed. To account for the imperfect slice profile, the signal is a sum of the signals along the slice profile.

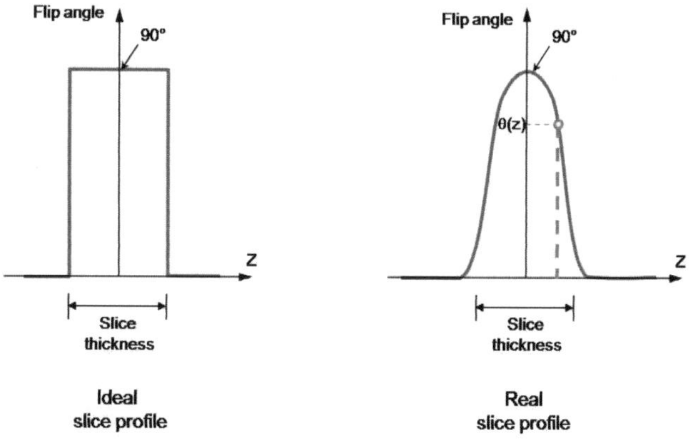

Figure 11. Ideal and realistic slice profile corresponding with a 90° slice selective RF pulse.

3.1.4. The Look-Locker sequence

A fast sequence to acquire an R1 map is the Look-Locker sequence [38] of which the RF sequence is shown in figure 12a. The Look-Locker sequence is similar to the IRGE sequence but the 90° RF pulse is replaced by a train of smaller α-pulses. Each α-pulse transfers some longitudinal magnetization into the transverse plane and gives way to a gradient echo by applying appropriate gradients. Also between each α-pulse a small phase encoding gradient is applied so that each gradient echo results in a new k-space line. This means that for each inversion excitation part of k-space is sampled instead of only one k-space line in the case of the IRGE sequence. It can be shown that the relaxation weighting of the different k-space lines results in broadening of the point-spread-function (PSF) thus limiting the intrinsic resolution. The signal from the nth readout pulse is given by

$$S = S_{sc}(\tau,T1,\alpha,N)\cdot\left(1-D(\tau,T1,\alpha,N)\cdot\exp(-R1^{*}\cdot n\cdot\tau)\right) \tag{13}$$

with $S_{sc}(\tau,T_1,\alpha,N)$ a scaling factor and $D(\tau,T_1,\alpha,N)$ a dynamic range parameter. τ is the inter-readout-pulse interval and N is the total number of echoes. $R1^*$ is the enhanced relaxation rate given by

$$R1^* = R1 - \frac{\ln(\cos\alpha)}{\tau} \tag{14}$$

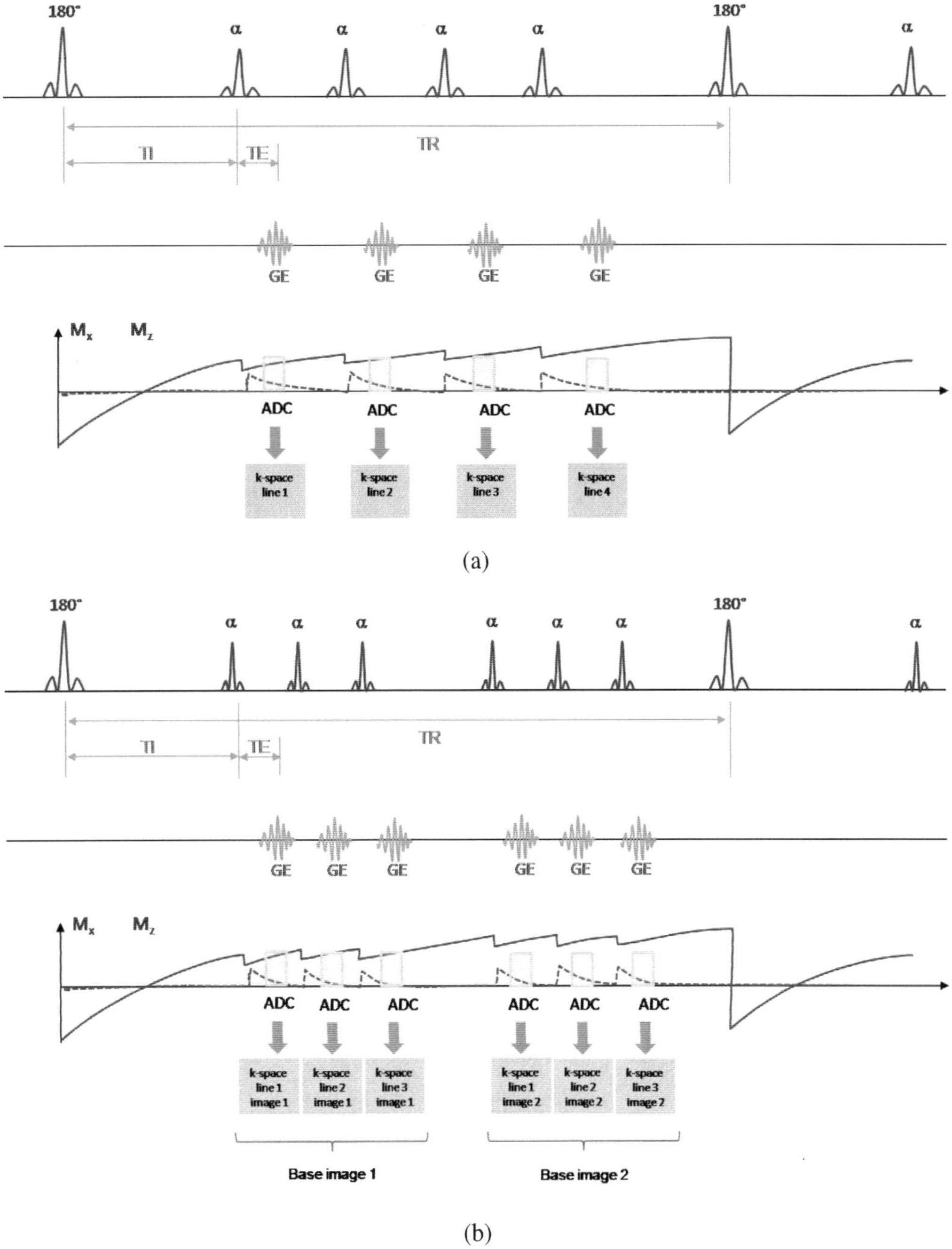

Figure 12. Look-Locker sequence (a) and TOMROP sequence (b) with magnetization history of the longitudinal component (M_z) and the transverse component (M_x).

To avoid severe artefacts as a result of the relaxation between the different k-space lines, a k-space ordering scheme is often proposed. In this scheme the central k-space lines (close to $k_y = 0$) are from echoes recorded immediately after the first α-pulse. A variant on the Look-Locker sequence is the TOMROP (T1 by multiple readout pulses) sequence (figure 12b). In this sequence the gradient echoes are grouped and interleaved with some recovery time to allow T1 relaxation to take place. Each group of gradient echoes is used in a separate T1 weighted image. The different T1-weighted images are then used to reconstruct an R1 map.

The Look-Locker sequence is particularly sensitive to RF pulse imperfections. Also changes in the flip angle distribution within the image (B1 field inhomogeneities) will affect the signal. A B1 inhomogeneity map can also be acquired when using equations 13-14. More complex analytical expressions for the signal may be derived. In addition more sophisticated numerical algorithms may be required to solve for the relaxation rate R1.

3.1.5. Steady-state free precession sequences (SSFP)

If a sequence of RF pulse excitations is given interleaved with a time TR, the signal will come in a steady state. In between two successive pulses the frequency encoding gradients can be placed in such a way that two echoes are obtained (figure 13).
It can be shown that the signal amplitude of both echoes is given by [36]

$$S_I = S_0 \tan\left(\frac{\alpha}{2}\right)\cdot\left(1-\left(\frac{(\exp(-R1\cdot TR)-\cos\alpha)(1-\exp(-2\cdot R2\cdot TR))}{\sqrt{p^2-q^2}}\right)\right) \qquad (15)$$

and

$$S_I = S_0 \tan\left(\frac{\alpha}{2}\right)\cdot\left(1-\left(\frac{(1-\exp(-R1\cdot TR)\cdot\cos\alpha)(1-\exp(-2\cdot R2\cdot TR))}{\sqrt{p^2-q^2}}\right)\right) \qquad (16)$$

with

$$\begin{aligned} p &= 1-\exp(-R1\cdot TR)\cdot\cos\alpha-\exp(-2\cdot R2\cdot TR)\cdot(\exp(-R1\cdot TR)-\cos\alpha) \\ q &= \exp(-R2\cdot TR)\cdot(1-\exp(-R1\cdot TR))(1+\cos\alpha) \end{aligned} \qquad (17)$$

The signal for both echoes as a function of α and TR is shown in figure 14a and b respectively.

After recording images with various repetition times TR and flip angles α, R1 and R2 maps can be obtained by fitting the pixel intensities in the base images against equations 15-17.

If the repetition time TR is much greater than the spin-spin relaxation time T2 (i.e. TR >> T2), the signal amplitude becomes independent of R2.

$$S_I = S_0\cdot\sin\alpha\cdot\left(\frac{1-\exp(-R1\cdot TR)}{1-\exp(-R1\cdot TR)\cdot\cos\alpha}\right) \qquad \text{TR} \gg \text{T2} \qquad (18)$$

In the latter case, the flip angle can be varied to extract R1.

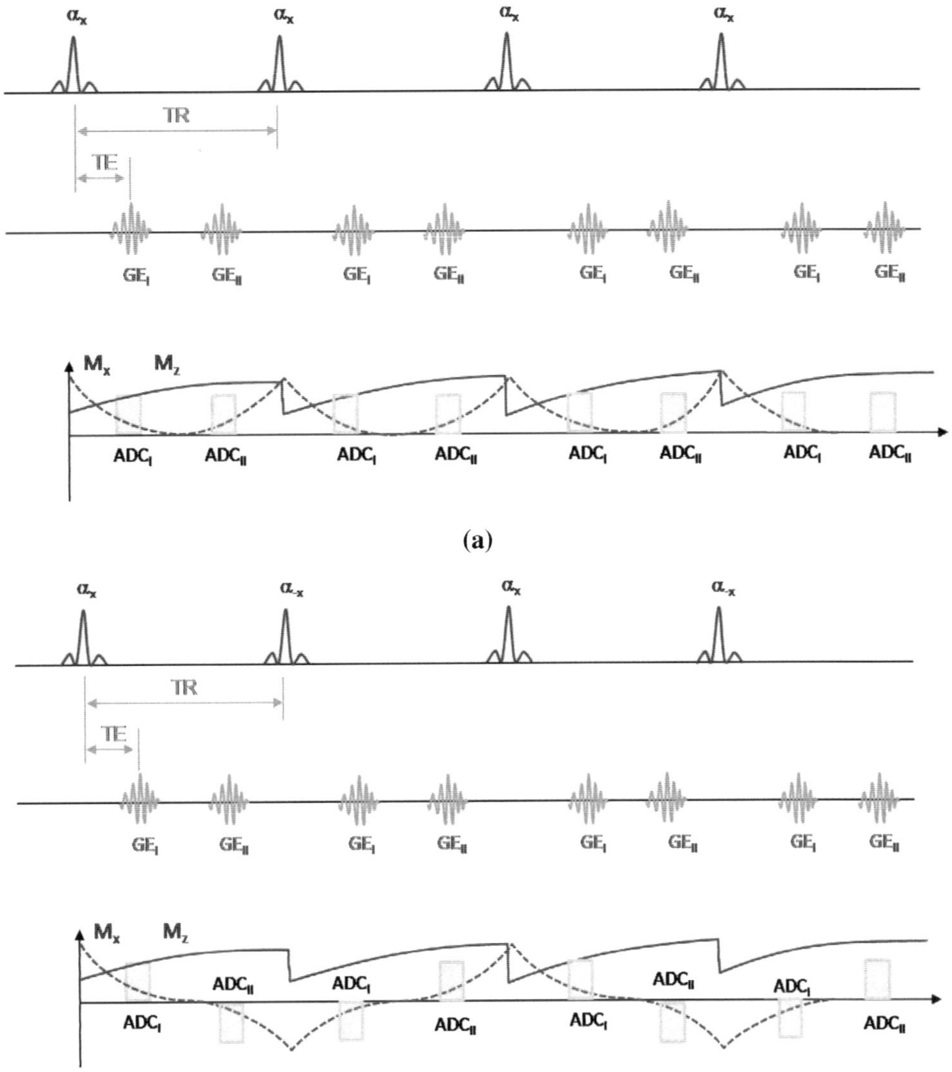

(b)

Figure 13. Two variants of the steady-state free precession (SSFP) sequence. Depending on the phase of the flip angle sequence the sign of the SSFP signals are either all positive (a) or alternated in polarity every two echoes (b). The magnitude of the SSFP signal is the same in both cases.

Under this condition, if we take two measurement with different flip angles of which the first is 90° and the second is α the R1 can be calculated from:

$$R1 = \frac{1}{TR} \cdot \ln\left(\frac{S(\alpha) \cdot \cos\alpha}{S(\alpha) - S(90°) \cdot \sin\alpha} \right) \qquad TR >> T2 \qquad (19)$$

A special case of SSFP is when the net gradient area is zero at any of the three gradient axes during one TR interval (between two successive RF excitations). This sequence is called balanced SSFP (figure 15b). In this case, only one echo is obtained.

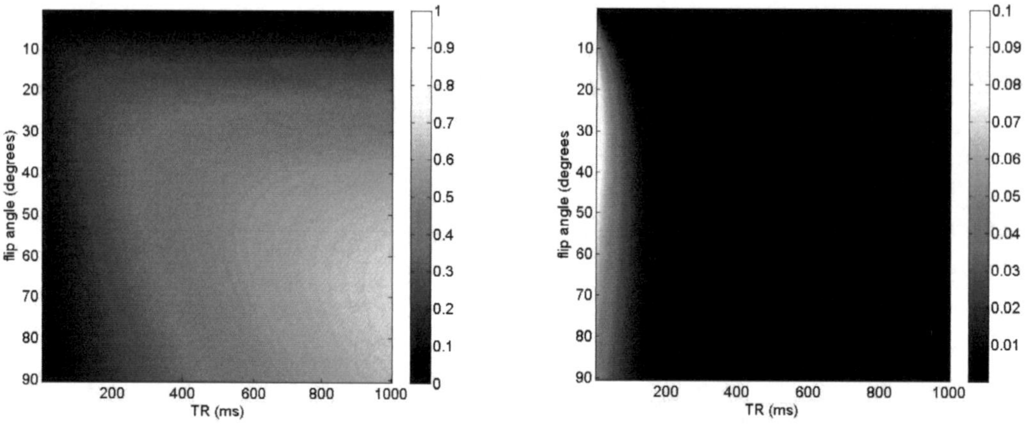

(a) **(b)**

Figure 14. Signal S_I (a) and S_{II} (b) in the SSFP sequence for various flip angles α and repetition times TR for a sample with R1 = 1 s^{-1} and an R2 = 10 s^{-1}.

The signal amplitude of the echo in balanced SSFP is given by [36]:

$$S = S_0 \cdot \sin\alpha \cdot \left(\frac{(1-\exp(-R1\cdot TR))\cdot \exp(-R2\cdot TE)}{1-(\exp(-R1\cdot TR)+\exp(-R2\cdot TR))\cdot\cos\alpha+\exp(-(R1+R2)\cdot TR)} \right) \tag{20}$$

for no sign alternation between successive RF pulses (i.e. α_x-α_x-α_x-…) and

$$S = S_0 \cdot \sin\alpha \cdot \left(\frac{(1-\exp(-R1\cdot TR))\cdot \exp(-R2\cdot TE)}{1-(\exp(-R1\cdot TR)-\exp(-R2\cdot TR))\cdot\cos\alpha-\exp(-(R1+R2)\cdot TR)} \right) \tag{21}$$

when sign alternation is used between successive RF pulses (i.e. α_x-α_{-x}-α_x-…).

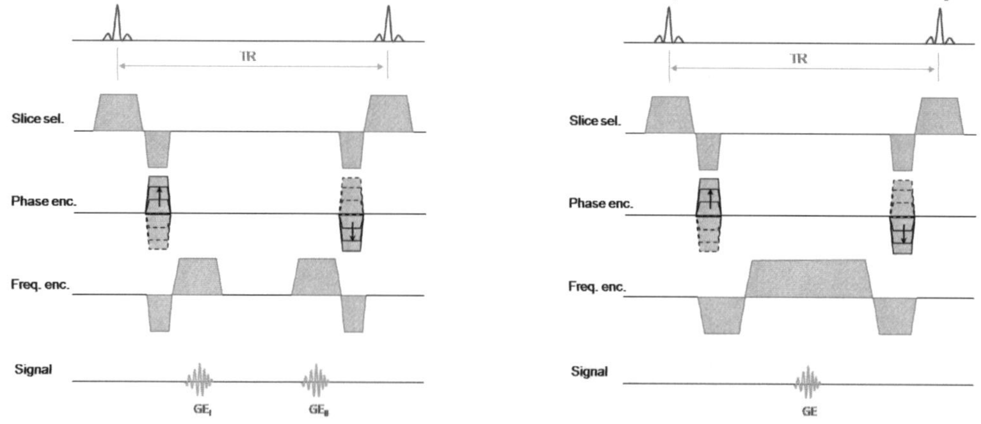

(a) **(b)**

Figure 15. Non-balanced SSFP (a) yielding two echoes and balanced SSFP (b) yielding only one echo. The difference lies in the frequency encoding gradient which is completely refocused within one TR in the case of the balanced SSFP sequence.

3.1.6. Very fast acquisition sequences

Theoretically, very fast acquisition schemes such as echo-planar imaging (EPI), gradient and spin-echo acquisition (GRASE) and half-Fourier acquired single shot turbo spin-echo (HASTE) can be used in combination with a T1 magnetization preparation sequence. For T1 mapping, an inversion recovery preparation can be applied. Also the Look-Locker sequence can be used in which an EPI readout is performed for each α-pulse. In this case, all the base images for a complete T1 map can be acquired within 3 s.

However, due to the very low bandwidth in the phase encoding direction, considerable image distortion can be produced in regions with off-resonance effects, such as field inhomogeneity, magnetic susceptibility variations, long time eddy currents and concomitant magnetic fields. Post-processing correction schemes have been proposed for EPI image distortions.

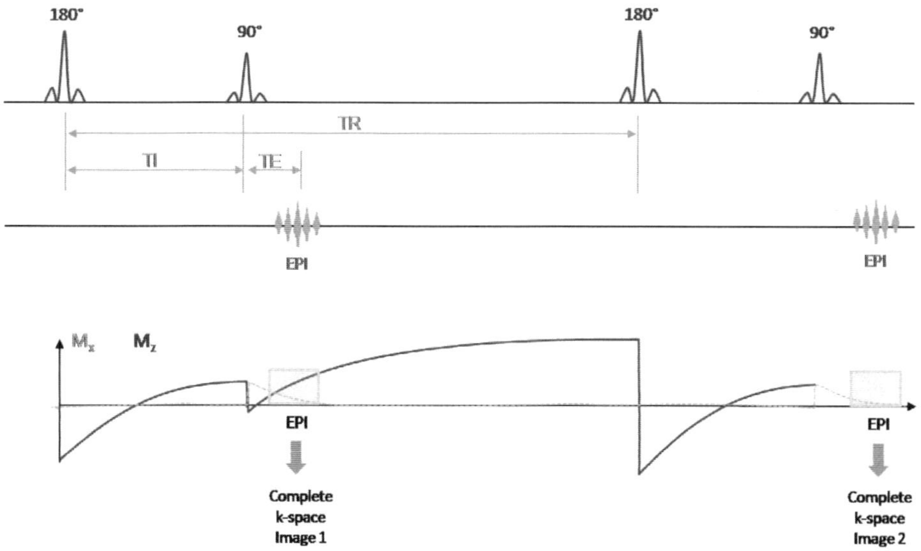

Figure 16. Inversion-recovery sequence with Echo Planar Imaging (EPI) readout. The magnetization history of the longitudinal component (M_z) and the transverse component (M_x) are also shown.

Because for gel dosimetry, high spatial accuracy is a prerequisite for the accuracy of the final dose maps, such very fast acquisition sequences may not be the best choice. Very fast acquisition may be used in combination with other k-space sampling schemes (such as spiral and projection acquisition). These sequences may perform much better in terms of spatial accuracy but care should be taken as with such acquisition schemes, artefacts will be smeared out over the entire image. This may compromise the dosimetric accuracy. Also the point-spread-function can be deteriorated significantly.

3.2. Quantitative R2 imaging sequences

3.2.1. The spin-echo sequence

The spin-echo has intrinsic T2 weighting. Changing the echo time, TE while keeping the repetition time, TR constant will result in a different T2 weighted spin-echo signal (figure 17). From two images recorded with different echo time an R2 map can be calculated. The R2 value in each pixel of the R2 map can thus be easily calculated from the corresponding pixel intensities in the different T2 weighted base images according to

$$R2 = \frac{1}{TE_2 - TE_1} \cdot \ln\left(\frac{S(TE_1)}{S(TE_2)}\right) \qquad (22)$$

with TE_1 and TE_2 the echo times for both images and $S(TE_1)$ and $S(TE_2)$ the pixel intensity in corresponding pixels.

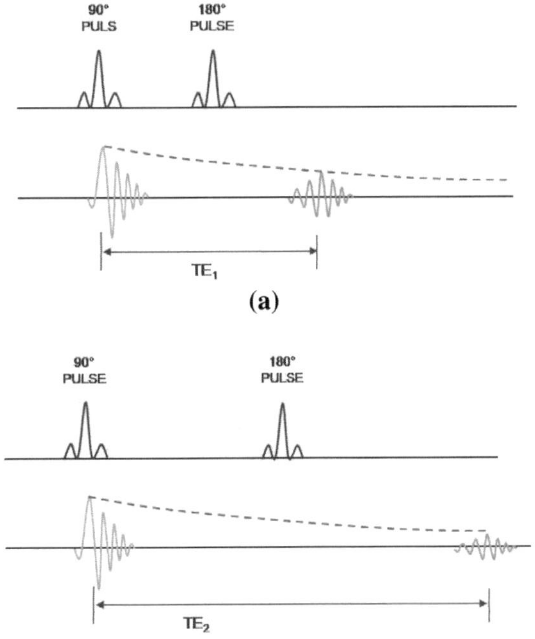

(a)

$$S(TE_1) = S_0 \cdot \exp(-R2 \cdot TE_1)$$

(b)

$$S(TE_2) = S_0 \cdot \exp(-R2 \cdot TE_2)$$

$$\Rightarrow R2 = \frac{1}{TE_2 - TE_1} \cdot \ln\left(\frac{S(TE_1)}{S(TE_2)}\right)$$

Figure 17. A spin-echo experiment repeated twice with two different echo times yields enough information to reconstruct an R2 map.

The advantage of the basic spin-echo sequence is that it is readily available on all scanners and that it is easy to understand and extract the R2 relaxation rate (equation 22).

Any displacement of both images with respect to each other (for example caused by eddy currents or motion of the object) can result in errors in the calculated R2 map. Although no object motion can be expected in the case of gel dosimeters, eddy current effects have been detected [39].

Another disadvantage of single spin-echo sequences concerns the acquisition time. If only a few images are recorded. There is a lot of dead time in the sequence (required for T1 recovery and T2 relaxation) but which is not been used for signal acquisition. This results in long measurement times to obtain sufficient signal-to-noise (thus dose resolution) in the dose maps or by other words a poor intrinsic precision (see further). It can be noted that multiple slices can be recorded in the time after the spin-echo and before the next (90°) excitation pulse (i.e. the time for T1 signal recovery). However for the longer echo time (figure 17b) this time period is significantly smaller.

3.2.2. Fast spin-echo sequence (turbo spin-echo sequence)

The acquisition for one slice can be accelerated by acquiring multiple echoes after each excitation pulse. Each echo is incrementally phase encoded for different phase lines in the k-space. The echo time in the calculation of the R2 map is replaced by an effective echo time. This effective echo time depends on the k-space phase ordering scheme. The number of echoes acquired within one repetition time in the fast spin-echo sequence is called the "turbo factor". The problem with this acquisition scheme is that the different k-space lines have a different T2-weighting. This leads to a distortion of the point-spread-function (PSF) (see further) which compromises the intrinsic spatial resolution in the images []. With many echoes, this distortion of the PSF may become rather significant.

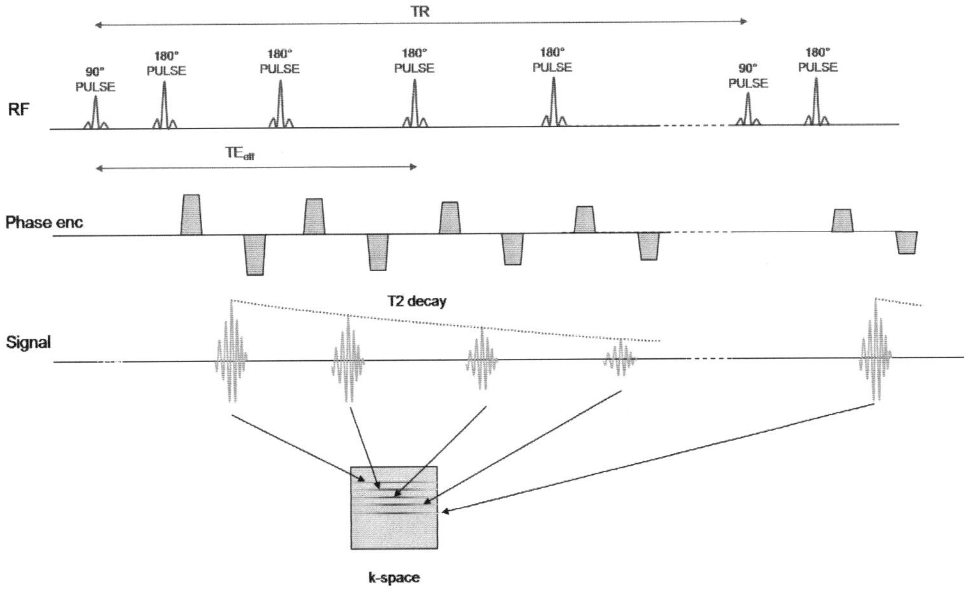

Figure 18. Fast spin-echo sequence yielding several echoes within one repetition time. All echoes within one repetition time have the same phase encoding. Each echo is stored in a separate k-space matrix. Finally, as many base images are obtained as there are echoes.

Different k-space line ordering schemes may help in minimizing the distortion of the PSF. In this case, the effective echo time is sometime difficult to determine. The fast spin-echo sequence is often regarded as a faster way to record R2 maps. However, when more than one slice is recorded, which is most often the case for 3D gel dosimetry, a multiple spin-echo sequence may be faster.

3.2.3. The multiple spin-echo sequence

In the multiple spin-echo sequence a train of 180° RF pulses is applied. In contrast to the fast spin-echo sequence, the spin-echo signals have the same phase encoding and are stored in separate k-space images which after FFT result in different T2-weighted base images.

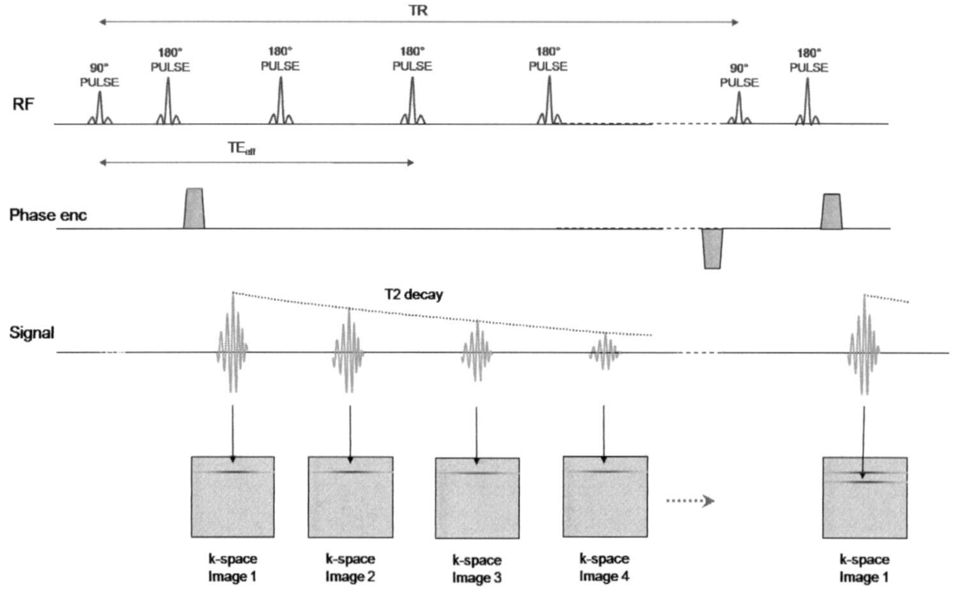

Figure 19. Multiple spin-echo sequence yielding several echoes within one repetition time. All echoes within one repetition time have the same phase encoding. Each echo is stored in a separate k-space matrix. Finally, as many base images are obtained as there are echoes.

A Levenberg-Marquardt χ^2-minimization algorithm can be used to fit the R2 value from the base images. Also with the multiple spin-echo sequence, multiple slices can be recorded interleaved. In that case, after each echo train a new slice is excited (with a 90° RF pulse) within the same repetition time.. It is shown that the number of echoes (or the echo time spacing) can be optimized with respect to dose precision for a certain number of slices [40]. The number of echoes in imaging sequences provided by the manufacturer is often restricted. In that case, the echo time spacing between the echoes can be optimized [41]. It is shown that heating of the phantom may be induced by the large number of 180° refocusing pulses. To compensate for temperature related imaging artefacts, a central k-space ordering scheme is preferred [42]. Also, it is shown that the interplay of stimulated echoes and varying eddy currents induced may induce imaging artefacts. To prevent these effects, a gradient train can be applied to bring the eddy currents in a steady state before the onset of the sequence [39]. Although the concept of fitting an R2 map from the different base images compensates theoretically for B1-field inhomogeneities, the R2 map may still be susceptible to secondary effects caused by the interplay of stimulated echoes and the B1 field inhomogeneities. Several strategies for compensating the remaining inhomogeneities in the R2 maps have been proposed [43, 44].

3.2.4. Single-shot free precession sequences

As demonstrated in section 3.1.5 the SSFP signal is weighted with both R1 and R2. With a repetition time TR much smaller than T1 and a flip angle $\alpha = 90°$ the ratio of the two signals in the non-balanced SSFP sequence becomes

$$\frac{S_{II}}{S_I} = \exp\left(-2 \cdot R2 \cdot (TR - \Delta)\right) \qquad TR \ll T1 \qquad (23)$$

with Δ the time between the second echo and the next 90° excitation pulse.

3.3. Quantitative magnetization transfer imaging sequences

Polymer gel dosimeters contain both synthetic and bio-polymers which create a macromolecular proton pool. These protons give rise to a broadened spectral peak (as a result of their short T2). Upon polymerisation, the proton pool increases which results in a larger fraction of the broad spectral component. This makes it possible to image polymer gel dosimeters using magnetization transfer (MT) based contrast [45].
Using an off-resonance RF pulse (figure 20), part of the macromolecular proton magnetization can be saturated ('demagnetized'). Due to magnetization transfer between the macromolecular proton pool and the liquid proton pool the loss of magnetization of the macromolecular proton pool will be reflected in a decrease in magnetization of the liquid proton pool. As a result, an MR signal decrease will be observed. For a quantitative interpretation of magnetization transfer between the different proton pools, the reader is referred to more specialized literature [46-48]

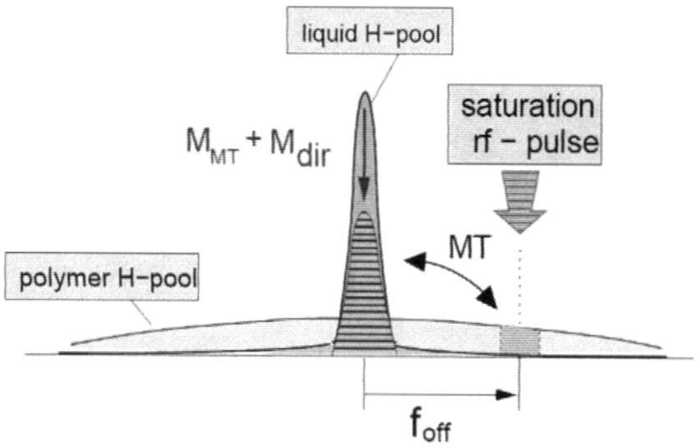

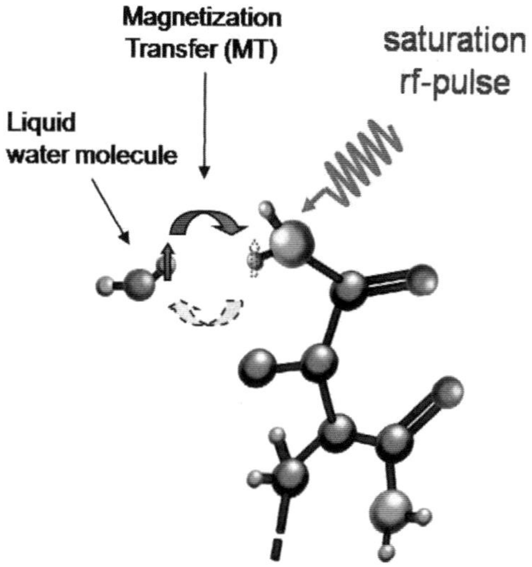

Figure 20. Principle of magnetization transfer. Short T2 (long R2) proton pools (such as from (bio)polymers and lipids) result in broadened spectral peaks. If an off-resonance RF pulse is applied, a fraction of these protons will be demagnetized. Due to the transfer of magnetization from the 'liquid' proton pool to the demagnetized protons the signal from the liquid proton pool will be decreased.

In the case of low-density foam gel dosimeters, R2 mapping is no longer possible due to the signal loss caused by diffusion weighting in the microscopic background magnetic field gradients that originate from susceptibility differences between the gas bubbles and the gel fraction [49]. It has been shown that quantitative magnetization transfer imaging at the other hand is well suited for these lung equivalent gel dosimeters [50].

3.3.1. MT prepared sequence

A pulsed MT prepared spin-echo sequence is shown in figure 21. The sequence consists of a magnetization preparation part and a readout part. The magnetization preparation consists of a series of off-resonance MT saturation pulses. The number (N_{sat}), frequency offset (f_{off}) and intensity of the MT saturation pulses can be varied by the user. In theory, a single MT pulse with a long duration and significant amplitude can be used. This is recognized as the initial continuous wave MT sequence. Most scanners however, do not allow such long RF pulses because of RF energy restrictions (patient safety limits and coil protection). The readout part can be any basic imaging sequence scheme such as spin-echo, fast spin-echo, gradient echo or EPI. A quantitative MT ratio image can be obtained by acquiring images with different MT saturation pulse intensities. Also the offset frequency can be varied.

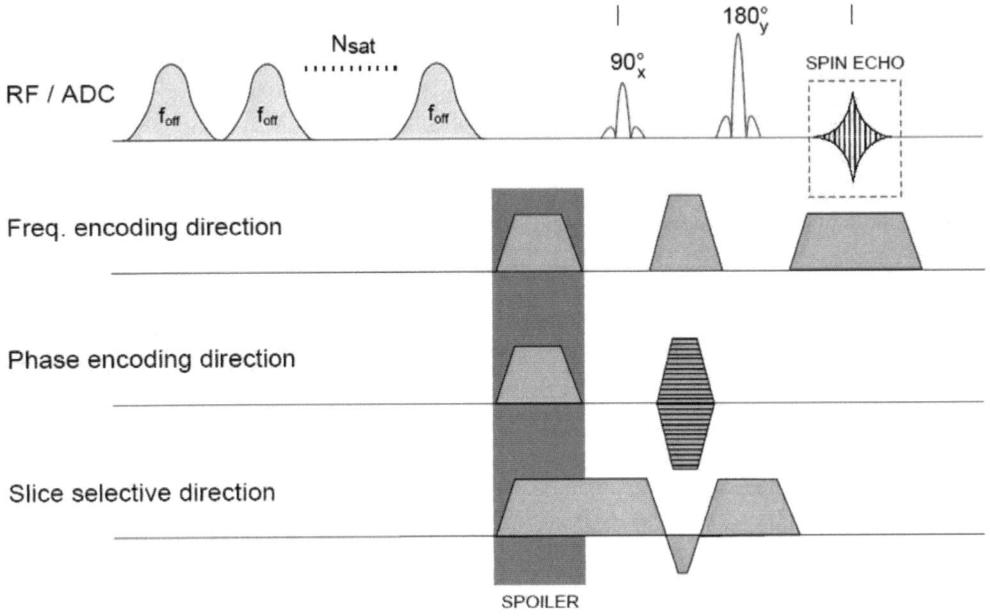

Figure 21. Magnetization transfer imaging sequence with MT magnetization preparation consisting of a number (N_{sat}) of off-resonance MT saturation pulses, spoiler gradients and a spin-echo readout sequence. The number, intensity and frequency offset of the MT saturation pulses can be varied.

The observed magnetization in the MT experiment is M_{SAT}. The observed relative signal decrease is called the magnetization transfer ratio MTR and is defined as

$$MTR = \frac{M_0 - M_{SAT}}{M_0} \tag{24}$$

with M_0 as the equilibrium magnetization, which is obtained by scanning without MT saturation pulses. The MTR consists of two contributions: M_{dir}, the direct effect contribution which is due to the saturation of the water proton pool, and *M*MT, the true magnetization transfer contribution due to magnetization transfer between the water proton pool and the polymer proton pool

$$MTR = \frac{M_{dir}}{M_0} + \frac{M_{MT}}{M_0} \tag{25}$$

The direct effect M_{dir} is experimentally determined from the signal intensity in a water sample M_{H2O}. As a result, the true magnetization transfer ratio, $MT=M_{MT}/M_0$ is given by

$$MT = \frac{M_{MT}}{M_0} = MTR - \frac{M_{dir}}{M_0} = \frac{M_{H_2O} - M_{SAT}}{M_0} \tag{26}$$

An MT offset frequency dispersion plot can be obtained by recording the signal intensity as a function of the MT saturation pulse frequency offset (figure 22b). An optimum offset frequency can be determined from this plot.

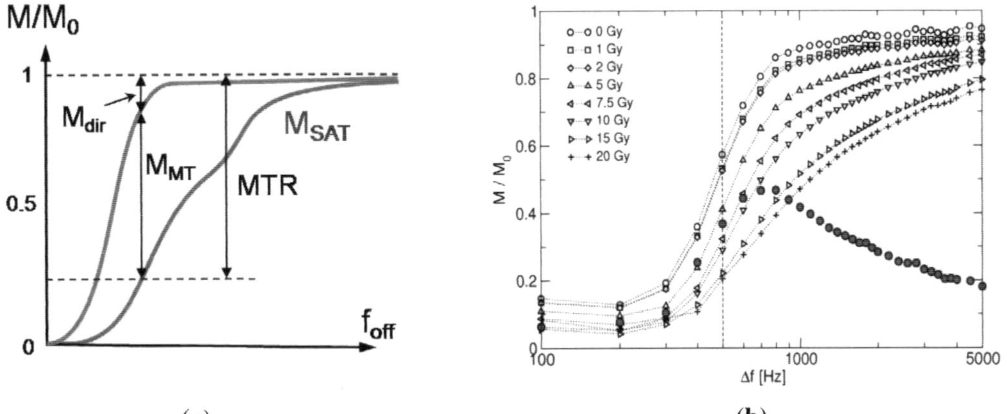

Figure 22. The observed relative signal decrease MTR is due to both direct saturation of the water protons M_{dir} and to magnetization transfer between the water proton pool and the polymer proton pool M_{MT} (a). A spectral MT measurement of a low-density methacrylic acid polymer gel foam (b). The red dots represent the MT difference between a sample irradiated to 20 Gy and an unirradiated sample.

The correlation between the number of MT saturation pulses and the MT weighting can be analytically derived [51].

3.3.2. Pulsed magnetization transfer steady state imaging

As for R1 and R2 mapping, also for MT imaging, a steady state sequence can be used [52]. A schematic drawing of the pulsed MT steady state sequence is shown in figure 23. The pulse sequence consists of three distinct elements: An off-resonance MT saturation pulse (with duration Δ, frequency offset f_{off} and amplitude θ) succeeded by a spoiler gradient, a small RF pulse with flip angle α which produces transverse magnetization for the liquid pool but has negligible effect on the semisolid pool, and finally a repetition time TR, which incorporates spatial encoding (gradients not shown in figure 23) and evolution of image contrast.

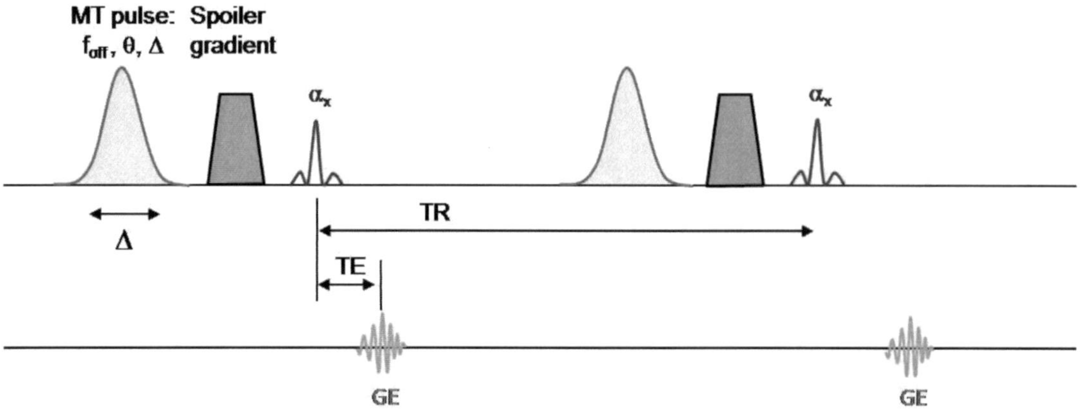

Figure 23. The pulsed magnetization transfer steady state imaging sequence consisting of MT pulses with spoiler gradient (with total time duration T) and a small flip angle excitation (flip angle α).

Image contrast is determined by five parameters: the MT pulse amplitude θ and duration Δ, the MT pulse offset frequency f_{off} and the spoiled gradient echo parameters α and TR. The correlation between the true magnetization transfer ratio, MT, and the two pool model can be described by a numerical model [53].

A different steady state sequence has been proposed by Gochberg et al [54] in which a train of RF pulses is used that selectively invert the magnetization of the free water protons while the macromolecular proton pool is minimally disturbed.

3.3.3. Stimulated echo preparation sequence

In describing the magnetization transfer experiment, a two compartmental model (figure 24) consisting of both a liquid (A) and a macromolecular proton pool (B) can be considered. This model is similar to a pharmacokinetic dilution model. If in pool A the protons are labelled (filled circles), magnetisation transfer (described by the exchange rates k_{AB} and k_{BA}) causes redistribution of the labelled protons. Also R1 relaxation takes place in both pools decreasing the labelled protons ($R1_A$ and $R1_B$). The process can be modelled by differential equations describing the change in magnetization in both pools

$$\begin{aligned} \frac{dM_A(t)}{dt} &= -M_A(t)\cdot(R1_A + k_{AB}) + M_B(t)\cdot k_{BA} \\ \frac{dM_B(t)}{dt} &= -M_B(t)\cdot(R1_B + k_{BA}) + M_A(t)\cdot k_{AB} \end{aligned} \quad (27)$$

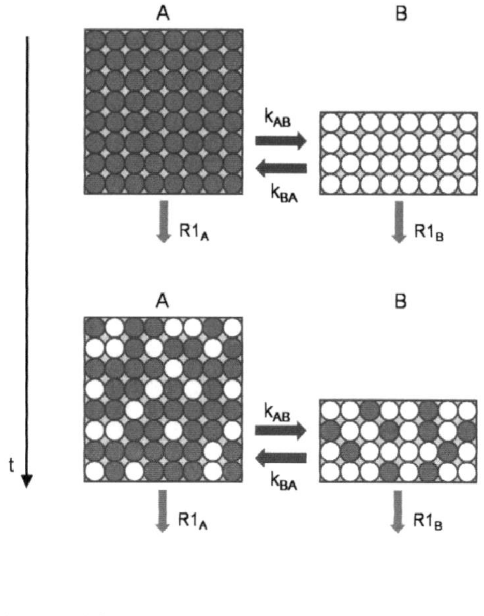

Figure 24. Evolution with time of a two compartment model consisting of 2 proton pools A and B with respective exchange rates k_{AB} and k_{BA} and R1 relaxation rates $R1_A$ and $R1_B$.

where $M_A(t)$ and $M_B(t)$ refer to the free and bound pool magnetization of the labelled spins, respectively. In the stimulated echo preparation sequence (figure 25a), labelling of the protons in the liquid proton pool (A) is performed by a spin preparation consisting of a 90° RF pulse followed by a modulation gradient and a second 90° RF pulse that flips the prepared transverse magnetization back in the longitudinal direction. The spins in the bound proton pool are not affected by this preparation because the transverse magnetization will vanish during a typical pulse separation time of a few milliseconds due to the short T2 decay of the bound protons. Solving the differential equation 27 with the boundary condition $M_B(t=0) = 0$, yields a bi-exponential decay of the magnetization in the free proton pool.

$$M_A(t) = M_0 \cdot \left(C_1 \cdot \exp(-\lambda_1 \cdot t) + C_2 \cdot \exp(-\lambda_2 \cdot t)\right) \tag{28}$$

λ_1 is a "fast" rate responsible for the quick approach to a steady state between both pools while λ_2 is a "slow" rate is roughly comparable to R1. The time t in equation 28 is the time after the sin preparation (i.e. time after the second 90° RF pulse). The different coefficients can be fitted from a biexponential fit to a stimulated echo dataset obtained with different mixing times. From these coefficients the fraction of bound proton pool at total proton pool can be derived.

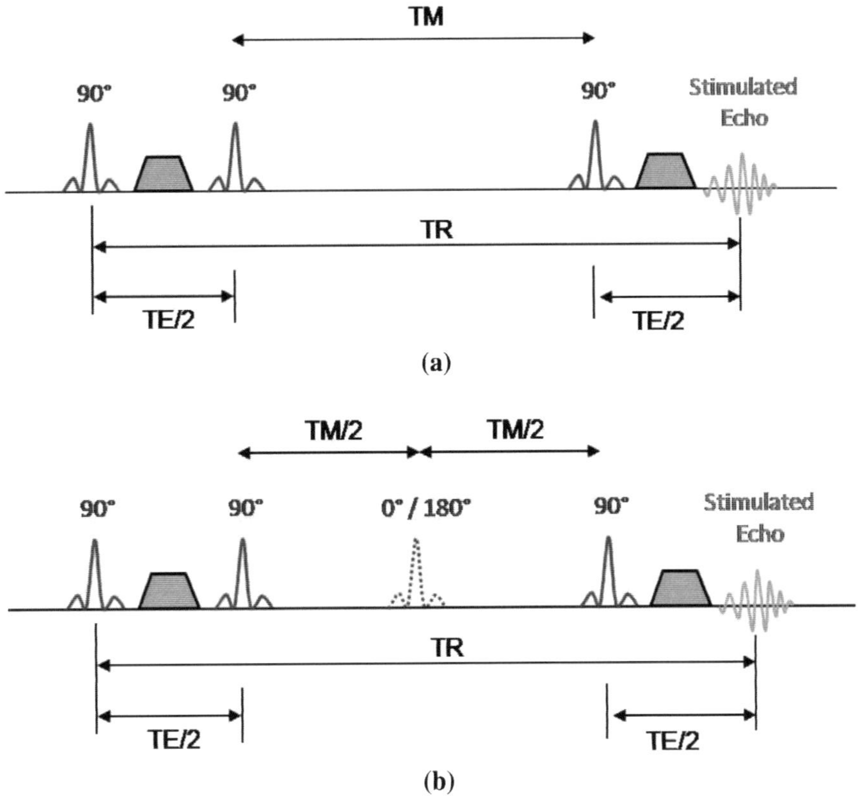

Figure 25. Stimulated echo preparation sequence for labelling spins.

A modified version of the stimulated echo sequence requiring only two measurements to extract the pool fractions is shown in figure 25b. This sequence is run twice: Once with a 180° RF pulse during the mixing time (time between the second and third 90° RF pulse) and once without the 180° RF pulse [55]. With a non-selective 180° RF pulse (1 ms) placed in the centre of the mixing period which changes the sign of the phase of the free-pool spins only, an imbalance is achieved. After the 180° RF pulse, the originally labelled spins in pool B will be diluted by the spins in pool A, which have opposite phase. Conversely, the spins in pool A, with opposite phase, will be diluted by the originally labelled spins from pool B. The pool fraction can be determined from both measurements.

4. How to evaluate the performance of the imaging sequence?

4.1. Accuracy and imaging artefacts

Whatever imaging technique is applied to map a dose-effective contrast (R1, R2, MT, optical absorbance, X-ray absorbance, ultrasound attenuation, dielectric constant, etc.) there are two essential criteria that should be accomplished in the imaging technique: First of all, the images should be accurate both in space as in dose (i.e. not contain systematic errors) and secondly, the stochastic deviations on each voxel should be as low as possible within a reasonable measurement time and for a certain image resolution. An interesting parameter to evaluate the stochastic deviations is the minimum detectable dose difference, defined as the dose resolution [56].

The target figure of accuracy that is aimed in gel dosimetry for high-precision radiotherapy is about 2–3% of the maximum dose in regions of homogeneous dose and a spatial error of less than about 2 mm in regions of high-dose gradients. However, in a conventional MR scanner, several imaging artefacts may cause errors in the final dose map. These errors may be classified in dose inaccuracies or in deformations of the dose maps. Studies of these different artefacts have resulted in different compensation strategies.

In treating imaging artefacts, we will mainly focus on T1 and T2 measurements but many of the artefacts that occur in these measurements may also appear in images obtained with other MR sequences. The spatial accuracy of the dose maps is affected by *geometrical distortions*. Eddy currents, main magnetic field inhomogeneities and gradient inhomogeneities cause machine related geometrical distortions, while susceptibility variations and chemical shift variations in the imaging volume cause object related geometrical distortions. Also *dose inaccuracies* can be machine related such as these caused by eddy currents, stimulated echoes, inhomogeneities in the B1-field and standing radiofrequency (B1) waves or they can be object related such as temperature drift and molecular self-diffusion. Table 1 gives an overview of the most important MRI artefacts that may compromise the reliability of gel dosimetry.

Table 1. Overview of important artifacts that may occur in MRI gel dosimetry classified by two criteria. The abbreviation as most often used in the literature is also given.

Geometrical distortions		Dose inaccuracies	
Machine related	Object related	Machine related	Object related
Field inhomogeneity $B_0(x,y,z)$	Susceptibility differences $\Delta\chi(x,y,z)$	Eddy currents $\Delta B_0(t)$, $\Delta g_{x,y,z}(t)$	Temperature Drift $T(t)$
Gradient non-linearity $g(x,y,z)$	Chemical shifts $\sigma(x,y,z)$	Stimulated echoes $M_{xy}(TE)$	Molecular self-diffusion $D(x,y,z)$
Eddy currents $\Delta B_0(t)$, $\Delta g_{x,y,z}(t)$		B1 field Inhomogeneity $B_1(x,y,z)$	
		Slice profile $M_{xy}(z)$	
		Standing waves $B1(x,y,z)$	

Journal of Physics: Conference Series **164** (2009) 012033 doi:10.1088/1742-6596/164/1/012033

4.1.1. Image non-uniformities

Manifestation

A parametric MR image of a uniform sample may demonstrate areas of non-uniform signal intensity. For gel dosimetry purposes, non-uniformities result in areas of over- and under-dosages if not corrected for.

Causes

Image non-uniformities will mainly occur in images recorded with a surface coil acting as a transmitter. A realistic transmitting RF coil produces a non-uniform RF field also called B1-field. Some imaging sequences are more sensitive to B1-field inhomogeneities than others.

In T1 imaging (for Fricke gel dosimetry), severe non-uniformities were found when a spin-echo sequence was used with different repetition times (TR) [57]. These non-uniformities are attributed to a non-uniform B1 field in combination with inherent differences in T1 relaxation for different relaxation times. These non-uniformities were found to be considerably less pronounced when an inversion recovery sequence was used with different inversion times (TI).

In T2 imaging for monomer/polymer gel dosimetry using a multi-echo sequence in combination with a circularly polarized (CP) transmit/receive head coil, the R2 map was found to be uniform over an area of 120 cm in the centre of the coil, while the R2 values (apparent dose values) decreased considerably near the edges of the coil [43] (figure 26d). Using a body coil, the B1-field is far more uniform (figure 26a) resulting in a much more uniform R2 map (figure 26c). However, measuring R2 using the body coil both as transmitter and receiver is at the cost of signal-to-noise (SNR).

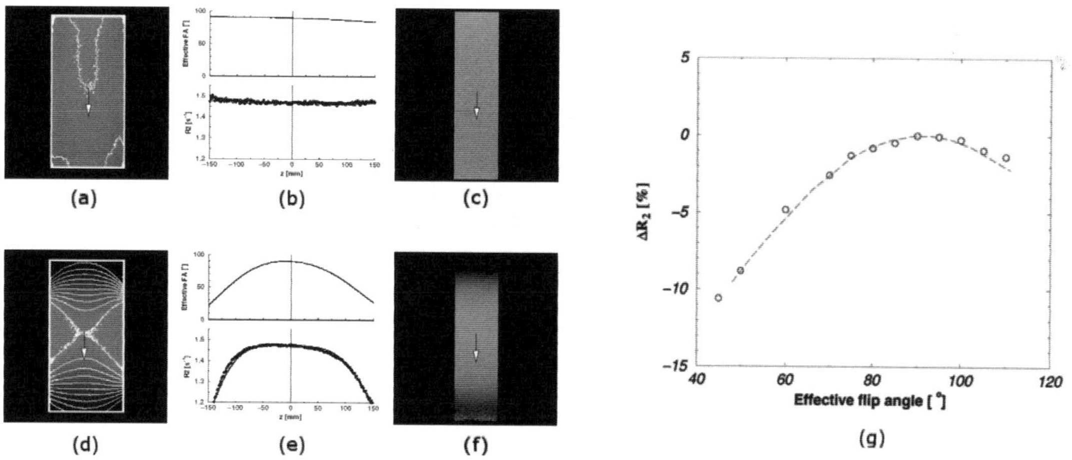

Figure 26. B1 field map (effective flip angle) acquired with a body coil (a) and with a transmit/receive head coil (d). R2 images of a blank non-irradiated polymer gel dosimeter acquired with the body coil and head coil are shown in (c) and (f) showing the effect of the inhomogeneous B1-field for the head coil. The field-of-view (FOV) is 320 x 320 mm^2. Corresponding longitudinal profiles through the center are also shown (b and e). The plot in figure (g) gives the relation between the change in R2 and the flip angle for a nominal flip angle of 90°.

The plot of change in R2 versus effective flip angle provided in figure 26g is dependent on the rf pulses used in the multiple spin-echo sequence. This relation can be theoretically derived using numerical simulations and can be experimentally derived by acquiring the signal of a homogeneous (blank) phantom in the centre of the coil for different excitation and refocusing flip angles [43].

There are several sources of non-uniformities in spin-echo images. These sources include B1-field inhomogeneity, RF standing waves, skin effect at high field strengths, crosstalk between multiple slices, mistuning of the head coil, eddy currents and static field inhomogeneities [58]. Most of these non-uniformities scale with the image intensity in the base images. As a result, the artefacts are cancelled out after fitting the quantitative MR parameter (T1, T2 or MT). However, some of these artefacts may not vanish completely in the parametric images if some other sequence specific mechanisms interfere. In clinical MR-scanners, the non-uniformities in the T1 and T2 images are mainly caused by B1 field inhomogeneities. It is well known that the B1 field generated by a radio-frequency coil is not completely homogeneous. B1 field inhomogeneities are attributed to RF amplifier distortions [59], digitization of the RF pulse [60], RF coil geometry [61-63] and penetration in the scanned object [64,65].

In the quantitative T1 spin-echo experiment, the incomplete excitation in some parts of the phantom due to B1-field inhomogeneity makes that the restoration of the longitudinal component of the magnetization is different for different repetition times. It is clearly seen that the inhomogeneous signal intensity pattern varies over the different T1 weighted base images [57]. Also in our T2 studies, the non-uniformities are mainly caused by an inhomogeneous B1-field in combination with stimulated echoes. As a result of the inhomogeneous B1-field, the flip angle of the nuclear magnetic moments after an excitation or a refocusing pulse will not be equal throughout the scanned volume. Thus the history of spin magnetization during a multiple spin-echo sequence will depend on the position of the nuclear moments in the slice and therefore the deviation in the R2 value will depend on the position [42]. Computer simulations solving the Bloch equations for the multi-echo sequence predict the correlation of the measured R2 with the local B1 field. The simulations are completely in accordance with the experimental findings. By mapping the B1 field it is now possible to predict how the R2 deviations will vary with slice position [42].

Another source of non-uniformities is an inhomogeneous temperature drift of the gel phantom during scanning. Due to RF power absorbed by the gel, a temperature rise in the order of 1-3 °C is not uncommon. As the outer boundaries of the phantom are surrounded by air (mostly in flow), temperature differences in the phantom will build up [43]. As the T2 of the dosimeter gel is temperature dependent, dose errors in the order of 3% to 10% (relative to the maximum dose) can be expected.

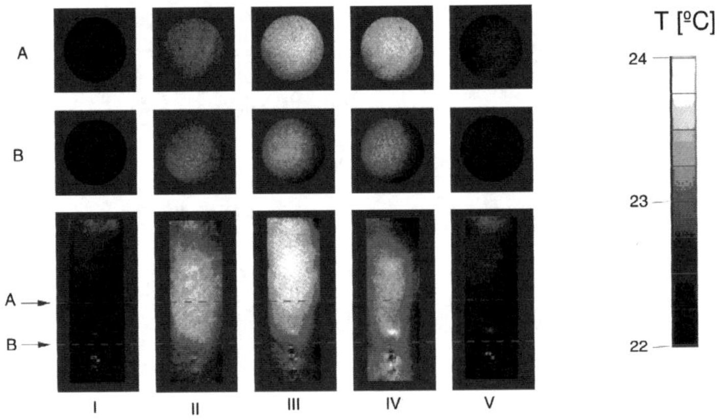

Figure 27. Temperature images recorded with the proton resonance frequency method at several time points during a multi-echo sequence. The interval between two successive maps is approx. 4 hours. The sequence was stopped after the recording of image IV. The scale is in degrees Celsius.

Compensation strategies

It has already been mentioned that T1 images can be obtained by an alternative spin inversion recovery sequence [57] if the uniformity in the T1 images exceeds 2-3% with the spin-echo sequence. The B1 field can be measured by several ways and can be used to correct the R1 and R2 images. However, in that case, also the influence of the B1 field upon the R1 and R2 image has to be known. An analytical expression cannot be derived easily in the case of a multiple spin-echo sequence because of the rapid (exponential) increase in magnetization pathways after a few imperfect RF pulses. Computer simulations solving the Bloch equations may help in deriving this correlation. A more practical, but theoretically less founded method is to measure the R2 distribution in a homogeneous phantom (for example, the gel phantom before irradiation) and using this image-set as a template to correct the resulting R2 images. It is obvious that these post-processing corrections are founded on the fact that the position of the phantom can be determined accurately. In applying this strategy, quality control of the reproducibility of the non-uniformities and the dependency of the non-uniformities on phantom shape should be performed on a regular basis. A more direct way of minimizing B1-field related non-uniformities is by using the body coil as a transmitter at the cost of a lower signal-to-noise ratio. Some scanners are equipped with receive-only coils that can be used in combination with the body coil acting as transmitter. This results in a more homogeneous B1-field while still preserving a good signal-to-noise ratio. The temperature drift related non-uniformities can be disposed of by centric k-space filling [42].

As non-uniformities may lead to severe dose errors and are very likely to occur to some extend in three dimensional scanning of the gel phantom, it is of ultimate importance that the uniformity of the parametric images is measured in advance and that scanning is performed using adequate compensation strategies to keep the non-uniformity level below acceptable levels.

4.1.2. Image distortions

Manifestation

The geometrical shape of the phantom is deformed in the dose maps (parametric images). These deformations may have an effect on the global image or may occur only locally in part of an image, especially around inclusions of other structures in the phantom. Some structures of the phantom may be displaced with respect to others. Important to note is that displacements and scaling of the whole phantom may occur both in the image plane as out of the image plane.

Causes

Low magnetic field scanners may suffer more from machine related image distortions than cryogenic magnets. The inclusion of other structures such as air cavities and low-density materials may result in object related distortions in the vicinity of these structures. Slice displacements may occur in scanners with non-active shielded gradients.

Spatial encoding of the recorded MR-signal is based upon the correlation of magnetic field strength and the frequency of the rf pulses, described by the Larmor equation (equation 1). By switching linear magnetic field gradients in three orthogonal directions during the MR-sequence, an encoding in three orthogonal directions is obtained. The magnetic field gradients, produced by gradient coils, may deviate from the assumed linearity to some extent. Furthermore, the main magnetic field, B_0, is not completely homogeneous. This results in a deformation of the recorded image. Similar encoding errors may occur during phase encoding and slice selection .

To measure the magnitude of deformations from non-linear gradients and main magnetic field inhomogeneities several phantoms have been proposed [66-68]. In observing in-plane distortions a pin-cushion phantom is often used. To account for errors in the construction of the phantom, the phantom is first scanned by use of CT. By overlaying the images indicating the positions of the tubes of the MR images on the CT images a distortion map can be derived. Apart from these static deviations in the magnetic field and gradients, the magnetic field may also deviate from the theoretical expected shape due to time-varying magnetic field components. These time-dependent magnetic field deviations are caused by "eddy currents". The eddy currents are induced through switching of the imaging gradients giving way to an electromotive force that acts on the cryostat and metal casings. Several techniques to measure and analyze eddy currents in clinical scanners have been described [39, 69-76]. The eddy current induced magnetic field is mostly described by a first order approximation composed of a global offset of the main magnetic field, $\Delta B_0(t)$ and a change in the magnetic field gradients, $\Delta g_i(t)$ (i = x,y,z) [69]. The eddy current related spatial encoding errors are mainly due to eddy current fields that are present during frequency encoding and slice selection. Depending on the moment in the sequence that eddy currents are present, they will cause slice shifts or slice tilting in different directions. To measure slice displacements, a pyramidal phantom can be used [39]. A special designed quality control phantom consisting of a perspex plate with holes drilled in special directions and filled with gel can also visualize slice warping (see further).

To extend gel dosimetry to phantoms that include air cavities (for example to investigate effects of electronic disequilibrium), materials with different electron densities are inserted in the gel phantom. These materials very often have a different magnetic susceptibility. This will result in susceptibility related distortions in the base images and in the final parametric images. The distortion is inversely proportional to the receiver bandwidth. Also the inclusion of materials with a different chemical composition (organic liquids, fats) in the gel may result in a (chemical) shift of these objects in the frequency encoding direction.

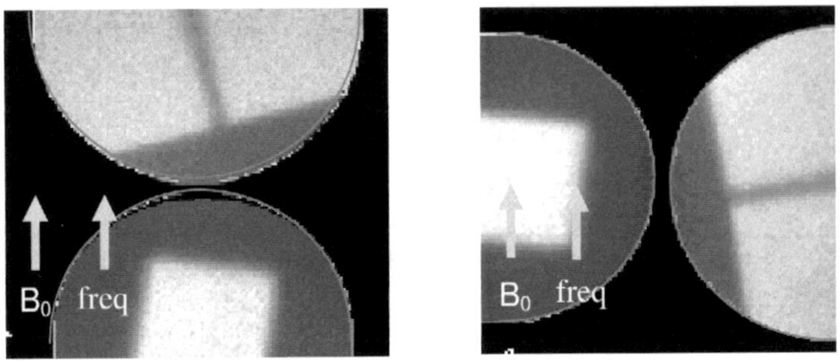

(a) **(b)**

Figure 28. Two cylindrical gel phantoms scanned in different directions illustrate susceptibility related deformations at the interface between both recipients. The circular lines are drawn on the image to show the actual boundary of the phantoms.

An example of a susceptibility related artefact is shown in figure 28a. This example was obtained by coincidence: In order to save some measurement time, two gel phantoms were placed in the scanner at the same time. The magnetic field and the frequency encoding direction are oriented upward in the images. It can be clearly seen that the phantoms are distorted near the interface between the phantoms. When the phantoms are oriented perpendicular to the magnetic field, the artefact disappears. Also changing the frequency and phase encoding direction will make the distortion disappear. Another example of susceptibility artefacts caused by a guiding catheter of a brachy source can be found elsewhere [77]. In this example the magnetic field distortion is shown as imaged and calculated numerically.

Compensation strategies

External (fiducial) markers on the phantom are often used to allow image fusion with computer planning or other dosimetry techniques. These external markers may give a first indication of deformations or slice displacements. A scaling error of the phantom can also be detected by use of image fusion/matching software.

To measure machine related distortions, use can be made of a dedicated quality control phantom (as discussed further). A distortion map can be derived and used as a correction matrix for the dose map as has been done for MR-guided stereotactic neurosurgery [68].

Object related geometrical distortions may be compensated by first measuring or simulating the local magnetic field distortions caused by susceptibility differences and chemical shift artefacts. These magnetic field maps can then be used to construct a correction template which can be used to correct the parametric images [78]. Another method to correct for local magnetic field distortions is by a technique called 'view-angle-tilting' [79]. This method has the advantage that no post-processing is needed but has the disadvantage that the point-spread function is broadened which may lead to an erroneous interpretation of the spatial resolution. Important to emphasize is that the artefacts act on a pixel-related scale instead of a geometrical scale. Increasing the resolution will decrease the artefact on a geometrical scale. Another important parameter is the receiver bandwidth. The distortions are inversely proportional to the receiver bandwidth. However increasing the bandwidth will decrease the signal-to-noise ratio.

4.1.3. Dose errors

Manifestation
Dose errors cannot be observed directly by inspection of the parametric images or dose maps but may be discovered by comparison of the gel measured dose values against dose values measured with other dosimetry technique. In some cases, the dose response may be dependent on the scanning orientation and other sequence parameters such as the field-of-view (FOV). The whole dose response can be higher or lower with a constant offset or the dose sensitivity can be altered.

Causes
With the use of a multiple spin-echo sequence the dose-response curve can be dependent upon scanning orientation and image parameters (field-of-view). These effects may be more pronounced in scanners with non-active shielded gradients and in applications where high gradient strengths and short echo times are used. It is well known that the R2 values of the gel are temperature dependent [10,19]. A temperature increase during scanning may result in a dose-underestimation. If the gels are not thermally equilibrated before scanning, temperature changes may cause severe dose errors. The chance for dose errors to occur is higher if scanning of the calibration tubes is performed in different circumstances and imaging parameters than the gel phantom.

The effect of eddy currents on image deformation has already been discussed in previous paragraph. The magnetic fields induced by eddy currents experience a certain decay time. The succession of many imaging gradients in a multi-echo sequence may lead to a continuous increase of magnetic field during the start of the sequence. From measurements of eddy current created by gradient trains containing different numbers of gradients it is found that saturation in the eddy current induced magnetic field offset is obtained after 10 gradients [39]. Through computer simulations (solving the Bloch equation numerically) it is proven that the continuous change in the eddy current induced magnetic field offset during the first 10 echoes leads to a deterioration of the slice profiles which results in a change of the measured T2-decay curve. As the eddy currents (and especially the magnetic field offset) are dependent on the imaging direction, the disturbance of the excitation history of stimulated echoes is different and therefore also the measured T2 values.

Partial volume effects may also lead to a misinterpretation of the dose in each pixel. It has been shown that partial volume effects can cause severe dose discrepancies in pixels adjacent to a point source as occurring in a brachytherapy experiment [77]. Also outer-volume effects may result in dose errors. In-plane outer volume effects are related to the point spread function while in the slice selective direction outer volume effects may be expected from imperfect slice profiles. In multi-slice imaging care has to be taken in excitation of the slices. Even if excitation of the different slices is performed interleaved, cross-talk between slices may lead to inter-slice variations.

Compensation strategies

The addition of a gradient train may bring the eddy currents in a steady state regime. It is shown experimentally and through computer simulations that by this preparation gradient train, the slice profiles are shifted in the slice-encoding direction but that the shape of the slice profiles is preserved. It is advisable for other scanners to measure the eddy current fields for different numbers of gradients in order to determine the required number of gradients to bring the eddy currents in a steady state regime.

Another source of dose errors is related to a temperature drift in the phantom during scanning. Although precautions are made in letting the gel phantom equilibrate at the room temperature of the scanner, a temperature increase may still occur during scanning by absorption of the RF energy of the excitation and refocusing pulses. In figure 29, the temperature in the gel phantom during a long scanning experiment is shown.
Note that a temperature increase is present during T2 scanning but not during temperature mapping (FLASH-sequence) illustrating the high-energy transmittance during the multiple spin-echo sequence. The sequence can be made significantly less sensitive to temperature drift by using a centric k-space ordering scheme instead of a standard linear k-space ordering scheme [42].

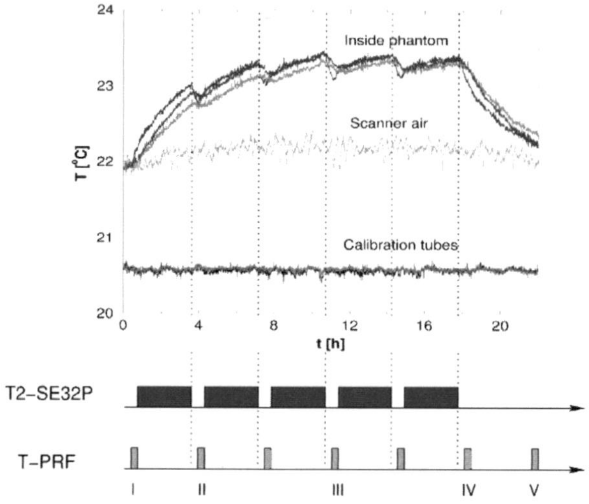

Figure 29. Temperature course during a long-term scanning experiment of a cylindrical gel phantom. The temperature is recorded by use of a fluoroptic thermometry system with 6 operating channels (probes). The temperature distribution is measured using an MRI temperature mapping using the proton resonance shift method. The measurement history is also provided in a corresponding timetable.

Partial volume effects can be minimized by using special designed slice-selective RF pulses and by increasing the time between excitation of adjacent slices.

Some websites that provide lists of other imaging artefacts that may occur in clinical imaging applications are provided in the references [60-62].

4.2. Precision and acquisition time

The overall dosimetric precision is governed by variations in the several operations that take place in the dosimetry experiment. The first step in a gel dosimetry experiment is weighing the chemicals. Stochastic variations in the weighting will result in variations in the measured dose-related value (R2, MT, OD) as the dose-response is determined by the chemical composition. It is found that other manufacturing conditions may also have an influence on the dose-response such as the temperature during fabrication. Stochastic variations in the controlled temperature will therefore also lead to variations in the measured dose-related value. Also during irradiation there are different sources of stochastic variable contributions that determine the overall dosimetric precision such as variations in the dose delivery, variations in the temperature during irradiation and stochastic variations in the positioning of the calibration phantoms. Any form of scanning the gel dosimeter will introduce thermal detector noise. The noise contribution is determined by some scan parameters. Often, the scan parameters can be optimized in order to achieve an optimal figure of precision.

The concept of dose-resolution was introduced to evaluate *the intrinsic dosimetric precision* in terms of dose sensitivity and scanning signal-to-noise [56]. The dose resolution, written as D_Δ^p, is defined as the minimal detectable dose difference within a given level of confidence, p. The dose resolution is related to the standard deviation on dose σ_D by the equation

$$D_{\Delta}^{p} = k_p \cdot \sqrt{2} \cdot \sigma_D \tag{29}$$

For a 95% confidence level the dose resolution becomes $D_{\Delta}^{95\%} = 2.77 \cdot \sigma_D$.

In most radiation dosimetry experiments, gel dosimeters are used in a relative manner in the sense that the dosimeter is exposed to the same treatment as the patient but with a different total radiation dose. The total dose delivered to the dosimeter is scaled to cover the active dose range of the dosimeter. In this context, it is preferable to use the concept of dose resolution relative to the operating dose range, here defined as relative dose resolution:

$$D_{\Delta\%}^{p} = \frac{D_{\Delta}^{p}}{(D_{\max} - D_{\min})} = \sqrt{2} \cdot k_p \cdot \left(\frac{\sigma_D}{D_{\max} - D_{\min}} \right) \tag{30}$$

If the dose maps are derived from quantitative NMR-Ψ maps, it can be shown that the relative dose resolution ($D_{\Delta\%}^{p}$) is equal to the relative Ψ resolution ($\Psi_{\Delta\%}^{p}$) which is defined in a similar way:

$$D_{\Delta\%}^{p} = \sqrt{2} \cdot k_p \cdot \left(\frac{\sigma_D}{D_{\max} - D_{\min}} \right) = \sqrt{2} \cdot k_p \cdot \left(\frac{\sigma_\Psi}{\Psi_{\max} - \Psi_{\min}} \right) = \psi_{\Delta\%}^{p} \tag{31}$$

with Ψ either R1, R2, MT or any other contrast. This also applies to non-MRI contrasts.

It should be emphasized that dose-resolution as defined here does not include stochastic variations in chemical concentrations, in dose delivery or in the calibration procedure. For that reason, dose-resolution can be considered as an intrinsic lower limit of dosimetric precision. It is a misconception that the dose-resolution would be a parameter that is only related to the type of gel dosimeter. In some publications, the concept of dose resolution has been misused as the criteria to compare different types of gel dosimeters. This is misleading as these studies report on dose resolutions obtained with sub-optimal scanning parameters. A completely different outcome would be obtained when different scanning parameters would be used. The concept of dose resolution however is very practical to optimize the intrinsic NMR sequence in terms of intrinsic precision [40,41]. In optimizing the NMR sequence, it is also important to take into account the number of slices that are required for the 3D dosimetry application as the optimization may also be dependent on the number of slices.

A dose-Ψ curve is used to calibrate the Ψ map [41] (again with Ψ either R1, R2, MT or any other contrast). The uncertainty on the dose value σ_{D*} extracted from a linear dose-Ψ plot with equation $\Psi = \Psi_0 + \alpha \cdot D$ is given by

$$\sigma_{D^*} = \frac{\sigma_c}{\alpha} \cdot \sqrt{\frac{(D^* - \bar{D})^2}{\sum_{i=1}^{N_{cal}} (D_i - \bar{D})^2} + \frac{1}{N_{cal}}} \tag{32}$$

with σ_c the standard deviation on the property Ψ in the calibration points [41]. This value is derived from the standard deviation in a region of interest of the calibration vials σ_{ROI}. If N_{ROI} is the number of points in the region of interest (ROI), the standard deviation on the calibration point is given by $\sigma_c = \frac{\sigma_{ROI}}{\sqrt{N_{ROI}}}$.

D^* is the estimated dose, $\bar{D}$ is the mean dose of all dose values in the calibration plot (i.e. $\bar{D} = \frac{1}{N_{cal}} \sum_{i=1}^{N_{cal}} D_i$ with D_i the dose in the ith calibration point and N_{cal} the number of calibration points.

In order to make an equitable comparison between different imaging modalities, a figure of intrinsic precision on dose readout (IP_D) can be defined which is independent of spatial resolution and measurement time. The concept of intrinsic precision on dose readout (IP_D) can be used to compare the precision obtained with different imaging modalities such as MRI, optical CT and X-ray CT.

$$IP_D = \frac{1}{D^p_{\Delta\%} \cdot \Delta V \cdot \sqrt{t_{meas}}} \qquad (33)$$

with ΔV the real imaging voxel size and t_{meas} the measurement time. In comparing different imaging modalities it is also important to compare the IP_D for the same scanning volume. A large IP_D is the most optimal.

5. How to optimize my MRI readout protocol?

Here we provide some guidelines for optimizing your MRI readout protocol. Many of the elements mentioned here, also apply to other imaging modalities.

1. *Choose the contrast mechanism and imaging sequence that is suitable for your gel system.*
   Any gel system may have a different dose sensitivity with respect to different MR properties (such as R1, R2 and MT). For stable gel dosimeters, the choice of contrast mechanism and imaging sequence depends on the optimum intrinsic precision on dose readout that can be obtained. It should be mentioned that at a first glimpse a particular contrast and sequence may appear to have a better dose sensitivity, however it should be emphasized that also the measurement time, spatial resolution and total imaging volume should be taken into account when evaluating the overall precision on readout (equation 33).
   In some cases however, fast scanning is required to avoid deterioration of the measured dose distribution by instability of the gel system (such as in the case of diffusion of contrast carrying chemicals). In these cases, the accuracy requirement overrules the search for maximum precision.

2. *Evaluate the imaging sequence on quality control phantoms of which the MR properties (R1, R2, MT) are comparable to the gel dosimeter that you plan to use.*
   A thorough investigation of all the possible imaging artefacts (table 1) may be a tedious job. However, by scanning a blank gel phantom (an unirradiated phantom) the homogeneity in the dose maps can be investigated. The blank phantom should be scanned with the same scanning parameters and RF coil as in the normal gel dosimetry protocol. Also post-processing should be performed in the same manner. Alternatively, a liquid filled phantom can be used, but care should be taken with convection currents of the liquid that may occur in the phantom.

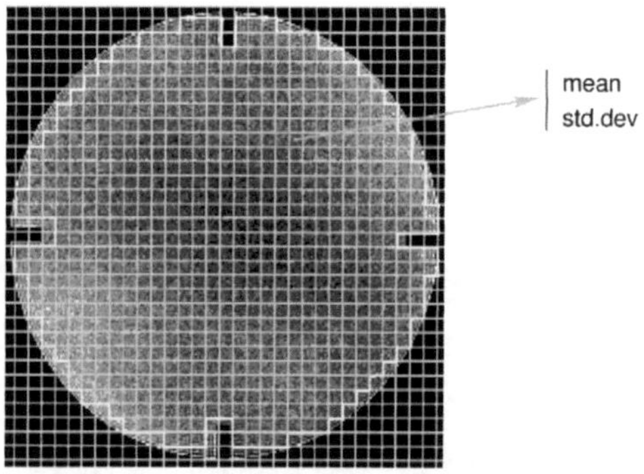

Figure 30. Subdivision of the image in square regions for statistical measurements

These currents may lead to additional motion artifacts that may be interpreted as non-uniformities. By subdividing the image in square regions of 7-by-7 pixels a distinction can be made between the stochastic and structural deviations. The stochastic deviation (noise) is obtained by calculating the average of the standard deviations in all square regions within the phantom. The structural deviation is obtained by calculating the standard deviation of the mean intensity values of all square regions.

Other parameters of interest are the slice profile, the actual spatial resolution, image deformations and slice warping. To investigate these properties, use can be made of a dedicated quality control phantom (figure 31).

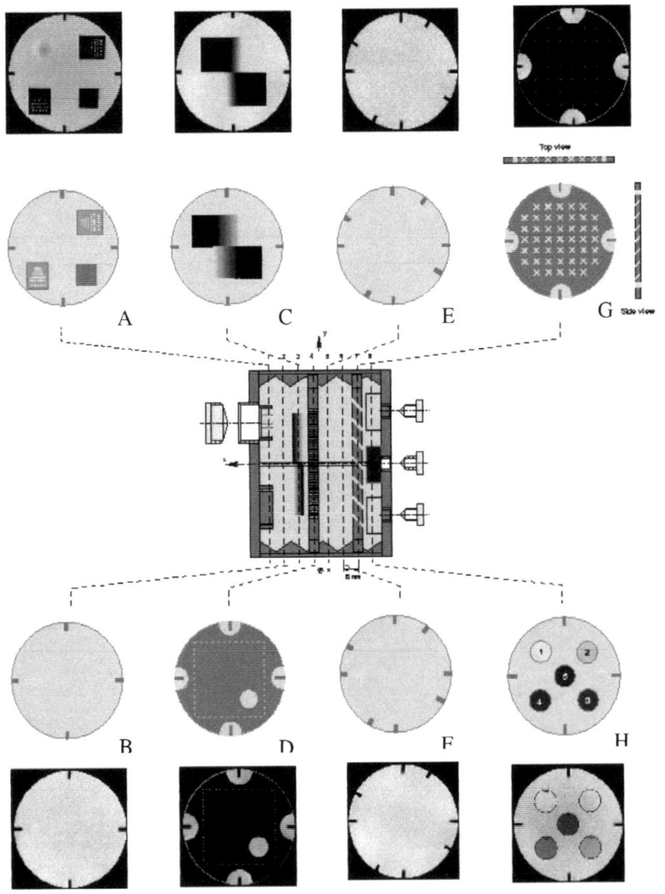

Figure 31. A multi-purpose quality control phantom of which eight image slices through the phantom provide a check of respectively (from left to right and top to bottom): image resolution (A), slice profile (C), signal-to-noise ratio (E), slice warp (G), ghost artifacts (B), pincushion distortion (D), quadrature errors (F) and susceptibility and chemical shift artifacts (H). With the drawing of the different sections, some images as obtained at an Expert 1T Siemens scanner (Ghent University hospital) are provided. For a more comprehensive description of the phantom see ref. 83.

3. *Optimize your sequence in terms of dose precision*
Imaging parameters (TE, TR, α, etc.) can be optimized with respect to the dose precision. The relative dose resolution L can be derived theoretically by analytical or by numerical means through a Monte Carlo random generation of signal noise in the base signal. Theoretical derivations for the optimization of echo time spacing and number of echoes for R2 measurements can be found in the literature [40,41]. Similar derivations can be made for other imaging sequences. Alternatively, the dose resolution can be determined experimentally for different imaging parameters.

4. *Optimize the calibration procedure*
   Enough calibration points should be taken. Both the number of calibration points and the dose precision of the calibration points determine the precision of the calibration (equation 32). The accuracy of the calibration procedure should be investigated by validation of the gel measured dose map of a known dose distribution. To minimize effects of chemical instability and temperature variations it is advisable to scan calibration vials together with the volumetric gel dosimeter phantom.

5. *Validate your sequence on a well defined test case of which dose profiles can be compared against some "gold standard".*
   This is an important step that is often neglected. By comparing dose profiles obtained with a golden standard against the gel measured dose profiles, an idea on the overall accuracy is obtained. It should be noted however that through this procedure not only the scanning protocol is validated, but also the fabrication and irradiation of the gel dosimeter (figure 32). Any error that occurs in the latter procedures may appear as a deviation between the gel measured dose distribution and the dose measured by the golden standard.

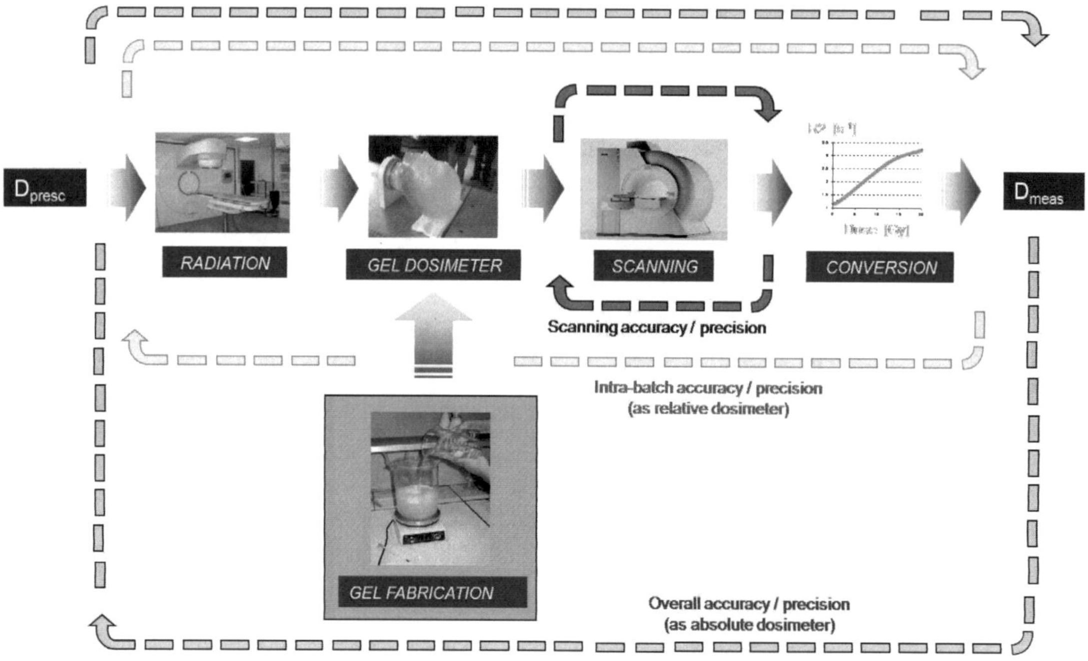

Figure 32. Gel dosimetry is performed in different stages. Errors can occur at each of the stages, leading to a decrease in the overall precision and accuracy. The scanning accuracy and precision can be determined by scanning a phantom with a predefined distribution of MR contrast (R1, R2, MT) several times.

References

[1] Gore J C, Kang Y S and Schulz R J 1984 Measurement of radiation dose distributions by nuclear magnetic resonance (NMR) imaging *Phys. Med. Biol.* **29** 1189–97

[2] Fricke H and Morse S 1927 The chemical action of Roentgen rays on dilute ferrosulphate solutions measure of dose *Am. J. Roentgenol. Radium Therapy Nucl. Med.* **18** 430–2

[3] Olsson L E, Petersson S, Ahlgren L and Mattsson S 1989 Ferrous sulphate gels for determination of absorbed dose distributions using MRI technique: basic studies *Phys. Med. Biol.* **34** 43–52

[4] Baldock C, Harris P J, Piercy A R and Healy B 2001b Experimental determination of the diffusion coefficient in two-dimensions in ferrous sulphate gels using the finite element method. *Australas Phys Eng Sci Med* **24** 19-30

[5] Podgorsak M B and Schreiner L J 1992 Nuclear magnetic relaxation characterization of irradiated Fricke solution *Med. Phys.* **19** 87–95

[6] Duzenli C, Sloboda R and Robinson D 1994 A spin-spin relaxation rate investigation of the gelatin ferrous sulphate NMR dosimeter *Phys. Med. Biol.* **39** 1577–92

[7] Audet C and Schreiner L J 1997 Multiple-site fast exchange model for spin-latttice relaxation in the Fricke-gelatin dosimeter *Med. Phys.* **24** 201–9

[8] Alexander P 1954 The degradation of solid polymethylmethacrylate by ionizing radiation *Proceedings of the royal society of london series A - Mathematical and physical sciences* **223** 392

[9] Kennan R P, Richardson K A, Zhong J, Maryanski M J and Gore J C 1996 The effects of cross-link density and chemical exchange on magnetization transfer in polyacrylamide gels *J Magn Reson B* **110** 267-77

[10] Maryanski M J, Gore J C, Kennan R P and Schulz R J 1993 NMR relaxation enhancement in gels polymerized and cross-linked by ionizing radiation: A new approach to 3D dosimetry by MRI *Magn. Reson. Imaging* **11** 253–8

[11] Zimmerman J R and Brittin W E 1957 Nuclear magnetic resonance studies in multiple phase systems: lifetime of a water molecule in an adsorbing phase on silica gel *J. Phys. C: Solid State Phys.* **61**

[12] Audet C 1995 NMR dose response of the gels used for 3-D radiation dosimetry by magnetic resonance imaging (*Ph.D. thesis, McGill University, Montreal, Quebec*)

[13] De Deene Y, Hanselaer P, De Wagter C, Achten E and De Neve W 2000 An investigation of the chemical stability of a monomer/polymer gel dosimeter *Phys. Med. Biol.* **45** 859–78

[14] Lepage M, Whittaker A K, Rintoul L, Back S A J and Baldock C 2001 The relationship between radiation-induced chemical processes and transverse relaxation times in polymer gel dosimeters *Phys. Med. Biol.* **46** 1061–74

[15] Kennan R P, Richardson K A, Zhong J, Maryanski M J and Gore J C 1996 The effects of crosslink density and chemical exchange on magnetization transfer in polyacrylamide gels *J. Magn. Reson. – series B* **110** 267–77

[16] Gochberg D F, Fong P M and Gore J C 2001 Studies of magnetization transfer and relaxation in irradiated polymer gels – interpretation of MRI-based dosimetry *Phys. Med. Biol.* **46** 799–811

[17] Lepage M, McMahon K, Galloway G J, De Deene Y, Back S A J and Baldock C 2002 Magnetization transfer imaging for polymer gel dosimetry *Phys. Med. Biol.* **47** 1881–90

[18] Murphy P S, Cosgrove V P, Schwarz A, Webb S and Leach M O 2000 Proton spectroscopic imaging of polyacrylamide gel dosimeters for absolute radiation dosimetry *Phys. Med. Biol.* **45** 835–45

[19] De Deene Y, De Wagter C, Van Duyse B, Derycke S, De Neve W and Achten E 1998 Three-dimensional dosimetry using polymer gel and magnetic resonance imaging applied to the

verification of conformal radiation therapy in head-and-neck cancer *Radiother. Oncol.* **48** 283–91

[20] Haraldsson P, Bäck S A J, Magnusson P and Olsson L E 2000 Dose response characteristics and basic dose distribution data for a polymerization-based dosemeter gel evaluated using MR *Br. J. Radiol.* **73** 58–65

[21] De Deene Y, De Wagter C, Van Duyse B, Derycke S, Mersseman B, De Gersem W, Voet T, Achten E and De Neve W 2000 Validation of MR-based polymer gel dosimetry as a preclinical three-dimensional verification tool in conformal radiotherapy. *Magn. Reson. Med.* **43** 116–25

[22] Pappas E, Seimenis I, Angelopoulos A, Georgolopoulou P, Kamariotaki-Paparigopoulou M, Maris T, Sakelliou L, Sandilos P and Vlachos L 2001 Narrow stereotactic beam profile measurements using N-vinylpyrrolidone based polymer gels and magnetic resonance imaging *Phys. Med. Biol.* **46** 783–97

[23] Ibbott G S, Maryanski M J, Eastman P, Holcomb S D, Zhang Y, Avison R G, Sanders M and Gore J C 1997 Three-dimensional visualization and measurement of conformal dose distributions using magnetic resonance imaging of BANG polymer gel dosimeters *Int. J. Radiat. Oncol. Biol. Phys.* **38** 1097–103

[24] Oldham M, Baustert I, Lord C, Smith T A D, McJury M, Warrington A P, Leach M O and Webb S 1998 An investigation into the dosimetry of a nine-field tomotherapy irradiation using BANG-gel dosimetry *Phys. Med. Biol.* **43** 1113–32

[25] Meeks S L, Bova F J, Maryanski M J, Kendrick L A, Ranade M K, Buatti J M and Friedman W A 1999 Image registration of BANG gel dose maps for quantitative dosimetry verification. *Int. J. Radiat. Oncol. Biol. Phys.* **43** 1141–53

[26] Cosgrove V P, Murphy P S, McJury M, Adams E J, Warrington A P, Leach M O and Webb S 2000 The reproducibility of polyacrylamide gel dosimetry applied to stereotactic conformal radiotherapy *Phys. Med. Biol.* **45** 1195–210

[27] Ramm U, Weber U, Bock M, Kramer M, Bankamp A, Damrau M, Thilmann C, Bottcher H D, Schad L R and Kraft G 2000 Three-dimensional BANG gel dosimetry in conformal carbon ion radiotherapy *Phys. Med. Biol.* **45** N95–102

[28] Farajollahi A R, Bonnett D E, Tattam D and Green S 2000 The potential use of polymer gel dosimetry in boron neutron capture therapy *Phys. Med. Biol.* **45** N9-N14

[29] Grebe G, Pfaender M, Roll M and Luedemann L 2001 Dynamic arc radiosurgery and radiotherapy: Commissioning and verification of dose distributions *Int. J. Radiat. Oncol. Biol. Phys.* **49** 1451–60

[30] De Deene Y 2006 On the accuracy and precision of gel dosimetry *J. Phys: Conf Series* **56** 72-85

[31] Slichter C P 1978 Principles of magnetic resonance, Springer-Verlag Berlin Heidelberg

[32] Levitt M H 2002 Spin Dynamics: Basics of nuclear magnetic resonance, John Wiley and Sons, West Sussex, England.

[33] Hornak J P 2007 The basics of MRI [web document] http://www.cis.rit.edu/htbooks/mri/

[34] Magnetic Resonance Technology – Information Portal 2008 MR-Tip [web document] http://www.mr-tip.com/

[35] Campus Medica 2007 e-MRI, Magnetic Resonance Imaging physics and technique course on the web [web document] http://www.e-mri.org/

[36] Bernstein M A, King K F and Zhou X J 2004 Handbook of MRI pulse sequences, Elsevier Academic Press, Burlington, Massachutes, USA

[37] Deoni S C L, Rutt B K and Peters T M 2003 Rapid combined T1 and T2 mapping using gradient recalled acquisition in the steady state *Magn. Reson. Med.* **49** 515-26

[38] Look D C and Locker D R 1970 Time saving in measurement of NMR and EPR relaxation time, *Rev Sci Instrum* **41**, 250–251

[39] De Deene Y, De Wagter C, De Neve W, Achten E 2000, Artefacts in multi-echo T2 imaging for high-precision gel dosimetry: I. Analysis and compensation of eddy currents *Phys Med Biol*

45 1807-23
[40] De Deene Y and Baldock C 2002 Optimization of multiple spin-echo sequences for 3D polymer gel dosimetry *Phys Med Biol* **47** 3117-41
[41] De Deene Y, Van deWalle R, Achten E and DeWagter C 1998 Mathematical analysis and experimental investigation of noise in quantitative magnetic resonance imaging applied in polymer gel dosimetry *Signal Processing* **70** 85-101
[42] De Deene Y and DeWagter C 2001 Artefacts in multi-echo T2 imaging for high-precision gel dosimetry: III. Effects of temperature drift during scanning *Phys Med Biol* **46**, 2697-711.
[43] De Deene Y, De Wagter C, De Neve W 2000, Artefacts in multi-echo T2 imaging for high-precision gel dosimetry: II. Analysis of B1 field inhomogeneity *Phys Med Biol* **45** 1825-39.
[44] Lepage M, Tofts P S, Bck S A J, Jayasekera P M and Baldock C 2001 Simple methods for the correction of T2 maps of phantoms *Magn Reson Med* **46** 1123-9
[45] Lepage M, McMahon K, Galloway G J, De Deene Y, Back S A J and Baldock C 2002 Magnetization transfer imaging for polymer gel dosimetry *Phys Med Biol* **47** 1881-90
[46] Henkelman R M, Huang X, Xiang Q, Stanisz G J, Swanson S D and Bronskill M J 1993 Quantitative interpretation of magnetization transfer *Magn Reson Med* **29** 759-66
[47] Gochberg D F, Kennan R P, Maryanski M J and Gore J C 1998 The role of specific side groups and pH in magnetization transfer in polymers *J Magn Reson* **131** 191-8
[48] Gochberg D F, Fong P M and Gore J C 2001 Studies of magnetization transfer and relaxation in irradiated polymer gels – interpretation of MRI-based dosimetry *Phys Med Biol* **46** 799-811
[49] Baete S H, De Deene Y, Masschaele B and De Neve W 2008, Microstructural analysis of foam by use of NMR R2 dispersion *J Magn Reson,* In press, doi:10.1016/j.jmr.2008.05.010
[50] De Deene Y, Vergote K, Claeys C and De Wagter C 2006 Three dimensional radiation dosimetry in lung-equivalent regions by use of a radiation sensitive gel foam: Proof of principle *Med Phys* **33**, 2586-97
[51] Helms G and Hagberg G A 2004 Pulsed saturation of the standard two-pool model for magnetization transfer. Part I: The steady state *Concepts Magn Reson - Part A* **21** 37-49
[52] Graham S J and Henkelman R M 1999 Pulsed magnetization transfer imaging: evolution of technique *Radiology* **212** 903-910.
[53] Sled J G and Pike G B 2000 Quantitative interpretation of magnetization transfer in spoiled gradient echo MRI sequences *J Magn Reson* **145** 24-36
[54] Gochberg D F, Kennan R P, Robson M D and Gore J C 1999 Quantitative Imaging of magnetization transfer using multiple selective pulses *Magn Reson Med* **41** 1065-72
[55] Ropele S, Seifert T, Enzinger C and Fazekas F 2003 Method for quantitative imaging of the macromolecular ^{1}H fraction in tissues *Magn Reson Med* **49** 864-71
[56] Baldock C, Lepage M, Back S A J, Murry P J, Jayasekera P M, Porter D and Kron T 2001 Dose resolution in radiotherapy polymer gel dosimetry: effect of echo spacing in MRI pulse sequence *Phys Med Biol* **46** 449-60
[57] Magnusson P, Back S A J and Olsson L E 1999 MRI image plane nonuniformity in evaluation of ferrous sulphate dosimeter gel (FeGel) by means of T1-relaxation time. *Magn Reson Imaging.* **17**, 1357-70
[58] Simmons A, Tofts P S, Barker G J and Arridge S R 1994 Sources of Intensity Nonuniformity in Spin Echo Images at 1.5T *Magn Reson Med.* **32**, 121-8
[59] Chan F, Pauly J and Macovski A 1992 Effects of RF amplifier distortion on selective excitation and their correction by prewarping. *Magn Reson Med.* **23**, 224-38
[60] Slotboom J, Creyghton J H N, Korbee D, Mehlkopf A F, Bovee W 1993 Spatially selective RF pulses and the effects of digitization on their performance *Magn Reson Med* **30** 732-40
[61] Collins C M, Li S, Yang Q X and Smith M B 1997 A method for accurate calculation of B1 fields in three dimensions, effects of shield geometry on field strength and homogeneity in the birdcage coil *J Magn Reson* **125** 233-41
[62] Foo T K F, Hayes C E, Kang Y 1986 An analytical model for the design of RF resonators for

MR body imaging *Magn Reson Med.* **3** 707-21

[63] Li S, Yang Q X and Smith M B 1994 RF coil optimization: Evaluation of B1 field homogeneity using field histograms and finite element calculations *Magn Reson Imaging* **12** 1079-87

[64] Scott G C, Joy M L G, Armstrong R L and Henkelman R M 1992 RF current density imaging in homogeneous media *Magn Reson Med* **28** 186-201

[65] Bottomley P A and Andrew E R 1978 RF magnetic field penetration, phase shift and power dissipation in biological tissue: Implications for NMR imaging *Phys Med Biol* **23** 630-43

[66] Price R R, Axel L, Morgan T, Newman R, Perman W, Schneiders N, Selikson M, Wood M and Thomas S R 1990 Quality assurance methods and phantoms for magnetic resonance imaging: report of AAPM nuclear magnetic resonance Task Group No. 1. *Med Phys* **17**, 287-95

[67] Michiels J, Bosmans H, Pelgrims P, Vandermeulen D, Gybels J, Marchal G and Suetens P 1994 On the problem of geometric distortion in magnetic resonance images for stereotactic neurosurgery *Magn. Reson. Imaging* **12** 749-65

[68] Schad L R, Ehricke H, Wowra B, Layer G, Engenhart R, Kauczor H, Zabel H, Brix G and Lorenz W J 1992 Correction of spatial distortion in magnetic resonance angiography for radiosurgical treatment planning of cerebral arteriovenous malformations *Magn Reson Imaging* **10** 609-21

[69] Jehenson P, Westphal M and Schuff N 1990 Analytical method for the compensation of eddy-current effects induced by pulsed magnetic field gradients in NMR systems *J Magn Reson* **90** 264-78

[70] Ahn C B and Cho Z H 1991 Analysis of the eddy-current induced artifacts and the temporal compensation in NMRI *IEEE Trans Med Imag* **10** 47-52

[71] Wysong R E and Lowe I J 1991 A simple method of measuring gradient induced eddy currents to set compensation networks *Magn Reson Med* **17** 149-63

[72] Ahn C B and Cho Z H 1991 Analysis of eddy currents in nuclear magnetic resonance imaging *Magn Reson Med* **17** 149-63

[73] Liu Q, Hughes D G and Allen P S 1994 Quantitative characterization of the eddy current fields in a 40-cm bore superconducting magnet *Magn Reson Med* **31** 73-6

[74] Boesch Ch, Gruetter R and Martin E 1991 Temporal and spatial analysis of fields generated by eddy current in superconducting magnets *Magn Reson Med* **20** 268-84

[75] Robertson S, Hughes D G, Liu Q and Allen P S 1992 Analysis of the temporal and spatial dependence of the eddy current fields in a 40-cm bore magnet *Magn Reson Med* **25** 158-66

[76] Schiff J, Rotem H, Stokar S and Kaplan N 1994 New efficient Eddy-Field-Mapping Procedure (FAME) *J Magn Reson, Series B* **104** 73-6

[77] De Deene Y, Reynaerdt N and De Wagter C 2001 On the accuracy of monomer/polymer gel dosimetry in the proximity of a high-dose-rate 192Ir source *Phys. Med. Biol.* **46** 2801-25

[78] Park H W, Ro Y M and Cho Z H 1988 Measurement of the magnetic susceptibility effect in high-field NMR imaging *Phys. Med. Biol.* **33** 339-49

[79] Cho Z H, Kim D J and Kim Y K 1988 Total inhomogeneity correction including chemical shifts and susceptibility effect in high-field NMR imaging *Med. Phys.* **15** 7-11

[80] Gregory C D 1997 Carl's Roost: MRI artefact gallery [web document] http://chickscope.beckman.uiuc.edu/roosts/carl/artifacts.html

[81] Patola W and Coulter B 1997 MRI artifacts [web document] http://www1.stpaulshosp.bc.ca/stpaulsstuff/MRartifacts.html

[82] Ballinger J R 1996 MRI artifacts [web document] http://www.mritutor.org/mritutor/artifact.htm

[83] De Deene Y, Fundamentals of MRI measurements for gel dosimetry 2004 *J Phys Conf Series* **3** 87-114

Journal of Physics: Conference Series **164** (2009) 012034 doi:10.1088/1742-6596/164/1/012034

Preliminary evaluation of optical CT scanning versus MRI for nPAG gel dosimetry: the Ghent experience

Jan Vandecasteele[1] and Yves De Deene[1,2]

[1]Research group QMRI, Department for experimental cancer research, nuclear medicine and radiotherpy (ECNURAD), Ghent University Hospital, De Pintelaan 185, 9000 Gent, Belgium

[2]Medical image and signal processing (MEDISIP), Ghent University Hospital, De Pintelaan 185, 9000 Gent, Belgium

E-mail: Jan.Vandecasteele@ugent.be

Abstract The aim of this study was to evaluate fast laser-scanning optical CT versus MRI for an nPAG gel dosimeter in terms of accuracy and precision. Three small cylindrical volumetric gel phantoms were fabricated and irradiated with photon beams. The gel dosimeters were scanned with an MR scanner and an in house developed laser scanning optical CT scanner. A comparison between MRI and optical CT scanning was performed based on the reconstructed images.

Preliminary results show a fair correspondence in the MRI acquired and optical CT acquired dose maps. Still, ringing artifacts contaminate the reconstructed optical CT images. These may be related to sub-pixel misalignments between the blank projection and the acquired transmission projection of the gel phantom. Another artifact may be caused by refraction near the edges of the field. Further optimisation of our optical CT scanner is required to obtain the same accuracy as with MRI. To make a comparison between the two imaging modalities in terms of precision, the intrinsic dose precision on readout (IP_D) was calculated which is independent of spatial resolution and acquisition time. It is shown that optical CT has a better intrinsic dose precision.

1. Introduction

In the last decade radiotherapy has focused on shaping the radiation dose distribution according to the shape of the tumour, what is called conformal radiotherapy (IMRT, IMAT, radiosurgery, tomotherapy). As a consequence there are steep dose gradients in three dimensions that pose a serious dosimetric challenge. Several dosimetric techniques are able to record the dose in one or two dimensions. Gel dosimetry is the only technique able to measure the absorbed dose distribution of a complete treatment in three dimensions and integrated over time. During the last years, gel dosimetry has evolved into a more reliable technique. This is due to extensive research into the chemical and physical characteristics of the gel system. Gel dosimeters have been used to validate conformal radiotherapy treatments [1]. A strong advantage of gel dosimetry above other techniques is that the gel dosimeter follows the complete treatment pathway as the patient during his or her treatment.

MRI is the most extensively studied imaging modality for gel dosimetry. Still only a few radiotherapy centres use this valuable dosimeter as an overall control for complex radiotherapy treatment verification. This can be partially attributed to the fabrication procedure, which requires a chemical laboratory. The greatest difficulty to implement gel dosimetry in the first place is the limited access to MR scanners. Also the expertise to get good quantitative images is not always readily available.

As an attempt to make gel dosimetry more accessible, optical computed tomography (optical CT) was developed and being researched for 12 years now. Different generations of optical CT scanners were proposed in the past [2].

In this study, an optical CT scanner was constructed on site which is able to measure 3D polymer gels. The aim was to evaluate the performance in terms of accuracy and precision of the novel optical CT scanning versus MRI for an nPAG gel dosimeter.

2. Materials and methods

2.1. Gel fabrication

The gel consists of a hydrogel matrix of 7% (w/w) gelatin from porcine skin (bloom 300, Sigma-Aldrich, Bornem, Belgium) in which (3% (w/w) N,N'-methylene-bis-acrylamide (Bis) and 3% (w/w) acrylamide (Aam) (both monomers were obtained from Sigma-Aldrich and are of electrophoresis grade) are dissolved. An antioxidant (tetrakis hydroxymethyl phosphonium chloride) is added to the mixture in a concentration of 3mM. The monomers and gelatin were dissolved into equal amounts of the total water volume (87% (w/w)). The monomer solution was heated to 60 °C until everything was dissolved. The gelatin was dissolved in water at room temperature and allowed to swell. Next the solution was heated up to 45 °C, at which the gelatin dissolves. Subsequently, both solutions were cooled down to 30 °C before mixing. After mixing, the antioxidant was added and the solution was thoroughly stirred for 30 s. The gel was transferred into three cylindrical transparent volumetric phantoms and sixteen test tubes that served as calibration vials.

2.2. Gel irradiation

The irradiation was performed using a clinical linear accelerator (SL-18, Elekta, Stockholm, Sweden). Each calibration vial (R: 0.8 cm, h: 10 cm, glass thickness: 0.3 cm) was irradiated separately in a cubic water tank ($22x22x22\ cm^3$) with 6 MV photons at a reference depth of 5 cm and with a source-to-surface distance (SSD) of 95 cm. The field size was $10x10\ cm^2$.

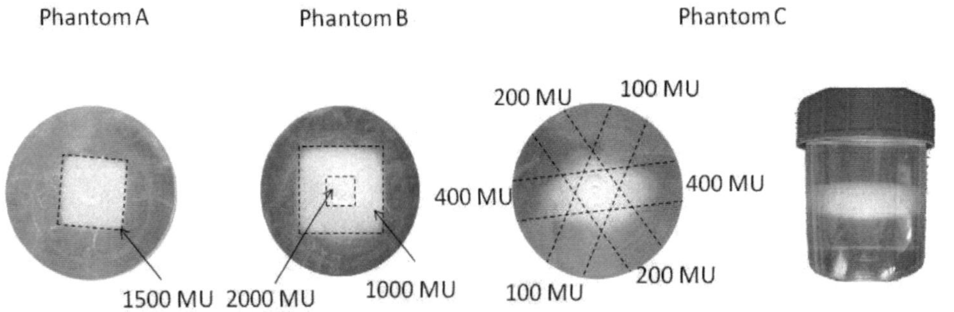

Figure 1. Photographic overview of the three cylindrical phantoms. A radiation beam with a fluence of 1500 MU was given to phantom A. Phantom B was irradiated with two centered square beams of $3x3\ cm^2$ and $1x1\ cm^2$ respectively, both with a fluence of 1000 MU. Phantom C was irradiated along 6 beam directions. The fluences are shown for each incident beam. The last photograph is a side view of phantom C. The black dotted line represents the field shapes.

The dose rate for all volumetric phantoms was 4 Gy min^{-1}. The axis of the calibration vials was perpendicular to the axis of the beam. Using this method, irradiation doses of 0 Gy to 30 Gy with increment of 2 Gy were delivered in the calibration vials.
The three small cylindrical volumetric gel phantoms (R: 2.6 cm, h: 6.7 cm) (figure 1) were irradiated with 6MV photon beams at an SSD of 100 cm. The field size used to irradiate phantom A was 2x2 cm². A radiation beam with a fluence of 1500 MU was given. Phantom B was irradiated with two centered square beams of 3x3 cm² and 1x1 cm² respectively, both with a fluence of 1000 MU. The axis of both phantoms A and B was parallel to the axis of the beam. For phantom C, the field size was 1.5x1.5 cm². Six beam directions were chosen (0°, 60°, 120°, 180°, 240°, 300°) with fluences of respectively 100 MU, 200 MU, 400 MU, 100 MU, 200 MU, 400 MU. The axis of phantoms C was perpendicular to the axis of the beam.

2.3. MR measurements and data processing

Irradiated gels together with the calibration vials were scanned in a Siemens SP 1.5 T whole body scanner using a standard transmit/receive RF head coil. A multiple spin-echo sequence (32 equidistant spin echoes) with a CPMG encoding scheme was used. Imaging parameters were TE = 40-1280 ms, TR = 5000 ms, FOV = 150 mm and MS = 256x256.
In house Matlab (The MathWorks, Natick, USA) code was employed to calculate R2 images from the set of base images. A calibration curve, extracted from the calibration vials, was used to convert the R2 maps of the volumetric gel phantoms to dose maps.

2.4. Optical CT scanner

2.4.1. General overview
An optical CT scanner (figure 2) was constructed in house (OPTOSCAN). We use a 2.0 mW, 632.8 nm HeNe-laser (JDS Uniphase Model 1122p, CA,USA) with a beam diameter of 0.63 mm (1/e² points). Before reaching a galvanomirror, the laser beam is guided through two pinhole collimators to sharpen the laser beam size. The galvanomirror reflects the laser beam to a plano-convex lens (Melles-Griot, Albuquerque, USA) which is positioned at focus distance (450mm) from the galvanomirror.

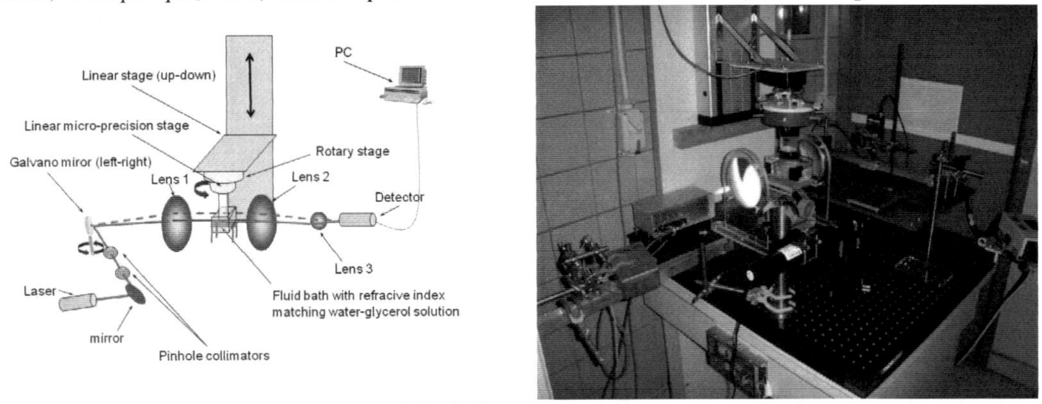

Figure 2. Left: Schematic basic design of the in house optical laser scanner. The laser beam is pointed at a rotating galvanomirror. Before reaching this galvanomirror the laser beam is guided through two pinhole collimators. The galvanomirror reflects the laser beam towards a plano-convex lens. The laser beam leaving the plano-convex lens travels in parallel lines through the gel dosimeter until it passes another plano-convex lens which guides the laser beam towards the photodetector. Right: Photograph of the OPTOSCAN optical CT scanner.

The laser beam leaving the plano-convex lens travels in parallel lines through the gel dosimeter until it passes another plano-convex lens which guides the laser beam towards the photodetector. The

photodetector is a large area balanced photoreceiver (Thorlabs, model 2307, Munich, Germany). In between the lenses a small fluid bath (14.2x7.5x 5.5 cm^3) is placed. The refractive index matching procedure is performed automatically based on a dry and wet off axis refraction measurement and using an automated pump system with a refractive index matching fluid (water-glycerol solution).
Matlab (The MathWorks) code was developed in house to control the optical scanner, construct sinograms and optical density images.

2.4.2. Optical CT scanning protocol
One hundred projections of 128 mm were recorded per slice with a phantom rotation increment of 1.8°. 128 spatial increments were acquired per projection yielding a profile resolution of 1mm. The same region was scanned both optically and with MRI.

2.4.3. Post processing
The measured projections per rotational increment were aligned in the sinogram using an in house Matlab (The MathWorks) script before reconstructing the optical density map. The optical density was calculated using equation 1

$$OD' = \log_{10}\left(\frac{(SB-SD)}{(S(x,\theta,z)-SD)}\right) \quad (1)$$

were SB is the mean of a blank projection, SD is the dark current of the photodetector and $S(x,\theta,z)$ represents the signal at position x, angle Θ and vertical slice position z.

2.4.4. Comparison between MRI and optical CT
The optical CT measured OD images were scaled to the MRI measured dose maps using a bilinear interpolation. The optical density maps were normalized to the MRI measured dose maps using a conversion factor that is derived from the maximum dose regions in the three volumetric phantoms. In order to make a fair comparison of precision between both imaging modalities, the intrinsic dosimetric precision on readout (equation 2) was calculated, which is independent of spatial resolution and acquisition time.

$$IP_D = \frac{1}{D^{p}_{\Delta\%} \cdot \Delta V \cdot \sqrt{t_{meas}}} \quad (2)$$

ΔV is the voxel volume, t_{meas} is the total acquisition time and $D^{p}_{\Delta\%}$ is the relative dose resolution (equation 3)

$$D^{95\%}_{\Delta\%} = \sqrt{2} \cdot k_p \cdot \frac{\sigma_D}{D_{max} - D_{min}} \quad (3)$$

with k_p the coverage factor for a 95% confidence interval and σ_D the standard deviation on dose. D_{max} and D_{min} are the maximum and minimum doses in the image.

To align the resulting images of MRI and optical measurements a rotation was carried out using a bilinear interpolation. A subtraction image was calculated between MRI and optical CT data for every phantom. Cross-profiles through both MRI and optical CT derived dose maps were averaged in the perpendicular direction over 5mm wide.

3. Results

Figure 3 shows the MRI and optical CT derived dose maps. The conversion factor derived from the three phantoms did not differ by more than 1.3%. The OD images were normalized to dose using the average of the three conversion factors.

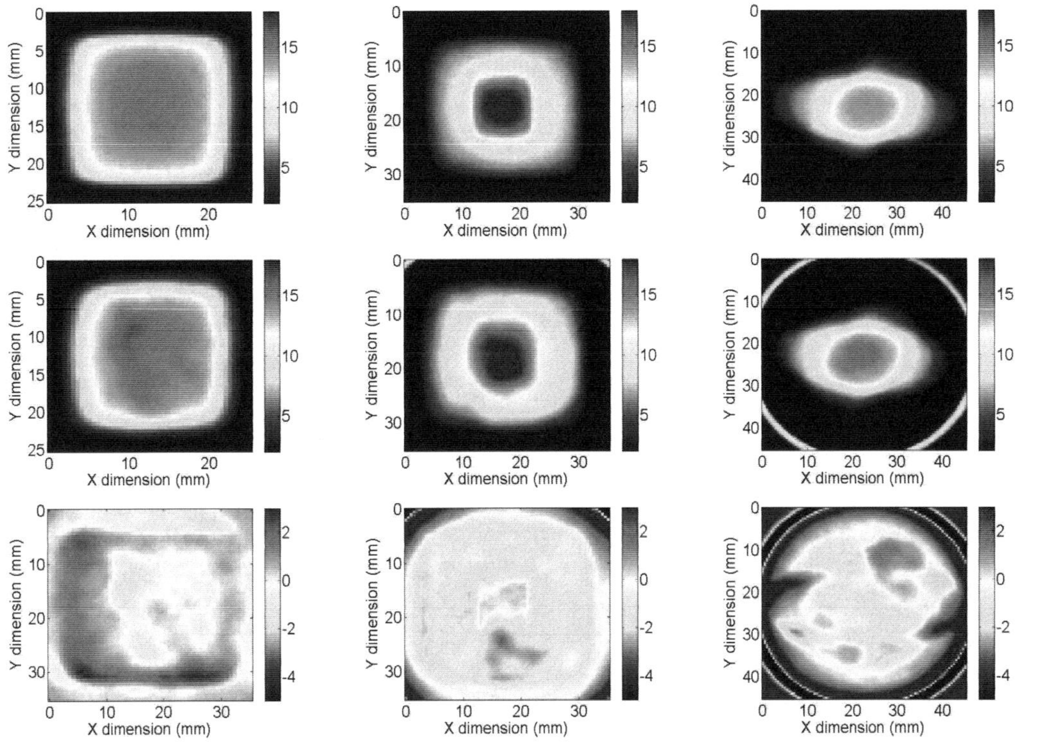

Figure 3. MRI acquired (top row) and optical CT acquired (second row) dose maps for the tree volumetric phantoms (from left to right: phantoms A, B and C). The bottom row shows the corresponding difference maps. Colorbar is expresses dose in Gy.

For further comparison, cross-profiles were also taken along the center of the irradiated fields for phantom A and phantom B (figure 4) .

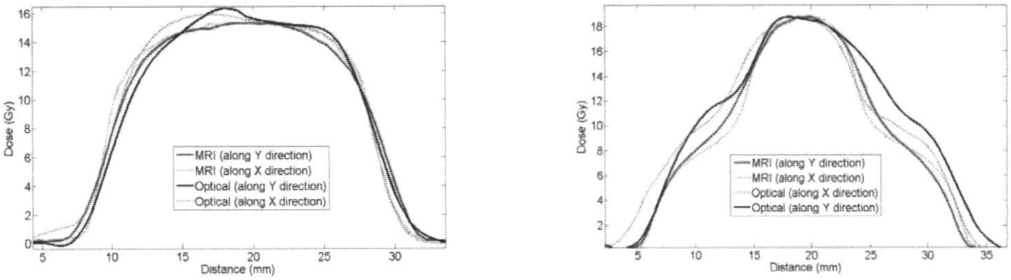

Figure 4. Lateral cross-profiles through phantoms A (left) and B (right) in both orthogonal directions.

The acquisition time for MRI data was 255 min and for optical CT was 20 min. Acquisition and performance parameters are listed in table 1.

	MRI	Optical CT
Voxel size (mm^3)	1.72 (0.586 x 0.586 x 5.00)	1.27 (1.00 x 1.00 x 1.27)
Minimum R2 (t^{-1})/ minimum OD (-)	1.18	-5.92 10^{-3}
Maximum R2 (t^{-1})/ maximum OD (-)	2.38	1.43 10^{-2}
$\square_D$	5.96 10^{-3}	1.09 10^{-4}
$D^{95\%}_{\square\%}$ (%)	1.38	1.49
IP_D ($mm^{-3}s^{-1/2}$)	3.42 10^{-1}	1.83

Table 1. Acquisition and performance parameters for MRI and optical CT measurements

4. Discussion

To investigate the overall performance of an in house built fast-laser scanning optical CT scanner, a comparison with a 'gold standard' is required. In this study, we have used an nPAG gel dosimeter and MRI to obtain an 'idea' of the accuracy and precision of the optical CT scan technique. A fair quantitative correspondence was found between MRI measured dose maps and optical CT measured dose maps. However, from figure 3 and figure 4 it can be seen that some imaging artifacts in the optical CT images still compromise the accuracy.
A dose hotspot is clearly visible inside the square field of phantom A. This may be related to some residual ringing artifacts. The field lines are less sharply delineated in the optical CT images which can also be seen in the subtraction image (figure 3) were the biggest deviation occurs near the edges of the irradiated field. We also observed refraction of the laser beam when grazing the edge of the field. This is probably due to a change in refractive index between the two structures (polymerized and non polymerized gel). This artifact is more pronounced at higher doses which was the case for phantom B.
Phantom C shows the potential of scanning more complex dose distributions. The optical CT data has a high degree of alignment with the MRI data within the irradiated field.
Still further work is needed to develop an adequate calibration procedure for optical CT. This could be done using calibration films or with calibration phantoms.

For the intrinsic dose precision on readout (IP_D), it is shown that optical CT scanning has a five times better intrinsic dose precision than MRI.

5. Conclusions

A fast laser scanning optical CT scanner was constructed and optimized at the Ghent university. To make an evaluation of the overall accuracy and precision of the optical CT scanner, a comparison was needed with a 'gold standard'. Therefore an nPAG gel dosimeter was used to make the comparison between MRI and optical CT. Preliminary results show a fair correspondence in the derived dose maps for MRI and optical CT. Still, ringing artifacts contaminate the reconstructed optical CT images. These may be related to sub-pixel misalignments between the blank projection and the acquired transmission projection of the gel phantom. Another artifact may be caused by refraction near the edges of the field. Further optimisation of our optical CT scanner is required to obtain the same accuracy as with MRI. To make a comparison between the two imaging modalities in terms of precision, the intrinsic dose precision on readout (IP_D) was calculated. It is shown that optical CT has a better intrinsic dose precision.

References

[1] De Deene Y, De Wagter C et al, 1998 Three-dimensional dosimetry using polymer gel and magnetic resonance imaging applied to the verification of conformal radiation therapy in head-and-neck cancer. *Radiother. Oncol.* 48 283–91
[2] Oldham M 2004 Optical-CT scanning of polymer gels *J. Phys.: Conf. Ser.* 3 122-135

Relaxometry Changes in a Gel Dosimetry Phantom due to Continued RF Exposure

Gary P. Liney, Mark J. Godber, Andrew D. Wilson, John W. Goodby, LindsayW. Turnbull.

Centre for MR Investigations, Hull Royal Infirmary, Hull, England & Imagel Ltd, York Science Park, York, England

g.p.liney@hull.ac.uk

Abstract. This study investigates the potential alteration in relaxation times in phantoms used for MRI gel-dosimetry due to their continued RF exposure from the MRI scanner used in the measurement process. The work quantifies these changes and establishes a tolerance for the image acquisition time as well as mapping the spatial distribution of these effects.

1. Introduction.

As MRI-gel dosimetry becomes increasingly used in the field of radiotherapy verification, the accuracy of the technique needs to be thoroughly established. There are many MRI-dependent factors that have the potential to introduce errors into the final dose map. Of these, the effect of temperature changes within the gel from the heating caused by RF absorption during the imaging procedure itself has been the least addressed. The dosimetric term for RF absorption is known as specific absorption rate (SAR) and has a quadratic dependence on field strength and flip angle and is dependent on other factors such as coil geometry. In order to map the dose deposited within the gel the dose-R_2 characteristic needs to be determined. This requires the acquisition of multiple echo imaging sequences in order to perform the necessary transverse relaxation time measurements, and requires the use of multiple RF pulses. Often repeat scanning in one flask is used to provide in situ calibration and/or for highly detailed plan verification. This leads to a RF intensive procedure that has the potential to introduce significant heating effects in the phantom. Changes in temperature will alter the background relaxation value and in turn affect the calibration efficacy leading to errors in the final dose calculation. As the diameter of the phantom becomes comparable with (half) the resonant wavelength of the system, the RF absorption becomes more efficient and temperature increases are likely to be an important consideration. At 3.0 Tesla the resonant wavelength is 26 cm. This study attempts to establish the tolerance for continued exposure to RF and map any spatial effects that exist during a typical scanning procedure.

2. Materials & Methods.

All imaging was performed on a 3.0 Tesla GE Signa scanner using a commercially available linear RF head coil. A two-litre flask of approximately 12 cm in diameter containing a MAGIC-type gel medium (Imagel Ltd, York, UK) and widely used for gel-dosimetry was imaged. This flask had previously been left overnight to reach ambient scan-room temperature of approximately 18 °C. The imaging acquisition consisted of two dual-echo fast spin-echo (FSE) sequences with TEs equal to 30,105 & 60,180 ms and a TR of 4 s. This protocol has been used in previous dosimetry studies at our Centre. Axial images were acquired through the gel container with 5 mm slice thickness and 1 mm gap with an in-plane resolution of 0.9 mm. The gel temperature was measured using an MR-compatible fluoroscopic probe and extension (Lambda photometrics, UK). A needle and sheath was used to introduce the probe into the gel for ease and accuracy of positioning and then withdrawn outside the bore to leave the probe in place. Measurements were taken at the edges and centre of the phantom during the acquisitions. Data-logging software was used to continually record temperature. Scanner bore temperature was also recorded before and after scanning. A total of seven repeat acquisitions (two signal averages each) were successively collected over a total period of 72 minutes. R_2 was estimated at the end of each sequence in small regions-of-interest using a semi-weighted log fit to the data. In addition, pixel-by-pixel maps of ΔR_2 were also produced by subtracting each subsequent repeat measurement from the first acquisition using in-house developed software (Dosemap2003, MATLAB) in order to visually monitor heating changes.

3. Results

Bore temperature remained constant throughout the examination. Within the gel itself, temperature changes of up to 1.4 °C were observed at the edges of the flask and a concomitant decrease of R_2 by 4.0 %. Figure 1 plots the temperature distribution (blue) during the course of scanning together with changes in R_2 measurements (pink) in a region at the edge of the flask. In contrast, the centre of the flask demonstrated smaller changes with a temperature increase and R_2 decrease of 0.3 °C and 2.6 % respectively.

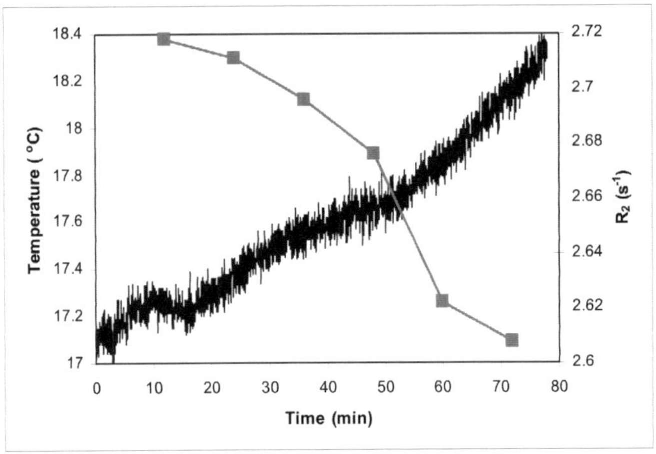

Figure 1: (above) A plot of temperature changes (in blue) and R_2 values (in pink) in a region at the edge of the gel flask during continued RF exposure of up to 72 minutes.

Figure 2 shows maps of ΔR_2 calculated by subtracting each subsequent acquisition from the first R_2 map. These maps (left to right and top to bottom) demonstrate increasing values of ΔR_2 (i.e. a continuing decrease in R_2) from which the increase in temperature can be inferred. There is a clear trend for the heating to dominate at the periphery of the flask with a much smaller and more gradual increase towards the centre. The increased temperature at the edges also appears asymmetrical.

Using previous data from our earlier calibration studies, the maximum temperature increases observed here would equate to between a 0.8 % and 5.4 % error in the estimated dose. This is clearly dependent on the characteristic (sensitivity) of the gel material.

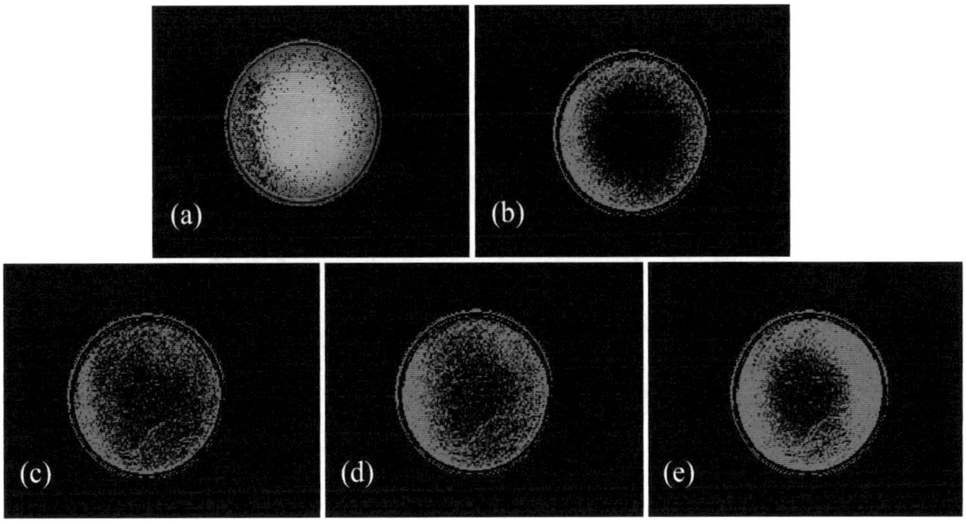

Figure 2: (above) Pixel-by-pixel maps of ΔR_2 every 12 minutes (a to e) show the evolution of the relaxation rate decrease due to temperature changes (T_2-weighted image of gel phantom also shown in first map).

4. Discussion.

This study establishes that temperature is unlikely to cause a significant component of the final dose error if scanning is restricted to less than 36 min based on previous data from our gel dosimetry studies. However, scanning beyond this time creates a significant heating effect at the edge of the dosimeter and alters the relaxation time of the gel in this region of sufficient degree that this could limit the absolute accuracy of the technique. It is of note that this temperature distribution is not uniform and will be dependent on the flask size, and be specific to the imaging sequence and RF component of the system in use. The maps of ΔR_2 are an effective way of mapping these RF heating changes and may be useful for investigating other sequences and systems in a similar way to establish the tolerance of these effects.

Journal of Physics: Conference Series **164** (2009) 012036 doi:10.1088/1742-6596/164/1/012036

Response Verification of Dose Rate and Time Dependence of PAGAT Polymer Gel Dosimeters By Photon Beams Using Magnetic Resonance Imaging.

Azadbakht B[1], Hadad K[1], Zahmatkesh M.H[2].

[1]Shiraz university,
[2]Institute of novine medical radiation

Corresponding author : e-mail : azad_bakhtiar@yahoo.com

Abstract : The purpose of this study was to evaluate dependence of PAGAT polymer gel dosimeter 1/T2 on different post time imaging as well as on different mean dose rates for a standard clinically used Co-60 therapy unit and an electa linear accelerator. Using MRI, the formulation to give the maximum change in the transverse relaxation rate R2(1/T2) was determined to be 4.5% N,N'-methylen-bis-acrylamide(bis), 4.5% acrylamid(AA), 5% gelatine, 5 mM tetrakis (hydroxymethyl) phosphonium chloride (THPC), 0.01 mM hydroquinone (HQ) and 86% HPLC(Water).When the preparation of final polymer gel solution is completed, it is transferred into phantoms and allowed to set by storage in a refrigerator at about 4°C. The sensitivity of the dosimeter was represented by the slope of calibration curve in the linear region measured for each modality. A calibration curve (in the linear region) based on 16 dosimeters (15 irradiated and one background) was obtained for 1.25 MV photon beam. To determine the calibration curve of the PAGAT polymer gel dosimeter, there are two linear responses between 2-10Gy and 10-30Gy and the R2-dose sensitivity showed stability with imaging post time after 38 days. Dose rate of dependence was studied in 6 MV photon beam with the use of dose rates 80, 160, 240, 320, 400 and 480 cGy/min. Evaluation of dosimeters were performed on Siemens Symphony, Germany 1.5T Scanner in the head coil. A multi echo sequence with 32 equidistant echoes was used for the evaluation of irradiated polymer gel dosimeters. The parameters of the sequence were as follows: TR 3000ms, TE 20ms, Slice Thickness 4 mm and FOV 256mm. No trend in polymer-gel dosimeter 1/T2 dependence was found on mean dose rate for photon beams.

Key words: polymer gels, PAGAT gel ,dose rate, timestability, MRI.

1. Introduction

In 1984, magnetic resonance imaging (MRI) demonstrated great potential in visulazing three dimensional (3D) dose distributions of ferric or ferrous sulphate gel dosimeters [1]. Subsequently, studies were undertaken to investigate the feasibility of using Ferric gel as a 3D dosimetry system in radiation oncology [2]. The major limitation in Ferric gel dosimetry is that it suffers from blurness of dose with time which is due to the migration of ferrous and ferric ions in gel matrix, known as diffusion [3]. In 1993, a polymer gel dosimeter was developed that maintained spatial information following irradiation which could be visualized using MRI [4]. Polymer gel dosimetry is a technique that has the ability to map absorbed radiation dose distribution in three dimensions (3D) with high spatial resolution. Polymer gel dosimeters offer a number of advantages over traditional dosimeters such as ionization chambers, thermoluminescent dosimeter (TLD) and radiographic film. These advantages include independence of radiation direction, radiological soft tissue equivalence, integration of dose for a number of sequential treatment fields, and perhaps most significantly, evaluation of a complete volume at once [5-6]. In 2001, the first normoxic gels were suggested that could be produced, stored and irradiated in a normal condition. Magnetic Resonance Imaging (MRI) has been most extensively used for the evaluation of absorbed dose distributions in polymer gel dosimeters. In the MRI evaluation of polymer gel dosimeters, changes in T2 is a result of physical density changes of irradiated polymer gel dosimeters. Many factors such as polymer gel composition, temperature variation during irradiation, type and energy of radiation, dose rate, temperature during MRI evaluation, time between irradiation to MRI evaluation, and strength of magnetic field have been studied by different authors [4-7-8-9]. All these factors can potentially affect polymer gel dosimeter response and significantly influence measured results. Consequently, it is important to evaluate and quantify each individual factor. This study has been focused on evaluation of the dependence of PAGAT polymer gel dosimeter response on post time and dose rates of the photon beams in a Co-60 therapy unit and an electa linear accelerator.

2. Materials and methods

2.1. PAGAT preparation

The PAGAT polymer gel formulation by % mass consisted of 4.5% N,N'-methylen-bis-acrylamide (bis), 4.5% acrylamid (AA), 5% gelatine, 5 mM tetrakis (hydroxymethyl) phosphonium chloride (THPC), 0.01 mM hydroquinone (HQ) and 86% HPLC(Water) [10]. All components were mixed on the bench top under a fume hood. The gelatine was added to the ultra-pure de-ionized water and left to soak for 12 min, followed by heating to 48C° using an electrical heating plate controlled by a thermostat. Once the gelatine completely dissolved the heat was turned off and the cross-linking agent, bis was added and stirred untile dissolved. Once the bis was completely dissolved the AA was added and stirred untile dissolved. Using pipettes, various concentration of the polymerization inhibitor HQ and the THPC anti-oxidant were combined with the polymer gel solution. When the preparation of final polymer gel solution is completed, it is transferred into phantoms and allowed to set by storage in a refrigerator at about 4C° [10].

Table 1 lists the component with different percent weight in normoxic PAGAT polymer gel dosimeter.

Table 1. Different chemicals and percent weight of PAGAT gel

Component	Percent Weight
Gelatine(300Bloom)	**5%**
N,N'-methylen-bis- acrylamide(bis)	**4.5%**
Acrylamide(AA)	**4.5%**
Tetrakis-phosphonium chloride(THPC)	**5mM**
Hydroquinone(HQ)	**0.01mM**
HPLC(Water)	**86%**

2.2. Irradiation

Irradiation of vials was performed using photon beams by Co-60 therapy unit and an electa linear accelerator with SSD = 80cm, field size of 20x 20cm^2 and the depth was selected at 5cm. to determine the sensitivity of PAGAT polymer gel dosimeter with different energies (e.g. 4, 6 and 18

MV), irradiation of vials was performed using an electa linear accelerator with SSD =80cm, field size= 20x20cm^2,dose rate=400 $\frac{cGy}{min}$ and the depth was selected at 5cm. The optimal post-manufacture irradiation was determined to be 1 day.

2.3. Imaging

Before imaging, all polymer gel dosimeters were transferred to a temperature controlled MRI scanning room to equilibrate to room temperature. The PAGAT polymer gel dosimeters were imaged in a Siemens Symphony 1.5 Tesla clinical MRI scanner using a head coil. T2 weighted imaging was performed using a standard Siemens 32-echo pulse sequence with TE of 20 ms, TR of 3000 ms, slice thickness of 4 mm, FOV of 256 mm. The optimal post imaging times was determined to be 1 day. The images were transferred to a personnel computer where T2 and R2 maps were computed using modified radiotherapy gel dosimetry image processing software coded in MATLAB (The Math Works, Inc).The mean T2 value of each vial was plotted as a function of dose with the quasi-linear section being evaluated for R2-dose sensitivity
table 2 lists the protocol of magnetic resonance imaging (MRI) was used in PAGAT polymer gel dosimeter.

Table 2. The protocol of magnetic resonance imaging (MRI)

PARAMETERS	
Field of view(FOV) {mm}	**256**
Marrix size(MS)	512×512
Slice Thickness(d) {mm}	**4**
Repetition Time(TR) {ms}	**3000**
Echo Time(TE) {ms}	**20**
Number of Slices	**1,2,3,4**
Number of Echoes	**32**
Total Measurement Time {min}	**25-30**
mm}{Resolution	**0.5**
Band With $\left\{ Hz/_{Pixel} \right\}$	**130**

3. Results

3.1. R2-dose Sensitivity of PAGAT polymer gel dosimeter

PAGAT gels with optimum value of ingredient was manufactured and irradiated to different doses. As it can be seen in figure 1, PAGAT has a linear response up to 30Gy. the response of the PAGAT gel is very similar in the lower dose region and The R2-dose response for doses less than 2Gy is not exact. The R2-dose response of the PAGAT polymer gel dosimeter is linear between 10-30Gy and 2-10Gy. Figure 1
shows that PAGAT polymer gel has a dynamic range of at least S.ISSS$^{1'1}$ for doses up to 30Gy compared with less than 2.55$^{1''1}$ for doses up to 30Gy in the preliminary study of the PAGAT polymer gel dosimeter.

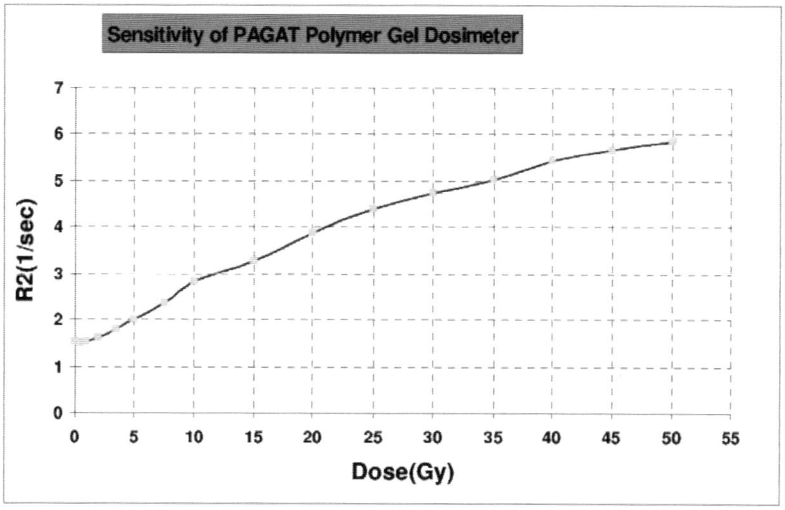

Figure 1. Sensitivity of PAGAT with different range of doses

Table 3 lists the Sensitivity of PAGAT polymer gel dosimeter with different range of doses.

Table 3. Sensitivity of PAGAT with different range of doses

dose(Gy)	R2- dose sensitivity ($S^{-1}Gy^{-1}$)	Correlation coefficient
0-2	0.0456	0.6175
2-10	0.1512	0.9949
10-30	0.0983	0.9939
30-50	0.0577	0.9825

3.2. R2-dose Response of PAGAT polymer gel dosimeter with Post Time

The R2-dose response of the PAGAT polymer gel dosimeter is linear between 10-30Gy doses. Figure 2 shows the R2-dose Response with Post Time (e.g. 1, 8, 15, 29 and 38 days). In this study the R2-dose response was linear up to 30Gy with R2-dose sensitivities of 0.0905, 0.1037, 0.1023, 0.0907 and 0.123 *S^Gy''l* when imaged at 1, 8, 15, 29 and 38 days post-irradiation respectively. The R2-dose sensitivity showed stability with imaging post time after 38 days.

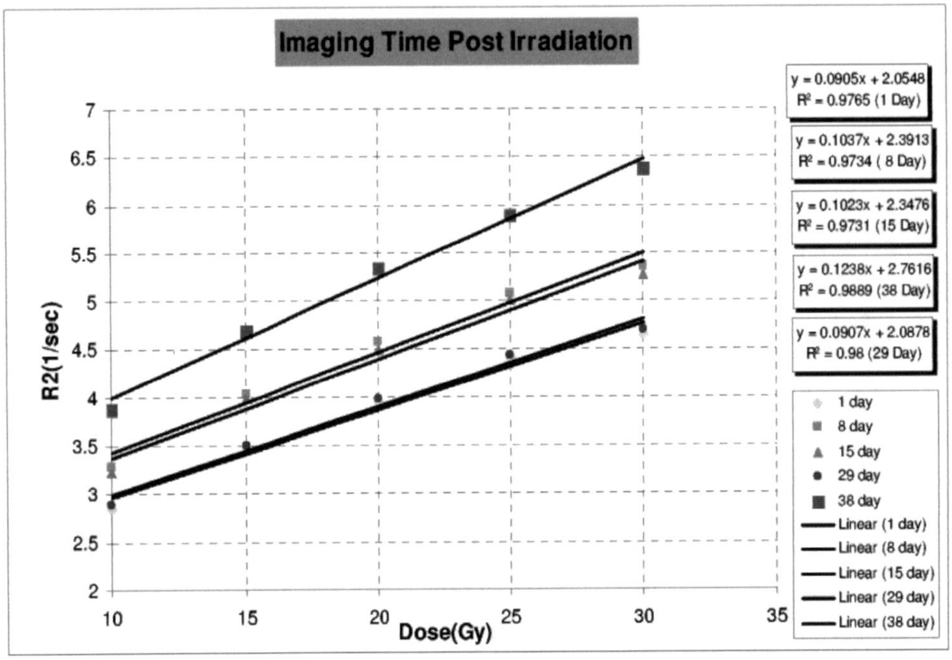

Figure 2. R2-dose response curve of the PAGAT polymer gel dosimeter evaluated at 1, 8, 15, 29 and 38 days post-irradiation.

Table 4 lists the R2-dose sensitivity and correlation coefficients for the five post-irradiation imaging times.
Table 4 indicate the PAGAT had reached steady-state by 38 days post-irradiation, therefore the R2-dose sensitivity of PAGAT polymer gel between 10Gy and 30Gy wa stable. This study has shown that the normoxic PAGAT polymer gel dosimeter has the properties of a dosimetric tool, which can be used in clinical radiotherapy.

Table 4. R2-dose sensitivity and correlation coefficient of PAGAT polymer gel between 10Gy and 30Gy.

Imaging time post Irradiation(day)	R2- dose sensitivity ($S^{-1}Gy^{-1}$)	Correlation coefficient
1	0.0905	0.9765
8	0.1037	0.9734
15	*0.1023*	*0.9731*
29	0.0907	0.98
38	0.123	0.9889

Figure 3 shows the R2-dose sensitivities of 0.0905, 0.1037, 0.1023, 0.0907 and 0.123 $s^{-1}Gy^{-1}$ when imaged at 1, 8, 15, 29 and 38 days post-irradiation respectively therefore the R2-dose sensitivity of PAGAT polymer gel dosimeter had reached steady-state by 38 days post-irradiation

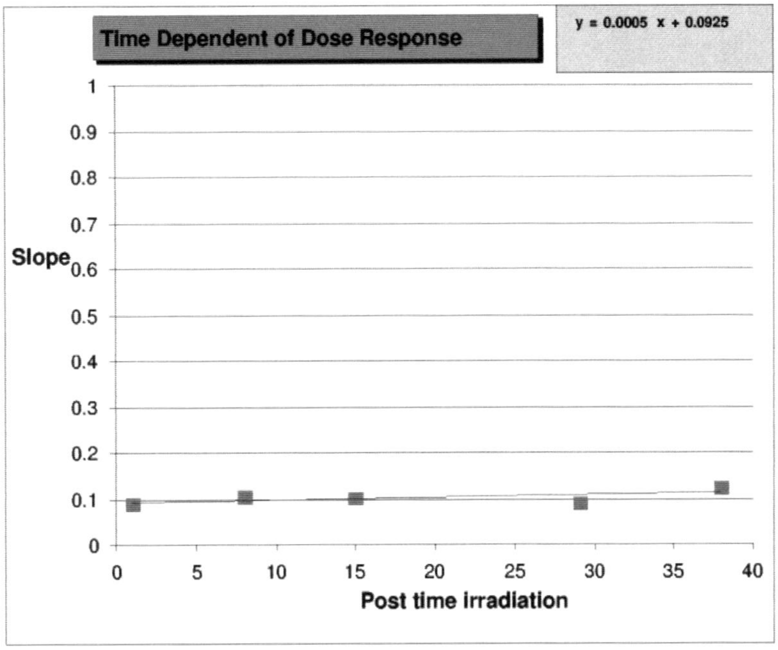

Figure 3. Time dependent of dose response

3.3. Verification Dose rate dependence of PAGAT polymer gel dosimeter

Dose rate dependence for PAGAT gel was verified in figure (4) with a 6 MV photon beam 20Gy dose was delivered to the gel phantoms with dose rates varying from 80, 160, 240, 320, 400 and 480cGy/min. A line was fitted to the data which shows no significant dependence of R2 to dose rate.

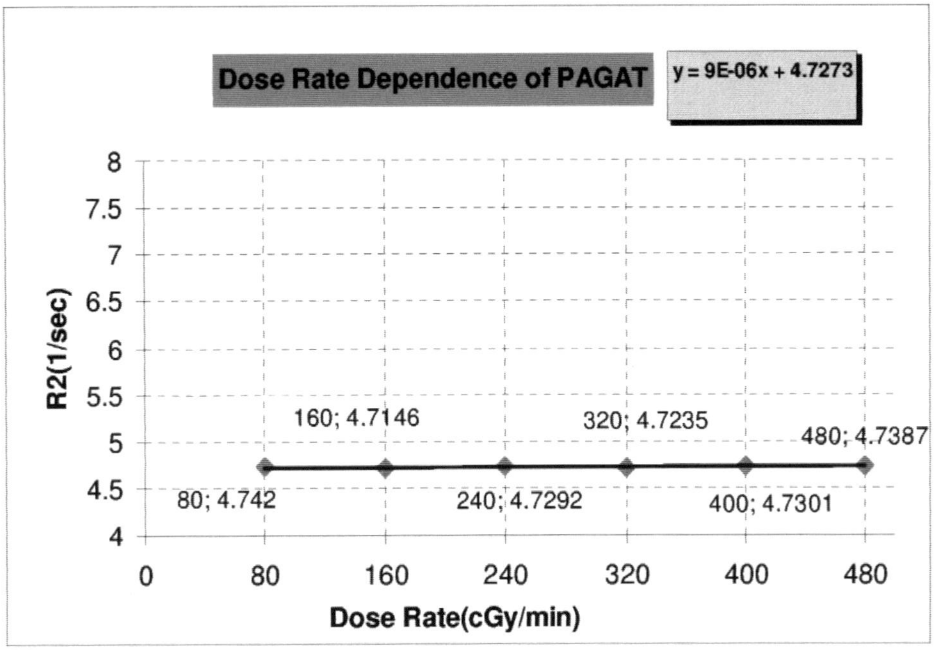

Figure 4. Verification R2-dose response of PAGAT polymer gel dosimeter with different dose rates

figure (5) shows the PAGAT polymer gel dosimeter sensitivity with different dose rates, therefore no significant dose rate effects in PAGAT polymer gel have been observed using NMR evaluation when dose rate is changed from 80cGy/min to 480cGy/min.

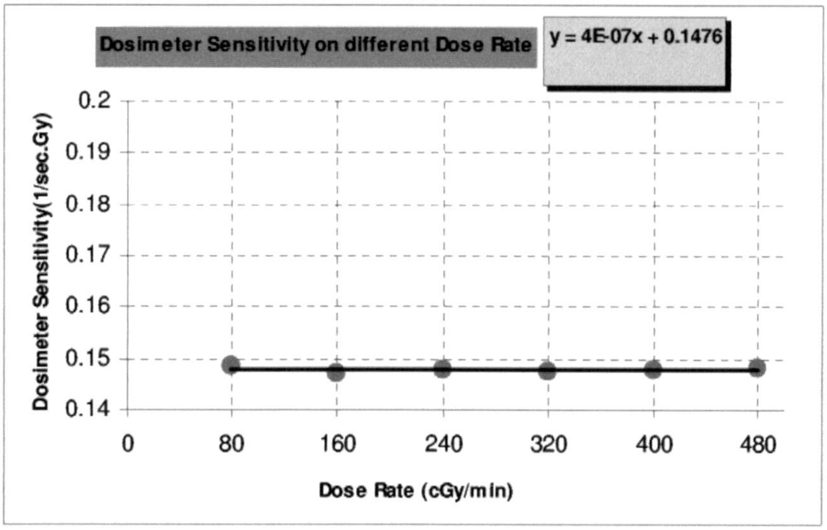

Figure5. Sensitivity of **PAGAT** Polymer gel dosimeter with different dose rates

4. Discussion

According to the best knowledge of the authors only a few studies were published to evaluate dose rate dependence for the polymer gel dosimeters. Ibbott et al. which worked on BANG polymer gel, concluded that the polymer gel dose response is independent of both dose rate and beam modality and used the polymer gel calibration curve determined with 6 MV x rays for evaluation of the gel dosimeter irradiated with ^{60}Co without any limitations or approximations [11]. Similarly De Wagter et al. which worked on polymer gel stated in their study that polymer gel responds practically independently of beam quality [12]. Maryanski et al., used central axis percentage depth dose of 6 MV x-ray and 15 MeV electron beams measured in the BANG-2 gel dosimeter (composed of: 5% Gelatin, 3% N,N'-methylene-bis-acrylamide(BIS), 3% Acrylic acid, 1% NaOH & 88% Water) and compared them with diode measurements in water. They concluded that there is no energy as well as dose rate dependence of gel dosimeter for electron beams in the range 2-15MeV and dose rates in the range 20-400*MU* min'[1] [8]. They found that gel response is independent to the dose rate. This is similar to our work however the ingredients in BANG-2 and PAGAT are totally different.

5. conclusions

The R2-dose response of the PAGAT formulation determined in this study was found to have a linear range up to 30Gy. the response of the PAGAT gel is very similar in the lower dose region and The R2-dose response for doses less than 2Gy is not exact.

Journal of Physics: Conference Series **164** (2009) 012036 doi:10.1088/1742-6596/164/1/012036

The R2-dose response of the PAG AT polymer gel dosimeter is linear between 10 to 30 Gy with R2-dose sensitivities of 0.0905, 0.1037, 0.1023, 0.0907 and 0.123 $s\sim^{1}Gy\text{-}^{1}$ when imaged at 1, 8, 15, 29 and 38 days post-irradiation respectively, therefore The R2-dose sensitivity showed stability with imaging post time after 38 days. To verification of dose rate dependence, different phantom of PAGAT gels was irradiated to 20Gy of doses by 6MV photons with different dose rates (e.g. 80, 160, 240, 320, 400 and 480cGy/min). therefore no significant dose rate effects in PAGAT polymer gel have been observed using NMR evaluation when dose rate is changed from 80cGy/min to 480cGy/min.

To avoid potential problems with different dosimeter response in different physical conditions one should perform calibration of polymer gel dosimeter and exposure of test phantom under the same or very similar physical conditions. PAGAT polymer gel dosimeter displayed good dose rate and time stability responses, which are important consideration when developing polymer gel dosimeters. The PAGAT polymer gel dosimeter in this study exhibited the essential characteristics required for clinical radiotherapy dosimetry. PAGAT polymer gel offers simplification in to the routine clinical radiotherapy environment.

References

[1] J.C. Gore, Y.S. Kang, and R.J. Schulz, "Measurement of radiation dose distributions by nuclear magnetic resonance (NMR) imaging". Phys. Med. Biol. **29,** 1189-97(1984).

[2] L. E. Olsson, A. Fransson, A. Ericsson, and S. Mattsson, "MR imaging of absorbed dose distributions for radiotherapy using ferrous sulphate gels". Phys. Med. Biol. 35,1623-31(1990).

[3] R. J. Schulz, and J. C. Gore, "Reported in Imaging of 3D dose Distributions by NMR National Cancer Institute Grant Application". 2RO1CA46605-04. (1990).

[4] M. J. Maryanski, J. C. Gore, R. Kennan, and R. J. Schulz, "NMR relaxation enhancement in gels polymerized and cross linked by ionizing radiation: a new approach to 3D dosimetry by MRI (Magn. Reson. Imag.)". **11,** 253-8 (1993).

[5] De Deene Y., "Essential characteristics of polymer gel dosimeters". Third International Conference Radiotherapy Gel Dosimetry. Journal of Physics: Conference Series 3 (2004) 34-57.

[6] Brindha S., Vennin A.J., Hill B., Baldock C., "Experimental study of attenuation properties of normoxic polymer gel dosimeters". Phys. Med. Biol. 49 (2004) N353-N361.

[7] M. J. Maryanski, R. J. Schulz, G. S. Ibbott, J. C. Gatenby, J. Xie, D. Horton, and J. C. Gore, "Magnetic resonance imaging of radiation dose distributions using o polymer gel dosimeter". Phys. Med. Biol. 39, 1437-1455 (1994).

[8] M. J. Maryanski, G. S. Ibbott, P. Estman, R. J. Schulz, and J. C. Gore, "Radiation therapy dosimetry using magnetic resonance imaging of polymer gels". Med. Phys. 23, 699-705 (1996).

[9] R. P. Kennan, K. J. Richardson, J. Zhong, M. J. Maryanski, and J. C. Gore, "The effects of cross-link density and chemical exchange on magnetization transfer in polyacrylamide gels". J. Magn. Reson., Ser. 6110,267-277(1996).

[10] Yenning A.J., Hill B., Brindha S., Healy B.J., and Baldock C., "Investigation of the PAGAT polymer gel dosimeter using magnetic resonance imaging". **Phys. Med. Biol.** 50 (2005) 3875-3888.

[11] G. S. Ibbott, M. J. Maryanski, P. Eastman, S. D. Holcomb, Y. Zhang, R. G. Avison, M. Sanders, and J. C. Gore, "Three-dimensional visualization and measurrment of conformal dose distributions using magnetic resonance imaging of BANG polymer gel dosimeters" Int. j. Radiat. Oncol, Biol., Phys., 38, 1097-1103(1997).

[12] C. De Wagter, Y. De Deene, B. Mersseman, and W. De Neve, "Comparison between polymer gel and film dosimetry in anthropomorphic phantoms" Abstract No. 21, 15th Annual ESTRO Meeting, Vienna, Austria, 23-26 September(1996).

Verification of Percentage depth dose of MAGICA polymer gel dosimeter with electron beams

Adinehvand K., Aghamiri M.R., Zahmatkesh M.H., (Ph.D.),Akhlaghpor Sh.

Borujerd Azad Univ.,Borujerd, Iran.
Division of Radiation Medicine, Shahid Beheshti Univ., Tehran, Iran
Novin Radiation Medicine Institute, Tehran, Iran.
Novin Radiation Medicine Institute, Tehran, Iran.

K_adinehvand@yahoo.com

Abstract: In this work investigation of the normoxic MAGICA polymer gel dosimeter has been undertaken. Using MRI, the formulation to give the maximum change in the transverse relaxation rate R2 was determined to be 8% gelatin, 0.5% agarose, 9% methacrylic acid, 0.0352% ascorbic acid, 0.0015% CuSO4.5H2O, 0.002% hydroquinone and 82.3% HPLC(Water).When the preparation of final polymer gel solution is completed, it is transferred into phantoms and allowed to set by storage in a refrigerator at about $4C^{\circ}$. The optimal post-manufacture irradiation and post imaging times were both determined to be 1 day. The R2-dose response was linear to 4000cGy. The response of the MAGICA gel is very similar in the lower dose region and The R2-dose response for doses less than 250cGy is not exact. The R2-dose response of the MAGICA polymer gel dosimeter is linear between 500 to 1750 cGy with R2-dose sensitivities of 0.0020 and 0.0023 $S^{-1}cGy^{-1}$ when imaged at 1and 8 days post-irradiation respectively. In this study, the percentage depth dose (PDD) of MAGICA polymer gel dosimeter is determined and different phantoms of MAGICA gels was irradiated to 1000cGy of doses by 6, 8, 12 and 18MeV electrons and The maximum percentage depth dose (PDD) is located at the depths of 0.5, 1, 2 and 3cm.

Key words : polymer gels, MAGICA gel, magnetic resonance imaging (MRI)

1. Introduction

In 1984, magnetic resonance imaging (MRI) demonstrated great potential in visulazing three dimensional (3D) dose distributions of ferric or ferrous sulphate gel dosimeters [1]. Subsequently, studies were undertaken to investigate the feasibility of using Ferric gel as a 3D dosimetry system in radiation oncology [2]. The major limitation in Ferric gel dosimetry is that it suffers from blurness of dose with time which is due to the migration of ferrous and ferric ions in gel matrix, known as diffusion [3]. In 1993, a polymer gel dosimeter was developed that maintained spatial information following irradiation which could be visualized using MRI [4].

In 2001, the first normoxic gels were suggested that could be produced, stored and irradiated in a normal condition. The polymer gel dosimeter, known as MAGIC, was based the polymerization of Methacrylic Acid (MAA) infused with copper (II) sulphate and the anti oxidant ascorbic acid in a gel matrix [5]. Although MAGIC gel has found its place in gel dosimetry for its feasibility to be manufactured in normal condition in the presence of oxygen, it suffers lack of stability in the casts and molding. Subsequently, number of normoxic polymer gel dosimeters were investigated with adding some Agarose to the MAGIC formulation known as MAGICA, to optimize the gel characteristics including stability, dynamic range, reproducibility and increasing stiffness [6]. Magnetic Resonance Imaging (MRI) has been most extensively used for the evaluation of absorbed dose distributions in polymer gel dosimeters. In the MRI evaluation of polymer gel dosimeters, changes in T2 is a result of physical density changes of irradiated polymer gel dosimeters. This study has been focused on evaluation of the dependence of MAGICA polymer gel dosimeter Percentage depth dose on different energies of the electron beams in a Medical Linear accelerator.

2. Materials and methods

2.1. MAGICA preparation

A MAGICA polymer gel dosimeter was prepared. All chemicals (gelatin, ascorbic acid, CuSO4.5H2O, hedroquanine and methacrylic acid) were provided by Sigma Aldridge and Flucka with experimental grade. HPLC water was obtained from Novin Medical Radiation Institute in Tehran. The preparation of the gel was carried out a similar procedure as described by Fong et al. (2001) with slight difference due to the presence of Agarose in MAGICA formulation. First, water was divided into 5 flasks of varying sizes beforehand, ready for dissolving each substance. Gelatin was added in to about 60% of the total HPLC de-ionized water. Two electrical heating plates provided with magnetic stirring and thermostatic control were used to heat the solutions. Gelatin was allowed to swell for about an hour and then the solution was stirred and heated to about 50 $^{\circ}C$ until a clear solution was obtained, ensuring all gelatin has been dissolved. When the gelatin solution temperature reached near 40 $^{\circ}C$, Agarose is added to about 30% of warm water which has been heated up to 50 $^{\circ}C$ beforehand. Agarose solution was stirred and heated to about 90 $^{\circ}C$ at which Agarose was thoroughly dissolved. At this time gelatin solution should have reached near 50 $^{\circ}C$. Both solutions were allowed to cool. The gelatin solution was larger in volume compared to the Agarose solution, thus Agarose solution cools faster in spite of its higher temperature. However, the cooling rates can be adjusted with respect to each other by proper adjustment of the heating plates. When both solutions cooled to an equal temperature about 47 $^{\circ}C$, Agarose solution was added to the gelatin solution and stirring continued. Stirring never stops before the end of fabrication. At 45 $^{\circ}C$, Hydroquinone which has been solved in about 5% of water was added to the mixture. The remaining 5% of water were divided into two portions and in each portion Ascorbic acid (AA) and Copper (II) sulphate were dissolved after being weighed. These two chemicals which together play the role of oxygen scavenger were added to the mixture when temperature declined to about 37 $^{\circ}C$. Methacrylic acid(MAA) was added at the same temperature and the stirring was continued. The amount of MAA for all gel fabrications was 9% of the total weight of gel except in one experiment in which more MAA was used. The gel was then decanted into test tubes or poured into the phantoms and left in a typical refrigerator at about 4 $^{\circ}C$ to set. Gel phantoms and calibration tubes were not irradiated in the first 24 hours after being manufactured. All irradiations were performed this period [5-6].

2.2. Irradiation

Irradiation of vials was performed using electron beams by Elekta linear accelerator with SSD = 100cm, field size of $20\times 20cm^2$ and the depth was selected at 1cm. To determine the Percentage depth dose of MAGICAT polymer gel dosimeter with different energies 6, 8,12 and 18 MeV and dose rate= $400\frac{cGy}{\min}$. The optimal post-manufacture irradiation was determined to be 1 day [7].

2.3. Imaging

Before imaging, all polymer gel dosimeters were transferred to a temperature controlled MRI scanning room to equilibrate to room temperature. The MAGICA polymer gel dosimeters were imaged in a Siemens Symphony 1.5 Tesla clinical MRI scanner using a head coil. T2 weighted imaging was performed using a standard Siemens 32-echo pulse sequence with TE of 20 ms, TR of 3000 ms, slice thickness of 4 mm, FOV of 256 mm. The optimal post imaging times was determined to be 1 day. The images were transferred to a personnel computer where T2 and R2 maps were computed using modified radiotherapy gel dosimetry image processing software coded in MATLAB (The Math Works, Inc).The mean T2 value of each vial was plotted as a function of dose with the quasi-linear section being evaluated for R2-dose sensitivity. Table 1 lists the protocol of magnetic resonance imaging (MRI) was used in MAGICA polymer gel dosimeter [7].

3. Results

3.1. Calibration Curve of MAGICA polymer gel dosimeter

MAGICA gels with optimum value of ingredient was manufactured and irradiated to different doses. As it can be seen in figure 1, MAGICA has a linear response to 4000cGy. The response of the MAGICA gel is very similar in the lower dose region and The R2-dose response for doses less than 250cGy is not exact. The R2-dose response of the MAGICA polymer gel dosimeter is linear between 500-1750cGy and 1750-4000cGy. Figure 1 shows Calibration Curve of MAGICA polymer gel dosimeter. Table 2 lists the Sensitivity of MAGICA polymer gel dosimeter with different range of doses.

3.2. Percentage depth dose(PDD) of MAGICA polymer gel dosimeter

In this study, the percentage depth dose (PDD) of MAGICA polymer gel dosimeter is determined and different electrons of MAGICA gels was irradiated to 1000cGy of doses by 6, 8, 12, and 18MeV electrons and The maximum percentage depth dose (PDD) is located at the depths of 0.5, 1, 2 and 3 cm. Figure 2 shows at depth of 8cm, the percentage depth dose for 6, 8, 12 and 18MeV electrons is determined, 56.11%, 49.13%, 53.79% and 61.42% respectively.

Table 1. The protocol of magnetic resonance imaging (MRI)

PARAMETERS		
Field of View(FOV)	[mm]	256
Matrix Size(MS)		512×512
Slice Thickness	[mm]	4
Repetition Time(TR)	[ms]	3000
Echo Time(TE)	[ms]	20
Number of Slices		1,2,3,4
Number of Echoes		32
Total Measurement Time	[min]	25-30
Resolution	[mm]	0.5
Band With $[Hz/Pixel]$		130

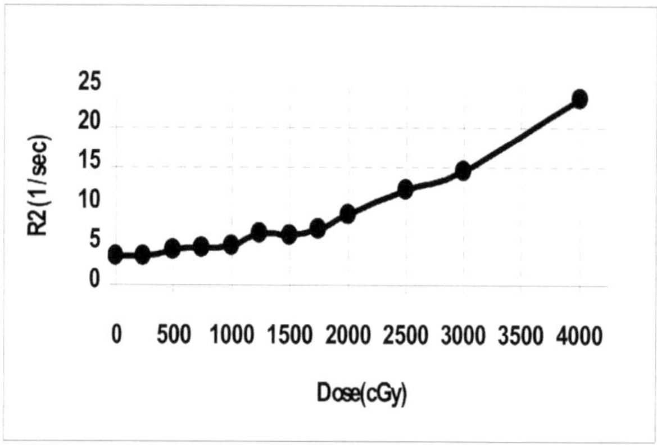

Figure 1. Sensitivity of MAGICA with different range of doses

Table 2. Sensitivity of MAGICA with different range of doses

dose(cGy)	R2- dose sensitivity (S-1cGy-1)
500-1750	0.002
1750-4000	0.0072

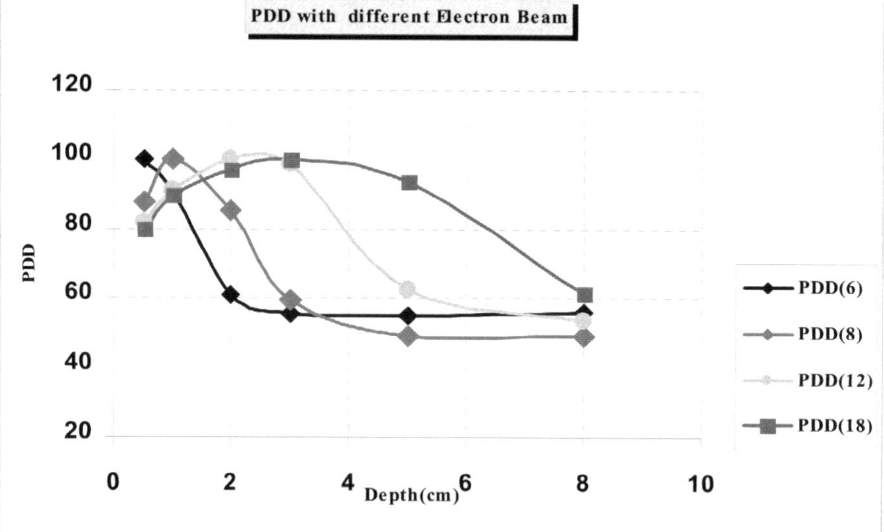

Figure 2. Verification of Percent Depth Dose (PDD) in MAGICA polymer gel dosimeter.

Journal of Physics: Conference Series **164** (2009) 012037 doi:10.1088/1742-6596/164/1/012037

4. Conclusion

For calibration curve of MAGICA polymer gel dosimeter formulation determined in this study was found to have a linear range to 4000cGy. the response of the MAGICA gel is very similar in the lower dose region and the R2-dose response for doses less than 250cGy is not exact. The R2-dose response of the MAGICA polymer gel dosimeter is linear between 500 to 1750 cGy with R2-dose sensitivities of0.0020 and 0.0023 $S^{-1}cGy^{-1}$ when imaged at 1 and 8 days post-irradiation respectively, therefore The R2-dose sensitivity showed stability with imaging post time after 8 days.

MAGICA gels were irradiated to 1000cGy of doses by 6, 8, 12 and 18MeV electron beams and the maximum percentage depth dose (PDD) is located at the depths of 0.5, 1, 2 and 3cm. At depth of 8cm, the percentage depth dose for 6, 8, 12 and 18MeV electrons is determined, 56.11%, 49.13%, 53.79% and 61.42% respectively. The MAGICA polymer gel dosimeter in this study exhibited the essential characteristics required for clinical radiotherapy dosimetry. MAGICA polymer gel offers simplification in to the routine clinical radiotherapy environment.

References

[1] J.C. Gore, Y.S. Kang, and R.J. Schulz, "Measurement of radiation dose distributions by nuclear magnetic resonance (NMR) imaging". Phys. Med. Biol. 29, 1189-97 (1984).
[2] L. E. Olsson, A. Fransson, A. Ericsson, and S. Mattsson, "MR imaging of absorbed dose distributions for radiotherapy using ferrous sulphate gels". Phys. Med. Biol. 35, 1623-31 (1990).
[3] R. J. Schulz, and J. C. Gore, "Reported in Imaging of 3D dose Distributions by NMR National Cancer Institute Grant Application". 2RO1CA46605-04. (1990).
[4] M. J. Maryanski, J. C. Gore, R. Kennan, and R. J. Schulz, "NMR relaxation enhancement in gels polymerized and cross linked by ionizing radiation: a new approach to 3D dosimetry by MRI (Magn. Reson. Imag.)". 11, 253-8 (1993).
[5] P. M. Fong, D. C. Keil, M. Does, and C. J. Gore, "Polymer gels for magnetic resonance imaging of radiation dose distributions at normal room atmosphere". Phys. Med. Biol. 46, 3105-3113 (2001).
[6] M. H Zahmatkesh, R. Kousari, Sh. Akhlaghpour, and S. A .Bagheri, "MRI gel dosimetry with Methacrylic acid., Ascorbic acid., Hydroquinone and Copper in Agarose (MAGICA) gel". Preliminary Proceedinges of DOSGEL 2004. Sep 13-16. 2004. Ghent. Belgium.(2004).
[7] Adinehvand K., "Verification Response of MAGICA Gel Dosimeter in Beams Electron with Technique MRI" . MS. Thesis. Department of Nuclear Engineering . Shahid Beheshti University. Tehran. Iran. (2007).

An overview of polymer gel dosimetry using x-ray CT

A. Jirasek[1] and M. Hilts[2]

1. Dept. of Physics and Astronomy, University of Victoria, Victoria, BC, CANADA

2. BC Cancer Agency, Vancouver Island Centre, Victoria, BC, CANADA

mhilts@bccancer.bc.ca

1. Introduction

The use of x-ray computed tomography (CT) for the imaging of irradiated polymer gels was first introduced at the 1st international conference on radiotherapy gel dosimetry, DOSGEL99, and in a subsequent paper [1,2]. These works showed dose dependent contrast in CT images of irradiated polyacrylamide (PAG) gel, quantified a CT dose response for PAG gel and thereby established the potential of CT as an alternative to MRI and optical CT for gel read-out. The advantages of CT read-out are practical ones: accessibility of CT to clinical radiation therapy environments, ease and speed of image acquisition, and a CT dose response that is stable and relatively insensitive to environmental factors. However, significant drawbacks include a very low sensitivity of CT number (H) to dose (~1%) and the fact that the act of imaging with CT will impart additional dose to the gel. Recent work has focused on means of understanding and minimizing these drawbacks.

The review paper presented at Dosgel 2006 [3] provided a thorough summary of the state of knowledge in the field at that time, and covered the following topics: the fundamental x-ray CT properties of polymer gel, dose response characteristics, CT imaging considerations, image filtering for noise reduction and CT gel dosimetry applications. This review serves as an "update" to the 2006 review paper and highlights new work that has been completed since this time. The material of the 2006 review will be repeated here only in very brief summary, and readers are referred to the 2006 review paper for details. The new work discussed includes: (a) normoxic polyacrylamide gel dosimetry; (b) x-ray CT dose in CT gel dosimetry and (c) image filtering for improved dose resolution.

2. A brief review of X-ray CT imaging of polymer gel dosimeter basics

2.1. Fundamentals

A small density change occurs in polymer gel upon irradiation which thereby changes the gel's x-ray attenuation (μ) and therefore the CT number (N_{CT}) and results in dose dependent contrast observed in CT images of irradiated gel. This change in density with dose has been measured experimentally for both PAG and MAGIC gels [4,5]. A typical CT dose response is mono-exponential with a region that can be approximated as linear at lower doses (the "quasi-linear" region, frequently used for relative dosimetry). An example (PAG gel, 6%T, 50%C) is shown below (Figure 1). This figure also illustrates the excellent reproducibility in CT dose response.

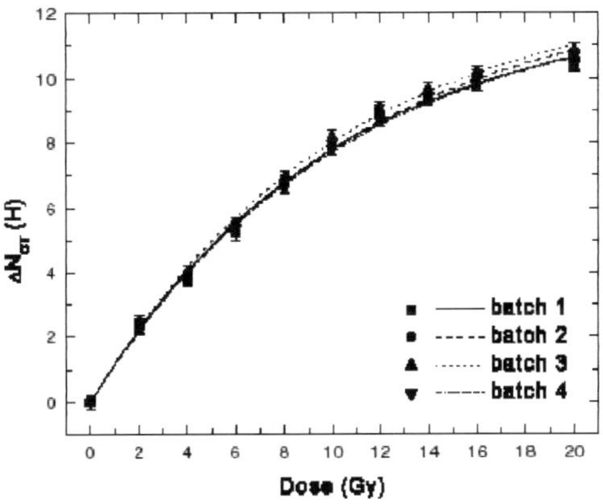

Figure 1: A typical CT dose response for polymer gel (formulation shown is PAG gel, 6%T, 50%C). Note the excellent dose response reproducibility. From [3].

Gel formulation can affect the shape and dose range of CT dose response, as well as the extent (and in some cases existence) of a quasi-linear region [6-13]. A model has been developed which provides an understanding of how composition might affect the density change [6]. For a given dose, the measured gel density change (or change in N_{CT}) is the result of two factors: 1) the amount of polymer formed and 2) an intrinsic density change that occurs on conversion of monomer to polymer. For PAG gel, the intrinsic density change was found to vary with the type of polymer formed and hence with the relative fraction of cross-linker in the system. The amount of polymer formed increased linearly with the total fraction of monomer (%T). Hence scope exists for optimizing gel formulation to CT imaging.

2.2. CT imaging optimization

Selection of CT technique (kV, mAs) does not affect the sensitivity of polymer gel dose response, however tube temperature does have an effect and a CT scanner should be fully warmed up prior to imaging [1]. Scan technique does have a tremendous effect on image noise and hence dose resolution [10,14 - 16]. Table 1, below, summarizes the effect selectable imaging parameters have on image noise. Increasing kV, mAs, slice thickness, number of averages and pixel dimension (post-processing) all serve to reduce image noise. However, these means of noise reduction all come at a price. In selecting a protocol for CT imaging polymer gel one must consider requirements for dose resolution (i.e. image noise), imaging time and spatial resolution. The difficulty lies in the conflict between achieving low noise (high CT scanning technique, large slice thickness and pixel size) and achieving both high spatial resolution (thin slices and small pixel size) and short imaging times (fewer slices imaged, low scan technique and therefore reduced load on the x-ray tube) [15]. Compromise will always be required and the "optimum" imaging protocol will depend on each application.

Table 1: Factors affecting CT image noise. From [3].

Factor affecting image noise (symbol)	Relationship to image noise ($\sigma_{N_{CT}}$)
Phantom diameter (d)	$\sigma_{N_{CT}} \propto e^{d}$
Tube voltage (kV)	$\sigma_{N_{CT}} \propto (kV)^{-1.4}$
Tube current (mA)	$\sigma_{N_{CT}} \propto (mA)^{-0.5}$
Slice scan time (s)	$\sigma_{N_{CT}} \propto s^{-0.5}$
Number of averages (NAX)	$\sigma_{N_{CT}} \propto (NAX)^{-0.5}$
Pixel dimension (w)	$\sigma_{N_{CT}} \propto e^{w}$ ($or\ w^{-0.65}$)
Slice thickness (h)	$\sigma_{N_{CT}} \propto h^{-0.5}$

3. Normoxic polyacrylamide gel dosimetry

Traditionally, CT polymer gel dosimetry has been performed with anoxic polyacrylamide gel [1,7]. Recently, the technique has been extended to normoxic polyacrylamide gel (nPAG, [12]). Normoxic polyacrylamide x-ray CT gel dosimetry provides a reproducible, stable platform from which to undertake 3D dosimetry. Figure 2 illustrates the reproducibility in slope of a series of irradiated nPAG gels imaged with CT. Also shown are the comparable slopes for anoxic PAG. As can be seen, adding antioxidant, in this case tetrakis (hydroxymethyl) phosphonium chloride (THPC), reduces the overall dose sensitivity of the gel. nPAG CT gel sensitivity has been quoted at ~0.36 +/- 0.03 H/Gy [12, 9].

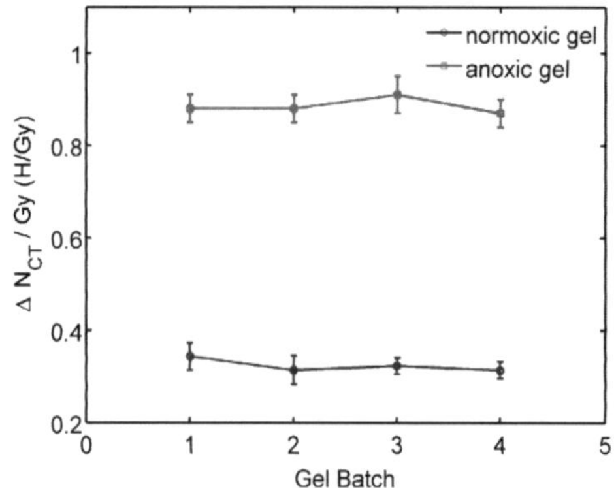

Figure 2: Dose response reproducibility of nPAG with x-ray CT imaging. Also shown is the dose response reproducibility of anoxic PAG. From [12].

Recently, efforts have been made to improve the overall dose sensitivity of polymer gels for readout with x-ray CT imaging. Koeva *et al* have investigated the use of bis-acrylamide co-solvents as a means of increasing the overall solubility limit of bis-acrylamide in the gel and hence increasing the dose sensitivity of the gel [17]. With no co-solvent, the solubility of bis-acrylamide in polymer gel is ~3% by weight. With bis-acrylamide co-solvent, the solubility limit can be increased to 10% (using isopropanol as co-solvent, [18]). Figure 3 illustrates the effects of adding increasing amounts of isopropanol to a N-isopropyl acrylamide gel containing 10% total monomer. Shown is the clear increase in dose sensitivity of polymer gel dosimetery using x-ray CT imaging. Bis-acrylamide co-solvent for increased dose response sensitivity has not, to date, been fully explored and offers excellent potential for improving the dose sensitivity performance of CT polymer gel dosimetry.

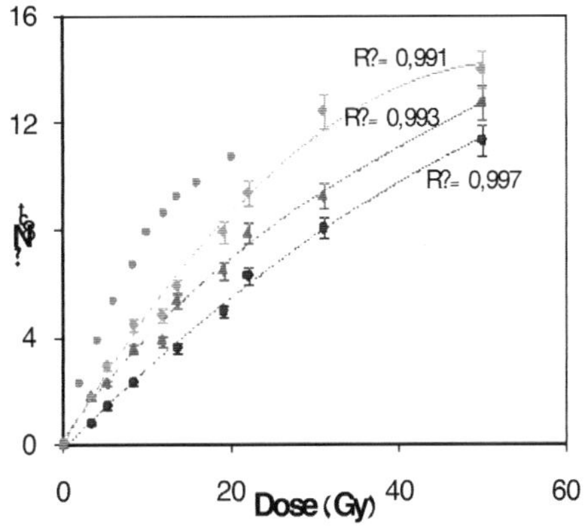

Figure 3: X-ray CT dose response for 10%, 50% C NIPAM/Bis dosimeters containing 0 (●), 10 (▲) and 30 (♦) wt% of isopropanol. An anoxic PAG dosimeter (6% T 50% C) (●) is included for comparison. From [17].

4. X-ray CT dose in imaged polymer gel dosimeters

A potential issue with utilising x-ray CT imaging with polymer gel dosimetry is the CT dose delivered to the gel during imaging, creating further polymerisation within the gel. CT dose inducing additional gel polymerisation can be particularly problematic in imaging protocols requiring a large number of image averages per slice. The effect of CT dose on gel polymerisation has been studied recently by [19]. Using a combination of CT dose measurements (with an ionisation chamber), CT imaging and Raman spectroscopic measurements, the CT-dose induced change in CT number for a range of imaging protocols and gel container sizes was reported. Table 2 shows the delivered CT dose and induced CT number (ΔN_{CT}) in a range of imaging protocols. This study was performed for a polyacrylamide gel manufactured with tetrakis (hydroxymethyl) phosphonium chloride (THPC) as antioxidant, with a reported gel sensitivity of 0.36 +/- 0.03 H/Gy.

As can be seen from table 2, induced ΔN_{CT} for all imaging protocols remains below the minimum detectable ΔN_{CT} of 0.2 H. For imaging protocols requiring large image averages (e.g. 64 averages) per slice, the ΔN_{CT} can exceed the minimum detectable limit (e.g. 0.5 H) and hence care must be taken to ensure that the gel is inactive prior to imaging. Of final note, the current literature contains results for only one type of gel (normoxic polyacrylamide gel). Recent efforts in the development of gels with increased sensitivity for CT imaging [17] will alter these results to some extent. Hence, future gel formulations will need to be tested for CT dose contamination.

Table 2: CT dose and induced ΔN_{CT} for a range of imaging protocols. From [19].

Imaging protocol	Phantom	kVp	mA s	Slice (mm)	Dose/image (cGy)	# Image averages	ΔN_{CT}
Volumetric (CTDI)	16 cm diam. (Center)	120	200	2	3.38±0.14	16	0.105±0.001
				5	3.4±0.1	16	0.105±0.001
				10	3.4±0.1	16	0.107±0.001
	16 cm diam. (Edge)	120	200	2	3.5±0.1	16	0.109±0.001
				5	3.5±0.1	16	0.110±0.001
				10	3.6±0.1	16	0.111±0.001
	16 cm diam. (Center)	140	200	2	4.4±0.2	16	0.136±0.002
				5	4.4±0.2	16	0.136±0.002
				10	4.5±0.2	16	0.138±0.002
	16 cm diam. (Edge)	140	200	2	4.5±0.2	16	0.139±0.002
				5	4.5±0.2	16	0.139±0.002
				10	4.6±0.2	16	0.141±0.002
Single slice (PD)	16 cm diam. (Center)	120	200	2	0.70±0.05	32	0.0430±0.0002
				5	0.91±0.06	32	0.0560±0.0004
				10	1.07±0.07	32	0.066±0.001
	16 cm diam. (Edge)	120	200	2	1.6±0.1	32	0.099±0.001
				5	1.7±0.1	32	0.107±0.001
				10	2.1±0.2	32	0.133±0.002
	16 cm diam. (Center)	140	200	2	0.83±0.06	32	0.0510±0.0003
				5	1.07±0.07	32	0.066±0.001
				10	1.23±0.09	32	0.076±0.001
	16 cm diam. (Edge)	140	200	2	1.5±0.1	32	0.095±0.001
				5	2.0±0.1	32	0.125±0.002
				10	2.1±0.2	32	0.130±0.002
Calibration (PD)	Calibration Gel vials	140	200	10	2.3±0.2	16	0.0720±0.0001

5. Image filtering for signal enhancement in CT imaging or polymer gel dosimeters

A range of image filtering methods have been proposed for image enhancement in x-ray CT imaging of polymer gel dosimeters. We here review two of the most promising methods proposed to date, adaptive mean filtering and two point maximum entropy regularisation of CT images.

5.1. Adaptive filtering

Adaptive mean filtering utilizes local image statistics, within a given filter window (or kernel size, K), to determine the power of the filter. The filter is expressed as

$$f(x,y) = g(x,y) - \frac{\sigma_N^2}{\sigma_L^2}\left[g(x,y) - m_L\right] \quad (1)$$

where f(x,y) is the filtered image, g(x,y) is the original unfiltered image, σ^2_N is the local variance (within the filter window), σ^2_L is the image noise, m_L is the local mean value of the image. Adaptive mean, sometimes called Weiner, filtering is typically performed using kernel sized between 3x3 and 11x11. Although larger kernels typically produce smoother images, these larger kernels also tend to compromise sharp gradients in the image. Hence, a tradeoff occurs between smoothness and image fidelity.

An interesting characteristic of the Adaptive filter is that it is well suited to multiple passes over the same image. Hence, not only can the kernel size be adjusted to a specific need, the number of iterations over the image can also be tuned. The effects of kernel size and number of iterations was studied recently for a range of image SNR [20]. As an example, figure 4 illustrates the results of Weiner filtering a noisy "U"-shaped synthetic CT image with varying kernel sizes and a range of total number of iterations. As can be seen from figure 4, increasing kernel size and/or the number of iterations has a positive effect on noise reduction in filtered images. At some point, moving to excessively high kernel size or number of iterations becomes counterproductive, as image distortion (blurring) becomes unacceptably high. Hence, care must be taken in choosing the kernel and iteration number.

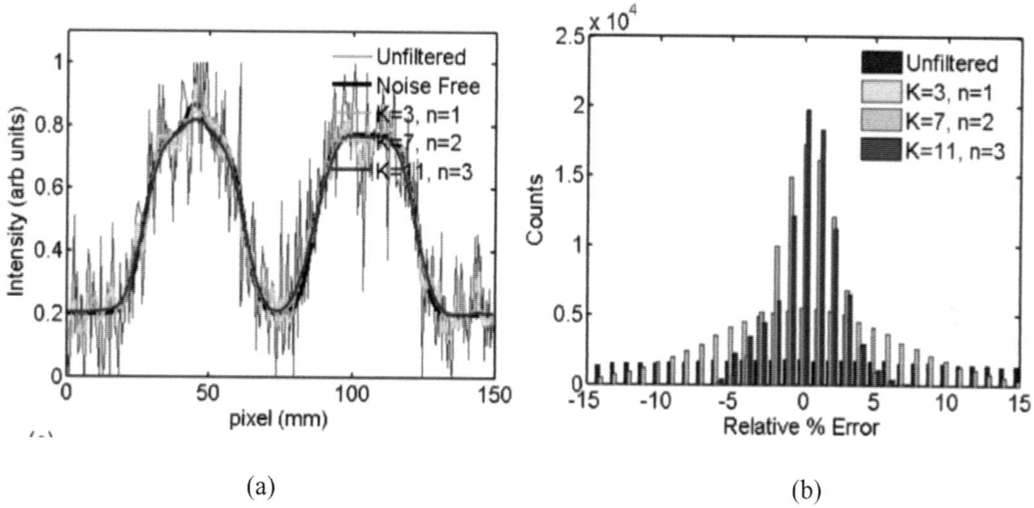

Figure 4: (a) Profiles through a "U" shaped image using kernal sizes if K=3, 7, and 11 and between 1 and 3 filter iterations. (b) Error histograms between filtered and unfiltered images. From [20].

A second example utilizes a more realistic, although still synthetic, "CT gel image" of a prostate conformal therapy dose distribution. Figure 5 illustrates the effects of Weiner filtering on a noise filled conformal prostate dose distribution. Most impressive perhaps is the large improvement in dose-area histograms afforded by Adaptive mean filtering. In this case a relatively large kernel (K=11) and moderate iterations (n=2) was used. This combination of parameters was made due to the reasonably low dose gradients in the image, hence allowing for a large kernel.

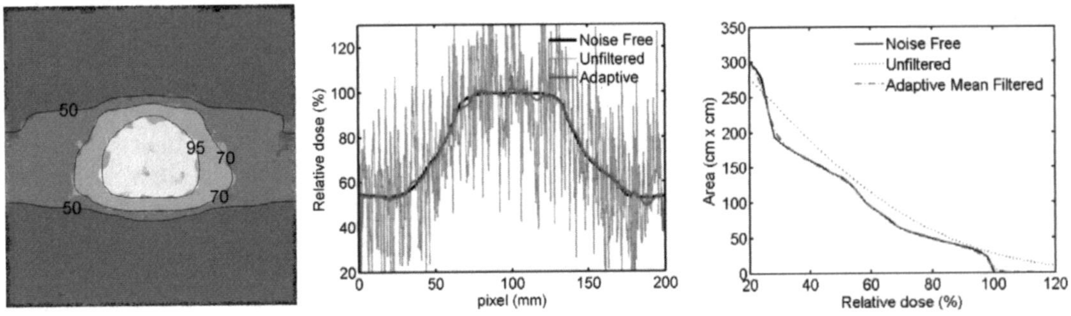

(a) (b) (c)

Figure 5: Adaptive mean filtering (K=11, n=2) for a conformal prostate "gel" image. (a) noise free image with filtered image contour overlay, (b) profiles through noise-free, filtered and unfiltered images, and (c) dose area histograms of noise-free, filtered and unfiltered images. From [20].

5.2. Two point maximum entropy

A second method for noise reduction in CT gel images has been investigated recently [21]. Two point maximum entropy method (TPMEM) regularisation is a powerful technique that, although more complex than Adaptive filtering, works particularly well on low SNR data and also allows for a large amount of flexibility in the amount of smoothing undertaken on the image. The aim of TPMEM is to minimise the function

$$T=-S+\lambda\chi^2 \quad (2)$$

where S is the entropy function, χ^2 is the chi square function, and λ is a Lagrange multiplier which mediates the relative contributions of smoothness (S) and fidelity (χ^2) in the resultant image. The entropy function is given by

$$S=\sum -\left(p\ln p+q\ln q\right) \quad (3)$$

where NP is the number of points in the input data and p and q are fractional probabilities in any two adjacent data points,

$$p=\frac{x_i}{\left(x_i+x_{i+1}\right)} \qquad q=\frac{x_{i+1}}{\left(x_i+x_{i+1}\right)} \quad (4)$$

The χ^2 function is calculated over all point and between the recovered (smoothed) and initial dataset (image). In general, the routine is run until a stopping criterion of $\chi^2 = NP$ is reached. However, this stopping criterion is to some extent flexible and hence can be written as

$$\chi^2=X\cdot NP \quad (5)$$

where X is a tuneable parameter, typically between X = 0.6 – 1.4. This offers a very unique capability to tune the resultant image to either greater smoothness or greater fidelity, depending on the desired outcome. Furthermore, this is not attained via a typical windowing method where greater smoothness is attained through a larger filter window and, hence, at the potential expense of image gradient degradation.

The algorithm for implementation of TPMEM for 1D (vector) data can be written as:

i. Initiate initial recovered spectrum to uniform values (e.g. choose each x_i as mean value of all d_i). Choose minimum and maximum limits for λ (e.g. $10^{-9} < \lambda < 10$).

ii. Set $\lambda = \lambda_{low}$. Set initial vector to flat estimate (i.e. use (i)). Minimise equation (1) using a conjugate gradient decent method. Record resulting χ^2.

iii. Set $\lambda = \lambda_{high}$. Set initial vector to flat estimate (i.e. use (i)). Minimise equation (1) using a conjugate gradient decent method. Record resulting χ^2 .

iv. Bisect λ. ($\lambda = (\lambda_{low} + \lambda_{high})/2$) using a root bisection method. Set input vector to output of (ii) or (iii), depending on which has lower χ^2. Minimise equation (1) using a conjugate gradient decent method. Record resulting χ^2.

v. Is $\chi^2 = X \cdot NP$ within tolerance (e.g. 5% of NP)?

   a) If no and if $\chi^2 > NP$, set $\lambda_{high} = \lambda$. Go to (iv).

   b) If no and if $\chi^2 < NP$ set $\lambda_{low} = \lambda$. Go to (iv).

   c) If yes, exit and record data.

The algorithm terminates when either $\chi^2 = X \cdot NP$ within a specified tolerance, or when the root-bisection accuracy has been exceeded. To extend the algorithm to 2D (image) data, The algorithm is extended to:

(i) Initiate initial recovered image to uniform values (e.g. choose each x_i as mean value of all d_i). Choose minimum and maximum limits for λ (e.g. $10^{-9} < \lambda < 10$).

(ii) Set $\lambda = \lambda_{low}$.

(iii) Set input vector to row 1 of image. Minimise equation (1).

(iv) Set input vector to column 1 of image. Minimise equation (1).

(v) Set input vector to row 2 of image. Minimise equation (1).

(vi) Set input vector to column 2 of image. Minimise equation (1). ...

(vii) Set input vector to row N of image. Minimise equation (1).

(viii) Set input vector to column N of image. Minimise equation (1).

(ix) Calculate and record χ^2 for entire image.

(x) Set $\lambda = \lambda_{high}$. Repeat (iii)–(ix).

(xi) Bisect λ. ($\lambda = (\lambda_{low} + \lambda_{high})/2$). Set input image to output of λ_{low} or λ_{high} minimisation, depending on which has lower χ^2 . Minimise image, (iii)–(ix).

(xii) Is $\chi^2 = X \cdot NP$ within tolerance (e.g. 5% of NP)?

a) If no and if $\chi^2 > X \cdot NP$, set $\lambda_{high} = \lambda$. Go to (iii).

b) If no and if $\chi^2 < X \cdot NP$ set $\lambda_{low} = \lambda$. Go to (iii).

c) If yes, exit and record data.

Note that this implementation of the algorithm is designed for square (NxN) images. NxM images can easily be accommodated through padding to NxN.

From equation 2 and the algorithms presented above, it can be seen that TPMEM does not rely on a window, pre se, to achieve smoothing and that, hence, window size does not determine the amount of smoothing to be undertaken. Rather, the technique relies on an interplay between two functions, one dedicated to smoothing the data (the entropy function) and the other dedicated to maintaining image fidelity (the χ^2 function). Furthermore, this interplay can be finely tuned through the X parameter in the stopping criterion, hence a user defined balance can be obtained for smoothness vs fidelity in any given image.

Figure 6 illustrates the TPMEM method applied to the same U-shaped pattern as in figure 4. Different amounts of smoothing were generated through the application of a range of X values in the stopping criterion. As can be seen from figure 6, larger X values lead to larger amounts of smoothing. Although not as dramatic as in the Adaptive mean filtering case, some sharp contrast definition can be lost in this method if too large an X parameter is chosen in the routine.

It should be noted that TPMEM comes with a larger "overhead" than Adaptive filtering. That is, whereas Adaptive filtering can be executed in Matlab in very few lines of code (the Weiner filter is built into Matlab), TPMEM requires some user generated code to enable the algorithm to run effectively. However, the tradeoff for the added complexity is that TPMEM works exceedingly well on very low SNR data, which can be beneficial for some x-ray CT imaging protocols. Hence, choosing between Adaptive filtering and TPMEM should be based on the quality of the given data.

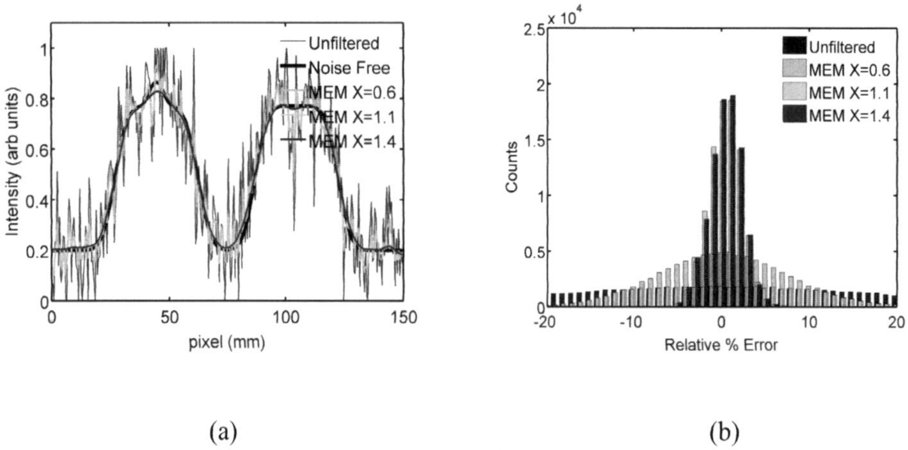

Figure 6: TPMEM filtering of a U-shaped patter. (a) profiles through unfiltered, noise-free, and TPMEM filtered images. X values range between X=0.6 to X=1.4; (b) error histograms showing improvement in image fidelity for filtered images. From [21].

6. Conclusions

Recent work in x-ray CT polymer gel dosimetry has concentrated on (i) moving to normoxic gel formulations, (ii) understanding the effects of CT dose on polymer gel dose distribution integrity, (iii) image filtering for dose resolution improvement and, recently, (iv) new gel formulations for improved x-ray CT dose response sensitivity. Results indicate that CT imaging remains an attractive possibility for polymer gel dosimetry. Further work on improved formulations for CT polymer gel dosimetry is currently underway and stands to offer increased dose response sensitivity, hence bringing the technique closer to MRI and optical CT in its clinical utility.

Acknowledgments
Funding from the Canadian Foundation for Innovation, the BC Knowledge Development Fund, the Natural Sciences and Engineering Research Council of Canada, and the BC Cancer Agency is acknowledged.

References

1. Hilts, M., Audet, C., Duzenli, C. and Jirasek, A. (2000) Polymer gel dosimetry using x-ray computed tomography: a feasibility study. *Phys Med Biol*, **45**, 2559-2571.
2. Hilts, M., Audet, C., Duzenli, C. and Robar, J. (1999) In Schreiner, L. J. and Audet, C. (eds.), *Proceedings of the 1st international workshop on radiation gel dosimetry.* Canadian College of Physicists in Medicine and Canadian Organization of Medical Physicists, Lexington, KY, pp. 190-192.
3. M. Hilts. X-ray computed tomography imaging of polymer gel dosimeters. *Journal of Physics: Conference Series*, 56:95–107, 2006.
4. Trapp, J.V., Michael, G., de Deene, Y. and Baldock, C. (2002) Attenuation of diagnostic energy photons by polymer gel dosimeters. *Phys Med Biol*, **47**, 4247-4258.
5. Mather, M.L., De Deene, Y., Whittaker, A.K., Simon, G.P., Rutgers, R. and Baldock, C. (2002) Investigation of ultrasonic properties of PAG and MAGIC polymer gel dosimeters. *Phys Med Biol*, **47**, 4397-4409.
6. Hilts, M., Jirasek, A. and Duzenli, C. (2004) Effects of gel composition on the radiation induced density change in PAG polymer gel dosimeters: a model and experimental investigations. *Phys Med Biol*, **49**, 2477-2490.
7. Trapp, J.V., Back, S.A., Lepage, M., Michael, G. and Baldock, C. (2001) An experimental study of the dose response of polymer gel dosimeters imaged with x-ray computed tomography. *Phys Med Biol*, **46**, 2939-2951.
8. Fong, P.M., Keil, D.C., Does, M.D. and Gore, J.C. (2001) Polymer gels for magnetic resonance imaging of radiation dose distributions at normal room atmosphere. *Phys. Med. Biol.*, **46**, 3105-3113.
9. Brindha, S., Venning, A., Hill, B. and Baldock, C. (2004) Experimental study of attenuation properties of normoxic polymer gel dosimeters. *Phys Med Biol*, **49**, N353-N361.
10. Hill, B., Venning, A. and Baldock, C. (2005) The dose response of normoxic polymer gel dosimeters measured using X-ray CT. *British J. Radiology*, **78**, 623-630.
11. Brindha, S., Venning, A., Hill, B. and Baldock, C. (2004) Radiological attenuation properties of normoxic polymer gel dosimeters. *J Phys Conference Series*, **3**, 196-199.
12. Jirasek, A., Hilts, M., Shaw, C. and Baxter, P. (2006) Investigation of tetrakis hydroxymethyl phosphonium chloride as an antioxidant for use in x-ray computed tomography polyacrylamide gel dosimetry. *Physics in Medicine and Biology*, **51**, 1891.
13. Venning, A., Brindha, S., Hill, B. and Baldock, C. (2004) Preliminary study of a normoxic PAG gel dosimeter with tetrakis (hydroxymethyl) phyosphonium chloride as an anti-oxidant. *J Phys Conference Series*, **3**, 155- 158.
14. Trapp, J., Michael, G. and Baldock, C. (2001) In Baldock, C. and De Deene, Y. (eds.), *Proc. 2nd International Conf. on Radiotherapy Gel Dosimetry.* Queensland University of Technology, Brisbane, AUS.
15. Hilts, M., Jirasek, A. and Duzenli, C. (2005) Technical considerations for implementation of x-ray CT polymer gel dosimetry *Phys Med Biol*, **50**, 1727 - 1745.

16. Hilts, M. and Duzenli, C. (2004) Image noise in x-ray CT polymer gel dosimetry. *J Phys Conference Series*, **3**, 252-256.
17. V.I. Koeva, K. B. McAuley, A.Jirasek, L.J. Schreiner , (2008). Preliminary investigation of the NMR, optical and x-ray CT dose-response of polymer gel dosimeters with cosolvents and increased crosslinker levels. DosGel 2008 Proceedings.
18. V.I. Koeva, E.S. Csaszar, R.J. Senden, K.B. McAuley, L.J. Schreiner, (2008a). Polymer gel dosimeters with increased solubility: a preliminary investigation of the NMR and optical dose-response using different crosslinkers and co-solvents. Macromol. Symp. 261, 157–166.
19. P. Baxter, A. Jirasek, and M. Hilts, (2007). X-ray CT dose in normoxic polyacrylamide gel dosimetry. Med. Phy, 34(6), 1934–43.
20. M. Hilts and A. Jirasek. Adaptive mean filtering for noise reduction in ct polymer gel dosimetry, (2008). Med. Phys, 35(1), 344 - 55.
21. Jirasek, A., Matthews, Q., Hilts, M., Schulze, G., Blades, M. and Turner, R. (2006) Investigation of a 2D two-point maximum entropy regularization method for signal-to-noise ratio enhancement: application to CT polymer gel dosimetry. *Phys Med Biol*, **51**, 2599-2617.

Journal of Physics: Conference Series **164** (2009) 012039 doi:10.1088/1742-6596/164/1/012039

Imaging of dose distributions using polymer gels based on radiation induced changes in stiffness

Remo A Crescenti[1], Jeffrey C Bamber[1], Assad A Oberai[2], Paul E Barbone[3], Joseph P Richter[2], Carlos Rivas[3], Nigel L Bush[1] and Steve Webb[1]

[1] Joint Department of Physics, Institute of Cancer Research and Royal Marsden NHS Foundation Trust, Downs Road, Sutton, Surrey, SM2 5TP, UK
[2] Department of Mechanical, Aerospace and Nuclear Engineering, Rensselaer Polytechnic Institute, 5048 JEC, 110 8th Street, Troy, NY, USA
[3] Department of Aerospace and Mechanical Engineering, Boston University, 110 Cummington Street, Boston, MA 02215, USA

E-mail: remo.crescenti@icr.ac.uk
Abstract. Previously, dose determination based on radiation induced stiffness difference measurements has received no or very little attention. Here, a preliminary evaluation of a combined system for dosimetry based on radiation sensitive gels, ultrasonic elastography and a plane strain inverse algorithm is presented. A block of gel was irradiated along one of its axes producing stiff rod-like regions. The dose distribution found with quantitative ultrasound elastography was compared with a reference dose distribution measured with magnetic resonance imaging. In these early results, the high dose areas were clearly detected, while noise in the ultrasound measurement and strong regularisation in the inverse computing introduced shape distortions, noise in the dose estimates and problems estimating the correct dose contrast. Improvements in the experimental setup and inverse computing are possible, for example by acquisition of transversal ultrasound data, which could essentially reduce the noise and restrict direct influence of the experimental boundary condition on the dose estimation by providing additional information for inverse computing. Based on the preliminary results and the potential for improvement it is concluded that further investigations should follow to establish the potential of the rapidly developing field of elastography for measuring radiation dose based on radiation induced changes in stiffness.

1. Motivation and background

In radiation sensitive polymer gels, monomers polymerize and the molecular structure changes [1]. To date, for dose determination, predominantly irradiation-induced changes of the magnetic properties have been investigated, but this requires magnetic resonance imaging (MRI) [e.g. 2], which is expensive. A quick, cheap and easy-to-use alternative for measuring a dose-dependent property of these gels would be appreciated. Optical imaging may stand as an alternative, if it is used in combination with radiation sensitive plastics [3]. Indeed, optical transmission tomography systems are commercially available, but at considerable cost. With many ultrasound systems already in daily use in clinics, ultrasound methods could provide the benefits of an easy-to-use low cost alternative to MRI. One method that can take advantage of widely available ultrasound equipment is quantitative elastography, which combines the detection of strain distribution within a deformed object with an inverse problem solving algorithm to convert the measurement of displacement or strain into an estimated stiffness distribution. This is potentially useful in gel dosimetry because the elastic properties of polymer gels change under irradiation [4]. This paper describes a preliminary evaluation of the potential of quantitative elastography for dose imaging. For simplicity, a plane strain inverse experimental setup and algorithm were employed, estimating dose only in a plane.

2. Material and methods

A deliberately tall (y-axis) and narrow (x-axis) block of radiation sensitive polymer gel (4 × 10 × 10 cm^3 along x, y and z-axis, respectively) was produced by adding ultrasonic scatterers to the recipe by Fong et al [5]. It was then irradiated with a linear accelerator (Elekta SL series with RT desktop software) applying three small 10 × 8 mm^2 fields to produce a stiffness distribution that was complex in the x-y-plane but uniform along the z-axis, forming three rod-shaped regions along the z-axis. The output factor of the linear accelerator to calculate the radiation dose was not known for such small fields but was estimated to be approximately 18 Gy from measurements with diamond detectors. The block of gel was placed in a mechanical frame (figure 1), which allowed expansion in the x-direction but prevented z-direction motion. The block was compressed in the y-direction at a strain rate of 1 % per s up to a maximum strain of 6 %, while radiofrequency (RF) ultrasonic echo data were acquired at 14 Hz (Z.one™ with an L10-5 linear array, Zonare Medical Systems Inc.) in repeated compressions in the x-y-plane at three different z-positions. This was accomplished with the sample and the mechanical frame sitting in an oil bath. The oil decreased ultrasonic reflections from the boundaries of the gel, attenuated sound outside the sample and provided slippery boundaries between the gel and the compressor plates. Cross-correlation was applied to track the movement of the echo pattern in a small region within a reference RF-frame to its new position in the RF-frame that was acquired at a later time-point, and the process repeated for other reference regions to build up a local displacement map for a region of interest (ROI) in the centre of the scanned plane. An average was then taken of such displacement maps from the different planes. From the average displacement map, the stiffness distribution was determined using an iterative inverse algorithm [6] using the finite element method, in which the material in the ROI was modelled as an incompressible, linear elastic solid.

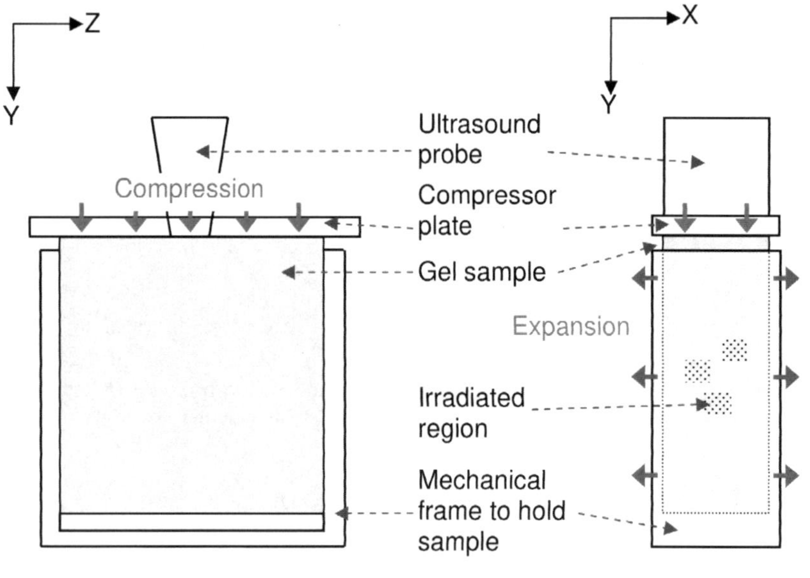

Figure 1: The radiosensitive gel was positioned in a frame that only allowed movement (compression or expansion) in the x-y-plane. The gel had been irradiated with three small fields through the x-y-face to create three rod-like, stiff, high dose regions along the z-axis.

Zero traction boundaries were chosen, as the tall and narrow shape of the sample was designed so that friction at the physical top or bottom boundary would not introduce additional stress in the ROI. For coping with the measurement noise and for turning this ill-posed inverse mathematical problem into a well conditioned problem, total variation diminishing (TVD) regularization was used; the regularization parameters were adjusted so that the algorithm produced an estimation with an homogeneously stiff background (an 'a priory' assumption, as it was not irradiated). For simplification, only the two-dimensional reconstruction described above has been accomplished. The plane strain experimental setup described in Figure 1 (uniform stiffness distribution along z and compression and expansion only in the x-y-plane) was chosen for such computation.

As a reference dose distribution and because the dose to the gel was only estimated, magnetic resonance imaging (MRI) of the same x-y-planes was employed using a 1.5 T scanner (Philips Medical Systems, Best, The Netherlands). Also with MRI, it was established that the dose distribution was indeed uniform in z-direction and that there was no major oxygen contamination that had inhibited the radiation induced polymerisation. T_2-weigthed multiple-spin-echo sequences with 8 echo times from 20 to 160 ms, and a repetition time of 2 s were used to acquire data with a 1×1 mm^2 in-plane resolution and 5 mm slice thickness.

Independent calibration was performed for both MRI and quantitative elastography, each with four smaller cylindrical samples of gel, to relate Young's modulus and the transversal relaxation rate (R_2) with dose, using MRI and a standard mechanical testing device (Instron™). For this purpose, two samples were homogeneously irradiated to 20.8 Gy and two were left non-irradiated.

Between production, irradiation, MRI and elastography measurement, the gels were stored in a fridge at approximately 9 °C. For each experiment, all gels were warmed to room temperature in a water bath for a sufficiently long time to reach temperature equilibrium.

3. Results

Figures 2 and 3 show the calibration results for the compression experiment and the MRI-measurement, respectively. For both modalities, the standard deviation from a series of compressions or R_2 measurements was computed. However such error bars are smaller than the symbols in the figure. The derived dose images are given in figures 4 and 5, again for both

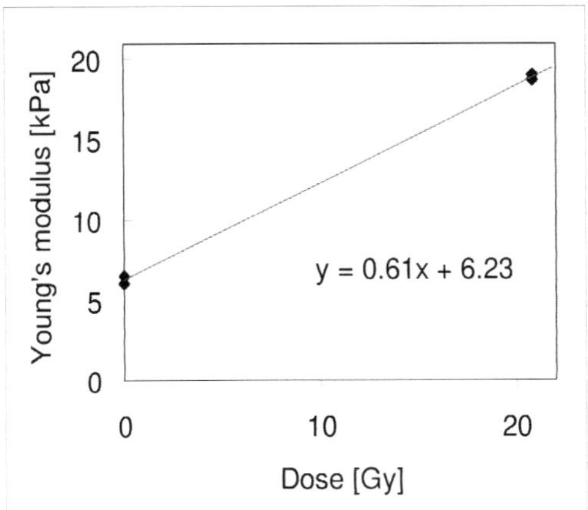

Figure 2: Young's modulus versus dose relationship from mechanical measurements of the four homogeneously irradiated gels.

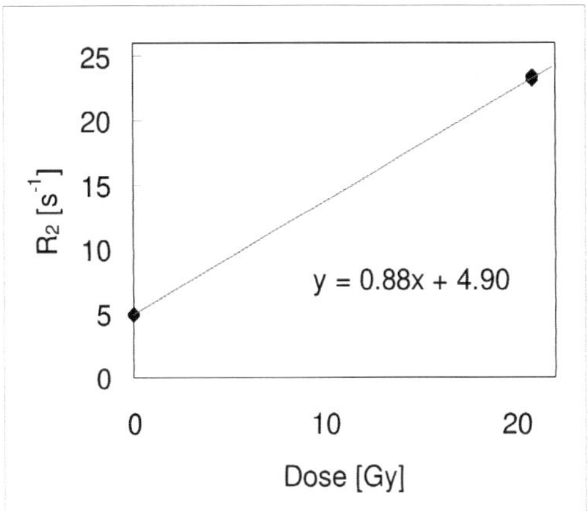

Figure 3: Transversal relaxation rate versus dose relationship from MRI measurements of the four homogeneously irradiated gels.

Journal of Physics: Conference Series **164** (2009) 012039 doi:10.1088/1742-6596/164/1/012039

imaging modalities. For the inverse computation, the regularization parameter was adjusted to best filter the noise in the measurement while conserving real stiffness contrast and edge sharpness.

4. Discussion

All three high dose areas were detected with good spatial accuracy (figure 4), however with strong shape distortions. The reason for the shape distortions is not known, but is thought to be associated with the combined effect of the TVD regularization and measurement noise due to ultrasound beam attenuation and reflection from the sides of the sample that was just 2 mm wider than the ultrasound transducer. Both effects decreased the signal to noise ratio in the ultrasound measurement. The smoothing characteristics of the regularization probably also caused the incorrect estimation of the dose contrast in the high dose areas, whilst the measurement noise also influenced the dose estimation in the background, especially at the sides. However, the top and bottom boundary regions were artefact-free, which indicates (as expected) that friction, if present at the gel-compressor plate boundary, did not introduce lateral (x-direction) stress in the region of interest.

Improvement of the experiment is possible, and is indeed planned for future work, for example by changing the sample geometry to minimise the reflections from the edges or by acquisition of ultrasound data from more than one direction to estimate lateral as well as axial displacement. Also the inverse algorithm is to be adapted to incorporate the additional lateral information, which will allow estimates for boundary conditions to be computed, rather than assuming that they are traction free. It is expected that such improvements will produce data with better signal-to-noise ratio while preserving real stiffness and hence dose gradients, and reduce or eliminate the influence of the boundary conditions.

Overall the technique shows potential if further improvement can resolve the difficulties described above. Although the method is currently being studied using experiments with controlled two-dimensional dose distributions it could be expanded to three dimensions (3D) by

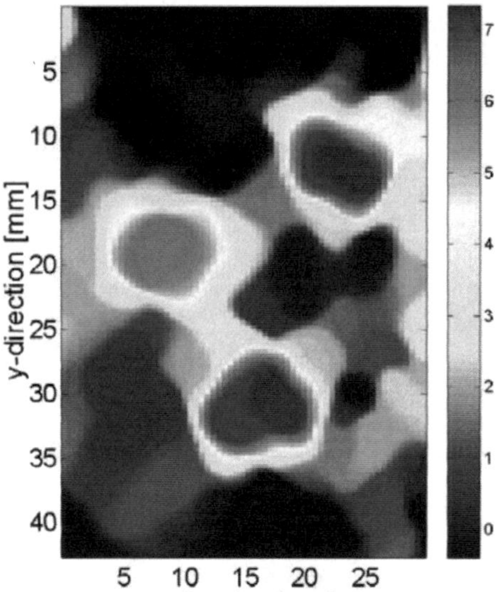

Figure 4: Dose distribution (in Gy) from ultrasound elastography and inverse algorithm.

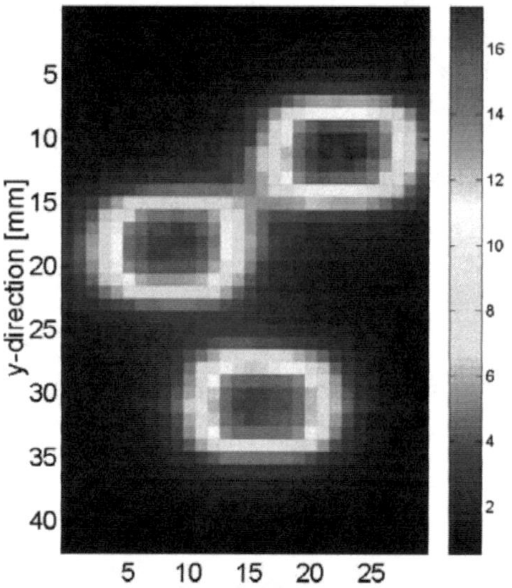

Figure 5: Dose distribution (in Gy) from MRI.

using a 3D model and by acquiring 3D image data with, for example, two-dimensional ultrasonic transducers. Such technology is currently under development and should enable arbitrary three-dimensional dose distributions to be determined.

5. Conclusion

We have visualized irradiation-induced changes of the elasticity in a polymer gel using ultrasound elastography and a plane strain inverse algorithm. Based on these preliminary results, and the known potential of further improvement, it can be anticipated that accurate determination of 3D dose distributions may eventually be possible.

Acknowledgement

This project is funded by the Institute of Cancer Research and the URP program at the Rensselaer Polytechnic Institute. PEB acknowledges the support of the NSF.

References

[1] Baldock C, Legape M, Rintoul L, Murry P and Whittaker A K 1999 Investigation of polymerisation of radiation dosimetry polymer gels DosGel'99, Schreiner L J (ed) (Lexington, KY: Canadian Organization of Medical Physicists) 99-105.

[2] De Deene Y, Vergote K, Claeys C and De Wagter C 2006 The fundamental radiation properties of normoxic polymer gel dosimeters: a comparison between a methacrylic acid based gel and Acrylamide based gels *Phys. Med. Biol.* **51** 653-673.

[3] Adamovics J, Jordan K and Dietrich J 2006 PRESAGE™ – Development and optimization studies of a 3D radiochromic plastic dosimeter. DOSGEL 2006, Lepage M, Jirasek A and Schreiner LJ (ed) (Sherbrooke, QC: Université de Sherbrooke) 105-107.

[4] Crescenti R, Bamber J, Bush N and Webb S 2006 Read-out of radiosensitive gels with ultrasound elastography: technique description and preliminary studies. DOSGEL 2006, Lepage M, Jirasek A and Schreiner LJ (ed) (Sherbrooke, QC: Université de Sherbrooke)234-236

[5] Fong P M, Keil D C, Does M D and Gore J C 2001 Polymer gels for magnetic resonance imaging of radiation dose distributions at normal room atmosphere *Phys. Med. Biol.* **46** 3105-3113.

[6] Oberai A A, Gokhale N H, Doyley M M, Bamber J C 2004 Evaluation of the Adjoint Equation Based Algorithm for Elasticity Imaging *Phys. Med. Biol.* **49** 2955-2974.

Normoxic Gel dosimetry using multislice X-ray CT: Preliminary study

N. Gopishankar , S. Vivekanandhan , S. Thulkar , R.K Bisht , V. Subramani, M.A. Laviraj , K. JothyBasu , S. SenthilKumaran, S.S. Kale , G.K Rath, B.S Sharma

All India Institute of Medical Sciences, New Delhi, India

Email : gshankar1974@yahoo.com

1. Introduction

In recent years there has been significant increase in the utility of X-ray CT for gel dosimetry due to easy accessibility of CT in clinical radiation therapy. Among various gel dosimeters available Normoxic polymer gels, is a new class of dosimeter that can be manufactured on a bench-top in presence of oxygen. Several antioxidants are available of which THPC was found to be very effective in oxygen scavenging capabilities. Several authors have shown that X-ray CT imaging is capable of providing dose resolutions approaching those of MRI gel within shorter imaging time (Hilts et al 2005). The dose information extracted from polymer gels are by radiation induced density change, which provides contrast in CT images. The focus of this paper is to outline considerations for implementing CT polymer gel dosimetry, specifically phantom design, CT imaging protocol, image averaging, and gel dose response. Image processing may be considered to be composed of two independent stages: acquisition stage/detection stage and processing/display stage. The most commonly used 'hands-on' form of digital image processing in medical imaging is 'image enhancement'. In this work, PAGAT Normoxic gel was investigated with a multislice X-ray CT. We establish the dose response reproducibility of PAGAT dosimeter when imaged with X-Ray CT for a range of irradiation times post gel manufacture (23h). Various factors influencing image noise for CT based gel dosimetry was studied and ways to improve were outlined. We also investigated the implementation of digital image filtering techniques on gel CT images to reduce image noise and improve dose resolution using several spatial filters.

2. Materials and methods

Siemens Volume Zoom CT scanner was used for all CT imaging. PET (Polyethylene Terephthalate) containers were used and not glass containers as the latter can produce extreme artefacts. Scan parameters (as show in Table1) were varied from set of parameters to access dependence of noise on each parameter. From each set of scan parameters, two images were taken to remove artefacts by background subtraction for noise study. Standard deviation in CT number (σNCT) was measured from pixel region at centre of final images. FOV (130x130mm) was used for dose response experiments. For 512x512 matrix size in CT images, scanned pixel dimension was 0.25mm. For image averaging study FOV of (380x380mm) was used. Image processing and analysis were done using Matlab software.

Scan parameter	Values
Tube voltage (kV)	80, 120, 140
Tube current (mA)	80, 125, 250, 300, 350, 400
Slice Scan time (s)	1
Reconstruction Algorithm	B10, B30, B41, B60, B80
Slice thickness (mm)	0.5, 1, 2.5, 5, 8
Field of View (mm)	50, 130, 250, 380, 400, 478

Table 1

A. Factors influencing image noise

A1. kV & mAs study

Increasing x-ray tube voltage (kV), x-ray tube current (mA) decreases image noise as shown in figure 1 and 2.

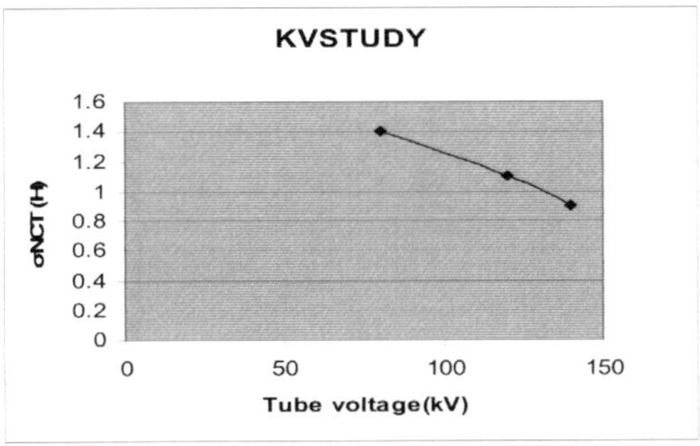

Fig 1

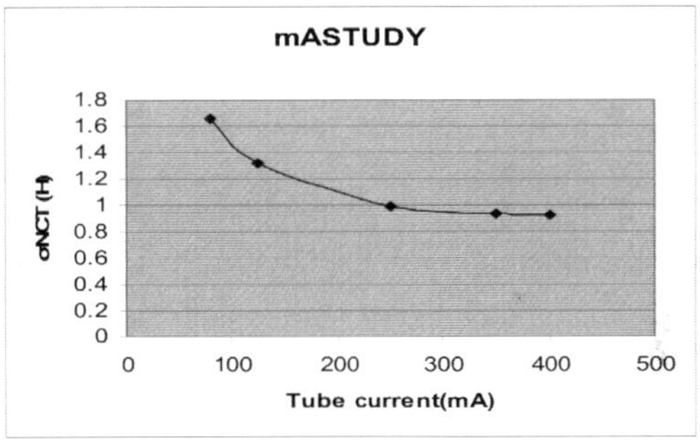

Fig 2

Also previously it has proven that increasing tube voltage has a greater effect on image noise than does an equal change in tube current (Hilts et al. 2005).

A2. Effect of Phantom Diameter

The effect of phantom diameter on image noise was investigated by imaging water filled PET containers with different diameters. CT image noise increases with phantom size. Similar work was done with glass container previously (Hilts et al. 2005). Irrespective of phantom material used, CT image noise tends to increase with phantom size.

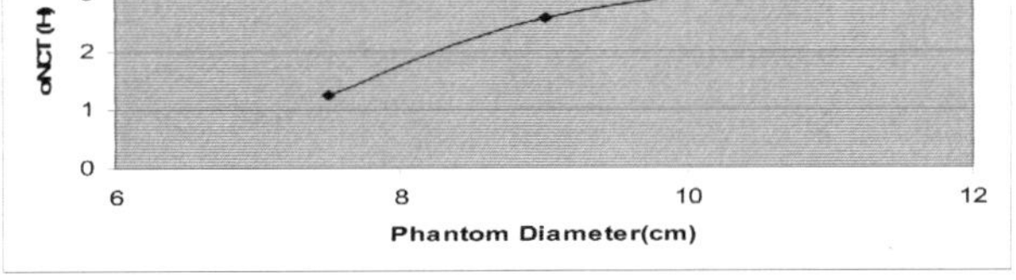

Fig 3

A3. Image averaging

The ability of a gel dosimetry read-out technique to produce spatially uniform images is crucial for accurate gel dosimetry (Hilts et al 2005). Mean and standard deviation for 36 spatially distinct 21x21 pixel ROI's in a single image, 16 imageaverages and 32 image averages are shown in figure 4, 5 and 6 respectively.

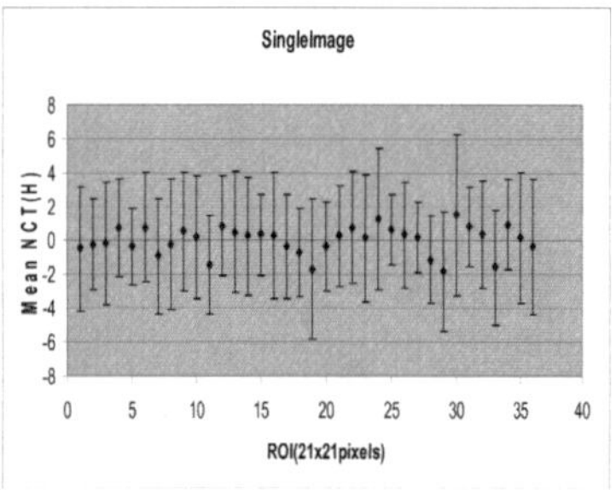

Fig 4

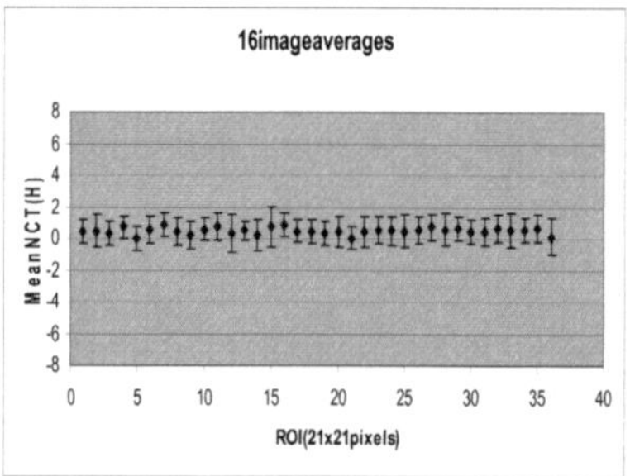

Fig 5

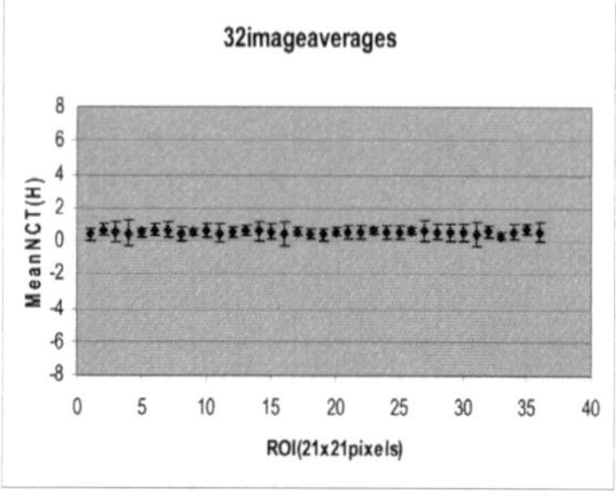

Fig 6

A4. Reconstruction Algorithm

The reconstruction algorithm has a significant influence on the image noise. The algorithm (B30 by Siemens) appears suitable for most gel dosimetry applications. The effect of different algorithms available on Siemens scanner on image noise is shown in figure 7.

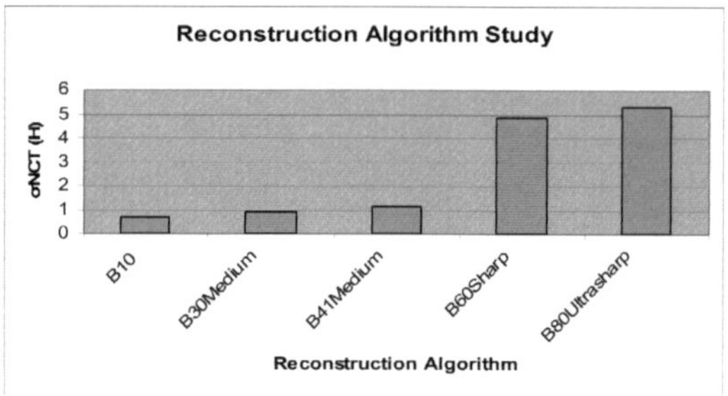

Fig 7

A5. Slice thickness and FOV (Spatial resolution)

Increasing slice thickness and FOV decreases image noise. A comparative study was done for different tube voltage and tube current combination for different slice thickness. The standard deviation was found to be minimal for combination of 140kV/140mAs.

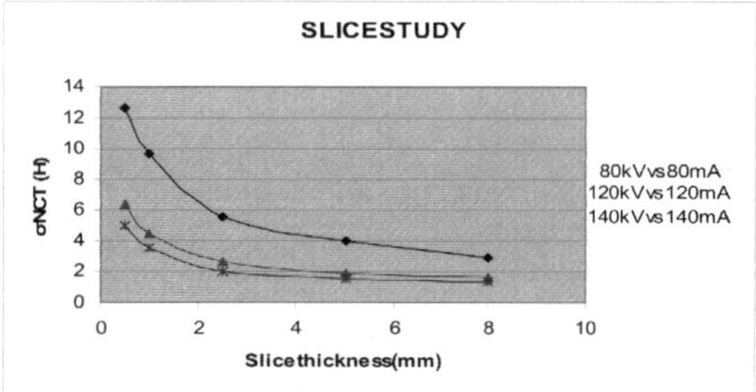

Fig 8

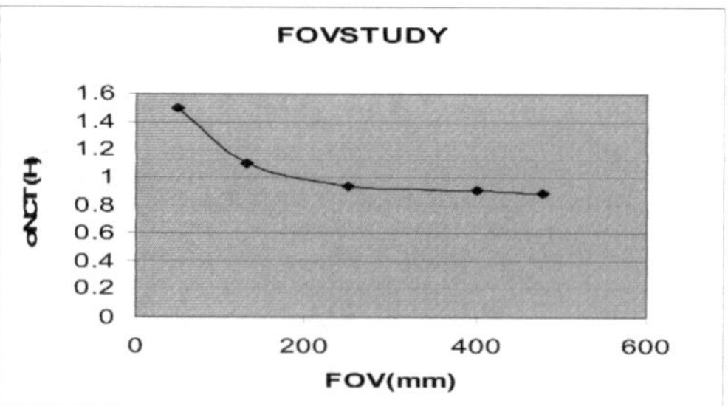

Fig 9

B. Gel Preparation

Normoxic gel was prepared under normal atmospheric conditions. The components used for the preparation of the dosimeter were 6% gelatin (300 Bloom from Porcine skin), 3% Acrylamide, 3% N, N' methylene-bis-acrylamide (BIS), 88% distilled water and 10mM tetrakis(hydroxy methyl) phosphonium chloride (THP). All the components were purchased from Sigma Aldrich, India. For preparation of PAGAT gel, 6% of gelatin by total weight of the dosimeter was mixed with 88% distilled water and allowed to soak for 45minutes before heating to 50°C in specially designed water bath with temperature control (thermostat), also crossed checked with a thermometer. Waterbath was designed in-house specifically to prepare 10 to 15 Liters of gel. The container with gelatin was heated to 50°C and simultaneously stirred with overhead stirrer. BIS (3%) was added and stirred thoroughly in the gel solution till it dissolved completely for 30minutes. 3% Acrylamide was added subsequently and the whole content was stirred for an hour. Once a clear solution was obtained 10mM of THPC was added and stirred and the container was removed from waterbath. A final volume of 4.5 l of gel was prepared. The gel was immediately poured into the three PET containers (*each with 1mm thickness and 1500mL capacity*) till screw top and tightly sealed with clingfilm and container lid. One container was used for calibration, one for clinical beam exposure and one for background subtraction purpose. The gel containers were immediately covered with black plastic cover to avoid exposure to daylight that might cause photopolymerization and then were refrigerated at 5°C.

C. Gel Irradiation

Irradiations were done approximately 23h post manufacture on Varian Clinac2300CD linear accelerator (Varian Assoc., Palo Alto, CA). The containers were brought to the room temperature of linac room before exposure to radiation. The PET containers (*10cm dia, 18.5cm height*) for exposure were left in the linac room for 1 hr. The PET (1) was irradiated to doses from 0Gy to 14Gy at Dmax by large flask geometry method (Taylor et al. 2007). The PET (2) was irradiated with four intersecting 4 x 4cm^2 6MV field doses of 2, 5, 7, and 11Gy at the depth of maximum dose. The PET (3) container was used for background subtraction. Reference marks were made on the containers for scanning reproducibility. After irradiation the gel containers were exposed to atmosphere one day after irradiation in order to make gels inactive and thus prevent further polymerization.

D. Gel imaging

CT imaging of irradiated gel was performed using Siemens Volume Zoom CT scanner. Irradiated and unirradiated gel used for background subtraction were positioned and oriented identically within scanner bore using reference marking. Acquisition parameters for dose response experiments were chosen from overall scanner parameters in table 1 and as follows: tube voltage 140kV, tube current 200mA, scan time 1sec, FOV 130x130, image thickness 2.5mm, Reconstruction algorithm B30 medium. Chosen FOV well encompassed cylindrical phantom. Images were obtained at 21 $\pm$ 1 °C. Before scanning gel containers were left in CT room for an hour to attain the scanning room temperature. Number of images averaged for dose response study was 15 images per scan position to reduce heat loading and imaging time without much loss in SNR. Artefacts were removed by subtracting an identical but unirradiated average background gel image from irradiated gel image.

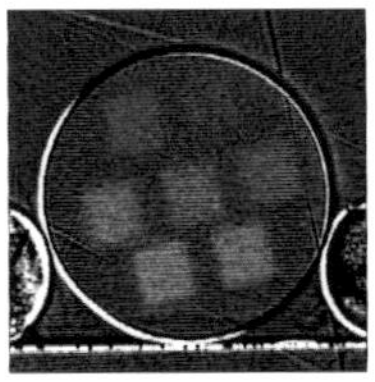

Fig 10

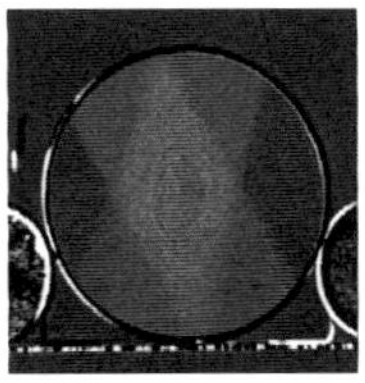

Fig 11

E. ΔN_{CT} – dose response

ΔN_{CT} – dose responses showed a linear region for doses upto 12Gy followed by slight decrease in response. The dose sensitivity of (4.68 $\pm$ 0.3) x 10 $^{-3}$ HcGy^{-1}was obtained 24hrs post irradiation. The mechanism for increase of ΔN_{CT} with dose must be related to increase in linear attenuation coefficient of the gel relative to that of water (Hilts et al. 2004). Overall the monoexponential fit was found to be better than the linear fit.

F. Reproducibility

The ΔN_{CT} – dose response reproducibility for given gel was determined over several imaging sessions on different days. The average slope for post irradiation hrs 24, 53, 145, 288, 436 was found to be (4.96 $\pm$ 0.3) x 10 $^{-3}$ HcGy^{-1} for linear fit (0 to 12Gy). Jirasek et al 2006 as shown that time of PAGAT gel CT scanning post gel irradiation does not alter resultant gel dose response characteristics. However we observed a slightly deviation in slope, intercept values. This may be due to reason that imaging requires proper temperature control system for improving the temporal stability (De Deene et al 2007).

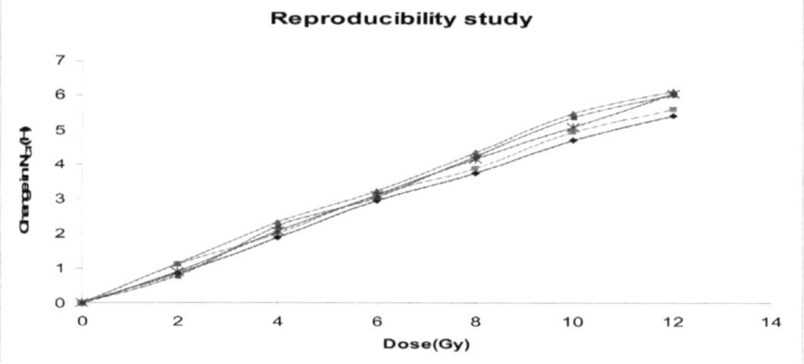

Fig 12

G. Influence of gel temperature during imaging

The effect of gel temperature during imaging on ΔN_{CT} – dose response was investigated a) immediately after removing from refrigerator b) after attaining room temperature on same day. The response remained linear for both conditions. The sensitivities of the ΔN_{CT} – dose response however differed and are given by:

$\Delta N_{CT}(H)$ = (6.85 $\pm$1.1) x 10^{-3} dose (cGy), condition 1 = immediate scanning

$\Delta N_{CT}(H)$ = (5.12 $\pm$0.2) x 10^{-3} dose (cGy), condition 2 = scanning at room temperature

L. Digital filtering

Digital image filtering is an effective method in reducing image noise while maintaining accurate spatial dose information. Generally it is performed in either frequency domain or spatial domain. The evaluated filters in this study are: mean (amean, harmean, charmean, gmean), median, midpoint, wiener2, alpha-trimmedmean, adaptivemedian. For this study slope of linear region (0 to 12Gy) from calibration gel was used to calculate σ_D. Measured noise $\sigma N_{CT}(H)$ from region of interest in the gel (25pixels) exposed with clinical beams and calculated values of σ_D and dose resolution ($D\Delta^{95\%}$) were analyzed for dose regions in gel irradiated for simulated clinical treatment. The response had dose region of slope 216$\pm$14 cGy/H (R^2 = 0.9972) and $D\Delta^{95\%}$ of 498 cGy.

3. Results and Discussion

The Siemens VolumeZoom x-ray CT scanner was optimized to evaluate PAGAT Normoxic gel dosimeters. In the image averaging study the number of images required to reduce noise was found to be 32. Further increase in number of images would certainly reduce noise but at the cost of tube loading and imaging time. In the phantom size study phantoms with smaller dimensions were effective in reducing $\sigma N_{CT}(H)$. The noise decreases with increase in kV, mAs, slice thickness, and field of view. But scan parameters for dose response study were chosen similar to a previous study in order to get comparative results The average slope was found to be $(4.96 \pm 0.3) \times 10^{-3}$ $HcGy^{-1}$ for the linear region. But full response (0 to 14Gy) is well modeled by a monoexponential fit than a linear fit. The large flask method used for gel calibration is straightforward and less prone to setup errors and is commendable for X-Ray CT gel dosimetry. In the stability study dose response was linear but with slight upward shift in the response curve. Proper temperature controlling devices are required to control the gel temperature to produce good results. The dose resolution in PAGAT CT images can be improved by filtering with spatial filters for reducing noise (Hilts et al 2004). Meanfilter and wiener2filter gave similar filtering results. In our study dose resolution highly improved by applying alpha-trimmed filter with 7x7pixel mask which showed a $D\Delta^{95\%}$ of 233cGy. In case of midpoint filter the σN_{CT} values increased with increase in mask size.

4. Conclusion

In summary gel dosimetry with multislice CT scanners significantly improves the quality of imaging and specifically reduces the imaging time. Also in general density changes occurring in irradiated polymer gel due to X-ray CT imaging are minute and therefore image contrast is very low. Hence choosing optimal scanning parameters and post-image acquisition data processing like image filtering is vital for CT gel dosimetry. Future study will include the study of other filters such as SUSAN and TPMEM for noise reduction in CT gel dosimetry.

References

[1] Trapp J V et al, Phys. Med. Bio, 46, 2939-2951 (2001).
[2] Hilts M et al, Med. Phys, 31, 39-49 (2004).
[3] Hilts M et al, Med. Phys, 50, 1727-1745 (2005).
[4] Jirasek A et al, Phys. Med. Bio, 51, 2599-2617 (2006)
[5] Taylor M L et al, Phys. Med. Bio, 52, 3991-4005 (2007).
[6] De Deene et al, Phys. Med. Bio, 52, 2719-2728 (2007).

Journal of Physics: Conference Series **164** (2009) 012041 doi:10.1088/1742-6596/164/1/012041

Recent advances in non gel tissue equivalent dosimeters

Kevin Jordan

London Regional Cancer Program, London Health Sciences Centre, London, Ontario, Canada

Email: kevin.jordan@lhsc.on.ca

1. Introduction

This review highlights some of the developments in radiation dosimetry, other than gel dosimetry, that are relevant to radiotherapy and have been reported since the previous DOSGEL meeting in 2006. The results included in this presentation were obtained primarily from searches of the PubMed database.

Two advantages of gel dosimetry are the tissue or water equivalence of the dose response and the inherent 3D geometry of the gel phantoms. However, in practice ionization chambers and film have been the most common dosimeters for clinical radiation dosimetry. Ionization chambers are easy to use and provide precise measurements. Protocols have been developed so that uniform results can be attained throughout the world. Silver halide film provides excellent spatial resolution and with careful processing can provide reproducible results. However, ionization chambers are limited by spatial resolution and detector perturbations of local dose gradients. Silver halide film's major drawback is the difference in energy response relative to water or tissue which becomes measurable in regions of high dose gradients and where energies less than 200 keV are present. As the complexity of delivered radiotherapy treatments increases verification at many distinct points becomes essential because many smaller fields are used to deposit the dose. This leads to a situation where dose delivered from the beam penumbral regions is a significant fraction of the total dose. Accurate measurements in penumbral regions, where the dose gradient is high and the energy spectrum is rapidly changing require small tissue or water equivalent dosimeters to avoid dose inaccuracies. If the beam is incorporated into a treatment planning system then errors would be propagated in subsequent calculations.

In this review highlights of advances for point (0D) and planar (2D) dosimeters are presented. Dosimeters that are near water or tissue equivalent are more relevant for direct comparisons with 3D radiation sensitive gels and plastics. This equivalence and sufficient dose sensitivities for accurate measurements in the clinically relevant range of 0.01 to 20 Gy range were the criteria for selecting material. Many of the papers are related to small field dosimetry where much of the dose is deposited in high gradient penumbral regions. Often the minor differences in dosimetry measurements in large fields can be clearly observed in the high gradient regions. Conversely, dosimeter systems that are accurate for small field dosimetry should also be the most accurate for larger fields. In the case of 3D dosimetry, the case of small fields with high gradients is also a reasonable place to compare the performance of different dosimetry systems. For example which items result in more accurate dosimetry: scattering versus transparent materials, low diffusion versus no diffusion materials, optical CT versus MRI scanners, laser versus light emitting diode optical CT scanners? It is interesting to recognize that most of the issues related to film and film densitometers are the same items that require evaluation for accurate 3D optical CT scanning.

2. Point detectors

Intensity modulated radiotherapy (IMRT) is becoming standard practice because multi-leaf collimators and IMRT capable planning programs are now standard technologies for new equipment purchases. Key to IMRT commissioning is accurate dosimetry for small radiation fields or beamlets. Small, water equivalent dosimeters are ideal for these measurements. Some of the examples listed could also be configured for linear, planar or volume dosimetry. For example a crystalline material could be distributed in a line, sheet or fill a container.

2.1 Ion chambers

Ion chambers are the local standard dosimeter in nearly all radiotherapy centers. They are calibrated against a national standard. Ion chambers give precise measurements and their accuracy is tied to the dosimetry protocol employed. The accuracy of relative dosimeters is directly tied to the accuracy of the ion chamber calculated dose.

Wang and Rogers have compared Monte Carlo calculations of replacement factors, P_{repl} (or p(cav)*P(dis) in IAEA notation), with recent measurements and found closer agreement than with values recommended in current protocols [1]. The magnitude of the discrepancies can be up to 0.5%. They recommend updating the parameters in current protocols to reflect these more accurate values, which reduce variations between measurement and calculations. As the differences between ion chamber and other dosimeter systems approach the 2% level the question becomes which measurement is more accurate.

Farmer type axial chambers and parallel plate chambers perturb the dose distribution that would be deposited if the chamber is replaced with water or tissue equivalent material. For this reason, smaller chambers such as the Pin-Point micro-ionization chamber were developed. While providing higher spatial resolution and few perturbations these smaller chambers also have lower sensitivity. Recently, to provide better agreement with higher performance detectors such as diamond detectors the PTW chamber was modified to an aluminum electrode (type PTW 31014). Profiles for MV photon beams measured with the Pin-Point chamber were 0.6 mm larger than those recorded with EDR2 film [2]. Stem and polarity effects up to 0.5% were measured for larger fields, requiring corrections to obtain accurate measurements.

2.2 Solids (powders, crystals)

In general, powders and crystalline materials can act as dosimeters by generating metastable species such as molecular radicals and populating excitation traps. If these materials are primarily organic then dose response will be essentially water or tissue equivalent. Quantitative detection can involve microwave photon absorption with electron paramagnetic resonance spectrometers or luminescence stimulated thermally, optically or chemically. Alanine / EPR dosimetry is one of the most thoroughly investigated powder dosimeters, but its dose sensitivity is low for the 0.01 to 20 Gy range. However, comparisons with XV film, mini-TLD and ionization chamber demonstrated agreement with 1 mm diameter by 3 mm long paraffin-alanine rods [3].

Lithium formate monohydrate with EPR dosimetry has been investigated due to increased sensitivity compared to alanine [4]. Dose response in the 0.5 to 3.5 Gy range was reported and a thorough uncertainty analysis determined a 2.5% uncertainty for doses greater than 3 Gy. The work suggested further increases in sensitivity may be achieved from similar materials.

2.3 Optically stimulated luminescence (OSL)

Carbon doped aluminum oxide (Al_2O_3:C) bound to a polystyrene sheet forms a 2D OSL film dosimeter. Following irradiation the recorded dose is read by recording the blue luminescence stimulated by the absorption green light. Upconversion is a very sensitive optical readout mechanism

Journal of Physics: Conference Series **164** (2009) 012041 doi:10.1088/1742-6596/164/1/012041

since photomultiplier tube phosphors that are much more sensitive to blue light are available. This OSL material is now commercially available from Landauer Inc. in the form of 2 mm diameter dots (Nanodot dosimeter) or 15 cm strips. The dots can be read with an OSL reader (Microstar, Landauer) or returned to the manufacturer for batch readings. The performance of the commercial dosimeters and readers revealed ~2% variations in response for prescreened detectors at doses > 0.1 Gy and a dose uncertainty of 2%(at 2σ) [5]. No fractionation effects, energy or dose rate dependence was found. A thorough characterization of the earlier commercial dosimeter (InLight/OSL Dot) and the Microstar reader was recently reported [6]. This system has several advantages over conventional thermal luminescence products: it is much smaller, does not require compressed gas, there are no annealing ovens, samples can be read repeatedly to improve precision, samples can be irradiated repeatedly up to a total dose of approximately 20 Gy before a simple optical annealing is required to restore initial dose response and the system is less expensive. Jursinic reported: repeat reading variations of 0.6%, following a 10 minute fade interval, sensitivity was constant up to 20 Gy, dose response nearly linear to 2 Gy and then becoming supralinear, energy, dose per pulse independence for the typical ranges of a linear medical accelerator. Also, temperature and angular independence simplified usage for in vivo dosimetry. These results indicate that OSL systems will be very practical for quantitative megavoltage dosimetry. A second study evaluated the same product but readings were performed by the manufacturer typically two weeks post irradiation [7]. The deviations in measurements were typically lower for this study possibly due to a larger number of samples and to higher precision in the Landauer commercial reader. The results clearly show a 4% decrease in energy response from 6 MV to 18MeV. The work also demonstrated the relative dose response of OSL, GafChromic EBT and Kodak X-Omat V films 5 cm outside the beam edge for 6 and 18 MV photon beams. The energy responses indicate an over response relative to water or tissue for the low energy components of the scattered photons. Linear response was up to 2 Gy was also reported in this study. Authors noted that OSL film may be useful for 2D dosimetry once appropriate readers are developed.

3. Two dimensional detectors

The optical density of films is measured with several types of densitometers and these instruments need to be independently calibrated for accurate dosimetry. As clinics evolve into digital imaging, the highly developed automatic film processors for silver halide film are quickly disappearing and new facilities do not include space for this technology. For this reason there has been a dramatic shift towards the usage of radiochromic film that is self developing. Also, a new version of radiochromic film GAFChromic EBT (external beam therapy), was recently introduced that has higher dose sensitivity than previous versions.

3.1 Plastic scintillator sheet

In principle, plastic scintillator sheets can be adapted for two-dimensional water equivalent dosimetry. The characteristics of plastics can be precisely tuned for physical and electronic density providing near ideal agreement with tissue or water dose responses. Plastics are inexpensive, chemically inert and can be formed into arbitrary shapes and volumes. Several approaches have been demonstrated to account for the light not generated by absorbed dose in the scintillator detector. While most of the research has concentrated on fiber based scintillator systems there are a few reports concerning 2D measurements. In the previous review the work of Petric et al was reported. In that study the scintillator plate was sandwiched between water phantoms [8]. A similar geometry has been investigated by Frelin et al with the added variation of replacing the water with polystyrene [9]. The polystyrene introduces substantial increases in phantom Cerenkov, fluorescence, absorption and optical aberrations. Also, a checkerboard plate was included to allow measurement if interference light spatially. Filtering to correct for the effect of this grid array led to measureable artifacts in the processed images. The work clearly shows the added complexity of trying to discriminate the

Journal of Physics: Conference Series **164** (2009) 012041 doi:10.1088/1742-6596/164/1/012041

scintillator from phantom signals. The optical problems described in this study are analogous to effects present in broad beam optical CT scanning.

3.2 Radiochromic film

Radiochromic films from International Speciality Products known as GAFChromic films have been widely used as tissue equivalent 2D dosimeters. A comprehensive reference list is provided on the manufacturer's website, www.ispcorp.com . EBT is the most sensitive film to date and is suitable for measurements in the 0.1 to 8 Gy range. The film contains two active layers for a total thickness of 0.229 mm and 0.3 % Li and 0.3% Cl appear to be the key sensitizing additives. Cheung *et al* have also investigated the post-irradiation darkening of EBT film and recommended that quantitative measurments include a wait time of at least 6 hours [10]. Otherwise a calibration film with the same post-irradation time should be referenced.

The higher sensitivity radiochromic films have also have higher energy absorption spectra and film scanners optimized for the older films will be of limited value in reading the newer film types. Laser densitometers at 633 nm are well matched to the main absorption peak. Soares has published a comparison of the common radiochromic films measured with several densitometers [11]. The film uniformity was measured to be 6% peak to peak and highlight the importance of film orientation for scanning to minimize polarization effects. Calculations of mass-energy absorption coefficients relative to water predict an over response at energies less than 100 keV. The absorption spectrum as a function of dose was evaluated in detail and implications for quantitative dosimetry clearly addressed by Devic et al [12]. The spectrum was decomposed into 8 peaks and the dose response of each peak was presented. The authors recommend that 20 nm bandwith light sources centred at 583 and 635 nm would be efficient for effective densitometry. The red band would provide higher sensitivity at low doses and the yellow band would be useful for higher doses. This paper discusses many issues that are also relevant to the design of optical CT scanners. It would be useful for clinical physicists if the same scanner could be optimized for both EBT films and the 3D radiochromic plastics and gels. Possibly more radiochromic 3D materials will be designed to have spectra with peaks at similar wavelengths to EBT film. Film orientation was also examined by Butson et al and they found up to a 15% variation in response at 50 cGy with film orientation [13]. This value decreased to 4% at 3 Gy but was a 3 fold larger affect that for previous versions of GAFChromic film.

Flatbed document scanners have been widely evaluated for quantitative 2D film densitometry based on the high spatial resolution achievable with a very low cost commercial product. However, these devices are generally not well matched to the spectrum of EBT film and scatter introduces significant artifacts. Because of these limitations users should be cautious when using such devices for quantitative dosimetry. Many groups have reported corrections to improve the performance of these devices but scanners optimized for EBT film will need to be developed before the full potential of this film is realized. Very effective corrections can be obtained by scanning films previously irradiated to known steps in dose [14,15]. Because EBT film is nearly independent of energy the dose calibration curves generated at one field size are valid for other field sizes, unlike silver based Kodak X-Omat V film [16]. This simplifies calibration procedures. Another advantage of an energy independent film is that measured beam penumbras are smaller and more accurate than for silver films [17]. The measured 80%-20% penumbra for a 6 MV, 10 cm x 10 cm field with EBT film was 0.2 mm and 0.6 mm smaller relative to EDR2 and XV films respectively. These differences can introduce measurable errors when modeling penumbras in treatment planning systems for IMRT calculations. Further documentation of the performance of EBT film with the Epson 1680 scanner reported a 1.3% standard deviation at 2.3 Gy in water by using at least two independent film measurements in the center of the scanner [18]. A study of dose rate, depth, field size, fractionation and electron beam energy demonstrated that a single energy, single depth and single cone size is sufficient for electron beam dose response calibration to an accuracy of +/- 4% [19]. Higher accuracy requires individual beam energy calibrations.

Attempts at adding coloured and polarized filters to commercial flatbed scanners has provided additional data suggesting that customized scanners for EBT film are required to realize optimum dosimetry [20]. The limitations of flatbed scanners for quantitative measurements were further documented by demonstrating variations in response along the CCD array and large film orientation effects and heating of film during repeated scans [21,22]. An early design of an LED based scanning spot densitometer optimized for EBT film has demonstrated sub millimeter spatial resolution, with no scatter or interference or rotation artifacts [23]. More expensive film scanners, designed for digitization of colour film slides have been evaluated and are promising if the red and green LED sources can be changed to match the EBT film absorption peaks [24]. Possibly these instruments could also be modified for high resolution optical CT scanning of smaller objects within the available 9 cm x 6 cm aperture.

The high spatial resolution of film allows certain dosimetry measurements to be easily performed. For example, accurate skin dose measurements at a 70 micron depth have been demonstrated with EBT film [25]. X-ray beams with energies less than 1 MV are under investigation for specialized treatments that require small penumbras such as stereotactic radiosurgery of small intracranial lesions and for radiotherapy of small animals. EBT film with digital microscope readout had been demonstrated as an effective method to measure penumbras to within 30 microns [26]. Finally, EBT film with an Epson 1680 scanner has been verified as a clinically acceptable replacement of radiographic film for QA of stereotactic radiotherapy procedures [27].

4. Accurate 3D positioning phantoms

EBT film has another practical advantage in being insensitive to immersion in water because of the laminated construction. Silver based films must be protected from high humidity and emulsions cannot be wetted. Often air pockets get trapped in light and water tight custom film jackets and lead to dosimetry artifacts. Water phantoms are versatile for performing 3D dosimetry since dosimeters can be placed at the optimum locations within the phantom. Pallota et al developed a water phantom suitable for arc therapy and IMRT and demonstrated usage of EBT and laminated EDR2 film [28]. Gels could easily be inserted into this phantom as well.

Finally, motion tracking during treatment delivery continues to be developed for routine radiotherapy. One approach to QA dosimetry is to have 3D motion control for the dosimetry phantom. A three stage linear motion movable phantom has been developed and characterized [29]. The moveable stages have 100 mm of motion and 0.1 mm accuracy with a maximum error in absolute position of 0.56 mm. Such accuracy will be necessary for high resolution 3D dosimetry when motion is involved.

5. Summary

In the last two years OSL dosimetry and EBT film have been introduced into clinical usage. The commercial OSL system from Landauer is much simpler to implement than thermal luminescence dosimetry systems. The performance is sufficient for quantitative MV dosimetry. EBT film has quickly become established in clinics using flatbed document scanners for readout. The true performance of this film requires scanners and densitometers that optimize the light sources to the main peaks at 583 and 639 nm and minimize the artifacts related to interference, scatter and film orientation. Possibly a single scanner will be developed to perform both EBT film dosimetry and optical CT scanning and more 3D radiation sensitive materials will be developed that can be probed at these same wavelengths.

References

[1] Wang L L and Rogers D W 2008 Calculation of the replacement correction factors for ion chambers in megavoltage beams by Monte Carlo simulation *Med. Phys.* 35 1747-1755

[2] Agostinelli S, Garelli S, Piergentili M and Foppiano F 2008 Response to high-energy photons of PTW31014 PinPoint ion chamber with a central aluminum electrode *Med. Phys.* **35** 3293-3301

[3] Abrego F C, Calcina C S, de Almeida A, de Almeida C E and Baffa O 2007 Relative output factor and beam profile measurements of small radiation fields with an L-alanine/K-band EPR minidosimeter Med. Phys. **34** 1573-1582

[4] Gustafsson H, Lund E and Olsson S 2008 Lithium formate EPR dosimetry for verifications of planned dose distributions prior to intensity-modulated radiation therapy Phys. Med. Biol. **53** 4667-4682

[5] Miller S D and Murphy M K 2007 Technical performance of the Luxel Al_2O_3:C optically stimulated luminescence dosimeter element at radiation oncology and nuclear accident dose levels Radiat. Prot. Dosimetry **123** 435-442

[6] Jursinic P A 2007 Characterization of optically stimulated luminescent dosimeters, OSLDs, for clinical dosimetric measurements Med. Phys. **34** 4594-4604

[7] Schembri V and Heijmen B J 2007 Optically stimulated luminescence (OSL) of carbon-doped aluminium oxide (Al_2O_3:C) for film dosimetry in radiotherapy Med. Phys. **34** 2113-2118

[8] Petric M P, Robar J L and Clark B G 2006 Development and characterization of a tissue equivalent plastic scintillator based dosimetry system *Med. Phys.* **33** 96-105

[9] Frelin A-M, Fontbonne J-M, Ban G, Colin J, Labalme M, Batalla A, Vela A, Boher P, Braud M and Leroux T 2008 The DosiMap, a new 2D scintillating dosimeter for IMRT quality assurance: Characterization of two Čerenkov discrimination methods Med. Phys. **35** 1651-1662

[10] Cheung T, Butson M J and Yu P K N 2005 Post-irradiation colouration of Gafchromic EBT radiochromic film *Phys. Med. Biol.* **50** N281-285

[11] Soares C G 2006 New developments in radiochromic film dosimetry Radiat. Prot. Dosim. **120** 100-106

[12] Devic S, Tomic N, Pang Z, Seuntjens J, Podgorsak E B and Soares C G 2007 Absorption spectroscopy of EBT model Gafchromic film Med. Phys. **34** 112-118

[13] Butson M J, Cheung T and Yu P K 2006 Scanning orientation effects on Gafchromic EBT film Australas. Phys. Eng. Sci. Med. **29** 281-284

[14] Saur S and Frengen J 2008 Gafchromic EBT film dosimetry with flatbed CCD scanner: A novel background correction method and full dose uncertainty analysis Med. Phys. **35** 3094-3101

[15] Menegotti L, Delana A and Martignano A 2008 Radiochromic film dosimetry with flatbed scanners: A fast and accurate method for dose calibration and uniformity correction with single film exposure Med. Phys. **35** 3073-3085

[16] Cheung T, Butson M J and Yu P K N 2006 Independence of calibration curves for EBT Gafchromic films of the size of high-energy X-ray fields Appl. Radiat. Isot. **64** 1027-1030

[17] Cheung T, Buston M J and Yu P K N 2006 Measurement of high energy x-ray beam penumbra with Gafchromic™ EBT radiochromic film Med. Phys. **33** 2912-2914

[18] Van Battum L J, Hoffmans D, Piersma H and Heukelom S 2008 Accurate dosimetry with Gafchromic EBT film of a 6 MV photon beam in water: what level is achievable? Med. Phys. **35** 704-716

[19] Su F-C, Liu Y. Stathakis S. Shi C, Esquivel and Papanikolaou 2007 Dosimetry characteristics of Gafchromic® EBT film responding to therapeutic electron beams Appl. Radiat. Isot **65** 1187-1192

[20] Kamomae T, Miyabe Y, Nakata M, Yano S, Okada T, Takakura T, Fujimoto T, Koshida K and Kikuchi Y 2007 Sensitivity characteristic that uses optical filters in Gafchromic radiation

dosimetry Nippon Hoshasen Gijutsu Gakkai Zasshi **63** 1253-1256
[21] Devic S, Wang Y-Z, Tomic N and Podgorsak E B 2006 Sensitivity of linear CCD array based film scanners used for film dosimetry Med. Phys. **33** 3993-3996
[22] Lynch B D, Kozelka J, Ranade M K, Li J G, Simon W E and Dempsey J F 2006 Important considerations for radiochromic film dosimetry with flatbed CCD scanners and EBT Gafchromic film Med. Phys. **33** 4551-4556
[23] Ranade M K, Li J G, Dubose R S, Kozelka J, Simon W E and Dempsey J F 2008 A prototype quantitative film scanner for radiochromic film dosimetry Med. Phys. **35** 473-479
[24] Hupe O and Brunzendorf J 2006 A novel method of radiochromic film dosimetry using a color scanner Med. Phys. **33** 4085-4094
[25] Devic S, Seuntjens J, Abdel-Rahman W, Evans M, Olivares M, Podgorsak E B, Vuong T and Soares C 2006 Accurate skin dose measurements using radiochromic film in clinical applications Med. Phys. **33** 1116-1124
[26] Keller B M, Peressotti C and Pignol J-P 2008 Optical imaging analysis of microscopic radiation dose gradients in Gafchromic EBT film using a digital microscope Med. Phys. **35** 3740-3747
[27] Sturtewagen E, Fuss M, Paelinck L, De Wagter C and Georg D 2008 Multi-demensional verification of stereotactic radiotherapy for uveal melanoma using radiochromic EBT film Z. Med. Phys. **18** 7-36
[28] Pallotta S, Marrazzo L and Bucciolini M 2007 Design and implementation of a water phantom for IMRT, arc therapy and tomotherapy dose distribution measurements Med. Phys. 34 3724-3731
[29] Nakayama H, Mizowaki T, Narita Y, Kawada N, Takahashi K, Mihara K and Hiraoka M 2008 Development of a three-dimensionally movable phantom system for dosimetric verifications Med. Phys. 35 1643-1650

RADBALL™: a new departure for 3-D dosimetry

Simon J Doran[1,2], Steven J Stanley[3], Paul M Jenneson[4] , Erwan Prott[5] and John Adamovics[6]

[1] CRUK Clinical MR Research Group, Institute of Cancer Research, Sutton, UK
[2] Department of Physics, University of Surrey, Guildford, UK
[3] Nexia Solutions, Hinton House, Risley, Warrington, WA3 6AS, UK
[4] HMS Sultan, Military Road, Gosport, UK
[5] address currently unavailable
[5] Dept of Chemistry and Biology, Rider University, Lawrenceville, NJ, USA

Corresponding authors: Simon.Doran@icr.ac.uk, Steven.J.Stanley@nexiasolutions.com

Abstract. This paper describes a new device, RADBALL™ for mapping environmental radiation fields, as found in the area of nuclear decontamination. The system consists of a specially shaped PRESAGE™ dosimeter, which sits inside a custom-designed lead collimator. This is imaged using optical CT to yield data from which the position of either point sources or extended objects may be reconstructed. The principle of the technique is explained, simulations and preliminary data are given and the current design of the dosimeter and collimator are presented.

1. Introduction

To date, the primary use of spatially resolved 3-D radiation dosimetry has been within the medical community, for simulating the dose deposited during radiotherapy treatments. However, the concentration of the 3-D dosimetry literature in this area and, consequently, the relatively restricted set of journals in which the material is published, have meant that potentially important applications in other industrial fields have not yet been explored.

As the cost of oil continues to rise inexorably and amid the current fears over energy security, the UK and other countries around the world are currently experiencing a renaissance in the nuclear industry. Despite this, a continuing area of public concern is the decommissioning of nuclear facilities at the end of their lifetime. Although many problems of decontamination have now been solved, there is still a need for 3-D position-sensitive detectors of radiation for the location of unidentified sources. Typical environments in which such devices would be useful are glove boxes, active cells and other confined spaces, in which the level of radiation might be high or access might be limited. Whilst the concept of 4π Compton imaging, using sold state detectors, is well known and a number of experimental implementations have been demonstrated in the research literature [1,2], such systems may not be suitable for use within the nuclear industry. Ideally, a device should be cheap, robust, stable against extremes of temperature and remotely deployable, without the need for connecting wires or intervention during the measurement process.

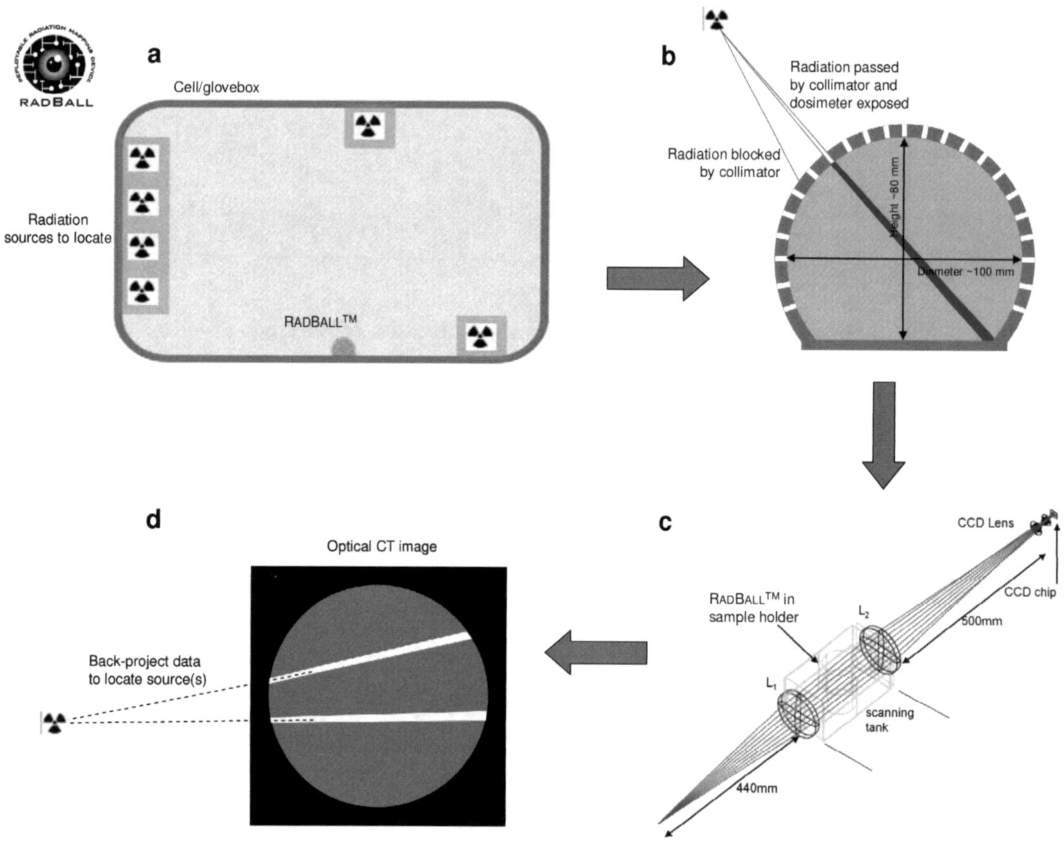

Figure 1: RadBall™ concept: (a) deployment of sample; (b) exposure; (c) optical tomography scanning of the sample; (d) reconstruction of the object position by reverse ray-tracing/back-projection.

This paper introduces RADBALL™, a solution to the problem described above. Subsequent sections describe the concept, give the results of simulations to test the idea, present preliminary data and show pictures of the current device.

2. Concept

The PRESAGE™ dosimeter formulation [3] is well known for being physically robust and is stable during a prolonged irradiation period and at elevated temperatures. It exhibits negligible chromophore diffusion and has good post-irradiation storage properties. It has a linear response over a wide range of doses (up to several hundred Gy) for the energies of incident radiation so far tested. Unlike polymer gels, PRESAGE™ appears, in tests to date, to be entirely insensitive to oxygen. It is thus an ideal material for applications in the nuclear industry.

The RadBall™ concept is illustrated in Figure 1. RadBall™ consists of the combination of an integrating radiation detector, made of PRESAGE™ with a custom built lead collimator. The device is deployed in the area which is to be mapped and radiation from the source passes through the holes in the collimator to expose the PRESAGE™. If the source is discrete, then easily identifiable tracks will appear in the sample, whereas extended sources will give a more diffuse pattern. After optical CT

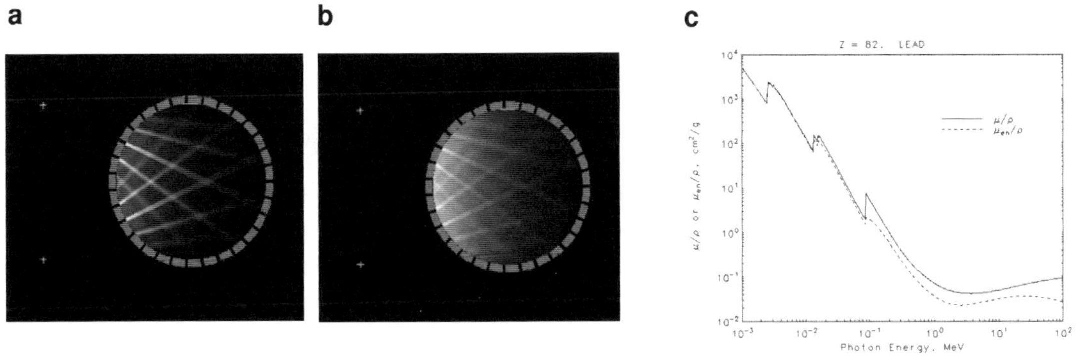

Figure 2: Simulation of one plane through the collimator for two different energies (a) 433 keV; (b) 1.2 MeV. Note the reduction in image contrast as the 5 mm thickness of lead gives lower shielding at increased energies. (c) variation of the linear attenuation coefficient of lead at different energies, showing the there is significant potential of optimising the device for different radioactive contaminants.

scanning, a map of the sources is reconstructed. For a set of discrete sources, this can be performed simply by tracing the distinct lines in the optical CT images back to their intersection point. If one is dealing with an extended source, then more sophisticated reconstructions are necessary.

3. Simulations and development of the lead collimator

In accordance with the basic principle for source location, as outlined above, software was written in IDL (ITT Visual Information Solutions) to simulate a lead collimator of the type proposed. A variety of case studies were simulated, demonstrating that it is feasible to identify distinct “tracks” in the dose distribution corresponding to radiation paths through the holes in the lead collimator. These simulations were further used to estimate a suitable size and separation for the holes in the collimator and inform the manufacturing process.

A particular concern was the need for a compromise between having the lead shell as thick as possible to provide the maximum shielding for regions away from the tracks, whilst maximising the solid angle subtended by each hole. In order to ensure that the device is sensitive to radiation from a variety of angles and thus operates like an ideal “pinhole camera”, the holes in the lead shell should be thin. However, if the collimator thickness is not great enough then its attenuation is small and we lose image contrast between the track regions and the nominally unirradiated regions. This compromise is energy-dependent, as the linear attenuation coefficient varies very rapidly in the region of interest (hundreds of keV – several MeV).

Figure 2 illustrates these points, using the source configuration adopted for a hypothetical irradiation by two photon-emitting point sources. These are positioned 5 cm perpendicularly on either side of a radial line from the centre of the sphere, at a distance of 10 cm from the sphere centre. Holes of diameter 2 mm and separation 10 mm are drilled in the lead shell. Figure 1(a) illustrates the case of a relatively low energy photon at 433 keV (μ = 2.18 cm^{-1}), whilst Figure 1(b) is the corresponding case for ^{60}Co (E = 1.2 MeV, μ = 0.635 cm^{-1}). The reduced image contrast in 1(b) is evident, but the images would still be suitable for source location. Moreover, by choosing materials other than lead for the absorbing collimator one can obtain increased values of linear attenuation coefficient.

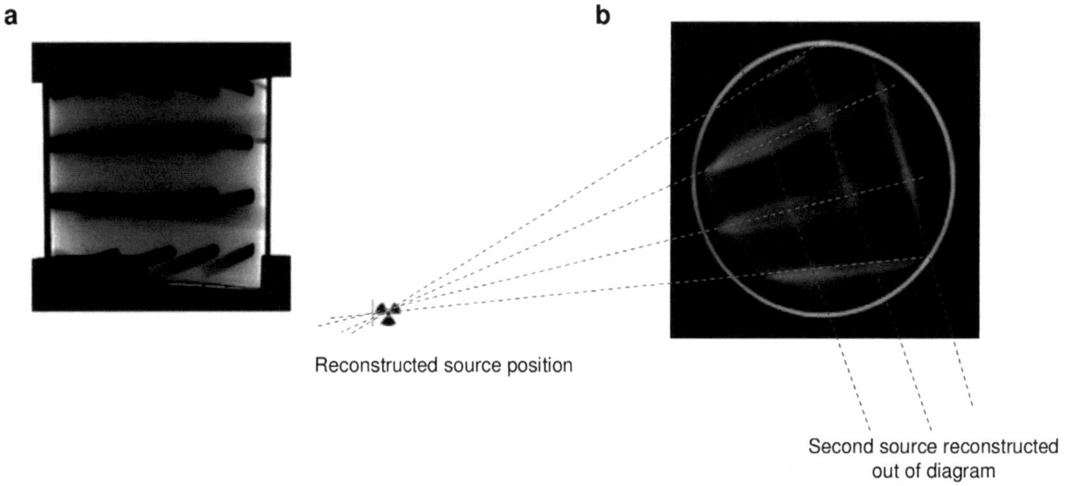

Figure 3: The preliminary experiment: (a) sample projection image; (b) oblique slice through reconstructed 3-D dataset. Images shown are from the CCD scanner

4. Preliminary experiments

Prior to the manufacture of the complex hemispherical dosimeter and collimator, the concept of RADBALL™ was tested using a cylindrical PRESAGE™ sample of diameter 6 cm and height 6 cm. This was surrounded by a lead sheet of thickness 2 mm, in which a simple grid of 4 mm holes was drilled (four rows of 13 holes). It was placed in front of an X-ray tube, operating at 80 keV and 250 μA at two different distances, 10 and 20 cm with orientations approximately 90° apart. The dose rate delivered was 410 μGy s^{-1} at 10 cm and 110 μGy s^{-1} at 20 cm. In each case, the dose delivered was approximately 46 Gy, chosen in order to gain clear visualisation of the radiation tracks. It should be remembered that at this low energy, the linear attenuation coefficient is much higher than at therapeutic energies and the dose deposited decays very rapidly as it passes through the sample. This explains why it was possible to use as little as 2 mm lead, whilst the production version of RADBALL™ uses 5 mm. The samples were imaged on the CCD-based scanner at the University of Surrey, using methods previously reported [4].

Figure 3 shows the results of the experiment. Clear tracks are seen both in the projection images and the reconstructed slices. Notice that the tracks do not appear to span the entire diameter of the dosimeter. There are three reasons for this: (i) At the proximal end of the track wall artefacts appear. In this case, since we are interested in knowing the exact distance of the source from the sample edge, it is helpful *not* to eliminate this from the image (which could be achieved by performing the standard processing method of data correction by a scan of the unirradiated dosimeter). (ii) At the distal end of the track, the signal has been strongly attenuated (see above). (iii) The two sets of tracks are not exactly co-planar and so a single slice image cannot show them both completely.

By considering two well-chosen planes, the position of the radiation sources may be reconstructed. An initial estimate, based on a "by eye" tracing (using the MacOS X medical visualisation application OsiriX for on-image measurements), gave 9.9 and 20.9 cm respectively as the distances from the dosimeter surface to the two sources respectively, in good agreement with the actual values. For this initial manual measurement, the errors in these measurements were estimated to be approximately 5%. However, further work is needed to develop automated software to perform this task, together with an objective estimate of the measurement error.

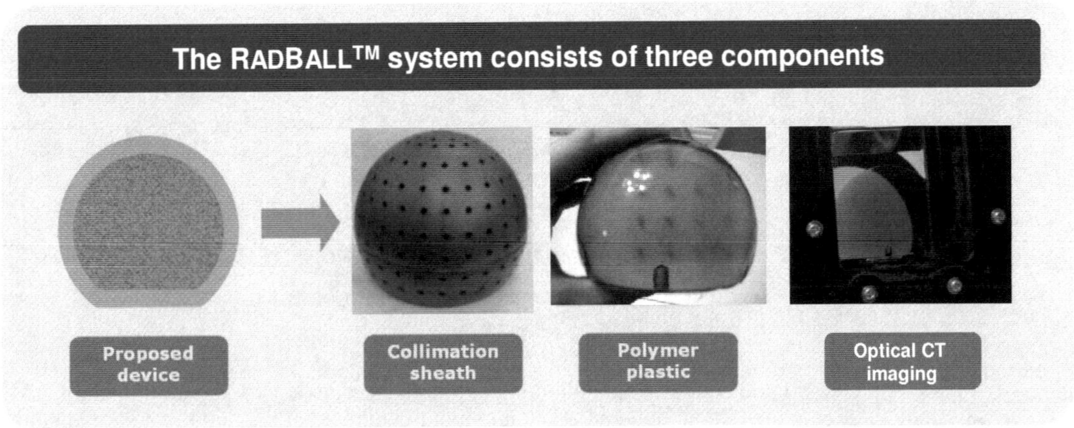

Figure 4: The current system, showing the collimation sheath, sample and the sample sitting in the optical CT imager.

5. Completed RADBALL™ system and future work

Figure 4 shows the completed collimator, together with the custom-manufactured PRESAGE™ insert. Initial irradiations are underway. One issue that remains to be resolved is the length of time that the samples must currently be left *in situ* to receive enough dose to obtain successful images. This is being addressed by investigation of different formulations of PRESAGE™. It should be noted that one advantage of this industrial application over the medical dosimetry scenario is that there is no requirement for the dosimeter to be tissue equivalent. This allows a different set of elemental ratios to be introduced in the PRESAGE™ composition and hence may lead to dosimeters that are optimal for the requirements of industrial dosimetry.

References

[1] Lehner C E *et al.*, 4π Compton imaging using a 3-D position-sensitive CdZnTe detector via weighted list-mode maximum likelihood, *IEEE Tran. Nucl. Sci.* **51**(4), 1618–1624 (2004)

[2] Zhang F and He Z, New readout electronics for 3-D position sensitive CdZnTe/HgI_2 detector arrays, *IEEE Tran. Nucl. Sci.* **53**(5), 3021–3027 (2006)

[3] Adamovics J and Maryanski M, Characterisation of PRESAGE™: a new 3-D radiochromic solid polymer dosimeter for ionising radiation, *Rad. Prot. Dosimetry* **120**(4), 107–112 (2006)

[4] Krstajic N and Doran S J, Focusing optics of a parallel beam CCD optical tomography apparatus for 3D radiation gel dosimetry, *Phys. Med. Biol.* **51**, 2055–2075 (2006)

Journal of Physics: Conference Series **164** (2009) 012043 doi:10.1088/1742-6596/164/1/012043

An attempt to determine the saturation dose for PRESAGE^{TM}

Shamsa Al-Nowais[1] , Andrew Nisbet[1,2] , John Adamovics[3] and Simon J Doran[4,1]

[1]Department of Physics, University of Surrey, Guildford, UK
[3] Royal Surrey County Hosptial, Guildford, UK
[3] Dept of Chemistry and Biology, Rider University, Lawrenceville, NJ, USA
[4] CRUK Clinical MR Research Group, Institute of Cancer Research, Sutton, UK

Corresponding author e-mail: Simon.Doran@icr.ac.uk

Abstract. This brief work-in-progress outlines two methods that we have attempted for determining the dose at which the linear relation between optical density of a PRESAGE^{TM} dosimeter and the dose deposited breaks down. Both methods were equally successful in mapping the linear relation up to an optical density of approximately 6.25 cm^{-1} (absorbance 2..5), but no saturation was found in this region.

1. Introduction

It is well known that gel dosimeters exhibit saturation effects. In most cases, the "response" to increasing dose (be it NMR R_2, Hounsfield number, ultrasound properties or optical absorption) starts off as a linear relation and then "rolls off", with a lower and lower response for additional dose above a certain threshold. Although different gel formulations have different saturation doses, these generally lie in the region of tens of Gy. We were keen to know at what dose saturation occurs for PRESAGE^{TM} [1], as this may have a bearing on our ongoing studies of the interaction of this dosimeter with protons. The results of this study, albeit provisional, are presented here on the basis that other researchers may find them interesting (and perhaps surprising).

2. Materials and methods

10 cuvettes containing PRESAGE^{TM}, of optical path length 4 mm, were supplied by the manufacturer (Heuris Pharma, Skillman, NJ). These were irradiated using a 6 MV Varian linac at the Royal Surrey County Hospital. A build-up region was provided by covering the samples with a slab of solid water, such that D_{max} was reached at the surface of the cuvettes, and a bolus was placed around the samples laterally. The field size was 200 × 200 mm^2 and the cuvettes were irradiated together, abutting each other, to ensure as uniform a dose as possible to the batch. The dose given was fractionated and after each fraction was given, one cuvette was removed, to yield the following nominal doses in the ten cuvettes: 40, 60, 80, 100, 120, 140, 160, 180, 240 and 280 Gy.

Each cuvette was measured in two different ways. Firstly, the cuvette was placed in a standard double-beam spectrophotometer (Camspec, Cambridge, UK) together with an unirradiated cuvette. Secondly, the cuvette was placed in the beam of the fast laser scanner described in [2]. The standard acquisition was modified by making the field-of-view (FOV) extremely small, such that the entire FOV was contained within a sinle cuvette. This second experiment was performed on the basis that it might prove possible to extend the range of optical densities measured.

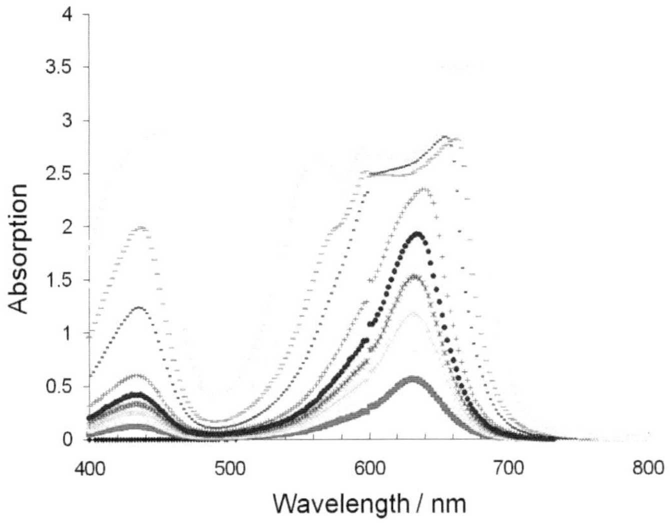

Figure 1: Spectra of PRESAGE™ cuvettes irradiated to increasing doses. Note that, since a double-beam spectrophotometer was used, the baseline spectrum of the compound is subtracted, leaving only the changes that occur with increasing dose.

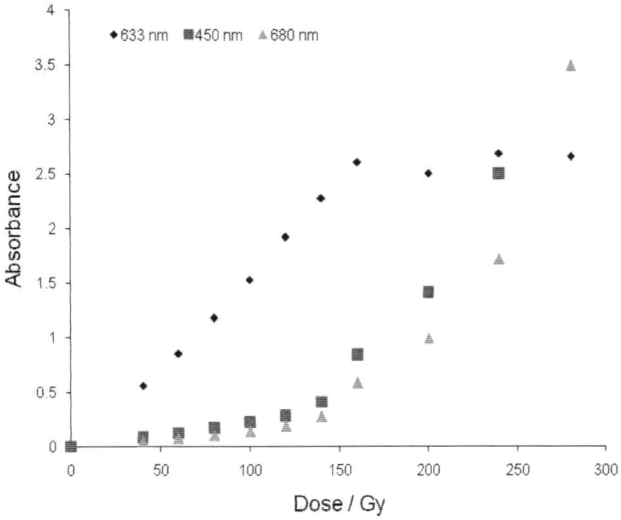

Figure 2: Absorbance of the PRESAGE™ samples as a function of dose for three different wavelengths (633, 450 and 680 nm). Notice that, although the data at 450 and 680 nm fall within the measureable range of the spectrophotometer for a larger number of cuvettes, the non-linear relation of the peak height with dose would make it much more difficult to determine the saturation threshold.

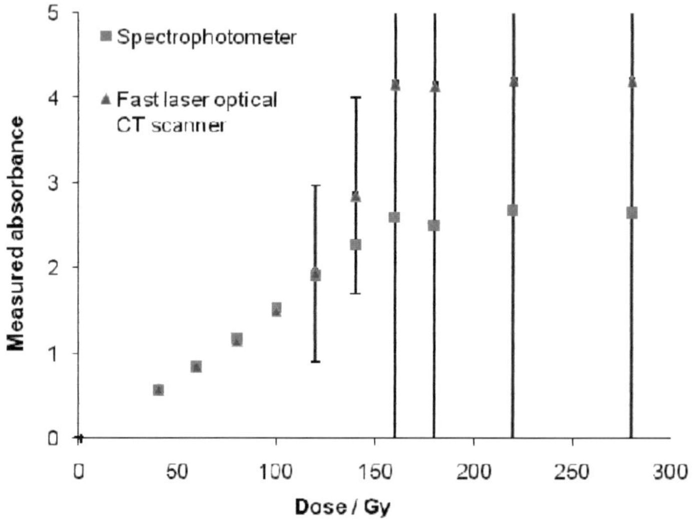

Figure 3: Comparison of the absorbance measurements made by the spectrophotometer and the laser optical CT scanner. The error bars are for the laser scanner and are too small to plot for doses less than 100 Gy. They are calculated from the variation in data seen across the nominally uniform images.

3. Results and discussion

Figure 1 shows the spectra recorded from the ten cuvettes and the unirradiated sample. It will immediately be obvious that the spectrophotometer records the spectra of the first six samples faithfully, but fails for samples 7–10 in the approximate region 600–680 nm — precisely the region in which we are interested. The upper limit of reliable measurement at 633 nm (the HeNe laser wavelength for which PRESAGE™ was designed) is an absorbance of about 2.6. Since the sample is 4 mm thick, this corresponds to an optical density of the PRESAGE™ of approximately 6.5 cm^{-1}, which is exceedingly large.

Initially, it was hoped to use regions of the spectrum slightly away from the 633 nm peak, which do not (optically) saturate the spectrometer in order to measure the (chemical) saturation of the PRESAGE™ samples. However, it will be seen from Figure 2 that the other regions in the absorption spectrum show a strongly non-linear dependence on dose and are thus unsuitable for this purpose.

Figure 3 compares the results of the laser optical CT scanner to those of the spectrophotometer. Once the absorbance of the unirradiated sample and the fixed absorbance of the scanner's lenses has been taken into account, the two sets of data overly each other almost perfectly up to 120 Gy (absorbance 1.94). It is no surprise that the error in absorbance increases dramatically above 100 Gy. Standard propagation of errors tells us that the error in log x is proportional to $1/x$. This means that the size of the error will rise by an order of magnitude with every unit increase in absorbance (all other things being equal).

It should be possible to better adapt the laser scanner for the task of measuring absorbance, on a cuvette sized object, by modifying the position of the sample and photodiode. Some light is currently lost through stray light reflections. Based on the specification of the photoreceiver used, it is anticipated that significantly higher values of absorbance should be measureable.

4. Conclusion

A linear relation was found between dose and optical response between 0 and 150 Gy (absorbance 2.6), the limit of reliable measurement of our spectrophotometer. No detectable saturation effect was observed. The fast laser scanner gave almost identical results to the spectrophotometer in the dose range 0 – 120 Gy (absorbance 0 – 1.94). Further improvements to the measurement capabilities of the latter instrument should be easy to achieve.

References

[1] Adamovics J and Maryanski M, Characterisation of PRESAGE™: a new 3-D radiochromic solid polymer dosimeter for ionising radiation, Rad. Prot. Dosimetry 120(4), 107–112 (2006)

[2] Krstajic, N and Doran S J, Fast laser scanning optical-CT apparatus for 3D radiation dosimetry. Phys. Med. Biol., 52(11) N257-N263 (2007).

Micelle hydrogels for three-dimensional dose verification

Babic S, Battista J and Jordan K

London Regional Cancer Program at London Health Sciences Centre, Department of Physics and Engineering, 790 Commissioners Road East, London, Ontario, Canada, N6A 4L6

The University of Western Ontario, Department of Medical Biophysics, Medical Sciences Bldg., London, Ontario, Canada, N6A 5C1

Steven.Babic@lhsc.on.ca

Abstract Gelatin hydrogels form a transparent and colourless matrix for polymerization or chromic reactions initiated by absorption of ionizing radiation. Generally, hydrogel chemistries have been limited to water soluble reactants. Work to adapt a water insoluble colourless leuco dye to coloured dye conversion reaction in hydrogels, led to the idea that micelles (i.e. tiny aggregates of surfactant molecules) may provide the necessary polar and nonpolar hybrid environment. Both leucomalachite green and leuco crystal violet radiochromic gels have been developed as three-dimensional (3-D) radiochromic dosimeters for optical computed tomography (CT) scanners. It has been found that the post-irradiation diffusion rates strongly correlate with the solubility of the leuco dyes. Since the crystal violet dye is more soluble in the micelle than in the surrounding water, the dose distribution degrades at the slower rate of micelle diffusion, thus yielding stable images of dose. A dosimetric characterization of leucomalachite green and leuco crystal violet gels, respectively, reveals that tissue equivalent micelle hydrogels are promising dosimeters for radiation therapy 3-D dose verification.

1. Introduction

The simple optical absorption spectrum of a gel containing only a single dye species is desirable for quantitative optical computed tomography (CT) scanning. Ferrous xylenol-orange gels exhibit a composite of at least four species that absorb light, with their yields being dose dependent [1]. Polymerization gels have a distribution of molecular weights that complicate quantitative analysis of attenuation profiles. Colourless, transparent hydrogels are spectrally ideal for three-dimensional (3-D) radiation dosimetry with optical CT readout. The lack of initial absorbance allows gels with optical pathlengths of greater than 20 cm to be efficiently scanned. Many radiochromic chemistries have been investigated for radiation dosimetry. One popular approach has involved conversion of a colourless material or solution to a coloured one. Most of the chemistries have involved non-aqueous systems. This work reports on the successful adaptation of a colourless leuco triphenylmethane dye to coloured dye conversion reaction using micelles (i.e. aggregates of surfactant molecules) which provide the required hybrid environment to dissolve these dyes into a hydrogel. Two different leuco dyes were investigated: leucomalachite green (LMG) and leuco crystal violet (LCV). It has been found that the post-irradiation diffusion rates correlate well with the solubility of leuco dyes [2]. Since the crystal violet dye is more soluble in the micelle than in the surrounding water, it was hypothesized that the dose distribution will degrade at the slower rate of micelle diffusion, thus yielding stable images of dose. In some situations, radiochromic hydrogels may be preferable for use over transparent 3-D polyurethane plastics (PRESAGE™) which also contain a leuco dye and have a similar optical performance to gels [3], since the radiochromic plastics are likely to limit dosimeter size due to the cost of the raw materials and complexity of manufacturing large volume dosimeters.

2. Methods and Materials

2.1 Hydrogel preparation

A systematic examination of surfactants revealed that Triton X-100 provided adequate solubility to keep both LMG dye and LCV dye dissolved in 4% gelatin hydrogels. LCV micelle gel samples were prepared by dissolving LCV dye (Sigma-Aldrich), surfactant Triton X-100 (Sigma-Aldrich) and trichloroacetic acid (Mallinckrodt) into a solution of 4% by mass 300 bloom porcine gelatin (Sigma G-2500) and distilled water. LMG micelle gel samples were prepared in a similar way except that LMG dye (Sigma-Aldrich) was used in place of LCV dye. The reactant concentrations were varied to determine the optimal gel sensitivity defined as the change in optical attenuation coefficient per unit radiation dose for both hydrogel samples, respectively.

Upon preparation, the aqueous radiochromic gelatin solutions were poured into 1 cm polymethymethacrylate (PMMA) cuvettes and allowed to gel overnight in a refrigerator at 5°C. Additionally, from the same batch, bulk gel samples were prepared in 1 L polyethylene terephthalte (PETE) cylindrical jars.

2.2 Irradiation procedure

The next day post gel manufacture, cuvettes were returned to room temperature (20°C). For the dosimetric characterization experiments, gel-filled cuvettes were irradiated in a water equivalent plastic cuvette holder using parallel-opposed-pair cobalt-60 beams (average of 1.25 MeV γ-rays) to doses between 0 – 30 Gy. The bulk 1L gel samples were irradiated with a 12 MeV electron beam (Varian Clinac 2100C, 100 cm source to surface distance, 6x6 cm^2 field size) to a dose of 30 Gy at depth of 30 mm. For the diffusion rate experiments, a parallel-opposed beam irradiation with a 6 MV X-ray beam (Varian Clinac 2100C) was applied only to one half of a cuvette filled with LCV gels made with and without surfactant Triton X-100 and LMG gels, respectively. A lead block was used to shield half of the cuvette in order to generate a step-shaped edge profile.

2.3 Hydrogel optical transmission measurements

An absorption spectrum over the visible wavelength region (400 – 650 nm) was initially measured for a gel cuvette sample irradiated to varying doses in order to determine the optimal readout wavelength. At 20 minutes post-irradiation, optical transmission was measured through individual gel-filled cuvettes and referenced to a water-filled cuvette at 590 nm for LCV gels and 633 nm for LMG gels, using an absorption spectrophotometer (Perkin-Elmer 139). The change in the attenuation coefficient, $\Delta\mu$, per unit dose interval was then measured with respect to the non-irradiated gel sample.

The bulk 1L gel samples were scanned 20 minutes post-irradiation using a commercial cone-beam optical CT scanner (Vista™, Modus Medical Devices Inc.) at a central wavelength frequency of 590 nm and 633 nm for LCV and LMG gels, respectively. 512 projections were collected over 360° with each projection averaged to yield a reconstructed pixel size of 0.5 mm. From the optical CT reconstruction, the central-axis change in optical attenuation with depth in the respective hydrogel was fit against the TG-21 corrected ion chamber central-axis depth dose data measured for the electron beam in water.

2.4 Hydrogel diffusion rate measurements

The change in optical density measurements through 1 cm pathlength were collected using a focused laser beam (λ = 594, spot size ~ 0.2 mm) along the length of each cuvette to sample the step-shaped edge profile across the unirradiated to irradiated regions. Optical transmission profiles were acquired from times of 0.33 to 23.5 hours after irradiation in order to determine the respective hydrogel diffusion coefficients (mm^2 hr^{-1}).

3. Results and Discussion

The change in the LCV gel absorption spectrum with increasing absorbed dose exhibited a peak near 600 nm. The spectral data for the LMG gel showed a distinct peak at 633 nm. These results suggest that readily available yellow and red light sources (i.e. laser, light emitting diode) are an optimal choice for optical attenuation measurements with LCV and LMG gels, respectively.

The most radiation sensitive LMG gel formulation contained: 4% gelatin, 6 mM Triton X-100, 15 mM trichloroacetic acid and 0.1 mM leucomalachite green dye. For the LCV gel, the optimal formulation was: 4% gelatin, 4 mM Triton X-100, 30 mM trichloroacetic acid and 1 mM leuco crystal violet dye. With these reactant concentrations, the maximum dose response for the LCV and LMG gels, respectively, was 0.012 $cm^{-1}Gy^{-1}$ at 590 nm and 0.005 $cm^{-1}Gy^{-1}$ at 633 nm. For both hydrogels, the dose response was linear over the studied dose range of 0 – 30 Gy.

The optical density profiles along the length of a half-irradiated cuvette filled with LMG gel is shown as a function of time in figure 1. Table 1 summarizes the measured diffusion rates for LCV gels made with and without surfactant, as well, LMG gels. LMG gels had a diffusion coefficient equal to 0.31 ± 0.01 $mm^2 hr^{-1}$.

Radiochromic Gel	Diffusion Coefficient (mm^2 hr^{-1})
LMG	0.31 ± 0.01
LCV (no surfactant)	0.33 ± 0.02
LCV (with surfactant)	0.036 ± 0.001

Table 1 Measured diffusion coefficients for LMG and LCV gels with and without surfactant, respectively.

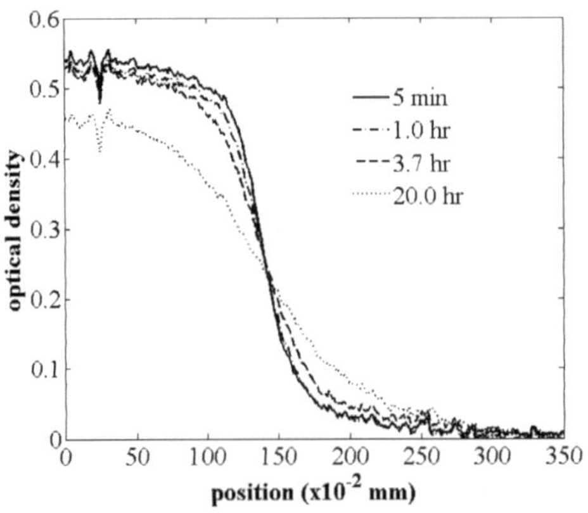

Figure 1 Optical density line profiles through the length of a half-irradiated LMG gel-filled cuvette at different time points.

This value was comparable to the diffusion rate of LCV gels made without surfactant Triton X-100 which was found to be 0.33 ± 0.02 $mm^2\ hr^{-1}$. Adding Triton X-100 lowered the diffusion coefficient of LCV gels by a factor of 10 to 0.036 ± 0.001 $mm^2\ hr^{-1}$. The latter result validates the hypothesis that the diffusion rate of the crystal violet dye is coupled to that of the micelle thereby increasing the post-irradiation dose stability. The low diffusion rate coupled with near water and tissue equivalency (see tables 2 and 3), suggests that radiochromic micelle gels may be a promising alternative for quantitative 3-D gel dosimetry. Presently, the lower dose sensitivity (~10-20 times less than the ferrous xylenol-orange gel) is an inferior property of these gels but it is anticipated that we can overcome this limitation with additional chemical additives (i.e. halocarbons) or purifying the dyes by recrystallization [4].

Material	**Elemental Composition (weight fractions)**									
	H	**C**	**N**	**O**	**Na**	**Mg**	**P**	**S**	**Cl**	**K**
LCV gel	0.100	0.124	0.024	0.729					0.022	
LMG gel	0.102	0.110	0.025	0.752					0.011	
Water	0.112			0.888						
Muscle (ICRU)	0.102	0.123	0.035	0.729	0.001	0.0002	0.002	0.005		0.003

Table 2 The elemental composition (in weight fractions) of the two different radiochromic micelle gels compared to water and striated muscle.

Material	**Physical density ($kg\ m^{-3}$)**	**Electron density ($x\ 10^{29}$ elec m^{-3})**	**Electron density relative to water**	**$^*Z_{eff}$ (photoelectric)**	**$^{**}Z_{eff}$ (pair production)**
LCV gel	1005 ± 5	3.3682	1.0076	7.46	6.61
LMG gel	1015 ± 5	3.3692	1.0079	7.43	6.60
Water	1000	3.3428	1.000	7.42	6.60
Muscle, striated	1040	3.4450	1.031	7.46	6.54

Table 3 The physical density, electron density and effective atomic number (Z_{eff}) of LCV and LMG gels compared to water and muscle. The exponent in the Z_{eff} calculation was *2.94 [5] and **1.00 [6].

Figure 2 shows a post-irradiation transmission image (camera view) of the LMG gel irradiated with a 12 MeV electron beam (a) and the corresponding cone-beam optical CT reconstructed image (b). A comparison of the central axis LMG gel attenuation coefficients normalized at depth of maximum dose (d_{max}) with TG21-corrected ion chamber data, demonstrated agreement only in the buildup region (figure 3a). Beyond d_{max}, the LMG gel over-responded but it was found that this spatial error could be eliminated by pre-exposing the LMG gels to a uniform 'priming' dose of 5 Gy. The reason for this is the subject of future work. LCV gels on the other hand, exposed to the same 12 MeV electron beam, showed agreement with ion chamber data at all depths (figure 3b) thereby confirming energy and dose-rate independence.

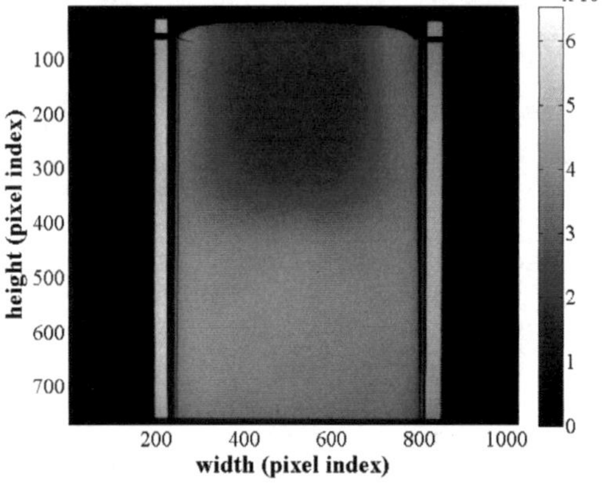

Figure 2a Post-irradiation transmission image (camera view) of a LMG gel irradiated with a 12 MeV electron beam (6x6 cm^2 field size).

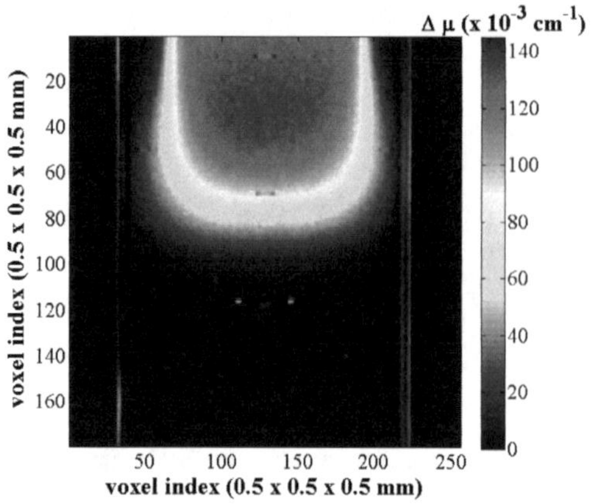

Figure 2b The corresponding reconstructed cone-beam optical CT image.

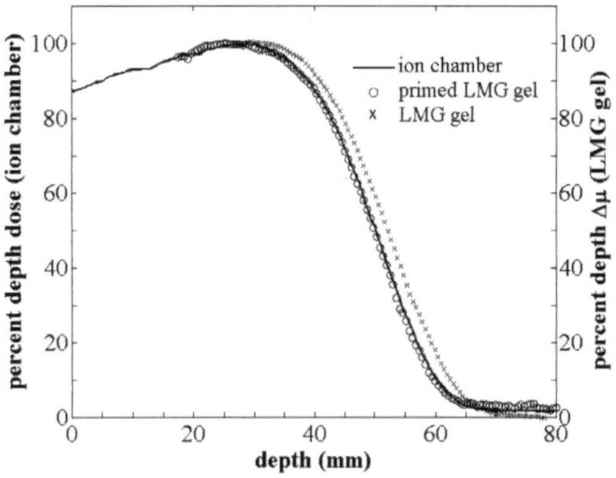

Figure 3a A comparison of the 12 MeV electron beam percent depth dose ion chamber data with the normalized central axis attenuation coefficient in LMG gel (with and without a 5 Gy 'priming' dose) at 633 nm.

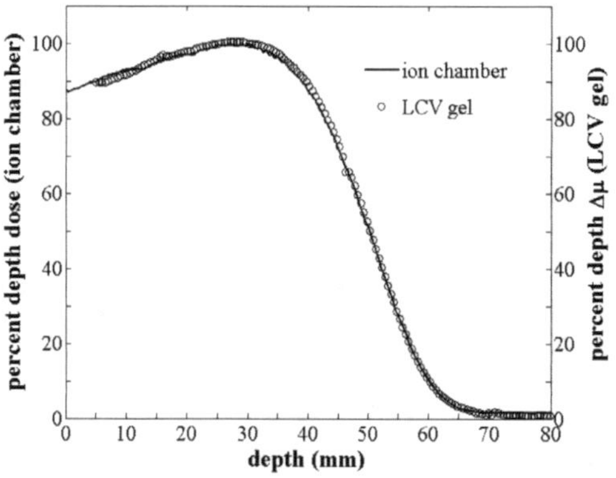

Figure 3b A comparison of the 12 MeV electron beam percent depth dose ion chamber data with the normalized central axis attenuation coefficient in LCV gel at 590 nm.

4. Conclusions

By using micelles, leuco dyes can be dissolved to provide transparent, colourless hydrogels. Radiochromic LCV micelle gels in particular show little diffusion effects and a dose response that is linear, energy and dose rate independent. Micelle hydrogels may be an alternative to radiochromic plastics for 3-D radiation therapy dosimetry.

5. References

[1] Babic S, Battista J, and Jordan K 2008 An apparent threshold dose response in ferrous xylenol-orange gel dosimeters when scanned with a yellow light source *Phys. Med. Biol.* **53** 1637 – 50

[2] De S, Girigoswami A, and Mandal S 2002 Enhanced fluorescence of triphenylmethane dyes in aqueous surfactant solutions at supramicellar concentrations – effect of added electrolyte. *Spectrochimica Acta Part A* **58** 2547-55

[3] Guo P, Adamovics J, and Oldham M 2006 Characterization of a new radiochromic three-dimensional dosimeter *Med. Phys.* **33** (5) 1338-45
[4] Miyaji T, Tokita S, Tachikawa T, and Azuma C 2001 Development of a radiation dosimeter consisting of leuco crystal violet and a small amount of halocarbons *J. of Photopolymer Sci & Tech* **14** (2) 225-26
[5] McCullough E and Holmes T 1985 Acceptance testing computerized radiation therapy treatment planning systems: Direct utilization of CT scan data *Med. Phys.* **12** (2) 237-41
[6] Murty R C 1965 Effective atomic numbers of heterogeneous materials *Nature* **207** 398-99

Acknowledgements
We thank Dr. John Adamovics, PhD, for his insightful discussion on leuco crystal violet dye. Support from the London Regional Cancer Program Endowment fund, Varian Research fund and the Ontario Consortium for Image Guided Therapy & Surgery for gel dosimetry and optical CT, is gratefully appreciated.

Journal of Physics: Conference Series **164** (2009) 012045 doi:10.1088/1742-6596/164/1/012045

A stable black-refractive-index-matching liquid for optical CT scanning of hydrogels

Kevin Jordan and Jerry Battista

London Regional Cancer Program, London Health Sciences Centre and Department of Medical Biophysics, University of Western Ontario, London, ON, Canada

kevin.jordan@lhsc.on.ca

Abstract. The calibration of optical CT scanners often involves referencing of transmission values to surrounding refractive index matching liquid. Often these liquids suffer from wavelength dependencies or can degrade during the course of a scan. In this paper, we describe a black liquid that serves as a stable reference for attenuation measurements. An optically stable, refractive index matching liquid for optical CT scanning of hydrogels was developed. The solution contains: water, glycerol, surfactant and carbon black. Sequential transmission measurements demonstrated less than 5% variation in transmission during a two week storage period at room temperature. Black solutions are expected to minimize spectral beam hardening artefacts associated with polychromatic light sources.

1. Introduction

Refractive index matching liquids are valuable for optical computed tomography (CT) scanning. Their use minimizes refraction and reflection of the imaging light and associated image artefacts. Radiation sensitive hydrogels and plastics (Presage^{TM}) have refractive indices in the ranges 1.34-1.37 and 1.52-1.54, respectively. This report focuses on optimizing the refractive index matching reference liquids for optical CT scanning of hydrogels. Currently, commercial optical CT scanners are equipped with either monochromatic lasers or narrow bandwidth light emitting diodes (LED). Often coloured compounds are added to the refractive index matching liquid in order to control the range of optical transmission by making the attenuation coefficient in the liquid similar to that of nonirradiated gel. Effectively this approach nulls out the base attenuation of the nonirradiated gel. However, if the absorption spectra of the added coloured compound changes over the spectral range of the LED source, transmission values through the hydrogel may be negatively affected. Specifically radiochromic gels such as ferrous-xylenol orange, crystal violet and malachite green have absorption peaks between 570 and 630 nm. Unless the light source is matched precisely to these absorption peaks, there is the potential for producing spectral hardening artefacts in the transmission CT images. This spectral hardening problem may occur with LED light sources with spectral bandwidths that span a range where the absorption spectrum of the samples is changing.

Water is the best choice for the base liquid since the gels are primarily water and the refractive indices are very similar. Water is also, convenient, nontoxic, inflammable, inexpensive and it has a low vapour pressure. Often, glycerol or propylene glycol are added to water in order to match the hydrogel refractive index. Other solutions based on dissolved NaCl and sucrose can be readily prepared with appropriate refractive indices. However, during experiments liquids often come into contact with other surfaces. Sucrose solutions are sticky as they dry and NaCl solutions are corrosive to many metals. Often the CT scanner aquarium contains machined plastic materials. Solvents such as alcohols can lead to hazing and fracturing of these machined surfaces and premature failure. The solutions are open to the atmosphere and scanning is often performed in an enclosed room where volatile, flammable solvents such as ethanol could be a safety hazard. Dust from the air accumulates in the liquids and periodic filtration is required to restore optical clarity of the liquid. Many coloured materials will adhere to paper filters leading to an increased transmission following filtration. Generally if mixtures of dyes are used then one dye will preferentially diminish during filtration leading to a change in their composite transmission spectrum. Still other types of coloured compounds are sensitive to pH and their spectra change as the pH is lowered by dissolution of carbon dioxide. Dissolved oxygen often plays a direct role in gradual degradation of coloured materials. In practice, we have found that patent blue violet and nigrosin dyes dissolved in water or propylene glycol aqueous solutions are reasonably stable over periods of days. However, these solutions form floating gelatinous particles over weeks in propylene glycol solutions. Nigrosin slowly adheres to plastic surfaces lowering the solution concentration. Gelatinous moulds, algae and bacteria also grow in these solutions unless preservatives are added. In brief, stable and user friendly liquids are necessary for reference liquids in quantitative optical CT scanning.

Black solutions are optimum for scanning with non-laser CT scanners since spectral hardening is limited only to the hydrogel. Black ink solutions have been reported in evaluation of laser CT scanners(1). Previous investigations, within our lab tested several black inks but most were determined to be mixtures of several dyes, with crystal violet as a common ingredient. This work, evaluated surfactants as a method of preparing black, transparent, aqueous solutions with refractive indices matched to hydrogels. This initial study has identified a promising approach to preparing stable solutions that will simplify and improve the calibration and accuracy of optical CT scanning of hydrogels by wavelength-independent stable referencing.

2. Materials and methods

Crude carbon black was prepared by collecting soot from a candle flame onto metal foil. The carbon black was dissolved in a dilute aqueous solution containing liquid dish soap (Sunlight, Lever Ponds). The clear solution was brownish grey in colour. Several black inks were dissolved in similar soap solutions but only the drawing ink from Staedtler Mars had a similar visible colour. The absorption spectra of these two samples were recorded with a spectrometer (Hitachi Perkin-Elmer model 139). Subsequent studies were performed with this ink assuming that carbon black is the principle coloured ingredient in this product.

Triton X-100 (Sigma) a non-ionic surfactant and a commercial detergent ('Simple Green® house and siding cleaner', Sunshine Makers Inc, product code 18201) which contains non-ionic surfactants (polyethyleneglycol ether derivatives) were chosen since both were clear and colourless. Combinations of solutions containing 10% glycerine or propylene glycol with either 2% Triton X-100 or 'Simple Green®' were prepared. Ink was added dropwise to obtain solutions with an estimated linear attenuation coefficient of 2 cm^{-1}. These samples were much darker than would be typically required for optical CT scanning. Samples were observed for precipitate formation over several days and then filtered with 5 micron paper to test for adhesion to paper fibers.

Plastic cuvettes were filled with the samples containing 10% glycerine and sealed with wax paper. These samples were stored at 295 K. Transmission images were recorded through the cuvettes over a period of 500 hours with an Apogee AltaU47+ 16 bit CCD camera. The reference image was air for this experiment.

3. Results

Figure 1 shows the visible attenuation spectra of aqueous dishsoap solutions (Sunlight) contaning carbon black from candle flame soot and black drawing ink. Note that the spectrometer filter switches at 600 and 650 nm, resulting in added structure to the smoothly increasing transmission spectrum. These spectra are similar those reported data for carbon black dispersed in mineral oil[(2)].

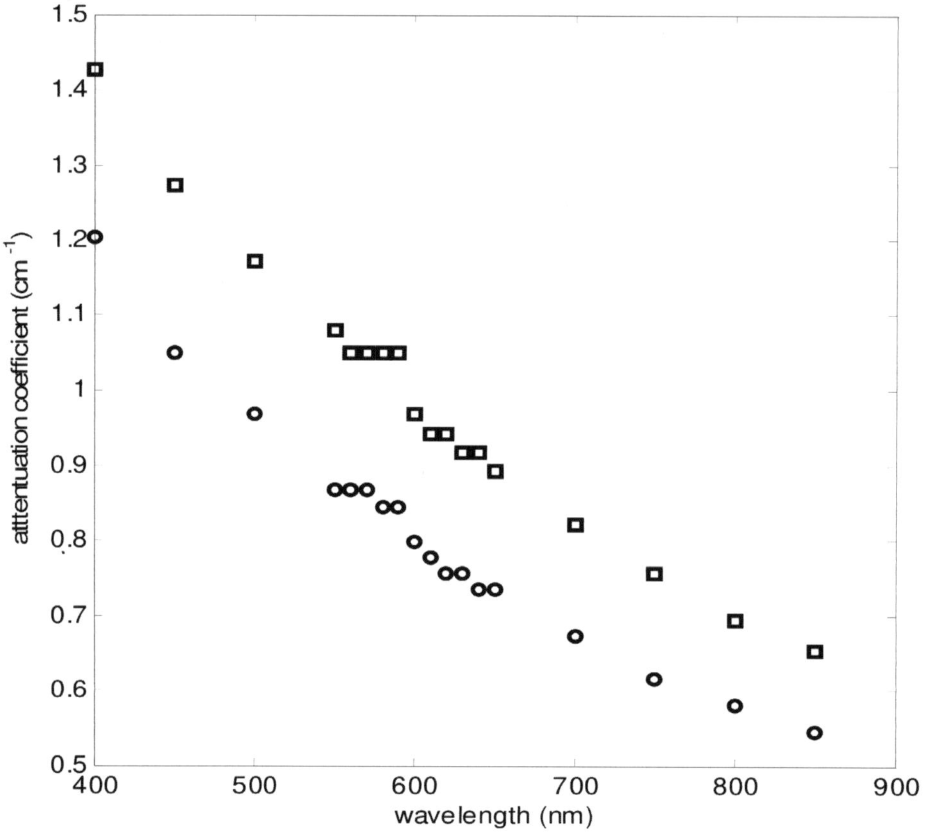

Figure 1. Transmission spectra for aqueous soap solutions of carbon black: candle flame soot (circle) and black drawing ink, Staedtler Mars (square). The similarity if the spectra confirm that Staedtler Mars ink is acceptable. The discontinuity at 600 nm is due to the switching filters in the spectrometer.

Figure 2 is a transmission image through cuvettes containing water and the black solutions with the surfactants Triton X-100 and Simple Green® house and siding cleaner. Mean values of a region of interest (2x10 mm) through the centres of the cuvettes referenced to air were recorded as a function of time. Water provided a reference liquid for this three week experiment. Table 1 lists the sample transmission as a function of time. The Triton X-100 sample transmission increased continuously. After four days a slight amount of black precipitate had settled in the cuvette and the sample was beginning to scatter light more than the Simple Green® solution. By 21 days, the Simple Green® sample was also developing a measurably greater transmission and transparency was beginning to degrade. Both samples were supporting mould or algae growth. Future experiments will identify a suitable anti-microbial agent that is effective and does not significantly alter the sample absorption spectrum.

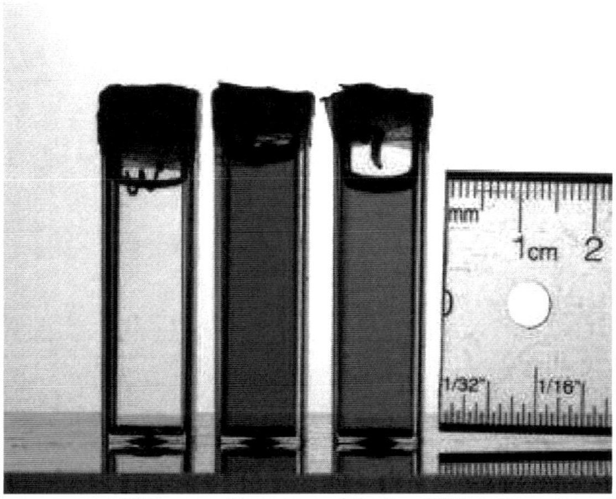

Figure 2. Transmission image at 590 nm of carbon black, 10% glycerine 2% surfactant aqueous solutions: water (left), Triton X-100 (middle), 'Simple Green®' (right) in 1 cm pathlength cuvettes.

Transmission (air reference)			
Time(hours)	H_2O	Triton X-100	"Simple green"
48	.936	.296	.384
72	.936	.305	.387
98	.923	.327	.383
144	.908	.374	.393
216	.917	.412	.398
360	.924	.475	.405
528	.923	.547	.424

Table 1. Transmission of carbon black solutions as a function of storage time at 22°C.

4. Conclusions

Carbon black is a relatively inert material and stable transparent aqueous solution with refractive index of 1.36 have been successfully prepared with a surfactant. For this example the commercial product 'Simple Green® house and siding cleaner' which contains predominantly a non-ionic surfactant was used. A solution with a transmission of 40% per centimetre demonstrated a stable transmission for approximately 400 hours. It is anticipated that storage times for these solutions can be dramatically increased with the addition of appropriate antimicrobial agents. These stable black solutions will simplify quantitative optical CT scanning by minimizing the potential spectral changes for narrow band light sources and providing accurate attenuation coefficients from LED type scanners similar to those obtained with laser scanners.

Acknowledgements: This work was supported by the London Regional Cancer Program Endowment Fund and the Ontario Research and Development Challenge fund (ORDCF-OCITS project).

References

[1] Islam K T, Dempsey J F, Ranade M K, Maryanski M J and Low D A, 2003 Initial evaulation of commercial optical CT-based 3D gel dosimeter, Med Phys 30, 2159-69

[2] Cataldo F, 2002 An investigation on the optical properties of carbon black, fullerite, and other carbonaceous materials in relation to the spectrum of interstellar extinction of light, Full., Nanotubes, & Carb. Nanostructures 10, 155-79

A transparent black non-diffusing micelle gel for optical CT performance evaluation phantoms

Kevin Jordan and Jerry Battista

London Regional Cancer Program, London Health Sciences Centre and Department of Medical Biophysics, University of Western Ontario, London, ON, Canada

kevin.jordan@lhsc.on.ca

Abstract. Performance evaluation of optical CT scanners requires a set of phantoms with well known optical and geometric properties. Gels are often used and tinted with colouring agents but these suffer from wavelength dependencies, diffusion and degradation over time. In this paper we describe a new approach to creating suitable test phantoms using micelles. Adding surfactants to gelatin hydrogels allows materials insoluble in water to become suspended. Carbon black nanoparticles were dissolved into a transparent hydrogel consisting of 4% gelatine and 0.2% Triton X-100. The lack of macroscopic diffusion of the black particles was demonstrated by recording transmission images over 500 hours.

1. Introduction

Optical computed tomography (CT) scanners have been developed for three-dimensional (3D) quantitative measurement of optical attenuation in transparent materials. For dosimetry purposes, polymerization gels form white, translucent images and radiochromic gels form coloured transparent images following the absorption of radiation. In testing their performance, inserts of known optical density are often used. Transparent, blue, finger, hydrogel phantoms were reported by Oldham et al[1]. Bosi et al[2] developed a stable white, translucent phantom. These phantoms were applied to the characterization of optical contrast resolution in CT scanners. Diffusion and colour stability limited the practicality of phantoms when prepared simply with food dyes.

Several optical CT geometries and several light sources have been reported. The light sources include; monochromatic coherent lasers, narrow bandwidth incoherent light emitting diodes, low-pressure sodium lamps and mercury arc lamps. Direct scanner comparisons of performance are complicated by the fact that the spectra of the non-laser scanners change as the light passes through the coloured materials. This problem is analogous to that of 'beam hardening' in x-ray CT. A transparent hydrogel which absorbs light, independent of wavelength, would therefore be very useful for performance evaluation of optical CT scanners with light sources of different spectra. Neutral density filters are commercially available for evaluating and calibration of traditional (2D) densitometers. These filters are typically made by evaporating different thicknesses of aluminum onto glass windows. Black materials evenly absorb light across the entire visible spectrum. Often several dyes or compounds are added to form a black appearing material, such as black ink. Some of the truly black dye molecules are relatively toxic and should be avoided. Carbon black is essentially a standard material for making black substances such as paints, plastics, papers, inks and printer toners. Carbon black is a nanoparticle material, relatively inert and inexpensive. However, carbon black is insoluble in water limiting its value for preparing transparent black hydrogel 'finger' phantoms. Recent investigations have demonstrated the potential of exploiting radiochromic reactions in hydrogels containing surfactants Jordan[(3)]. In this study, initial results are presented where a surfactant has been employed to form a transparent black, non-diffusing stable hydrogel that can be cast in phantoms for optical CT performance evaluation. The phantom can be used to create sharp optical edges used in spatial resolution tests, or 'finger' phantoms used in contrast resolution and attenuation calibration tests.

2. Methods and materials

Carbon black from two sources was investigated. Initially carbon black was produced by placing a metal foil over a candle flame and depositing a black soot film. The second source of carbon black was drawing ink (Staedtler Mars). Both samples were dissolved in Triton X100 aqueous solutions and the absorption spectra were recorded with a visible absorption spectrometer (Hitachi-Perkin Elmer model 139). Subsequent gels were prepared using the black ink. Hydrogels were prepared by dissolving porcine gelatin (300 Bloom, Sigma) 4% by mass in water at 45°C with Triton X-100 (Sigma) 0.25% by mass. The solution was filtered and poured into transparent colourless plastic(PETE) jars. One of the jars also had hollow polymethylmethacrylate rods of 25.4, 12.6 and 4.8 mm outer diameters placed parallel to the rotation symmetry axis. The rods were coated with a thin film of petroleum-based grease and the bottom ends sealed. The 'finger' phantom and reference were stored at 4°C in a refrigerator for 24 hours before filling the finger cavities. Plastic 1cm optical cuvettes were half filled with gelatin solution and set aside to gel to form optical edges. The next day, the remaining stock gelatin was melted and darkened by adding drop wise an aqueous solution containing Triton X100 and ink. The end seals were gently removed from the tubes in the finger phantom and the dark gel solution (~25°C) was added as the tubes were withdrawn gently from the cool gel phantom. Also the upper cuvette halves were filled with darkened gel. Cuvette samples were stored both at 22°C and 4°C for subsequent measurements.

Transmission images were recorded through the cuvettes as a function of time with a 16 bit CCD camera (Apogee Alta- U47+). For these studies, the reference image was air. A diffuse light source was constructed by back-illuminating a white plastic sheet with a yellow Luxeon III LED spectrally filtered with 590 nm bandpass filter and an OG570 longpass filter. The finger phantom and uniform reference were both laser scanned at 594 nm, acquiring 225 CT projections over 180 degrees and reconstructed at 0.5 mm pixel size with filtered backprojection (Matlab V7.4, Mathworks algorithm iradon).

3. Results

The visible absorption spectra for solutions containing the candle soot and black ink were similar indicating the primary pigment in this brand of drawing ink was carbon black. Figure 1 is a transmission image of the gel-filled cuvettes. Note the meniscus that formed when the cuvettes were half-filled. Transmission images were calculated by dividing the cuvette images by the corresponding air image. Vertical profiles through the centre of the half-black cuvette stored at 295K are plotted for 20 and 500 hours post sample preparation in Figure 2. Note the profiles are essentially identical, demonstrating that the gel is extremely stable with respect to both optical attenuation and geometry. After two weeks mould growth was observed in the 22°C samples of black gel. The experiment was stopped at 500 hours when the mould colonies started to decrease visible light transmission. Figure 3, is a reconstructed optical laser CT slice through the 'finger' phantom. The full width half maximum diameters of the reconstructed fingers were of 25.3, 12.5 and 5.0 mm in agreement with the tube outer diameters 25.4, 12.6 and 4.8 mm. Note that the streaking artifacts present are due to the seams in the plastic jar. Also, darker regions are evident within the fingers. This non-uniformity was due to pouring the black gel in steps as the tubes were withdrawn in increments of a few cm. Finger uniformity is also very sensitive to the gel temperatures. Casting warmer black gels can provide more uniform fingers at the expense of degrading the boundary sharpness due to side melting.

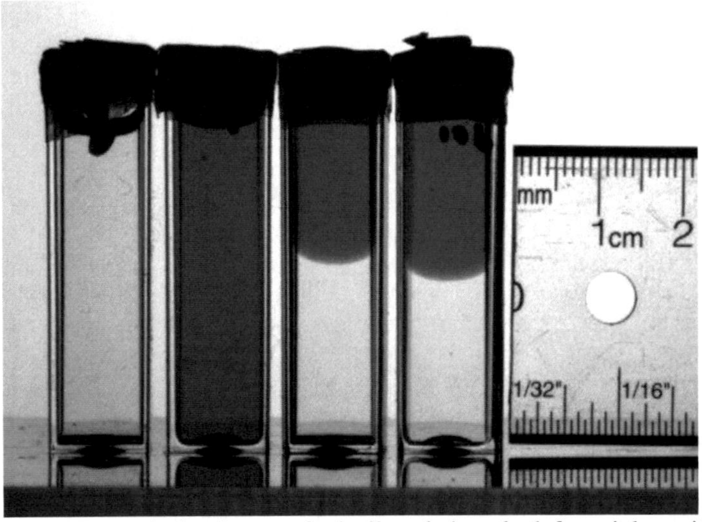

Figure 1. Transmission image of micelle gelatin gels: left to right, uniform micelle gel, uniform black gel, gradient gels stored at 22°C and 4°C. Vertical profile of 22°C sample plotted in Figure 2.

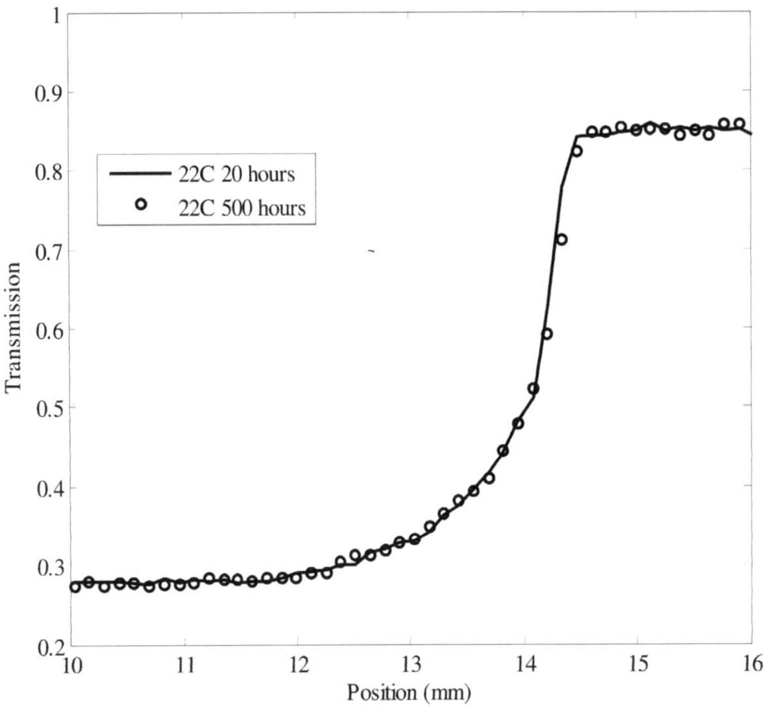

Figure 2. Plot of vertical profile across meniscus edge in Figure 1 of 22°C gel at 20 and 500 hours post preparation.

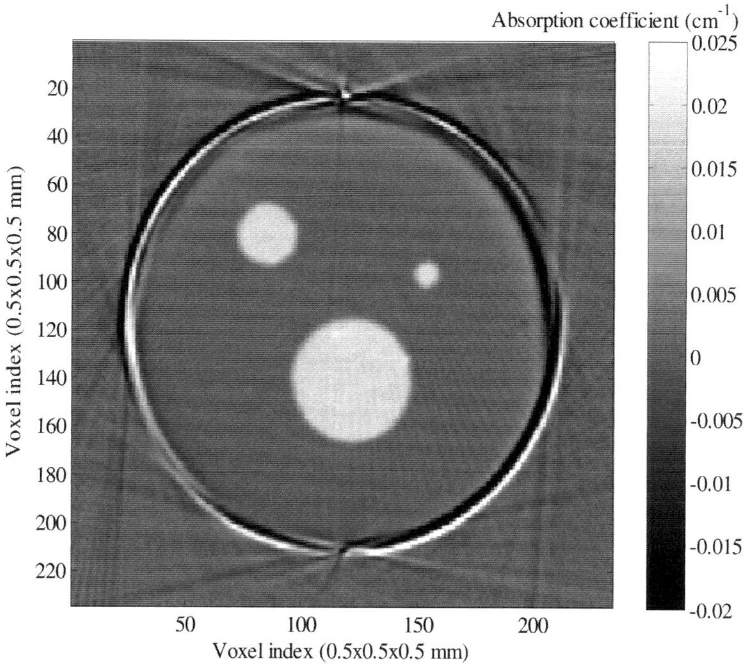

Figure 3. Laser CT reconstructed slice of black micelle finger contrast phantom. Streaking artifacts are due to seams in the plastic jar.

4. Conclusions

Black, non-diffusing, gel for finger phantom manufacture can be prepared by using a surfactant to suspend carbon black nanoparticles in micelle gelatin hydrogels. Triton X100 is one example of a nonionic surfactant that adds minimal scatter to 4% gelatin hydrogels and provides a stable optical attenuation over several weeks. These properties allow the production of calibration and test phantoms that can used with both monochromatic laser CT scanners and broader bandwidth incoherent LED based scanners. Because the gel is black, absorption is nearly independent of wavelength and spectral hardening artifacts are avoided. Future work will investigate the use of micelle gel hosts to prepare imaging performance phantoms for both X-ray CT and magnetic resonance imaging using appropriate contrast agents.

Acknowledgements: This work was supported by the London Regional Cancer Program Endowment Fund and the Ontario Research and Development Challenge fund(ORDLF – OCITS project).

References

1. Oldham M, Siewerdsen J H, Kumar S, Wong J and Jaffray D A, 2003 Optical-CT gel dosimetry I: Basic investigations, Med Phys 30 623-34A

2. Bosi S, Pourandokht N, Puran A, Davies J and Baldock C, 2007 Initial investigation of a novel light-scattering gel phantom for evaluation of optical CT scanners for radiotherapy gel dosimetry, Phys Med Biol 52 2893-2903

3. Jordan K, Avaakumov N and Battista J, 2007 Preliminary investigations of a leucomalachite green hydrogel for three-dimensional radiation dosimetry, Med Phys 34, 2485 (abstract)

New 3D radiochromic gel dosimeters with inhibited diffusion

Jaroslav Šolc[1] and Václav Spěváček[2]

[1]Czech Metrology Institute - Inspectorate for Ionizing Radiation, Radiová 1, CZ 10227 Prague 10, Czech Republic

[2]Czech Technical University in Prague, Faculty of Nuclear Sciences and Physical Engineering, Břehová 7, 115 19 Prague 1, Czech Republic

jsolc@cmi.cz

Abstract. Two new radiochromic gels for 3D dosimetry, both with inhibited diffusion, are described. The first dosimeter is based on the radiation-induced creation of a dye Turnbull blue. The dose sensitivity of the gel measured in terms of spectrophotometric absorbance is $(5.0 \pm 0.1)\text{x}10^{-3}$ $Gy^{-1}cm^{-1}$ (1σ) and the background value 24 h after gel preparation is 0.14 cm^{-1}. The diffusion coefficient of the gel stored at 24°C is less than $4\text{x}10^{-3}$ mm^2h^{-1} (1σ). The second dosimeter makes use of the radiation-induced formation of insoluble mercurous chloride which creates a fog inside an originally bright gel. The dose sensitivity of the dosimeter is $(3.8 \pm 0.2)\text{x}10^{-2}$ $Gy^{-1}cm^{-1}$ (1σ) and its background value 24 h after preparation is 0.18 cm^{-1}. The diffusion coefficient is less than $2\text{x}10^{-4}$ mm^2h^{-1} (1σ).

1. Introduction

Gel dosimeters are integral chemical dosimetry systems with the unique ability to record and retain information about a spatial distribution of dose. Radiochromic gels which change color during irradiation represent one class of these dosimeters. The main advantages of radiochromic gels include the possibility to evaluate their response using relatively cheap optical methods, e.g. cone-beam optical computed tomography (CT) [1]. So far, a Fricke-gel dosimeter with xylenol orange ion indicator incorporated into a gel matrix (FXG dosimeter, [2]) has come into widespread use only, but its disadvantages are the fast spontaneous increase of the background value and the significant feathering of dose patterns owing to ferric ions diffusion [3]. This work describes basic characteristics of two newly developed radiochromic gels with inhibited diffusion of dose patterns.

2. Materials and methods

2.1. Dosimeter based on Turnbull blue

The first new radiochromic gel dosimeter is based on the radiation-induced creation of an insoluble dye Turnbull blue, $K[Fe^{II}Fe^{III}(CN)_6]$ (the TB dosimeter, see Figure 1, [4]). Basically, the TB dosimeter is composed of a gel matrix, potassium ferricyanide, and a ferric compound dissolved in an acidic medium. During irradiation ferric ions (Fe^{3+}) are reduced to ferrous ions (Fe^{2+}) which subsequently interact with potassium ferricyanide forming Turnbull blue.

Various compositions and concentrations of compounds were tested in order to achieve high sensitivity and low background simultaneously. Three versions of the gel are suggested (see Table 1).

Table 1: Composition of all three versions of the TB dosimeter

Compound		Version 1	Version 2	Version 3
gelatin (300 bloom)	-	3 % w/w	5 % w/w	-
agarose (phytagel, [5])	-	-	-	0.4 % w/w
potassium ferricyanide	$K_3Fe(CN)_6$	3 mM	1.5 mM	1.5 mM
ferric chloride	$FeCl_3.6H_2O$	1.5 mM	-	0.45 mM
ferric ammonium citrate	$C_6H_8O_7.FeNH_3$	-	1.5 mM	1.5 mM
sulphuric acid	H_2SO_4	-	15 mM	-

Figure 1: Samples of unirradiated (left) and irradiated (25 Gy – center; 100 Gy – right) TB dosimeter

2.1.1. *Gel preparation.* Gelatin (or agarose) was completely dissolved in distilled water (half of the total gel volume) at ~50°C (or at ~80°C in case of agarose). All other compounds were prepared separately as solutions in distilled water. It was difficult to dissolve ferric chloride hence its solution was heated and stirred while hydrochloric acid was being slowly added until the precipitate disappeared. The final mixing of the gel was carried out in semi-darkness. Gelatin (or agarose) was cooled down to 35-40°C and during its intensive stirring, the solutions were added successively into the gel as follows: ferricyanide and ferric chloride (version 1), or ferricyanide, sulphuric acid, and ferric ammonium citrate (version 2), or ferricyanide, ferric ammonium citrate, and ferric chloride (version 3). The gel was filled up with distilled water to the final volume, then separated into vials, and refrigerated. No special atmosphere is needed but it is desirable to use compounds in a high degree of purity.

2.2. Dosimeter based on mercurous chloride

The second presented radiochromic gel dosimeter is based on formation of mercurous chloride, Hg_2Cl_2 (the Hg dosimeter, see Figure 2, [6]). Hg_2Cl_2 is an insoluble compound created by radiation-induced reduction of Hg^{2+} contained in acidic solution of mercuric chloride, $HgCl_2$. Composition of the Hg dosimeter is shown in Table 2.

Table 2: Composition of the Hg dosimeter

Compound		Concentration
gelatin (300 bloom)	-	3 % w/w
mercuric chloride	$HgCl_2$	50 mM
oxalic acid	$(COOH)_2.2H_2O$	65 mM

Figure 2: Samples of unirradiated (right) and irradiated (30 Gy, left) Hg dosimeter

2.2.1. *Gel preparation.* Gelatin was completely dissolved in distilled water (half of the total gel volume) at ~50°C and all other compounds were prepared separately as solutions in distilled water. In addition, a ~0.5 M solution of sodium hydroxide, NaOH, was prepared for adjustments of the pH of the gel. This solution was subsequently used for increasing the pH of oxalic acid to the value of ~2.7. The final mixing of the gel was again carried out in semi-darkness. Gelatin was cooled down to 35-40°C and during its intensive stirring, at first oxalic acid and then mercuric chloride were added into the gel. While still stirring, the pH of the gel was slowly adjusted to the value of 3.6. In the end, the gel was filled up with distilled water to the final volume, then separated into vials, and refrigerated.

2.3. Measurement of dosimetric characteristics

Several basic dosimetric characteristics of new gels were determined: the dose sensitivity, the background value, the gel ageing, and the diffusion rates. The dose sensitivity was calculated as a slope of dependence of the spectrophotometric absorbance on dose; the background value was given as absorbance of an unirradiated gel sample, and the gel ageing was determined by measuring absorbance of an unirradiated gel sample during several days after manufacture. Gel samples were prepared into short spectrophotometric plastic optical cuvettes (1x1x4.5 cm^3 inner dimensions) and their absorbance was evaluated by a spectrophotometer Helios β. The method of measuring the diffusion rate was as follows: gel samples in long plastic columns (1x15x1 cm^3) were irradiated over half their length to form an edge spread function (ESF). The rate of change of the ESF over a period of one month was determined from profiles of intensity of grey taken from black-and-white photographs [7]. The intensity profiles were fitted with a curve approximating the error function and subsequently the diffusion coefficients were calculated according to equations presented in [8].

3. Results and discussion

3.1. TB dosimeter

3.1.1. *Gel composition.* Optimization of composition of the gel was focused mainly to suppression of increased background. It was found that gelatin negatively influences the background values hence its concentration was hold as low as possible. In the version 2 ferric chloride was replaced by ferric ammonium sulfate which led to slight decrease of the gel sensitivity but also to significant decrease of background. However, this dosimeter is highly light-sensitive and should be always stored in a dark place. In the version 3 gelatin was replaced with agarose which is able to solidify in much lower concentration. Unfortunately, even during intensive stirring agarose was lumping when the acidic solution of $FeCl_3$ was being added to the gel. That is why the concentration of $FeCl_3$ had to be lowered, but for maintaining the sensitivity ferric ammonium citrate was added into the gel. This version of the TB dosimeter shows the lowest background and the slowest ageing although its manufacture is a bit complicated due to a high melting temperature of agarose.

3.1.2. *Dosimetric characteristics.* Absorption spectra of the TB dosimeter are shown in Figure 3. Peak of the absorption band of Turnbull blue corresponds to 690 nm therefore all values of absorbance were measured at this wavelength.

The dose dependence of the TB dosimeter version 3 is presented in Figure 4. The dependence is linear from 0 Gy up to at least 400 Gy. The dose sensitivity is equal to $(5.0 \pm 0.1)x10^{-3}$ $Gy^{-1}cm^{-1}$ (1σ) and the background value 24 h after gel preparation is 0.14 cm^{-1} and 0.19 cm^{-1} if stored in a dark place at 5°C and 24°C, respectively. Ageing of a gel stored at three different conditions is shown in Figure 5. The gel is very stable if stored in a refrigerator but, however, the background values are significantly increasing if the gel is stored in the light.

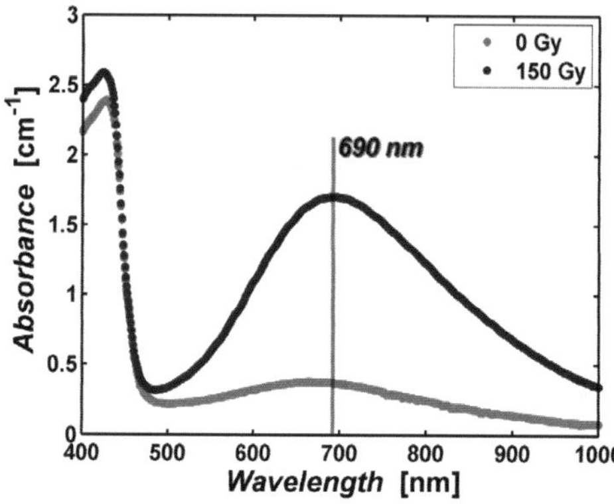

Figure 3: Absorption spectra of the TB dosimeter. An unirradiated sample (red) and a sample irradiated to 150 Gy (blue).

The diffusion coefficients were measured for gels stored in a dark place at 5°C and 24°C and in both cases the absolute value was lower than its uncertainty. It was found that the diffusion coefficient is less than $2x10^{-3}$ mm^2h^{-1} (1σ) and less than $4x10^{-3}$ mm^2h^{-1} (1σ) at 5°C and 24°C, respectively, hence the diffusion is practically negligible. This is caused by the nature of molecules of Turnbull blue which create huge clusters preventing the diffusion through the gel matrix.

Characteristics of all versions of the TB dosimeter are summed up in Table 3.

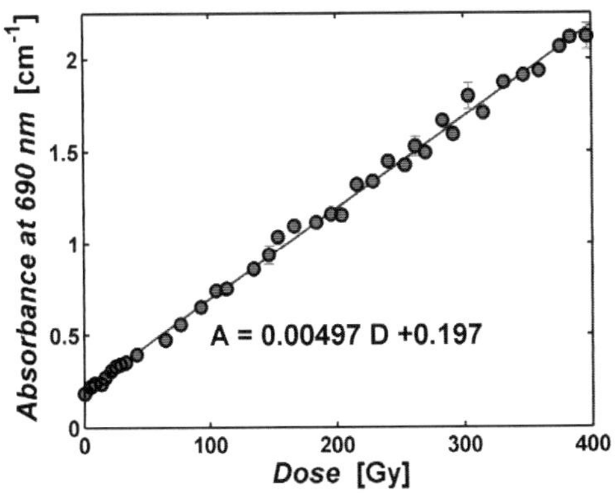

Figure 4: The dose dependence and the sensitivity of the TB dosimeter version 3 measured at the wavelength of 690 nm. Data dispersion is caused by slightly different light-history of each gel sample due to very long (3 days) irradiation.

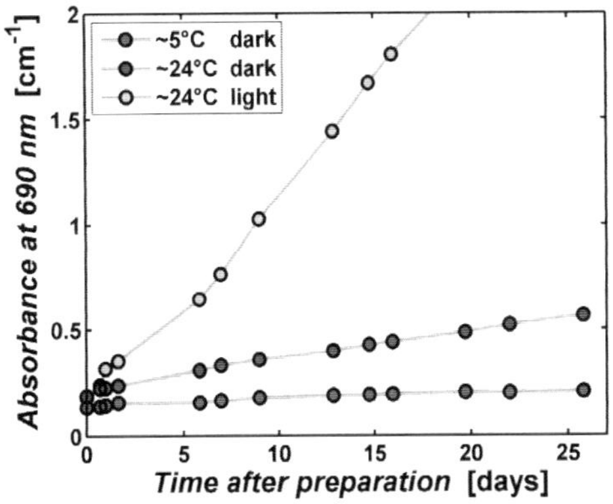

Figure 5: The ageing of three unirradiated samples of the TB dosimeter version 3 stored in a refrigerator (red), in a dark place at 24°C (blue), and in the indirect light at 24°C (green).

3.2. Hg dosimeter

3.2.1. *Gel composition.* It was observed that the dose sensitivity of the gel depends on its pH and it was found that an optimal value of pH is ~3.6. Unfortunately, from the point of view of the user the gel contains toxic mercury and so after use it should be treated as a toxic waste.

3.2.2. *Dosimetric characteristics.* Absorption spectra of irradiated gels are presented in Figure 6. Hg_2Cl_2 emerging during irradiation creates insoluble particles with dimensions comparable to the wavelength of light therefore the gel scatters the light resulting in an increase of absorbance at all wavelengths. The consequence is that this gel cannot be evaluated with the cone-beam optical CT. Unfortunately, the gel gives no response on magnetic resonance imaging too.

The dose response is linear and the gel sensitivity is highest at the wavelength of 850 nm where the linearity of the dose response extends to at least 30 Gy and the sensitivity reaches $(3.8 \pm 0.2)\times10^{-2}$ $Gy^{-1}cm^{-1}$ (1σ; see Figure 7). If stored in a dark place at 24°C, the background value is very low: 0.04, 0.08, and 0.18 cm^{-1} 1, 5, and 24 h after gel manufacture, respectively. The ageing of the gel measured at 850 nm is very slight if the gel is stored in a refrigerator but increases if stored at a higher temperature or in the light.

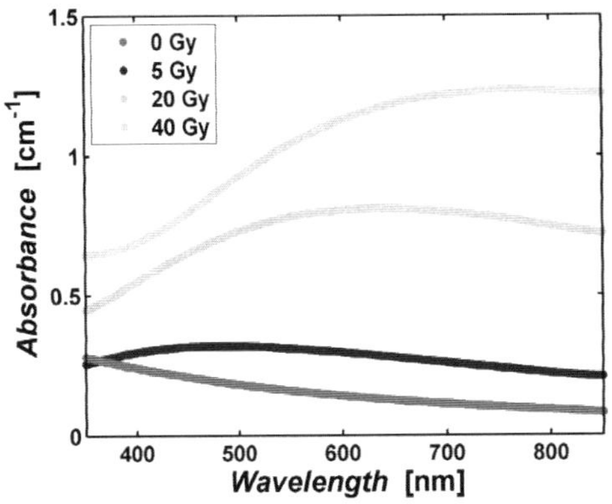

Figure 6: Absorption spectra of the Hg dosimeter. An unirradiated sample (red) and samples irradiated to 5 Gy (blue), 20 Gy (green), and 40 Gy (cyan).

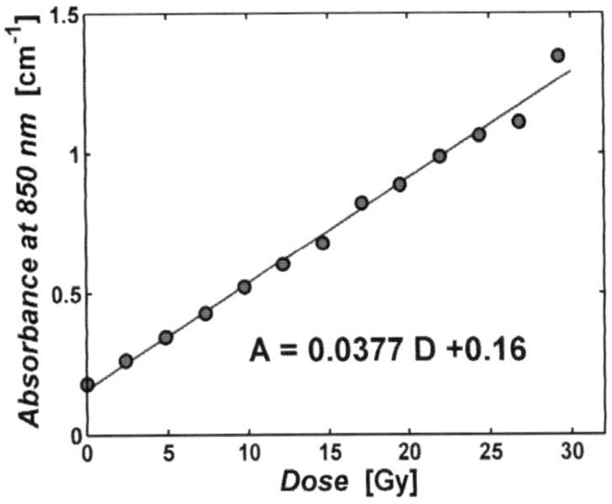

Figure 7: The dose dependence and the sensitivity of the Hg dosimeter measured at 850 nm.

The diffusion rate was determined for a gel stored in a dark place at 24°C only. Again, it was found that the absolute value of the diffusion coefficient is lower than its uncertainty because solid particles of Hg_2Cl_2 are much larger than the pores in the gel and cannot move through. The result is that the diffusion coefficient of the Hg dosimeter is lower than $2x10^{-4}$ mm^2h^{-1} (1σ).

All characteristics of the Hg dosimeter are summed up in Table 3.

Table 3; *the FXG dosimeter is presented for comparison (composition used: 5 % w/w gelatin, 0.1 mM xylenol orange, 0.5 mM ferrous ammonium sulphate, 25 mM sulphuric acid)

Dosimeter	Sensitivity ($Gy^{-1}cm^{-1}$; 1σ)	Background value 24 h after manuf. at 5°C/24°C (cm^{-1})	Linearity of the dose response	Diffusion coeff. at 5°C/24°C ($mm^2{\cdot}h^{-1}$; 1σ)	Ageing at 5°C	Light sensitivity
FXG*	$(7.6 \pm 0.1)x10^{-2}$ (585 nm)	0.25/0.35 (585 nm)	3 - 30 Gy	(0.60 ± 0.02) / (0.74 ± 0.03)	medium	medium
TB dos. ver.1	$(6.3 \pm 0.3)x10^{-3}$ (690 nm)	0.65/0.75 (690 nm)	0 – min.400 Gy	$< 2x10^{-3}$ / $< 4x10^{-3}$	medium	medium
TB dos. ver.2	$(4.4 \pm 0.2)x10^{-3}$ (690 nm)	0.17/0.64 (690 nm)	0 – min.400 Gy	$< 2x10^{-3}$ / $< 4x10^{-3}$	slow	high
TB dos. ver.3	$(5.0 \pm 0.1)x10^{-3}$ (690 nm)	0.14/0.19 (690 nm)	0 – min.400 Gy	$< 2x10^{-3}$ / $< 4x10^{-3}$	very slow	higher
Hg dos.	$(3.8 \pm 0.1)x10^{-2}$ (850 nm)	0.15/0.18 (850 nm)	0 – min.30 Gy	$< 2x10^{-4}$ / $< 2x10^{-4}$	slow	low

4. Conclusions

Two new radiochromic gel dosimeters with inhibited diffusion were described. The dosimeter based on Turnbull blue is suitable for cone-beam optical CT evaluation and could be used in high dose measurements with steep gradients, e.g. dose distribution measurements around brachytherapy sources. The second gel, the dosimeter based on formation of mercurous chloride, possesses high dose sensitivity and could be of use, for example, in measurements of beam profiles, depth dose curves, etc.

References

[1] Wolodzko J.G.; Marsden C. and Appleby A. *CCD imaging for optical tomography of gel radiation dosimeters.* Med. Phys., Vol.26, No.11 (1999), p.2508-13

[2] Bero M.A.; Gilboy W.B.; Glover P.M. and El-masri H.M. *Tissue-equivalent gel for non-invasive spatial radiation dose measurements.* NIM-B, Vol.166, No. (2000), p.820-825

[3] Rae I.D.W.; Willemse C.A.; Lotter M.G.; Engelbrecht J.S. and Swarts J.C. *Chelator effect on ion diffusion in ferrous-sulfate-doped gelatin gel dosimeters as analyzed by MRI.* Med. Phys., Vol.23, No.1 (1996), p.15-23

[4] Balog, J. et al. *Radiation chemical formation and precipitation of Turnbull blue from Fe(III)-oxalate.* Rad.Phys.Chem., 1999, vol.55, p.483-487

[5] Phytagel: www.sigmaaldrich.com/catalog/search/ProductDetail/SIGMA/P8169 (20.5.2008)

[6] Oza P.M. and Rao P.S. *Mercuric chloride as a low level gamma-ray dosimeter: radiolytic and photolytic constants at various pH.* Z. Phys. Chem., No.57 (1968), p.265-8

[7] Rae W. et al. *Chelator effect on ion diffusion in ferrous-sulfate-doped gelatin gel dosimeters as analyzed by MRI.* Med.Phys., January 1996, vol.23, no.1, p.15-23

[8] Kron T. et al. *Fast T1 Imaging of Dual Gel Samples for Diffusion Measurements in NMR Dosimetry Gels.* Magnetic Resonance Imaging, 1997, vol.15, no.2, p.211-221.

In-situ radiation dosimetry based on Radio-Fluorogenic Co-Polymerization

John M. Warman, Leonard H. Luthjens, Matthijs P. de Haas

Technical University of Delft, RID, Mekelweg 15, 2629JB Delft, The Netherlands

J.M.Warman@TUDelft.NL

Abstract. A fluorimetric method of radiation dosimetry is presented for which the intensity of the fluorescence of a (tissue equivalent) medium is linearly dependent on accumulated dose from a few Gray up to kiloGrays. The method is based on radio-fluorogenic co-polymerization (RFCP) in which a normally very weakly fluorescent molecule becomes highly fluorescent when incorporated into a (radiation-initiated) growing polymer chain. The method is illustrated with results of *in-situ* measurements within the chamber of a cobalt-60 irradiator. It is proposed that RFCP could form the basis for fluorimetric multi-dimensional dose imaging.

1. Introduction

Many methods for measuring the energy deposited in materials by high-energy radiation have been developed over the years based on a variety of physico-chemical effects resulting from the interaction of radiation with matter. Of these only a few are capable of monitoring *in-situ* and in real-time the dose rate or the accumulated dose after a given exposure. The most commercially viable and generally accepted methods used at present are based on the current generated by "ionizing" radiation in gaseous (Geiger counter) or solid (semiconductor) media. Optical methods of *in-situ* monitoring have been proposed based on the radio-luminescence or optically-stimulated luminescence of (doped) crystals or glasses. However. the latter have not, as yet, been generally accepted as suitable alternatives within the field of radiotherapy dosimetry. The increasing refinement of radiotherapeutic techniques and the necessity of better definition and *in-vivo* control of the geometrical dose distribution profiles has made it essential (and possibly legally imperative) that suitable dosimetry methods are available.

In this article we present a new method of monitoring *in-situ* and in real-time the accumulated dose within a small (<< 1 cc) volume of a medium which is close to tissue equivalent in its attenuation properties. In addition, the method requires no electrical wiring or applied voltages, depending for the the transmission of information from the monitoring site on optical fibers. The method is based on the process of radio-fluorogenic co-polymerization (RFCP), first demonstrated by one of the present authors in 1997 [1]. In RFCP, polymerization of a bulk monomeric medium is initiated by radiation. Also present in the medium is a small concentration of a compound which is normally non-fluorescent but which on co-polymerization into the growing polymer chains becomes fluorescent. The resulting intensity of the fluorescence is then proportional to the degree of polymerization which in turn is proportional to the yield of initiating free-radicals produced by the radiation, and hence to the radiation dose.

The RFCP method has been used in the past to study radiation-induced polymerization of methylmethacrylate with particular attention given to the "gel effect" [2], in which autoacceleration of polymerization takes place at relatively high total doses (kiloGrays). It was realized more recently that the effect could be used in the low dose regime to monitor accumulated dose. Demonstration of this potential is the subject of the present article. The fact that the fluorescent product is a high molecular weight polymer indicates that RFCP could be applied to obtain fixed multidimensional fluorescent images of dose distribution using semirigid films or bulk gel phantoms.

2. Experimental

The overall experimental set-up, using optical fibre transmission components, is illustrated in Fig. 1.

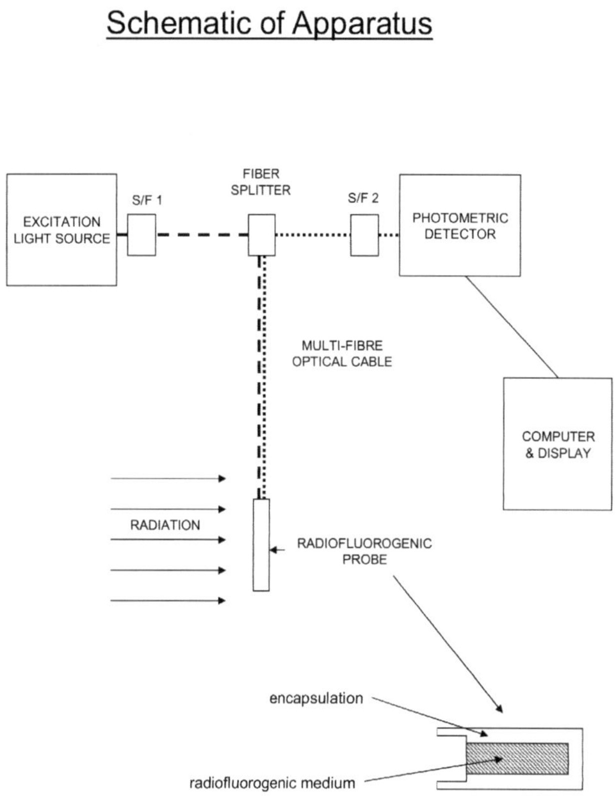

FIGURE 1

The apparatus consists of an "excitation light source" (ELS); a shutter/optical-filter holder (S/F1); a fiber splitter (S); the "radio-fluorogenic probe" (RFP); a second shutter/optical-filter holder (S/F2); a "spectrophotometric detector" (SPD); and computer hardware and software for analysis and storage of the photometer output. The ELS used was either a pulsed N_2 laser (LTB Lasertechnik Berlin, MSG 800) or a CW 360 nm light-emitting diode (Avalight-LED360, Avantes). A UG11 long-wavelength cut-off filter was used in S/F1 to remove the visible wavelength tail of the LED. The SPD was a miniature spectrophotometer capable of spectral resolution from 200 to 900 nm (Ocean Optics SD2000 with data analysis software provided). If considered necessary a GG420 short-wavelength cut-off filter was used in S/F2 to remove any residual light from the excitation source. The optical cable connecting to the probe consisted of a central 200 μm fiber core surrounded by 6 200 μm fibers

(Avantes FCR-7UV200-2). At the splitter, the fibers were separated into two cables with the inner fiber going to the SPD and the outer fibers going to the ELS.

The radio-fluorogenic probe consisted of a 20 mm long poly-tetrafluoroethylene (PTFE) closed cylinder of 8 mm OD and 4 mm ID which could be tightly fitted onto the end of the optical cable as shown in Figure 1. This was filled with the radio-fluorogenic solution which, in the experiments presented here, consisted of methylmethacrylate (MMA, Merk) as bulk solvent which had been passed over a DHR4 hydroquinone-stabilizer removal column, with a *ca* millimolar concentration (Optical density at 360 nm *ca* 2 per cm*)* of maleimido-Fluoroprobe (MFP, for molecular structure see Figure 2). The MFP was synthesised by the group of prof. J.W. Verhoeven. (University of Amsterdam). The solution was deaerated by bubbling with nitrogen within a continuously nitrogen-flushed glove box. The solution was transferred to the probe cylinder and the cylinder was attached to the end of the fiber cable within the glove box.

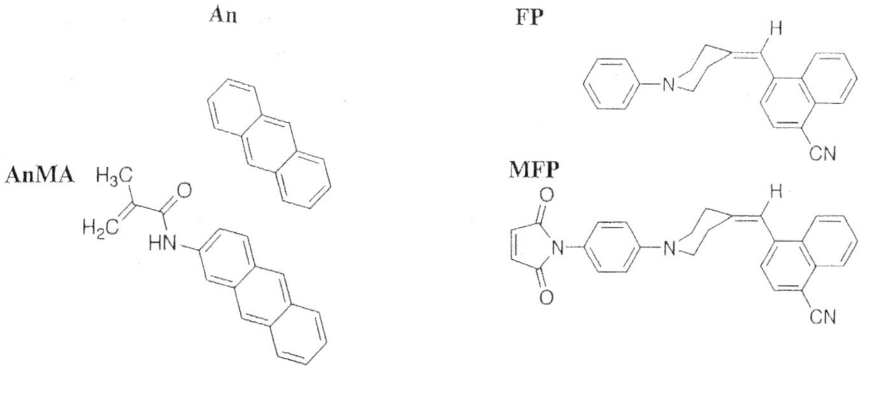

Figure 2. The molecular structures of two of the fluorogenic probe molecules used in RFCP studies; AnMA and MFP (present work) together with their fluorescent moieties anthracene (An) and Fluoroprobe (FP).

The attached probe was placed in the centre of the raised irradiation chamber of a Gammacell 200 cobalt-60 irradiator (Atomic Energy of Canada Ltd). Measurement of the fluorescence emanating from the probe was made prior to and after lowering the chamber into the irradiation zone. The dose rate of *ca* 5 Gy/min was accurately known and continuously corrected for the natural decay of ^{60}Co. Using cylindrical lead attenuators the dose rate could be reduced by factors of 4.15 and 14.6.

3. Results and Discussion

In Figure 3 is shown the spectrally-resolved light emanating from the radio-fluorogenic probe before (lowest trace) and after lowering the probe into the irradiation zone. The dominant feature, with an emission maximum at 530 nm (150 channel numbers in the figure), is the fluorescence of the FP chromaphore which increases with time due to the RFCP effect. The fluorescence spectrum prior to irradiation, corresponds to a quantum yield of less than 0.01. The sharply-peaked short-wavelength emission is residual light from the excitation source. The fluorescent component alone was obtained by taking the integral from 450 to 600 nm.

The integrated fluorescence was automatically recorded every 30 seconds using an integration time of 10 seconds. The results, from which the initial, preirradiation value has been automatically subtracted, are shown in Figure 4 for the 3 dose rates used. It should be pointed out that these three plots were obtained sequentially using the same probe by raising and lowering the chamber and inserting the required lead attenuator.

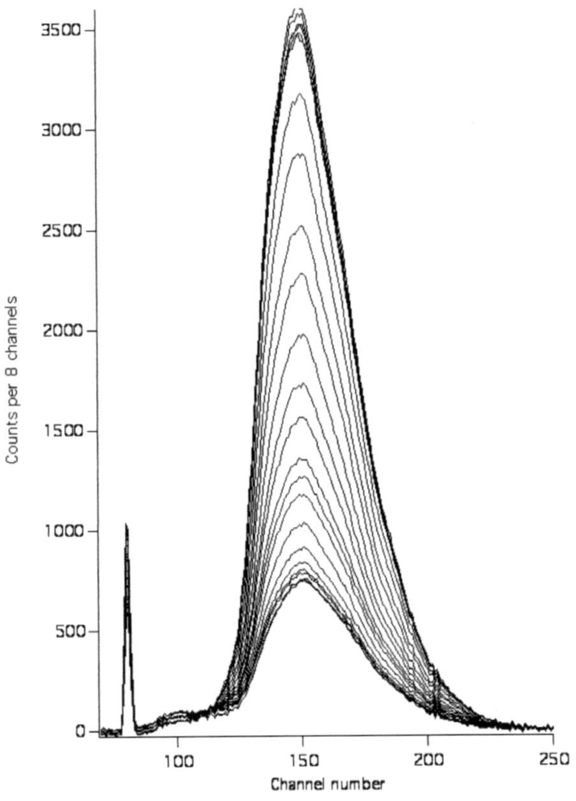

Figure 3. The increase in the spectrally resolved fluorescence emanating from a radiofluorogenic probe containing a solution of MFP in MMA as a function of time after lowering into the irradiation zone of the cobalt-60 source. The maximum in the fluorescence at *ca* 150 channel numbers corresponds to a wavelength of 530 nm. The maximum irradiation time was 90 minutes corresponding to an accumulated dose of *ca* 400 Gy. Several spectra were taken after irradiation was stopped to illustrate the lack of post-irradiation growth. The sharp, peaked spectral feature at short wavelengths is remnant light from the excitation source.

As can be seen in Figure 4, the fluorescence obeys a very good linear dependence on exposure time (or dose) for the three dose rates used. Importantly the data show no indication of an initial delay which would result from the presence of trace amounts of oxygen in the solution. For the highest dose rate the linearity extends up to a total accumulated dose of 160 Gy. In other measurements, the linearity has in fact been found to extend up to at least 1.6 kGy.

As expected from the nature of the underlying polymerization process, the slopes of the traces in Figure 4 show a very good linear dependence on the square root of the dose rate. This results from the fact that the ultimate degree of polymerization is determined by radical-radical recombination [3]. While a linear dose rate dependence might be considered to be preferable a perfect square root dependence should not be too difficult to take into account in quantifying data, and has the advantage that the method is most sensitive to the lowest relative dose rates.

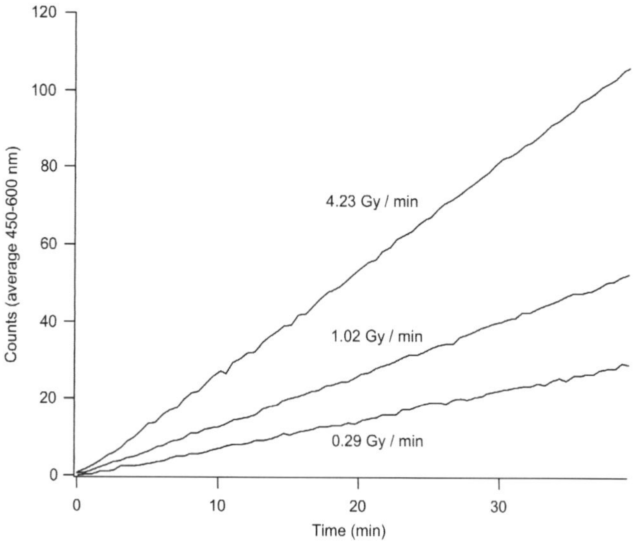

Figure 4. The dependence of the increase in the integrated fluorescence on irradiation time for the three dose rates given using a radiofluorogenic solution of MFP in MMA.

4. Conclusions

We have shown that radio-fluorogenic co-polymerization, RFCP, can provide a method of monitoring *in-situ* and in real-time (seconds) the accumulated radiation dose within a small volume (<< 1 cc) using a probe which is tissue equivalent and biologically compatible. In addition, the method requires no electric wiring or applied voltages making it more suitable for potential *in vivo* applications than existing methods. A linear dependence on dose from a few Gray up to kiloGrays has been found for a methylmethacrylate based solution, allowing multiple use of the same probe. Extension of the sensitivity to lower or higher dose ranges could be achieved by the choice of other polymerizable monomers and fluorogenic probe molecules. The fact that the fluorescent product of RFPC is a polymer indicates that this process could be applied to multi-dimensional dose imaging using semi-rigid, gel matrices.

References

[1] J. M. Warman, R. D. Abellon, H. J. Verhey, J. W. Verhoeven, J. W. Hofstraat, Phys. Chem. B 101 (1997) 4913.

[2] J. M. Warman, R. D. Abellon, L. H. Luthjens, J. W. A. Suykerbuyk, H. J. Verhey, J. W. Verhoeven, Nucl. Instr. and Meth. in Phys. Res. B 151 (1999) 361; M. S. Frahn, L. H. Luthjens, J. M. Warman, Polymer, 44 (2003) 7933; M. S. Frahn, J. M. Warman, R. D. Abellon, L. H. Luthjens, Rad. Phys. Chem. 60 (2001) 433; M. S. Frahn, R. D. Abellon, L. H. Luthjens, M. J. W. Vermeulen, J. M. Warman, Nucl. Instr. and Meth. in Phys. Res. B 208 (2003) 405.

[3] L. H. Luthjens, M. S. Frahn, R. D. Abellon, M. L. Hom, J. M. Warman, Res. Chem. Intermed. 27 (2001) 765.

Journal of Physics: Conference Series **164** (2009) 012049 doi:10.1088/1742-6596/164/1/012049

Clinical applications of gel dosimeters

Sven ÅJ Bäck

Department of Medical Radiation Physics, Lund University, Malmö University Hospital, Sweden

Sven.Back@med.lu.se

Abstract. This short communication raises some issues regarding the use of gel dosimeters in a clinical setting. Further, the need for improvements to attract future interest is discussed. The gel dosimetry method is unique owing to its 3D properties. This is not fully utilised in the present publications. New methods to evaluate and compare dose distributions are needed to further develop the dosimeter gel systems. Undoubtedly there is a continuous need for 3D dosimetry as radiation therapy applications become more and more complex.

1. Introduction

The potential for a dose image read-out of a radio-sensitive gel volume and thereby the possibility to undertake true 3D dosimetry in clinical applications has attracted a great interest. As stated more than 15-years ago, gel dosimetry is a totally non-invasive and non-destructive method, since the dosimeter gel phantom itself forms the detector- a feature not shared with any dosimetric system [1].

This short communication raises some issues regarding the use of gel dosimeters in a clinical setting. Further, the need for improvements to attract future interest is discussed. For a complete review please refer to previous DOSGEL proceedings.

2. Development of the gel dosimeter method for clinical applications

The development of gel dosimetry must be twofold, basic investigations and feasibility studies to find the suitable applications where gel dosimetry puts extra value into the quality assurance procedure (Figure 1, from reference [2]).

Today several new applications that would benefit from 3D dosimetry can be added to right in Figure 1. One example is dynamic arc therapy, which could include variations in gantry angles, MLC movement, MLC speed, and dose rate. Further, various gated treatments that involve the problem of a moving target and deliberately interrupted beams introduce delicate problems for conventional detector methods.

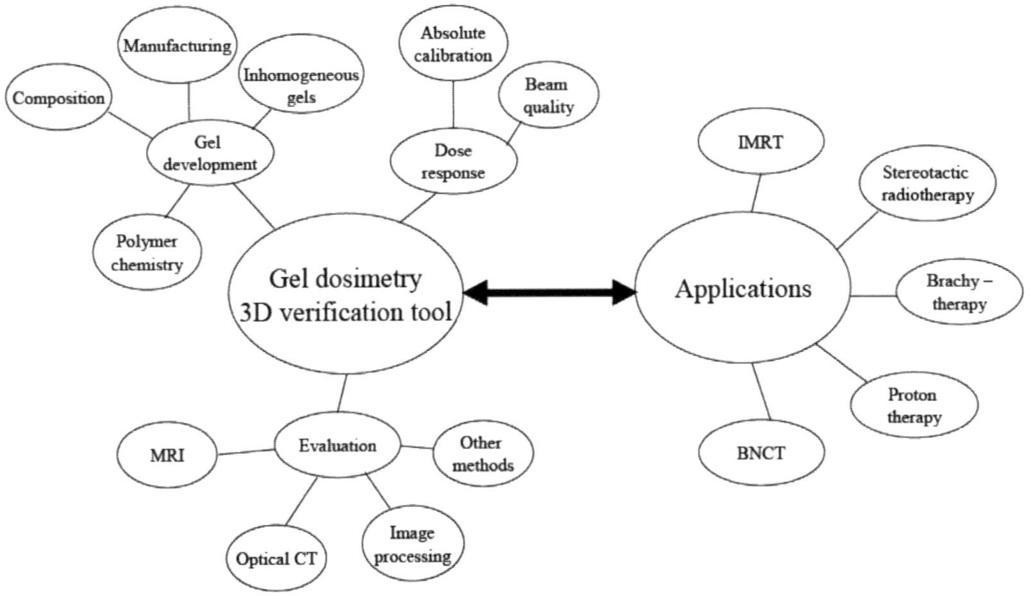

Figure 1. In order for gel dosimetry to become a standard tool for 3D verification in radiotherapy a number of areas should be investigated (left). There are a number of clinical applications for which such a system could be highly suitable (right).

It is difficult and often unfair to compare the uncertainty of a volumetric dosimetry method with traditional detectors. However, there is a lack of benchmarking studies to help the users understand what can be achieved when using gel dosimetry. As with all detectors there are inherent limitations and uncertainties. Despite this, a gel measurement could be of great value even with a relatively poor standard uncertainty. This can be accepted as the combined attractive features of the method, as the potential for high-resolution dosimetry with an excellent geometric integrity and volumetric dose determinations, add new measured information regarding the evaluated clinical applications.

To put the development into perspective two studies from our group will serve as examples. One of the first quantitative studies where gel data was compared with treatment planning data in a clinical application (Gamma Knife) was presented in 1992 [3] (Figure 2).

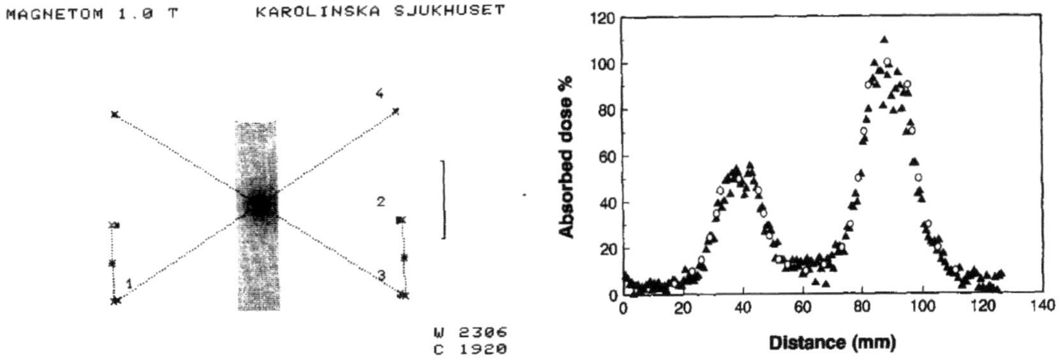

Figure 2. MR image (left) and relative dose profiles (right) through the isocenter in the Gamma Knife application from 1992 (o calculated, Δ measured).

From this study the size of the target and the geometric positioning in 2D could be determined. However, one can note the large scatter in the data points. More recent stereotactic data include volumetric evaluations as comparisons of calculated and measure dose volume histograms, gamma distributions, isodose volumes and traditional dose profiles in three dimensions [4] (Figure 3).

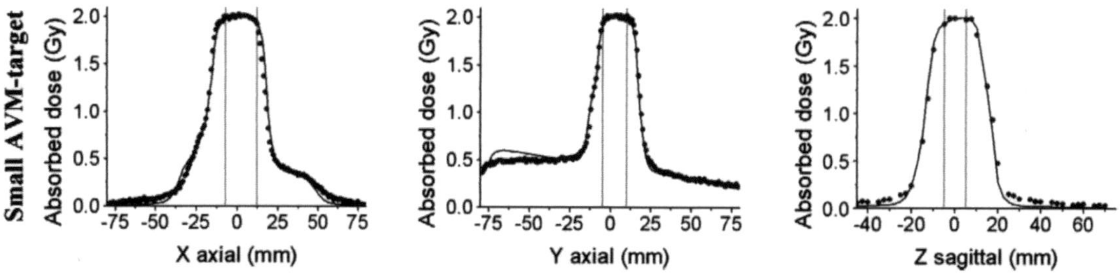

Figure 3. Absorbed dose profiles through the isocenter for a stereotactic application from 2008 (— calculated, • measured). The X axis shows the lateral, Y the anterior-posterior and Z the cranial-caudal direction. The gray vertical lines in each plot mark the extension of the planning target volume (PTV). Zero is the isocenter position on each plot.

In addition to the obvious improvement of the data uncertainty, the introduction of an independent method to normalize the gel data allows for an absolute absorbed dose comparison (in Gy). Gel dosimetry cannot be expected to alone solve all dosimetry problems and this is particularly true if the uncertainty level should be preserved. The combination of several dosimetry methods is needed to accomplish a safe and effective process when for example commissioning new treatment techniques.

3. Future aspects

Undoubtedly there is a continuous need for 3D dosimetry as radiation therapy applications become more and more complex. The publication rate and the citation frequency (gel dosimetry/Web of Science (ISI)) indicate that the research area has been at a relatively high and constant level during the last recent years (Figure 4).

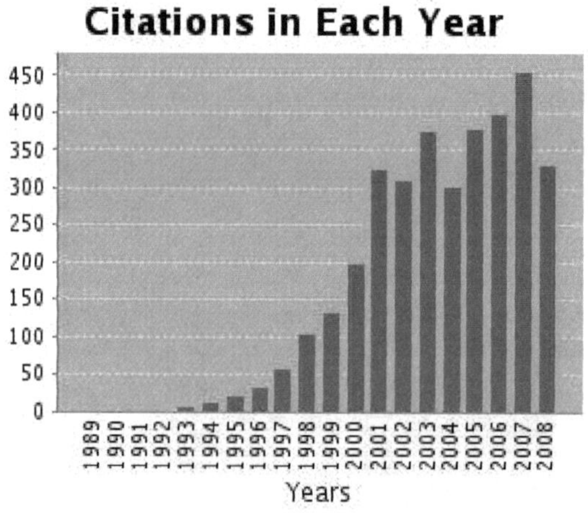

Figure 4. Citation frequency of publications matching the search criteria "gel dosimetry"(Web of Science (ISI), September 2008).

To continue to attract the interest of clinical active medical physicists and researchers there are some important issues to be aware of when using gel dosimetry in clinical applications:

- Use the gel systems in relevant situations where the 3D properties are of importance.
- Combine with traditional detector systems.
- Published "good practice" strategies should be followed.
- Be aware of, or when possible avoid, the limitations of the specific gel system used (dose rate, fractionation, LET etc).

Some of the important future aspects to be addressed by the gel dosimetry community include:

- Development of new 3D dose/volume comparison methods.
- Development of new improved, non-toxic gel compositions.
- Reporting of relevant uncertainty estimations (c.f. [5] and references therein).
- The use of anthropomorphic phantoms.

References

[1] Day MJ, 1990, Radiation dosimetry using nuclear magnetic resonance: an introductory review, Phys. Med. Biol. 35, 1605-9.
[2] Gustavsson H, Radiotherapy Gel Dosimetry; Development and Application of Normoxic Polymer Gels, Ph.D. Thesis, Department of Medical Radiation Physics, Lund University, Malmö University Hospital, Malmö 2004.
[3] Olsson LE, Arndt J, Fransson A, Nordell B, 1992, Three-dimensionell dose mapping from gamma knife treatment using a dosimeter gel and MR-imaging, Radiother. Oncol. 24, 82-86.
[4] Björeland A, Lindvall P, Karlsson A, Gustavsson H, Bäck SÅJ, Karlsson M, Bergenheim T, 2008, Liquid ionization chamber calibrated gel dosimetry in conformal stereotactic radiotherapy of brain lesions, Acta Oncol. 47, 1099 –109.
[5] Karlsson A, Characterization and Clinical Application of Normoxic Polymer Gel in Radiation Terapy Dosimetry, Ph.D. Thesis, Department of Medical Radiation Physics, Lund University, Malmö University Hospital, Malmö 2007.

Preliminary investigation of PAGAT polymer gel radionuclide dosimetry of Tc-99m

Kelly Braun[1], Dale Bailey[1,2], Brendan Hill[1,3], Clive Baldock[1]

[1]Institute of Medical Physics, School of Physics, University of Sydney;
[2]Department of Nuclear Medicine, Royal North Shore Hospital, Sydney;
[3]Medical Physics Section, Department of Biomedical Engineering, Canberra Hospital, Canberra, Australia.

Email: c.baldock@physics.usyd.edu.au

Abstract PAGAT polymer gel was investigated as a suitable dosimeter materials for measuring absorbed dose from the unsealed source radionuclide Tc-99m. Differing amounts of Tc-99m over the range of 25-5000 MBq were introduced into a normoxic polymer gel mixture (PAGAT) in sealed nitrogen-filled P6 glass vials. After irradiation the gels were evaluated using MRI more than 48 hours after preparation to allow for radioactive decay. The dose delivered to the vial was also calculated empirically. R2 versus total activity curves were obtained over a number of experiments and these were used to evaluate the relationship between the amount of gel polymerization and the dose deposited by the radionuclide. A linear response up to 1000 MBq (corresponding to 20Gy) was displayed and was still behaving monotonically at 5000 MBq. Polymer gels offer the potential to measure radiation dose three-dimensionally using MRI.

1. Introduction

Treatments involving the use of ionizing radiation such as radiotherapy and targeted radionuclide therapy play an essential role in medicine and are becoming ever more commonplace with the development of new therapies and technologies. With these treatments comes the need for dosimetry in order to measure the amount of absorbed radiation dose to ensure patient health and safety and to reduce the health risks that can result from an overexposure to radiation. Dosimetry is important both for having the ability to reproduce and compare treatments and techniques and for knowing the limit beyond which administered radiation would cause harmful side effects. A variety of dosimetry techniques exist which can provide precise absorbed dose measurements in one or two dimensions. However, while one and two dimensional dosimeters can provide accurate and precise dose measurements, they have limitations such as in the dosimetry of high dose gradients or in the ability to provide adequate spatial resolution in three dimensions. Polymer gel dosimetry however can provide true three-dimensional absorbed dose information [1,2,3,4].

Gel dosimeters have primarily been studied in the context of clinical radiotherapy in order to evaluate dose distributions from external radiation beams. The most common gel formulations are known as polymer gels, which consist of monomers that polymerize upon irradiation. These dosimeters can be employed to verify absorbed dose estimates from treatment planning systems and to analyze dose gradients created by multileaf collimator systems. Polymer gel dosimeters have the potential to increase the accuracy of radiotherapy treatments and to aid in the development of more personalized dosimetry for treatment plans [1,2,3,4].

In nuclear medicine, radiopharmaceuticals, pharmaceuticals tagged with radioactive isotopes, are administered to the patient orally, intravenously, or through inhalation. The radiopharmaceutical partakes in a certain physiological pathway and is taken up by the organ of interest. Once it is administered, the radionuclide causes photons and/or charged particles to be emitted within the body. Gamma cameras, SPECT or PET cameras are used in order to detect the emergent photons. For diagnostic imaging, only trace amounts of radiopharmaceuticals are administered. In targeted radionuclide therapy, higher doses of radiopharmaceuticals are used in order to treat and eradicate tumours. Dosimetry for radionuclide therapy and diagnostic studies currently relies on imaging in order to map the distribution of the radiopharmaceutical in the body. The images obtained are quantified in order to measure the activity of the radiopharmaceutical in a certain part of the body at a given time in which corrections must be applied in order to finally calculate the absorbed dose distribution.

There is a clear need for a standardized method of radionuclide absorbed dose calculations. Dosimetry in nuclear medicine may be performed in different ways by different clinics or may not be performed at all because of the time required for such calculations and the potentially large uncertainties associated with the calculated doses. The lack of standardized methodology thus means that treatments delivered in different clinics may not be consistent. While accurate absorbed dose calculations have been developed for radiotherapy and have been adopted worldwide, radionuclide absorbed dose calculations used clinically are still approximations at best. The integration of polymer gel dosimeters into nuclear medicine radionuclide dosimetry will potentially provide an accurate method of calculating absorbed dose and aid in the standardization of dose calculation methods. This is coupled with the recent introduction of combined multi-modality SPECT/CT and PET/CT scanners thus providing anatomical electron density maps (the CT scan) and the distribution of the radiopharmaceutical, affording new possibilities for determining individualised dose plans for internal radionuclide therapy.

Absorbed radiation dose from radionuclides are more difficult to measure than the dose from an external beam of radiation. Current absorbed dose measurements in radiotherapy measure the absorbed dose to water from a beam of radiation and use correction factors to account for inconsistencies and inhomogeneities associated with the measurement. Doses are measured over the period during which the radiation beam is turned on. Conversely, absorbed dose measurements in nuclear medicine must take into account the fact that the radionuclide is located within the body and causes a decaying absorbed dose to accumulate. Measurements must therefore be taken over the entire period of time in which the radionuclide is active within the body. Depending on the physical and biological half-lives of the radionuclide, the absorbed dose will be deposited over a period of hours or days.

Promising results from polymer gel radiotherapy studies sparked an interest in investigating the use of gel dosimeters for internal dosimetry in nuclear medicine [5,6,7]. The previous radionuclide dosimetry studies using gel dosimetry for radionuclide therapy used MRI to evaluate the radionuclide dose deposited in a MAGIC polymer gel dosimeters [8]. The studies found a relationship between dose and the amount of polymerization, indicated by change in R2.

The current study aimed to investigate the dosimetry of Technetium-99m (Tc-99m) in conjunction with the PAGAT normoxic polymer gel dosimeter [9] for the purposes of diagnostic nuclear medicine dosimetry. The radiation dose from different activities of the radionuclide was measured by analyzing the radiation-induced change in the gels with MRI.

2. Materials and Methods

The formulation used for the PAGAT gel was according to Venning et al. [9]. The gelatin was added to deionized water and allowed to soak for fifteen minutes. The solution was then heated to 50°C and stirred with a magnetic stir bar in order for the gelatin to dissolve in the water. Once the temperature of the solution had reached 50°C, the temperature was held constant. The gel solution was continuously stirred by the stir bar throughout the course of the experiment in order to thoroughly mix the other ingredients into the gel. When the solution became clear in colour, indicating that the gelatin had dissolved, the acrylamide was added. The bis-acrylamide was then added once the acrylamide had mixed into the gel, followed by the hydroquinone. The solution was left to stir for five minutes, during which time activities of Tc-99m were added to P-6 nitrogen-filled vials [10,11] via a syringe in up to 5000 MBq. Tetrakis was then added to the gel and stirred for a further two minutes before the gel was added to the Tc-99m in the vials via a syringe. When the gel was added to each vial, the tip of the syringe needle was pushed into the Tc-99m solution to help ensure that the Tc-99m would mix evenly into the gel. After removing the empty syringes, the vial was shaken and stored behind a lead shield. As each of the needles became contaminated with Tc-99m, the needle was replaced before adding gel to other vials in order to avoid contaminating samples with activity from other samples.

The irradiated vials were analyzed within 48 hours post-irradiation using a Siemens Vision 1.5T clinical MRI scanner using the same imaging protocol as used by Venning et al. [9]. T2 weighted imaging was performed using a multiple spin-echo pulse sequence. 64 equidistant spin-echoes were used along with a TE of 20ms and a TR of 5000ms. For this preliminary study, the vials were imaged using two 5mm slices. The first slice was close to the bottom of the vial and therefore might suffer from overlap with the glass. The second slice was closer to the top of the gel and further from the base of the vial. A region of interest was drawn inside the gel in order to avoid inconsistencies in the data from the effects of the glass edges of the vial. The values obtained for each sample were the mean R2 and the associated standard deviation.

Medical Internal Radiation Dose (MIRD) calculations [12,13,14,15] were undertaken to determine the accumulated dose from the Tc-99m to the polymer gel.

3. Results

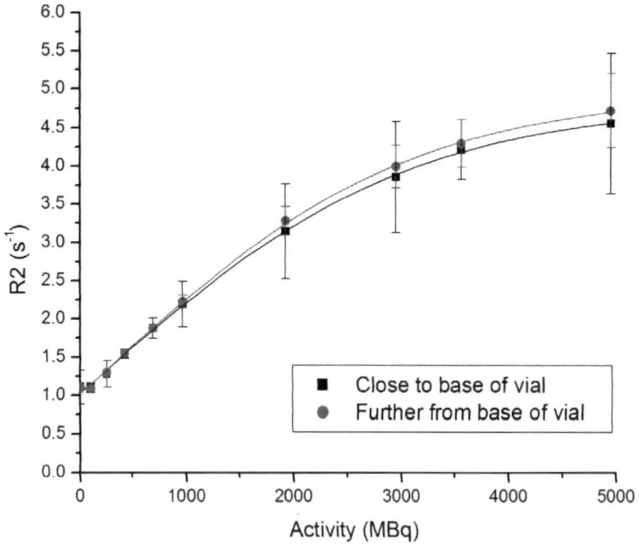

Figure 1 R_2 plotted against activity.

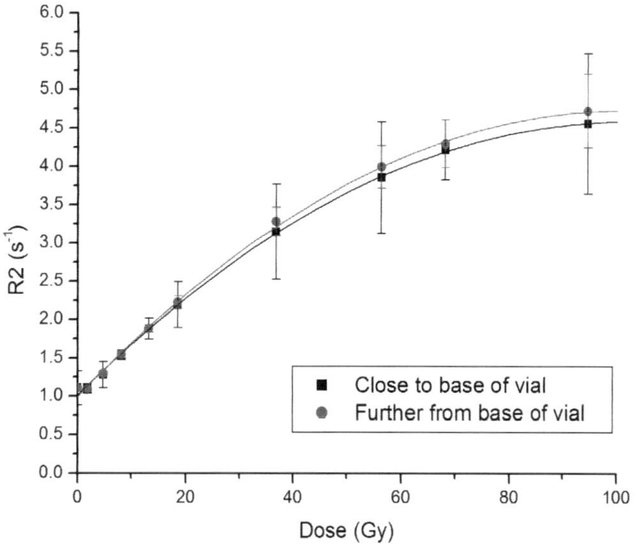

Figure 2 R_2 plotted against dose.

3. Discussion

This preliminary study has shown that gel dosimeters have the potential for measuring the absorbed dose from Tc-99m radionuclide. Not only do the gels respond to Tc-99m, but they polymerize as a function of the activity of the added radionuclide. PAGAT gels would be potentially good dosimeters for use in nuclear medicine. The gels are not prone to diffusion and therefore the results are a permanent record of the intergrated radionuclide dose. PAGAT polymer gels are radiological tissue equivalent.

Several improvements can be made in order to facilitate future experiments. Higher activities such as 7000MBq and 9000MBq should be used in order to provide a better idea of the curve characteristics and the activity at which the gel saturates. More work could be done to determine which syringe size is best for adding gel to the vials or cuvettes in order to ensure the reproducibility of the experiment and the accuracy of the added gel volume. More gel should be added to the P-6 vials in future experiments in order to facilitate MRI analysis. The gel was only divided up into two MRI slices because of the short depth of the gel in the vial. The first slice was close to the bottom of the vial and therefore might suffer from overlap with the glass. Adding more gel would allow for more MRI slices to be taken and thus provide a more accurate analysis of gel polymerization. If more gel is added to the vial, however, more of the radionuclide must also be added.

Future studies will be translated to the use of gels in three-dimensional anthropomorphic phantoms to more closely relate the administered activity to the absorbed dose in the body. The implications of the implementation of gels into nuclear medicine include determining breast dose from sentinel lymph node injections, local dose and the dose to surrounding tissues from radionuclide therapy treatments, or dose from other treatments such as radiation synovectomy.

The use of polymer gel dosimeters to measure the absorbed dose from Tc-99m in diagnostic nuclear medicine has not previously been investigated and is a new method of calculating the absorbed dose from the radionuclide over the length of time that the radionuclide remains active within the patient. This new methodology provides the potential to further personalize treatments and to more accurately calculate dose and future work will examine its utility with other radionuclides.

References

[1] *DOSGEL 1999 Proceedings of the 1st International Workshop on Radiation Therapy GelDosimetry* (Lexington, USA) Eds L J Schreiner and C Audet

[2] *DOSGEL 2001 Proceedings of the 2nd International Conference on Radiotherapy GelDosimetry* (Brisbane, Australia) Eds C Baldock and Y De Deene

[3] *DOSGEL 2004 Proceedings of the 3rd International Conference on Radiotherapy Gel Dosimetry* (Ghent, Belgium) Eds C Baldock and Y De Deene

[4] *DOSGEL 2006 Proceedings of the 4th International Conference on Radiotherapy Gel Dosimetry* (Sherbrooke, Canada) Eds L J Schreiner, A Jirasek and M Lepage

[5] Courbon F, Love P, Chittenden S, Flux G, Ravel P, Cook G 2006 Preparation and use of I-131 MAGIC gel as a dosimeter for targeted radionuclide therapy *Cancer Biotherapy & Radiopharmaceuticals* **21** 427-436

[6] Gear J I, Flux G D, Charles-Edwards E, Partridge M, Cook G, Ott R J 2006 The application of polymer gel dosimeters to dosimetry for targeted radionuclide therapy *Phys. Med. Biol.* **51** 3503–3516

[7] Gear J I, Charles-Edwards E, Partridge M, Flux G D 2007 A quality-control method for SPECT-based dosimetry in targeted radionuclide therapy *Cancer Biotherapy & Radiopharmaceuticals* **22** 166-174

[8] Fong P M, Keil D C, Does M D, Gore J C. Polymer gels for magnetic resonance imaging of radiation dose distributions at normal room atmosphere 2001 *Phys. Med. Biol.* **46** 3105-3113

[9] Venning A J, Hill B, Brindha S, Baldock C 2005 Investigation of PAGAT normoxic polymer gel dosimeter using magnetic resonance imaging *Phys. Med. Biol.* **50** 3875–3888

[10] Baldock C, Burford R P, Billingham N, Wagner G S, Patval S, Badawi R D, Keevil S F 1998 Experimental procedure for the manufacture and calibration of polyacrylamide gel (PAG) for magnetic resonance imaging (MRI) radiation dosimetry *Phys. Med. Biol.* **43** 695-702

[11] Michael G J, Henderson C J, Nitschke K, Baldock C 1999 Effects of glass and backscatter on measurement of absorbed dose in polymer gel dosimetry *Phys. Med. Biol.* **45** N133-N138

[12] Ljungberg M, Strand S E, King M A 1998 *Monte Carlo Calculations in Nuclear Medicine* IOP Publishing Ltd (London)

[13] Loevinger R, Budinger T F, Watson E E 1998 *MIRD primer for absorbed dose calculations* The Society of Nuclear Medicine (New York)

[14] International Commission on Radiation Units and Measurements (ICRU) 2002 The MIRD Schema. Absorbed dose specification in nuclear medicine. ICRU Report 67. *J. ICRU* **2** 35-39

[15] Dillman L T 1969 Radionuclide decay schemes and nuclear parameters for use in radiation-dose estimation. MIRD Pamphlet No.4. *J. Nucl. Med.* **10** Suppl 2 5-32

On the use of VIP Gel dosimetry in HDR brachytherapy

L Petrokokkinos[1,2*], A Moutsatsos[2], P Karaiskos[2], V Kouridou[3], E Pantelis[4], P Papagiannis[2], and I Seimenis[5]

[1]Nuclear & Particle Physics Section, Physics Department, University of Athens, Panepistimioupoli Ilissia, 157 71, Athens, Greece

[2]Medical Physics Laboratory, Medical School, University of Athens, Mikras Asias 75 (Goudi), 115 27, Athens, Greece

[3]Anticancer Hospital of Athens "St. Savvas", L.Alexandras 171, 115 22, Athens, Greece

[4]CyberKnife centre, Iatropolis - Magnitiki Tomografia, Ethnikis Antistaseos 54-56, Chalandri, 152 31, Athens, Greece

[5]Medical Diagnostic Centre Ayios Therissos, 92 Troodous Ave., Strovolos, Nicosia, Cyprus

*Corresponding author. e-mail: lpetrok@phys.uoa.gr

Abstract An experimental procedure is discussed with regard to its potential in ^{192}Ir HDR brachytherapy dosimetry. Two samples of VIP normoxic gel formulation are used; one for gel response calibration and the other for acquiring experimental data. Using the same irradiation method for both calibration and experimental purposes (an ^{192}Ir HDR brachytherapy source) and treating the two samples identically (i.e. the two samples are prepared, irradiated and scanned at the same time and stored together at all times) leads to total dose uncertainties comparable to those of other well established dosimetry methods over a significant dose range (~7Gy-40Gy). In this dose range, the described procedure can be used to either acquire absolute dosimetry results for the characterisation of new ^{192}Ir HDR brachytherapy sources, or to facilitate the planning of relative dosimetry experiments for the verification of calculations by new generation treatment planning systems that are currently phasing in, in complex 3D dose distributions involving inhomogeneities and finite medium geometries.

1. Introduction

Polymer gel dosimetry, the only inherently three dimensional (3D) dosimetry method currently available, has proved effective for the verification of complex dose distributions, especially in the case where steep dose gradients are involved such as those encountered in brachytherapy or stereotactic radiosurgery. In the majority of polymer gel dosimetry studies, however, results are reported in the form of relative dose distributions, i.e. normalized to a point in the dosimeter's volume where dose is known by means of calculation or the use of an "absolute" dosimeter. Polymer gels are not inherently absolute dosimeters in the sense that measurements (irrespective of the imaging modality employed) are not directly translated to dose based on fundamental physical processes and constants, but rather using a calibration procedure. Nonetheless, polymer gel dosimetry could be employed to report absolute dosimetric data as well as relative dosimetry results, with an uncertainty comparable to other well established dosimetric methods. This is discussed in this work for ^{192}Ir High Dose Rate (HDR) brachytherapy dosimetry using a normoxic gel with appropriate calibration and experimental procedures, and Magnetic Resonance Imaging (MRI) as the readout method.

2. Materials and methods

2.1. Gel preparation and irradiation

A recently introduced normoxic polymer gel formulation referred to as VIP [1, 2, 3], was used for the purpose of this work. This formulation is based on VIPAR gel formulation [4 and aims at retaining the wide dose range of VIPAR response [5] while extending it in the low dose region [2, 3], and eliminating the deoxygenation procedure to achieve speed, ease of manufacture and increased reproducibility of the dosimeter's characteristics.

N,N'-methylenebisacrylamide (4%), *N*-Vinylpyrrolidone (8%), gelatin (7.5%), Copper Sulfate (0.0008%) and Ascorbic Acid (0.007%) were diluted in hyperpure water (resistivity > 18 M□·cm) as described elsewhere [2, 3]. Two 100 ml borosilicate glass bottles were used as gel containers with a closed end glass catheter fitted through the cap to ensure impermeability to oxygen through the irradiation catheter. The gel filled bottles were tightly sealed and left overnight in a cool, dark place for the gel to solidify before irradiation.

The brachytherapy irradiation was carried out using a VariSource ^{192}Ir (5 mm active core length) HDR remote afterloader (Varian Oncology Systems, Palo Alto, USA) programmed for each bottle to irradiate in a single dwell position for 283.5 s, with a source kerma strength (S_k) at the time of irradiaton of 12861.42 cGy·cm^2/h (with a 3% uncertainty). Taking into account that 11.15 Gy was delivered at 1 cm distance along the transversal source axis.

The irradiated gel samples were scanned after 2 days, providing ample time for the polymerization and crosslinking procedures to evolve and for achieving thermal equilibrium of the gel substance.

2.2. Magnetic Resonance readout and data acquisition

The irradiated gel samples were scanned in the same session using a 1.5 T whole body Philips Gyroscan NT MR imager (Philips Medical Systems, Best, The Netherlands) and a quadrature RF receiver coil. Both bottles were placed in the center of the receiver coil and as close as possible to one another to ensure minimal RF field inhomogeneities. A 3D Carr-Purcell-Meiboom-Gill (CPMG) imaging pulse sequence (20 equidistant echoes, TE_1=40 ms, TR=2000 ms) was employed. Axial slices were reconstructed with an in-plane imaging pixel of (0.73864 x 0.73864) mm^2 and 0.75 mm slice thickness to enhance spatial resolution and minimize averaging effects in the steep dose gradient regions close to the HDR source.

The echo base images were exported in DICOM format and a 3D relaxation time (T_2) matrix was reconstructed through the application of a least squares mono-exponential fitting routine of pixel signal intensity versus echo time for each pixel in all echo base images. The relaxation rate (R_2=1/T_2) 3D matrix was used to calibrate the dosimeter's response over dose, by matching gel response data to corresponding doses at various distances along the source's transverse axis.

2.3. Calibration procedure and associated uncertainty

The reference dosimetric data used for calibration were Monte Carlo (MC) calculated relative dosimetry results for the HDR source along its transverse axis [6] coupled with the source S_k at the time of irradiation. The MC data set presents an uncertainty of less than 1%, while S_k uncertainty is

3%. Thus, the reference dose data set presents an overall uncertainty of 3% since S_k uncertainty is the dominant factor.

The spatial matching of the delivered dose distribution to the gel response distribution is crucial to an accurate calibration procedure. This is accomplished with excellent accuracy (<0.05 mm) by taking advantage of the source's cylindrical symmetry. Due to this symmetry the registered signal around the source's center presents a point and an axis of symmetry which can be pinpointed in a reference coordinates system by the use of image processing algorithms in the three dimensional dataset. The source center is calculated to be the center of mass of the signal distribution in the irradiated gel volume, and its central axis as the line connecting the centers of the ellipsoidal distributions in sequential axial slices.

For distances (r_i) spanning from 3 mm from the source's center up to 22 mm along its transverse axis, with a step of 0.74 mm, the signal residing on a circle of radius r_i around the source was averaged ($R_{2i}=R_2mean(r_i)$) and its standard deviation ($\square R_{2i}= \square R_2mean(r_i)$) was calculated to be used as a meaningful measure of the uncertainty of the mean due to random effects associated with the measuring procedure and the determination of the source's center. In the implementation of the above procedure the spacing of data points on each circle's circumference was chosen to be comparable to the pixel size.

The dosimeter's response values (R_{2i}) to different dose levels (reference dose values, D_i) were matched to form a dose-response curve (figure 1). The VIP response is linear from 2.5 Gy up to 33 Gy, with sublinearity gradually setting in at higher doses. For brachytherapy dosimetry purposes a wide range of doses is necessitated. Thus a second degree polynomial function ($R_2= a_1D^2+a_2D+a_3$) was fitted to the response data ($R_{2i}=f(D_i)$) to make use of a larger dose range of VIP gel response. The coefficients of the fitting function were determined by weighted least squares regression, since each point in the dose-response data-set is known with different uncertainty, as clearly seen in figure 1. The weights used were equal to the inverse squared standard deviation for each point ($w_i=1/\square R_{2i}^2$).

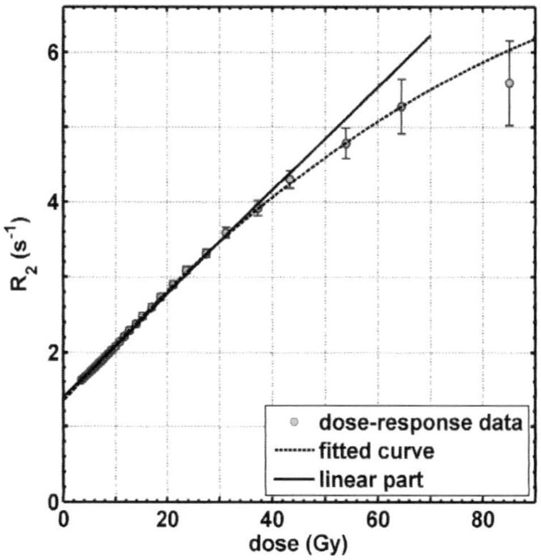

Figure **1**: The dosimeter's dose-response data. A curve is fitted to obtain a calibration function. The linear response part of the curve is denoted

The constants of the fitting function were found to be: $a_1=-(0.00028\pm0.00004)$ $Gy^{-2}\cdot s^{-1}$, $a_2=(0.0792\pm0.0018)$ $(Gy\cdot s)^{-1}$ and $a_3=(1.341\pm0.011)$ s^{-1}.

The above described calibration function is used to obtain doses (D_{exp}) in an experiment which produces gel response measurements (R_{2meas}). The sources of uncertainty to be considered in experimental dosimetry using this calibration procedure are the uncertainties of the fitting coefficients and the uncertainty of the reference doses (u_{Dref}). Thus the calibration uncertainty of a measured dose would be:

$$u_{cal} = \sqrt{\sum\left(\frac{\partial cal}{\partial a_i}\cdot\delta a_i\right)^2 + (0.03\cdot D)^2} \quad (1)$$

since $u_{Dref}=(D/Sk)u_{Sk}=0.03D$ (2).

2.4. Experimental dosimetry uncertainty

Besides calibration uncertainty, experimental dosimetry results are subject to the uncertainty induced by the use of the polymer gel – MRI method to register the signal distribution to be translated to the corresponding dose distribution around the HDR source. Thus the uncertainty of the measured dose shall include the uncertainty of the measured R_2 values (u_{meas}) which incorporates all uncertainty factors related to noise and discrepancies induced to the final measurement due to MR scanning, radiation and chemical factors [7]. The cumulative effect of all these factors, which end up affecting

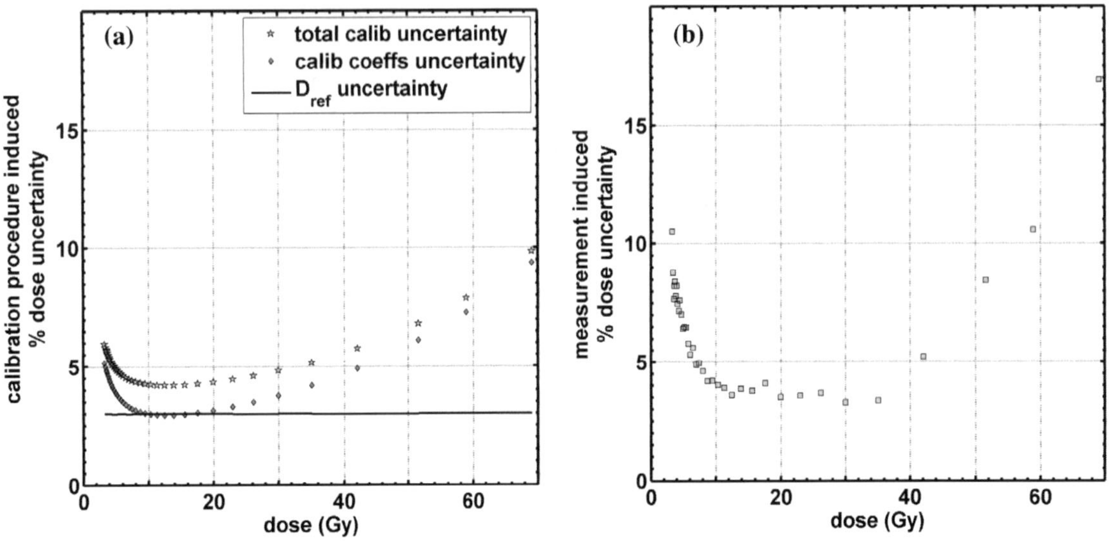

Figure **2**: Percent dose uncertainties due to (a) calibration procedure and (b) measurement of the irradiated gel dosimeter. The two calibration procedure uncertainty's constituents, the coefficients uncertainty (ua_i) and the reference dose uncertainties (u_{Dref}) are plotted separately (a)

the final measurement in a statistical way, can be quantified by the standard deviation of the measured R_2 values which is easily calculated due to the cylindrical symmetry in single source dwell position irradiations as described above. Moreover, the use of normoxic gel of the same preparation batch for both calibration and measurement bottles, combined with an experimental procedure that secures the same thermal history and the same irradiation and scanning conditions for both samples, cancels out the effect of the majority of potential systematic uncertainties and ensures that the effect of those remaining is kept minimal and relatively smaller than the uncertainty already accounted for by the combination of u_{cal} and u_{meas}.

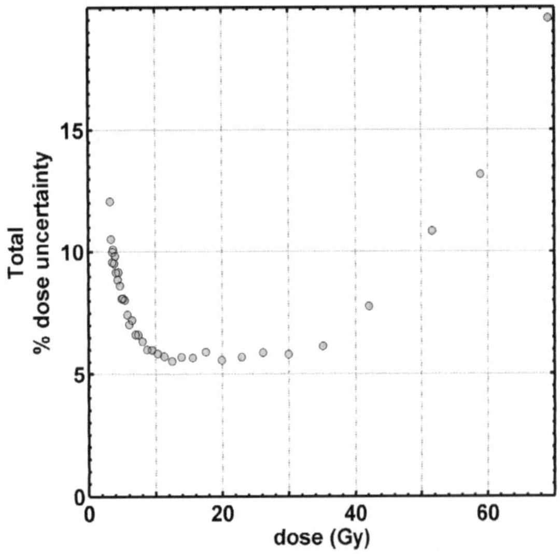

Figure **3**: Total % uncertainty in the measured source's dose resulting from corresponding calibration and measurement uncertainties

3. Results & discussion

Figure 2 presents the relative percent uncertainties induced in experimental brachytherapy dosimetry by (a) the calibration procedure and (b) the measurement process. The effect of the reference dose uncertainty is constant while the coefficients' effect is comparable to that of u_{Dref} for a range of doses spanning from 5 Gy up to about 35 Gy, summing up to a total dose uncertainty due to the calibration procedure below 5% in this range (figure 2a). The uncertainty in the measured dose due to the measurement solely (i.e. apart from the calibration procedure) spans in a much wider range of dose percentages (figure 2b), whereas the general trend is similar to that shown in figure 2a: a region of more or less low uncertainty (below 4%) outside of which the uncertainty increases rapidly.

The total uncertainty of measured dose is presented in figure 3. In the range from 7 Gy up to 37 Gy (which corresponds to a distance range

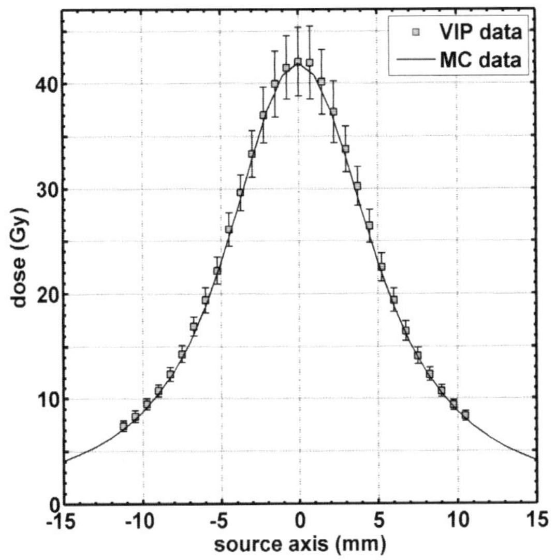

Figure **4**: MC and experimental dose profiles along a line parallel to the source axis at a 5 mm distance

from 13 mm down to 5.4 mm along the source's transversal axis) the uncertainty is lower than 6%, the suggested limit of uncertainty for an experimental dosimetry method used to characterize a new HDR source [8]. Thus the VIP polymer gel-based method proposed herein can be used for the dosimetric characterization of an ^{192}Ir HDR source, preferably using two different experimental gel bottles irradiated so as to cover the distance range of interest around the source with acceptable accuracy.

In order to demonstrate this potential the polymer gel measured dose along a line parallel to the source long axis at 5 mm distance is compared to the corresponding MC-based, reference dose distribution in figure 4. The gel measured values emanate from the second gel bottle, with the experimental dose values being computed using the calibration data of the first gel bottle. In this figure it can be seen that although significant uncertainties occur in a few, relatively high dose points, the experimental dose distribution is in excellent agreement with the reference dose distribution since this distribution is deduced by experimental dose results averaged according to cylindrical symmetry, and most important, concurrent measurement of a dose distribution (an inherent advantage of the polymer gel method) constricts the effect of individual point uncertainties.

An experimental procedure similar to that proposed herein has been successfully employed with a different gel formulation for the dosimetry of low energy/low dose rate brachytherapy sources [9, 10]. In such applications however not only inter-batch but also intra-batch variability is an issue due to the low source S_k that leads to prolonged irradiation times. In HDR brachytherapy dosimetry, the small irradiation times determine a time frame that allows for the response of a polymer gel dosimeter, especially a normoxic one, to be accurately evaluated. It should also be noted that the optimum experimental dose range, in terms of dose uncertainty, could be extended if the calibration source S_k was known with improved accuracy.

4. Conclusions

Irradiation of a gel sample with a single HDR source dwell position allows for a meaningful estimation of precision through repeatability (i.e. the agreement of concurrent measurements carried out under the same conditions of measurement) by exploiting the cylindrical symmetry of delivered dose and corresponding measured response. An experimental procedure that treats normoxic gel filled bottles for calibration and experimental purposes identically, in terms of storage, irradiation and scanning conditions, ensures minimal effect of systematic uncertainties. The total dose uncertainties of the method were calculated and found comparable to those related to other well established dosimetry methods, with the added advantage of concurrent measurements in three dimensions in a single experiment and for an adequate dose range. This dose range can be used for the dosimetric characterization of new HDR source designs that have been recently proposed, as well as for the verification of complex three dimensional dose distributions in brachytherapy.

References

[1] Pantelis E, Antypas C, Petrokokkinos L, Karaiskos P, Papagiannis P, Kozicki M, Georgiou E, Sakelliou L and Seimenis I 2008 Dosimetric characterization of CyberKnife radiosurgical

photon beams using polymer Gels *Med Phys* **35**, 2312-20

[2] Petrokokkinos L, Kozicki M, Pantelis E, Seimenis I, Karaiskos P, Antypas C, Fijuth J and Sakelliou L 2007 Characterization of a new polymer gel for radiosurgery dosimetry using Magnetic Resonance Imaging *4th International Conference on Imaging Technologies in Biomedical Sciences (ITBS 2007)*

[3] Kozicki M, Maras P, Rybka K, Biegański T, Kadłubowski S and Petrokokkinos L 2007 On the development of the VIPAR polymer gel dosimeter for three-dimensional dose measurements *Macromol. Symp. 2007*, 345–352

[4] Pappas E, Maris T, Angelopoulos A, Paparigopoulou M, Sakelliou L, Sandilos P, Voyiatzi S and Vlachos L 1999 A new polymer gel for magnetic resonance imaging (MRI) radiation dosimetry *Phys. Med. Biol.* **44** 2677-2684

[5] Kipouros P, Pappas E, Baras P, Hatzipanayoti D, Karaiskos P, Sakelliou L, Sandilos P and Seimenis I 2001 Wide dynamic dose range of VIPAR polymer gel dosimetry *Phys. Med. Biol.* **46** 2143-2159

[6] A. Angelopoulos, P. Baras, L. Sakelliou, P. Karaiskos, and P. Sandilos 2000 Monte Carlo dosimetry of a new 192Ir high dose rate brachytherapy source *Med. Phys.* **27**, 2521-2527

[7] DeDeene Y 2006 On the accuracy and precision of gel dosimetry *J. Phys.: Conf. Ser.* **56** 72-85

[8] Li Z, Das RK, DeWerd LA, Ibbot GS, Meigooni AS, Perez-Calatayud J, Rivard MJ, Sloboda RS and Williamson JF 2007 Dosimetric prerequisites for routine clinical use of photon emitting brachytherapy sources with average energy higher than 50 keV *Med. Phys.* **34**, 37-40

[9] Pantelis E, Lymperopoulou G, Papagiannis P, Sakelliou L, Stiliaris S, Sandilos P, Seimenis I, Kozicki M and Rosiak J M 2005 Polymer gel dosimetry close to an 125I interstitial brachytherapy seed *Phys. Med. Biol.* **50,** 4371–84

[10] Papagiannis P, Pantelis E, Georgiou E, Karaiskos P, Angelopoulos A, Sakelliou L, Stiliaris S, Baltas D and Seimenis I 2006 Polymer gel dosimetry for the TG-43 dosimetric characterization of a new ^{125}I interstitial brachytherapy seed *Phys. Med. Biol.* **51**, 2101-11

RapidArc™ treatment verification using polymer gel dosimetry

Sofie Ceberg[1], Helen Gustavsson[1], Stine Korreman[2], Joakim Medin[2], Flemming Kjær-Kristoffersen[2] and Sven ÅJ Bäck[1]

[1]Department of Medical Radiation Physics, Lund University, Malmö University Hospital, Sweden
[2]Department of Radiation Oncology, Rigshospitalet, Copenhagen University Hospital, Denmark

Sofie.ceberg@med.lu.se

Abstract. The aim of this study was to verify a novel volumetric arc therapy technique, RapidArc™. Polymer gel dosimetry system was used to measure the advanced inhomogeneous 3D dose distribution produced using the technique RapidArc™. A preclinical installation of the novel beam delivery approach was set up on a linear accelerator at Rigshospitalet in Copenhagen. A prostate treatment plan was delivered to a 1.3 l nPAG gel phantom using one single arc rotation from 200 to 160 degrees, and a target dose of 3.3 Gy. Magnetic resonance imaging of the gel was carried out using the 1.5 T scanner and MATLAB was used for image processing and 3D rendering. The difference in relative absorbed dose between the treatment planning system (TPS) and gel measurement was calculated voxel by voxel within the 80% and the 95% isodose volume, respectively. Measurements agreed well with the TPS within the treated volume. Within both isodose volumes 90% of the voxels showed a deviation less or equal to 5%. This study shows that the 3D gel dosimetry system is a useful tool for dose verification of advanced treatment delivery techniques.

1. Introduction

RapidArc™ is a novel radiation therapy technique where the treatment is delivered during one single rotation (up to 360 degrees) of the linear accelerator gantry. This type of volumetric arc therapy (VMAT) is made possible by simultaneously varying the MLC positions, the rotation speed of the gantry, and the dose rate during the treatment [1]. RapidArc™ delivery aims to shorten the treatment time and to produce a better or equal target-volume coverage and sparing of the normal tissue compared with multiple-field IMRT. The first treatment of a cancer patient with this technique was carried out very recently and few scientific articles regarding the use of RapidArc™ has been published [2]. The advanced inhomogeneous 3D dose distribution produced by RapidArc™ is difficult to verify with conventional means, using single dose point measurements or 2D detectors arrays. The 3D gel dosimetry has the advantage to be able to obtain the absorbed dose in the entire irradiated volume, since the gel integrates the absorbed dose and that the response is independent of incident radiation direction [3]. In addition, the high-resolution gel system functions both as a phantom and a detector since the gel composition is near soft tissue equivalent [4]. A preclinical installation of the RapidArc™ beam delivery approach was set up on a linear accelerator (Clinac iX, Varian) at Rigshospitalet in Copenhagen. The purpose of this installation was to perform measurements to verify the correctness of the doses delivered. The polymer gel system was one of several detector systems used in this investigation [5].

2. Material and methods

2.1. Gel preparation

In this study the normoxic polymer gel system nPAG was used. It is based on 3% w/w acrylamide (electrophoresis grade, ≥99%, powder, Sigma Aldrich) and 3% w/w *N,N'*-methylenebisacrylamide (electrophoresis grade, ≥98%, powder, Sigma Aldrich). Gelatine (swine skin, 300 bloom, Sigma Aldrich) was used as the matrix substance and tetrakis(hydroxymethyl)-phosphonium chloride (techn. ~80% in water, Sigma Aldrich) as an oxygen scavenger. The remaining constituent was ultra pure deionized water (resistivity > 18.2 MΩ cm). The gelatine and water were mixed in room temperature and then heated to 45°C until the gelatine was completely dissolved, which took about 1 hour. The monomers were added before cooling the mixture to 35°C. Ten minutes before the gel was poured into bottles and vials the oxygen scavenger was added. The gels were prepared under normal levels of oxygen and the solutions were stirred continuously through the entire mixing procedure. The phantoms were left to set in the dark at room temperature for about 24 hours before irradiation.

2.2. Treatment planning and RapidArc™ dose delivery

As a base for the treatment planning, CT-images of the gel phantom were acquired using a spiral CT scanner (Sensation Open, Siemens Medical Systems). The slice thickness was 3 mm. The treatment planning system (beta version RapidArc™ optimiser, Varian Medical Systems) was used to generate the RapidArc™ 18 MV treatment plan, with an arc rotation from 200 to 160 degrees, and a target dose of 3.3 Gy. The output varied between 200 and 600 MU/min during the 320 degree rotation. A 1.3 l gel phantom was irradiated using a Clinac iX linear accelerator (Varian Medical Systems) (figure 1).

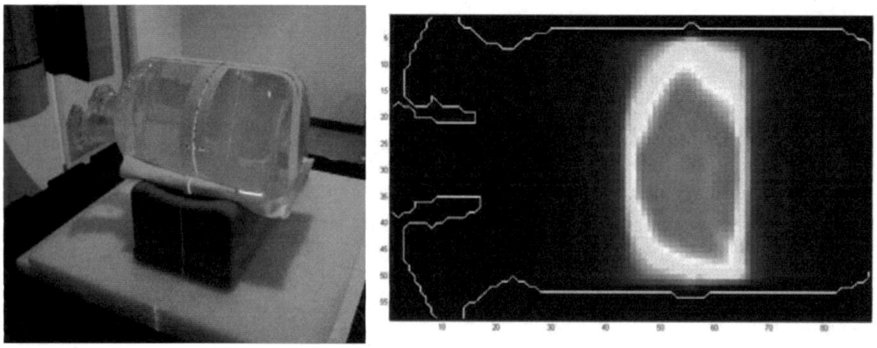

Figure 1. (left) The gel phantom aligned with the set-up lasers at the linear accelerator, and (right) a intensity image of a horizontal slice through the gel phantom with the measured relative absorbed dose (as calculated from the MRI R2 image).

2.3. Gel R2 response linearity validation

The dose response was evaluated by means of R2 measurements as a function of absorbed dose and investigated for absorbed doses between 0 and 8 Gy. To ensure homogenous absorbed dose to the gel vials, they were placed at 3 cm depth in a 30x30x30 cm^3 cubic water phantom and irradiated using a 20 x 20 cm^2 18 MV photon beam. The linear accelerator was set to deliver 600 MU/min. The R2-data of the irradiated 1.3 l gel phantom was converted to relative absorbed dose using background subtraction and normalization in a region of homogenous dose.

2.4. Magnetic resonance imaging and image processing

Approximately 24 h after irradiation, magnetic resonance imaging of the gel was carried out using the 1.5 T MRI unit (Siemens Symphony, Siemens Medical Systems), and the same protocol and set up as described elsewhere [6]. The images were acquired using a 32-echo multi spin echo sequence with an inter-echo spacing of 25 ms. The repetition time was 4000 ms and the voxel size was 1 x 1 x 3 mm^3. In-house developed software was used for T2-calculation [7], and MATLAB 7.4.0 was used for image processing and 3D rendering. The raw data was smoothed with a 3x3x3 box-filter.

3. Results and discussion

The experimental gel dosimetry measurements carried out in this study were used for the verification of the absorbed dose delivery using a novel volumetric arc therapy technique (RapidArc™). The 3D R2-data of the irradiated gel phantom was converted to 3D relative absorbed dose using background subtraction and normalization in a region of homogenous dose close to the isocentre. The relative evaluation was justified since a approximately linear dose response covering the used dose range was observed (figure 2).

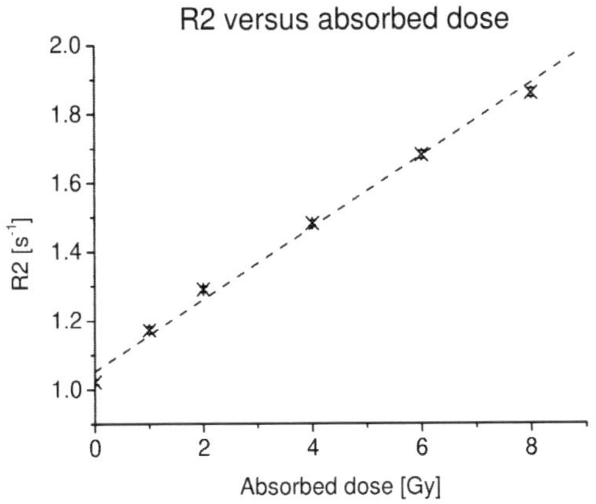

Figure 2. The R2 value as a function of the absorbed dose. The r^2 value of the linear fit equals 0.996.

The treatment plan data and the measured relative absorbed dose volume were aligned by matching the inner surface of the glass bottles in the isocentre planes. The relative dose distributions in both volumes were normalized to a volume of 1 x 10 x 10 voxels in the centre of the isocentre slice. The dose differences were calculated for the volume within the 95% isodose surface corresponding to the treated volume [8]. The volume within the 80% isodose surface was analysed in the same way. The relative dose difference between the TPS and the gel measurement were compared in the isocentre slices. The 80% and 95% isodose curves shows a very good agreement and the difference image shows deviations within ±3% in absorbed dose (figure 3). The white isodoses illustrates the 80% (3 pt line) and 95% (2 pt line) measured relative absorbed dose levels respectively. The black isodoses illustrates the same levels for the absorbed doses as calculated by the TPS. To illustrate the good agreement the 80% isodose surfaces were calculated and projected in 3D view (figure 4). The difference in relative absorbed dose between the TPS and gel measurement was calculated voxel by voxel for the 112 071 voxels within the 80% isodose volume and 76 877 voxels within the 95% isodose volume. The mean difference and the standard uncertainty was (0.9±3.1) % and (1.4±2.8) % for the 80% and 95% isodose volumes, respectively (figure 6). Within both isodose volumes 90% of the voxels showed a deviation less or equal to 5%. Outside the analysed volumes some large deviations were detected, especially close to the glass walls. A factor potentially contributing to the deviations could be the uncertainty in the absorbed dose calculations by the TPS outside the field and close to the surface. Since the dose rate variation increases with the distance from isocentre due to the gantry rotation, an additional explanation to the discrepancies could be the dose rate dependency of the polymer gel system [9]. Measurements with other detector systems are desirable since these deviations in the low dose volume could depend on either or both the TPS and the gel measurements.

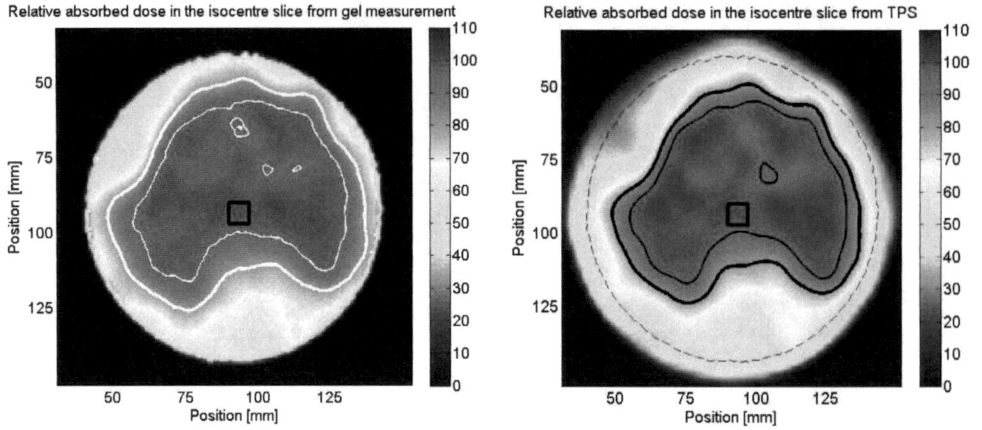

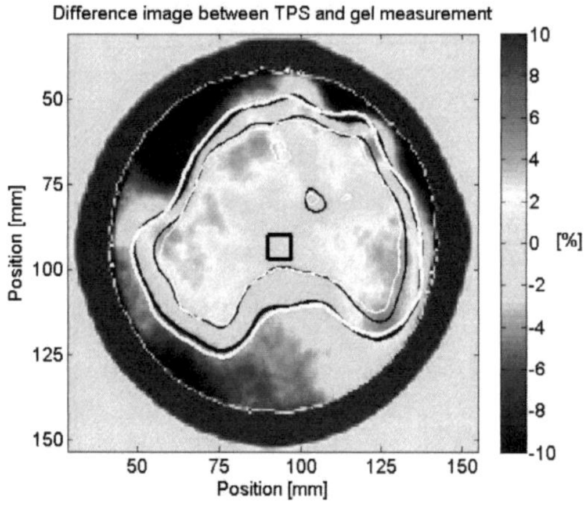

Figure 3. (left) Relative absorbed dose in the isocentre slice from the gel measurement and (upper right) the treatment planning system. The broken line illustrates the inner contour of the glass bottle. (lower right) The difference image between the two upper images given in percentage relative difference. The brown ring corresponds to the 6.7 mm thick glass wall of the gel bottle, which did not give any signal during magnetic resonance imaging. The black square illustrates the normalisation region.

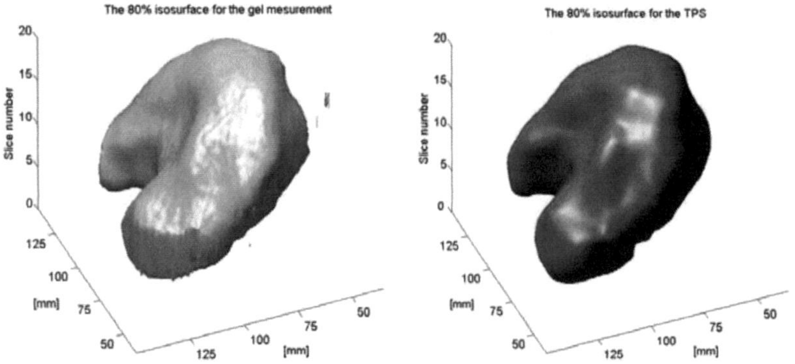

Figure 4. (left) The 80% absorbed dose volume of the gel. (right) The 80% absorbed dose volume according to TPS.

5. Conclusion

A 3D gel dosimetry system was used for a dosimetric verification of a RapidArc™ treatment. Measurements agreed well with the treatment planning system within the treated volume. Within analysed isodose volumes of 80% and 95% respectively, 90% of the voxels showed a deviation less or equal to 5%. Discrepancies between the calculated and measured relative absorbed dose in the low dose volume could be attributed to both calculations and measurements, and a comparison with additional detector systems would be of great interest. This study shows that gel dosimetry is a very useful tool for dose verification of advanced treatment delivery techniques such as RapidArc™.

Figure 5. (left) An overlay of the 80% isosurfaces from the TPS (red) and the gel measurement (green). Difference images are presented for (upper right) one slice above isocentre and (lower right) one slice below.

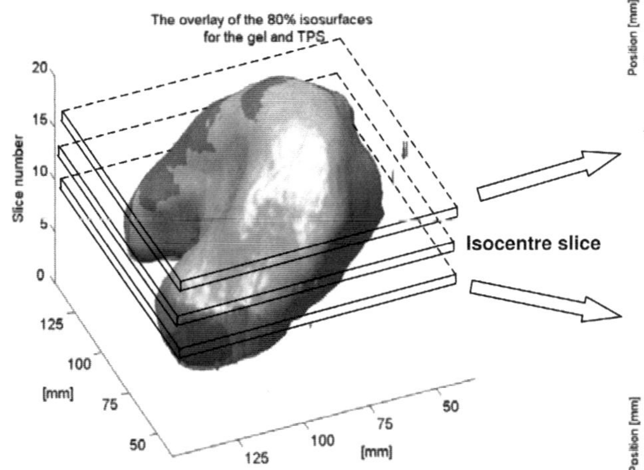

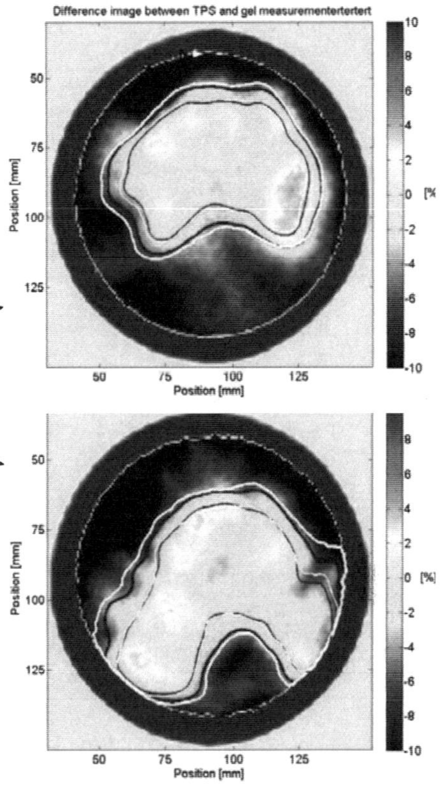

Figure 6.
The distribution of the deviations between the TPS and the gel measurement. (left) The 112 071 voxels corresponds to the 80%-absorbed dose volume. (right) The 76 877 voxels corresponds to the 95%-absorbed dose volume.

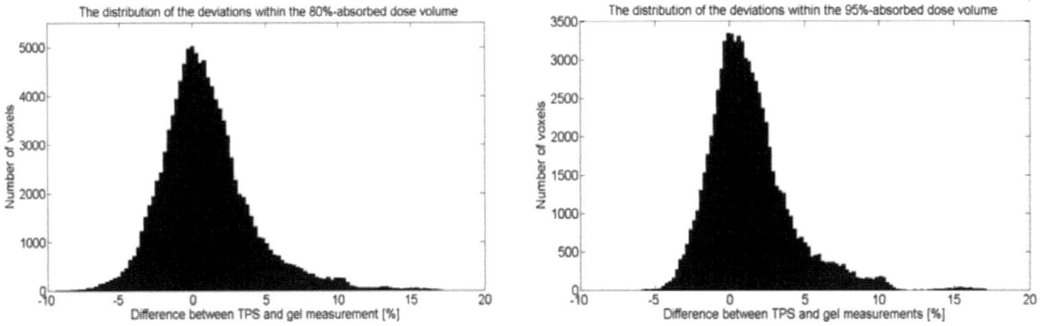

References

[1] Otto K. Volumetric modulated arc therapy: IMRT in a single gantry arc. Med Phys 2008 Jan;35(1):310-7.

[2] Kjær-Kristoffersen F, Ohlhues L, Medin J and Korreman S. RapidArc volumetric modulated therapy planning for prostate cancer patients. Accepted to Acta Oncologica 2008.

[3] Olsson L E, Bäck S Å J, Magnusson P and Haraldsson P 1998 3D dosimetry using gels and MRI In: Imaging in radiation therapy (Madison, WI, USA: Medical Physics Publishing) 475-504.

[4] Keall P and Baldock C 1999 A theoretical study of the radiological properties and water equivalence of fricke and polymer gels used for radiation dosimetry. Australas. Phys. Eng. Sci. Med. 22 85-91.

[5] Korreman S, Medin J and Kjær-Kristoffersen F. Dosimetric verification of RapidArc treatment delivery. Accepted to Acta Oncologica 2008.

[6] Karlsson A, Gustavsson H, Månsson S, McAuley KB, Bäck SA. Dose integration characteristics in normoxic polymer gel dosimetry investigated using sequential beam irradiation. Phys Med Biol. 2007.

[7] Karlsson A, Gustavsson H, Olsson L and Bäck S. A. (2003) Image Processing Software for 3D Dose Evaluation in Gel Dosimetry. World Congress on Medical Physics and Biomedical Engineering.

[8] ICRU Report 62 (1999): Prescribing, Recording and Reporting Photon Beam Therapy.

[9] De Deene Y, Vergote K, Claeys C and De Wagter C. The fundamental radiation properties of normoxic polymer gel dosimeters: a comparison between a methacrylic acid based gel and acrylamide based gels. Phys. Med. Biol. 2006 51 653-673.

Journal of Physics: Conference Series **164** (2009) 012053 doi:10.1088/1742-6596/164/1/012053

Gamma Knife relative dosimetry using VIP polymer gel and EBT radiochromic films.

A Moutsatsos[1], L Petrokokkinos[1,2], K Zourari[1], P Papagiannis[1], P Karaiskos[1,3], K Dardoufas[3], J Damilakis[4], I Seimenis[5] and E Georgiou[1]

[1] Medical Physics Laboratory, Medical School, University of Athens, 75 Mikras Asias, 115 27, Athens, Greece

[2] Nuclear and Particle Physics Section, Physics Department, University of Athens, Panepistimioupolis, Ilisia, 157 71 Athens, Greece

[3] Radiotherapy Department, Hygeia Hospital, Kiffisias Avenue & 4 Erythrou Stavrou, Marousi, 151 23 Athens, Greece

[4] Department of Medical Physics, University of Crete, Medical School, Stavrakia, Iraklion, Crete, Greece

[5] Medical Diagnostic Center Ayios Therissos, 92 Troodous Avenue, Strovolos, Nicosia, Cyprus

Corresponding author: A Moutsatsos (armouts@phys.uoa.gr)

Abstract. The VIP polymer gel–MRI method and EBT Gafchromic films were employed to obtain relative dosimetry results for the Gamma Knife (GK) radiation fields of 4 mm and 18 mm nominal diameter. Results are compared to the corresponding calculations of GammaPlan Treatment Planning System (TPS) in the form of 1D profiles and 2D distributions. Measured and planned relative dosimetry datasets are found in close agreement within experimental uncertainties. A corresponding agreement is shown for Dose Volume Histogram (DVH) results that are available only through the application of the polymer gel method.

1. Introduction

The introduction of 3D conformal radiotherapy techniques - including Gamma Knife (GK) Stereotactic RadioSurgery (SRS) - into clinical practice, has stressed the importance of experimental verification of the complex dose distributions delivered. Currently, film dosimetry is a well-established method utilized to obtain experimental relative dosimetry data for commissioning and Quality Assurance (QA) purposes in such radiation therapy modalities. Polymer gel-MRI dosimetry studies have also been reported in the literature for SRS applications [1-8]. This work compares relative dosimetry results of the 4mm and 18mm GK radiation fields measured using a new normoxic gel dosimeter (VIP) [8] and a radiochromic film of improved dosimetric characteristics (EBT) [9], and exploits the inherent 3D dose integrating character of the polymer gel method to verify the planned Dose Volume Histogram (DVH) for each delivered dose distribution.

2. Materials and methods

The VIP gel dosimeter has already been successfully employed in SRS dosimetry [8]. Its composition (w/v: 8% NVP, 4% MBA, 7.5% gelatin, 0.0008% $CuSO_4$ and 0.007% ascorbic acid) comprises the result of an effort to maintain the wide dose response range of its "anoxic" predecessor (VIPAR [10]) while decreasing the lower limit of dose detection and simplifying its manufacturing procedure. Following preparation the gel solution was poured into two cylindrical glass vials of 95 mm height and 50 mm outer diameter that were hermitically sealed and stored overnight at room temperature to solidify. The first vial was employed in the GK irradiation while the second – to which a flexible, closed end catheter of 1.5mm external diameter was introduced through an appropriate hole drilled into its cap – was used for irradiation with a microSelectron ^{192}Ir HDR remote afterloading unit for calibration purposes [10].

The phantom and experimental procedure followed for the GK irradiation is described in detail in a previous work [6] and ensures that the vial's longitudinal axis is aligned to the z-axis of the unit. The GammaPlan Treatment Planning System (TPS) with the output factor values proposed by the vendor was utilized to plan and deliver the same dose of 30 Gy at two points along the z axis spaced 25 mm apart, employing the 4mm and 18mm helmet collimator respectively (see Fig. 1(a)). The calibration irradiation was performed on the same day by programming the mHDR-v2 ^{192}Ir source to deliver 10 Gy at 1 cm distance along the transverse source bisector in a single dwell position.

The readout of the two irradiated gel vials was performed 3 days post irradiation via Magnetic Resonance Imaging (MRI) on a 1.5 Tesla Philips Gyroscan NT MR scanner (Philips Medical Systems, Nederland BV), employing a volume selective, Carr-Purcell-Meiboom-Gill (CPMG), 18-echo pulse sequence (with an initial echo time (TE) of 40 ms, further 40 ms increments, and a repetition time (TR) of 2300 ms). A built-in quadrature radio-frequency (RF) body coil and a RF head coil were used for proton excitation and signal detection, respectively. The gel vials were placed at the center of the receiver coil in order to minimize RF field inhomogeneity effects. The acquisition and reconstruction inplane resolution was (0.54 x 0.54) mm^2, while the slice thickness was set to 0.75 mm resulting to a reconstructed voxel size of (0.54 x 0.54 x 0.75) mm^3. This voxel size compromises the contradicting demands for submillimeter spatial dose-measurements resolution [11] and acceptable statistical noise.

After discarding the first echo image due to imperfections in the signal decay curve, a T_2 map was calculated for each reconstructed axial (xy plane) slice using a mono-exponential fitting routine of pixel signal intensity versus echo time on a pixel by pixel basis. A 3D T_2 matrix was then constructed for each vial by the acquired T_2 maps and converted to the corresponding R_2 ($=1/T_2$) relaxation rate matrix that allows for the reconstruction of the scanned volume in any chosen oblique plane. Dose response data of the VIP polymer gel formulation were derived from the brachytherapy vial [10] and a linear fit was performed in the dose region of 2.5-33 Gy ($R_2 = (1.608\pm0.004)s^{-1}+(0.0820\pm0.0004)s^{-1}Gy^{-1} \times D$). For the GK vial, custom-written routines were utilized to determine the coordinates of the irradiation center of each delivered dose distribution, by exploiting the 3D symmetry of the GK radiation fields in space with respect to this point.

EBT radiochromic films have been introduced for radiotherapy dosimetry exhibiting high sensitivity and fine spatial resolution, without problems associated with energy dependence of response and water non-equivalence [9]. Two film sheets from the same batch were used; one for calibration and one for GK irradiation purposes. For the GK irradiation the same phantom and experimental procedure as with the VIP gel irradiation were employed. The gap for the accommodation of the gel vial was filled with two custom-made Plexiglas inserts and a (95 x 50) mm^2 film piece was aligned to the central coronal plane (xz plane) of the GK unit. Two single shots of 5 Gy maximum dose, spaced 35 mm apart were planned and delivered employing the 4 mm and the 18 mm helmet collimator respectively (see Fig. 1(b)). For calibration purposes, twenty two, (2.5 x 2.5) cm^2 precut EBT film pieces were irradiated using a Siemens Oncor 6 MV LINAC to deliver doses ranging from 10 to 2000 cGy. The films were placed perpendicular to the central axis of a (10 x 10) cm^2 collimated photon beam at 10 cm depth (SSD = 100 cm) in a solid water phantom of appropriate size to ensure full scatter conditions.

The scanning of all EBT films was performed 24 hours after irradiation to preclude postirradiation coloration effects, using a Microtek ScanWizard Pro (ScanMaker 9800XL) flatbed optical scanner in transmission mode. All films were placed in the same predetermined area of the scanner bed, maintaining the same orientation. 48-bit, RGB mode images were acquired with a resolution of 150 dpi (pixel size = 0.169 mm), and only the red color channel of the image was utilized and further processed. The mean pixel value (PV) of each calibration film piece was assigned to the corresponding delivered dose and a fifth degree polynomial fit was applied.

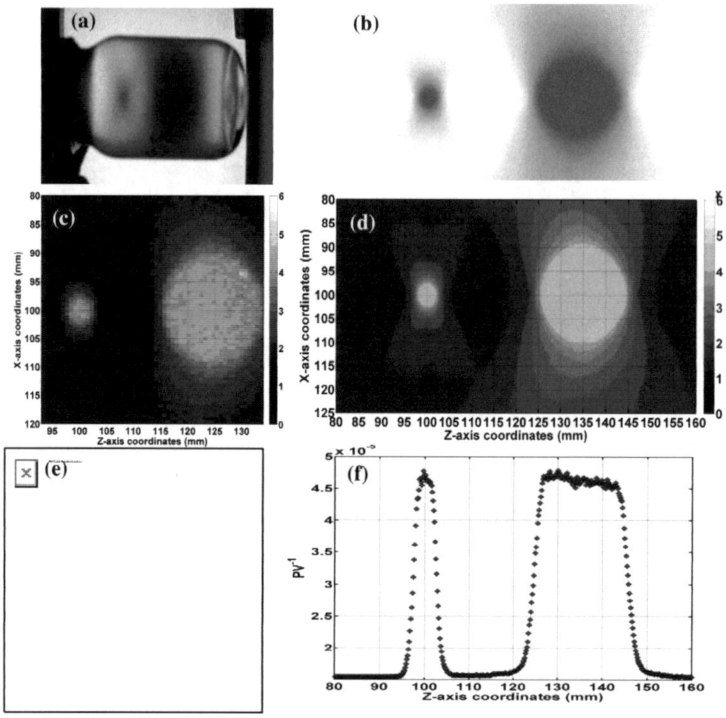

Figure 1. (a),(b) Photographs of the irradiated VIP gel and EBT film, respectively. (c),(d) The central coronal R_2 map of the gel and the PV^{-1} map of the film, respectively. (e),(f) The corresponding R_2 and PV^{-1} profiles along the z-axis

For the GK irradiated film, the aforementioned custom-written routines were used (in 2D) to determine the coordinates of the irradiation center of each delivered radiation field. Contrary to the 3D gel data, relative dosimetry results derived from the 2D film data are sensitive to potential misalignments of the film plane and the central coronal (xz) plane of the unit during irradiation (i.e. rotations about the x or z axes). The accuracy of the experimental procedure employed [6] however, ensures that no significant misalignment (i.e. capable of affecting relative dosimetry measurements) took place (see also Fig. 1(d) and 1(f)).

In order to obtain experimental, relative dosimetry results comparable to corresponding TPS calculations, a normalization of the measured data to the dose level at the irradiation center of each GK field is required. In an effort to suppress individual point uncertainties, this dose level was calculated by sampling a spherical volume (for gel data) or a circular surface (for film data) centered on the reference point using linear interpolation and estimating the mean dose value. The density of the interpolated points was similar to the corresponding grid of measurements, while the size of the sampled volume (or surface) was determined considering TPS calculations for the corresponding dose distribution in order to preclude averaging effects.

3. Results and Discussion

The planned distance between the two GK fields delivered to each dosimeter was verified within 0.3 mm (24.75±0.02 mm for the gel and 34.88±0.13 mm for the film; see also Fig. 1(c)-1(e)). In view of the GK unit spatial dose delivery accuracy (0.5 mm) this is an excellent agreement that negates any considerable misalignment of the film plane and the central coronal plane of the planned dose distributions during film irradiation.

Figure 2 presents the measured and TPS calculated relative dose profiles along the z-axis. The spatial coordinates of the irradiation center of each field were assigned to (x,y,z) = (100,100,100) following the common GK unit convention. In this figure, gel results under 2.5 Gy (i.e. ~ 8.5%) are not presented since this dose region lies below the method's lower limit of detection, while gel measurements beyond z = 109 mm are not available for the 18mm field since the vial's length was not enough to accommodate the full delivered distribution (see figures 1(a) and 1(c)). Considering uncertainties attributed to single voxel (or pixel) results (i.e R_2 or PV), an overall good matching of the three data sets is observed. It should also be noted that the TPS profiles are normalized to the dose level at the unit centre point (UCP), while experimental results are normalized to the dose level at the irradiation center of each field. According to the unit specifications, this should not deviate from the mechanical UCP more than 0.5 mm which can therefore be assumed as the maximum systematic uncertainty in the comparison between experimental results and TPS calculations.

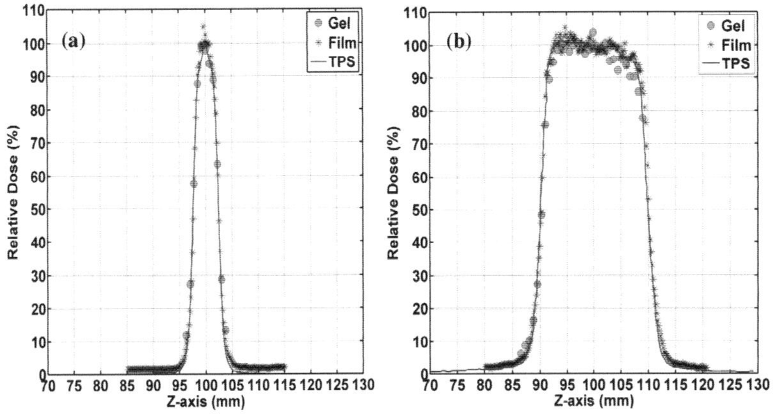

Figure 2. Experimental and TPS calculated relative dose profiles along the z-axis for the 4mm (a) and the 18mm (b) GK fields.

Figures 3(a)-(b) present the comparison of measured and planned 2D relative dose distributions on the central coronal plane. In these figures, the TPS calculated and the corresponding experimental relative iso-dose lines are superimposed on the gel's central coronal relative dose map. The observed deviation of measured and calculated data is attributed to the combined effect of experimental uncertainties - owing mainly to the uncertainty in the definition of the dose normalization level – and the total (geometrical and dose delivery) tolerances of the GK unit. Overall, the calculated and gel measured 2D relative dose sets present no systematic offsets and a distance-to-agreement better than 1 MR imaging pixel (0.5 mm). The same applies for film results, with the exception of a worse distance-

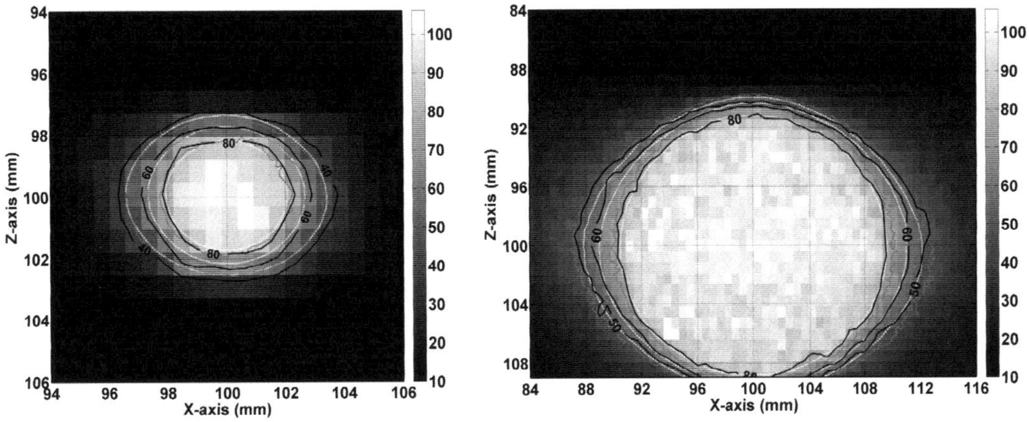

Figure 3. A comparison of measured (VIP gel: black solid lines, EBT film: red dashed lines) and TPS calculated (yellow dashed lines) relative iso-dose lines for the 4 mm (a) and 18 mm (b) helmet collimator fields, presented on the central coronal plane and superimposed on the gel's corresponding relative dose map.

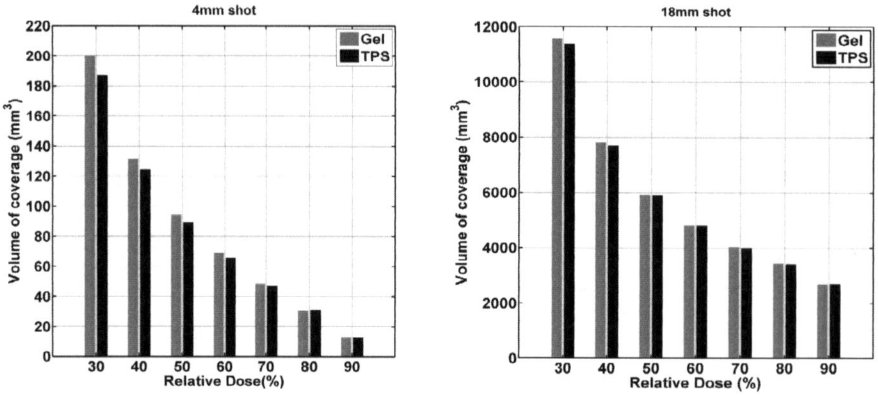

Figure 4. Comparison of the VIP gel measured and TPS calculated Dose Volume Histograms for the 4mm and 18mm GK fields .

to-agreement (< 1 mm) observed for the 50% and the 60% relative dose levels at the 18mm field.

Figure 4 compares DVH values calculated by gel relative dosimetry results (sum of voxels to which relative dose values above a certain level are recorded) and corresponding TPS calculations for the GK fields involved in this work. For the 18mm field, experimental results and TPS calculations are in close agreement (better than 1.5%). Gel measurements for the 4mm field however, seem to systematically overestimate DVH values corresponding up to the 70% relative dose level. This is partially attributed to the uncertainty in the definition of the dose normalization level, which has a more pronounced effect on the 4mm helmet dose distribution due to its confined spatial extent. Small changes of this dose level, providing an acceptable distance-to-agreement between the measured and planned dose distribution, yield significant changes to the corresponding DVH results for the 4mm GK field (e.g. a 2% change of the dose normalization level results to a change of measured DVH values up to 5%). Moreover, the observed deviations between experimental and TPS calculated results (with the maximum being 6.5% for the 30% relative dose level) are practically comparable to the system tolerances considering the 0.5 mm spatial dose delivery accuracy that the vendor claims.

4. Conclusions

VIP polymer gel-MRI and EBT radiochromic film are of equal potential for 1D and 2D relative dosimetry in GK SRS applications. However, only polymer gel dosimetry can offer experimental dose verification in 3D, a feature that could justify the added effort and skill required by this method.

Acknowledgment

This work was partially supported by the Greek Ministry of Health (Central Council of Health).

References

[1] Ertl A, Berg A, Zehetmayer M and Frigo P. High-resolution dose profile studies based on MR imaging with polymer BANG^{TM} gels in stereotactic radiation techniques. Magn. Reson. Imaging **18** 343–9 (2000)

[2] Mack A, Scheib S G, Major J, Gianolini S, Pazmandi G, Feist H, Czempiel H and Kreiner H-J. Precision dosimetry for narrow photon beams used in radiosurgery-determination of Gamma Knife output factors. Med. Phys. **29** 2080–9 (2002)

[3] Scheib S G, Schenkel Y and Gianolini S. Absolute dose verifications in small photon fields using BANG^{TM} gel. 3rd Int. Conf. on Radiotherapy Gel Dosimetry J. Phys.: Conf. Series 3 228–31 (2004)

[4] Novotny J,Dvorak P, Spevacek V, Tintera J,Novotny J, Cechak T and Liscak R. Quality control of the stereotactic radiosurgery procedure with the polymer-gel dosimetry. Radiother. Oncol. **63** 223–30 (2002)

[5] Scheib S G and Gianolini. Three-dimensional dose verification using BANG gel: a clinical example. J. Neurosurg. **97** 582–7 S (2002)

[6] Karaiskos P, Petrokokkinos L, Tatsis E, Angelopoulos A, Baras P, Kozicki M, Papagiannis P, Rosiak J, Sakelliou L, Sandilos P, Vlachos L. Dose verification of single shot gamma knife applications using VIPAR polymer gel and MRI. Phys Med Biol 2005; **50**:1235-50

[7] Papagiannis P, Karaiskos P, Kozicki M, Rosiak J, Sakelliou L, Sandilos P, Seimenis I, Torrens M. Three-dimensional dose verification of the clinical application of gamma knife stereotactic radiosurgery using polymer gel and MRI. Phys Med Biol 2005; **50**:1979-90

[8] E. Pantelis, C. Antypas, L Petrokokkinos, P. Karaiskos, P. Papagiannis, M. kozicki, E. Georgiou, L. Sakelliou and I. Seimenis. Dosimetric characterization of CyberKnife radiosurgical photon beams using polymer gels. Med. Phys. **35**, 2312-2320 (2008)

[9] M. Todorovic, M. Fischer, F. Cremers, E. Thom, and R. Schmidt. Evaluation of GafChromic EBT prototype B for external beam dose verification. Med. Phys. **33**, 1321–1328 (2006)

[10] Kipouros P, Pappas E, Baras P, Hatzipanayoti D, Karaiskos P, Sakelliou L, Sandilos P, Seimenis I. Wide dynamic dose range of VIPAR polymer gel dosimetry. Phys. Med. Biol 2001; **46**:2143-59

[11] A. Berg, M. Pernkopf, C. Waldhäusl, W. Schmidt, and E. Moser. High resolution MR based polymer dosimetry versus film densitometry: A systematic study based on the modulation transfer function approach. Phys. Med. Biol. 2004; **49**:4087–4108

Stereotactic radiosurgery photon field profile dosimetry using conventional dosimeters and polymer gel dosimetry. Analysis and inter-comparison.

E. Pappas[1,3], T. G. Maris[1], S. Manolopoulos[2], F. Zacharopoulou[1], A. Papadakis[1], S. Green[2] and C. Wojnecki

[1] Department of Medical Physics, Faculty of Medicine, University of Crete, 711 10 Stavrakia-Heraklion, Crete, Greece

[2] Queen Elizabeth Medical Centre, University Hospital Birmingham NHS Trust, Birmingham B15 2TH, UK

[3] Medical Physics Department, Anticancer-Oncologic Hospital of Athens 'Saint Savvas', 171 Alexandras Ave., 115 22 Athens, Greece

Corresponding author's e-mail address: epappas@edu.med.uoc.gr

Abstract. Small photon fields are increasingly used in modern radiotherapy and especially in IMRT and SRS/SRT treatments. Accurate beam profile measurements of such beams are crucial for a precise and effective treatment. In this work four different dosimetric methods have been used for profile measurements of three small 6 MV circular fields having diameters of 7.5, 15.0 and 30.0 mm. A small sensitive volume air ion chamber, a diamond detector, a novel silicon-diode array and Vinyl-Pyrrolidone based polymer gel dosimetry. The results of this work reveal the well-known disadvantages and/or problems of the conventional dosimeters for this kind of measurements and support that polymer gel dosimetry may overcome these problems. Conclusively, it is estimated that polymer gels could play an important role towards the minimization of the total SRS/SRT treatment error that is related with small field profile measurements.

1. Introduction

Narrow high-energy photon beams are increasingly used in modern radiotherapy (RT) and especially in Intensity Modulated RT (IMRT - beamlets) and Stereotactic Radiosurgery – RT (SRS/SRT) applications. Accurate measurements of the dose characteristics of such beams (percentage depth doses, beam profiles, output factors) are required as input to the Treatment Planning Systems in order to apply these modern RT techniques with high precision and in a trusted manner. Uncertainties related especially to profile measurements, result to certain clinical impact in 3D-CRT and in SRS/SRT.[1,2] Broadening of the measured penumbra, usually an effect related to the size of the detector (volume averaging), result in over-irradiation of the adjacent Organs at Risk (OAR) since greater margins are needed to fully encompass the PTV.[1] However, small photon beam measurements are problematic because of the high dose gradients that are present and the non-existence of lateral electronic equilibrium. Moreover, the dosimeters used for such measurements should ideally exhibit certain characteristics such as dose-rate, directional and energy independence, tissue equivalency, high spatial resolution measurements, small sensitive volume, and without beam perturbation and positioning problems.

The aim of this work is to analyze and investigate the agreement between profile measurements of 6 MV circular SRS beams, provided by the BrainLab collimator system, performed using a PTW PinPoint ion chamber, a PTW diamond detector, a novel silicon diode array and polymer gel dosimetry. The whole process results to a better 'knowledge' of the true profiles.

2. Materials and methods

2.1 Source and the SRT system

All irradiations were performed using a Varian C600 Clinac 6 MV x-ray beam. The linear accelerator is adapted for stereotactic treatments via an additional tertiary collimator system by BrainLAB. The latter uses brass shelled lead collimators with a conical aperture of various sizes ranging from 7.5 to 35 mm.

2.2 Detectors

A linear diode array (DOSI) was fabricated on a single piece of high resistivity (1-10 kΩ cm) 300 μm thick n-type silicon by means of ion implantation. The diode pitch is 250 μm providing with sufficient resolving power for sub-millimetre measurements. The sensitive area of the device is 0.25x32 mm^2. For the DOSI measurements RW3 solid water slabs of various thickness were used.

A small waterproof diamond detector (type 60003 by PTW Freiburg) was also used. This is made from natural diamond with a sensitive volume of less than 6 mm^3, area between 3 and 15 mm^2 and thickness smaller than 0.4 mm (nominal value).

In addition, a small ionisation chamber specifically developed for small field dosimetry measurements (PinPoint 31014 by PTW Freiburg) has been used. It is waterproof and fully guarded, with a sensitive volume of 15 mm^3.

2.2.1 Polymer gels

The polymer gels prepared and used in this study were Normoxic – VinylPyrrolidone (VPL) based gels. The gel manufacturing process is described analytically in Papadakis *et al.*[3]

The calibration gel vials were irradiated to doses up to 25 Gy while the experimental gel vial was irradiated at 3 different sites, using 3 different SRS circular collimators: 7.5, 15.0 and 30.0 mm diameter, delivering 25 Gy at the depth of maximum dose (d_{max}) in each separate beam.

The irradiated gels were scanned with a 1.5 T Siemens Vision Plus MR Imager (Siemens medical Systems, Erlangen, Germany) using a slice selective 16-echo Carr-Purcell-Meiboom-Gill pulse sequence with an initial echo time of 40 ms with further 40 ms increments and a repetition time of 5 s. The slice thickness was selected to be 5 mm while the in-plane image resolution was 0.5 mm. This resolution size is directly related to the methods sensitive size (or 'gel detector size'). The value of 0.5 mm is adequate for not introducing volume averaging effects. Nine signal acquisitions were performed resulting in a high SNR (~ 100) T2 map, derived using the imagers' software.

2.2.1.1. Polymer gel data analysis

Central pixel: Within the T2-map (Figure 1), a pixel that corresponds to the beams' central axis (central pixel) is firstly defined for each separate irradiation. This task was achieved by superimposing random iso-T2 lines on the irradiated area of the T2-map, and subsequently by fitting corresponding circles (Figure 2). The calculated centers of these circles were found to lie within, or to exist in the vicinity, of a unique pixel with dimensions 0.5 x 0.5 mm^2, the central pixel.

Profiles: As long as a linear dose-R2 response exists, the background subtracted spin-spin relaxation rate (R2net) profile along a line that intersects the central pixel, equals to the off-axis ratio of the studied irradiation area. The direction of the line can be arbitrary and profiles using a number of arbitrary lines can be averaged in order to reduce the statistical noise and derive a profile that characterizes the whole beam. However, in order to use *all* the available gel data for the measurement of a profile that characterizes a certain small field (and not just a number of random line profiles passing through the central pixel), the following process can be followed: A large number of circles centered at the irradiated area's *central pixel* can be used (three of them are presented in Figure 3 for the 30.0 mm collimator irradiated area). The average R2net value of the pixels that lie in the periphery of each separate circle (normalized to the central pixel R2net value) correspond to the off-axis ratio for a distance from the central axis equal to the circle radius. The normalization point in all cases is the mean value that correspond to the central pixel and its eight surrounding pixels.

3. Results and discussion

Polymer gel calibration test tubes measurement and analysis revealed that a linear dose – R2 response exists (up to 25 Gy). Therefore, it can be stated that for the polymer gel batch produced and used in this study, relative R2net measurements equal relative dose measurements.

In Figure 3 are presented all the profile measurements performed in this study for the 7.5 mm, 15 mm and 30 mm collimator. The gel-data plotted uncertainties correspond to the standard deviation of the measurements used for each separate Off Axis Distance. The normalization point was at the beams' central axis, for each separate beam. Polymer gels resulted in high spatial resolution and high spatial density dose measurements without positioning difficulties or electron transport alteration effects. Each gel data-point correspond to the mean value of all pixels lying on a circle with a radius equal to the Off Axis Distance while the profile measurements performed with the other dosimeters correspond to a random direction that is perpendicular to the central axis (and ideally intersects with it). A general satisfying agreement between datasets for each separate collimator is observed. However, it is easily observed that the PinPoint ion chamber broadens the measured penumbra. The diamond detector and the DOSI provides rather accurate results but they result to a slight pseudosharpening of the profiles possibly due to their high density. Moreover, for the 7.5 mm cone the diamond detector provides a rather underestimated FWHM. This effect is probably related with set-up errors that are very common in this kind of measurements.

In Table I are presented the 80-20 % penumbra widths of all the studied beams.

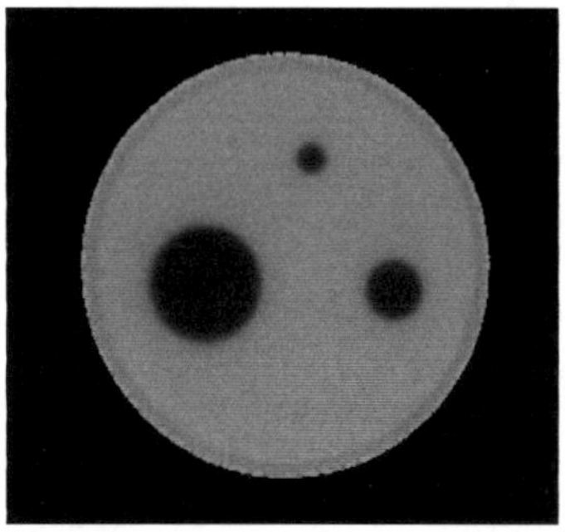

Figure 1: T2-map of the experimental gel-vial irradiated using the 7.5, 15.0 and 30.0 mm collimators

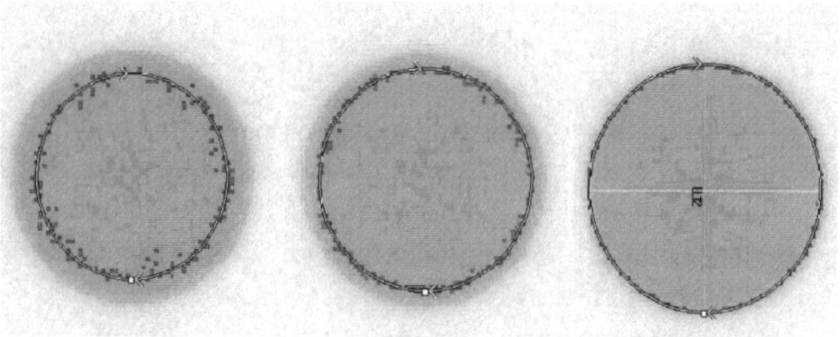

Figure 2: Three separate random iso-T2 pixels superimposed on the T2-map (30.0 mm irradiated area – magnified). The centers of the fitted circles lie within or adjacent to a unique pixel: the central pixel.The mean value of the iso-T2 pixels that lie on a fitted circle correspond to the T2 value at an Off Axis Distance equal to the circles' radius

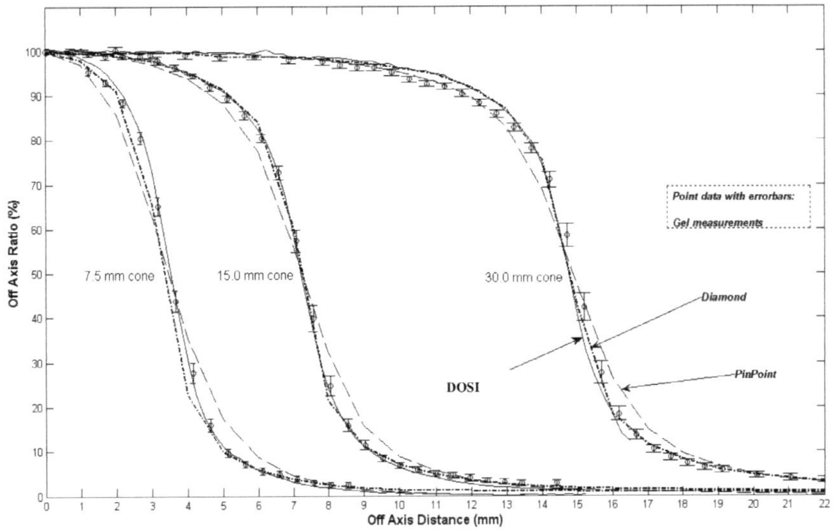

Figure 3: Beam profiles of all studied small fields using all dosimetric techniques used in this study.

TABLE I 80-20% (mm)

Beam Diameter	**Diamond**	**PinPoint**	**DOSI**	**Gel Dosimetry**
7.5 mm	1.5	2.5	1.6	1.7
15.0 mm	1.8	2.9	1.9	2.1
30.0 mm	2.2	3.2	2.2	2.5

The measured 80-20% penumbra widths of all the studied beams

4. Conclusions

The findings of this work support the argument that the detector size and composition together with positioning difficulties are the main problems related to small field file measurements. These problems can be addressed by the use of polymer gel dosimetry. Since small field profile measurement uncertainty contributes directly to the total treatment error, it is believed that polymer gels could have an important role towards the minimization of this error and therefore towards an improved clinical outcome of modern SRS/SRT treatments.

References

[1] F. Garcia-Vicente, M. J. Bejar, L. Perez, and J. J. Torres, "Clinical impact of the detector size effect in 3D-CRT," Radiother. Oncol. **74**, 315-322 (2005).

[2] W. U. Laub and T. Wong, "The volume effect of detectors in the dosimetry of small fields used in IMRT," Med. Phys. **30**, 341-347 (2003).

[3] A. E. Papadakis, T. G. Maris, F. Zacharopoulou, E. Pappas, G. Zacharakis, and J. Damilakis, "An evaluation of the dosimetric performance characteristics of N-vinylpyrrolidone-based polymer gels," Phys. Med. Biol. **52**, 5069-5083 (2007).

Journal of Physics: Conference Series **164** (2009) 012055 doi:10.1088/1742-6596/164/1/012055

Carbon beam dosimetry using VIP polymer gel and MRI

I Kantemiris[1*], L Petrokokkinos[1,2], A Angelopoulos[1], N Bassler[3,4,5], I Seimenis[6], P Karaiskos[2]

[1]Nuclear and Particle Physics Section, Physics Department, University of Athens, Panepistimioupoli, Ilissia, 157 71 Athens, Greece

[2]Medical Physics Laboratory, Medical School, University of Athens, Greece

[3]Department of Medical Physics in Radiation Oncology, German Cancer Research Center (DKFZ), Germany

[4] Heidelberg University Hospital, Heidelberg, Germany

[5]Department of Experimental Clinical Oncology, Aarhus University Hospital, Aarhus, Denmark

[6]Medical Diagnostic Center "Ayios Therissos, Strovolos, Nicosia, Cyprus

*Corresponding author. E-mail: ikantem@phys.uoa.gr

Abstract. VIP polymer gel dosimeter was used for Carbon ion beam dosimetry using a 150 MeV/n beam with 10 Gy plateau dose and a SOBP irradiation scheme with 5 Gy Bragg peak dose. The results show a decrease by 8 mm in the expected from Monte Carlo simulation range in water, suggesting that the dosimeter is non water equivalent. However VIP shows efficiency close to 1 in the plateau region and significantly reduced efficiency in the peak. On the other hand the SOBP results yield an efficiency close to 1 in the SOBP implying that the dose response of the VIP dosimeter may not be solely related to LET.

1. Introduction

To date, at least three ion-beam radiotherapy facilities worldwide are in clinical use. The main advantage that heavy ion beam irradiation presents in comparison to conventional (using photon or electron radiation) radiotherapeutic methods is the refinement of dose deposition with depth due to the Bragg peak, which spares healthy tissue before and after the target, as well as the extremely higher biological effectiveness of ion radiation in comparison to photon radiation.

However, the dosimetry of such modalities presents a scientific and practical challenge, mainly due to the higher LET which changes with energy and depth.

In this work, preliminary results of VIP polymer gel dosimeter's response to Carbon ion irradiation are presented along with corresponding Monte Carlo simulation data in order to investigate the dosimeter's properties in high LET radiation beams.

2. Materials and methods

2.1. Polymer gel

For this study's purposes the VIP polymer gel dosimeter, a normoxic formulation recently introduced by our group [1, 2], was used. It consists of 4% (w/v) *N,N'*-methylenebisacrylamide, 8% (w/v) *N*-Vinylpyrrolidone and 7.5% (w/v) gelatin, while 0.0008% (w/v) Copper Sulfate and 0.007% (w/v) Ascorbic Acid were added to act as oxygen scavengers. Hyper pure water (resistivity > 18 MΩ·cm) was used to prepare the solution in normal atmospheric conditions. The manufacturing process is described in detail in previous works [1, 2, 3]. The VIP's elemental composition and density are presented in table 1.

Table 1: Elemental composition and density of the VIP formulation

Density (1.0313g/cm^3)

Elements	**w/w (%)**
Carbon	11.6534
Hydrogen	10.4521
Oxygen	74.9725
Nitrogen	2.9215
Copper	3.06E-04
Sulphur	1.54E-04

Two 100 ml borosilicate glass bottles were completely filled up with the prepared gel solution and subsequently were tightly sealed using Parafilm and Teflon tape, to ensure non penetration of oxygen. The bottles were left overnight in room temperature for the gel to solidify and subsequently were shipped to the GSI (Darmstadt, Germany) installation for irradiation.

2.2. Irradiation

The GSI Carbon ion radiotherapy unit was used for irradiation purposes. A 150 MeV/n monoenergetic beam was employed for the first bottle's irradiation, using a square field of (2x2)cm^2 size, through the bottle's cap, so as to avoid interactions and secondary particles generation in the borosilicate glass walls. The irradiation of the first bottle (first irradiation) was planned to deliver 10 Gy physical plateau dose in water using the 150 MeV/n monoenergetic beam, with the central beam axis parallel to the bottle's main axis.

The second bottle was irradiated using a multi-energetic spectrum beam, along the bottle's longitudinal axis, with a (1x1) cm^2 square field, planned to deliver 5 Gy in a spread out Bragg peak (SOBP) residing between 3 cm and 6 cm depth in water (second irradiation).

2.3. Magnetic Resonance Imaging

Following irradiation, the bottles were immediately shipped to the MRI facility. They were scanned using a 1.5 T magnet and a quadrature RF receiver coil, five days post irradiation. Both bottles were placed in the receiver coil's centre to ensure minimal RF field inhomogeneities and were scanned using a three dimensional multiple spin–echo CPMG (Carr-Purcell-Meiboom-Gill) imaging pulse sequence (32 equidistant echoes, TE_1=40 ms, TR=2000ms). Coronal slices were reconstructed with a (0.73864 x 0.73864) mm^2 pixel size and a slice thickness of 0.75 mm.

The three dimensional data were fitted on a voxel by voxel basis using a least squares mono-exponential fitting routine of signal intensity versus echo time, for all the acquired echo base images, thus obtaining a three dimensional relaxation time (T_2) matrix, out of which a relaxation rate ($R_2=1/T_2$) matrix was calculated, a quantity known to be proportional to the polymer gel recorded dose for photon beam irradiation.

2.4. Monte Carlo simulation

The FLUKA [4, 5] general purpose Monte Carlo code was employed to simulate the carbon ion beam incident to the VIP gel dosimeter, also serving as a phantom. The monoenergetic beam's field size and energy as well as the VIP's elemental composition and density, presented in Table 1, were used to exactly simulate the experimental setup of the first irradiation. An additional simulation was performed in water, for reasons of comparison, since the planned dose distributions (10 Gy in the plateau region) were performed assuming the gel as water.

For optimum dose calculations, the option HADROTHE was used in the DEFAULT card in order to take advantage of the low particle transport threshold and small fraction of kinetic energy lost in each step. The dose was firstly scored via USRBIN card in a 3-D matrix with a voxel (0.74x0.74x0.75) mm^3 similar to the spatial resolution of the MR scanninng. Due to the small voxel the uncertainty in dose is high (~10%), thus a second detector was also defined via USRBIN. This detector is consisted of cylindrical voxels with 0.5 mm height, along the incident beam's axis and 1 cm radius so as to acquire an acceptable statistical uncertainty (<1%) in a short time frame (<20h).

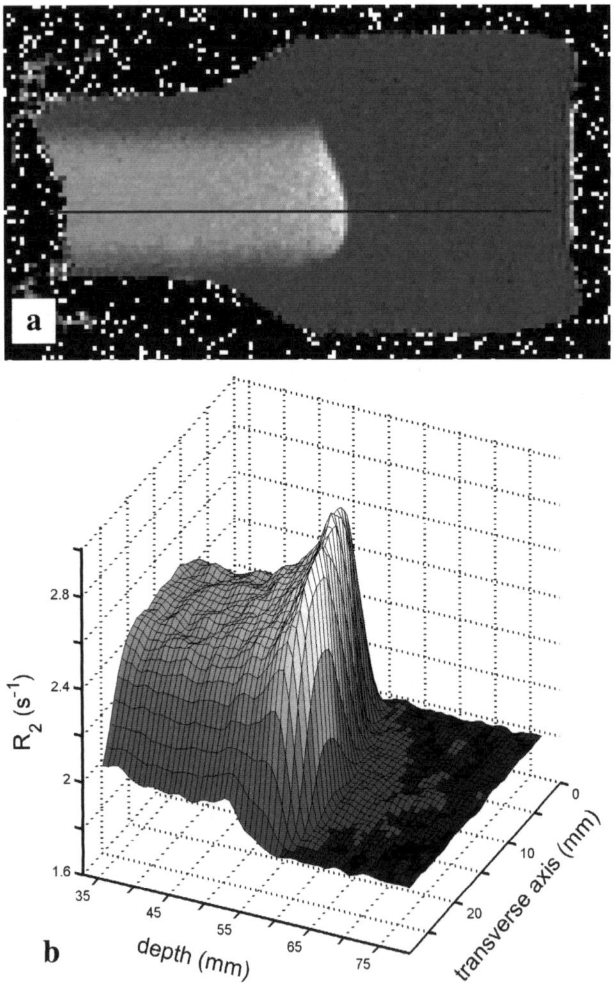

Figure 1: A coronal R_2 map of the gel of the first irradiation, residing in the centre of the Carbon-ions beam on top (a).A surface plot of VIP's response along this plane on bottom (b). The height of the plotted surface in each point represents the corresponding R_2 value. The surface of the irradiated gel is obviously malformed; however the well recorded Bragg peak region follows closely the surface shape.

3. Results and discussion

In Figure 1(a) the central coronal R_2 map of the gel of the first irradiation, which corresponds to the 150 MeV/n monoenergetic beam, is presented. As it can be clearly observed by the R_2 map, gel's surface is malformed. This can be attributed to the fact that the VIP substance was not completely solidified prior to shipment. A small gap appeared below the bottle's cap during solidification and gel shrinkage and since the package could have been left with the side down, the surface of the gel solidified in such a non flat fashion. However this non-flatness is accurately recorded in the R2 map thus revealing the ability of the gel to accurately record the Bragg peak distance of the carbon ion beam.

The general characteristics of measured R_2 distribution on the coronal plane of figure 1(a), can also be observed in figure 1(b). Although the surface of the irradiated gel is malformed the recorded Bragg peak region follows closely the surface shape.

In figure 2(a) an axial R_2 map, in the depth of 37 mm, is presented. The rectangular shape of the incident beam is well depicted. The beam's Full Width at Half Maximum, shown in figure 2(b) and measured across a transverse to the central beam axis profile, was found to be 20 mm, in excellent agreement with corresponding Monte Carlo calculations. Moreover, the shape and size of the recorded field were found to be the same at various depths and similar to those expected and verified by the Monte Carlo simulation.

Figure 3(a) presents the response of the dosimeter with depth along the line drawn on the coronal R_2 map of figure 1(a). The shape of this response is similar to the one expected with the plateau, the Bragg peak found at a depth of 51 mm from the gel surface and the tail region directly after the Bragg peak. However, according to Monte Carlo simulations in water, the Bragg peak was expected to be at a depth of 59 mm. As pointed out in figure 3(b), the observed difference is due to the different energy deposition process with depth between the gel substance and water. Thus carbon ions beam imparts the maximum energy in a lower depth in the VIP volume compared to water. Monte Carlo calculations revealed that the above differences are mainly due to differences in the elemental composition between the VIP gel and water (6 of the 8mm difference in the Bragg peak depth between polymer gel and water are due to differences in composition) and secondly due to differences in density (2 of the 8mm difference in the Bragg peak depth between polymer gel and water are due to differences in density). Thus VIP polymer gel dosimeter, and polymer gels in general, is not water equivalent phantom material for carbon ion beam dosimetry. However, results presented in Figure 3 suggest that the polymer gel method can provide relative accurate dosimetric measurements for the plateau region and the depth of the Bragg peak (taking into account its density and composition) and moreover, can record the expected tail after the Bragg peak (figures 1b, 3a) which is attributed to beam's and target's fragmentation due to nuclear reactions. The plateau value recorded by VIP, as seen in figure 3a is about 2.23 s^{-1}. Taking into account the typical dose–response characteristics of the VIP polymer gel [2] with photon beams, the dosimeter presents a linear response from ~3 Gy up to ~38 Gy. In this dose region the typical linear response function for photons is:

$$R_2 = 1.4(s^{-1})+0.08(Gy\cdot s)^{-1}\cdot D \qquad (1)$$

while the unirradiated gel signal exhibits $R_2(0)=1.6s^{-1}$.

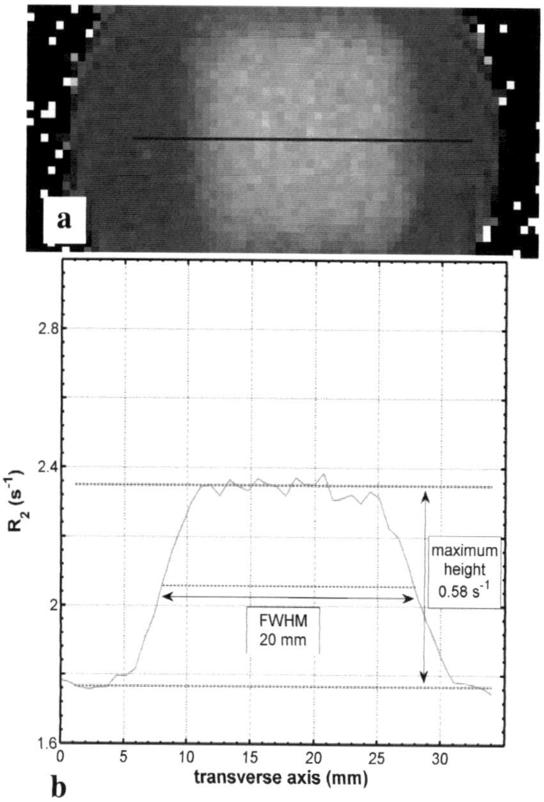

Figure 2: An axial R_2 map, transverse to the beam's central axis at 37 mm depth, measuring from the gel surface, is presented (a). The response profile presented on b is along the black line on the axial R_2 map. The "background" and "maximum" signal are clearly marked so as to figure out the beam's FWHM, which is found to be 20 mm.

According to this response curve for photons, the recorded plateau signal corresponds to a 10.38 Gy dose value, in good agreement with corresponding Monte Carlo calculations (10.83 Gy). This indicates that the VIP gel dosimeter presents a response similar to photons in the plateau region of the Carbon ions beam (relative to photons response efficiency close to 1) where the LET is of the order of 50 keV/μm. Comparing this value with an efficiency around 0.4 reported by Ramm et al for BANG-3 dosimeters and around 0.63 for BANG-1 dosimeters [6], the VIP gel seems to present an improved relative to photons efficiency in the plateau region. However the recorded value in the Bragg peak is about 2.92 s^{-1}, a value which corresponds to a dose of 19 Gy by using the dose-response curve for photons. This yields a relative to photon response efficiency of about 0.29 which is mainly attributed to the high LET (over 150 keV/μm) of the Carbon beam in the Bragg peak. It is also noted that the recorded value at Bragg peak may be significantly affected from volume averaging effects because of the extremely low width of the peak.

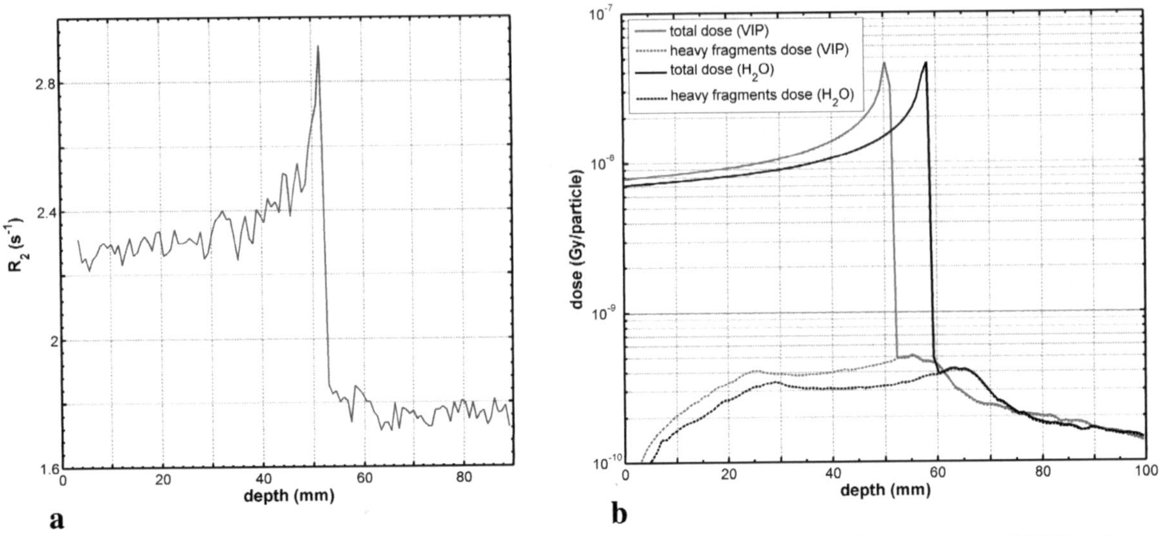

Figure 3: The dose-depth curve as calculated by Monte Carlo simulation, both for water and VIP polymer gel. The figure is presented in logarithmic dose scale so as to show the non-primary heavy particles contribution.

The second bottle irradiation scheme involved doses (~ 5Gy) just over the detection limit of VIP for photons (~3 Gy). However an average signal of 1.87 s^{-1} for the SOBP was recorded between 3 cm and 6 cm depth from gel surface, while the $R_2(0)$ value for the bottle was found 1.76 s^{-1}. According to the photon response curve of equation (1) the SOBP response value corresponds to 5.88 Gy, thus obtaining a relative to photon response efficiency close to 1, within experimental uncertainties. Although this dose recorded in the SOBP is close to the VIP's dose detection limit, the aforementioned finding suggests that the low efficiency of the VIP dosimeter in high LET – high dose regions (Bragg peak of figure 3a) may be also related to the specific dose characteristics of the gel dosimeter (i.e. dose range, dose response and in general the specific monomers used which determine these characteristics).

References

[1] M Kozicki, P Maras, K Rybka, T Biegański, S Kadłubowski and L Petrokokkinos "On the development of the VIPAR polymer gel dosimeter for three-dimensional dose measurements" Macromol. Symp. 2007, 345–352

[2] L Petrokokkinos, M Kozicki, E Pantelis, I Seimenis, P Karaiskos, C Antypas, J Fijuth and L Sakelliou "Characterization of a new polymer gel for radiosurgery dosimetry using Magnetic Resonance Imaging" 4th International Conference on Imaging Technologies in Biomedical Sciences (ITBS 2007)

[3] E Pantelis, C Antypas, L Petrokokkinos, P Karaiskos, P Papagiannis, M Kozicki, E Georgiou, L Sakelliou and I Seimenis "Dosimetric characterization of CyberKnife radiosurgical photon beams using polymer Gels" Med Phys **35** (2008), 2312-20

[4] A Fasso, A Ferrari, J Ranft, and P R Sala, "FLUKA: a multi-particle transport code", CERN-2005-10 (2005), INFN/TC_05/11, SLAC-R-773

[5] A Fasso, A Ferrari, S Roesler, P R Sala, G Battistoni, F Cerutti, E Gadioli, M V Garzelli, F Ballarini, A Ottolenghi, A Empl and J Ranft, "The physics models of FLUKA: status and recent developments", Computing in High Energy and Nuclear Physics 2003 Conference (CHEP2003), La Jolla, CA, USA, March 24-28, 2003, (paper MOMT005), eConf C0303241 (2003), arXiv:hep-ph/0306267

[6] U Ramm, U Weber, M Bock, M Kramer, A Bankamp, M Damrau,C Thilmann, H D Bottcher, L R Schad and G Kraf "Three-dimensional BANG™ gel dosimetry in conformal carbon ion radiotherapy" Med Phys **45** (2000), N95-N102

3-D dosimetric evaluation of 2.5 mm HD120® multileaf system for intensity modulated stereotactic radiosurgery using optical CT based polymer gel dosimetry

Cheng-Shie Wuu, Jack Kessel, and Y. Xu

Department of Radiation Oncology, Columbia University, New York, NY, USA

Email : csw6@columbia.edu

Abstract : A Trilogy TX equipped with a 2.5 mm HD120® multileaf collimator system is available for the treatment of radiosurgery and IMRT. In this study, we evaluated the 3-D dosimetric impact of leaf width on an IMRT radiosurgery plan by comparing the target coverage and the dose gradient around the target, produced from both a 2.5 mm HD120® high-definition MLC system and a 5mm-leaf-width millennium 120 MLC system, using an optical CT based polymer gel dosimetry system. The 2.5 mm MLC improves target conformity and surrounding tissue sparing when compared to that of 5 mm MLC.

1. Introduction

Radiosurgery, either cranial or extracranial, is a procedure that precisely delivers large radiation doses to tumors and other relevant anatomical targets in a single fraction or in a small number of fractions (typically up to five). The goal of a non-invasive stereotactic radiosugery (SRS) is to destroy, or make inactive, the target without damaging surrounding healthy tissue and without involving traditional surgery. The desirable conditions of dose distribution in SRS are: a). The shape of the isosurface defined by the prescription dose conforms to the outline of the target; b). Nonuniform dose distribution throughout the target volume and the location of the hot spot are clinically acceptable; and c). A rapid falloff of dose from the target volume to surrounding healthy tissue following the criteria that are clinically important. Therefore, the success of stereotactic radiosugery requires precise targeting and rapid dose falloff in surrounding normal tissues. The ability to deliver an accurate desired dose to the target requires precise volume targeting. To achieve this, an accurate patient positioning, immobilization, and compensation for organ motion is essential. Recent developments in geometric accuracy of linear accelerator and immobilization devices as well as imaging-guided radiotherapy systems have improved the spatial accuracy when treating patient with stereotactic radiosurgery. Various linear accelerator radiosurgery techniques include multiple non-coplanar arc fields using circular cones, multiple static conformal beams using either blocks or multileaf collimator (MLC), and intensity modulated fields using dynamic MLC. There have been reports on dosimetric comparisons among these radiosurgery techniques [1-3]. In general, via modulation of the beam intensity of each field, the use of IMRT for radiosurgery has the features of good dose conformity and steep dose gradients for organ sparing. In addition, inverse treatment planning for IMRT replaces the traditional forward planning for other radiosurgery techniques, which potentially can reduce the treatment planning time for complex cases [4] While it is still not clear as to which technique is superior for SRS, this study was focused on the dosimetric impact of leaf width, using a Varian Trilogy TX vs. a Varian 21 EX linear accelerator. The Trilogy TX has a 2.5 mm HD120® high-definition multileaf collimator system and the 21EX has a 5 mm millennium multileaf collimator system. We evaluated the 3-D dosimetric impact of leaf width on an IMRT radiosurgery plan by comparing the target coverage and the dose gradient around the target, produced from both a 2.5 mm HD120® high-definition MLC system and a 5mm-leaf-width millennium 120 MLC system, using an optical CT based polymer gel dosimetry system.

A 3-D polymer gel dosimeter, with various techniques for imaging the dose distribution, has been applied to verify the complex 3-D dose distribution [5-9]. Our previous studies have shown that our optical CT based polymer gel dosimetry can be used for dose verification in 3DCRT, IMRT, and SRS. In this study, we investigated the detailed dosimetric impact on the stereotactic radiosurgery using a 2.5 mm HD120 multileaf system for a 2.5 cc small brain lesion.

2. Materials and Methods

In this study, a polymer gel dosimeter and an optical CT scanner have been employed to implement 3-D dose distribution measurements. A plastic cylinder of 15 cm diameter and 14 cm height, filled with BANG®3 polymer gels (MGS Research, Inc., Madison, CT) and modified to optimal dose-response characteristics, was used for dose measurement. A slice thickness of 2.5 mm without spacing was used for CT simulation on both the patient and the gel cylinder. The DICOM CT images were then sent to the treatment planning system through the network. The Varian Eclipse treatment planning system was used to design the IMRT radiosurgery plan for a patient with a 2.5 cc small brain tumor treated with 6 MV photon beams.

To test the impact of leaf-width on the target coverage and the dose fall-off in the immediate vicinity of the target, a 2-mm margin outside the target was outlined to form a volume of avoidance. Dose-volume constraints for the optimization process include a) 100% of the target volume being covered by 100% of the prescription dose, b) The maximum dose to the target is limited to 110%, and c) $\leq$ 30% of the avoidance volume should receive $\geq$60% of the prescription dose. The inverse planning was first performed for the patient with the Trilogy TX HD120®. The dose calculation algorithm in the treatment planning system is based on the AAA model with an inhomogeneity correction. In the process of selecting the best treatment plan for the purpose of dosimetric comparison between 2.5mm and 5mm MLCs, one strategy was to give the highest priority to target coverage and dose distribution conformity while simultaneously minimizing the dose to the volume of avoidance. The other approach was to give a higher priority to the dose constraint for the OAR (avoidance volume, in this case) as long as 100% of the target receives an acceptable minimum dose (e.g. 100% of the target volume receiving $\geq$95% of the prescription dose). For instance, this can happen when the dose to the organ at risk (OAR) is the limiting factor for the treatment plan. For the gel study we selected the latter as the criteria for planning, as the difference in target coverage between two MLC systems can be verified from the gel measurements. Once the "best" plan was identified, the same fluence map file was used to recalculate the dose distribution using the 21EX 5 mm millennium 120 MLC.

The leave sequence files and monitor units for the patient plans from both 2.5 mm HD 120 and 5mm millennium 120 were then transferred to the gel phantom to generate a hybrid phantom plan. The gel irradiation was performed under the same set-up geometry as used in the hybrid phantom plan. The set-up geometry of the gel measurement was verified with source-to-surface distance from irradiation gantry angles.

Two plastic cylinders of 15 cm diameter and 14 cm height, filled with BANG polymer gel, were used to verify the 3-D dose distributions from two MLC systems. One gel was irradiated with the plan generated from Trilogy TX with a 2.5 mm MLC system. One gel was irradiated with the plan generated from 21EX with a 5mm MLC system.

In order to correlate the optical density response of the gel with radiation dose, the same type of cylinder containing the same batch of gel was irradiated with a 16 MeV electron beam, with 6 x 6 cm cone, to a known dose at the depth of dmax. The optical density at a specific depth and the PDD table of the electron beam can be used to obtain the optical density dose response of the gel. The gel sample was scanned with 1 mm pixel resolution using a commercial optical CT scanner, OCTOPUS^{TM} (MGS Research Inc., Madison, CT). Details of the operating principles, 3-D dose reconstruction, and instrument architecture of the optical scanner were described previously [6].

3. Results and Discussion

The gel samples for the hybrid phantom irradiation and the gel sample for the calibration of optical density dose response were from the same batch. The optical density versus dose was found to be linear in the dose range of 2-20 Gy. Figure 1 shows the dose-volume histograms (DVHs) for the target volume and avoidance volume, generated from both Trilogy TX (2.5 mm MLC) and 21EX (5 mm MLC) with the same fluence map files. As shown in Fig.1a, the highest priority during the treatment planning was given to target coverage while trying to minimize the dose to the avoidance volume. While 98.5% of the target volume received ≥100% of the prescription dose from both machines, the avoidance volume consistently received more dose at all dose levels from 21EX plan than from Trilogy. Furthermore, the target dose from 21EX is also less uniform. Fig. 1b shows the plan comparison with highest priority given to the avoidance volume (OAR). When limiting 30% of the avoidance volume to receive ≥60% of the prescription dose, only 87% of the target volume can be covered by the prescription dose from 21EX (5mm MLC). Monk et al. [3] reported that, based on the treatment planning study of 14 clinical cases, the micro-MLC (3 mm) can improve both PTV conformity and surrounding tissue sparing when compared to that of a standard linear accelerator using 5 mm MLC for intracranial stereotactic radiotherapy. However, the comparison presented in Figure 1 and that from Monk et al. [3] are purely based on a treatment planning study. For such a small lesion (2.5 cc) in this study, one may question the clinical importance and the dosimetric accuracy for the above difference observed between 2.5 mm MLC and 5 mm MLC. Putting aside the clinical implication, the gel dosimeter was used to investigate the dosimetry accuracy in the treatment planning system as well as the MLC comparison.

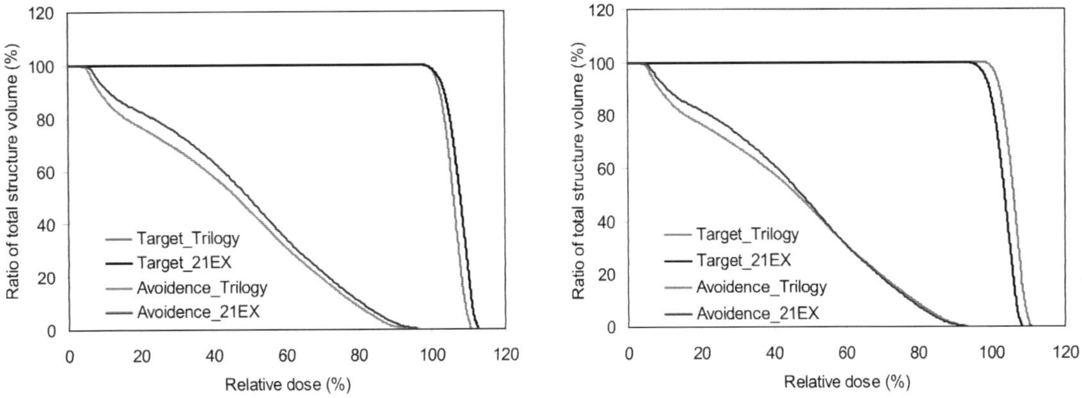

a) b)

Figure1. Comparison of the DVHs from Eclipse treatment planning system for Trilogy and 21EX.

Figure 2 shows dose distribution at the central slice in axial plane between the gel measurements and the treatment planning calculations. Both measured and calculated dose distributions are in reasonable agreement for both machines. However, the isodose lines from the gel measurements show more variation than those from the planning system calculation, and this trend is more significant for the Trilogy TX 2.5 mm MLC. These discrepancies may be partly attributed to the fact that the calculation grid for the planning system is 2.5 mm yet the resolution of gel measurements is 1 mm, as well as the Trilogy TX having a smaller leaf width. As shown in Fig. 1b, the target coverage is lower for the plan generated with the 21EX 5mm MLC.

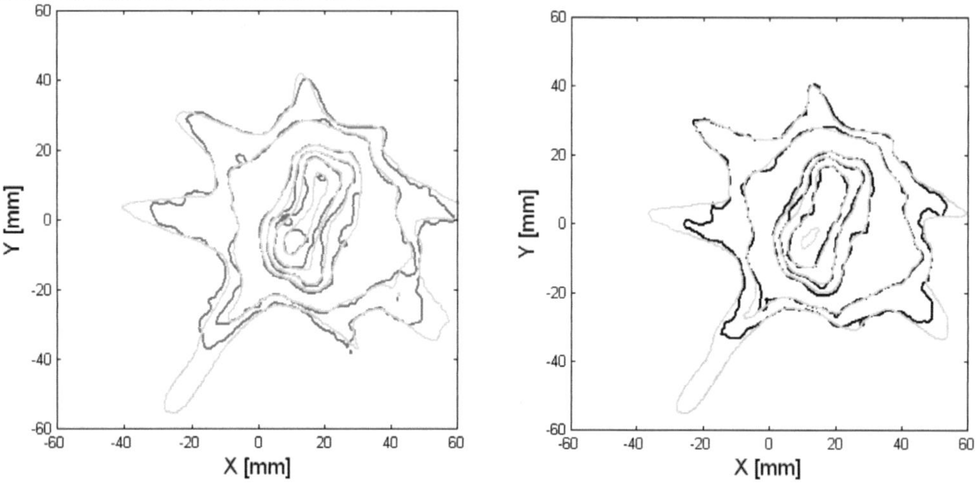

Figure 2. Comparison of the isodose lines (110, 100, 90, 80, 50, and 30 percent) from the Eclipse planning system (green lines) and the gel experiments for the central transverse slice. Trilogy experiment: red lines; 21EX experiment: black lines.

Figure 3 displays the dose distribution from gel measurements in the axial plane at 3 mm inferior from the central slice, where the target is still well covered by the Trilogy TX 2.5 mm MLC; whereas with the 21EX 5mm MLC the target coverage is appreciably lower. It was noted that the target volume in study changes significantly from this slice to the next inferior slice. The 5 mm leaf width from 21EX could not provide the desired dose conformity for such a drastic change in target shape. As a result, both the target volume and the avoidance volume received a higher dose compared to those from the Trilogy TX, as demonstrated in Figure 1b. The other alternative was to renormalize the dose distribution in order to achieve the dose constraint for the avoidance volume but at the same time miss part the target coverage, as shown in Fig. 1a. Another desirable dosimetric feature is the dose falloff outside the target area. The dose distributions at 15 mm inferior to the central slice are shown at Figure 4. The steeper dose falloff from the 2.5 mm MLC than that of the 21EX can be partly attributed to the smaller penumbra for the 2.5 mm HD 120.

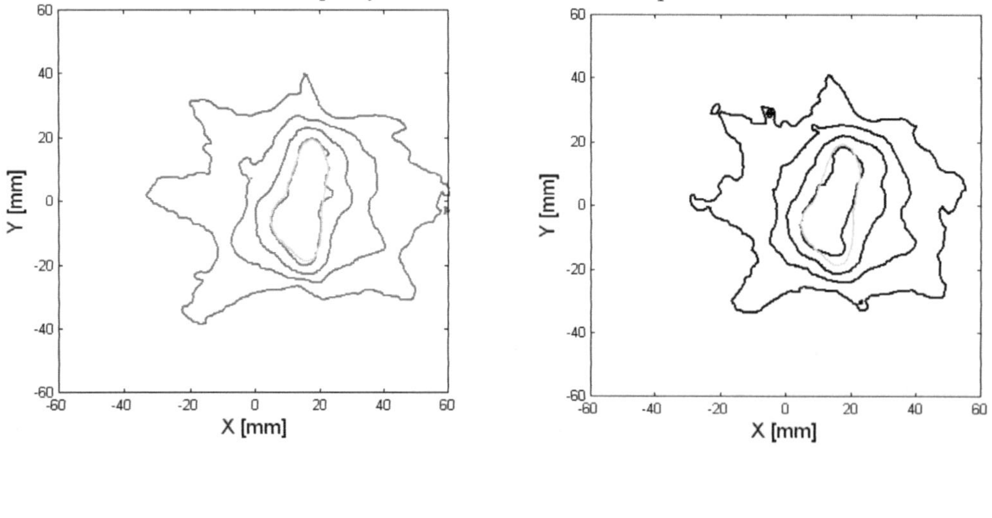

Figure 3. Isodose distribution (100, 80, 60, and 40 percent) from the gel experiments for the transverse slice that is 3 mm inferior to the central slice. Trilogy experiment: red lines; 21EX experiment:black lines. The green line is the PTV on this slice.

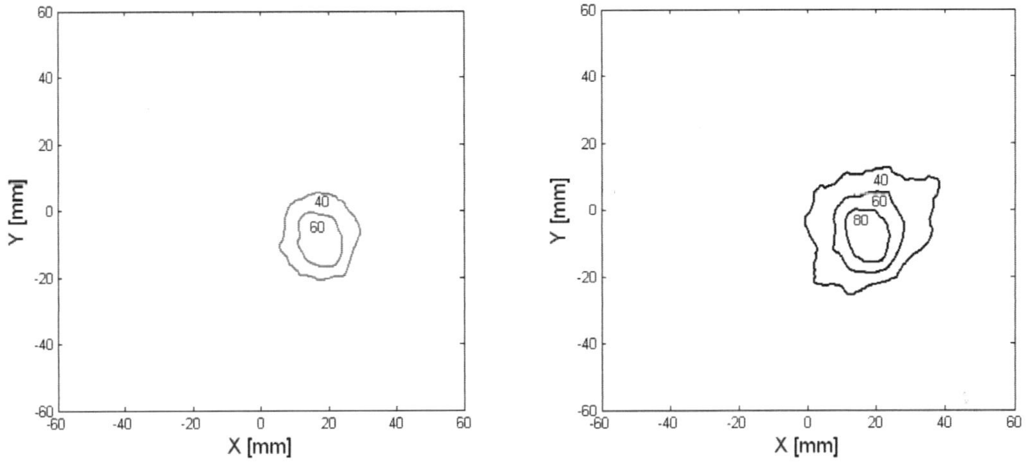

Figure 4. Isodose distribution (80, 60, and 40 percent) from the gel experiments for the transverse slice that is 15 mm inferior to the central slice. Trilogy experiment: red lines; 21EX experiment: black lines.

4. Conclusion

We have evaluated the dosimetric impact of leaf width on an IMRT radiosurgery plan by comparing the target coverage and the dose gradient around the target, produced from both a 2.5 mm HD120® high-definition MLC system on the Trilogy TX and a 5mm-leaf-width millennium 120 MLC system at 21EX. The 3-D dose distributions from the treatment planning were verified using an optical CT based polymer gel dosimetry. The 2.5 mm MLC improves target conformity and surrounding tissue sparing when compared to that of 5 mm MLC.

References

[1] Hamilton RJ, Kuchnir FT, Sweeney P, et al. Comparison of static conformal field with multiple non-coplanar arc techniques for stereotactic radiosurgery or stereotactic radiotherapy. Int J Radiat Oncol Biol Phys 1995;33:1221-1228.

[2] Kubo HD, Pappas CTE, Wilder RB. A comparison of arc-based and static mini-multileaf collimator-based radiosurgery treatment plans. Radiother Oncol 1997;45:89-93.

[3] Monk JE, Perks JR, Doughty D, et al. Comparison of a micro-multileaf collimator with a 5mm-leaf-width collimator for intracranial stereotactic radiotherapy. Int J Radiat Oncol Biol Phys 2003;57:1443-1449.

[4] Dvorak P, Georg D, Bogner J, et al. Impact of IMRT and leaf width on stereotactic body radiotherapy of liver and lung lesions. Int J Radiat Oncol Biol Phys 2005;61:1572-1581.

[5] De Deene Y, De Wagter C, Van Duyse B, et al. Three dimensional dosimetry using polymer gel and magnetic resonance imaging applied to the verification of conformal radiation therapy in head and neck cancer. Radiother Oncol 1998;48:283-291.

[6] Xu Y, Wuu CS, and Maryanski MJ. Determining optimal gel sensitivity in optical CT scanning of gel dosimeters. Med. Phys. 2003:30: 2257-2263.

[7] Xu Y, Wuu CS, and Maryanski MJ. Performance of optical CT scanning of polymer gels as a tool for 3D dose verification. Med. Phys. 2004;31:3024-3033.

[8] McJury M et al. Radition dosimetry using polymer gels: Methods and application. Br. J. Radiol. 2000;73:919-929.
[9] Wuu CS and Xu Y. Three-dimensional dose verification for intensity modulated radiation therapy using optical CT based polymer gel dosimetry. Med. Phys. 2006:33:1412-1419.

Journal of Physics: Conference Series **164** (2009) 012057 doi:10.1088/1742-6596/164/1/012057

Phantom for Moving Organ Dosimetry with Gel

Paul Ravindran[1], Anurupa Mahata[2], Ebenezer Suman Babu[1]

[1]Department of Radiation Oncology, Christian Medical College, Vellore India

[2]Department of Radiation Oncology, Indo-American Cancer Hospital and Research Institute, Hyderabad, India

paul@cmcvellore.ac.in

Abstract. The displacements caused by the cardiac and respiratory motions cause smearing of the dose distribution that defeats the purpose of high precision radiotherapy. A phontom that holds a gel cylinder and radiochromic film, was designed and developed to simulate the respiratory motion in the superior and inferior directions. The effect of lung movement on dose distribution was studied by exposing gel as well as a radiochromic film using the phantom. The results obtained with Gel was comparable to those obtained with the radiochromic films.

1. Introduction

Precision in delivery of radiotherapy has been successfully addressed due to the development of radiation delivery techniques such as 3D-CRT, IMRT and IGRT. However organ motions such as respiratory and cardiac motions and in bladder, prostate and bowel negate the clinical benefits of these highly conformal radiotherapy techniques. Though techniques have been developed to treat the organs that have predictable movements such as respiratory and cardiac motions, these are not widely practised as they need sophisticated equipment and skilled personnel to perform these treatments. The respiratory and cardiac motions, though predictable, cause the largest displacements in intra- thoracic tumors. The magnitude of displacement varies markedly with patient, position and size of the tumor and respiratory pattern of the patient. These organ motions during treatment results in an increase in the treated volume and the delivered dose distribution does not match the intended dose distribution. A few authors have performed dosimetry of these moving targets with ion chamber, diode arrays and gel dosimetry [1,2]. A few organ motion phantoms have been developed to study the effect of dose distribution due to moving organs [3]. The purpose of this study is to develop an organ motion phantom that could simulate the tumor motion in the lung and investigate the resultant dose distribution due to the longitudinal motion of the lung with Radiochromic Fe Gel and with GafChromic films

2. Material and Methods

2.1. The Thorax Phantom:

A thorax shaped phantom made of acrylic filled with water was designed and developed to study the dosimetry due to organ motion. The phantom has two cylindrical inserts that represent the lungs and one of them is attached to a movement system that can simulate the longitudinal respiratory motion of the lung with varying frequencies and amplitude. The lung cylinder has a frame insert that can hold a GafChromic film and a cylindrical insert for Fe gel.

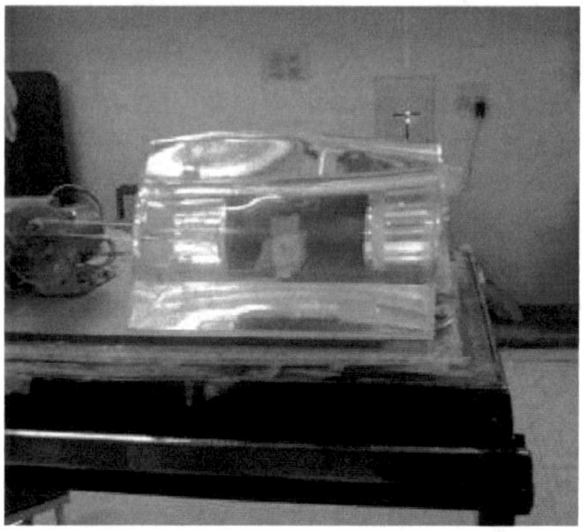

Figure 1. Thorax phantom developed for dosimetry with organ motion

Figure 2. Inserts for dosimetry with GafChromic Film and Gel

The thorax phantom developed is shown in Figure 1 and the inserts for the film and Gel are shown in Figure 2. The inserts are connected to a stepper motor with a free wheel arrangement that enables the longitudinal motion. The respiratory motion in the phantom could be altered to simulate the respiratory movement of the patient. The amplitude of the movement is decided by the angle of rotation of the wheel and the frequency is decided by the timer that passes the pulses for driving the stepper motor. The stepper motor is driven by a PC and an in-house software developed in Visual Basic 6.

2.2. Measurement with GafChromic films:

GafChromic films were mounted on the lung insert and placed in the thorax phantom. The films were exposed to a $4cm^2$ field first without any respiratory movement and then with varying frequencies of the respiratory motion. The GafChromic film was calibrated and read using a flat bed transmission scanner. The films were read four hours after exposure. The dose distribution obtained with no movement of the lung insert was compared with the dose distribution obtained with various respiratory movements.

2.3 Measurement with Fe Gel:

The experiment was repeated with Fe Gel. The Gel was prepared by adding 4% Gelatin to 50 mM H_2SO_4, 0.5 mM Ferrous Ammonium Sulphate and 0.05mM Xylenol Orange. This was prepared in the acrylic cylindrical insert made for the thorax phantom. The Gel was exposed initially with no respiratory movement and then with varying frequencies and amplitudes of respiratory motions.

The exposed gels were read with optical cone beam CT scanner developed in-house. The cylindrical insert with the Fe Gel was scanned before exposure and then 30 minutes after exposure. The projection data were acquired for every degree for 194 degrees and reconstructed using the software developed for the optical cone beam CT scanner.

3. Results and Discussion:

The developed phantom has longitudinal motion and can hold film and cylinder containing the Gel. The frequency and the amplitude of longitudinal motion can be controlled with the computer. The dosimetry performed with both the film and the gel showed significant smearing of the dose distribution with the respiratory movement. No variation was observed in the dose distribution with varying frequencies for both film and the gel. Increasing the amplitude of the movement resulted in further smearing of the dose distribution and also the volume receiving the prescribed dose reduced significantly.

4. Conclusion:

A phantom has been successfully developed to study the effect of dose due organ motion with Gel and film. The dose distributions obtained with Fe Gel was compared with those obtained with the Gafchromic film and found to match reasonably well. This study shows that Gel dosimetry could be used to study the 3D dosimetry of moving organs.

References

[1] X. Allen Li,a_ Christopher Stepaniak, and Elizabeth Gore, "Technical and dosimetric aspects of respiratory gating usinga pressure-sensor motion monitoring system"; Medical Physics, Vol. 33, 145-154

[2] Sofie Månsson, Anna Karlsson, Helen Gustavsson1, Johan Christensson and Sven Å J, "Dosimetric verification of breathing adapted radiotherapy using polymer gel", Journal of Physics: Conference Series 56 (2006) 300–303 4th International Conference on Radiotherapy Gel Dosimetry

[3] Nioutsikou E, Seppenwoolde Y, Symonds-Tayler JR, Heijmen B, Evans P, Webb S. "Dosimetric investigation of lung tumor motion compensation with a robotic respiratory tracking system: an experimental study"; Med Phys. 2008 Apr;35(4):1232-40

Journal of Physics: Conference Series **164** (2009) 012058 doi:10.1088/1742-6596/164/1/012058

Investigating the Feasibility of 3D Dosimetry in the RPC IMRT H&N Phantom

H S Sakhalkar, D Sterling, [*]J Adamovics, [#]G Ibbott and M Oldham

Department of Radiation Oncology Physics, Duke University Medical Center, Durham, NC

*Department of Chemistry and Biology, Rider University, Lawrenceville, NJ

#Department of Radiation Physics, M. D. Anderson Cancer Center, Houston, Tx

mark.oldham@duke.edu

Abstract. An urgent requirement for 3D dosimetry has been recognized because of high failure rate (~25%) in RPC credentialing, which relies on point and 2D dose measurements. Comprehensive 3D dosimetry is likely to resolve more errors and improve IMRT quality assurance. This work presents an investigation of the feasibility of PRESAGE/optical-CT 3D dosimetry in the Radiologic Physics Center (RPC) IMRT H&N phantom. The RPC H&N phantom (with standard and PRESAGE dosimetry inserts alternately) was irradiated with the same IMRT plan. The TLD and EBT film measurement data from standard insert irradiation was provided by RPC. The 3D dose measurement data from PRESAGE insert irradiation was readout using the OCTOPUS™ 5X optical-CT scanner at Duke. TLD, EBT and PRESAGE dose measurements were inter-compared with Eclipse calculations to evaluate consistency of planning and delivery. Results showed that the TLD point dose measurements agreed with Eclipse calculations to within 5% dose-difference. Relative dose comparison between Eclipse dose, EBT dose and PRESAGE dose was conducted using profiles and gamma comparisons (4% dose-difference and 4 mm distance-to-agreement). Profiles showed good agreement between measurement and calculation except along steep dose gradient regions where Eclipse modelling might be inaccurate. Gamma comparisons showed that the measurement and calculation showed good agreement (>96%) if edge artefacts in measurements are ignored. In conclusion, the PRESAGE/optical-CT dosimetry system was found to be feasible as an independent dosimetry tool in the RPC IMRT H&N phantom.

1. Introduction

The Radiological Physics Center (RPC) Head and Neck (H&N) Intensity Modulated Radiation Therapy (IMRT) credentialing test is routinely taken by institutions that wish to participate in the H&N clinical trial protocols[1]. The dosimetry for credentialing comprises point and 2D dose measurements using TLDs and EBT Gafchromic film respectively. A lenient passing criterion of 7% dose-difference and 4 mm distance-to-agreement is used for comparison of measurement with planning calculations. It has been reported that 25% of the institutions fail the credentialing test at first attempt in spite of 2D measurements and generous verification criterial[2]. The standard of IMRT quality assurance could be improved if comprehensive 3D dosimetry is utilized.

The potential of the PRESAGE™/optical-CT system as a practical, 3D, dosimetric quality assurance tool in the clinic has been demonstrated[3]. This system has many favorable characteristics for optical-CT readout including stability in clinical settings, linear radiochromic (light absorbing not light-scattering) response to radiation and non-requirement of external containers[3,4]. In this study, the first implementation of the PRESAGE™/optical-CT 3D dosimetry system for 3D dosimetry in the RPC IMRT H&N phantom is demonstrated.

2. Materials and Methods

A picture of the RPC H&N phantom with the standard insert (henceforth called RPC insert) and a modified insert (for PRESAGE™ 3D dosimeter) is shown in **fig 1**. The components of the RPC insert are: an EBT film in the central axial plane, two EBT films in the sagittal plane, a primary planning target volume (PTV) structure, a secondary PTV structure, an organ at risk (OAR) structure and 8 TLDs dispersed in the three structures. The structures have different densities than the rest of the standard insert material such that they are clearly identified in an x-ray CT scan.

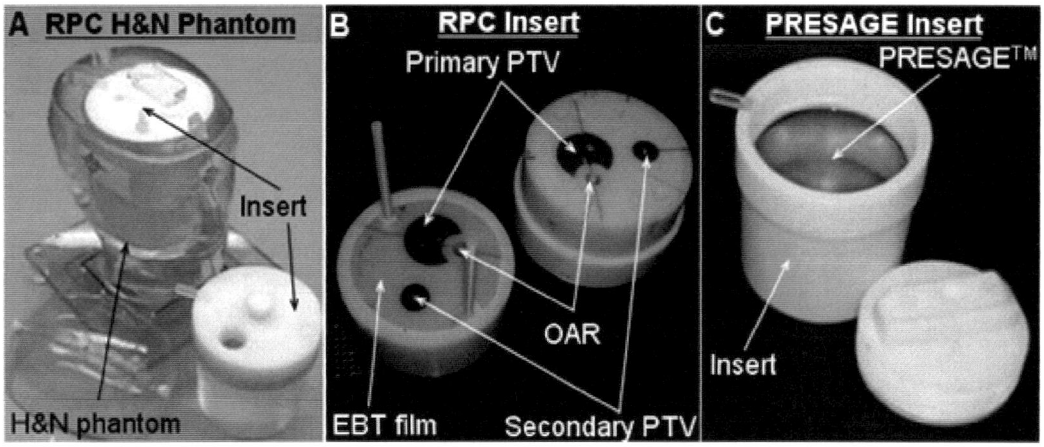

Fig 1. RPC H&N phantom and inserts. A picture of the phantom with insert is shown in **A**. The central cross-section of the standard RPC insert is shown in **B**. The PRESAGE insert is shown in **C**.

Treatment planning and delivery for the H&N phantom (with standard insert as well as PRESAGE insert) followed the same procedure used for an actual patient treatment. After x-ray CT scans, an IMRT treatment plan was designed based on the constraints provided by the RPC. The only difference between the treatment plans for the RPC insert and PRESAGE insert was that the prescription dose was reduced to 4Gy for the PRESAGE insert from 6.6 Gy for the RPC insert. The reduction in prescription dose was required for optimal OD change in the PRESAGE dosimeter and did not change the relative dose distribution (see **fig 2**) and/or relative fluence. MapCheck IMRT QA was employed to check consistency of relative fluence between the RPC insert plan and PRESAGE insert plan. A gamma comparison with a 0.3% dose-difference and 0.3 mm distance-to-agreement criteria showed that the agreement was >98% for all beams. This facilitated direct comparison of relative dose distributions from PRESAGE™ measurement (PRESAGE dose), treatment planning system calculation (Eclipse dose) and EBT film measurement (EBT dose).

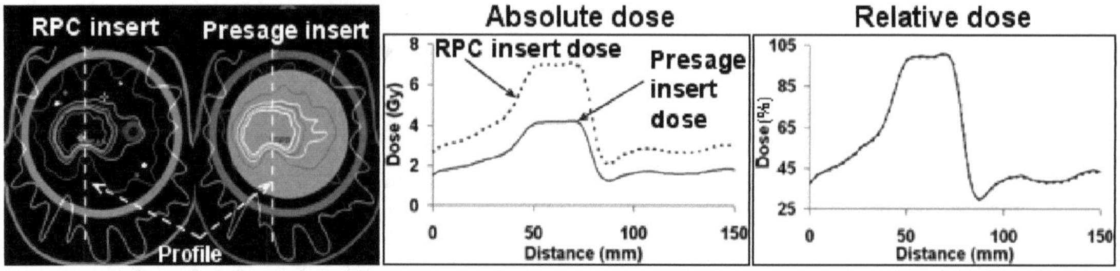

Fig 2. Reduction in prescription dose did not change relative dose distribution in RPC insert and PRESAGE insert.

After irradiation, the RPC standard insert was sent back to the RPC for dose readout from the TLDs and the axial EBT film whereas the PRESAGE dosimeter was scanned at Duke using the OCTOPUS^{TM} 5X optical-CT scanner. The OCTOPUS^{TM} 5X scanner is an upgraded version of the OCTOPUS scanner described in Guo et al[3] with a 5x fast scanning capability. The 3D scan consisted of 20 axial slices (inter-slice spacing of 4 mm) where each slice comprised 600 linear projection scans (image matrix of 168 pixels with 1 mm pixel spacing) acquired at 0.6^0 angular increments. The 3D dose cube was reconstructed with a spatial resolution of 1 mm x 1mm x 4mm using an in house MATLAB code. After dose readout, the absolute point dose measurements form TLDs were compared with Eclipse calculations. The relative PRESAGE dose, Eclipse dose and EBT dose distributions were also compared for 2D and 3D analysis. Dose profiles and gamma maps were used for comparative analysis.

3. Results and Discussion

Successful RPC credentialing requires that the TLD measurements in the primary PTV and secondary PTV are within 7% of Eclipse calculations. Results showed that the TLD measurements comfortably passed the RPC threshold in primary PTV and secondary PTV where dose-difference was within 5% of Eclipse calculations. In the OAR, the pass criterion is that the distance-to-agreement (DTA) should be within 4 mm. TLD results showed that the DTA agreement was within 1mm thereby passing the credentialing test.

Fig 3 shows a central slice from PRESAGE dose measurement. Improvements in optics and acquisition technique resulted in lower noise (reduced to 2%) and reduced edge artefact (< 4 mm). The low noise and significantly reduced edge artefact facilitated highly accurate and precise dose measurement even at low dose levels and to within 4 mm of the edge.

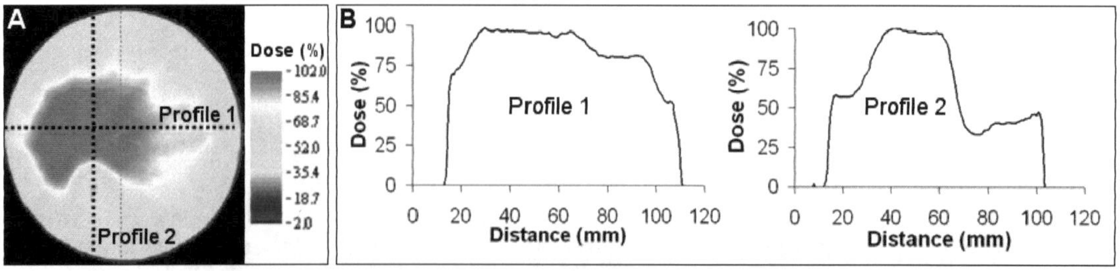

Fig 3. PRESAGE dose measurement. A) Central slice. **B)** Dose profiles along dotted lines in A show reduced noise and edge artefact.

Dose profile and gamma comparison between PRESAGE dose, EBT dose and Eclipse dose in the central axial plane is shown in **fig 4**. Profiles showed good agreement in the high dose region. In general, PRESAGE dose measurement showed better noise characteristics than EBT film dose measurement. Some differences were observed in dose gradient regions and near the edge. The differences in dose gradient regions were because of modeling errors of Eclipse while differences near the edge were because of edge artefacts in PRESAGE and EBT dose measurement. Gamma comparisons showed that all dose comparisons showed full agreement if a 4% dose-difference and 4 mm DTA criteria was used and if the outer 4 mm ring (edge artefact) is ignored. Similar agreement was observed for a full 3D gamma comparison between PRESAGE measurement and Eclipse calculation demonstrating the feasibility of accurate 3D dosimetry in the RPC IMRT H&N phantom.

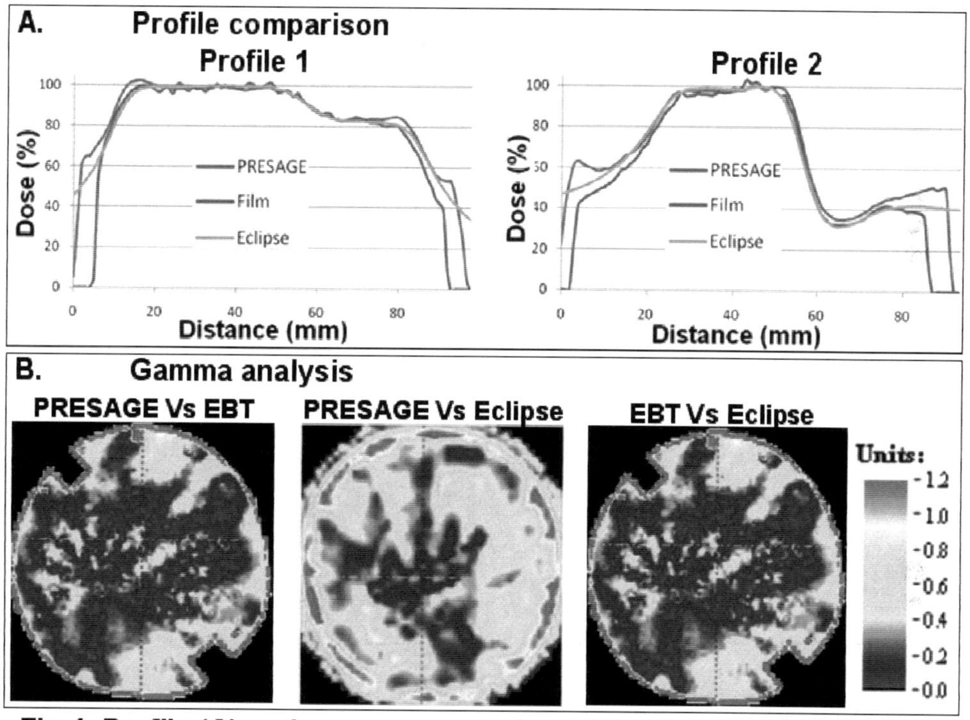

Fig 4. Profile (A) and gamma comparison (B) between PRESAGE dose, EBT dose and Eclipse dose

4. Conclusions

Significant failure rates for participating institutions have been reported in the 2D RPC credentialing tests even though generous verification criteria are used. Comprehensive 3D dosimetry (e.g. PRESAGE/optical-CT dosimetry system) has the potential to improve IMRT quality assurance and resolve more errors. This work is the first implementation of the PRESAGE/optical-CT dosimetry system for 3D dosimetry in the RPC IMRT H&N phantom. Noise level of PRESAGE measurement (2%) was significantly improved and was lesser than that of EBT film measurement used in RPC credentialing. Profile comparisons showed encouraging agreement of PRESAGE measurement with EBT film measurement and Eclipse calculations. Gamma map comparisons showed that if edge artefacts (significantly reduced to < 4mm) were ignored, PRESAGE measurement showed full agreement with EBT film measurement and Eclipse dose calculations to within 4% dose-difference and 4 mm distance-to-agreement constraints. In general, results showed that the PRESAGE/optical-CT system was a feasible tool for 3D dosimetry in the RPC H&N phantom and it has the potential to extend the RPC credentialing process to three dimensions.

References

1. Molineu, A., *et al.* Design and implementation of an anthropomorphic quality assurance phantom for intensity-modulated radiation therapy for the Radiation Therapy Oncology Group. *Int J Radiat Oncol Biol Phys* **63**, 577-583 (2005).
2. Radiological Physics Center Report #129. Report to the AAPM Therapy Physics Committee 2008.
3. Guo, P., Adamovics, J. & Oldham, M. A practical three-dimensional dosimetry system for radiation therapy. *Med Phys* **33**, 3962-3972 (2006).
4. Guo, P.Y., Adamovics, J.A. & Oldham, M. Characterization of a new radiochromic three-dimensional dosimeter. *Med Phys* **33**, 1338-1345 (2006).

Journal of Physics: Conference Series **164** (2009) 012059 doi:10.1088/1742-6596/164/1/012059

An Investigation into the Robustness of Optical-CT Dosimetry of a Radiochromic Dosimeter Compatible with the RPC Head-and-Neck Phantom

H S Sakhalkar, J Adamovics, G Ibbott and M Oldham

Department of Radiation Oncology Physics, Duke University Medical Center, Durham, NC

*Department of Chemistry and Biology, Rider University, Lawrenceville, NJ

#Department of Radiation Physics, M. D. Anderson Cancer Center, Houston, Tx

mark.oldham@duke.edu

Abstract. The potential of the PRESAGE™/Optical-CT system as a comprehensive 3D dosimetry tool has been demonstrated. The current study focused on detailed characterization of robustness (intra-dosimeter uniformity and temporal stability) and reproducibility (inter-dosimeter reproducibility) of PRESAGE™ inserts compatible with the RPC H&N phantom. In addition, the accuracy and precision of PRESAGE dose measurement was also evaluated. Four identical PRESAGE™ dosimeters (10cm diameter and 7cm height cylinders) were irradiated with the same rotationally symmetric treatment plan using a Varian accelerator. The treatment plan was designed to rigorously evaluate robustness and reproducibility for multiple dose levels and in 3D. All dosimeters were scanned by optical-CT at daily intervals to study temporal stability. Dose comparisons were made between PRESAGE, ECLIPSE, and independent measurement with EBT film at a select depth. The use of improved optics and acquisition technique yielded substantially higher quality 3D dosimetry data from PRESAGE than has been achieved previously (noise reduced to ~1%, accuracy to within 3%). Data analysis showed excellent intra-dosimeter uniformity, temporal stability and inter-dosimeter reproducibility of relative radiochromic response. In general, the PRESAGE™ dose-distribution was found to agree better with EBT (~99% pass rate) than with ECLIPSE calculations (~92% pass rate) especially in penumbral regions for a 3% dose-difference and 3 mm distance-to-agreement evaluation criteria. The results demonstrate excellent robustness and reproducibility of the PRESAGE™ for relative 3D-dosimetry and represent a significant step towards incorporation in the RadOnc-clinic (e.g. integration with RPC phantom).

1. Introduction

PRESAGE™ is a relatively novel radiochromic 3D dosimeter, which has many favorable characteristics for dose readout using optical-CT[1,2,3]. Most studies have focused on comparing planned dose distribution with a one-time measurement of dose[2,3]. Comprehensive 3D characterization of the robustness and reproducibility of PRESAGE™ is still pending. In this study, experiments were specifically conducted to evaluate robustness (intra-dosimeter uniformity and temporal stability of response) and reproducibility (inter-dosimeter) of radiochromic response for multiple dose levels and in 3D. Some robustness and reproducibility characteristics were evaluated previously but were limited to single profile measurements[3]. Robustness and reproducibility were evaluated by irradiating multiple 3D dosimeters with an identical treatment plan, tracking the readout at periodic time intervals and inter-comparison of the readouts. The accuracy and precision of dosimetry was also characterized by comparison with treatment planning system (TPS) calculations and with EBT film measurements in the central plane. The treatment plan was carefully designed to facilitate rigorous investigation of robustness, reproducibility, precision and accuracy for multiple dose levels and in 3D (see methods). The PRESAGE dosimeters were designed to fit as inserts in the RPC H&N phantom to test the viability of the PRESAGE/optical-CT dosimetry system as a clinical tool (e.g. integration with the RPC IMRT credentialing test)

2. Materials and Methods

2.1. Experiment to evaluate robustness, reproducibility, accuracy and precision

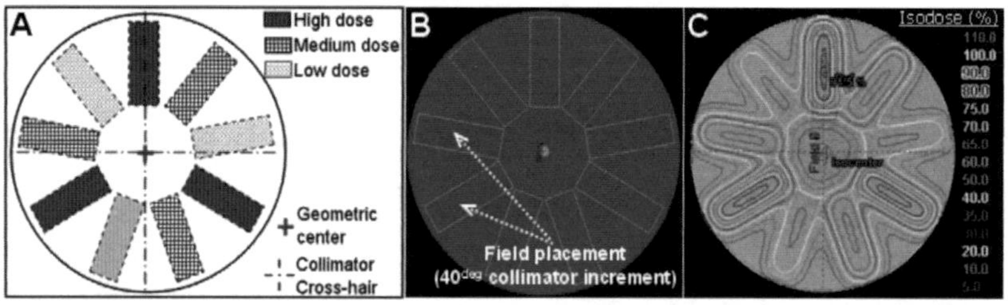

Fig 1. Rotationally symmetric treatment plan. A) Pictorial representation of planning and delivery. **B)** Frontal view from CT scan with field overlay showing placement of the nine fields in a rotationally symmetric pattern. **C)** Rotationally symmetric dose distribution.

Four identical PRESAGE™ 3D dosimeters were irradiated with an identical treatment plan. The dosimeters (10 cm diameter) were from the same batch of pre-mold mixture. The PRESAGE inserts had an effective Z number of 8.3, a physical density of 1.07g/cm3, and a CT number of ~180.

An x-ray CT scan was acquired and imported into an Eclipse workstation for planning. For the CT scan, the PRESAGE™ inserts were placed with their flat circular face on the treatment couch such that circular sections were available in the frontal plane of reconstruction (**fig 1**). The gantry was positioned at 180^0 and nine rotationally symmetric beams (collimator angular increments of 40^0 starting at 0^0) were placed. The field size for all beams was 1x3 cm^2. A dose of 4Gy (100%) was prescribed at the center of the first beam (0^0 collimator) and at a depth of 2 cm along the central axis of the PRESAGE™ insert. The second (40^0 collimator) and third beams (80^0 collimator) were weighted to 72% and 32% of the first beam respectively. The remaining beams were a repetition of the first three beams. The overall dose distribution in any axial slice was highly modulated and consisted of three identical sectors of dose distribution i.e. rotationally symmetric. Each sector consisted of dose delivered by a high dose (100%), a medium dose (72%) and a low dose (32%) beam. Along the long axis of PRESAGE inserts, there was only a slight variation of dose distribution. The significance of this treatment plan is that: i) it facilitated rigorous investigation of intra dosimeter uniformity of radiochromic response for multiple dose levels because of rotational symmetry of distribution, ii) it facilitated rigorous investigation of temporal variation and inter-dosimeter reproducibility for multiple dose levels and in 3D and iii) it enabled accurate EBT film dosimetry (insensitive to slight errors in film placement) for independent comparison in the axial plane. All PRESAGE inserts were irradiated with the same treatment plan but on different days. Accuracy and symmetry of delivery was ascertained by a quality assurance test whereby the collimator jaws were retracted, the cross-hair was aligned with the geometrical center of the circular face of dosimeters and the collimator was rotated to test the alignment of cross-hair with the geometrical center at all angular increments.

The OCTOPUS™ 5X optical-CT scanner (MGS Research Inc, Madison, CT) was used for dose readout from PRESAGE™ inserts. The OCTOPUS™ 5X optical-CT scanner is an upgrade over previously used apparatus[2]. The upgrade facilitated ~5X faster scanning because of changes in motor configurations. Each 3D scan consisted of 15 slices with inter-slice spacing of 5 mm. Each slice comprised 600 linear projection scans with rotational increments of 0.6^0 commensurate with the Nyquist sampling criteria. The time required to scan each slice was ~7-8 minutes. Every projection had a matrix size of 168 mm (168 pixels with 1 mm spacing). An in house MATLAB-based (Mathworks, Natic, MA) code was used to reconstruct the 3D distribution of radiochromic response (optical density; units: /meter) with a resolution of 1 mm x 1 mm x 5 mm. The PRESAGE inserts were scanned at periodic time intervals of 24 hrs post irradiation to evaluate temporal variation. The relative dose distribution was proportional to the change in OD and is called PRESAGE dose. After optical-CT scanning, one of the dosimeters was cut centrally and an EBT film was inserted at the cut-plane for independent verification of dose distribution. A calibration curve was acquired on the day of irradiations to enable conversion to dose distribution (EBT dose)[2]. The EPSON® Perfection 4990 flatbed scanner was used for scanning EBT films and its use has been described previously.

2.2. Data analysis

Data analysis was done in DoseQA (www.3cognition.com) or using in house routines in MATLAB. The regions receiving identical dose should elicit identical radiochromic response (change in optical density (OD)). This was the basis for the evaluation of intra-dosimeter uniformity of response and was facilitated by the rotationally symmetric treatment plan and delivery. A 3D PRESAGE dose was compared with itself but rotated by 120^0 and 240^0. Dose profiles and gamma maps (2% dose difference and 1 mm distance-to-agreement (DTA) criteria) were used in comparative analysis. The 2% dose difference (accounting for highly modulated dose distribution) and 1 mm DTA criteria (accounting for setup error) were very stringent as compared to conventional clinical criteria.

The 3D radiochromic response of a dosimeter was compared with 3D response at 24 hour intervals post-irradiation for upto 90 hrs to evaluate temporal stability. The radiochromic response showed a consistent creep (increase of OD) with time and this creep was evaluated for linearity with time. Each point in the 3D distribution was fit to a linear parametric model represented by; $y = m\ x\ t + C$; using linear least squared regression. Here, m (rate of net OD change or creep rate; units: /meter/hr) and C (initial OD; units: /meter) are the parameters of the model; y (estimate of net OD at any time t; units: /meter) is the dependent variable and t (time; units: hr) is the independent variable. The R^2 value for regression at each data point was used to evaluate linearity of fit. The creep-rate (% change of OD / hr) is a measure of consistent aging of the radiochromic response at all dose levels and was used to evaluate linearity of temporal variation of OD.

Inter-dosimeter reproducibility was evaluated by comparing 3D dose distributions from the four different dosimeters. The evaluation consisted of profile comparisons and gamma evaluations. A criterion of 2% dose-difference (accounting for highly modulated distribution) and 2 mm distance-to-agreement (accounting for inter-dosimeter setup error) was used in gamma evaluations.

The accuracy and precision of PRESAGE dose was evaluated by comparison with independent EBT dose measurement in the central plane and with 3D calculations from the Eclipse treatment planning system (henceforth termed Eclipse dose). Dose profile comparisons and gamma evaluations were used. Gamma criteria of 3% dose-difference and 3 mm DTA was used in dose comparisons between independent datasets.

3. Results and Discussion

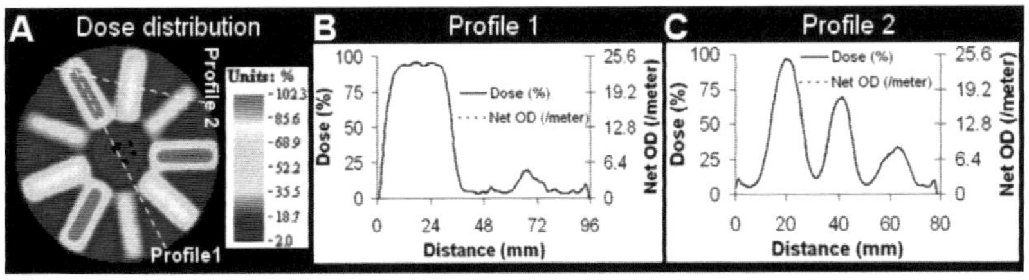

Fig 2. PRESAGE dose in central plane. A) Rotationally symmetric dose-distribution in central plane. **B&C)** Dose profiles along dotted lines in **A** show reduced noise and edge-artefact.

Improved optics and data-acquisition technique facilitated reconstruction of dose-distribution with reduced noise (1%-2%) and reduced edge artefact (~3mm from edge). **Fig 2** shows dose distribution in the central slice and profiles along dotted lines which confirm reduced noise. Such high quality data facilitated rigorous evaluation of robustness and reproducibility characteristics of PRESAGE.

Evaluation of intra-dosimeter uniformity is shown in **fig 3**. The 3D readout from a PRESAGE dosimeter was compared with the same readouts, but rotated by 120^0 and 240^0 respectively. This was possible because of the rotationally symmetric treatment and delivery. Flat and penumbral dose profiles showed close agreement for all dose levels. These profiles were chosen because any differences in distribution are likely to show up because of steep dose gradients. Gamma comparisons showed excellent (>98% pass rate) agreement in 3D confirming excellent intra dosimeter uniformity of radiochromic response in 3D.

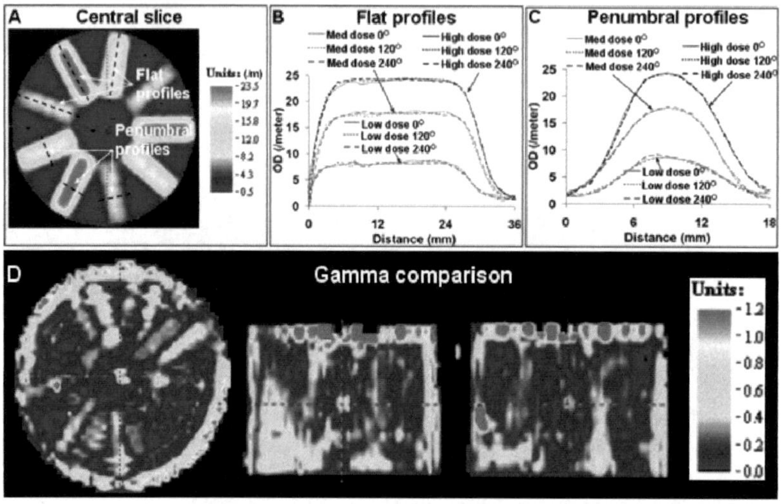

Fig 3. Intra-dosimeter uniformity. PRESAGE distribution was registered with the same distribution rotated by 120⁰ and 240⁰. Such comparison was possible because of radially symmetric distribution. A) Central slice radiochromic response showing the position of profiles. **B&C)** Dose profiles along dotted lines in **A** show encouraging intra-dosimeter uniformity. **D)** Gamma comparison (2% dose-difference and 1 mm DTA) shows encouraging agreement in 3D.

Temporal variation of radiochromic response was evaluated by periodic tracking for ~90 hrs post-irradiation and results are shown in **fig 4**. Results showed that the dosimeters turned darker with time i.e. they showed temporal creep (increase in OD) post irradiation (**fig 4b**). The creep was linear with time as seen by the excellent fit (R^2 > 0.98) to the linear regression model for all data points in the central slice. The images in **fig 4** are representative for all slices. The creep rate was calculated from the parameters of the model and was estimated to be ~ 0.54% per hour. The results demonstrate excellent relative temporal stability.

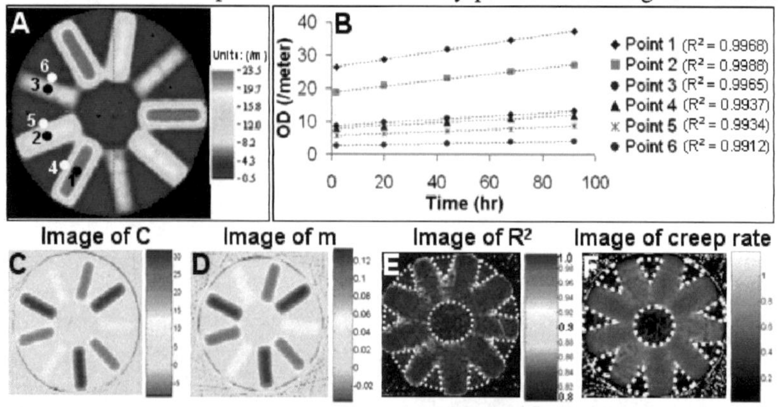

Fig 4. Temporal stability of radiochromic response. A) Central slice radiochromic response. **B)** Temporal variation for six representative points which are shown in A. **C, D, E & F)** The linear regression model for each point in central slice. C&D are images of parameters C and m. E is an image of the R^2 value of linear fit. F is an image of the creep rate calculated from the parameters of the fit.

Inter dosimeter reproducibility was evaluated by inter-comparison of PRESAGE dose from multiple dosimeters and was found to be excellent as shown in **fig 5**. The relative dose distributions showed excellent agreement (>98% gamma pass rate) to within 2% dose difference and within 2 mm DTA.

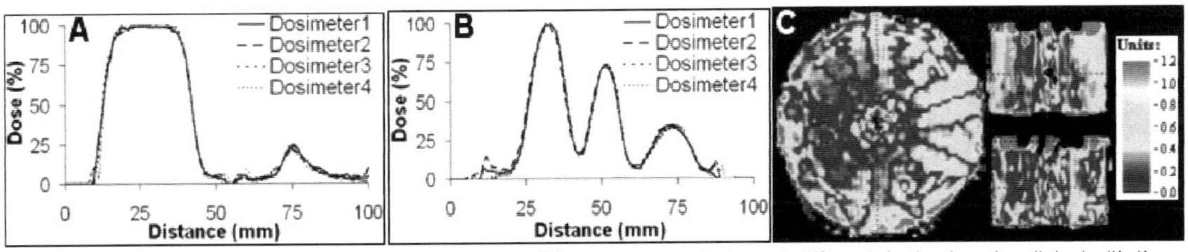

Fig. 5 Inter-dosimeter reproducibility of PRESAGE™. Measurements from 4 different dosimeters irradiated with the same treatment plan showed excellent agreement to within 2% dose-difference and 2mm distance to agreement as seen in the profile plots of **A** and **B** and representative 3D gamma comparison map between dose readout from two different dosimeters.

Comparison of PRESAGE dose with EBT dose and Eclipse dose in the central slice is shown in **fig 6**. PRESAGE dose clearly agreed better with EBT dose than with Eclipse dose, particularly in penumbral regions. Excellent agreement of PRESAGE dose and EBT dose (3% dose-difference and 3 mm DTA) confirmed that PRESAGE was an accurate relative dosimeter.

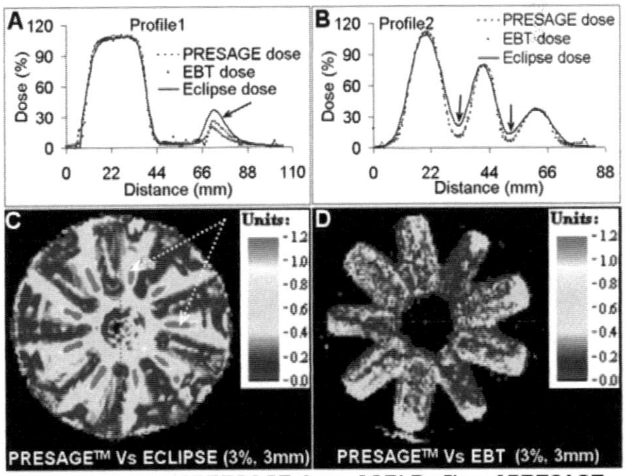

Fig 6. Accuracy of PRESAGE dose. A&B) Profiles of PRESAGE, EBT and Eclipse dose. C) Gamma comparison between PRESAGE and Eclipse dose. D) Gamma comparison between PRESAGE dose and EBT dose. In general, PRESAGE dose agreed better with EBT dose than with Eclipse dose, particularly in penumbrae.

4. Conclusions

This study presents the first rigorous investigations into the robustness, reproducibility, precision and accuracy of 3D dosimetry in PRESAGE™ inserts compatible with the RPC H&N IMRT credentialing phantom. Improvements in optics and acquisition technique produced very high quality (noise ~ 2%) 3D data even at low dose levels. Results indicated that the PRESAGE™ inserts were extremely robust for relative 3D dosimetry as indicated by excellent intra-dosimeter consistency (within 2%) of radiochromic response and stability of relative dose distribution for > 90 hrs post-irradiation. Reproducibility of response for different 3D inserts was also evaluated and was highly encouraging (within 2%) at all dose levels. Accuracy of relative dose measurement was evaluated by comparison with independent EBT film measurement and was within 2% even along the penumbra and at low dose levels. The excellent quality of 3D dose measurement that was achieved with the PRESAGE™/OCTOPUS™ dosimetry system (comparable to EBT film) confirms its viability as a clinical 3D dosimetry tool (e.g. integration with RPC IMRT credentialing).

References

1. Adamovics, J. & Maryanski, M.J. Characterisation of PRESAGE: A new 3-D radiochromic solid polymer dosemeter for ionising radiation. *Radiat Prot Dosimetry* **120**, 107-112 (2006).
2. Guo, P., Adamovics, J. & Oldham, M. A practical three-dimensional dosimetry system for radiation therapy. *Med Phys* **33**, 3962-3972 (2006).
3. Guo, P.Y., Adamovics, J.A. & Oldham, M. Characterization of a new radiochromic three-dimensional dosimeter. *Med Phys* **33**, 1338-1345 (2006).

On the role of polymer gels in the dosimetry of small photon fields used in radiotherapy

Evangelos Pappas[1,2]

[1] Department of Medical Physics, Faculty of Medicine, University of Crete, 711 10 Stavrakia-Heraklion, Crete, Greece

[2] Medical Physics Department, Anticancer-Oncologic Hospital of Athens 'Saint Savvas', 171 Alexandras Ave., 115 22 Athens, Greece

email : epappas@edu.med.uoc.gr

1. Introduction

Small photon fields, having dimensions from 30 mm down to 3 mm, are increasingly used in modern radiotherapy techniques and especially in Stereotactic radiosurgery / radiotherapy (SRS / SRT) and in intensity modulated radiation therapy (IMRT - beam segments). These small photon fields are used in gamma-knife, X-knife, Cyberknife, helical Tomotherapy and general in LINAC - IMRT treatments using conventional or micro-MLCs.

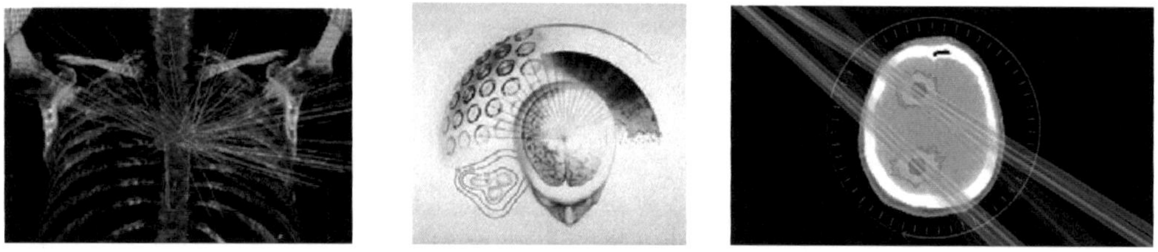

Figure 1. Cyberknife beams Gamma – knife beams Tomotherapy beams

The technological advances that allowed the use of such modern techniques for adequate encompassing of complex target volumes with the desired iso-dose surfaces were not followed by the corresponding necessary advances in dosimetry. The international dosimetry protocols (e.g. AAPM TG51, IAEA TRS 398) are mainly referred to photon field sizes from about 4 cm up to 40 cm.

The commissioning and periodic quality assurance of a radiotherapy system that utilizes small photon fields require accurate measurements of percentage depth doses (pdd), output factors and beam profiles. Large dosimetric errors during treatment and target-loss and/or Organs

At Risk over- or under-irradiation can often happen due to the increased uncertainty in measuring these parameters.

Small photon field dosimetry is very demanding due to:

a) the high dose gradients that are present in narrow photon beams,
b) the non-existence of lateral electronic equilibrium.

Moreover, the dosimeters used for such measurements should ideally exhibit certain characteristics:

a) they should be tissue equivalent (and not perturb the radiation beam)
b) they should *not* exhibit dose rate dependence of response (especially for profile measurements, since dose rate varies across the beam)
c) they should *not* have directional dependence of response (especially for pdd and profile measurements, since the angular distribution of the electrons and scattered photons varies with distance from the beam central axis and with depth)
d) they should *not* have energy dependence of response. (Usually dosimeters are calibrated at broad beams that exhibit a different (softer) photon spectrum compared to narrow beams)
e) the spatial resolution should be high (in order to perform accurate measurements at high dose gradient regions – In narrow beams the dose gradient can be higher than 40% / mm)
f) the sensitive volume of the dosimeter should be as small as possible in order to avoid volume averaging effects (that lead, for example, to penumbra broadening), but in the same time the detector sensitivity should be adequate for precise measurements (keep the statistical noise low)
g) they should *not* perturb the radiation beam (and result to electron transport alterations)
h) finally positioning (set-up) problems are always present (eg it is difficult to match the scanning path of the detectors' effective point of measurement with a direction that intersects the central axis of a narrow beam, when measuring profiles).

It is a fact that no single detector fulfills all the mentioned requirements. It has been proposed (Podgorsak *et al* 1999) that the use of a variety of detectors for small beam dosimetry is a good practice. In the literature have been presented small photon field dose characteristics measurements using air and liquid ionization chambers, radiographic and Gafchromic films, diamond detectors, plastic scintillators, TLDs, MOSFETs, radiophotoluminescence glass plates, polymer gels and silicon diodes (Buccilioni *et al* 2003, Westermark *et al* 2000, Robar and Clark 1999, Vach *et al* 2001, Martens *et al* 2000). The advantages and disadvantages of each dosimeter have been well reported.

The most important problems, among those listed above, present in almost all dosimetric systems used for narrow photon beam measurements are: a) dosimeter perturbation effects, b) volume averaging and c) positioning (set-up) errors (Das *et al* 2008, McKerracher and Twaites, 1999, 2006). It is believed that these problems can be addressed using polymer gel dosimetry which is a dosimetric method where the dosimeter and the phantom are identical and practically water equivalent, the spatial resolution of the measurements can be as low as sub-mm in three-dimensions and the derived data can be well manipulated for overcoming averaging and set-up errors (Pappas *et al* 2005, 2006).

A combination of narrow-beam data measurements performed using: a) conventional dosimeters and b) polymer gel dosimetry, can lead to a final trustable beam dataset that can be used safely as input in a Treatment Planning System and for quality assurance purposes.

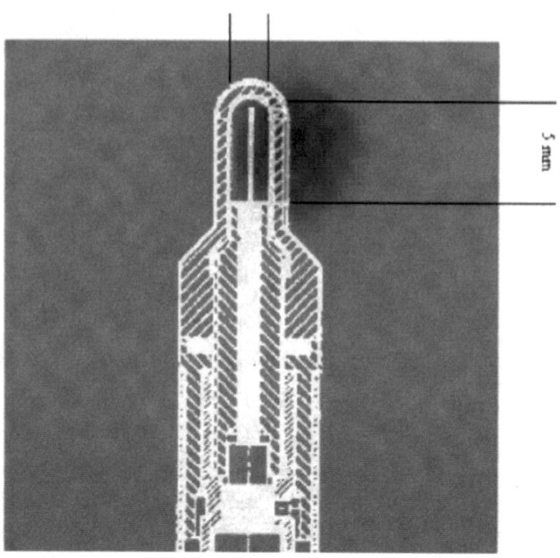

Figure 2. Positioning problems, volume averaging and electron transport alterations errors are the most important problems in small field dosimetry using conventional dosimeters. In this figure one can see a schematic representation of the longitudinal cross section of the PTW-PinPoint ion chamber, superimposed on a T2 map of an irradiated polymer gel with a 5 mm X-knife circular collimator (6 MV x-ray beam))

Below, are briefly presented: a) basic principles of polymer gel data manipulation methods for overcoming set-up and volume averaging errors and b) actual pdd, OF, and profile measurements of gamma-knife, X-knife and Cyberknife beams.

2. Manipulation of polymer gel data for overcoming positioning (set-up) and volume averaging problems

2.1 Overcoming set-up errors

The scanning path of the effective point of measurement of a conventional dosimeter, when for example one tries to measure the profile of a narrow beam, should cross the beam central axis. Figure 2 schematically shows that this is a difficult task. Small set-up errors can lead to erroneous results. Polymer gel dosimetry data can be well manipulated so that the central axis signal can be well defined, avoiding in this way positioning errors.

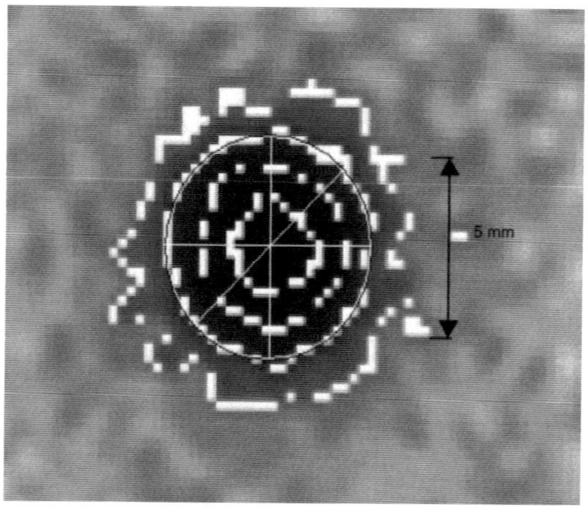

Figure 3. 0.25 mm resolution axial *T2*-map (perpendicular to the beams axis) of an irradiated polymer gel using a 5 mm circular collimator (6 MV x-rays). The four superimposed quasi-circular groups of white pixels correspond to four different random iso-*T2* values. The circle that fits one of them is also presented along with its center. All corresponding circle-centers lie within, or exist in the vicinity, of a unique pixel defined as the 'central pixel'. This pixel is presumed to belong to the beam's central axis

Within a *T2*-map (scanning level perpendicular to the beams central axis) of a polymer gel irradiated with a small circular photon field, a single pixel exists that corresponds to the beams' central axis (central pixel). This pixel can be determined by superimposing random iso-*T2* lines on the irradiated area of the *T2*-map, and subsequently by fitting corresponding circles as can be seen in the Figure 3. The calculated centers of these circles ideally lie within, or exist in the vicinity, of a unique pixel, hence, the central pixel. Accordingly, the important set-up errors present in small beam profile measurements can be eliminated.

2,2 Overcoming volume averaging errors

Volume averaging errors lead to beam penumbra broadening (that results to Organs At Risk over-irradiation) and to Output Factor underestimation (that results to higher delivered dose that that described). These volume averaging errors can be well estimated and eliminated by manipulating polymer gel data.

Estimating volume averaging errors in beam profiles measurements

By analyzing the *T2*-map of a 5 mm beam irradiated gel, the spin-spin relaxation rate *R2*-line profile that intercepts the *central pixel* (defined in the previous paragraph) can be converted to a % dose-line profile. As long as the delivered doses lie within the dose range where a linear dose – *R2* response exists, the background subtracted *R2* [*R2*net=*R2*(D)-*R2*(0)] can be normalized to the corresponding value of the central pixel. In this way, a beam profile that corresponds to a gel-detector size equal to the *T2*-map pixel size can be determined.

The same procedure can be followed if a square AxA mm^2 signal integration area is used for 'scanning' the *T2*-map along a linear direction that intercepts the central pixel (Figure 4). In Pappas *et al* (2006), 'A', was chosen to be 0.5, 1.5, 2.5, 3.5, 4.5, 5.5 and 7.5 mm respectively. Thus, the same *true* profile (*Pt*) can be measured using a variety of gel-detector sizes ('A'), each resulting in a corresponding *measured* profile (Pm_A).

Figure 4. A square signal integration area AxA is superimposed on the *T2*-map of the irradiated polymer gel (A=3.5 mm in this figure). The same *T2*-map is presented sevenfold along with the same signal integration area at seven different positions along its 'scanning' path which crosses the 'central pixel'.

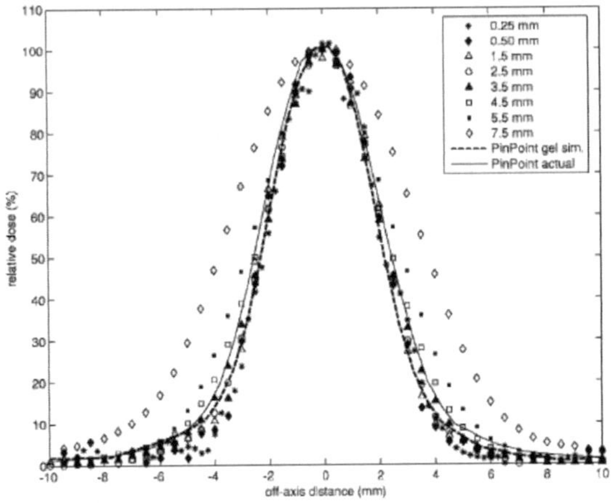

Figure 5. Beam profiles of the 5 mm diameter 6 MV stereotactic beam derived using polymer gel dosimetry with a variety of signal integration areas (or gel-detector sizes). In the same figure are also presented: a) actual corresponding PinPoint measurements and b) gel-derived measurements performed using a signal integration volume that approximates that of the PinPoint ion chamber

The knowledge of the dependence of the beam profile on the detector size used for measuring it, can lead to an accurate measurement of the true profile by extrapolating the measurements to zero detector size. In practice, a detector size equal to ~ 0.5 mm is adequate for not introducing volume averaging effects. By using an integration area equal to that of a known dosimeter, the volume averaging error introduced by this dosimeter can be estimated. Therefore a convolution kernel for this dosimeter can be extracted and then used for de-convolving actual dosimeter measurements and eliminating volume averaging errors. However, possible electron transport alterations will still affect the results. Since it is not the aim of this brief report to analyze deeper this issue, the reader is referred to Pappas *et al* (2006) for a more analytical presentation.

2.3 Estimating volume averaging errors in Output Factor measurements

By manipulating polymer gel data using a variety of signal integration areas one can also evaluate the volume effect to the Output Factor measurement. A small detector size (corresponding to a small integration area) results to relatively accurate but not very precise measurements. On the other hand, as the integration area is increasing (having the central pixel always at its center) the measurement becomes less accurate due to volume averaging. In Figure 6 are presented the OF measurements of a 5 mm diameter collimator (6 MV x-rays), performed using a variety of integration areas.

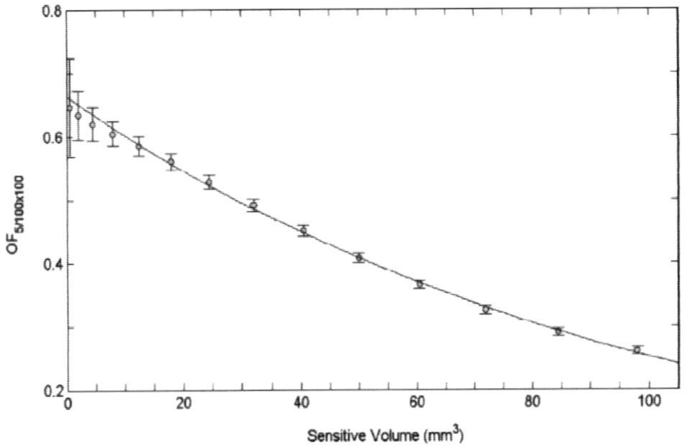

Figure 6. OF measurements of a 5 mm diameter X-knife collimator (6 MV x-rays), performed using a variety of integration areas using polymer gel data

In the case where high dose gradients exist in 3 dimensions, one can use the full 3D data set from polymer gel dosimetry. In figure 7 is presented the percentage of OF underestimation due to volume averaging for the 4 mm and 8 mm Gamma-knife helmets.

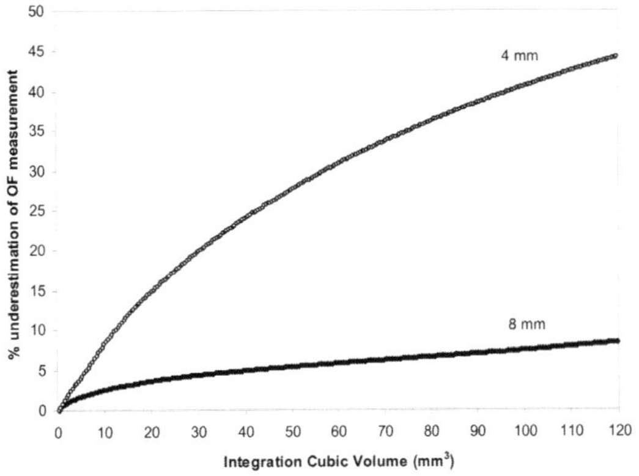

Figure 7. Percentage of OF underestimation due to volume averaging for the 4 mm and 8 mm Gamma-knife helmets

2.4 Pdds, profile and OF measurements

In the final section of this report are briefly presented pdd, profile and OF polymer gel derived measurements of small photon fields used in modern radiotherapy. Corresponding measurements performed using conventional detectors are also presented for comparison.

Pdds

The relatively low dose gradients along the central axis of a small field, makes the pdd measurements the less problematic among the small field data characteristics. In the figure below are presented polymer gel pdd measurements of the 5 and 10 mm diameter photon beams and corresponding conventional and PinPoint measurements.

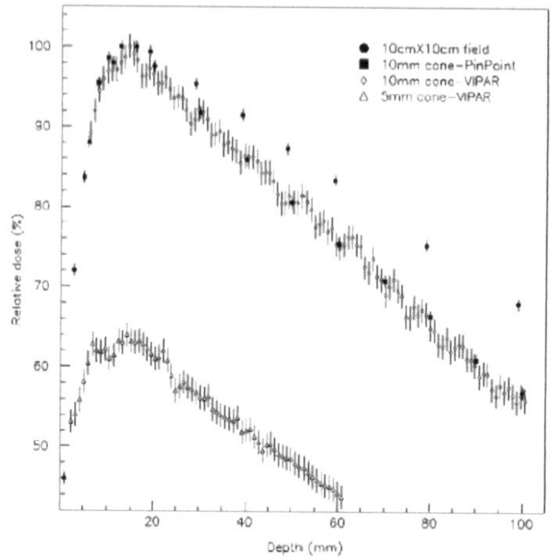

Figure 8. Pdd measurements of the 5 and 10 mm diameter photon beams (6 MV x-rays) performed using polymer gel dosimetry. PinPoint measurements for the 10 mm beam and for a 10 cm X 10 cm field are also presented

Profiles

By analyzing the *T2*-map of a narrow beam irradiated gel, the *R*2net profile along a line that intersects the central pixel, equals to the off-axis ratio of the irradiation area. The direction of the line can be arbitrary and profiles using a number of arbitrary lines can be averaged in order to reduce the statistical noise and derive a profile that characterizes the whole beam. However, in order to use *all* the available gel data for the measurement of a profile that characterizes a certain small field (and not just a number of random line profiles passing through the central pixel), the following process can be followed. A large number of circles centered at the irradiated area's *central pixel* can be used (three of them are presented in Figure 9 for a 30 mm collimator irradiated area). The average *R*2net value of the pixels that lie in the periphery of each separate circle (normalized to the central pixel R2net value) correspond to the off-axis ratio for a distance from the central axis equal to the circle radius. The normalization point in all cases can be the mean value that correspond to the central pixel and its eight surrounding pixels.

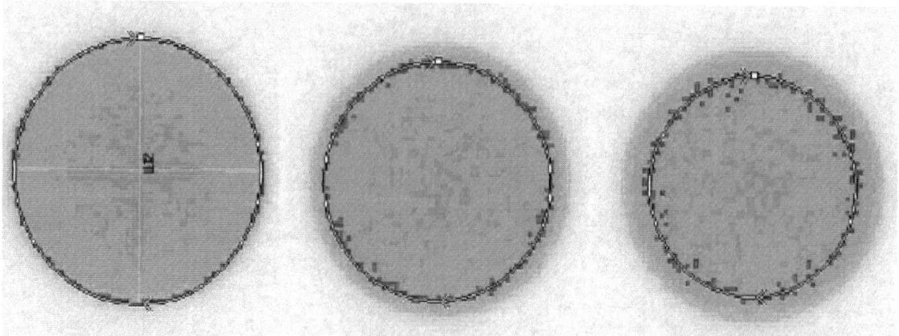

Figure 9. Three snapshots that correspond to three different random iso-*T2* pixels superimposed on the *T2*-map of the irradiated gel (30.0 mm irradiated area – magnified). The centers of the fitted circles lie within or adjacent to, a unique pixel: the central pixel. The mean value of the iso-*T2* pixels that lie on a fitted circle correspond to the *T2* value at a Off Axis Distance equal to the circles' radius.

Figures 10,11 and 12 present Cyberknife (5, 7.5 and 10 mm beams), gamma-knife (8 mm collimator helmet) and Novalis (7.5, 15 and 30 mm collimators) profile measurements performed with polymer gel dosimetry. Corresponding measurements performed using diode detectors, PinPoint air ion chamber, films, diamond detector or diode array detector are also presented for reasons of comparison

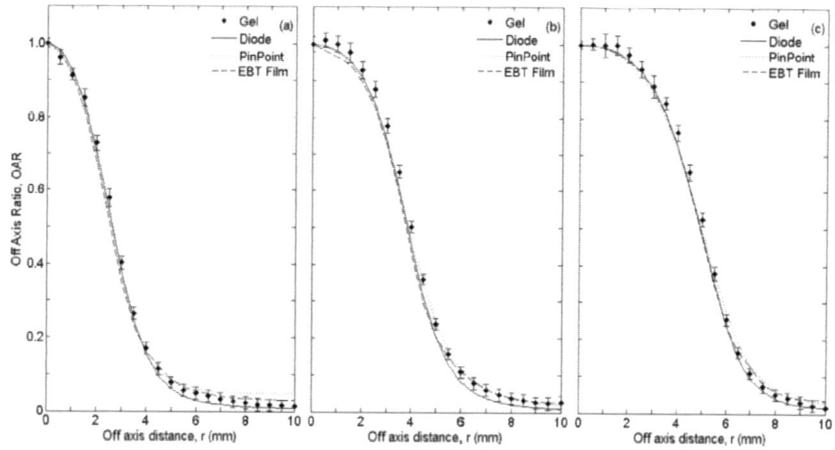

Figure 10. Polymer gel, diode, PinPoint and film measured Off Axis Measurements for the Cyberknife beams of 5, 7.5 and 10 mm diameter beams (from Pantelis *et al* 2008)

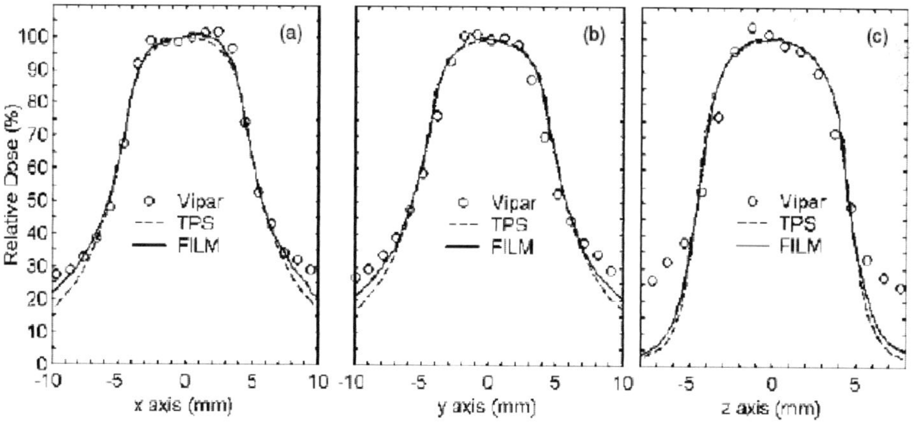

Figure 11. Gamma-knife 8 mm collimator helmet relative dose profiles. Corresponding Gammaplan calculations as well as radiochromic film measurements are also presented for comparison (from Karaiskos *et al* 2005)

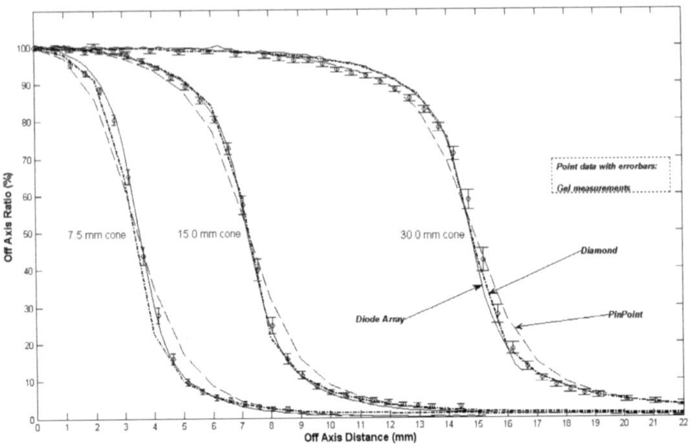

Figure 12. Beam profiles of 7.5, 15 and 30 mm Novalis beams. The solid line corresponds to the diode array (DOSI), dash-dot to diamond and dashed line to the PinPoint measurements respectively. Polymer gel data points are *not* connected with a line (Pappas *et al* 2008 in press)

Output Factors

Cyberknife output factors for the 5, 7.5 and 10 mm beams were measured equal to 0.702, 0.872 and 0.929 respectively, using polymer gel dosimetry.

X-knife (6 MV) output factor for the 5 mm beam was measured equal to 0.66, using polymer gel dosimetry.

Corresponding OF measurements using conventional detectors and Monte Carlo simulations are also available in the literature (Pantelis *et al* 2008, Pappas *et al* 2005).

3. Conclusions

Volume averaging problems, set-up errors and detector perturbation effects are the main problems present in small photon field dosimetry. Polymer gel dosimetry characteristics allow the elimination of these problems. Polymer gels can and should become a useful tool in clinical practice for small field data acquisition and periodic quality assurance in modern radiotherapy techniques where small photon fields are used.

References

- M. Buccilioni *et al,* Med. Phys. **30**, 2149-2154 (2003).
- J. Das *et al,* Med. Phys. **35**, 206-215 (2008).
- P. Karaiskos *et al,* Phys. Med. Biol. **50**, 1235-1250 (2005).
- C. Martens *et al,* Phys. Med. Biol. **45**, 2519-2530 (2000).
- C. McKerracher, and D. I. Thwaites, Phys. Med. Biol. **44**, 2143-2160 (1999).
- C. McKerracher and D. I. Thwaites, Radiother. Oncol. **79**, 348-351 (2006).
- E. Pantelis *et al,* Med. Phys. **35**, 2312-2320 (2008)
- E. Pappas *et al,* Phys. Med. Biol. **46**, 783-97 (2001).
- E. Pappas *et al,* Med. Phys. **32**, 1513-1520 (2005).
- E. Pappas *et al,* Med. Phys. **33**, 3700-3710 (2006).
- E. Pappas *et al*, Med. Phys. (*in press*).

- E B Podgorsak, Neurosurg. Clin. North Am. **3**, 9-34 (1992).
- J. L. Robar and B. G. Clark, Med. Phys. **26**, 2144-2150 (1999).
- Y. W. Vach *et al,* Med. Phys. **28**, 303-309 (2001).
- M. Westermark *et al,* Phys. Med. Biol. **45**, 685-702 (2000).

Journal of Physics: Conference Series **164** (2009) 012061 doi:10.1088/1742-6596/164/1/012061

VIPARnd - GeVero® tool in planning of TPS scheduled brain tumour radiotherapy

Marek Kozicki[1,*], Piotr Maras[2], Krzysztof Rybka[3,4], Tadeusz Biegański[3,4,5]

[1]Institute of Architecture of Textiles, Faculty of Material Technologies and Textile Design, Technical University of Lodz, Zeromskiego 116, 90-924 Lodz, Poland

[2]Medical Physics Department, Copernicus Hospital, Pabianicka 62, 93-513 Lodz, Poland

[3]Department of Medical Imaging Techniques, Medical University of Lodz, Kosciuszki 4, 90-419 Lodz, Poland

[4]Polish Mother Memorial Hospital – Research Institute, Department of Diagnostic Imaging, Rzgowska 281/289, 93-338 Lodz, Poland

[5]Department of Paediatric Radiology, Medical University of Lodz, Kosciuszki 4, 90-419 Lodz, Poland

*Corresponding author email: mkozicki@mitr.p.lodz.pl

Abstract. In this paper, VIPARnd - GeVero® tool is presented for the first time in an application to a brain tumour radiotherapy. Whereas usefulness of VIPARnd polymer gel in various radiotherapy techniques has recently been confirmed, GeVero® software for calculation of MRI polymer gel data and comparison with TPS dose distribution simulation is now examined. The results demonstrate satisfactory agreement between polymer gel dosimetry-MRI and TPS dose distributions and prove helpfulness of the software and VIPARnd polymer gel in radiotherapy dosimetry. It is also believed that the software facilitates data processing and therefore should be of further support in po-gel dosimetry studies.

1. Introduction

Polymer gel dosimetry has been established as a powerful method for three-dimensional high resolution measurements of absorbed radiation dose distribution and therefore it has been widely examined in radiotherapy dosimetry applications. As radiation detectors, polymer gels are used that are physical hydrogel matrixes containing radiation sensitive vinyl monomer ingredients. If exposed to ionizing radiation, radical polymerisation and crosslinking takes place leading to formation of large structures scattering visible light. This feature allows for naked eye observation of radiation beams tracks after complex irradiation.

In seventies of the last century the direct relation between polymer molecular weight, crosslinking, polymer chains entanglements and nuclear magnetic resonance relaxation times described professor Charlesby [1,2]. Then, various systems were studied in order to establish relation between radical polymerisation and crosslinking of aqueous vinyl monomer solutions and NMR relaxation times, e.g. ionizing radiation induced acrylamide polymerisation and formation of hydrogels was analyzed by NMR measurements [3]. It was also possible to derive a method of sol-gel analysis for irradiated aqueous solutions of vinyl substrates and follow polymerisation process in the systems via T_2 measurements [4]. The latter studies resulted in several polymer gel compositions based on poly(ethylene glycol) diacrylate and *N,N'*-methylenebisacrylamide of acronym PABIG [5,6] as well as *N*-vinylpyrrolidone and *N,N'*-methylenebisacrylamide. The latter was also analyzed elsewhere and the gel was named VIPAR [7]. Improvement of VIPAR polymer gel composition was a subject of profound studies thereafter and have lead to compositions of much lower dose threshold [8] as well as an addition of normoxic ingredients proposed elsewhere [9] resulted in a composition named VIPARnd of simpler and faster preparation scheme [10,11]. Description of another normoxic *N*-vinylpyrrolidone based polymer gel can be found elsewhere [12].

Currently, magnetic resonance imaging measurement of T2 relaxation time is the most common technique allowing precise and detail analysis of changes inside polymer gels. However, polymer gel MRI data analysis may be lengthy, and requires significant familiarity of DICOM structures, if not accompanied by specific software. Therefore, fast, user friendly software for polymer gels examination was developed in our group. GeVero® software equipped with VIPARnd polymer gel is believed to substantially ease 3D radiation dose recording and data processing. Consequently, in this work we present the first data on application VIPARnd - GeVero® tool in planning of a brain tumor treatment.

2. Materials and methods

2.1. Polymer gel dosimeter

The VIPARnd polymer gel dosimeter (*N*-vinylpyrrolidone 8% w/v, *N,N'*-methylenebisacrylamide 4% (w/v), gelatine 7.5% (w/v), copper sulphate ($CuSO_4 x 5H_2O$) and L-ascorbic acid of 0.0008% (w/v) and 0.007% (w/v), respectively) was prepared according to the procedure described earlier [10]. Briefly, *N,N'*-methylenebisacrylamide was dissolved in deionised water under stirring and heating below 50°C. Afterwards, gelatine was added and completely dissolved before the solution was cooled down to a temperature of about 33°C in order to add *N*-vinylpyrrolidone. Finally, ascorbic acid and $CuSO_4$ x $5H_2O$ were added and the composition was mixed carefully.

Two types of vials for irradiation and magnetic resonance imaging were used. A batch of 0.8 dm^3 VIPARnd was prepared to fill in fourteen 50 cm^3 long calibration vials. However, 2.5 dm^3 of the same polymer gel was prepared for treatment planning verification experiment. In the latter case, a large vial corresponding to the average volume of brain was used.

Notwithstanding the fact that VIPARnd belongs to so called normoxic range of polymer gel dosimeters, all vials were tightly closed to provide stable conditions for polymerization reactions of monomers and prevent the air from inflowing.

2.2. Irradiation

The former scheme of irradiation of the calibration vials was adopted [10]. All vials were irradiated 24 hours post preparation to ensure that the solidification of gelatine is completed. Each 50 cm^3 VIPARnd calibration vial was centrally placed inside cubic phantom of RW3 plates (18x18x18 cm^3 cube). Prior to irradiation, however, the phantom with VIPARnd vial was scanned (CT, Toshiba, 1 mm step) and 3D images were employed by treatment planning system (TPS, Eclipse External Beam Planning, Anisotropic Analytical Algorithm, AAA, for photons) to calculate three bands of different maximal dose with 5 cm spacing between the centers of consecutive 3.5 cm long bands. The irradiation was performed with the aid of Clinac 600CD, 6MV beam (Varian Medical Systems, Palo Alto, CA; a four-field box technique was applied to irradiate each band). The absorbed dose range was equal 0 – 38 Gy (300 MU min^{-1}) (Fig. 1).

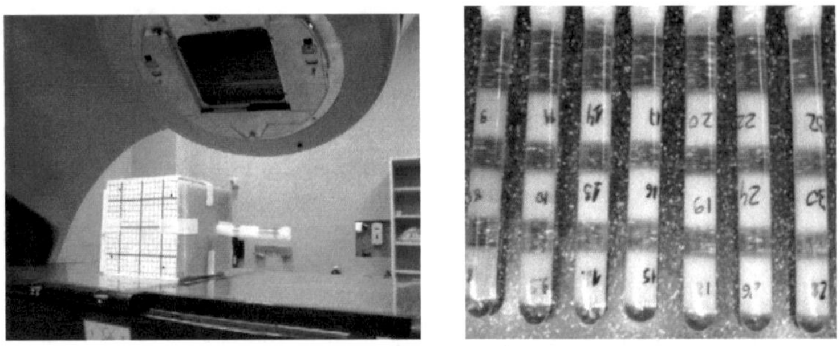

Figure 1. RW3 phantom with VIPARnd calibration vial (left) and the effect of the irradiation of calibration gels (right) (Clinac 600, 300 MU min^{-1}).

CT images of a patient brain tumor were used for radiotherapy planning with the aid of Eclipse External Beam Planning (Anisotropic Analytical Algorithm, AAA, for photons). This system generated a plan of irradiation of the 2.5 dm^3 VIPARnd vial. Further, the irradiation plan was put into operation with the aid of Clinac 600 (300 MU min^{-1}, prescribed dose - 36Gy). Results are shown in Figure 2.

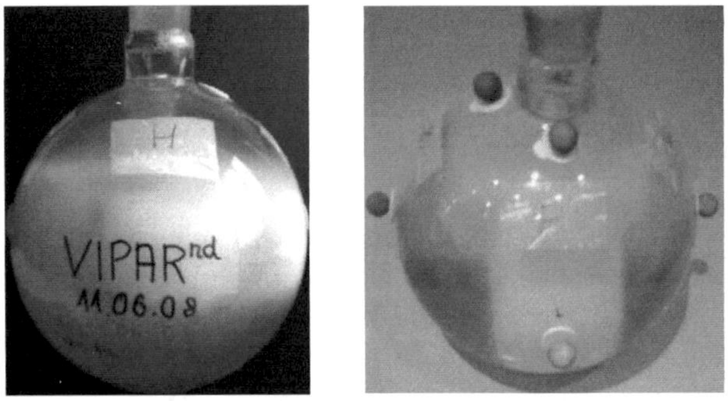

Figure 2. The 2.5 dm^3 VIPARnd vial after irradiation with Clinac 600 (300 MU min^{-1}) according to brain tumor plan delivered by Eclipse External Beam Planning.

2.3. Magnetic Resonance Imaging

All vials of VIPARnd underwent MR imaging (Picker Edge 1.5 T) 24 hours post irradiation. Previous protocol was applied [13,14], however in this study, TE where equal 30, 60, 90, 120 ms, echo averaging was set to 1, 2, 4 and 8, 256x256 matrix, 0.78 mm pixel spacing (for the calibration samples). The parameters of MR scanning of the 2.5 dm^3 vial were as follow: TE 30, 60, 90, 120 ms, echo averaging was set to 1, pixel spacing 0.74 mm; slice thickness 1.5 mm; 256x256 matrix. Four echo-based images for each slice of both type vials were obtained. The pixel intensities of the four echo-based images were further analyzed with the aid of GeVero® software.

2.4. Dose distribution calculations

Magnetic resonance DICOM images of irradiated 50 cm^3 and 2.5 dm^3 vials were processed by GeVero® software described elsewhere [15]. The software calculates R2 relaxation rate and absorbed dose distribution in an irradiated gel. Calibration option is one of the features of the software as well as processing of TPS data of a planned dose distribution, here in VIPARnd polymer gel. Moreover, TPS and VIPARnd - MRI data were compared: gamma index, isodoses, dose profiles, histograms.

3. Results and discussion

Figure 3 shows calibration R2 dependence on absorbed dose for VIPARnd gel at different MR setup.

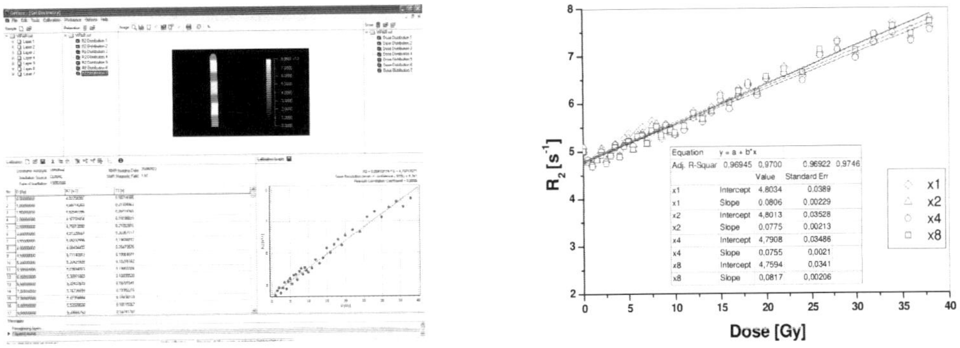

Figure 3. Calibration curves for VIPARnd calculated by GeVero® for different MR signal averaging (1, 2, 4, 8).

The data revealed no significant difference in dose sensitivity and R2(0) parameters, which were important for further calculations of dose distribution in 2.5 dm^3 vial. Therefore, the large gel vial was MR scanned without signal averaging option that significantly reduced the time of measurement. It is also worth mentioning that dose sensitivity value of the gel was found to be similar to the one reported before [10].

In Figure 4, the comparison of TPS calculations and MR measurements for VIPARnd is visualised for selected XY dose plane with the aid of GeVero®. Quantitative and qualitative analysis in all planes (XY, XZ, YZ) and 108 layers revealed good matching of dose distributions in vast majority of the gel layers. Mismatching of dose distributions was also discerned especially close to the glass walls of the gel vial. The discrepancies are to be scrutinised in due time in order to find satisfactory explanation of this phenomenon.

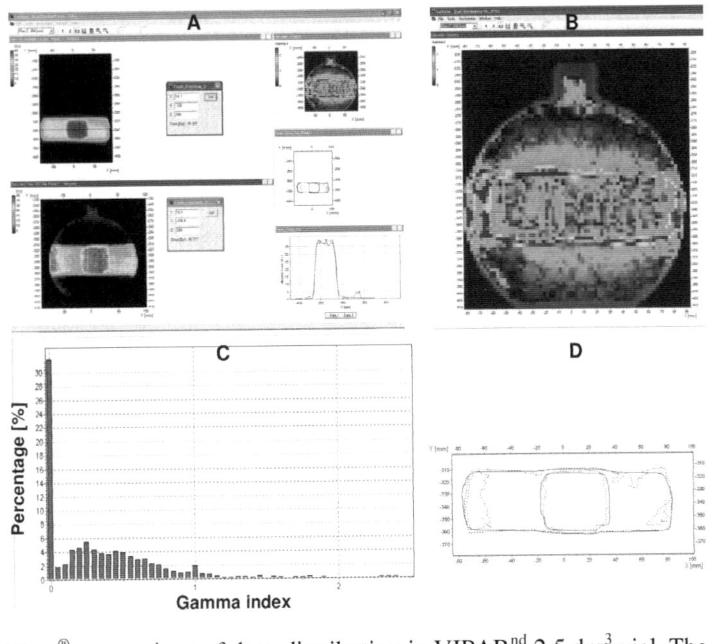

Figure 4. A - GeVero® comparison of dose distribution in VIPARnd 2.5 dm^3 vial. The dose distribution data calculated with Eclipse External Beam Planning system and the polymer gel-MRI method (A). B – Gamma index; C – Gamma index histogram; D – isodoses superimposition. The data is presented for selected cross-section of VIPARnd in XY plane.

4. Conclusions
In this work VIPARnd-GeVero® was presented as prospective tool for user-friendly comparison of calculations against measurements of 3D absorbed dose distributions. The worked out software facilitated typically laborious calculations of polymer gel dosimetry data. A good agreement was found between TPS irradiation plan and the polymer gel-MRI dose distribution.

5. Acknowledgements
Polish Ministry of Science and Higher Education is acknowledged for support of this study (project no N205 048 32/2930).

References
[1] Charlesby A., The use of pulsed NMR technique in the measurement of radiation effects in polymer, *Radiat. Phys. Chem.*, **14**, 919-930 (1979)

[2] Charlesby A., Käfer P., Folland R., Study of very high molecular weight polyethylene using pulsed NMR techniques, *Radiat. Phys. Chem.*, **11**, 83-91 (1978)

[3] Rosiak J.M., Burczak K., Pękala W., Piślewski N., Idziak S., Charlesby A. Studies of polymerisation and crosslinking of aqueous acrylamide, *Radiat. Phys. Chem.*, **32**, 793-796 (1988)

[4] Kozicki M., Radiation-induced reactions of divinyl monomers and their application in polymer gel dosimetry, PhD thesis at the Institute of Applied Radiation Chemistry, Technical University of Lodz, Poland (2004)

[5] Kozicki M., Rosiak J.M. "Basic properties of a new PABIG polymer gel dosimeter for dose distribution assessment in radiotherapy treatment", 2nd International Symposium on Reactive Polymers in Inhomogeneous Systems, in Melts and at Interfaces, September 28 – October 1, 2003, Dresden, Germany, book of abstract, S3/33

[6] Pantelis E., Lymperopoulou G., Papagiannis P., Sakelliou L., Stiliaris S., Sandilos P., Seimenis I., Kozicki M. and Rosiak J.M., Polymer gel dosimetry close to an 125I interstitial brachytherapy seed, *Phys. Med. Biol.* **50,** 4371-4384 (2005)

[7] Pappas E., Moris T., Angelopoulos A., Paparigopoulou M., Sakelliou L., Sandilos P., Voyiatzi S., Vlachos L., A new polymer gel for magnetic resonance imaging (MRI) radiation dosimetry, *Phys. Med. Biol.*, **44**, 2677-2684 (1999)

[8] Kozicki M., Petrokokkinos L., Papagiannis P., Sakelliou L., Angelopoulos A., Pappas E., Rosiak J.M., Study of VIPAR gel dosimeter dose range improvement, Medical Physics Proceedings of the jointly held Congresses ICMP 2005, BTM 2005, September 14-17, Nuremberg, Germany, *Biomedizinische Technik*, **50**, 1368-1369 (2005)

[9] Fong P.M., Keil D.C., Does M.D., Gore J.C., Polymer gel for magnetic resonance imaging of radiation dose distributions at normal room atmosphere, *Phys. Med. Biol.*, **46**, 3105-3113 (2001)

[10] Kozicki M., Maras P., Rybka K., Bieganski T., Kadłubowski S., Petrokokkinos L. On the development of the VIPAR polymer gel dosimeter for three-dimensional dose measurements, *Macromol. Symp.*, **254**, 345–352 (2007)

[11] Pantelis E., Antypas C., Petrokokkinos L., Karaiskos P., Papagiannis P., Kozicki M., Georgiou E., Sakelliou L. and Seimenis I., Dosimetric characterization of CyberKnife radiosurgical photon beams using polymer gels, *Med. Phys.*, **35**, 2312-2320 (2008)

[12] Papadakis A.E., Maris T.G., Zacharopoulou F., Pappas E., Zacharakis G. and Damilakis J., An evaluation of the dosimetric performance characteristics of N-vinylpyrrolidone-based polymer gels, *Phys. Med. Biol.*, **52**, 5069–5083 (2007)

[13] Kozicki M., Rybka K., Bieganski T., Petrokokkinos L., Angelopoulos A., 23rd Annual Scientific Meeting of the European Society for Magnetic Resonance in and Biology,

Proceedings September 21–23, Warsaw, Poland p. 107 (2006)

[14] Petrokokkinos L., Kozicki M., Baras P., Angelopoulos A., Papagiannis P., Rybka K., Fijuth J., Bieganski T., Implementing polymer gel dosimetry in clinics with basic MRI installations, *Biomedizinische Technik*, **50**, 1061–1062 (2005)

[15] Kozicki M., Maras P., Jankowski J., Karwowski A.C., Polymer gel - TPS radiotherapy dosimetry GeVero® software for ionizing radiation absorbed dose 3D distribution calculations, submitted to DOSGEL 2008

Journal of Physics: Conference Series **164** (2009) 012062 doi:10.1088/1742-6596/164/1/012062

Polymer gel - TPS radiotherapy dosimetry GeVero® software for ionizing radiation absorbed dose 3D distribution calculations

Marek Kozicki[1,*], Piotr Maras[2], Jacek Jankowski[3] and Andrzej C. Karwowski[1]

[1]Institute of Architecture of Textiles, Faculty of Material Technologies and Textile Design, Technical University of Lodz, Zeromskiego 116, 90-924 Lodz, Poland

[2]Medical Physics Department, Copernicus Hospital, Pabianicka 62, 93-513 Lodz, Poland

[3]Department of Strength of Materials and Structures, Faculty of Mechanical Engineering, Technical University of Lodz, Stefanowskiego 1/15, 90-924 Lodz, Poland

*Corresponding author email: mkozicki@mitr.p.lodz.pl

Abstract. Implementation of polymer gel dosimetry in radiotherapy departments calls for: easily manufactured gel dosimeters of required physical-chemical properties, set-up procedures of irradiation, adaptation of three-dimensional scanning procedures and instruments as well as fast tool for calculation of 3D absorbed dose distribution in the polymer gel dosimeters and comparison with another treatment planning system calculated dose distribution. These challenges resulted in several propositions in polymer gel dosimetry area. In this work, however, a summary of results on construction of polymer gel dosimetry software facilitating usually laborious 3D dose distributions data processing is provided.

1. Introduction

The polymer gel dosimetry is meant for support of three-dimensional dosimetry in radiotherapy departments. It employs polymer gel dosimeters consisting of gel matrix with monomers inside, which convert to crosslinked structures if exposed to ionizing radiation. If, for example, PABIG [1,2] or VIPAR [3-7] polymer gels are irradiated, water radiolysis radical products are formed: $^{\bullet}OH$, $H^{\bullet}$, e^{-}_{aq}. The products effectively react with monomer molecules causing monomer radicals formation (initiation step of polymerisation). These radicals are capable of reacting with other monomer molecules leading to formation of a growing polymer chains (propagation step). The propagation step characterises by the increase in molecular weight of polymer and its radius of gyration. The propagation step is very efficient in case of typical monomers of polymer gel dosimeters. This step is followed by termination where dead, non-radical polymer chains are formed. Growing or final chains of polymer can react with monomers, especially difunctional molecules, leading to formation of three-dimensional polymer network (crosslinking reactions).

The products of radical polymerisation and crosslinking can be analyzed with the aid of nuclear magnetic resonance (NMR) relaxation times measurements, since as it was reported earlier [8], their molecular weight influences on NMR relaxation times. Therefore, the technique of magnetic resonance imaging is currently the most commonly used in assessment of radiation-induced changes in polymer gel dosimeters.

The scheme of polymer gel dosimetry application requires irradiation of a gel dosimeter according to a pre-defined irradiation plan. Measurement with the aid of e.g. magnetic resonance imaging and calculation of dose distribution inside polymer gel dosimeter are the following steps. The latter may require significant qualifications if three-dimensional polymer gel data matrix is to be analyzed and compared with a treatment planning system (TPS) calculated dose distribution. In order to simplify this step, we have undertaken to construct software allowing for user friendly readout and processing of polymer gel dosimeter magnetic resonance data as well as comparison with corresponding TPS dose distribution simulation (Fig. 1). Functions of GeVero® software (gel dosimetry verification of treatment planning system software) are illustrated below.

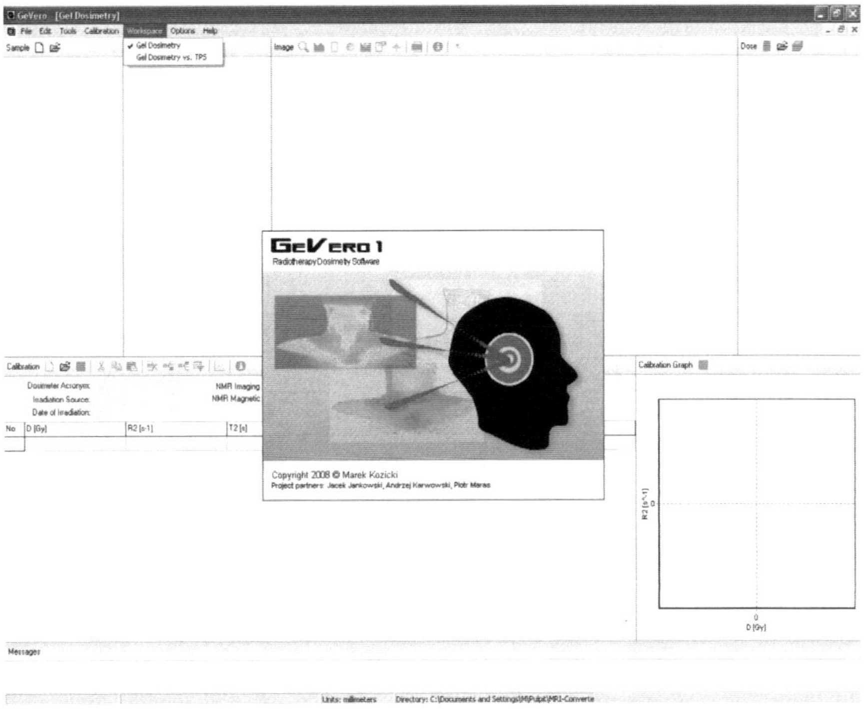

Figure 1. GeVero® radiotherapy dosimetry software when starts.

2. Results

GeVero® radiotherapy dosimetry software takes a mathematical approach to calculate 3D radiation dose distribution in polymer gel dosimeters and compares data with 3D radiation dose distribution calculated with the aid of Treatment Planning Systems (TPS). The software was designed to operate in bi-window mode and a user can chose whether to calculate dose distribution in a polymer gel dosimeter or compare and contrast polymer gel-MRI with TPS data.

The main options of the software include:

1. Importing of DICOM data of Magnetic Resonance Imaging (MRI) echo images (or T2 images) of a polymer gel dosimeter; automatic ordering of echo images and assigning them to the appropriate layers of the scanned polymer gel dosimeter.

2. Calculation of spin-spin T2 MR relaxation time for each layer of the polymer gel dosimeter.

Input data of signal value and echo time corresponding to echo images of each layer is taken on pixel by pixel basis in order to calculate the function of $y = y0 + a*\exp(-R2*t)$, where y0 and a are constants and $R2=1/T2$ – is spin-spin relaxation rate (T2 – relaxation time [s]), y – signal value, t – echo time [ms]. Taking logarithm of both sides of the above equation allows for determination of R2 and a values.

This procedure leads to R2 layers of a polymer gel, which further may be used either for calibration curve parameters or dose distribution calculations.

3. The calibration option for construction a calibration curve: $R2[s^{-1}]$ = Dose sensitivity $[Gy^{-1}s^{-1}]$ * Dose [Gy] + R(0) $[s^{-1}]$.

If calibration option is chosen, R2 matrixes for calibration samples must be calculated first. Then, it is possible to create a new calibration table that includes information on absorbed dose by polymer gel samples and corresponding R2 values. Additionally, the table stores all necessary characteristics of a gel dosimeter, such as irradiation and scanning date, irradiation source features and irradiation conditions, composition of a polymer gel, MR scanning protocol, etc.

GeVero® calculates the dose sensitivity parameter and R2(0) – relaxation rate at 0Gy, on the basis of the least squares method. The calibration curve equation is attached to the calibration table together with dose resolution information for a polymer gel dosimeter (Fig. 2).

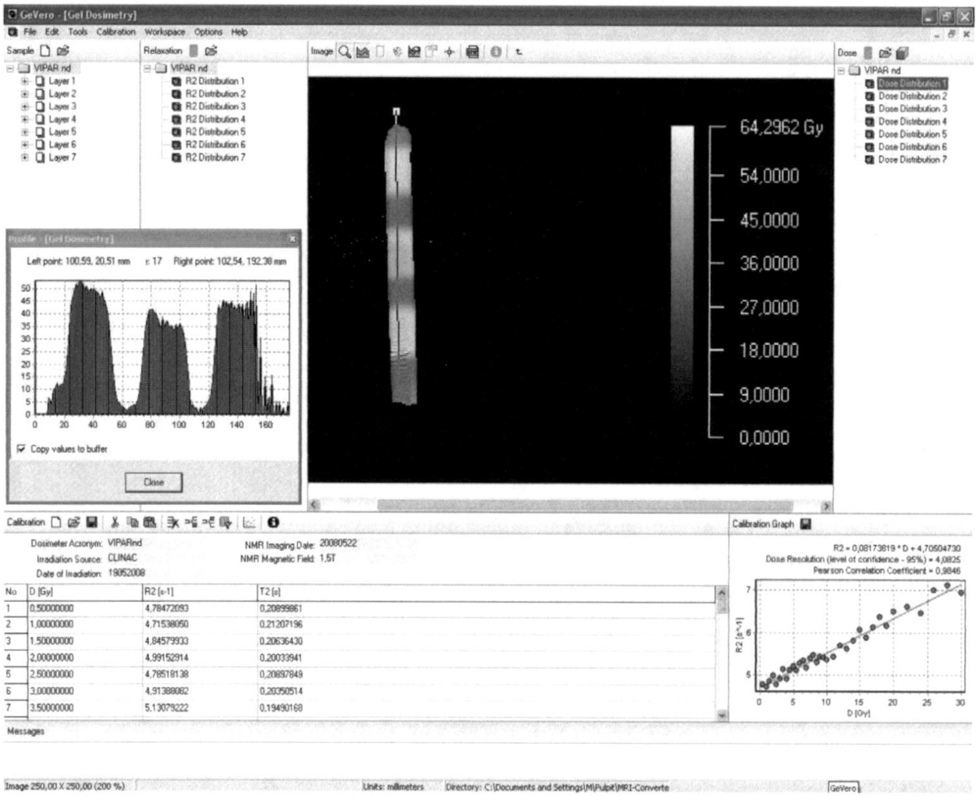

Figure 2. GeVero® when calculates R2 and dose distribution in long 50 cm^3 vials of $VIPAR^{nd}$ as well as a trial calculation of $VIPAR^{nd}$ calibration curve.

If dose distribution needs to be calculated, the dose option should be used. A user is asked to type conversion parameters for a particular polymer gel, which are R2(0) [1/s] and a [1/(Gy*s)] – dose sensitivity. Those parameters should be searched in appropriate calibration table files. Afterwards, dose distribution is calculated. Further, vectorial transformation and isodose centre selection is performed for the dose distribution 2D matrixes in order to create 3D dose matrix, if comparison with TPS dose distribution simulation is considered.

4. Importing TPS dose distribution data, dose distribution visualization in 3D, comparing and contrasting polymer gel dosimetry vs. TPS data (gamma parameter, histograms, isodoses, 2D dose difference).

This part of the software permits an import and separate analysis of polymer gel dose distribution data or treatment planning system dose distribution (import of DICOM RT, e.g. Eclipse system). Comparative analysis of data coming both form calculation methods is also feasible. In each case of separate data analysis, absolute and relative dose profiles as well as histograms drawing can be performed. When both data of dose distributions are imported, one can compare and contrast them thanks to the option of superimposition the data on each other and calculation of isodoses (qualitative comparison) as well as dose profiles or gamma index on the basis of Low et al approach [9] (quantitative comparison). Subtraction of 2D dose distributions of two different data is also possible.

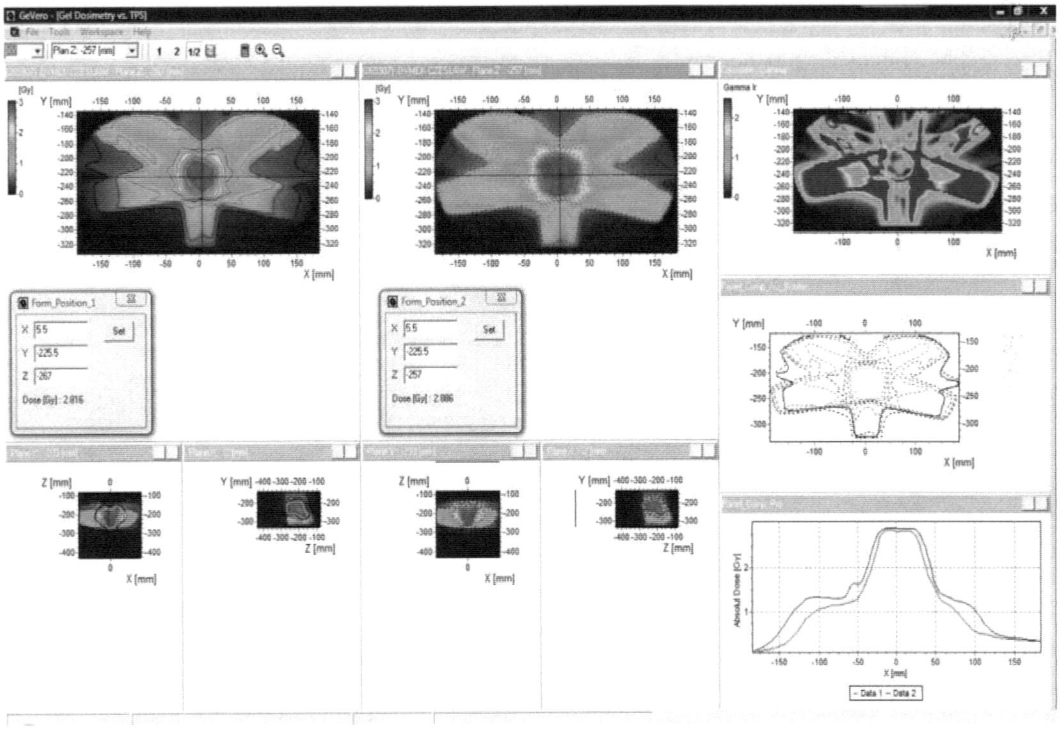

Figure 3. GeVero® when compares two data of various planning systems via gamma index (upper right), isodoses (middle right) and absolute dose profiles (lower right) superimpositions. The six graphs counting form left correspond to two data set on the planes of XY, ZX, YZ. Visualization of separate single data for detailed inspection is feasible.

3. Conclusion

GeVero® radiotherapy dosimetry software eases and shortens time of calculations of dose distributions in polymer gel dosimeters. In combination with PABIG^{nx} and VIPARnd compositions [5,6] of simplified preparation procedures it is believed to have been complementary solution towards facilitation of routine application of the gel dosimetry technique. Currently, further studies on function of GeVero®-VIPARnd/PABIG^{nx} in radiotherapy dosimetry are ongoing.

Acknowledgement
Polish Ministry of Science and Higher Education is acknowledged for support of this study (project no N205 048 32/2930).

References

[1] Ulanski P., Janik I., Kadlubowski S., Kozicki M., Kujawa P., Pietrzak M., Stasica P. and Rosiak J.M., Polymeric Biomaterials Synthesized by Radiation technique - Current Studies at IARC Poland, *Polym. Adv. Technol.*, **13**, 951-95 (2002)

[2] Pantelis E., Lymperopoulou G., Papagiannis P., Sakelliou L., Stiliaris S., Sandilos P., Seimenis I., Kozicki M. and Rosiak J M, Polymer gel dosimetry close to an 125I interstitial brachytherapy seed, *Phys. Med. Biol.* **50**, 4371-4384 (2005)

[3] Pappas E., Moris T., Angelopoulos A., Paparigopoulou M., Sakelliou L., Sandilos P., Voyiatzi S., Vlachos L. A new polymer gel for magnetic resonance imaging (MRI) radiation dosimetry, *Phys. Med. Biol.*, **44**, 2677-2684 (1999)

[4] Kozicki M., Rybka K., Bieganski T., Petrokokkinos L., Angelopoulos A., Adaptation of basic MRI protocols for the implementation of polymer gel dosimetry in clinical setting, 23rd Annual Scientific Meeting of the European Society for Magnetic Resonance in Medicine and Biology (ESMRMB 2006), e-Poster no. 663 on-line available (http://posters.webges.com/esmrmb/epos), Proceedings p. 107, September 21-23, 2006, Warsaw, Poland

[5] Kozicki M., Maras P., Rybka K., Bieganski T., Kadłubowski S., Petrokokkinos L. On the development of the VIPAR polymer gel dosimeter for three-dimensional dose measurements, *Macromol. Symp.*, **254**, 345–352 (2007a)

[6] Kozicki M., Rosiak J.M., Sakelliou L., Angelopoulos A., „Gel dosimeters for UV light and ionizing radiation dose distribution spatial measurements", Polish Patent Application, P-383256 (June 2007b)

[7] Pantelis E., Antypas C., Petrokokkinos L., Karaiskos P., Papagiannis P., Kozicki M., Georgiou E., Sakelliou L. and Seimenis I., Dosimetric characterization of CyberKnife radiosurgical photon beams using polymer gels, *Med. Phys.*, **35**, 2312-2320 (2008)

[8] Charlesby A., The use of pulsed NMR technique in the measurement of radiation effects in polymer, *Radiat. Phys. Chem.*, **14**, 919-930 (1979)

[9] Low D.A., Harms W.B., Mutic S., Purdy J.A., A technique for the quantitative evaluation of dose distributions, *Med. Phys.*, **25**, 656-661 (1998)

Journal of Physics: Conference Series **164** (2009) 012063 doi:10.1088/1742-6596/164/1/012063

Fricke gel layer dosimeters in BNCT: recent applications

G Bartesaghi[1], M Carrara[2], G Gambarini[1], A. Negri[1]

[1]Department of Physics of the University and INFN, Milan, Italy

[2]Department of Medical Physics, National Cancer Institute of Milan, Italy

E-mail: giacomo.bartesaghi@unimi.it

Abstract. Radiochromic gel dosimeters in form of layers based on Fricke solution have shown to be a valuable method for in-phantom dose measurements in a mixed BNCT radiation field. Using dosimeters with different isotopic composition it is possible to determine bidimensional distribution of the different dose components. In the last years, the method has been continuously improved, permitting to achieve reliable dose images with areas until 14x16cm^2, also in epithermal columns with a relevant spectral component of fast neutrons.

1. Dosimetry in BNCT

Boron neutron capture therapy (BNCT) is an experimental technique whose aim is to selectively release heavy charged particles in the cancer cells exploiting the high capture cross section of ^{10}B. A BNCT treatment is composed by two steps: at first, a boron carrier capable to concentrate itself mainly in the tumour cells is administered to the patient; then, at a proper time after the drug administration, the patient (or the organ) to be treated is exposed to a thermal or epithermal neutron field. ^{10}B nuclides capture neutrons with high probability (σ = 3837b for 25meV neutrons), leading to one of the following reactions:

$$^{10}B + n \rightarrow {}^{4}He + {}^{7}Li + \gamma\ (0{,}48\ MeV) + 2{,}31\ MeV\ (94\%)$$
$$^{10}B + n \rightarrow {}^{4}He + {}^{7}Li + 2{,}79\ MeV\ (6\%)$$

The ^{4}He and ^{7}Li particles release almost all their kinetic energy in the malignant cells, since they have a combined average range of 12-13μm, comparable to the linear dimensions of a mammalian cell; moreover, being heavy charged particles with a total average energy of 2.34MeV, they have an elevated linear energy transfer (LET) resulting in a high radiobiological effectiveness (RBE), providing an effective tumour–seeking boron carrier.

BNCT is a complex technique that in the last years has obtained encouraging results, however without passing completely the toxicity and efficacy trials; it needs multidisciplinary efforts in order to become competitive with the traditional radiotherapy based on electrons and photons. The crucial point is the normal tissue sparing, achievable developing highly selective boron carriers and studying suitable neutron sources.

Different kinds of neutron sources for BNCT are under study, but until now only nuclear reactors can provide a neutron field with acceptable characteristics of energy spectrum, flux and background gamma radiation. In particular, epithermal neutron beams, that is with a relevant component of neutrons with energy between 0.5eV and 10keV, are convenient for BNCT trials of deep tumour treatments. However, this kind of sources have unavoidable components of background photons and fast neutrons (>10keV) damaging indistinctly all the tissues.

Differently from the conventional radiotherapies, the tissues exposed to a BNCT radiation field are traversed by four distinct radiation components, each one with a different LET and causing the absorption of dose with different RBE. The so-called therapeutic dose is due to boron neutron capture. Fast neutrons, interacting with nuclei, release their kinetic energy mainly to the hydrogen nuclei in tissues, causing recoil protons. Hydrogen can capture thermal neutrons, generating 2.2MeV photons; other gamma rays come from the column structure. Finally, also nitrogen can capture neutrons, emitting a proton with the $^{14}N(n,p)^{14}C$ reaction.

As far as dosimetry is concerned, it is important to determine beside the therapeutic dose also the dose due to the undesired radiation components. In fact, the neutron fluence has to be sufficiently high in order to achieve the tumor control; but, on the other hand, the maximum admitted fluence is established on the basis of the dose in healthy tissues, which has to be within tolerance limits [1].

Usually, when the effect on a patient of a BNCT treatment has to be foreseen, the boron and the nitrogen doses are estimated by means of the kerma approach considering the in-tissue thermal fluence; the last is mainly evaluated performing in-phantom activation foils measurements [2].

Dose measurements in BNCT are challenging. The most common detectors for gamma and fast neutron dose evaluation are paired ionization chambers and thermoluminescence detectors (TLD) but, because of a number of unpredictable factors, such dosimeters can hardly reach an accuracy below 8-10% (one standard deviation). It is therefore recommended to perform both the gamma and fast neutron dose measurements exploiting more than one type of experimental technique.

Fricke gel dosimeters in form of layers have proved to give a valuable method to determine in-phantom absorbed dose spatial distributions. Differently from ionization chambers and TLDs that perform point measurements, these gel dosimeters can provide bidimensional images of the different absorbed dose components. The measurement shown in this paper have been carried out at the epithermal column of the LVR-15 reactor at Řež (CZ) [3]. Utilizing Fricke gel dosimeters with different isotopic composition, it was possible to obtain distinctly images of boron, photon and fast dose components.

2. Fricke gel dosimeters for BNCT

Fricke gel dosimeters in form of layer are laboratory-made radiochromic gels with the following standard composition: ferrous sulphate [1mM $Fe(NH_4)_2(SO4)_2{\cdot}6H_2O)$]; sulphuric acid [25mM H_2SO_4]; porcine skin [3% of the final weight] as gelling agent and Xylenol Orange [0.165mM $C_{31}H_{27}N_2Na_5O_{13}S$] as metal ion indicator. The dosimeters can be framed in different shape and size, usually with a thickness of 3mm, and are inserted in a phantom having a good tissue equivalence for neutrons. As far as the neutron transport in phantom is concerned, the non-perturbative effect due to dosimeters of such thickness with different isotopic composition of such thickness has been demonstrated by means of Monte Carlo (MC) simulations.

The dosimeters are imaged before and after the irradiation with a charged coupled device (CCD) camera. The optical transmittance at 585nm is acquired by detecting grey levels (GL) matrix. A proper software has been developed in order to perform a pixel-to-pixel elaboration and obtain the difference of optical density [□(OD)]. By means of calibration with a reference photon source, the □(OD) images are converted into dose images.

Exploiting the GL images of dosimeters with different compositions irradiated in the same position of the phantom, elaboration algorithms have been developed in order to provide dose images due to the different radiation components [4,5].

A couple of standard and boron added (usually with 40ppm of ^{10}B) dosimeters is used to determine both gamma and boron dose distributions; while the method to separate the gamma and fast neutron doses is based on a couple of standard and heavy-water-made dosimeters.

In the last years, the Fricke-gel-layer experimental technique has been continuously improved. A particular care has been devoted to the improvement of the method reliability; in particular, in one of the experimental configurations it has been estimated that an uncertainty of 6-7% in the total dose evaluation resulted in an error of around 30% of the determined boron dose.

Moreover, since in BNCT the amount of absorbed dose in tissues far away from the target volume is not negligible, larger dosimeters have been assembled (with areas until 16x14cm^2). Dosimeters with such areas show problems of thickness homogeneity which have been solved by performing transmittance images before the irradiation also at 430nm and utilizing properly developed algorithms [6].

The data shown in this paper have been obtained utilizing rectangular dosimeters with areas of 12x6cm^2 and 16x14cm^2 (the sensitive areas are about 11x5cm^2 and 15x13cm^2, respectively). The layers were inserted either in a BNCT standard (50x50x25cm^3) water phantom or in a cylindrical (d = 16cm, h = 14cm) water equivalent phantom, having dimensions comparable to a human head. Fig.1 shows, as an example, o couple of standard and 40ppm boron 12x6cm^2 dosimeters placed on the planar light source; these dosimeters have been irradiated for 90' in the cylindrical phantom.

Figure 1 Gel dosimeters (12x6cm^2) placed on the planar light source for transmittance imaging detection after a 90' irradiation. In the boron added dosimeter (above) the nuance due to dose absorption is more apparent that in the standard one (below).

3. Measurements with Fricke gel layers

In order to determine boron dose distributions, it is necessary to irradiate a standard Fricke dosimeters and a boron added one in the same position in the phantom. Fig. 2 shows the Δ(OD) images of a couple of 16x14cm^2 dosimeters exposed in the cylindrical phantom; the two dosimeters have been irradiated for 2 hours at the nominal power of the reactor. The obtained Δ(OD) images are proportional to the absorbed dose, and are shown with some percentage isodose curves pointed out with colors.

Applying to these data a coefficient obtained by means of a calibration performed with a photon source, it is possible to convert them into the so-called "gamma-equivalent dose" distributions. The total absorbed dose is clearly due to the different components of the radiation field, and applying to the Δ(OD) images a coefficient obtained with a gamma calibration, the different sensitivity of Fricke gel to radiation having different LET is neglected. The gamma-equivalent dose distributions are presented as 3d plots in Fig. 3.

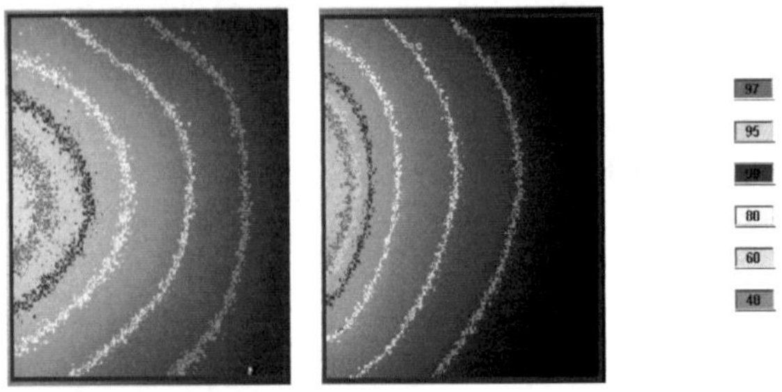

Figure 2 Δ(OD) images of a couple of 16x14cm^2 gel layers, exposed for 2h in the water phantom; a standard (on the left) and a 40ppm boron added (on the right) dosimeter are shown. The neutron beam collimator was positioned on the left side of the images. Some percentage isodose curves are highlighted.

Since a standard Fricke gel exposed in an epithermal column radiation field is sensitive only to the gamma and the fast neutron components, while a boron added dosimeter absorbs also the dose due to the boron captures, it is possible to obtain the boron dose distribution by subtraction, taking into account the relative sensitivity of the dosimeters to photons and to the heavy charged particles produced by the boron capture. The boron dose distribution achieved from data shown in Fig. 3 is depicted in Fig. 4. The trend of the boron dose distribution is related to the thermal neutron fluence in the phantom; for example, the presence of a dose maximum on the central beam axis at around 2 cm of depth is due to the thermalization of the epithermal neutron beam traversing the phantom material.

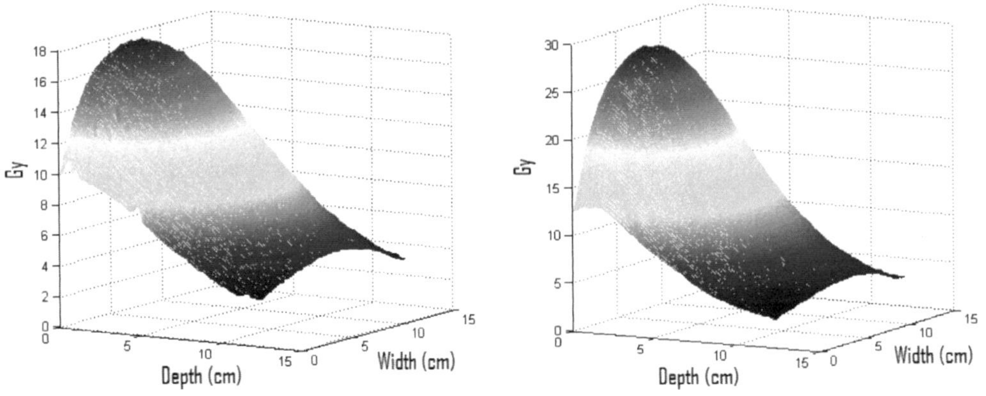

(a) (b)

Figure 3 Gamma-equivalent absorbed dose distributions in a standard (a) and a 40ppm boron added dosimeter (b) obtained by images shown in Fig. 2. Note the different scales on z-axis.

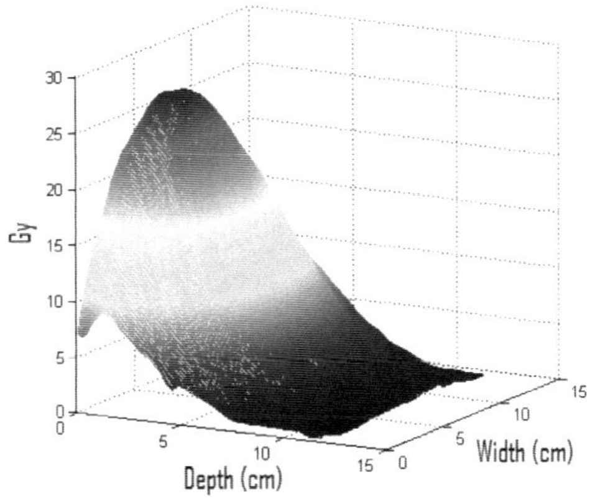

Figure 4 Boron dose distribution imaged in the water phantom.

The separation of the photon and the fast neutron doses was carried out applying a suitable procedure, which includes the use of a standard and a heavy-water-made dosimeter and proper MC simulations in order to calculate the ratio between the delivered energy of recoil protons and deuterons [7]. As far as the treatment planning is concerned, it is crucial to evaluate accurately the absorbed dose due to fast neutrons, since this radiation component deposits its energy non selectively in tissue and has a high radiobiological effectiveness.

A couple of 12x6cm^2 dosimeters was inserted in the cylindrical phantom and exposed for 2 hours. Fig. 5 shows the on-axis profiles; the photon dose was measured also with thermoluminescence dosimeters (TLD), while MC simulations have been carried out in order to calculate the relative fast neutron absorbed dose. Both TLD and MC data are plotted together with the gel dosimeter data for comparison (MC data are normalized at the point at 2.25cm of depth), showing a satisfying agreement.

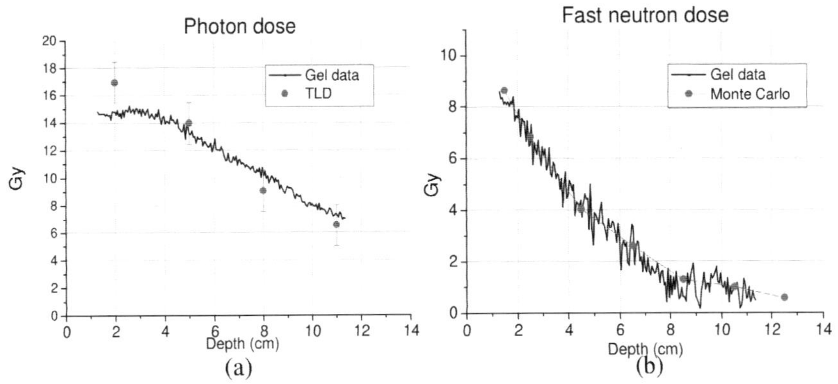

Figure 5 Photon (a) and fast neutron (b) dose on-axis profiles. The dosimeters have been exposed for 2 hours in the cylindrical phantom. Fricke gel data are compared with TLD measurements and MC calculations.

Journal of Physics: Conference Series **164** (2009) 012063 doi:10.1088/1742-6596/164/1/012063

4. Conclusions and acknowledgements

Fricke gel dosimeters in form of layers have proved to perform valid measurements of the different dose components in tissue or tissue-equivalent material exposed to a reactor BNCT epithermal beam.

Recent improvements allow to acquire dose images of areas until 13x15cm^2 and to provide a reliable separation not only of boron and photon doses but of fast neutron absorbed dose too.

This work has been carried out thanks to the collaboration of the BNCT team of the Nuclear Research Institute of Řež (CZ) and was supported by the National Institute of Nuclear Physics (INFN), Italy.

References

[1] Voorbrak W P, Recommendation for the dosimetry of Boron Neutron Capture Therapy, NRG 2003

[2] Current status of neutron capture therapy, IAEA, Vienna, 2001

[3] Burian J, Flibor S, Marek M, Rejchrt J, Viererbl L and Tomandl I, Physics for BNCT, *J. Phys: Conf. Ser.*, **41** (2006) 174-186

[4] Gambarini G, Colli V, Gay S, Petrovich C, Pirola L and Rosi G, In-phantom imaging of all dose components in boron neutron capture therapy by means of gel dosimeters, *Appl. Rad. Isot.*, **61** (2004) 759-763

[5] Gambarini G, Carrara M, Colli V, Gay S and Tomatis S, Further developments and applications of layer gel dosimetry, *J. Phys: Conf. Ser.*, **3** (2004) 213-216

[6] Carrara M, Gambarini G, Bartesaghi G, Fallai C and Negri A, A procedure to mathematically amend possible thickness disuniformities in gel-layer dosimetry, *submitted to J. Phys: Conf. Ser.* Work presented at this conference.

[7] Bartesaghi G, Burian J, Carrara M, Gambarini G, Marek M, Negri A, Pirola L and Viererbl L, Fast neutron dose evaluation in BNCT with Fricke gel detectors, *submitted to IEEE Trans. Nuc. Sc.*

Detection of Ultraviolet Radiation Using Tissue Equivalent Radiochromic Gel materials

M. A. Bero* and I. Abukassem

National Radiation Metrology Laboratory (NRML), Protection and Safety Department, Atomic Energy Commission, Damascus P.O.Box 6091, Syria.

mbero@aec.org.sy

Abstract. Ferrous Xylenol-orange Gelatin gel (FXG) is known to be sensitive to ionising radiation such as γ and X-rays. The effect of ionising radiation is to produce an increase in the absorption over a wide region of the visible spectrum, which is proportional to the absorbed dose. This study demonstrates that FXG gel is sensitive to ultraviolet radiation and therefore it could functions as UV detector. Short exposure to UV radiation produces linear increase in absorption measured at 550nm, however high doses of UV cause the ion indicator colour to fad away in a manner proportional to the incident UV energy. Light absorbance increase at the rate of 1.1% per minute of irradiation was monitored. The exposure level at which the detector has linear response is comparable to the natural summer UV radiation. Evaluating the UV ability to pass through tissue equivalent gel materials shows that most of the UV gets absorbed in the first 5mm of the gel materials, which demonstrate the damaging effects of this radiation type on human skin and eyes. It was concluded that FXG gel dosimeter has the potential to offer a simple, passive ultraviolet radiation detector with sensitivity suitable to measure and visualises the natural sunlight UV exposure directly by watching the materials colour changes.

1. Introduction:

Ultraviolet light is the section of electromagnetic radiation that lies between visible light and X-rays. Human eyes are sensitive to wavelengths of between 790nm (red) and 400nm (violet), while soft X-rays have wavelengths that are less than a few nano-meters [3]. The ultraviolet spectrum is generally subdivided to the following spectral bands: UVC (100-280 nm), UVB (280-315 nm), and UVA (315-400 nm) [1]. UVA is the main component of the terrestrial ultraviolet spectrum, and is important in the generation of photochemical smog and also in the fading of plastics, paints and fabrics colors. Only 1% of solar radiation lies within the UVB band, and most of this is absorbed by ozone [2]. However, small changes in ozone cause large changes in UVB radiation reaching the Earth's surface. UVC is totally absorbed by ozone and other gases and does not reach the Earth's surface. Time of day, season, latitude and altitude affect the level of terrestrial UV as the light attenuation is greater the further it has to travel through the atmosphere. Between 20% and 30% of the total daily UV amount is received during two hours around within the local midday at mid latitudes in summer. One kilometer increase in altitude will increase the UV flux by 6%. Similarly, the increases of cloud cover and high levels of pollution in the atmosphere reduce the amount of UV incident at the surface.

* Corresponding Author: e-mail: scientific@aec.org.sy, (M. A. Bero), mbero@aec.org.sy, Fax: 00963116112289

The sun is responsible for the development and continued existence of life on Earth, but the ultraviolet content of sunlight causes a deleterious effect on biological systems. In some regions on the earth, skin tumor is the most common human cancer. It is estimated that one out of three people might develop a skin cancer in their lifetime [2]. Non-melanoma skin cancers are attributed to cumulative UV exposure and malignant melanomas are thought to be linked to infrequent, intense exposures, especially sunburn in childhood [3]. Cancers typically take one of three forms corresponding to the three major types of skin cells: basal cells, squamous sells and melanocytes [4]. Cancer of the melanocytes, malignant melanoma, is the most lethal variety but also the least common. Despite the fact that UVC is the most energetic part of the UV spectrum, no UVC reaches the Earth's surface and, hence, UVB is the most damaging light to biological systems [5]. The most effective wavelength of sunlight for the production of erythema is 300nm (UVB), rapidly falling to just 1% of the effectiveness at 320nm [3] and in general, UVB is approximately one thousand times more effective at producing sunburn than UVA. The production of vitamin D by sunlight incident on skin is the only thoroughly established beneficial effect of UV radiation [6]. From spring until autumn, 15 minutes of sun exposure for the hand, face and arms between 9am and 4pm is adequate to provide the vitamin D requirement. Also, many common devices and industrial processes involve the emission of UV radiation, including arc discharges and bactericidal lamps.

2. Detector materials and preparation:

The materials used to produce detection medium sensitive to UV radiation are the following. Pure water triply distilled and deionised makes about 95% of the detector, 5% by weight of gelatin powder $(C_{17}H_{32}N_5O_6)_x$ with high gel strength indicator; analytical grade ferrous ammonium sulfate hexahydrate $Fe(NH_4)_2(SO_4)_2.6H_2O$ with a 0.5 mM concentration, 0.1mM of the xylenol orange-sodium salt ion indicator, $C_{31}H_{28}N_2O_{13}SNa_4$; and finally a small amount 50×10^{-3}N of concentrated sulfuric acid H_2SO_4. The radiochromic gel is prepared according to a procedure described in [7]. Ultraviolet effects are determined through quantitative evaluation of induced changes in gel material. Gels samples are put in standard size 1cm light pass length UV grade quartz cuvettes. Samples which have different thicknesses were also made in order to study the percentage absorption of UV radiation energy.

2.1. Irradiation:

An ultraviolet (UV) source simulating the solar radiation is used to irradiate the FXG gel samples. It is also known as sun simulator, and reproduces UVA and UVB spectrum equivalent to natural sunlight. The ground level of solar radiation reaching the earth's surface varies significantly with atmospheric condition, location, time of the day, earth/sun distance, and solar activity. Standard spectra have been developed to provide a basis for standardization of theoretical evaluation of the effects of solar radiation.

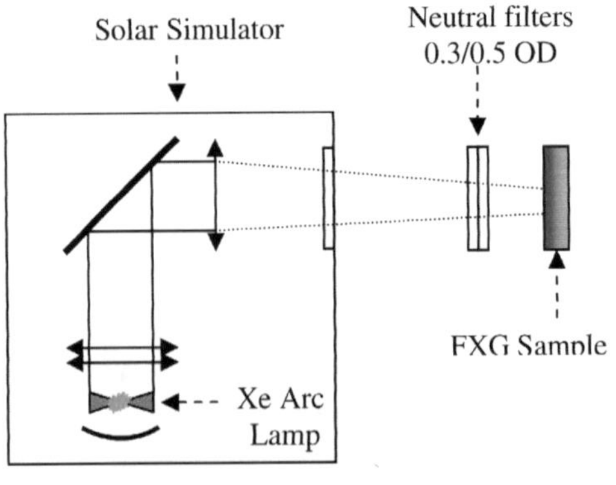

Figure 1: UV irradiation setup.

Xenon arc lamp and power supply, Model XPS-200, (Solar Light Company, 721 Oak Lane, Philadelphia, PA 19126. USA.) is used as UV radiation source. The sun simulator UVA beam is reduced by applying suitable filters to the natural sunlight level of 55 ± 5 W/m^2 [8], measured with Optometer P9710 (Gigahertz-Optik GmbH, Postfach 1445, 82170 Puchheim, Germany). Figure 1 gives a schematic representation of the irradiation experimental setup.

2.2. Spectrophotometry analysis of samples

Optical measurements were obtained from a SPECORD-210 double beam UV-visible spectrophotometer. A baseline was set with empty quartz cuvettes in both beams in order to establish a zero reading, and then cuvette filled with FXG gel was placed in the analyzing beam in order to get the changes in the gel absorption. The absorption was measured in the wavelengths range from 350nm to 650nm with 1nm intervals and a peak centered at about 550nm is found suitable for quantitative measurements. Quartz cuvettes are transparent to UV therefore it can be used for the irradiation and optical measurement of gel samples. The gel itself absorbs UV radiation much better than water; for example, UVB has the ability to penetrate about 17 cm of water with reduction of only 30% in its intensity [8].

3. Results and Discussion:

3.1. The Action of Radiation on FXG Gel

The Fricke solution depends on the radiolysis of water giving number of free radicals which are active species that will interact with the ferrous ions and cause them to oxidize. Similarly, the aerated FXG system relies on H˙, OH. and H_2O_2 to oxidize the ferrous ions [7]. The energy of UV light is such that it lies between ionizing and non-ionizing radiation for water. However, it is found that the ionizing events that can produce highly reactive radicals inside FXG materials are not the only processes taking place in the gel, but the excitation collisions also serve as an intermediary species within the interactions between radiation and the sensitive chemical system. Furthermore, it may also be the case that vibrationally excited (rather than electronically excited) molecules may instigate chemical reactions [7]. The production of ferric ions is not strictly dependent on the radiolysis of water. Instead, the ferric ions are likely to be produced indirectly by oxidizing events that are happing in the water by the absorbed energy of UV radiation. Figure 2 shows the optical absorbance measured for FXG sample irradiated with 105 W/m2 artificial UVA radiations.

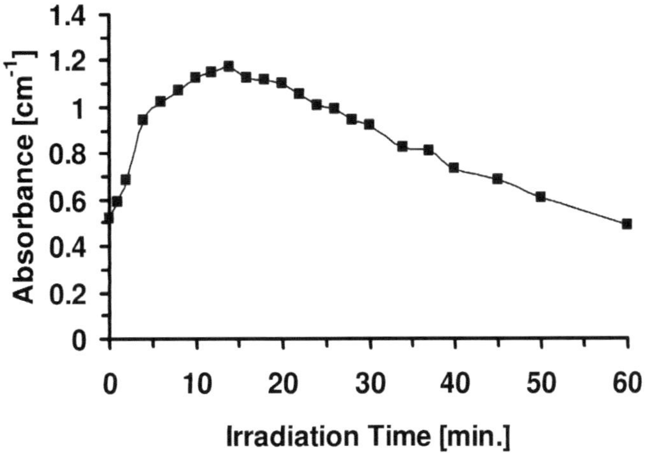

Figure 2: The optical absorbance of 1cm thick FXG samples under 105 W/m^2 UVA irradiation.

3.2. Ferrous Xylenol orange Gel (FXG) Dosimeter

By exploiting the visible colour change from orange to purple, FXG samples optical absorption measurements can be made to quantify the production of ferric ions. The ferric ions, produced by the radiation action, form complex with xylenol, but the ferrous ions do not, therefore the changes in ferric ion concentration results in proportional changes in the absorption of light. The conversion of ferrous to ferric ions is utilized to provide a practical dosimeter; the changes in the light absorption measured at different wavelengths for irradiated gel samples can be quantified and related to the delivered radiation doses. The optical response of the gel system (FXG) was studied at the wavelength 550nm. Before irradiation the FXG gel is orange, but once the gel is irradiated, the absorption at the 550 nm increases and the gel color becomes purple. By increasing the UVA irradiation (E_e) W/m^2 after filtering the beam to the measured natural sunlight level light absorption found increases linearly over the range 0 up to 15 minutes at exposure rate of 55 W/m^2 to UVA radiation, see figure 2.

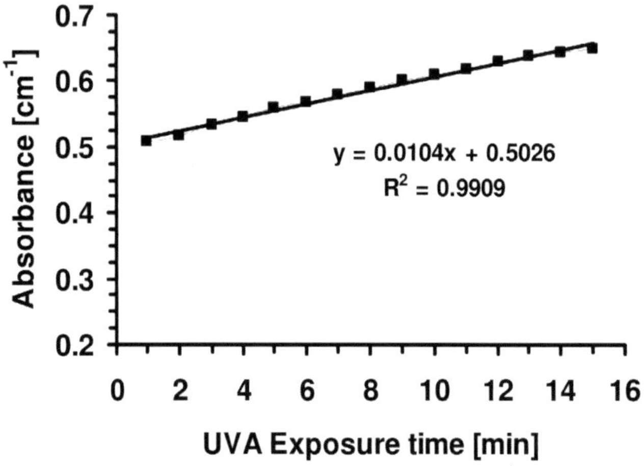

Figure 3: FXG response to UVA measured with standard size quartz cuvettes at 550 nm.

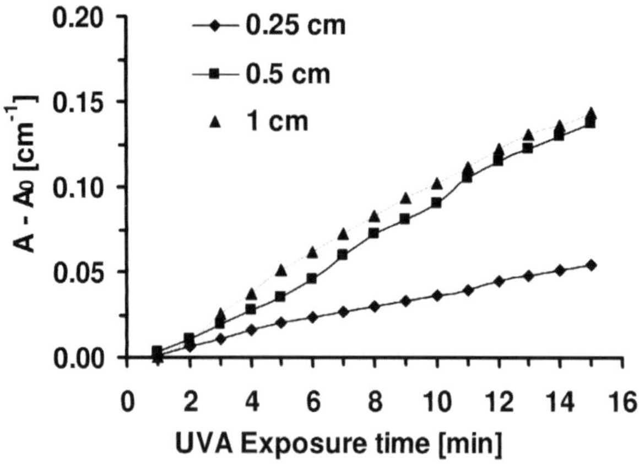

Figure 4: Absorption variation for three FXG layers of different thickness after illuminating the absorption of the unexposed samples.

Figure 3, illustrates the UVA effects on three FXG samples with different thicknesses. These samples were studied in order to evaluate its penetration ability and to demonstrate the hazards caused by this type of radiation and its damaging effects on human skin. It shows that major portion of UVA radiation get absorbed in 5 mm thickness of FXG gel.

4. Conclusions

The results show that exposing FXG gel to ultraviolet radiation gives a measurable and reproducible effect. The increase in absorption at 585nm has been shown to be linearly dependent on the energy incident on the gel and, consequently, the gel can potentially measure the ultraviolet exposure and visualize its harmful effects directly. Finally, the UV exposure of the population has a prominent profile, due to the grave implications of its link to skin cancer. The public awareness of UV is very high, and burn times are even routinely included in weather forecasts during the summer. Hence, there could be a demand for a passive UV detector made from very simple chemical ingredients and which has the attractive property of changing from orange to purple with increased exposure.

Acknowledgments

The authors would like to thank Prof. I. Othman, AECS general director, for his support and interest in this work. Thanks also to Dr. M.S. Al-Masri for useful discussions of the project.

References

[1] Commission Internationale de l'Eclairage (CIE), 1987, Vocabulaire International de l'Eclairage, Publication **50** pp. 845.

[2] Diffey, BL, 2004, Climate change, ozone depletion and the impact on ultraviolet exposure of human skin, *Phys. Med. Biol.* **49** R1–R11.

[3] Diffey, BL, 1991, Solar ultraviolet radiation effects on biological systems, *Phys. Med. Biol.* **36** 299-328.

[4] Leffell, DJ and Brash, DE, 1996, Sunlight and Skin Cancer, *Scientific American Magazine*.

[5] Lee, J., 1998, Our Health in Flux, Inside Science Number **116,** *NewScientist* **12** pp. 4.

[6] International Commission on Non-Ionizing Radiation Protection, 2007, Protecting Workers from Ultraviolet Radiation Protection, Editors: Paolo Vecchia, et.al., ICNIRP, Germany.

[7] BERO, MA, et al., 2000, Tissue-equivalent gel for non-invasive spatial radiation dose measurements, *Nucl. Instrum. Meth.* **B 166–167** 820–5.

[8] Othman, I, et al., 1997, Seasonal Variations of Solar UVB and UVA in Syria, *Health Physics* **72**, 49-52.

[9] Ching-Shen, S and Shyi-Ming, Y, 1996, UV Attenuation Coefficient in Water Determined by Thermoluminescence Detector, *Radiation Measurements* **26**, 83-6.

Journal of Physics: Conference Series **164** (2009) 012065 doi:10.1088/1742-6596/164/1/012065

Designing phantom for head-and-neck treatment verification: feasibility tests with bone and bone equivalent material incorporated into polymer gel

Jitka Šemnická[1,2], Václav Spěváček[1], Tomáš Veselský[1], Ondřej Konček[2,3], Josef Novotný Jr.[4]

[1]Czech Technical University in Prague, Faculty of Nuclear Sciences and Physical Engineering, Břehová 7, Prague 1, 115 19, Czech Republic

[2]Na Homolce Hospital, Department of Medical Physics, Roentgenova 2, Prague 5, 150 30, Czech Republic

[3]Motol Hospital, Department of Radiotherapy and Oncology, Prague 5, V Úvalu 84, Czech Republic

[4]University of Pittsburg, Center for Image-Guided Neurosurgery, 200 Lothrop Street, Pittsburgh, USA

jitka.semnicka@homolka.cz

Abstract. This work presents different approaches to manufacture of polyacrylamide gel (PAG) with bone inhomogeneity in an attempt to construct an authentic phantom for verification of head-and-neck irradiation specially for quantifying the absorbed dose in the spine. A special phantom was designed which includes air tube simulating thorax and accommodates model of spinal canal. Several methods of bone implantation into PAG were tested. The results indicate that proper mechanical surface cleaning of the bones eliminates the detrimental effects of chemicals used for bone conservation, while sufficient nitrogen flushing during the manufacture procedure effectively lowers the amount of oxygen present in the bone pores.

1. Introduction

In the last decade, polymer gels proved to be a very useful tool for three-dimensional visualization and measurement of absorbed dose distribution in a variety of medical applications – stereotactic radiosurgery (SRS) [1], intensity modulated radiotherapy (IMRT) [2], brachytherapy [3] or breathing adapted radiotherapy (BART) [4]. Most gel authors sought to model the real world situation using appropriate shapes and dimensions of gel phantoms (e.g. pelvic phantom for prostate IMRT verification, etc...); however, the flexibility of gel dosimeters enables to manufacture even more authentic phantoms, incorporating different kinds of inhomogeneities either of natural origin, simulating bones [5, 6] and cavities [6, 7], or artificial implants and clips [8]. These kinds of heterogeneous phantoms along with a convenient three-dimensional read-out method provide the only experimental way to quantify the absorbed dose with sufficient spatial resolution in regions of electronic disequilibrium [5]. The manufacture of heterogeneous phantoms is more complicated than pure gel phantoms since the implanted heterogeneity must be chemically stable and oxygen free (if normoxic gel is not used) to eliminate fatal errors in measurement results interpretation [5]. Moreover, when magnetic resonance imaging (MRI) is used to read-out the response of the irradiated gel, a special requirement is imposed on the MRI compatibility of the implanted heterogeneity [8] and the gel container.

This work aims to investigate the possibility of implanting real bone samples and bone substitutes into a gel sample. Attention is paid to the head-and-neck treatment verification as this location presents a complex site with irregular surface and different heterogeneous organs (e.g. spinal cord, vertebrae, larynx, etc.). A dedicated glass phantom was designed to simulate the shape of the head-and-neck region. It can accommodate vertebrae (real or bone-like material) and thus simulate the spinal canal filled with the radiosensitive gel.

The following text summarizes practical experience and complications encountered when implanting real bones into a polymer gel.

2. Materials and Methods

2.1. Head-and-neck phantom design

A glass hollow bust-shaped container was designed with maximum wall thickness four millimeters. The container fits into a polystyrene bed, which contains contrast markers (tubes with copper sulphate solution or copper wires) to facilitate co-registration between different sets of transversal image data. The phantom can be supplied with vertebrae or its substitutes (Figure 1).

During the first MRI scanning session the phantom was tested for artifacts. A real set of neck vertebrae was positioned inside the container, which was filled with gelatine to fix the bones in one position (Figure 2).

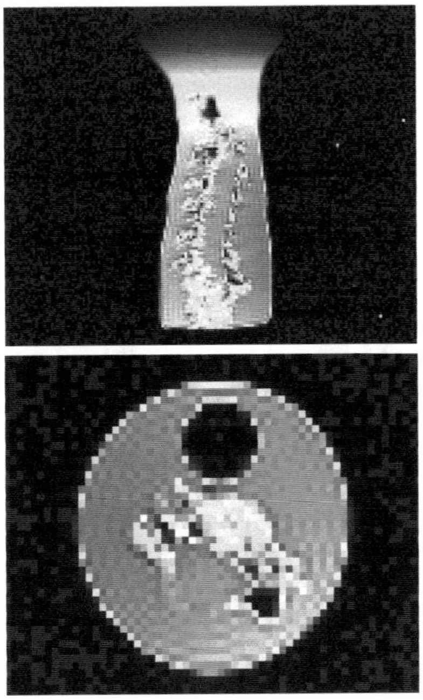

Figure 1: sagittal and axial T2 images of gelatine filled H&N phantom with vertebrae inserted. The pixel size 2 mm, slice thickness 5 mm, 1 acquisition. Note the spinal canal in the sagittal image and the air tube in the axial image simulating the throat.

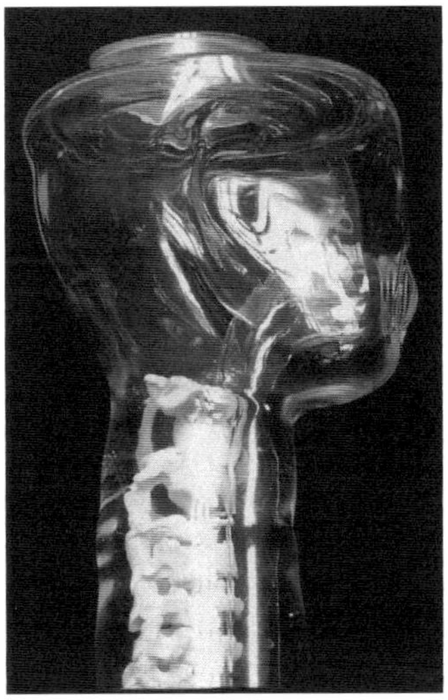

Figure 2: Head-and-neck phantom (H&N)

An alternative bone equivalent material [9] (relative electron density to water 1.514, mass density 1.609 g/cm^3, used for electron density/HU calibration in radiotherapy planning) was also tested for MRI compatibility, but shaped bone substitutes are not available yet.

2.2. Gel composition and preparation

The composition of a polyacrylamide gel (PAG) used in this work is given in Table 1. The gel preparation followed a well-established procedure using a perspex nitrogen filled glove box [10]. All gel containers were washed with distilled water prior to filling to avoid gel contamination.

Component	*Weight fraction*	*Manufacturer*
Gelatine, swine skin	5	SIGMA Aldrich
Acrylamide	3	Riedel-de Haen, Seelze
N,N'-methylene-bisacrylamide	3	APPLICHEM, Darmstadt
Distilled water	89	

Table 1: PAG composition, weight fraction in [%]

2.3. Implanting bone samples into gels
To investigate the chemical influence of a bone on the gel dosimeter (no previous information was available on the chemicals used to conserve the bone samples) a small piece of bone (the same batch as the vertebrae) was implanted in a glass flask filled with PAG during the standard manufacture procedure (in the nitrogen flushed perspex box). This sample wasn't irradiated due to vast polymerization which occurred within few hours from preparation.

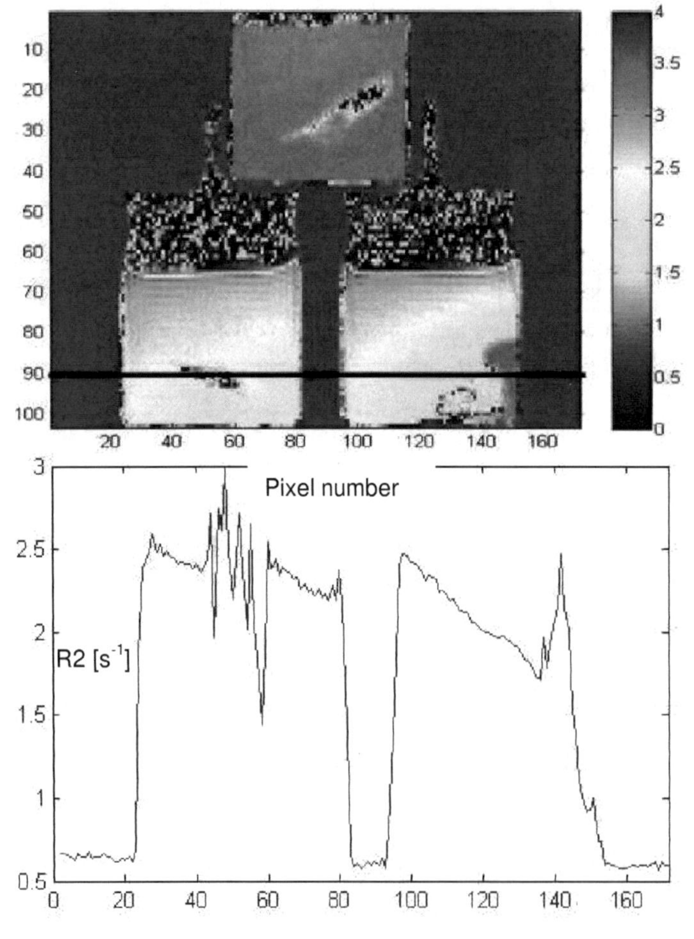

Figure 3: R2 map [s^{-1}](left bottom: mechanically cleaned bone; 11,5 Gy, right bottom: bone cleaned with acetone 11,5 Gy, top: mechanically cleaned bone; unirradiated) and corresponding R2 profile (smoothed) of gel samples with bone implants; coronal orientation, slice thickness 5 mm, pixel size 1mm, 2 acquisitions

During the second test, three samples of bones were used. Two out of the three were mechanically cleaned and washed with distilled water. The third piece was additionally cleaned with acetone and then washed with distilled water as well. The bone samples were immersed in a nitrogen atmosphere (approximately 1 hour) to flush oxygen from its pores prior to the positioning into the glass containers.

2.4. Gel irradiation
The PAG samples (the second test) with implanted bones were homogenously irradiated with a Gammacell 220 unit (MDS Nordion, Canada) with the dose rate of 7.75 Gy/h five days from the

manufacture procedure (stored in refrigerator). Two samples with bones, one cleaned mechanically and the one cleaned with acetone received 11.5 Gy. The remaining sample (mechanically cleaned) was left unirradiated as a reference.

2.5. MRI scanning protocol and data processing

The three PAG samples (the second test) were scanned 12 hours from irradiation. The same MRI scanning sequence was used for the PAG samples and the H&N phantom: 2D T2 weighted multiple spin echo sequence (16 echoes), with equidistant TE 22.5 ms, TR 2000 ms. Number of signal averaging, slice orientation, pixel size and slice thickness varied and are indicated alongside of each particular figure. The scanning was performed using Siemens Magnetom Expert (Siemens, Germany) 1T scanner in the head coil and the T2 map was reconstructed using the scanner's manufacturer software (Numaris) since no significant difference was observed in comparison with least-squares fit of echo images performed aside; the first echo was excluded.

T2 maps and profiles were further processed in Matlab 6.5 (The Mathworks, USA) equipped with the Image Processing Toolbox.

3. Results and discussion

Phantom imaging revealed that no significant artifacts are present in the images. The major complication could be caused by susceptibility effects close to the phantom walls, which would be more pronounced, if stronger magnetic field was used. Glass wall also presents the major drawback from the point of view of designing an authentic phantom.

In the first experiment, with the bone implant without any surface cleaning, the polymerization process was noticeable within few hours after gel preparation around the bone implant apparently induced by unknown chemicals from the bone (the sample was not irradiated).

Results of the second experiment with bones cleaned mechanically and with acetone are summarized in Figure 3. It shows the R2 map with the three bone samples and corresponding R2 profile.

In the second experiment, the first container with mechanically cleaned bone (11.5 Gy) shows reduction of response around the bone most probably caused by remaining oxygen in its pores. The acetone-cleaned sample exhibits stronger reduction in R2 probably due to traces of acetone. The last sample (mechanically cleaned, unirradiated) shows that polymerization takes place only on the surface of the implanted bone, which implies that the mechanical surface cleaning reduced the effect of unknown chemicals (compared to the first trial with bone implanted without any additional manipulation).

The results indicate that the way of reducing oxygen levels in bone pores by sufficient nitrogen flushing can be successful when it is performed for longer time interval. PAG compositions enriched with tetrakis(hydroxymethyl)phospohonium chloride (THPC) were not tested due to its reduced sensitivity compared to the standard PAG. A different way how to overcome complications with real bone implants could be to coat them in some chemically resistant substance or to use bone substitutes with similar parameters as a real bone (electron density and mass density) in order to model similar interaction processes of radiation [5, 6, 7].

4. Conslusions

Further research in bone equivalent materials and real bone implants is ongoing.

5. Acknowledgments

This work is supported by IGS2008 (The Czech Technical University in Prague).

References

[1] Novotny J., Dvorak P., Spevacek V., et al. Quality control of the stereotactic radiosurgery procedure with polymer-gel dosimetry. Radiother. Oncology 2002, 63(2):223-30

[2] Gusstavsson H., Karlsson A., Back S.A.J., et al. MAGIC-type polymer gel for three-dimensional dosimetry: Intensity-modulated radiation therapy verification. Med. Phys. 2003, 30(6):1264-1271

[3] Pappagiannis P., Karaiskos P., Georgiou E. On the use of high dose rate ^{192}Ir ^{169}Yb sources with the MammoSite® Radiation therapy system. Med. Phys. 2007, 34(9):3614-3619

[4] Månsson S., Karlsson A., Gustavsson H. Preliminary proceedings of DOSGEL 2006, 7 – 10 August 2006, 291-293

[5] Hepworth S.J., McJury M., Oldham M., et al. Dose mapping of inhomogeneities positioned in radiosensitive polymer gels. NIMA 422, 1999, 756-760

[6] Wanatabe Y., Mooij R., Perera G.M., et al. Heterogenity phantoms for visualization of 3D dose distributions by MRI-based polymer gel dosimetry. Med. Phys. 2004, 31(5):975-984

[7] Isbakan F., Ülgen Y., Bilge H., et al. Gamma Knife 3-D dose distribution near the area of tissue inhomogeneities by normoxic gel dosimetry. Med. Phys. 2007, 34(5), 1623-1630

[8] Geso M., Ackerly T., Brown S., et al. Determination of dosimetric perturbations caused by aneurysm clip in stereotactic radiosurgery using gel phantoms and EBT-Gafrochromic films. Med. Phys. 2008, 35(2):744-752

[9] www.agmecolt.cz

[10] Novotny J., Spevacek V., Dvorak P. Energy and dose rate dependence of BANG-2 polymer gel dosimeter. Med. Phys. 2001, 28(11):2379-2386

Evaluation of Dose Delivery Accuracy of Gamma Knife Using MRI Polymer Gel Dosimeter in an Inhomogeneous Phantom

A. Pourfallah T.[1], Riahi Alam N.[1*], Allahverdi M.[1], Ay M.[1,2], Zahmatkesh M.[3]

[1]Department of Medical Physics and Biomedical Engineering, School of Medicine, Tehran University of Medical Sciences (TUMS),Tehran-Iran

[2]Research Center for Science and Technology in Medicine, Tehran University of Medical Sciences (TUMS), Tehran-Iran

[3]Novin Medical Radiation Center, Tehran-Iran

*Corresponding author : e-mail: riahinad@sina.tums.ac.ir

Abstract : Polymer gel dosimetry is still the only dosimetry method for directly measuring three-dimensional dose distributions. MRI Polymer gel dosimeters are tissue equivalent and can act as a phantom material. Because of high dose response sensitivity, the MRI was chosen as readout device.
In this study dose profiles calculated with treatment-planning software (LGP) and measurements with the MR polymer gel dosimeter for single-shot irradiations were compared.
A custom-built 16 cm diameter spherical plexiglas head phantom was used in this study. Inside the phantom, there is a cubic cutout for insertion of gel phantoms and another cutout for inserting the inhomogeneities. The phantoms were scanned with a 1.5T MRI (Siemens syngo MR 2004A 4VA25A) scanner. The multiple spin-echo sequence with 32 echoes was used for the MRI scans. Calibration relations between the spin-spin relaxation rate and the absorbed dose were obtained by using small cylindrical vials, which were filled with the PAGAT polymer gel from the same batch as for the spherical phantom.
1D and 2D data obtained using gel dosimeter for homogeneous and inhomogeneous phantoms were compared with dose obtained using LGP calculation.
The distance between relative isodose curves obtained for homogeneous phantom and heterogeneous phantoms exceed the accepted total positioning error (>±2mm).
The findings of this study indicate that dose measurement using PAGAT gel dosimeter can be used for verifying dose delivering accuracy in GK unit in presence of inhomogeneities.

Key words: Gamma Knife (GK), PAGAT Gel Dosimeter, Inhomogeneity, Planning System (LGP)

1. Introduction

Stereotactic Gamma-Knife (GK) radiosurgery plays an important role in managing small intracranial brain lesions (of volume typically less than 25 cm^3) (1, 2).
Currently, polymer gel dosimetry is still the only dosimetry method for directly measuring three-dimensional dose distributions. Polymer gel dosimeters are tissue equivalent and can act as a phantom material (3, 4).
In this study the PAGAT (PAG And THPC anti-oxidant) gel dosimeter was used to investigate the effects of air and PTFE inhomogeneities on dose delivery accuracy of Gamma Knife systems.
Some studies examined the shot placement accuracy of Gamma Knife units by measuring dose distribution or measuring the distance between the location of the maximum dose and the mechanical center (5-7).
Comparison of 1D line profiles and 2D dose distributions between measurements and LGP calculations was undertaken by some investigators (7-11). However, investigation of dose distribution in a single shut GK treatment in presence of air and PTFE inhomogeneities is the purpose of this study.

2. Materials and methods

In this study PAGAT polymer gel dosimeter was fabricated according to composition proposed by Venning et al (2005) (12) who noted Using MRI, the formulation to give the maximum change in the transverse relaxation rate $R2$ was determined to be 4.5% *N,N'*–methylene-bis-acrylamide (bis), 4.5% acrylamide (AA), 5% gelatine, 5 mM THPC, 0.01 mM hydroquinone (HQ) and 86% H2O.
For fabricating the gel dosimeter, the De Deene et al (2006) (13) proposed method have been used.
Phantom in this study was a 16cm spherical plexiglas in which there is a cubic cutout for inserting the gel vials (4×4×4 cm^3) and another cutout (4×4×3 cm^3) for inserting the air and/or a bone equivalent material (Poly-tetra-fluoro-ethylene (PTFE), with density of 2.2 gm/cm^3). Designing the phantom and it's cutouts for inserting the inhomogeneities were performed using EGSnrc code (unpublished results).
Figure 2 shows the phantom placed in a Gamma Knife unit (model C) for irradiation.
The calibration tubes were irradiated using the Theratron Co-60 machine using a special container in which the calibration vials could be located horizontally. The calibration vials were irradiated from 0 to 50 Gy with steps of 2.5 and 5 (i.e. 0, 2.5, 5, 10, 15, 20, 25, 30, 35, 40,45 and 50). Post-manufacture irradiation time was 24 hours.

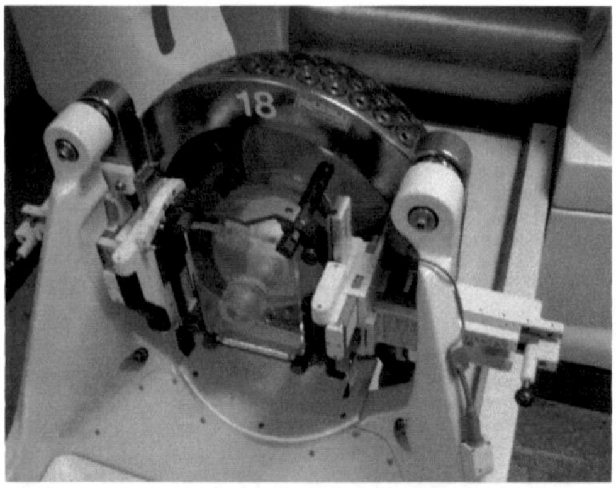

Figure 1. Phantom placed in a Gamma Knife unit (model C) for irradiation.

The evaluation of the polymerized dosimeter was performed on a Siemens (syngo MR 2004A 4VA25A) 1.5 T scanner in the transmitter/receiver head coil. A multi-echo sequence with 32 echoes was used for the evaluation of irradiated polymer-gel dosimeters. The parameters of the sequence were as follows: TR 3000 ms, TE 22–640 ms, slice thickness 1 mm, FOV 128 mm, matrix size 256×256, pixel size 0.5×0.5 mm^2, one acquisitions. The *R*2 and dose maps were computed using modified radiotherapy gel dosimetry image processing software coded in MatLab.

3. Results and Discussion

Calibration data for the PAGAT gel batch used in this work were derived by the analysis of axial T2 maps of the calibration gel tubes 24 hours post-irradiation and a cubic and quadratic fit was performed on R2 values of PAGAT(A) and (2), in the dose region of 0–30 Gy, and 5-40, respectively. Post-irradiation imaging time was also 24 houres.
After discarding the first echo of the 32-echo train, a single T2 (spin–spin relaxation time) map was automatically derived for each reconstructed slice. These maps were exported from the scanner in DICOM-3 format and then imported into MatLab (14) to construct a 2D T2 matrix which was subsequently converted to an R2 (= 1/T2) relaxation rate matrix. The *R*2 matrix was subsequently converted into a relative dose matrix normalized to the maximum prescribed dose of 40 Gy.
Figure 2 compares the relative dose profiles along the X axis of the experimental reference coordinate system for the PAGAT gel irradiated with the 18 mm collimator helmet between homogeneous and GP prediction. Dose difference between these profiles is in average 2.7%±1.9%, and in maximum is 6%.

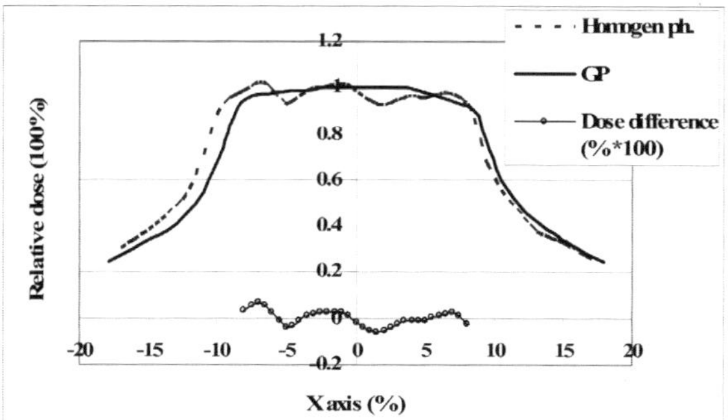

Figure 2. Relative dose profiles along the X axis of the experimental reference coordinate system for the irradiation with the 18 mm collimator helmet, obtained using homogeneous phantom (- - -), and GP prediction (—). Dose difference is also depicted (–◦–).

Figure 3 compares the relative dose profiles along the X axis of the experimental reference coordinate system for the PAGAT gel irradiated with the 18 mm collimator helmet between homogeneous and heterogeneous phantoms.

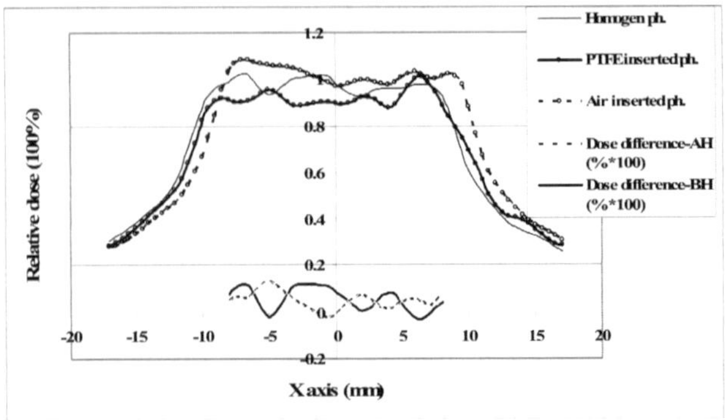

Figure 3. Relative dose profiles along the X axis of the experimental reference coordinate system for the irradiation with the 18 mm collimator helmet, obtained using homogeneous phantom (—), PTFE inserted phantom (-◦-), and air inserted phantom (-◦-◦-).Dose difference is also depicted, between air inserted phantom and homogeneous phantom (- - -), and PTFE inserted phantom (—).Dose difference between homogeneous phantom and air inserted phantom is in average 5.2%±3.3%, and dose difference between homogeneous phantom and PTFE inserted phantom is in average 6.1%±3.9%.

Dose difference between homogeneous phantom and air inserted phantom is in average 5.2%±3.3% and in maximum 13.2%, and dose difference between homogeneous phantom and PTFE inserted phantom is in average 6.1%±3.9% and in maximum 12%.
Figure 4 shows the comparative relative dose distributions for the irradiation with the 18 mm collimator helmet between homogeneous and PTFE inserted phantom and also between homogeneous and air inserted phantom.

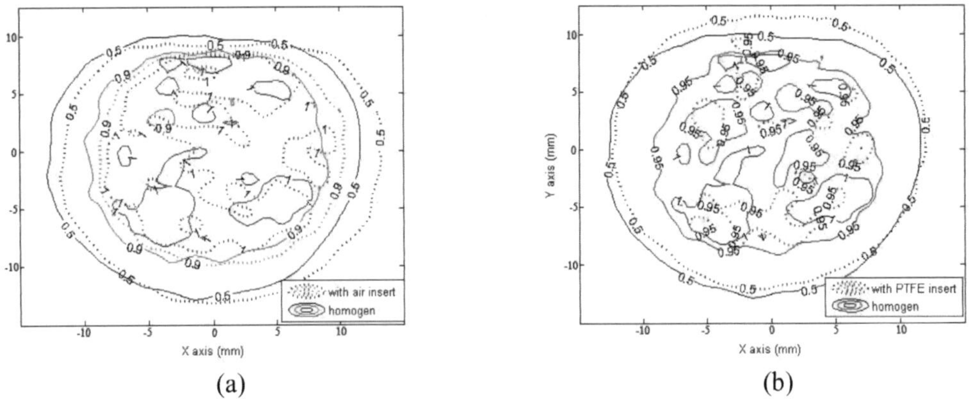

Figure 4. Comparative relative dose distributions for the irradiation with the 18 mm collimator helmet between homogeneous and air inserted phantom (a), and also homogeneous and PTFE inserted phantom (b). Distance between 100%, 90%, and also 50% isodose lines in maximum exceed the 2mm.

As depicted in figure 4, distance between 100%, 90%, and also 50% isodose lines in maximum exceed the 2mm. In addition, in PTFE inserted phantom relative isodose at most reach 95% and in air inserted phantom it exceed 105%.
The total positioning error, based on the surface contouring accuracy, MRI fiducial correspondence and overall positioning accuracy of the Gamma Knife is ±2 mm, similar to the

error expected for patient treatment (15), so, the distance between relative isodose curves exceed the total positioning error.

4. Conclusion

The discrepancies observed between the results obtained for heterogeneous and homogeneous phantoms suggest that GP predictions must be corrected in order to take care of the air- and bone-tissue inhomogeneities.
In this respect it is worthwhile to mention that an air inhomogeneity that could be assumed as maxillary or frontal sinuses gives rise to modifications of the dose profiles and dose distribution.

References

[1] Leksell L. The stereotaxic method and radiosurgery of the brain. *Acta Chir Scand* 1951;102:316–319
[2] Mack A, Scheib SG, Major J, et al. Precision dosimetry for narrow photon beams used in radiosurgery-determination of Gamma Knife output factors. *Med Phys* 2002;29:2080–2089
[3] Fong P, Keil D, Does M, Gore J. Polymer gels for magnetic resonance imaging of radiation dose distributions at normal room atmosphere. *Phys Med Biol* 2001;46:3105–3113
[4] Venning A, Nitschke K, Keall P, Baldock C. Radiological properties of normoxic polymer gel dosimeters. *Med Phys* 2005;32:1047–1053
[5] Ertl A, Saringer W, Heimberger K, Kindl P. assurance for the Leksell gamma unit: Considering magnetic resonance image-distortion and delineation failure in the targeting of the internal auditory canal. *Med Phys* 1999;26:166–170
[6] Guo WY, Chu WC, Wu MC. An evaluation of the accuracy of magnetic-resonance-guided Gamma Knife surgery. *Stereotact Funct Neurosurg* 1996;;66:85–92
[7] Watanabe Y, Akimitsu T, Hirokawa Y, Mooij RB, Perera GM. Evaluation of dose delivery accuracy of Gamma Knife by polymer gel dosimetry. *J Appl Clin Med Phys* 2005;6:133-142
[8] Novotny J, Dvorak P, Spevacek V. Quality control of the stereotactic radiosurgery procedure with the polymer-gel dosimetry. *Radiother Oncol* 2002;63:223–230
[9] Ertl A, Berg A, Zehetmayer M, Frigo P. High-resolution dose profile studies based on MR imaging with polymer BANG(TM) gels in stereotactic radiation techniques *Magn Reson Imag* 2000;18:343–349
[10] Novotny JJ, Novotny J, Spevacek V. Application of polymer gel dosimetry in gamma knife radiosurgery. *J Neurosurg* 2002;7:556–562.
[11] Scheib SG, Gianolini S. Three-dimensional dose verification using BANG gel: A clinical example. .. *J Neurosurg* 2002;97:582–587
[12] Venning AJ, Hill B, Brindha S, Healy BJ, Baldock C. Investigation of the PAGAT polymer gel dosimeter using magnetic resonance imaging. *Phys Med Biol* 2005;50:3875-3888
[13] De Deene Y, Vergote K, Claeys C, De Wagter C. The fundamental radiation properties of normoxic polymer gel dosimeters: a comparison between a methacrylic acid based gel and acrylamide based gels. *Phys Med Biol* 2006;51:653-673
[14] MathWorks. Image Processing Toolbox 5. *MATLAB v 7* 2007
[15] Moskvin V, Timmerman R, DesRosiers C, et al. Monte Carlo simulation of the Leksell gamma Knife (R): II. Effects of heterogeneous versus homogeneous media for stereotactic radiosurgery. *Phys Med Biol* 2004;49:4879-4895

Journal of Physics: Conference Series **164** (2009) 012067 doi:10.1088/1742-6596/164/1/012067

MAGIC with formaldehyde applied to dosimetry of HDR Brachytherapy source

Marques, T. [1], **Fernandes, J.** [1], **Barbi, G.** [2], **Nicolucci, P.** [1], **Baffa, O.** [1]

[1]Departamento de Física e Matemática, FFCLRP, Universidade de São Paulo, Av. Bandeirantes 3900, 14040-901 – Ribeirão Preto, SP, Brazil.

[2]Hospital das Clinicas da Faculdade de Medicina de Ribeirão Preto, Av. Bandeirantes 3900, 14040-900 – Ribeirão Preto, SP, Brazil.

tatiana_mp@terra.com.br

Abstract. The use of polymer gel dosimeters in brachytherapy can allow the determination of three-dimensional dose distributions in large volumes and with high spatial resolution if an adequate calibration process is performed. One of the major issues in these experiments is the polymer gel response dependence on dose rate when high dose rate sources are used and the doses in the vicinity of the sources are to be determinated. In this study, the response of a modified MAGIC polymer gel with formaldehyde around an Iridium-192 HDR brachytherapy source is presented. Experimental results obtained with this polymer gel were compared with ionization chamber measurements and with Monte Carlo simulation with PENELOPE. A maximum difference of 3.10% was found between gel dose measurements and Monte Carlo simulation at a radial distance of 18 mm from the source. The results obtained show that the gel's response is strongly influenced by dose rate and that a different calibration should be used for the vicinity of the source and for regions of lower dose rates. The results obtained in this study show that, provided the proper calibration is performed, MAGIC with formaldehyde can be successfully used to accurate determinate dose distributions form high dose rate brachytherapy sources.

1. Introduction

Polymer gel dosimetry offers a great potential to measure complex dose distributions in three dimensions, with the ability of measuring high dose gradients. This special characteristic of gel dosimetry can be applied on high dose rate (HDR) brachytherapy, solving some of the most important problems in this dosimetry: the lack of spatial resolution and non-tissue equivalence of conventional dosimeters used like thermoluminescent dosimeters, diodes and films [1].

The usefulness of gel dosimetry for external beam radiation therapy has been shown by several studies [2-4], but its application for HDR brachytherapy studies is conditioned to the understanding of some specific imaging and radiochemical influences that can result in significant dose errors [5]. Measurements in the vicinity of Brachytherapy sources are susceptible to oxygen permeability from the catheter wall that leads to an inhibition of polymerization and to magnetic susceptibility artefacts during imaging of the gel phantom. Another source of errors can be the diffusion of monomers during irradiation causing overshoots in regions of high dose and high dose gradients.

In this study we used MAGIC gel with formaldehyde to measure dose distributions of a 192-Iridium brachytherapy source while some simple procedures were used to avoid the above cited problems. We have also compared our results with Monte Carlo Simulation performed using PENELOPE code with an implementation that considers the extended source geometry used and a voxel model to obtain volumetric dosimetry.

2. Materials and methods

2.1. *Gel preparation*

The gel preparation was described elsewhere [6]. Part of the gel was poured into six cylindrical glass tubes routinely used for blood sample collection (BD Vacutainer® with 5ml volume, 12mm diameter and 75 mm height, closed with a 20mm hermetic stopper inserted in a plastic cover) and the rest was poured into an acrylic rectangular recipient that is impervious to oxygen and is normally used to make and visualize cell cultures (CellStar® 50 ml volume).

2.2. *Gel irradiation*

The GammaMed Plus 232 192-Ir source used to irradiation has an active cylindrical volume of 0.6 mm diameter and 3.5 mm height encapsulated in stainless steel that is welded to a steel cable. An esophageal catheter of 1.6 mm radius was used to guide the source. The source activity was 5.43 Ci (200.91 GBq) by the time the experimental procedures were performed.

A cubic box of 30 cm height filled with water was used to accommodate the recipients containing the polymer gel during irradiation. The rectangular acrylic recipient was irradiated with the catheter held on its surface while the source was scheduled to stop at the center of the gel volume, as shown in figure 1.a. In this case, a continuous distribution of dose rates reaching the dosimeter gel is achieved, the portions of the gel closer to the source position receiving higher dose rates. The glass tubes were positioned in an acrylic fixture to keep them aligned alongside with the center of the catheter during irradiation, as shown in figure 1.b. Five tubes have been irradiated (one at a time), the first one being 16 mm away from the catheter and the next ones in 10 mm distances from each other. In this case, the polymer gel was in a region of lower dose gradient. In water measurements were also made in several radial distances from the source with a plane parallel ionization chamber, *Markus* type, of 0.05 cm^3 volume, as shown in figure 1.c.

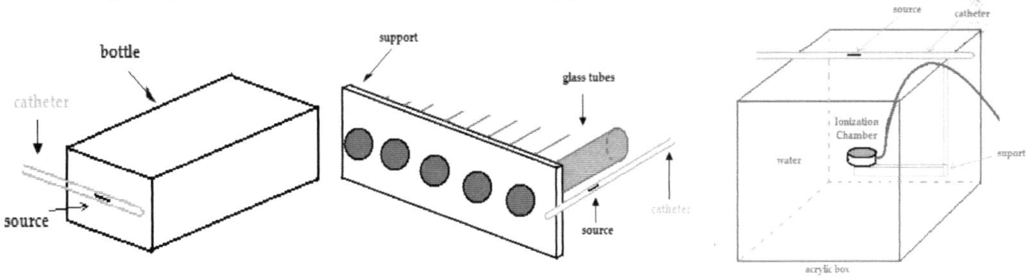

(1.a.) *(1.b.)* *(1.c.)*

Figure 1 – Irradiation setup: (1.a) Acrylic recipient - high dose gradient; (1.b) glass tubes - low dose gradient; (1.c) ionization chamber.

2.3. *MRI imaging*

MR images were acquired one day after the irradiation to allow enough time for gel reaction completion and achievement of thermal equilibrium with the MRI scanner room temperature. The images were acquired using a 1.5 T scanner (Siemens, Magneton Vision) with a head coil and multi spin echo sequences with 16 echo times multiples of 22.5 ms, a repetition time of 4000 ms and a matrix size of 256 × 256 pixels. The slice thickness was 2mm and the FOV was 150 mm. Two acquisitions were averaged for each scan. The transverse relaxation rate R2 (=1/T2) was calculated by fitting the image signal intensities versus the echo time on a pixels wise base in a specific program developed by our group in MatLab® 6.5 (Mathworks Inc).

2.4. *Monte Carlo simulation.*

The Monte Carlo simulations were run with PENELOPE code system. A user code was written to describe faithfully the geometry of the source and simulate all anisotropic properties of the brachytherapy extended source. Beta radiation of 240, 536 and 672 keV and an average energy of 376 keV for photons were used to represent the source's spectra. The simulated cubic water phantom dimensions (20 cm) were assumed to be sufficient for obtaining full scattering conditions for the relevant range of distances.

A 3D voxel-dosimetry routine was implemented to record the energy deposition values in the volume along the simulation.

3. Results and Discussions

The R2 *versus* depth curves obtained with the two experimental configurations (figures 1.a and 1.b) are plotted in figure 2.

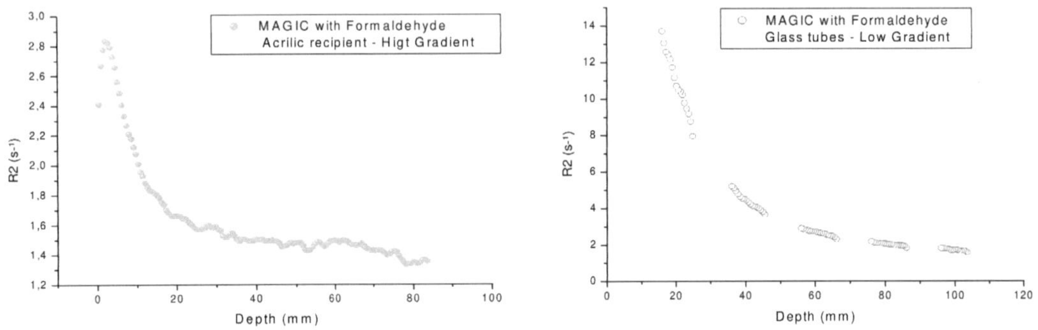

(2.a.) *(2.b.)*

Figure 2 – R2 versus depth curves for MAGIC with formaldehyde irradiated in 192- Iridium HDR brachytherapy source: (2.a.) Acrylic recipient; (2.b.) Glass tubes.

Different exposure times for each of the settings were used in order to achieve a significant signal in each configuration. Exposure times of 40.2 and 1331.6 s were used in the region of high and low gradients, respectively, in order to avoid saturation due to the radiation-induced macromolecule breakings in the higher gradient region or lack of signal in the region of lower gradient.

Figures 3.a and 3.b show the fitted curves used to correlate the R2 with dose for the acrylic recipient (high dose region) and for the glass tubes (lower dose regions), respectively. The calibration curve for the glass tubes using the same exposure time as used for the acrylic recipient resulted in the same calibration curve shown in figure 3.b.

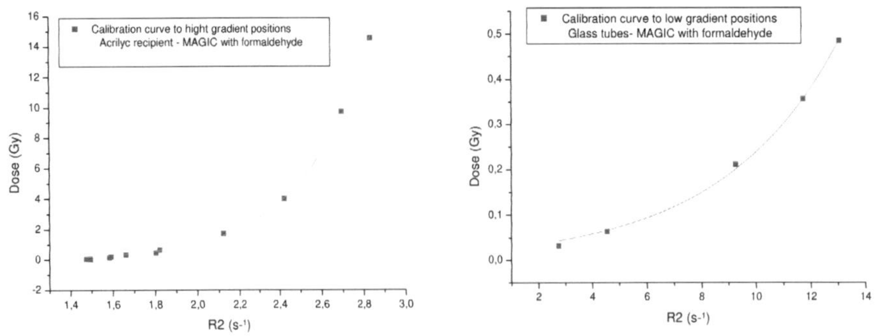

(3.a.) *(3.b.)*

Figure 3 – Calibration curves used in high and low gradient regions to correlate MAGIC gel with formaldehyde R2 MRI signal with dose: (3.a.) Acrylic recipient, with corresponding fitting $D=0.00216.e^{(1/0.32093).R2}$; *(3.b) Glass tubes, with corresponding fitting* $D=0.02306.\ e^{(1/4.27832).R2}$.

Figure 4.a presents three depth dose curves relative to radial distances from the center of the source obtained with gel data (acrylic recipient and glass tubes) and the ionization chamber. In figure 4.b. the doses are normalized to the dose at 2.0 mm away from the center of the active volume of the source (point just outside the source's capsule) to be compared with PENELOPE simulated data.

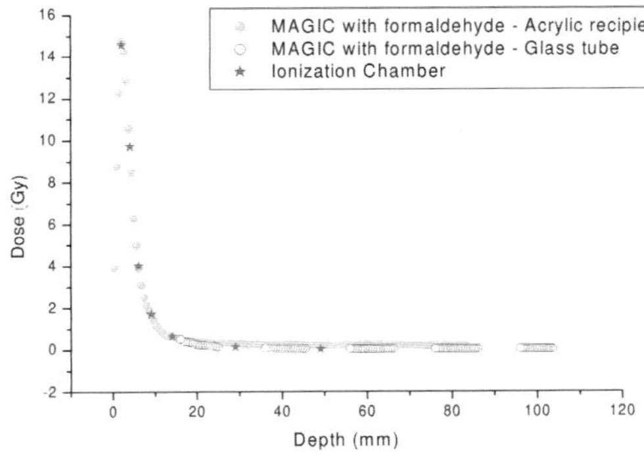

(4.a.)

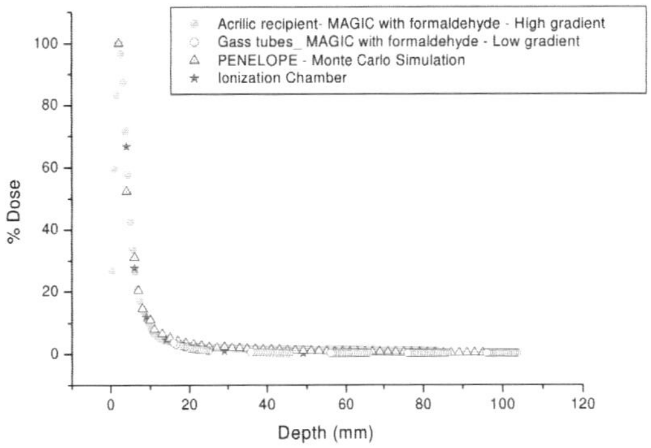

(4.b.)

Figure 4 – Depth dose curves obtained with MAGIC with formaldehyde polymer gel, ionization chamber and PENELOPE Monte Carlo simulation code in radial distances from the center of the source: (4.a.) Magic gel with formaldehyde in the acrylic recipient and in the glass tubes and ionization chamber data; (4.b.) Magic gel with formaldehyde in the acrylic recipient and in the glass tubes and ionization chamber data normalized to dose in 2.0 mm away from the center of the active source volume.

PENELOPE Monte Carlo simulation and MAGIC with formaldehyde polymer gel data has an average concordance of 99.20% in dose percentages measured with the HDR brachytherapy 192-Iridim source. A maximum difference of 3.10% between simulation and experimental dose obtained with gel appears at 18 mm from the center of the source for the acrylic recipient (at this same distance the difference between simulation and data for glass tubes is of 2.56%). The results suggest that a calibration recipient including both high and low dose rates, such as the acrylic recipient used, provides an accurate calibration curve for all therapeutic relevant distances.

R2 distributions, dose maps and dose gradient maps for the acrylic recipient and the glass tubes are shown in figure 5. The left side of this figure (5.a, 5.c, 5.e) shows results for the acrylic recipient, relative to high gradient regions, and the right side of the figure (5.b, 5.d, 5.f) shows results for the glass tubes, relative to lower gradient regions.

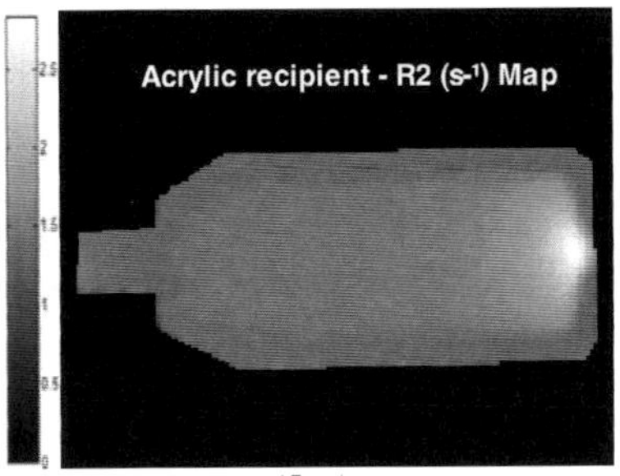

(5.a.)

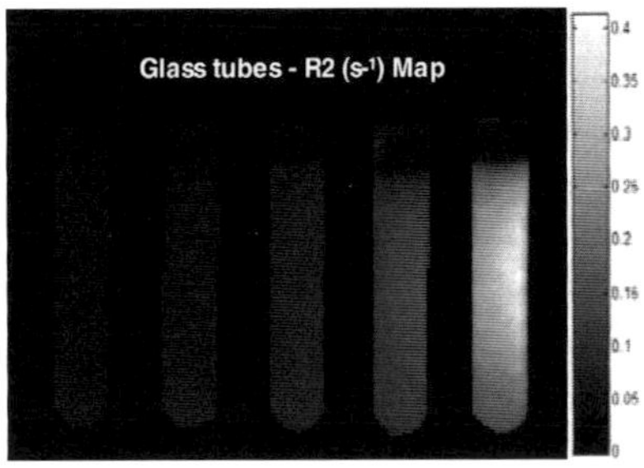

(5.b.)

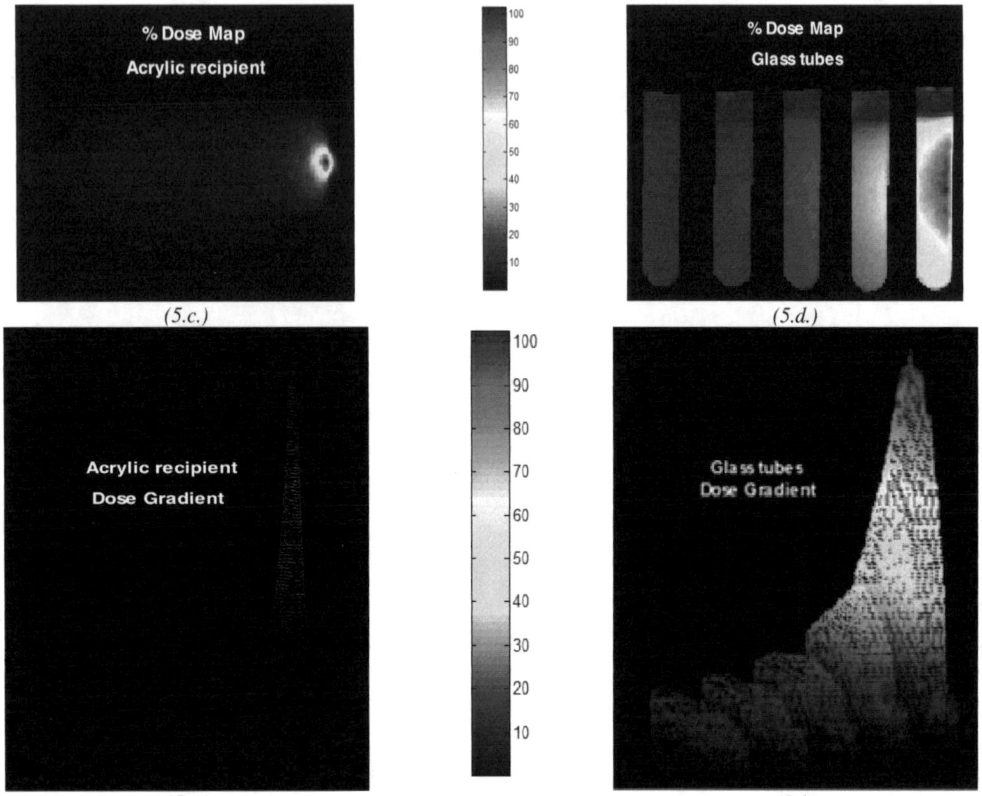

Figure 5 – R2, dose and dose gradient maps obtained from MRI image analysis: (5.a. and 5.b.) R2 in high-dose rate region and lower-dose rate regions respectively; (5.c. and 5.d.) Normalized dose in high-dose rate region and lower-dose rate regions respectively; (5.e. and 5.f.) Dose gradient 3D plots in high-dose rate region and lower-dose rate regions respectively.

4. Conclusions

MAGIC with formaldehyde has suitable features for use in HDR 192-Iridium brachytherapy source three-dimensional dosimetry such as high spatial resolution. This work shows that a proper gel calibration can provide accurate dose measurements in regions of high and low dose rates, allowing it to express dose rate gradients with high dose resolution.

The comparison of depth dose data obtained with gel, ionization chamber and Monte Carlo simulation showed that simulation data can be used as reference to polymer gel studies since it provides similar spatial resolution as the gel dosimeter.

Acknowlegdgments

This work was partially supported by CNPq, CAPES and FAPESP. The technical support from Brunello, L Rocha, H Ferreira, H Oliveira and JL Aziani is also appreciated.

References

[1] Ballestera F., Hernandez C., Perez-calatayud J., Lliso F.: "Monte Carlo calculation of dose rate distributions around 192Ir wires", Med. Phys. 24(8), 1221-1228, 1997.

[2] Oldham M., Guo P., Gluckman G., Adamovics J.: "IMRT verification using radiochromic/optical – CT dosimetry system", DosGel_2006 – Abstract Book, 144-144, 2006.

[3] Bayreder C., Schön R., Wieland M., Georg D., Moser E., Berg A.: "The spatial resolution in

dosimetry with normoxic polymer-gels investigated with dose modulation transfer approach", Med. Phys. 35(5), 1756-1769 ,2008.

[4] De Deene Y., Pittomvils G., Visalatchi S.: "The influence of cooling rate on the accuracy of normoxic polymer gel dosimeters", Phys. Med. Biol. 52, 2719-2728, 2007.

[5] De Deene Y., Reynaert N., Wagter C.: "On the accuracy of monomer/polymer gel dosimetry in the proximety of a high-dose-rate 192Ir source", Phys. Med. Biol. 46, 2801-2825, 2001.

[6] Fernandes J., Pastorello B., Araujo ., Baffa O.: "Formaldehyde increases MAGIC gel dosimeter melting point and sensitivity", Phys. Med. Biol. 53, N53-N58, 2008.

AUTHOR INDEX

AUTHOR INDEX

Institute of Physics
Dirac House, Temple Back
Bristol BS1 6BE UK

CURRAN ASSOCIATES INC
proceedings.com